J	Japan
KIU	1000 international units
L	Luxembourg
LD_{50}	lethal dose 50%
LDA	lithium diisopropylamide
Leu	leucine
LHRH	luteinizing hormone-releasing hormone
liq.	liquid
LUX	Luxembourg
lyo.	lyophilizate
Lys	lysine
M	mouse
Me	CH_3 (in structural formulas)
Met	Methionine
MEX	Mexico
MF	molecular formula
Mf	mouse, female
ml	milliliter (cubic centimeter)
Mm	mouse, male
MTBE	methyl *tert*-butyl ether
MW	molecular weight
N	Norway
NBS	*N*-bromosuccinimide
NZ	New Zealand
ophth.	ophthalmic
P	Portugal
p.	page
p.o.	per os (orally)
prior.	priority (patent)
Pro	proline
psi	pounds per square inch
q. v.	quod vide (which see)
R	rat
Rf	rat, female
Rm	rat, male
RN	Chemical Abstracts Registry Number Note: CA Registry numbers are subject to change without notice
s. r.	sustained release
s.c.	subcutaneously
Ser	serine
SF	Finland
SIE	saccharose inhibitory unit
sol.	solution
suppos.	suppositories

susp.	suspension
tabl.	tablets
TEA	triethylamine
TEBA	triethylbenzylammonium chloride
temp.	temperature
TEA	triethylamine
THF	tetrahydrofuran
Thr	threonine
Tos	tosyl, 4-methylphenylsulfonyl
TosOH	4-methylbenzenesulfonic acid
Trp	tryptophan
Tyr	tyrosine
USA	United States of America
Val	valine
wfm	withdrawn from the market (for commercial products)
YU	Yugoslavia
Z	benzyloxycarbonyl

Country codes in patent descriptions

AT	Austria
AU	Australia
BE	Belgium
CA	Canada
CH	Switzerland
DAS	Deutsche Auslegungsschrift
DD	German Democratic Republic
DE	Federal Republic of Germany
DOS	Deutsche Offenlegungsschrift
DRP	Deutsches Reichspatent
EP	European Patent
ES	Spain
GB	Great Britain
HU	Hungary
IL	Israel
IN	India
NL	Netherlands
IT	Italia
JP	Japan
LU	Luxemburg
SE	Sweden
SU	Soviet Union
US	United States of America
WO	World Patent
ZA	South Africa

121

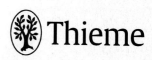

Pharmaceutical Substances

Syntheses, Patents, Applications

A–M

Axel Kleemann and
Jürgen Engel

Bernhard Kutscher and
Dietmar Reichert

4th Edition

Thieme

Stuttgart · New York

Prof. Dr. Axel Kleemann
Amselstr. 2
D-63454 Hanau

Dr. Dietmar Reichert
Degussa-Hüls
FC-FEA1
Postfach 13 45
D-63403 Hanau

Prof. Dr. Jürgen Engel
ASTA Medica AG
Weismüllerstr. 45
D-60314 Frankfurt

Prof. Dr. Bernhard Kutscher
ASTA Medica AG
Weismüllerstr. 45
D-60314 Frankfurt

Die Deutsche Bibliothek –
CIP-Einheitsaufnahme

Kleemann, Axel:
Pharmaceutical substances : syntheses, patents,
applications / Axel Kleemann and
Jürgen Engel. Bernhard Kutscher and
Dietmar Reichert. – 4th ed. – Stuttgart ;
New York : Thieme, 2001
  Einheitssacht.: Pharmazeutische Wirkstoffe
<engl.>

© 2001 Georg Thieme Verlag
Rüdigerstraße 14
D-70469 Stuttgart
http://www.thieme.de
electronic documents and indexes:
FIZ CHEMIE BERLIN
Franklinstr. 11
D-10587 Berlin
http: //www.fiz-chemie.de

Coverdesign: Thieme Marketing

Printed in Germany
by Konrad Triltsch, 97199 Ochsenfurt-Hohestadt

ISBN 3-13-558404-6 (GTV, Stuttgart)
ISBN 1-58890-031-2 (TNY, New York)

1 2 3 4 5 6

Table of Contents

Preface to the fourth edition

The third edition of this book was a great success. Within less than 18 months it was more or less sold out, the printed version as well as the electronic one. So the publishers plagued the authors to prepare the next edition. Again, our "junior authors", Prof. Dr. Bernhard Kutscher and Dr. Dietmar Reichert, were doing an excellent job in elaborating the manuscript of about 100 new drug monographs. Furthermore we took the chance to carry out numerous additions and corrections. As a consequence of this extension it was no longer possible to stay with a "one volume edition" and we now have the fourth edition with two volumes.

This fourth edition contains 2267 active pharmaceutical ingredients ("API's") including those which were launched only recently. No fundamental changes were made in comparison to the third edition, which means that the preface, the acknowledgements and the introduction to the third edition are also still valid for the new edition.

We are very grateful to the many readers who kindly provided us with proposals for corrections. Again, the authors are deeply indebted to Dr. Hans-Georg Scharnow and his excellent team from the FIZ CHEMIE BERLIN, for their contributions to the preparation process of this book. We also would like to thank Dr. Elisabeth Hillen and Dr. Guido Herrmann from the Georg Thieme Verlag for the good cooperation. Finally the authors are very grateful to the colleagues of the experienced "ASTA Medica team", who were already mentioned in the "Acknowledgements" to the third edition. In the meantime Dr. Knut Eis joined this team.

Frankfurt a. M., Autumn 2000   Axel Kleemann
Jürgen Engel

Preface to the third edition

For many years, the second edition of "Pharmazeutische Wirkstoffe" has been out of print and colleagues and friends have asked and plagued us to publish a new edition. Everyone who has ever prepared a similar reference book knows how much energy and enthusiasm has to be put into such a task besides the "daily job" in industry. After some hesitation we finally decided to prepare the new edition, in English. Two fortunate circumstances made things a lot easier.

Firstly, Mr. Willi Plein brought us into contact with Dr. H.-G. Scharnow of the FIZ CHEMIE BERLIN[1], who was interested in preparing the new edition as an electronic version by making use of their specialised know-how and self-developed software. With this technology we were able to transform greater parts of the second edition, especially the formula drawings, into the new electronic version – thereby establishing an innovative method of book production, making future new editions much easier to prepare, and also enabling the publisher to offer a CD-ROM version. The FIZ CHEMIE team was successful in attracting the BMBF[2] for this interesting project and luckily received generous financial support.

Secondly, we persuaded the young colleagues Prof. Dr. Bernhard Kutscher and Dr. Dietmar Reichert to take responsibility as "junior authors" for this book. About a quarter of the workload for this volume was done by them and furthermore they have the capability and talent to contribute to future editions – if hopefully necessary.

The purpose and object of this book are to establish a link between INN's[3], structure, synthesis and production processes, patent (and literature) situation, medical use and trade names of important pharmaceuticals.

This volume contains a collection of 2171 active pharmaceutical ingredients, which are listed alphabetically according to their INN's. Often there are still other Generic Names in use, for example BAN[4], DCF[5], USAN[6], which are added in brackets under the INN's and are listed in the index. In addition to the foregoing editions, we have added the molecular formulas and molecular weights as well as the CAS Registry Numbers[7] and – where available – the ATC Code Numbers[8], the EINECS numbers[9], and data for

acute toxicology as well as pharmaceutical dosage forms. In the last section of each monograph, the trade names in the six most important markets are listed. The trade names were taken from the relevant reference books on pharmaceutical specialities[10] [11] [12].

Within the 16 years since the second edition, respectively 11 years since the supplement volumes were published, many changes have occurred. Companies disappeared or were acquired and the trade names of many products were changed. Some products were withdrawn from the market, either as mono drugs or combinations. In such cases, we added "w f m" behind the trade name, when we felt that we should not eliminate

---

[1] FIZ CHEMIE BERLIN, Fachinformationszentrum Chemie GmbH, Franklinstraße 11, D-10587 Berlin

[2] Bundesministerium für Bildung, Wissenschaft, Forschung und Technologie, Bonn

[3] INN = International Nonproprietary Names; synonymous with "Generic Names"

[4] BAN = British Approved Name

[5] DCF = Dénomination Commune Française

[6] USAN = United States Adopted Name

[7] CAS = Chemical Abstracts Service

[8] ATC = Anatomic Therapeutic Chemical (Classification of Drugs)
   a) ATC Index, WHO Collaborating Center for Drug Statistics Methodology, Oslo 1997
   b) European Drug Index, N. F. Miller, R. P. Dessing, 4th ed., Deutscher Apotheker Verlag Stuttgart 1997

[9] EINECS = European Inventory of Existing Chemical Substances

[10] "WDI, World Drug Index" – database, Derwent Information, 1998

[11] "MDDR"-database, Drug Data Report, Prous Science, MDL, San Leandro, 1998

[12] Speciality lists for
   D: "Rote Liste 2000", Hrsg. Bundesverband der Pharmazeutischen Industrie e.V., Frankfurt/M.
   "List of Pharmaceutical Substances", 10. edition, ABDATA Pharma-Daten-Service, Eschborn, March 1997
   F: "Dictionnaire VIDAL 1998" OUP, Paris
   GB: "MIMS" (Monthly Index of Medical Specialities), Haymarket Publishing Services Ltd., London 1998
   I: "L'Informatore Farmaceutico", 58. edit., Organizzazione Editoriale Medico Farmaceutico srl, Milano 1998
   J: a) "Japta List" 1987, Japanese Drug Directory, 3rd ed., Tokyo 1987
   b) Drugs in Japan/Ethical Drugs 1970, 10, ed. by Japan Pharmaceutical Information Center, Tokyo, Japan
   USA: "PDR, Physicians Desk Reference", 52. edition, 1998; Medical Economics Comp., Montvale, N. J.

the monograph or its former trade name because the drug is still in use in other countries or in numerous combinations – even if not listed here – or is of historical interest. The given trade names are normally representing the mono drug product. In some cases, we listed the year of market introduction behind the name of the company.

It was the intention of the authors to present the synthesis routes in broad details. In many cases, we describe different synthesis routes, especially for the economically most important drugs – but also when we are not completely sure which particular route is applied technically. The practicability of such a reference book is to a great extent dependent on clearly arranged and in-depth indexes. The reader will find appropriate and comprehensive indexes for

1. Trade Names
2. Intermediates
3. Enzymes, Microorganisms, Plants, Animal Tissues
4. Substance classes.

As many abbreviations are used in this book, a separate "list of abbreviations" has been added, see front and back endpapers. Concerning abbreviations for chemicals and reagents we adhered to the "Standard List of Abbreviations", published in J. Org. Chem., Vol. **68**, No. 1, 1998, p. 19A.

We have had many discussions regarding the inclusion of the newer biopharmaceuticals ("Pharmaproteins") which are produced by recombinant DNA methods (e.g., Interferons, human Insulin, Erythropoietin etc.). After several trials, we dropped this plan because we could not find a

concept consistent with the original character of the book, which predominantly contains synthetic drugs as well as some antibiotics made by conventional fermentation and some plant ingredients that are produced by extraction. A borderline case exists with many synthethic peptides. It is often practically impossible to find out the applied technical synthesis process among the manifold routes described in the literature. So, we more or less stayed with the peptide drugs that were already included in the 2nd edition and supplement volume.

The book is designed to serve the needs of not only the specialists in drug synthesis but also of a broad usergroup in the chemical, intermediates and pharmaceutical industry, pharmacists, universities and other research institutes. So, it is important for everybody who has to deal with pharmaceuticals or their raw materials and intermediates and would like to have a quick survey. For the analytical chemist, knowledge about the synthesis route of an active ingredient is essential in order to recognize contaminations with intermediates or possible byproducts. Often, also the allergologist has a need for information regarding the synthesis of a drug in order to localize allergic byproducts or intermediates. The synthetic chemist will be provided with stimulating suggestions for his work.

We hope that this book will prove to be a standard reference for everyone who is interested in pharmaceuticals.

Frankfurt a. M., Autumn 1998   Axel Kleemann
                                 Jürgen Engel

Acknowledgements

Preparing a reference book of this kind is by no means feasible without the contributions of numerous dedicated colleagues who bring in their special expertise and knowledge.

The authors are deeply indebted to Dr. Hans-Georg Scharnow, Ingo Adamczyk, Gerhard Fabian, Martin Holberg, Gudrun Lippmann, Ulrich Quandt, Karin Raatsch and Dr. Katrin Seemeyer from FIZ CHEMIE BERLIN, for providing their computer know-how and software expertise. This FIZ CHEMIE team has taken a decisive part in the preparation of the book, also in reminding the authors regularly of their obligation to deliver the manuscripts on time. It was always a reliable, fruitful and pleasant cooperation. We also wish to extend our gratefulness to the BMBF (Bundesministerium für Bildung, Wissenschaft, Forschung und Technologie) for generous support of this project.

It is the authors special desire to express their gratitude to the many co-workers of the Chemical Research Department of ASTA Medica AG, who willingly agreed to assist and support the process of manuscript preparation and correction. For their assistance in patent and literature searching, we wish to thank Dr. Ilona Fleischhauer, Dieter Heiliger, Christa Krehmer, Thomas Kynast, Horst Mai, Dr. Ulf Niemeyer, Wolfgang Paul, Horst Rauer, Jürgen Schael and Norbert Schulmeyer. Important contributions in the search for and correction of trade names have been made by Dr. Thomas Arnold, Dr. Gerhard Nößner, Dr. Klaus Paulini and Dr. Heinz Weinberger and the trade name searching process was greatly supported by Dr. Patricia Charpentier, Dr. Stefan Eils, Anja Heddrich, Dr. Gilbert Müller, Sandra Nemetz, Sven-Oliver Schmidt, Anita Storch and Silvia Werner.

Important advice concerning the pharmaceutical specialities on the Japanese market was provided by our Japanese colleague Nobuo Kumagai, whom we wish to thank sincerely.

A word of thanks has also to be addressed to Dr. Elisabeth Hillen and Dr. Rolf Hoppe from the Georg Thieme Verlag for the always good cooperation and acceptance of many special suggestions concerning the production of this book and its CD-ROM version.

Finally, we would like to express our gratitude to the many readers and users of the foregoing editions who took the time to provide us with comments, corrections and suggestions.

The authors

Introduction

This reference book describes the production/isolation processes of 2267 active pharmaceutical substances (including the syntheses of their intermediates) that are or have been marketed. In order to illustrate what particular information can be drawn from the book a typical monograph is depicted below and labelled. It is self-explanatory.

The frequently-used abbreviations are explained in a separate list, q.v. see front and back endpapers.

With respect to the names of reagents and intermediates in the course of the syntheses the authors tried to use the names found in the catalogues of the commercial providers of fine chemicals, e.g. Sigma-Aldrich, otherwise the Chemical Abstracts Names are given.

In addition to the respective patents and publications in journals several standard reference books for Pharmaceuticals and Fine Chemicals were used and sometimes also cited. The most important sources are given here:

- The Merck Index, 12 ed., Merck & Co., Inc.; NJ 1996
  abbrev.: Merck Index
- Z. Budesinsky and M. Protiva, Synthetische Arzneimittel, Akademie-Verlag, Berlin 1961
  abbrev.: Budesinsky-Protiva
- G. Ehrhart, H. Ruschig, Arzneimittel, vols. 1–5, Verlag Chemie, Weinheim 1972
  abbrev.: Ehrhart-Ruschig

- Index Nominum, international drug directory, ed. by Swiss Pharmaceutical Society, 1992/93
- A. Kleemann, E. Lindner, J. Engel, Arzneimittel-Fortschritte 1972–1985, VCH Verlagsgesellschaft mbH, Weinheim 1987
- D. Lednicer, L. A. Mitscher, The Organic Chemistry of Drug Synthesis, vols. 1–6, John Wiley & Sons, New York 1977–1999
- Ullmanns Encyclopädie der technischen Chemie, 3. and 4. edition, Verlag Chemie Weinheim;
- Ullmann's Encyclopedia of Industrial Chemistry, 5. ed., vols. A1–A28, B1–B8, VCH Verlagsgesellschft mbH, Weinheim 1985–1997 abbrev.: Ullmanns Encycl. Techn. Chem.
- Ullmann's Encyclopedia of Industrial Chemistry, 6. ed., on CD-ROM, 1999 and 2000
- M. Negwer, Organic-chemical drugs and their synonyms, Akademie-Verlag, Berlin 1994
- USP Dictionary of USAN and International Drug Names, US Pharmacopeia, Rockville, MD 1998

The acute toxicity data were in most cases taken from different data banks or other secondary sources, so no guarantee can be given for validity.

For information regarding pharmaceutical dosage forms see preface, footnote 12.

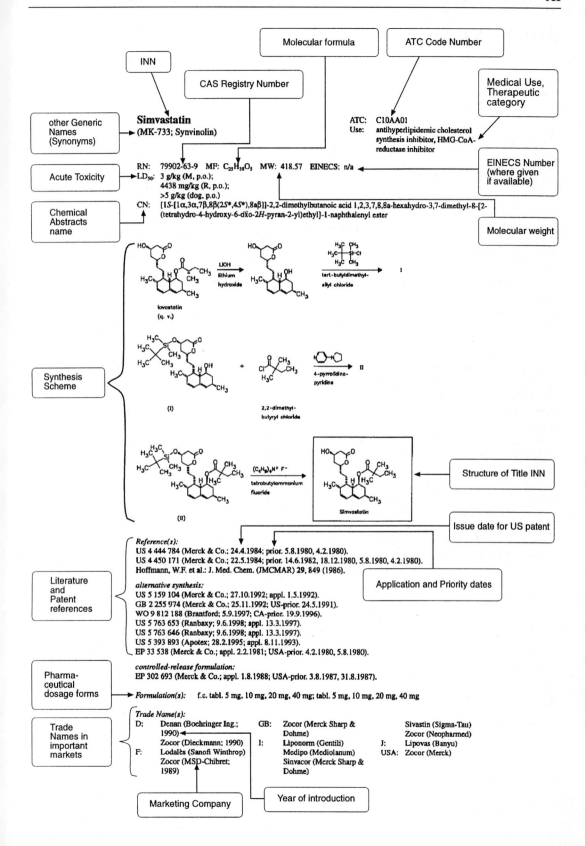

Molecular formula

ATC Code Number

INN

CAS Registry Number

Medical Use, Therapeutic category

other Generic Names (Synonyms)

Simvastatin
(MK-733; Synvinolin)

ATC: C10AA01
Use: antihyperlipidemic cholesterol synthesis inhibitor, HMG-CoA-reductase inhibitor

Acute Toxicity

RN: 79902-63-9 MF: $C_{25}H_{38}O_5$ MW: 418.57 EINECS: n/a
LD$_{50}$: 3 g/kg (M, p.o.);
4438 mg/kg (R, p.o.);
>5 g/kg (dog, p.o.)

EINECS Number (where given if available)

Chemical Abstracts name

CN: [1S-[1α,3α,7β,8β(2S*,4S*),8aβ]]-2,2-dimethylbutanoic acid 1,2,3,7,8,8a-hexahydro-3,7-dimethyl-8-[2-(tetrahydro-4-hydroxy-6-oxo-2H-pyran-2-yl)ethyl]-1-naphthalenyl ester

Molecular weight

Synthesis Scheme

lovastatin (q. v.)

LiOH lithium hydroxide

tert-butyldimethyl-silyl chloride

I

(I)

2,2-dimethyl-butyryl chloride

4-pyrrolidino-pyridine

II

(II)

$(C_4H_9)_6N^+ F^-$ tetrabutylammonium fluoride

Simvastatin

Structure of Title INN

Issue date for US patent

Literature and Patent references

Reference(s):
US 4 444 784 (Merck & Co.; 24.4.1984; prior. 5.8.1980, 4.2.1980).
US 4 450 171 (Merck & Co.; 22.5.1984; prior. 14.6.1982, 18.12.1980, 5.8.1980, 4.2.1980).
Hoffmann, W.F. et al.: J. Med. Chem. (JMCMAR) 29, 849 (1986).

alternative synthesis:
US 5 159 104 (Merck & Co.; 27.10.1992; appl. 1.5.1992).
GB 2 255 974 (Merck & Co.; 25.11.1992; US-prior. 24.5.1991).
WO 9 812 188 (Brantford; 5.9.1997; CA-prior. 19.9.1996).
US 5 763 653 (Ranbaxy; 9.6.1998; appl. 13.3.1997).
US 5 763 646 (Ranbaxy; 9.6.1998; appl. 13.3.1997).
US 5 393 893 (Apotex; 28.2.1995; appl. 8.11.1993).
EP 33 538 (Merck & Co.; appl. 2.2.1981; USA-prior. 4.2.1980, 5.8.1980).

Application and Priority dates

Pharmaceutical dosage forms

controlled-release formulation:
EP 302 693 (Merck & Co.; appl. 1.8.1988; USA-prior. 3.8.1987, 31.8.1987).

Formulation(s): f.c. tabl. 5 mg, 10 mg, 20 mg, 40 mg; tabl. 5 mg, 10 mg, 20 mg, 40 mg

Trade Names in important markets

Trade Name(s):

D:	Denan (Boehringer Ing.; 1990)	GB: Zocor (Merck Sharp & Dohme)	Sivastin (Sigma-Tau)
	Zocor (Dieckmann; 1990)	I: Liponorm (Gentili)	Zocor (Neopharmed)
F:	Lodalès (Sanofi Winthrop)	Medipo (Mediolanum)	J: Lipovas (Banyu)
	Zocor (MSD-Chibret; 1989)	Sinvacor (Merck Sharp & Dohme)	USA: Zocor (Merck)

Marketing Company

Year of introduction

Alphabetical List of Drug Monographs

Amidotrizoic acid 86
Amifostine 87
Amikacin 88
Amiloride 89
Amineptine 90
Aminocaproic acid 91
Aminoglutethimide 92
Aminophenazone 92
Aminophylline 2012
Aminoprofen 1203
Aminopromazine 93
Aminopyrine 92
p-Aminosalicylic acid 94
Aminosalylum 94
Aminosidine 1564
L-2-Aminosuccinic acid 140
Amiodarone 94
Amiphenazole 95
Amitriptyline 96
Amitriptylinoxide 97
Amixetrine 98
Amlexanox 99
Amlodipine 99
Ammoidin 1289
Amobarbital 101
Amodiaquine 102
Amorolfine 102
Amosulalol 103
Amoxapine 104
Amoxicillin 105
Amoxycillin 105
Amphetaminil 108
Amphotericin B 108
Ampicillin 109
Ampiroxicam 111
Amprenavir 112
Amrinone 115
Amsacrine 116
Amylobarbitone 101
Anagestone acetate 117
Anagrelide hydrochloride 118
Anastrozole 120
Ancitabine 121
Ancrod 121
Androstanolone 122
Anethole 122
Anethole trithione 123
Aneurine 2014
Angiotensinamide 124
Anileridine 126
Aniracetam 127
Anisindione 128
Anisotropine methylbromide 1477
Antazoline 128
Anthralin 687
Antrafenine 129
Apalcillin 130
Apazone 157
APD 1551
Aplonidine 131
Apomorphine 130
Apraclonidine 131
Aprindine 132
Aprobarbital 133
Aprotinine 133
Aquocobalamin 1028
ara C 569

Aranidipine 134
Arginine aspartate 135
Arginine pidolate 136
Arginine pyroglutamate 136
Aritromicina 166
Arotinolol 136
AS-4370 1369
5-ASA 1259
Ascorbic acid 137
Aseptamide 992
Asparaginase 139
L-Asparaginase 139
Aspartame 139
L-Aspartic acid 140
Aspirin 28
Aspoxicillin 141
Astemizole 143
Astromicin 144
AT-877 830
AT-4140 1889
Atebrin 1243
Atenolol 145
Atorvastatin calcium 146
Atracurium besilate 150
Atropine 151
Atropine methonitrate 152
Atropinmethylnitrat 152
AU-7801 1203
Auranofin 153
Avastatin 410
Axerophthol 1802
Ay 416 1406
Azacitidine 154
Azacort 581
Azacortid 870
Azacosterol 155
Azacyclonol 156
6-Azamianserin 1343
Azapetine 157
Azapropazone 157
Azatadine 158
Azathioprine 159
Azelaic acid 160
Azelastine 161
Azidamfenicol 162
Azidoamphenicol 162
Azidocillin 163
Azidothymidine 2197
Azimilide hydrochloride 163
Azintamide 165
Azithromycin 166
Azlocillin 167
Azolinic acid 480
Azosemide 168
AZT 2197
Azthreonam 169
Aztreonam 169

B

B 8510-29 1554
BA-168 1180
DA-598 DR 920
Bacampicillin 172
Bacitracin 172
Baclofen 174
Balsalazide sodium 175

Bambuterol 176
Bamethan 176
Bamifylline 177
Bamipine 178
Barbexaclone 179
Barbital 179
Barnidipine 180
Batroxobin 181
Bay-12-8039 1372
Bay-g-5421 5
Bay-K 5552 1449
Bay-m-1099 1334
Bay-o-9867 482
Bay u 3405 1784
Bay-W-6228 410
BBP 230
BCH-790 1136
BCNU 354
BDF-5895 1375
Beclamide 182
Beclobrate 182
Beclometasone 183
Beclomethasone 183
Befunolol 185
Bekanamycin 186
Bemegride 187
Bemetizide 187
Benactyzine 188
Benaprizine 188
Benapryzine 188
Benazepril 189
Benciclano 190
Bencyclane 190
Bendacort 191
Bendacortone 191
Bendazac 191
Bendrofluazide 192
Bendroflumethiazide 192
Benethamine Penicilline 204
Benexate 193
Benfluorex 194
Benfotiamine 195
Benfurodil hemisuccinate 195
Benidipine 196
Benmoxin 197
Benorilate 197
Benorilato 197
Benorylate 197
Benoxaprofen 198
Benoxinate 1528
Benperidol 199
Benproperine 199
Benserazide 200
Bentiamine 201
Bentiromide 202
Benzalkonium chloride 203
Benzapril 189
Benzarone 204
Benzathine benzylpenicillin 204
Benzatropine 205
Benzbromarone 206
Benzethonium chloride 206
Benzhexol 2111
Benzilonium bromide 207
Benzilpenicillin 204
Benziodarone 208
Benzocaine 208

Benzoctamine 209
Benzoesäurebenzylester 215
Benzonatate 209
Benzothiazide 213
Benzoyl peroxide 210
Benzoylsulfanilamide 1916
Benzoylthiamine disulfide 252
Benzperidol 199
Benzphetamine 211
Benzquinamide 211
Benzthiazide 213
Benztiazide 213
Benztropine 205
Benzydamine 213
Benzyl alcohol 214
Benzyl benzoate 215
Benzyl mustard oil 215
Benzyl nicotinate 1432
Benzylpenicillin 216
Bephenium hydroxynaphtho-ate 217
Bepridil 217
Beractant 1952
Betacarotene 218
Betacarotin 218
Betahistine 222
Betainchloralum 529
Betaine aspartate 222
Betaine hydrate 223
Betamethasone 224
Betamethasone acetate 228
Betamethasone adamantoate 229
Betamethasone benzoate 229
Betamethasone butyrate pro-pionate 230
Betamethasone dipropionate 231
Betamethasone divalerate 232
Betamethasone phosphate 232
Betamethasone valerate 233
Betanidine 234
Betaxolol 235
Betazole 236
Bethanechol chloride 237
Bethanidine 234
Bevantolol 238
Bevonium methylsulfate 239
Bevonium metilsulfate 239
Bezafibrate 239
B-HT-920 1966
BIBR 277 1978
Bibrocathin 240
Bibrocathol 240
BIBR 277SE 1978
Bicalutamide 241
Bietamiverine 242
Bietaserpine 242
Bifluranol 243
Bifonazole 244
Bifonazolum 244
Dindazac 191
Binedaline 245
Binodaline 245
Biolf-62 954
Biotin 246

Mofezolac 1355
Molindone 1356
Molsidomine 1357
Mometasone furoate 1358
Monakolin-K 1198
Monobenzone 1359
Montelukast sodium 1359
Moperone 1362
Mopidamol 1363
Mopiperone 1362
Moracizine 1364
Morclofone 1365
Morfazinamida 1366
Moricizine 1364
Morinamide 1366
Moroxydine 1366
Morphine 1367
Morphium 1367
Mosapramine 1368
Mosapride citrate 1369
Moxalactam 1143
Moxaverine 1370
Moxestrol 1371
Moxifloxacin hydrochloride 1372
Moxisylyte 1374
Moxonidine 1375
MPC-1304 134
MS 932 39
MST 16 1881
Mupirocin 1376
Muroctasin 1834
Muzolimine 1377
Mycophenolate mofetil 1378
Mycophenolic acid 1379
Mydecamycin 1329
Myrtecaine 1379

N

N-22 1355
N-696 2039
Nabilone 1381
Nabumetone 1382
N'-acetylsulfanilamide 1918
Nadolol 1383
Nadoxolol 1384
Nafamostat 1385
Nafamstat 1385
Nafcillin 1386
Nafronyl 1386
Naftidrofuryl 1386
Naftifine 1387
Naftifungin 1387
Naftopidil 1388
Nalbuphine 1389
Nalidixic acid 1390
Nalmefene 1390
Nalmetrene (base) 1390
Nalorphine 1391
Naloxone 1392
Naltrexone 1393
Nandrolone 1394
Nandrolone decanoate 1395
Nandrolone hexyloxyphenyl-propionate 1396

Nandrolone phenpropionate 1396
Nandrolone phenylpropionate 1396
Nandrolone undecanoate 1397
Nandrolone undecylate 1397
Naphazoline 1398
Naproxen 1399
Naproxen piperazine 1657
Naratriptan 1403
Natamycin 1405
Nateglinide 1406
Natrium-picosulfat 1884
Nazasetron 1407
Nebivolol 1408
Nedaplatin 1409
Nedocromil 1410
Nefazodone hydrochloride 1411
Nefopam 1413
Nelfinavir mesylate 1414
Nemonapride 1418
Neomycin B 941
Neostigmine methylsulfate 1419
Neticonazole hydrochloride 1419
Netilmicin 1420
Nevirapine 1421
Niacin 1432
Niacinamide 1431
Nialamide 1423
Niaprazine 1424
Nicardipine 1425
Nicergoline 1426
Niceritrol 1427
Nicethamide 1440
Niclosamide 1428
Nicoclonate 1428
Nicofuranose 1429
Nicorandil 1430
Nicotafuryl 1430
Nicotinamide 1431
Nicotinic acid 1432
Nicotinic acid benzyl ester 1432
Nicotinyl alcohol 1433
Nicotinyl cyclandelate 1325
Nicoumalone 13
Nifedipine 1434
Nifenalol 1434
Nifenazone 1435
Niflumic acid 1436
Nifluril 1436
Nifuratel 1437
Nifuroxazide 1437
Nifurprazine 1438
Nifurtimox 1439
Nifurtoinol 1439
Nifurzide 1440
NIH-10365 1390
Nikethamide 1440
Nilutamide 1441
Nilvadipine 1442
Nimesulide 1443
Nimetazepam 1444
Nimodipine 1445
Nimorazole 1446

Nimustine 1447
Nipradilol 1447
Nipradolol 1447
Niprodipine 1442
Niridazole 1448
Nisoldipine 1449
Nitalapram 488
Nitrazepam 1450
Nitrefazole 1451
Nitrendipine 1452
Nitrimidazine 1446
Nitrofural 1453
Nitrofurantoin 1453
Nitrofurazone 1453
Nitroxoline 1454
Nizatidine 1455
Nizofenone 1456
NKT-01 989
NND 318 1141
NO-05-0328 2026
2'-NOG 954
Nomegestrol acetate 1457
Nomifensine 1458
Nomurtide 1834
Nonoxinol 9 1459
Nonoxynol 9 1459
Nopoxamine 1379
5'-Noranhydrovinblastine 2177
Norbolethone 1459
Norboletone 1459
Nordazepam 1460
Nordiazepam 1460
DL-Norephedrine 1624
Norethandrolone 1461
Norethindrone 1461
Norethindrone acetate 1462
Norethisterone 1461
Norethisterone acetate 1462
Norethisterone enanthate 1463
Norethynodrel 1464
Noretynodrel 1464
Norfenefrine 1465
Norfloxacin 1466
Norgestimate 1467
Norgestrel 1468
D-Norgestrel 1169
Norgestrienone 1469
Norisoephedrine 1471
Normethadone 1470
Normethandrolone 1296
Normethandrone 1296
Normolaxol 1470
D-Norpseudoephedrine 1471
p-Norsynephrine 1478
Nortriptyline 1472
Novobiocin 1473
Novonal 2154
Noxiptiline 1474
Noxitiolinum 1475
Noxythiolin 1475
Noxytiolin 1475
NSC-66847 2005
NSC-125973 1545
NSC-266046 1510
NSC-296961 87
NSC-356894 989

NSC-609669 2080
NSC-616348 1096
NSC-628503 690
NSC-249008/352122 2120
NSC-105014-F 492
NY-198 1182
Nylidrine 290
Nystatin 1475
NZ-105 735

O

Obidoxime chloride 1477
Octatropine methylbromide 1477
Octopamine 1478
Octotiamine 1479
Oestradiol 779
Oestradiolbenzoat 781
Oestradiol-17-cyclopentyl-propionat 782
Oestradiolundecanoat 784
Oestron 787
Ofloxacin 1479
(S)-Ofloxacin 1165
1-OHP 1510
OKY-046 1543
Olanzapine 1480
Oleandomycin 1481
Oleum tropaeoli 215
Olprinone hydrochloride 1482
Olsalazine sodium 1483
Omapatrilat 1484
Omatropina 1016
Omeprazole 1486
Omoconazole nitrate 1488
Ondansetron 1489
ONO-1078 1682
OPC-8212 2163
OPC-13013 470
OPC-17116 979
Opipramol 1491
Orazamide 1491
Orciprenaline 1492
ORF-11676 1390
Org-3770 1343
Org-9426 1829
Orgotein 1493
Orientomycin 563
Orlipastat 1494
Orlistat 1494
Ornipresina 1501
Ornipressin 1501
Orotic acid 1502
Orphenadrine 1504
Orpressin 1501
Oseltamivir 1504
Otilonium bromide 1506
Ouabain 1908
Ox-373 479
Oxaceprol 1507
Oxacillin 1508
Oxaflozane 1508
Oxaflumazine 1509
Oxaliplatin 1510
Oxametacin 1510

S-1991 1866
S-3341 1815
S-6059 1143
6315-S 864
S-9490 1600
SA-96 277
Saccharin 1847
Salacetamide 1848
Salazosulfapyridine 1849
Salbutamol 1849
Salicylamide 1851
Salicylate de choline 455
Salicylazosulfapyridine 1849
Salicylic acid 1852
Salmaterol 1853
Salmeterol 1853
Salsalate 1854
SAM 40
Saquinavir 1854
Saralasin acetate 1858
SB-1 763
SB-75 418
SB-205312 1682
SC-58635 409
Sch-16134 1765
Sch-21420 1097
Sch-29851 1188
Sch-33844 1892
Sch-34117 592
Sch-60936 763
Schizophyllan 1880
Scopolamine 1859
Scopolamine butyl bromide 311
(−)sddc 1136
SDT-DJN 608 1406
SDZ 212-713 1826
SDZ-205502 1768
Secbutabarbital 1860
Secbutobarbitone 1860
Secnidazole 1860
Secobarbital 1861
Secretin 1862
SED-9490 (as erbumine) 1600
Selegiline 1863
Seratrodast 1864
Sertaconazole 1865
Sertindole 1866
Sertraline 1868
Serum-Tryptase 858
Setastine 1869
Setiptiline 1870
Sevoflurane 1871
SF-86327 1991
SH-401 1048
SHB 331 960
Sibutramine hydrochloride 1872
Sildenafil 1873
Silibinin 1876
Simethicone 665
Simfibrate 1876
Simvastatin 1877
Sisomicin 1878
β-Sitosterin 1879
β-Sitosterol 1879
Sizofiran 1880

SK & F-96022 1554
SK & F-101468 1837
SK & F-101468A 1837
SK & F-S 104864-A 2080
SKB 108566 758
SKF 108566 758
SKF 82526-J 844
SL-85.0324 1349
SL-80-0750-23N 2207
SM-7338 1257
SM-3997 (as citrate) 1971
SN-305 1133
SND-919Y 1679
Sobrerol 1881
Sobuzoxane 1881
Sodium aurothiomalate 1882
Sodium Butabarbital 1860
Sodium dioctyl sulfosuccinate 1882
Sodium picosulfate 1884
Sodium picosulphate 1884
Sofalcone 1884
Sorbitol 1885
Sorivudine 1886
Sotalol 1888
Sparfloxacin 1889
Spectinomycin 1890
Spiperone 1891
Spiramycin 1891
Spirapril 1892
Spironolactone 1895
Spiroylsäure 1852
Spirsäure 1852
Spizofurone 1897
SPM-925 1353
SQ-28555 937
SR-41319 2040
SR-47436 1095
SR-41319B 2040
SR-25990C 526
SRI-62320 921
SS-717 1419
ST-1396 409
Stallimycin 1897
Stanazol 1899
Stanolone 122
Stanozolol 1899
Stavudine 1900
Stepronin 1904
Streptokinase 1905
Streptomycin 1905
Streptoniazid 1906
Streptonicozid 1906
Streptozocin 1907
Streptozotocin 1907
g-Strophanthin 1908
k-Strophanthin 1909
k-Strophanthin-α 1910
k-Strophanthin-β 1909
k-Strophanthin-β + k-Strophanthoside 1909
k-Strophanthoside 1909
g-Strophantoside 1908
Styramate 1911
SU-88 1884
SU 101 1148
Succinylcholine chloride 1952

Succinylsulfathiazole 1911
Sucralfate 1912
Sufentanil 1913
Sulbactam 1913
Sulbenicillin 1914
Sulbentine 1915
Sulconazole 1915
Sulfabenzamide 1916
Sulfacarbamide 1917
Sulfacetamide 1918
Sulfachlorpyridazine 1919
Sulfacitine 1919
Sulfacytine 1919
Sulfadiazine 1920
Sulfadicramide 1921
Sulfadimethoxine 1922
Sulfadoxine 1923
Sulfaethidole 1924
Sulfafurazole 1924
Sulfaguanidine 1925
Sulfaguanole 1926
Sulfalene 1926
Sulfaloxic acid 1927
Sulfamerazine 1928
Sulfameter 1932
Sulfamethizole 1929
Sulfamethoxazole 1930
Sulfamethoxypyridazine 1931
Sulfametopyrazine 1926
Sulfametoxydiazine 1932
Sulfametrole 1933
Sulfamoxole 1933
Sulfanilamide 1934
Sulfanylurea 1917
Sulfaperin 1935
Sulfaphenazole 1936
Sulfaproxyline 1937
Sulfasalazine 1849
Sulfathiazole 1938
Sulfatolamide 1204
Sulfinpyrazone 1938
Sulfisomidine 1939
Sulfisoxazole 1924
Sulfisoxazole Acetyl 29
Sulforidazine 1940
Sulformethoxine 1923
Sulformetoxinum 1923
Sulforthomidine 1923
Sulfoxone sodium 1941
Sulindac 1941
Sulmetozin 1942
Suloctidil 1943
Sulphabenzamide 1916
Sulphacetamide 1918
Sulphadiazine 1920
Sulphafurazole 1924
Sulphaguanidine 1925
Sulphaloxate 1927
Sulphaloxic Acid 1927
Sulphamethizole 1929
Sulphamethoxazole 1930
Sulphamethoxypyridazine 1931
Sulphamoxole 1933
Sulphanilamide 1934
Sulphaphenazole 1936
Sulphasalazine 1849
Sulphasomidine 1939

Sulphathiazole 1938
Sulphaurea 1917
Sulphinpyrazone 1938
Sulpiride 1944
Sultamicillin 1945
Sulthiame 1946
Sultiame 1946
Sultopride 1947
Sultroponium 1948
Sumatriptan 1948
SUN-1165 1635
SUN 5555 828
Suplatast tosilate 1950
Suprofen 1951
Surfactant TA 1952
Suxamethonium chloride 1952
Suxibuzone 1953
Synephrine 1953
Synvinolin 1877
Syrosingopine 1954

T

TA-058 141
TA-064 587
TA-167 1980
TA-870 689
TA-903 193
TA-2711 727
TA-6366 1051
TA-8704 689
Tacalcitol 1956
Tacrine 1957
Tacrolimus 1958
Talampicillin 1964
Talinolol 1965
Talipexole 1966
Tamoxifen 1967
Tamsulosin hydrochloride 1969
Tandospirone 1971
Tazanolast 1972
Tazarotene 1973
3 TC 1136
TC-80 1094
TCV-116 329
Teciptiline 1870
Teclothiazide 1974
TECZA 2107
Tegafur 1974
Tegafur-Uracil 1976
Teichomycin 1976
Teicoplanin 1976
Telmesteine 1978
Telmisartan 1978
Temafloxacin 1980
Temazepam 1981
Temocapril 1981
Temocillin 1984
Teniposide 1987
Tenoglicine 1904
Tenonitrozole 1988
Tenoxicam 1988
Tenylidone 1989
Teofyllamin 2012
Terazosin 1990

ATC
Anatomical Therapeutic Chemical Classification

In 1981, the WHO[1] Regional Office for Europe recommended the *Anatomical Therapeutic Chemical* (ATC) classification system. In the ATC system drugs are classified in groups at 5 different levels. The drugs are divided into main groups (1st level), with two therapeutic/pharmacological subgroups (2nd and 3rd levels). Level 4 is a therapeutic/ pharmacological/chemical subgroup and level 5 is the chemical substance.

The complete classification of *simvastatin* illustrates the structure of the code:

C	Cardiovascular System
C10	Serum Lipid Reducing Agents
C10A	Cholesterol And Triglyceride Reducers
C10AA	HMG CoA reductase inhibitors
C10AA01	Simvastatin

The ATC classification system was originally based on the same main principles as the *Anatomical Classification* (AC-system)[2] developed by the *European Pharmaceutical Market Research Association* (EPhMRA) and the *Pharmaceutical Business Intelligence and Research Group* (PBIRG)[3].

---

[1] WHO Collaborating Centre for Drug Statistics Methodology, c/o Norsk Medisinaldepot AS, P. O. Box 100, Veltvet, N-0518 Oslo, Norway, Telephone: (47)22169810/22169801, Telefax: (47)22169818
[2] http://www.ephmra.org/6_001
[3] Formerly called International Pharmaceutical Market Research Group (IPMRG)

A01AB	Antiinfectives for local oral treatment		A03AX	Other synthetic anticholinergic agents
A01AC	Corticosteroids for local oral treatment		A03BA	Belladonna alkaloids, tertiary amines
A01AD	Other agents for local oral treatment		A03BB	Belladonna alkaloids semisynthetic, quaternary ammonium compounds
A02	Antacids, drugs for treatment of peptic ulcer and flatulence		A03CA	Synthetic anticholinergic agents in combination with psycholeptics
A02A	Antacids		A03DA	Synthetic anticholinergic agents in combination with analgesics
A02AB	Aluminum compounds		A03E	Antispasmodics and anticholinergics in combination with other drugs
A02AD	Combinations and complexes of aluminium, calcium and magnesium compounds		A03FA	Propulsives
A02B	Drugs for treatment of peptic ulcer		A04	Antiemetics and antinauseants
A02BA	H_2-receptor antagonists		A04A	Antiemetics and antinauseants
A02BB	Prostaglandins		A04AA	Serotonin (5HT3) antagonists
A02BC	Proton pump inhibitors		A04AD	Other antiemetics
A02BD	Proton pump inhibitors		A05	Bile and liver therapy
A02BO	Antiulcerants		A05A1	Bile therapy
A02BX	Other drugs for treatment of peptic ulcer		A05AA	Bile acid preparations
A03	Antispasmodic and anticholinergic agents and propulsives		A05AB	Preparations for biliary tract therapy
A03A	Synthetic antispasmodic and anticholinergic agents		A05AX	Other drugs for bile therapy
			A05B	Liver therapy, lipotropics
A03AA	Synthetic anticholinergics, esters with tertiary amino group		A05BA	Liver therapy
			A06A	Laxatives
A03AB	Synthetic anticholinergics, quaternary ammonium compounds		A06AB	Contact laxatives
			A06AD	Osmotically acting laxatives
A03AC	Synthetic antispasmodics, amides with tertiary amines		A06AG	Enemas

A07	Antidiarrheals, intestinal antiinflammatory and antiinfective agents	B04AA	Cholesterol- and triglyceride reducers
A07AA	Antibiotics	B04AB	HMG Co-A-reductase inhibitors
A07AB	Suflonamides	B04AC	Fibrates
A07AC	Imidazole derivatives	B05A	Blood and related products
A07AX	Other intestinal antiinfectives	B05BB	Solutions affecting the electrolyte balance
A07DA	Antipropulsives	B05CA	Antiinfectives
A07EA	Corticosteroids for local use	B05CX	Other irrigating solutions
A07EB	Antiallergic agents, excl. corticosteroids	B05XX	Other i.v. solution additives
A07EC	Aminosalicylic acid and similar agents	B06A	Other hematological agents
A07XA	All other antidiarrhoeals	B06AA	Enzymes
A08A	Antiobesity preparations, excl. diet products	C01	Cardiac therapy
A08AA	Centrally acting antiobesity products	C01A	Cardiac glycosides
A08AB	Peripherally acting antiobesity products	C01AA	Digitalis glycosides
A09A	digestives, incl. Enzymes	C01AB	Scilla glycosides
A09AB	Acid preparations	C01AC	Strophantus glycosides
A10	Drugs use in diabetes	C01AX	Other cardiac glycosides
A10BA	Biguanides	C01B	Antiarrhythmics, class I and II
A10BB	Sulfonamides, urea derivatives	C01BA	Antiarrhythmics, class IA
A10BC	Sulfonamides (heterocyclic)	C01BB	Antiarrhythmics, class IB
A10BF	glycosidase inhibitors	C01BC	Antiarrhythmics, class IC
A10BG	Thiazolinediones	C01BD	Antiarrhythmics, class III
A10BX	Other oral blood glucose lowering drugs	C01BG	Other class I antiarrhythmics
A10XA	Aldose reductase inhibitors	C01CA	Adrenergic and dopaminergic agents
A11	Vitamins	C01CE	Phosphodiesterase inhibitors
A11CA	Vitamin A, plain	C01CX	Other cardiac stimulants
A11CC	Vitamin D and analogues	C01D	Vasodilators used in cardiac diseases
A11DA	Vitamin B_1, plain	C01DA	Organic nitrates
A11DB	Vitamin B_1 in comb. with vitamin B_6 and/or vitamin B_{12}	C01DB	Quinolone vasodilators
		C01DX	Other vasodilators used in cardiac diseases
A11GA	Ascorbic acid (vit C), plain	C01EB	Other cardiac preparations
A11HA	Other plain vitamin preparations	C02	Antihypertensives
A11JC	Vitamins, other combinations	C02AA	Rauwolfia alkaloids
A12AX	Calcium, combinations with other drugs	C02AB	Methyldopa
A12BA	Potassium	C02AC	Imidazoline receptor agonists
A13A	Tonics	C02BB	Secondary and tertiary amines
A14	Anabolic agents for systemic use	C02CA	α-adrenoceptor blocking agents
A14A	Anabolic steroids	C02CB	α- and β-adrenoceptor blocking agent
A14AA	Androstan derivatives	C02CC	Guanidine derivatives
A14AB	Estren derivatives	C02DA	Thiazide derivatives
A14B	Other anabolic agents	C02DB	Hydrazinophthalazine derivatives
A16AA	Amino acids and derivatives	C02DC	Pyrimidine derivatives
A16AX	Various alimentary tract and metabolism products	C02DE	Calcium channel blockers
		C02DG	Guanidine derivatives
B01AA	Vitamin K antagonists	C02KB	Tyrosine hydroxylase inhibitors
B01AB	Heparin group	C02KC	MAO inhibitors
B01AC	Platelet aggregation inhibitors excl. heparin	C02KD	Serotonin antagonists
B01AD	Enzymes	C02L	Antihypertensives and diuretics in combination
B01AX	Other antithrombotic agents	C02LA	Rauwolfia alkaloids and diuretics in combination
B02AA	amino acids		
B02AB	Proteinase inhibitors	C02LX	Other antihypertensives and diuretics
B02BA	Vitamin K	C03	Diuretics
B02BC	Local hemostatics	C03AA	Thiazides, plain
B02BX	Other systemic hemostatics	C03AX	Thiazides, combinations with other drugs
B03BA	Vitamin B_{12} (cyanocobalamin and derivatives)	C03BA	Sulfonamides, plain
		C03BD	Xanthine derivatives
		C03BX	Other low-ceiling diuretics
		C03CA	Sulfonamides, plain
B03BB	Folic acid and derivatives	C03CC	Aryloxyacetic acid derivatives

C03CD	Pyrazolone derivatives
C03CX	Other high-ceiling diuretics
C03DA	Aldosterone antagonists
C03DB	Other potassium-sparing agents
C03E	Diuretics and potassium-sparing agents in combination
C03EA	Low-ceiling diuretics and potassium-sparing agents
C04	Peripheral vasodilators
C04A	Peripheral vasodilators
C04AA	2-amino-1-phenylethanol derivatives
C04AB	Imidazoline derivatives
C04AC	Nicotinic acid and derivatives
C04AD	Purine derivatives
C04AE	Ergot alkaloids
C04AX	Other peripheral vasodilators
C05	Vasoprotectives
C05AA	Products containing corticosteroids
C05AD	Products containing local anesthetics
C05AX	Other antihemorrhoidals for topical use
C05BA	Preparations with heparin for topical use
C05BB	Sclerosing agents for local injection
C05BX	Other sclerosing agents
C05C	Capillary stabilizing agents
C05CA	Bioflavonoids
C05CX	Other capillary stabilizing agents
C07A	Beta blocking agents
C07AA	β blocking agents, non-selective
C07AB	β blocking agents, selective
C07AG	α and β blocking agents
C07BA	β blocking agents, non-selective, and thiazides
C07BB	β blocking agents, selective, and thiazides
C07DA	β blocking agents, non-selective, thiazides and other diuretics
C07EA	β blocking agents, non-selective, and vasodilators
C08CA	Dihydropyridine derivatives
C08CX	Other selective calcium channel blockers with mainly vascular effects
C08DA	Phenylalkylamine derivatives
C08DB	Benzothiazepine derivatives
C08EA	Phenylalkylamine derivatives
C08EX	Other non-selective calcium channel blockers
C09	Agents acting on the Renin-Angiotensin System
C09A	ACE inhibitors, plain
C09AA	ACE inhibitors, plain
C09BA	ACE inhibitors and diuretics
C09CA	Angiotensin II antagonists, plain
C10A	Cholesterol and triglyceride reducers
C10AA	HMG CoA reductase inhibitors
C10AB	Fibrates
C10AD	Nicotinic acid and derivatives
C10AX	Other cholesterol and triglyceride reducers
D01	Antifungal for dermatological use
D01A	Antifungals for topical use
D01AA	Antibiotics
D01AC	Imidazole derivatives
D01AE	Other antifungals for topical use
D01BA	Antifungals for systemic use
D02B	Protectives against UV-radiation
D02BB	Protectives against UV-radiation for systemic use
D03	Preparations for treatment of wounds and ulcers
D03A	Cicatrizants
D03AX	Other cicatrizants
D04	Antipruritics, incl. antihistamines, anesthetics, etc.
D04AA	Antihistamines for topical use
D04AB	Anesthetics for topical use
D05	Antipsoriatics
D05AC	Antracen derivatives
D05AD	Psoralens for topical use
D05AX	Other antipsoriatics for topical use
D05B	Antipsoriatics for systemic use
D05BA	Psoralens for systemic use
D05BB	Retinoids for treatment of psoriasis
D06A	Antibiotics for topical use
D06AA	Tetracycline and derivatives
D06AX	Other antibiotics for topical use
D06BA	Sulfonamides
D06BB	Antivirals
D06BX	Other chemotherapeutics
D07A	Corticosteroids, plain
D07AA	Corticosteroids, weak (group I)
D07AB	Corticosteroids, moderately potent (group II)
D07AC	Corticosteroids, potent (group III)
D07AD	Corticosteroids, very potent (group IV)
D07BB	Corticosteroids, moderately potent, combinations with antiseptics
D07BC	Corticosteroids, potent, combinations with antiseptics
D07CB	Corticosteroids, moderately potent, combinations with antibiotics
D07CC	Corticosteroids, potent, combinations with antibiotics
D07XA	Corticosteroids, weak, other combinations
D07XB	Corticosteroids, moderately potent, other combinations
D07XC	Corticosteroids, potent, other combinations
D08	Antiseptics and disinfectants
D08A	Antiseptics and disinfectants
D08AA	Acridine derivatives
D08AC	Biguanides and amidines
D08AE	Phenol and derivatives
D08AF	Furan derivatives
D08AG	Iodine products
D08AH	Quinoline derivatives
D08AJ	Quaternary ammonium compounds
D08AK	Mercurial products
D08AX	Other antiseptic and disinfectants
D09AA	Ointment dressings with antiinfectives
D10AA	Corticosteroids, combinations for treatment of acne

D10AB	Preparations containing sulfur
D10AD	Retinoids for topical use in acne
D10AE	Peroxides
D10AF	Antiinfectives for treatment of acne
D10AX	Other anti-acne preparations for topical use
D10BA	Retinoids for treatment of acne
D11AA	Antihidrotics
D11AE	Androgens for topical use
D11AX	Other dermatologicals
G01AA	Antibiotics
G01AB	Arsenic compounds
G01AC	Quinoline derivatives
G01AE	Sulfonamides
G01AF	Imidazole derivatives
G01AG	Triazole derivatives
G01AX	Other antiinfectives and antiseptics
G02AB	Ergot alkaloids
G02AD	Prostaglandins
G02B	Contraceptives for topical use
G02BB	Intravaginal contraceptives
G02CA	Sympathomimetics, labour repressants
G02CB	Prolactine inhibitors
G02CC	Antiinflammatory products for vaginal administration
G03	Sex hormones and modulators of the genital system
G03A	Hormonal contraceptives for systemic use
G03AA	Progestogens and estrogens, fixed combinations
G03AB	Progestogens and estrogens, sequential preparations
G03AC	Progestogens
G03B	Androgens
G03BA	3-oxoandrosten-4 derivatives
G03BB	5-androstanon-3 derivatives
G03C	Estrogens
G03CA	Natural and semisynthetic estrogens, plain
G03CB	Synthetic estrogens, plain
G03CC	Estrogens, combinations with other drugs
G03D	Progestogens
G03DA	Pregnen-4 derivatives
G03DB	Pregnadien derivatives
G03DC	Estren derivatives
G03EA	Androgens and estrogens
G03EB	Androgen, progestogen and estrogen in combination
G03EK	Androgens and female sex hormones
G03FA	Progestogens and estrogens, fixed combinations
G03FB	Progestogens and estrogens, sequential preparations
G03GB	Ovulation stimulants, synthetic
G03H	Antiandrogens
G03HA	Antiandrogens, plain preparations
G03HB	Antiandrogens and estrogens
G03X	Other sex hormones and modulators of the genital system
G03XA	Antigonadotropins and similar agents
G03XB	Antiprogestogens

G04A	Urinary antiseptives and antiinfectives
G04AA	Methenamine preparations
G04AB	Quinolone derivatives (excl. J01M)
G04AC	Nitrofuran derivatives
G04AG	Other urinary antiseptics and antiinfectives
G04BD	Urinary antispasmodics
G04BE	Papaverine and derivatives
G04BX	Other urologicals
G04C	Drugs used in benign prostatic hypertrophy
G04CA	α-adrenoreceptor antagonists
G04CB	Testosterone-5-α-reductase inhibitors
H01BA	Vasopressin and analogues
H01BB	Oxytocin and derivatives
H01C	Hypothalamic hormones
H01CA	Gonadotropin-releasing hormones
H02AA	Mineralocortocoides
H02AB	Glucocorticoides
H02CA	Anticorticosteroids
H03AA	Thyroid hormones
H03BA	Thiouracils
H03BB	Sulfur-containing imidazole derivatives
H03CA	Iodine therapy
J01	Antibacterials for systemic use
J01AA	Tetracyclines
J01BA	Amphenicols
J01C	β-lactam antibacterials, penicillins
J01CA	Penicillin with extended spectrum
J01CE	β-lactamase sensitive penicillins
J01CF	β-lactamase resistent penicillins
J01CG	β-lactamase inhibitors
J01CR	Combinations of penicillins, incl. beta-lactamase inhibitors
J01DA	Cephalosporins and related substances
J01DF	Monobactams
J01DH	Carbapenems
J01E	Sulfonamides and trimethoprim
J01EA	Trimethoprim and derivatives
J01EB	Short-acting sulfonamides
J01EC	Intermediate-acting sulfonamides
J01ED	Long-acting sulfonamides
J01FA	Macrolides
J01FF	Lincosamides
J01G	Aminoglycoside antibacterials
J01GA	Streptomycins
J01GB	Other aminoglycosides
J01HA	Penicillinase sensitive penicillins
J01KD	Aminoglycoside antibiotics
J01M	Quinolone antibacterials
J01MA	Fluoroquinolones
J01MB	Other quinolones excl. G04AB
J01XA	Glycopeptide antibacterials
J01XB	Polymyxins
J01XD	Imidazole derivatives
J01XX	Other antibacterials
J02AA	Antibiotics
J02AB	Imidazole derivatives
J02AC	Triazole derivatives
J02AX	Other antimycotics for systemic use
J04A	Drugs for treatment of tuberculosis

J04AA	Aminosalicylic acid and derivatives
J04AB	Antibiotics
J04AC	Hydrazides
J04AD	Thiocarbamide derivatives
J04AK	Other drugs for treatment of tuberculosis
J04BA	Drugs for treatment of lepra
J05	Antivirals for systemic use
J05A	Direct acting antivirales
J05AB	Nucleosides and nucleotides
J05AC	Cyclic amines
J05AE	HIV-proteinase inhibitors
J05AF	Nucleosides and nucleotides excl. reverse transcriptase inhibitors
J05AG	Non-nucleoside reverse transcriptase inhibitors
J05AX	Other antivirals
L01	Antineoplastic agents
L01AA	Nitrogen mustard analogues
L01AB	Alkyl sulfonates
L01AC	Ethylene imines
L01AD	Nitrosoureas
L01AX	Other alkylating agents
L01BA	Folic acid analogues
L01BB	Purine analogues
L01BC	Pyrimidine analogues
L01C	Plant alkaloids and other natural products
L01CA	Vinca alkaloids and analogues
L01CB	Podophyllotoxin derivatives
L01CD	Taxanes
L01DA	Actinomycines
L01DB	Anthracyclines and related substances
L01DC	Other cytotoxic antibiotics
L01XA	Platinum compounds
L01XB	Methylhydrazines
L01XX	Other antineoplastic agents
L02A	Hormones and related agents
L02AA	Estrogens
L02AB	Progestogens
L02AE	Gonadotropin releasing hormone analogues
L02AX	Other hormones
L02B	Hormone antagonists and related agents
L02BA	Anti-estrogens
L02BB	Anti-androgens
L02BG	Enzyme inhibitors
L03A	Immunostimulating agents
L03AX	Other immunostimulating agents
L04	Immunosuppressive agents
L04AA	Selective immunosuppressive agents
L04AX	Other immunosuppressive agents
M01	Antiinflammatory and antirheumatic products
M01A	Antiinflammatory and antirheumatic agents, non steroids
M01AA	Butylpyrazolidines
M01AB	Acetic acid derivatives and related substances
M01AC	Oxicams
M01AE	Propionic acid derivatives
M01AG	Fenamates
M01AH	Cycloxygenase-2-inhibitors
M01AX	Other antiinflammatory and antirheumatic agents, non-steroids
M01BA	Antiinflammatory and antirheumatic agents in combination with corticosteroids
M01CB	Gold preparations
M01CC	Penicillamine and similar agents
M02A	Topical products for joint and muscular pain
M02AA	Antiinflammatory preparations, non-steroids for topical use
M02AC	Preparations with salicylic acid derivatives
M02AX	Other topical products for joint and muscular pain
M03	Muscle relaxants
M03A	Muscle relaxants, peripherally acting agents
M03AA	Curare alkaloids
M03AB	Choline derivatives
M03AC	Other quaternary ammonium compounds
M03B	Muscle relaxants, centrally acting agents
M03BA	Carbamic acid esters
M03BB	Oxazol, thiazine, triazine derivatives
M03BC	Ethers, chemically close to antihistamines
M03BX	Other centrally acting agents
M03CA	Dantrolene and derivatives
M04	Antigout preparations
M04AA	Preparations inhibiting uric acid production
M04AB	Preparations increasing uric acid excretion
M05BA	Bisphosphonates
M05BX	Other drugs affecting mineralization
M09AB	Enzymes
N01A	Anesthetics, general
N01AA	Ethers
N01AB	Halogenated hydrocarbons
N01AF	Barbiturates, plain
N01AH	Opioid anesthetics
N01AX	Other general anesthetics
N01B	Anesthetics, local
N01BA	Esters of aminobenzoic acid
N01BB	Amides
N01BC	Esters of benzoic acid
N01BX	Other local anesthetics
N02	Analgesics
N02A	Opioids
N02AA	Natural opium alkaloids
N02AB	Phenylpiperidine derivatives
N02AC	Diphenylpropylamine derivatives
N02AD	Benzomorphan derivatives
N02AE	Oripavine derivatives
N02AF	Morphinan derivatives
N02AX	Other opioids
N02B	Other analgesics and antipyretics
N02BA	Salicylic acid and derivatives
N02BB	Pyrazolones
N02BE	Anilides
N02BG	Other analgesics and antipyretics
N02CA	Ergot alkaloids
N02CB	Corticosteroid derivatives
N02CC	Selective 5HT1-receptor agonists

N02CX	Other antimigraine preparations
N03AA	Barbiturates and derivatives
N03AB	Hydantoin derivatives
N03AC	Oxazolidine derivatives
N03AD	Succinimide derivatives
N03AE	Benzodiazepine derivatives
N03AF	Carboxamide derivatives
N03AG	Fatty acid derivatives
N03AX	Other antiepileptics
N04	Anti-parkinson drugs
N04A	Anticholinergic agents
N04AA	Tertiary amines
N04AB	Ethers chemically close to antihistamines
N04AC	Ethers of tropine or tropine derivatives
N04B	Dopaminergic agents
N04BA	Dopa and dopa derivatives
N04BB	Adamantane derivatives
N04BC	Dopamine agonists
N04BD	Monoamine oxidase Type B inhibitors
N05	Psycholeptics
N05A	Antipsychotics
N05AA	Phenothiazines with aliphatic side-chain
N05AB	Phenothiazines with piperazine structure
N05AC	Phenothiazines with piperidine structure
N05AD	Butyrophenone derivatives
N05AE	Indole derivatives
N05AF	Thioxanthene derivatives
N05AG	Diphenylbutylpiperidine derivatives
N05AH	Diazepines and oxazepines
N05AK	Neuroleptics, in tardive dyskinesia
N05AL	Benzamides
N05AX	Other antipsychotics
N05B	Anxiolytics
N05BA	Benzodiazepine derivatives
N05BB	Diphenylmethane derivatives
N05BC	Carbamates
N05BD	Dibenzo-bicyclo-octadiene derivatives
N05BE	Azaspirodecanedione derivatives
N05BX	Other anxiolytics
N05C	Hypnotics and sedatives
N05CA	Barbiturates, plain
N05CC	Aldehydes and derivatives
N05CD	Benzodiazepine derivatives
N05CE	Piperidinedione derivatives
N05CF	Cyclopyrrolones
N05CG	Imidazopyridines
N05CM	Other hypnotics and sedatives
N06	Psychoanaleptics
N06A	Antidepressants
N06AA	Non-selective monoamine reuptake inhibitors
N06AB	Selective serotonin reuptake inhibitors
N06AE	Monocyclic derivatives
N06AF	Monoamine oxidase inhibitors, nonselective
N06AG	Monoamine oxidase type A inhibitors
N06AX	Other antidepressants
N06B	Psychostimulants and nootropics
N06BA	Centrally acting sympathomimetics
N06BC	Xanthine derivatives
N06BX	Other psychostimulants and nootropics
N07	Other nervous system drugs
N07A	Parasympathomimetics
N07AA	Anticholinesterases
N07AB	Choline esters
N07AX	Other parasympathomimetics
N07CA	Antivertigo preparations
N07X	Other nervous system drugs
N07XX	Other nervous system drugs
P01AA	Hydroxyquinoline derivatives
P01AB	Nitroimidazole derivatives
P01AC	Dichloroacetamide derivatives
P01AX	Other agents against amoebiasis and other protozoal diseases
P01BA	Aminoquinolines
P01BB	Biguanides
P01BC	Quinine alkaloids
P01BD	Diaminopyrimidines
P01BX	Other antimalarials
P01C	Agents against Leishmaniasis and Trypanosomiasis
P01CC	Nitrofurane derivatives
P01CD	Arsenic compounds
P01CX	Other agents against leishmaniasis and trypanosomiasis
P02	Anthelmintics
P02BA	Quinoline derivatives and related substances
P02BX	Other antitrematodal agents
P02CA	Benzimidazole derivatives
P02CB	Piperazine and derivatives
P02CC	Tetrahydropyrimidine derivatives
P02CE	Imidazothiazole derivatives
P02CX	Other antinematodals
P02DA	Salicylic acid derivatives
P02DX	Other anticestodals
P03A	Ectoparasiticides, incl. scabicides
P03AA	Sulphur containing products
P03AX	Other ectoparasiticides, incl. scabicides
R01A	Decongestants and other nasal preparations for topical use
R01AA	Sympathomimetics, plain
R01AB	Sympathomimetics, combinations excl. corticosteroids
R01AC	Antiallergic agents, excl. corticosteroids
R01AD	Corticosteroids
R01AX	Other nasal preparations
R01BA	Sympathomimetics
R02AA	Antiseptics
R02AB	Antibiotics
R02AD	Anesthetics, local
R03	Anti-asthmatics
R03A	Adrenergics, inhalants
R03AA	α and β-adrenoceptor agonists
R03AB	Non-selective β-adrenoceptor agonists
R03AC	Selective β-2-adrenoceptor agonists
R03AK	Adrenergics and other anti-asthmatics
R03B	Other anti-asthmatics, inhalants
R03BA	Glucocorticoids

R03BB	Anticholinergics
R03BC	Antiallergic agents, excl. corticosteroids
R03BX	Other anti-asthmatics, inhalants
R03CA	α and β-adrenoceptor agonists
R03CB	Non-selective β-adrenoceptor agonists
R03CC	Selective β-2-adrenoceptor agonists
R03D	Other anti-asthmatics for systemic use
R03DA	Xanthines
R03DC	Leukotriene receptor antagonists
R03DX	Other anti-asthmatics for systemic use
R05CA	Expectorants
R05CB	Mucolytics
R05DA	Opium alkaloids and derivatives
R05DB	Other cough suppressants
R06	Antihistamines for systemic use
R06A	Antihistamines for systemic use
R06AA	Aminoalkyl ethers
R06AB	Substituted alkylamines
R06AC	Substituted ethylene diamines
R06AD	Phenothiazine derivatives
R06AE	Piperazine derivatives
R06AX	Other antihistamines for systemic use
R07A	Other respiratory system products
R07AA	Lung surfactants
R07AB	Respiratory stimulants
S01AA	Antibiotics
S01AB	Sulfonamides
S01AD	Antivirals
S01AX	Other antiinfectives
S01B	Antiinflammatory agents
S01BA	Corticosteroids, plain
S01BC	Antiinflammatory agents, non-steroids
S01CA	Corticosteroids and antiinfectives in combination
S01CB	Corticosteroids, antiinfectives and mydriatics in combination
S01EA	Sympathomimetics in glaucoma therapy
S01EB	Parasympathomimetics
S01EC	Carbonic anhydrase inhibitors
S01ED	β blocking agents
S01EX	Other antiglaucoma preparations
S01FA	Anticholinergics
S01FB	Sympathomimetics excl. antiglaucoma preparations
S01GA	Sympathomimetics used as decongestants
S01GX	Other antiallergics
S01HA	Local anesthetics
S01JA	Colouring agents
S01KK	Other surgical acids
S01XA	Other ophthalmologicals
S02	otologicals
S02AA	Antiinfectives
S02BA	Corticosteroids
S02CA	Corticosteroids and antiinfectives in combination
S02DA	Analgesics and anesthetics
S03	Ophthalmological and otological preparations
S03AA	Antiinfectives
S03BA	Corticosteroids
S03CA	Corticosteroids and antiinfectives in combination
V03A	All other therapeutic products
V03AA	Drugs for treatment of chronic alcoholism
V03AB	Antidotes
V03AC	Iron chelating agents
V03AF	Detoxifying agents for antineoplastic treatment
V03AG	Drugs for treatment of hypercalcemia
V03AH	Drugs for treatment of hypoglycemia
V04CA	Tests for diabetes
V04CC	Tests for bile duct patency
V04CD	Tests for pituitary function
V04CG	Tests for gastric secretion
V04CJ	Tests for thyreoidea function
V04CK	Tests for pancreatic function
V08	Contrast media
V08AA	Watersoluble, nephrotropic, high osmolar X-ray contrast media
V08AB	Watersoluble, nephrotropic, low osmolar X-ray contrast media
V08AC	Watersoluble, hepatotropic X-ray contrast media
V08AD	Non-watersoluble X-ray contrast media
V09D	Hepatic and reticulo endothelial system

Abacavir
(1592U89)

ATC: J05AF06
Use: antiviral, anti HIV, reverse transcriptase inhibitor

RN: 136470-78-5   MF: $C_{14}H_{18}N_6O$ MW: 286.34
CN: (1S,4R)-4-[2-Amino-6-(cyclopropylamino)-9H-purin-9-yl]-2-cyclopentene-1-methanol

succinate
RN: 168146-84-7   MF: $C_{14}H_{18}N_6O \cdot C_4H_6O$ MW: 356.43
sulfate
RN: 188062-50-2   MF: $C_{14}H_{18}N_6O \cdot 1/2H_2SO_4$ MW: 670.76

(a)

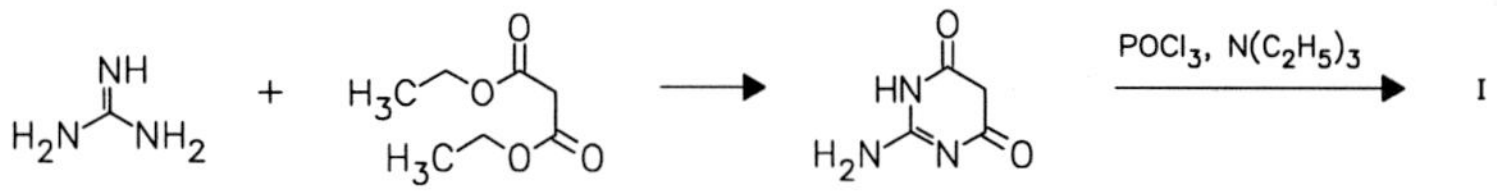

guanidine + diethyl malonate → 2-amino-4,6-pyrimidinedione

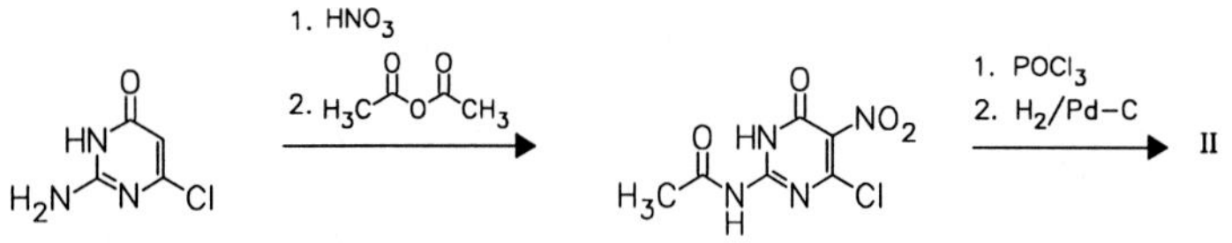

$POCl_3$, $N(C_2H_5)_3$ → I

2-amino-6-chloro-4(3H)-pyrimidinone (I)

1. HNO_3
2. H_3C—O—CH₃ (acetic anhydride)

2-acetylamino-6-chloro-5-nitro-4(3H)-pyrimidinone

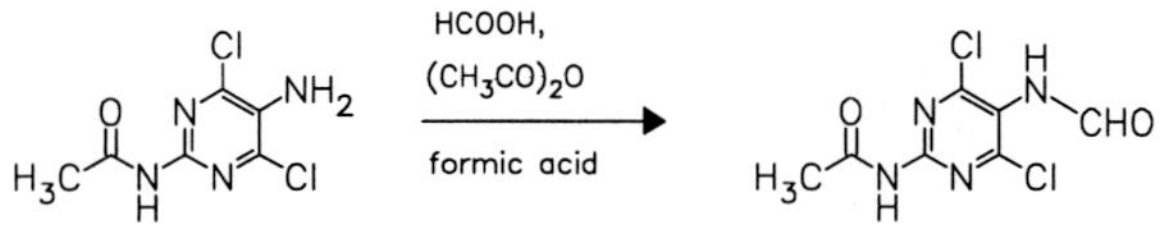

1. $POCl_3$
2. H_2/Pd–C → II

N-(5-amino-4,6-dichloropyrimidin-2-yl)acetamide (II)

HCOOH, $(CH_3CO)_2O$
formic acid

N-[4,6-dichloro-5-(formylamino)-2-pyrimidinyl]acetamide (III)

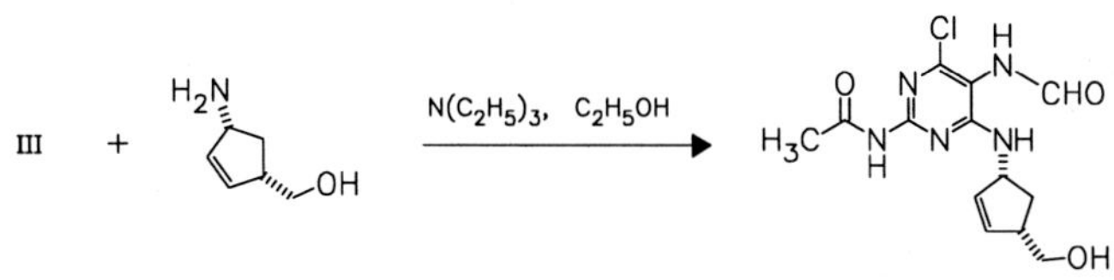

III + (1S-cis)-4-amino-2-cyclopentene-1-methanol (IV)

$N(C_2H_5)_3$, C_2H_5OH

(V)

V 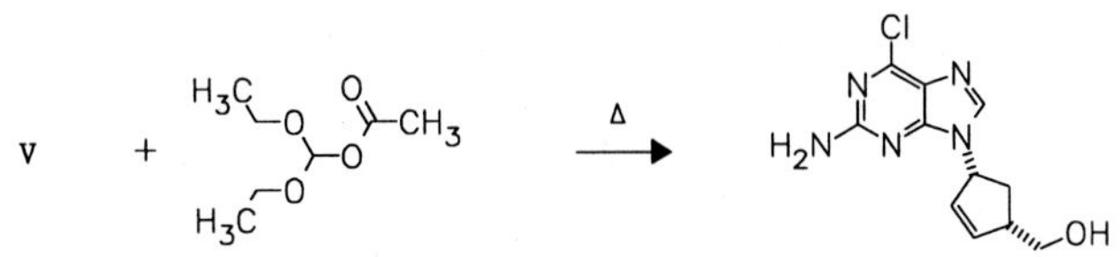 + acetic acid diethoxymethyl ester (VI)

Δ

VII

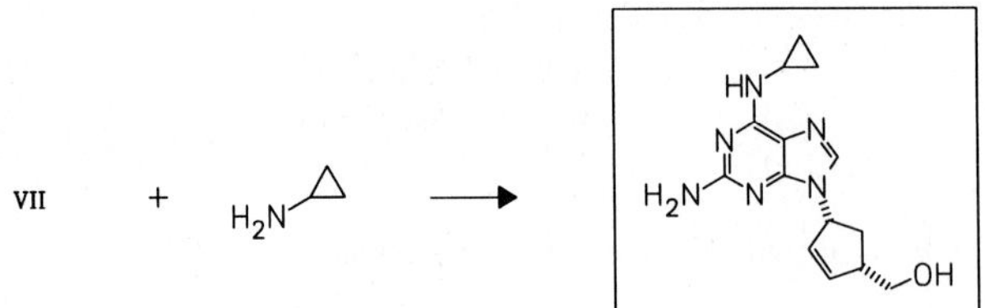

VII   +   H₂N—◁   →   Abacavir

cyclopropyl-
amine (VIII)

(aa) synthesis of (1S-cis)-4-amino-2-cyclopentene-1-methanol (IV)

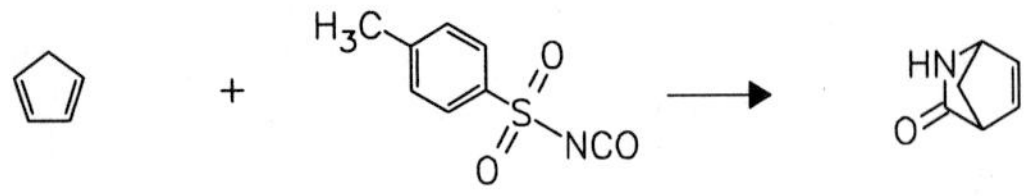

◇   +   H₃C—⬡—S(=O)(=O)—NCO   →   2-azabicyclo-

cyclo-   4-methylbenzene-   2-azabicyclo-
pentadiene   sulfonyl isocyanate   [2.2.1]hept-5-en-
                                              3-one (IX)

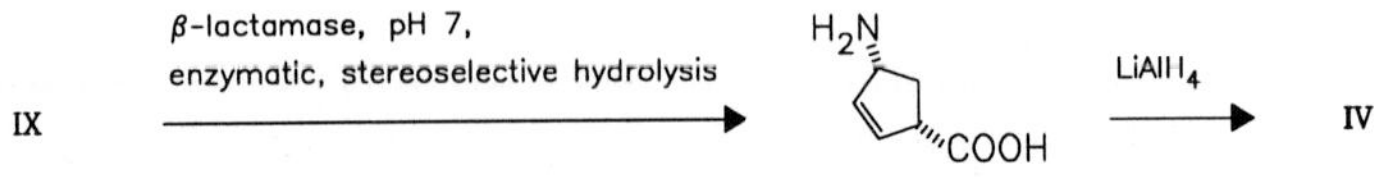

IX   —(β-lactamase, pH 7, enzymatic, stereoselective hydrolysis)→   cis-4-amino-2-cyclopentene-1-carboxylic acid   —(LiAlI₄)→   IV

cis-4-amino-2-
cyclopentene-1-
carboxylic acid

(b)

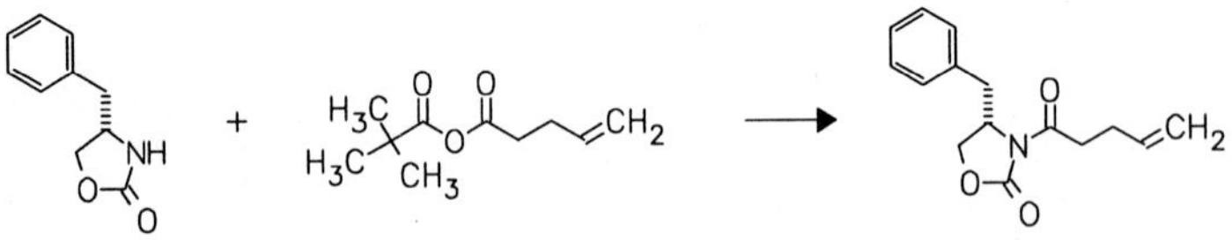

4(S)-benzyl-   +   4-pentenoic pivalic   →   (X)
oxazolidin-2-one   anhydride

X   +   acrolein   —(Bu₂BOTf, dibutylboryl triflate)→   (XI)

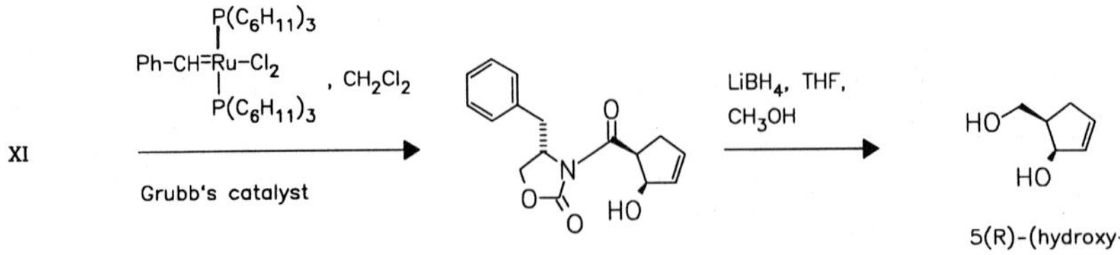

XI   —(Ph-CH=Ru-Cl₂, P(C₆H₁₁)₃, P(C₆H₁₁)₃, CH₂Cl₂, Grubb's catalyst)→   —(LiBH₄, THF, CH₃OH)→   5(R)-(hydroxymethyl)-2-cyclopenten-1(R)-ol (XII)

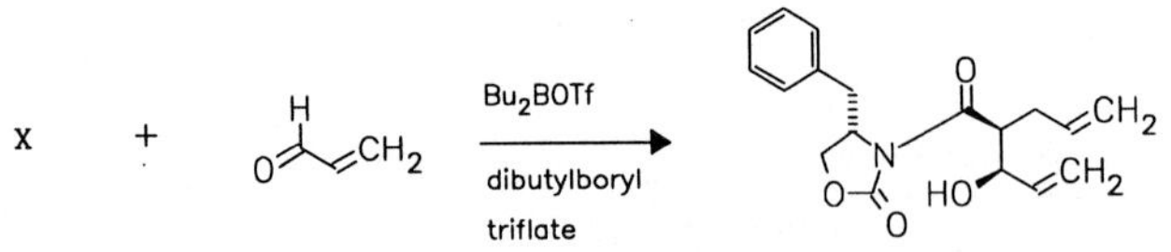

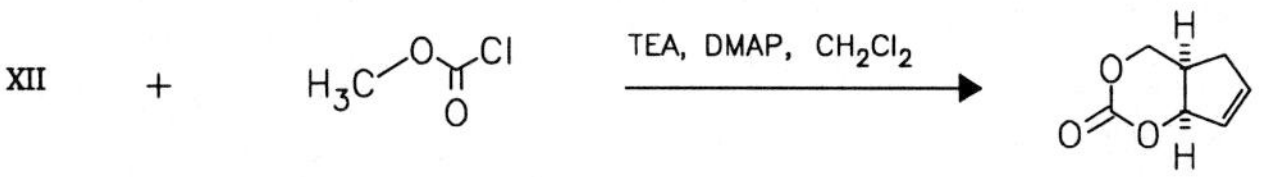

XII + H₃C—O—C(=O)—Cl   TEA, DMAP, CH₂Cl₂ → (XIII)

methyl
chloroformate

XIII + 2-amino-6-chloropurine (cf. famciclovir)   NaH, THF, DMSO, Pd(PPh₃)₄   tetrakis(triphenyl-phosphine)palladium → VII   VIII → Abacavir

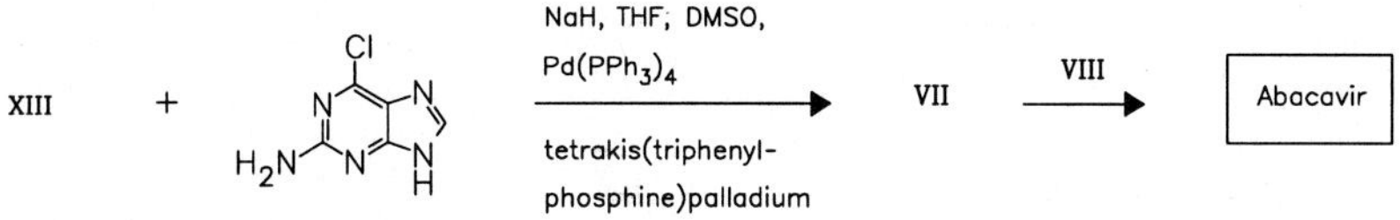

c

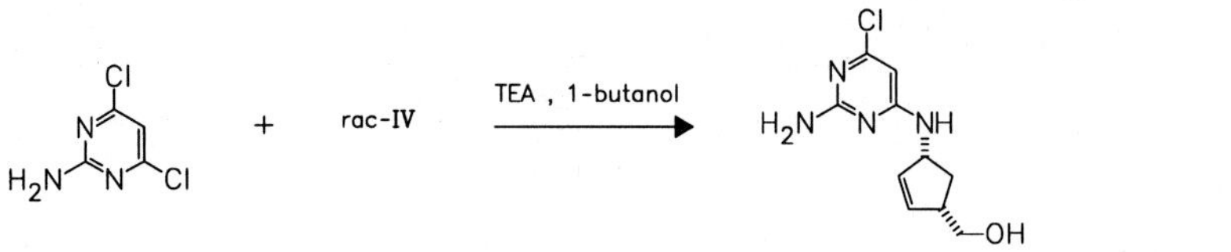

2-amino-4,6-dichloropyrimidine + rac-IV   TEA , 1-butanol → (±)-cis-4-[(2-amino-4-chloro-6-pyrimidinyl)-amino]-2-cyclopentene-1-methanol (XIV)

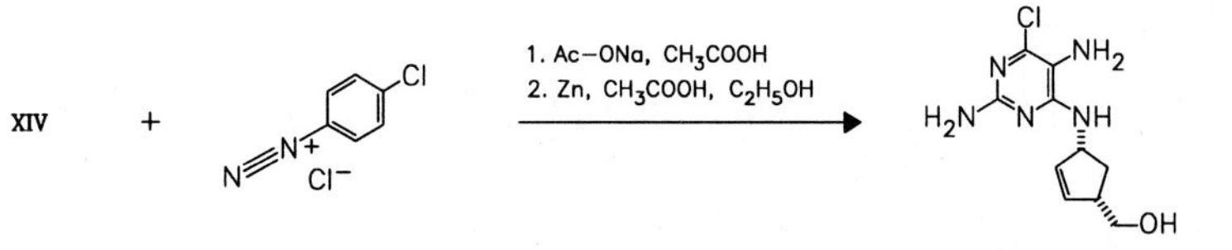

XIV + 4-chlorobenzene-diazonium chloride   1. Ac—ONa, CH₃COOH   2. Zn, CH₃COOH, C₂H₅OH → (±)-cis-4-[(2,5-diamino-4-chloro-6-pyrimidinyl)-amino]-2-cyclopentene-1-methanol (XV)

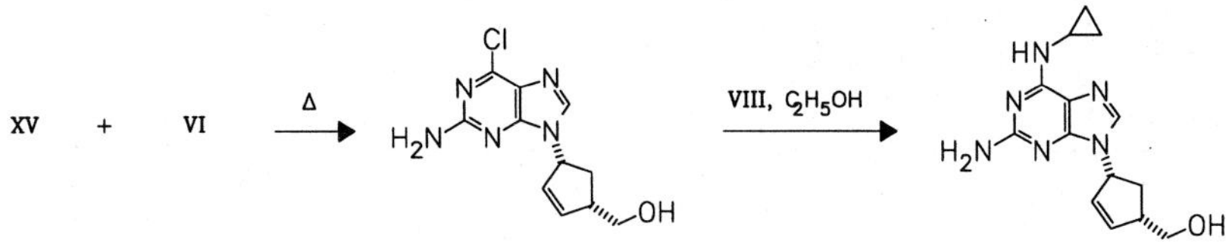

XV + VI   Δ →   VIII, C₂H₅OH → (XVI)

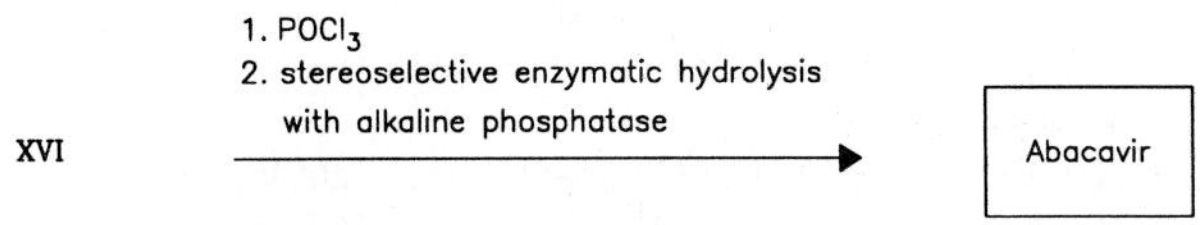

XVI   1. POCl₃   2. stereoselective enzymatic hydrolysis with alkaline phosphatase → Abacavir

(d)

2-amino-6-
chloropurine

VIII

2-amino-6-
(cyclopropyl-
amino)purine

XIII, CsCO₃,
Pd(PPh₃)₄, DMSO

cesium carbonate,
tetrakis(triphenyl-
phosphine)palladium

Abacavir

Reference(s):
a EP 434 450 (Wellcome Found.; 26.6.1991; appl. 21.12.1990; USA-prior. 22.12.1989).
   Crimmins, M.T. et al.: J. Org. Chem. (JOCEAH) **61** 4192 (1996).
aa EP 424 064 (Enzymatix; appl. 24.4.1991; GB-prior. 16.10.1989).
b Olivo, H.F. et al.: J. Chem. Soc., Perkin Trans. 1 (JCPRB4) **1998**, 391.
c US 5 034 394 (Welcome Found.; 23.7.1991; appl. 22.12.1989; GB-prior. 27.6.1988).
d WO 9 924 431 (Glaxo; appl. 12.11.1998; WO-prior. 12.11.1997).

alternative syntheses:
EP 878 548 (Lonza; appl. 13.5.1998; CH-prior. 13.5.1997).

condensation of pyrimidines *with* cyclopentylamine IV:
Vince, R.; Hua, M.: J. Med. Chem. (JMCMAR) **33** (1), 17 (1990).
EP 349 242 (Wellcome Found.; appl. 26.6.1989; GB-prior. 27.6.1988).
EP 366 385 (Wellcome Found.; appl. 23.10.1989; GB-prior. 24.10.1988).
Grumam, A. et al.: Tetrahedron Lett. (TELEAY) **36** (42), 7767 (1995).
JP 1 022 853 (Asahi Glass Co.; appl. 17.7.1987).

alternative preparation of 4-amino-2-cyclopentene-1-methanol:
EP 926 131 (Lonza; appl. 24.11.1998; CH-prior. 27.11.1997).
WO 9 745 529 (Lonza; appl. 30.5.1997; CH-prior. 30.5.1996).

abacavir succinate *as antiviral agent*:
WO 9 606 844 (Wellcome; 7.3.1996; appl. 25.8.1995; GB-prior. 26.8.1994).

synergistic combinations for treatment of HIV infection:
WO 9 630 025 (Wellcome; 3.10.1996; appl. 28.3.1996; GB-prior. 30.3.1995).

Formulation(s):   oral sol. 20 mg/ml; tabl. 300 mg (as sulfate)

Trade Name(s):
D: Ziagen (Glaxo Wellcome;   USA: Ziagen (Glaxo Wellcome)
      1999)

Abciximab

(7E3; C7E3; C7E3 Fab; C7E3-F(ab')2)

ATC: B01AC13
Use: platelet antiaggregation inhibitor,
      antianginal, GPIIb/IIIa-receptor
      antagonist

RN: 143653-53-6   MF: unspecified   MW: unspecified
CN: immunoglobulin G (human-mouse monoclonal c7E3 clone p7E3V$_H$hCγ_4 Fab fragment antihuman
      glycoprotein IIb/IIIa receptor), disulfide with human-mouse monoclonal c7E3 clone p7E3V$_K$hC$_K$ light
      chain

Reference(s):
Gold, H.K. et al.: Circulation Suppl. (CISUAQ) **80**(4) (1989), Abst. 1063.

Formulation(s):   vial 10 mg/5 ml

Trade Name(s):
D: ReoPro (Lilly)   GB: Reopro (Lilly)
F: ReoPro (Lilly)   USA: ReoPro (Lilly)

Acamprosate calcium

ATC: V03AA
Use: alcohol deterrent

RN: 77337-73-6   MF: $C_{10}H_{20}CaN_2O_8S_2$ MW: 400.49   EINECS: 278-665-3
LD$_{50}$: >10 g/kg (M, p.o.)
CN: 3-(acetylamino)-1-propanesulfonic acid calcium salt (2:1)

free acid
RN: 77337-76-9   MF: $C_5H_{11}NO_4S$ MW: 181.21   EINECS: 278-667-4

3-amino-1-   3-aminopropane-   Acamprosate calcium
propanol   1-sulfonic acid

Reference(s):
DE 3 019 350 (Lab. Meram; appl. 21.5.1980; F-prior. 23.5.1979).

synthesis of 3-aminopropane-1-sulfonic acid:
JP 46 002 012 (Kowa; appl. 19.1.1971).
Fujii, A. et al.: J. Med. Chem. (JMCMAR) **18**, 502 (1975).
WO 8 400 958 (Mitsui; appl. 15.3.1984; J-prior. 7.9.1982, 19.7.1983, 8.9.1982).

Formulation(s):   tabl. 333 mg

Trade Name(s):
D: Campral (Lipha)   F: Aotal (Meram)   GB: Campral (Lipha)

Acarbose
(Bay-g-5421)

ATC: A10BF01
Use: antidiabetic, α-glucosidase inhibitor,
         hypoglycemic

RN: 56180-94-0   MF: $C_{25}H_{43}NO_{18}$ MW: 645.61   EINECS: 260-030-7
LD$_{50}$: >500.000 SIE/kg (M, i.v.); >1000.000 SIE/kg (M, p.o.);
         478.000 SIE/kg (R, i.v.); >1000.000 SIE/kg (R, p.o.)
         65.000 SIE = 1g (SIE = saccharase inhibitory units)
CN: [1S-(1α,4α,5β,6α)]-O-4,6-dideoxy-4-[[4,5,6-trihydroxy-3-(hydroxymethyl)-2-cyclohexen-1-yl]amino]-
        α-D-glucopyranosyl(1→4)-O-α-D-glucopyranosyl-(1→4)-D-glucose

Acarbose

Fermentation of *Actinoplanes* SE50/110.

Reference(s):
US 4 062 950 (Bayer; 13.12.1977; D-prior. 22.9.1973).
DOS 2 347 782 (Bayer; appl. 21.9.1973).
Schmidt, D.D. et al.: Naturwissenschaften (NATWAY) **64**, 535 (1977).

total synthesis:
Ogawa, S.; Shibata, Y.: Chem. Commun. (CCOMA8) **1988**, 605.

review:
Tschesche, H. in Arzneimittel, Fortschritte 1972-1985 (Ed. A. Kleemann, E. Lindner, J. Engel), p. 87, VCH
Verlagsgesellschaft, Weinheim 1987.

Formulation(s):   tabl. 50 mg, 100 mg

Trade Name(s):
D:	Glucobay (Bayer; 1990)	GB:	Glucobay (Bayer)	USA:	Precose (Bayer)
F:	Glucor (Bayer)	J:	Glucobay (Bayer)		

Acebutolol

ATC: C07AB04; C07BB04
Use: β-adrenergic receptor blocker

RN: 37517-30-9   MF: $C_{18}H_{28}N_2O_4$ MW: 336.43   EINECS: 253-539-0
LD_{50}: 75.2 mg/kg (M, i.v.);
        4 mg/kg (dog, i.v.)
CN: (±)-*N*-[3-acetyl-4-[2-hydroxy-3-[(1-methylethyl)amino]propoxy]phenyl]butanamide

(*R*)-base
RN: 68107-81-3   MF: $C_{18}H_{28}N_2O_4$ MW: 336.43
(*S*)-base
RN: 68107-82-4   MF: $C_{18}H_{28}N_2O_4$ MW: 336.43
(*RS*)-monohydrochloride
RN: 34381-68-5   MF: $C_{18}H_{28}N_2O_4 \cdot HCl$ MW: 372.89   EINECS: 251-980-3
LD_{50}: 185 mg/kg (M, i.p.); 53 mg/kg (M, i.v.); 4050 mg/kg (M, p.o.); 291 mg/kg (M, s.c.);
        222 mg/kg (R, i.p.); 103 mg/kg (R, i.v.); 6620 mg/kg (R, p.o.); 1310 mg/kg (R, s.c.);
        41 mg/kg (rabbit, i.v.); 296 mg/kg (rabbit, p.o.)

butyric anhydride   4-aminophenol   4-butyramidophenol   acetyl chloride

O-acetyl-4-butyramidophenol (I)

2-acetyl-4-butyramidophenol

O-(2-oxiranylmethyl)-
2-acetyl-4-butyramidophenol (II)

Acebutolol

Reference(s):
GB 1 247 384 (May & Baker; appl. 22.12.1967).
DAS 1 815 808 (May & Baker; appl. 19.12.1968; GB-prior. 22.12.1967, 14.5.1968, 2.8.1968).
US 3 726 919 (May & Baker; appl. 19.12.1968; GB-prior. 22.12.1967, 14.5.1968, 2.8.1968).
US 3 857 952 (May & Baker; appl. 3.8.1972).

preparation of 4-butyramidophenol:
Kuhn; Koehler; Koehler: Hoppe-Seyler's Z. Physiol. Chem. (HSZPAZ) **247**, 197, 216 (1937).
Verma, K.K.; Tyagi, P.: Anal. Chem. (ANCHAM) **56** (12), 2157 (1984).
US 2 824 838 (Esso Research & Eng. Co.; 25.2.1958; appl. 13.1.1955).

Formulation(s): amp. 25 mg; tabl. 200 mg, 400 mg (as hydrochloride)

Trade Name(s):

D:	Prent (Bayer; 1977)		Sectral (Rhône-Poulenc	J:	Acetanol (Rhodia; 1984)
	Sali-Prent (Bayer; 1982)-		Rorer; 1975)		Sectral (Kanebo; 1981)
	comb.	I:	Acecor (SPA)	USA:	Sectral (Wyeth-Ayerst;
	Tredalat (Bayer)-comb.		Alol (SIT)		1985)
F:	Sectral (Specia; 1976)		Prent (Bayropharm; 1981)		
GB:	Secadrex (Rhône-Poulenc		Sectral (Rhône-Poulenc		
	Rorer; 1982)-comb.		Rorer; 1980)		

Acecarbromal

(Acetylcarbromal; Acetcarbromal)

ATC: N05CM
Use: sedative, hypnotic

RN: 77-66-7 MF: $C_9H_{15}BrN_2O_3$ MW: 279.13 EINECS: 201-047-1
LD_{50}: 1600 mg/kg (M, p.o.)
CN: *N*-[(acetylamino)carbonyl]-2-bromo-2-ethylbutanamide

carbromal
(q. v.)

acetic anhydride

Acecarbromal

Reference(s):
DRP 225 710 (Bayer; 1910).

alternative syntheses:
DRP 286 760 (Bayer; 1913).
DRP 327 129 (Bayer; 1917).

Formulation(s):   drg. 100 mg

Trade Name(s):
D: Abasin (Bayer); wfm   USA: Carbased (Mallard); wfm
      Afrodor (Farco-Pharma)   Sedamyl (Riker); wfm

Aceclidine

ATC: S01EB08; S01EB58
Use: antiglaucoma, miotic

RN: 827-61-2   MF: $C_9H_{15}NO_2$ MW: 169.22   EINECS: 212-574-1
LD$_{50}$:   78 mg/kg (M, i.p.); 36 mg/kg (M, i.v.); 165 mg/kg (M, p.o.); 102 mg/kg (M, s.c.);
       45 mg/kg (R, i.v.); 225 mg/kg (R, s.c.)
CN: 1-azabicyclo[2.2.2]octan-3-ol acetate (ester)

hydrochloride
RN: 6109-70-2   MF: $C_9H_{15}NO_2 \cdot HCl$ MW: 205.69   EINECS: 228-071-5
LD$_{50}$:   27 mg/kg (M, i.v.); 165 mg/kg (M, p.o.);
       45 mg/kg (R, i.v.)
salicylate (1:1)
RN: 6821-59-6   MF: $C_9H_{15}NO_2 \cdot C_7H_6O_3$ MW: 307.35
LD$_{50}$:   113 mg/kg (M, s.c.)

3-hydroxy-
quinuclidine
(cf. clidinium
bromide synthesis)

acetic anhydride

Aceclidine

Reference(s):
US 2 648 667 (Roche; 1953; prior. 1951).
Grob, C.A. et al.: Helv. Chim. Acta (HCACAV) **40**, 2170 (1957).

Formulation(s):   eye drops 200 mg (as hydrochloride), 20 mg

Trade Name(s):
D: Glaucotat (Chibret)   Glaucostat (Merck Sharp &
F: Glaucadrine (Merck Sharp   Dohme-Chibret)
      & Dohme-Chibret)-comb.   I: Glaunorm (Farmigea)

Aceclofenac

ATC: M01AB16
Use: non-steroidal anti-inflammatory,
        analgesic, antipyretic, prostaglandin
        synthesis inhibitor

RN: 89796-99-6   MF: $C_{16}H_{13}Cl_2NO_4$ MW: 354.19
LD$_{50}$:   121 mg/kg (M, p.o.)
CN: 2-[(2,6-dichlorophenyl)amino]benzeneacetic acid carboxymethyl ester

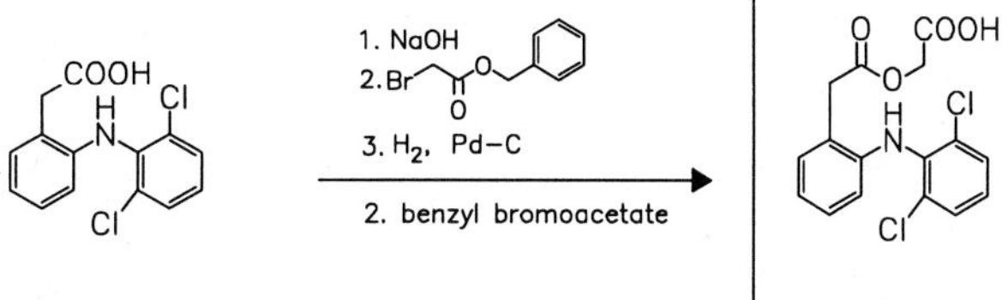

diclofenac
(q. v.)

Aceclofenac

Reference(s):
EP 119 932 (Prodes; appl. 19.3.1984; E-prior. 21.3.1983).
US 4 548 952 (Prodes; 22.10.1985; appl. 15.3.1984; E-prior. 21.3.1983).

alternative synthesis:
ES 2 020 146 (Prodesfarma; appl. 29.5.1990).

Formulation(s):   cream 1.5 %; vial 150 mg; tabl. 100 mg

Trade Name(s):
GB: Preservex (Bristol-Myers
       Squibb; 1992)

Acediasulfone

ATC: S02
Use: antibacterial, cytotoxic agent

RN: 80-03-5   MF: $C_{14}H_{14}N_2O_4S$ MW: 306.34   EINECS: 201-243-7
CN: N-[4-[(4-aminophenyl)sulfonyl]phenyl]glycine

monosodium salt
RN: 127-60-6   MF: $C_{14}H_{13}N_2NaO_4S$ MW: 328.32   EINECS: 204-852-6

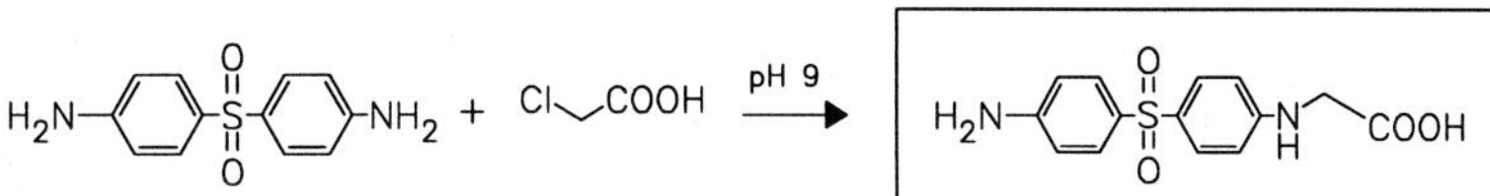

dapsone
(q. v.)

chloroacetic
acid

Acediasulfone

Reference(s):
CH 254 803 (Cilag; appl. 1946).
CH 278 482 (Cilag; appl. 1949).
US 2 589 211 (Parke Davis; 1952; appl. 1948).
US 2 454 835 (Parke Davis; 1948; prior. 1943).
US 2 751 382 (Cilag; 1956; D-prior. 6.7.1953).

Trade Name(s):
D: Ciloprin (Cilag-Chemie)-
      comb.; wfm

Acefylline

ATC: R03B
Use: cardiotonic, diuretic, antispasmodic, bronchodilator

RN: 652-37-9 MF: $C_9H_{10}N_4O_4$ MW: 238.20 EINECS: 211-490-2
LD$_{50}$: 1180 mg/kg (M, i.p.); 2733 mg/kg (M, p.o.)
CN: 1,2,3,6-tetrahydro-1,3-dimethyl-2,6-dioxo-7H-purine-7-acetic acid

theophylline + chloroacetic acid → Acefylline (I)

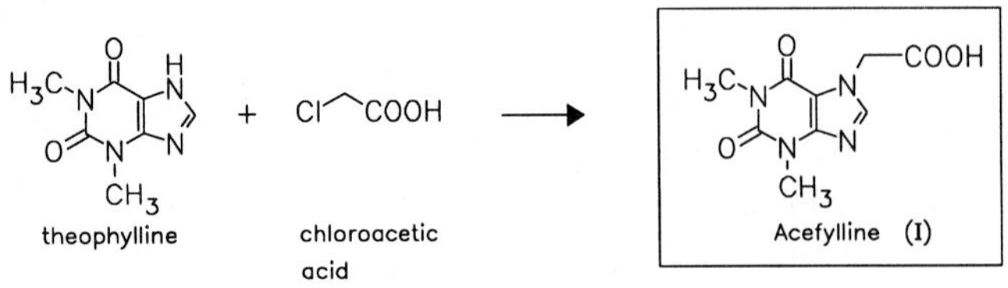

Acepifylline

RN: 18833-13-1 MF: $C_9H_{10}N_4O_4 \cdot xC_4H_{10}N_2$ MW: unspecified EINECS: 242-614-3
CN: 1,2,3,6-tetrahydro-1,3-dimethyl-2,6-dioxo-7H-purine-7-acetic acid compd. with piperazine

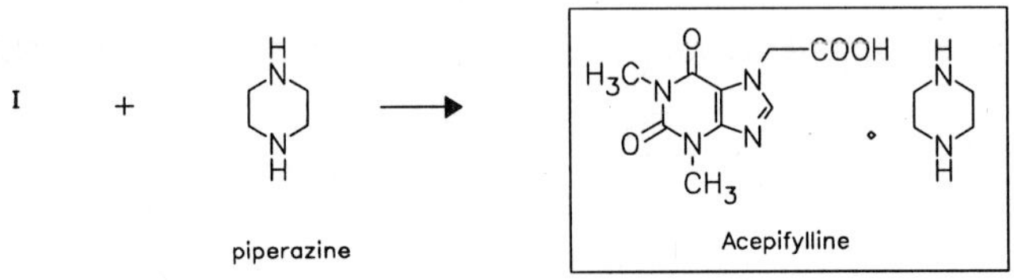

I + piperazine → Acepifylline

Acefylline heptaminol

RN: 59989-20-7 MF: $C_9H_{10}N_4O_3 \cdot C_8H_{19}NO$ MW: 367.45 EINECS: 262-012-4
CN: 1,2,3,6-tetrahydro-1,3-dimethyl-2,6-dioxo-7H-purine-7-acetic acid compd. with 6-amino-2-methyl-2-heptaminol (1:1)

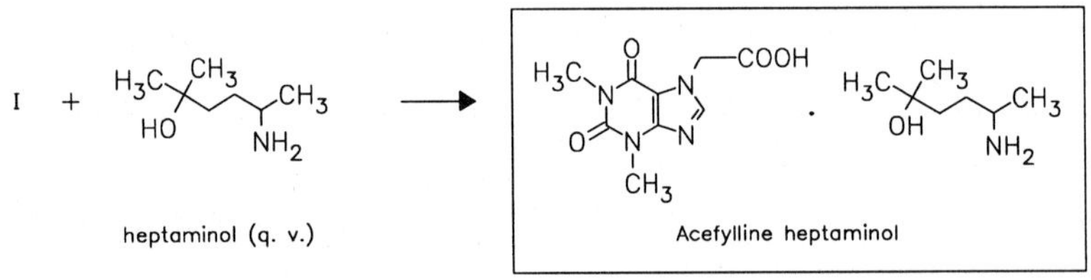

I + heptaminol (q. v.) → Acefylline heptaminol

Reference(s):
Blaisse, J.: Bull. Soc. Chim. Fr. (BSCFAS) **1949**, 769.

Formulation(s): amp. 500 mg/200 ml; drg. 250 mg; suppos. 500 mg; tabl. 250 mg (acefylline); drg. 250 mg; inj. 0.5 g; suppos. 0.5-1 g

Trade Name(s):
D: Etaphydel (Delalande; as acepifylline); wfm

F: Sureptil (Synthélabo; as acefylline-heptaminol)-comb.

GB: Etophylate (Delalande; as acepifylline); wfm
I: Sureptil (Delalande Isnardi)-comb.

Aceglutamide aluminum

ATC: A02AB; N06B
Use: peptic ulcer therapeutic

RN: 12607-92-0   MF: $C_{35}H_{59}Al_3N_{10}O_{24}$ MW: 1084.85
LD_{50}:   460 mg/kg (M, i.v.); 13.1 g/kg (M, p.o.);
    400 mg/kg (R, i.v.); >14.5 g/kg (R, p.o.)
CN: pentakis(N^2-acetyl-L-glutaminato)tetrahydroxytrialuminum

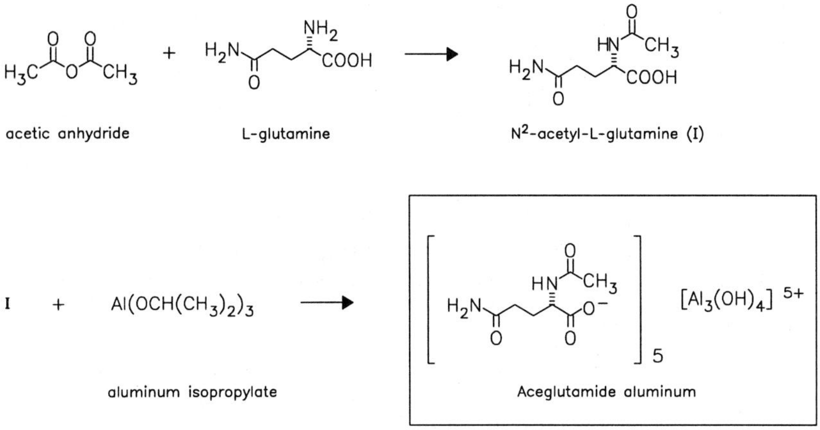

acetic anhydride   L-glutamine N^2-acetyl-L-glutamine (I)

I   + $Al(OCH(CH_3)_2)_3$

aluminum isopropylate   Aceglutamide aluminum

Reference(s):
DOS 2 127 176 (Kyowa Hakko; appl. 1.6.1971; J-prior. 5.6.1970).
US 3 787 466 (Kyowa Hakko; 22.1.1974; J-prior. 5.6.1970).

preparation of N^2-acetyl-L-glutamine:
Reddy, A.V; Ravindranath, B.: Synth. Commun. (SYNCAV) **22** (2), 257 (1992).
Synge: Biochem. J. (BIJOAK) **33**, 673 (1939).

Formulation(s):   gran. 700 mg

Trade Name(s):
J: Glumal (Kyowa Hakko)

Acemetacin

ATC: M01AB11
Use: non-steroidal anti-inflammatory

RN: 53164-05-9   MF: $C_{21}H_{18}ClNO_6$ MW: 415.83   EINECS: 258-403-4
LD_{50}:   55 mg/kg (Mm, p.o.); 18.42mg/kg (Mf, p.o.);
    24.2 mg/kg (Rm, p.o.); 30.1 mg/kg (Rf, p.o.)
CN: 1-(4-chlorobenzoyl)-5-methoxy-2-methyl-1*H*-indole-3-acetic acid carboxymethyl ester

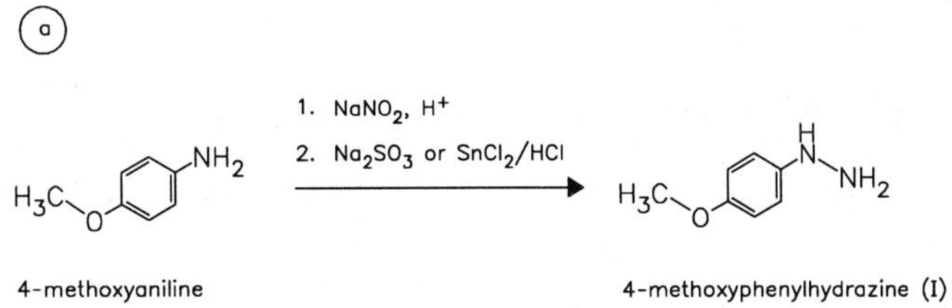

1. $NaNO_2$, H^+
2. Na_2SO_3 or $SnCl_2$/HCl

4-methoxyaniline   4-methoxyphenylhydrazine (I)

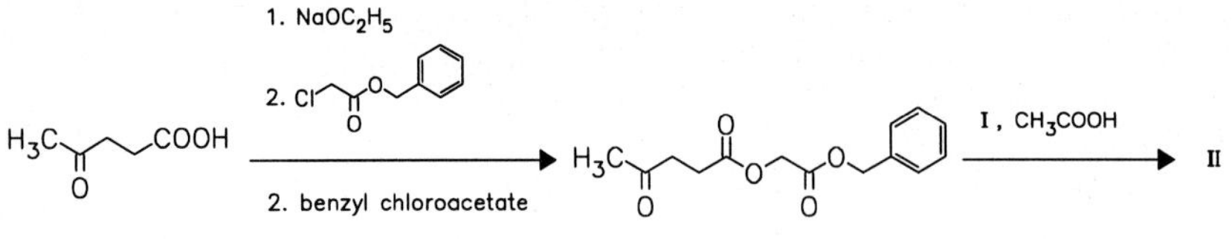

levulinic acid

benzyl levulinoyloxyacetate

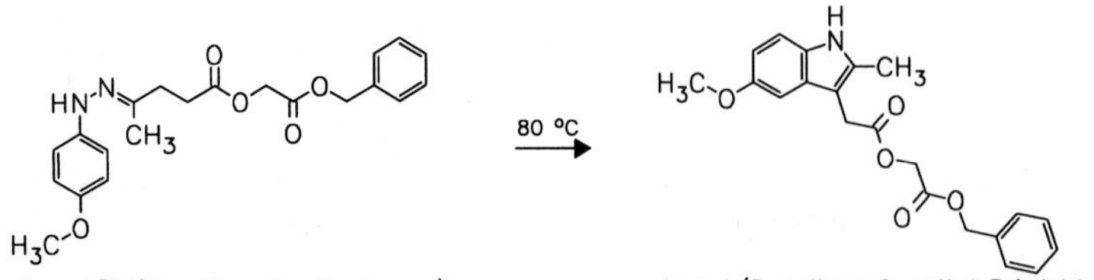

benzyl [2-(4-methoxyphenylhydrazono)-
valeryloxy]acetate (II)

benzyl (5-methoxy-2-methyl-3-indolyl-
acetoxy)acetate (III)

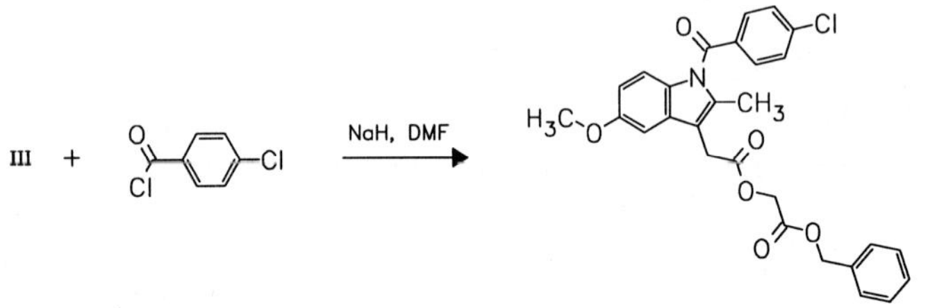

4-chloro-
benzoyl chloride

benzyl [1-(4-chlorobenzoyl)-5-methoxy-
2-methyl-3-indolylacetoxy]acetate (IV)

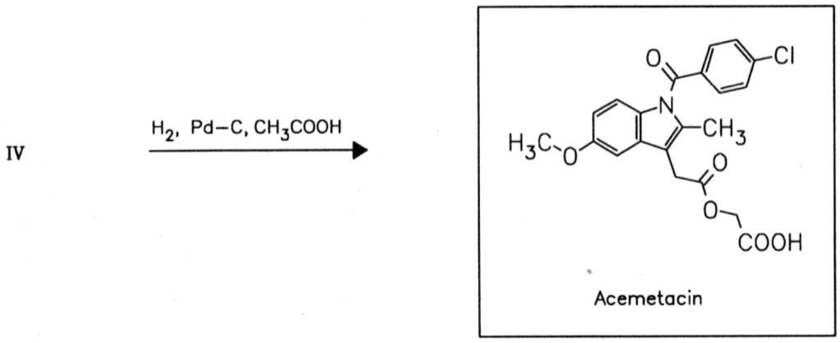

Acemetacin

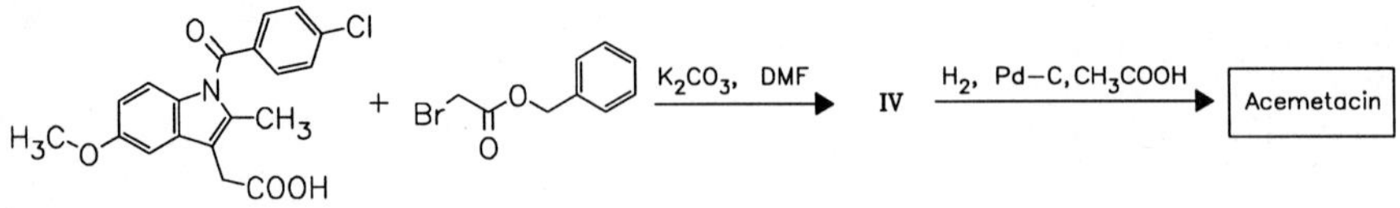

indometacin
(q. v.)

benzyl bromo-
acetate

Acemetacin

Reference(s):
DOS 2 234 651 (Tropon; appl. 14.7.1972).
FR 2 192 828 (Tropon; appl. 13.7.1973; D-prior. 14.7.1972).
US 3 910 952 (Troponwerke Dinklage; 7.10.1975; appl. 28.6.1973; D-prior. 14.7.1972).

preparation of 4-methoxyphenylhydrazine *from* 4-methoxyaniline (*p*-anisidine):
Lee, A.-R. et al.: J. Heterocycl. Chem. (JHTCAD) **32** (1), 1-12 (1995).
Clade, D.W. et al.: J. Chem. Soc., Perkin Trans. 2 (JCPKBH) , 909-916 (1982).
DE 70 459 (Riedel; 12.11.1891).
Altschul: Ber. Dtsch. Chem. Ges. (BDCGAS) **25**, 1849 (1892).

preparation of benzyl levulinoyloxyacetate:
Boltze, K.-H.; Brendler, O.; Jacobi, H.; Opitz, W.; Raddatz, S. et al.: Arzneim.-Forsch. (ARZNAD) **30** (8a), 1314-1325 (1980).

Formulation(s):   cps. 30 mg, 60 mg; s. r. cps. 90 mg

Trade Name(s):
D: Rantudil (Bayer; 1980)   I: Acemix (Bioprogress)   J: Rantudil (Kowa; 1984)
GB: Emflex (Merck)   Solar (Bioindustria)

Acenocoumarol

(Acenocumarin; Nicoumalone)

ATC: B01AA07
Use: anticoagulant

RN: 152-72-7   MF: $C_{19}H_{15}NO_6$ MW: 353.33   EINECS: 205-807-3
LD$_{50}$:   115 mg/kg (M, i.p.); 1470 mg/kg (M, p.o.);
513 mg/kg (R, p.o.)
CN: 4-hydroxy-3-[1-(4-nitrophenyl)-3-oxobutyl]-2*H*-1-benzopyran-2-one

methyl   acetic anhydride   methyl acetyl-   4-hydroxy-
salicylate   salicylate   coumarin (I)
(wintergreen oil)

4-nitro-   Acenocoumarol
benzalacetone

Reference(s):
US 2 648 862 (Geigy; 1953; CH-prior. 1950).

Formulation(s):   tabl. 1 mg, 4 mg

Trade Name(s):
D: Sintrom (Geigy); wfm   GB: Sinthrome (Geigy)   J: Sintrom (Ciba-Geigy)
F: Sintrom (Novartis)   I: Sintrom (Novartis)   USA: Sintrom (Geigy); wfm

Acepromazine

ATC: N05AA04
Use: neuroleptic, anti-emetic, tranquilizer

RN: 61-00-7   MF: $C_{19}H_{22}N_2OS$ MW: 326.46   EINECS: 200-496-0
LD$_{50}$: 59 mg/kg (M, i.v.)
CN: 1-[10-[3-(dimethylamino)propyl]-10H-phenothiazin-2-yl]ethanone

maleate (1:1)
RN: 3598-37-6   MF: $C_{19}H_{22}N_2OS \cdot C_4H_4O_4$ MW: 442.54   EINECS: 222-748-9
LD$_{50}$: 65 mg/kg (M, i.v.);
        95 mg/kg (R, i.v.); 400 mg/kg (R, p.o.)

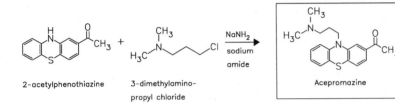

2-acetylphenothiazine   3-dimethylamino-   Acepromazine
                             propyl chloride

Reference(s):
DE 1 049 865 (Bayer; appl. 7.9.1955).
Schmitt, J. et al.: Bull. Soc. Chim. Fr. (BSCFAS) **1957**, 938, 1474.

Formulation(s):   drops 1 mg/10 drops; syrup 2.5 mg; tabl. 10 mg (as maleate)

Trade Name(s):
F: Noctran (Menarini)-comb.   J: Plebal (Fujinaga-Sankyo)-
                                           comb.

Aceprometazine

ATC: N05AA
Use: neuroleptic, antitussive

RN: 13461-01-3   MF: $C_{19}H_{22}N_2OS$ MW: 326.46   EINECS: 236-661-9
LD$_{50}$: 517 mg/kg (M, p.o.)
CN: 1-[10-[2-(dimethylamino)propyl]-10H-phenothiazin-2-yl]ethanone

maleate
RN: 7455-18-7   MF: $C_{19}H_{22}N_2OS \cdot C_4H_4O_4$ MW: 442.54

2-acetylphenothiazine   2-dimethylamino-   Aceprometazine
                             propyl chloride

Reference(s):
DE 1 049 865 (Bayer; appl. 7.9.1955).

Formulation(s):   tabl. 13.55 mg (as maleate in combination with 400 mg meprobramate)

Trade Name(s):

D: Clindorm (Midy)-comb.
F: Mépronizine (Sanofi)-comb.
Noctran (Menarini)-comb.
J: Noctran (Clin-Midy-Sanofi); wfm

Acetarsol

(Acetarsone)

ATC: A07AX02; G01AB01; P01CD02
Use: antiprotozoal (trichomonas)

RN: 97-44-9 MF: $C_8H_{10}AsNO_5$ MW: 275.09 EINECS: 202-582-3
LD_{50}: 180 mg/kg (M, i.v.); 4 mg/kg (M, p.o.)
CN: [3-(acetylamino)-4-hydroxyphenyl]arsonic acid

monosodium salt
RN: 5892-48-8 MF: $C_8H_9AsNNaO_5$ MW: 297.07 EINECS: 227-573-1

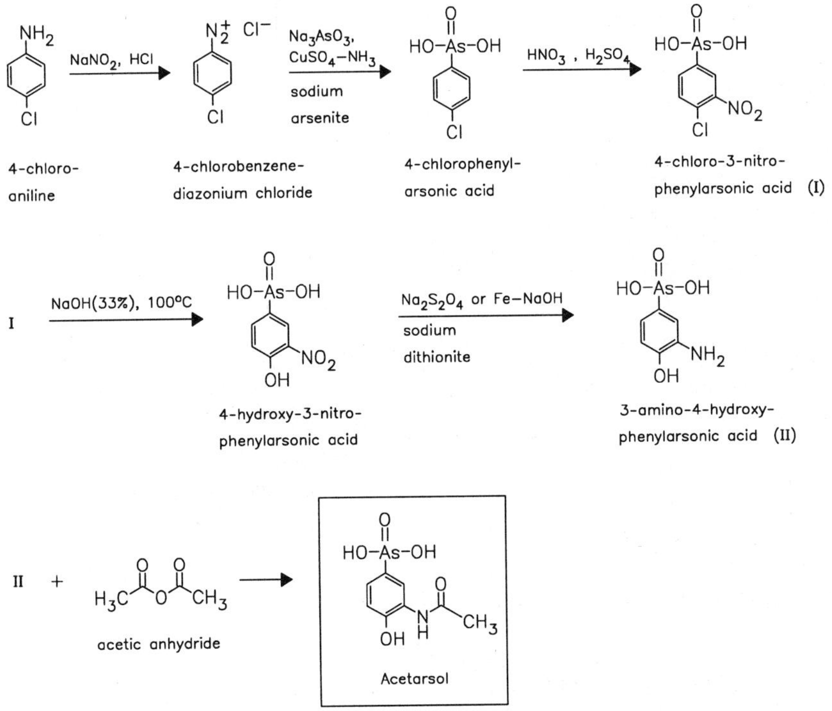

Reference(s):
Raiziss, G.W.; Gavron, J.L.: J. Am. Chem. Soc. (JACSAT) **43**, 583 (1921).
Raiziss, G.W.; Fisher, B.C.: J. Am. Chem. Soc. (JACSAT) **48**, 1323 (1926).
DRP 250 264 (H. Bart; appl. 1910).
DRP 245 536 (Hoechst; appl. 1911).
DRP 224 953 (Hoechst; appl. 1909).

Formulation(s): collutorium (mouth wash) 0.5 mg/100 g

Trade Name(s):

F: Arpha collutoire (Fournier)-comb.; wfm
Collargent acétarsol (Sarbach)-comb.; wfm
Gynoplix (Doms-Adrian); wfm
Gynoplix (Théraplix)-comb.; wfm
Humex collutoire (Fournier)-comb.; wfm
Humex Fournier collutoire (Fournier)-comb.; wfm

Polygynax (Innothéra)-
comb.; wfm
Polygynax Virgo
(Innothéra)-comb.; wfm
Pyorex (Bailly-Speab)-
comb.; wfm

Sanogyl (Pharmascience)-
comb.; wfm
Sanogyl (Vilette); wfm
GB: Pyorex (Bengue)-comb.;
wfm
S. V. C. (May & Baker)

I: Gynoplix (Vaillant)
J: Neo Osvarsan (Banyu)
Osvarsan (Banyu)

Acetazolamide
(Acetazoleamide)

ATC: S01EC01
Use: diuretic

RN: 59-66-5 MF: $C_4H_6N_4O_3S_2$ MW: 222.25 EINECS: 200-440-5
LD_{50}: 1175 mg/kg (M, i.p.); >3000 mg/kg (M, i.v.); 4300 mg/kg (M, p.o.); >3000 mg/kg (M, s.c.);
2750 mg/kg (R, i.p.);
>1500mg/kg (g. p., s.c.)
>2000 mg/kg (dog, i.v.);
CN: N-[5-(aminosulfonyl)-1,3,4-thiadiazol-2-yl]acetamide

NH_4^+ $^-$S—CN + H_2N—NH_2 ⟶ (hydrazine-1,2-bis-(thiocarboxamide)) ⟶ COCl$_2$ ⟶ 2-amino-5-mercapto-1,3,4-thiadiazole (I)

ammonium
rhodanide

hydrazine

hydrazine-1,2-bis-
(thiocarboxamide)

2-amino-5-mercapto-
1,3,4-thiadiazole (I)

I + acetic anhydride ⟶ 2-acetylamino-5-mercapto-1,3,4-thiadiazole ⟶ Cl$_2$ chlorine ⟶ 2-acetylamino-1,3,4-thiadiazole-5-sulfonyl chloride (II)

II ⟶ NH$_3$ ⟶ Acetazolamide

Reference(s):
US 2 554 816 (American Cyanamid; 1951; prior. 1950).
Roblin, R.O.; Clapp, J.W.: J. Am. Chem. Soc. (JACSAT) **72**, 4890 (1950).

similar process:
US 2 980 679 (Omikron-Gagliardi; 18.4.1961; I-prior. 4.4.1957).

Formulation(s):   amp. 500 mg; cream 10 %; lyo. 500 mg; powder 500 mg; s. r. cps. 500 mg; tabl. 125 mg,
250 mg

Trade Name(s):
D: Diamox (Lederle)
Diuramid (medpharm)
Glaupax (CIBA Vision)
F: Défiltran (Labs. Jumer)
Diamox (Théraplix)
GB: Diamox (Storz)

Diamox Sustets (Lederle);
wfm
I: Diamox (Cyanamid)
J: Acetamox (Santen)
Atenezol (Tsuruhara)
Diamox (Lederle-Takeda)

Diamox S. R. (Lederle-
Takeda)
Didoc (Sawai)
Donmox (Hotta)
Zohnox (Konto)
USA: Diamox (Lederle)

Acetiamine

ATC: A11
Use: vitamin B_1-derivative, neurotropic analgesic

RN: 299-89-8   MF: $C_{16}H_{22}N_4O_4S$ MW: 366.44
CN: ethanethioic acid S-[1-[2-(acetyloxy)ethyl]-2-[[(4-amino-2-methyl-5-pyrimidinyl)methyl]formylamino]-1-propenyl] ester

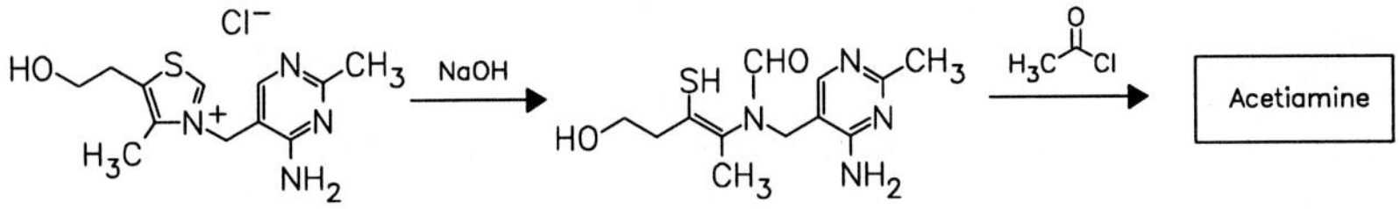

thiamine (I)
(q. v.)

(II)

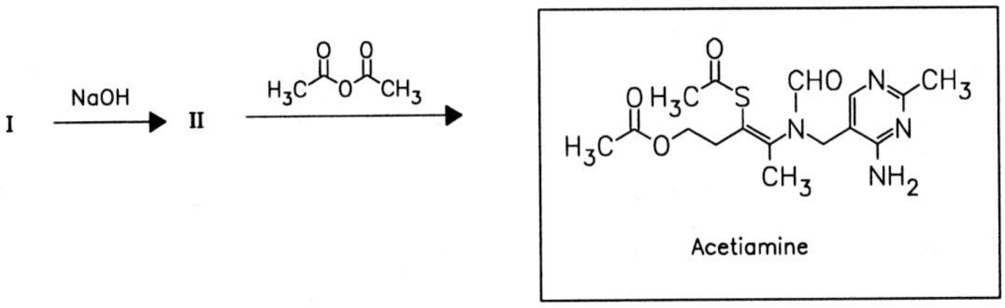

Acetiamine

Acetiamine

Reference(s):
US 2 752 348 (Takeda; 1956; J-prior. 1952).
Matsukawa, T.; Kawasaki, H.: Yakugaku Zasshi (YKKZAJ) **23**, 705 (1953).
Gauthier, B. et al.: Ann. Pharm. Fr. (APFRAD) **21**, 655 (1963).

Formulation(s):   drg. 50 mg

Trade Name(s):
D: Thianeurone (Rhône-Poulenc); wfm
F: Algo-Névriton (Pharmuka); wfm

Acetohexamide
(Cyclamide)

ATC: A10BB31
Use: antidiabetic

RN: 968-81-0   MF: $C_{15}H_{20}N_2O_4S$ MW: 324.40   EINECS: 213-530-4
LD_{50}: >2500 mg/kg (M, p.o.);
      5g/kg (R, p.o.)
CN: 4-acetyl-N-[(cyclohexylamino)carbonyl]benzenesulfonamide

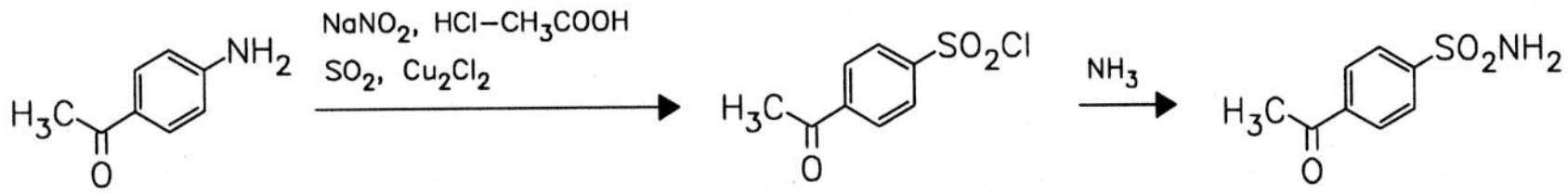

4-amino-acetophenone

4-acetylphenyl-sulfonyl chloride

4-acetylbenzene-sulfonamide (I)

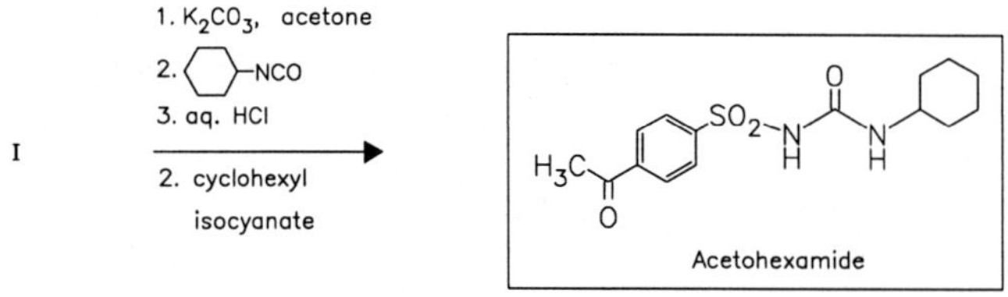

1. K$_2$CO$_3$, acetone

2. ⬡-NCO

3. aq. HCl
_____
2. cyclohexyl
isocyanate

I →

H$_3$C—⬡—SO$_2$—NH—C(=O)—NH—⬡ cyclohexyl

Acetohexamide

Reference(s):
US 3 320 312 (Lilly; 16.5.1967; prior. 28.4.1960).
DE 1 177 631 (Lilly; appl. 21.4.1961; USA-prior. 28.4.1960).
DE 1 135 891 (Hoechst; appl. 30.6.1960).

Formulation(s):   tabl. 250 mg, 500 mg

Trade Name(s):
GB: Dimelor (Lilly); wfm   J: Dimelin (Shionogi)
I: Dimelor (Lilly); wfm   USA: Dymelor (Lilly)

Acetophenazine

ATC: N05AB07
Use: neuroleptic, antipsychotic

RN: 2751-68-0   MF: C$_{23}$H$_{29}$N$_3$O$_2$S   MW: 411.57
CN: 1-[10-[3-[4-(2-hydroxyethyl)-1-piperazinyl]propyl]-10*H*-phenothiazin-2-yl]ethanone

maleate (1:2)
RN: 5714-00-1   MF: C$_{23}$H$_{29}$N$_3$O$_2$S · 2C$_4$H$_4$O$_4$ MW: 643.71   EINECS: 227-202-3
LD$_{50}$:   71 mg/kg (M, i.v.);
        60 mg/kg (R, i.p.); 39 mg/kg (R, i.v.); 415 mg/kg (R, p.o.)

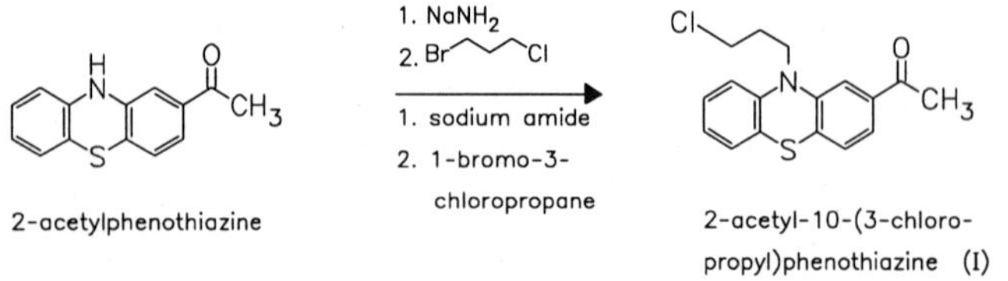

1. NaNH$_2$
2. Br⁀⁀Cl
_____
1. sodium amide
2. 1-bromo-3-
chloropropane

2-acetylphenothiazine   →   2-acetyl-10-(3-chloro-propyl)phenothiazine   (I)

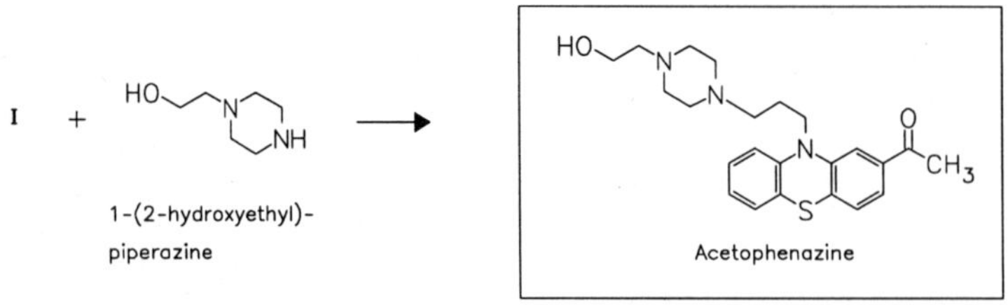

I +   1-(2-hydroxyethyl)-piperazine   →   Acetophenazine

Reference(s):
US 2 985 654 (Schering Corp.; 23.5.1961; prior. 21.9.1956).

Formulation(s):   tabl. 20 mg (as dimaleate)

Trade Name(s):
USA: Tindal (Schering); wfm

Acetorphan
(Racecadotril)

ATC: A07XA04
Use: antisecretory, enkephalinaseinhibitor

RN: 81110-73-8 MF: $C_{21}H_{23}NO_4S$ MW: 385.48
CN: (±)-*N*-[2-[(Acetylthio)methyl]-1-oxo-3-phenylpropyl]glycine phenylmethyl ester

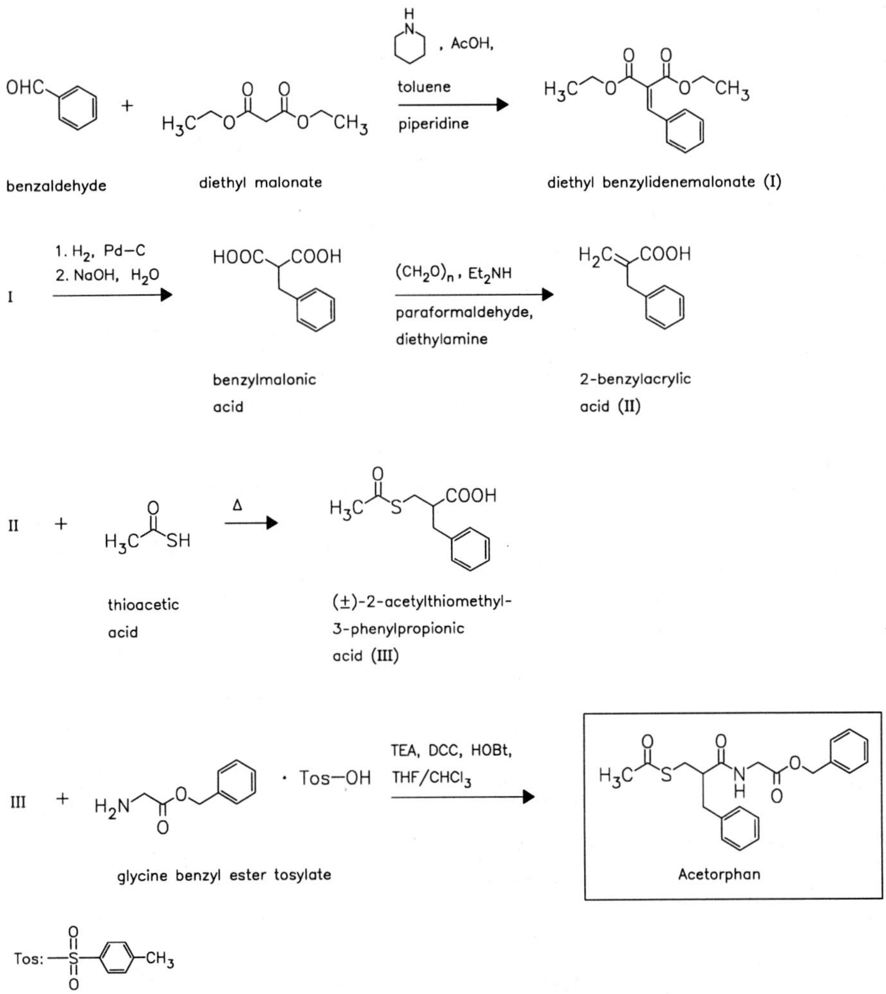

benzaldehyde
diethyl malonate
AcOH,
toluene
piperidine
diethyl benzylidenemalonate (I)
1. H2, Pd–C
2. NaOH, H2O
benzylmalonic acid
(CH2O)n, Et2NH
paraformaldehyde, diethylamine
2-benzylacrylic acid (II)
thioacetic acid
(±)-2-acetylthiomethyl-3-phenylpropionic acid (III)
· Tos–OH
TEA, DCC, HOBt, THF/CHCl3
glycine benzyl ester tosylate
Acetorphan
Tos:

Reference(s):
EP 38 758 (Roques, B. et al.; appl. 17.4.1981; F-prior. 17.4.1980).
EP 729 936 (Soc. Civile Bioprojet; appl. 1.3.1996; F-prior. 3.3.1995).

synthesis of III:
Mannich, C.; Ritsert, K.: Ber. Dtsch. Chem. Ges. (BDCGAS) **57**, 1116 (1924).

Formulation(s): cps. 100 mg

Trade Name(s):
F: Tiorfan (Bioprojet; 1993)

Acetrizoic acid

ATC: V08AA07
Use: X-ray contrast medium

RN: 85-36-9   MF: $C_9H_6I_3NO_3$ MW: 556.86   EINECS: 201-600-7
LD_{50}:   8000 mg/kg (M, i.v.); 20 g/kg (M, p.o.)
CN: 3-(acetylamino)-2,4,6-triiodobenzoic acid

meglumine salt (1:1)
RN: 22154-43-4   MF: $C_9H_6I_3NO_3 \cdot C_7H_{17}NO_5$ MW: 752.08
LD_{50}:   10.1 g/kg (M, i.v.)
sodium salt
RN: 129-63-5   MF: $C_9H_5I_3NNaO_3$ MW: 578.85   EINECS: 204-956-1
LD_{50}:   12156 mg/kg (M, i.m.); 7800 mg/kg (M, i.v.);
          6400 mg/kg (R, i.v.);
          5200 mg/kg (rabbit, i.v.);
          5600 mg/kg (cat, i.v.);
          6300 mg/kg (dog, i.v)

benzoic acid → 3-nitrobenzoic acid → 3-aminobenzoic acid → 3-amino-2,4,6-triiodobenzoic acid (I)

Acetrizoic acid

Reference(s):
US 2 611 786 (Mallinckrodt; 1952; appl. 1950; prior. 21.7.1948).
Wallingford et al.: J. Am. Chem. Soc. (JACSAT) **74**, 4365 (1952).

3-amino-2,4,6-triiodobenzoic acid:
Kretzer: Ber. Dtsch. Chem. Ges. (BDCGAS) **30**, 1944 (1897).

Formulation(s):   vial. 250 mg/ml, 500 mg/ml

Trade Name(s):
F: Vasurix (Guerbet); wfm   J: Diaginol (Banyu); wfm   Pyelokon-R
GB: Diaginol (May & Baker);   USA: Cystocon (Mallinckrodt);   (Mallinckrodt); wfm
      wfm   wfm   Salpix (Ortho); wfm

Acetylcholine chloride

ATC: S01EB09
Use: parasympathomimetic, miotic,
       vasodilator (peripheral)

RN: 60-31-1   MF: $C_7H_{16}ClNO_2$ MW: 181.66   EINECS: 200-468-8
LD_{50}:   10 mg/kg (M, i.v.); 3 g/kg (M, p.o.);
          22 mg/kg (R, i.v.); 2500 mg/kg (R, p.o.)
CN: 2-(acetyloxy)-N,N,N-trimethylethanaminium chloride

hydroxide
RN: 56-13-3   MF: $C_7H_{17}NO_3$ MW: 163.22
bromide
RN: 66-23-9   MF: $C_7H_{16}BrNO_2$ MW: 226.11   EINECS: 200-622-4
LD_{50}: 170 mg/kg (M, s.c.)

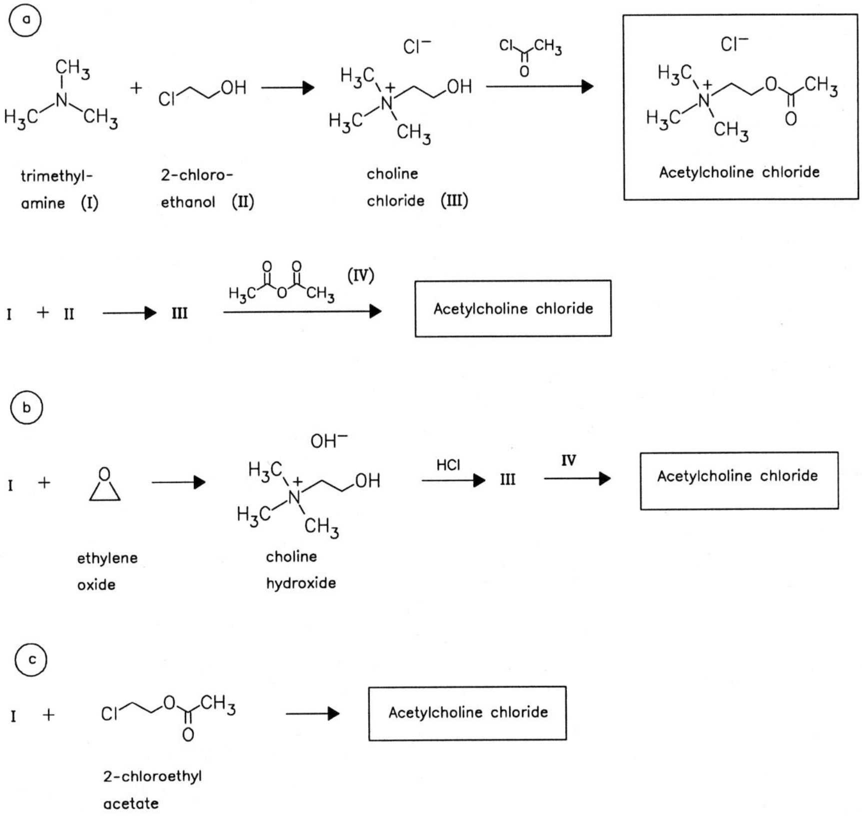

a)

trimethyl-   2-chloro-   choline   Acetylcholine chloride
amine (I)   ethanol (II)   chloride (III)

I + II ⟶ III ⟶ Acetylcholine chloride

b)

I + [ethylene oxide] ⟶ choline hydroxide ⟶ (HCl) III ⟶ (IV) Acetylcholine chloride

ethylene
oxide

choline
hydroxide

c)

I + [2-chloroethyl acetate] ⟶ Acetylcholine chloride

2-chloroethyl
acetate

Reference(s):
Baeyer, A. v.: Justus Liebigs Ann. Chem. (JLACBF) **142**, 235 (1867).
Nothnagel: Arch. Pharm. (Weinheim, Ger.) (ARPMAS) **232**, 265 (1894).
Fourneau, E.; Page, H.J.: Bull. Soc. Chim. Fr. (BSCFAS) [4] **15**, 544 (1914).
DE 801 210 (BASF; appl. 1948).
US 1 957 443 (Merck & Co.; 1934; appl. 1931).
US 2 012 268 (Merck & Co.; 1935; appl. 1931).
US 2 013 536 (Merck & Co.; 1935; appl. 1931).

Formulation(s):   amp. 20 mg; eye drops 1 %

Trade Name(s):
D: Miochol-E (CIBA Vision)   J: Acetylcholine (Roche)
I: Farmigea acetilcolina   Neucholin-A (Zeria); wfm
      (Farmigea); wfm   Ovisot (Daiichi); wfm

Acetylcysteine

ATC: R05CB01; S01XA08; V03AB23
Use: mucolytic agent

RN: 616-91-1   MF: $C_5H_9NO_3S$ MW: 163.20   EINECS: 210-498-3
LD_{50}:   400 mg/kg (M, i.p.); 3800 mg/kg (M, i.v.); 7888 mg/kg (M, p.o.);
        1140 mg/kg (R, i.v.); 5050 mg/kg (R, p.o.);
        700 mg/kg (dog, i.p.); 700 mg/kg (dog, i.v.); >1 g/kg (dog, p.o.)
CN: N-acetyl-L-cysteine

monosodium salt
RN: 19542-74-6   MF: $C_5H_8NNaO_3S$ MW: 185.18   EINECS: 243-143-6
LD_{50}:   3800 mg/kg (M, i.v.);
        2559 mg/kg (R,i.v.)
monoammonium salt
RN: 50807-78-8   MF: $C_5H_9NO_3S \cdot H_3N$ MW: 180.23

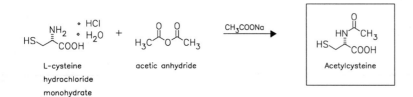

L-cysteine hydrochloride monohydrate   acetic anhydride   Acetylcysteine

Reference(s):
US 3 091 569 (Mead Johnson;. 28.5.1963; appl. 26.8.1960).
US 3 184 505 (Mead Johnson; 18.5.1965; appl. 18.6.1962).
Smith, H.A.; Gorin, G.: J. Org. Chem. (JOCEAH) **26**, 820 (1961).

ammonium salt (mucolysis of bronchial mucus by nebulization):
DOS 2 305 271 (Bristol-Myers; appl. 2.2.1973; USA-prior. 3.2.1972).

Formulation(s):   amp. 300 mg (as monosodium salt); cps. 200 mg; eff. tabl. 100 mg, 200 mg, 600 mg; f. c. tabl.
        100 mg, 200 mg, 600 mg; gran. 10 mg, 100 mg, 200 mg, 600 mg; lyo. for syrup 100 mg; syrup
        200 mg/10 ml; tabl. 100 mg, 200 mg, 600 mg

Trade Name(s):
D: ACC (Hexal)
    Acemuc (betapharm)
    Fluimucil-100/-200
    (Zambon)
    Rinofluimucil (Inpharzam)-
    comb.
    numerous combination and
    generic preparations
F: Broncoclar (Oberlin)
    Codotussyl (Whitehall)
    Euronac (Europhta)
    Exomuc (Bouchara)

Fluimucil (Zambon)
Fluimucil Antibiotic 750
(Zambon)
Genac (Génévrier)
Mucolator (Abbott)
Mucomyst (Bristol-Myers
Squibb)
Mucothiol (SCAT)
Rhinofluimucil (Débat)-
comb.
Solmucol (Génévrier)
Tixair (Byk)

GB: Ilube (Alcon)-comb.
    Parvolex (Evans)
I: Brunac (Bruschettini)
    Fluimucil (Zambon)
    Mucisol (Deca)
    Rinofluimucil (Zambon)-
    comb.
J: Acetein (Senju)
    Mucofilin Sol. (Eisai)
USA: Mucosit (Dey)

Acetyldigitoxin

ATC: C01AA01
Use: cardiotonic, cardiac glycoside

RN: 1111-39-3   MF: $C_{43}H_{66}O_{14}$ MW: 806.99   EINECS: 214-178-4
LD_{50}:   >30 mg/kg (g. p., p.o.);
　　　514 µg/kg (cat, i.v.); 250 µg/kg (cat, p.o.)
CN: (3β,5β)-3-[(O-3-acetyl-2,6-dideoxy-β-D-*ribo*-hexopyranosyl-(1→4)-O-2,6-dideoxy-β-D-*ribo*-hexopyranosyl-(1→4)-2,6-dideoxy-β-D-*ribo*-hexopyranosyl)oxy]-14-hydroxycard-20(22)-enolide

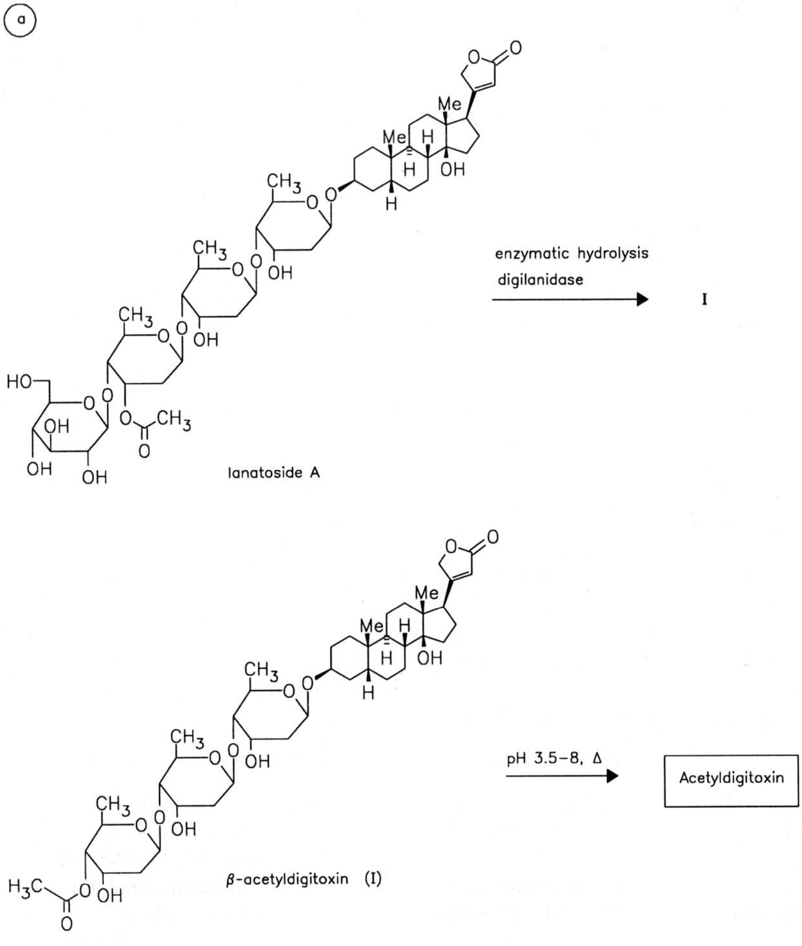

enzymatic hydrolysis
digilanidase ⟶ I

lanatoside A

pH 3.5–8, Δ ⟶ Acetyldigitoxin

β-acetyldigitoxin (I)

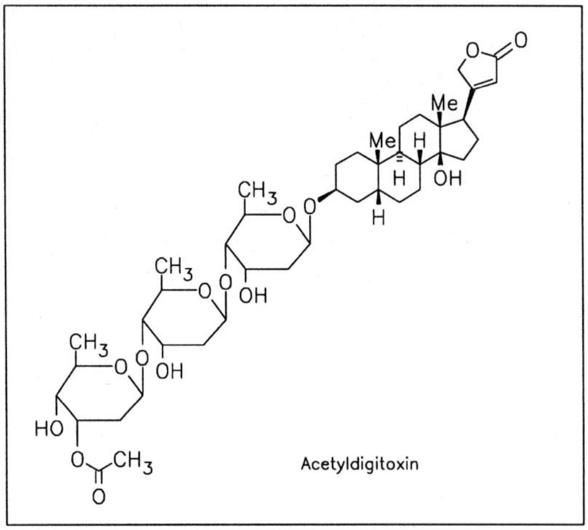

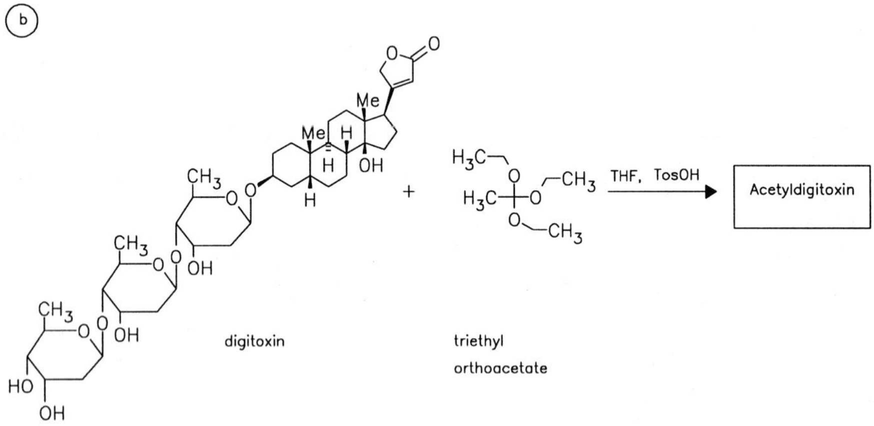

Acetyldigitoxin

b

digitoxin

triethyl
orthoacetate

THF, TosOH → Acetyldigitoxin

Reference(s):
a Stoll, A. et al.: Helv. Chim. Acta (HCACAV) **34**, 397 (1951).
   Gisvold, O.: J. Pharm. Sci. (JPMSAE) **61**, 1320 (1972).
   HU 155 716 (Richter Gedeon; appl. 20.1.1968).
   DE 925 047 (Sandoz; appl. 1954; CH-prior. 1952).
b DE 2 010 422 (Boehringer Ing.; appl. 5.3.1970).

alternative synthesis:
DE 2 206 737 (Boehringer Mannh.; appl. 12.2.1972) (α-Acetyldigoxin, q. v.).

Formulation(s):   tabl. 0.2 mg

Trade Name(s):
D: Acylanid (Sandoz); wfm   F: Acylanid (Sandoz); wfm   USA: Acylanid (Sandoz); wfm

α-Acetyldigoxin

ATC: C01AA02
Use: cardiotonic, cardiac glycoside

RN: 5511-98-8   MF: $C_{43}H_{66}O_{15}$ MW: 822.99   EINECS: 226-855-1
LD$_{50}$: 3300 µg/kg (g. p., p.o.);
       200 µg/kg (cat, p.o.)
CN: (3β,5β,12β)-3-[(O-3-O-acetyl-2,6-dideoxy-β-D-*ribo*-hexopyranosyl-(1→4)-O-2,6-dideoxy-β-D-*ribo*-
       hexopyranosyl-(1→4)-2,6-dideoxy-β-D-*ribo*-hexopyranosyl)oxy]-12,14-dihydroxycard-20(22)-enolide

ⓐ

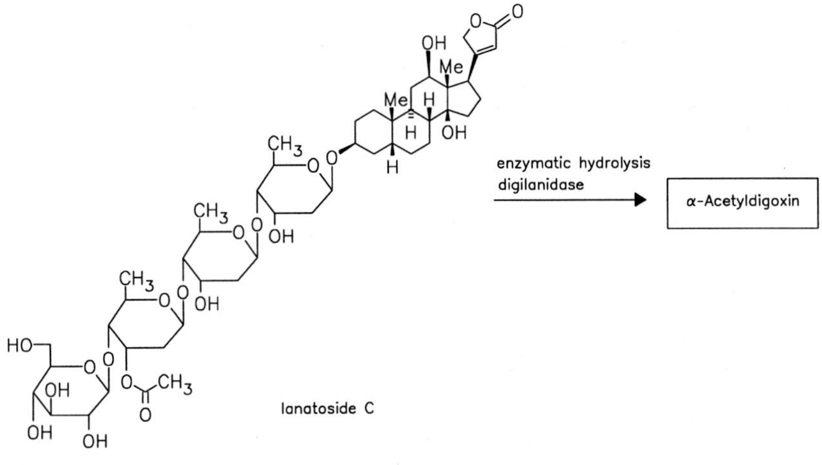

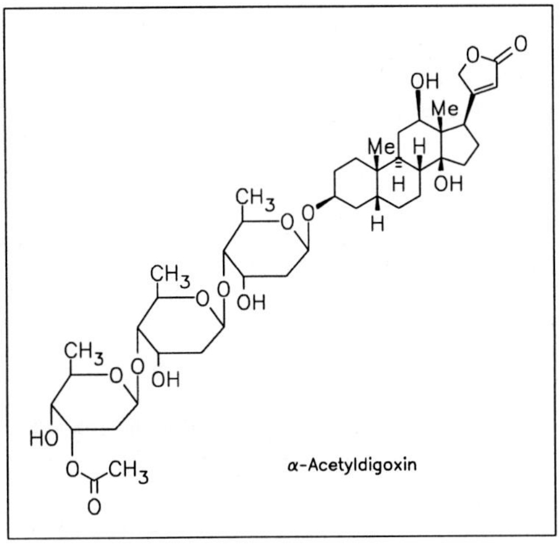

lanatoside C

enzymatic hydrolysis
digilanidase

⟶ α-Acetyldigoxin

α-Acetyldigoxin

ⓑ

digilanide

(lanatoside A, B, C)

enzymatic hydrolysis ⟶

acetyldigitoxin
+
acetyldigoxin
+
acetylgitoxin

partition of the methanol
solution between
H_2O and $CHCl_3$ ⟶

α-Acetyldigoxin

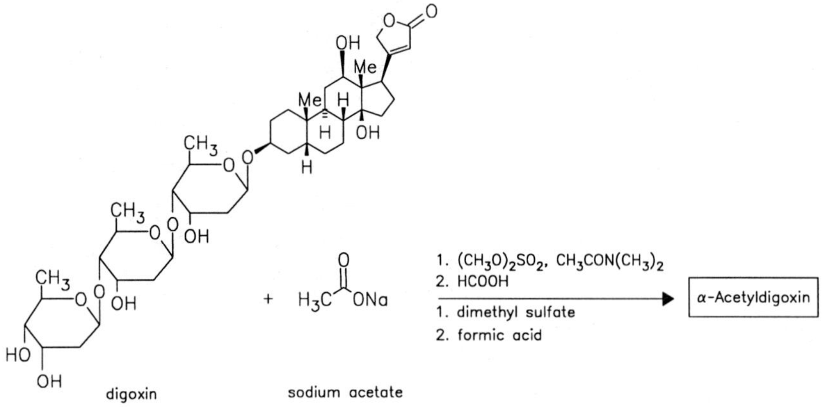

Reference(s):
a   Fieser, L.F.; Fieser, M.: Steroide, p. 801, Verlag Chemie, Weinheim 1961.
b   GB 1 162 614 (Heilmittelwerke Wien; appl. 1.2.1968; A-prior. 7.2.1967).
    Gisvold, O.: J. Pharm. Sci. (JPMSAE) **61**, 1320 (1972).

alternative syntheses:
DE 2 010 422 (Boehringer Ing.; appl. 5.3.1970). (Acetyldigitoxin, q. v.).
Rietbrock, N.; Kuhlmann, J.: Naunyn-Schmiedeberg's Arch. Pharmacol. (NSAPCC) **279**, 413 (1973).

Formulation(s):   sol. 0.5 mg/ml; tabl. 0.25 mg, 0.2 mg

Trade Name(s):

D:	Card-Hydergin (Sandoz)-comb.; wfm	Lanadigin (Promonta); wfm		Nitro-Sandolanid (Sandoz)-comb.; wfm
	Digi-Complamin (Beecham-Wülfing)-comb.; wfm	Lanadigin EL (Promonta); wfm	F:	Sandolanid (Sandoz) Acygoxine (Sandoz); wfm
		Lanadigin + Theophyllin (Promonta)-comb.; wfm	I:	Cedigossina (Sandoz)

β-Acetyldigoxin

ATC: C01AA02
Use: cardiotonic, cardiac glycoside

RN: 5355-48-6   MF: $C_{43}H_{66}O_{15}$ MW: 822.99   EINECS: 226-337-5
LD$_{50}$:   2400 µg/kg (g. p., p.o.);
        422 µg/kg (dog, p.o.)
CN: (3β,5β,12β)-3-[(O-4-O-acetyl-2,6-dideoxy-β-D-ribo-hexopyranosyl-(1→4)-O-2,6-dideoxy-β-D-ribo-hexopyranosyl-(1→4)-2,6-dideoxy-β-D-ribo-hexopyranosyl)oxy]-12,14-dihydroxycard-20(22)-enolide

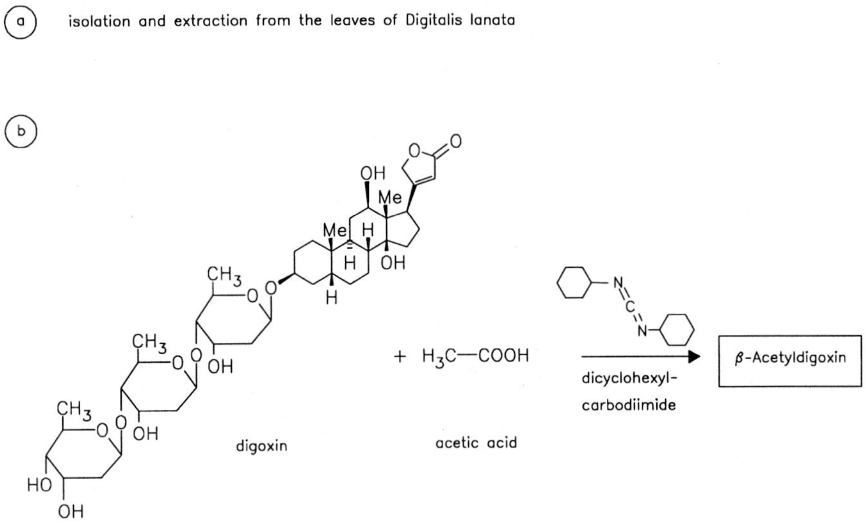

(a) isolation and extraction from the leaves of Digitalis lanata

(b)

digoxin   +   H₃C—COOH   acetic acid   dicyclohexyl-carbodiimide   →   β-Acetyldigoxin

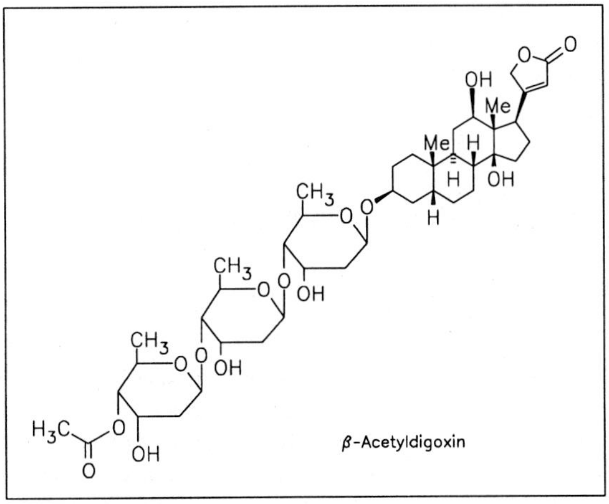

β-Acetyldigoxin

Reference(s):

a   Hopponen, R.E.; Gisvold, O.: J. Am. Pharm. Assoc. (JPHAA3) **41**, 146 (1952).
    Rangaswami, S. et al.: Indian J. Pharm. (IJPAAO) **17**, 253 (1955).
b   HU 7 147 (Richter Gedeon; appl. 5.6.1972).

alternative syntheses:
Haberland, G.: Arzneim.-Forsch. (ARZNAD) **15**, 481 (1965).
Graf, E.; Pfaff, J.: Arch. Pharm. (Weinheim, Ger.) (ARPMAS) **307**, 943 (1974).
DOS 2 826 532 (LEK tovarna farm.; appl. 16.6.1978; YU-prior. 22.6.1977).

medical use:
DOS 1 921 307 (Boehringer Ing.; appl. 25.4.1969) addition to DOS 1 767 553.

Formulation(s):   tabl. 0.1 mg, 0.2 mg

Trade Name(s):
D: Beta-Acetyldigoxin
      (ratiopharm)
      Beta-Acetyldigoxin R.A.N.
      = glycotop (R.A.N.)

      Beta-Acetyldigoxin-
      ratiopharm 0,1 mg/0,2 mg
      (ratiopharm)
      Digostada 0.2/-mite
      (Stadapharm)
      Digostade (Stada)

      Digotab (ASTA Medica
      AWD)
      Digox (ct-Arzneimittel)
      Digoxin-Didier
      (Hormosan)
      Gladixol (Corax)

Kardiamed (Medice)
Novodigal (Beiersdorf)
Stillacor (Wolff)   I:

numerous combination
preparations
Beta-Acigoxia (Inverni
della Beffa); wfm

Cardioreg (Nattermann);
wfm

Acetylsalicylic acid

(Acidum acetylsalicylicum; Aspirin)

ATC: A01AD05; B01AC06; M01BA03;
        N02BA01; N02BA51
Use: analgesic, antipyretic, antirheumatic,
        platelet aggregation inhibitor

RN: 50-78-2   MF: $C_9H_8O_4$ MW: 180.16   EINECS: 200-064-1
LD_{50}: 280 mg/kg (M, i.p.); 250 mg/kg (M, p.o.); 1520 mg/kg (M, s.c.);
       340 mg/kg (R, i.p.); 200 mg/kg (R, p.o.); 1600 mg/kg (R, s.c.);
       1075 mg/kg (g. p., p.o.);
       1010 mg/kg (rabbit, p.o.);
       681 mg/kg (dog, i.v.); 700 mg/kg (dog, p.o.)
CN: 2-(acetyloxy)benzoic acid

aluminum salt
RN: 147-31-9   MF: $C_{27}H_{21}AlO_{12}$ MW: 564.44
calcium salt
RN: 69-46-5   MF: $C_{18}H_{14}CaO_8$ MW: 398.38   EINECS: 200-707-6
lithium salt
RN: 552-98-7   MF: $C_9H_7LiO_4$ MW: 186.09   EINECS: 209-029-5
sodium salt
RN: 493-53-8   MF: $C_9H_7NaO_4$ MW: 202.14   EINECS: 207-777-7
LD_{50}: 730 mg/kg (M, i.p.);
       1450 mg/kg (R, i.p.)
magnesium salt
RN: 132-49-0   MF: $C_{18}H_{14}MgO_8$ MW: 382.61   EINECS: 205-062-4
LD_{50}: 620 mg/kg (M, s.c.)
lysine salt (1:1)
RN: 62952-06-1   MF: $C_9H_8O_4 \cdot C_6H_{14}N_2O_2$ MW: 326.35
LD_{50}: 950 mg/kg (M, i.v.); 3270 mg/kg (M, p.o.);
       1525 mg/kg (R, i.v.); 4350 mg/kg (R, p.o.)

salicylic acid   +   acetic anhydride   →   Acetylsalicylic acid

Reference(s):
Ullmanns Encykl. Tech. Chem., 3. Aufl., Vol. **13**, 90.
US 3 235 583 (Norwich Pharmacal; 15.2.1966; appl. 22.7.1964).

acetylation in presence of pyridine *for avoidance of formation of* acetylsalicylic anhydride *and*
acetylsalicylsalicylic acid:
DOS 2 635 540 (A. L. de Week, H. Bundgaard; appl. 6.8.1976).

acetylation in presence of H_2SO_4:
US 2 731 492 (J. Kamlet; 1956; appl. 1954).

crystallization:
US 2 890 240 (Monsanto; 1959; appl. 1957).

aluminum salts:
DRP 585 986 (Chinoin; appl. 1931; H.-prior. 1931).
US 2 698 332 (Reheis Comp.; 1954; appl. 1951).
US 2 918 485 (Keystone Chemurgic Corp.; 1959; appl. 1955).
GB 888 666 (Hardman & Holden; appl. 1959).

aluminum acetylsalicylate glutaminate:
DOS 2 909 829 (Kyowa Hakko; appl. 13.3.1979; J-prior. 13.3.1978).

Formulation(s):   cps. 325 mg, 500 mg, suppos. 125 mg, 150 mg, 300 mg; tabl. 50 mg, 75 mg, 100 mg, 300 mg, 500 mg

Trade Name(s):

D: Alka-Seltzer (Bayer)
Aspirin (Bayer; 1899)
Aspisol (Bayer; as DL-lysine salt)
Aspro (Roche Nicholas)
ASS Dura (durachemie)
ASS-ratiopharm (ratiopharm)
Godamed (Pfleger)
Micristin (OPW)
Miniasal (OPW)
Romigal (Romogal-Werk)
Santasal (Merckle)
Togal (Togal)
numerous combination preparations

F: Actron (Bayer)-comb.
Afebry (Galephar)-comb.
Alka-Seltzer (Bayer)-comb.
Antigrippine (SmithKline Beecham)-comb.
Aspégic 500 (Synthélabo; as lysine salt)
Aspirine Bayér (Bayer)
Aspirine duRhône (Bayer)
Aspirine pH 8 (3M Santé)
Aspirine Upsa (UPSA)
Aspirine Upsa Vitamine C (UPSA)-comb.

Aspirine Vitamine C (Oberlin)-comb.
Aspirisucre (Arkomedika)
Aspro (Nicholas)
Catalgine (Schwarz)
Claragine (Nicholas)
Kardégic (Synthélabo)
Rhonal (Théraplix)
Sargépirine (ASTA Médica)
Solupsan (UPSA)
numerous combination preparations

GB: Angettes (Bristol-Myers)
Aspan (Hoechst)-comb.
Aspirin (Bayer)
Caprin (Sinclair)
Disprin CO (Reckitt & Colman)
Nu-Seals Aspirin (Lilly)
Post MI (Ashbourne)
numerous combination preparations

I: Ac Acsal (Formulario Naz.; Tariff. Nazionale; Scfm; Iema; Farmacologico Milanese)
Acesal (Geymonat)

Alupir (Farmacologico Milanese; as aluminum salt)
Aspergum (Farmades)
Aspirina (Bayer)
Aspirinetta (Bayer)
Aspro (Roche)
Bufferin (Bristol-Myers Squibb)
Cemirit (Bayer)
Endydol (Guidotti)
Kilios (Carlo Erba)
numerous combination preparations

J: generic preparations
USA: Acuprin (Richwood)
Ecotrin (SmithKline Beecham Consumer)
Equagesic (Wyeth-Ayerst)
Fiorinal (Novartis)
Halfprin (Kramer)
Norgeric (3M)
Percotan (Endo)
Roboxisal (Robins)

Acetylsulfafurazole
(Acetylsulfisoxazole; Sulfisoxazole Acetyl)

ATC: S01AB
Use: antibacterial

RN: 80-74-0 MF: $C_{13}H_{15}N_3O_4S$ MW: 309.35 EINECS: 201-305-3
CN: *N*-[(4-aminophenyl)sulfonyl]-*N*-(3,4-dimethyl-5-isoxazolyl)acetamide

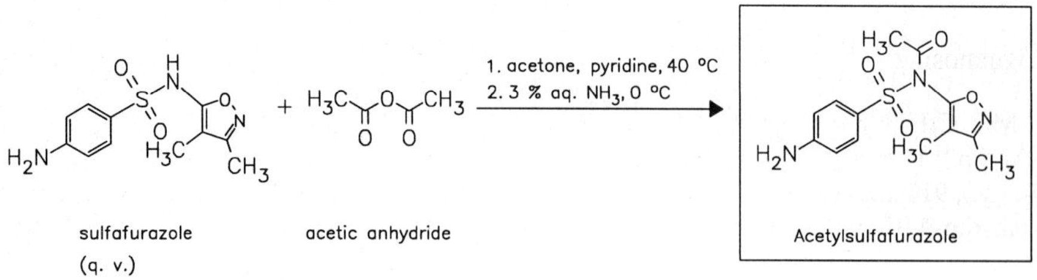

sulfafurazole   acetic anhydride   Acetylsulfafurazole
(q. v.)

Reference(s):
US 2 721 200 (Roche; 1955; appl. 1953).

Formulation(s):   susp. 500 mg/5 ml

Trade Name(s):
USA: Eryzole (Alra)   Pediazole (Ross)

Acexamic acid
(Acide acexamique)

ATC: D03A
Use: antifibrinolytic

RN: 57-08-9   MF: $C_8H_{15}NO_3$ MW: 173.21   EINECS: 200-310-8
CN: 6-(acetylamino)hexanoic acid

sodium salt
RN: 7234-48-2   MF: $C_8H_{14}NNaO_3$ MW: 195.19   EINECS: 230-635-0

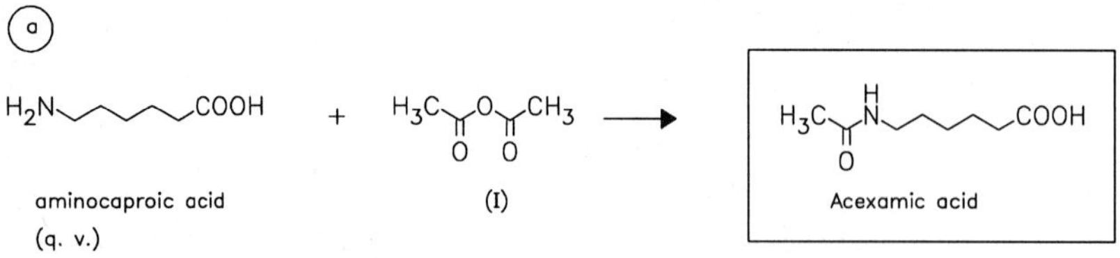

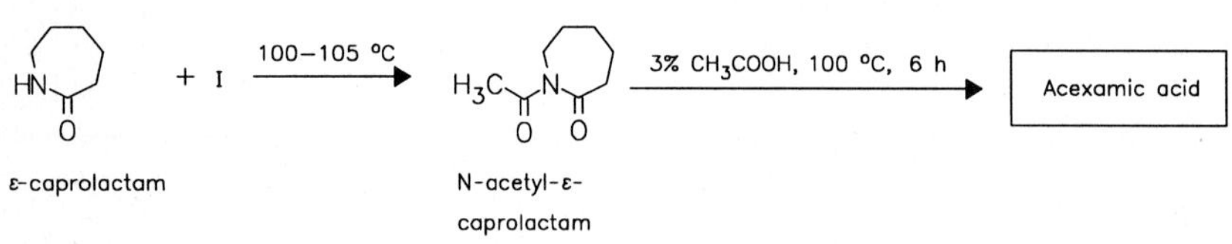

aminocaproic acid   (I)   Acexamic acid
(q. v.)

ε-caprolactam   N-acetyl-ε-   Acexamic acid
                       caprolactam

Reference(s):
Offe, H.A.: Z. Naturforsch., B: Anorg. Chem., Org. Chem., Biochem., Biophys., Biol. (ZENBAX) **2**, 182 (1947).
FR-M 2 332 (Rowa; appl. 1963).

Formulation(s):   amp. 5 g (as sodium salt); cps. 300 mg (as zinc salt); ointment 5 % (as sodium salt);
                     susp. 300 mg (as zinc salt)

Trade Name(s):
F: Plasténan (Isopharm)   Plasténan Néomycine   I: Plastenan (Italfarmaco);
                                     (Isopharm)-comb.   wfm

Aciclovir
(Acyclovir; Acycloguanosine)

ATC: D06BB03; J05AB01; S01AD03
Use: antiviral

RN: 59277-89-3 MF: $C_8H_{11}N_5O_3$ MW: 225.21 EINECS: 261-685-1
LD$_{50}$: 1000 mg/kg (M, i.p.); 1118 mg/kg (M, i.v.); >10000 mg/kg (M, p.o.); 1118 mg/kg (M, s.c.);
860 mg/kg (R, i.p.); 910 mg/kg (R, i.v.); >20000 mg/kg (R, p.o.); 620 mg/kg (R, s.c.)
CN: 2-amino-1,9-dihydro-9-[(2-hydroxyethoxy)methyl]-6H-purin-6-one

monosodium salt
RN: 69657-51-8 MF: $C_8H_{10}N_5NaO_3$ MW: 247.19
LD$_{50}$: 999 mg/kg (M, i.p.); >10000 mg/kg (M, p.o.);
1210 mg/kg (R, i.p.); >600 mg/kg (R, i.v.); >20000 mg/kg (R, p.o.); 650 mg/kg (R, s.c.)

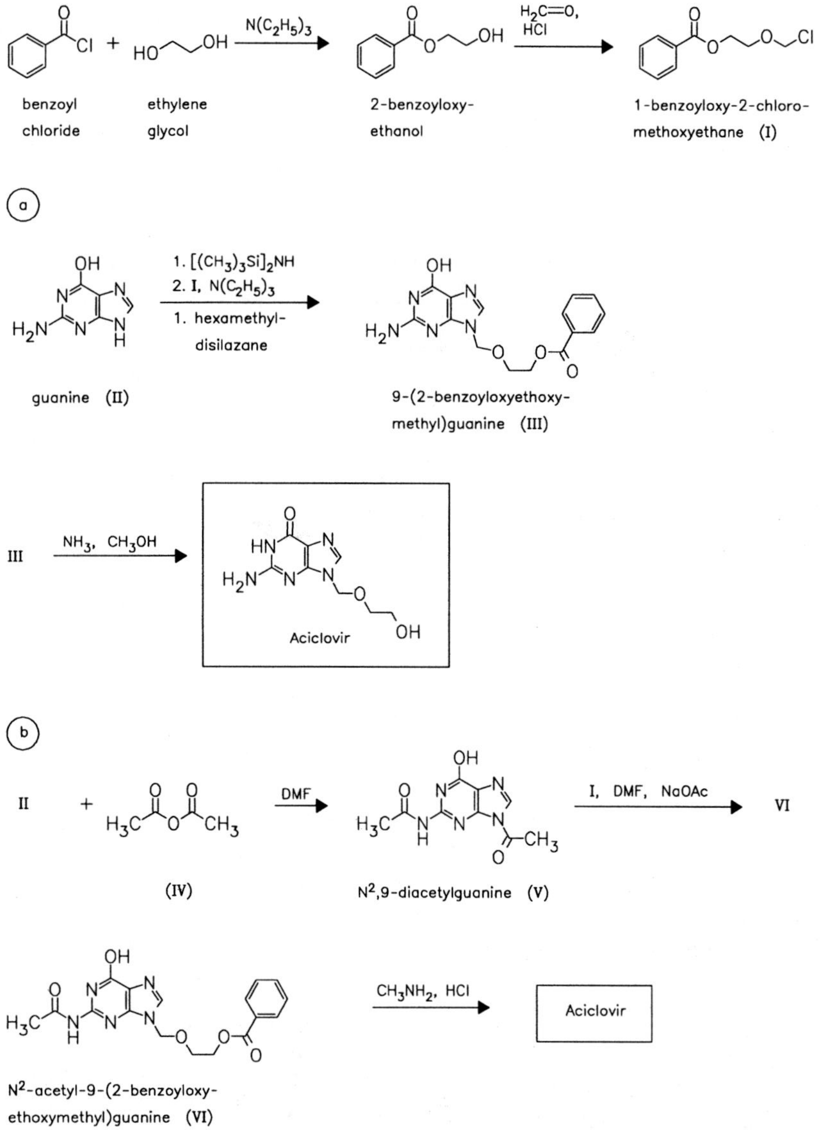

benzoyl chloride + ethylene glycol → 2-benzoyloxy-ethanol → 1-benzoyloxy-2-chloro-methoxyethane (I)

a

guanine (II) → 9-(2-benzoyloxyethoxy-methyl)guanine (III)

III → Aciclovir

b

II + (IV) → N²,9-diacetylguanine (V) → VI

N²-acetyl-9-(2-benzoyloxy-ethoxymethyl)guanine (VI) → Aciclovir

ⓒ

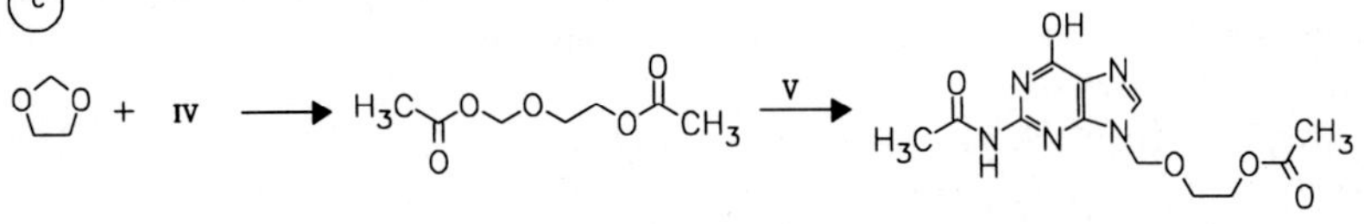

1,3-dioxo-
lane (VII)

2-acetoxyethyl
acetoxymethyl ether (VIII)

"diacetylaciclovir" (IX)

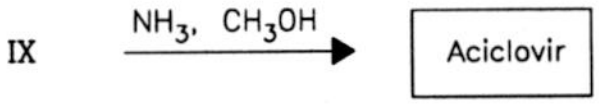

IX $\xrightarrow{\text{NH}_3, \text{ CH}_3\text{OH}}$ Aciclovir

ⓓ

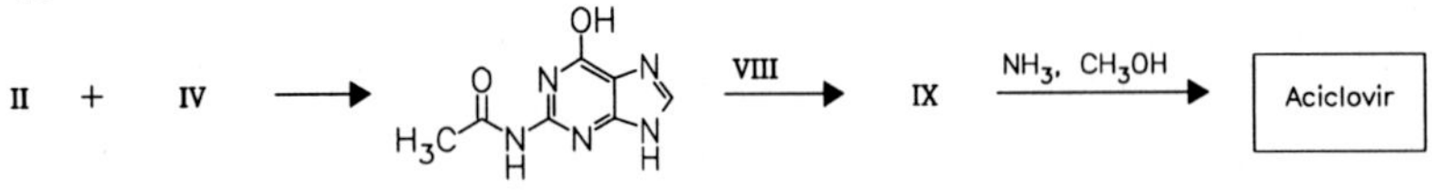

II + IV $\longrightarrow$

N²-acetylguanine

$\xrightarrow{\text{VIII}}$ IX $\xrightarrow{\text{NH}_3, \text{ CH}_3\text{OH}}$ Aciclovir

ⓔ

2,6-dichloropurine

$+$ I $\xrightarrow[\text{DMF}]{\text{N(C}_2\text{H}_5)_3,}$

$\xrightarrow{\begin{array}{l}\text{1. NH}_3, \text{CH}_3\text{OH}\\\text{2. NaNO}_2, \text{ AcOH}\\\text{3. NH}_3, \text{CH}_3\text{OH}\end{array}}$ Aciclovir

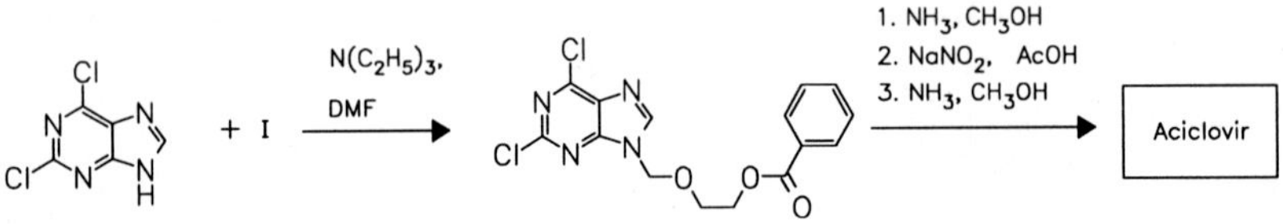

ⓕ

II +

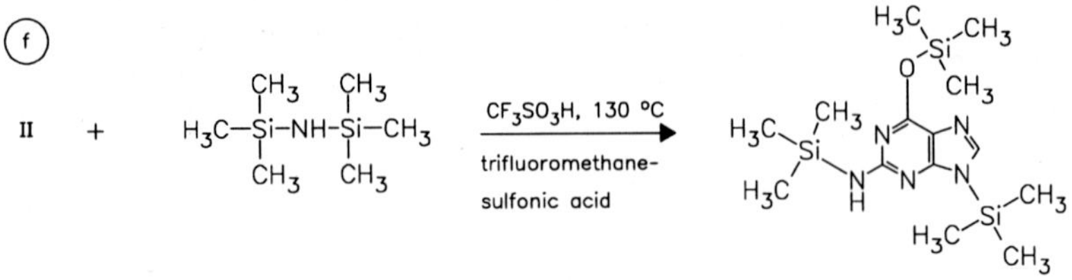

hexamethyl-
disilazane

$\xrightarrow[\text{sulfonic acid}]{\text{CF}_3\text{SO}_3\text{H, 130 °C} \atop \text{trifluoromethane-}}$

(X)

X + VII $\xrightarrow{\begin{array}{l}\text{1. }\Delta\\\text{2. H}_2\text{O}\end{array}}$

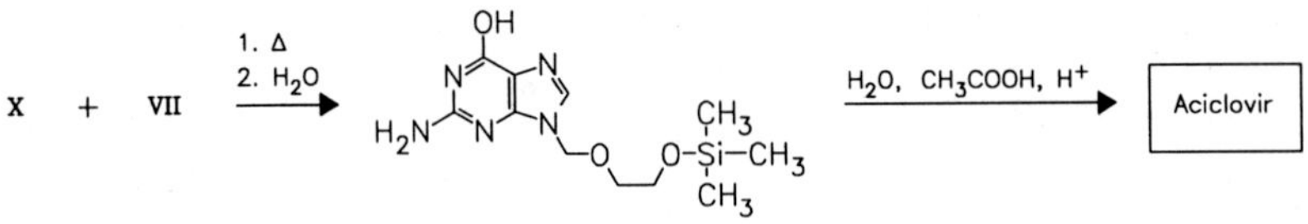

9-(2-trimethylsilyloxy-
ethoxymethyl)guanine

$\xrightarrow{\text{H}_2\text{O, CH}_3\text{COOH, H}^+}$ Aciclovir

ⓖ

V + VII $\xrightarrow{\text{tosyl chloride}}$

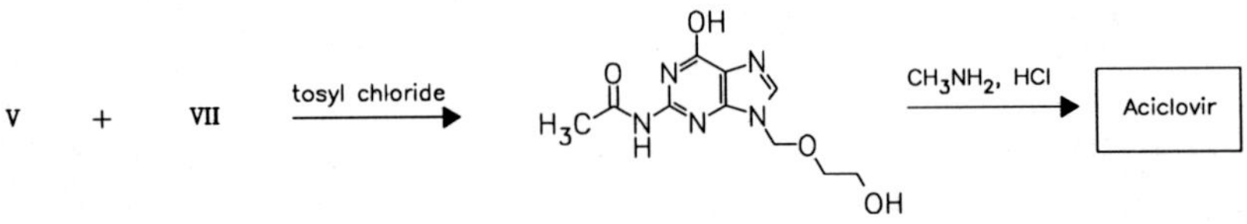

"monoacetylaciclovir"

$\xrightarrow{\text{CH}_3\text{NH}_2, \text{ HCl}}$ Aciclovir

Reference(s):

Schaeffer, H.J. et al.: Nature (London) (NATUAS) **272**, 583 (1978).
DE 2 539 963 (Wellcome; appl. 2.9.1975; GB-prior. 2.9.1974).
US 4 199 574 (Wellcome; 22.4.1980; GB-prior. 2.9.1974).
GB 1 523 865 (Burroughs Wellcome; GB-prior. 2.9.1974).
c   GB 1 567 671 (Wellcome; appl. 26.8.1977; USA-prior. 27.8.1976).
    Matsumoto, H. et al.: Chem. Pharm. Bull. (CPBTAL) **36**, 1153 (1988).
f   EP 709 385 (Roche; appl. 13.7.1995; USA-prior. 26.7.1994, 27.4.1995).

alternative synthesis from 4-aminoimidazole-5-carboxamide:
WO 9 011 283 (GEA Farm.; 4.10.1990; DK-prior. 20.3.1989).

alternative synthesis via formylguanine:
WO 9 507 281 (Recordati; appl. 3.2.1994; I-prior. 10.9.1993).

synthesis using 1,3-dioxolane:
US 5 567 816 (Syntex; appl. 27.4.1995; USA-prior. 27.7.1994).

improved procedures:
DE 19 536 164 (Boehringer Ingelheim; D-prior. 28.9.1995).
WO 9 724 357 (Mallinckrodt; appl. 17.12.1996; USA-prior. 28.12.1995).
DE 19 604 101 (B. Lehmann; 6.2.1996).
EP 806 425 (Lupin Lab.; EP-prior. 9.4.1996).
US 5 792 868 (Ajinomoto; appl. 18.3.1994; J-prior. 18.9.1991).

Formulation(s):   cps. 200 mg; cream 50 mg/g; eye ointment 30 mg/g; susp. 8 %; tabl. 200 mg, 400 mg, 800 mg;
vial 250 mg, 500 mg

Trade Name(s):

D:	Zovirax (Glaxo Wellcome; 1983)	Alovir (Foletto)		Sifiviral (SIFI)
		Avirase (Lampugnani)		Zovirax (Wellcome; 1984)
F:	Activir (Warner-Lambert)	Avyclor (Bioprogress)	J:	Zovirax (Seimitomo-
	Zovirax (Wellcome; 1983)	Cycloviran (Sigma-Tau)		Wellcome; 1985)
GB:	Herpetad (Boehringer Ing.)	Dravyr (Drug Research)	USA:	Zovirax (Glaxo Wellcome;
	Zovirax (Glaxo Wellcome; 1981)	Efrivir (Aesculapius-Bs)		1985)
		Esavir (Boniscontro &		
I:	Aciviran (Ripari-Gero)	Gazzone)		
	Acyvir (Delalande Isnardi)	Neviran (Coli)		

Acipimox

ATC: C10AD06
Use: antihyperlipoproteinemic

RN: 51037-30-0   MF: $C_6H_6N_2O_3$ MW: 154.13   EINECS: 256-928-3
LD_{50}: 3500 mg/kg (M, p.o.)
CN: 5-methylpyrazinecarboxylic acid 4-oxide

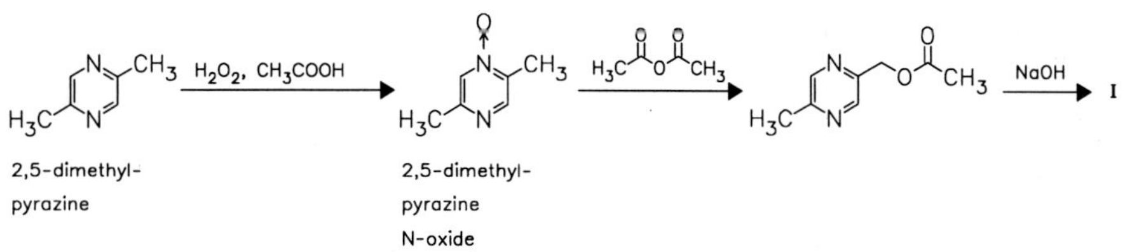

2,5–dimethyl–
pyrazine

2,5–dimethyl–
pyrazine
N–oxide

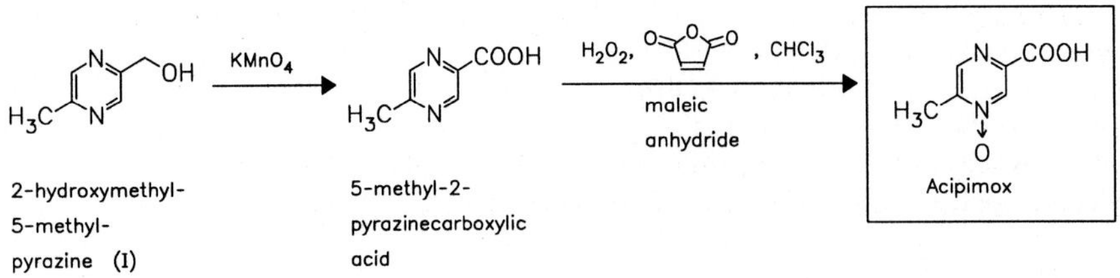

2-hydroxymethyl-
5-methyl-
pyrazine (1)

5-methyl-2-
pyrazinecarboxylic
acid

Acipimox

Reference(s):
US 4 002 750 (Carlo Erba; 11.1.1977; I-prior. 28.4.1972).
US 4 051 245 (Carlo Erba; 27.9.1977; I-prior. 28.4.1972).
DOS 2 319 834 (Carlo Erba; appl. 18.4.1973; I-prior. 28.4.1972).
GB 1 361 967 (Carlo Erba; appl. 12.4.1973; I-prior. 28.4.1972).
Brubrogi, V. et al.: Eur. J. Med. Chem. (EJMCA5) **15**, 157 (1980).

5-methyl-2-pyrazinecarboxylic acid:
Pitré, D. et al.: Chem. Ber. (CHBEAM) **99**, 364 (1966).

2-hydroxymethyl-5-methylpyrazine:
Klein, B. et al.: J. Org. Chem. (JOCEAH) **26**, 129 (1961).

Formulation(s):   cps. 25 mg, 250 mg

Trade Name(s):
I: Olbetam (Pharmacia &
       Upjohn; 1985)

Aclarubicin
(Aclacinomycin A)

ATC: L01DB04
Use: antineoplastic

RN: 57576-44-0   MF: $C_{42}H_{53}NO_{15}$ MW: 811.88   EINECS: 260-824-3
LD_{50}:   22.6 mg/kg (M, i.p.); 33.7 mg/kg (M, i.v.)
CN: [1*R*-(1α,2β,4β)]-2-ethyl-1,2,3,4,6,11-hexahydro-2,5,7-trihydroxy-6,11-dioxo-4-[[2,3,6-trideoxy-4-*O*-
       [2,6-dideoxy-4-*O*-[(2*R*-*trans*)-tetrahydro-6-methyl-5-oxo-2*H*-pyran-2-yl]-α-L-*lyxo*-hexopyranosyl]-3-
       (dimethylamino)-α-L-*lyxo*-hexopyranosyl]oxy]-1-naphthacenecarboxylic acid methyl ester

hydrochloride
RN: 75443-99-1   MF: $C_{42}H_{53}NO_{15} \cdot HCl$ MW: 848.34

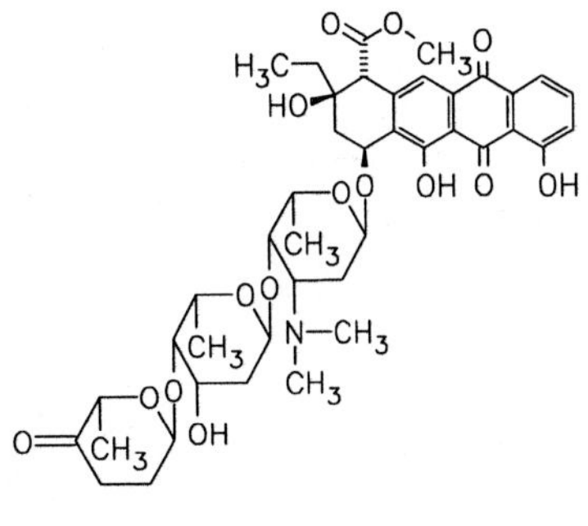

Aclarubicin

By fermentation of *Streptomyces galilaeus* MA 144-M1 (ATCC 3113); separation of aclacinomycin A and B by column chromatography.

Reference(s):
DOS 2 532 568 (Zaidanhojin Biseibutsu Kagaku Kenkykai; appl. 21.1.1975; J-prior. 27.7.1974).
US 3 988 315 (Zaidanhojin Biseibutsu Kagaku Kenkykai, 26.10.1976; J-prior. 27.7.1974).

Formulation(s):   powder 20 mg; vial 20 mg (as hydrochloride)

Trade Name(s):
D: Aclaplastin (medac)   F: Aclacinomycine (Roger   J: Aclacinon (Sanraku)
                                          Bellon); wfm

Aclatonium napadisilate

ATC: A03AB
Use: antispasmodic, cholinergic

RN: 55077-30-0   MF: $C_{10}H_{20}NO_4 \cdot 1/2C_{10}H_6O_6S_2$ MW: 722.83
LD$_{50}$:   41.9 mg/kg (M, i.v.); 15 g/kg (M, p.o.);
       46 mg/kg (R, i.v.); >13.9 g/kg (R, p.o.);
       >10 g/kg (dog, p.o.)
CN: 2-[2-(acetyloxy)-1-oxopropoxy]-*N,N,N*-trimethylethanaminium 1,5-naphthalenedisulfonate (2:1)

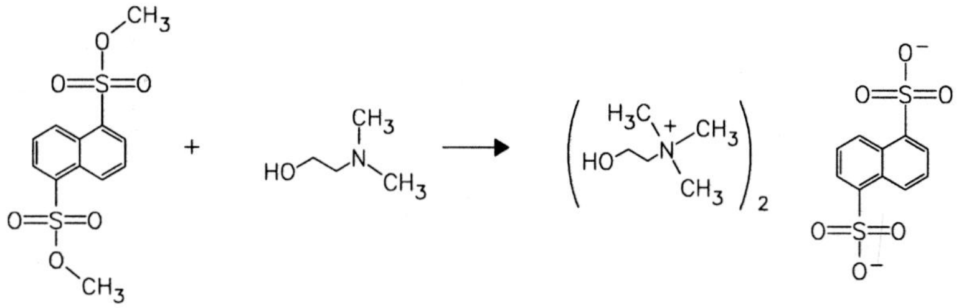

dimethyl   2-dimethyl-   (I)
naphthalene-   aminoethanol
1,5-disulfonate

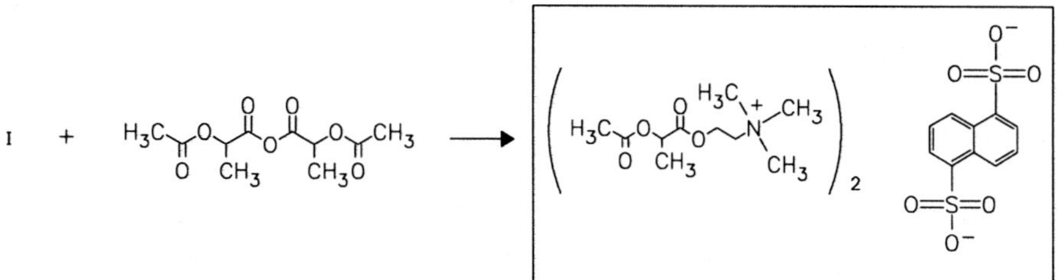

(2-acetoxypropionic) anhydride   Aclatonium napadisilate

Reference(s):
DE 2 425 983 (Toyama; appl. 30.5.1974; J-prior. 12.6.1973).
US 3 903 137 (Toyama; 2.9.1975; J-prior. 12.6.1973, 20.6.1973).

Formulation(s):   cps. 25 mg, 50 mg

Trade Name(s):
J: Abovis (Toyama; 1981)

Acriflavinium chloride

(Acriflavine hydrochloride)

ATC: R02AA13
Use: antiseptic, chemotherapeutic (local infections)

RN: 8063-24-9 MF: $C_{14}H_{14}ClN_3 \cdot C_{13}H_{11}N_3 \cdot 3HCl$ MW: 578.38
CN: 3,6-diamino-10-methylacridinium chloride monohydrochloride mixt. with 3,6-acridinediamine dihydrochloride

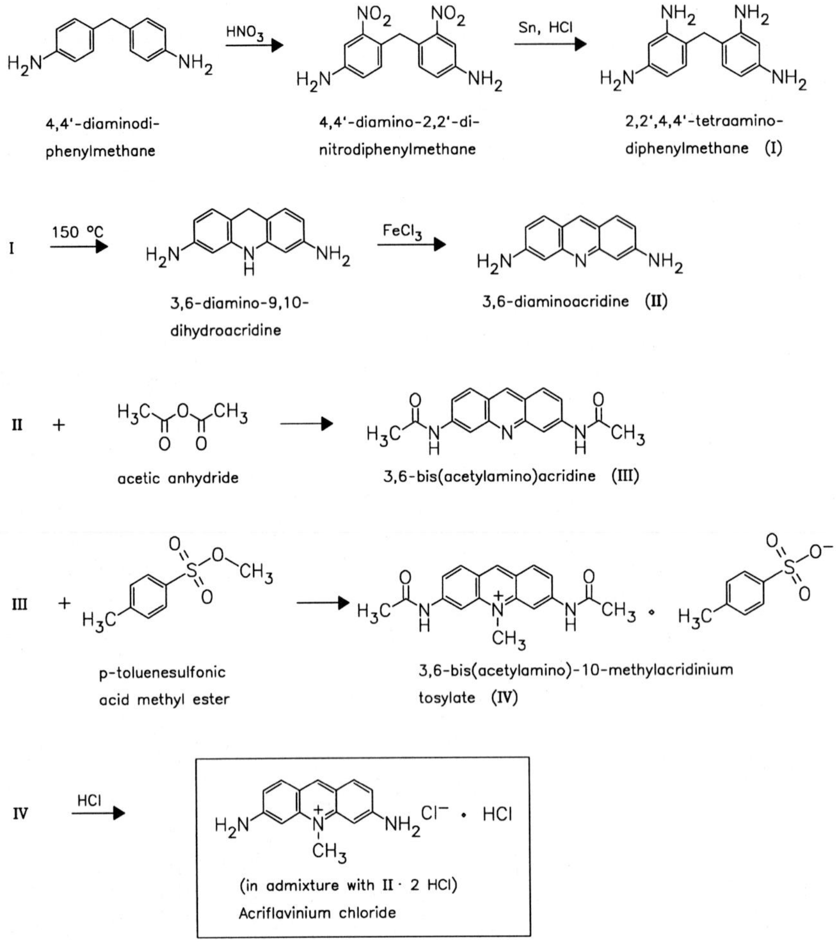
HNO3
Sn, HCl
4,4'-diaminodiphenylmethane
4,4'-diamino-2,2'-dinitrodiphenylmethane
2,2',4,4'-tetraaminodiphenylmethane (I)
I
150 °C
FeCl3
3,6-diamino-9,10-dihydroacridine
3,6-diaminoacridine (II)
II +
acetic anhydride
3,6-bis(acetylamino)acridine (III)
III +
p-toluenesulfonic acid methyl ester
3,6-bis(acetylamino)-10-methylacridinium tosylate (IV)
IV
HCl
(in admixture with II · 2 HCl)
Acriflavinium chloride

Reference(s):
FR 686 606 (I. G. Farben; 1929).

Formulation(s): sol. 150 mg/100 g; tabl. 0.15 mg (comb. with 5 mg benzocaine)

Trade Name(s):
D: Nordapanin (Michallik)-comb.
F: Chromargon (M. Richard)-comb.
J: Isravin (Takeda); wfm

Acrivastine

(BW-825C)

ATC: R06AX18
Use: non-sedative antihistaminic (for treatment of allergic rhinitis, urticaria)

RN: 87848-99-5 MF: $C_{22}H_{24}N_2O_2$ MW: 348.45
CN: (*E,E*)-3-[6-[1-(4-methylphenyl)-3-(1-pyrrolidinyl)-1-propenyl]-2-pyridinyl]-2-propenoic acid

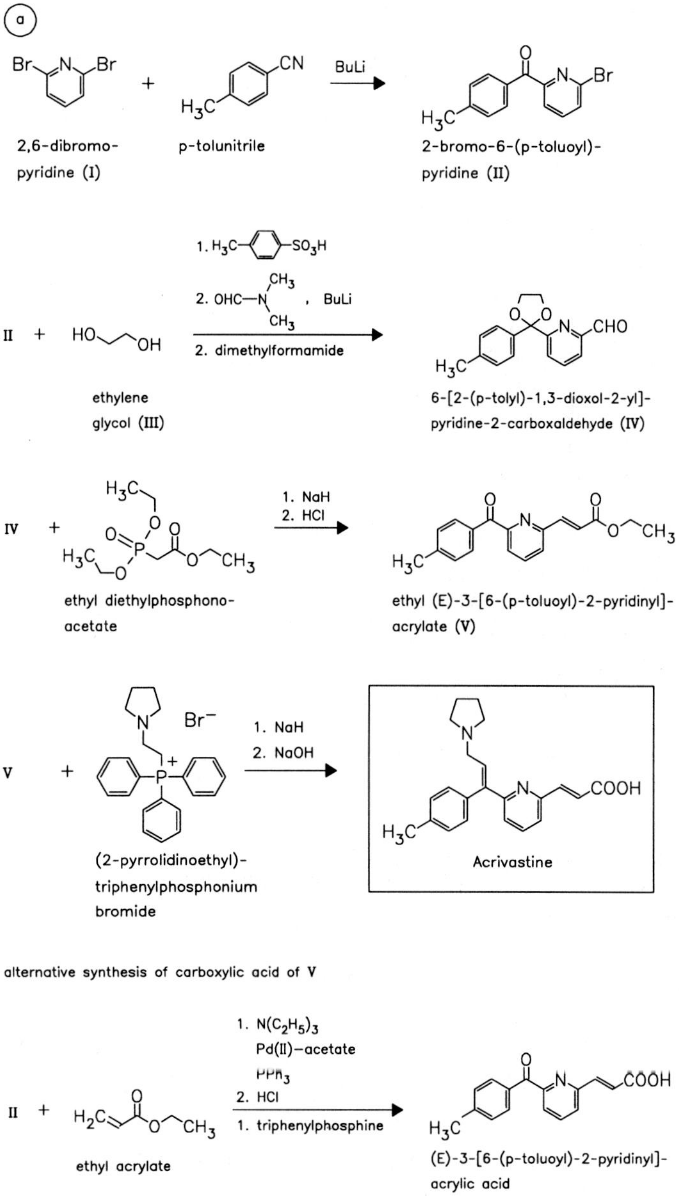
a
BuLi
2,6-dibromo-pyridine (I)
p-tolunitrile
2-bromo-6-(p-toluoyl)-pyridine (II)
II + HO OH
1. H3C SO3H
2. OHC—N(CH3)2 , BuLi
2. dimethylformamide
ethylene glycol (III)
6-[2-(p-tolyl)-1,3-dioxol-2-yl]-pyridine-2-carboxaldehyde (IV)
IV +
1. NaH
2. HCl
ethyl diethylphosphono-acetate
ethyl (E)-3-[6-(p-toluoyl)-2-pyridinyl]-acrylate (V)
V +
Br⁻
1. NaH
2. NaOH
(2-pyrrolidinoethyl)-triphenylphosphonium bromide
Acrivastine
alternative synthesis of carboxylic acid of V
II +
1. N(C2H5)3
Pd(II)-acetate
PPh3
2. HCl
1. triphenylphosphine
ethyl acrylate
(E)-3-[6-(p-toluoyl)-2-pyridinyl]-acrylic acid

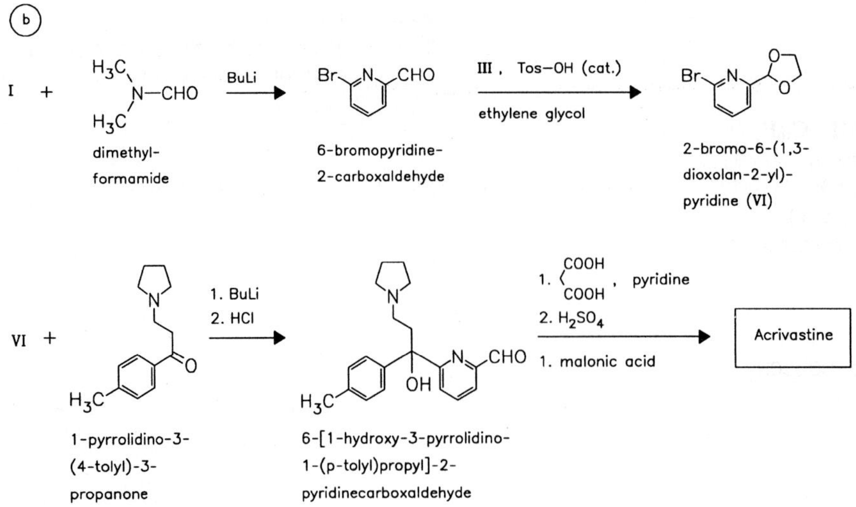

ⓑ

I + dimethyl-formamide →[BuLi] 6-bromopyridine-2-carboxaldehyde →[III, Tos—OH (cat.); ethylene glycol] 2-bromo-6-(1,3-dioxolan-2-yl)-pyridine (VI)

VI + 1-pyrrolidino-3-(4-tolyl)-3-propanone →[1. BuLi; 2. HCl] 6-[1-hydroxy-3-pyrrolidino-1-(p-tolyl)propyl]-2-pyridinecarboxaldehyde →[1. (COOH, COOH), pyridine; 2. H₂SO₄; 1. malonic acid] Acrivastine

Reference(s):

EP 85 959 (Wellcome; appl. 3.2.1983; GB-prior. 4.2.1983).
US 4 501 893 (Burroughs Wellcome; 26.2.1985; GB-prior. 4.2.1982).
US 4 562 258 (Burroughs Wellcome; 31.12.1985; GB-prior. 4.2.1982).
US 4 650 807 (Burroughs Wellcome; 17.3.1987; GB-prior. 4.2.1982).
US 4 657 918 (Burroughs Wellcome; 14.4.1987; GB-prior. 4.2.1982).
EP 249 950 (Wellcome; appl. 3.2.1983; GB-prior. 4.2.1982, 18.10.1982).

preparation of 2,6-dibromopyridine:
Nakagawa, H. et al.: Chem. Pharm. Bull. (CPBTAL) **46** (10), 1656-1657 (1998).
Malinowski, M., Kczmarek, L.: Synthesis (SYNTBF) **11** 1013-1015 (1987).
den Hertog; Wibaut: Recl. Trav. Chim. Pays-Bas (RTCPA3) **51** 940, 947 (1932).
McElvain; Goese: J. Am. Chem. Soc. (JACSAT) **65** 2227, 2230 (1943).

preparation of 2-bromo-6-(1,3-dioxolan-2-yl)pyridine:
Davies, S.R. et al.: J. Organomet. Chem. (JORCAI) **550** (1-2), 29 (1998).
Niemitz, J.: Synth. Commun. (SYNCAV) **11** (4), 273 (1981)
Heirtzler, F.R.; Neuberger, N.; Zehnder, Margareta; Constable, E.G.: Liebigs Ann./Recl. (LIARFV) (2), 297-302 (1997)

preparation of 6-bromopyridine-6-carboxaldehyde:
Meth-Cohn, O.; Jiang, H.: J. Chem. Soc., Perkin Trans. 1 (JCPRB4) **22**, 3737 (1998).
Uenishi, J.; Nishiwaki, K.; Hata, S., Nakamura, K.: Tetrahedron Lett. (TELEAY) **35** (43), 7973 (1994).
Ashimori, A. et al.: Chem. Pharm. Bull. (CPBTAL) **38** (9), 2446 (1990).

preparation of 1-pyrrolidino-3-(4-tolyl)-3-propanone *via Mannich-condensation from p*-methylacetophenone:
Adamson et al.: J. Chem. Soc. (JCSOA9) 312, 322 (1958).
Huang, Y.; Hall, I: Pharmazie (PHARAT) **51** (4), 199-206 (1996).

Formulation(s):   cps. 8 mg; syrup 4 mg

Trade Name(s):

GB:	Benadryl (Warner-Lambert Consumer)	Semprex (Calmic; 1988)	USA: Semprex-D (Medeva)
		I:	Semprex (Wellcome)

Actarit
(MS 932)

ATC: M01
Use: analgesic (non-opoid), antirheumatic, immunomodulator, antiarthritic

RN: 18699-02-0   MF: $C_{10}H_{11}NO_3$ MW: 193.20   EINECS: 242-511-3
LD$_{50}$: 14.7 g/kg (M, p.o.);
        14.8 g/kg (R, p.o.);
        >6.05 g/kg (dog, p.o.)
CN: 4-(acetylamino)benzeneacetic acid

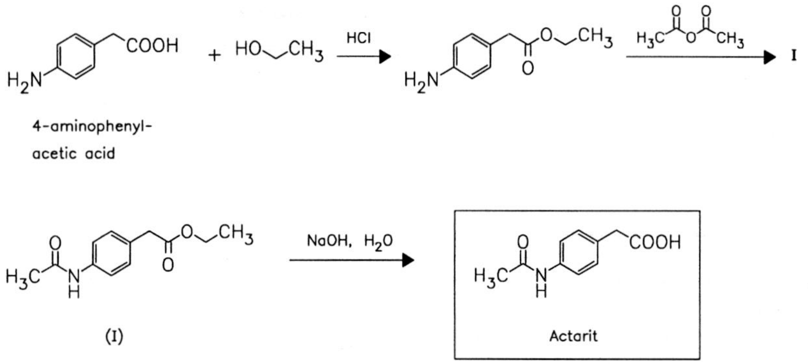

4-aminophenyl-
acetic acid

(I)   Actarit

Reference(s):
DE 3 317 107 (Mitsubishi Chem. Ind.; appl. 24.11.1983; J-prior. 11.5.1982).
EP 94 599 (Mitsubishi Chem. Ind.; appl. 23.11.1983; J-prior. 11.5.1982).
Yoshida, H. et al.: Int. J. Immunother. (IJIMET) **3**(4), 261 (1987).

Formulation(s):   tabl. 100 mg

Trade Name(s):
J: Mover (Mitsubishi Chem./   Orcl (Nippon Shinyaku;
      Nikken Chem.)   1994)

Actinoquinol
(Etoquinol)

ATC: D02B
Use: light protection agent

RN: 15301-40-3   MF: $C_{11}H_{11}NO_4S$ MW: 253.28   EINECS: 239-334-9
CN: 8-ethoxy-5-quinolinesulfonic acid

sodium salt
RN: 7246-07-3   MF: $C_{11}H_{10}NNaO_4S$ MW: 275.26   EINECS: 230-651-8

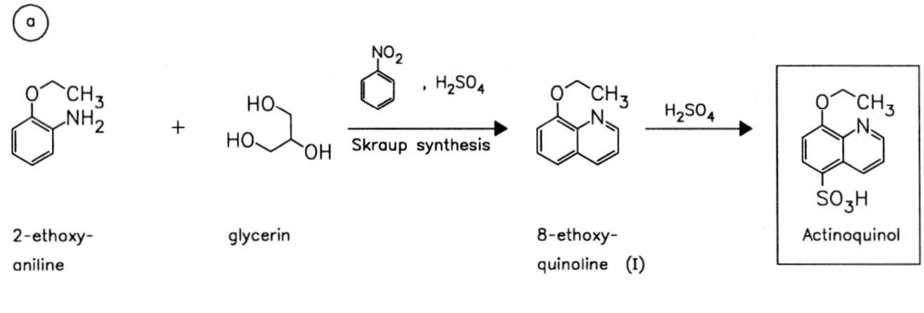

2-ethoxy-   glycerin   8-ethoxy-   Actinoquinol
aniline   quinoline (I)

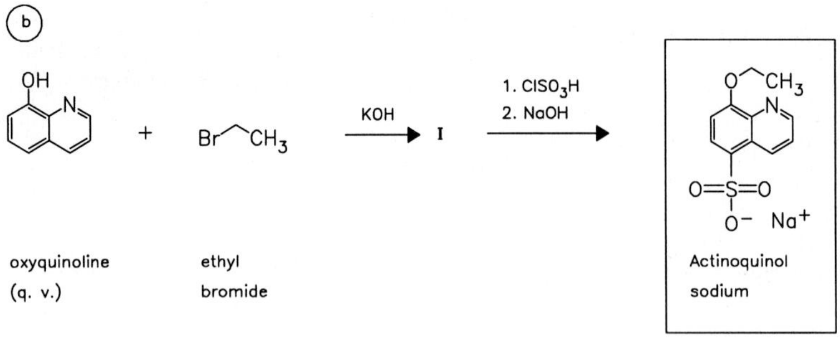

ⓑ

oxyquinoline   ethyl   Actinoquinol
(q. v.)   bromide   sodium

Reference(s):
Ghosh, T.N.; Roy, A.C.: J. Indian Chem. Soc. (JICSAH) **22**, 39 (1945).

Formulation(s):   eye drops 0.3 %

Trade Name(s):
D: dura Ultra (durachemie)-   Idrilsine (Winzer)-comb.   I: Fotofil (Intes)-comb.
      comb.   Tele-Stulln (Stulln)-comb.

Ademetionine

(Adenosylmethionine; Methioninyl adenylate; SAM)

ATC: A16AA02
Use: antirheumatic (degenerative arthropathy)

RN: 29908-03-0   MF: $C_{15}H_{22}N_6O_5S$ MW: 398.44   EINECS: 249-946-8
CN: 5'-[[(3S)-3-amino-3-carboxypropyl]methylsulfonio]-5'-deoxyadenosine inner salt

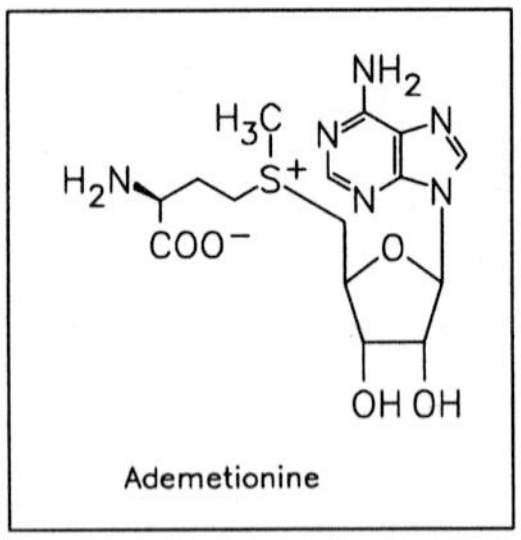

Ademetionine

Preparation by fermentation of *Saccharomyces cerevisiae* (baker yeast) with addition of L- or DL-methionine, lyse of cells with ethyl acetate and purification by ion-exchange chromatography.

Reference(s):
fermentation and isolation:
Schlenk: Enzymologia (ENZYAS) **29**, 283 (1965).
DE 1 803 978 (Boehringer Mannh.; appl. 18.10.1968).
US 3 962 034 (Ajinomoto; 8.6.1976; J.-prior. 27.11.1973).
DOS 3 231 569 (Nippon Zeon; appl. 25.8.1982).
DOS 3 304 468 (Nippon Zeon; appl. 9.2.1983; J.-prior. 25.2.1982, 26.2.1982).
DOS 3 329 218 (Nippon Zeon; appl. 12.8.1983; J.-prior. 13.8.1982).

stable salts:
4-toluenesulfonates:
DOS 2 336 401 (Errekappa Euroterapici; appl. 17.7.1973; I-prior. 2.8.1972).
US 3 893 999 (Bioresearch; 8.7.1975; I-prior. 2.8.1972).

4-toluenesulfonate sulfates:
US 3 954 726 (Bioresearch; 4.5.1976; I-prior. 27.6.1973; 24.5.1974).

other sulfonates:
DOS 2 530 898 (Bioresearch; appl. 10.7.1975; I-prior. 12.7.1974).
US 4 057 686 (Bioresearch; 8.11.1977; I-prior. 12.7.1974).
US 4 465 672 (Bioresearch; 14.8.1984; I-prior. 24.8.1981).
EP 72 980 (Bioresearch; appl. 12.8.1982; I-prior. 24.8.1981).
EP 162 323 (Bioresearch; appl. 25.4.1985; I-prior. 16.5.1984).
EP 162 324 (Bioresearch; appl. 25.4.1985; I-prior. 16.5.1984).

other salts:
EP 73 376 (Bioresearch; appl. 12.8.1982; I-prior. 24.8.1981).
EP 74 555 (Bioresearch; appl. 30.8.1982; I-prior. 11.9.1981).
EP 108 817 (Kanegafuchi; appl. 6.11.1982).
EP 141 462 (Tecofar; appl. 19.10.1984; I-prior. 26.10.1983).

formulations:
injection form:
EP 136 463 (Bioresearch; appl. 1.8.1984; I-prior. 24.8.1983).

gastric juice resistant form:
EP 136 464 (Bioresearch; appl. 1.8.1984; I-prior. 24.8.1983).

Formulation(s):   amp. 384 mg; tabl. 384 mg (as bisulfate)

Trade Name(s):

D:	Gumbaral (ASTA Medica AWD)	Ergen (San Carlo)	Turin (San Carlo)
		Samyr (Bioresearch)	
I:	Donamet (Knoll)	Transmetil (Bioresearch)	

Adiphenine

ATC: A03AA
Use: antispasmodic, anticholinergic

RN: 64-95-9   MF: $C_{20}H_{25}NO_2$ MW: 311.43   EINECS: 200-599-0
LD_{50}:   182 mg/kg (M, i.p.); 21.5 mg/kg (M, i.v.); 600 mg/kg (M, p.o.); 400 mg/kg (M, s.c.);
      27 mg/kg (R, i.v.);
      35 mg/kg (dog, i.v.);
      30 mg/kg (rabbit, i.v.)
CN: α-phenylbenzeneacetic acid 2-(diethylamino)ethylester

hydrochloride
RN: 50-42-0   MF: $C_{20}H_{25}NO_2 \cdot HCl$ MW: 347.89   EINECS: 200-036-9
LD_{50}:   185 mg/kg (M, i.p.); 500 mg/kg (M, p.o.); 650 mg/kg (M, s.c.);
      250 mg/kg (R, i.p.); 17.3 mg/kg (R, i.v.)

diphenylacetic acid → diphenylacetyl chloride → (with 2-diethylamino-ethanol) → Adiphenine

Reference(s):
DE 626 539 (Ciba; 1934).

Formulation(s):   drg. 20 mg, 25 mg; suppos. 40 mg, 50 mg

Adipiodone

(Iodipamide)

ATC: V08AC04
Use: X-ray contrast medium

RN: 606-17-7   MF: $C_{20}H_{14}I_6N_2O_6$ MW: 1139.76   EINECS: 210-105-5
LD$_{50}$: 2440 mg/kg (M, i.v.)
CN: 3,3'-[(1,6-dioxo-1,6-hexanediyl)diimino]bis[2,4,6-triiodobenzoic acid]

disodium salt
RN: 2618-26-0   MF: $C_{20}H_{12}I_6N_2Na_2O_6$ MW: 1183.73   EINECS: 220-049-3
LD$_{50}$: 3400 mg/kg (R, i.v.)
meglumine salt (1:2)
RN: 3521-84-4   MF: $C_{20}H_{14}I_6N_2O_6 \cdot 2C_7H_{17}NO_5$ MW: 1530.19   EINECS: 222-534-5
LD$_{50}$: 3195 mg/kg (M, i.v.);
       5000 mg/kg (R, i.v.); 1921 mg/kg (R, parenteral);
       1446 mg/kg (rabbit, parenteral);
       1200 mg/kg (dog, i.v.)

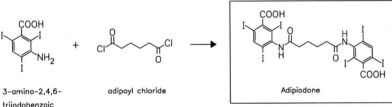

3-amino-2,4,6-
triiodobenzoic
acid
(cf. acetrizoic
acid synthesis)

adipoyl chloride

Adipiodone

Reference(s):
US 2 776 241 (Schering AG; 1957; D-prior. 1952).
DE 936 928 (Schering AG; appl. 1952).
DE 962 698 (Schering AG; appl. 1952).
DE 962 699 (Schering AG; appl. 1953).
DE 1 006 428 (Schering AG; appl. 1955).

starting material:
Kretzer, H.: Ber. Dtsch. Chem. Ges. (BDCGAS) 30, 1944 (1897).

Formulation(s):   amp. 20 ml with 300 mg meglumine salt/ml

Adrafinil
(CRL-40028)

ATC: N06BX17
Use: α-adrenergic agonist (for symptomatic treatment of vigilance and depressive manifestations), antidepressant

RN: 63547-13-7   MF: $C_{15}H_{15}NO_3S$ MW: 289.36   EINECS: 264-303-1
LD_{50}: >2048 mg/kg (M, i.p.); 1950 mg/kg (M, p.o.)
CN: 2-[(diphenylmethyl)sulfinyl]-N-hydroxyacetamide

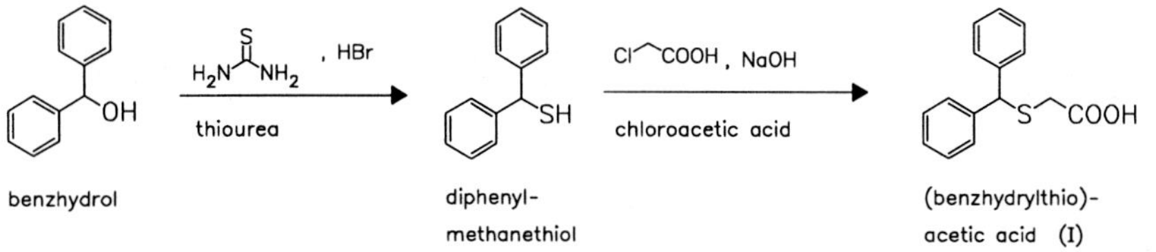

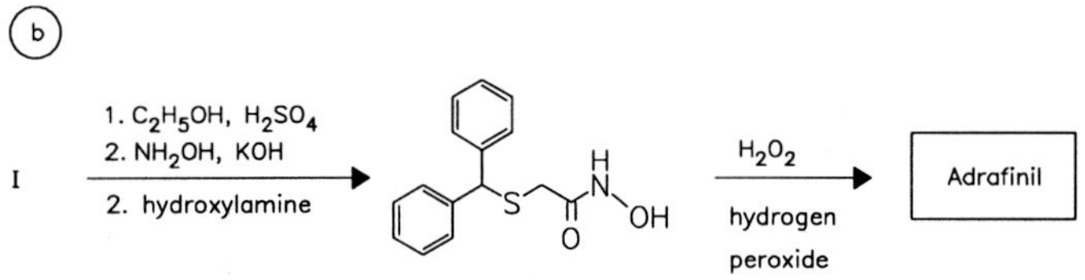

benzhydrol → [thiourea, HBr] → diphenyl-methanethiol → [chloroacetic acid, ClCH2COOH, NaOH] → (benzhydrylthio)-acetic acid (I)

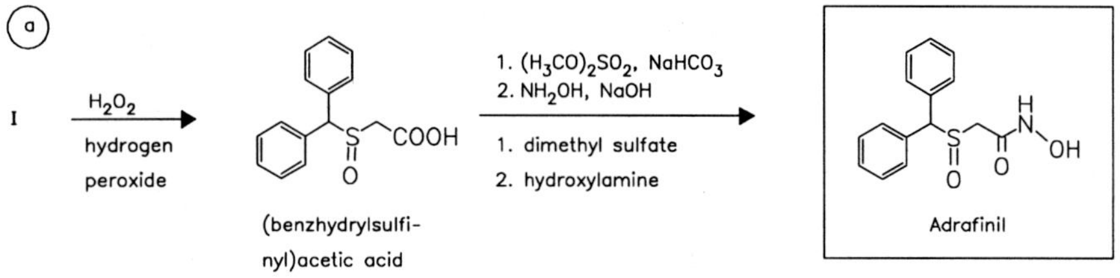

a)

I → [H2O2 hydrogen peroxide] → (benzhydrylsulfi-nyl)acetic acid → [1. (H3CO)2SO2, NaHCO3; 2. NH2OH, NaOH / 1. dimethyl sulfate; 2. hydroxylamine] → Adrafinil

b)

I → [1. C2H5OH, H2SO4; 2. NH2OH, KOH / 2. hydroxylamine] → [H2O2 hydrogen peroxide] → Adrafinil

Reference(s):
DOS 2 642 511 (Lab. Lafon; appl. 22.9.1976; GB-prior. 2.10.1975).
US 4 066 686 (Lab. Lafon; 3.1.1978; GB-prior. 2.10.1975).
US 4 098 824 (Lab. Lafon; 4.7.1978; GB-prior. 2.10.1975).

Formulation(s):   cps. 300 mg

Trade Name(s):
F: Olmifon (Lafon; 1985)

Adrenalone

ATC: A01AD06; B02BC05
Use: sympathomimetic, vasoconstrictor, hemostyptic

RN: 99-45-6   MF: $C_9H_{11}NO_3$ MW: 181.19   EINECS: 202-756-9
LD_{50}: 275 mg/kg (M, i.v.)
CN: 1-(3,4-dihydroxyphenyl)-2-(methylamino)ethanone

hydrochloride
RN: 62-13-5   MF: $C_9H_{11}NO_3 \cdot HCl$ MW: 217.65   EINECS: 200-525-7
LD_{50}: 902 mg/kg (M, i.p.)

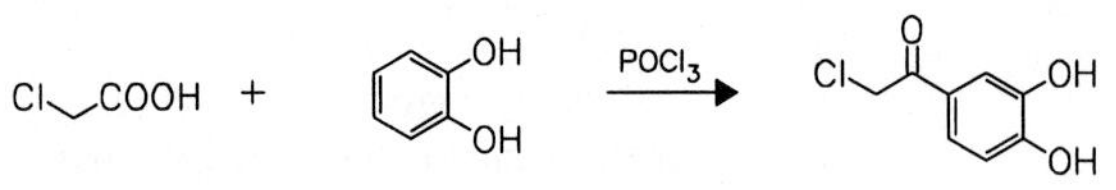

chloroacetic acid   catechol   2-chloro-3',4'-dihydroxy-acetophenone (I)

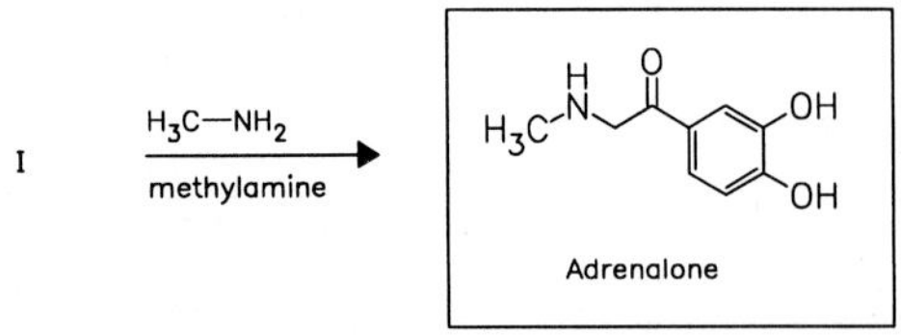

Adrenalone

Reference(s):
DRP 152 814 (Hoechst; 1903).

Formulation(s):   60 mg/stick

Trade Name(s):
D: Stryphnasal (Sertürner)   F: Adrénalone Tétracaine   Hémorrodine (Rocher)-
                                          Guillon (Pharmascience)-   comb.; wfm
                                          comb.; wfm

Afloqualone

ATC: M03A
Use: muscle relaxant

RN: 56287-74-2   MF: C$_{16}$H$_{14}$FN$_3$O   MW: 283.31
LD$_{50}$:   397 mg/kg (M, p.o.);
         249 mg/kg (R, p.o.)
CN: 6-amino-2-(fluoromethyl)-3-(2-methylphenyl)-4(3H)-quinazolinone

hydrochloride
RN: 56287-75-3   MF: C$_{16}$H$_{14}$FN$_3$O · xHCl   MW: unspecified

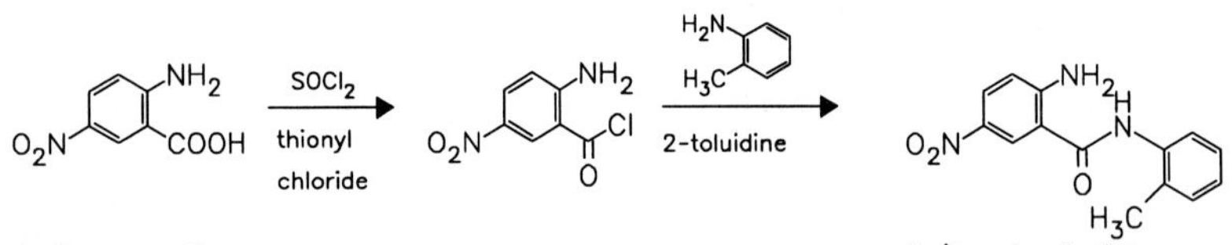

5-nitroanthranilic acid   N-(2-amino-5-nitro-benzoyl)-2-toluidine

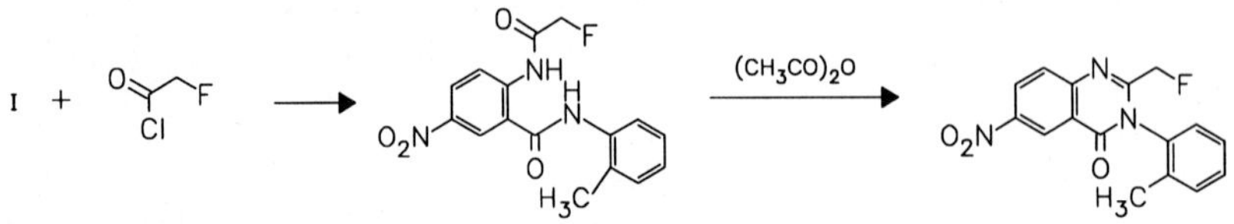

fluoroacetyl chloride   (II)

II $\xrightarrow{\text{H}_2/\text{Pd--C or SnCl}_2}$

Afloqualone

Reference(s):
DOS 2 449 113 (Tanabe; appl. 15.10.1974; J-prior. 15.10.1973).
US 3 966 731 (Tanabe; 29.6.1976; J-prior. 15.10.1973).
Tani, J. et al.: J. Med. Chem. (JMCMAR) **22**, 95 (1979).

Formulation(s):   tabl. 20 mg

Trade Name(s):
J: Aflospan (Kyowa)   Arofuto (Tanabe; 1983)

Ajmaline
(Rauwolfine)

ATC: C01BA05
Use: antiarrhythmic

RN: 4360-12-7   MF: $C_{20}H_{26}N_2O_2$ MW: 326.44   EINECS: 224-439-4
LD_{50}:   75 mg/kg (M, i.p.); 21 mg/kg (M, i.v.); 255 mg/kg (M, p.o.); 180 mg/kg (M, s.c.);
       94 mg/kg (R, i.p.); 26 mg/kg (R, i.v.); 360 mg/kg (R, p.o.); 216 mg/kg (R, s.c.)
CN: (17R,21α)-ajmalan-17,21-diol

monohydrochloride
RN: 4410-48-4   MF: $C_{20}H_{26}N_2O_2 \cdot HCl$ MW: 362.90   EINECS: 224-562-3
LD_{50}:   105 mg/kg (M, i.p.); 26 mg/kg (M, i.v.); 205 mg/kg (M, p.o.);
       86 mg/kg (R, i.p.); 19.3 mg/kg (R, i.v.); 290 mg/kg (R, p.o.);
       135 mg/kg (g. p., p.o.)

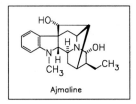

Ajmaline

By extraction from the pulverized roots of *Rauwolfia serpentina* (L.) Beuth.

Reference(s):
Siddiqui, S.; Siddiqui, R.H.: J. Indian Chem. Soc. (JICSAH) **8**, 667 (1931); **9**, 539 (1932); **12**, 37 (1935).

Formulation(s):   amp. 2 mg/2 ml, 50 mg/2 ml, 10 mg/10 ml, 50 mg/10 ml

Trade Name(s):
D: Gilurytmal (Solvay   F: Cardiorythmine (Servier);   Ritmosedina (Inverni della
       Arzneimittel)   wfm   Beffa)-comb.
       Tachmalin (ASTA Medica   Dipaxan (Innothéra)-   J: Gilurytmal (Giulini-Tokyo
       AWD)   comb.; wfm   Tanabe)
                              I: Aritmina (UCM)

Alacepril
(DU-1219)

ATC: C09A
Use: antihypertensive (ACE inhibitor),
       metabolizes partly to captopril

RN: 74258-86-9   MF: C$_{20}$H$_{26}$N$_2$O$_5$S   MW: 406.50
LD$_{50}$: >5 g/kg (M, p.o.);
        >5 g/kg (R, p.o.)
CN: (S)-N-[1-[3-(acetylthio)-2-methyl-1-oxopropyl]-L-prolyl]-L-phenylalanine

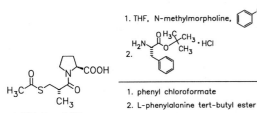

1-[(S)-3-acetylthio-
2-methylpropanoyl]-
L-proline
(cf. captopril synthesis)

1. THF, N-methylmorpholine, [phenyl chloroformate], −15 °C
2. [L-phenylalanine tert-butyl ester · HCl]

1. phenyl chloroformate
2. L-phenylalanine tert-butyl ester
   hydrochloride

→ I

anisole
CF$_3$COOH

(I)

Alacepril

Reference(s):
US 4 248 883 (Dainippon Pharmac. Co.; 3.2.1981; J-prior. 6.7.1978).
EP 7 477 (Dainippon Pharmac Co.; appl. 3.7.1979; J-prior. 6.7.1978).

pharmacology:
Takeyama, K. et al.: Arzneim.-Forsch. (ARZNAD) **35**, 1502 (1985).

metabolism:
Matsumoto, K. et al.: Arzneim.-Forsch. (ARZNAD) **36**, 40 (1986).

Formulation(s):   tabl. 12.5 mg, 25 mg

Trade Name(s):
J: Cetapril (Dainippon; 1988)

L-Alanine

Use: non-essential proteinogenic amino
       acid (part of infusion solutions)

RN: 56-41-7   MF: C$_3$H$_7$NO$_2$ MW: 89.09   EINECS: 200-273-8
CN: L-alanine

(a)   extraction from protein hydrolyzates by ion-exchange chromatography

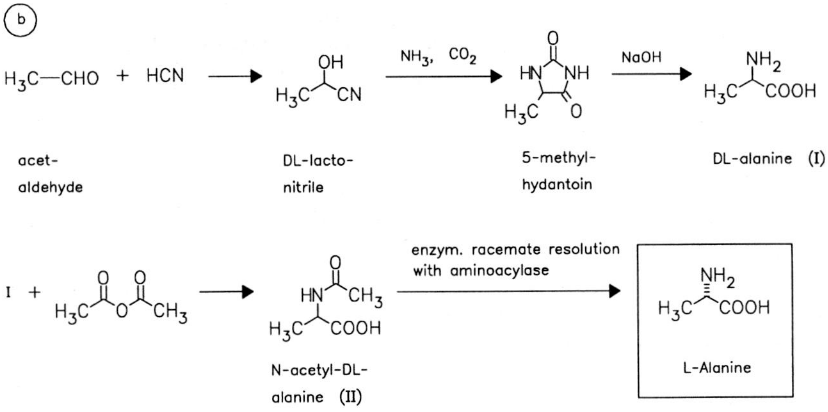

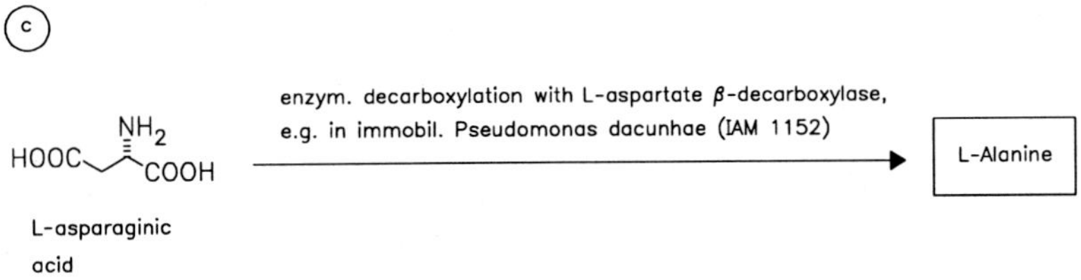

(b)

| acet-aldehyde | DL-lacto-nitrile | 5-methyl-hydantoin | DL-alanine (I) |

N-acetyl-DL-alanine (II)   L-Alanine

(c)

L-asparaginic acid

Reference(s):

review:
Ullmann's Encyclopedia of Industrial Chemistry, 5th Ed., Vol. **A2**, 69.
Kaneko, T.; Izumi, Y.; Chibata, I.; Itoh, T.: Synthetic Production and Utilization of Amino Acids, Kodansha Ltd. and John Wiley & Sons, Tokyo, New York, p. 62 (1974).
  c US 3 898 128 (Tanabe; 5.8.1975; J-prior. 20.11.1972).
     Yamamoto, K. et al.: Biotechnol. Bioeng. (BIBIAU) **22**, 2045 (1980).

Formulation(s):   tabl. 400 mg

Trade Name(s):
F: Theraplix (Abufene)

Alatrofloxacin mesilate
(CP 116517; CP 116517-27)

ATC: J01MA
Use: antibacterial, prodrug of trovafloxacin

RN: 146961-77-5   MF: $C_{26}H_{25}F_3N_6O_5 \cdot CH_4O_3S$ MW: 654.62
CN: L-Alanyl-*N*-[(1α,5α,6α)-3-[6-carboxy-8-(2,4-difluorophenyl)-3-fluoro-5,8-dihydro-5-oxo-1,8-naphthyridin-2-yl]-3-azabicyclo[3.1.0]hex-6-yl]-L-alaninamide monomethanesulfonate

base
RN: 146961-76-4   MF: $C_{26}H_{25}F_3N_6O_5$ MW: 558.52

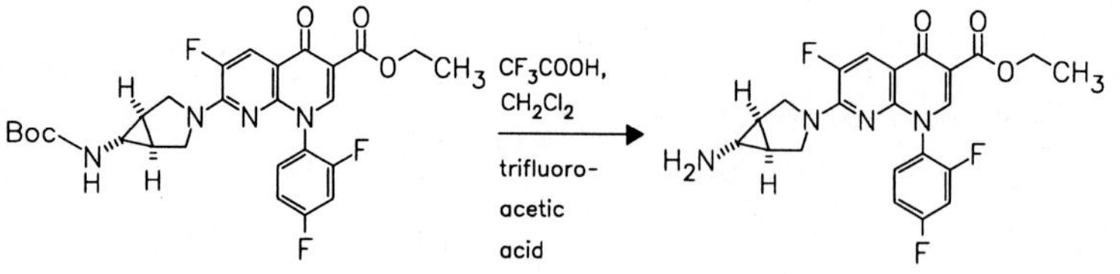
CF3COOH,
CH2Cl2
trifluoro-
acetic
acid

(cf. trovafloxacin mesylate) (I)

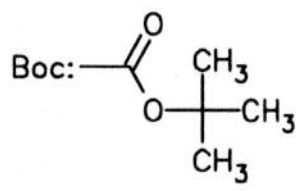
Boc:

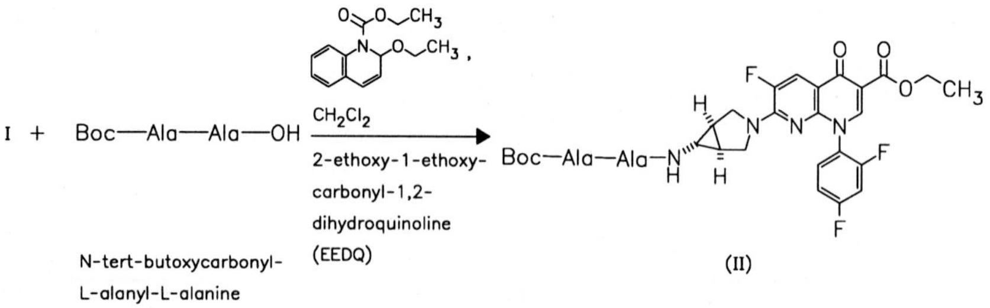
I +
Boc—Ala—Ala—OH
CH2Cl2
2-ethoxy-1-ethoxy-
carbonyl-1,2-
dihydroquinoline
(EEDQ)
N-tert-butoxycarbonyl-
L-alanyl-L-alanine
Boc—Ala—Ala—N
(II)

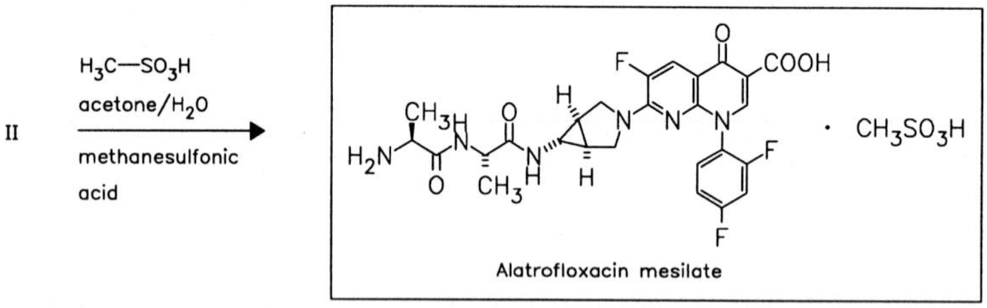
II
H3C—SO3H
acetone/H2O
methanesulfonic
acid
· CH3SO3H
Alatrofloxacin mesilate

Reference(s):
US 5 164 402 (Pfizer; 17.11.1992; appl. 4.2.1991; WO-prior. 16.8.1989).
WO 9 700 268 (Pfizer; appl. 27.3.1996; USA-prior. 15.6.1995).

Formulation(s): vials 200 mg/40 ml, 300 mg/60 ml (5 mg/ml) (as mesilate)

Trade Name(s):

D:	TROVAN (Pfizer); wfm	GB:	Turvel (Pfizer); wfm	USA:	Trovan (Pfizer); wfm
F:	Turvel (Pfizer); wfm	I:	Turvel (Pfizer); wfm		

Alclofenac

ATC: M01AB06
Use: analgesic, antipyretic, anti-inflammatory

RN: 22131-79-9 MF: $C_{11}H_{11}ClO_3$ MW: 226.66 EINECS: 244-795-4
LD_{50}: 508 mg/kg (M, i.p.); 1100mg/kg (M, p.o.);
465 mg/kg (R, i.p.); 1050 mg/kg (R, p.o.)
CN: 3-chloro-4-(2-propenyloxy)benzeneacetic acid

sodium salt
RN: 24049-18-1 MF: $C_{11}H_{10}ClNaO_3$ MW: 248.64
LD_{50}: 530 mg/kg (R, i.p.); 1050 mg/kg (R, p.o.)

a

H_2C~Br + allyl bromide (I) 2-chlorophenol $\xrightarrow{K_2CO_3}$ 2-chlorophenyl allyl ether (II)

II + HCHO formaldehyde $\xrightarrow{HCl, As_2O_3}$ 4-allyloxy-3-chloro-benzyl chloride $\xrightarrow[\text{sodium cyanide}]{NaCN, DMSO}$ III

4-allyloxy-3-chloro-benzyl cyanide (III) $\xrightarrow{KOH, C_2H_5OH}$ Alclofenac

b

I + 3-chloro-4-hydroxy-benzaldehyde $\longrightarrow$ 4-allyloxy-3-chloro-benzaldehyde (IV)

IV + formaldehyde dimethyl mercaptal S-oxide $\xrightarrow[\text{ammonium hydroxide}]{\text{benzyltrimethyl-}}$ 1-(2-methylsulfinyl-2-methyl-thiovinyl)-4-allyloxy-3-chlorobenzene $\xrightarrow{HCl}$ Alclofenac

Reference(s):
a BE 704 368 (Madan; appl. 27.9.1967).
BE 718 930 (Madan; appl. 1.8.1968; prior. 27.9.1967).
GB 1 174 535 (Madan; appl. 28.8.1968; B-prior. 27.9.1967, 1.8.1968).
b GB 1 504 828 (Sagami; appl. 26.11.1976; J-prior. 1.12.1975).

lysine salt:
DOS 2 711 964 (Biochefarm; appl. 18.3.1977).

Formulation(s):   amp. 833 mg; tabl. 1 g, 500 mg

Trade Name(s):
D: Neoston (Beiersdorf); wfm   I: Rentenac (Tosi); wfm   J;   Allopydin (Chugai)
GB: Prinalgin (Berk); wfm   Zumaril (Sidus); wfm   Epinal (Mitsubishi Yuka)

Alclometasone dipropionate

ATC: D07AB; S01BA
Use: topical steroidal anti-inflammatory (glucocorticoid)

RN: 66734-13-2   MF: $C_{28}H_{37}ClO_7$ MW: 521.05   EINECS: 266-464-3
LD$_{50}$:   2506 mg/kg (M, s.c.);
3593 mg/kg (R, s.c.)
CN: (7α,11β,16α)-7-chloro-11-hydroxy-16-methyl-17,21-bis(1-oxopropoxy)pregna-1,4-diene-3,20-dione

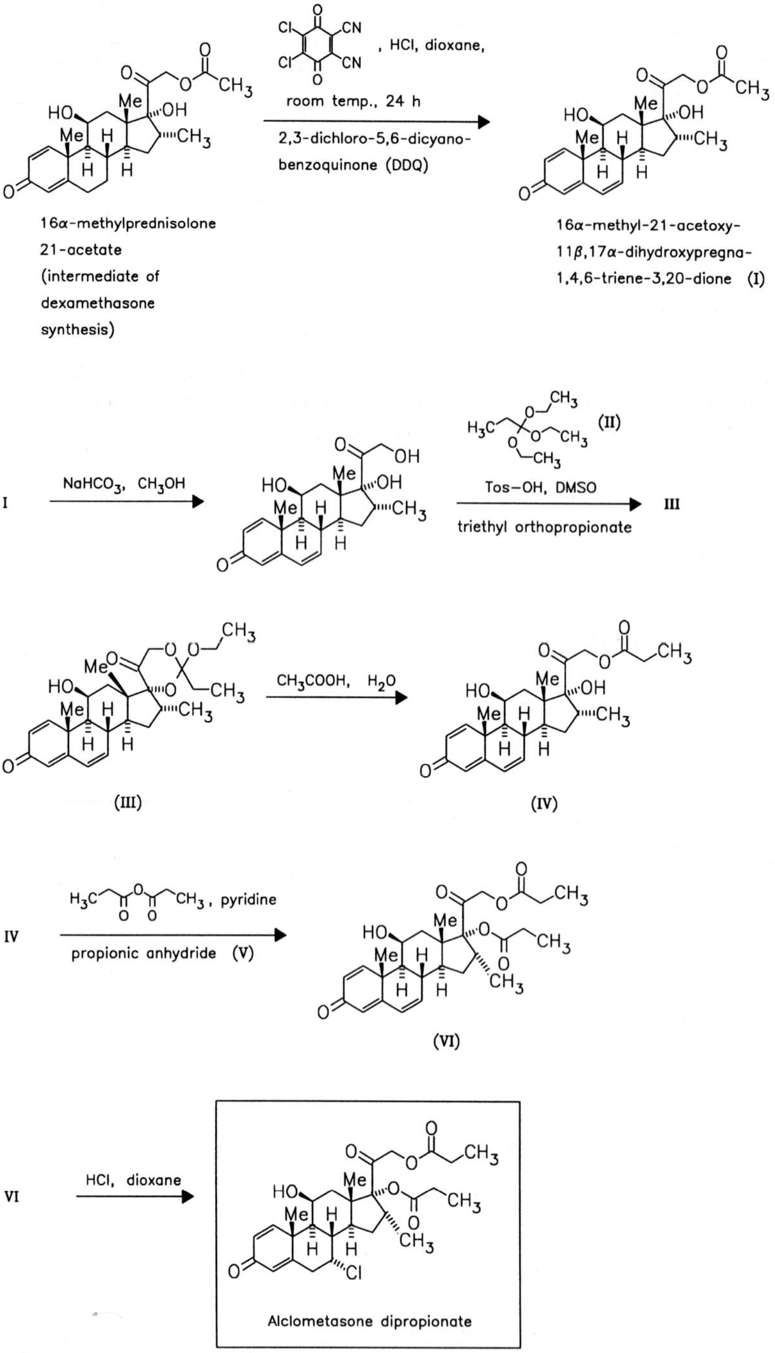

16α-methylprednisolone
21-acetate
(intermediate of
dexamethasone
synthesis)

, HCl, dioxane,

room temp., 24 h

2,3-dichloro-5,6-dicyano-
benzoquinone (DDQ)

16α-methyl-21-acetoxy-
11β,17α-dihydroxypregna-
1,4,6-triene-3,20-dione (I)

I

NaHCO₃, CH₃OH

(II)

Tos—OH, DMSO

triethyl orthopropionate

III

(III)

CH₃COOH, H₂O

(IV)

IV

H₃C⌐O⌐CH₃, pyridine

propionic anhydride (V)

(VI)

VI

HCl, dioxane

Alclometasone dipropionate

Reference(s):
Shue, H.-J.; Green, M.J.: J. Med. Chem. (JMCMAR) **23**, 430 (1980).
US 4 124 707 (Schering Corp.; 7.11.1978; prior. 12.12.1976, 7.11.1977).
US 4 076 708 (Schering Corp.; 28.2.1978; prior. 22.12.1976).
DOS 2 756 550 (Scherico; appl. 19.12.1977; USA-prior. 22.12.1976).

Formulation(s):   cream and ointment 0.5 mg/1 g

Trade Name(s):

D:	Delonal (Essex Pharma; 1985)	GB:	Modrasone (Dominion; 1986)	J:	Almeta (Shionogi)
F:	Aclosone (Schering-Plough)	I:	Legederm (Schering-Plough; 1988)	USA:	Aclovate (Glaxo Wellcome; 1986)

Alcuronium chloride

ATC: M03AA01
Use: muscle relaxant

RN: 15180-03-7 MF: $C_{44}H_{50}Cl_2N_4O_2$ MW: 737.82 EINECS: 239-229-8

LD_{50}: 610 µg/kg (M, i.p.); 240 µg/kg (M, i.v.); 38500 µg/kg (M, p.o.); 610 µg/kg (M, s.c.); 270 µg/kg (R, i.p.); 27600 µg/kg (R, p.o.); 280 µg/kg (R, s.c.)

CN: [1*R*-(1α,3a*S**,10α,11aβ,12α,14a*S**,19aα,20bα,21α,22aβ,23*E*,26*E*)]-2,3,11,11a,13,14,22,22a-octahydro-23,26-bis(2-hydroxyethylidene)-1,12-di-2-propenyl-10*H*,19a*H*,20b*H*,21*H*-1,21:10,12-diethano-dipyrrolo[3,2-*f*:3',2'-*f*'][1,5]diazocino[3,2,1-*jk*:7,6,5-*j'k'*]dicarbazolium dichloride

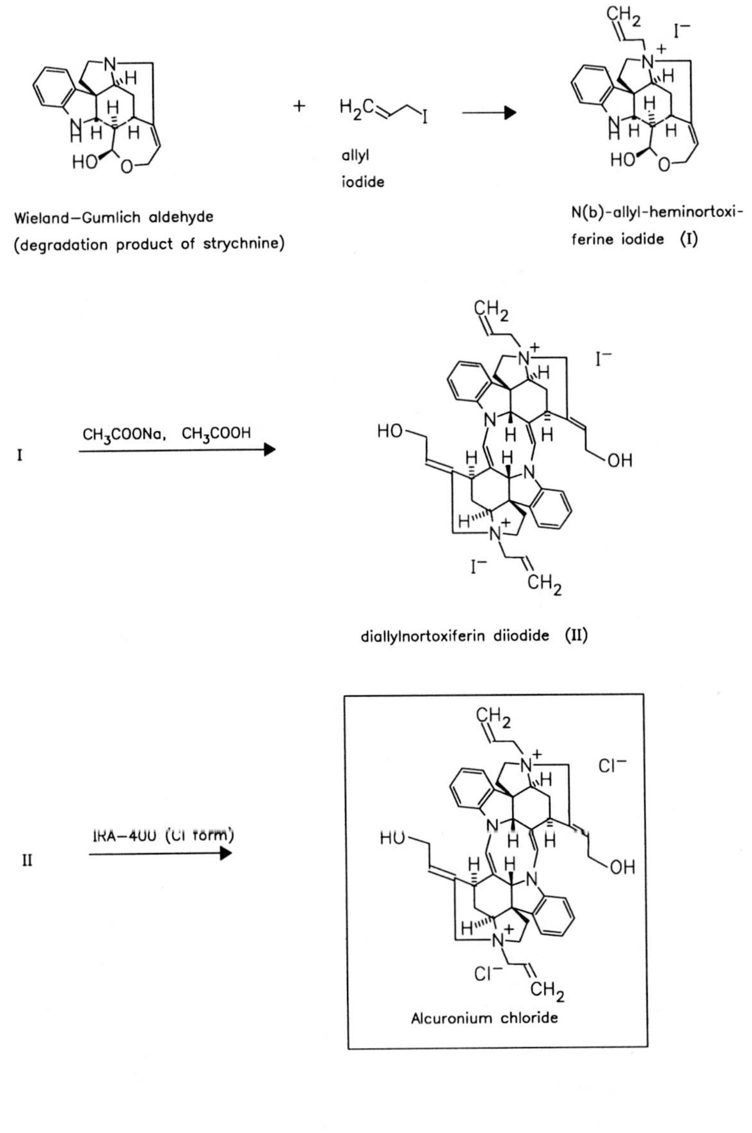
allyl
iodide
Wieland–Gumlich aldehyde
(degradation product of strychnine)
N(b)-allyl-heminortoxiferine iodide (I)
I
CH3COONa, CH3COOH
diallylnortoxiferin diiodide (II)
II
IRA-400 (Cl form)
Alcuronium chloride

Reference(s):
US 3 080 373 (Roche; 5.3.1963; F-prior. 29.8.1960).
Karrer, P. et al.: Angew. Chem. (ANCEAD) **70**, 644 (1958).

Formulation(s):   amp. 5 ml, 10 ml (1 mg/ml); inj. sol. 10 mg/2 ml

Trade Name(s):

D:	Alloferin Amp. (Roche)	GB:	Alloferin (Roche); wfm	USA:	Alloferin (Roche); wfm
F:	Alloférine (Roche); wfm	J:	Dialferin (Roche)		

Aldosterone

ATC: H02AA01
Use: mineralocorticoid

RN: 52-39-1   MF: $C_{21}H_{28}O_5$ MW: 360.45   EINECS: 200-139-9
CN: (11β)-11,21-dihydroxy-3,20-dioxopregn-4-en-18-al

21-O-acetylcorticosterone

NOCl, pyridine
nitrosyl chloride

21-O-acetyl-11-O-nitrosyl-
corticosterone (I)

I   hν, toluene, N₂ →

21-O-acetylaldosterone 18-oxime

NaNO₂, CH₃COOH →   II

21-O-acetylaldosterone (II)

basic saponification →

Aldosterone

Reference(s):
Barton, D.H.R.; Beaton, J.M.: J. Am. Chem. Soc. (JACSAT), **82**, 2641 (1960).

starting material:
The Merck Index, 2513 (Rahway 1976).

alternative syntheses:
US 3 002 972 (Ciba; 3.10.1961; appl. 28.11.1958; CH-prior. 5.12.1957).
US 3 014 029 (Ciba; 19.12.1961; appl. 16.6.1959; CH-prior. 18.6.1958).
US 3 049 539 (Wisconsin Alumni Res. Found.; 14.8.1962; appl. 29.7.1957).
Wettstein, A. et al.: Helv. Chim. Acta (HCACAV) **44**, 502 (1961).
Reichstein, T. et al.: Helv. Chim. Acta (HCACAV) **38**, 1432 (1957).

review:
Fieser, L.F.; Fieser, M.: Steroide p. 766 ff, Verlag Chemie, Weinheim 1961.

total synthesis:
Johnson, P.S. et al.: J. Am. Chem. Soc. (JACSAT) **80**, 2585 (1958).
Blickenstaff, R.T.; Ghosh, A.C.; Wolf, G.C.: Total Synthesis of Steroids (Organic Chemistry Vol. **30**) p. 187 ff, Academic Press, New York, London 1974.

Formulation(s):   tabl. 500 mg, 750 mg

Trade Name(s):
D: Aldocorten (Ciba); wfm   USA: Aldocortin (Burroughs
GB: Aldocorten (Ciba); wfm   Wellcome); wfm
I: Sinsurrene Forte (Parke   Electrocortin (Ciba-Geigy);
      Davis)-comb.   wfm

Alendronate sodium

ATC: M05BA04
Use: treatment of osteoporosis

RN: 121268-17-5   MF: $C_4H_{12}NNaO_7P_2 \cdot 3H_2O$ MW: 325.12
LD_{50}:   >4 g/kg (dog, p. o.)
CN: (4-Amino-1-hydroxybutylidene)bis[phosphonic acid] monosodium salt trihydrate

acid
RN: 66376-36-1   MF: $C_4H_{13}NO_7P_2$ MW: 249.10
anhydrous monosodium salt
RN: 129318-43-0   MF: $C_4H_{12}NNaO_7P_2$ MW: 271.08

γ-aminobutyric acid

(I)

Alendronate sodium

Reference(s):
WO 9 506 052 (Merck & Co.; USA-prior. 25.8.1993).
WO 9 533 756 (Merck & Co.; appl. 2.6.1995; USA-prior. 6.6.1994).
US 5 510 517 (Merck & Co.; 2.3.1995; USA-prior. 25.8.1993).
DE 3 016 289 (Henkel KG; D prior. 28.4.1980).
BE 896 453 (Ist. Gentili s. p. a.; appl. 14.4.1983; I-prior. 15.4.1982, 16.2.1983).
BE 903 513 (Ist. Gentili s. p. a.; appl. 25.10.1985; I-prior. 29.10.1984).
EP 494 844 (Ist. Gentili s. p. a.; appl. 2.1.1992; I-prior. 8.1.1991).
US 4 621 077 (Ist. Gentili s. p. a.; 8.6.1984; I-prior. 15.4.1982).
US 5 019 651 (Merck & Co.; 27.12.1991; USA-prior. 20.6.1990).
US 4 922 007 (Merck & Co.; 1.5.1990; USA-prior. 9.6.1989).

alternative process for the production of alendronate:
WO 9 834 940 (Apotex Inc.; CA-prior. 11.2.1997).

Formulation(s):   amp. 5 mg, 10 mg; tabl. 5 mg, 10 mg, 40 mg (as sodium salt)

Trade Name(s):

D: Fosamax (Merck Sharp & Dohme)
GB: Fosamax (Merck Sharp & Dohme)
I: Adronat (Neopharmed)

Alendros (Gentili)
Dronal (Sigmatau)
Fosamax (Merck Sharp & Dohme)
J: Onclast (Banyu)

Teiroc (Teijin)
USA: Fosamax (Merck Sharp & Dohme; 1993)

Alfacalcidol

(1α-Hydroxycholecalciferol; 1α-Hydroxy-vitamin D₃)

ATC: A11CC03
Use: calcium metabolism regulator, vitamin D-derivative

RN: 41294-56-8 MF: $C_{27}H_{44}O_2$ MW: 400.65 EINECS: 255-297-1
LD$_{50}$: 440 g/kg (M, p.o.);
340 g/kg (R, p.o.)
CN: (1α,3β,5Z,7E)-9,10-secocholesta-5,7,10(19)-triene-1,3-diol

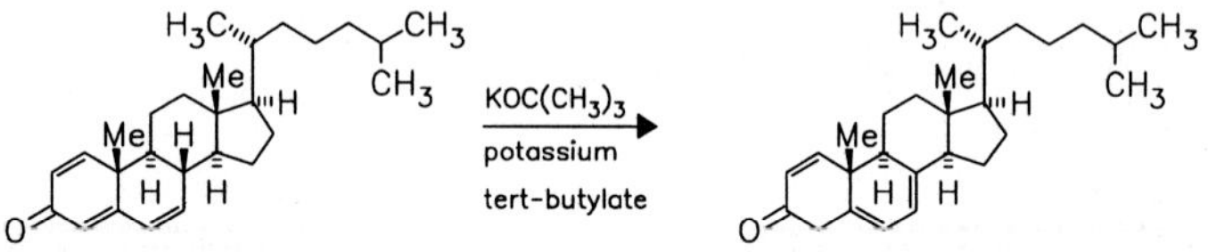

3-oxo-1,4,6-cholestatriene

KOC(CH₃)₃
potassium tert-butylate

3-oxo-1,5,7-cholestatriene (I)

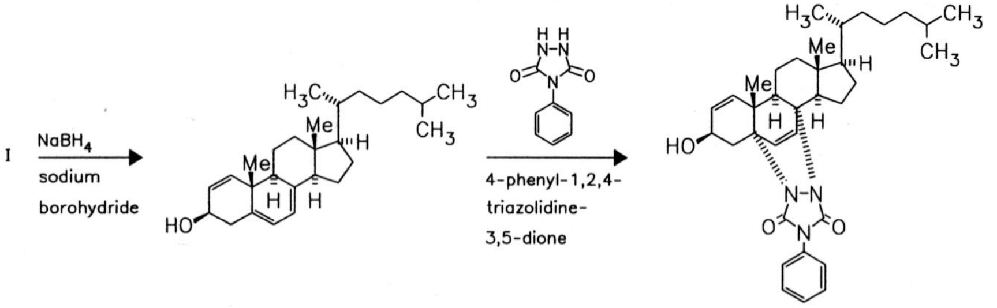

I → NaBH₄ sodium borohydride

1,5,7-cholestatrien-3β-ol

4-phenyl-1,2,4-triazolidine-3,5-dione

(II)

II → 3-chloroper-benzoic acid

LiAlH₄, THF
lithium aluminum hydride

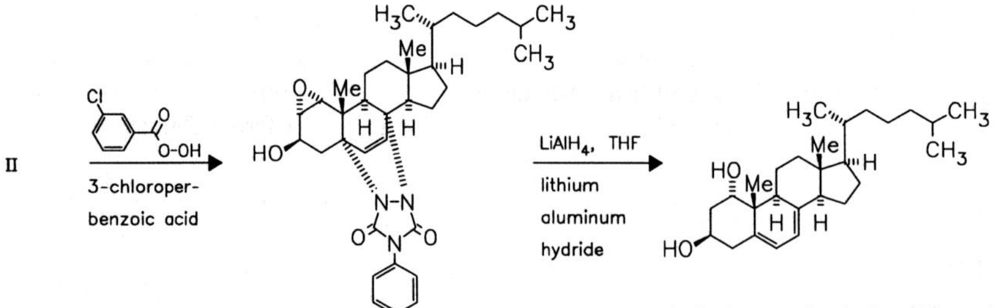

1α-hydroxyprovitamin D₃ (III)

Alfacalcidol

Reference(s):
US 3 929 770 (Wisconsin Alumni Res.; 30.12. 1975; J-prior. 3.12.1973).

alternative syntheses:
Holick, M.F. et al.: Science (Washington, D.C.) (SCIEAS) **180**, 190 (1973).
Barton, D.H.R. et al.: J. Am. Chem. Soc. (JACSAT) **95**, 2748 (1973).
Fürst, A. et al.: Helv. Chim. Acta (HCACAV) **56**, 1708 (1973).
US 3 966 777 (Yeda Res. & Devel.; 29.6.1976; IL-prior. 22.10.1974).
DOS 2 259 661 (Wisconsin Alumni Res.; appl. 1.12.1972; USA-prior. 2.12.1971).
BE 877 356 (Wisconsin Alumni Res.; appl. 28.6.1979; USA-prior. 15.1.1979, 21.5.1979).
GB 1 553 321 (Merck & Co.; valid from 30.6.1977; USA-prior. 1.7.1976).
DOS 2 923 953 (Upjohn; appl. 13.6.1979; USA-prior. 19.6.1978).

total synthesis:
Harrison, R.G. et al.: Tetrahedron Lett. (TELEAY) **1973**, 3649.

synthesis of intermediates:
US 4 046 760 (Merck & Co., 6.9.1977; prior. 1.7.1976).

pharmaceutical formulation:
JP-appl. 78 136 512 (Chugai; appl. 28.4.1977).
US 4 164 569 (Chugai; 14.8.1979; J-prior. 8.4.1977).

use as anti-inflammatory:
FR 2 389 377 (J. Brohult, appl. 6.5.1977).

Formulation(s):   amp. 0.001 mg, 0.002 mg; cps. 0.001 mg, 0.0025 mg, 1 mg; inj. 2 µg/ml

Trade Name(s):

D:	Bondiol (Gry)		One Alpha (Leo)		Diseon (Smith Kline &
	Eins Alpha (Leo)	I:	Dediol (Rhône-Poulenc		French)
F:	Un-Alfa (Leo)		Rorer)	J:	Alfarol (Chugai)
GB:	Alfa D (Berk)				

Alfadolone acetate
(Alphadolone acetate)

ATC: N01A
Use: anesthetic (intravenous)

RN: 23930-37-2   MF: $C_{23}H_{34}O_5$ MW: 390.52   EINECS: 245-942-5
LD_{50}: >30 mg/kg (rabbit, i.v.)
CN: (3α,5α)-21-(acetyloxy)-3-hydroxypregnane-11,20-dione

alfadolone
RN: 14107-37-0   MF: $C_{21}H_{32}O_4$ MW: 348.48   EINECS: 237-961-2
LD_{50}: 59 mg/kg (M, i.v.)

(a)

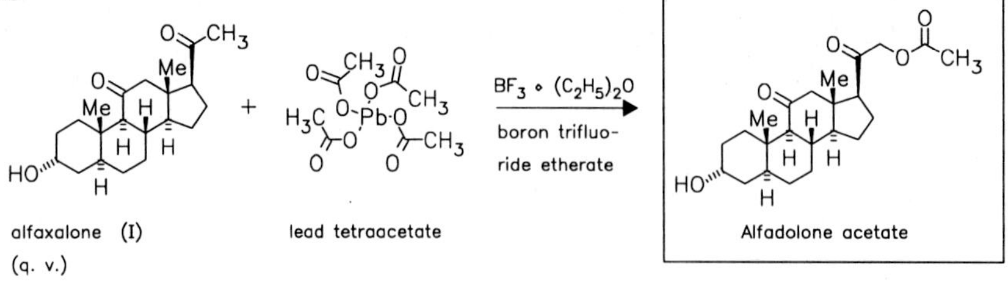

alfaxalone (I)   lead tetraacetate   Alfadolone acetate
(q. v.)

(b)

21-bromoalfaxalone

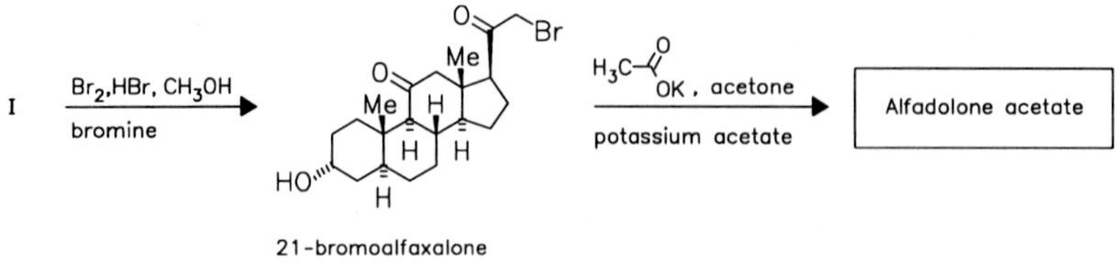

Reference(s):
DE 2 030 402 (Glaxo; appl. 19.6.1970; GB-prior. 20.6.1969; 11.6.1970).
ZA 703 861 (Glaxo; appl. 8.6.1970; GB-prior. 20.6.1969)
(alternative synthesis).
Browne, P.A.; Kirk, D.N.: J. Chem. Soc. (JCSOA9) **1969**, 1653.

Formulation(s):   amp. 0.5 mg/ml

Trade Name(s):
D: Aurantex (Glaxo)-comb.;   GB: Althesin (Glaxo)-comb.;
       wfm   wfm
F: Alfatesine (Glaxo)-comb.;   I: Althesin (Glaxo)-comb.;
       wfm   wfm

Alfaxalone
(Alphaxalone)

ATC: N01AX05
Use: anesthetic (intravenous)

RN: 23930-19-0   MF: $C_{21}H_{32}O_3$ MW: 332.48
LD$_{50}$:   430 mg/kg (M, i.p.); 36.9 mg/kg (M, i.v.); 880 mg/kg (M, p.o.); 5220 mg/kg (M, s.c.);
       116 mg/kg (R, i.p.); 19.4 mg/kg (R, i.v.); 297 mg/kg (R, p.o.); >2200 mg/kg (R, s.c.);
       9.36 mg/kg (rabbit, i.v.)
CN: (3α,5α)-3-hydroxypregnane-11,20-dione

acetate
RN: 51267-69-7   MF: $C_{23}H_{34}O_4$ MW: 374.52

(a)

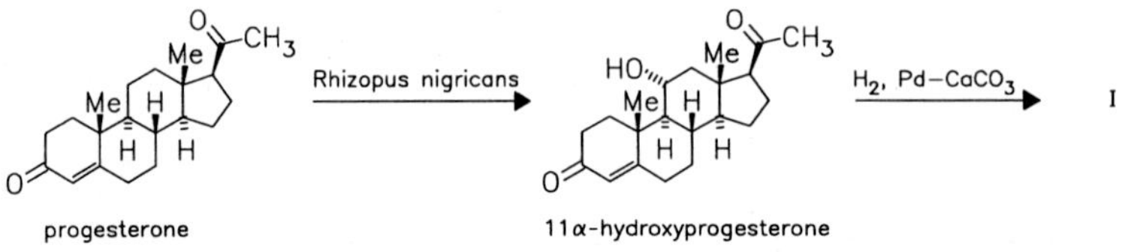

progesterone   11α-hydroxyprogesterone

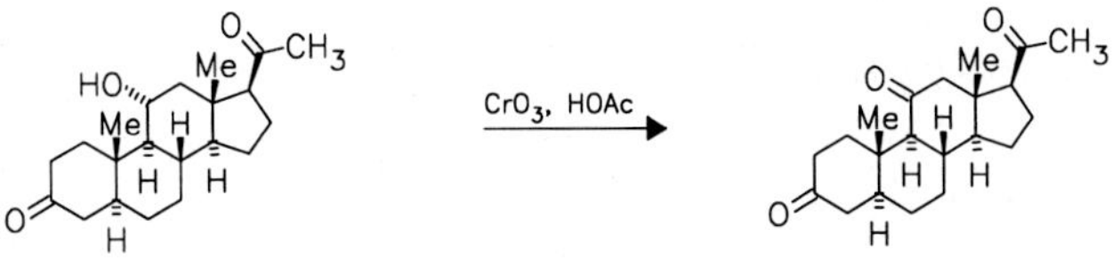

11α-hydroxy-5α-pregnane-
3,20-dione (I)

CrO₃, HOAc →

3,11,20-trioxo-5α-
pregnane (II)

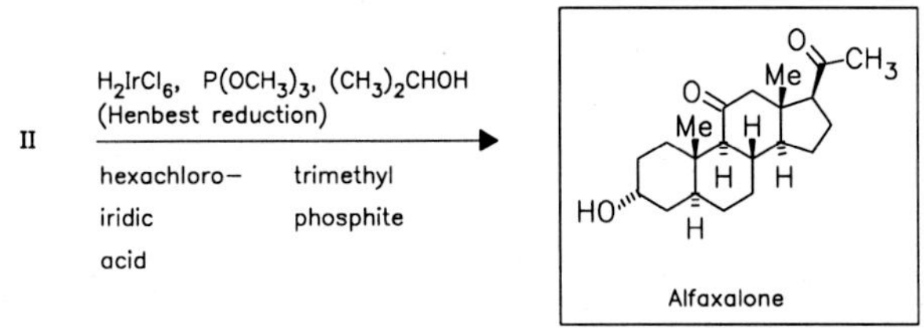

II

H₂IrCl₆, P(OCH₃)₃, (CH₃)₂CHOH
(Henbest reduction)

hexachloro-   trimethyl
iridic   phosphite
acid

→

Alfaxalone

(b)

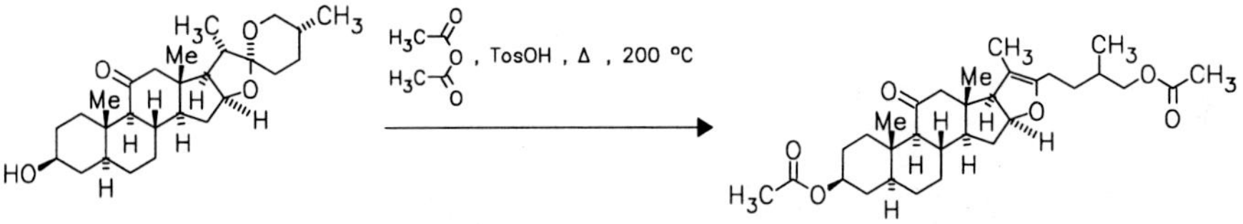

11-oxotigogenin
(from hecogenin)

$H_3C-C(O)-O-C(O)-CH_3$, TosOH , Δ , 200 °C →

3β,26-diacetoxy-5α-furost-
20(22)-en-11-one (III)

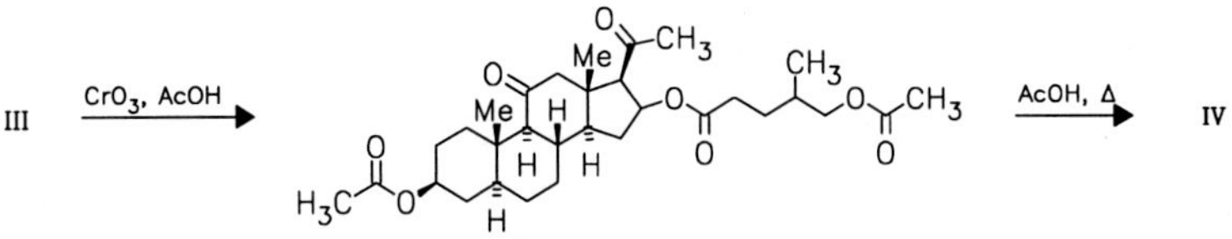

III   CrO₃, AcOH →

3β-acetoxy-16-(5-acetoxy-4-methyl-
pentanoyl)-5α-pregnane-11,20-dione

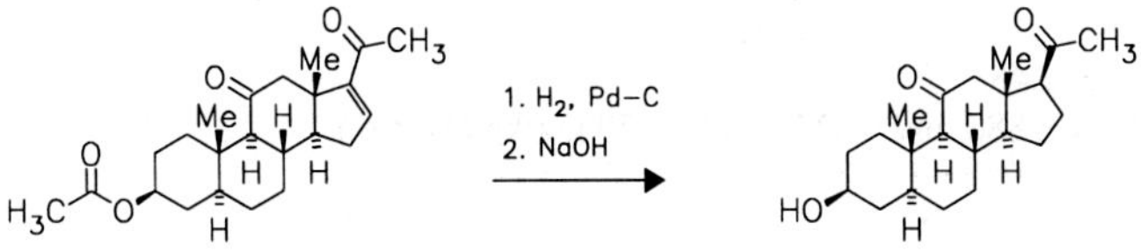

AcOH, Δ →   IV

3β-acetoxy-11,20-dioxo-
16-pregnene (IV)

1. H₂, Pd-C
2. NaOH →

11,20-dioxo-3β-hydroxy-
5α-pregnane (V)

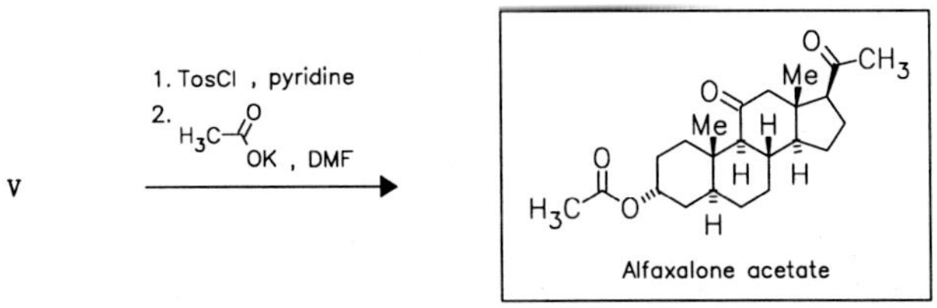

V

1. TosCl , pyridine
2. $H_3C-C(O)-OK$, DMF
→

Alfaxalone acetate

Reference(s):
a   Browne, P.A.; Kirk, D.N.: J. Chem. Soc. C (JSOOAX) **1969**, 1653.
b   Nagata, W. et al.: Helv. Chim. Acta (HCACAV) **42**, 1399 (1959).

medical use:
DE 2 030 402 (Glaxo; appl. 19.6.1970; GB-prior. 20.6.1969, 11.6.1970).

Formulation(s):　amp. 5 ml, 10 ml, 0.3 %

Trade Name(s):
J:　　Alphadione (Shin Nihon
　　　Jitsugyo)

Alfentanil

ATC:　N01AH02
Use:　analgesic, short-time anesthetic (for basal narcosis)

RN:　71195-58-9　MF: C$_{21}$H$_{32}$N$_6$O$_3$　MW: 416.53
CN:　N-[1-[2-(4-ethyl-4,5-dihydro-5-oxo-1H-tetrazol-1-yl)ethyl]-4-(methoxymethyl)-4-piperidinyl]-N-phenylpropanamide

monohydrochloride
RN:　69049-06-5　MF: C$_{21}$H$_{32}$N$_6$O$_3$ · HCl　MW: 452.99　EINECS: 273-846-3
monohydrochloride monohydrate
RN:　70879-28-6　MF: C$_{21}$H$_{32}$N$_6$O$_3$ · HCl · H$_2$O　MW: 471.00

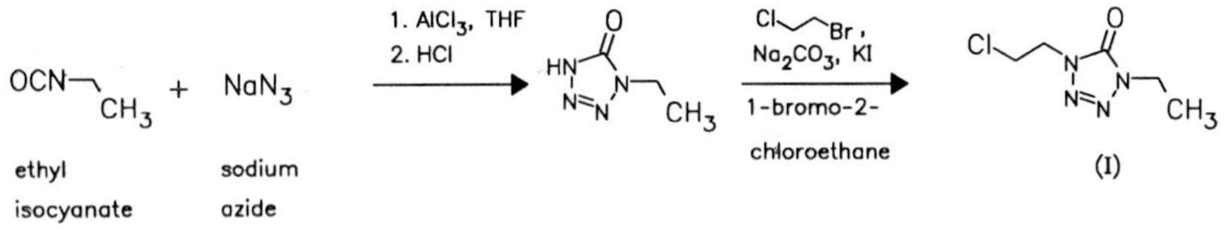

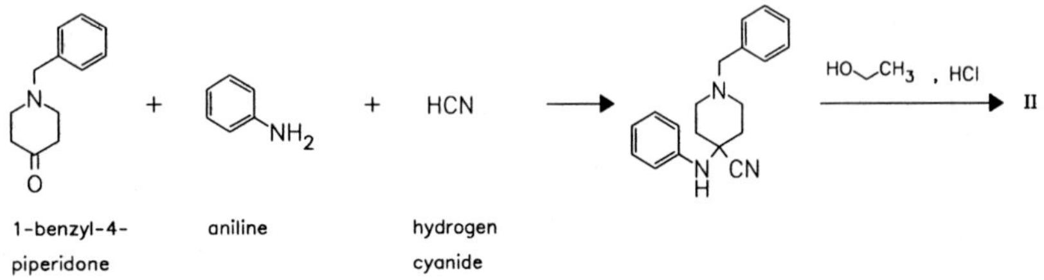

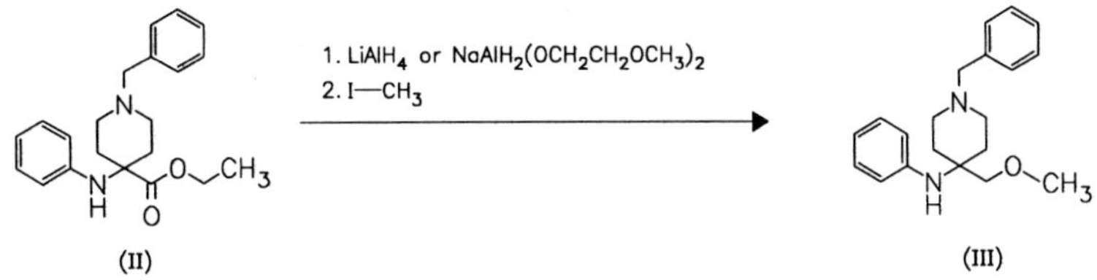

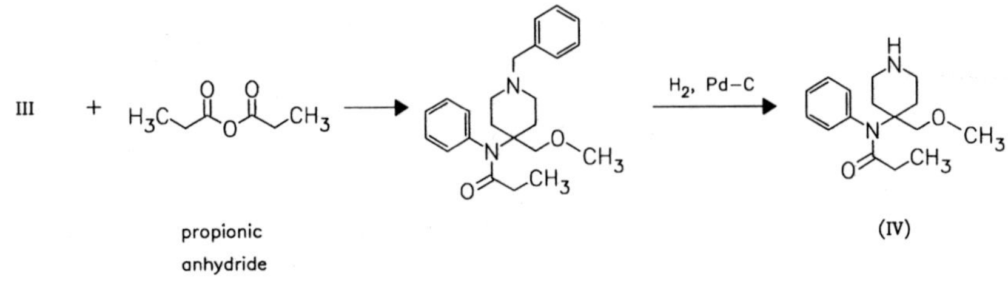

ethyl isocyanate　+　sodium azide　→　→　1-bromo-2-chloroethane　→　(I)

1-benzyl-4-piperidone　+　aniline　+　HCN　→　→　II

(II)　→　(III)

III　+　propionic anhydride　→　H$_2$, Pd–C　→　(IV)

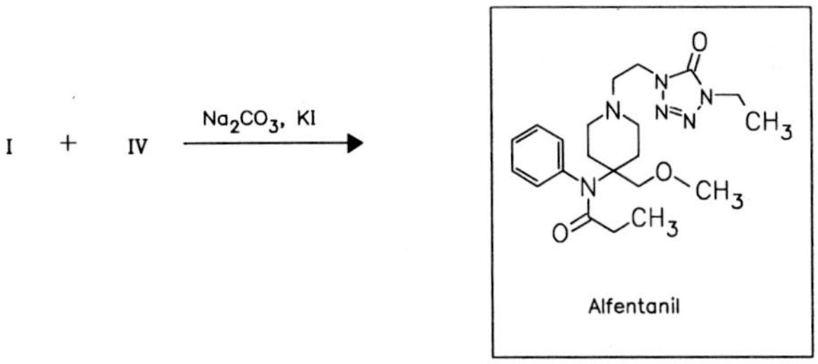
I + IV
Na2CO3, KI
CH3
CH3
CH3
Alfentanil

Reference(s):
GB 1 598 872 (Janssen; appl. 3.5.1978; USA-prior. 5.5.1977).
DOS 2 819 873 (Janssen; appl. 5.5.1978; USA-prior. 5.5.1977, 13.3.1978).
US 4 167 574 (Janssen; 11.9.1979; appl. 25.10.1978; prior. 13.3.1978).

Formulation(s): amp. 500 µg/ml; inj. sol. 1 mg/2 ml, 5 mg/10 ml; intensive care inj. 5 mg/ml

Trade Name(s):

D:	Rapifen (Janssen-Cilag; 1983)	GB:	Rapifen (Janssen-Cilag; 1983)
F:	Rapifen (Janssen-Cilag)	USA:	Alfenta (Janssen; 1987)

Alfuzosin

ATC: C02CA; G04CB01
Use: antihypertensive, α_1-antagonist, treatment of benign prostatic hypertrophy

RN: 81403-80-7 MF: $C_{19}H_{27}N_5O_4$ MW: 389.46
CN: (±)-*N*-[3-[(4-amino-6,7-dimethoxy-2-quinazolinyl)methylamino]propyl]tetrahydro-2-furancarboxamide

monohydrochloride
RN: 81403-68-1 MF: $C_{19}H_{27}N_5O_4 \cdot HCl$ MW: 425.92

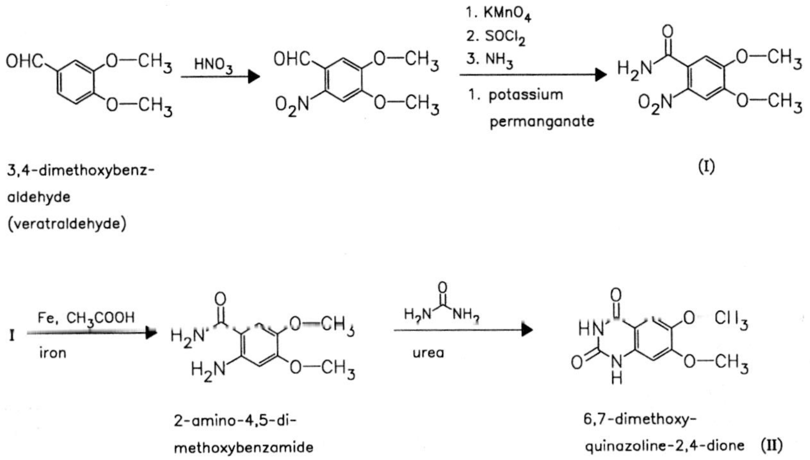
OHC
O—CH3
O—CH3
HNO3
O2N
1. KMnO4
2. SOCl2
3. NH3
1. potassium permanganate
H2N
3,4-dimethoxybenz-aldehyde (veratraldehyde)
(I)
I
Fe, CH3COOH
iron
H2N
urea
HN
2-amino-4,5-dimethoxybenzamide
6,7-dimethoxy-quinazoline-2,4-dione (II)

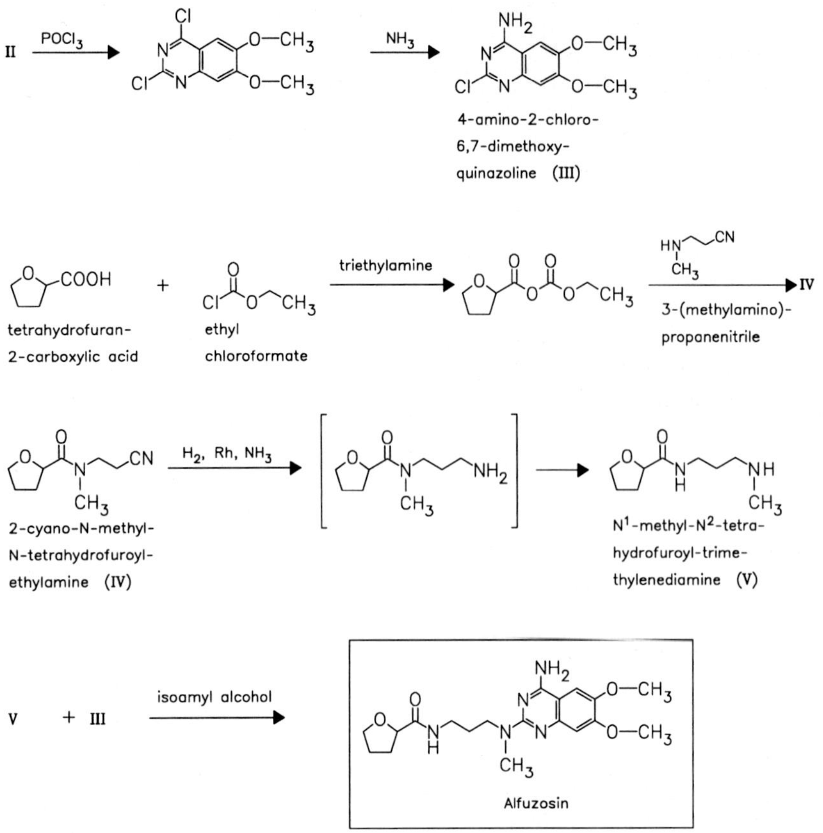

Reference(s):

US 4 315 007 (Synthelabo; 9.2.1982; F-prior. 6.2.1978, 29.12.1978).

DE 290 445 (Synthelabo; appl. 16.8.1979; F-prior. 6.2.1978, 29.12.1978).

Manoury, P.M. et al.: J. Med. Chem. (JMCMAR) **29**, 19 (1986).

synthesis of 6,7-dimethoxyquinazoline-2,4-dione:

Althuis, T.H.; Hess, H.J.: J. Med. Chem. (JMCMAR) **20**, 146 (1977).

Formulation(s):   tabl. 2.5 mg (as hydrochloride)

Trade Name(s):

D:	Urion (Byk Gulden)		Xatral (Synthélabo; 1989)	Xatral (Synthelabo)
	Uroxatral (Synthelabo)	GB:	Xatral (Lorex)	
F:	Urion (Zambon)	I:	Mittoval (Schering)	

Algestone acetophenide

(Alfasone acetophenide; Alphasone acetophenide)

ATC: D10AX; G03DA

Use: antiacne, progestogen

RN: 24356-94-3   MF: C$_{29}$H$_{36}$O$_4$ MW: 448.60   EINECS: 246-195-8

CN: [16α(R)]-16,17-[(1-phenylethylidene)bis(oxy)]pregn-4-ene-3,20-dione

algestone

RN: 595-77-7   MF: C$_{21}$H$_{30}$O$_4$ MW: 346.47   EINECS: 209-869-2

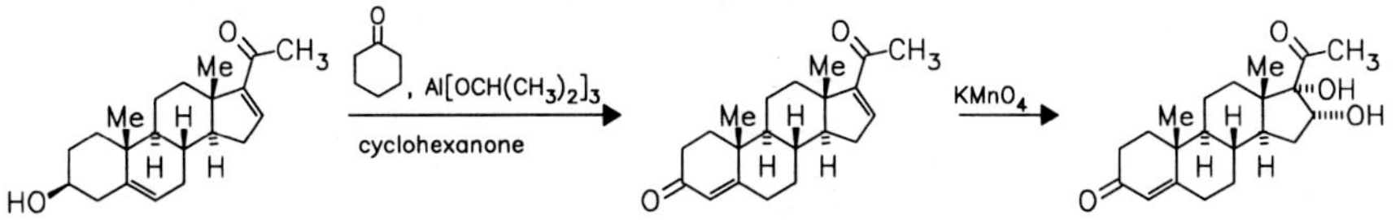

16-dehydropregnenolone

16-dehydroprogesterone

16α,17α-dihydroxy-
progesterone (I)

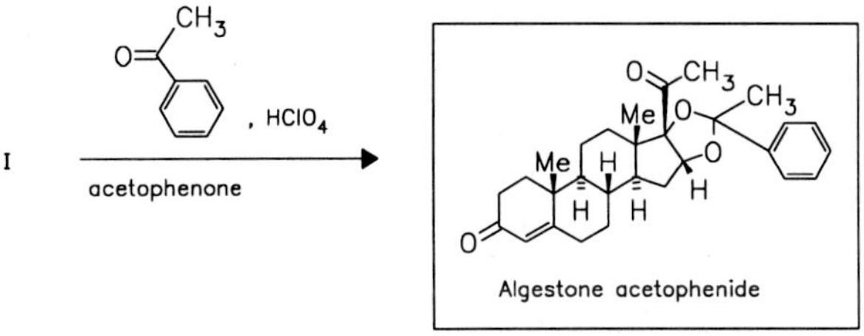

Algestone acetophenide

Reference(s):
DE 1 125 423 (Olin Mathieson; appl. 1959; USA-prior. 1958).
Fried, J. et al.: Chem. Ind. (London) (CHINAG) **1961**, 465.

alternative synthesis:
US 3 008 958 (Olin Mathieson; 1961; prior. 1961).

synthesis of intermediates:
US 2 727 909 (Searle; 1955; prior. 1954).
US 3 165 541 (Olin Mathieson; 12.1.1965; prior. 20.5.1963).
Cooley, G. et al.: J. Chem. Soc. (JCSOA9) **1955**, 4373.
Inhoffen, H.H. et al.: Chem. Ber. (CHBEAM) **87**, 593 (1954).
Hydorn, A.E. et al.: Steroids (STEDAM) **3**, 493 (1964).

injection solution:
US 3 164 520 (Olin Mathieson; 5.1.1965; prior. 29.10.1962).

medical use as contraceptive:
GB 1 060 632 (Olin Mathieson; appl. 16.8.1963; USA-prior. 11.11.1962).

Formulation(s):   cream 2 %

Trade Name(s):
I: Neolutin Depos. (Medici)

Alibendol

ATC: C10A; A03
Use: antispasmodic, choleretic,
cholekinetic

RN: 26750-81-2   MF: $C_{13}H_{17}NO_4$ MW: 251.28   EINECS: 247-960-9
LD$_{50}$:   >3000 mg/kg (M, p.o.); >2000 mg/kg (M, s.c.)
CN: 2-hydroxy-*N*-(2-hydroxyethyl)-3-methoxy-5-(2-propenyl)benzamide

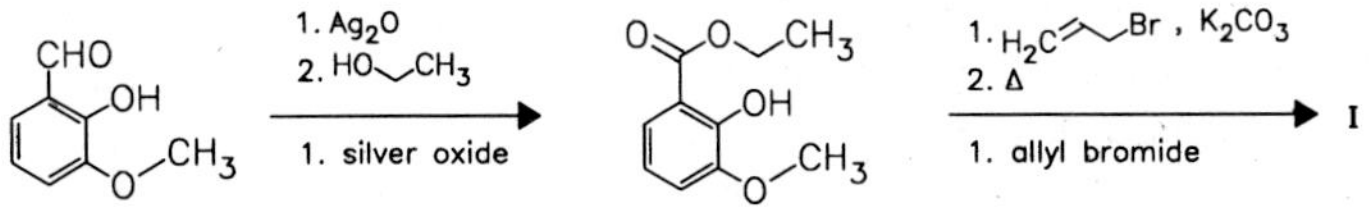

2-hydroxy-3-
methoxy-
benzaldehyde

ethyl 2-hydroxy-
3-methoxybenzoate

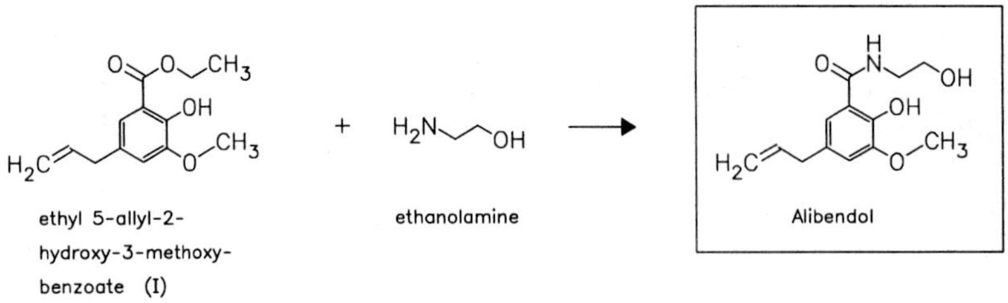

H2C
OH
CH3
H2N
OH
ethyl 5-allyl-2-hydroxy-3-methoxybenzoate (I)
ethanolamine
Alibendol

Reference(s):
DE 1 768 615 (Roussel-Uclaf; appl. 1968; F-prior. 1967).
Clemence, F. et al.: Chim. Ther. (CHTPBA) **5**, 188 (1970).

Formulation(s): tabl. 100 mg

Trade Name(s):
F: Cebera (Irex)

Alimemazine
(Trimeprazine)

ATC: R06AD01
Use: antihistaminic, psychosedative

RN: 84-96-8 MF: $C_{18}H_{22}N_2S$ MW: 298.45 EINECS: 201-577-3
LD_{50}: 33 mg/kg (M, i.v.); 300 mg/kg (M, p.o.);
35 mg/kg (R, i.v.); 210 mg/kg (R, p.o.)
CN: *N*,*N*,β-trimethyl-10*H*-phenothiazine-10-propanamine

tartrate (2:1)
RN: 4330-99-8 MF: $C_{18}H_{22}N_2S \cdot 1/2C_4H_6O_6$ MW: 746.99 EINECS: 224-368-9
LD_{50}: 33 mg/kg (M, i.v.); 300 mg/kg (M, p.o.);
35 mg/kg (R, i.v.); 210 mg/kg (R, p.o.)

NaNH2
sodium amide
phenothiazine
3-dimethylamino-2-methylpropyl chloride
Alimemazine

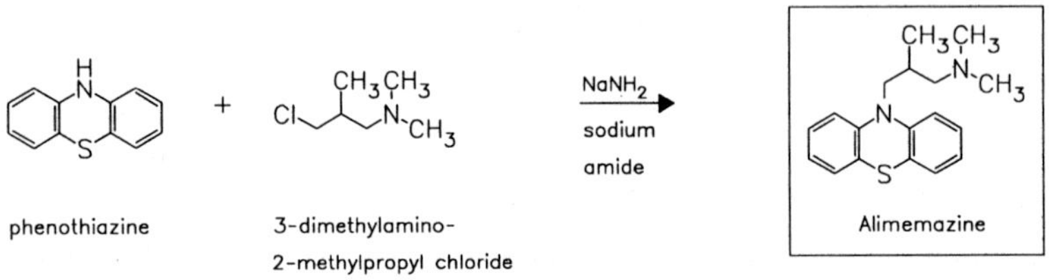

Reference(s):
US 2 837 518 (Rhône-Poulenc; 1958; F-prior. 1954).
DE 1 034 639 (Rhône-Poulenc; appl. 1955; GB-prior. 1954 and 1955).

Formulation(s): drops 40 mg; tabl. 2.5 mg, 5 mg (as tartrate)

Trade Name(s):
D: Repeltin (Bayer)
F: Théralène (Evans Medical)
Théralène Pectoral (Evans Medical)-comb.
GB: Vallergan (Rhône-Poulenc Rorer; as tartrate)
I: in comb. with prednisolone
J: Alimezine (Daiichi)

Alizapride

ATC: A03FA05; A04AD
Use: anti-emetic, neuroleptic

RN: 59338-93-1   MF: C$_{16}$H$_{21}$N$_5$O$_2$ MW: 315.38   EINECS: 261-710-6
LD$_{50}$:   92.7 mg/kg (M, i.v.)
CN: 6-methoxy-*N*-[[1-(2-propenyl)-2-pyrrolidinyl]methyl]-1*H*-benzotriazole-5-carboxamide

p-amino-
salicylic acid
(q. v.)

dimethyl
sulfate

methyl 4-amino-
2-methoxybenzoate

1. HNO$_3$
2. H$_2$, Raney—Ni

methyl 4,5-diamino-
2-methoxybenzoate   (I)

NaNO$_2$, HCl
sodium
nitrite

methyl 6-methoxy-
benzotriazole-5-
carboxylate

1-allyl-2-
aminomethyl-
pyrrolidine

Alizapride

Reference(s):
DE 2 500 919 (Delagrange; appl. 11.1.1975).
US 4 039 672 (Delagrange; 2.8.1977; D-prior. 11.1.1975).

synthesis of methyl 4-amino-2-methoxybenzoate:
DOS 1 966 212 (Yamanouchi; appl. 29.12.1969; J-prior. 2.12.1968, 9.12.1968, 4.4.1969).

Formulation(s):   amp. 50 mg/2 ml; drinking amp. 360 mg; suppos. 50 mg; tabl. 50 mg

Trade Name(s):
D: Vergentan (Synthelabo)   I: Limican (Synthelabo)
F: Plitican (Synthélabo)   Nausilen (Baldacci)

Allantoin

ATC: D03; D05
Use: wound remedy, antipsoriatic,
adstringent, web stimulant,
keratolytic, antacid

RN: 97-59-6   MF: C$_4$H$_6$N$_4$O$_3$ MW: 158.12   EINECS: 202-592-8
CN: (2,5-dioxo-4-imidazolidinyl)urea

Alcloxa

RN: 1317-25-5   MF: C$_4$H$_9$Al$_2$ClN$_4$O$_7$ MW: 314.55   EINECS: 215-262-3
LD$_{50}$:   >8 g/kg (M,R, p.o.)
CN: chloro[(2,5-dioxo-4-imidazolidinyl)uretato]tetrahydroxyaluminum

Aldioxa

RN: 5579-81-7   MF: C$_4$H$_7$AlN$_4$O$_5$ MW: 218.11   EINECS: 226-964-4
LD$_{50}$:   >8 g/kg (M, p.o.)
CN: [(2,5-dioxo-4-imidazolidinyl)ureato]dihydroxyaluminum

urea   glyoxylic   Allantoin
              acid

Alcloxa   Aldioxa

Reference(s):
DOS 1 939 924 (BASF; appl. 6.8.1969).

from glyoxal via "in situ"-glyoxylic acid:
DOS 2 714 938 (Akad. d. Wiss. der DDR; appl. 2.4.1977; DDR-prior. 29.10.1976).

from chloral hydrate via "in situ"-glyoxylic acid:
DOS 2 717 698 (Akad. d. Wiss. der DDR; appl. 21.4.1977; DDR-prior. 29.10.1976).

by oxidation of uric acid with PbO, *or* H$_2$O$_2$ *or* potassium permanganate:
Org. Synth. (ORSYAT) **13** 1 (1933).

by oxidation of glycoluril *with* H$_2$O$_2$:
Biltz, H.; Schiemann, G.: J. Prakt. Chem. (JPCEAO) **113**, 92 (1926).
US 2 802 011 (Carbogen Corp.; 1957; appl. 1956).

by condensation of glyoxylic acid esters or glyoxylic acid acetal esters with urea:
US 2 158 098 (Merck & Co.; 1939; appl. 1937).

Formulation(s):   cream 0.2 %; ointment 2 %; powder 0.5 %; tabl. 100 mg

Trade Name(s):
D: more than 70 combination
        preparations
        allantoin
        Brand- und Wundgel (Eu
        Rho Arznei)-comb.
        Contractubex Gel (Merz &
        Co.)-comb.
        Ellsurex (Galderma)-comb.
        Essaven (Nattermann)-
        comb.
        HAEMO-Exhirud (Sanofi
        Winthrop)-comb.
        Hydro Cordes (Block Drug
        Company; Ichthyol)-comb.

Lipo Cordes (Block Drug
Company)-comb.
Psoralon (Hermal)-comb.
Psoriasis-Salbe M
(Balneopharm)
Ulcurilen (Spitzner)-comb.
alcloxa
Ansudor (Basotherm)-
comb.
aldioxa
Ansudor (Basotherm)-
comb.
Dexa-Mederma Akne
(Merz & Co.)-comb.

Elmedal (Thiemann)-comb.
Mederima (Merz & Co.)-
comb.
ZeaSorb Puder (Stiefel)-
comb.
F: *alcloxa*
        Ulfon (Lafon)-comb.
        aldioxa
        Ulfon (Lafon)-comb.
GB: *allantoin*
        Actinac (Hoechst)-comb.
        with chloramphenicol and
        hydrocortisone

	Alphosyl (Stafford-Miller)-comb.	Antiacne Samil (Samil)-comb.	Cervex (Medics)-comb.; wfm

Alphosyl (Stafford-Miller)-comb.

Aphosyl HC (Stafford-Miller)-comb. with hydrocortisone

Dermalex (Sanofi Winthrop)-comb. with squalene and hexachlorophane

I: *allantoin*

Alphosyle (Poli)-comb.

Antiacne Samil (Samil)-comb.

Apsor pomata (IDI Farmaceutici)-comb.

J: *aldioxa*

Aldioxa (Isei)

Chlokale (Sawai)

USA: *allantoin*

Alphosyl (Reed & Carnrick)-comb.; wfm

Bahnex (Maxsil)-comb.; wfm

Cervex (Medics)-comb.; wfm

Cutemol Creme (Summers); wfm

Herpecin-L (Campbell)-comb.; wfm

Sufamal (Milex)-comb.; wfm

Vagilia (Lemmon)-comb.; wfm

Allobarbital

(Allobarbitone)

ATC: N05CA21
Use: hypnotic, sedative

RN: 52-43-7   MF: $C_{10}H_{12}N_2O_3$ MW: 208.22   EINECS: 200-140-4
LD$_{50}$:   218 mg/kg (M, i.v.)
CN: 5,5-di-2-propenyl-2,4,6(1*H*,3*H*,5*H*)-pyrimidinetrione

allyl bromide   barbituric acid   Allobarbital

Reference(s):
DRP 268 158 (Ciba; appl. 1911).
DRP 526 854 (Hoffmann-La Roche; appl. 1930).

Formulation(s):   tabl. 30 mg, 100 mg, 300 mg

Trade Name(s):
D: Toximer (Merckle)-comb.; wfm

F: Spasmo-Cibalgine (Ciba)-comb.; wfm

I: Allobarb (Tariff. Integrativo)

USA: Diadol (Durst); wfm

Allopurinol

ATC: M04AA01
Use: uricosuric agent

RN: 315-30-0   MF: $C_5H_4N_4O$ MW: 136.11   EINECS: 206-250-9
LD$_{50}$:   >1 g/kg (M, p.o.)
CN: 1,5-dihydro-4*H*-pyrazolo[3,4-*d*]pyrimidin-4-one

a

ethyl cyanoacetate + triethyl ortho-formate → ethyl ethoxymethylene-cyanoacetate (I)

I $\xrightarrow{\text{H}_2\text{N}-\text{NH}_2}{\text{hydrazine}}$ ethyl 5-amino-pyrazole-4-carboxylate $\xrightarrow{\text{H}_2\text{N}-\text{CHO}}{\text{formamide}}$ Allopurinol

b

cyanoacetamide + formamidine hydrochloride → 3-amino-2-cyano-acrylamide (II)

II $\xrightarrow{\text{H}_2\text{N}-\text{NH}_2}{\text{hydrazine}}$ 5-aminopyrazole-4-carboxamide $\xrightarrow{\text{H}_2\text{N}-\text{CHO}}{\text{formamide}}$ Allopurinol

Reference(s):
a   US 2 868 803 (Ciba; 13.1.1959; CH-prior. 10.2.1956).
    US 3 624 205 (Burroughs Wellcome; 30.11.1971; USA-prior. 25.4.1967).
b   DAS 1 720 024 (Wellcome Found; appl. 12.7.1967; GB-prior. 14.7.1966).

similar process:
DAS 1 904 894 (Wellcome Found; appl. 31.1.1969; GB-prior. 2.2.1968).
US 4 146 713 (Burroughs Wellcome; 27.3.1979; GB-prior. 2.2.1968).

alternative syntheses:
US 3 474 098 (Burroughs Wellcome; 21.10.1969; prior. 29.3.1956).
DAS 2 224 382 (Henning Berlin; appl. 18.5.1972).
DE 1 118 221 (Wellcome Found; appl. 4.8.1956; GB-prior. 10.8.1955).
DAS 1 814 082 (Wellcome Found; appl. 11.12.1968).
DAS 1 950 075 (Henning Berlin; appl. 3.10.1969).
DOS 2 018 345 (Delmar Chemicals; appl. 16.4.1970; GB-prior. 17.4.1969).

combination with benzbromarone:
GB 1 493 237 (Henning Berlin; appl. 11.5.1976; D-prior. 10.12.1975).

Formulation(s):   tabl. 100 mg, 200 mg, 300 mg

Trade Name(s):
D: Allo-300-Tablinen (ct-Arzneimittel)   Allomaron (Nattermann)-comb.   Allo-Puren 100/-300 (Isis Puren)
Bleminol (gepepharm)

Cellidrin (Henning)
dura Al 300 (durachemie)
Foligan (Henning Berlin)
Remid 100/-300 (TAD)
Suspendol (Merckle)
Uribenz 300 (R.A.N.)
Uripurinol 100/300
(Azupharma)
Urosin (Boehringer
Mannh.)

Zyloric (Glaxo Wellcome;
1966)
combination preparations
F: Zyloric (Glaxo Wellcome;
1968)
GB: Zyloric (Glaxo Wellcome;
1966)
I: Allopuri (Formulario Naz.)
Allurit (RBS Pharma)

Allurit (Rhône-Poulenc
Rorer)
Uricemil (ICT)
Uricodue (IFI)-comb.
Zyloric (Wellcome; 1969)
J: Zyloric (Tanabe; 1969)
USA: Zyloprim (Glaxo
Wellcome; 1966)

Allylestrenol
(Allyloestrenol)

ATC: G03DC01
Use: progestogen

RN: 432-60-0 MF: C$_{21}$H$_{32}$O MW: 300.49 EINECS: 207-082-9
LD$_{50}$: >640 mg/kg (M, p.o.)
CN: (17β)-17-(2-propenyl)estr-4-en-17-ol

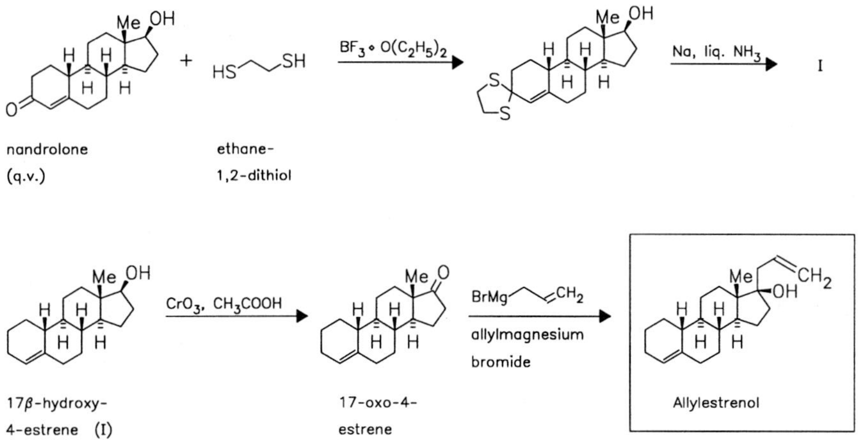

nandrolone
(q.v.)

ethane-
1,2-dithiol

17β-hydroxy-
4-estrene (I)

17-oxo-4-
estrene

allylmagnesium
bromide

Allylestrenol

Reference(s):
GB 841 411 (Organon; appl. 2.4.1958; NL-prior. 10.4.1957).

alternative syntheses:
GB 875 549 (Organon; appl. 31.12.1959; NL-prior. 13.1.1959).
US 2 878 267 (Organon; appl. 16.4.1958; NL-prior. 1.5.1957).

Formulation(s): tabl. 5 mg

Trade Name(s):
D: Gestanon (Organon); wfm
GB: Gestanin (Organon); wfm
I: Gestanon (Organon Italia)
J: Gestanon (Sankyo)

Alminoprofen

ATC: M01AE16
Use: non-steroidal anti-inflammatory,
analgesic

RN: 39718-89-3 MF: C$_{13}$H$_{17}$NO$_2$ MW: 219.28 EINECS: 254-604-6
LD$_{50}$: 2400 mg/kg (M, p.o.)
CN: α-methyl-4-[(2-methyl-2-propenyl)amino]benzeneacetic acid

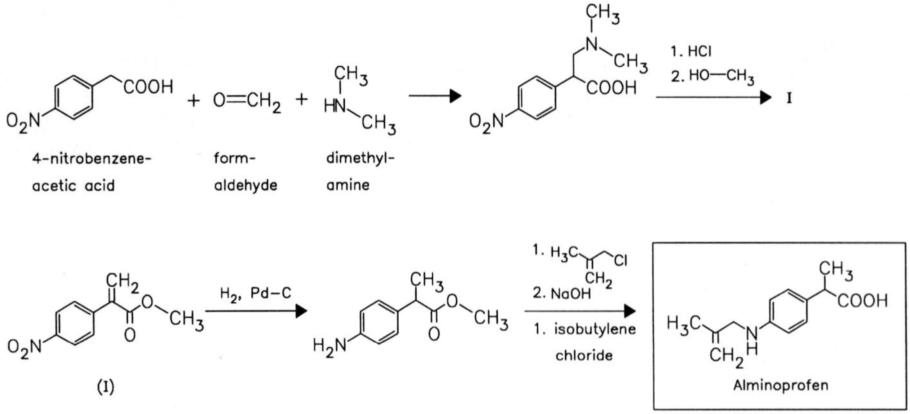

4-nitrobenzene-   form-   dimethyl-
acetic acid   aldehyde   amine

(I)

Alminoprofen

Reference(s):
Dumaitre, B. et al.: Eur. J. Med. Chem. (EJMCA5) **14**, 207 (1979).

alternative synthesis:
FR 2 289 180 (Lab. Bouchara; appl. 17.5.1971).

Formulation(s):   tabl. 150 mg, 300 mg

Trade Name(s):
F: Minalfène (Bouchara)   J: Minalfen (Fujirebio)

Almitrine

ATC: R07AB07
Use: analeptic, respiratory stimulant

RN: 27469-53-0   MF: $C_{26}H_{29}F_2N_7$ MW: 477.56   EINECS: 248-475-5
CN: 6-[4-[bis(4-fluorophenyl)methyl]-1-piperazinyl]-*N,N'*-di-2-propenyl-1,3,5-triazine-2,4-diamine

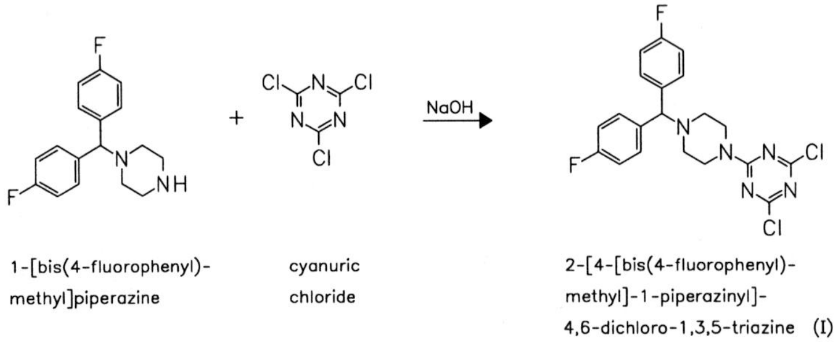

1-[bis(4-fluorophenyl)-   cyanuric   2-[4-[bis(4-fluorophenyl)-
methyl]piperazine   chloride   methyl]-1-piperazinyl]-
4,6-dichloro-1,3,5-triazine (I)

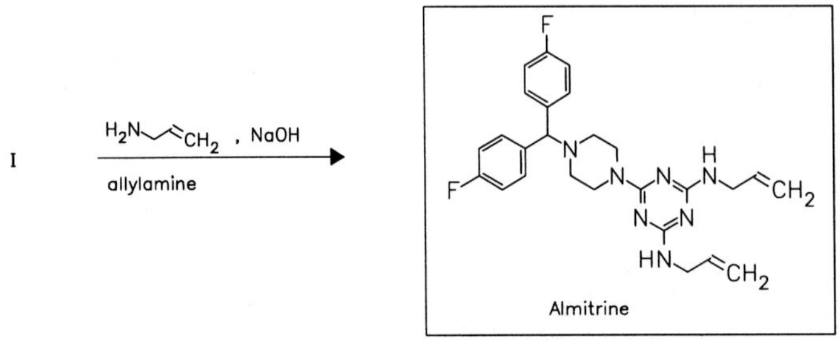

Almitrine

Reference(s):
FR 2 019 646 (Science Union; appl. 22.9.1969; GB-prior. 2.10.1968).
DOS 1 947 332 (Science Union; appl. 18.9.1969; GB-prior. 2.10.1968).
US 3 647 794 (Science Union; 7.3.1972; GB-prior. 2.10.1968).
GB 1 256 513 (Science Union; appl. 2.10.1968; valid from 30.9.1969).

Formulation(s):   f. c. tabl. 50 mg; vial 15 mg/5 ml; tabl. 50 mg

Trade Name(s):
D: Vectarion (Servier; 1984)   F: Duxil (Therval Médical;   Vectarion (Euthérapie;
                                       1979)-comb.   1983)

Aloxiprin

ATC: B01AC15; N02BA02
Use: analgesic

RN: 9014-67-9   MF: unspecified   MW: unspecified
CN: aloxiprin

polymeric condensation product of aluminum oxide and acetylsalicylic acid

3 Al[OCH(CH$_3$)$_2$]$_3$ +   5   (COOH ... O–C–CH$_3$, O)   →   Al$_3$O$_2$-[C$_6$H$_4$(OCOCH$_3$)-COO]$_5$

aluminum isopropylate   acetylsalicylic acid   Aloxiprin
                             (q. v.)

Reference(s):
Cummings, A.J. et al.: J. Pharm. Pharmacol. (JPPMAB) **15**, 56 (1963).

Formulation(s):   tabl. 400 mg, 450 mg, 600 mg

Trade Name(s):
GB: Palaprin (Nicholas); wfm   Palaprin forte (Nicholas);
                                         wfm

Alphaprodine
(Alfaprodina)

ATC: N02AB
Use: analgesic

RN: 77-20-3   MF: C$_{16}$H$_{23}$NO$_2$ MW: 261.37   EINECS: 201-011-5
CN: *cis*-1,3-dimethyl-4-phenyl-4-piperidinol propanoate (ester)

hydrochloride
RN: 561-78-4   MF: C$_{16}$H$_{23}$NO$_2$ · HCl   MW: 297.83
LD$_{50}$:   32 mg/kg (M, i.v.);
         25 mg/kg (R, i.v.); 90 mg/kg (R, p.o.);
         36.2 mg/kg (dog, i.v.)

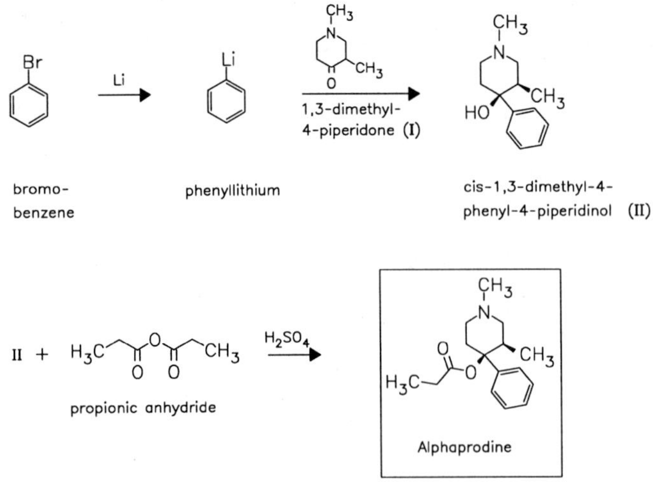

bromo-
benzene

phenyllithium

1,3-dimethyl-
4-piperidone (I)

cis-1,3-dimethyl-4-
phenyl-4-piperidinol (II)

II + H₃C (propionic anhydride) $\xrightarrow{H_2SO_4}$

propionic anhydride

Alphaprodine

Reference(s):
US 2 498 433 (Hoffmann-La Roche; 1950; prior. 1946).

starting material:

1,3-dimethyl-4-piperidone:
Howton: J. Org. Chem. (JOCEAH) **10**, 277 (1945).

Formulation(s):   amp. 4 %, 6 %

Trade Name(s):
USA: Nisentil (Roche); wfm

Alpidem

ATC: N05B
Use: anxiolytic, ω_1-agonist

RN: 82626-01-5 MF: $C_{21}H_{23}Cl_2N_3O$ MW: 404.34
CN: 6-chloro-2-(4-chlorophenyl)-N,N-dipropylimidazo[1,2-a]pyridine-3-acetamide

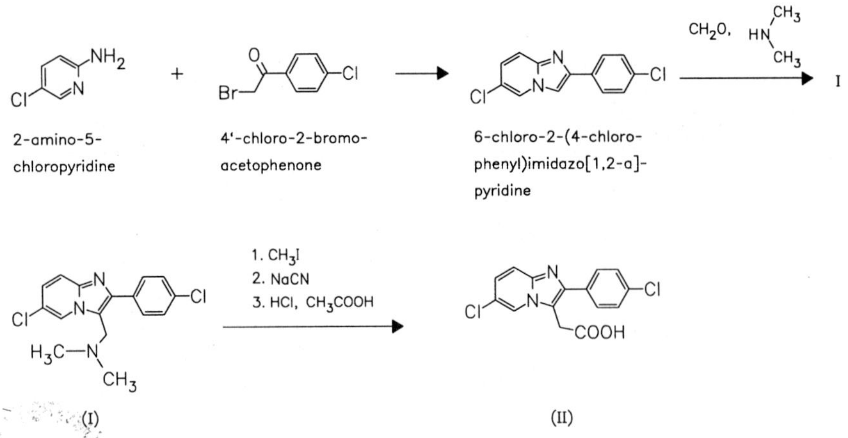

2-amino-5-
chloropyridine

4'-chloro-2-bromo-
acetophenone

6-chloro-2-(4-chloro-
phenyl)imidazo[1,2-a]-
pyridine

1. CH₃I
2. NaCN
3. HCl, CH₃COOH

(I)

(II)

II 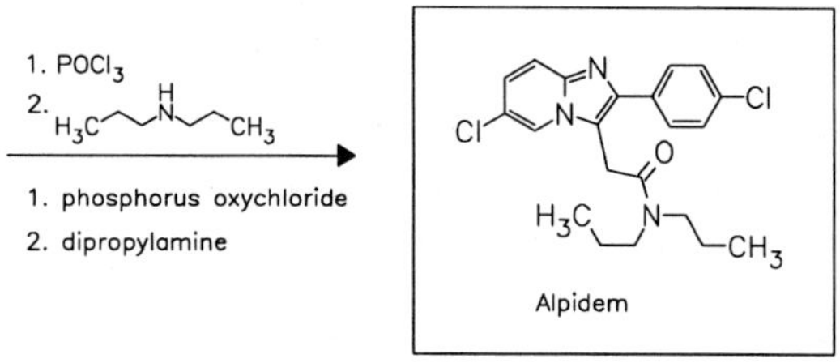
1. POCl3
2. H3C N CH3
1. phosphorus oxychloride
2. dipropylamine
Alpidem

Reference(s):
EP 50 563 (Synthelabo; appl. 15.10.1981; F-prior. 22.10.1980).
US 4 382 938 (Synthelabo; 10.5.1983; F-prior. 22.10.1980).
US 4 460 592 (Synthelabo; 17.7.1984; F-prior. 22.10.1980).

Formulation(s): tabl. 50 mg

Trade Name(s):
F: Anaxyl (Synthélabo; 1991);
wfm

Alprazolam

ATC: N05BA12
Use: tranquilizer

RN: 28981-97-7 MF: $C_{17}H_{13}ClN_4$ MW: 308.77 EINECS: 249-349-2
LD_{50}: 770 mg/kg (M, p.o.);
1220 mg/kg (R, p.o.)
CN: 8-chloro-1-methyl-6-phenyl-4*H*-[1,2,4]triazolo[4,3-*a*][1,4]benzodiazepine

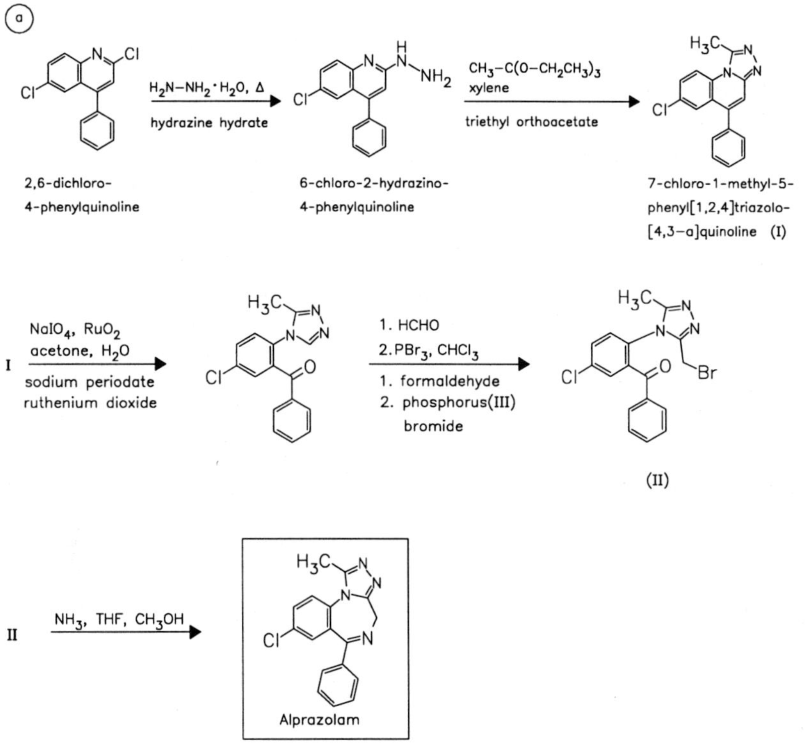
a
H2N–NH2 · H2O, Δ
hydrazine hydrate
CH3–C(O–CH2CH3)3
xylene
triethyl orthoacetate
2,6-dichloro-
4-phenylquinoline
6-chloro-2-hydrazino-
4-phenylquinoline
7-chloro-1-methyl-5-
phenyl[1,2,4]triazolo-
[4,3–a]quinoline (I)
I
NaIO4, RuO2
acetone, H2O
sodium periodate
ruthenium dioxide
1. HCHO
2. PBr3, CHCl3
1. formaldehyde
2. phosphorus(III)
bromide
(II)
II
NH3, THF, CH3OH
Alprazolam

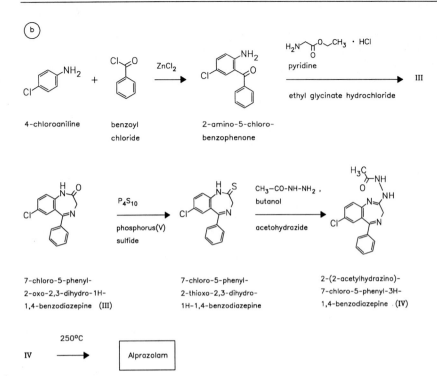

ⓑ

4-chloroaniline + benzoyl chloride →(ZnCl₂) 2-amino-5-chloro-benzophenone

H_2N...O...CH_3 · HCl
pyridine
ethyl glycinate hydrochloride
→ III

7-chloro-5-phenyl-
2-oxo-2,3-dihydro-1H-
1,4-benzodiazepine (III)

P_4S_{10}
phosphorus(V) sulfide
→

7-chloro-5-phenyl-
2-thioxo-2,3-dihydro-
1H-1,4-benzodiazepine

$CH_3-CO-NH-NH_2$,
butanol
acetohydrazide
→

2-(2-acetylhydrazino)-
7-chloro-5-phenyl-3H-
1,4-benzodiazepine (IV)

IV —(250°C)→ | Alprazolam |

Reference(s):

US 3 987 052 (Upjohn; 19.10.1976; appl. 29.10.1969; USA-prior. 17.3.1969).
US 3 980 789 (Upjohn; 14.9.1976; appl. 19.6.1972; USA-prior. 29.3.1971).
DE 1 955 349 (Takeda; D-prior. 4.11.1969).
GB 1 298 364 (Upjohn; GB-prior. 27.10.1969).
a DOS 2 203 782 (Upjohn; appl. 27.1.1972; USA-prior. 9.2.1971).
   US 3 709 898 (Upjohn; 9.1.1973; prior. 9.2.1971).
   US 3 781 289 (Upjohn; 25.12.1973; prior. 11.5.1972).
b DOS 2 012 190 (Upjohn; appl. 14.3.1970; USA-prior. 17.3.1969).

Formulation(s):   tabl. 0.25 mg, 0.5 mg, 1 mg, 1 g

Trade Name(s):
D: Cassadan 0,25/0,5/1 (ASTA
      Medica AWD)
      Tafil 0,5/1,0 Tabletten
      (Pharmacia & Upjohn;
      1984)
      Xanax (Pharmacia &
      Upjohn)
F: Xanax (Upjohn; 1984)
GB: Xanax (Pharmacia &
      Upjohn; 1983)
I: Frontal (UCM)
      Mialin (Biomedica
      Foscama)
      Valeans (Valeas)
      Xanax (Upjohn; 1985)
J: Constan (Takeda; 1984)
      Solanax (Upjohn-
      Sumitomo; 1984)
USA: Xanax (Pharmacia &
      Upjohn; 1981)

Alprenolol

ATC: C07AA01
Use: beta blocking agent

RN: 13655-52-2   MF: $C_{15}H_{23}NO_2$ MW: 249.35   EINECS: 237-140-9
LD₅₀: 20 mg/kg (M, i.v.)
CN: 1-[(1-methylethyl)amino]-3-[2-(2-propenyl)phenoxy]-2-propanol

hydrochloride

RN: 13707-88-5   MF: $C_{15}H_{23}NO_2 \cdot HCl$ MW: 285.82   EINECS: 237-244-4

LD_{50}:   29 mg/kg (M, i.v.); 184 mg/kg (M, p.o.);
       17 mg/kg (R, i.v.); 590 mg/kg (R, p.o.);
       18 mg/kg (dog, i.v.); 383 mg/kg (dog, p.o.)

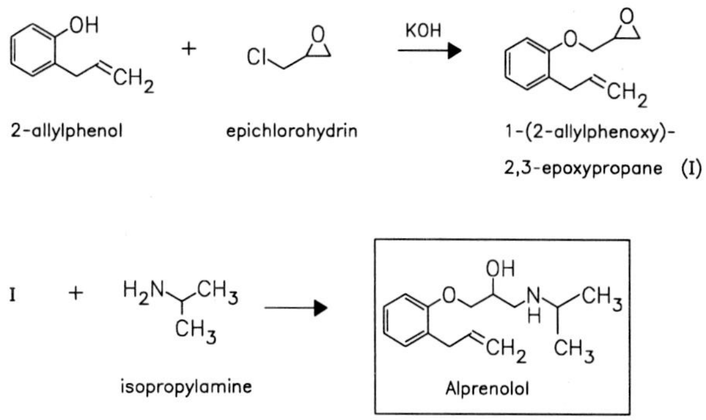

2-allylphenol   epichlorohydrin   1-(2-allylphenoxy)-
                                          2,3-epoxypropane (I)

isopropylamine   Alprenolol

Reference(s):

US 3 466 376 (AB Hässle; 9.9.1969; prior. 18.1.1966, 17.6.1966).
Brandström, A.: Acta Pharm. Suec. (APSXAS) **1966**, 303.

2-allylphenol *by rearrangement of* allyl phenyl ether:
DOS 2 746 002 (Firestone; appl. 13.10.1977; USA-prior. 18.10.1976).

Formulation(s):   cps. 10 mg, 20 mg, 40 mg, 50 mg; lyo. for inf. 42.6 mg; tabl. 200 mg

Trade Name(s):

D: Aptin-Duriles (Astra)   J: Apllobal (Fujisawa; as
                                          hydrochloride)

Altizide

(Althiazide)

ATC: C03EA01; C03EA04
Use: diuretic, antihypertensive

RN: 5588-16-9   MF: $C_{11}H_{14}ClN_3O_4S_3$ MW: 383.90   EINECS: 226-994-8

CN: 6-chloro-3,4-dihydro-3-[(2-propenylthio)methyl]-2H-1,2,4-benzothiadiazine-7-sulfonamide 1,1-dioxide

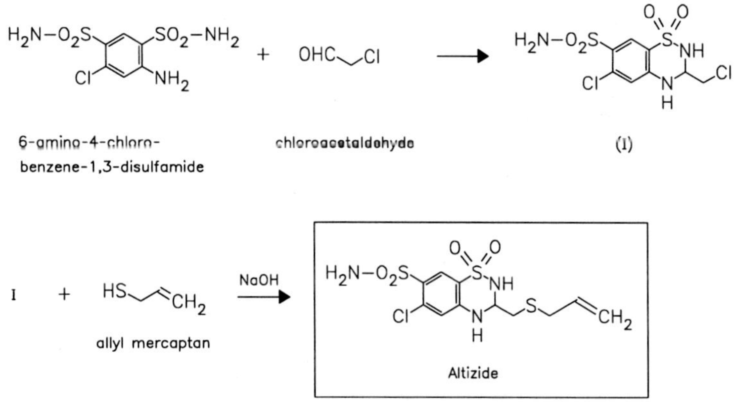

6-amino-4-chloro-   chloroacetaldehyde   (I)
benzene-1,3-disulfamide

allyl mercaptan

Altizide

Reference(s):

GB 902 658 (Pfizer; appl. 10.1.1961; USA-prior. 27.9.1960).

Formulation(s):   cps. 0.25 mg, 0.5 mg; drops 1 mg; sol. 0.1 mg/ml; tabl. 0.25 mg, 0.5 mg, 1 mg, 2 mg

Trade Name(s):

F: Aldactazine (Monsanto)-comb.
Practazin (Cardel)-comb.

Prinactizide (Dakota)-comb.
Spiroctazine (Boehringer Mannh.)-comb.

I: Aldatense (SPA)-comb.; wfm

USA: Aldactazide (Searle)-comb.; wfm

Altretamine
(Hexamethylmelamine)

ATC: L01XX03
Use: antineoplastic

RN: 645-05-6   MF: $C_9H_{18}N_6$ MW: 210.29   EINECS: 211-428-4
LD_{50}:   350 mg/kg (R, p.o.)
CN: N,N,N',N',N'',N''-hexamethyl-1,3,5-triazine-2,4,6-triamine

hydrochloride
RN: 15468-34-5   MF: $C_9H_{18}N_6 \cdot xHCl$ MW: unspecified
LD_{50}:   100 mg/kg (M, i.v.)

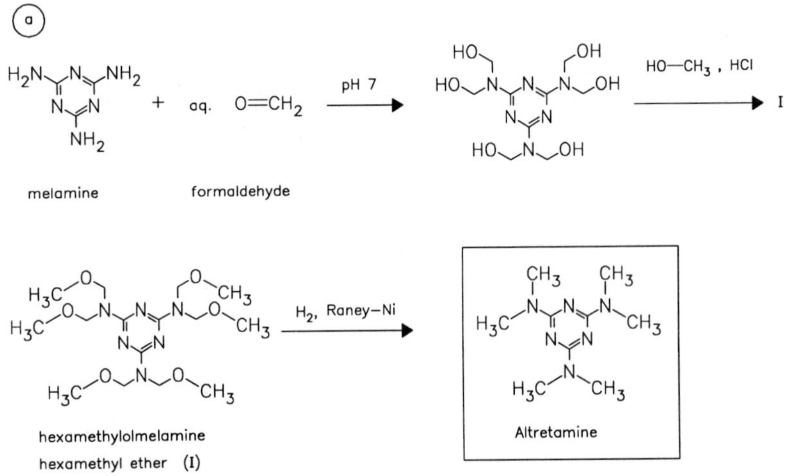

melamine   formaldehyde

hexamethylolmelamine
hexamethyl ether   (I)

Altretamine

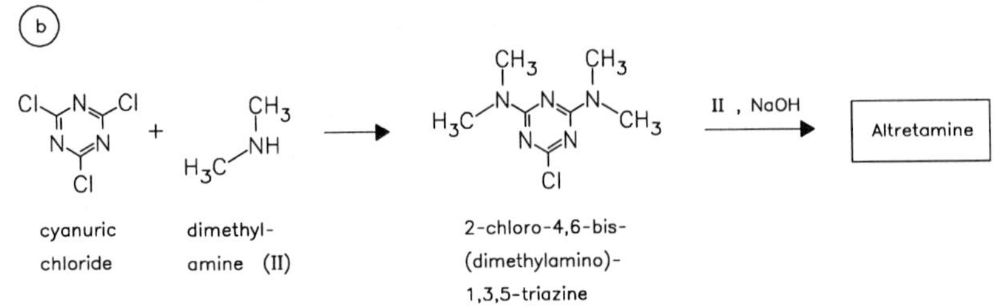

cyanuric chloride   dimethyl-amine   (II)

2-chloro-4,6-bis-(dimethylamino)-1,3,5-triazine

Altretamine

Reference(s):

a   DE 1 240 870 (Cassella; appl. 17.11.1965).
b   Gunduz, T.: Commun. Fac. Sci. Univ. Ankara, Ser. B: Chim. (CAKBA9) **15**, 69 (1968).
      Cumber, A.J.; Ross, W.C.J.: Chem.-Biol. Interact. (CBINA8) **17**, 349 (1977).

synthesis of hexamethylolmelamine hexamethyl ether:
Gams, A. et al.: Helv. Chim. Acta (HCACAV) **24**, 302 (1941).
US 3 322 762 (Pittsburgh Plate Glass; 30.5.1967; prior. 27.2.1962; 8.4.1964).

Formulation(s):   cps. 50 mg, 100 mg

Trade Name(s):
D: Hexamethylmelamin   F: Hexastat (Roger Bellon);   I: Hexastat (Rhône-Poulenc
       (Rhône-Poulenc); wfm   wfm   Rorer)
                                   GB: Hexalen (Speywood)   USA: Hexalen (U.S. Bioscience)

Alufibrate
(Aluminium clofibrate)

ATC: C01AB03
Use: cholesterol depressant

RN: 24818-79-9   MF: $C_{20}H_{21}AlCl_2O_7$ MW: 471.27   EINECS: 246-477-0
CN: bis[2-(4-chlorophenoxy-κO)-2-methylpropanoato-κO]hydroxyaluminum

clofibric acid   aluminum   Alufibrate
(cf. clofibrate   ethylate
synthesis)

Reference(s):
GB 860 303 (ICI; appl. 20.6.1958).

Formulation(s):   tabl. 500 mg

Trade Name(s):
D: Atherolipin (Schwarz);   F: Athérolip (Millot-Solac);
       wfm   wfm

Aluminum nicotinate

ATC: C10AD04
Use: antihyperlipidemic, vasodilator
         (peripheral)

RN: 1976-28-9   MF: $C_{18}H_{12}AlN_3O_6$ MW: 393.29   EINECS: 217-832-7
CN: 3-pyridinecarboxylic acid aluminum salt

nicotinic acid   aluminum   Aluminum nicotinate
                        hydroxide

Reference(s):
US 2 970 082 (Walker Labs.; 31.1.1961; appl. 7.10.1958).

Formulation(s):   tabl. 125 mg

Trade Name(s):
USA: Nicalex (Merrell-National);
         wfm

Alverine
(Dipropyline; Fenpropamine)

ATC: A03AX08
Use: antispasmodic

RN: 150-59-4   MF: $C_{20}H_{27}N$ MW: 281.44   EINECS: 205-763-5
CN: *N*-ethyl-*N*-(3-phenylpropyl)benzenepropanamine

citrate (1:1)
RN: 5560-59-8   MF: $C_{20}H_{27}N \cdot C_6H_8O_7$ MW: 473.57   EINECS: 226-929-3

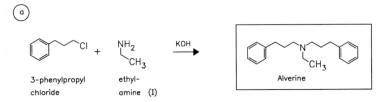

a

3-phenylpropyl   ethyl-
chloride   amine   (I)   Alverine

b

cinnamaldehyde

Reference(s):
a   Külz, F. et al.: Ber. Dtsch. Chem. Ges. (BDCGAS) **72**, 2161 (1939).
b   Stühmer, W.; Elbrächter, E.-A.: Arch. Pharm. Ber. Dtsch. Pharm. Ges. (APBDAJ) **287**, 139 (1954).

Formulation(s):   inj. sol. 40 mg/2 ml; suppos. 80 mg; tabl. 40 mg

Trade Name(s):
D: Spasmocol (Norgine)-
        comb.; wfm
F: Hepatoum (Hepatoum)-
        comb.
        Météospasmyl (Mayoly-
        Spindler)-comb.

Schoum comprimés
(Pharmysiène)-comb.
Spasmavérine (Théraplix)
Spasmavérine suppos.
(Théraplix)-comb.
GB: Alvercol (Norgine; as
        citrate)-comb.

Spasmonal (Norgine; as
citrate)
I: Profenil (Ipti); wfm
        Spasmaverine (Roger
        Bellon); wfm
USA: Spacolin (Philips Roxane);
        wfm

Amantadine

ATC: J05AC; N04BB01
Use: antiparkinsonian, antiviral

RN: 768-94-5   MF: $C_{10}H_{17}N$ MW: 151.25   EINECS: 212-201-2
LD_{50}:   700 mg/kg (M, p.o.);
         900 mg/kg (R, p.o.)
CN: tricyclo[3.3.1.1^{3,7}]decan-1-amine

hydrochloride
RN: 665-66-7   MF: $C_{10}H_{17}N \cdot HCl$ MW: 187.71   EINECS: 211-560-2
LD$_{50}$:   95 mg/kg (M, i.v.); 700 mg/kg (M, p.o.);
         90 mg/kg (R, i.v.); 800 mg/kg (R, p.o.);
         37 mg/kg (dog, i.v.)
sulfate (2:1)
RN: 31377-23-8   MF: $C_{10}H_{17}N \cdot 1/2H_2SO_4$ MW: 400.58   EINECS: 250-604-5

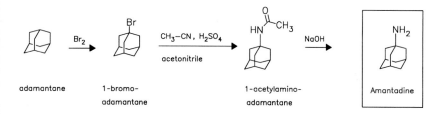

| adamantane | 1-bromo-
adamantane | | 1-acetylamino-
adamantane | Amantadine |

Reference(s):
Stetter, H. et al.: Chem. Ber. (CHBEAM) **93**, 226 (1960).
US 3 310 469 (Du Pont; 21.3.1967; prior. 28.8.1961, 15.4.1963, 22.10.1963).

synthesis from adamantane, HCN *and* H_2SO_4:
US 3 152 180 (Studiengesellschaft Kohle; 6.10.1964, D-prior. 25.8.1960).

combination with molindone *(antidepressant):*
US 4 148 896 (Du Pont; 10.4.1979; appl. 22.2.1978).

Formulation(s):   f. c. tabl. 100 mg, 150 mg; cps. 100 mg; amp. 200 mg/500 ml (as sulfate); syrup 50 mg/5 ml

Trade Name(s):
D: Amantadin (ratiopharm)   GB: Symmetrel (Geigy; as   J: Symmetrel (Fujisawa-
       PK-Merz (Merz & Co.)   hydrochloride)   Novartis)
F: Mantadix (Du Pont)   I: Mantadan (Boehringer   USA: Symmetrel (Endo)
                                         Ing.)

Ambazone

ATC: R02AA01
Use: antiseptic, disinfectant (oral and
       pharyngeal chemotherapeutic),
       antineoplastic

RN: 539-21-9   MF: $C_8H_{11}N_7S$ MW: 237.29   EINECS: 208-713-0
LD$_{50}$:   1 g/kg (M, p.o.);
         750 mg/kg (R, p.o.)
CN: 2-[4-[(aminoiminomethyl)hydrazono]-2,5-cyclohexadien-1-ylidene]hydrazinecarbothioamide

monohydrate
RN: 6011-12-7   MF: $C_8H_{11}N_7S \cdot H_2O$ MW: 255.31

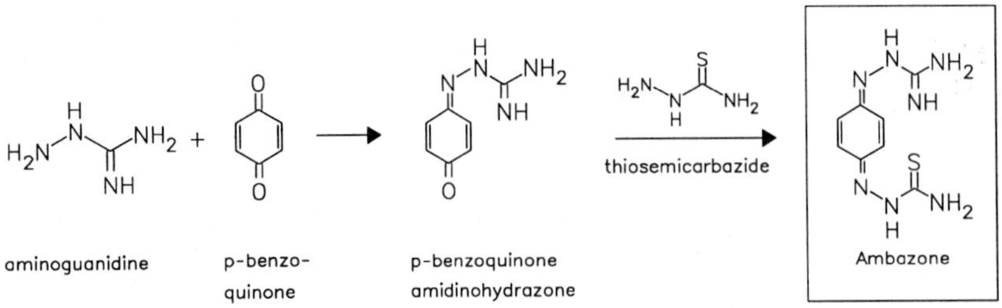

aminoguanidine · p-benzo-quinone · p-benzoquinone amidinohydrazone · thiosemicarbazide · Ambazone

Reference(s):
DE 965 723 (Bayer; appl. 1953).

Formulation(s):   tabl. 10 mg, 100 mg

Trade Name(s):
D: Iversal (Bayer); wfm   GB: Iversal (Bayer); wfm
F: Iversal (Bayer); wfm   I: Primar (Bayer); wfm

Ambenonium chloride

ATC: N07AA30
Use: cholinesterase inhibitor

RN: 115-79-7   MF: $C_{28}H_{42}Cl_4N_4O_2$ MW: 608.48   EINECS: 204-107-5
LD$_{50}$:   1510 g/kg (M, i.v.); 145 mg/kg (M, p.o.);
       2720 g/kg (R, i.v.); 18.5 mg/kg (R, p.o.)
CN: N,N'-[(1,2-dioxo-1,2-ethanediyl)bis(imino-2,1-ethanediyl)]bis[2-chloro-N,N-diethylbenzenemethanaminium] dichloride

hydroxide
RN: 470-78-0   MF: $C_{28}H_{44}Cl_2N_4O_4$ MW: 571.59

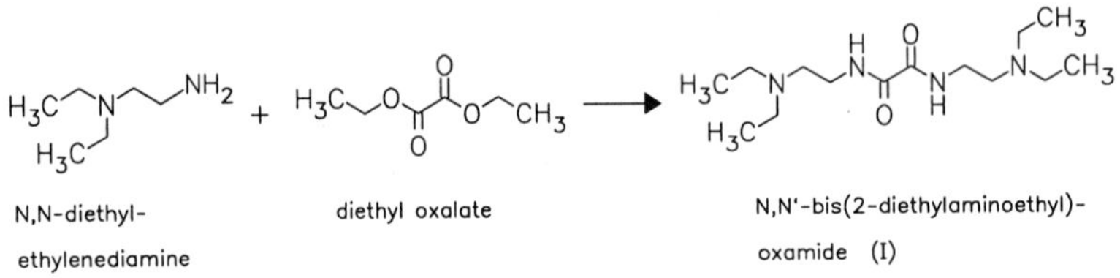

N,N-diethyl-ethylenediamine · diethyl oxalate · N,N'-bis(2-diethylaminoethyl)-oxamide (I)

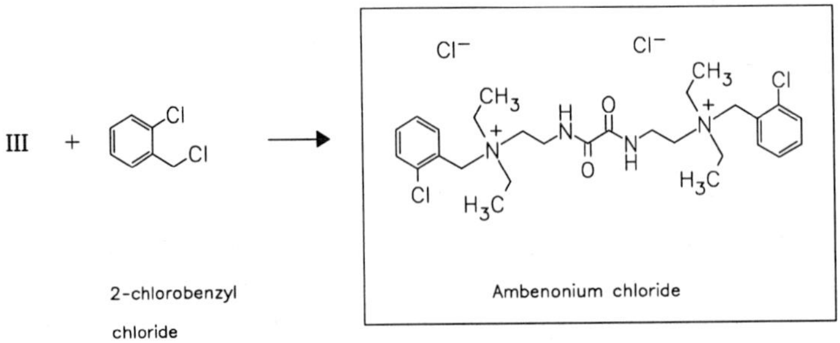

III + 2-chlorobenzyl chloride → Ambenonium chloride

Reference(s):
DE 1 024 517 (Sterling Drug; appl. 1954; USA-prior. 1953).
US 3 096 373 (Sterling Drug; 2.7.1963; appl. 1956).
Phillips, A.P.: J. Am. Chem. Soc. (JACSAT) **73**, 5822 (1951).

oxamide intermediate:
US 2 438 200 (Du Pont; 1948; appl. 1946).

Formulation(s):   cps. 10 mg; tabl. 10 mg

Trade Name(s):

D:	Mytelase Tabletten (Winthrop); wfm	GB:	Mytelase (Winthrop); wfm	J:	Mytelase (Winthrop-Nippon Shoji)
F:	Mytélase (Sanofi Winthrop)	I:	Mytelase (Winthrop); wfm	USA:	Mytelase (Winthrop); wfm

---

Ambroxol

ATC: R05CB
Use: expectorant

RN: 18683-91-5   MF: $C_{13}H_{18}Br_2N_2O$ MW: 378.11   EINECS: 242-500-3
LD_{50}:   138 mg/kg (M, i.v.); 2720 mg/kg (M, p.o.);
13.4 g/kg (R, p.o.)
CN: *trans*-4-[[(2-amino-3,5-dibromophenyl)methyl]amino]cyclohexanol

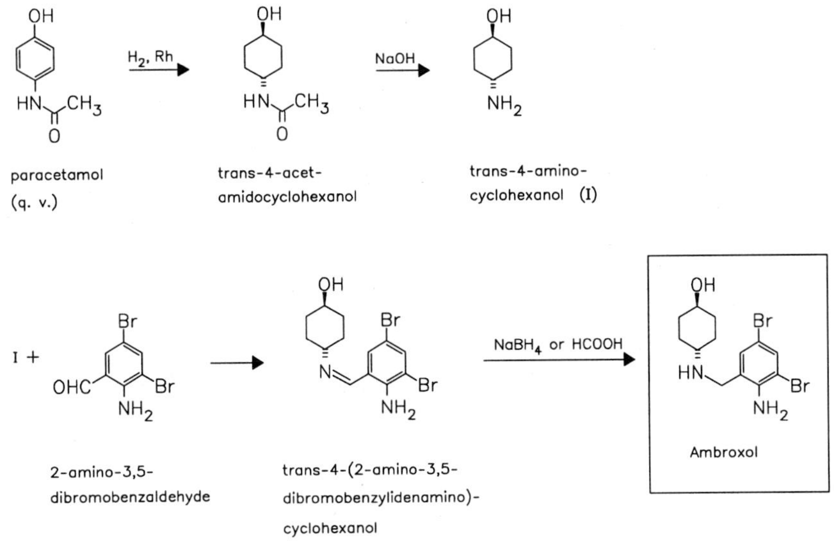

paracetamol (q. v.)

trans-4-acet-amidocyclohexanol

trans-4-amino-cyclohexanol (I)

2-amino-3,5-dibromobenzaldehyde

trans-4-(2-amino-3,5-dibromobenzylidenamino)-cyclohexanol

Ambroxol

Reference(s):

GB 1 178 034 (Boehringer Ingelh.; appl. 10.5.1967; D-prior. 10.5.1966).
US 3 536 713 (Boehringer Ingelh.; 27.10.1970; appl. 10.5.1967; S-prior. 10.5.1966).
DE 1 593 579 (Thomae; appl. 10.5.1966).
DOS 2 218 647 (Thomae; appl. 18.4.1972).
DOS 2 223 193 (Thomae; appl. 12.5.1972).
Keck, J.: Justus Liebigs Ann. Chem. (JLACBF) **707**, 107 (1967).

Formulation(s):   amp. 15 mg; cps. 75 mg; drops 7.5 mg, 30 mg; eff. tabl. 30 mg, 60 mg; f. c. tabl. 30 mg, 60 mg; inhalation sol. 7,5 mg; inj. 1000 mg; syrup 15 mg, 30 mg; tabl. 30 mg, 60 mg (as hydrochloride)

Trade Name(s):

D:	Ambril (Glaxo Wellcome)	frenopect (Hefa Pharma)	Mucosolvan (Boehringer Ing.; 1979)
	Bronchopront (Mack, Illert.)	Lindoxyl (Lindopharm)	
		Mucoclear (Mundipharma)	Mucotablin-Tropfen (Sanorania)
	duramucal (durachemie)	Mucophlogat (Azuchemie)	

Ambuside

ATC: C02L
Use: diuretic, antihypertensive

RN: 3754-19-6   MF: $C_{13}H_{16}ClN_3O_5S_2$ MW: 393.87   EINECS: 223-158-4
CN: 4-chloro-6-[(3-hydroxy-2-butenylidene)amino]-N^1-2-propenyl-1,3-benzenedisulfonamide

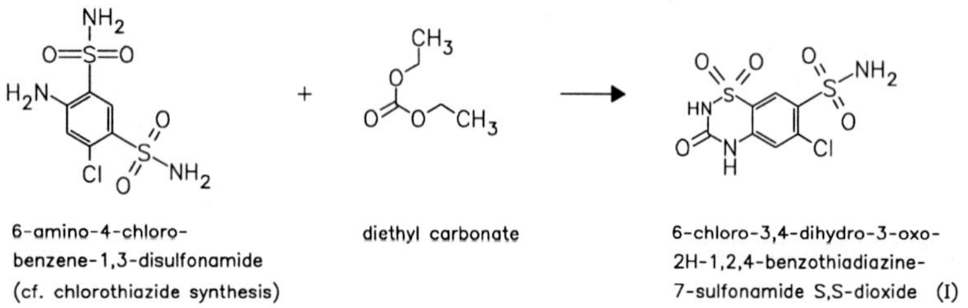

6-amino-4-chloro-benzene-1,3-disulfonamide
(cf. chlorothiazide synthesis)

diethyl carbonate

6-chloro-3,4-dihydro-3-oxo-2H-1,2,4-benzothiadiazine-7-sulfonamide S,S-dioxide   (I)

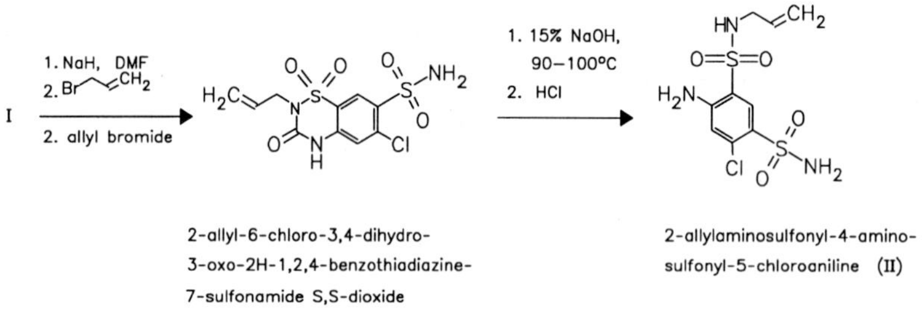

1. NaH, DMF
2. Br—CH₂

2. allyl bromide

2-allyl-6-chloro-3,4-dihydro-3-oxo-2H-1,2,4-benzothiadiazine-7-sulfonamide S,S-dioxide

1. 15% NaOH, 90–100°C
2. HCl

2-allylaminosulfonyl-4-aminosulfonyl-5-chloroaniline   (II)

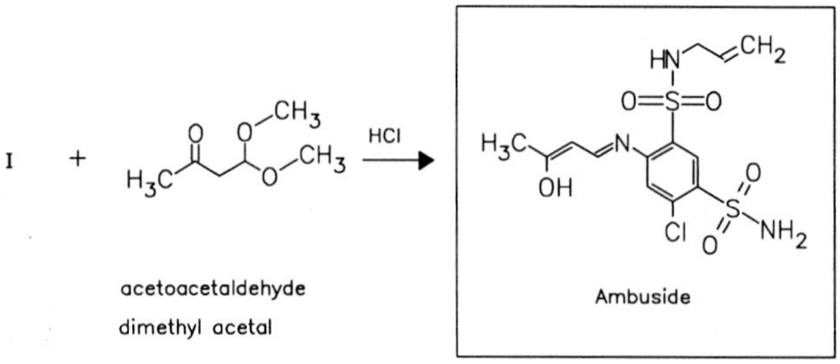

acetoacetaldehyde dimethyl acetal

Ambuside

Reference(s):
US 3 188 329 (Colgate-Palmolive; 8.6.1965; appl. 10.4.1962).

intermediates:
Close, W.J. et al.: J. Am. Chem. Soc. (JACSAT) **82**, 1132 (1960).

Trade Name(s):
F: Hydrion (Robert et Carrière); wfm

Amcinonide
(Triamcinolone acetate cyclopentanoide)

ATC: D07AC11; H02AB
Use: topical glucocorticoid

RN: 51022-69-6   MF: $C_{28}H_{35}FO_7$ MW: 502.58   EINECS: 256-915-2
LD$_{50}$:   >5 g/kg (M, p.o.);
    >2 g/kg (R, p.o.)
CN: (11β,16α)-21-(acetyloxy)-16,17-[cyclopentylidenebis(oxy)]-9-fluoro-11-hydroxypregna-1,4-diene-3,20-dione

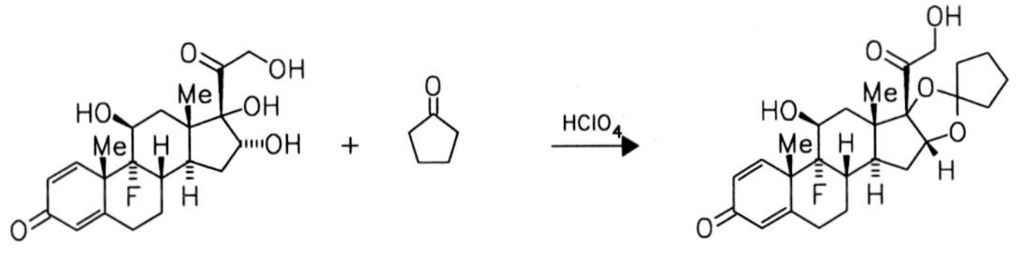

triamcinolone   cyclopentanone   triamcinolone cyclopentanonide (I)
(q. v.)

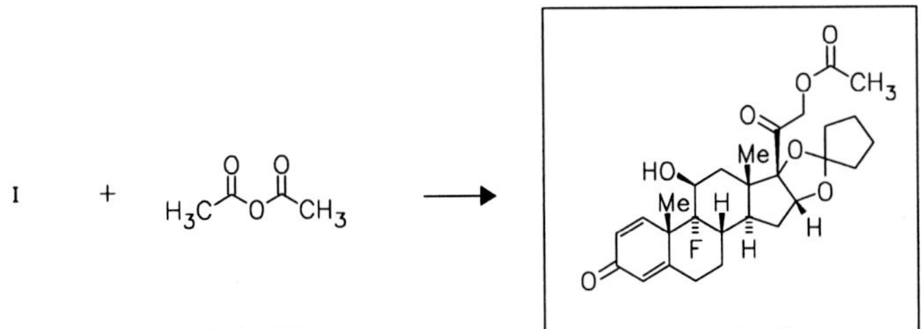

acetanhydride   Amcinonide

Reference(s):
GB 1 442 925 (American Cyanamid; USA-prior. 17.8.1973).
DOS 2 437 847 (American Cyanamid; appl. 6.8.1974; USA-prior. 17.8.1973).
BE 818 929 (American Cyanamid; appl. 16.8.1974; USA-prior. 17.8.1973).
US 4 158 055 (American Cyanamid; 12.6.1979; USA-prior. 6.6.1975).

Formulation(s):   cream 0.1 %; lotion 0.1 %; ointment 0.1 %

Trade Name(s):
D: Amciderm (Hermal/Merck; 1985)
F: Penticort (Wyeth-Lederle; 1980)
I: Amcinil (Crosara)
J: Visderm (Lederle; 1982)
Penticort Neomycine (Wyeth-Lederle)-comb.
USA: Cyclocort (Fujisawa; 1979)

Amezinium metilsulfate

ATC: C01CA00
Use: selective noradrenergic antihypotensive

RN: 30578-37-1   MF: $C_{11}H_{12}N_3O \cdot CH_3O_4S$ MW: 313.33   EINECS: 250-248-0
LD$_{50}$:   28 mg/kg (M, i.v.); 1330 mg/kg (M, p.o.);
    24 mg/kg (R, i.v.); 1410 mg/kg (R, p.o.);
    60 mg/kg (dog, i.v.); 100 mg/kg (dog, p.o.)
CN: 4-amino-6-methoxy-1-phenylpyridazinium methyl sulfate

chloride
RN: 51410-15-2 MF: $C_{11}H_{12}ClN_3O$ MW: 237.69

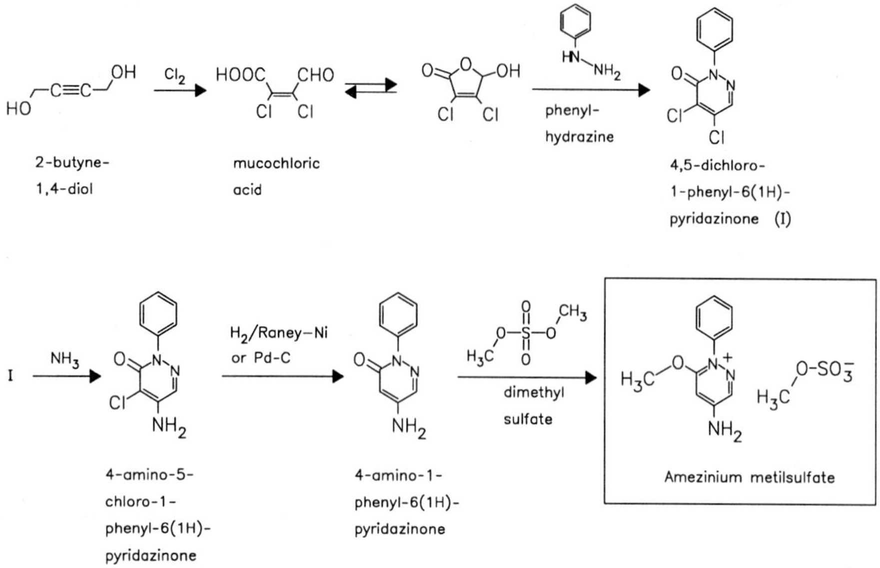
Cl2
HOOC CHO
Cl Cl
phenyl-
hydrazine
2-butyne-
1,4-diol
mucochloric
acid
4,5-dichloro-
1-phenyl-6(1H)-
pyridazinone (I)
NH3
H2/Raney-Ni
or Pd-C
dimethyl
sulfate
4-amino-5-
chloro-1-
phenyl-6(1H)-
pyridazinone
4-amino-1-
phenyl-6(1H)-
pyridazinone
Amezinium metilsulfate

Reference(s):
Reicheneder, F. et al.: Arzneim.-Forsch. (ARZNAD) **31** (II), 1529 (1981).
DE 1 912 941 (BASF; appl. 14.3.1969).
DOS 2 139 687 (BASF; appl. 7.8.1971).
DOS 2 211 662 (BASF; appl. 10.3.1972).
DOS 3 114 496 (BASF; appl. 10.4.1981).
EP 63 267 (BASF; appl. 31.3.1982; D-prior. 10.4.1981).

precursors:
DE 2 100 685 (BASF; appl. 8.1.1971).

Formulation(s): amp. 5 mg; tabl. 10 mg

Trade Name(s):
D: Regulton (Knoll) Supratonin (Grünenthal)

Amfebutamone
(Bupropion)

ATC: N06AE
Use: antidepressant

RN: 34911-55-2 MF: $C_{13}H_{18}ClNO$ MW: 239.75
LD_{50}: 544 mg/kg (M, p.o.)
CN: (±)-1-(3-chlorophenyl)-2-[(1,1-dimethylethyl)amino]-1-propanone

hydrochloride
RN: 31677-93-7 MF: $C_{13}H_{18}ClNO \cdot HCl$ MW: 276.21 EINECS: 250-759-9
LD_{50}: 230 mg/kg (M, i.p.); 575 mg/kg (M, p.o.);
210 mg/kg (R, i.p.); 600 mg/kg (R, p.o.)

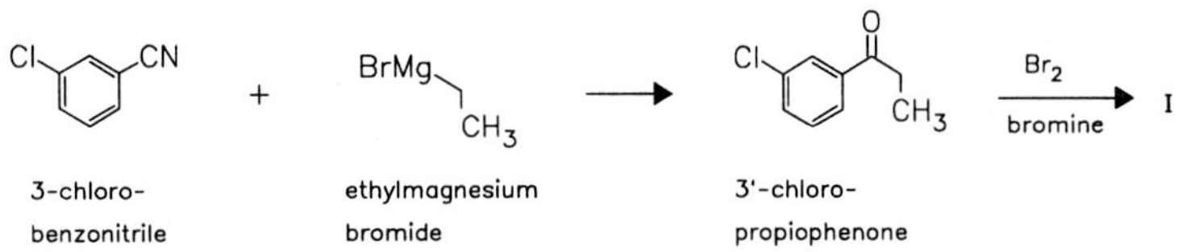

3-chloro-
benzonitrile

ethylmagnesium
bromide

3'-chloro-
propiophenone

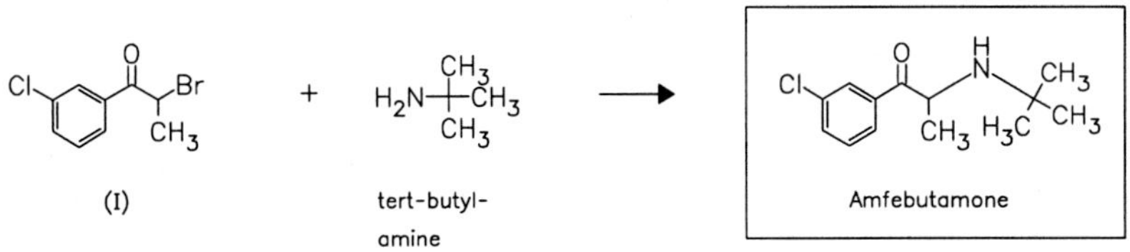

(I)

tert-butyl-
amine

Amfebutamone

Reference(s):

DOS 2 059 618 (Wellcome; appl. 3.12.1970; GB-prior. 4.12.1969).
DOS 2 064 934 (Wellcome; appl. 3.12.1970; GB-prior. 4.12.1969).
CA 977 778 (Wellcome; appl. 15.11.1970).

Formulation(s):   s. r. tabl. 100 mg, 150 mg (as hydrochloride); tabl. 75 mg, 100 mg

Trade Name(s):

USA: Wellbutrin (Glaxo   Zyban (Glaxo Wellcome)
      Wellcome)

Amfenac sodium

ATC: M01AB
Use: non-steroidal anti-inflammatory,
       analgesic

RN: 61941-56-8   MF: $C_{15}H_{12}NNaO_3$ MW: 277.26
LD$_{50}$: 550 mg/kg (M, i.v.); 615 mg/kg (M, p.o.);
       277 mg/kg (R, i.v.); 311 mg/kg (R, p.o.)
CN: 2-amino-3-benzoylbenzeneacetic acid monosodium salt

monohydrate
RN: 61618-27-7   MF: $C_{15}H_{12}NNaO_3 \cdot H_2O$ MW: 295.27
amfenac
RN: 51579-82-9   MF: $C_{15}H_{13}NO_3$ MW: 255.27
LD$_{50}$: 615 mg/kg (M, p.o.);
       311 mg/kg (R, p.o.)

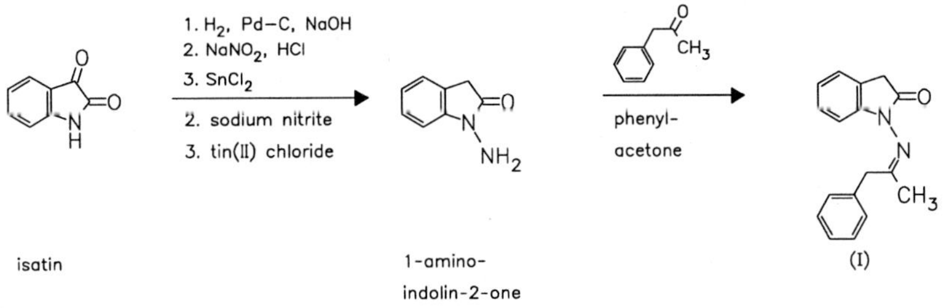

isatin

1. H_2, Pd–C, NaOH
2. NaNO$_2$, HCl
3. SnCl$_2$

2. sodium nitrite
3. tin(II) chloride

1-amino-
indolin-2-one

phenyl-
acetone

(I)

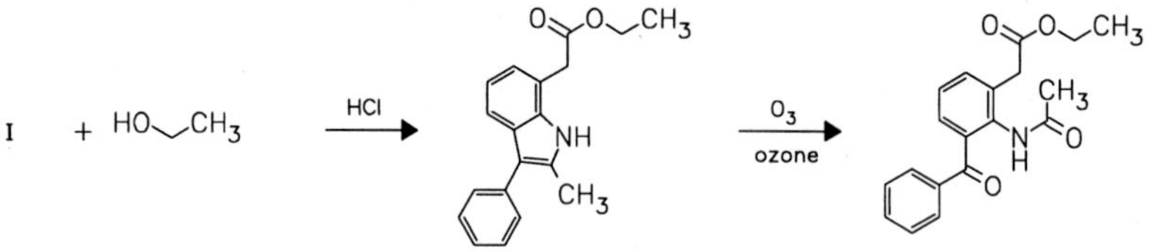

(II)

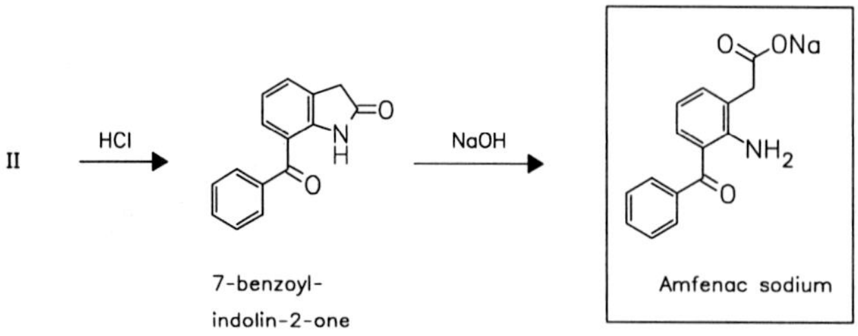

7-benzoyl-
indolin-2-one

Amfenac sodium

Reference(s):
DOS 2 324 768 (Robins; appl. 16.5.1973; USA-prior. 17.5.1972).
US 4 045 576 (Robins; USA-prior. 17.5.1972)
Welstead, W.J. et al.: J. Med. Chem. (JMCMAR) **22**, 1074 (1979).

1-aminoindolin-2-one:
Lora Tamayo, M. et al.: Org. Prep. Proced. Int. (OPPIAK) **8**, 45 (1976).

Formulation(s):　　tabl. 5 mg

Trade Name(s):
J:　　Fenazox (Meiji Seika)

Amfepramone
(Diethylpropion)

ATC:　A08AA03
Use:　appetite depressant

RN:　90-84-6　MF: $C_{13}H_{19}NO$　MW: 205.30　EINECS: 202-019-1
LD$_{50}$:　160 mg/kg (M, p.o.);
　　　>400 mg/kg (R, p.o.)
CN:　2-(diethylamino)-1-phenyl-1-propanone

hydrochloride
RN:　134-80-5　MF: $C_{13}H_{19}NO \cdot HCl$　MW: 241.76　EINECS: 205-156-5
LD$_{50}$:　50 mg/kg (M, i.v.); 385 mg/kg (M, p.o.);
　　　400 mg/kg (R, p.o.)

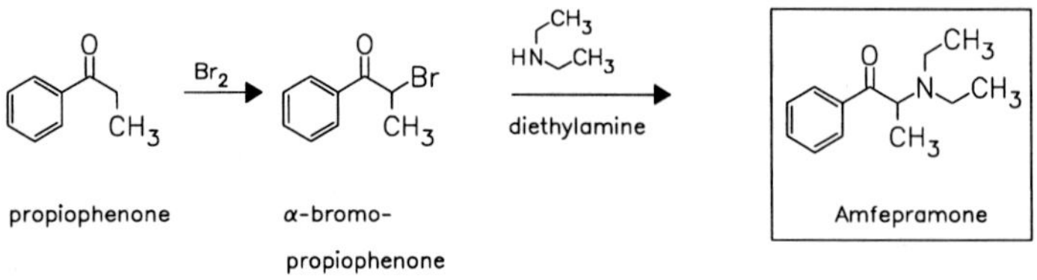

propiophenone

α-bromo-
propiophenone

diethylamine

Amfepramone

Reference(s):
US 3 001 910 (Temmler-Werke; 26.9.1961; D-prior. 16.4.1958).

Formulation(s):   cps. 25 mg, 75 mg; s. r. cps. 375 mg; s. r. tabl. 75 mg; tabl. 25 mg, 75 mg

Trade Name(s):

D:	Regenon retard (Temmler)	GB:	Apisate (Wyeth)-comb.;	USA:	Tenuate (Merrell-National);
	Tenuate (Synomed)		wfm		wfm
F:	Modératan (Théranol-		Tenuate (Merrell); wfm		Tepanil (Riker); wfm
	Deglaude)		Tenuate Dospan (Merrell);		
	Préfamone (Dexo)		wfm		
	Tenuate-Dospan (Marion	I:	Linea Valeas (Valeas)		
	Merrell)		Tenuate Dospan (Lepetit)		

Amidephrine mesilate
(Amidefrine mesilate)

ATC: R03A
Use: rhinological therapeutic,
vasoconstrictor, sympathomimetic

RN: 1421-68-7   MF: $C_{10}H_{16}N_2O_3S \cdot CH_4O_3S$ MW: 340.42
LD_{50}:   190 mg/kg (M, i.v.); 2284 mg/kg (M, p.o.);
13 mg/kg (R, p.o);
1400 g/kg (dog, i.v.)
CN: (+)-*N*-[3-[1-hydroxy-2-(methylamino)ethyl]phenyl]methanesulfonamide monomethanesulfonate

amidephrine
RN: 37571-84-9   MF: $C_{10}H_{16}N_2O_3S$ MW: 244.32

methanesulfonyl chloride + 3-aminoaceto-phenone → pyridine → 3-methylsulfonylamino-acetophenone (I)

I → Br_2 → α-bromo-3-methyl-sulfonylaminoacetophenone → N-benzyl-methylamine → α-benzylmethylamino-3-methylsulfonylaminoacetophenone (II)

II → 1. H_2, Pd-C 2. CH_3SO_3H → Amidephrine mesilate

Reference(s):
FR-M 3 027 (Mead Johnson; appl. 23.1.1963; USA-prior. 24.1.1962, 14.12.1962).

Formulation(s):   sol. 0.1 %

Trade Name(s):
GB: Dricol (Bristol); wfm

Amidotrizoic acid
(Diatrizoic acid)

ATC: V08AA01
Use: X-ray contrast medium

RN: 117-96-4   MF: $C_{11}H_9I_3N_2O_4$ MW: 613.92   EINECS: 204-223-6
LD_{50}:   8900 mg/kg (M, i.v.);
    >12.3 g/kg (R, i.v.)
CN: 3,5-bis(acetylamino)-2,4,6-triiodobenzoic acid

monosodium salt
RN: 737-31-5   MF: $C_{11}H_8I_3N_2NaO_4$ MW: 635.90   EINECS: 212-004-1
LD_{50}:   14 g/kg (M, i.v.); >7 g/kg (M,R, p.o.);
    11.4 g/kg (R, i.v.);
    13.2 g/kg (dog, i.v.)
meglumine salt
RN: 8064-12-8   MF: $C_{11}H_8I_3N_2NaO_4 \cdot C_{11}H_9I_3N_2O_4 \cdot C_7H_{17}NO_5$ MW: 1445.03
LD_{50}:   11.5 g/kg (M, i.v.);
    29.2 mg/kg (R, i.v.)

3,5-dinitrobenzoic acid → 3,5-diaminobenzoic acid → 3,5-diamino-2,4,6-triiodobenzoic acid (I)

acetic anhydride → Amidotrizoic acid

Reference(s):
Larsen, A.A. et al.: J. Am. Chem. Soc. (JACSAT) **78**, 3210 (1956).
GB 748 319 (Schering AG; appl. 1954; D-prior. 1953).
GB 782 313 (Mallinckrodt; appl. 1955; USA-prior. 1954).
US 3 076 024 (Sterling Drug; 29.1.1963; appl. 19.2.1954).
DE 1 260 477 (Schering AG; appl. 1954; USA-prior. 1953).

salts with amino acids:
DAS 2 261 584 (Dr. F. Köhler Chemie; appl. 15.12.1972).

Formulation(s):   amp. 0.65 g/ml; inj. sol. 31 %-73 %

Trade Name(s):

D:	Angiografin (Schering)	GB: Gastrografin (Schering Chemicals); wfm
	Gastrografin (Schering)	Hypaque (Winthrop); wfm
	Peritrast (Köhler; as lysine salt)	Urografin (Schering Chemicals); wfm
	Urografin (Schering)	
	Urovison (Schering)	I: Gastrografin (Schering)-comb.
F:	Angiografine (Schering)	Selectografin (Schering)-comb.
	Gastrografine (Schering)	
	Radiosélectan (Schering)	

J: Urografin (Schering-Nichidoku Yakuhin)
USA: Cardiografin (Squibb); wfm
    Cystografin (Squibb); wfm
    Gastrografin (Squibb); wfm
    Hypaque-Cysto (Winthrop); wfm
    Hypaque-Diu (Winthrop); wfm

Hypaque Sodium
(Winthrop); wfm
Meglumine Diatrizoate
(Squibb); wfm

Reno-M-30 (Squibb); wfm
Reno-M-60 (Squibb); wfm
Reno-M-DIP (Squibb);
wfm

Renovist (Squibb); wfm
Sinografin (Squibb)-comb.
with adipiodon; wfm

Amifostine

(Ethiophos; Gammaphos; NSC-296961; WR 2721)

ATC: V03AF05
Use: mucolytic agent, radioprotector,
reduction of cisplatin induced renal
toxicity

RN: 20537-88-6   MF: $C_5H_{15}N_2O_3PS$ MW: 214.23
LD$_{50}$:   557 mg/kg (M, i.v.); 842 mg/kg (M, p.o.);
826 mg/kg (R, p.o.)
CN: 2-[(3-aminopropyl)amino]ethanethiol dihydrogen phosphate (ester)

2-(3-aminopropylamino)ethyl
bromide dihydrobromide

trisodium thiophosphate
dodecahydrate

Amifostine

Reference(s):
DD 289 448 (Amt für Atomsicherheit; appl. 29.7.1982; DDR-prior. 29.7.1982).
DD 289 449 (Amt für Atomsicherheit; appl. 29.7.1983; DDR-prior. 29.7.1983).

composition having improved stability:
WO 9 403 179 (US Bioscience; appl. 30.7.1993; USA-prior. 31.7.1992).

preparation of monohydrate:
JP 54 046 722 (Yamanouchi; appl. 12.4.1979; J-prior. 21.9.1977).

preparation via 2-(3-aminopropylamino)ethyl bromide:
SU 751 030 (Kortun; 30.6.1981; SU-prior. 4.1.1979).

use for protection during radio- and chemotherapy:
US 5 298 499 (Res. Triangle Inst.; appl. 5.7.1991; USA-prior. 5.7.1991).
WO 8 907 942 (US Bioscience; appl. 21.2.1989; USA-prior. 23.2.1988).
US 5 167 947 (Southwest Res. Inst.; appl. 26.10.1989; USA-prior. 26.10.1989).
US 3 892 824 (Southern Res. Inst.; appl. 16.12.1988; USA-prior. 16.12.1988).

use for reducing side effects with azidothymidine:
WO 9 014 007 (US Bioscience; appl. 9.5.1990; USA-prior. 24.5.1989).

use for prevention of cytostatic alopecia:
DE 3 509 071 (ASTA-Werke; appl. 14.3.1985; D-prior. 29.3.1984).

Formulation(s):   amp. 500 mg; vial 500 mg dry substance for inj.

Trade Name(s):
D: Ethyol (Essex Pharma;   GB: Ethyol (Schering-Plough)
1995)

Amikacin

ATC: D06AX12; J01GB06; S01AA21
Use: aminoglycoside antibiotic

RN: 37517-28-5   MF: $C_{22}H_{43}N_5O_{13}$ MW: 585.61   EINECS: 253-538-5
LD_{50}:   280 mg/kg (M, i.v.); >6 g/kg (M, p.o.)
CN: (S)-O-3-amino-3-deoxy-α-D-glucopyranosyl-(1→6)-O-[6-amino-6-deoxy-α-D-glucopyranosyl-(1→4)]-N^1-(4-amino-2-hydroxy-1-oxobutyl)-2-deoxy-D-streptamine

sulfate (1:2)
RN: 39831-55-5   MF: $C_{22}H_{43}N_5O_{13} \cdot 2H_2SO_4$ MW: 781.76   EINECS: 254-648-6
LD_{50}:   181 mg/kg (M, i.v.); >10.679 g/kg (M, p.o.);
    234 mg/kg (R, i.v.); >4 g/kg (R, p.o.);
    383 mg/kg (dog, i.v.)

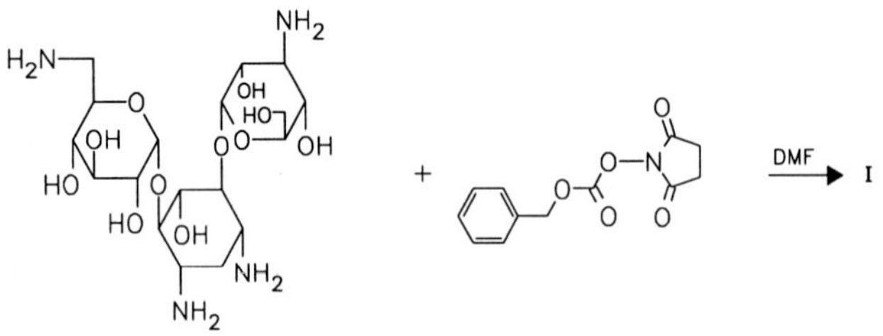

kanamycin A
(q. v.)

N-(benzyloxycarbonyl-
oxy)succinimide

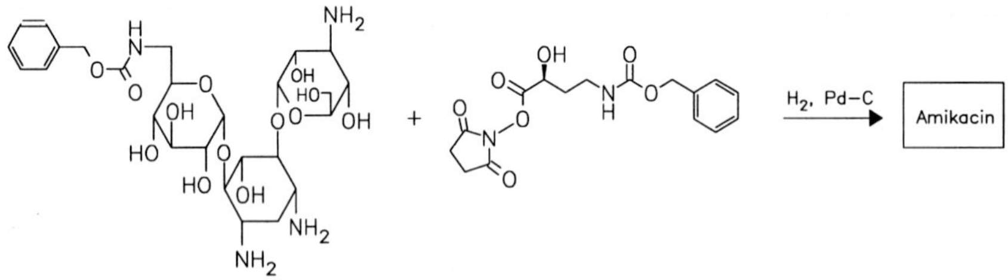

N^6-(benzyloxycarbonyl)kanamycin A   (I)

L(-)-γ-benzyloxycarbonylamino-
α-hydroxybutyric acid
succinimido ester

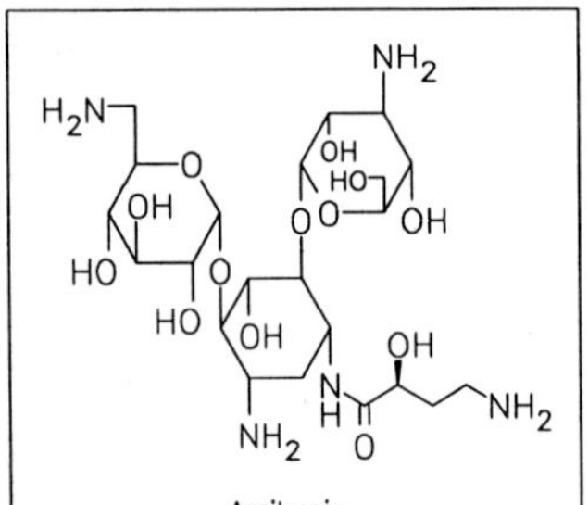

Amikacin

Reference(s):
GB 1 401 221 (Bristol Myers; appl. 13.7.1972; USA-prior. 13.7.1971).
DE 2 234 315 (Bristol-Myers; appl. 12.7.1972; USA-prior. 27.1.1972, 13.7.1971).
US 3 781 268 (Bristol-Myers; 25.12.1973; prior. 27.1.1972, 13.7.1971).
Kawaguchi, H. et al.: J. Antibiot. (JANTAJ) **25**, 695 (1972).

alternative syntheses:
NL 7 401 517 (Bristol-Myers; appl. 4.2.1974; USA-prior. 7.2.1973).
NL 7 414 668 (Bristol-Myers; appl. 11.11.1974; USA-prior. 14.11.1973, 23.5.1974).
US 3 974 137 (Bristol-Myers; 10.8.1976; prior. 23.5.1974).
DOS 2 432 644 (Takeda; appl. 8.7.1974; J-prior. 12.7.1973).
DOS 2 716 533 (Pfizer; appl. 14.4.1977; GB-prior. 14.4.1976).
DOS 2 818 822 (Bristol-Myers; appl. 28.4.1978; USA-prior. 28.4.1977, 20.3.1978).
DOS 2 818 992 (Bristol-Myers; appl. 28.4.1978; USA-prior. 28.4.1977; 20.3.1978).

disulfate pentahydrate:
FR 2 308 373 (Bristol-Myers; appl. 22.3.1976; USA-prior. 23.4.1975).

review:
Kawaguchi, H.; Hiroshi: Drug Action Drug Resist. Bact. (DADRBY) **2**, 45 (1975).

Formulation(s):   cream 2.5 %, 5 %; eye drops 0.3 %, 0.5 %; gel 5.5; vial 100 mg/2 ml, 250 mg/2 ml,
                     500 mg/2 ml

Trade Name(s):

D:	Biklin (Bristol-Myers Squibb; 1976)	Chemacin (CT) Likacin (Lisapharma; 1981)	Pierami (Pierrel; 1980) Sifamic (SIFI)
F:	Amiklin (Bristol-Myers Squibb)	Lukadin (San Carlo)	J: Amikacin Sulfate (Banyu) Biklin (Banyu-Bristol-
GB:	Amikin (Bristol-Myers Squibb; 1976)	Migracin (SmithKline Beecham)	Myers Squibb)
I:	Amicasil (Biotekfarma) Bb-k8 (Bristol; 1978)	Mikavir (Salus Research; 1986)	USA: Amikin (BMS; 1976)

Amiloride

ATC: C03DB01
Use: diuretic, antihypertonic

RN: 2609-46-3   MF: $C_6H_8ClN_7O$ MW: 229.63   EINECS: 220-024-7
CN: 3,5-diamino-*N*-(aminoiminomethyl)-6-chloropyrazinecarboxamide

monohydrochloride
RN: 2016-88-8   MF: $C_6H_8ClN_7O \cdot HCl$ MW: 266.09   EINECS: 217-958-2

glyoxal   5,6-diaminouracil   lumazine   3-aminopyrazine-2-carboxylic acid (I)

methanol   methyl 3-amino-pyrazine-2-carboxylate   methyl 6-chloro-3,5-diaminopyrazine-2-carboxylate (II)

II + guanidine → Amiloride

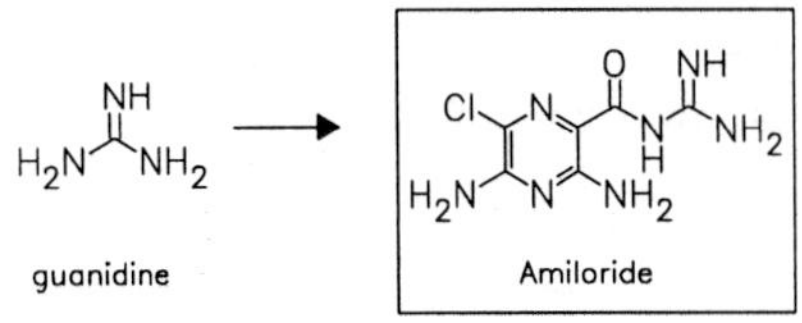

Reference(s):

DE 1 470 053 (Mercle & Co.; appl. 28.10.1963; USA-prior. 30.10.1962).
US 3 313 813 (Merck & Co.; 11.4.1967; prior. 30.10.1962, 7.10.1963).
GB 1 066 855 (Merck & Co.; appl. 24.10.1963; USA-prior. 30.10.1962, 7.10.1963).
Bicking, J.B. et al.: J. Med. Chem. (JMCMAR) **8**, 638 (1965).
Cragoe, E.J. et al.: J. Med. Chem. (JMCMAR) **10**, 66 (1967).

improved method for 5,6-diaminouracil:
DOS 2 831 037 (Lonza; appl. 14.7.1978; CH-prior. 20.7.1977).

combination with etacrynic acid:
US 3 781 430 (Merck & Co.; 25.12.1973; prior. 30.10.1962, 7.10.1963, 7.2.1966, 18.2.1969, 21.12.1971).

Formulation(s):   tabl. 2.5 mg, 5 mg, 10 mg in comb. with hydrochlorothiazide (as hydrochloride)

Trade Name(s):

D: Amiduret (Trommsdorff; 1985)-comb.
Diaphal (Pierre Fabre Pharma)-comb.
Diursan (TAD)-comb.
Esmalorid (Merck)-comb.
Moducrin (MSD; 1978)-comb.
Moduretik, -mite (Du Pont Pharma; 1973)-comb.
Rhefluin, -mite (Kytta-Siegfried)-comb.

F: Logiréne (Pharmacia & Upjohn SA)-comb.
Modamide (Merck Sharp & Dohme; 1973)

Moducren (Merck Sharp & Dohme-Chibret; 1979)-comb.
Modurétic (Merck Sharp & Dohme; 1973)-comb.

GB: Amilco (Baker Norton; 1983)-comb. with hydrochlorothiazole
Burinex A (Leo)-comb.
FruCo (Baker Norton)-comb.
Frumil (Rhône-Poulenc Rorer; 1983)-comb.
Kalten (Zeneca; 1985)-comb.

Lasoride (Hoechst; 1987)-comb.
Moducren (Morson; 1981)-comb.
Moduret-25 (Du Pont; 1984)-comb.
Moduretic (Du Pont; 1970)
Navispare (Novartis)-comb.

I: Moduretic (Merck Sharp & Dohme; 1975)-comb.

USA: Midamor (Merck Sharp & Dohme; 1981)
Moduretic (Merck Sharp & Dohme; 1981)-comb.

Amineptine

ATC: N06AA19
Use: psychoanaleptic, CNS stimulant

RN: 57574-09-1   MF: $C_{22}H_{27}NO_2$ MW: 337.46   EINECS: 260-818-0
LD$_{50}$:   115 mg/kg (M, i.p.)
CN: 7-[(10,11-dihydro-5*H*-dibenzo[*a,d*]cyclohepten-5-yl)amino]heptanoic acid

hydrochloride
RN: 30272-08-3   MF: $C_{22}H_{27}NO_2 \cdot HCl$ MW: 373.92   EINECS: 250-107-3
LD$_{50}$:   405 mg/kg (M, p.o.)

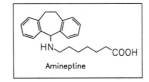

5-oxo-10,11-dihydro-
5H-dibenzo[a,d]cyclo-
heptene

5-hydroxy-10,11-
dihydro-5H-di-
benzo[a,d]cycloheptene

5-chloro-10,11-dihydro-5H-
dibenzo[a,d]cycloheptene (I)

I + H_2N ... ethyl 7-aminoheptanoate → Amineptine

Amineptine

Reference(s):
DOS 2 011 806 (Science Union; appl. 12.3.1970; GB-P. 27.3.1969).
US 3 758 528 (Science Union; 11.9.1973; appl. 13.3.1970).
US 3 821 249 (Science Union; 28.6.1974; prior. 13.3.1970, 30.10.1972).

Formulation(s):   tabl. 100 mg (as hydrochloride)

Trade Name(s):
F: Survector (Euthérapic;   I: Maneon (Poli; 1983)
      1978); wfm 1999   Survector (Stroder; 1983)

Aminocaproic acid
(Acide aminocaproique; Epsilcapramin)

ATC: B02AA01
Use: antifibrinolytic, plasmin inhibitor

RN: 60-32-2 MF: $C_6H_{13}NO_2$ MW: 131.18 EINECS: 200-469-3
LD_{50}: 4900 mg/kg (M, i.v.); 14.3 g/kg (M, p.o.);
       3300 mg/kg (R, i.v.);
       >7 g/kg (dog, p.o.)
CN: 6-aminohexanoic acid

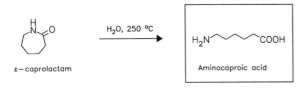

ε–caprolactam   Aminocaproic acid

Reference(s):
US 2 453 234 (American Enka Corp.; 1948; NL-prior. 1946).

Formulation(s):   inj. flask 250 mg/ml; syrup 25 %; tabl. 500 mg

Trade Name(s):
D: Epsilon-Aminocapronsäure   F: Hexalense (Leurquin)   Resplamin (Kyorin)
      "Roche" (Roche); wfm   I: Caprolisin (Malesci)   USA: Amicar (Immunex)
      Epsilon-Tachostypan   J: Capusumine (Nichiiko)
      (Hormon-Chemie)-comb.;   Hemotin (Hokuriku)
      wfm   Ipsilon (Daiichi)

Aminoglutethimide

ATC: J04AA01
Use: antineoplastic (aromatase inhibitor)

RN: 125-84-8 MF: $C_{13}H_{16}N_2O_2$ MW: 232.28 EINECS: 204-756-4
LD_{50}: 625 mg/kg (M, i.p.)
CN: 3-(4-aminophenyl)-3-ethyl-2,6-piperidinedione

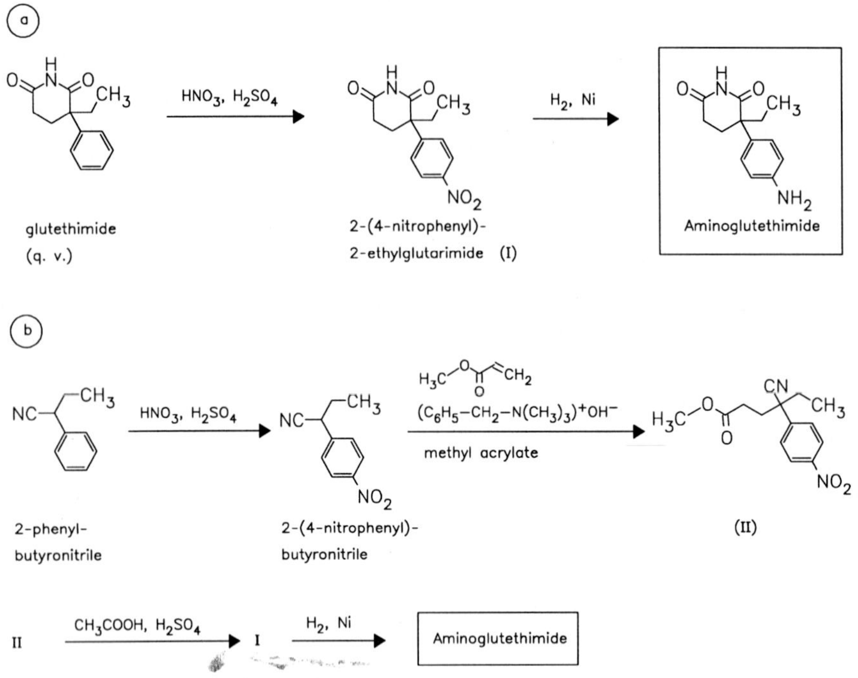

a
HNO3, H2SO4
H2, Ni
glutethimide
(q. v.)
2-(4-nitrophenyl)-
2-ethylglutarimide (I)
Aminoglutethimide
b
HNO3, H2SO4
(C6H5-CH2-N(CH3)3)+OH-
methyl acrylate
2-phenyl-
butyronitrile
2-(4-nitrophenyl)-
butyronitrile
(II)
II
CH3COOH, H2SO4
I
H2, Ni
Aminoglutethimide

Reference(s):
US 2 848 455 (Ciba; 1958; CH-prior. 1955).

racemate resolution:
Finch, N. et al.: Experientia (EXPEAM) **31**, 1002 (1975).

Formulation(s): tabl. 250 mg

Trade Name(s):

D:	Orimeten (Novartis Pharma)	F:	Orimétène (Novartis)	USA:	Cytadren (Novartis)
	Rodazol (Novartis Pharma)	GB:	Orimeten (Novartis)		
		I:	Orimeten (Novartis)		

Aminophenazone

(Amidophenazon; Amidopyrin; Aminopyrine)

ATC: N02BB03
Use: analgesic, antipyretic, anti-inflammatory

RN: 58-15-1 MF: $C_{13}H_{17}N_3O$ MW: 231.30 EINECS: 200-365-8
LD_{50}: 78 mg/kg (M, i.v.); 350 mg/kg (M, p.o.);
98 mg/kg (R, i.v.); 285 mg/kg (R, p.o.);
121 mg/kg (dog, i.v.); 220 mg/kg (dog, p.o.)
CN: 4-(dimethylamino)-1,2-dihydro-1,5-dimethyl-2-phenyl-3*H*-pyrazol-3-one

ascorbate
RN: 23635-43-0   MF: $C_{13}H_{17}N_3O \cdot C_6H_8O_6$ MW: 407.42

ethyl acetoacetate   +   phenylhydrazine   →   3-methyl-1-phenyl-5-Δ^3-pyrazolone   dimethyl sulfate (I) →   2,3-dimethyl-1-phenyl-5-Δ^3-pyrazolone (II)

II   NaNO₂ →   2,3-dimethyl-4-nitroso-1-phenyl-5-Δ^3-pyrazolone   NaHSO₃ →   4-amino-2,3-dimethyl-1-phenyl-5-Δ^3-pyrazolone   I →   Aminophenazone

Reference(s):
DRP 193 632 (E. Scheitlin; 1907).
Ehrhart, Ruschig **I**, 171.

Formulation(s):   suppos. 200 mg, 500 mg; tabl. 100 mg, 300 mg

Trade Name(s):
D: Compretten (Cascan); wfm
      Dimametten (Hormosan); wfm

I: Pyramidon (Hoechst); wfm
      Farmidone (Farmitalia)
      Fugantil (Ghimas)

      numerous combination preparations
J: Neophyllin (Nippon Eisai)

Aminopromazine
(Proquamezine)

Use: antispasmodic

RN: 58-37-7   MF: $C_{19}H_{25}N_3S$ MW: 327.50   EINECS: 200-378-9
CN: *N,N,N',N'*-tetramethyl-3-(10*H*-phenothiazin-10-yl)-1,2-propanediamine

fumarate (2:1)
RN: 3688-62-8   MF: $C_{19}H_{25}N_3S \cdot 1/2C_4H_4O_4$ MW: 771.06   EINECS: 222-987-9

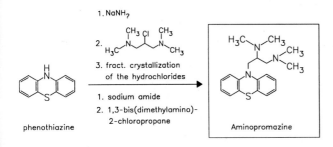

phenothiazine
1. NaNH₂
2. 1,3-bis(dimethylamino)-2-chloropropane
3. fract. crystallization of the hydrochlorides
1. sodium amide
Aminopromazine

Reference(s):
GB 800 635 (Rhône-Poulenc; appl. 1954).
DE 1 034 637 (Rhône-Poulenc; appl. 1955; F-prior. 1954).

Trade Name(s):
D: Lorusil (Bayer); wfm   F: Lispamol (Specia); wfm

---

p-Aminosalicylic acid
(Aminosalylum; PAS)

ATC: J04AA01
Use: tuberculostatic

RN: 65-49-6   MF: $C_7H_7NO_3$ MW: 153.14   EINECS: 200-613-5
LD$_{50}$: 3898 mg/kg (M, i.v.); 4 g/kg (M, p.o.)
CN: 4-amino-2-hydroxybenzoic acid

calcium salt (2:1)
RN: 133-15-3   MF: $C_{14}H_{12}CaN_2O_6$ MW: 344.34   EINECS: 205-095-4
LD$_{50}$: 6500 mg/kg (M, p.o.)
monosodium salt
RN: 133-10-8   MF: $C_7H_6NNaO_3$ MW: 175.12   EINECS: 205-091-2
LD$_{50}$: 3380 mg/kg (M, i.v.); 6900 mg/kg (M, p.o.);
        8 g/kg (R, p.o.)

3-amino-   carbon   p-Aminosalicylic acid
phenol   dioxide

$KHCO_3$, 5-10 atm

Reference(s):
US 2 540 104 (Parke Davis; 1951; prior. 1949).

purification:
US 2 844 625 (Miles, 1958; appl. 1954).

Formulation(s):   vial 13.49 g (as monosodium salt)

Trade Name(s):
D: Pas-Fatol N (Fatol)   GB: Asacol (SmithKline   Salf-Pas (Salf; as sodium
F: B-PAS (Salvoxyl-Wander);   Beecham)   salt)
       wfm   Pentasal (Yamanouchi)   J: PAS Calcium (Sumitomo);
       PAS Elbiol   Salofalk (Thames)   wfm
       (Pharmacotechnie); wfm   I: Eupasal sodico (Bieffe   Sanpas Cal. (Sankyo); wfm
                                             Medital; as sodium salt) USA: Paser (Jacobus)

Amiodarone

ATC: C01BD01
Use: antiarrhythmic

RN: 1951-25-3   MF: $C_{25}H_{29}I_2NO_3$ MW: 645.32   EINECS: 217-772-1
LD$_{50}$: 178 mg/kg (M, i.v.); >4 g/kg (M, p.o.)
CN: (2-butyl-3-benzofuranyl)[4-[2-(diethylamino)ethoxy]-3,5-diiodophenyl]methanone

hydrochloride
RN: 19774-82-4   MF: $C_{25}H_{29}I_2NO_3 \cdot HCl$ MW: 681.78

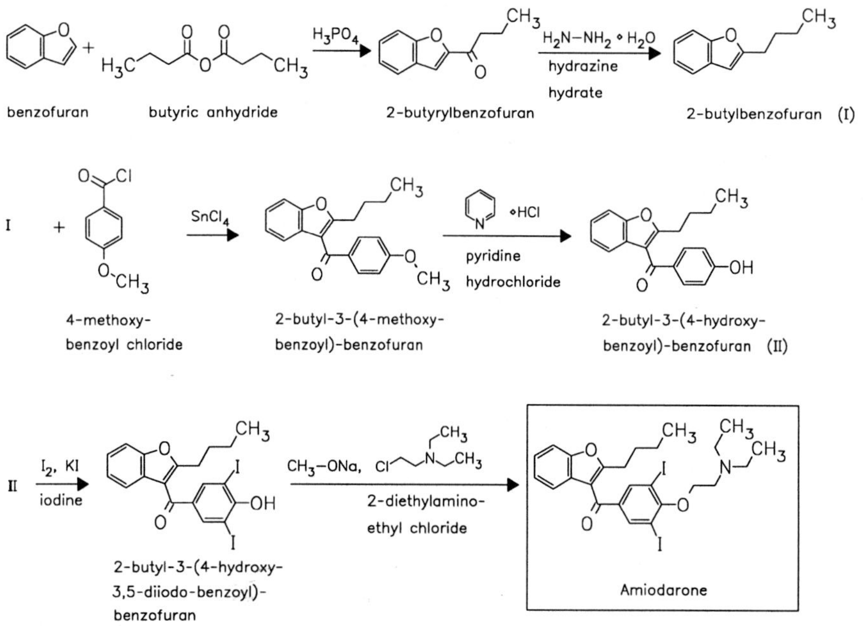

benzofuran   butyric anhydride   2-butyrylbenzofuran   2-butylbenzofuran (I)

4-methoxy-benzoyl chloride   2-butyl-3-(4-methoxy-benzoyl)-benzofuran   2-butyl-3-(4-hydroxy-benzoyl)-benzofuran (II)

2-butyl-3-(4-hydroxy-3,5-diiodo-benzoyl)-benzofuran   Amiodarone

Reference(s):
FR 1 339 389 (Labaz; appl. 22.11.1962).
US 3 248 401 (Labaz; 26.4.1966; prior. 24.11.1961).

2-butylbenzofuran:
Buu-Hoï, N.P. et al.: J. Chem. Soc. (JCSOA9) **1964**, 173.

Formulation(s):   inj. sol. 150 mg/3ml; tabl. 200 mg

Trade Name(s):

D:	Cordarex (Sanofi Winthrop)	F:	Cordarone (Sanofi Winthrop)	I:	Amiodar (Midy) Cordarone (Sigma-Tau)
	Tachydaron (ASTA Medica AWD)	GB:	Cordarone X (Sanofi Winthrop)	USA:	Cordarone (Wyeth-Ayerst; as hydrochloride)

Amiphenazole

ATC: R07A
Use: respiratory stimulant, morphine antagonist, antidote (barbiturate poisonings)

RN: 490-55-1   MF: $C_9H_9N_3S$ MW: 191.26   EINECS: 207-713-8
LD_{50}: 400 mg/kg (M, p.o.)
CN: 5-phenyl-2,4-thiazolediamine

monohydrochloride
RN: 942-31-4   MF: $C_9H_9N_3S \cdot HCl$ MW: 227.72   EINECS: 213-389-9
LD_{50}: 372 mg/kg (M, p.o.)

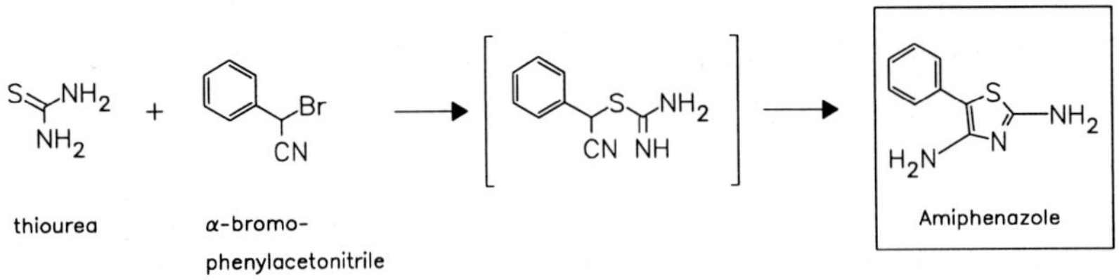

thiourea

α-bromo-
phenylacetonitrile

Amiphenazole

Reference(s):

Davis, W. et al.: J. Chem. Soc. (JSCOA9) **1955**, 3491.

Chase, B.H. et al.: J. Chem. Soc. (JSCOA9) **1955**, 4443.

Formulation(s):   inj. flask 150 mg

Trade Name(s):

D: Daptazile 100 (Nicholas);   Daptazile Injektion   GB: Daptazole (Nicholas); wfm
      wfm   (Nicholas); wfm

Amitriptyline

ATC: N06AA09
Use: antidepressant

RN: 50-48-6   MF: $C_{20}H_{23}N$ MW: 277.41   EINECS: 200-041-6

LD_{50}:   16 mg/kg (M, i.v.); 140 mg/kg (M, p.o.);
       320 mg/kg (R, p.o.)

CN: 3-(10,11-dihydro-5H-dibenzo[a,d]cyclohepten-5-ylidene)-N,N-dimethyl-1-propanamine

hydrochloride

RN: 549-18-8   MF: $C_{20}H_{23}N \cdot HCl$ MW: 313.87   EINECS: 208-964-6

LD_{50}:   21 mg/kg (M, i.v.); 140 mg/kg (M, p.o.);
       14 mg/kg (R, i.v.); 240 mg/kg (R, p.o.);
       >27 mg/kg (dog, i.v.)

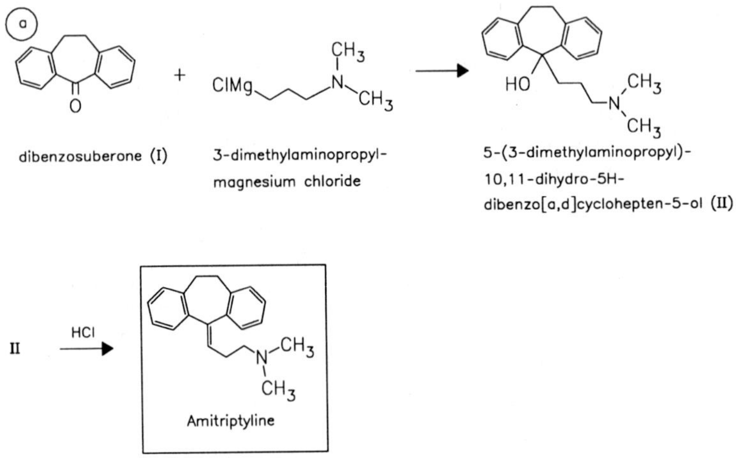

dibenzosuberone (I)   3-dimethylaminopropyl-
                       magnesium chloride

5-(3-dimethylaminopropyl)-
10,11-dihydro-5H-
dibenzo[a,d]cyclohepten-5-ol (II)

Amitriptyline

(b)

I + BrMg cyclopropyl

cyclopropyl-
magnesium
bromide

HO

5-cyclopropyl-10,11-
dihydro-5H-dibenzo-
[a,d]cyclohepten-5-ol

HBr, CH₃COOH   →   III

Br

5-(3-bromopropylidene)-
10,11-dihydro-5H-
dibenzo[a,d]cycloheptene (III)

+ HN CH₃ / CH₃

dimethylamine

→   Amitriptyline

Reference(s):
a GB 858 187 (Hoffmann-La Roche; appl. 24.3.1959; CH-prior. 3.4.1958).
   DE 1 109 166 (Hoffmann-La Roche; appl. 16.3.1959; CH-prior. 3.4.1958).
   BE 584 061 (Merck & Co.; appl. 27.10.1959; USA-prior. 31.10.1958).
   BE 609 095 (Kefalas A/S; appl. 12.10.1961; DK-prior. 12.10.1960).
b Hoffsommer, R.D. et al.: J. Org. Chem. (JOCEAH) **27**, 4134 (1962).

alternative synthesis:
DAS 1 468 138 (Kefalas; appl. 12.3.1963; GB-prior. 23.3.1962, 9.11.1962).
US 3 205 264 (Merck & Co.; 7.9.1965; appl. 15.6.1962).

Formulation(s):   amp. 56.6 mg; f. c. tabl. 10 mg, 25 mg, 50 mg; drg. 11.32 mg, 28.3 mg; drops 40 mg/1 ml;
                      inj. 50 mg/2 ml; tabl. 25 mg, 50 mg (as hydrochloride)

Trade Name(s):

D:	Amineurin (Neuro Hexal)	I:	Adepril (Lepetit)	J:	Tryptanol (Merck-Banyu;
	Limbatril (ICN)		Amilit-ifi (IFI)		as hydrochloride)
	Saroten (Bayer Vital)		Amitript (Formulario Naz.)	USA:	Elavil (Zeneca; as
F:	Elavil (Merck Sharp &		Diapatol (Teofarma)-comb.		hydrochloride)
	Dohme-Chibret)		Laroxyl (Roche)		Etrafon (Schering)
	Laroxyl (Roche)		Limbitryl (Roche)-comb.		Limbitrol (Roche Products;
GB:	Lentizol (Parke Davis)		Sedans (Ganassini)-comb.		as hydrochloride)
	Triptafen (Goldshield)-		Triptizol (Merck Sharp &		Triavil (Merck; as
	comb.		Dohme)		hydrochloride)
	Tryptizol (Morson)		combination preparations		generics

Amitriptylinoxide

ATC: N06AA09
Use: antidepressant

RN;   4317-14-0   MF: C₂₀H₂₃NO   MW; 293,41
LD₅₀: 320 mg/kg (M, i.p.); 87 mg/kg (M, i.v.); 330 mg/kg (M, p.o.);
        120 mg/kg (R, i.p.); 25 mg/kg (R, i.v.); 1800 mg/kg (R, p.o.);
        330-460 mg/kg (rabbit, p.o.);
        330 mg/kg (dog, p.o.)
CN: 3-(10,11-dihydro-5H-dibenzo[a,d]cyclohepten-5-ylidene)-N,N-dimethyl-1-propanamine N-oxide

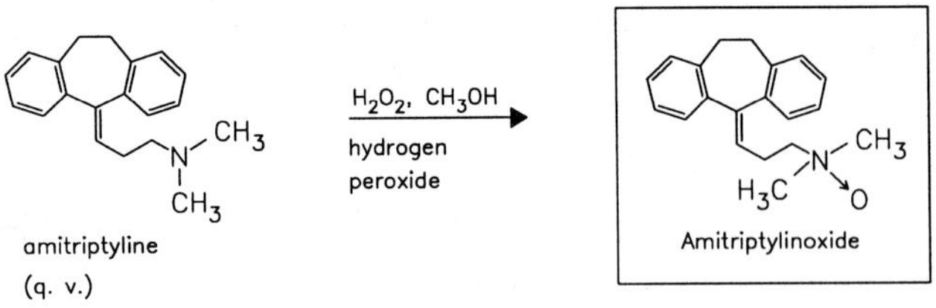

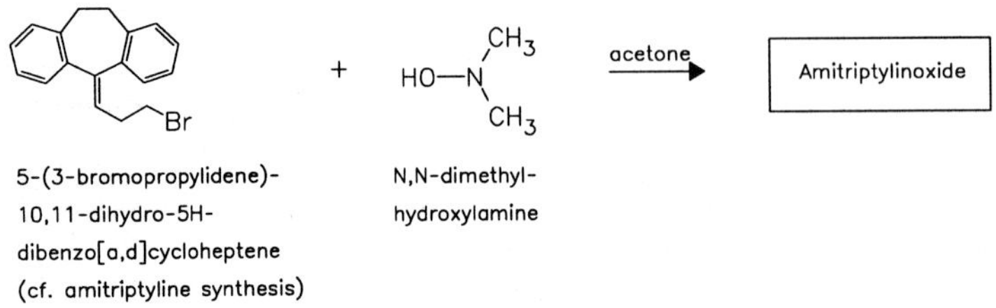

amitriptyline
(q. v.)

H₂O₂, CH₃OH
hydrogen
peroxide

Amitriptylinoxide

5-(3-bromopropylidene)-
10,11-dihydro-5H-
dibenzo[a,d]cycloheptene
(cf. amitriptyline synthesis)

+

N,N-dimethyl-
hydroxylamine

acetone

Amitriptylinoxide

Reference(s):
DE 1 243 180 (Dumex; appl. 15.2.1964; GB-prior. 20.2.1963).
FR-M 3 222 (Dumex; appl. 20.2.1964; GB-prior. 20.2.1963).
NL-appl. 6 511 947 (Merck & Co., appl. 14.9.1965; USA-prior. 14.9.1964).

Formulation(s):   tabl. 30 mg, 60 mg, 90 mg, 120 mg

Trade Name(s):
D: Equilibrin (Rhône-Poulenc
      Rorer)

Amixetrine

ATC: N06A; R03BB
Use: anticholinergic, antidepressant,
        antispasmodic

RN: 24622-72-8   MF: $C_{17}H_{27}NO$ MW: 261.41
CN: 1-[2-(3-methylbutoxy)-2-phenylethyl]pyrrolidine

hydrochloride
RN: 24622-52-4   MF: $C_{17}H_{27}NO \cdot HCl$ MW: 297.87   EINECS: 246-365-1

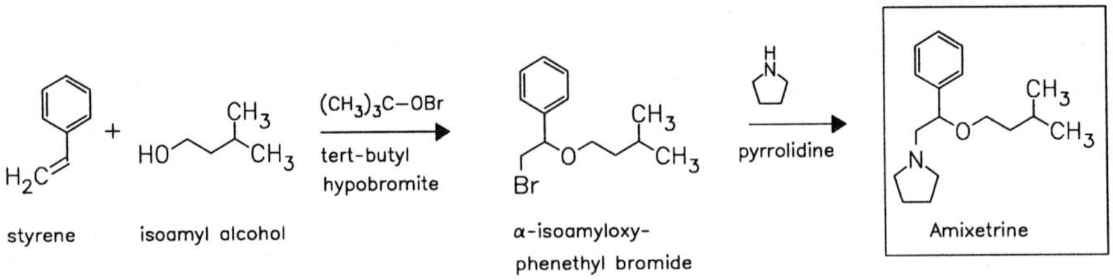

styrene   isoamyl alcohol

$(CH_3)_3C-OBr$
tert-butyl
hypobromite

α-isoamyloxy-
phenethyl bromide

pyrrolidine

Amixetrine

Reference(s):
DOS 1 811 767 (Mauvernay; appl. 29.11.1968; F-prior. 15.12.1967).

Formulation(s):   tabl. 50 mg

Trade Name(s):
F: Somagest (Riom); wfm

Amlexanox
(AA-673)

ATC: R03DX01; R06AX
Use: antiallergic, antiasthmatic

RN: 68302-57-8   MF: $C_{16}H_{14}N_2O_4$ MW: 298.30
LD_{50}: 2320 mg/kg (M, p.o.);
         10 g/kg (R, p.o.)
CN: 2-amino-7-(1-methylethyl)-5-oxo-5H-[1]benzopyrano[2,3-b]pyridine-3-carboxylic acid

2-hydroxy-5-isopropyl-   dimethyl-   6-isopropyl-4-oxo-
acetophenone   formamide   4H-1-benzopyran-3-
                                                carboxaldehyde (I)

                           6-isopropyl-4-oxo-   2-amino-6-isopropyl-
                           4H-1-benzopyran-   4-oxo-4H-1-benzopyran-
                           3-carbonitrile   3-carboxaldehyde (II)

ethyl cyanoacetate   Amlexanox

Reference(s):
DOS 2 809 720 (Takeda; appl. 7.3.1978; J-prior. 8.3.1977, 20.12.1977).
US 4 143 042 (Takeda; 6.3.1979; J-prior. 8.3.1977, 20.12.1977).
US 4 255 576 (Takeda; 10.3.1981; J-prior. 8.3.1977, 10.12.1977).
US 4 299 963 (Takeda; 10.11.1981; J-prior. 8.3.1977, 10.12.1977).
Nohara, A. et al.: J. Med. Chem. (JMCMAR) 28, 559 (1985).

synthesis of 6-isopropyl-4H-1-benzopyran-3-carbonitrile:
US 3 896 114 (Takeda Chemical Ind.; appl. 22.7.1975; J-prior. 12.4.1972, 14.4.1972).
DE 2 317 899 (Takeda Chemical Ind.; appl. 25.10.1973; J-prior. 12.4.1972).

Formulation(s):   cream 5 %; tabl. 100 mg

Trade Name(s):
J: Solfa (Takeda; 1989)   USA: Aphthasol (Block Drug
                                             Company)

Amlodipine

ATC: C02DE; C08CA01
Use: calcium antagonist, antianginal,
        antihypertensive

RN: 88150-42-9   MF: $C_{20}H_{25}ClN_2O_5$ MW: 408.88
CN: 2-[(2-aminoethoxy)methyl]-4-(2-chlorophenyl)-1,4-dihydro-6-methyl-3,5-pyridinedicarboxylic acid 3-
       ethyl 5-methyl ester

maleate (1:1)
RN: 88150-47-4   MF: $C_{20}H_{25}ClN_2O_5 \cdot C_4H_4O_4$ MW: 524.95

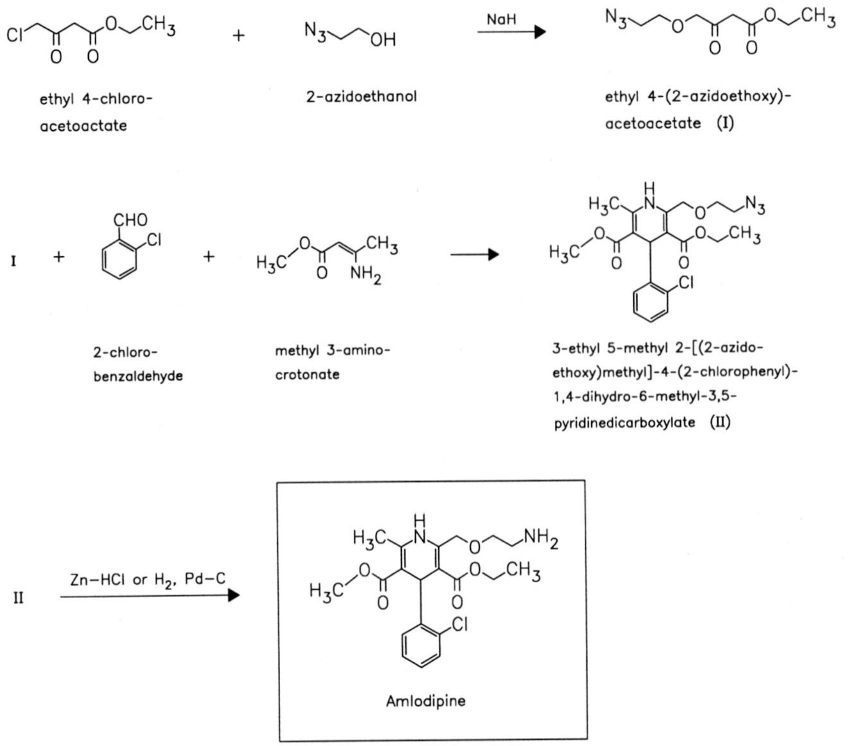

ethyl 4-chloro-
acetoacetate

2-azidoethanol

ethyl 4-(2-azidoethoxy)-
acetoacetate (I)

2-chloro-
benzaldehyde

methyl 3-amino-
crotonate

3-ethyl 5-methyl 2-[(2-azido-
ethoxy)methyl]-4-(2-chlorophenyl)-
1,4-dihydro-6-methyl-3,5-
pyridinedicarboxylate (II)

Amlodipine

Reference(s):
EP 89 167 (Pfizer; appl. 8.3.1983; GB-prior. 11.3.1982).
EP 599 220 (Lek; appl. 19.11.1993; SI-prior. 26.11.1992).
CA 2 188 071 (Apotex; appl. 17.10.1996; NZ-prior. 1.11.1995).

besylate salt:
EP 244 944 (Pfizer; appl. 31.3.1987; GB-prior. 4.4.1986).

racemate resolution:
EP 331 315 (Pfizer; appl. 16.2.1989; GB-prior. 27.2.1988).
Arrowsmith, J.E. et al.: J. Med. Chem. (JMCMAR) **29**, 1696 (1986).

combination with ACE-inhibitors:
WO 9 628 185 (Pfizer; appl. 26.2.1996; USA-prior. 16.3.1995).

Formulation(s):   cps. 5 mg, 15 mg, 20 mg; tabl. 2.5 mg, 5 mg, 10 mg

Trade Name(s):
D: Norvasc (Mack, Illert; Pfizer)
F: Amlor (Pfizer)
GB: Istin (Pfizer; 1990)

I: Antacal (Errekappa Euroter.; 1991)
Monopina (Bioindustria; 1991)
Norvasc (Pfizer; 1990)

J: Amlodin (Sumitomo)
Norvasc (Pfizer)
USA: Lotrel (Novartis)
Norvasc (Pfizer; 1991)

Amobarbital
(Amylobarbitone)

ATC: N05CA02
Use: hypnotic

RN: 57-43-2 MF: $C_{11}H_{18}N_2O_3$ MW: 226.28 EINECS: 200-330-7
LD$_{50}$: 345 mg/kg (M, p.o.);
250 mg/kg (R, p.o.);
58 mg/kg (dog, i.v.)
CN: 5-ethyl-5-(3-methylbutyl)-2,4,6(1H,3H,5H)-pyrimidinetrione

monosodium salt
RN: 64-43-7 MF: $C_{11}H_{17}N_2NaO_3$ MW: 248.26 EINECS: 200-584-9
LD$_{50}$: 505 mg/kg (M, p.o.);
128 mg/kg (R, i.v.); 275 mg/kg (R, p.o.);
75 mg/kg (dog, i.v.); 99 mg/kg (dog, p.o.)

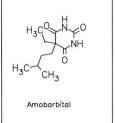

Reference(s):
GB 191 008 (E. Layraud; 1922; F-prior. 1921).
US 1 856 792 (Eli Lilly; 1932; prior. 1929).

Formulation(s): tabl. 15 mg, 30 mg, 50 mg, 100 mg

Trade Name(s):

D:	Ansudoral (Basotherm)-comb.; wfm	Météoxane (Gallier)-comb.; wfm	GB: Amytal (Flynn)
	Jalonac (Röhm Pharma)-comb.; wfm	Nardyl (Vernin)-comb.; wfm	Sodium Amytal (Flynn) Tuinal (Flynn)-comb.
	Metrotonin (Temmler)-comb.; wfm	Noctadiol (Millot-Solac)-comb.; wfm	I: Amobarb (Tariff. Integrativo)
	Stadadorm Tabl. (Stada); wfm	Supponoctal (Houdé)-comb.; wfm	J: Amytal (Yamanouchi) Isomytal (Nippon
F:	Binoctal (Houdé)-comb.; wfm	Tensophoril (Synlab)-comb.; wfm	Shinyaku) USA: Amytal (Lilly)
	Carlytène amobarbital (Dedieu)-comb.; wfm	Viscéralgine comprimés (Riom)-comb.; wfm	Amytal Sodium (Lilly) Tuinal (Lilly)

Amodiaquine

ATC: P01BA06
Use: antimalarial

RN: 86-42-0   MF: $C_{20}H_{22}ClN_3O$ MW: 355.87   EINECS: 201-669-3
LD_{50}: 550 mg/kg (M, p.o.)
CN: 4-[(7-chloro-4-quinolinyl)amino]-2-[(diethylamino)methyl]phenol

dihydrochloride dihydrate
RN: 69-44-3   MF: $C_{20}H_{22}ClN_3O \cdot 2HCl \cdot 2H_2O$ MW: 464.82   EINECS: 200-706-0

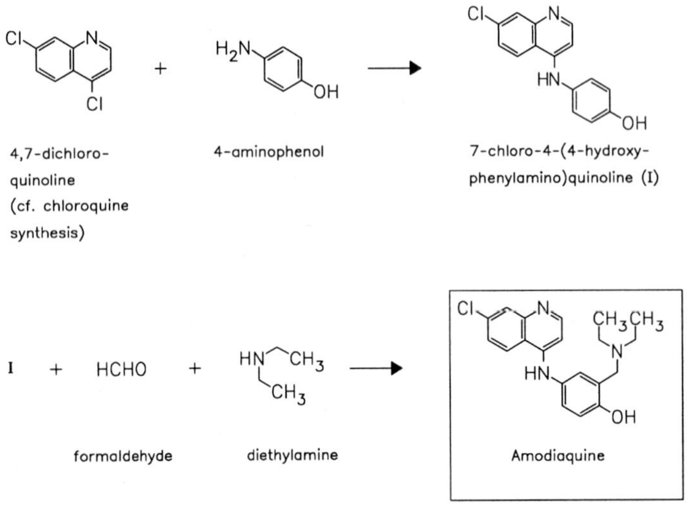

4,7-dichloro-   4-aminophenol   7-chloro-4-(4-hydroxy-
quinoline   phenylamino)quinoline (I)
(cf. chloroquine
synthesis)

formaldehyde   diethylamine   Amodiaquine

Reference(s):
US 2 474 821 (Parke Davis; 1949; prior. 1945).
Burckhalter, J.F. et al.: J. Am. Chem. Soc. (JACSAT) **68**, 1894 (1946).

Formulation(s):   tabl. 200 mg (as dihydrochloride dihydrate)

Trade Name(s):
F: Flavoquine (Roussel   GB: Camoquin (Parke Davis);   USA: Camoquin (Parke Davis);
      Diamant)   wfm   wfm

Amorolfine

ATC: D01AE16
Use: topical antimycotic

RN: 78613-35-1   MF: $C_{21}H_{35}NO$ MW: 317.52
CN: cis-(±)-4-[3-[4-(1,1-dimethylpropyl)phenyl]-2-methylpropyl]-2,6-dimethylmorpholine

hydrochloride
RN: 78613-38-4   MF: $C_{21}H_{35}NO \cdot HCl$ MW: 353.98

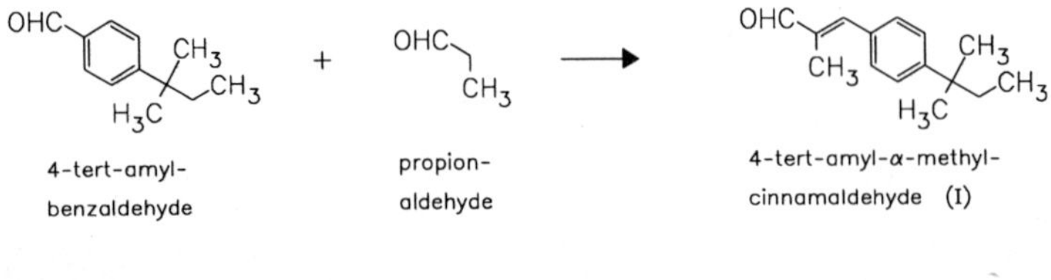

4-tert-amyl-   propion-   4-tert-amyl-α-methyl-
benzaldehyde   aldehyde   cinnamaldehyde (I)

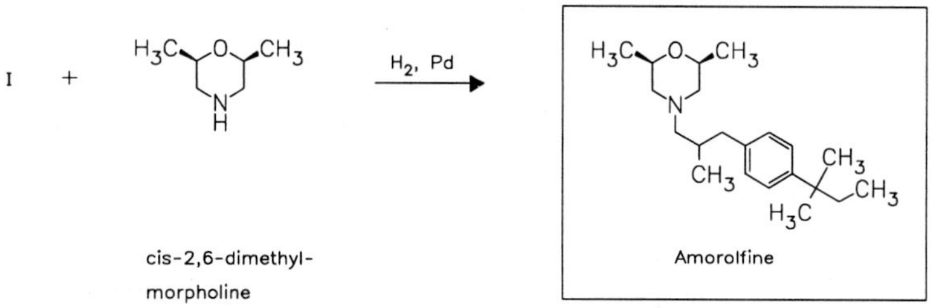

I   +

cis-2,6-dimethyl-
morpholine

H₂, Pd

Amorolfine

Reference(s):
DE 2 752 135 (Hoffmann-La Roche; appl. 22.11.1976).
EP 24 334 (Hoffmann-La Roche; appl. 7.8.1980; CH-prior. 17.8.1979, 29.5.1980).

antimycotic nail varnish:
EP 389 778 (Hoffmann-La Roche; appl. 15.2.1990; CH-prior. 9.11.1989, 24.2.1989).

Formulation(s):   cream 0.25 %, sol. 5 %

Trade Name(s):
D: Loceryl (Roche)   GB: Loceryl (Roche; 1992 as   J: Pekiron (Kyorin)
F: Loceryl (Roche)   hydrochloride)

Amosulalol
(YM-09538)

ATC: C02CB
Use: α- and β-adrenoceptor blocker,
             antihypertensive

RN: 85320-68-9   MF: $C_{18}H_{24}N_2O_5S$ MW: 380.47
CN: (±)-5-[1-hydroxy-2-[[2-(2-methoxyphenoxy)ethyl]amino]ethyl]-2-methylbenzenesulfonamide

monohydrochloride
RN: 70958-86-0   MF: $C_{18}H_{24}N_2O_5S \cdot HCl$ MW: 416.93

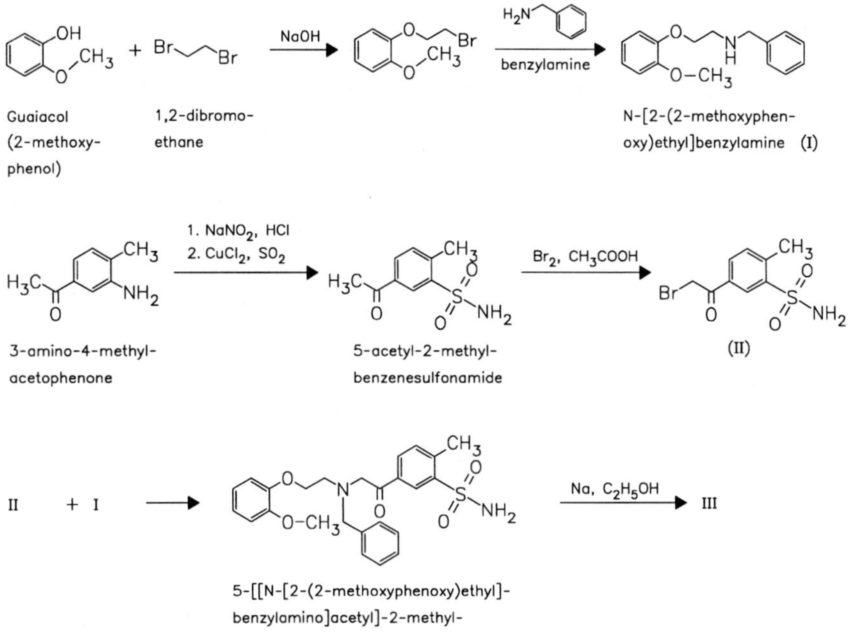

Guaiacol
(2-methoxy-
phenol)

1,2-dibromo-
ethane

N-[2-(2-methoxyphen-
oxy)ethyl]benzylamine (I)

3-amino-4-methyl-
acetophenone

5-acetyl-2-methyl-
benzenesulfonamide

(II)

II   +   I

5-[[N-[2-(2-methoxyphenoxy)ethyl]-
benzylamino]acetyl]-2-methyl-
benzenesulfonamide

Na, C₂H₅OH   III

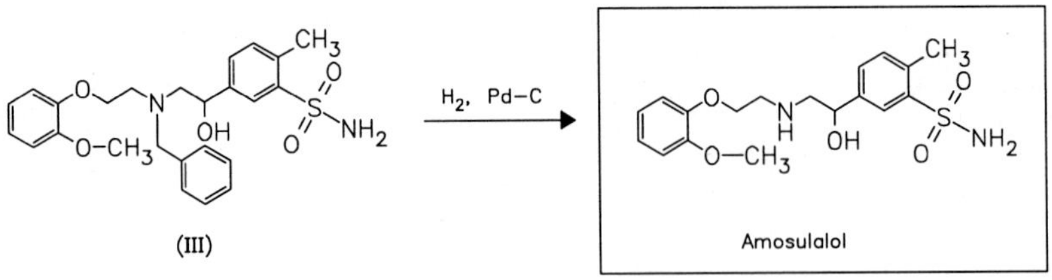

(III)   →   Amosulalol

H₂, Pd–C

Reference(s):
DOS 2 843 016 (Yamanouchi; appl. 3.10.1978; J-prior. 12.10.1977, 26.10.1977, 23.12.1977, 21.6.1978).
GB 2 006 772 (Yamanouchi; appl. 12.10.1978; J-prior. 12.10.1976, 26.10.1977, 23.12.1977, 21.6.1978).

synthesis of I:
Augstein, J. et al.: J. Med. Chem. (JMCMAR) **8**, 365 (1965).

synthesis of II:
EP 162 404 (Seitetsu Kagaku; appl. 14.5.1985; J-prior. 15.5.1984, 18.9.1984, 3.4.1985).

synthesis of ¹⁴C-amosulalol:
Arima, H.; Tamazawa, K.: J. Labelled Compd. Radiopharm. (JLCRD4) **20**, 803 (1983).

Formulation(s):   tabl. 10 mg

Trade Name(s):
J: Lowgan (Yamanouchi;
          1988 as hydrochloride)

Amoxapine

ATC: N06AA17
Use: antidepressant

RN: 14028-44-5   MF: $C_{17}H_{16}ClN_3O$ MW: 313.79   EINECS: 237-867-1
LD$_{50}$:   122 mg/kg (M, i.p.); 112 mg/kg (M, p.o.)
CN: 2-chloro-11-(1-piperazinyl)dibenzo[*b,f*][1,4]oxazepine

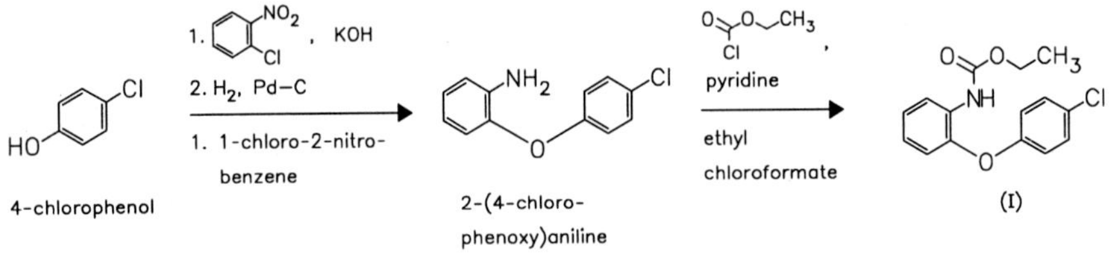

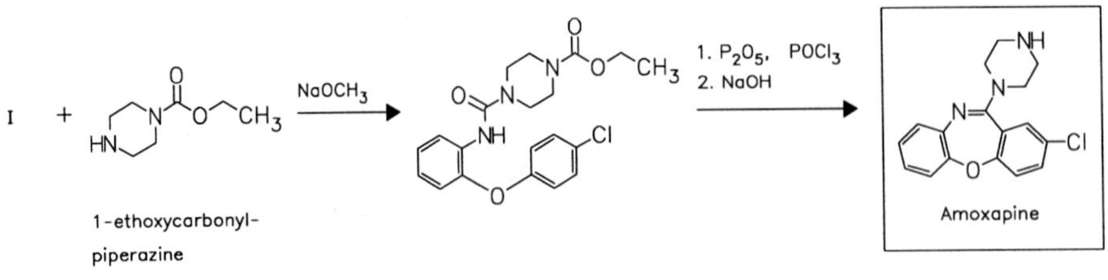

4-chlorophenol

1. 1-chloro-2-nitro-
   benzene

2-(4-chloro-
phenoxy)aniline

(I)

1-ethoxycarbonyl-
piperazine

Amoxapine

Reference(s):
US 3 681 357 (American Cyanamide; 16.5.1972; prior. 20.5.1966).
US 3 444 169 (American Cyanamide; 13.5.1969; prior. 17.1.1966).
GB 1 177 956 (American Cyanamide; prior. 23.12.1966).
GB 1 192 812 (American Cyanamide; USA-prior. 20.5.1966).
DE 1 645 954 (American Cyanamide; appl. 17.1.1967; USA-prior. 17.1.1966).
GB 1 157 957 (American Cyanamide; prior. 15.9.1965).
US 3 663 696 (American Cyanamide; 16.5.1972; prior. 28.2.1964, 20.5.1966, 22.7.1970).
Schmutz, J. et al.: Helv. Chim. Acta (HCACAV) **50**, 245 (1967).
Schmutz, J. et al.: Chim. Ther. (CHTPBA) **2**, 424 (1967).

preparation of 2-(4-chlorophenoxy)aniline:
DE 216 642 (Bayer; 1908).
Wassmundt, F.W.; Pedemonte, R.P.: J. Org. Chem. (JOCEAH) **60** (16), 4991 (1995).

Formulation(s):   sol. 5 %; sol. 5 %; tabl. 25 mg, 50 mg, 100 mg, 150 mg

Trade Name(s):
F: Défanyl (Wyeth-Lederle)   J: Amoxan (Lederle; 1981)   USA: Asendin (Lederle Labs.;
GB: Asendis (Wyeth)   1980)

Amoxicillin
(Amoxycillin)

ATC: J01CA04
Use: antibiotic

RN: 26787-78-0   MF: $C_{16}H_{19}N_3O_5S$ MW: 365.41   EINECS: 248-003-8
LD_{50}:   >25 g/kg (M, p.o.);
      >15 g/kg (R, p.o.)
CN: [2S-[2α,5α,6β(S*)]]-6-[[amino(4-hydroxyphenyl)acetyl]amino]-3,3-dimethyl-7-oxo-4-thia-1-
      azabicyclo[3.2.0]heptane-2-carboxylic acid

sodium D(-)-α-(4-hydroxy-
phenyl)-α-(2-methoxy-
carbonyl-1-methylethenyl-
amino)acetate
(DANE salt; cf.
ampicillin, method (c))

ethyl chloroformate

N-methylmorpholine → I

D-α-(4-hydroxyphenyl)-α-(2-
methoxycarbonyl-1-methyl-
ethenylamino)acetic acid
anhydride with monoethyl
carbonate (I)

6-amino-
penicillanic acid (II)

1. (CH₃)₃SiCl, N(C₂H₅)₃,
2. H⁺, pH 1.1-1.2

1. trimethyl-
chlorosilane

→ Amoxicillin

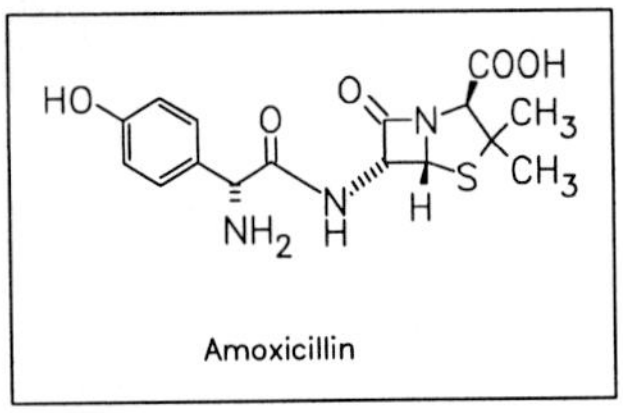

Amoxicillin

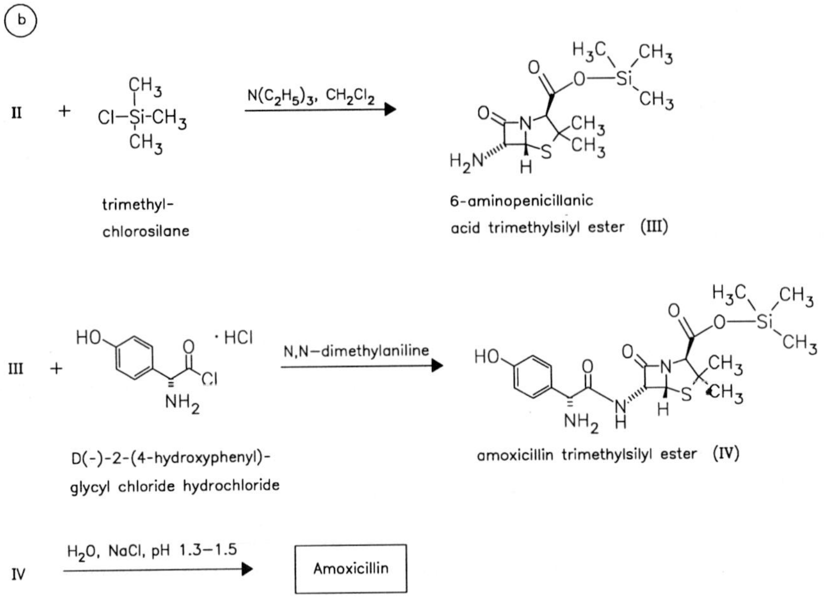

b

II + trimethyl-chlorosilane → 6-aminopenicillanic acid trimethylsilyl ester (III)

III + D(-)-2-(4-hydroxyphenyl)-glycyl chloride hydrochloride → amoxicillin trimethylsilyl ester (IV)

IV → Amoxicillin

Reference(s):
"*racemic amoxicillin*":
US 3 674 776 (Beecham; 4.7.1972; prior. 23.8.1968).
GB 1 241 844 (Beecham; appl. 18.8.1969; prior. 23.8.1968).
DE 1 942 693 (Beecham; appl. 18.8.1969; GB-prior. 23.8.1968).
GB 978 178 (Beecham; appl. 2.11.1962; valid from 25.10.1963).
US 3 192 198 (Beecham; 29.6.1965; GB-prior. 2.11.1962).

amoxicillin:
Long, A.A.W. et al.: J. Chem. Soc. C (JSOOAX) **1971**, 1920.
US 3 674 776 (Beecham; 4.7.1972; appl. 18.8.1969; GB-prior. 23.8.1968).
DOS 1 942 693 (Beecham; appl. 21.8.1969; GB-prior. 23.8.1968).
GB 1 241 844 (Beecham; appl. 23.8.1968; valid from 20.8.1969).
a   US 4 128 547 (Gist-Brocades; 5.12.1978; NL-prior. 6.9.1977).
     GB 1 339 605 (Beecham; appl. 1.4.1971; valid from 28.3.1972).
     preparation of "DANE salt":
     DE 2 400 489 (Upjohn Co.; appl. 5.1.1974; USA-prior. 12.1.1973).
     US 3 904 606 (Upjohn Co.; prior. 12.1.1973).
     Dane, E. et al.: Angew. Chem. (ANCEAD) **76**, 342 (1964).
     Dane, E. et al.: Chem. Ber. (CHBEAM) **98**, 789 (1965).
b   DAS 2 611 286 (Bristol-Myers; appl. 17.3.1976; USA-prior. 17.3.1975).
     preparation of D(-)-2-(4-hydroxyphenyl)glycyl chloride hydrochloride:
     CA 1 024 507 (Bristol Myers Co.; appl. 16.1.1974; USA-prior. 18.1.1973).

alternative syntheses:
US 4 053 360 (Bristol-Myers; 11.10.1977; GB-prior. 5.6.1974, 19.3.1975).
DOS 2 454 841 (Archifar; appl. 19.11.1974; I-prior. 17.5.1974).
DOS 2 755 903 (Dobfar; appl. 15.12.1977; I-prior. 16.12.1976).
GB 1 535 291 (Bristol-Myers; appl. 5.3.1976; USA-prior. 17.3.1975).
US 4 098 796 (Novo; 4.7.1978; appl. 7.6.1976).
BE 867 414 (Antibioticos S.A.; appl. 24.5.1978; E-prior. 4.6.1977).

microbiologic acylation of 6-APA with methyl D-α-(4-hydroxyphenyl)-glycinate hydrochloride *by means of Aphanocladium aranearum* (ATCC 20453):
US 4 073 687 (Shionogi; 14.2.1978; J-prior. 14.5.1975).

sodium salt:
GB 1 543 317 (Beecham; valid from 4.8.1976; prior. 27.9.1975).
DOS 2 729 112 (Beecham; appl. 28.6.1977; GB-prior. 7.7.1976).

trihydrate:
DAS 2 611 286 (Bristol-Myers; appl. 17.3.1976; USA-prior. 17.3.1975).
DOS 2 732 528 (Bristol-Myers; appl. 19.7.1977; GB-prior. 20.7.1976).

water soluble salts with arginine *or* lysine:
GB 1 504 767 (Beecham; valid from 23.8.1976; prior. 2.7.1975, 30.9.1975; 3.11.1975).
GB 1 539 510 (Beecham; valid from 23.8.1976; prior. 23.8.1975, 30.9.1975, 3.11.1975).

"amorphous" amoxicillin:
DAS 2 112 634 (Beecham; appl. 16.3.1971; GB-prior. 16.3.1970).

formulation for injection solutions:
GB 1 532 993 (Beecham; appl. 7.3.1975; valid from 9.2.1976).

O-acetylamoxicillin:
US 4 053 360 (Bristol-Myers; 11.10.1977; GB-prior. 5.6.1974, 19.3.1975).

Formulation(s):   syrup 500 mg/5 ml, 2.5 %, 5 %, 10 %; tabl. 500 mg, 750 mg, 1 g

Trade Name(s):

D: Amagesan (Pharbita)
    Amoxi-Diolan (Engelhard)
    Amoxillat (Azupharma)
    Amoxypen (Grünenthal)
    Augmentan (SmithKline
    Beecham; 1982)-comb.
    Clamoxyl (SmithKline
    Beecham; 1974)
    dura AX (durachemie)
    Flanamox (Wolff)
    Sigamopen (Kytta-
    Siegfried)
F: Agram (Inava)
    Amodex (Bouchara)
    Amophar (Dakota)
    Amoxine (Negma)
    Augmentin (SmithKline
    Beecham; 1984)-comb.
    Bactox (Innotech
    International)
    Bristamox (Bristol-Myers
    Squibb)
    Ciblor (Inava)-comb.

    Clamoxyl (SmithKline
    Beecham; 1974)
    Flemoxine (Yamanouchi
    Pharma)
    Gramidil (EG Labo)
    Hiconcil (Bristol-Myers
    Squibb)
    Zamocilline (Zambon)
GB: Amoran (Eastern)
    Amoxil (Bencard; 1972)
    Augmentin (SmithKline
    Beecham; 1984)-comb.
    Galenamox (Galen)
I: Alfamox (Alfa
    Wassermann)
    Am-73 (Medici)
    Amoflux (Lampugnani)
    Amox (Salus Research)
    Amoxina (Magis)
    Amoxipen (Metapharma)
    Cabermox (Caber)
    Ibiamox (IBI; as trihydrate)
    Isimoxin (ISI)
    Mopen (Firma)

    Pamocil (Farma Uno)
    Simoxil (Herdel)
    Sintopen (Mitim)
    Velamox (SmithKline
    Beecham)
    Zimox (Carlo Erba)
    generics and numerous
    combination preparations
J: Amolin (Takeda)
    Clamoxyl (SmithKline
    Beecham; 1975)
    Delacillin (Sankyo)
    Efpenix (Toyo Jozo)
    Hiconcil (Bristol)
    Himinomax (Kaken)
    Pacetocin (Kyowa)
    Sawacillin (Fujisawa)
    Widecillin (Meiji)
    generics
USA: Amoxil (SmithKline
    Beecham; 1974)
    Wymox (Wyeth-Ayerst;
    1978)

Amphetaminil
(Amfetaminil)

Use: psychotonic

RN: 17590-01-1   MF: $C_{17}H_{18}N_2$ MW: 250.35   EINECS: 241-560-8
LD$_{50}$:   182 mg/kg (M, p.o.);
37.6 mg/kg (R, p.o.)
CN: α-[(1-methyl-2-phenylethyl)amino]benzeneacetonitrile

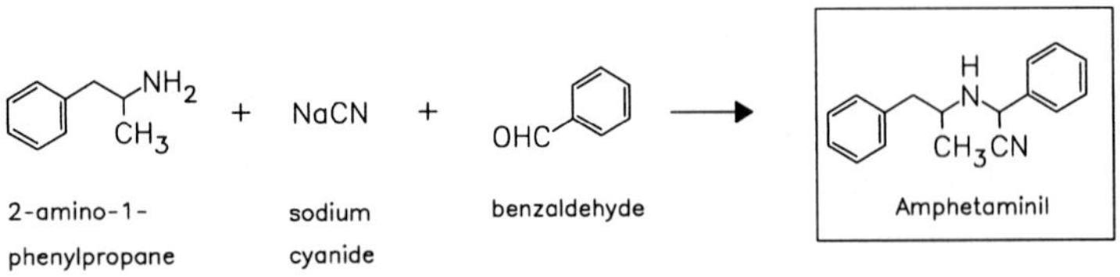

2-amino-1-
phenylpropane

sodium
cyanide

benzaldehyde

Amphetaminil

Reference(s):
AT 223 606 (Dr. H. Voigt; appl. 25.4.1961; valid from 15.3.1962).
Klosa, J.: J. Prakt. Chem. (JPCEAO) 20, 275 (1963).

Formulation(s):   amp. 20 mg, 60 mg

Trade Name(s):
D: AN 1 (Voigt); wfm   Ton-O$_2$ (Voigt)-comb.; wfm   Vit-O$_2$ (Voigt)-comb.; wfm

Amphotericin B

ATC: A01AB04; G01AA03; J02AA01
Use: fungicidal antibiotic

RN: 1397-89-3   MF: $C_{47}H_{73}NO_{17}$ MW: 924.09   EINECS: 215-742-2
LD$_{50}$:   1200 µg/kg (M, i.v.); >8 g/kg (M, p.o.);
1600 µg/kg (R, i.v.); >5 g/kg (R, p.o.);
6 mg/kg (dog, i.v.)
CN: [1R-
(1R*,3S*,5R*,6R*,9R*,11R*,15S*,16R*,17R*,18S*,19E,21E,23E,25E,27E,29E,31E,33R*,35S*,36R*,37
S*)]-33-[(3-amino-3,6-dideoxy-β-D-mannopyranosyl)oxy]-1,3,5,6,9,11,17,37-octahydroxy-15,16,18-
trimethyl-13-oxo-14,39-dioxabicyclo[33.3.1]nonatriaconta-19,21,23,25,27,29,31-heptaene-36-carboxylic
acid

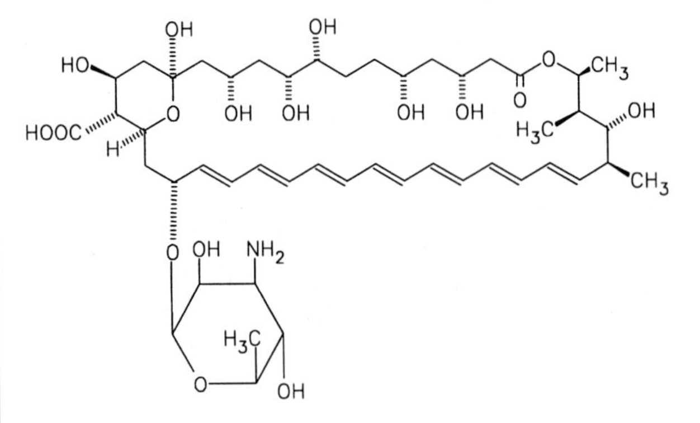

Amphotericin B

Fermentatively from *Streptomyces nodosus*.

Reference(s):
US 2 908 611 (Olin Mathieson; 1959; prior. 1954).

Formulation(s):   caramels 10 mg; cream 30 mg/g; ointment 30 mg/1 g; powder 50 mg; susp. 100 mg, 500 mg; tabl. 10 mg, 100 mg; liposome-encapsulated amphotericin B in a complex with dimyristoyl phosphatidylcholine and dimyristoyl phosphatidylglycerol, vials 20 ml

Trade Name(s):

D:	AmBisome (NeXstar; 1999)		Fungizone (Squibb)
	Ampho-Moronal (Bristol-Myers Squibb)	GB:	Abelcet (Liposome Co.)
			Ambisone (NeXstar)
			Amphicol (Zeneca)
	Amphotericin B zur Infusion (Bristol-Myers Squibb)		Fungilin r (Squibb)
			Fungizone (Squibb)
	Mysteclin (Bristol-Myers Squibb)-comb.	I:	Fungilin (Mead Johnson)
			Fungizone (Bristol-Myers Squibb)
F:	Amphocycline (Bristol-Myers Squibb)-comb.	J:	Fungizone (Bristol-Myers Squibb-Sankyo)

USA: Abelect Injection (Liposome Co.)
Amphotec for Injection (Sequus)
Fungizone (Bristol-Myers Squibb, Oncology/Immunology)

Ampicillin

ATC: J01CA01; S01AA19
Use: antibiotic

RN: 69-53-4   MF: $C_{16}H_{19}N_3O_4S$ MW: 349.41   EINECS: 200-709-7
LD_{50}: 4600 mg/kg (M, i.v.); >5 g/kg (M, p.o.);
         6200 mg/kg (R, i.v.)
CN: [2S-[2α,5α,6β(S^*)]]-6-[(aminophenylacetyl)amino]-3,3-dimethyl-7-oxo-4-thia-1-azabicyclo[3.2.0]heptane-2-carboxylic acid

trihydrate
RN: 7177-48-2   MF: $C_{16}H_{19}N_3O_4S \cdot 3H_2O$ MW: 403.46
LD_{50}: 15.2 g/kg (M, p.o.);
         10 g/kg (R, p.o.)
monosodium salt
RN: 69-52-3   MF: $C_{16}H_{18}N_3NaO_4S$ MW: 371.39   EINECS: 200-708-1
LD_{50}: >5314 mg/kg (M,R, p.o.)
monopotassium salt
RN: 23277-71-6   MF: $C_{16}H_{18}KN_3O_4S$ MW: 387.50   EINECS: 245-550-4

D(-)-Cbo-phenylglycine   ethyl chloro-formate (I)   D-Cbo-phenylglycine anhydride with monoethyl carbonate (II)

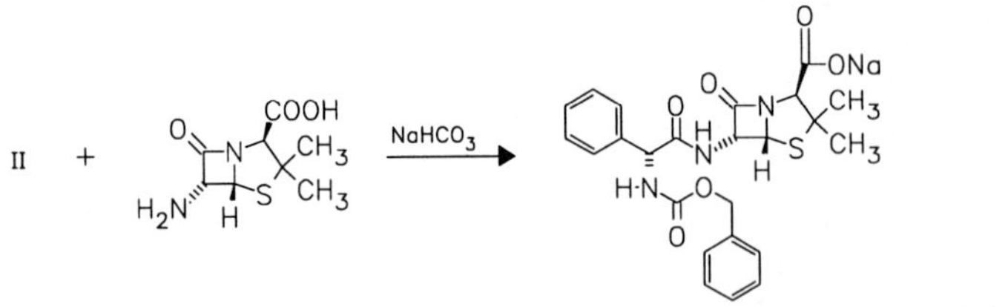

II   +

6-amino-
penicillanic acid   (III)

Cbo-ampicillin sodium salt   (IV)

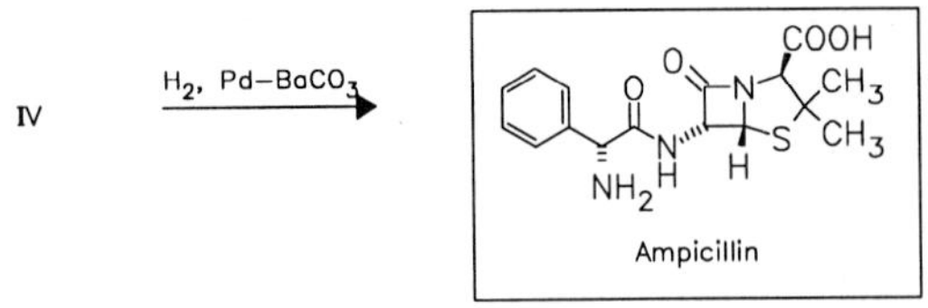

IV

Ampicillin

b

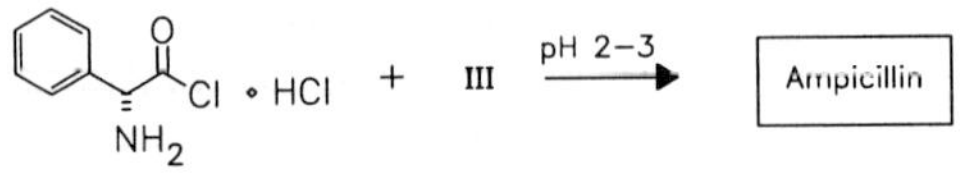

D(-)-α-phenylglycine
chloride hydrochloride

pH 2-3

Ampicillin

c

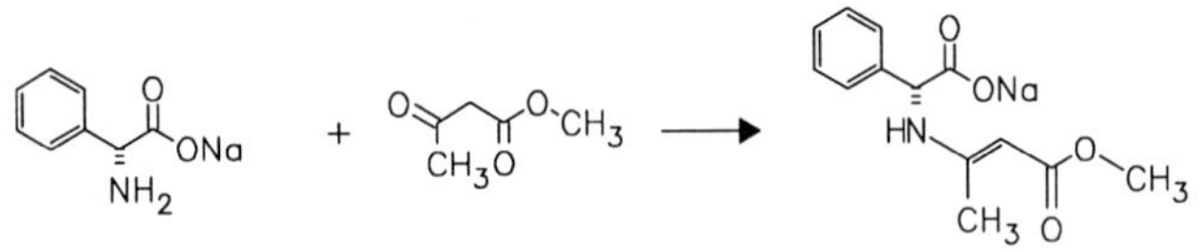

D(-)-phenylglycine
sodium salt

methyl
acetoacetate

N-(2-methoxycarbonyl-1-methyl-
ethenyl)-D(-)-phenylglycine sodium salt   (V)

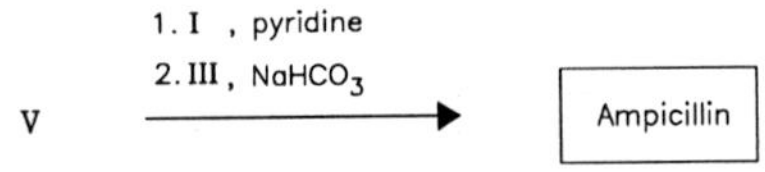

1. I , pyridine
2. III , NaHCO₃

V

Ampicillin

Reference(s):
a   GB 893 049 (Beecham; appl. 6.10.1958, 12.5.1959).
    GB 902 703 (Beecham; valid from 19.5.1961; prior. 25.8.1960).
    US 2 985 648 (Beecham; 23.5.1961; GB-prior. 6.10.1958).
    DAS 1 139 844 (Beecham; appl. 6.10.1959; GB-prior. 6.10.1958, 12.5.1960).
    DE 1 156 078 (Beecham; appl. 29.5.1961; GB-prior. 25.8.1960).
b   US 3 140 282 (Bristol-Myers; 7.7.1964; appl. 5.3.1962).
c   GB 991 586 (Beecham; appl. 28.2.1963, 3.12.1963; valid from 13.2.1964).

alternative syntheses:
DE 1 168 910 (Beecham; appl. 3.7.1962; GB-prior. 21.7.1961).
US 3 144 445 (American Home; 11.8.1964; appl. 26.12.1962).
DAS 1 445 506 (Bristol-Myers; appl. 24.10.1963; USA-prior. 29.10.1962).
DAS 1 545 534 (Astra; appl. 4.3.1965; S-prior. 6.3.1964).
DAS 2 029 195 (Yamanouchi; appl. 13.6.1970; J-prior. 16.6.1969).
DAS 1 800 698 (American Home Products; appl. 2.10.1968; USA-prior. 2.10.1967).
DOS 2 755 903 (Dobfar; appl. 15.12.1977; I-prior. 16.12.1976).

enzymatic and microbiological methods:
US 3 079 307 (Bayer; 26.2.1963; D-prior. 7.10.1961).
DE 1 966 521 (Kyowa Hakko; appl. 9.9.1969; J-prior. 18.9.1968, 8.10.1968).
DAS 1 967 074 (Kyowa Hakko; appl. 9.9.1969; J-prior. 18.9.1968, 8.10.1968).
DAS 2 050 983 (Kyowa Hakko; appl. 16.10.1970; J-prior. 16.10.1969).
US 4 073 687 (Shionogi; 14.2.1978; J-prior. 14.5.1975).

ampicillin salts:
DE 1 197 460 (Bayer; appl. 4.9.1962).
DOS 1 670 111 (Bristol-Myers; appl. 16.7.1966).
DE 1 670 191 (Beecham; appl. 24.2.1967; GB-prior. 3.3.1966).
DAS 1 795 129 (Beecham; appl. 14.8.1968; USA-prior. 18.8.1967).
DE 1 903 388 (American Home Products; appl. 23.1.1969; USA-prior. 23.1.1968).
DAS 2 623 835 (Boehringer Ing.; appl. 28.5.1976).

trihydrate:
US 3 157 640 (Bristol-Myers; 17.11.1964; appl. 21.3.1963).

Formulation(s):   amp. 0.5 g, 1 g, 2 g, 5 g; lyo. 532 mg, 1060 mg, 2128 mg, 5320 mg

Trade Name(s):

D: Binotal (Grünenthal)
     Jenampin (Jenapharm)
     Unacid (Pfizer)
F: Ampicilline (Arkodex;
     Panpharma)
     Proampi (Stafford-Miller)
     Totapen (Bristol-Myers
     Squibb)
     Unacim (Jouveinal)-comb.
GB: Ampiclox (Beecham)
     Magnapen (Beecham)-
     comb.
     Penbritin (Beecham)
I: Ampici (Formulario Naz.)
     Ampicillina (Pierrel)
     Ampilisa (Lisapharma)
     Ampilux (Allergan)
     Ampiplus Simplex
     (Menarini)

Amplital (Farmitalia)
Amplizer (OFF)
Citicil (CT)
Ibimicyn (IBI)
Lampocillina Orale (Salus
Research)
Pentrexyl (Bristol It. Sud)
Platocillina (Crosara)
generics and numerous
combination preparations
J: Acucillin (Fuji)-comb.
     Adobacillin (Tobishi)
     Amipenix (Toyo Jozo)
     Bionacillin (Takata)
     Bonapicillin (Taiho)
     Cilleral (Bristre-Banyu)
     Combipenix (Toyo Jozo)-
     comb.
     Domicillin (Marupi)

Isocillin (Kanto)
Ohtecin (Kyowa)
Penbritin (Beecham-
Fujisawa)
Penimic (SS Seiyaku)
Pentrex (Banyu)
Pharcillin (Toyo Pharmar)
Solcillin (Takeda)-comb.
Synpenin (Sankyo)
Tokiocillin (Isei)
Totacillin (Beecham)
Totaclox (Beecham)-comb.
Viccillin (Meiji)
USA: Amcill (Parke Davis)
     Omnipen (Wyeth-Ayerst)
     Unosyn for Injection
     (Pfizer)
     generics

Ampiroxicam
(CP-65703)

ATC: M01
Use: anti-inflammatory, prodrug of
       piroxicam

RN: 99464-64-9   MF: $C_{20}H_{21}N_3O_7S$ MW: 447.47
LD_{50}:   747 mg/kg (R, p.o.)
CN: carbonic acid ethyl 1-[[2-methyl-3-[(2-pyridinyl)aminocarbonyl]-2H-1,2-benzothiazin-4-yl]oxy]ethyl
      ester S,S-dioxide

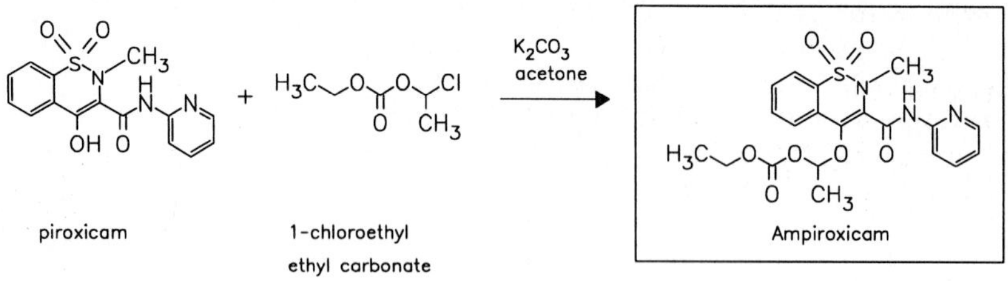

piroxicam

1-chloroethyl
ethyl carbonate

Ampiroxicam

Reference(s):
EP 147 177 (Pfizer Inc.; appl. 19.12.1984; USA-prior. 21.12.1983).

topical preparations:
JP 07 316 075 (Pola Kasei Kogyo; appl. 26.5.1994; J-prior. 26.5.1994).

Formulation(s):   cps. 13.5 mg, 27 mg

Trade Name(s):
J: Flucam (Pfizer/Toyama;   Nacyl (Toyama)
      1994)

Amprenavir
(KUX 478; UX 478; 141W94)

ATC: J05AE05
Use: antiviral, HIV protease inhibitor

RN: 161814-49-9   MF: $C_{25}H_{35}N_3O_6S$ MW: 505.64
CN: [(1S,2R)-3-[[(4-Aminophenyl)sulfonyl](2-methylpropyl)amino]-2-hydroxy-1-
       (phenylmethyl)propyl]carbamic acid (3S)-tetrahydro-3-furanyl ester

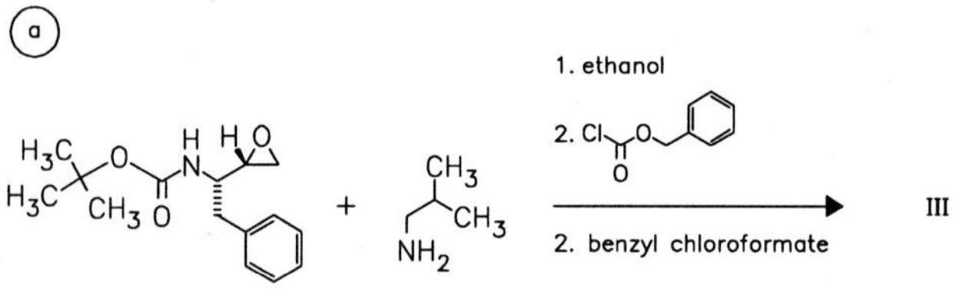

3(S)-tert-butoxy-
carbonylamino-1,2(S)-
epoxy-4-phenyl-
butane (I)
(cf. saquinavir
synthesis)

isobutyl-
amine (II)

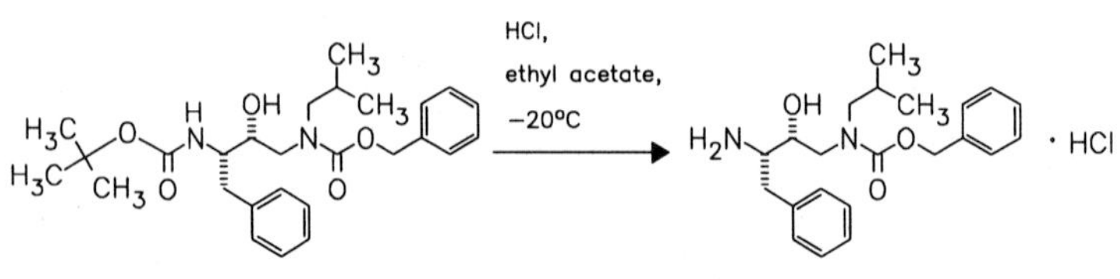

III

(III)

(IV)

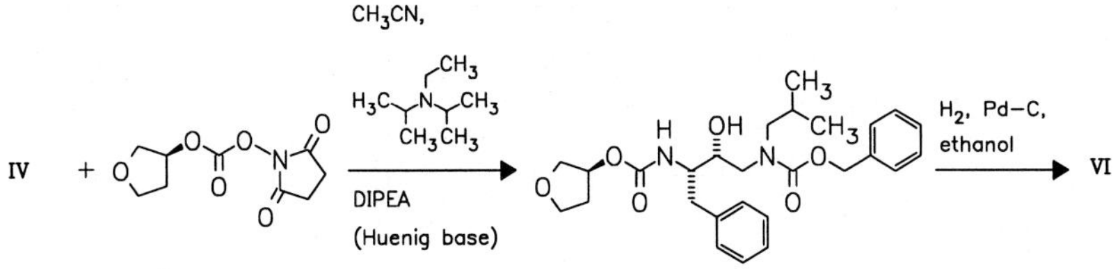

IV + succinimido (S)-3-tetrahydro-furyl carbonate (V)

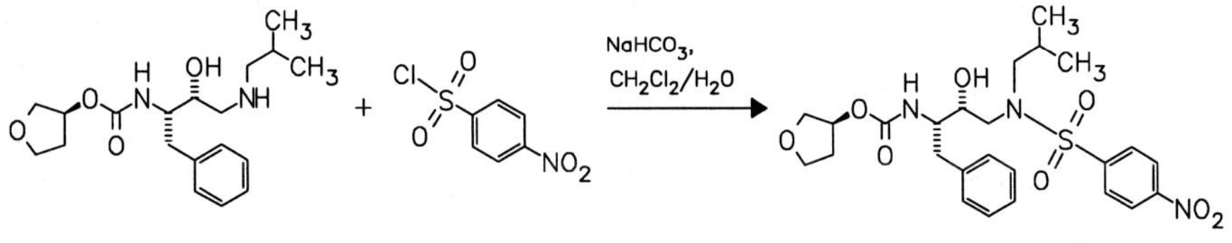

CH₃CN, DIPEA (Huenig base) → VI

H₂, Pd–C, ethanol → VI

(VI) + p-nitrobenzene-sulfonyl chloride (VII)

NaHCO₃, CH₂Cl₂/H₂O → (VIII)

VIII

H₂, Pd–C, ethyl acetate →

Amprenavir

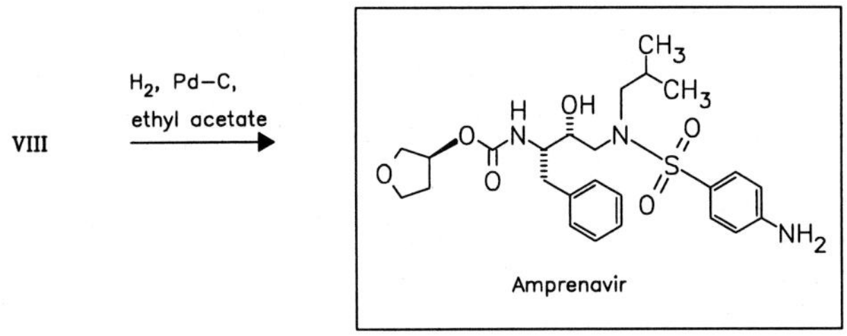

(aa)   synthesis of V

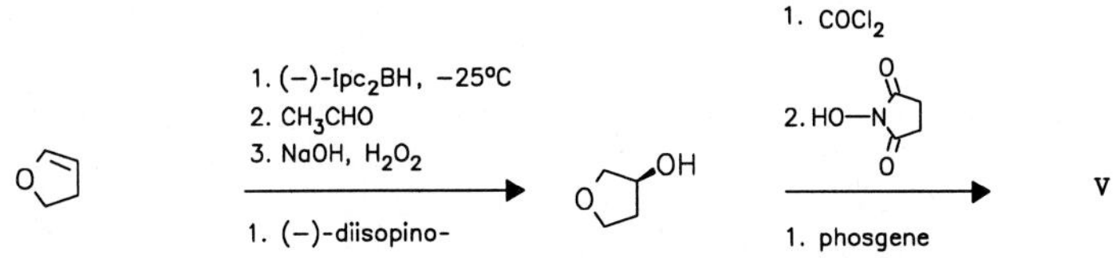

2,3-dihydro-furan

1. (−)-Ipc₂BH, −25°C
2. CH₃CHO
3. NaOH, H₂O₂

1. (−)-diisopino-camphenylborane

(S)-(+)-3-hydroxy-tetrahydro-furan

1. COCl₂
2. HO—N

1. phosgene
2. N-hydroxy-succinimide

V

(−)-Ipc₂BH:

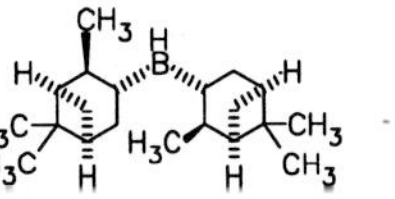

(b)

I   +   II

ethanol →

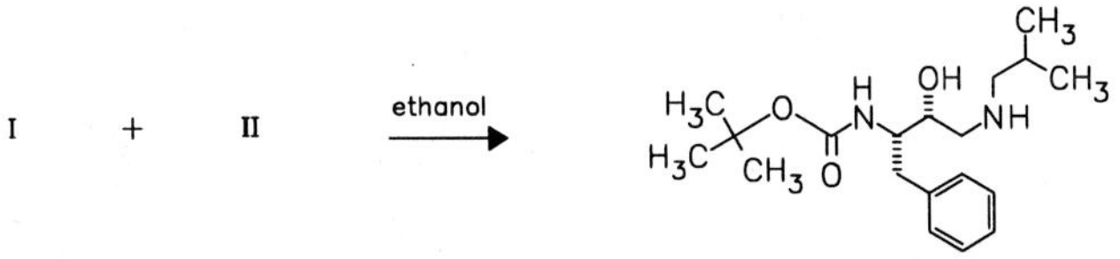

(IX)

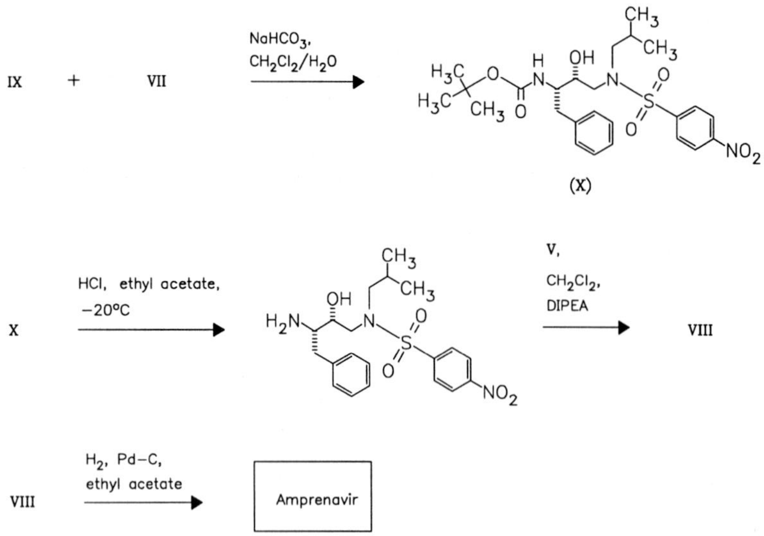

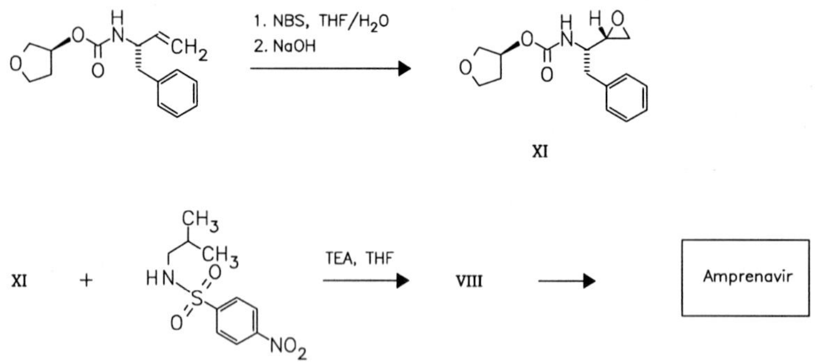

Reference(s):

a   WO 9 405 639 (Vertex Pharm.; appl. 7.9.1993; USA-prior. 8.9.1992).
aa Brown, H.C. et al.: J. Am. Chem. Soc. (JACSAT) **108**, 2049-2054 (1986).
b   WO 9 633 184 (Vertex Pharm.; appl. 18.4.1996; USA-prior. 19.4.1995; 8.9.1992).
c   JP 09 124 630 (Kissei Pharm.; appl. 26.10.1995).

nanocrystalline formulations:
WO 9 902 665 (Nanosystems; appl. 9.7.1998; USA-prior. 9.7.1997).

stable crystal polymorphs:
WO 9 857 648 (Vertex Pharm.; appl. 16.6.1998; USA-prior. 16.6.1997).

novel crystal form V:
WO 9 856 781 (Glaxo; 17.12.1998; appl. 11.6.1998; USA-prior. 13.6.1997).

combination with AZT:
WO 9 720 554 (Vertex Pharm.; 12.6.1997; appl. 5.12.1996; USA-prior. 5.12.1995).

crystallization of amprenavir:
JP 09 071 575 (Kissei Pharm.; appl. 7.9.1995).

Formulation(s):   cps. 50 mg, 150mg, oral sol. 15 mg/ml

Trade Name(s):
USA: Agenerase (Glaxo
       Wellcome; 1999)

Amrinone

ATC: C01CE01
Use: cardiotonic (positive inotropic effect)

RN: 60719-84-8   MF: $C_{10}H_9N_3O$ MW: 187.20   EINECS: 262-390-0
LD$_{50}$:   150 mg/kg (M, i.v.); 288 mg/kg (M, p.o.);
       75 mg/kg (R, i.v.); 102 mg/kg (R, p.o.)
CN: 5-amino[3,4'-bipyridin]-6(1H)-one

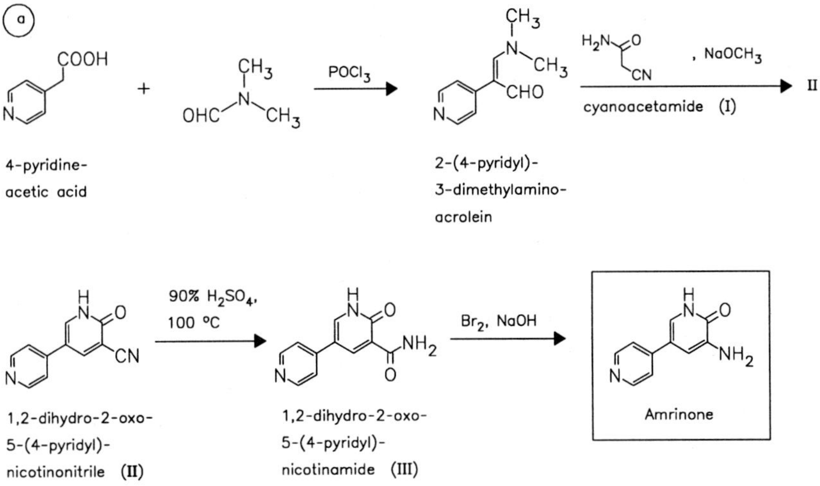

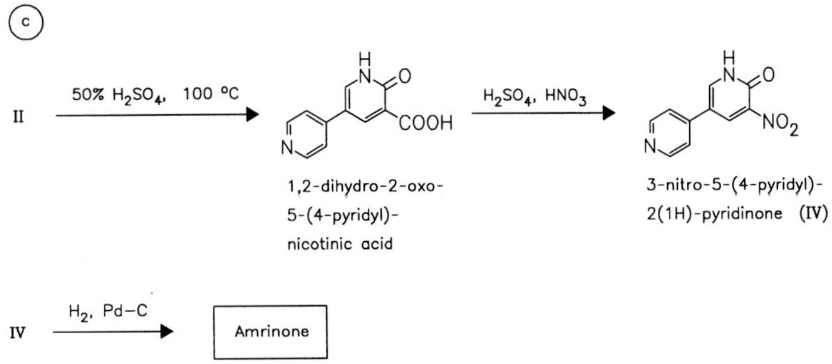

4-pyridine-
acetic acid

2-(4-pyridyl)-
3-dimethylamino-
acrolein

cyanoacetamide (I)

1,2-dihydro-2-oxo-
5-(4-pyridyl)-
nicotinonitrile (II)

1,2-dihydro-2-oxo-
5-(4-pyridyl)-
nicotinamide (III)

Amrinone

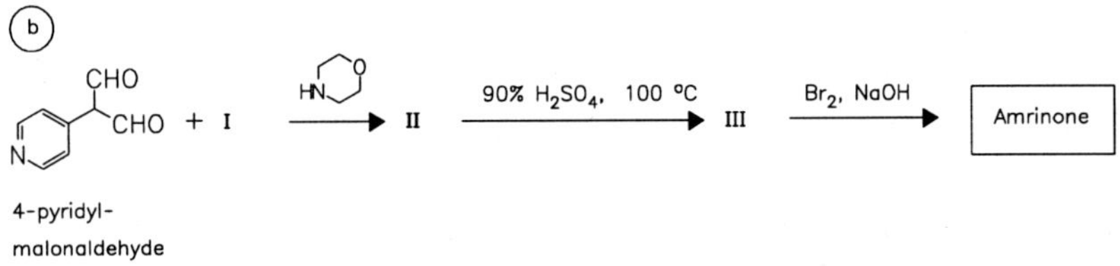

4-pyridyl-
malonaldehyde

Amrinone

1,2-dihydro-2-oxo-
5-(4-pyridyl)-
nicotinic acid

3-nitro-5-(4-pyridyl)-
2(1H)-pyridinone (IV)

IV $\xrightarrow{\text{H}_2, \text{ Pd-C}}$ Amrinone

Reference(s):
US 4 072 746 (Sterling Drug; 7.2.1978; appl. 21.7.1976; prior. 14.10.1975).
GB 1 512 129 (Sterling Drug; appl. 28.9.1976; USA-prior. 14.10.1975).
DE 2 646 469 (Sterling Drug; appl. 21.7.1976; USA-prior. 14.10.1975).
US 4 004 012 (Sterling Drug; 18.1.1977; appl. 14.10.1975).

improved method analogous to **a**:
GB 2 070 008 (Sterling Drug; appl. 20.2.1981; USA-prior. 26.2.1980).

preparation of 4-pyridineacetic acid *from* 4-acetylpyridine:
Katritzky: J. Chem. Soc. (JCSOA9) **1955**, 2586, 2592.

preparation of 4-pyridyl-malondialdehyde *from* 4-methylpyridine *and* DMF:
Niedrich,H.; Heyne,H.-U.; Schroetter,E.; Jaensch,H.-J.; Heidrich,H.-J. et al.: Pharmazie (PHARAT) **41**(3), 173 (1986).

Formulation(s):   amp. 5 mg/ml, 100 mg

Trade Name(s):

D:	Wincoram (Sanofi Winthrop; 1984)	I:	Inocor (Maggioni-Winthrop)		Vesistol (Inverni della Beffa)

Amsacrine

(m-AMSA)

ATC: L01XX01
Use: antineoplastic

RN: 51264-14-3   MF: $C_{21}H_{19}N_3O_3S$ MW: 393.47   EINECS: 257-094-3
LD$_{50}$:   33.7 mg/kg (M, i.v.); 53.42 mg/kg (M, p.o.);
        6.25 mg/kg (dog, i.v.); 50 mg/kg (dog, p.o.)
CN: N-[4-(9-acridinylamino)-3-methoxyphenyl]methanesulfonamide

monohydrochloride
RN: 54301-15-4   MF: $C_{21}H_{19}N_3O_3S \cdot HCl$ MW: 429.93
LD$_{50}$:   60 mg/kg (M, i.p.)

2-chloro-
benzoic acid   aniline   2-anilino-
                               benzoic acid   9-chloro-
                                                        acridine (I)

3-methoxy-4-
nitroaniline   methane-
                      sulfonyl
                      chloride (II)   (III)

2-methoxy-4-
nitroaniline

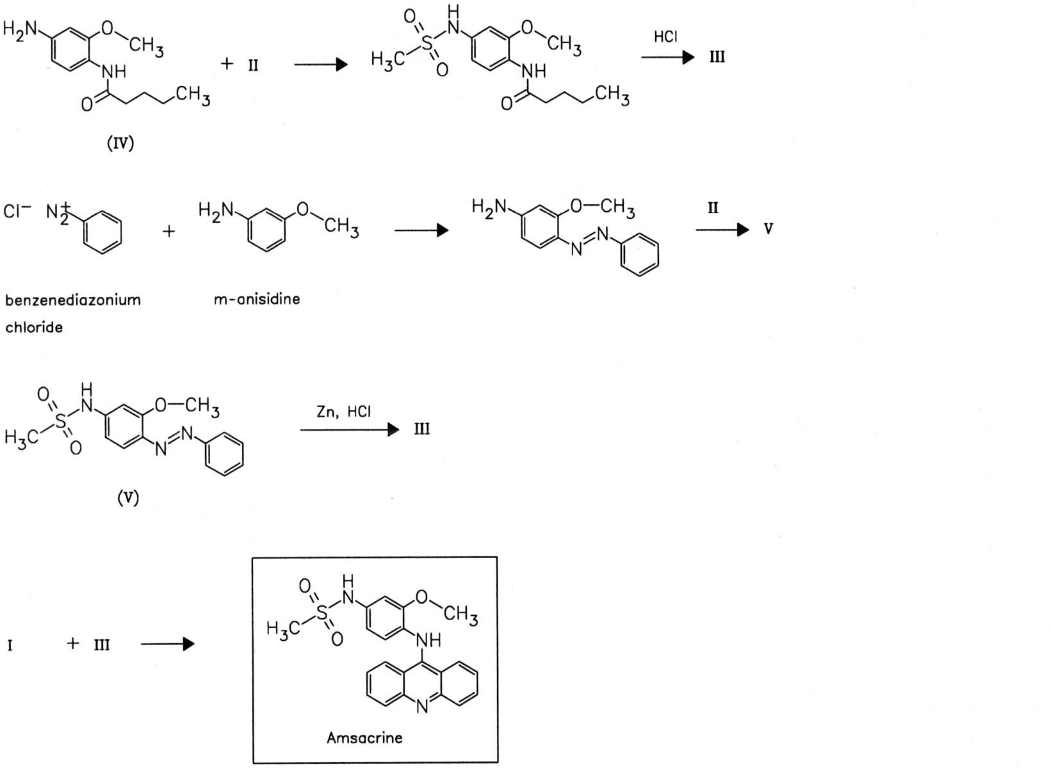

(IV)

benzenediazonium chloride   m-anisidine

(V)

I + III →

Amsacrine

Reference(s):

Cain, B.F. et al.: J. Med. Chem. (JMCMAR) **18**, 1110 (1975); **20**, 987 (1977).
Denny, W.A. et al.: J. Med. Chem. (JMCMAR) **21**, 5 (1978).
Rewcastle, G.W. et al.: J. Med. Chem. (JMCMAR) **25**, 1231 (1982).

preparation of III from benzenediazonium chloride:
Wilson, W.R. et al.: J. Med. Chem. (JMCMAR) **32**, 23 (1989).

Formulation(s):   amp. 85 mg/1.7 ml, 75 mg/1.5 ml

Trade Name(s):
D: Amsidyl (Gödecke)   F: Inocor (Sanofi Winthrop)   GB: Amsidine (Goldshield)

Anagestone acetate

ATC: G03DA
Use: progestogen

RN: 3137-73-3   MF: $C_{24}H_{36}O_3$ MW: 372.55   EINECS: 221-535-8
CN: (6α)-17-(acetyloxy)-6-methylpregn-4-en-20-one

anagestone
RN: 2740-52-5   MF: $C_{22}H_{34}O_2$ MW: 330.51

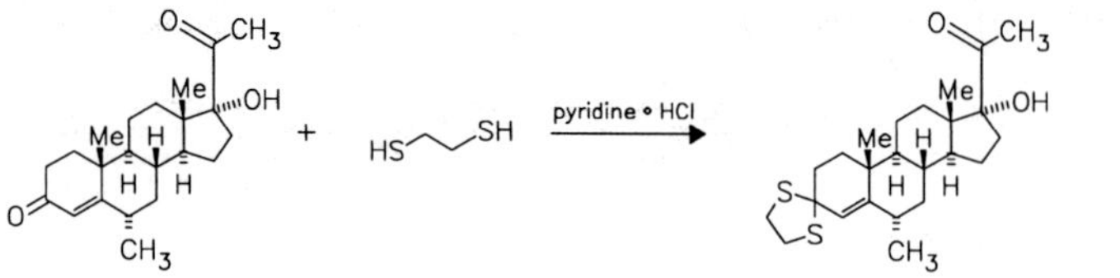

medroxyprogesterone (q.v.) — 1,2-ethane-dithiol — 3,3-ethylenedithio-17-hydroxy-6α-methyl-4-pregnen-20-one (I)

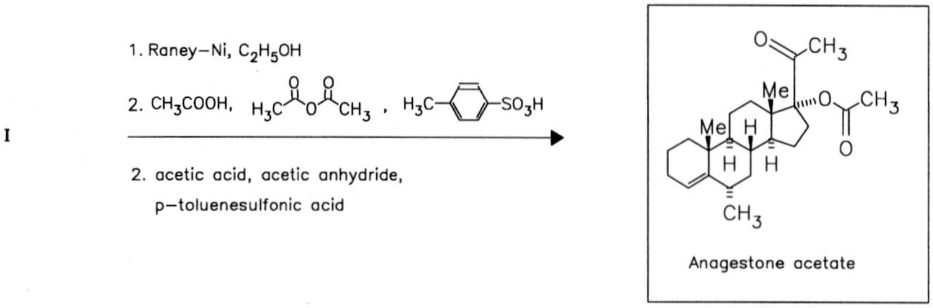

I
1. Raney–Ni, C_2H_5OH
2. CH_3COOH, $H_3C-CO-O-CO-CH_3$, $H_3C-C_6H_4-SO_3H$
2. acetic acid, acetic anhydride, p–toluenesulfonic acid
Anagestone acetate

Reference(s):
BE 624 370 (Ortho; appl. 31.10.1962; USA-prior. 6.6.1962, 24.7.1961; F-prior. 23.7.1962).

Trade Name(s):
USA: Anatropin (Ortho); wfm

Anagrelide hydrochloride
(BL-4162A; BMY-26538-01)

ATC: B01AC14
Use: antithrombotic, phosphodiesterase III (PDEIII)-inhibitor that reduces platelet counts

RN: 58579-51-4 MF: $C_{10}H_7Cl_2N_3O \cdot HCl$ MW: 292.55
CN: 6,7-Dichloro-1,5-dihydroimidazo[2,1-*b*]quinazolin-2(3*H*)-one hydrochloride

base
RN: 68475-42-3 MF: $C_{10}H_7Cl_2N_3O$ MW: 256.09

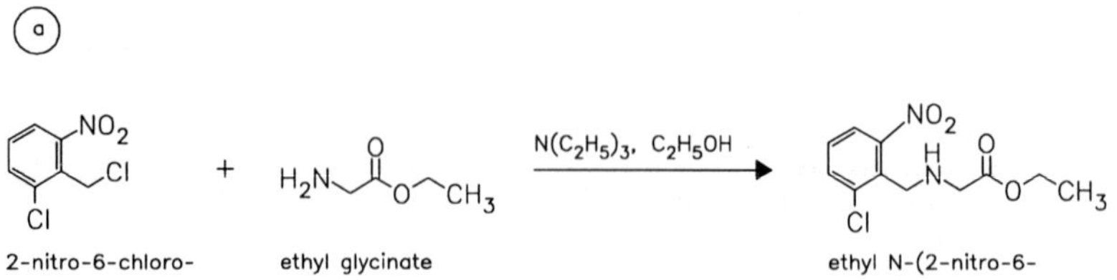

a
$N(C_2H_5)_3$, C_2H_5OH

2-nitro-6-chloro-benzyl chloride — ethyl glycinate — ethyl N-(2-nitro-6-chlorobenzyl)glycinate (I)

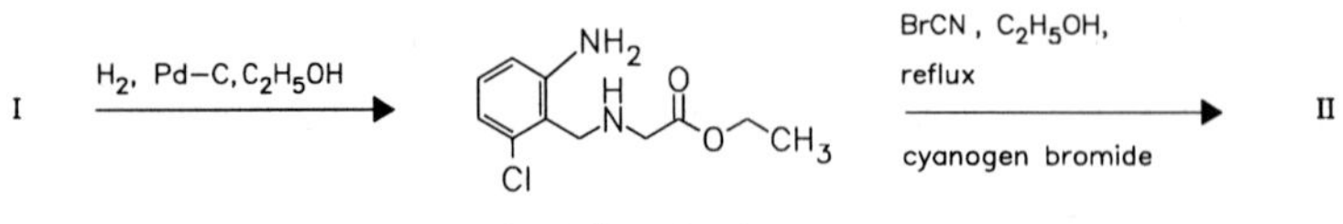

I
H_2, Pd–C, C_2H_5OH
BrCN, C_2H_5OH, reflux
cyanogen bromide
II

ethyl N-(2-amino-6-chlorobenzyl)glycinate

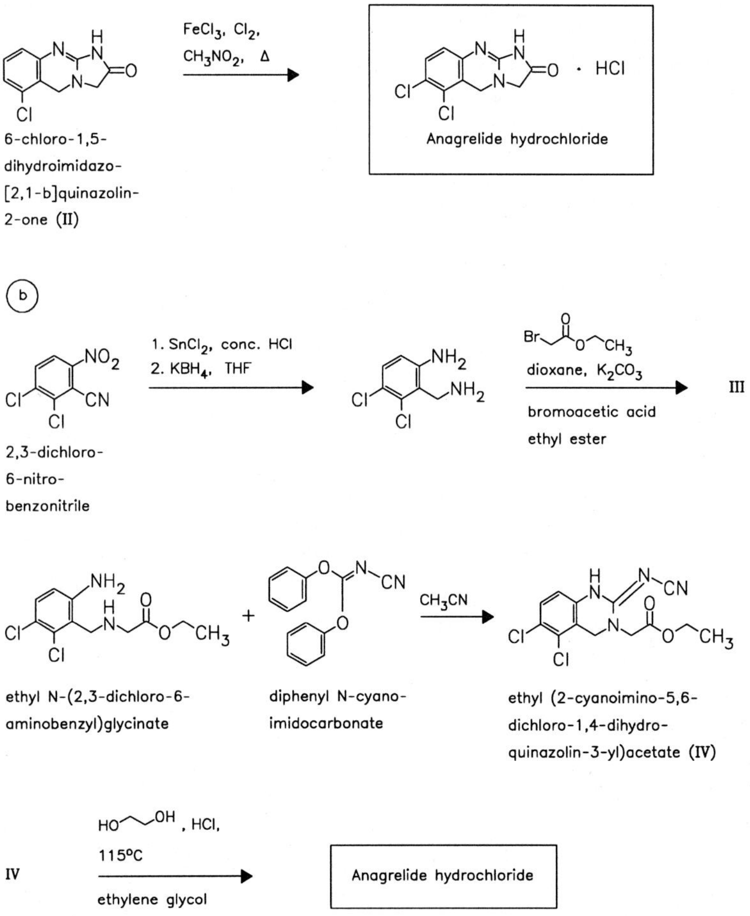
FeCl3, Cl2, CH3NO2, Δ
· HCl
6-chloro-1,5-dihydroimidazo-[2,1-b]quinazolin-2-one (II)
Anagrelide hydrochloride
b
1. SnCl2, conc. HCl
2. KBH4, THF
dioxane, K2CO3
bromoacetic acid ethyl ester
III
2,3-dichloro-6-nitro-benzonitrile
CH3CN
ethyl N-(2,3-dichloro-6-aminobenzyl)glycinate
diphenyl N-cyano-imidocarbonate
ethyl (2-cyanoimino-5,6-dichloro-1,4-dihydro-quinazolin-3-yl)acetate (IV)
IV
, HCl, 115°C
ethylene glycol
Anagrelide hydrochloride

alternative synthesis of anagrelide precursor III

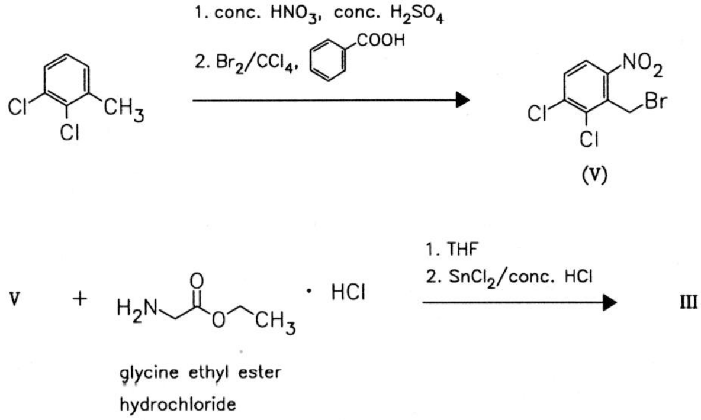
1. conc. HNO3, conc. H2SO4
2. Br2/CCl4,
(V)
V +
· HCl
1. THF
2. SnCl2/conc. HCl
III
glycine ethyl ester hydrochloride

Reference(s):

a US 3 932 407 (Bristol Myers Co.; 13.1.1976; USA-prior. 4.2.1972).
alternative cyclization:
US 4 208 521 (Bristol Myers Co.; 17.6.1980; USA-prior. 31.7.1978).
alternative syntheses from 5-chloroisatine *or* 1,2,3-trichlorobenzene:
US 4 146 718 (Bristol Myers Co.; 27.3.1979; USA-prior. 10.4.1978).

b EP 514 917 (EGIS Gyogyszergyar; appl. 22.5.1992; HU-prior. 22.5.1991).
Trinka, P.; Reiter, J.: J. Prakt. Chem./Chem.-Ztg. (JPCCEM) **338** (8), 750 (1997).

synthesis of ethyl *N*-(2,3-dichloro-6-aminobenzyl)glycinate:
HU 60 998 (Egis Gyogyszergyar; HU-prior. 22.5.1991)
Trinka, P.; Slegel, P.; Reiter, J.: J. Prakt. Chem./Chem.-Ztg. (JPCCEM) **338** (7), 675 (1996).
synthesis of 2,3-dichloro-6-nitrobenzonitrile *from* 1,2,3-trichlorobenzene:
Trinka, P.; Berecz, G.; Reiter, J.: J. Prakt. Chem./Chem.-Ztg. (JPCCEM) **338** (7), 679 (1996).

synthesis of anagrelide *precursor*
EP 778 258 (Roberts Lab.; appl. 8.3.1996; USA-prior. 4.12.1995).

pharmaceutical compositions:
US 4 357 330 (Bristol-Myers Co.; 2.11.1982; USA-prior. 30.7.1981).

Formulation(s):   cps. 0.5 mg, 1 mg (as hydrochloride hydrate)

Trade Name(s):
USA: Agrylin (Roberts
          Pharmaceutical; 1998)

Anastrozole
(ICI-D1033; ZD-1033)

ATC: L02BG03
Use: antineoplastic, non-steroidal
          aromatase inhibitor

RN: 120511-73-1   MF: $C_{17}H_{19}N_5$ MW: 293.37
CN: α,α,α',α'-tetramethyl-5-(1*H*-1,2,4-triazol-1-ylmethyl)-1,3-benzenediacetonitrile

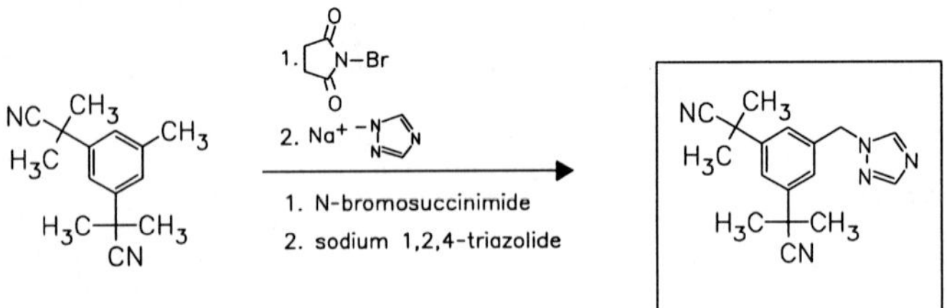

2,2'-(5-methyl-
m-phenylene)bis-
(2-methyl-
propionitrile)

Reference(s):
EP 296 749 (ICI; appl. 14.6.1988; GB-prior. 16.6.1987).

combination with 5α-reductase inhibitors:
WO 9 218 132 (Merck & Co.; appl. 4.6.1992; USA-prior. 17.4.1991).

Formulation(s):   tabl. 1 mg

Trade Name(s):
D: Arimidex (Zeneca)   GB: Arimidex (Zeneca)
F: Arimidex (Zeneca)   USA: Arimidex (Zeneca)

Ancitabine
(Cyclooxytidine)

ATC: L01BC
Use: antineoplastic

RN: 31698-14-3   MF: $C_9H_{11}N_3O_4$ MW: 225.20
LD_{50}: 800 mg/kg (M, i.v.); 3400 mg/kg (M, p.o.);
820 mg/kg (R, i.v.); >7 g/kg (R, p.o.);
CN: [2R-(2α,3β,3aβ,9aβ)]-2,3,3a,9a-tetrahydro-3-hydroxy-6-imino-6H-furo[2',3':4,5]oxazolo[3,2-a]pyrimidine-2-methanol

monohydrochloride
RN: 10212-25-6   MF: $C_9H_{11}N_3O_4 \cdot HCl$ MW: 261.67   EINECS: 233-515-6
LD_{50}: 800 mg/kg (M, i.v.); >7 g/kg (M, p.o.);
820 mg/kg (R, i.v.); >7 g/kg (R, p.o.);
344 mg/kg (dog, i.v.)

cytidine → Ancitabine
POCl₃, H₂O, CH₃COOC₂H₅

Reference(s):
Kanai, T. et al.: Chem. Pharm. Bull. (CPBTAL) **18**, 2569 (1970).

alternative syntheses:
Walwick, E.R. et al.: Proc. Chem. Soc., London (PCSLAW) **1959**, 84.
Doerr, L.L.; Fox, J.J.: J. Org. Chem. (JOCEAH) **31**, 1465 (1967).
Ruyle, W.V.; Shenn, T.Y.: J. Med. Chem. (JMCMAR) **10**, 331 (1967).
Kugawa, K.K.; Ichino, M.: Tetrahedron Lett. (TELEAY) **1970**, 867.
The Merck Index, 11th Ed., 663 (Rahway 1991).

Formulation(s):   amp. 10 mg, 500 mg (as hydrochloride)

Trade Name(s):
J: Cyclo-C (Kohjin; as
hydrochloride)

Ancrod

ATC: C04A
Use: anticoagulant, fibrinolytic

RN: 9046-56-4   MF: unspecified   MW: unspecified   EINECS: 232-933-6
CN: proteinase, agkistrodon serine

Fibrinolytic effecting protease enzyme with glycoprotein structure; relative mol mass ca. 30000. Isolation from the poison secretion (venom) of *Agkistrodon rhodostoma* (malayan pit viper) with chromatographic purification.

Reference(s):
US 3 657 416 (Natl. Res. Dev. Corp., London; 18.4.1982; GB-prior. 21.2.1964).

Formulation(s):   amp. 70 iu.

Trade Name(s):
D: Arwin (Knoll); wfm   GB: Arvin (Armour); wfm

Androstanolone

(Stanolone)

ATC: G03BB02
Use: androgen

RN: 521-18-6   MF: C$_{19}$H$_{30}$O$_2$ MW: 290.45   EINECS: 208-307-3
CN: (5α,17β)-17-hydroxyandrostan-3-one

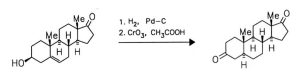

androstenolone   3,17-androstanedione   (I)

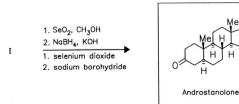

Androstanolone

Reference(s):
US 2 927 921 (Schering; 8.3.1960; prior. 19.5.1954, 24.1.1952).

alternative syntheses:
Butenandt, A. et al.: Chem. Ber. (CHBEAM) **68**, 2097 (1935).
Ruzicka, L. et al.: Helv. Chim. Acta (HCACAV) **20**, 1557 (1937); **24**, 1151 (1941).

Formulation(s):   amp. 2 %, 5 %; gel 2.5 %; tabl. 5 mg, 25 mg

Trade Name(s):
D: Ophthovitol (Dr. Winzer)-   F: Andractim (Besins-   J: Apeton (Fujisawa); wfm
      comb.; wfm   Iscovesco)   USA: Neodrel (Pfizer); wfm
                                   I: Anabolex (Samil); wfm

Anethole

ATC: A16AX02
Use: expectorant, carminative, aroma

RN: 4180-23-8   MF: C$_{10}$H$_{12}$O   MW: 148.21   EINECS: 224-052-0
CN: (E)-1-methoxy-4-(1-propenyl)benzene

a Isolation from essential oils, e. g. anise oil (80-90 %), staranise oil (>90 %), fennel oil (up to 80 %).
b From American sulfatterpentinol.

c Synthetic:

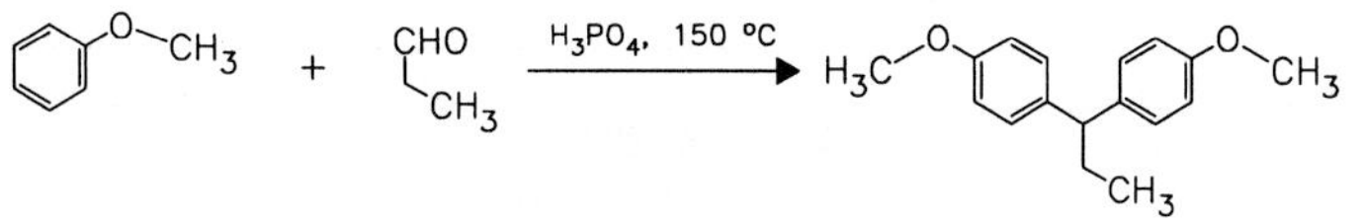

anisole + propion-aldehyde → 1,1-bis(4-methoxyphenyl)-propane (I)

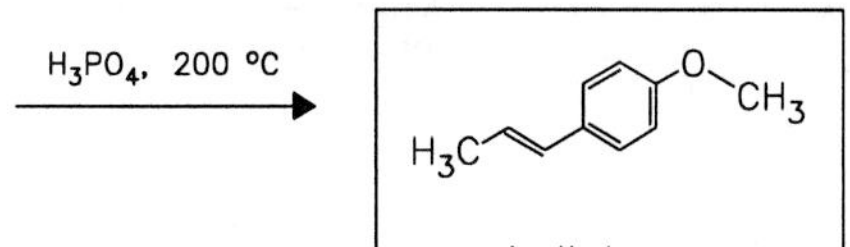

Anethole

Reference(s):
review:
Ullmanns Encykl. Tech. Chem., 4. Aufl., Vol. **20**, 241.
DE 2 418 974 (Haarmann & Reimer; appl. 19.4.1974).

Formulation(s):   cps. 75 mg; sol. 4 g/100 g

Trade Name(s):
D: Pinimenthol (Spitzner)-   Rowatinex (Rowa-   GB: Rowatinex (Rowa)-comb.
     comb.   Wagner)-comb.

Anethole trithione

ATC: A16AX02
Use: choleretic

RN: 532-11-6   MF: $C_{10}H_8OS_3$ MW: 240.37   EINECS: 208-528-5
LD$_{50}$:   1480 mg/kg (M, p.o.)
CN: 5-(4-methoxyphenyl)-3*H*-1,2-dithiole-3-thione

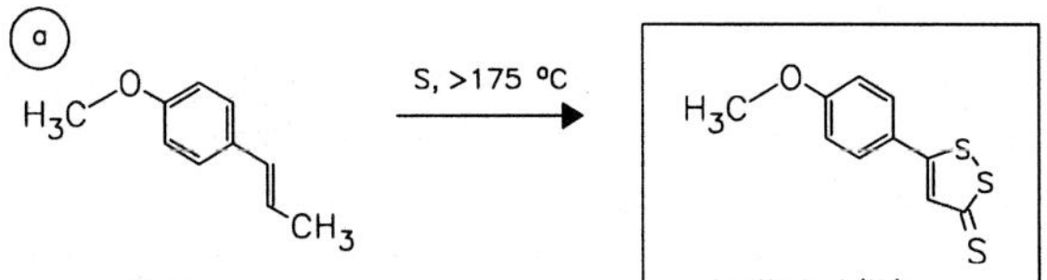

anethole → Anethole trithione

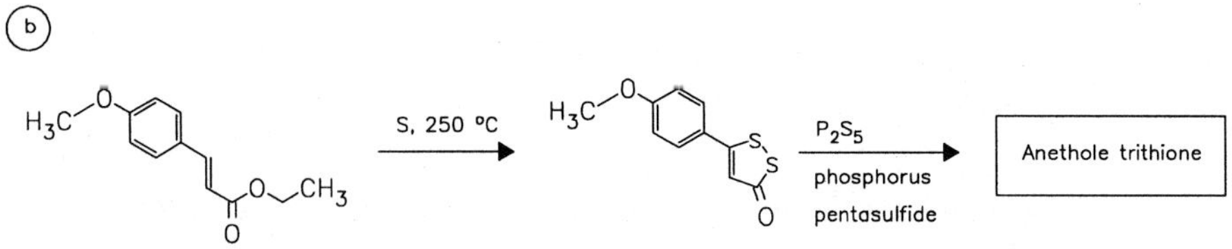

ethyl 4-methoxycinnamate → 5-(4-methoxyphenyl)-1,2-dithiol-3-one → Anethole trithione

Reference(s):
a   DE 855 865 (B. Böttcher; appl. 1942).
    DE 869 799 (B. Böttcher; appl. 1940).
b   DE 874 447 (B. Böttcher; appl. 1944).
    Schmidt, U. et al.: Justus Liebigs Ann. Chem. (JLACBF) **631**, 129 (1960).

Formulation(s):   cps. 4 mg, 75 mg; sol. 4 g/100 g

Trade Name(s):
D: Mucinol (Sanofi Winthrop)   Liverin (Sir)-comb.; wfm   J: Felviten (Nippon
F: Sulfarlem (Solvay Pharma)   Sulfalerm (Farmades); wfm   Shinyaku)
I: Liverin (Perkins)-comb.;   Sulfalerm (Sir); wfm
      wfm

Angiotensinamide

ATC: C01CX06
Use: hypertensive

RN: 53-73-6   MF: $C_{49}H_{70}N_{14}O_{11}$ MW: 1031.19   EINECS: 200-182-3
CN: 1-L-asparagine-5-L-valineangiotensin II

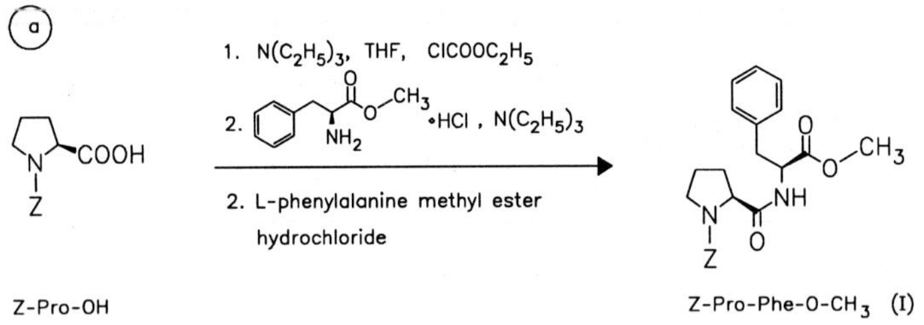

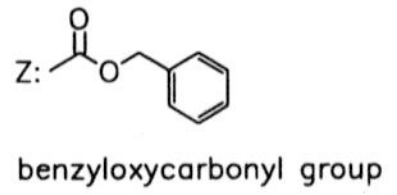

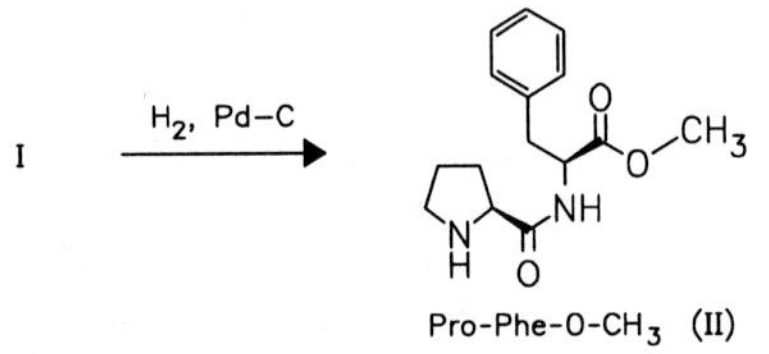

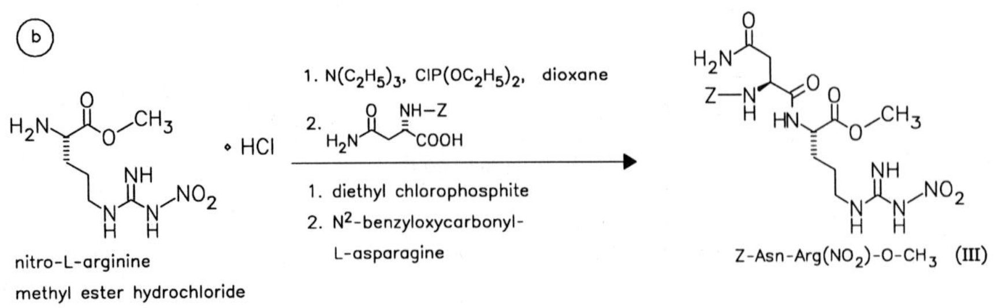

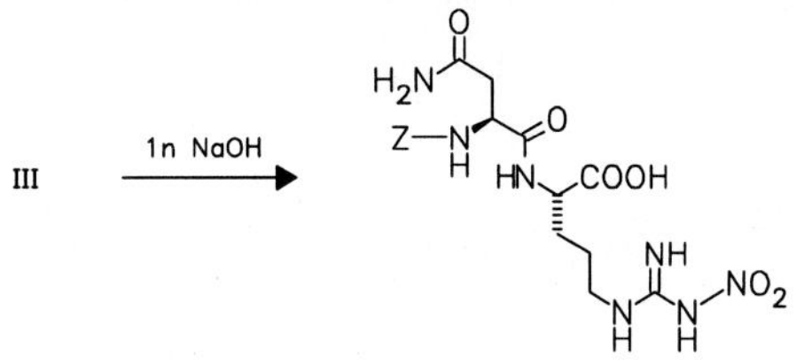
III
1n NaOH

Z-Asn-Arg(NO_2)-OH (IV)

c

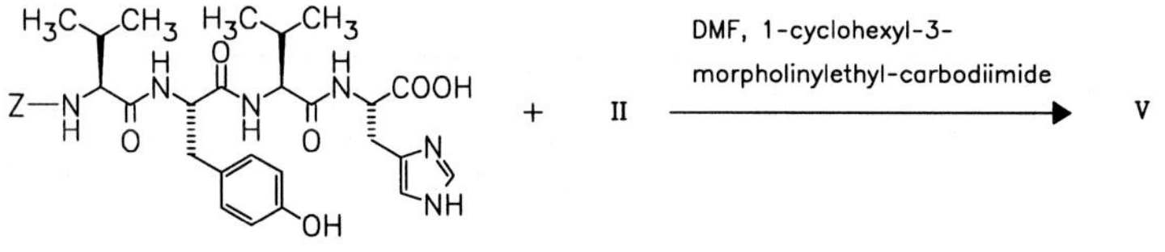
+ II
DMF, 1-cyclohexyl-3-morpholinylethyl-carbodiimide
V

N-benzyloxycarbonyl-L-valyl-L-tyrosyl-L-valyl-L-histidine

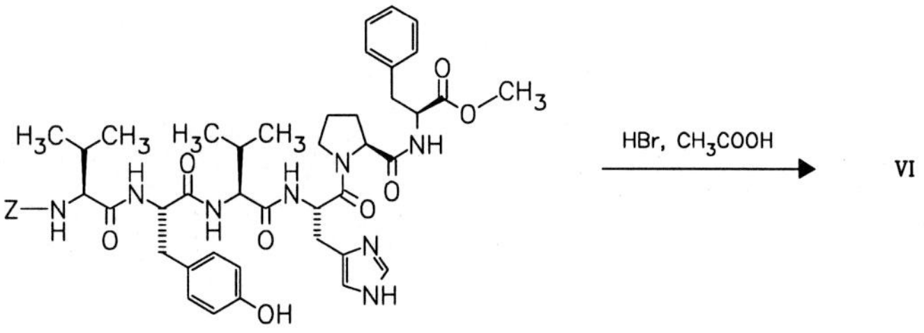
HBr, CH_3COOH
VI

Z-Val-Tyr-Val-His-Pro-Phe-O-CH_3 (V)

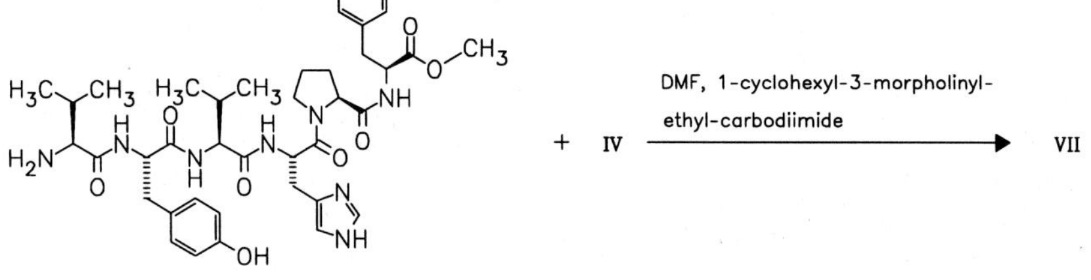
+ IV
DMF, 1-cyclohexyl-3-morpholinyl-ethyl-carbodiimide
VII

H-Val-Tyr-Val-His-Pro-Phe-O-CH_3 (VI)

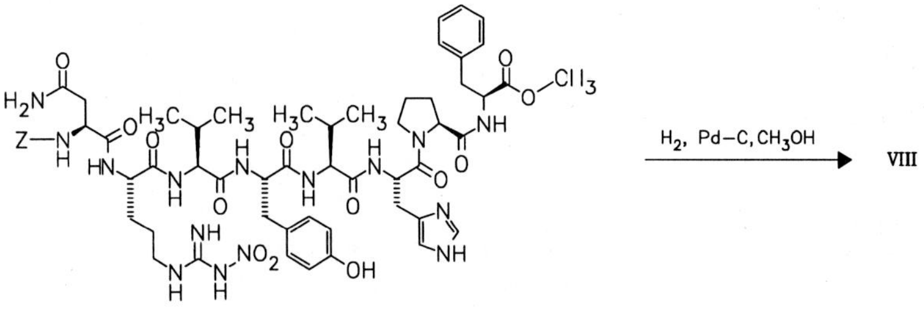
H_2, Pd–C, CH_3OH
VIII

Z-Asn-Arg(NO_2)-Val-Tyr-Val-His-Pro-Phe-O-CH_3 (VII)

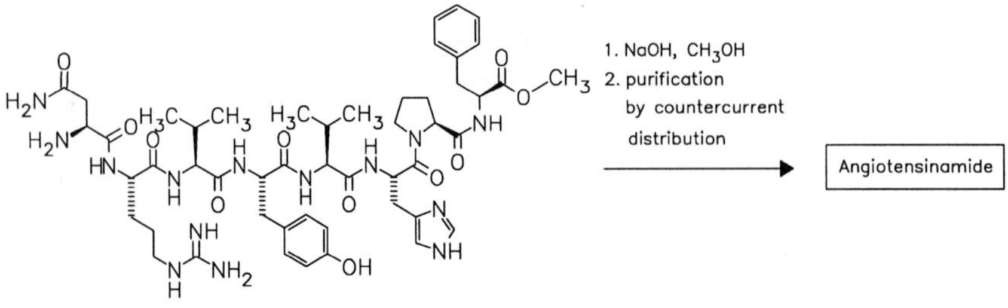

1. NaOH, CH$_3$OH
2. purification by countercurrent distribution

Angiotensinamide

H–Asn–Arg–Val–Tyr–Val–His–Pro–Phe–O–CH$_3$ (VIII)

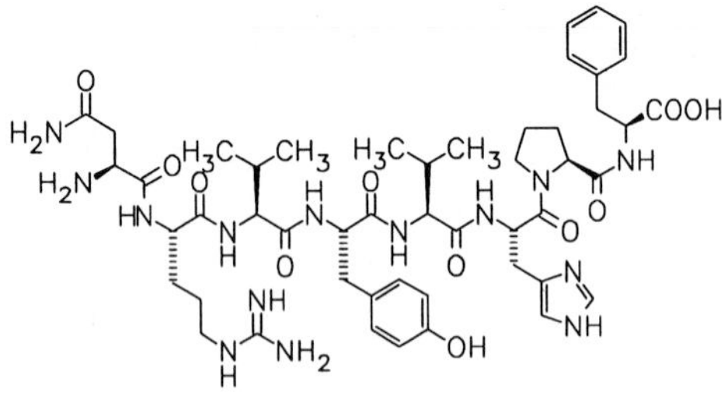

H–Asn–Arg–Val–Tyr–Val–His–Pro–Phe–OH

Angiotensinamide

Reference(s):
DE 1 125 942 (Ciba; appl. 2.9.1957; CH-prior. 6.9.1956, 8.2.1957, 6.3.1957, 31.7.1957).

Formulation(s):   amp. 2.5 mg

Trade Name(s):
D: Hypertensin (Ciba); wfm

Hypertensin CIBA (Ciba); wfm

GB: Hypertensin CIBA (Ciba); wfm

Anileridine

ATC: N01AH05; N01AX
Use: analgesic

RN: 144-14-9   MF: C$_{22}$H$_{28}$N$_2$O$_2$ MW: 352.48
CN: 1-[2-(4-aminophenyl)ethyl]-4-phenyl-4-piperidinecarboxylic acid ethyl ester

dihydrochloride
RN: 126-12-5   MF: C$_{22}$H$_{28}$N$_2$O$_2$ · 2HCl   MW: 425.40   EINECS: 204-770-0
LD$_{50}$:   22 mg/kg (M, i.v.); 229 mg/kg (M, p.o.);
        175 mg/kg (R, p.o.)
phosphate (1:1)
RN: 4268-37-5   MF: C$_{22}$H$_{28}$N$_2$O$_2$ · H$_3$PO$_4$ MW: 450.47

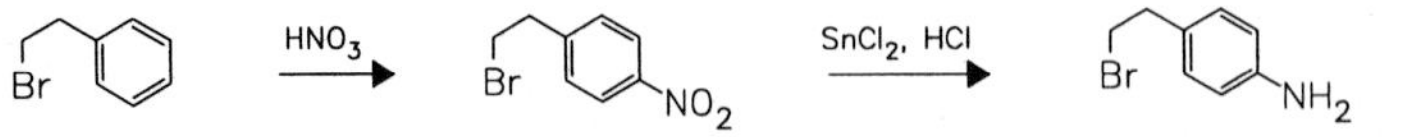

2-phenylethyl bromide

HNO$_3$

2-(4-nitrophenyl)- ethyl bromide

SnCl$_2$, HCl

2-(4-aminophenyl)- ethyl bromide (I)

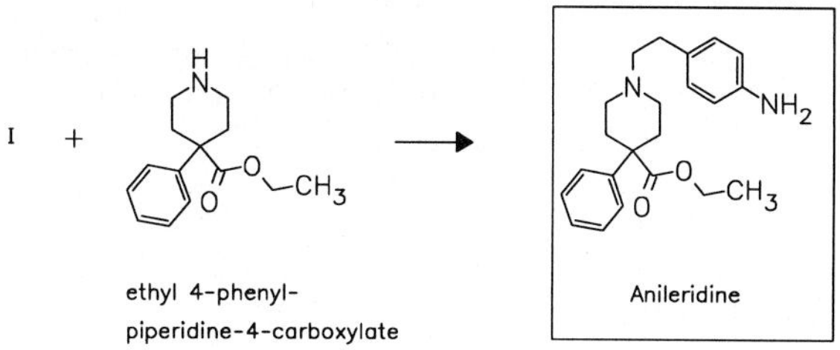
I
+
H
N
O
CH3
O
ethyl 4-phenyl-
piperidine-4-carboxylate
N
NH2
O
CH3
O
Anileridine

Reference(s):
US 2 966 490 (Merck & Co.; 27.12.1960; prior. 26.5.1955).

Formulation(s): amp. 25 mg; tabl. 25 mg

Trade Name(s):
USA: Leritine (Merck Sharp & Dohme); wfm

Aniracetam
(Ro-13-5057)

ATC: N06BX11
Use: nootropic (against senile dementia and cerebral insufficiency), cognition enhancer

RN: 72432-10-1 MF: $C_{12}H_{13}NO_3$ MW: 219.24
LD_{50}: >100 mg/kg (M, i.v.); >5000 mg/kg (M, p.o.)
>50 mg/kg (R, i.v.); 4500 mg/kg (R, p.o.)
CN: 1-(4-methoxybenzoyl)-2-pyrrolidinone

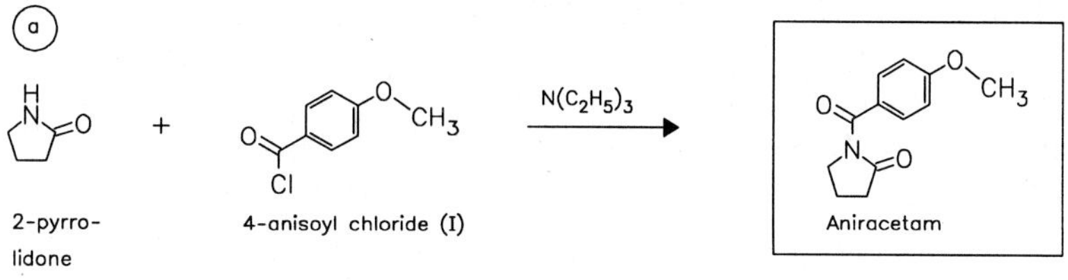
a
H
N
O
+
O
CH3
O
Cl
2-pyrro-
lidone
4-anisoyl chloride (I)
N(C2H5)3
O
CH3
O
N
O
Aniracetam

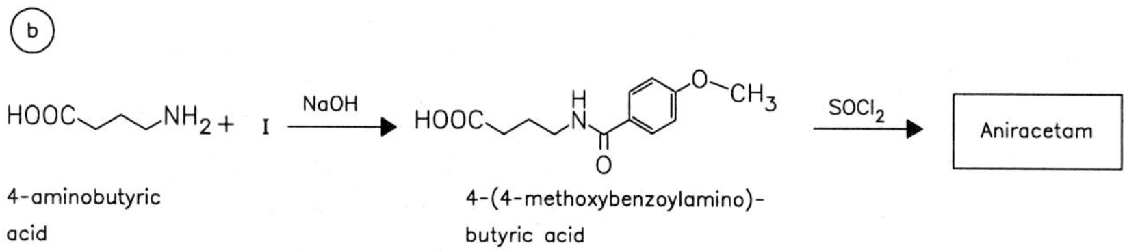
b
HOOC
NH2
+
I
NaOH
HOOC
H
N
O
O
CH3
4-aminobutyric
acid
4-(4-methoxybenzoylamino)-
butyric acid
SOCl2
Aniracetam

Reference(s):
EP 5 143 (Hoffmann-La Roche, Sparamedica; appl. 9.2.1979; CH-prior. 10.2.1978, 22.11.1978).
EP 44 088 (Hoffmann-La Roche; appl. 9.2.1979; CH-prior. 10.2.1978, 22.11.1978).
(also alternative synthesis).

medical use for treatment of claudicatio intermittens:
EP 243 336 (UCB; appl. 10.4.1987; GB-prior. 14.4.1986).

Formulation(s): powder 1.5 g; tabl. 100 mg, 200 mg, 750 mg

Trade Name(s):

I:	Ampamet (Menarini)	Reset (Biomedica	J:	Draganon (Nippon Roche)
	Draganon (Roche; 1992)	Foscama)		Sarpul (Toyama Chem.)

Anisindione

Use: anticoagulant

RN: 117-37-3 MF: $C_{16}H_{12}O_3$ MW: 252.27 EINECS: 204-186-6
LD_{50}: 300 mg/kg (M, p.o.)
CN: 2-(4-methoxyphenyl)-1*H*-indene-1,3(2*H*)-dione

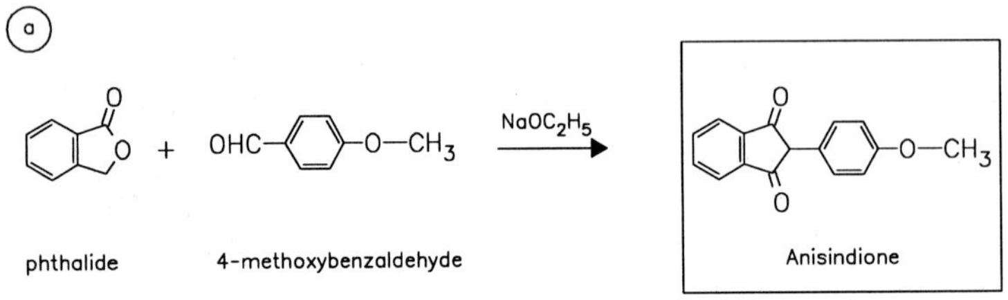

a
NaOC2H5
phthalide
4-methoxybenzaldehyde
Anisindione

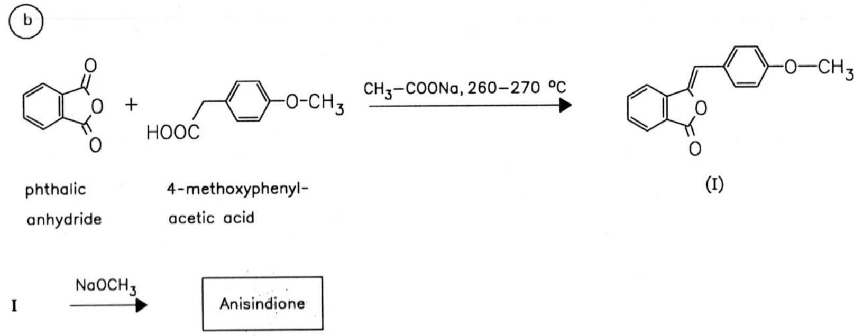

b
CH3–COONa, 260–270 °C
phthalic anhydride
4-methoxyphenyl-acetic acid
(I)
I
NaOCH3
Anisindione

Reference(s):
US 2 899 358 (Schering Corp.; 11.8.1959; prior. 23.2.1956).

Formulation(s): tabl. 75 mg, 100 mg, 300 mg

Trade Name(s):

F:	Midone (Cétrane); wfm	Unidone (Unilabo); wfm	USA:	Miradon (Schering)

Antazoline

ATC: R01AC04; R06AX05
Use: antihistaminic

RN: 91-75-8 MF: $C_{17}H_{19}N_3$ MW: 265.36 EINECS: 202-094-0
LD_{50}: 61 mg/kg (M, i.v.); 398 mg/kg (M, p.o.)
CN: 4,5-dihydro-*N*-phenyl-*N*-(phenylmethyl)-1*H*-imidazole-2-methanamine

monohydrochloride
RN: 2508-72-7 MF: $C_{17}H_{19}N_3 \cdot HCl$ MW: 301.82 EINECS: 219-719-8
LD_{50}: 30 mg/kg (dog, i.v.)
sulfate (1:1)
RN: 24359-81-7 MF: $C_{17}H_{19}N_3 \cdot H_2SO_4$ MW: 363.44
monomesylate
RN: 3131-32-6 MF: $C_{17}H_{19}N_3 \cdot CH_4O_3S$ MW: 361.47 EINECS: 221-523-2

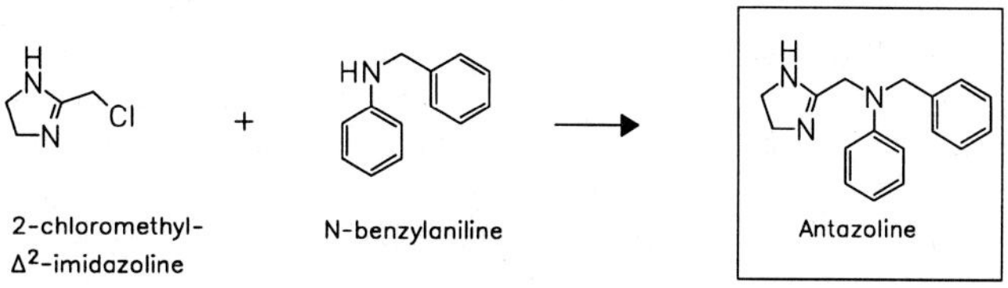

2-chloromethyl-
Δ²-imidazoline   N-benzylaniline   Antazoline

Reference(s):
US 2 449 241 (Ciba; 1948; CH-prior. 1944).

Formulation(s):   eye drops 0.15 mg/ml, 0.5 mg/ml, 5 mg/ml

Trade Name(s):

D:	Allergopos (Ursapharm)-comb.	F:	Alcolène (Alcon)-comb.; wfm		Zincoimidazyl (Allergan)-comb.
	Antistin-Privin (CIBA Vision)-comb.	GB:	Otrivine-Antistin (CIBA Vision)-comb.	USA:	Arithmin (Lannett); wfm Azolone (Smith, Miller & Patch); wfm
	Ophtalmin (Winzer)-comb. Spersallerg (CIBA Vision)-comb.	I:	Antistin Privina (Novartis)-comb. Eubetal (SIFI)-comb.		

Antrafenine

ATC: S02DA
Use: analgesic, anti-inflammatory

RN: 55300-29-3   MF: $C_{30}H_{26}F_6N_4O_2$ MW: 588.55
LD$_{50}$:   4 g/kg (M, p.o.)
CN: 2-[[7-(trifluoromethyl)-4-quinolinyl]amino]benzoic acid 2-[4-[3-(trifluoromethyl)phenyl]-1-piperazinyl]ethyl ester

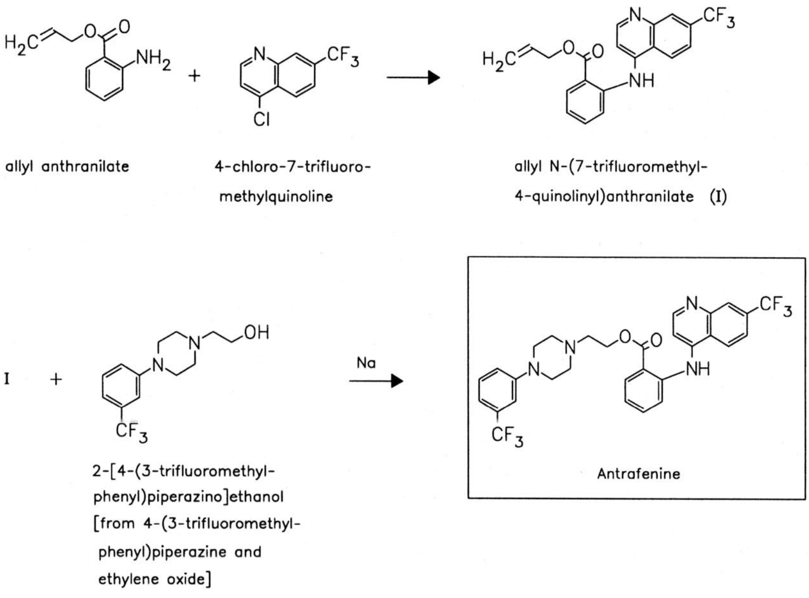

allyl anthranilate   4-chloro-7-trifluoro-
                            methylquinoline   allyl N-(7-trifluoromethyl-
                                                     4-quinolinyl)anthranilate   (I)

2-[4-(3-trifluoromethyl-
phenyl)piperazino]ethanol
[from 4-(3-trifluoromethyl-
phenyl)piperazine and
ethylene oxide]   Antrafenine

Reference(s):
DOS 2 415 982 (Synthelabo; appl. 2.4.1974; F-prior. 6.4.1973, 9.5.1973, 17.12.1973).

Apalcillin

ATC:　　J01CA
Use:　　semisynthetic β-lactam antibiotic

RN:　　63469-19-2　MF: $C_{25}H_{23}N_5O_6S$　MW: 521.55
LD_{50}:　1300 mg/kg (M, i.v.)
CN:　　[2S-[2α,5α,6β(S*)]]-6-[[[[(4-hydroxy-1,5-naphthyridin-3-yl)carbonyl]amino]phenylacetyl]amino]-3,3-dimethyl-7-oxo-4-thia-1-azabicyclo[3.2.0]heptane-2-carboxylic acid

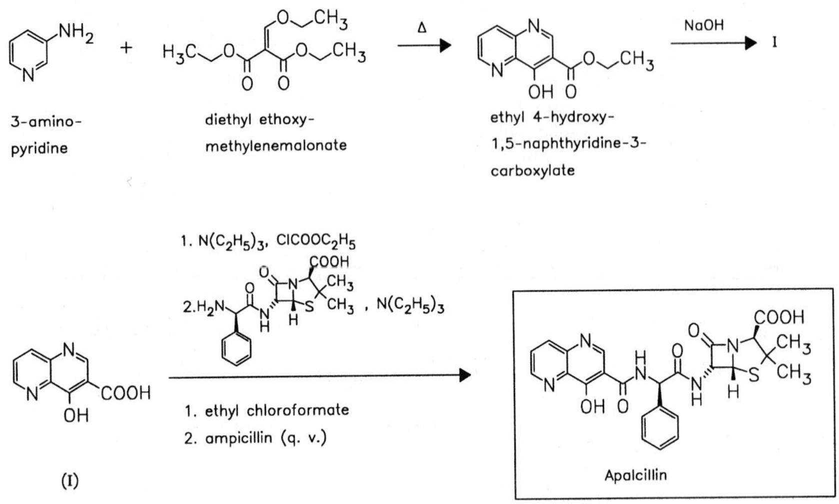

3-amino-
pyridine

diethyl ethoxy-
methylenemalonate

ethyl 4-hydroxy-
1,5-naphthyridine-3-
carboxylate

(I)

1. ethyl chloroformate
2. ampicillin (q. v.)

Apalcillin

Reference(s):
US 3 864 329 (Sumitomo; 4.2.1975; J-prior. 29.12.1970).
US 4 005 075 (Sumitomo; 25.1.1977; J-prior. 5.4.1973).
DOS 2 416 449 (Sumitomo; appl. 4.4.1974; J-prior. 5.4.1973).
US 3 945 995 (Sumitomo; 23.3.1976; J-prior. 5.4.1973).

Formulation(s):　　lyo. 1042 mg, 3126 mg

Apomorphine

ATC:　　N04BC07
Use:　　emetic, expectorant

RN:　　58-00-4　MF: $C_{17}H_{17}NO_2$　MW: 267.33　EINECS: 200-360-0
LD_{50}:　56 mg/kg (M, i.v.); >100 mg/kg (M, p.o.)
CN:　　(R)-5,6,6a,7-tetrahydro-6-methyl-4H-dibenzo[de,g]quinoline-10,11-diol

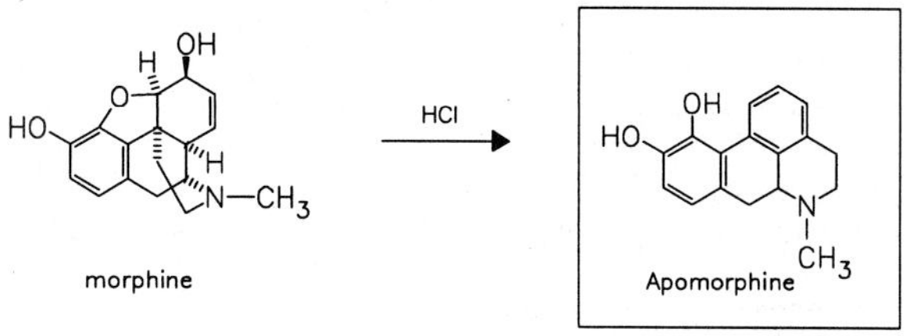

morphine → Apomorphine

Reference(s):
Small, L. et al.: J. Org. Chem. (JOCEAH) **5**, 334 (1940).

Formulation(s):   amp. 10 mg; inj. sol. 10 mg/1 ml, 5 mg/1 ml; tabl. 3 mg

Trade Name(s):
F: Apokinon (Aguettant)   I: Apomor (Tariff.
GB: Britagel (Britannia; as   Integrativo; as
      hydrochloride)   hydrochloride)

Apraclonidine
(Aplonidine)

ATC: S01EA03
Use: selective α_2-agonist (for postsurgical control of intraocular pressure)

RN: 66711-21-5   MF: $C_9H_{10}Cl_2N_4$ MW: 245.11
CN: 2,6-dichloro-N^1-2-imidazolidinylidene-1,4-benzenediamine

monohydrochloride
RN: 73218-79-8   MF: $C_9H_{10}Cl_2N_4 \cdot HCl$ MW: 281.57
LD_{50}:   6 mg/kg (M, i.v.); 3 mg/kg (M, p.o.);
            9 mg/kg (R, i.v.); 38 mg/kg (R, p.o.)
dihydrochloride
RN: 73217-88-6   MF: $C_9H_{10}Cl_2N_4 \cdot 2HCl$ MW: 318.04

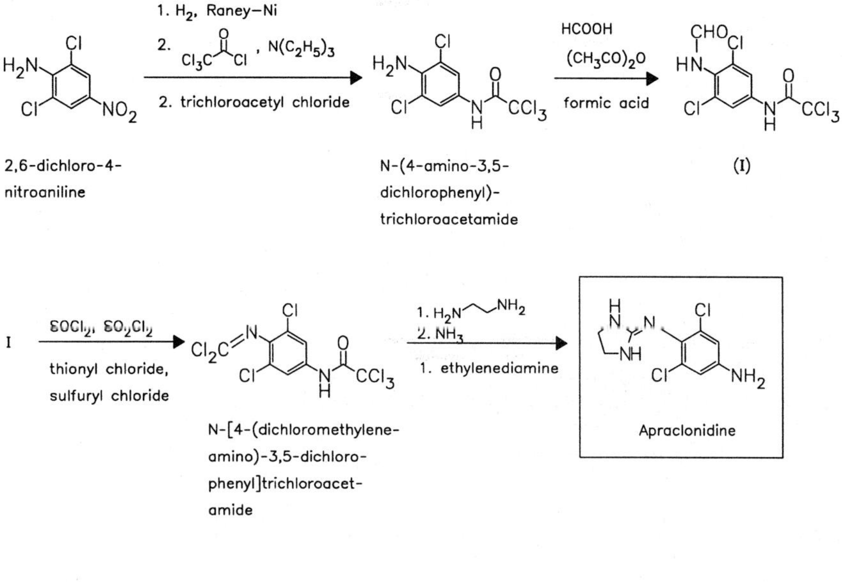

2,6-dichloro-4-nitroaniline

N-(4-amino-3,5-dichlorophenyl)-trichloroacetamide

(I)

N-[4-(dichloromethylene-amino)-3,5-dichloro-phenyl]trichloroacet-amide

Apraclonidine

Reference(s):
EP 81 924 (Alcon; appl. 19.11.1982; USA-prior. 20.11.1981).
EP 81 923 (Alcon; appl. 19.11.1982; USA-prior. 20.11.1981).
US 4 461 904 (Alcon; 24.7.1984; prior. 20.11.1981).
Ronot, B.; Leclerc, G.: Bull. Soc. Chim. Fr. (BSCFAS) **Pt. 2**, 520 (1979).

combination with β-receptor antagonist:
EP 365 662 (Alcon; 26.4.1989; USA-prior. 26.4.1988).

preparation of 2,6-dichloro-4-nitroaniline:
Goldschmidt; Strohmenger: Ber. Dtsch. Chem. Ges. (BDCGAS) **55**, 2455 (1922).
Pausadeer; Scroggie: Aust. J. Chem. (AJCHAS) **12**, 430, 432 (1959).
Fluerscheim: J. Chem. Soc. (JCSOA9) **93** 1774 (1908).
Datta; Müller: J. Am. Chem. Soc. (JACSAT) **41**, 2036 (1919).
Koerner: Gazz. Chim. Ital. (GCITA9) **4**, 376 (1874).
Kohn; Pfeifer: Monatsh. Chem. (MOCMB7) **48**, 236 (1927).

Formulation(s):   eye drops 0.5 %, 1 %; ophthalmic sol. 10 mg/ml

Trade Name(s):

D:	Iopidine (Alcon)	GB:	Iopidine (Alcon)
F:	Iopidine (Alcon)		

USA: Iopidine (Alcon; 1988); wfm

Aprindine

ATC: C01BB04
Use: antiarrhythmic

RN: 37640-71-4   MF: $C_{22}H_{30}N_2$ MW: 322.50
LD_{50}:   274 mg/kg (M, p.o.)
CN: *N*-(2,3-dihydro-1*H*-inden-2-yl)-*N'*,*N'*-diethyl-*N*-phenyl-1,3-propanediamine

monohydrochloride
RN: 33237-74-0   MF: $C_{22}H_{30}N_2 \cdot HCl$ MW: 358.96   EINECS: 251-418-7
LD_{50}:   17.1 mg/kg (M, i.v.); 262 mg/kg (M, p.o.);
     16.6 mg/kg (R, i.v.); 525 mg/kg (R, p.o.)

2-indanone   aniline (I)   N-(2-indanylidene)-aniline   2-anilinoindane (II)

2-indanyl methanesulfonate

1. NaNH₂
2. Cl CH₃ ... CH₃

2. 3-diethylamino-propyl chloride

Aprindine

Reference(s):
DE 2 060 721 (Christiaens S. A.; appl. 10.12.1970; GB-prior. 19.12.1968, 26.11.1970).

Formulation(s):   cps. 50 mg; inj. sol. 200 mg/20 ml

Trade Name(s):
D: Amidonal (PCR   F: Fiboran (Nycomed)
     Arzneimittel)   J: Aspenone (Mitsui)

Aprobarbital

ATC: N05CA05
Use: hypnotic, sedative

RN: 77-02-1   MF: $C_{10}H_{14}N_2O_3$ MW: 210.23   EINECS: 200-997-4
LD_{50}:   200 mg/kg (M, i.p.); 350 mg/kg (M, s.c.)
CN: 5-(1-methylethyl)-5-(2-propenyl)-2,4,6(1*H*,3*H*,5*H*)-pyrimidinetrione

monosodium salt
RN: 125-88-2   MF: $C_{10}H_{13}N_2NaO_3$ MW: 232.22   EINECS: 204-760-6
LD_{50}:   85 mg/kg (R, i.p.)

5-isopropyl-   allyl bromide   Aprobarbital
barbituric acid

Reference(s):
US 1 444 802 (Hoffmann-La Roche; 1923; appl. 1921).

Formulation(s):   elixir 40 mg; tabl. 20 mg, 40 mg, 80 mg

Trade Name(s):
D: Allional (Hoffmann-La   Nervisal (Lappe)-comb.;   Vita-Dor (Steigerwald)-
     Roche)-comb.; wfm   wfm   comb.; wfm
     Mandotrilan-"forte"   Nervolitan (Kettelhack)-   USA: Alurate (Roche); wfm
     (Henk)-comb.; wfm   comb.; wfm
     Nervinum Stada (Stada)-   Resedorm (Lappe)-comb.;
     comb.; wfm   wfm

Aprotinine
(Trasylol; Triazinin; Zymofren)

ATC: B02AB
Use: proteinase inhibitor, kallikrein
       inhibitor

RN: 9087-70-1   MF: $C_{284}H_{432}N_{84}O_{79}S_7$ MW: 6511.55   EINECS: 232-994-9
LD_{50}:   >50 ml/kg (M, i.p.); >50 ml/kg (M, s.c.);
       >40 ml/kg (R, i.p.); >40 ml/kg (R, s.c.)
CN: trypsin inhibitor (ox pancreas basic)

antagosan
RN: 9050-74-2   MF: $C_{284}H_{432}N_{84}O_{79}S_7$ MW: 6511.55
iniprol
RN: 11004-21-0   MF: $C_{284}H_{432}N_{84}O_{79}S_7$ MW: 6511.55

ox pancreas basic
RN: 12407-79-3   MF: $C_{284}H_{432}N_{84}O_{79}S_7$ MW: 6511.55
ox pancreas basic reduced
RN: 11061-94-2   MF: $C_{284}H_{438}N_{84}O_{79}S_7$ MW: 6517.60

By extraction of animal lymph glands, parotid glands, pancreas, liver, milt and blood serum with diluted acetic acid-ethanol-mixtures upon removal of fat and proteins.

Reference(s):
US 2 890 986 (Bayer; 16.6.1959; D-prior. 29.5.1954).

Formulation(s):   amp. 200000 KIU; inj. sol. 100000 KIU/10 ml, 500000 KIU/50 ml

Trade Name(s):
D: Antagosan (Hoechst)   Biscol (Lab. Français du   Trasylol (Bayer)
      Beriplast (Centeon   Fractionement et des   J: Trasylol (Bayer-Yoshitomi;
      Pharma)-comb.   Biotechnologies)-comb.   as solution)
      Tissucol (Immuno)   Trasylol (Bayer-Pharma)   USA: Trasylol Injection (Bayer)
      Trasylol (Bayer Vital)   I: Antagosan (Behring)
F: Antagosan (Hoechst   Fase (Astra-Simes)
      Houdé)   Kir Richter (Lepetit)

Aranidipine
(MPC-1304)

ATC: C04
Use: antihypertensive, calcium channel
        blocker

RN: 86780-90-7   MF: $C_{19}H_{20}N_2O_7$ MW: 388.38
LD$_{50}$: 143 mg/kg (M, p.o.);
         1459 mg/kg (R, p.o.);
         3333 mg/kg (dog, p.o.)
CN: 1,4-dihydro-2,6-dimethyl-4-(2-nitrophenyl)-3,5-pyridinedicarboxylic acid methyl 2-oxopropyl ester

diketene   +   2,2-(ethylene-   →[NaH, benzene]   2,2-(ethylenedioxy)-
                 dioxy)-1-propanol   propyl acetoacetate   (I)

I   +   2-nitro-   →[piperidine, acetic acid]   2,2-(ethylenedioxy)-
          benzaldehyde   propyl 2-(2-nitro-
                                                     benzylidene)acetoacetate   (II)

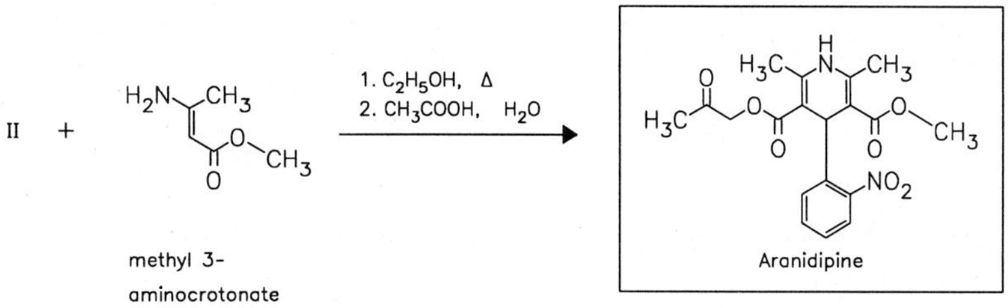

methyl 3-
aminocrotonate

Aranidipine

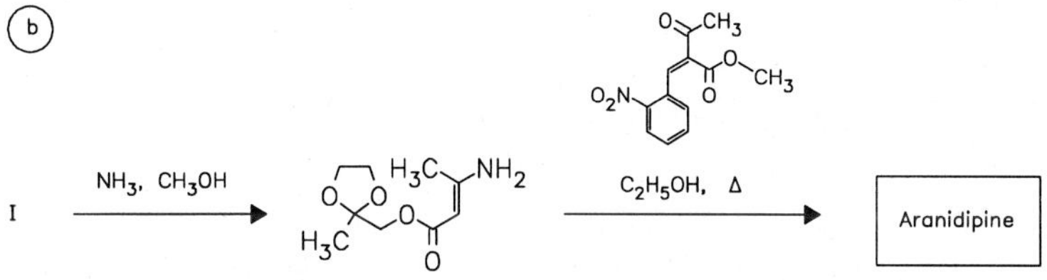

(b)

Aranidipine

Reference(s):

a   Ohno, S. et al.: Chem. Pharm. Bull. (CPBTAL) **34**(4), 1589-1606 (1986).
b   FR 2 514 761 (Maruko Seiyaku; appl. 19.10.1982; J-prior. 19.10.1981).

topical ophthalmic formulation:
WO 9 323 082 (Alcon Lab.; appl. 12.5.1993; USA-prior. 13.5.1992).

Formulation(s):   gran. 20 mg/g (2 %)

Trade Name(s):
J: Bec (Maruko; Bristol-   Sapresta (Taiho)
       Myers Squibb)

Arginine aspartate

ATC: A13A
Use: liver dysfunction therapeutic, tonic

RN: 7675-83-4   MF: $C_6H_{14}N_4O_2$ MW: 174.20   EINECS: 231-656-8
CN: L-aspartic acid compd. with L-arginine (1:1)

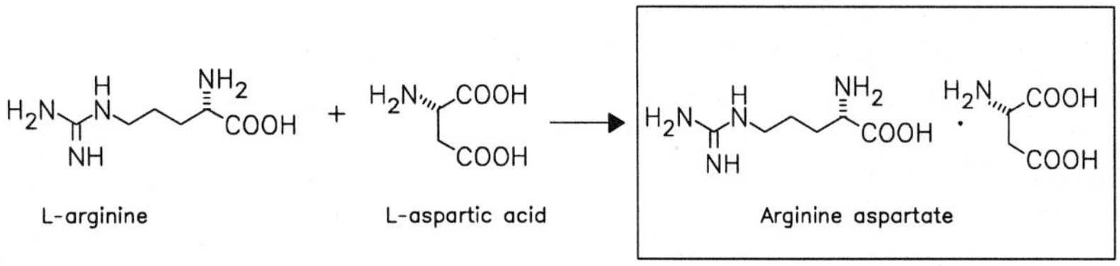

L-arginine   L-aspartic acid   Arginine aspartate

Very pure preparations are obtained by treatment of L-aspartic acid loaded strong basic anion-exchange resins with an aqueous solution of L-arginine hydrochloride.

Reference(s):
DAS 1 518 033 (Mundipharma; appl. 17.9.1965; F-prior. 14.1.1965).

Formulation(s):   tabl. 1 g; sol. 1 g/10 ml, 1 g/5 ml; tabl. 500 mg, 1 g

Trade Name(s):
D: Argihepar (Chephasaar);   F: Sargenor (ASTA Medica)   I: Glutargin (Terapeutico
       wfm   M.R.)-comb.

Sargenor (ASTA Medica)

Arginine pidolate
(Arginine pyroglutamate)

ATC: A13A
Use: tonic, cerebrostimulant

RN: 64855-91-0   MF: $C_6H_{14}N_4O_2 \cdot C_5H_7NO_3$ MW: 303.32   EINECS: 265-253-3
CN: 5-oxo-proline compd. with L-arginine (1:1)

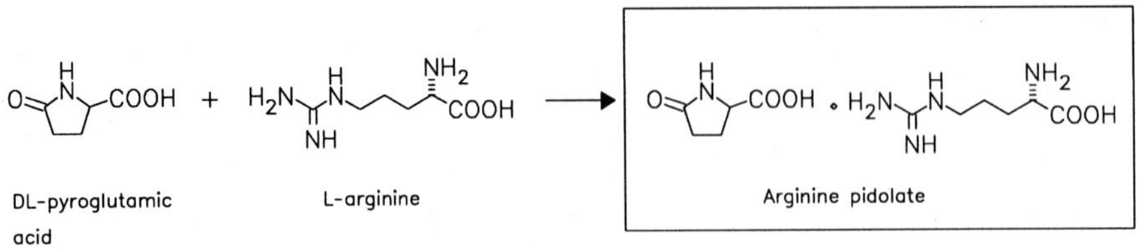

DL-pyroglutamic acid   L-arginine

Arginine pidolate

Reference(s):
DAS 2 416 339 (Manetti Roberts; appl. 4.4.1974; I-prior. 4.4.1973).
GB 1 421 089 (Manetti Roberts; appl. 27.3.1974; I-prior. 4.4.1973).
Provenzano, P.M. et al.: Arzneim.-Forsch. (ARZNAD) **27**, 1553 (1977).

use as sexual tonic:
DOS 3 125 512 (Manetti Roberts; appl. 29.6.1981; I-prior. 30.6.1980).

Formulation(s):   lyo. 500 mg, 1 g; tabl. 500 mg

Trade Name(s):
I: Adiuvant (Manetti Roberts)   Detoxergon Polvere   Neoiodarsolo (Baldacci)-
     Detoxergon Fiale   (Baldacci)-comb.   comb.
     (Baldacci)-comb.

Arotinolol
(S-596)

ATC: C07AA
Use: β-adrenoceptor blocker,
        antihypertensive, antianginal

RN: 52560-77-7   MF: $C_{15}H_{21}N_3O_2S_3$ MW: 371.55
LD_{50}: >360 mg/kg (M, i.p.); 86 mg/kg (M, i.v.); >5000 mg/kg (M, p.o.)
CN: 5-[2-[[3-[(1,1-dimethylethyl)amino]-2-hydroxypropyl]thio]-4-thiazolyl]-2-thiophenecarboxamide

hydrochloride
RN: 80416-73-5   MF: $C_{15}H_{21}N_3O_2S_3 \cdot xHCl$ MW: unspecified

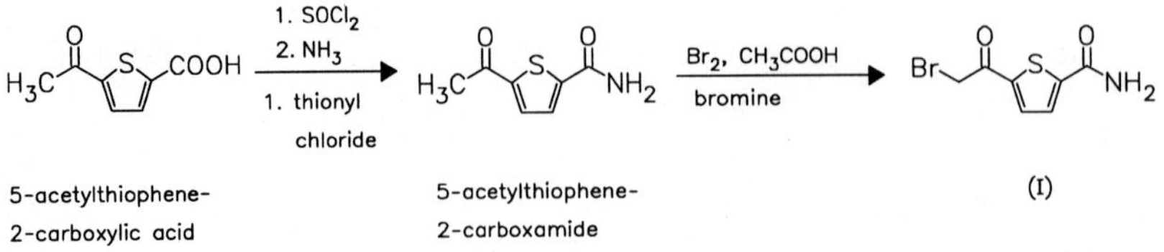

5-acetylthiophene-   5-acetylthiophene-   (I)
2-carboxylic acid   2-carboxamide

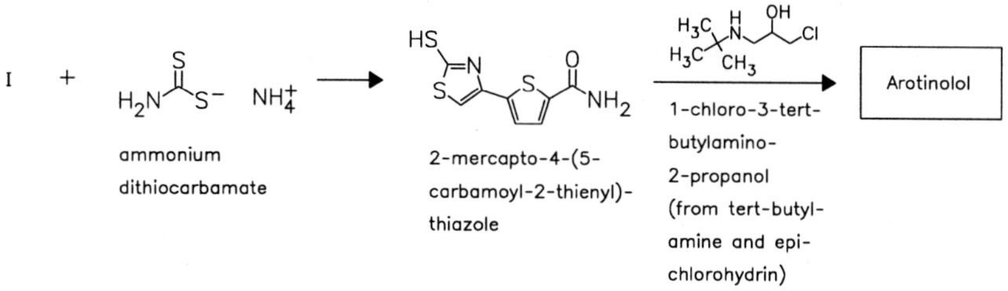
I
ammonium
dithiocarbamate
2-mercapto-4-(5-
carbamoyl-2-thienyl)-
thiazole
1-chloro-3-tert-
butylamino-
2-propanol
(from tert-butyl-
amine and epi-
chlorohydrin)
Arotinolol

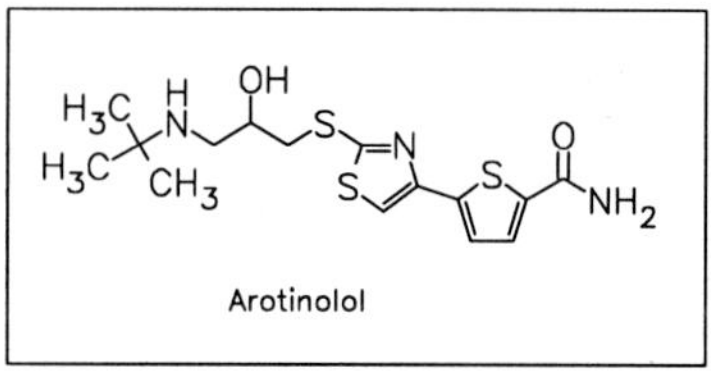
Arotinolol

Reference(s):
DOS 2 341 753 (Sumitomo; appl. 17.8.1973; J-prior. 17.8.1972, 5.4.1973).
US 3 932 400 (Sumitomo; 13.1.1976; J-prior. 17.8.1972, 5.4.1973).
Hara, Y. et al.: J. Pharm. Sci. (JPMSAE) **67**, 1334 (1978).

Formulation(s): tabl. 10 mg

Trade Name(s):
J: Almarl (Sumitomo; 1986)

Ascorbic acid

(Acide ascorbique; Vitamin C)

ATC: A11GA01
Use: antiscorbutical vitamin, antioxidant

RN: 50-81-7 MF: $C_6H_8O_6$ MW: 176.12 EINECS: 200-066-2
LD_{50}: 518 mg/kg (M, i.v.); 3367 mg/kg (M, p.o.);
>4 g/kg (R, i.v.); 11.9 g/kg (R, p.o.)
CN: L-ascorbic acid

monopotassium salt
RN: 15421-15-5 MF: $C_6H_7KO_6$ MW: 214.21 EINECS: 239-432-1
monosodium salt
RN: 134-03-2 MF: $C_6H_7NaO_6$ MW: 198.11 EINECS: 205-126-1
calcium salt (2:1)
RN: 5743-27-1 MF: $C_{12}H_{14}CaO_{12}$ MW: 390.31 EINECS: 227-261-5
magnesium salt (2:1)
RN: 15431-40-0 MF: $C_{12}H_{14}MgO_{12}$ MW: 374.54 EINECS: 239-442-6
Fe(II) salt (2:1)
RN: 24808-52-4 MF: $C_{12}H_{14}FeO_{12}$ MW: 406.08

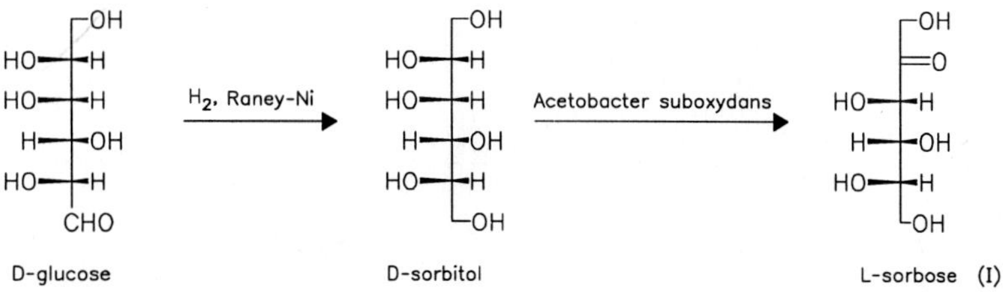
H2, Raney-Ni
Acetobacter suboxydans
D-glucose
D-sorbitol
L-sorbose (I)

acetone   diacetone-L-sorbose   diacetone-2-oxo-
                                               L-gulonic acid (II)

2-oxo-L-
gulonic acid   Ascorbic acid

Reference(s):
Reichstein, T.; Grüssner, A.: Helv. Chim. Acta (HCACAV) **17**, 311 (1934).
Ullmanns Encykl. Tech. Chem., 3. Aufl., Vol. **18**, 223 ff.
Kirk-Othmer Encycl. Chem. Technol., 2. Ed. (15SWA8), Vol. **2**, (1963-1971), p. 747 ff.

Formulation(s):   amp. 562 mg; eff. tabl. 1 g; powder 1 g; tabl. 50 mg, 200 mg, 500 mg

Trade Name(s):
D: Ascorvit (Jenapharm)
      Cebion (Merck)
      Cetebe (SmithKline
      Beecham)
      Cevitt (Hermes)
      Vitamin C Phytopharma
      (OTW)
      numerous combination
      preparations
F: Ascofer (Gerda; as iron-
      salt)
      Laroscorbine (Roche
      Nicholas)
      Midy Vitamine C
      (SmithKline Beecham)
      Vitascorbol (Théraplix)
      generics (as salts) and
      numerous combination
      preparations
GB: Ferfolic SV (Sinclair)-
      comb.
      Redoxon (Roche
      Consumer)
I: Acidylina (Ist. Italiano
      Ferm.)

Agrumina (Also)
Agruvit (Lepetit)
Ascomed (Ripari-Gero)
Ascorgil (Biomedica
Foscama)
Aster C (Corvi)
Bio-Ci (Ceccarelli)
C-Lisa (Lisapharma)
C-Tard (Eurand-Mi)
Cebion (Bracco)
Cecon (Abbott)
Cevit (Italfarmaco)
Duo-C (Geymonat)
Idro-C (Blue Cross)
Lemonvit (Molteni)
Redoxon (Roche)
Vicifte (Iacopo Monico)
Vicitina (CT)
Vitamina C Vca
(Bergamon)
Vitamina C Vita
(Synthelabo)
combination preparations
J: Ascoyl (Shionogi); wfm
      Hicee (Takeda); wfm

Viscorin (Daiichi); wfm
Vitacimin (Takeda); wfm
numerous combination
preparations
USA: ACES Antioxidant Soft
      Gels (Carlson)
      Ce-Vi-Sol (Mead Johnson)
      Cevi-Bid (Geriatric
      Pharm.)
      Chromagen (Savage)
      CitraDerm (Pedinol)
      Ferancee (Stuart)
      Fero-Folic-500 (Abbott)
      Fero-Grad-500 (Abbott)
      Fetrin (Lunsco)
      Irospam (Fielding)
      Materna (Lederle Labs.)
      Trinsicon (UCB)
      Vi-Daylin ADC (Ross)
      Vitron-C (Fisons)
      numerous combination
      preparations

Asparaginase

(L-Asparaginase)

ATC: L01XX; L01XX02
Use: antineoplastic

RN: 9015-68-3   MF: unspecified   MW: unspecified   EINECS: 232-765-3
LD$_{50}$:   136 g/kg (M, i.v.);
        7568 mg/kg (R, i.v.);
        227 mg/kg (dog, i.v.)
CN: asparaginase

L-Asparagine-amidohydrolase.
Relative mol mass 133000 ± 5000.
By separation from bacterial culture such as *Escherichia coli, Serratia marcescens, Erwinia aroideae, Erwinia atroseptica, Erwinia carotovora.*

Reference(s):
Ho et al.: J. Biol. Chem. (JBCHA3) **245**, 3708 (1970).
DAS 1 642 615 (Bayer; appl. 27.12.1967).
DAS 1 942 833 (Secret. of State for Social Services, London; appl. 22.8.1969; GB-prior. 23.8.1968).
DAS 1 942 900 (Secret. of State for Social Services, London; appl. 22.8.1969; GB-prior. 23.8.1968).

Formulation(s):   inj. powder 10000 iu/2.5 ml

Trade Name(s):
D: Erwinase (Ipsen Pharma)   I: Crasnitin (Bayer); wfm   USA: Elspar (Merck)
F: Kidrolase (Bellon)   J: Leunase (Kyowa Hakko)

Aspartame

Use: sweetener (pharmaceutical agent)

RN: 22839-47-0   MF: C$_{14}$H$_{18}$N$_2$O$_5$ MW: 294.31   EINECS: 245-261-3
CN: *N*-L-α-aspartyl-L-phenylalanine 1-methyl ester

L-aspartic acid → L-aspartic anhydride hydrochloride → Aspartame

L-phenylalanine methyl ester

(along with aspartame are formed up to 20% β-isomer, separated by crystallization)

Reference(s):
DE 1 692 768 (Searle & Co.; prior. 16.2.1968).
Mazur, R.H. et al.: J. Am. Chem. Soc. (JACSAT) **91**, 264 (1969).
Ariyoshi, Y. et al.: Bull. Chem. Soc. Jpn. (BCSJA8) **46**, 1893 (1973).
DE 2 104 620 (Ajinomoto; appl. 1.2.1971; J-prior. 31.1.1970, 23.2.1970, 26.6.1970).
DAS 2 152 111 (Ajinomoto; appl. 19.10.1971; J-prior. 26.10.1970).
DAS 2 233 535 (Ajinomoto; appl. 7.7.1972; J-prior. 9.7.1972).
US 3 492 131 (Searle; 27.1.1970; appl. 18.4.1966).

alternative synthesis:
US 4 238 392 (Pfizer; 9.12.1980; appl. 29.10.1979).
US 4 173 562 (Monsanto; 6.11.1979; prior. 27.12.1976, 31.3.1978).
EP 143 881 (Gema; appl. 6.7.1984; CH-prior. 7.9.1983).

fermentative preparation from L-aspartic acid *and* L-phenylalanine methyl ester:
US 4 506 011 (Toyo Soda; 19.3.1985; J-prior. 5.9.1981; 14.10.1981, 18.1.1982).

purification:
US 3 798 207 (Ajinomoto; 19.3.1974; J-prior. 26.10.1970).

Formulation(s):   eff. tabl. 20 mg; tabl. 18 mg

Trade Name(s):

D:	Canderel (Wander)	Bil Aspatame dolaf		Hermesetas Gold
F:	Candérel (Monsanto; 1980)	(Pietrasanta)		(Milanfarma)
I:	Aspardolce Dolafic	Dietoman aspartame		Snel Miel (Fea)
	(Ganassini)	(Sterling Midy)		Suaviter (Boehringer
	Asparel Dietason	Dolcor aspartame		Mannh.)
	(Formenti)	(Gazzoni)		Tac Aspartame (Also)
	Aspartina (Ilex)	Futura aspartame	J:	Pal-Sweet (Ajinomoto)
		(Farmacologico Milanese)		

L-Aspartic acid
(L-2-Aminosuccinic acid; Acide aspartique)

Use: non-essential proteinogenic amino acid (for infusion solutions and as salt former)

RN: 56-84-8   MF: $C_4H_7NO_4$ MW: 133.10   EINECS: 200-291-6
LD_{50}:   6 g/kg (M, i.p.)
CN: L-aspartic acid

monopotassium salt
RN: 1115-63-5   MF: $C_4H_6KNO_4$ MW: 171.19   EINECS: 214-226-4
monosodium salt
RN: 3792-50-5   MF: $C_4H_6NNaO_4$ MW: 155.09   EINECS: 223-264-0
magnesium salt (2:1)
RN: 2068-80-6   MF: $C_8H_{12}MgN_2O_8$ MW: 288.50   EINECS: 218-191-6
magnesium salt (2:1) tetrahydrate
RN: 7018-07-7   MF: $C_8H_{12}MgN_2O_8 \cdot 4H_2O$ MW: 360.56

ammonium fumarate   L-Aspartic acid

Reference(s):
with immovable aspartase:
Tosa, T. et al.: Biotechnol. Bioeng. (BIBIAU) **15**, 69 (1973).

with immovable E. coli (ATCC 11303):

Sato, T. et al.: Biotechnol. Bioeng. (BIBIAU) **17**, 1797 (1975).
US 3 791 926 (Tanabe; 12.2.1974; J.-prior. 28.10.1971).
US 4 138 292 (Tanabe, 6.2.1979; J-prior. 2.7.1976).
Fusee, M.C. et al.: Appl. Environ. Microbiol. (AEMIDF) **42**, 672 (1981).
US 4 436 813 (Purification Engineering; 13.3.1984; appl. 16.3.1982).
US 4 560 653 (Grace; 24.12.1985; appl. 6.6.1983).
EP 110 422 (Tanabe; appl. 2.12.1983; J.-prior. 3.12.1982).

Formulation(s):   many different formulations

Trade Name(s):

D:	Eubiol (Chephasaar) numerous combination preparations	F:	Mégamag (Mayoly-Spindler; as magnesium salt) Sargenon (ASTA Medica; with arginine)		numerous combination preparations
				I:	Oral K (Sclavo)-comb. Polase (Wyeth)-comb.

Aspoxicillin

(TA-058)

ATC: J01

Use: semisynthetic penicillin (for parenteral administration), derivative of amoxicillin

RN: 63358-49-6   MF: $C_{21}H_{27}N_5O_7S$ MW: 493.54

LD_{50}:   >10 g/kg (M, i.v.)

CN: [2*S*-(2α,5α,6β)]-*N*-methyl-D-asparaginyl-*N*-(2-carboxy-3,3-dimethyl-7-oxo-4-thia-1-azabicyclo[3.2.0]hept-6-yl)-D-2-(4-hydroxyphenyl)glycinamide

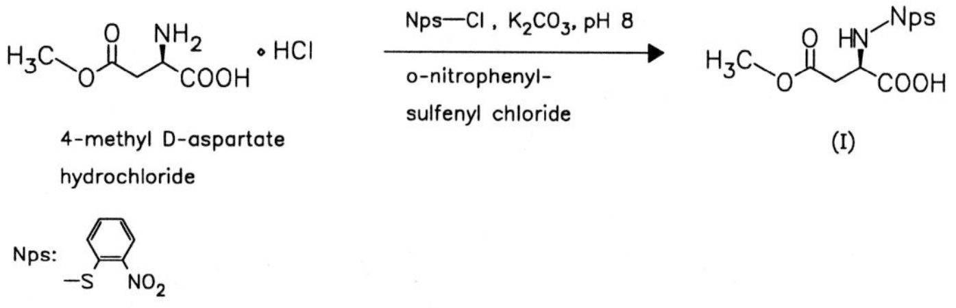

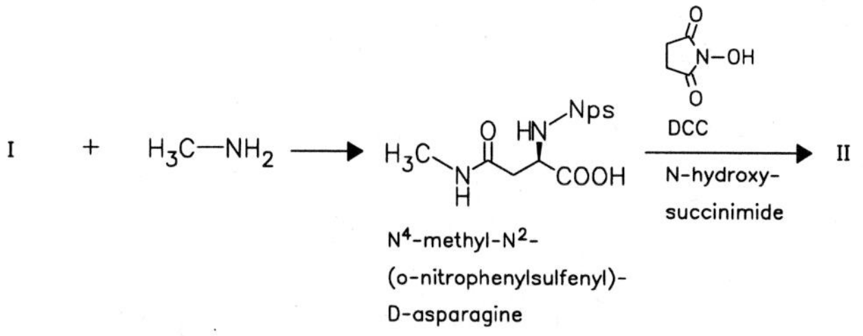

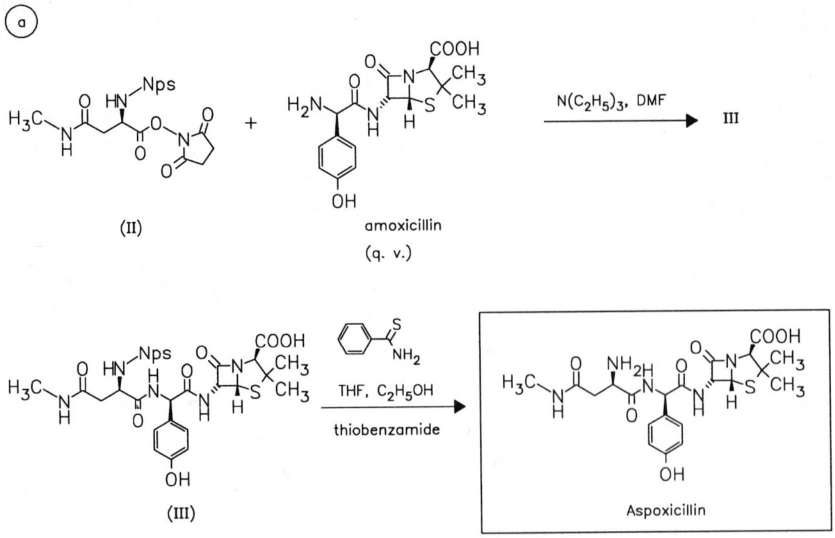

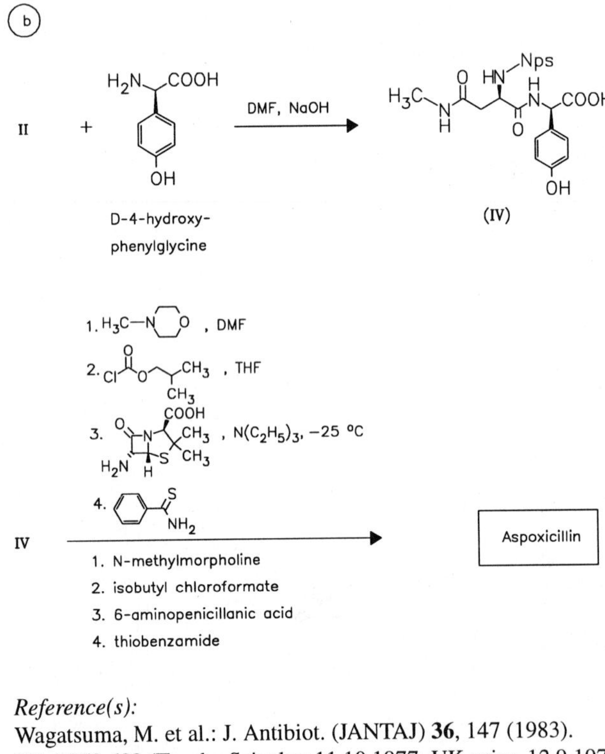

a

H_3C ... (II) + amoxicillin (q. v.) $\xrightarrow{N(C_2H_5)_3,\ DMF}$ III

(III) + thiobenzamide $\xrightarrow{THF,\ C_2H_5OH}$ Aspoxicillin

b

II + D-4-hydroxy-phenylglycine $\xrightarrow{DMF,\ NaOH}$ (IV)

1. $H_3C-N\!\!\bigcirc\!\!O$, DMF

2. $Cl-C(O)-O-CH(CH_3)CH_3$, THF

3. 6-aminopenicillanic acid , $N(C_2H_5)_3$, $-25\ ^{\circ}C$

4. thiobenzamide

IV $\longrightarrow$ Aspoxicillin

1. N-methylmorpholine
2. isobutyl chloroformate
3. 6-aminopenicillanic acid
4. thiobenzamide

Reference(s):

Wagatsuma, M. et al.: J. Antibiot. (JANTAJ) **36**, 147 (1983).
US 4 053 609 (Tanabe Seiyaku; 11.10.1977; UK-prior. 12.9.1975; J-prior. 27.12.1975, 29.12.1975).
GB 1 533 413 (Tanabe Seiyaku; appl. 12.9.1975 and 16.8.1976).
GB 1 533 414 (Tanabe Seiyaku; appl. 3.12.1976; J-prior. 27.12.1975).
DOS 2 638 067 (Tanabe Seiyaku; appl. 24.8.1976; GB-prior. 12.9.1975, 27.12.1975, 29.12.1975).

purification:
US 4 313 875 (Tanabe Seiyaku; 2.2.1982; J-prior. 11.9.1979).
EP 25 233 (Tanabe Seiyaku; appl. 10.9.1980; J-prior. 11.9.1979).

trihydrate:
US 4 866 170 (Tanabe Seiyaku; 12.9.1989; J-prior. 24.9.1986).
EP 261 823 (Tanabe Seiyaku; appl. 3.9.1987; J-prior. 24.9.1986).

lyophilized preparation:
US 4 966 899 (Tanabe Seiyaku; 30.10.1990; J-prior. 14.1.1987).

Formulation(s): powder in vial 1 g, 2 g

Trade Name(s):
J: Doyle (Tanabe; 1987)

Astemizole

(R 43512)

ATC: R06AX11
Use: antihistaminic, antiallergic

RN: 68844-77-9 MF: $C_{28}H_{31}FN_4O$ MW: 458.58 EINECS: 272-441-9
LD_{50}: 35 mg/kg (M, i.v.); 2560 mg/kg (M, p.o.);
28.2 mg/kg (R, i.v.); >2560 mg/kg (R, p.o.);
21.8 mg/kg (dog, i.v.); >320 mg/kg (dog, p.o.)
CN: 1-[(4-fluorophenyl)methyl]-*N*-[1-[2-(4-methoxyphenyl)ethyl]-4-piperidinyl]-1*H*-benzimidazol-2-amine

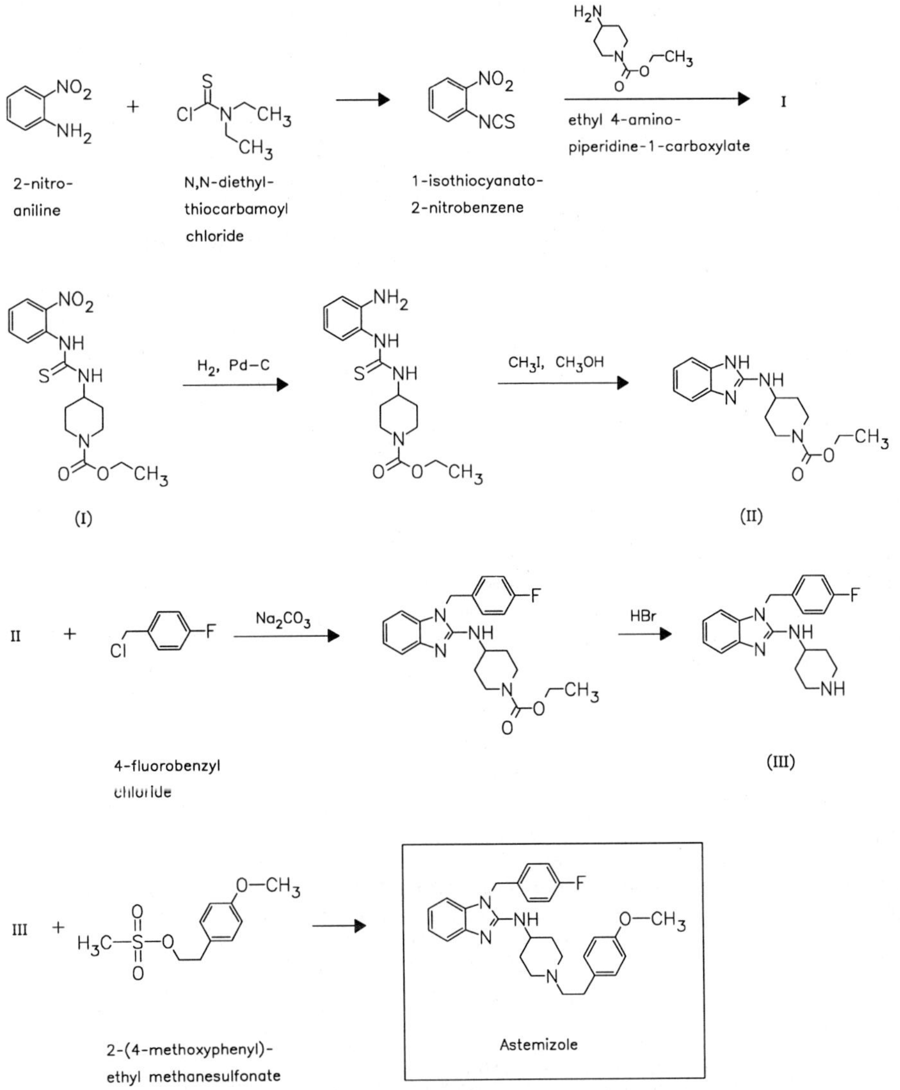
2-nitro-
aniline
N,N-diethyl-
thiocarbamoyl
chloride
1-isothiocyanato-
2-nitrobenzene
ethyl 4-amino-
piperidine-1-carboxylate
I
(I)
H2, Pd-C
CH3I, CH3OH
(II)
II
Na2CO3
4-fluorobenzyl
chloride
HBr
(III)
III
2-(4-methoxyphenyl)-
ethyl methanesulfonate
Astemizole

Reference(s):
US 4 219 559 (Janssen; 26.8.1980; prior. 10.1.1979).
EP 5 318 (Janssen; appl. 30.3.1979; USA-prior. 10.1.1979; 3.4.1978).

synthesis of 1-isothiocyanato-2-nitrobenzene:
Sayigh, A.A.R.: J. Org. Chem. (JOCEAH) **30**, 2465 (1965).

*synthesis of N,N-*diethylthiocarbamoyl chloride:
Goerdeler, J.; Luedke, H.: Chem. Ber. (CHBEAM) **103**, 3393 (1970).
v. Braun: Ber. Dtsch. Chem. Ges. (BDCGAS) **36**, 2274 (1903).
Billeter: Ber. Dtsch. Chem. Ges. (BDCGAS) **26**, 1686 (1893).
Goshorn et al.: Org. Synth. (ORSYAT) **35**, 55 (1955).
US 2 466 276 (Sharples Chemicals Inc.; 5.4.1949; appl. 2.2.1946).

Formulation(s):   drops 2 mg/ml; susp. 30 ml (0.2 %); tabl. 10 mg

Trade Name(s):

D:	Hismanal (Janssen-Cilag; 1985)	GB:	Hismanal (Janssen-Cilag; 1983)	J:	Hismanal (Mochida)
F:	Hismanal (Janssen-Cilag; 1986)	I:	Hismanal (Janssen; 1987) Histamen (Polifarma)	USA:	Hismanal (Janssen)

Astromicin

(Fortimicin A)

ATC: J01G
Use: aminoglycoside antibiotic

RN: 55779-06-1   MF: $C_{17}H_{35}N_5O_6$ MW: 405.50
LD$_{50}$:   380 mg/kg (M, i.v.); 400 mg/kg (M, p.o.)
CN: 4-amino-1-[(aminoacetyl)methylamino]-1,4-dideoxy-3-*O*-(2,6-diamino-2,3,4,6,7-pentadeoxy-β-L-*lyxo*-heptopyranosyl)-6-*O*-methyl-L-*chiro*-inositol

sulfate (1:2)
RN: 72275-67-3   MF: $C_{17}H_{35}N_5O_6 \cdot 2H_2O_4S$ MW: 601.65
LD$_{50}$:   94 mg/kg (M, i.v.); 13.6 g/kg (M, p.o.);
86 mg/kg (R, i.v.); >10 g/kg (R, p.o.);
214 mg/kg (dog, i.v.)

Astromicin

Preparation by fermentation of *Micromonospora olivoasterospora* FERM-P1560 (identical with *Micromonospora sp.* MK-70; ATCC 31009 and ATCC 31010) and isolation/purification on ion-exchanger and column chromatography.

Reference(s):
Nara, T. et al.: J. Antibiot. (JANTAJ) **30**, 533 (1977).
Okachi, R. et al.: J. Antibiot. (JANTAJ) **30**, 541 (1977).
US 3 976 768 (Abbott; 24.8.1976; appl. 22.7.1974; J.-prior. 23.7.1973).
GB 1 473 356 (Abbott; appl. 22.7.1974; J.-prior. 23.7.1973).
FR 2 238 502 (Kyowa Hakko; appl. 22.7.1974; J.-prior. 23.7.1973).
DE 2 435 160

structure:
Egan, R.S. et al.: J. Antibiot. (JANTAJ) **30**, 552 (1977).

Formulation(s): amp. 200 mg

Trade Name(s):
J: Fortimicin (Kyowa Hakko; 1985)

Atenolol

ATC: C07AA; C07AB03
Use: antiadrenergic (β-receptor), antihypertensive

RN: 29122-68-7 MF: $C_{14}H_{22}N_2O_3$ MW: 266.34 EINECS: 249-451-7
LD_{50}: >57 mg/kg (M, i.v.); 2 g/kg (M, p.o.);
77 mg/kg (R, i.v.); 2 g/kg (R, p.o.)
CN: 4-[2-hydroxy-3-[(1-methylethyl)amino]propoxy]benzeneacetamide

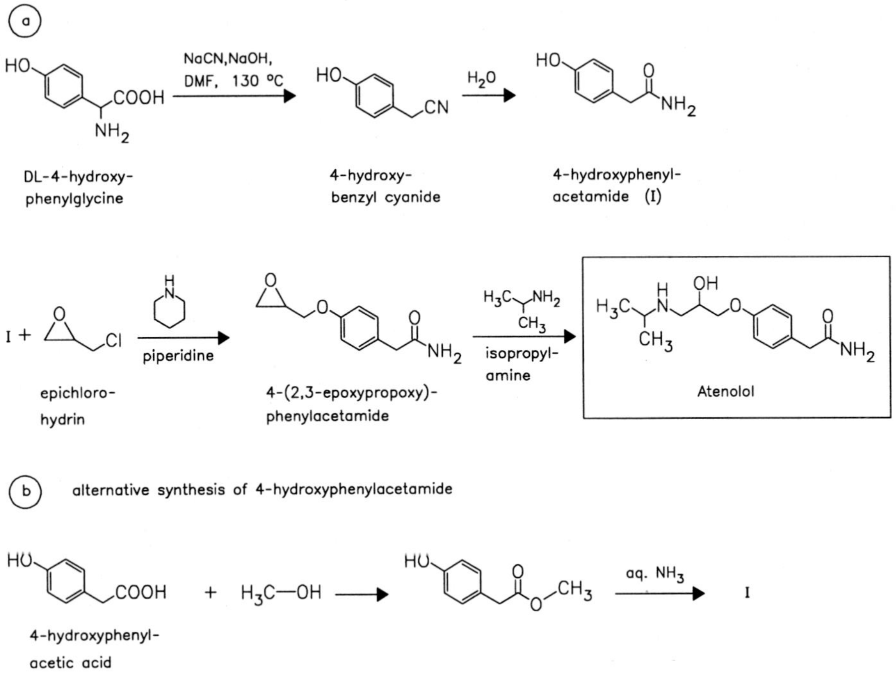
a
NaCN,NaOH, DMF, 130 °C
H2O
DL-4-hydroxyphenylglycine
4-hydroxybenzyl cyanide
4-hydroxyphenylacetamide (I)
piperidine
isopropylamine
epichlorohydrin
4-(2,3-epoxypropoxy)-phenylacetamide
Atenolol
b
alternative synthesis of 4-hydroxyphenylacetamide
aq. NH3
4-hydroxyphenylacetic acid

Reference(s):
US 3 934 032 (ICI; 20.1.1976; prior. 10.3.1972).
US 3 663 607 (ICI; 16.5.1972; GB-prior. 21.2.1969).
US 3 836 671 (ICI; 17.9.1974; GB-prior. 21.2.1969, 24.9.1969, 18.11.1970 and 19.11.1970).
DOS 2 007 751 (ICI; appl. 19.2.1970; GB-prior. 21.2.1969 and 24.9.1969).
GB 1 285 038 (ICI; appl. 21.2.1969; valid from 24.9.1969).

alternative synthesis:
GB 1 391 444 (ICI; appl. 13.7.1971; valid from 19.6.1972).

4-hydroxybenzyl cyanide:
GB 1 522 477 (ICI; appl. 13.8.1974; valid from 11.11.1975).
US 4 154 757 (ICI; 15.5.1979; appl. 22.5.1978).

Formulation(s):   amp. 5 mg/10 ml; f. c. tabl. 25 mg, 50 mg, 100 mg

Trade Name(s):

D:	Atebeta (betapharm)		Bêta-Adalate (Bayer)-		Tenoret-50 (Zeneca)-comb.
	duratenol (durachemie)		comb.		Tenoretic (Zeneca)-comb.
	Falitonsin (ASTA Medica		Tenordate (Zeneca)-comb.		Tenormin (Zeneca; 1976)
	AWD)		Ténormine (Zeneca; 1979)		Totamol (CP Pharm.)
	Tenormin (Zeneca; 1976)	GB:	Beta-adalat (Bayer)-comb.	J:	Tenormin (ICI-Sumitomo
	Tri-Normin (Zeneca; 1984)		Kalten (Zeneca)-comb.		Chem.; 1984)
F:	Betatop (EG Labo)		Tenben (Galen)-comb.	USA:	Tenoretic (Zeneca; 1984)
			Tenif (Zeneca)-comb.		Tenormin (Zeneca; 1981)

Atorvastatin calcium
(CI-981; YM-548)

ATC: C10AA05
Use: hyperlipidemic, HMG-CoA-
reductase inhibitor

RN: 134523-03-8   MF: $C_{66}H_{68}CaF_2N_4O_{10}$ MW: 1155.36
CN: [R-(R^*,R^*)]-2-(4-fluorophenyl)-β,δ-dihydroxy-5-(1-methylethyl)-3-phenyl-4-[(phenylamino)carbonyl]-
1H-pyrrole-1-heptanoic acid calcium salt (2:1)

free acid
RN: 134523-00-5   MF: $C_{33}H_{35}FN_2O_5$ MW: 558.65

2-isobutyryl-
acetanilide

benz-
aldehyde

2-benzylidene-4-
methyl-3-oxo-N-
phenylpentanamide

4-fluorobenzaldehyde,
thiazolium salt

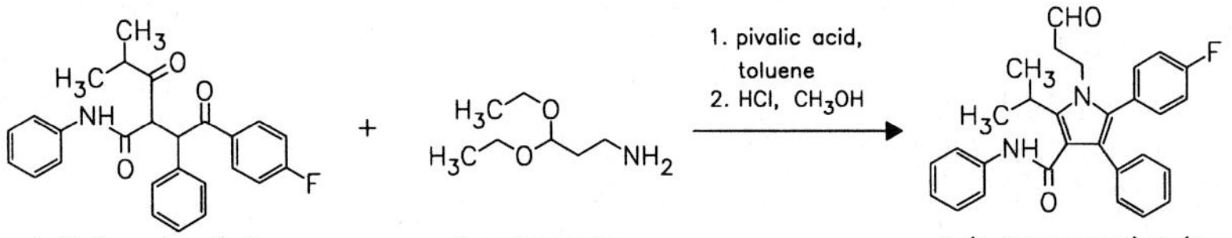

4-(4-fluorophenyl)-2-isobutyryl-3-phenyl-4-oxo-N-phenyl-butyramide (I)

3-aminopropion-aldehyde diethyl acetal

5-(4-fluorophenyl)-2-(1-methylethyl)-1-(3-oxo-propyl)-N,4-diphenyl-1H-pyrrole-3-carbox-amide (II)

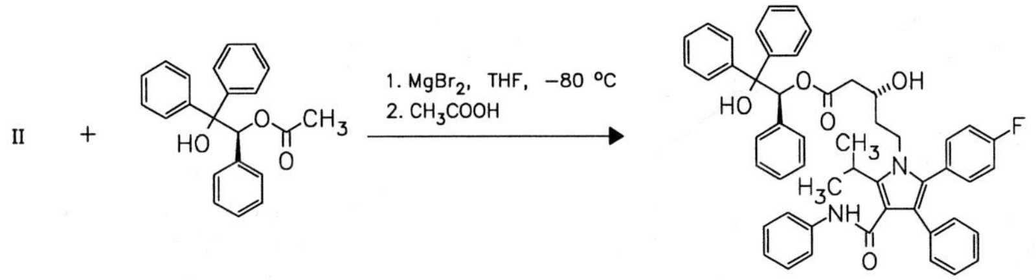

(S)-(+)-2-acetoxy-1,1,2-triphenyl-ethanol

(III)

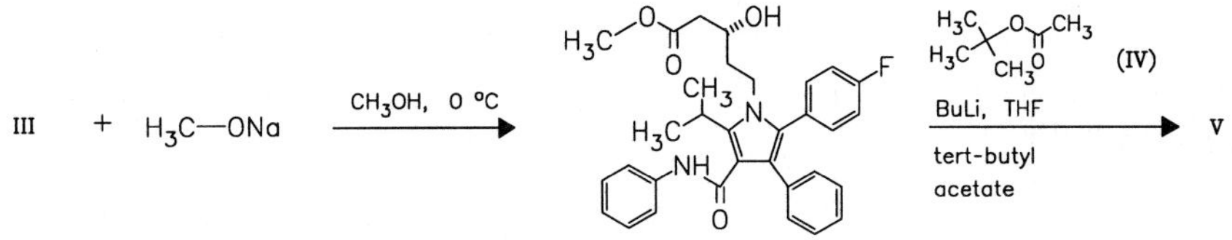

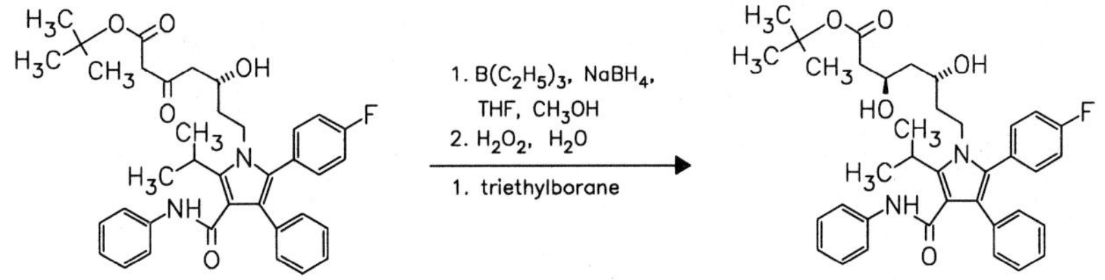

(V)

(VI)

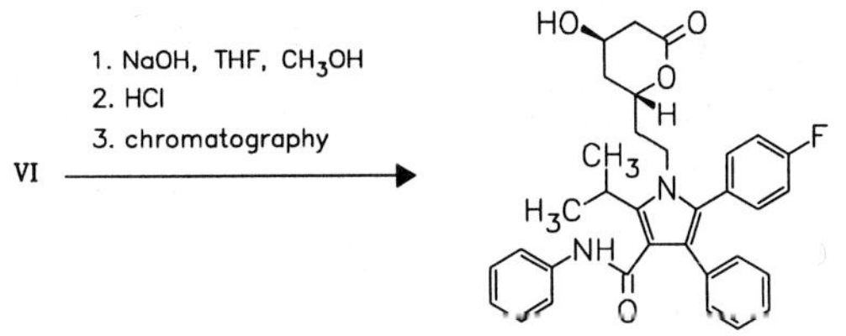

(2R-trans)-5-(4-fluoro-phenyl)-2-(1-methylethyl)-N,4-diphenyl-1-[2-(tetra-hydro-4-hydroxy-6-oxo-2H-pyran-2-yl)ethyl]-1H-pyrrole-3-carboxamide (VII)

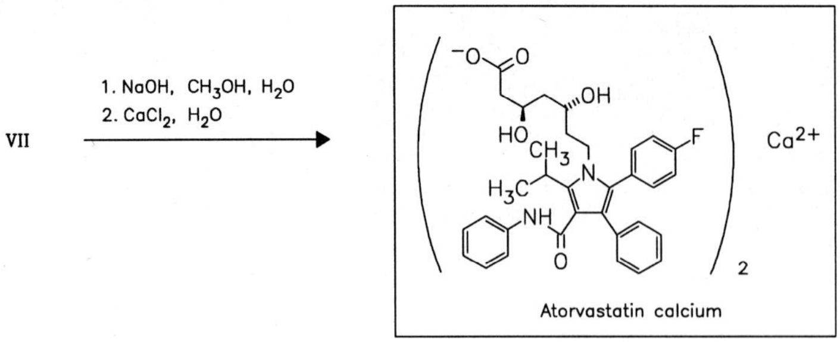

VII
$\xrightarrow{\text{1. NaOH, CH}_3\text{OH, H}_2\text{O}}$
2. CaCl$_2$, H$_2$O

Atorvastatin calcium

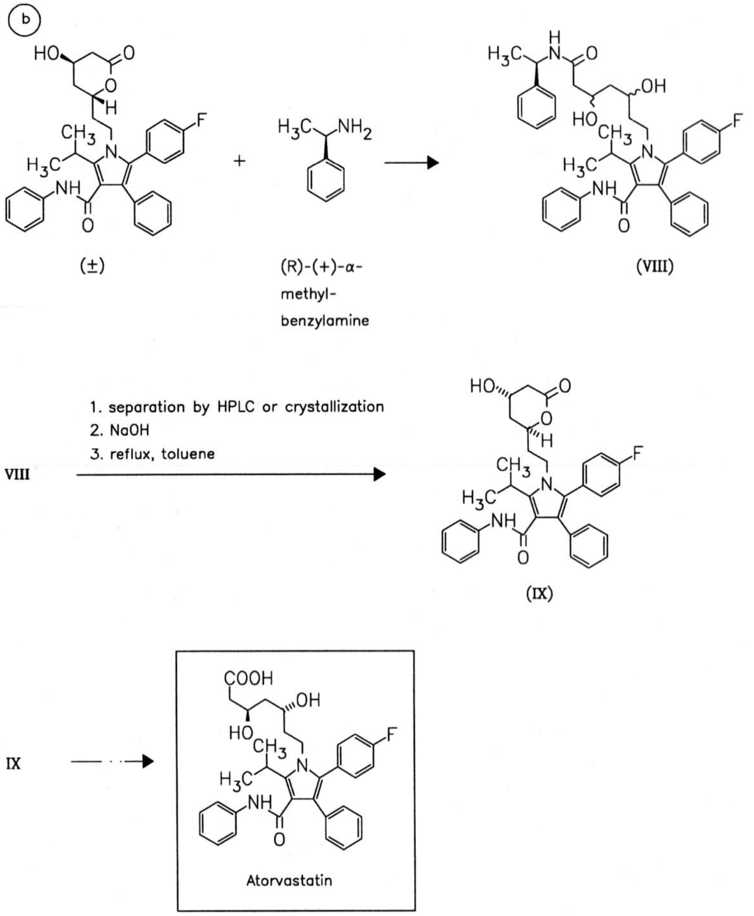

(b)

(±)   +   (R)-(+)-α-methyl-benzylamine $\rightarrow$ (VIII)

VIII
$\xrightarrow{\text{1. separation by HPLC or crystallization}}$
2. NaOH
3. reflux, toluene

(IX)

IX $\dashrightarrow$

Atorvastatin

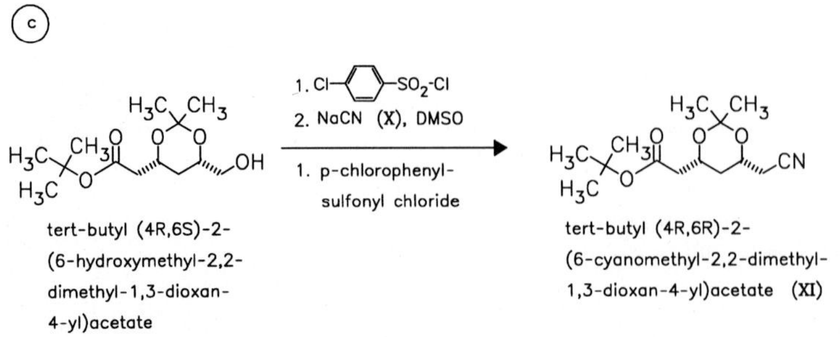

(c)

$\xrightarrow{\text{1. Cl—}\bigcirc\text{—SO}_2\text{-Cl}}$
2. NaCN (X), DMSO

1. p-chlorophenyl-sulfonyl chloride

tert-butyl (4R,6S)-2-(6-hydroxymethyl-2,2-dimethyl-1,3-dioxan-4-yl)acetate

tert-butyl (4R,6R)-2-(6-cyanomethyl-2,2-dimethyl-1,3-dioxan-4-yl)acetate (XI)

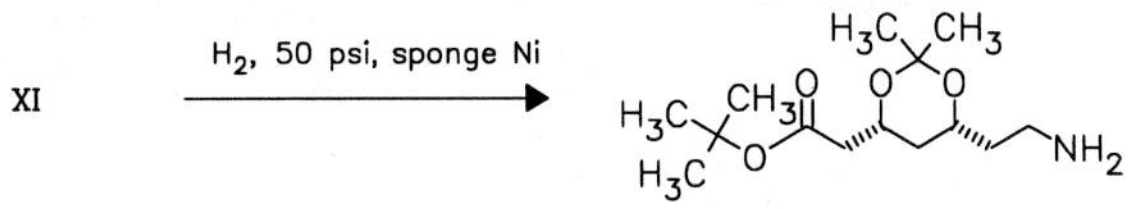

XI $\xrightarrow{\text{H}_2,\ 50\ \text{psi, sponge Ni}}$

tert-butyl (4R,6R)-2-
[6-(2-aminoethyl)-2,2-
dimethyl-1,3-dioxan-4-
yl]acetate (XII)

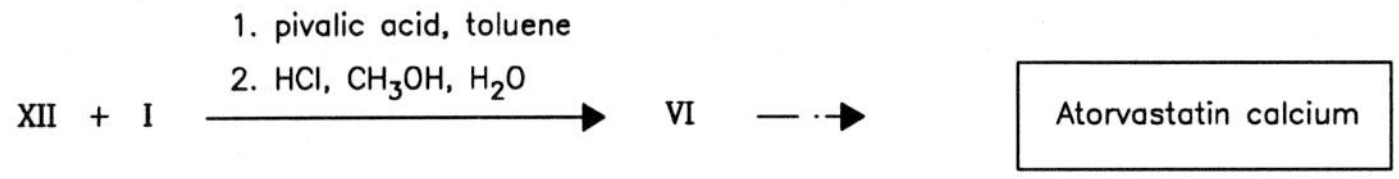

XII + I $\xrightarrow[\text{2. HCl, CH}_3\text{OH, H}_2\text{O}]{\text{1. pivalic acid, toluene}}$ VI — ·→ │ Atorvastatin calcium │

(ca) alternative synthesis of intermediate XI:

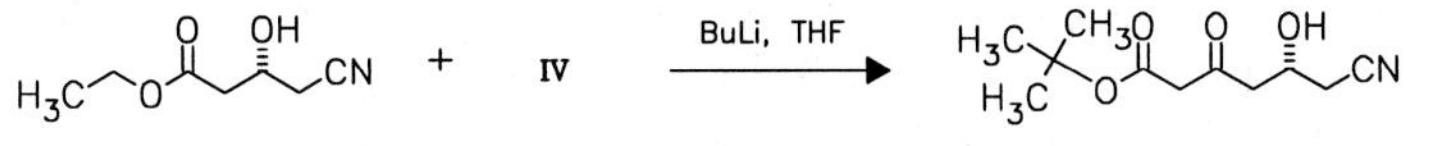

ethyl 3(R)-hydroxy-
4-cyanobutyrate (XIII)

+ IV $\xrightarrow{\text{BuLi, THF}}$

tert-butyl 5(R)-hydroxy-
6-cyano-3-oxohexanoate (XIV)

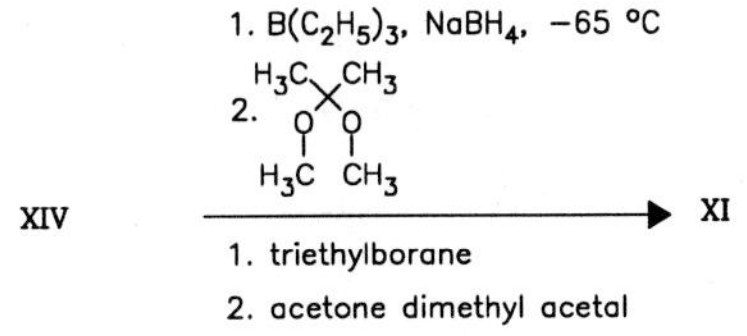

XIV $\xrightarrow[\substack{\text{1. triethylborane} \\ \text{2. acetone dimethyl acetal}}]{\substack{\text{1. B(C}_2\text{H}_5)_3,\ \text{NaBH}_4,\ -65\ ^\circ\text{C} \\ \text{2.}}}$ XI

(cb) synthesis of the starting material XIII:

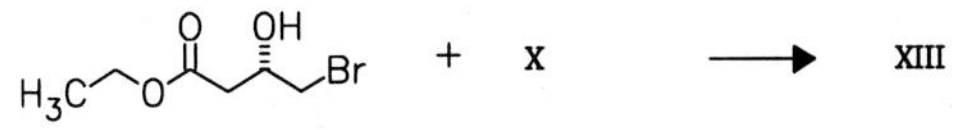

ethyl 4-bromo-3(S)-
hydroxybutanoate

+ X $\longrightarrow$ XIII

(d) alternative synthesis of intermediate II:

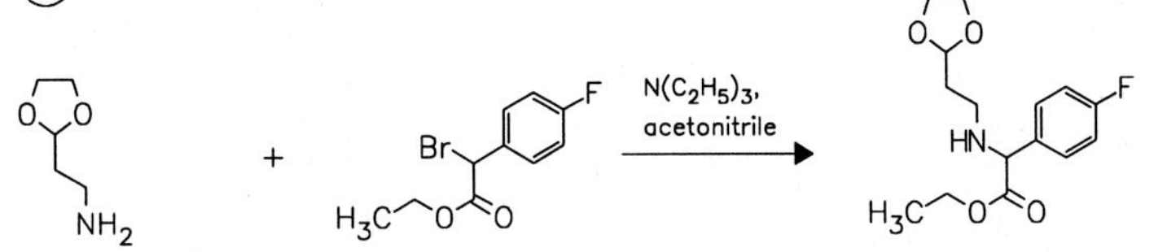

2-(1,3-dioxolan-
2-yl)ethylamine

+

ethyl 2-bromo-
2-(4-fluorophenyl)-
acetate

$\xrightarrow{\substack{\text{N(C}_2\text{H}_5)_3, \\ \text{acetonitrile}}}$

ethyl 2-[2-(1,3-dioxolan-
2-yl)ethylamino]-2-(4-
fluorophenyl)acetate (XV)

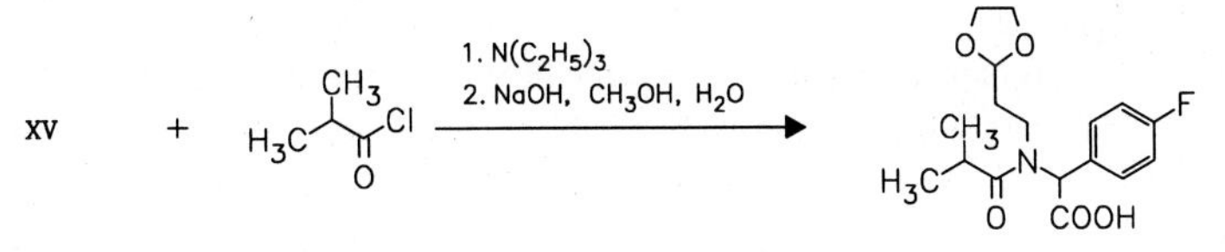

XV + $\underset{\text{chloride}}{\underset{\text{isobutyryl}}{}}$ H$_3$C CH$_3$ Cl $\xrightarrow[\text{2. NaOH, CH}_3\text{OH, H}_2\text{O}]{\text{1. N(C}_2\text{H}_5)_3}$

(XVI)

XVI   +

N,3-diphenyl-
propynamide

DCC, 90 °C,
(CH₃CO)₂O

HCl   →   II

1-[2-(1,3-dioxolan-2-
yl)ethyl]-5-(4-fluoro-
phenyl)-2-(1-methyl-
ethyl)-N,4-diphenyl-
1H-pyrrole-3-carb-
oxamide

Reference(s):
a,b US 4 681 893 (Warner-Lambert; appl. 21.7.1987; USA-prior. 30.5.1986).
EP 409 281 (Warner-Lambert; appl. 23.1.1991; USA-prior. 21.7.1989, 26.2.1991).
EP 680 320 (Warner-Lambert; appl. 8.11.1995; USA-prior. 19.1.1993).
c Naeminga, T. et al.: Tetrahedron Lett. (TELEAY) **33**, 2279-2282 (1992).
WO 9 703 960 (Warner-Lambert; appl. 6.2.1997; USA-prior. 17.7.1995).
ca Baumann, K.L. et al.: Tetrahedron Lett. (TELEAY) **33**, 2283-2284 (1992).
cb Isbell, H. et al.: Carbohydr. Res. (CRBRAT) **72**, 301-304 (1972).
d Roth, B.D. et al.: J. Med. Chem. (JMCMAR) **34**, 357-366 (1991).
preparation of N,3-diphenylpropynamide:
Gadwhal, S. et al.: Indian J. Chem., Sect. B: Org. Chem. Incl. Med. Chem. (IJSBDB) **37B** (8), 725-727
(1998).

preparation of intermediates:
WO 9 932 434 (Warner-Lambert; appl. 2.12.1998; USA-prior. 19.12.1997).
WO 9 957 109 (Kaneka Corp.; appl. 28.4.1999; J-prior. 30.4.1998).
WO 9 804 543 (Warner-Lambert; appl. 1.7.1997; USA-prior. 29.7.1996).
US 5 155 251 (Warner-Lambert; appl. 13.10.1992; 11.10.1991).
US 5 103 024 (Warner-Lambert; prior. 17.10.1990).

new crystalline forms of atorvastatin:
WO 9 703 959 (Warner-Lambert; appl. 8.7.1996; USA-prior. 17.7.1995).
WO 9 703 958 (Warner-Lambert; appl. 6.2.1997; USA-prior. 17.7.1995).

stable oral formulation:
WO 9 416 693 (Warner-Lambert; appl. 4.8.1994; USA-prior. 19.1.1993).

Formulation(s):   tabl. 10 mg, 20 mg, 40 mg

Trade Name(s):
D: Sotis (Parke Davis/   J: Lipitor (Warner-Lambert)   Xavator (Parke Davis)
       Gödecke; Mack, Illert.)   Torvast (Pfizer)   USA: Lipitor (Parke Davis;
GB: Lipitor (Parke Davis)   Tozalip (Guidotti)   Pfizer)

Atracurium besilate

ATC: M03AC04
Use: skeletal muscle relaxant

RN: 64228-81-5   MF: C₅₃H₇₂N₂O₁₂ · 2C₆H₅O₃S   MW: 1243.50   EINECS: 264-743-4
CN: 2,2'-[1,5-pentanediylbis[oxy(3-oxo-3,1-propanediyl)]]bis[1-[(3,4-dimethoxyphenyl)methyl]-1,2,3,4-
        tetrahydro-6,7-dimethoxy-2-methylisoquinolinium] dibenzenesulfonate

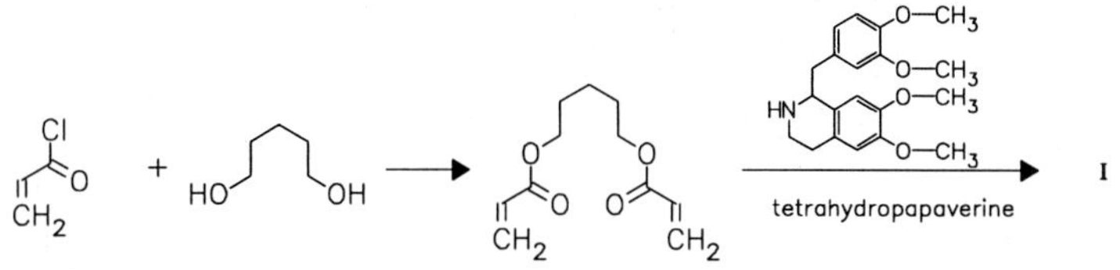

acryloyl chloride

pentane-1,5-diol

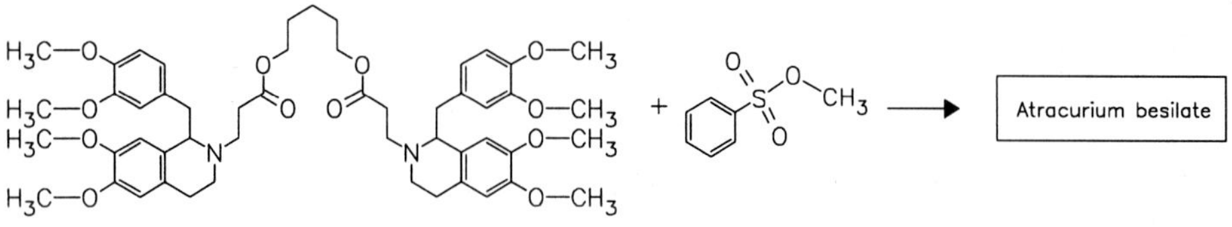

tetrahydropapaverine

(I)

methyl benzene-sulfonate

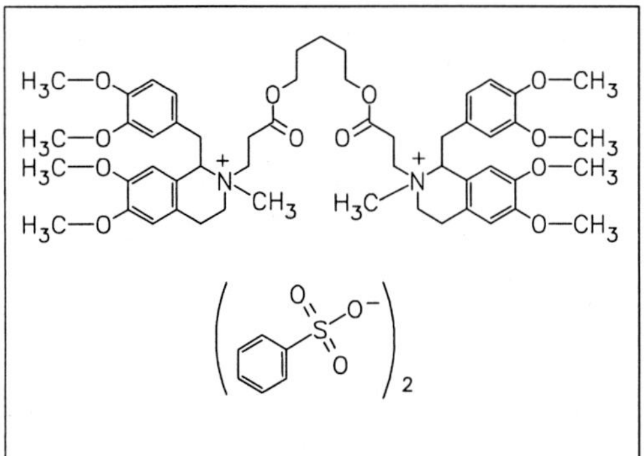

Atracurium besilate

Atracurium besilate

Reference(s):

DOS 2 655 883 (Wellcome; appl. 9.12.1976; GB-prior. 10.12.1975, 29.10.1976).
US 4 179 507 (Wellcome; 18.12.1979, GB-prior. 10.12.1975, 29.10.1976).
Stenkale, J.B. et al.: Eur. J. Med. Chem. (EJMCA5) **16**, 515 (1981).

Formulation(s): amp. 25 mg/2.5 ml, 50 mg/5 ml

Trade Name(s):

D:	Tracrium (Glaxo Wellcome; 1987)	F:	Tracrium (Glaxo Wellcome)	I:	Tracrium (Wellcome)	
		GB:	Tracrium (Wellcome; 1982)	USA:	Tracrium (Glaxo Wellcome; 1983)	

Atropine

(DL-Hyoscyamine)

ATC: A03BA01; S01FA01
Use: anticholinergic, mydriatic, antispasmodic

RN: 51-55-8 MF: $C_{17}H_{23}NO_3$ MW: 289.38 EINECS: 200-104-8
LD$_{50}$: 30 mg/kg (M, i.v.); 75 mg/kg (M, p.o.);
73 mg/kg (R, i.v.); 500 mg/kg (R, p.o.)
CN: *endo*-(±)-α-(hydroxymethyl)benzeneacetic acid 8-methyl-8-azabicyclo[3.2.1]oct-3-yl ester

borate (1:1)
RN: 51460-78-7   MF: $C_{17}H_{23}NO_3 \cdot H_3BO_3$ MW: 351.21
sulfate (2:1)
RN: 55-48-1   MF: $C_{17}H_{23}NO_3 \cdot 1/2H_2SO_4$ MW: 676.83   EINECS: 200-235-0
LD_{50}:   31 mg/kg (M, i.v.); 468 mg/kg (M, p.o.);
         37 mg/kg (R, i.v.); 600 mg/kg (R, p.o.);
         60 mg/kg (dog, i.v.)

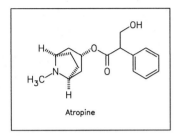

Atropine

By extraction of Solanacean drugs, especially *Atropa belladonna*, *Hyoscyamus niger* or other species. On careful extraction L-hyoscyamine is obtained first, which can be racemized to atropine by addition of alkali in ethanolic solution.

Reference(s):
Ullmanns Encykl. Tech. Chem., 3. Aufl., Vol. **3**, 201 f.
Ullmanns Encykl. Tech. Chem., 4. Aufl., Vol. **7**, 151.

Formulation(s):   amp. for inj. 100 mg; eye drops 10 mg; inj. sol. 0.25 mg, 0.5 mg, 1 mg, 2 mg; tabl. 0.5 mg

Trade Name(s):
D: Angiocardyl (Rhenomed)
       Atropin in der Ophtiole
       (Mann)
       Atropinol Augentropfen
       (Winzer)
       Atropin POS (Ursapharm)
       Borotropin Augentropfen
       (Winzer)
       Cansat (Sanofi Winthrop)
       generics and combination
       preparations
F: Atropine Aguettant
       (Aguettant)
       Atropine Lavoisier (Caix et
       du Marais)
       Atropine Martinet (CIBA
       Vision Ophthalmics)

GB: Chibro-Atropine (Merck
       Sharp & Dohme-Chibret)
       generics and numerous
       combination preparations
       Lomotil (Searle)-comb.
I: Atropina Aolfato (Scfm)
       Atropina Lux (Allergan; as
       sulfate)
       Atropi S (Formulario Naz.;
       Tariff. Nazionale; Bieffe
       Medital; Bioindustria;
       Collalto; Farge; Galenica
       (Senese); Jacobo Monicol;
       Ogna; Salf)
       Atrop S (Sifra)
       Atro S (Farge)
       Liotropina (SIFI; as sulfate)

       generics (Farmigea; Scfm)
       and combination
       preparations
J: generics
USA: Arco-Lase Plus (Arco)
       Atrohist Plus (Medeva)
       Bellatal (Richwood)
       Donnatal (Robins)
       Enlon-Plus (Ohmeda)
       Larox (Geneva)
       Lomotil (Searle)
       Motofen (Carnrick)
       Prosed/DS (Star)
       Urised (PolyMedica)

Atropine methonitrate
(Atropinmethylnitrat; Methylatropine Nitrate;
   Methonitrate d'atropine)

ATC: A03BB02
Use: anticholinergic, mydriatic,
       antispasmodic

RN: 52-88-0   MF: $C_{18}H_{26}N_2O_6$ MW: 366.41   EINECS: 200-156-1
LD_{50}:   9300 µg/kg (M, i.v.); 1320 mg/kg (M, p.o.);
         1902 mg/kg (R, p.o.)
CN: *endo*-(±)-3-(3-hydroxy-1-oxo-2-phenylpropoxy)-8,8-dimethyl-8-azoniabicyclo[3.2.1]octane nitrate

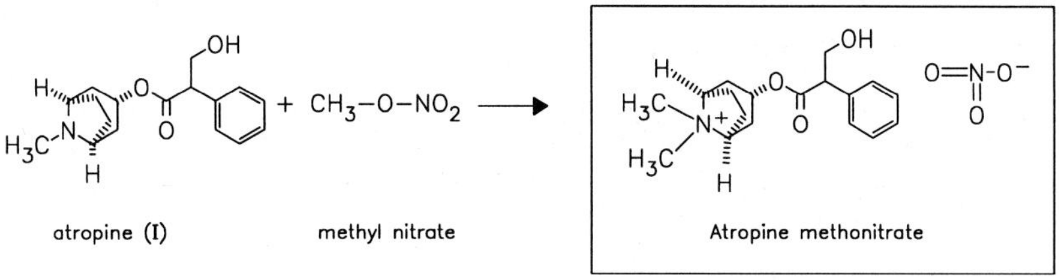
atropine (I)
methyl nitrate
Atropine methonitrate

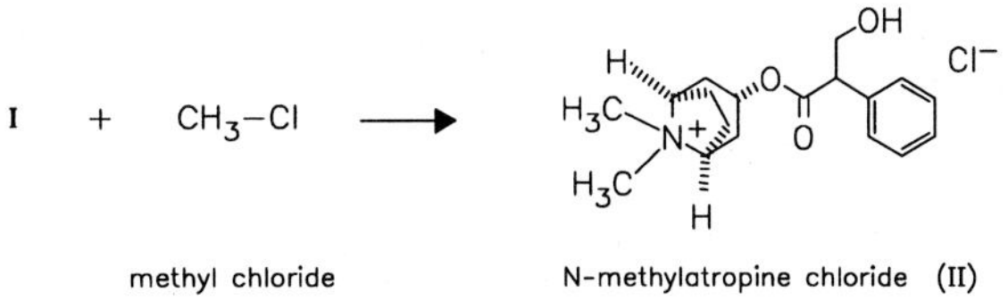
methyl chloride
N-methylatropine chloride (II)

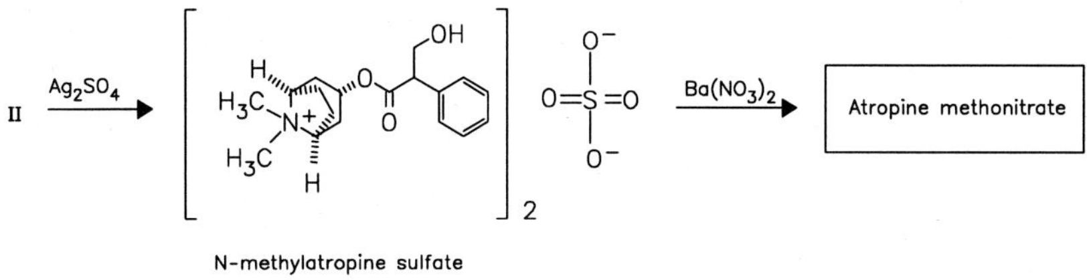
Ag2SO4
Ba(NO3)2
Atropine methonitrate
N-methylatropine sulfate

Reference(s):

DRP 137 622 (Bayer; 1901).
DRP 138 443 (Bayer; 1901).

Formulation(s): drops

Trade Name(s):

D: Afdosa (Hefa-Frenon)-comb.; wfm
Afpred (Hefa-Frenon)-comb.; wfm
Ansudoral (Basotherm)-comb.; wfm
Bronchovydrin Inhalationslösung (Searle-Endopharm)-comb.; wfm
Brox (Redel)-comb.; wfm
Myocardetten (Byk Gulden)-comb.; wfm
Myocardon (Byk Gulden)-comb.; wfm
Perphyllon (Homburg)-comb.; wfm
Tonaton (Luitpold)-comb.; wfm

USA: Festalan (Hoechst-Roussel)-comb.

Auranofin

ATC: M01CB03
Use: rheumatoid arthrosis therapeutic

RN: 34031-32-8 MF: $C_{20}H_{34}AuO_9PS$ MW: 678.52 EINECS: 251-801-9
LD_{50}: 310 mg/kg (M, p.o.);
265 mg/kg (R, p.o.)
CN: (2,3,4,6-tetra-*O*-acetyl-1-thio-β-D-glucopyranosato-*S*)(triethylphosphine)gold

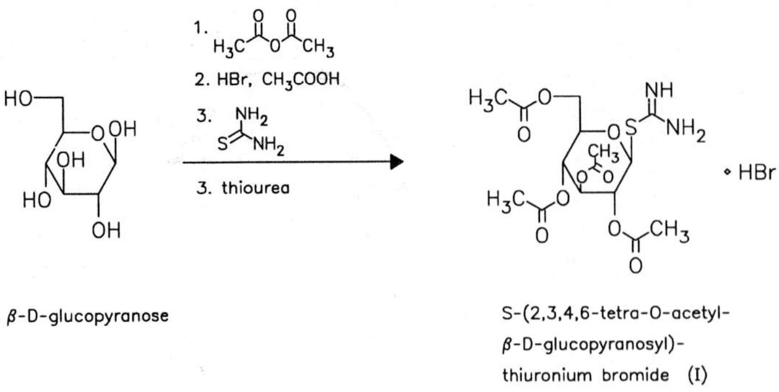

β-D-glucopyranose

S-(2,3,4,6-tetra-O-acetyl-
β-D-glucopyranosyl)-
thiuronium bromide (I)

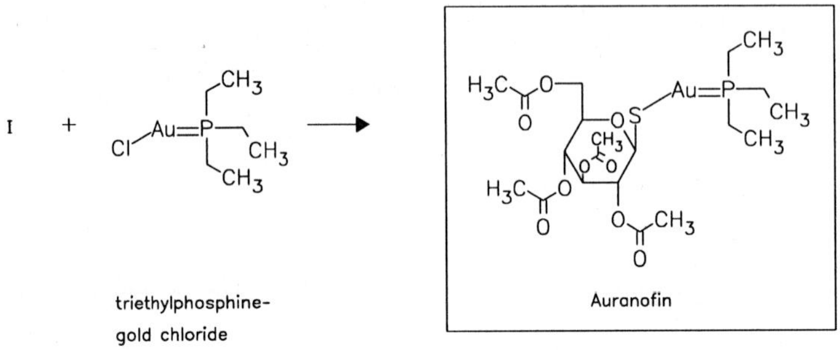

triethylphosphine-
gold chloride

Auranofin

Reference(s):

US 3 635 945 (Smith Kline & French; 18.1.1972; prior. 28.10.1969).
DE 2 051 495 (Smith Kline & French; appl. 20.10.1970; USA-prior. 28.10.1969).
US 3 708 579 (Smith Kline & French; 2.1.1973; prior. 28.10.1969, 1.10.1971).
Sutton, B.M. et al.: J. Med. Chem. (JMCMAR) **15**, 1095 (1972).

synthesis of S-(2,3,4,6-tetra-O-acetyl-β-D-glucopyranosyl)thiuronium bromide:
Bommer, W.A.; Kahn, J.R.: J. Am. Chem. Soc. (JACSAT) **73**, 2241 (1951).
DOS 2 215 653 (Konishiroku; appl. 30.3.1972).
Horton, D.: Methods Carbohydr. Chem. (MCACAL) **3**, 435 (1963).

Formulation(s):   f. c. tabl. 3 mg

Trade Name(s):

D:	Ridaura (Yamanouchi; 1982)	GB:	Ridaura (Yamanouchi; 1987)	J: Ridaura (Fujisawa; 1986)
F:	Ridauran (Robapharm)	I:	Ridaura (Smith Kline & French; 1984)	USA: Ridaura (SmithKline Beecham; 1985)

Azacitidine

ATC: L01BC
Use: antineoplastic

RN: 320-67-2   MF: $C_8H_{12}N_4O_5$ MW: 244.21   EINECS: 206-280-2
LD_{50}: 1159 mg/kg (M, i.v.); 572.3 mg/kg (M, p.o.)
CN: 4-amino-1-β-D-ribofuranosyl-1,3,5-triazin-2(1*H*)-one

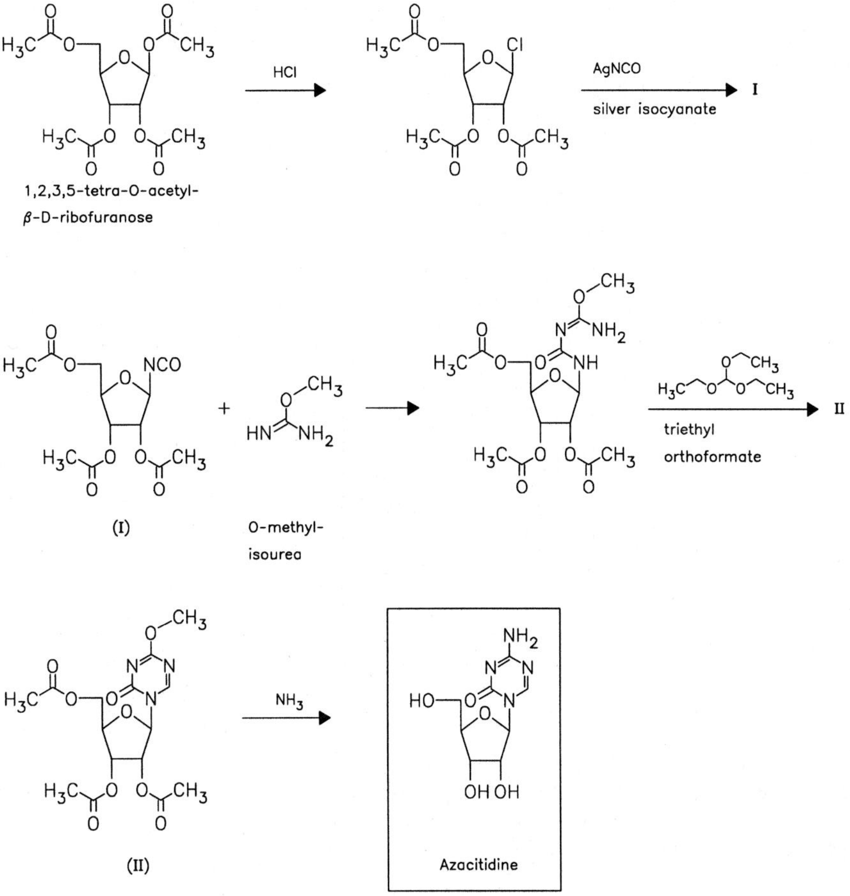
1,2,3,5-tetra-O-acetyl-
β-D-ribofuranose
HCl
AgNCO
silver isocyanate
I
(I)
O-methyl-
isourea
triethyl
orthoformate
II
(II)
NH3
Azacitidine

Reference(s):
Piskala, A.; Sorm, F.: Collect. Czech. Chem. Commun. (CCCCAK) **29**, 2060 (1964).
US 3 350 388 (F. Sorm, A. Piskala; 1967; prior. 1965).

formation from Streptoverticillium ladakanus:
Hanka, L.J. et al.: Antimicrob. Agents Chemother. (AACHAX) **1966**, 619.

isolation and structure elucidation:
Bergy, M.E.; Herr, R.R.: Antimicrob. Agents Chemother. (AACHAX) **1966**, 625.

Trade Name(s):
USA: Mylosar (Upjohn); wfm

Azacosterol

(Diazasterol)

Use: cholesterol depressant

RN: 313-05-3 MF: $C_{25}H_{44}N_2O$ MW: 388.64
LD_{50}: 90 mg/kg (M, p.o.)
CN: (3β,17β)-17-[[3-(dimethylamino)propyl]methylamino]androst-5-en-3-ol

dihydrochloride
RN: 1249-84-9 MF: $C_{25}H_{44}N_2O \cdot 2HCl$ MW: 461.56
LD_{50}: 380 mg/kg (M, p.o.);
470 mg/kg (R, p.o.)

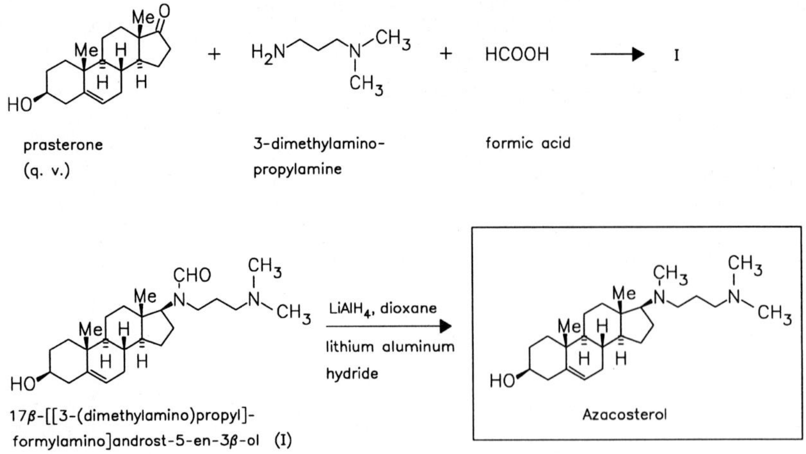
+
+
HCOOH
I
prasterone
(q. v.)
3-dimethylamino-
propylamine
formic acid
LiAlH4, dioxane
lithium aluminum
hydride
17β-[[3-(dimethylamino)propyl]-
formylamino]androst-5-en-3β-ol (I)
Azacosterol

Reference(s):
US 3 084 156 (Searle; 2.4.1963; prior. 30.11.1961, 28.3.1961).
Counsell, R.E. et al.: J. Med. Pharm. Chem. (JMPCAS) **5**, 1224 (1962).

Trade Name(s):
USA: Ornitrol (Searle); wfm

Azacyclonol

Use: anxiolytic

RN: 115-46-8 MF: $C_{18}H_{21}NO$ MW: 267.37 EINECS: 204-092-5
LD_{50}: 177 mg/kg (M, i.v.); 650 mg/kg (M, p.o.)
CN: α,α-diphenyl-4-piperidinemethanol

hydrochloride
RN: 1798-50-1 MF: $C_{18}H_{21}NO \cdot HCl$ MW: 303.83 EINECS: 217-284-9
LD_{50}: 121 mg/kg (M, i.v.); 650 mg/kg (M, p.o.)

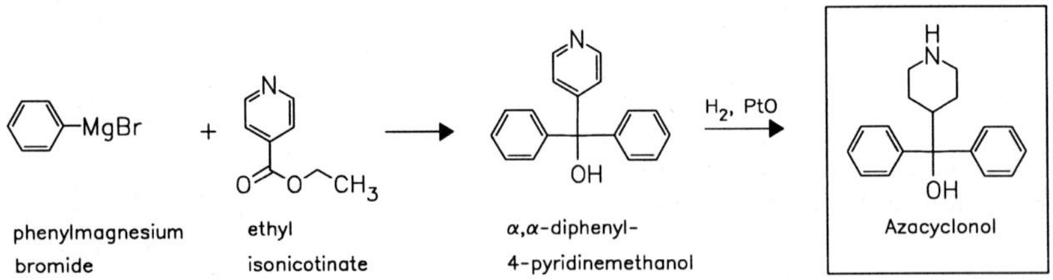
+
H2, PtO
phenylmagnesium
bromide
ethyl
isonicotinate
α,α-diphenyl-
4-pyridinemethanol
Azacyclonol

Reference(s):
US 2 804 422 (Merrell; 1957; prior. 1954).

Formulation(s): amp. 5 mg/ml (as hydrochloride); tabl. 20 mg

Trade Name(s):
F: Frenquel (Merrell) J: Frenquel (Shionogi)

Azapetine

ATC: C04AX30
Use: sympatholytic, vasodilator

RN: 146-36-1   MF: $C_{17}H_{17}N$ MW: 235.33   EINECS: 205-667-3
LD_{50}:   27 mg/kg (M, i.v.); 460 mg/kg (M, p.o.);
     50 mg/kg (dog, i.v.)
CN: 6,7-dihydro-6-(2-propenyl)-5H-dibenz[c,e]azepine

phosphate (1:1)
RN: 130-83-6   MF: $C_{17}H_{17}N \cdot H_3PO_4$ MW: 333.32   EINECS: 204-997-5
LD_{50}:   26 mg/kg (M, i.v.); 460 mg/kg (M, p.o.);
     50 mg/kg (dog, i.v.)

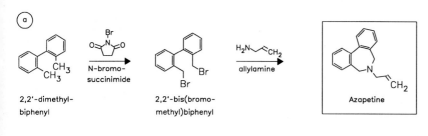

Reference(s):
a   US 2 619 484 (Hoffmann-La Roche; 1952; appl. 1950).
b   US 2 693 465 (Hoffmann-La Roche; 1954; appl. 1953).

Formulation(s):   tabl. 25 mg

Trade Name(s):
D: Ilidar (Roche); wfm

Azapropazone
(Apazone; Cinnopropazone)

ATC: M01AX04
Use: anti-inflammatory, analgesic

RN: 13539-59-8   MF: $C_{16}H_{20}N_4O_2$ MW: 300.36   EINECS: 236-913-8
LD_{50}:   680 mg/kg (M, i.v.); 1080 mg/kg (M, p.o.);
     660 mg/kg (R, i.v.); 1800 mg/kg (R, p.o.)
CN: 5-(dimethylamino)-9-methyl-2-propyl-1H-pyrazolo[1,2-a][1,2,4]benzotriazine-1,3(2H)-dione

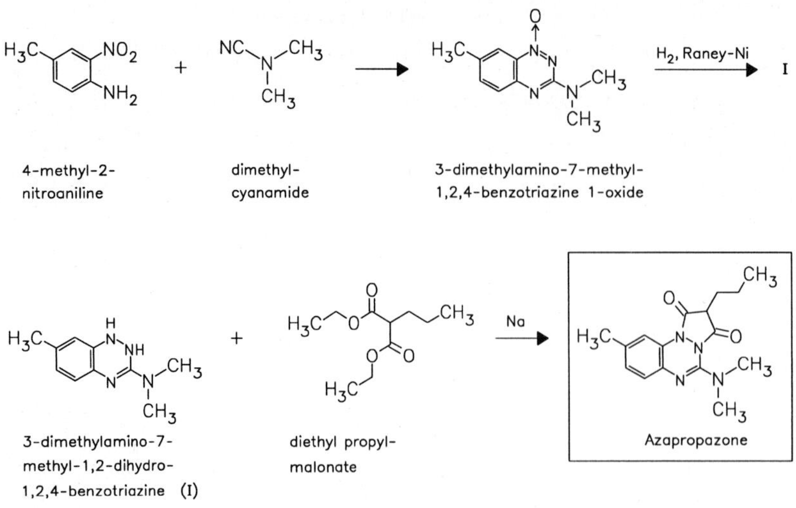

H2, Raney-Ni
4-methyl-2-nitroaniline
dimethylcyanamide
3-dimethylamino-7-methyl-1,2,4-benzotriazine 1-oxide
Na
3-dimethylamino-7-methyl-1,2-dihydro-1,2,4-benzotriazine (I)
diethyl propylmalonate
Azapropazone

Reference(s):

US 3 349 088 (Siegfried AG; 24.10.1967; CH-prior. 22.10.1963).
Mixich, G.: Helv. Chim. Acta (HCACAV) **51**, 532 (1968).

Formulation(s): cps. 150 mg, 200 mg, 300 mg; tabl. 600 mg

Trade Name(s):

D:	Tolyprin (Du Pont Pharma)	GB:	Rheumox (Wyeth)	J:	Sinnamin (Nippon Chemiphar)
F:	Prolixan (J. Logeais); wfm	I:	Prolixan (Malesci); wfm		

Azatadine

ATC: R06AX09
Use: antihistaminic

RN: 3964-81-6 MF: $C_{20}H_{22}N_2$ MW: 290.41
CN: 6,11-dihydro-11-(1-methyl-4-piperidinylidene)-5*H*-benzo[5,6]cyclohepta[1,2-*b*]pyridine

dimaleate

RN: 3978-86-7 MF: $C_{20}H_{22}N_2 \cdot 2C_4H_4O_4$ MW: 522.55 EINECS: 223-615-8
LD_{50}: 165 mg/kg (M, p.o.);
440 mg/kg (R, p.o.)

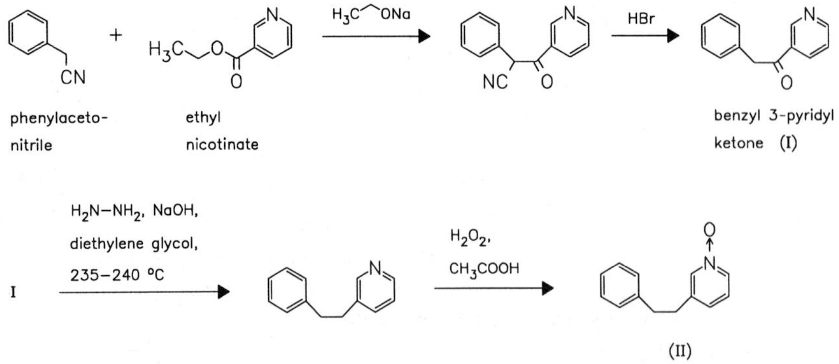

H3C ONa
HBr
phenylacetonitrile
ethyl nicotinate
benzyl 3-pyridyl ketone (I)
H2N–NH2, NaOH, diethylene glycol, 235–240 °C
H2O2, CH3COOH
(II)

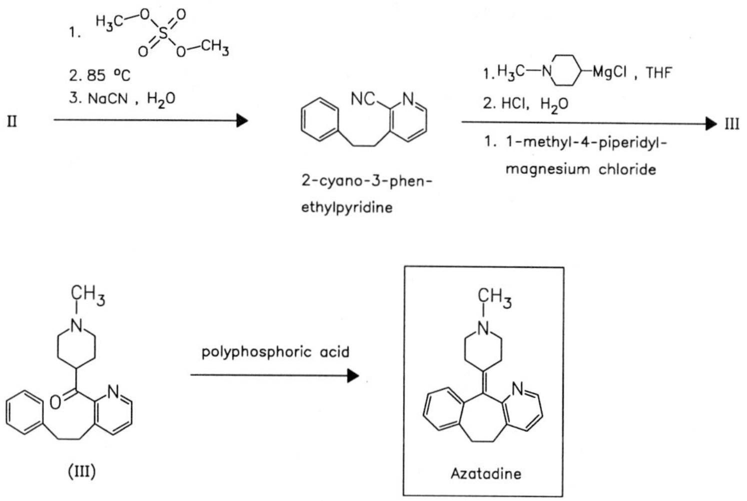

2-cyano-3-phen-
ethylpyridine

1. 1-methyl-4-piperidyl-
magnesium chloride

(III)

polyphosphoric acid

Azatadine

Reference(s):

US 3 301 863 (Schering Corp.; 31.1.1967; prior. 24.4.1963, 13.12.1963, 21.12.1964, 18.3.1965).
US 3 326 924 (Schering Corp.; 20.6.1967; prior. 24.4.1963, 13.12.1963).
US 3 357 986 (Schering Corp.; 12.12.1967; prior. 24.4.1963, 13.12.1963, 19.9.1966).
US 3 366 635 (Schering Corp.; 30.1.1968; prior. 24.4.1963, 13.12.1963).
US 3 419 565 (Schering Corp.; 31.12.1968; prior. 24.4.1963, 19.9.1966).

improved process for 2-cyano-3-phenethylpyridine:
US 4 954 632 (SmithKline Beecham Corp.; 4.9.1990; prior. 2.12.1987, 10.2.1989).
Villani, F.J. et al.: J. Med. Chem. (JMCMAR) 15, 750 (1972).

Formulation(s):   syrup 0.5 mg (as dimaleate); tabl. 1 mg (azatadine maleate)

Trade Name(s):
GB: Optimine (Schering-   USA: Trinalin (Key Pharm.)-
       Plough)   comb.

Azathioprine

ATC: L01BB; L04AX01
Use: antineoplastic, immunosuppressive

RN: 446-86-6   MF: $C_9H_7N_7O_2S$ MW: 277.27   EINECS: 207-175-4
LD$_{50}$: 1389 mg/kg (M, p.o.);
       535 mg/kg (R, p.o.)
CN: 6-[(1-methyl-4-nitro-1H-imidazol-5-yl)thio]-1H-purine

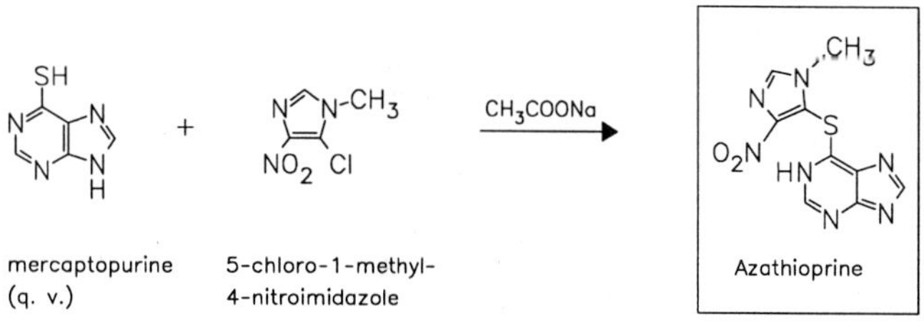

mercaptopurine   5-chloro-1-methyl-   Azathioprine
(q. v.)   4-nitroimidazole

Reference(s):
US 3 056 785 (Burroughs Wellcome; appl. 2.10.1962; prior. 21.3.1960).

Formulation(s): amp. 50 mg; f. c. tabl. 50 mg, 25 mg; lyo. 54.1 mg

Trade Name(s):

D:	Azamedac (medac)	F:	Imurel (Glaxo Wellcome)	I:	Imuran (Wellcome)
	Imurek (Glaxo Wellcome)	GB:	Azamune (Penn)	J:	Imuran (Tanabe)
	Zytrim (Isis Puren)		Imuran (Glaxo Wellcome)	USA:	Imuran (Glaxo Wellcome)

Azelaic acid

ATC: D10AX03; D11AX
Use: topical treatment of hyperpigmentary disorders and skin cancers, acne therapeutic

RN: 123-99-9 MF: $C_9H_{16}O_4$ MW: 188.22 EINECS: 204-669-1
LD_{50}: >5 g/kg (R, p.o.)
CN: nonanedioic acid

disodium salt
RN: 17265-13-3 MF: $C_9H_{14}Na_2O_4$ MW: 232.19 EINECS: 241-298-4
calcium salt (1:1)
RN: 14488-58-5 MF: $C_9H_{14}CaO_4$ MW: 226.29

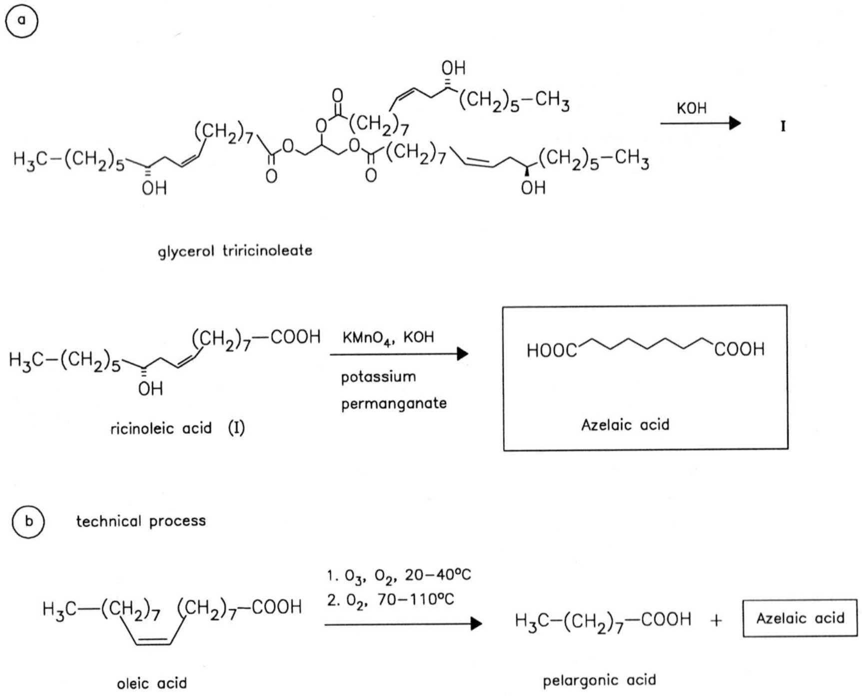

glycerol triricinoleate

ricinoleic acid (I)

Azelaic acid

technical process

oleic acid

pelargonic acid

Azelaic acid

Reference(s):
a Hill, J.W.; McEwen, W.L.: Org. Synth. (ORSYAT), Coll. Vol. **2**, 53 (1943).
b Ullmann's Encyclopedia of Industrial Chemistry, 5th Ed., Vol. **A8**, 526.

alternative synthesis:
US 3 402 108 (Emery; 17.9.1968; appl. 7.7.1966).
US 3 810 937 (V.P. Kuceski; 14.5.1974; appl. 15.9.1970).
JP 56 169 640 (Nippon Oil; appl. 31.5.1980).
JP 58 140 038 (Kuraray; appl. 16.2.1982).
DOS 2 035 558 (Degussa; appl. 17.7.1970).
DOS 2 052 815 (Degussa; appl. 28.10.1970).
DOS 2 106 307 (Degussa; appl. 10.2.1971).
DOS 2 106 913 (Degussa; appl. 13.2.1971).
DOS 2 316 203 (Henkel; appl. 31.3.1973).

topical treatment:
US 4 818 768 (Schering AG; 4.4.1989; appl. 29.1.1982; I-prior. 19.4.1977, 30.12.1977).

Formulation(s):   cream 200 mg/g (20 %)

Trade Name(s):
D: Skinoren (Schering; 1988)   GB: Skinoren (Schering Health   USA: Azelex (Allergan)
F: Skinoren (Schering)   Care)

Azelastine

ATC: R01AC03; R03D; R06AX19;
          S01GX07
Use: antiasthmatic, antiallergic,
          antihistaminic

RN: 58581-89-8   MF: $C_{22}H_{24}ClN_3O$ MW: 381.91
LD_{50}: 36 mg/kg (R, i.v.); 130 mg/kg (R, p.o.)
CN: 4-[(4-chlorophenyl)methyl]-2-(hexahydro-1-methyl-1*H*-azepin-4-yl)-1(2*H*)-phthalazinone

hydrochloride
RN: 37932-96-0   MF: $C_{22}H_{24}ClN_3O \cdot xHCl$ MW: unspecified   EINECS: 253-720-4

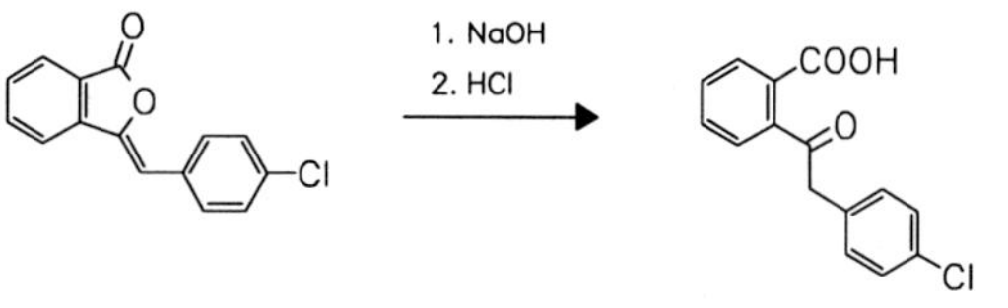

3-(4-chlorobenzylidene)-
phthalide

2-(4-chlorophenyl-
acetyl)benzoic
acid (I)

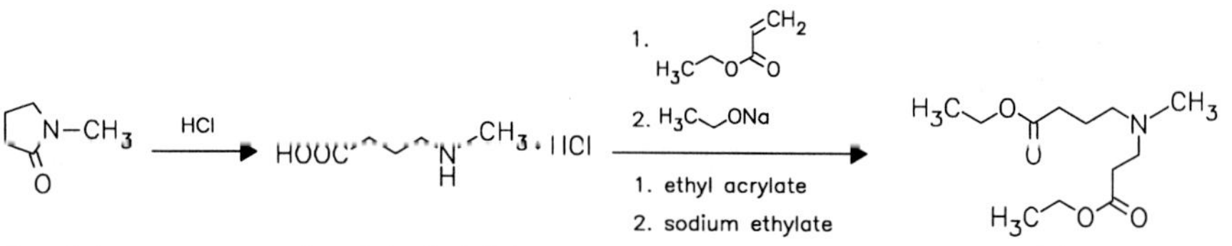

1-methyl-
pyrrolidin-
2-one

4-methylaminobutyric
acid hydrochloride

1. ethyl acrylate
2. sodium ethylate

(II)

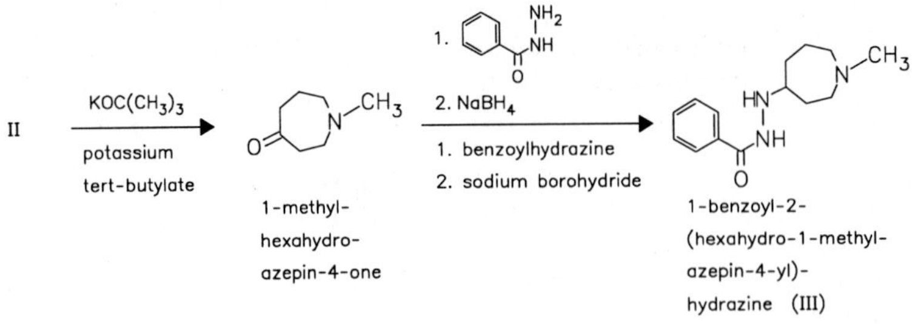

II → KOC(CH₃)₃ potassium tert-butylate → 1-methyl-hexahydro-azepin-4-one

1. benzoylhydrazine
2. sodium borohydride

1-benzoyl-2-(hexahydro-1-methyl-azepin-4-yl)-hydrazine (III)

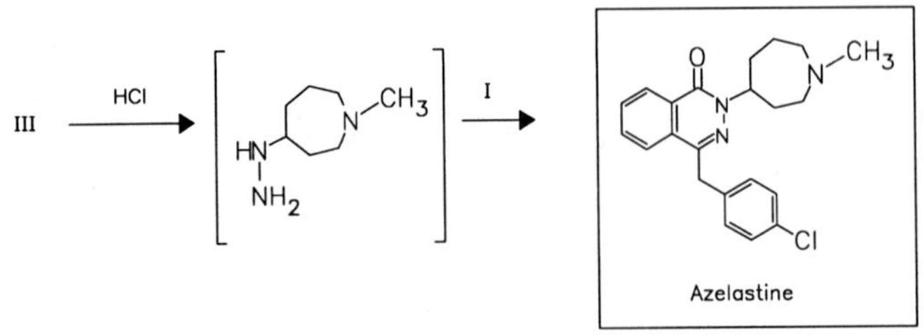

III → HCl → [...] → I → Azelastine

Reference(s):
DE 2 164 058 (ASTA-Werke; appl. 23.12.1971).
US 3 813 384 (ASTA-Werke; 28.5.1974; CH-prior. 22.1.1971).
EP 316 633 (ASTA Medica AG; appl. 27.10.1988; D-prior. 13.11.1987).

Formulation(s):   f. c. tabl. 1 mg, 2 mg, 4 mg; nasal spray (as hydrochloride, 0.2 mg/puff)

Trade Name(s):

D:	Allergodil (ASTA Medica; 1992)	
		Radetazin (Arzneimittelwerk Dresden; 1992)
	F:	Allergodil (ASTA Medica)

GB: Rhinolast (ASTA Medica; 1991)
J: Azeptin (Eisai; 1986)
USA: Astelin (Wallace)

Azidamfenicol
(Azidoamphenicol)

ATC: J01BA
Use: antibiotic

RN: 13838-08-9   MF: C₁₁H₁₃N₅O₅   MW: 295.26   EINECS: 237-552-9
CN: [R-(R*,R*)]-2-azido-N-[2-hydroxy-1-(hydroxymethyl)-2-(4-nitrophenyl)ethyl]acetamide

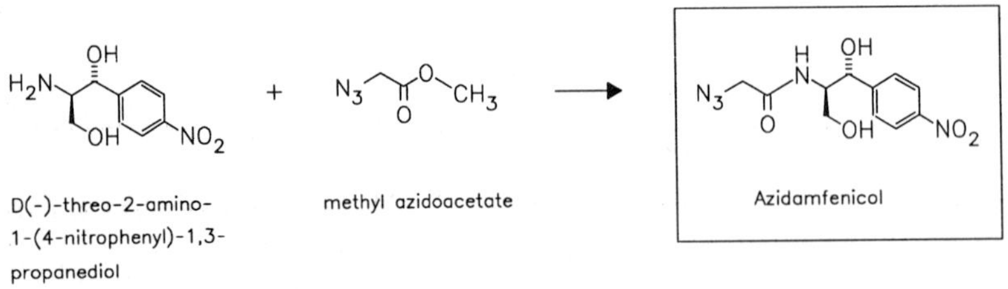

D(-)-threo-2-amino-1-(4-nitrophenyl)-1,3-propanediol

methyl azidoacetate

Azidamfenicol

Reference(s):
US 2 882 275 (Bayer; 14.4.1959, D-prior. 28.1.1955).

Formulation(s):   eye drops 10 mg/ml (1 %)

Azidocillin

ATC: J01CE04; J01HA
Use: antibiotic

RN: 17243-38-8   MF: $C_{16}H_{17}N_5O_4S$ MW: 375.41   EINECS: 241-278-5
CN: [2S-[2α,5α,6β(S*)]]-6-[(azidophenylacetyl)amino]-3,3-dimethyl-7-oxo-4-thia-1-
       azabicyclo[3.2.0]heptane-2-carboxylic acid

monopotassium salt
RN: 22647-32-1   MF: $C_{16}H_{16}KN_5O_4S$ MW: 413.50
sodium salt
RN: 35334-12-4   MF: $C_{16}H_{16}N_5NaO_4S$ MW: 397.39

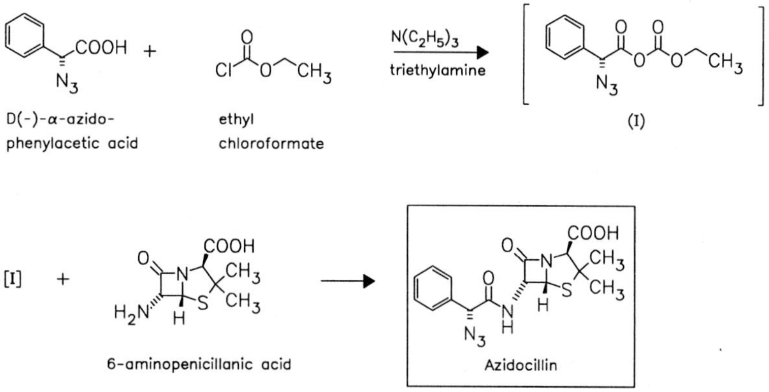

D(-)-α-azido-   ethyl   (I)
phenylacetic acid   chloroformate

6-aminopenicillanic acid   Azidocillin

Reference(s):
US 3 293 242 (Beecham; 20.12.1966; GB-prior. 21.7.1961).
DE 1 168 910 (Beecham; appl. 3.7.1962; GB-prior. 21.7.1961).
GB 940 488 (Beecham; appl. 21.7.1961; valid from 23.7.1962).

Formulation(s):   gran. 250 mg; syrup 250 mg; tabl. 750 mg (as sodium salt)

Azimilide hydrochloride

Use: class III antiarrhythmic agent

RN: 149888-94-8   MF: $C_{23}H_{28}ClN_5O_3 \cdot 2HCl$ MW: 530.88
CN: 1-[[[5-(4-Chlorophenyl)-2-furanyl]methylene]amino]-3-[4-(4-methyl-1-piperazinyl)butyl]-2,4-
       imidazolidinedione dihydrochloride

base
RN: 149908-53-2   MF: $C_{23}H_{28}ClN_5O_3$ MW: 457.96

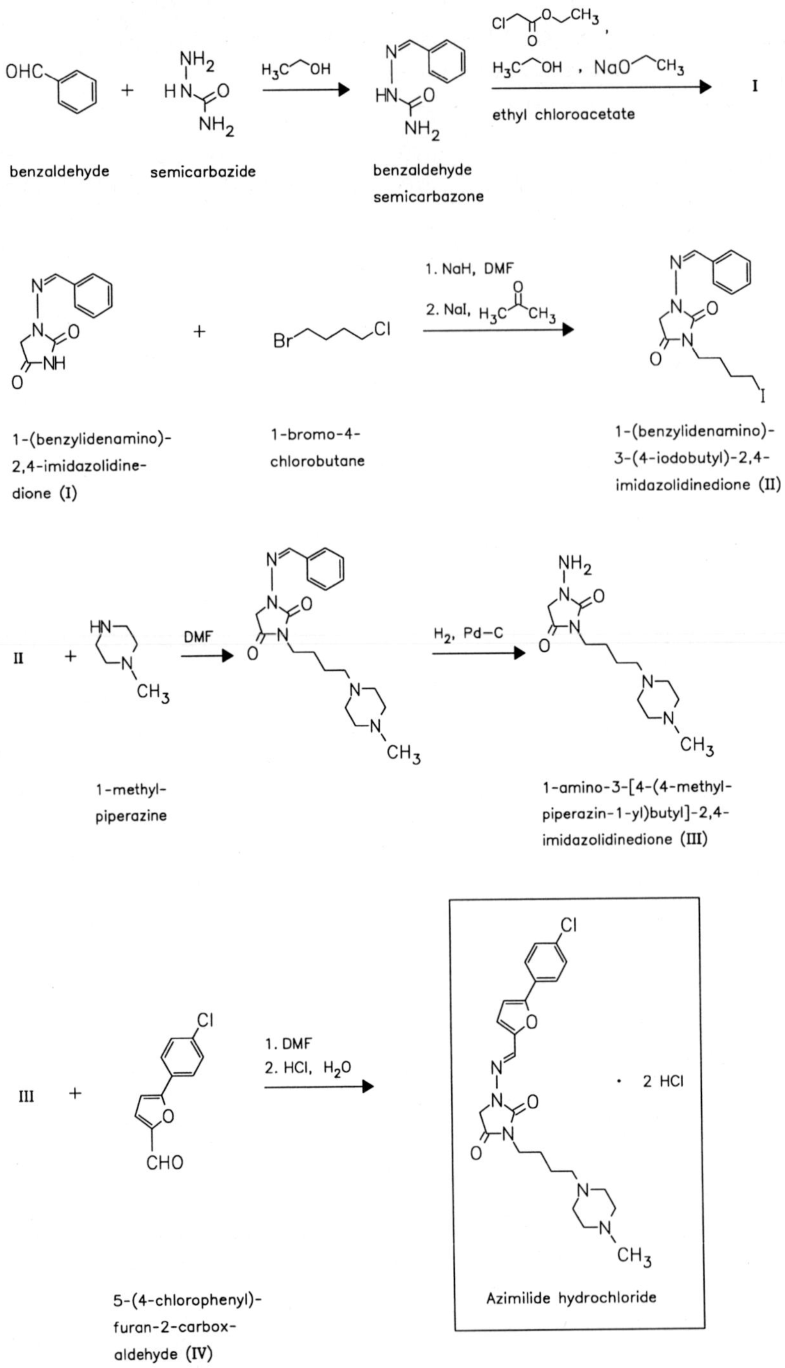

benzaldehyde   semicarbazide

benzaldehyde
semicarbazone

ethyl chloroacetate

I

1-(benzylidenamino)-
2,4-imidazolidine-
dione (I)

1-bromo-4-
chlorobutane

1. NaH, DMF

2. NaI, H₃C—CO—CH₃

1-(benzylidenamino)-
3-(4-iodobutyl)-2,4-
imidazolidinedione (II)

II   +

DMF

H₂, Pd–C

1-methyl-
piperazine

1-amino-3-[4-(4-methyl-
piperazin-1-yl)butyl]-2,4-
imidazolidinedione (III)

III   +

1. DMF
2. HCl, H₂O

· 2 HCl

5-(4-chlorophenyl)-
furan-2-carbox-
aldehyde (IV)

Azimilide hydrochloride

preparation of intermediate IV

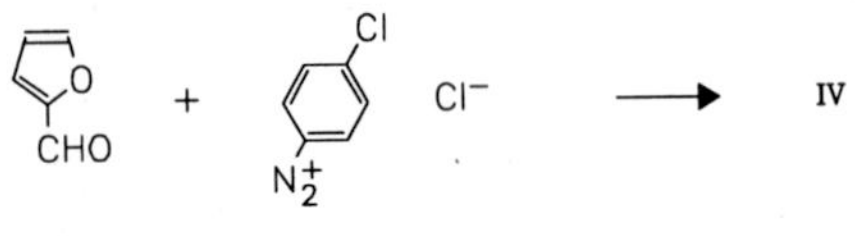

IV

Reference(s):
WO 9 304 061 (Procter and Gamble Co.; appl. 10.8.1992; USA-prior. 14.8.1991).

preparation of 1-benzylidenaminoimidazoline-2,4-dione:
Jack: J. Pharm. Pharmacol. (JPPMAP) **11**, Suppl. 108, 112 (1959).

preparation of 5-(4-chlorophenyl)furan-2-carboxaldehyde:
Pong, S.F.; Pelosi, S.S.; Wessels, F.L.; Yu, C.-N.; Burns, H.: Arzneim.-Forsch. (ARZNAD) **33** (10), 1411 (1983).

Trade Name(s):
USA: Stedicor (Procter &
            Gamble; 1999)

Azintamide

ATC: A03E; A05A1; A05AX
Use: choleretic

RN: 1830-32-6   MF: $C_{10}H_{14}ClN_3OS$ MW: 259.76   EINECS: 217-384-2
LD$_{50}$:   1150 mg/kg (M, p.o.);
            1550 mg/kg (R, p.o.)
CN: 2-[(6-chloro-3-pyridazinyl)thio]-*N,N*-diethylacetamide

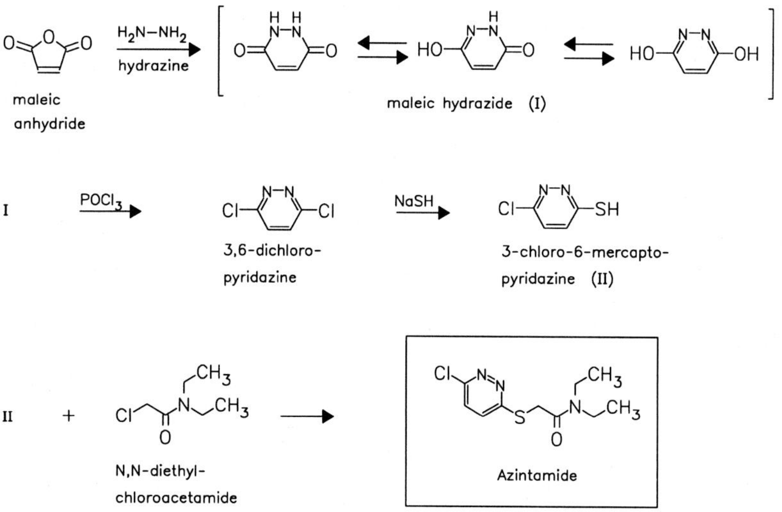

Reference(s):
DE 1 188 604 (Lentia; appl. 17.11.1961).
BE 624 848 (Österr. Stickstoffwerke; appl. 14.11.1962; A-prior. 16.11.1961).
Stormann, H.: Arzneim.-Forsch. (ARZNAD) **14**, 266 (1964).

Formulation(s).   drg. 100 mg

Trade Name(s):
D: Oragallin (Truw)-comb.

Azithromycin
(Aritromicina)

ATC: J01FA10
Use: macrolide antibiotic

RN: 83905-01-5   MF: $C_{38}H_{72}N_2O_{12}$ MW: 749.00
LD_{50}:   1200 mg/kg (M, i.p.); 825 mg/kg (M, i.p.); 3 g/kg (M, p.o.);
>2 g/kg (R, p.o.)
CN: [2R-(2R*,3S*,4R*,5R*,8R*,10R*,11R*,12S*,13S*,14R*)]-13-[(2,6-dideoxy-3-C-methyl-3-O-methyl-α-
L-ribo-hexopyranosyl)oxy]-2-ethyl-3,4,10-trihydroxy-3,5,6,8,10,12,14-heptamethyl-11-[[3,4,6-trideoxy-
3-(dimethylamino)-β-D-xylo-hexopyranosyl]oxy]-1-oxa-6-azacyclopentadecan-15-one

monohydrochloride
RN: 90581-30-9   MF: $C_{38}H_{72}N_2O_{12} \cdot HCl$ MW: 785.46

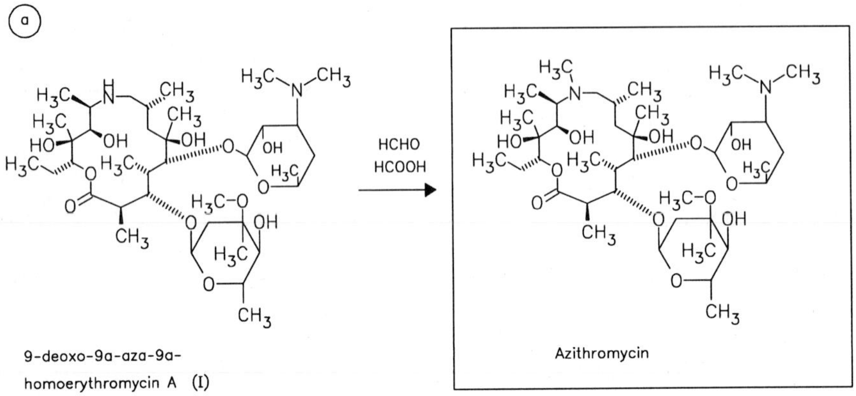

9-deoxo-9a-aza-9a-
homoerythromycin A   (I)
(from erythromycin A, q. v.)

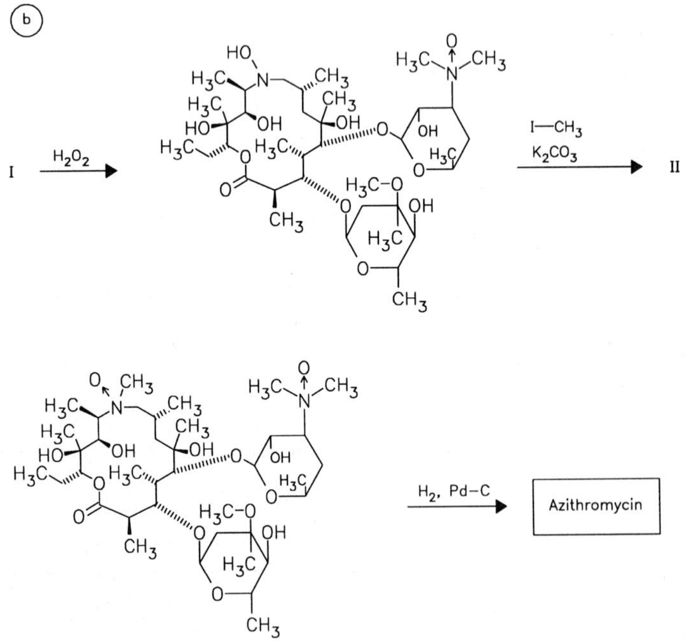

Azithromycin

(II)

Reference(s):
a DOS 3 140 449 (Pliva; appl. 12.10.1981; YU-prior. 6.3.1981).
  US 4 517 359 (Pliva; 14.5.1985; appl. 22.9.1981; YU-prior. 6.3.1981).
b EP 101 186 (Pliva; appl. 14.7.1983; USA-prior. 19.7.1982, 15.11.1982).
  US 4 474 768 (Pliva; 2.10.1984; prior. 19.7.1982, 15.11.1982).
  Djokic, S. et al.: J. Antibiot. (JANTAJ) **40**, 1006 (1987).

stable, non-hygroscopic dihydrate:
EP 298 650 (Pfizer; appl. 28.6.1988).

medical use for treatment of protozoal infections:
US 4 963 531 (Pfizer; 16.10.1990; prior. 16.8.1988, 10.9.1987).

Formulation(s):   cps. 250 mg; susp. 200 mg (as dihydrate)

Trade Name(s):
D: Zithromax (Mack)   GB: Zithromax (Richborough;   USA: Zithromax (Pfizer; as
F: Zithromax (Pfizer)   1991)   dihydrate)

Azlocillin

ATC: J01CA09
Use: antibiotic

RN: 37091-66-0   MF: $C_{20}H_{23}N_5O_6S$ MW: 461.50   EINECS: 253-348-2
CN: [2S-[2α,5α,6β(S*)]]-3,3-dimethyl-7-oxo-6-[[[[(2-oxo-1-
       imidazolidinyl)carbonyl]amino]phenylacetyl]amino]-4-thia-1-azabicyclo[3.2.0]heptane-2-carboxylic acid

monosodium salt
RN: 37091-65-9   MF: $C_{20}H_{22}N_5NaO_6S$ MW: 483.48   EINECS: 253-347-7
LD_{50}:   5065 mg/kg (M, i.v.);
       1793 mg/kg (R, i.v.)

2-imida-   phosgene   1-chloroformyl-   N-(2-oxoimidazolidinocarbonyl)-
zolidinone   imidazolidinone   D-phenylglycine (I)

6-aminopenicillanic acid   Azlocillin

Reference(s):
FR 2 100 682 (Bayer; appl. 25.5.1971; D-prior. 25.5.1970).
US 3 933 795 (Bayer; 20.1.1976; D-prior. 25.5.1970).
DOS 2 104 579 (Bayer; appl. 1.2.1971).
US 3 978 223 (Bayer; 20.1.1976; D-prior. 25.5.1970).
DE 2 025 415 (Bayer; prior. 25.5.1970).

combination with other semisynthetic penicillins:
DOS 2 737 673 (Bayer; appl. 20.8.1977).

Formulation(s):   amp. 2 g/20 ml, 5 g, 750 ml; inf. powder 500 mg, 1 g, 2 g, 5 g; lyo. 524 mg, 1048 mg,
2096 mg, 4192 mg, 5240 mg

Trade Name(s):

D: Securopen (Bayer; 1977)	F: Securopen (Bayer-Pharma); wfm	GB: Securopen (Bayer; 1980) I: Securopen (Bayer; 1985)

Azosemide

ATC: C03CA
Use: diuretic

RN: 27589-33-9   MF: $C_{12}H_{11}ClN_6O_2S_2$ MW: 370.85   EINECS: 248-549-7
LD_{50}:   138 mg/kg (M, i.v.); 6350 mg/kg (M, p.o.);
252 mg/kg (R, i.v.); 2545 mg/kg (R, p.o.)
CN: 2-chloro-5-(1H-tetrazol-5-yl)-4-[(2-thienylmethyl)amino]benzenesulfonamide

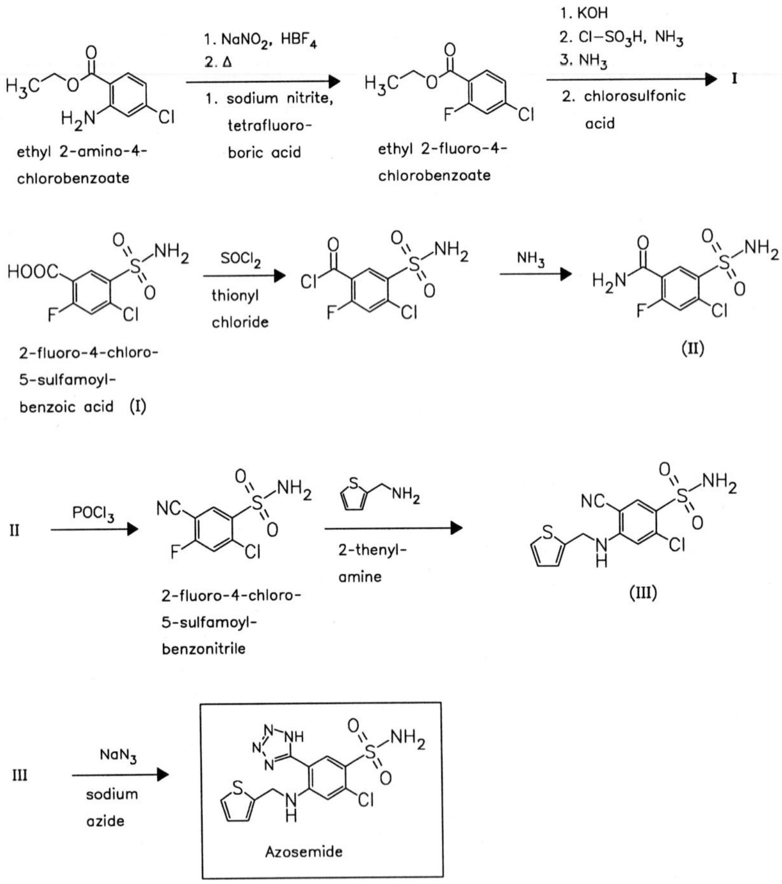

Reference(s):
DOS 1 815 922 (Boehringer Mannh.; appl. 20.12.1968).
US 3 665 002 (Boehringer Mannh.; 23.5.1972; D-prior. 20.12.1968).

alternative synthesis:
DOS 3 034 664 (Boehringer Mannh.; appl. 13.9.1980).

synthesis of 2-fluoro-4-chloro-5-sulfamoylbenzoic acid:
Sturm, K. et al.: Chem. Ber. (CHBEAM) **99**, 328 (1966).

combination preparations:
DOS 2 423 550 (Boehringer Mannh.; appl. 15.5.1974).
DOS 2 423 606 (Boehringer Mannh.; appl. 15.5.1974).
DOS 2 556 001 (Boehringer Mannh.; appl. 2.12.1975).

Formulation(s):   f. c. tabl. 80 mg

Trade Name(s):
D: Luret (Sanofi Winthrop)   J: Diart (Sanwa Kagaku)

Aztreonam

(Azthreonam)

ATC: J01DF01
Use: synthetic monobactam antibiotic

RN: 78110-38-0   MF: $C_{13}H_{17}N_5O_8S_2$ MW: 435.44   EINECS: 278-839-9
LD$_{50}$:   1963 mg/kg (M, i.v.); >10 g/kg (M, p.o.);
        2001 mg/kg (R, i.v.); >10 g/kg (R, p.o.)
CN: [2S-[2α,3β(Z)]]-2-[[[1-(2-amino-4-thiazolyl)-2-[(2-methyl-4-oxo-1-sulfo-3-azetidinyl)amino]-2-oxoethylidene]amino]oxy]-2-methylpropanoic acid

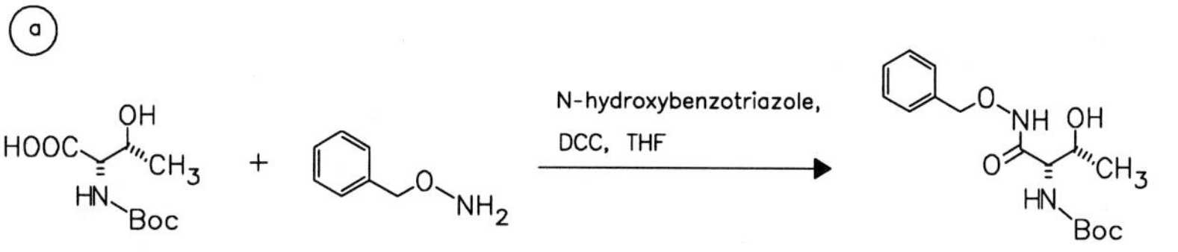

N-tert-butyloxy-carbonyl-L-threonine

O-benzyl-hydroxylamine

(I)

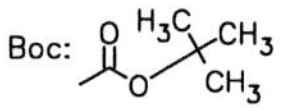

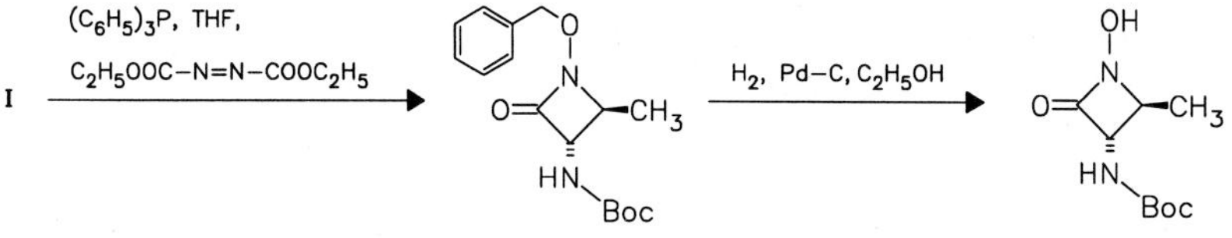

(3S-trans)-1-benzyl-oxy-3-tert-butyl-oxycarbonylamino-4-methyl-2-azetidinone

(II)

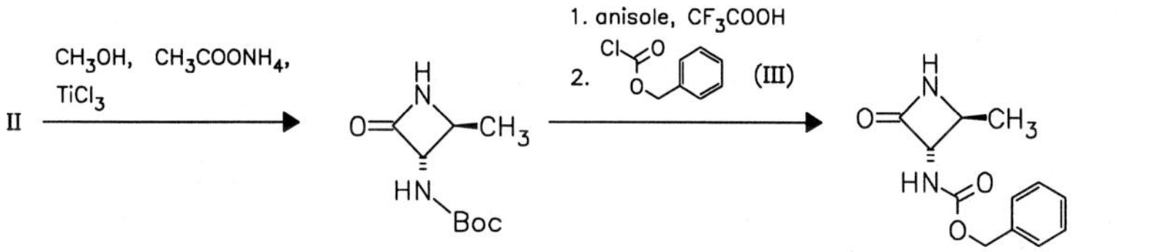

(3S-trans)-3-benzyl-oxycarbonylamino-4-methyl-2-azetidinone (IV)

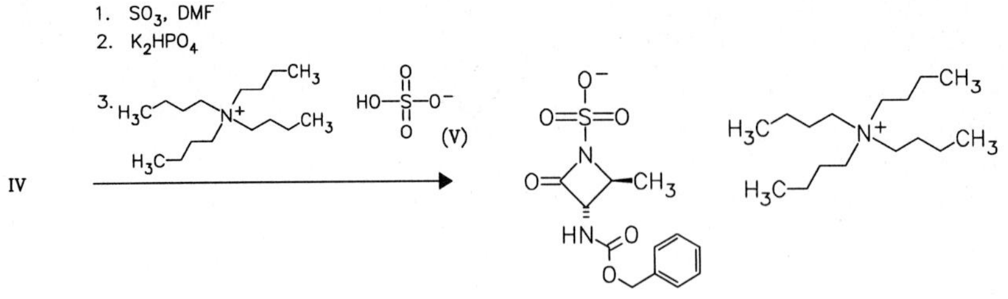

1. SO₃, DMF
2. K₂HPO₄

3. (V)

IV → (VI)

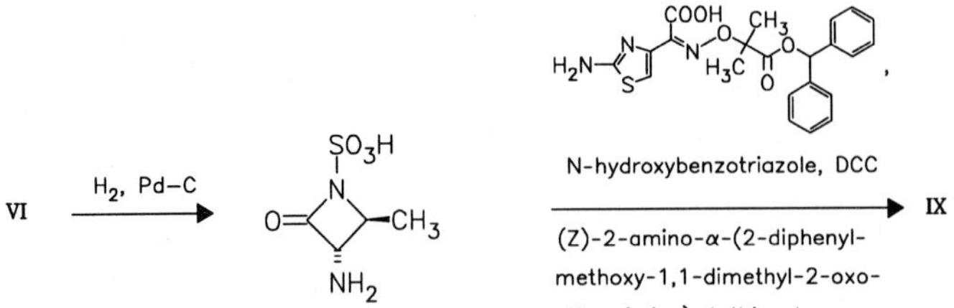

VI → (3S-trans)-3-amino-4-methyl-2-oxo-1-azetidine-sulfonic acid (VII)

H₂, Pd–C

N-hydroxybenzotriazole, DCC

(Z)-2-amino-α-(2-diphenyl-methoxy-1,1-dimethyl-2-oxo-ethoxyimino)-4-thiazole-acetic acid (VIII)

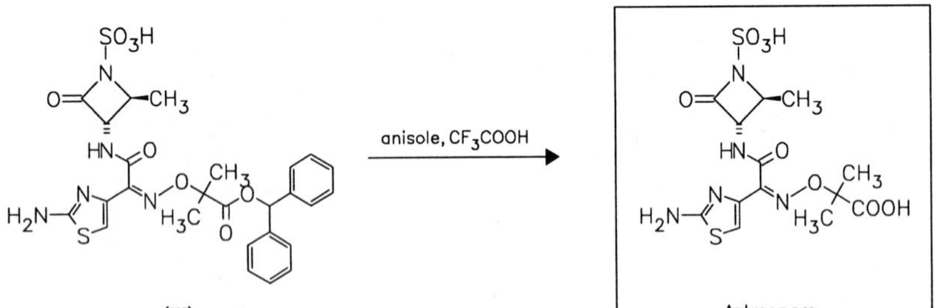

→ IX

(IX) → anisole, CF₃COOH → Aztreonam

b

HOOC—C(OH)(CH₃)NH₂ L-threonine

+ H₃C—OH

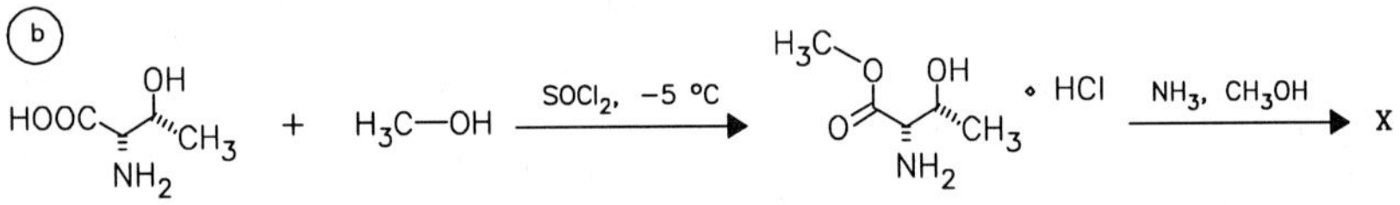

SOCl₂, –5 °C →

L-threonine methyl ester hydrochloride · HCl

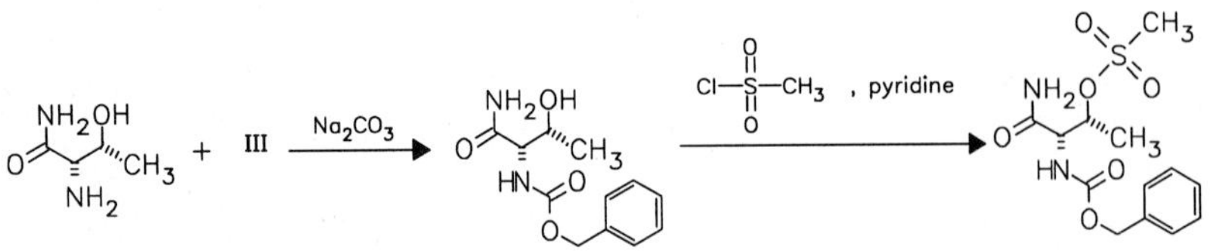

NH₃, CH₃OH → X

L-threonin-amide (X) + III → Na₂CO₃ → N²-benzyloxy-carbonyl-L-threoninamide

Cl—S(=O)₂—CH₃ , pyridine → (XI)

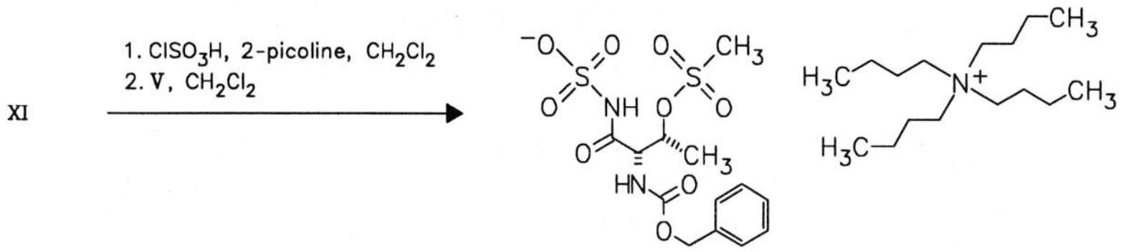

1. ClSO$_3$H, 2-picoline, CH$_2$Cl$_2$
2. V, CH$_2$Cl$_2$

XI →

N^1-sulfo-N^2-benzyloxycarbonyl-O-methylsulfonyl-
L-threoninamide tetrabutylammonium salt   (XII)

XII $\xrightarrow{\text{K}_2\text{CO}_3, \ \text{C}_2\text{H}_4\text{Cl}_2, \ \text{H}_2\text{O}}$ VI $\xrightarrow{\text{H}_2, \ \text{Pd}-\text{C}, \ \text{C}_2\text{H}_5\text{OH}}$ VII $\xrightarrow[\text{N-hydroxybenzotriazole, DCC}]{\text{VIII,}}$ IX

IX $\xrightarrow{\text{anisole, CF}_3\text{COOH}}$ | Aztreonam |

Reference(s):
US 4 386 034 (Squibb; 31.5.1983; prior. 10.2.1982).
US 4 529 698 (Squibb; 16.7.1985; prior. 5.11.1984).
US 4 625 022 (Squibb; prior. 25.11.1986; 2.2.1981).
DOS 3 104 145 (Squibb; appl. 6.2.1981; USA-prior. 29.8.1980).
GB 2 071 650 (Squibb; appl. 6.2.1981; USA-prior. 7.2.1980, 29.9.1980).

Formulation(s):   amp. 2 g; inj. powder 500 mg, 1 g, 2 g; lyo. for inf. 2 g; vial 1 g/3 ml

Trade Name(s):
D:	Azactam (Bristol-Myers Squibb; 1985)	GB:	Azactam (Bristol-Myers Squibb; 1986)	J:	Azactam (Squibb; 1987)
F:	Azactam (Sanofi Winthrop)	I:	Azactam (Squibb; 1984)	USA:	Azactam (Bristol-Myers Squibb; 1987)

Bacampicillin

ATC: J01CA06
Use: antibiotic (broad spectrum penicillin)

RN: 50972-17-3 MF: $C_{21}H_{27}N_3O_7S$ MW: 465.53
CN: [2S-[2α,5α,6β(S*)]]-6-[(aminophenylacetyl)amino]-3,3-dimethyl-7-oxo-4-thia-1-azabicyclo[3.2.0]heptane-2-carboxylic acid 1-[(ethoxycarbonyl)oxy]ethyl ester

monohydrochloride
RN: 37661-08-8 MF: $C_{21}H_{27}N_3O_7S \cdot HCl$ MW: 501.99 EINECS: 253-580-4
LD$_{50}$: 184 mg/kg (M, i.v.); 8529 mg/kg (M, p.o.);
176 mg/kg (R, i.v.); 10 g/kg (R, p.o.)

azidocillin sodium salt
(q. v.)

1-chloroethyl
ethyl carbonate

6-(D-α-azidophenylacetamido)-
penicillanic acid 1-ethoxycarbonyl-
oxyethyl ester (I)

Bacampicillin

Reference(s):
DAS 2 144 457 (Astra; appl. 4.9.1971; S-prior. 17.9.1970, 20.11.1970).
US 3 873 521 (Astra; 25.3.1975; S-prior. 17.9.1970; 20.11.1970).
US 3 939 270 (Astra; 17.2.1976; S-prior. 17.9.1970; 20.11.1970).

Formulation(s):   f. c. tabl. 400 g, 800 mg; susp. 125 mg (as hydrochloride)

Trade Name(s):
D: Ambacamp (Pharmacia & Upjohn; 1981)
Penglobe (Astra; 1977)
F: Bacampicine (Pharmacia & Upjohn)
Penglobe (Lematte et Boinot)
GB: Ambaxin (Upjohn; 1980)
I: Ambaxin (Upjohn)
Amplibac (Schwarz)
Bacacil (Pfizer)
Penglobe (Bracco)
J: Bacacil (Pfizer Taito)
Penglobe (Yoshitomi)
USA: Spectrobid (Pfizer; 1981)

Bacitracin

ATC: D06AX05; R02AB04
Use: polypeptide antibiotic (mainly topical application)

RN: 1405-87-4 MF: $C_{66}H_{103}N_{17}O_{16}S$ MW: 1422.72 EINECS: 215-786-2
LD$_{50}$: 360 mg/kg (M, i.v.); >3750 mg/kg (M, p.o.)
CN: bacitracin

bacitracin A

RN: 22601-59-8   MF: $C_{66}H_{103}N_{17}O_{16}S$ MW: 1422.72   EINECS: 245-115-9

Bacitracin A

"Bacitracin" is submitted as mixture of bacitracin A with other bacitracins.
From culture of *Bacillus subtilis* and purification on ion-exchangers.

Reference(s):
US 2 498 165 (US-Secret. of War; 1950; appl. 1946).
US 2 828 246 (Commercial Solvents Corp.; 1958; appl. 1956).

purification:
US 2 457 887 (Ben Venue Labs.; 1949; appl. 1947).
US 2 609 324 (Commercial Solvents Corp.; 1952; appl. 1949).
US 2 774 712 (S. B. Penick & Co.; 1956; appl. 1955).
US 2 776 240 (Commercial Solvents Corp.; 1957; appl. 1954).
US 2 834 711 (Commercial Solvents Corp.; 1958; appl. 1956).
US 2 915 432 (Merck & Co.; 1959; appl. 1955).
US 2 960 437 (Pfizer; 1960; appl. 1955).
US 3 795 663 (Commercial Solvents Corp.; 5.3.1974; appl. 1.5.1972).
US 4 101 539 (IMC Chemical; 18.7.1978; appl. 17.10.1977).

complexes with nickel salts:
US 2 903 357 (Grain Processing Corp.; 1959; appl. 1958).

Na-bacitracin methanesulfonate:
US 3 205 137 (Warner-Lambert; 7.9.1965; appl. 19.3.1963).

complexes with zinc, cobalt *or* manganese sulfate *resp.* sulfonates*:*
US 3 384 631 (Spofa; 21.5.1968; appl. 23.6.1965; CSSR-prior. 26.6.1964).

complexes with metal methanesulfinates:
US 3 441 646 (Commercial Solvents; 29.4.1969; appl. 22.1.1965).

complex with calcium or magnesium alkylbenzenesulfonates:
US 3 891 615 (Commercial Solvents; 24.6.1975; appl. 25.10.1973).

Formulation(s):   amp. 50000 iu; vial 5000 iu; nasal ointment 300 iu; ointment 300 iu, 500 iu; powder 300 iu.

Trade Name(s):
D: Anginomycin (MIP
       Pharma)
       Batrax (Gewo)-comb.
       Bivacyn (medphano)
       Cicatrex (Glaxo
       Wellcome)-comb.

       Frubienzym (Boehringer
       Ing.)-comb.
       Nebacetin (Yamanouchi)-
       comb.
       Neobac (Dermapharm)
       Neotracin (CIBA Vision)

       Polyspectran (Alcon)-
       comb.
       Prednitracin (CIBA
       Vision)-comb.
       Tonsilase (Media)-comb.
F: Bacicoline (Merck Sharp &
       Dohme-Chibret)-comb.

| | Collunovar (Synthélabo)-comb. | | generics | | USA: | Betadine (Purdue |
| | comb. | GB: | Cicatrin (Glaxo | | | Frederick) |

Collunovar (Synthélabo)-comb.
Lysopaine ORL (Boehringer Ing.)-comb.
Maxilase Bacitracine (Sanofi Winthrop)-comb.
Oropivalone (Jouveinal)-comb.
Pimafucort (Beytout)-comb.

GB: Cicatrin (Glaxo Wellcome)-comb.
Polyfax (Dominion)-comb.
I: Bimixin (Lusofarmaco)-comb.
Enterostop (Teafarma)-comb.
Orobicin (Fulton)-comb.
J: Bacitracin (Ono)

generics

USA: Betadine (Purdue Frederick)
Cortisporin (Burroughs Wellcome)-comb.
Neosporin (Glaxo Wellcome; Warner-Lambert)
Polysporin (Warner-Lambert)
generics

Baclofen

ATC: M03BX01
Use: muscle relaxant (antispasmodic)

RN: 1134-47-0   MF: $C_{10}H_{12}ClNO_2$ MW: 213.66   EINECS: 214-486-9
LD$_{50}$:   31 mg/kg (M, i.v.); 200 mg/kg (M, p.o.);
    78 mg/kg (R, i.v.); 145 mg/kg (R, p.o.)
CN: β-(aminomethyl)-4-chlorobenzenepropanoic acid

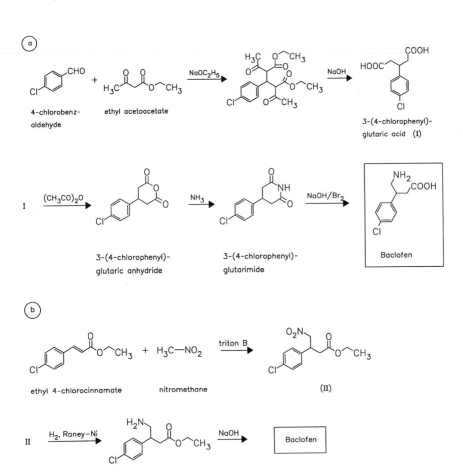

Reference(s):
a   US 3 471 548 (Ciba; 7.10.1969; CH-prior. 9.6.1963; 22.5.1964).
    US 3 634 428 (Ciba; 11.1.1972; CH-prior. 9.7.1963, 22.5.1964).
b   JP 45 016 692 (Uchimaru, F. et al.; Daiichi Seiyku; appl. 10.6.1970); C.A. (CHABA8) **73**, 77617w (1970).

combination with neuroleptics:
US 3 947 579 (Nelson Research & Dev.; 30.3.1976; appl. 3.6.1974).
US 3 978 216 (Nelson Research & Dev.; 31.8.1976; prior. 3.6.1974, 16.7.1975).
US 4 138 484 (Nelson Research & Dev.; 6.2.1979; prior. 3.6.1974, 16.7.1975, 16.8.1976, 25.7.1977).

Formulation(s):   inj. sol. 0.05 mg/1 ml, 10 mg/20 ml, 10 mg/5 ml; intrathecal inj. 50 µg/ml, 0.05 mg/ml, 10 mg/20 ml, 10 mg/5 ml; liquid 5 mg/5 ml; tabl. 5 mg, 10 mg, 25 mg

Trade Name(s):
D:	Lebic (Isis Puren)	GB:	Lioresal (Novartis)	Lioresal (Novartis)
	Lioresal (Novartis Pharma)	I:	Lioresal (Ciba)	
F:	Liorésal (Novartis)	J:	Gabalon (Daiichi)	

Balsalazide sodium
(BX-661-A)

ATC: D08
Use: anti-inflammatory

RN: 82101-18-6   MF: $C_{17}H_{13}N_3Na_2O_6$ MW: 401.29
CN: 5-[[4-[[(2-carboxyethyl)amino]carbonyl]phenyl]azo]-2-hydroxybenzoic acid disodium salt

(*E*)-free acid
RN: 80573-04-2   MF: $C_{17}H_{15}N_3O_6$ MW: 357.32

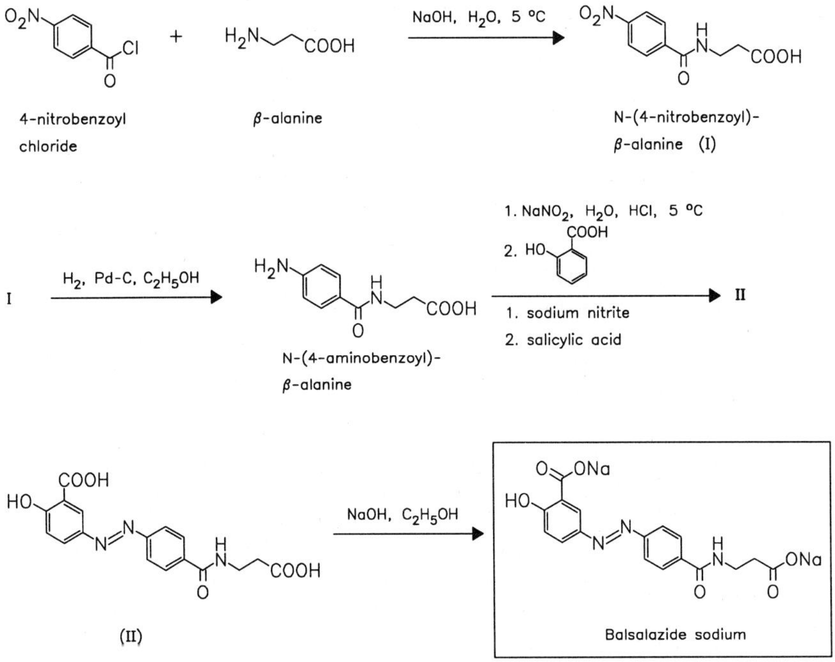

4-nitrobenzoyl chloride

β-alanine

N-(4-nitrobenzoyl)-β-alanine (I)

N-(4-aminobenzoyl)-β-alanine

(II)

Balsalazide sodium

Reference(s):
DE 3 128 819 (Biorex Lab.; appl. 21.7.1981; GB-prior. 21.7.1980, 7.7.1981).

Formulation(s):   cps. 750 mg (as disodium salt)

Trade Name(s):
GB: Colazide (Astra/manuf. by
       Salix)

Bambuterol
(KWD-2183)

ATC: R03CC12
Use: β₂-receptor agonist, orally active lipophilic terbutaline ester prodrug, long lasting bronchodilator

RN: 81732-65-2   MF: $C_{18}H_{29}N_3O_5$ MW: 367.45
CN: dimethylcarbamic acid 5-[2-[(1,1-dimethylethyl)amino]-1-hydroxyethyl]-1,3-phenylene ester

monohydrochloride
RN: 81732-46-9   MF: $C_{18}H_{29}N_3O_5 \cdot HCl$ MW: 403.91

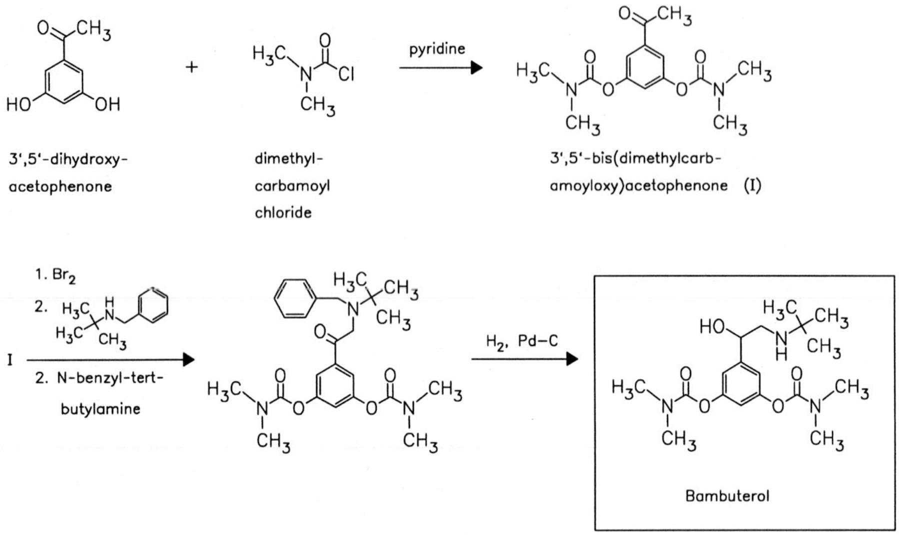

3',5'-dihydroxy-acetophenone   dimethyl-carbamoyl chloride   3',5'-bis(dimethylcarbamoyloxy)acetophenone (I)

Bambuterol

Reference(s):
EP 43 807 (Draco; appl. 30.6.1981; GB-prior. 9.7.1980, 29.5.1981).
DOS 3 163 871 (Draco; appl. 23.1.1981; GB-prior. 9.7.1980, 29.5.1981).
(alternative synthesis given).

Formulation(s):   tabl. 10 mg, 20 mg; sol. 0.1%

Trade Name(s):
D: Bambec (Astra)

Bamethan
(Butylnorsynephrine; Butyloctopamine)

ATC: C04AA31
Use: sympathomimetic, vasodilator

RN: 3703-79-5   MF: $C_{12}H_{19}NO_2$ MW: 209.29   EINECS: 223-043-9
LD₅₀: 72 mg/kg (M, i.v.); 562 mg/kg (M, p.o.);
       80 mg/kg (R, i.v.)
CN: α-[(butylamino)methyl]-4-hydroxybenzenemethanol

sulfate (2:1)
RN: 5716-20-1   MF: $C_{12}H_{19}NO_2 \cdot 1/2H_2SO_4$ MW: 516.66   EINECS: 227-214-9
LD₅₀: 72 mg/kg (M, i.v.); 1600 mg/kg (M, p.o.);
       >1500 mg/kg (R, p.o.)

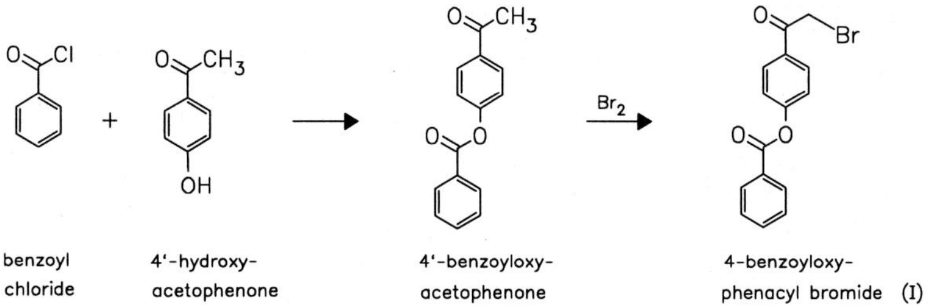

| benzoyl chloride | 4'-hydroxy-acetophenone | 4'-benzoyloxy-acetophenone | 4-benzoyloxy-phenacyl bromide (I) |

| butylamine | N-(4-benzoyloxy-phenacyl)butylamine | Bamethan |

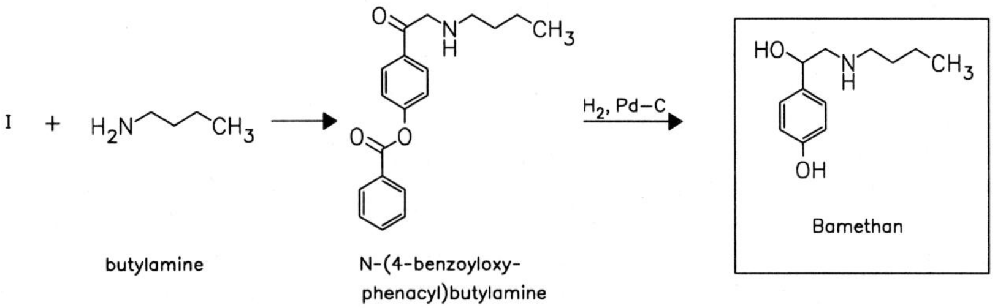

Reference(s):

Corrigan, J.R. et al.: J. Am. Chem. Soc. (JACSAT) **67**, 1894 (1945).

Formulation(s):   gel 1.5 g/100 g

Trade Name(s):

D:	Emasex (Eurim Pharma)-comb.	GB:	Vasculit (Boehringer Ing.); wfm	Pericardin (Santen-Yamanouchi)
	Heweven (Hevert)	I:	Vasculat (Boehringer Ing.;	Simpelate (Seiko Eiyo)
	Medigel (Medice)-comb.		as sulfate)	Valtolmin (Sanwa)
	Theo-Hexanicit (Astra/ Promed)	J:	Bloodbin (Nakataki)	Vasculat (Tanabe; as sulfate)
	Vasoforte N Kapseln (Krugmann)		Butibatol (Hishiyama) Butosin (Kobayashi)	Vasolat (Kanto)
F:	Escinogel (Doms-Adrian)-comb.		Cyclate (Hokuriku) Garmin (Fuso) Pan Line (Maruishi)	Vasolen (Toho) Vasstol (Nichiiko) Yonomol A (Sawai)

Bamifylline

ATC: R03BA; R03DA08
Use: bronchodilator, coronary vasodilator

RN: 2016-63-9   MF: $C_{20}H_{27}N_5O_3$ MW: 385.47
CN: 7-[2-[ethyl(2-hydroxyethyl)amino]ethyl]-3,7-dihydro-1,3-dimethyl-8-(phenylmethyl)-1H-purine-2,6-dione

monohydrochloride
RN: 20684-06-4   MF: $C_{20}H_{27}N_5O_3 \cdot HCl$ MW: 421.93   EINECS: 243-967-6

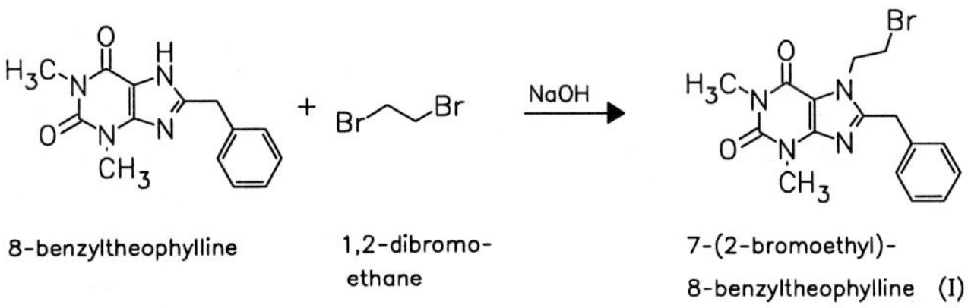

| 8-benzyltheophylline | 1,2-dibromo-ethane | 7-(2-bromoethyl)-8-benzyltheophylline (I) |

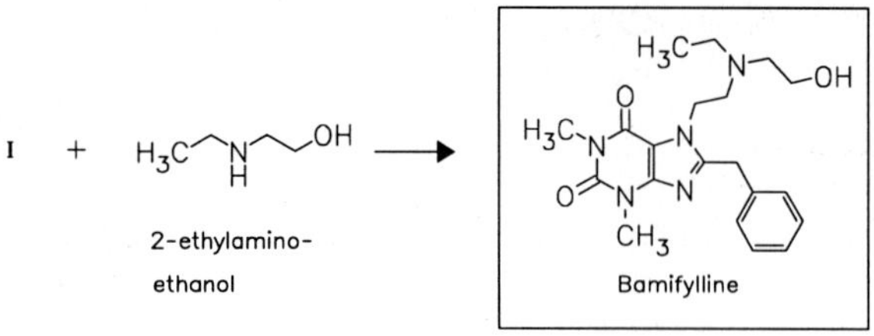

I   +   H₃C〜N(H)〜OH
2-ethylamino-
ethanol

Bamifylline

Reference(s):
BE 602 888 (A. Christiaens S.A.; appl. 21.4.1961; GB-prior. 22.4.1960).

Formulation(s):   inj. sol. 300 mg/5 ml; suppos. 250 mg, 750 mg; tabl. 300 mg

Trade Name(s):
D: Trentadil (Fresenius); wfm
F: Trentadil (Evans Medical)
      Trentadil (Sedaph); wfm

Trentadil injectable (Evans
Medical)
GB: Trentadil (Armour); wfm

I: Bamifix (Chiesi)
      Briafil (Alfa Wassermann)

Bamipine

ATC: D04AA15; R06AX01
Use: antihistaminic

RN: 4945-47-5   MF: C₁₉H₂₄N₂   MW: 280.42   EINECS: 225-587-2
LD₅₀: 250 mg/kg (M, p.o.)
CN: 1-methyl-N-phenyl-N-(phenylmethyl)-4-piperidinamine

monohydrochloride
RN: 1229-69-2   MF: C₁₉H₂₄N₂ · HCl   MW: 316.88
LD₅₀: 60 mg/kg (M, i.v.); 750 mg/kg (M, p.o.);
       460 mg/kg (R, p.o.);
       189 mg/kg (dog, p.o.)
lactate (1:1)
RN: 61670-09-5   MF: C₁₉H₂₄N₂ · C₃H₆O₃   MW: 370.49   EINECS: 262-887-2

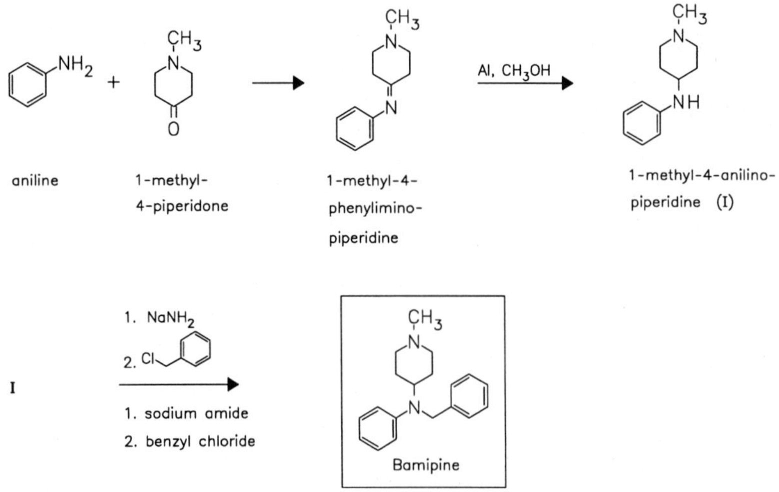

aniline

1-methyl-
4-piperidone

1-methyl-4-
phenylimino-
piperidine

Al, CH₃OH

1-methyl-4-anilino-
piperidine   (I)

I

1. NaNH₂
2. Cl〜

1. sodium amide
2. benzyl chloride

Bamipine

Reference(s):
US 2 683 714 (Knoll AG; 1954; D-prior. 1949).

Formulation(s):   cream 20 mg; drg. 20 mg; f. c. tabl. 50 mg; gel 20 mg

Trade Name(s):
D: Bamipin (ratiopharm)   F: Taumidrine (Knoll); wfm
      Soventol (Knoll)   I: Soventol (Knoll); wfm

Barbexaclone

ATC: N03AA04
Use: antiepileptic

RN: 4388-82-3   MF: $C_{12}H_{12}N_2O_3$ MW: 232.24   EINECS: 224-504-7
LD$_{50}$:   334 mg/kg (M, p.o.);
      306 mg/kg (R, p.o.)
CN: 5-ethyl-5-phenyl-2,4,6(1*H*,3*H*,5*H*)-pyrimidinetrione, compd. with (*S*)-*N*,α-dimethylcyclohexaneethanamine (1:1)

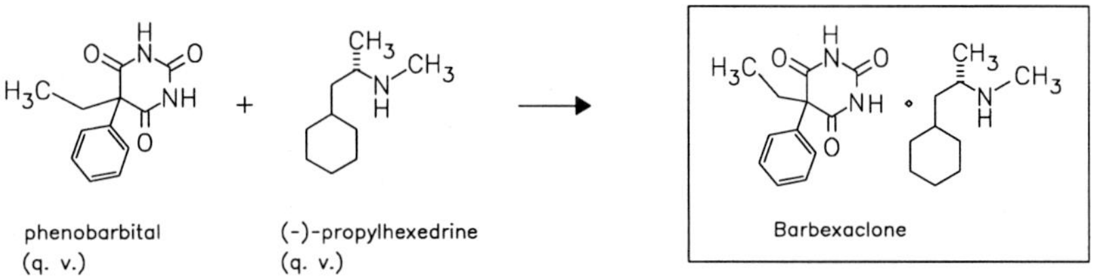

phenobarbital   (-)-propylhexedrine   Barbexaclone
(q. v.)   (q. v.)

Reference(s):
DE 1 120 452 (Knoll; appl. 16.4.1960).

Formulation(s):   drg. 100 mg, 25 mg

Trade Name(s):
D: Maliasin (Knoll)   I: Maliasin (Ravizza)

Barbital

ATC: N05CA04
Use: hypnotic

RN: 57-44-3   MF: $C_8H_{12}N_2O_3$ MW: 184.20   EINECS: 200-331-2
LD$_{50}$:   600 mg/kg (M, p.o.)
CN: 5,5-diethyl-2,4,6(1*H*,3*H*,5*H*)-pyrimidinetrione

monosodium salt
RN: 144-02-5   MF: $C_8H_{11}N_2NaO_3$ MW: 206.18   EINECS: 205-613-9
LD$_{50}$:   830 mg/kg (M, i.v.); 800 mg/kg (M, p.o.);
      280 mg/kg (R, i.v.); 600 mg/kg (R, p.o.)

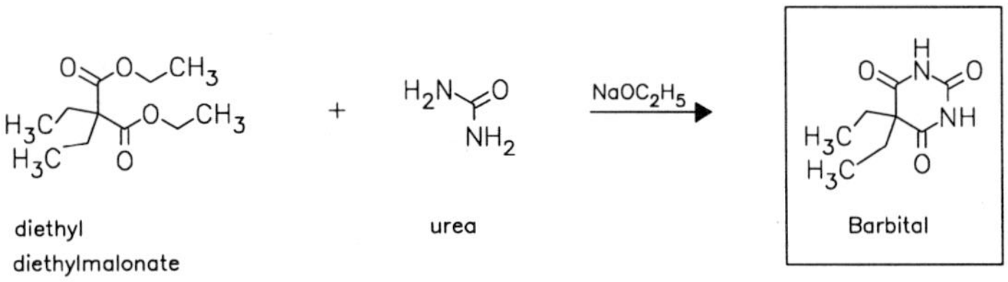

diethyl   urea   Barbital
diethylmalonate

Reference(s):
Fischer; Dilthey: Justus Liebigs Ann. Chem. (JLACBF) **335**, 338 (1904).

Formulation(s): tabl. 250 mg, 500 mg

Trade Name(s):
D: Barbimetten (Hormosan); F: combination preparations; I: Barbitt (Tariff. Integrativo)
wfm wfm Veronidia (Vaillant); wfm

Barnidipine

(Mepirodipine hydrochloride)

ATC: C08CA12
Use: antihypertensive agent, long active calcium antagonist

RN: 104757-53-1 MF: $C_{27}H_{29}N_3O_6 \cdot HCl$ MW: 528.01
CN: [S-(R^*,R^*)]-1,4-Dihydro-2,6-dimethyl-4-(3-nitrophenyl)-3,5-pyridinedicarboxylic acid methyl 1-(phenylmethyl)-3-pyrrolidinyl ester hydrochloride

free base
RN: 104713-75-9 MF: $C_{27}H_{29}N_3O_6$ MW: 491.54
racemate
RN: 71863-55-3 MF: $C_{27}H_{29}N_3O_6$ MW: 491.54

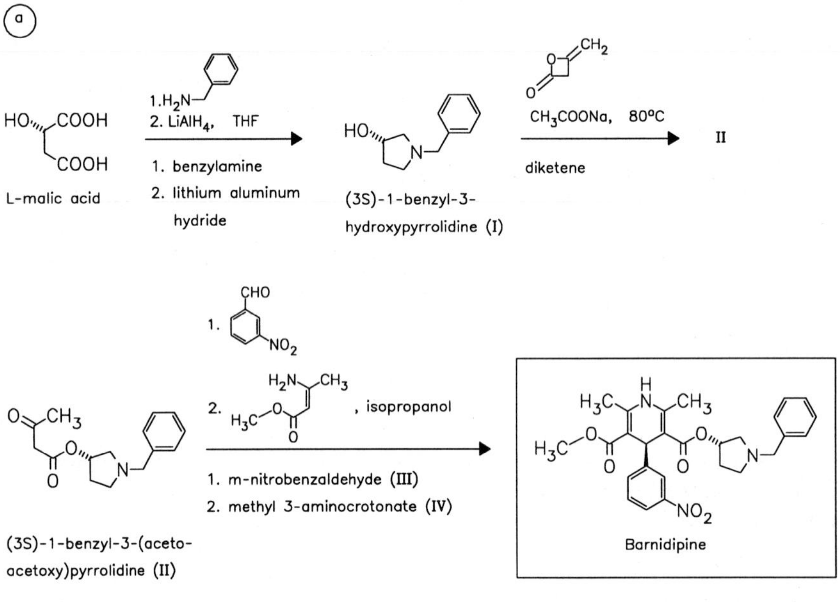

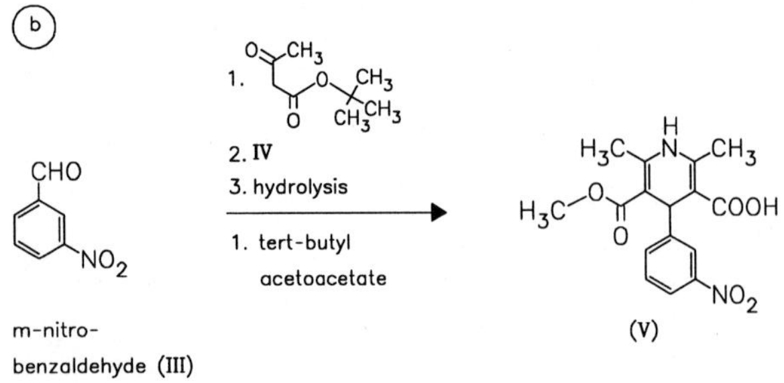

(a)

HO‚‚‚COOH
COOH
L-malic acid

1. H₂N—
2. LiAlH₄, THF

1. benzylamine
2. lithium aluminum hydride

(3S)-1-benzyl-3-hydroxypyrrolidine (I)

CH₃COONa, 80°C
diketene

II

(3S)-1-benzyl-3-(aceto-acetoxy)pyrrolidine (II)

1. m-nitrobenzaldehyde (III)
2. methyl 3-aminocrotonate (IV)

, isopropanol

Barnidipine

(b)

CHO
NO₂
m-nitro-benzaldehyde (III)

1. tert-butyl acetoacetate
2. IV
3. hydrolysis

(V)

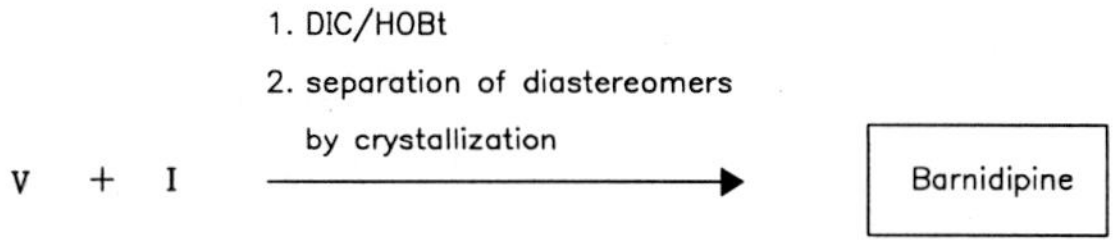

1. DIC/HOBt

2. separation of diastereomers

   by crystallization

V + I ⟶ | Barnidipine |

Reference(s):

a DE 2 904 552 (Yamanouchi Pharm.; appl. 7.2.1979; J-prior. 14.2.1978).
b CN 85 107 590 (Faming Zhuanli Sheqing Gonhai S.; appl. 11.10.1985; J-prior. 24.1.1985).

alternative syntheses:
Hirose, Y.; Kariya, K.; Sasaki, I.; Kuronom Y; Achiwa, K.: Tetrahedron Lett. (TELEAY) 34 (37), 5915 (1993).
JP 6 279 409 (Mercian Corp.; J-prior. 26.3.1993).
JP 7 070 066 (Amano Pharma Co.; prior. 3.9.1993).

alternative synthesis of optically active 1-benzyl-3-hydroxypyrrolidine:
JP 9 263 578 (Koei Chemical Co.; appl. 29.3.1996).

X-ray structure and synthesis of all enantiomers:
Tamazawa, K et al.: J. Med. Chem. (JMCMAR) 29 (12), 2504 (1986)

Formulation(s):   tabl. 5 mg, 10 mg, 15 mg (as hydrochloride)

Trade Name(s):
J: Hypoca R (Yamanouchi;
      1992)

Batroxobin

ATC: B02BX03
Use: anticoagulant, fibrinolytic

RN: 9039-61-6 MF: unspecified MW: unspecified EINECS: 232-918-4
LD_{50}: 384 μg/kg (M, i.v.);
       210 μg/kg (R, i.v.);
       380 μg/kg (dog, i.v.)
CN: bothrops atrox serine proteinase

Fibrinolytic effecting protease enzyme from the poison secretion (venom) of Bothrops atrox with glycoprotein structure. It has thrombin similarly endopeptidase activity.
Purification by chromatographic methods.

Reference(s):
US 3 849 252 (Pentapharm; 19.11.1974; CH-prior. 18.1.1971).
DOS 2 201 993 (Pentapharm; appl. 17.1.1972; CH-prior. 18.1.1971).

Formulation(s):   amp. 20 iu.

Trade Name(s):
I: Botropase (Ravizza)   J: Defibrase (Tobishi-
      Reptilase (Lepetit)   Fujisawa)

Beclamide

ATC: N03AX30
Use: antiepileptic, anticonvulsant

RN: 501-68-8   MF: C$_{10}$H$_{12}$ClNO   MW: 197.67   EINECS: 207-927-1
LD$_{50}$: 1 g/kg (M, p.o.);
        770 mg/kg (R, i.v.); 3200 mg/kg (R, p.o.)
CN: 3-chloro-N-(phenylmethyl)propanamide

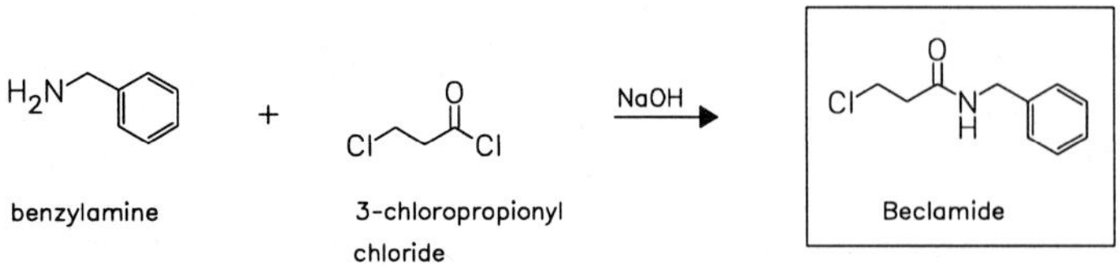

benzylamine   3-chloropropionyl   Beclamide
                     chloride

Reference(s):
US 2 569 288 (American Cyanamid; 1951; prior. 1949).

Formulation(s):   drg. 330 mg, 500 mg

Trade Name(s):
D: Neuracen (Promonta); wfm   F: Posédrine (Aron); wfm   I: Posedrine (Aron); wfm
      Posedrin (Promonta); wfm   GB: Nydrane (Rona); wfm

Beclobrate

ATC: B04AC
Use: hyperlipidemic

RN: 55937-99-0   MF: C$_{20}$H$_{23}$ClO$_3$ MW: 346.85   EINECS: 259-912-4
CN: (±)-2-[4-[(4-chlorophenyl)methyl]phenoxy]-2-methylbutanoic acid ethyl ester

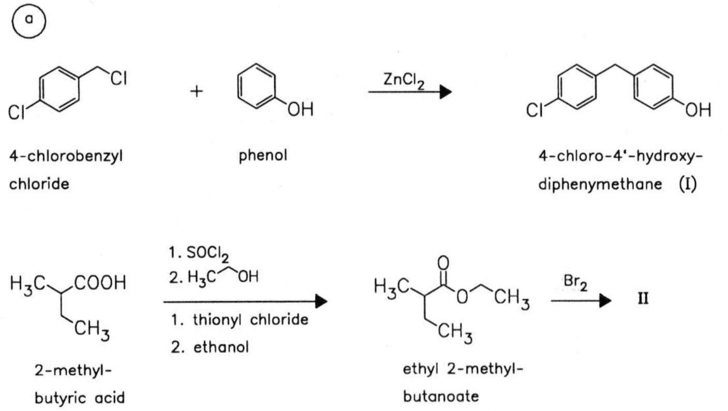

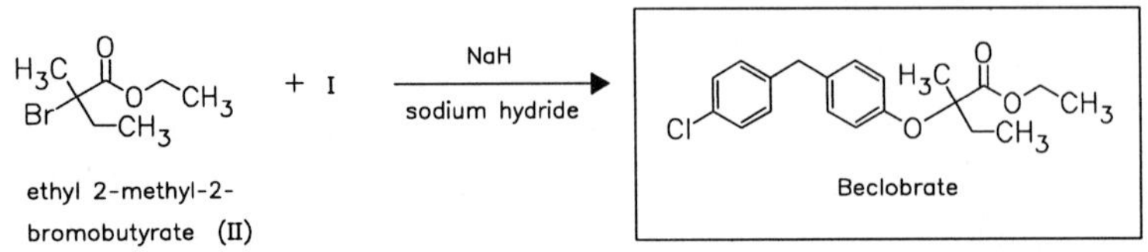

4-chlorobenzyl   phenol   4-chloro-4'-hydroxy-
chloride   diphenymethane (I)

2-methyl-   1. thionyl chloride   ethyl 2-methyl-
butyric acid   2. ethanol   butanoate

ethyl 2-methyl-2-   Beclobrate
bromobutyrate (II)

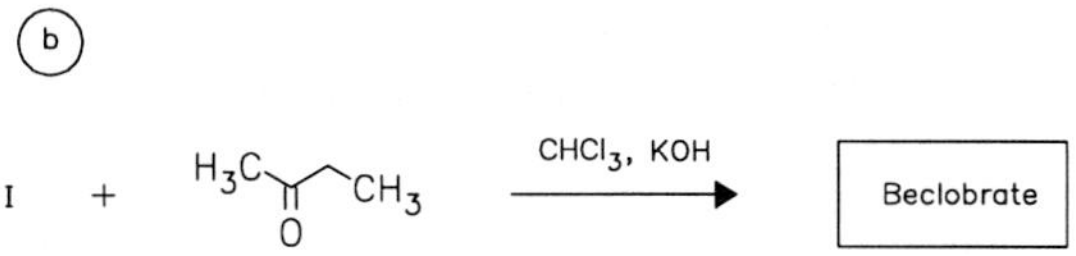

b)

I + H₃C—C(O)—CH₂—CH₃   → [Beclobrate]

ethyl methyl
ketone

(CHCl₃, KOH above arrow)

Reference(s):
DOS 2 461 069 (Siegfried; appl. 23.12.1974; CH-prior. 27.12.1973, 28.3.1974, 3.10.1974, 18.11.1974).
BE 823 904 (Siegfried; appl. 18.11.1974; CH-prior. 27.12.1973).
US 4 153 803 (Siegfried; 8.5.1979; CH-prior. 27.12.1973, 28.3.1974, 18.11.1974).
Thiele, K. et al.: Arzneim.-Forsch. (ARZNAD) **29**, 711 (1979).

synthesis of I:
Klarmann, E. et al.: J. Am. Chem. Soc. (JACSAT) **54**, 3315 (1932).
Huston, R.C. et al.: J. Am. Chem. Soc. (JACSAT) **55**, 4639 (1933).

synthesis of ethyl 2-methylbutanoate:
Gardner, R.: J. Chem. Soc. (JCSOA9) **1938**, 53.

Formulation(s):   tabl. 100 mg

Trade Name(s):
CH: Beclipur (Siegfried; 1988)   Beclosclerin (Siegfried;   Turec (Zyma; 1988)
                                         1988)

Beclometasone
(Beclomethasone)

ATC: A07EA07; D07AC15; R01AD01;
        R03BA01
Use: glucocorticoid

RN: 4419-39-0   MF: C₂₂H₂₉ClO₅   MW: 408.92   EINECS: 224-585-9
CN: (11β,16β)-9-chloro-11,17,21-trihydroxy-16-methylpregna-1,4-diene-3,20-dione

dipropionate
RN: 5534-09-8   MF: C₂₈H₃₇ClO₇   MW: 521.05   EINECS: 226-886-0
LD₅₀: >5 g/kg (M, p.o.);
       >3.75 g/kg (R, p.o.)

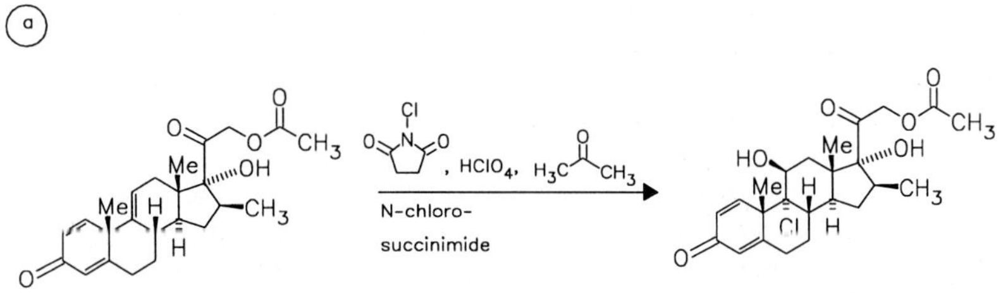

a)

21-acetoxy-17-hydroxy-16β-
methylpregna-1,4,9(11)-triene-
3,20-dione
(intermediate in syntheses
of betamethasone, q.v.)

beclometasone 21-acetate   (I)

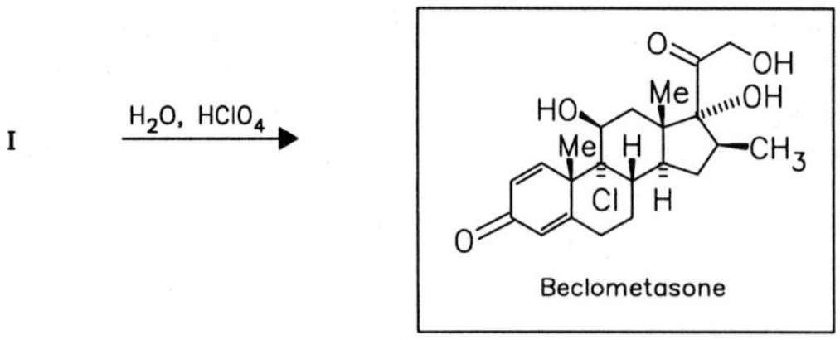

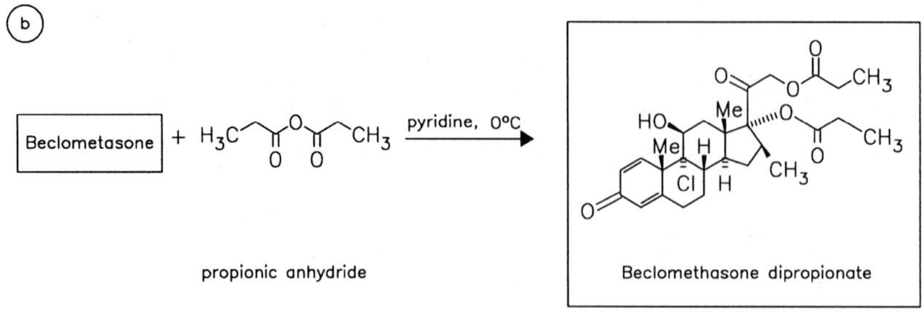

I $\xrightarrow{\text{H}_2\text{O, HClO}_4}$

Beclometasone

(b)

Beclometasone + H_3C propionic anhydride $\xrightarrow{\text{pyridine, 0°C}}$ Beclomethasone dipropionate

Reference(s):

a GB 912 378 (Merck & Co.; appl. 3.6.1959; USA-prior. 19.6.1958).
   GB 912 379 (Merck & Co.; appl. 3.6.1959; USA-prior. 19.6.1958).
   alternative synthesis:
   GB 901 093 (Scherico; appl. 22.7.1958; USA-prior. 22.7.1957).
   US 4 041 055 (Upjohn; 9.8.1977, appl. 17.11.1975).
b BE 649 170 (Glaxo; appl. 11.6.1964; GB-prior. 11.6.1963).
   FR 2 274 309 (Plurichemie; appl. 27.3.1975; P-prior. 27.3.1974, 10.3.1975).

medical use:
DOS 2 320 111 (Allen & Hanburys; appl. 19.4.1973; GB-prior. 20.4.1972).

Formulation(s):   cream 0.025 %; dose aerosol (0.05 µg, 0.25 µg/puff); nasal spray (0.05µg/puff); powder inhaler

Trade Name(s):

D: AeroBec Autohaler (3M
      Medica/ASTA Medica
      AWD)
      Beclorhinol (Lindopharm)
      Becloturmant (Desitin)
      Beconase (Glaxo)
      Beconase Aquosum
      (Glaxo)
      Beconase Dosier-Spray
      (Glaxo)
      Sanasthmax (Glaxo)
      Sanasthmyl (Glaxo)
      Sanasthmyl Dosier-Aerosol
      Rotadish (Glaxo)
      Viarox (Byk Essex)
      Viarox (Byk Gulden)
      Viarox Dosier-Aerosol
      (Byk Essex)
F: Beclojet (Promedica)
      Béconase (Glaxo
      Wellcome)

      Bécotide (Glaxo
      Wellcome)
      Prolair Autohaler (3M
      Santé)
      Rhinirex (Irex)
      Spir (Inava)
GB: Aerobec (3M)
      Asmabec (Evans)
      Beclazone (Baker Norton)
      Becloforte (Allen &
      Hanburys)
      Becodisks (Allen &
      Hanburys)
      Beconase (Allen &
      Hanburys)
      Becotide (Allen &
      Hanburys)
      Filair (3M)
      Propaderm (Glaxo
      Wellcome)
I: Becotide (Glaxo)
      Bronco-Turbinal (Valeas)

      Cleniderm (Chiesi)
      Clenigen (Chiesi)-comb.
      Clenil (Chiesi)
      Clenil spray (Chiesi)
      Inalone (Lampugnani)
      Menaderm (Menarini)
      Proctisone (Chiesi)-comb.
      Propaderm (Demcan)
      Rino-Clenil (Chiesi)
      Turbinal (Valeas)
J: Aldecin (Schering-Plough)
      Becloderm (Kobayashi)
      Beconase (Glaxo)
      Becotide (Nippon Glaxo)
      Becotide (Glaxo)
      Belg (Kowa)
      Betozon (Ohta)
      Betozon (Ohta Seiyaku)
      Entyderma (Taiyo)
      Hibisterin (Nippon Zoki)
      Korbutone (Nippon Glaxo)
      Mulunet (Tatsumi)

Propaderm (Shin Nihon;
Jitsugyo-Glaxo Fuji; as
dipropionate)
Rhinocort (Fujisawa)
Salcoat (Fujisawa)
Soluroid (Nikken)

USA: Beclovent (Glaxo
Wellcome)
Beconase (Glaxo
Wellcome)
Beconase (Glaxo
Wellcome; as dipropionate)

Vancenase (Schering-
Plough; as dipropionate)
Vanceril (Schering; as
dipropionate)

Befunolol

ATC: C07AA; S01ED06
Use: β-adrenoceptor blocker

RN: 39552-01-7   MF: $C_{16}H_{21}NO_4$ MW: 291.35
LD_{50}: 100-105 mg/kg (M, i.v.)
CN: (±)-1-[7-[2-hydroxy-3-[(1-methylethyl)amino]propoxy]-2-benzofuranyl]ethanone

hydrochloride
RN: 39543-79-8   MF: $C_{16}H_{21}NO_4 \cdot HCl$ MW: 327.81
LD_{50}: 65 mg/kg (M, i.v.); 950 mg/kg (M, p.o.);
922 mg/kg (R, p.o.)

2-hydroxy-3-
methoxy-
benzaldehyde

chloro-
acetone

2-acetyl-7-
methoxybenzo-
furan

1. hydrobromic acid
2. epichlorohydrin, piperidine

(±)-2-acetyl-7-glycidyl-
oxybenzofuran (I)

isopropyl-
amine

Befunolol

Reference(s):
DOS 2 223 184 (Kakenyaku Kako; appl. 12.5.1972; J-prior. 13.5.1971, 14.7.1971, 28.10.1971, 6.1.1972).
US 3 853 923 (Kakenyaku Kako; 10.12.1974; J-prior. 13.5.1971, 14.7.1971, 28.10.1971, 6.1.1972).
US 4 056 626 (Kakenyaku Kako; 1.11.1977; J-prior. 6.1.1972; 13.5.1971).

synthesis of 2-acetyl-7-methoxybenzofuran:
Bergel et al.: J. Chem. Soc. (JCSOA9), **1944**, 261.

Formulation(s):   eye drops 2.5 mg/ml, 5 mg/ml

Trade Name(s):
D: Glauconex (Alcon; 1984)
F: Bentos (CIBA Vision
Ophthalmics; 1987)

I: Betaclan (Angelini)
J: Bentos (Kaken; as
hydrochloride; 1983)

Bekanamycin
(Kanamycin B)

ATC: A07AA; J01KD; S01AA
Use: aminoglycoside antibiotic

RN: 4696-76-8   MF: $C_{18}H_{37}N_5O_{10}$ MW: 483.52   EINECS: 225-170-5
LD$_{50}$: 136 mg/kg (M, i.v.)
CN: O-3-amino-3-deoxy-α-D-glucopyranosyl-(1→6)-O-[2,6-diamino-2,6-dideoxy-α-D-glucopyranosyl-(1→4)]-2-deoxy-D-streptamine

sulfate (1:1)
RN: 29701-07-3   MF: $C_{18}H_{37}N_5O_{10} \cdot H_2SO_4$ MW: 581.60
LD$_{50}$: 112 mg/kg (M, i.v.);
         141 mg/kg (R, i.v.); >10 g/kg (R, p.o.)

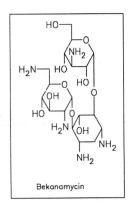

Bekanamycin

Fermentation of *Streptomyces kanamyceticus* (ATCC 12853) and precipitation with sodium dodecylphenylsulfonate.

Reference(s):
DAS 1 115 413 (H. Umezawa; appl. 1958; USA-prior. 1957).
US 2 967 177 (Bristol-Myers; 1961; prior. 1958).
US 3 032 547 (Merck & Co., 15.1.1962; prior. 12.9.1958).

alternative synthesis:
US 2 931 798 (H. Umezawa et al.; 1960; J-prior. 1956).
US 2 936 307 (Bristol-Myers; 1960; prior. 1957).

total synthesis:
Umezawa, S. et al.: Bull. Chem. Soc. Jpn. (BCSJA8) **42**, 537 (1969).

structure:
Ito, T.: J. Antibiot., Ser. A (JAJAAA) **17**, 189 (1964).

review:
Wakazawa, T. et al.: J. Antibiot., Ser. A (JAJAAA) **14A**, 180, 187 (1961).

Formulation(s):   cps. 250 mg; gran. 250 mg; powder; susp. 200 mg

Trade Name(s):
I: Kanendos (Crinos; as
      sulfate)
      Micomplex (Schiapparelli
      Searle)-comb.

      Visumicina (Merck Sharp
      & Dohme)-comb.
J: Kanendomycin (Meiji
      Seika; as sulfate)

Bemegride

ATC: R07AB05
Use: antidote for barbiturate poisoning, analeptic

RN: 64-65-3 MF: $C_8H_{13}NO_2$ MW: 155.20 EINECS: 200-588-0
LD_{50}: 16 mg/kg (M, i.v.); 41 mg/kg (M, p.o.);
16 mg/kg (R, i.v.)
CN: 4-ethyl-4-methyl-2,6-piperidinedione

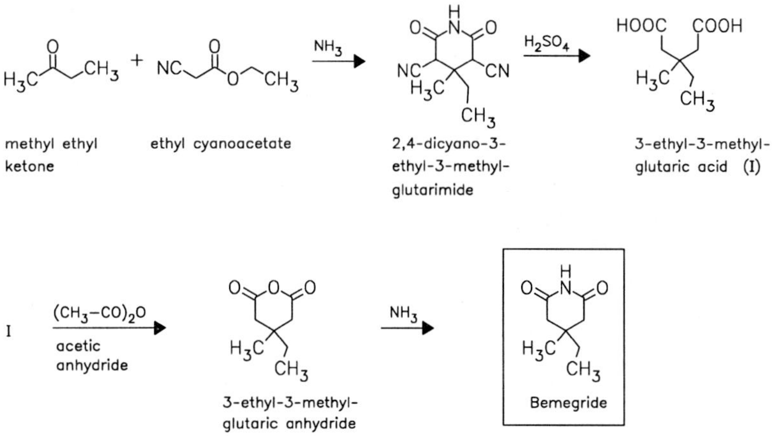
methyl ethyl ketone
ethyl cyanoacetate
NH3
2,4-dicyano-3-ethyl-3-methylglutarimide
H2SO4
3-ethyl-3-methylglutaric acid (I)
I
(CH3-CO)2O
acetic anhydride
3-ethyl-3-methylglutaric anhydride
NH3
Bemegride

Reference(s):
Benica, W.S.; Wilson, C.H.O.: J. Am. Pharm. Assoc. (JPHAA3) **39**, 451, 454 (1950).

Formulation(s): amp. 5 mg/ml (5 %, 10 %)

Trade Name(s):
D: Eukraton (Nordmark); wfm
F: Mégimide (Aspros-Nicholas); wfm
GB: Megimide (Nicholas); wfm
J: Antibarbi (Tanabe)

Bemetizide

ATC: C03E
Use: diuretic

RN: 1824-52-8 MF: $C_{15}H_{16}ClN_3O_4S_2$ MW: 401.90 EINECS: 217-357-5
LD_{50}: 345 mg/kg (M, i.v.); >5 g/kg (M, p.o.);
>5 g/kg (R, p.o.)
CN: 6-chloro-3,4-dihydro-3-(1-phenylethyl)-2*H*-1,2,4-benzothiadiazine-7-sulfonamide 1,1-dioxide

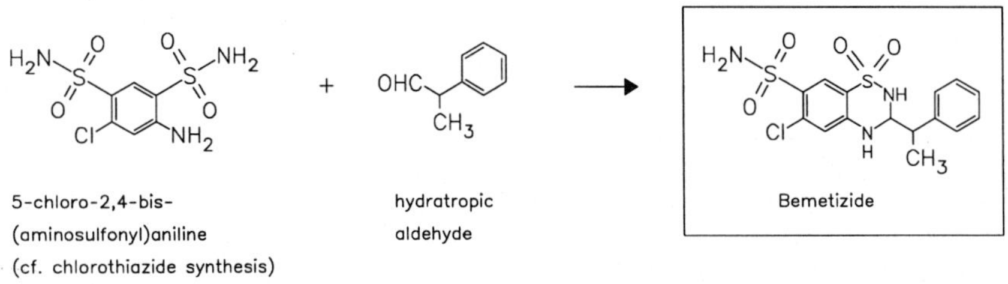
5-chloro-2,4-bis-(aminosulfonyl)aniline
(cf. chlorothiazide synthesis)
hydratropic aldehyde
Bemetizide

Reference(s):
AT 230 382 (Dr. H. Voigt; appl. 8.3.1961; D-prior. 23.2.1961).
Topliss, J.G. et al.: J. Org. Chem. (JOCEAH) **26**, 3842 (1961).
Jacobi, H.; Fontaine, R.: Arzneim.-Forsch. (ARZNAD) **16**, 1186, 1332 (1966).

Formulation(s):   drg. 10 mg, 20 mg

Trade Name(s):
D: Dehydro sanol (Sanol)-   Diucomb (Melusin)-comb.   F: Tensigradyl (Oberval)-
       comb.   comb.; wfm

Benactyzine

ATC: N04A
Use: ataractic, neuroleptic, anticholinergic

RN: 302-40-9   MF: $C_{20}H_{25}NO_3$ MW: 327.42   EINECS: 206-123-8
LD_{50}:   100 mg/kg (M, i.p.); 159 mg/kg (M, s.c.);
       135 mg/kg (R, i.m.)
CN: α-hydroxy-α-phenylbenzeneacetic acid 2-(diethylamino)ethyl ester

hydrochloride
RN: 57-37-4   MF: $C_{20}H_{25}NO_3 \cdot HCl$ MW: 363.89   EINECS: 200-324-4
LD_{50}:   14.3 mg/kg (M, i.v.); 160 mg/kg (M, p.o.);
       184 mg/kg (R, p.o.)

ethyl benzilate   2-diethylamino-   Benactyzine
                         ethanol

Reference(s):
US 2 394 770 (American Cyanamid; 1946; prior. 1942).

Formulation(s):   amp. 2 mg/ml, 0.3 %; tabl. 1 mg

Trade Name(s):
D: Brondiletten (Albert-   I: Pre Ciclo (Ibis)-comb.;   J: Morcain (Tatsumi); wfm
       Roussel)-comb.; wfm   wfm   Parpon (Santen); wfm
       Perasthman (Polypharm)-   Sirenitas (Benvegna)-
       comb.   comb.; wfm

Benaprizine
(Benapryzine)

ATC: N04
Use: antiparkinsonian

RN: 22487-42-9   MF: $C_{21}H_{27}NO_3$ MW: 341.45
CN: α-hydroxy-α-phenylbenzeneacetic acid 2-(ethylpropylamino)ethyl ester

hydrochloride
RN: 3202-55-9   MF: $C_{21}H_{27}NO_3 \cdot HCl$ MW: 377.91
LD_{50}:   500 mg/kg (M, p.o.)

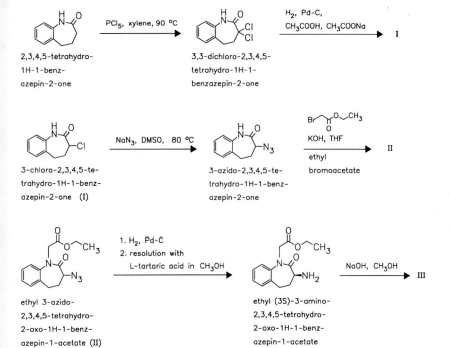

methyl benzilate   **2-(ethylpropylamino)- ethanol**   **Benaprizine**

Reference(s):
US 3 746 743 (Beecham; 17.7.1973; GB-prior. 22.8.1963).

Formulation(s):   tabl. 10 mg (as hydrochloride), 50 mg

Trade Name(s):
GB: Brizin (Beecham); wfm   I: Zinadril (Smith Kline
                                            Beecham)

Benazepril

(Benzapril)

ATC: C09AA07
Use: antihypertensive (ACE inhibitor)

RN: 86541-75-5   MF: $C_{24}H_{28}N_2O_5$ MW: 424.50
CN: [S-(R^*,R^*)]-3-[[1-(ethoxycarbonyl)-3-phenylpropyl]amino]-2,3,4,5-tetrahydro-2-oxo-1H-1-benzazepine-1-acetic acid

monohydrochloride
RN: 86541-74-4   MF: $C_{24}H_{28}N_2O_5 \cdot$ HCl   MW: 460.96

2,3,4,5-tetrahydro- 1H-1-benz- azepin-2-one

PCl₅, xylene, 90 °C

3,3-dichloro-2,3,4,5- tetrahydro-1H-1- benzazepin-2-one

H₂, Pd-C, CH₃COOH, CH₃COONa   I

3-chloro-2,3,4,5-te- trahydro-1H-1-benz- azepin-2-one (I)

NaN₃, DMSO, 80 °C

3-azido-2,3,4,5-te- trahydro-1H-1-benz- azepin-2-one

KOH, THF   II

ethyl bromoacetate

ethyl 3-azido- 2,3,4,5-tetrahydro- 2-oxo-1H-1-benz- azepin-1-acetate (II)

1. H₂, Pd-C
2. resolution with L-tartaric acid in CH₃OH

ethyl (3S)-3-amino- 2,3,4,5-tetrahydro- 2-oxo-1H-1-benz- azepin-1-acetate

NaOH, CH₃OH   III

(3S)-3-amino-
1-(carboxymethyl)-
2,3,4,5-tetrahydro-
1H-1-benzazepin-
2-one (III)

ethyl 2-oxo-
4-phenylbutyrate

Benazepril

Reference(s):
Watthey, J.W.H. et al.: J. Med. Chem. (JMCMAR) **28**, 1511 (1985).
US 4 410 520 (Ciba-Geigy; 18.10.1983; prior. 11.8.1981, 9.11.1981, 19.7.1982).
EP 72 352 (Ciba-Geigy; appl. 5.8.1982; USA-prior. 11.8.1981, 9.11.1981).
EP 206 993 (Ciba-Geigy; appl. 9.6.1986; CH-prior. 13.6.1985)

Formulation(s):   f. c. tabl. 5 mg, 10 mg, 20 mg (as hydrochloride)

Trade Name(s):

D:	Cibacen (Novartis Pharma)		Cibadrex (Novartis)-comb.	J:	Cibacen (Novartis; as
	Cibadrex (Novartis	I:	Cibadrex (Ciba-Geigy)-		hydrochloride)
	Pharma)-comb.		comb.	USA:	Lotensin (Ciba)
F:	Briazide (Pierre Fabre)-		Tensanil (Zyma)		Lotensin (Ciba)-comb. with
	comb.		Zinadur (Smith Kline		Hydrochlorothiazide
	Briem (Pierre Fabre)		Beecham)-comb.		Lotrel (Ciba)-comb. with
	Cibacére (Novartis)				Amlodipine

Bencyclane
(Benciclano)

ATC: C04AX11
Use: antispasmodic, vasodilator

RN: 2179-37-5   MF: $C_{19}H_{31}NO$ MW: 289.46   EINECS: 218-547-0
CN: *N,N*-dimethyl-3-[[1-(phenylmethyl)cycloheptyl]oxy]-1-propanamine

fumarate (1:1)
RN: 14286-84-1   MF: $C_{19}H_{31}NO \cdot C_4H_4O_4$ MW: 405.54   EINECS: 238-204-9
LD_{50}:   45 mg/kg (M, i.v.); 446 mg/kg (M, p.o.);
        41 mg/kg (R, i.v.); 414 mg/kg (R, p.o.)

benzylmag-
nesium chloride

cyclo-
heptanone

1-benzyl-
cycloheptanol

1. sodium amide
2. 3-dimethylamino-
propyl chloride

Bencyclane

Reference(s):
HU 151 865 (Egyesült Gyogyszer; appl. 18.8.1963).

Formulation(s):   drg. 75 mg, 100 mg

Trade Name(s):

D: Card-Fludilat (Thiemann)-   I: Angiociclan (Organon
      comb. with digoxin   Italia)
      Fludilat (Thiemann)   F: Novo-Card-Fludilat   J: Halidor (Sumitomo; as
                                     (Thiemann)-comb.   fumarate)
                                     Fludilat (Organon); wfm

Bendacort

(Bendacortone)

ATC: D07XA
Use: glucocorticoid

RN: 53716-43-1   MF: $C_{37}H_{42}N_2O_7$ MW: 626.75   EINECS: 258-710-3
CN: (11β)-11,17-dihydroxy-21-[[[[1-(phenylmethyl)-1H-indazol-3-yl]oxy]acetyl]oxy]pregn-4-ene-3,20-dione

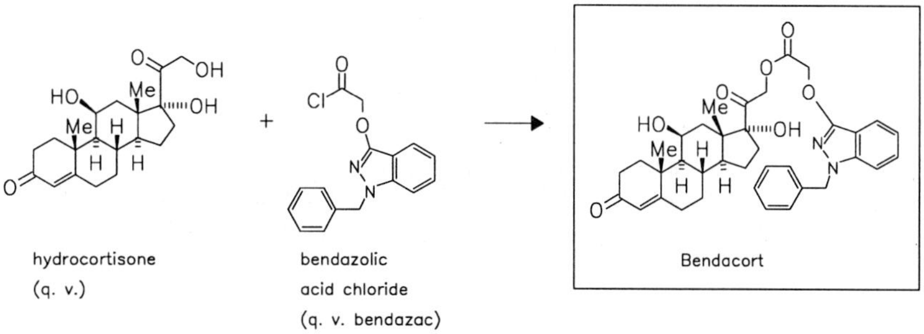

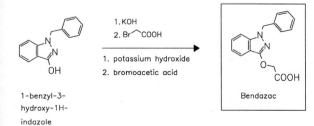

hydrocortisone   bendazolic   Bendacort
(q. v.)   acid chloride
                            (q. v. bendazac)

Reference(s):
DOS 2 601 367 (Angelini; appl. 15.1.1976; I-prior. 13.2.1975).

Formulation(s):   cream 3 %; ointment 3 %

Trade Name(s):
I: Versacort (Angelini)

Bendazac

(Bindazac; Acido bendazolico)

ATC: M02AA11; S01BC07
Use: anti-inflammatory

RN: 20187-55-7   MF: $C_{16}H_{14}N_2O_3$ MW: 282.30   EINECS: 243-569-2
LD$_{50}$:   380 mg/kg (M, i.v.); 1105 mg/kg (M, p.o.);
      304 mg/kg (R, i.v.); 1200 mg/kg (R, p.o.)
CN: [[1-(phenylmethyl)-1H-indazol-3-yl]oxy]acetic acid

sodium salt
RN: 23255-99-4   MF: $C_{16}H_{13}N_2NaO_3$ MW: 304.28   EINECS: 245-528-4

1. KOH
2. Br⌒COOH

1. potassium hydroxide
2. bromoacetic acid

1-benzyl-3-
hydroxy-1H-
indazole

Bendazac

Reference(s):
US 3 470 194 (Angelini Francesco; 30.9.1969; I-prior. 29.8.1966).

starting material:
Palazzo, G. et al.: J. Med. Chem. (JMCMAR) **9**, 38 (1966).

use in ointments, lotions etc.:
US 3 470 298 (Angelini Francesco; 30.9.1969; prior. 29.1.1969, 24.5.1968, 3.1.1967).
(Bendacort, q. v.)

Formulation(s):   cream 1 %, 3 %; ointment 1 %, 3 %

Trade Name(s):
I: Bendaline (Angelini)-   Versus (Angelini)
      comb. with lysine   J: Zildasac (Chugai)

Bendroflumethiazide
(Bendrofluazide)

ATC: C03AA01
Use: diuretic, antihypertensive

RN: 73-48-3   MF: $C_{15}H_{14}F_3N_3O_4S_2$ MW: 421.42   EINECS: 200-800-1
LD$_{50}$:   395 mg/kg (M, i.v.); >10 g/kg (M, p.o.)
CN: 3,4-dihydro-3-(phenylmethyl)-6-(trifluoromethyl)-2*H*-1,2,4-benzothiadiazine-7-sulfonamide 1,1-dioxide

3-trifluoromethyl-
aniline

4-amino-6-trifluoro-
methyl-1,3-benzene-
disulfochloride

2,4-diaminosulfonyl-
5-trifluoromethylaniline (I)

phenylacetyl
chloride

Bendroflumethiazide

β-ethoxystyrene

Bendroflumethiazide

Reference(s):
US 3 265 573 (Squibb; 9.8.1966, appl. 27.7.1962).
US 3 392 168 (Lovens Kemiske Fabrik; 9.7.1968; GB-prior. 13.8.1958).
Holdrege, C.T. et al.: J. Am. Chem. Soc. (JACSAT) **81**, 4807 (1959).

Formulation(s):   cps. 1.25 mg, 2.5 mg; tabl. 2.5 mg, 5 mg

Trade Name(s):
D: Docidrazin (Rhein-Pharma;   Dociretic (Thiemann)-   Pertenso (Fournier
      Zeneca)-comb.   comb.   Pharma)-comb.

Repicin (Boehringer Ing.)-comb.

Sali-Aldopur, - forte (Hormosan)-comb.

Sotaziden (Bristol-Myers Squibb)

Spirostada comp. -forte (Stadapharm)-comb.

Tensoflux (Hennig)-comb.

F: Naturine (Leo)

Precyclan-Leo (Leo)-comb.

Tensionorme (Leo)-comb.

GB: Aprinox (Knoll)

Corgaretic (Sanofi Winthrop)-comb.

Inderetic (Zeneca)-comb.

Inderex (Zeneca)-comb.

Neo-Naclex (Goldshield)

Prestim (Leo)-comb.

Tenben (Galen)-comb.

I: Idrexin-Na (Vermont); wfm

Menserene (Squibb)-comb.; wfm

Notens (Farge); wfm

Polidiuril (Bios); wfm

Salural (Icb); wfm

Sodiuretic (Squibb); wfm

J: Centyl (Leo-Sankyo)

Benexate

(TA-903)

ATC: A02BX

Use: cytoprotective agent (for treatment of gastric ulcer), chymotrypsin inhibitor

RN: 78718-52-2   MF: $C_{23}H_{27}N_3O_4$ MW: 409.49

LD_{50}:   7600 mg/kg (M, p.o.);
8010 mg/kg (R, p.o.)

CN: *trans*-2-[[[4-[[(aminoiminomethyl)amino]methyl]cyclohexyl]carbonyl]oxy]benzoic acid phenylmethyl ester

monohydrochloride

RN: 78718-25-9   MF: $C_{23}H_{27}N_3O_4 \cdot HCl$ MW: 445.95

monotosylate

RN: 82576-86-1   MF: $C_{23}H_{27}N_3O_4 \cdot C_7H_8O_3S$ MW: 581.69

monohydrochloride, clathrate with β-cyclodextrin (1:1)

RN: 86157-91-7   MF: $C_{23}H_{27}N_3O_4 \cdot HCl \cdot C_{42}H_{70}O_{35}$ MW: 1580.93

tranexamic acid (q. v.)

S-methyl-thiouronium chloride

trans-4-(guanidinomethyl)-cyclohexanecarboxylic acid (I)

1. DCC
2.

I   +

benzyl salicylate

1. dicyclohexylcarbodiimide
2. β-cyclodextrin

Benexate

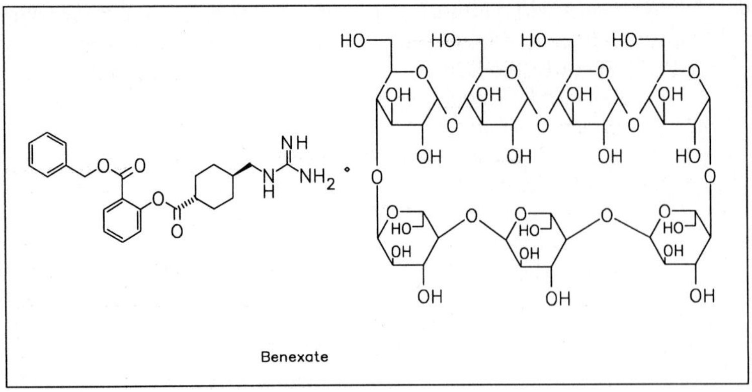

Benexate

Reference(s):
DE 3 035 086 (Nippon Chemiphar; appl. 17.9.1980; J-prior. 20.9.1979, 26.12.1979).
US 4 348 410 (Nippon Chemiphar, Teikoku Chem.; 7.9.1982; J-prior. 20.9.1979, 26.12.1979).

preparation of the clathrate with β-cyclodextrin:
EP 78 599 (Teikoku Chem.; appl. 27.8.1982; J-prior. 1.9.1981).

alternative synthesis:
JP 57 035 556 (Nippon Chemiphar; 26.2.1982; prior. 8.8.1980).
JP 88 051 146 (Nippon Chemiphar; 13.10.1988; prior. 8.8.1980).
Satoh, T. et al.: Chem. Pharm. Bull. (CPBTAL) **33**, 647 (1985).

Formulation(s):   cps. 200 mg

Trade Name(s):
J: Loumiel (Teikoku; as   Ulgut (Shionogi)
      hydrochloride β-
      cyclodextrin clathrate)

Benfluorex

ATC: B04AA; C10AX04
Use: appetite depressant

RN: 23602-78-0   MF: $C_{19}H_{20}F_3NO_2$ MW: 351.37   EINECS: 245-777-9
LD$_{50}$: 2300 mg/kg (M, p.o.)
CN: 2-[[1-methyl-2-[3-(trifluoromethyl)phenyl]ethyl]amino]ethanol benzoate (ester)

hydrochloride
RN: 23642-66-2   MF: $C_{19}H_{20}F_3NO_2 \cdot HCl$ MW: 387.83   EINECS: 245-801-8
LD$_{50}$: 108 mg/kg (M, i.p.)

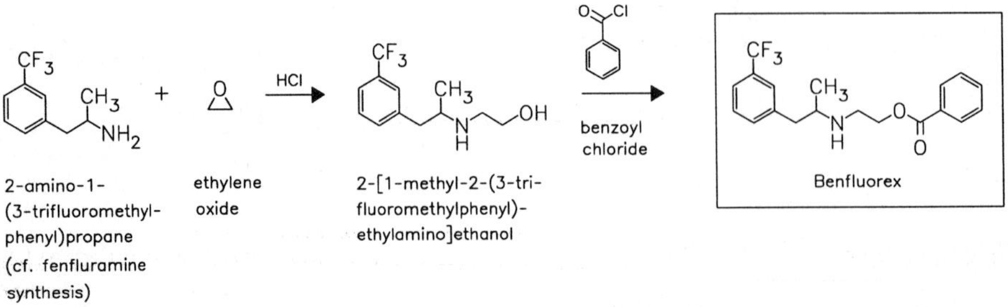

2-amino-1-   ethylene   2-[1-methyl-2-(3-tri-   Benfluorex
(3-trifluoromethyl- oxide   fluoromethylphenyl)-
phenyl)propane   ethylamino]ethanol
(cf. fenfluramine
synthesis)

benzoyl
chloride

Reference(s):
DE 1 593 991 (Science Union; appl. 14.4.1967; GB-prior. 15.4.1966).
FR 1 517 587 (Science Union; appl. 5.4.1967; GB-prior. 15.4.1966).
FR-M 6 564 (Science Union; appl. 3.7.1967; GB-prior. 15.4.1966).
US 3 607 909 (Science Union; 21.9.1971; GB-prior. 15.4.1966).

Formulation(s):   drg. 150 mg; tabl. 150 mg

Trade Name(s):
F: Mediator (Biopharma)   I: Mediaxal (Servier)   Minolip (Master Pharma)

Benfotiamine

ATC: A11DB
Use: neurotropic analgesic

RN: 22457-89-2   MF: $C_{19}H_{23}N_4O_6PS$ MW: 466.46   EINECS: 245-013-4
LD$_{50}$:   2200 mg/kg (M, i.v.); 15 g/kg (M, p.o.)
CN: benzenecarbothioic acid S-[2-[[(4-amino-2-methyl-5-pyrimidinyl)methyl]formylamino]-1-[2-(phosphonooxy)ethyl]-1-propenyl] ester

thiamine
(q. v.)

1. orthophosphoric acid
2. benzoyl chloride

Benfotiamine

Reference(s):
DE 1 130 811 (Sankyo Kabushiki Kaisha; appl. 14.4.1960; J-prior. 14.4.1959, 17.10.1959, 3.12.1959).

Formulation(s):   tabl. 40 mg, 50 mg, 100 mg, 300 mg

Trade Name(s):
D: Milgamma (Wörwag)-   Vitalgesic (Clin-Midy)-   I: Tridodilan (Roussel)-comb.
       comb.   comb.; wfm   J: Biotamin (Sankyo)
       Milneuron (Wörwag)-   Vitanevril (Clin-Comar-
       comb.   Byla); wfm
F: Vitalgesic (Clin-Comar-   Vitanevril (Clin-Midy);
       Byla)-comb.; wfm   wfm

Benfurodil hemisuccinate

ATC: C01D
Use: cardiotonic, vasodilator

RN: 3447-95-8   MF: $C_{19}H_{18}O_7$ MW: 358.35   EINECS: 222-367-8
LD$_{50}$:   520 mg/kg (M, p.o.)
CN: butanedioic acid mono[1-[5-(2,5-dihydro-5-oxo-3-furanyl)-3-methyl-2-benzofuranyl]ethyl] ester

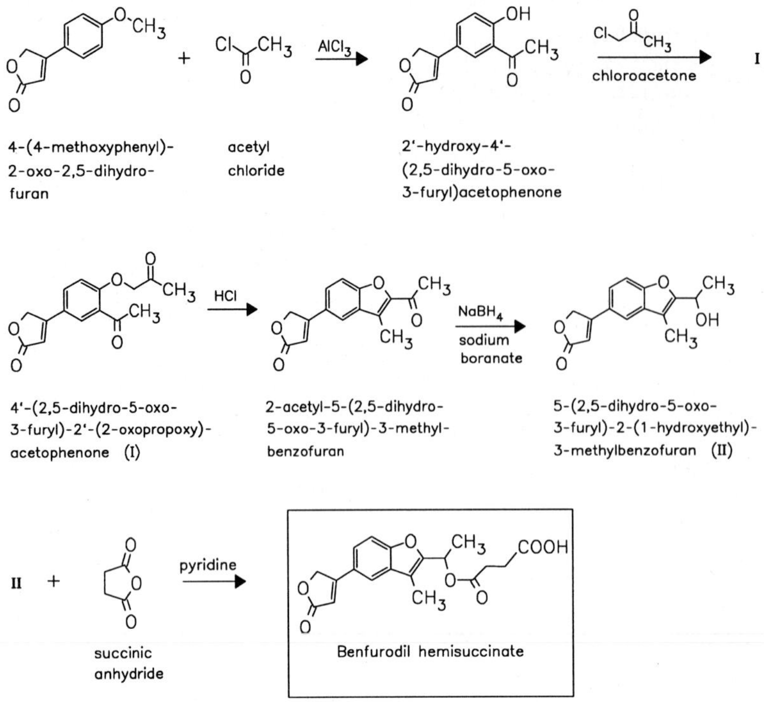

AlCl3
chloroacetone
I
4-(4-methoxyphenyl)-2-oxo-2,5-dihydrofuran
acetyl chloride
2'-hydroxy-4'-(2,5-dihydro-5-oxo-3-furyl)acetophenone
HCl
NaBH4
sodium boranate
4'-(2,5-dihydro-5-oxo-3-furyl)-2'-(2-oxopropoxy)-acetophenone (I)
2-acetyl-5-(2,5-dihydro-5-oxo-3-furyl)-3-methylbenzofuran
5-(2,5-dihydro-5-oxo-3-furyl)-2-(1-hydroxyethyl)-3-methylbenzofuran (II)
II
pyridine
succinic anhydride
Benfurodil hemisuccinate

Reference(s):
FR 1 408 721 (Clin-Byla; appl. 7.2.1964).
US 3 355 463 (Clin-Byla; 28.11.1967; F-prior. 7.2.1964).

Formulation(s): amp. 2.5 %/2 ml; tabl. 150 mg

Trade Name(s):
F: Eucilat (Clin-Comar-Byla);
wfm

Benidipine
(KW-3049)

ATC: C02DE
Use: calcium antagonist, antihypertensive, antianginal

RN: 105979-17-7 MF: $C_{28}H_{31}N_3O_6$ MW: 505.57
CN: (*R**,*R**)-(±)-1,4-dihydro-2,6-dimethyl-4-(3-nitrophenyl)-3,5-pyridinedicarboxylic acid methyl 1-(phenylmethyl)-3-piperidinyl ester

monohydrochloride
RN: 91599-74-5 MF: $C_{28}H_{31}N_3O_6 \cdot HCl$ MW: 542.03
LD_{50}: 21.5 mg/kg (M, i.p.); 2.5 mg/kg (M, i.v.); 322 mg/kg (M, p.o.); 33.5 mg/kg (M, s.c.);
15.1 mg/kg (R, i.p.); 4.4 mg/kg (R, i.v.); 87.6. mg/kg (R, p.o.); 276 mg/kg (R, s.c.)

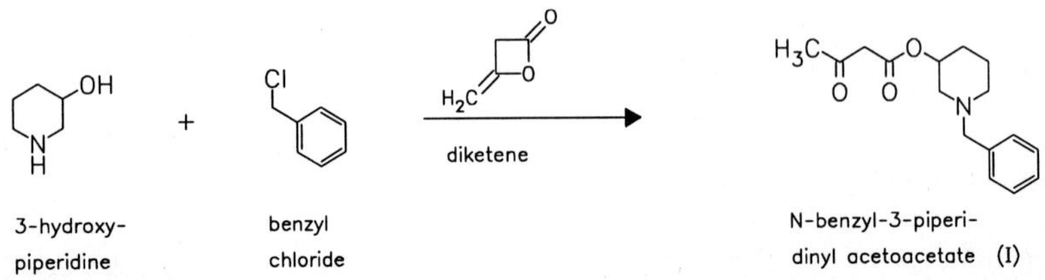

diketene
3-hydroxypiperidine
benzyl chloride
N-benzyl-3-piperidinyl acetoacetate (I)

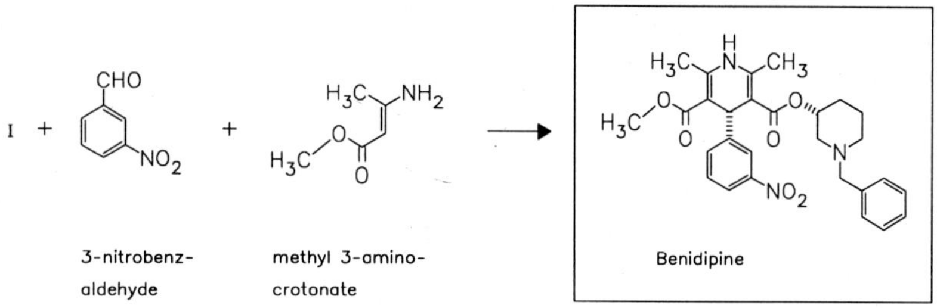

3-nitrobenz-   methyl 3-amino-   Benidipine
aldehyde   crotonate

Reference(s):
EP 63 365 (Kyowa Hakko; appl. 15.4.1982; J-prior. 17.4.1981).

alternative synthesis:
EP 106 275 (Kyowa Hakko; appl. 5.10.1983; J-prior. 15.10.1982, 27.1.1983; 3.6.1983).

Formulation(s):   tabl. 2 mg, 4 mg, 8 mg

Trade Name(s):
J: Coniel (Kyowa Hakko;
        1991)

Benmoxin

ATC: N06A
Use: antidepressant

RN: 7654-03-7   MF: $C_{15}H_{16}N_2O$ MW: 240.31   EINECS: 231-619-6
LD$_{50}$:   250 mg/kg (M, p.o.);
        675 mg/kg (R, p.o.)
CN: benzoic acid 2-(1-phenylethyl)hydrazide

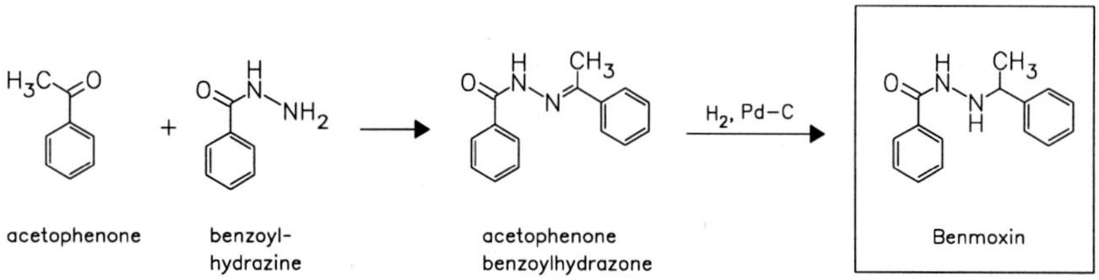

acetophenone   benzoyl-   acetophenone   Benmoxin
                  hydrazine   benzoylhydrazone

Reference(s):
GB 919 491 (ICI; appl. 1958; valid from 1959).
FR 1 314 362 (ICI; appl. 1959; GB-prior. 1958).

Trade Name(s):
F: Neuralex (Millot); wfm

Benorilate

(Benorylate; Benorilato)

ATC: N02BA10
Use: analgesic, antirheumatic

RN: 5003-48-5   MF: $C_{17}H_{15}NO_5$ MW: 313.31   EINECS: 225-674-5
LD$_{50}$:   1551 mg/kg (M, p.o.);
        3500 mg/kg (R, p.o.)
CN: 2-(acetyloxy)benzoic acid 4-(acetylamino)phenyl ester

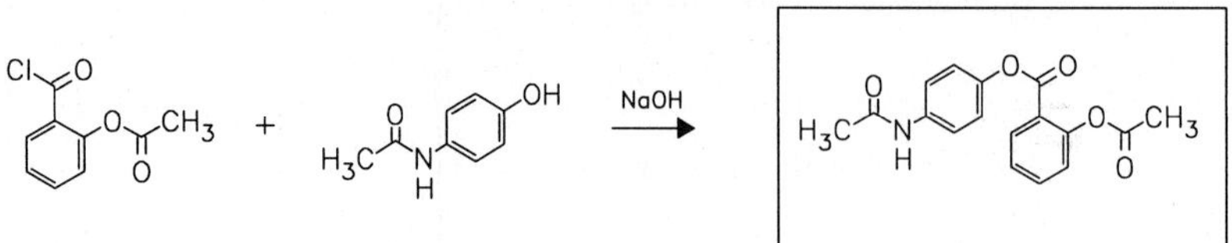

O-acetyl-
salicyloyl chloride

paracetamol
(q. v.)

Benorilate

Reference(s):
US 3 431 293 (Sterling Drug; 4.3.1969; GB-prior. 9.4.1964).
FR 1 436 870 (Sterwin; appl. 8.4.1965; GB-prior. 9.4.1964).

Formulation(s):   gran. 2 g; powder 2 g; susp. 2 g, 400 mg; tabl. 750 mg

Trade Name(s):
D: Benortan (Winthrop); wfm   Salipran (Evans Medical)   I: Bentum (Zambon); wfm
F: Benortan (Winthrop); wfm   GB: Benoral (Sanofi Winthrop)   Winolate (Winthrop); wfm

---

Benoxaprofen

ATC: M01AE06
Use: non-steroidal anti-inflammatory,
analgesic

RN: 51234-28-7   MF: $C_{16}H_{12}ClNO_3$ MW: 301.73   EINECS: 257-069-7
LD_{50}:   800 mg/kg (M, p.o.);
118 mg/kg (R, p.o.)
CN: 2-(4-chlorophenyl)-α-methyl-5-benzoxazoleacetic acid

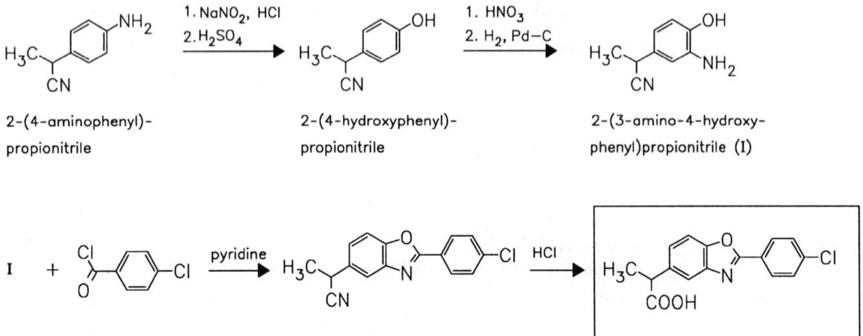

2-(4-aminophenyl)-
propionitrile

2-(4-hydroxyphenyl)-
propionitrile

2-(3-amino-4-hydroxy-
phenyl)propionitrile (I)

4-chlorobenzoyl
chloride

2-(4-chlorophenyl)-α-
methyl-5-benzoxazole-
acetonitrile

Benoxaprofen

Reference(s):
Dunwell, D.W. et al.: J. Med. Chem. (JMCMAR) **81**, 53 (1975).
DOS 2 324 443 (Lilly; appl. 15.5.1973; GB-prior. 18.5.1972).

Trade Name(s):
D: Coxigon (Lilly); wfm   F: Inflamid (Eli Lilly); wfm   GB: Opren (Dista); wfm

Benperidol
(Benzperidol)

ATC: N05AD07
Use: neuroleptic

RN: 2062-84-2 MF: $C_{22}H_{24}FN_3O_2$ MW: 381.45 EINECS: 218-172-2
LD$_{50}$: 20 mg/kg (M, i.v.); 432 mg/kg (M, p.o.);
21 mg/kg (R, i.v.)
CN: 1-[1-[4-(4-fluorophenyl)-4-oxobutyl]-4-piperidinyl]-1,3-dihydro-2*H*-benzimidazol-2-one

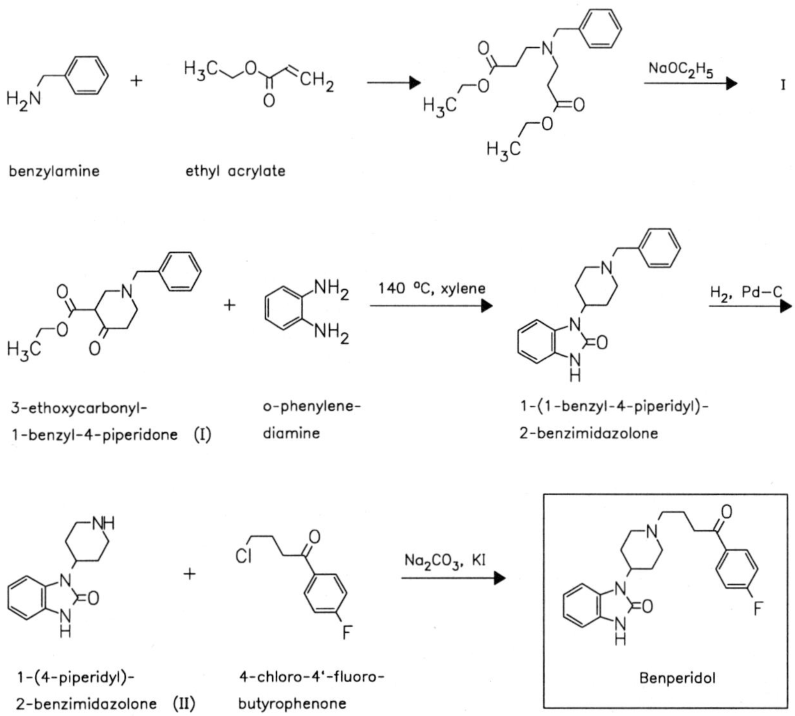

benzylamine
ethyl acrylate
NaOC2H5
I
3-ethoxycarbonyl-
1-benzyl-4-piperidone (I)
o-phenylene-
diamine
140 °C, xylene
1-(1-benzyl-4-piperidyl)-
2-benzimidazolone
H2, Pd-C
II
1-(4-piperidyl)-
2-benzimidazolone (II)
4-chloro-4'-fluoro-
butyrophenone
Na2CO3, KI
Benperidol

Reference(s):
GB 989 755 (Janssen; appl. 24.12.1962; USA-prior. 22.12.1961).
US 3 161 645 (Janssen; 15.12.1964; prior. 22.12.1961).
DE 1 470 120 (Janssen; appl. 19.12.1962; USA-prior. 22.12.1961).

Formulation(s): amp. 2 mg; drops 2 mg; tabl. 0.25 mg, 2 mg; 5 mg; 10 mg

Trade Name(s):

D:	Glianimon (Bayer Vital)		Frenactil (Clin-Midy); wfm
F:	Frenactil (Clin-Comar-Byla); wfm	GB:	Anquil (Janssen-Cilag)
		I:	Psicoben (Ravizza); wfm

Benproperine

ATC: R05DB02
Use: antitussive

RN: 2156-27-6 MF: $C_{21}H_{27}NO$ MW: 309.45
LD$_{50}$: 1087 mg/kg (M, p.o.)
CN: 1-[1-methyl-2-[2-(phenylmethyl)phenoxy]ethyl]piperidine

dihydrogen phosphate
RN: 19428-14-9 MF: $C_{21}H_{27}NO \cdot H_3PO_4$ MW: 407.45 EINECS: 243-050-0
LD_{50}: 32 mg/kg (M, i.v.); 1100 mg/kg (M, p.o.)

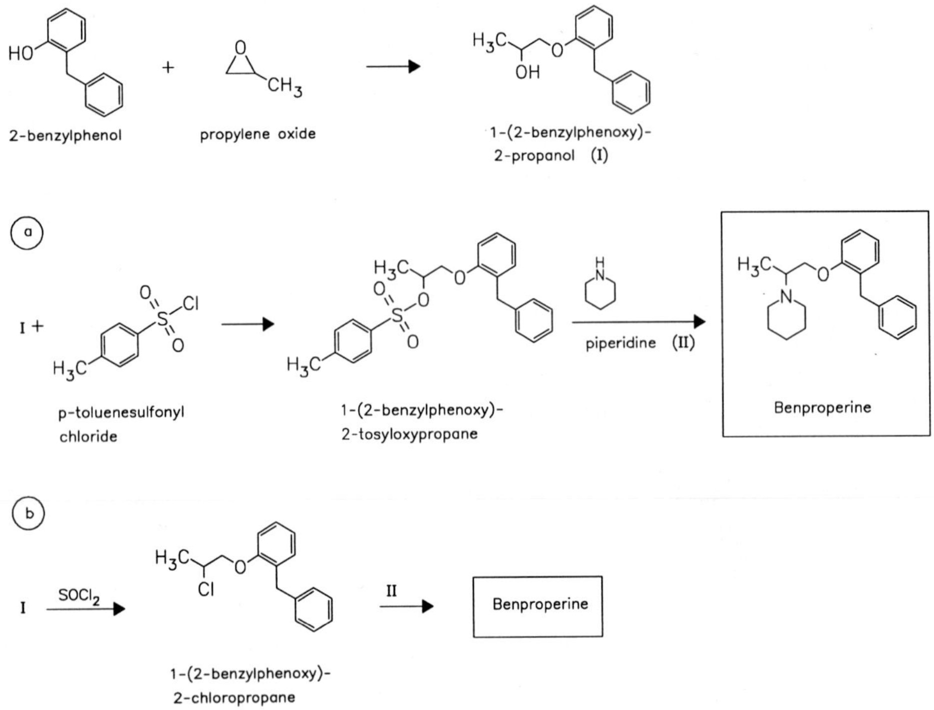

HO
+
O
CH3
H3C
O
OH
2-benzylphenol
propylene oxide
1-(2-benzylphenoxy)-
2-propanol (I)
a
I +
O
Cl
S
O
H3C
O
O
S
O
N
H
piperidine (II)
N
Benproperine
p-toluenesulfonyl
chloride
1-(2-benzylphenoxy)-
2-tosyloxypropane
b
I
SOCl2
Cl
II
Benproperine
1-(2-benzylphenoxy)-
2-chloropropane

Reference(s):
DAS 1 420 955 (Pharmacia; appl. 24.4.1961; DK-prior. 28.4.1960).
US 3 117 059 (Pharmacia; 7.1.1964; DK-prior. 28.4.1960).

Formulation(s): drg. 33 mg; susp. 15 mg; syrup 24.4 mg

Trade Name(s):

D:	Tussafug (Robugen)	I:	Blascorid Sosp. (Guidotti; as embonate)	J:	Flaveric (Taito Pfizer; as phosphate)

Benserazide

ATC: N04BA02
Use: antiparkinsonian (in combination with levodopa), decarboxylase inhibitor

RN: 322-35-0 MF: $C_{10}H_{15}N_3O_5$ MW: 257.25
CN: DL-serine 2-[(2,3,4-trihydroxyphenyl)methyl]hydrazide

monohydrochloride
RN: 14919-77-8 MF: $C_{10}H_{15}N_3O_5 \cdot HCl$ MW: 293.71 EINECS: 238-991-9
LD_{50}: 5 g/kg (M, p.o.);
5300 mg/kg (R, p.o.)

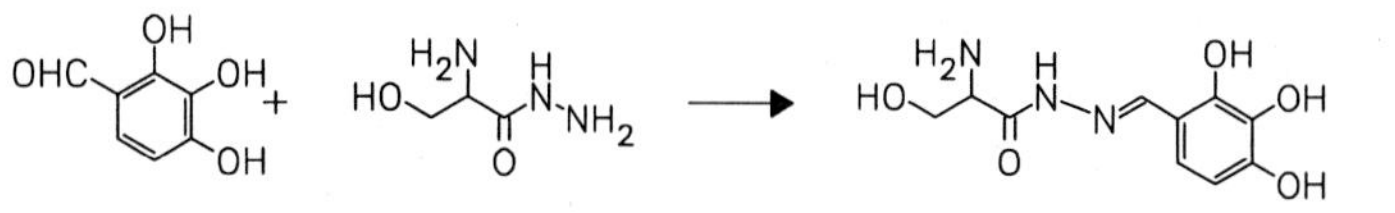

2,3,4-trihydroxy-
benzaldehyde

DL-serine
hydrazide

N-(DL-seryl)-2,3,4-
trihydroxybenzaldehyde hydrazone   (I)

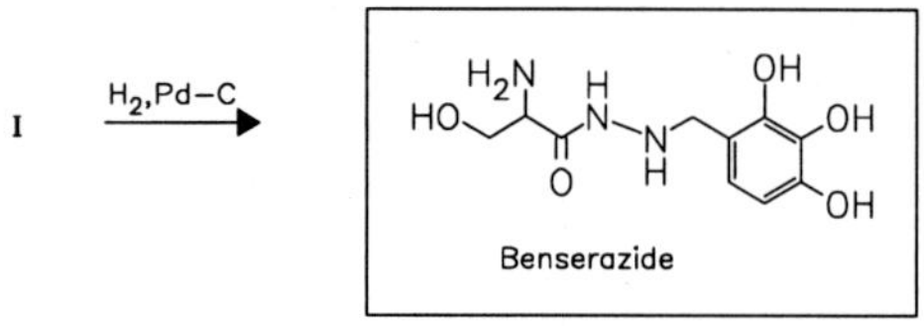

Benserazide

Reference(s):

DE 1 165 607 (Roche; appl. 8.5.1962; CH-prior. 16.6.1961).
US 3 178 476 (Roche; 13.4.1965; CH-prior. 16.6.1961).

L-*form:*

US 3 557 292 (Roche; 19.1.1971; appl. 16.8.1968).
DE 1 941 284 (Roche; appl. 13.8.1969; CH-prior. 16.8.1968).
DAS 1 966 821 (Roche; appl. 13.8.1969; CH-prior. 16.8.1968).

Formulation(s):   cps. 12.5 mg, 14.25 mg, 25 mg, 28.5 mg, 50 mg; dispersible tabl. 12.5 mg 25 mg; s. r. cps.
28.5 mg; tabl. 28.5 mg, 57 mg

Trade Name(s):

D:	Madopar (Roche)-comb. with levodopa	I:	Madopar (Roche)-comb.; wfm
F:	Modopar (Roche)-comb. with levodopa		Madopar (Roche)-comb. with levodopa; wfm
GB:	Madopar (Roche)-comb. with levodopa	J:	EC-doparl (Kyowa Hakko)-comb. with levodopa

Madopair (Roche)-comb.
with levodopa
Neodopasol (Daiichi)-
comb. with levodopa

Bentiamine

(Dibenthiamine; Dibenzoylthiamine)

ATC: A11
Use: vitamin B$_1$-derivative, neurotropic
analgesic

RN: 299-88-7   MF: $C_{26}H_{26}N_4O_4S$ MW: 490.58   EINECS: 206-084-7
LD$_{50}$:   7480 mg/kg (M, p.o.)
CN: benzenecarbothioic acid S-[2-[[(4-amino-2-methyl-5-pyrimidinyl)methyl]formylamino]-1-[2-
(benzoyloxy)ethyl]-1-propenyl] ester

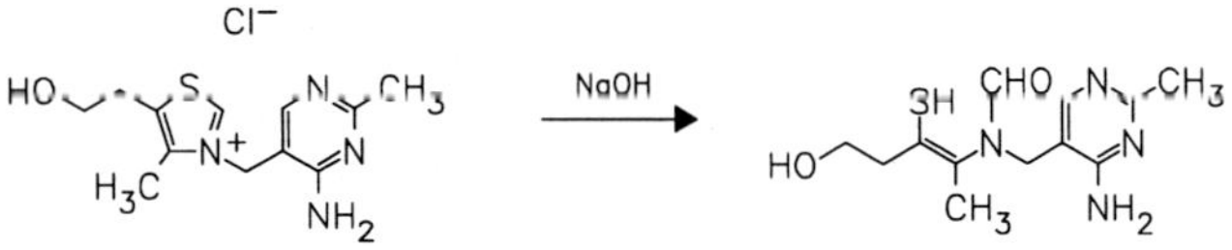

thiamine
(q. v.)

N-(4-amino-2-methylpyrimidin-
5-ylmethyl)-N-(4-hydroxy-1-
methyl-2-mercaptobut-1-enyl)-
formamide (I)

I + benzoyl chloride → Bentiamine

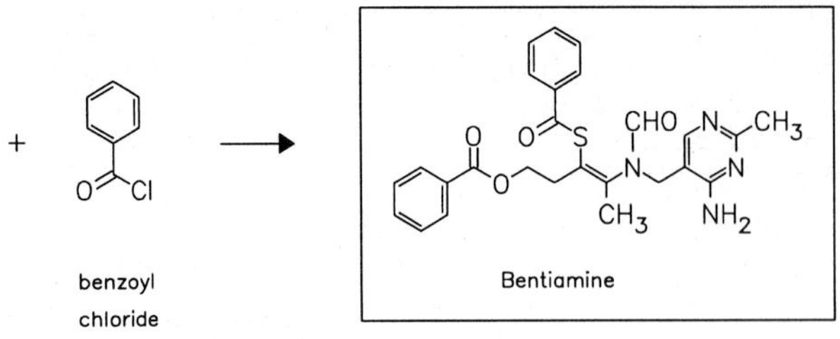

Reference(s):

US 2 752 348 (Takeda; 1956; J-prior. 1952).

Matsukawa, T.; Kawasaki, H.: Yakugaku Zasshi (YKKZAJ) **23**, 705 (1953).

Trade Name(s):

D: only combination
      preparations; wfm

Bentiromide

ATC: V04CK03
Use: pancreas function diagnostic

RN: 37106-97-1   MF: $C_{23}H_{20}N_2O_5$ MW: 404.42   EINECS: 253-349-8

LD_{50}:   1020 mg/kg (M, i.v.); >6 g/kg (M, p.o.);
       485 mg/kg (R, i.v.); >6 g/kg (R, p.o.)

CN: (*S*)-4-[[2-(benzoylamino)-3-(4-hydroxyphenyl)-1-oxopropyl]amino]benzoic acid

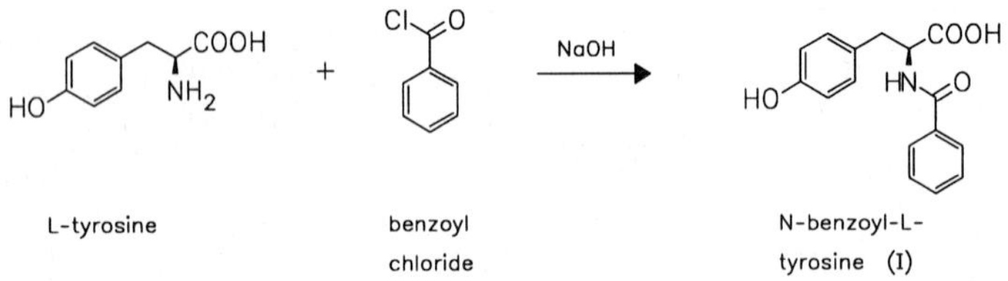

L-tyrosine   +   benzoyl chloride   NaOH →   N-benzoyl-L-tyrosine (I)

I

1. 4-methylmorpholine
2. ethyl chloroformate
3. 4-aminobenzoic acid

Bentiromide

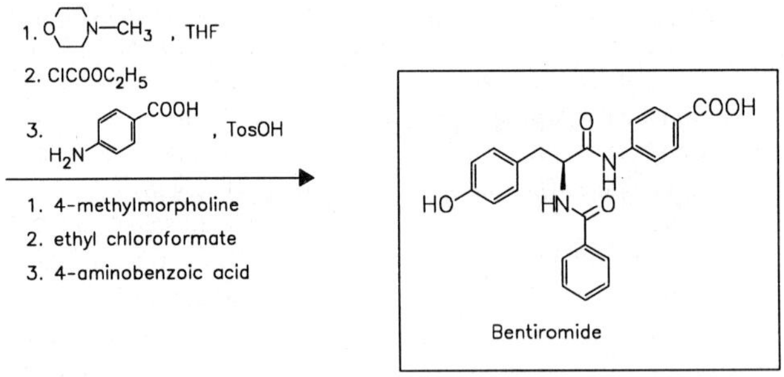

Reference(s):

Benneville, P.L. de et al.: J. Med. Chem. (JMCMAR) **15**, 1098 (1972).

US 3 745 212 (Rohm & Haas; 10.7.1973; appl. 19.11.1970).

DE 2 156 835 (Rohm & Haas; appl. 16.11.1971; USA-prior. 19.11.1970).

Formulation(s):   sol. 500 mg/10 ml; tabl. 333 mg

Trade Name(s):

D: PFT Roche (Roche); wfm   J: PFD (Eisai); wfm   USA: Chymex (Adria); wfm

Benzalkonium chloride

ATC: D08AJ01; D09AA11; R02AA16
Use: antiseptic, cation active tenside

RN: 8001-54-5   MF: unspecified   MW: unspecified
CN: benzalkonium chloride

mixture of
fatty alcohols
(by hydrogenation of
the mixture of coconut
fatty acids)

R: $C_8H_{17} - C_{18}H_{37}$

dimethylamine

mixture of
N,N-dimethyl-
alkylamines (I)

benzyl
chloride

Benzalkonium chloride

Reference(s):
Ehrhart, Ruschig **IV**, 50.
Guyer et al.: Helv. Chim. Acta (HCACAV) **20**, 1462 (1937).
Ralston, A.W. et al.: J. Am. Chem. Soc. (JACSAT) **69**, 2095 (1947).

Formulation(s):   nail lacquer 1 oz.; sol. 1 oz.

Trade Name(s):

D: Baktonium (Bode)
       Laudamonium (Henkel)
       Lysoform-Killovon
       (Lysoform)
       Sagrotan Med (Schülke &
       Mayr)
       and 100 more combination
       preparations
F: Biseptine (Nicholas)-comb.
       Chlorure de benzalkonium
       Théramex (Théramex)
       Kenalcol (Bristol-Myers
       Squibb)
       Pharmatex (Innothéra)
       Rhinoflumicin (Zambon)-
       comb.
       Sparaplaie Na (Médicine
       Végetale)
GB: Bradosol (Novartis)

       Conotrane (Yamanouchi)-
       comb.
       Dermol (Dermal)-comb.
       Drapolene (Warner-
       Lambert)-comb.
       Emulsiderm (Dermal)-
       comb.
       Ionil T (Alcon)-comb.
       Oilatum Plus (Stiefel)-
       comb.
       Timodine (Reckitt &
       Colman)-comb.
I: Alfac (Bracco)
       Alfafluorone
       (Biotekfarma)-comb.
       Atisteril (Ati)
       Benzal (Tariff. Nazionale)
       Citralkon (Schiapparelli
       Salute)

       Citrosil (Glaxo)
       Dil Mill (SIT)
       Herbagola propoli (Grica
       Chemical)
       Lacribase Saluzine
       (Allergan)
       Quatersal (Ascor)
       Sapocitrosil (Glaxo)
       Steramina "G" (Formenti)
       Streptosil (Boehringer Ing.)
       Video bagno (Farmila)
       Video gocce (Farmila)
       Vittoria Lazione
       (Ottolenghi)
J: Osvan (Daigo-Takeda)
USA: Amino-Cerv (Milex)-comb.
       Ony-Clear (Pedinol)
       Zephiran (Winthrop-Breon)

Benzarone

ATC: C05CX
Use: antihemorrhagic, antispasmodic, vein therapeutic

RN: 1477-19-6   MF: $C_{17}H_{14}O_3$ MW: 266.30   EINECS: 216-026-2
LD$_{50}$: >12 g/kg (M, p.o.);
>12 g/kg (R, p.o.)
CN: (2-ethyl-3-benzofuranyl)(4-hydroxyphenyl)methanone

salicyl-   chloroacetone   2-acetyl-   2-ethylbenzofuran   (I)
aldehyde   benzofuran

4-methoxy-   2-ethyl-3-(4-methoxy-   Benzarone
benzoyl chloride   benzoyl)benzofuran

Reference(s):
DE 1 076 702 (Labaz; appl. 20.12.1957; B-prior. 21.12.1956).
US 3 012 042 (Labaz; 5.12.1961; B-prior. 21.12.1956).

alternative synthesis of 2-acetylbenzofuran *(from* benzofuran *and* acetic anhydride/H_3PO_4):
Buu-Hoï, N.P.: J. Chem. Soc. (JCSOA9) **1964**, 173.

Formulation(s):   tabl. 100 mg

Trade Name(s):
D: Fragivix (Sanol); wfm   F: Derol (Labaz)-comb. with   I: Fragivix (Sigma-Tau); wfm
      Vasoc (Lindopharm); wfm   lidocaine; wfm   Venagil (Logifarm); wfm
                                           Fragivix (Labaz); wfm   Venagil (Scalari); wfm

---

Benzathine benzylpenicillin

(Benethamine Penicilline; Benzilpenicillin; Penicillin G Benzathine)

ATC: J01CE08
Use: depot antibiotic

RN: 1538-09-6   MF: $C_{16}H_{18}N_2O_4S \cdot 1/2C_{16}H_{20}N_2$ MW: 909.14   EINECS: 216-260-5
LD$_{50}$: 2 g/kg (M, p.o.)
CN: [2S-(2α,5α,6β)]-3,3-dimethyl-7-oxo-6-[(phenylacetyl)amino]-4-thia-1-azabicyclo[3.2.0]heptane-2-carboxylic acid compd. with N,N'-bis(phenylmethyl)-1,2-ethanediamine (2:1)

ethylene-   benzaldehyde   1,2-bis(benzyliden-
diamine   amino)ethane

N,N'-dibenzyl-   benzylpenicillin   Benzathine benzylpenicillin
ethylenediamine   (I)

Reference(s):
US 2 627 491 (Wyeth; 1953; prior. 1950).

Formulation(s):   gel 0.1 g/100 g

Trade Name(s):
D: Depotpen (Dauelsberg)-   F: Extencilline (Specia)   Wycillina A. P. (Carlo
      comb.   GB: Penidural (Wyeth); wfm   Erba)
      Sulfa-Tardocillin (Bayer)-   I: Benzil B (Formulario Naz.)   J: Bicillin (Banyu)
      comb.   Tri-Wycillina A. P. (Carlo   USA: Bicillin (Wyeth); wfm
      Tardocillin (Bayer)   Erba)-comb.   Permapen (Pfizer); wfm

Benzatropine
(Benztropine)

ATC: N04AC01
Use: parasympatholytic, antiparkinsonian

RN: 86-13-5   MF: C$_{21}$H$_{25}$NO   MW: 307.44
LD$_{50}$:   25 mg/kg (M, i.v.)
CN: endo-3-(diphenylmethoxy)-8-methyl-8-azabicyclo[3.2.1]octane

mesylate
RN: 132-17-2   MF: C$_{21}$H$_{25}$NO · CH$_4$O$_3$S   MW: 403.54   EINECS: 205-048-8
LD$_{50}$:   24 mg/kg (M, i.v.); 91 mg/kg (M, p.o.);
        940 mg/kg (R, p.o.)

diphenyl-   tropine   Benzatropine
diazomethane

Reference(s):
US 2 595 405 (Merck & Co.; 1952; prior. 1949).

Formulation(s):   amp. 2 mg; tabl. 0.5 mg, 1mg, 2 mg

Trade Name(s):
D: Cogentinol (Astra)   GB: Cogentin (Merck Sharp &   USA: Cogentin (Merck Sharp &
F: Cogentine (Merck Sharp &   Dohme; as mesylate)   Dohme; as mesylate)
      Dohme); wfm

Benzbromarone

ATC: M04AB; N04AC01
Use: uricosuric agent

RN: 3562-84-3　MF: $C_{17}H_{12}Br_2O_3$　MW: 424.09　EINECS: 222-630-7
LD_{50}: 77 mg/kg (M, i.v.); 618 mg/kg (M, p.o.);
248 mg/kg (R, p.o.)
CN: (3,5-dibromo-4-hydroxyphenyl)(2-ethyl-3-benzofuranyl)methanone

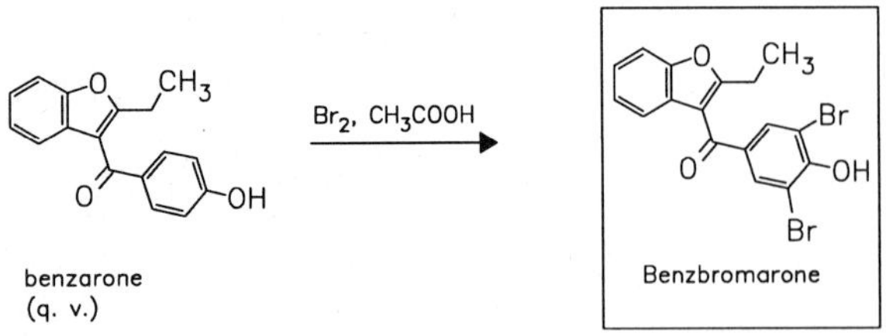

benzarone
(q. v.)

Br_2, CH_3COOH

Benzbromarone

Reference(s):
DE 1 080 144 (Labaz; appl. 20.12.1957; B-prior. 21.12.1956).
US 3 012 042 (Labaz; 5.12.1961; B-prior. 21.12.1956).

combination with allopurinol:
GB 1 493 237 (Henning Berlin; appl. 11.5.1976; D-prior. 10.12.1975).

Formulation(s):　f. c. tabl. 20 mg

Trade Name(s):
D: Acifugan (Henning Berlin)-comb.
Allomaron (Nattermann)-comb.

Azubromaron (Azupharma)
Harpagin (Merz & Co.)
Narcaricin (Heumann)
F: Désuric (Sanofi Winthrop)

I: Desuric (Sigma-Tau); wfm
J: Urinorm (Torii)

Benzethonium chloride

ATC: R02AA09
Use: disinfectant, antiseptic

RN: 121-54-0　MF: $C_{27}H_{42}ClNO_2$　MW: 448.09　EINECS: 204-479-9
LD_{50}: 30 mg/kg (M, i.v.); 338 mg/kg (M, p.o.);
19 mg/kg (R, i.v.); 368 mg/kg (R, p.o.)
CN: N,N-dimethyl-N-[2-[2-[4-(1,1,3,3-tetramethylbutyl)phenoxy]ethoxy]ethyl]benzenemethanaminium chloride

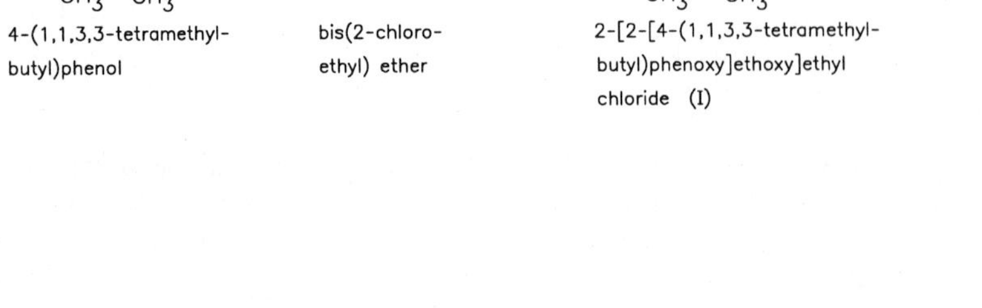

4-(1,1,3,3-tetramethyl-butyl)phenol

bis(2-chloro-ethyl) ether

NaOH

2-[2-[4-(1,1,3,3-tetramethyl-butyl)phenoxy]ethoxy]ethyl
chloride　(I)

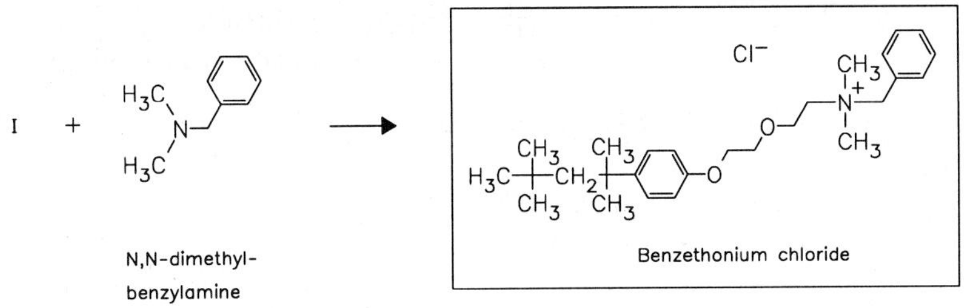

N,N-dimethyl-
benzylamine

Benzethonium chloride

Reference(s):

US 2 115 250 (Rohm & Haas; 1938; appl. 1936).
US 2 170 111 (Rohm & Haas; 1939; appl. 1936).
US 2 229 024 (Rohm & Haas; 1941; appl. 1939).

Formulation(s):   many different formulations

Trade Name(s):

D:	Brand- und Wundgel (Medica)-comb.		Ta-Ro-Cap (Soekami)-comb.; wfm		Ribex Gola (Formenti) Sterilix (Formenti)
F:	Alcolène (Alcon)-comb.; wfm		Vasol (Fumouze)-comb.; wfm	J:	Hyamine-T (Sankyo) Neostelin-Green (Bayer-Nihonshika)
	Ineka (Soekami)-comb.; wfm	GB:	Emko (Syntex)-comb.		
		I:	Air Sanitzer (Chifa)		

Benzilonium bromide

ATC: A03AB
Use: anticholinergic

RN: 1050-48-2   MF: $C_{22}H_{28}BrNO_3$ MW: 434.37   EINECS: 213-885-5
LD_{50}:   11.2 mg/kg (M, i.v.); 363 mg/kg (M, p.o.);
        760 mg/kg (R, p.o.)
CN: 1,1-diethyl-3-[(hydroxydiphenylacetyl)oxy]pyrrolidinium bromide

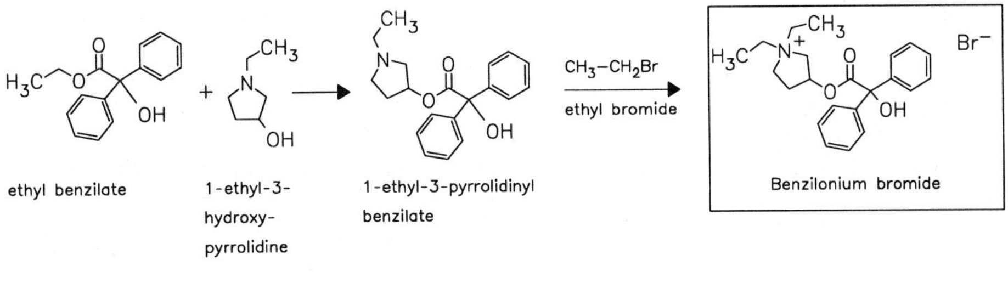

ethyl benzilate   1-ethyl-3-hydroxy-pyrrolidine   1-ethyl-3-pyrrolidinyl benzilate   ethyl bromide   Benzilonium bromide

Reference(s):

GB 821 436 (Parke Davis; appl. 22.2.1956).
DE 1 136 338 (Parke Davis; appl. 12.2.1957; GB-prior. 22.2.1956, 29.1.1957).

Formulation(s):   cps. 10 mg; tabl. 10 mg

Trade Name(s):

D:	Minelcin (Parke Davis); wfm	F:	Portyn (Parke Davis); wfm	J:	Portyn (Parke Davis-Sankyo)
		GB:	Portyn (Parke Davis); wfm		

Benziodarone

ATC: C01DA; C01DX04
Use: coronary vasodilator, uricosuric agent

RN: 68-90-6 MF: $C_{17}H_{12}I_2O_3$ MW: 518.09 EINECS: 200-695-2
LD_{50}: 450 mg/kg (M, p.o.)
CN: (2-ethyl-3-benzofuranyl)(4-hydroxy-3,5-diiodophenyl)methanone

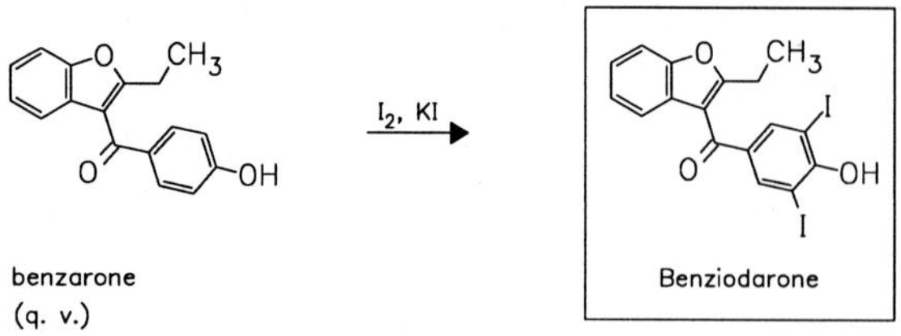

benzarone
(q. v.)

Benziodarone

Reference(s):
GB 836 272 (Labaz; appl. 17.12.1957; B-prior. 21.12.1956).

Formulation(s): cps. in comb. with allopurinol

Trade Name(s):
F: Ampliuril pH (Labaz); wfm   Amplivix (Labaz); wfm   I: Uricodue (IFI)-comb.

Benzocaine
(Ethoforme)

ATC: C05AD03; D04AB04; N01BA05;
      R02AD01
Use: local anesthetic

RN: 94-09-7 MF: $C_9H_{11}NO_2$ MW: 165.19 EINECS: 202-303-5
LD_{50}: 216 mg/kg (M, i.p.)
CN: 4-aminobenzoic acid ethyl ester

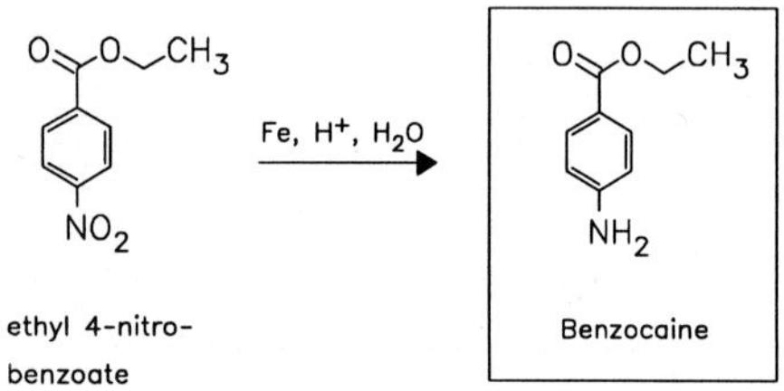

ethyl 4-nitro-
benzoate

Benzocaine

Reference(s):
Org. Synth. (ORSYAT) **8**, 66 (1928).

Formulation(s): cream 100 mg; ointment 5 %, 10 %, 20 %; pills 4 mg, 8 mg, 20 mg; powder 60 mg; suppos.
                   100 mg

Trade Name(s):
D: Anaesthesin (Ritsert)   I: Anes Par (Tariff.   Cetacaine (Cetylite)
      Flavamed (Berlin-Chemie)   Integrativo)   Hurricaine (Beutlich)
      Subcutin (Ritsert)   Gengivarium (Kemyos)   Tympagesic (Savage)
      Zahnerol (Janssen)   USA: Americaine (Medeva)
GB: generics   Auralgan (Wyeth-Ayerst)

Benzoctamine

ATC: N05BD01
Use: psychosedative, tranquilizer

RN: 17243-39-9 MF: C$_{18}$H$_{19}$N MW: 249.36
LD$_{50}$: 30 mg/kg (M, i.v.); 280 mg/kg (M, p.o.);
36 mg/kg (R, i.v.); 600 mg/kg (R, p.o.);
>10 mg/kg (dog, i.v.); >200 mg/kg (dog, p.o.)
CN: N-methyl-9,10-ethanoanthracene-9(10H)-methanamine

hydrochloride
RN: 10085-81-1 MF: C$_{18}$H$_{19}$N · HCl MW: 285.82 EINECS: 233-216-0
LD$_{50}$: 26 mg/kg (R, i.v.); 700 mg/kg (R, p.o.)

anthracene + N-methyl-formanilide → (POCl$_3$, 90–95 °C) anthracene-9-carboxaldehyde → (H$_2$C=CH$_2$, DMF, 170 °C, ethylene) I

9,10-dihydro-9,10-ethanoanthracene-9-carboxaldehyde (I) →
1. H$_3$C—NH$_2$
2. H$_2$, Raney–Ni
1. methylamine
2. hydrogenation → Benzoctamine

Reference(s):
Wilhelm, M.; Schmidt, P.: Helv. Chim. Acta (HCACAV) **52**, 1385 (1969).
BE 610 863 (Ciba; appl. 28.11.1961; CH-prior. 29.11.1960, 10.10.1961).
US 3 399 201 (Ciba; 27.8.1968; CH-prior. 29.11.1960, 10.10.1961, 1.11.1963, 23.12.1964, 24.11.1965, 10.12.1965).
DE 1 228 605 (Ciba; appl. 24.11.1961; CH-prior. 29.11.1960, 10.10.1961).

Formulation(s):   syrup 2 mg/2 ml; tabl. 5 mg, 10 mg

Trade Name(s):
D: Tacitin (Ciba); wfm   F: Tacitine (Ciba); wfm   GB: Tacitin (Ciba); wfm

---

Benzonatate

ATC: R05DB01
Use: antitussive

RN: 104-31-4 MF: C$_{30}$H$_{53}$NO$_{11}$ MW: 603.75 EINECS: 203-194-7
LD$_{50}$: 9 mg/kg (M, i.v.); 400 mg/kg (M, p.o.)
CN: 4-(butylamino)benzoic acid 3,6,9,12,15,18,21,24,27-nonaoxaoctacos-1-yl ester

ethyl 4-butyl-
aminobenzoate

nonaethylene glycol
monomethyl ether

Benzonatate

Reference(s):
US 2 714 608 (Ciba; 1955; CH-prior. 1950).
US 2 714 609 (Ciba; 1955; CH-prior. 1950).

Formulation(s):   cps. 100 mg; perls 100 mg

Trade Name(s):
USA: Tessalon (Forest)

Benzoyl peroxide
(Peroxide de benzoyle)

ATC: D10AE01
Use: keratolytic, antiseptic

RN: 94-36-0   MF: $C_{14}H_{10}O_4$ MW: 242.23   EINECS: 202-327-6
LD_{50}: 5700 mg/kg (M, p.o.);
      7710 mg/kg (R, p.o.)
CN: dibenzoyl peroxide

benzoyl
chloride

Benzoyl peroxide

Reference(s):
Ullmanns Encykl. Tech. Chem., 4. Aufl., Vol. **17**, 671.

stabilization of aqueous formulations with sodium dioctylsulfosuccinate:
US 4 387 107 (Dermik Labs.; 7.6.1983; prior. 25.7.1979, 16.12.1980).

alternative formulations:
US 3 535 422 (Stiefel Labs.; 20.10.1970; prior. 30.3.1966, 11.3.1968).
US 4 056 611 (Stiefel Labs.; 1.11.1977; appl. 16.4.1973).
US 4 545 990 (L'Oreal; 8.10.1985; appl. 21.11.1983; LU-prior. 22.11.1982).

combination with salicylic acid:
US 4 318 907 (Westwood; 9.3.1982; appl. 4.4.1978).
US 4 355 028 (Westwood; 19.10.1982; appl. 30.4.1981).

Formulation(s):   cps. 100 mg; gel 5 %, 10 %

Trade Name(s):
D: Abmederm (gepepharm)   Benzaknen (Galderma)   Sanoxit (Galderma)
      Akne-Aid-Lotion (Stiefel)   Benzoxyl 20 Lotion   Scherogel (Asche)
      Aknefug-oxid (Wolff)   (Stiefel)   Ultra Clearasil (Wick
      Akneroxid (Hermal)   Pan Oxyl (Stiefel)   Pharma)

F: Cutacnyl (Galderma)
     Eclaran (Pierre Fabre)
     Effacné (Roche-Posay)
     Pannogel (Labs. CS)
     Panoxyl (Stiefel)
GB: Acnezide (Galderma)
     Acnidazil (Janssen-Cilag)-
     comb.
     Benzamycin (Bioglan)-
     comb.
     Nericur (Schering)
     Panoxyl (Stiefel)

     Quinoderm (Quinoderm)-
     comb.
     Quinoped (Quinoderm)-
     comb.
I: Acnidazil (Fisons
     Italchimici)-comb.
     Benoxid (Brocades)
     Benzac (Galderma)
     Benzoil Peros (Formulario
     Naz.)
     Benzomix (Savoma)
     Fatroxid (Fatro)
     Reloxyl (Rdc)

USA: Benzac (Galderma)
     Benzagel (Dermik)
     Benzamycin (Dermik)
     Benzashave (Medicis)
     Brevoxyl (Stiefel)
     Desquam-E (Westwood-
     Squibb)
     Desquam-X (Westwood-
     Squibb)
     PanOxyl (Stiefel)
     Triaz (Medicis)
     Vanoxide-HC (Dermik)

Benzphetamine

Use: appetite depressant

RN: 156-08-1   MF: $C_{17}H_{21}N$ MW: 239.36
LD_{50}:   227 mg/kg (M, p.o.);
       160 mg/kg (R, p.o.)
CN: (+)-N,α-dimethyl-N-(phenylmethyl)benzeneethanamine

hydrochloride
RN: 5411-22-3   MF: $C_{17}H_{21}N \cdot HCl$ MW: 275.82   EINECS: 226-489-2

(+)-deoxyephedrine   benzyl chloride   Benzphetamine

Reference(s):
US 2 789 138 (Upjohn; 1957; prior. 1952).

Formulation(s):   tabl. 25 mg, 50 mg

Trade Name(s):
F: Inapetyl (Upjohn); wfm   GB: Didrex (Upjohn); wfm

Benzquinamide

ATC: N05AK
Use: anti-emetic, tranquilizer

RN: 63-12-7   MF: $C_{22}H_{32}N_2O_5$ MW: 404.51
CN: 2-(acetyloxy)-N,N-diethyl-1,3,4,6,7,11b-hexahydro-9,10-dimethoxy-2H-benzo[a]quinolizine-3-
     carboxamide

2-(3,4-dimethoxy-
phenyl)ethylamine   diethyl malonate   ethyl N-(3,4-dimethoxy-
phenethyl)malonamate (I)

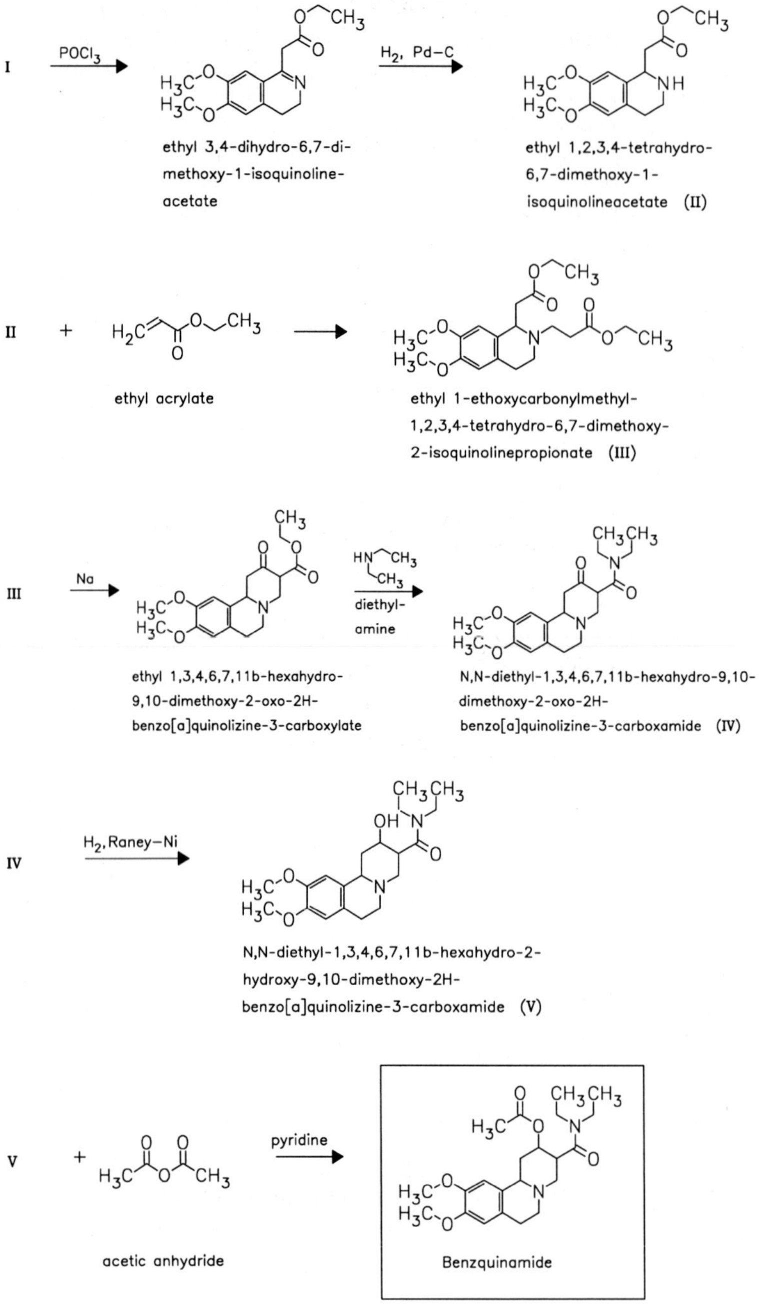

I   POCl$_3$ →

ethyl 3,4-dihydro-6,7-di-
methoxy-1-isoquinoline-
acetate

H$_2$, Pd–C →

ethyl 1,2,3,4-tetrahydro-
6,7-dimethoxy-1-
isoquinolineacetate   (II)

II   +   ethyl acrylate →

ethyl 1-ethoxycarbonylmethyl-
1,2,3,4-tetrahydro-6,7-dimethoxy-
2-isoquinolinepropionate   (III)

III   Na →

ethyl 1,3,4,6,7,11b-hexahydro-
9,10-dimethoxy-2-oxo-2H-
benzo[a]quinolizine-3-carboxylate

diethyl-
amine →

N,N-diethyl-1,3,4,6,7,11b-hexahydro-9,10-
dimethoxy-2-oxo-2H-
benzo[a]quinolizine-3-carboxamide   (IV)

IV   H$_2$, Raney–Ni →

N,N-diethyl-1,3,4,6,7,11b-hexahydro-2-
hydroxy-9,10-dimethoxy-2H-
benzo[a]quinolizine-3-carboxamide   (V)

V   +   acetic anhydride   pyridine →

Benzquinamide

Reference(s):
US 3 053 845 (Pfizer; appl. 29.8.1961).
US 3 055 894 (Pfizer; appl. 9.3.1960).
BE 621 895 (Pfizer; appl. 29.8.1962; USA-prior. 9.3.1960, 29.8.1961).
DE 1 303 628 (Pfizer; appl. 30.5.1962; USA-prior. 29.8.1961, 6.9.1961).

starting material:
Brossi, A. et al.: Helv. Chim. Acta (HCACAV) **41**, 119 (1958).

Formulation(s):   amp. 50 mg

Benzthiazide
(Benzothiazide; Benztiazide)

ATC: C03
Use: diuretic, antihypertensive

RN: 91-33-8 MF: $C_{15}H_{14}ClN_3O_4S_3$ MW: 431.95 EINECS: 202-061-0
LD$_{50}$: 410 mg/kg (M, i.v.); >5 g/kg (M, p.o.);
     422 mg/kg (R, i.v.); >10 g/kg (R, p.o.);
     >5 g/kg (dog, p.o.)
CN: 6-chloro-3-[[(phenylmethyl)thio]methyl]-2*H*-1,2,4-benzothiadiazine-7-sulfonamide 1,1-dioxide

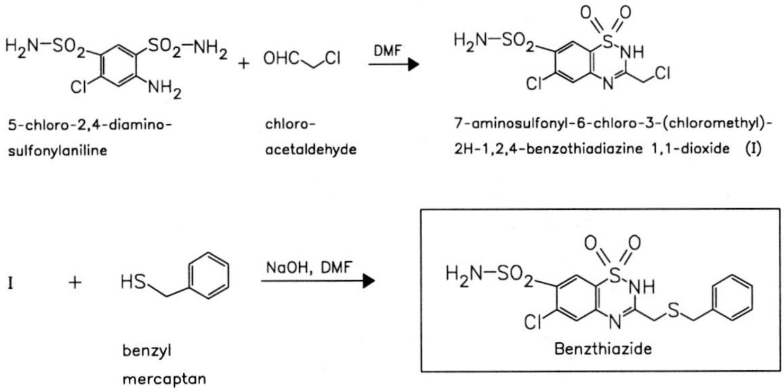

5-chloro-2,4-diamino-
sulfonylaniline

chloro-
acetaldehyde

7-aminosulfonyl-6-chloro-3-(chloromethyl)-
2H-1,2,4-benzothiadiazine 1,1-dioxide (I)

benzyl
mercaptan

Benzthiazide

Reference(s):
US 3 111 517 (Pfizer; 19.11.1963).

Formulation(s):   cps. 25 mg

Benzydamine

ATC: A01AD02; G02CC03; M01AX07;
     M02AA05; M02AX
Use: analgesic, antipyretic, anti-
     inflammatory

RN: 642-72-8 MF: $C_{19}H_{23}N_3O$ MW: 309.41 EINECS: 211-388-8
LD$_{50}$: 25 mg/kg (M, i.v.); 460 mg/kg (M, p.o.),
     950 mg/kg (R, p.o.)
CN: *N,N*-dimethyl-3-[[1-(phenylmethyl)-1*H*-indazol-3-yl]oxy]-1-propanamine

monohydrochloride
RN: 132-69-4 MF: $C_{19}H_{23}N_3O \cdot HCl$ MW: 345.87 EINECS: 205-076-0
LD$_{50}$: 33 mg/kg (M, i.v.); 440 mg/kg (M, p.o.);
     43.5 mg/kg (R, i.v.); 740 mg/kg (R, p.o.)

methyl   benzyl   methyl 2-benzyl-   1-benzyl-3-hydroxy-
anthranilate   chloride   aminobenzoate   1H-indazole (I)

1. NaOCH₃
2. Cl⌒⌒N-CH₃
        CH₃

1. sodium methylate
2. 3-dimethylamino-
   propyl chloride

Benzydamine

Reference(s):
FR 1 382 855 (Angelini Francesco; appl. 21.2.1964; I-prior. 9.8.1963).

Formulation(s):   amp. 25 mg; cps. 50 mg; cream 30 mg; drg. 50 mg; drops 50 mg; liquid 1.5 mg;
                     powder 500 mg (as hydrochloride)

Trade Name(s):
D: Tantum (Solvay   Multum (Lampugnani)   Benzyrin (Yoshitomi)
      Arzneimittel)   Saniflor (Esseti)   Enzamin (Kowa)
F: Opalgyne (Innothéra)   Tantum Biotic (Angelini)-   Epirotin (Nakataki)
GB: Difflam (3M; as   comb. with tetracycline   Lilizin (Beppu)
      hydrochloride)   Verax (Tosi-Novara)   Riripen (Daiichi)
I: Afloben (Esseti)   numerous combination   Salyzoron (Hishiyama)
      Berzirin (Fater)   preparations   Sanal (Sana)
      Ginesal (Farmigea)   J: Antol (Seiko Eiyo)
      Leucorsan (Zilliken)-comb.   Benzidan (Nikken)

Benzyl alcohol

(Alcoholum benzylicum; Phenylcarbinolum)

ATC: R02AD
Use: disinfectant, local anesthetic

RN: 100-51-6   MF: C_7H_8O MW: 108.14   EINECS: 202-859-9
LD_{50}:   324 mg/kg (M, i.v.); 1360 mg/kg (M, p.o.);
       53 mg/kg (R, i.v.); 1230 mg/kg (R, p.o.)
CN: benzenemethanol

benzyl   aq. NaOH or Na₂CO₃, Δ   Benzyl alcohol
chloride

Reference(s):
Ullmanns Encykl. Tech. Chem., 4. Aufl., Vol. **8**, 437.

Formulation(s):   amp. 1 %, 2 %; cream 1 %; sol. 1 g/100 g

Trade Name(s):

D: Spitacid (Henkel)-comb.   GB: Pabrinex (Link)-comb.   Foille (Delalande Isnardi)-
     numerous combination   Sudocrem (Tosara)-comb.   comb.
     preparations   I: Borocaina (Schiapparelli)

Benzyl benzoate

(Benzoesäurebenzylester)

ATC: P03AX01
Use: scabicide, pharmaceutic agent

RN: 120-51-4   MF: $C_{14}H_{12}O_2$ MW: 212.25   EINECS: 204-402-9
LD$_{50}$:   1400 µL/kg (M, p.o.);
       1700 µL/kg (R, p.o.);
       >22440 mg/kg (dog, p.o.)
CN: benzoic acid phenylmethyl ester

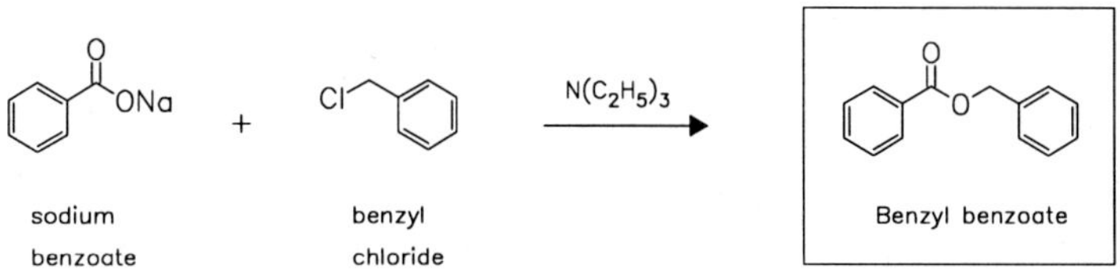

sodium   benzyl   Benzyl benzoate
benzoate   chloride

Reference(s):
Tharp, I.D. et al.: Ind. Eng. Chem. (IECHAD) **39**, 1300 (1947).

Formulation(s):   emulsion 250 mg

Trade Name(s):

D: Acarosan (Allergopharma)   Anusol HC (Warner-   I: Antiscabbia Candioli al
     Antiscabiosum Mago KG   Lambert)-comb.   D.D.T. terap. (Candioli)-
     (Strathmann)   Ascabiol (Rhône-Poulenc   comb.
F: Ascabiol (Evans Medical)   Rorer)   Benz Be (Formulario Naz.;
GB: Anugesic HC (Parke   Tariff. Integrativo)
     Davis)-comb.

Benzyl mustard oil

(Oleum tropaeoli)

ATC: S01AA
Use: antibiotic

RN: 622-78-6   MF: C_8H_7NS MW: 149.22   EINECS: 210-753-9
LD$_{50}$:   150 mg/kg (M, s.c.)
CN: (isothiocyanatomethyl)benzene

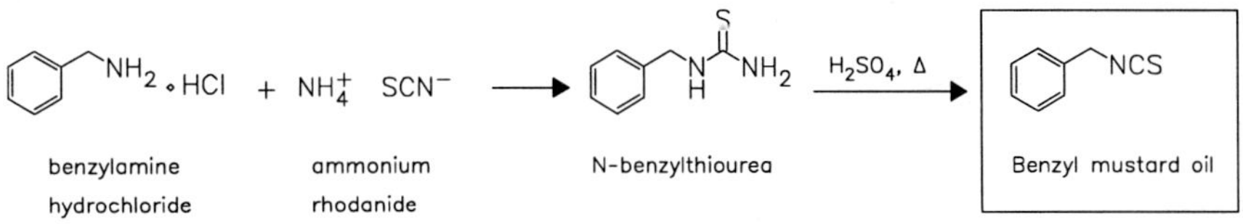

benzylamine   ammonium   N-benzylthiourea   Benzyl mustard oil
hydrochloride   rhodanide

Reference(s):
Ullmanns Encykl. Tech. Chem., 4. Aufl., Vol. **23**, 156.

Formulation(s):   cps. 14.4 mg

Benzylpenicillin
(Penicillin G)

ATC: J01CE01; J01HA; S01AA14
Use: antibiotic

RN: 61-33-6   MF: $C_{16}H_{18}N_2O_4S$ MW: 334.40   EINECS: 200-506-3
LD$_{50}$:   329 mg/kg (M, i.v.); >5 g/kg (M, p.o.);
      8 g/kg (R, p.o.)
CN: [2S-(2α,5α,6β)]-3,3-dimethyl-7-oxo-6-[(phenylacetyl)amino]-4-thia-1-azabicyclo[3.2.0]heptane-2-
      carboxylic acid

monosodium salt
RN: 69-57-8   MF: $C_{16}H_{17}N_2NaO_4S$ MW: 356.38   EINECS: 200-710-2
LD$_{50}$:   1500 mg/kg (M, i.v.); >4 g/kg (M, p.o.);
      3020 mg/kg (R, i.v.); 6916 mg/kg (R, p.o.)
monopotassium salt
RN: 113-98-4   MF: $C_{16}H_{17}KN_2O_4S$ MW: 372.49   EINECS: 204-038-0
LD$_{50}$:   240 mg/kg (M, i.v.); 6257 mg/kg (M, p.o.);
      243 mg/kg (R, i.v.); 8900 mg/kg (R, p.o.)

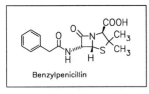

Benzylpenicillin

From fermentation solutions of *Penicillium notatum* Westling or *Penicillium chrysogenum* Thom by addition of
phenylacetic acid as precursor.

Reference(s):
Ehrhart, Ruschig **IV**, 286 ff.

Formulation(s):   eff. tabl. 653.6 mg; f. c. tabl. 392.2 mg, 653.6 mg, 982.32 mg; lyo. for syrup 1986.59 mg

Bephenium hydroxynaphthoate

ATC: P02CX02
Use: anthelmintic

RN: 3818-50-6   MF: $C_{17}H_{22}NO \cdot C_{11}H_7O_3$ MW: 443.54   EINECS: 223-306-8
CN: N,N-dimethyl-N-(2-phenoxyethyl)benzenemethanaminium 3-hydroxy-2-naphthoate (1:1)

sodium phenolate   2-(dimethylamino)-ethyl chloride   N-(2-phenoxyethyl)-dimethylamine   benzyl chloride   I

bephenium chloride (I)   sodium 3-hydroxy-2-naphthoate   Bephenium hydroxynaphthoate

Reference(s):
US 2 918 401 (Borroughs Wellcome; 22.12.1959; GB-prior. 29.3.1956).
DE 1 117 600 (Wellcome Found.; appl. 21.3.1957; GB-prior. 29.3.1956, 24.1.1957).

Formulation(s):   gran. 2.5 g, 4.33 g; powder 5 g

Trade Name(s):
D: Alcopar (Wellcome); wfm   J: Alcopar-P (Wellcome-   USA: Alcopara (Borroughs
F: Alcopar (Wellcome); wfm   Tanabe)   Wellcome); wfm
GB: Alcopar (Wellcome); wfm

Bepridil

ATC: C02DE; C08EA02
Use: calcium channel blocker, antianginal

RN: 64706-54-3   MF: $C_{24}H_{34}N_2O$ MW: 366.55   EINECS: 256-384-7
LD$_{50}$: 1955 mg/kg (M, p.o.); 23,5 mg/kg (M, i.v.)
CN: α-[(2-methylpropoxy)methyl]-N-phenyl-N-(phenylmethyl)-1-pyrrolidineethanamine

monohydrochloride
RN: 68099-86-5   MF: $C_{24}H_{34}N_2O \cdot HCl$ MW: 403.01   EINECS: 268-472-2
monohydrochloride monohydrate
RN: 74764-40-2   MF: $C_{24}H_{34}N_2O \cdot HCl \cdot H_2O$ MW: 421.03
LD$_{50}$: 23.5 mg/kg (M, i.v.); 1955 mg/kg (M, p.o.);
        >21.3 mg/kg (R, i.v.); 6850 mg/kg (R, p.o.)
(+)-form
RN: 110143-74-3   MF: $C_{24}H_{34}N_2O$ MW: 366.55
(–)-form
RN: 110143-75-4   MF: $C_{24}H_{34}N_2O$ MW: 366.55
(±)-form
RN: 89035-90-5   MF: $C_{24}H_{34}N_2O$ MW: 366.55

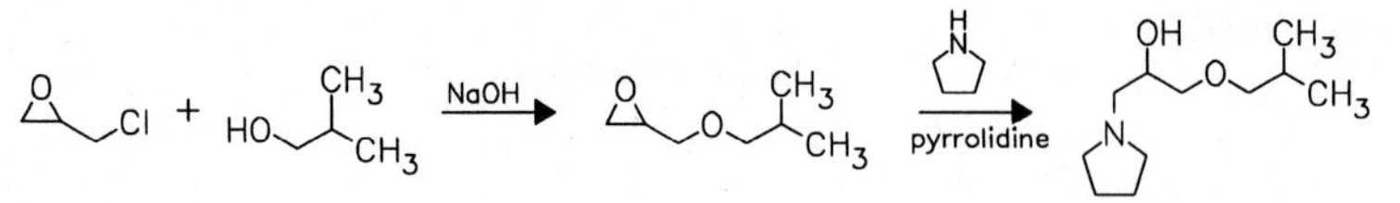

epichloro-   isobutanol   glycide isobutyl ether   1-(2-hydroxy-3-iso-
hydrin   butoxypropyl)pyrrolidine (I)

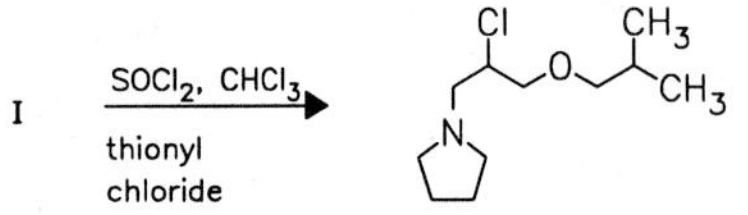

1-(2-chloro-3-iso-
butoxypropyl)pyrrolidine (II)

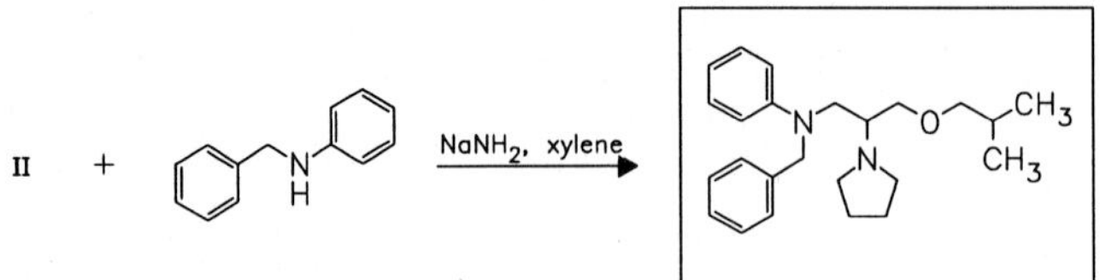

N-benzylaniline   Bepridil

Reference(s):

DOS 2 310 918 (CERM; appl. 5.3.1973; F-prior. 6.3.1972).
DE 2 802 864 (CERM; appl. 13.1.1978; F-prior. 25.1.1977).
US 3 962 238 (CERM; 8.6.1976; appl. 27.2.1973; F-prior. 6.3.1972).
GB 1 377 327 (CERM; appl. 27.2.1973; F-prior. 6.3.1972).
GB 1 595 031 (CERM; appl. 13.1.1978; F-prior. 25.1.1977).

Formulation(s):   tabl. 100 mg

Trade Name(s):

F: Cordium (Riom; 1981)   J: Bepricor (Nippon Organon;   USA: Vascor (Ortho-McNeil; as
                                            Sankyo; as hydrochloride   hydrochloride)
                                            hydrate)

Betacarotene

(β-Carotene; Betacarotin; β-Carotin)

ATC: D02BB01
Use: provitamin A

RN: 7235-40-7   MF: $C_{40}H_{56}$ MW: 536.89   EINECS: 230-636-6
CN: (all-*E*)-1,1'-(3,7,12,16-tetramethyl-1,3,5,7,9,11,13,15,17-octadecanonaene-1,18-diyl)bis[2,6,6-
       trimethylcyclohexene]

1 Roche:

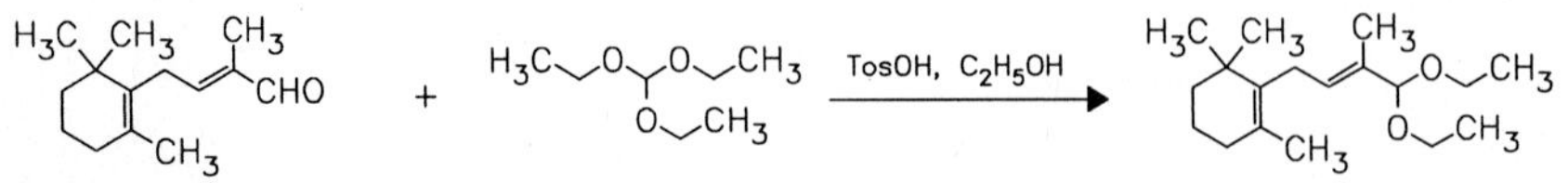

"aldehyde C_{14}"   triethyl   (II)
(cf. retinol synthesis)   orthoformate (I)

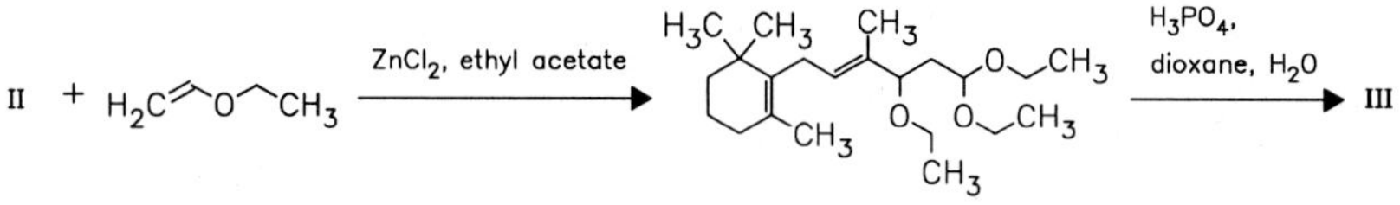

ethyl vinyl ether

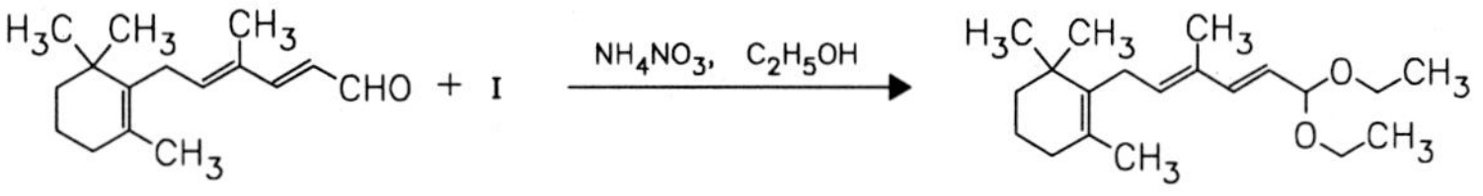

(III) (IV)

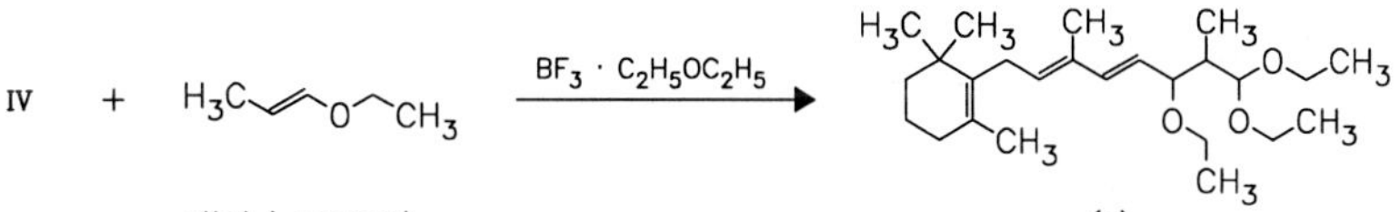

ethyl 1-propenyl ether

(V)

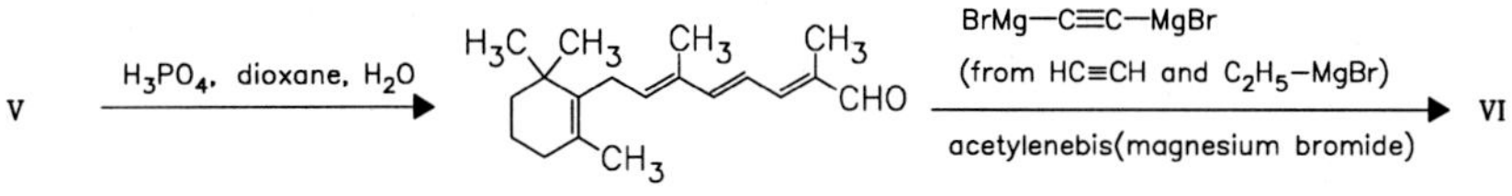

"β-aldehyde C_{19}"

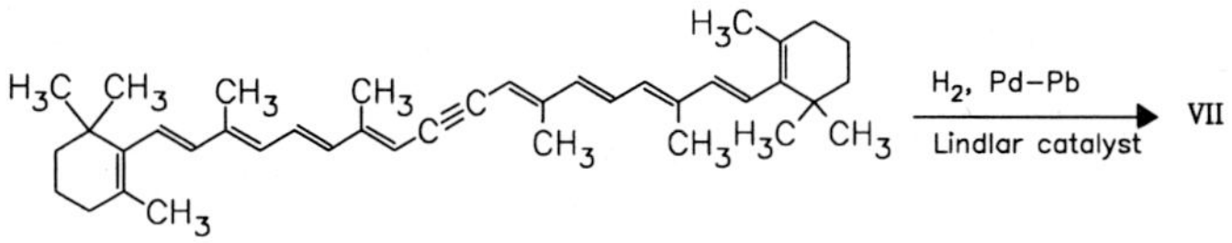

15-dehydro-β-carotene (VI)

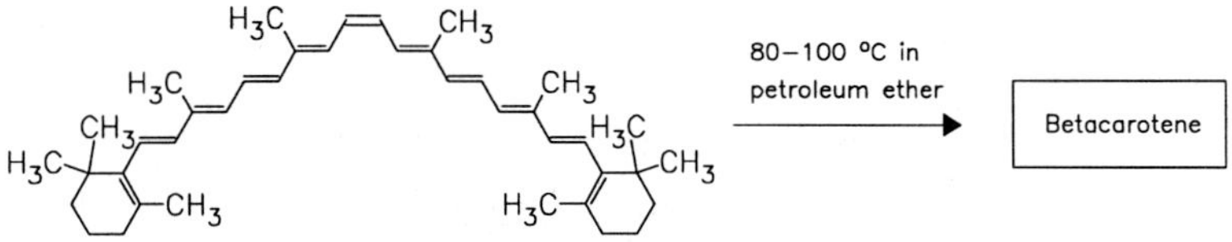

15-cis-β-carotene (VII)

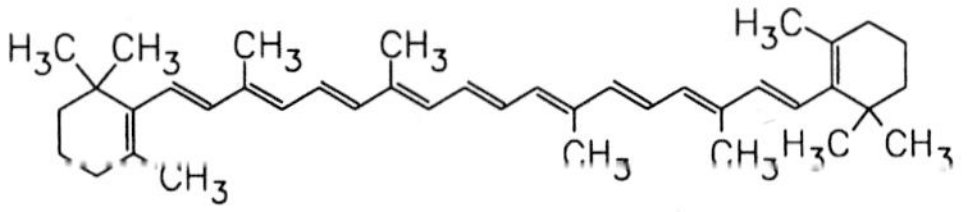

Betacarotene

2 BASF:

a

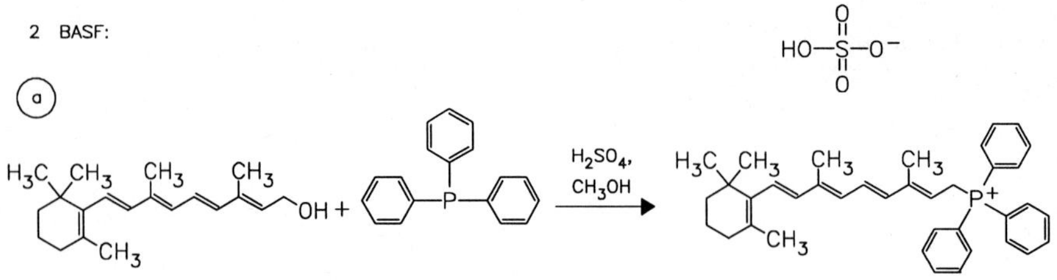

retinol (VIII)   triphenyl-   (IX)
(q. v.)   phosphine

IX

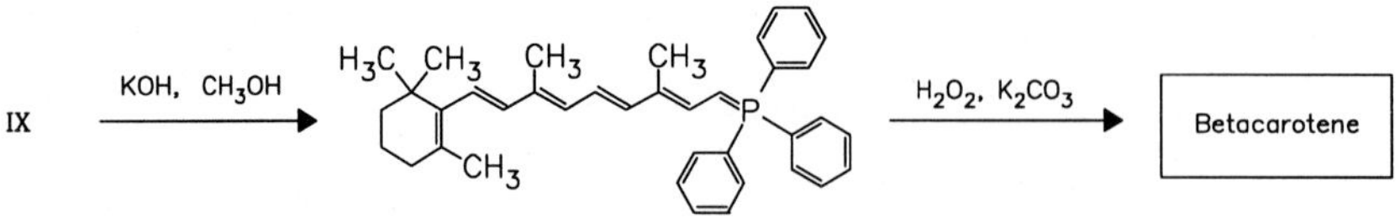

(X)

b

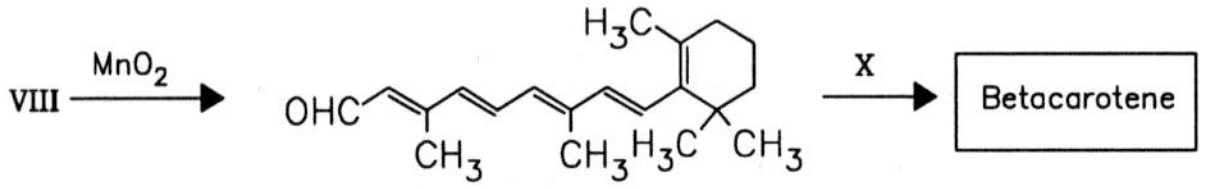

VIII

retinal

X → Betacarotene

c

1,4-dibromo-   trimethyl   tetramethyl 2-butene-
2-butene   phosphite   1,4-diylbisphosphonate

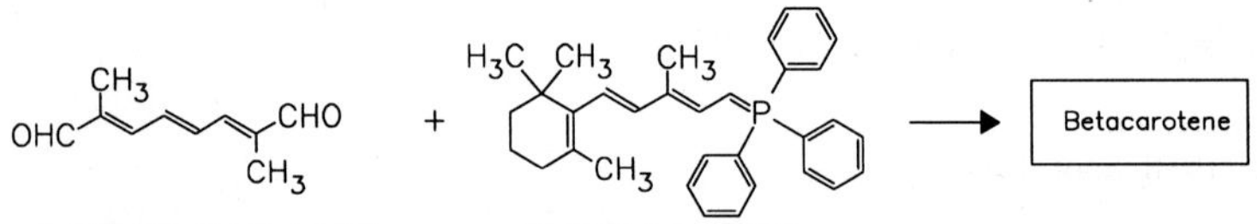

1. NaOCH₃
2. H₃C—O...CH₃
3. H₂O/H⁺

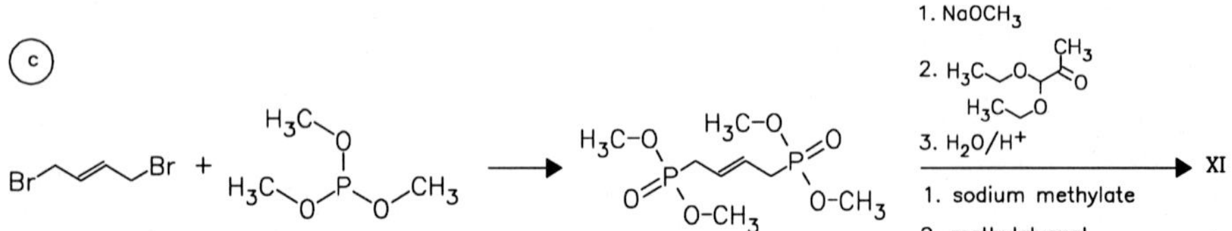

1. sodium methylate
2. methylglyoxal
   diethyl acetal

→ XI

(all-E)-2,7-dimethyl-2,4,6-   triphenylphosphinylide
octatrienedial (XI)   from vinyl-β-ionol
                               (cf. retinol synthesis)

→ Betacarotene

3 Rhone-Poulenc:

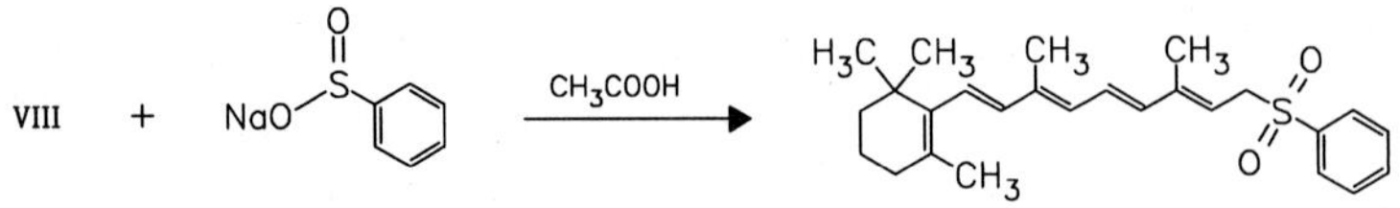

VIII +   NaO—S⟨phenyl⟩   CH₃COOH

sodium   retinyl phenyl sulfone (XII)
benzenesulfinate

retinyl chloride

"sulfone C$_{40}$" (XIII)

Reference(s):
review:
Ullmanns Encykl. Tech. Chem., 4. Aufl., Vol. **23**, 633 ff.

1   Isler, O. et al.: Helv. Chim. Acta (HCACAV) **39**, 249 (1956).
    Isler, O.: Angew. Chem. (ANCEAD) **68**, 547 (1956).
    DE 855 399 (Roche; appl. 26.5.1950).
    DE 858 095 (Roche; appl. 1.10.1950).
    DE 953 073 (Roche; appl. 23.5.1954; CH-prior. 29.6.1953).
    DE 953 074 (Roche; appl. 5.6.1954; CH-prior. 1.7.1953).
    isomerization to all-trans-form:
    US 3 367 985 (Roche; 6.2.1968; appl. 18.4.1966).
    DE 2 440 747 (Roche; appl. 26.8.1974; USA-prior. 29.8.1973).

2 *review:*
Pommer, H.: Angew. Chem. (ANCEAD) **72**, 911 (1960).
Pommer, H.: Angew. Chem. (ANCEAD) **89**, 437 (1977).
a   DE 2 505 869 (BASF; appl. 12.2.1975).
b   DE 1 068 709 (BASF; appl. 6.6.1958).
    DE 1 158 505 (BASF; appl. 23.5.1962).
c   DE 954 247 (BASF; appl. 20.10.1954).
    DE 1 068 705 (BASF; appl. 22.3.1958).
    DE 1 068 703 (BASF; appl. 14.3.1958).
    "C$_{10}$-dialdehyde":
    DE 1 092 472 (BASF; appl. 2.10.1958).
3   DE 2 224 606 (Rhône-Poulenc; appl. 19.5.1972; F-prior. 19.5.1971).

isolation from carrots and similar material:
US 2 848 508 (H. M. Harnett et al.; 1958; appl. 1954).

fermentative production:
US 2 959 521 (Grain Processing Corp.; 1960; appl. 1959).
US 2 959 522 (Grain Processing Corp.; 1960; appl. 1959).
US 3 001 912 (Commercial Solvents Corp.; 1961; appl. 1958).
US 3 128 236 (Grain Processing Corp.; 1964; appl. 1961).

Formulation(s):   cps. 25 mg

Trade Name(s):

D:	Bella Carotin (3M Medica)	Bétasellen (Arkopharma)-		Phénoro Roche (Roche)-
	Carotaben (Hermal)	comb.		comb.
	combination preparations	Dijrarel 100 (Leurquin)-	I:	Fotoretin (Farmila)-comb.
F:	Azinc complexe	comb.		Mirtilene (SIFI)-comb.
	(Arkopharma)-comb.		USA:	Aces (Carlson)

Betahistine

ATC: C04AX; N07CA01
Use: diaminooxydase inhibitor

RN: 5638-76-6   MF: $C_8H_{12}N_2$ MW: 136.20   EINECS: 227-086-4
LD$_{50}$:   2920 mg/kg (M, p.o.);
       6110 mg/kg (R, p.o.)
CN: N-methyl-2-pyridineethanamine

dihydrochloride
RN: 5579-84-0   MF: $C_8H_{12}N_2 \cdot 2HCl$ MW: 209.12   EINECS: 226-966-5
dimesylate
RN: 54856-23-4   MF: $C_8H_{12}N_2 \cdot 2CH_4O_3S$ MW: 328.41   EINECS: 259-377-7
LD$_{50}$:   505 mg/kg (M, i.v.); 500 mg/kg (M, p.o.);
       604 mg/kg (R, i.v.); 3030 mg/kg (R, p.o.)

2-picoline   paraform-   2-(2-pyridyl)-   2-(2-bromoethyl)-   Betahistine
                aldehyde   ethanol   pyridine

Reference(s):
Löffler, K.: Ber. Dtsch. Chem. Ges. (BDCGAS) **37**, 161 (1904).
Walter, L.A. et al.: J. Am. Chem. Soc. (JACSAT) **63**, 2771 (1941).

Formulation(s):   drops 1.25 % (as dihydrochloride); s. r. tabl. 20 mg; tabl. 6 mg, 12 mg (as dimesylate), 8 mg,
                     16 mg (as dihydrochloride)

Trade Name(s):
D: Aequamen (Promonta   Vasomotal (Solvay   GB: Serc (Solvay; as
      Lundbeck)   Arzneimittel)   hydrochloride)
      Betavert (Henning)   F: Extovyl (Marion Merrell) I: Microser (Formenti)
      Melopat (Pharmasal)   Lectil (Bouchara)   J: Merislon (Eisai)
      Ribrain (Searle-   Serc (Solvay Pharma)   USA: Serc (Unimed); wfm
      Endopharm; Yamanouchi)

Betaine aspartate

ATC: A05BA; A09AB; A12BA
Use: liver therapeutic, stomach therapeutic

RN: 52921-08-1   MF: $C_5H_{11}NO_2 \cdot C_4H_6NO_4$ MW: 249.24   EINECS: 258-258-7
CN: 1-carboxy-N,N,N-trimethylmethanaminium hydrogen L-aspartate

betaine   L-aspartic acid   Betaine aspartate

Reference(s):
FR 1 356 945 (M. R. Cote; appl. 5.12.1962; MC-prior. 14.12.1961).
FR-M 2 462 (Albert Rolland; appl. 9.10.1962).

Formulation(s):   amp. 2 g/dose;sol. 10 ml

Trade Name(s):

F: Somatyl (Anphar-Rolland);
      wfm
      Somatyl (L'Hépatrol); wfm
I: Betaina Manzoni
      (Manzoni)-comb.
      Betascor (Manetti
      Roberts)-comb.
      Bios Liver (Ausonia)-
      comb.

Ciatox (Ibirn)-comb.
Citroepatina (Roussel-
Maestretti)-comb.
Eparbolic (Carlo Erba)-
comb.
Equipar (Lampugnani)-
comb.
Glicobil (Medici Domus)-
comb.

Glution (Boniscontro &
Gazzone)-comb.
Inobetin (Boniscontro &
Gazzone)-comb.
Kloref (Samil)-comb.
Somatyl (Prophin)-comb.

Betaine hydrate

ATC: A09AB02
Use: liver therapeutic, gastric therapeutic

RN: 590-47-6   MF: $C_5H_{13}NO_3$ MW: 135.16   EINECS: 209-684-7
CN: 1-carboxy-*N,N,N*-trimethylmethanaminium hydroxide inner salt

hydrochloride
RN: 590-46-5   MF: $C_5H_{12}ClNO_2$ MW: 153.61   EINECS: 209-683-1
dihydrogen citrate (1:1)
RN: 17671-50-0   MF: $C_6H_7O_7 \cdot C_5H_{12}NO_2$ MW: 309.27   EINECS: 241-648-6

trimethyl-
amine

sodium
chloroacetate

betaine chloride (I)

basic ion exchanger (e. g. IRA–410)

Betaine hydrate

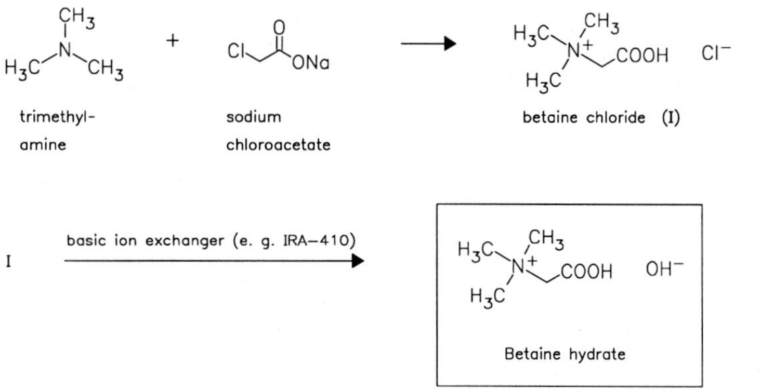

(b) by-product of beet-sugar production; isolation by acidic precipitation or by ion-exchange
methods from the mash

Reference(s):
Stoltzenberg, H.: Z. Physiol. Chem. (ZPCHA5) **92**, 445 (1914).
a   DRP 269 701 (AG für Anilin-Fabrikation; appl. 1912).
     US 2 800 502 (Internat. Minerals & Chem. Corp.; 1957; appl. 1953).
b   US 1 685 758 (D. K. Tressler; 1928; appl. 1925).

Formulation(s):   gran. 400 mg

Trade Name(s):
D: Flacar (Schwabe)-comb.   F: Citrarginine (Laphal)-
                                       comb.

Citrate de bétaïne Beaufour
(Beaufour)

Citrate de bétaïneeffervescent Upsa (UPSA)-comb. Gastrobul (Guerbet)-comb. Hépagrume (Synthélabo)- comb. Ornitaïne (Schwarz)-comb.	GB: I:	Kloref (Cox; as hydrochloride)-comb. Betaina Manzoni (Gaymonat; as citrate) Citroepatina (Roussel)- comb.	J:	Somatyl (Teofarma; as aspartate) Apellet-BT (Ono)-comb. Molmagen (Toa Yakuhin- Torii)-comb.

Betamethasone

ATC: A07EA04; D07AC01; C05AA05;
D07XC01; H02AB01; R01AD06;
R03BA04; S01BA06; S01CB04;
S03CA06

Use: glucocorticoid

RN: 378-44-9   MF: $C_{22}H_{29}FO_5$ MW: 392.47   EINECS: 206-825-4
LD$_{50}$: >4.5 g/kg (M, p.o.)
CN: (11β,16β)-9-fluoro-11,17,21-trihydroxy-16-methylpregna-1,4-diene-3,20-dione

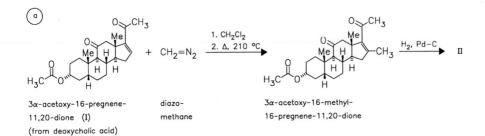

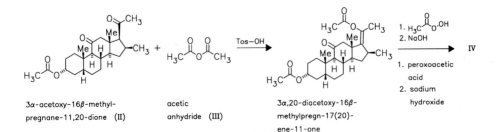

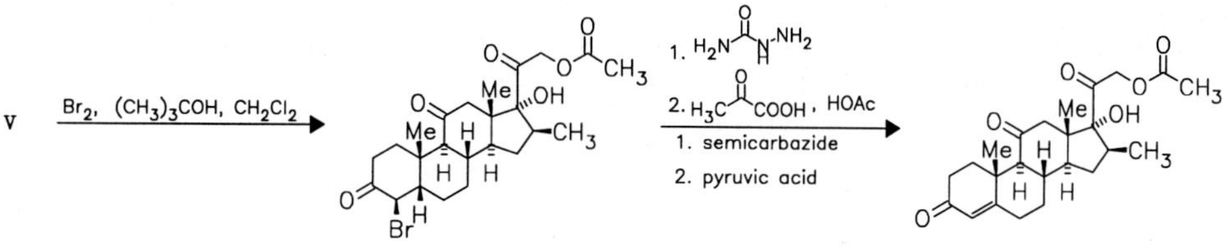

V →[Br₂, (CH₃)₃COH, CH₂Cl₂]

4β-bromo-17α,21-di-
hydroxy-16β-methyl-
pregnane-3,11,20-trione 21-acetate

1. H₂N-NH-NH₂
2. H₃C-CO-COOH, HOAc
1. semicarbazide
2. pyruvic acid

16β-methylcortisone
21-acetate (VI)

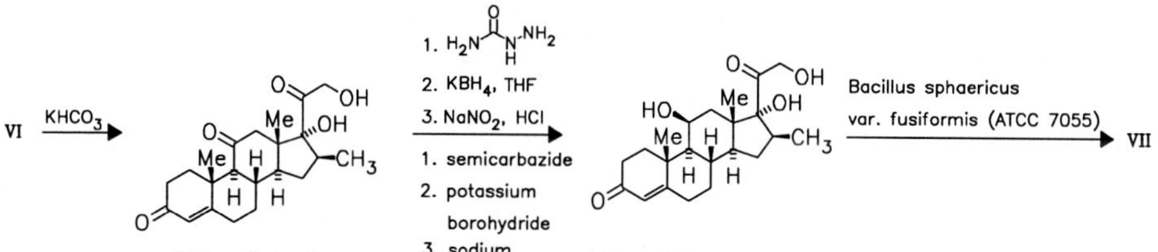

VI →[KHCO₃]

16β-methylcortisone

1. H₂N-NH-NH₂
2. KBH₄, THF
3. NaNO₂, HCl
1. semicarbazide
2. potassium borohydride
3. sodium nitrite

16β-methylhydrocortisone

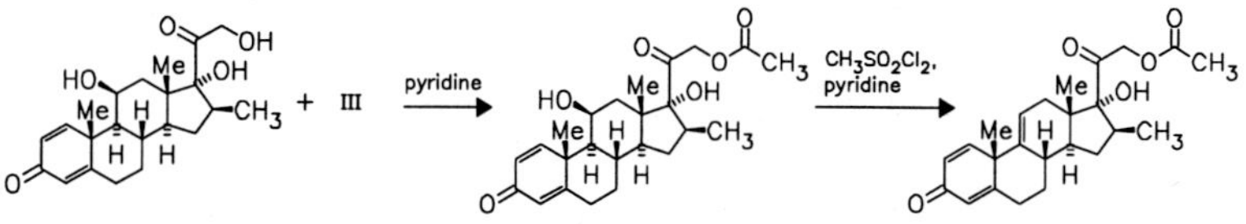

Bacillus sphaericus
var. fusiformis (ATCC 7055) → VII

16β-methylprednisolone (VII) + III →[pyridine]

16β-methylprednisolone
21-acetate

→[CH₃SO₂Cl₂, pyridine]

16β-methyl-17α,21-dihydroxy-
1,4,9(11)-pregnatriene-
3,20-dione 21-acetate (VIII)

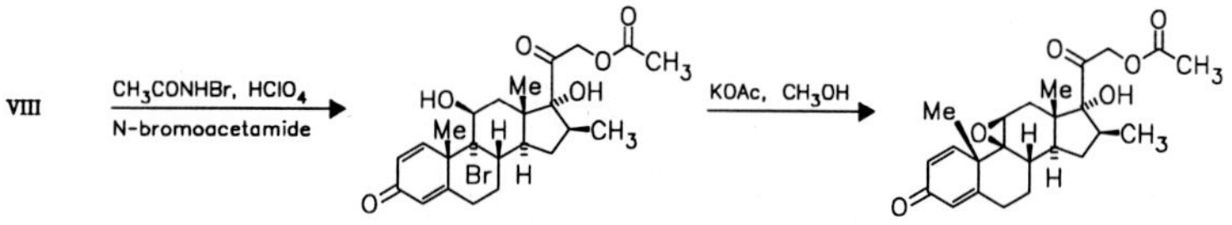

VIII →[CH₃CONHBr, HClO₄ / N-bromoacetamide]

9α-bromo-16β-methyl-
prednisolone 21-acetate (IX)

→[KOAc, CH₃OH]

9β,11β-epoxy-17α,21-dihydroxy-
16β-methyl-1,4-pregnadiene-
3,20-dione 21-acetate (X)

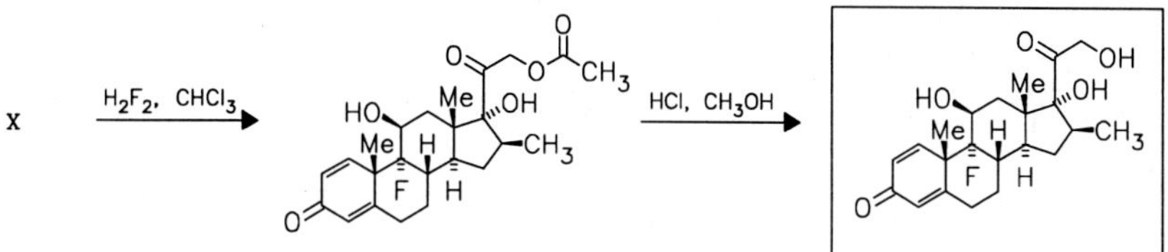

X →[H₂F₂, CHCl₃]

betamethasone acetate (XI)

→[HCl, CH₃OH]

Betamethasone

(b)

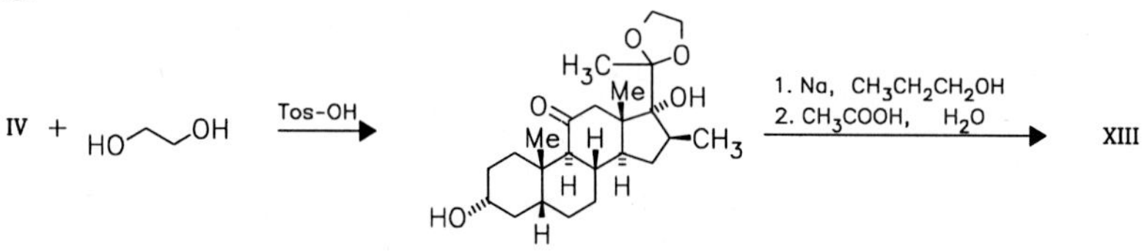

IV + HO-CH₂-CH₂-OH (ethylene glycol (XII)) →[Tos-OH]

3α,17α-dihydroxy-16β-
methyl-5β-pregnane-11,20-
dione 20-ethylene acetal

1. Na, CH₃CH₂CH₂OH
2. CH₃COOH, H₂O → XIII

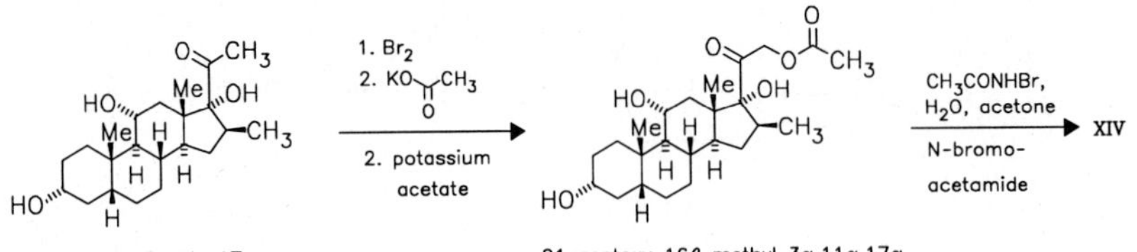

16β−methyl−3α,11α,17α−
trihydroxy−5β−pregnan−
20−one (XIII)

1. Br₂
2. KO—C—CH₃
2. potassium
acetate

21−acetoxy−16β−methyl−3α,11α,17α−
trihydroxy−5β−pregnan−20−one

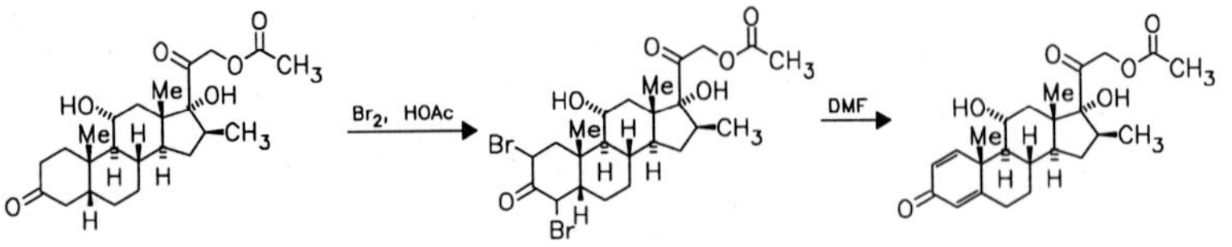

CH₃CONHBr,
H₂O, acetone
N−bromo−
acetamide

XIV

21−acetoxy−11α,17α−dihydroxy−
16β−methyl−5β−pregnane−
3,20−dione (XIV)

Br₂, HOAc

21−acetoxy−2,4−dibromo−
11α,17α−dihydroxy−16β−
methyl−5β−pregnane−3,20−dione

DMF

21−acetoxy−11α,17α−dihydroxy−
16β−methylpregna−1,4−diene−
3,20−dione (XV)

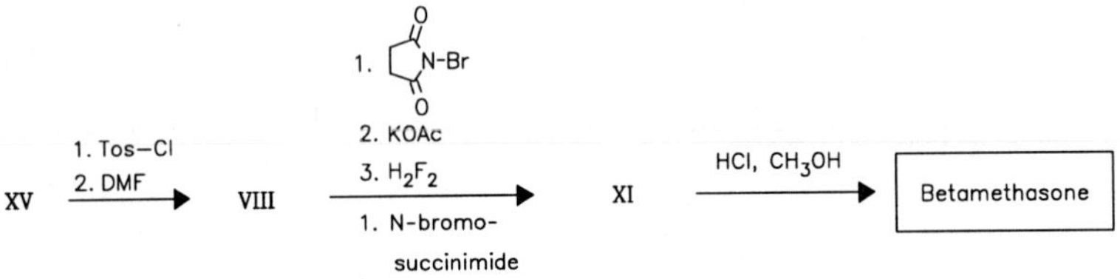

XV

1. Tos—Cl
2. DMF

VIII

1. N−Br
2. KOAc
3. H₂F₂
1. N−bromo−
succinimide

XI

HCl, CH₃OH

Betamethasone

(c)

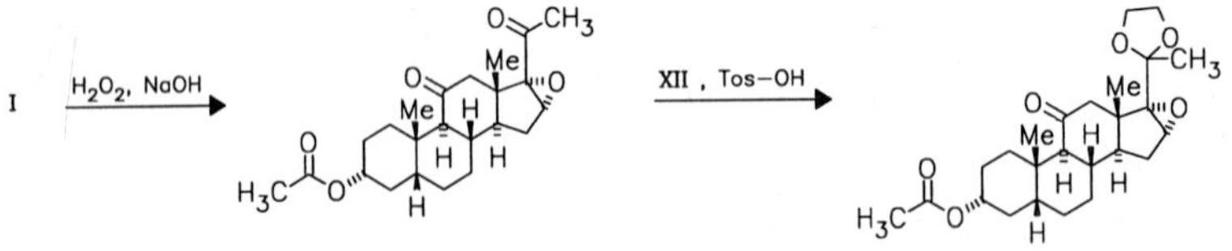

I

H₂O₂, NaOH

3α−acetoxy−11,20−dioxo−
16α,17α−epoxy−5β−pregnane

XII , Tos−OH

3α−acetoxy−16α,17α−epoxy−
5β−pregnane−11,20−dione
21−ethylene acetal (XVI)

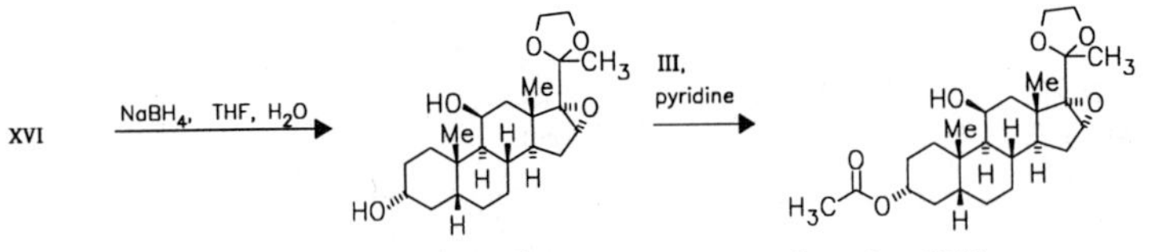

XVI

NaBH₄, THF, H₂O

20,20−ethylenedioxy−
16α,17α−epoxy−5β−
pregnane−3α,11β−diol

III,
pyridine

3α−acetoxy−20,20−
ethylenedioxy−16α,17α−
epoxy−5β−pregnan−11β−ol (XVII)

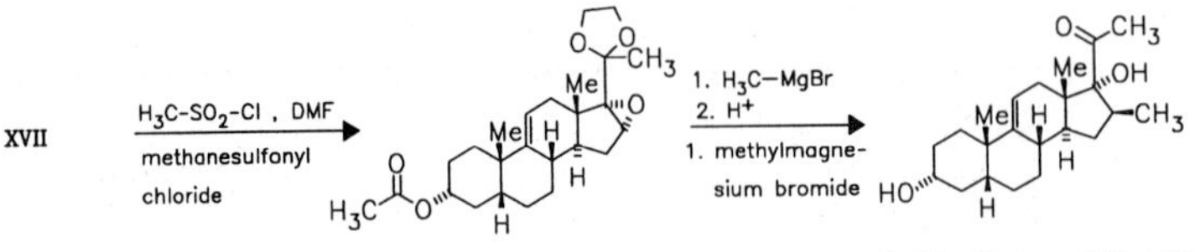

XVII

H₃C−SO₂−Cl , DMF
methanesulfonyl
chloride

3α−acetoxy−20,20−
ethylenedioxy−16α,17α−
epoxy−5β−pregn−9(11)−ene

1. H₃C—MgBr
2. H⁺
1. methylmagne−
sium bromide

3α,17α−dihydroxy−16β−methyl−
5β−pregn−9(11)−en−
20−one (XVIII)

XVIII $\xrightarrow[\text{chromium(VI)}]{\text{CrO}_3, \text{H}_2\text{SO}_4, \text{ acetone}}$ $\xrightarrow[\text{bromine}]{\text{Br}_2, \text{CH}_3\text{COOH}}$

17α-hydroxy-16β-methyl-
5β-pregn-9(11)-ene-3,20-dione

2β,4β-dibromo-17α-hydroxy-
16β-methyl-5β-pregn-9(11)-ene-
3,20-dione (XIX)

XIX $\xrightarrow[\substack{\text{lithium} \quad \text{lithium} \\ \text{carbonate} \quad \text{bromide}}]{\text{Li}_2\text{CO}_3, \text{ LiBr}, \text{ DMF}}$ $\xrightarrow[\text{iodine}]{\text{I}_2, \text{ Ca(OH)}_2, \text{ CaCl}_2}$ XX

17α-hydroxy-16β-methyl-
pregna-1,4,9(11)-triene-
3,20-dione

$\xrightarrow[\text{acetone}]{\substack{\text{H}_3\text{C—COOK} \\ \text{CH}_3\text{COOH}}}$ VIII $\xrightarrow[\substack{\text{N-bromo-} \\ \text{succinimide}}]{\text{N-Br}, \text{ HClO}_4}$ IX

17α-hydroxy-21-iodo-16β-
methylpregna-1,4,9(11)-
triene-3,20-dione (XX)

IX $\xrightarrow[]{\substack{1. \text{CH}_3\text{COONa}, \quad \text{CH}_3\text{OH} \\ 2. \text{CH}_3\text{COOH}, \quad \text{pyridine}}}$ X $\xrightarrow[\text{hydrogen fluoride}]{\text{H}_2\text{F}_2, \text{ THF}, \text{ CHCl}_3}$ XI $\xrightarrow[]{\text{NaOCH}_3, \text{ CH}_3\text{OH}}$ Betamethasone

Reference(s):
a US 3 164 618 (Schering Corp.; 5.1.1965; prior. 23.7.1957, 8.5.1958).
b Oliveto, E.P. et al.: J. Am. Chem. Soc. (JACSAT) **80**, 4428 (1958).
   Oliveto, E.P. et al.: J. Am. Chem. Soc. (JACSAT) **80**, 6687 (1958).
c US 3 104 246 (Roussel-Uclaf; 17.9.1963; appl. 26.7.1962; F-prior. 18.8.1961).
   Julian, P.L. et al.: J. Am. Chem. Soc. (JACSAT) **77**, 4601 (1955).

alternative syntheses:
US 3 053 865 (Merck & Co.; 11.9.1962; prior. 19.3.1958, 1.3.1960).
Taub, D. et al.: J. Am. Chem. Soc. (JACSAT) **80**, 4435 (1958); **28**, 4012 (1960).
US 4 041 055 (Upjohn; 9.8.1977; appl. 17.1.1975).

Formulation(s):   syrup 0.6 mg/5 ml; tabl. 0.5 mg, 0.6 mg, 1 mg

Trade Name(s):
D: Beta-Creme (Lichtenstein)
      Betagalen (Pharmagalen)
      Betam-Ophtal (Winzer)
      Beta-Stulln (Pharma Stulln)
      Betnesol (Glaxo Wellcome)
      Betnesol-V (Glaxo
      Wellcome/Cascan)
      Celestamine N (Essex
      Pharma)
      Celestan (Essex Pharma)
      Cordes Beta (Ichthyol)

      Diprosis (Essex Pharma)
      Diprosone (Essex Pharma)
      Euvaderm (Parke Davis)
F: Betnesalic (Glaxo
      Wellcome)-comb.
      Betnesol (Glaxo Wellcome)
      Betneval (Glaxo Wellcome)
      Betneval néomycine (Glaxo
      Wellcome)-comb.
      Célestamine (Schering-
      Plough)-comb.

      Célestène (Schering-
      Plough)
      Célestoderm (Schering-
      Plough)
      Diprolène (Schering-
      Plough)
      Diprosalic (Schering-
      Plough)-comb.
      Diprosept (Schering-
      Plough)-comb.

Diprosone néomycine (Schering-Plough)-comb.
Gentasone (Schering-Plough)-comb.

GB: Betnelan (Evans)
Betnesol (Evans)
Vista-Metasone (Martindale)

I: Alfaflor (Intes)-comb.
Apsor pom. derm. (IDI)-comb.
Beben (Parke Davis; as benzoate)
Bentelan (Glaxo; as phosphate)
Beta (IDI; as valeroacetate)
Betabioptal (Farmila)-comb.
Betameta (Formulario Naz.; as dipropionate)-comb.
Biorinil (Farmila)-comb.

Brumeton coll. (Bruschettini)-comb.
Celestoderm (Schering-Plough; as valerate)
Celestone (Schering-Plough)
Deltavagin (Farma-Biagini)-comb.
Dermatar (IDI)-comb.
Diproform (Schering-Plough)-comb.
Diprogenta (Sca)-comb.
Diprorecto (Schering-Plough)-comb.
Diprosalic (Schering-Plough)-comb.
Diprosone (Schering-Plough; as dipropionate)
Ecoval (Glaxo; as valerate)
Eubetal (SIFI)-comb.
Fluororinil (Farmila)-comb.
Gentalyn Beta (Schering-Plough)-comb.

Micutrin Beta crema (Schiapparelli Searle)-comb.
Minisone (IDI)
Stranoval pom. derm. (Teofarma)
Viobeta (IDI)-comb.
Visublefarite sosp. oft. (Merck Sharp & Dohme)-comb.
Visumetazone Antib. (Merck Sharp & Dohme)-comb.
several combination preparations

J: Betamamallet (Showa Yakuhin)
Betametha (Dojin)
Betnelan (Daiichi)
Dabbeta (Zenyaku)
Rinderon (Shionogi)
Rinesteron (Fuso)

USA: Celestone (Schering)

Betamethasone acetate

ATC: H02AB
Use: glucocorticoid

RN: 987-24-6   MF: $C_{24}H_{31}FO_6$ MW: 434.50   EINECS: 213-578-6
CN: (11β,16β)-21-(acetyloxy)-9-fluoro-11,17-dihydroxy-16-methylpregna-1,4-diene-3,20-dione

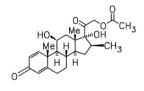

16β-methylprednisolone 21-acetate (from meprednisone acetate)

1. CH_3-SO_2-Cl, pyridine
2. $CH_3-CO-NH-Br$, dioxane
3. $KO-COCH_3$, CH_3OH
4. H_2F_2, $CHCl_3$

1. methanesulfonyl chloride
2. N-bromoacetamide
3. potassium acetate
4. hydrogen fluoride

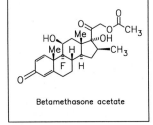

Betamethasone acetate

Reference(s):
US 3 164 618 (Schering Corp., 5.1.1965; prior. 8.5.1958, 23.7.1957).

additional literature:
betamethasone, q. v.

Formulation(s):   amp. 3 mg/ml, 3 mg/ml (in combination with betamethasone dihydrogen phosphate)

Trade Name(s):
D: Celestan Depot (Essex Pharma)-comb.
F: Betafluorene (Lepetit); wfm

Célestàne chronodose (Schering-Plough)-comb.; wfm

I: Celestone Cronodose (Schering-Plough)-comb.
USA: Celestone Soluspan (Schering)-comb.

Betamethasone adamantoate

ATC: H02AB
Use: glucocorticoid

RN: 40242-27-1 MF: C$_{33}$H$_{43}$FO$_6$ MW: 554.70 EINECS: 254-855-1
CN: (11β,16α)-9-fluoro-11,17-dihydroxy-16-methyl-21-[(tricyclo[3.3.1.1^{3,7}]dec-1-ylcarbonyl)oxy]pregna-1,4-diene-3,20-dione

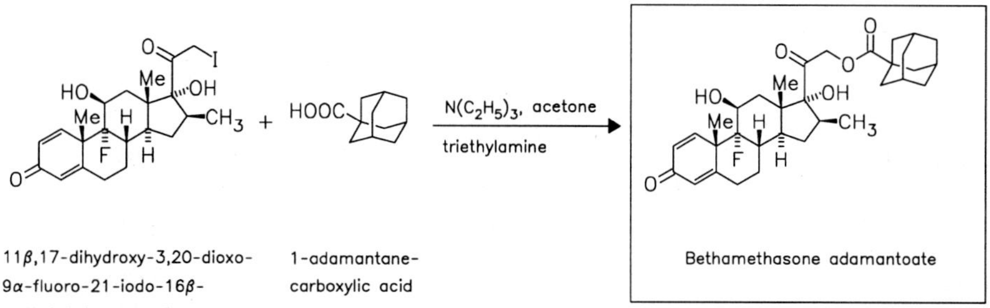

11β,17-dihydroxy-3,20-dioxo-
9α-fluoro-21-iodo-16β-
methyl-1,4-pregnadiene
(from betamethasone)

1-adamantane-
carboxylic acid

Bethamethasone adamantoate

Reference(s):
DOS 2 232 827 (Glaxo; appl. 4.7.1972; GB-prior. 5.7.1971).
(also alternative syntheses).

Trade Name(s):
GB: Betsovet (Glaxo); wfm

Betamethasone benzoate

ATC: D07AC
Use: glucocorticoid

RN: 22298-29-9 MF: C$_{29}$H$_{33}$FO$_6$ MW: 496.58 EINECS: 244-897-9
CN: (11β,16β)-17-(benzoyloxy)-9-fluoro-11,21-dihydroxy-16-methylpregna-1,4-diene-3,20-dione

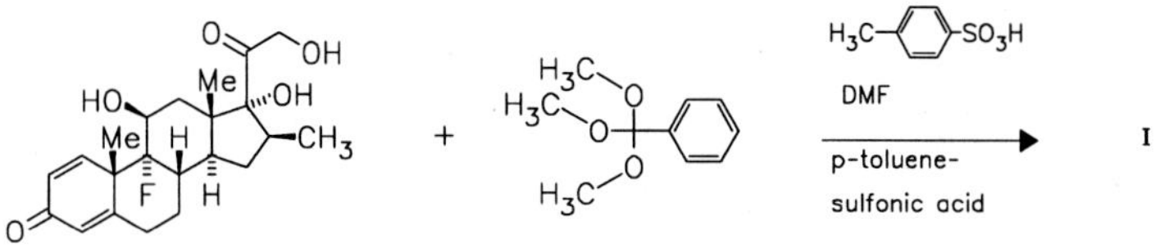

betamethasone
(q. v.)

trimethyl
orthobenzoate

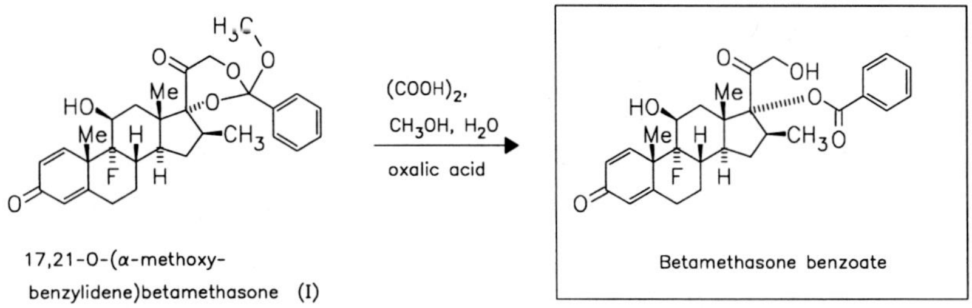

17,21-0-(α-methoxy-
benzylidene)betamethasone (I)

Betamethasone benzoate

Reference(s):
US 3 529 060 (Warner-Lambert; 15.9.1970; I-prior. 1.3.1967).
Ercoli, A. et al.: J. Med. Chem. (JMCMAR) **15**, 783 (1972).

alternative synthesis:
DOS 2 340 591 (Glaxo; appl. 10.8.1973; GB-prior. 11.8.1972).

pharmaceutical formulation:
US 3 749 773 (Warner-Lambert; 31.7.1973; prior. 25.2.1971).

Formulation(s):   cream 1 g/0.25 mg, 1 g/1 mg; gel 1 g/1 mg; lotion 0.1 %; ointment 0.1 %

Trade Name(s):
D:	Euvaderm (Parke Davis)	I:	Beben crema derm. (Parke Davis)		combination preparations
GB:	Bebate (Warner); wfm			J:	Asakin (Mikasa)
			Beben Sid (Parke Davis)		

Betamethasone butyrate propionate
(BBP; TO-186)

Use: topical anti-inflammatory, steroidal agent

RN: 5534-02-1   MF: $C_{29}H_{39}FO_7$ MW: 518.62
CN: (11β,16β)-9-fluoro-11-hydroxy-16-methyl-17-(1-oxobutoxy)-21-(1-oxopropoxy)pregna-1,4-diene-3,20-dione

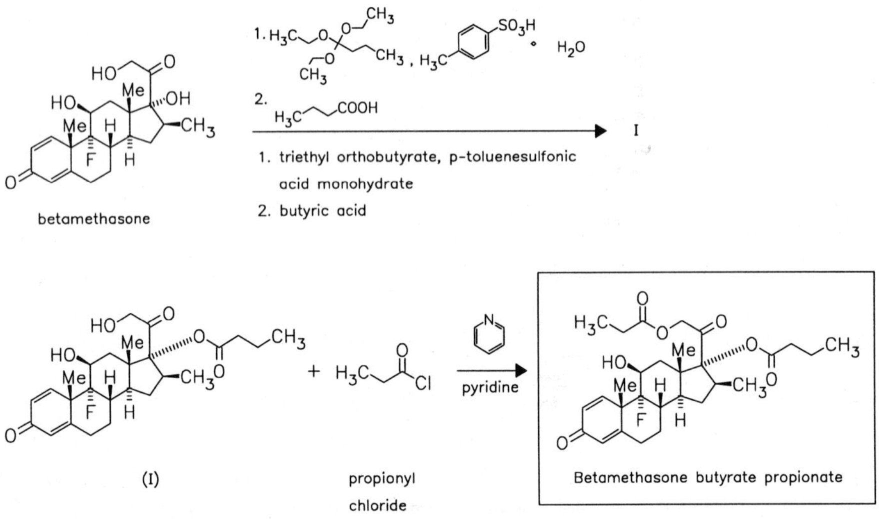

betamethasone

1. triethyl orthobutyrate, p-toluenesulfonic acid monohydrate
2. butyric acid

(I)   propionyl chloride   Betamethasone butyrate propionate

Reference(s):
Imai, S. et al.: Clin. Rep. **24**(11), 113 (1990).
Shue, H.-J. et al.: J. Med. Chem. (JMCMAR) **23**(4), 430 (1980).

Trade Name(s):
J: Antebate (Torii)

Betamethasone dipropionate

ATC: D07AC; D07BC; D07CC; H02AB
Use: glucocorticoid

RN: 5593-20-4 MF: $C_{28}H_{37}FO_7$ MW: 504.60 EINECS: 227-005-2
LD$_{50}$: >5 g/kg (M, p.o.);
>4 g/kg (R, p.o.)
CN: (11β,16β)-9-fluoro-11-hydroxy-16-methyl-17,21-bis(1-oxopropoxy)pregna-1,4-diene-3,20-dione

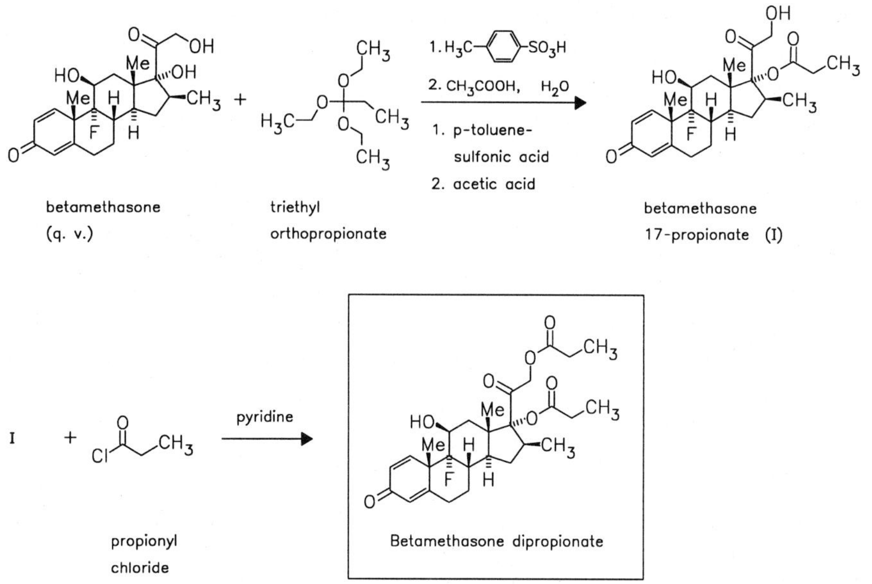

betamethasone (q. v.) + triethyl orthopropionate

1. H_3C—⟨⟩—SO_3H
2. CH_3COOH, H_2O
1. p-toluene-sulfonic acid
2. acetic acid

betamethasone 17-propionate (I)

I + propionyl chloride —pyridine→ Betamethasone dipropionate

Reference(s):
US 3 312 591 (Glaxo; 4.4.1967; GB-prior. 10.5.1963, 28.1.1964).
US 3 312 590 (Glaxo; 4.4.1967; GB-prior. 11.6.1963, 28.1.1964).
DE 1 443 957 (Glaxo; appl. 10.6.1964; GB-prior. 11.6.1963, 28.1.1964).

review:
Ferrante, M.C.; Rudy, B.C.: Anal. Profiles Drug Subst. (APDSB7) **6**, 43 (1977).

Formulation(s): aerosol 0.1 %; amp. 5 mg/ml; cream 0.05 %; ointment 0.05 %

Trade Name(s):

D: Diprogenta (Essex Pharma)-comb.
Diprosalic (Essex Pharma)-comb.
Diprosis (Essex Pharma)
Diprosone (Essex Pharma)
Diprosone depot (Essex Pharma)-comb.

F: Diproléne (Schering-Plough)
Diprosalic (Schering-Plough)-comb.
Diprosept (Schering-Plough)-comb.

Diprosone (Schering-Plough)
Diprosone Neomycin (Schering-Plough)-comb.
Diprostène (Schering-Plough)-comb.

GB: Diprosalic (Schering-Plough)-comb.
Diprosone (Schering-Plough)

I: Betameta Diprop (Formulario Naz.)
Diprosone (Schering-Plough)

numerous combination preparations

J: Dermosol-DP (Iwaki)
Diprocel (Schering-Plough)
Etynderon-DP (Taiyo)
Floderon (Ohta)
Ijilone-DP (Maeda)
Rinderon-DP (Shionogi)

USA: Diprolene (Schering)
Diprosone (Schering)
Lotrisane (Schering)

Betamethasone divalerate

ATC: D07AC
Use: glucocorticoid

RN: 38196-44-0   MF: $C_{32}H_{45}FO_7$ MW: 560.70   EINECS: 253-820-8
CN: (11β,16β)-9-fluoro-11-hydroxy-16-methyl-17,21-bis[(1-oxopentyl)oxy]pregna-1,4-diene-3,20-dione

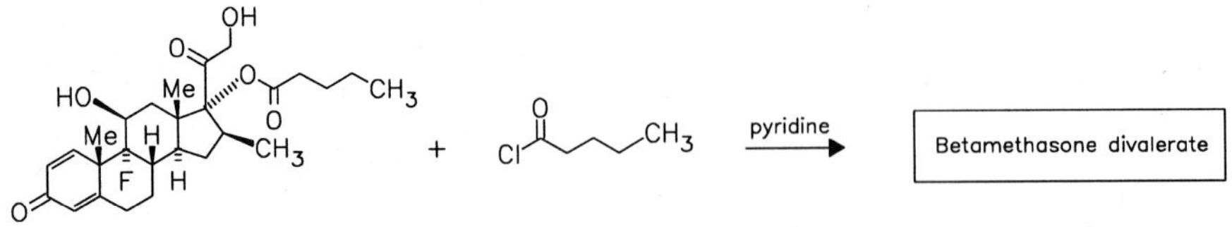

betamethasone valerate   valeryl chloride
(q. v.)

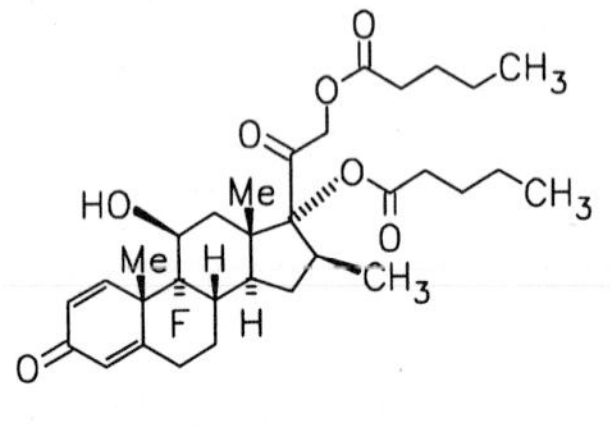

Betamethasone divalerate

Reference(s):
US 3 312 591 (Glaxo; 4.4.1967; GB-prior. 10.5.1963, 28.1.1964).
US 3 312 590 (Glaxo; 4.4.1967; GB-prior. 11.6.1963, 28.1.1964).
DE 1 443 957 (Glaxo; 10.6.1964; GB-prior. 11.6.1963, 28.1.1964).
cf. also betamethasone dipropionate.

Formulation(s):   cream 0.1 %; lotion 0.1 %; ointment 0.1 %; rectal ointment 0.05 %

Trade Name(s):
I: Betadival (Fardeco); wfm   Diprosone Creme (Essex);
                                        wfm

Betamethasone phosphate

ATC: H02AB; D07AC
Use: glucocorticoid

RN: 360-63-4   MF: $C_{22}H_{30}FO_8P$ MW: 472.45   EINECS: 206-636-7
LD$_{50}$:   700 mg/kg (M, i.p.)
CN: (11β,16β)-9-fluoro-11,17-dihydroxy-16-methyl-21-(phosphonooxy)pregna-1,4-diene-3,20-dione

disodium salt
RN: 151-73-5   MF: $C_{22}H_{28}FNa_2O_8P$ MW: 516.41   EINECS: 205-797-0
LD$_{50}$:   1304 mg/kg (M, i.v.); 1607 mg/kg (M, p.o.);
              1276 mg/kg (R, i.v.); 1877 mg/kg (R, p.o.)

1. CH₃–SO₂Cl, pyridine
2. NaI, acetone
3. AgH₂PO₄, NaOH
4. H⁺

1. methanesulfonyl chloride
2. sodium iodide
3. silver dihydrogen phosphate

betamethasone
(q. v.)

Betamethasone phosphate

Reference(s):
GB 913 941 (Merck & Co.; valid from 1959; USA-prior. 1958).

alternative syntheses:
US 2 939 873 (Merck & Co.; 1960; prior. 1959).
DOS 2 225 658 (I. Villax; appl. 14.12.1972; P-prior. 5.6.1971).
DE 1 134 075 (Merck AG; appl. 1959).

aqueous solution stabilized by 1-mercapto-2,3-propanediol:
DE 2 021 446 (Gruppo Lepetit; appl. 2.5.1970; I-prior. 7.5.1969).

Formulation(s):   amp. 2.63 mg/ml, 5.3 mg/ml; sol. 6.6 mg/100 g

Trade Name(s):

D:	Betnesol Past. (Glaxo Wellcome)	Célestène (Schering-Plough)		Vista-Methasone (Daniels)-comb.
	Betnesol Rekt. (Glaxo Wellcome/Cascan)	Célestène Chronodose (Schering-Plough)-comb.	I:	Bentelan (Glaxo)
	Betnesol WL (Glaxo Wellcome/Cascan)	Diprostène (Schering-Plough)-comb.		Celestone Ar. and im (Schering-Plough)
	Celestan depot (Essex Pharma)-comb.	Gentasone (Schering-Plough)	J:	Barbesolone (Nihon Tenganyaku)
	Diprosone depot (Essex Pharma)-comb.	GB: Betnesol (Glaxo)		Betnesol (Daiichi)
F:	Betnesol (Glaxo Wellcome)	Betnesol N (Glaxo)-comb.		Linolosal (Wakamoto)
		Vista-Methasone (Daniels)		Linosal (Wakamoto)
				Rinderon (Shionogi)
				Sanbetason (Santen)

Betamethasone valerate

ATC: D07AC
Use: glucocorticoid

RN: 2152-44-5   MF: C₂₇H₃₇FO₆   MW: 476.59   EINECS: 218-439-3
LD₅₀: >3 g/kg (M, p.o.);
        >3 g/kg (R, p.o.)
CN: (11β,16β)-9-fluoro-11,21-dihydroxy-16-methyl-17-[(1-oxopentyl)oxy]pregna-1,4-diene-3,20-dione

1. H₃C—⟨ ⟩—SO₃H
2. H₂SO₄, CH₃OH, H₂O

1. p-toluene-sulfonic acid
2. sulfuric acid

betamethasone
(q. v.)

trimethyl orthovalerate

Betamethasone valerate

Reference(s):
US 3 312 590 (Glaxo; 4.4.1967; GB-prior. 11.6.1963, 28.1.1964).
US 3 312 591 (Glaxo; 4.4.1967; GB-prior. 10.5.1963, 28.1.1964).

alternative synthesis:
DOS 2 055 221 (Lab. Chim. Farm. Blasina; appl. 10.11.1970).
DOS 2 340 591 (Glaxo; appl. 10.8.1973; GB-prior. 11.8.1972).
DOS 2 431 377 (Lark; appl. 29.6.1974; I-prior. 4.1.1974).

dermatological use:
ZA 7 700 678 (S. Fourie et al.; appl. 7.2.1977).
FR-M 5 399 (P. Temime; appl. 14.10.1965).
BE 829 197 (L. Grosjean; appl. 16.5.1975).

Formulation(s):   cream 0.1 %; lotion 0.1 %; ointment 0.1 %; tabl. 0.1 mg

Trade Name(s):

D: Betamethason Wolff
        (Wolff)
      Betnesol V, -"mite" (Glaxo
        Wellcome/Cascan)-comb.
      Celestan V, -"mite", -
        crinale (Essex Pharma)
      Celestan V mit Neomycin
        (Essex Pharma)-comb.
      Celestan V mit Sulmycin
        (Essex Pharma)-comb.
      Cordes Beta (Ichthyol)
      Sulmycin (Essex Pharma)-
        comb.
F: Betnesalic (Glaxo
        Wellcome)-comb.
      Betneval (Glaxo Wellcome)
      Betneval Néomycin (Glaxo
        Wellcome)-comb.
      Célestoderm (Schering-
        Plough)
      Célestoderm Relais
        (Schering-Plough)
GB: Betacap (Dermal)

      Betnovate (Glaxo
        Wellcome)
      Betnovate Rectal (Glaxo
        Wellcome)-comb.
      Bettamousse (Evans)
      Fucibet (Leo)-comb.
I: Celestoderm-V (Schering-
        Plough)
      Dermovaleas (Valeas)
      Ecoval (Glaxo)-comb.
      Ecoval-70 (Glaxo)
J: Ain V (Kobayashi)
      Asdesolon (Maruishi)
      Bectmiran (Towa)
      Betaclin (Sawai)
      Betnevate (Glaxo-Daiichi)
      Betnevate N (Daiichi)-
        comb.
      Calamiraderon V (Fukuchi)
      Cordel (Taisho)
      Dermitt (Mitgamitsu
        Mitsui)
      Dermosol (Iwaki)

      Hormeton (Tobishi)
      Hormezon (Tobishi Jakuhin
        Kogyo)
      Ijilone V (Maeda Kyowa;
        Ahishin)
      Keligroll (Kaigai Horita)
      Muhibeta V (Ikeda
        Mohando)
      Muhibeta V (Nippon Shoji)
      Nolcart (Tatsumi)
      Otumazon (Fukuchi)
      Rapoletin (Zeria)
      Rinderon-V (Shionogi)
      Rinderon V (Shionogi)-
        comb.
      Rinderon VA (Shionogi)-
        comb.
      Rinderon VG (Shionogi)-
        comb.
      Tochiprobetasone (Shinsei
        Kowa)
USA: Beta-Val (Teva)

Betanidine

(Bethanidine)

ATC: C02CC01
Use: antihypertensive

RN: 55-73-2   MF: $C_{10}H_{15}N_3$ MW: 177.25
LD_{50}:   16.307 mg/kg (M, i.v.)
CN: N,N'-dimethyl-N''-(phenylmethyl)guanidine

sulfate (2:1)
RN: 114-85-2   MF: $C_{10}H_{15}N_3 \cdot 1/2H_2SO_4$ MW: 452.58   EINECS: 204-056-9
LD_{50}:   12 mg/kg (M, i.v.); 520 mg/kg (M, p.o.);
        20 mg/kg (R, i.v.)

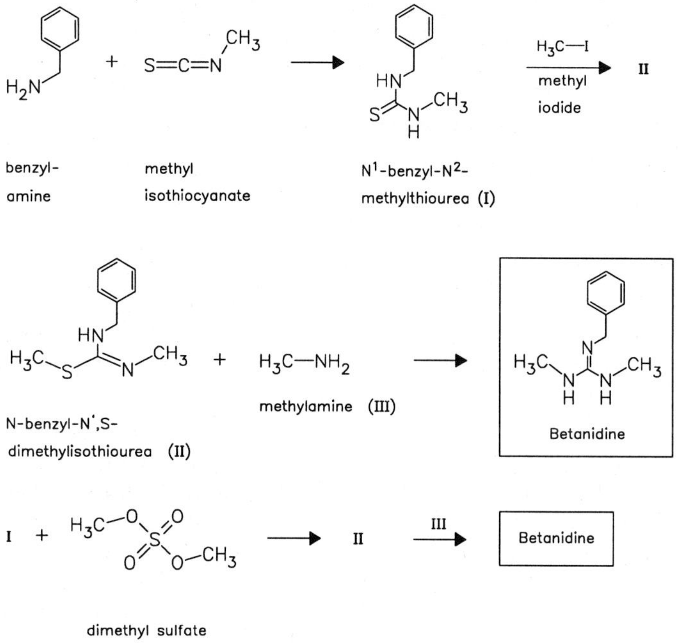
benzyl-
amine
methyl
isothiocyanate
N[1]-benzyl-N[2]-
methylthiourea (I)
methyl
iodide
II
N-benzyl-N',S-
dimethylisothiourea (II)
methylamine (III)
Betanidine
I
II
III
Betanidine
dimethyl sulfate

Reference(s):
GB 973 882 (Wellcome Found.; appl. 15.12.1960; prior. 23.12.1959).

alternative synthesis:
DAS 1 568 057 (GEA; appl. 9.12.1966).

Formulation(s): tabl. 10 mg, 50 mg

Trade Name(s):

F:	Esbatal (Wellcome); wfm	I:	Esbatal (Wellcome); wfm	Hypersin (Zeria)
GB:	Bendogen (Lagap); wfm	J:	Benzoxine (Sanwa)	
	Esbatal (Calmic); wfm		Betaindol (Tanabe)	

Betaxolol

ATC: C07AB05; S01ED02
Use: selective β-adrenoceptor blocker, antihypertensive

RN: 63659-18-7 MF: $C_{18}H_{29}NO_3$ MW: 307.43
LD_{50}: 37 mg/kg (M, i.v.); 944 mg/kg (M, p.o.)
CN: (±)-1-[4-[2-(cyclopropylmethoxy)ethyl]phenoxy]-3-[(1-methylethyl)amino]-2-propanol

hydrochloride
RN: 63659-19-8 MF: $C_{18}H_{29}NO_3 \cdot HCl$ MW: 343.90 EINECS: 264-384-3
LD_{50}: 37 mg/kg (M, i.v.); 48 mg/kg (M, p.o.);
27.4 mg/kg (R, i.v.); 998 mg/kg (R, p.o.);
30 mg/kg (dog, p.o.)

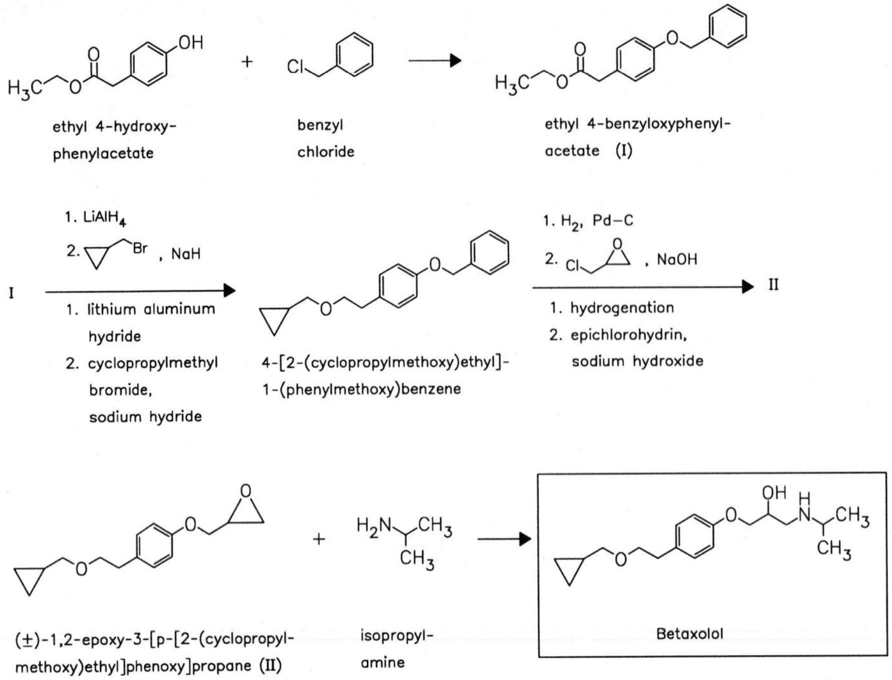

ethyl 4-hydroxy-
phenylacetate

benzyl
chloride

ethyl 4-benzyloxyphenyl-
acetate (I)

1. LiAlH₄

2. cyclopropylmethyl Br , NaH

I →

1. lithium aluminum
hydride

2. cyclopropylmethyl
bromide,
sodium hydride

4-[2-(cyclopropylmethoxy)ethyl]-
1-(phenylmethoxy)benzene

1. H₂, Pd–C

2. Cl epichlorohydrin , NaOH   II

1. hydrogenation

2. epichlorohydrin,
sodium hydroxide

(±)-1,2-epoxy-3-[p-[2-(cyclopropylmethoxy)ethyl]phenoxy]propane (II)

+   H₂N isopropyl-
amine

Betaxolol

Reference(s):

DOS 2 649 605 (Synthelabo; appl. 29.10.1976; F-prior. 6.11.1975).
US 4 252 984 (Synthelabo; 24.2.1981; appl. 20.10.1976; F-prior. 6.11.1975).
US 4 311 708 (Synthelabo; 24.2.1981, F-prior. 6.11.1975).
US 4 342 783 (Synthelabo; 3.8.1983; prior. 30.6.1980).

Formulation(s):   eye drops 0.25 %, 0.5 %; f. c. tabl. 20 mg; tabl. 10 mg, 20 mg, 25 mg (as hydrochloride)

Trade Name(s):

D: Betoptima (Alcon; 1985)
      Kerlone (Synthelabo; 1984)
F: Betoptic (Alcon; 1987)

      Kerlone (Robert et
      Carrière; Synthélabo/
      Schwarz; 1983)
GB: Betoptic (Alcon; 1986)
      Kerlone (Lorex; 1984)

I: Betoptic coll. (Alcon;
      1986)
      Kerlon (Synthelabo; 1987)
USA: Betoptic (Alcon; 1985)
      Kerlone (Searle)

Betazole

(Ametazole)

ATC: V04CG02
Use: gastric acid diagnostic, gastric acid
        stimulant

RN: 105-20-4   MF: C₅H₉N₃   MW: 111.15   EINECS: 203-278-3
CN: 1*H*-pyrazole-3-ethanamine

dihydrochloride

RN: 138-92-1   MF: C₅H₉N₃ · 2HCl   MW: 184.07   EINECS: 205-345-2
LD₅₀: 803 mg/kg (M, i.v.); 860 mg/kg (M, p.o.)

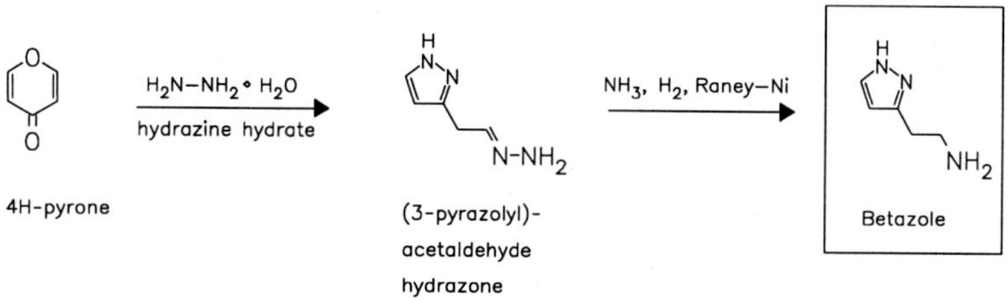

4H-pyrone → (3-pyrazolyl)-acetaldehyde hydrazone → Betazole

Reference(s):
US 2 785 177 (Eli Lilly; 12.3.1957; prior. 7.1.1952).

Formulation(s):   amp. 50 mg (5 %, as dihydrochloride)

Trade Name(s):
D: Betazole "Lilly"; wfm   J: Histimin (Shionogi)
GB: Histalog (Lilly); wfm   USA: Histalog (Lilly); wfm

Bethanechol chloride

ATC: N07AB02
Use: parasympathomimetic

RN: 590-63-6   MF: $C_7H_{17}ClN_2O_2$ MW: 196.68   EINECS: 209-686-8
LD_{50}:   10 mg/kg (M, i.v.); 250 mg/kg (M, p.o.);
        21 mg/kg (R, i.v.); 1500 mg/kg (R, p.o.)
CN: 2-[(aminocarbonyl)oxy]-*N,N,N*-trimethyl-1-propanaminium chloride

β-methylcholine chloride   phosgene (I)   →   Bethanechol chloride

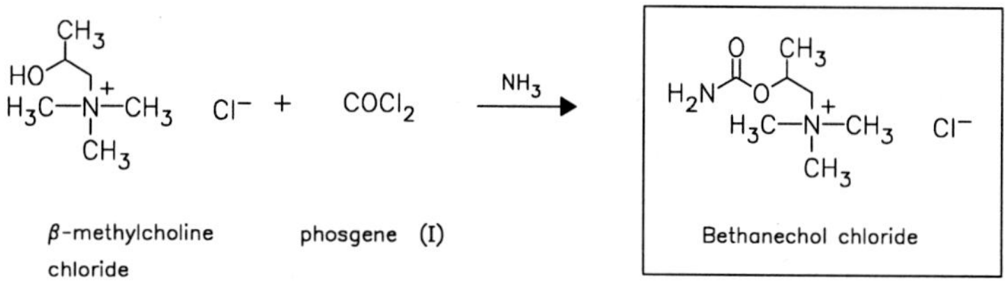

1-chloro-2-propanol   + I   →   2-(aminocarbonyl-oxy)propyl chloride   →   Bethanechol chloride

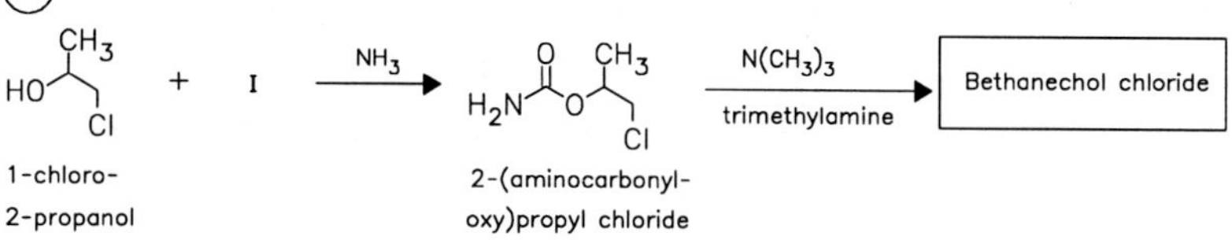

Reference(s):
a   US 2 322 375 (Merck & Co.; 1943; prior. 1940).
b   US 1 894 162 (O. Dahner, C. Diehl; 1933; D-prior. 1930).

Formulation(s):   amp. 5 mg; tabl. 5 mg, 10 mg, 25 mg, 50 mg

Trade Name(s):
GB: Myotonine (Glenwood)   J: Besacolin (Eisai)   Perista (Nissin)
I: Urecholine (Merck Sharp   Bethachorol (Nichiiko)   USA: Urecholine (Merck)
       & Dohme)   Paracholin (Kanto)

Bevantolol

ATC: C07AB06
Use: long acting cardioselective β_1-adrenoceptor blocker

RN: 59170-23-9   MF: $C_{20}H_{27}NO_4$ MW: 345.44
LD$_{50}$:   419 mg/kg (M, p.o.);
       38 mg/kg (R, i.v.)
CN: 1-[[2-(3,4-dimethoxyphenyl)ethyl]amino]-3-(3-methylphenoxy)-2-propanol

hydrochloride
RN: 42864-78-8   MF: $C_{20}H_{27}NO_4 \cdot HCl$ MW: 381.90
LD$_{50}$:   419 mg/kg (M, p.o.);
       25.1 mg/kg (R, i.v.); 460 mg/kg (R, p.o.)

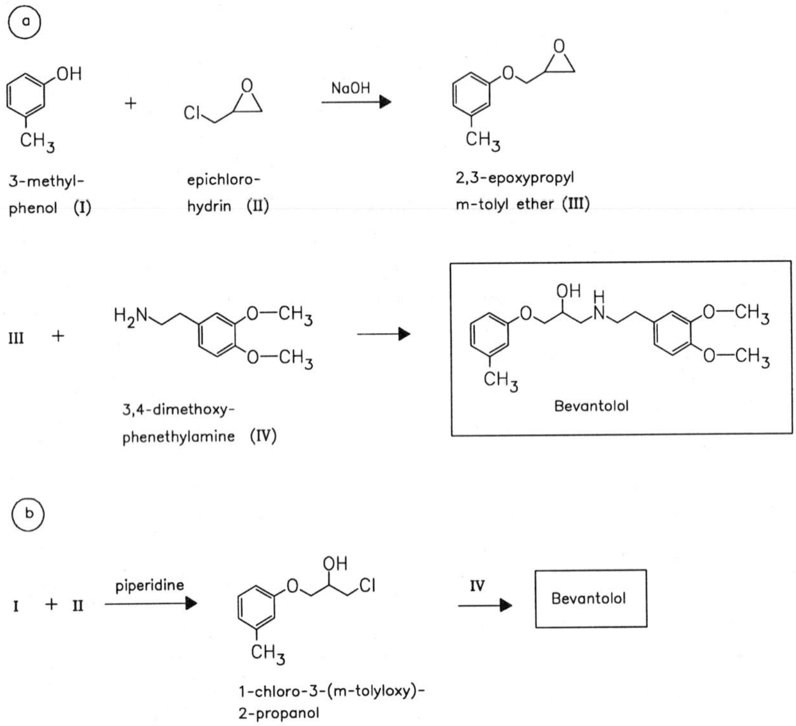

3-methyl-phenol (I)   epichloro-hydrin (II)   2,3-epoxypropyl m-tolyl ether (III)

3,4-dimethoxy-phenethylamine (IV)   Bevantolol

1-chloro-3-(m-tolyloxy)-2-propanol   Bevantolol

Reference(s):
DE 2 259 489 (Parke Davis; appl. 5.12.1972; USA-prior. 14.12.1971).
US 3 857 891 (Parke Davis; 31.12.1974; appl. 14.2.1971).
US 3 929 856 (Parke Davis; 30.12.1975; appl. 3-9-1974; prior. 3.9.1974, 14.12.1971).
Crowther, A.F. et al.: J. Med. Chem. (JMCMAR) **12**, 638 (1979).
Hoetle, M.L. et al.: J. Med. Chem. (JMCMAR) **18**, 148 (1975).

Formulation(s):   tabl. 100 mg, 200 mg

Trade Name(s):
J: Calvan (Nippon
      Chemiphar; Torii; as
      hydrochloride)

USA: Vantol (Parke Davis; as
        hydrochloride); wfm

Bevonium metilsulfate

(Bevonium methylsulfate; Piribenzil; Pyribenzil)

ATC: A03AB13
Use: anticholinergic, antispasmodic

RN: 5205-82-3   MF: $C_{22}H_{28}NO_3 \cdot CH_3O_4S$ MW: 465.57   EINECS: 226-001-8
LD$_{50}$:   17.4 mg/kg (M, i.v.); 1360 mg/kg (M, p.o.);
        26 mg/kg (R, i.v.); 5080 mg/kg (R, p.o.);
        1 g/kg (dog, p.o.)
CN: 2-[[(hydroxydiphenylacetyl)oxy]methyl]-1,1-dimethylpiperidinium methyl sulfate

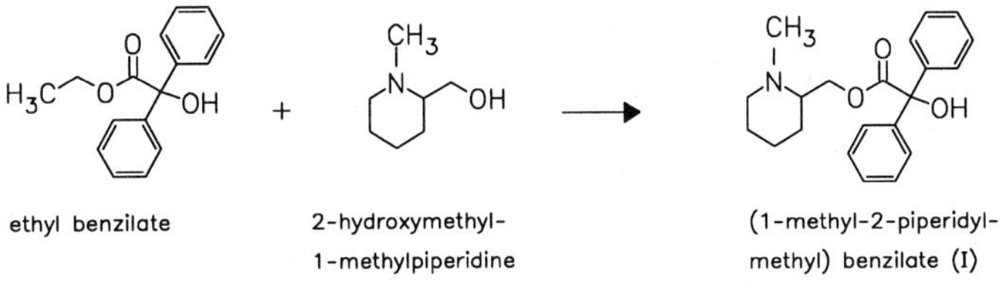

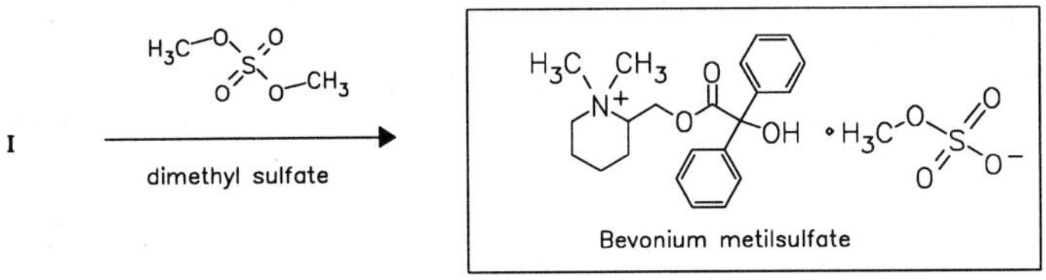

ethyl benzilate   2-hydroxymethyl-   (1-methyl-2-piperidyl-
                         1-methylpiperidine   methyl) benzilate (I)

Bevonium metilsulfate

Reference(s):
BE 616 951 (Grünenthal; appl. 26.4.1962; D-prior. 29.4.1961).

piribenzil:
DE 1 188 081 (Grünenthal; appl. 19.2.1960).

Formulation(s):   amp. 10 mg (0.25 %); tabl. 50 mg

Trade Name(s):
D: Acabel (Grünenthal); wfm   J: Acabel (Dainippon)

Bezafibrate

ATC: B04AA; C01AB02
Use: antiarteriosclerotic
        (antihyperlipidemic)

RN: 41859-67-0   MF: $C_{19}H_{20}ClNO_4$ MW: 361.83   EINECS: 255-567-9
LD$_{50}$:   723 mg/kg (M, p.o.);
        1082 mg/kg (R, p.o.)
CN: 2-[4-[2-[(4-chlorobenzoyl)amino]ethyl]phenoxy]-2-methylpropanoic acid

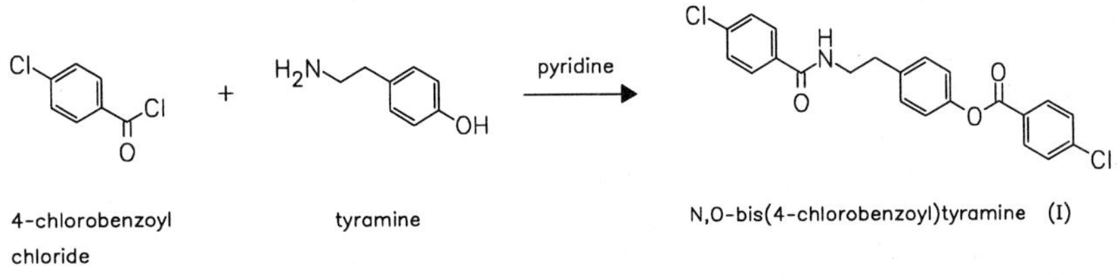

4-chlorobenzoyl   tyramine   N,O-bis(4-chlorobenzoyl)tyramine   (I)
chloride

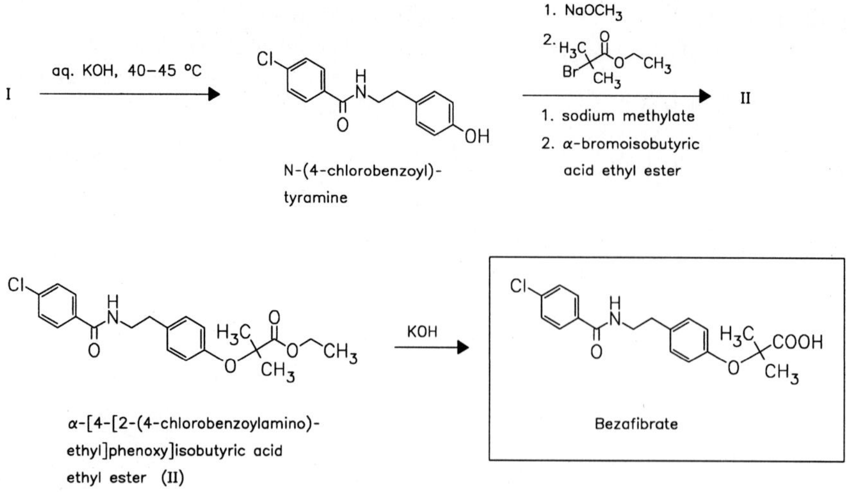

N-(4-chlorobenzoyl)-
tyramine

1. NaOCH₃
2. α-bromoisobutyric
acid ethyl ester

1. sodium methylate
2. α-bromoisobutyric
acid ethyl ester

α-[4-[2-(4-chlorobenzoylamino)-
ethyl]phenoxy]isobutyric acid
ethyl ester (II)

KOH

Bezafibrate

Reference(s):
DOS 2 149 070 (Boehringer Mannh.; appl. 1.10.1971).
FR-appl. 2 154 739 (Boehringer Mannh.; appl. 29.9.1972; D-prior. 1.10.1971, 22.6.1972).

Formulation(s):   drg. 200 mg; f. c. tabl. 200 mg; s. r. drg. 400 mg; tabl. 200 mg

Trade Name(s):
D: Azufibrate (Azupharma)   Pegradin (Berlin-Chemie)   GB: Bezalip (Bristol-Myers
     Befibrate (Henning)   Sklerofibrate (Merckle)   Squibb)
     Bezacur (Hexal)   F: Béfizal (Boehringer   I: Bezalip (Boehringer
     Cedur (Boehringer Mannh.)   Mannh.)   Mannh.)
     Lipox (TAD)

Bibrocathol

(Bibrocathin; Bismucatebrol)

ATC: S01AX05
Use: antiseptic

RN: 6915-57-7   MF: $C_6HBiBr_4O_3$ MW: 649.67   EINECS: 230-023-3
CN: 4,5,6,7-tetrabromo-2-hydroxy-1,3,2-benzodioxabismole

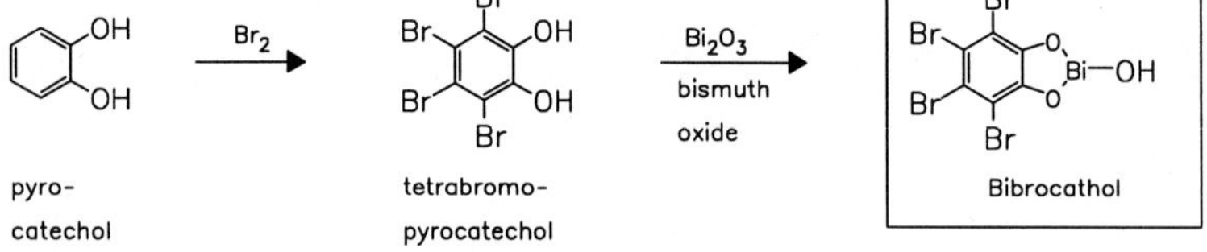

pyro-
catechol

Br₂

tetrabromo-
pyrocatechol

Bi₂O₃
bismuth
oxide

Bibrocathol

Reference(s):
DRP 207 544 (Chem. Fabrik von Heyden; appl. 1908).
Hundrup: Arch. Pharm. Chemi (APCEAR) **54**, 537 (1947).

Formulation(s):   eye ointment 1 %, 2 %, 3 %, 5 %

Trade Name(s):
D: Noviform (CIBA Vision)   Novifort (Dispersa)-comb.   Posiformin (Ursapharm)

Bicalutamide
(ICI-176334)

ATC: L02BB03
Use: non-steroidal antiandrogen,
        antineoplastic, anti(prostate)cancer

RN: 90357-06-5   MF: $C_{18}H_{14}F_4N_2O_4S$ MW: 430.38
CN: (±)-N-[4-cyano-3-(trifluoromethyl)phenyl]-3-[(4-fluorophenyl)sulfonyl]-2-hydroxy-2-methylpropanamide

R-enantiomer
RN: 113299-40-4   MF: $C_{18}H_{14}F_4N_2O_4S$ MW: 430.38
S-enantiomer
RN: 113299-38-0   MF: $C_{18}H_{14}F_4N_2O_4S$ MW: 430.38

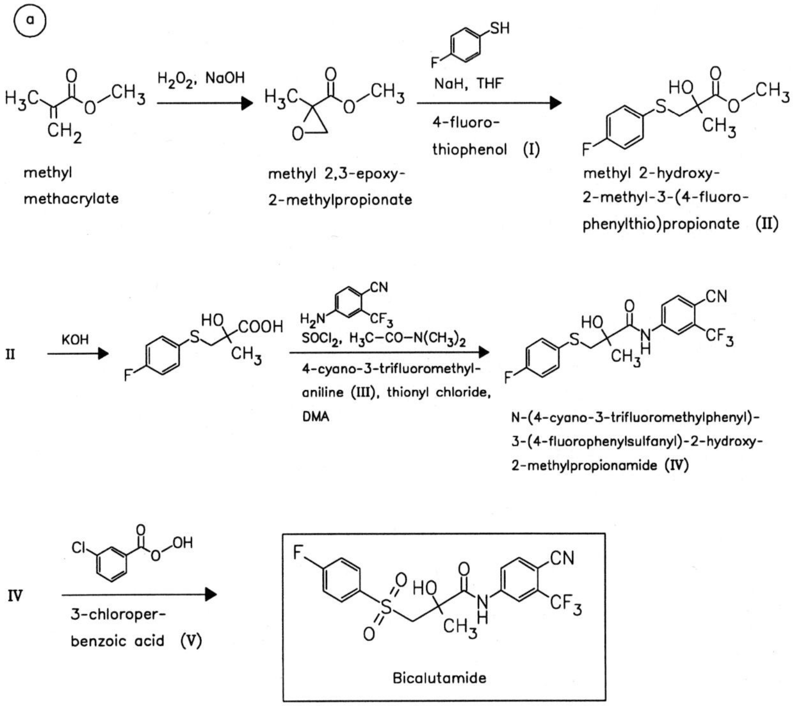

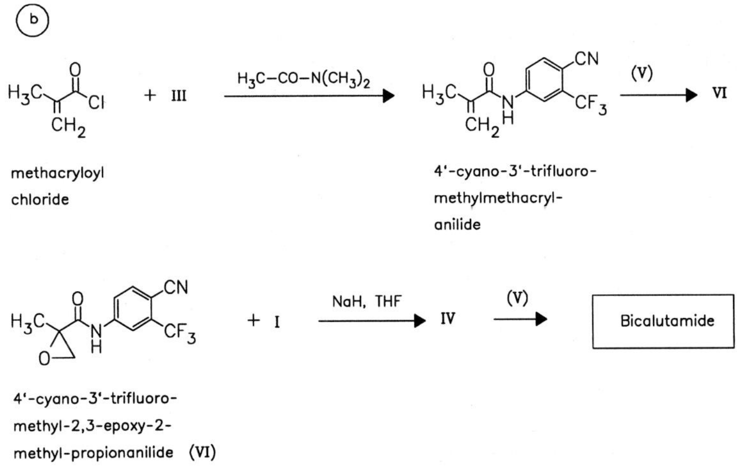

Reference(s):
EP 100 172 (ICI; appl. 8.7.1983; UK-prior. 23.7.1982).

active enantiomer (R-(–)-bicalutamide) for treating e. g. prostate cancer, acne:
WO 9 519 770 (Sepracor Inc.; appl. 27.7.1995; USA-prior. 21.1.1994).

combination with progesterone antagonists:
DE 4 318 371 (Schering AG; 1.12.1994; D-prior. 28.5.1993).

combination with sex steroid biosynthesis inhibitors:
WO 9 100 733 (Endorecherche Inc.; 24.1.1994; USA-prior. 7.7.1989).

Formulation(s):   tabl. 50 mg

Trade Name(s):
D: Casodex (Zeneca)   GB: Casodex (Zeneca)   USA: Casodex (Zeneca)

Bietamiverine
(Dietamiverin)

ATC: A03AA
Use: antispasmodic

RN: 479-81-2   MF: $C_{19}H_{30}N_2O_2$ MW: 318.46   EINECS: 207-538-7
CN: α-phenyl-1-piperidineacetic acid 2-(diethylamino)ethyl ester

dihydrochloride
RN: 2691-46-5   MF: $C_{19}H_{30}N_2O_2 \cdot 2HCl$ MW: 391.38   EINECS: 220-262-1
LD_{50}: 55 mg/kg (M, i.v.); 1247 mg/kg (M, p.o.)

α-chlorophenyl-   2-diethylamino-
acetyl chloride   ethanol

2-phenyl-2-chloroacetic acid
2-(diethylamino)ethyl ester (I)

I

piperidine

Bietamiverine

Reference(s):
DE 859 892 (Nordmark; appl. 1950).

Trade Name(s):
D: Spasmaparid (Nordmark);   J: Sparine A (Tokyo Tanabe)
      wfm

Bietaserpine

ATC: C02AA07
Use: antihypertensive

RN: 53-18-9   MF: $C_{39}H_{53}N_3O_9$ MW: 707.87   EINECS: 200-165-0
CN: (3β,16β,17α,18β,20α)-1-[2-(diethylamino)ethyl]-11,17-dimethoxy-18-[(3,4,5-
       trimethoxybenzoyl)oxy]yohimban-16-carboxylic acid methyl ester

bitartrate (1:1)
RN: 1111-44-0   MF: $C_{39}H_{53}N_3O_9 \cdot C_4H_6O_6$ MW: 857.95   EINECS: 214-180-5

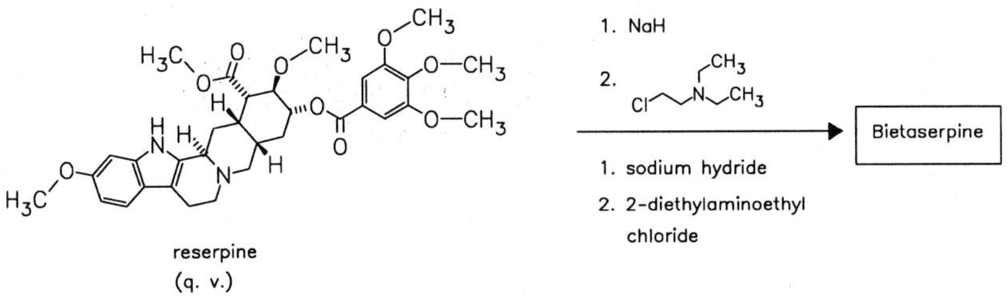

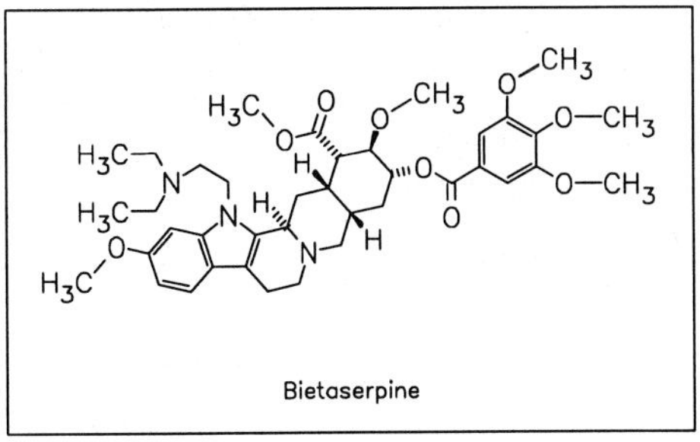

reserpine
(q. v.)

1. NaH
2.
1. sodium hydride
2. 2-diethylaminoethyl
   chloride

→ Bietaserpine

Bietaserpine

Reference(s):
FR 1 256 524 (Dautreville et Lebas et A. Buzas; appl. 13.2.1959).
FR-M 102 (Soc. Nogentaise de Prod. Chim. et A. Buzas; appl. 3.8.1960).

Trade Name(s):
F: Tensibar (Lefranca); wfm   I: Pleiantensin simplex
                                           (Guidotti); wfm

Bifluranol

ATC: G03HA
Use: antiandrogen, treatment of benign
        prostatic hypertrophy

RN: 34633-34-6   MF: $C_{17}H_{18}F_2O_2$ MW: 292.33
CN: (R*,S*)-4,4'-(1-ethyl-2-methyl-1,2-ethanediyl)bis[2-fluorophenol]

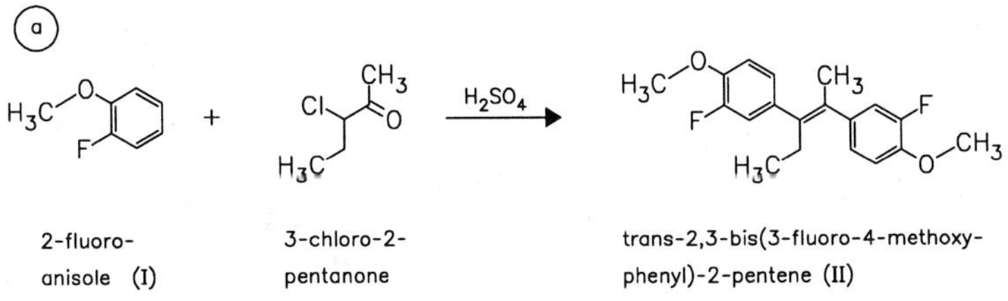

2-fluoro-
anisole (I)

3-chloro-2-
pentanone

H_2SO_4 →

trans-2,3-bis(3-fluoro-4-methoxy-
phenyl)-2-pentene (II)

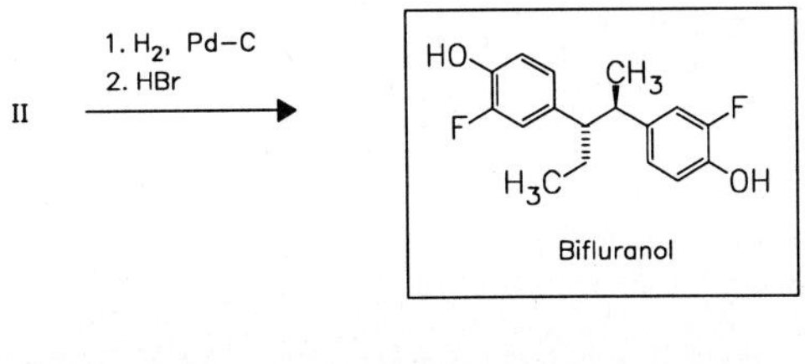

1. H_2, Pd–C
2. HBr

II →

Bifluranol

(b)

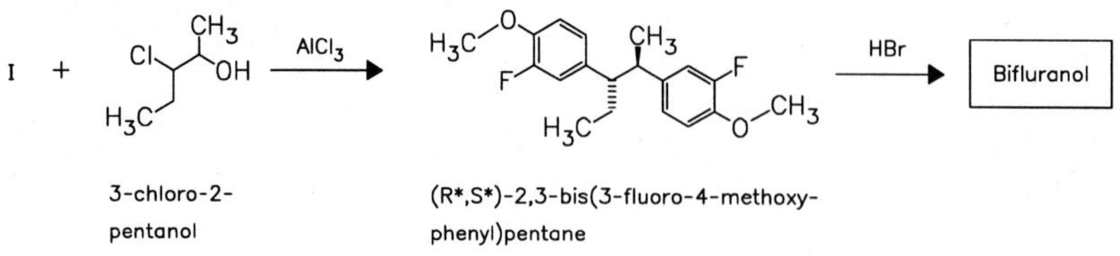

3-chloro-2-
pentanol

(R*,S*)-2,3-bis(3-fluoro-4-methoxy-
phenyl)pentane

Reference(s):
DE 2 110 428 (Biorex; appl. 4.3.1971; GB-prior. 16.3.1970).
US 4 051 263 (Biorex; 27.9.l977; GB-prior. 16.3.1970).

Formulation(s):   amp.

Trade Name(s):
GB: Prostarex (Biorex); wfm

Bifonazole
(Bifonazolum)

ATC: D01AC10
Use: topical antimycotic (inhibitor of
ergosterin biosynthesis in yeasts and
dermatophytes)

RN: 60628-96-8   MF: $C_{22}H_{18}N_2$ MW: 310.40   EINECS: 262-336-6
LD$_{50}$:   57 mg/kg (M, i.v.); 2629 mg/kg (M, p.o.);
63 mg/kg (R, i.v.); 1463 mg/kg (R, p.o.);
>500 mg/kg (dog, p.o.)
CN: 1-([1,1'-biphenyl]-4-ylphenylmethyl)-1H-imidazole

monohydrochloride
RN: 60629-09-6   MF: $C_{22}H_{18}N_2 \cdot HCl$ MW: 346.86
sulfate
RN: 60629-08-5   MF: $C_{22}H_{18}N_2 \cdot xH_2O_4S$ MW: unspecified

(a)

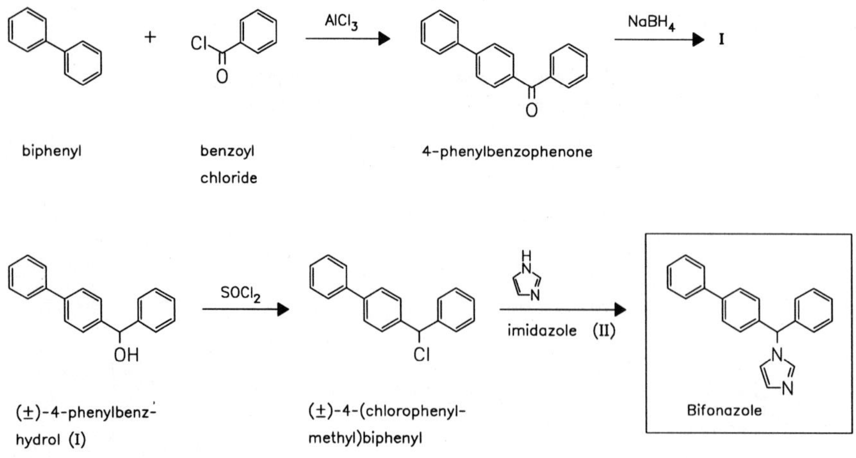

biphenyl

benzoyl
chloride

4-phenylbenzophenone

(±)-4-phenylbenz-
hydrol (I)

(±)-4-(chlorophenyl-
methyl)biphenyl

Bifonazole

(b)

II →(SOCl₂, CH₃CN)→ [1,1'-sulfinyl-bisimidazole] →(I)→ Bifonazole

Reference(s):
DOS 2 461 406 (Bayer; appl. 5.12.1975; USA-prior. 24.12.1974).
US 4 118 487 (Bayer; 3.11.1978; appl. 5.12.1975; prior. 24.12.1974).

effective mechanism:
Berg, D. et al.: Arzneim.-Forsch. (ARZNAD) **34** (I), 139 (1984).

Formulation(s):   cream 10 mg (1 %); gel 10 mg; lotion 1 %; powder 10 mg (1 %); sol. 10 mg (1 %)

Trade Name(s):
D: Bifomyk (Hexal)   F: Amycor (Lipha Santé;   I: Azolmen (Menarini; 1987)
      Bifon (Dermapharm)   1987)   Bifazol (Bayropharm;
      Mycospor (Bayer; 1983)   Amycor onychoset (Lipha   1986)
                                      Santé)-comb.

Binedaline
(Binodaline)

ATC: N06AB
Use: antidepressant

RN: 60662-16-0   MF: $C_{19}H_{23}N_3$ MW: 293.41
LD_{50}: 54 mg/kg (M, i.v.); 770 mg/kg (M, p.o.);
       27 mg/kg (R, i.v.)
CN: N,N,N'-trimethyl-N'-(3-phenyl-$1H$-indol-1-yl)-1,2-ethanediamine

monohydrochloride
RN: 57647-35-5   MF: $C_{19}H_{23}N_3 \cdot HCl$ MW: 329.88   EINECS: 260-877-2
LD_{50}: 54 mg/kg (M, i.v.); 760 mg/kg (M, p.o.);
       26 mg/kg (R, i.v.); 1160 mg/kg (R, p.o.);
       >20 mg/kg (dog, i.v.)

| 2-amino-benzophenone | + | H₃C—MgBr methylmagnesium bromide | →(H₂O/H⁺)→ | 2-(1-phenylvinyl)-aniline | →(1. NaNO₂, HCl 2. NH₃)→ | 4-phenyl-cinnoline (I) |

I →(H₂, Pd–Al₂O₃)→ 1,4-dihydro-4-phenyl-cinnoline →(H₃C—COOH, H₂O)→ N-(3-phenylindol-1-yl)-acetamide (II)

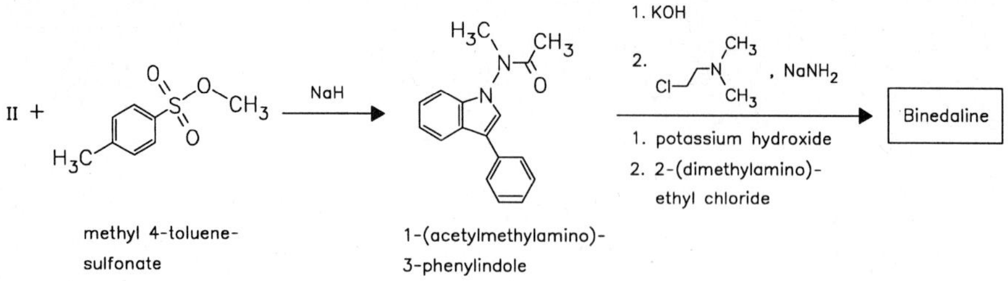

II +
NaH
1. KOH
2. , NaNH2
1. potassium hydroxide
2. 2-(dimethylamino)-
ethyl chloride
Binedaline
methyl 4-toluene-
sulfonate
1-(acetylmethylamino)-
3-phenylindole

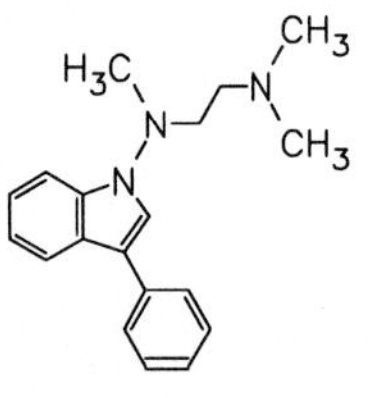

Binedaline

Reference(s):

DOS 2 512 702 (Siegfried AG; appl. 22.3.1975; CH-prior. 29.3.1974).
US 4 204 998 (Siegfried AG; 27.5.1980; CH-prior. 29.3.1974).
Schatz, F. et al.: Arzneim.-Forsch. (ARZNAD) **30**, 919 (1980).

synthesis of 1,4-dihydro-4-phenylcinnoline:
Simpson, J.C.F. et al.: J. Chem. Soc. (JCSOA9) **1945**, 646.
Scheifele, H.J. Jr. et al.: Org. Synth. (ORSYAT) **32**, 8 (1952).
Sternbach, L.H. et al.: J. Org. Chem. (JOCEAH) **26**, 4488 (1961).

Formulation(s): tabl. 25.5 mg

Trade Name(s):
J: Ixprim (Roussel-Uclaf)

Biotin

(Vitamin B_7; Vitamin H)

ATC: A11HA05
Use: growth factor, vitamin

RN: 58-85-5 MF: $C_{10}H_{16}N_2O_3S$ MW: 244.32 EINECS: 200-399-3
CN: [3a*S*-(3aα,4β,6aα)]-hexahydro-2-oxo-1*H*-thieno[3,4-*d*]imidazole-4-pentanoic acid

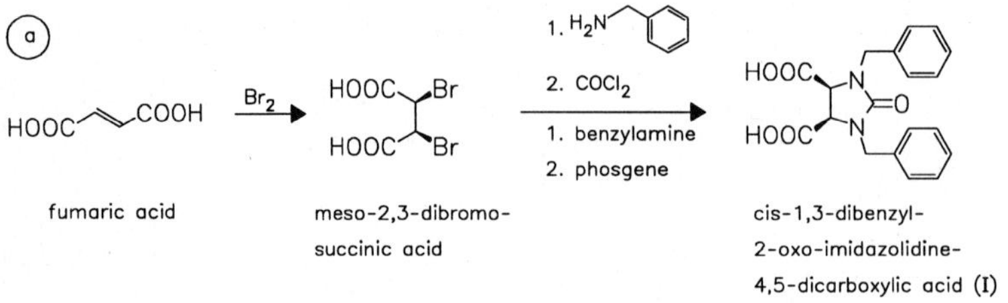

a
Br2
1. H2N
2. COCl2
1. benzylamine
2. phosgene
fumaric acid
meso-2,3-dibromo-
succinic acid
cis-1,3-dibenzyl-
2-oxo-imidazolidine-
4,5-dicarboxylic acid (I)

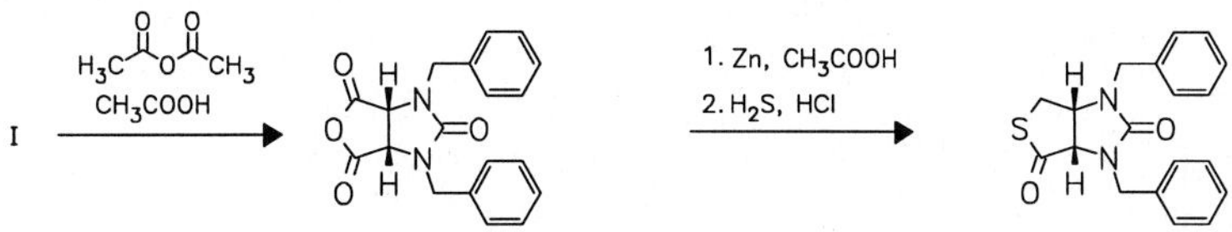
I
CH_3COOH
1. Zn, CH_3COOH
2. H_2S, HCl

cis-1,3-dibenzyl-perhydrofuro[3,4-d]-imidazole-2,4,6-trione (II)

cis-1,3-dibenzyl-perhydrothieno[3,4-d]-imidazole-2,4-dione (III)

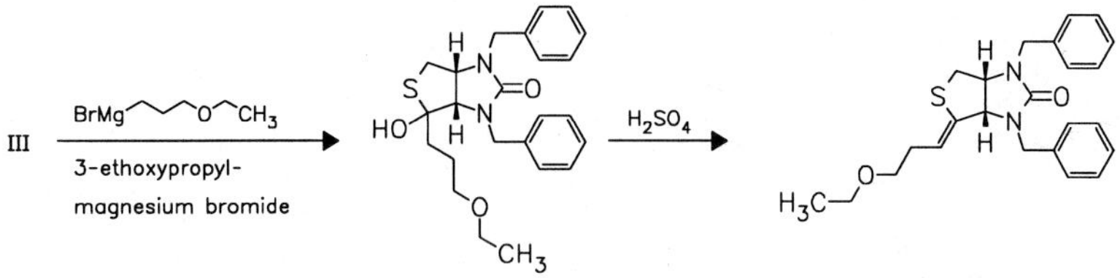
III
3-ethoxypropyl-magnesium bromide
H_2SO_4

1,3-dibenzyl-4-(3-ethoxy-propyl)-4-hydroxy-cis-perhydrothieno[3,4-d]-imidazol-2-one

1,3-dibenzyl-4-(3-ethoxy-propylidene)-cis-perhydro-thieno[3,4-d]imidazol-2-one (IV)

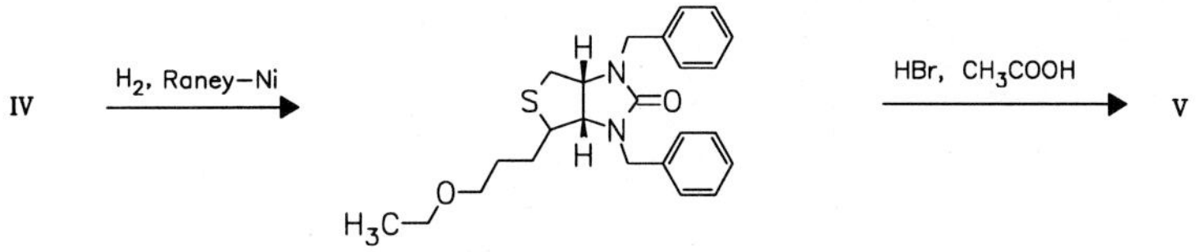
IV
H_2, Raney–Ni
HBr, CH_3COOH
V

1,3-dibenzyl-4-(3-ethoxypropyl)-cis-perhydrothieno[3,4-d]-imidazol-2-one

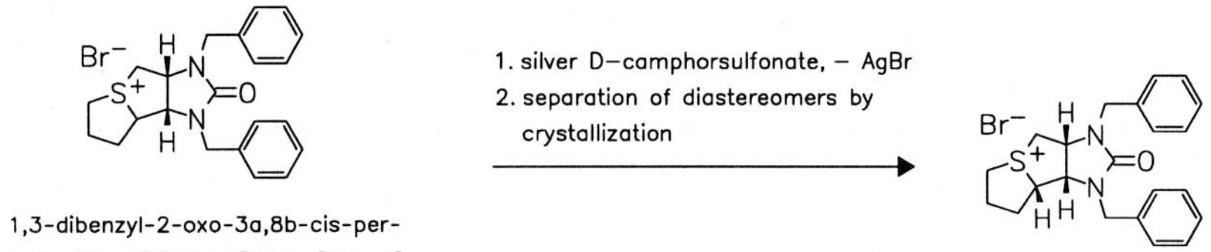
1. silver D–camphorsulfonate, – AgBr
2. separation of diastereomers by crystallization

1,3-dibenzyl-2-oxo-3a,8b-cis-per-hydrothieno[1',2':1,2]thieno[3,4–d]-imidazolium bromide (V)

(VI)

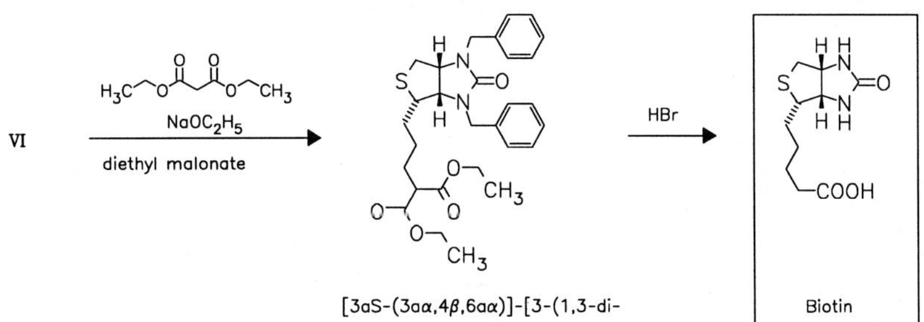
VI
$NaOC_2H_5$
diethyl malonate
HBr
Biotin

[3aS-(3aα,4β,6aα)]-[3-(1,3-di-benzyl-2-oxo-perhydrothieno-[3,4-d]imidazol-4-yl)propyl]-malonic acid diethyl ester

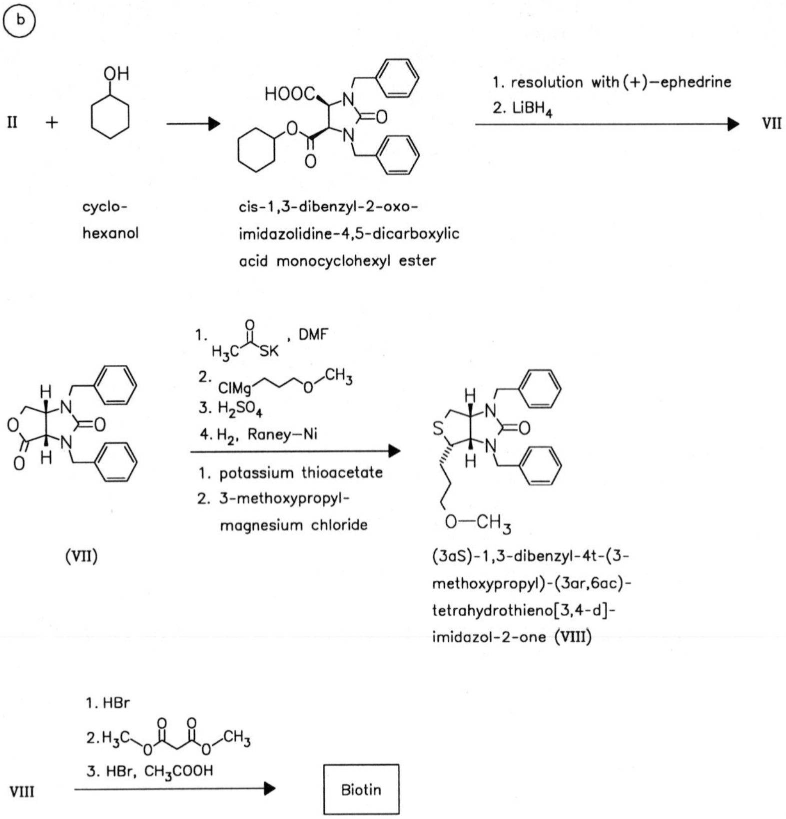

ⓑ

II + cyclo-hexanol → cis-1,3-dibenzyl-2-oxo-imidazolidine-4,5-dicarboxylic acid monocyclohexyl ester

1. resolution with (+)−ephedrine
2. LiBH₄ → VII

(VII)

1. H_3C-C(O)-SK , DMF
2. ClMg-CH₂CH₂CH₂-O-CH₃
3. H_2SO_4
4. H_2, Raney−Ni

1. potassium thioacetate
2. 3-methoxypropyl-magnesium chloride

(3aS)-1,3-dibenzyl-4t-(3-methoxypropyl)-(3ar,6ac)-tetrahydrothieno[3,4-d]-imidazol-2-one (VIII)

VIII

1. HBr
2. H_3C-O-C(O)-CH₂-C(O)-O-CH₃
3. HBr, CH_3COOH

→ Biotin

Reference(s):

US 2 489 232 (Roche; 1949; appl. 1946).
US 2 489 233 (Roche; 1949; appl. 1947).
US 2 489 234 (Roche; 1949; appl. 1947).
US 2 489 235 (Roche; 1949; appl. 1947).
US 2 489 236 (Roche; 1949; appl. 1947).
US 2 489 237 (Roche; 1949; appl. 1948).
US 2 489 238 (Roche; 1949; appl. 1948).
US 2 519 720 (Roche; 1950; appl. 1948).
US 3 740 416 (Roche; 1973; CH-prior. 29.11.1969).

newer syntheses:
DAS 2 331 244 (Sumitomo; appl. 19.6.1973; J-prior. 22.6.1972, 23.3.1973).
DAS 2 534 962 (Teikoku; appl. 5.8.1975; J-prior. 5.8.1974, 6.8.1974, 8.8.1974).
DOS 2 730 341 (Roche; appl. 5.7.1977; USA-prior. 12.7.1976).
DOS 2 807 200 (Roche; appl. 20.2.1978; USA-prior. 23.2.1977).
US 4 054 740 (Roche; 18.10.1977; prior. 24.12.1974, 5.9.1975).
US 4 130 712 (Roche; 19.12.1978; prior. 12.7.1976, 17.6.1977).
US 4 130 713 (Roche; 19.12.1978; prior. 5.8.1977).
Lavielle, S. et al.: J. Am. Chem. Soc. (JACSAT) **100**, 1558 (1978).

Formulation(s):   amp. 0.5 mg, 5 mg; cps. 0.06 mg, 0.1 mg; drg. 0.15 mg, 0.5 mg; tabl. 5 mg, 10 mg

Trade Name(s):

D:	Bio-H-Tin (Engelfried & Bartel)	Deacura (Dermapharm)	Polybion (Merck)-comb.
	Brodermatin (Engelfried & Bartel)	Mediobiotin (Medopharm)	Rombellin (Simons)
		Multibionta (Merck)-comb.	numerous combination
		Piorin (Roche Nicholas)	preparations

F: Alvityl (Solvay Pharma)-
      comb.
      Azedavit (Whitehall)-
      comb.
      Azinc complexe
      (Arkopharma)-comb.
      Berocca (Nicholas)
      Biotine (Roche)
      Cernévit (Baxtersa/Clintel
      Parentéral)-comb.

      Élévit Vitamine B9
      (Nicholas)-comb.
      Lofenalac (Bristol-Myers
      Squibb)-comb.
      Plènyl (Oberlin)-comb.
      Soluvit (Pharmacia &
      Upjohn)-comb.
      Supradyne (Roche
      Nicholas)-comb.
      Survitine (Roche
      Nicholas)-comb.

      Vivamyne (Whitehall)-
      comb.
      generics
GB: Ketovite (Paines & Byrne)-
      comb.
I: Biodermatin (Lafare)
      Diathymil (Dermalife)
J: Havita (Kakenyaku)
USA: Mega-B (Arco)
      Megadose (Arco)
      combination preparations

Biperidene

ATC: N04AA
Use: antiparkinsonian

RN: 514-65-8   MF: $C_{21}H_{29}NO$ MW: 311.47   EINECS: 208-184-6
LD_{50}:   56 mg/kg (M, i.v.); 530 mg/kg (M, p.o.);
       750 mg/kg (R, p.o.);
       340 mg/kg (dog, p.o.)
CN: α-bicyclo[2.2.1]hept-5-en-2-yl-α-phenyl-1-piperidinepropanol

hydrochloride
RN: 1235-82-1   MF: $C_{21}H_{29}NO \cdot HCl$ MW: 347.93   EINECS: 214-976-2
LD_{50}:   56 mg/kg (M, i.v.); 530 mg/kg (M, p.o.);
       750 mg/kg (R, p.o.);
       340 mg/kg (dog, p.o.)
lactate (1:1)
RN: 7085-45-2   MF: $C_{21}H_{29}NO \cdot C_3H_6O_3$ MW: 401.55   EINECS: 230-388-9
LD_{50}:   61 mg/kg (M, i.v.)

acetophenone   paraform-aldehyde   piperidine   3-piperidino-propiophenone (I)

bicyclo[2.2.1]-hept-5-en-2-yl-magnesium chloride   Biperidene

cyclo-pentadiene   methyl vinyl ketone   2-acetyl-5-norbornene   2-(3-piperidinopropionyl)-5-norbornene (II)

II   +   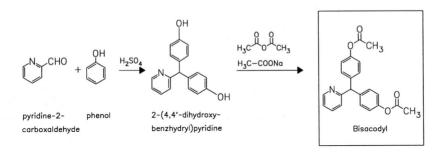 ⟶   Biperidene

MgBr

phenylmagnesium
bromide

Reference(s):
US 2 789 110 (Knoll; 1957; D-prior. 1953).
DE 1 005 067 (Knoll; appl. 1953).

Formulation(s):   amp. 5 mg/ml; powder 1 %; s. r. drg. 4 mg; tabl. 2 mg

Trade Name(s):
D: Akineton (Knoll)   GB: Akineton (Abbott); wfm   J: Akineton (Dainippon)
      Desiperiden (Desitin)   I: Akineton (Ravizza; as   Tasmofin (Yoshitomi)
      Norakin (Neuro Hexal)   chloride)   Tasmolin (Yoshitomi)
F: Akineton retard (Knoll)   Akineton (Knoll; as lactate) USA: Akineton (Knoll Labs.)

Bisacodyl

ATC: A06AB02; A06AG02
Use: laxative

RN: 603-50-9   MF: $C_{22}H_{19}NO_4$ MW: 361.40   EINECS: 210-044-4
LD_{50}:   17.5 g/kg (M, p.o.);
      4.32 g/kg (R, p.o.);
      >15 g/kg (dog, p.o.)
CN: 4,4'-(2-pyridinylmethylene)bis[phenol] diacetate (ester)

pyridine-2-   phenol   2-(4,4'-dihydroxy-   Bisacodyl
carboxaldehyde   benzhydryl)pyridine

Reference(s):
DE 951 987 (Thomae; appl. 1952).
US 2 764 590 (Thomae; 1956; D-prior. 1952).

alternative synthesis:
DE 951 988 (Thomae; appl. 1952).

Formulation(s):   drg. 5 mg; suppos. 10 mg; tabl. 5 mg

Trade Name(s):
D: Agaroletten (Warner-   Drix (Hermes)   Laxbene (Merckle)
      Lambert)   Dulcolax (Boehringer Ing.)   Laxoberal (Boehringer
      Bekunis (roha)   Florisan (Boehringer Ing.)   Ing.)
      Bisco-Zifron extra stark   Laxagetten (ct-   Mandrolax (Dolorgiet)
      (Biscova)   Arzneimittel)   Marienbader (RIAM)
      Biscu (Biscova)   Laxanin (Schwarzhaupt)   Mediolax (Medice)
      Darmol (Omegin)   Laxanin N (Schwarzhaupt)   Pyrilax (Berlin-Chemie)

Stadalax (Stada Chemie)
Tempolax (Hommel)
Tirgon (Woelm)
Vinco (OTW)
Vinco-Abführperlen
(OTW)-comb.
numerous combination
preparations
F: Contalax (3M Santé)
Dulcolax (Boehringer Ing.)
Pilules Dupuis
(Synthélabo)-comb.

GB: Dulcolax (Boehringer Ing.)
I: Alaxa (Angelini)
Dulcolax (Fher)
Fisiolax (Manetti Roberts)-
comb.
Normalene
(Montefarmaco)
J: Anan (Ono)
Biomit (Sampo)-comb.
Cathalin (Hokuriku)-comb.
Ethanis (Taisho)-comb.
Prépacol (Guerbet)-comb.

Lax (Kanto)-comb.
Satolax-10 (Sato)
Telemin Soft (Funai)-comb.
Vemas (Nippon Zoki)-
comb.
Vencoll (Maruko)-comb.
USA: Dulcolax (Novartis
Consumer)
Evac-Q-Kwik (Savage)
Fleet Prep Kits (Fleet)

Bisantrene

(CL-216942)

ATC: L01
Use: intercalanting antineoplastic (against
adult acute non-lymphocytic
leucemia)

RN: 78186-34-2   MF: $C_{22}H_{22}N_8$ MW: 398.47
LD$_{50}$:   245 mg/kg (M, route unreported)
CN: 9,10-anthracenedicarboxaldehyde bis(4,5-dihydro-1H-imidazol-2-ylhydrazone)

dihydrochloride

RN: 71439-68-4   MF: $C_{22}H_{22}N_8 \cdot$ 2HCl   MW: 471.40

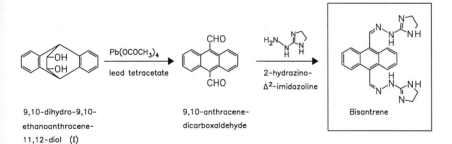

anthracene   vinylene   9,10,11,12-tetrahydro-
carbonate   9,10-[4,5][1,3]dioxolo-
anthracen-14-one

9,10-dihydro-9,10-   9,10-anthracene-   Bisantrene
ethanoanthracene-   dicarboxaldehyde
11,12-diol (I)

KOH, C_2H_5OH → I

Pb(OCOCH$_3$)$_4$
lead tetracetate

H_2N hydrazino
2-hydrazino-
Δ^2-imidazoline

Reference(s):
DOS 2 850 822 (American Cyanamid; appl. 23.11.1978; USA-prior. 28.11.1977, 5.5.1978, 19.9.1978,
2.10.1978).
US 4 187 373 (American Cyanamid; 5.2.1980; appl. 2.10.1978).
Murdock, K.L. et al.: J. Med. Chem. (JMCMAR) **25**, 505 (1982).

Formulation(s):   vial 50 mg, 250 mg, 500 mg

Trade Name(s):
F: Zantrene (Lederle; 1990 as   USA: Cyabin (Lederle; as
dihydrochloride); wfm   dihydrochloride); wfm

Bisbentiamine
(Benzoylthiamine disulfide)

ATC: A11
Use: neurotropic analgesic, vitamin B_1-derivative

RN: 2667-89-2   MF: $C_{38}H_{42}N_8O_6S_2$ MW: 770.94   EINECS: 220-206-6
LD$_{50}$: 194 mg/kg (M, i.v.); 9 g/kg (M, p.o.)
CN: N,N'-[dithiobis[2-[2-(benzoyloxy)ethyl]-1-methyl-2,1-ethenediyl]]-bis[N-[(4-amino-2-methyl-5-pyrimidinyl)methyl]formamide]

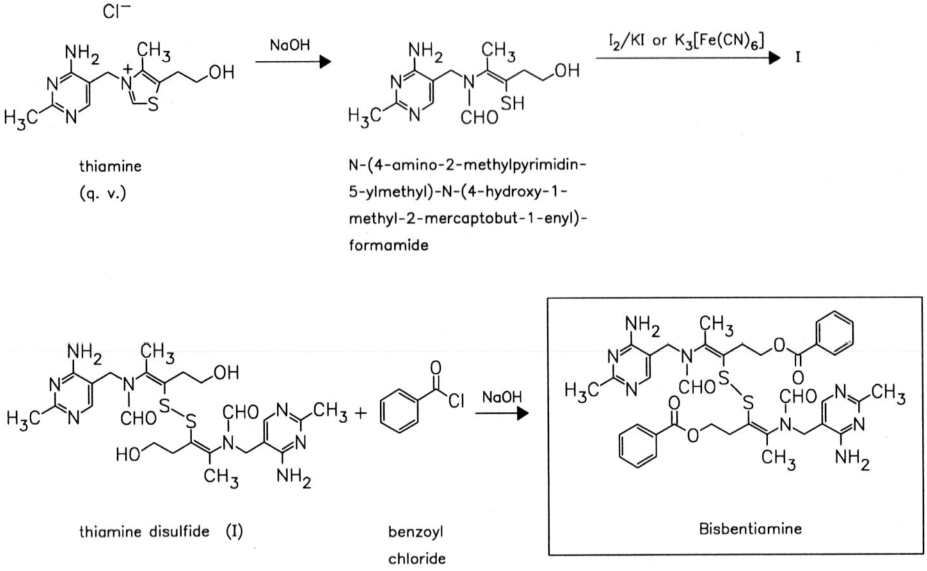

thiamine
(q. v.)

N-(4-amino-2-methylpyrimidin-
5-ylmethyl)-N-(4-hydroxy-1-
methyl-2-mercaptobut-1-enyl)-
formamide

thiamine disulfide (I)

benzoyl
chloride

Bisbentiamine

Reference(s):
US 3 109 000 (Tanabe; 1963; J-prior. 1960).
GB 922 444 (Tanabe; appl. 1961; J-prior. 1960).

similar method:
DOS 1 954 519 (Hitachi; appl. 29.10.1969).

Formulation(s):   cps. 50 mg

Trade Name(s):
D: Neuro-Fortamin (Asche)-   J: Beston (Tanabe)
      comb.; wfm

Bisoprolol

ATC: C07AB07
Use: beta blocking agent

RN: 66722-44-9   MF: $C_{18}H_{31}NO_4$ MW: 325.45
CN: (±)-1-[4-[[2-(1-methylethoxy)ethoxy]methyl]phenoxy]-3-[(1-methylethyl)amino]-2-propanol

fumarate
RN: 104344-23-2   MF: $C_{18}H_{31}NO_4 \cdot 1/2C_4H_4O_4$ MW: 766.97

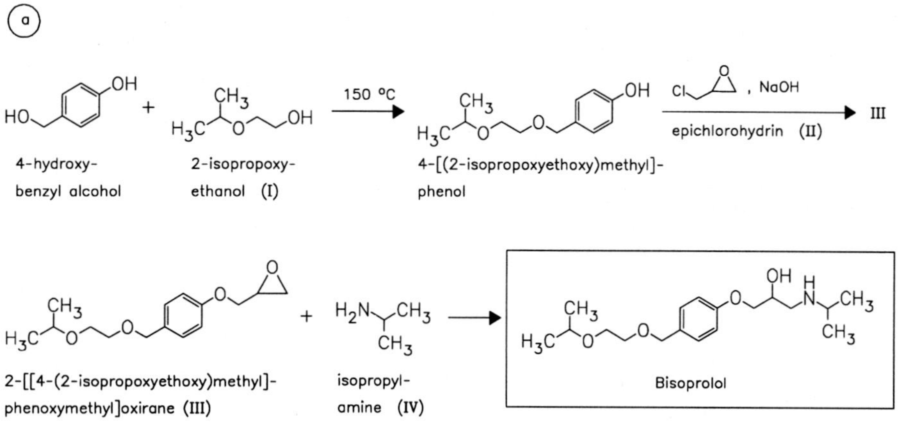

a

4-hydroxy-
benzyl alcohol

2-isopropoxy-
ethanol (I)

4-[(2-isopropoxyethoxy)methyl]-
phenol

epichlorohydrin (II)

III

2-[[4-(2-isopropoxyethoxy)methyl]-
phenoxymethyl]oxirane (III)

isopropyl-
amine (IV)

Bisoprolol

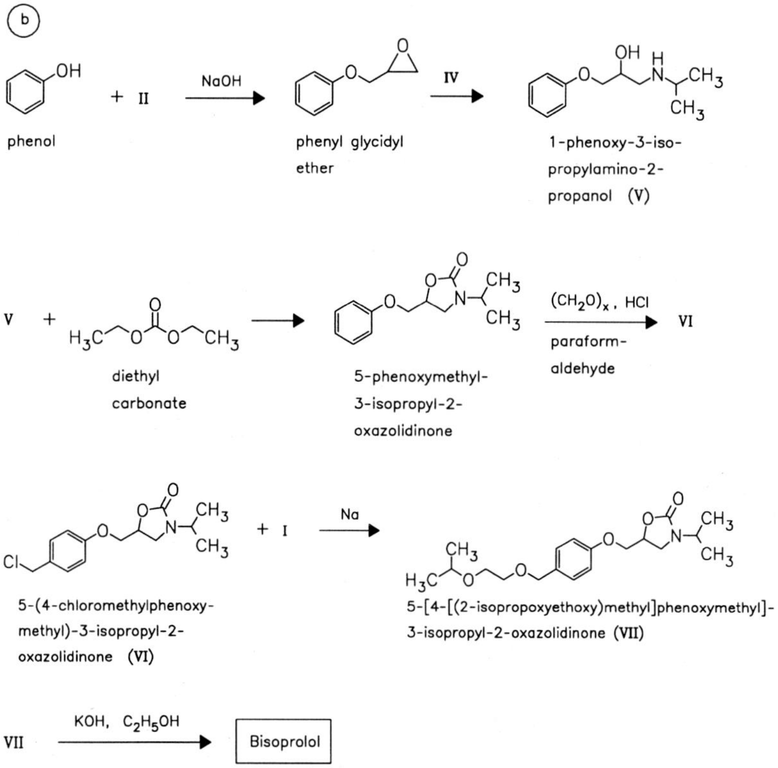

b

phenol

+ II

phenyl glycidyl
ether

IV

1-phenoxy-3-iso-
propylamino-2-
propanol (V)

V +

diethyl
carbonate

5-phenoxymethyl-
3-isopropyl-2-
oxazolidinone

paraform-
aldehyde

VI

5-(4-chloromethylphenoxy-
methyl)-3-isopropyl-2-
oxazolidinone (VI)

+ I

Na

5-[4-[(2-isopropoxyethoxy)methyl]phenoxymethyl]-
3-isopropyl-2-oxazolidinone (VII)

VII

KOH, C₂H₅OH

Bisoprolol

Reference(s):

Harting, J. et al.: Arzneim.-Forsch. (ARZNAD) **36**, 200 (1986).

a DOS 2 645 710 (Merck Patent GmbH; appl. 9.10.1976).
US 4 258 062 (Merck Patent GmbH; 24.3.1981; appl. 30.5.1979; D-prior. 9.10.1976).

b DOS 3 205 457 (Merck Patent GmbH; appl. 16.2.1982).

Formulation(s):   f. c. tabl. 10 mg; f. c. tabl. 5 mg, 10 mg (as fumarate)

Trade Name(s):

D: Bisobloc (Azupharma)   F: Détensiel (Lipha Santé;   GB: Emcor (Merck)
Concor (Merck; 1986)   1987)   Monocor (Wyeth)
Fondril (Procter & Gamble)   Soprol (Wyeth-Lederle;   Monozide 10 (Wyeth)-
                                    1988)   comb.

I:	Concor (Bracco)	USA:	Zebeta (Lederle)	Ziac (Lederle)-comb.

Bitolterol

ATC: R03AC17
Use: selective β_2-adrenoceptor agonist, bronchodilator

RN: 30392-40-6 MF: $C_{28}H_{31}NO_5$ MW: 461.56
CN: 4-methylbenzoic acid 4-[2-[(1,1-dimethylethyl)amino]-1-hydroxyethyl]-1,2-phenylene ester

mesylate
RN: 30392-41-7 MF: $C_{28}H_{31}NO_5 \cdot CH_4O_3S$ MW: 557.66 EINECS: 250-177-5
LD_{50}: 31.4 mg/kg (M, i.v.); 4116 mg/kg (M, p.o.);
44 mg/kg (R, i.v.); >6221 mg/kg (R, p.o.)

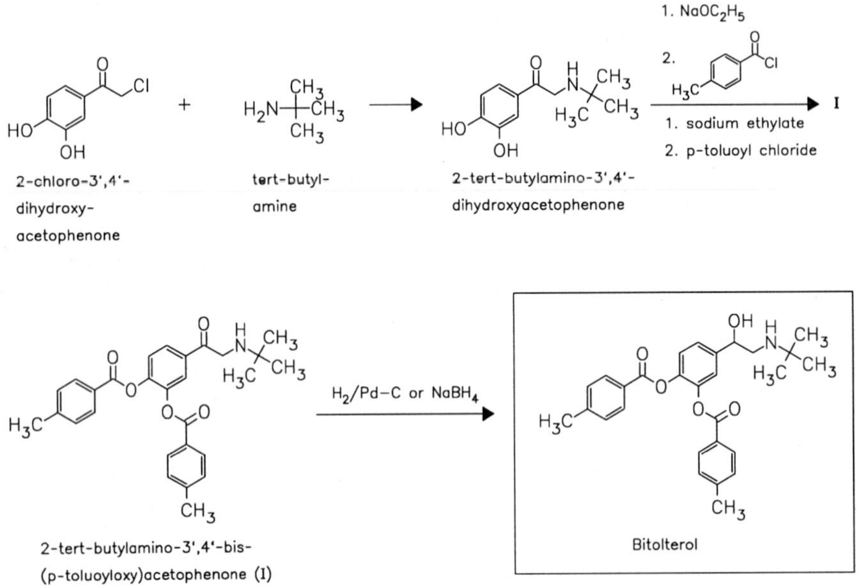

1. NaOC2H5
2.
1. sodium ethylate
2. p-toluoyl chloride
I
2-chloro-3',4'-dihydroxy-acetophenone
tert-butyl-amine
2-tert-butylamino-3',4'-dihydroxyacetophenone
H2/Pd-C or NaBH4
2-tert-butylamino-3',4'-bis-(p-toluoyloxy)acetophenone (I)
Bitolterol

Reference(s):
DOS 2 015 573 (Sterling Drug; appl. 1.4.1970; USA-prior. 1.4.1969).
Corrigan, J.R. et al.: J. Am. Chem. Soc. (JACSAT) **71**, 530 (1949).
Fuller, B.F. et al.: J. Med. Chem. (JMCMAR) **19**, 834 (1976).

Formulation(s): aerosol 10 ml (0.8 %); tabl. 4 mg

Trade Name(s):

I:	Asmalene (Firma) Tolbet (Corvi)	J:	Effectin (Shionogi-Winthrop; as mesylate)

Bleomycin

ATC: L01DC01
Use: antineoplastic (peptide antibiotic)

RN: 11116-31-7 MF: $C_{55}H_{84}N_{17}O_{21}S_3$ MW: 1415.57 EINECS: 234-356-5
LD_{50}: 100 mg/kg (M, i.v.)
CN: N^1-[3-(dimethylsulfonio)propyl]bleomycinamide

50 % Bleomycin A$_2$, 20 % Bleomycin B$_2$.

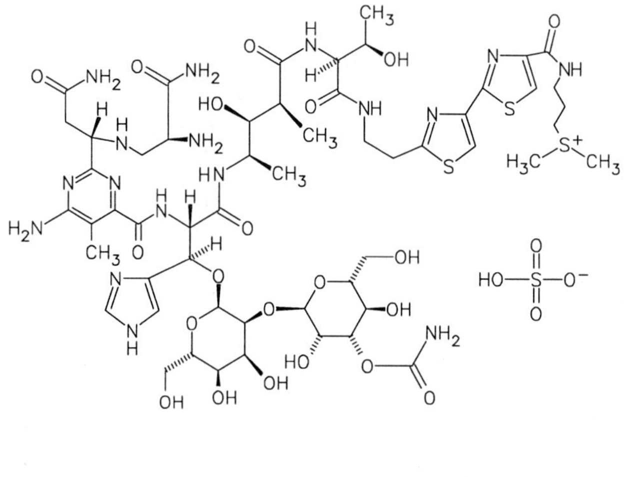

Bleomycin A$_2$

From culture of *Streptomyces verticillus* by ion-exchange adsorption and column chromatographic purification (on alumina) via the copper complex.

Reference(s):
DE 1 217 549 (Zaidan Hojin Biseibutsu Kagaku Kenkyu Kai = Microbial Chemistry Research Foundation; Tokyo; appl. 5.3.1964; J-prior. 5.3.1963).

Formulation(s):   amp. 15 mg (as sulfate)

Trade Name(s):

D:	BLEO-cell (cell pharm) Bleomycinum-Mack (Mack, Illert.)	I:	Bleomicina (Rhône-Poulenc Rorer) generics
F:	Bleomycine Roger Bellon (Roger Bellon) generics	J:	Bleo (Nippon Kayaku; as hydrochloride)

Bleo S (Nippon Kayaku; as sulfate)

USA: Blenoxane (Bristol-Myers Squibb Oncology/ Immunology; as sulfate)

Bluensomycin

ATC: A07AA
Use: antibiotic

RN: 11011-72-6   MF: C$_{21}$H$_{39}$N$_5$O$_{14}$ MW: 585.56
LD$_{50}$:   2250 mg/kg (M, i.v.);
     >2500 mg/kg (R, p.o.)
CN: *O*-2-deoxy-2-(methylamino)-α-L-glucopyranosyl(1→2)-*O*-5-deoxy-3-*C*-(hydroxymethyl)-α-L-lyxofuranosyl-(1→2)-1-[(aminoiminomethyl)amino]-1-deoxy-D-*scyllo*-inositol 5-carbamate

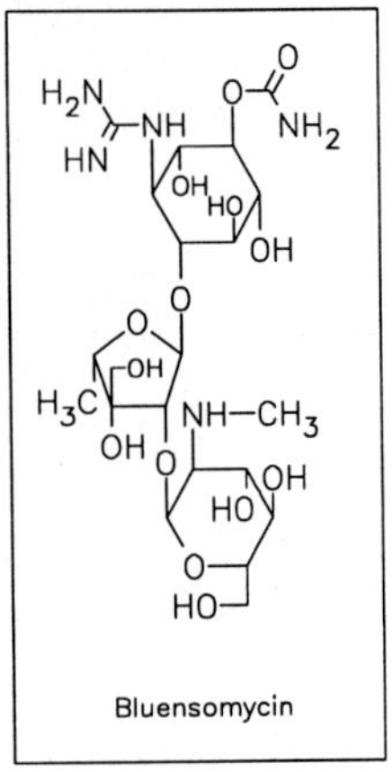

Bluensomycin

From fermentation solutions of *Streptomyces bluensis* NRRL 2876.

Reference(s):

Mason, O.J. et al.: Antimicrob. Agents Chemother. (AACHAX) **1963**, 607.

Bergy, M.E. et al.: Antimicrob. Agents Chemother. (AACHAX) **1963**, 614.

DAS 1 183 631 (Upjohn; appl. 19.7.1962; USA-prior. 7.8.1961).

structure:

Bannister, B.; Argoudelis, A.D.: J. Am. Chem. Soc. (JACSAT) **85**, 119, 234 (1963).

McGilveray, I.J.; Rinehart, U.L.: J. Am. Chem. Soc. (JACSAT) **87**, 4003 (1965).

Trade Name(s):

USA: Bluensomycin "Upjohn"
          (Upjohn); wfm

Bolasterone

ATC: G03BA
Use: anabolic

RN: 1605-89-6   MF: $C_{21}H_{32}O_2$ MW: 316.49   EINECS: 216-519-2

CN: (7α,17β)-17-hydroxy-7,17-dimethylandrost-4-en-3-one

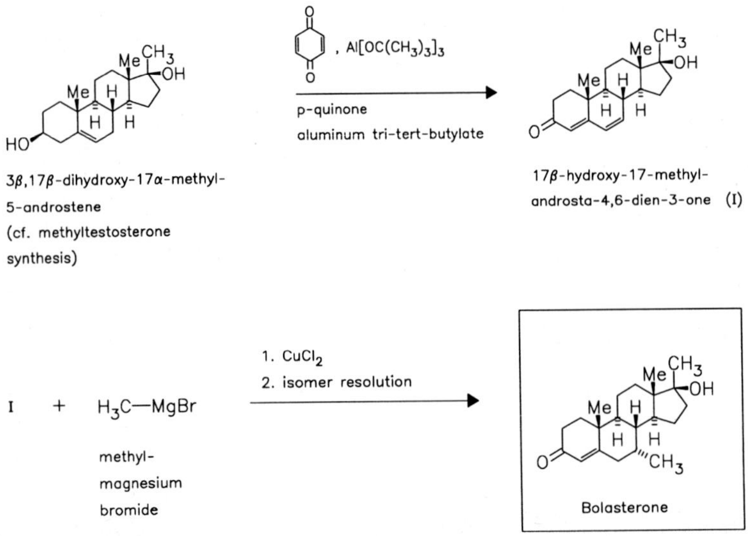

3β,17β-dihydroxy-17α-methyl-
5-androstene
(cf. methyltestosterone
synthesis)

p-quinone
aluminum tri-tert-butylate

17β-hydroxy-17-methyl-
androsta-4,6-dien-3-one   (I)

I   +   H₃C—MgBr

methyl-
magnesium
bromide

1. CuCl₂
2. isomer resolution

Bolasterone

Reference(s):
US 3 341 557 (Upjohn; 12.9.1967; prior. 5.6.1961, 6.11.1960, 6.6.1958).
Campbell, J.A.; Babcock, J.C.: J. Am. Chem. Soc. (JACSAT) **81**, 4069 (1959).

Trade Name(s):
USA: Myagen (Upjohn); wfm

Boldenone undecenylate

ATC: G03B
Use: anabolic

RN: 13103-34-9   MF: $C_{30}H_{44}O_3$ MW: 452.68   EINECS: 236-024-5
CN: (17β)-17-[(1-oxo-10-undecenyl)oxy]androsta-1,4-dien-3-one

boldenone
RN: 846-48-0   MF: $C_{19}H_{26}O_2$ MW: 286.42   EINECS: 212-686-0

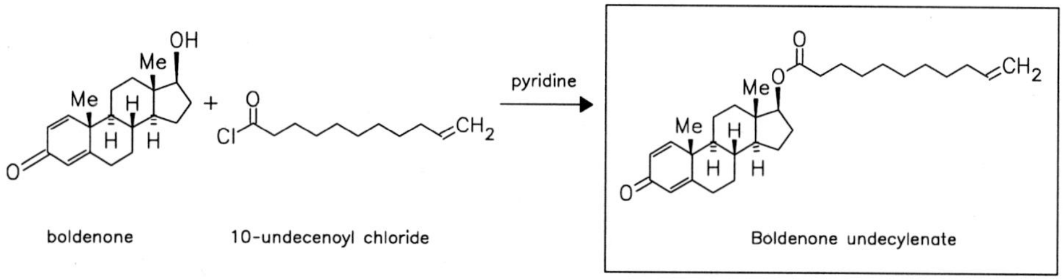

boldenone   10-undecenoyl chloride   Boldenone undecylenate

Reference(s):
BE 623 277 (Merck AG; appl. 5.10.1962; D-prior. 5.10.1961).

starting material:
CA 803 490 (Upjohn; appl. 1956; USA-prior. 1955).
GB 922 525 (Loevens Kemiske Fabrik; valid from 6.11.1961; prior. 9.11.1960).
US 2 837 464 (Schering; 1958; prior. 1955).
US 2 875 196 (Olin Mathieson; 1959; prior. 1956, 1955).
Meystre, Ch. et al.: Helv. Chim. Acta (HCACAV) **39**, 734 (1956).

Trade Name(s):
D: Vebonol (Ciba); wfm   USA: Parenabol (Ciba); wfm

Bopindolol

ATC: C07AA17
Use: β-adrenoceptor antagonist,
       antihypertensive

RN: 62658-63-3   MF: $C_{23}H_{28}N_2O_3$ MW: 380.49
LD_{50}:   17 mg/kg (M, i.v.)
CN: (±)-1-[(1,1-dimethylethyl)amino]-3-[(2-methyl-1H-indol-4-yl)oxy]-2-propanolbenzoate (ester)

(E)-2-butenedioate (1:1)
RN: 62658-64-4   MF: $C_{23}H_{28}N_2O_3 \cdot C_4H_4O_4$ MW: 496.56
LD_{50}:   17mg/kg (M, i.v.)

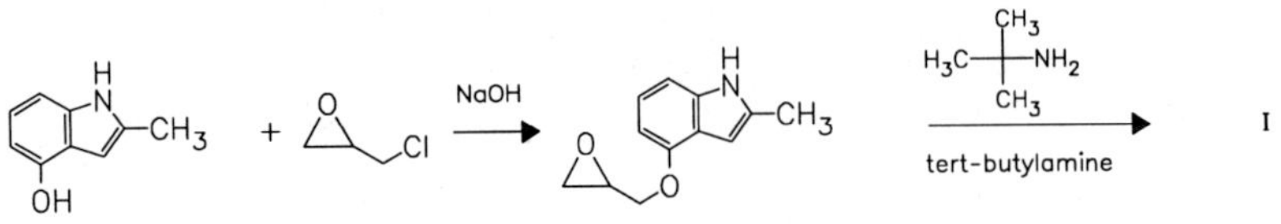

4-hydroxy-2-methylindole (cf. mepindolol synthesis)   epichlorohydrin   2-methyl-4-oxiranylmethoxyindole

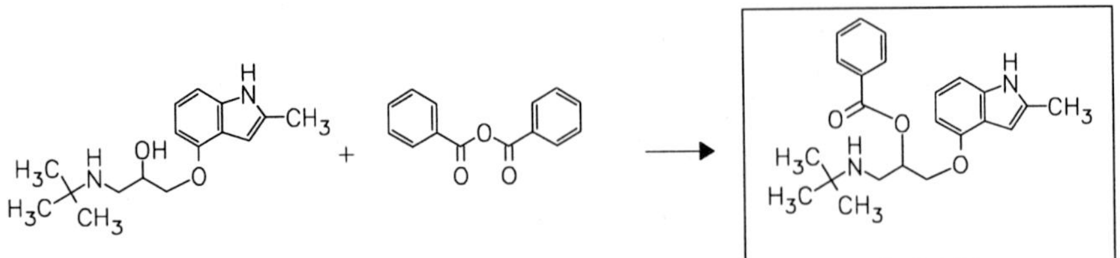

4-(3-tert-butylamino-2-hydroxypropoxy)-2-methylindole (I)   benzoic anhydride   Bopindolol

Reference(s):
DOS 2 635 209 (Sandoz; appl. 5.8.1976; CH-prior. 15.8.1975).
GB 1 575 509 (Sandoz; appl. 13.8.1976; CH-prior. 15.8.1975).
GB 1 575 510 (Sandoz; appl. 13.8.1976; CH-prior. 15.8.1975).

Formulation(s):   tabl. 1 mg

Trade Name(s):
D: Wandonorm (Novartis   J: Sandonorm (Novartis; as
     Pharma; 1989 as hydrogen   malonate)
     malonate)

Bornaprine

ATC: N04AA11
Use: antiparkinsonian

RN: 20448-86-6   MF: $C_{21}H_{31}NO_2$ MW: 329.48
LD$_{50}$:   26 mg/kg (M, i.v.)
CN: 2-phenylbicyclo[2.2.1]heptane-2-carboxylic acid 3-(diethylamino)propyl ester

hydrochloride
RN: 26908-91-8   MF: $C_{21}H_{31}NO_2 \cdot HCl$ MW: 365.95   EINECS: 248-100-5

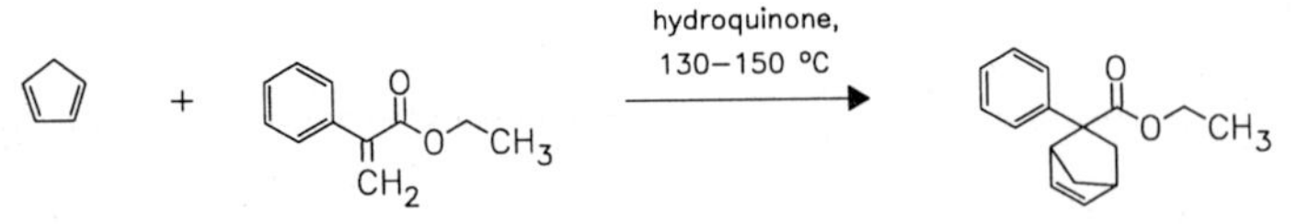

cyclopentadiene   ethyl 2-phenylacrylate   ethyl 2-phenylbicyclo[2.2.1]-5-heptene-2-carboxylate (I)

I
H2, Raney-Ni, 100 at
KOH
ethyl 2-phenylbicyclo[2.2.1]heptane-2-carboxylate
2-phenylbicyclo-[2.2.1]heptane-2-carboxylic acid (II)

II
SOCl2
3-diethylamino-1-propanol
Bornaprine

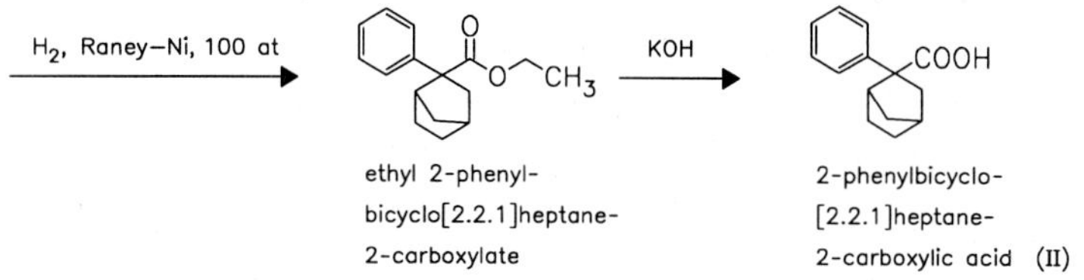

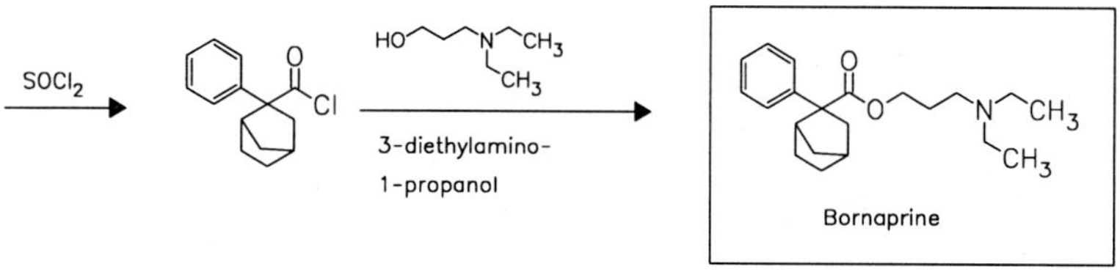

Reference(s):
DE 1 044 809 (Knoll; appl. 16.6.1956).

Formulation(s): tabl. 4 mg (as hydrochloride)

Trade Name(s):
D: Sormodren (Knoll) I: Sormodren (Ravizza)

Brimonidine

(UK-14304; UK-14304-08; AGN-190342LF (tartrate))

ATC: N07
Use: antihypertensive, α_2-receptor antagonist

RN: 59803-98-4 MF: $C_{11}H_{10}BrN_5$ MW: 292.14
LD_{50}: 160 mg/kg (M, p.o.)
CN: 5-bromo-*N*-(4,5-dihydro-1*H*-imidazol-2-yl)-6-quinoxalinamine

tartrate (1:1)
RN: 70359-46-5 MF: $C_{11}H_{10}BrN_5 \cdot C_4H_6O_6$ MW: 442.23

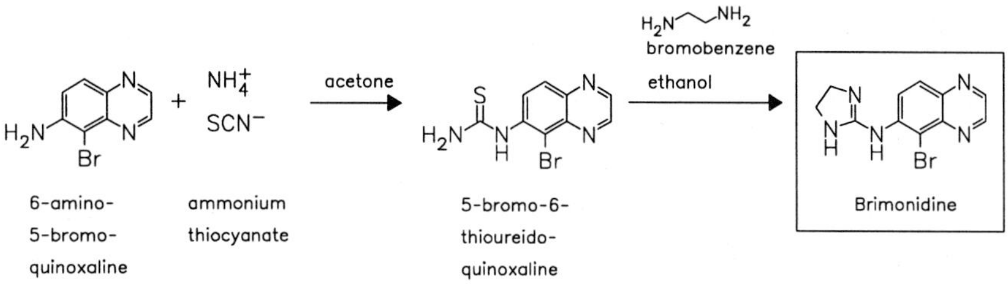

acetone
bromobenzene
ethanol
6-amino-5-bromoquinoxaline
ammonium thiocyanate
5-bromo-6-thioureidoquinoxaline
Brimonidine

Reference(s):
DE 2 538 620 (Pfizer; appl. 29.8.1975; GB-prior. 6.9.1974).

use:
WO 9 510 280 (Allergan; appl. 19.9.1994; USA-prior. 13.10.1993).
WO 9 701 339 (Allergan; appl. 17.6.1996; USA-prior. 28.6.1995).
WO 9 635 424 (Allergan; appl. 5.9.1996; USA-prior. 12.5.1995).

combinations:
WO 9 613 267 (Allergan; appl. 20.10.1995; USA-prior. 27.10.1994).
US 5 215 991 (Allergan; appl. 20.12.1990; USA-prior. 26.1.1990).

suspension formulations for controlled delivery:
WO 9 211 871 (Allergan; appl. 17.12.1991; USA-prior. 27.12.1990).
WO 9 205 770 (Allergan; appl. 10.9.1991; USA-prior. 27.9.1990).

Formulation(s): eye drops 0.2 %

Trade Name(s):

GB:	Alphagan (Allergan; as tartrate)	USA:	Alphagan (Allergan; as tartrate)

Brinzolamide
(AL-4862)

ATC: S01EC04
Use: antiglaucoma, topical carbonic anhydrase inhibitor

RN: 138890-62-7 MF: $C_{12}H_{21}N_3O_5S_3$ MW: 383.51
CN: (4*R*)-4-(Ethylamino)-3,4-dihydro-2-(3-methoxypropyl)-2*H*-thieno[3,2-*e*]-1,2-thiazine-6-sulfonamide 1,1-dioxide

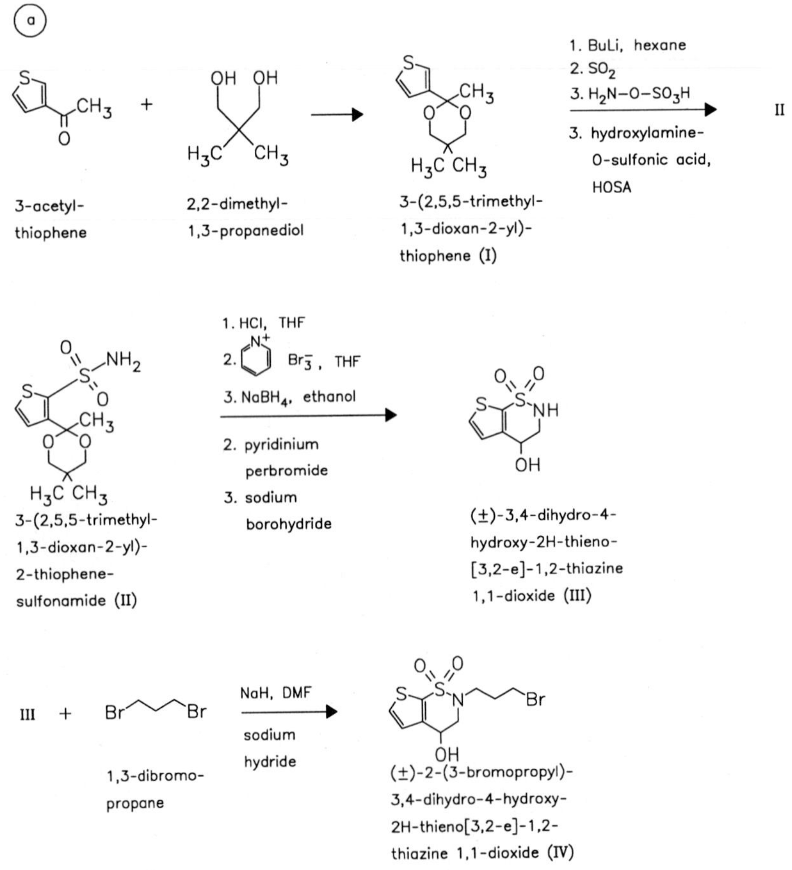
a
3-acetyl-thiophene
2,2-dimethyl-1,3-propanediol
3-(2,5,5-trimethyl-1,3-dioxan-2-yl)-thiophene (I)
1. BuLi, hexane
2. SO2
3. H2N-O-SO3H
3. hydroxylamine-O-sulfonic acid, HOSA
II
3-(2,5,5-trimethyl-1,3-dioxan-2-yl)-2-thiophene-sulfonamide (II)
1. HCl, THF
2. Br3⁻, THF
3. NaBH4, ethanol
2. pyridinium perbromide
3. sodium borohydride
(±)-3,4-dihydro-4-hydroxy-2H-thieno-[3,2-e]-1,2-thiazine 1,1-dioxide (III)
III
1,3-dibromo-propane
NaH, DMF
sodium hydride
(±)-2-(3-bromopropyl)-3,4-dihydro-4-hydroxy-2H-thieno[3,2-e]-1,2-thiazine 1,1-dioxide (IV)

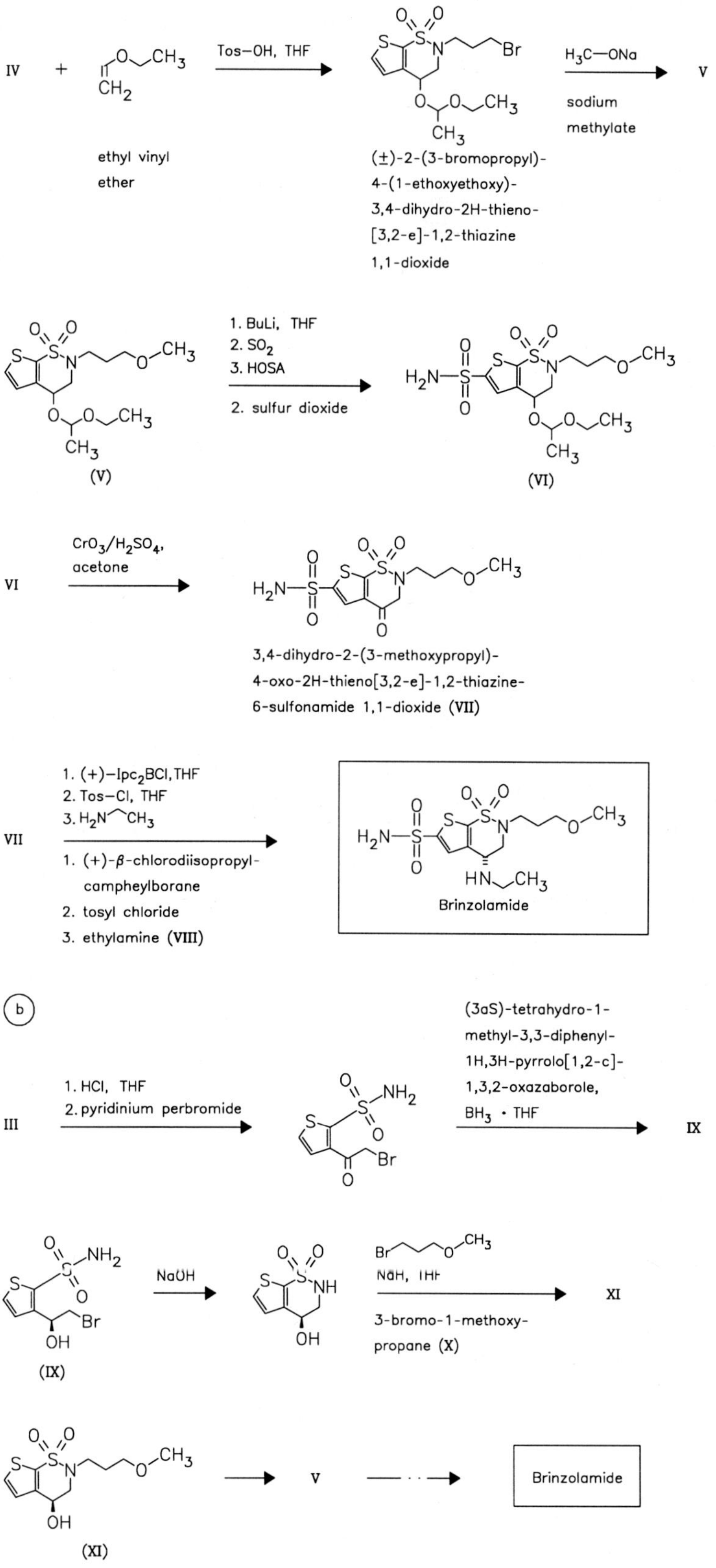

IV + ethyl vinyl ether →(Tos–OH, THF) (±)-2-(3-bromopropyl)-4-(1-ethoxyethoxy)-3,4-dihydro-2H-thieno-[3,2-e]-1,2-thiazine 1,1-dioxide →(H₃C—ONa, sodium methylate) V

(V) →(1. BuLi, THF 2. SO₂ 3. HOSA 2. sulfur dioxide) (VI)

VI →(CrO₃/H₂SO₄, acetone) 3,4-dihydro-2-(3-methoxypropyl)-4-oxo-2H-thieno[3,2-e]-1,2-thiazine-6-sulfonamide 1,1-dioxide (VII)

VII →(1. (+)–Ipc₂BCl, THF 2. Tos–Cl, THF 3. H₂N‿CH₃ 1. (+)-β-chlorodiisopropyl-campheylborane 2. tosyl chloride 3. ethylamine (VIII)) Brinzolamide

b

III →(1. HCl, THF 2. pyridinium perbromide) →((3aS)-tetrahydro-1-methyl-3,3-diphenyl-1H,3H-pyrrolo[1,2-c]-1,3,2-oxazaborole, BH₃ · THF) IX

(IX) →(NaOH) →(Br‿‿O‿CH₃ / NaH, THF 3-bromo-1-methoxy-propane (X)) XI

(XI) → V — · · → Brinzolamide

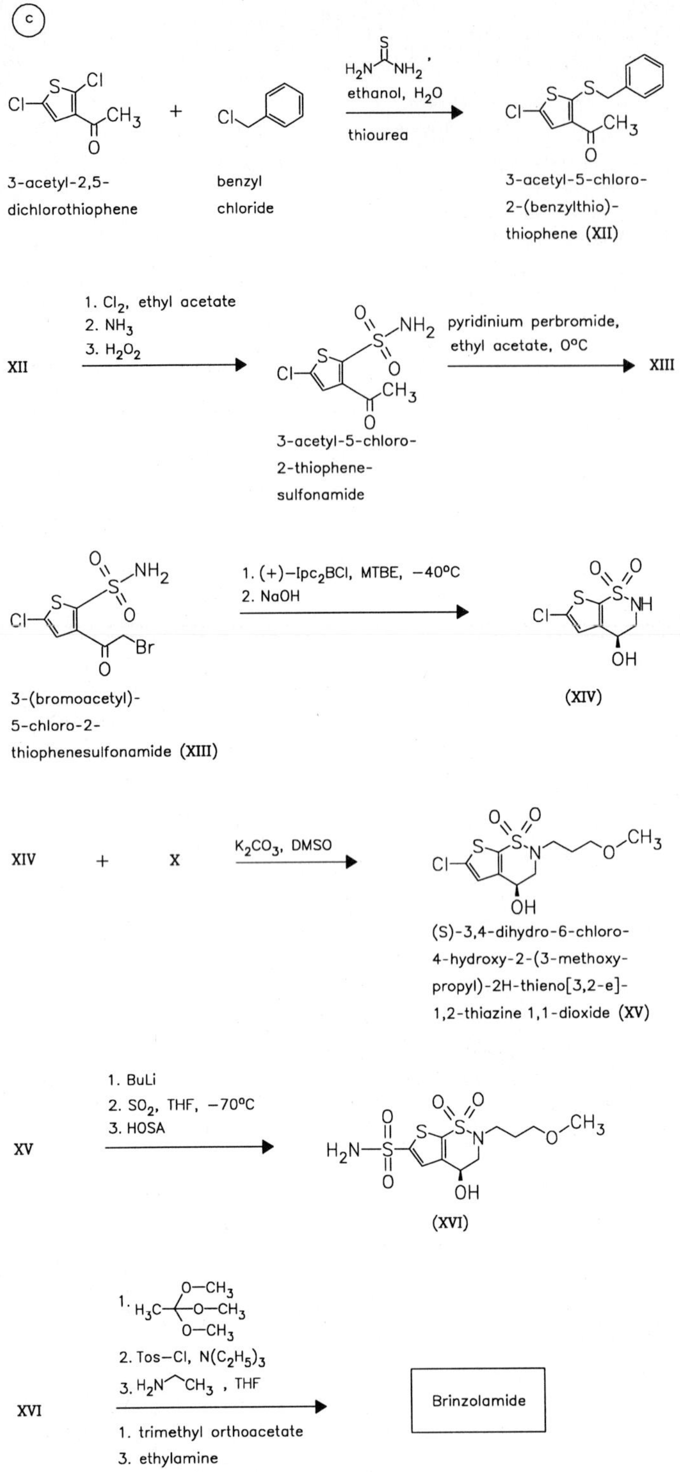

ⓒ

3-acetyl-2,5-
dichlorothiophene

benzyl
chloride

3-acetyl-5-chloro-
2-(benzylthio)-
thiophene (XII)

XII

1. Cl$_2$, ethyl acetate
2. NH$_3$
3. H$_2$O$_2$

pyridinium perbromide,
ethyl acetate, 0°C

XIII

3-acetyl-5-chloro-
2-thiophene-
sulfonamide

3-(bromoacetyl)-
5-chloro-2-
thiophenesulfonamide (XIII)

1. (+)-Ipc$_2$BCl, MTBE, −40°C
2. NaOH

(XIV)

XIV   +   X

K$_2$CO$_3$, DMSO

(S)-3,4-dihydro-6-chloro-
4-hydroxy-2-(3-methoxy-
propyl)-2H-thieno[3,2-e]-
1,2-thiazine 1,1-dioxide (XV)

XV

1. BuLi
2. SO$_2$, THF, −70°C
3. HOSA

(XVI)

XVI

1. H$_3$C—C(O—CH$_3$)(O—CH$_3$)(O—CH$_3$)
2. Tos—Cl, N(C$_2$H$_5$)$_3$
3. H$_2$N—CH$_3$, THF

1. trimethyl orthoacetate
3. ethylamine

Brinzolamide

Reference(s):
a   US 5 378 703 (Alcon; 3.1.1995; USA-prior. 9.4.1990).
b   US 5 470 973 (Alcon; 28.11.1995; USA-prior. 3.10.1994).
c   Conrow, R.E. et al.: Org. Process Res. Dev. (OPRDFK) **3** 114-120 (1999).

ophthalmic compositions with prostaglandins:
WO 9 853 809 (Merck + Co.; 3.12.1998; appl. 26.5.1998; USA-prior. 30.5.1997).
WO 9 819 680 (Alcon; appl. 5.9.1997; USA-prior. 1.11.1996).

process for manufacturing ophthalmic suspensions:
WO 9 825 620 (Alcon; appl. 5.9.1997; USA-prior. 11.12.1996).

pharmaceutical compositions:
WO 9 702 825 (Alcon; appl. 12.7.1995).
WO 9 115 486 (Alcon; appl. 3.4.1991; USA-prior. 9.4.1990).

Formulation(s):   ophth. susp. 1% in dispensers (2.5, 5, 10 and 15 ml)

Trade Name(s):
USA: Azopt (Alcon; 1998)

Brodimoprim
(Ro-10-5970)

ATC: J01EA02
Use: antibacterial

RN: 56518-41-3   MF: C$_{13}$H$_{15}$BrN$_4$O$_2$ MW: 339.19   EINECS: 260-238-8
CN: 5-[(4-Bromo-3,5-dimethoxyphenyl)methyl]-2,4-pyrimidinediamine

hydrochloride
RN: 56518-40-2   MF: C$_{13}$H$_{15}$BrN$_4$O$_2$ · HCl   MW: 375.65

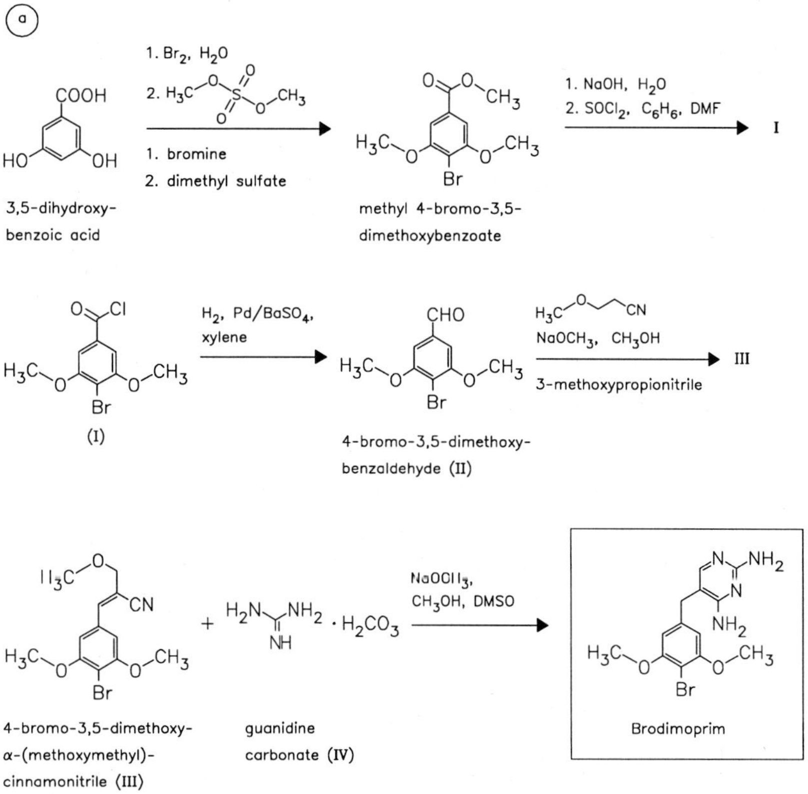

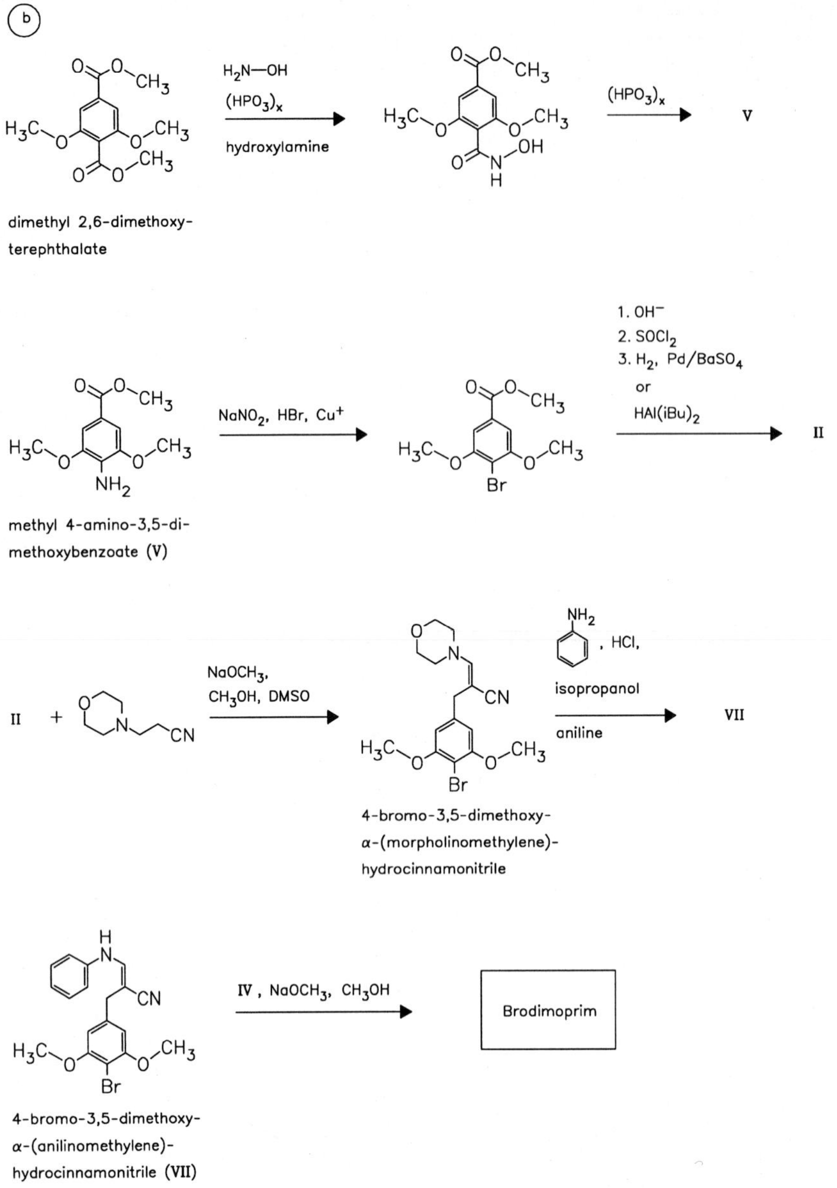

b

dimethyl 2,6-dimethoxy-
terephthalate

methyl 4-amino-3,5-di-
methoxybenzoate (V)

4-bromo-3,5-dimethoxy-
α-(morpholinomethylene)-
hydrocinnamonitrile

4-bromo-3,5-dimethoxy-
α-(anilinomethylene)-
hydrocinnamonitrile (VII)

Reference(s):

a   CA 1 017 743 (Hoffmann-La Roche; CH-prior. 8.11.1973).
    DE 2 452 889 (Hoffmann-La Roche; appl. 7.11.1974; CH-prior. 8.11.1973).
b   Kompis, I., Wick. A.: Helv. Chim. Acta (HCACAV) **60** (8), 3025 (1977).

alternative preparation of 4-bromo-3,5-dimethoxybenzaldehyde:
Barfknecht, C.F.; Nichols, D.E.: J. Med. Chem. (JMCMAR) **14**, 370 (1971).

Formulation(s):   gran. 200 mg; susp. 1%, 50 mg; tabl. 200 mg

Trade Name(s):
I: Hyprim (Fisons)   Unitrim (Fisons; 1993)

Bromazepam

ATC: N05BA08
Use: tranquilizer

RN: 1812-30-2   MF: $C_{14}H_{10}BrN_3O$ MW: 316.16   EINECS: 217-322-4
LD_{50}:   879 mg/kg (M, p.o.);
          1950 mg/kg (R, p.o.)
CN: 7-bromo-1,3-dihydro-5-(2-pyridinyl)-2H-1,4-benzodiazepine-2-one

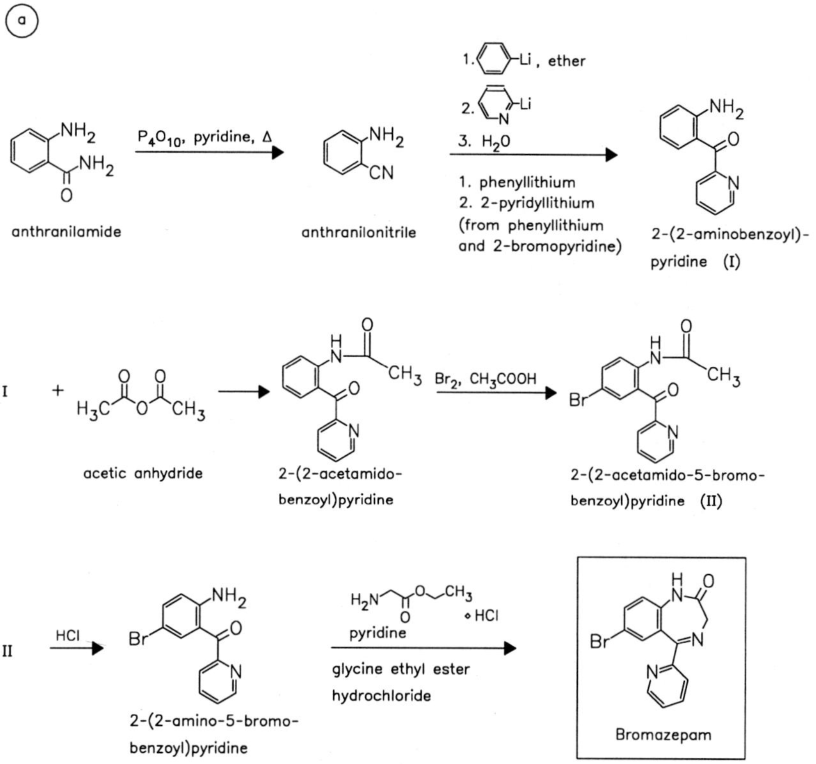

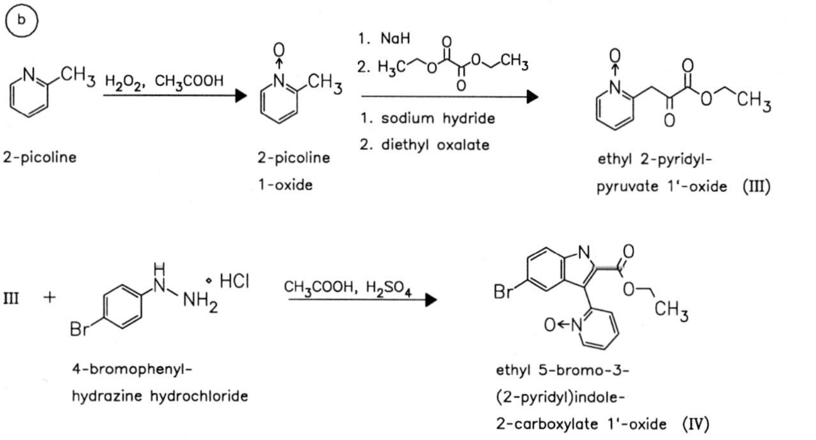

a

anthranilamide → P$_4$O$_{10}$, pyridine, Δ → anthranilonitrile

1. phenyllithium
2. 2-pyridyllithium
(from phenyllithium and 2-bromopyridine)

2-(2-aminobenzoyl)-pyridine (I)

I + acetic anhydride → 2-(2-acetamido-benzoyl)pyridine → Br$_2$, CH$_3$COOH → 2-(2-acetamido-5-bromo-benzoyl)pyridine (II)

II → HCl → 2-(2-amino-5-bromo-benzoyl)pyridine

glycine ethyl ester hydrochloride, pyridine → Bromazepam

b

2-picoline → H$_2$O$_2$, CH$_3$COOH → 2-picoline 1-oxide

1. NaH
2. H$_3$C-O...O-CH$_3$

1. sodium hydride
2. diethyl oxalate

ethyl 2-pyridyl-pyruvate 1'-oxide (III)

III + 4-bromophenyl-hydrazine hydrochloride → CH$_3$COOH, H$_2$SO$_4$ → ethyl 5-bromo-3-(2-pyridyl)indole-2-carboxylate 1'-oxide (IV)

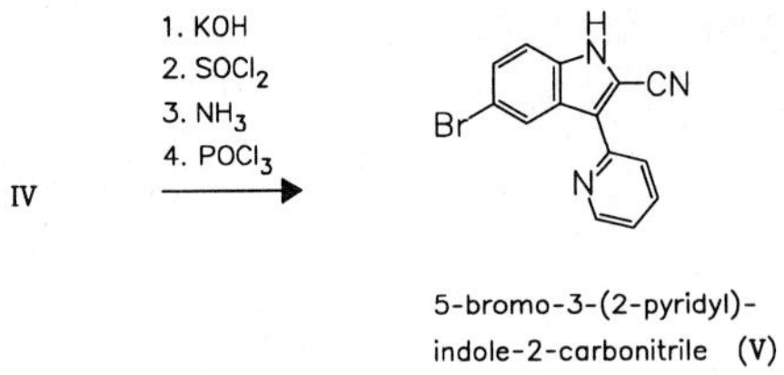

1. KOH
2. SOCl₂
3. NH₃
4. POCl₃

IV →

5-bromo-3-(2-pyridyl)-
indole-2-carbonitrile (V)

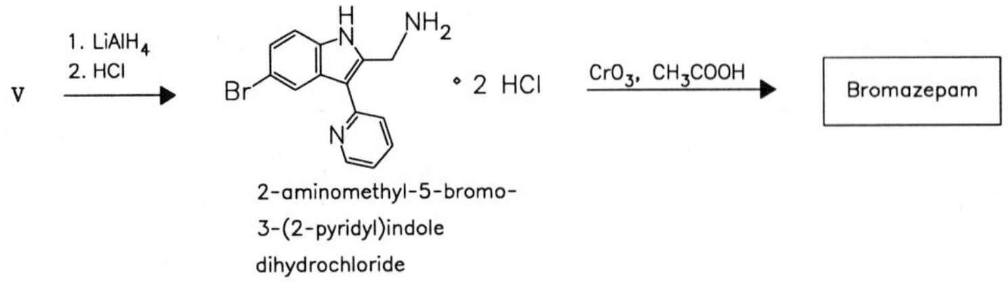

1. LiAlH₄
2. HCl

V →

· 2 HCl   CrO₃, CH₃COOH →   Bromazepam

2-aminomethyl-5-bromo-
3-(2-pyridyl)indole
dihydrochloride

Reference(s):

a US 3 100 770 (Roche; 13.8.1963; appl. 11.8.1961).
US 3 182 065 (Roche; 4.5.1965; appl. 9.4.1964; prior. 19.4.1963).
US 3 182 066 (Roche; 4.5.1965; appl. 9.4.1964; prior. 19.4.1963).
US 3 182 067 (Roche; 4.5.1965; appl. 9.4.1964).
Fryer, R.I. et al.: J. Pharm. Sci. (JPMSAE) **53**, 264 (1964).
modified methods:
DAS 2 233 483 (Roche; appl. 7.7.1972; GB-prior. 8.7.1971, 7.10.1971).
DOS 2 252 378 (Roche; appl. 25.10.1972; CH-prior. 18.11.1971).
alternative synthesis of 2-(2-amino-5-bromobenzoyl)pyridine:
DAS 2 256 614 (Roche; appl. 17.11.1972).
b DAS 1 813 241 (Roche; appl. 6.12.1968; J-prior. 8.12.1967, 9.12.1967, 12.12.1967, 25.4.1968).

combination with sulpiride:
DAS 2 342 214 (Roche; appl. 21.8.1973; CH-prior. 21.9.1972).

Formulation(s):   tabl. 3 mg, 6 mg

Trade Name(s):

D: Bromazenil (Neuro Hexal)   neo OPT (Optimed)   I: Compendium (Polifarma)
    Bromazepam (Heumann)   Normoc (Merckle)   Lexotan (Roche)
    Durazanil (durachemie)   F: Anxyrex (Irex)   J: Lexotan (Nippon Roche)
    Gityl (Krewel Meuselbach)   Lexomil Roche (Roche)   USA: Lectopam (Roche); wfm
    Lexotanil (Roche)   GB: Lexotan (Roche)

Bromazine

(Bromdiphenhydramine)

ATC: R06AA01
Use: antihistaminic

RN: 118-23-0 MF: C₁₇H₂₀BrNO MW: 334.26 EINECS: 204-238-8
CN: 2-[(4-bromophenyl)phenylmethoxy]-*N,N*-dimethylethanamine

hydrochloride
RN: 1808-12-4 MF: C₁₇H₂₀BrNO · HCl MW: 370.72 EINECS: 217-310-9
LD₅₀: 63 mg/kg (M, i.v.); 366 mg/kg (M, p.o.);
       55 mg/kg (R, i.v.); 602 mg/kg (R, p.o.);
       21 mg/kg (dog, i.v.)

| 4-bromo-benzhydrol | | 4-bromobenz-hydryl bromide | | Bromazine |

Reaction scheme: 4-bromo-benzhydrol → (PBr₃, CCl₄) → 4-bromobenzhydryl bromide → (2-dimethylamino-ethanol) → Bromazine

Reference(s):
GB 670 622 (Parke Davis; appl. 1948; CH-prior. 1947).

Formulation(s): cps. 25 mg

Trade Name(s):
D: Ambodryl (Parke Davis); wfm   I: Ambodryl (Parke Davis); wfm
F: Ambodryl (Parke Davis); wfm   USA: Ambodryl (Parke Davis); wfm

Bromelain
(Bromelin)

ATC: B06AA11; J01AA
Use: anti-inflammatory, antineoplastic

RN: 9001-00-7   MF: unspecified   MW: unspecified   EINECS: 232-572-4
LD$_{50}$: 30 mg/kg (M, i.v.); >10 g/kg (M, p.o.);
>10 g/kg (R, p.o.)
CN: bromelain, juice

A concentrate of proteolytic enzymes derived from *Ananas comosus* Merr.
proteolytic enzyme (glycoprotein)
relative molecular mass ≡ 33000
By extraction from pineapple stems with water and precipitation with acetone or ammonium sulfate.

Reference(s):
Heinicke, R.M.: Science (Washington, D.C.) (SCIEAS) **118**, 753 (1953).
US 3 002 891 (Pineapple Research Inst. Hawai; 3.10.1961; appl. 12.12.1958).

purification:
US 2 950 227 (Schering AG; 1960; prior. 1956, 1959).

Formulation(s): drg. 4.5 mg, 8 mg, 20 mg, 40 mg, 45 mg, 90 mg; tabl. 500 F.I.P.-E.

Trade Name(s):
D: Bromelain 200 (Ursapharm)
Enzym-Wied (Wiedemann)-comb.
Floradix (Salushaus)-comb.
Mulsal (Mucos)-comb.

Phlogenzym (Mucos)
Proteozym (Wiedemann)
Traumanase (Nattermann)
Wobenzym (Mucos)-comb.
F: Extranase (Rottapharm)

Tetranase (Rottapharm)-comb.
I: Ananase (Rottapharm)
J: Kimotab (Mochida)
USA: Ananase (Rorer); wfm

Bromfenac sodium
(AHR-10282)

ATC: N02
Use: anti-inflammatory

RN: 91714-93-1 MF: C$_{15}$H$_{11}$BrNNaO$_3$ MW: 356.15
CN: 2-amino-3-(4-bromobenzoyl)benzeneacetic acid monosodium salt

sesquihydrate
RN: 120638-55-3 MF: C$_{15}$H$_{11}$BrNNaO$_3$ · 3/2H$_2$O MW: 766.35
free acid
RN: 91714-94-2 MF: C$_{15}$H$_{12}$BrNO$_3$ MW: 334.17

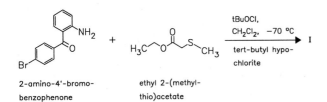

2-amino-4'-bromo-
benzophenone

ethyl 2-(methyl-
thio)acetate

tBuOCl,
CH$_2$Cl$_2$, −70 °C
tert-butyl hypo-
chlorite

I

7-(4-bromobenzoyl)-
3-(methylthio)-2,3-
dihydro-1H-
indol-2-one (I)

1. Raney-Ni, THF
2. NaOH

Bromfenac sodium

Reference(s):
US 4 126 635 (Robins Co.; appl. 15.4.1977; USA-prior. 17.5.1972, 25.4.1973).
US 4 568 695 (Robins Co.; USA-prior. 7.12.1983).
Welsh, D.A. et al.: J. Med. Chem. (JMCMAR) **27**, 1379-1388 (1984).

Formulation(s):   cps. 25 mg (as sodium salt)

Trade Name(s):
USA: Duract (Wyeth-Ayerst)

Bromhexine
(Bromexina)

ATC: R05CB02
Use: expectorant

RN: 3572-43-8 MF: C$_{14}$H$_{20}$Br$_2$N$_2$ MW: 376.14 EINECS: 222-684-1
CN: 2-amino-3,5-dibromo-N-cyclohexyl-N-methylbenzenemethanamine

monohydrochloride
RN: 611-75-6 MF: C$_{14}$H$_{20}$Br$_2$N$_2$ · HCl MW: 412.60 EINECS: 210-280-8
LD$_{50}$: 44 mg/kg (M, i.v.); 3 g/kg (M, p.o.);
      6 g/kg (R, p.o.)

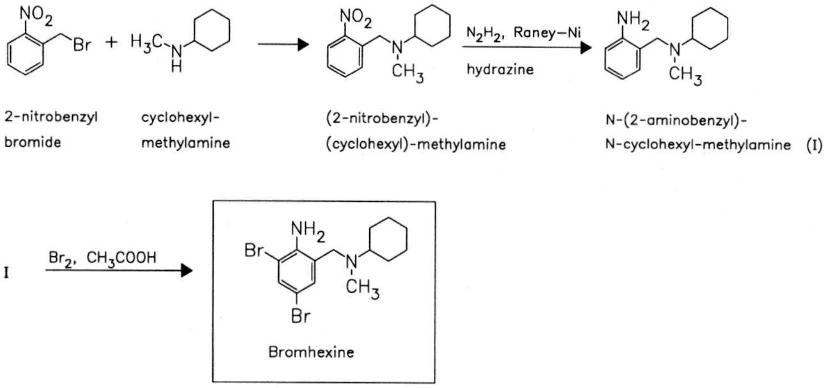

2-nitrobenzyl   cyclohexyl-   (2-nitrobenzyl)-   N-(2-aminobenzyl)-
bromide   methylamine   (cyclohexyl)-methylamine   N-cyclohexyl-methylamine   (I)

Bromhexine

Reference(s):

DE 1 169 939 (Thomae; appl. 20.11.1961).
US 3 336 308 (Boehringer Ing.; 15.8.1967; D-prior. 14.10.1963).
Keck, J.: Justus Liebigs Ann. Chem. (JLACBF) **662**, 171 (1963).
Engelhorn, R.; Püschmann, S.: Arzneim.-Forsch. (ARZNAD) **13**, 464 (1963).
Arch, F.: Arzneim.-Forsch. (ARZNAD) **13**, 480 (1963).

alternative syntheses:

DAS 2 311 637 (Thomae; appl. 9.3.1973).
DAS 2 365 624 (Thomae; appl. 27.3.1973; J-prior. 30.3.1972, 4.7.1972).
DAS 2 315 310 (Thomae; appl. 27.3.1973; J-prior. 30.3.1972, 4.7.1972).
DAS 2 443 712 (Thomae; appl. 12.9.1974).
DOS 2 633 518 (Egyt; appl. 26.7.1976; H-prior. 28.10.1975).
DOS 2 412 119 (Huhtamäki; appl. 13.3.1974; SF-prior. 15.3.1973, 2.7.1973, 9.1.1974, 8.2.1974).

use as mucous membrane local anesthetic:
DOS 2 729 786 (Thomae; appl. 1.7.1977).

Formulation(s):   amp. 8 mg/4 ml; drg. 8 mg, 12 mg; drops 8 mg/4 ml; syrup 4 mg (as hydrochloride);
    tabl. 4 mg, 8 mg, 10 mg, 20 mg

Trade Name(s):

D:	Aparsonin (Merckle)		Customed (Chefaro)	I:	Bertabronc (Berta-Mi)-

D: Aparsonin (Merckle)   Customed (Chefaro)   I: Bertabronc (Berta-Mi)-
    Berotec solvens   Lubrirhin (Alcon)   comb.
    (Boehringer Ing.)-comb.   Omniapharm (Merckle)   Bisolvon (Boehringer Ing.)
    Bisolvomycin (Boehringer   Synergomycin (Abbott)   Broncokin (Geymonat)
    Ing.)-comb.   F: Bisolvon (Boehringer Ing.)   Tauglicolo (SIT)-comb.
    Bisolvon (Boehringer Ing.)   GB: Alupent (Boehringer Ing.)-   combination preparations
    Bisolvonat (Boehringer   comb.   J: Bisolvon (Tanabe; as
    Ing.)-comb.   hydrochloride)

Bromindione

(Bromophenindione; Brophenadione)

ATC: M04
Use: anticoagulant

RN: 1146-98-1   MF: $C_{15}H_9BrO_2$ MW: 301.14
LD$_{50}$:   200 mg/kg (M, p.o.)
CN: 2-(4-bromophenyl)-1*H*-indene-1,3(2*H*)-dione

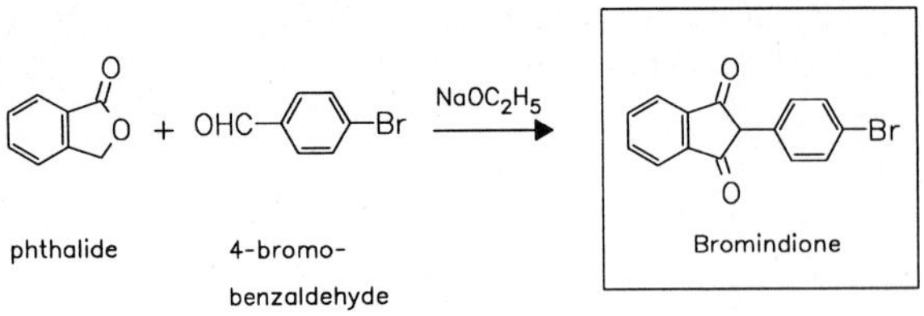

phthalide + 4-bromo-benzaldehyde → Bromindione

Reference(s):
US 2 847 474 (USV; 1958; appl. 1954).
cf. also anisindione

Formulation(s): amp. 8 mg/4 ml

Trade Name(s):
F: Fluidane (Metadier-Tours);
     wfm

Bromisoval

ATC: N05CM03
Use: sedative, slightly hypnotic

RN: 496-67-3   MF: $C_6H_{11}BrN_2O_2$ MW: 223.07   EINECS: 207-825-7
LD_{50}:   2 g/kg (M, p.o.);
        1 g/kg (R, p.o.)
CN: *N*-(aminocarbonyl)-2-bromo-3-methylbutanamide

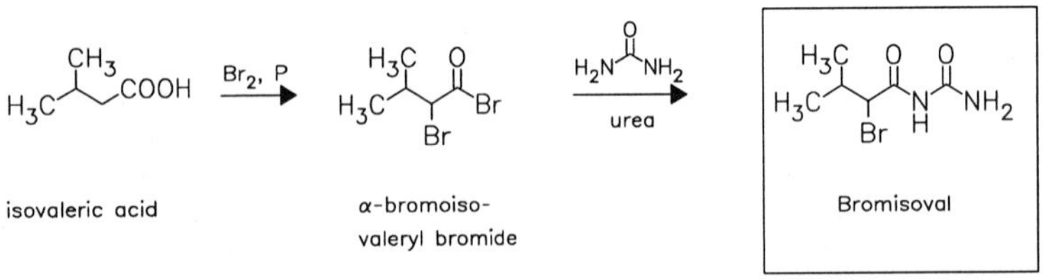

isovaleric acid → α-bromoiso-valeryl bromide → Bromisoval

Reference(s):
DRP 185 962 (Knoll; 1906).

Formulation(s): drg. 20 mg, 50 mg

Trade Name(s):
D: Brom-Nervacit (Herbert)-
     comb.; wfm
     Bromural (Knoll)-comb.;
     wfm
     Diffucord, -N (Dolorgiet)-
     comb.; wfm
     Rebuso Tabletten
     (Ravensberg)-comb.; wfm

Sekundal (Woelm)-comb.;
wfm
Steno-Valocordin
(Promonta)-comb.; wfm
Tempidorm N (Roland)-
comb.; wfm
Valocordin (Promonta)-
comb.; wfm

Ventrivert Tabletten
(Dolorgiet)-comb.; wfm
F: Beneural (Chantereau);
     wfm
J: Brovarin (Nippon
     Shinyaku)
USA: Bromural (Knoll); wfm

Bromocriptine
(2-Bromoergocryptine)

ATC: G02CB01; N04BC01
Use: prolactin inhibitor

RN: 25614-03-3   MF: $C_{32}H_{40}BrN_5O_5$ MW: 654.61   EINECS: 247-128-5
LD_{50}: >800 mg/kg (M, p.o.);
72 mg/kg (R, i.v.)
CN: (5'α)-2-bromo-12'-hydroxy-2'-(1-methylethyl)-5'-(2-methylpropyl)ergotaman-3',6',18-trione

monomesylate
RN: 22260-51-1   MF: $C_{32}H_{40}BrN_5O_5 \cdot CH_4O_3S$ MW: 750.71   EINECS: 244-881-1
LD_{50}: 189 mg/kg (M, i.v.); 2502 mg/kg (M, p.o.);
10.5 mg/kg (R, i.v.); >2 g/kg (R, p.o.)

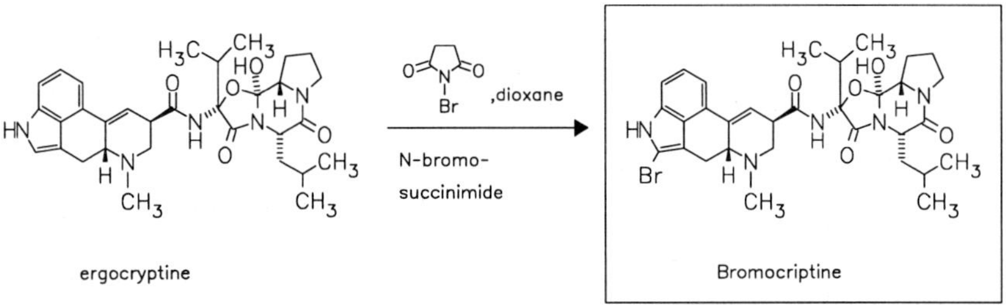

ergocryptine

N-bromo-succinimide

Bromocriptine

Reference(s):
US 3 752 814 (Sandoz; 14.8.1973; CH-prior. 31.5.1968).
DAS 1 926 045 (Sandoz; appl. 22.5.1969; CH-prior. 31.5.1968).

medical use:
US 3 752 888 (Sandoz; 14.8.1973; CH-prior. 31.5.1968).

nasal formulation:
DOS 2 802 113 (Sandoz; appl. 19.1.1978).

Formulation(s):   cps. 5 mg, 10 mg; tabl. 2.5 mg (as mesylate)

Trade Name(s):
D: Kirim (Hormosan)
      Pravidel (Novartis Pharma; 1977)
F: Bromo-Kin (Irex)

      Parlodel (Novartis; 1978)
GB: Parlodel (Novartis; 1976)
I: Parlodel (Sandoz; 1979)
      Serocryptin (Serono)

J: Parlodel (Novartis; 1979)
USA: Parlodel (Novartis; 1976)

Bromopride

ATC: A03FA04
Use: anti-emetic, gastric therapeutic

RN: 4093-35-0   MF: $C_{14}H_{22}BrN_3O_2$ MW: 344.25   EINECS: 223-842-2
LD_{50}: 310 mg/kg (M, p.o.);
545 mg/kg (R, p.o.)
CN: 4-amino-5-bromo-N-[2-(diethylamino)ethyl]-2-methoxybenzamide

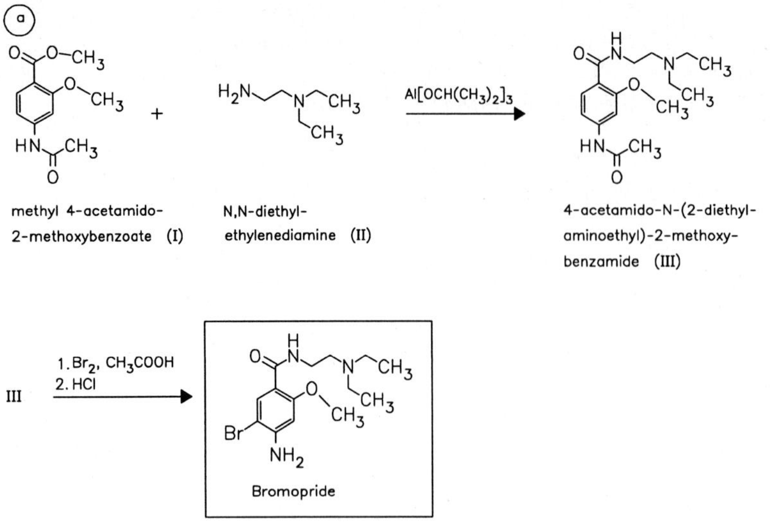

a

methyl 4-acetamido-
2-methoxybenzoate (I)

N,N-diethyl-
ethylenediamine (II)

4-acetamido-N-(2-diethyl-
aminoethyl)-2-methoxy-
benzamide (III)

III $\xrightarrow{\text{1. Br}_2\text{, CH}_3\text{COOH} \atop \text{2. HCl}}$

Bromopride

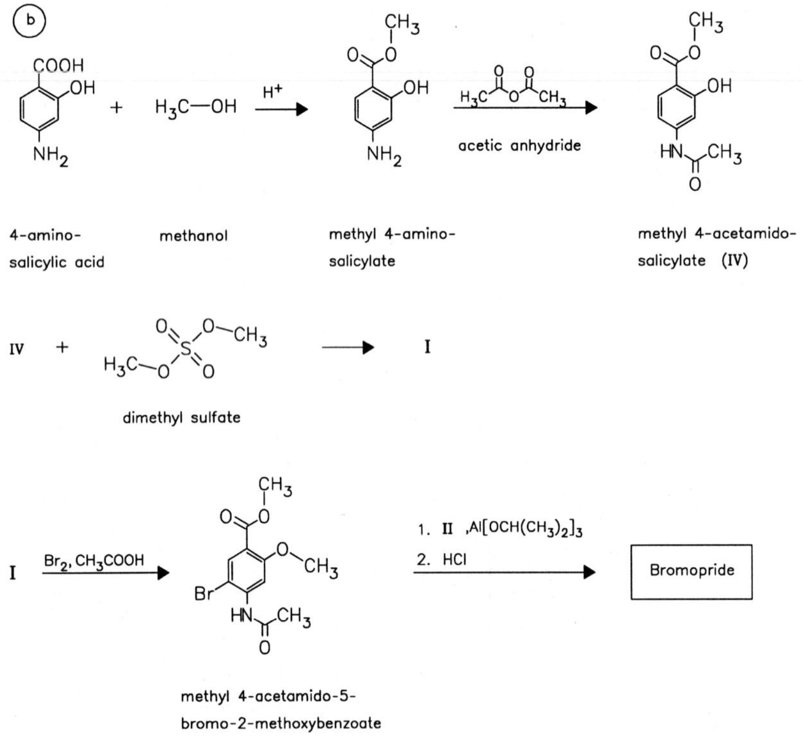

b

4-amino-
salicylic acid

methanol

methyl 4-amino-
salicylate

acetic anhydride

methyl 4-acetamido-
salicylate (IV)

IV + ⟶ I

dimethyl sulfate

I $\xrightarrow{\text{Br}_2\text{, CH}_3\text{COOH}}$

methyl 4-acetamido-5-
bromo-2-methoxybenzoate

$\xrightarrow{\text{1. II ,Al[OCH(CH}_3)_2]_3 \atop \text{2. HCl}}$ Bromopride

Reference(s):

US 3 177 252 (Soc. d'Etudes Scientifiques et Industrielles de l'Ile-de-France; 6.4. 1965; F-prior. 25.7.1961).

US 3 219 528 (Soc. d'Etudes Scientifiques et Industrielles de l'Ile-de-France; 23.11.1965; F-prior. 5.8.1960, 4.11.1960, 25.7.1961).

US 3 357 978 (Soc. d'Etudes Scientifiques et Industrielles de l'Ile-de-France; 12.12.1967; F-prior. 5.3.1963).

DE 1 233 877 (Soc. d'Etudes Scientifiques et Industrielles de l'Ile-de-France; appl. 14.7.1962; F-prior. 25.7.1961).

alternative synthesis:
DAS 2 102 848 (Delmar; appl. 21.1.1971; USA-prior. 21.1.1970).
DAS 2 119 724 (Teikoku Hormone Mfg.; appl. 22.4.1971; J-prior. 24.4.1970).
DAS 2 162 917 (Soc. d'Etudes Scientifiques et Industrielles de l'Ile-de-France; appl. 17.12.1971; J-prior. 21.12.1970).
DAS 2 166 118 (Teikoku Hormone Mfg.; appl. 22.4.1971; J-prior. 24.4.1970).
DOS 2 435 222 (Soc. d'Etudes Scientifiques et Industrielles de l'Ile-de-France; appl. 22.7.1974; J-prior. 24.7.1973).

Formulation(s):   amp. 10 mg; cps. 10 mg; drops 13.3 mg

Trade Name(s):
D: Cascapride (Merck)   I: Opridan (Locatelli; as   Valopride (Synthelabo)
                                               dihydrochloride
                                               monohydrate)

Bromperidol

ATC: N05AD06
Use: antipsychotic, neuroleptic

RN: 10457-90-6   MF: $C_{21}H_{23}BrFNO_2$ MW: 420.32   EINECS: 233-943-3
LD$_{50}$:   18.9 mg/kg (M, i.v.); 174 mg/kg (M, p.o.);
         10 mg/kg (R, i.v.); 359 mg/kg (R, p.o.)
CN: 4-[4-(4-bromophenyl)-4-hydroxy-1-piperidinyl]-1-(4-fluorophenyl)-1-butanone

monohydrochloride
RN: 59453-24-6   MF: $C_{21}H_{23}BrFNO_2 \cdot HCl$ MW: 456.78
LD$_{50}$:   18.9 mg/kg (R, i.v.); 174 mg/kg (R, p.o.)

fluorobenzene   4-chlorobutyryl   4-chloro-4'-fluoro-
                    chloride   butyrophenone (**I**)
                                             (cf. haloperidol)

4-bromo-α-methyl-   form-   4-(4-bromophenyl)-
styrene   aldehyde   1,2,3,6-tetrahydro-
                                                                 pyridine (**II**)

4-(4-bromophenyl)-
4-hydroxypiperidine   Bromperidol

Reference(s):
US 3 438 991 (Janssen; appl. 15.4.1969; GB-prior. 18.11.1959).
DE 1 289 845 (Janssen; appl. 18.4.1959; GB-prior. 22.4.1958).
Niemegeers, C.J.E.; Janssen, P.A.J.: Arzneim.-Forsch. (ARZNAD) **24**, 45 (1974).

Formulation(s):   amp. 5 mg/ml; drops 2 mg/ml; tabl. 1 mg, 5 mg, 10 mg

Trade Name(s):
D: Impromen (Janssen; 1983)   I: Impromen (Formenti;   J: Impromen (Yoshitomi;
      Tesoprel (Organon; 1984)   1989)   1986)

Brompheniramine

ATC: R06AB01
Use: antihistaminic

RN: 86-22-6   MF: $C_{16}H_{19}BrN_2$ MW: 319.25   EINECS: 201-657-8
CN: γ-(4-bromophenyl)-*N,N*-dimethyl-2-pyridinepropanamine

maleate (1:1)
RN: 980-71-2   MF: $C_{16}H_{19}BrN_2 \cdot C_4H_4O_4$ MW: 435.32   EINECS: 213-562-9
LD_{50}:   318 mg/kg (R, p.o.)

4-bromo-   2-chloro-   (4-bromophenyl)-   2-(4-bromophenyl)-4-
benzyl   pyridine   (2-pyridyl)-   dimethylamino-2-(2-
cyanide   acetonitrile   pyridyl)butyronitrile (I)

Brompheniramine

Reference(s):
US 2 567 245 (Schering Corp.; 1951; prior. 1948).
US 2 676 964 (Schering Corp.; 1954; prior. 1950).

Formulation(s):   elixir 4 mg, 2 mg/5 ml; cps. 12 mg

Trade Name(s):
D: Dimegan (Kreussler)   Rupton Chronules (Dexo)-   USA: Bromfed (Muro; as
F: Chronotrophir (Sanofi   comb.   maleate)
      Winthrop)   GB: Dimotane (Wyeth)   Dallergy (Laser; as
      Dimégan (Dexo)   Dimotane Plus (Wyeth)-   maleate)
      Dimetane Expectorant   comb.   Dimetane (Robins)
      (Whitehall)   I: Ilvin (Bracco; as maleate)   Ladrame (ECR; as maleate)
      Martigène (CIBA Vision J: Bromrun (Hokuriku)   Poly-Histine (Sanofi; as
      Ophthalmics)-comb.   maleate)

Respahist (Respa; as maleate) | Rondec (Dura; as maleate) | Ultrabrom (We; as maleate)

Broparestrol

ATC: G03
Use: estrogen (synthetic)

RN: 479-68-5 MF: $C_{22}H_{19}Br$ MW: 363.30 EINECS: 207-537-1
CN: 1-(2-bromo-1,2-diphenylethenyl)-4-ethylbenzene

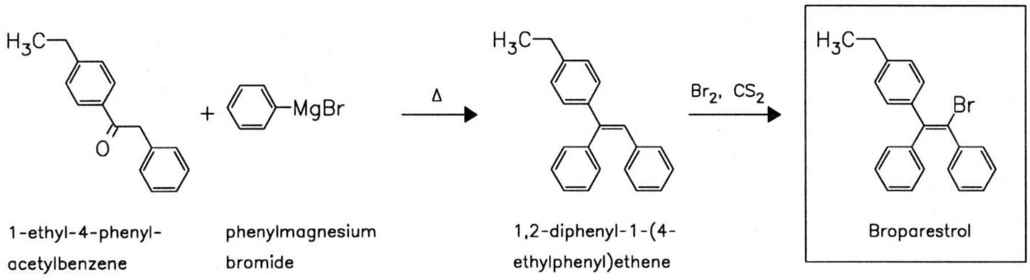

H3C
MgBr
Δ
Br2, CS2
H3C
H3C
Br
O
1-ethyl-4-phenyl-acetylbenzene
phenylmagnesium bromide
1,2-diphenyl-1-(4-ethylphenyl)ethene
Broparestrol

Reference(s):
Dvolaitzky, M.; Jacques, J.: Bull. Soc. Chim. Biol. (BSCIA3) **40**, 939 (1958).

Formulation(s): cream 10 %; emulsion 5 %

Trade Name(s):
F: Acnestrol (Devimy)-comb.; wfm | Longestrol (Laroche Navarron); wfm
I: Acnestrol (Scharper); wfm

Brotizolam

ATC: N05CD09
Use: tranquilizer, hypnotic

RN: 57801-81-7 MF: $C_{15}H_{10}BrClN_4S$ MW: 393.70 EINECS: 260-964-5
LD_{50}: 920 mg/kg (M, i.p.); >10000 mg/kg (M, p.o.);
1000 mg/kg (R, i.p.); >10000 mg/kg (R, p.o.)
CN: 2-bromo-4-(2-chlorophenyl)-9-methyl-6*H*-thieno[3,2-*f*][1,2,4]triazolo[4,3-*a*][1,4]diazepine

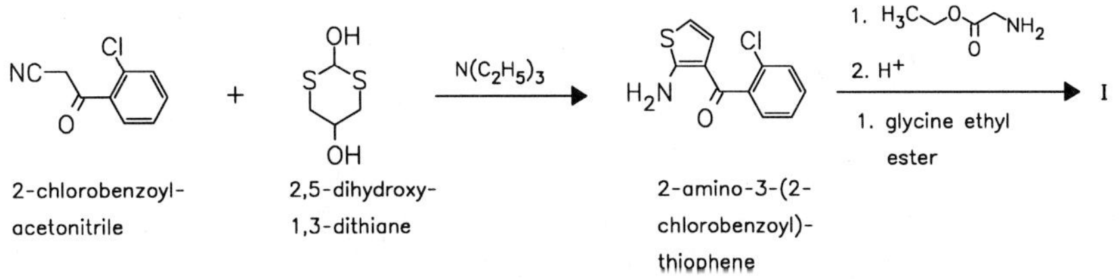

Cl
NC
O
OH
S
S
OH
N(C2H5)3
S
Cl
H2N
O
1. H3C O NH2 O
2. H+
1. glycine ethyl ester
I
2-chlorobenzoyl-acetonitrile
2,5-dihydroxy-1,3-dithiane
2-amino-3-(2-chlorobenzoyl)-thiophene

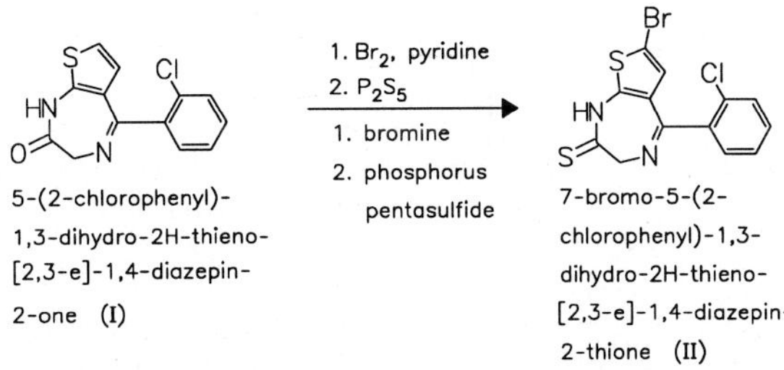

S
Cl
HN
O
N
1. Br2, pyridine
2. P2S5
1. bromine
2. phosphorus pentasulfide
Br
S
Cl
HN
S
N
5-(2-chlorophenyl)-1,3-dihydro-2H-thieno-[2,3-e]-1,4-diazepin-2-one (I)
7-bromo-5-(2-chlorophenyl)-1,3-dihydro-2H-thieno-[2,3-e]-1,4-diazepin-2-thione (II)

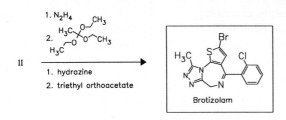

II
1. hydrazine
2. triethyl orthoacetate

Brotizolam

Reference(s):
DOS 2 410 030 (Boehringer Ing.; appl. 2.3.1974).
Weber, K.H. et al.: Arzneim.-Forsch. (ARZNAD) **36**, 518 (1986).

alternative synthesis:
DOS 2 503 235 (Boehringer Ing.; appl. 27.1.1975).
DOS 2 533 924 (Boehringer Ing.; appl. 30.7.1975).

synthesis of 5-(2-chlorophenyl)-1,3-dihydro-2H-thieno[2,3-e]-1,4-diazepin-2-one:
DOS 2 221 623 (Hoffmann-La Roche; appl. 3.5.1972; CH-prior. 14.5.1971).
Gewald, K.: Chem. Ber. (CHBEAM) **98**; 3571 (1965).

Formulation(s):   tabl. 0.25 mg

Trade Name(s):
D: Lendormin (Boehringer   I: Lendormin (Boehringer   J: Lendormin (Nippon
    Ing.)   Ing.)   Boehringer)

Broxyquinoline

ATC: A07AX01; G01AC06; P01AA01
Use: intestinal antiseptic

RN: 521-74-4   MF: $C_9H_5Br_2NO$ MW: 302.95   EINECS: 208-317-8
LD_{50}: 7420 mg/kg (M, p.o.)
CN: 5,6-dibromo-8-quinolinol

oxyquinoline
(q. v.)

Br_2 →

Broxyquinoline

Reference(s):
Bedall, K; Fischer, O. et al.: Ber. Dtsch. Chem. Ges. (BDCGAS) **14**, 1367 (1881).
Zinnei; Fiedler: Arch. Pharm. Ber. Dtsch. Pharm. Ges. (APBDAJ) **291**, 493 (1958).
DOS 2 515 476 (Chem. Fabrik Kalk; appl. 9.4.1975).

Formulation(s):   ointment 1.5 %

Trade Name(s):
D: Dysentrocym (Sanol)-   Intestopan (Sandoz)-comb.;   F: Colipar (Ucépha); wfm
    comb.; wfm   wfm   Entercine (Robapharm);
    Fenilor Lutschtabletten   Sandoin/-C (Sandoz)-   wfm
    (UCB); wfm   comb.; wfm

Intestopan (Sandoz)-comb.;
wfm

Norquinol (Norgan)-comb.;
wfm

Bucillamine

(DE-019; SA-96; Tiobutarit)

ATC: M01CC02
Use: immunomodulator, treatment of
       rheumatoid arthritis

RN: 65002-17-7   MF: C$_7$H$_{13}$NO$_3$S$_2$ MW: 223.32
LD$_{50}$: 2285 mg/kg (M, i.p.); 989 mg/kg (M, i.v.)
CN: N-(2-mercapto-2-methyl-1-oxopropyl)-L-cysteine

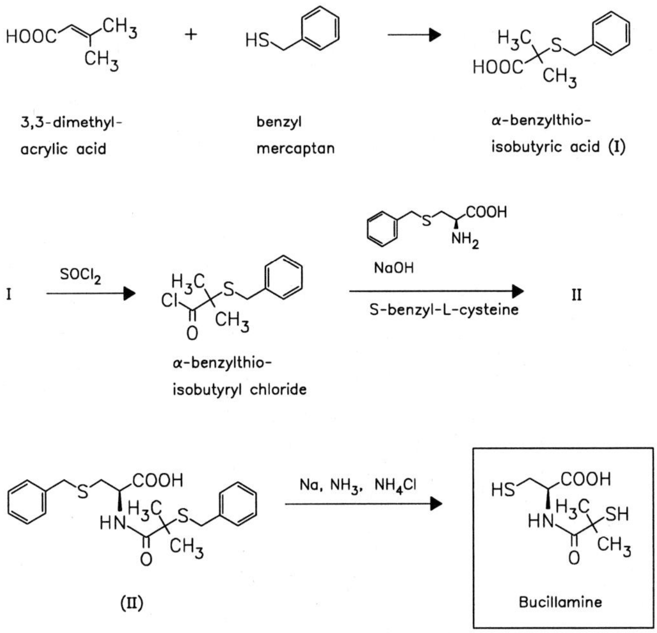

3,3-dimethyl-
acrylic acid

benzyl
mercaptan

α-benzylthio-
isobutyric acid (I)

S-benzyl-L-cysteine

α-benzylthio-
isobutyryl chloride

(II)

Bucillamine

Reference(s):
US 4 137 420 (Santen; 30.1.1979; J-prior. 8.3.1976).
DE 2 709 820 (Santen; appl. 7.3.1977; J-prior. 8.3.1976).

medical use as mucolytic:
US 4 305 958 (Santen; 15.12.1981; J-prior. 8.3.1976).

Formulation(s):   tabl. 100 mg (sugar coated)

Trade Name(s):
J: Rimatil (Santen)

Bucladesine sodium

ATC: C01CE04
Use: cardiotonic, phosphodiesterase
       inhibitor, positive inotropic acting
       drug

RN: 16980-89-5   MF: C$_{18}$H$_{23}$N$_5$NaO$_8$P   MW: 491.37   EINECS: 241-059-4
LD$_{50}$: 543 mg/kg (M, i.v.); >5 g/kg (M, p.o.);
       448 mg/kg (R, i.v.); >5 g/kg (R, p.o.)
CN: N-(1-oxybutyl)adenosine cyclic 3',5'-(hydrogen phosphate) 2'-butanoate monosodium salt

bucladesine
RN: 362-74-3 MF: $C_{18}H_{24}N_5O_8P$ MW: 469.39 EINECS: 206-649-8

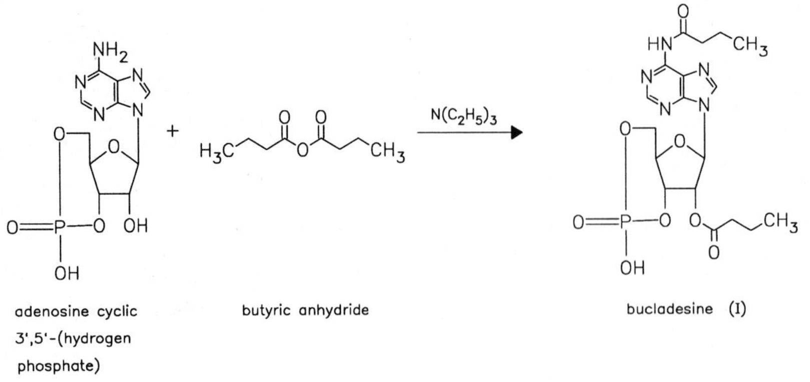
NH2
N(C2H5)3
HN
CH3
H3C
ONa
adenosine cyclic 3',5'-(hydrogen phosphate)
butyric anhydride
bucladesine (I)

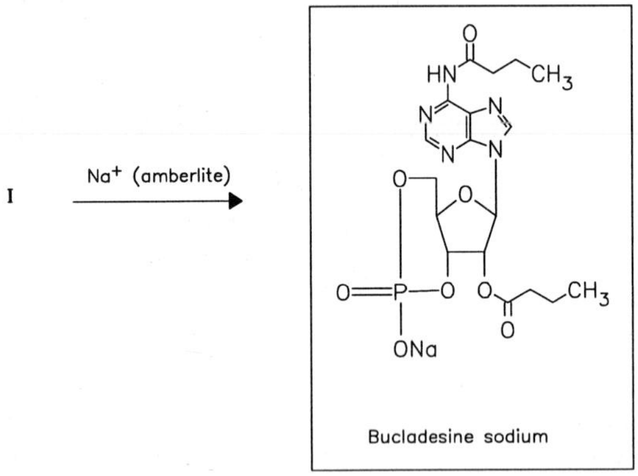
I
Na+ (amberlite)
Bucladesine sodium

Reference(s):
JP 5 195 096 (Daiichi Seiyaku; appl. 14.2.1975).
JP 51 113 896 (Daiichi Seiyaku; appl. 31.3.1975).
JP 5 239 699 (Daiichi Seiyaku; appl. 26.9.1975).

Formulation(s): amp. 0.05 mg, 0.2 mg; tabl. 200 mg, 400 mg

Trade Name(s):
J: Actocin (Daiichi)

Buclizine
(Histabutizine)

ATC: R06AE01
Use: antiallergic, antihistaminic

RN: 82-95-1 MF: $C_{28}H_{33}ClN_2$ MW: 433.04 EINECS: 201-448-1
CN: 1-[(4-chlorophenyl)phenylmethyl]-4-[[4-(1,1-dimethylethyl)phenyl]methyl]piperazine

dihydrochloride
RN: 129-74-8 MF: $C_{28}H_{33}ClN_2 \cdot 2HCl$ MW: 505.96 EINECS: 204-962-4
LD_{50}: 2100 mg/kg (M, p.o.)

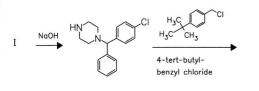

4-chlorobenz-
hydryl chloride

ethyl piperazine-
N-carboxylate

ethyl 4-(4-chlorobenz-
hydryl)piperazine-1-carboxylate (I)

I → NaOH

1-(4-chlorobenz-
hydryl)piperazine

4-tert-butyl-
benzyl chloride

Buclizine

Reference(s):
DE 964 048 (H. Morren; appl. 1952; B-prior. 1951).

Formulation(s):   tabl. 25 mg

Trade Name(s):
D: Migralave (Temmler)-   F: Aphilan (Darcy)   GB: Migraleve (Pfizer
       comb. with paracetamol   Consumer)-comb.

Buclosamide

ATC: D01AE12
Use: fungicide

RN: 575-74-6   MF: C$_{11}$H$_{14}$ClNO$_2$ MW: 227.69   EINECS: 209-390-9
CN: *N*-butyl-4-chloro-2-hydroxybenzamide

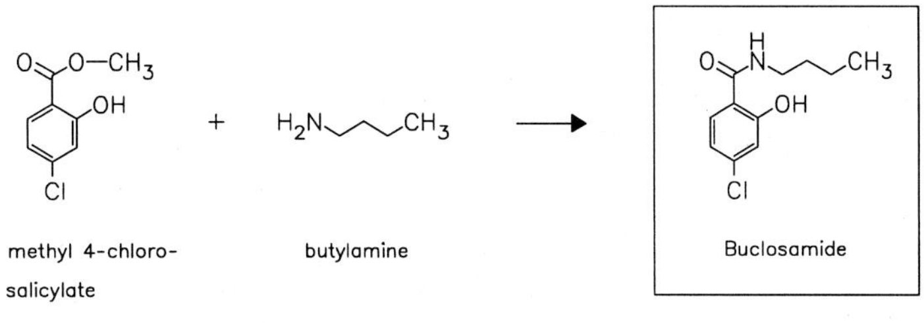

methyl 4-chloro-
salicylate

butylamine

Buclosamide

Reference(s):
US 2 923 737 (Hoechst; 2.2.1960; D-prior. 26.1.1956).

Formulation(s):   ointment 10 g/100 g; sol. 10 g/100 ml

Trade Name(s):
D: Jadit (Hoechst)-comb.;   Jadit-Hydrocortisone   I: Jadit (Hoechst)-comb.;
       wfm   (Hoechst)-comb.; wfm   wfm
F: Jadit (Hoechst); wfm

Bucloxic acid

(Acide bucloxique)

ATC: N02
Use: anti-inflammatory

RN: 32808-51-8   MF: C$_{16}$H$_{19}$ClO$_3$ MW: 294.78   EINECS: 251-231-0
LD$_{50}$:   852 mg/kg (M, p.o.);
             120 mg/kg (R, p.o.)
CN: 3-chloro-4-cyclohexyl-γ-oxobenzenebutanoic acid

calcium salt
RN: 32808-53-0   MF: C$_{32}$H$_{36}$CaCl$_2$O$_6$ MW: 627.62   EINECS: 251-232-6
LD$_{50}$:   1700 mg/kg (M, p.o.);
             175 mg/kg (R, p.o.)

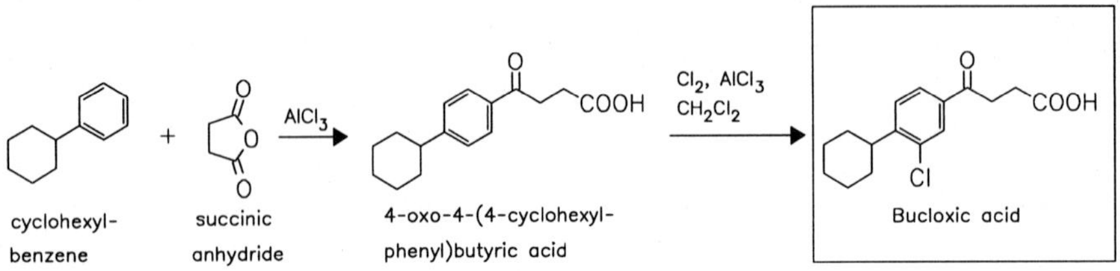

cyclohexyl-   succinic   4-oxo-4-(4-cyclohexyl-   Bucloxic acid
benzene   anhydride   phenyl)butyric acid

Reference(s):
DE 2 021 445 (Clin-Byla; appl. 2.5.1970; F-prior. 12.5.1969).
GB 1 315 542 (Clin-Byla; appl. 7.5.1970; F-prior. 12.5.1969).

Trade Name(s):
F: Esfar (Midy); wfm

Bucumolol

ATC: C07A
Use: β-adrenoceptor blocker

RN: 58409-59-9   MF: C$_{17}$H$_{23}$NO$_4$ MW: 305.37
LD$_{50}$:   31 mg/kg (M, i.v.); 680 mg/kg (M, p.o.)
CN: 8-[3-[(1,1-dimethylethyl)amino]-2-hydroxypropoxy]-5-methyl-2H-1-benzopyran-2-one

hydrochloride
RN: 30073-40-6   MF: C$_{17}$H$_{23}$NO$_4$ · HCl   MW: 341.84

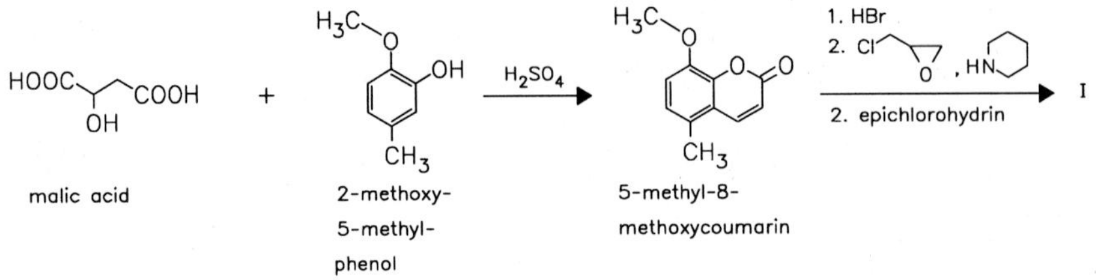

malic acid   2-methoxy-   5-methyl-8-
              5-methyl-   methoxycoumarin
              phenol

5-methyl-8-(3-chloro-
2-hydroxypropoxy)-
coumarin (I)

tert-butyl-
amine

Bucumolol

Reference(s):
DOS 2 021 958 (Sankyo; appl. 27.4.1970; J-prior. 28.4.1969).
US 3 663 570 (Sankyo; 16.5.1972; J-prior. 28.4.1969; 27.10.1969).
Sato, Y. et al.: Chem. Pharm. Bull. (CPBTAL) **20**, 905 (1972).

Formulation(s):   tabl. 5 mg, 10 mg

Trade Name(s):
J: Bucumarol (Sankyo)

Budesonide

ATC: A07EA06; D07AC09; H02AB16;
        R01AD05; R03AB; R03BA02
Use: topical glucocorticoid, antiasthmatic

RN: 51333-22-3   MF: $C_{25}H_{34}O_6$ MW: 430.54   EINECS: 257-139-7
LD_{50}:   124 mg/kg (M, i.v.); 4750 mg/kg (M, p.o.);
        98.9 mg/kg (R, i.v.); >3200 mg/kg (R, p.o.)
CN: (11β,16α)-16,17-[butylidenebis(oxy)]-11,21-dihydroxypregna-1,4-diene-3,20-dione

16α-hydroxyprednisolone
(cf. desonide synthesis)

butyraldehyde

Budesonide

Reference(s):
US 3 929 768 (Bofors; 30.12.1975; appl. 14.5.1973; S-prior. 19.5.1972).
DOS 2 323 215 (Bofors; appl. 19.5.1973; S-prior. 19.5.1972).
US 3 983 233 (Bofors; prior. 14.5.1973).
US 4 835 145 (Sicor; 30.5.1989; I-prior. 11.6.1984, 2.1.1987).

separation of diastereomers:
DOS 2 323 216 (Bofors; appl. 19.5.1973; S-prior. 19.5.1972).

Formulation(s):   aerosol 0,2 mg/puff; cream 0.025 %; nasal aerosol 0.05 mg/puff; ointment 0.025 mg;
        pumpspray 0.05 mg/puff; susp. 0.5 mg/2 ml, 1 mg/2 ml

Trade Name(s):
D: Benosid (Farmasan)   Entocort (Astra)   F: Pulmicort (Astra)
      Bronchocux (TAD)   Pulmicort (Astra/pharma-   GB: Entocort CR (Astra)
      Budecort (Klinge)   stern; 1983)   Pulmicort (Astra; 1983)
      Budegat (Fatol)   Respicort (Mundipharma)

Rhinocort Aqua (Astra; 1984)

I: Bidien (IDI)

J: Prefenid lipocrema (Brocades)
Budeson (Fujisawa)

USA: Pulmicort (Astra)
Rhinocort (Astra)

Budipine
(BY-701)

ATC: N04AA
Use: antiparkinsonian

RN: 57982-78-2 MF: C$_{21}$H$_{27}$N MW: 293.45 EINECS: 261-062-4
LD$_{50}$: 33 mg/kg (M, i.v.); 120 mg/kg (M, p.o.);
28 mg/kg (R, i.v.); 165 mg/kg (R, p.o.)
CN: 1-(1,1-dimethylethyl)-4,4-diphenylpiperidine

hydrochloride
RN: 63661-61-0 MF: C$_{21}$H$_{27}$N · HCl MW: 329.92 EINECS: 264-388-5

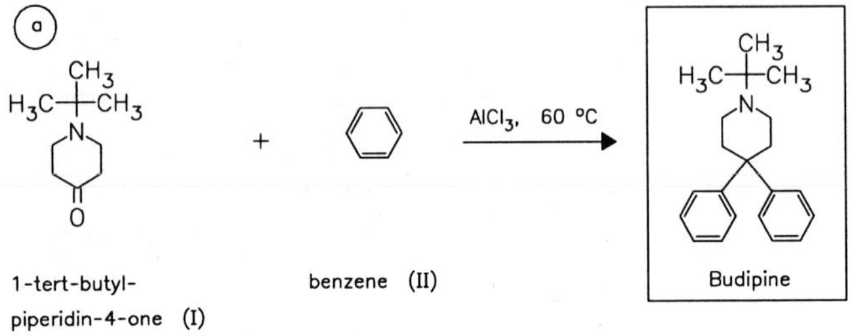

1-tert-butyl-
piperidin-4-one (I)

benzene (II)

Budipine

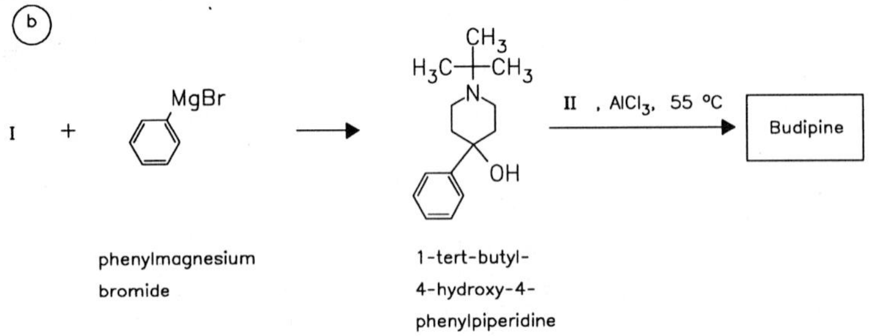

phenylmagnesium
bromide

1-tert-butyl-
4-hydroxy-4-
phenylpiperidine

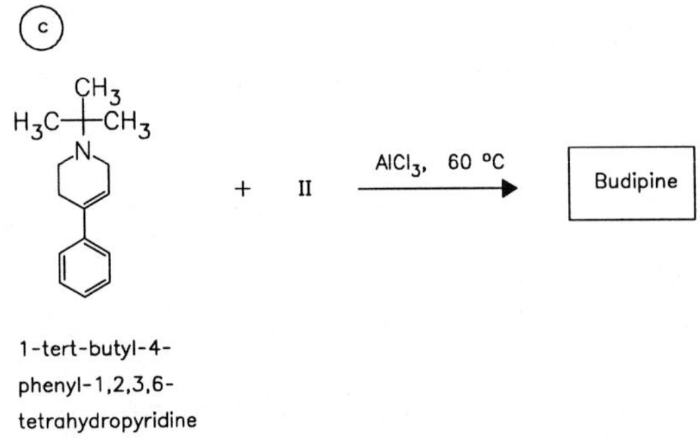

Budipine

1-tert-butyl-4-
phenyl-1,2,3,6-
tetrahydropyridine

Budipine

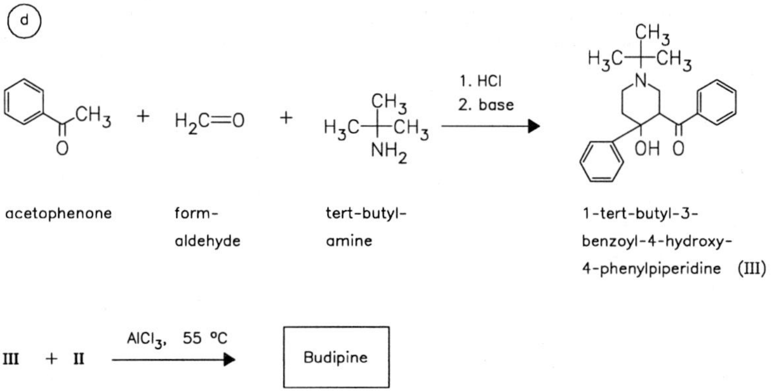

(d)

| acetophenone | form-aldehyde | tert-butyl-amine | 1-tert-butyl-3-benzoyl-4-hydroxy-4-phenylpiperidine (III) |

III + II $\xrightarrow{\text{AlCl}_3,\ 55\ °C}$ Budipine

Reference(s):

a-d Schaefer, H. et al.: Arzneim.-Forsch. (ARZNAD) **34**, 233-240 (1984).
   DE 2 825 322 (Byk Gulden; appl. 11.1.1979; LU-prior. 30.6.1977).

Formulation(s):   tabl. 10 mg, 20 mg, 30 mg (as hydrochloride)

Trade Name(s):
D: Parkinsan (Promonta
      Lundbeck)

Budralazine

ATC: C02DB
Use: antihypertensive

RN: 36798-79-5   MF: $C_{14}H_{16}N_4$ MW: 240.31
LD$_{50}$:   4020 mg/kg (M, i.p.); 1820 mg/kg (M, p.o.);
      3570 mg/kg (R, i.p.); 620 mg/kg (R, p.o.)
CN: 1(2*H*)-phthalazinone (1,3-dimethyl-2-butenylidene)hydrazone

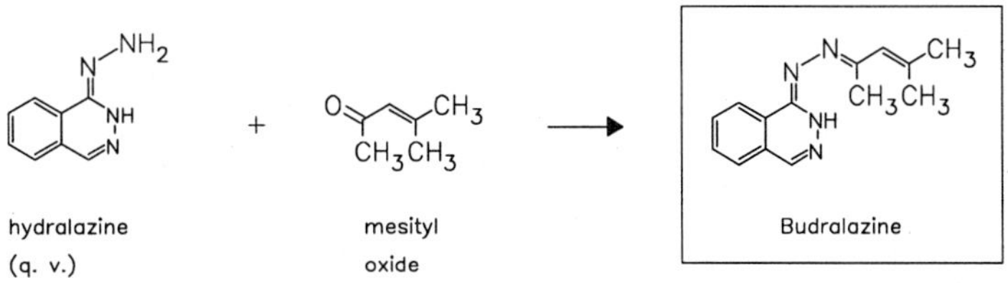

| hydralazine (q. v.) | mesityl oxide | Budralazine |

Reference(s):
Ueno, K. et al.: Chem. Pharm. Bull. (CPBTAL) **24**, 1068 (1976).
DOS 2 145 359 (Daiichi Seiyaku; appl. 13.9.1971; J-prior. 14.9.1970).
US 3 840 539 (Daiichi Seiyaku; 8.10.1974; appl. 2.9.1971; J-prior. 14.9.1970).

Formulation(s):   gran. 1 %; tabl. 30 mg, 60 mg

Trade Name(s):
J: Buterazine (Daiichi; 1983)

Bufetolol

(Bufetrol)

ATC: C07AA
Use: beta blocking agent

RN: 53684-49-4 MF: $C_{18}H_{29}NO_4$ MW: 323.43
CN: 1-[(1,1-dimethylethyl)amino]-3-[2-[(tetrahydro-2-furanyl)methoxy]phenoxy]-2-propanol

hydrochloride
RN: 35108-88-4 MF: $C_{18}H_{29}NO_4 \cdot HCl$ MW: 359.89 EINECS: 252-369-4
LD_{50}: 50.3 mg/kg (M, i.v.); 402 mg/kg (M, p.o.);
59.4 mg/kg (R, i.v.); 1088 mg/kg (R, p.o.)

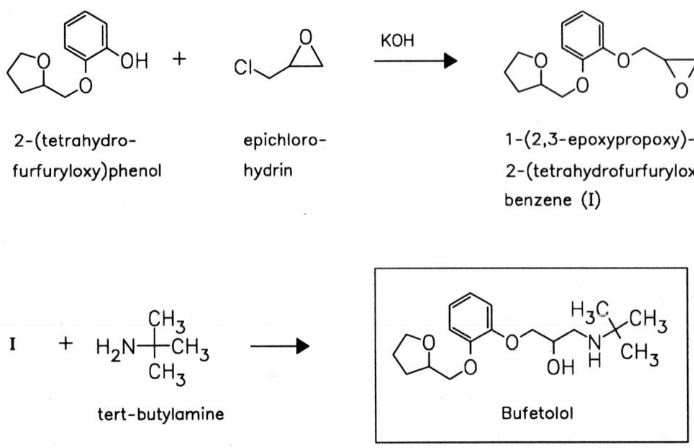

KOH
2-(tetrahydro-furfuryloxy)phenol
epichloro-hydrin
1-(2,3-epoxypropoxy)-2-(tetrahydrofurfuryloxy)-benzene (I)
I
tert-butylamine
Bufetolol

Reference(s):
DOS 2 024 001 (Yoshitomi; appl. 15.5.1970; J-prior. 16.5.1969, 2.10.1969, 3.4.1970).
US 3 723 476 (Yoshitomi; 27.3.1973; J-prior. 16.5.1969, 2.10.1969, 3.4.1970).

Formulation(s): f. c. tabl. 5 mg, 10 mg (as hydrochloride)

Trade Name(s):
I: Adobiol (Menarini)
J: Adobiol (Yoshitomi; as hydrochloride)

Bufexamac

ATC: M01AB17
Use: anti-inflammatory

RN: 2438-72-4 MF: $C_{12}H_{17}NO_3$ MW: 223.27 EINECS: 219-451-1
LD_{50}: 8 g/kg (M, p.o.);
3370 mg/kg (R, p.o.)
CN: 4-butoxy-*N*-hydroxybenzeneacetamide

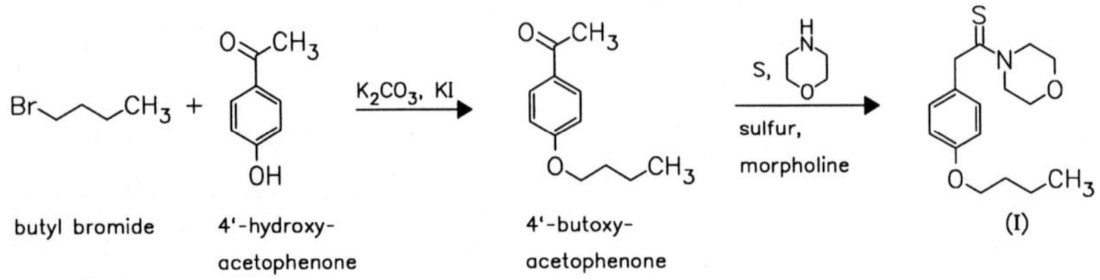

K_2CO_3, KI
S,
sulfur, morpholine
butyl bromide
4'-hydroxy-acetophenone
4'-butoxy-acetophenone
(I)

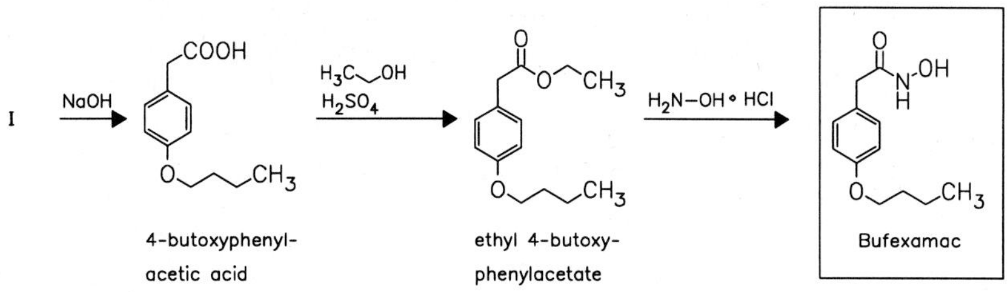

4-butoxyphenyl-
acetic acid

ethyl 4-butoxy-
phenylacetate

Bufexamac

Reference(s):

Buu-Hoï, N.P. et al.: C. R. Hebd. Seances Acad. Sci. (COREAF) **261**, 2259 (1965).
BE 661 226 (Madan; appl. 17.3.1965).
US 3 479 396 (Madan; 18.11.1969; B-prior. 5.6.1964, 17.3.1965).
DAS 1 768 406 (Madan; appl. 1.6.1965; B-prior. 5.6.1964, 17.3.1965).

Formulation(s):   ointment 50 mg/g; suppos. 250 mg

Trade Name(s):

D:	Bufederm (Pharmagalen)		Parfenac (Novalis; Lederle;		Calmaderm (Whitehall)
	Duradermal (durachemie)		1976)		Parfenac (Whitehall; 1975)
	Ekzemase (Azupharma)		Proctoparf (Novalis; 1984)-	I:	Parfenal (Cyanamid)
	Jomax (Hexal)		comb.		Viafen (Zyma)
	Malipuran (Heumann)	F:	Bufal (Pierre Fabre)	J:	Anderm (Lederle)

Buflomedil

ATC: C04AX20
Use: vasodilator, antispasmodic

RN: 55837-25-7   MF: $C_{17}H_{25}NO_4$ MW: 307.39   EINECS: 259-851-3
CN: 4-(1-pyrrolidinyl)-1-(2,4,6-trimethoxyphenyl)-1-butanone

hydrochloride
RN: 35543-24-9   MF: $C_{17}H_{25}NO_4 \cdot HCl$ MW: 343.85   EINECS: 252-611-9
LD_{50}:   40 mg/kg (M, i.v.); 275 mg/kg (M, p.o.);
        58.5 mg/kg (R, i.v.); 410 mg/kg (R, p.o.)

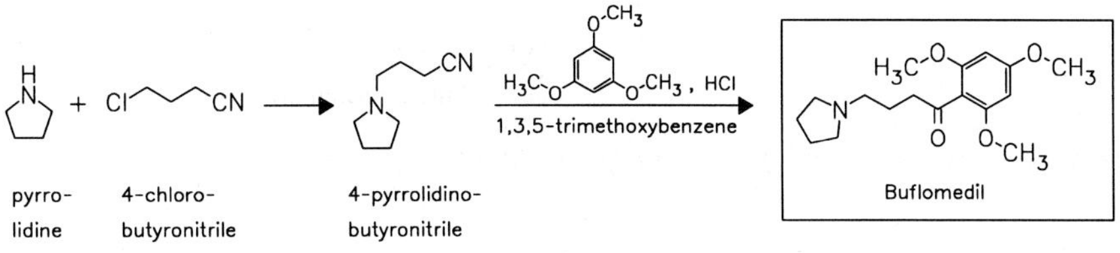

pyrro-
lidine

4-chloro-
butyronitrile

4-pyrrolidino-
butyronitrile

1,3,5-trimethoxybenzene

Buflomedil

Reference(s):

GB 1 325 192 (Orsymonde; appl. 6.5.1970; valid from 6.5.1971).
DE 2 122 144 (Orsymonde; appl. 3.5.1971; GB-prior. 6.5.1970).
US 3 895 030 (Orsymonde; 15.7.1975; appl. 5.5.1971; GB-prior. 6.5.1970).

Formulation(s):   amp. 0.4 g/40 ml, 0.4 g/120 ml, 50 mg/5 ml; s. r. tabl. 600 mg; tabl. 150 mg, 300 mg (as
        hydrochloride)

Trade Name(s):

D:	Bufedil (Abbott; 1982)		Loftyl (Abbott)		Buflan (Pierrel; 1982)
	Buflo (AbZ-Pharma)	F:	Fonzylane (Lafon; 1976)		Buflocit (CT)
	Defluina (Nattermann)	I:	Bufene (Ist. Chim. Inter.)		Buflofar (Farge)

Emoflux (Metapharma)	Irrodan (Biomedica	Medil (Crosara)
Flomed (Pulitzer)	Foscama)	Perfudan (Piam)
Flupress (Drug Research)	Loftyl (Abbott; 1982)	Pirxane (Lisapharma)

Bumadizone

ATC: M01AB07
Use: anti-inflammatory, antipyretic

RN: 3583-64-0   MF: $C_{19}H_{22}N_2O_3$ MW: 326.40   EINECS: 222-710-1
LD$_{50}$:   1350 mg/kg (M, p.o.)
CN: butylpropanedioic acid mono(1,2-diphenylhydrazide)

calcium salt (2:1)
RN: 34461-73-9   MF: $C_{38}H_{42}CaN_4O_6$ MW: 690.85   EINECS: 252-048-9
LD$_{50}$:   1500 mg/kg (M, p.o.);
        750 mg/kg (R, p.o.)

phenylbutazone
(q. v.)

NaOH, NaCl, H_2O

Bumadizone

Reference(s):
US 3 455 999 (Geigy; 15.7.1969; CH-prior. 7.6.1963).
DE 1 235 936 (Geigy; appl. 5.6.1964; CH-prior. 7.6.1963).
DE 2 055 845 (Byk Gulden; appl. 13.11.1970).

Formulation(s):   tabl. 110 mg

Trade Name(s):
D: Eumotol (Byk Gulden);   Rheumatol (Tosse); wfm   I: Eumotol (Byk Gulden);
      wfm   F: Eumotol (Valpan); wfm   wfm

Bumetanide

ATC: C03CA02
Use: diuretic

RN: 28395-03-1   MF: $C_{17}H_{20}N_2O_5S$ MW: 364.42   EINECS: 249-004-6
LD$_{50}$:   >200 mg/kg (M, i.v.); 4624 mg/kg (M, p.o.);
        >200 mg/kg (R, i.v.); >6 g/kg (R, p.o.)
CN: 3-(aminosulfonyl)-5-(butylamino)-4-phenoxybenzoic acid

4-chloro-
benzoic acid

ClSO$_3$H

4-chloro-3-
(chlorosulfonyl)-
benzoic acid

HNO$_3$

4-chloro-3-(chlorosulfonyl)-
5-nitrobenzoic acid

NH$_3$

5-sulfamoyl-4-chloro-
3-nitrobenzoic acid   (I)

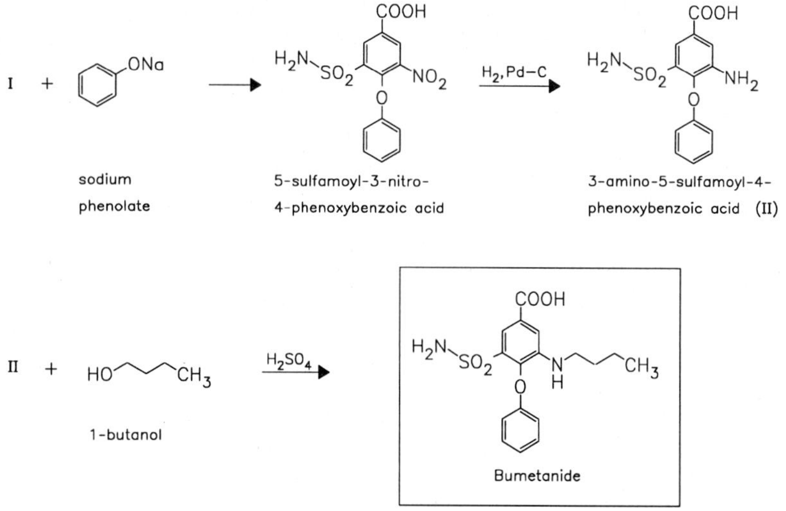

I + sodium phenolate → 5-sulfamoyl-3-nitro-4-phenoxybenzoic acid $\xrightarrow{H_2, Pd-C}$ 3-amino-5-sulfamoyl-4-phenoxybenzoic acid (II)

II + 1-butanol $\xrightarrow{H_2SO_4}$ Bumetanide

Reference(s):

GB 1 249 490 (Loevens Kemiske Fabr.; valid from 22.12.1969; prior. 24.12.1968, 18.6.1969, 29.7.1969).
DOS 1 964 503 (Loevens Kemiske Fabr.; appl. 23.12.1969; GB-prior. 24.12.1968, 18.6.1969, 29.7.1969).
DE 1 964 504 (Loevens Kemiske Fabr.; appl. 23.12.1969; GB-prior. 24.12.1968, 18.6.1969, 29.7.1969; USA-prior. 24.7.1969).
DAS 1 966 878 (Loevens Kemiske Fabr.; appl. 23.12.1969; GB-prior. 24.12.1968, 18.6.1969, 29.7.1969).
US 3 634 583 (Loevens Kemiske Fabr.; 11.1.1972; appl. 24.7.1969).
US 3 806 534 (Leo Pharm.; 23.4.1974; appl. 22.12.1969; GB-prior. 24.12.1968).

Formulation(s): amp. 1 mg/2 ml, 5 mg/10 ml; tabl. 1 mg

Trade Name(s):

D:	Burinex (Leo)	Burinex A (Leo)-comb.	I:	Fontego (Polifarma)
F:	Burinex (Leo; 1987)	with amiloride	J:	Lunetoron (Sankyo)
	Lixil (Leo); wfm	Burinex K (Leo)-comb.	USA:	Bumex (Roche; 1983)
GB:	Burinex (Leo; 1973)	with potassium chloride		

Bunamiodyl
(Buniodyl)

ATC: V08
Use: X-ray contrast medium

RN: 1233-53-0 MF: $C_{15}H_{16}I_3NO_3$ MW: 639.01
CN: 2-[[2,4,6-triiodo-3-[(1-oxobutyl)amino]phenyl]methylene]butanoic acid

monosodium salt
RN: 1923-76-8 MF: $C_{15}H_{15}I_3NNaO_3$ MW: 660.99
LD$_{50}$: 418 mg/kg (M, i.v.); 2690 mg/kg (M, p.o.);
480 mg/kg (R, i.v.); 2800 mg/kg (R, p.o.)

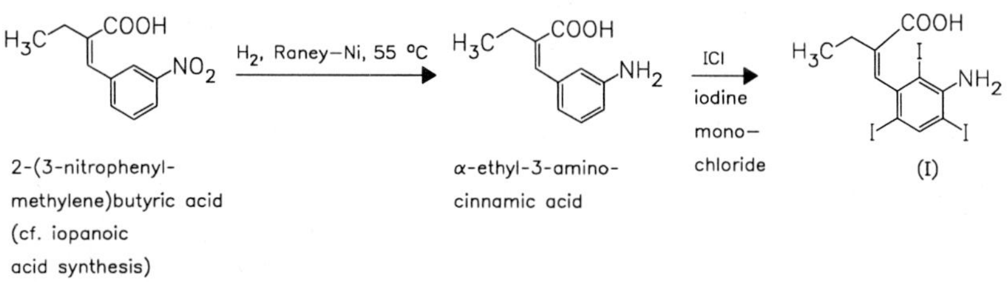

2-(3-nitrophenyl-methylene)butyric acid (cf. iopanoic acid synthesis) $\xrightarrow{H_2, \text{ Raney-Ni, 55 °C}}$ α-ethyl-3-amino-cinnamic acid $\xrightarrow[\text{iodine mono-chloride}]{ICl}$ (I)

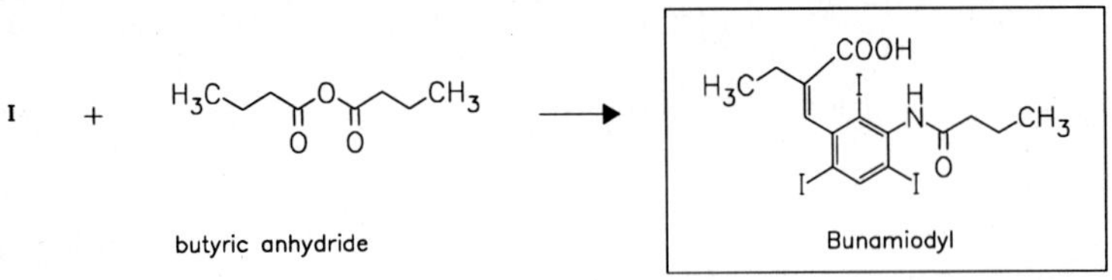

butyric anhydride

Bunamiodyl

Reference(s):
Cassebaum, H.; Dierbach, K.: Pharmazie (PHARAT) **16**, 392 (1961).

Formulation(s): sol. 4.5 g

Trade Name(s):
D: Orabilix (Hefa-Frenon); F: Orabilix (Guerbet); wfm
wfm J: Orabilix (Kodama); wfm

Bunazosin

ATC: C02
Use: antihypertensive

RN: 80755-51-7 MF: $C_{19}H_{27}N_5O_3$ MW: 373.46
CN: 1-(4-amino-6,7-dimethoxy-2-quinazolinyl)hexahydro-4-(1-oxobutyl)-1H-1,4-diazepine

monohydrochloride
RN: 52712-76-2 MF: $C_{19}H_{27}N_5O_3 \cdot HCl$ MW: 409.92
LD$_{50}$: 57 mg/kg (M, i.v.); 1201 mg/kg (M, p.o.);
50 mg/kg (R, i.v.); 980 mg/kg (R, p.o.)

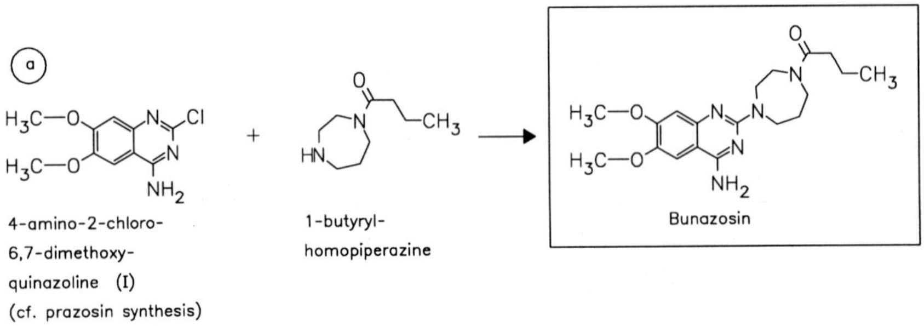

4-amino-2-chloro-
6,7-dimethoxy-
quinazoline (I)
(cf. prazosin synthesis)

1-butyryl-
homopiperazine

Bunazosin

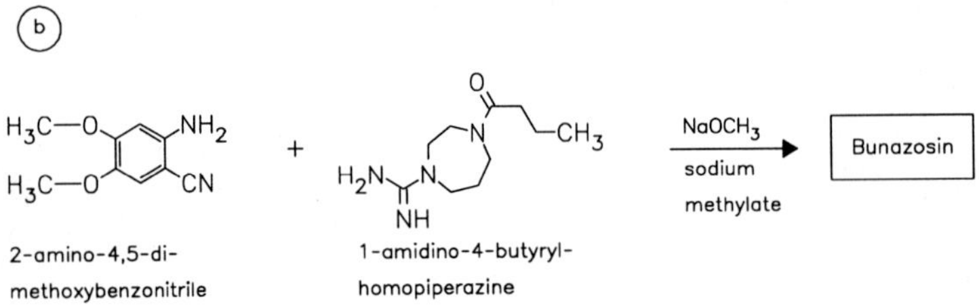

2-amino-4,5-di-
methoxybenzonitrile

1-amidino-4-butyryl-
homopiperazine

NaOCH₃
sodium
methylate

Bunazosin

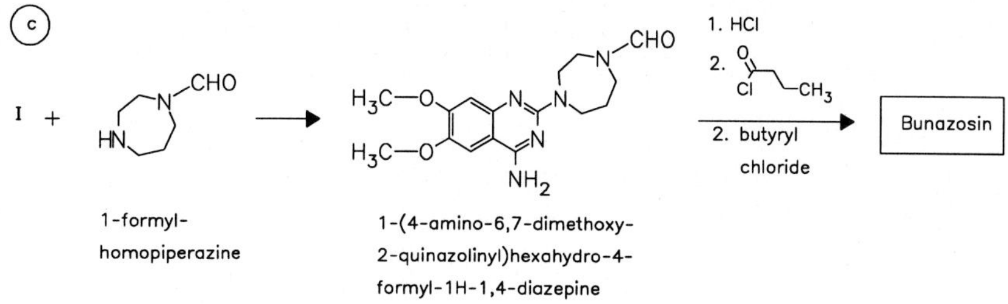

1-formyl-homopiperazine

1-(4-amino-6,7-dimethoxy-2-quinazolinyl)hexahydro-4-formyl-1H-1,4-diazepine

Reference(s):

a   JP 7 682 285 (Eisai; appl. 16.4.1974).
b   JP 75 140 474 (Eisai; appl. 16.4.1974).
c   DOS 2 354 389 (Eisai; appl. 30.10.1973; J-prior. 30.10.1972).
    US 3 920 636 (Eisai; 18.11.1975; appl. 29.10.1973; J-prior. 30.10.1972).

Formulation(s):   gran. 0.5 %; tabl. 0.5 mg, 1 mg, 3 mg

Trade Name(s):
J: Detantol (Eisai; 1985)

Bunitrolol

ATC: C07AA
Use: beta blocking agent

RN: 34915-68-9   MF: $C_{14}H_{20}N_2O_2$ MW: 248.33
LD$_{50}$:   46 mg/kg (M, i.v.)
CN: 2-[3-[(1,1-dimethylethyl)amino]-2-hydroxypropoxy]benzonitrile

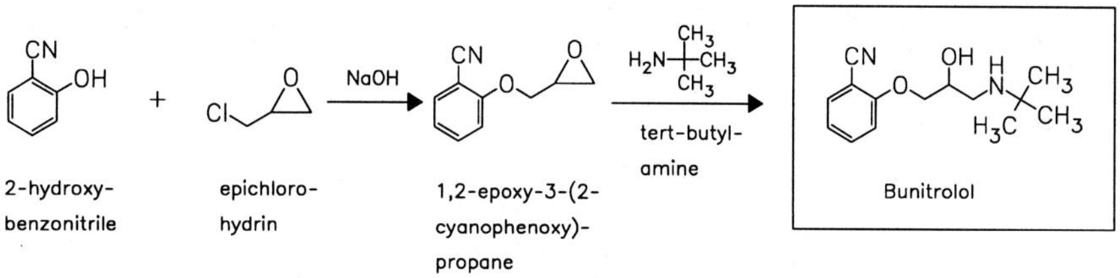

2-hydroxy-benzonitrile

epichlorohydrin

1,2-epoxy-3-(2-cyanophenoxy)-propane

tert-butyl-amine

Bunitrolol

Reference(s):
DAS 1 593 782 (Boehringer Ing.; appl. 15.6.1967).
US 3 541 130 (Boehringer Ing.; 17.11.1970; D-prior. 6.2.1967, 15.6.1967, 25.7.1967).
US 3 868 460 (Boehringer Ing.; 25.2.1975; appl. 23.10.1973; D-prior. 6.2.1967).

alternative synthesis:
DOS 2 503 222 (Boehringer Ing.; appl. 27.1.1975).

Formulation(s):   tabl. 5 mg (as hydrochloride)

Trade Name(s):
D: Stresson (Boehringer Ing.;   I: Betrilol (Boehringer Ing.);   J: Betrilol (Boehringer; 1983)
        1977); wfm   wfm

Buphenine

(Nylidrine)

ATC: C04AA02; G02CA02
Use: vasodilator, sympathomimetic

RN: 447-41-6 MF: $C_{19}H_{25}NO_2$ MW: 299.41 EINECS: 207-182-2
CN: 4-hydroxy-α-[1-[(1-methyl-3-phenylpropyl)amino]ethyl]benzenemethanol

hydrochloride
RN: 849-55-8 MF: $C_{19}H_{25}NO_2 \cdot HCl$ MW: 335.88 EINECS: 212-701-0
LD_{50}: 40 mg/kg (M, i.v.); 250 mg/kg (M, p.o.);
37.4 mg/kg (R, i.v.); >4800 mg/kg (R, p.o.)

a

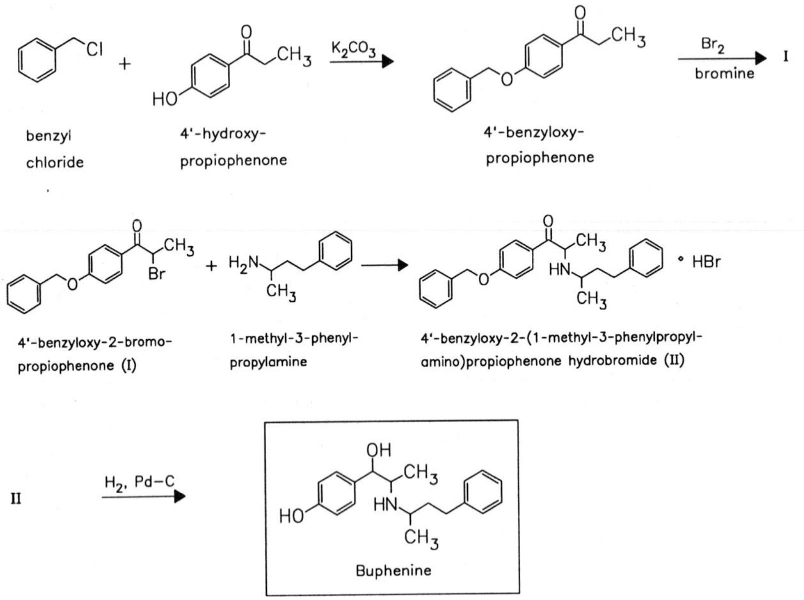
K2CO3
Br2
bromine
I
benzyl chloride
4'-hydroxy-propiophenone
4'-benzyloxy-propiophenone
4'-benzyloxy-2-bromo-propiophenone (I)
1-methyl-3-phenyl-propylamine
4'-benzyloxy-2-(1-methyl-3-phenylpropyl-amino)propiophenone hydrobromide (II)
II
H2, Pd-C
Buphenine

b

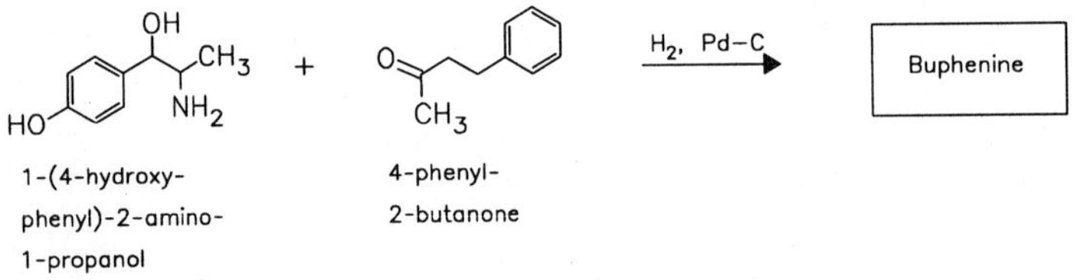
H2, Pd-C
Buphenine
1-(4-hydroxy-phenyl)-2-amino-1-propanol
4-phenyl-2-butanone

Reference(s):

a US 2 661 373 (F. Külz, C. Schöpf; 1953; prior. 1953).
DE 815 043 (Troponwerke; 1948).
DAS 1 182 245 (Philips; appl. 19.1.1962; NL-prior. 23.1.1961).

b US 2 661 372 (Troponwerke; 1953; prior. 1949).

Formulation(s): amp. 5 mg; drops 4 mg; tabl. 6 mg

Trade Name(s):

D: Apoplectal (Klinge)-comb.
opino N gel (biomo; as hydrochloride)

F: Ophtadil (Chauvin)-comb.

Phlébogel (Lipha Santé)-comb.	I:	Opino (Bayropharm)-comb.; wfm	USA: Adrin (Major); wfm Arlidin (USV); wfm

Bupheniode

ATC: C02
Use: antihypertensive, vasodilator

RN: 22103-14-6   MF: $C_{19}H_{23}I_2NO_2$ MW: 551.21   EINECS: 244-781-8
LD_{50}:   >600 mg/kg (M, i.p.); >2 g/kg (M, p.o.)
CN: 4-hydroxy-3,5-diiodo-α-[1-[(1-methyl-3-phenylpropyl)amino]ethyl]benzenemethanol

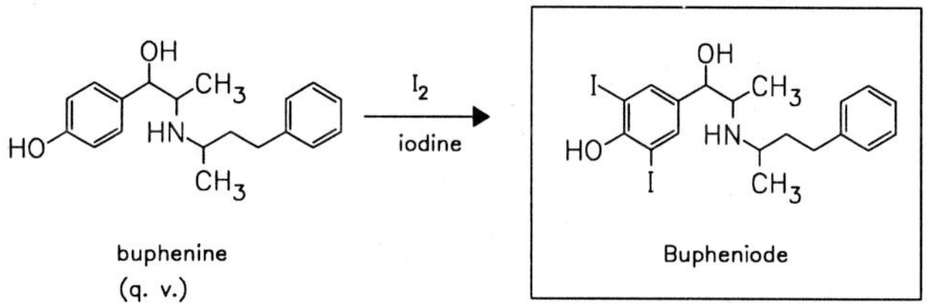

buphenine
(q. v.)

Bupheniode

Reference(s):
ZA 680 046 (Houdé; appl. 29.12.1967; F-prior. 10.1.1967, 21.12.1967).

Formulation(s):   amp. 4 mg, 6 mg; tabl. 4 mg, 6 mg

Trade Name(s):
F: Proclival (Houdé); wfm

Bupivacaine
(Marcain)

ATC: N01BB01
Use: local anesthetic

RN: 2180-92-9   MF: $C_{18}H_{28}N_2O$ MW: 288.44   EINECS: 218-553-3
LD_{50}:   7100 μg/kg (M, i.v.);
      5600 μg/kg (R, i.v.)
CN: 1-butyl-*N*-(2,6-dimethylphenyl)-2-piperidinecarboxamide

monohydrochloride
RN: 18010-40-7   MF: $C_{18}H_{28}N_2O \cdot HCl$ MW: 324.90   EINECS: 241-917-8

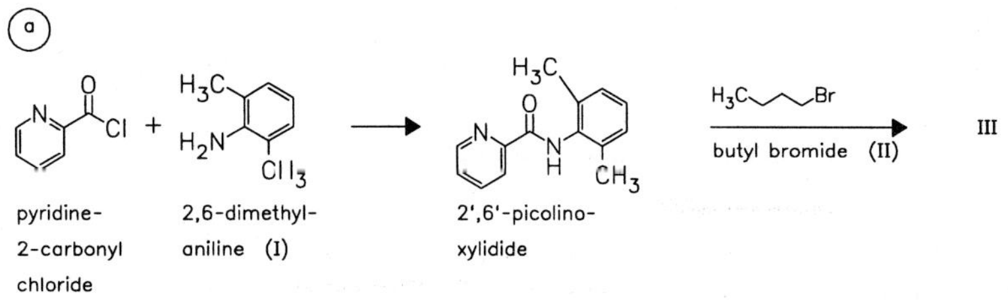

pyridine-
2-carbonyl
chloride

2,6-dimethyl-
aniline   (I)

2',6'-picolino-
xylidide

butyl bromide   (II)

III

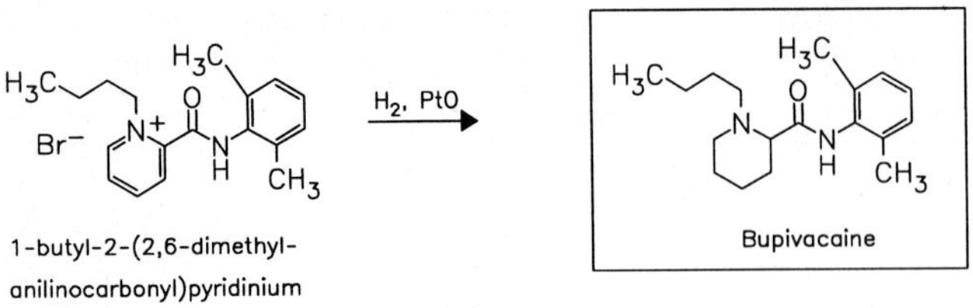

1-butyl-2-(2,6-dimethyl-
anilinocarbonyl)pyridinium
bromide (III)

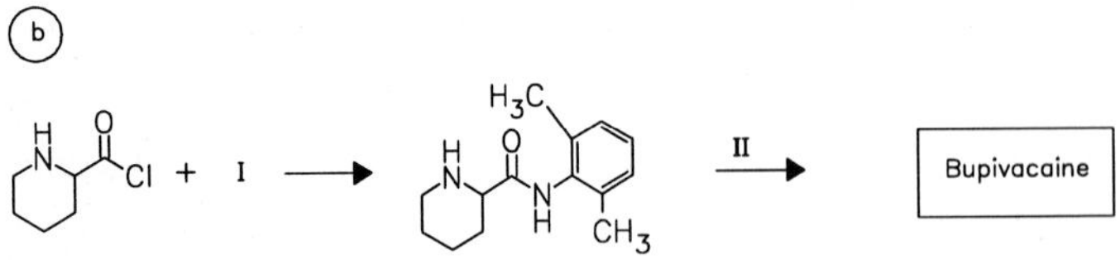

Bupivacaine

b)

pipecolinoyl
chloride

piperidine-2-carboxylic
acid (2,6-dimethyl-
anilide)

Bupivacaine

Reference(s):
DE 1 161 900 (AB Bofors; appl. 19.7.1955; S-prior. 6.4.1955).
DE 1 169 941 (AB Bofors; appl. 19.7.1955; S-prior. 28.4.1955).
GB 869 978 (AB Bofors; appl. 13.2.1959; S-prior. 13.3.1958).
Ekenstam, B. af et al.: Acta Chem. Scand. (ACHSE7) **11**, 1183 (1957).

alternative syntheses:
US 2 792 399 (AB Bofors; 1957; S-prior. 1954).
US 2 955 111 (AB Bofors; 1960; appl. 1957).

Formulation(s):   amp. 0.25 %, 0.5 %; inj. flask 0.25 %, 0.5 %, 0.75 %

Trade Name(s):
D:	Bupivacain (Rhône-Poulenc Rorer)	GB:	Marcain (Astra)		Marcaina iperbarica (Pierrel)
	Carbostesin (Astra)		Marcain with Adrenaline (Astra)-comb.	J:	Marcain (Yoshimoti-Takeda; as hydrochloride)
	Dolanaest (Strathmann)	I:	Bupiforan (Bieffe Medital)		
F:	Marcaïne (Astra)		Marcaina (Pierrel)	USA:	Sensocraine (Astra; as hydrochloride)
	Marcaïne adrénaline (Astra)-comb.		Marcaina adrenalina (Pierrel)-comb.		

Bupranolol

ATC: C07AA19
Use: beta blocking agent

RN: 14556-46-8   MF: $C_{14}H_{22}ClNO_2$ MW: 271.79
LD$_{50}$: 45 mg/kg (M, i.v.)
CN: 1-(2-chloro-5-methylphenoxy)-3-[(1,1-dimethylethyl)amino]-2-propanol

hydrochloride
RN: 15148-80-8   MF: $C_{14}H_{22}ClNO_2 \cdot HCl$ MW: 308.25   EINECS: 239-208-3
LD$_{50}$: 39.3 mg/kg (M, i.v.); 329 mg/kg (M, p.o.);
        15.3 mg/kg (R, i.v.); 518 mg/kg (R, p.o.);
        438 mg/kg (dog, p.o.)

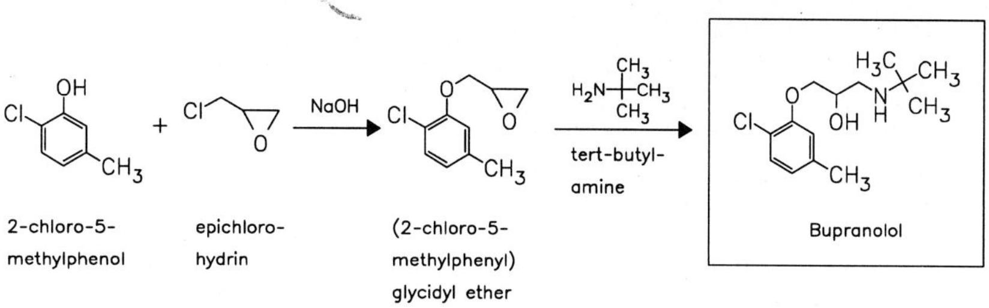

2-chloro-5-
methylphenol

epichloro-
hydrin

(2-chloro-5-
methylphenyl)
glycidyl ether

tert-butyl-
amine

Bupranolol

Reference(s):
DE 1 236 523 (Sanol-Arzneimittel; appl. 15.2.1962).
US 3 309 406 (Sanol; 14.3.1967; appl. 24.3.1965).

Formulation(s):   tabl. 50 mg, 100 mg, 200 mg

Trade Name(s):
D: Betadrenol (Schwarz)   J: Bupranolol Hydrochloride   Looser (Kaken; as
F: Bétadran (J. Logeais); wfm   (Shin Nihon Jitsugyo; as   hydrochloride)
I: Betadrenol (Schwarz)   hydrochloride)

Buprenorphine

ATC: N02AE01
Use: analgesic

RN: 52485-79-7   MF: $C_{29}H_{41}NO_4$ MW: 467.65   EINECS: 257-950-6
LD_{50}:   24 mg/kg (M, i.v.); 260 mg/kg (M, p.o.);
        31 mg/kg (R, i.v.)
CN: [5α,7α(S)]-17-(cyclopropylmethyl)-α-(1,1-dimethylethyl)-4,5-epoxy-18,19-dihydro-3-hydroxy-6-
        methoxy-α-methyl-6,14-ethenomorphinan-7-methanol

hydrochloride
RN: 53152-21-9   MF: $C_{29}H_{41}NO_4 \cdot HCl$ MW: 504.11   EINECS: 258-396-8

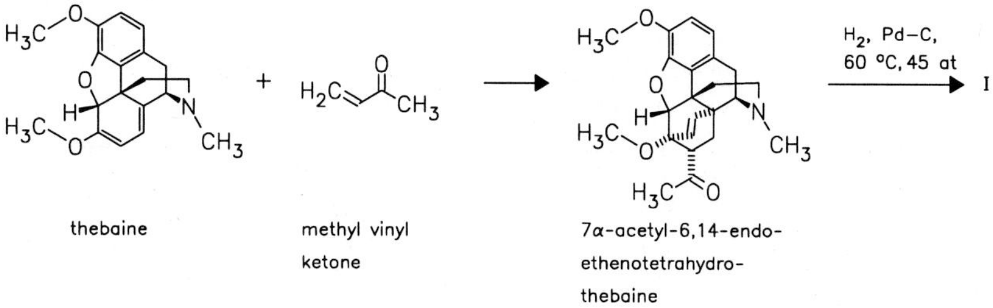

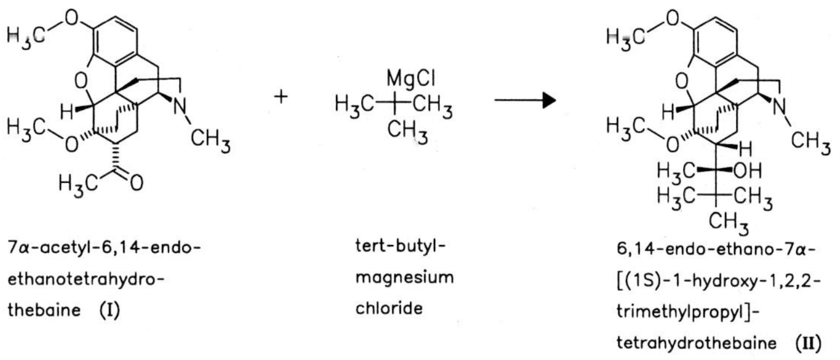

thebaine

methyl vinyl
ketone

7α-acetyl-6,14-endo-
ethenotetrahydro-
thebaine

H_2, Pd–C,
60 °C, 45 at   I

7α-acetyl-6,14-endo-
ethenotetrahydro-
thebaine **(I)**

tert-butyl-
magnesium
chloride

6,14-endo-ethano-7α-
[(1S)-1-hydroxy-1,2,2-
trimethylpropyl]-
tetrahydrothebaine **(II)**

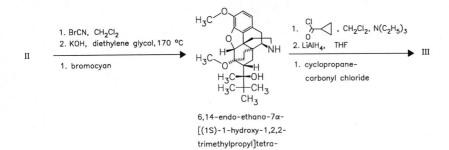

II →
1. BrCN, CH₂Cl₂
2. KOH, diethylene glycol, 170 °C

1. bromocyan

6,14-endo-ethano-7α-
[(1S)-1-hydroxy-1,2,2-
trimethylpropyl]tetra-
hydronorthebaine

1. cyclopropane-
carbonyl chloride

1. $\overset{Cl}{\underset{O}{\text{C}}}$◁ , CH₂Cl₂, N(C₂H₅)₃
2. LiAlH₄, THF → III

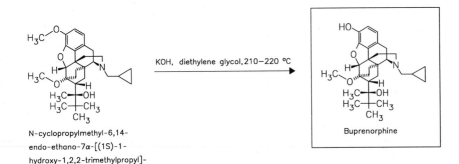

KOH, diethylene glycol, 210–220 °C →

N-cyclopropylmethyl-6,14-
endo-ethano-7α-[(1S)-1-
hydroxy-1,2,2-trimethylpropyl]-
tetrahydronorthebaine (III)

Buprenorphine

Reference(s):
DE 1 620 206 (Reckitt & Colman; appl. 15.6.1966; GB-prior. 15.6.1965).
US 3 433 791 (Reckitt & Sons Ltd; 18.3.1969; GB-prior. 15.6.1965).

formulation with naloxone:
EP 144 243 (Reckitt & Colman; appl. 5.12.1984; GB-prior. 6.12.1983).

Formulation(s):   amp. 0.3 mg/ml; sublingual tabl. 0.4 μg 200 μg (as hydrochloride)

Trade Name(s):
D: Temgesic (Roche; 1981)
F: Subutex (Schering-Plough)
      Temgésic (Schering-
      Plough; 1987)

GB: Temgesic (Reckitt &
       Colman; 1978)
I: Temgesic (Boehringer
       Mannh.)

J: Lepetan (Otsuka; 1984)
USA: Buprenex (Reckitt &
       Colman; 1985)

Buserelin

ATC: L02AE01
Use: synthetic nonapeptide agonist analog
        of gonadorelin (LH-RH), gonad
        stimulating principle for treatment of
        hormone sensitive prostatic
        carcinoma and endometriosis

RN: 57982-77-1   MF: C₆₀H₈₆N₁₆O₁₃   MW: 1239.45   EINECS: 261-061-9
CN: 6-[O-(1,1-dimethylethyl)-D-serine]-9-(N-ethyl-L-prolinamide)-10-deglycinamideluteinizing hormone-
       releasing factor (pig)

monoacetate
RN: 68630-75-1   MF: C₆₀H₈₆N₁₆O₁₃ · C₂H₄O₂   MW: 1299.50
LD₅₀: 56 mg/kg (M, i.v.); >1 g/kg (M, p.o.);
       36 mg/kg (R, i.v.); >400 mg/kg (R, p.o.)

diacetate

RN: 59179-42-9   MF: $C_{60}H_{86}N_{16}O_{13} \cdot 2C_2H_4O_2$ MW: 1359.55

5-oxo-Pro[1]	His[2]	Trp[3]	Ser[4]	Tyr[5]	D-Ser[6]	Leu[7]	Arg[8]	Pro[9]
							Z—ONSu H—	NHEt
							Z—	NHEt
					But / Z—OTcp H—			NHEt
			Z—OBt H—OH	Bzl / Z—	But			NHEt
			Z—	Bzl / —OH H—	But			NHEt
		—N$_2$H$_3$ Z—		Bzl	But			NHEt
		—N$_3$ H—			But			NHEt
					But			NHEt

abbreviations:

OBt:	3-hydroxy-4-oxo-3,4-dihydro-1,2,3-benzotriazinyl ester
Z:	benzyloxycarbonyl
Bzl:	benzyl ether
OTcp:	2,4,5-trichlorophenyl ester
ONSu:	N-hydroxysuccinimidyl ester
N$_3$:	azide
N$_2$H$_3$:	hydrazide
Bu^t:	tert-butyl ether

Buserelin

Reference(s):
DE 2 438 350 (Hoechst; appl. 9.8.1974).
US 4 024 248 (Hoechst; 17.5.1977; D-prior. 9.8.1974).

alternative synthetic methods:
DE 2 905 502 (Hoechst; appl. 14.2.1979).

parenteral depot formulations:
1) *microcapsules with* poly-D-(–)-3-hydroxybutyric acid *as carrier:*
DE 3 428 372 (Hoechst; appl. 1.8.1984).
EP 172 422 (Hoechst; appl. 20.7.1985; D-prior. 1.8.1984).
EP 262 583 (Hoechst; appl. 24.9.1987; D-prior. 2.10.1986, 13.12.1986).

2) *with biodegradable poly(hydroxyalkyl)aminodicarboxylic acid derivatives:*
EP 274 127 (Hoechst; appl. 29.12.1987; D-prior. 3.1.1987).

medical use as contraceptive:
DOS 2 735 515 (Hoechst; appl. 6.8.1977).
EP 764 (Hoechst; appl. 1.8.1978; D-prior. 6.8.1977).

Formulation(s): nasal spray 10 mg/10 ml; sol. for s. c. amp. 15 mg/10 g; sol. 5.5 mg/5.5 ml for s. c. inj. with 6.6 mg buserelin acetate on polyglycolide matrix

Trade Name(s):

D:	Profact (Hoechst; 1984)		Suprefact (Hoechst Houdé;	I:	Suprefact (Hoechst Italia)
	Suprecur (Hoechst)		1986)	J:	Suprecur (Hoechst Japan)
F:	Bigonist (Cassenne)	GB:	Suprecur (Shire)		

Buspirone

ATC: N05BE01
Use: tranquilizer

RN: 36505-84-7 MF: $C_{21}H_{31}N_5O_2$ MW: 385.51 EINECS: 253-072-2
LD$_{50}$: 136 mg/kg (R, i.p.)
CN: 8-[4-[4-(2-pyrimidinyl)-1-piperazinyl]butyl]-8-azaspiro[4.5]decane-7,9-dione

monohydrochloride
RN: 33386-08-2 MF: $C_{21}H_{31}N_5O_2 \cdot HCl$ MW: 421.97 EINECS: 251-489-4
LD$_{50}$: 655 mg/kg (M, p.o.);
196 mg/kg (R, p.o.);
586 mg/kg (dog, p.o.)

1-(2-pyrimidyl)-piperazine + 4-chloro-butyronitrile →(Na$_2$CO$_3$) 4-(2-pyrimidyl)-1-(3-cyanopropyl)piperazine →(H$_2$, Raney–Ni) I

1-(2-pyrimidyl)-4-(4-aminobutyl)piperazine (I) + 8-oxaspiro[4.5]-decane-7,9-dione → Buspirone

Reference(s):
DOS 2 057 845 (Bristol-Myers; appl. 24.11.1970; USA-prior. 24.11.1969).
US 3 976 776 (Mead Johnson; 24.8.1976; prior. 24.11.1969).
Wu, Y.H. et al.: J. Med. Chem. (JMCMAR) **15**, 477 (1972).
US 3 907 801 (Mead Johnson; 23.9.1975; prior. 24.11.1969).
US 3 717 634 (Mead Johnson; 20.2.1973; prior. 24.11.1969).

Formulation(s): tabl. 5 mg, 10 mg

Trade Name(s):

D:	Bespar (Bristol-Myers; 1985)	F:	Buspar (Bristol-Myers Squibb; 1988)	GB:	Buspar (Bristol-Myers Squibb; 1987)

I: Axoren (Glaxo Wellcome)   Buspimen (Menarini)   USA: Buspar (Bristol-Myers
      Buspar (Bristol It. Sud)   Squibb; 1986)

Busulfan

ATC: L01AB01
Use: antineoplastic

RN: 55-98-1   MF: $C_6H_{14}O_6S_2$ MW: 246.30   EINECS: 200-250-2
LD_{50}:   110 mg/kg (M, p.o.);
      1800 μg/kg (R, i.v.)
CN: 1,4-butanediol dimethanesulfonate

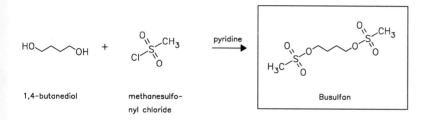

1,4-butanediol   methanesulfo-   Busulfan
                      nyl chloride

Reference(s):
GB 700 677 (Wellcome Found.; appl. 1950).
US 2 917 432 (Burroughs Wellcome; 15.12.1959; prior. 5.10.1954).

Formulation(s):   tabl. 0.5 mg, 2 mg

Trade Name(s):
D: Myleran (Glaxo Wellcome)   GB: Myleran (Glaxo Wellcome)   J: Mablin (Takeda)
F: Misulban (Techni-Pharma);   I: Misulban (Nuovo ISM)   USA: Myleran (Glaxo Wellcome)
      wfm   Myleran (Wellcome)

Butacaine

ATC: D04AB
Use: local anesthetic

RN: 149-16-6   MF: $C_{18}H_{30}N_2O_2$ MW: 306.45   EINECS: 205-734-7
CN: 3-(dibutylamino)-1-propanol 4-aminobenzoate (ester)

monohydrochloride
RN: 5892-15-9   MF: $C_{18}H_{30}N_2O_2 \cdot HCl$ MW: 342.91   EINECS: 227-568-4
LD_{50}:   21 mg/kg (M, i.v.)
sulfate (2:1)
RN: 149-15-5   MF: $C_{18}H_{30}N_2O_2 \cdot 1/2H_2SO_4$ MW: 710.98   EINECS: 205-733-1
LD_{50}:   12 mg/kg (M, i.v.)

4-nitro-   3-dibutylamino-   (I)
benzoyl   1-propanol
chloride

I →(H₂, Raney-Ni)

Butacaine

Reference(s):
US 1 358 751 (Abbott; 1920; appl. 1920).
US 1 676 470 (Abbott; 1928; GB-prior. 1921).

preparation of 3-dibutylamino-1-propanol *from* allylalcohol *and* dibutylamine:
US 2 437 984 (Abbott; 1948; appl. 1945).

butacaine-pamoate:
DAS 2 401 605 (Rocador; appl. 14.1.1974; E-prior. 18.1.1973).

Formulation(s): cps. 50 mg

Trade Name(s):
F: Relaxoddi (Leurquin)-
   comb.
USA: Butyn Metaphen (Abbott);
   wfm
Butyn Sulfate (Abbott);
   wfm

Butalamine

ATC: C04AX23
Use: vasodilator

RN: 22131-35-7 MF: C₁₈H₂₈N₄O MW: 316.45 EINECS: 244-794-9
CN: *N,N*-dibutyl-*N'*-(3-phenyl-1,2,4-oxadiazol-5-yl)-1,2-ethanediamine

hydrochloride
RN: 28875-47-0 MF: C₁₈H₂₈N₄O · xHCl MW: unspecified EINECS: 249-279-2

5-amino-3-phenyl-
1,2,4-oxadiazole
(cf. imolamine
synthesis)

1. NaNH₂
2. (2-dibutylamino-ethyl chloride)

1. sodium amide
2. 2-dibutylamino-
   ethyl chloride

Butalamine

Reference(s):
DAS 1 445 409 (J.M.D. Aron-Samuel, J.J. Sterne; appl. 6.7.1962; GB-prior. 11.7.1961, 12.6.1962).
US 3 338 899 (Aron-Samuel; 29.8.1967; prior. 9.7.1962).

Formulation(s): f. c. tabl. 40 mg, 80 mg

Trade Name(s):
D: Adrevil (Novartis)
F: Oxadiléne (Leurquin)-
   comb.
I: Surheme (Aron)
   Surheme (Lipha); wfm
   Surheme (Spemsa); wfm

Butalbital

(Allylbarbituric acid)

ATC: N05C
Use: sedative

RN: 77-26-9   MF: $C_{11}H_{16}N_2O_3$ MW: 224.26   EINECS: 201-017-8
LD$_{50}$: 160 mg/kg (R, s.c.)
CN: 5-(2-methylpropyl)-5-(2-propenyl)-2,4,6(1H,3H,5H)-pyrimidinetrione

isobutyl   diethyl   diethyl isobutyl-
bromide   malonate   malonate

5-isobutyl-   allyl   Butalbital
barbituric acid (I)   bromide

Reference(s):
Volwiler, E.H.: J. Am. Chem. Soc. (JACSAT) **47**, 2236 (1925).

Formulation(s):   f. c. tabl. 300 mg

Trade Name(s):

D:	Aequiton (Südmedica)-comb.	Bupap (FCR)	Pacaps (Lunsco)
	Optalidon (Sandoz)-comb.	Esgic (Forest)-comb.	Phrenilin (Carnrick)
F:	Optalidon (Sandoz)-comb.	Fioricet (Novartis)	Repan (Everett)
I:	Optalidon (Sandoz)-comb.	Fiorinal (Novartis)-comb.	Sedapap (Merz)
USA:	Anolor (Blansett)	Fiortal w/Codeine (Genera)	Tenake (Seatrace)
	Axocet (Savage)	Medigesic (U.S. Pharmaceutical)	

Butamirate

ATC: R05DB13
Use: antitussive

RN: 18109-80-3   MF: $C_{18}H_{29}NO_3$ MW: 307.43   EINECS: 242-005-2
CN: α-ethylbenzeneacetic acid 2-[2-(diethylamino)ethoxy]ethyl ester

citrate (1:1)
RN: 18109-81-4   MF: $C_{18}H_{29}NO_3 \cdot C_6H_8O_7$ MW: 499.56   EINECS: 242-006-8
LD$_{50}$: 47.2 mg/kg (M, i.v.); 865 mg/kg (M, p.o.);
37.2 mg/kg (R, i.v.); 4164 mg/kg (R, p.o.)

2-phenyl-   2-(2-diethylamino-   Butamirate
butyryl chloride   ethoxy)ethanol

Reference(s):
DE 1 151 515 (Hommel AG; appl. 9.3.1960; CH-prior. 12.3.1959).
US 3 349 114 (Hommel AG; 24.10.1967; appl. 17.5.1963).

Formulation(s):   drops 30 mg; syrup 1.772 mg; suppos. 20 mg; syrup 10.65 mg

Trade Name(s):
D: Pertix-Hommel (Hommel)   Sinecod (Zyma)   Sinecod (Zyma)
      Sinecod (Karlspharma)   I: Butiran (Ecobi)

Butanilicaine

ATC: N01BB05
Use: local anesthetic

RN: 3785-21-5   MF: $C_{13}H_{19}ClN_2O$ MW: 254.76
CN: 2-(butylamino)-N-(2-chloro-6-methylphenyl)acetamide

monohydrochloride
RN: 6027-28-7   MF: $C_{13}H_{19}ClN_2O \cdot HCl$ MW: 291.22   EINECS: 227-893-1
LD_{50}:   30 mg/kg (M, i.v.)
phosphate (1:1)
RN: 2081-65-4   MF: $C_{13}H_{19}ClN_2O \cdot H_3PO_4$ MW: 352.76   EINECS: 218-211-3

chloroacetyl   2-chloro-6-   2-chloro-N-(2-   Butanilicaine
chloride   methylaniline   chloro-6-methyl-
                                   phenyl)acetamide

Reference(s):
DE 939 633 (Hoechst; 1953).
DE 1 005 075 (Hoechst; 1952).

process variant:
DE 1 009 633 (Hoechst; 1953).

Formulation(s):   amp. 51 mg/1.7 ml; vial 1 % sol.

Trade Name(s):
D: Hostacain (Hoechst); wfm   J: Hostacain (Hoechst); wfm

Butaperazine

ATC: N05AB09
Use: neuroleptic

RN: 653-03-2   MF: $C_{24}H_{31}N_3OS$ MW: 409.60   EINECS: 211-493-9
LD_{50}:   67 mg/kg (M, i.v.);
        413 mg/kg (R, p.o.)
CN: 1-[10-[3-(4-methyl-1-piperazinyl)propyl]-10H-phenothiazin-2-yl]-1-butanone

diphosphate
RN: 7389-45-9   MF: $C_{24}H_{31}N_3OS \cdot 2H_3PO_4$ MW: 605.59   EINECS: 230-972-3

dimaleate
RN: 1063-55-4 MF: $C_{24}H_{31}N_3OS \cdot 2C_4H_4O_4$ MW: 641.74 EINECS: 213-900-5
LD$_{50}$: 17.6 mg/kg (M, i.v.); 296 mg/kg (M, p.o.);
63 mg/kg (R, i.v.); 264 mg/kg (R, p.o.);
>50.7 mg/kg (dog, i.v.)

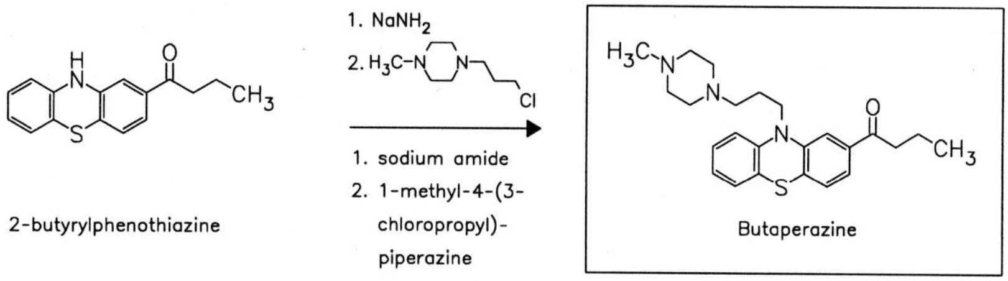

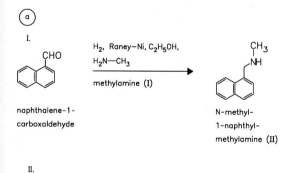

2-butyrylphenothiazine

1. NaNH$_2$
2. H$_3$C-N N-CH$_2$CH$_2$CH$_2$-Cl

1. sodium amide
2. 1-methyl-4-(3-
chloropropyl)-
piperazine

Butaperazine

Reference(s):
DE 1 120 451 (Bayer; appl. 30.5.1956).
US 2 985 654 (Schering Corp.; 23.5.1961; appl. 21.9.1956).

Formulation(s): tabl. 0.1 mg, 0.25 mg, 0.5 mg

Trade Name(s):
D: Östrogynal (Asche)-comb.; USA: Repoise (Robins); wfm
wfm

Butenafine

Use: antifungal for topical use

RN: 101828-21-1 MF: $C_{23}H_{27}N$ MW: 317.48
CN: *N*-[[4-(1,1-Dimethylethyl)phenyl]methyl]-*N*-methyl-1-naphthalenemethanamine

hydrochloride
RN: 101827-46-7 MF: $C_{23}H_{27}N \cdot HCl$ MW: 353.94

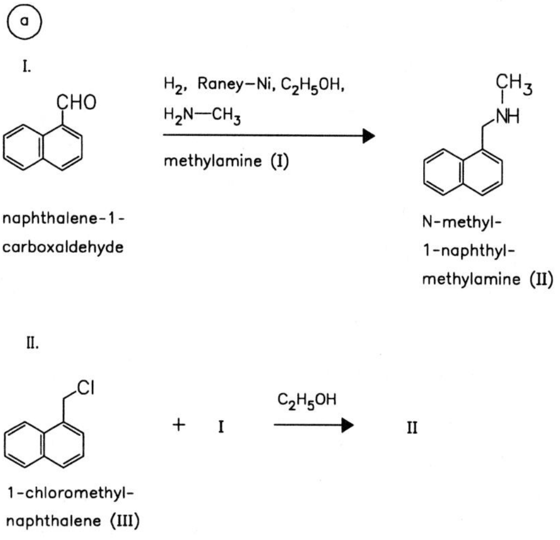

(a)

I.

naphthalene-1-
carboxaldehyde

H$_2$, Raney–Ni, C$_2$H$_5$OH,
H$_2$N—CH$_3$

methylamine (I)

N-methyl-
1-naphthyl-
methylamine (II)

II.

1-chloromethyl-
naphthalene (III)

+ I $\xrightarrow{C_2H_5OH}$ II

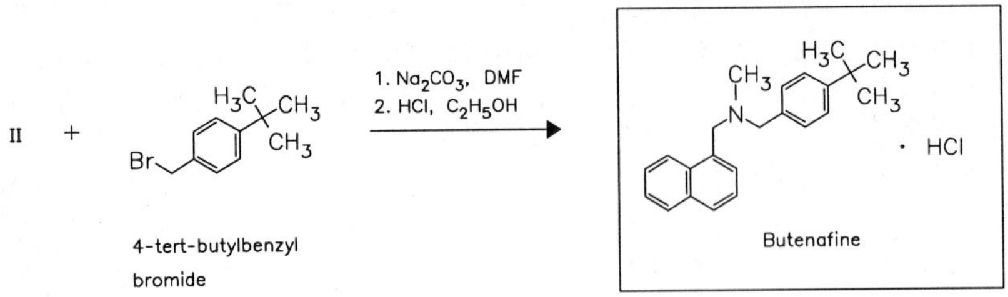

II +

4-tert-butylbenzyl
bromide

1. Na₂CO₃, DMF
2. HCl, C₂H₅OH

Butenafine · HCl

b

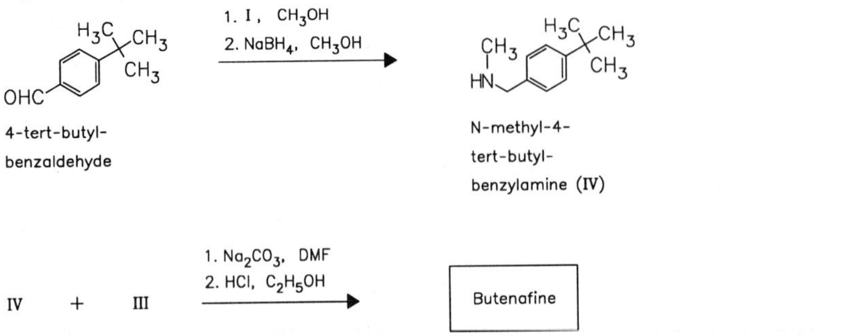

4-tert-butyl-
benzaldehyde

1. I, CH₃OH
2. NaBH₄, CH₃OH

N-methyl-4-
tert-butyl-
benzylamine (IV)

IV + III

1. Na₂CO₃, DMF
2. HCl, C₂H₅OH

Butenafine

c

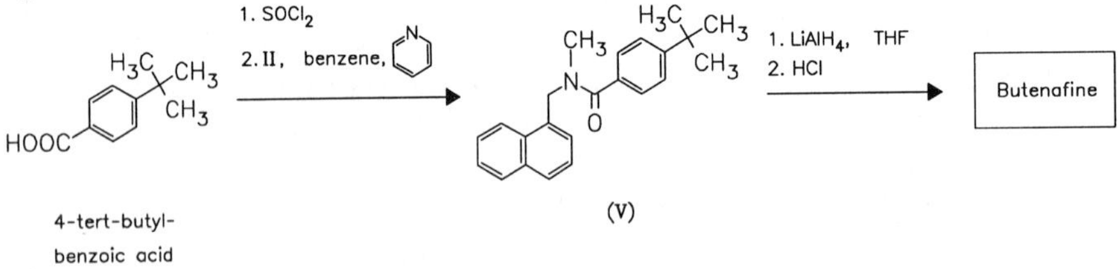

4-tert-butyl-
benzoic acid

1. SOCl₂
2. II, benzene,

(V)

1. LiAlH₄, THF
2. HCl

Butenafine

d

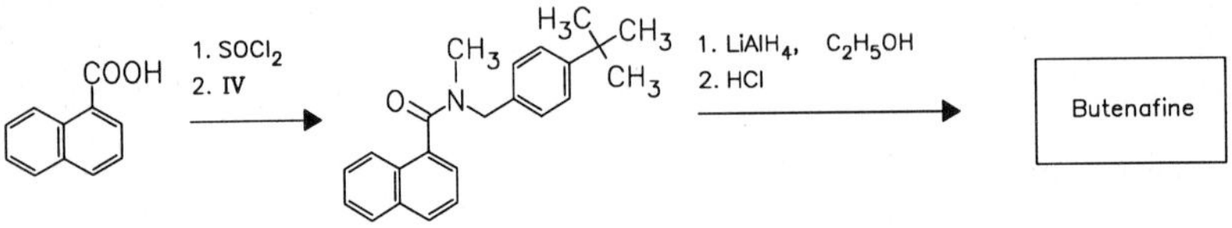

1. SOCl₂
2. IV

1. LiAlH₄, C₂H₅OH
2. HCl

Butenafine

Reference(s):
a   EP 221.781 (Mitsui Toatsu Chem.; appl. 31.10.1986; J-prior. 1.11.1985).
b   JP 03 200 747 (Kokai Tokkyo Koho; appl. 28.12.1989; J-prior. 2.9.1991).
c,d EP 164 697 (Kaken Pharmaceutical Co.; appl. 6.6.1985; J-prior. 9.6.1984).

preparation of N-methyl-1-naphthylmethylamine:
Dalm, Zoller: Helv. Chim. Acta (HCACAV) **35** 1348, 1353 (1952).
Elslager, E.F; Johnson, J.L.; Werbel, L.M.: J. Med. Chem. (JMCMAR) **24** (2), 140 (1981).
Baltzly, I.: J. Am. Chem. Soc. (JACSAT) **65** 1984 (1943)
Lutz et al.: J. Org. Chem. (JOCEAH) **12** 760 (1947)

Formulation(s):   cream 1%; sol. 1% (als hydrochloride)

Trade Name(s):
J: Mentax (Kaken; 1992)   Volley (Hisamitsu)   USA: Mentax (Bertek Pharms.)

Butetamate
(Butethamate)

ATC: S01FA
Use: antispasmodic

RN: 14007-64-8   MF: $C_{16}H_{25}NO_2$ MW: 263.38   EINECS: 237-817-9
CN: α-ethylbenzeneacetic acid 2-(diethylamino)ethyl ester

citrate
RN: 13900-12-4   MF: $C_{16}H_{25}NO_2 \cdot xC_6H_8O_7$ MW: unspecified   EINECS: 237-671-6

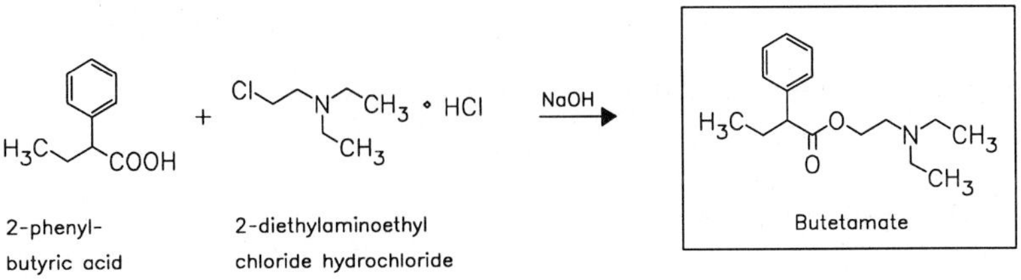

2-phenyl-   2-diethylaminoethyl
butyric acid   chloride hydrochloride

Butetamate

Reference(s):
CH 291 375 (Hommel; appl. 1950).
Engelhardt, A.: Arzneim.-Forsch. (ARZNAD) **11**, 217 (1957).

Formulation(s):   sol. 14.5 mg/5 ml

Trade Name(s):
D: Baldicap (Giulini)-comb.;   numerous combination   I: Pertix (Bonomelli Farm.)-
      wfm   preparations; wfm   comb.; wfm
      Pertix-Hommel Liquidum   GB: Cam (Rybar); wfm
      (Hommel); wfm

Butethamine

ATC: N01B
Use: local anesthetic

RN: 2090-89-3   MF: $C_{13}H_{20}N_2O_2$ MW: 236.32
CN: 2-[(2-methylpropyl)amino]ethanol 4-aminobenzoate (ester)

monohydrochloride
RN: 553-68-4   MF: $C_{13}H_{20}N_2O_2 \cdot HCl$ MW: 272.78
LD_{50}:   36 mg/kg (M, i.v.);
      28 mg/kg (R, i.v.)

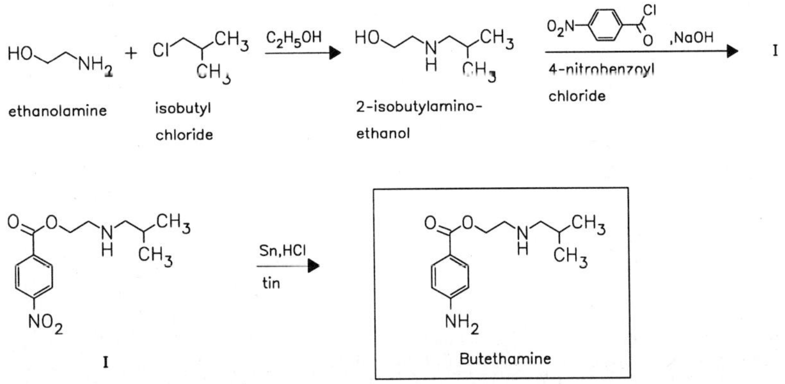

ethanolamine   isobutyl   2-isobutylamino-   4-nitrobenzoyl
                  chloride   ethanol   chloride

Butethamine

Reference(s):
US 2 139 818 (Novocol Chem.; 1938; prior. 1935).

Formulation(s): amp.

Trade Name(s):
USA: Dentocaine (Amer. Chem.); wfm
Monocaine formate (Novocol); wfm
Monocaine hydrochloride (Philadelphia Labs.); wfm

Butibufen

ATC: M02A
Use: non-steroidal anti-inflammatory

RN: 55837-18-8 MF: $C_{14}H_{20}O_2$ MW: 220.31 EINECS: 259-849-2
LD_{50}: 810 mg/kg (M, p.o.);
1600 mg/kg (R, p.o.)
CN: α-ethyl-4-(2-methylpropyl)benzeneacetic acid

sodium salt
RN: 60682-24-8 MF: $C_{14}H_{19}NaO_2$ MW: 242.29 EINECS: 262-374-3

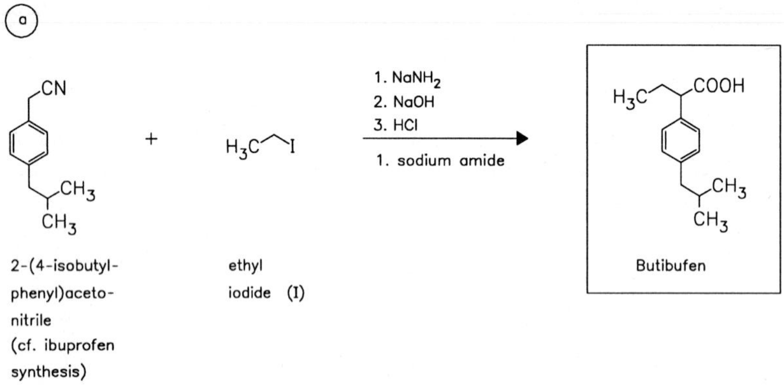
a
CN
CH3
CH3
+
H3C I
1. NaNH2
2. NaOH
3. HCl
1. sodium amide
H3C
COOH
CH3
CH3
2-(4-isobutyl-phenyl)aceto-nitrile (cf. ibuprofen synthesis)
ethyl iodide (I)
Butibufen

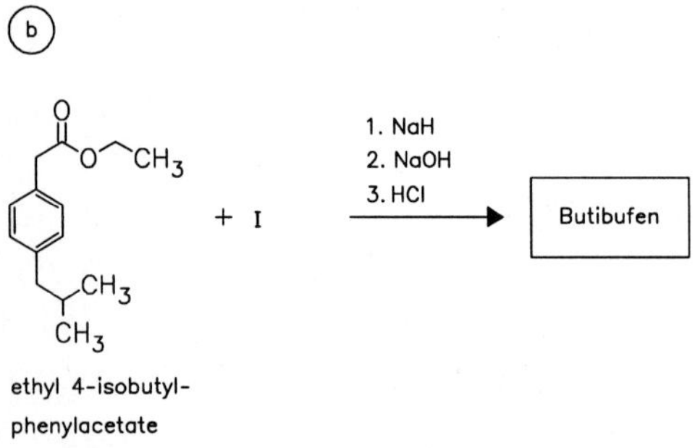
b
O
O
CH3
CH3
CH3
+ I
1. NaH
2. NaOH
3. HCl
Butibufen
ethyl 4-isobutyl-phenylacetate

Reference(s):
DE 2 505 813 (Juste; appl. 12.2.1975).
CH 573 891 (Juste; appl. 16.6.1975).
US 4 031 243 (Juste; 21.6.1977; appl. 25.2.1975).

alternative synthesis:
EP 184 573 (Sanofi; appl. 28.11.1985; F-prior. 29.11.1985).

Formulation(s): cps. 350 mg; sachets 500 mg; suppos. 500 mg; tabl. 500 mg

Trade Name(s):
E: Mijal (Juste; 1992)

Butizide
(Buthiazide; Thiabutazide)

ATC: C03E
Use: diuretic, antihypertensive

RN: 2043-38-1 MF: $C_{11}H_{16}ClN_3O_4S_2$ MW: 353.85 EINECS: 218-048-8
CN: 6-chloro-3,4-dihydro-3-(2-methylpropyl)-2*H*-1,2,4-benzothiadiazine-7-sulfonamide 1,1-dioxide

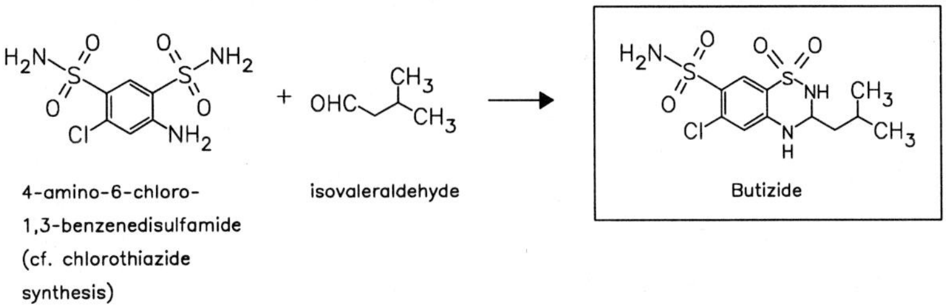

4-amino-6-chloro-1,3-benzenedisulfamide (cf. chlorothiazide synthesis)
isovaleraldehyde
Butizide

Reference(s):
GB 861 367 (Ciba; appl. 2.3.1959; USA-prior. 9.4.1958, 9.6.1958, 29.7.1958, 29.9.1958).
Werner, I.H. et al.: J. Am. Chem. Soc. (JACSAT) **82**, 1161 (1960).

Formulation(s): cps. 50 mg; drg. 50 mg

Trade Name(s):
D: Aldactone 50-Saltucin (Boehringer Mannh.)-comb.
Modenol (Boehringer Mannh.)-comb.
Saltucin (Boehringer Mannh.)
Torrat (Boehringer Mannh.)-comb.
Tri-Torrat (Boehringer Mannh.)-comb.
F: Eunéphran (Servier); wfm
I: Kadiur (Boots Italia)-comb.
Saludopin (SIT)-comb.

Butoconazole

ATC: G01AF15
Use: topical antifungal

RN: 64872-76-0 MF: $C_{19}H_{17}Cl_3N_2S$ MW: 411.78
LD_{50}: >1600 mg/kg (M, i.p.); >3200 mg/kg (M, p.o.);
940 mg/kg (R, i.p.)
CN: (±)-1-[4-(4-chlorophenyl)-2-[(2,6-dichlorophenyl)thio]butyl]-1*H*-imidazole

mononitrate
RN: 64872-77-1 MF: $C_{19}H_{17}Cl_3N_2S \cdot HNO_3$ MW: 474.80
LD_{50}: >3200 mg/kg (M, p.o.);
1720 mg/kg (R, p.o.)

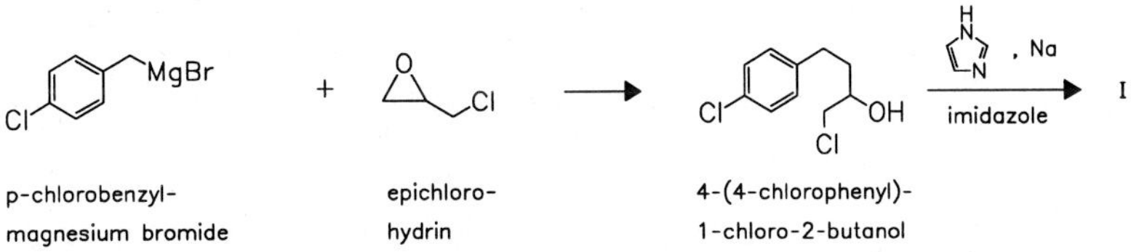

MgBr
Cl
OH
imidazole
Na
I
p-chlorobenzyl-magnesium bromide
epichlorohydrin
4-(4-chlorophenyl)-1-chloro-2-butanol

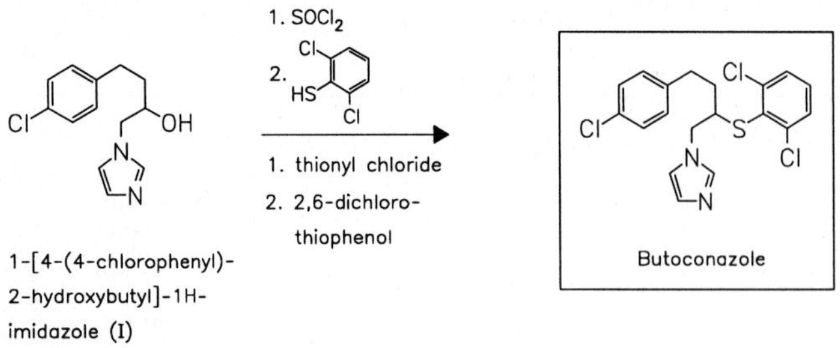

1-[4-(4-chlorophenyl)-
2-hydroxybutyl]-1H-
imidazole (I)

1. SOCl₂

2. [structure: Cl, HS, Cl]

1. thionyl chloride
2. 2,6-dichloro-
   thiophenol

Butoconazole

Reference(s):

Walker, K.A.M. et al.: J. Med. Chem. (JMCMAR) **21**, 840 (1978).

US 4 078 071 (Syntex; USA-prior. 28.7.1975).

DOS 2 800 755

Formulation(s):   vaginal cream 2 %

Trade Name(s):

F: Gynomyk (Cassenne)   USA: Femstat (Syntex; 1986);
                                   wfm

Butoctamide

Use: hypnotic, antineoplastic

RN: 32838-26-9   MF: C₁₂H₂₅NO₂   MW: 215.34

LD₅₀: 476 mg/kg (M, i.p.); 2000 mg/kg (M, p.o.)

CN: N-(2-ethylhexyl)-3-hydroxybutanamide

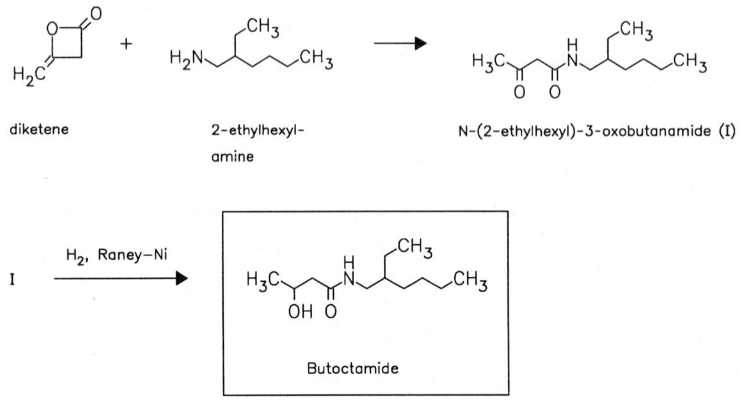

[reaction scheme]

diketene   +   2-ethylhexyl-
                  amine
                         →   N-(2-ethylhexyl)-3-oxobutanamide (I)

I   →[H₂, Raney–Ni]   Butoctamide

Reference(s):

DOS 1 768 445 (Lion Hamigaki; appl. 15.5.1968; J-prior. 15.5.1967).

US 3 639 457 (A. Sakuma et al.; 1.2.1972; J-prior. 15.5.1967).

Formulation(s):   600 mg

Trade Name(s):

J: Listomine (Lion; as
      hemisuccinate)

Butofilolol

Use: β-adrenoceptor blocker

RN: 64552-17-6 MF: $C_{17}H_{26}FNO_3$ MW: 311.40
CN: (±)-1-[2-[3-[(1,1-dimethylethyl)amino]-2-hydroxypropoxy]-5-fluorophenyl]-1-butanone

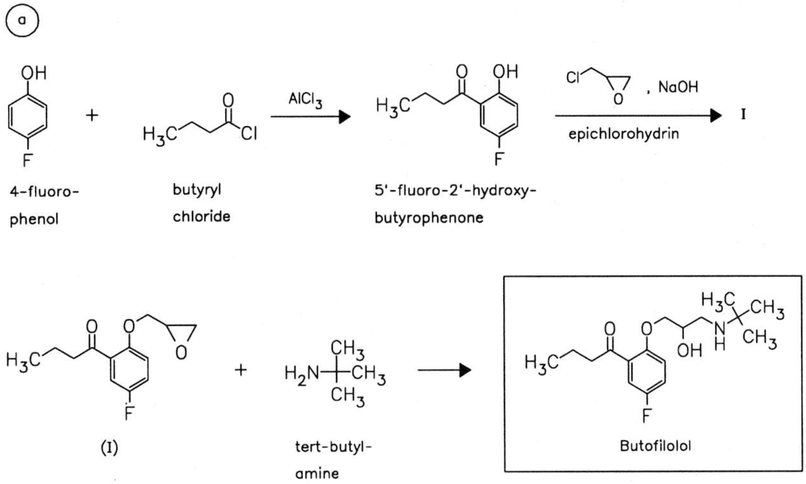
a
4-fluoro-phenol
butyryl chloride
AlCl3
5'-fluoro-2'-hydroxy-butyrophenone
NaOH
epichlorohydrin
I
(I)
tert-butyl-amine
Butofilolol

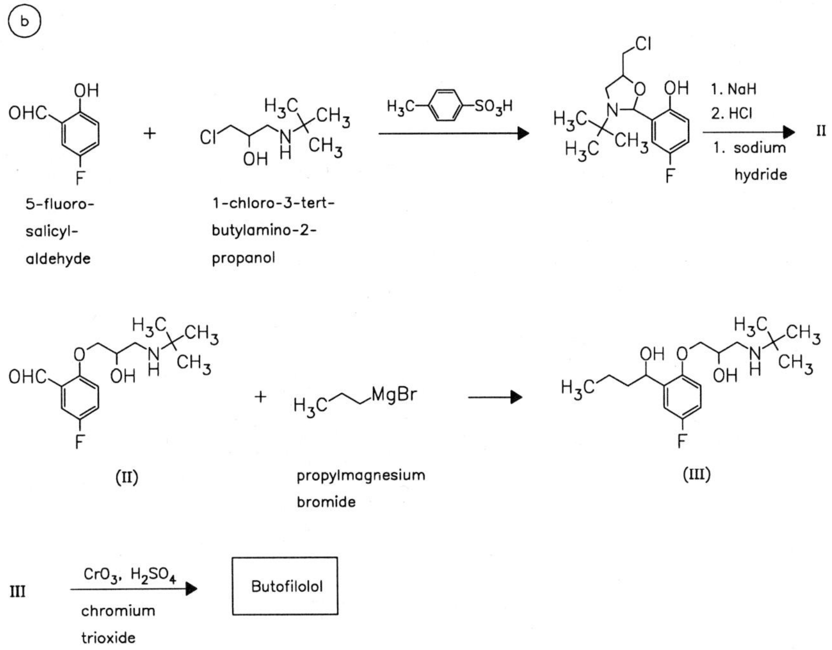
b
5-fluoro-salicyl-aldehyde
1-chloro-3-tert-butylamino-2-propanol
1. NaH
2. HCl
1. sodium hydride
II
(II)
propylmagnesium bromide
(III)
III
CrO3, H2SO4
chromium trioxide
Butofilolol

Reference(s):
DOS 2 528 147 (CM Industries; appl. 24.6.1975; GB-prior. 28.6.1974).

Trade Name(s):
F: Cafide (Lab. Labaz); wfm

Butorphanol

ATC: N02AF01
Use: analgesic

RN: 42408-82-2 MF: $C_{21}H_{29}NO_2$ MW: 327.47 EINECS: 255-808-8
CN: 17-(cyclobutylmethyl)morphinan-3,14-diol

tartrate (1:1)
RN: 58786-99-5 MF: $C_{21}H_{29}NO_2 \cdot C_4H_6O_6$ MW: 477.55 EINECS: 261-443-5
LD_{50}: 36 mg/kg (M, i.v.); 395 mg/kg (M, p.o.);
17 mg/kg (R, i.v.); 315 mg/kg (R, p.o.);
10 mg/kg (dog, i.v.); >50 mg/kg (dog, p.o.)

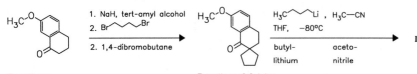

1. NaH, tert-amyl alcohol
2. 1,4-dibromobutane
7-methoxy-tetralone
7-methoxy-2,2-tetra-methylenetetralone
THF, −80°C
butyl-lithium
aceto-nitrile
I
LiAlH4, THF
HCl, (C2H5)2O
pyridine, CH2Cl2
ethyl chloroformate
II
(I)
3-chloroper-benzoic acid
(II)
(III)
III
NaH, tert-amyl alcohol
POCl3, pyridine
N-ethoxycarbonyl-14-hydroxy-3-methoxyisomorphinan
N-ethoxycarbonyl-3-methoxy-8,14-didehydromorphinan (IV)
IV
KOH
pyridine
cyclobutane-carbonyl chloride
3-methoxy-8,14-didehydromorphinan
(V)

(VI)

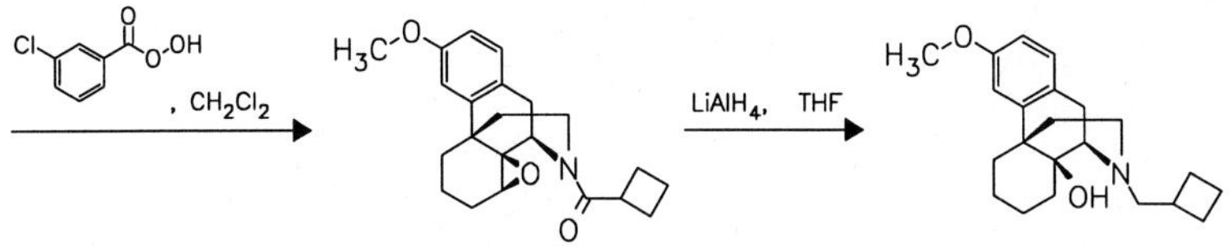

Butorphanol

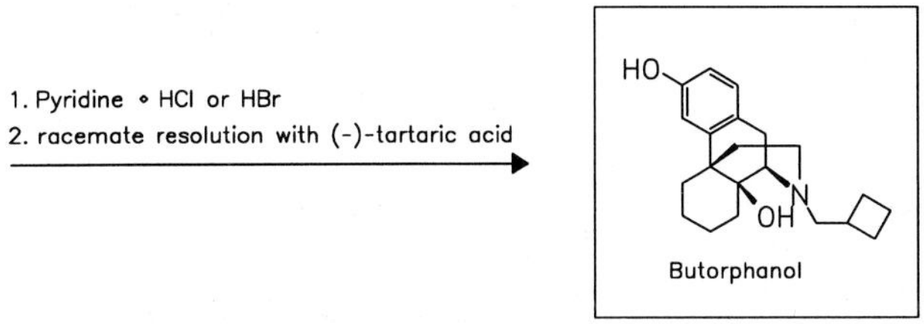

Reference(s):
US 3 775 414 (Bristol-Myers; 27.11.1973; appl. 10.5.1972).
US 3 819 635 (Bristol-Myers; 25.6.1974; prior. 8.9.1971, 13.1.1972).
DOS 2 243 961 (Bristol-Myers; appl. 7.9.1972; USA-prior. 8.9.1971, 13.1.1972).
DOS 2 265 255 (Bristol-Myers; appl. 7.9.1972; USA-prior. 8.9.1971, 13.1.1972).
DOS 2 265 256 (Bristol-Myers; appl. 7.9.1972; USA-prior. 8.9.1971, 13.1.1972).
US 3 980 641 (Bristol-Myers; 14.9.1976; appl. 31.7.1975).
Monkovic, J. et al.: J. Am. Chem. Soc. (JACSAT) **95**, 7910 (1973).

alternative syntheses:
from 7-methoxy-2-tetralone:
US 4 017 497 (Bristol-Myers; 12.4.1977; appl. 18.11.1975).

from (–)-1-(4-methoxybenzyl)-2-methyl-1,2,3,4,5,6,7,8-octahydroisoquinoline:
US 4 115 389 (Bristol-Myers; 19.9.1978; appl. 2.5.1977).

Formulation(s):   nasal spray 15 mg/2.5 ml

Trade Name(s):
I: Stadole (Bristol Europe;   J: Stadol (Bristol; 1986)   USA: Stadol (Bristol-Myers
       1984); wfm   Squibb; 1978)

Butriptyline

ATC: N06AA15
Use: antidepressant

RN: 35941-65-2   MF: $C_{21}H_{27}N$ MW: 293.45
CN: (±)-10,11-dihydro-*N,N*,β-trimethyl-5*H*-dibenzo[*a,d*]cycloheptene-5-propanamine

hydrochloride
RN: 5585-73-9   MF: $C_{21}H_{27}N \cdot HCl$ MW: 329.92   EINECS: 226-983-8
LD$_{50}$: 48 mg/kg (M, i.v.); 345 mg/kg (M, p.o.);
       700 mg/kg (R, p.o.)

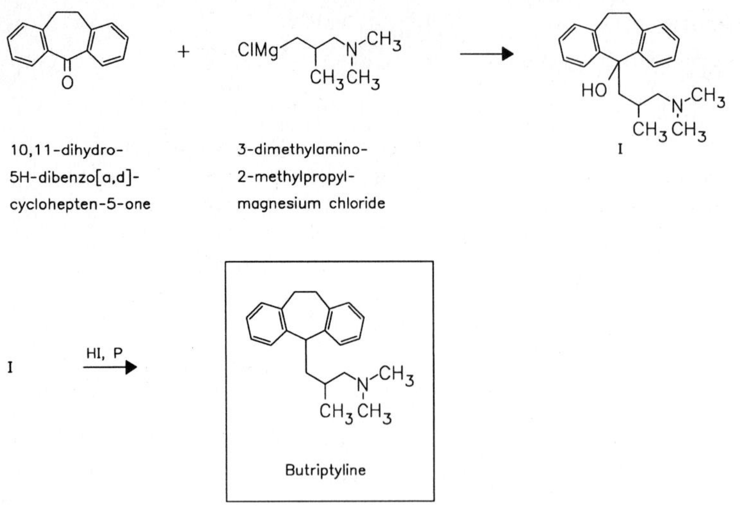

10,11-dihydro-
5H-dibenzo[a,d]-
cyclohepten-5-one

3-dimethylamino-
2-methylpropyl-
magnesium chloride

I

I $\xrightarrow{\text{HI, P}}$

Butriptyline

Reference(s):
BE 613 750 (Ayerst; appl. 9.2.1962; CDN-prior. 10.2.1961).
US 3 409 640 (Schering Corp.; 5.11.1968; appl. 22.7.1959).

Formulation(s):   tabl. 25 mg (as hydrochloride)

Trade Name(s):
GB: Evadyne (Ayerst); wfm   I: Evadene (Wyeth-Ayerst)

Butropium bromide

Use: antispasmodic

RN: 29025-14-7   MF: $C_{28}H_{38}BrNO_4$ MW: 532.52   EINECS: 249-375-4
LD$_{50}$:   6400 µg/kg (M, i.v.); 1500 mg/kg (M, p.o.);
21 mg/kg (R, i.v.)
CN: [3(S)-endo]-8-[(4-butoxyphenyl)methyl]-3-(3-hydroxy-1-oxo-2-phenylpropoxy)-8-methyl-8-
azoniabicyclo[3.2.1]octane bromide

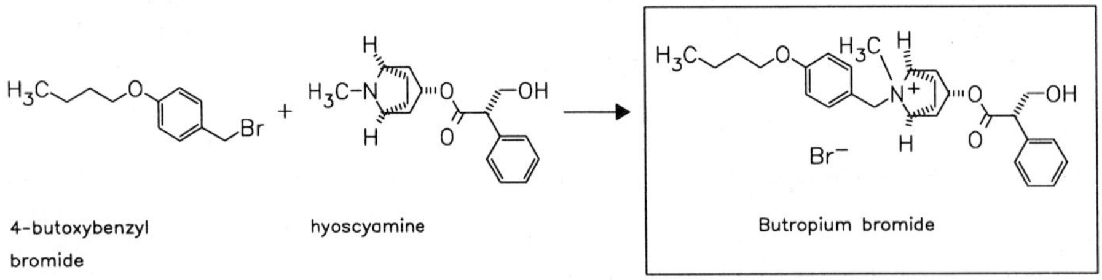

4-butoxybenzyl
bromide

hyoscyamine

Butropium bromide

Reference(s):
DOS 1 950 378 (Eisai Kabushiki Kaisha; appl. 6.10.1969; J-prior. 18.2.1969).
US 3 696 110 (Eisai Kabushiki Kaisha; 3.10.1972; J-prior. 18.2.1969).

Formulation(s):   amp. 4 mg

Trade Name(s):
J: Coliopan (Eisai; 1974)

Butylscopolammonium bromide

(Butylscopolamine bromide; Scopolamine butyl bromide; Hyoscin butyl bromide)

ATC: A03BB01
Use: antispasmodic

RN: 149-64-4   MF: $C_{21}H_{30}BrNO_4$ MW: 440.38   EINECS: 205-744-1
LD$_{50}$:   10.3 mg/kg (M, i.v.); 1170 mg/kg (M, p.o.);
          24 mg/kg (R, i.v.); 1040 mg/kg (R, p.o.)
CN: [7(S)-(1α,2β,4β,5α,7β)]-9-butyl-7-(3-hydroxy-1-oxo-2-phenylpropoxy)-9-methyl-3-oxa-9-azoniatricyclo[3.3.1.0^{2,4}]nonane bromide

butyl bromide   (−)-scopolamine   Butylscopolammonium bromide

Reference(s):
DE 856 890 (Boehringer Ing.; appl. 1950).

Formulation(s):   amp. 20 mg/ml; drg. 10 mg; f. c. tabl. 10 mg; suppos. 7.5 mg, 10 mg; tabl. 20 mg;
                   vial 200 mg/10 ml

Trade Name(s):
D: Buscopan (Boehringer Ing.)-comb.
F: Génoscopolamine (Amino)
      Scopoderm (Novartis)
GB: Buscopan (Boehringer Ing.)
I: Buscopan (Boehringer Ing.)
      Buscopan composto (Boehringer Ing.)-comb.
      Tranquo-Buscopan (Boehringer Ing.)-comb.
J: Antispasmin (Green Cross)
      Bubusco-S (Sawai)

Buscopan (Boehringer-Tanabe)
Buscoridin (Kanebo)
Buscote (Kotani)
Buspon (Toyo Pharmar)
Butibol (Towa)
Butylpan (Hokuriku)
Butymide (Ohta)
Butysco (Kobayashi)
Diaste-M (Fukuchi-Fujizoki)
Hyoscomin (Vitacain)

Hyospan (Toiyo)
Moryspan (Beppu)
Reladan (Isei)
Scobro (Ono)
Scobron (Mohan)
Scobutylamin (Horii)
Scordin-B (Ono)
Scorpan (Kanto)
Sparicon (Yamanouchi)
Spasmopan (Nichiiko)
Stibron (Iwaki)

Cabergoline

ATC: G02CB03
Use: dopamine D_2 receptor antagonist, prolactin inhibitor for prevention or suppression of puerperal lactation

RN: 81409-90-7 MF: $C_{26}H_{37}N_5O_2$ MW: 451.62
CN: (8β)-N-[3-(Dimethylamino)propyl]-N-[(ethylamino)carbonyl]-6-(2-propenyl)ergoline-8-carboxamide

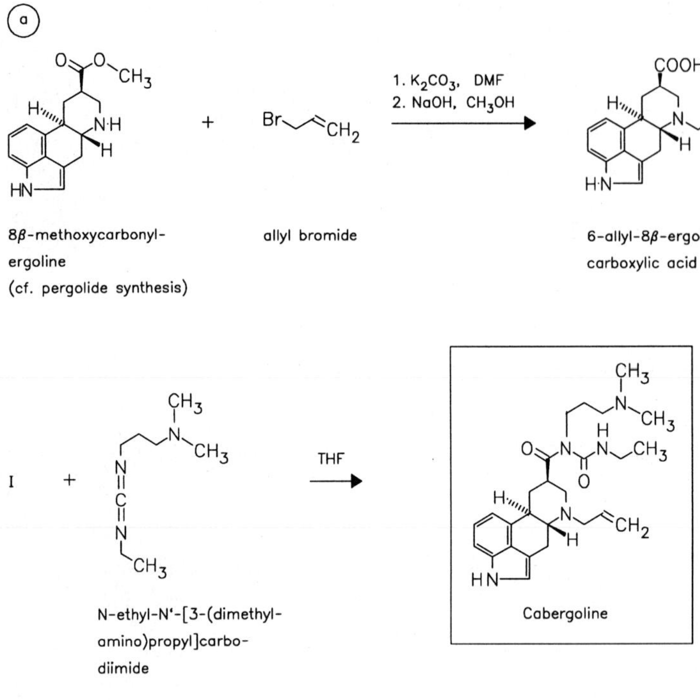

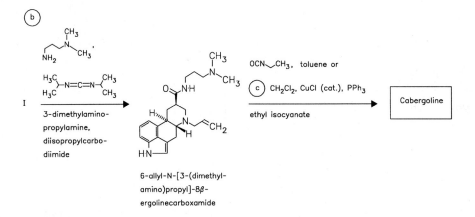

a

8β-methoxycarbonyl-ergoline
(cf. pergolide synthesis)

+ allyl bromide

1. K_2CO_3, DMF
2. NaOH, CH_3OH

6-allyl-8β-ergoline-carboxylic acid (I)

I + N-ethyl-N'-[3-(dimethyl-amino)propyl]carbo-diimide

THF

Cabergoline

b

I

3-dimethylamino-propylamine, diisopropylcarbo-diimide

6-allyl-N-[3-(dimethyl-amino)propyl]-8β-ergolinecarboxamide

$OCN\text{---}CH_3$, toluene or
c CH_2Cl_2, CuCl (cat.), PPh_3

ethyl isocyanate

Cabergoline

Reference(s):
a GB 2 074 566 (Farmitalia Carlo Erba S.p.A.; appl. 31.3.1981; GB-prior. 3.4.1980).
b US 4 526 892 (Farmitalia Carlo Erba S.p.A.; 2.7.1985; USA-prior. 3.3.1981).
    BE 894 060 (Farmitalia Carlo Erba S.p.A.; appl. 9.8.1982; GB-prior. 11.8.1981).
b,c WO 9 318 034 (Farmitalia Carlo Erba S.p.A.; appl. 15.2.1993; GB-prior. 12.3.1982).
    Candiani; Cabri, W.; Zarini, F.; Bedeschi, A.: Synlett (SYNLES) **1995** (6), 605.

synthesis and nidation inhibitory activity of a new class of ergoline derivatives:
Brambillà, E.; Disalle, E.; Briatico, G.; Mantegani, S.; Temperilli, A.: Eur. J. Med. Chem. (EJMCA5) **24**, 421 (1989)

Formulation(s):   tabl. 0.5 mg, 1 mg, 2 mg, 4 mg

Trade Name(s):

D:	CABASERIL (Pharmacia & Upjohn)	GB:	Cabaser (Pharmacia & Upjohn)	I:	Dostinex (Pharmacia & Upjohn)
	Dostinex (Pharmacia & Upjohn)		Dostinex (Pharmacia & Upjohn)	USA:	Dostinex (Pharmacia & Upjohn)

Cadexomer iodine

ATC: D03AX01; D08AG
Use: antiseptic for treatment of decubitus and venous leg ulcers

RN: 94820-09-4   MF: unspecified   MW: unspecified
LD_{50}:   >2 g/kg (R, i.p.); >2 g/kg (R, s.c.)
CN: cadexomer iodine

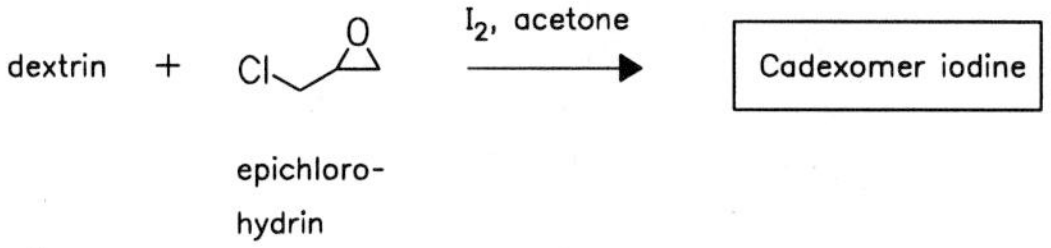

dextrin + epichlorohydrin $\xrightarrow{I_2,\ acetone}$ Cadexomer iodine

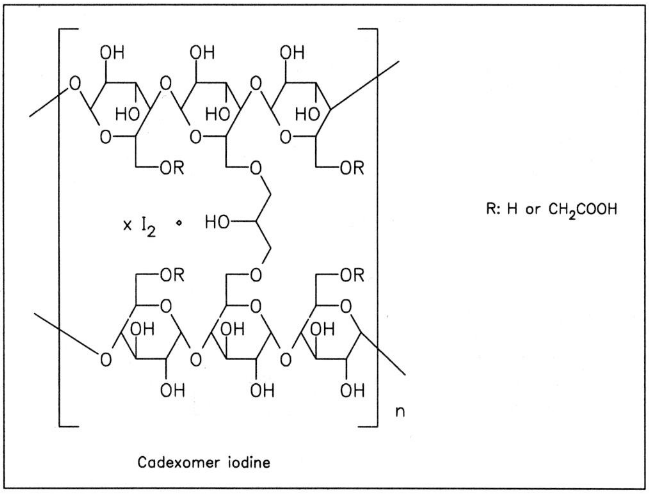

R: H or CH_2COOH

Cadexomer iodine

Reference(s):
DE 2 533 159 (A. O. Johannson; appl. 24.7.1975).
US 4 010 259 (A. O. Johannson; 1.3.1977; appl. 17.7.1975).
FR 2 320 112 (A. O. Johannson; appl. 5.8.1975).

Formulation(s):   dry sterile powder, micropellets, sachet 3 g, 1 % bioavailable iodine

Trade Name(s):

D:	Iodosorb (Strathmann)	F:	Iodosorb (Millot; 1984); wfm

Cadralazine

ATC: C02DB04
Use: antihypertensive, vasodilator

RN: 64241-34-5   MF: $C_{12}H_{21}N_5O_3$ MW: 283.33
LD$_{50}$:   700 mg/kg (M, i.p.);
269 mg/kg (R, i.v.); 2060 mg/kg (R, p.o.)
CN: 2-[6-[ethyl(2-hydroxypropyl)amino]-3-pyridazinyl]hydrazinecarboxylic acid ethyl ester

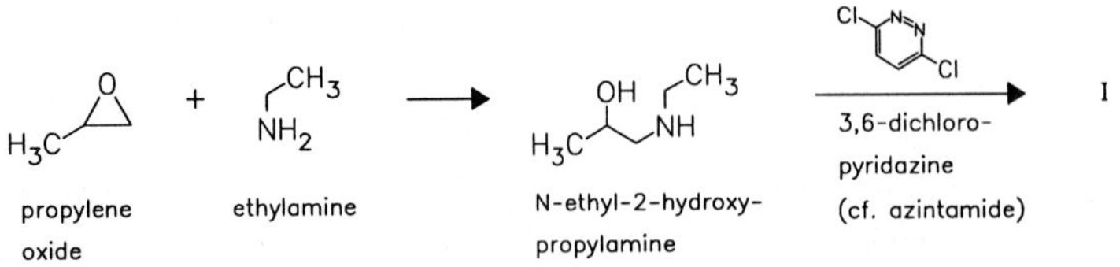

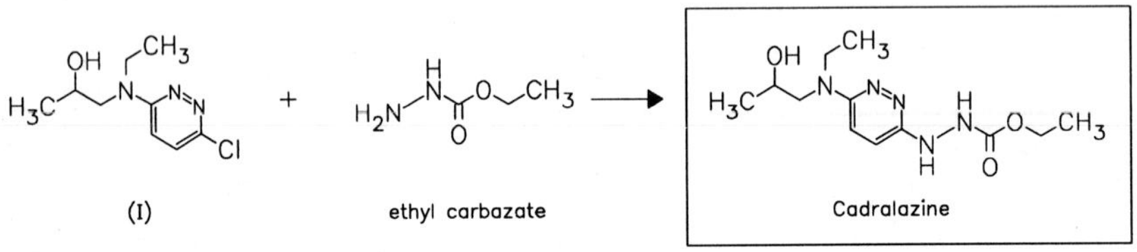

propylene oxide   ethylamine   N-ethyl-2-hydroxy-propylamine   3,6-dichloro-pyridazine (cf. azintamide)

(I)   ethyl carbazate   Cadralazine

Reference(s):
US 4 002 753 (I.S.F.; 11.1.1977; I-prior. 7.3.1973).

alternative syntheses:
US 4 575 552 (I.S.F.; 11.3.1986; I-prior. 28.4.1983).
US 4 632 982 (I.S.F.; 30.12.1986; I-prior. 28.4.1983).
US 4 757 142 (I.S.F.; 12.7.1988; I-prior. 13.5.1985).
cf. also synthesis of pildralazine

Formulation(s):   cps. 10 mg, 15 mg, 20 mg

Trade Name(s):
I: Cadraten (SmithKline   Cadrilan (Novartis)
Beecham)   J: Cadral (Novartis)

Cafaminol

Use: rhinological therapeutic

(Mecoffaminum; Methylcoffanolamine)

RN: 30924-31-3   MF: $C_{11}H_{17}N_5O_3$ MW: 267.29   EINECS: 250-390-3
LD$_{50}$:   700 mg/kg (M, s.c.)
CN: 3,7-dihydro-8-[(2-hydroxyethyl)methylamino]-1,3,7-trimethyl-1H-purine-2,6-dione

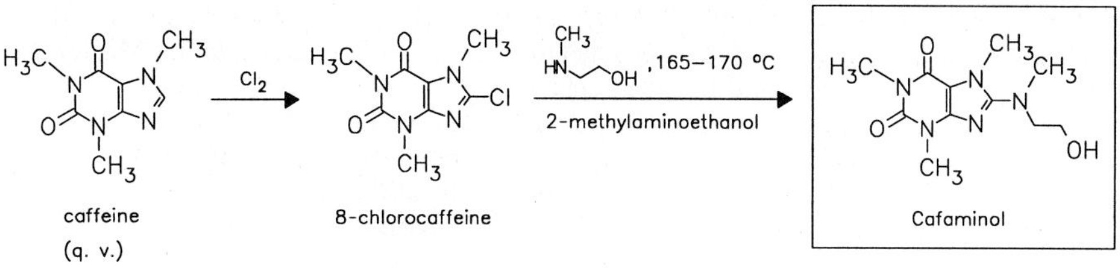

caffeine
(q. v.)

8-chlorocaffeine

Cafaminol

Reference(s):
DE 1 085 530 (J. Klosa; appl. 15.8.1958).
US 3 094 531 (Delmar Chemicals; appl. 30.4.1959).

Formulation(s):   drg. 50 mg

Trade Name(s):
D: Rhinoptil (Promonta); wfm   I: Katasma balsamico
                                          (Bruschettini)

Cafedrine

ATC: C01CA21
Use: circulatory analeptic

RN: 58166-83-9   MF: $C_{18}H_{23}N_5O_3$ MW: 357.41
CN: 3,7-dihydro-7-[2-[(2-hydroxy-1-methyl-2-phenylethyl)amino]ethyl]-1,3-dimethyl-1H-purine-2,6-dione

[R-(R*,S*)]-cafedrine
RN: 78396-34-6   MF: $C_{18}H_{23}N_5O_3$ MW: 357.41
[R-(R*,S*)]-monohydrochloride
RN: 3039-97-2   MF: $C_{18}H_{23}N_5O_3 \cdot HCl$ MW: 393.88   EINECS: 221-243-0
LD$_{50}$:   525 mg/kg (M, i.p.)

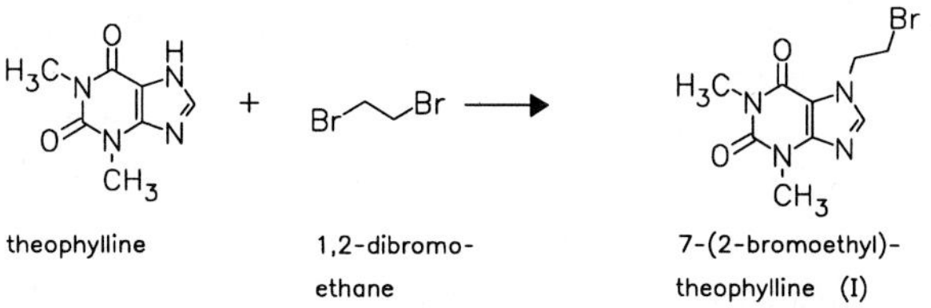

theophylline

1,2-dibromo-
ethane

7-(2-bromoethyl)-
theophylline   (I)

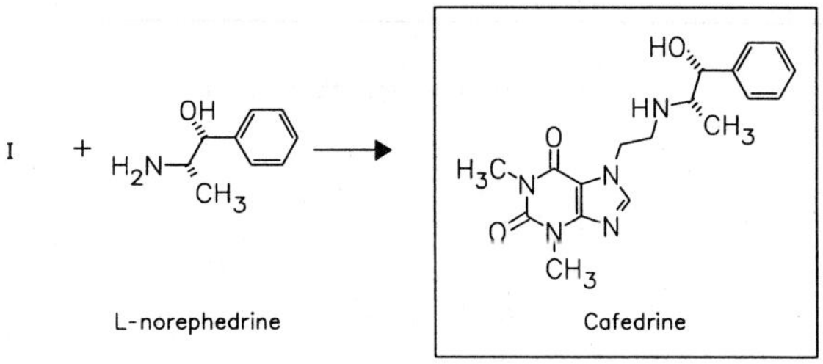

L-norephedrine

Cafedrine

Reference(s):
DE 1 095 285 (Degussa; appl. 25.9.1956).
US 3 029 239 (Degussa; 10.4.1962; D-prior. 17.4.1954).

Formulation(s):   amp. 200 mg; f. c. tabl. 100 mg

Caffeine

(Caféine; Coffein)

ATC: N06BC01
Use: analeptic, diuretic

RN: 58-08-2   MF: $C_8H_{10}N_4O_2$ MW: 194.19   EINECS: 200-362-1
LD_{50}:   62 mg/kg (M, i.v.); 127 mg/kg (M, p.o.);
       105 mg/kg (R, i.v.); 192 mg/kg (R, p.o.);
       140 mg/kg (dog, p.o.)
CN: 3,7-dihydro-1,3,7-trimethyl-1*H*-purine-2,6-dione

theophylline
(q. v.)   dimethyl sulfate   Caffeine

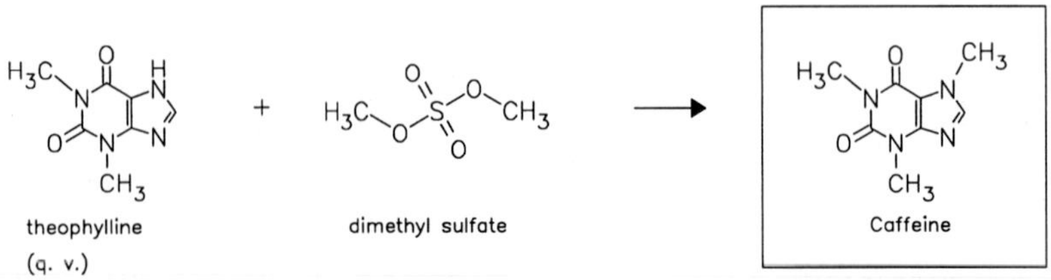

Reference(s):
DE 834 105 (Boehringer Ing.; appl. 1949).

Formulation(s):   tabl. 200 mg

Caffeine acetyltryptophanate

(A 50; Coftrinum)

ATC: N06BC01
Use: psychotonic

RN: 60364-24-1   MF: $C_{13}H_{14}N_2O_3 \cdot C_8H_{10}N_4O_2$ MW: 440.46
CN: 1-acetyl-L-tryptophan compd. with 3,7-dihydro-1,3,7-trimethyl-1*H*-purine-2,6-dione (1:1)

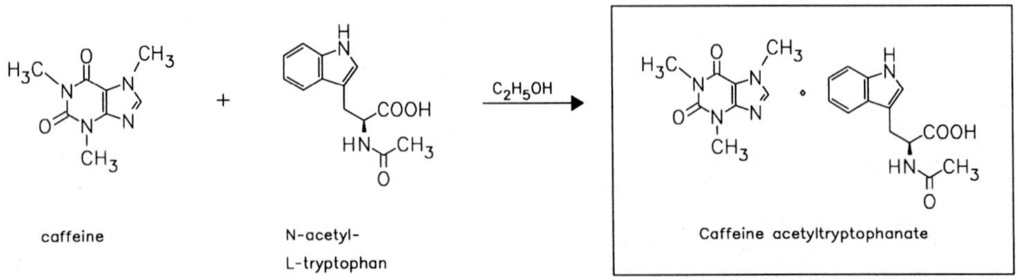

caffeine   N-acetyl-   Caffeine acetyltryptophanate
                  L-tryptophan

Reference(s):
FR-M 1 759 (A. E. C.; appl. 22.2.1962).

Trade Name(s):
F: Adrifane (Adrian, Paris);
         wfm

Calcifediol

ATC: A11CC06
Use: calcium regulator

RN: 19356-17-3   MF: $C_{27}H_{44}O_2$ MW: 400.65   EINECS: 242-990-9
CN: (3β,5Z,7E)-9,10-secocholesta-5,7,10(19)-triene-3,25-diol

25-hydroxycholesterol 3-acetate   acetic anhydride   (I)

1. 1,3-dibromo-5,5-
   dimethylhydantoin
2. trimethyl phosphite
3. lithium aluminum
   hydride

(II)   (III)

Calcifediol

1. 1,3-dibromo-5,5-
   dimethylhydantoin
2. trimethyl phosphite
3. methylmagnesium
   iodide

II $\xrightarrow{h\nu, \Delta}$ Calcifediol

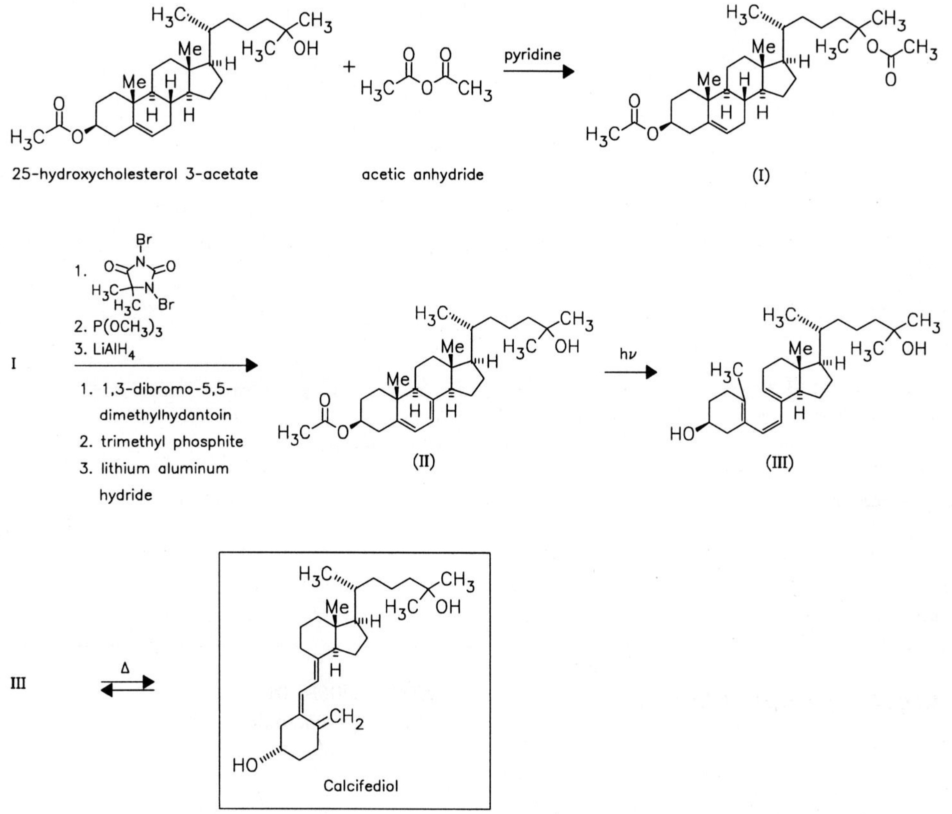

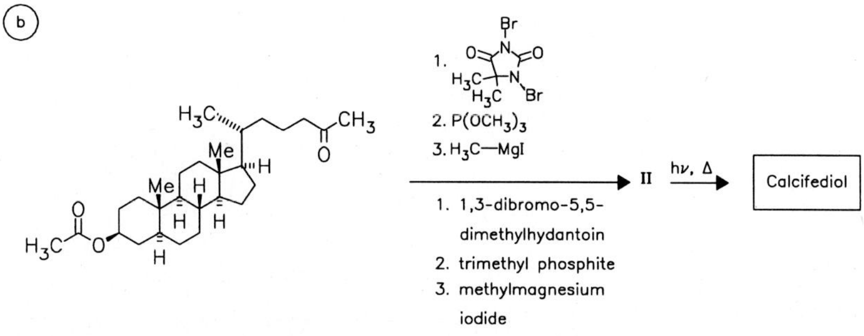

Reference(s):

a,b Blunt, J.W.; DeLuca, H.F.: Biochemistry (BICHAW) **8**, 671 (1969).

DeLuca, H.F.: Am. J. Clin. Nutr. (AJCNAC) **22**, 412 (1969).

Halkes, S.J.; Vliet, N.P. van: Recl. Trav. Chim. Pays-Bas (RTCPA3) **88**, 1080 (1969).

alternative syntheses:

Sodano, Ch. S.: Vitamins, Synthesis, Production and Use, p. 131, 159 (New Jersey 1979).

US 4 001 096 (Upjohn; 4.1.1977; prior. 21.2.1975).

US 3 833 622 (Upjohn; 3.9.1974; prior. 17.3.1969).

structure and isolation:

DeLuca, H.F.: Am. J. Clin. Nutr. (AJCNAC) **22**, 412 (1969).

Formulation(s):   drops 0.15 mg/ml, 0.45 mg/ml

Trade Name(s):

D:	Dedrogyl (Albert-Roussel, Hoechst)	F:	Dédrogyl (Roussel) Un-Alfa (Léo)	I:	Didrogyl (Roussel)
				USA:	Calderol (Organon)

Calcipotriol

ATC: A11CC; D05AX02
Use: antipsoriatic, topical vitamin D_3-analog

RN: 112828-00-9   MF: $C_{27}H_{40}O_3$ MW: 412.61
CN: (1α,3β,5Z,7E,22E,24S)-24-cyclopropyl-9,10-secochola-5,7,10(19),22-tetraene-1,3,24-triol

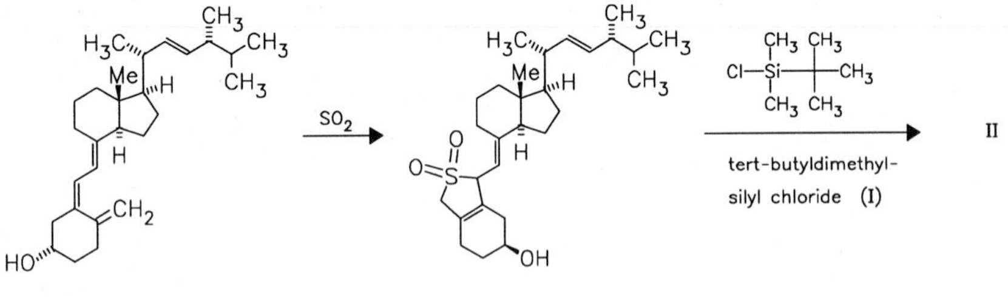

vitamin D_2

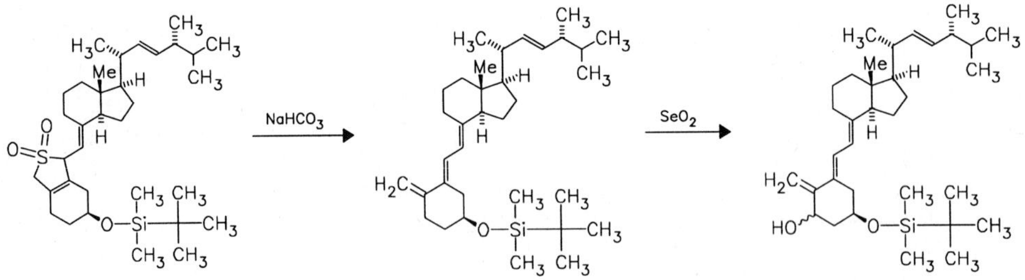

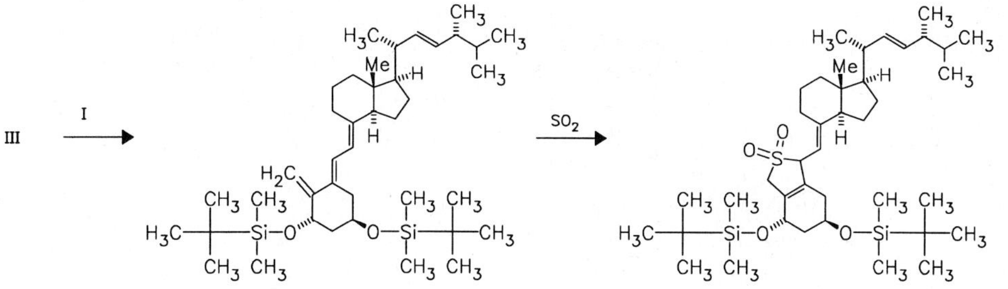

III $\xrightarrow{I}$ (IV) $\xrightarrow{SO_2}$

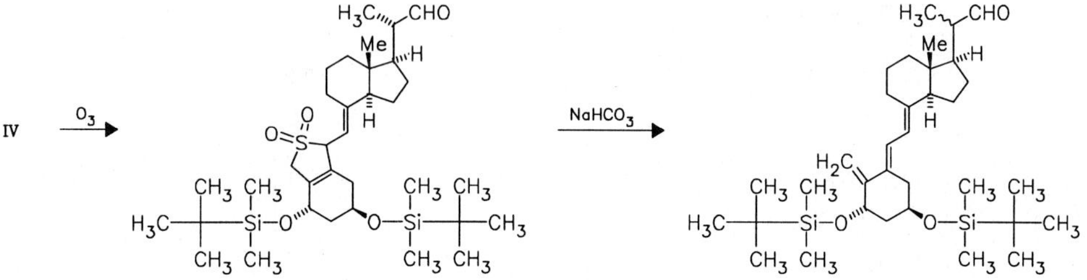

IV $\xrightarrow{O_3}$ $\xrightarrow{NaHCO_3}$ (V)

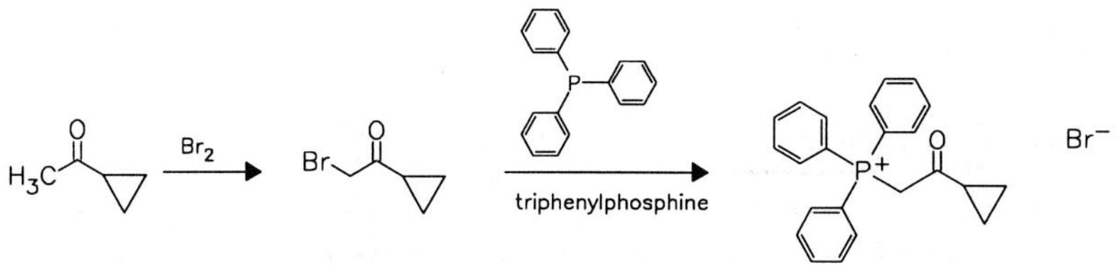

acetylcyclo-
propane $\xrightarrow{Br_2}$ $\xrightarrow{\text{triphenylphosphine}}$ (VI)

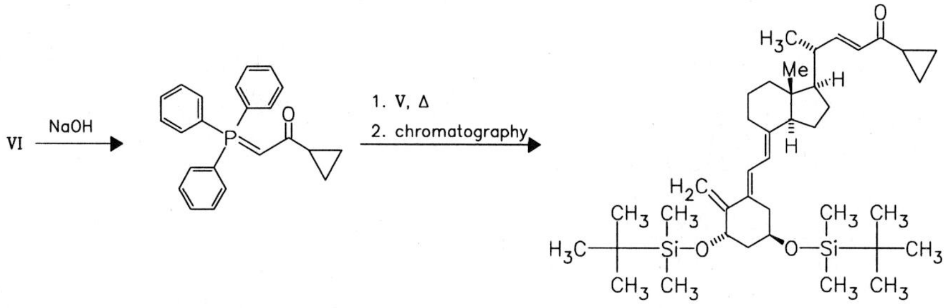

VI $\xrightarrow{NaOH}$ (cyclopropylcarbo-
nylmethylene)tri-
phenylphosphorane $\xrightarrow[\text{2. chromatography}]{\text{1. V, }\Delta}$ (VII)

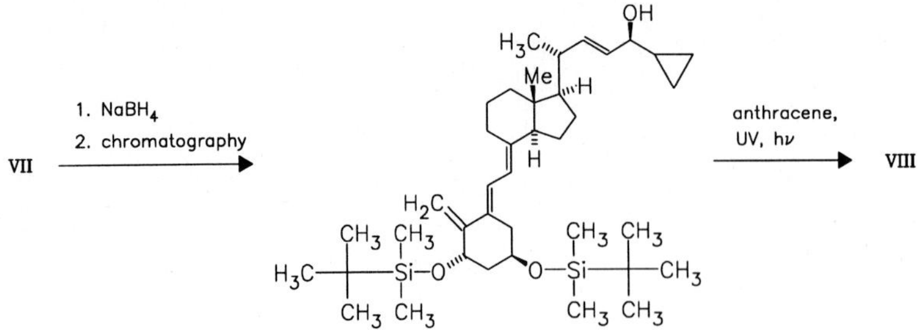

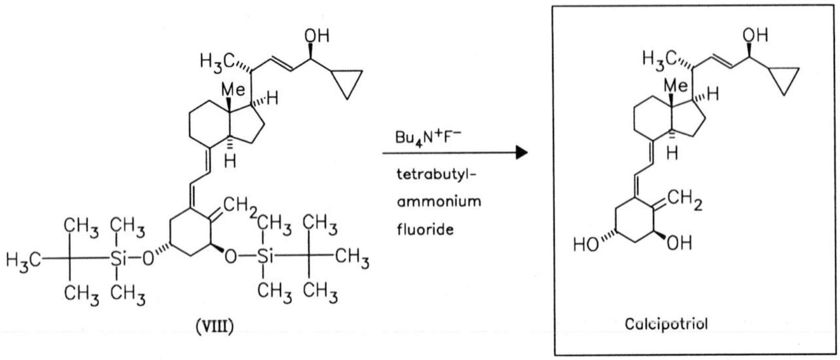

Reference(s):
EP 227 826 (Leo; appl. 14.7.1986; GB-prior. 2.8.1985).
WO 8 700 834 (Leo; appl. 14.7.1986; GB-prior. 2.8.1985).
Calverley, M.J.: Tetrahedron (TETRAB) **43**, 4609 (1987).

Formulation(s):   ointment 50 µg/g

Trade Name(s):
D: Daivonex (Leo)   F: Daivonex (Léo)   I: Daivonex (Formenti)
        Psorcutan (Schering AG)   GB: Daivonex (Leo; 1991)   Psorcutan (Schering)

Calcitriol
(1α,25-Dihydroxy-vitamin D$_3$)

ATC: A11CC04; D05AX03
Use: calcium regulator

RN: 32222-06-3   MF: C$_{27}$H$_{44}$O$_3$ MW: 416.65   EINECS: 250-963-8
LD$_{50}$:   1350 µg/kg (M, p.o.);
        105 µg/kg (R, i.v.); 620 µg/kg (R, p.o.)
CN: (1α,3β,5Z,7E)-9,10-secocholesta-5,7,10(19)-triene-1,3,25-triol

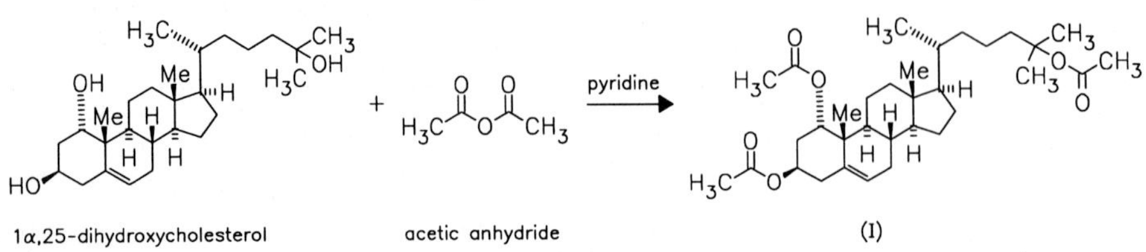

1α,25-dihydroxycholesterol   acetic anhydride   (I)

(II)

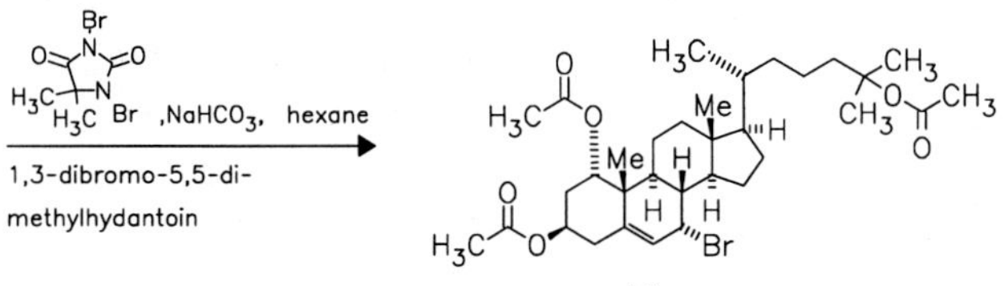

(III)

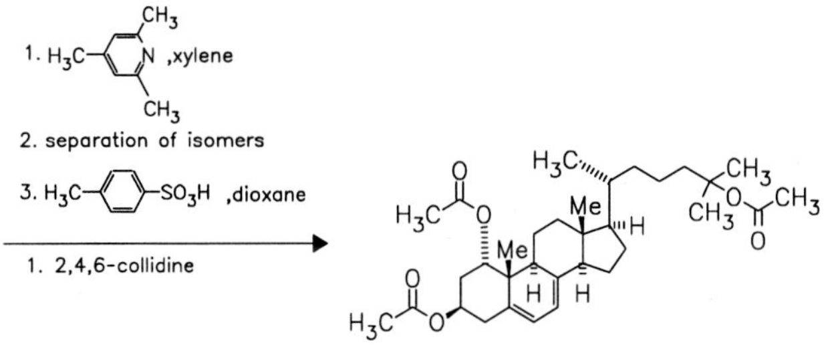

Calcitriol

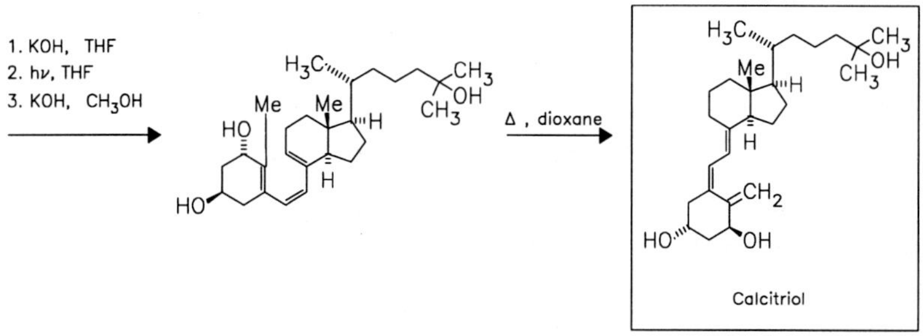

Reference(s):
US 3 993 675 (Hoffmann-La Roche Inc.; 23.11.1976; prior. 12.11.1973, 24.2.1975).

alternative synthesis:
DOS 2 754 759 (Chugai Seiyaku; appl. 8.12.1977; J-prior. 8.12.1976).
Semmler, E.J. et al.: Tetrahedron Lett. (TELEAY) **1972**, 4147.
Barton, D.R. et al.: J. Chem. Soc., Chem. Commun. (JCCCAT) **1974**, 203.

synthesis of 1α,25-dihydroxycholesterol:
DOS 2 453 648 (Hoffmann-La Roche; appl. 12.11.1974; USA-prior. 18.11.1973).

Formulation(s):   amp. 1 µg/ml, 2 µg/ml; cps. 0.25 µg, 0.5 µg

Trade Name(s):
D: Rocaltrol (Roche)   I: Calcijex (Abbott)   USA: Calcijex (Abbott)
F: Rocaltrol (Roche)   Rocaltrol (Roche)   Rocaltrol (Roche Labs.)
GB: Rocaltrol (Roche)   J: Rocaltrol (Roche-Kyorin)

Calcium dobesilate

(Dobesilate de calcium)

ATC: C05BX01
Use: hemostatic (capillary protective)

RN: 20123-80-2   MF: $C_{12}H_{10}CaO_{10}S_2$ MW: 418.41   EINECS: 243-531-5
LD$_{50}$: 775 mg/kg (M, i.v.); 7549 mg/kg (M, p.o.);
       7061 mg/kg (R, p.o.)
CN: 2,5-dihydroxybenzenesulfonic acid calcium salt (2:1)

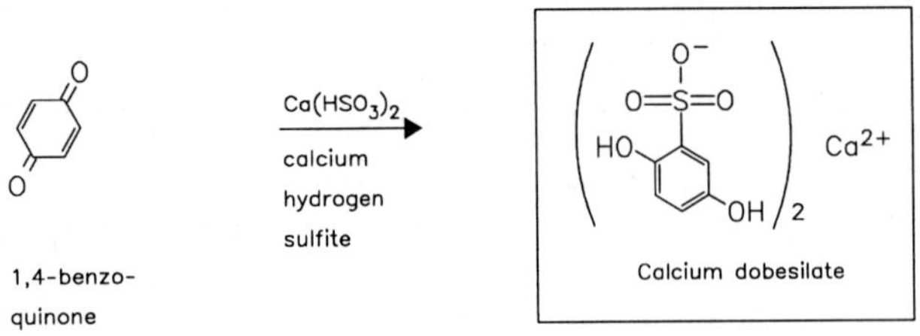

1,4-benzo-
quinone

Calcium dobesilate

Reference(s):
US 3 509 207 (Lab. Om; 28.4.1970; CH-prior. 20.1.1966).

Formulation(s):   cps. 250 mg; tabl. 250 mg

Trade Name(s):
D:	Dexium (Synthelabo)	I:	Dobesifar (Farmila)	Doxium (Delalande
	Dobica (OPW)		Doxiproct-Plus (Delalande	Isnardi)
F:	Doxium (Synthélabo)		Isnardi)-comb.	

Calcium hopantenate

ATC: N06B
Use: cerebral activator

RN: 17097-76-6   MF: $C_{20}H_{36}CaN_2O_{10}$ MW: 504.59
CN: (R)-4-[(2,4-dihydroxy-3,3-dimethyl-1-oxobutyl)amino]butanoic acid calcium salt (2:1)

hopantenic acid
RN: 18679-90-8   MF: $C_{10}H_{19}NO_5$ MW: 233.26
LD_{50}:   2250 mg/kg (M, i.p.); 5720 mg/kg (M, route unreported)

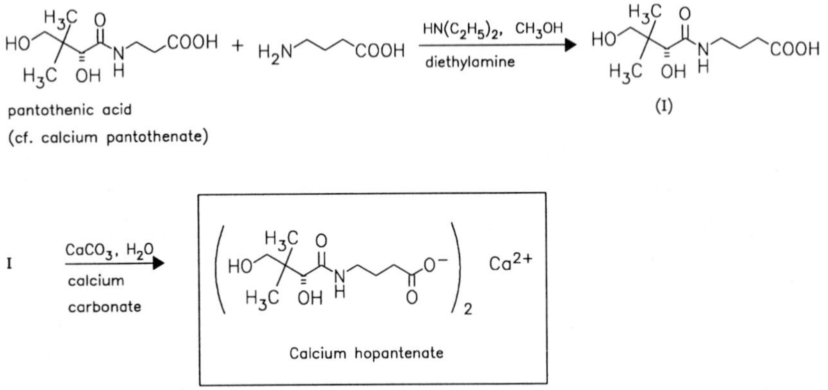

pantothenic acid
(cf. calcium pantothenate)

(I)

Calcium hopantenate

Reference(s):
Kopelevich, V.M. et al.: Khim. Farm. Zh. (KHFZAN) **5**, 21 (1971).
JP 26 189 (64) (Takeda; appl. 23.10.1962).

alternative syntheses:
McFall Desha, C.; Fuerst, R.: Biochim. Biophys. Acta (BBACAQ) **86**, 33 (1964).
JP 732 (66) (Tanabe; appl. 25.2.1964).

review:
Nishizawa, Y.; Kodama, T.: Proc. Jpn. Acad. (PCACAW) **42**, 841 (1966).

Trade Name(s):
J: Hopate (Tanabe)

Calcium pantothenate
(Vitamin B$_5$)

ATC: A11HA31; D03AX04
Use: growth factor

RN: 137-08-6   MF: $C_{18}H_{32}CaN_2O_{10}$ MW: 476.54   EINECS: 205-278-9
LD$_{50}$:   1443 mg/kg (M, i.p.); 2490 mg/kg (M, route unreported); 2500 mg/kg (M, s.c.);
      3500 mg/kg (R, s.c.)
CN: (*R*)-*N*-(2,4-dihydroxy-3,3-dimethyl-1-oxobutyl)-β-alanine calcium salt (2:1)

Pantothenic acid

RN: 79-83-4   MF: $C_9H_{17}NO_5$ MW: 219.24   EINECS: 201-229-0
LD$_{50}$:   910 mg/kg (M, i.v.); 10 g/kg (M, p.o.);
      830 mg/kg (R, i.v.); >10 g/kg (R, p.o.)
CN: (*R*)-*N*-(2,4-dihydroxy-3,3-dimethyl-1-oxobutyl)-β-alanine
monosodium salt
RN: 867-81-2   MF: $C_9H_{16}NNaO_5$ MW: 241.22   EINECS: 212-768-6

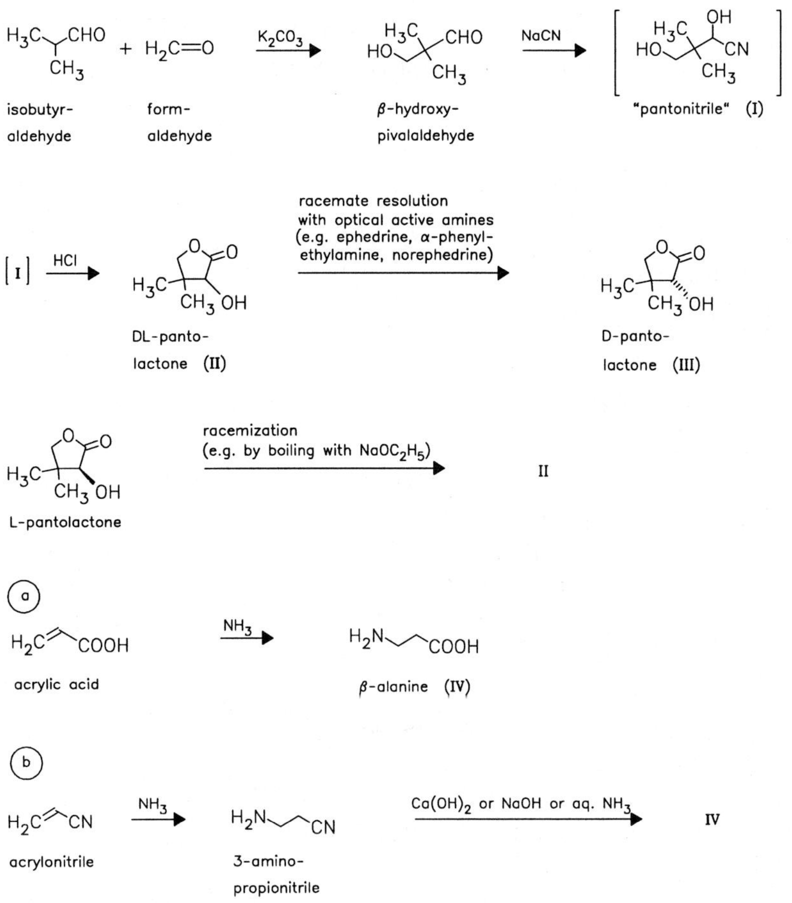

β-alanine
calcium salt

Calcium pantothenate

Reference(s):
DL-pantolactone:
Glaser: Monatsh. Chem. (MOCMB7) **25**, 46 (1904).
Stiller et al.: J. Am. Chem. Soc. (JACSAT) **62**, 1785 (1940).
Reichstein, Grüssner: Helv. Chim. Acta (HCACAV) **23**, 650 (1940).
Carter; Ney: J. Am. Chem. Soc. (JACSAT) **63**, 312 (1941).
US 2 552 530 (Upjohn; 1958; appl. 1954).
US 2 863 878 (Union Carbide; 1958; appl. 1954).
GB 857 128 (Nopco; appl. 1958; USA-prior. 1958).
US 2 967 869 (Nopco; 1961; appl. 1958).
US 3 024 250 (Nopco; 1962; appl. 1958).
DOS 2 758 883 (BASF; appl. 30.12.1977).
US 4 082 775 (Soc. Chim. des Charbonnages; 4.4.1978; F-prior. 7.7.1975).
GB 1 490 680 (Soc. Chim. des Charbonnages; appl. 21.6.1976; F-prior. 7.7.1975).
US 4 095 952 (VEB Jenapharm; 20.6.1978; prior. 19.10.1972, 4.6.1974, 16.3.1976, 15.10.1976).

extraction of pantolactone from aqueous solutions with methyl tert-butyl ether:
DOS 2 809 179 (BASF; appl. 3.3.1978).

D-pantolactone:
racemate resolution with ephedrine:
US 2 460 239 (Nopco; 1949; appl. 1945).
US 2 460 240 (Nopco; 1949; appl. 1945).

with L-(+)-1-(4-nitrophenyl)-2-aminopropane-1,3-diol:
DD 16 982 (R. Ring; appl. 1957).
DD 32 628 (W. Braune et al.; appl. 25.3.1963).

with 1-α-phenylethylamine:
US 3 185 710 (Nopco; 25.5.1965; appl. 6.9.1961).

with d-3-aminomethylpinane:
GB 1 495 162 (BASF; appl. 29.1.1975; D-prior. 30.1.1974, 9.11.1974).

with d-norephedrine *and derivatives:*
DAS 2 558 508 (Alps; appl. 24.12.1975; J-prior. 19.2.1975).
US 4 045 450 (Alps; 30.8.1977; J-prior. 19.2.1975).

by fractional crystallization of ammonium pantoate:
FR 1 522 111 (Fuji Chemical; appl. 9.5.1967; J-prior. 10.5.1966).

of guanidinium pantoate:
DOS 2 838 689 (A. E. C.; appl. 5.9.1978; F-prior. 5.9.1977).

of lithium pantoate:
US 4 115 443 (VEB Jenapharm; 19.9.1978; prior. 17.10.1973, 10.1.1975, 25.3.1976).
FR-appl. 2 231 638 (VEB Jenapharm; appl. 31.5.1974; DDR-prior. 4.6.1973).

racemization of L-pantolactone:
US 2 976 298 (Nopco; 1961; appl. 1958).
US 2 434 061 (Merck & Co.; 1948; appl. 1945).
US 2 463 734 (Nopco; 1949; appl. 1945).
US 2 967 869 (Nopco; 1961; appl. 1958).

β-alanine:
a US 2 376 334 (Univ. of California; 1945; appl. 1941).
  DAS 2 232 090 (Tokyo Fine Chem.; appl. 30.6.1972).
b US 2 336 067 (Lederle; 1943; appl. 1942).
  US 2 377 401 (Lederle; 1945; appl. 1942).
  US 2 461 842 (Sharpies Chemicals; 1949; appl. 1943).
  US 2 819 303 (Nopco; 1958; appl. 1953).
  US 2 935 524 (Nopco; 1960; appl. 1957).
  DE 1 084 730 (Degussa; appl. 1959).
  US 2 956 080 (Merck & Co.; 1960; appl. 1953).
  DAS 2 223 236 (VEB Jenapharm; appl. 12.5.1972; DDR-prior. 4.6.1971).
  DAS 2 232 090 (Tokyo Fine Chemical; appl. 30.6.1972).

calcium pantothenate:
DE 875 359 (Roche; appl. 1941; CH-prior. 1940).
DE 873 089 (E. Merck AG; appl. 1941; USA-prior. 1940).
GB 571 915 (Lederle; appl. 1943; USA-prior. 1942).
US 2 809 213 (Chemlek Labs.; 1957; appl. 1954).
DAS 1 041 967 (Pfizer; appl. 1954; USA-prior. 1953).
US 2 957 025 (Pfizer; 1960; appl. 1953).
US 2 780 645 (Comm. Solvents Corp.; 1957; appl. 1954).
US 2 935 528 (Nopco; 1960; appl. 1957).

purification:
US 2 390 499 (Lederle; 1945; appl. 1942).
US 2 496 363 (Merck & Co.; 1950; appl. 1948).
US 2 957 025 (Pfizer; 1960; appl. 1953).
GB 1 511 216 (Diamond Shamrock; appl. 1.2.1977).
DOS 2 708 016 (Diamond Shamrock; appl. 24.2.1977).

Formulation(s):   cps. 6 mg, 10 mg, 50 mg; ophthalmic ointment 25 mg/g

Trade Name(s):

D:	Kerato Biciron (S & K Pharma)		numerous combination preparations	Lasonil H (Bayer)-comb.
				J: Panto (Daiichi)
	numerous combination preparations	GB:	combination preparations	numerous combination preparations
		I:	Fisiolax (Manetti Roberts)-	
F:	Modane (RPR Cooper)-		comb.	USA: Mega-B (Arco)
	comb.		Lasaproct (Bayer)-comb.	

Calusterone

ATC: G03BA; L02A
Use: androgen, antineoplastic (mamma carcinoma)

RN: 17021-26-0   MF: $C_{21}H_{32}O_2$ MW: 316.49
CN: (7β,17β)-17-hydroxy-7,17-dimethylandrost-4-en-3-one

3β,17β-diacetoxy-17α-methyl-   methyl-   (I)
7-oxo-5-androstene   lithium

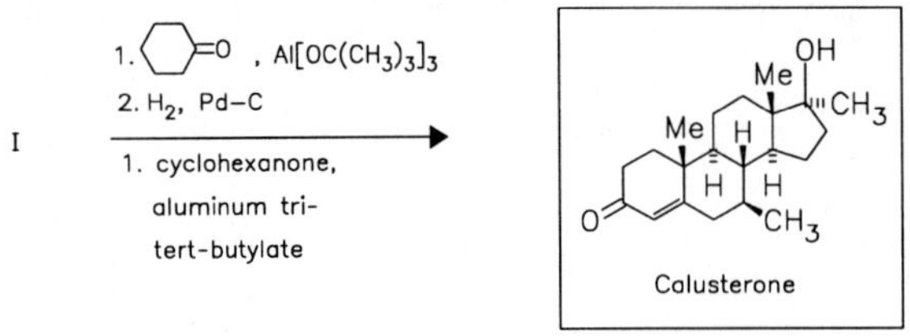

Calusterone

Reference(s):
US 3 029 263 (Upjohn; 10.4.1962; prior. 22.12.1958, 6.6.1958)
(synthesis of starting material is also described).

alternative synthesis:
US 3 341 557 (Upjohn; 12.9.1967; prior. 5.6.1961, 6.11.1960, 6.6.1958).

Formulation(s):   tabl. 50 mg

Trade Name(s):
USA: Methosarb (Upjohn); wfm

Camazepam

ATC: N05BA15
Use: sedative, tranquilizer

RN: 36104-80-0   MF: $C_{19}H_{18}ClN_3O_3$ MW: 371.82   EINECS: 252-866-6
LD$_{50}$:   970 mg/kg (M, p.o.);
    >4 g/kg (R, p.o.)
CN: dimethylcarbamic acid 7-chloro-2,3-dihydro-1-methyl-2-oxo-5-phenyl-1H-1,4-benzodiazepin-3-yl ester

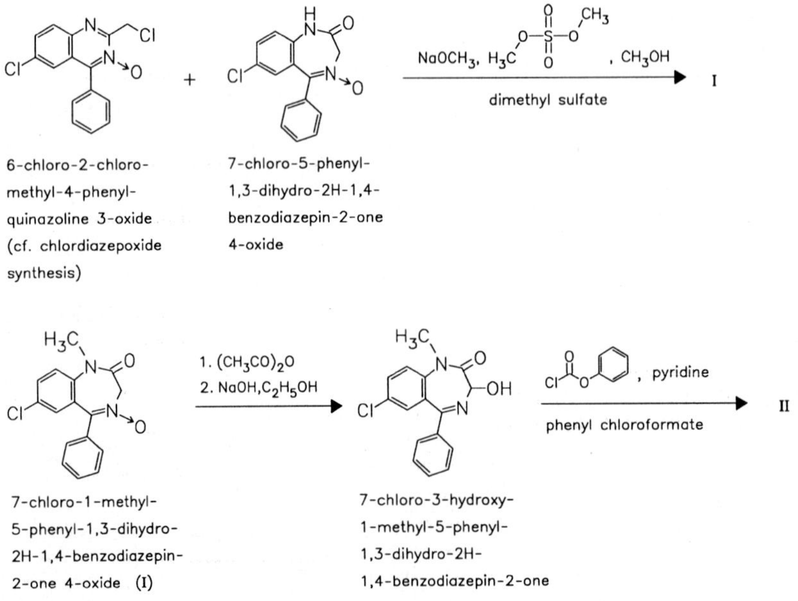

6-chloro-2-chloro-
methyl-4-phenyl-
quinazoline 3-oxide
(cf. chlordiazepoxide
synthesis)

7-chloro-5-phenyl-
1,3-dihydro-2H-1,4-
benzodiazepin-2-one
4-oxide

dimethyl sulfate

7-chloro-1-methyl-
5-phenyl-1,3-dihydro-
2H-1,4-benzodiazepin-
2-one 4-oxide (I)

7-chloro-3-hydroxy-
1-methyl-5-phenyl-
1,3-dihydro-2H-
1,4-benzodiazepin-2-one

phenyl chloroformate

(II)   dimethylamine   Camazepam

Reference(s):
DOS 2 142 181 (Siphar; appl. 23.8.1971; CH-prior. 24.8.1970).
US 3 799 920 (Siphar; 26.3.1974; CH-prior. 24.8.1970).
US 3 867 529 (Siphar; 18.2.1975; CH-prior. 24.8.1970).

alternative synthesis (reaction of the 3-hydroxy-compd. with dimethylcarbamoyl chloride):
DOS 2 558 015 (Siphar; appl. 22.12.1975; CH-prior. 6.3.1975).

precursors:
GB 972 968 (Roche; appl. 9.12.1960; USA-prior. 10.12.1959, 15.1.1960, 26.4.1960, 27.6.1960).
Sternbach, L.H.; Reeder, E.: J. Org. Chem. (JOCEAH) **26**, 4936 (1961).
Bell, S.C.; Childress, S.J.: J. Org. Chem. (JOCEAH) **27**, 562, 1691 (1962).

Formulation(s):   drg. 10 mg, 20 mg

Trade Name(s):
D: Albego (Boehringer Ing.);   I: Albego (Simes); wfm
       wfm   Limpidon (Crinos); wfm

Camostat

ATC: B02AB04
Use: trypsin inhibitor (for treatment of chronic pancreatitis)

RN: 59721-28-7   MF: $C_{20}H_{22}N_4O_5$ MW: 398.42
CN: 4-[[4-[(aminoiminomethyl)amino]benzoyl]oxy]benzeneacetic acid 2-(dimethylamino)-2-oxoethyl ester

monomesylate
RN: 59721-29-8   MF: $C_{20}H_{22}N_4O_5 \cdot CH_4O_3S$ MW: 494.53
LD_{50}:   200 mg/kg (M, i.v.); 3 g/kg (M, p.o.);
       152 mg/kg (R, i.v.); 3 g/kg (R, p.o.)

4-guanidino-   4-guanidinobenzoyl
benzoic acid   chloride hydrochloride (I)

(4-hydroxyphenyl)-   2-bromo-N,N-   N,N-dimethylcarbamoylmethyl
acetic acid   dimethyl-   (4-hydroxyphenyl)acetate (II)
                       acetamide

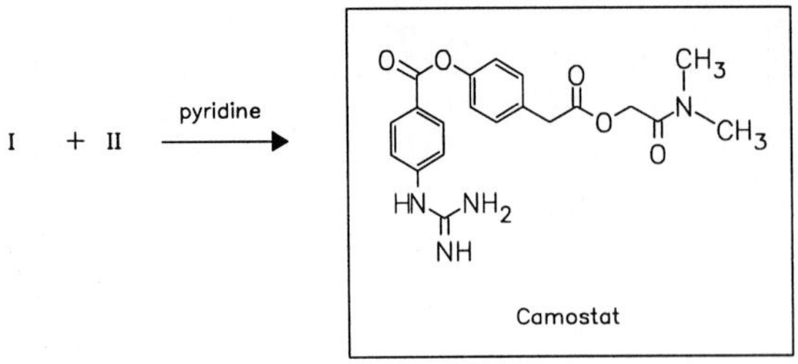

Camostat

Reference(s):
DOS 2 548 886 (Ono Pharmac.; appl. 31.10.1975; J-prior. 1.11.1974, 17.12.1974, 27.5.1975).
US 4 021 472 (Ono Pharmac.; 3.5.1977; J-prior. 1.11.1974, 17.12.1974, 27.5.1975).
GB 1 472 700 (Ono Pharmac.; appl. 23.10.1975; J-prior. 1.11.1974; 17.12.1974, 27.5.1975).
FR 2 289 181 (Ono Pharmac.; appl. 30.10.1975; J-prior. 1.11.1974, 17.12.1974, 27.5.1975).

Formulation(s): gran. 200 mg

Trade Name(s):
J: Foipan (Ono; 1985)

Camphotamide
(Camphetamide)

ATC: N06
Use: analeptic

RN: 4876-45-3 MF: $C_{11}H_{17}N_2O \cdot C_{10}H_{15}O_4S$ MW: 424.56 EINECS: 225-484-2
LD$_{50}$: 422 mg/kg (M, i.v.)
CN: 3-[(diethylamino)carbonyl]-1-methylpyridinium salt with 4,7,7-trimethyl-3-oxobicyclo[2.2.1]heptane-2-sulfonic acid (1:1)

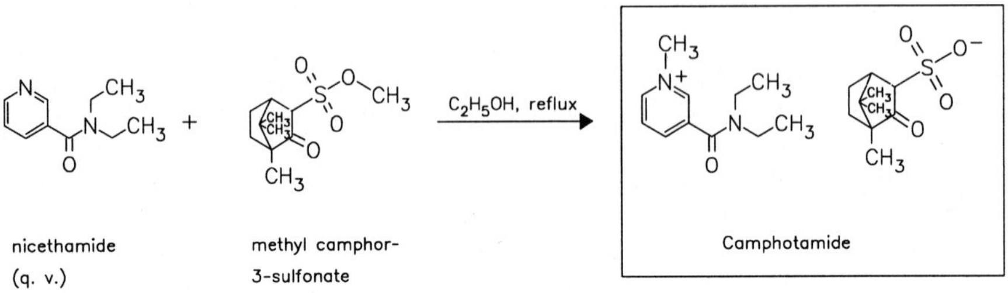

nicethamide
(q. v.)

methyl camphor-
3-sulfonate

Camphotamide

Reference(s):
FR 812 032 (Soc. Franc. de Rech. Biochimiques; appl. 1936).

Trade Name(s):
F: Tonicorine (Lematte et
Boinot); wfm

Camylofin
(Acamylophenin)

ATC: A03AA03
Use: antispasmodic

RN: 54-30-8 MF: $C_{19}H_{32}N_2O_2$ MW: 320.48 EINECS: 200-202-0
LD$_{50}$: 760 mg/kg (M, p.o.) 2
CN: α-[[2-(diethylamino)ethyl]amino]benzeneacetic acid 3-methylbutyl ester

dihydrochloride

RN: 5892-41-1   MF: $C_{19}H_{32}N_2O_2 \cdot 2HCl$ MW: 393.40   EINECS: 227-571-0
LD$_{50}$:   49.2 mg/kg (M, i.v.); 760 mg/kg (M, p.o.);
        >.15 g/kg (R, p.o.)

phenylglycine   2-diethylaminoethyl   Camylofin
isopentyl ester   chloride hydrochloride

Reference(s):
DE 842 206 (ASTA; appl. 1950).

Formulation(s):   amp. 24 mg/ml; drg. 60 mg; suppos. 40 mg

Trade Name(s):
D: Avacan (ASTA Medica);   Spasmo-Urolong   Avacan (Uji); wfm
      wfm   (Thiemann)-comb.; wfm   Rugo (Hokuriku); wfm
      Avafortan (ASTA Medica)-   Ullus Apotheker Vetter
      comb.; wfm   (Vetter)-comb.; wfm
      Avamigran (Degussa   F: Avafortan (Lucien)-comb.
      Pharma/ASTA)-comb.;   I: Avacan (Schering); wfm
      wfm   J: Adopon (Kowa); wfm

Candesartan cilexetil
(TCV-116)

Use: antihypertensive, angiotensin II
        antagonist

RN: 145040-37-5   MF: $C_{33}H_{34}N_6O_6$ MW: 610.67
CN: (±)-2-ethoxy-1-[[2'-(1*H*-tetrazol-5-yl)[1,1'-biphenyl]-4-yl]methyl]-1*H*-benzimidazole-7-carboxylic acid
      1-[[(cyclohexyloxy)carbonyl]oxy]ethyl ester

3-nitrophthalic   ethyl 2-chloro-
acid   formyl-3-nitro-
                                             benzoate

ethyl 2-(tert-butoxy-   4-(2-cyanophenyl)-
carbonylamino)-3-   benzyl bromide
nitrobenzoate (I)   (cf. losartan
                               synthesis)

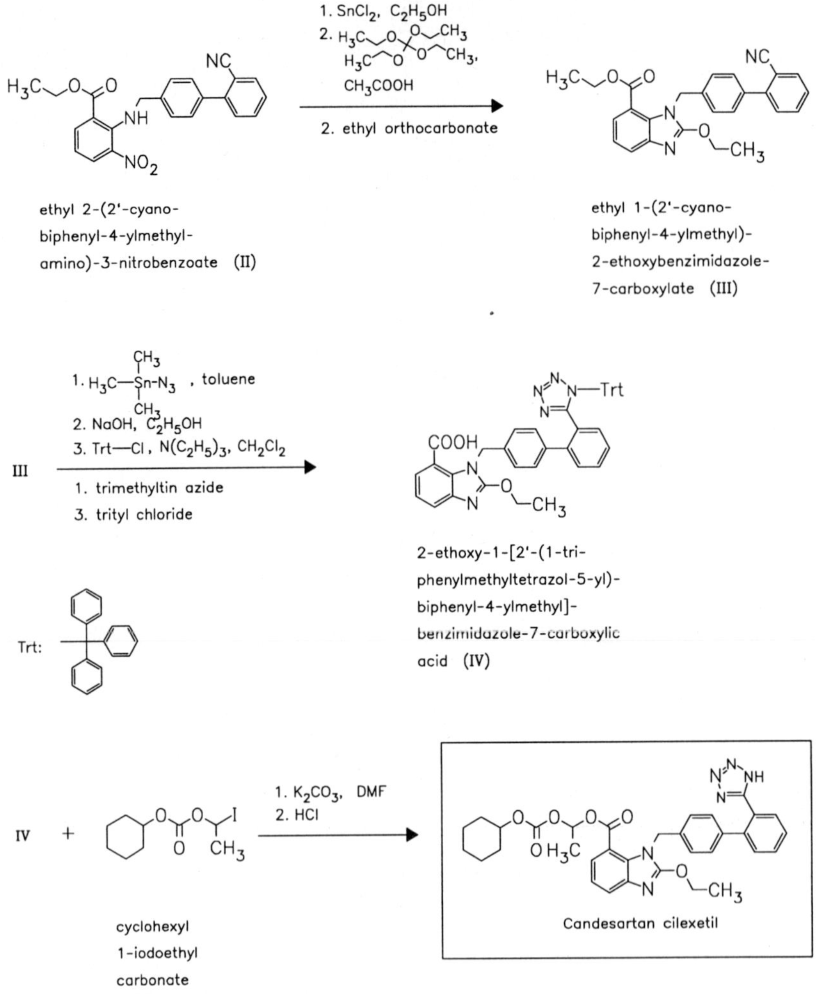
1. $SnCl_2$, C_2H_5OH
2. CH_3COOH
2. ethyl orthocarbonate
ethyl 2-(2'-cyano-biphenyl-4-ylmethyl-amino)-3-nitrobenzoate (II)
ethyl 1-(2'-cyano-biphenyl-4-ylmethyl)-2-ethoxybenzimidazole-7-carboxylate (III)
III
1. toluene
2. NaOH, C_2H_5OH
3. Trt—Cl, $N(C_2H_5)_3$, CH_2Cl_2
1. trimethyltin azide
3. trityl chloride
Trt:
2-ethoxy-1-[2'-(1-triphenylmethyltetrazol-5-yl)-biphenyl-4-ylmethyl]-benzimidazole-7-carboxylic acid (IV)
IV +
1. K_2CO_3, DMF
2. HCl
cyclohexyl 1-iodoethyl carbonate
Candesartan cilexetil

Reference(s):

EP 459 136 (Takeda Chem. Ind.; appl. 19.4.1991; J-prior. 27.4.1990, 30.5.1990, 1.10.1990).

Formulation(s): tabl. 4 mg, 8 mg, 16 mg

Trade Name(s):

D:	Atacand (Astra/Promed)	GB:	Amias (Astra; Takeda)
	Blopress (Takeda)	USA:	Atacand (Astra)

Canthaxanthin

ATC: S01JA
Use: photoprotector, dye stuff

RN: 514-78-3 MF: $C_{40}H_{52}O_2$ MW: 564.85 EINECS: 208-187-2
LD_{50}: 10 g/kg (M, p.o.)
CN: β,β-carotene-4,4'-dione

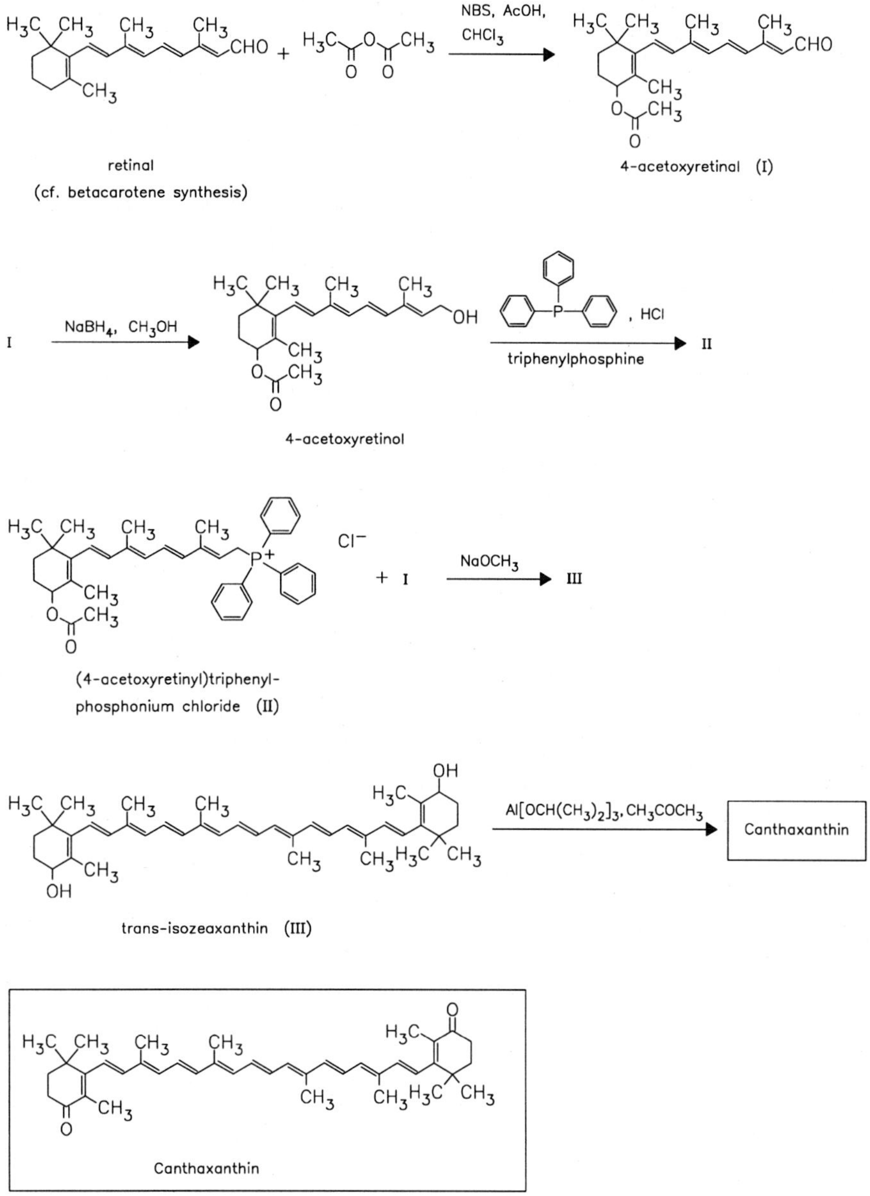
NBS, AcOH, CHCl3
retinal
(cf. betacarotene synthesis)
4-acetoxyretinal (I)
NaBH4, CH3OH
4-acetoxyretinol
triphenylphosphine
, HCl
(4-acetoxyretinyl)triphenylphosphonium chloride (II)
NaOCH3
trans-isozeaxanthin (III)
Al[OCH(CH3)2]3, CH3COCH3
Canthaxanthin
Canthaxanthin

Reference(s):

US 3 311 656 (Roche; 28.3.1967; appl. 12.5.1964).

Surmatis, J.D. et al.: Helv. Chim. Acta (HCACAV) **53**, 974 (1970).

alternative syntheses:

DOS 2 037 935 (Roche; appl. 30.7.1970; CH-prior. 1.8.1969).

US 4 000 198 (Roche; 28.12.1976; appl. 9.6.1975).

DOS 2 625 259 (Roche; appl. 4.6.1976; USA-prior. 9.6.1975).

Formulation(s): gel 10 mg/600 mg; 15 mg/900 mg

Trade Name(s):
F: Phenoro "Roche" (Prod.
Roche S.A.R.L.)-comb.
with betacarotene

Capecitabine
(Ro-09-1978)

ATC: L01BC06
Use: anticancer, orally active prodrug of
doxifluridine

RN: 154361-50-9 MF: $C_{15}H_{22}FN_3O_6$ MW: 359.35
CN: 5'-Deoxy-5-fluoro-N-[(pentyloxy)carbonyl]cytidine

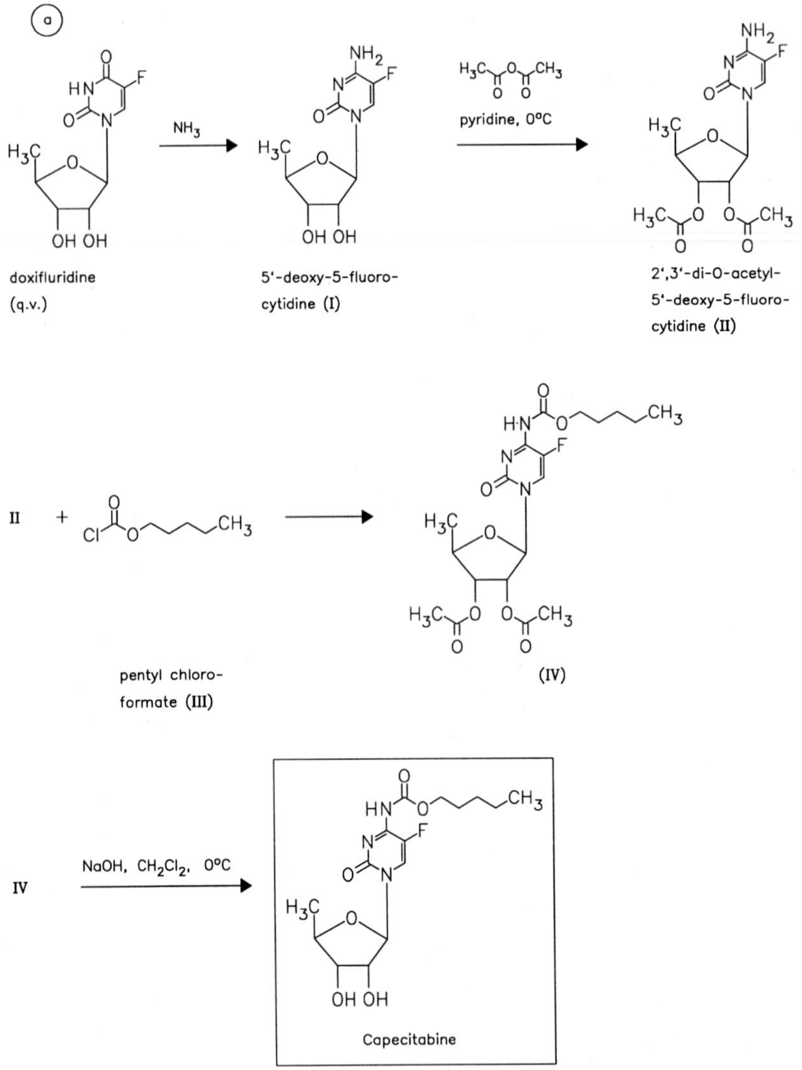

doxifluridine
(q.v.)

5'-deoxy-5-fluoro-
cytidine (I)

2',3'-di-O-acetyl-
5'-deoxy-5-fluoro-
cytidine (II)

II +

pentyl chloro-
formate (III)

(IV)

IV

NaOH, CH₂Cl₂, 0°C

Capecitabine

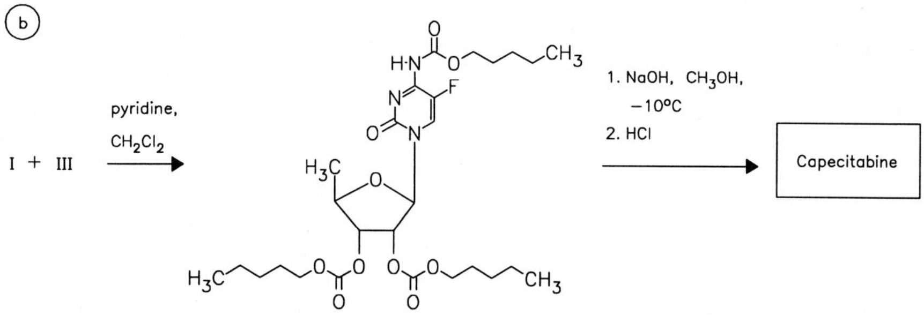

(b)

I + III →[pyridine, CH$_2$Cl$_2$] →[1. NaOH, CH$_3$OH, −10°C 2. HCl] Capecitabine

alternative preparation of intermediate II

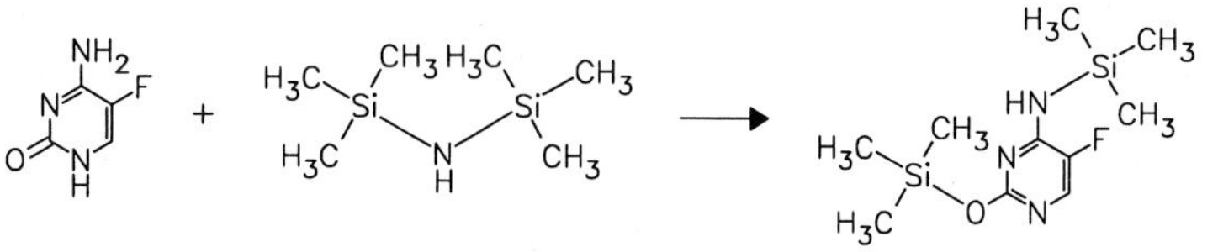

5-fluoro-
cytosine

hexamethyldisilazane

5-fluoro-2-0,4-N-bis-
(trimethylsilyl)cytosine (V)

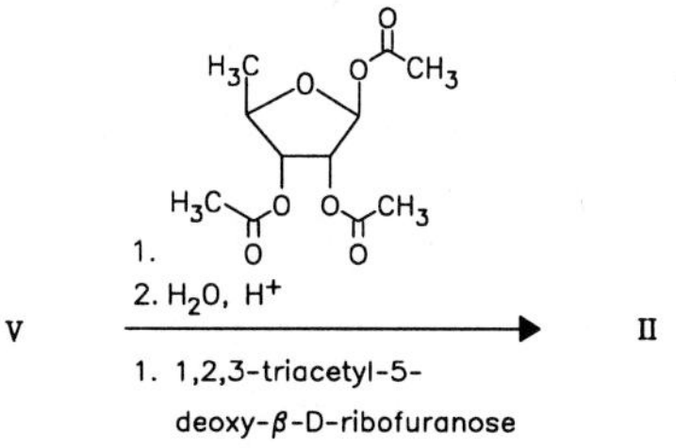

V →[1. 1,2,3-triacetyl-5-deoxy-β-D-ribofuranose 2. H$_2$O, H$^+$] II

1. 1,2,3-triacetyl-5-
deoxy-β-D-ribofuranose

Reference(s):
a   EP 602 454 (Hoffmann-La Roche; appl. 1.12.1993; EP-prior. 18.12.1992)
    US 5 472 949 (Hoffmann-La Roche; 5.12.1995; EP-prior. 18.12.1992).
b   US 5 476 932 (Hoffmann-La Roche; 19.12.1995; USA-prior. 26.8.1994)

compositions of interleukin *and pyrimidine nucleosides:*
WO 9 637 214 (Hoffmann-La Roche; appl. 15.5.1996; EP-prior. 26.5.1995)

Formulation(s):   tabl. 150 mg, 500 mg

Trade Name(s):
USA: Xeloda (Roche; 1998)

Capreomycin

(Caprenomycin)

ATC: J04AB30
Use: tuberculostatic, peptide antibiotic

RN: 11003-38-6   MF: C$_{25}$H$_{44}$N$_{14}$O$_8$ MW: 668.72
LD$_{50}$:   238 mg/kg (M, i.v.)
CN: capreomycin (mixture of capreomycin IB, IA, IIA and IIB)

sulfate
RN: 1405-37-4   MF: H$_2$SO$_4$ · unspecified   MW: unspecified   EINECS: 215-776-8
LD$_{50}$:   250 mg/kg (M, i.v.);
    325 mg/kg (R, i.v.)

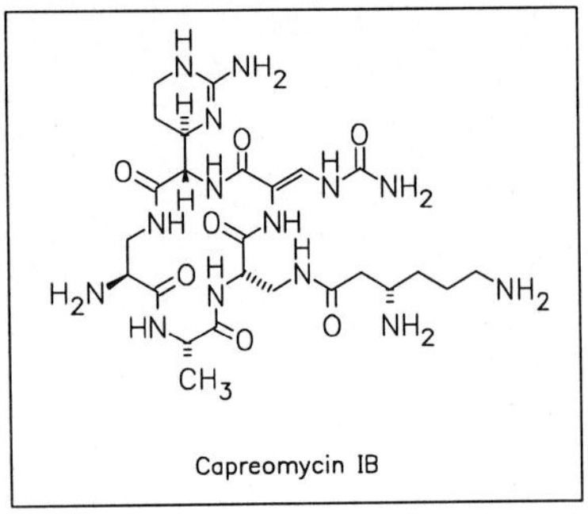

Capreomycin IB

From culture of *Streptomyces capreolus* by ion-exchange adsorption.

Reference(s):
US 3 143 468 (Eli Lilly; 4.8.1964; appl. 25.5.1962; prior. 2.11.1959).

Formulation(s):   vial 1 g

Trade Name(s):
D: Ogostal (Lilly); wfm   J: Capastat (Shionogi; as
F: Capastat (Lilly); wfm   sulfate)
GB: Capastat (King; as sulfate) USA: Capastat (Dura)

Captodiame

ATC: N05BB02
Use: psychoregulant, sedative

RN: 486-17-9   MF: $C_{21}H_{29}NS_2$ MW: 359.60   EINECS: 207-629-1
LD$_{50}$: 72 mg/kg (M, i.v.); 1630 mg/kg (M, p.o.);
    3800 mg/kg (R, p.o.)
CN: 2-[[[4-(butylthio)phenyl]phenylmethyl]thio]-*N,N*-dimethylethanamine

hydrochloride
RN: 904-04-1   MF: $C_{21}H_{29}NS_2 \cdot HCl$ MW: 396.06   EINECS: 212-992-4
LD$_{50}$: 44 mg/kg (M, i.v.)

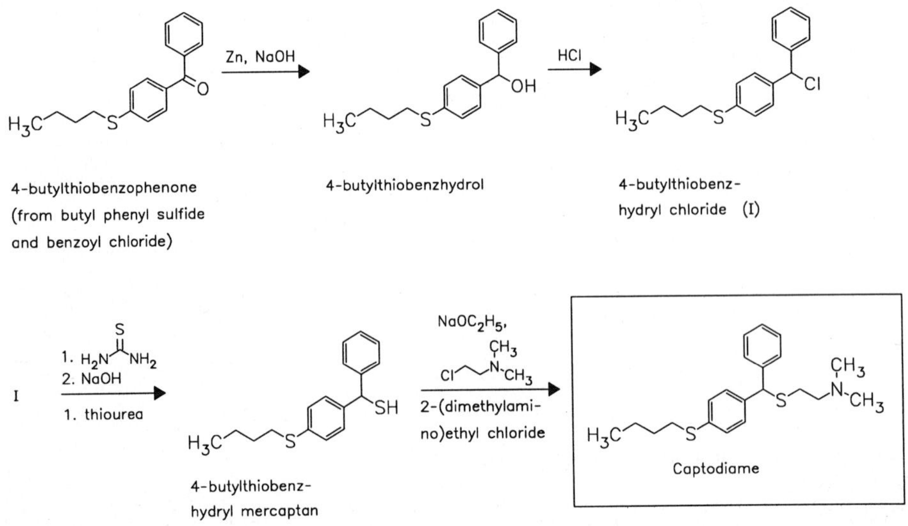

4-butylthiobenzophenone
(from butyl phenyl sulfide
and benzoyl chloride)

4-butylthiobenzhydrol

4-butylthiobenz-
hydryl chloride (I)

4-butylthiobenz-
hydryl mercaptan

Captodiame

Reference(s):
US 2 830 088 (O.H. Hubner, P.V. Petersen; 1958; DK-prior. 1952).

Formulation(s):   tabl. 50 mg

Trade Name(s):
F: Covatine (Bailly)

Captopril

ATC: C09AA01
Use: antihypertensive (ACE inhibitor)

RN: 62571-86-2   MF: $C_9H_{15}NO_3S$ MW: 217.29   EINECS: 263-607-1
LD$_{50}$:   663 mg/kg (M, i.v.); 2500 mg/kg (M, p.o.);
        554 mg/kg (R, i.v.); 4245 mg/kg (R, p.o.)
CN: (S)-1-(3-mercapto-2-methyl-1-oxopropyl)-L-proline

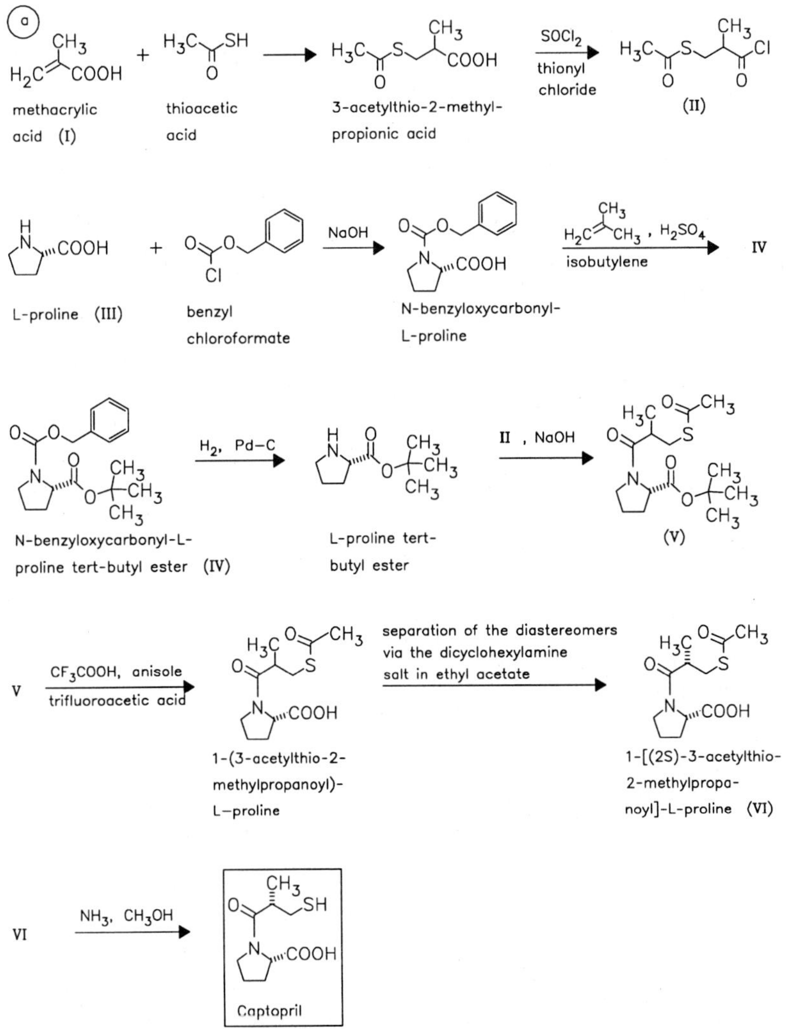

methacrylic acid (I) + thioacetic acid → 3-acetylthio-2-methyl-propionic acid → (SOCl₂, thionyl chloride) → (II)

L-proline (III) + benzyl chloroformate → (NaOH) → N-benzyloxycarbonyl-L-proline → (H₂C=CH₃ isobutylene, H₂SO₄) → IV

N-benzyloxycarbonyl-L-proline tert-butyl ester (IV) → (H₂, Pd-C) → L-proline tert-butyl ester → (II, NaOH) → (V)

V → (CF₃COOH, anisole trifluoroacetic acid) → 1-(3-acetylthio-2-methylpropanoyl)-L-proline → separation of the diastereomers via the dicyclohexylamine salt in ethyl acetate → 1-[(2S)-3-acetylthio-2-methylpropanoyl]-L-proline (VI)

VI → (NH₃, CH₃OH) → Captopril

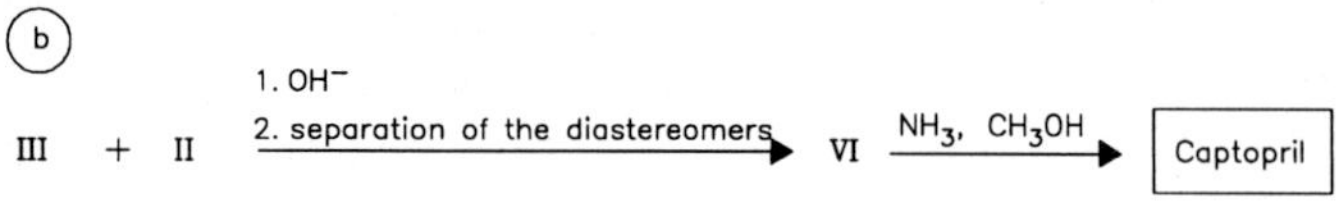

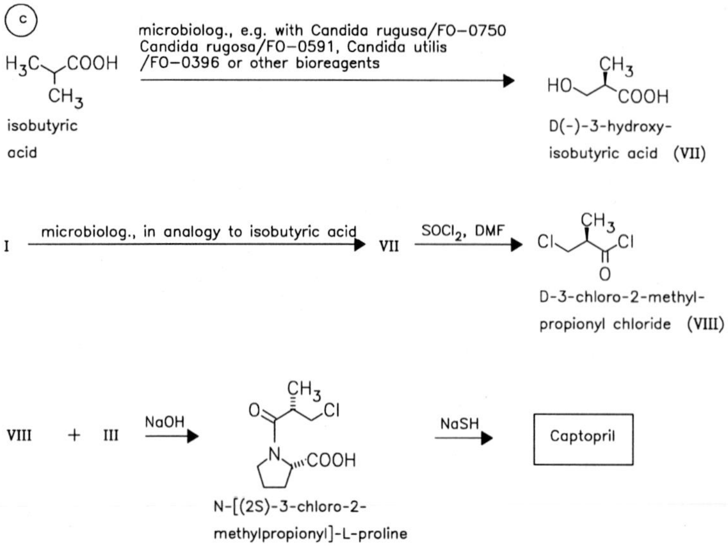

b

III + II →[1. OH⁻ / 2. separation of the diastereomers] VI →[NH₃, CH₃OH] Captopril

c

H₃C-CH(CH₃)-COOH (isobutyric acid) →[microbiolog., e.g. with Candida rugosa/FO–0750 Candida rugosa/FO–0591, Candida utilis /FO–0396 or other bioreagents] HO-CH(CH₃)-COOH D(-)-3-hydroxy-isobutyric acid (VII)

I →[microbiolog., in analogy to isobutyric acid] VII →[SOCl₂, DMF] Cl-CH₂-CH(CH₃)-CO-Cl D-3-chloro-2-methyl-propionyl chloride (VIII)

VIII + III →[NaOH] N-[(2S)-3-chloro-2-methylpropionyl]-L-proline →[NaSH] Captopril

Reference(s):

a,b DOS 2 703 828 (Squibb; appl. 31.1.1977; USA-prior. 13.2.1976; 21.6.1976, 22.12.1976).
US 4 046 889 (Squibb; 6.9.1977; appl. 13.2.1976).
US 4 105 776 (Squibb; 8.8.1978; prior. 13.2.1976, 21.6.1976, 22.12.1976).
US 4 154 840 (Squibb; 15.5.1979; prior. 13.2.1976, 21.6.1976, 22.12.1976, 9.3.1977).
US 4 154 935 (Squibb; 15.5.1979; prior. 21.2.1978, 1.9.1978).
c GB 2 065 643 (Kanegafuchi; appl. 1.12.1980; J-prior. 13.12.1979, 28.12.1979, 8.3.1980).
US 4 460 780 (Kanegafuchi; 17.7.1984; J-prior. 20.1.1982).

microbiological production of D-(–)-3-hydroxyisobutyric acid:
US 4 310 635 (Kanegafuchi; 12.1.1982; J-prior. 6.11.1979, 14.2.1980, 7.7.1980).

similar methods with D- *or* DL-3-halogeno-2-methylpropionic acids *as intermediates:*
Nam, D.H. et al.: J. Pharm. Sci. (JPMSAE) **73**, 1843 (1984).
DE 3 049 273 (Egyt; appl. 29.12.1980; HU-prior. 29.12.1979).
US 4 332 725 (Egyt; 1.6.1982; HU-prior. 29.12.1979).
GB 2 066 252 (Egyt; appl. 29.12.1980; HU-prior. 29.12.1979).
US 4 399 144 (Wyeth; 16.8.1983; GB-prior. 30.4.1980).

alternative syntheses:
via 3-mercapto-2-D-methylpropionic acid:
US 4 384 139 (Kanegafuchi; 17.5.1983; J-prior. 20.8.1980).
GB 2 082 174 (Kanegafuchi; appl. 7.8.1981; J-prior. 20.8.1980).

via 3-acylthio-2-D-methylpropionic acids:
EP 8 831 (Océ-Andeno; appl. 31.8.1979; NL-prior. 7.9.1978).

racemate resolution of DL-3-acylthio-2-methylpropionic acids:
EP 35 811 (Océ-Andeno; appl. 26.2.1981; NL-prior. 6.3.1980).
US 4 346 045 (Océ-Andeno; 24.8.1982; NL-prior. 6.3.1980).
US 4 297 282 (Sumitomo; 27.10.1981; J-prior. 2.3.1979, 13.3.1979, 28.6.1979, 25.7.1979).

racemization of L-3-acylthio-2-methylpropionic acids:
US 4 411 836 (Sumitomo; 25.10.1983; J-prior. 13.3.1979).

purification of captopril *(removal of disulfide):*
US 4 332 726 (Squibb; 1.6.1982; appl. 25.8.1980).

medical use for treatment of glaucoma:
EP 99 239 (Squibb; appl. 6.7.1983; USA-prior. 6.7.1982).

combination with diuretics:
DOS 2 854 316 (Squibb; appl. 15.12.1978; USA-prior. 27.12.1977).

controlled-release formulation:
US 4 505 890 (Squibb; 19.5.1985; appl. 30.6.1983).

Formulation(s):   tabl. 6.25 mg, 12.5 mg, 25 mg, 50 mg

Trade Name(s):

D:	Capozide (Bristol-Myers Squibb; 1984)-comb. with hydrochlorothiazide Lopirin (Bristol-Myers Squibb; 1981) Tensobon (Schwarz; 1983)		Lopril (Bristol-Myers Squibb; 1982)	Aceplus (Bristol-Myers Squibb)-comb. Acepress (Guidotti) Capoten (Bristol-Myers Squibb; 1981)
		GB:	Acepril (Ashbourne) Azecide (Ashbourne)-comb.	
F:	Captéa (Bellon)-comb. Captolane (Bellon; 1984) Ecazide (Bristol-Myers Squibb)-comb.		Capoten (Bristol-Myers Squibb; 1981) Capozide (Bristol-Myers Squibb)-comb.	J: Captoril (Sankyo; 1983) USA: Capoten (Bristol-Myers Squibb; 1981) Capozide (Bristol-Myers Squibb; 1986)
		I:	Acediur (Guidotti)-comb.	

Carazolol

ATC: C07AA
Use: non-selective β-adrenoceptor blocker, antihypertensive, antianginal

RN: 57775-29-8   MF: $C_{18}H_{22}N_2O_2$ MW: 298.39   EINECS: 260-945-1
CN: 1-(9*H*-carbazol-4-yloxy)-3-[(1-methylethyl)amino]-2-propanol

4-hydroxy-carbazole + epichloro-hydrin → NaOH, dioxane → 4-(2,3-epoxypropoxy)-carbazole (I)

I + isopropyl-amine (C_2H_5OH) → Carazolol

Reference(s):
DOS 2 240 599 (Boehringer Mannh.; appl. 18.8.1972).
GB 1 369 580 (Boehringer Mannh.; valid from 9.10.1974; D-prior. 18.8.1972).

synthesis of 4-hydroxycarbazole:
DOS 2 928 483 (Boehringer Mannh., appl. 14.7.1979).

Formulation(s):   tabl. 5 mg

Trade Name(s):
D: Conducton (Klinge)

Carbachol

(Carbacholine)

ATC: N07AB01; S01EB02
Use: parasympathomimetic

RN: 51-83-2   MF: $C_6H_{15}ClN_2O_2$ MW: 182.65   EINECS: 200-127-3
LD_{50}: 300 µg/kg (M, i.v.); 15 mg/kg (M, p.o.);
       100 µg/kg (R, i.v.); 40 mg/kg (R, p.o.)
CN: 2-[(aminocarbonyl)oxy]-*N,N,N*-trimethylethanaminium chloride

2-chloro-   phosgene   2-chloroethyl   2-chloroethyl
ethanol   chloroformate   carbamate (I)

trimethyl-   Carbachol
amine

Reference(s):
DRP 539 329 (E. Merck AG; appl. 1930).
DRP 553 148 (E. Merck AG; appl. 1930).
DRP 590 311 (E. Merck AG; appl. 1932).
Hayworth, R.D. et al.: J. Chem. Soc. (JCSOA9) **1947** 176.

alternative synthesis from choline chloride:
US 2 374 367 (Merck & Co.; 1945; prior. 1943).

Formulation(s):   amp. 0.25 mg; tabl. 2 mg

Trade Name(s):

D:	Carbamann (Mann)	F:	Iricoline (Lematte et	I:	Mios (Intes)-comb.
	Doryl (Merck)		Boinot); wfm	J:	Calpinol (Tanabe)
	Isopto-Carbachol (Alcon)		Isopto Carbachol (Alcon);	USA:	Isopto Carbachol Solut.
	Jesytryl (Chauvin		wfm		(Alcon)
	ankerpharm)	GB:	Isopto-Carbachol (Alcon)-		
			comb.		

Carbamazepine

ATC: N03AF01; N03AX
Use: antiepileptic, anticonvulsant

RN: 298-46-4   MF: $C_{15}H_{12}N_2O$ MW: 236.27   EINECS: 206-062-7
LD_{50}: 529 mg/kg (M, p.o.);
       1957 mg/kg (R, p.o.);
       5620 mg/kg (dog, p.o.)
CN: 5*H*-dibenz[*b,f*]azepine-5-carboxamide

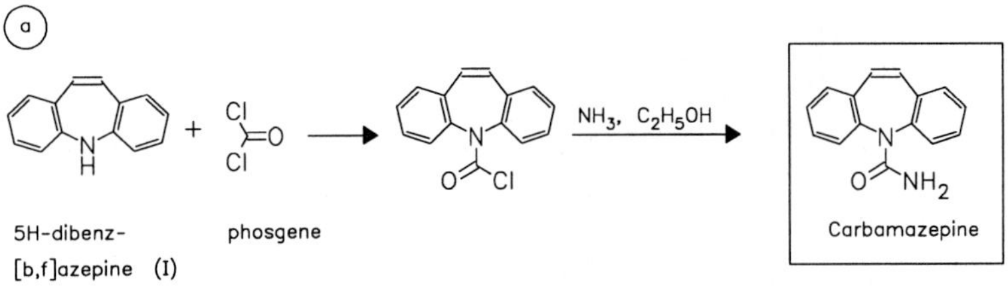

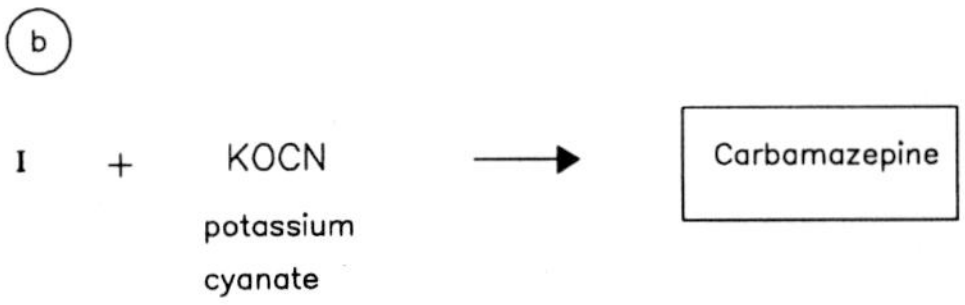

(a)

5H-dibenz-   phosgene
[b,f]azepine (I)

Carbamazepine

(b)

I   +   KOCN   ⟶   Carbamazepine

        potassium
        cyanate

Reference(s):
US 2 948 718 (Geigy; 9.8.1960; CH-prior. 20.12.1957).

alternative synthesis:
DD 133 052 (R. Müller; appl. 8.9.1977).

Formulation(s):   s. r. tabl. 200 mg, 400 mg; susp. 100 mg; tabl. 200 mg

Trade Name(s):

D:	Finlepsin (ASTA Medica		Timonil (Desitin)	J:	Tegretol (Fujisawa)
	AWD; Boehringer Mannh.)	F:	Tegrétol (Novartis)	USA:	Epitol (Teva)
	Sirtal (Merck Generika)	GB:	Tegretol (Novartis)		Tegretol (Novartis)
	Tegretal (Novartis Pharma)	I:	Tegretol (Novartis)		

N-Carbamoyl-L-aspartic acid calcium salt
(Calcii carbaspartas)

Use: psychoenergetic, tranquilizer

RN: 16649-79-9   MF: $C_5H_6CaN_2O_5$ MW: 214.19   EINECS: 240-698-6
CN: *N*-(aminocarbonyl)-L-aspartic acid calcium salt (1:1)

N-carbamoyl-L-aspartic acid
RN: 13184-27-5   MF: $C_5H_8N_2O_5$ MW: 176.13   EINECS: 236-134-3
LD_{50}:   >1 g/kg (M, p.o.)

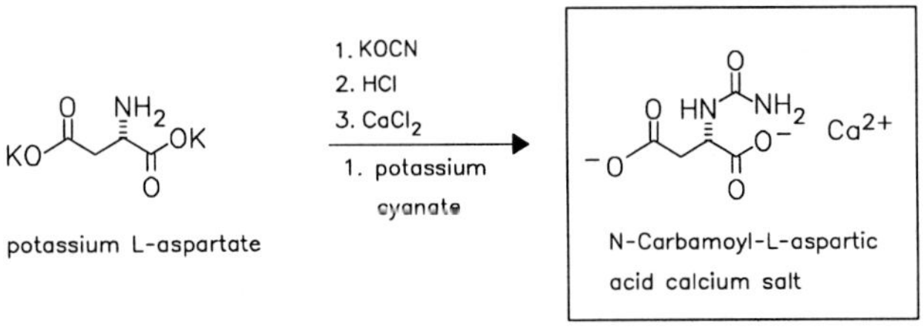

potassium L-aspartate

1. KOCN
2. HCl
3. $CaCl_2$
1. potassium
  cyanate

N-Carbamoyl-L-aspartic
acid calcium salt

Reference(s):
FR-M 6 376 (Roussel-Uclaf; appl. 18.4.1967).

Carbasalate calcium

(Calcium carbaspirin)

ATC: B01AC08; N02BA15
Use: analgesic

RN: 5749-67-7   MF: $C_{18}H_{14}CaO_8 \cdot CH_4N_2O$ MW: 458.44   EINECS: 227-273-0
CN: 2-(acetyloxy)benzoic acid calcium salt compd. with urea (1:1)

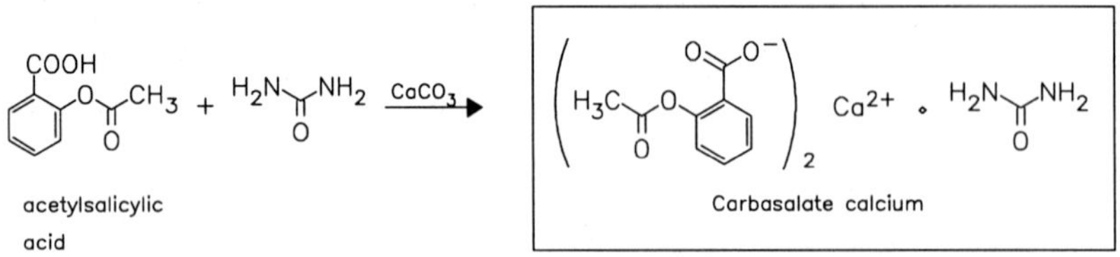

acetylsalicylic acid

Carbasalate calcium

Reference(s):
Parrott, E.L.: J. Pharm. Sci. (JPMSAE) **51**, 897 (1962).

calcium acetylsalicylate:
US 2 003 374 (Lee Labs.; 1935; appl. 1932).

Formulation(s):   tabl. 500 mg

Carbazochrome

ATC: B02BX02
Use: antihemorrhagic, hemostatic

RN: 69-81-8   MF: $C_{10}H_{12}N_4O_3$ MW: 236.23   EINECS: 200-717-0
LD$_{50}$:   >35.832 g/kg (M, p.o.);
              >17.280 g/kg (R, p.o.)
CN: 2-(1,2,3,6-tetrahydro-3-hydroxy-1-methyl-6-oxo-5H-indol-5-ylidene)hydrazinecarboxamide

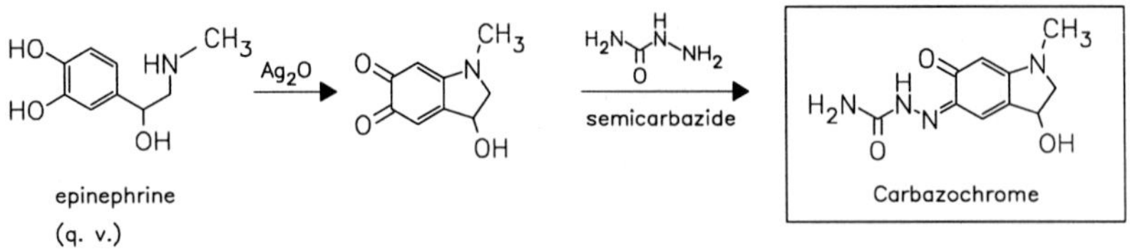

epinephrine
(q. v.)

semicarbazide

Carbazochrome

Reference(s):
US 2 506 294 (Soc. Belge de l'Azote et des Prod. Chim.; 1950; B-prior. 1943).
GB 806 908 (Labaz; appl. 1957; USA-prior. 1956).
US 3 244 591 (Endo Labs.; 5.4.1966; appl. 10.8.1960).

oxidation of adrenaline with persulfate:
DOS 2 713 652 (Nippon Gohsei; appl. 28.3.1977; J-prior. 31.3.1976).

Formulation(s):   inj. sol. 1.5 mg/3.6 ml, 50 mg/10 ml; tabl. 2.5 mg, 10 mg

Trade Name(s):

D: Adrenoxyl (Sanofi
      Winthrop)
F: Adrénoxyl (Labaz); wfm
      Bivenon (Lab. Français de
      Thérapeutique); wfm
I: Fleboside (Synthelabo)-
      comb.
J: Adcal (Nissin)
      Adedolon (Sanwa)
      Adenaron (Kowa)
      Adnamin (Kanto)

Adona (Tanabe)
Adonamin (Kanto)
Adorzon (Hokuriku)
Adostill-AC (Dojin Iyaku)
Adozon (Kyorin)
Adrechros (Toho Iyaku)
Adrezon (Ono)
Blochel (Mochida)
Carbazon (Hokuriku)
Carbinate (Fuji Zoki)
Carnamid (Kanebo)

Chichina (Fuso)
Donaseven (Kini Yakult
Seizo)
Kealain (Funai)
Ohproton (Ohta)
Olynate (Sanwa)
Perichron (Toho Yakuhin)
Shiketsumin (Ohta)
Tazin (Grelan)

USA: Adrenosem (Beecham-
        Massengill); wfm

Carbenicillin

ATC: J01CA03
Use: antibiotic

RN: 4697-36-3   MF: $C_{17}H_{18}N_2O_6S$ MW: 378.41   EINECS: 225-171-0
LD_{50}: 2363 mg/kg (M, i.v.)
CN: [2S-(2α,5α,6β)]-6-[(carboxyphenylacetyl)amino]-3,3-dimethyl-7-oxo-4-thia-1-azabicyclo[3.2.0]heptane-
       2-carboxylic acid

disodium salt
RN: 4800-94-6   MF: $C_{17}H_{16}N_2Na_2O_6S$ MW: 422.37   EINECS: 225-360-8
LD_{50}: 4500 mg/kg (M, i.v.); >12 g/kg (M, p.o.);
       6800 mg/kg (R, i.v.); >10 g/kg (R, p.o.)
monopotassium salt
RN: 17230-86-3   MF: $C_{17}H_{17}KN_2O_6S$ MW: 416.50   EINECS: 241-269-6

phenylmalonic acid   6-aminopenicil-   carbenicillin benzyl ester   (I)
benzyl ester chloride   lanic acid

Carbenicillin

Reference(s):
US 3 142 673 (Pfizer; 28.7.1964; appl. 31.3.1961).
US 3 282 926 (Beecham; 1.11.1966; GB-prior. 23.4.1963).
US 3 492 291 (Beecham; 27.1.1970; GB-prior. 23.4.1963).
DE 1 295 558 (Beecham; appl. 23.4.1964; GB-prior. 23.4.1963).
GB 1 004 670 (Beecham; appl. 23.4.1963; valid from 20.4.1964).
GB 1 197 973 (Beecham; appl. 18.4.1967).
DAS 1 770 225 (Beecham; appl. 18.4.1968; GB-prior. 18.4.1967).

from phenylmalonic acid monochloride:
DAS 2 244 556 (Pfizer; appl. 11.9.1972; USA-prior. 1.10.1971).

alternative syntheses:
DE 1 931 097 (Koninkl. Nederland. Gisten Spiritusfabriek; appl. 19.6.1969; NL-prior. 19.6.1968).
DE 1 966 702 (Koninkl. Nederland. Gisten Spiritusfabriek; appl. 19.6.1969; NL-prior. 19.6.1968).
DOS 2 622 456 (Bayer; appl. 20.5.1976).

Formulation(s):   tabl. 382 mg

Trade Name(s):

D:	Anabactyl (Beecham); wfm	I:	Geopen (Pfizer; as sodium salt)	Pyopen (Beecham-Massengill); wfm
	Carindapen (Pfizer); wfm			
	Microcillin (Bayer); wfm	J:	Gripenin (Fujisawa; as sodium salt)	
F:	Pyopen (Beecham-Sévigné); wfm	USA:	Geopen (Roerig); wfm	

Carbenoxolone

ATC: A02BX01
Use: peptic ulcer therapeutic

RN: 5697-56-3   MF: $C_{34}H_{50}O_7$ MW: 570.77   EINECS: 227-174-2
LD_{50}:   290 mg/kg (M, i.v.); 1400 mg/kg (M, p.o.);
2450 mg/kg (R, p.o.);
371 mg/kg (dog, i.v.)
CN: (3β,20β)-3-(3-carboxy-1-oxopropoxy)-11-oxoolean-12-en-29-oic acid

disodium salt
RN: 7421-40-1   MF: $C_{34}H_{48}Na_2O_7$ MW: 614.73   EINECS: 231-044-0
LD_{50}:   198 mg/kg (M, i.v.);
2450 mg/kg (R, p.o.);
371 mg/kg (dog, i.v.); 3900 mg/kg (dog, p.o.)

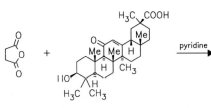

succinic anhydride   glycyrrhetic acid   Carbenoxolone

Reference(s):
DE 1 076 684 (Biorex; appl. 1.7.1958; GB-prior. 16.7.1957).
US 3 070 623 (Biorex; 25.12.1962; GB-prior. 16.7.1957).
US 3 262 851 (Biorex; 26.7.1966; GB-prior. 16.7.1957).

Formulation(s):   tabl. 50 mg

Trade Name(s):

D: Biogastrone/-Duodenal
        Degussa(Homburg); wfm
        Neogel (Homburg); wfm
        Ulcus-Tablinen
        (Beiersdorf-Tablinen); wfm
        Ulcus-Tablinen
        (Sanorania); wfm

F: Duogastrone (Merrell);
        wfm

GB: Bioral (SmithKline
        Beecham)
        Pyrogastrone (Sanofi
        Winthrop)-comb.

I: Gastrausil (ISF); wfm

        Gastrausil (Searle); wfm
        Megast (Searle)-comb.;
        wfm

J: Biogastrone (Richardson-
        Merrell-Shionogi)

Carbidopa

ATC: N04BA02
Use: decarboxylase inhibitor (at levodopa
          therapy)

RN: 28860-95-9   MF: $C_{10}H_{14}N_2O_4$ MW: 226.23   EINECS: 249-271-9
LD$_{50}$:   468 mg/kg (M, i.p.);
        2804 mg/kg (R, i.p.)
CN: (S)-α-hydrazino-3,4-dihydroxy-alphamethylbenzenepropanoic acid

monohydrate
RN: 38821-49-7   MF: $C_{10}H_{14}N_2O_4 \cdot H_2O$ MW: 244.25

(3,4-dimethoxy-
phenyl)acetone

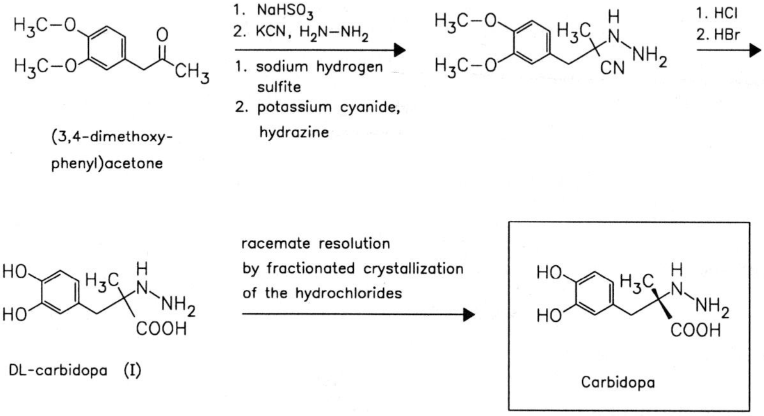

1. NaHSO₃
2. KCN, H₂N–NH₂

1. sodium hydrogen
   sulfite
2. potassium cyanide,
   hydrazine

1. HCl
2. HBr

I

DL-carbidopa (I)

racemate resolution
by fractionated crystallization
of the hydrochlorides

Carbidopa

L-N-acetyl-3-(3,4-dimethoxy-
phenyl)-2-methylalanine
(cf. methyldopa synthesis)

NaNO₂, HCl

Zn, HCl

II

HCl, 120 °C or HBr

Carbidopa

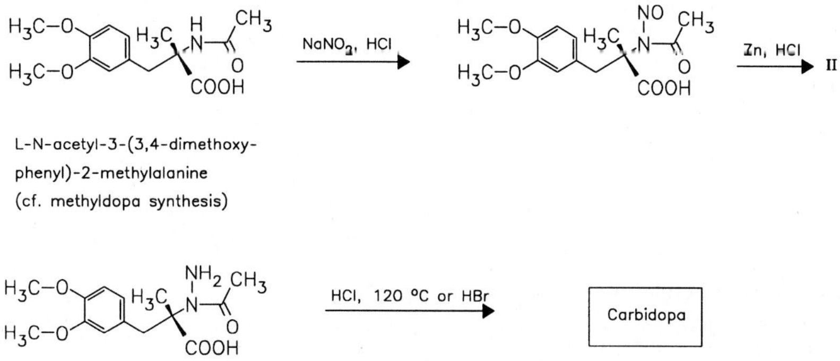

(c)

L-3-(3,4-dimethoxy-
phenyl)-2-methylalanine

potassium
cyanate

L-3-(3,4-dimethoxy-
phenyl)-2-hydrazino-
2-methylalanine (III)

Reference(s):
DL-carbidopa:
US 3 462 536 (Merck & Co.; 19.8.1969; prior. 28.7.1960 and 29.6.1961).
GB 940 596 (Merck & Co.; appl. 17.7.1961; USA-prior. 28.7.1960).

carbidopa:
DOS 2 062 285 (Merck & Co.; appl. 17.12.1970; USA-prior. 18.12.1969, 5.2.1970, 24.2.1970, 25.3.1970).
DOS 2 062 332 (Merck & Co.; appl. 17.12.1970; USA-prior. 18.12.1969, 5.2.1970, 24.2.1970, 25.3.1970).
Karady, S. et al.: J. Org. Chem. (JOCEAH) **36**, 1946, 1949 (1971).

alternative synthesis from methyldopa:
US 3 781 415 (Merck & Co.; 25.12.1973; appl. 9.9.1971; prior. 18.6.1969).
US 3 830 827 (Merck & Co.; 20.8.1974; appl. 7.9.1972; prior. 18.6.1969).

*combination with m-*tyrosine:
US 3 839 585 (Merck & Co.; 1.10.1974; appl. 30.4.1973; prior. 5.8.1970).

combination with benzimidazolyl- and benzoxazolylalanines:
US 4 069 333 (Merck & Co.; 17.1.1978; appl. 8.2.1977; prior. 13.2.1976).

combination with other antihypertensives:
US 4 086 354 (Merck & Co.; 25.4.1978; prior. 13.2.1976, 8.2.1977).

combination with hydralazine:
US 4 055 645 (Merck & Co.; 25.10.1977; appl. 13.2.1976).

Formulation(s):   s. r. tabl. 27 mg, 54 mg; tabl. 27 mg

Trade Name(s):

D:	isicom (Desitin)	GB:	Sinemet (Du Pont)-comb.		Neodopaston (Sankyo)-
	Nacom (Du Pont Pharma)-		with levodopa		comb. with levodopa
	comb. with levodopa	I:	Sinemet (Du Pont Pharma	USA:	Atamet (Athena)
	Striaton (Knoll)		Italia)-comb. with levodopa		Sinemet (Du Pont)
F:	Sinemet (Du Pont Pharma)-	J:	Menesit (Merck-Banyu)-		
	comb. with levodopa		comb. with levodopa		

Carbimazole

ATC: H03BB01
Use: antithyroid drug

RN: 22232-54-8   MF: C$_7$H$_{10}$N$_2$O$_2$S   MW: 186.24   EINECS: 244-854-4
CN: 2,3-dihydro-3-methyl-2-thioxo-1*H*-imidazole-1-carboxylic acid ethyl ester

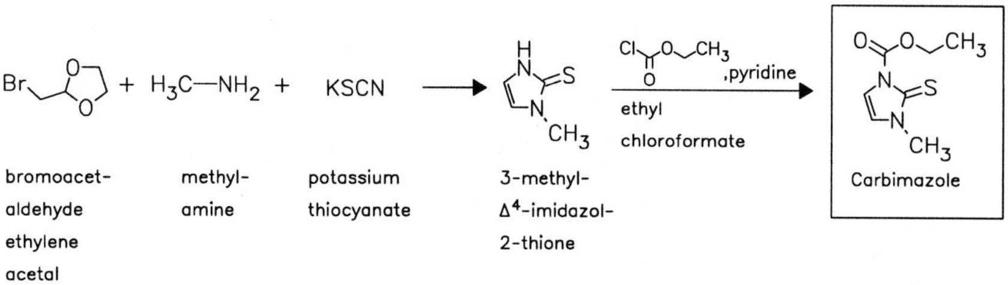

Br — [ethylene acetal] + H₃C—NH₂ + KSCN → 3-methyl-Δ⁴-imidazol-2-thione → Carbimazole

| bromoacet-aldehyde ethylene acetal | methyl-amine | potassium thiocyanate | 3-methyl-Δ⁴-imidazol-2-thione | Carbimazole |

Reference(s):

US 2 671 088 (Nat. Res. Dev. Corp.; 1954; GB-prior. 1951).
US 2 815 349 (Nat. Res. Dev. Corp.; 1957; GB-prior. 1956).
Baker, J.A.: J. Chem. Soc. (JCSOA9) **1958**, 2387.

Formulation(s):   tabl. 5 mg, 10 mg

Trade Name(s):
D: Neo-Thyreostat (Herbrand)   I: Carbotiroid (Borromeo);
F: Néo-Mercazole (Nicholas)   wfm
GB: Neo-Mercazole (Roche)   Neo-Tireol (Granata); wfm

Carbinoxamine

ATC: R06AA08
Use: antihistaminic

RN: 486-16-8   MF: $C_{16}H_{19}ClN_2O$ MW: 290.79   EINECS: 207-628-6
LD_{50}:   18 mg/kg (M, i.v.)
CN: 2-[(4-chlorophenyl)-2-pyridinylmethoxy]-*N,N*-dimethylethanamine

maleate (1:1)
RN: 3505-38-2   MF: $C_{16}H_{19}ClN_2O \cdot C_4H_4O_4$ MW: 406.87   EINECS: 222-498-0
LD_{50}:   32 mg/kg (M, i.v.); 162 mg/kg (M, p.o.)

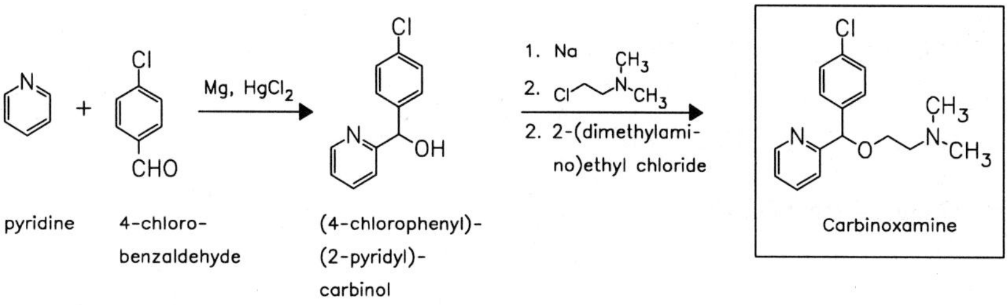

| pyridine | 4-chloro-benzaldehyde | (4-chlorophenyl)-(2-pyridyl)-carbinol | | Carbinoxamine |

1. Na
2. 2-(dimethylamino)ethyl chloride

Reference(s):

US 2 606 195 (Merrell; 1952; prior. 1947).

alternative synthesis:
US 2 800 485 (McNeil; 1957; appl. 1955).

Formulation(s):   tabl. 2 mg

Trade Name(s):
D: Polistin T-Caps   F: Allergefon (Lafon)   Torfam (Abbott)-comb.
     (Trommsdorff)   Humex Fournier gélule   J: Chlorcap Nyscap (S. S.
     Rhinopront (Mack, Illert.)   (Urgo)-comb.   Pharm.; as maleate)
     Rhinotussal (Mack, Illert.)   I: Rondec (Abbott)-comb.   Hislosine (Toho)

USA: Biohist (Wakefield)
        Rondec (Dura; as maleate)

Carbocisteine
(Carboxymethylcysteine)

ATC: R05CB03
Use: secretolytic, mucolytic agent

RN: 638-23-3   MF: $C_5H_9NO_4S$ MW: 179.20   EINECS: 211-327-5
LD_{50}:   8400 mg/kg (M, p.o.);
        >15 g/kg (R, p.o.)
CN: S-(carboxymethyl)-L-cysteine

chloroacetic   L-cysteine   Carbocisteine
acid

Reference(s):
FR 1 288 907 (Rech. et Propagande Scientifiques; appl. 15.2.1961).

preparation from L-cystine:
DAS 2 647 094 (Degussa; appl. 19.10.1976).
US 4 129 593 (Degussa; 12.12.1978; D-prior. 19.10.1976).

Formulation(s):   cps. 375 mg; syrup 280 mg

Trade Name(s):
D: Mucopront (Mack, Illert.)
      Sedotussin (Rodleben;
      UCB; Vedim)
      Transbronchin (ASTA
      Medica AWD)
F: Bronchathiol (Martin-
      Johnson & Johnson-MSD)
      Bronchocyst (SmithKline
      Beecham)
      Bronchokod
      (Biogalénique)
      Broncloclar (Oberlin)
      Broncorinol (Roche
      Nicholas)
      Bronkirex (Irex)

Cadotussyl (Whitehall)
Drill Expectorant (Pierre
Fabre)
Fluditec (Innotech
International)
Fluvic (Pierre Fabre)
Médibronc (Elerté)
Muciclar (Parke Davis)
Mucotrophir (Sanofi
Winthrop)
Pectasan (RPR Cooper)
Rhinathiol (Joullié)
GB: Mucodyne (Rhône-Poulenc
       Rorer)
I: Carbocit (CT)

Fluifort (Dompé)
Lisil (KBR)
Lisomucil (Synthelabo)
Mucocis (Crosara)
Mucojet (Polifarma)
Mucolase (Lampugnani)
Mucosol (Tosi-Novara)
Mucotreis (Ecobi)
Polimucil (Poli)-comb.
Reomucil (Astra-Simes)
Solfomucil (Locatelli)
Solucis (Magis)
Superthiol (Francia Farm.)
J: Mucodyne (Kyorin)

Carbocromen
(Carbochromen; Chromonar)

ATC: C01DX05
Use: coronary vasodilator

RN: 804-10-4   MF: $C_{20}H_{27}NO_5$ MW: 361.44   EINECS: 212-356-6
LD_{50}:   35.5 mg/kg (M, i.v.); 6300 mg/kg (M, p.o.)
CN: [[3-[2-(diethylamino)ethyl]-4-methyl-2-oxo-2H-1-benzopyran-7-yl]oxy]acetic acid ethyl ester

hydrochloride
RN: 655-35-6   MF: $C_{20}H_{27}NO_5 \cdot HCl$ MW: 397.90   EINECS: 211-511-5
LD_{50}:   34 mg/kg (M, i.v.); 6300 mg/kg (M, p.o.);
        8 g/kg (R, p.o.)

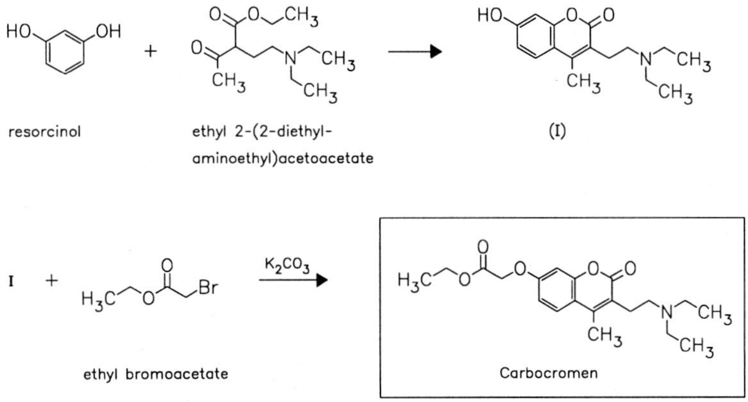

resorcinol

ethyl 2-(2-diethyl-
aminoethyl)acetoacetate

(I)

I +   ethyl bromoacetate $\xrightarrow{K_2CO_3}$ Carbocromen

Reference(s):

BE 621 327 (Cassella; appl. 10.8.1962; D-prior. 12.8.1961).
DE 1 210 883 (Cassella; appl. 9.11.1961).
US 3 282 938 (Cassella; 1.11.1966; D-prior. 12.8.1961, 9.11.1961, 26.1.1962).

Formulation(s):   cps. 75 mg, 150 mg; tabl. 450 mg

Trade Name(s):

D: Intensain (Hoechst)   I: Cardiocap (Miba; as
F: Intensain (Diamant); wfm   hydrochloride)
      Sédo-Intensain (Diamant)-   J: Intensain (Takeda; as
      comb.; wfm   hydrochloride)

Carboplatin

(CBDCA; Paraplatin)

ATC: L01XA02
Use: antineoplastic

RN: 41575-94-4   MF: $C_6H_{12}N_2O_4Pt$ MW: 371.25   EINECS: 255-446-0
LD$_{50}$:   150 mg/kg (M, i.p.); 140 mg/kg (M, i.v.);
       85 mg/kg (R, i.v.)
CN: (*SP*-4-2)-diammine[1,1-cyclobutanedi(carboxylato-κ*O*)(2-)]platinum

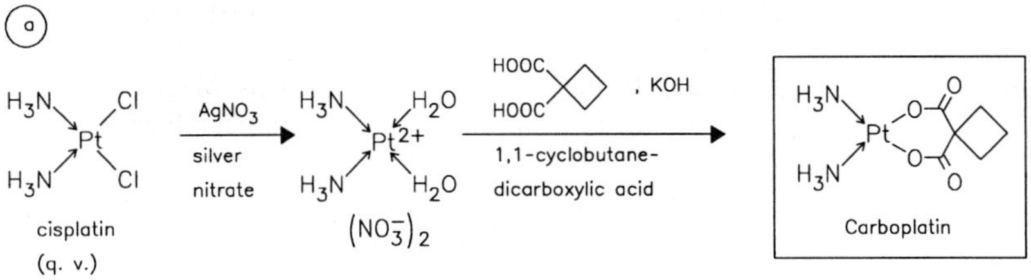

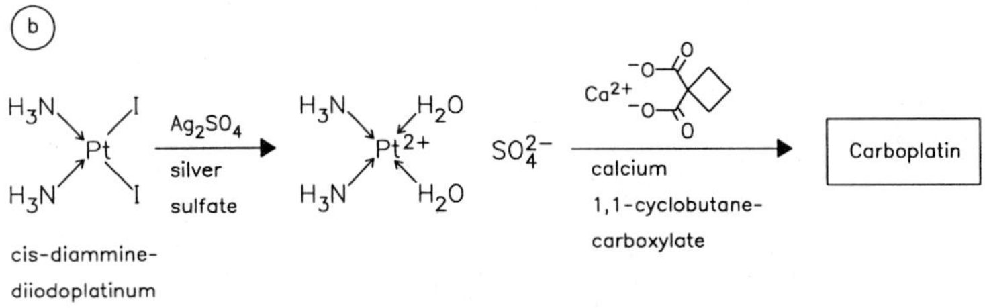

Reference(s):
a   US 4 140 707 (Research Corp.; 20.2.1979; prior. 8.6.1972).
    DE 2 329 485 (Research Corp.; appl. 8.6.1973; USA-prior. 8.6.1972).
    GB 1 380 228 (Research Corp.; Complete specification 8.1.1975; USA-prior. 8.6.1972).
b   Harrison, R.C. et al.: Inorg. Chim. Acta (ICHAA3) **46**, L15 (1980).

Formulation(s):   vial 50 mg/5 ml, 150 mg/15 ml, 450 mg/45 ml

Trade Name(s):

D:	Carboplat (Bristol-Myers Squibb)	GB:	Paraplatin (Bristol-Myers Squibb; 1985)
	Ribocarbo (ribosepharm)	I:	Paraplatin (Bristol It. Sud)
F:	Paraplatine (Bristol-Myers Squibb)	J:	Paraplatin (Bristol-Myers Squibb)

USA: Paraplatin (Bristol-Myers Squibb Oncology/ Immunology)

Carboquone
(Carbazilquinone)

ATC: L01AC03
Use: antineoplastic

RN: 24279-91-2   MF: $C_{15}H_{19}N_3O_5$ MW: 321.33
LD_{50}:   5430 µg/kg (M, i.v.); 28.6 mg/kg (M, p.o.);
          3620 µg/kg (R, i.v.); 27.3 mg/kg (R, p.o.)
CN: 2-[2-[(aminocarbonyl)oxy]-1-methoxyethyl]-3,6-bis(1-aziridinyl)-5-methyl-2,5-cyclohexadiene-1,4-dione

2-(2-aminocarbonyl-
oxy-1-methoxyethyl)-
5-methylhydroquinone
dimethyl ether

2-(2-aminocarbonyl-
oxy-1-methoxyethyl)-
5-methyl-1,4-
benzoquinone

Carboquone

Reference(s):
DOS 1 905 224 (Sankyo; appl. 28.1.1969; J-prior. 29.1.1968, 28.12.1968).
Nakao, H. et al.: Ann. Sankyo Res. Lab. (SKKNAJ) **27**, 1 (1976).

Formulation(s):   amp. 1 mg; tabl. 0.5 mg

Trade Name(s):
J: Carbazilquinone (Sankyo)   Esquinone (Sankyo)

Carbromal

ATC: N05CM04
Use: sedative, hypnotic

RN: 77-65-6   MF: $C_7H_{13}BrN_2O_2$ MW: 237.10   EINECS: 201-046-6
LD_{50}:   464 mg/kg (M, p.o.);
          427 mg/kg (R, i.v.); 316 mg/kg (R, p.o.)
CN: *N*-(aminocarbonyl)-2-bromo-2-ethylbutanamide

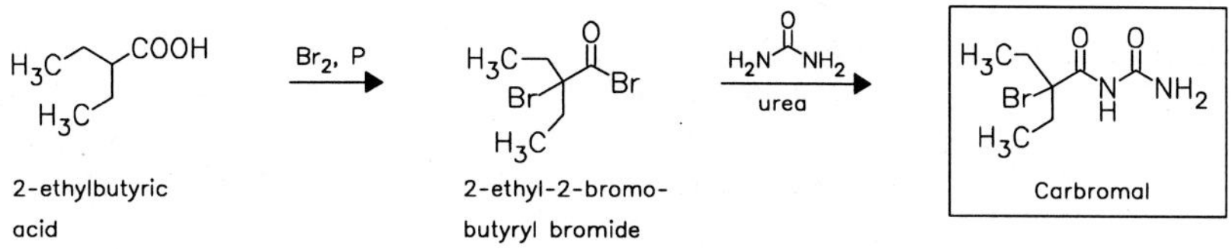

2-ethylbutyric acid → 2-ethyl-2-bromo-butyryl bromide → Carbromal

Reference(s):
DRP 225 710 (Bayer; 1909).

Formulation(s):   drg. 250 mg

Trade Name(s):

D: Adalin (Bayer); wfm
    Addisomnol (Synochem);
    wfm
    Mirfudorm (Diabetylin);
    wfm

    Mirfurdorm (Merckle);
    wfm
    Staurodorm Neu
    (Dolorgiet)-comb.; wfm

F: Divalentyl (Promedica)-
    comb.; wfm
    Dormopan (Bayer-Pharma)

I: Bonares (ISF)-comb.; wfm
    Contradol Merz (SIT); wfm

Carbutamide

(Butylcarbamide; Glybutamide)

ATC: A10BB06
Use: antidiabetic

RN: 339-43-5   MF: $C_{11}H_{17}N_3O_3S$ MW: 271.34   EINECS: 206-424-4
LD$_{50}$:   1920 mg/kg (M, i.v.); 2800 mg/kg (M, p.o.);
            980 mg/kg (R, i.v.); 7800 mg/kg (R, p.o.)
CN: 4-amino-*N*-[(butylamino)carbonyl]benzenesulfonamide

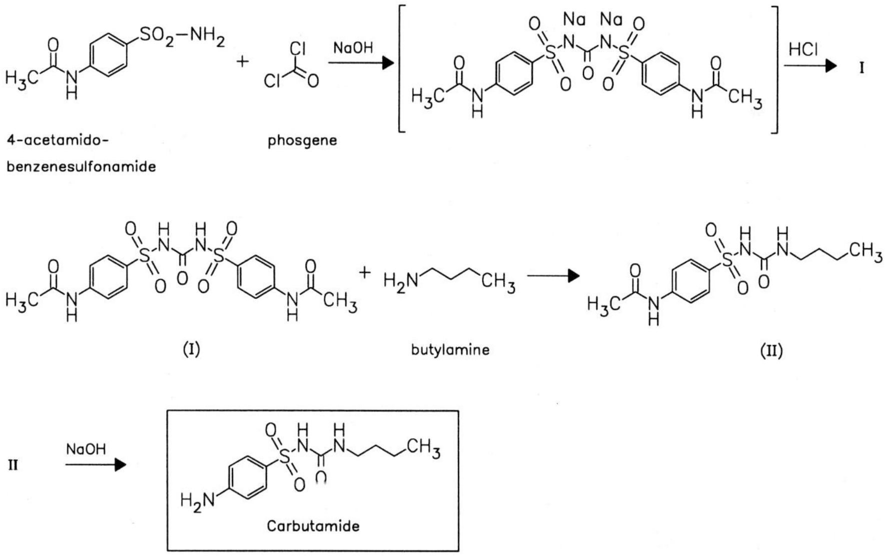

4-acetamido-benzenesulfonamide + phosgene → NaOH → (I) → HCl → I

(I) + butylamine (H_2N-CH$_3$) → (II)

II → NaOH → Carbutamide

Reference(s):
DE 1 117 103 (Boehringer Mannh.; appl. 1953).
US 2 907 692 (Boehringer Mannh.; 6.10.1959; D-prior. 11.2.1953).
Haack, E.: Arzneim.-Forsch. (ARZNAD) **8**, 444 (1958).

Formulation(s):   tabl. 0.5 g

Trade Name(s):

D: Dia-Tablinen (Sanorania); wfm
Invenol (Hoechst); wfm
Nadisan (Boehringer Mannh.); wfm

F: Glucidoral (Servier)

I: Diabetoplex (Vaillant); wfm
Insoral (Valeas); wfm
Invenol (Hoechst); wfm

J: Mellitos (Oho)
Rovan (Sumitomo)

Carbuterol

ATC: R03AC10; R03CC10
Use: selective β-adrenoceptor agonist, bronchodilator

RN: 34866-47-2 MF: $C_{13}H_{21}N_3O_3$ MW: 267.33 EINECS: 252-257-5
LD_{50}: 38 mg/kg (M, i.v.); 3134 mg/kg (M, p.o.);
77,2 mg/kg (R, i.v.)
CN: [5-[2-[(1,1-dimethylethyl)amino]-1-hydroxyethyl]-2-hydroxyphenyl]urea

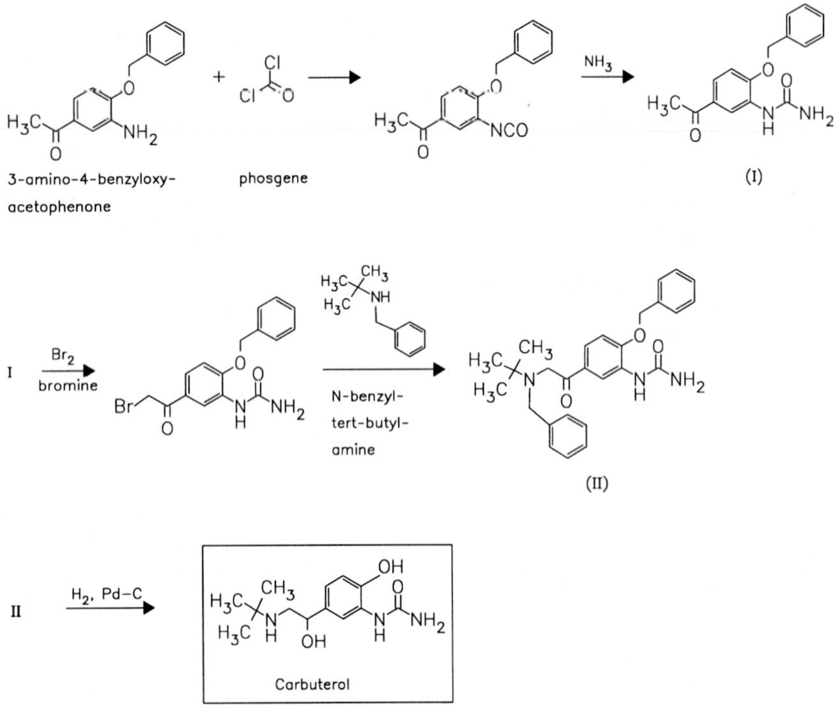
3-amino-4-benzyloxy-acetophenone
phosgene
NH_3
(I)
I
Br_2
bromine
N-benzyl-tert-butyl-amine
(II)
II
H_2, Pd-C
Carbuterol

Reference(s):

US 3 763 232 (Smith Kline & French; 2.10.1973; prior. 17.2.1970, 11.1.1971).
DOS 2 106 620 (Smith Kline & French; appl. 12.2.1971; USA-prior. 17.2.1970).
US 3 917 847 (Smith Klinc & Frcnch; 4.11.1975; prior. 11.1.1971, 17.2.1970).
Kaiser, C. et al.: J. Med. Chem. (JMCMAR) **17**, 49 (1974).

Formulation(s): aerosol 0.1 mg/puff; sol. 1 mg/0.8 ml; tabl. 2 mg

Trade Name(s):

D: Pirem (Gödecke); wfm
Pirem (Gödecke/Sasse); wfm

I: Bronsecur (Parke Davis)
Dynavent (Piam)

Carfecillin

ATC: G01AA08; J01CA
Use: antibiotic

RN: 27025-49-6 MF: C$_{23}$H$_{22}$N$_2$O$_6$S MW: 454.50 EINECS: 248-171-2
LD$_{50}$: 728 mg/kg (M, i.v.); 3924 mg/kg (M, p.o.)
CN: [2S-(2α,5α,6β)]-6-[(1,3-dioxo-3-phenoxy-2-phenylpropyl)amino]-3,3-dimethyl-7-oxo-4-thia-1-azabicyclo[3.2.0]heptane-2-carboxylic acid

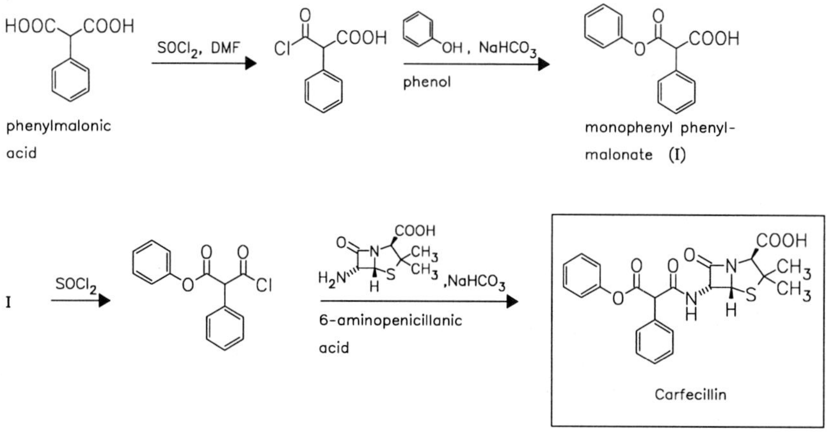

phenylmalonic acid

SOCl$_2$, DMF

phenol, OH, NaHCO$_3$

monophenyl phenyl-malonate (I)

I SOCl$_2$

6-aminopenicillanic acid, H$_2$N, NaHCO$_3$

Carfecillin

Reference(s):
US 3 853 849 (Beecham; 10.12.1974; prior. 2.11.1967 and 29.5.1969).
US 3 881 013 (Beecham; 29.4.1975; GB-prior. 5.11.1966 and 27.1.1967).

Formulation(s): tabl. 500 mg

Trade Name(s):
I: Uricillina (IBI) J: Gripenin-O (Fujisawa)
    Urocarf (SPA; as sodium   Uticillin (SmithKline
    salt)   Beecham)

Carfenazine
(Carphenazine)

ATC: N05AK
Use: neuroleptic

RN: 2622-30-2 MF: C$_{24}$H$_{31}$N$_3$O$_2$S MW: 425.60 EINECS: 220-072-9
CN: 1-[10-[3-[4-(2-hydroxyethyl)-1-piperazinyl]propyl]-10H-phenothiazin-2-yl]-1-propanone

dimaleate
RN: 2975-34-0 MF: C$_{24}$H$_{31}$N$_3$O$_2$S · 2C$_4$H$_4$O$_4$ MW: 657.74 EINECS: 221-019-2
LD$_{50}$: 42 mg/kg (M, i.v.); 156 mg/kg (M, p.o.);
      162 mg/kg (R, p.o.)

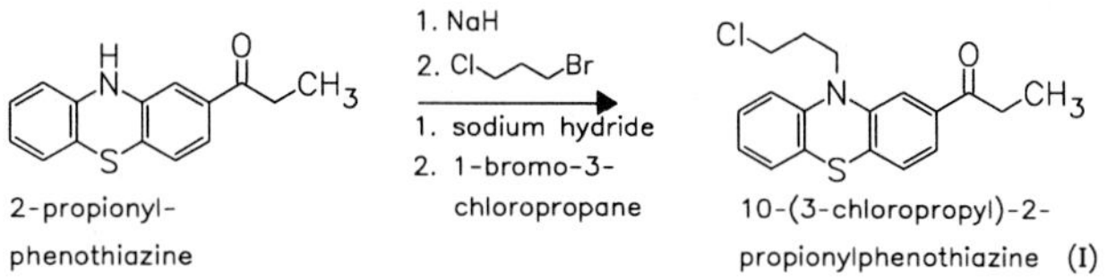

2-propionyl-phenothiazine

1. NaH
2. Cl—Br

1. sodium hydride
2. 1-bromo-3-chloropropane

10-(3-chloropropyl)-2-propionylphenothiazine (I)

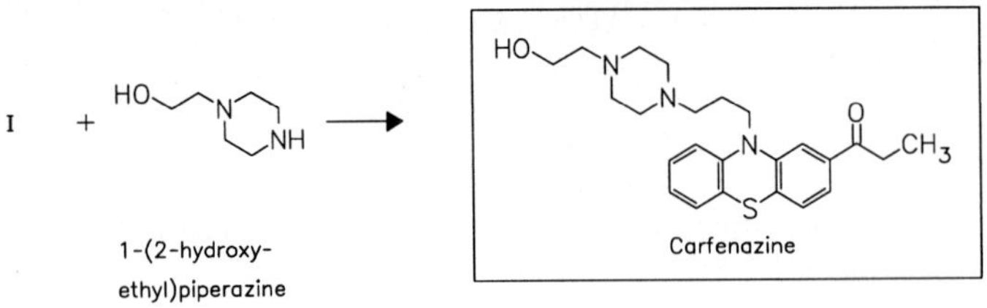

I + 1-(2-hydroxy-ethyl)piperazine → Carfenazine

Reference(s):
US 2 985 654 (Schering Corp.; 1961; appl. 1956).
US 3 023 146 (American Home; 27.2.1962; appl. 6.6.1960; prior. 3.6.1959).

Formulation(s):   tabl. 25 mg, 400 mg

Trade Name(s):
USA: Proketazine (Wyeth); wfm

Carindacillin
(Indanylcarbenicilline; Carbenicillin Indanyl Sodium)

ATC: J01CA05
Use: antibiotic

RN: 35531-88-5   MF: $C_{26}H_{26}N_2O_6S$ MW: 494.57
LD$_{50}$:   3600 mg/kg (M, p.o.);
2 g/kg (R, p.o.);
>500 mg/kg (dog, p.o.)
CN: [2S-(2α,5α,6β)]-6-[[3-[(2,3-dihydro-1H-inden-5-yl)oxy]-1,3-dioxo-2-phenylpropyl]amino]-3,3-dimethyl-7-oxo-4-thia-1-azabicyclo[3.2.0]heptane-2-carboxylic acid

monosodium salt
RN: 26605-69-6   MF: $C_{26}H_{25}N_2NaO_6S$ MW: 516.55   EINECS: 247-845-3
LD$_{50}$:   210 mg/kg (M, i.v.); 4400 mg/kg (M, p.o.);
295 mg/kg (R, i.v.); 4450 mg/kg (R, p.o.);
>15.3 mg/kg (dog, p.o.)

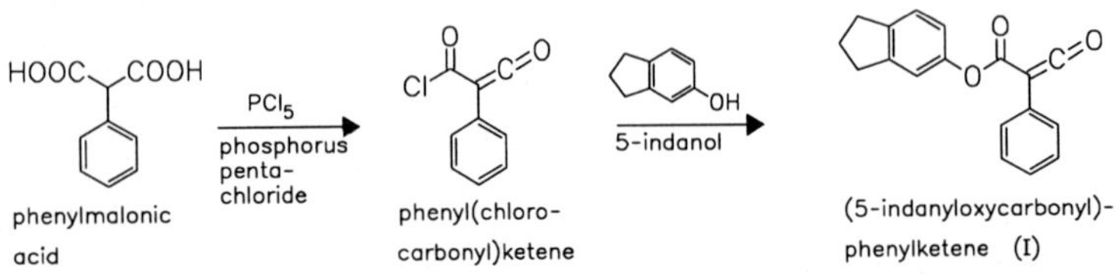

phenylmalonic acid → (PCl₅ phosphorus pentachloride) → phenyl(chloro-carbonyl)ketene → (5-indanol) → (5-indanyloxycarbonyl)-phenylketene (I)

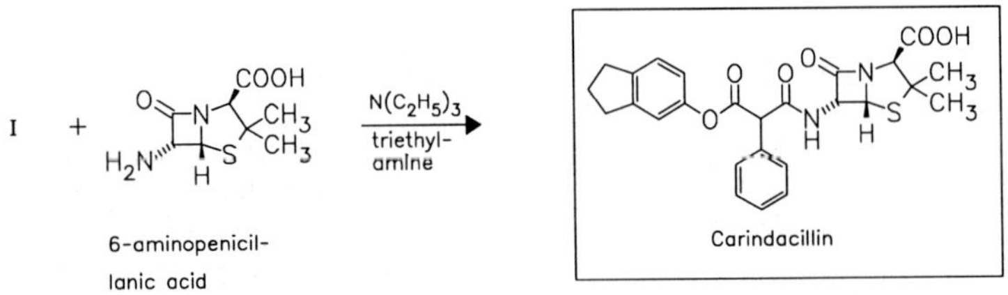

I + 6-aminopenicillanic acid → (N(C₂H₅)₃ triethylamine) → Carindacillin

Reference(s):
US 3 557 090 (Pfizer; 19.1.1971; appl. 5.1.1968).
US 3 574 189 (Pfizer; 6.4.1971; appl. 5.1.1968).
US 3 679 801 (Pfizer; 25.7.1972; prior. 5.1.1968, 4.6.1969, 19.5.1970).
DAS 1 967 024 (Pfizer; appl. 3.1.1969; USA-prior. 5.1.1968).

alternative synthesis:
DOS 1 959 569 (Pfizer; appl. 27.11.1969; USA-prior. 23.1.1969).

Formulation(s):   tabl. 500 mg

Trade Name(s):
D: Carindapen (Pfizer; 1973);   J: Geopen-U (Taito Pfizer;
    wfm   1976)
I: Geopen orale (Pfizer; 1973)   USA: Geocillin (Pfizer; 1972)

Carisoprodol

ATC: M03BA02
Use: muscle relaxant

RN: 78-44-4   MF: $C_{12}H_{24}N_2O_4$ MW: 260.33   EINECS: 201-118-7
LD$_{50}$:   165 mg/kg (M, i.v.); 1800 mg/kg (M, p.o.);
    450 mg/kg (R, i.v.); 1320 mg/kg (R, p.o.)
CN: (1-methylethyl)carbamic acid 2-[[(aminocarbonyl)oxy]methyl]-2-methylpentyl ester

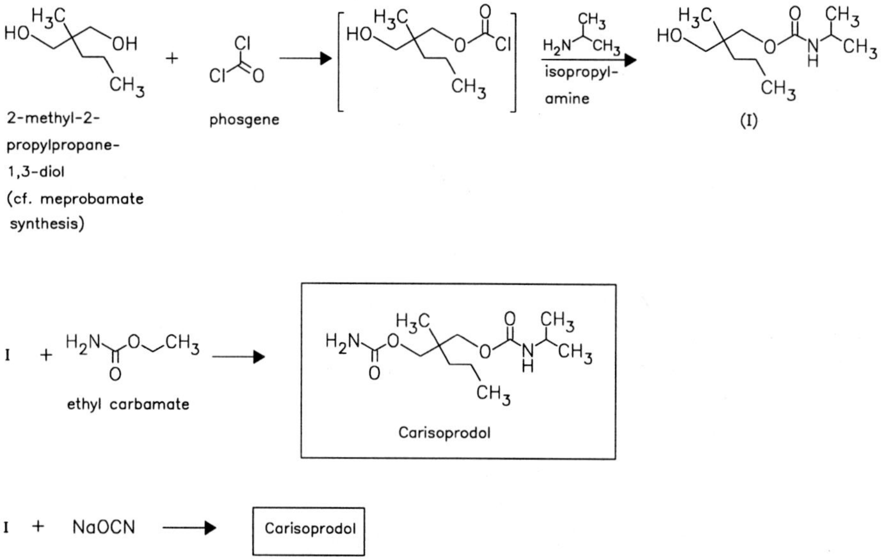

2-methyl-2-
propylpropane-
1,3-diol
(cf. meprobamate
synthesis)

phosgene

isopropyl-
amine

(I)

ethyl carbamate

Carisoprodol

I + NaOCN ⟶ Carisoprodol

Reference(s):
US 2 937 119 (Carter Products; 17.5.1960; prior. 11.6.1959).

Formulation(s):   tabl. 350 mg

Trade Name(s):
D: Sanoma (Heilit)   Flexartal (Clin-Midy); wfm   numerous combination
F: Flexagit (Clin-Midy)-   GB: Carisoma (Pharmax)   preparations
    comb.; wfm   I: Flexidone (Pierrel)-comb.   J: Myobutazolidin (Ciba-
    Flexalgit (Clin-Comar-   Soma Complex   Geigy-Fujisawa)-comb.
    Byla)-comb.; wfm   (Teofarma)-comb.   Somanil (Banyu)
    Flexartal (Clin-Comar-   Teknadone (Teknofarma)-   USA: Soma (Wallace)
    Byla); wfm   comb.

Carmofur
(HCFU)

ATC: L01BC04
Use: antineoplastic, orally active fluorouracil derivative

RN: 61422-45-5   MF: $C_{11}H_{16}FN_3O_3$ MW: 257.27
LD$_{50}$: 1129 mg/kg (M, p.o.);
        268 mg/kg (R, p.o.);
        65 mg/kg (dog, p.o.)
CN: 5-fluoro-N-hexyl-3,4-dihydro-2,4-dioxo-1(2H)-pyrimidinecarboxamide

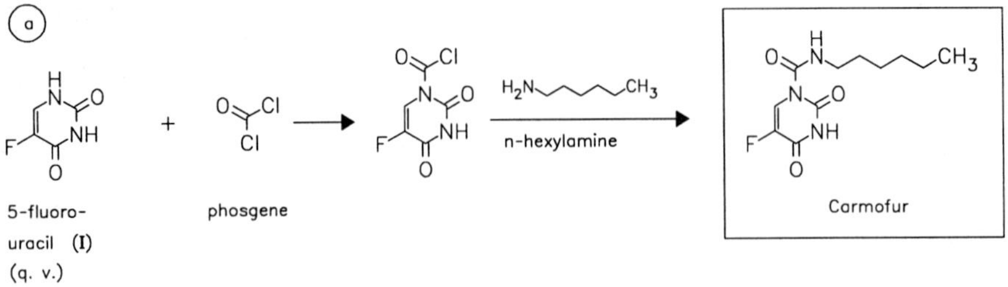

5-fluoro-
uracil (I)
(q. v.)

phosgene

n-hexylamine

Carmofur

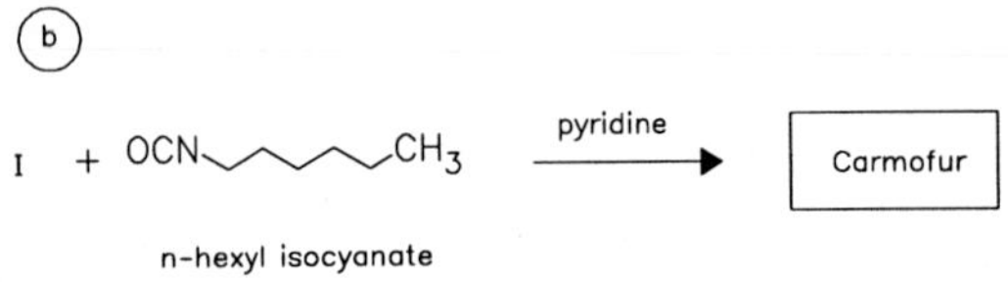

I + OCN⌒⌒⌒CH$_3$ —pyridine→   Carmofur

n-hexyl isocyanate

Reference(s):
a   JP 53/098 977 (Mitsui; appl. 2.8.1977).
b   DOS 2 639 135 (Mitsui; appl. 31.8.1976; USA-prior. 5.11.1975).
    US 4 071 519 (Mitsui; 31.1.1978; prior. 5.11.1975).

Formulation(s):   tabl. 100 mg

Trade Name(s):
J: Mifurol (Mitsui; 1981)   Yamaful (Yamanouchi; 1981)

Carmustine
(BCNU)

ATC: L01AD01
Use: antineoplastic

RN: 154-93-8   MF: $C_5H_9Cl_2N_3O_2$ MW: 214.05   EINECS: 205-838-2
LD$_{50}$: 26 mg/kg (M, i.p.); 45 mg/kg (M, i.v.); 19 mg/kg (M, p.o.); 24 mg/kg (M, s.c.);
        13.8 mg/kg (R, i.v.); 20 mg/kg (R, p.o.)
CN: N,N'-bis(2-chloroethyl)-N-nitrosourea

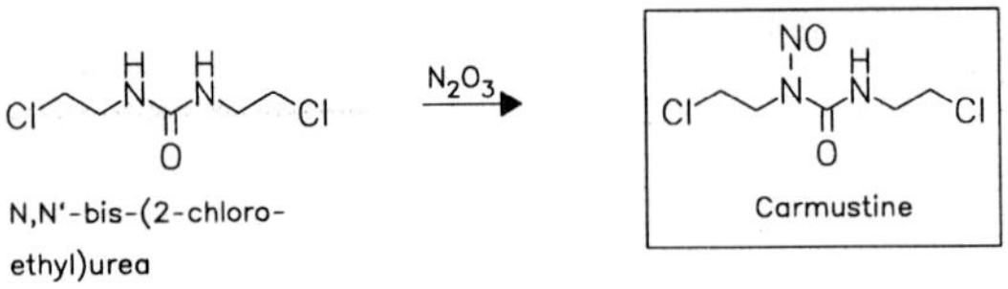

N,N'-bis-(2-chloro-
ethyl)urea

Carmustine

Reference(s):
DOS 2 528 365 (The Government of US; appl. 25.6.1975; USA-prior. 13.11.1974).

nitrosation with NaNO$_2$:
Johnston, T.P. et al.: J. Med. Chem. (JMCMAR) **6**, 669 (1963).

synthesis of N,N'-bis-(2-chloroethyl)urea:
Bastian, H.: Justus Liebigs Ann. Chem. (JLACBF) **566**, 210 (1950).

review:
Carter, S.K. et al.: "Advances in Cancer Research" (Ed. G. Klein, S. Weinhouse) **16**, 273 (1972).

Formulation(s):   tabl. 7.7 mg

Trade Name(s):

D:	Carmubris (Bristol-Myers Squibb)	GB:	BICNU (Bristol-Myers Squibb)
F:	BICNU (Bristol-Myers Squibb)	I:	Nitrumon (Astra-Simes)

USA: BICNU (Bristol-Myers Squibb)
Gliadel (Rhône-Poulenc Rorer)

Carnitine
(Levocarnitine)

ATC: A12AX; A14B; A11JC
Use: appetite stimulant, antiarrhythmic, cardiomyopathy therapeutic

RN: 541-15-1 MF: C$_7$H$_{15}$NO$_3$ MW: 161.20
LD$_{50}$: 9 g/kg (M, s.c.);
      7 g/kg (dog, route unreported)
CN: (R)-3-carboxy-2-hydroxy-N,N,N-trimethyl-1-propanaminium hydroxide inner salt

L-hydrochloride
RN: 6645-46-1 MF: C$_7$H$_{15}$NO$_3$ · HCl MW: 197.66 EINECS: 229-663-6
DL-carnitine
RN: 406-76-8 MF: C$_7$H$_{15}$NO$_3$ MW: 161.20 EINECS: 206-976-6
DL-hydrochloride
RN: 461-05-2 MF: C$_7$H$_{15}$NO$_3$ · HCl MW: 197.66 EINECS: 207-309-1
LD$_{50}$: 6 g/kg (M, s.c.);
      10 g/kg (R, s.c.)

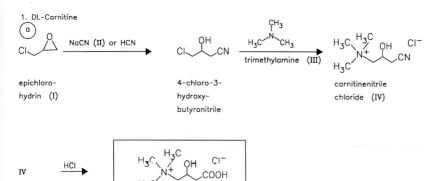

1. DL-Carnitine

epichlorohydrin (I) → NaCN (II) or HCN → 4-chloro-3-hydroxy-butyronitrile → trimethylamine (III) → carnitinenitrile chloride (IV)

IV → HCl → DL-Carnitine hydrochloride

(b)

III   +   I $\longrightarrow$ (2,3-epoxypropyl)-trimethylammonium chloride $\xrightarrow{\text{II}}$ IV $\xrightarrow{\text{HCl}}$ DL-Carnitine hydrochloride

(2,3-epoxypropyl)-
trimethylammonium
chloride

(c)

III   +   ethyl 4-chloro-acetoacetate $\xrightarrow{\text{NaOC}_2\text{H}_5}$ $\xrightarrow{\text{H}_2, \text{ Pt-C}}$ V

ethyl 4-chloro-
acetoacetate

carnitine ethyl ester chloride   (V) $\xrightarrow{\text{HCl}}$ DL-Carnitine hydrochloride

2. L-Carnitine

(a)

IV $\xrightarrow{\text{H}_2\text{O}_2}$ DL-carnitinamide chloride $\xrightarrow{\text{OH}^-, \text{ anion exchanger}}$ DL-carnitinamide hydroxide   (VI)

DL-carnitinamide chloride   DL-carnitinamide hydroxide   (VI)

VI   +   D-camphoric acid $\xrightarrow{\text{isopropanol}}$ L-carnitinamide D-camphorate $\xrightarrow{\text{HCl}}$ VII

D-camphoric acid   L-carnitinamide D-camphorate

L-carnitinamide chloride   (VII) $\xrightarrow[\text{2. oxalic acid}]{\text{1. H}_2\text{O}, \ \Delta}$ L-Carnitine hydrochloride

(b)

DL-carnitine $\xrightarrow{\text{H}_2\text{SO}_4, \ \Delta}$ crotonobetaine   (VIII) $\xrightarrow[\text{044 K74}]{\substack{\text{microbiologically} \\ \text{e.g. with E.coli}}}$ L-Carnitine

DL-carnitine   crotonobetaine   (VIII)   L-Carnitine

(c)

VIII $\xrightarrow{\substack{\text{microbiologically} \\ \text{e.g. with Acinetobacter Iwoffi (ATCC 9036)} \\ \text{or Proteus mirabilis (ATCCF 15290)}}}$ L-Carnitine

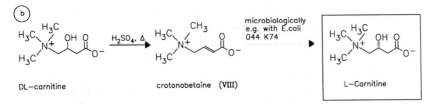

Reference(s):

1a US 3 135 788 (Nihon Zoki Seiyaku; 2.6.1964; J-prior. 28.9.1959).
   hydrolysis of carnitinenitrile chloride *with conc.* HCl:
   DAS 1 090 676 (Labaz; appl. 24.10.1958).
b US 4 070 394 (Ethyl Corp.; 24.1.1978; appl. 11.3.1977; prior. 23.1.1976).
c CH 588 433 (Lonza; appl. 25.9.1974).
   similar process from γ-chloroacetanilide:
   CH 589 604 (Lonza; appl. 26.4.1974).
2a DOS 2 927 672 (C. Cavazza; appl. 9.7.1979; I-prior. 10.7.1978).
   US 4 254 053 (C. Cavazza; 3.3.1981; I-prior. 10.7.1978).
   electrolytic methods for release of base:
   DOS 3 342 713 (Sigma-Tau; appl. 25.11.1983; I-prior. 25.11.1982).
   US 4 521 285 (Sigma-Tau; 4.6.1985; I-prior. 25.11.1982).
b EP 148 132 (Sigma-Tau; appl. 31.10.1984; DDR-prior. 3.11.1983).
c EP 122 794 (Ajinomoto; appl. 13.4.1984; J-prior. 13.4.1983).

enzymatic methods from γ-butyrobetaine *and* 2-ketoglutaric acid *with* γ-butyrobetaine hydroxylase *from Neurospora crassa:*
GB 2 078 742 (Sigma-Tau; appl. 23.6.1981; I-prior. 24.6.1980).

synthesis from D-mannitol:
US 4 413 142 (Anic; 1.11.1983; I-prior. 18.3.1981).

use as antiarrhythmic:
US 3 830 931 (S. L. De Felice; 20.8.1974; appl. 6.11.1972).
US 3 968 241 (S. L. De Felice; 6.7.1976; prior. 6.11.1972, 2.7.1974).

parenteral use for improvement of myocard function:
US 4 075 352 (S. L. De Felice; 21.2.1978; appl. 28.4.1976).

use for reduction of cardiotoxicity of cytostatics, e. g. daunomycin:
US 4 320 110 (S. L. De Felice; 16.3.1982; appl. 4.10.1979).
US 4 400 371 (S. L. De Felice; 23.8.1983; appl. 12.5.1981).

use as appetite stimulant:
US 3 810 994 (Ethyl Corp.; 14.5.1974; appl. 1.6.1972).

additive to parenteral feeding:
DOS 3 032 300 (A. Lohninger, Wien; appl. 27.8.1980).
US 4 320 145 (C. Cavazza; 16.3.1982; I-prior. 5.10.1979).
EP 59 775 (Leopold & Co.; appl. 1.6.1981; YU-prior. 9.6.1980).

use as antihyperlipidemic:
US 4 315 944 (Sigma-Tau; 16.2.1982; I-prior. 21.9.1979).

use as geriatric for improvement of mental ability:
US 4 474 812 (Sigma-Tau; 2.10.1984; I-prior. 29.10.1982).

treatment of lung diseases:
DE 2 360 332 (Otsuka; appl. 4.12.1973; J-prior. 7.12.1972).

Formulation(s):   drinking sol. 1 g/10 ml; inj. sol. 1g/5 ml; syrup 1 g/3.3 ml

Trade Name(s):

D: Biocarn (Medice)
     L-Carn (Sigma-Tau)
F: Lévoearnil (Sigma-Tau)
I: Anetin (Ibirn)
     Biocarnil (Gentili)
     Briocor (Farge)
     Cardimet (Errekappa
     Euroter.)

Cardiogen (Chemil)
Carnitene (Sigma-Tau)
Carnitolo (Recordati
Farma)
Carnitop (Virginia Farmac.)
Carnovis (Duncan)
Carnum (Firma)
Carrier (Chiesi)

Carvit (AGIPS)
Eucar (Salus Research)
Eucarnil (Pulitzer)
Kernit (CT)
L-Carnitina Coli (Coli)
Lefcar (Glaxo)
Levocarvit (Mitim)
Medocarnitin (Medosan)

Metina (Pierrel)
Miocardin (Magis)   J:
Miocor (Ecobi)
Miotonal (Caber)

Transfert (Piam)
Abedine (Nippon Zoki)
Entomin (Maruko)
Monocamin (Tanabe)

USA: L-Carnitine (Tyson)
       Carnitor (Sigma-Tau)

Caroxazone

ATC: N06A
Use: antidepressant

RN: 18464-39-6   MF: $C_{10}H_{10}N_2O_3$ MW: 206.20   EINECS: 242-345-1
LD$_{50}$:   728 mg/kg (M, p.o.)
CN: 2-oxo-2H-1,3-benzoxazine-3(4H)-acetamide

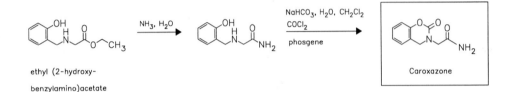

ethyl (2-hydroxy-
benzylamino)acetate

Caroxazone

Reference(s):
CH 586 687 (Farmitalia; appl. 26.4.1974).
ZA 742 435 (Farmitalia; appl. 17.4.1974).

alternative synthesis:
US 3 427 313 (Farmitalia; 11.2.1969; I-prior. 23.12.1965, 14.9.1966).
Bernardi, L. et al.: Experientia (EXPEAM) **24**, 774 (1968).

Trade Name(s):
I: Timostenil (Carlo Erba);
       wfm

Timostenil (Farmitalia);
wfm

Carpipramine
(Carbadipimidine)

ATC: N06B
Use: antidepressant

RN: 5942-95-0   MF: $C_{28}H_{38}N_4O$ MW: 446.64   EINECS: 227-700-0
LD$_{50}$:   28 mg/kg (M, i.v.); 2180 mg/kg (M, p.o.);
       37 mg/kg (R, i.v.); 1025 mg/kg (R, p.o.)
CN: 1'-[3-(10,11-dihydro-5H-dibenz[b,f]azepin-5-yl)propyl][1,4'-bipiperidine]-4'-carboxamide

dihydrochloride monohydrate
RN: 7075-03-8   MF: $C_{28}H_{38}N_4O \cdot 2HCl \cdot H_2O$ MW: 537.58   EINECS: 230-372-1
LD$_{50}$:   136 mg/kg (M, i.p.); 28 mg/kg (M, i.v.); 2180 mg/kg (M, p.o.);
       76 mg/kg (R, i.p.); 37 mg/kg (R, i.v.); 1025 mg/kg (R, p.o.);
       18 mg/kg (rabbit, i.v.)
maleate (1:1)
RN: 100482-23-3   MF: $C_{28}H_{38}N_4O \cdot C_4H_4O_4$ MW: 562.71
LD$_{50}$:   147 mg/kg (M, i.p.); 2055 mg/kg (M, p.o.);
       169 mg/kg (R, i.p.)

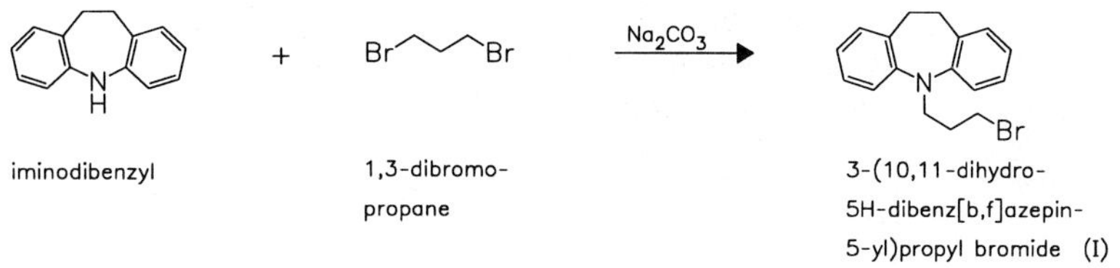

iminodibenzyl + 1,3-dibromo-propane → Na₂CO₃ → 3-(10,11-dihydro-5H-dibenz[b,f]azepin-5-yl)propyl bromide (I)

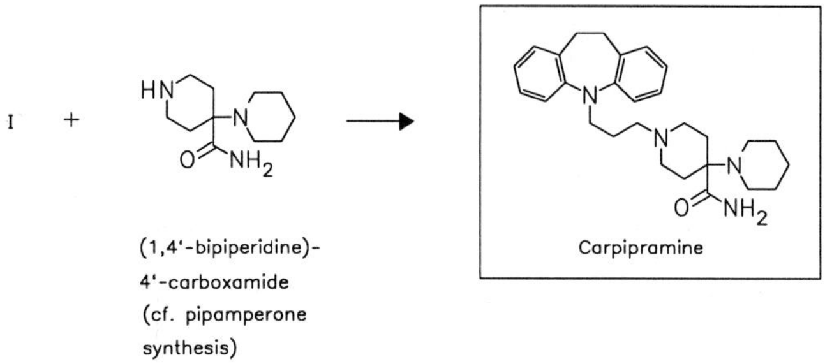

I + (1,4'-bipiperidine)-4'-carboxamide (cf. pipamperone synthesis) → Carpipramine

Reference(s):

JP 66 006 572 (Yoshitomi; appl. 29.6.1963).

Nakanishi, M. et al.: J. Med. Chem. (JMCMAR) **13**, 644 (1970).

medical use as anxiolytic, hypnotic:

EP 374 042 (Rhône-Poulenc; appl. 13.12.1989; F-prior. 16.12.1988).

Formulation(s):   powder 10 %; tabl. 25 mg, 50 mg

Trade Name(s):

F: Prazinil (Pierre Fabre)   J: Defekton (Yoshitomi)

Carprofen

ATC: M01AE

Use: non-steroidal anti-inflammatory

RN: 53716-49-7   MF: $C_{15}H_{12}ClNO_2$ MW: 273.72   EINECS: 258-712-4

LD_{50}:   400 mg/kg (M, p.o.)

CN: (±)-6-chloro-α-methyl-9H-carbazole-2-acetic acid

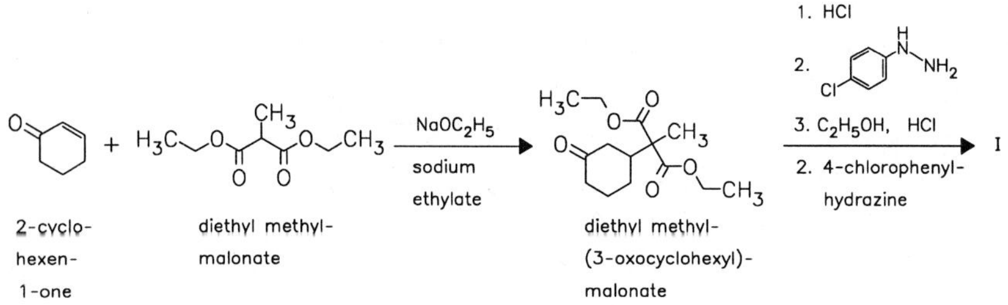

2-cyclo-hexen-1-one + diethyl methyl-malonate → NaOC₂H₅ sodium ethylate → diethyl methyl-(3-oxocyclohexyl)-malonate →
1. HCl
2. 4-chlorophenyl-hydrazine
3. C₂H₅OH, HCl → I

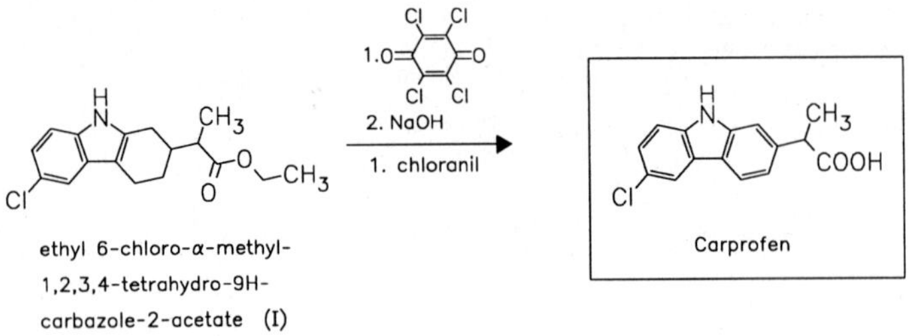

ethyl 6-chloro-α-methyl-
1,2,3,4-tetrahydro-9H-
carbazole-2-acetate (I)

Carprofen

Reference(s):
US 3 896 145 (Hoffmann-La Roche; 22.7.1975; prior. 17.5.1973, 24.7.1972).

Formulation(s):   tabl. 150 mg

Trade Name(s):
USA: Rimadyl (Roche); wfm

Carteolol

ATC: C07AA15; S01ED05
Use: beta blocking agent

RN: 51781-06-7   MF: $C_{16}H_{24}N_2O_3$ MW: 292.38
LD$_{50}$: 810 mg/kg (M, p.o.);
        830 mg/kg (dog, p.o.)
CN: 5-[3-[(1,1-dimethylethyl)amino]-2-hydroxypropoxy]-3,4-dihydro-2(1H)-quinolinone

monohydrochloride
RN: 51781-21-6   MF: $C_{16}H_{24}N_2O_3 \cdot HCl$ MW: 328.84   EINECS: 257-415-7
LD$_{50}$: 54.5 mg/kg (M, i.v.); 810 mg/kg (M, p.o.);
        153 mg/kg (R, i.v.); 1330 mg/kg (R, p.o.);
        830 mg/kg (dog, p.o.)

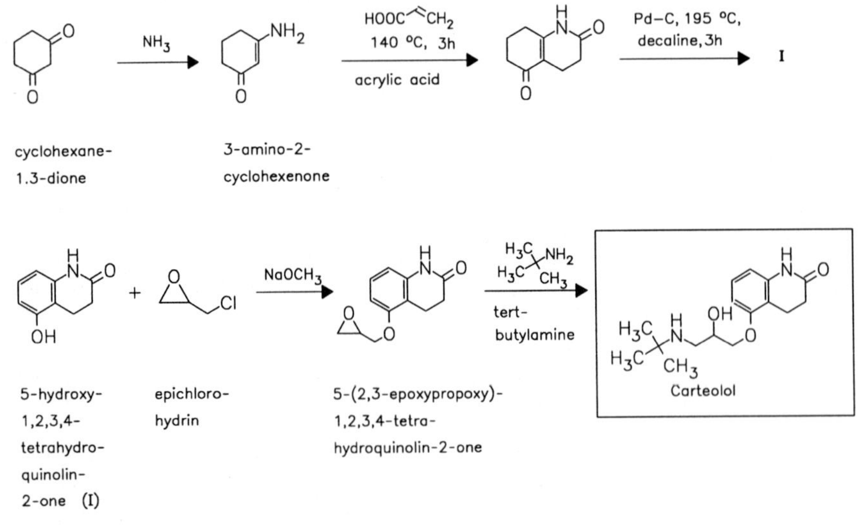

cyclohexane-
1.3-dione

3-amino-2-
cyclohexenone

5-hydroxy-
1,2,3,4-
tetrahydro-
quinolin-
2-one (I)

epichloro-
hydrin

5-(2,3-epoxypropoxy)-
1,2,3,4-tetra-
hydroquinolin-2-one

Carteolol

Reference(s):
Winkler, W.: Arzneim.-Forsch. (ARZNAD) **33**, 279 (1983).
DOS 2 302 027 (Otsuka; appl. 16.1.1973; J-prior. 13.4.1972).
US 3 910 924 (Otsuka; 7.10.1975; appl. 19.1.1973; J-prior. 13.4.1972).

synthesis of intermediate 5-hydroxy-1,2,3,4-tetrahydroquinolin-2-one:
Shono, T. et al.: J. Org. Chem. (JOCEAH) **46**, 3719 (1981).

Formulation(s):   eye drops 1 %, 2 %, tabl. 2.5 mg, 5 mg, 20 mg (as hydrochloride)

Trade Name(s):
D: Arteoptic (CIBA Vision;   Cartéol (Chauvin; 1985)   I: Carteol (SIFI; 1987)
    1984)   Mikelan (Lipha Santé)   J: Mikelan (Otsuka; 1980)
    Endak (Madaus; 1982)   GB: Teoptic (CIBA Vision;   USA: Cartrol (Abbott)
F: Carpilo (Chauvin)-comb.   1986)

Carticaine

ATC: N01B; N01BB08; N01BB58
Use: local anesthetic

RN: 23964-58-1   MF: $C_{13}H_{20}N_2O_3S$ MW: 284.38
CN: 4-methyl-3-[[1-oxo-2-(propylamino)propyl]amino]-2-thiophenecarboxylic acid methyl ester

monohydrochloride
RN: 23964-57-0   MF: $C_{13}H_{20}N_2O_3S \cdot HCl$ MW: 320.84   EINECS: 245-957-7
LD_{50}: 37 mg/kg (M, i.v.)

methyl 3-amino-   2-bromo-
4-methylthiophene-   propionyl
2-carboxylate   chloride

Carticaine

Reference(s):
DAS 1 643 325 (Hoechst; appl. 7.7.1967).
US 3 855 243 (Hoechst; 17.12.1974; D-prior. 7.7.1967).

Formulation(s):   amp. 10mg/ml, 20 mg/ml, 40 mg/ml, 50 mg/ml

Trade Name(s):
D: Ultracain (Hoechst)   F: Alphacaïne (SPAD)-comb.

Carumonam
(AMA-1080; Ro-17-2301)

ATC: S01AA
Use: antibacterial (monobactam antibiotic)

RN: 87638 04 8   MF: $C_{12}H_{14}N_6O_{10}S_2$ MW: 466.41
CN: [2S-[2α,3α(Z)]]-[[[2-[2-[[(aminocarbonyl)oxy]methyl]-4-oxo-1-sulfo-3-azetidinyl]amino]-1-(2-amino-4-thiazolyl)-2-oxoethylidene]amino]oxy]acetic acid

disodium salt
RN: 86832-68-0   MF: $C_{12}H_{12}N_6Na_2O_{10}S_2$ MW: 510.37

(a) azetidinone intermediate:

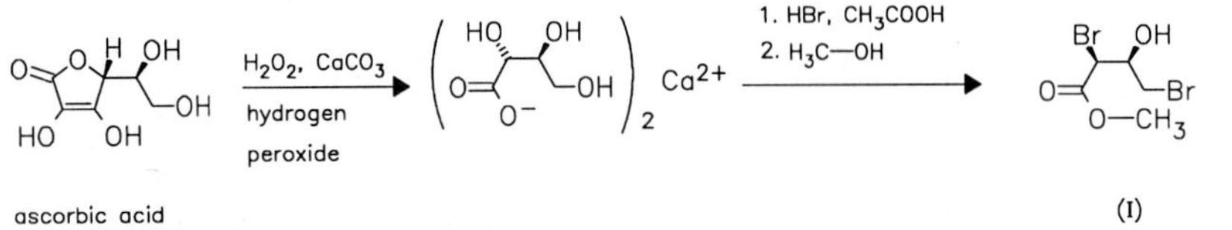

ascorbic acid

(I)

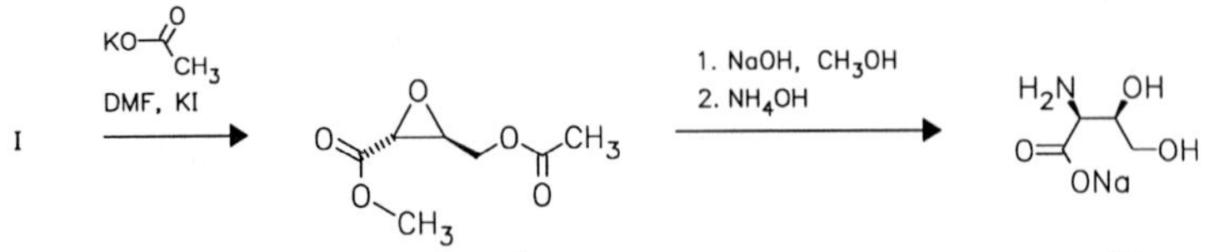

methyl (2R,3S)-4-
acetoxy-2,3-epoxy-
butanoate

(II)

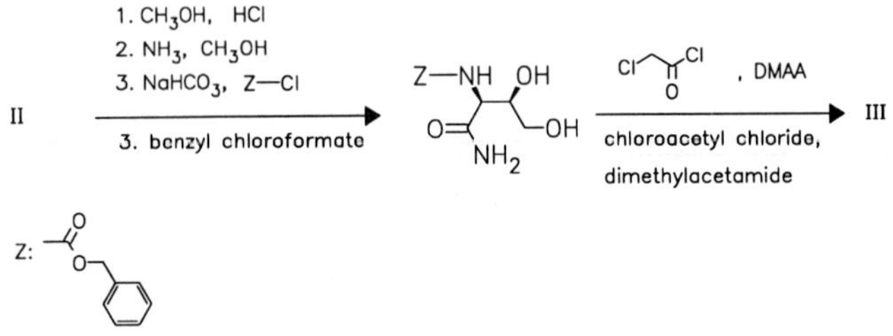

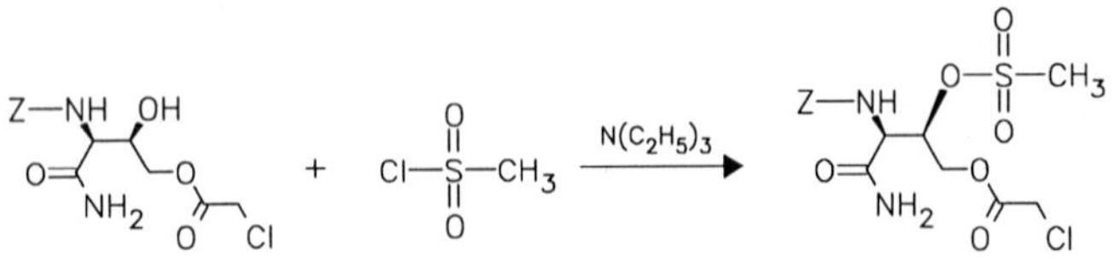

Z:

(III)   methanesulfo-
nyl chloride

(IV)

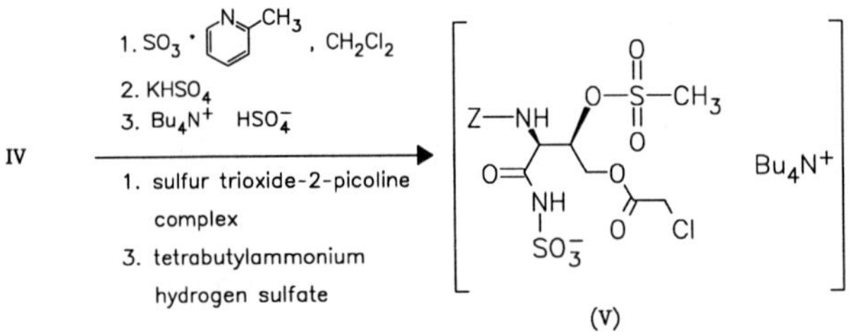

(V)

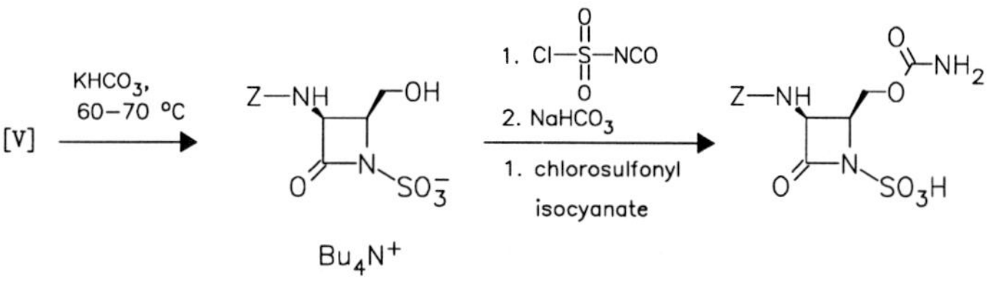

b side chain:

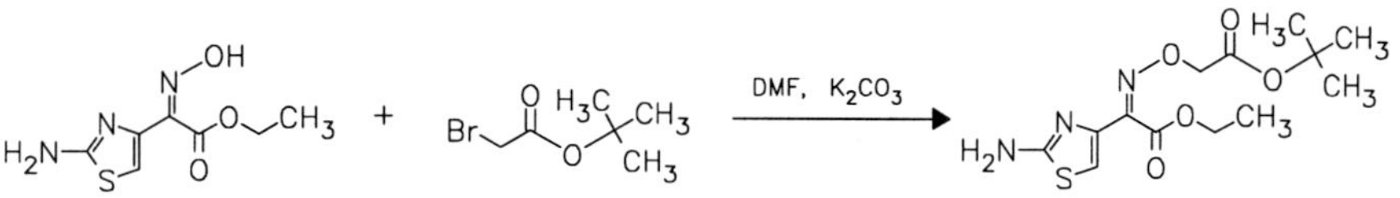

ethyl 2-(2-amino-4-thiazolyl)-2(Z)-hydroxyiminoacetate

tert-butyl bromoacetate

(VII)

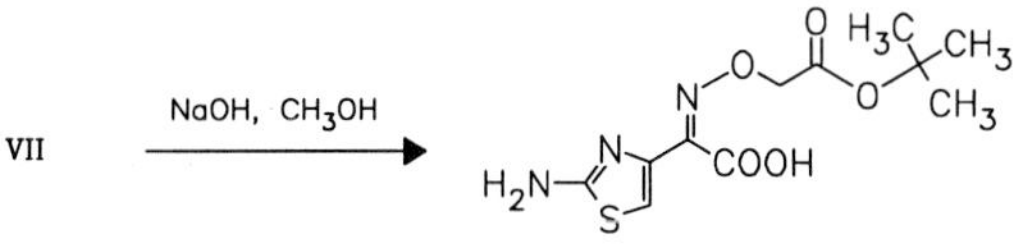

(VIII)

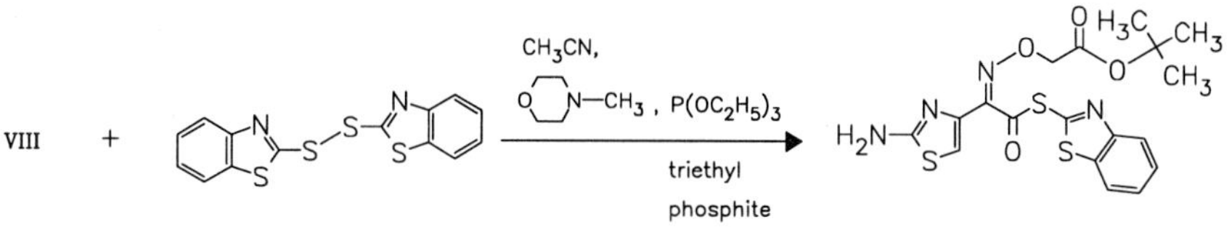

2,2'-dithiobis-(benzothiazole)

(IX)

c final product:

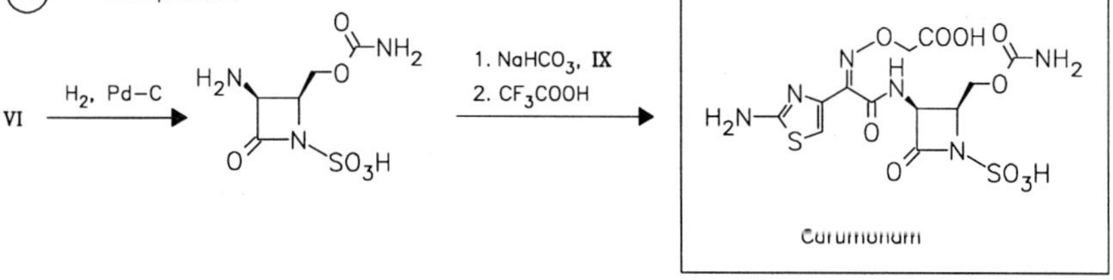

Carumonam

Reference(s):

Kishimoto, S. et al.: J. Antibiot. (JANTAJ) **36**, 1421 (1983).

Sendai, M. et al.: J. Antibiot. (JANTAJ) **38**, 346 (1985).

US 4 572 801 (Takeda; 25.2.1986; PCT-prior. 30.4.1981, 21.8.1981, 24.9.1981; J-prior. 30.4.1982, 31.5.1982; USA-appl. 3.12.1981, 5.8.1982, 31.5.1983).

special route according to **a** *for VI:*

Manchand, P.S. et al.: J. Org. Chem. (JOCEAH) **53**, 5507 (1988).

alternative route for VI:
Hashigushi, S. et al.: Heterocycles (HTCYAM) **24**, 2273 (1986).

further synthetic routes for carumonam *and its intermediates:*
US 4 673 739 (Takeda; 16.6.1987; PCT-prior. 5.12.1980, 30.4.1981, 21.8.1981, 24.9.1981; J-prior. 30.4.1982, 31.5.1982; USA-appl. 3.12.1981, 5.8.1982, 31.5.1983, 18.9.1985).
US 4 675 397 (Takeda 23.6.1987; PCT-prior. 5.12.1980, 30.4.1981, 21.8.1981, 24.9.1981; J-prior. 30.4.1982; USA-appl. 3.12.1981, 5.8.1982) - 446 pages.
US 4 782 147 (Takeda; 1.11.1988; PCT-prior. 5.12.1980, 30.4.1981, 21.8.1981, 24.9.1981; J-prior. 31.5.1982; USA-appl. 3.12.1981, 31.5.1983) - 504 pages.
US 4 502 994 (Hoffmann-La Roche; 5.3.1985; appl. 9.12.1982).
US 4 652 651 (Hoffmann-La Roche; 24.3.1987; prior. 31.5.1983, 14.4.1986).
US 4 663 469 (Hoffmann-La Roche; 5.5.1987; prior. 9.12.1982, 10.12.1984).
EP 96 297 (Hoffmann-La Roche; appl. 25.5.1983; CH-prior. 3.6.1982, 25.4.1983).
EP 185 221 (Hoffmann-La Roche; appl. 25.11.1985; CH-prior. 19.12.1984).

Formulation(s):   (disodium salt) vial 0.5 g (i.m. and i.v. inj.), 1 g (i.v. inj.)

Trade Name(s):
D: Amasulin (Takeda); wfm

Carvedilol
(BM-14190)

ATC: C07AG02; C07EA
Use: non-selective β_1-adrenoceptor blocker with vasodilating activity

RN: 72956-09-3   MF: $C_{24}H_{26}N_2O_4$ MW: 406.48
LD$_{50}$:   364 mg/kg (M, i.p.); 27 mg/kg (M, i.v.);
      769 mg/kg (R, i.p.); 25 mg/kg (R, i.v.);
      >1 g/kg (dog, p.o.)
CN: 1-(9*H*-carbazol-4-yloxy)-3-[[2-(2-methoxyphenoxy)ethyl]amino]-2-propanol

4-hydroxy-carbazole + epichloro-hydrin → NaOH → 4-(2,3-epoxypropoxy)-carbazole (I)

I + 2-(2-methoxyphen-oxy)ethylamine → diglyme → Carvedilol

Reference(s):
DOS 2 815 926 (Boehringer Mannh.; appl. 13.4.1978).
EP 4 920 (Boehringer Mannh.; appl. 4-7-1979; D-prior. 13.4.1978).

synthesis of enantiomers:
EP 127 099 (Boehringer Mannh.; appl. 19.5.1984; D-prior. 26.5.1983).

Formulation(s):   tabl. 6.25 mg, 12.5 mg, 25 mg, 50 mg

Trade Name(s):

D: Dilatrend (Boehringer Mannh.; SmithKline Beecham; 1991)
Querto (Byk Gulden)
F: Kredex (Boehringer Mannh.; SmithKline Beecham)
GB: Eucardic (Boehringer Mannh. VK)
I: Carvipress (Gentili)
Dilatrend (Boehringer Mannh.)
J: Artist (Daiichi Seiyaku)
USA: Coreg (SmithKline Beecham)

Carzenide

ATC: M01AE01
Use: antispasmodic, diuretic (carboanhydrase inhibitor)

RN: 138-41-0 MF: $C_7H_7NO_4S$ MW: 201.20 EINECS: 205-327-4
LD_{50}: >1 g/kg (M, i.p.);
350 mg/kg (R, i.p.)
CN: 4-(aminosulfonyl)benzoic acid

monosodium salt
RN: 6101-29-7 MF: $C_7H_6NNaO_4S$ MW: 223.18

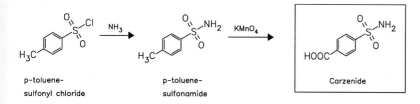

NH3
KMnO4
p-toluene-sulfonyl chloride
p-toluene-sulfonamide
Carzenide

By-product of saccharin production.

Reference(s):
DRP 64 624 (Dr. C. Fahlberg; appl. 1891).

Formulation(s): f. c. tabl. 200 mg

Trade Name(s):
D: Dismenol (Simons)-comb.

Cefacetrile
(Cephacetrile)

ATC: J01DA34
Use: antibiotic

RN: 10206-21-0 MF: $C_{13}H_{13}N_3O_6S$ MW: 339.33 EINECS: 233-508-8
CN: (6*R*-*trans*)-3-[(acetyloxy)methyl]-7-[(cyanoacetyl)amino]-8-oxo-5-thia-1-azabicyclo[4.2.0]oct-2-ene-2-carboxylic acid

monosodium salt
RN: 23239-41-0 MF: $C_{13}H_{12}N_3NaO_6S$ MW: 361.31 EINECS: 245-513-2
LD_{50}: 3700 mg/kg (M, i.v.); 19 g/kg (M, p.o.);
3100 mg/kg (R, i.v.); 15.1 g/kg (R, p.o.)

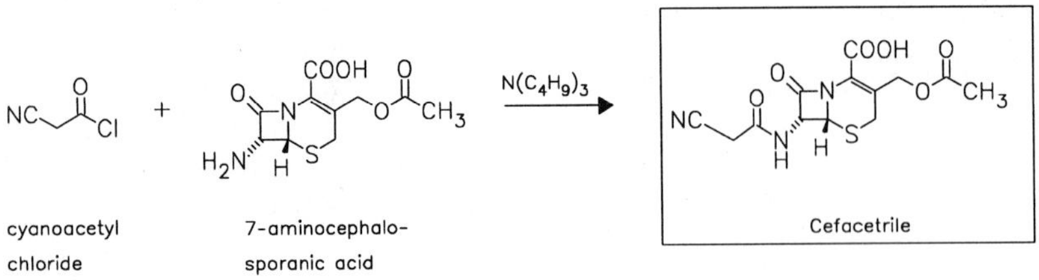

cyanoacetyl
chloride

7-aminocephalo-
sporanic acid

Cefacetrile

Reference(s):
DAS 1 670 324 (Ciba-Geigy; appl. 8.1.1966; CH-prior. 18.1.1965, 1.4.1965, 10.5.1965, 20.10.1965).
US 3 483 197 (Ciba; 9.12.1969; CH-prior. 18.1.1965, 1.4.1965, 10.5.1965, 20.10.1965).
NL-appl. 6 600 586 (Ciba; appl. 17.1.1966; CH-prior. 18.1.1965, 1.4.1965, 10.5.1965, 20.10.1965).

acylation with mixed anhydrides of cyanoacetic acid:
DOS 2 730 580 (Pierrel S.p.A.; appl. 6.7.1977; GB-prior. 10.7.1976).

acylation via 1,3,2-dioxaboranyl-derivatives:
DOS 2 755 902 (Dobfar; appl. 15.12.1977; I-prior. 16.12.1976).

sodium salt:
US 4 061 853 (Ciba-Geigy; 6.12.1977; CH-prior. 9.12.1975).

Formulation(s):   vial 1 g/5 ml

Trade Name(s):

D:	Celospor (Ciba/ Grünenthal); wfm	I:	Celospor (Novartis; as sodium salt)
F:	Celospor (Ciba); wfm	J:	Celospor (Novartis)

Celtol (Takeda)

Cefaclor

ATC: J01DA08
Use: antibiotic

RN: 53994-73-3   MF: $C_{15}H_{14}ClN_3O_4S$ MW: 367.81   EINECS: 258-909-5
LD_{50}:   >20 g/kg (M, p.o.);
    >20 g/kg (R, p.o.)
CN: [6R-[6α,7β(R*)]]-7-[(aminophenylacetyl)amino]-3-chloro-8-oxo-5-thia-1-azabicyclo[4.2.0]oct-2-ene-2-carboxylic acid

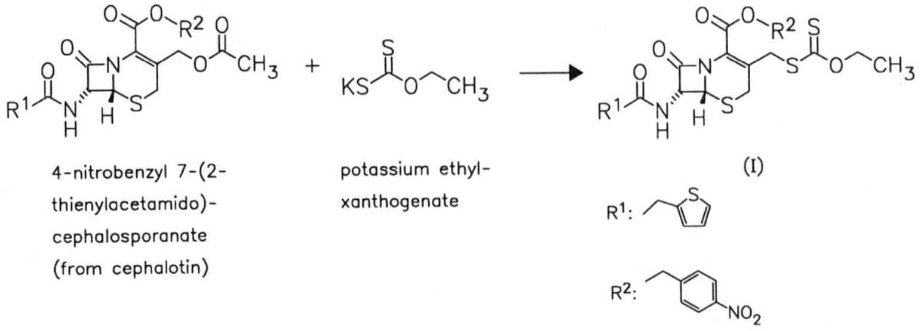

4-nitrobenzyl 7-(2-
thienylacetamido)-
cephalosporanate
(from cephalotin)

potassium ethyl-
xanthogenate

(I)

R^1:

R^2:

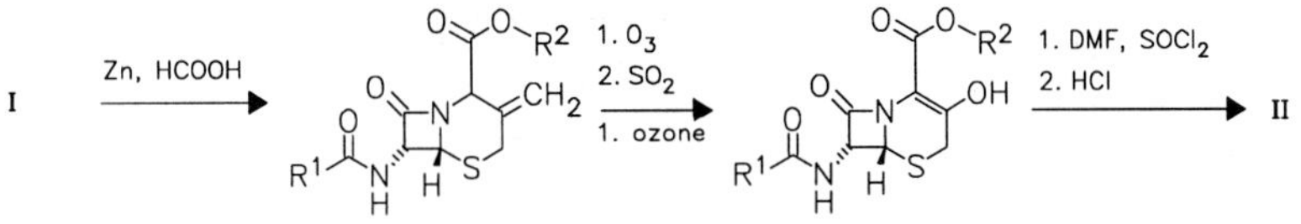

4-nitrobenzyl
3-methylene-7-(2-
thienylacetamido)-
cepham-4-carb-
oxylate

4-nitrobenzyl 3-
hydroxy-7-(2-thienyl-
acetamido)-3-
cephem-4-carb-
oxylate

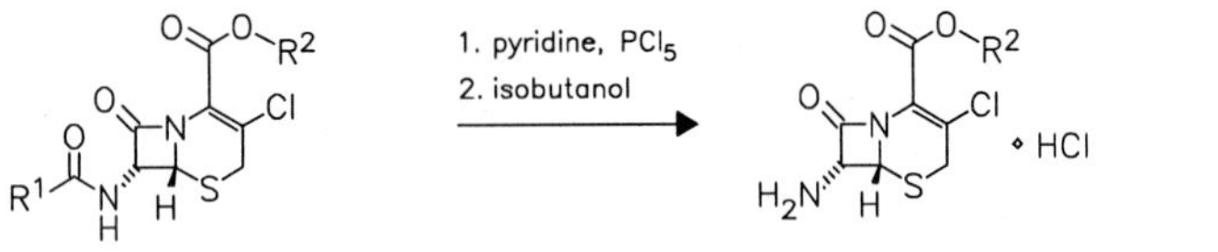

4-nitrobenzyl 3-chloro-
7-(2-thienylacetamido)-
3-cephem-4-carboxylate (II)

4-nitrobenzyl 7-amino-3-chloro-
3-cephem-4-carboxylate
hydrochloride (III)

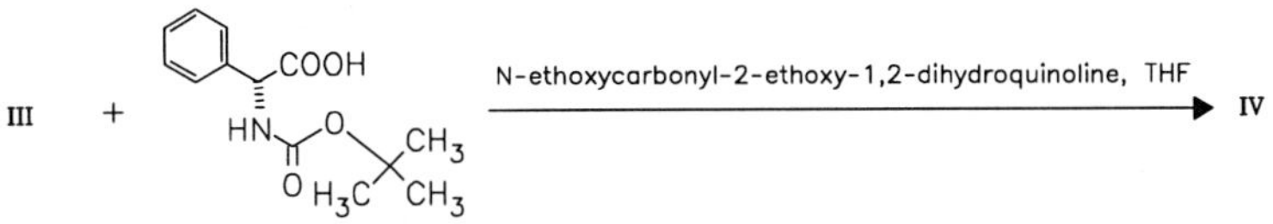

N-tert-butoxycarbonyl—
D-α-phenylglycine

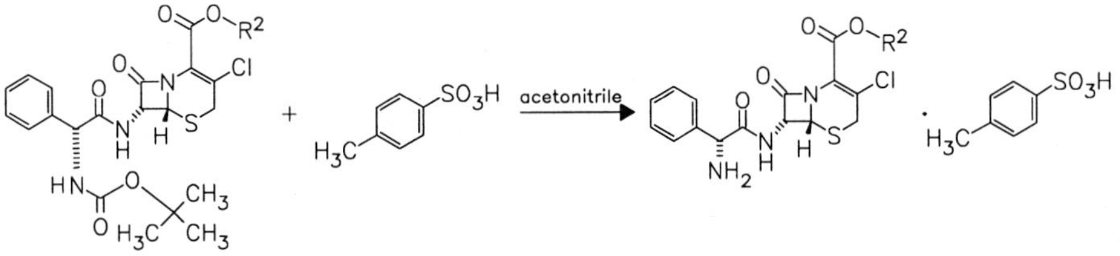

(IV)

(V)

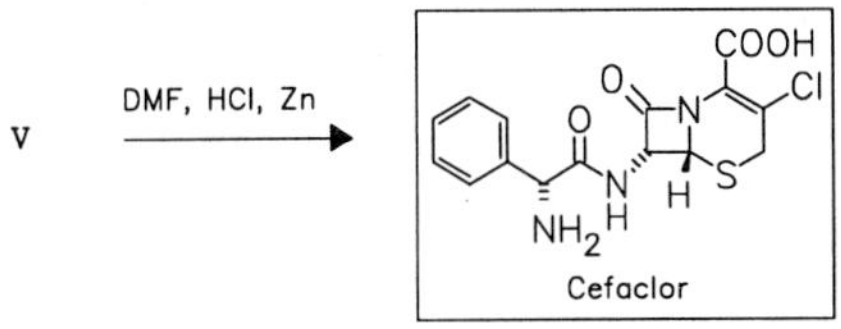

Cefaclor

Reference(s):
US 3 925 372 (Lilly; 9.12.1975; prior. 23.2.1973, 1.4.1974).
DOS 2 408 698 (Lilly; appl. 22.2.1974; USA-prior. 23.2.1973).
Chauvette, R.R.; Pennington, P.A.: J. Med. Chem. (JMCMAR) **18**, 403 (1975).

3-halogenocephem precursors:
DOS 2 408 686 (Lilly; appl. 22.2.1974; USA-prior. 23.2.1973).
US 4 115 643 (Lilly; 19.9.1978; prior. 16.8.1976, 8.8.1977).

3-hydroxycephem intermediates:
US 3 917 587 (Lilly; 4.11.1975; appl. 28.11.1972).

3-methylenecephem intermediates:
US 3 932 393 (Lilly; 13.1.1976; appl. 25.2.1971).
US 4 075 203 (Lilly; 21.2.1978; appl. 16.6.1976).

3-chlorocephem intermediates:
US 3 962 227 (Lilly; 8.6.1976; prior. 23.2.1973, 1.4.1974).
US 4 064 343 (Lilly; 20.12.1977; prior. 23.2.1973, 1.4.1974, 9.2.1976).

Formulation(s):　cps. 250 mg, 500 mg; gran. 125 mg, 250 mg; s. r. tabl. 375 mg, 500 mg; syrup 125 mg/ml,
250 mg/ml

Trade Name(s):
D:　Kefspor (ASTA Medica　　　Sigacefal (Kytta-Siegfried)　J:　Kefral (Shionogi; 1982)
　　AWD)　　　　　　　　F:　Alfatil (Lilly; 1981)　　USA:　Ceclor (Lilly; 1979)
　　Muco Panoral (Lilly)-　　GB:　Distaclor MR (Lilly; 1979)　　Ceclor CD (Dura)
　　comb.　　　　　　　　　Keftid (Galen)
　　Panoral (Lilly; 1979)　　I:　Panacef (Lilly)

Cefadroxil

ATC:　J01DA09
Use:　antibiotic

RN:　50370-12-2　MF: $C_{16}H_{17}N_3O_5S$　MW: 363.39　EINECS: 256-555-6
LD_{50}:　>1.5 g/kg (M, i.v.); >10 g/kg (M, p.o.);
　　　>1 g/kg (R, i.v.); >10 g/kg (R, p.o.);
　　　>2 g/kg (dog, p.o.)
CN:　[6R-[6α,7β(R^*)]]-7-[[amino(4-hydroxyphenyl)acetyl]amino]-3-methyl-8-oxo-5-thia-1-
azabicyclo[4.2.0]oct-2-ene-2-carboxylic acid

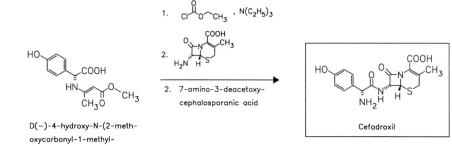

D(−)-4-hydroxy-N-(2-meth-
oxycarbonyl-1-methyl-
ethenyl)phenylglycine

2. 7-amino-3-deacetoxy-
cephalosporanic acid

Cefadroxil

Reference(s):
DE 1 795 292 (Bristol-Myers; appl. 5.9.1968; USA-prior. 5.9.1967).
US 3 489 752 (Bristol-Myers; 13.1.1970; appl. 5.9.1967).
GB 1 240 687 (Bristol-Myers; appl. 5.9.1968; USA-prior. 5.9.1967).
US 3 985 741 (Bristol-Myers; 12.10.1976; prior. 15.9.1972, 18.10.1974).
GB 1 532 682 (Bristol-Myers; appl. 27.4.1976; valid from 7.3.1977).

crystalline monohydrate:
US 4 160 863 (Bristol-Myers; 10.7.1979; prior. 7.4.1977, 2.2.1978).
DOS 2 718 741 (Bristol-Myers; appl. 27.4.1977; GB-prior. 27.4.1976, 7.3.1977).

Formulation(s):　cps. 500 mg; oral susp. 125 mg/5 ml, 250 mg/5 ml, 500 mg/5 ml; tabl. 1 g

Trade Name(s):
D:　Bidocef (Bristol-Myers　　　Cedrox (Hexal)　　　　　Grüncef (Bristol-Myers
　　Squibb; 1980)　　　　　　　　　　　　　　　　Squibb; Grünenthal)

F: Oracefal (Bristol-Myers Squibb; 1977)
GB: Baxan (Bristol-Myers Squibb; 1982)
I: Cefadril (AGIPS)
Ceoxil (Magis)
Cephos (CT)
Crenodyn (Ist. Italiano Ferm.)
Oradroxil (Lampugnani)
J: Sedral (Banyu; 1982)
Sumacef (Bristol Banyu; 1982)
USA: Duricef (Bristol-Myers Squibb; 1978)
Ultracef (Bristol; 1981)

Cefalexin
(Cephalexin)

ATC: J01DA01
Use: antibiotic

RN: 15686-71-2 MF: $C_{16}H_{17}N_3O_4S$ MW: 347.40 EINECS: 239-773-6
LD_{50}: 1495 mg/kg (M, p.o.);
>20 g/kg (R, p.o.)
CN: [6*R*-[6α,7β(*R**)]]-7-[(aminophenylacetyl)amino]-3-methyl-8-oxo-5-thia-1-azabicyclo[4.2.0]oct-2-ene-2-carboxylic acid

monohydrate
RN: 23325-78-2 MF: $C_{16}H_{17}N_3O_4S \cdot H_2O$ MW: 365.41

(a)

(b)

(c)

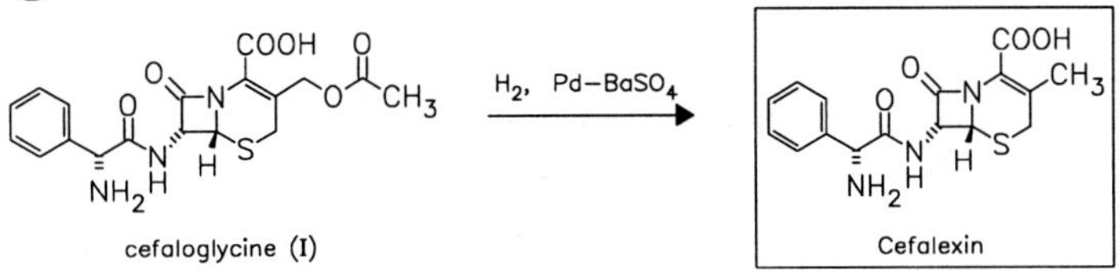

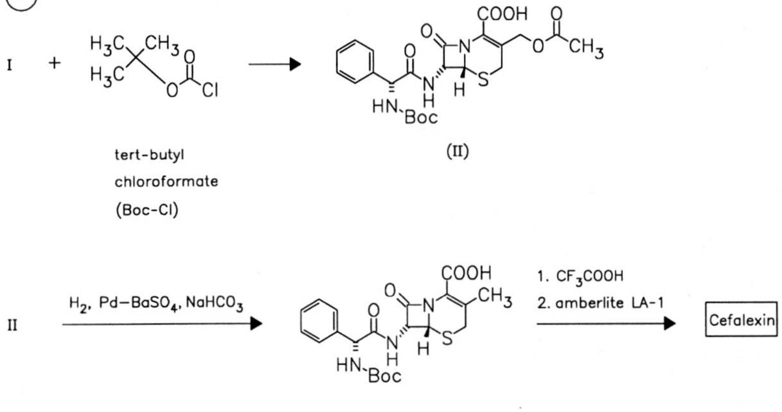

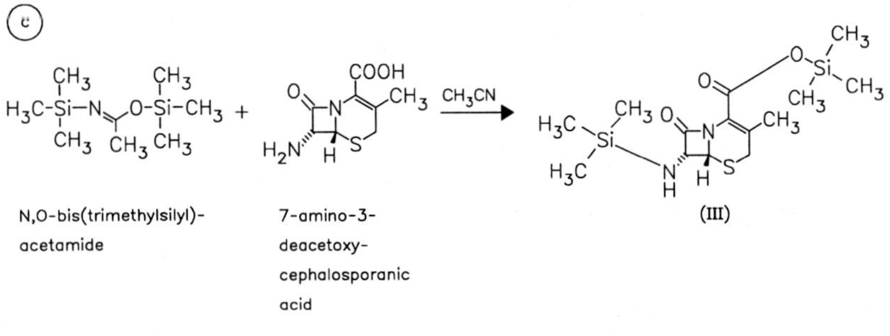

III   +

1. CH_3CN, $(H_3C)_2N-CH_2-C_6H_5$, $Cl-COOCH_3$
2. CH_3OH

1. benzyldimethylamine

→   Cefalexin

D(−)-methyl 3-(α-
carboxybenzylamino)-
crotonate sodium salt

Reference(s):
DE 1 670 625 (Lilly; appl. 28.3.1967; USA-prior. 14.9.1966).
US 3 507 861 (Lilly; 21.4.1970; prior. 31.7.1962, 14.9.1966).
a,b Ryan, C.W. et al.: J. Med. Chem. (JMCMAR) **12**, 310 (1968).
    FR 1 524 225 (Eli Lilly; appl. 23.3.1967; USA-prior. 14.9.1966).
    GB 1 174 335 (Eli Lilly; appl. 7.3.1967).
c   DOS 1 942 454 (Lilly; appl. 20.8.1969; USA-prior. 23.8.1968).
    GB 1 459 807 (Proter S.p.A.; appl. 27.5.1975).

purification:
US 3 634 416 (Glaxo; 11.1.1972; GB-prior. 26.3.1969).
US 3 676 437 (Glaxo; 11.7.1972, GB-prior. 26.9.1969).

alternative syntheses (also ring extension of penicillin sulfoxide esters):
GB 1 204 394 (Eli Lilly; appl. 8.5.1968; USA-prior. 8.5.1967).
US 3 502 663 (Eli Lilly; 24.3.1970; appl. 21.4.1969).
US 3 671 449 (Lilly; 20.6.1972; prior. 23.8.1968, 19.8.1970).
DAS 2 012 955 (Eli Lilly; appl. 18.3.1970; USA-prior. 18.3.1969).
DOS 2 117 377 (Bristol-Myers; appl. 8.4.1971; USA-prior. 10.4.1970, 5.10.1970).
DOS 2 127 225 (Yamanouchi; appl. 2.6.1971; J-prior. 12.6.1970, 15.6.1970).
DAS 2 241 091 (Toyo Jozo; appl. 21.8.1972; J-prior. 20.8.1971, 14.1.1972).
DAS 2 242 684 (Lilly; appl. 30.8.1972; GB-prior. 11.9.1971).
US 3 946 002 (Eli Lilly; 23.3.1976; appl. 11.7.1974).
DOS 2 728 578 (Lilly; appl. 24.6.1977; USA-prior. 1.7.1976).
Chauvette, R.R. et al.: J. Org. Chem. (JOCEAH) **36**, 1259 (1971).

acylation via 1,3,2-dioxaboranyl-derivatives:
DOS 2 755 902 (Dobfar; appl. 15.12.1977; I-prior. 16.12.1976).

microbiological acylation:
US 4 073 687 (Shionogi; 14.2.1978; J-prior. 14.5.1975).

crystalline monohydrate:
US 3 531 481 (Lilly; 29.9.1970; prior. 21.4.1969).
US 3 655 656 (Lilly; 11.4.1972; prior. 21.4.1969, 4.6.1970).

salts with sulfonic acids:
US 3 676 434 (Lilly; 11.7.1972; prior. 29.7.1970).

retard preparation:
GB 1 543 543 (Shionogi; appl. 11.5.1977; J-prior. 13.5.1976).

Formulation(s):   cps. 500 mg; f. c. tabl. 500 mg, 1000 mg; gran. 125 mg, 250 mg; vial 1 g/4 ml; susp. 250 mg/5
    ml; syrup 50 mg/ml, 250 mg/ml

Trade Name(s):
D: Ceporexin (Glaxo
    Wellcome; Hoechst; 1973)
    Oracef (Lilly; 1971)
F: Cefacet (Norgine)
    Ceporexine (Glaxo
    Wellcome)

    Keforal (Lilly; 1970)
GB: Ceporex (Glaxo Wellcome;
    1969)
    Keflex (Lilly; 1985)
I: Cefalexi (Formulario Naz.;
    Lifepharma)

    Cefalexina (Marco Viti)
    Ceporex (Glaxo)
    Foce (Medici)-comb.
    Fosfolexin (Lifepharma)-
    comb.
    Lafarin (Lafare)

Pivacef (Firma)
Zetacef (Menarini)
J: Cephalomax (Daisan)
Cephazal (Hokuriku)
Cepol (Torii)
CEX (Glaxo)
Ciponium (Nippon
Kayaku)
Derantel (Nippon
Chemiphar)

Garasin (Wakamoto)
Iwalexin (Iwaki)
Keflex (Shionogi)
Larixin (Toyama)
Madlexin (Meiji)
Mamalexin (Showa)
Mepilacin-DS (Kanto Ishi)
Ohlexin (Ohta)
Oracocin (Tobishi)
Oroxin (Otsuka)

Rinesal (Kissei)
Salitex (Banyu)
Segoramin (Takata)
Sencephalin (Takeda)
Suciralin (Mohan)
Syncl (Toyo Jozo)
Taicelexin (Taiyo)
Tokiolexin (Isei)
Xakl (SS Seiyaku)

USA: Keflex (Dista; 1971)

Cefaloglycin

(Cephaloglycin)

ATC: J01DA
Use: antibiotic

RN: 3577-01-3   MF: $C_{18}H_{19}N_3O_6S$ MW: 405.43   EINECS: 222-696-7
LD_{50}:   >10 g/kg (M, p.o.);
>10 g/kg (R, p.o.)
CN: [6R-(6α,7β(R^*))]-3-[(acetyloxy)methyl]-7-[(aminophenylacetyl)amino]-8-oxo-5-thia-1-
azabicyclo[4.2.0]oct-2-ene-2-carboxylic acid

dihydrate
RN: 22202-75-1   MF: $C_{18}H_{19}N_3O_6S \cdot 2H_2O$ MW: 441.46

N-Cbo-D-phenyl-
glycine

1. isobutyl chloroformate
2. 7-aminocephalosporanic
acid

N-Cbo-cefaloglycin (I)

Cefaloglycin

Cbo:

Reference(s):
GB 985 747 (Eli Lilly; appl. 22.8.1962; USA-prior. 11.9.1961).
US 3 497 505 (Eli Lilly; 24.2.1970; appl. 24.10.1966).
GB 1 017 624 (Merck & Co.; appl. 10.1.1963; USA-prior. 16.1.1962).

acylation via silyl-derivatives:
DOS 1 942 454 (Lilly; appl. 20.8.1969; USA-prior. 23.8.1968).

microbiological acylation:
US 4 073 687 (Shionogi; 14.2.1978; J-prior. 14.5.1975).

Formulation(s):   cps. 250 mg, 500 mg

Cefaloridine
(Cephaloridine)

ATC: J01DA02
Use: antibiotic

RN: 50-59-9   MF: $C_{19}H_{17}N_3O_4S_2$ MW: 415.49   EINECS: 200-052-6
LD_{50}:   2200 mg/kg (M, i.v.); >20 g/kg (M, p.o.);
     1065 mg/kg (R, i.v.); 2500 mg/kg (R, p.o.)
CN: (6R-trans)-1-[[2-carboxy-8-oxo-7-[(2-thienylacetyl)amino]-5-thia-1-azabicyclo[4.2.0]oct-2-en-3-yl]methyl]pyridinium hydroxide inner salt

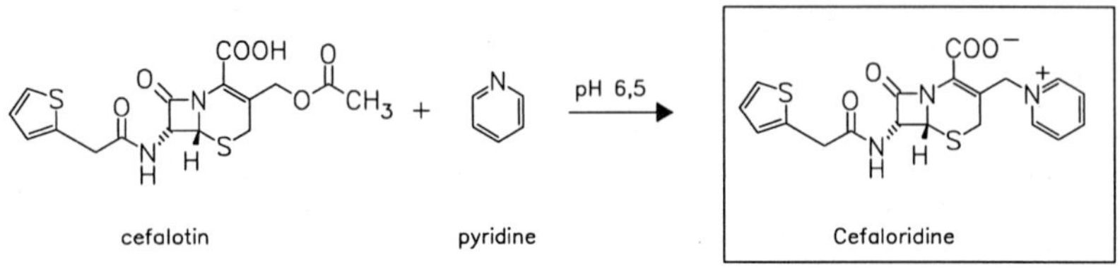

cefalotin   pyridine   Cefaloridine

Reference(s):
GB 1 030 630 (Glaxo; appl. 14.12.1962).
DE 1 445 828 (Glaxo; appl. 14.12.1963; GB-prior. 14.12.1962, 2.12.1963).
FR 1 384 197 (Glaxo; appl. 13.12.1963; GB-prior. 14.12.1962, 2.12.1963).
DAS 1 670 599 (Lilly; appl. 17.1.1966; USA-prior. 5.3.1965).
DAS 1 795 581 (Glaxo; appl. 4.11.1964; GB-prior. 13.7.1964, 29.9.1964).
DE 1 795 610 (Glaxo; appl. 4.11.1964; GB-prior. 4.11.1963, 13.7.1964, 29.9.1964).

Formulation(s):   amp. 250 mg/2 ml, 500 mg/3 ml, 1 g/4 ml

Cefalotin
(Cephalotin)

ATC: J01DA03
Use: antibiotic

RN: 153-61-7   MF: $C_{16}H_{16}N_2O_6S_2$ MW: 396.44   EINECS: 205-815-7
LD_{50}:   4990 mg/kg (M, i.v.);
     >5 g/kg (R, i.v.)
CN: (6R-trans)-3-[(acetyloxy)methyl]-8-oxo-7-[(2-thienylacetyl)amino]-5-thia-1-azabicyclo[4.2.0]oct-2-ene-2-carboxylic acid

monosodium salt
RN: 58-71-9   MF: $C_{16}H_{15}N_2NaO_6S_2$ MW: 418.43   EINECS: 200-394-6
LD_{50}:   4800 mg/kg (M, i.v.);
     5600 mg/kg (R, i.v.); >10 g/kg (R, p.o.)

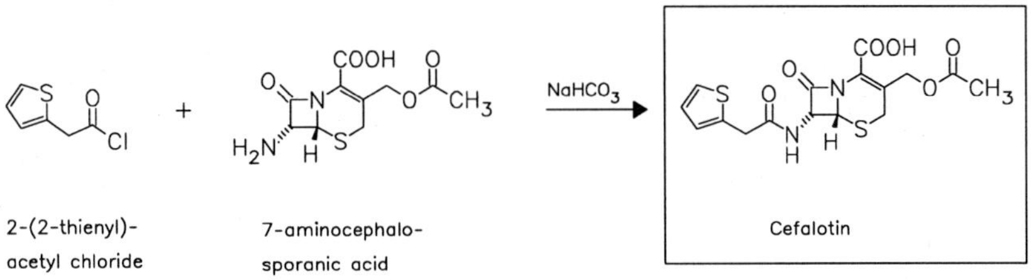

2-(2-thienyl)-
acetyl chloride

7-aminocephalo-
sporanic acid

Cefalotin

Reference(s):

DE 1 445 684 (Eli Lilly; appl. 4.6.1962; USA-prior. 8.6.1961).
BE 618 663 (Eli Lilly; appl. 7.6.1962; USA-prior. 8.6.1961).
DAS 1 670 641 (Lilly; appl. 23.11.1967; USA-prior. 23.11.1966).
DOS 2 730 579 (Pierrel S.p.A.; appl. 6.7.1977; GB-prior. 10.7.1976).

acylation via silyl-derivatives of 7-aminocephalosporanic acid:
DOS 1 942 454 (Lilly; appl. 20.8.1969; USA-prior. 23.8.1968).

acylation via 1,3,2-dioxaboranyl-derivatives:
DOS 2 755 902 (Dobfar; appl. 15.12.1977; I-prior. 16.12.1976).

total synthesis:
Ratcliffe, R.W.; Christensen, G.B.: Tetrahedron Lett. (TELEAY) **1973**, 4649.

"easily soluble form" for parenteral application by freeze-drying:
US 4 029 655 (Lilly; 14.6.1977; appl. 11.4.1975).
US 4 132 848 (Lilly; 2.1.1979; prior. 3.11.1977).
DOS 2 752 442 (Lilly; appl. 24.11.1977).

crystalline sterile preparation for parenteral application:
US 4 029 655 (Lilly; 14.6.1977; appl. 11.4.1975).

Formulation(s):   amp. 500 mg, 1 g, 2 g, 4 g (as sodium salt)

Trade Name(s):

D:	Cepovenin (Hoechst; 1973); wfm	J:	Cephation (Meiji)	Sucira N (Mohan)

D: Cepovenin (Hoechst;
     1973); wfm
F: Céfalotine (Panpharma)
     Kéflin (Lilly; 1966)
     generics
I: Cefalo (Formulario Naz.)
     Keflin (Lilly)

J: Cephation (Meiji)
     CET (Glaxo)
     Coaxin (Tobishi)
     Keflin (Shionogi Lilly)
     Resting (Ono)
     Sodium Cephalotin (Green
     Cross)

     Sucira N (Mohan)
     Synclotin (Toyo Jozo)
     Toricelocin (Torii)
USA: Keflin (Lilly; 1975); wfm
     Seffin Neutral (Glaxo;
     1984); wfm

Cefamandole

ATC: J01DA07
Use: antibiotic

RN: 34444-01-4   MF: $C_{18}H_{18}N_6O_5S_2$ MW: 462.51   EINECS: 252-030-0
CN: [6R-[6α,7β(R*)]]-7-[(hydroxyphenylacetyl)amino]-3-[[(1-methyl-1H-tetrazol-5-yl)thio]methyl]-8-oxo-5-thia-1-azabicyclo[4.2.0]oct-2-ene-2-carboxylic acid

formate monosodium salt (nafate)
RN: 42540-40-9   MF: $C_{19}H_{17}N_6NaO_6S_2$ MW: 512.50   EINECS: 255-877-4
LD$_{50}$:   3915 mg/kg (M, i.v.);
      2562 mg/kg (R, i.v.)

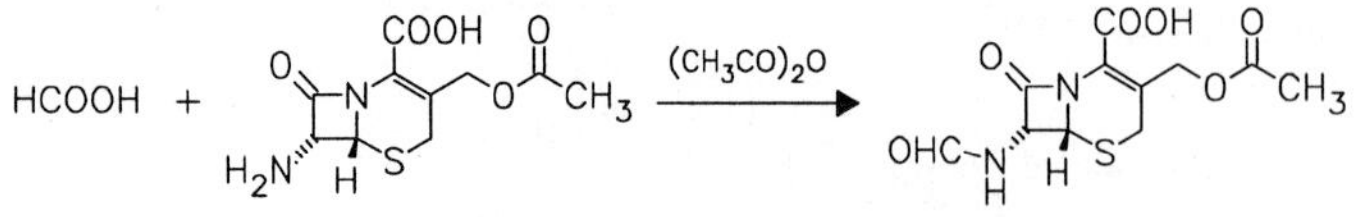

formic acid

7-aminocephalosporanic acid

7-formamidocephalosporanic acid (I)

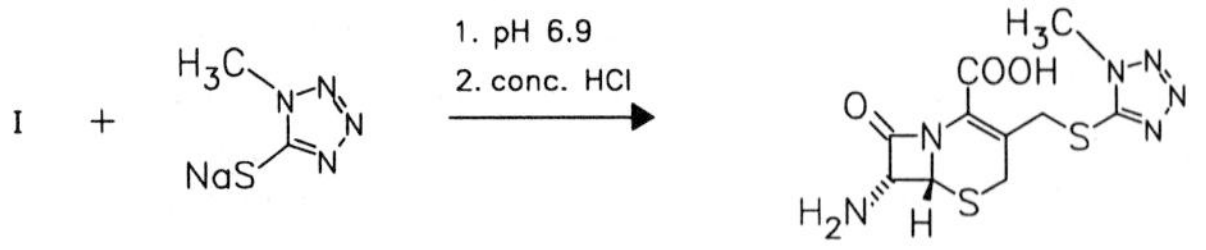

1-methyl-1H-tetrazole-5-thiol sodium salt

7-amino-3-(1-methyl-tetrazol-5-ylthiomethyl)-3-cephem-4-carboxylic acid (II)

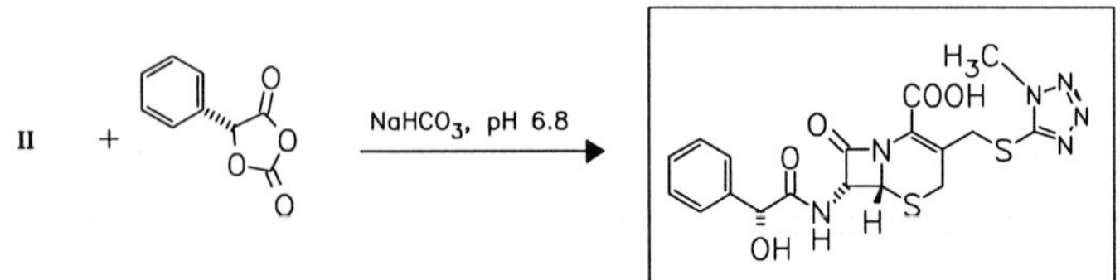

D-anhydro-O-carboxymandelic acid (from D(−)-mandelic acid and phosgene)

Cefamandole

Reference(s):

US 3 641 021 (Lilly; 8.2.1972; appl. 18.4.1969).
DE 2 018 600 (Lilly; appl. 17.4.1970; USA-prior. 18.4.1969).
DAS 2 065 621 (Lilly; appl. 17.4.1970; USA-prior. 18.4.1969).
US 3 840 531 (Lilly; 8.10.1974; appl. 21.3.1972).
US 3 903 278 (Smith Kline Corp.; 2.9.1975; prior. 4.11.1971).
DOS 2 730 579 (Pierrel S.p.A.; appl. 6.7.1977; GB-prior. 10.7.1976).

preparation and/or purification via the trimethylsilyl-derivatives:
DOS 2 711 095 (Lilly; appl. 14.3.1977; USA-prior. 17.3.1976).

purification:
US 4 115 644 (Lilly; 19.9.1978; appl. 19.9.1978).
DOS 2 839 670 (Lilly; appl. 12.9.1978; USA-prior. 19.9.1977).

crystalline sodium salt:
US 4 054 738 (Lilly; 18.10.1977; appl. 22.12.1975).
US 4 168 376 (Lilly; 18.9.1979; appl. 5.6.1978).

lithium salt:
GB 1 546 757 (Lilly; appl. 10.4.1975; valid from 7.4.1976).

O-formyl-derivative:
US 3 928 592 (Lilly; 23.12.1975; appl. 21.2.1974).
GB 1 493 676 (Lilly; appl. 20.2.1975; USA-prior. 22.2.1974).
GB 1 546 898 (Lilly; appl. 7.4.1976; USA-prior. 11.4.1975).
DOS 2 506 622 (Lilly; appl. 17.2.1975; USA-prior. 22.2.1974).

*crystalline sodium salt of O-*formylcefamandole:*
US 4 006 138 (Lilly; 1.2.1977; appl. 11.4.1975).

complex of cefamandole sodium *with* 1,4-dioxane *and* water:
US 3 947 414 (Lilly; 30.3.1976; appl. 23.12.1974).

complex of cefamandole sodium *with* ethyl L-(–)-lactate:
US 3 947 415 (Lilly; 30.3.1976; appl. 23.12.1974).

Formulation(s):   vial 0.5 g, 1 g, 2 g (as nafate)

Trade Name(s):

D:	Mandokef (Lilly; 1977)	Cefaseptolo (Miba)			Mandolsan (San Carlo)
F:	Kefandol (Lilly)	Cefiran (Pierrel)			Neocefal (Metapharma)
GB:	Kefadol (Dista; 1978)	Cemado (Francia Farm.)			Septomandolo (IPA)
I:	Bergacef (Bergamon)	Fado (Caber)	J:		Kefadole (Shionogi)
	Cedol (Eurofarmaco)	Lampomandol (AGIPS)	USA:		Mandol (Lilly; 1978)
	Cefam (Magis)	Mancef (Lafare)			
	Cefamen (Menarini)	Mandokef (Lilly)			

Cefapirin

(Cephapirin; Cefaprin)

ATC: J01DA30
Use: β-lactam antibiotic

RN: 21593-23-7   MF: $C_{17}H_{17}N_3O_6S_2$ MW: 423.47   EINECS: 244-466-5
LD$_{50}$:   >760 mg/kg (M, i.v.); 26.1 g/kg (M, p.o.);
6048 mg/kg (R, i.v.); 16.356 g/kg (R, p.o.)
CN: (6R-*trans*)-3-[(acetyloxy)methyl]-8-oxo-7-[[(4-pyridinylthio)acetyl]amino]-5-thia-1-azabicyclo[4.2.0]oct-2-ene-2-carboxylic acid

monosodium salt
RN: 24356-60-3   MF: $C_{17}H_{16}N_3NaO_6S_2$ MW: 445.45   EINECS: 246-194-2
LD$_{50}$:   4600 mg/kg (M, i.v.); 26.1 g/kg (M, p.o.);
4580 mg/kg (R, i.v.); 16.4 g/kg (R, p.o.);
2500 mg/kg (dog, i.v.)

4-chloro-pyridine + thioglycolic acid → 4-pyridylthio-acetic acid → (I)

I + 7-aminocephalosporanic acid (II) → Cefapirin

II + bromoacetyl bromide → 4-mercaptopyridine → Cefapirin

Reference(s):

Crast, L.B. et al.: J. Med. Chem. (JMCMAR) **16**, 1413 (1973).
US 3 422 100 (Bristol-Myers; 14.1.1969; appl. 2.5.1967; prior. 5.1.1967).
US 3 503 967 (Bristol-Myers; 31.3.1970; appl. 26.8.1968).
US 3 578 661 (Bristol-Myers; 11.5.1971; appl. 2.6.1969).
DE 1 670 301 (Bristol-Myers; appl. 5.1.1968; USA-prior. 5.1.1967).

acylation via 1,3,2-dioxaboranyl-derivatives:
DOS 2 755 902 (Dobfar; appl. 15.12.1977; I-prior. 16.12.1976).

salts with amino acids:
FR-appl. 2 479 228 (Dobfar; appl. 25.3.1981; I-prior. 1.4.1980).

Formulation(s):   vial 0.5 g, 1 g, 2 g, 4 g (as sodium salt)

Trade Name(s):

D:	Bristocef (Bristol; 1974); wfm	Brisporin (Bristol It. Sud); wfm	Cefatrexyl (Nihon Bristol) Cepotril (Tobishi-Kaken)
F:	Cefaloject (Bristol-Myers Squibb; 1974)	Piricef (CT); wfm	Ceropirin (Nichiiko) Taicelepirin (Taiyo)
		J: Antibalin (Nippon Chemiphar)	Vacian (Kantoishi)
I:	Ambrocef (Lusofarmaco); wfm	Cefarin (Fuji)	USA: Cefadyl (Bristol; 1974)

Cefatrizine

ATC: J01DA21
Use: β-lactam antibiotic

RN: 51627-14-6   MF: $C_{18}H_{18}N_6O_5S_2$ MW: 462.51   EINECS: 257-324-2
CN: [6R-[6α,7β(R*)]]-7-[[amino(4-hydroxyphenyl)acetyl]amino]-8-oxo-3-[(1H-1,2,3-triazol-4-ylthio)methyl]-5-thia-1-azabicyclo[4.2.0]oct-2-ene-2-carboxylic acid

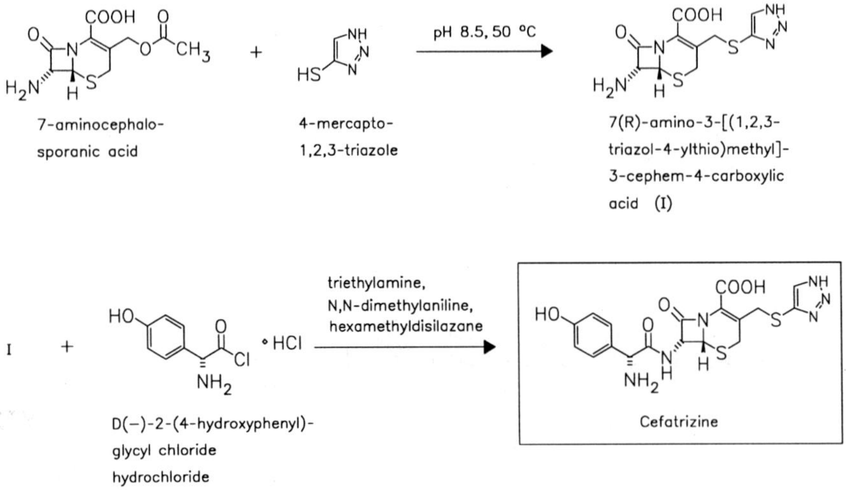

7-aminocephalo-
sporanic acid

4-mercapto-
1,2,3-triazole

7(R)-amino-3-[(1,2,3-
triazol-4-ylthio)methyl]-
3-cephem-4-carboxylic
acid   (I)

D(−)-2-(4-hydroxyphenyl)-
glycyl chloride
hydrochloride

Cefatrizine

Reference(s):

US 3 899 394 (Bristol-Myers; 12.8.1975; prior. 26.12.1972).
US 3 867 380 (SmithKline Corp.; 18.2.1975; prior. 17.12.1970, 18.2.1971, 14.6.1972).
DOS 2 364 192 (Bristol-Myers; appl. 21.12.1973; USA-prior. 26.12.1972).
DAS 2 622 985 (Bristol-Myers; appl. 21.5.1976; USA-prior. 23.5.1975).
US 3 970 651 (Bristol-Myers; 20.7.1976; prior. 7.1.1974, 18.12.1974).
US 3 985 747 (Bristol-Myers; 12.10.1976; appl. 24.5.1974).

acylation via 1,3,2-dioxaboranyl derivatives:
DOS 2 755 902 (Dobfar; appl. 15.12.1977; I-prior. 16.12.1976).

Formulation(s):   cps. 125 mg, 250 mg, 500 mg; susp. 5 %; syrup 125 mg, 250 mg

Trade Name(s):

F:	Céfaperos (Bristol-Myers	Lampotrix (Leben's)	Trixilan (Pulitzer)
	Squibb; 1987)	Latocef (Delsaz &	Trizina (Francia Farm.)
I:	Cefatrix (tekfarma bkf)	Filippini)	Zanitrin (Bristol It. Sud)
	Cefotrizin (Firma)	Miracef (Tosi-Novara)	Zinaf (Crosara)
	Cetrazil (Herdel)	Novacef (Locatelli)	Zitrix (Metapharma)
	Faretrizin (Lafare)	Orotrix (San Carlo)	J: Bricef (Bristol)
	Ipatrizina (IPA)	Tamyl (Fisons Italchimici)	Cepticol (Banyu; 1980)
	Kefoxina (CT)	Tricef (Eurofarmaco)	
	Ketrizin (Esseti)	Trixidine (ASTA Medica)	

Cefazedone

ATC: J01DA15
Use: antibiotic

RN: 56187-47-4   MF: $C_{18}H_{15}Cl_2N_5O_5S_3$ MW: 548.45
CN: (6R-trans)-7-[[(3,5-dichloro-4-oxo-1(4H)-pyridinyl)acetyl]amino]-3-[[(5-methyl-1,3,4-thiadiazol-2-yl)thio]methyl]-8-oxo-5-thia-1-azabicyclo[4.2.0]oct-2-ene-2-carboxylic acid

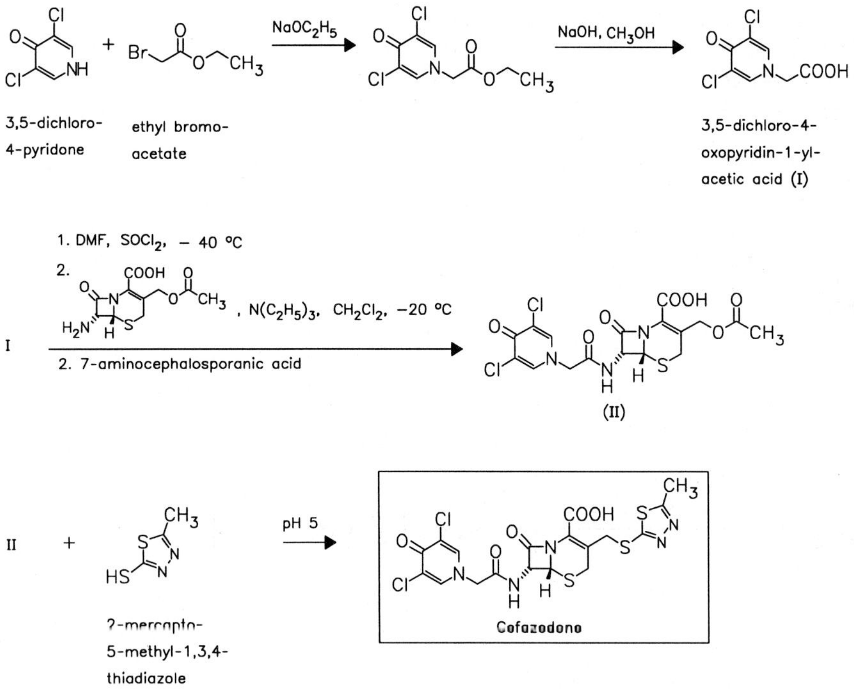

3,5-dichloro-4-pyridone   ethyl bromo-acetate   3,5-dichloro-4-oxopyridin-1-yl-acetic acid (I)

2. 7-aminocephalosporanic acid

(II)

2-mercapto-5-methyl-1,3,4-thiadiazole   Cefazedone

Reference(s):
Gericke, R.; Rogalski, W.: Arzneim.-Forsch. (ARZNAD) **29** (I), 362 (1979).
DOS 2 427 224 (E. Merck; appl. 6.6.1974).
DOS 2 345 402 (E. Merck; appl. 8.9.1973).
DOS 2 621 011 (E. Merck; appl. 12.5.1976).
GB 1 436 989 (E. Merck; appl. 5.9.1974; D-prior. 8.9.1973, 6.6.1974).
US 4 153 693 (E. Merck; 8.5.1979; D-prior. 8.9.1973, 6.6.1974).
GB 1 539 158 (E. Merck; appl. 11.5.1977; D-prior. 12.5.1976).

Formulation(s):   vial 1 g, 2 g

Trade Name(s):
D: Refosporin (E. Merck);
      wfm

Cefazolin
(Cephazolin)

ATC: J01DA04
Use: antibiotic

RN: 25953-19-9   MF: $C_{14}H_{14}N_8O_4S_3$ MW: 454.52   EINECS: 247-362-8
LD$_{50}$:   3 g/kg (M, i.v.)
CN: (6R-*trans*)-3-[[(5-methyl-1,3,4-thiadiazol-2-yl)thio]methyl]-8-oxo-7-[(1*H*-tetrazol-1-ylacetyl)amino]-5-thia-1-azabicyclo[4.2.0]oct-2-ene-2-carboxylic acid

monosodium salt
RN: 27164-46-1   MF: $C_{14}H_{13}N_8NaO_4S_3$ MW: 476.50   EINECS: 248-278-4
LD$_{50}$:   3900 mg/kg (M, i.v.); >11 g/kg (M, p.o.);
      2760 mg/kg (R, i.v.); >11 g/kg (R, p.o.);
      2200 mg/kg (dog, i.v.)

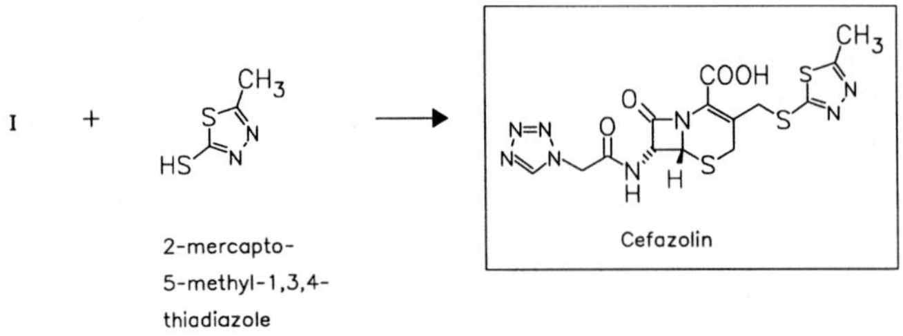

Reference(s):
US 3 516 997 (Fujisawa; 23.6.1970; appl. 12.4.1968; J-prior. 15.4.1967, 24.10.1967, 28.10.1967).
DE 1 170 168 (Fujisawa; appl. 10.4.1968; J-prior. 14.4.1967).

corresponding:
GB 1 206 305 (Fujisawa; appl. 11.4.1968; J-prior. 15.4.1967).
NL 6 805 179 (Fujisawa; appl. 11.4.1968; J-prior. 15.4.1967).
DOS 1 953 861 (Fujisawa; appl. 25.10.1969).

alternative syntheses:
DOS 2 055 796 (Fujisawa; appl. 13.11.1970; J-prior. 17.11.1969).
DOS 2 540 374 (Lilly; appl. 10.9.1975; USA-prior. 12.9.1974).

acylation via 1,3,2-dioxaboranyl-derivatives:
DOS 2 755 902 (Dobfar; appl. 15.12.1977; I-prior. 16.12.1976).

purification:
US 4 115 645 (Lilly; 19.9.1978; appl. 10.5.1977).

sodium salt:
DOS 2 752 443 (Lilly; appl. 24.11.1977; USA-prior. 24.11.1976).

sodium salt monohydrate:
US 4 104 470 (Lilly; 1.8.1978; appl. 3.6.1977).

rapidly soluble spray dried sodium salt:
US 4 146 971 (Lilly; 3.4.1979; prior. 24.11.1976, 14.12.1977).

suspension for parenteral application:
GB 1 546 479 (Lilly; appl. 23.4.1976; USA-prior. 28.4.1975).

Formulation(s):   vial 250 mg, 500 mg, 1 g, 2 g (as sodium salt)

Trade Name(s):

D:	Elzogram (Lilly; 1974)	Cefabiozim (IPA)		Zolisint (Locatelli)
	Gramaxin (Roche; 1974)	Cefamezin (Carlo Erba)	J:	Cefamezin (Fujisawa;
F:	Cefacidal (Bristol-Myers	Cefazil (Delsaz &		1971)
	Squibb; 1976)	Filippini)	USA:	Ancef (SmithKline
	Kefzol (Lilly; 1976)	Cromezin (Crosara)		Beecham; 1973)
GB:	Kefzol (Lilly; 1974)	Firmacef (Firma)		Kefzol (Lilly; 1973)
I:	Acef (Eurofarmaco)	Recef (Farma Uno)		
	Biazolina (Ist. Italiano	Totacef (Bristol It. Sud)		
	Ferm.)	Zolin (San Carlo)		

Cefbuperazone

ATC: J01DA
Use: β-lactam antibiotic

RN: 76610-84-9   MF: $C_{22}H_{29}N_9O_9S_2$ MW: 627.66
CN: [6R-[6α,7α,7(2R*,3S*)]]-7-[[2-[[(4-ethyl-2,3-dioxo-1-piperazinyl)carbonyl]amino]-3-hydroxy-1-oxobutyl]amino]-7-methoxy-3-[[(1-methyl-1H-tetrazol-5-yl)thio]methyl]-8-oxo-5-thia-1-azabicyclo[4.2.0]oct-2-ene-2-carboxylic acid

1. (CH$_3$)$_3$SiCl, N(C$_2$H$_5$)$_3$

2. H$_3$C , N(C$_2$H$_5$)$_3$ · HCl

H$_2$N COOH
H$_3$C OH

D-threonine

1. trimethylchlorosilane
2. 2,3-dioxo-4-ethyl-1-piperazine-
   carbonyl chloride
   (cf. piperacillin synthesis)

2(R)-(2,3-dioxo-4-ethyl-
1-piperazinecarboxamido)-
3(S)-hydroxybutyric
acid (I)

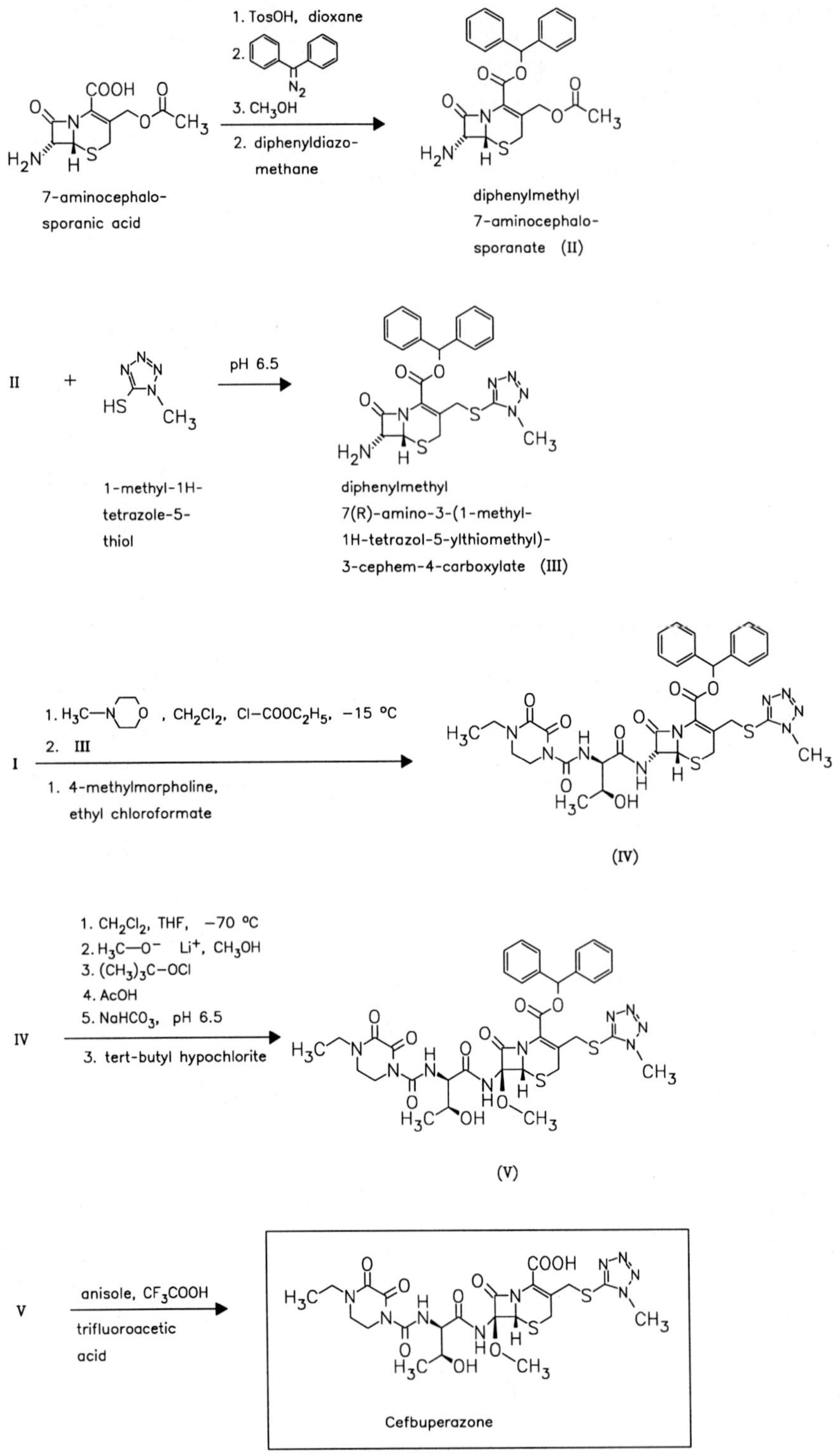

7-aminocephalo-
sporanic acid

diphenylmethyl
7-aminocephalo-
sporanate (II)

1. TosOH, dioxane
2. [diphenyldiazomethane structure] N₂
3. CH₃OH
2. diphenyldiazo-
methane

II +

1-methyl-1H-
tetrazole-5-
thiol

pH 6.5

diphenylmethyl
7(R)-amino-3-(1-methyl-
1H-tetrazol-5-ylthiomethyl)-
3-cephem-4-carboxylate (III)

1. H₃C—N O , CH₂Cl₂, Cl—COOC₂H₅, −15 °C
2. III

I

1. 4-methylmorpholine,
ethyl chloroformate

(IV)

1. CH₂Cl₂, THF, −70 °C
2. H₃C—O⁻ Li⁺, CH₃OH
3. (CH₃)₃C—OCl
4. AcOH
5. NaHCO₃, pH 6.5

IV

3. tert-butyl hypochlorite

(V)

V

anisole, CF₃COOH

trifluoroacetic
acid

Cefbuperazone

Reference(s):
DOS 2 939 747 (Toyama; appl. 1.10.1979; J-prior. 23.4.1979, 7.8.1979).
FR 2 455 051 (Toyama; appl. 4.10.1979; J-prior. 23.4.1979, 7.8.1979).
US 4 263 292 (Toyama; 21.4.1981; J-prior. 13.6.1978, 23.4.1979, 7.8.1979).
GB 2 048 241 (Toyama; appl. 26.9.1979; J-prior. 23.4.1979, 7.8.1979).

Formulation(s):   vial 500 mg, 1 g

Trade Name(s):
J: Keiperazon (Kaken; 1985)   Tomiporan (Toyama; 1985)

Cefditoren pivoxil
(ME-1207)

ATC: S01AA
Use: cephalosporin

RN: 117467-28-4   MF: $C_{25}H_{28}N_6O_7S_3$ MW: 620.73
CN: [6R-[3(Z),6α,7β(Z)]]-7-[[(2-amino-4-thiazolyl)(methoxyimino)acetyl]amino]-3-[2-(4-methyl-5-thiazolyl)ethenyl]-8-oxo-5-thia-1-azabicyclo[4.2.0]oct-2-ene-2-carboxylic acid (2,2-dimethyl-1-oxopropoxy)methyl ester

[6R-[3(Z),6α,7β(E)]]-form
RN: 104145-87-1   MF: $C_{25}H_{28}N_6O_7S_3$ MW: 620.73

4-methoxybenzyl 3-(chloro-methyl)-7(R)-(phenylacetamido)-3-cephem-4-carboxylate

1. $P(C_6H_5)_3$, NaI, acetone
2. (CHO thiazole) , NaHCO₃, CH₂Cl₂

1. triphenylphosphine, sodium iodide
2. 4-methylthiazole-5-carboxaldehyde

→ I

(I)

PCl₅, pyridine
−30 °C
phosphorus pentachloride in pyridine

(II)

II +

2-(methoxyimino)-2-(2-tritylamino-thiazol-4-yl)acetic acid

1. POCl₃
2. F₃C—COOH, anisole
2. trifluoroacetic acid in anisole

→ III

(III) · F₃C—COOH

1. NaHCO₃, H₂O
2. Diaion HP20 chromatography

→ IV

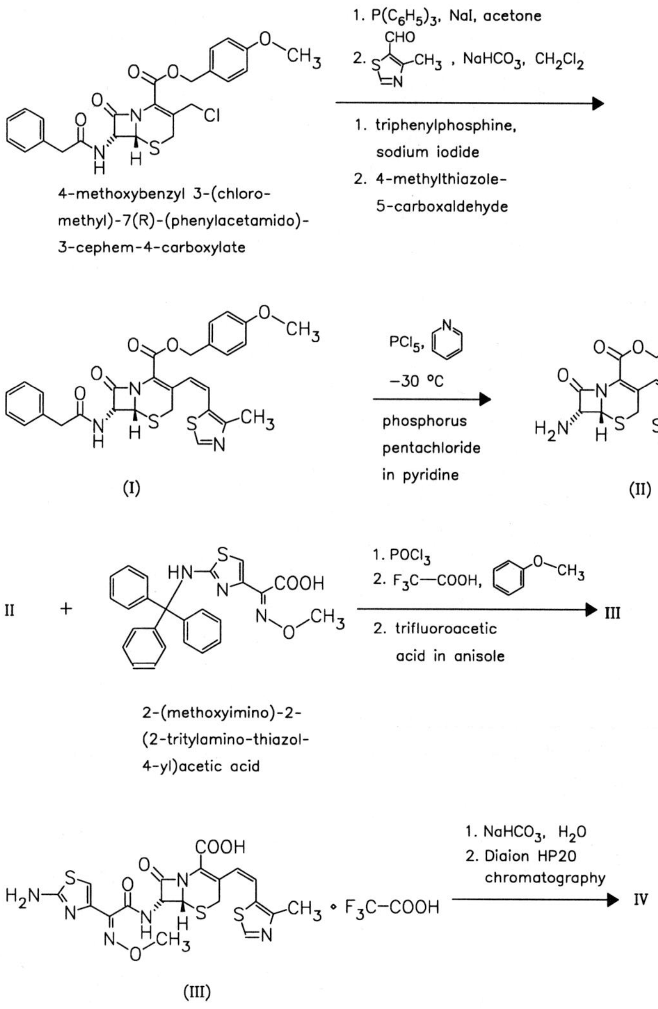

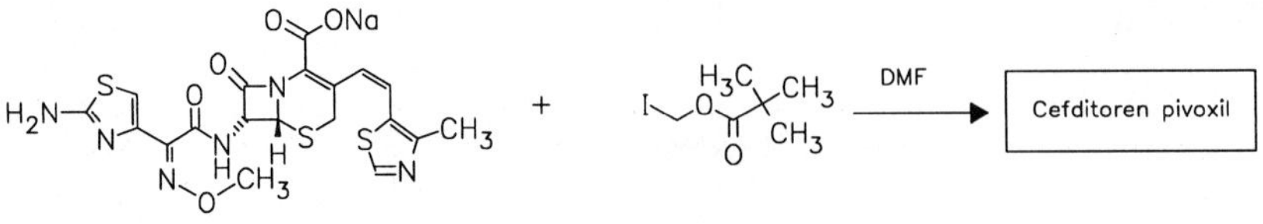

(IV)   iodomethyl
                        pivalate

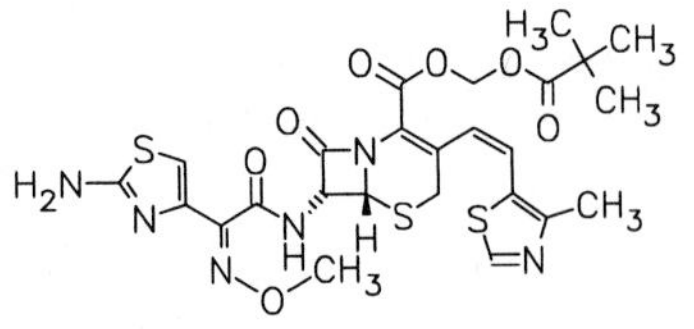

Cefditoren pivoxil

Reference(s):
synthesis:
EP 175 610 (Meiji Seika Kaisha; appl. 26.3.1986; J-prior. 7.9.1984, 18.7.1985).
Sakagami, K. ct al.: J. Antibiot. (JANTAJ) **43**(8), 1047 (1990).
Sakagami, K. et al.: Chem. Pharm. Bull. (CPBTAL) **39**(9), 2433 (1992).

pharmaceutical compositions:
EP 339 465 (Meiji Seika Kaisha; appl. 2.11.1989; J-prior. 19.4.1988).
EP 629 404 (Meiji Seika Kaisha; appl. 21.12.1994; J-prior. 16.6.1993).

Formulation(s):   gran. 100 mg; tabl. 100 mg

Trade Name(s):
J: Meiact (Meiji Seika)

Cefixime
(CL-284635; FK-027; FR-17027)

ATC: J01DA23
Use: semisynthetic third generation
        cephem antibiotic (for oral
        administration), high β-lactamase
        stability

RN: 79350-37-1   MF: $C_{16}H_{15}N_5O_7S_2$ MW: 453.46
LD$_{50}$: 4420 mg/kg (M, i.v.); >10 g/kg (M, p.o.);
        6990 mg/kg (R, i.v.); >10 g/kg (R, p.o.);
        >3200 mg/kg (dog, i.v.); >600 mg/kg (dog, p.o.)
CN: [6R-[6α,7β(Z)]]-7-[[(2-amino-4-thiazolyl)[(carboxymethoxy)imino]acetyl]amino]-3-ethenyl-8-oxo-5-
        thia-1-azabicyclo[4.2.0]oct-2-ene-2-carboxylic acid

trihydrate
RN: 125110-14-7   MF: $C_{16}H_{15}N_5O_7S_2 \cdot 3H_2O$ MW: 507.50

Synthesis of intermediates:

(1)

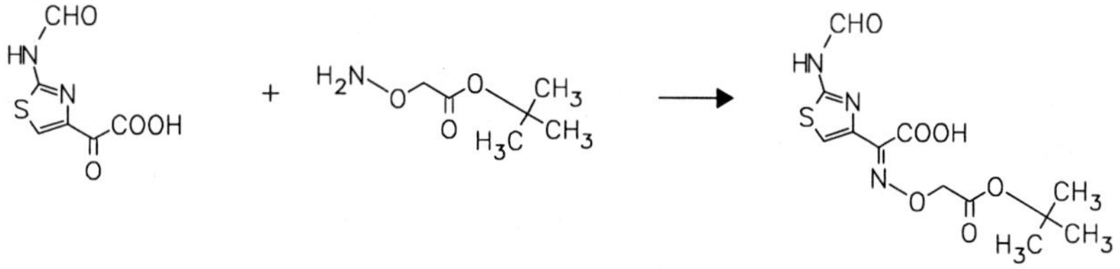

(2-formamido-
thiazol-4-yl)-
oxoacetic acid

tert-butyl
aminooxyacetate

(Z)-2-(tert-butoxycarbonyl-
methoxyimino)-2-(2-form-
amidothiazol-4-yl)-
acetic acid (I)

(2)

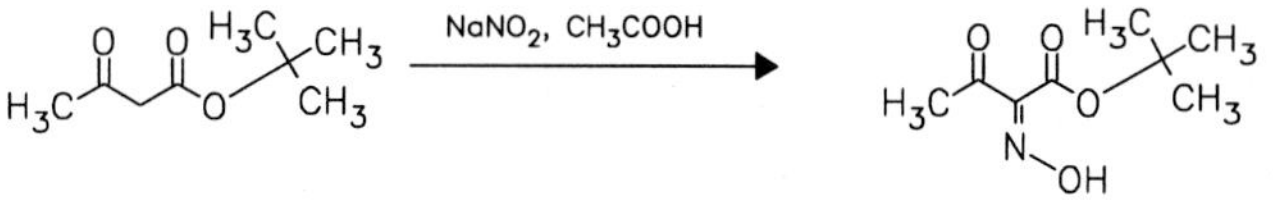

tert-butyl
acetoacetate

NaNO₂, CH₃COOH

tert-butyl 2-hydroxy-
imino-3-oxobutyrate (II)

1. acetyl acetate, DMF, K₂CO₃

2. Cl—CH₃ , 7.5 h

II

2. methyl chloroacetate

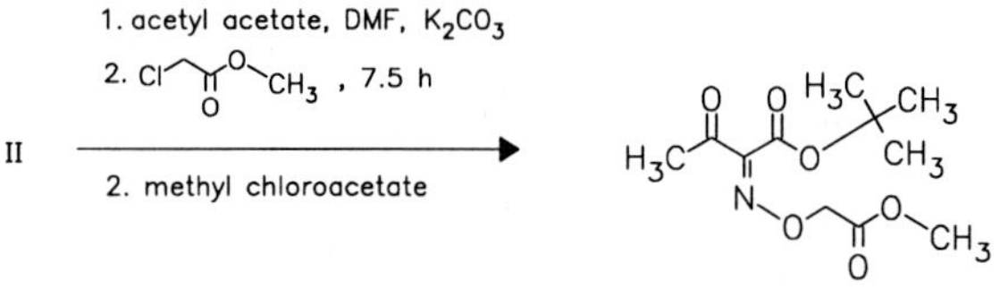

(Z)-tert-butyl 2-(methoxy-
carbonylmethoxyimino)-
3-oxobutyrate (III)

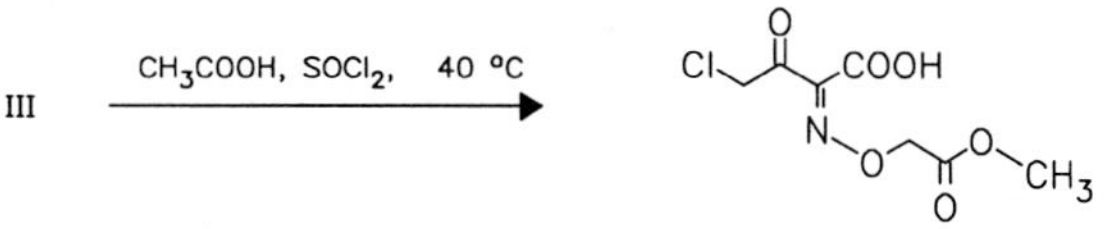

III

CH₃COOH, SOCl₂,   40 °C

(Z)-4-chloro-2-(methoxy-
carbonylmethoxyimino)-
3-oxobutyric acid (IV)

(3)

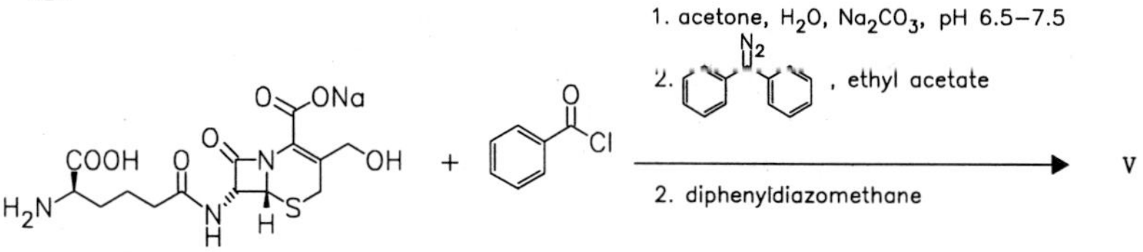

deacetylcephalosporin C
sodium salt

benzoyl
chloride

1. acetone, H₂O, Na₂CO₃, pH 6.5–7.5

2. (diphenyl diazo compound) , ethyl acetate

2. diphenyldiazomethane

V

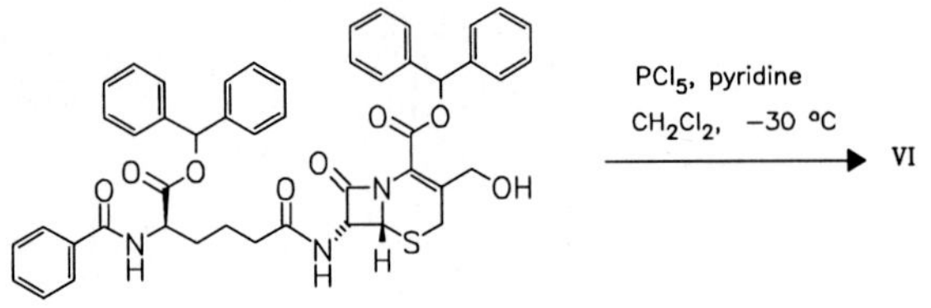

diphenylmethyl 7-[5-benzamido-
5-(diphenylmethoxycarbonyl)-
pentanamido]-3-hydroxymethyl-
3-cephem-4-carboxylate (V)

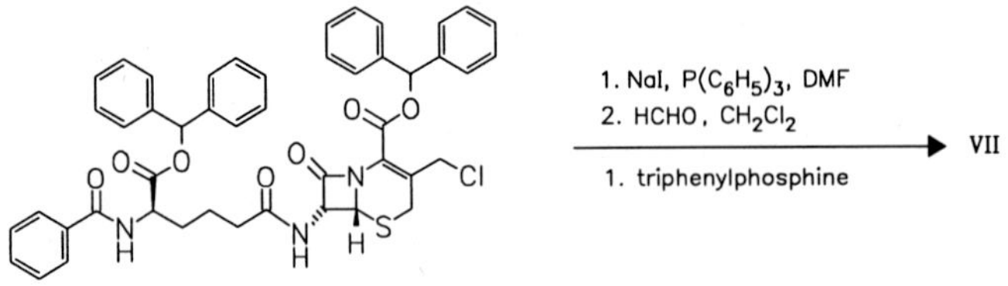

PCl$_5$, pyridine
CH$_2$Cl$_2$, −30 °C
⟶ VI

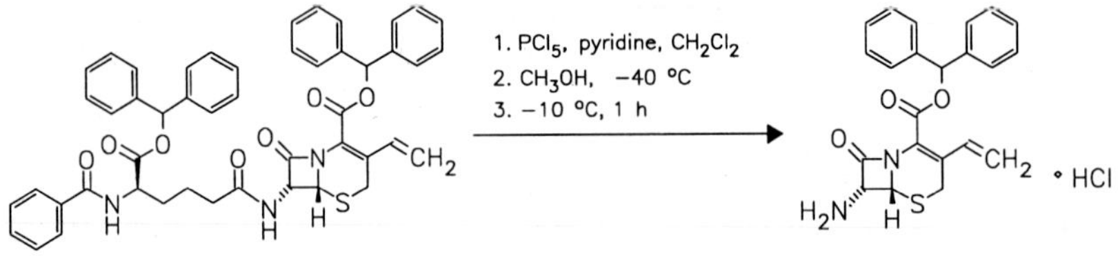

(VI)

1. NaI, P(C$_6$H$_5$)$_3$, DMF
2. HCHO, CH$_2$Cl$_2$
⟶ VII

1. triphenylphosphine

(VII)

1. PCl$_5$, pyridine, CH$_2$Cl$_2$
2. CH$_3$OH, −40 °C
3. −10 °C, 1 h
⟶

diphenylmethyl 7-amino-
3-vinyl-3-cephem-4-
carboxylate hydrochloride (VIII)

④

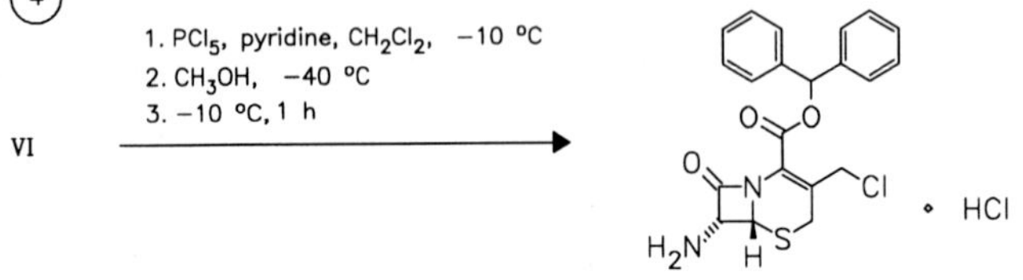

VI

1. PCl$_5$, pyridine, CH$_2$Cl$_2$, −10 °C
2. CH$_3$OH, −40 °C
3. −10 °C, 1 h
⟶

diphenymethyl
7-amino-3-chloro-
methyl-3-cephem-
4-carboxylate
hydrochloride (IX)

synthesis of Cefixime:

ⓐ

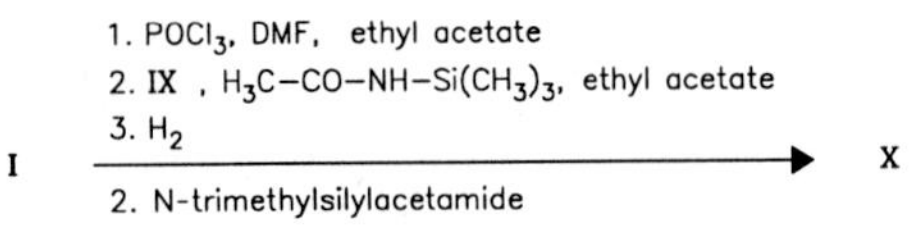

I

1. POCl$_3$, DMF, ethyl acetate
2. IX , H$_3$C−CO−NH−Si(CH$_3$)$_3$, ethyl acetate
3. H$_2$

2. N-trimethylsilylacetamide
⟶ X

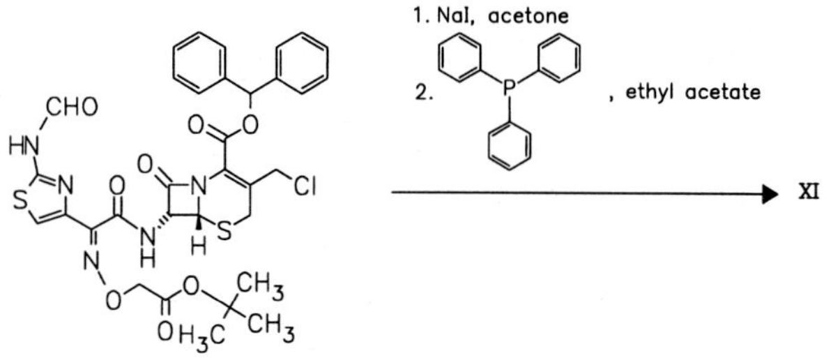
1. NaI, acetone
2. , ethyl acetate
XI

(X)

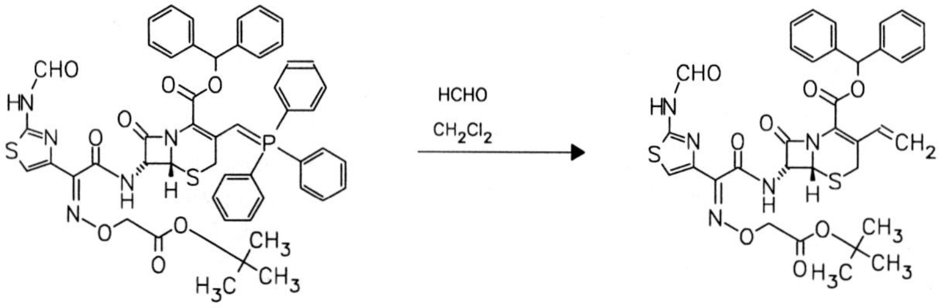
HCHO
CH_2Cl_2

(XI) (XII)

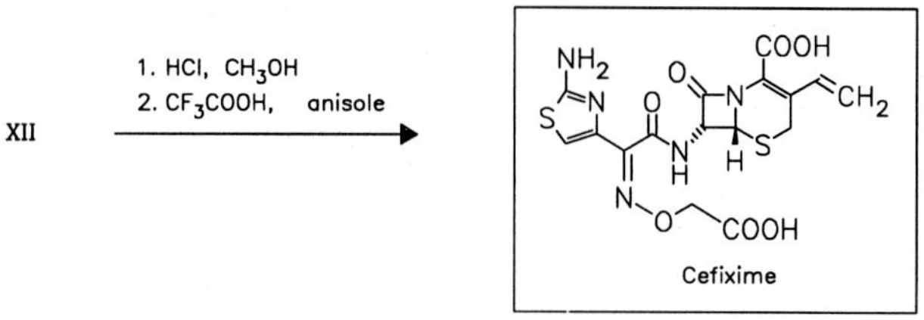
XII
1. HCl, CH_3OH
2. CF_3COOH, anisole
Cefixime

b

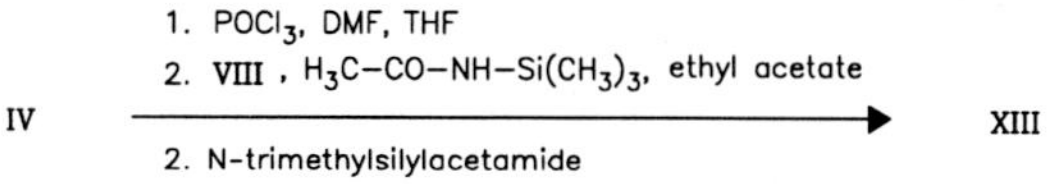
IV
1. $POCl_3$, DMF, THF
2. VIII, $H_3C-CO-NH-Si(CH_3)_3$, ethyl acetate
2. N-trimethylsilylacetamide
XIII

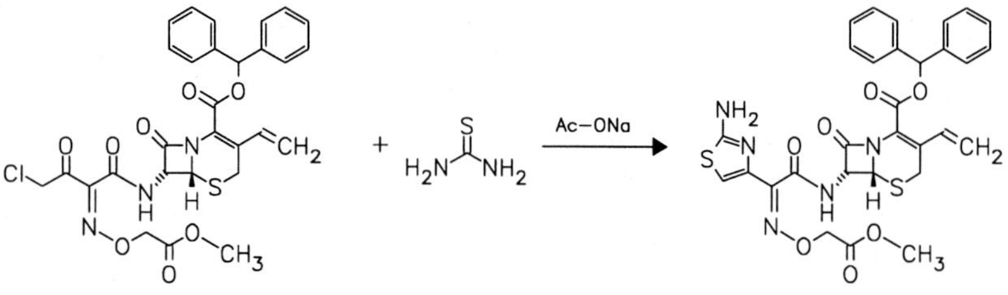
Ac–ONa

(XIII) thiourea (XIV)

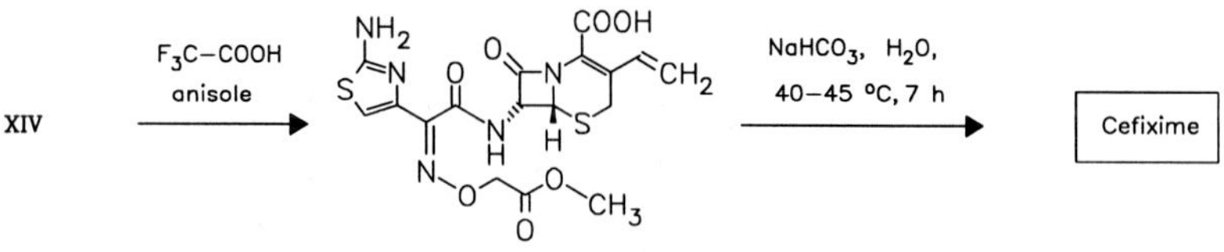
XIV
$F_3C-COOH$
anisole
$NaHCO_3$, H_2O,
40–45 °C, 7 h
Cefixime

Reference(s):
alternative synthesis routes for VIII starting from 7-aminocephalosporanic acid *are also described in the cited literature:*
Yamanaka, H. et al.: J. Antibiot. (JANTAJ) **38**, 1738 (1985).
Yamanaka, H. et al.: J. Antibiot. (JANTAJ) **39**, 101 (1986).
Kawabata, K. et al.: J. Antibiot. (JANTAJ) **39**, 405 (1986).
US 4 409 214 (Fujisawa; 11.10.1983; UK-prior. 19.11.1979, 8.2.1980, 21.4.1980, 14.7.1980).
US 4 423 213 (Fujisawa; 27.12.1983; UK-prior. 19.11.1979, 8.2.1980, 21.4.1980, 14.7.1980).
US 4 487 927 (Fujisawa; 11.12.1984; UK-prior. 19.11.1979, 8.2.1980, 21.4.1980, 14.7.1980).
US 4 585 860 (Fujisawa; 29.4.1984; UK-prior. 19.11.1979, 8.2.1980, 21.4.1980, 14.7.1980).
EP 30 630 (Fujisawa; appl. 15.11.1980; UK-prior. 19.11.1979, 8.2.1980, 21.4.1980, 14.7.1980).

Formulation(s):   cps. 100 mg, 200 mg; fine gran. 50 mg/g; f. c. tabl. 200 mg, 400 mg; oral susp. 100 mg/5 ml; syrup 100 mg/5 ml; supplied as trihydrate in all formulations

Trade Name(s):

D:	Cephoral (Merck; 1991)	F:	Oroken (Bellon)		Unixime (Firma)
	Suprax (Klinge)	I:	Cefixoral (Menarini)	J:	Cefspan (Fujisawa; 1987)
	Uro-Cephoral (Merck)		Suprax (Cyanamid)	USA:	Suprax (Lederle Labs.)

Cefmenoxime

ATC: J01DA16
Use: β-lactam antibiotic (cefalosporin derivative)

RN: 65085-01-0   MF: $C_{16}H_{17}N_9O_5S_3$ MW: 511.57
CN: [6R-[6α,7β(Z)]]-7-[[(2-amino-4-thiazolyl)(methoxyimino)acetyl]amino]-3-[[(1-methyl-1*H*-tetrazol-5-yl)thio]methyl]-8-oxo-5-thia-1-azabicyclo[4.2.0]oct-2-ene-2-carboxylic acid

hydrochloride (2:1)
RN: 75738-58-8   MF: $C_{16}H_{17}N_9O_5S_3 \cdot 1/2HCl$ MW: 1059.60   EINECS: 278-299-4
LD$_{50}$:   7830 mg/kg (M, i.v.); 17.54 g/kg (M, p.o.);
          2680 mg/kg (R, i.v.); >20 g/kg (R, p.o.)

2-(2-chloroacetamido-
4-thiazolyl)-2-meth-
oxyiminoacetyl
chloride
(cf. ceftriaxone
synthesis)

7(R)-amino-3-(1-methyl-
1H-tetrazol-5-ylthio-
methyl)-3-cephem-4-
carboxylic acid
(cf. cefamandole
synthesis)

dimethyl-
acetamide

I

. C₂H₅OH, THF

thiourea

(I)

Cefmenoxime

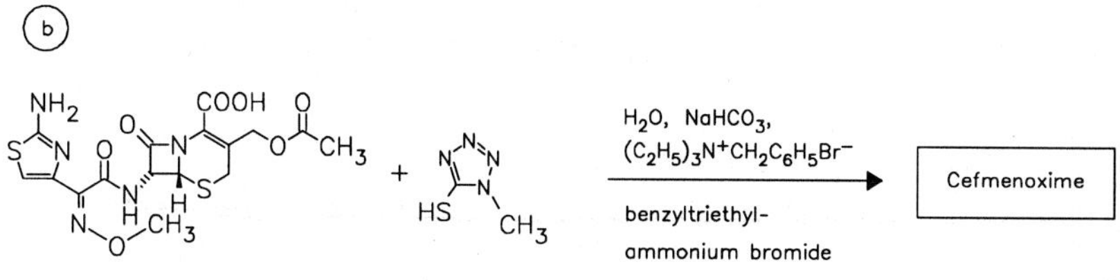

b

cefotaxime
(q. v.)

1-methyl-1H-
tetrazole-5-
thiol

H₂O, NaHCO₃,
(C₂H₅)₃N⁺CH₂C₆H₅Br⁻

benzyltriethyl-
ammonium bromide

Cefmenoxime

Reference(s):

Ochiai, M. et al.: Chem. Pharm. Bull. (CPBTAL) **25**, 3115 (1977).
DOS 2 556 736 (Takeda; appl. 17.12.1975; J-prior. 19.12.1974; GB-prior. 9.6.1975).
US 4 098 888 (Takeda; 4.7.1978; J-prior. 1974; GB-prior. 9.6.1975).
DOS 2 715 385 (Takeda; appl. 6.4.1977; J-prior. 14.4.1976; 8.9.1976).

Formulation(s):   vial 500 mg, 1 g, 2 g (as hydrochloride)

Trade Name(s):
D: Tacef (Takeda; 1983)   F: Cemix (Takeda); wfm   J: Bestcall (Takeda; 1983)

Cefoperazone

ATC: J01DA32
Use: antibiotic

RN: 62893-19-0   MF: $C_{25}H_{27}N_9O_8S_2$ MW: 645.68   EINECS: 263-749-4
CN: [6R-[6α,7β(R*)]]-7-[[[[(4-ethyl-2,3-dioxo-1-piperazinyl)carbonyl]amino](4-
hydroxyphenyl)acetyl]amino]-3-[[(1-methyl-1H-tetrazol-5-yl)thio]methyl]-8-oxo-5-thia-1-
azabicyclo[4.2.0]oct-2-ene-2-carboxylic acid

sodium salt
RN: 62893-20-3   MF: $C_{25}H_{26}N_9NaO_8S_2$ MW: 667.66

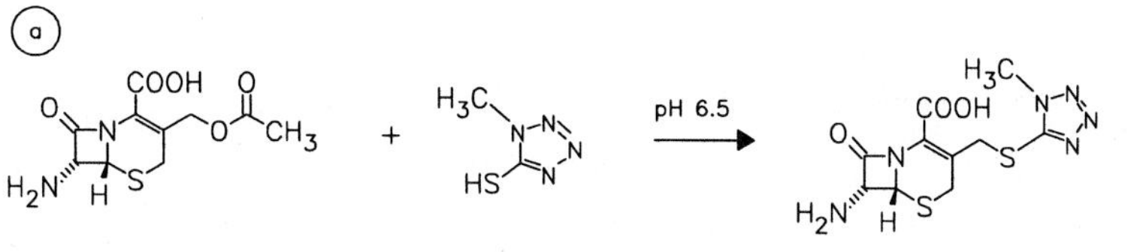

a

7-aminocephalo-
sporanic acid

1-methyl-1H-
tetrazole-5-thiol

pH 6.5

7-amino-3-(1-methyl-
tetrazol-5-ylthiomethyl)-
3-cephem-4-carboxylic acid (I)
(cf. cefamandole synthesis)

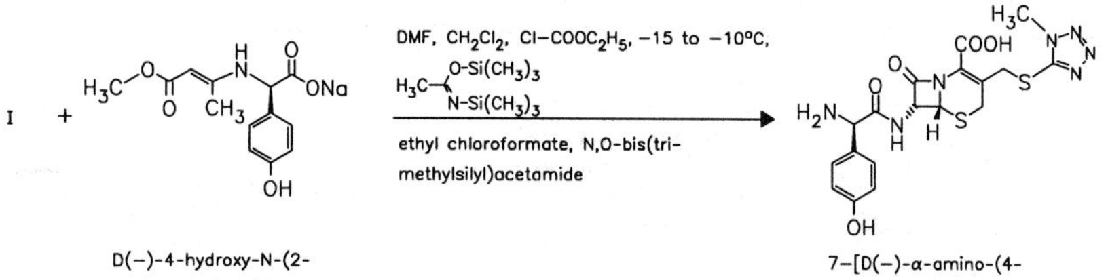

I +

DMF, CH₂Cl₂, Cl-COOC₂H₅, −15 to −10°C,

ethyl chloroformate, N,O-bis(trimethylsilyl)acetamide

D(−)-4-hydroxy-N-(2-
methoxycarbonyl-1-
methylethenyl)phenyl-
glycine sodium salt

7-[D(−)-α-amino-(4-
hydroxyphenyl)acetamido]-
3-(1-methyltetrazol-5-yl-
thiomethyl)-3-cephem-
4-carboxylic acid (II)

II + 2,3-dioxo-4-ethyl-1-piperazinecarbonyl chloride (III) (cf. piperacillin synthesis) → (K$_2$CO$_3$, 0–5°C) → Cefoperazone

b)

III + D(–)-4-hydroxy-phenylglycine sodium salt → (N(C$_2$H$_5$)$_3$) → D(–)-α-(2,3-dioxo-4-ethyl-1-piperazino-carbonylamino)-4-hydroxyphenylacetic acid (IV)

IV →

1. CH$_2$Cl$_2$, DMF, N,N-dimethylaniline, –15°C, Cl–COOC$_2$H$_5$
2. I, CH$_3$CN, H$_3$C–C(O–Si(CH$_3$)$_3$)=N–Si(CH$_3$)$_3$, –20°C

1. ethyl chloroformate
2. 7-amino-3-[5-(1-methyltetrazolyl)thiomethyl]-3-cephem-4-carboxylic acid, N,O-bis(trimethylsilyl)acetamide

→ Cefoperazone

Reference(s):

DE 2 600 880 (Toyama; D-prior. 12.1.1976).

US 4 410 522 (Toyama; 18.10.1983; J-prior. 9.5.1974).

US 4 110 327 (Toyama; 29.8.1978; J-prior. 9.5.1974, 13.5.1974, 31.5.1974, 13.8.1974, 26.9.1974, 13.12.1974, 27.3.1975).

DOS 2 519 400 (Toyama; appl. 30.4.1975; J-prior. 9.5.1974, 13.5.1974, 31.5.1974, 13.8.1974, 26.9.1974, 13.12.1974, 27.3.1975).

GB 1 508 062 (Toyama; appl. 28.4.1975; J-prior. 9.5.1974, 13.5.1974, 31.5.1974, 24.7.1974, 7.8.1974, 13.8.1974, 26.9.1974, 12.10.1974, 28.10.1974, 6.12.1974, 13.12.1974, 17.2.1975, 26.3.1975, 27.3.1975).

GB 1 508 071 (Toyama; appl. 19.1.1976).

N,N-dimethylacetamide adducts:

DOS 2 841 706 (Toyama; appl. 25.9.1978; J-prior. 27.9.1977).

Formulation(s):　vial 250 mg, 500 mg, 1 g, 2 g (as sodium salt)

Trade Name(s):

D:　Cefobis (Pfizer; 1981)
F:　Céfobis (Pfizer); wfm
I:　Bioperazone (Leben's)
　　Cefazone (Locatelli)
　　Cefobid (Pfizer)
　　Cefogram (Metapharma)

Cefoneg (Tosi-Novara)
Cefoper (Menarini)
Cefosint (Crosara)
Dardum (Lisapharma)
Farecef (Lafare)
Ipazone (IPA)

Kefazon (Esseti)
Mediper (Medici)
Novobiocyl (Francia Farm.)
Perocef (Pulitzer)
Prontokef (Master Pharma)
Tomabef (Salus Research)

Zoncef (AGIPS)   J: Cefobid (Toyama/Pfizer;   Cefoperazin (Toyama)
                                1981)   USA: Cefobid (Pfizer; 1982)

Cefotaxime

ATC: J01DA10
Use: antibiotic

RN: 63527-52-6   MF: $C_{16}H_{17}N_5O_7S_2$ MW: 455.47   EINECS: 264-299-1
CN: [6R-[6α,7β(Z)]]-3-[(acetyloxy)methyl]-7-[[(2-amino-4-thiazolyl)(methoxyimino)acetyl]amino]-8-oxo-5-
       thia-1-azabicyclo[4.2.0]oct-2-ene-2-carboxylic acid

monosodium salt
RN: 64485-93-4   MF: $C_{16}H_{16}N_5NaO_7S_2$ MW: 477.45
LD$_{50}$: 6845 mg/kg (M, i.v.); >20 g/kg (M, p.o.);
       7 g/kg (R, i.v.); >20 g/kg (R, p.o.);
       >1.5 g/kg (dog, i.v.)

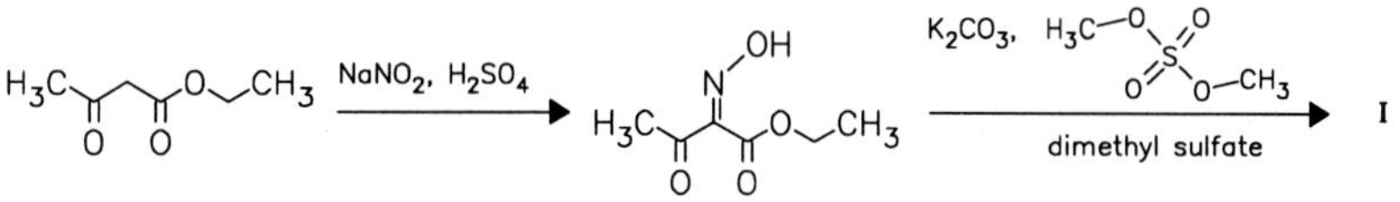

ethyl acetoacetate

ethyl 2-(hydroxy-
imino)acetoacetate

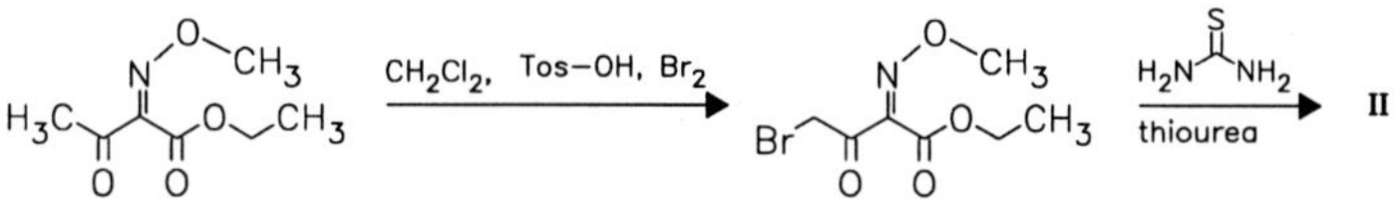

ethyl 2-(methoxy-
imino)acetoacetate (I)

ethyl 4-bromo-2-
(methoxyimino)-
acetoacetate

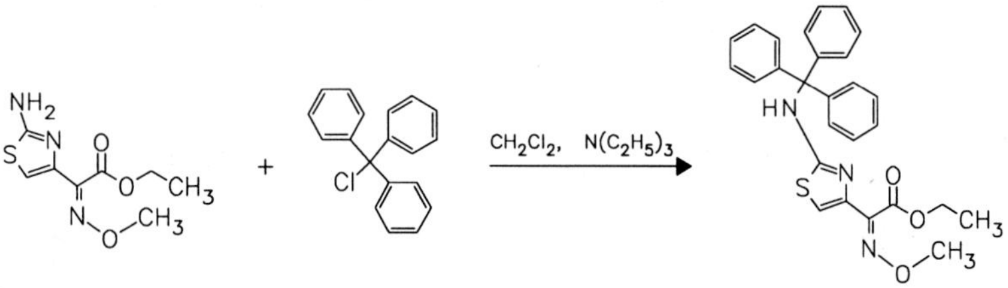

ethyl 2-(2-amino-
4-thiazolyl)-2-(meth-
oxyimino)acetate (II)

trityl
chloride

ethyl 2-(methoxy-
imino)-2-[2-(trityl-
amino)-4-thiazolyl]-
acetate (III)

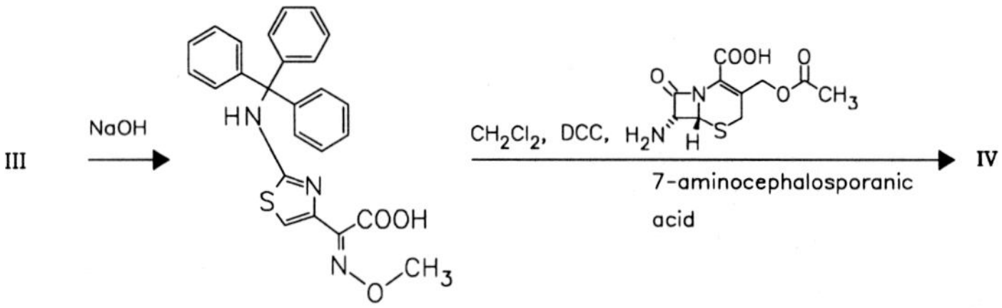

7-aminocephalosporanic
acid

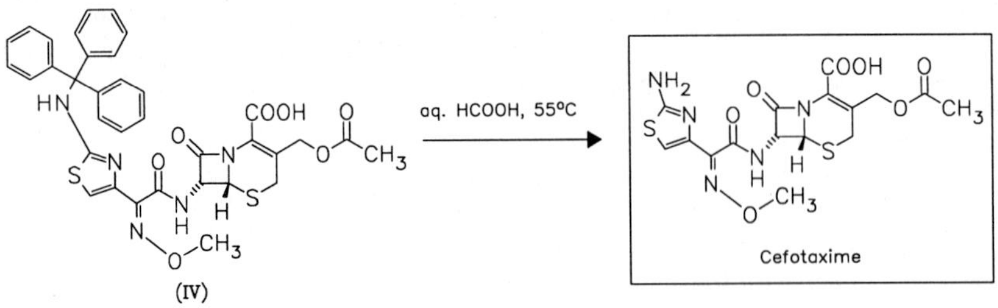

(IV) → Cefotaxime

aq. HCOOH, 55°C

Reference(s):

DOS 2 702 501 (Roussel-Uclaf; appl. 21.1.1977; F-prior. 23.1.1976, 11.6.1976, 18.8.1976).
US 4 152 432 (Roussel-Uclaf; 1.5.1979; F-prior. 23.1.1976).

sodium salt:
DAS 2 708 439 (Hoechst; appl. 26.2.1977).

Formulation(s):   vial 250 mg, 500 mg, 1 g, 2 g (as sodium salt)

Trade Name(s):
D: Claforan (Hoechst; 1980)   Zariviz (Hoechst Italia Sud)   USA: Claforan (Hoechst Marion
F: Claforan (Hoechst)   J: Cefotax (Roussel-Chugai;   Roussel)
GB: Claforan (Roussel; 1981)   1981)
I: Claforan (Roussel)   Claforan (Hoechst; 1981)

Cefotetan

ATC: J01DA14
Use: β-lactam antibiotic (cefalosporin derivative)

RN: 69712-56-7   MF: $C_{17}H_{17}N_7O_8S_4$ MW: 575.63   EINECS: 274-093-3
LD_{50}:   4990 mg/kg (M, i.v.); >10 g/kg (M, p.o.);
       5 g/kg (R, i.v.); >10 g/kg (R, p.o.);
       >6 g/kg (dog, i.v.)
CN: [6R-(6α,7α)]-7-[[[4-(2-amino-1-carboxy-2-oxoethylidene)-1,3-dithietan-2-yl]carbonyl]amino]-7-methoxy-3-[[(1-methyl-1H-tetrazol-5-yl)thio]methyl]-8-oxo-5-thia-1-azabicyclo[4.2.0]oct-2-ene-2-carboxylic acid

disodium salt
RN: 74356-00-6   MF: $C_{17}H_{15}N_7Na_2O_8S_4$ MW: 619.59   EINECS: 277-834-9
LD_{50}:   4990 mg/kg (M, i.v.); >10 g/kg (M, p.o.);
       6790 mg/kg (R, i.v.); >10 g/kg (R, p.o.);
       >6 g/kg (dog, i.v.)

methyl malonamate   + CS₂   NaH →   → NaOH →   4-carboxy-3-hydroxy-5-mercaptoisothiazole trisodium salt (I)

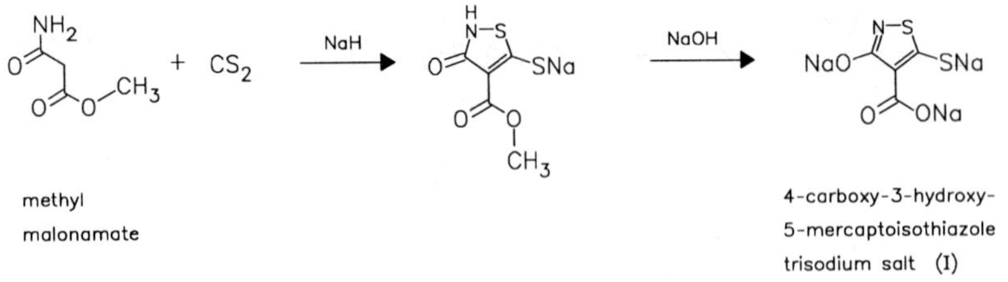

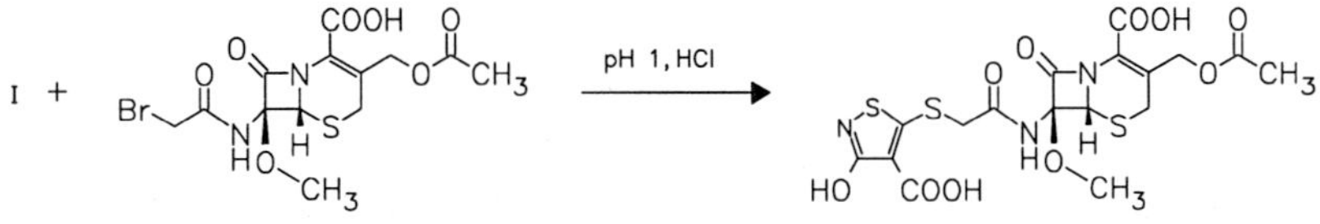

I + 7(S)-bromoacetamido-
7-methoxycephalo-
sporanic acid

pH 1, HCl → (II)

II + 1-methyl-1H-
tetrazole-5-
thiol

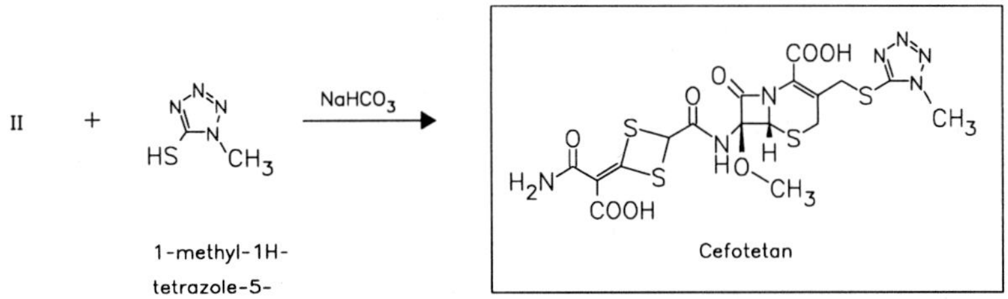

NaHCO₃ → Cefotetan

Reference(s):

DOS 2 824 559 (Yamanouchi; appl. 5.6.1978; J.-prior. 10.6.1977).
US 4 263 432 (Yamanouchi; 21.4.1981; appl. 7.6.1978; J-prior. 28.7.1977).

Formulation(s):   vial 250 mg, 500 mg, 1 g, 2 g (as sodium salt)

Trade Name(s):
D: Apatef (ICI; 1985); wfm   I: Apatef (Zeneca)   J: Yamatetan (Yamanouchi)
F: Apacef (Zeneca)   Cepan (IBI)   USA: Cefotan (Stuart; 1986)

Cefotiam

ATC: J01DA19
Use: antibiotic

RN: 61622-34-2   MF: $C_{18}H_{23}N_9O_4S_3$ MW: 525.64
LD$_{50}$:   3840 mg/kg (M, i.v.)
CN: (6R-trans)-7-[[(2-amino-4-thiazolyl)acetyl]amino]-3-[[[1-[2-(dimethylamino)ethyl]-1H-tetrazol-5-yl]thio]methyl]-8-oxo-5-thia-1-azabicyclo[4.2.0]oct-2-ene-2-carboxylic acid

dihydrochloride
RN: 66309-69-1   MF: $C_{18}H_{23}N_9O_4S_3 \cdot 2HCl$ MW: 598.56

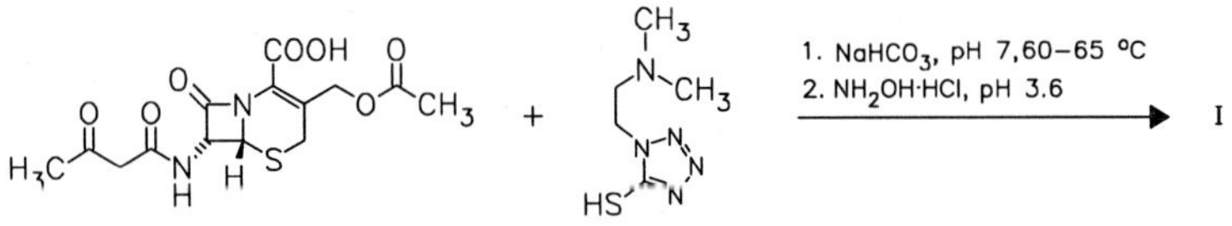

7-acetoacetamidocephalosporanic
acid

1-(2-dimethyl-
aminoethyl)-
1H-tetrazole-
5-thiol

+

1. NaHCO₃, pH 7,60–65 °C
2. NH₂OH·HCl, pH 3.6
→   I

(I)

2-amino-4-
thiazoleacetic
acid hydrochloride

Cefotiam

Reference(s):
Tsushima, S. et al.: Chem. Pharm. Bull. (CPBTAL) **27**, 696 (1979).
DOS 2 461 478 (Takeda; appl. 24.12.1974; J-prior. 25.12.1973).
DAS 2 462 736 (Takeda; appl. 24.12.1974; J-prior. 25.12.1973).
US 4 080 498 (Takeda; 21.3.1978; appl. 20.12.1974; J-prior. 25.12.1973).
DE 2 738 711 (Takeda; appl. 27.8.1977; J-prior. 31.8.1976).
US 4 146 710 (Takeda; 27.3.1979; 29.8.1977; J-prior. 31.8.1976).

intermediates:
DOS 2 607 064 (Takeda; appl. 21.2.1976; J-prior. 24.2.1975).

Formulation(s):   vial 500 mg, 1 g, 2 g (as dihydrochloride); tabl. 100 mg, 200 mg

Trade Name(s):
D: Spizef (Takeda; 1982)   Texodil (Cassenne)   J: Pansporin T (Takeda; 1981)
F: Taketiam (Takeda)   I: Sporidyn (Zoja)   Sporidyn (Cyanamid)

Cefoxitin

ATC: J01DA05
Use: antibiotic

RN: 35607-66-0   MF: $C_{16}H_{17}N_3O_7S_2$ MW: 427.46   EINECS: 252-641-2
LD$_{50}$:   4970 mg/kg (M, i.v.); >10 g/kg (M, p.o.);
       8580 mg/kg (R, i.v.); >10 g/kg (R, p.o.);
       >10 g/kg (dog, i.v.)
CN: (6R-cis)-3-[[(aminocarbonyl)oxy]methyl]-7-methoxy-8-oxo-7-[(2-thienylacetyl)amino]-5-thia-1-
       azabicyclo[4.2.0]oct-2-ene-2-carboxylic acid

monosodium salt
RN: 33564-30-6   MF: $C_{16}H_{16}N_3NaO_7S_2$ MW: 449.44   EINECS: 251-574-6
LD$_{50}$:   4970 mg/kg (M, i.v.); >10 g/kg (M, p.o.);
       8580 mg/kg (R, i.v.); >10 g/kg (R, p.o.);
       10 g/kg (dog, i.v.)

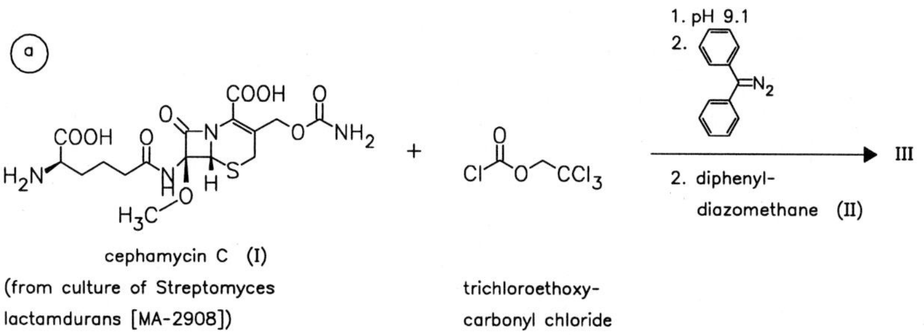

a

cephamycin C (I)
(from culture of Streptomyces
lactamdurans [MA-2908])

trichloroethoxy-
carbonyl chloride

1. pH 9.1
2.
2. diphenyl-
   diazomethane (II)

III

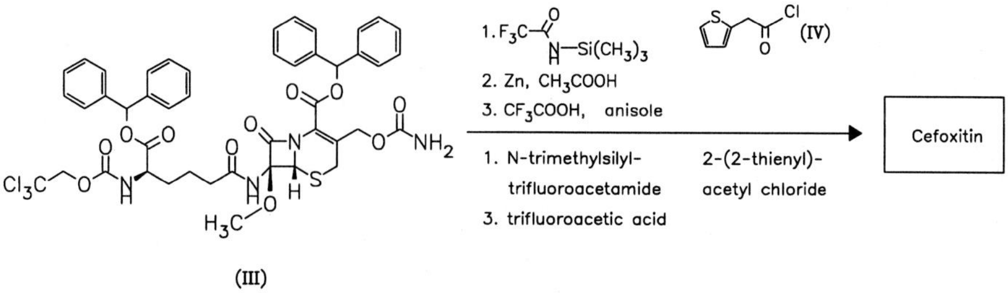

(III)

1. F$_3$C—N—Si(CH$_3$)$_3$
2. Zn, CH$_3$COOH
3. CF$_3$COOH, anisole

1. N-trimethylsilyl-
   trifluoroacetamide
2.
3. trifluoroacetic acid

2-(2-thienyl)-
acetyl chloride

(IV)

Cefoxitin

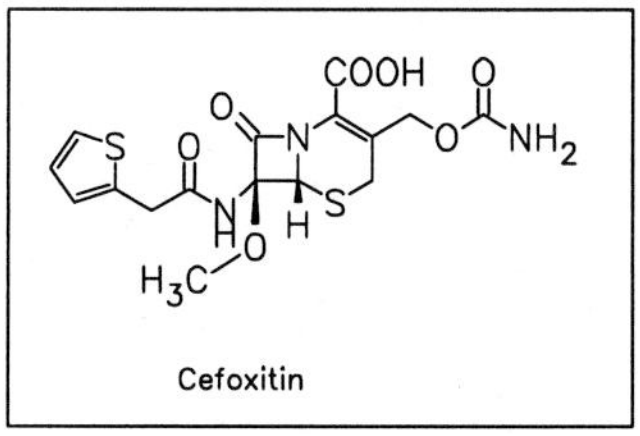

Cefoxitin

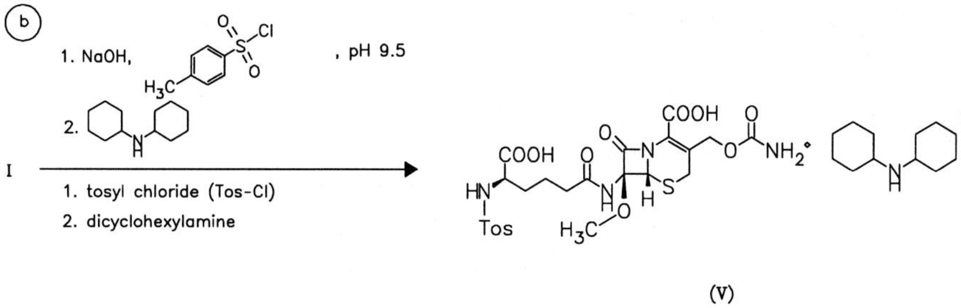

b

1. NaOH,
   Tos-Cl, pH 9.5
2.
1. tosyl chloride (Tos-Cl)
2. dicyclohexylamine

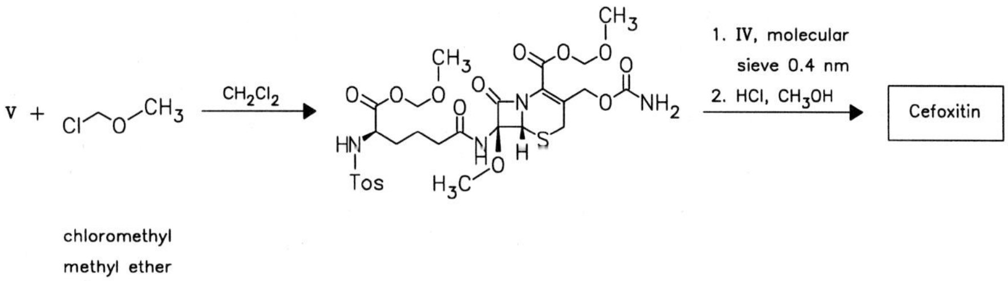

I

(V)

v + chloromethyl methyl ether

CH$_2$Cl$_2$

1. IV, molecular
   sieve 0.4 nm
2. HCl, CH$_3$OH

Cefoxitin

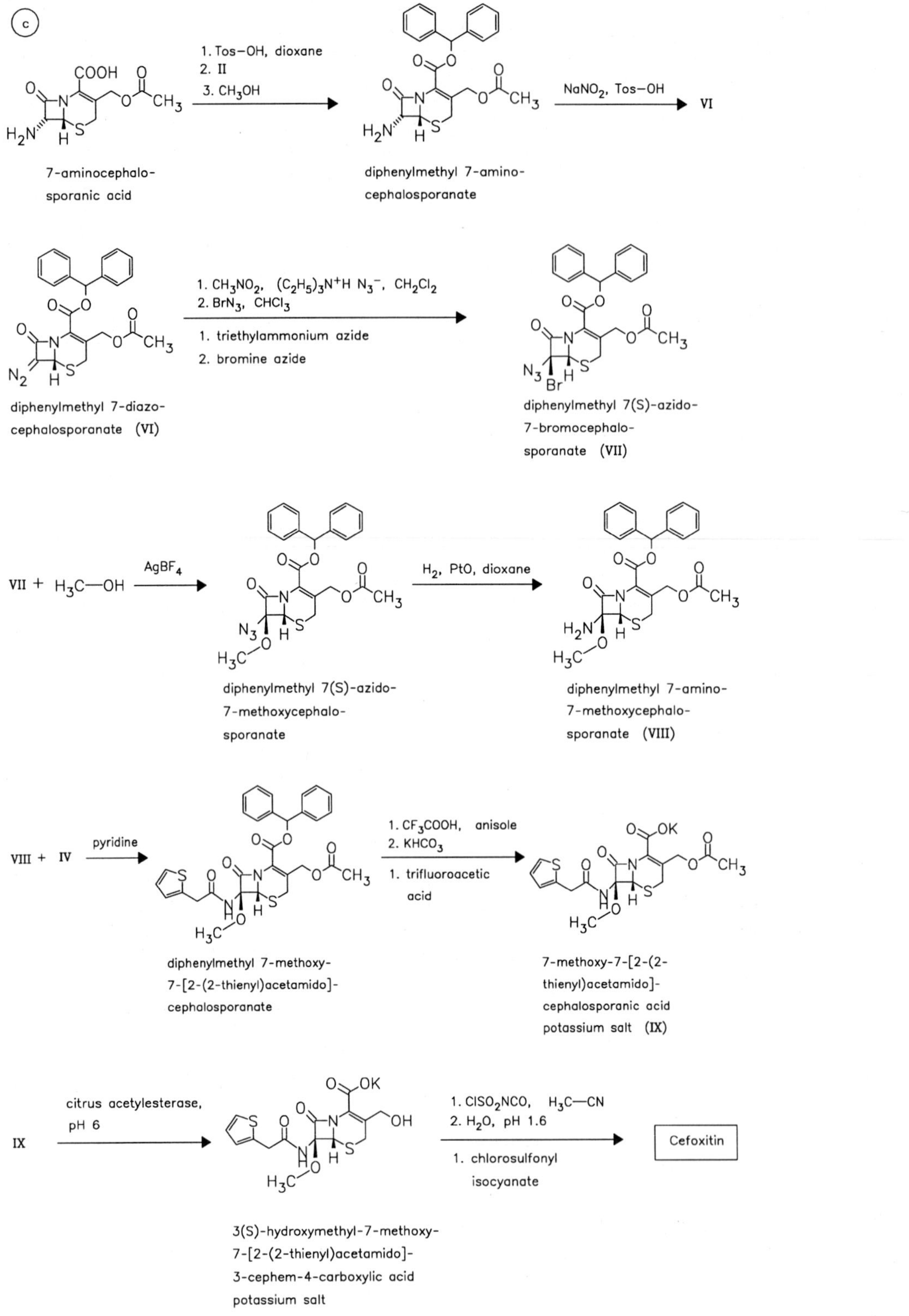

7-aminocephalo-
sporanic acid

1. Tos–OH, dioxane
2. II
3. CH₃OH

diphenylmethyl 7-amino-
cephalosporanate

NaNO₂, Tos–OH   VI

diphenylmethyl 7-diazo-
cephalosporanate (VI)

1. CH₃NO₂, (C₂H₅)₃N⁺H N₃⁻, CH₂Cl₂
2. BrN₃, CHCl₃

1. triethylammonium azide
2. bromine azide

diphenylmethyl 7(S)-azido-
7-bromocephalo-
sporanate (VII)

VII + H₃C—OH $\xrightarrow{AgBF_4}$ diphenylmethyl 7(S)-azido-
7-methoxycephalo-
sporanate

$\xrightarrow{H_2, PtO, dioxane}$ diphenylmethyl 7-amino-
7-methoxycephalo-
sporanate (VIII)

VIII + IV $\xrightarrow{pyridine}$ diphenylmethyl 7-methoxy-
7-[2-(2-thienyl)acetamido]-
cephalosporanate

1. CF₃COOH, anisole
2. KHCO₃

1. trifluoroacetic
acid

7-methoxy-7-[2-(2-
thienyl)acetamido]-
cephalosporanic acid
potassium salt (IX)

IX $\xrightarrow{citrus\ acetylesterase,\ pH\ 6}$ 3(S)-hydroxymethyl-7-methoxy-
7-[2-(2-thienyl)acetamido]-
3-cephem-4-carboxylic acid
potassium salt

1. ClSO₂NCO, H₃C—CN
2. H₂O, pH 1.6

1. chlorosulfonyl
isocyanate

Cefoxitin

Reference(s):
a,c US 4 297 488 (Merck & Co. 27.10.1981; appl. 2.6.1971; GB-prior. 16.6.1970).
 DOS 2 129 675 (Merck & Co.; appl. 15.6.1971; GB-prior. 16.6.1970).
 DOS 2 143 331 (Merck & Co.; appl. 15.6.1971; GB-prior. 16.6.1970).
 DOS 2 203 653 (Merck & Co.; appl. 26.1.1972; GB-prior. 27.1.1971).
 US 3 775 410 (Merck & Co.; 27.11.1973; appl. 29.11.1971).
 US 3 780 033 (Merck & Co.; 18.12.1973; appl. 29.11.1971).
 DOS 2 258 278 (Merck & Co.; appl. 28.11.1972; USA-prior. 29.11.1971).
 US 3 843 641 (Merck & Co.; 22.10.1974; prior. 29.11.1971).
b DOS 2 456 528 (Merck & Co.; appl. 29.11.1974; USA-prior. 30.11.1973).
c DE 2 318 829 (Merck & Co.; appl. 13.4.1973; USA-prior. 14.4.1972).
 DOS 2 365 582 (Merck & Co.; appl. 13.4.1973; USA-prior. 14.4.1972).

azido-intermediates:
DOS 2 365 406 (Merck & Co.; appl. 13.4.1973; USA-prior. 14.4.1972).

7-amino-7-methoxycephalosporanic acid esters:
DOS 2 365 456 (Merck & Co.; appl. 13.4.1973; USA-prior. 14.4.1972).

from 7-amino-7-methoxypenicillanic acid derivatives:
DAS 2 229 246 (Merck & Co.; appl. 15.6.1972; USA-prior. 18.6.1971, 13.12.1971).

fermentative preparation of cephamycin C:
US 3 914 157 (Merck & Co.; 21.10.1975; prior. 13.3.1970, 30.6.1970, 1.12.1971, 12.2.1973).
US 3 962 224 (Merck & Co.; 8.6.1976; prior. 14.4.1972, 30.6.1972, 10.10.1972, 5.3.1973).
GB 1 515 809 (Merck & Co.; appl. 7.9.1976; USA-prior. 21.11.1975).
US 4 137 405 (Merck & Co.; 30.1.1979; appl. 28.7.1977).

7-(5-amino-5-carboxypentanoylamino)-7-methoxy-3-hydroxymethyl-3-cephem-4-carboxylic acid

from cephamycin A *or* B:
DAS 2 509 337 (Meiji Seika Kaisha; appl. 4.3.1975; J-prior. 11.3.1974).

common synthetic methods for 7-methoxycephalosporine:
Hiraoka, T. et al.: Heterocycles (HTCYAM) **8**, 719 (1977).

Formulation(s): vial 1 g, 2 g (as sodium salt)

Trade Name(s):

D:	Mefoxitin (MSD; 1978)	I:	Betacef (Firma)		Tifox (Select Pharma)
F:	Mefoxin (Merck Sharp & Dohme-Chibret)		Cefociclin (Ist. Italiano Ferm.)	J:	Cenomycin (Daiichi) Merxin (Merck-Banyu)
GB:	Mefoxin (Merck Sharp & Dohme; 1978)		Mefoxin (Merck Sharp & Dohme)	USA:	Mefoxin (Merck; 1978)

Cefpiramide

ATC: J01DA27
Use: β-lactam antibiotic

RN: 70797-11-4 MF: $C_{25}H_{24}N_8O_7S_2$ MW: 612.65
CN: [6R-[6α,7β(R*)]]-7-[[[[(4-hydroxy-6-methyl-3-pyridinyl)carbonyl]amino](4-hydroxyphenyl)acetyl]amino]-3-[[(1-methyl-1H-tetrazol-5-yl)thio]methyl]-8-oxo-5-thia-1-azabicyclo[4.2.0]oct-2-ene-2-carboxylic acid

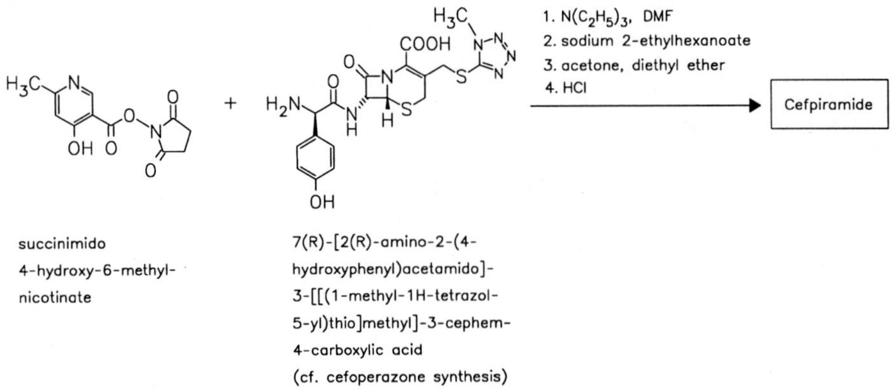

H3C
COOH
1. N(C2H5)3, DMF
2. sodium 2-ethylhexanoate
3. acetone, diethyl ether
4. HCl
Cefpiramide
OH
succinimido
4-hydroxy-6-methyl-
nicotinate
7(R)-[2(R)-amino-2-(4-
hydroxyphenyl)acetamido]-
3-[[(1-methyl-1H-tetrazol-
5-yl)thio]methyl]-3-cephem-
4-carboxylic acid
(cf. cefoperazone synthesis)

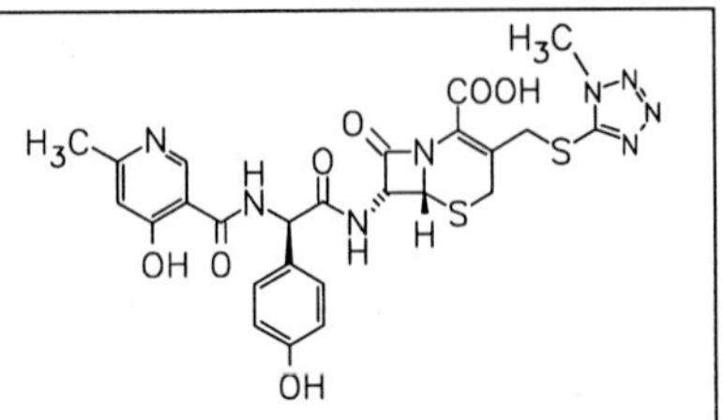
Cefpiramide

Reference(s):

DOS 2 539 664 (Sumitomo; appl. 5.9.1975; J-prior. 6.9.1974, 19.9.1974, 20.3.1975).
US 4 156 724 (Sumitomo; 29.5.1979; appl. 8.9.1975; J-prior. 6.9.1974, 19.9.1974, 20.3.1975).
US 4 160 087 (Sumitomo; 3.7.1979; appl. 10.5.1977; J-prior. 6.9.1974, 19.9.1974, 20.3.1975).
GB 1 510 730 (Sumitomo; appl. 5.9.1975; J-prior. 6.9.1974, 19.9.1974, 20.3.1975).
FR 2 283 688 (Sumitomo; appl. 5.9.1975; J-prior. 6.9.1974, 19.9.1974, 20.3.1975).

Formulation(s): vial 1 g (as sodium salt)

Trade Name(s):

J: Sepatren (Sumitomo) Suncefal (Yamanouchi)

cis-Cefprozil

ATC: J01DA41
Use: broad-spectrum cephalosporin (orally active)

RN: 92665-29-7 MF: $C_{18}H_{19}N_3O_5S$ MW: 389.43
CN: [6*R*-[3(*Z*),6α,7β(*R**)]]-7-[[Amino(4-hydroxyphenyl)acetyl]amino]-8-oxo-3-(1-propenyl)-5-thia-1-azabicyclo[4.2.0]oct-2-ene-2-carboxylic acid

monohydrate

RN: 121123-17-9 MF: $C_{18}H_{19}N_3O_5S \cdot H_2O$ MW: 407.45

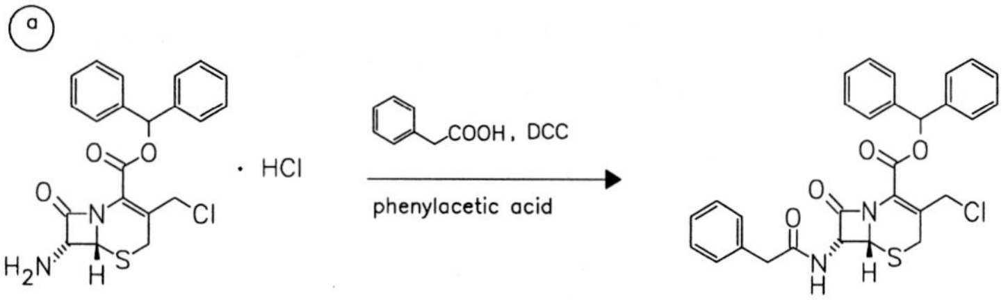

diphenylmethyl 7-amino-
3-chloromethyl-3-cephem-
4-carboxylate hydrochloride (I)
(cf. cefixime synthesis)

phenylacetic acid

(II)

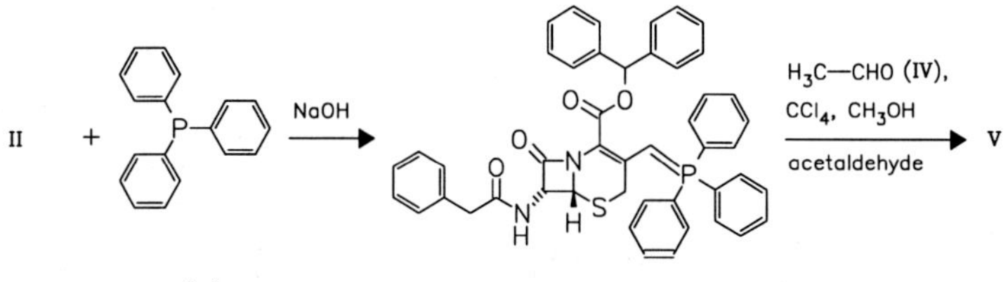

(III)

diphenylmethyl 7-phenyl-
acetamido-3-[(triphenylphos-
phoranylidene)methyl]-3-
cephem-4-carboxylate

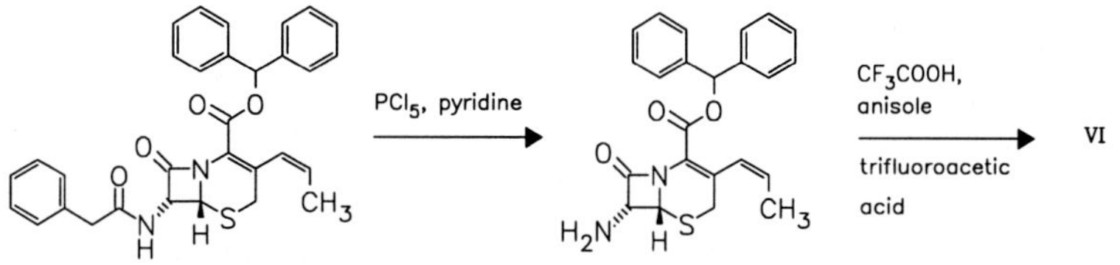

diphenylmethyl 7-phenyl-
acetamido-3-[(Z)-1-propenyl]-
3-cephem-4-carboxylate (V)

diphenylmethyl
7-amino-3-[(Z)-1-
propenyl]-3-cephem-
4-carboxylate

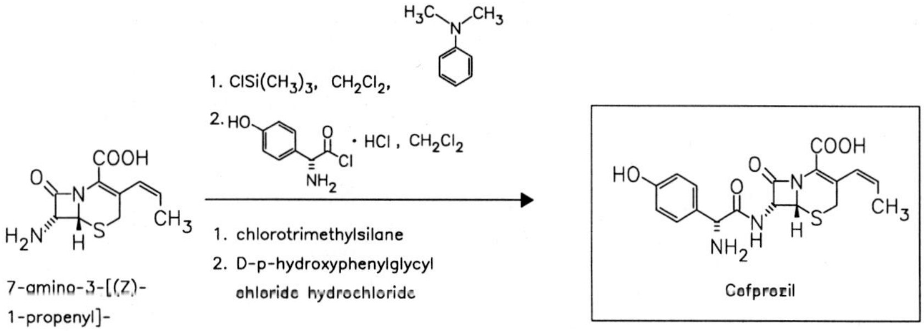

trifluoroacetic
acid

7-amino-3-[(Z)-
1-propenyl]-
3-cephem-4-
carboxylic acid (VI)

1. chlorotrimethylsilane
2. D-p-hydroxyphenylglycyl
   chloride hydrochloride

Cefprozil

b

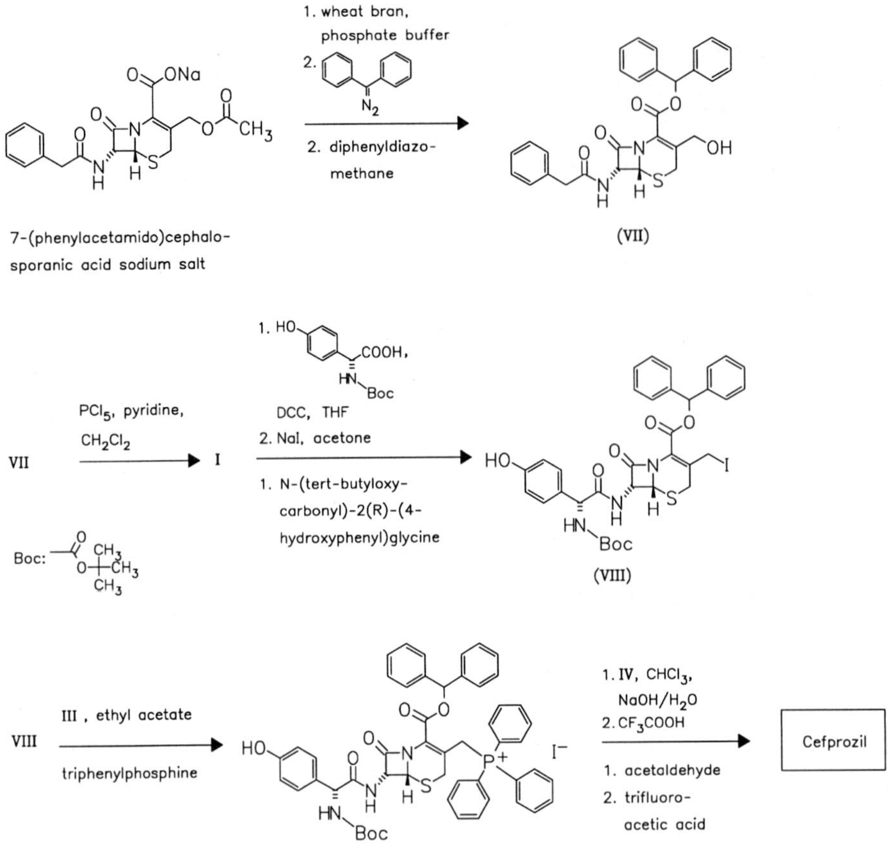

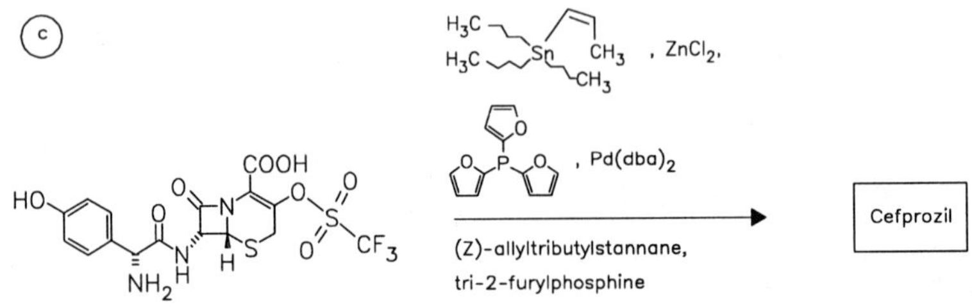

7-(phenylacetamido)cephalo-
sporanic acid sodium salt

(VII)

(VIII)

Cefprozil

c

Cefprozil

Reference(s):
a   US 4 694 079 (Bristol-Myers Squibb. Co.; 15.9.1987; USA-prior. 29.7.1985).
b   DE 3 402 642 (Bristol-Myers Squibb. Co.; appl. 26.1.1984; USA-prior. 28.1.1983).
c   US 4 870 168 (Bristol-Myers Squibb & Co.; 29.11.1989; USA-prior. 26.2.1987).

Formulation(s):   oral susp. 125 mg/5 ml, 250 mg/5ml; tabl. 250 mg, 500 mg

Trade Name(s):
USA: Cefzil (Bristol-Myers
        Squibb; 1992)

Cefradine
(Cephradine)

ATC: J01DA31
Use: antibiotic

RN: 38821-53-3   MF: $C_{16}H_{19}N_3O_4S$ MW: 349.41   EINECS: 254-137-8
LD$_{50}$: 3539 mg/kg (M, i.v.); 3549 mg/kg (M, p.o.);
      >2500 mg/kg (R, i.v.); >12 g/kg (R, p.o.)
CN: [6R-[6α,7β(R*)]]-7-[(amino-1,4-cyclohexadien-1-ylacetyl)amino]-3-methyl-8-oxo-5-thia-1-azabicyclo[4.2.0]oct-2-ene-2-carboxylic acid

dihydrate
RN: 31828-50-9   MF: $C_{16}H_{19}N_3O_4S \cdot 2H_2O$ MW: 385.44
LD$_{50}$: 3 g/kg (M, i.v.); 5 g/kg (M, p.o.);
      >8.5 g/kg (R, p.o.)

D(−)−α-phenyl-glycine

D-2-(1,4-cyclo-hexadienyl)glycine

methyl acetoacetate

(I)

ethyl chloroformate

7-amino-3-deacetoxy-cephalosporanic acid

triethylamine

Cefradine

Reference(s):
US 3 485 819 (Squibb; 23.12.1969; appl. 2.7.1968).
DAS 1 931 722 (Squibb; appl. 23.6.1969; USA-prior. 2.7.1968).

acylation via 1,3,2-dioxaboranyl-derivatives:
DOS 2 755 902 (Dobfar; appl. 15.12.1977; I-prior. 16.12.1976).

microbiological acylation with Aphanocladium aranearum (ATCC 20453):
US 4 073 687 (Shionogi; 14.2.1978; J-prior. 14.5.1975).

Formulation(s):   cps. 250 mg, 500 mg; tabl. 1 g; vial 250 mg, 500 mg, 1 g

Trade Name(s):

D:	Eskacef (SK Dauelsberg; 1977); wfm	GB:	Velosef (Bristol-Myers Squibb)
	Sefril (Heyden; 1973); wfm	I:	Cefrabiotic (Leben's)
F:	Cefirex (Irex)		Citicef (CT)
	Doncef (Pharma 2000)		Ecosporina (Ecobi)
	Kelsef (Jumer)		Lisacef (Lisapharma)
	Zadyl (Thera France)	J:	Cefro (Sankyo)
	Zeefra (Doms-Adrian)		Dicefalin (Nikon Squibb)

USA: Anspor (Smith Kline & French; 1974); wfm
Cephradine (Lederle Standard)
Velosef (Squibb; 1974); wfm

Cefsulodin

ATC: J01DA12
Use: β-lactam antibiotic

RN: 62587-73-9   MF: $C_{22}H_{20}N_4O_8S_2$ MW: 532.55
CN: [6R-[6α,7β(R*)]]-4-(aminocarbonyl)-1-[[2-carboxy-8-oxo-7-[(phenylsulfoacetyl)amino]-5-thia-1-azabicyclo[4.2.0]oct-2-en-3-yl]methyl]pyridinium hydroxide inner salt

monosodium salt
RN: 52152-93-9   MF: $C_{22}H_{19}N_4NaO_8S_2$ MW: 554.54   EINECS: 257-692-4
LD$_{50}$:   3780 mg/kg (M, i.v.); >15 g/kg (M, p.o.);
3030 µg/kg (R, i.v.); >15 g/kg (R, p.o.);
>15 g/kg (dog, p.o.)

(R)-α-sulfophenyl-acetyl chloride   7-aminocephalo-sporanic acid

(I)   isonicotin-amide   Cefsulodin

Reference(s):
Nomura, H. et al.: J. Med. Chem. (JMCMAR) **17**, 1312 (1974).
US 4 065 619 (Takeda; 27.12.1977; J-prior. 17.7.1971, 22.10.1971).
DE 2 234 280 (Takeda; appl. 12.7.1972; J-prior. 17.7.1971, 22.10.1971).
FR 2 146 313 (Takeda; appl. 17.7.1972; J-prior. 17.7.1971, 22.10.1971).
GB 1 387 656 (Takeda; appl. 17.7.1972; J-prior. 17.7.1971, 22.10.1971).

Formulation(s):   vial 500 mg, 1 g, 2 g (as sodium salt)

Trade Name(s):
D: Pseudocef (Takeda; 1981)   GB: Monaspor (Ciba; 1982);   J: Takesulin (Takeda; 1981)
F: Pyocefal (Takeda)   wfm

Ceftazidime

ATC: J01DA11
Use: β-lactam antibiotic (cefalosporin derivative)

RN: 72558-82-8   MF: $C_{22}H_{22}N_6O_7S_2$ MW: 546.59   EINECS: 276-715-9
LD$_{50}$:   6300 mg/kg (M, i.v.); >20 g/kg (M, p.o.);
5800 mg/kg (R, i.v.); >20 g/kg (R, p.o.)
CN: [6R-[6α,7β(Z)]]-1-[[7-[[(2-amino-4-thiazolyl)[(1-carboxy-1-methylethoxy)imino]acetyl]amino]-2-carboxy-8-oxo-5-thia-1-azabicyclo[4.2.0]oct-2-en-3-yl]methyl]pyridinium hydroxide inner salt

dihydrochloride
RN: 73547-70-3   MF: $C_{22}H_{22}N_6O_7S_2 \cdot 2HCl$ MW: 619.51

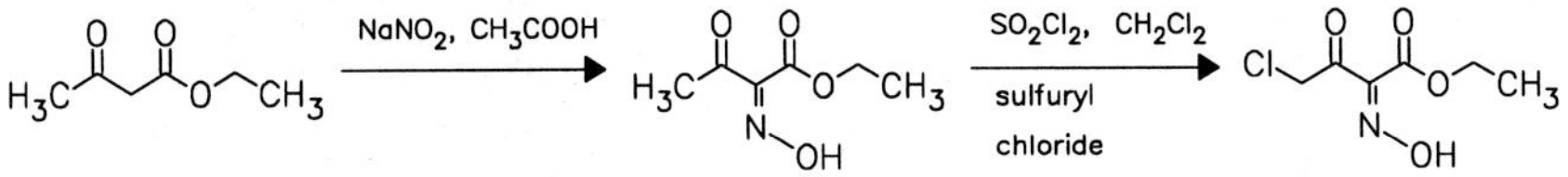

ethyl acetoacetate

ethyl 2-hydroxy-
iminoacetoacetate

ethyl 4-chloro-2-
hydroxyimino-
acetoacetate (I)

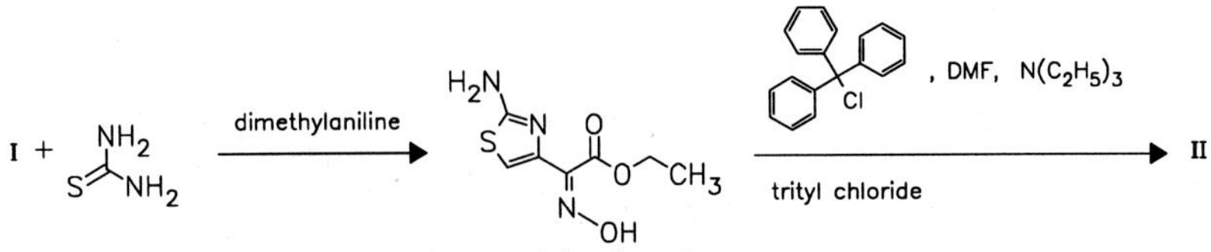

I + thiourea

ethyl (Z)-2-(2-amino-
thiazol-4-yl)-2-hydroxy-
iminoacetate

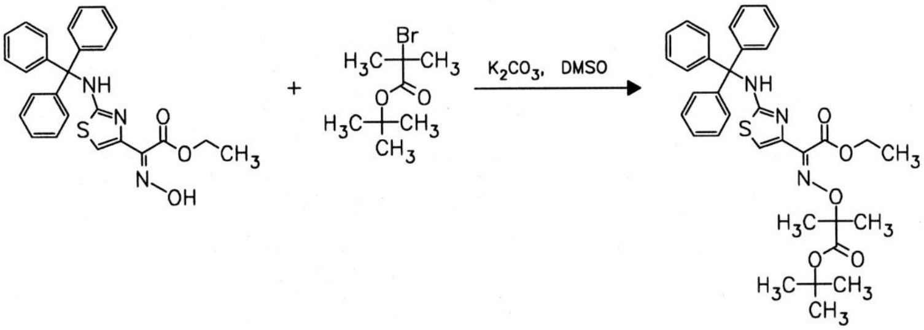

trityl chloride

II

ethyl (Z)-2-hydroxyimino-
2-(2-tritylaminothiazol-4-
yl)acetate (II)

tert-butyl
2-bromo-2-
methyl-
propionate

ethyl (Z)-2-(1-tert-butoxy-
carbonyl-1-methylethoxyimino)-
2-(2-tritylaminothiazol-4-
yl)acetate (III)

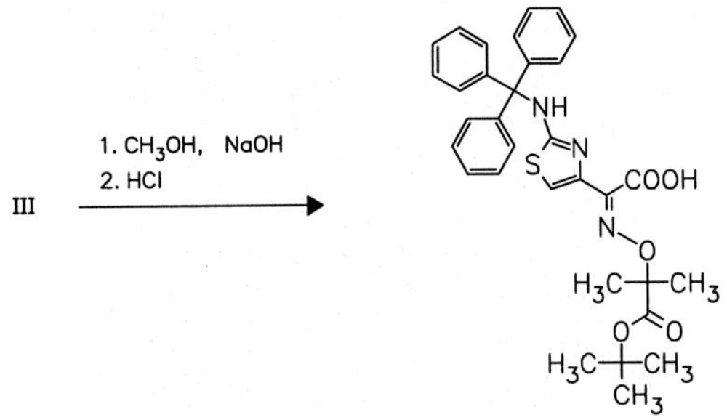

III

1. CH$_3$OH, NaOH
2. HCl

(Z)-2-(1-tert-butoxycarbonyl-
1-methylethoxyimino)-2-
(2-tritylaminothiazol-4-yl)-
acetic acid (IV)

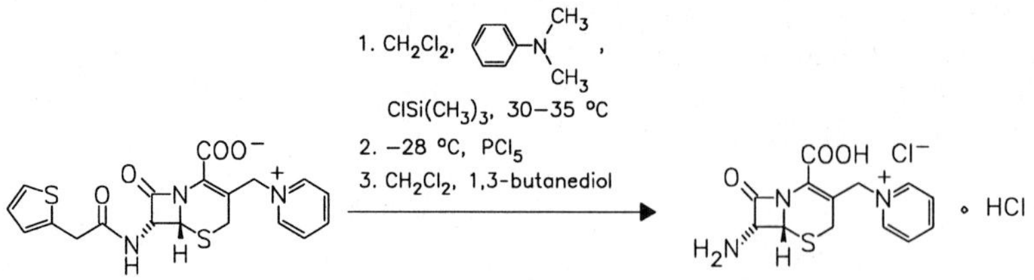

cefaloridine
(q. v.)

1. CH₂Cl₂, [structure with CH₃, N, CH₃],
ClSi(CH₃)₃, 30−35 °C
2. −28 °C, PCl₅
3. CH₂Cl₂, 1,3-butanediol

7(R)−amino−3−(1−
pyridiniomethyl)−3−
cephem−4−carboxylic
acid chloride
monohydrochloride (V)

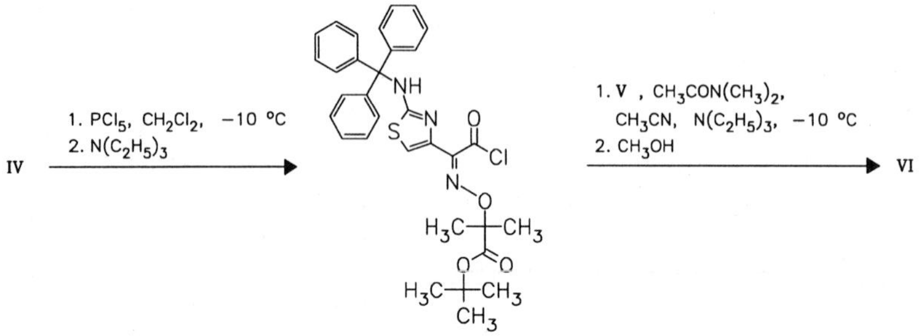

IV

1. PCl₅, CH₂Cl₂, −10 °C
2. N(C₂H₅)₃

1. V , CH₃CON(CH₃)₂,
CH₃CN, N(C₂H₅)₃, −10 °C
2. CH₃OH

VI

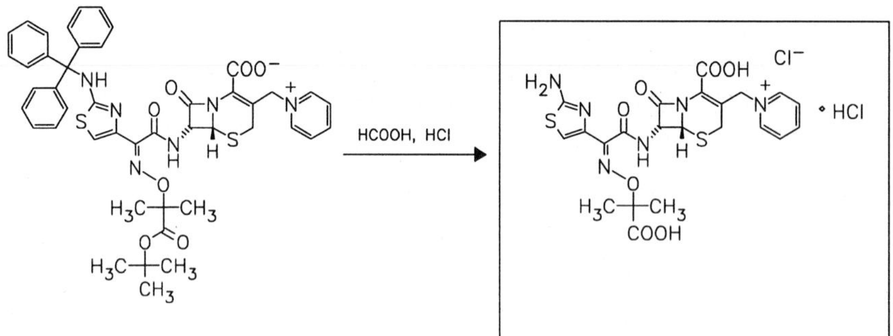

(VI)

HCOOH, HCl

Ceftazidime dihydrochloride

Reference(s):
DOS 2 921 316 (Glaxo; appl. 25.5.1979; GB-prior. 26.5.1978).
US 4 258 041 (Glaxo; 24.3.1981; GB-prior. 26.5.1978).
GB 2 025 398 (Glaxo; appl. 25.5.1979; GB-prior. 26.5.1978).
US 4 525 587 (Eli Lilly; 25.6.1985; prior. 27.12.1982, 3.2.1984).

intermediate IV:
US 4 497 956 (Glaxo; 5.2.1985; GB-prior. 13.11.1981).

acid chloride of IV:
EP 101 148 (Glaxo; appl. 28.4.1983; GB-prior. 29.4.1982).

intermediate V:
EP 135 258 (Eli Lilly; appl. 18.6.1984; USA-prior. 20.6.1983).
EP 70 706 (Glaxo; appl. 16.7.1982; GB-prior. 17.7.1981).

salts and crystal modifications:
crystalline dihydrochloride:
US 4 467 086 (Glaxo; 21.8.1984; GB-prior. 2.10.1979).

pentahydrate:
DOS 3 037 102 (Glaxo; appl. 1.10.1980; GB-prior. 2.10.1979).
GB 2 063 871 (Glaxo; appl. 1.10.1980; GB-prior. 2.10.1979).
US 4 329 453 (Glaxo; 11.5.1982; appl. 9.9.1980; GB-prior. 2.10.1979).

sesquihydrate:
DOS 3 313 816 (Hoechst; appl. 16.4.1983).
EP 122 584 (Hoechst; appl. 10.4.1984; D-prior. 16.4.1983).

anhydrous crystal modification:
DOS 3 313 818 (Hoechst; appl. 16.4.1983).
EP 122 585 (Hoechst; appl. 10.4.1984; D-prior. 16.4.1983).

pharmaceutical formulations:
DOS 3 332 616 (Glaxo; appl. 9.9.1983; GB-prior. 10.9.1982).

Formulation(s):   vial 250 mg, 500 mg, 1 g, 2 g, 3 g

Trade Name(s):

D:	Fortum (Cascan-Glaxo; 1984)		Kefadim (Lilly)		Starcef (Firma)
		I:	Ceftim (Glaxo Allen)	J:	Modacin (Shin Nihon)
F:	Fortum (Glaxo Wellcome)		Glazidim (Glaxo)	USA:	Ceptaz (Glaxo)
GB:	Fortum (Glaxo Wellcome; 1983)		Panzid (Duncan)		Fortaz (Glaxo; 1985)
			Spectrum (Sigma-Tau)		Tazdime (Lilly; 1985)

Ceftezole

ATC: J01DA36
Use: antibiotic

RN: 26973-24-0   MF: $C_{13}H_{12}N_8O_4S_3$ MW: 440.49
CN: (6R-trans)-8-oxo-7-[(1H-tetrazol-1-ylacetyl)amino]-3-[(1,3,4-thiadiazol-2-ylthio)methyl]-5-thia-1-azabicyclo[4.2.0]oct-2-ene-2-carboxylic acid

Reference(s):
DE 1 770 168 (Fujisawa; appl. 10.4.1968; J-prior. 15.4.1967, 23.10.1967, 28.10.1967).
US 3 516 997 (Fujisawa; 23.6.1970; J-prior. 15.4.1967, 24.10.1967, 28.10.1967).
GB 1 206 305 (Fujisawa; appl. 11.4.1968; J-prior. 15.4.1967, 24.10.1967, 28.10.1967).

combination with penicillins:
DOS 2 508 443 (Fujisawa; appl. 27.2.1975; J-prior. 28.2.1974, 27.3.1974).

Formulation(s):   vial 250 mg, 500 mg, 1 g, 2 g (as sodium salt)

Trade Name(s):
I: Alomen (Benedetti)   J: Celoslin (Fujisawa)   Falomesin (Chugai)

Ceftizoxime

ATC: J01DA22
Use: β-lactam antibiotic (cefalosporin derivative)

RN: 68401-81-0   MF: $C_{13}H_{13}N_5O_5S_2$ MW: 383.41
CN: [6R-[6α,7β(Z)]]-7-[[(2-amino-4-thiazolyl)(methoxyimino)acetyl]amino]-8-oxo-5-thia-1-azabicyclo[4.2.0]oct-2-ene-2-carboxylic acid

monohydrochloride
RN: 68401-80-9   MF: $C_{13}H_{13}N_5O_5S_2 \cdot HCl$ MW: 419.87
monosodium salt
RN: 68401-82-1   MF: $C_{13}H_{12}N_5NaO_5S_2$ MW: 405.39
LD_{50}:   5150 mg/kg (M, i.v.); >10 g/kg (M, p.o.);
    5570 mg/kg (R, i.v.); >10 g/kg (R, p.o.)

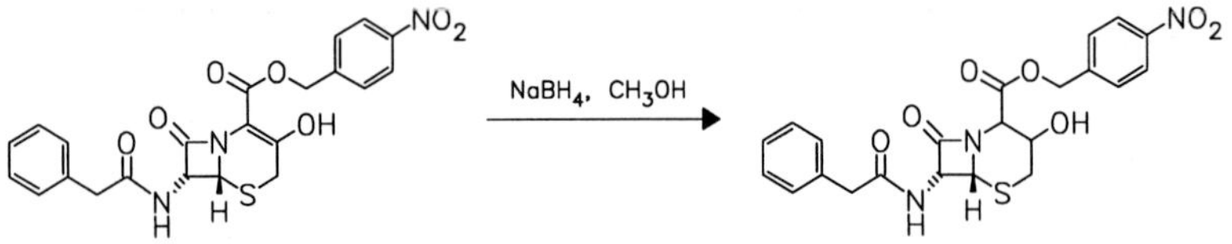

4-nitrobenzyl 3-hydroxy-7(R)-
phenylacetamido-3-cephem-4-
carboxylate
(cf. cefaclor synthesis)

4-nitrobenzyl 3-hydroxy-7(R)-
phenylacetamidocepham-4-
carboxylate (I)

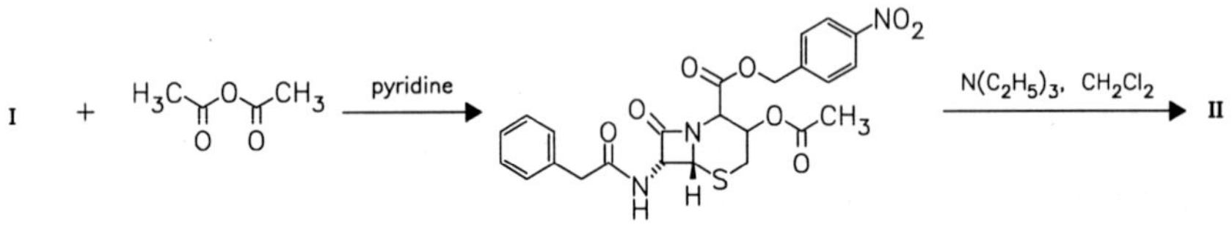

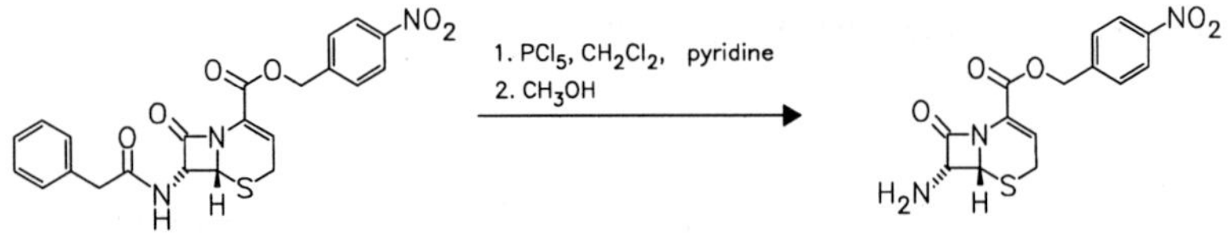

4-nitrobenzyl 7(R)-phenyl-
acetamido-3-cephem-4-
carboxylate (II)

4-nitrobenzyl 7(R)-amino-
3-cephem-4-carboxylate (III)

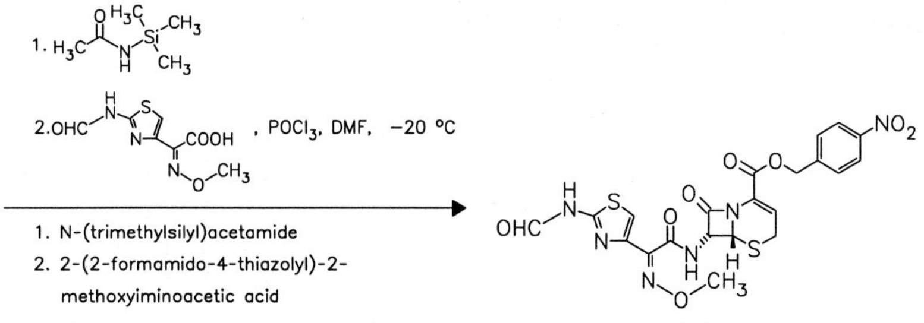

1. N-(trimethylsilyl)acetamide
2. 2-(2-formamido-4-thiazolyl)-2-methoxyiminoacetic acid

(IV)

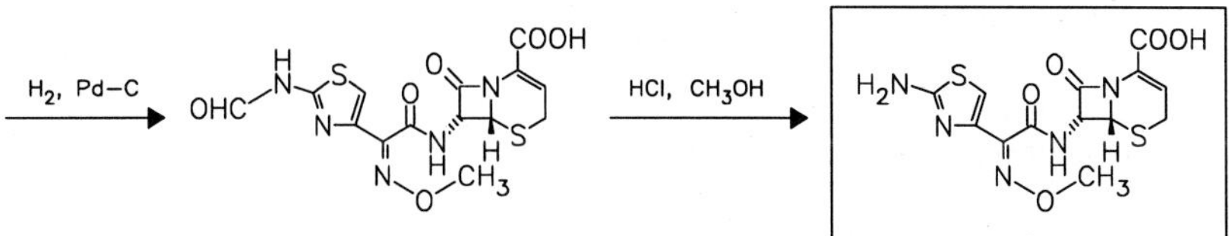

Ceftizoxime

Reference(s):
US 4 427 674 (Fujisawa; 24.1.1984; GB-prior. 14.3.1977, 12.7.1977, 11.10.1977, 3.1.1978).
US 4 463 002 (Fujisawa; 31.7.1984; J-prior. 21.5.1981).

Formulation(s):   vial 0.25 g, 0.5 g, 1 g, 2 g (as sodium salt)

Trade Name(s):
D: Ceftix (Roche; 1983)   GB: Cefizox (Wellcome); wfm   J: Epocelin (Fujisawa)
F: Cefizox (Bellon)   I: Eposerin (Farmitalia)   USA: Cefizox (Fujisawa; 1983)

Ceftriaxone

ATC: J01DA13
Use: β-lactam antibiotic (cefalosporin derivative)

RN: 73384-59-5   MF: $C_{18}H_{18}N_8O_7S_3$ MW: 554.59   EINECS: 277-405-6
CN: [6R-[6α,7β(Z)]]-7-[[(2-amino-4-thiazolyl)(methoxyimino)acetyl]amino]-8-oxo-3-[[(1,2,5,6-tetrahydro-2-methyl-5,6-dioxo-1,2,4-triazin-3-yl)thio]methyl]-5-thia-1-azabicyclo[4.2.0]oct-2-ene-2-carboxylic acid

disodium salt
RN: 74578-69-1   MF: $C_{18}H_{16}N_8Na_2O_7S_3$ MW: 598.55   EINECS: 277-930-0
LD$_{50}$:   2200 mg/kg (M, i.v.); >10 g/kg (M, p.o.);
              1900 mg/kg (R, i.v.); >10 g/kg (R, p.o.)

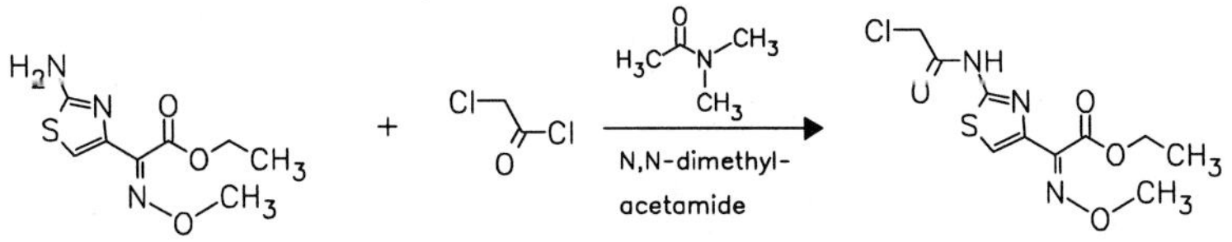

ethyl 2-(2-amino-
4-thiazolyl)-2-(meth-
oxyimino)acetate
(cf. cefotaxime
synthesis)

chloro-
acetyl
chloride

ethyl 2-(2-chloroacet-
amido-4-thiazolyl)-2-
methoxyiminoacetate (I)

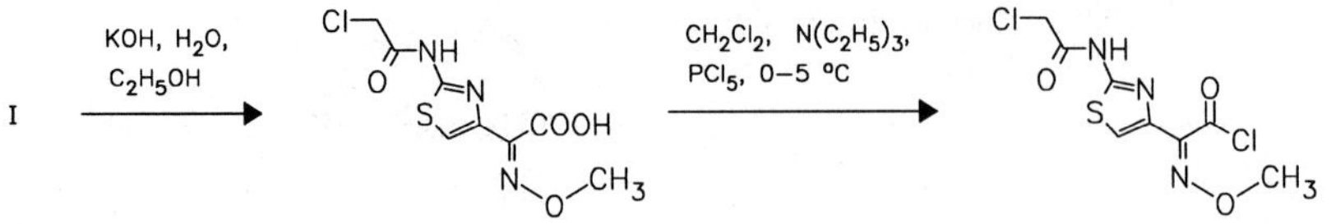

2-(2-chloroacetamido-
4-thiazolyl)-2-methoxy-
iminoacetyl chloride (II)

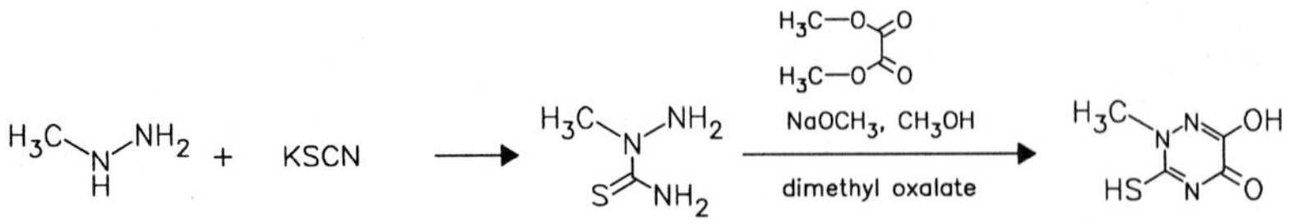

methyl-
hydrazine

potassium
rhodanide

N-amino-
N-methyl-
thiourea

2,5-dihydro-6-
hydroxy-2-methyl-
3-mercapto-5-oxo-
1,2,4-triazine (III)

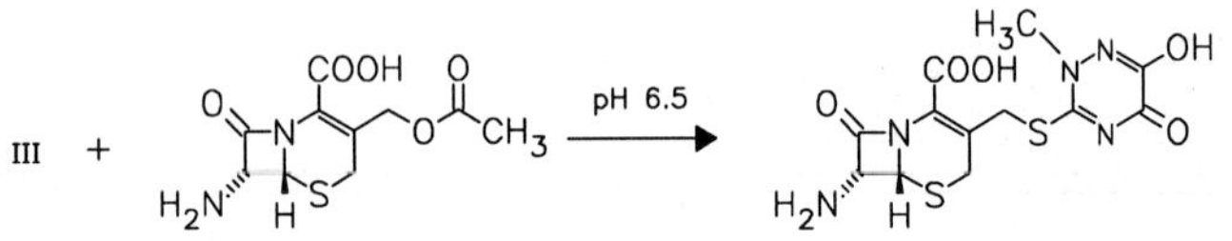

7-aminocephalo-
sporanic acid

7(R)-amino-3-[[(2,5-dihydro-
6-hydroxy-2-methyl-5-oxo-
1,2,4-triazin-3-yl)thio]methyl]-
3-cephem-4-carboxylic acid (IV)

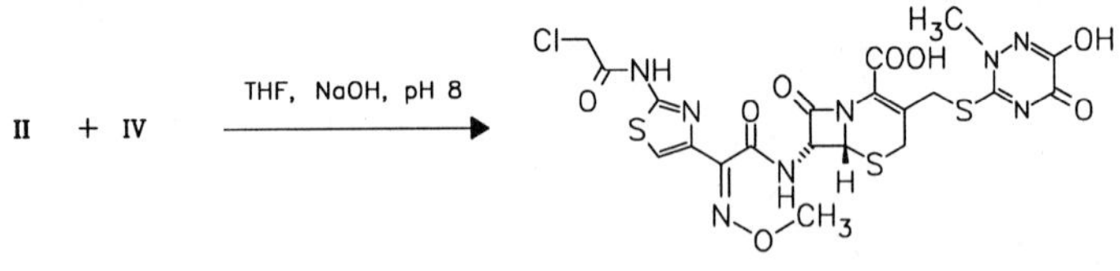

(v)

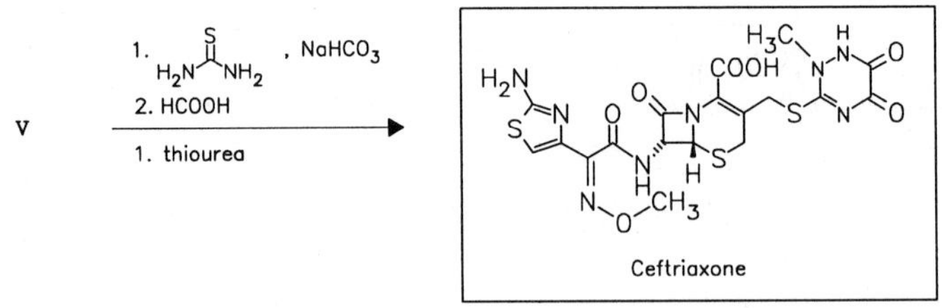

Ceftriaxone

Reference(s):
DOS 2 922 036 (Roche; appl. 30.5.1979; CH-prior. 30.5.1978, 8.3.1979).
US 4 327 210 (Roche; 27.4.1982; appl. 24.11.1978; CH-prior. 30.5.1978).

alternative synthesis from cefotaxime *and* 2,5-dihydro-6-hydroxy-2-methyl-3-mercapto-5-oxo-1,2,4-triazine:
US 4 431 804 (Roche; 14.2.1984; CH-prior. 17.2.1981).

Formulation(s): — inj. powder 250 mg, 1 g, 2 g

Trade Name(s):
D: Rocephin (Roche; 1983)   F: Rocéphine (Roche)   GB: Rocephin (Roche)

I: Rocefin (Roche)   USA: Rocefin (Roche Labs.;
J: Rocephin (Roche)   1985)

Cefuroxime

ATC: J01DA06
Use: antibiotic

RN: 55268-75-2   MF: $C_{16}H_{16}N_4O_8S$ MW: 424.39   EINECS: 259-560-1
LD$_{50}$: 10.4 g/kg (M, i.v.); >10 g/kg (M, p.o.);
        >8 g/kg (R, i.v.); 10 g/kg (R, p.o.)
CN: [6R-[6α,7β(Z)]]-3-[[(aminocarbonyl)oxy]methyl]-7-[[2-furanyl(methoxyimino)acetyl]amino]-8-oxo-5-thia-1-azabicyclo[4.2.0]oct-2-ene-2-carboxylic acid

sodium salt
RN: 56238-63-2   MF: $C_{16}H_{15}N_4NaO_8S$ MW: 446.37

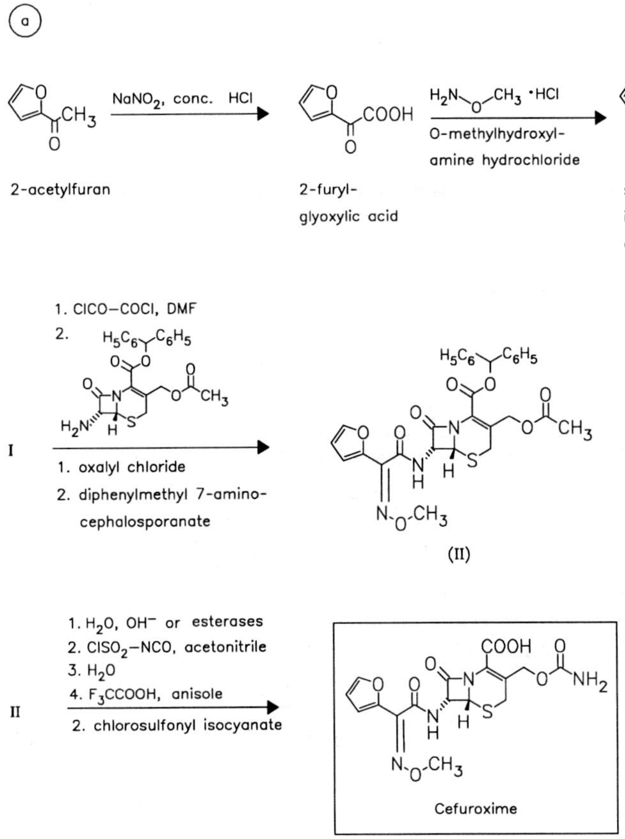

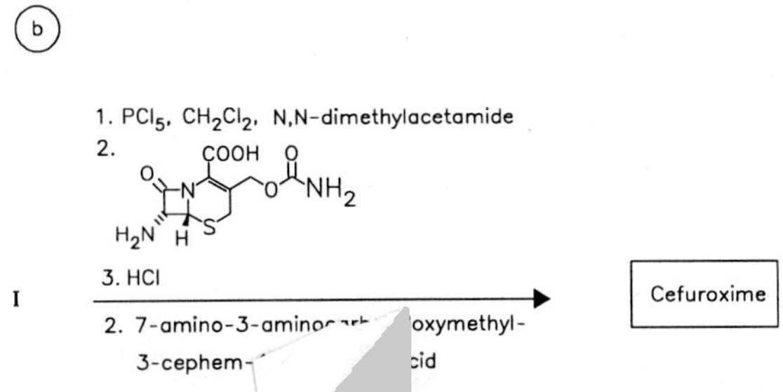

2-acetylfuran

2-furyl-
glyoxylic acid

O-methylhydroxyl-
amine hydrochloride

syn-2-methoxy-
imino-2-(2-furyl)-
acetic acid (I)

(II)

Cefuroxime

Cefuroxime

Reference(s):
GB 1 453 049 (Glaxo; appl. 21.8.1973; valid from 13.8.1974).
DAS 2 439 880 (Glaxo; appl. 20.8.1974; GB-prior. 21.8.1973).
DOS 2 462 376 (Glaxo; appl. 20.8.1974; GB-prior. 21.8.1973).
DOS 2 204 060 (Glaxo; appl. 28.1.1972; GB-prior. 29.1.1971, 1.10.1971 and 14.1.1972).
DOS 2 223 375 (Glaxo; appl. 12.5.1972; GB-prior. 14.5.1971 and 1.10.1971).
DOS 2 265 234 (Glaxo; appl. 12.5.1972; GB-prior. 14.5.1971 and 1.10.1971).
US 3 966 717 (Glaxo; 29.6.1976; GB-prior. 14.5.1971, 1.10.1971, 21.8.1973).
DOS 2 439 880 (Glaxo; appl. 20.8.1974; GB-prior. 21.8.1973).
US 3 971 778 (Glaxo; 27.7.1976; GB-prior. 25.10.1972).
US 3 974 153 (Glaxo; 10.8.1976; GB-prior. 14.5.1971, 1.10.1971, 21.8.1973).

crystalline sodium salt:
DOS 2 901 730 (Glaxo; appl. 17.1.1979; GB-prior. 17.1.1978).

syn-2-methoxyimino-2-(2-furyl)acetic acid, *resp.* -acetyl chloride:
US 4 017 515 (Glaxo; 12.4.1977; GB-prior. 14.5.1971, 1.10.1971, 12.5.1972 and 25.10.1972).

2-furylglyoxylic acid:
GB 1 503 649 (Glaxo; appl. 28.6.1974; valid from 27.6.1975).
US 4 013 680 (Glaxo; 22.3.1977; prior. 18.6.1975).

L-*lysine salt of* cefuroxime:
US 4 128 715 (Glaxo; 5.12.1978; GB-prior. 28.4.1976).
DOS 2 718 730 (Glaxo; appl. 27.4.1977; GB-prior. 28.4.1976).

alternative methods for 3-hydroxymethylcephalosporin derivatives:
DE 1 545 915 (Glaxo; appl. 29.10.1965; GB-prior. 30.10.1964, 27.1.1965, 19.10.1965).
DOS 2 745 219 (Glaxo; appl. 7.10.1977; GB-prior. 8.10.1976).
DAS 1 795 777 (Glaxo; appl. 29.10.1965; GB-prior. 30.10.1964, 27.1.1965, 19.10.1965).
GB 1 474 519 (Glaxo; appl. 14.5.1973; valid from 6.5.1974).

alternative methods for 3-carbamoyloxymethyl-cephalosporin-derivatives from the corresponding 3-hydroxymethyl-derivatives (enzymatic with O-transcarbamoylase):
US 4 075 061 (Glaxo; 21.2.1978; GB-prior. 19.2.1976).
US 4 164 447 (Glaxo; 14.8.1979; GB-prior. 19.2.1976).
US 4 164 447 (Glaxo; 14.8.1979; GB-prior. 19.2.1976).
(also reaction of the hydroxymethyl-compd. with chlorosulfonyl isocyanate corresponding at the cefoxitin-synthesis, q. v.).

Formulation(s):   amp. 250 mg, 500 mg, 750 mg, 1 g, 1.5 g/20 ml, 1-5 g/40 ml; gran. 125 mg (as sodium salt); tabl. 125 mg, 250 mg, 500 mg

Trade Name(s):

D:	Elobact (Glaxo Wellcome/ Cascan)	Cefamar (Firma)	Kefox (CT)
	Zinacef (Glaxo Wellcome; Hoechst; 1977)	Cefoprim (Esseti)	Kesint (Mendelejeff)
		Cefumax (Locatelli)	Lafurex (Lafare)
		Cefur (Eurofarmaco)	Lamposporin (Leben's)
	Zinnat (Glaxo Wellcome)	Cefurex (Salus Research)	Medoxim (Medici)
F:	Cepazine (Sanofi Winthrop)	Cefurin (Magis)	Polixima (Sifarma)
		Coliofossim (Coli)	Supero (Francia Farm.)
	Zinnat (Glaxo Wellcome)	Curoxim (Glaxo)	J: Oracef (Shin Nihon-Glaxo)
GB:	Zinacef (Glaxo Wellcome)	Deltacef (Pulitzer)	USA: Kefurox (Glaxo Wellcome;
I:	Biociclin (Delsaz & Filippini)	Duxima (Ecobi)	1986)
		Gibicef (Metapharma)	Zinacef (Glaxo Wellcome;
	Biofurex (KBR)	Ipacef (IPA)	1983)
	Bioxima (Kemyos)	Itorex (Biotekfarma)	

Celecoxib

(SC-58635; YM-177)

ATC: M01AH01
Use: anti-inflammatory, cyclooxygenase-2 inhibitor

RN: 169590-42-5   MF: $C_{17}H_{14}F_3N_3O_2S$ MW: 381.38
CN: 4-[5-(4-Methylphenyl)-3-(trifluoromethyl)-1H-pyrazol-1-yl]benzenesulfonamide

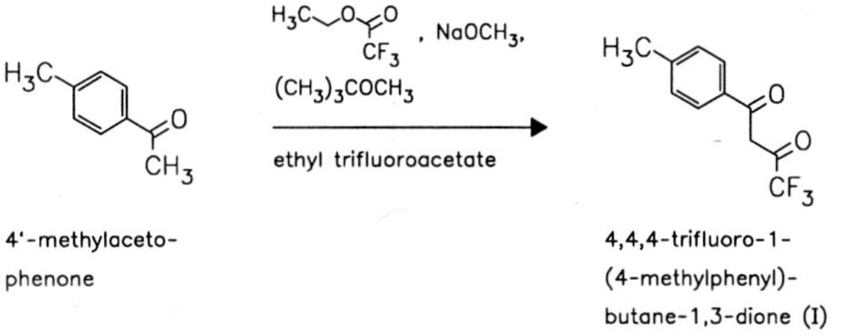

4'-methylaceto-phenone

ethyl trifluoroacetate

4,4,4-trifluoro-1-(4-methylphenyl)-butane-1,3-dione (I)

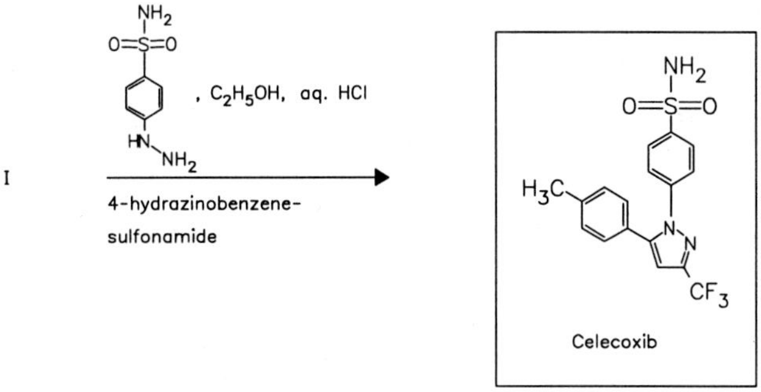

4-hydrazinobenzene-sulfonamide

Celecoxib

Reference(s):
WO 9 515 316 (Searle & Co.; appl. 14.11.1994; USA-prior. 30.11.1993, 6.4.1994).
WO 9 637 476 (Searle & Co.; appl. 23.5.1996; USA-prior. 25.5.1995).
US 5 892 053 (Searle & Co.; 6.4.1999; USA-prior. 25.5.1995).
Penning, T.D. et al.: J. Med. Chem. (JMCMAR) **40** (9), 1347 (1997).
De Vleeschauwer, M.; Gauthier, J.Y.: Synlett (SYNLES) **1997** (4), 375.

Formulation(s):   cps. 100 mg, 200 mg

Trade Name(s):
USA: Celebrex (Pfizer; Searle; 1999)

Celiprolol

(ST-1396)

ATC: C07AB08
Use: cardioselective β-receptor antagonist

RN: 56980-93-9   MF: $C_{20}H_{33}N_3O_4$ MW: 379.50   EINECS: 260-497-7
CN: N'-[3-acetyl-4-[3-[(1,1-dimethylethyl)amino]-2-hydroxypropoxy]phenyl]-N,N-diethylurea

monohydrochloride
RN: 57470-78-7   MF: $C_{20}H_{33}N_3O_4 \cdot HCl$ MW: 415.96   EINECS: 260-752-2
LD_{50}:   42.3 mg/kg (M, i.v.); 1362 mg/kg (M, p.o.);
        68.3 mg/kg (R, i.v.); 2157 mg/kg (R, p.o.)

3-acetyl-4-
hydroxyaniline

diethylcarbamoyl
chloride

N,N-diethyl-N'-(3-acetyl-
4-hydroxyphenyl)urea (I)

I

1. epichlorohydrin
2. tert-butylamine

Celiprolol

Reference(s):
DOS 2 458 624 (Lentia; appl. 11.12.1974; A-prior. 20.12.1973, 19.11.1974, 20.11.1974, 25.11.1974).
US 4 034 009 (Chemie Linz 5.7.1977; appl. 17.12.1974; A-prior. 20.12.1973).

purification:
EP 229 947 (Lentia; appl. 2.12.1986; D-prior. 13.12.1985).

synthesis of enantiomers:
EP 135 162 (Chemie Linz; appl. 17.8.1984; D-prior. 19.8.1983).
EP 155 518 (Chemie Linz; appl. 20.2.1985; D-prior. 21.3.1984).

ophthalmic formulation:
EP 366 765 (Alcon; appl. 26.4.1989; USA-prior. 26.4.1988).
EP 109 561 (Rorer; appl. 20.10.1983; USA-prior. 27.10.1982).

sustained release formulation:
EP 285 871 (Lentia; appl. 16.3.1988; D-prior. 10.4.1987).
EP 268 813 (Lentia; appl. 16.10.1987; D-prior. 24.10.1986).

Formulation(s):   f. c. tabl. 100 mg, 200 mg (as hydrochloride)

Trade Name(s):
D: Celipro Lich (Lichtenstein)   F: Célectol (Bellon)
      Selectol (Pharmacia &   GB: Celectol (Rhône-Poulenc
      Upjohn; 1986)   Rorer)

Cerivastatin sodium
(Avastatin; Bay-W-6228; Rivastatin)

Use: hyperlipidemic, HMG-CoA-
        reductase inhibitor,
        antihypercholesterolemic agent

RN: 143201-11-0   MF: $C_{26}H_{33}FNNaO_5$ MW: 481.54
CN: [S-[R*,S*-(E)]]-7-[4-(4-fluorophenyl)-5-(methoxymethyl)-2,6-bis(1-methylethyl)-3-pyridinyl]-3,5-
       dihydroxy-6-heptenoic acid monosodium salt

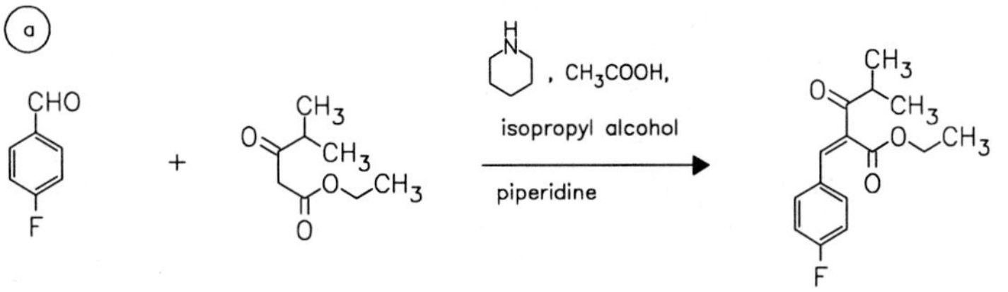

4-fluorobenz-
aldehyde

ethyl
isobutyrylacetate

ethyl 2-isobutyryl-
3-(4-fluorophenyl)-
acrylate (I)

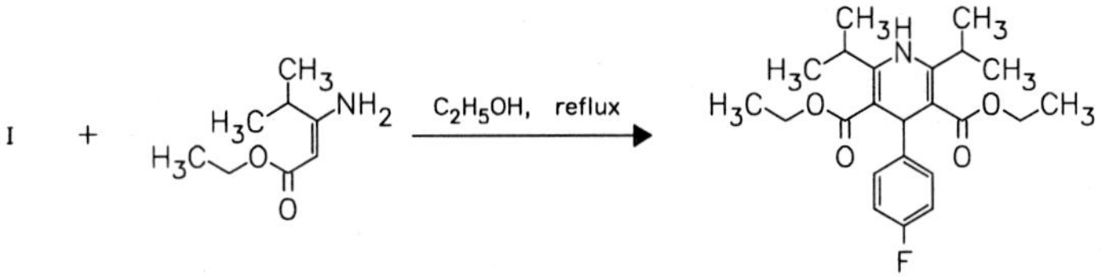

ethyl 3-amino-
4-methyl-2-
pentenoate

diethyl 1,4-dihydro-2,6-
diisopropyl-4-(4-fluorophenyl)-
pyridine-3,5-dicarboxylate (II)

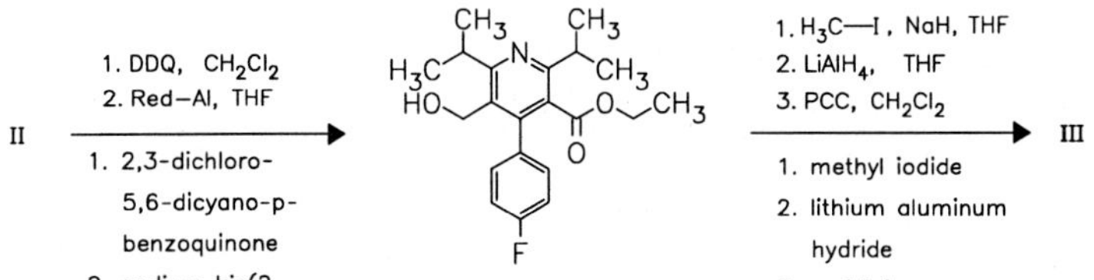

1. DDQ, CH$_2$Cl$_2$
2. Red-Al, THF

1. 2,3-dichloro-
5,6-dicyano-p-
benzoquinone
2. sodium bis(2-
methoxyethoxy)-
dihydridoaluminate

ethyl 2,6-diisopropyl-4-
(4-fluorophenyl)-5-hydroxy-
methylpyridine-3-carboxylate

1. H$_3$C—I, NaH, THF
2. LiAlH$_4$, THF
3. PCC, CH$_2$Cl$_2$

1. methyl iodide
2. lithium aluminum
hydride
3. pyridinium
chlorochromate

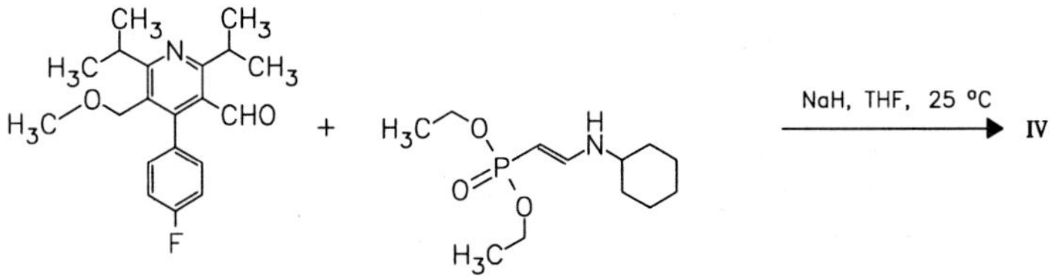

2,6-diisopropyl-
4-(4-fluorophenyl)-
5-methoxymethyl-
3-pyridine-
carboxaldehyde (III)

diethyl 2-
(cyclohexylamino)-
vinylphosphonate

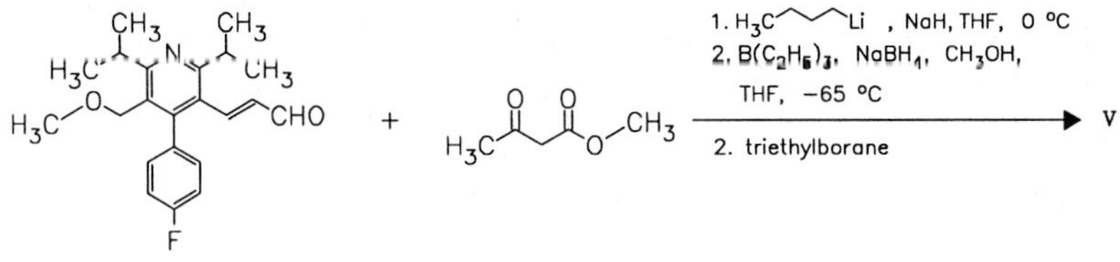

NaH, THF, 25 °C

IV

(IV)

methyl
acetoacetate

1. H$_3$C$\diagup\diagdown\diagup$Li , NaH, THF, 0 °C
2. B(C$_2$H$_5$)$_3$, NaBH$_4$, CH$_3$OH,
    THF, −65 °C

2. triethylborane

V

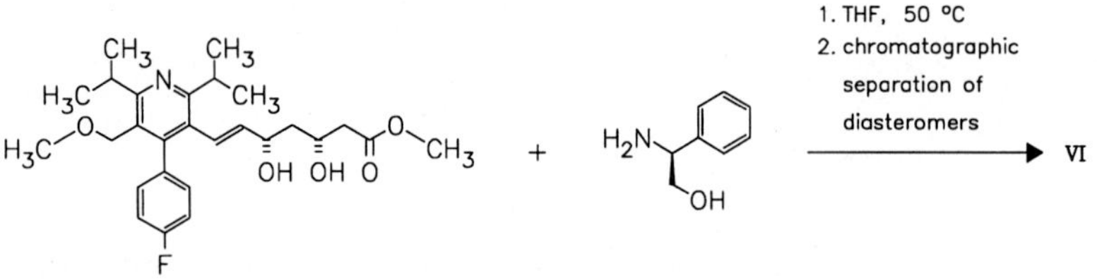

methyl erythro-(E)-7-[2,6-diisopropyl-
4-(4-fluorophenyl)-5-methoxymethyl-
3-pyridyl]-3,5-dihydroxy-hept-6-enoate (V)

(+)-(S)-phenyl-
glycinol

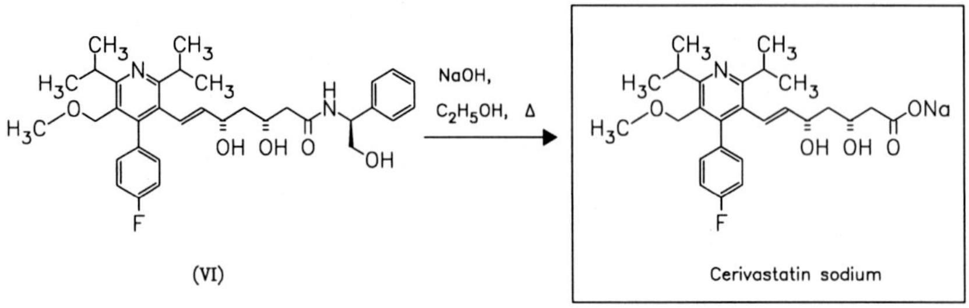

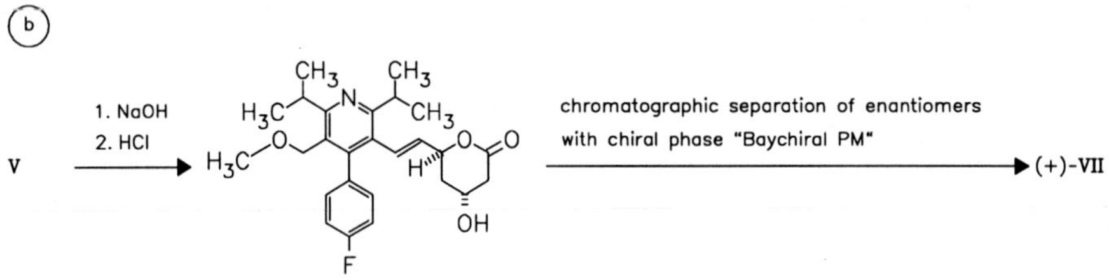

1. THF, 50 °C
2. chromatographic
   separation of
   diasteromers

VI

(VI)

NaOH,
C$_2$H$_5$OH, Δ

Cerivastatin sodium

(b)

V

1. NaOH
2. HCl

(±)-(VII)

chromatographic separation of enantiomers
with chiral phase "Baychiral PM"

(+)-VII

(+)-(VII)

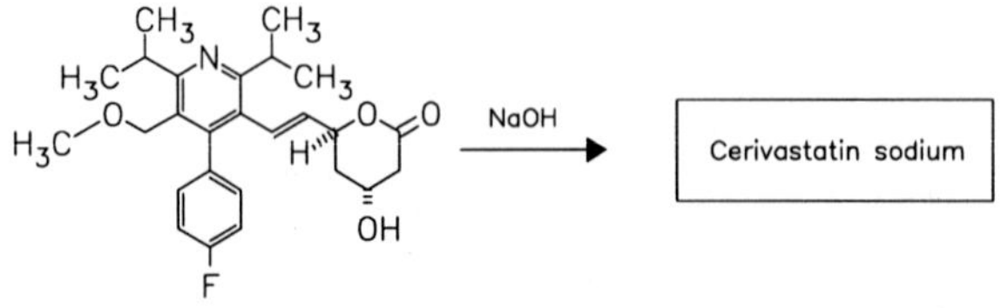

NaOH

Cerivastatin sodium

Reference(s):
a   DE 4 040 026 (Bayer; appl. 14.12.1990).
    EP 325 130 (Bayer AG; appl. 9.1.1989; D-prior. 20.1.1988).
    EP 491 226 (Bayer AG; appl. 3.12.1991; D-prior. 14.12.1990).
    AU 9 189 615 (Bayer AG; appl. 11.12.1991; D-prior. 14.12.1990).
    (R)-(+)-α-phenethylamine *can be used instead of S*-(+)-phenylglycinol.
b   Drugs Future (DRFUD4) **19**, 537-541 (1994).

Formulation(s):   tabl. 0.1 mg, 0.2 mg, 0.3 mg

Trade Name(s):
D: Lipobay (Bayer)   GB: Lipobay (Bayer)   USA: Baycol (Bayer)

Ceruletide

(Caerulein)

ATC: V04CC04; V04CK
Use: diagnostic (for pancreatic function),
        stimulant of gastric secretory

RN: 17650-98-5   MF: $C_{58}H_{73}N_{13}O_{21}S_2$ MW: 1352.42
LD_{50}: 1012 mg/kg (M, i.v.)
CN: 5-oxo-L-prolyl-L-glutaminyl-L-α-aspartyl-O-sulfo-L-tyrosyl-L-threonylglycyl-L-tryptophyl-L-methionyl-
       L-α-aspartyl-L-phenylalanin amide

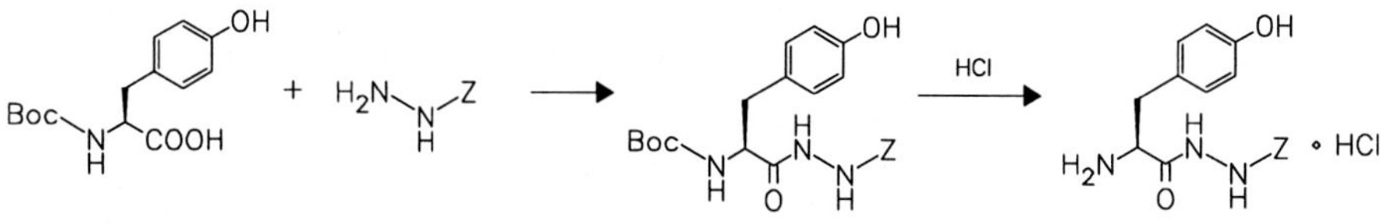

Boc-Tyr   Boc-Tyr-NH-NH-Z   Tyr-NH-NH-Z · HCl (I)

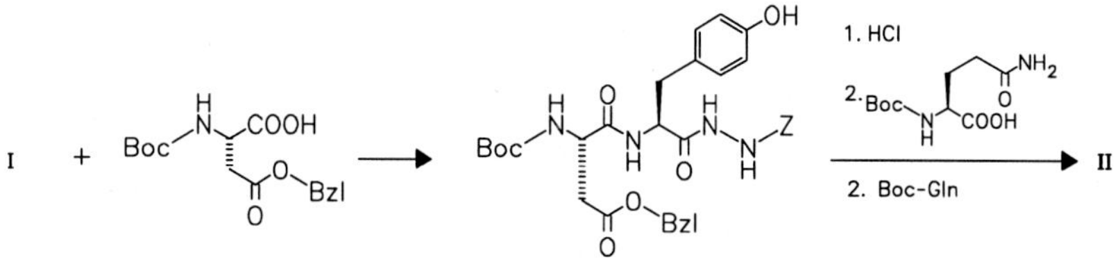

Boc-Asp(OBzl)   Boc-Asp(OBzl)-Tyr-NH-NH-Z

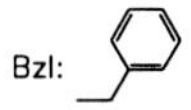

Bzl:

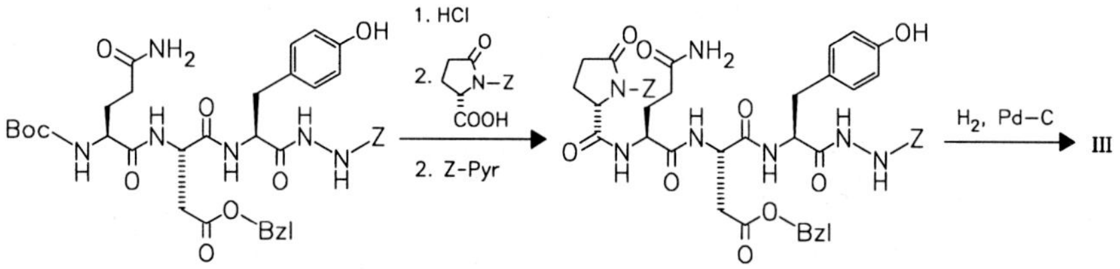

Boc-Gln-Asp(OBzl)-Tyr-NH-NH-Z (II)   Z-Pyr-Gln-Asp(OBzl)-Tyr-NH-NH-Z

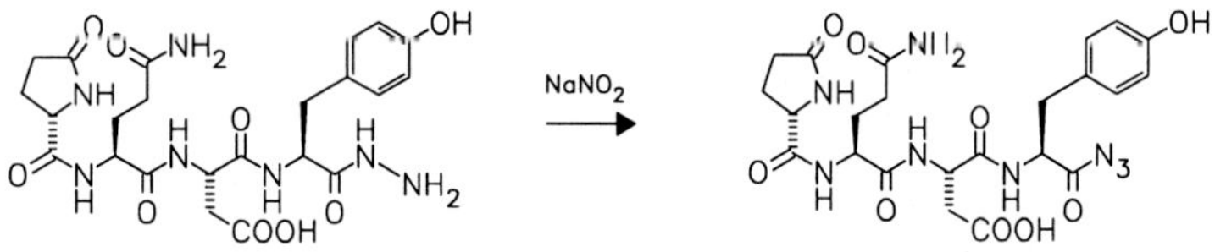

Pyr-Gln-Asp-Tyr-NH-NH₂ (III)   Pyr-Gln-Asp-Tyr-N₃ (IV)

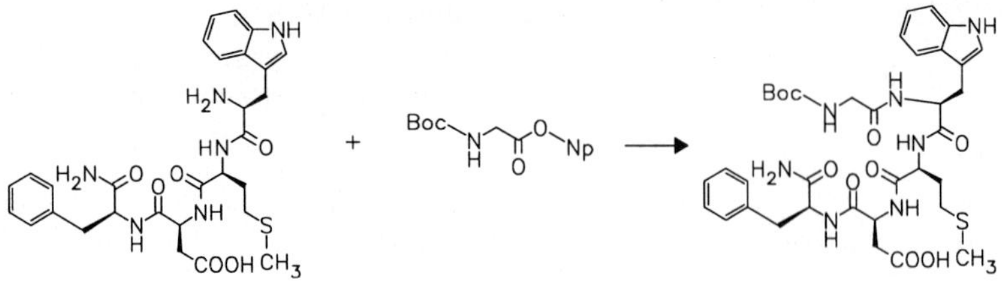

Trp-Met-Asp-Phe-NH$_2$ + Boc-Gly-O-Np → Boc-Gly-Trp-Met-Asp-Phe-NH$_2$ (V)

Np:

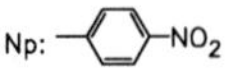

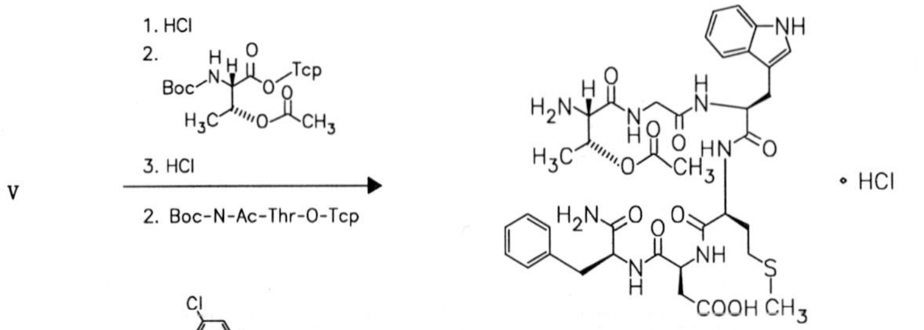

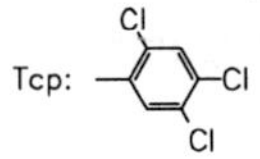

Ac-Thr-Gly-Trp-Met-Asp-Phe-NH$_2$ · HCl (VI)

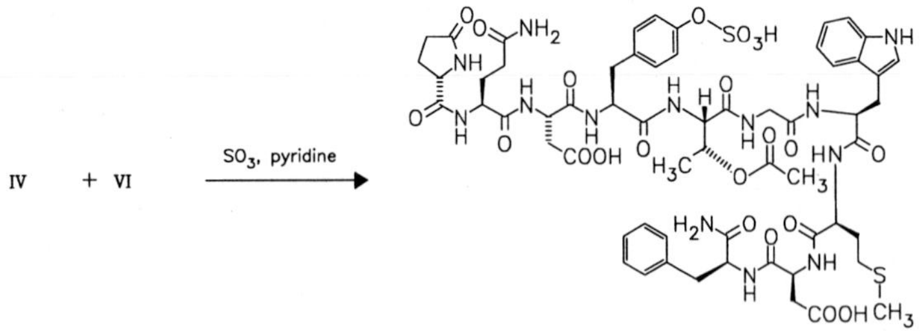

Tcp:

Pyr-Gln-Asp-Tyr(SO$_3$H)-Thr(Ac)-Gly-Trp-Met-Asp-Phe-NH$_2$ (VII)

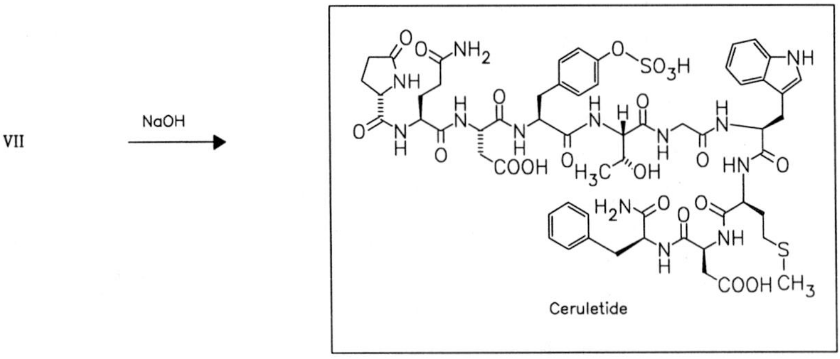

Ceruletide

Reference(s):

DE 1 643 504 (Soc. Farmaceutici Italia; appl. 6.4.1972; I-prior. 9.8.1966).
US 3 472 832 (Soc. Farmaceutici Italia; appl. 9.8.1966; I-prior. 9.8.1966).
Bernardi, L. et al.: Experientia (EXPEAM) **23**, 700 (1967).

structure and isolation from Hyla caerulea:
Anastasi, A. et al.: Experientia (EXPEAM) **23**, 699 (1967).

Formulation(s): amp. 5 µg/ml, 40 µg/2 ml; vial 20 µg, 30 µg, 40 µg

Trade Name(s):
D: Takus (Pharmacia &   J: Ceosunin (Kyowa Hakko)
    Upjohn)

Cetalkonium chloride

ATC: S01AA
Use: antiseptic, bactericide

RN: 122-18-9 MF: $C_{25}H_{46}ClN$ MW: 396.10 EINECS: 204-526-3
CN: *N*-hexadecyl-*N,N*-dimethylbenzenemethanaminium chloride

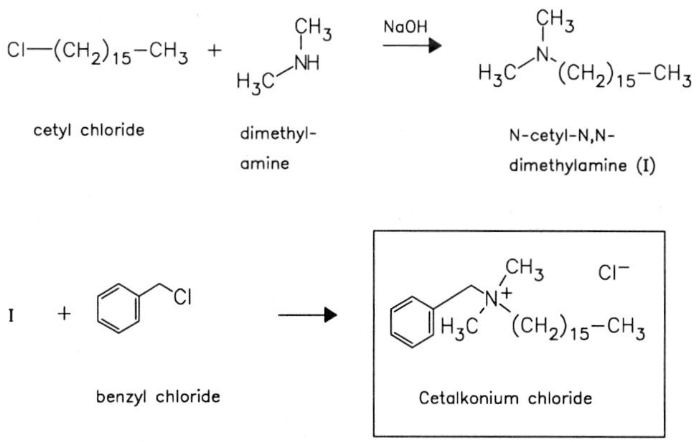

cetyl chloride dimethyl- N-cetyl-N,N-
               amine   dimethylamine (I)

benzyl chloride Cetalkonium chloride

Reference(s):
FR 771 746 (I. G. Farben; 1934).

Formulation(s): sol. 13 g/100 g, 130 mg

Trade Name(s):
D: Baktonium (Bode)   F: Pansoral (Pierre Fabre   GB: Bonjela (Reckitt &
    Mundisal (Mundipharma)-   Santé)-comb.   Colman)-comb.
    comb.   J: Lazal (Shionogi)

Cethexonium bromide

ATC: D08AX; R02AA20
Use: antiseptic

RN: 1794-74-7 MF: $C_{24}H_{50}BrNO$ MW: 448.57
CN: *N*-hexadecyl-2-hydroxy-*N,N*-dimethylcyclohexanaminium bromide

chloride
RN: 58703-78-9 MF: $C_{24}H_{50}ClNO$ MW: 404.12

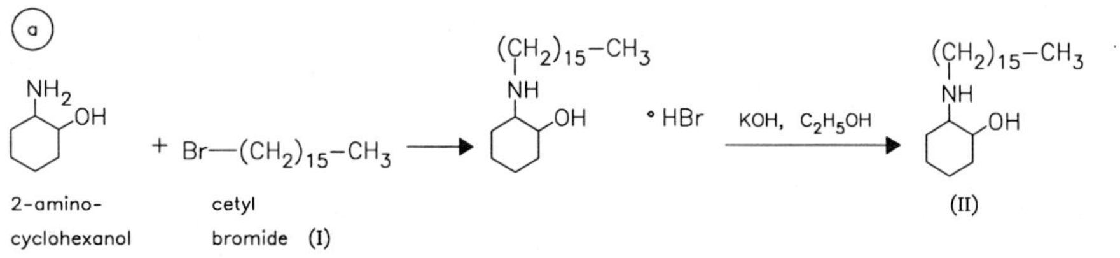

2-amino-   cetyl
cyclohexanol   bromide (I)   (II)

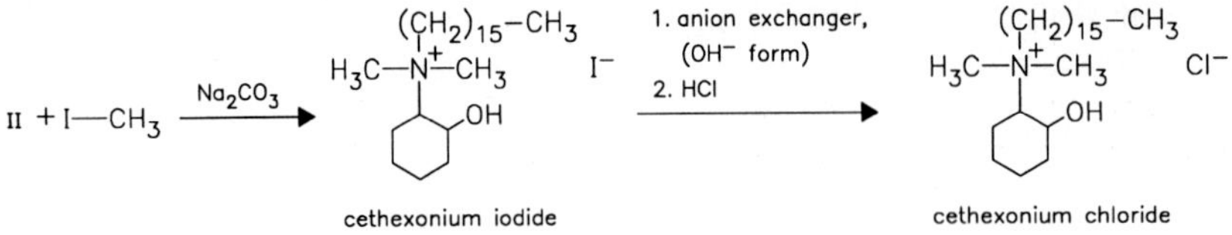

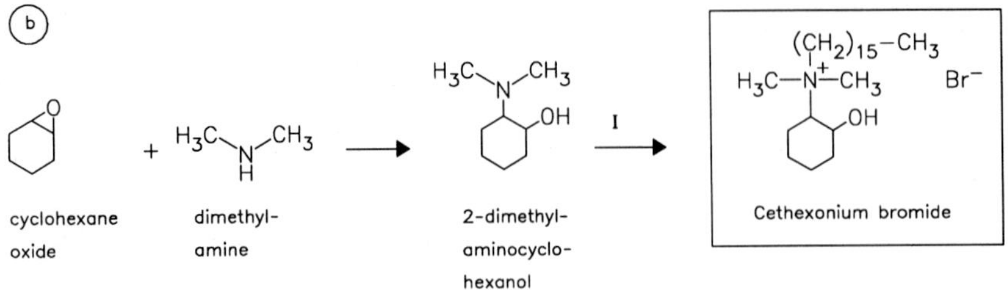

II + I—CH₃ →(Na₂CO₃)

cethexonium iodide

1. anion exchanger, (OH⁻ form)
2. HCl
→

cethexonium chloride

(b)

cyclohexane
oxide

+ dimethyl-
amine

→ 2-dimethyl-
aminocyclo-
hexanol

I →

Cethexonium bromide

Reference(s):
Winternitz, F. et al.: Bull. Soc. Chim. Biol. (BSCIA3) **33**, 369 (1951).

Formulation(s):   collutorium 0.025 g/100 ml, 0.1 mg/0.4 ml, 0.3 g/100 ml; eye drops 0.025 % (bromide);
ointment 1 g/100 g; powder 1.5 g/100 g; sol 50 mg/100 ml

Trade Name(s):
F: Biocidan (Menarini)

Cetiedil

ATC: C04AX26
Use: vasodilator (peripheral)

RN: 14176-10-4   MF: C₂₀H₃₁NO₂S   MW: 349.54   EINECS: 238-028-2
LD₅₀: 1726 mg/kg (M, p.o.)
CN: α-cyclohexyl-3-thiopheneacetic acid 2-(hexahydro-1H-azepin-1-yl)ethyl ester

citrate (1:1)
RN: 16286-69-4   MF: C₂₀H₃₁NO₂S · C₆H₈O₇   MW: 541.66   EINECS: 240-381-2

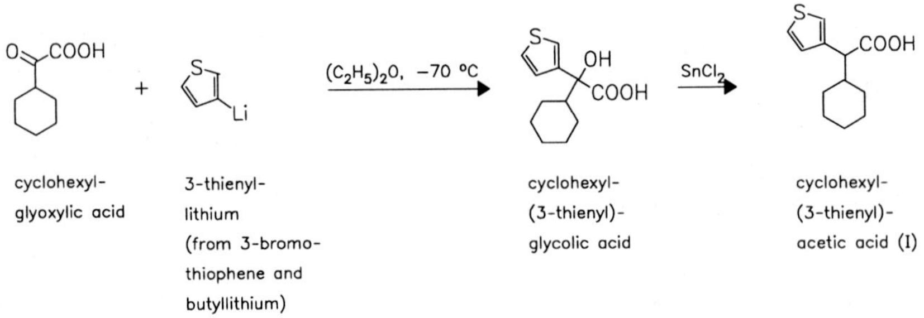

cyclohexyl-
glyoxylic acid

+ 3-thienyl-
lithium
(from 3-bromo-
thiophene and
butyllithium)

(C₂H₅)₂O, −70 °C →

cyclohexyl-
(3-thienyl)-
glycolic acid

SnCl₂ →

cyclohexyl-
(3-thienyl)-
acetic acid (I)

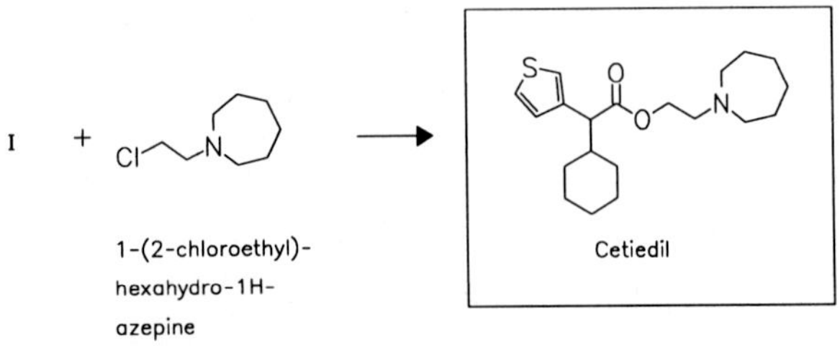

I + 1-(2-chloroethyl)-
hexahydro-1H-
azepine

→ Cetiedil

Reference(s):
FR 1 460 571 (Innothéra; appl. 10.6.1965).
FR-M 5 504 (Innothéra; appl. 10.6.1965).
Robba, M.; Guen, Y. Le: Chim. Ther. (CHTPBA) **1967** (No. 2), 120.

synthesis of starting materials:
Robba, M.; Guen, Y. Le: Chim. Ther. (CHTPBA) **1966** (No. 4), 238.
FR-appl. 2 260 575 (Innothéra; appl. 11.2.1974).
FR-appl. 2 260 576 (Innothéra; appl. 11.2.1974).

synthesis from 3-thienylacetonitrile:
US 4 108 865 (Labaz; 22.8.1978; prior. 29.8.1974, 1.3.1976).

Formulation(s):   cps. 100 mg (as citrate)

Trade Name(s):
F: Stratene (Gerda)   Vasocet (Cipharm)   I: Stratene (Sigma-Tau); wfm

---

Cetirizine

ATC: R06AE07
Use: non-sedative antihistaminic

RN: 83881-51-0   MF: $C_{21}H_{25}ClN_2O_3$ MW: 388.90
CN: (±)-[2-[4-[(4-chlorophenyl)phenylmethyl]-1-piperazinyl]ethoxy]acetic acid

dihydrochloride
RN: 83881-52-1   MF: $C_{21}H_{25}ClN_2O_3 \cdot 2HCl$ MW: 461.82
LD$_{50}$:   365 mg/kg (R, p.o.);
        >320 mg/kg (dog, p.o.)

4-chlorobenzhydryl chloride   N-carbethoxy-piperazine   →(Na$_2$CO$_3$)   ethyl 4-(4-chlorobenzhydryl)-piperazine-1-carboxylate   →(HCl)   1-(4-chlorodiphenyl-methyl)piperazine (I)

I   →(methyl 2-(2-chloroethoxy)-acetate , sodium carbonate , Na$_2$CO$_3$)   methyl 2-[2-[4-(4-chlorodiphenylmethyl)-1-piperazinyl]ethoxy]acetate   →(KOH)   Cetirizine

Reference(s):
EP 58 146 (UCB; appl. 5.2.1982; GB-prior. 6.2.1981, 8.4.1981).

alternative synthesis (also enantiomers):
GB 2 225 321 (UCB; appl. 23.11.1988).
EP 801 064 (UCB; appl. 9.4.1997; BE-prior. 10.4.1996).
WO 9 737 982 (UCB; appl. 28.3.1997; BE-prior. 4.10.1996).
WO 9 802 425 (Apotex; appl. 11.7.1997; CA-prior. 11.7.1996).

synthesis of 1-(4-chlorodiphenylmethyl)piperazine:
US 2 819 269 (Abbott; 1958).
HU 17 343 (Richter Gedeon; appl. 26.5.1977).
US 2 709 169 (UCB; 1952).

Formulation(s):   drops 10 mg; sol. 0.1 %; tabl. 10 mg (as dihydrochloride)

Trade Name(s):
D: Zyrtec (UCB; Rodleben;   GB: Zirtek (UCB)   USA: Zyrtec (Pfizer; as
      Vedim)   I: Formistin (Formenti; 1990)   hydrochloride)
F: Virlix (Synthélabo)   Virlix (Chemil)
      Zyrtec (UCB)   Zirtec (UCB; 1990)

Cetrimonium bromide
(Cetrimide)

ATC: R02AA17
Use: antiseptic

RN: 57-09-0   MF: $C_{19}H_{42}BrN$ MW: 364.46   EINECS: 200-311-3
LD$_{50}$:   32 mg/kg (M, i.v.);
      44 mg/kg (R, i.v.); 410 mg/kg (R, p.o.)
CN: *N,N,N*-trimethyl-1-hexadecanaminium bromide

hydroxide
RN: 505-86-2   MF: $C_{19}H_{43}NO$ MW: 301.56   EINECS: 208-022-4

cetyl bromide   trimethylamine   Cetrimonium bromide

Reference(s):
Shelton, R.S. et al.: J. Am. Chem. Soc. (JACSAT) **68**, 753 (1946).

Formulation(s):   sol. 117 mg/100 g; tabl. 4 mg

Trade Name(s):
D: Lemocin (Novartis   Buccawalter (SmithKline   GB: Ceanel Conc. (Quinoderm)-
      Consumer Health)   Beecham)-comb.   comb.
      Xylastesin (Espe)   Cétavlon (Zeneca)   Cetavlex (Zeneca)
      numerous combination   Dérinox (Thérabel Lucien)-   I: Cetavlon (Zeneca)
      preparations   comb.   Xylonor (Ogna)-comb.
F: Aseptit (Riom)   Rectoquotane (Evans   J: Cetavlon (Sumitomo
                                  Medical)-comb.   Chem.)

Cetrorelix
(SB-75; D-20761)

ATC: H01CC02
Use: LHRH-antagonist

RN: 120287-85-6   MF: $C_{70}H_{92}ClN_{17}O_{14}$ MW: 1431.06
CN: *N*-Acetyl-3-(2-naphthalenyl)-D-alanyl-4-chloro-D-phenylalanyl-3-(3-pyridinyl)-D-alanyl-L-seryl-L-
      tyrosyl-*N*^{5}-(aminocarbonyl)-D-ornithyl-L-leucyl-L-arginyl-L-prolyl-D-alaninamide

acetate
RN: 145672-81-7   MF: $C_{70}H_{92}ClN_{17}O_{14} \cdot xC_2H_4O_2$ MW: unspecified
diacetate
RN: 130143-01-0   MF: $C_{70}H_{92}ClN_{17}O_{14} \cdot 2C_2H_4O_2$ MW: 1551.17
trifluoroacetate
RN: 130289-71-3   MF: $C_{70}H_{92}ClN_{17}O_{14} \cdot C_2HF_3O_2$ MW: 1545.09
pamoate
RN: 132741-85-6   MF: $C_{70}H_{92}ClN_{17}O_{14} \cdot C_{23}H_{16}O_6$ MW: 1819.44

(a) solid-phase synthesis:

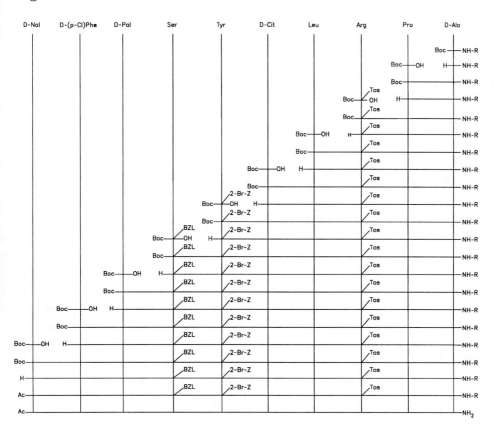

Cit:	citrulline (N^5-carbamoylornithine)
Pal:	3-pyridylalanine
Nal:	2-naphthylalanine
(p-Cl)Phe:	4-chlorophenylalanine
Boc:	tert-butoxycarbonyl
BZL:	benzyl
2-Br-Z:	2-bromobenzyloxycarbonyl
Ac:	acetyl
Tos:	tosyl

R: polystyrene-divinylbenzene (MBHA-resin)

After cleavage from the resin the crude peptide is dissolved
in aqueous acetic acid and purified via prep. HPLC.

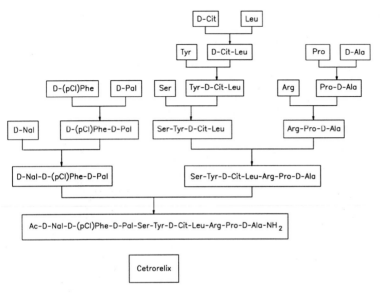

Cetrorelix

b classical, liquid-phase synthesis:

```
                              ┌───────┐ ┌─────┐
                              │ D-Cit │ │ Leu │
                              └───────┘ └─────┘
                                  │         │
                              ┌─────┐ ┌──────────┐                ┌─────┐ ┌───────┐
                              │ Tyr │ │ D-Cit-Leu│                │ Pro │ │ D-Ala │
                              └─────┘ └──────────┘                └─────┘ └───────┘
                                  │         │                         │       │
┌───────────┐ ┌───────┐     ┌─────┐ ┌──────────────┐           ┌─────┐ ┌───────────┐
│ D-(pCl)Phe│ │ D-Pal │     │ Ser │ │ Tyr-D-Cit-Leu│           │ Arg │ │ Pro-D-Ala │
└───────────┘ └───────┘     └─────┘ └──────────────┘           └─────┘ └───────────┘
        │         │              │         │                        │        │
┌───────┐ ┌──────────────┐   ┌──────────────────┐              ┌──────────────────┐
│ D-Nal │ │ D-(pCl)Phe-D-Pal│ │ Ser-Tyr-D-Cit-Leu│            │ Arg-Pro-D-Ala    │
└───────┘ └──────────────┘   └──────────────────┘              └──────────────────┘
     │           │                      │                                │
┌──────────────────────────┐      ┌────────────────────────────────────────┐
│ D-Nal-D-(pCl)Phe-D-Pal   │      │ Ser-Tyr-D-Cit-Leu-Arg-Pro-D-Ala        │
└──────────────────────────┘      └────────────────────────────────────────┘
            │                                        │
    ┌──────────────────────────────────────────────────────────────────────────┐
    │ Ac-D-Nal-D-(pCl)Phe-D-Pal-Ser-Tyr-D-Cit-Leu-Arg-Pro-D-Ala-NH₂            │
    └──────────────────────────────────────────────────────────────────────────┘
                              │
                      ┌──────────────┐
                      │  Cetrorelix  │
                      └──────────────┘
```

abbreviations see method a

Reference(s):
a   Bajusz, S. et al.: Int J. Pept. Protein Res. (IJPPC3) **32**, 425 (1988).
    EP 299 402 (ASTA Medica; appl. 11.7.1988; USA-prior. 17.7.1987).
a,b Kleemann, A. et al.: Proc. Akabori Conf.: Ger.-Jpn. Symp. Pept. Chem., 4th, 1991, 96-101.
b   Kunz, F.R. et al.: Proc. Akabori Conf.: Ger.-Jpn. Symp. Pept. Chem., 5th, 1994, 15-16.

long-acting injection suspension with pamoate salt:
US 773 032 (ASTA Medica; 10.6.1996; D-prior. 9.12.1993).

sterile acetate formulation:
EP 611 572 (ASTA Medica; D-prior. 19.2.1993).

use in fertility control:
EP 788 799 (ASTA Medica; USA-prior. 7.2.1996).

use for BPH or prostate cancer:
WO 9 810 781 (ASTA Medica; USA-prior. 12.9.1996).

Formulation(s):   vial 0.25 mg, 3 mg (as acetate)

Trade Name(s):
D: Cetrotide (ASTA Medica   GB: Cetrotide (ASTA Medica)
     AWD; 1999)

Cetylpyridinium chloride

ATC: B05CA01; D08AJ03; D09AA07;
      R02AA06
Use: disinfectant, antiseptic

RN: 123-03-5   MF: $C_{21}H_{38}ClN$ MW: 340.00   EINECS: 204-593-9
LD_{50}: 10 mg/kg (M, i.v.); 108 mg/kg (M, p.o.);
     200 mg/kg (R, p.o.)
CN: 1-hexadecylpyridinium chloride

Reference(s):
Budesinsky-Protiva, 531-532.

Formulation(s):   eff. tabl. 1.5 mg, 3 mg; lozenge 1.4 mg; sol. 5 mg/10 ml (0.01 %, 0.05 %); tabl. 2 mg

Trade Name(s):
D: Dobendan (Cassella-med)   Merocaine (Seton)-comb.   Penaten (Johnson &
     Formamint N (Beecham-   Merocet (Seton)   Johnson)
     Wülfing)   I: Borocaina (Schiapparelli   Ragaden (Ganassini)
     Tyrosolvetten (Byk Gulden;   Salute)   Vidermina (Ganassini)
     Roland)-comb.   Fluprim (Roche)   numerous combination
     numerous combination   Gola (Sella)   preparations
     preparations   Golagamma (Avantgarde)   J: Colgen 123 (Kowa)-comb.
F: Alodont (Warner-Lambert)-   Neocepacol (Lepetit)   Pabron Troche (Taisho)
     comb.   Neocoricidin (Schering-   Suprol (Iwaki)
     Broncorinol (Roche   Plough)   USA: Cepacol (Lakeside)-comb.;
     Nicholas)-comb.   Neoformitrol (Sandoz)   wfm
     Cétylyre (Oberlin)   Neogola (Sella)   Cobrex (Reid-Rowell)-
GB: Calgel (Warner-Lambert)   Noalcool (Sella)   comb.; wfm

Chenodeoxycholic acid

(Chenodiol; Acide chenodéoxycholique;
   Chenodesoxycholsäure)

ATC: A05AA01
Use: choleretic, anticholethithogenic
      dissolution of cholesterol gallstones

RN: 474-25-9   MF: $C_{24}H_{40}O_4$ MW: 392.58   EINECS: 207-481-8
LD_{50}: 100 mg/kg (M, i.v.); 3 g/kg (M, p.o.);
     106 mg/kg (R, i.v.); 4 g/kg (R, p.o.);
     >1 g/kg (dog, p.o.)
CN: (3α,5β,7α)-3,7-dihydroxycholan-24-oic acid

cholic acid + HO—CH$_3$ →[H_2SO_4] methyl cholate →[pyridine] I

methyl 3α,7α-diacetoxy-
12α-hydroxycholanate (I)

→[$Na_2Cr_2O_7$, CH_3COOH]

methyl 3α,7α-diacetoxy-
12-oxocholanate (II)

II →[1. $H_2N-NH_2 \cdot H_2O$, $HO\frown OH$, 200 °C; 2. HCl / 1. hydrazine hydrate]

Chenodeoxycholic acid

Reference(s):

Fieser, L.F.; Rajagopalan, S.: J. Am. Chem. Soc. (JACSAT) **72**, 5530 (1950).

Hauser, E. et al.: Helv. Chim. Acta (HCACAV) **43**, 1595 (1960).

Hofmann, A.F.: Acta Chem. Scand. (ACHSE7) **17**, 173 (1963).

Sato, Y.; Ikekawa, N.: J. Org. Chem. (JOCEAH) **24**, 1367 (1959).

purification:

DOS 2 302 744 (Union International; appl. 20.1.1973; GB-prior. 20.1.1972).

DE 2 404 102 (Schering AG; appl. 25.1.1974).

DOS 2 613 346 (Diamalt; appl. 29.3.1976).

US 4 163 017 (Diamalt; 31.7.1979; D-prior. 29.3.1976).

JP-appl. 52 153 955 (Tokyo Tanabe; appl. 18.6.1976).

isolation from animal bile:

US 3 919 266 (Intellectual Property Dev. Corp.; 11.11.1975; prior. 21.9.1972, 19.11.1973).

US 4 014 908 (Intellectual Property Dev. Corp.; 29.3.1977; prior. 21.9.1972, 19.11.1973, 7.5.1974, 30.5.1974, 20.2.1976).

US 4 072 695 (Intellectual Property Dev. Corp.; 7.2.1978; prior. 21.9.1972, 19.11.1973, 7.5.1974, 30.5.1974, 20.2.1976, 17.9.1976).

combination with hymecromone:

DOS 2 700 085 (Lipha; appl. 4.1.1977; F-prior. 13.7.1976).

Formulation(s): cps. 250 mg; tabl. 250 mg

Trade Name(s):

D:	Chenofalk (Falk)	Hekbilin (Strathmann)-	GB: Chendol (CP Pharm.)
	Cholit-Ursan (Fresenius-	comb.	Chenofalk (Thames)
	Praxis)	Ursofalk (Falk)-comb.	I: Chenofalk (Interfalk)
		F: Chénodex (Hoechst Houdé)	Chenossil (Midy)

J: Fluibil (Zambon Italia)   Cholasa (Tokyo Tanabe)   USA: Chenix (Reid-Rowell);
      Chenochol (Yamanouchi)   wfm

Chloral hydrate

ATC: N05CC01
Use: hypnotic, sedative

RN: 302-17-0   MF: $C_2H_3Cl_3O_2$ MW: 165.40   EINECS: 206-117-5
LD_{50}:   530 mg/kg (M, i.v.); 1100 mg/kg (M, p.o.);
       479 mg/kg (R, p.o.)
CN: 2,2,2-trichloro-1,1-ethanediol

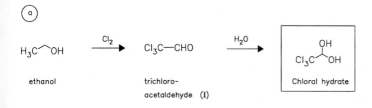

Reference(s):
Ullmanns Encykl. Tech. Chem., 4. Aufl., Vol. **9**, 377.

Formulation(s):   cps. 250 mg, 500 mg

Trade Name(s):
D: Chloraldurat (Pohl)   GB: Welldorm elixir (S & N)   USA: Noctec (Squibb); wfm
F: numerous combination   I: Cloral (Tariff. Nazionale)   generica
      preparations   J: Escre (SS)

Chloralodol

(Chlorhexadol)

ATC: N05CC02
Use: hypnotic

RN: 3563-58-4   MF: $C_8H_{15}Cl_3O_3$ MW: 265.56   EINECS: 222-634-9
CN: 2-methyl-4-(2,2,2-trichloro-1-hydroxyethoxy)-2-pentanol

Reference(s):
US 2 931 838 (Det Danske Med.-& Kem.-Komp.; 5.4.1960; DK-prior. 8.12.1956).

Formulation(s):   tabl. 400 mg, 800 mg

Trade Name(s):
GB: Medodorm (Medo); wfm

Chlorambucil

ATC: L01AA02
Use: antineoplastic

RN: 305-03-3   MF: $C_{14}H_{19}Cl_2NO_2$ MW: 304.22   EINECS: 206-162-0
LD$_{50}$:   80 mg/kg (M, p.o.);
76 mg/kg (R, p.o.)
CN: 4-[bis(2-chloroethyl)amino]benzenebutanoic acid

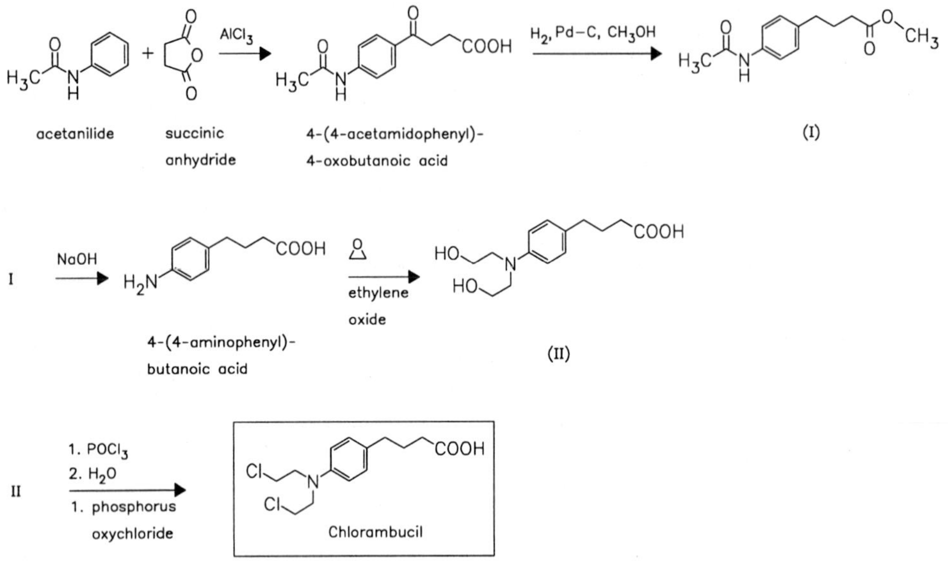

acetanilide   succinic   4-(4-acetamidophenyl)-
anhydride   4-oxobutanoic acid
(I)

4-(4-aminophenyl)-   ethylene
butanoic acid   oxide
(II)

1. POCl$_3$
2. H$_2$O
II   ——————→
1. phosphorus
oxychloride   Chlorambucil

Reference(s):
US 3 046 301 (Borroughs Wellcome; 24.7.1962; prior. 29.10.1959).

Formulation(s):   drg. 2 mg; tabl. 2 mg, 5 mg

Trade Name(s):

D:	Leukeran (Glaxo Wellcome)	GB:	Leukeran (Glaxo Wellcome)		Linfolysin (Nuovo ISM)
F:	Chloraminophéne (Techni-Pharma)	I:	Leukeran (Glaxo Wellcome)	USA:	Leukeran (Glaxo Wellcome)

Chloramphenicol

ATC: D06AX02; D10AF03; G01AA05;
J01BA01; S01AA01; S02AA01;
S03AA08
Use: antibiotic

RN: 56-75-7   MF: $C_{11}H_{12}Cl_2N_2O_5$ MW: 323.13   EINECS: 200-287-4
LD$_{50}$:   110 mg/kg (M, i.v.); 1500 mg/kg (M, p.o.);
171 mg/kg (R, i.v.); 2500 mg/kg (R, p.o.)
CN: [R-(R*,R*)]-2,2-dichloro-N-[2-hydroxy-1-(hydroxymethyl)-2-(4-nitrophenyl)ethyl]acetamide

a

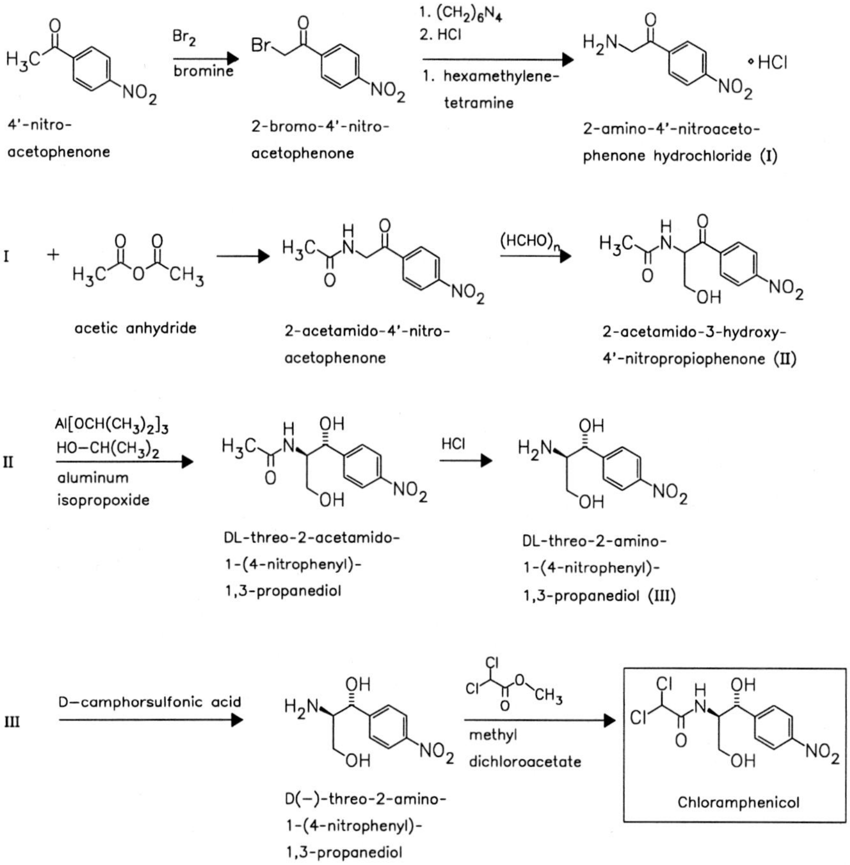

4'-nitro-
acetophenone

2-bromo-4'-nitro-
acetophenone

2-amino-4'-nitroaceto-
phenone hydrochloride (I)

I +

acetic anhydride

2-acetamido-4'-nitro-
acetophenone

2-acetamido-3-hydroxy-
4'-nitropropiophenone (II)

II

aluminum
isopropoxide

DL-threo-2-acetamido-
1-(4-nitrophenyl)-
1,3-propanediol

DL-threo-2-amino-
1-(4-nitrophenyl)-
1,3-propanediol (III)

III

D–camphorsulfonic acid

D(−)-threo-2-amino-
1-(4-nitrophenyl)-
1,3-propanediol

methyl
dichloroacetate

Chloramphenicol

b

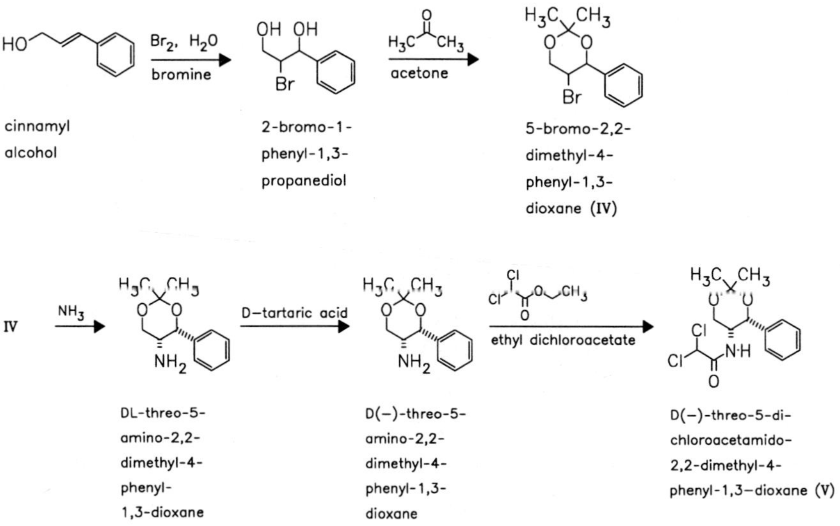

cinnamyl
alcohol

2-bromo-1-
phenyl-1,3-
propanediol

acetone

5-bromo-2,2-
dimethyl-4-
phenyl-1,3-
dioxane (IV)

IV

D-tartaric acid

ethyl dichloroacetate

DL-threo-5-
amino-2,2-
dimethyl-4-
phenyl-
1,3-dioxane

D(−)-threo-5-
amino-2,2-
dimethyl-4-
phenyl-1,3-
dioxane

D(−)-threo-5-di-
chloroacetamido-
2,2-dimethyl-4-
phenyl-1,3-dioxane (V)

D(−)-threo-2-dichloro-
acetamido-1-(4-nitro-
phenyl)-1,3-propane-
diol dinitrate

Reference(s):
Ehrhart-Ruschig **IV**, 398 ff.
a Long, L.M.; Troutman, H.D.: J. Am. Chem. Soc. (JACSAT) **71**, 2469, 2473 (1949).
US 2 483 871 (Parke Davis; 1949; appl. 1948).
US 2 483 884 (Parke Davis; 1949; appl. 1948).
US 2 483 885 (Parke Davis; 1949; appl. 1949).
US 2 483 892 (Parke Davis; 1949; appl. 1948).
US 2 687 434 (Parke Davis; 1954; appl. 1953).
US 2 651 661 (Monsanto; 1953; appl. 1950).
US 2 786 870 (Parke Davis; 1957; appl. 1954).
Rebstock, M.C. et al.: J. Am. Chem. Soc. (JACSAT) **71**, 2458-2468 (1949).
b BE 539 991 (Boehringer Mannh.; appl. 1955; D-prior. 1954).
DE 1 016 718 (Boehringer Mannh.; appl. 1953).
BE 558 378 (Boehringer Mannh.; appl. 14.6.1957; D-prior. 27.6.1956, 22.12.1956).

alternative synthesis (from benzaldehyde *and* nitromethane *or* O-nitroethanol *via* 2-nitro-1-phenyl-1,3-propanediol):
DE 862 302 (Parke Davis; appl. 1949; USA-prior. 1948).
DE 1 064 937 (Boehringer Mannh.; appl. 1957).
DOS 2 708 301 (Egyt; appl. 25.2.1977; H-prior. 25.2.1976).

O-3-monophosphate:
DAS 1 668 961 (Roussel-Uclaf; appl. 20.2.1968; I-prior. 20.2.1967, 18.5.1967).

Formulation(s):   amp. 1 g (as hydrogen succinate sodium salt); cps. 250 mg, 500 mg; ear drops 5 g/100 ml,
50 mg/g; eye drops 5 mg, 10 mg; ointment 1 % (10 mg/g)

Trade Name(s):

D:	Aquamycetin (Winzer)	GB: Chloromycetin
	Chloramphenicol-PW	(Goldshield)
	(Pharma Wernigerode)	Kemicetine Succinate
	Chloramsaar (Chephasaar)	(Pharmacia & Upjohn)
	Oleomycetin (Winzer)	Minims Chloramphenicol
	Paraxin (Boehringer	(Chauvin)
	Mannh.)	SNO Phenicol (Chauvin)
	Thilocanfol C (Alcon)	I: Chemicetina (Carlo Erba)
	numerous combination	Chloromycetin (Parke
	preparations	Davis)
F:	Cébédexacol (Chauvin)-	Cloram (Formulario Naz.)
	comb.	Minims (Smith & Nephew)
	Cébénicol (Chauvin)	Mycetin (Farmigea)
	numerous combination	Sificetina (SIFI)
	preparations	Vitamfenicolo (Allergan)

numerous salts and
combination preparations
J: Antacin (Sumitomo)
Chloromycetin (Sankyo)
Kemicetine (Fujisawa)
Myclocin (Takeda)
Paraxin (Yamanouchi)
Synthomycetine (Otsuka)
numerous generics and
combination preparations
USA: Elase-Chloromycetin
(Fujisawa)

Chlorazanil

ATC: C03
Use: diuretic

RN: 500-42-5   MF: C$_9$H$_8$ClN$_5$ MW: 221.65   EINECS: 207-904-6
LD$_{50}$:   300 mg/kg (M, p.o.);
16 mg/kg (R, i.v.)
CN: N-(4-chlorophenyl)-1,3,5-triazine-2,4-diamine

monohydrochloride
RN: 2019-25-2   MF: C$_9$H$_8$ClN$_5$ · HCl   MW: 258.11   EINECS: 217-962-4

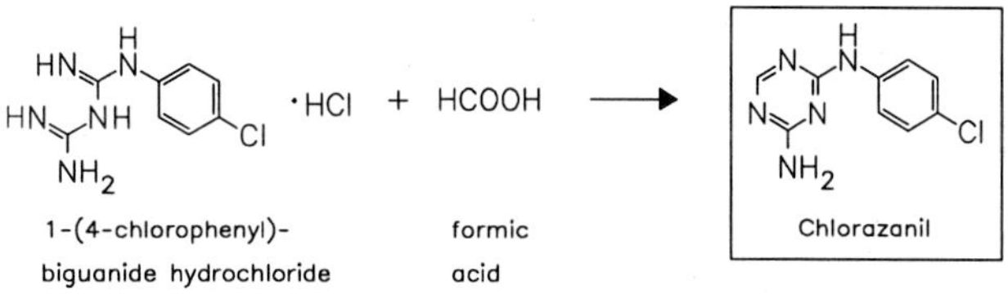

1-(4-chlorophenyl)-   formic   Chlorazanil
biguanide hydrochloride   acid

Reference(s):
DE 1 008 303 (Heumann & Co.; appl. 1955).

Formulation(s):   tabl. 150 mg

Trade Name(s):
D: Orpidan-150 (Heumann);
wfm

Chlorbenzoxamine

ATC: A03AX03
Use: anticholinergic

RN: 522-18-9   MF: C$_{27}$H$_{31}$ClN$_2$O   MW: 435.01   EINECS: 208-323-0
CN: 1-[2-[(2-chlorophenyl)phenylmethoxy]ethyl]-4-[(2-methylphenyl)methyl]piperazine

dihydrochloride
RN: 5576-62-5   MF: C$_{27}$H$_{31}$ClN$_2$O · 2HCl   MW: 507.93   EINECS: 226-951-3
LD$_{50}$:   1400 mg/kg (M, p.o.);
66 mg/kg (R, i.v.); 3350 mg/kg (R, p.o.)

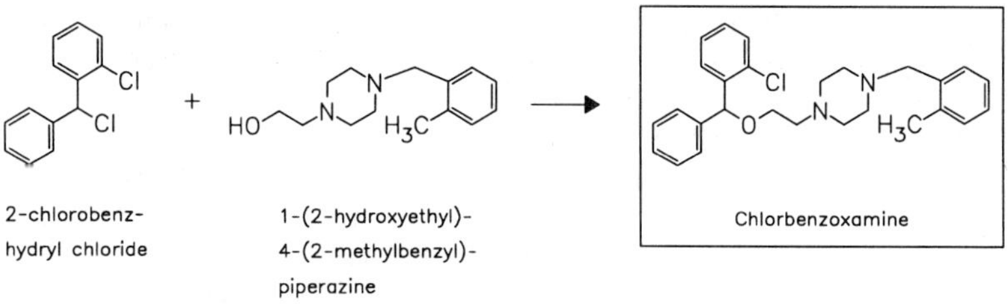

2-chlorobenz-   1-(2-hydroxyethyl)-   Chlorbenzoxamine
hydryl chloride   4-(2-methylbenzyl)-
                     piperazine

Reference(s):
BE 549 420 (H. Morren; appl. 10.7.1956).

Formulation(s):   tabl. 30 mg (as dihydrochloride)

Chlorcyclizine
(Histachlorazine)

ATC: R06AE04
Use: antihistaminic

RN: 82-93-9   MF: $C_{18}H_{21}ClN_2$ MW: 300.83   EINECS: 201-446-0
CN: 1-[(4-chlorphenyl)phenylmethyl]-4-methylpiperazine

monohydrochloride
RN: 14362-31-3   MF: $C_{18}H_{21}ClN_2 \cdot HCl$ MW: 337.29
dihydrochloride
RN: 129-71-5   MF: $C_{18}H_{21}ClN_2 \cdot 2HCl$ MW: 373.76

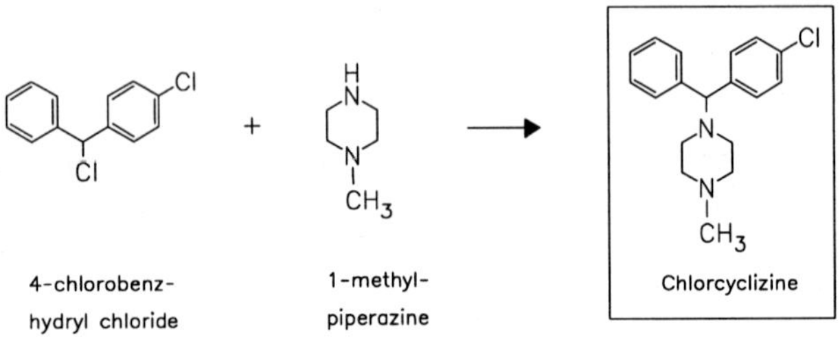

4-chlorobenz-   1-methyl-   Chlorcyclizine
hydryl chloride   piperazine

Reference(s):
US 2 630 435 (Burroughs Wellcome; 1953; prior. 1948).

Formulation(s):   tabl. 50 mg (as hydrochloride)

Chlordiazepoxide

ATC: N05BA02
Use: tranquilizer

RN: 58-25-3   MF: $C_{16}H_{14}ClN_3O$ MW: 299.76   EINECS: 200-371-0
LD_{50}:   95 mg/kg (M, i.v.); 200 mg/kg (M, p.o.);
       165 mg/kg (R, i.v.); 392 mg/kg (R, p.o.)
CN: 7-chloro-N-methyl-5-phenyl-3H-1,4-benzodiazepin-2-amine 4-oxide

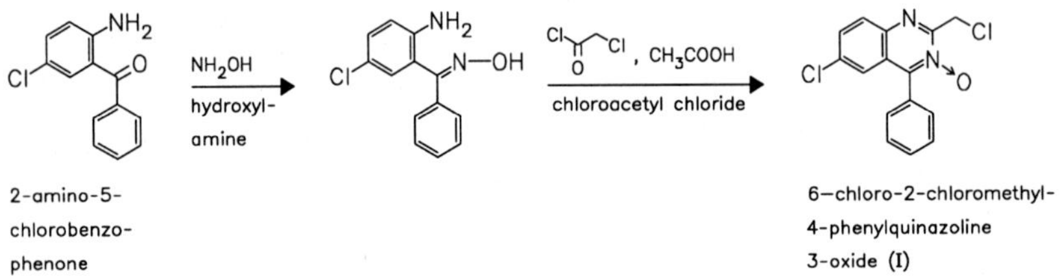

2-amino-5-   6-chloro-2-chloromethyl-
chlorobenzo-   4-phenylquinazoline
phenone   3-oxide (I)

I + H₂N—CH₃ → Chlordiazepoxide

methylamine

Reference(s):
US 2 893 992 (Hoffmann-La Roche; 7.7.1959; prior. 15.5.1958).
DE 1 096 363 (Hoffmann-La Roche; appl. 24.4.1959; USA-prior. 15.5.1958).
Sternbach, L.H. et al.: J. Org. Chem. (JOCEAH) **26**, 1111 (1961).

Formulation(s):   drg. 10 mg; f. c. tabl. 5 mg; tabl. 5 mg, 10 mg, 25 mg

Trade Name(s):

D:	Limbatril (Roche)-comb.	Psicofar (Terapeutico)	USA: Librax (Roche Products; as
	Multum (Rosen Pharma)	Reliberan (Geymonat)	hydrochloride)
	Radepur (ASTA Medica	Sedans (Ganassini)-comb.	Librium (Roche Products;
	AWD)	Seren Vita (Synthelabo)	as hydrochloride)
F:	Librax (Roche)-comb.	J: Balance (Yamanouchi)	Limbitrol (Roche Products)
GB:	Librium (Roche)	Contol (Takeda)	generics
I:	Diapatol (Teofarma)-comb.	Sophiamin (Kyowa	
	Librium (Roche)	Yakuhin)	
	Limbitryl (Roche)-comb.	Trakipearl (Hishiyama)	

Chlorhexidine

ATC: A01AB03; B05CA02; D08AC02;
                D09AA12; R02AA05; S01AX09;
                S02AA09; S03AA04
Use: antiseptic

RN: 55-56-1   MF: $C_{22}H_{30}Cl_2N_{10}$ MW: 505.46   EINECS: 200-238-7
LD_{50}: 24 mg/kg (M, i.v.); 2515 mg/kg (M, p.o.);
        21 mg/kg (R, i.v.); 9200 μL/kg (R, p.o.)
CN: *N,N''*-bis(4-chlorophenyl)-3,12-diimino-2,4,11,13-tetraazatetradecanediimidamide

diacetate
RN: 56-95-1   MF: $C_{22}H_{30}Cl_2N_{10} \cdot 2C_2H_4O_2$ MW: 625.56   EINECS: 200-302-4
LD_{50}: 25 mg/kg (M, i.v.); 2 g/kg (M, p.o.)
dihydrochloride
RN: 3697-42-5   MF: $C_{22}H_{30}Cl_2N_{10} \cdot 2HCl$ MW: 578.38   EINECS: 223-026-6
LD_{50}: >5 g/kg (M, s.c.)
di-D-gluconate
RN: 18472-51-0   MF: $C_{22}H_{30}Cl_2N_{10} \cdot 2C_6H_{12}O_7$ MW: 897.77   EINECS: 242-354-0
LD_{50}: 12.9 mg/kg (M, i.v.); 1260 mg/kg (M, p.o.);
        24.2 mg/kg (R, i.v.); 2 g/kg (R, p.o.)

a)

hexamethylenebis-
(dicyanodiamide)

4-chloroaniline
hydrochloride

130–140 °C

Chlorhexidine

b)

hexamethylenediamine
dihydrochloride

(4–chlorophenyl)–
dicyanodiamide

150–160 °C

Chlorhexidine

Reference(s):
GB 705 838 (ICI; appl. 1951; valid from 1952).

Formulation(s):   gel 1 g/100 g; powder 1 g/100 g; sol. 0.1 g/100 g, 0.2 g/100 g, 1 g/50 ml (as digluconate)

Trade Name(s):

D: Chlorhexamed (Blend-a-
        med)
        Chlorhexidindigluconat
        (Engelhard)
        Corsodyl (SmithKline
        Beecham OTC Medicines)
        Frubilurgyl (Boehringer
        Mannh.)
        Hansamed (Beiersdorf)
        numerous combination
        preparations
F: Antalyre (Boehringer Ing.;
        as gluconate)-comb.
        Collunovar (Dexo; as
        gluconate)
        Collupressine (Synthélabo;
        as gluconate)-comb.
        Collustan (Oberlin; as
        digluconate)-comb.
        Corsadyl (SmithKline
        Beecham; as digluconate)
        Cytéal (Sinbio; as
        gluconate)-comb.
        Dacryne (Johnson &
        Johnson)
        Diseptine (Nicholas; as
        gluconate)-comb.
        Eludril (Inava; as
        digluconate)-comb.

Hibidil (Zeneca; as
digluconate)
Hibiscrub (Zeneca; as
digluconate)
Hibitane (Zeneca; as
gluconate)
Merfene (Novartis; as
gluconate)
Plurexid (Evans; as
gluconate)
Prexidine (Pred; as
gluconate)
Sepéal (Sinbio; as
digluconate)
Thiovalone (Eurorga; as
diacetate)-comb.
GB: Bactigras (Smith &
        Nephew)
        Chlorhexitulle (Hoechst)
        Corsodyl (SmithKline
        Beecham)
        numerous preparations
I: Clorex (Formulario Naz.)
        Contact (Vaas; as
        hydrochloride)
        Corsodyl (SmithKline
        Beecham)
        Effetre (Farma3 Medicalex)
        Hansamed (Beiersdorf)

Hibidil (Zeneca)
Hibiscrup (Zeneca)
Hibitane (Zeneca)
Lenixil (Eurospital)
Neomercurocromo (SIT)
Neoxene (Ecobi)
Odontoxina (Ipfi)
Oramil (Ganassini)
Plak Out (Byk Gulden)
Sanoral (Kemiprogress)
Savlodil (Zeneca)-comb.
Savlol (Zeneca)-comb.
Vaxidina (Vaas)
Vidermina (Ganassini)
J: Hexadol (Green Cross)
        Hibiscrub (ICI)
        Hibitane Digluconate
        (Sumitomo)
        Maskin (Maruishi)
        Pabron (Taisho Seiyaku)
        White Gol (Tamagawa
        Eizai)
        White Rive (Eisai)
USA: Betasept (Purdue
        Frederick; as gluconate)
        Hibiclens (Zeneca; as
        gluconate)
        Hibistat (Zeneca; as
        gluconate)

Peridex (Procter &
Gamble; as gluconate)

Periogard (Colgate Oral; as
gluconate)

Chlormadinone acetate

ATC: G03D
Use: progestogen

RN: 302-22-7   MF: $C_{23}H_{29}ClO_4$ MW: 404.93   EINECS: 206-118-0
LD$_{50}$: >2 g/kg (M, i.v.); >15 g/kg (M, p.o.);
        >10 g/kg (R, p.o.)
CN: 17-(acetyloxy)-6-chloropregna-4,6-diene-3,20-dione

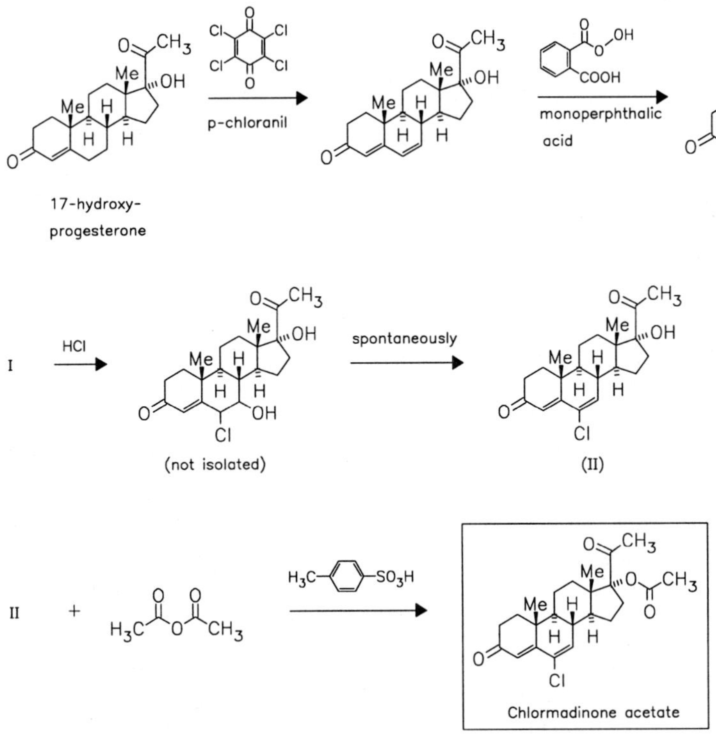

17-hydroxy-
progesterone

p-chloranil

monoperphthalic
acid

(I)

I

HCl

spontaneously

(not isolated)

(II)

II   +

Chlormadinone acetate

Reference(s):
DE 1 075 114 (E. Merck AG; appl. 29.4.1958).
Brückner, K. et al.: Chem. Ber. (CHBEAM) **94**, 1225 (1961).

Formulation(s):   tabl. 2 mg, 5 mg

Trade Name(s):
D: Chlormadinon (Jenapharm)
       Gestafortin (Merck)
       Gestamestrol (Hermal-
       Chemie)-comb.

       Menova (Merck)-comb.
       Neo-Eunomin
       (Grünenthal)-comb.
F: Lutéran (Solymés)

I: Fisiosequil (Recordati);
       wfm
J: Lutoral (Shionogi)

Chlormerodrin

ATC: C03
Use: diuretic

RN: 62-37-3   MF: $C_5H_{11}ClHgN_2O_2$ MW: 367.20   EINECS: 200-530-4
LD$_{50}$:   215 mg/kg (M, p.o.);
150 mg/kg (R, p.o.)
CN: [3-[(aminocarbonyl)amino]-2-methoxypropyl-C^1,O^3]chloromercury

allylurea   Chlormerodrin

Reference(s):
US 2 635 982 (Lakeside Labs.; 1953; prior. 1951).

Formulation(s):   amp.; tabl. 18 mg

Trade Name(s):
USA: Neohydrin (Lakeside); wfm

Chlormezanone

ATC: M03BB02
Use: muscle relaxant

RN: 80-77-3   MF: $C_{11}H_{12}ClNO_3S$ MW: 273.74   EINECS: 201-307-4
LD$_{50}$:   600 mg/kg (M, p.o.);
605 mg/kg (R, p.o.);
500 mg/kg (dog, p.o.)
CN: 2-(4-chlorophenyl)tetrahydro-3-methyl-4H-1,3-thiazin-4-one 1,1-dioxide

| 3-mercapto-propionic acid | methyl-amine | 4-chloro-benzaldehyde | 2-(4-chlorophenyl)-3-methyltetra-hydro-1,3-thiazin-4-one | Chlormezanone |

Reference(s):
GB 815 203 (Sterling Drug; appl. 3.7.1957; USA-prior. 20.7.1956).
Surrey, A.R. et al.: J. Am. Chem. Soc. (JACSAT) **80**, 3469, 3471 (1958).

Formulation(s):   suppos. 200 mg; tabl. 100 mg, 200 mg, 400 mg

Trade Name(s):
D: Muskel Trancopal (Winthrop); wfm
Muskel Trancopal comp. (Winthrop)-comb. with paracetamol; wfm

Muskel Trancopal cum codeino (Winthrop)-comb. with paracetamol and codeine phosphate; wfm
F: Alinam (Lucien); wfm

Supotran (Winthrop); wfm
Trancogésic (Winthrop)-comb. with aspirine; wfm
Trancopal (Winthrop); wfm
GB: Trancopal (Winthrop); wfm

Trancoprin (Winthrop)-
comb. with aspirine; wfm
I: Clormetadone (Nuovo
Cons. Sanit. Naz.)-comb.
Condol (Maggioni-
Winthrop)-comb.

Eblimon (Guidotti)-comb.
numerous combination
preparations
J: Myolespen (Dojin Iyaku)
Relizon (Mochida)
Trancopal (Daiichi)

Transanate (Teikoku
Hormone)
USA: Trancopal (Sanofi)

Chlormidazole

(Clomidazolum)

ATC: D01AC
Use: fungistatic, antifungal

RN: 3689-76-7   MF: $C_{15}H_{13}ClN_2$ MW: 256.74   EINECS: 222-998-9
CN: 1-[(4-chlorophenyl)methyl]-2-methyl-1H-benzimidazole

hydrochloride
RN: 54118-67-1   MF: $C_{15}H_{13}ClN_2 \cdot HCl$ MW: 293.20

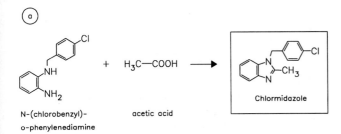

N-(chlorobenzyl)-   acetic acid
o-phenylenediamine

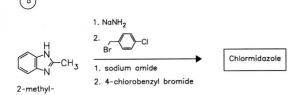

2-methyl-
benzimidazole

1. NaNH₂
1. sodium amide
2. 4-chlorobenzyl bromide

Chlormidazole

Reference(s):
US 2 876 233 (Grünenthal; 3.3.1959; prior. 29.10.1956).

Formulation(s):   cream 5 %; ointment 5 % (as hydrochloride)

Trade Name(s):
D: Myco-Jellin (Grünenthal)-
comb. with fluocinolone
acetonide, wfm

Polycid N (Grünenthal)-
comb.; wfm

Chlorobutanol

ATC: A04AD04
Use: hypnotic, anesthetic

RN: 57-15-8   MF: $C_4H_7Cl_3O$ MW: 177.46   EINECS: 200-317-6
CN: 1,1,1-trichloro-2-methyl-2-propanol

acetone   +   chloroform $\xrightarrow{\text{KOH}}$ Chlorobutanol

Reference(s):
Budesinsky-Protiva, 235.
Willgerodt, C.: Ber. Dtsch. Chem. Ges. (BDCGAS) **14**, 2451 (1881).
US 2 462 389 (Socony-Vac Oil; 1949; prior. 1946).

Formulation(s):   sol. 250 mg/100 ml

Trade Name(s):

D: Givalex (Norgine)

F: Alodont (Warner-Lambert)-
comb.
Angispray (Monot)-comb.
Balsamorhinol (Janssen)-
comb.
Ciella (RPR Cooper)-comb.
Eludril (Inava)-comb.
Givalex (Nagine Pharma)-
comb.

Liquifilm (Allergan)-comb.
Optrex (Etris)-comb.

GB: Cerumol (L.A.B.)-comb.
Eludril (Chefaro)-comb.
Monphytol (L.A.B.)-comb.

I: Abiostil (Deca)-comb.
Antipulmina (Lisapharma)
Cerumenex (ASTA
Medica)-comb.

Clorobutanolo (Tariff.
Integrativo)
Corizzina (SIT)-comb.
Desalfa (Intes)-comb.
Fialetta odontalg. Knapp
(Montefarmaco)-comb.
Oftalzina (SIT)-comb.
Respiro (Pierrel)-comb.
Rinoleina (Granelli)-comb.

Chloroprednisone acetate

ATC: H02AB; D07AB
Use: topical glucocorticoid

RN: 14066-79-6   MF: $C_{23}H_{27}ClO_6$ MW: 434.92   EINECS: 237-919-3
CN: (6α)-21-(acetyloxy)-6-chloro-17-hydroxypregna-1,4-diene-3,11,20-trione

6α-chlorocortisone $\xrightarrow{\text{SeO}_2, \ (H_3C)_3C-OH}$ Chloroprednisone acetate

Reference(s):
DE 1 079 042 (Syntex; appl. 1958; MEX-prior. 1957).
FR-M 666 (Syntex; appl. 20.9.1960).

alternative synthesis:
US 3 130 211 (Upjohn; 21.4.1964; prior. 1957, 1958).

pharmaceutical formulation:
GB 955 891 (Organon; valid from 1962; NL-prior. 1961).

Formulation(s):   cream; ointment

Trade Name(s):
USA: Adremycin (Organon);   Topilan (Syntex); wfm
      wfm

Chloroprocaine

ATC: N01BA04
Use: local anesthetic

RN: 133-16-4   MF: $C_{13}H_{19}ClN_2O_2$ MW: 270.76
CN: 4-amino-2-chlorobenzoic acid 2-(diethylamino)ethyl ester

monohydrochloride
RN: 3858-89-7   MF: $C_{13}H_{19}ClN_2O_2 \cdot HCl$ MW: 307.22   EINECS: 223-371-2
LD_{50}:   266 mg/kg (M, i.p.); 700 mg/kg (M, s.c.)

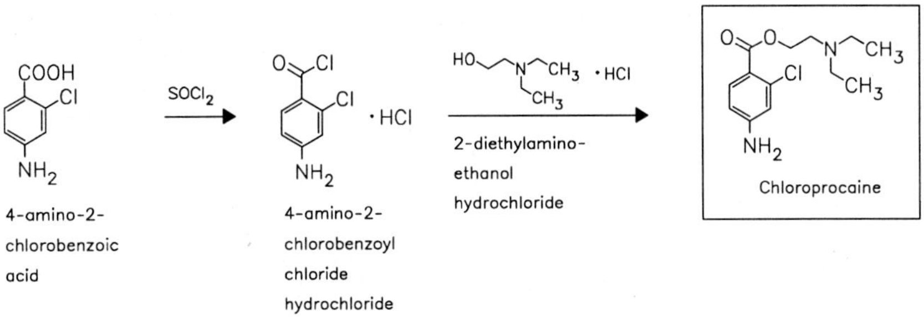

4-amino-2-chlorobenzoic acid → 4-amino-2-chlorobenzoyl chloride hydrochloride → (2-diethylamino-ethanol hydrochloride) → Chloroprocaine

Reference(s):
US 2 460 139 (Wallace & Tiernan; 1949; appl. 1945).

Formulation(s):   multiple-dose vial 1 %, 2 %; single-dose vial 2 %, 3 % (as hydrochloride)

Trade Name(s):
J: Piocaine (Teikoku Kagaku-   USA: Nesacaine (Astra)
    Nagase)   Nesacaine (Pennwalt)

Chloropyramine
(Halopyramine; Chlortripelennamine)

ATC: D04AA09; R06AC03
Use: antihistaminic

RN: 59-32-5   MF: $C_{16}H_{20}ClN_3$ MW: 289.81   EINECS: 200-421-1
LD_{50}:   24.1 mg/kg (M, i.v.); 354 mg/kg (M, p.o.);
      32.5 mg/kg (R, i.v.); 920 mg/kg (R, p.o.)
CN: N-[(4-chlorophenyl)methyl]-N',N'-dimethyl-N-2-pyridinyl-1,2-ethanediamine

monohydrochloride
RN: 6170-42-9   MF: $C_{16}H_{20}ClN_3 \cdot HCl$ MW: 326.27   EINECS: 228-216-2

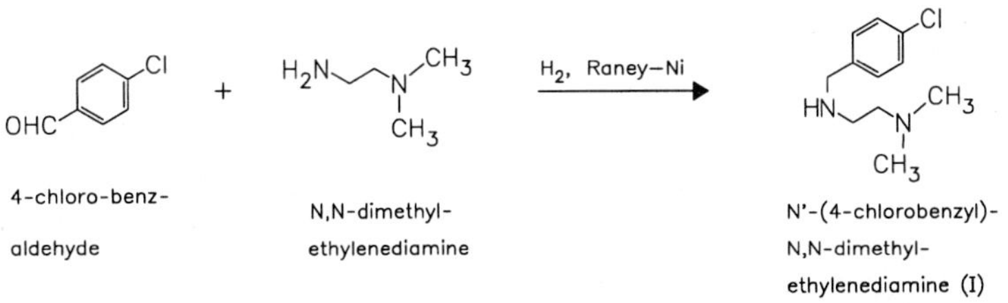

4-chloro-benz-aldehyde + N,N-dimethyl-ethylenediamine → (H₂, Raney-Ni) → N'-(4-chlorobenzyl)-N,N-dimethyl-ethylenediamine (I)

I + [2-bromo-pyridine] → [Chloropyramine]

2-bromo-
pyridine

Chloropyramine

Reference(s):
US 2 569 314 (American Cyanamid; 1951; appl. 1947).
Vaughan, J.R. et al.: J. Org. Chem. (JOCEAH) **14**, 228 (1949).

Formulation(s):   amp. 20 mg; cream 1 %; tabl. 25 mg

Trade Name(s):
D: Synpen (Geigy); wfm   I: Sinopen (Geigy); wfm

Chloropyrilene
(Chlorthenylpyramine)

ATC: R06AC
Use: antihistaminic

RN: 148-65-2   MF: $C_{14}H_{18}ClN_3S$ MW: 295.84
LD_{50}: 105 mg/kg (M, i.p.)
CN: N-[(5-chloro-2-thienyl)methyl]-N',N'-dimethyl-N-2-pyridinyl-1,2-ethanediamine

citrate (1:1)
RN: 148-64-1   MF: $C_{14}H_{18}ClN_3S \cdot C_6H_8O_7$ MW: 487.96   EINECS: 205-720-0
monohydrochloride
RN: 135-35-3   MF: $C_{14}H_{18}ClN_3S \cdot HCl$ MW: 332.30
LD_{50}: 438 mg/kg (M, p.o.)

[5-chloro-2-chloromethyl-thiophene] + [N,N-dimethyl-N'-(2-pyridyl)ethylene-diamine] → NaNH₂, toluene / sodium amide → [Chloropyrilene]

5-chloro-2-
chloromethyl-
thiophene

N,N-dimethyl-N'-
(2-pyridyl)ethylene-
diamine

NaNH₂, toluene
sodium amide

Chloropyrilene

Reference(s):
(cf. thenyldiamine, methapyrilene)
US 2 581 868 (Monsanto; 1952; prior. 1946).
Clapp, R.C. et al.: J. Org. Chem. (JOCEAH) **14**, 216 (1949).
Clapp, R.C. et al.: J. Am. Chem. Soc. (JACSAT) **69**, 1549 (1947).

Formulation(s):   tabl. 25 mg

Trade Name(s):
I: Brevirina (Prodatti Erma)-   Panta-Valeas (Valeas); wfm   Tagathen (Lederle; as
      comb.; wfm   USA: Tagathen (Lederle); wfm   citrate); wfm

Chloroquine

ATC: P01BA01
Use: antirheumatic, antimalarial

RN: 54-05-7 MF: C$_{18}$H$_{26}$ClN$_3$ MW: 319.88 EINECS: 200-191-2
LD$_{50}$: 21.6 mg/kg (M, i.v.); 311 mg/kg (M, p.o.);
60 mg/kg (R, i.v.); 330 mg/kg (R, p.o.)
CN: N^4-(7-chloro-4-quinolinyl)-N^1,N^1-diethyl-1,4-pentanediamine

diphosphate
RN: 50-63-5 MF: C$_{18}$H$_{26}$ClN$_3$ · 2H$_3$PO$_4$ MW: 515.87 EINECS: 200-055-2
LD$_{50}$: 500 mg/kg (M, p.o.)
sulfate (1:1)
RN: 132-73-0 MF: C$_{18}$H$_{26}$ClN$_3$ · H$_2$SO$_4$ MW: 417.96 EINECS: 205-077-6
sulfate (1:1) monohydrate
RN: 6823-83-2 MF: C$_{18}$H$_{26}$ClN$_3$ · H$_2$O$_4$S · H$_2$O MW: 435.97
dihydrochloride
RN: 3545-67-3 MF: C$_{18}$H$_{26}$ClN$_3$ · 2HCl MW: 392.80 EINECS: 222-592-1
2,5-dihydroxybenzoate
RN: 16510-14-8 MF: C$_{18}$H$_{26}$ClN$_3$ · xC$_7$H$_6$O$_4$ MW: unspecified EINECS: 240-578-3
diorotate
RN: 16301-30-7 MF: C$_{18}$H$_{26}$ClN$_3$ · 2C$_5$H$_4$N$_2$O$_4$ MW: 632.07 EINECS: 240-389-6
LD$_{50}$: 1130 mg/kg (M, p.o.)

starting products:
1. 4.7-Dichloroquinoline

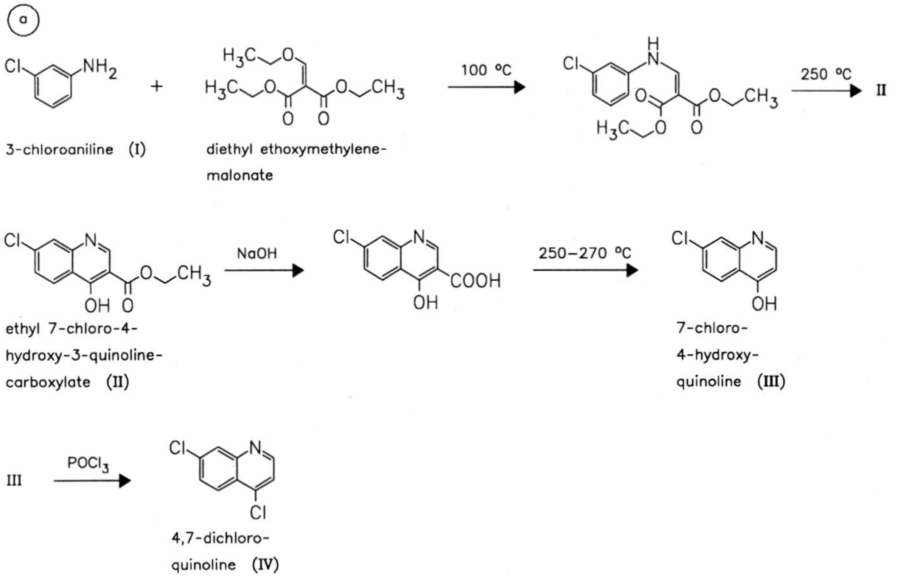

a

3-chloroaniline (I) + diethyl ethoxymethylene-malonate → 100 °C → 250 °C → II

ethyl 7-chloro-4-hydroxy-3-quinoline-carboxylate (II) → NaOH → 250–270 °C → 7-chloro-4-hydroxy-quinoline (III)

III → POCl$_3$ → 4,7-dichloro-quinoline (IV)

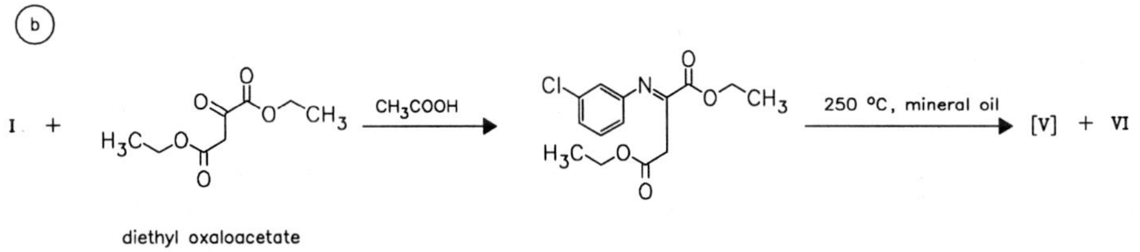

b

I + diethyl oxaloacetate → CH$_3$COOH → 250 °C, mineral oil → [V] + VI

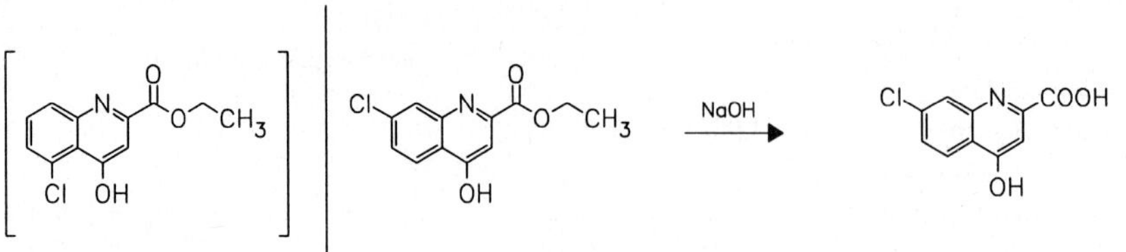

ethyl 5-chloro-4-
hydroxy-2-quinoline-
carboxylate (V)
(removal by crystallization
from acetic acid)

ethyl 7-chloro-4-
hydroxy-2-quinoline-
carboxylate (VI)

(VII)

VII $\xrightarrow{\text{270 °C, mineral oil}}$ III $\xrightarrow{\text{POCl}_3}$ IV

c

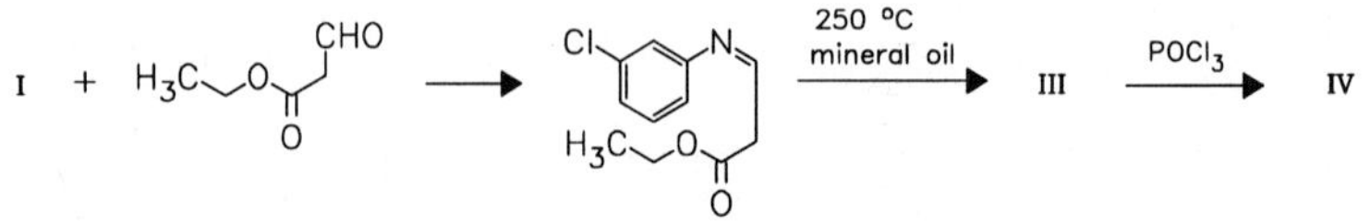

I + H₃C formyl-acetate ethyl formyl-
acetate

$\xrightarrow{\text{250 °C}\atop\text{mineral oil}}$ III $\xrightarrow{\text{POCl}_3}$ IV

2. Novoldiamine

a

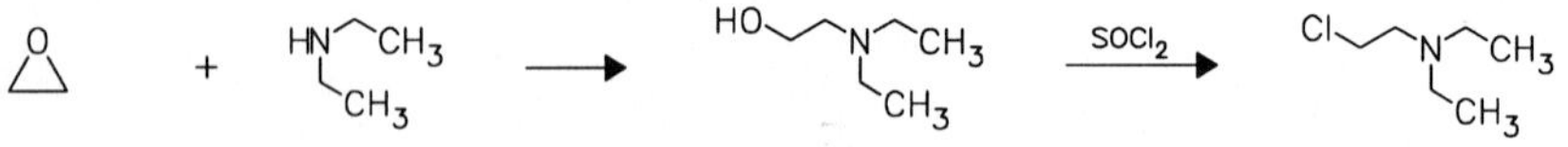

ethylene
oxide (VIII)

diethyl
amine (IX)

2-diethylamino-
ethanol $\xrightarrow{\text{SOCl}_2}$

2-diethylamino-
ethyl chloride (X)

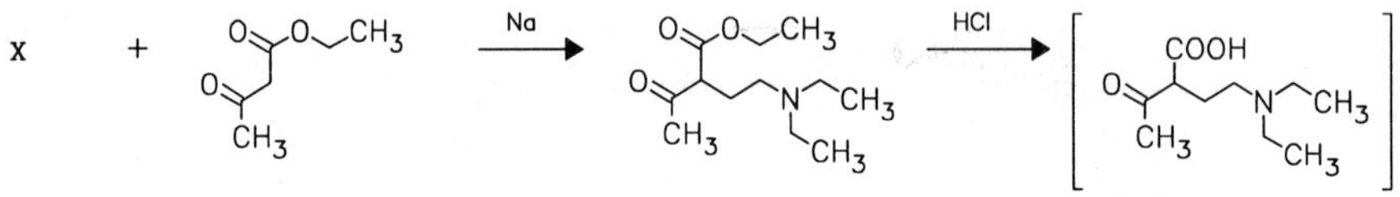

X + ethyl
acetoacetate (XI) $\xrightarrow{\text{Na}}$ $\xrightarrow{\text{HCl}}$ (XII)

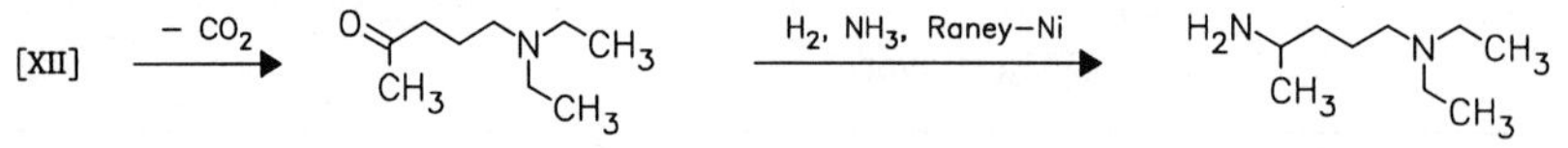

[XII] $\xrightarrow{-\text{CO}_2}$ 1-diethylamino-
4-pentanone (XIII) $\xrightarrow{\text{H}_2, \text{NH}_3, \text{Raney-Ni}}$ novoldiamine (XIV)

b

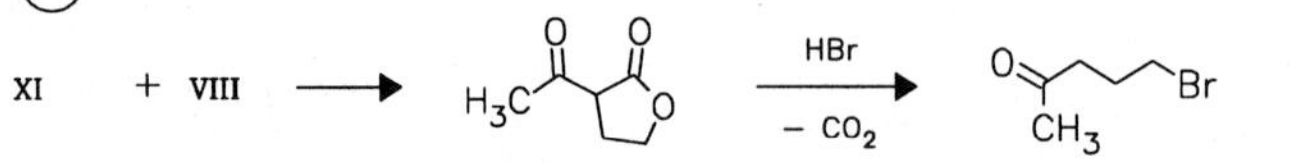

XI + VIII $\longrightarrow$ 2-acetyl-
butyrolactone $\xrightarrow[-\text{CO}_2]{\text{HBr}}$ 1-bromo-
4-pentanone (XV)

XV + IX $\longrightarrow$ XIII $\xrightarrow{\text{H}_2, \text{NH}_3, \text{Raney-Ni}}$ XIV final product:

Chloroquine

IV   +   XIV $\xrightarrow{180°C}$

Chloroquine

Reference(s):
US 2 233 970 (Winthrop; 1941; D-prior. 1937).
DRP 683 692 (I. G. Farben; appl. 1937).
Drake, N.L. et al.: J. Am. Chem. Soc. (JACSAT) **68**, 1214 (1946).
1a Price, C.C.; Roberts, R.M.: J. Am. Chem. Soc. (JACSAT) **68**, 1204 (1946).
   DD 53 065 (S. Schwarz et al.; appl. 1966).
b   Surrey, A.R.; Hammer, H.F.: J. Am. Chem. Soc. (JACSAT) **68**, 113 (1946).
c   US 2 478 125 (American Cyanamid; 1949; appl. 1944).
2a DRP 486 079 (I. G. Farben; appl. 1924).
b   Elderfield, R.C. et al.: J. Am. Chem. Soc. (JACSAT) **68**, 1579 (1946).

alternative syntheses of novoldiamine:
US 2 365 825 (Monsanto; 1944; appl. 1942).
GB 1 157 637 (Sterling Drug; appl. 1966; USA-prior. 1965).

aminating hydrogenation of novolketone, *continuous method:*
DOS 2 923 472 (Bayer; appl. 9.6.1979).

alternative synthesis of 4,7-dichloroquinoline *from* 3-chloroaniline *and acrylic acid ester:*
FR 1 514 280 (Roussel-Uclaf; appl. 10.1.1967).
EP 56 765 (Rhône-Poulenc; appl. 15.1.1982; F-prior. 16.1.1981).

alternative synthesis of chloroquine *from* 7-chloro-4-oxo-1,2,3,4-tetrahydroquinoline *and* novoldiamine:
EP 56 766 (Rhône-Poulenc; appl. 15.1.1982; F-prior. 16.1.1981).

chlorination of 7-chloro-4-hydroxyquinoline *with* benzotrichloride:
DOS 3 112 415 (Dynamit Nobel; appl. 28.3.1981).

Formulation(s):   amp. 250 mg/5 ml; syrup 15 mg; tabl. 50 mg, 155 mg, 300 mg (as phosphate)

Trade Name(s):

D:	Resochin (Bayer Vital)	GB:	Avloclor (Zeneca)	USA: Aralen (Sanofi; as
F:	Nivaquine (Rhône-Poulenc		Nivaquine (Rhône-Poulenc	hydrochloride)
	Rorer Specia)		Rorer)	Aralen (Sanofi; as
	Savarine (Zeneca Pharma)-	I:	Cloroc (Formulario Naz.)	phosphate)
	comb.		Clorochina (Bayer)	

Chlorothiazide

ATC: C03AA04
Use: diuretic

RN: 58-94-6   MF: $C_7H_6ClN_3O_4S_2$ MW: 295.73   EINECS: 200-404-9
LD_{50}:   940 mg/kg (M, i.v.); 8 g/kg (M, p.o.);
    200 mg/kg (R, i.v.); 10 g/kg (R, p.o.);
    1 g/kg (dog, i.v.)
CN: 6-chloro-2*H*-1,2,4-benzothiadiazine-7-sulfonamide 1,1-dioxide

sodium salt
RN: 7085-44-1   MF: $C_7H_5ClN_3NaO_4S_2$ MW: 317.71

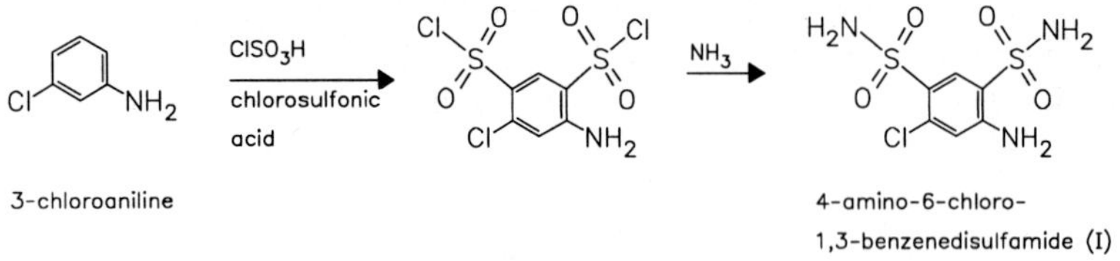

3-chloroaniline

4-amino-6-chloro-
1,3-benzenedisulfamide (I)

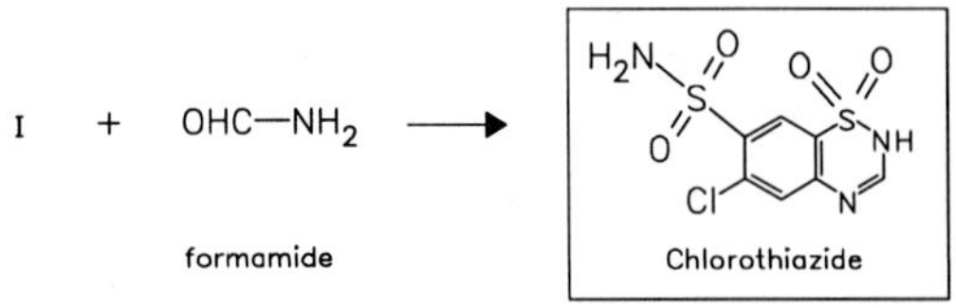

I   +   OHC—NH₂   ⟶

formamide

Chlorothiazide

Reference(s):
US 2 809 194 (Merck & Co.; 8.10.1957; prior. 2.5.1956).
US 2 937 169 (Merck & Co.; 17.5.1960; prior. 25.9.1958).
Novello, E.C.; Sprague, J.M.: J. Am. Chem. Soc. (JACSAT) **79**, 2028 (1957).

alternative synthesis of 4-amino-6-chloro-1,3-benzenedisulfamide *(chlorosulfonation of* 1,3-dichlorobenzene *and subsequent reaction with* ammonia*):*
DE 1 119 290 (Hoechst; appl. 7.11.1959).

Formulation(s):   amp. 500 mg/20 ml (as sodium salt); tabl. 250 mg, 500 mg

Trade Name(s):

D:	Chlotride (Sharp & Dohme); wfm	I:	Clotride (Merck Sharp & Dohme); wfm	USA: Aldochlor (Merck)
F:	Diupreskal (Théraplix)-comb.; wfm		Saluren (Croce Bianca); wfm	Diupres (Merck)
	Diurilix (Théraplix)-comb.; wfm	J:	Aldoclor (Merck Sharp & Dohme)-comb. with methyldopa	Diuril (Merck; as sodium salt)
GB:	Saluric (Merck Sharp & Dohme)		Chlotride (Merck-Banyu)	generics

---

Chlorotrianisene

ATC: G03CA06
Use: synthetic estrogen

RN: 569-57-3   MF: C₂₃H₂₁ClO₃   MW: 380.87   EINECS: 209-318-6
CN: 1,1',1''-(1-chloro-1-ethenyl-2-ylidene)tris[4-methoxybenzene]

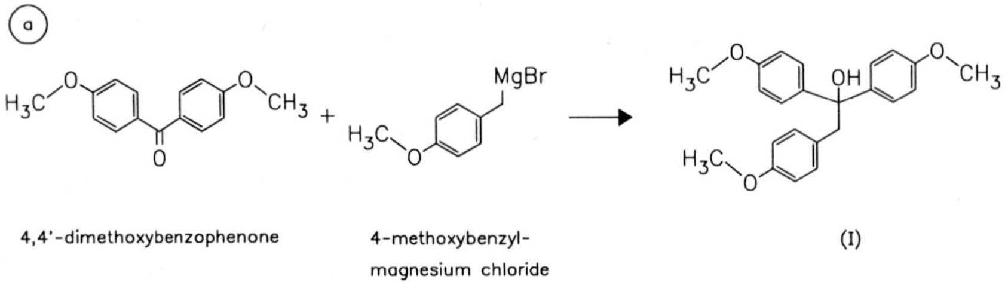

4,4'-dimethoxybenzophenone

4-methoxybenzyl-
magnesium chloride

(I)

Chlorotrianisene

b

I $\xrightarrow{Cl_2}$ Chlorotrianisene

Reference(s):
a   US 2 430 891 (Merrell; 1947; prior. 1941).
b   BE 561 508 (ICI; valid from 1943; prior. 1942).

Formulation(s):   cps. 12 mg, 24 mg, 72 mg

Trade Name(s):

D:	Merbentul (Marion Merrell); wfm		Tace-FN (Merrell-Toraude); wfm	USA:	Chlotride (Merck & Co.); wfm
F:	Tace (Merrell Dow); wfm	GB:	Tace (Merrell); wfm		Diuril (Merck & Co.); wfm
		I:	Anisene (Farmila); wfm		Tace (Merrell Dow); wfm

Chloroxylenol

(Parachlorometaxylenol)

ATC: D08AE05
Use: antiseptic

RN: 88-04-0   MF: C$_8$H$_9$ClO   MW: 156.61   EINECS: 201-793-8
LD$_{50}$:   1 g/kg (M, p.o.);
          3830 mg/kg (R, p.o.)
CN: 4-chloro-3,5-dimethylphenol

3,5-dimethyl-
phenol

$\xrightarrow[\text{sulfuryl chloride}]{\text{Cl}_2 \text{ or SO}_2\text{Cl}_2}$

Chloroxylenol

Reference(s):
US 2 350 677 (W. Wiggins Cocker; 1944; GB-prior. 1939).

Formulation(s):   cream 0.33 g/100 g; powder 0.33 g/100 g (combination); sol. 1 g/100 g (combination)

Trade Name(s):

D:	Bacillotox (Bode)-comb. Gehwol (Gerlach)-comb.	I:	Dettol (Manetti Roberts); wfm		Cortic (Everett) Zoto HC (Horizon)
GB:	Rinstead (Schering-Plough)-comb. Zeasorb (Stiefel)-comb.		Foille (Delalande Isnardi)-comb.		
		USA:	Cortane-B (Blansett)		

Chlorphenamine
(Chlorpheniramine)

ATC: R06AB04
Use: antihistaminic

RN: 132-22-9   MF: $C_{16}H_{19}ClN_2$ MW: 274.80   EINECS: 205-054-0
LD$_{50}$:   20 mg/kg (M, i.v.); 121 mg/kg (M, p.o.);
         118 mg/kg (R, p.o.)
CN: γ-(4-chlorophenyl)-N,N-dimethyl-2-pyridinepropanamine

maleate (1:1)
RN: 113-92-8   MF: $C_{16}H_{19}ClN_2 \cdot C_4H_4O_4$ MW: 390.87   EINECS: 204-037-5
LD$_{50}$:   26.1 mg/kg (M, i.v.); 130 mg/kg (M, p.o.);
         306 mg/kg (R, p.o.);
         97.6 mg/kg (dog, i.v.)

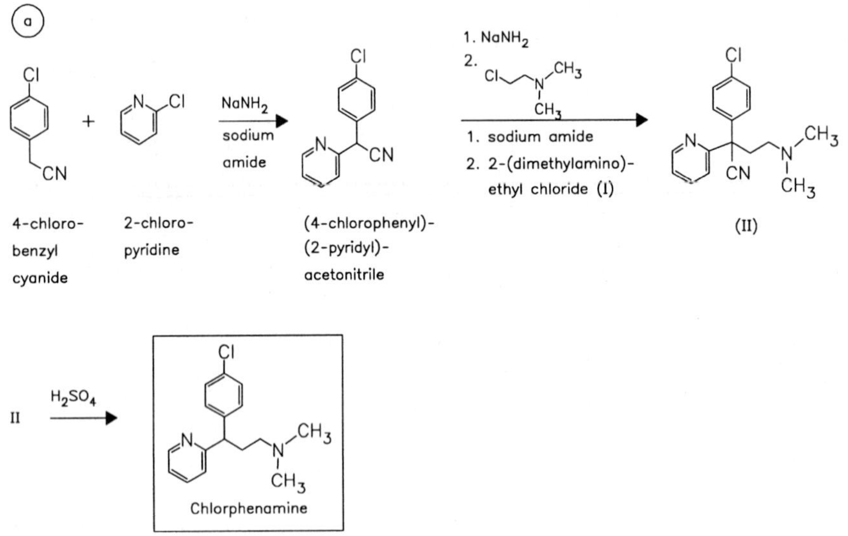

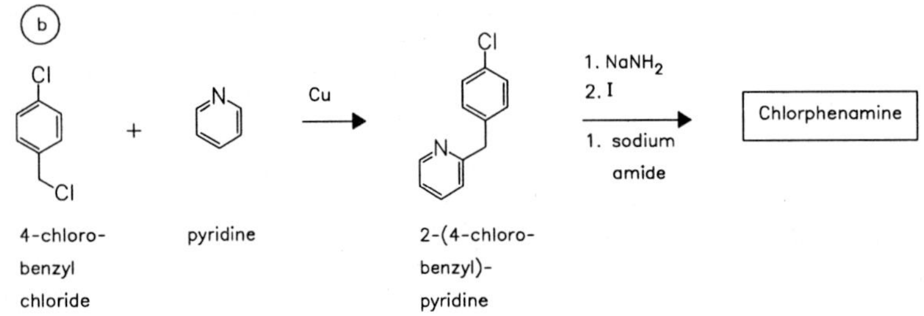

4-chloro-   2-chloro-   (4-chlorophenyl)-
benzyl   pyridine   (2-pyridyl)-
cyanide   acetonitrile

(II)

Chlorphenamine

4-chloro-   pyridine   2-(4-chloro-
benzyl   benzyl)-
chloride   pyridine

Chlorphenamine

Reference(s):
US 2 567 245 (Schering Corp.; 1951; prior. 1948).
US 2 676 964 (Schering Corp.; 1954; prior. 1950).

Formulation(s):   amp. 10 mg; cps. 2.5 mg, 4 mg, 8 mg (as maleate); syrup 3 mg/15 ml

Trade Name(s):
D: Balkis (Dolorgiet)-comb.
      Codicaps (Thiemann)-
      comb.

      Contac (SmithKline
      Beecham)
      Grippostad (Stada)

F:

      Sedotussin (Rodleben,
      UCB, Vedim)
      Arpha (Fournier SCA)-
      comb.

Bronchalène (Martin)-
comb.
Hexapneumine (Doms-
Adrian)-comb.
Hyrvalan (Monot)-comb.
Pneumopan (SmithKline
Beecham)-comb.
Poroncorinol (Roche
Nicholas)-comb.
Rhinofebral (Martin)-
comb.
Rumicine (Schering-
Plough)-comb.
Sup-Rhinite (SmithKline
Beecham)
GB: Galepsend (Galen; as
maleate)-comb.
Haymine (Pharmax; as
maleate)-comb.
Piriton (Stafford-Miller)
I: Fienamina (Recordati)-
comb.
Lentostamin (SIT)
Neocoricidin (Schering-
Plough)-comb.
Rectocoricidin (Schering-
Plough)-comb.
Trimeton (Schering-
Plough)
combination preparations
J: Allergin (Sankyo)
Atalis-D (Kanto-Isei)
Bismilla (Fuso)
Chlodamin (Maruko)
Chlor-Trimeton (Schering)
Lekrica (Yoshitomi)

Neorestamin (Kowa)
Poracemin (Horita)
USA: Ah Chew (We; as maleate)
Ana-Kit (Bayer Allergy; as
maleate)
Anaplex (ECR; as maleate)
Atrohist (Medeva; as
maleate)
Atrohist (Medeva; as
tannate)
Brexin (Savage; as maleate)
Codimal (Schwarz; as
maleate)
Co-Pyronil (Dista; as
maleate)
Cura-Vent/DA (Dura; as
maleate)
D.A. II (Dura; as maleate)
Dallergyl (Laser; as
maleate)
Donatussin (Laser; as
maleate)
Endal (Forest; as maleate)
Extendryl (Fleming; as
maleate)
Fedahist (Schwarz; as
maleate)
Histussin (Sanofi; as
maleate)
Hycamine (Endo; as
maleate)
Hydrocodone
(Pharmaceutical
Associates; as maleate)
Kronofec (Ferndale; as
maleate)

Mescolor (Horizon; as
maleate)
Nalex-A (Blansett; as
maleate)
ND (Seatrace; as maleate)
Notamine (Carnrick; as
maleate)
Omnihist (We; as maleate)
Ornace (SmithKline
Beecham; as maleate)
Pediacof (Sanofi; as
maleate)
Protid (Lunsco; as maleate)
Rescon (Ion; as maleate)
Respa ARM (Respa; as
maleate)
Rynaton (Wallace; as
tannate)
Rynatuss (Wallace; as
tannate)
Sinulin (Cernick; as
maleate)
Sinutas Sinus Allergy MS
(Warner-Lambert; as
maleate)
Tamafed (Horizon; as
tannate)
Triotann (Duramed; as
tannate)
Tuss (Seatrace; as maleate)
Tussar (Rhône-Poulenc
Rorer; as maleate)
Tussend (Monarch; as
maleate)
Tylenol (McNeil; as
maleate)

Chlorphenesin

ATC: D01AE07
Use: antifungal

RN: 104-29-0 MF: C$_9$H$_{11}$ClO$_3$ MW: 202.64 EINECS: 203-192-6
LD$_{50}$: 911 mg/kg (M, s.c.)
CN: 3-(4-chlorophenoxy)-1,2-propanediol

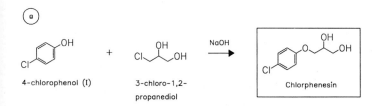

4-chlorophenol (I)   3-chloro-1,2-
                          propanediol   Chlorphenesin

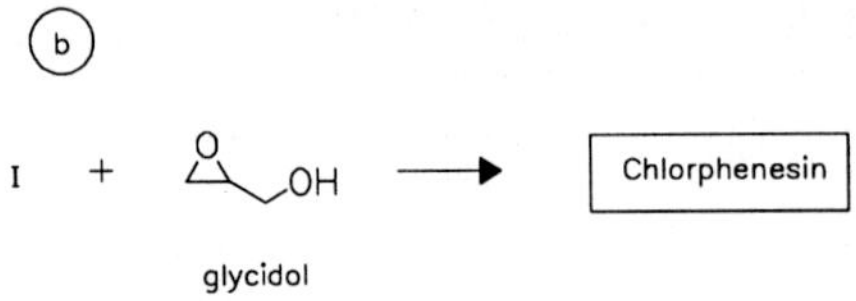

(b)

I + glycidol → Chlorphenesin

Reference(s):
GB 628 497 (British Drug Houses; appl. 1948).

Formulation(s):   cream 10 mg/1 g; vaginal suppos. 10 mg

Trade Name(s):
D: Soorphenesin (Kade)   Miol Cream
      Soorphenesin H (Kade)-   (Comprehensive)-comb.;
      comb.   wfm
GB: Aero-Mycil (Duncan,   Mycil (Duncan, Flockhart);
      Flockhart); wfm   wfm

Chlorphenesin carbamate

ATC: D01AE07
Use: analgesic, muscle relaxant,
        tranquilizer

RN: 886-74-8   MF: $C_{10}H_{12}ClNO_4$ MW: 245.66   EINECS: 212-954-7
LD$_{50}$:   239 mg/kg (M, i.v.); 807 mg/kg (M, p.o.);
       236 mg/kg (R, i.v.); 744 mg/kg (R, p.o.)
CN: 3-(4-chlorophenoxy)-1,2-propanediol 1-carbamate

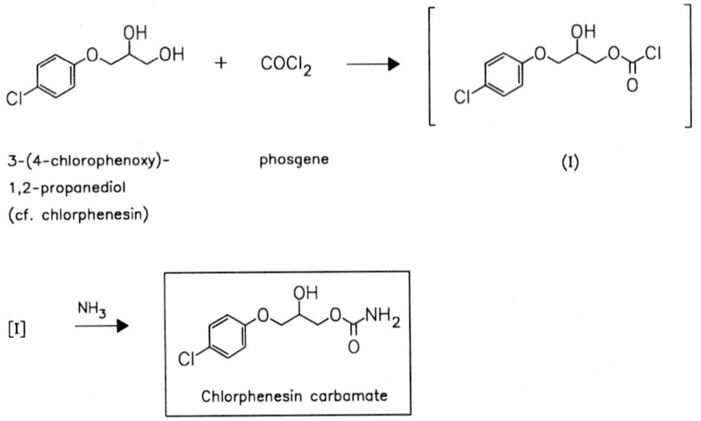

3-(4-chlorophenoxy)-   phosgene   (I)
1,2-propanediol
(cf. chlorphenesin)

Chlorphenesin carbamate

Reference(s):
US 3 161 567 (Upjohn; 15.12.1964; prior. 29.5.1963; medical use).
US 3 214 336 (Upjohn; 26.10.1965; prior. 26.8.1960).

Formulation(s):   tabl. 400 mg

Trade Name(s):
USA: Maolate (Upjohn); wfm

Chlorphenoxamine

ATC: D04AA34; R06AA06
Use: antihistaminic

RN: 77-38-3 MF: $C_{18}H_{22}ClNO$ MW: 303.83
LD_{50}: 376 mg/kg (M, p.o.)
CN: 2-[1-(4-chlorophenyl)-1-phenylethoxy]-*N,N*-dimethylethanamine

hydrochloride
RN: 562-09-4 MF: $C_{18}H_{22}ClNO \cdot HCl$ MW: 340.29 EINECS: 209-227-1
LD_{50}: 44 mg/kg (M, i.v.); 345 mg/kg (M, p.o.);
     1 g/kg (R, p.o.);
     30.8 mg/kg (dog, i.v.)

4-chloro-
benzophenone

methyl-
magnesium
chloride

1-(4-chlorophenyl)-
1-phenylethanol

1. sodium amide
2. 2-(dimethylamino)-
   ethyl chloride

Chlorphenoxamine

Reference(s):
US 2 785 202 (ASTA-Werke; 12.3.1957; D-prior. 1952).
DE 1 009 193 (ASTA-Werke; appl. 1955).

Formulation(s):   cream 15 mg/g; drg. 20 mg, 30 mg (combination); gel 15 mg/g; suppos. 24 mg, 60 mg;
               tabl. 20 mg

Trade Name(s):
D: Rodavan (ASTA Medica)-
       comb.
       Systral (ASTA Medica)

   Systral (ASTA Medica)-
   comb.
F: Systral (Lucien); wfm

GB: Clorevan (Evans); wfm
J: Systral (Kyorin)
USA: Phenoxene (Dow); wfm

Chlorpromazine

ATC: N05AA01
Use: antipsychotic, neuroleptic,
      psychosedative

RN: 50-53-3 MF: $C_{17}H_{19}ClN_2S$ MW: 318.87 EINECS: 200-045-8
LD_{50}: 16 mg/kg (M, i.v.); 135 mg/kg (M, p.o.);
     23 mg/kg (R, i.v.); 142 mg/kg (R, p.o.);
     30 mg/kg (dog, i.v.)
CN: 2-chloro *N,N* dimethyl 10*H* phenothiazine 10 propanamine

monohydrochloride
RN: 69-09-0 MF: $C_{17}H_{19}ClN_2S \cdot HCl$ MW: 355.33 EINECS: 200-701-3
LD_{50}: 20 mg/kg (M, i.v.); 135 mg/kg (M, p.o.);
     25 mg/kg (R, i.v.); 145 mg/kg (R, p.o.)

2-chloro-
phenothiazine

3-dimethylamino-
propyl chloride

Chlorpromazine

Reference(s):
US 2 645 640 (Rhône-Poulenc; 1953; F-prior. 1950).
DE 910 301 (Rhône-Poulenc; appl. 1951; F-prior. 1950).

Formulation(s):   amp. 25 mg/ml, 50 mg/2 ml; drops 20 mg/ml; suppos. 25 mg, 100 mg; syrup 10 mg/5 ml; tabl.
                   10 mg, 25 mg, 50 mg, 100 mg, 200 mg (as hydrochloride)

Trade Name(s):
D: Propaphenin (Rodleben)   Prozin (Lusofarmaco)   Promexin (Meiji)
F: Largactil (Rhône-Poulenc   J: Acemin (Sanko)   Wintermin (Shionogi)
       Rorer Specia)   Contomin (Yoshitomi)   USA: Thorazine (SmithKline
GB: Largactil (Rhône-Poulenc   Copormin (Kaken)   Beecham)
       Rorer)   Doimazin (Nippon   Thorazine (SmithKline
I: Clorpr (Formulario Naz.;   Shinyaku)   Beecham; as
       Biologici Italia; Sifra)   Epokuhl (Kyowa)   hydrochloride)
       Largactil (Rhône-Poulenc   Ishitomin (Kanto)
       Rorer)   Norcozine (Iwaki)

Chlorpropamide

ATC: A10BB02
Use: antidiabetic

RN: 94-20-2   MF: $C_{10}H_{13}ClN_2O_3S$ MW: 276.74   EINECS: 202-314-5
LD$_{50}$:   500 mg/kg (M, i.v.); 1100 mg/kg (M, p.o.);
       590 mg/kg (R, i.v.); 2150 mg/kg (R, p.o.)
CN: 4-chloro-N-[(propylamino)carbonyl]benzenesulfonamide

4-chlorobenzene-
sulfonamide

propyl isocyanate

Chlorpropamide

Reference(s):
US 3 013 072 (Pfizer; 1961; prior. 1958).
US 3 349 124 (Pfizer; 24.10.1967; prior. 20.5.1957).
Ruschig, H. et al.: Arzneim.-Forsch. (ARZNAD) 8, 448 (1958).

Formulation(s):   tabl. 100 mg, 250 mg

Trade Name(s):
D: Chloronase (Hoechst); wfm   I: Clorprop (Formulario Naz.)   J: Abemide (Kobayashi
       Diabetoral (Boehringer   Diabemide (Guidotti)   Kako)
       Mannh.); wfm   Diabexan (Crosara)   Arodoc-C (Sawai)
F: Diabinèse (Pfizer)   Pleiamide (Guidotti)-comb.   Chloronase (Hoechst)
GB: Diabinese (Pfizer)   Diabinese (Taito Pfizer)

Diamide (Kanto)	Shuabate (Toyama)	USA: Diabinese (Pfizer)
Mellitos C (Ono)	Toyomelin (Toyo Jozo)	

Chlorprothixene

ATC: N05AF03
Use: neuroleptic

RN: 113-59-7 MF: $C_{18}H_{18}ClNS$ MW: 315.87 EINECS: 204-032-8
LD_{50}: 36 mg/kg (M, i.v.); 50.1 mg/kg (M, p.o.);
200 mg/kg (R, p.o.)
CN: (*Z*)-3-(2-chloro-9*H*-thioxanthen-9-ylidene)-*N*,*N*-dimethyl-1-propanamine

hydrochloride
RN: 6469-93-8 MF: $C_{18}H_{18}ClNS \cdot HCl$ MW: 352.33 EINECS: 229-289-3
LD_{50}: 42.4 mg/kg (M, i.v.); 242 mg/kg (M, p.o.)
acetate
RN: 58889-16-0 MF: $C_{18}H_{18}ClNS \cdot C_2H_4O_2$ MW: 375.92

(a)

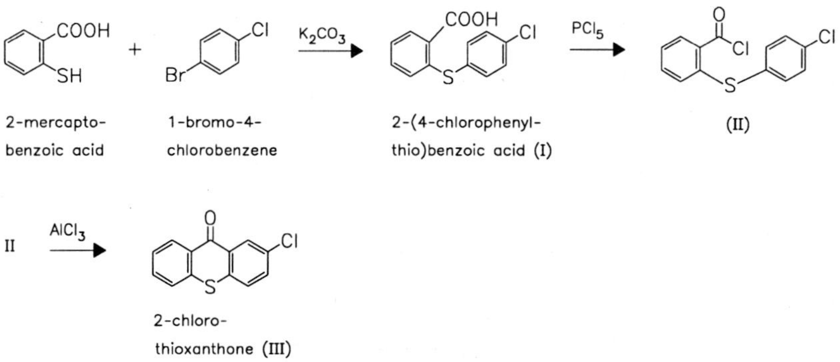
COOH
SH
Br
Cl
K2CO3
PCl5
S
2-mercapto-benzoic acid
1-bromo-4-chlorobenzene
2-(4-chlorophenyl-thio)benzoic acid (I)
(II)
II
AlCl3
2-chloro-thioxanthone (III)

(b)

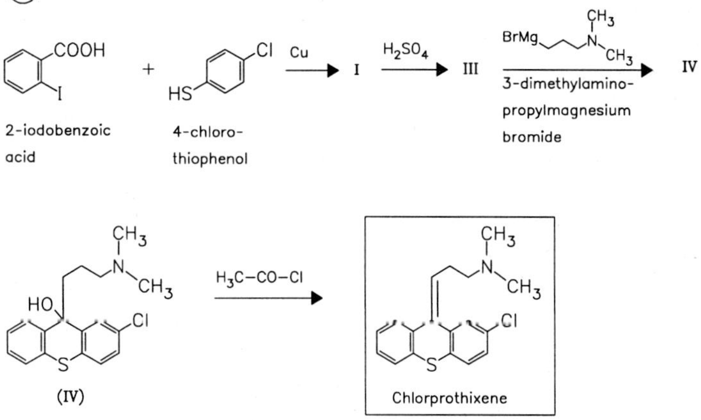
COOH
I
HS
Cl
Cu
I
H2SO4
III
BrMg
CH3
N
IV
3-dimethylamino-propylmagnesium bromide
2-iodobenzoic acid
4-chloro-thiophenol
HO
S
(IV)
H3C–CO–Cl
Chlorprothixene

(c)

allylmagnesium
bromide

2-chloro-9-(2-
propen-1-ylidene)-
thioxanthene (V)

V + CH₃ 100 °C
   HN   ────→ [Chlorprothixene]
     CH₃

dimethyl-
amine

Reference(s):
a DE 1 044 103 (Hoffmann-La Roche; appl. 22.5.1957; CH-prior. 12.6.1956, 29.6.1956, 5.7.1956).
   CH 349 617 (Hoffmann-La Roche; appl. 29.6.1956).
   BE 558 171 (Hoffmann-La Roche; appl. 6.6.1957; CH-prior. 12.6.1956, 29.6.1956, 5.7.1956).
b US 2 951 082 (Merck & Co.; 30.8.1960; prior. 9.7.1956).
c US 3 116 291 (Kefalas; 31.12.1963; DK-prior. 4.12.1958).
   DE 1 168 446 (Kefalas; appl. 1959; DK-prior. 1958).
   DE 1 418 517 (Kefalas; appl. 1959; DK-prior. 1958).

separation of isomers:
US 3 115 502 (Roche; 24.12.1963; CH-prior. 19.6.1959).

alternative synthesis:
DE 1 162 382 (Kefalas; appl. 1959; DK-prior. 1958).

isomerization:
DE 1 190 955 (Roche; appl. 1960; CH-prior. 1959).

review:
Bonricino, G.E. et al.: J. Org. Chem. (JOCEAH) **26**, 2383 (1961).

alternative synthesis:
DOS 1 918 739 (Egyesült; appl. 12.4.1969; H-prior. 12.4.1968).

Formulation(s):   amp. 50 mg/ml; drg. 15 mg, 50 mg; f. c. tabl. 15 mg, 50 mg, 100 mg; liquid 40 mg;
                     sol. 20 mg/ml; susp. 20 mg/ml

Trade Name(s):
D: Truxal (Promonta   GB: Taractan (Roche); wfm   Tra-Quilan (Eisai)
      Lundbeck)   I: Taractan (Roche); wfm   Truxal (Toyama)
F: Taractan (Roche); wfm   J: Chlothixen (Yoshitomi)   USA: Taractan (Roche); wfm

Chlorquinaldol
(Clorquinaldol)

ATC: D08AH02; G01AC03; P01AA04;
        R02AA11
Use: antiseptic, antifungal

RN: 72-80-0   MF: C₁₀H₇Cl₂NO   MW: 228.08   EINECS: 200-789-3
LD₅₀: 660 mg/kg (R, p.o.);
       2250 mg/kg (dog, p.o.)
CN: 5,7-dichloro-2-methyl-8-quinolinol

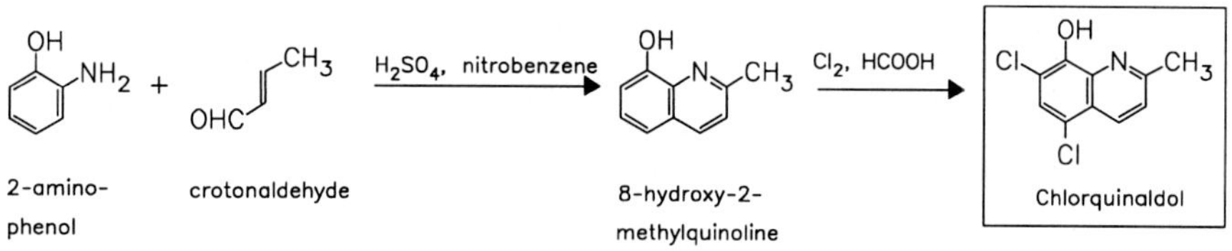

2-amino-   crotonaldehyde   8-hydroxy-2-   Chlorquinaldol
phenol   methylquinoline

Reference(s):

US 2 411 670 (Geigy; 1946; CH-prior. 1942).

Bourquin, J.-P. et al.: Arch. Pharm. Ber. Dtsch. Pharm. Ges. (APBDAJ) **295**, 383 (1962).

Formulation(s):   cream 10 mg, 130 mg

Trade Name(s):

D: Nerisona (Schering)-comb.   Siogène (Geigy); wfm   J: Lonjee (Sampo)-comb.
      Proctaspre (Henning)-   Sterosan (Geigy); wfm   Rub-All T (Toyama)-comb.
      comb.   GB: Lacoid C (Yamanouchi)-   Siosteran (Ciba-Geigy-
F: Gynothérax (Bouchard);   comb.   Fujisawa)
      wfm   I: Eczecur (Schering)-comb.
      Nérisone (Schering)-comb.;   Impetex (Roche)-comb.
      wfm   Norisona (Schering)-comb.

Chlortalidone

(Chlorthalidone)

ATC: C03BA04
Use: diuretic, antihypertensive

RN: 77-36-1   MF: $C_{14}H_{11}ClN_2O_4S$ MW: 338.77   EINECS: 201-022-5

LD_{50}:   >5 g/kg (M, p.o.);
       >5 g/kg (R, p.o.)

CN: 2-chloro-5-(2,3-dihydro-1-hydroxy-3-oxo-1H-isoindol-1-yl)benzenesulfonamide

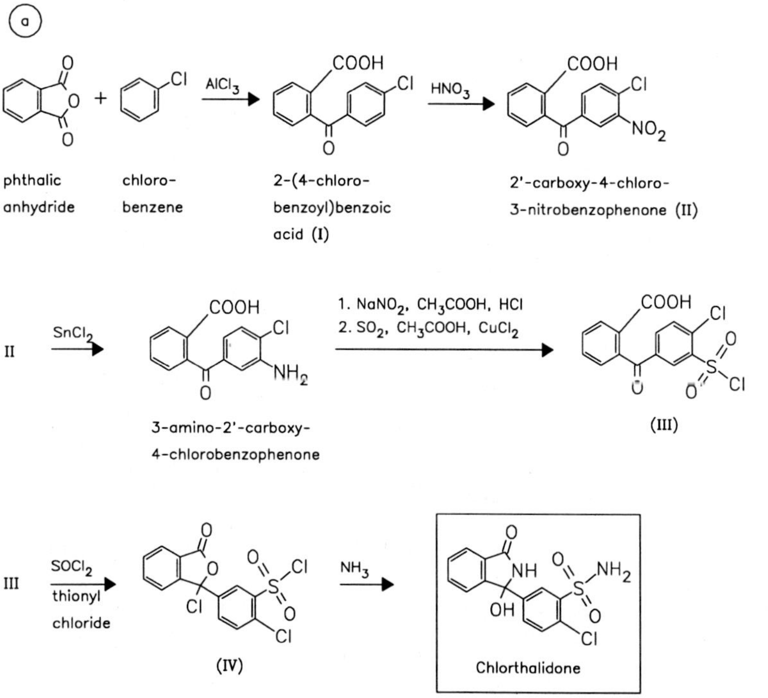

phthalic   chloro-   2-(4-chloro-   2'-carboxy-4-chloro-
anhydride   benzene   benzoyl)benzoic   3-nitrobenzophenone (II)
                                   acid (I)

                3-amino-2'-carboxy-   (III)
                4-chlorobenzophenone

                (IV)   Chlorthalidone

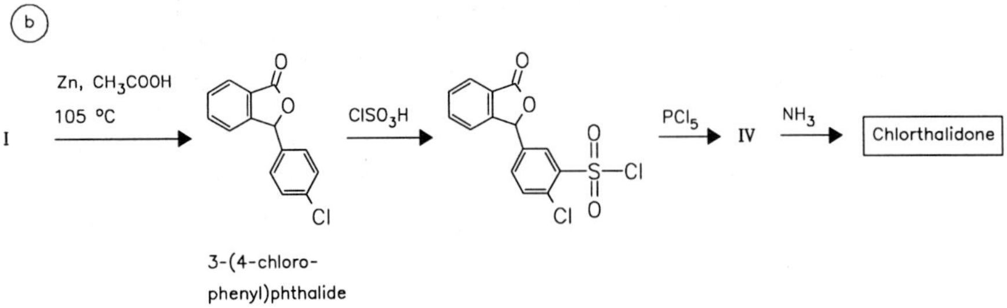

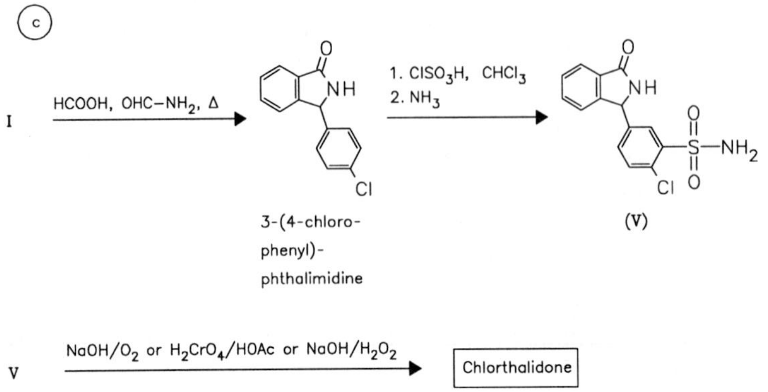

b)

Zn, CH$_3$COOH
105 °C

I → [3-(4-chloro-phenyl)phthalide] → ClSO$_3$H → [structure with S–Cl] → PCl$_5$ → IV → NH$_3$ → [Chlorthalidone]

3-(4-chloro-
phenyl)phthalide

c)

I → HCOOH, OHC–NH$_2$, Δ → [3-(4-chloro-phenyl)-phthalimidine] → 1. ClSO$_3$H, CHCl$_3$ 2. NH$_3$ → [structure (V) with S–NH$_2$]

3-(4-chloro-
phenyl)-
phthalimidine

(V)

V → NaOH/O$_2$ or H$_2$CrO$_4$/HOAc or NaOH/H$_2$O$_2$ → [Chlorthalidone]

Reference(s):

a US 3 055 904 (Geigy; 25.9.1962; CH-prior. 4.11.1957).
  Graf, W. et al.: Helv. Chim. Acta (HCACAV) **42**, 1085 (1959).
b US 4 188 330 (Dow; 12.2.1980; appl. 10.10.1978).
c EP 51 215 (USV; appl. 22.10.1981; USA-prior. 31.10.1980).
  EP 51 217 (USV; appl. 22.10.1981; USA-prior. 31.10.1980).

water soluble dispersions:
EP 125 420 (Boehringer Ing.; appl. 15.3.1984; USA-prior. 16.3.1983).

Formulation(s):   tabl. 25 mg, 50 mg, 100 mg

Trade Name(s):

D: Combipresan (Boehringer
    Ing.)-comb.
    Darebon (Novartis
    Pharma)-comb.
    Diu-Atenolol Verla (Verla)
    Hydro-Long-Tablinen
    (Sanorania)
    Hygroton (Novartis
    Pharma)
    Prelis (Novartis Pharma)-
    comb.
    Teneretic (Zeneca)-comb.
    Trasitensin (Novartis
    Pharma)-comb.
    Trepress (Novartis
    Pharma)-comb.
    TRI-Horm (Zeneca)-comb.
    combination preparations

F: Hygroton (Novartis)
    Logroton (Novartis)-comb.
    Trasitensine (Novartis)-
    comb.

GB: Hygroton (Novartis)
    Kalspare (Dominion)-
    comb.
    Tenoret 50 (Zeneca)-comb.
    Tenoretic (Zeneca)-comb.

I: Ataclor (Crosara)-comb.
    Atenigron (Mitim)-comb.
    Biotens (Kemyos
    Biomedical Research)-
    comb.
    Carmian (Lifepharma)-
    comb.
    Combipresan (Boehringer
    Ing.)-comb.

    Diube (SIT)-comb.
    Diurolab (Leben's)-comb.
    Eupres Mite (Schiapparelli
    Searle)-comb.
    Igroseles (Carlo Erba)-
    comb.
    Igroton (Novartis)
    Igroton-Lopresor
    (Novartis)-comb.
    Igroton Reserpina
    (Novartis)-comb.
    Target (Lisapharma)-comb.
    Tenolone (Lusofarmaco)-
    comb.
    Tenoretic (Zeneca)-comb.
    Zambesil (Gentili)

J: Hybasedock (Sawai)
    Hygroton (Ciba-Geigy)

USA: Combipres (Boehringer   Hygroton (Rhône-Poulenc   Tenoretic (Zeneca)
      Ing.)   Rorer)   Thalitone (Monarch)

Chlortetracycline

ATC: A01AB21; D06AA02; J01AA03;
      S01AA02
Use: antibiotic

RN: 57-62-5   MF: $C_{22}H_{23}ClN_2O_8$ MW: 478.89   EINECS: 200-341-7
LD_{50}: 134 mg/kg (M, i.v.); 1500 mg/kg (M, p.o.);
      118 mg/kg (R, i.v.);
      150 mg/kg (dog, i.v.); 750 mg/kg (dog, p.o.)
CN: [4S-(4α,4aα,5aα,6β,12aα)]-7-chloro-4-(dimethylamino)-1,4,4a,5,5a,6,11,12a-octahydro-3,6,10,12,12a-
      pentahydroxy-6-methyl-1,11-dioxo-2-naphthacenecarboxamide

monohydrochloride
RN: 64-72-2   MF: $C_{22}H_{23}ClN_2O_8 \cdot HCl$ MW: 515.35   EINECS: 200-591-7
LD_{50}: 100 mg/kg (M, i.v.); 2314 mg/kg (M, p.o.);
      100 mg/kg (R, i.v.)

Chlortetracycline

From fermentation solutions of *Streptomyces aureofaciens*.

Reference(s):
US 2 482 055 (American Cyanamid; 1949; prior. 1948).
US 2 609 329 (American Cyanamid; 1949; prior. 1948).
US 2 899 422 (American Cyanamid; 1959; prior. 1956).
US 2 987 449 (American Cyanamid; 6.6.1961; prior. 23.2.1960).
US 3 050 446 (American Cyanamid; 21.8.1962; prior. 28.7.1960).
Duggar, B.M.: Ann. N. Y. Acad. Sci. (ANYAA9) **51**, 175 (1948).

Formulation(s):   cream 10 mg/g, 30 mg/g, 3 %; eye ointment 10 mg/g (1 %); ointment 30 mg/10 g (3 %); pastes
                    30 mg; pessaries 100 mg (as hydrochloride)

Trade Name(s):
D: Aureodelf (Lederle)-comb.   GB: Aureocort (Wyeth)-comb.   Aureomix (SIT)-comb.
      Aureomycin (Lederle)   Aureomycin (Wyeth)   J: Aureomycin (Lederle)
F: Auréomycine (Specia);   Deteclo (Wyeth)-comb.   USA: Aureomycin (Lederle);
      wfm   I: Aureocort (Cyanamid)-   wfm
      Tri-antibiotique Chibret   comb.
      (Chibret)-comb.; wfm   Aureomicina (Cyanamid)

Chlorthenoxazine

ATC: N02B
Use: anti-inflammatory, antipyretic, analgesic

RN: 132-89-8   MF: $C_{10}H_{10}ClNO_2$ MW: 211.65   EINECS: 205-082-3
LD_{50}:   11.155 g/kg (M, p.o.);
10 g/kg (R, p.o.)
CN: 2-(2-chloroethyl)-2,3-dihydro-4H-1,3-benzoxazin-4-one

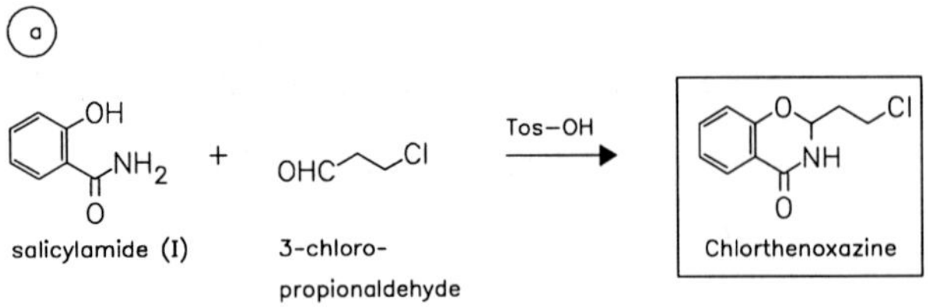

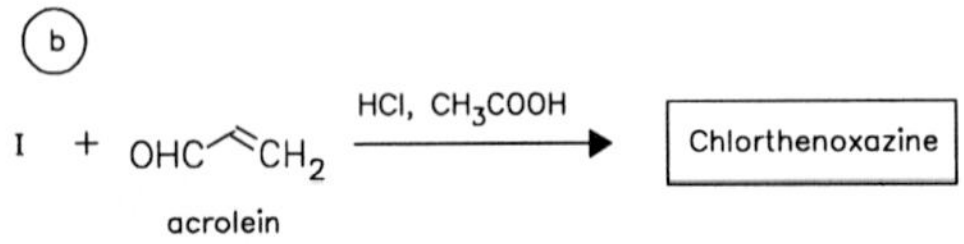

salicylamide (I)   3-chloro-propionaldehyde   Chlorthenoxazine

acrolein   Chlorthenoxazine

Reference(s):
a   DE 1 021 848 (Thomae; appl. 1955).
b   DE 1 028 999 (Thomae; appl. 1956; addition to DE 1 021 848).

Formulation(s):   tabl. 200 mg

Trade Name(s):
D: Cimporhin (Tomae)-comb.; wfm
Fiobrol (Geigy)-comb.; wfm
I: Atossipirina (Borromeo)-comb.; wfm

Betix (Saba); wfm
Megapir (Biotrading)-comb.; wfm
Ossazin (Sealari); wfm
Ossazone (Brocchieri); wfm

Ossipirina (Radiumfarma); wfm
Oxal (Saita); wfm
Reugaril (Farber-Ref); wfm
Reulin (Isola-Ibi); wfm
Reumital (Farge); wfm

Chlorzoxazone

ATC: M03BB03
Use: muscle relaxant

RN: 95-25-0   MF: $C_7H_4ClNO_2$ MW: 169.57   EINECS: 202-403-9
LD_{50}:   440 mg/kg (M, p.o.);
763 mg/kg (R, p.o.)
CN: 5-chloro-2(3H)-benzoxazolone

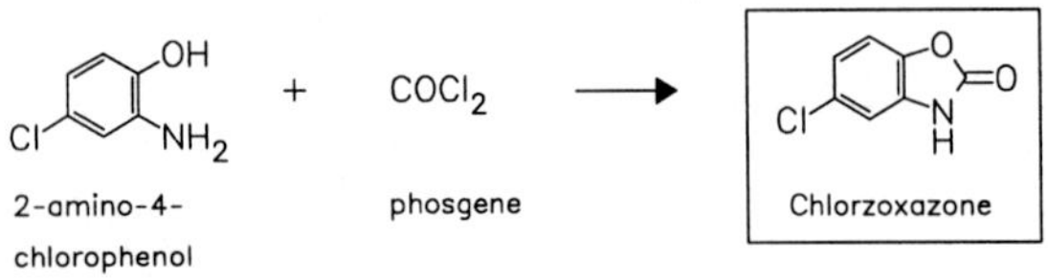

2-amino-4-chlorophenol   phosgene   Chlorzoxazone

Reference(s):
US 2 895 877 (McNeil; 21.7.1959; prior. 30.7.1956).

Formulation(s):   tabl. 250 mg, 500 mg

Trade Name(s):

D:	Paraflex (Cilag-Chemie)-comb.; wfm		Deltapyrin (Kodama)-comb.		Salinalon (Nippon Kayaku)-comb.
I:	Biomioran (Bioindustria); wfm		Framenco (Fuso)		Solaxin (Eisai)
	Paraflex (Cilag-Chemie); wfm		Kiricoron (Sampo)-comb.		Sorazin (Toho)
			Mesin (Yamanouchi)		Trancrol (Mohan)
J:	Chlozoxine (Sanko)		Nichirakishin (Nichiiko)	USA:	Parafon Forte (Ortho-McNeil Pharmceutical)-comb.
	Chroxin (Kanto)		Pathorysin (Kowa Yakuhin)		
			Rheumadex Comp. (Nakataki)-comb.		

Cholestyramine

(Colestyramine)

Use: antipruritic at biliary congestion

RN: 11041-12-6   MF: unspecified   MW: unspecified   EINECS: 234-270-8
LD$_{50}$:   >7.5 g/kg (M, p.o.);
>4 g/kg (R, p.o.)
CN: cholestyramine

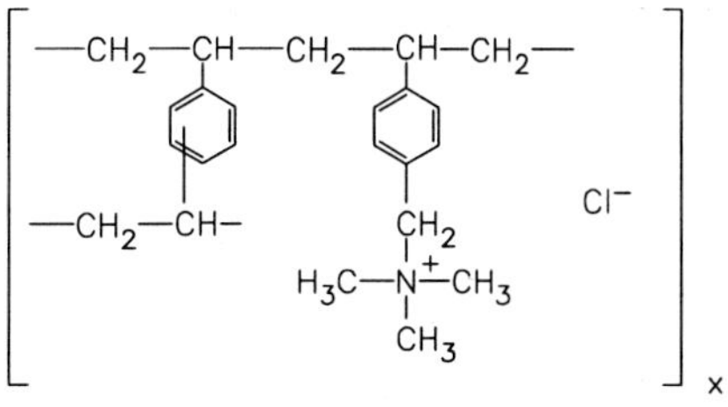

Cholestyramine

Chloromethylation of styrene-divinylbenzene-mixing polymerizate and following reaction with trimethylamine.

Reference(s):
"medical use"
US 3 383 281 (Merck & Co.; 14.5.1968; appl. 22.9.1961; prior. 15.7.1958).

Formulation(s):   eff. tabl. 2 g; gran. 4 g; powder 4 g

Trade Name(s):

D:	Lipocol (Merz & Co.)	GB:	Questran (Bristol-Myers Squibb)	USA:	LoCholest (Warner Chilcott Professional Products)
	Quantalan (Bristol-Myers Squibb)				
	Vasocan (Felgenträger)	I:	Cholestrol (Formenti)		Questran (Bristol-Myers Squibb)
F:	Questran (Allard; Bristol-Myers Squibb)		Questran (Bristol It. Sud; as hydrochloride)		

Choline chloride

ATC: A05B
Use: choleretic

RN: 67-48-1   MF: $C_5H_{14}ClNO$ MW: 139.63   EINECS: 200-655-4
LD$_{50}$:   53 mg/kg (M, i.v.); 3900 mg/kg (M, p.o.);
3400 mg/kg (R, p.o.)
CN: 2-hydroxy-*N,N,N*-trimethylethanaminium chloride

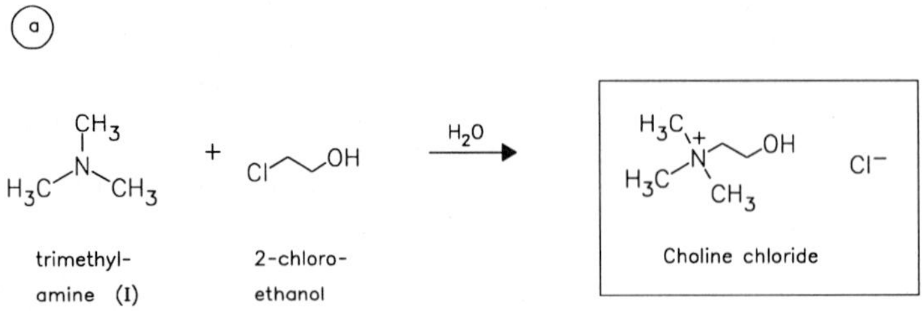

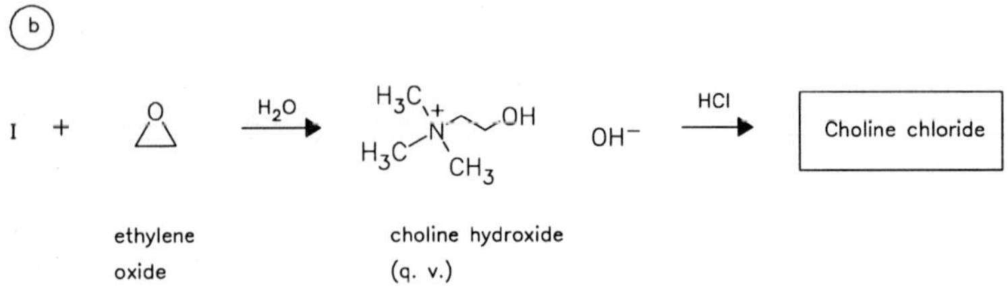

trimethyl-   2-chloro-   Choline chloride
amine (I)   ethanol

ethylene   choline hydroxide
oxide   (q. v.)

Reference(s):
Ullmanns Encykl. Tech. Chem., 4. Aufl., Vol. **9**, 586.
US 2 623 901 (Nopco; 1952; appl. 1950).

Formulation(s):   emulsion 400 mg/5 ml

Trade Name(s):
D: Geriatrie-Mulsin (Mucos)-   I: Betotal (Carlo Erba)-comb.   numerous combination
comb.   Colina Cloruro (Tariff.   preparations
F: Desintex-Choline (M.   Integrativo)
Richard)-comb.

Choline dihydrogen citrate

ATC: C04AX; M03AB
Use: lipotropic

RN: 77-91-8   MF: $C_6H_7O_7 \cdot C_5H_{14}NO$ MW: 295.29   EINECS: 201-068-6
LD$_{50}$:   >4800 mg/kg (M, i.v.); >4800 mg/kg (M, p.o.);
>4800 mg/kg (R, i.v.); >4800 mg/kg (R, p.o.)
CN: 2-hydroxy-*N,N,N*-trimethylethanaminium salt with 2-hydroxy-1,2,3-propanetricarboxylic acid (1:1)

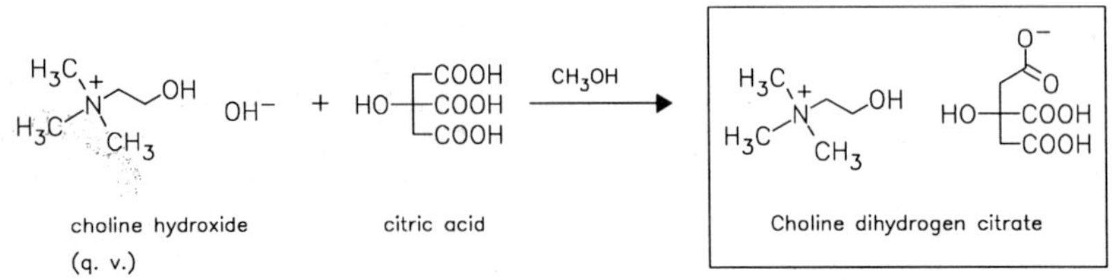

choline hydroxide   citric acid   Choline dihydrogen citrate
(q. v.)

Reference(s):
US 2 870 198 (Nopco; 1959; appl. 1954).

Formulation(s):   amp. 300 mg/ml

Trade Name(s):

D: Neurotropan (Phönix)-
    comb.
F: Citrocholine (Thérica)-
    comb.
    Hepacholine Sortriol
    (Synthélabo)-comb.

Hépagrume (Synthélabo)-
comb.
Kalicitrine (Promedica)-
comb.
Romarine-choline
(Aérocid)-comb.

I: Ipocol (Arnaldi)-comb.;
    wfm
    Liverin (Perkins)-comb.;
    wfm
    Rybutol (Bergamon)-
    comb.; wfm

Choline hydroxide

Use: parasympathomimetic

RN: 123-41-1   MF: $C_5H_{15}NO_2$ MW: 121.18   EINECS: 204-625-1
LD$_{50}$: 21.4 mg/kg (M, i.v.)
CN: 2-hydroxy-*N,N,N*-trimethylethanaminium hydroxide

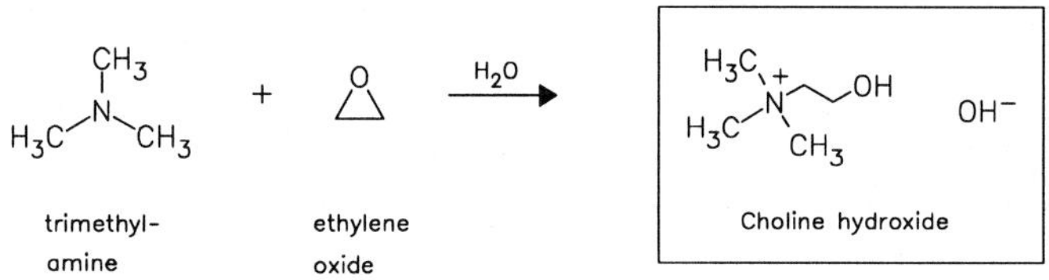

trimethyl-   ethylene   Choline hydroxide
amine   oxide

Intermediate for choline salts.

Reference(s):
Ullmanns Encykl. Tech. Chem., 4. Aufl., Vol. **9**, 586.
GB 379 260 (F. Körner; appl. 1932; D-prior. 1931).
DRP 655 882 (Prod. Aminés S. A., Brüssel; appl. 1931; B-prior. 1931).
Renshaw, R.R.: J. Am. Chem. Soc. (JACSAT) **32**, 128 (1910).
US 2 774 759 (American Cyanamid; 1956; appl. 1955).

alternative synthesis from trimethylamine *and* 2-chloroethanol:
DE 801 210 (BASF; appl. 1948).
US 2 623 901 (Nopco; 1952; appl. 1950).

Trade Name(s):
USA: Choline/Inoritol Tablets   Lipo-C (Legere); wfm
      (Solgar); wfm

Choline salicylate
(Salicylate de choline)

ATC: N02BA03
Use: analgesic, anti-inflammatory,
        antipyretic

RN: 2016-36-6   MF: $C_5H_{14}NO \cdot C_7H_5O_3$ MW: 241.29   EINECS: 217-948-8
LD$_{50}$: 2690 mg/kg (M, p.o.)
CN: 2-hydroxy-*N,N,N*-trimethylethanaminium salicylate (1:1)

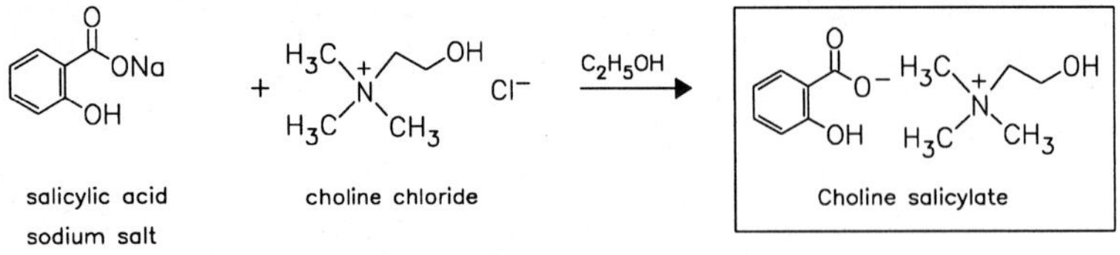

salicylic acid
sodium salt

choline chloride

Choline salicylate

Reference(s):
US 3 069 321 (Labs. for Pharmac. Dev.; 18.12.1962; appl. 4.4.1960).
BE 583 513 (Mundipharma; appl. 12.10.1959).

Formulation(s):   drops 200 mg/ml; gel 87.1 mg/g; sol. 500 mg/100 ml

Trade Name(s):
D: Audax (Mundipharma)
      Givalex (Norgine)-comb.
      Mundisal (Mundipharma)-
      comb.

F: Givalex (Norgine Pharma)-
      comb.
      Pansoral (Inava)-comb.

GB: Bonjela (Reckitt &
      Colman)-comb.

I: Salicol (Sais); wfm
J: Satibon (Grelan)
USA: Trilisate (Purdue Frederick)

Choline stearate

ATC: C05
Use: anti-inflammatory, liver therapeutic

RN: 60154-01-0   MF: $C_{18}H_{35}O_2 \cdot C_5H_{14}NO$ MW: 387.65
CN: 2-hydroxy-*N,N,N*-trimethylethanaminium octadecanoate (salt)

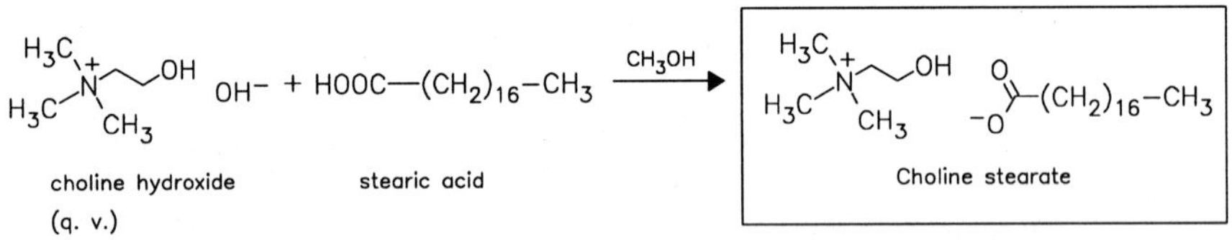

choline hydroxide
(q. v.)

stearic acid

Choline stearate

Reference(s):
US 2 774 759 (American Cyanamid; 1956; appl. 1955).

Formulation(s):   ointment 2.95 g/100 g

Trade Name(s):
D: Chomelanum (Schur)

Choline theophyllinate

(Cholinophylline; Oxtriphylline; Oxytrimethylline)

ATC: R03DA02
Use: bronchodilator

RN: 4499-40-5   MF: $C_7H_7N_4O_2 \cdot C_5H_{14}NO$ MW: 283.33   EINECS: 224-798-7
CN: 2-hydroxy-*N,N,N*-trimethylethanaminium, salt with 3,7-dihydro-1,3-dimethyl-1*H*-purine-2,6-dione (1:1)

choline
hydrogen carbonate

theophylline

Choline theophyllinate

Reference(s):
US 2 776 287 (Nepera; 1957; appl. 1954).

Formulation(s):   f. c. tabl. 200 mg; s. r. tabl. 400 mg, 600 mg

Trade Name(s):
D: Euspirax (Asche)
GB: Choledyl (Warner); wfm
          Sabidal (Zyma); wfm
I: Sclerofillina (Medici
          Domus); wfm

Teofilcolina (Salfa); wfm
Teofilcolina sedativa
(Salfa)-comb.; wfm

Theophyl-Choline
(Perkins)-comb. with
theophyllineacetate; wfm
J: Ishicolin (Kanto-Isei)
          Theocolin (Eisai)

Chymopapain

ATC: M09AB01
Use: intervertebral disk damages
            therapeutic

RN: 9001-09-6   MF: unspecified   MW: unspecified   EINECS: 232-580-8
LD$_{50}$:   42.3 mg/kg (M, i.v.);
          36.1 mg/kg (R, i.v.)
CN: chymopapain

Proteolytic enzyme from the latex of *Carica papaya* with an approximate molecular weight of 27000. It is differentiated from papain in electrophoresis behavior, in solubility and in substrate specifity. Isolation by acidify of papaya-latex with HCl, salting out with NaCl and following chromatographic purification. The formulation contains L-cysteine as reducing agent.

Reference(s):
Jansen, E.F.; Balls, A.K.: J. Biol. Chem. (JBCHA3) **137**, 459 (1941).
US 2 313 875 (E. F. Jansen, A.K. Balls; 1943; appl. 1940).
US 3 558 433 (Baxter Labs.; 26.1.1971; appl. 7.11.1967).

medical use:
US 4 439 423 (Smith Labs.; 27.3.1984; appl. 13.5.1981).
US 3 320 131 (Baxter Labs.; 1967; prior. 1963, 1964).

Formulation(s):   vial 4 iu, 5 iu, 10 iu/1000 iu.

Trade Name(s):
D: Discase (Travenol); wfm
F: Chymodiactine (Knoll)

USA: Chymodiactin (Smith);
            wfm

α-Chymotrypsin
(Alphachymotrypsin)

ATC: B06AA04; S01KX01
Use: anti-inflammatory, proteolytic

RN: 9004-07-3 MF: unspecified MW: unspecified EINECS: 232-671-2
LD_{50}: 89 mg/kg (M, i.v.); >6 g/kg (M, p.o.);
84 mg/kg (R, i.v.); >4 g/kg (R, p.o.)
CN: chymotrypsin

Isolation from homogenized bovine pancreas by
1. extraction with 0,25 normal H_2SO_4.
2. Fractionated ammonium sulfate precipitaion of α-chymotrypsinogen (further fractions contain deoxyribonuclease, chymotrypsinogen B, ribonuclease, trypsinogen).
3. Activation of α-chymotrypsinogen by dissolution in 0,005 normal HCl, standardization to 0,1 molar $CaCl_2$ and 0,1 molar borate buffer pH 8.0; separation of inactive precipitate after 24 h; precipitation of Ca^{2+} as sulfate.
4. Fractionated ammonium sulfate precipitation (twice).
5. Crystallization from borat buffer at pH 8.0 (twice).
6. Desalting by gel chromatography or dialysis.
7. Sterile filtration.
8. Lyophilization.

Reference(s):
Ullmanns Encykl. Tech. Chem., 4. Aufl., Vol. **10**, 536.

properties, review:
Niemann, C.: Science (Washington, D.C.) (SCIEAS) **143**, 1287 (1964).

Formulation(s):   amp. ca. 5 mg/ 5 ml; ointment ca. 5 mg/30 g

Trade Name(s):
D: Alpha-Chymocutan (Strathmann)
    Alpha-Chymotrase (Strathmann)
    Enzym-Wied (Wiedemann)-comb.
    Wobe-Mugos (Mucos)-comb.
F: Alphachymotrypsine Choay (Sanofi Winthrop)
    Alphacutanée (Leurquin)

GB: Cirkan (Sinbio)-comb.
    Chymar (Armour); wfm
    Chymocyclar (Armour); wfm
    Chymoral (Armour)-comb.; wfm
    Deanase (Consolidated Chemicals); wfm
I: Ribociclina (Puropharma)-comb.

    Zonulasi (SmithKline Beecham)
J: Chymoral (Tokyo Tanabe)
    Chymotase (Mochida)
    Chymozym (Teikoku Hormone)
    Kimopsin (Eisai)
    Zonolysine (Mochida)
USA: Orenzyme (Merrell Dow); wfm

Cianidanol
((+)-Catechin; (+)-Catechol; Cianidol; Cyanidanol; Cyanidol; Dexcyanidanol)

ATC: V09D
Use: liver therapeutic (inhibition of lipide peroxidation)

RN: 154-23-4 MF: $C_{15}H_{14}O_6$ MW: 290.27 EINECS: 205-825-1
CN: (2R-*trans*)-2-(3,4-dihydroxyphenyl)-3,4-dihydro-2H-1-benzopyran-3,5,7-triol

Cianidanol

Ingredient of various plants and trees ("catechu" from *Uncaria gambir* and *Acacia catechu*), obtained by extraction with water or ethyl acetate.

Reference(s):
Freudenberg, K. et al.: Ber. Dtsch. Chem. Ges. (BDCGAS) **54**, 1204 (1921).
Freudenberg, K. et al.: Ber. Dtsch. Chem. Ges. (BDCGAS) **55**, 1737 (1922).
Freudenberg, K. et al.: Justus Liebigs Ann. Chem. (JLACBF) **444**, 135 (1925).

absolute configuration:
Hardegger, E. et al.: Helv. Chim. Acta (HCACAV) **40**, 1819 (1957).

new crystal modifications:
US 4 515 804 (Zyma; 7.5.1985; GB-prior. 24.2.1982).

salts with basic amino acids:
US 4 285 964 (Continental Pharma; 25.8.1981; appl. 30.8.1979).
GB 2 057 437 (Continental Pharma; appl. 19.8.1980; USA-prior. 30.8.1979).
US 4 507 314 (Midit, Soc. Fiduciaire; 26.3.1985; appl. 20.7.1983).

Formulation(s):   tabl. 750 mg

Trade Name(s):
D: Catergen (Zyma); wfm   Catergen (Zyma); wfm   Transepar (Dompé); wfm
F: Catergène (Zyma); wfm   DrenoliveR (Biochimica   J: Catergen (Kanebo-Sankyo)
I: Ausoliver (Ausonia); wfm   Zanardi); wfm

Cibenzoline

ATC: C01BG07
Use: class I antiarrhythmic

RN: 53267-01-9   MF: C$_{18}$H$_{18}$N$_2$ MW: 262.36   EINECS: 258-453-7
CN: 2-(2,2-diphenylcyclopropyl)-4,5-dihydro-1*H*-imidazole

1. N$_2$H$_4$
2. HgO or with H$_3$C—C(O)—O-OH

1. hydrazine
2. mercury oxide or with peracetic acid

H$_2$C=CN
acrylonitrile

I

benzo-
phenone

diphenyldiazo-
methane

+ ethylenediamine   p-toluenesulfonate   →   Cibenzoline

1-cyano-2,2-
diphenyl-
cyclopropane   (I)

Reference(s):
DOS 2 359 795 (Hexachimie; appl. 30.9.1973; GB-prior. 30.11.1972, 6.2.1973).
DOS 2 359 816 (Hexachimie; appl. 30.9.1973; GB-prior. 30.11.1972, 6.2.1973, 2.8.1973).
US 3 903 104 (Hexachimie; 9.1975; GB-prior. 30.11.1972, 6.2.1973, 2.8.1973).
US 3 905 993 (Hexachimie; 16.9.1975; GP-prior. 30.11.1972, 6.2.1973).

synthesis of diphenyldiazomethane:
Staudinger, H. et al.: Chem. Ber. (CHBEAM) **49** (1916), 1932
Adamson, J.R. et al.: J. Chem. Soc., Perkin Trans. 1 (JCPRB4) **1975**, 2030.

Formulation(s):   cps. 130 mg; tabl. 130 mg; vial 100 mg

Trade Name(s):
F: Cipralan (UPSA; 1985)   Exacor (Monsanto)

Ciclacillin

(Cyclacillin)

ATC: J01CA
Use: antibiotic

RN: 3485-14-1   MF: $C_{15}H_{23}N_3O_4S$ MW: 341.43   EINECS: 222-470-8
LD$_{50}$:   5010 mg/kg (M, p.o.);
          5010 mg/kg (R, p.o.);
          2500 mg/kg (dog, p.o.)
CN: [2S-(2α,5α,6β)]-6-[[(1-aminocyclohexyl)carbonyl]amino]-3,3-dimethyl-7-oxo-4-thia-1-azabicyclo[3.2.0]heptane-2-carboxylic acid

1-aza-3-oxa-
spiro[4.5]decane-
2,4-dione

6-amino-
penicillanic
acid

Ciclacillin

Reference(s):
US 3 194 802 (American Home; 13.7.1965; appl. 7.2.1963; prior. 26.2.1962).
US 3 553 201 (American Home; 5.1.1971; appl. 3.10.1967; prior. 13.5.1966).

alternative synthesis:
DOS 2 755 903 (Dobfar; appl. 15.12.1977; I-prior. 16.12.1976).

enzymatic:
DAS 2 050 982 (Kyowa Hakko; appl. 16.10.1970; J-prior. 24.10.1969).

via silyl-derivatives:
US 3 478 018 (American Home; 11.11.1969; appl. 2.10.1967).

Formulation(s):   gran. 10 %; tabl. 250 mg, 500 mg

Trade Name(s):

D:	Ultracillin (Grünenthal); wfm	Teejel (Napp)-comb.; wfm	Vastcillin (Takeda)
		Bionacillin-C (Takeda)	Wyvital (Wyeth)
GB:	Calthor (Ayerst); wfm	Citosarin (Toyo Jozo)	USA: Cyclapen (Wyeth); wfm

J: placed before Bionacillin-C (Takeda)

Cicletanine

(Cycletanide)

ATC: C03BX03
Use: diuretic, antihypertensive

RN: 89943-82-8   MF: $C_{14}H_{12}ClNO_2$ MW: 261.71
LD_{50}:   4500 mg/kg (M, p.o.);
        5000 mg/kg (R, p.o.)
CN: (±)-3-(4-chlorophenyl)-1,3-dihydro-6-methylfuro[3,4-c]pyridin-7-ol

hydrochloride
RN: 82747-56-6   MF: $C_{14}H_{12}ClNO_2 \cdot HCl$ MW: 298.17

pyridoxine   acetone   3,4$^\alpha$-O-iso-
(q. v.)   propylidene-
                                                    pyridoxine

2,2,8-trimethyl-5-   4-chlorophenyl-   Cicletanine
formyl-4H-pyrido-   magnesium
[3,4-d]-1,3-dioxane (I) bromide

Reference(s):
DOS 3 204 596 (Soc. de Conseils de Recherche et d'Appl. Sci.; appl. 10.2.1982; GB-prior. 10.2.1981).
US 4 383 998 (Soc. de Conseils de Recherche et d'Appl. Sci.; 17.5.1983; GB-prior. 10.2.1981).

synthesis of 2,2,8-trimethyl-5-formyl-4*H*-pyrido[3,4-*d*]-1,3-dioxane:
Koryntyk, W.; Wiedemann, W.: J. Org. Chem. (JOCEAH) **27**, 2531 (1962).
Koryntyk, W.; Kris, E.J.; Singh, R.P.: J. Org. Chem. (JOCEAH) **29**, 574 (1964).
Sattsangi, P.D.; Argoudelis, C.J.: J. Org. Chem. (JOCEAH) **33**, 1337 (1968).

Formulation(s):   cps. 50 mg, 100 mg (hydrochloride)

Trade Name(s):
D: Justar (Intersan; 1989)   F: Tenstaten (Ipsen-Beaufour;
                                               1988)

Ciclometasone

ATC: D07AB; H02AB
Use: glucocorticoid

RN: 86022-88-0   MF: $C_{32}H_{44}ClNO_7$ MW: 590.16   EINECS: 289-141-9
CN: [11β,16β,21(*trans*)]-21-[[[4-[(acetylamino)methyl]cyclohexyl]carbonyl]oxy]-9-chloro-11,17-dihydroxy-
        16-methylpregna-1,4-diene-3,20-dione

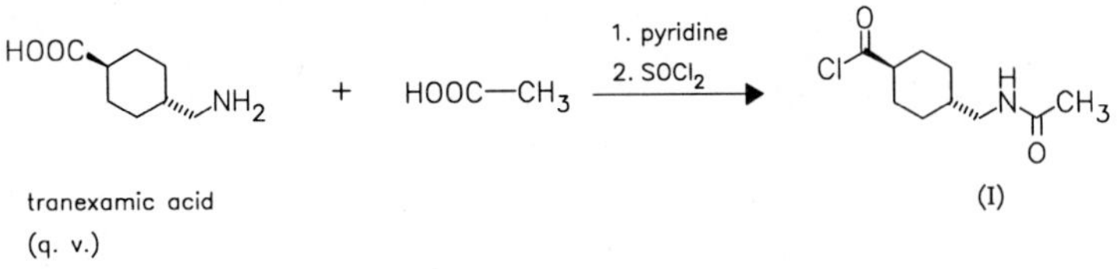

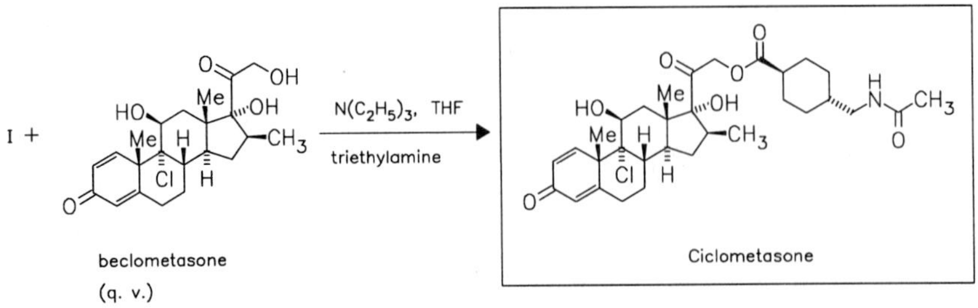

tranexamic acid
(q. v.)

(I)

beclometasone
(q. v.)

Ciclometasone

Reference(s):
FR 2 280 384 (Rorer; appl. 1.8.1974).

synthesis of 4-aminomethylcyclohexanecarboxylic acid:
Levine, M.; Sedlecky, R.: J. Org. Chem. (JOCEAH) **24**, 115 (1959).

Trade Name(s):
I: Cycloderm (Rottapharm)   Telecort Sray (Rottapharm)

Ciclonium bromide

ATC: A03DA04
Use: antispasmodic, anticholinergic

RN: 29546-59-6   MF: $C_{22}H_{34}BrNO$ MW: 408.42   EINECS: 249-687-0
LD$_{50}$:   400 mg/kg (M, p.o.);
          1030 mg/kg (R, p.o.)
CN: 2-(1-bicyclo[2.2.1]hept-5-en-2-yl-1-phenylethoxy)-*N,N*-diethyl-*N*-methylethanaminium bromide

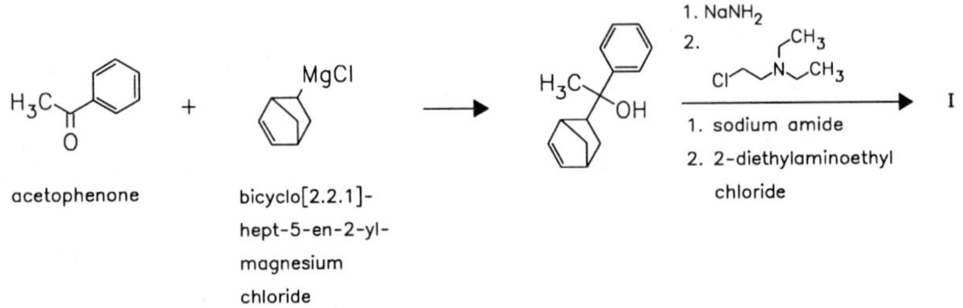

acetophenone   bicyclo[2.2.1]-
                hept-5-en-2-yl-
                magnesium
                chloride

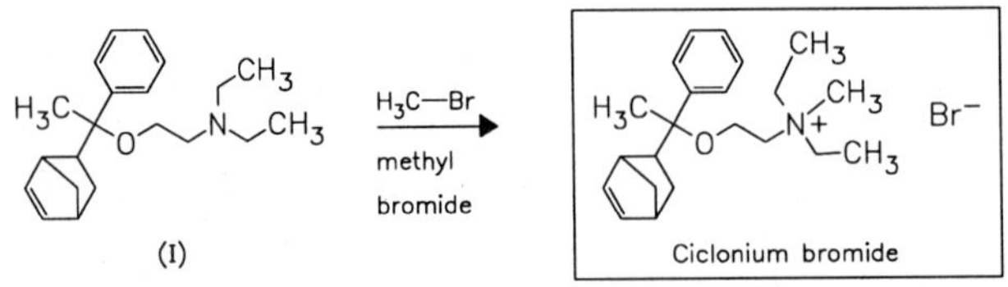

1. sodium amide
2. 2-diethylaminoethyl
   chloride

(I)   methyl
       bromide

Ciclonium bromide

Reference(s):
DE 1 052 982 (ASTA; appl. 29.6.1957).

Formulation(s):   amp. 10 mg/2 ml

Ciclopirox

ATC: D01AE14; G01AX12
Use: antifungal

RN: 29342-05-0   MF: $C_{12}H_{17}NO_2$ MW: 207.27   EINECS: 249-577-2
CN: 6-cyclohexyl-1-hydroxy-4-methyl-2(1*H*)-pyridinone

ciclopirox olamine
RN: 41621-49-2   MF: $C_{12}H_{17}NO_2 \cdot C_2H_7NO$ MW: 268.36   EINECS: 255-464-9
LD_{50}: 71 mg/kg (M, i.v.); 1740 mg/kg (M, p.o.);
      72 mg/kg (R, i.v.); 2350 mg/kg (R, p.o.)

6-cyclohexyl-
4-methyl-2-
pyrone

hydroxylamine, 2-amino-
pyridine

Ciclopirox

methyl 5-cyclohexyl-
3-methyl-5-oxo-2-
pentenoate (mixture
with the 3-pentenoate;
from methyl 4-methyl-
2-butenoate)

hydroxylamine
hydrochloride

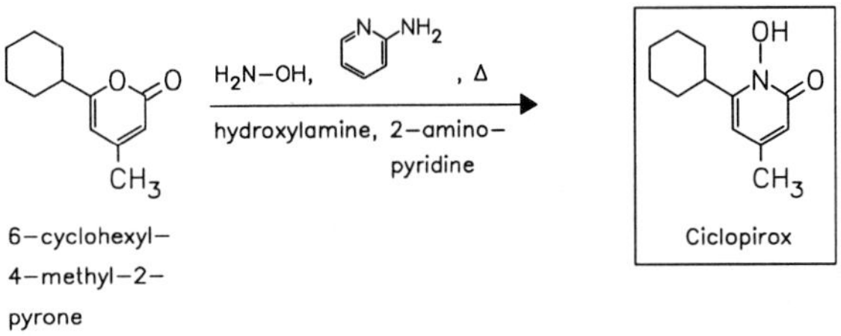

methyl 5-cyclohexyl-
3-methyl-5-hydroxyimino-
2-pentenoate (I)

Ciclopirox olamine

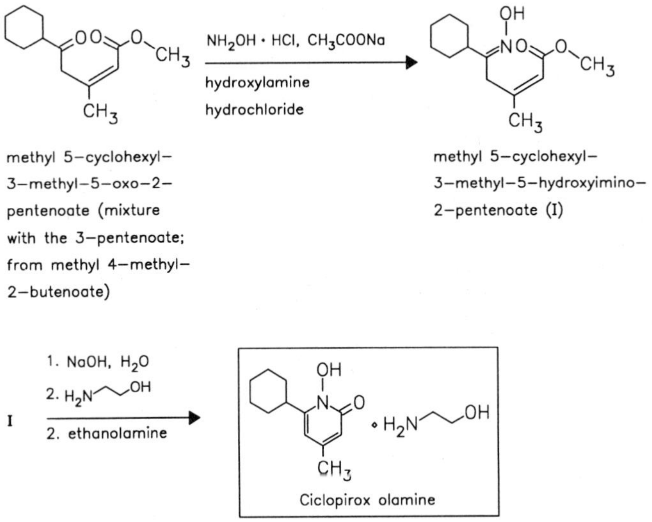

Reference(s):
a   US 3 883 545 (Hoechst AG; 13.5.1975; appl. 16.11.1971; prior. 22.12.1972).
    US 3 972 888 (Hoechst AG; 3.8.1976; D-prior. 25.3.1972).
b   ZA 696 039 (Hoechst AG; appl. 12.8.1969; D-prior. 31.8.1968).
    DE 1 795 270 (Hoechst AG; appl. 31.8.1968).
    DOS 2 214 608 (Hoechst AG; appl. 25.3.1972).

Formulation(s):   cream 1 %; powder 1 % (as olamine); sol. 10 mg/ml (as ciclopirox); vaginal cream 1 % (as olamine)

Trade Name(s):

D: Batrafen (Hoechst; 1981)
      Nagel-Batrafen (Hoechst)
F: Mycoster (Pierre Fabre; 1986)
I: Batrafen (Hoechst)

      Brumixol (Bruschettini)
      Dafnegin (Poli)
      Miclast (Pierre Fabre Phar.)
      Miclast (Lifepharma)
      Micomicen (Synthelabo)

      Micoxolamina (Delalande Isnardi)
J: Batrafen (Hoechst)
USA: Loprox (Hoechst Marion Roussel; 1983)

Ciclosporin

(Cyclosporin A)

ATC: L04AA01
Use: immunosuppressive

RN: 59865-13-3   MF: $C_{62}H_{111}N_{11}O_{12}$ MW: 1202.64
CN: [R-[R*,S*-(E)]]-cyclic(L-alanyl-D-alanyl-N-methyl-L-leucyl-N-methyl-L-leucyl-N-methyl-L-valyl-3-hydroxy-N,4-dimethyl-L-2-amino-6-octenoyl-L-α-aminobutyryl-N-methylglycyl-N-methyl-L-leucyl-L-valyl-N-methyl-L-leucyl)

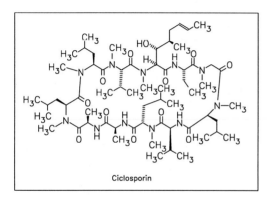

Ciclosporin

Cyclic peptide from 11 amino acids. Preparation by fermentation of *Tolypocladium inflatum* Gams with addition of DL-α-aminobutyric acid to the fermentation medium. Isolation by homogenization of mycelium, extraction with 90 % methanol and column chromatographic purification.

Reference(s):
US 4 117 118 (Sandoz; 26.9.1978; prior. 29.11.1974, 15.8.1975, 9.9.1976; CH-prior. 9.4.1976).
DE 2 455 859 (Sandoz; appl. 26.11.1974; CH-prior. 6.12.1973, 21.10.1974).
Rüegger, A. et at.: Helv. Chim. Acta (HCACAV) **59**, 1075 (1976).
Kobel, H.; Traber, R.: Eur. J. Appl. Microbiol. Biotechnol. (EJABDD) **14**, 237 (1982).

structure:
Petcher, T.J. et al.: Helv. Chim. Acta (HCACAV) **59**, 1480 (1976).

total syntheses:
Wenger, R.M.: Angew. Chem. (ANCEAD) **97**, 88 (1985).
EP 34 567 (Sandoz; appl. 13.2.1981; CH-prior. 14.2.1980).

oral formulation:
US 5 766 629 (SangStat Med. Corp.; 16.6.1998; prior. 25.8.1995, 21.3.1996).

Formulation(s):   cps. 25 mg, 50 mg, 100 mg; inj. sol. 250 mg/5 ml; oral sol. 100 mg/ml; sol. 100 mg

Trade Name(s):

D: Sandimmun (Novartis; 1983)

F: Sandimmun (Novartis)
GB: Neoral (Novartis)

      Sandimmun (Novartis; 1983)

I: Sandimmun (Sandoz)   USA: Neoral (Novartis)
J: Sandimmun (Novartis;   Sandimmune (Novartis;
        1986)   1983)

Cicloxilic acid

ATC: A05AX; A06AB
Use: choleretic, hepatic protectant

RN: 57808-63-6   MF: $C_{13}H_{16}O_3$ MW: 220.27   EINECS: 260-966-6
LD_{50}: 2095 mg/kg (M, p.o.);
        1570 mg/kg (R, p.o.)
CN: cis-2-hydroxy-2-phenylcyclohexanecarboxylic acid

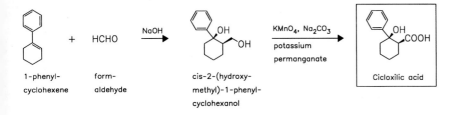

1-phenyl-   form-   cis-2-(hydroxy-   Cicloxilic acid
cyclohexene   aldehyde   methyl)-1-phenyl-
                                      cyclohexanol

Reference(s):
BE 848 143 (Guidotti Int.; appl. 9.12.1976; I-prior. 12.11.1975).

alternative synthesis and use as choleretic:
US 3 700 775 (L. Turbanti; 24.10.1972; I-prior. 29.4.1966).

stereochemistry:
Turbanti, L. et al.: Arzneim.-Forsch. (ARZNAD) **28** (II), 1449 (1978).

Formulation(s):   amp. 60 mg; drg. 40 mg

Trade Name(s):
I: Plecton (Guidotti)   Pleiabil (Guidotti)-comb.

Cicrotoic acid

Use: choleretic

(Acide cicrotoique)

RN: 25229-42-9   MF: $C_{10}H_{16}O_2$ MW: 168.24   EINECS: 246-739-4
LD_{50}: 1925 mg/kg (M, p.o.);
        2900 mg/kg (R, p.o.)
CN: 3-cyclohexyl-2-butenoic acid

acetylcyclo-   ethyl   ethyl 3-cyclohexyl-
hexane   bromoacetate   3-hydroxybutanoate

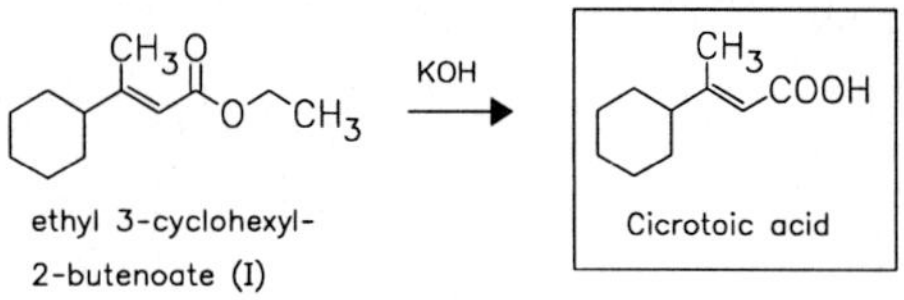

ethyl 3-cyclohexyl-
2-butenoate (I)

Cicrotoic acid

Reference(s):
FR-M 4 665 (A. E. C. Soc. de Chim. Organ. et Biol.; appl. 3.5.1965).
Young et al.: J. Org. Chem. (JOCEAH) **28**, 928 (1963).

Formulation(s):   cps. 250 mg

Trade Name(s):
F: Accroibile (Adrian-
      Marinier); wfm

Cidofovir
(HPMPC; GS-504; GS-0504)

ATC: J05AB12
Use: antiviral

RN: 113852-37-2   MF: C$_8$H$_{14}$N$_3$O$_6$P   MW: 279.19
CN: (*S*)-[[2-(4-amino-2-oxo-1(2*H*)-pyrimidinyl)-1-(hydroxymethyl)ethoxy]methyl]phosphonic acid

dihydrate
RN: 149394-66-1   MF: C$_8$H$_{14}$N$_3$O$_6$P · 2H$_2$O   MW: 315.22

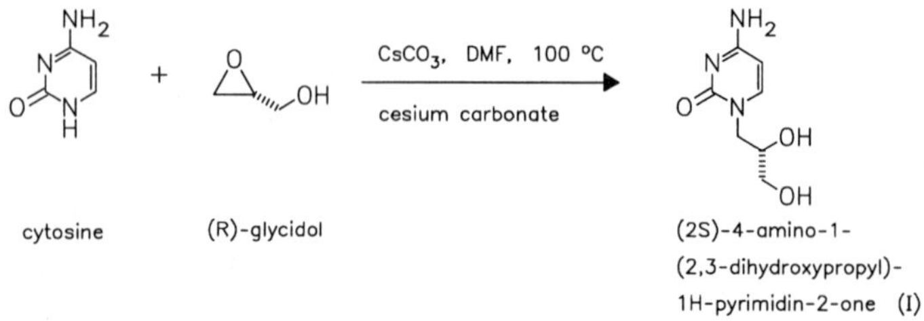

cytosine   (R)-glycidol

(2S)-4-amino-1-
(2,3-dihydroxypropyl)-
1H-pyrimidin-2-one   (I)

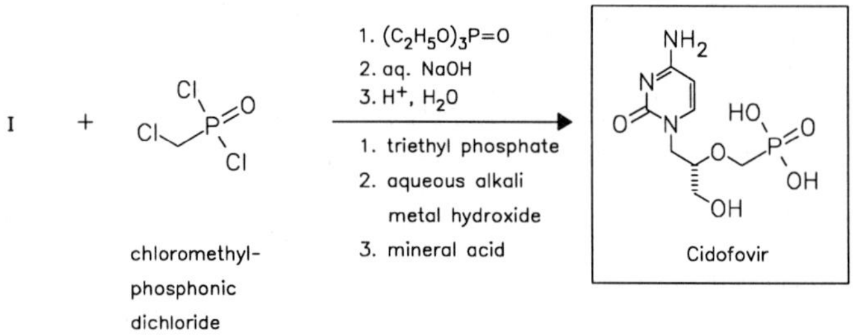

I +

chloromethyl-
phosphonic
dichloride

1. (C$_2$H$_5$O)$_3$P=O
2. aq. NaOH
3. H$^+$, H$_2$O

1. triethyl phosphate
2. aqueous alkali
   metal hydroxide
3. mineral acid

Cidofovir

Reference(s):
EP 253 412 (Ceskoslovenska Akademie Ved., Czech., appl. 20.1.1988; CS-prior. 18.7.1986).
WO 9 624 355 (Astra; appl. 15.8.1996; WO-prior. 6.2.1995).
WO 9 713 528 (Dumex-Alpharma; appl. 17.4.1997; prior. 12.10.1995).

synthesis of (2S)-4-amino-1-(2,3-dihydroxypropyl)-1*H*-pyrimidin-2-one:
Holy, A.: Collect. Czech. Chem. Commun. (CCCCAK) **58** (3), 649 (1993).
Holy, A.: Collect. Czech. Chem. Commun. (CCCCAK) **43**, 2054 (1978).
Martin, J.C. et al.: Nucleosides Nucleotides (NUNUD5) **8** (5-6), 923 (1989).

Formulation(s):   vial 375 mg (75 mg/ml anhydrous) for i.v. infusion

Trade Name(s):
D: VISTIDE (Pharmacia &   USA: Vistide (Gilead Science)
      Upjohn)

Cilastatin

ATC: J01DH51
Use: dehydropeptidase inhibitor (for combination with imipenem)

RN: 82009-34-5   MF: $C_{16}H_{26}N_2O_5S$ MW: 358.46   EINECS: 279-875-8
LD$_{50}$:   8 g/kg (M, route unreported);
      8 g/kg (R, route unreported)
CN: [R-[R*,S*-(Z)]]-7-[(2-amino-2-carboxyethyl)thio]-2-[[(2,2-dimethylcyclopropyl)carbonyl]amino]-2-heptenoic acid

monosodium salt
RN: 81129-83-1   MF: $C_{16}H_{25}N_2NaO_5S$ MW: 380.44   EINECS: 279-694-4
LD$_{50}$:   6786 mg/kg (M, i.v.); >10 g/kg (M, p.o.);
      5027 mg/kg (R, i.v.); >10 g/kg (R, p.o.)

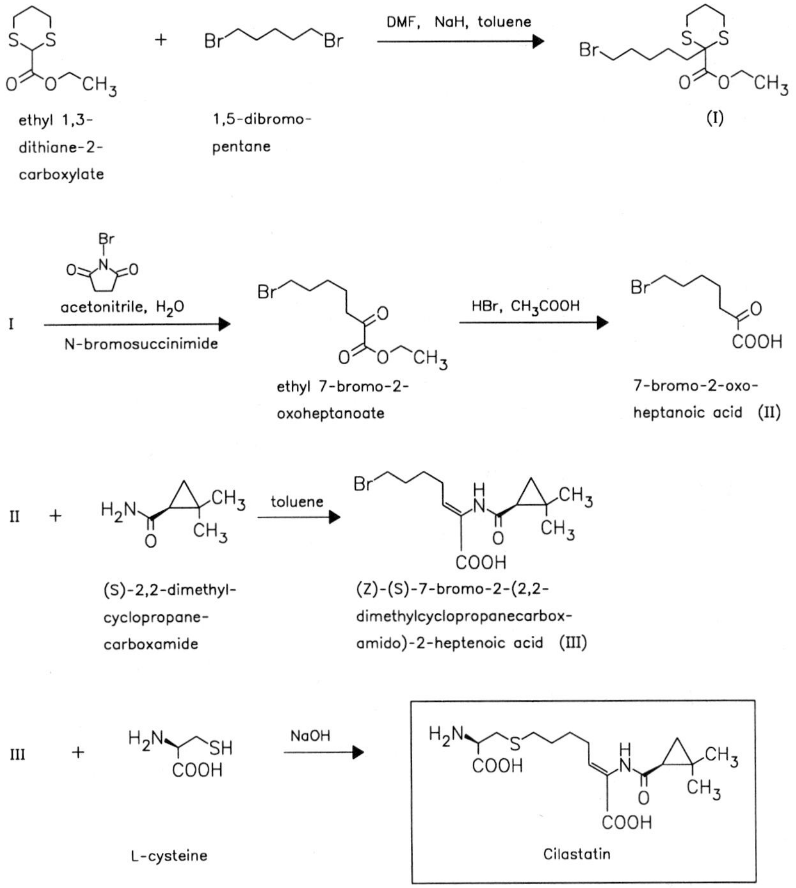

ethyl 1,3-dithiane-2-carboxylate   +   1,5-dibromo-pentane   →[DMF, NaH, toluene]   (I)

I   →[acetonitrile, H$_2$O / N-bromosuccinimide]   ethyl 7-bromo-2-oxoheptanoate   →[HBr, CH$_3$COOH]   7-bromo-2-oxo-heptanoic acid (II)

II   +   (S)-2,2-dimethyl-cyclopropane-carboxamide   →[toluene]   (Z)-(S)-7-bromo-2-(2,2-dimethylcyclopropanecarbox-amido)-2-heptenoic acid (III)

III   +   L-cysteine   →[NaOH]   Cilastatin

Reference(s):
EP 10 573 (Merk & Co.; appl. 24.7.1979; USA-prior. 24.7.1978).
EP 48 301 (Merck & Co.; appl. 24.9.1980).

Formulation(s):   amp. 250 mg, 500 mg, 750 mg (as sodium salt)

Trade Name(s):

D:	Zienam (MSD; 1985)- comb. with imipenem	I:	Imipem (Neopharmed)- comb.		Tienam (MSD)-comb.
F:	Tienam (Merck Sharp & Dohme-Chibret)-comb.		Tenacid (Sigma-Tau)- comb.	J:	Tienam (Banyu; 1987)- comb.
				USA:	Primaxin (Merck; 1985)

Cilazapril

ATC: C09AA08
Use: antihypertensive (ACE inhibitor)

RN: 88768-40-5   MF: $C_{22}H_{31}N_3O_5$ MW: 417.51
CN: [1S-[1α,9α(R*)]]-9-[[1-(ethoxycarbonyl)-3-phenylpropyl]amino]octahydro-10-oxo-6H-pyridazino[1,2-a][1,2]diazepine-1-carboxylic acid

monohydrate
RN: 92077-78-6   MF: $C_{22}H_{31}N_3O_5 \cdot H_2O$ MW: 435.52

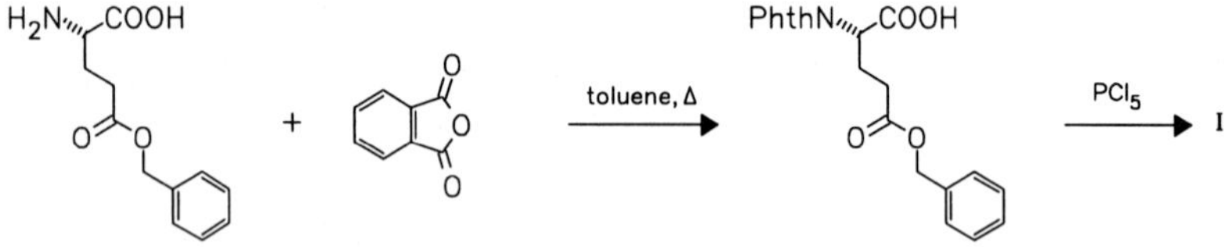

γ-benzyl
L-glutamate

phthalic
anhydride

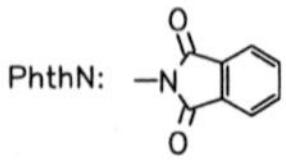

PhthN:

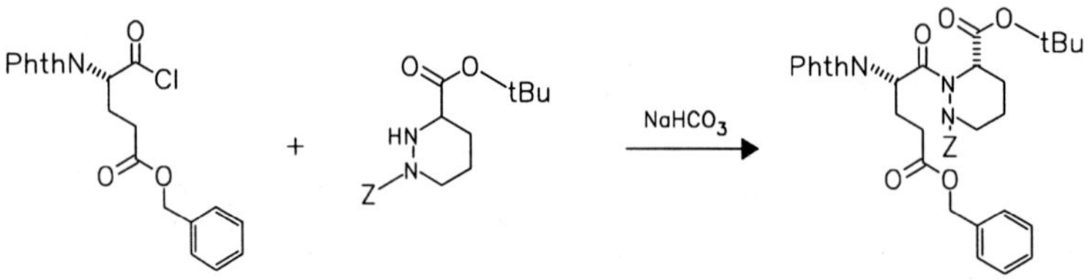

(I)

tert-butyl 1-(benzyl-
oxycarbonyl)-hexa-
hydro-3-pyridazine-
carboxylate

(II)

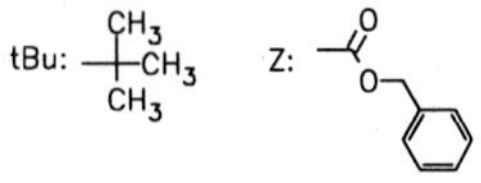

tBu: Z:

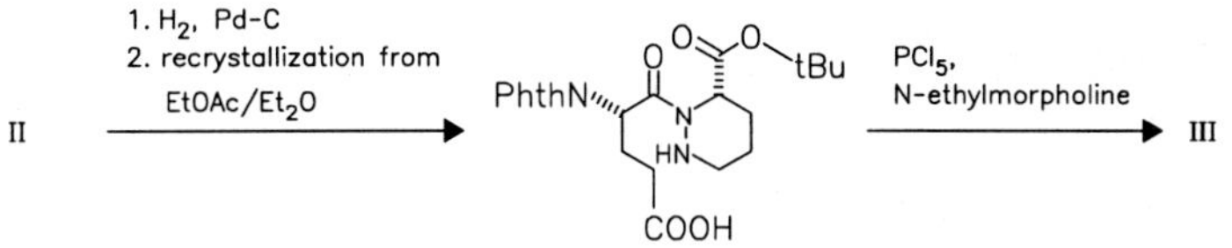

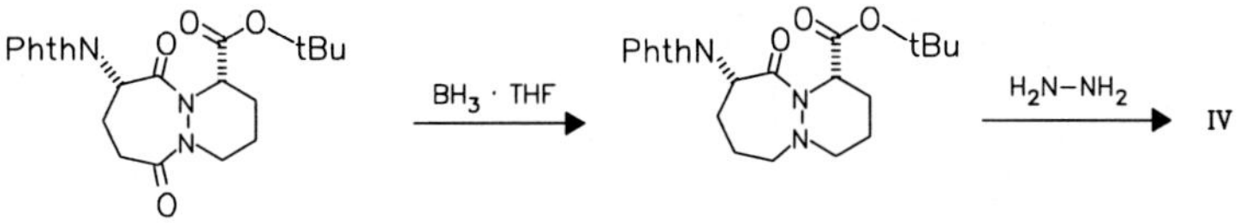

tert-butyl octahydro-
6,10-dioxo-9(S)-phthal-
imido-6H-pyridazo[1,2-a]-
[1,2]diazepine-1(S)-
carboxylate (III)

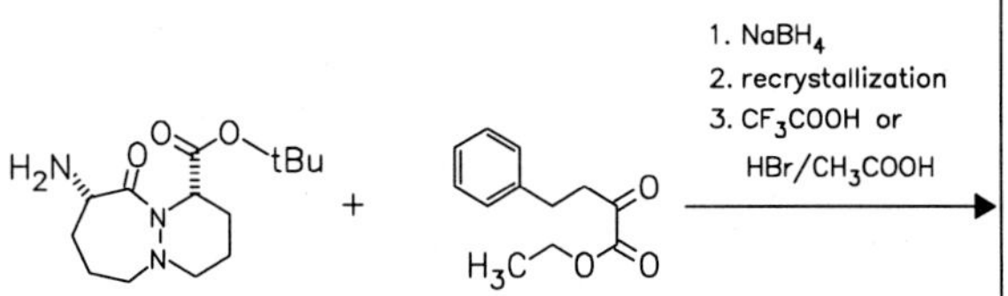

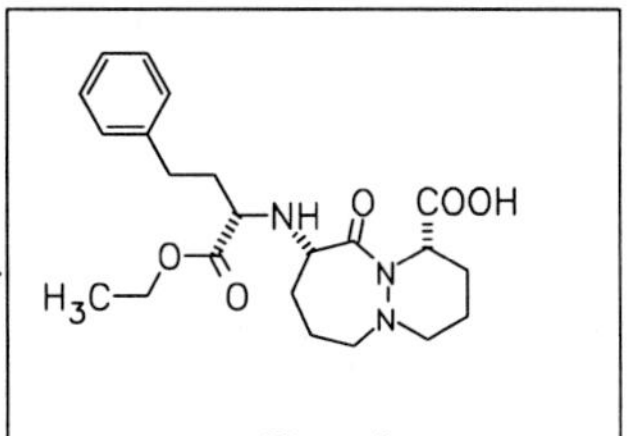

tert-butyl 9(S)-amino-
octahydro-10-oxo-6H-pyri-
dazo[1,2-a][1,2]diazepine-
1(S)-carboxylate (IV)

Cilazapril

Reference(s):
US 4 512 924 (Hoffmann-La Roche; 23.4.1985; GB-prior. 12.5.1982, 28.2.1983).
US 4 658 024 (Hoffmann-La Roche; 14.4.1987; GB-prior. 12.5.1982).
Attwood, M.R. et al.: FEBS Lett. (FEBLAL) **165**, 201 (1984).

tert-butyl 1-(benzyloxycarbonyl)-hexahydro-3-pyridazinecarboxylate:
Hassall, C.H. et al.: J. Chem. Soc., Perkin Trans. 1 (JCPRB4), 1451 (1979).

Formulation(s):   f. c. tabl. 0.5 mg, 1 mg, 2.5 mg, 5 mg

Trade Name(s):
D: Dynorm (Merck/Roche)   GB: Vascace (Roche)   Initiss (Carlo Erba)
F: Justor (Jacques Logeais)   I: Inibace (Roche)

Cilnidipine
(FRC-8653)

Use: antihypertensive, calcium antagonist

RN: 132203-70-4   MF: $C_{27}H_{28}N_2O_7$ MW: 492.53
LD$_{50}$:   >5 g/kg (M, p.o.);
          4412 mg/kg (R, p.o.);
          >2 g/kg (dog, p.o.)
CN: (E)-(±)-1,4-dihydro-2,6-dimethyl-4-(3-nitrophenyl)-3,5-pyridinedicarboxylic acid 2-methoxyethyl 3-
       phenyl-2-propenyl ester

unspecified stereochemistry
RN: 102106-21-8   MF: $C_{27}H_{28}N_2O_7$ MW: 492.53

(+)-enantiomer
RN: 132338-87-5   MF: $C_{27}H_{28}N_2O_7$ MW: 492.53
(−)-enantiomer
RN: 132295-21-7   MF: $C_{27}H_{28}N_2O_7$ MW: 492.53

2-methoxyethyl   cinnamyl 2-(3-   Cilnidipine
3-aminocrotonate   nitrobenzylidene)-
                        acetoacetate

Reference(s):
EP 161 877 (Fujirebio; appl. 2.5.1985; J-prior. 4.5.1984, 20.6.1984).
Drugs Future (DRFUD4) **21**(3), 249-253 (1996).

Formulation(s):   tabl. 5 mg, 10 mg

Trade Name(s):
J: Atelec (Ajinomoto/Nippon-   Cinalong (Fujirebio)   Ciscard (Nippon
        HMR)   Boehringer Ing.)

Cilostazol
(OPC-13013)

ATC: B01AC
Use: platelet aggregation inhibitor,
        cerebral vasodilating activity

RN: 73963-72-1   MF: $C_{20}H_{27}N_5O_2$ MW: 369.47
LD_{50}: >5 g/kg (M, p.o.);
        >5 g/kg (R, p.o.);
        >2 g/kg (dog, p.o.)
CN: 6-[4-(1-cyclohexyl-1H-tetrazol-5-yl)butoxy]-3,4-dihydro-2(1H)-quinolinone

cyclohexyl-   5-chloropenta-   N-(5-chloropen-
amine   noyl chloride   tanoyl)cyclohexyl-
                                             amine

1-cyclohexyl-5-(4-   6-hydroxy-3,4-   1,5-diaza-   Cilostazol
chlorobutyl)tetrazole (I)   dihydrocarbo-   bicyclo-
                            styril   [5.4.0]undec-
                            (cf. carteolol   5-ene
                            synthesis)

Reference(s):
DOS 2 934 747 (Otsuka; appl. 28.8.1979; J-prior. 1.9.1978).
US 4 277 479 (Otsuka; 7.7.1981; J-prior. 1.9.1978).
Nishi, T. et al.: Chem. Pharm. Bull. (CPBTAL) **31**, 1151 (1983).

medical use for treatment of nephritis:
JP 2 178 227 (Otsuka; appl. 28.12.1988).

medical use for treatment of Raynaud's syndrome:
JP 2 178 226 (Otsuka, appl. 28.12.1988).

Formulation(s):   tabl. 50 mg, 100 mg

Trade Name(s):
J: Pletaal (Otsuka)   Retal (Otsuka; 1988)

Cimetidine

ATC: A02BA01
Use: peptic ulcer therapeutic (H_2-receptor antagonist)

RN: 51481-61-9   MF: $C_{10}H_{16}N_6S$ MW: 252.35   EINECS: 257-232-2
LD_{50}:   150 mg/kg (M, i.v.); 2550 mg/kg (M, p.o.);
        106 mg/kg (R, i.v.); 5 g/kg (R, p.o.);
        206 mg/kg (dog, i.v.); 2600 mg/kg (dog, p.o.)
CN: *N*-cyano-*N'*-methyl-*N''*-[2-[[(5-methyl-1*H*-imidazol-4-yl)methyl]thio]ethyl]guanidine

ethyl 2-chloro-   formamide   ethyl 5-methyl-   4-hydroxymethyl-
acetoacetate   imidazole-4-   5-methylimidazole
                                       carboxylate   hydrochloride (I)

cysteamine   4-[(2-aminoethylthio)-   dimethyl cyano-
hydrochloride   methyl]-5-methyl-   carboimidodithioate
                            imidazole dihydrochloride   (from cyanamide, KOH,
                                                        carbon disulfide and
                                                        dimethyl sulfate)

(II)   Cimetidine

Reference(s):
US 3 894 151 (Smith Kline & French; 8.7.1975; GB-prior. 20.4.1972).
US 4 000 302 (Smith Kline & French; 28.12.1976; GB-prior. 20.4.1972).
DOS 2 320 131 (Smith Kline & French; appl. 19.4.1973; GB-prior. 20.4.1972) – medical use.
DOS 2 344 779 (Smith Kline & French; appl. 5.9.1973; GB-prior. 5.9.1972 and 8.2.1973).
US 3 950 333 (Smith Kline & French; 13.4.1976; appl. 14.3.1974; prior. 29.2.1972 and 20.9.1972).

cimetidine "A":
DOS 2 742 531 (Smith Kline & French; appl. 21.9.1977; GB-prior. 21.9.1976, 24.1.1977).
GB 1 543 238 (Smith Kline & French; appl. 21.9.1976, 24.1.1977, 20.9.1977; valid from 13.12.1977).

precursors and alternative methods:
DOS 2 637 670 (Smith Kline & French; appl. 20.8.1976; USA-prior. 20.8.1975 and 27.5.1976).
FR 2 321 490 (Smith Kline & French; appl. 16.8.1976; USA-prior. 20.8.1975 and 27.5.1976).
GB 1 338 169 (Smith Kline & French; appl. 9.3.1971 and 22.7.1971; valid from 9.3.1972).
DAS 2 211 454 (Smith Kline & French; appl. 9.3.1972; GB-prior. 9.3.1971 and 22.7.1971).
US 4 018 931 (Smith Kline & French; 19.4.1977; appl. 4.12.1975; prior. 29.2.1972, 20.9.1972 and 14.3.1974).
US 4 013 678 (Smith Kline & French; 22.3.1977; GB-prior. 2.9.1974).
US 3 984 293 (Smith Kline & French; 5.10.1976; prior. 2.9.1974).
BE 853 954 (Smith Kline Corp. GB appl. 26.4.1977; USA-prior. 22.2.1977).
US 4 063 023 (Smith Kline & French; 13.12.1977; prior. 20.8.1975).
DOS 2 649 059 (Smith Kline; appl. 28.10.1976; USA-prior. 29.10.1975).
DOS 2 718 715 (Smith Kline; appl. 27.4.1977; USA-prior. 22.2.1977).
US 4 049 672 (Smith Kline & French; 20.9.1977; appl. 17.3.1976; prior. 29.2.1972, 20.9.1972, 14.3.1974).
US 4 104 472 (Smith Kline & French; 1.8.1978; prior. 9.2.1977, 24.5.1977).
US 4 163 858 (Smith Kline; 7.8.1979; prior. 9.2.1977, 24.5.1977, 8.3.1978).
DOS 2 805 221 (Smith Kline; appl. 8.2.1978; USA-prior. 9.2.1977, 24.5.1977).
DOS 2 814 355 (BASF; appl. 3.4.1978).
DOS 2 855 836 (Lab. Om; appl. 22.12.1978; CH-prior. 28.12.1977, 7.12.1978).
FR-appl. 2 386 525 (Ricorvi; appl. 17.10.1977; E-prior. 6.4.1977).

X-ray structure:
Hädicke, E. et al.: Chem. Ber. (CHBEAM) **111**, 3222 (1978).

combination with conventional antihistaminics:
US 4 104 382 (Smith Kline & French; 1.8.1978; prior. 9.4.1973, 16.4.1975, 27.9.1976).

Formulation(s):   amp. 200 mg/2 ml, 400 mg/4 ml, 1000 mg/10 ml; eff. tabl. 400 mg, 800 mg; f. c. and tabl. 200 mg, 400 mg, 800 mg

Trade Name(s):

D: Altramet (ASTA Medica AWD)
    Azucimet (Anipharma)
    Tagamet (SmithKline Beecham; 1977)
F: Stomédine (SmithKline Beecham)
    Tagamet (SmithKline Beecham; 1977)
GB: Algitec (SmithKline Beecham)-comb.
    Dyspamet (SmithKline Beecham)
    Galenamet (Galen)
    Tagamet (SmithKline Beecham; 1976)

    Zita (Eastern)
I: Biomag (Pulitzer)
    Brumetidina (Bruschettini)
    Citimid (CT)
    Dina (San Carlo)
    Eureceptor (Zambon)
    Gastromet (Bayropharm)
    Neo Gastransil (Schiapparelli Searle)
    Notul (Mendelejeff; as hydrochloride)
    Stomet (Allergan)
    Tagamet (Smith Kline & French; 1977)
    Tametin (SmithKline Beecham)

    Temic (Farma Uno)
    Ulcedin (AGIPS)
    Ulcestop (Metapharma; as hydrochloride)
    Ulcodina (Locatelli)
    Ulcofalk (Interfalk)
    Ulcomedina (Leben's)
    Ulis (Lafare)
    Vagolisal (Biotekfarma)
J: Tagamet (SKF-Fujisawa; 1982)
USA: Tagamet (SmithKline Beecham; 1977)

Cimetropium bromide

ATC: A03BB05
Use: anticholinergic, antispasmodic

RN: 51598-60-8   MF: $C_{21}H_{28}BrNO_4$ MW: 438.36
CN: [7(S)-(1α,2β,4β,5α,7β)]-9-(cyclopropylmethyl)-7-(3-hydroxy-1-oxo-2-phenylpropoxy)-9-methyl-3-oxa-9-azoniatricyclo[3.3.1.0^{2,4}]nonane bromide

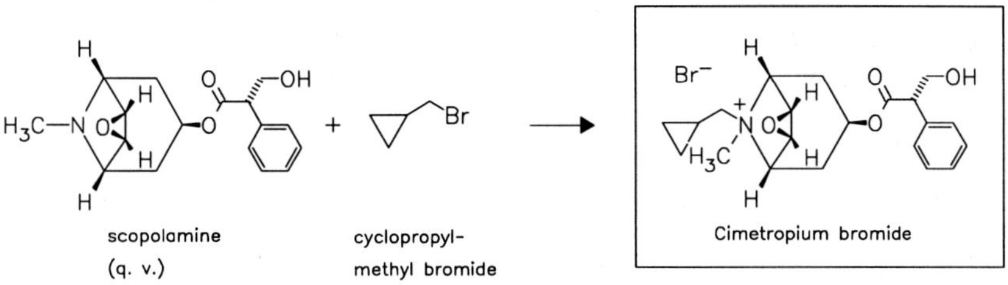

scopolamine (q. v.) + cyclopropyl-methyl bromide → Cimetropium bromide

Reference(s):

US 3 853 886 (De Angeli; 10.12.1974; appl. 13.4.1973; GB-prior. 18.4.1972).
US 3 952 108 (De Angeli; 20.4.1976, GB-prior. 18.4.1972).
DOS 2 316 728 (De Angeli; appl. 4.4.1973; GB-prior. 18.4.1972).

Formulation(s): amp. 5 mg/ml; suppos. 50 mg; syrup 1 %; tabl. 50 mg

Trade Name(s):
I: Alginor (Boehringer Ing.; 1985)

Cinchocaine
(Dibucaine)

ATC: C05AD04; D04AB02; N01BB06; S01HA06
Use: local anesthetic

RN: 85-79-0 MF: $C_{20}H_{29}N_3O_2$ MW: 343.47 EINECS: 201-632-1
LD$_{50}$: 24.5 mg/kg (M, i.p.); 28.5 mg/kg (M, s.c.)
CN: 2-butoxy-*N*-[2-(diethylamino)ethyl]-4-quinolinecarboxamide

monohydrochloride
RN: 61-12-1 MF: $C_{20}H_{29}N_3O_2 \cdot HCl$ MW: 379.93 EINECS: 200-498-1
LD$_{50}$: 3800 µg/kg (M, i.v.);
52 mg/kg (R, i.v.)

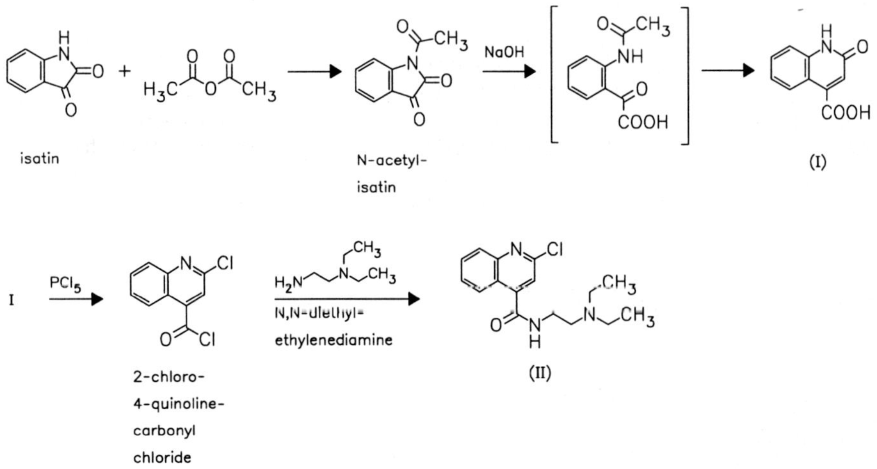

isatin + (acetic anhydride) → N-acetyl-isatin → (NaOH) → [ ] → (I)

I → (PCl$_5$) → 2-chloro-4-quinoline-carbonyl chloride → (N,N-diethyl-ethylenediamine) → (II)

II + HO⌒⌒CH₃ →(Na) [Cinchocaine structure]

1-butanol   Cinchocaine

Reference(s):
DRP 537 104 (Ciba; appl. 1926).
US 1 825 623 (Ciba; 1931; D-prior. 1926).
Miescher, K.: Helv. Chim. Acta (HCACAV) **15**, 163 (1932).

Formulation(s):   amp. 6 mg/3 ml (as hydrochloride); rectal ointment 5 mg/100 g; suppos. 1 mg

Trade Name(s):

D:	Anumedin (Kade)-comb.		Ultrapoct (Schering)-comb.	I:
	Butazolidin (Novartis	F:	Deliproct (Schering)-comb.	
	Pharma)-comb.		Ultraproct (Schering)-	
	Dolo-Posterine (Kade)		comb.	
	Faktu (Byk Gulden;	GB:	Nupercainal (Novartis)	J:
	Roland)-comb.		Proctosedyl (Hoechst)-	
	Otobacid (Asche)-comb.		comb.	
	Procto-Kaban (Asche)-		Scheriproct (Schering)-	
	comb.		comb.	USA:
	Protospre (Hennig)-comb.		Ultraproct (Schering)-	
	Scheriproct (Schering)-		comb.	
	comb.		Uniroid (Unigreg)-comb.	

I: Algolisina (Celsius)-comb.
      Nupercainal (Ciba); wfm
      Ultraproct (Schering)-
      comb.
J: Nupercain (Ciba-Geigy-
      Takeda)
      Percamin (Teikoku
      Kagaku-Nagase)
USA: Nupercaine (Ciba); wfm

Cineole

(Cajeputol; Eucalyptol)

ATC: R01AX; R05CA
Use: antiseptic, expectorant

RN: 470-82-6   MF: C₁₀H₁₈O   MW: 154.25   EINECS: 207-431-5
LD₅₀: 2480 mg/kg (R, p.o.)
CN: 1,3,3-trimethyl-2-oxabicyclo[2.2.2]octane

Cineole

Principal ingredient of eucalyptus oils, isolated after separation of remaining terpenes with sulfuric acid.

Reference(s):
Ullmanns Encykl. Tech. Chem., 4. Aufl., Vol. **22**, 542.
DRP 499 732 (Rhein. Kampfer-Fabrik; appl. 1928).
US 2 090 620 (Newport Industries; 1937; appl. 1936).

Formulation(s):   cps. 100 mg; sol. 2 g/100 g, 15 g/100 g

Trade Name(s):

D:	Denesol (Doerernkamp)-	Eufimenth (Lichtenstein)-	Pinimenthol (Spitzner)-
	comb.	comb.	comb.

Rowachol (Rowa-Wagner)-
comb.
Rowatinex (Rowa-
Wagner)-comb.
Soledum (Cassella-med)
Transpulmin (ASTA
Medica)
Wick VapoRup (Wick
Pharma)

numerous combination
preparations
GB: Rowachol (Rowa)-comb.
Rowatinex (Rowa)-comb.
I: Alc Ment Cmp (Formulario
Naz.)-comb.
Balsamic (Formulario
Naz.)-comb.
Brochenolo Balsamo
(Midy)-comb.

Calyptol (Rhône-Poulenc
Rorer)
Eucal (Tariff. Nazionale)
Eucalipt (Tariff. Nazionale)
Rinostil (Deca)-comb.
numerous combination
preparations
USA: Listerine Antiseptic
(Warner-Lambert)

Cinepazet

(Cinepazate)

ATC: C01DX14
Use: vasodilator, antianginal

RN: 23887-41-4   MF: $C_{20}H_{28}N_2O_6$ MW: 392.45   EINECS: 245-927-3
LD_{50}:   1300 mg/kg (M, p.o.); 300 mg/kg (M, i.v.)
CN: 4-[1-oxo-3-(3,4,5-trimethoxyphenyl)-2-propenyl]-1-piperazineacetic acid ethyl ester

maleate (1:1)
RN: 50679-07-7   MF: $C_{20}H_{28}N_2O_6 \cdot C_4H_4O_4$ MW: 508.52   EINECS: 256-709-2

3,4,5-trimethoxy-
cinnamoyl chloride

ethyl 1-piperazine-
acetate

Cinepazet

Reference(s):
DAS 1 795 402 (Delalande; appl. 26.9.1968; GB-prior. 29.9.1967).

Formulation(s):   tabl. 300 mg

Trade Name(s):
F: Vascoril (Delalande); wfm   I: Vascoril (Delalande); wfm

Cinepazide

ATC: C04AX27
Use: vasodilator (peripheral)

RN: 23887-46-9   MF: $C_{22}H_{31}N_3O_5$ MW: 417.51   EINECS: 245-928-9
CN: 1-[2-oxo-2-(1-pyrrolidinyl)ethyl]-4-[1-oxo-3-(3,4,5-trimethoxyphenyl)-2-propenyl]piperazine

maleate (1:1)
RN: 26328-04-1   MF: $C_{22}H_{31}N_3O_5 \cdot C_4H_4O_4$ MW: 533.58   EINECS: 247-613-1
LD_{50}:   617 mg/kg (M, i.v.); 1000 mg/kg (M, p.o.);
414 mg/kg (R, i.v.); 1310 mg/kg (R, p.o.)

3,4,5-trimethoxy-
cinnamoyl chloride

piperazinoacetic
acid pyrrolidide

Cinepazide

Reference(s):
DE 1 915 795 (Delalande; appl. 27.3.1969; GB-prior. 3.4.1968).
DOS 2 043 350 (Delalande; appl. 1.9.1970; F-prior. 17.10.1969).
US 3 634 411 (Delalande; 11.1.1972; GB-prior. 3.4.1968).

Formulation(s):   amp. 80 mg/2 ml; tabl. 200 mg (as maleate)

Trade Name(s):
F: Vasodistal (Delalande;
      1974); wfm
I: Vasodistal (Delalande;
      1978); wfm
J: Anapazin (Zenyaku)
      Bilbvarde (Yoshindo)
      Brendil (Daiichi; 1981)
      Brentomine (Daito Koeki
      Nichiiko)

Brepanael (Hotta)
Cinema (Choseido)
Ekarusin (Seiko Eiyo)
Madesol (Sanwa)
Mishiline (Mishiyama)
Neubcat (Nippon Shoji)
Prosmet (Sawai)
Schulandere (Tsuruhara)
Scorojile (Kotobuki)

Sebdeel (Mohan)
Sylpinale (Teikoku
Kagaku)
Tatsumedil (Tatsumi)
Tineup (Maruko)
Vasodeniel (MF-Taiyo)

Cinitapride
(LAS-17177)

ATC: A04
Use: gastrointestinal

RN: 66564-14-5   MF: $C_{21}H_{30}N_4O_4$ MW: 402.50
CN: 4-amino-N-[1-(3-cyclohexen-1-ylmethyl)-4-piperidinyl]-2-ethoxy-5-nitrobenzamide

fumarate (1:1)
RN: 67135-13-1   MF: $C_{21}H_{30}N_4O_4 \cdot C_4H_4O_4$ MW: 518.57
tartrate
RN: 96623-56-2   MF: $C_{21}H_{30}N_4O_4 \cdot xC_4H_6O_6$ MW: unspecified
LD_{50}: >450 mg/kg (M, R, p.o.)

2-ethoxy-4-nitro-
benzoic acid

4-amino-2-ethoxy-
5-nitrobenzoic acid   (I)

4-amino-1-(3-
cyclohexen-1-yl-
methyl)piperidine (II)

ethyl chloroformate

Cinitapride

synthesis of II

1. Li, NH3
2. H2, Pd
II

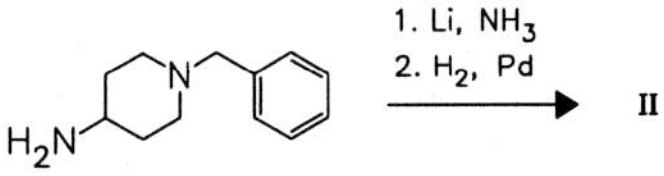

4-amino-1-benzyl-
piperidine

Reference(s):
GB 1 574 419 (Anphar; appl. 3.9.1980; GB-prior. 16.11.1976).
CH 628 886 (Anphar; appl. 31.3.1982; CH-prior. 1.1.1978).

synthesis of I:
Goldstein, H.; Brochon, R.: Helv. Chim. Acta (HCACAV) **32**, 2334 (1949).

*synthesis of II/*cinitapride:
DE 2 706 038 (A. Gallardo; appl. 12.2.1977; GB-prior. 17.2.1976).

alternative synthesis of cinitapride:
ES 2 001 458 (Fordonal; appl. 16.5.1988; E-prior. 12.12.1986).

Formulation(s): sol. 0.2 mg/ml (as tartrate); tabl. 1 mg

Trade Name(s):
E: Cidine (Almirall; 1990)

Cinmetacin

ATC: M01AB
Use: non-steroidal anti-inflammatory

RN: 20168-99-4 MF: $C_{21}H_{19}NO_4$ MW: 349.39 EINECS: 243-555-6
LD_{50}: 360 mg/kg (M, i.p.); 750 mg/kg (M, p.o.);
590 mg/kg (R, i.p.); 1020 mg/kg (R, p.o.)
CN: 5-methoxy-2-methyl-1-(1-oxo-3-phenyl-2-propenyl)-1*H*-indole-3-acetic acid

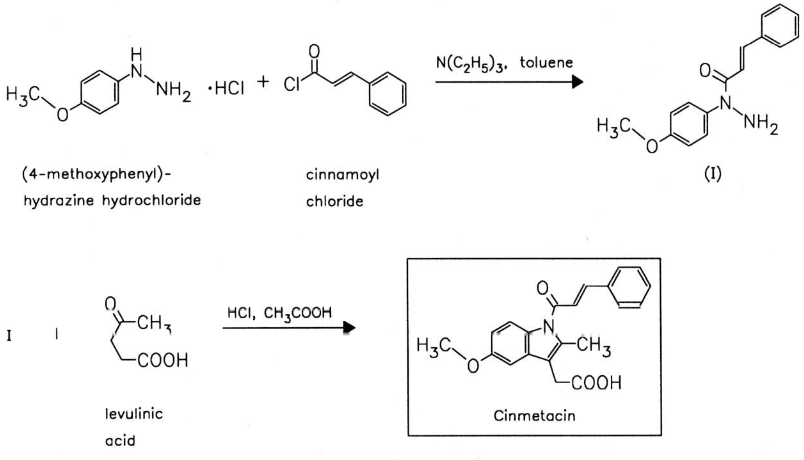

·HCl
N(C2H5)3, toluene
(4-methoxyphenyl)-
hydrazine hydrochloride
cinnamoyl
chloride
(I)
I
HCl, CH3COOH
levulinic
acid
Cinmetacin

Reference(s):
US 3 576 800 (Sumitomo; 27.4.1971; J-prior. 12.5.1966, 27.6.1966, 30.6.1966, 8.7.1966, 1.8.1966, 19.8.1966, 15.12.1966, 16.12.1966, 20.12.1966, 6.1.1967, 7.1.1967, 16.1.1967, 17.1.1967).
ZA 672 683 (Sumitomo; appl. 12.4.1967; J-prior. 12.5.1966, 27.6.1966, 30.6.1977).
Yamamoto, H.; Nakao, M.: J. Med. Chem. (JMCMAR) **12**, 176 (1969).

Formulation(s):   cps 300 mg; suppos. 375 mg, 750 mg

Trade Name(s):
I: Cindomet (Chiesi); wfm   J: Indolacin (Sumitomo)

Cinnamedrine
(Cinnamylephedrine)

ATC: N02
Use: uterine antispasmodic, treatment of
        menstrua syndrom

RN: 90-86-8   MF: C$_{19}$H$_{23}$NO   MW: 281.40   EINECS: 202-021-2
CN: α-[1-[methyl(3-phenyl-2-propenyl)amino]ethyl]benzenemethanol

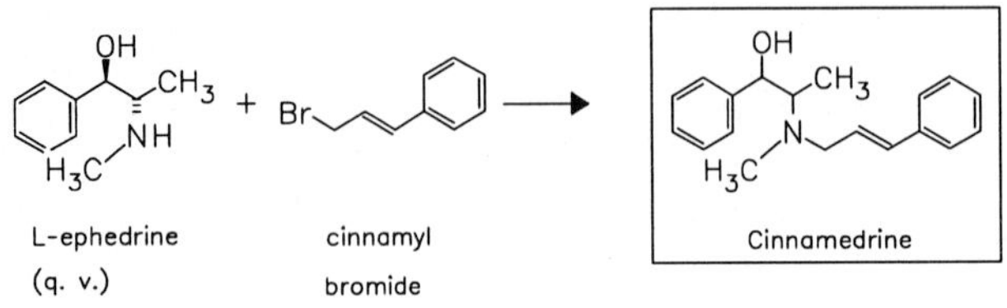

L-ephedrine   cinnamyl   Cinnamedrine
(q. v.)   bromide

Reference(s):
US 1 959 392 (Winthrop; 1934; D-prior. 1930).
Welsh, L.H.; Kennan, G.L.: J. Am. Pharm. Assoc. (JPHAA3) **30**, 123 (1941).

Formulation(s):   tabl. 14.9 mg in combination with aspirine, coffeine

Trade Name(s):
USA: Midol (Glenbrook)-comb.;
        wfm

Cinnarizine

ATC: N07CA02
Use: antihistaminic, vasodilator

RN: 298-57-7   MF: C$_{26}$H$_{28}$N$_2$ MW: 368.52   EINECS: 206-064-8
LD$_{50}$:   22 mg/kg (M, i.v.); >4500 mg/kg (M, p.o.);
        24 mg/kg (R, i.v.); >6500 mg/kg (R, p.o.);
        >500 mg/kg (dog, p.o.)
CN: 1-(diphenylmethyl)-4-(3-phenyl-2-propenyl)piperazine

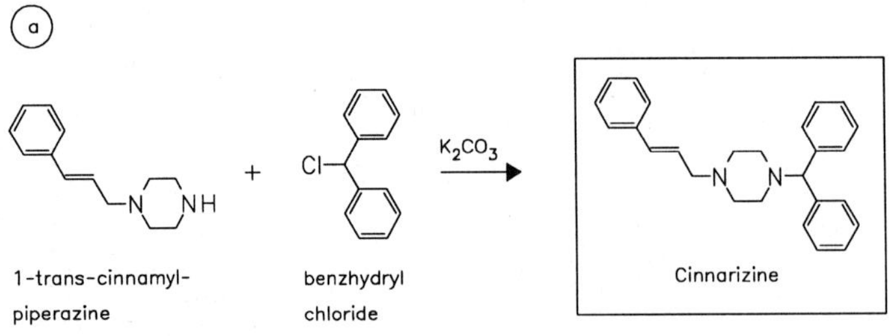

1-trans-cinnamyl-   benzhydryl   Cinnarizine
piperazine   chloride

(b)

cinnamyl
chloride

1-benzhydryl-
piperazine

K_2CO_3 → Cinnarizine

Reference(s):
US 2 882 271 (Janssen; 14.8.1959; NL-prior. 20.4.1956).
DE 1 086 235 (Janssen; appl. 10.4.1957; NL-prior. 20.4.1956).

combination with dihydroergotamine:
DOS 2 820 937 (Dolorgiet; appl. 12.5.1978).

Formulation(s): cps. 75 mg; tabl. 20 mg, 25 mg, 75 mg

Trade Name(s):

D:	Arlevert (Henning)-comb.		Toliman (Corvi)
	Cinnacet (Sanofi Winthrop)	J:	Annarizine (Sioe)
	Cinnarizin forte R.A.N.		Aplactan (Eisai)
	(R.A.N.)		Aplexal (Taiyo Yakuko
F:	Sureptil (Synthélabo)-		Takayama)
	comb.		Apomiterl (Teizo)
GB:	Stugeron (Janssen-Cilag)		Apsatan (Wakamoto)
I:	Cinazyn (Fisons		Cerebalan (Tobishi)
	Italchimici)		Corathiem (Ohta)
	Stugeron (Janssen)		Denapol (Teisan)
	Sureptil (Delalande		Eglen (Tatsumi)
	Isnardi)-comb.		Hirdsyn (Fuso)

Izaberizin (Toho)
Katoseran (Hishiyama)
Milactan (Miwa)
Processine (Sankyo)
Roin (Maruishi)
Salarizine (Iwaki)
Sapratol (Daigo-Takeda)
Sedatromin (Takeda)
Sigmal (Fuji Zoki)
Siptazin (Isei)
Spaderizine (Kotobuki)
Tolesmin (Sato)

Cinolazepam

(Ox-373)

ATC: N05CD13
Use: hypnotic benzodiazepine

RN: 75696-02-5 MF: $C_{18}H_{13}ClFN_3O_2$ MW: 357.77
LD_{50}: 3.5 g/kg (R, p.o.)
      3.5 g/kg (M, p.o)
CN: (±)-7-Chloro-5-(2-fluorophenyl)-2,3-dihydro-3-hydroxy-2-oxo-1H-1,4-benzodiazepine-1-propanenitrile

2-amino-5-chloro-
2'-fluorobenzophenone
(cf. flunitrazepam
synthesis,
doxefazepam synthesis)

7-chloro-5-
(2-fluorophenyl)-
1,3-dihydro-2H-
1,4-benzo-
diazepin-2-one

(I)

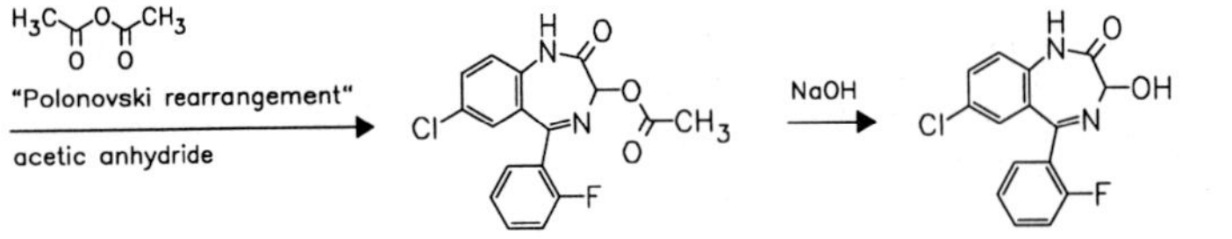

(±)-7-chloro-3-hydroxy-
5-(2-fluorophenyl)-
1,3-dihydro-2H-1,4-
benzodiazepin-2-one (II)

Cinolazepam

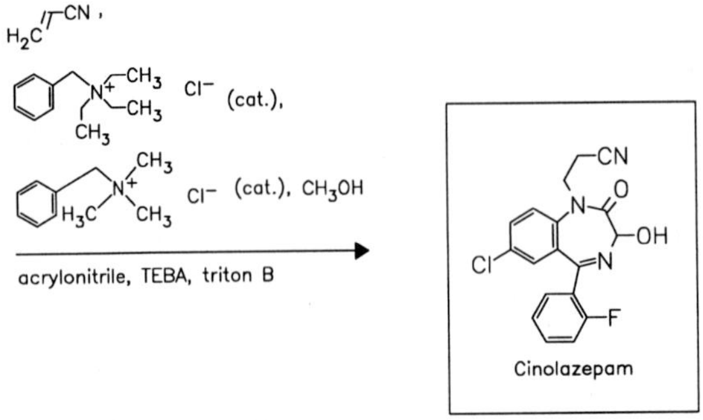

Reference(s):
DE 2 950 235 (Gerot Pharmazeutika; appl. 23.12.1979; A-prior. 18.12.1978).

synthesis of intermediate II:
Earley, J.V.; Fryer, R.I.; Winter, D.; Sternbach, L.H.: J. Med. Chem. (JMCMAR) **11** (4), 774 (1968).

Formulation(s):   tabl. 40 mg

Trade Name(s):
A: Geroderm (Gerot; 1993)

Cinoxacin

(Acidum azolinicum; Azolinic acid)

ATC: G04AB05
Use: antibacterial (treatment of urinary
tract infections)

RN: 28657-80-9   MF: $C_{12}H_{10}N_2O_5$ MW: 262.22   EINECS: 249-133-8
LD_{50}:   900 mg/kg (R, i.v.); 4160 mg/kg (R, p.o.)
CN: 1-ethyl-1,4-dihydro-4-oxo[1,3]dioxolo[4,5-g]cinnoline-3-carboxylic acid

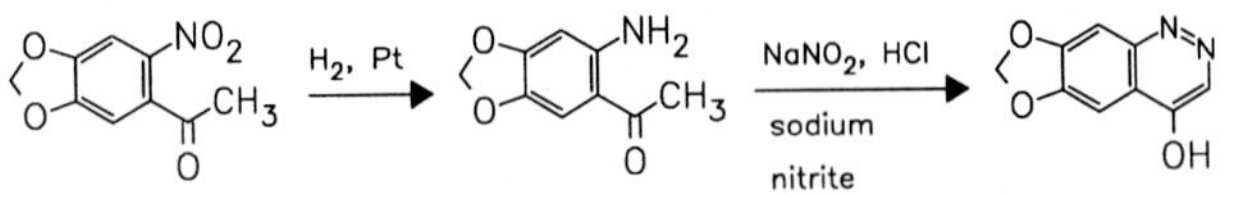

4,5-methylene-
dioxy-2-nitro-
acetophenone

2-amino-4,5-
methylenedioxy-
acetophenone

4-hydroxy-
6,7-methylene-
dioxycinnoline (I)

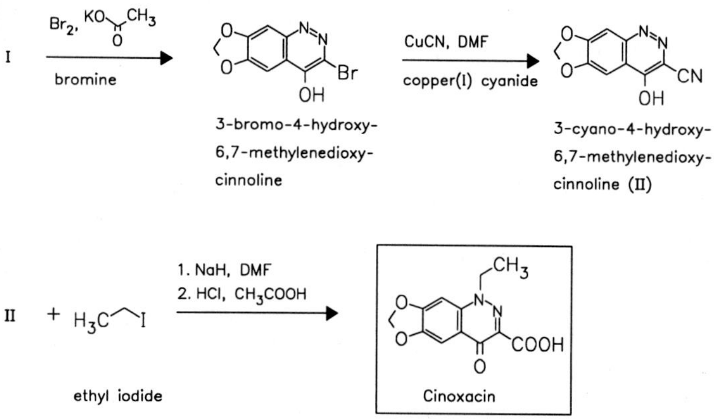

3-bromo-4-hydroxy-
6,7-methylenedioxy-
cinnoline

3-cyano-4-hydroxy-
6,7-methylenedioxy-
cinnoline (II)

ethyl iodide   Cinoxacin

Reference(s):
US 3 669 965 (Eli Lilly; 13.6.1972; prior. 29.12.1969).
DOS 2 005 104 (Eli Lilly; appl. 4.2.1970).

Formulation(s):   cps. 250 mg, 500 mg

Trade Name(s):

D:	Cinoxacin (Rosen Pharma)		Noxigram (Firma)	J:	Cinobact (Shionogi)
GB:	Cinobac (Lilly; 1979)		Uronorm (Alfa	USA:	Cinobac (Dista; 1981);
I:	Cinobac (Lilly)		Wassermann)		wfm
	Nossacin (Corvi)		Uroxacin (Malesci)		

Ciprofibrate

ATC: B04AC; C01AB08
Use: antihyperlipidemic, clofibrate derivative

RN: 52214-84-3   MF: $C_{13}H_{14}Cl_2O_3$ MW: 289.16   EINECS: 257-744-6
CN: 2-[4-(2,2-dichlorocyclopropyl)phenoxy]-2-methylpropanoic acid

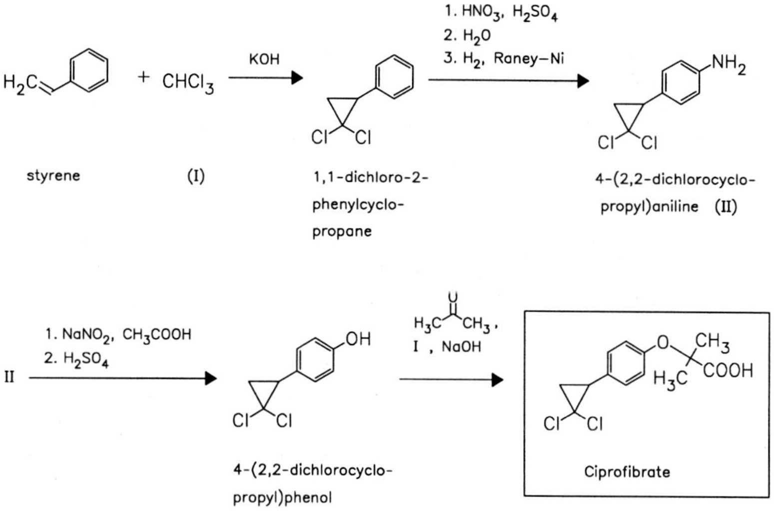

styrene   (I)   1,1-dichloro-2-
phenylcyclo-
propane

4-(2,2-dichlorocyclo-
propyl)aniline   (II)

4-(2,2-dichlorocyclo-
propyl)phenol

Ciprofibrate

Reference(s):
US 3 948 973 (Sterling Drug; 6.4.1976; prior. 29.8.1972).
DOS 2 343 606 (Sterling Drug; appl. 29.8.1973; USA-prior. 29.8.1972).

synthesis of 4-(2,2-dichlorocyclopropyl)aniline:
Nefedov, O.M.; Shafran, R.N.: Zh. Org. Khim. (ZORKAE) **1974,** 477.
C.A. (CHABA8) **80**, 145626o (1974).

Formulation(s): cps. 100 mg, tabl. 100 mg

Trade Name(s):
F: Lipanor (Sanofi Winthrop; 1985)
GB: Modalim (Sanofi Winthrop)

Ciprofloxacin

(Bay-o-9867)

ATC: J01MA02; S03AA07
Use: antibacterial

RN: 85721-33-1 MF: $C_{17}H_{18}FN_3O_3$ MW: 331.35
LD_{50}: 122 mg/kg (M, i.v.); 5 g/kg (M, p.o.);
207 mg/kg (R, i.v.); >2 g/kg (R, p.o.)
CN: 1-cyclopropyl-6-fluoro-1,4-dihydro-4-oxo-7-(1-piperazinyl)-3-quinolinecarboxylic acid

monohydrate
RN: 113078-43-6 MF: $C_{17}H_{18}FN_3O_3 \cdot H_2O$ MW: 349.36
monohydrochloride
RN: 93107-08-5 MF: $C_{17}H_{18}FN_3O_3 \cdot HCl$ MW: 367.81
hydrochloride
RN: 86483-48-9 MF: $C_{17}H_{18}FN_3O_3 \cdot xHCl$ MW: unspecified
LD_{50}: 258 mg/kg (M, i.v.); >5 g/kg (M, p.o.);
300 mg/kg (R, i.v.); >5 g/kg (R, p.o.)
lactate (1:1)
RN: 97867-33-9 MF: $C_{17}H_{18}FN_3O_3 \cdot C_3H_6O_3$ MW: 421.43

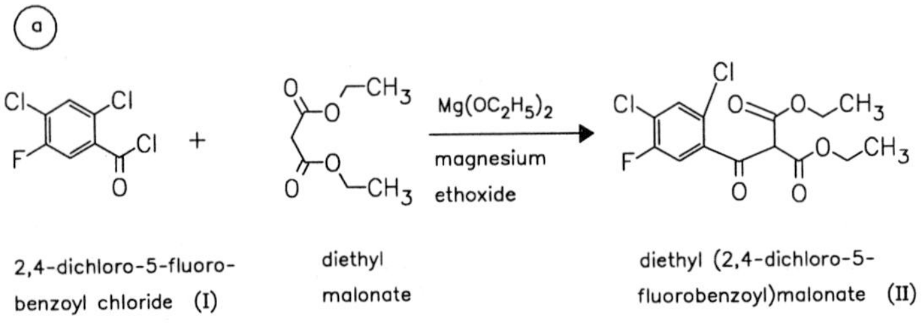

a
Mg(OC2H5)2
magnesium ethoxide
2,4-dichloro-5-fluoro-benzoyl chloride (I)
diethyl malonate
diethyl (2,4-dichloro-5-fluorobenzoyl)malonate (II)

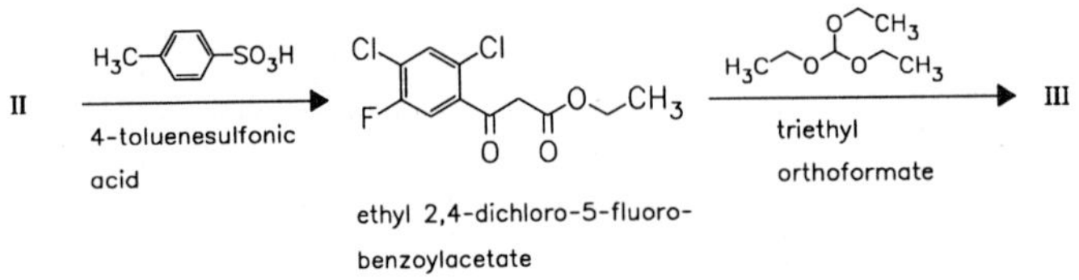

II
4-toluenesulfonic acid
ethyl 2,4-dichloro-5-fluoro-benzoylacetate
triethyl orthoformate
III

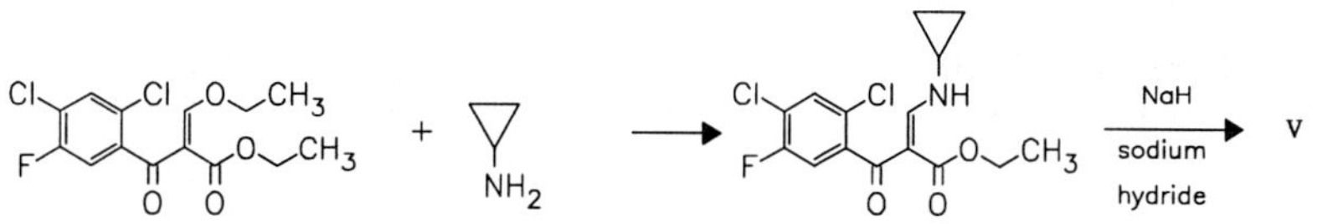

ethyl 2-(2,4-dichloro-5-fluorobenzoyl)-3-ethoxyacrylate (III)

cyclopropyl-amine (IV)

ethyl 3-cyclopropylamino-2-(2,4-dichloro-5-fluoro-benzoyl)acrylate

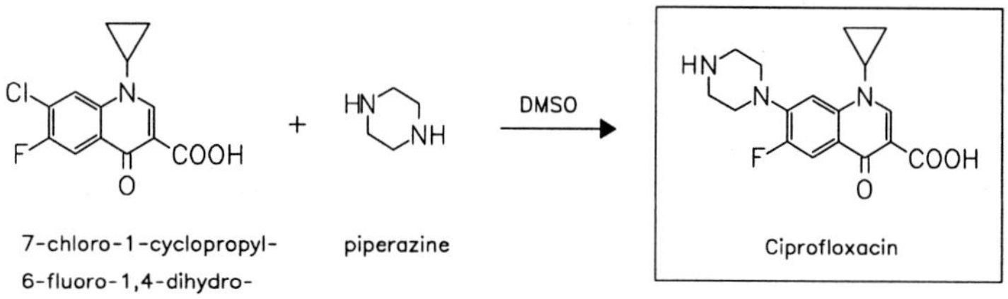

7-chloro-1-cyclopropyl-6-fluoro-1,4-dihydro-4-oxoquinoline-3-carboxylic acid (V)

piperazine

Ciprofloxacin

b

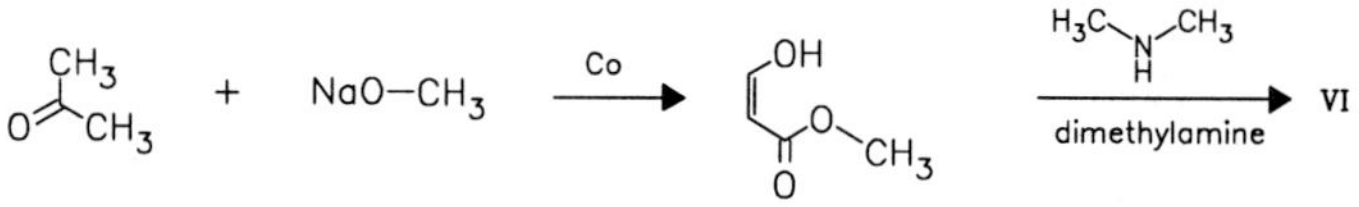

acetone

sodium methylate

methyl 3-hydroxyacrylate

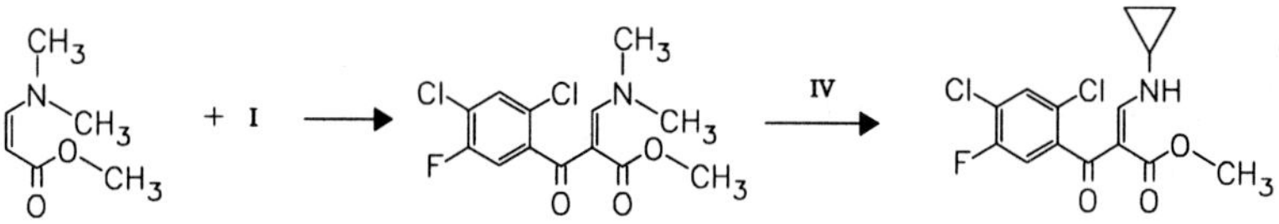

methyl 3-di-methylamino-acrylate (VI)

(VII)

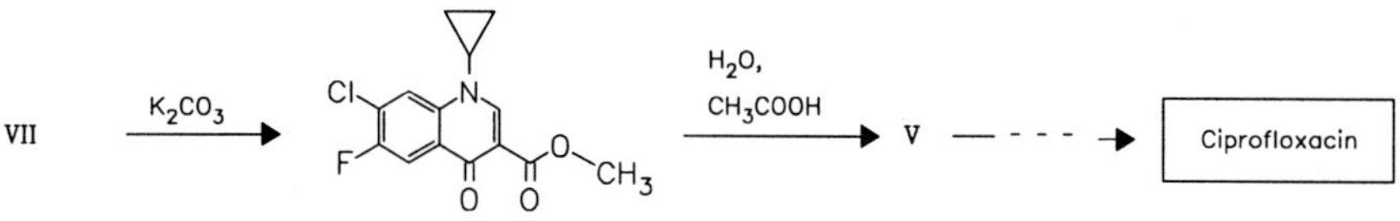

Reference(s):
EP 49 355 (Bayer AG; appl. 21.8.1981; D-prior. 3.9.1980).
US 4 670 444 (Bayer AG; 2.6.1987; D-prior. 3.9.1980).
DE 3 273 892
DOS 3 142 854 (Bayer AG; appl. 29.10.1981).
US 4 620 007 (Bayer AG; 28.10.1986; D-prior. 3.9.1980, 29.10.1981).
Grohe, K.; Heitzer, H.: Liebigs Ann. Chem. (LACHDL) **1987**, 29.
EP 657 448 (Bayer AG; appl. 28.11.1994; D-prior. 10.12.1993).

Formulation(s):   amp. 100 mg/10 ml, 200 mg/200 ml, 400 mg/400 ml; eye drops 3 mg/3 ml; tabl. 100 mg, 200 mg, 250 mg, 500 mg, 750 mg; vial 100 mg/50 ml, 200 mg/100 ml (as hydrochloride)

D: Ciloxan (Alcon)   F: Ciflox (Bayer)   Flociprin (IBI; 1989)
      Ciprobay (Bayer Vital;   GB: Ciloxan (Alcon)   USA: Ciloxan (Alcon)
      1987)   Ciproxin (Bayer; 1987)   Cipro (Bayer; 1987)
      Uniflox (Bayer)   I: Ciproxin (Bayer; 1989)

Cisapride

ATC: A03FA02
Use: gastrokinetic, promotility

RN: 81098-60-4   MF: C$_{23}$H$_{29}$ClFN$_3$O$_4$ MW: 465.95   EINECS: 279-689-7
CN: (±)-cis-4-Amino-5-chloro-2-methoxy-N-[1-[3-(4-fluorophenoxy)propyl]-3-methoxy-4-
       piperidyl]benzamide

(+)-tartrate
RN: 189888-25-3   MF: C$_{23}$H$_{29}$ClFN$_3$O$_4$ · C$_4$H$_6$O$_6$ MW: 616.04

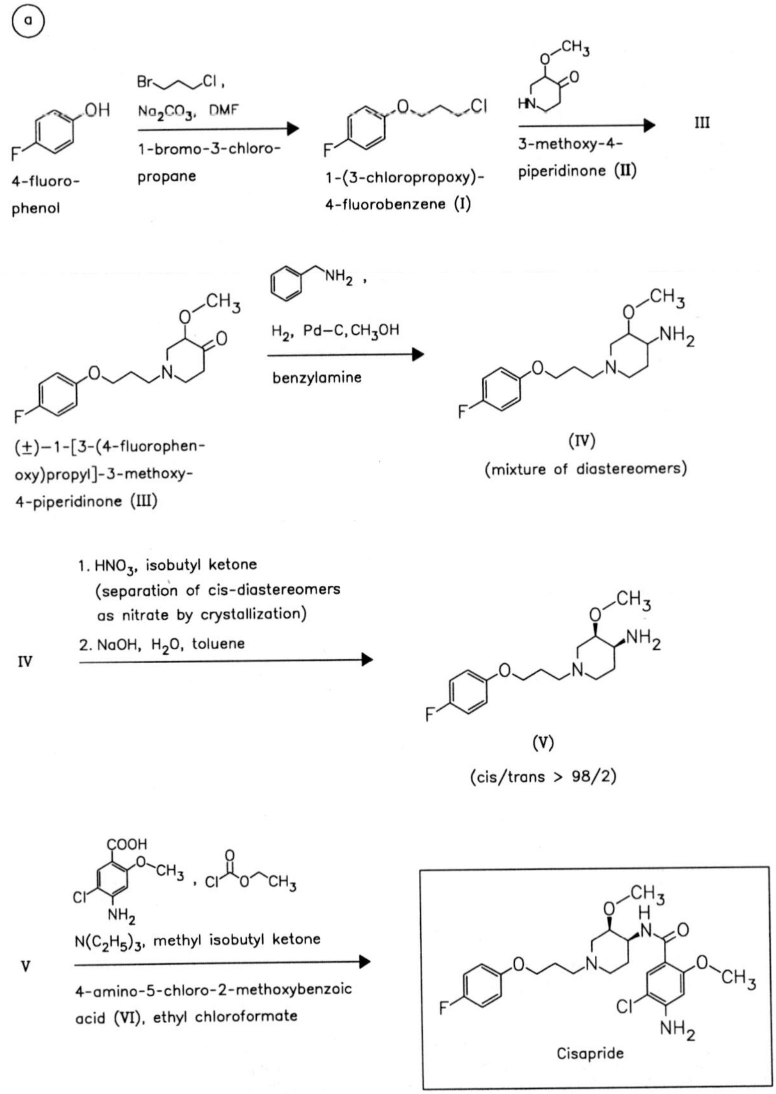

Cisapride

(b)

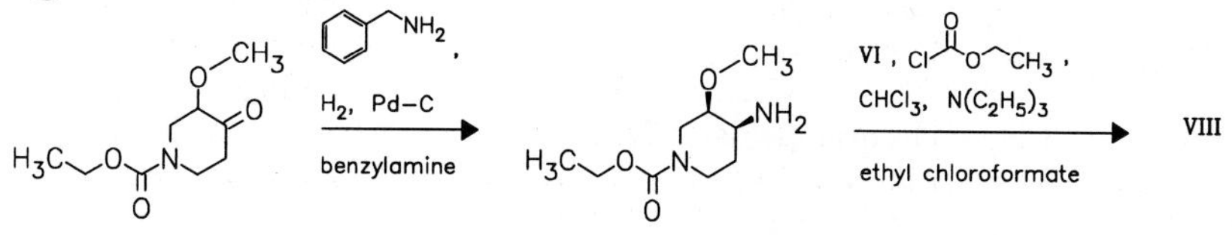
H2, Pd-C
benzylamine
VI, CHCl3, N(C2H5)3
ethyl chloroformate
VIII

ethyl (±)-3-methoxy-4-oxo-1-piperidinecarboxylate (VII)

ethyl (±)-cis-4-amino-3-methoxy-1-piperidinecarboxylate

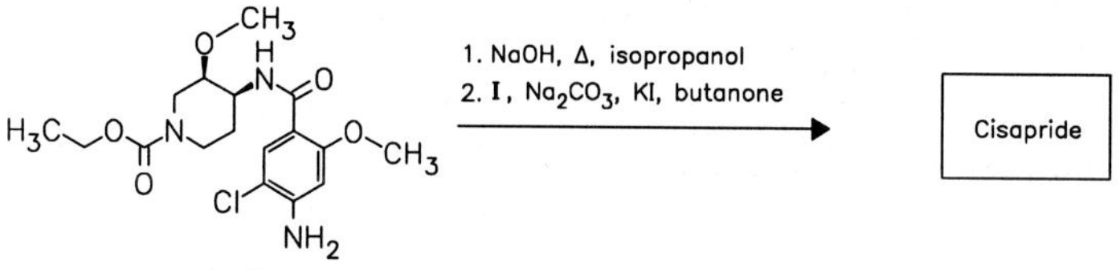
1. NaOH, Δ, isopropanol
2. I, Na2CO3, KI, butanone
Cisapride

(VIII)

preparation of 3-methoxy-4-piperidinone (II):

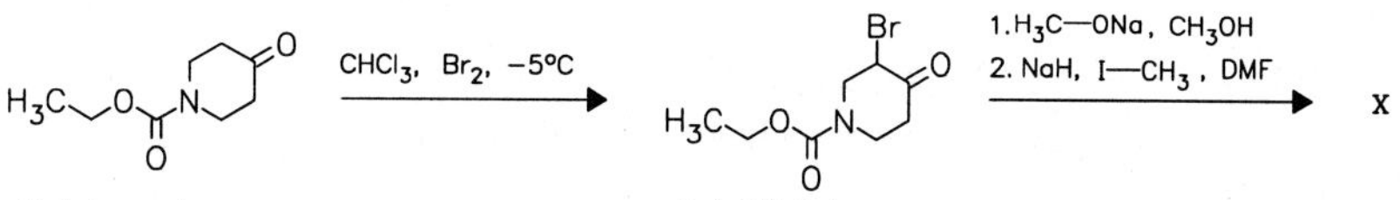
CHCl3, Br2, −5°C
1. H3C—ONa, CH3OH
2. NaH, I—CH3, DMF
X

ethyl 4-oxo-1-piperidinecarboxylate (IX)

ethyl (±)-3-bromo-4-oxo-1-piperidinecarboxylate

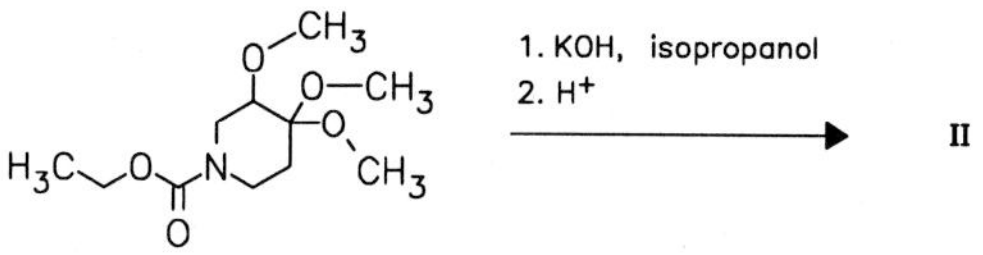
1. KOH, isopropanol
2. H+
II

ethyl (±)-3,4,4-trimethoxy-1-piperidinecarboxylate (X)

preparation of ethyl (±)-3-methoxy-4-oxo-1-piperidinecarboxylate (VII):

(ba)

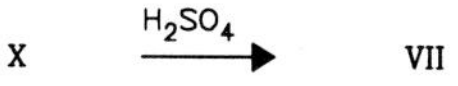
X
H2SO4
VII

(bb)

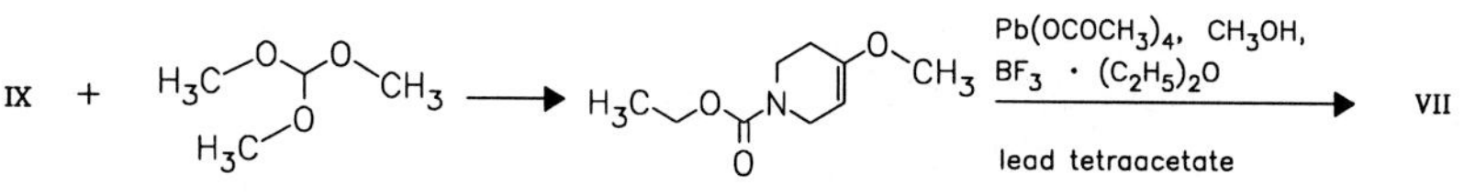
IX +
Pb(OCOCH3)4, CH3OH, BF3 · (C2H5)2O
lead tetraacetate
VII

trimethoxymethane

Reference(s):

a WO 9 816 511 (Janssen Pharmaceuticals; appl. 9.10.1997; EP-prior. 15.10.1996).
a,b EP 76 530 (Janssen Pharmaceuticals; 13.4.1983; USA-prior. 1.10.1981).
bb Singh, V.S.; Singh, C.; Dikshit, D.K.: Synth. Commun. (SYNCAV) **28** (1), 45 (1998).

Formulation(s):   f. c. cps. 400 mg; susp. 1 mg/ml (as hydrate); tabl. 5 mg, 10 mg, 20 mg, 25 mg, 50 mg, 100 mg.

Trade Name(s):

D:	Alimix (Janssen-Cilag)	GB:	Prepulsid (Janssen-Cilag)
	Propulsin (Janssen-Cilag)	I:	Alimix (Cilag)
F:	Prepulsid (Janssen-Cilag)		Cipril (Fisons)

Prepulsid (Janssen-Cilag)
USA: Propulsid (Janssen; 1997)

Cisatracurium besylate
(51W89; 51W)

ATC: M03AC11
Use: neuromuscular blocker

RN: 96946-42-8   MF: $C_{53}H_{72}N_2O_{12} \cdot 2C_6H_5O_3S$ MW: 1243.50
CN: [1R-[1α,2α(1'R*,2'R*)]]-2,2'-[1,5-pentanediylbis[oxy(3-oxo-3,1-propanediyl)]]bis[1-[(3,4-dimethoxyphenyl)methyl]-1,2,3,4-tetrahydro-6,7-dimethoxy-2-methylisoquinolinium] dibenzenesulfonate

cation
RN: 96946-41-7   MF: $C_{53}H_{72}N_2O_{12}$ MW: 929.16

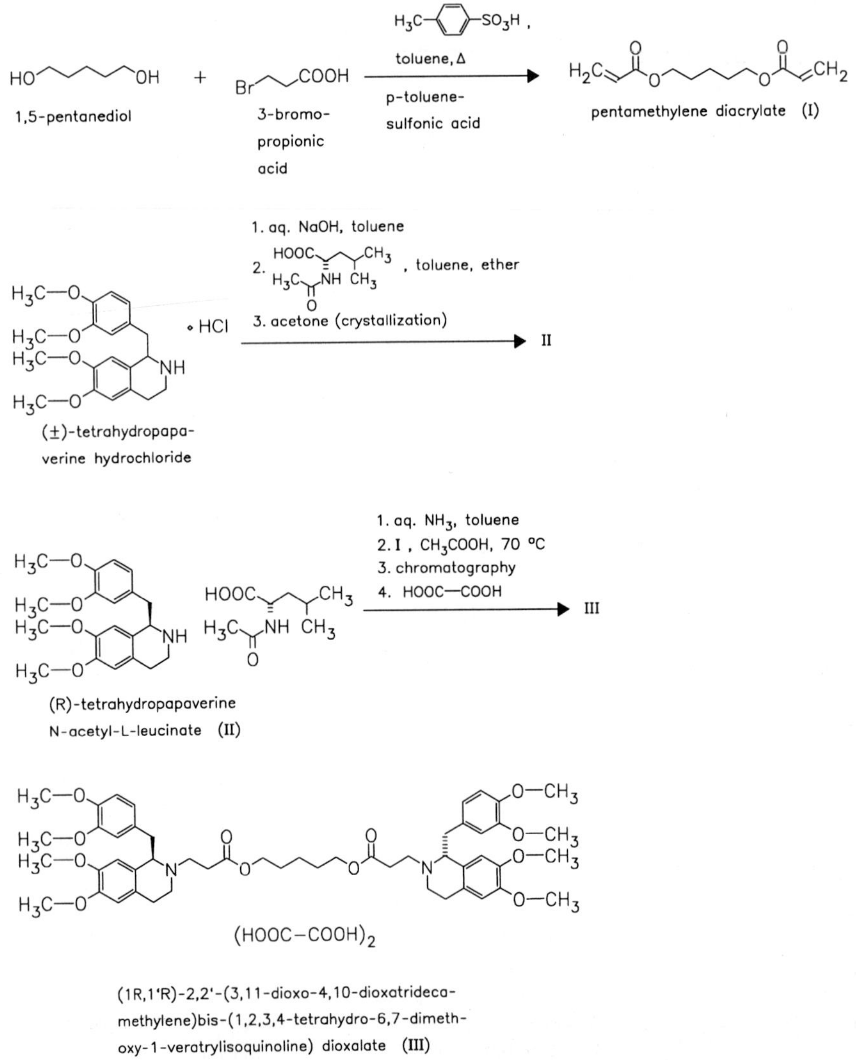

1,5-pentanediol   +   3-bromo-propionic acid → pentamethylene diacrylate (I)

(±)-tetrahydropapa-verine hydrochloride → II

(R)-tetrahydropapaverine N-acetyl-L-leucinate (II) → III

(1R,1'R)-2,2'-(3,11-dioxo-4,10-dioxatrideca-methylene)bis-(1,2,3,4-tetrahydro-6,7-dimeth-oxy-1-veratrylisoquinoline) dioxalate (III)

Cisatracurium besylate

Reference(s):
WO 9 200 965 (Wellcome Foundation; appl. 23.1.1992; GB-prior. 13.7.1990).
US 5 453 510 (Burroughs Wellcome Co.; appl. 26.9.1995; GB-prior. 13.7.1990; USA-prior. 12.7.1991).
Boyd, A.H. et al.: Br. J. Anaesth. (BJANAD) **74** (4), 400 (1995).

Formulation(s):   amp. (inj.) 2 mg/ml (25 ml, 10 ml, 2.5 ml)

Trade Name(s):
D: Nimbex (Glaxo Wellcome;   GB: Nimbex (Glaxo Wellcome;   J: Ciprxan (Bayer)
      Zeneca)   as besylate)   USA: Nimbex (Glaxo Wellcome)

Cisplatin

ATC: L01XA01
Use: antineoplastic

RN: 15663-27-1   MF: $Cl_2H_6N_2Pt$ MW: 300.05   EINECS: 239-733-8
LD$_{50}$:   3.4 mg/kg (R, i.v.);
       9.7 mg/kg (g. p., i.v.)
CN: diamminedichloroplatinum (*SP*-4-2)

| $K_2[PtCl_6]$ | $\xrightarrow[\text{hydrazine}]{N_2H_4}$ | $K_2[PtCl_4]$ | $\xrightarrow{NH_3,\ NH_4Cl}$ | Cisplatin |

potassium   potassium   Cisplatin
hexachloro−   tetrachloro-
platinate(IV)   platinate(II)

Reference(s):
US 4 273 755 (MPD Techn.; 16.6.1981; prior. 16.8.1979).
EP 30 782 (MPD Techn.; USA-prior. 16.8.1979).
DE 3 305 248 (Degussa AG; D prior. 16.2.1983).
Kaufmann, G.B. et al.: Inorg. Synth. (INSYA3) **7**, 239 (1963).
Rosenberg, B. et al.: Nature (London) (NATUAS) **222**, 385 (1969).

injectable solution:
DOS 2 906 700 (Bristol-Myers; appl. 21.2.1979; USA-prior. 30.5.1978).

Formulation(s):   vial (lyo.) 10 mg, 25 mg, 50 mg; vial (sol.) 10 mg/20 ml, 50 mg/100 ml, 100 mg/200 ml

Trade Name(s):
D: Cisplatin Azupharma   Cisplatin-Lösung (ASTA   Cisplatin medac (medac)
      (Azupharma)   Medica AWD)

Platiblastin (Pharmacia & Upjohn)
Platinex (Bristol-Myers Squibb; 1979)
generics and combination preparations

F: Cisplatine Dakota (Dakota)
Cisplatine Lilly (Lilly)
Cisplatyl (Rhône-Poulenc Rorer Bellon)

GB: Neoplatin (Mead Johnson; 1979); wfm
Platinex (Bristol-Myers); wfm
Platosin (Nordic); wfm

I: Citoplatino (Rhône-Poulenc Rorer)
Platamine (Farmitalia)
Platinex (Bristol It. Sud)

Pronto Platamine (Farmitalia)
generics and combination preparations

J: Briplatin (Bristol Squibb)
Randa (Nippon Kayaku)

USA: Platinol (Bristol-Myers Squibb; 1978)

Citalopram
(Nitalapram; LU 10171; ZD-211)

ATC: N06AB04
Use: antidepressant, selective serotonin-uptake inhibitor

RN: 59729-33-8   MF: $C_{20}H_{21}FN_2O$ MW: 324.40   EINECS: 261-891-1
CN: 1-[3-(dimethylamino)propyl]-1-(4-fluorophenyl)-1,3-dihydro-5-isobenzofurancarbonitrile

monohydrobromide
RN: 59729-32-7   MF: $C_{20}H_{21}FN_2O \cdot HBr$ MW: 405.31   EINECS: 261-890-6
monohydrochloride
RN: 85118-27-0   MF: $C_{20}H_{21}FN_2O \cdot HCl$ MW: 360.86   EINECS: 285-680-9
fumarate
RN: 107190-73-8   MF: $C_{20}H_{21}FN_2O \cdot xC_4H_4O_4$ MW: unspecified

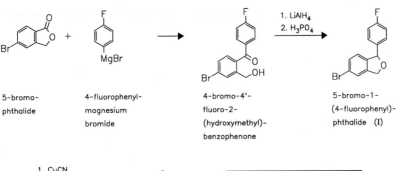

5-bromo-phthalide

4-fluorophenyl-magnesium bromide

4-bromo-4'-fluoro-2-(hydroxymethyl)-benzophenone

5-bromo-1-(4-fluorophenyl)-phthalide (I)

1. cuprous cyanide
2. 3-dimethylaminopropyl chloride

Citalopram

Reference(s):
DE 2 657 013 (Kefalas; appl. 16.12.1976; GB-prior. 14.1.1976).
US 4 136 193 (Kefalas; 7.1.1977; appl. 23.1.1979; GB-prior. 14.1.1976).
Bigler, A.J. et al.: Eur. J. Med. Chem. (EJMCA5) **12**, 289 (1977).

alternative synthesis:
EP 171 943 (Lundbeck; appl. 19.7.1985; GB-prior. 6.8.1984).
WO 9 819 511 (Lundbeck; appl. 10.11.1997; WO-prior. 10.11.1997).
WO 9 819 512 (Lundbeck; WO-prior. 10.12.1997).
WO 9 819 513 (Lundbeck; DK-prior. 8.7.1997).

synthesis of enantiomers:
EP 347 066 (Lundbeck; appl. 1.6.1989; GB-prior. 14.6.1988).

preparation of 5-bromophthalide:
Levy; Stephen: J. Chem. Soc. (JCSOA9) **1931**, 867, 870.

Formulation(s):   f. c. tabl. 20 mg, 40 mg (as hydrobromide)

Trade Name(s):

D:	Cipramil (Promonta Lundbeck)	GB:	Cipramil (Lundbeck; as hydrochloride)	USA:	Celexa (Forest)

Citicoline

ATC: N06BX06
Use: cerebrostimulant, antiparkinsonian, lipometabolism coenzyme (lecithin- and plasmalogen biosynthesis)

RN: 987-78-0   MF: $C_{14}H_{26}N_4O_{11}P_2$ MW: 488.33   EINECS: 213-580-7
LD_{50}:   4600 mg/kg (M, i.v.); 27.142 g/kg (M, p.o.);
     2973 mg/kg (R, i.v.); 18.501 g/kg (R, p.o.)
CN: cytidine-5'-(trihydrogen diphosphate) mono[2-(trimethylammonio)ethyl] ester hydroxide inner salt

phosphorylcholine chloride   morpholine   dicyclohexylcarbodiimide   (I)

cytidine-5'-phosphate tributylamine salt   Citicoline

(b)   fermentatively from cytidylic acid, choline phosphate and glucose in presence of alkali phosphates, magnesium sulfate by means of microorganisms, which produce fructose-1,6-bisphosphate

Reference(s):
Kennedy, E.P.: J. Biol. Chem. (JBCHA3) **222**, 185 (1956).
a   JP-appl. 7 004 747 (Takeda; appl. 18.12.1967).
    similar processes:
    JP-appl. 6 540 ('64) (Takeda; appl. 11.5.1960).
    JP-appl. 6 541 ('64) (Takeda; appl. 23.8.1960).
    JP-appl. 13 024 ('60) (Takeda; appl. 9.9.1960).
    JP-appl. 1 384 ('67) (Takeda; appl. 22.8.1963).
    JP-appl. 7 004 505 (Toho; appl. 22.9.1967).

crystalline monohydrate:
DOS 2 019 308 (Takeda; appl. 22.4.1970; J-prior. 24.4.1969).
b   DOS 2 054 785 (Asahi; appl. 6.11.1970; J-prior. 26.11.1969).

similar process:
DOS 2 037 988 (Kyowa Hakko; appl. 30.7.1970; J-prior. 4.8.1969).

Formulation(s):   amp. 250 mg/2 ml, 500 mg/4 ml, 1000 mg/8 ml (as sodium salt)

Trade Name(s):

F:	Rexort (Takeda)	Kemodyn (Esseti)	Daicoline (Daisan)
I:	Anticolin (Farge)	Logan (Ist. Chim. Inter.)	Dereb (Ohta)
	Brassel (Schiapparelli	Neurex (Salus Research)	Emicholine-F (Dojin)
	Searle)	Neuroton (Nuovo Cons.	Emilian (Beppu)
	Cebroton (Sancarlo)	Sanit. Naz.)	Ensign (Yamanouchi)
	Cidifos (Neopharmed)	Nicholin (Cyanamid)	Erholen (Nichiiko)
	Cidilin (Errekappa	Nicolsint (Leben's)	Haibrain (Ono)
	Euroter.)	Polineural (Biotekfarma)	Hornbest (Hoei)
	Citicolin (Piam)	Sinkron (Ripari-Gero)	Intelon (Takata)
	Citifar (Lafare)	Sintoclar (Pulitzer)	Nicholin (Takeda)
	Citsav (Savio IBN)	J: Andes (Nippon Kayaku)	Plube (Mochida)
	Difosfocin (Magis)	Ceregut (Kodama)	Recognan (Toyo Jozo)
	Encclin (Crosara)	Colite (Nippon Chemiphar)	Rupis (Vitacain)
	Flussorex (Lampugnani)	Corenalin (Kaken)	Suncholin (Mohan)
	Gerolin (CT)	Cyscholin (Kanto)	

Citiolone
(Acetylhomocysteine thiolactone)

ATC: A05BA04
Use: liver therapeutic

RN: 1195-16-0   MF: $C_6H_9NO_2S$ MW: 159.21   EINECS: 214-793-8
LD_{50}:   1200 mg/kg (M, i.v.)
CN: *N*-(tetrahydro-2-oxo-3-thienyl)acetamide

N-acetyl-DL-methionine   → Na/liq. NH$_3$, − 40 °C → N-acetylhomocysteine sodium salt (I)

[I] → HCl → Citiolone

Reference(s):
DE 1 134 683 (Degussa; appl. 16.3.1961).

Formulation(s):   cps. 200 mg, 400 mg; gran. 200 mg; suppos. 250 mg, 500 mg

Trade Name(s):
D: Contralum Ultra (Hermal)-
      comb.; wfm

Hepa-Merz (Merz)-comb.;
wfm

Hepasteril B. compositum, forte (Fresenius)-comb.; wfm Mederma (Merz)-comb.; wfm	Reducdyn (Nordmark)-comb.; wfm Sterofundin-CH (Braun Melsungen); wfm Tutofusin LC (Pfrimmer); wfm	F: Thioncycline (Merrell)-comb.; wfm Thioxidréne (Bottu); wfm I: Citiolase (Roussel)

Citrulline

ATC: V03AB99
Use: liver therapeutic

RN: 372-75-8   MF: $C_6H_{13}N_3O_3$ MW: 175.19   EINECS: 206-759-6
CN: N^5-(aminocarbonyl)-L-ornithine

malate (1:1)
RN: 70796-17-7   MF: $C_6H_{13}N_3O_3 \cdot C_4H_4O_4$ MW: 291.26

①

1. NaOH
2. CuO
3. H_2S

L-arginine hydrochloride   Citrulline

② by fermentation
a from Saccharomyces genus
b from ornithine
c from Arthrobacter

Reference(s):
1   Fox, S.W.: J. Biol. Chem. (JBCHA3) **123**, 687 (1938).
2a JP 52 143 288 (Kyowa; appl. 20.5.1976).
2b JP 50 148 588 (Miura; appl. 23.5.1974).
2c JP 53 075 387 (Kyowa; appl. 13.12.1976).

alternative syntheses:
JP 122 48/67 (Ajinomoto; appl. 11.9.1965).
JP 117 58/68 (Kyowa; appl. 15.11.1965).
Fox, S.W. et al.: J. Org. Chem. (JOCEAH) **6**, 410 (1941).

crystallization:
JP 7 100 174 (Ajinomoto; appl. 20.11.1968).

isolation from Citrullus vulgaris Schrad.:
Wada, M.; Biochem. Z. (BIZEA2) **224**, 420 (1930)

use as liver therapeutic:
FR-M 4 182 (Inst. de Recherche Sci.; appl. 9.3.1965).
FR-M 5 594 (Dimaphar; appl. 1.7.1966).
FR-M 5 703 (Lab. Carriere Carron; appl. 30.8.1965).
FR-M 6 305 (Dimaphar; appl. 15.12.1966).

use as digestant:
FR-M 5 695 (Lab. Carriere Carron; appl. 29.8.1966).

citrulline fumarate:
FR-M 6 306 (Dimaphar; appl. 15.12.1966).

citrulline maleate:
FR-M 6 443 (Dimaphar; appl. 21.4.1967).

Formulation(s):   amp. 60 mg/15 ml; drg. 25 mg, 100 mg

Trade Name(s):
D: Polilevo (Taurus Pharma)-
       comb.
F: Azonutril (Pharmacia &
       Upjohn)-comb.
       Epuram (Pharmafarm)-
       comb.

Perifago (Pharmacia &
Upjohn)
Stimol (Biocodex; as
malate)
I: Biotassina (UCM)-comb.

Citruplexina (Synthelabo)-
comb.
Ipoazotal (SIT)-comb.

Cladribine
(NSC-105014-F; RWJ-26251; 2-CdA)

ATC: L01BB04
Use: antineoplastic

RN: 4291-63-8   MF: $C_{10}H_{12}ClN_5O_3$ MW: 285.69
CN: 2-chloro-2'-deoxyadenosine

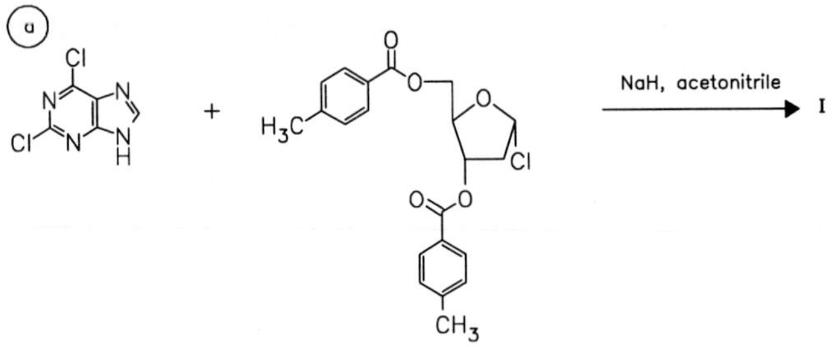

2,6-dichloro-
purine

2-deoxy-3,5-di-O-p-
toluoyl-α-D-erythro-
pentofuranosyl chloride

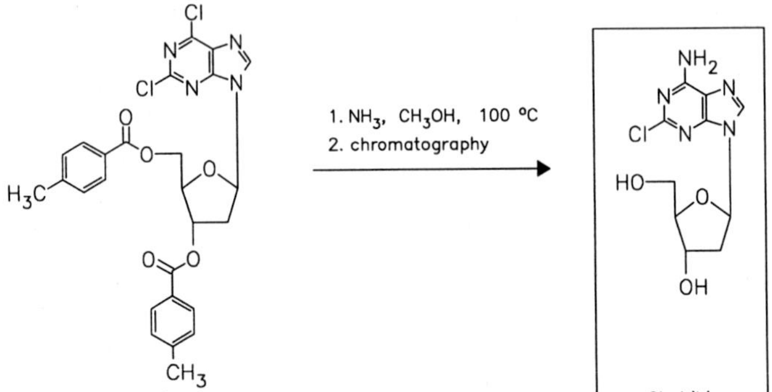

NaH, acetonitrile → I

2,6-dichloro-9-(2-deoxy-
3,5-di-O-p-toluoyl-β-D-
erythro-pentofuranosyl)-
purine (I)

1. NH₃, CH₃OH, 100 °C
2. chromatography

Cladribine

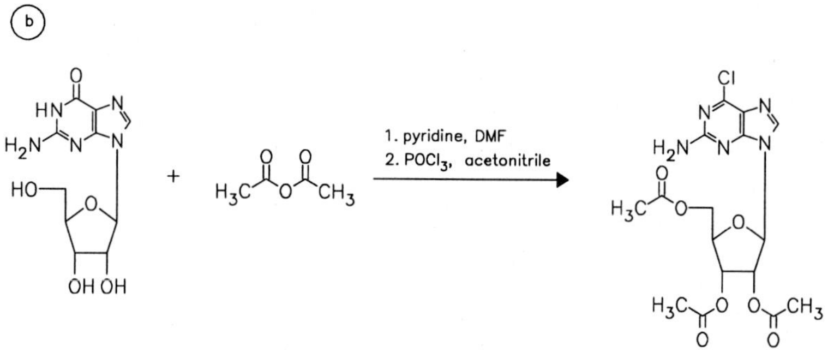

b

guanosine

9-(2,3,5-tri-O-acetyl-
β-D-ribofuranosyl)-2-
amino-6-chloropurine (II)

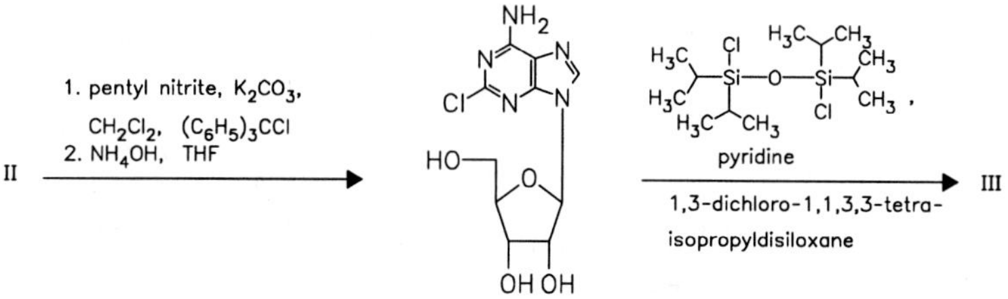

1. pyridine, DMF
2. POCl₃, acetonitrile

1. pentyl nitrite, K₂CO₃,
   CH₂Cl₂, (C₆H₅)₃CCl
2. NH₄OH, THF

II

2-chloro-
adenosine

1,3-dichloro-1,1,3,3-tetra-
isopropyldisiloxane

pyridine

III

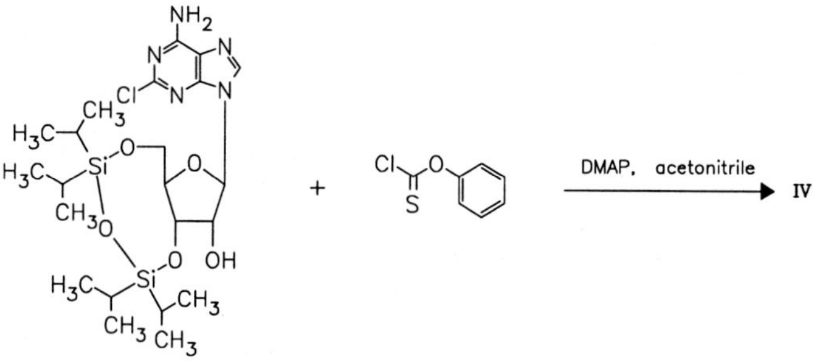

(III)

O-phenyl chloro-
thioformate

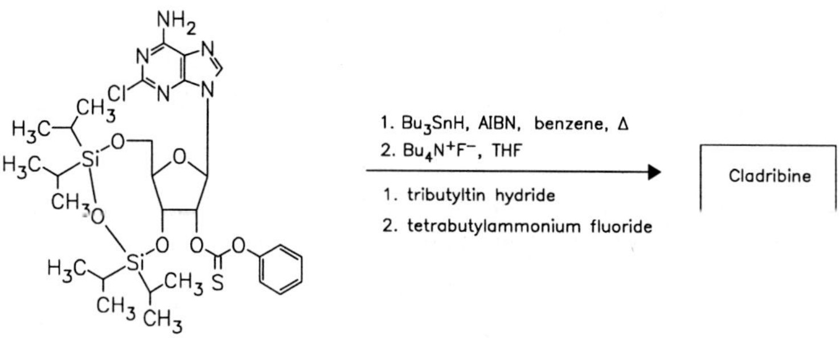

DMAP, acetonitrile

IV

1. Bu₃SnH, AIBN, benzene, Δ
2. Bu₄N⁺F⁻, THF

1. tributyltin hydride
2. tetrabutylammonium fluoride

Cladribine

2-chloro-2'-O-phenoxy-
thiocarbonyl-3',5'-O-
(tetraisopropyldisiloxanylene)-
adenosine (IV)

Reference(s):
a Kazimierczuk, Z. et al.: J. Am. Chem. Soc. (JACSAT) **106**, 6379-6382 (1984).
   EP 173 059 (Univ. Brigham Young; appl. 17.7.1985; USA-prior. 6.8.1984, 15.1.1987).
   Christensen, L.F. et al.: J. Med. Chem. (JMCMAR) **15**, 735 (1972).
b US 5 208 327 (Ortho Pharm. Corp.; appl. 16.4.1992; USA-prior. 18.12.1991).

compositions for treatment of rheumatoid arthritis:
US 5 310 732 (Scripps Res. Inst.; appl. 19.2.1992; USA-prior. 3.2.1986).

Formulation(s):   inj. sol. 10 mg/10 ml

Trade Name(s):
D: Leustatin (Janssen-Cilag)   GB: Leustat (Janssen-Cilag)
F: Leustatine (Janssen-Cilag)   USA: Leustatin (Ortho Biotech)

Clavulanic acid

ATC: J01CR02
Use: β-lactamase inhibitor

RN: 58001-44-8   MF: $C_8H_9NO_5$ MW: 199.16   EINECS: 261-069-2
LD_{50}:   4 g/kg (M, i.v.); 4526 mg/kg (M, p.o.);
      7936 mg/kg (R, p.o.)
CN: [2R-(2α,3Z,5α)]-3-(2-hydroxyethylidene)-7-oxo-4-oxa-1-azabicyclo[3.2.0]heptane-2-carboxylic acid

monosodium salt
RN: 57943-81-4   MF: $C_8H_8NNaO_5$ MW: 221.14   EINECS: 261-032-0
LD_{50}:   4 g/kg (M, i.p.); 4500 mg/kg (M, s.c.)

Clavulanic acid

From cultures of *Streptomyces clavuligerus*.

Reference(s):
US 4 529 720 (Beecham; 16.7.1985; GB-prior. 2.4.1974).
US 4 367 175 (Glaxo; 4.1.1983; GB-prior. 7.2.1975).
GB 1 508 977 (Beecham; appl. 11.4.1975; GB-prior. 20.4.1974, 21.6.1974, 9.10.1974, 11.12.1974).
DOS 2 517 316 (Beecham; appl. 18.4.1975; GB-prior. 20.4.1974, 21.6.1974, 9.10.1974, 11.12.1974).
DE 2 560 074 (Beecham; appl. 18.4.1975; GB-prior. 20.4.1974, 21.6.1974, 9.10.1974, 11.12.1974).

pure salts (e. g. Na-, Li- and other salts):
US 4 490 294 (Beecham; 25.12.1984; GB-prior. 7.2.1975, 17.3.1975).
US 4 490 295 (Beecham; 25.12.1984; GB-prior. 7.2.1975, 17.3.1975).
GB 1 543 563 (Glaxo; appl. 7.2.1975, 17.3.1975; Compl. Spect. 6.2.1976).

tert-butylamine salt:
EP 26 044 (Beecham; appl. 15.8.1980; GB-prior. 24.8.1979).
US 4 454 069 (Beecham; 12.6.1984; GB-prior. 24.8.1979).

various salts:
US 4 367 175 (Glaxo; 4.1.1983; GB-prior. 7.2.1975, 17.3.1975).

esters:
GB 1 508 978 (Beecham; appl. 11.4.1975; GB-prior. 20.4.1974, 21.6.1974, 9.10.1974, 11.12.1974).

formulation with amoxicillin:
EP 8 905 (Beecham; appl. 21.8.1979; GB-prior. 6.9.1978).
US 4 301 149 (Beecham; 17.11.1981; GB-prior. 11.10.1977).
EP 49 061 (Beecham; appl. 6.9.1981; GB-prior. 27.9.1980).
EP 52 962 (Beecham; appl. 2.11.1981; GB-prior. 20.11.1980).
GB 2 084 016 (Beecham; GB-prior. 27.9.1980).

formulation with penicillins and cephalosporins:
DOS 2 559 411 (Beecham; appl. 18.4.1975; GB-prior. 20.4.1974, 21.6.1974, 9.10.1974, 11.12.1974).

Formulation(s):   drops 12.5 mg/ml; f. c. tabl. 125 mg; tabl. 125 mg; vial 0.1 g, 0.2 g, 0.275 g, 0.6 g, 1.2 g, 2.2 g
(as potassium salt)-comb. with amoxicillin

Trade Name(s):

D: Augmentan (SmithKline Beecham; 1982)-comb. with amoxicillin
Betabactyl (SmithKline Beecham)-comb. with ticarcillin
F: Augmentin (SmithKline Beecham; 1984)
Ciblor (Inava)
Claventin (SmithKline Beecham)
GB: Augmentin (SmithKline Beecham; 1984)-comb. with amoxicillin

Timentin (SmithKline Beecham)-comb.
I: Augmentin (SmithKLine B. Farm.)-comb.
Clavucar (Smith Kline & French)-comb. with ticarcilline
Clavulin (Carlo Erba)-comb.
Neoduplamox (Smith Kline & French)-comb.
Timentin (SmithKline Beecham)-comb. with ticarcilline

all combination preparations with amoxicillin
J: Augmentin (Beecham-Meiji)-comb. with amoxicillin
USA: Augmentin (SmithKline Beecham; 1984)-comb. with amoxicillin
Timentin (SmithKline Beecham)

Clebopride

ATC: A03FA06; A04AD
Use: anti-emetic, specific antagonist of peripheral and central dopamine receptors, reversible MAO-inhibitor

RN: 55905-53-8   MF: $C_{20}H_{24}ClN_3O_2$ MW: 373.88   EINECS: 259-885-9
LD$_{50}$: 260 mg/kg (M, i.m.); 40 mg/kg (M, i.p.); 51 mg/kg (M, i.v.); 490 mg/kg (M, p.o.); 350 mg/kg (M, s.c.);
1450 mg/kg (R, i.m.); 155 mg/kg (R, i.p.); 39 mg/kg (R, i.v.); 2540 mg/kg (R, p.o.); 4850 mg/kg (R, s.c.)
CN: 4-amino-5-chloro-2-methoxy-*N*-[1-(phenylmethyl)-4-piperidinyl]benzamide

monohydrochloride
RN: 57645-39-3   MF: $C_{20}H_{24}ClN_3O_2 \cdot HCl$ MW: 410.35
LD$_{50}$: >1 g/kg (M, p.o.)
malate (1:1)
RN: 57645-91-7   MF: $C_{20}H_{24}ClN_3O_2 \cdot C_4H_6O_5$ MW: 507.97   EINECS: 260-874-6
LD$_{50}$: 51 mg/kg (M, i.v.); 490 mg/kg (M, p.o.);
39 mg/kg (R, i.v.); 2540 mg/kg (R, p.o.);
>800 mg/kg (dog, p.o.)

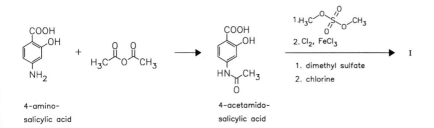

4-amino-
salicylic acid

4-acetamido-
salicylic acid

1. dimethyl sulfate
2. chlorine

methy 2-methoxy-
4-acetamido-5-
chlorobenzoate (I)

2-methoxy-4-
amino-5-chloro-
benzoic acid (II)

1-benzyl-
piperidine-4-one

1-benzyl-4-amino-
piperidine (III)

Clebopride

Reference(s):
DE 2 513 136 (Anphar; appl. 21.3.1975; GB-prior. 21.3.1974).
US 4 138 492 (Anphar; 6.2.1979; appl. 17.3.1975; GB-prior. 21.3.1974).
Prieto, J. et al.: J. Pharm. Pharmacol. (JPPMAB) **29**, 147 (1977).

alternative synthesis:
JP 63 295 558 (Asahi; appl. 26.5.1987).
JP 63 295 557 (Asahi; appl. 26.5.1987).

synthesis of intermediates:
JP 63 295 559 (Asahi; appl. 26.5.1987).

transdermal patch:
EP 303 445 (Fordonal; appl. 9.8.1988; J-prior. 13.8.1987).

Formulation(s):   amp. 1 mg; sol. 0.5 mg; syrup 0.5 mg; tabl. 0.25 mg, 0.5 mg (as hydrogen maleate)

Trade Name(s):
I: Cleprid (Recordati; 1987)   J: Amicos (Banyu; 1985)
       Motilex (Guidotti)   Clast (Meiji)

Clemastine
(Meclastine)

ATC: D04AA14; R06AA04
Use: antiallergic, antihistaminic

RN: 15686-51-8   MF: $C_{21}H_{26}ClNO$ MW: 343.90
CN: [R-(R*,R*)]-2-[2-[1-(4-chlorophenyl)-1-phenylethoxy]ethyl]-1-methylpyrrolidine

hydrogen fumarate (1:1)
RN: 14976-57-9   MF: $C_{21}H_{26}ClNO \cdot C_4H_4O_4$ MW: 459.97   EINECS: 239-055-2
LD_{50}:   43 mg/kg (M, i.v.); 730 mg/kg (M, p.o.);
       82 mg/kg (R, i.v.); 3550 mg/kg (R, p.o.)

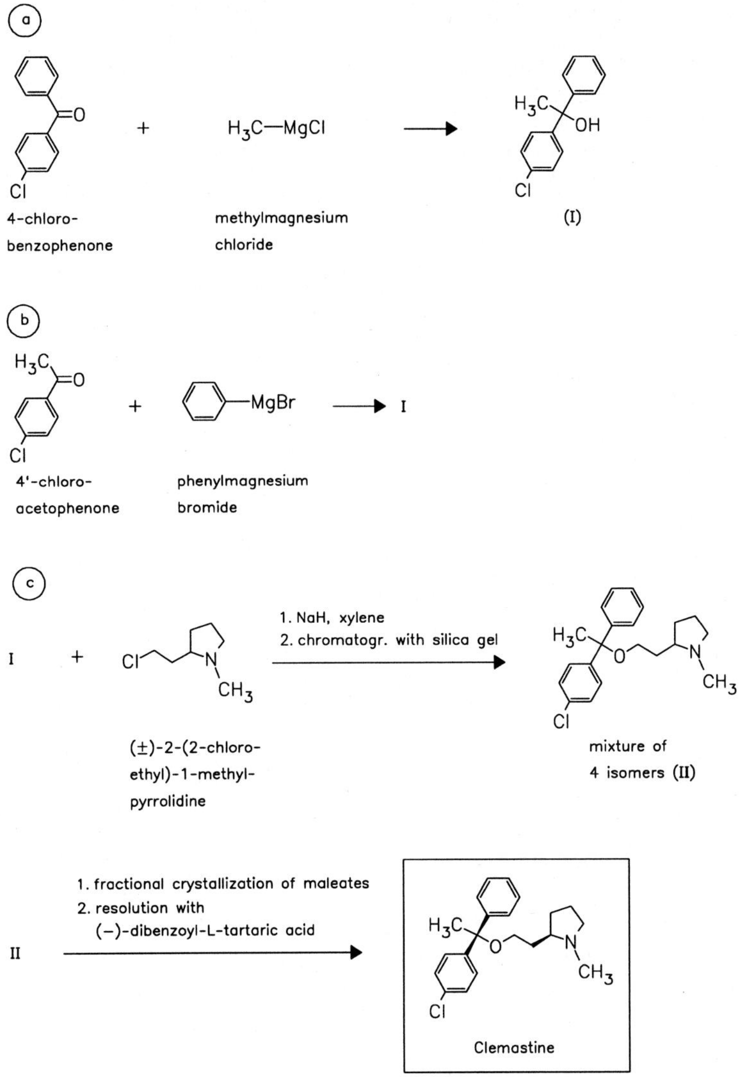

a
O
$H_3C—MgCl$
H_3C
OH
Cl
4-chloro-
benzophenone
methylmagnesium
chloride
(I)
b
H_3C
O
MgBr
I
Cl
4'-chloro-
acetophenone
phenylmagnesium
bromide
c
I
Cl
N
CH_3
1. NaH, xylene
2. chromatogr. with silica gel
H_3C
O
N
CH_3
Cl
(±)-2-(2-chloro-
ethyl)-1-methyl-
pyrrolidine
mixture of
4 isomers (II)
II
1. fractional crystallization of maleates
2. resolution with
(−)-dibenzoyl-L-tartaric acid
H_3C
O
N
CH_3
Cl
Clemastine

Reference(s):

Ebnöther, A.; Weber, H.-P.: Helv. Chim. Acta (HCACAV) **59**, 2462 (1976).
GB 942 152 (Sandoz; appl. 14.12.1960; CH-prior. 19.1.1960, 3.8.1960, 27.9.1960).
FR-M 1 313 (Sandoz; appl. 13.7.1961).

preparation of 2-(2-chloroethyl)-1-methylpyrrolidine *enantiomers:*
Vernier, J.M. et al.: J. Med. Chem. (JMCMAR) **42** (10), 1684 (1999).

Formulation(s): amp. 2 mg/5 ml; gel 300 mg/g (as hydrogen fumarate); syrup 0.5 mg/10 ml; tabl. 1 mg

Trade Name(s):

D: Corto-Tavegil (Novartis Pharma)-comb.
Tavegil (Novartis Consumer Health)
F: Tavégil (Sandoz); wfm
GB: Tavegil (Novartis; as hydrogen fumarate)
I: Tavegil (Sandoz)
J: Alagyl (Sawai)
Alusas (Fuso)
Anhistan (Nippon Zoki)
Antriptin (Nippon Yakuhin)
Batomu (Zensei)
Benanzyl (Isei)
Chlonaryl (Ohta)
Clemanyl (Kyoritsu Yamagata)
Fuluminol (Tatsumi)
Fumalestine (Hishiyama)
Fumartin (Torii)
Histamedine (Mohan)
Inbestan (Maruko)
Kinotomin (Toa Eiyo)

Lacretin (Tokyo Tanabe)
Lecasol (Kaken)
Maikohis (Nihon Yakuhin)
Mallermin-F (Taiyo
Yakuko)
Marsthine (Towa)
Masletene (Shioe)

Natarilan (Nippon
Chemiphar)
Piloral (Nippon Kayaku)
Raseltin (Maruishi)
Reconin (Toyama)
Romien (Fuji Zoki)
Tavegyl (Sandoz-Sankyo)

Telgin G (Takata)
Trabest (Hoei)
Xolamin (Sanko)
generics and combination
preparations
USA: Tavist (Dorsey); wfm
Travist (Sandoz); wfm

Clemizole

ATC: R06A
Use: antihistaminic, antiallergic

RN: 442-52-4   MF: $C_{19}H_{20}ClN_3$ MW: 325.84   EINECS: 207-133-5
CN: 1-[(4-chlorophenyl)methyl]-2-(1-pyrrolidinylmethyl)-1H-benzimidazole

monohydrochloride
RN: 1163-36-6   MF: $C_{19}H_{20}ClN_3 \cdot HCl$ MW: 362.30   EINECS: 214-605-4
LD$_{50}$:   75 mg/kg (M, i.v.); 837 mg/kg (M, p.o.);
74 mg/kg (R, i.v.); 1950 mg/kg (R, p.o.)

1-chloro-
2-nitro-
benzene

4-chloro-
benzylamine

chloroacetyl
chloride

(I)

pyrrolidine

Clemizole

Reference(s):
US 2 689 853 (Schering AG; 1954; D-prior. 1950).

alternative syntheses:
DE 980 644 (Schering AG; appl. 1950).
DE 901 649 (Schering AG; appl. 1951).

Formulation(s):   cream 10 mg/40 g; suppos. 5 mg

Trade Name(s):
D: Megacillin (Grünenthal)-
comb. with penicillin; wfm
Scheriproct (Scherax)-
comb.; wfm
Ultraproct (Scherax)-
comb.; wfm

F: Deliproct (Schering)-
comb.; wfm
Ultralan (Schering); wfm
Ultraproct (Schering)-
comb.; wfm
GB: Scheriproct (Schering)-
comb.; wfm

Ultraproct (Schering)-
comb.; wfm
I: Ultraproct (Schering)-
comb.
J: Histacur (Nichidoku)

Clenbuterol

ATC: R03AC14; R03CC13
Use: bronchodilator

RN: 37148-27-9 MF: $C_{12}H_{18}Cl_2N_2O$ MW: 277.20   EINECS: 253-366-0
LD_{50}: 27.6 mg/kg (M, i.v.)
CN: 4-amino-3,5-dichloro-α-[[(1,1-dimethylethyl)amino]methyl]benzenemethanol

hydrochloride
RN: 21898-19-1   MF: $C_{12}H_{18}Cl_2N_2O \cdot HCl$ MW: 313.66

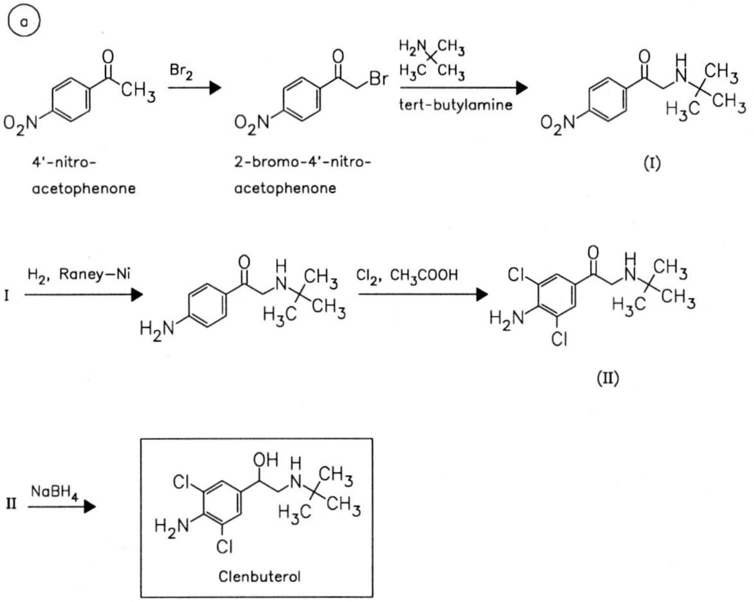

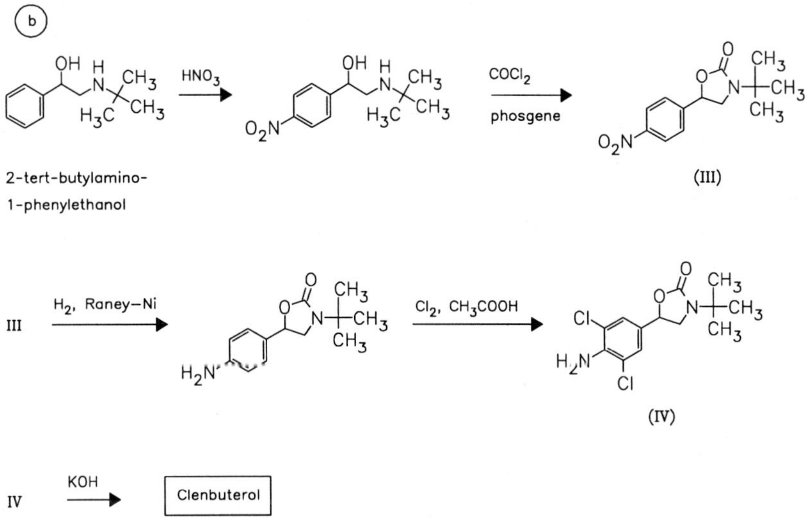

(I)

(II)

Clenbuterol

2-tert-butylamino-
1-phenylethanol

(III)

(IV)

IV →(KOH)→ Clenbuterol

Reference(s):
Keck, J. et al.: Arzneim.-Forsch. (ARZNAD) **22**, 861 (1972).
a DOS 1 793 416 (Thomae; appl. 5.9.1967).
　BE 704 213 (Thomae; appl. 22.9.1967; D-prior. 22.9.1966, 15.2.1967, 2.6.1967).
　US 3 536 712 (Boehringer Ing.; 27.10.1970; D-prior. 22.9.1966, 15.2.1967, 2.6.1967).
b DOS 2 157 040 (Thomae; appl. 17.11.1971).
　DE 1 543 928 (Thomae; appl. 22.9.1966).

alternative syntheses:
DAS 2 354 959 (Thomae; appl. 2.11.1973).

Formulation(s): 　drops 0.059 mg/ml, 15 mg/2 ml; syrup 0.005 mg/5 ml; tabl. 0.01 mg, 0.02 mg (as
　　　　　　　　　hydrochloride)

Trade Name(s):

D:	Contraspasmin (ASTA Medica AWD) Spiropent (Boehringer Ing.) Spasmo-Mucosolvan (Boehringer Ing.)-comb.	I:	Broncodil (Leben's) Clenasma (Biomedica Foscama) Contrasmina (Falqui) Monores (Valeas)	Prontovent (Salus Research) Spiropent (Boehringer Ing.)
				J:	Spiropent (Teijin-Kissei)

Clidanac

ATC:　M01AB
Use:　non-steroidal anti-inflammatory, antipyretic

RN:　34148-01-1　MF: $C_{16}H_{19}ClO_2$　MW: 278.78
LD_{50}:　41 mg/kg (R, p.o.)
CN:　6-chloro-5-cyclohexyl-2,3-dihydro-1*H*-indene-1-carboxylic acid

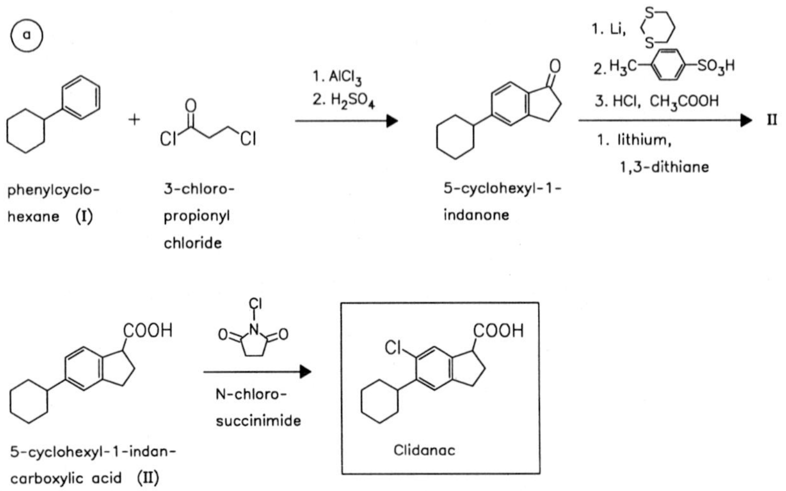

phenylcyclo-　　3-chloro-　　　　　5-cyclohexyl-1-
hexane　(I)　　propionyl　　　　indanone
　　　　　　　chloride

5-cyclohexyl-1-indan-
carboxylic acid　(II)

Clidanac

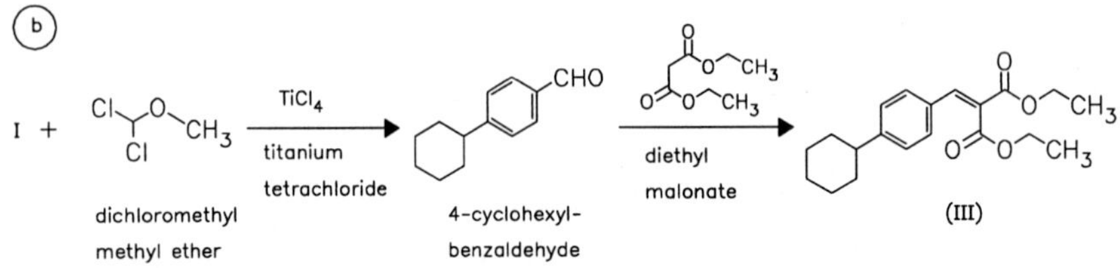

I +　dichloromethyl　　TiCl₄　　4-cyclohexyl-　　diethyl　　　(III)
　　methyl ether　　titanium　　benzaldehyde　　malonate
　　　　　　　　tetrachloride

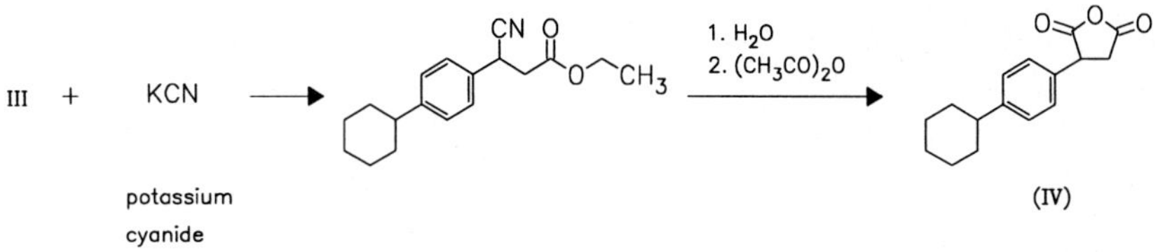

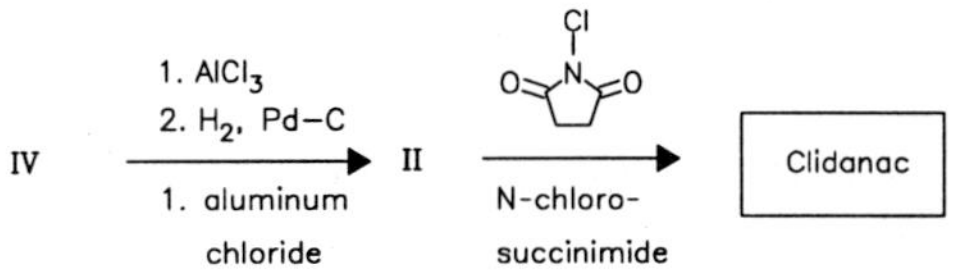

III + KCN

potassium
cyanide

(IV)

IV →
1. AlCl₃
2. H₂, Pd–C
1. aluminum
   chloride
II
N–chloro–
succinimide
Clidanac

Reference(s):
a   Juby, P.F. et al.: J. Med. Chem. (JMCMAR) **15**, 1297 (1972).
    (alternative synthesis described)
b   DOS 2 004 038 (Bristol-Myers; appl. 29.1.1970; USA-prior. 31.1.1969).
    US 3 565 943 (Bristol-Myers; 23.2.1971; prior. 17.9.1969, 31.1.1969).
    US 3 663 627 (Bristol-Myers; 16.5.1972; prior. 1.6.1970).

alternative synthesis:
DOS 2 330 856 (Takeda; appl. 16.6.1973; J-prior. 19.6.1972, 21.11.1972).

Formulation(s):   tabl. 15 mg

Trade Name(s):
J: Britai (Bristol-Banyu)   Indanol (Takeda)

Clidinium bromide

ATC: A03CA02
Use: anticholinergic

RN: 3485-62-9   MF: C₂₂H₂₆BrNO₃   MW: 432.36   EINECS: 222-471-3
LD₅₀: 16 mg/kg (M, i.v.); 492 mg/kg (M, p.o.);
       26 mg/kg (dog, i.v.)
CN: 3-[(hydroxydiphenylacetyl)oxy]-1-methyl-1-azoniabicyclo[2.2.2]octane bromide

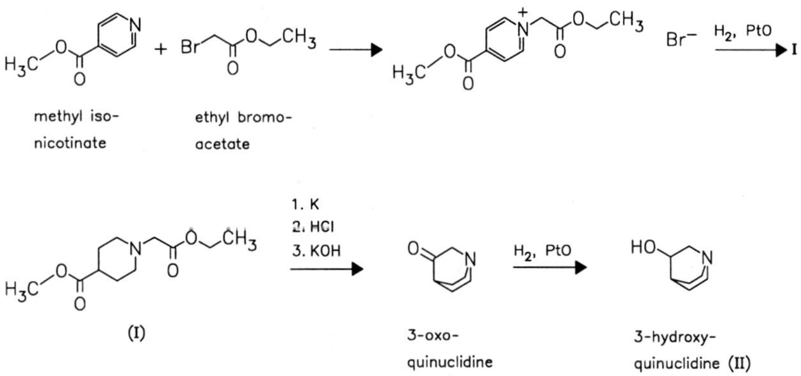

methyl iso-
nicotinate

ethyl bromo-
acetate

(I)

1. K
2. HCl
3. KOH

3-oxo-
quinuclidine

H₂, PtO

3-hydroxy-
quinuclidine (II)

benzilic
chloride

Clidinium bromide

Reference(s):
US 2 648 667 (Hoffmann-La Roche; 1955; prior. 1951).

Formulation(s):   drg. 2.5 mg

Trade Name(s):
D: Librax (Roche)-comb. with   GB: Libraxin (Roche)-comb.   I: Librax (Roche)-comb. with
       chlorodiazepoxide; wfm   with chlorodiazepoxide;   chlorodiazepoxide
F: Librax (Roche)-comb. with   wfm   USA: Librax (Roche)-comb. with
       chlorodiazepoxide   chlorodiazepoxide

Clindamycin

ATC: D10AF01; G01AA10; J01FF01
Use: antibiotic

RN: 18323-44-9   MF: $C_{18}H_{33}ClN_2O_5S$ MW: 424.99   EINECS: 242-209-1
LD$_{50}$:   2618 mg/kg (R, s.c.)
CN: (2S-*trans*)-methyl 7-chloro-6,7,8-trideoxy-6-[[(1-methyl-4-propyl-2-pyrrolidinyl)carbonyl]amino]-1-thio-
       L-*threo*-α-D-*galacto*-octopyranoside

monohydrochloride
RN: 21462-39-5   MF: $C_{18}H_{33}ClN_2O_5S \cdot HCl$ MW: 461.45   EINECS: 244-398-6
LD$_{50}$:   245 mg/kg (M, i.v.); 2539 mg/kg (M, p.o.);
       2193 mg/kg (R, p.o.)

lincomycin $(H_5C_6)_3P/H_3C-CN/Cl_2$
                              Rydon reagent

Clindamycin

Reference(s):
US 3 496 163.
DE 1 795 740
US 3 418 414 (Upjohn; 24.12.1968; appl. 31.8.1966).
US 3 475 407 (Upjohn; 28.10.1969; appl. 22.12.1967).
US 3 509 127 (Upjohn; 28.4.1970; appl. 30.4.1968).

use as antimalarial:
US 3 627 887 (Upjohn; 14.12.1971; appl. 17.10.1969).

Rydon reagent:
Landauer, S.R.; Rydon, H.N.: J. Chem. Soc. (JCSOA9) **1953**, 2224.
Coe, D.G. et al.: J. Chem. Soc. (JCSOA9) **1954**, 2281.
Rydon, H.N.; Tonge, B.L.: J. Chem. Soc. (JCSOA9) **1956**, 3043.

Formulation(s):   amp. 300 mg/2 ml, 600 mg/4 ml, 900 mg/6 ml; cps. 150 mg, 300 mg; gel 10 mg/g;
                       sol. 10 mg/ml (as hydrochloride or phosphate); vaginal cream 20 mg

Trade Name(s):

D:	Sobelin (Upjohn; 1968)	Clinamycina (Savoma)	USA: Cleocin (Pharmacia &
	generics	Dalacin C (Pharmacia &	Upjohn; 1970)
F:	Dalacine (Pharmacia &	Upjohn; 1969)	Clinda-Derm (Paddock)
	Upjohn; 1972)	I: Dalacin C (Upjohn)	Clindets Pledgets (Stiefel)
GB:	Cleocin (Upjohn)	J: Dalacin (Upjohn; 1971)	

Clinofibrate

ATC: B04AC
Use: antihyperlipidemic, clofibrate
         derivative

RN: 30299-08-2   MF: $C_{28}H_{36}O_6$ MW: 468.59
LD$_{50}$: 255 mg/kg (M, i.p.); 1800 mg/kg (M, p.o.); 410 mg/kg (M, s.c.);
         205 mg/kg (R, i.p.); >4000 mg/kg (R, p.o.); 2200 mg/kg (R, s.c.)
CN: 2,2'-[cyclohexylidenebis(4,1-phenyleneoxy)]bis[2-methylbutanoic acid]

cyclo-   phenol   1,1-bis(4-hydroxy-
hexanone   phenyl)cyclohexane (I)

2-butanone   Clinofibrate

Reference(s):
US 3 716 583 (Sumitomo; 13.2.1972; appl. 7.4.1970; J-prior. 16.4.1969).
US 3 821 404 (Sumitomo; 28.6.1974; J-prior. 16.4.1969).
DOS 2 017 331 (Sumitomo; appl. 10.4.1970; J-prior. 16.4.1969, 2.5.1969).

synthesis of 1,1-bis(4-hydroxyphenyl)cyclohexane:
DD 46 281 (G. Drefahl, E. Littmann; appl. 22.1.1962).

Formulation(s):   tabl. 100 mg, 200 mg

Trade Name(s):

J:	Deslipoze (Kowa Yakuhin)	Lipirate (Hishiyama)	Lipofibrate (Taiyo)
	Lipaderin (Uji)	Lipoclin (Sumitomo; 1981)	Prinmate (Sawai)

Clioquinol
(Chloroiodoquine; Iodochlorhydroxyquin)

ATC: D08AH30; D09AA10; G01AC02; P01AA02; S02AA05
Use: wound- and bowel-antiseptic

RN: 130-26-7　MF: C_9H_5ClINO　MW: 305.50　EINECS: 204-984-4
LD$_{50}$: 69 mg/kg (M, p.o.);
>5 g/kg (R, p.o.)
CN: 5-chloro-7-iodo-8-quinolinol

oxyquinoline
(q. v.)

5-chloro-8-
hydroxyquinoline

Clioquinol

Reference(s):
DRP 117 767 (Ciba; 1899).

Formulation(s):　cream 3 g/100 g; emulsion 0.5 g/100 g; ointment 3 g/100 g

Trade Name(s):
D:	Linola-sept (Wolff)		Synalar C (Zeneca)-comb.	Reticus (Farmila)-comb.
	Locacorten-Vioform		Vioform-hydrocortisone	Viobeta (IDI)-comb.
	(Novartis Pharma)-comb.		(Novartis)-comb.	Viocidina (IDI)
	Millicorten-Vioform	I:	Dermadex Chinol	J:	Emaform (Tanabe)
	(Novartis Pharma)-comb.		(SmithKline Beecham)-		Entero-Vioform (Ciba-
F:	Diprosept (Schering-		comb.		Geigy-Takeda)
	Plough)-comb.		Diproform (Schering-		Mexaform (Ciba-Geigy-
GB:	Betnovate-C (Glaxo		Plough)-comb.		Takeda)-comb.
	Wellcome)-comb.		Iodoclorossich TI (Tariff.	USA:	Racet (Lemmon)-comb.;
	Locorten-Vioform		Integrativo)		wfm
	(Novartis)-comb.		Locorten (Zyma)-comb.		Vioform (Ciba); wfm

Clobazam

ATC: N05BA09
Use: minor tranquilizer

RN: 22316-47-8　MF: $C_{16}H_{13}ClN_2O_2$　MW: 300.75　EINECS: 244-908-7
LD$_{50}$: 510 mg/kg (M, i.p.); 840 mg/kg (M, p.o.);
>2000 mg/kg (R, p.o.)
CN: 7-chloro-1-methyl-5-phenyl-1H-1,5-benzodiazepine-2,4(3H,5H)-dione

1,3-dichloro-
benzene

2,4-dichloro-
1-nitrobenzene

5-chloro-2-
nitrodiphenyl-
amine　(I)

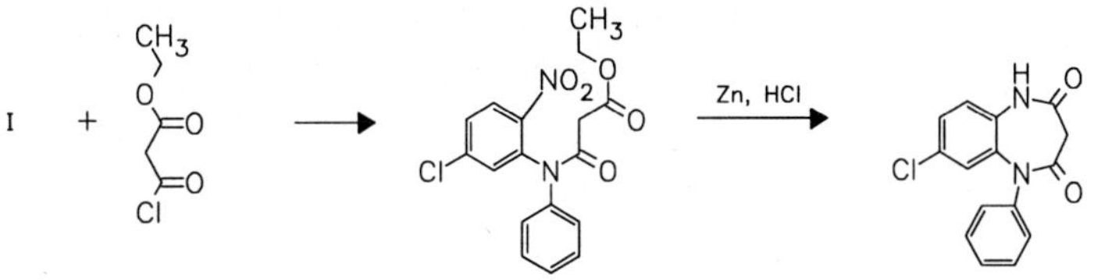

monoethyl
malonyl
chloride

1-demethyl-
clobazam (II)

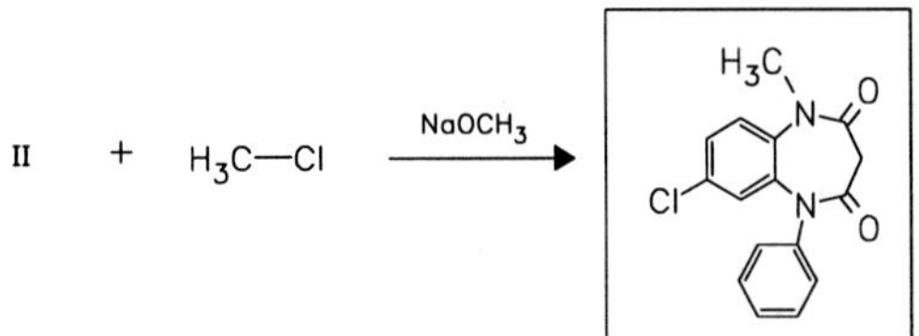

Clobazam

Reference(s):

DAS 1 793 837 (Roussel-Uclaf; appl. 14.12.1967; I-prior. 14.12.1966).
US 3 984 398 (Roussel-Uclaf; 5.10.1976; I-prior. 14.12.1966).
US 3 836 653 (Boehringer Ing.; 17.9.1974; D-prior. 7.2.1967, 18.1.1968).
DOS 1 670 190 (Boehringer Ing.; appl. 7.2.1967).
DOS 1 670 306 (Boehringer Ing.; appl. 18.1.1968).
Weber, K.H. et al.: Justus Liebigs Ann. Chem. (JLACBF) **756**, 128 (1972).

1-demethylclobazam:
DAS 1 668 634 (Roussel-Uclaf; appl. 14.12.1967; I-prior. 14.12.1966).

synthesis of 5-chloro-2-nitrodiphenylamine:
Laubenheimer: Ber. Dtsch. Chem. Ges. (BDCGAS) **9**, 771 (1876).

combination with nomifensine:
DOS 2 724 683 (Hoechst; appl. 1.6.1977).

Formulation(s):   cps. 10 mg; tabl. 10 mg, 20 mg

Trade Name(s):

D: Frisium (Hoechst; 1977)   GB: Frisium (Hoechst; 1979)
F: Urbanyl (Synthélabo;   I: Frisium (Hoechst Italia
       1975)   Sud)

Clobenoside

ATC: C05
Use: anti-inflammatory, vasoprotective

RN: 29899-95-4   MF: C$_{25}$H$_{32}$Cl$_2$O$_6$ MW: 499.43   EINECS: 249-940-5
CN: ethyl 5,6-bis-*O*-[(4-chlorophenyl)methyl]-3-*O*-propyl-D-glucofuranoside

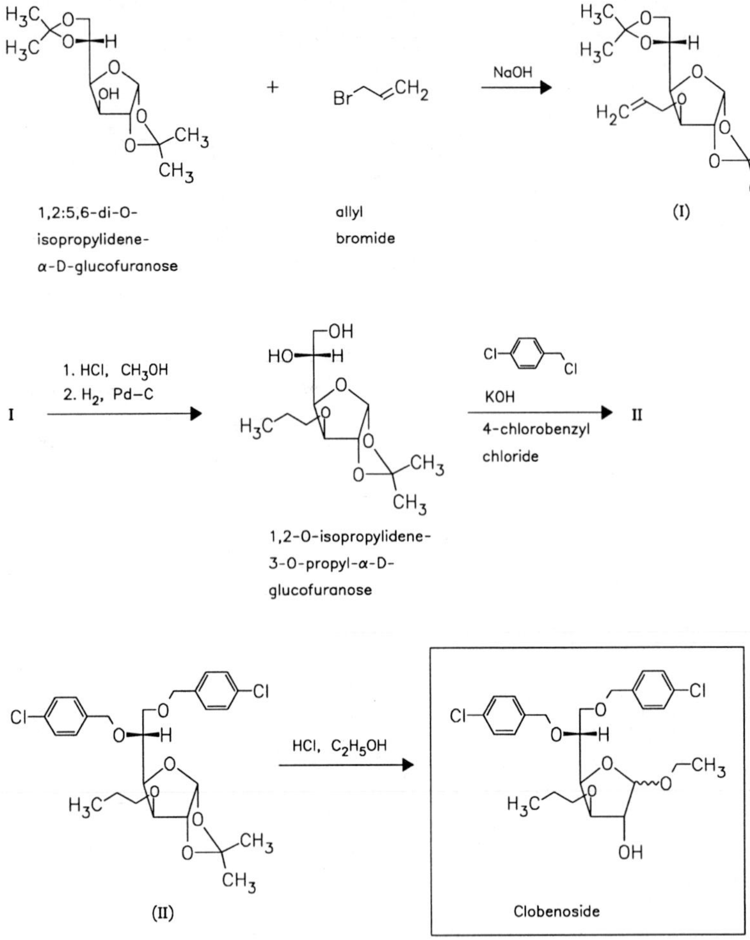

1,2:5,6-di-O-
isopropylidene-
α-D-glucofuranose

allyl
bromide

(I)

1. HCl, CH₃OH
2. H₂, Pd–C

1,2-O-isopropylidene-
3-O-propyl-α-D-
glucofuranose

4-chlorobenzyl
chloride

KOH

II

HCl, C₂H₅OH

(II)

Clobenoside

Reference(s):
DOS 1 793 338 (Ciba-Geigy; appl. 3.9.1968; CH-prior. 11.9.1967).
US 3 665 884 (Ciba-Geigy; 11.4.1972; CH-prior. 11.9.1967).
US 3 542 761 (Ciba-Geigy; 24.11.1970; appl. 25.4.1968; CH-prior. 11.9.1967).

synthesis of 1,2-O-isopropylidene-3-O-propyl-α-D-glucofuranose:
DOS 2 031 161 (Ciba; appl. 24.6.1970; CH-prior. 3.7.1969).
Cunningham, J. et al.: Tetrahedron Lett. (TELEAY) **19**, 1191 (1964).
Corbettand, W.M.; McKay, J.E.: J. Chem. Soc. (JCSOA9), 2930 (1961).

Formulation(s):   tabl. 200 mg, gel

Trade Name(s):
CH: Aglidin (Zyma)   Finocal (Zyma)
       Arvigol (Zyma)   Flogasol (Vifor)

Clobenzorex

ATC: A08AA08
Use: appetite depressant

RN: 13364-32-4   MF: C₁₆H₁₈ClN   MW: 259.78   EINECS: 236-434-4
CN: (+)-N-[(2-chlorophenyl)methyl]-α-methylbenzeneethanamine

hydrochloride
RN: 5843-53-8 MF: $C_{16}H_{18}ClN \cdot HCl$ MW: 296.24 EINECS: 227-434-5
LD$_{50}$: 103 mg/kg (M, i.p.);
103 mg/kg (R, i.p.)

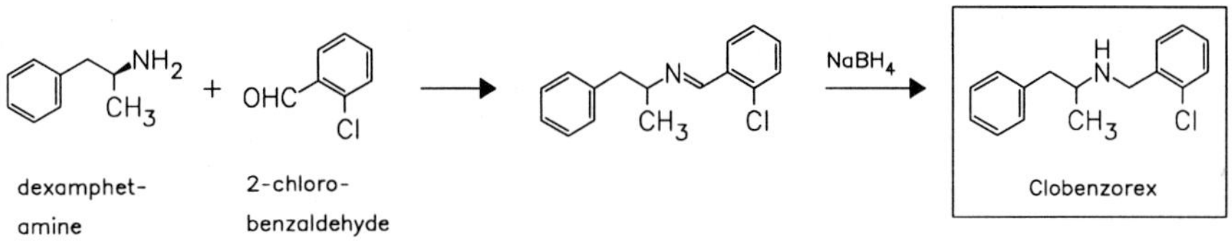

| dexamphet-
amine | 2-chloro-
benzaldehyde | | Clobenzorex |

Reference(s):
FR 1 429 306 (S. I. F. A.; appl. 23.11.1964).

Formulation(s): cps. 30 mg (as hydrochloride)

Trade Name(s):
F: Dinintel (Roussel Diamant;
as hydrochloride)

Clobenztropine
(Chlorobenztropine)

ATC: R06
Use: antihistaminic

RN: 5627-46-3 MF: $C_{21}H_{24}ClNO$ MW: 341.88
CN: 3-[(4-chlorophenyl)phenylmethoxy]-8-methyl-8-azabicyclo[3.2.1]octane

hydrochloride
RN: 14008-79-8 MF: $C_{21}H_{24}ClNO \cdot HCl$ MW: 378.34
LD$_{50}$: 174 mg/kg (M, p.o.);
364 mg/kg (R, p.o.)

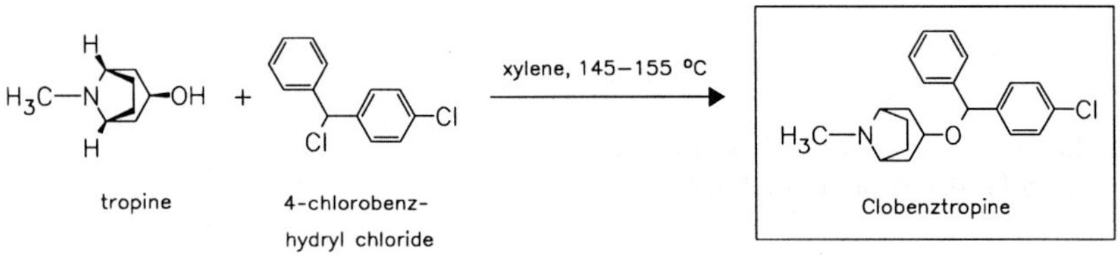

| tropine | 4-chlorobenz-
hydryl chloride | | Clobenztropine |

Reference(s):
US 2 782 200 (Schenley Labs.; 1957; appl. 1955).

alternative synthesis (with 4-chlorodiphenyldiazomethane).
US 2 799 680 (S. Fromer; 1957; appl. 1954).

Trade Name(s):
USA: Teprin (Endo); wfm

Clobetasol propionate

ATC: D07AD01
Use: topical glucocorticoid

RN: 25122-46-7   MF: $C_{25}H_{32}ClFO_5$ MW: 466.98   EINECS: 246-634-3
LD_{50}:   >3 g/kg (M, p.o.);
         >3 g/kg (R, p.o.)
CN: (11β,16β)-21-chloro-9-fluoro-16-methyl-17-(1-oxopropoxy)pregna-1,4-diene-3,20-dione

clobetasol
RN: 25122-41-2   MF: $C_{22}H_{28}ClFO_4$ MW: 410.91   EINECS: 246-633-8

betamethasone 17-propionate
(cf. betamethasone dipropionate
synthesis)

methanesulfonyl
chloride

(I)

I   LiCl,   , DMF
      H_3C CH_3
      lithium acetone
      chloride

Clobetasol propionate

Reference(s):
DE 1 902 340 (Glaxo; appl. 17.1.1969; GB-prior. 19.1.1968).
US 3 721 687 (Glaxo; 20.3.1973; GB-prior. 4.4.1968).

Formulation(s):   cream 0.05 %; ointment 0.05 %; sol. 0.5 mg/g (0.05 %)

Trade Name(s):
D: Dermoxin (Glaxo
       Wellcome/Cascan; 1976)
       Dermoxinale (Glaxo
       Wellcome/Cascan; 1977)
       Karison (Dermapharm)
F: Dermoval (Glaxo
       Wellcome)
GB: Dermovate (Glaxo
       Wellcome; 1973)

I: Clobesol (Glaxo
       Wellcome)
J: Betaleston (Nihon
       Yakuhin)
       Delspart (Kodama)
       Dermovate (Glaxo; 1979)
       Entyfluson (Taiyo)
       Glydil (Shinshin)
       Mahady (Wukamoto)

       Myalore (Ohta)
       Siodelbate (Tatsumi)
       Solvega (Hisamitsu)
USA: Cormax (Oclassen)
       Temovate (Glaxo
       Wellcome; 1986)

Clobetasone butyrate

ATC: D07AB01; S01BA09
Use: topical glucocorticoid

RN: 25122-57-0   MF: $C_{26}H_{32}ClFO_5$ MW: 478.99   EINECS: 258-953-5
LD_{50}:   >6 g/kg (M, p.o.);
         >6 g/kg (R, p.o.)
CN: (16β)-21-chloro-9-fluoro-16-methyl-17-(1-oxobutoxy)pregna-1,4-diene-3,11,20-trione

clobetasone

RN: 54063-32-0   MF: $C_{22}H_{26}ClFO_4$ MW: 408.90

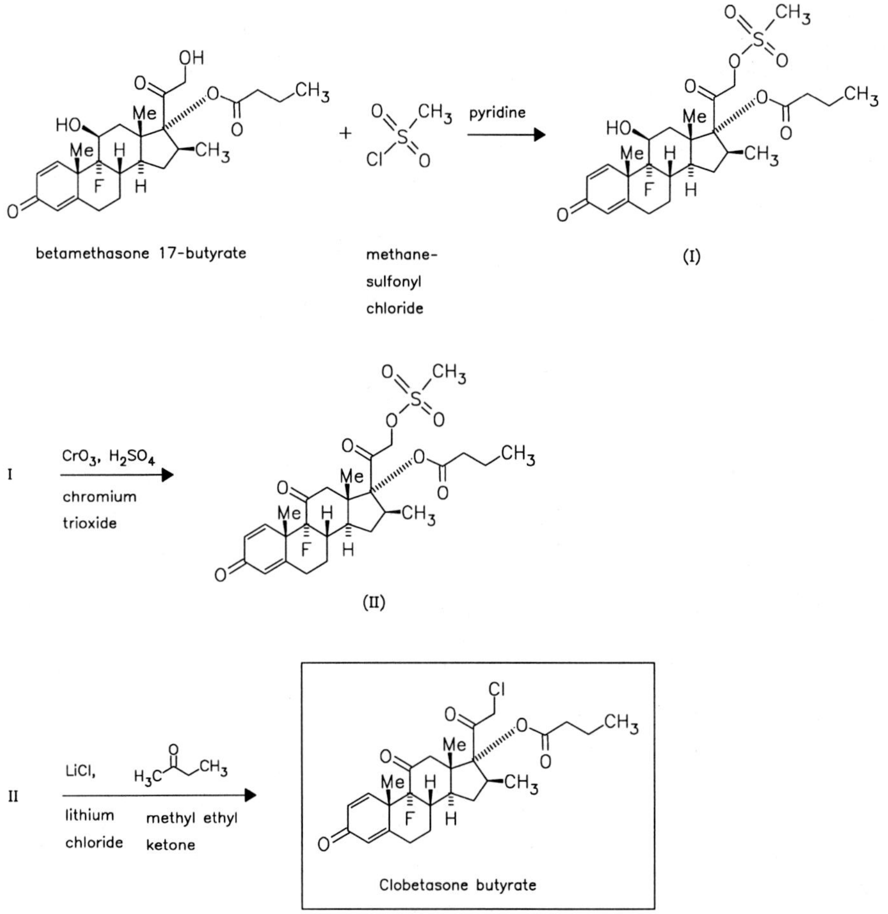

betamethasone 17-butyrate   methane-   (I)
                                 sulfonyl
                                 chloride

(II)

Clobetasone butyrate

Reference(s):
DE 1 902 340 (Glaxo; appl. 17.1.1969; GB-prior. 19.1.1968).
US 3 721 687 (Glaxo; 20.3.1973; GB-prior. 4.4.1968).
cf. synthesis of betamethasone-17-butyrate

Formulation(s):   eye drops 0.1 %; cream, ointment 0.5 mg/g (0.05 %)

Trade Name(s):

D: Emovate (Glaxo; 1980)
GB: Clobuvate (Dominion)
      Eumovate (Glaxo
      Wellcome; 1975)

      Trimovate (Glaxo
      Wellcome)-comb.
I: Eumovate (Glaxo; 1983)

      Visucloben (Merck Sharp
      & Dohme)
J: Kindavate (Glaxo; 1984)

Clobutinol

ATC: R05DB03
Use: antitussive

RN: 14860-49-2   MF: $C_{14}H_{22}ClNO$ MW: 255.79   EINECS: 238-926-4
LD_{50}:   53 mg/kg (M, i.v.); 334 mg/kg (M, p.o.);
        63 mg/kg (R, i.v.); 802 mg/kg (R, p.o.);
        45.3 mg/kg (dog, i.v.)
CN: 4-chloro-α-[2-(dimethylamino)-1-methylethyl]-α-methylbenzeneethanol

hydrochloride
RN: 1215-83-4   MF: $C_{14}H_{22}ClNO \cdot HCl$ MW: 292.25   EINECS: 214-931-7
LD_{50}:   40.9 mg/kg (M, i.v.); 334 mg/kg (M, p.o.);
        63 mg/kg (R, i.v.); 802 mg/kg (R, p.o.);
        45.3 mg/kg (dog, i.v.)

4-chlorobenzyl-
magnesium chloride

4-dimethylamino-
3-methyl-2-butanone

Clobutinol

Reference(s):
DE 1 146 068 (Thomae; appl. 21.3.1959).
DE 1 150 686 (Thomae; appl. 12.5.1960).
US 3 121 087 (Thomae; 11.2.1964; prior. 18.3.1960, 28.6.1961).
Engelhorn, R.: Arzneim.-Forsch. (ARZNAD) **10**, 794 (1960).

alternative synthesis:
DE 1 153 380 (Thomae; appl. 21.5.1959).

Formulation(s):   amp. 20 mg/2 ml; drg. 40 mg; cps. 40 mg, 80 mg; drops 40 mg/0.67 ml; syrup 40 mg/10 ml (as
        hydrochloride)

Trade Name(s):
D: Mentopin (Hermes)   Silomat (Boehringer Ing.)   Silomat compositum
      Nullatuss (Pharma   Tussamed (Hexal)   (Fher)-comb.
      Wernigerode)   F: Silomat (Boehringer Ing.)   J: Silomat (Morishita)
      Rotafuss (MIP Pharma)   I: Silomat (Fher)

Clocapramine

ATC: S01B
Use: thymoleptic

RN: 47739-98-0   MF: $C_{28}H_{37}ClN_4O$ MW: 481.08
CN: 1'-[3-(3-chloro-10,11-dihydro-5H-dibenz[b,f]azepin-5-yl)propyl][1,4'-bipiperidine]-4'-carboxamide

hydrochloride hydrate
RN: 28058-62-0   MF: $C_{28}H_{38}Cl_2N_4O \cdot HCl \cdot H_2O$ MW: 572.02

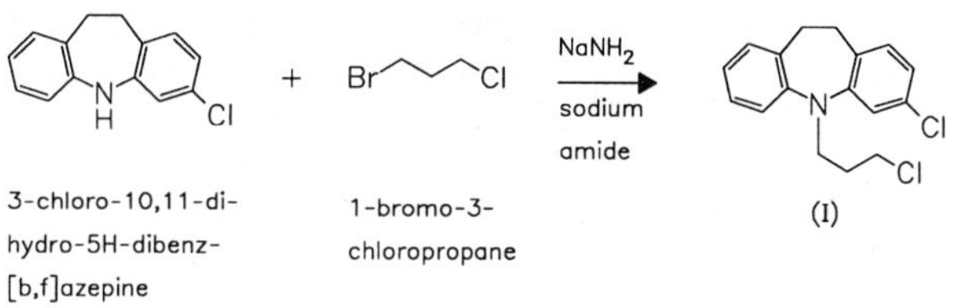

NaNH2
sodium
amide
3-chloro-10,11-dihydro-5H-dibenz[b,f]azepine
1-bromo-3-chloropropane
(I)

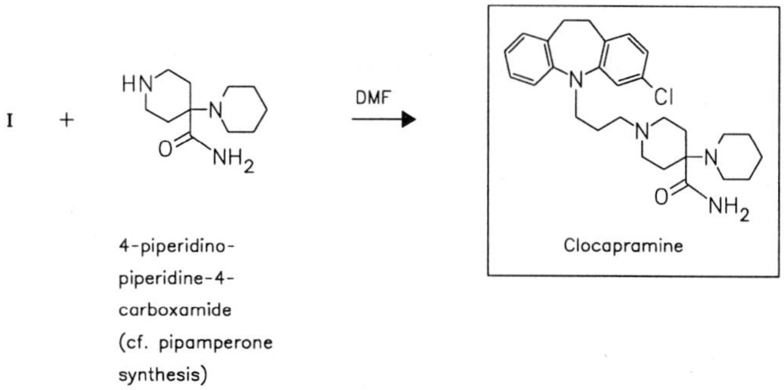

I +
DMF
4-piperidinopiperidine-4-carboxamide
(cf. pipamperone synthesis)
Clocapramine

Reference(s):

DOS 1 905 765 (Yoshitomi; appl. 6.2.1969; J-prior. 7.2.1968).
US 3 668 210 (Yoshitomi; 6.6.1972; J-prior. 7.2.1968).

Formulation(s): tabl. 10 mg, 25 mg (as hydrochloride)

Trade Name(s):

J: Clofekton (Yoshitomi-Takeda; 1974)

Clocortolone

ATC: D07AB21
Use: glucocorticoid

RN: 4828-27-7 MF: $C_{22}H_{28}ClFO_4$ MW: 410.91 EINECS: 225-406-7
CN: (6α,11β,16α)-9-chloro-6-fluoro-11,21-dihydroxy-16-methylpregna-1,4-diene-3,20-dione

pivalate
RN: 34097-16-0 MF: $C_{27}H_{36}ClFO_5$ MW: 495.03 EINECS: 251-826-5
caproate
RN: 4891-71-8 MF: $C_{28}H_{38}ClFO_5$ MW: 509.06 EINECS: 225-513-9

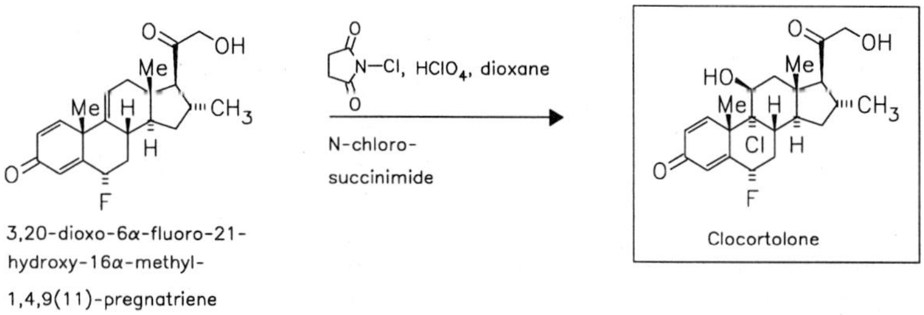

N–Cl, HClO4, dioxane
N-chloro-succinimide
3,20-dioxo-6α-fluoro-21-hydroxy-16α-methyl-1,4,9(11)-pregnatriene
Clocortolone

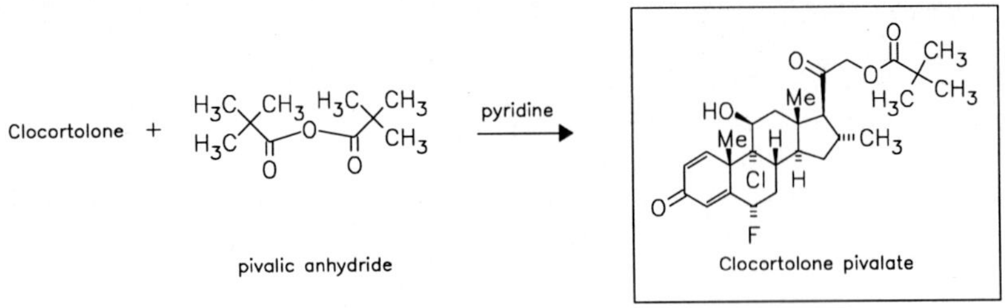

Clocortolone + pivalic anhydride →(pyridine) Clocortolone pivalate

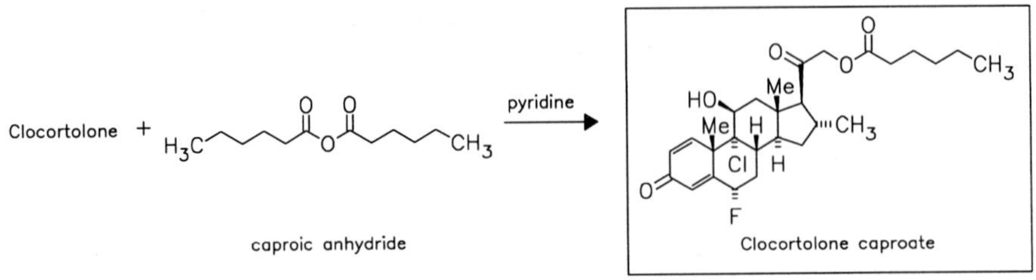

Clocortolone + caproic anhydride →(pyridine) Clocortolone caproate

Reference(s):
NL-appl. 6 412 708 (Schering AG; appl. 2.11.1964; D-prior. 9.11.1963).
FR 6 752 M (Schering AG; appl. 9.11.1966; D-prior. 9.11.1965).

synthesis of starting compound:
DOS 1 913 042 (Schering AG; appl. 11.3.1969).

alternative synthesis:
DOS 2 011 559 (Schering AG; appl. 7.3.1970).

Formulation(s):   cream 1 mg/g

Trade Name(s):
D: Crino-Kaban (Asche; as pivalate-comb.)
Kaban (Asche; as pivalate and capronate-comb.)

Kabanimat (Asche; as pivalate and capronate)
Procto Kaban (Asche; as capronate)-comb.

I: Cilder (Cilag; as pivalate); wfm
USA: Cloderm (Penederm; as pivalate)

Clodantoin
(Chlordantoin)

ATC: G01AX01
Use: fungicide

RN: 5588-20-5   MF: $C_{11}H_{17}Cl_3N_2O_2S$ MW: 347.69   EINECS: 226-995-3
LD$_{50}$: >1165 mg/kg (R, p.o.)
CN: 5-(1-ethylpentyl)-3-[(trichloromethyl)thio]-2,4-imidazolidinedione

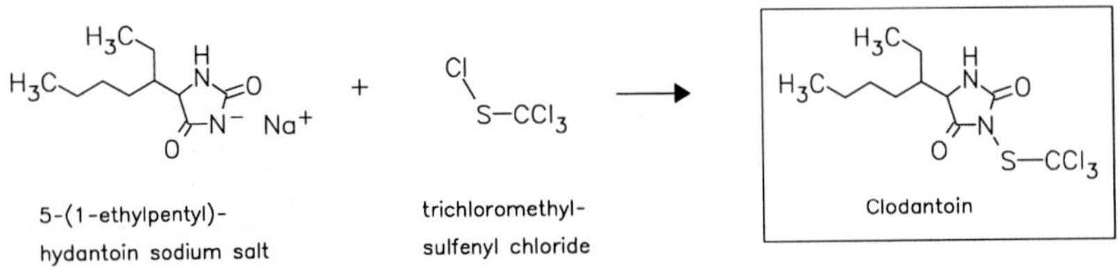

5-(1-ethylpentyl)-hydantoin sodium salt   +   trichloromethyl-sulfenyl chloride   →   Clodantoin

Reference(s):
US 2 553 770 (Standard Oil; 1951; prior. 1949).

Formulation(s):   cream; gel; powder

Trade Name(s):
GB: Sporostacin (Ortho); wfm   USA: Sporostacin (Ortho)-comb.;
J: Gynelan (Eisai)-comb.   wfm

Clodronate disodium
(Clodronic acid disodium salt)

ATC: M05BA02
Use: calcium metabolism regulator

RN: 22560-50-5   MF: $CH_2Cl_2Na_2O_6P_2$ MW: 288.86   EINECS: 245-078-9
CN: (dichloromethylene)bis(phosphonic acid) disodium salt

free acid
RN: 10596-23-3   MF: $CH_4Cl_2O_6P_2$ MW: 244.89   EINECS: 234-212-1

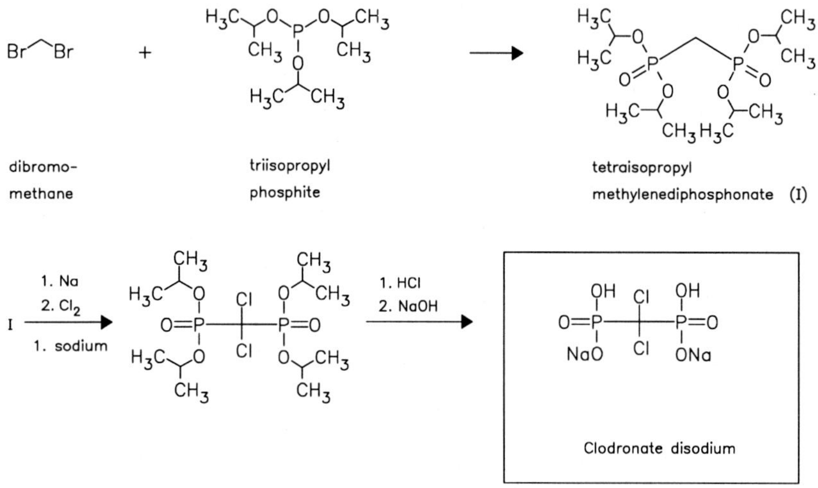

dibromo-methane   triisopropyl phosphite   tetraisopropyl methylenediphosphonate   (I)

Clodronate disodium

Reference(s):
DOS 1 467 655 (Procter & Gamble; appl. 17.3.1964; USA-prior. 18.3.1963).
DOS 1 793 768 (Procter & Gamble; appl. 17.3.1964; USA-prior. 18.3.1963).
US 3 404 178 (Procter & Gamble; 1.10.1968; appl. 18.3.1963, 7.10.1965).
US 3 422 021 (Procter & Gamble; 14.1.1969; appl. 18.3.1963).
Quimby, O.F. et al.: J. Org. Chem. (JOCEAH) **32**, 4111 (1967).

alternative synthesis:
McKenna, C.E. et al.: Phosphorus sulfur **37**, 1 (1998)

Formulation(s):   amp. 300 mg/5 ml, 300 mg/10 ml; cps. 400 mg; f. c. tabl. 520 mg, 800 mg

Trade Name(s):
D: Bonefos (Astra; medac)   F: Clastoban (Rorer; Roger   I: Clasteon (Gentili)
    Ostac (Roche; 1988)   Bellon)   Difosfonal (SPA)
                            Cytos (Roche)   Ossiten (Roche)

Clofedanol

(Chlophedianol)

ATC: R05DB10
Use: antitussive

RN: 791-35-5 MF: $C_{17}H_{20}ClNO$ MW: 289.81 EINECS: 212-340-9
LD_{50}: 70 mg/kg (M, i.v.); 300 mg/kg (M, p.o.)
CN: 2-chloro-α-[2-(dimethylamino)ethyl]-α-phenylbenzenemethanol

hydrochloride
RN: 511-13-7 MF: $C_{17}H_{20}ClNO \cdot HCl$ MW: 326.27 EINECS: 208-124-9
LD_{50}: 42 mg/kg (M, i.v.); 284 mg/kg (M, p.o.);
53 mg/kg (R, i.v.); 350 mg/kg (R, p.o.);
84 mg/kg (dog, p.o.)

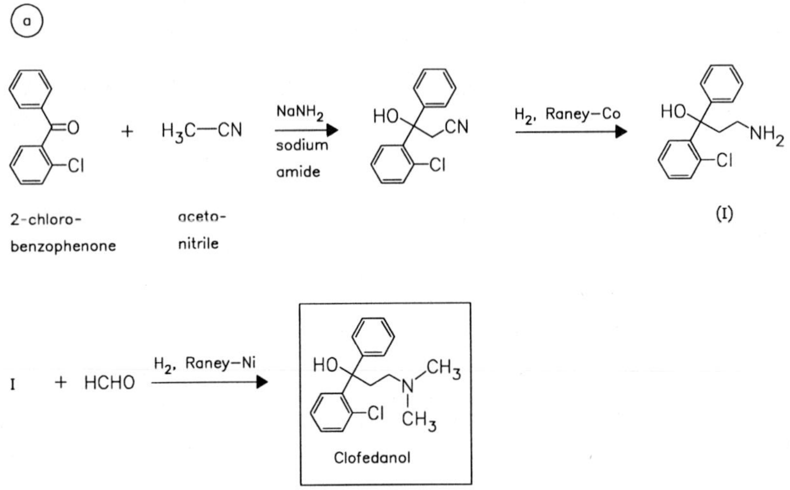
a
NaNH2
sodium
amide
H2, Raney-Co
2-chloro-
benzophenone
aceto-
nitrile
(I)
I + HCHO
H2, Raney-Ni
Clofedanol

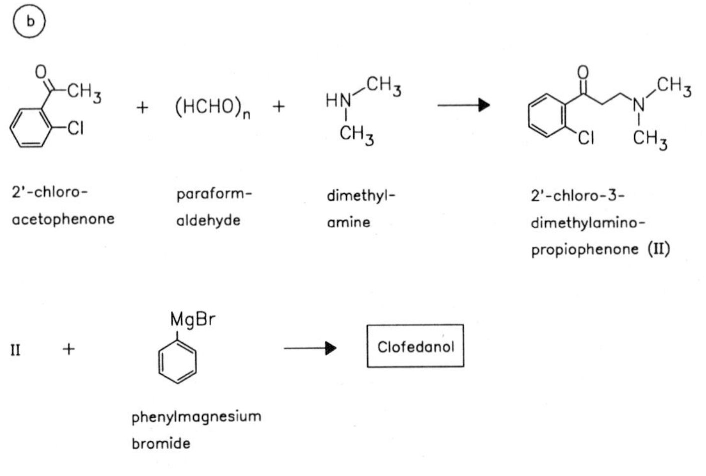
b
(HCHO)n
2'-chloro-
acetophenone
paraform-
aldehyde
dimethyl-
amine
2'-chloro-3-
dimethylamino-
propiophenone (II)
II +
MgBr
Clofedanol
phenylmagnesium
bromide

Reference(s):

DE 1 080 568 (Bayer; appl. 8.1.1958).
DE 1 083 277 (Bayer; appl. 19.3.1958).
US 3 031 377 (Bayer; 24.4.1962; appl. 26.11.1957).

Formulation(s): syrup 25 mg, 30 mg; tabl. 12.5 mg (as hydrochloride)

Trade Name(s):
D: Dicton (Dolorgiet)-comb.;   Pectolitan (Kettelhack-   I: Soltux (Corvi)-comb.
        wfm   Riker); wfm   J: Coldrin (Nippon Shinyaku)
                                     F: Tussiplégyl (Bayer); wfm   USA: Ulo (Riker); wfm

Clofexamide

ATC: N06B
Use: psychoanaleptic

RN: 1223-36-5   MF: $C_{14}H_{21}ClN_2O_2$ MW: 284.79   EINECS: 214-951-6
CN: 2-(4-chlorophenoxy)-N-[2-(diethylamino)ethyl]acetamide

4-chlorophenoxy-   N,N-diethyl-   Clofexamide
acetyl chloride   ethylenediamine

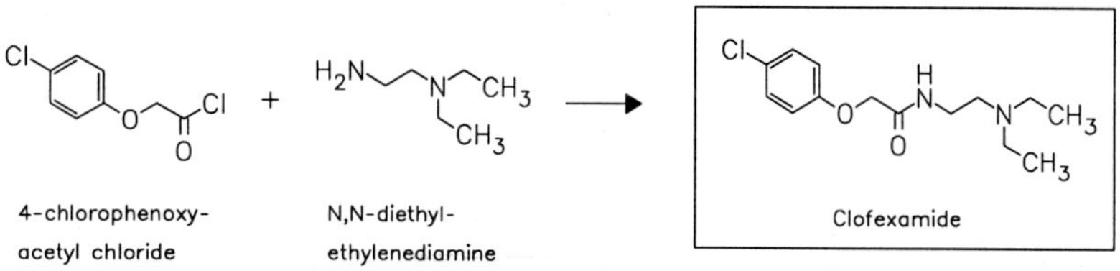

Reference(s):
GB 942 761 (Centre National de la Recherche Scientifique; appl. 8.4.1960; F-prior. 15.4.1959, 30.7.1959).

Formulation(s):   tabl. 50 mg, 400 mg

Trade Name(s):
F: Clofexan à la
        noramidopyrine (Anphar)-
        comb.; wfm

Clofezone

ATC: M01AA05; M02AA03
Use: anti-inflammatory, antirheumatic

RN: 17449-96-6   MF: $C_{19}H_{20}N_2O_2 \cdot C_{14}H_{21}ClN_2O_2$ MW: 593.17   EINECS: 241-466-7
LD_{50}: 1700 mg/kg (M, p.o.);
        1950 mg/kg (R, p.o.)
CN: 2-(4-chlorophenoxy)-N-[2-(diethylamino)ethyl]acetamide compd. with 4-butyl-1,2-diphenyl-3,5-
        pyrazolidinedione (1:1)

dihydrate
RN: 60104-29-2   MF: $C_{19}H_{20}N_2O_2 \cdot C_{14}H_{21}ClN_2O_2 \cdot 2H_2O$ MW: 629.20

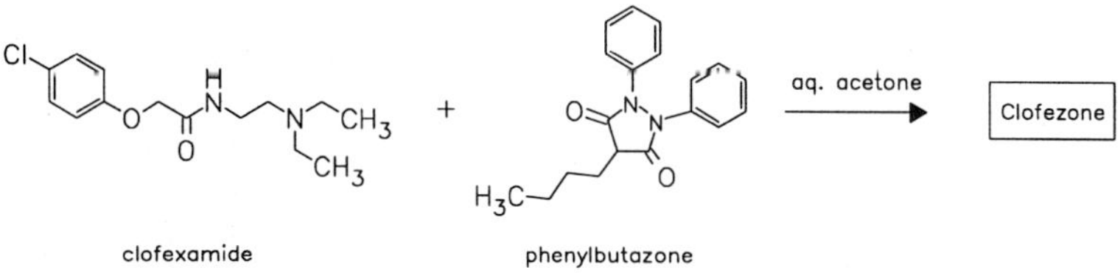

clofexamide   phenylbutazone

Clofezone

Reference(s):
US 3 491 190 (P. Rumpf, J.-E., G. Thuillier; 20.1.1970; F-prior. 8.9.1965).

Formulation(s):   cps. 200 mg, 400 mg; ointment 5 g/100 g; suppos. 400 mg

Trade Name(s):
D: Perclusone (Mack, Illert.);   F: Perclusone (Serb)-comb.   J: Panas (Grelan)
      wfm   I: Perclusone (Marxer); wfm

Clofibrate

ATC: C01AB01
Use: cholesterol depressant,
        antihyperlipidemic,
        antiarteriosclerotic

RN: 637-07-0   MF: $C_{12}H_{15}ClO_3$ MW: 242.70   EINECS: 211-277-4
CN: 2-(4-chlorophenoxy)-2-methylpropanoic acid ethyl ester

| 4-chloro-phenol | acetone | chloro-form | 2-(4-chloro-phenoxy)-2-methylpropanoic acid | Clofibrate |

Reference(s):
Julia, M. et al.: Bull. Soc. Chim. Fr. (BSCFAS) **1956**, 777.
US 3 262 850 (ICI; 26.7.1966; GB-prior. 20.6.1958).

Formulation(s):   cps. 250 mg, 500 mg

Trade Name(s):
D: Regelan N 500 (Zeneca)   Atemarol (Kowa Yakuhin)   Cholestol (Toho)
F: Lipavlon (Zeneca)   Athebrate (Kakenyaku)   Cholesrun (Hokuriku)
GB: Atromid S (Zeneca)   Atherolate (Fuji Zoki)   Clarol (Toyama)
I: Sinteroid (Crinos)   Atheromide (Ono)   Clobrate (Chugai)
J: Amotril (Sumitomo)   Atmol (Taisho)   Clobren (Morishita)
      Apoterin A (Seiko)   Atosterine (Kanto)   Clofbate (Mohan)
      Artehard (Nissin)   Auparton (Samva)   Climinon (Meiji)
      Ateculon (Nippon   Binograc (Zeria)   Deliva (Nippon Kayaku)
      Chemiphar)   Bresit (Toyo Jozo)   Hyclorate (Funay)
      Ateles (Tokyo Hosei)   Cholenal (Yamanouchi)   Hypocerol (Fuso)

Liprinal (Banyu)   Scrobin (Nikken)   USA: Atromid S (Wyeth-Ayerst)

Clofoctol

ATC: J01XX03; R07A
Use: antibacterial (in respiratory
       infections)

RN: 37693-01-9   MF: C$_{21}$H$_{26}$Cl$_2$O   MW: 365.34   EINECS: 253-632-6
LD$_{50}$:   >4000 mg/kg (R, p.o.)
CN: 2-[(2,4-dichlorophenyl)methyl]-4-(1,1,3,3-tetramethylbutyl)phenol

4-(1,1,3,3-tetra-   2,4-dichloro-   Clofoctol
methylbutyl)-   benzyl
phenol   chloride

Reference(s):
FR 1 602 455 (I. R. C. E. B. A.; appl. 21.8.1968; GB-prior. 31.8.1967).
US 3 830 852 (I. R. C. E. B. A.; 20.8.1974; F-prior. 18.8.1970).
DOS 2 140 765 (I. R. C. E. B. A.; appl. 13.8.1971; F-prior. 18.8.1970).

preparation of 4-(1,1,3,3-tetramethylbutyl)phenol:
US 2 726 270 (Dow Chem.; 1951).
DE 842 073 (Reichhold Chem. Inc.; 1950).
US 2 732 448 (California Research Corp.; 1953).
US 2 572 019 (DuPont de Nemours & Co.; 1950).
further patents are described before 1950.

Formulation(s):   suppos. 100 mg, 200 mg, 750 mg

Trade Name(s):
F: Octofène (Débat; 1978)   I: Gramplus (Chiesi)   Octofene (Roussel; 1985)

Clometacin

ATC: N02
Use: anti-inflammatory, analgesic

RN: 25803-14-9   MF: C$_{19}$H$_{16}$ClNO$_4$ MW: 357.79   EINECS: 247-271-3
LD$_{50}$:   1 g/kg (M, p.o.)
CN: 3-(4-chlorobenzoyl)-6-methoxy-2-methyl-1*H*-indole-1-acetic acid

4-methoxy-2-   nitroethane   1-(2-nitro-4-methoxy-
nitrobenzaldehyde   phenyl)-2-nitropropene

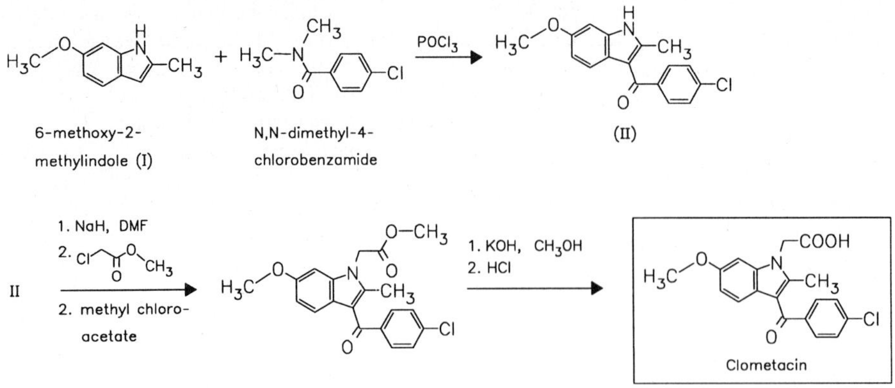

6-methoxy-2-
methylindole (I)

N,N-dimethyl-4-
chlorobenzamide

(II)

Clometacin

Reference(s):
DE 1 901 167 (Roussel-Uclaf; appl. 10.1.1969; F-prior. 11.1.1968, 10.4.1968, 10.9.1968, 11.9.1968, 10.12.1968).

Formulation(s):   tabl. 150 mg

Trade Name(s):
F: Dupéran (Cassenne); wfm

Clomethiazole

ATC: N05CM02
Use: anticonvulsant, hypnotic, sedative

RN: 533-45-9   MF: C_6H_8ClNS MW: 161.66   EINECS: 208-565-7
LD$_{50}$:   94 mg/kg (M, i.v.); 2110 mg/kg (M, p.o.)
CN: 5-(2-chloroethyl)-4-methylthiazole

ethanedisulfonate (2:1)
RN: 1867-58-9   MF: $C_6H_8ClNS \cdot 1/2C_2H_6O_6S_2$ MW: 513.51   EINECS: 217-483-0
LD$_{50}$:   150 mg/kg (M, i.v.); 835 mg/kg (M, p.o.)

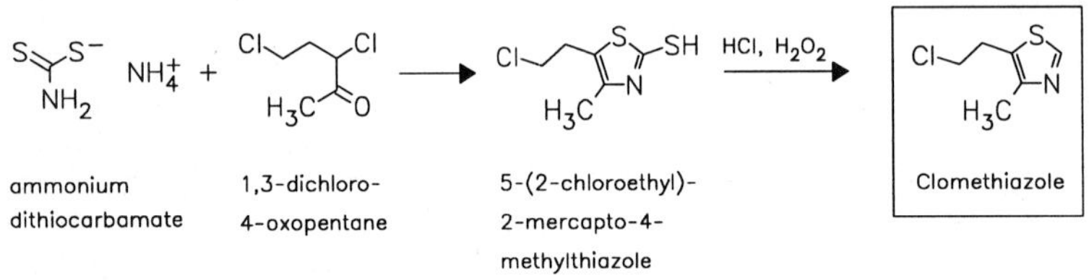

ammonium
dithiocarbamate

1,3-dichloro-
4-oxopentane

5-(2-chloroethyl)-
2-mercapto-4-
methylthiazole

Clomethiazole

Reference(s):
CH 200 248 (Roche; 1937).

sulfonate:
US 3 031 457 (R. Charonnat, J. Chareton, A. Boune; 24.4.1962; F-prior. 28.9.1955).

alternative syntheses:
Buchman, E.R.: J. Am. Chem. Soc. (JACSAT) **58**, 1803 (1936).
Sawa, Y.; Ishida, T.: Yakugaku Zasshi (YKKZAJ) **76**, 337 (1956).

Formulation(s):   amp. 0.8 % (as ethanedisulfonate); cps. 192 mg, 300 mg; syrup 320 mg/10 ml; tabl. 500 mg

Clometocillin
(Chlomethocillin)

ATC: J01CE07
Use: antibiotic

RN: 1926-49-4   MF: $C_{17}H_{18}Cl_2N_2O_5S$ MW: 433.31   EINECS: 217-657-6
CN: [2S-(2α,5α,6β)]-6-[[(3,4-dichlorophenyl)methoxyacetyl]amino]-3,3-dimethyl-7-oxo-4-thia-1-azabicyclo[3.2.0]heptane-2-carboxylic acid

monopotassium salt
RN: 15433-28-0   MF: $C_{17}H_{17}Cl_2KN_2O_5S$ MW: 471.40

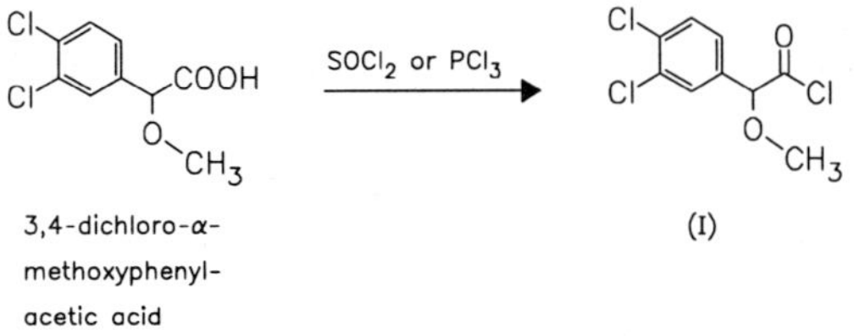

3,4-dichloro-α-
methoxyphenyl-
acetic acid

(I)

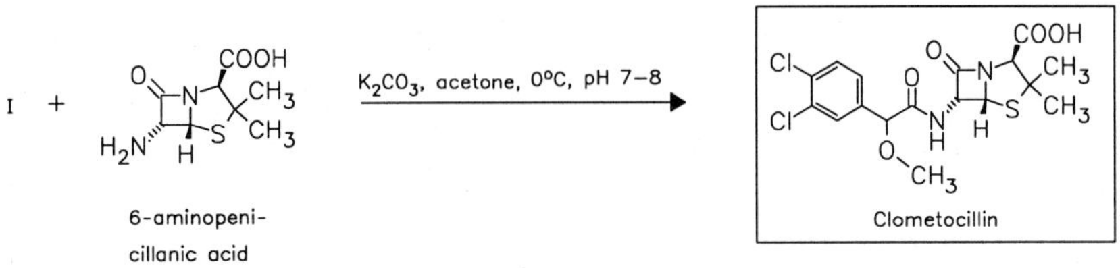

6-aminopeni-
cillanic acid

Clometocillin

Reference(s):
US 3 007 920 (Recherche Industrie Therapeutiques; 7.11.1961; GB-prior. 28.10.1960).

Formulation(s):   tabl. 500 mg

Trade Name(s):
F: Rixapen (Smith Kline &
        French); wfm

Clomifene
(Clomiphene)

ATC: G03GB02
Use: synthetic gonadotropin stimulant,
        antiestrogen

RN: 911-45-5   MF: $C_{26}H_{28}ClNO$ MW: 405.97   EINECS: 213-008-6
LD$_{50}$: 1700 mg/kg (M, p.o.)
CN: 2-[4-(2-chloro-1,2-diphenylethenyl)phenoxy]-N,N-diethylethanamine

citrate (1:1)
RN: 50-41-9   MF: $C_{26}H_{28}ClNO \cdot C_6H_8O_7$ MW: 598.09   EINECS: 200-035-3
LD$_{50}$: 1400 mg/kg (M, p.o.);
        5750 mg/kg (R, p.o.)

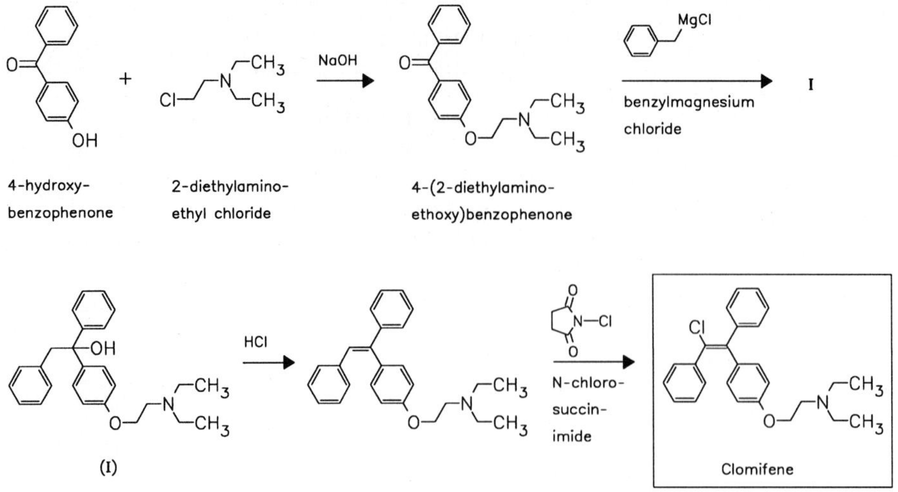

4-hydroxy-
benzophenone

2-diethylamino-
ethyl chloride

4-(2-diethylamino-
ethoxy)benzophenone

benzylmagnesium
chloride

(I)

N-chloro-
succin-
imide

Clomifene

Reference(s):
US 2 914 563 (Merrell; 24.11.1959; prior. 6.8.1957).

medical use:
BE 782 321 (Richardson-Merrell; appl. 19.4.1971).

Formulation(s):   cps. 50 mg; tabl. 50 mg (as citrate)

Trade Name(s):

D:	Clomhexal (Hexal)	GB:	Clomid (Hoechst)
	Dyneric (Henning Berlin)		Serophene (Hoechst)
	Pergotime (Serono)	I:	Clomid (Lepetit)
F:	Clomid (Marion Merrell)		Prolifen (Chiesi)
	Pergotime (Serono)	J:	Clomid (Shionogi)

Orifen (Iwaki)
USA: Clomid (Hoechst Marion
Roussel; as citrate)
Serophene (Serono)

Clomipramine

ATC: N06AA04
Use: antidepressant

RN: 303-49-1   MF: $C_{19}H_{23}ClN_2$ MW: 314.86   EINECS: 206-144-2
LD$_{50}$:   27 mg/kg (M, i.v.); 380 mg/kg (M, p.o.);
        613 mg/kg (R, p.o.)
CN: 3-chloro-10,11-dihydro-*N,N*-dimethyl-5*H*-dibenz[*b,f*]azepine-5-propanamine

monohydrochloride
RN: 17321-77-6   MF: $C_{19}H_{23}ClN_2 \cdot HCl$ MW: 351.32   EINECS: 241-344-3
LD$_{50}$:   22 mg/kg (M, i.v.); 470 mg/kg (M, p.o.);
        26 mg/kg (R, i.v.); 914 mg/kg (R, p.o.);
        32 mg/kg (dog, i.v.); 383 mg/kg (dog, p.o.)

ⓐ

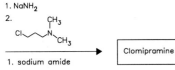

3-chloro-10,11-di-
hydro-5H-dibenz-
[b,f]azepine (I)

phosgene

160–210 °C

(II)

Clomipramine

ⓑ

1. NaNH₂
2.

I

1. sodium amide
2. 3-dimethylamino-
   propyl chloride

→ Clomipramine

Reference(s):
US 3 515 785 (Geigy; 2.6.1970; CH-prior. 6.12.1958; 12.1.1959).
DE 1 161 278 (Geigy; appl. 5.12.1959; CH-prior. 6.12.1958, 12.1.1959).
CH 371 799 (Geigy; appl. 6.12.1958).
Craig, P.N. et al.: J. Org. Chem. (JOCEAH) **26**, 135 (1961).

starting material:
DE 1 166 200 (Geigy; appl. 1959; CH-prior. 1958, 1959).
US 3 056 774 (Geigy; 1962; CH-prior. 1958).
US 3 056 776 (Geigy; 1962; CH-prior. 1959).

Formulation(s):   amp. 25 mg/2 ml; drg. 25 mg; f. c. tabl. 10 mg, 25 mg; s. r. tabl. 75 mg

Trade Name(s):
D: Anafranil (Novartis)   F: Anafranil (Novartis)   J: Anafranil (Fujisawa)
       Hydiphen (ASTA Medica   GB: Anafranil (Novartis)   USA: Anafranil (Novartis; as
       AWD)   I: Anafranil (Geigy)   hydrochloride)

Clomocycline
(Chlormethylencycline; Clomociclina)

ATC: J01AA11
Use: antibiotic

RN: 1181-54-0   MF: $C_{23}H_{25}ClN_2O_9$ MW: 508.91
LD$_{50}$:   115 mg/kg (M, i.v.); 2830 mg/kg (M, p.o.)
CN: [4S-(4α,4aα,5aα,6β,12aα)]-7-chloro-4-(dimethylamino)-1,4,4a,5,5a,6,11,12a-octahydro-3,6,10,12,12a-
       pentahydroxy-N-(hydroxymethyl)-6-methyl-1,11-dioxo-2-naphthacenecarboxamide

sodium salt
RN: 68-20-2   MF: $C_{23}H_{24}ClN_2NaO_9$ MW: 530.89

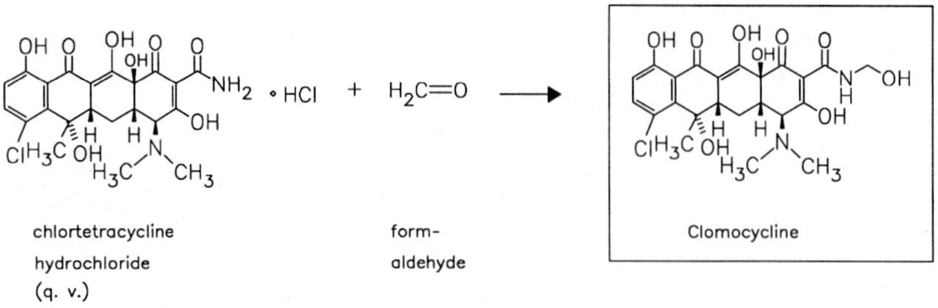

chlortetracycline   form-   Clomocycline
hydrochloride   aldehyde
(q. v.)

Reference(s):
BE 628 142 (Leo Ind. Chim. Farm. S.p.A.; appl. 7.2.1963).

Formulation(s):   cps. 170 mg (as sodium salt)

Trade Name(s):
GB: Megaclor (Pharmax); wfm   I: Megaclor (Pharmax); wfm

Clonazepam

ATC: N03AE01
Use: anticonvulsant

RN: 1622-61-3   MF: C$_{15}$H$_{10}$ClN$_3$O$_3$ MW: 315.72   EINECS: 216-596-2
CN: 5-(2-chlorophenyl)-1,3-dihydro-7-nitro-2H-1,4-benzodiazepin-2-one

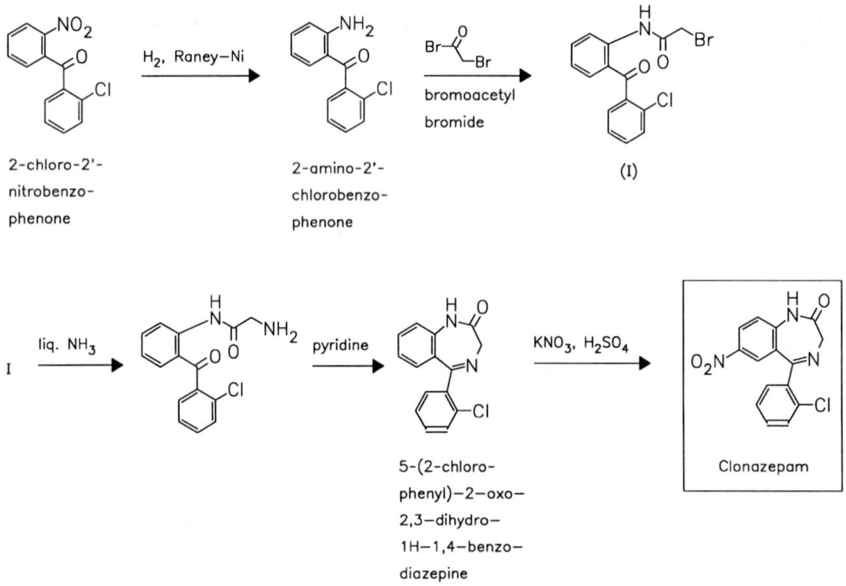

2-chloro-2'-   2-amino-2'-   (I)
nitrobenzo-   chlorobenzo-
phenone   phenone

                           5-(2-chloro-   Clonazepam
                           phenyl)-2-oxo-
                           2,3-dihydro-
                           1H-1,4-benzo-
                           diazepine

Reference(s):
Sternbach, L.H. et al.: J. Med. Chem. (JMCMAR) **6**, 261 (1963).
US 3 116 203 (Hoffmann-La Roche; 31.12.1963; prior. 14.3.1962).
US 3 123 529 (Hoffmann-La Roche; 3.3.1964; prior. 9.3.1962).
US 3 121 114 (Roche; 11.2.1964; CH-prior. 2.12.1960).
US 3 203 990 (Roche; 31.8.1965; prior. 27.6.1960, 20.4.1961, 21.3.1962).
US 3 335 181 (Roche; 8.8.1967; appl. 17.4.1964).

Formulation(s):   amp. 1 mg/ml; sol. 2.5 mg/ml; tabl. 0.25 mg, 0.5 mg, 1 mg, 2 mg

Trade Name(s):
D: Antalepsin (ASTA Medica   F: Rivotril (Roche)   USA: Klonopin (Roche Labs.)
       AWD)   GB: Rivotril (Roche)
       Rivotril (Roche)   I: Rivotril (Roche)

Clonidine

ATC: C02AC01; N02CX02; S01EA04
Use: antihypertensive

RN: 4205-90-7   MF: $C_9H_9Cl_2N_3$ MW: 230.10   EINECS: 224-119-4
CN: N-(2,6-dichlorophenyl)-4,5-dihydro-1H-imidazol-2-amine

monohydrochloride
RN: 4205-91-8   MF: $C_9H_9Cl_2N_3 \cdot HCl$ MW: 266.56   EINECS: 224-121-5
LD_{50}:   17.6 mg/kg (M, i.v.); 135 mg/kg (M, p.o.);
       29 mg/kg (R, i.v.); 126 mg/kg (R, p.o.);
       6 mg/kg (dog, i.v.); 30 mg/kg (dog, p.o.)

2,6-dichloro-   ammonium   N-(2,6-di-   (I)
aniline   rhodanide   chlorophenyl)-
       thiourea

ethylene-   Clonidine
diamine

Reference(s):
DE 1 303 141 (Boehringer Ing.; appl. 9.10.1961).
US 3 202 660 (Boehringer Ing.; 24.8.1965; D-prior. 9.10.1961).
US 3 236 857 (Boehringer Ing.; 22.2.1966; D-prior. 9.10.1961, 4.10.1963).
BE 653 933 (Boehringer Ing.; appl. 2.10.1964; D-prior. 4.10.1963; 31.7.1964).
GB 1 016 514 (Boehringer Ing.; appl. 2.10.1962; D-prior. 9.10.1961).
GB 1 034 938 (Boehringer Ing.; appl. 28.9.1964; D-prior. 4.10.1963; addition to GB 1 016 514).

alternative syntheses:
DAS 1 770 874 (VEB Arzneimittelwerke Dresden; appl. 12.7.1968).
DAS 2 505 297 (Lentia; appl. 7.2.1975; A-prior. 5.4.1974).

Formulation(s):   eye drops 0.625 mg/ml, 1.25 mg/ml; inj. sol. 0.15 mg/1 ml, 0.75 mg/ml; s. r. cps. 0.25 mg;
       tabl. 0.075 mg, 0.15 mg, 0.3 mg

Trade Name(s):
D: Aruclonin (Chauvin   Dixarit (Boehringer Ing.)   F: Catapressan (Boehringer
       ankerpharm)   Haemiton (ASTA Medica   Ing.)
       Catapresan (Boehringer   AWD)   GB: Catapres (Boehringer Ing.)
       Ing.)   Haemiton (ASTA Medica   Dixarit (Boehringer Ing.)
       Combipresan (Boehringer   AWD)-comb.   I: Adesipress (Carlo Erba)
       Ing.)-comb.   Isoglaucon (Alcon)   Catapresan (Boehringer
       Dispaclonidin (CIBA   Mirfat (Merckle)   Ing.)-comb.
       Vision)   Paracefan (Boehringer Ing.)

Combipresan (Boehringer Ing.)	Isoglaucon (Boehringer Ing.; as hydrochloride)	J: Catapres (Tanabe)
		USA: Catapres (Boehringer Ing.)

Clopamide

ATC: C03BA03
Use: diuretic

RN: 636-54-4 MF: $C_{14}H_{20}ClN_3O_3S$ MW: 345.85 EINECS: 211-261-7
CN: cis-3-(aminosulfonyl)-4-chloro-N-(2,6-dimethyl-1-piperidinyl)benzamide

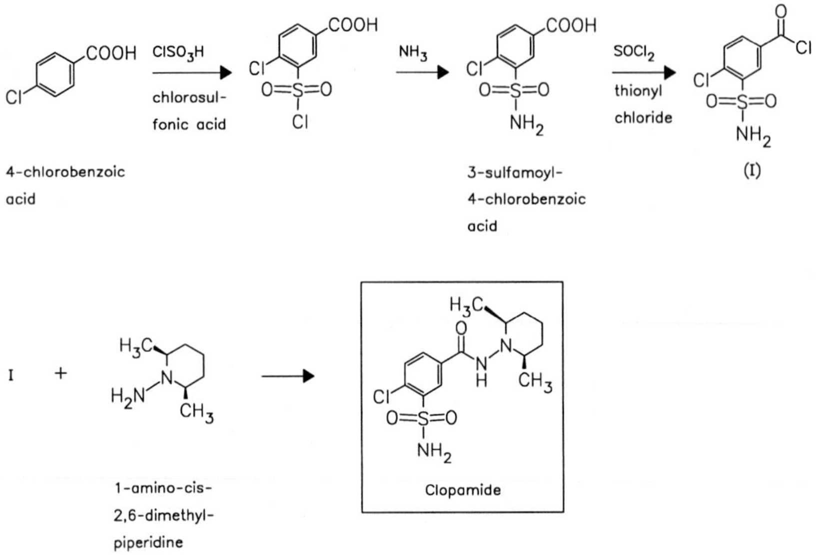

4-chlorobenzoic acid

3-sulfamoyl-4-chlorobenzoic acid

(I)

1-amino-cis-2,6-dimethyl-piperidine

Clopamide

Reference(s):
Jucker, E.; Lindenmann, A.: Helv. Chim. Acta (HCACAV) **45**, 2316 (1962).
CH 412 891 (Sandoz; appl. 6.6.1961; addition to CH 396 905).
CH 396 905 (Sandoz; appl. 9.11.1960).
CH 436 288 (Sandoz; appl. 11.6.1963).

combination with dihydroergocristine and reserpine:
US 3 567 828 (Sandoz; 2.3.1971; appl. 12.8.1968; CH-prior. 17.8.1967).
DAS 1 792 271 (Sandoz; appl. 13.8.1968; CH-prior. 17.8.1967).

Formulation(s): drg. 2.5 mg, 5 mg (comb. with reserpine); tabl. 20 mg

Trade Name(s):
D: Brinaldix (Novartis Pharma)
Briserin/mite (Novartis Pharma)-comb.

Viskaldix (Novartis Pharma)-comb.
F: Viskaldix (Sandoz)-comb.
GB: Viskaldix (Novartis)-comb.

I: Brinerdina (Sandoz)-comb.
USA: Brinaldix (Sandoz); wfm

Clopenthixol

ATC: N05AF02
Use: neuroleptic

RN: 982-24-1 MF: $C_{22}H_{25}ClN_2OS$ MW: 400.97 EINECS: 213-566-0
LD$_{50}$: 226 mg/kg (M, i.p.)
CN: 4-[3-(2-chloro-9H-thioxanthen-9-ylidene)propyl]-1-piperazineethanol

dihydrochloride
RN: 633-59-0   MF: $C_{22}H_{25}ClN_2OS \cdot 2HCl$ MW: 473.90   EINECS: 211-194-3
LD$_{50}$:   111 mg/kg (M, i.v.); 560 mg/kg (M, p.o.);
      125 mg/kg (R, i.v.); 660 mg/kg (R, p.o.)

chlorprothixen   1-(2-hydroxyethyl)-   Clopenthixol
               piperazine

Reference(s):
US 3 116 291 (Kefalas; 31.12.1963; DK-prior. 4.12.1958).
DE 1 231 254 (Kefalas; appl. 1960; DK-prior. 1959).
US 3 149 103 (Kefalas A/S; 15.9.1964; DK-prior. 14.7.1959).
GB 932 494 (Kefalas; appl. 3.12.1959; DK-prior. 4.12.1958, 14.8.1959).

alternative syntheses:
DE 1 443 983 (Roche; appl. 16.1.1962; CH-prior. 8.2.1961, 30.3.1961).
DE 1 795 506 (Kefalas; appl. 3.12.1959; DK-prior. 4.12.1958, 14.8.1959).
DOS 1 918 739 (Egyesült; appl. 12.4.1969; H-prior. 12.4.1968).

separation of isomers:
DAS 2 429 101 (Kefalas; appl. 18.6.1974; GB-prior. 25.6.1973).

Formulation(s):   amp. 10 mg/ml, 25 mg/ml; drg. 10 mg, 25 mg; tabl. 25 mg (as dihydrochloride)

Trade Name(s):
D: Ciatyl (Bayer Vital)   GB: Clopixol (Lundbeck)   I: Sordinol (Pierrel)

Cloperastine

ATC: R05DB21
Use: antitussive

RN: 3703-76-2   MF: $C_{20}H_{24}ClNO$ MW: 329.87   EINECS: 223-042-3
LD$_{50}$:   439 mg/kg (g.p., route unreported)
CN: 1-[2-[(4-chlorophenyl)phenylmethoxy]ethyl]piperidine

4-chloro-   1. sodium amide   Cloperastine
benzhydrol   2. 1-(2-chloroethyl)-
               piperidine

Reference(s):
GB 670 622 (Parke Davis; appl. 1948; CH-prior. 1947).

salt with 2-[(6-hydroxy[1,1'-biphenyl]-3-yl)carbonyl]benzoic acid:
GB 1 179 945 (Yoshitomi; appl. 7.7.1967; J-prior. 7.7.1966; 16.8.1966, 15.11.1966 and 1.3.1967).

Formulation(s):   syrup 0.2 %; tabl. 5 mg, 10 mg, 20 mg

Trade Name(s):

I:	Nitossil (Zyma)		Seki (Astra-Simes)
	Risoltuss (Magis)	J:	Hustazol (Yoshitomi)

Clopidogrel hydrogensulfate
(SR-25990C)

ATC: B01AC04
Use: platelet anti-aggregatory

RN: 120202-66-6 MF: $C_{16}H_{16}ClNO_2S \cdot H_2SO_4$ MW: 419.91
CN: (*S*)-.α-(2-Chlorophenyl)-6,7-dihydrothieno[3,2-*c*]pyridine-5(4*H*)-acetic acid methyl ester sulfate (1:1)

(±)-base
RN: 90055-48-4 MF: $C_{16}H_{16}ClNO_2S$ MW: 321.83
(+)-base
RN: 113665-84-2 MF: $C_{16}H_{16}ClNO_2S$ MW: 321.83

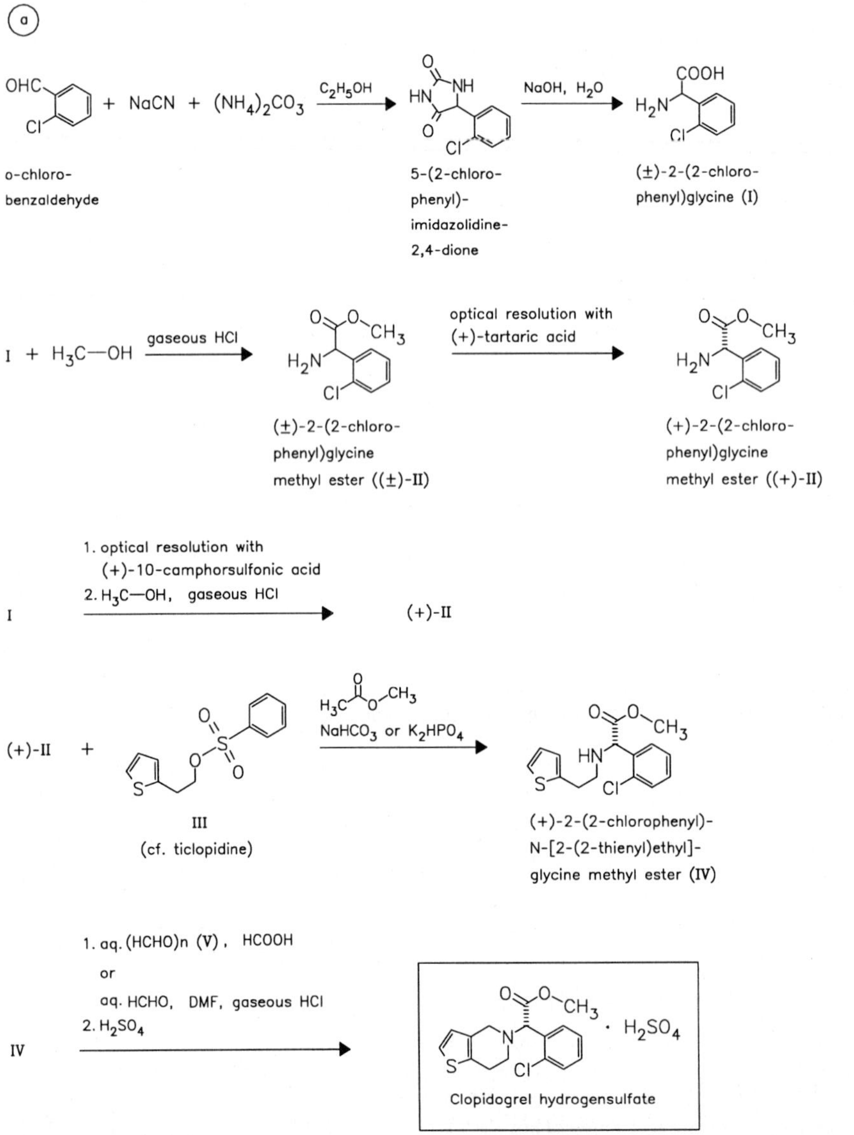

alternative route

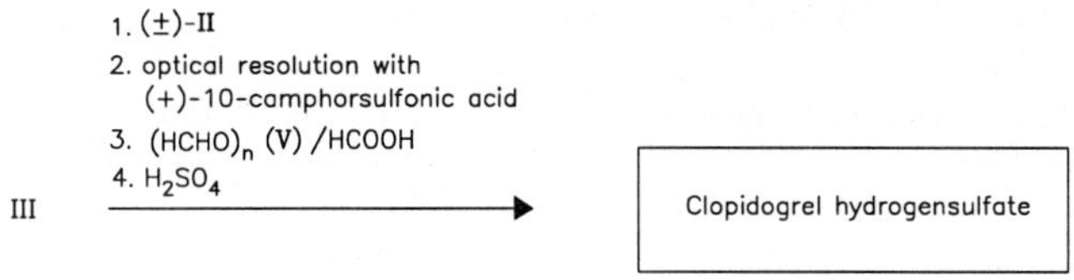

    1. (±)-II
    2. optical resolution with
       (+)-10-camphorsulfonic acid
    3. (HCHO)$_n$ (V) /HCOOH
    4. H$_2$SO$_4$

III   ⟶   Clopidogrel hydrogensulfate

synthesis of intermediate III

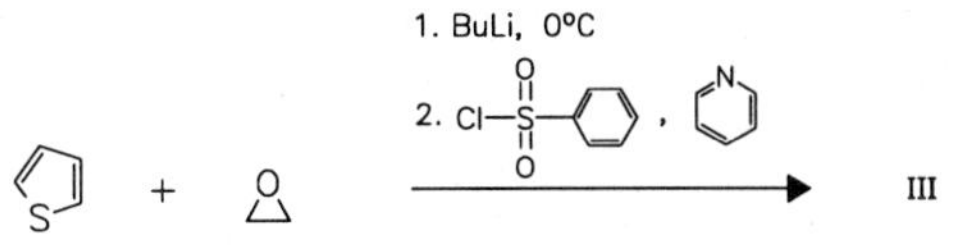

III

b

2-thiophene-
carboxaldehyde

2-thienyl-
acetaldehyde (VI)

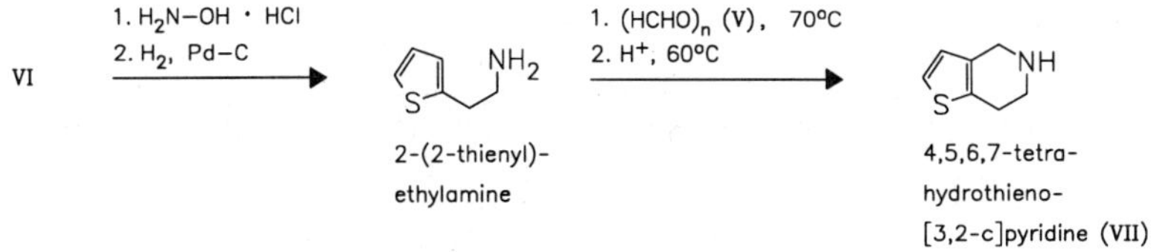

VI

2-(2-thienyl)-
ethylamine

4,5,6,7-tetra-
hydrothieno-
[3,2-c]pyridine (VII)

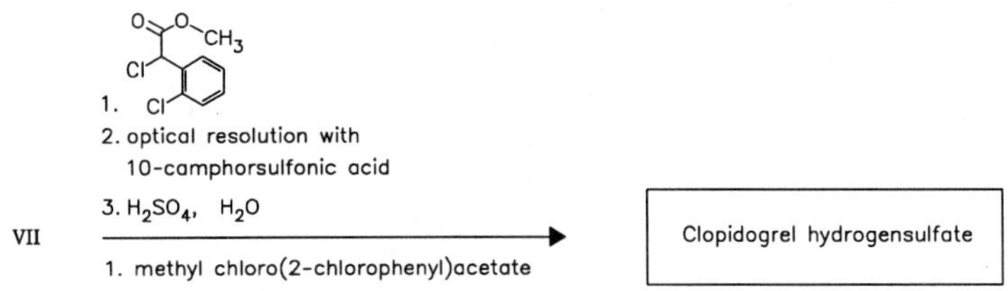

VII

1. methyl chloro(2-chlorophenyl)acetate

Clopidogrel hydrogensulfate

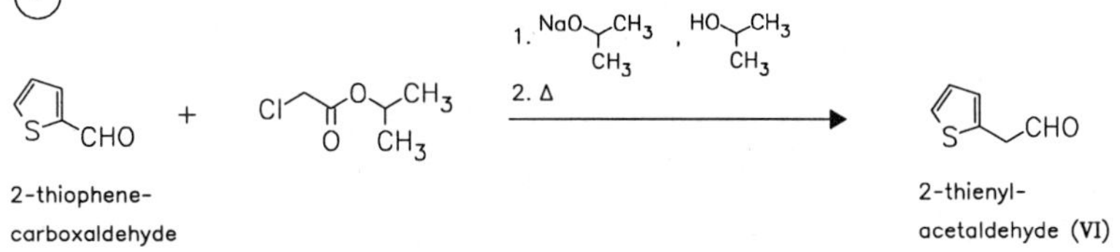

Reference(s):

a EP 99 802 (Sanofi; appl. 5.7.1983; F-prior. 13.7.1982).
   optical resolution of (±)-clopidogrel with (+)-10-camphorsulfonic acid:
   EP 281 459 (Elf Sanofi; appl. 16.2.1988; F-prior. 17.2.1987).
b EP 465 358 (Sanofi; appl. 3.7.1991; F-prior. 4.7.1990).

alternative route for preparing 2-thienylethylamine derivatives from thienylglycidic acid derivatives:
WO 9 839 322 (Sanofi; appl. 5.3.1998; F-prior. 5.3.1997).

pharmaceutical compositions:
WO 9 729 753 (Sanofi; appl. 17.2.1997; F-prior. 19.2.1996).
WO 9 717 064 (Sanofi; appl. 30.10.1996; F-prior. 3.11.1995).

synthesis of optical pure (2-chlorophenyl)glycine:
Garcia, M.J.; Azerad, R: Tetrahedron: Asymmetry (TASYE3) 8 (1), 85 (1997).

synthesis of (2-chlorophenyl)glycine:
Hayashi: Chem. Pharm. Bull. (CPBTAL) 7, 912, 1914 (1959).
Kobow, M.; Sprung, W.-D.; Schulz, E.: Pharmazie (PHARAT) 45 (7), 529 (1990).

Formulation(s): f. c. tabl. 75 mg (as hydrogen sulfate)

Trade Name(s):

D: Iscover (Bristol-Myers Squibb)
Plavix (Sanofi-Synthelabo; 1988)

GB: Plavix (Sanofi Winthrop)

USA: Plavix (Sanofi Pharm; Bristol-Myers Squibb; 1998)

Cloprednol

ATC: H02AB14
Use: glucocorticoid

RN: 5251-34-3 MF: $C_{21}H_{25}ClO_5$ MW: 392.88 EINECS: 226-052-6
CN: (11β)-6-chloro-11,17,21-trihydroxypregna-1,4,6-triene-3,20-dione

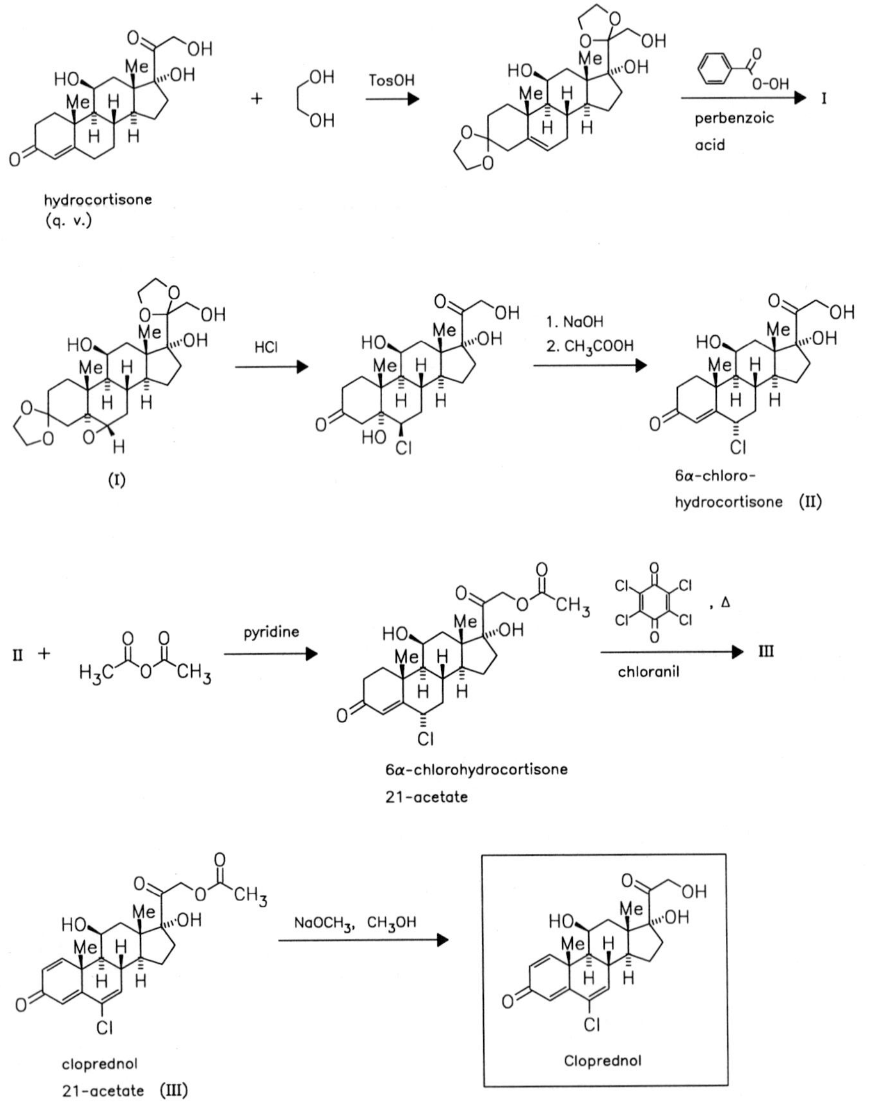
TosOH
perbenzoic acid
hydrocortisone (q. v.)
(I)
HCl
1. NaOH
2. CH3COOH
6α-chloro-hydrocortisone (II)
pyridine
chloranil
6α-chlorohydrocortisone 21-acetate
NaOCH3, CH3OH
cloprednol 21-acetate (III)
Cloprednol

Reference(s):

US 3 232 965 (Syntex; 1.2.1966; prior. 8.7.1957, 20.6.1958).
GB 890 835 (Syntex; appl. 20.6.1958; Mex.-prior. 22.6.1957, 20.7.1957).

alternative synthesis:
US 3 264 332 (Schering Corp.; 2.8.1966; prior. 7.1.1959).

Formulation(s):   tabl. 2.5 mg, 5 mg, 10 mg

Trade Name(s):
D: Syntestan (Syntex/Roche)   I: Cloradryn (Recordati)

Cloral betaine
(Betainchloralum; Chloral Betaine)

ATC: N05C
Use: hypnotic, sedative

RN: 2218-68-0   MF: $C_2H_3Cl_3O_2$ MW: 165.40   EINECS: 218-722-1
LD$_{50}$:   800 mg/kg (M, p.o.)
CN: 1-carboxy-*N,N,N*-trimethylmethanaminium inner salt compd. with 2,2,2-trichloro-1,1-ethanediol (1:1)

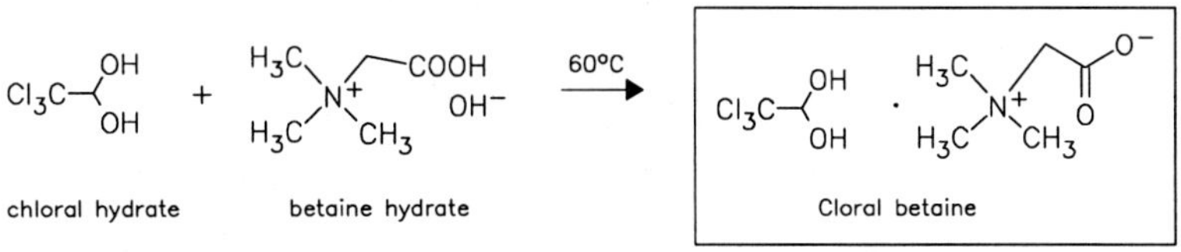

chloral hydrate   betaine hydrate   Cloral betaine

Reference(s):
US 3 028 420 (British Drug Houses; 3.4.1962).
GB 874 246 (British Drug Houses; appl. 26.6.1959; valid from 27.5.1960).

Trade Name(s):
USA: Beta-Chlor (Mead   Quinamm (Merrell-   Quinine sulfate (Perepac);
       Johnson); wfm   National); wfm   wfm

Clorexolone

ATC: C03BA12
Use: diuretic

RN: 2127-01-7   MF: $C_{14}H_{17}ClN_2O_3S$ MW: 328.82   EINECS: 218-342-6
LD$_{50}$:   230 mg/kg (M, i.v.); >6 g/kg (M, p.o.);
       120 mg/kg (R, i.v.); 6 g/kg (R, p.o.)
CN: 6-chloro-2-cyclohexyl-2,3-dihydro-3-oxo-1*H*-isoindole-5-sulfonamide

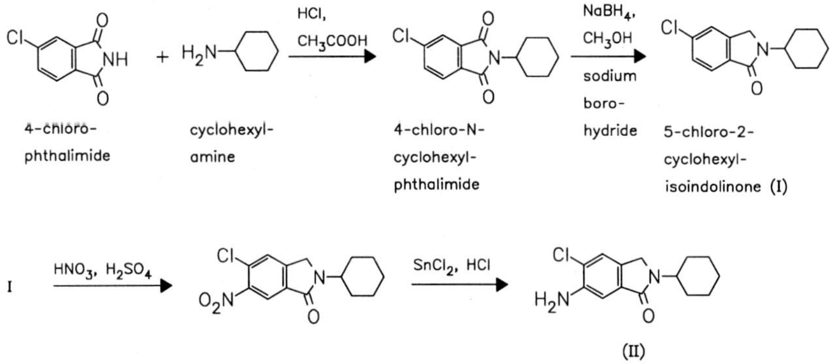

4-chloro-   cyclohexyl-   4-chloro-N-   hydride   5-chloro-2-
phthalimide   amine   cyclohexyl-   cyclohexyl-
                                        phthalimide   isoindolinone (I)

(II)

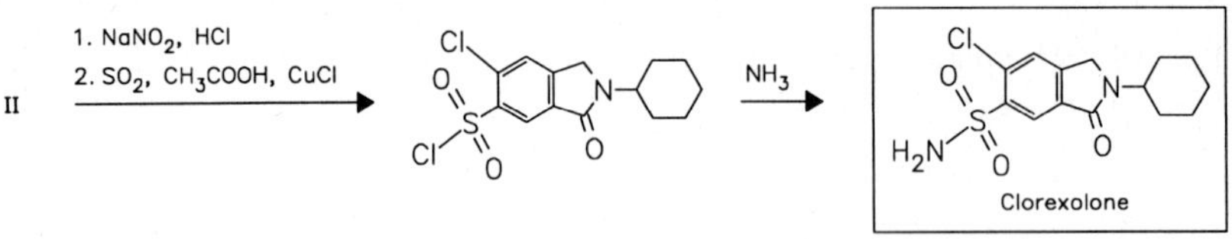

Clorexolone

Reference(s):
BE 620 654 (May & Baker; appl. 25.7.1962; GB-prior. 28.7.1961).

Formulation(s):   tabl. 10 mg, 25 mg

Trade Name(s):

F:	Flonatril (Specia); wfm	GB:	Nefrolan (May & Baker);
	Speciatensol (Specia)-		wfm
	comb.; wfm	J:	Nefrolan (Teikoku Zoki)

Cloricromen

(AD-6)

ATC: B01AC02; C01
Use: coronary vasodilator, antithrombotic

RN: 68206-94-0   MF: $C_{20}H_{26}ClNO_5$ MW: 395.88
LD_{50}:   10 mg/kg (R, i.v.); 1250 mg/kg (R, p.o.)
CN: [[8-chloro-3-[2-(diethylamino)ethyl]-4-methyl-2-oxo-2H-1-benzopyran-7-yl]oxy]acetic acid ethyl ester

hydrochloride
RN: 74697-28-2   MF: $C_{20}H_{26}ClNO_5 \cdot HCl$ MW: 432.34

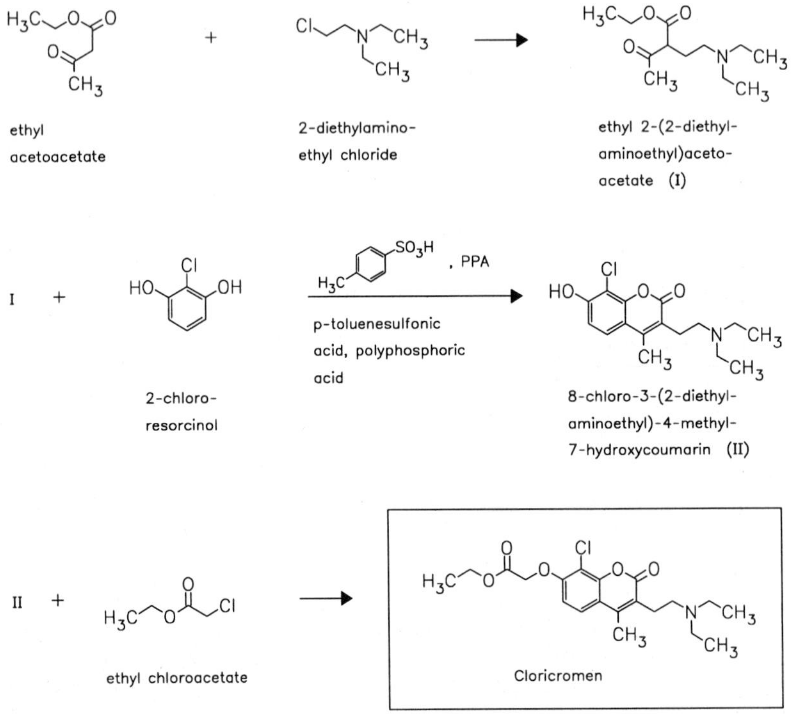

ethyl
acetoacetate

2-diethylamino-
ethyl chloride

ethyl 2-(2-diethyl-
aminoethyl)aceto-
acetate (I)

2-chloro-
resorcinol

p-toluenesulfonic
acid, polyphosphoric
acid

8-chloro-3-(2-diethyl-
aminoethyl)-4-methyl-
7-hydroxycoumarin (II)

ethyl chloroacetate

Cloricromen

Reference(s):
DOS 2 846 083 (Fidia; appl. 23.10.1978; I-prior. 17.11.1977).
US 4 296 039 (Fidia; 20.10.1981; I-prior. 17.11.1977).

synthesis of ethyl 2-(2-diethylaminoethyl)acetoacetate:
Weizmann, Ch.; Bergmann, E.; Sulzbacher, M.: J. Org. Chem. (JOCEAH) **15**, 918 (1950).

synthesis of 2-chlororesorcinol:
Schamp, N.: Bull. Soc. Chim. Belg. (BSCBAG) **73**, 35 (1946).
Wauzlick, H.V.; Mohrmann, S.: Chem. Ber. (CHBEAM) **96**, 2257 (1963).

Formulation(s):   amp. 30 mg/5 ml; cps. 100 mg; vial 30 mg

Trade Name(s):
I: Assogen (Metapharma)   Proendotel (Fidia; 1991)

Clorprenaline

ATC: R03
Use: bronchodilator

RN: 3811-25-4   MF: $C_{11}H_{16}ClNO$ MW: 213.71   EINECS: 223-291-8
CN: 2-chloro-α-[[(1-methylethyl)amino]methyl]benzenemethanol

hydrochloride monohydrate
RN: 5588-22-7   MF: $C_{11}H_{16}ClNO \cdot HCl \cdot H_2O$ MW: 268.18
LD_{50}:   54 mg/kg (M, i.v.); 298 mg/kg (M, p.o.);
        68 mg/kg (R, i.v.); 450 mg/kg (R, p.o.);
        >400 mg/kg (dog, p.o.)

2'-chloro-
acetophenone

2-bromo-
2'-chloro-
acetophenone

(I)

isopropylamine

Clorprenaline

Reference(s):
US 2 816 059 (Lilly; 1957; appl. 1956).

Formulation(s):   sol. 2 % (inhalation)

Trade Name(s):

J:	Aremans (Zensei)	Conselt (San-a)	Neoasutoma (Nihon
	Asthone (Eisai)	Cosmoline (Chemiphar)	Yakuhin)
	Bronocon (Wakamoto)	Fusca (Hoei)	Pentadoll (Showa)
	Clopinerin (Nippon Shoji)	Kalutein (Tatsumi)	Propran (Kobayashi Kako)
	Clorprenalin HCl (Kongo)		

Clortermine

ATC: N07
Use: appetite depressant

RN: 10389-73-8   MF: $C_{10}H_{14}ClN$ MW: 183.68
CN: 2-chloro-α,α-dimethylbenzeneethanamine

hydrochloride
RN: 10389-72-7   MF: $C_{10}H_{14}ClN \cdot HCl$ MW: 220.14
LD_{50}:   332 mg/kg (R, p.o.)

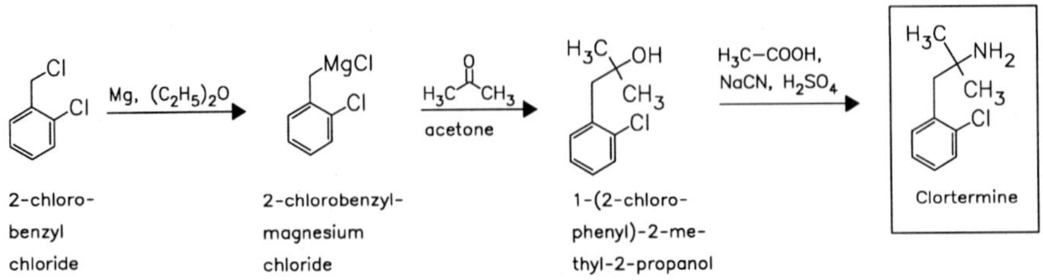

2-chloro-
benzyl
chloride

2-chlorobenzyl-
magnesium
chloride

1-(2-chloro-
phenyl)-2-me-
thyl-2-propanol

Clortermine

Reference(s):
US 3 415 937 (Ciba; 10.12.1968; appl. 1.12.1966; prior. 31.8.1964).

Formulation(s):   tabl. 50 mg

Trade Name(s):
USA: Voranil (USV); wfm

Clostebol acetate

ATC: D11AE
Use: anabolic

RN: 855-19-6   MF: $C_{21}H_{29}ClO_3$ MW: 364.91   EINECS: 212-720-4
CN: (17β)-17-(acetyloxy)-4-chloroandrost-4-en-3-one

clostebol
RN: 1093-58-9   MF: $C_{19}H_{27}ClO_2$ MW: 322.88   EINECS: 214-133-9

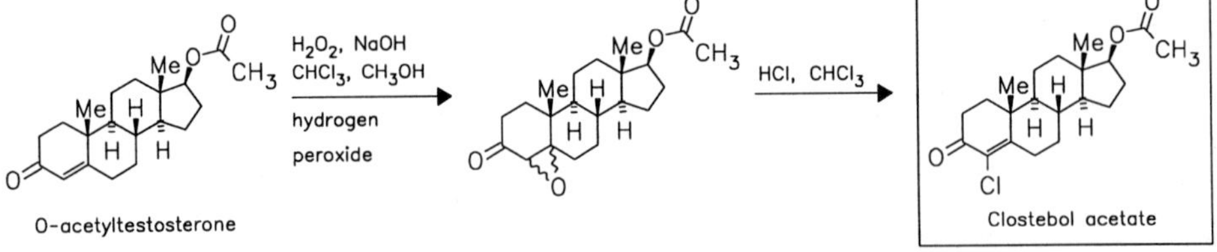

O-acetyltestosterone

Clostebol acetate

Reference(s):
US 2 953 582 (Soc. Farmaceutici Italia; 20.9.1960; appl. 26.10.1956; I-prior. 23.4.1956).
US 2 933 510 (Julian Labs.; 19.4.1960, Prior. 3.2.1955).

Formulation(s):   cream 0.5 %; sugar coated tabl. 15 mg; vial 10 mg/1.5 ml

Trade Name(s):
D: Megagrisevit (Pharmacia &
     Upjohn)

Megagrisevit N (Pharmacia
& Upjohn)-comb.

F: Trofoseptine (Boehringer
     Ing.)-comb.

I: Alfa-Trofodermine   Trofodermin (Farmitalia)-   J: Steranabol (Sumitomo)
      (Farmitalia)   comb. with neomycine

Clotiazepam

ATC: N05BA21
Use: anxiolytic, benzodiazepine analog

RN: 33671-46-4   MF: $C_{16}H_{15}ClN_2OS$ MW: 318.83   EINECS: 251-627-3
LD$_{50}$:   440 mg/kg (M, i.p.); 636 mg/kg (M, p.o.)
CN: 5-(2-chlorophenyl)-7-ethyl-1,3-dihydro-l-methyl-2H-thieno[2,3-e]-1,4-diazepin-2-one

(a)

3-(2-chloro-   ethyl glycinate   Clotiazepam
benzoyl)-5-ethyl-   hydrochloride (II)
2-methylamino-
thiophene (I)

(b)

N-benzyloxy-   (III)
carbonylglycyl
chloride

(c)

5-(2-chloro-
phenyl)-7-ethyl-
1,3-dihydro-2H-
thieno[2,3-e]-1,4-
diazepin-2-one

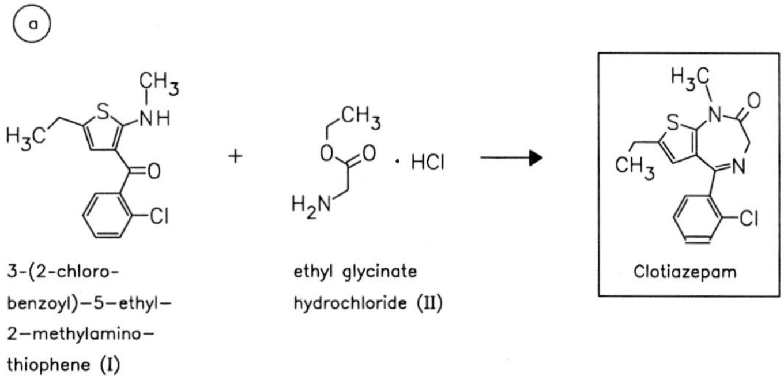

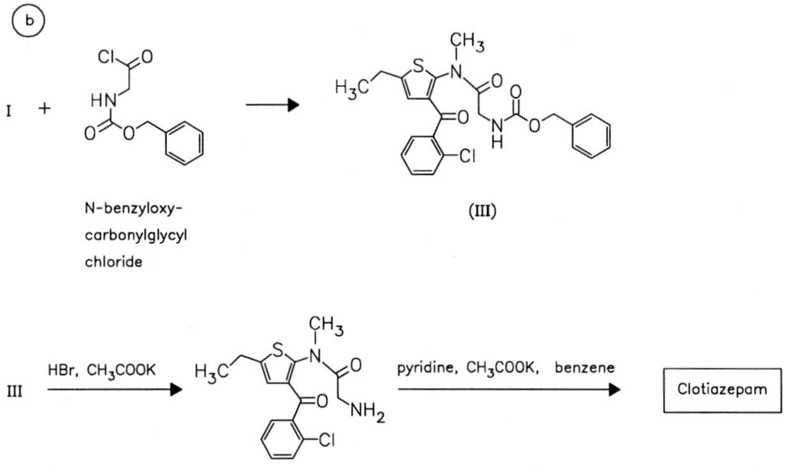

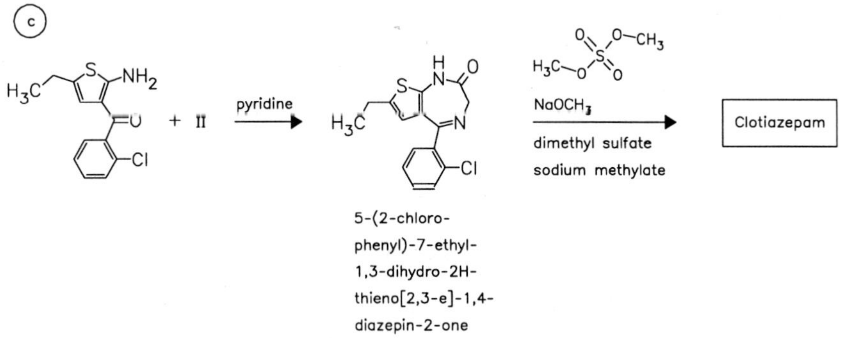

Reference(s):
DOS 2 107 356 (Yoshitomi; appl. 16.2.1971; J-prior. 17.2.1970, 23.2.1970, 7.3.1970, 25.6.1970, 31.7.1970).
US 3 849 405 (Yoshitomi; 19.11.1974; J-prior. 17.2.1970, 23.2.1970, 7.3.1970, 25.6.1970, 31.7.1970).

Formulation(s):   drops 1 %; tabl. 5 mg, 10 mg, 20 mg

Trade Name(s):
D: Trecalmo (Bayer Vital;   Tienor (Farmaka)   Reilyfter (Maruko)
      1979)   J: Emolex (Nichiiko)   Rize (Yoshitomi)
F: Vératran (Murat; 1984)   Isocline (Sawai)
I: Rizen (Puropharma; 1984)   Lieze (Yoshitomi)

Clotrimazole

ATC: A01AB18; D01AC01; G01AF02
Use: antifungal

RN: 23593-75-1   MF: $C_{22}H_{17}ClN_2$ MW: 344.85   EINECS: 245-764-8
LD$_{50}$:   761 mg/kg (M, p.o.);
       708 mg/kg (R, p.o.);
       >2 g/kg (dog, p.o.)
CN: 1-[(2-chlorophenyl)diphenylmethyl]-1H-imidazole

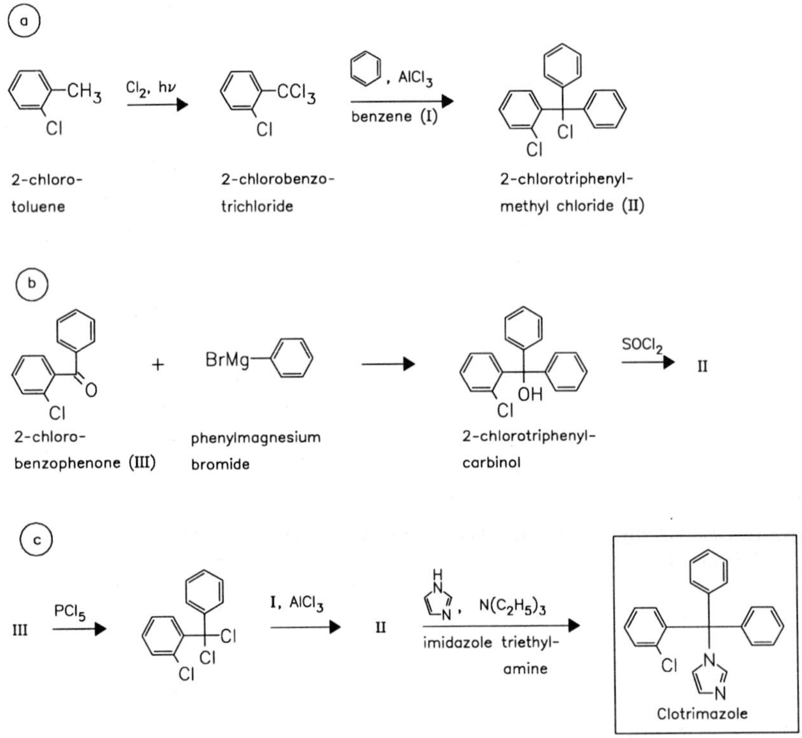

a)

2-chloro-   2-chlorobenzo-   2-chlorotriphenyl-
toluene   trichloride   methyl chloride (II)

b)

2-chloro-   phenylmagnesium   2-chlorotriphenyl-
benzophenone (III) bromide   carbinol

c)

imidazole triethyl-
amine

Clotrimazole

Reference(s):
DE 1 617 481 (Bayer; appl. 15.9.1967).
DAS 1 670 976 (Bayer; appl. 29.1.1968).
DE 1 670 977 (Bayer; D-prior. 29.1.1968).
US 3 660 577 (Bayer; 2.5.1972; D-prior. 15.9.1967).
US 3 705 172 (Bayer; 5.12.1972; D-prior. 15.9.1967).

mode of mechanism:
Berg, D. et al.: Arzneim.-Forsch. (ARZNAD) **34** (I), 139 (1984).

medical use at Herpes labialis:
US 4 438 129 (Pennwalt; 20.3.1984; appl. 27.9.1982).

special formulations:
DOS 3 321 043 (Bayer; appl. 10.6.1983).
EP 128 459 (Bayer; appl. 30.5.1984; D-prior. 10.6.1983).
EP 112 485 (Bayer; appl. 17.11.1983; D-prior. 25.11.1982).

combination with corticosteroids:
EP 49 468 (Schering Corp.; appl. 30.9.1981; USA-prior. 6.10.1980).

Formulation(s):   pessaries 100 mg, 200 mg, powder 10 mg/g (1 %); 500 mg; sol. 1 %, 10 mg; spray 10 mg/ml (1 %); topical cream 1 %, 10 mg; vaginal cream 10 mg, 20 mg/ml (2 %, 10 %); vaginal tabl. 100 mg, 200 mg, 500 mg

Trade Name(s):

D: Antifungal (Hexal)
Apocanda (esparma)
ARU Spray (Chauvin ankerpharm)
Azutrimazol (Azupharma)
Benzoderm (Athenstaedt)
Candazol (Apogepha)
Canesten (Bayer; 1973)
Canifug (Wolff)
Cloderm (Dermapharm)
Cutistad (Stada)
Durafungol (durachemie)
Gilt (Solvay Arzneimittel)
Gyno Canesten (Bayer Vital)
Holfungin (Holborn)
Jenamazol (Jenapharm)
Lobalacid (Kade)

Mono Baycuten (Bayropharm)
Mycofug (Hermal)
Myko Cordes (Ichthyol)
Mykofungin (Wyeth)
Mykohaug (betapharm)
Ovis (Warner-Lambert)
Pedisafe (BASF Generics)
Radical (Maurer)
Uromycol (Hayer)

F: Trimysten (Roger Bellon; 1978); wfm

GB: Canesten (Bayer; 1973)
Canesten HC (Bayer)-comb.
Lotriderm (Dominion)-comb.
Masnoderm (Dominion)

I: Antimicotico Same (Savoma)
Canesten (Bayropharm; 1973)
Desamix Effe (Savoma)-comb.
Gyno-Canesten (Bayropharm)
Meclon (Farmigea)

J: Empecid (Bayer-Takeda; 1976)
Tao (Toko-Fujisawa)

USA: Fungoid (Pedinol)
Lotrimin (Schering; 1976)
Lotrisone (Schering)
Mycelex (Bayer)

Cloxacillin

ATC: J01CF02
Use: antibiotic

RN: 61-72-3   MF: $C_{19}H_{18}ClN_3O_5S$ MW: 435.89   EINECS: 200-514-7
CN: [2S-(2α,5α,6β)]-6-[[[3-(2-chlorophenyl)-5-methyl-4-isoxazolyl]carbonyl]amino]-3,3-dimethyl-7-oxo-4-thia-1-azabicyclo[3.2.0]heptane-2-carboxylic acid

monosodium salt
RN: 642-78-4   MF: $C_{19}H_{17}ClN_3NaO_5S$ MW: 457.87   EINECS: 211-390-9
LD_{50}:   916 mg/kg (M, i.v.), 5 g/kg (M, p.o.);
1660 mg/kg (R, i.v.); 5 g/kg (R, p.o.)

monosodium salt monohydrate
RN: 7081-44-9   MF: $C_{19}H_{17}ClN_3NaO_5S \cdot H_2O$ MW: 475.89
LD_{50}:   1100 mg/kg (M, i.v.); 5 g/kg (M, p.o.);
1660 mg/kg (R, i.v.); 5 g/kg (R, p.o.)

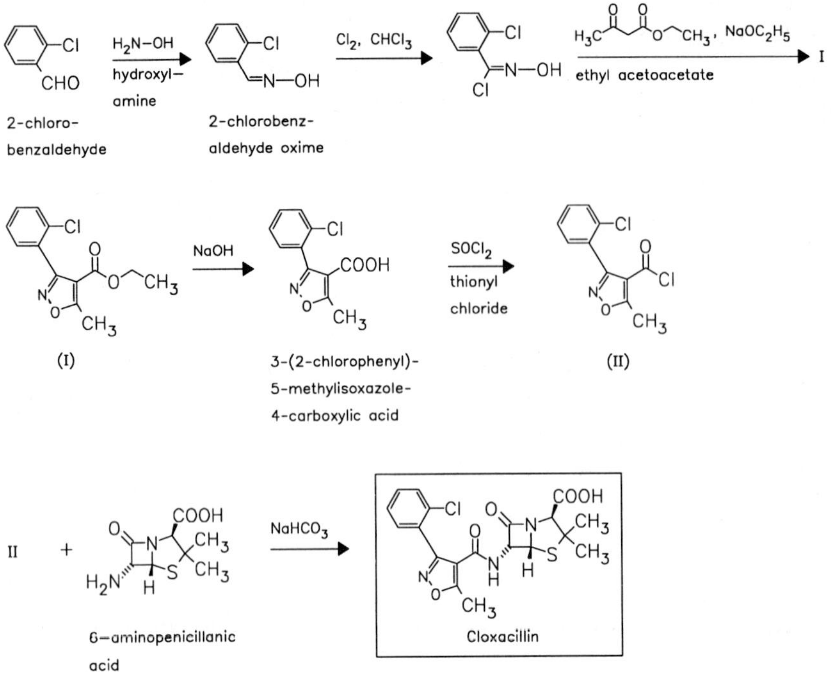

2-chloro-benzaldehyde

2-chlorobenzaldehyde oxime

(I)

3-(2-chlorophenyl)-5-methylisoxazole-4-carboxylic acid

(II)

6-aminopenicillanic acid

Cloxacillin

Reference(s):
US 2 996 501 (Beecham; 15.8.1961; GB-prior. 31.3.1960).
GB 905 778 (Beecham; appl. 31.3.1960; valid from 14.3.1961).
GB 958 478 (Beecham; appl. 28.2.1963; USA-prior. 13.3.1962).

Formulation(s): amp. 250 mg, 500 mg; cps. 250 mg, 500 mg; tabl. 250 mg, 500 mg (as sodium salt)

Trade Name(s):

D:	Ampiclox (Beecham)-comb. with ampicillin; wfm Pyoclox (Beecham)-comb. with carbenicillin; wfm Pyolox (Beecham)-comb. with carbenicillin; wfm	GB:	Ampiclox (SmithKline Beecham)-comb. with ampicillin Orbenin (Beecham); wfm		Methocillin-S (Meiji Seika) Orbenin (Beecham-Fujisawa) Prostaphlin (Bristre-Banyu) Solcillin C (Takeda)-comb.
		I:	Amplium (Sigma-Tau)-comb.		Totaclox (Beecham)-comb.
F:	Orbenine (SmithKline Beecham)		Cloxac (Formulario Naz.)	USA:	Cloxapen (Beecham); wfm
		J:	Acucillin (Fuji)-comb.		Tegopen (Bristol); wfm

Cloxazolam

ATC: N05BA22
Use: tranquilizer

RN: 24166-13-0 MF: $C_{17}H_{14}Cl_2N_2O_2$ MW: 349.22
LD$_{50}$: 2630 mg/kg (M, p.o.);
1780 mg/kg (R, p.o.)
CN: 10-chloro-11b-(2-chlorophenyl)-2,3,7,11b-tetrahydrooxazolo[3,2-*d*][1,4]benzodiazepin-6(5*H*)-one

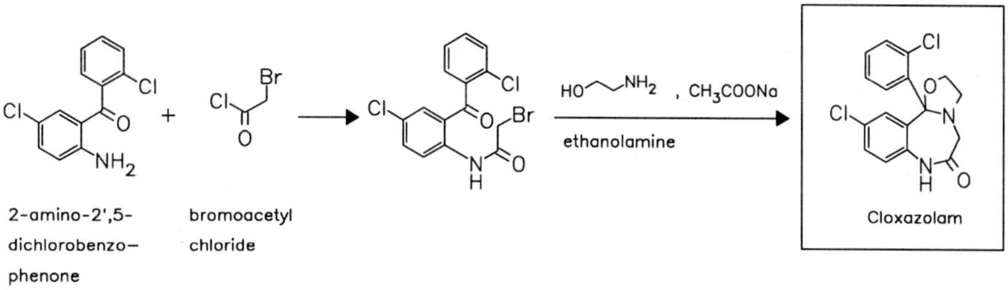
HO NH2 , CH3COONa
ethanolamine
2-amino-2',5-
dichlorobenzo-
phenone
bromoacetyl
chloride
Cloxazolam

Reference(s):
DOS 1 812 252 (Sankyo; appl. 26.11.1968; J-prior. 27.11.1967).
DOS 1 817 923 (Sankyo; appl. 26.11.1968; J-prior. 27.11.1967).

alternative synthesis:
DOS 1 954 065 (Sankyo; appl. 23.10.1969; J-prior. 24.10.1968, 17.4.1969).
US 3 696 094 (Sankyo; 3.10.1972; J-prior. 24.10.1968, 25.10.1968).
US 3 772 371 (Sankyo; 13.11.1973; J-prior. 27.11.1967).

review:
Schulte, E.: Dtsch. Apoth. Ztg. (DAZEA2) **115**, 1253, 1828 (1975).

Formulation(s): tabl. 1 mg, 2 mg

Trade Name(s):
J: Enadel (Taito Pfizer) Sepazon (Sankyo)

Clozapine

ATC: N05AH02
Use: neuroleptic

RN: 5786-21-0 MF: $C_{18}H_{19}ClN_4$ MW: 326.83 EINECS: 227-313-7
LD_{50}: 36.5 mg/kg (M, i.v.); 150 mg/kg (M, p.o.);
41.6 mg/kg (R, i.v.); 251 mg/kg (R, p.o.);
145 mg/kg (dog, p.o.)
CN: 8-chloro-11-(4-methyl-1-piperazinyl)-5*H*-dibenzo[*b*,*e*][1,4]diazepine

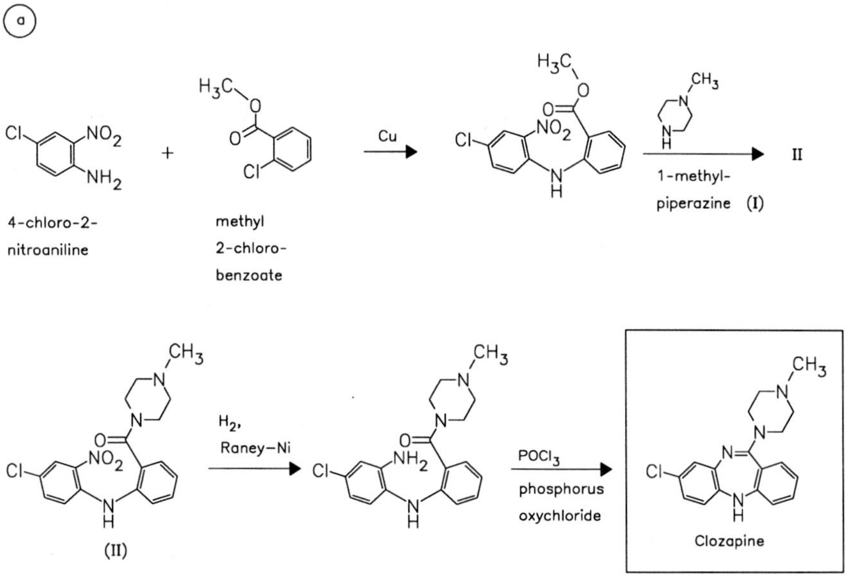
a
Cu
1-methyl-
piperazine (I)
II
4-chloro-2-
nitroaniline
methyl
2-chloro-
benzoate
(II)
H2,
Raney-Ni
POCl3
phosphorus
oxychloride
Clozapine

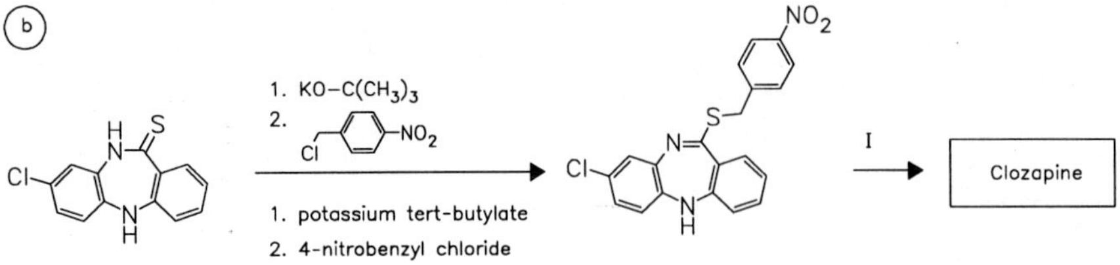

(b)

1. KO–C(CH₃)₃
2. 4-nitrobenzyl chloride

1. potassium tert-butylate
2. 4-nitrobenzyl chloride

I → Clozapine

8-chloro-11-thioxo-
10,11-dihydro-5H-
dibenzo[b,e][1,4]-
diazepine

Reference(s):

CH 404 677 (Dr. A. Wander; appl. 2.12.1960).
CH 398 620 (Dr. A. Wander; appl. 16.8.1960).
GB 980 853 (Dr. A. Wander; appl. 16.8.1961; CH-prior. 16.8.1960, 2.12.1960).
NL 147 426 (Dr. A. Wander; appl. 24.5.1963; CH-prior. 25.5.1962, 8.6.1962, 5.12.1962, 15.2.1963).
DE 1 280 879 (Wander; appl. 7.8.1961; CH-prior. 16.8.1960, 2.12.1960).
US 3 539 573 (Wander; 10.11.1970; CH-prior. 16.8.1060, 2.12.1960, 20.7.1961, 25.5.1962, 5.12.1962, 15.2.1963, 22.3.1967, 11.7.1967, 3.11.1967).
Hunziker, F. et al.: Helv. Chim. Acta (HCACAV) **50**, 1588 (1967).

Formulation(s):   inj. sol. 50 mg/2 ml; tabl. 25 mg, 100 mg

Trade Name(s):
D:　　Clozaril (Novartis)　　　　F:　　Leponex (Novartis)
　　　Leponex (Novartis Pharma)　USA:　Clozaril (Novartis)

Cobamamide
(Adenosylcobalamin; Coenzym B₁₂; Dibencozide)

ATC:　B03BA04
Use:　anabolic

RN:　13870-90-1　MF: C₇₂H₁₀₀CoN₁₈O₁₇P　MW: 1579.61　EINECS: 237-627-6
LD₅₀:　1 g/kg (M, i.v.)
CN:　cobinamide *Co*-(5'-deoxyadenosine-5') deriv. hydroxide dihydrogen phosphate (ester) inner salt 3'-ester with 5,6-dimethyl-1-α-D-ribofuranosyl-1*H*-benzimidazole

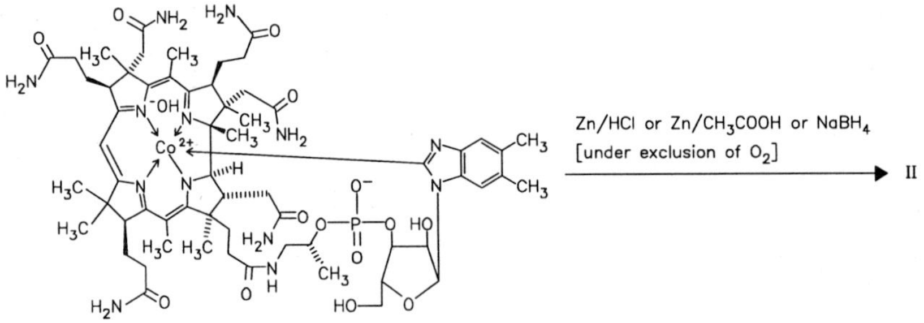

Zn/HCl or Zn/CH₃COOH or NaBH₄
[under exclusion of O₂]
→ II

hydroxocobalamin (I)
(q. v.)

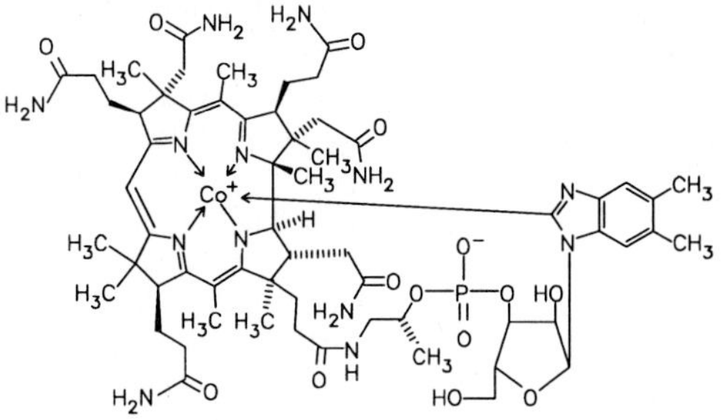

hydridocobalamin (II)

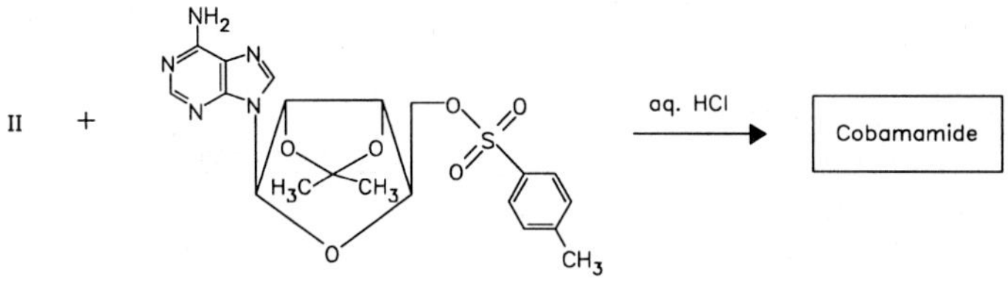

2',3'-O-isopropylidene-
5'-O-tosyladenosine

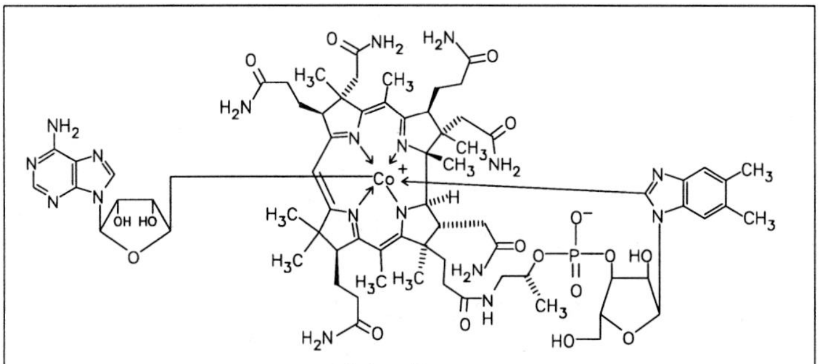

Cobamamide

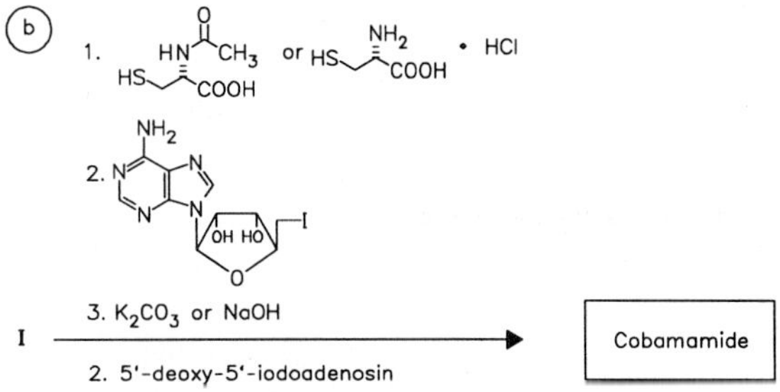

3. K₂CO₃ or NaOH

2. 5'-deoxy-5'-iodoadenosin

Reference(s):

a Bernhauer, K. et al.: Angew. Chem. (ANCEAD) **75**, 1145 (1963).
   US 3 213 082 (Glaxo; 19.10.1965; GB-prior. 11.12.1961).
b US 3 461 114 (Yamanouchi; 12.8.1969; J-prior. 1.10.1966).

Formulation(s):   cps. 0.25 mg, 1 mg; drops 30 mg; tabl. 0.25 mg, 1 mg, 2.5 mg

Trade Name(s):
D: Xobaline (Albert-Roussel);   wfm

F: Vibalgan (Doms-Adrian)-
      comb.
I: Amico (SIT)-comb.
      Aminozim (Pierrel)-comb.
      Anabasi (Zilliken)
      Calciozim (Pierrel)-comb.
      Calisvit (Menarini)-comb.
      Cobaforte (Roussel)
      Cocametina B12 (Sigma-
      Tau)-comb.
      Glutacomplex (Chemil)-
      comb.

      Hepafactor Complex
      (Sigma-Tau)-comb.
      Indusil (Recordati)
J: Actavix (Nippon Kayaku)
      Actimide (Tobishi)
      Ademide (Toyo Jozo)
      Cabarol (Daiko)
      Calomide (Yamanouchi)
      Cobaforte (Roussel-
      Chugai)
      Cobalan (Daiichi)
      Cobaltamin-S (Wakamoto)

      Cobamyde (Shiu Nihon
      Jitsugyo)
      Funacomide (Funai)
      Hokuramide (Hokuriku)
      Hycobal (Eisai)
      Hyrasedon (Sawai)
      Lasedmeide (Choseido)
      Metamide (Nakataki)
      Sabalamin (Sato-Santen)
      Satomid (Shinshin)
      generics and combination
      preparations

Cocarboxylase

ATC: A11DA
Use: enzyme against metabolic
        disturbance

RN: 154-87-0   MF: $C_{12}H_{19}ClN_4O_7P_2S$ MW: 460.77   EINECS: 205-836-1
LD$_{50}$:   >1 g/kg (M, i.m.);
      >500 mg/kg (R, i.m.)
CN: 3-[(4-amino-2-methyl-5-pyrimidinyl)methyl]-4-methyl-5-(4,6,6-trihydroxy-4,6-dioxido-3,5-dioxa-4,6-
      diphosphahex-1-yl)thiazolium chloride

Work up of the reaction mixtures on ion-exchangers.

Reference(s):
US 2 991 284 (E. Merck AG; 4.7.1961; D-prior. 28.9.1957).

Formulation(s):   vial 5.8 mg/10 mg, 20 mg, 50 mg, 100 mg

Trade Name(s):
D: Cernevit (Baxter)-comb.
F: Cernénit (Baxter)-comb.
      Plenyl (Oberlin)-comb.
      generics and combination
      preparations
I: Adenobeta (Salus
      Research)-comb.
      Adenoplex (Lepetit)-comb.
      Adenovit (Nuovo Cons.
      Sanit. Naz.)-comb.
      Benexol (Roche)-comb.
      Bivitasi (ISI)
      Firmavit (Firma)-comb.

      Fosforilasi (Polifarma)-
      comb.
      Neogeynevral (Geymonat)-
      comb.
      Piruvasi (Delalande
      Isnardi)-comb.
      salts and combination
      preparations
J: Bicholase (Fuso)
      Cocalbose (Fuji Zoki)
      Cocalox (Maruko)
      Co-Carten (Sanken)
      Coxylase (Funai)

      Carboxin (Toa Eiyo-
      Yamanouchi)
      Hiactose (Ohno)
      Metabolase (Takeda)
      Neo Alinachiol (Kanto)
      Nutrase (Kyorin)
      Paraboramin (Hoei)
      Proffit (Isei)
      Pyrolase (Chugai)
      Reborase (Kanto)
      Thiamilase (Hokuriku)

Codeine

ATC: R05DA04
Use: antitussive, narcotic, analgesic

RN: 76-57-3   MF: $C_{18}H_{21}NO_3$ MW: 299.37   EINECS: 200-969-1
LD_{50}:   54 mg/kg (M, i.v.); 250 mg/kg (M, p.o.);
75 mg/kg (R, i.v.); 427 mg/kg (R, p.o.);
69 mg/kg (dog, i.v.)
CN: (5α,6α)-7,8-didehydro-4,5-epoxy-3-methoxy-17-methylmorphinan-6-ol

hydrobromide
RN: 125-25-7   MF: $C_{18}H_{21}NO_3 \cdot HBr$ MW: 380.28   EINECS: 204-730-2
LD_{50}:   535 mg/kg (M, p.o.)
hydriodide
RN: 125-26-8   MF: $C_{18}H_{21}NO_3 \cdot HI$ MW: 427.28   EINECS: 204-731-8
phosphate (1:1)
RN: 52-28-8   MF: $C_{18}H_{21}NO_3 \cdot H_3PO_4$ MW: 397.36   EINECS: 200-137-8
LD_{50}:   62 mg/kg (M, i.v.); 237 mg/kg (M, p.o.);
54 mg/kg (R, i.v.); 85 mg/kg (R, p.o.);
97.8 mg/kg (dog, i.v.)

morphine + $H_3C—OH$ → Codeine

Reference(s):
Ehrhart-Ruschig **I**, 117-118.
DRP 247 180 (C. H. Boehringer; 1912).
Ullmanns Encykl. Tech. Chem., 3. Aufl., Vol. **3**, 232.

Formulation(s):   cps. 30 mg; drops 2.4 g/100 ml; suppos. 30 mg; syrup 0.117 g/100 g; tabl. 30 mg, 50 mg

Trade Name(s):
D: Bronchicum (Nattermann)
Codeinum phophoricum
Compretten (Glaxo
Wellcome/Cascan)
Codicept (Sanol)
Codipront (Mack, Illert.)
Dolomo (Klinge; as
phosphate)-comb.
Dolviran (Bayer Vital)-
comb.
Gelonida (Gödecke; as
phosphate)-comb.
Lonarid (Boehringer Ing.)-
comb.
Optipyrin (Thiemann)-
comb.
Spasmo-Cibalgin Comp.
(Novartis Pharma)-comb.

Treupel (ASTA Medica
AWD)
Tricodein (Soleo)
Tussipect (Beiersdorf-
Lilly)-comb.
numerous combination
preparations
F: numerous combination
preparations
GB: Aspar (Hoechst)-comb.
Codafen Continus (Napp;
as phosphate)-comb.
Migralere (Pfizer
Consumer; as phosphate)-
comb.
Solpadol (Sanofi Winthrop;
as phosphate)-comb.
Tylex (Schwarz; as
phosphate)-comb.

numerous combination
preparations
I: Bromocodeina (Menarini)
Codeinol (Saba)-comb.
Codipront (Bracco)-comb.
Lactocol (Ogna)-comb.
Hedrix Plan (Saba)-comb.
Senodin (Bristol-Myers
Squibb)-comb.
numerous combination
preparations
J: Codeine Phosphate
(Dainippon; Sankyo;
Shionogi; Takeda; Tanabe)
USA: Brontex (Procter &
Gamble; as phosphate)
Dimetane (Robins; as
phosphate)

| Nucofed (Monarch; as phosphate) | Robitussin (Robins; as phosphate) | numerous generics and combination preparations |

Colecalciferol
(Cholecalciferol; Vitamin D_3)

ATC: A11CC05
Use: antirachitic

RN: 67-97-0   MF: $C_{27}H_{44}O$ MW: 384.65   EINECS: 200-673-2
LD_{50}:   42.5 mg/kg (M, p.o.);
        42 mg/kg (R, p.o.);
        80 mg/kg (dog, p.o.)
CN: (3β,5Z,7E)-9,10-secocholesta-5,7,10(19)-trien-3-ol

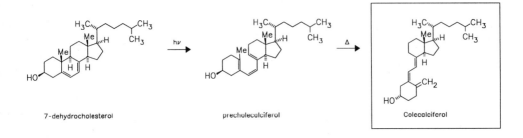

7-dehydrocholesterol   precholecalciferol   Colecalciferol

Reference(s):
Kirk-Othmer, Encycl. Chem. Technol., Vol. **21**, 549 ff.
Ullmanns Encykl. Tech. Chem., 3. Aufl., Vol. **18**, 236 ff. (synthesis of 7-dehydrocholesterol as described).

synthesis from 7-dehydrocholesterol ester:
US 3 661 939 (Nisshin Flour Milling; 9.5.1972; J-prior. 16.12.1969).

synthesis from 25-fluorocholesterol ester:
JP-appl. 540 46-768 (Teijin; appl. 19.9.1977).

crystalline vitamin D_3:
US 3 665 020 (Hoffmann-La Roche; 23.5.1972; CH-prior. 25.2.1969).

total synthesis:
Inhoffen, H.H.: Angew. Chem. (ANCEAD) **72**, 875 (1960).

Formulation(s):   amp. 1.25 mg/ml, 2.5 mg/ml; cps. 0.5 mg; drops 0.5 mg/ml; emulsion 6 mg/ml; tabl. 0.01 mg, 0.25 mg, 5 mg

Trade Name(s):
D: Dekristol (Jenapharm)
    D-Mulsin (Mucos)
    D-Tracetten (Albert-
    Roussel, Hoechst)
    D_3-Vicotrat (Heyl)
    Ospur D_3 (Henning)
    Provitina D_3 (Promonta)
    Vigantol (Merck)
    Vigantoletten (Merck)
    Vigorsan (Albert-Roussel,
    Hoechst)

    numerous combination
    preparations
F: Alvityl (Solvay)-comb.
    Cernévit (Baxter)-comb.
    Quotivit (Mayoly-
    Spindler)-comb.
    Survitine (Roche
    Nicholas)-comb.
    numerous combination
    preparations
GB: Octovit (Smith Kline &
    French)-comb.; wfm

I: Adiboran (Eurospital)-
    comb.
    Antilinf (Delalande
    Isnardi)-comb.
    Calciozim (Pierrel)-comb.
    Calisvit (Menarini)-comb.
    Haliborange (Eurospital)-
    comb.
    Iper D_3 (Zambon Italia)
    Tridelta (Ceccarelli)
J: Vitasol AD_3 + E (Tiger)-
    comb.

| USA: | Al-Vite (Drug Industries)- comb.; wfm | Ultra "A" & "D" (Nature's Bounty); wfm | Ultra "D"-Tabl. (Nature's Bounty); wfm |

Colfosceril palmitate

ATC: R07AA01
Use: synthetic lung surfactant, prophylactic treatment of respiratory distress syndrome

RN: 99732-49-7   MF: $C_{40}H_{80}NO_8P \cdot C_{16}H_{34}O \cdot [C_{14}H_{22}O \cdot C_2H_4O \cdot CH_2O]x$ MW: unspecified
CN: (R)-N,N,N-trimethyl-10-oxo-7-[(1-oxohexadecyl)oxy]-3,5,9-trioxa-4-phosphapentacosan-1-aminium-4-oxide inner salt, mixt. with formaldehyde polymer with oxirane and 4-(1,1,3,3-tetramethylbutyl)phenol and 1-hexadecanol

Colfosceril palmitate

Lyophilization of 1,2-dipalmitoyl-sn-3-glycerophosphorylcholine, n-hexadecan-1-ol, tyloxapol solution in 0.1 n NaCl.

Reference(s):
US 4 826 821 (The Regents of the Univ. of California; 2.5.1989; appl. 5.11.1986; prior. 26.6.1985, 17.10.1983).
EP 50 793 (The Regents of the Univ. of California; appl. 14.10.1981; USA-prior. 24.10.1980).

Formulation(s):   vial 108 mg (lyo.)

Trade Name(s):
D: Exosurf (Glaxo Wellcome)   I: Exosurf Neonatate (Glaxo   USA: Exosurf (Glaxo Wellcome;
GB: Exosurf Neonatal (Glaxo   Wellcome)   1991)
     Wellcome; 1991)

Colistin

(Colistin A + B; Polymyxin E)

ATC: A07AA10; J01XB01
Use: antibiotic (macrocyclic peptide)

RN: 1066-17-7   MF: unspecified   MW: unspecified   EINECS: 213-907-3
CN: colistin

sulfate ·
RN: 1264-72-8   MF: $H_2O_4S \cdot$ x unspecified   MW: unspecified   EINECS: 215-034-3
LD$_{50}$:   6 mg/kg (M, i.v.); 793 mg/kg (M, p.o.)
pentasodium mesylate
RN: 8068-28-8   MF: unspecified   MW: unspecified   EINECS: 232-516-9
LD$_{50}$:   222 mg (M, i.v.); >767 mg (M, p.o.);
      5450 mg/kg (R, p.o.)

Cyclopolypeptide antibiotic from *Aerobacillus colistinus*.

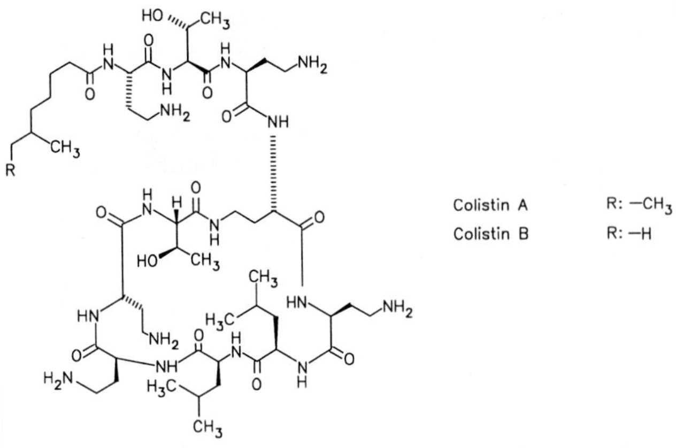

Colistin A   R: —CH₃
Colistin B   R: —H

Colistin

Reference(s):
Vogler, K.; Studer, R.O.: Experientia (EXPEAM) **22**, 345 (1966).

Formulation(s):   tabl. 24 mg, 95 mg; vial 33.3 mg

Trade Name(s):

D:	Diarönt (Chephasaar)
F:	Bacicoline (Merck Sharp & Dohme-Chibret)-comb.
	Colicort (Merck Sharp & Dohme-Chibret)-comb.

Colimycine (Bellon)
GB: Colomycin (Pharmax)
I: Colbiocin (SIFI)-comb.
     Colimicina (UCB)
J: Colimycin-S (Kaken)

Methacolimycin (Kaken)
USA: Coly-Mycin (Parke Davis; as sulfate)

Convallatoxin

ATC: C01AA
Use: cardiac glycoside, cardiotonic

RN: 508-75-8   MF: $C_{29}H_{42}O_{10}$ MW: 550.65   EINECS: 208-086-3
LD₅₀: 1 mg/kg (M, i.v.); >2 g/kg (M, p.o.);
       15.2 mg/kg (R, i.v.)
CN: (3β,5β)-3-[(6-deoxy-α-L-mannopyranosyl)oxy]-5,14-dihydroxy-19-oxocard-20(22)-enolide

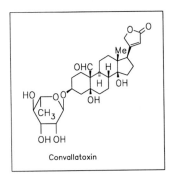

Convallatoxin

From *Convallaria majalis.*

Reference(s):
DRP 490 648 (Hoffmann-La Roche; appl. 1928; CH-prior. 1928).
Karrer, P.: Helv. Chim. Acta (HCACAV) **12**, 506 (1929).
SU 64 447 (F. D. Zilbert; appl. 1945).
PL 51 371 (Inst. Farmaceutyczny; appl. 17.5.1965).

partial synthesis:
Reichstein, T. et al.: Helv. Chim. Acta (HCACAV) **33**, 1541 (1950).
DD 19 239 (E. Lüdde; appl. 18.7.1960).
SU 198 319 (Kharkov Scientific Research Chemical-Pharmaceutical Institut; appl. 9.8.1965).

alternative syntheses:
The Merck Index, 2505 (Rahway 1990).
DOS 1 933 090 (Hoechst; appl. 30.6.1969).

Trade Name(s):

D:	Cor-Eusedon (Krewel); wfm	several combination products containing standardized *Convallaria majalis* extract.	I:	several combination products containing standardized *Convallaria majalis* extract.

Cortisone

ATC: H02AB10; S01BA03
Use: glucocorticoid

RN: 53-06-5   MF: $C_{21}H_{28}O_5$ MW: 360.45   EINECS: 200-162-4
LD_{50}:   230 mg/kg (M, i.p.)
CN: 17,21-dihydroxypregn-4-ene-3,11,20-trione

acetate
RN: 50-04-4   MF: $C_{23}H_{30}O_6$ MW: 402.49   EINECS: 200-006-5

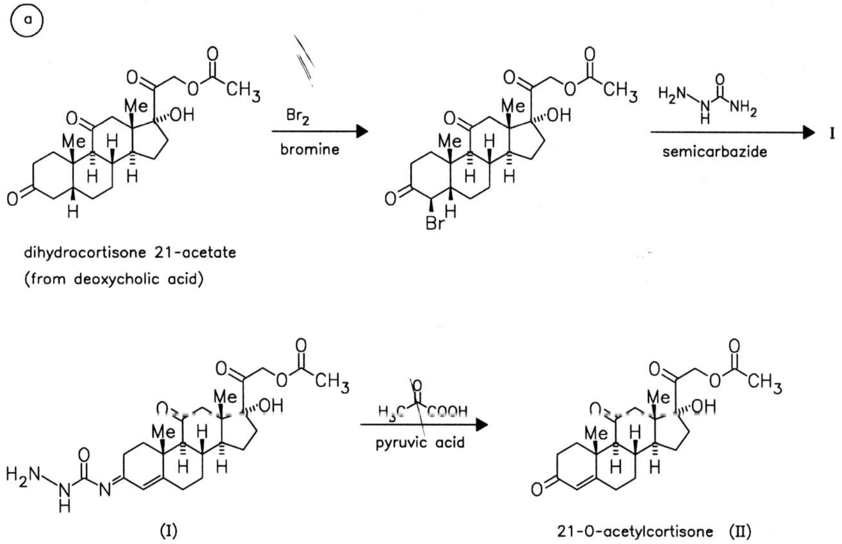

dihydrocortisone 21-acetate
(from deoxycholic acid)

21-0-acetylcortisone (II)

Cortisone

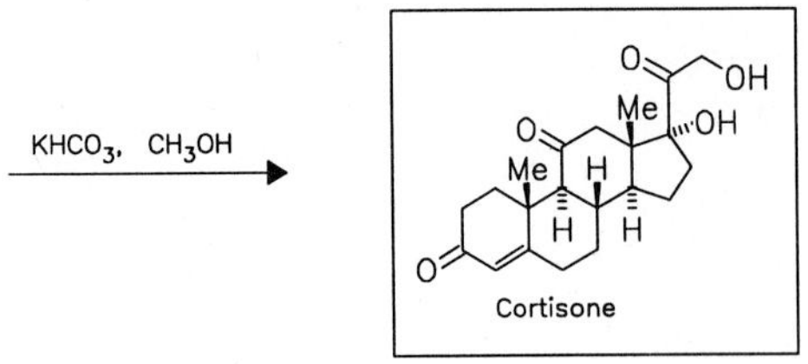

(b)

progesterone → (III)

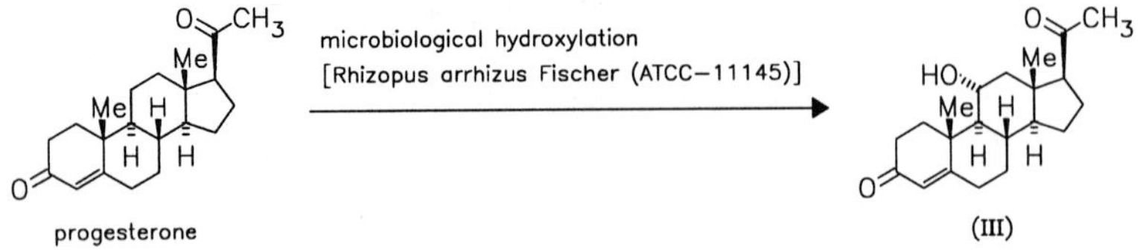

microbiological hydroxylation
[Rhizopus arrhizus Fischer (ATCC–11145)]

III → (IV)

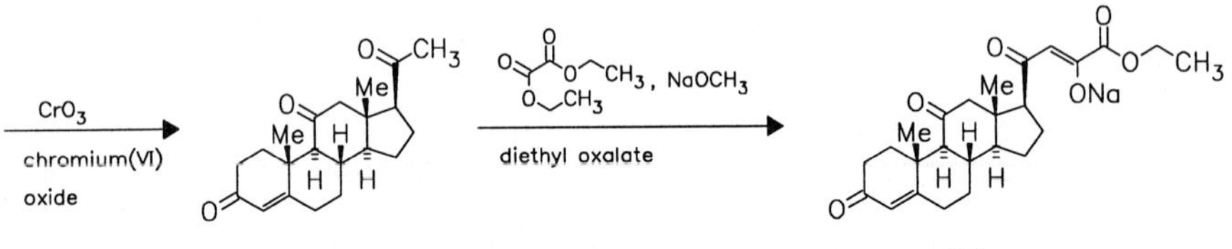

CrO₃
chromium(VI) oxide

diethyl oxalate

CH₃—O—CO—CO—O—CH₃, NaOCH₃

IV → (V)

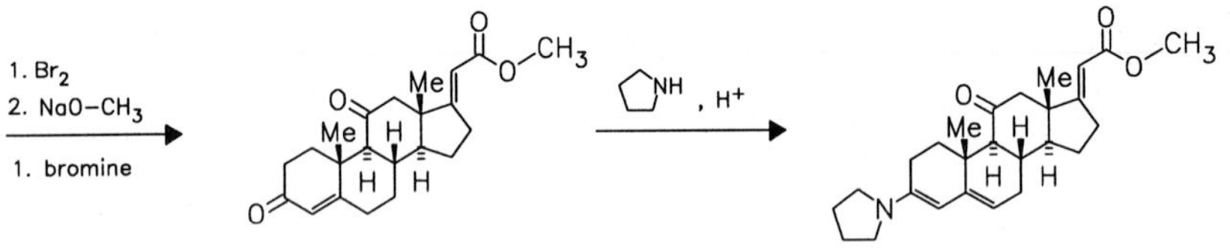

1. Br₂
2. NaO–CH₃
1. bromine

NH, H⁺

V → (VI)

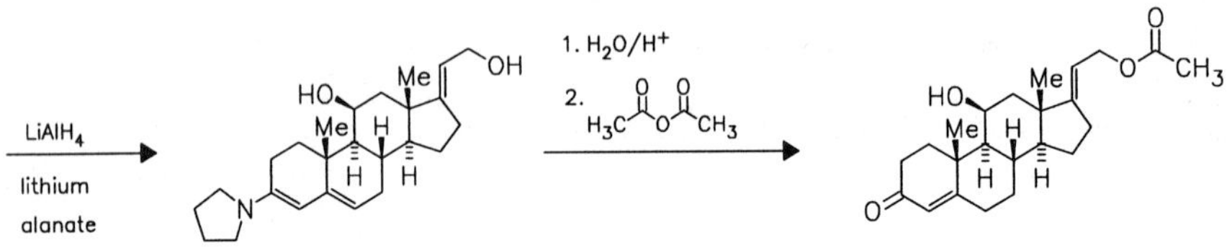

LiAlH₄
lithium alanate

1. H₂O/H⁺
2. H₃C—CO—O—CO—CH₃

VI → VII

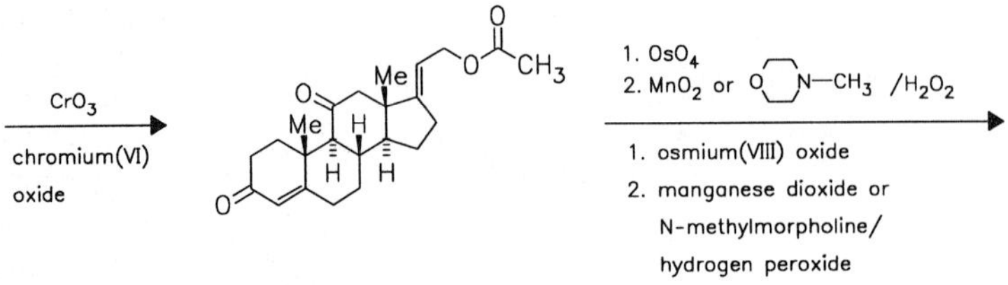

CrO₃
chromium(VI) oxide

1. OsO₄
2. MnO₂ or O=N—CH₃ /H₂O₂

1. osmium(VIII) oxide
2. manganese dioxide or
   N-methylmorpholine/
   hydrogen peroxide

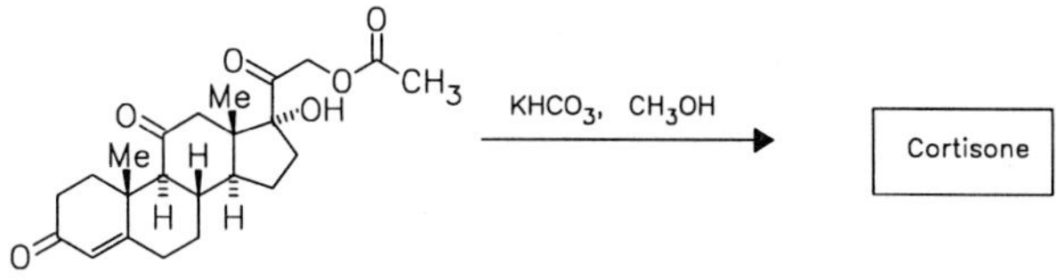

(VII)

Reference(s):

a   Applezweig, N.: Steroid Drugs, Vol. **1**, 14, 61 (New York, Toronto, London 1962).
Kendall, E.C. et al.: J. Biol. Chem. (JBCHA3) **166**, 345 (1946), **173**, 271 (1948).
synthesis of dihydrocortisone acetate:
Applezweig, N.: Steroid Drugs, Vol. **1**, 62 (New York, Toronto, London 1962).
The Merck Index, 2862 (Rahway 1976).

b   US 2 602 769 (Upjohn; 1952; prior. 1950).
US 2 769 823 (Upjohn; 1956; appl. 1954).
Applezweig, N.: Steroid Drugs, Vol. **1**, 59 (New York, Toronto, London 1962).

alternative syntheses:
FR 1 091 734 (Upjohn; appl. 1953; USA-prior. 1952).

cf. hydrocortisone *from* dehydropregnenolone acetate:
Ehrhart, Ruschig, **III**, 399.

from ergosterol *and* stigmasterol:
Rosenkranz, G.: Fortschr. Chem. Org. Naturst. (FCONAA) **10**, 274 (1953).

from hecogenin:
Applezweig, N.: Steroid Drugs, Vol. **1**, 66 (New York, Toronto, London 1962).

from sitosterol:
US 4 041 055 (Upjohn; 9.8.1977; appl. 17.11.1975).

total synthesis:
Fieser, L.F.; Fieser, M.: Steroide, 779 (Weinheim 1961).

review:
Fieser, L.F.; Fieser, M.: Steroide, 679 (Weinheim 1961).
Ullmanns Encykl. Tech. Chem., 4. Aufl., Vol. **13**, 50.

Formulation(s):   ointment 0.5 %, 1 %; tabl. 25 mg, 5 mg, 50 mg; vial 25 mg (2.5 mg/ml), 500 mg (50 mg/ml)
(as acetate)

Trade Name(s):

D:	Cortison Augensalbe Dr.	I:	Cortone Acetato (Merck	J:	Cortisone Acetat Sup.

D: Cortison Augensalbe Dr.
Winzer (Dr. Winzer)
Cortison Ciba (Novartis
Pharma)
F: Cortisme Roussel (Roussel)
GB: Cortisyl (Hoechst)

I: Cortone Acetato (Merck
Sharp & Dohme)
Dutimelan (Hoechst)-
comb.
generics

J: Cortisone Acetat Sup.
(Upjohn)
Cortone (Banyu)
Scheroson (Schering)
USA: Cortisone Acetate (Merck)
Cortone Acetat (Merck)

Cortivazol

ATC: H02AB17
Use: glucocorticoid

RN: 1110-40-3   MF: $C_{32}H_{38}N_2O_5$ MW: 530.67   EINECS: 214-175-8
CN: (11β,16α)-21-(acetyloxy)-11,17-dihydroxy-6,16-dimethyl-2'-phenyl-2'*H*-pregna-2,4,6-trieno[3,2-
c]pyrazol-20-one

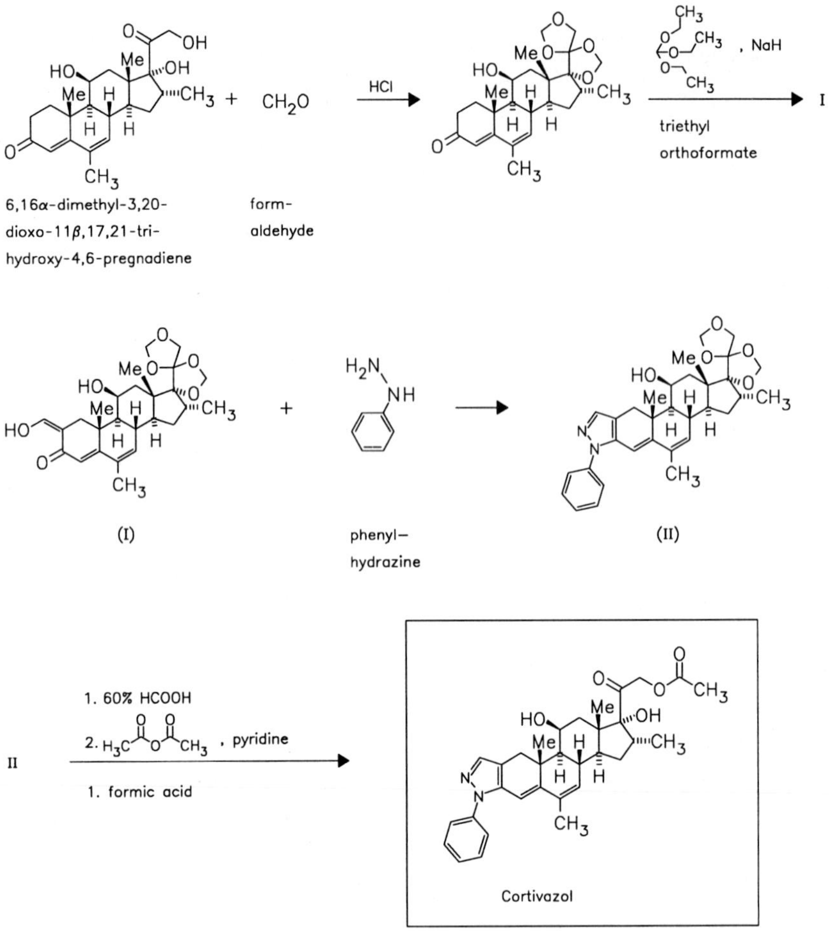

6,16α–dimethyl-3,20-
dioxo-11β,17,21-tri-
hydroxy-4,6-pregnadiene

form-
aldehyde

triethyl
orthoformate

(I)

phenyl–
hydrazine

(II)

1. 60% HCOOH

2. H_3C—C—O—C—CH$_3$, pyridine

II

1. formic acid

Cortivazol

Reference(s):
US 3 067 194 (Merck & Co.; 4.12.1962; prior. 1.12.1961, 4.11.1960).
US 3 300 483 (Merck & Co.; 24.1.1967; prior. 4.12.1962, 2.7.1962, 1.12.1961, 4.11.1960).
Fried, J.H. et al.: J. Am. Chem. Soc. (JACSAT) **85**, 236 (1963).

Formulation(s):   syringe 3.75 mg

Trade Name(s):
F: Altim (Roussel)

Creatinolfosfate
(Creatinol phosphate)

ATC: C01EB05
Use: cardiac preparation, cardiac stimulant

RN: 6903-79-3   MF: $C_4H_{12}N_3O_4P$ MW: 197.13   EINECS: 230-011-8
LD$_{50}$:   1200 mg/kg (M, i.v.); >5 g/kg (M, p.o.);
        1300 mg/kg (R, i.v.); >5 g/kg (R, p.o.)
CN: N-methyl-N-[2-(phosphonooxy)ethyl]guanidine

disodium salt
RN: 6903-80-6   MF: $C_4H_{10}N_3Na_2O_4P$ MW: 241.10

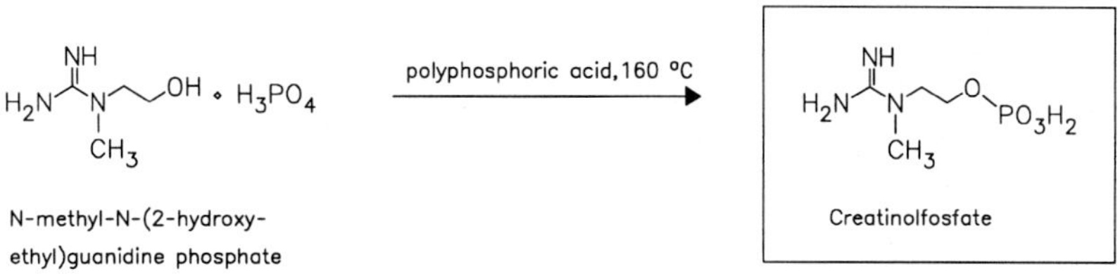

N-methyl-N-(2-hydroxy-
ethyl)guanidine phosphate

Creatinolfosfate

Reference(s):
DOS 2 550 430 (E. Allievi; appl. 13.11.1974; I-prior. 13.11.1974).

alternative syntheses:
FR-M 6 401 (Siphar; appl. 14.11.1966).
Ferrari, G.; Casagrande, C.: Farmaco, Ed. Sci. (FRPSAX) **20**, 879 (1965).

medical use as cardiac preparation:
DOS 2 144 584 (Siphar; appl. 6.9.1971; B-prior. 7.9.1970).

effervescent tablet:
BE 755 826 (Siphar; appl. 7.9.1970).

Formulation(s):   amp. 510 mg/4 ml; eff. gran. 500 mg/6 g; tabl. 250 mg

Trade Name(s):
I: Aplodan (Astra-Simes)

Croconazole
(Cloconazole)

ATC: D01A
Use: topical antifungal (for treatment of
candidiasis)

RN: 77175-51-0   MF: $C_{18}H_{15}ClN_2O$ MW: 310.78
CN: 1-[1-[2-[(3-chlorophenyl)methoxy]phenyl]ethenyl]-1*H*-imidazole

monohydrochloride
RN: 77174-66-4   MF: $C_{18}H_{15}ClN_2O \cdot HCl$ MW: 347.25
LD$_{50}$:   1150 mg/kg (M, p.o.);
2 g/kg (R, p.o.)

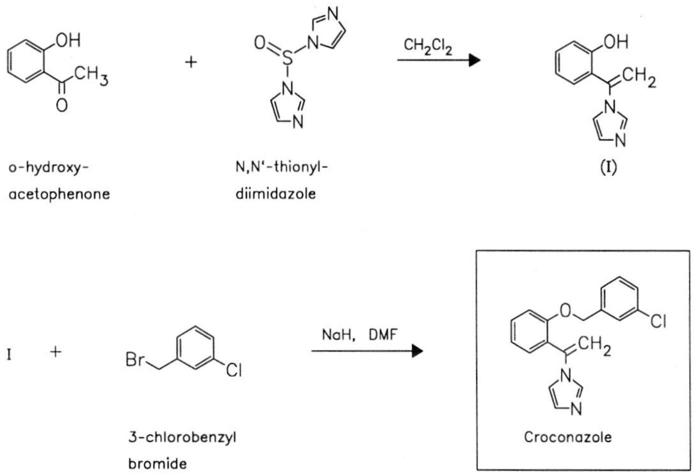

o-hydroxy-
acetophenone

N,N'-thionyl-
diimidazole

(I)

3-chlorobenzyl
bromide

Croconazole

Reference(s):
DOS 3 021 467 (Shionogi; appl. 6.6.1980; J-prior. 7.6.1979, 7.9.1979).
US 4 328 348 (Shionogi; 4.5.1982; J-prior. 7.6.1979, 7.9.1979).
US 4 463 011 (Shionogi; 31.7.1984; J-prior. 7.6.1979, 7.9.1979).
US 4 483 866 (Shionogi; 20.11.1984; J-prior. 7.6.1979, 7.9.1979).

Formulation(s):   cream 1 %; sol. 10 mg/g (1 %) (as hydrochloride)

Trade Name(s):
D: Pilzcin (Merz & Co.)   J: Pilzcin (Shionogi)

Cromoglicic acid
(Acidum cromoglicicum)

ATC: A07EB01; R01AC01; R03BC01;
          S01GX01
Use: antiallergic

RN: 16110-51-3   MF: $C_{23}H_{16}O_{11}$ MW: 468.37   EINECS: 240-279-8
LD_{50}:   >2.15 g/kg (R, p.o.)
CN: 5,5'-[(2-hydroxy-1,3-propanediyl)bis(oxy)]bis[4-oxo-4H-1-benzopyran-2-carboxylic acid]

disodium salt
RN: 15826-37-6   MF: $C_{23}H_{14}Na_2O_{11}$ MW: 512.33   EINECS: 239-926-7
LD_{50}:   3300 mg/kg (M, i.v.); >11 g/kg (M, p.o.);
      >4 g/kg (R, i.v.); >11 g/kg (R, p.o.);
      >1.6 g/kg (dog, i.v.); >4 g/kg (dog, p.o.)

2',6'-dihydroxy-acetophenone   epichloro-hydrin   diethyl oxalate

(I)

Cromoglicic acid

Reference(s):
DAS 1 543 579 (Fisons; appl. 23.3.1966; GB-prior. 25.3.1965, 9.12.1965, 17.12.1965).
GB 1 144 905 (Fisons; valid from 3.3.1966; prior. 25.3.1965, 9.12.1965, 17.12.1965).
US 3 419 578 (Fisons; 31.12.1968; GB-prior. 25.3.1965, 9.12.1965).
Barker, G. et al.: J. Med. Chem. (JMCMAR) **16**, 87 (1973).
US 3 671 625 (Fisons; 20.6.1972; GB-prior. 25.3.1965).
US 3 686 412 (Fitzmonrice et al.; 22.8.1972; GB-prior. 25.3.1965).
US 3 777 033 (Fisons; 4.12.1973; GB-prior. 25.3.1965).

disodium cromoglycate *with particular mass density:*
DOS 2 741 202 (Fisons; appl. 13.9.1977; GB-prior. 23.9.1976, 16.10.1976).

combination with anti-inflammatories:
US 4 066 756 (Fisons; 3.1.1978; GB-prior. 28.11.1975).
US 4 151 292 (Fisons; 24.4.1979; GB-prior. 25.1.1977).

Formulation(s):   aerosol 1 mg/0.05 ml; cps. 100 mg; gran. 100 mg, 200 mg; nasal spray 2.8 mg/0.14 ml, 20 mg/ ml; ophthalmic drops 10 mg/0.5 ml, 20 mg/ml (as disodium salt)

Trade Name(s):

D: Aarane (Rhône-Poulenc
        Rorer; 1983)-comb.
        Allergochrom (Ursapharm)
        Allergospasmin (ASTA
        Medica AWD; 1983)-comb.
        Colimune (Fisons)
        Colimune s 100/s 200
        (Fisons)
        Durachroman (durachemie)
        Intal (Fisons; Rhône-
        Poulenc Rorer; 1970)
        Lomupren (Fisons)
        Opticrom (Fisons)
        Pulbil (Klinge)
        Vividrin (Mann)
        generics
F: Cromedil (Europhta)
        Cromoptic (Chauvin)

        Intercron (Laphal)
        Lomudal (Rhône-Poulenc
        Rorer Specia)
        Lomusol (Rhône-Poulenc
        Rorer Specia)
        Nalcron (Rhône-Poulenc
        Rorer Specia)
        Opticron (Rhône-Poulenc
        Rorer Specia)
GB: Intal Syncroner (Rhône-
        Poulenc Rorer; 1968)
        Nalcrom (Rhône-Poulenc
        Rorer)
        Opticrom (Rhône-Poulenc
        Rorer)
        Rynacrom Spray (Rhône-
        Poulenc Rorer)

I: Cromantal (Nuovo Cons.
        Sanit. Naz.)
        Frenal (Schiapparelli
        Searle)
        Gastrofrenal (Schiapparelli
        Searle)
        Lomudal (Fisons)
        Nalcrom (Fisons)
        Rinofrenal (Schiapparelli
        Searle)-comb.
        Sificrom (SIFI)
        Visuglican (Merck Sharp &
        Dohme)-comb.
J: Intal (Fujisawa; 1971)
USA: Aarane (Syntex); wfm
        Intal (Fisons; 1973); wfm
        Nasalcrom (Fisons); wfm
        Opticrom (Fisons); wfm

Cropropamide

ATC: R07AB
Use: respiratory tonic

RN: 633-47-6   MF: $C_{13}H_{24}N_2O_2$ MW: 240.35   EINECS: 211-193-8
CN: *N*-[1-[(dimethylamino)carbonyl]propyl]-*N*-propyl-2-butenamide

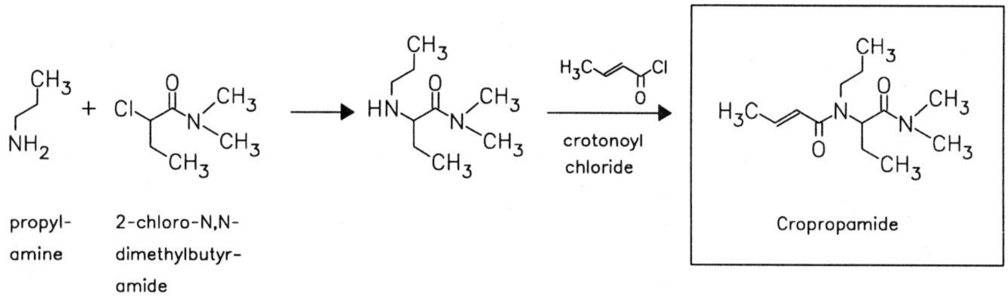

propyl-   2-chloro-N,N-
amine   dimethylbutyr-
                amide

crotonoyl
chloride

Cropropamide

Reference(s):
US 2 447 587 (Geigy; 1948; CH-prior. 1942).

Formulation(s):   drops 15 % (comb. with crotetamide)

Trade Name(s):

D: Micoren (Geigy)-comb.
        with crotetamide; wfm
F: Micorène (Ciba-Geigy)-
        comb. with crotetamide;
        wfm
GB: Micoren (Geigy)-comb.
        with crotetamide; wfm
I: Micoren (Geigy)-comb.
        with crotetamide; wfm

Crotamiton

ATC: P03A
Use: antipruritic, scabicide

RN: 483-63-6　MF: $C_{13}H_{17}NO$　MW: 203.29　EINECS: 207-596-3
LD$_{50}$: 1600 mg/kg (M, p.o.);
　　　1500 mg/kg (R, p.o.)
CN: N-ethyl-N-(2-methylphenyl)-2-butenamide

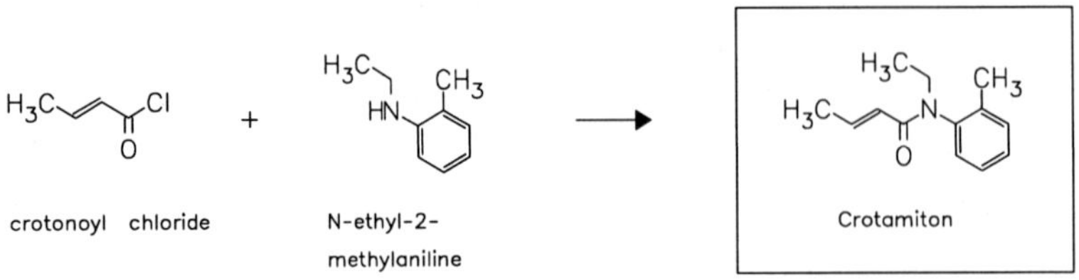

crotonoyl chloride　　N-ethyl-2-
　　　　　　　　　methylaniline　　　　　Crotamiton

Reference(s):
GB 615 137 (Geigy; appl. 1946).

Formulation(s): 　cream 0.1 g/g; gel 50 mg/100 g; ointment 100 mg/100 g; sol. 10 %

Trade Name(s):
D:	Crotamitex-Gel (gepepharm)	F:	Eurax (Zyma)	Eurax (Ciba-Geigy-Fujisawa)
	Euraxil (Novartis Consumer Health)	GB:	Eurax (Novartis Consumer)	
		I:	Eurax (Zyma)	USA: Eurax (Westwood-Squibb)
		J:	Dermarin (Taisho)	

Crotetamide

ATC: R07AB
Use: respiratory tonic

RN: 6168-76-9　MF: $C_{12}H_{22}N_2O_2$　MW: 226.32　EINECS: 228-208-9
CN: N-[1-[(dimethylamino)carbonyl]propyl]-N-ethyl-2-butenamide

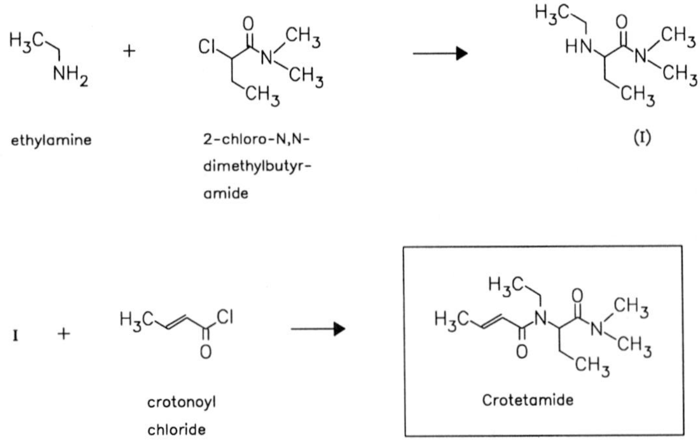

ethylamine　　　2-chloro-N,N-　　　　　　　　　(I)
　　　　　　　dimethylbutyr-
　　　　　　　amide

I　+　　crotonoyl　　　　　　Crotetamide
　　　　chloride

Reference(s):
US 2 447 587 (Geigy; 1948; CH-prior. 1942).

Formulation(s): 　drops 15 % (comb. with cropropamide)

Cyamemazine

(Cyamepromazine)

ATC: N05AA06
Use: neuroleptic, tranquilizer

RN: 3546-03-0   MF: $C_{19}H_{21}N_3S$ MW: 323.46   EINECS: 222-594-2
CN: 10-[3-(dimethylamino)-2-methylpropyl]-10H-phenothiazine-2-carbonitrile

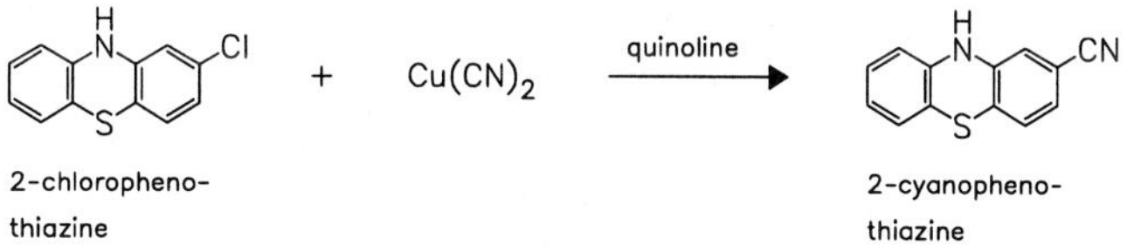

2-chloropheno-
thiazine

2-cyanopheno-
thiazine

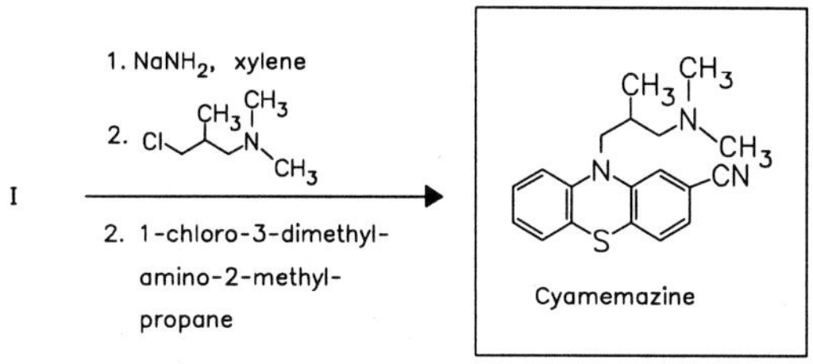

Cyamemazine

Reference(s):
US 2 877 224 (Rhône-Poulenc; 1959; F-prior. 1955).
DE 1 056 611 (Rhône-Poulenc; appl. 1956; F-prior. 1955).

Formulation(s):   amp. 50 mg/5 ml; drops 4 %; sol. 40 mg/ml; tabl. 25 mg, 100 mg

Cyanocobalamin

(Vitamin B_{12})

ATC: B03BA01
Use: antipernicious vitamin

RN: 68-19-9   MF: $C_{63}H_{88}CoN_{14}O_{14}P$ MW: 1355.39   EINECS: 200-680-0
CN: cobinamide cyanide hydroxide dihydrogen phosphate (ester) inner salt 3'-ester with 5,6-dimethyl-1-α-D-
       ribofuranosyl-1H-benzimidazole

Cyanocobalamin

By fermentation with *Streptomyces griseus, S. olivaceus, S. aureofaciens, Bacillus megatherium* or *Propionobacterium freudenreichii*. Molasses is used generally as fermentation medium, $CoCl_2$ and 5,6-dimethylbenzimidazole are added. Various adsorption and extraction methods are used for isolation from the fermentation liquors.

Reference(s):
review:
Ullmanns Encykl. Tech. Chem., 3. Aufl., Vol. **18**, 219.
Kirk-Othmer Encycl. Chem. Technol., 2nd Ed. (15SWA8), Vol. **21**, (1963-1971), 544.
Bernhauer, K. et al.: Angew. Chem. (ANCEAD) **75**, 1145 (1963).

fermentative preparation:
US 2 505 053 (Merck & Co.; 1950; appl. 1948).
US 2 530 416 (Merck & Co.; 1950; appl. 1949).
US 2 563 794 (Merck & Co.; 1951; appl. 1949).
US 2 582 589 (Abbott; 1952; appl. 1949).
US 2 595 499 (Merck & Co.; 1952; appl. 1948).
US 2 650 896 (Merck & Co.; 1953; appl. 1950).
US 2 703 302 (Merck & Co.; 1955; appl. 1952).
US 2 703 303 (Merck & Co.; 1955; prior. 1948).
DE 1 046 258 (Soc. Farmaceutici Italia; appl. 1956; I-prior. 1955).
DE 1 076 889 (Distillers; appl. 1958; GB-prior. 1957).
US 2 951 017 (Distillers; 1960; GB-prior. 1957).
US 3 000 793 (Merck & Co.; 1961; prior. 1955).
US 3 018 225 (Merck & Co.; 23.1.1962; prior. 1953).
DE 1 080 264 (Distillers; appl. 1958; GB-prior. 1957).
DE 1 091 705 (Roche; appl. 1959).
DE 1 109 317 (Roche; appl. 1959).
GB 1 451 694 (Richter Gedeon; appl. 25.10.1974; H-prior. 26.10.1973).
US 4 119 492 (Nippon Oil; 10.10.1978; J-prior. 5.2.1976).

yield increasing by addition of betaine to the nutritive medium:
US 3 000 793 (Merck & Co.; 1961; appl. 1957).
US 2 923 666 (Pabst Brewing Comp.; 1960; appl. 1954).

isolation from liver preparations:
US 2 594 314 (Merck & Co.; 1952; appl. 1948).
US 2 609 325 (Merck & Co.; 1952; appl. 1948).

purification and isolation:
US 2 607 717 (Merck & Co.; 1952; appl. 1949).
US 2 626 888 (Merck & Co.; 1953; appl. 1950).
US 2 628 186 (Research Corp.; 1953; appl. 1950).
US 3 057 851 (Armour; 9.10.1962; prior. 1955).

Formulation(s):   amp. 0.1 mg, 1 mg; drg. 1 mg; drops 0.05 mg; inj. flask 0.5 mg, 1 mg. 5 mg

Trade Name(s):

D: B$_{12}$ "Ankermann"
(Wörwag)
B$_{12}$-Horfervit (Arteva
Pharma)
B$_{12}$ Rotexmedica
(Rotexmedica)
B$_{12}$-Steigerwald
(Steigerwald)
B$_{12}$-Vicotrat (Heyl)
Biovital (Dr. Schieffer)-
comb.
Bryonon (Protina)-comb.
Cervevit (Baxter)-comb.
Cobidec (Warner-
Lambert)-comb.
Cytobion (Merck)
Dodecatol (Heyl)-comb.
Dolo-Neurobion
(Merckle)-comb.
Eryfer (Cassella-med)-
comb.
Eukalasan (Steigerwald)-
comb.
Hämo-Vibolex (Anphasaar)
Lophakomb (Lomapharm)
Multibionta (Merckle)-
comb.
Natabec (Warner-Lambert)
Neurotrat (Knoll)-comb.
Vicapan B$_{12}$ (Merckle)
Vitamin B 12 forte (Hevert)
Vitamin B$_{12}$ (OTW)

Vitamin B12
Injektionslösung
(Wiedemann)
Vitamin-B$_{12}$-ratiopharm
(ratiopharm)
numerous combination
preparations

F: Alvityl (Solvay)-comb.
Azedanit (Whitehall)-
comb.
B$_{12}$ Mille Delagrange
(Synthélabo)
Berocca (Nicholas)-comb.
Forvital (Whitehall)-comb.
Pharmaton (Boehringer
Ing.)-comb.
Soluvit (UCB)-comb.
Synergil (Dakota)
Vitamine B$_{12}$ Aguettant
(Aguettant)
Vitamine B$_{12}$ Lavoisier
(Chaix et du Marais)
Vivamyne (Whitehall)-
comb.
numerous combination
preparations

GB: Cytacon (Goldshield)
Cytamen (Evans)

I: Cobequin (Casarini)
Dobetin (Angelini)
Efargen (Teofarma)-comb.

Epargriseovit (Farmitalia)-
comb.
Eritrovit B12 (Lisapharma)
Mionevrasi (Boehringer
Mannh.)-comb.
Neoeparibiol (Ecobi)-
comb.
Reticulogen (Lilly)
Tonicum (SIT)-comb.
numerous combination
preparations

J: Actamin B$_{12}$ (Yashima)
Redisol (Merck-Banyu)
numerous combination
preparations

USA: Bevitamel (Westlake)
Chromagen (Savage)
Cyanocobalamin (Elkins-
Sinn)
Fetrin (Lunsco)
Hemocyte-F (U.S.
Pharmaceutical)
Mega-B (Arco)
Nascobal (Schwarz)
Niferex (Schwarz)
Nu-Iron-Plus Elixier
(Merz)
Rubramin PC (Squibb)
Trinsicon (UCB)
numerous combination
preparations

Cyclandelate

ATC: C04AX01
Use: antispasmodic

RN: 456-59-7   MF: C$_{17}$H$_{24}$O$_3$ MW: 276.38   EINECS: 207-271-6
LD$_{50}$:   >10 g/kg (M, p.o.);
5 g/kg (R, p.o.)
CN: α-hydroxybenzeneacetic acid 3,3,5-trimethylcyclohexyl ester

DL-mandelic
acid

3,3,5-trimethyl-
cyclohexanol

Cyclandelate

Reference(s):
US 2 707 193 (Brocades-Stheeman; 1955; NL-prior. 1949).

purification:
US 3 663 597 (American Home; 16.5.1972; appl. 5.5.1970).

Formulation(s):   cps. 400 mg; drg. 200 mg, 400 mg

Trade Name(s):

D:	Eucebral-N (Südmedica)-comb.	Hacosan (Sanko)
	Natil (3M Medica)	Hi-Cyclane Cap. (Tyama)
	Spasmocyclon (3M Medica)	Mandelic (Seiko)
		Marucyclan (Maruko)
F:	Cyclergine (Poirier)	Mitalon (Toyo Pharmar)
	Cyclospasmol (Yamanouchi)	Newcellan Cap. (Kowa)
	Novodil (Augot)	Saiclate (Morishita)
	Vascunormyl (Alcon)-comb.	Sancyclan (Santen)
GB:	Cyclobral (Norgine); wfm	Sepyron Cap. (Sankyo)
	Cyclospasmol (Brocades); wfm	Spadelate Cap. (Zeria)

I: Ciclospasmol (Brocades)
J: Anticen (Nippon Kayaku)
Aposelebin (Hokuriku)
Capilan (Takeda)
Capistar (Kowa Yakuhin)
Ceaclan (Mohan)
Cepidan (Meiji)
Circle-one (Funai)
Circulat (Kotani)
Cyclan (Ohta)
Cyclan-Cap. (Nichiiko)
Cyclansato (SS)
Cycleat Cap. (Hishiyama)
Cycralate (Kanto)

Venalal (Mochida)
Zirkulat (Nippon Shoji)
USA: Cyclospasmol (Ives); wfm
generics; wfm

Cyclizine

ATC: R06AE03
Use: antihistaminic, anti-emetic

RN: 82-92-8   MF: $C_{18}H_{22}N_2$ MW: 266.39   EINECS: 201-445-5
LD_{50}:   147 mg/kg (M, p.o.)
CN: 1-(diphenylmethyl)-4-methylpiperazine

monohydrochloride
RN: 303-25-3   MF: $C_{18}H_{22}N_2 \cdot HCl$ MW: 302.85   EINECS: 206-136-9
LD_{50}:   165 mg/kg (M, p.o.)
lactate (1:1)
RN: 5897-19-8   MF: $C_{18}H_{22}N_2 \cdot C_3H_6O_3$ MW: 356.47

benzhydryl   1-methyl-   Cyclizine
chloride   piperazine

Reference(s):
US 2 630 435 (Burroughs Wellcome; 1953; prior. 1948).

Formulation(s):   amp. 50 mg; suppos. 100 mg; tabl. 25 mg, 50 mg (as hydrochloride)

Trade Name(s):

D:	Migräne-Kranit spezial (Krewel)-comb.; wfm	
F:	Migwell (Glaxo Wellcome; as hydrochloride)-comb.	

GB: Diconal (Glaxo Wellcome)-comb.
Migril (Glaxo Wellcome)-comb.
Valoid (Glaxo Wellcome)

I: Marzine (Wellcome; as hydrochloride)
J: Cleamine (Kodama)-comb.
USA: Marezine (Burroughs Wellcome); wfm

Marezine (Burroughs
Wellcome; as
hydrochloride); wfm

Cyclobarbital
(Hexemal; Cyclobarbitone)

ATC: N05CA10
Use: hypnotic

RN: 52-31-3   MF: $C_{12}H_{16}N_2O_3$ MW: 236.27   EINECS: 200-138-3
LD$_{50}$:   840 mg/kg (M, p.o.)
CN: 5-(1-cyclohexen-1-yl)-5-ethyl-2,4,6(1H,3H,5H)-pyrimidinetrione

calcium salt
RN: 5897-20-1   MF: $C_{12}H_{16}N_2O_3 \cdot xCa$ MW: unspecified   EINECS: 227-590-4
calcium salt (2:1)
RN: 143-76-0   MF: $C_{24}H_{30}CaN_4O_6$ MW: 510.60   EINECS: 205-610-2

cyclo-   methyl   methyl   methyl 2-cyano-
hexanone   cyanoacetate   1-cyclohexenyl-   2-(1-cyclohexenyl)-
                              cyanoacetate   butyrate (I)

dicyano-
diamide   Cyclobarbital

Reference(s):
DRP 442 655 (Bayer; 1924).
GB 231 150 (Bayer; 1924).

Formulation(s):   cps. 75 mg; tabl. 100 mg, 200 mg (as calcium salt)

Trade Name(s):

D: Dormopan (Bayropharm)-
      comb.; wfm
      Gastripan (Merckle)-
      comb.; wfm
      Itridal (Homburg)-comb.;
      wfm
      Medinox (Pfleger)-comb.;
      wfm
      Phanodorm (Bayer); wfm

      Somnubene (Merckle)-
      comb.; wfm
      Somnupan C (Merckle);
      wfm
      Stodinox (Lorenz) comb.;
      wfm
      Tempidorm N (Roland)-
      comb.; wfm
      generics; wfm

F: Dormopan (Bayer-
      Pharma)-comb.; wfm
GB: Phanodorm (Winthrop);
      wfm
      Rapidal (Medo)
I: Cyclobarbitalum (Sale di
      calcio)
J: Adorm (Shionogi)

Cyclobenzaprine

ATC: M03BX08
Use: muscle relaxant, psychosedative

RN: 303-53-7   MF: $C_{20}H_{21}N$ MW: 275.40   EINECS: 206-145-8
LD_{50}:   36 mg/kg (M, i.v.); 250 mg/kg (M, p.o.)
CN: 3-(5H-dibenzo[a,d]cyclohepten-5-ylidene)-N,N-dimethyl-1-propanamine

hydrochloride
RN: 6202-23-9   MF: $C_{20}H_{21}N \cdot HCl$ MW: 311.86   EINECS: 228-264-4
LD_{50}:   36 mg/kg (M, i.v.); 250 mg/kg (M, p.o.)

dibenzo[a,d]cyclo-   3-dimethylamino-   (I)
hepten-5-one   propylmagnesium
                        chloride

Cyclobenzaprine

Reference(s):
US 3 272 864 (Merck & Co.; 13.9.1966; appl. 19.4.1962).
US 3 409 640 (Schering Corp.; 5.11.1968; appl. 22.7.1959).

medical use:
US 3 882 246 (Merck & Co.; 6.5.1975; prior. 31.1.1973, 21.5.1971, 9.4.1974).

Formulation(s):   cps. 10 mg; tabl. 10 mg, 30 mg (as hydrochloride)

Trade Name(s):
I: Flexiban (Neopharmed)   USA: Flexeril (Merck; as
                                            hydrochloride)

Cyclobutyrol

ATC: A05AX03
Use: choleretic

RN: 512-16-3   MF: $C_{10}H_{18}O_3$ MW: 186.25   EINECS: 208-138-5
LD_{50}:   2900 mg/kg (M, i.v.); >10 g/kg (M, p.o.);
            1760 mg/kg (R, i.v.); 4820 mg/kg (R, p.o.)
CN: α-ethyl-1-hydroxycyclohexaneacetic acid

monosodium salt
RN: 1130-23-0   MF: $C_{10}H_{17}NaO_3$ MW: 208.23   EINECS: 214-458-6
calcium salt
RN: 40043-69-4   MF: $C_{10}H_{18}O_3 \cdot xCa$ MW: unspecified
betaine salt (1:1)
RN: 23579-12-6   MF: $C_{10}H_{17}O_3 \cdot C_5H_{12}NO_2$ MW: 303.40   EINECS: 245-750-1

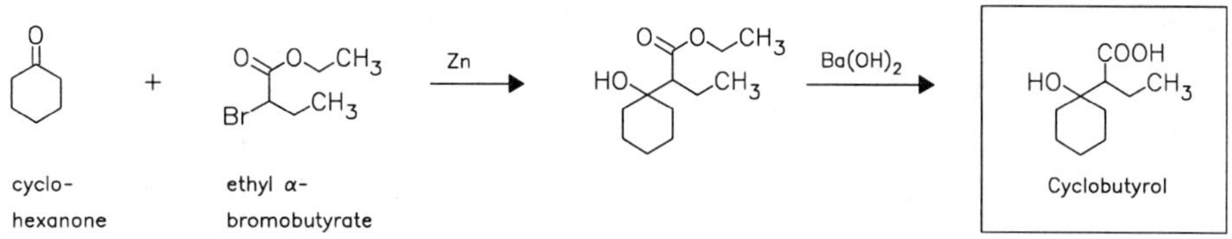
Zn
Ba(OH)2
cyclo-
hexanone
ethyl α-
bromobutyrate
Cyclobutyrol

Reference(s):
DE 1 094 254 (Lab. J. Logeais; appl. 14.2.1959; F-prior. 19.2.1958).
US 3 065 134 (Lab. J. Logeais; 20.11.1962; F-prior. 19.2.1958).
Maillard, J. et al.: Bull. Soc. Chim. Fr. (BSCFAS) **1958**, 244.

Formulation(s): amp. 200 mg; tabl. 250 mg (as sodium salt)

Trade Name(s):

D:	Benestan (Karlspharma)-comb.; wfm	I:	Epa-Bon (Sifarma)
	Trommgallol (Trommsdorff)-comb.; wfm	J:	Lipotrin (Eisai)
F:	Hébucol (J. Logeais)		Riphole N (Nichiiko)

Cyclofenil

ATC: G03GB01
Use: gonadotropin stimulant (against infertility)

RN: 2624-43-3 MF: $C_{23}H_{24}O_4$ MW: 364.44 EINECS: 220-089-1
LD_{50}: >12.5 g/kg (M, p.o.);
>12 g/kg (R, p.o.)
CN: 4-[[4-(acetyloxy)phenyl]cyclohexylidenemethyl]phenol acetate

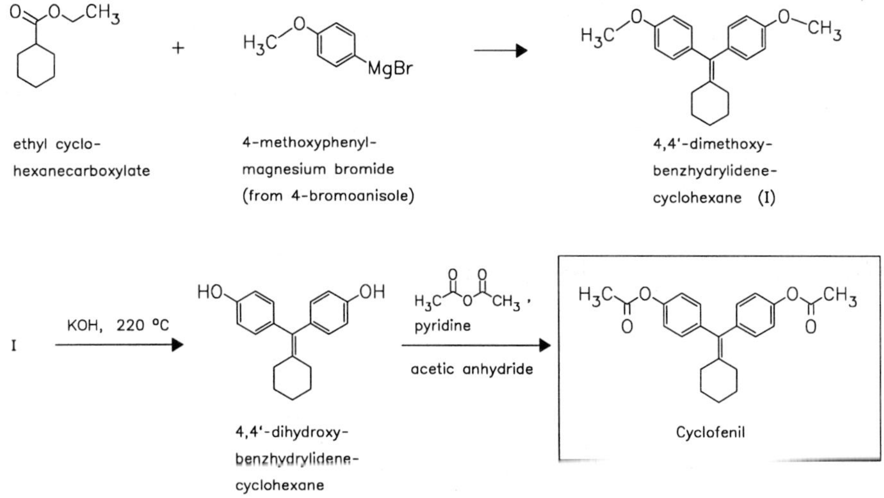
ethyl cyclo-
hexanecarboxylate
4-methoxyphenyl-
magnesium bromide
(from 4-bromoanisole)
4,4'-dimethoxy-
benzhydrylidene-
cyclohexane (I)
KOH, 220 °C
pyridine
acetic anhydride
4,4'-dihydroxy-
benzhydrylidene-
cyclohexane
Cyclofenil

Reference(s):
US 3 287 397 (K.G. Olsson et al.; 22.11.1966; GB-prior. 22.11.1960).

Formulation(s): tabl. 100 mg, 200 mg, 400 mg

Trade Name(s):

D:	Fertodur (Schering); wfm	GB:	Ondonit (Roussel); wfm	I:	Fertodur (Schering)
F:	Ondogyne (Roussel); wfm		Rehibin (Thames); wfm		Neoclym (Poli)

J: Sexovid (Teikoku Zoki)

Cyclomethycaine

ATC: S01HA
Use: local anesthetic

RN: 139-62-8   MF: $C_{22}H_{33}NO_3$ MW: 359.51
CN: 4-(cyclohexyloxy)benzoic acid 3-(2-methyl-1-piperidinyl)propyl ester

sulfate (2:1)
RN: 6202-05-7   MF: $C_{22}H_{33}NO_3 \cdot 1/2H_2SO_4$ MW: 817.10
sulfate (1:1)
RN: 50978-10-4   MF: $C_{22}H_{33}NO_3 \cdot H_2SO_4$ MW: 457.59

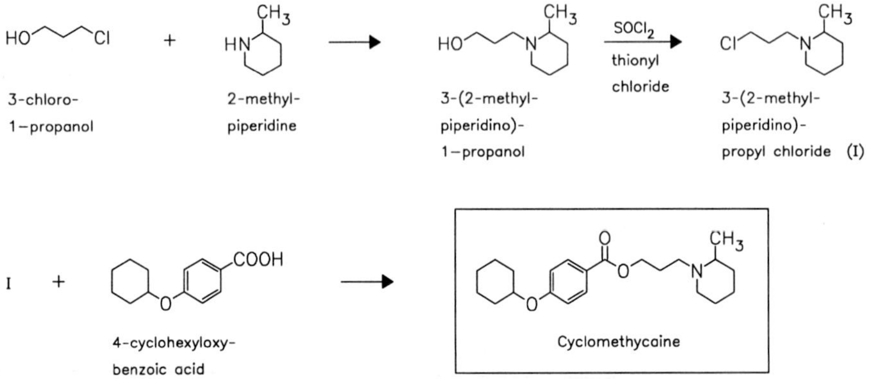

3-chloro-   2-methyl-   3-(2-methyl-   3-(2-methyl-
1-propanol   piperidine   piperidino)-   piperidino)-
                                             1-propanol   propyl chloride (I)

4-cyclohexyloxy-   Cyclomethycaine
benzoic acid

Reference(s):
US 2 439 818 (S. M. McElvain, T. P. Carney; 1948; appl. 1946).
McElvain, S.M.; Carney, T.P.: J. Am. Chem. Soc. (JACSAT) **68**, 2592 (1946).

Formulation(s):   cream 0.5 % - 1 %; ointment 0.5 - 1%; spray 0.25 %

Trade Name(s):
USA: Surfacaine (Lilly); wfm

Cyclopentamine

ATC: R01AA02
Use: sympathomimetic

RN: 102-45-4   MF: $C_9H_{19}N$ MW: 141.26
CN: N,α-dimethylcyclopentaneethanamine

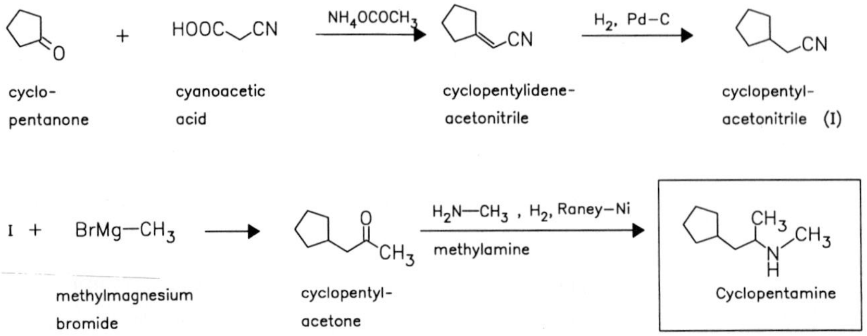

cyclo-   cyanoacetic   cyclopentylidene-   cyclopentyl-
pentanone   acid   acetonitrile   acetonitrile (I)

methylmagnesium   cyclopentyl-   Cyclopentamine
bromide   acetone

Reference(s):
US 2 520 015 (Eli Lilly; 1950; prior. 1948).

Formulation(s):   amp. 10 mg, 25 mg; sol.

Trade Name(s):
D:	Copyronilum (Lilly)- comb.; wfm	F:	Cyclonarol (Hépatrol); wfm		I: USA:	Copyronil (Lilly); wfm Clopane (Lilly); wfm

---

Cyclopenthiazide
(Cyclomethiazide)

ATC: C03AA07
Use: diuretic, antihypertensive

RN: 742-20-1   MF: $C_{13}H_{18}ClN_3O_4S_2$ MW: 379.89   EINECS: 212-012-5
LD$_{50}$:   232 mg/kg (M, i.v.); >1 g/kg (M, p.o.);
      142 mg/kg (R, i.v.); 1 g/kg (R, p.o.)
CN: 6-chloro-3-(cyclopentylmethyl)-3,4-dihydro-2H-1,2,4-benzothiadiazine-7-sulfonamide 1,1-dioxide

6-amino-4-chloro-
1,3-benzenedisulfamide
(cf. chlorothiazide
synthesis)

cyclopentyl-
acetaldehyde

Cyclopenthiazide

Reference(s):
BE 587 225 (Ciba; appl. 3.2.1960; USA-prior. 4.2.1959).
Whitehead, C.W. et al.: J. Org. Chem. (JOCEAH) **26**, 2814 (1961).

Formulation(s):   tabl. 0.25 mg, 0.5 mg

Trade Name(s):
D:	Navidrex (Ciba); wfm		Trasidrex (Novartis)-comb.	USA:	Navidrix (Ciba-Geigy);
GB:	Navidrex (Novartis)	J:	Navidrex (Ciba-Geigy-		wfm
	Navispare (Novartis)-comb.		Takeda)		

Cyclopentobarbital

ATC: N05CA
Use: hypnotic

RN: 76-68-6   MF: $C_{12}H_{14}N_2O_3$ MW: 234.26   EINECS: 200-979-6
LD$_{50}$:   90 mg/kg (R, i.p.)
CN: 5-(2-cyclopenten-1-yl)-5-(2-propenyl)-2,4,6(1H,3H,5H)-pyrimidinetrione

3-bromo-
cyclopentene

diethyl
malonate

diethyl 2-cyclo-
pentenylmalonate

allyl bromide

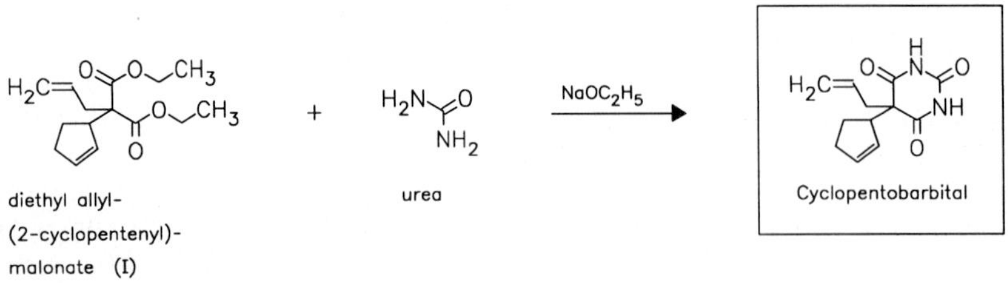

diethyl allyl-
(2-cyclopentenyl)-
malonate (I)

urea

Cyclopentobarbital

Reference(s):
DRP 589 947 (Comp. de Béthune; appl. 1930; F-prior. 1929).

Trade Name(s):
D: Cyclopal (Siegfried); wfm

Cyclopentolate

ATC: S01FA04
Use: antispasmodic, mydriatic

RN: 512-15-2   MF: $C_{17}H_{25}NO_3$ MW: 291.39   EINECS: 208-136-4
CN: α-(1-hydroxycyclopentyl)benzeneacetic acid 2-(dimethylamino)ethyl ester

hydrochloride
RN: 5870-29-1   MF: $C_{17}H_{25}NO_3 \cdot HCl$ MW: 327.85   EINECS: 227-521-8
LD_{50}: 84 mg/kg (M, i.v.); 960 mg/kg (M, p.o.);
        >4 g/kg (R, p.o.)

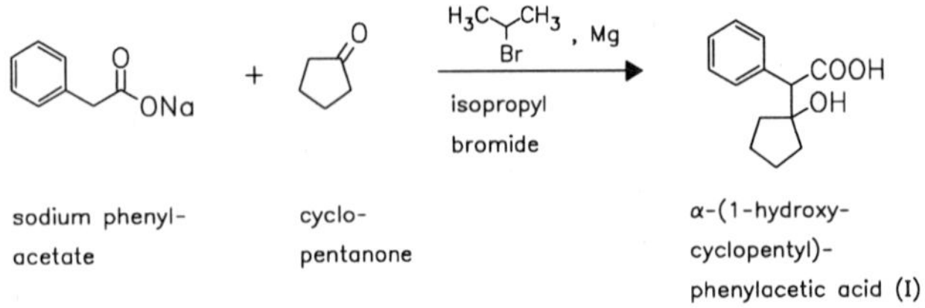

sodium phenyl-   cyclo-   α-(1-hydroxy-
acetate   pentanone   cyclopentyl)-
                                        phenylacetic acid (I)

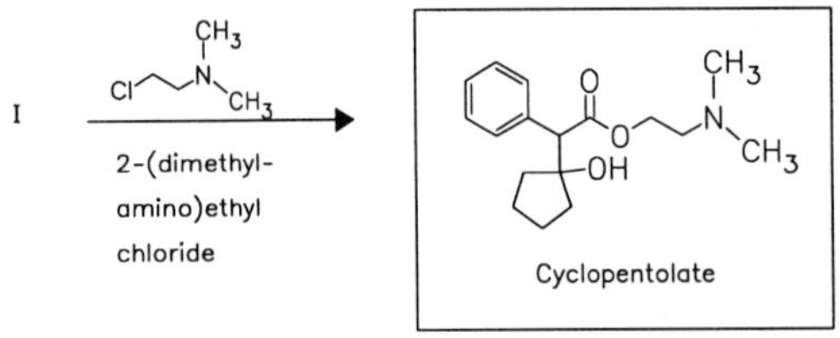

2-(dimethyl-
amino)ethyl
chloride

Cyclopentolate

Reference(s):
US 2 554 511 (Schieffelin & Co.; 1951; prior. 1949).

Formulation(s):   eye drops 5 mg (0.5 %, 1 %) (as hydrochloride)

Trade Name(s):
D: Cyclopentolat   GB: Minims Cyclopentolate   Cyclomydril (Alcon)-
        Augentropfen (Alcon)   (Chauvin)   comb.; wfm
        Zyklolat EDO (Mann)   Mydrilate (Boehringer Ing.) generics and combination
F: Skiacol (Alcon)   I: Ciclolux (Allergan)   preparations; wfm
                                   USA: Cyclogyl (Alcon); wfm

Cyclophosphamide

ATC: L01AA01
Use: antineoplastic

RN: 50-18-0 MF: $C_7H_{15}Cl_2N_2O_2P$ MW: 261.09 EINECS: 200-015-4
LD_{50}: 140 mg/kg (M, i.v.); 137 mg/kg (M, p.o.);
148 mg/kg (R, i.v.); 160 mg/kg (R, p.o.)
CN: *N,N*-bis(2-chloroethyl)tetrahydro-2*H*-1,3,2-oxazaphosphorin-2-amine 2-oxide

	bis(2-chloro-ethyl)amine	N,N-bis-(2-chloro-ethyl)phosphor-amidic dichloride	Cyclophosphamide

Reference(s):
DE 1 057 119 (ASTA-Werke; appl. 10.2.1956).
US 3 018 302 (ASTA-Werke; 23.1.1962; D-prior. 10.2.1956).

Formulation(s):   drg. 50 mg; f. c. tabl. 50 mg; vial 100 mg, 200 mg, 500 mg, 1000 mg

Trade Name(s):
D:	Cyclo-cell (cell pharm)	F:	Endoxan ASTA (ASTA Medica)	J:	Endoxan (Shionogi)
	Cyclostin (Pharmacia & Upjohn)			USA:	Cytoxan (Bristol-Myers Squibb)
		GB:	Endoxana (ASTA Medica)		
	Endoxan (ASTA Medica AWD)	I:	Endoxan-Asta (ASTA Medica)		

Cycloserine
(Orientomycin)

ATC: J04AB01
Use: antibiotic (tuberculostatic)

RN: 68-41-7 MF: $C_3H_6N_2O_2$ MW: 102.09 EINECS: 200-688-4
LD_{50}: 560 mg/kg (M, i.v.); 5290 mg/kg (M, p.o.);
>5 g/kg (R, p.o.);
>2 g/kg (dog, p.o.)
CN: (*R*)-4-amino-3-isoxazolidinone

hydrogen tartrate
RN: 17139-97-8 MF: $C_3H_6N_2O_2 \cdot C_4H_6O_6$ MW: 252.18

Ⓐ   from fermentation solutions of Streptomyces garyphalus, S. orchidaceus, S. lavendulae

Ⓑ

D-serine		D-serine methyl ester hydrochloride

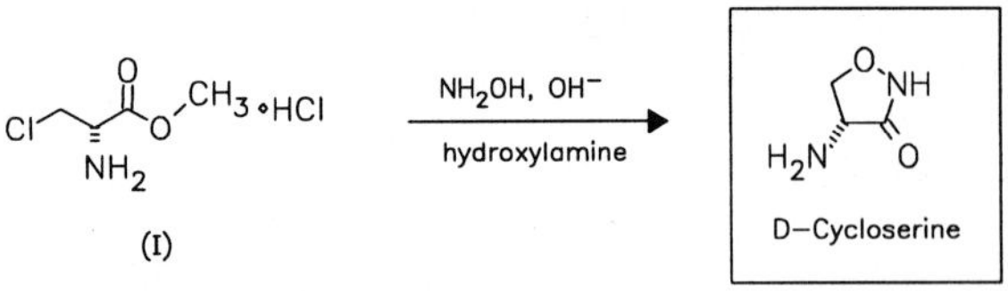

Reference(s):

a   US 2 773 878 (Pfizer; 1956; appl. 1952).
US 2 789 983 (Commercial Solvents; 1957; prior. 1954).
US 2 845 433 (Merck & Co.; 1958; appl. 1955).

b   Plattner, P.A. et al.: Helv. Chim. Acta (HCACAV) **40**, 1531 (1957).
Smrt, J. et al.: Experientia (EXPEAM) **13**, 291 (1957).

alternative syntheses:
US 2 772 280 (Merck & Co.; 1956; appl. 1954).
US 2 840 565 (Merck & Co.; 1958; appl. 1954).

Formulation(s):   cps. 250 mg; tabl. 250 mg

Trade Name(s):

D:	D-Cycloserin "Roche" (Roche); wfm	GB:	Cycloserine Roche (Roche); wfm		Orientmycin (Kayaku-Kakenyaku)

D: D-Cycloserin "Roche" (Roche); wfm

F: D-Cyclosérine Roche (Roche); wfm

GB: Cycloserine Roche (Roche); wfm

I: Ciclozer (Formulario Naz.)

J: Cyclomycin (Shionogi)

Orientmycin (Kayaku-Kakenyaku)

Seromycin (Lilly-Schionogi)

USA: Seromycin (Dura)

Cyclothiazide

ATC: C03AA09
Use: diuretic

RN: 2259-96-3   MF: $C_{14}H_{16}ClN_3O_4S_2$ MW: 389.88   EINECS: 218-859-7

LD$_{50}$:   >5 g/kg (M, p.o.);
       >5 g/kg (R, p.o.)

CN: 3-bicyclo[2.2.1]hept-5-en-2-yl-6-chloro-3,4-dihydro-2*H*-1,2,4-benzothiadiazine-7-sulfonamide 1,1-dioxide

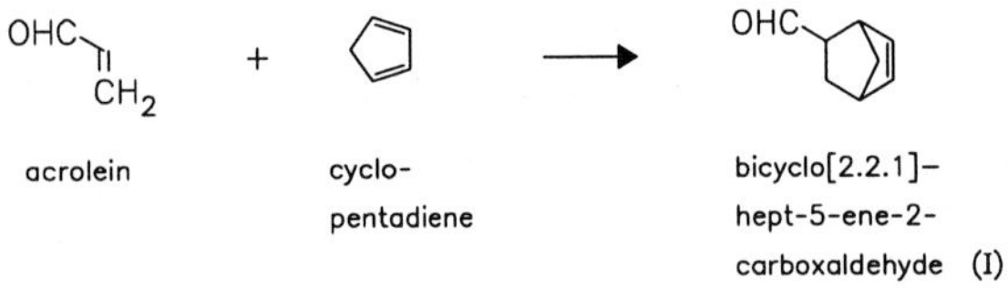

acrolein   cyclo-pentadiene   bicyclo[2.2.1]-hept-5-ene-2-carboxaldehyde (I)

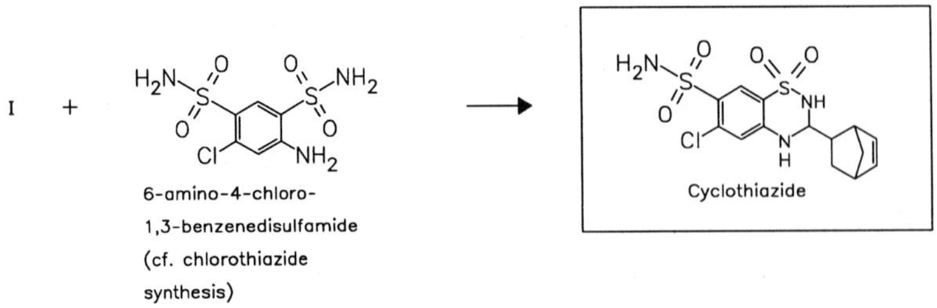

6-amino-4-chloro-1,3-benzenedisulfamide (cf. chlorothiazide synthesis)   Cyclothiazide

Reference(s):
US 3 275 625 (Boehringer Ing.; 27.9.1966; prior. 23.1.1961).
DE 1 125 938 (Thomae; appl. 12.2.1960).
GB 915 236 (Eli Lilly; appl. 25.7.1961; USA-prior. 31.10.1960).

Formulation(s):   tabl. 2.5 mg, 3 mg

Trade Name(s):

D:	Dimapres (Dieckmann)-comb.; wfm	F:	Cyclotériam (Roussel Diamant)-comb. with triamterene	J:	Valmiran (Boehringer-Tanabe)
				USA:	Anhydron (Lilly)

Cyclovalone

Use: digestant, choleretic

RN: 579-23-7   MF: $C_{22}H_{22}O_5$ MW: 366.41   EINECS: 209-438-9
LD_{50}:   56 mg/kg (M, i.v.)
CN: 2,6-bis[(4-hydroxy-3-methoxyphenyl)methylene]cyclohexanone

vanillin   cyclo-hexanone   Cyclovalone

Reference(s):
AT 180 258 (A. v. Waldheim Chem. Pharm. Fabrik; appl. 1953).
Rumpel, W.: Arch. Pharm. Ber. Dtsch. Pharm. Ges. (APBDAJ) **287**, 350 (1954).

Formulation(s):   gran. 0.66/100 g

Trade Name(s):

D:	Beveno (Fischer); wfm	F:	Vanilone (Iénapharm)	GB:	Vanisorbyl (Nicholas); wfm

Cycrimine

ATC: N04
Use: antiparkinsonian

RN: 77-39-4   MF: $C_{19}H_{29}NO$ MW: 287.45   EINECS: 201-024-6
CN: α-cyclopentyl-α-phenyl-1-piperidinepropanol

hydrochloride
RN: 126-02-3   MF: $C_{19}H_{29}NO \cdot HCl$ MW: 323.91   EINECS: 204-764-8
LD_{50}:   50 mg/kg (M, i.v.); 349 mg/kg (M, p.o.);
        628 mg/kg (R, p.o.)

aceto-phenone   paraform-aldehyde   piperidine   3-piperidino-propiophenone   cyclopentyl-magnesium bromide   Cycrimine

Reference(s):
US 2 680 115 (Winthrop-Stearns, 1954; prior. 1949).

Formulation(s):   tabl. 0.25 mg, 0.5 mg

Trade Name(s):
I: Pagitane (Lilly); wfm   USA: Pagitane (Lilly); wfm

Cynarine

ATC: A06AB20
Use: choleretic

RN: 1182-34-9   MF: $C_{25}H_{24}O_{12}$ MW: 516.46   EINECS: 214-655-7
LD_{50}: 1900 mg/kg (M, i.p.)
CN: (1α,3α,4α,5β)-1,4-bis[[3-(3,4-dihydroxyphenyl)-1-oxo-2-propenyl]oxy]-3,5-
        dihydroxycyclohexanecarboxylic acid

(a)   from Cynara scolymus (artichokes) leaves by extraction

(b)

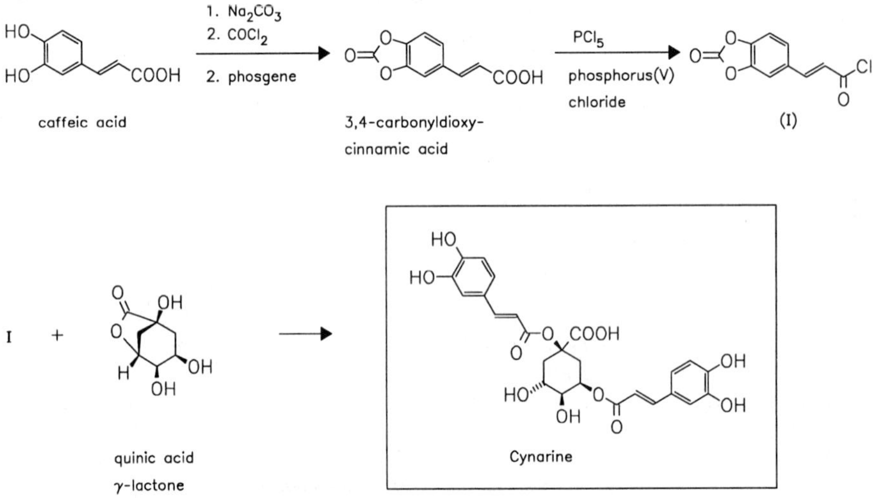

caffeic acid

3,4-carbonyldioxy-
cinnamic acid

(I)

quinic acid
γ-lactone

Cynarine

Reference(s):
a   US 2 863 909 (Farmitalia; 9.12.1958; I-prior. 28.5.1954).
b   US 3 100 224 (Farmitalia; 6.8.1963; I-prior. 28.5.1954).

synthesis of quinic acid lactone:
Wolinsky, J. et al.: J. Org. Chem. (JOCEAH) **29**, 3596 (1964).

Formulation(s):   tabl.

Trade Name(s):
D: Benestan (Karlspharma)-   Listrocol (Carlo Erba);   Methiocholin (Pfleger)-
        comb.; wfm   wfm   comb.; wfm

                                                                          J: Plemocil (Sumitomo)

Cyproheptadine

ATC: R06AX02
Use: antiallergic, appetite stimulant

RN: 129-03-3   MF: $C_{21}H_{21}N$ MW: 287.41   EINECS: 204-928-9
LD_{50}:   15 mg/kg (M, i.v.); 106 mg/kg (M, p.o.);
        295 mg/kg (R, p.o.)
CN: 4-(5*H*-dibenzo[*a,d*]cyclohepten-5-ylidene)-1-methylpiperidine

hydrochloride
RN: 969-33-5   MF: $C_{21}H_{21}N \cdot HCl$ MW: 323.87   EINECS: 213-535-1
LD_{50}:   23 mg/kg (M, i.v.); 69 mg/kg (M, p.o.);
        295 mg/kg (R, p.o.)

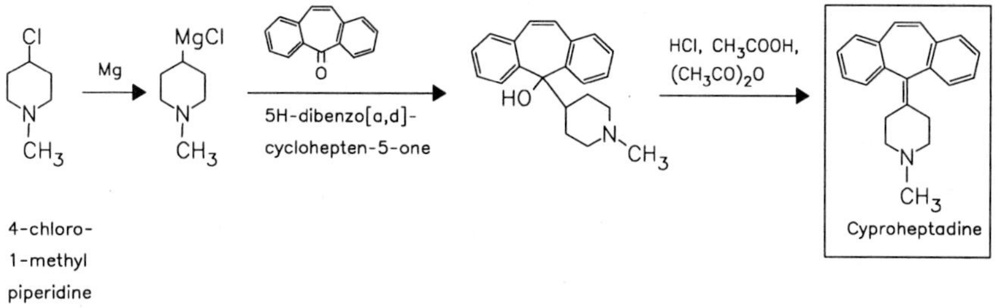

4-chloro-
1-methyl
piperidine

5H-dibenzo[a,d]-
cyclohepten-5-one

Cyproheptadine

Reference(s):
US 3 014 911 (Merck & Co.; 26.12.1961; prior. 13.7.1959).
Engelhardt, E.L. et al.: J. Med. Chem. (JMCMAR) **8**, 829 (1965).

Formulation(s):   syrup 2 mg/5 ml; tabl. 4 mg (as hydrochloride)

Trade Name(s):
D: Peritol (medphano)
F: Périactine (Merck Sharp &
        Dohme; as hydrochloride)
GB: Periactin (Merck Sharp &
        Dohme)
I: Periactin (Neopharmed)
J: Ifrasarl (Showa Shinyaku)
        Periactin (Merck-Banyu)
USA: Periactin (Merck; as
        hydrochloride)

Cyproterone acetate

ATC: L02BB; G03HB
Use: antiandrogen

RN: 427-51-0   MF: $C_{22}H_{27}ClO_3$ MW: 374.91   EINECS: 207-048-3
CN: (1β,2β)-17-(acetyloxy)-6-chloro-1,2-dihydro-3'*H*-cyclopropa[1,2]pregna-1,4,6-triene-3,20-dione

cyproterone
RN: 2098-66-0   MF: $C_{22}H_{27}ClO_3$ MW: 374.91

17-hydroxy-
progesterone

chloranil

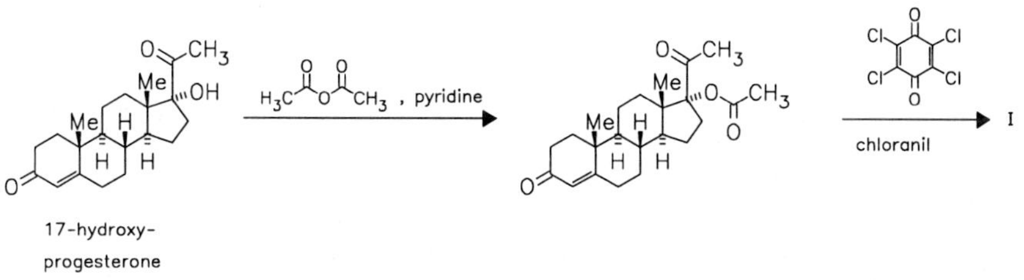

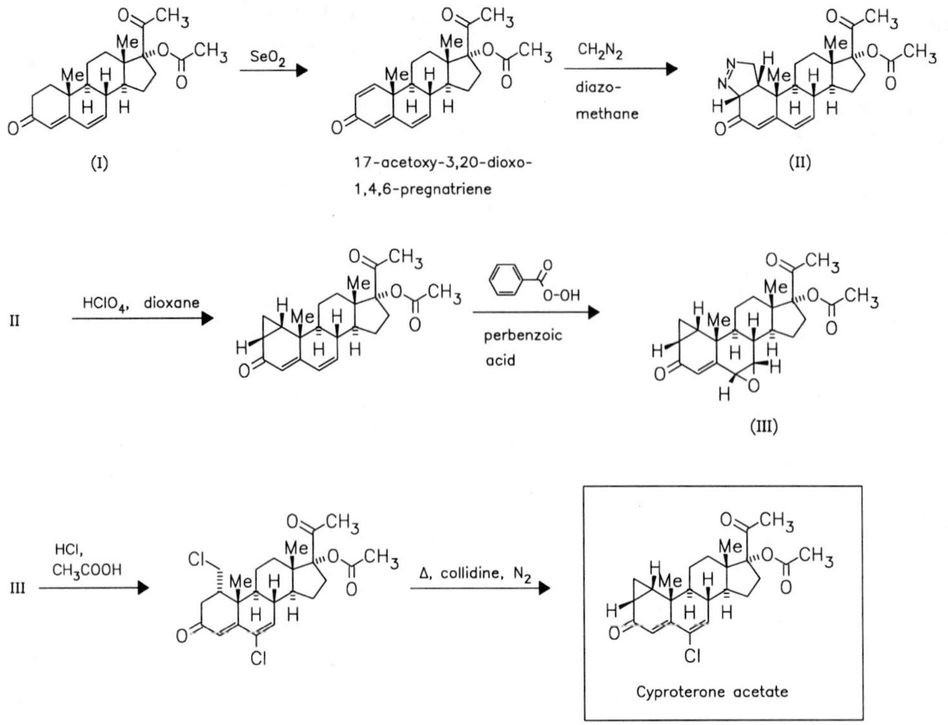

(I)

17-acetoxy-3,20-dioxo-
1,4,6-pregnatriene

(II)

(III)

Cyproterone acetate

Reference(s):
DE 1 072 991 (Schering AG; appl. 25.10.1958).
DE 1 096 353 (Schering AG; appl. 11.7.1959).
DE 1 158 966 (Schering AG; appl. 29.4.1961).
DE 1 189 991 (Schering AG; appl. 31.5.1963).
US 3 234 093 (Schering AG; 8.2.1966; appl. 24.4.1962; D-prior. 29.4.1961).

synthesis of 17-hydroxyprogesterone:
DE 1 119 266 (Schering AG; appl. 18.12.1957).
US 2 962 510 (Schering AG; 29.11.1960; appl. 9.12.1958; D-prior. 18.12.1957).

alternative synthesis:
DE 1 183 500 (Schering AG; appl. 12.10.1962).
DOS 3 331 824 (Schering AG; appl. 1.9.1983).
DOS 4 006 165 (Schering AG; appl. 25.2.1990).

review:
Wiechert, R.: Z. Naturforsch., B: Anorg. Chem., Org. Chem., Biochem., Biophys., Biol. (ZENBAX) 196, 944
(1964).

Formulation(s): amp. 300 mg/3 ml; tabl. 10 mg, 50 mg

Trade Name(s):
D:	Androcur (Schering)		Climéne (Schering)		Dianette (Schering)-comb.
	Climen (Schering)-comb.		Diane (Schering)-comb.	I:	Androcur (Schering)
	Diane (Schering)-comb.	GB:	Androcur (Schering)		Diane (Schering)-comb.
F:	Androcur (Schering)		Cyprostat (Schering)	J:	Androcur (Schering)

Cytarabine

(ara C; Cytosine arabinoside)

ATC: L01BC01
Use: antineoplastic, antiviral

RN: 147-94-4 MF: $C_9H_{13}N_3O_5$ MW: 243.22 EINECS: 205-705-9
LD$_{50}$: >7 g/kg (M, i.v.); 3150 mg/kg (M, p.o.);
>5 g/kg (R, i.v.); >5 g/kg (R, p.o.)
CN: 4-amino-1-β-D-arabinofuranosyl-2(1H)-pyrimidinone

monohydrochloride
RN: 69-74-9 MF: $C_9H_{13}N_3O_5 \cdot HCl$ MW: 279.68 EINECS: 200-713-9
LD$_{50}$: 826 mg/kg (M, p.o.);
>3.2 g/kg (R, p.o.);
172 mg/kg (dog, i.v.)

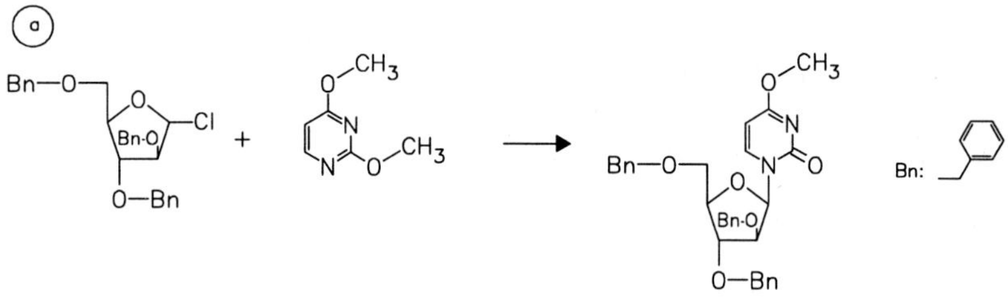

2,3,5-tri-O-benzyl-
D-arabinofuranosyl
chloride

2,4-dimethoxy-
pyrimidine

(I)

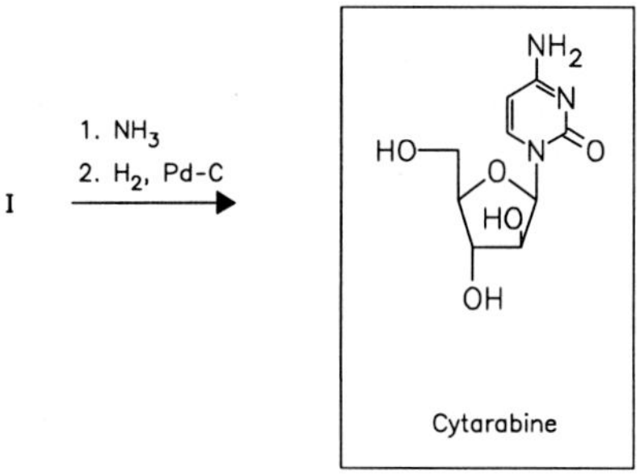

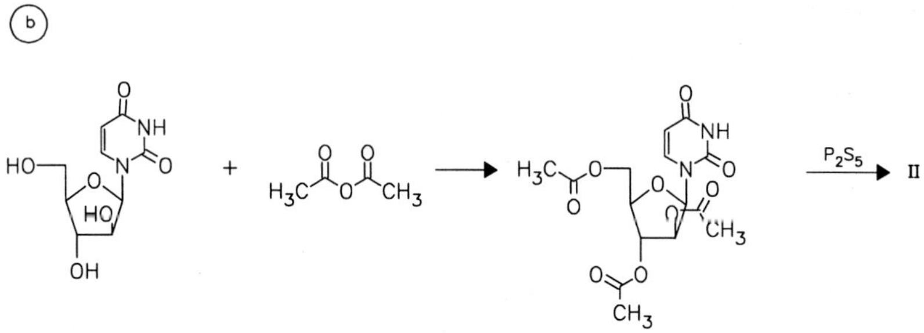

Bn:

1. NH$_3$
2. H$_2$, Pd-C

I →

Cytarabine

1-β-D-arabino-
furanosyluracil

acetic
anhydride

P$_2$S$_5$ → II

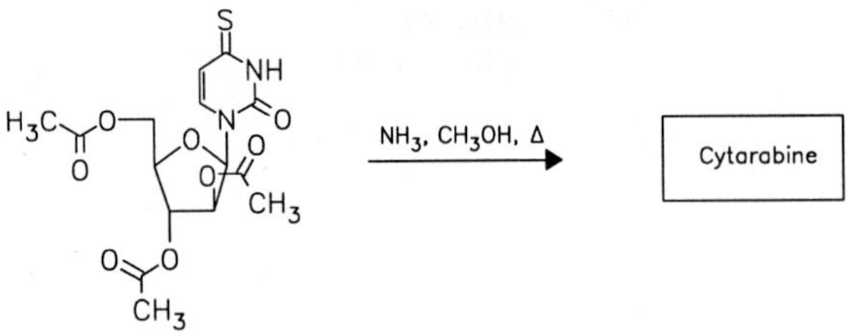

(II)

Reference(s):
NL-appl. 6 511 420 (Merck & Co.; appl. 1.9.1965; USA-prior. 2.9.1964).
US 3 116 282 (Upjohn; 1963, prior. 1959).

alternative synthesis:
Roberts, W.K.; Dekker, C.A.: J. Org. Chem. (JOCEAH) **32**, 816 (1967).
Fromageot, H.P.M.; Reese, C.B.: Tetrahedron Lett. (TELEAY) **1966**, 3499.
Claesen, C.A.A. et al.: Tetrahedron Lett. (TELEAY) **26**, 3859 (1985).

Formulation(s):   amp. 40 mg/2 ml, 100 mg/5 ml, 1 g/20 ml, 1 g/10 ml

Trade Name(s):

D:	Alexan (Mack)		Cytarbel (Rhône-Poulenc		Erpalfa (Intes)
	ARA-cell (cell pharm)		Roger Bellon)	J:	Cyclocide (Nippon
	Udicil (Pharmacia &	GB:	Cytosar (Pharmacia &		Shinyaku)
	Upjohn)		Upjohn)		Cytosar (Upjohn)
F:	Aracytine (Pharmacia &	I:	Alexan (Byk Gulden)	USA:	Cytosar-U (Pharmacia &
	Upjohn)		Aracytin (Upjohn)		Upjohn)

Dacarbazine
(DTIC)

ATC: L01XX13
Use: antineoplastic

RN: 4342-03-4   MF: $C_6H_{10}N_6O$ MW: 182.19   EINECS: 224-396-1
LD_{50}: 466 mg/kg (M, i.v.); 2032 mg/kg (M, p.o.);
      411 mg/kg (R, i.v.); 2147 mg/kg (R, p.o.)
CN: 5-(3,3-dimethyl-1-triazenyl)-1H-imidazole-4-carboxamide

5-amino-
imidazole-4-
carboxamide

4-carbamoyl-
5-diazonio-
N^1-imidazolide

Dacarbazine

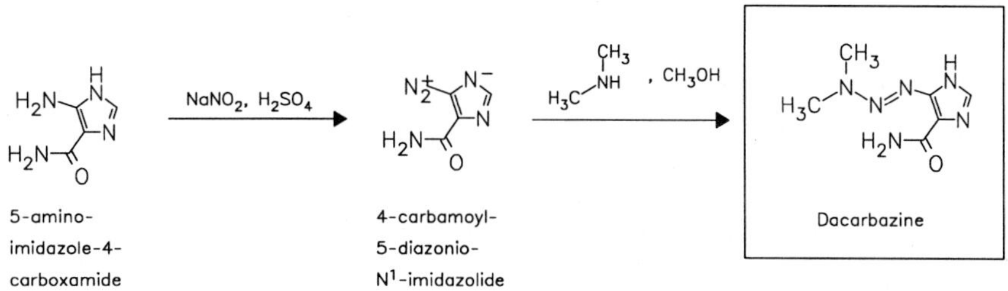

Reference(s):
Shealy, J.F. et al.: J. Org. Chem. (JOCEAH) **27**, 2150 (1962).

Formulation(s):   lyo. 100 mg, 200 mg

Trade Name(s):

D:	Detimedac (medac)	GB:	DTIC-DOME (Bayer)	USA:	DTIC-DOME (Bayer)
	D.T.I.C. 100/200 (Rhône-Poulenc)	I:	Deticene (Rhône-Poulenc Rorer)		
F:	Deticene (Rhône-Poulenc Rorer Bellon)	J:	Dacarbazine (Kyowa Hakko)		

Dactinomycin
(Actinomycin D; Meractinomycin)

ATC: L01DA01
Use: antibiotic, antineoplastic

RN: 50-76-0   MF: $C_{62}H_{86}N_{12}O_{16}$ MW: 1255.44   EINECS: 200-063-6
LD_{50}: 1025 µg/kg (M, i.v.); 13 mg/kg (M, p.o.);
      460 µg/kg (R, i.v.); 7200 µg/kg (R, p.o.)
CN: stereoisomer of N,N'-[(2-amino-4,6-dimethyl-3-oxo-3H-phenoxazine-1,9-diyl)bis[carbonylimino[2-(1-hydroxyethyl)-1-oxo-2,1-ethanediyl]imino[2-(1-methylethyl)-1-oxo-2,1-ethanediyl]-1,2-pyrrolidinediylcarbonyl(methylimino)(1-oxo-2,1-ethanediyl)]]bis[N-methyl-L-valine] di-ξ-lactone

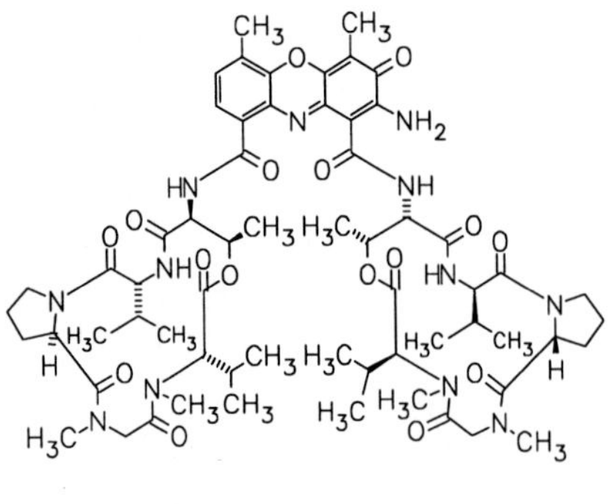

Dactinomycin

From cultures of *Actinomyces antibioticus* and chromatographic purification on Al$_2$O$_3$.

Reference(s):
US 2 378 876 (Merck & Co.; 1945; appl. 1941).

Formulation(s):   lyo. 0.5 mg

Trade Name(s):

D: Lyovac (Merck Sharp & Dohme)

I: Cosmegen (Merck Sharp & Dohme)

GB: Cosmegen Lyovac (Merck Sharp & Dohme)

USA: Cosmegen (Merck Sharp & Dohme)

Danazol

ATC: G03XA01
Use: antigonadotropin, anterior pituitary suppressant

RN: 17230-88-5   MF: C$_{22}$H$_{27}$NO$_2$ MW: 337.46   EINECS: 241-270-1
LD$_{50}$:   4830 mg/kg (M, p.o.);
        >17 g/kg (R, p.o.);
        >5 g/kg (dog, p.o.)
CN: (17α)-pregna-2,4-dien-20-yno[2,3-*d*]isoxazol-17-ol

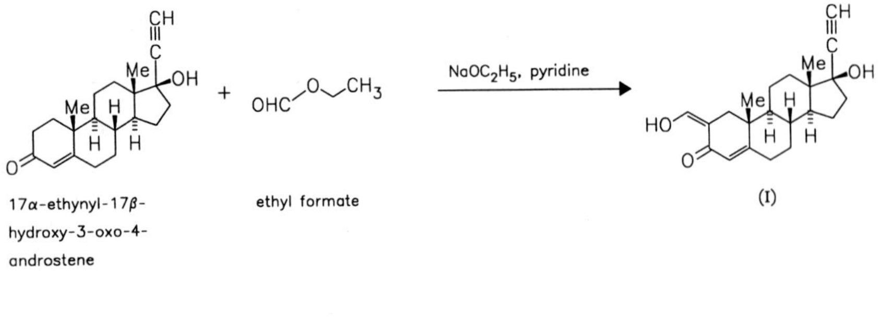

17α-ethynyl-17β-
hydroxy-3-oxo-4-
androstene

ethyl formate

NaOC$_2$H$_5$, pyridine

(I)

I
H2N–OH, Na–OCOCH3, CH3COOH
Danazol

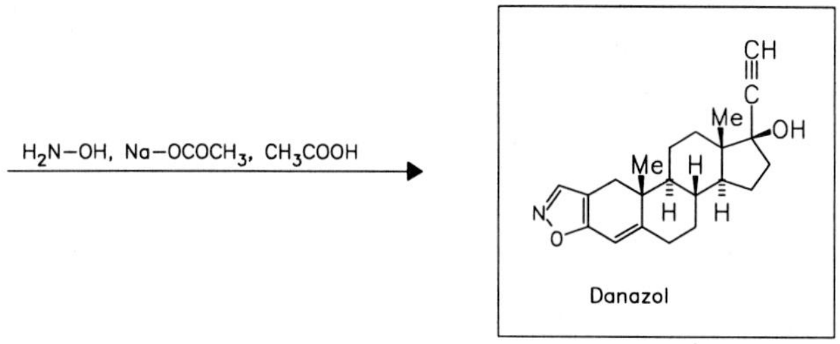

Reference(s):
GB 905 844 (Sterling Drug; valid from 1959; USA-prior. 1958).
US 3 135 743 (Sterling Drug; 2.6.1964; prior. 29.6.1960, 23.7.1958).
Pirkle, W.H. et al.: J. Med. Chem. (JMCMAR) **6**, 1 (1963).
Clinton, R.O. et al.: J. Am. Chem. Soc. (JACSAT) **83**, 1478 (1961).

Formulation(s): cps. 50 mg, 100 mg, 200 mg

Trade Name(s):

D:	Winobanin (Sanofi Winthrop)	GB:	Danol (Sanofi Winthrop)	J:	Bonzol (Tokyo Tanabe)
F:	Danatrol (Sanofi Winthrop)	I:	Danatrol (Maggioni-Winthrop)	USA:	Danocrine (Sanofi)

Dantrolene

ATC: M03CA01
Use: skeletal muscle relaxant, antispasmodic

RN: 7261-97-4 MF: $C_{14}H_{10}N_4O_5$ MW: 314.26 EINECS: 230-684-8
LD_{50}: >7 g/kg (M, i.v.)
CN: 1-[[[5-(4-nitrophenyl)-2-furanyl]methylene]amino]-2,4-imidazolidinedione

sodium salt hydrate (2:7)
RN: 24868-20-0 MF: $C_{14}H_9N_4NaO_5 \cdot 7/2H_2O$ MW: 798.58

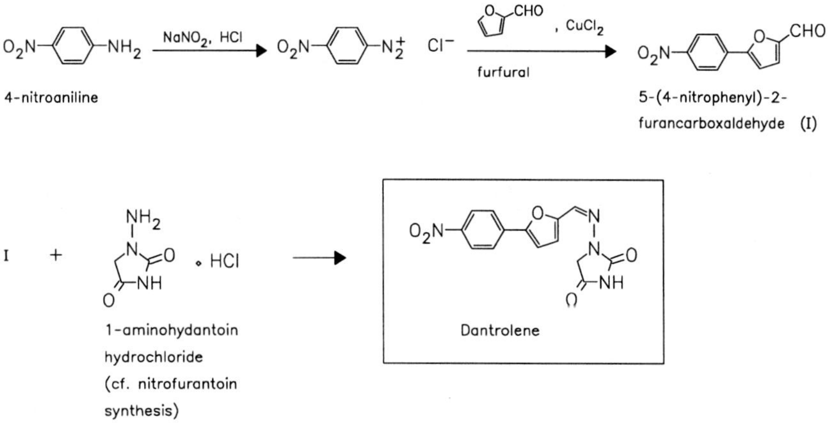

NaNO2, HCl
furfural
, CuCl2
4-nitroaniline
5-(4-nitrophenyl)-2-furancarboxaldehyde (I)
I
• HCl
1-aminohydantoin hydrochloride (cf. nitrofurantoin synthesis)
Dantrolene

Reference(s):
US 3 415 821 (Norwich Pharmacal Co.; 10.12.1968; appl. 7.9.1965).
Snyder, H.R. Jr. et al.: J. Med. Chem. (JMCMAR) **10**, 807 (1967).

use as antiarrhythmic:
EP 105 859 (Norwich Eaton; appl. 30.9.1983; USA-prior. 1.10.1982).

Formulation(s): amp. 20 mg (as sodium salt); cps. 25 mg, 50 mg (as sodium salt); susp. 5 mg/ml

Trade Name(s):

D:	Dantamacrin (Röhm Pharma)	F:	Dantrium (Lipha Santé Division Oberval; as sodium salt)	I:	Dantrium (Formenti)
	Dantrolen (Röhm Pharma)			J:	Dantrium (Yamanouchi)
		GB:	Dantrium (Procter & Gamble)	USA:	Dantrium (Procter & Gamble)

Dapiprazole

(AF-2139)

ATC: N05AX; S01EX02
Use: antipsychotic, antiglaucoma

RN: 72822-12-9 MF: $C_{19}H_{27}N_5$ MW: 325.46
LD_{50}: 260 mg/kg (M, i.p.)
CN: 5,6,7,8-tetrahydro-3-[2-[4-(2-methylphenyl)-1-piperazinyl]ethyl]-1,2,4-triazolo[4,3-a]pyridine

monohydrochloride
RN: 72822-13-0 MF: $C_{19}H_{27}N_5 \cdot HCl$ MW: 361.92

ethyl 3-chloro-
propionate

1-(o-tolyl)-
piperazine

ethyl 3-[4-(o-tolyl)-1-
piperazinyl]propionate (I)

3-[4-(o-tolyl)-1-
piperazinyl)]propionic
acid hydrazide

6-methoxy-2,3,4,5-
tetrahydropyridine

Dapiprazole

Reference(s):
DE 2 915 318 (Angelini; appl. 14.4.1979; I-prior. 18.4.1978).
US 4 307 095 (Angelini; 22.12.1981; prior. 29.3.1979, 29.8.1980; I-prior. 18.4.1978).
US 4 307 096 (Angelini; 22.12.1981; prior. 29.3.1979, 29.8.1980; I-prior. 18.4.1978).
US 4 325 952 (Angelini; 20.4.1982; prior. 29.3.1979, 29.8.1980; I-prior. 18.4.1978).
BE 877 161 (Angelini; appl. 21.6.1979).

ophthalmic composition:
EP 288 659 (Angelini; appl. 25.1.1988).
US 4 879 294 (Angelini; 7.11.1989; appl. 28.1.1988).

Formulation(s): eye drops 50 mg/10 ml (5 %) (as hydrochloride)

Dapsone

(DADPS; DDS; Diaphenylsulfone)

ATC: J04BA02
Use: chemotherapeutic (leprosy)

RN: 80-08-0   MF: $C_{12}H_{12}N_2O_2S$ MW: 248.31   EINECS: 201-248-4
LD_{50}:   225 mg/kg (M, i.v.); 250 mg/kg (M, p.o.);
          1 g/kg (R, p.o.)
CN: 4,4'-sulfonylbis[benzenamine]

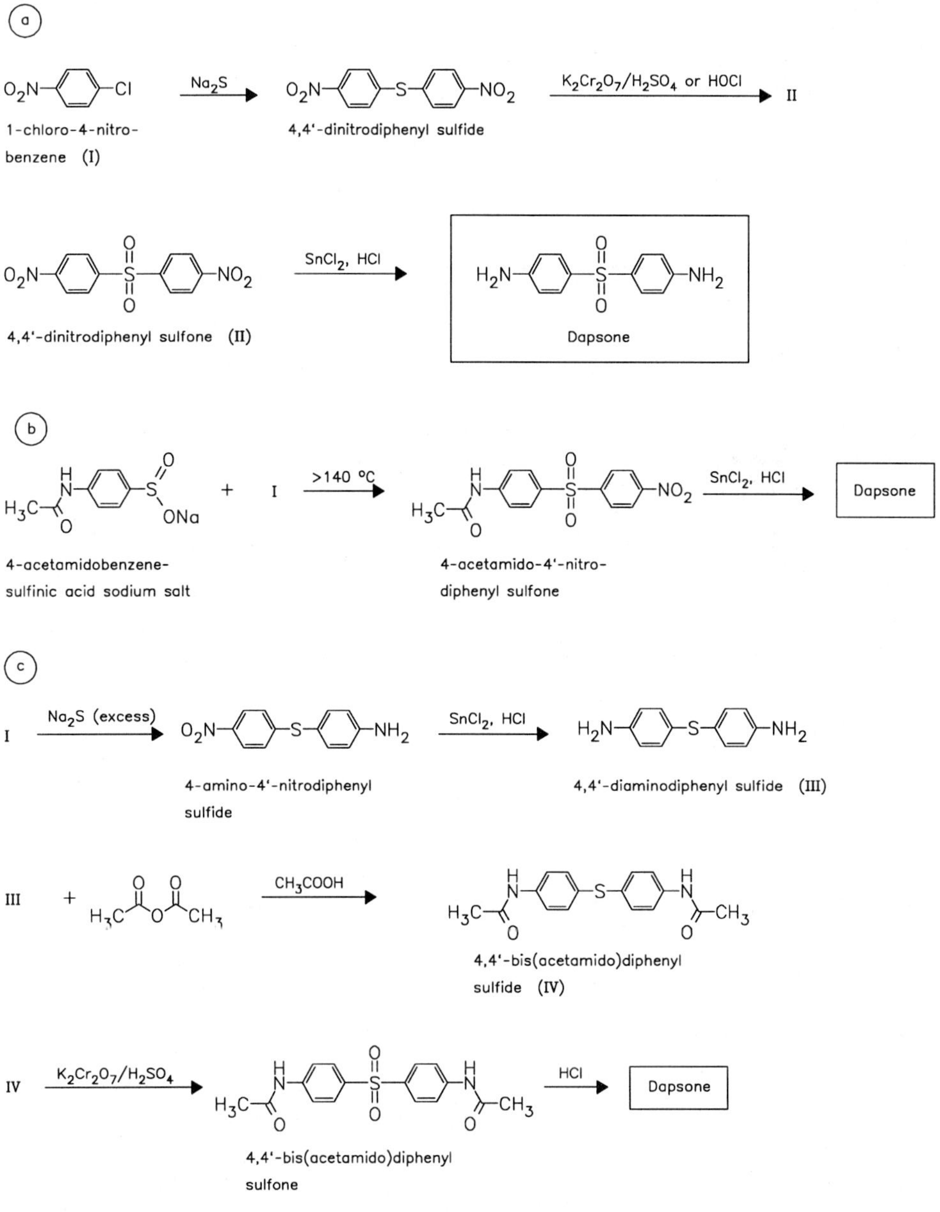

1-chloro-4-nitro-benzene (I)

4,4'-dinitrodiphenyl sulfide

4,4'-dinitrodiphenyl sulfone (II)

Dapsone

4-acetamidobenzene-sulfinic acid sodium salt

4-acetamido-4'-nitro-diphenyl sulfone

4-amino-4'-nitrodiphenyl sulfide

4,4'-diaminodiphenyl sulfide (III)

4,4'-bis(acetamido)diphenyl sulfide (IV)

4,4'-bis(acetamido)diphenyl sulfone

Reference(s):
a   Fromm, E.; Wittmann, J.: Ber. Dtsch. Chem. Ges. (BDCGAS) **41**, 2264 (1908).
    US 2 385 889 (Merck & Co.; 1945; appl. 1945).
b   Ferry, C.W. et al.: Org. Synth. (ORSYAT) **22**, 32 (1942).
    GB 510 127 (Schering AG; appl. 1938; D-prior. 1937).
    similar process:
    US 2 227 400 (American Cyanamide Co.; 1940; appl. 1939).
c   Raiziss, G.W. et al.: J. Am. Chem. Soc. (JACSAT) **61**, 2763 (1939).

S-oxidation with hydrogen peroxide:
Arendonk, A.M. Van; Kleiderer, E.C.: J. Am. Chem. Soc. (JACSAT) **62**, 3521 (1940).

preparation via 4,4'-dichlorodiphenyl sulfone:
GB 506 227 (I.G. Farben; appl. 1937).
FR 829 926 (I.G. Farben; appl. 1937; D-prior. 1936, 1937).
FR 844 220 (Lab. Franç. de Chimiothérapie et M. A. Girard; appl. 1938; D-prior. 1937).

Formulation(s):   tabl. 50 mg

Trade Name(s):

D:	Dapson-Fatol (Saarstickstoff-Fatol)	GB:	Maloprim (Glaxo Wellcome)-comb.	J: Protogen (Yoshitomi)
F:	Disulone (Rhône-Poulenc Rorer Specia)	I:	Avlosulfon (Ayerst-Usa); wfm	USA: generic

---

Daunorubicin
(Daunomycin)

ATC: L01DB02
Use: antineoplastic, anthracycline
       antibiotic

RN: 20830-81-3   MF: $C_{27}H_{29}NO_{10}$ MW: 527.53   EINECS: 244-069-7
LD_{50}:   5 mg/kg (M, i.p.); 29 mg/kg (M, i.v.);
       8 mg/kg (R, i.p.); 13 mg/kg (R, i.v.)
CN: (8S-cis)-8-acetyl-10-[(3-amino-2,3,6-trideoxy-α-L-*lyxo*-hexopyranosyl)oxy]-7,8,9,10-tetrahydro-6,8,11-
       trihydroxy-1-methoxy-5,12-naphthacenedione

hydrochloride
RN: 23541-50-6   MF: $C_{27}H_{29}NO_{10}$ · HCl   MW: 563.99   EINECS: 245-723-4
LD_{50}:   50 mg/kg (M, i.v.); 205 mg/kg (M, p.o.);
       14.3 mg/kg (R, i.v.); 290 mg/kg (R, p.o.)

Daunorubicin

Fermentation of *Streptomyces peucetius*.

Reference(s):
BE 639 897 (Farmitalia; appl. 14.11.1963).
Marco, A. Di et al.: Nature (London) (NATUAS) **201**, 706 (1964).

structure and stereochemistry:
Arcamone, F. et al.: J. Am. Chem. Soc. (JACSAT) **86**, 5334 (1964).
Iwamoto et al.: Tetrahedron Lett. (TELEAY) **1968**, 3891.
Arcamone, F. et al.: Gazz. Chim. Ital. (GCITA9) **100**, 949 (1970).

total synthesis:
Acton et al.: J. Med. Chem. (JMCMAR) **17**, 659 (1974).

alternative syntheses:
DOS 2 519 157 (Farmitalia; appl. 30.4.1975; GB-prior. 2.5.1974).
FR 2 183 710 (Farmitalia; appl. 6.5.1973; I-prior. 6.5.1972).
EP-appl. 100 075 (Sanraku-Ocean; appl. 22.7.1983; J-prior. 24.7.1982).

purification:
BE 898 506 (Farmitalia; appl. 20.12.1983; I-prior. 23.12.1982).

Formulation(s):   lyo. 20 mg

Trade Name(s):

D:	Daunoblastin (Carlo Erba)	GB:	Cerubidin (Rhône-Poulenc Rorer)	USA:	Cerubidine (Bedford; as hydrochloride)
F:	Cérubidine (Rhône-Poulenc Roger Bellon; as hydrochloride)		Daunoxome (NeXstar)		DaunoXome (NeXstar; as citrate)
		I:	Daunoblastina (Farmitalia)		
		J:	Daunomycin (Meiji Seika)		

Deanol acetamidobenzoate

(Deanoli acetabenzoas)

ATC: N06BX04
Use: stimulant

RN: 3635-74-3   MF: $C_9H_9NO_3 \cdot C_4H_{11}NO$ MW: 268.31   EINECS: 222-858-7
LD$_{50}$:   3918 mg/kg (M, p.o.)
CN: 4-(acetylamino)benzoic acid compd. with 2-(dimethylamino)ethanol (1:1)

2-dimethylamino-   4-acetamidobenzoic   Deanol acetamidobenzoate
ethanol   acid

Reference(s):
GB 879 259 (Riker; appl. 1957; USA-prior. 1956).

Formulation(s):   tabl. 100 mg

Trade Name(s):

D:	Deanol Riker (Kettelhack-Riker); wfm	F:	Diforène (Choay); wfm	USA:	Deaner (Riker); wfm
		I:	Pabenol (Gentili)		

Debrisoquin

ATC: C02CC04
Use: antihypertensive

RN: 1131-64-2   MF: $C_{10}H_{13}N_3$ MW: 175.24   EINECS: 214-470-1
CN: 3,4-dihydro-2(1*H*)-isoquinolinecarboximidamide

sulfate (2:1)

RN: 581-88-4   MF: $C_{10}H_{13}N_3 \cdot 1/2H_2SO_4$ MW: 448.55   EINECS: 209-472-4
LD_{50}:   31.7 mg/kg (M, i.v.); 235 mg/kg (M, p.o.);
        610 mg/kg (R, p.o.)

1,2,3,4-tetrahydro-   S-methylisothio-   Debrisoquin
isoquinoline   uronium sulfate

Reference(s):
BE 629 007 (Hoffmann-La Roche; appl. 28.2.1963; USA-prior. 6.3.1962, 18.12.1962).
DE 1 244 788 (Hoffmann-La Roche; appl. 25.2.1963; USA-prior. 6.3.1962, 18.12.1962).
Wenner, W.: J. Med. Chem. (JMCMAR) **8**, 125 (1965).

Formulation(s):   tabl. 10 mg

Trade Name(s):
GB: Declinax (Roche); wfm   USA: Declinax (Roche); wfm
       Declinax (Roche; as   Declinax (Roche; as
       sulfate); wfm   sulfate); wfm

Decamethonium bromide

ATC: M03
Use: muscle relaxant

RN: 541-22-0   MF: $C_{16}H_{38}Br_2N_2$ MW: 418.30   EINECS: 208-772-2
LD_{50}:   630 µg/kg (M, i.v.); 190 mg/kg (M, p.o.)
CN: *N,N,N,N',N',N'*-hexamethyl-1,10-decanediaminium dibromide

trimethyl-   1,10-dibromodecane   Decamethonium bromide
amine

Reference(s):
Blomquist, A.T. et al.: J. Am. Chem. Soc. (JACSAT) **81**, 678 (1959).

Formulation(s):   tabl. 0.25 mg, 0.5 mg

Trade Name(s):
USA: Syncurine (Burroughs
        Wellcome); wfm

Deferiprone

(L1; CGP-37391; CP20)

ATC: V03AC02
Use: metal antagonist

RN: 30652-11-0   MF: $C_7H_9NO_2$ MW: 139.15
LD$_{50}$: 2 g/kg (R, p.o.)
CN: 3-hydroxy-1,2-dimethyl-4(1H)-pyridinone

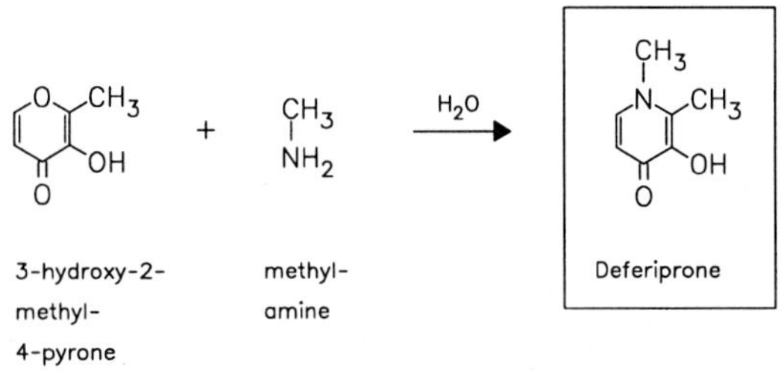

3-hydroxy-2-methyl-4-pyrone   +   methyl-amine   →   Deferiprone

Reference(s):
Dobbin, P.S. et al.: J. Med. Chem. (JMCMAR) **36**(17), 2448 (1993).
Konthogiorghes, G.J.; Sheppard, L.: Inorg. Chim. Acta (ICHAA3) **136**, 11 (1987).

clinical studies:
Vreugdenhil, G.; Swaak, G.; Kontoghiorghes, G.J.; VanEijk, H.G.: Lancet (LANCAO) **2**(8676), 1398-1399 (1989).

Formulation(s):   caps. 250 mg, 500 mg

Trade Name(s):
IND: Deferrum (Cangene)   Kelfer (Cipla)

Deferoxamine

(Desferrioxamine)

ATC: V03AC01
Use: iron complex former (for therapy of iron storage diseases)

RN: 70-51-9   MF: $C_{25}H_{48}N_6O_8$ MW: 560.69   EINECS: 200-738-5
LD$_{50}$: 250 mg/kg (M, i.v.); 1340 mg/kg (M, p.o.);
329 mg/kg (R, i.v.)
CN: N'-[5-[[4-[[5-(acetylhydroxyamino)pentyl]amino]-1,4-dioxobutyl]hydroxyamino]pentyl]-N-(5-aminopentyl)-N-hydroxybutanediamide

monomesylate
RN: 138-14-7   MF: $C_{25}H_{48}N_6O_8 \cdot CH_4O_3S$ MW: 656.80   EINECS: 205-314-3
LD$_{50}$: 273 mg/kg (M, i.v.); 15.2 g/kg (M, p.o.);
330 mg/kg (R, i.v.); 17.3 g/kg (R, p.o.)

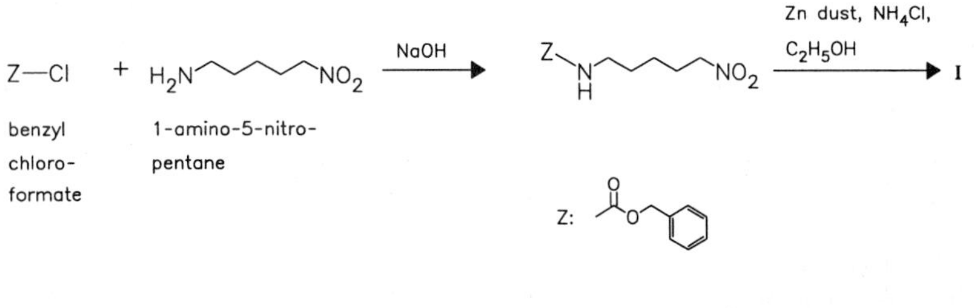

Z—Cl   + H_2N NO_2 →   Z NO_2 → I

benzyl chloro-formate   1-amino-5-nitro-pentane

Z:

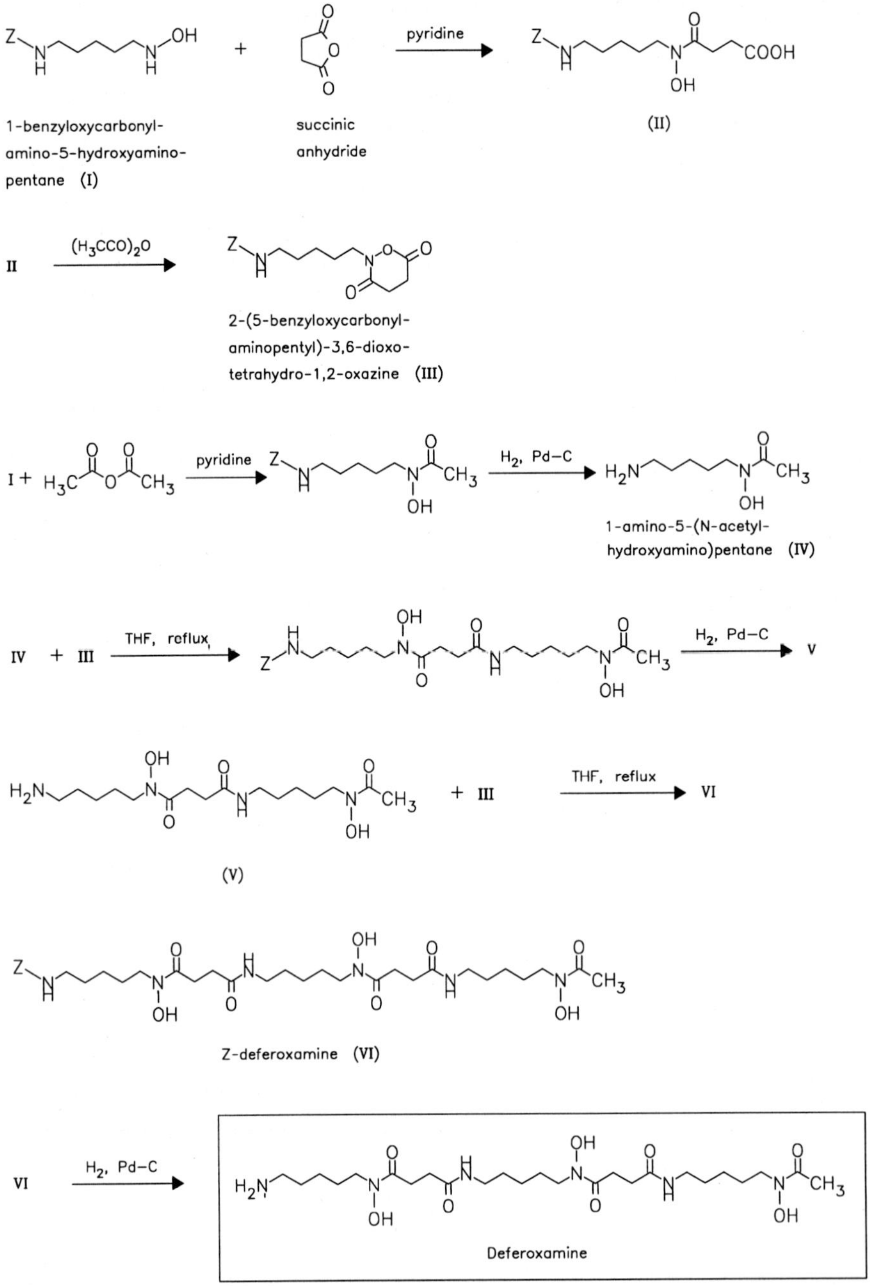

Z
OH
pyridine
COOH
OH
1-benzyloxycarbonyl-amino-5-hydroxyamino-pentane (I)
succinic anhydride
(II)
II
(H3CCO)2O
2-(5-benzyloxycarbonyl-aminopentyl)-3,6-dioxo-tetrahydro-1,2-oxazine (III)
I +
pyridine
H2, Pd-C
CH3
1-amino-5-(N-acetyl-hydroxyamino)pentane (IV)
IV + III
THF, reflux
V
+ III
VI
(V)
Z-deferoxamine (VI)
Deferoxamine

Reference(s):

isolation from metabolites of actinomyceten:

Bickel, H. et al.: Helv. Chim. Acta (HCACAV). **43**, 2118 (1960).

constitutional elucidation:

Bickel, H. et al.: Helv. Chim. Acta (HCACAV) **43**, 2129 (1960).

synthesis:

BE 609 053 (Ciba; appl. 11.10.1961; CH-prior. 11.10.1960, 23.11.1960, 7.4.1961, 26.4.1961, 29.6.1961, 10.8.1961, 11.8.1961).

BE 619 532 (Ciba; appl. 28.6.1962; CH-prior. 29.6.1961).

Prelog, V.; Walser, A.: Helv. Chim. Acta (HCACAV) **45**, 631 (1962).

Formulation(s):   amp. 500 mg; inj. powder 500 mg; lyo. 500 mg, 2 g

Trade Name(s):
D: Desferal (Ciba)   I: Desferal (Ciba-Geigy)   USA: Desferal (Novartis; as
F: Desféral (Ciba-Geigy)   J: Desferal (Novartis-Takeda)   mesylate)
GB: Desferal (Novartis)

Defibrotide

ATC: B01AX01; B06A
Use: antithrombotic, cholinergic channel
          modulator, stimulates fibrinolysis

RN: 83712-60-1   MF: unspecified   MW: unspecified
CN: defibrotide; polydeoxyribonucleotides from bovine lung

Extraction from mammalian organs with aqueous solution of Zn salts.

Reference(s):
DE 2 154 278 (Crinos; appl. 3.11.1971; I-prior. 3.11.1970).
DE 2 154 277 (Crinos; appl. 3.11.1971; I-prior. 3.11.1970).
US 3 829 567 (Crinos; 13.8.1974; I-prior. 3.11.1970).
US 3 899 481 (Crinos; 12.8.1975; I-prior. 3.11.1970).
EP 263 155 (Crinos; 10.4.1987; I-prior. 17.4.1986).

medical use for renal dialysis patients:
EP 317 766 (Crinos; appl. 20.10.1988; I-prior. 23.10.1987).

medical use for treatment of myocardial ischaemia:
EP 152 148 (Crinos; appl. 11.2.1985; I-prior. 16.2.1984).

medical use for treatment of peripheral arterial disease:
EP 137 543 (Crinos; appl. 7.9.1984; I-prior. 12.9.1983).

medical use for treatment of acute renal insufficiency:
EP 137 542 (Crinos; appl. 7.9.1984; I-prior. 12.9.1983).

Formulation(s):   vial 200 mg

Trade Name(s):
I: Dasovas (Carlo Erba)   Noravid (Roussel; 1986)   Prociclide (Crinos; 1986)

Deflazacort
(Azacort; Oxazacort)

ATC: H02AB13
Use: glucocorticoid, anti-inflammatory

RN: 14484-47-0   MF: $C_{25}H_{31}NO_6$ MW: 441.52   EINECS: 238-483-7
LD_{50}: 5200 mg/kg (M, p.o.)
CN: (11β,16β)-21-(acetyloxy)-11-hydroxy-2'-methyl-5'H-pregna-1,4-dieno[17,16-d]oxazole-3,20-dione

17α-azido-3β,16α-di-
acetoxy-5α-pregnane-
11,20-dione

1. H_2, PtO_2
2. HCl

3β-hydroxy-2'-methyl-5α,5'βH-
pregnano[17,16-d]oxazole-11,20-dione (I)

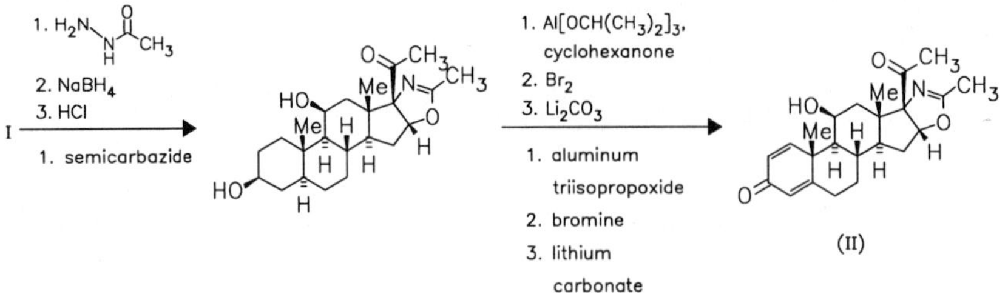

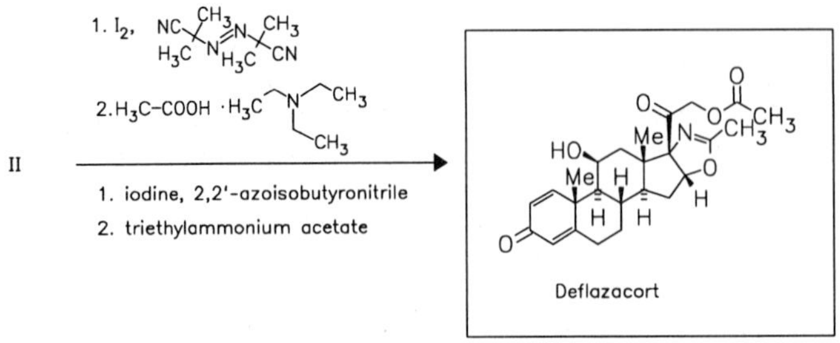

Deflazacort

Reference(s):

GB 1 077 393 (Lepetit; appl. 22.4.1965).
Nathansohn, G. et al.: J. Med. Chem. (JMCMAR) **10**, 799 (1967).

synthesis of 17α-azido-3β,16α-diacetoxy-5a-pregnane-11,20-dione:
Nathansohn, G. et al.: Gazz. Chim. Ital. (GCITA9) **35**, 1338 (1965).

alternative synthesis:

US 3 624 077 (Lepetit; 30.11.1971; GB-prior. 11.1.1966).
Nathansohn, G. et al.: Steroids (STEDAM) **13**, 383 (1969).

Formulation(s):   tabl. 6 mg, 30 mg

Trade Name(s):

D: Calcort (Albert-Roussel,   I: Deflan (Guidotti)
       Hoechst)   Flantadin (Lepetit)

Dehydrocholic acid

(Acide déhydrocholique)

ATC: A05
Use: choleretic, liver protective drug

RN: 81-23-2   MF: $C_{24}H_{34}O_5$ MW: 402.53   EINECS: 201-335-7
LD$_{50}$:   1492 mg/kg (M, i.v.); 3100 mg/kg (M, p.o.);
       750 mg/kg (R, i.v.); 4 g/kg (R, p.o.)
CN: (5β)-3,7,12-trioxocholan-24-oic acid

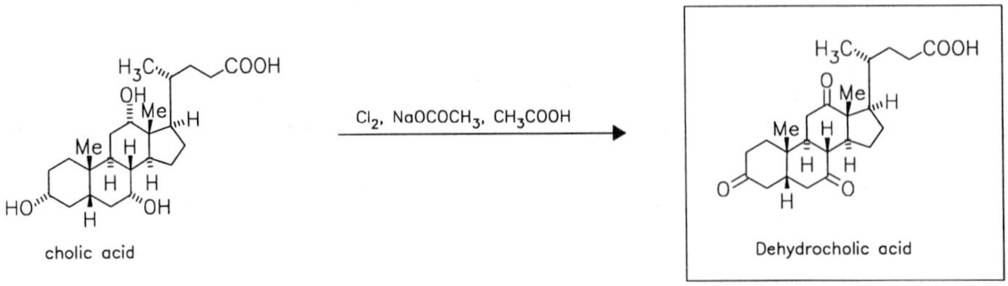

cholic acid

Cl_2, NaOCOCH$_3$, CH$_3$COOH

Dehydrocholic acid

Reference(s):
US 2 966 499 (Merck & Co.; 27.12.1960; prior. 9.4.1958).

Formulation(s):   amp. for i.v. and i.m. inj. 1 g/5 ml H$_2$O (as sodium salt); tabl. 250 mg

Trade Name(s):

D: Decholin (Cassella-Riedel)
    Eupond N (Ferring)
    Felacomp (Verla)-comb.
    numerous combination
    preparations
F: Dycholium (Théraplix);
    wfm
GB: Dehydrocholin (Duncan,
    Flockhart); wfm
I: Certobil (Metapharma)-
    comb.
    Debridat (Sigma-Tau)-
    comb.
    Heparbil (Montefarmaco)-
    comb.

    numerous combination
    preparations
J: Dehychol (Nippon Eiyo)
    Dehydrochol (Kanto;
    Sawai; Hokuriku)
    Hydrochol (Kyorin)
USA: Atrocholin (Glaxo); wfm
    Bilax (Drug Industries)-
    comb.; wfm
    Cholan (Pennwalt)-comb.;
    wfm
    Cholan-DH (Pennwalt);
    wfm
    Cholan-HMB (Pennwalt)-
    comb.; wfm

    Decholin (Miles); wfm
    Decholin Sodium (Dome);
    wfm
    Gastroenterase (Wallace)-
    comb.; wfm
    Hepahydrin (Great
    Southern); wfm
    Ketochol (Searle); wfm
    Neocholan (Dow); wfm
    Neolax (Central)-comb.;
    wfm
    Sodium Dehydrocholate
    (City Chem.); wfm
    Sodium Dehydrocholate
    (Endo); wfm

Delavirdine mesilate
(U-90152S)

ATC: J05AG02
Use: antiviral, HIV-1 reverse transcriptase
       inhibitor

RN: 147221-93-0   MF: C$_{22}$H$_{28}$N$_6$O$_3$S · CH$_4$O$_3$S   MW: 552.68
CN: 1-[3-[(1-Methylethyl)amino]-2-pyridinyl]-4-[[5-[(methylsulfonyl)amino]-1*H*-indol-2-yl]-
      carbonyl]piperazine monomethanesulfonate

base
RN: 136817-59-9   MF: C$_{22}$H$_{28}$N$_6$O$_3$S   MW: 456.57

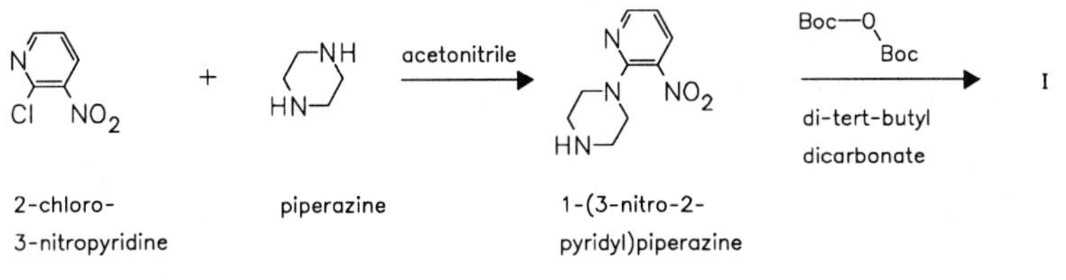

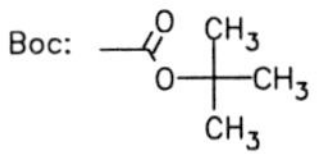

2-chloro-   piperazine   1-(3-nitro-2-
3-nitropyridine   pyridyl)piperazine

Boc:

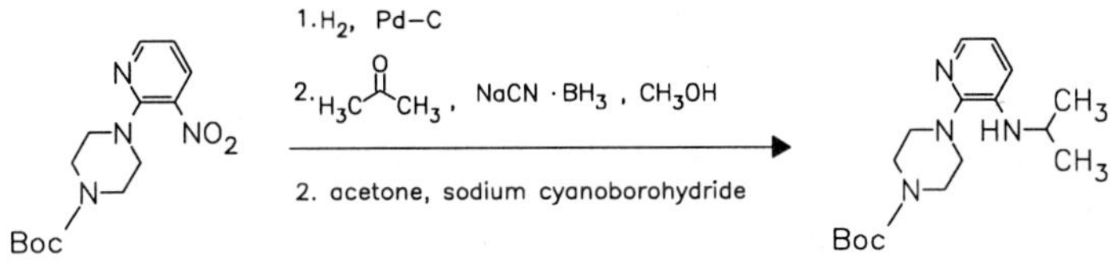

II   +   [structure: 5-nitroindole-2-carboxylic acid]

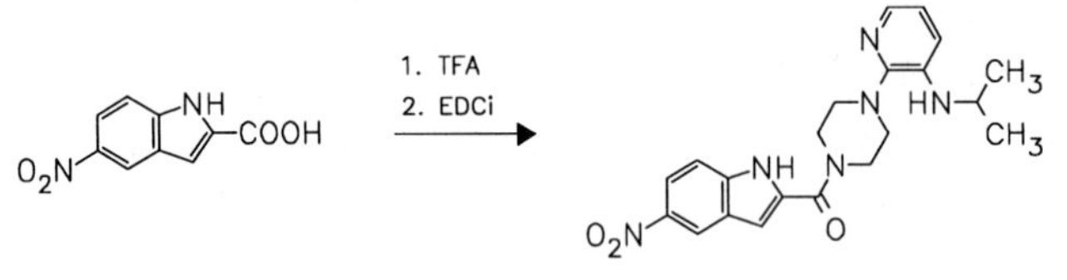

5-nitroindole-
2-carboxylic acid

1. TFA
2. EDCi

(III)

1. H₂, Pd–C

2. H₃C—S—Cl , pyridine, CH₂Cl₂

3. H₃C—SO₃H , acetonitrile

III

2. methanesulfonyl chloride

Delavirdine mesilate

· H₃C—SO₃H

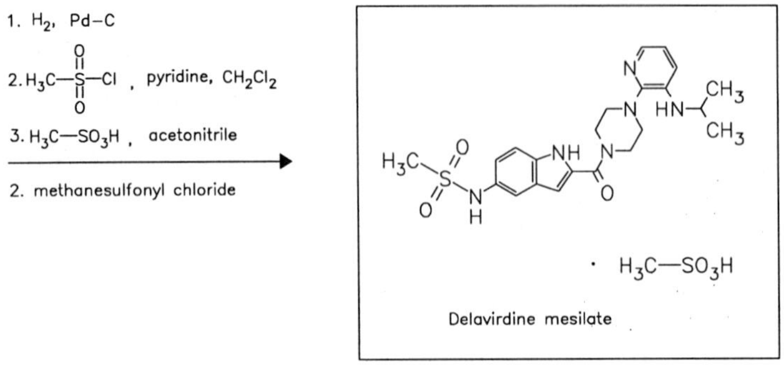

Reference(s):
Romero, D.L. et al.: J. Med. Chem. (JMCMAR) **36**, 1505 (1993).
WO 9 109 849 (Upjohn; USA-prior. 28.12.1989).
Pedersen, O.S.; Pedersen, E.B.: Synthesis (SYNTBF), **2000**, 479.

water clathrates:
WO 9 422 836 (Upjohn + Co.; appl. 15.3.1994; USA-prior. 26.3.1993).

novel crystal form:
WO 9 528 398 (Upjohn + Co.; appl. 1.3.1995; USA-prior. 15.4.1994).

combination with HIV-protease inhibitors:
WO 9 726 880 (Pharmacia + Upjohn; appl. 10.12.1996; USA-prior. 26.1.1996).
WO 9 616 675 (Rega Inst.; appl. 29.11.1995; USA-prior. 30.11.1994).

combination and use with other reverse transcriptase inhibitors:
WO 9 409 781 (Upjohn + Co.; appl. 10.9.1993; USA-prior. 28.10.1992).

Formulation(s):   tabl. 100 mg

Trade Name(s):
USA: Rescriptor (Pharmacia &
       Upjohn; 1997)

Demecarium bromide

ATC: S01EB04
Use: cholinesterase inhibitor

RN: 56-94-0   MF: C₃₂H₅₂Br₂N₄O₄   MW: 716.60   EINECS: 200-301-9
LD₅₀: 6490 µg/kg (M, p.o.)
CN: 3,3'-[1,10-decanediylbis[(methylimino)carbonyloxy]]bis[N,N,N-trimethylbenzenaminium] dibromide

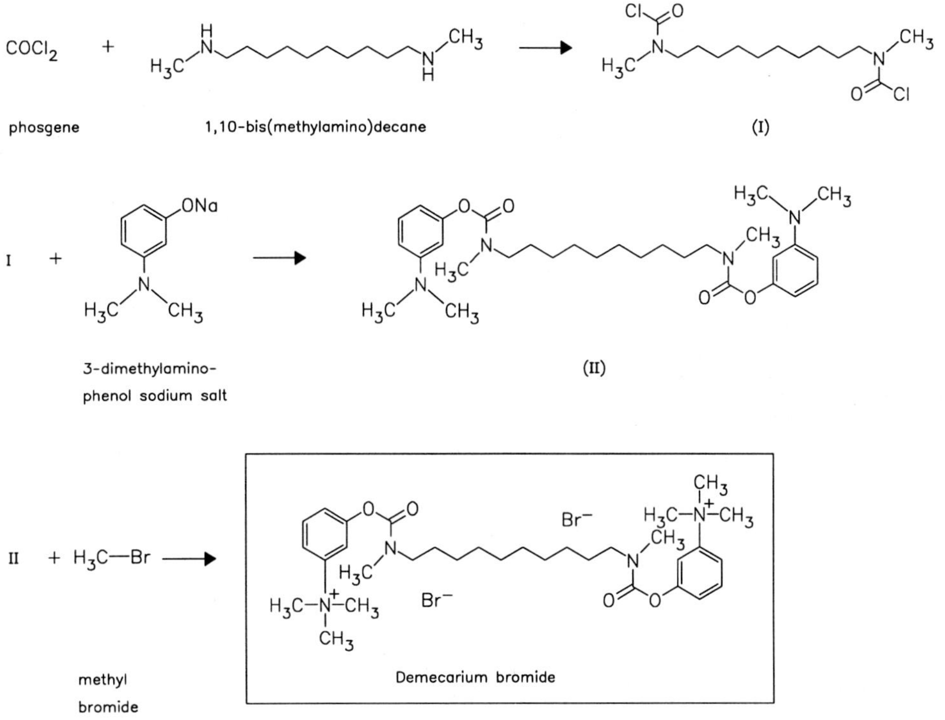

phosgene   1,10-bis(methylamino)decane   (I)

3-dimethylamino-
phenol sodium salt   (II)

II + H₃C—Br ⟶

methyl
bromide   Demecarium bromide

Reference(s):
US 2 789 891 (Österr. Stickstoffwerke; 1957; A-prior. 1954).

Formulation(s):   collyre 0.25 %, 0.5 %, 1 %

Trade Name(s):
D:	Tosmilen (Lentia); wfm	GB:	Tosmilen (Astra); wfm
F:	Tosmilène (Chibret); wfm		Tosmilen (Sinclair); wfm

J: Tosmilen (Chugai)
USA: Humorsol (Merck)

Demeclocycline
(Demethylchlortetracycline)

ATC: D06AA01; J01AA01
Use: antibiotic

RN: 127-33-3   MF: $C_{21}H_{21}ClN_2O_8$ MW: 464.86   EINECS: 204-834-8
LD$_{50}$:   79 mg/kg (M, i.v.);
>6.75 g/kg (R, p.o.)
CN: [4S-(4α,4aα,5aα,6β,12aα)]-7-chloro-4-(dimethylamino)-1,4,4a,5,5a,6,11,12a-octahydro-3,6,10,12,12a-pentahydroxy-1,11-dioxo-2-naphthacene carboxamide

monohydrochloride
RN: 64-73-3   MF: $C_{21}H_{21}ClN_2O_8 \cdot HCl$ MW: 501.32   EINECS: 200-592-2
LD$_{50}$:   275 mg/kg (M, i.v.); 2150 mg/kg (M, p.o.);
94 mg/kg (R, i.v.); 2372 mg/kg (R, p.o.)

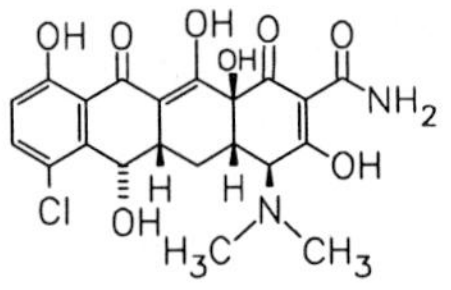

Demeclocycline

From fermentation solutions of a *Streptomyces aureofaciens* mutant.

Reference(s):

US 2 878 289 (American Cyanamid; 17.3.1959; prior. 28.5.1956).
US 3 012 946 (American Cyanamid; 12. 12. 1961; appl. 16.11.1960).
US 3 019 172 (American Cyanamid; 30.1.1962; appl. 14.3.1960).
US 3 050 446 (American Cyanamid; 21.8.1962; appl. 28.7.1960).
US 3 154 476 (Olin Mathieson; 27.10.1964; appl. 29.4.1963).
DE 1 041 213 (American Cyanamid; appl. 24.5.1957; USA-prior. 28.5.1956).
McCormick, J.R.D. et al.: J. Am. Chem. Soc. (JACSAT) **79**, 4561 (1957).

Formulation(s):   tabl. 150 mg, 300 mg (as hydrochloride)

Trade Name(s):

D: Demebronc (Lederle)-
      comb.
      Ledermycin (Novalis
      Arzn.)
      Lederstatin (Novalis
      Arzn.)-comb.
F: Ledermycine (Lederle-
      Novalis); wfm
GB: Deteclo (Wyeth)-comb.
      Ledermycin (Wyeth)
I: Demebronc (Cyanamid)-
      comb.
      Demetraciclina (Libral);
      wfm
      Detracin (Sierochimica);
      wfm

Detravis (Vis); wfm
Dimeral (Panther-Osfa
Chemie); wfm
Diuciclin (Benvegna); wfm
Elkamicina (Biotrading);
wfm
Fidocin (Farmaroma); wfm
Isodemetil (Isola-Ibi); wfm
Latomicina (Farber-Ref);
wfm
Ledermicina (Cyanamid)
Lesten (Serono)-comb.;
wfm
Magis-Ciclina (Tiber); wfm
Mirciclina (Francia Farm.);
wfm

Neo-Cromaciclin (Panther-
Osfa Chemie)
Oldem (Firma)-comb.
Tetradek (SIT)
Tollerclin (Scalari)
Varibiotic (Cyanamid)-
comb.
Veraciclina (AFI)
numerous combination
preparations
J: Demethylchlor Tetracycline
(Kaken)
Ledermycin (Lederle)
USA: Declomycin (Lederle
Labs.; as hydrochloride)

Demegestone

ATC: G03DB05
Use: progestogen

RN: 10116-22-0   MF: $C_{21}H_{28}O_2$ MW: 312.45   EINECS: 233-320-6
CN: 17-methyl-19-norpregna-4,9-diene-3,20-dione

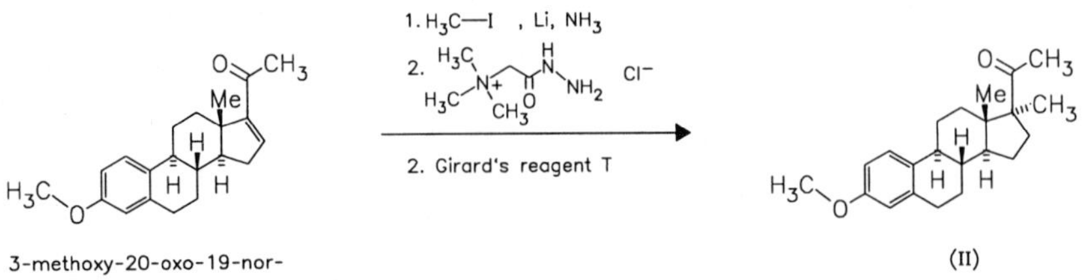

3-methoxy-20-oxo-19-nor-
pregna-1,3,5(10),16-tetraene   (I)
(from estrone 3-methyl ether)

(II)

Demegestone

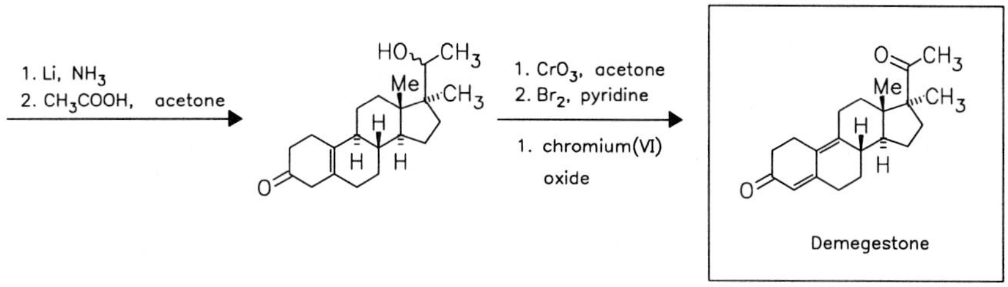

Reference(s):

US 3 453 267 (Roussel-Uclaf; 1.7.1969; F-prior. 31.12.1964, 25.2.1965, 24.3.1965, 14.6.1965, 3.9.1965, 17.9.1965).
US 3 547 959 (Roussel-Uclaf; 15.12.1970; F-prior. 27.12.1965).
Joly, R. et al.: Bull. Soc. Chim. Fr. (BSCFAS) **1973**, 2694.

starting material:
Burn, D.; Petrov, v.: J. Chem. Soc. (JCSOA9) **1962**, 364.

total synthesis:
Velluz, L. et al.: Tetrahedron (TETRAB), **1966**, Suppl. 8, part II, 495.

Formulation(s):   tabl. 500 mg

Trade Name(s):
F: Lutionex (Roussel
       Diamant)

Denopamine
(TA-064)

ATC: C01CA
Use: orally active cardiostimulant, β_1-
         receptor agonist

RN: 71771-90-9   MF: $C_{18}H_{23}NO_4$ MW: 317.39
LD_{50}:   198 mg/kg (M, i.v.); 5672 mg/kg (M, p.o.);
        9369 mg/kg (R, p.o.)
CN: (R)-α-[[[2-(3,4-dimethoxyphenyl)ethyl]amino]methyl]-4-hydroxybenzenemethanol

hydrochloride
RN: 64299-19-0   MF: $C_{18}H_{23}NO_4 \cdot HCl$ MW: 353.85

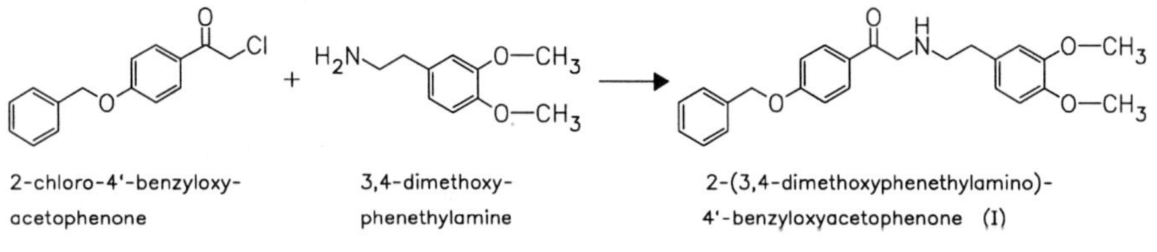

2-chloro-4'-benzyloxy-   3,4-dimethoxy-   2-(3,4-dimethoxyphenethylamino)-
acetophenone   phenethylamine   4'-benzyloxyacetophenone (I)

Denopamine

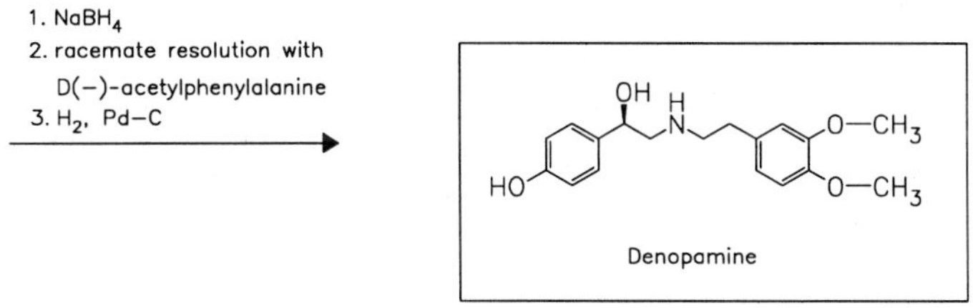

Reference(s):
DOS 2 542 881 (Tanabe; appl. 25.9.1975).
US 4 032 575 (Tanabe; 28.6.1977; appl. 1.10.1975).
US 4 072 759 (Tanabe; 7.2.1978; appl. 10.11.1976; prior. 1.10.1975).
Umino, N. et al.: Chem. Pharm. Bull. (CPBTAL) **27**, 1479 (1979).

enantioselective synthesis starting from optically active 4-hydroxyphenylglycine:
JP 85 009 702 (Tanabe; appl. 14.11.1977).
JP 85 009 703 (Tanabe; appl. 14.11.1977).
Ikezaki, M. et al.: Yakugaku Zasshi (YKKZAJ) **106**, 80 (1986).

Formulation(s):   gran. 5 %; tabl. 5 mg, 10 mg

Trade Name(s):
J: Kalgut (Tanabe; 1988)

Deptropine
(Dibenzheptropine)

ATC: R06AX16
Use: antihistaminic, anticholinergic

RN: 604-51-3   MF: $C_{23}H_{27}NO$ MW: 333.48   EINECS: 210-069-0
CN: *endo*-3-[(10,11-dihydro-5*H*-dibenzo[*a,d*]cyclohepten-5-yl)oxy]-8-methyl-8-azabicyclo[3.2.1]octane

citrate (1:1)
RN: 2169-75-7   MF: $C_{23}H_{27}NO \cdot C_6H_8O_7$ MW: 525.60   EINECS: 218-516-1
LD$_{50}$:   32 mg/kg (M, i.v.); 300 mg/kg (M, p.o.);
        445 mg/kg (R, p.o.);
        75 mg/kg (dog, p.o.)
methobromide
RN: 10139-98-7   MF: $C_{24}H_{30}BrNO$ MW: 428.41
LD$_{50}$:   1150 µg/kg (M, i.v.); 680 mg/kg (M, p.o.);
        1200 µg/kg (R, i.v.); 800 mg/kg (R, p.o.);
        71 mg/kg (dog, p.o.)
methiodide
RN: 38146-43-9   MF: $C_{24}H_{30}INO$ MW: 475.41

phthalic
anhydride

phenylacetic
acid

2-(2-phenylethyl)-
benzoic acid   (I)

I   →   polyphosphoric acid

5-oxo-10,11-di-
hydro-5H-dibenzo-
[a,d]cycloheptene

NaBH$_4$

5-hydroxy-10,11-
dihydro-5H-di-
benzo[a,d]cyclo-
heptene

SOCl$_2$ → II

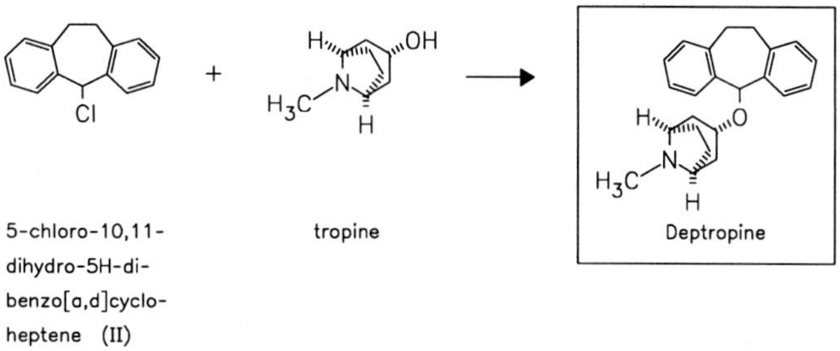

5-chloro-10,11-
dihydro-5H-di-
benzo[a,d]cyclo-
heptene (II)

tropine

Deptropine

Reference(s):
Stelt, C. van der et al.: J. Med. Pharm. Chem. (JMPCAS) **4**, 335 (1961).

Formulation(s):   tabl. 1 mg

Trade Name(s):
GB: Brontina (Brocades); wfm   Brontisol (Brocades)-   I: Brontin (Formenti)
                                                       comb.; wfm

Dequalinium chloride

ATC: D08AH01; G01AC05; R02AA02
Use: bacteriostatic, antifungal

RN: 522-51-0   MF: $C_{30}H_{40}Cl_2N_4$ MW: 527.58   EINECS: 208-330-9
LD_{50}: 1900 µg/kg (M, i.v.)
CN: 1,1'-(1,10-decanediyl)bis[4-amino-2-methylquinolinium] dichloride

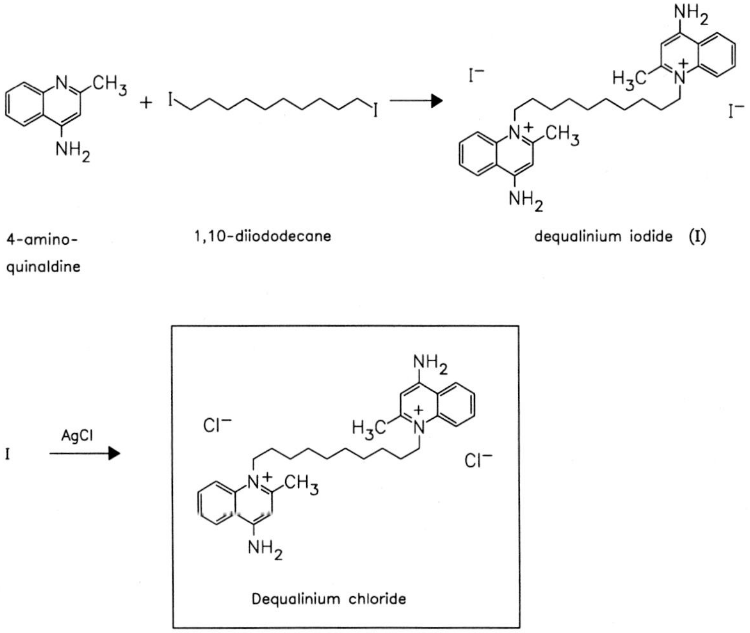

4-amino-
quinaldine

1,10-diiododecane

dequalinium iodide   (I)

Dequalinium chloride

Reference(s):
GB 745 956 (Allen & Hanburys; appl. 1953).

Formulation(s):   different tabl., creams, sol. and gels

Trade Name(s):

D: Dequafungan Hautspray (Kreussler)
Dequavagyn (Kreussler)
Eriosept (Kreussler)
Evazol (Ravensberg)
Maltyl (Merckle)
Optipect (Thiemann)
Phylletten (Arznei Müller-Rorer)
Soor-Gel (Engelhard)
Sorot (Ravensberg)
Tonsillol (Merckle)-comb.
numerous combination preparations

F: Humex Fournier Kinaldine (Labs. Urgo)-comb.
Oroseptol (SmithKline Beecham)-comb.

GB: Labosept (L.A.B.)

I: Aperdan (Tiber)-comb.
Decabis (Gazzoni)
Dequadin (Eurospital)
Dequadin (Importex)
Faringina (SIT)
Farmocillina (Zyma)
Golosan (Lifepharma)
Kinogen (Geymonat)-comb.
Lariquin (Manetti Roberts)
Osangin (Antonetto)
Rinospray (Midy)-comb.
Sterox (Granelli)-comb.
Transpulmina gola (Sigurtà)-comb.

J: Almani S (Tanabe)-comb.
Dequadin Lozenges (Torii)
Honkon-N (Zenyaku)-comb.
Neues Troch (Toyo Pharmar)
SP Troche (Meiji Seika)

Deserpidine

(Desmethoxyreserpine)

ATC: C02AA05
Use: neuroleptic

RN: 131-01-1 MF: $C_{32}H_{38}N_2O_8$ MW: 578.66 EINECS: 205-004-8

LD_{50}: 500 mg/kg (M, p.o.);
15 mg/kg (R, i.v.)

CN: (3β,16β,17α,18β,20α)-17-methoxy-18-[(3,4,5-trimethoxybenzoyl)oxy]yohimban-16-carboxylic acid methyl ester

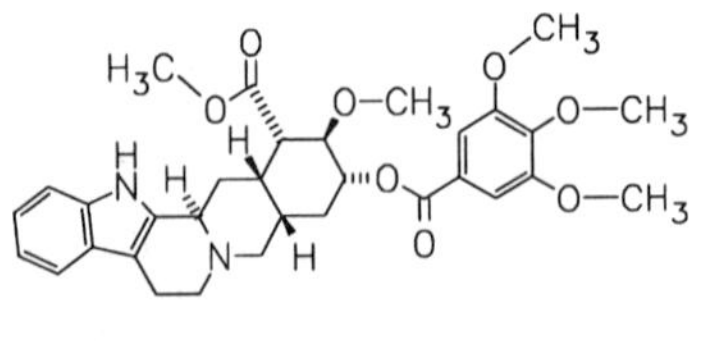

Deserpidine

a By extraction of *Rauwolfia serpentina* (L.) Beuth. roots, separation of in greater quantity available reserpine as heavy soluble thiocyanate and column chromatographic purification of the mother liquors.

b By extraction of *Rauwolfia canescens*, *R. hirsuta*, *R. tetraphylla*, *R. indecora, R. vomitoria* Afz. or *R.. cubana* roots and purification by fractional crystallization and/or column chromatography on Al_2O_3.

Reference(s):

a US 2 887 489 (Ciba; 1959; CH-prior. 1956).

b US 2 982 769 (Ciba; 1961; appl. 1955).

Formulation(s): tabl. 0.25 mg; tabl. 0.25 mg (comb. with 5 mg methyclothiazide)

Trade Name(s):

F: Enduronyl (Abbott)-comb.; wfm

GB: Enduronyl (Abbott)-comb.; wfm
Harmonyl (Abbott); wfm

USA: Enduronyl (Abbott)-comb.; wfm
Harmonyl (Abbott); wfm
Oreticyl (Abbott)-comb.; wfm

Desipramine

ATC: N06AA01
Use: antidepressant

RN: 50-47-5 MF: C$_{18}$H$_{22}$N$_2$ MW: 266.39 EINECS: 200-040-0
LD$_{50}$: 22 mg/kg (M, i.v.); 448 mg/kg (M, p.o.);
29 mg/kg (R, i.v.); 375 mg/kg (R, p.o.)
CN: 10,11-dihydro-N-methyl-5H-dibenz[b,f]azepine-5-propanamine

monohydrochloride
RN: 58-28-6 MF: C$_{18}$H$_{22}$N$_2$ · HCl MW: 302.85 EINECS: 200-373-1
LD$_{50}$: 37 mg/kg (M, i.v.); 315 mg/kg (M, p.o.);
19 mg/kg (R, i.v.); 871 mg/kg (R, p.o.);
25 mg/kg (dog, i.v.)

a

1. NaNH$_2$
2. Br⌒⌒Cl

1. sodium amide
2. 1-bromo-3-chloro-
propane

H$_3$C—NH$_2$
methylamine

10,11-dihydro-5H-
dibenz[b,f]azepine (I)

5-(3-chloropropyl)-
10,11-dihydro-5H-
dibenz[b,f]azepine

Desipramine

b

I +

3-(benzylmethyl-
amino)propyl chloride
(from 1-bromo-3-chloro-
propane and benzylmethylamine)

NaNH$_2$
sodium
amide

H$_2$, Pd–C

Desipramine

c

imipramine
(q. v.)

+

ethyl
chloroformate

KOH

Desipramine

Reference(s):
a FR-M 796 (Geigy; appl. 3.9.1960; CH-prior. 4.9.1959).
GB 908 788 (Geigy; appl. 1960; CH-prior. 1959).
DE 1 189 550 (Geigy; appl. 1960; CH-prior. 1959).
b US 3 454 698 (Colgate-Palmolive; 8.7.1969; prior. 25.5.1960).
US 3 454 554 (Colgate-Palmolive; 8.7.1969; prior. 25.5.1960).
c DE 1 288 599 (Geigy; appl. 13.3.1962; CH-prior. 14.3.1961).
DE 1 445 800 (Geigy; appl. 2.3.1962; CH-prior. 3.3.1961).

Formulation(s):   drg. 25 mg; tabl. 25 mg

Trade Name(s):

D: Pertofran (Novartis Pharma)
     Petylyl (ASTA Medica AWD)

F: Pertofran (Novartis; as hydrochloride)
GB: Pertofran (Geigy); wfm
I: Nortimil (Chiesi)

J: Pertofran (Fujisawa)
USA: Norpramin (Hoechst Marion Roussel; as hydrochloride)

Desloratadine

(Sch-34117)

Use: non-sedating antihistamine metabolite of loratadine

RN: 100643-71-8   MF: $C_{19}H_{19}ClN_2$ MW: 310.83
CN: 8-Chloro-6,11-dihydro-11-(4-piperidinylidene)-5H-benz[5,6]cyclohepta[1,2-b]pyridine

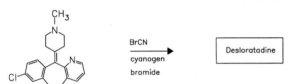

a

NaOH, C_2H_5OH

loratadine (q. v.)

Desloratadine

b

BrCN
cyanogen bromide

Desloratadine

8-chloroazatadine
(cf. loratadine)

Reference(s):
a,b WO 8 503 707 (Schering Corp.; appl. 8.2.1985; USA-prior. 15.2.1984).

polymorphs:
WO 9 901 450 (Schering Corp.; appl. 1.7.1998; USA-prior. 2.7.1997).

eye drops containing loratadine metabolites:
WO 9 848 803 (Schering-Plough K.K.; WO-prior. 25.4.1997).

treatment of allergic rhinitis and asthma with desloratadine:
WO 9 834 611 (Sepracor; appl. 10.2.1998; USA-prior. 11.2.1997).
WO 9 620 708 (Sepracor; appl. 11.12.1995; USA-prior. 30.12.1994).

transdermal dosage system:
DE 4 442 999 (Hexal Pharma; D-prior. 2.12.1994).

Trade Name(s):
USA: DCL (Schering-Plough; 2000)

Deslorelin

Use: GnRH-agonist (for treatment of precocious puberty)

RN: 57773-65-6   MF: $C_{64}H_{83}N_{17}O_{12}$ MW: 1282.48
CN: 6-D-tryptophan-9-(N-ethyl-L-prolinamide)-10-deglycinamide luteinizing hormone-releasing factor (pig)

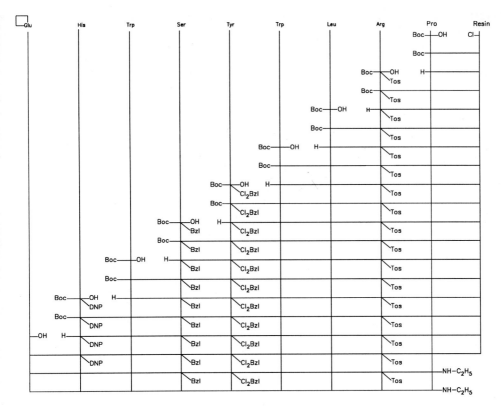

Glu	His	Trp	Ser	Tyr	Trp	Leu	Arg	Pro	Resin
								Boc—OH	Cl—
								Boc	
							Boc—OH / Tos	H	
							Boc / Tos		
						Boc—OH	H / Tos		
						Boc	Tos		
					Boc—OH	H	Tos		
					Boc		Tos		
				Boc—OH / Cl₂Bzl	H		Tos		
				Boc / Cl₂Bzl			Tos		
			Boc—OH / Bzl	H / Cl₂Bzl			Tos		
			Boc / Bzl	Cl₂Bzl			Tos		
		Boc—OH	H / Bzl	Cl₂Bzl			Tos		
		Boc	Bzl	Cl₂Bzl			Tos		
	Boc—OH / DNP	H	Bzl	Cl₂Bzl			Tos		
	Boc / DNP		Bzl	Cl₂Bzl			Tos		
Boc—OH	H / DNP		Bzl	Cl₂Bzl			Tos		
Boc	DNP		Bzl	Cl₂Bzl			Tos		
OH H	DNP		Bzl	Cl₂Bzl			Tos		—NH–C₂H₅
	DNP		Bzl	Cl₂Bzl			Tos		—NH–C₂H₅

Tos: Tosyl
Boc: tert-butoxycarbonyl
Cl₂Bzl: 2,4-dichlorobenzyl
Bzl: benzyl
DNP: 2,4-dinitrophenyl

Deslorelin

Reference(s):
DOS 2 830 629 (Salk Inst.; appl. 12.7.1978; USA-prior. 14.7.1977, 26.6.1978).
US 4 218 439 (Salk Inst.; 19.8.1980; appl. 26.6.1978; prior. 14.7.1977).

Formulation(s):   vial 500 µg

Trade Name(s):
GB: Somagard (Monmouth;   USA: Somagard (Roberts; 1990);
        1991); wfm   wfm

Desmopressin

ATC: H01BA02
Use: antidiuretic

RN: 16679-58-6   MF: $C_{46}H_{64}N_{14}O_{12}S_2$ MW: 1069.24   EINECS: 240-726-7
CN: 1-(3-mercaptopropanoic acid)-8-D-argininevasopressin

acetate (1:2)
RN: 16789-98-3   MF: $C_{46}H_{64}N_{14}O_{12}S_2 \cdot 2C_2H_4O_2$ MW: 1189.34

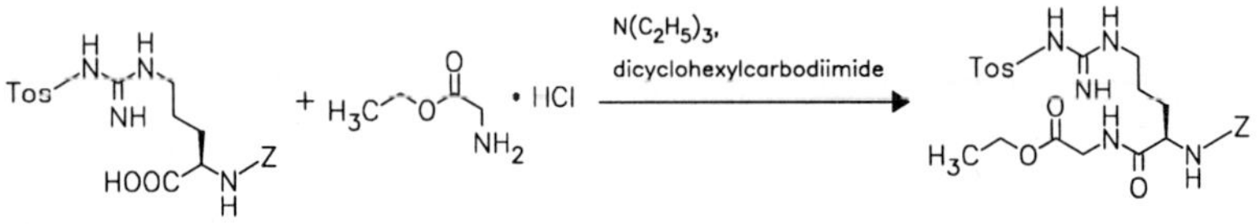

Z-D-Arg(Tos)-OH   Gly-OEt · HCl   Z-D-Arg(Tos)-Gly-OEt   (I)

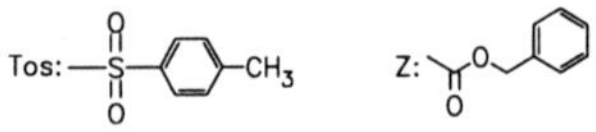

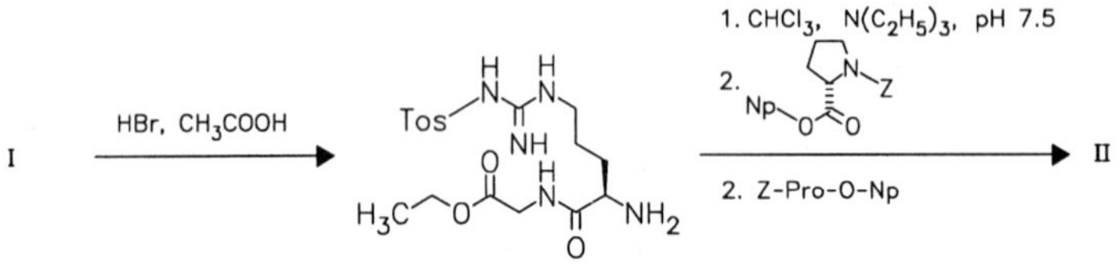

D-Arg(Tos)-Gly-OEt

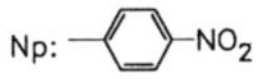

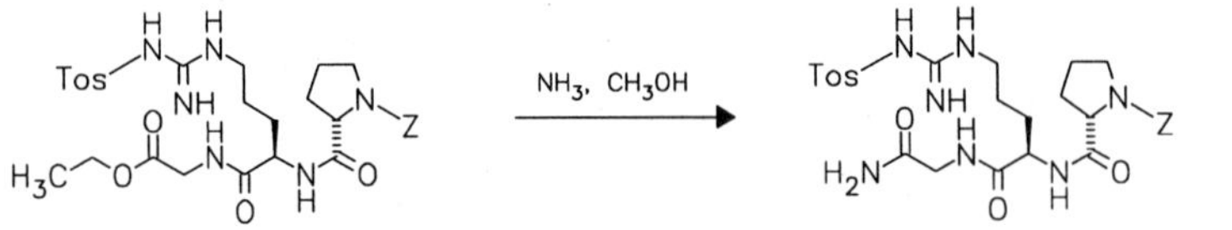

Z-Pro-D-Arg(Tos)-Gly-OEt   (II)   Z-Pro-D-Arg(Tos)-Gly-NH$_2$ (III)

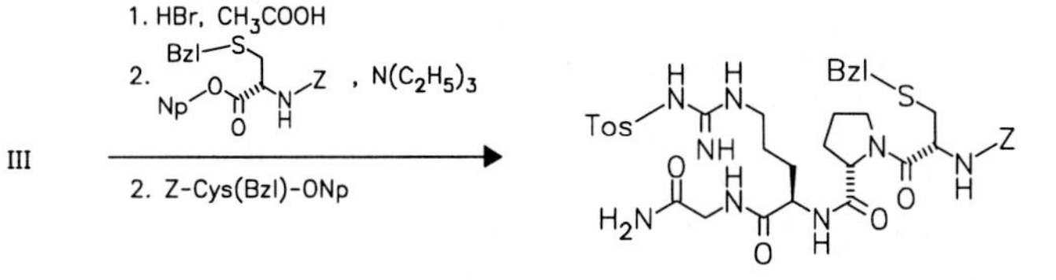
1. HBr, CH3COOH
2. , N(C2H5)3
III
2. Z-Cys(Bzl)-ONp

Z-Cys(Bzl)-Pro-D-Arg(Tos)-Gly-NH$_2$ (IV)

Bzl: 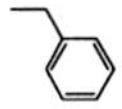

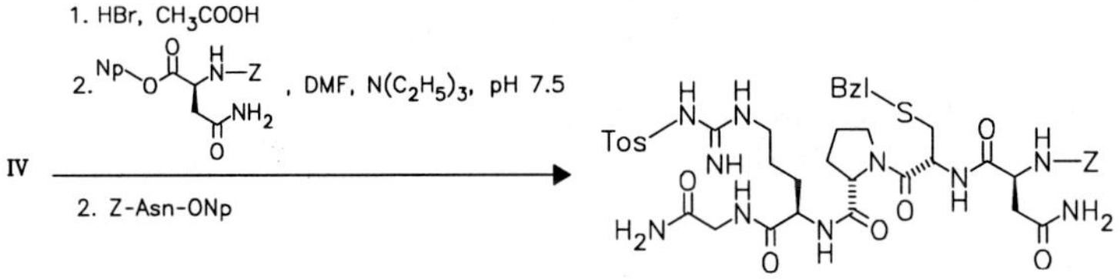
1. HBr, CH3COOH
2. , DMF, N(C2H5)3, pH 7.5
IV
2. Z-Asn-ONp

Z-Asn-Cys(Bzl)-Pro-D-Arg(Tos)-Gly-NH$_2$ (V)

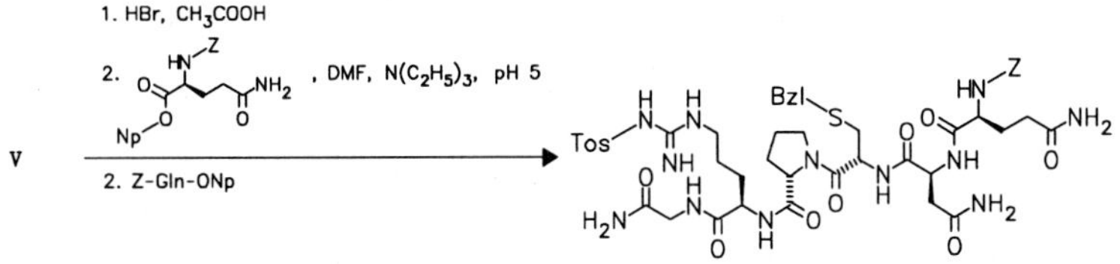
1. HBr, CH3COOH
2. , DMF, N(C2H5)3, pH 5
V
2. Z-Gln-ONp

Z-Gln-Asn-Cys(Bzl)-Pro-D-Arg(Tos)-Gly-NH$_2$ (VI)

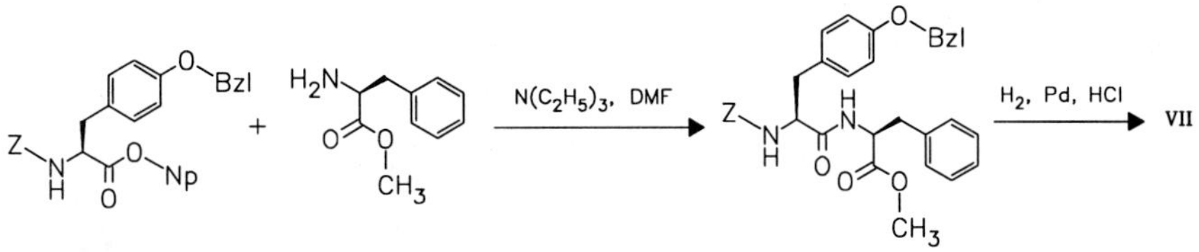
N(C2H5)3, DMF
H2, Pd, HCl
VII

Z-Tyr(Bzl)-ONp Phe-OMe Z-Tyr(Bzl)-Phe-OMe

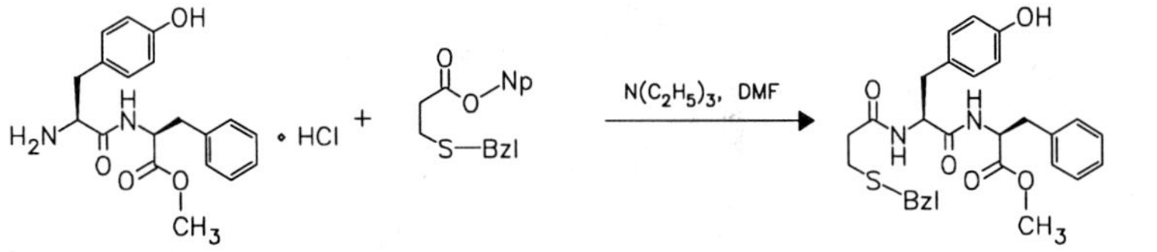
N(C2H5)3, DMF

Tyr-Phe-OMe · HCl Bzl-Mep-ONp Bzl-Mep-Tyr-Phe-OMe (VIII)

Mep: 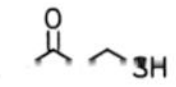

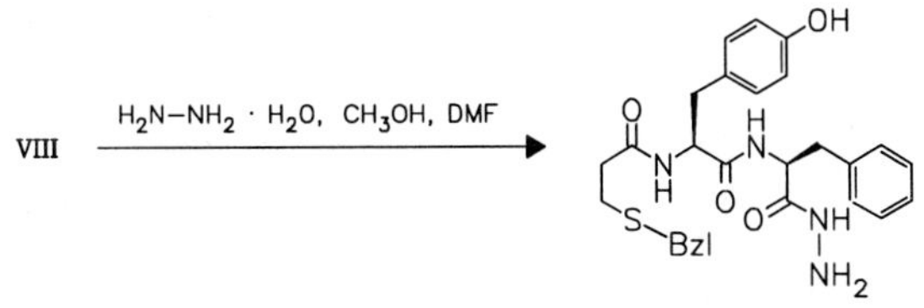
VIII
H2N–NH2 · H2O, CH3OH, DMF

Bzl-Mep-Tyr-Phe-NH-NH$_2$ (IX)

VI

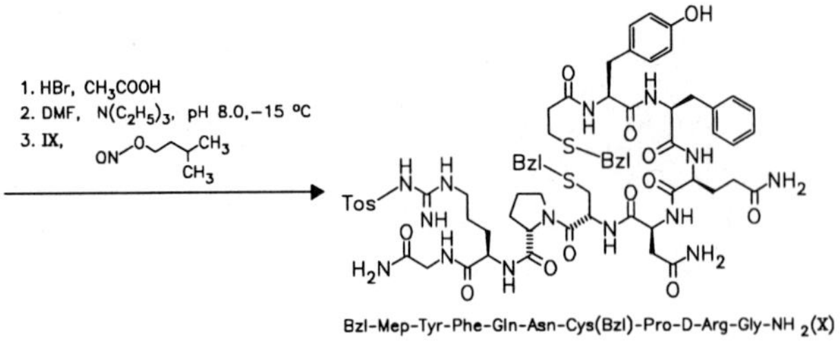
1. HBr, CH3COOH
2. DMF, N(C2H5)3, pH 8.0, –15 °C
3. IX,

Bzl-Mep-Tyr-Phe-Gln-Asn-Cys(Bzl)-Pro-D-Arg-Gly-NH $_2$(X)

X

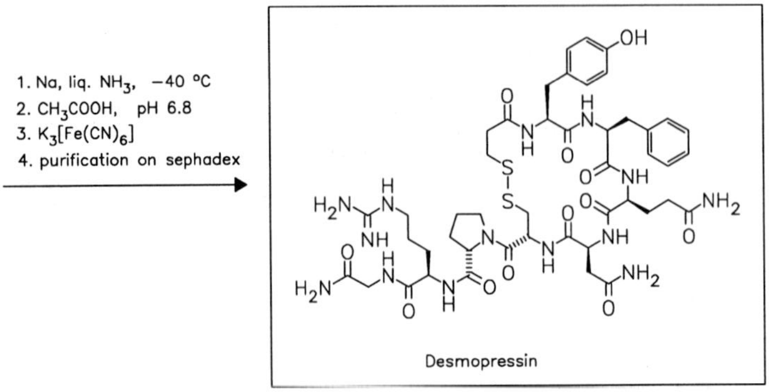
1. Na, liq. NH3, –40 °C
2. CH3COOH, pH 6.8
3. K3[Fe(CN)6]
4. purification on sephadex
Desmopressin

Reference(s):

US 3 454 549 (Sandoz; 8.7.1969; CH-prior. 17.7.1964).
US 3 497 491 (Ceskoslovenska Akad.; 24.2.1970; CS-prior. 15.9.1966).
Huguenin, R.L.; Boissonas, R.A.: Helv. Chim. Acta (HCACAV) **49**, 695 (1966).
DOS 2 723 453 (Ferring; appl. 24.5.1977; S-prior. 24.5.1976).
DOS 2 749 932 (Ferring; appl. 8.11.1977; S-prior. 12.11.1976).
GB 1 539 317 (Ferring; appl. 20.5.1977; S-prior. 24.5.1976).
GB 1 539 318 (Ferring; appl. 4.11.1977; S-prior. 12.11.1976).

Formulation(s): amp. 4 µg; doses spray 0.1 mg; tabl. 0.1 mg, 0.2 mg

Trade Name(s):

D: DDAVP (Ferring)
Minirin (Ferring)
F: Minirin (Ferring; as acetate)
GB: DDAVP (Ferring)
I: Minirin (Ferring)
J: Desmopressin (Kyowa Hakko)
USA: DDAVP (Rhône-Poulenc Rorer; as acetate)

Desogestrel

ATC: G03AA
Use: progestogen, oral contraceptive (in combination with ethinylestradiol)

RN: 54024-22-5 MF: $C_{22}H_{30}O$ MW: 310.48 EINECS: 258-929-4
CN: (17α)-13-ethyl-11-methylene-18,19-dinorpregn-4-en-20-yn-17-ol

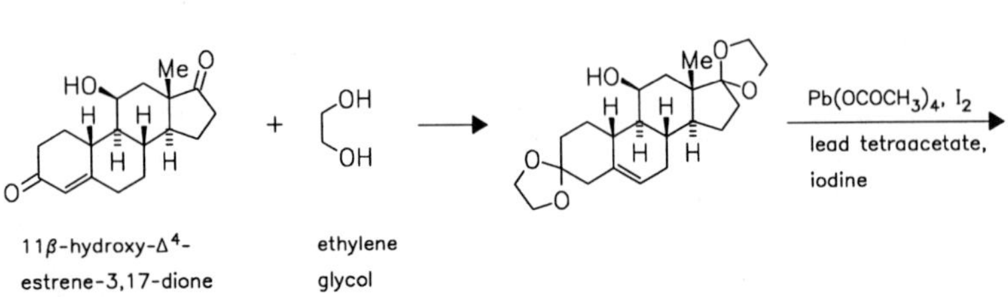
Pb(OCOCH3)4, I2
lead tetraacetate, iodine
11β-hydroxy-Δ4-estrene-3,17-dione
ethylene glycol

I

(I)   methylmagnesium bromide

1. N_2H_4
2. CrO_3

1. hydrazine
2. chromium(VI) oxide

→ II

(II)   methylenetriphe-nylphosphorane

HCl

11-methylene-18-methyl-Δ^4-estrene-3,17-dione

1. ⌐SH⌐SH
2. NaBH_4

1. 1,2-ethanedithiol
2. sodium borohydride

→ III

(III)

1. Na, NH_3
2. CrO_3

2. chromium(VI) oxide

HC≡CH , K, NH_3

acetylene

Desogestrel

Reference(s):
US 3 927 046 (Akzona; 16.12.1975; appl. 3.12.1973; NL-prior. 9.12.1972).
DE 2 361 120 (Organon; appl. 7.12.1973; NL-prior. 9.12.1972; 15.11.1973).
NL 7 411 607 (Akzo; appl. 2.9.1974).
Broek, A.S. van den et al.: Recl. Trav. Chim. Pays-Bas (RTCPA3) **94**, 35 (1975).

Formulation(s):   tabl. 150 µg (in combination with ethinylestradiol)

Trade Name(s):
D: Biviol (Nourypharma)-comb.
      Cyclosa (Nourypharma)-comb.
      Cydeane (Monsanto)-comb.
      Lovelle (Organon)-comb.
      Marvelon (Organon; 1981)-comb.

      Oviol (Nourypharma)-comb.
F: Cydeane (Monsanto)-comb.
      Mercilon (Organon)-comb.
      Varnoline (Organon; 1984)-comb.
GB: Marvelon (Organon)-comb.

      Mercilon (Organon; 1982)-comb.
I: Mercilon (Organon)-comb.
      Planum (Menarini)-comb.
      Practil (Organon)-comb.
      Securgin (Menarini)-comb.
USA: Desogen (Organon)
      Ortho-Cept (Ortho-McNeil Pharmaceutical)

Desonide
(Prednacinolone)

ATC: D07AB08; S01BA11
Use: topical glucocorticoid

RN: 638-94-8   MF: C_{24}H_{32}O_6   MW: 416.51   EINECS: 211-351-6
LD_{50}:   3710 mg/kg (M, p.o.)
CN: (11β,16α)-11,21-dihydroxy-16,17-[(1-methylethylidene)bis(oxy)]pregna-1,4-diene-3,20-dione

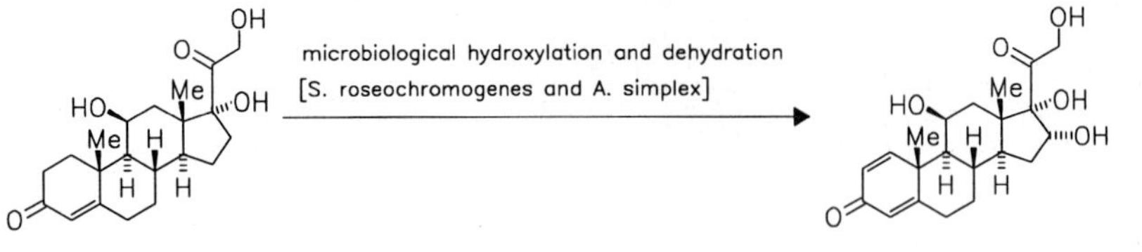

hydrocortisone
(q. v.)

microbiological hydroxylation and dehydration
[S. roseochromogenes and A. simplex]

16α-hydroxyprednisolone   (I)

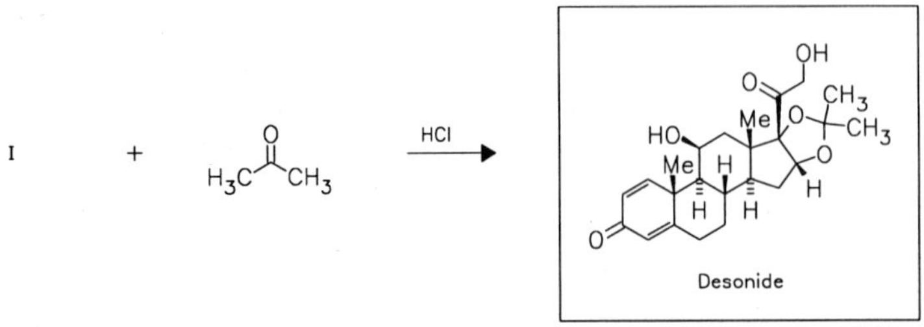

Desonide

Reference(s):

US 3 536 586 (Squibb; 27.10.1970; prior. 25.1.1968).
US 2 990 401 (American Cyanamid; 27.6.1961; prior. 18.6.1958, 11.3.1958).
Bernstein, S. et al.: J. Am. Chem. Soc. (JACSAT) **81**, 4573 (1959).

synthesis of hydrocortisone:
Allen, W.S.; Bernstein, S.: J. Am. Chem. Soc. (JACSAT) **78**, 1909 (1956).
Bernstein, S.: Recent Prog. Horm. Res. (RPHRA6) **14**, 1 (1958).

alternative synthesis:
US 3 549 498 (Squibb; 22.12.1970; prior. 2.4.1968).

Formulation(s):   cream 1 mg/g

Trade Name(s):

D:	Sterax (Galderma)	I:	Cloressidina
F:	Locapred (Pierre Fabre)		(Farmacologico Milanese)
	Locatrop (Pierre Fabre)		Desonix (Usar)-comb.
	Tridésonit (Dome-		PR 100 (Farmacologico
	Hollister-Stier)		Milanese)
GB:	Tridesilon (Lagap); wfm		Prenacid (SIFI)

Reticus (Farmila)
Reticus Antimicotico
(Farmila)-comb.
J: Tridesonit (Miles)
USA: Des Owen (Galderma)
Tridesilon (Bayer)

Desoximetasone
(Desoximethasone)

ATC: D07AC03; D07XC02
Use: topical glucocorticoid

RN: 382-67-2   MF: $C_{22}H_{29}FO_4$ MW: 376.47   EINECS: 206-845-3
CN: (11β,16α)-9-fluoro-11,21-dihydroxy-16-methylpregna-1,4-diene-3,20-dione

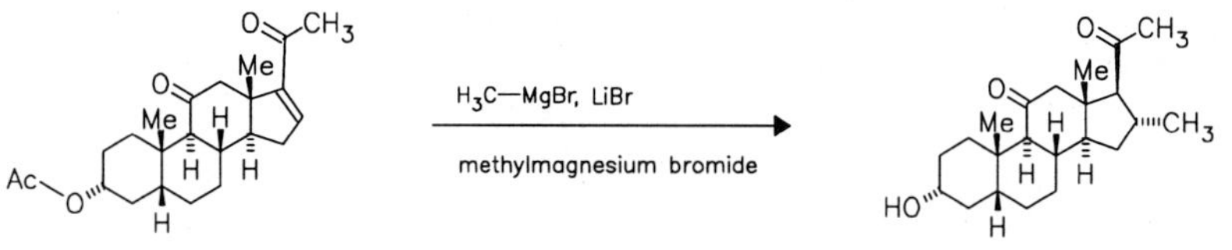

3α-acetoxy-16-pregnene-
11,20-dione
(from deoxycholic acid)

3α-hydroxy-16α-methyl-
pregnane-11,20-dione (I)

H₃C—MgBr, LiBr

methylmagnesium bromide

I

Br₂,
CH₃OH, CH₃COCl

NaO—Ac, DMF

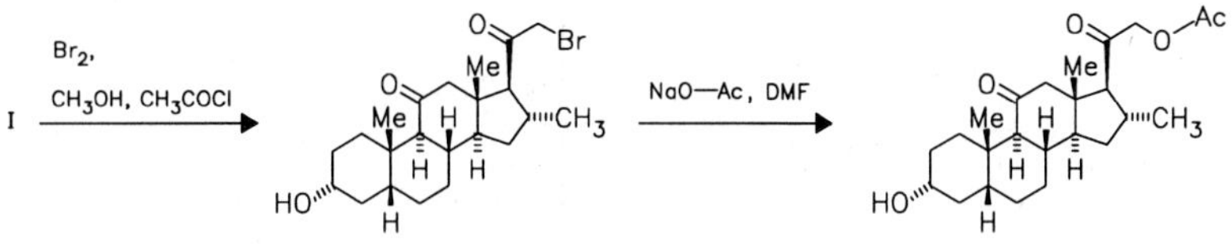

21-acetoxy-3α-hydroxy-
16α-methylpregnane-
11,20-dione (II)

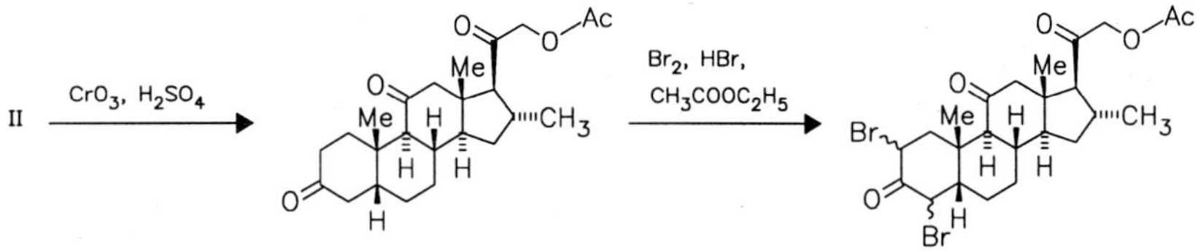

II

CrO₃, H₂SO₄

Br₂, HBr,
CH₃COOC₂H₅

21-acetoxy-16α-methyl-
pregnane-3,11,20-trione

(III)

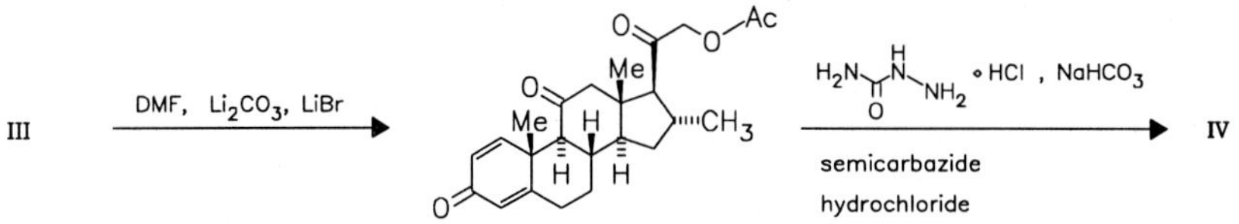

III

DMF, Li₂CO₃, LiBr

21-acetoxy-16α-methyl-
pregna-1,4-diene-3,11,20-
trione

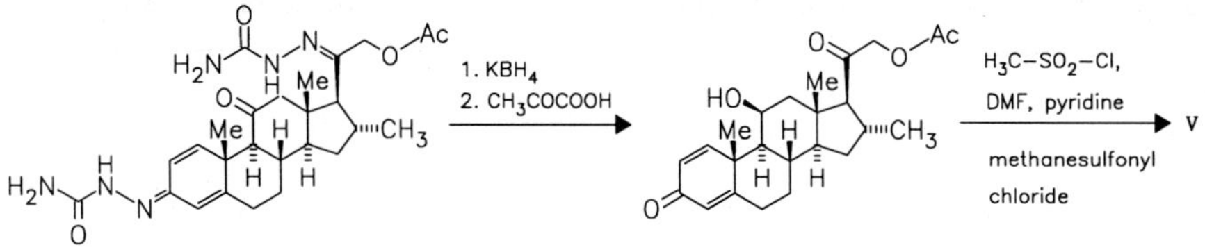

H₂N—C(=O)—NH—NH₂ • HCl , NaHCO₃

semicarbazide
hydrochloride

IV

(IV)

1. KBH₄
2. CH₃COCOOH

21-acetoxy-11β-hydroxy-
16-methyl-1,4-pregna-
diene-3,20-dione

H₃C—SO₂—Cl,

DMF, pyridine

methanesulfonyl
chloride

V

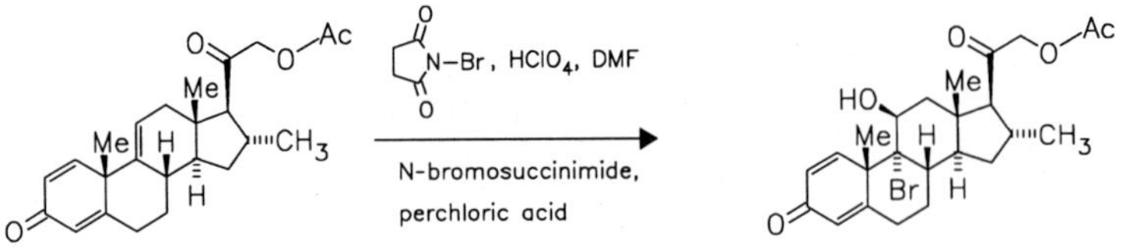

21-acetoxy-16α-methyl-
1,4,9(11)-pregnatriene-
3,20-dione   (V)

(VI)

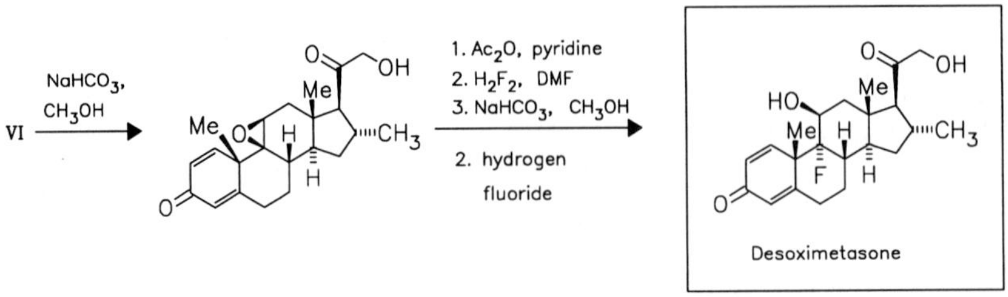

Desoximetasone

Reference(s):
US 3 099 654 (Roussel-Uclaf; 30.6.1963; F-prior. 17.8.1960).
DOS 1 159 441 (Roussel-Uclaf; appl. 4.8.1961; F-prior. 17.8.1960).
FR 1 296 544 (Roussel-Uclaf; appl. 17.8.1960).
Joly, R. et al.: Arzneim.-Forsch. (ARZNAD) **24**, 1 (1974).

synthesis of 21-acetoxy-11β-hydroxy-16α-methyl-1,4-pregnadien-3,20-dione:
DOS 1 205 096 (Roussel-Uclaf; appl. 12.5.1961; F-prior. 14.5.1960, 16.5.1960).

alternative synthesis:
BE 614 196 (Schering AG; appl. 21.2.1962; D-prior. 22.2.1961).
US 3 232 839 (Schering AG; 1.2.1966; D-prior. 22.2.1961, 27.6.1963).

Formulation(s):   cream 0.25 %, 0.05 %; lotion 0.25 %; ointment 0.35 %

Trade Name(s):

D:	Topisolon (Hoechst)	Topifram (Roussel
	Topisolon (Hoechst)-comb.	Diamant)-comb.
F:	Topicorte (Roussel	GB: Stiedex LP (Stiefel)
	Diamant)	I: Flubason (Hoechst)

USA: Topicort (Hoechst Marion
        Roussel)

Desoxycortone acetate

(Deoxycorticosterone acetate; Deoxycortone acetate)

ATC: H02AA03
Use: mineralocorticoid

RN: 56-47-3   MF: $C_{23}H_{32}O_4$ MW: 372.51   EINECS: 200-275-9
CN: 21-(acetyloxy)pregn-4-ene-3,20-dione

desoxycortone
RN: 64-85-7   MF: $C_{21}H_{30}O_3$ MW: 330.47   EINECS: 200-596-4
LD$_{50}$:   1 g/kg (M, route unreported)

a

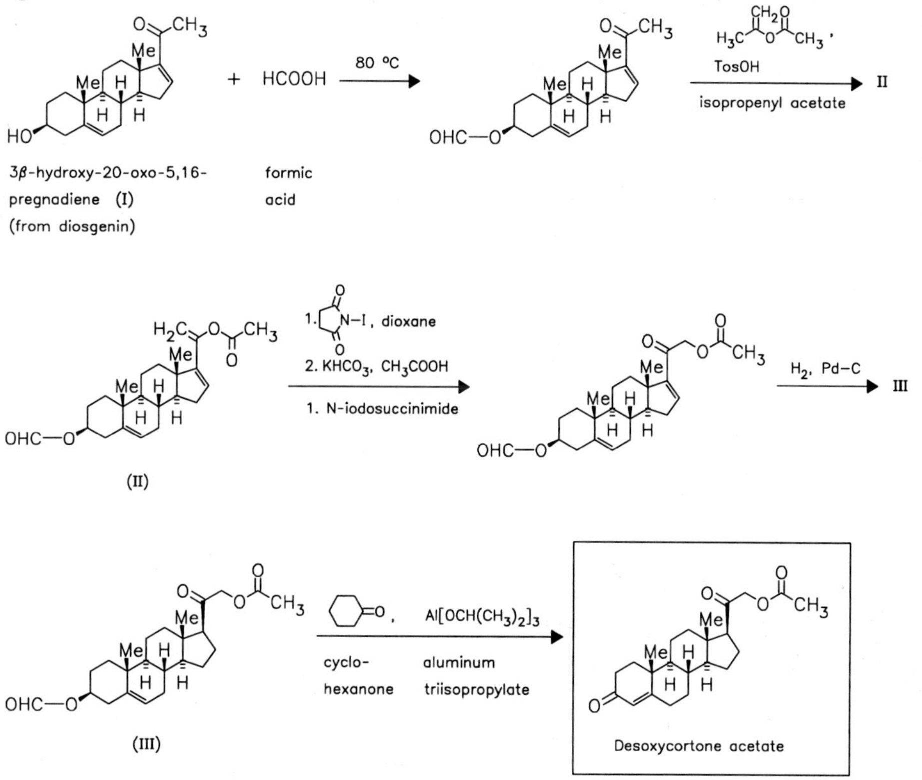

3β-hydroxy-20-oxo-5,16-pregnadiene (I) (from diosgenin) + formic acid → 80 °C → H₃C-acetate, TosOH, isopropenyl acetate → II

(II) → 1. N—I, dioxane; 2. KHCO₃, CH₃COOH (1. N-iodosuccinimide) → → H₂, Pd–C → III

(III) → cyclohexanone, aluminum triisopropylate Al[OCH(CH₃)₂]₃ → Desoxycortone acetate

b

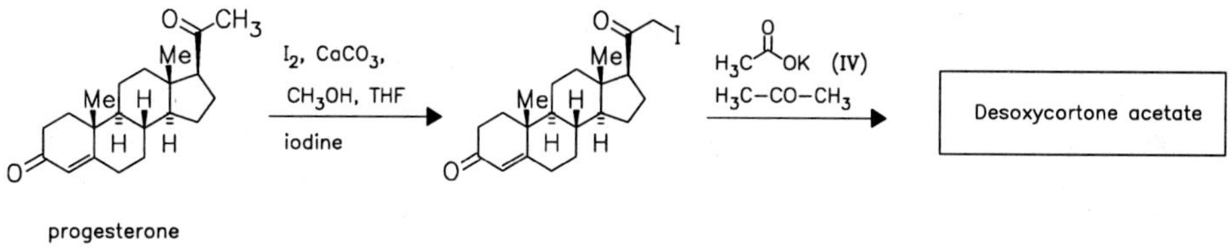

progesterone → I₂, CaCO₃, CH₃OH, THF (iodine) → H₃C-CO-OK (IV), H₃C–CO–CH₃ → Desoxycortone acetate

c

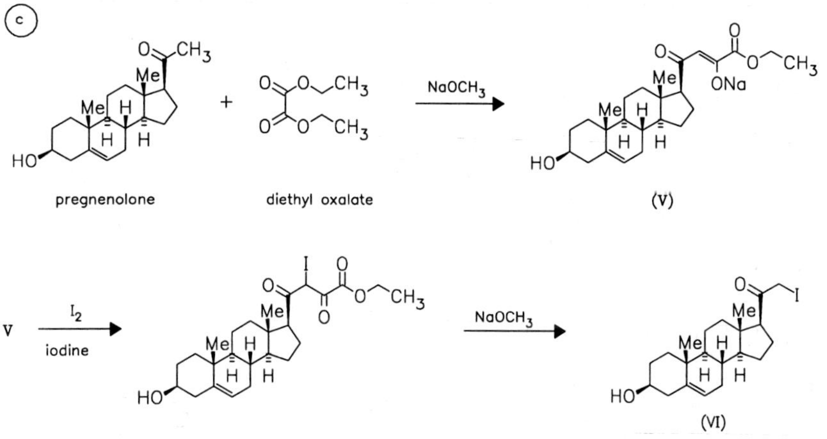

pregnenolone + diethyl oxalate → NaOCH₃ → (V)

V → I₂ (iodine) → → NaOCH₃ → (VI)

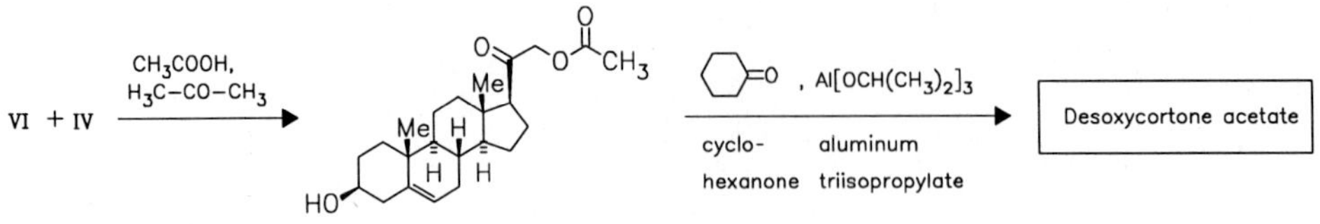

VI + IV → [CH₃COOH, H₃C-CO-CH₃] → [steroid structure] → cyclohexanone, aluminum triisopropylate, Al[OCH(CH₃)₂]₃ → **Desoxycortone acetate**

Reference(s):

a   Sondheimer, F. et al.: J. Am. Chem. Soc. (JACSAT) **79**, 5034 (1957).
   synthesis of I:
   Wall, M.E.: J. Am. Chem. Soc. (JACSAT) **77**, 5665 (1955).
b   Ringold, H.J.; Stork, G.: J. Am. Chem. Soc. (JACSAT) **80**, 250 (1958).
c   Ruschig, H.: Angew. Chem. (ANCEAD) **60**, 247 (1948).
   Ruschig, H.: Chem. Ber. (CHBEAM) **88**, 878 (1955).

alternative syntheses:
DE 871 153 (Hoechst; appl. 1937).
DE 875 353 (Schering AG; appl. 1938).
US 2 312 480 (Roche-Organon; 1943, CH-prior. 1937).
US 2 409 043 (Schering Corp.; 1946, D-prior. 1939).
US 2 470 903 (W. C. Ross; 1949; GB-prior. 1945).
Serini, A. et al.: Ber. Dtsch. Chem. Ges. (BDCGAS) **72**, 391 (1939).
Wilds, A.L.; Shunk, C.H.: J. Am. Chem. Soc. (JACSAT) **70**, 2427 (1948).

review:
Ullmanns Encykl. Tech. Chem., 4. Aufl., Vol. **13**, 52.

Formulation(s):   amp. 10 mg/ml

Trade Name(s):

D:	Docabolin (Nourypharma; as phenylpropionate-comb.); wfm	GB:	Percorten M Crystals (Ciba; as pivalate); wfm	Sinsurrene (Parke Davis)-comb.; wfm
F:	Syncortyl (Roussel Diamant)	I:	Cortiron (Schering); wfm Neodin (Lusofarmaco); wfm	J: Syncorta (Takeda) USA: Doca Acetate (Organon); wfm Percorten (Ciba); wfm

---

Detajmium bitartrate

ATC: C01B
Use: antiarrhythmic

RN: 53862-81-0   MF: $C_{27}H_{42}N_3O_3 \cdot C_4H_6O_6 \cdot H_2O$ MW: 624.75
LD₅₀: 6000 µg/kg (R, i.v.)
CN: (17R,21α)-4-[3-(diethylamino)-2-hydroxypropyl]-17,21-dihydroxyajmalanium salt with [R(R*,R*)]-2,3-dihydroxybutanedioic acid (1:1) monohydrate

tartrate (1:1)
RN: 33774-52-6   MF: $C_{27}H_{42}N_3O_3 \cdot C_4H_6O_6$ MW: 606.74
LD₅₀: 10 mg/kg (R, i.v.); 290 mg/kg (R, p.o.)

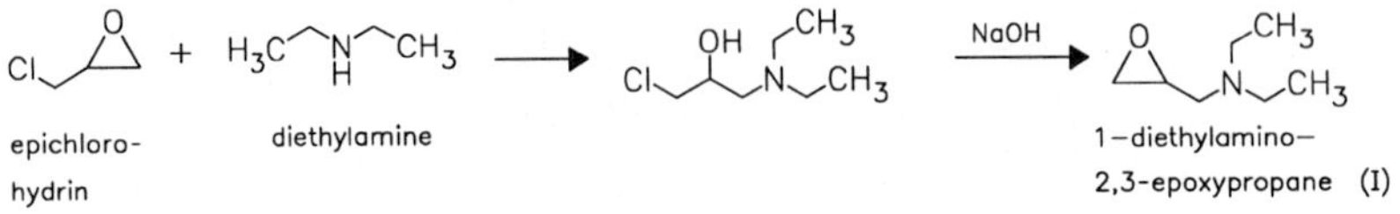

epichloro-hydrin + diethylamine → [intermediate] → NaOH → 1-diethylamino-2,3-epoxypropane (I)

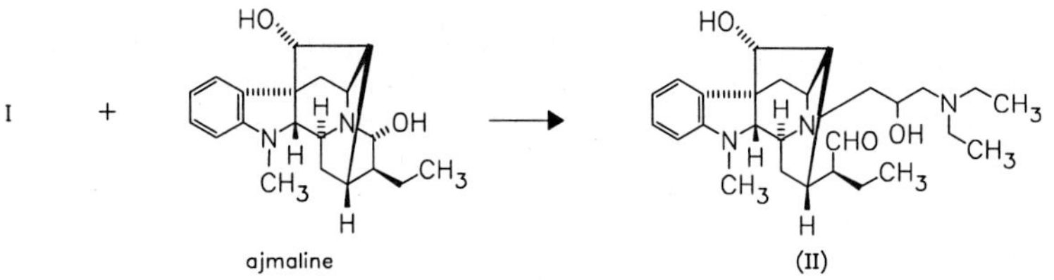

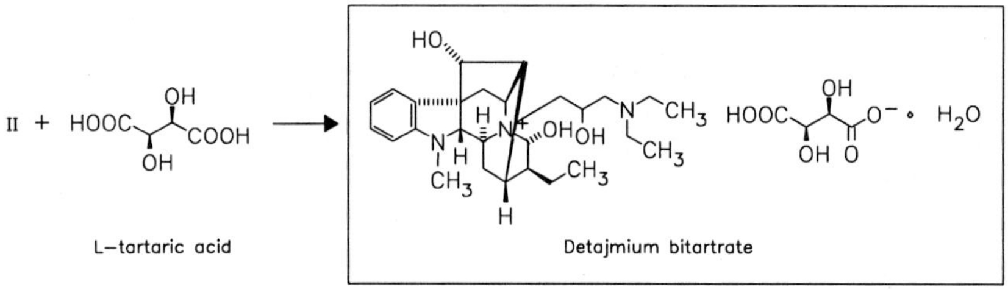

ajmaline → (II)

II + L-tartaric acid → Detajmium bitartrate

Reference(s):

DE 2 025 286 (VEB Arzneimittelwerk Dresden; appl. 23.5.1970; DDR-prior. 28.7.1969).
GB 1 244 597 (VEB Arzneimittelwerk Dresden; appl. 5.7.1970; DDR-prior. 28.7.1969).

Formulation(s):   sugar coated tabl. 25 mg

Trade Name(s):
D: Tachmalcor (ASTA Medica
      AWD)

Dexamethasone

(Dexametasone)

ATC: A01AC02; C05AA09; D07AB19;
        D07XB05; D10AA03; H02AB02;
        R01AD03; S01BA01; S01CB01;
        S02BA06; S03BA01

Use: glucocorticoid

RN: 50-02-2   MF: $C_{22}H_{29}FO_5$ MW: 392.47   EINECS: 200-003-9
LD$_{50}$:   >3 g/kg (R, p.o.)
CN: (11β,16α)-9-fluoro-11,17,21-trihydroxy-16-methylpregna-1,4-diene-3,20-dione

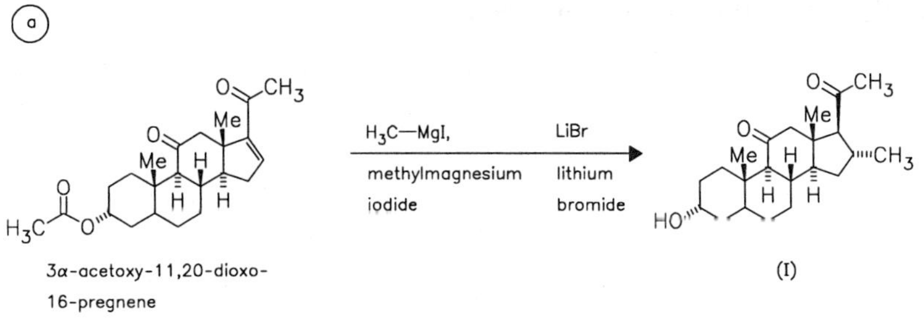

3α-acetoxy-11,20-dioxo-
16-pregnene

H_3C—MgI,   LiBr

methylmagnesium   lithium
iodide   bromide

(I)

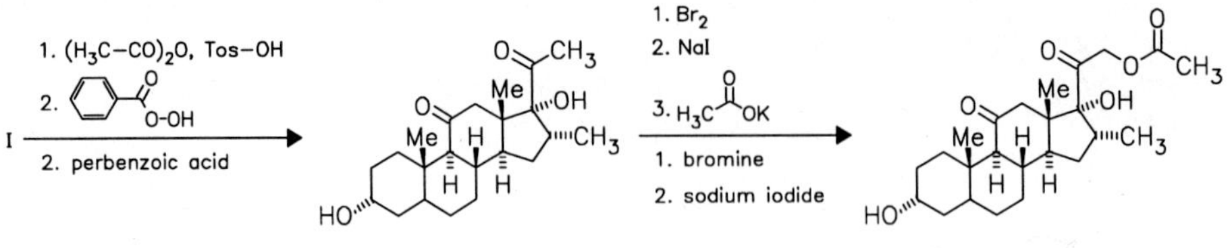

I

1. (H₃C—CO)₂O, Tos—OH

2. perbenzoic acid

→ (reaction gives intermediate with acetyl and OH groups)

1. Br₂
2. NaI
3. H₃C—CO—OK

1. bromine
2. sodium iodide

→ **(II)**

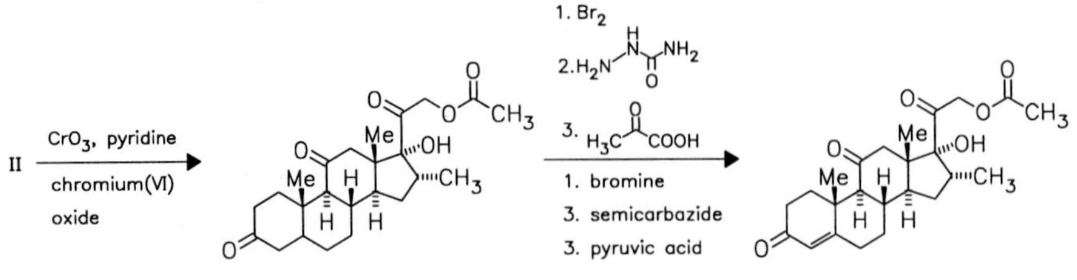

II

CrO₃, pyridine

chromium(VI) oxide

→

1. Br₂
2. H₂N—NH—CO—NH₂
3. H₃C—CO—COOH

1. bromine
3. semicarbazide
3. pyruvic acid

→ **(III)**

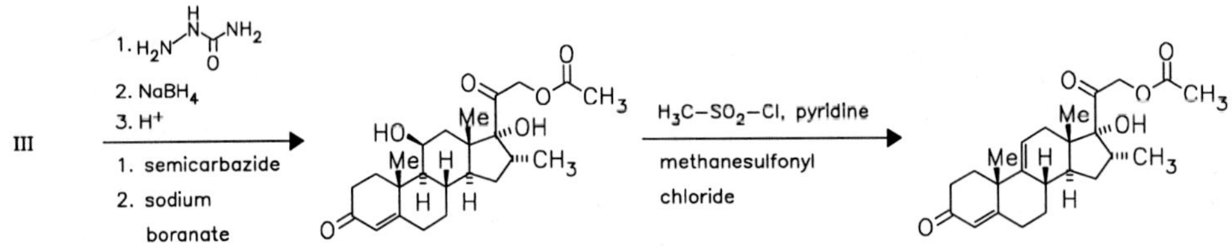

III

1. H₂N—NH—CO—NH₂
2. NaBH₄
3. H⁺

1. semicarbazide
2. sodium boranate

→

H₃C—SO₂—Cl, pyridine

methanesulfonyl chloride

→ **(IV)**

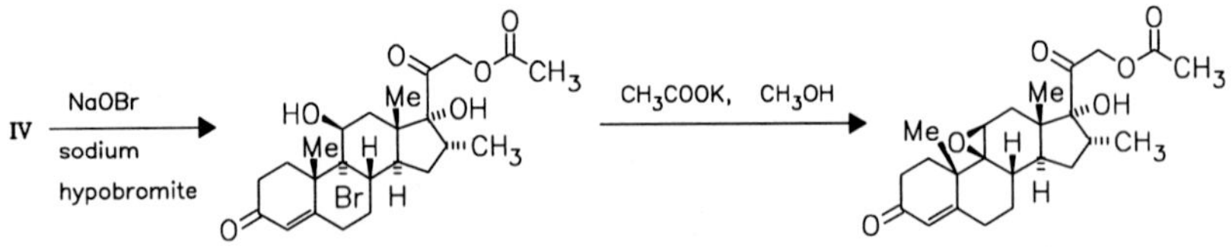

IV

NaOBr

sodium hypobromite

→

CH₃COOK, CH₃OH

→ **(V)**

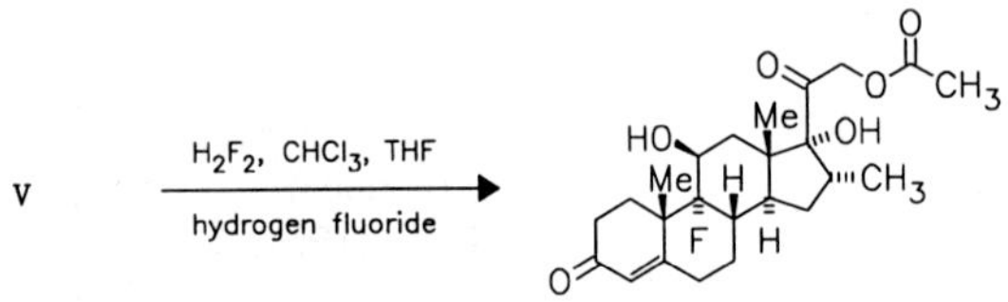

V

H₂F₂, CHCl₃, THF

hydrogen fluoride

→ **(VI)**

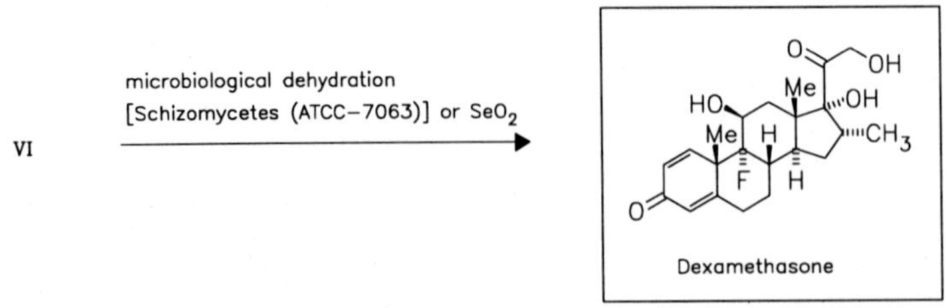

VI

microbiological dehydration
[Schizomycetes (ATCC–7063)] or SeO₂

→

Dexamethasone

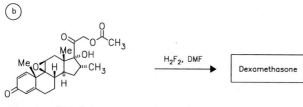

21-acetoxy-3,20-dioxo-
9β,11β-epoxy-17α-hydroxy-
16α-methyl-1,4-pregnadiene (VII)
(from 11-oxo-16-
dehydroprogesterone)

Reference(s):
a   Arth, G.E. et al.: J. Am. Chem. Soc. (JACSAT) **80**, 3160 (1958).
    DE 1 113 690 (Merck & Co.; appl. 22.2.1958; USA-prior. 27.2.1957).
    Applezweig, N.: Steroid Drugs, Vol. **1**, 72 (New York, Toronto, London 1962).
b   Oliveto, E.P. et al.: J. Am. Chem. Soc. (JACSAT) **80**, 4431 (1958).
    US 2 852 511 (Olin Mathieson; 1958; prior. 1953).
    US 3 007 923 (Lab. Franç. de Chimiothérapie, 7.11.1961; appl. 12.1.1960; F-prior. 22.1.1959).
    Applezweig, N.: Steroid Drugs, Vol. **1**, 74 (New York, Toronto, London 1962).

synthesis from tigogenin:
Ohta, T. et al.: Org. Process Res. Dev. (OPRDFK) **1**, 420 (1997).

synthesis of VII:
Oliveto, E.P. et al.: J. Am. Chem. Soc. (JACSAT) **80**, 4431 (1958).
Marker, R.E.; Crooks, H.: J. Am. Chem. Soc. (JACSAT) **64**, 1280 (1942).
GB 869 511 (Upjohn; appl. 24.4.1959; USA-prior. 26.5.1958).

alternative synthesis:
US 4 041 055 (Upjohn; 9.8.1977; appl. 17.11.1975).

review:
Ullmanns Encykl. Tech. Chem., 4. Aufl., Vol. **13**, 57.

Formulation(s):   aerosol 0.075 mg/per pump; amp. 5 mg/ml; eye drops 1 mg/ml; f. c. tabl. 0.5 mg, 0.75 mg,
                      1.5 mg; ointment 0.1 %; sol. 0.03 %; suppos. 2.2 mg

Trade Name(s):
D: afpred forte Dexa (Hefa-
        Frenon)
        Anemul (Medopharm)
        Baycuten (Bayropharm; as
        acetate)-comb.
        Chibro-Cadou (Chibret)
        Cortidexason Crinale,
        Salbe (Dermapharm)-
        comb.
        Cortisumman (Dr. Winzer)
        Corto-Tavegil (Novartis
        Pharma)-comb.
        Dexa-Allvoran (TAD)
        Dexa Biciron (Alcon)-
        comb.
        Dexagel (Mann)
        Dexa Loscon (Galderma)
        Dexamonozon (Medice)
        Dexamytex (Mann)-comb.

        Dexa-Philogout
        (Azupharma)
        Dexa Polyspectral (Alcon)
        Dexa-Rhinospray
        (Boehringer Ing.)
        Dexa-sine (Thilo)
        Duodexa N Salbe (Kade)
        Fortecortin (Merck)
        Isopto-Dex (Alcon)-comb.
        Lipotalan (Merckle)
        Localison (Dorsch)-comb.
        Millicorten (Novartis
        Pharma)-comb.
        Nystalocal (Nourypharma)-
        comb.
        Otobacid (Asche)-comb.
        Predni (Sanirania)
        Predni-F-Tablinen
        (Sanorania)

        Predni-F-Tablinen
        (Sanorania)
        Sokaral (Pharma-Allergan)
        Solutio Cordes (Ichthyol)
        Spersadex (CIBA Vision)
        Tuttozem (Strathmann)
        Tuttozem N (Mayo)
        various combination
        preparations and generics
F: Décadron (Merck Sharp &
        Dohme-Chibret)
        Dectancyl (Roussel
        Diamant; as acetate)
        Maxidex (Alcon)
        numerous combination
        preparations
GB: Decadron (Merck Sharp &
        Dohme)
        Maxidex (Alcon)-comb.

Maxitrol (Alcon)-comb.
Otomize (Stafford-Miller)
Sofradex (Florizel)-comb.

I: Antimicotico liquido/
pomata (IFI)-comb.
Aurizone (SIFI)-comb.
Decadron (Merck Sharp &
Dohme)
Desalark (Farmacologico
Milanese)
Desamix-neomicina
(Savoma)-comb.
Deseronil (Sca)
Fluorobioptal (Farmila)-
comb.
Lasoproct (Bayer)-comb.
Luxazone (Allergan)
Nasicortin (Bracco)-comb.
Neocortofen (Ripari-Gero)-
comb.
Rinedrone (Deca)-comb.
Tobradere (Alcon)-comb.
Visumetazone (Merck
Sharp & Dohme)

various combination
preparations

J: Alpermell (Nippon
Shinyaku)
Amumetazon (Choseido)
Aphtasolon (Showa Yakka)
Bisno-DS (Ohta)
Carulon (Yamanouchi)
Corson (Takeda)
Dab M (Zenyaku)
Decaderm (Banyu)
Decadron (Banyu)
Dectan (Nippon Roussel-
Chugai)
Dekisachosei (Choseido)
Delenar (Schering-
Shionogi)-comb.
Dersene (Ikeda)
Dethamedin (Ohta)
Dexa A (Shinsei Sawai)
Dexaltin (Nippon Kayaku)
Dexamamalet (Showa)
Dexame (Dojin)
Dexasone (Hokuriku)

Eurason D (Ciba-Geigy)
Metasolon (Shionogi)
Mitasone (Toyo Pharmar)
Orgadron (Organon-
Sankyo)
Rheumadex (Nakataki)-
comb.
Rheumatol (Sankyo)-comb.
Santeson (Santa)
Sawasone (Sawai)
Sunia-D Comp. (Zeria
Shinyaku)

USA: Dalalone (Forest; as
acetate)
Decadron (Merck)
Decadron (Merck; as
acetate)
Decaspray (Merck)
TobraDex (Alcon)
several combination
preparations and generics

Dexamethasone *tert*-butylacetate

ATC: A01AC; D07AB; R01AD
Use: glucocorticoid

RN: 24668-75-5   MF: $C_{28}H_{39}FO_6$ MW: 490.61   EINECS: 246-389-2
CN: (11β,16α)-21-(3,3-dimethyl-1-oxobutoxy)-9-fluoro-11,17-dihydroxy-16-methylpregna-1,4-diene-3,20-dione

dexamethasone
(q. v.)

tert-butylacetic
anhydride

Dexamethasone tert-butylacetate

Reference(s):
DOS 2 317 954 (Jelen. Zaklady Farm. Polfa; appl. 10.4.1973; PL-prior. 21.4.1972).

aerosol:
US 3 282 781 (Merck & Co. 1.11.1966; prior. 25.11.1960).

Formulation(s):   nasal drops 0.2 mg

Trade Name(s):
D: Dissiden (Allegopharma)-
comb.; wfm

Nasicortin (Merck)-comb.;
wfm

USA: Decadron T.B.A. (Merck
Sharp & Dohme); wfm

Dexamethasone 21-isonicotinate

ATC: D07AB; S01BA
Use: glucocorticoid

RN: 2265-64-7   MF: $C_{28}H_{32}FNO_6$ MW: 497.56   EINECS: 218-866-5
LD$_{50}$:   3470 mg/kg (M, p.o.);
3562 mg/kg (R, p.o.)
CN: (11β,16α)-9-fluoro-11,17-dihydroxy-16-methyl-21-[(4-pyridinylcarbonyl)oxy]pregna-1,4-diene-3,20-dione

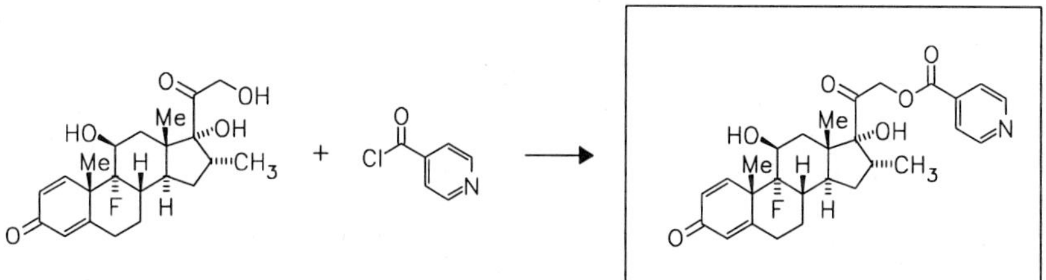

dexamethasone   isonicotinoyl   Dexamethasone 21-isonicotinate
(q. v.)   chloride

Reference(s):
ZA 623 489 (Thomae; appl. 1.8.1962; D-prior. 19.8.1961).

Formulation(s):   aerosol 0.125 mg/puff, 0.02 mg; eye drops 0.25 mg/ml; sol. 0.025 %

Trade Name(s):

D:	Auxiloson (Boehringer Ing.)	Dexa Biciron (Alcon)-comb.	F:	Auxisone (Boehringer Ing.)
	Corti Biciron (S & K Pharma)-comb.	Dexa Loscon (Galderma)	GB:	Dexa-Rhinaspray (Boehringer Ing.)-comb.
		Dexa-Rhinospray (Boehringer Ing.)-comb.	I:	Desalfa (Intes)-comb.

Dexamethasone 21-linolate

ATC: D07AB
Use: glucocorticoid

RN: 39026-39-6   MF: $C_{40}H_{59}FO_6$ MW: 654.90   EINECS: 254-254-4
CN: [11β,16α,21-(Z,Z)]-9-fluoro-11,17-dihydroxy-16-methyl-21-[(1-oxo-9,12-octadecadienyl)oxy]-pregna-1,4-diene-3,20-dione

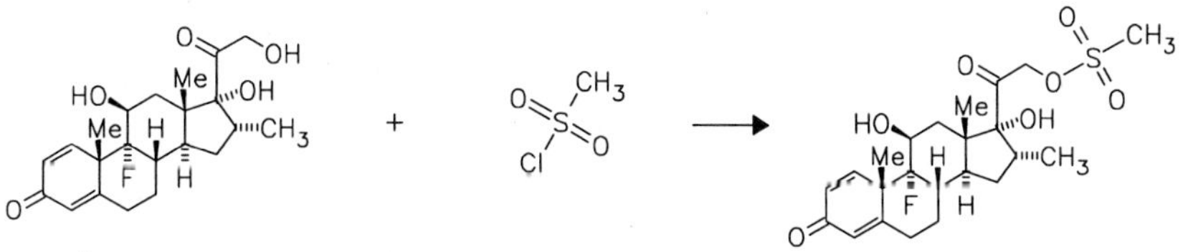

dexamethasone   methanesulfonyl   (I)
(q. v.)   chloride

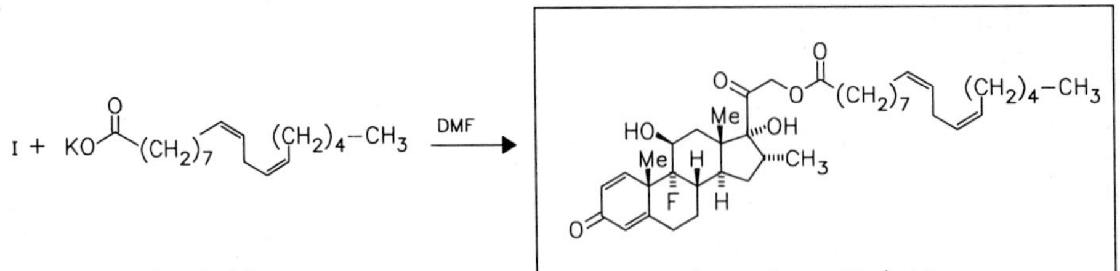

I + KO (CH₂)₇ ... (CH₂)₄-CH₃ potassium linolate — DMF →

Dexamethasone 21-linolate

Reference(s):
GB 1 292 785 (ISF; valid from 19.4.1971; I-prior. 17.10.1970).

Formulation(s):   cream 0.2 %; lotion 0.15 %

Trade Name(s):
I: Kanaderm 200 (Firma)-
      comb. with kanamycine
      Situalin (Puropharma)

Situalin Antibiotico
(Puropharma)-comb. with
bekanamycine

Dexamethasone phosphate

ATC: H02AB; S01BA; S03BA; D07AB
Use: glucocorticoid

RN: 312-93-6   MF: C₂₂H₃₀FO₈P   MW: 472.45   EINECS: 206-232-0
CN: (11β,16α)-9-fluoro-11,17-dihydroxy-16-methyl-21-(phosphonooxy)pregna-1,4-diene-3,20-dione

disodium salt
RN: 2392-39-4   MF: C₂₂H₂₈FNa₂O₈P   MW: 516.41   EINECS: 219-243-0
LD₅₀: 1800 mg/kg (M, p.o.)

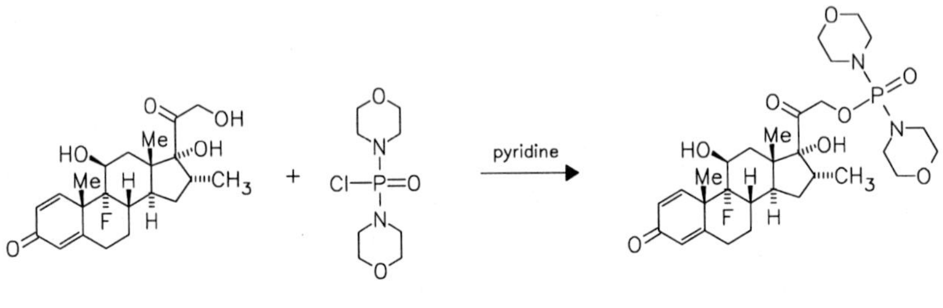

dexamethasone
(q. v.)

dimorpholino-
phosphinic
chloride

pyridine →

(I)

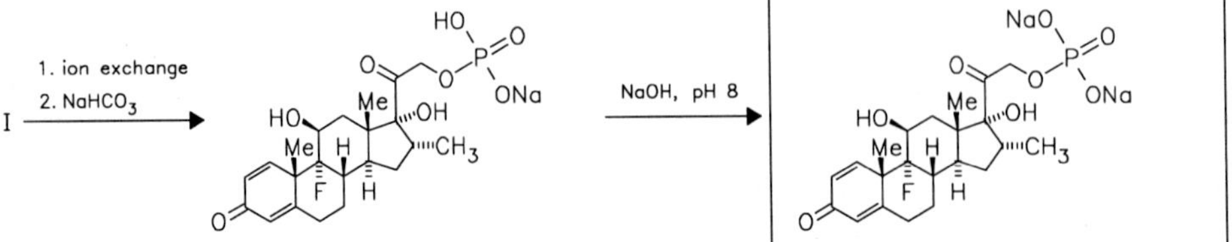

I — 1. ion exchange / 2. NaHCO₃ →

NaOH, pH 8 →

Dexamethasone phosphate

Reference(s):
DE 1 134 075 (Merck AG, appl. 26.11.1959).

alternative synthesis:
US 2 939 873 (Merck & Co.; 7.6.1960; appl. 26.1.1959; prior. 20.11.1957).
Jrmscher, K.: Chem. Ind. (London) (CHINAG) **1961**, 1035.

Formulation(s):   amp. 5 mg/ml, 48 mg/2 ml, 20 mg/5 ml, 8 mg/2 ml, 120 mg/5 ml; eye drops 1.1 mg/ml;
ointment 0.2 %

Trade Name(s):

D: Dexabene (Merckle)
    Dexa-Brachialin
    (Steigerwald)
    Dexa-Effekton (Brenner-
    Efeka)
    Spersadexolin (Dispersa)-
    comb.
    Totocortin (Winzer)
    various combination
    preparations and generics
F: Cébédex (Chauvin; as
    disodium salt)
    Cébédexacol (Chauvin; as
    disodium salt)-comb.

Chibro-Cardon (Merck
Sharp & Dohme-Chibret;
as monosodium salt)-comb.
Corticétine (Chauvin; as
monosodium salt)-comb.
Dexagrane (Leurquin; as
monosodium salt)
Frakidex (Chauvin; as
monosodium salt)-comb.
Soludécadron (Merck
Sharp & Dohme-Chibret;
as monosodium salt)
I: Decadron Fosfato (Merck
    Sharp & Dohme)
    Desalark (Farm. Mil.)

Eta-Biocortilen (SIFI)-
comb. with neomycine
Eta-Cortilen (SIFI)
Kanazone (SIT)-comb.
with kanamycine
Soldesam (Farm. Mil.)
J: Corson (Takeda)
    Decadron (Banyu)
    Donray (Kodama)
    Orgadrone (Sankyo)
    Solcort (Fuji)
    Teikason (Teika)
USA: Decadron Phosphate
    (Merck)
    Neo-Decadron Phosphate
    (Merck)-comb.

Dexamethasone pivalate
(Dexamethasone trimethylacetate)

ATC: D07AB
Use: glucocorticoid

RN: 1926-94-9   MF: $C_{27}H_{37}FO_6$ MW: 476.59   EINECS: 217-659-7
CN: (11β,16α)-21-(2,2-dimethyl-1-oxopropoxy)-9-fluoro-11,17-dihydroxy-16-methylpregna-1,4-diene-3,20-
dione

dexamethasone   pivaloyl   Dexamethasone pivalate
(q. v.)   chloride

Reference(s):
US 3 033 881 (Ciba; 8.5.1962; CH-prior. 4.7.1958).
CH 398 585 (Ciba; appl. 1956)

alternative syntheses:
ES 320 497 (Lab. M. Cuatrecasas; appl. 30.11.1965).
DOS 2 317 954 (Jelen. Zaklady Farm. Polfa; appl. 10.4.1973; PL-prior. 21.4.1972).

Formulation(s):   ointment 0.02 %

Dexamethasone valerate

ATC: D07AB
Use: glucocorticoid

RN: 33755-46-3   MF: $C_{27}H_{37}FO_6$ MW: 476.59   EINECS: 251-669-2
LD$_{50}$:   >3 g/kg (M, p.o.);
>3 g/kg (R, p.o.)
CN: (11β,16α)-9-fluoro-11,21-dihydroxy-16-methyl-17-[(1-oxopentyl)oxy]pregna-1,4-diene-3,20-dione

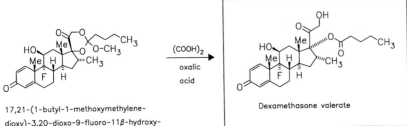

17,21-(1-butyl-1-methoxymethylene-
dioxy)-3,20-dioxo-9-fluoro-11β-hydroxy-
16α-methyl-1,4-pregnadiene
(cf. betamethasone valerate)

Dexamethasone valerate

Reference(s):
DOS 2 111 114 (Inst. Luso Farmaco; appl. 9.3.1971; I-prior. 14.3.1970).

alternative syntheses:
DOS 2 055 221 (Lab. Chim. Farm. Blasina; appl. 10.11.1970).

Formulation(s):   cream 0.1 %

Dexbrompheniramine

ATC: R06AB06
Use: antihistaminic

RN: 132-21-8   MF: $C_{16}H_{19}BrN_2$ MW: 319.25   EINECS: 205-053-5
CN: (S)-γ-(4-bromophenyl)-N,N-dimethyl-2-pyridinepropanamine

maleate (1:1)
RN: 2391-03-9   MF: $C_{16}H_{19}BrN_2 \cdot C_4H_4O_4$ MW: 435.32   EINECS: 219-236-2
LD$_{50}$:   25 mg/kg (M, i.v.); 176 mg/kg (M, p.o.);
191 mg/kg (R, p.o.)

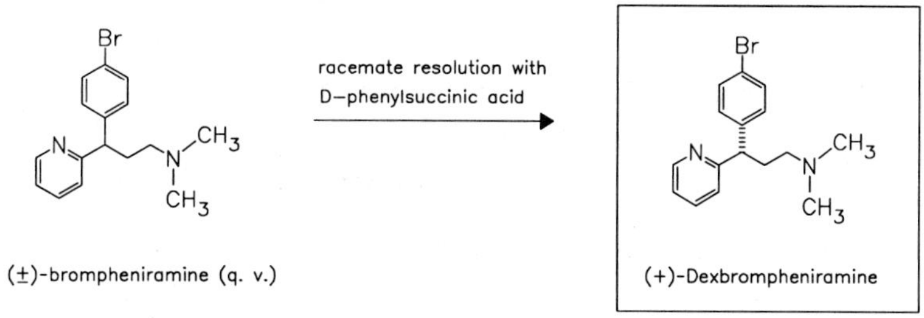

(±)-brompheniramine (q. v.) → racemate resolution with D—phenylsuccinic acid → (+)-Dexbrompheniramine

Reference(s):
US 3 030 371 (L. A. Walter; 17.4.1962; appl. 1958).
US 3 061 517 (Schering Corp.; 30.10.1962; prior. 16.2.1962).

Formulation(s): tabl. 2 mg

Trade Name(s):
USA: Disobrom (Geneva)

Dexchlorpheniramine

ATC: R06AB02
Use: antihistaminic

RN: 25523-97-1 MF: $C_{16}H_{19}ClN_2$ MW: 274.80 EINECS: 247-073-7
CN: (S)-γ-(4-chlorophenyl)-N,N-dimethyl-2-pyridinepropanamine

maleate (1:1)
RN: 2438-32-6 MF: $C_{16}H_{19}ClN_2 \cdot C_4H_4O_4$ MW: 390.87 EINECS: 219-450-6
LD_{50}: 28 mg/kg (M, i.v.); 189 mg/kg (M, p.o.);
267 mg/kg (R, p.o.);

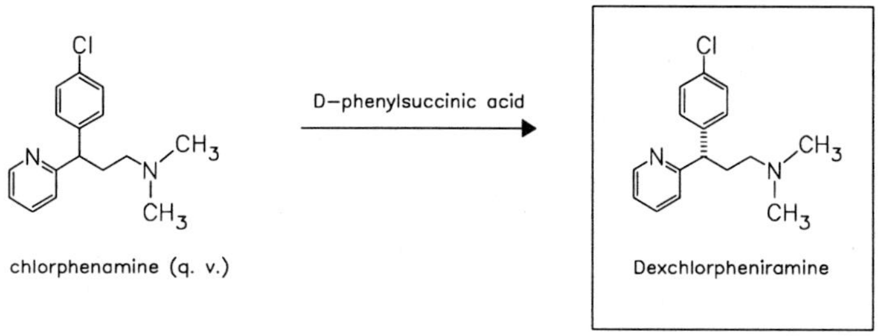

chlorphenamine (q. v.) → D—phenylsuccinic acid → Dexchlorpheniramine

Reference(s):
GB 834 984 (Scherico; appl. 4.7.1958; USA-prior. 4.3.1958).
US 3 030 371 (L. H. Walter; 17.4.1962; appl. 4.3.1958).
US 3 061 517 (Schering Corp.; 30.10.1962; prior. 16.2.1962).

Formulation(s): drg. 6 mg; tabl. 2 mg

Trade Name(s):
D: Celestamine (Essex)-comb.
Polaronil (Byk Essex)
F: Celestamine (Schering-
Plough; as meleate)

Polaramine (Schering-
Plough; as maleate)
I: Polaramin (Schering-
Plough)

J: Polaramine (Schering-
Shionogi)
USA: Baylarmine (Bay); wfm

Dexfenfluramine

ATC: A08AA04
Use: antiobesity, S-enantiomer of
fenfluramine

RN: 3239-44-9   MF: $C_{12}H_{16}F_3N$ MW: 231.26
LD$_{50}$:   115 mg/kg (R, p.o.)
CN: (S)-N-ethyl-α-methyl-3-(trifluoromethyl)benzeneethanamine

hydrochloride
RN: 3239-45-0   MF: $C_{12}H_{16}F_3N \cdot HCl$ MW: 267.72   EINECS: 221-806-0

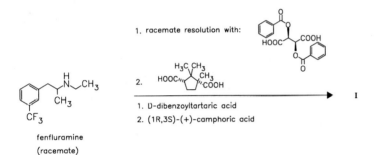

fenfluramine
(racemate)

(I)

Dexfenfluramine

Reference(s):
DE 1 293 774 (Sience-Union et Cie., Soc. Franç. de Recherche Médicale; appl. 22.6.1965; USA-prior.
27.7.1964).
GB 1 078 186 (Sience-Union et Cie., Soc. Franç. de Recherche Médicale; appl. 16.6.1965; USA-prior.
27.7.1964).

medical use as antidepressant:
EP 2 531 146 (R. J. Wurtman et al.; USA-prior. 16.6.1987).

medical use for intermittent carbohydrate craving:
EP 53 175 (J. R.Wurtman et al.; appl. 15.6.1981; USA-prior. 16.6.1980).

Formulation(s):   cps. 15 mg

Trade Name(s):
D:	Isomeride (as hydrochloride); wfm	GB:	Adifax (Servier; 1990 as hydrochloride); wfm	Isomeride (Servier; 1990 as hydrochloride)
F:	Isomeride (Ardix; as hydrochloride); wfm	I:	Glypolix (Servier; as hydrochloride)	USA: Redux (Wyeth-Ayerst)

Dexketoprofen trometamol

((S)-(+)-Ketoprofen; LM-1158 as acid)

ATC: M01AE17
Use: analgesic, anti-inflammatory

RN: 156604-79-4   MF: $C_{16}H_{14}O_3 \cdot C_4H_{11}NO_3$ MW: 375.42
CN: (S)-3-benzoyl-α-methylbenzeneacetic acid compd. with 2-amino-2-(hydroxymethyl)-1,3-propanediol (1:1)

dexketoprofen
RN: 22161-81-5   MF: $C_{16}H_{14}O_3$ MW: 254.29

1. C_2H_5OH, H^+
2. enzymatic hydrolysis of (S)-ester with esterase from ophiostoma or ceratocystis

(±)-ketoprofen (q. v.)

(S)-ketoprofen (I)

C_2H_5OH
H_2O

Dexketoprofen trometamol

Reference(s):
WO 9 411 332 (Lab. Menarini; appl. 9.11.1993; E-prior. 10.11.1992).
WO 9 420 449 (Dompé Farmac.; appl. 7.3.1994; I-prior. 9.3.1993).

enantio-selective synthesis of (+)-(S)-2-(3-benzoylphenyl)propionic acid:
Fadel, A.: Synlett. (SYNLES) **1**, 48 (1992).

stereoselective hydrolysis of ketoprofene esters using esterase:
US 5 912 164 (Lab. Menarini; appl. 9.5.1997; GB-prior. 3.3.1993; 3.3.1994; USA-prior. 31.8.1995; 5.9.1995).
WO 9 420 633 (Lab. Menarini; appl. 9.5.1997; GB-prior. 3.3.1993; 3.3.1994; USA-prior. 31.8.1995; 5.9.1995).
WO 9 304 189 (Lab. Menarini; appl. 19.8.1992; GB-prior. 22.8.1991).
Hernaiz, M.J.: J. Mol. Catal. A: Chem. (JMCCF2) **96** (3), 317 (1995).
Garcia, M.: Biotechnol. Lett. (BILED3) **19** (10), 999 (1997).
WO 9 015 146 (Rhône-Poulenc; appl. 1.6.1990; USA-prior. 5.6.1989).

Formulation(s):   tabl. 12.5 mg, 25 mg

Trade Name(s):
D: Enantyum (Lab. Menarini; 1998)

Dexpanthenol

(Pantothenyl alcohol; Panthenol)

ATC: A11HA30; D03AX03; S01XA12
Use: growth factor, wound remedy

RN: 81-13-0   MF: $C_9H_{19}NO_4$ MW: 205.25   EINECS: 201-327-3
LD_{50}: 7 g/kg (M, i.v.); 15 g/kg (M, p.o.)
CN: (R)-2,4-dihydroxy-N-(3-hydroxypropyl)-3,3-dimethylbutanamide

D(–)-2-hydroxy-
3,3-dimethyl-
butanolide

3-amino-1-
propanol

Dexpanthenol

Reference(s):
US 2 413 077 (Roche; 1946; CH-prior. 1942).

use as aerosol:
DAS 2 531 260 (Desitin-Werke; appl. 12.7.1975).

Formulation(s):   amp. 500 mg; emulsion 50 mg; eye and nasal ointment 50 mg; inj. sol. 500 mg/2 ml; nasal spray 50 mg; ointment 5 %, 50 mg; sol. 50 mg; tabl. 100 mg

Trade Name(s):

D:	Bepanthen Roche (Roche)	Ucee (Merck Produkte)	Hydrosol polyvitaminé
	Corveregel (Mann)	Urupan (Merckle)	(Roche)-comb.
	Cutemol (Medopharm)	generics	I: Bepanten (Roche)
	Dexpanthenol Heumann F:	Alvityl (Solvay Pharma)-	Pantenolo(Formulario
	(Heumann)	comb.	Naz.)
	Marolderm (Dermapharm)	Bécozyme injèctable	J: Pantene (Shionogi)
	Pan-Ophthal (Winzer)	(Roche)-comb.	Pantol (Toa Eiyo-
	Panthenol Drobena	Bepanthene (Roche)	Yamanouchi)
	(Drobena)	Cernévit (Clintec Nutrition USA:	Ilopan (Savage)
	Pelina (MIP Pharma)	Clinique)-comb.	Ilopan-Choline (Savage)

Dexrazoxane

(ICRF-187)

ATC: V03AB; V03AF02
Use: antineoplastic, protectant of anthracycline induced cardiotoxicity, (+)-enantiomer of razoxane (q. v.)

RN: 24584-09-6   MF: $C_{11}H_{16}N_4O_4$ MW: 268.27
CN: (S)-4,4'-(1-methyl-1,2-ethanediyl)bis[2,6-piperazinedione]

(S)-1,2-propane-
diamine

paraform-
aldehyde

(S)-N,N,N',N'-tetrakis-
(cyanomethyl)-1,2-
propanediamine   (I)

Dexrazoxane

Reference(s):
EP 330 381 (Erbamont; appl. 17.2.1989; USA-prior. 17.2.1988)

alternative synthesis:

EP 2 845 594 (Monsanto; appl. 22.3.1988; USA-prior. 23.3.1987).
DE 1 910 283 (National Research Development Corp.; appl. 28.2.1969; USA-prior. 2.7.1968).
GB 1 234 935 (National Research Development Corp.; appl. 3.7.1967).

Formulation(s):   lyo. for inf. 500 mg

Trade Name(s):
I: Cardioxane (Eurocetus)   USA: Zinecard (Pharmacia &
      Eucardion (Dompé Biotec)   Upjohn)

Dextromethorphan
(D-Methorphan)

ATC: R05DA09
Use: antitussive, analgesic

RN: 125-71-3   MF: $C_{18}H_{25}NO$ MW: 271.40   EINECS: 204-752-2
LD_{50}:   210 mg/kg (M, p.o.);
        16.286 mg/kg (R, i.v.); 116 mg/kg (R, p.o.)
CN: $(9\alpha,13\alpha,14\alpha)$-3-methoxy-17-methylmorphinan

hydrobromide
RN: 125-69-9   MF: $C_{18}H_{25}NO \cdot HBr$ MW: 352.32   EINECS: 204-750-1
LD_{50}:   34 mg/kg (M, i.v.); 165 mg/kg (M, p.o.);
        350 mg/kg (R, p.o.)
hydrobromide monohydrate
RN: 6700-34-1   MF: $C_{18}H_{25}NO \cdot HBr \cdot H_2O$ MW: 370.33

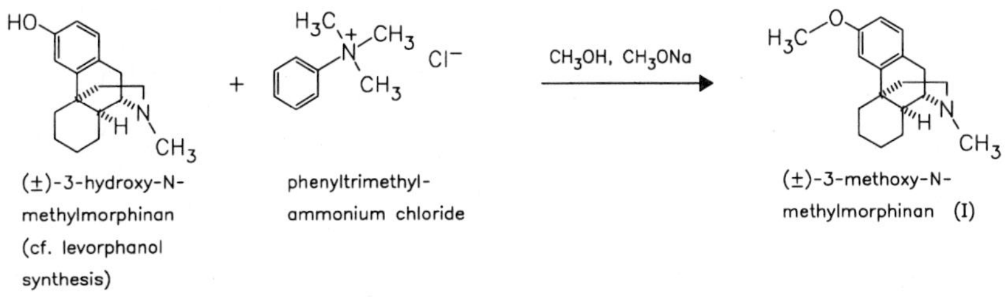

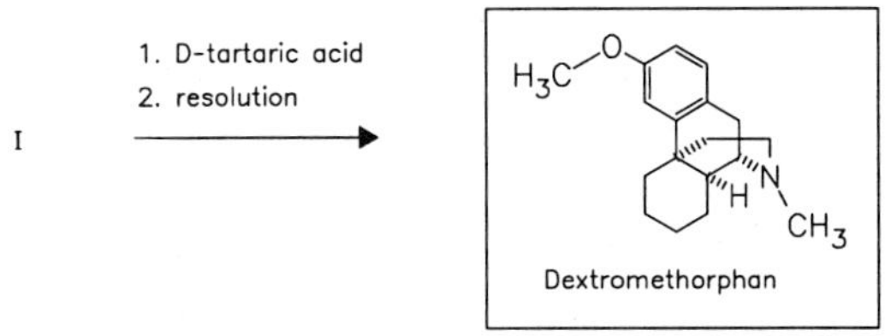

(±)-3-hydroxy-N-   phenyltrimethyl-   (±)-3-methoxy-N-
methylmorphinan   ammonium chloride   methylmorphinan   (I)
(cf. levorphanol
synthesis)

Dextromethorphan

Reference(s):
US 2 676 177 (Roche; 1954; CH-prior. 1949).
Schnider, O.; Grüssner, A.: Helv. Chim. Acta (HCACAV) **34**, 2211 (1951).

medical use as analgesic:
US 4 316 888 (Nelson Research; 23.2.1982; appl. 15.4.1980).
US 4 446 140 (Nelson Research; 1.5.1984; prior. 10.12.1981, 29.3.1982).

nasal use as antitussive:
US 4 454 140 (Roche; 12.6.1984; appl. 7.9.1982).

Formulation(s):   syrup 5 mg, 6.65 mg

Trade Name(s):

D: Arpha (Fournier Pharma)
NeoTussan (Novartis)
Robitussin plus
(Scheurich)-comb.
tuss (Rentschler)
Wick (Wick Pharma)

F: Nodex (Brothier; as
hydrobromide)
Nortussine (Norgine; as
hydrobromide)-comb.
Tuxium (Galephar; as
hydrobromide)

GB: Actifed Compound
(Wellcome)-comb.; wfm
Actifed Compound Linctus
(Wellcome)-comb.; wfm
Benafed (Parke Davis)-
comb.; wfm
Cosylan (Parke Davis);
wfm
Lotussin (Searle); wfm
Syrtussar (Armour); wfm

I: Actifed (Wellcome)-comb.
Aricodil (Malesci)-comb.
Balsatux (Edmond)-comb.
Bcchilar (Montefarmaco)
Benadryl Complex (Parke
Davis)-comb.
Broncal (SmithKline
Beecham)-comb.
Bronchenolo Tosse (Midy)
Broncobeta (Beta)-comb.
Broncodex (Pastor Farina)-
comb.

Canfodion (Gentili)
DextroB Afo (Afom)
Euci (Falqui)-comb.
Fluprim (Roche)
Ingro (Farmacologico
Milanese)-comb.
Iodozan (SmithKline
Beecham)-comb.
Neoborocillina
(Schiapparelli)-comb.
Ozopulmin (Geymonat)-
comb.
Resyl (Zyma)-comb.
Romilar (Roche)-comb.
Sanabronchiol (Kalda)
Sedotus Valda (Valda)
Torfan (Abbott)-comb.
Valatux (Farmacologico
Milanese)
Vicks Medinait (Procter &
Gamble)-comb.

J: Coughcon (Kyowa
Yakuhin-Santen)
Dextophan (Hishiyama)
Hihustan-M (Maruko)
Medicon (Shionogi)
Methorcon (Kyowa
Yakuhin)
Oricolon (Dojin)
Radeophan (Tokyo Tanabe)
Testamin (Toyama)

USA: Anatuss (Merz; as
hydrobromide)

Benylin (Warner-Lambert;
as hydrobromide)
Bromfed-DM (Muro; as
hydrobromide)
Codimal (Schwarz; as
hydrobromide)
Diabe-Tuss DM (Paddock;
as hydrobromide)
Dimetane-DX (Robins; as
hydrobromide)
Donatussin (Laser; as
hydrobromide)
Duratuss DM (UCB; as
hydrobromide)
Fenesin DM (Dura; as
hydrobromide)
Muco-Fen (Wakefield; as
hydrobromide)
Poly-Histine DM (Sanofi;
as hydrobromide)
Safe Tussin (Kramer; as
hydrobromide)
Syn-Rx DM (Medeva; as
hydrobromide)
Tussar DM (Rhône-
Poulenc Rorer; as
hydrobromide)-comb.
Tussi-Organidin (Wallace;
as hydrobromide)
Tylenol (McNeil; as
hydrobromide)
generics

Dextromoramide

ATC: N02AC01
Use: analgesic

RN: 357-56-2　MF: $C_{25}H_{32}N_2O_2$　MW: 392.54　EINECS: 206-613-1
LD$_{50}$: 21 mg/kg (M, i.v.); 168 mg/kg (M, p.o.);
13 mg/kg (R, i.v.); 71.8 mg/kg (R, p.o.)
CN: (S)-1-[3-methyl-4-(4-morpholinyl)-1-oxo-2,2-diphenylbutyl]pyrrolidine

bitartrate (1:1)
RN: 2922-44-3　MF: $C_{25}H_{32}N_2O_2 \cdot C_4H_6O_6$　MW: 542.63　EINECS: 220-870-7
LD$_{50}$: 71.8 mg/kg (R, p.o.)

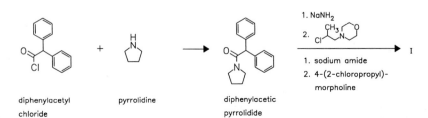

diphenylacetyl
chloride

pyrrolidine

diphenylacetic
pyrrolidide

1. NaNH$_2$
2.

1. sodium amide
2. 4-(2-chloropropyl)-
morpholine

I

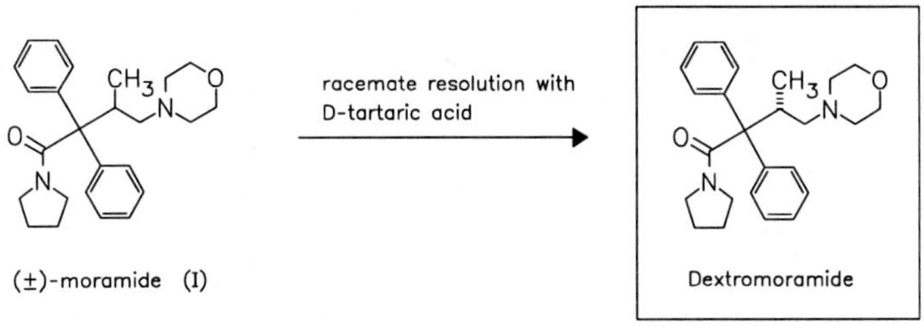

(±)-moramide (I)

Dextromoramide

Reference(s):
BE 544 757 (Janssen; appl. 5.2.1957; NL-prior. 9.2.1956).
DE 1 117 126 (Janssen; appl. 5.12.1956; NL-prior. 9.2.1956).
GB 822 055 (Janssen; appl. 23.10.1956; NL-prior. 9.2.1956).

Formulation(s):   suppos. 13.8 mg; tabl. 6.9 mg, 13.8 mg (as bitartrate)

Trade Name(s):
D: Jetrium (Hek); wfm   GB: Palfium (B.M. Pharm)
F: Palfium (Delalande)   I: Narcolo (Lusofarmaco)

Dextropropoxyphene

(Dextropropoxiphene; α-D-Propoxyphene; Propoxyphene)

ATC: N02AC04
Use: analgesic

RN: 469-62-5   MF: C$_{22}$H$_{29}$NO$_2$ MW: 339.48   EINECS: 207-420-5
LD$_{50}$:   25 mg/kg (M, i.v.); 140 mg/kg (M, p.o.);
           135 mg/kg (R, p.o.)
CN: [S-(R*,S*)]-α-[2-(dimethylamino)-1-methylethyl]-α-phenylbenzeneethanol propanoate (ester)

hydrochloride
RN: 1639-60-7   MF: C$_{22}$H$_{29}$NO$_2$ · HCl   MW: 375.94   EINECS: 216-683-5
LD$_{50}$:   28 mg/kg (M, i.v.); 282 mg/kg (M, p.o.);
           15 mg/kg (R, i.v.); 230 mg/kg (R, p.o.)
           29 mg/kg (dog, i.v.); 100 mg/kg (dog, p.o.)
napsylate (1:1) monohydrate
RN: 26570-10-5   MF: C$_{22}$H$_{29}$NO$_2$ · C$_{10}$H$_8$O$_3$S · H$_2$O   MW: 565.73
LD$_{50}$:   973 mg/kg (M, p.o.);
           485 mg/kg (R, p.o.); 990 mg/kg (Rf, p.o.)

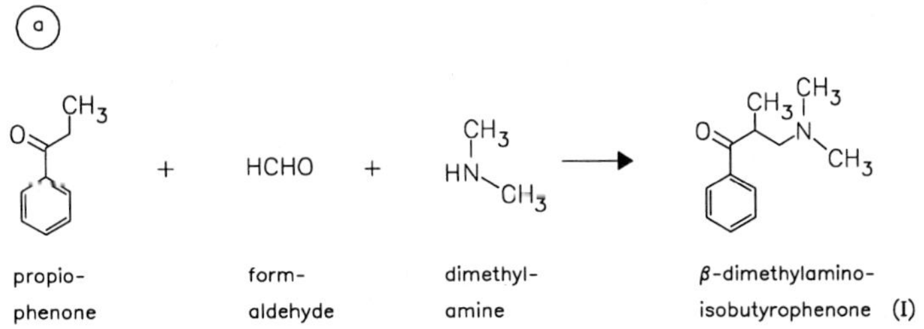

propio-   form-   dimethyl-   β-dimethylamino-
phenone   aldehyde   amine   isobutyrophenone (I)

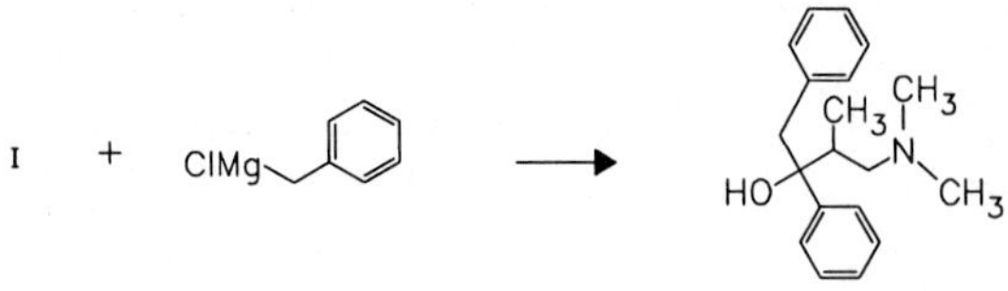

I + benzylmagnesium chloride (II) →

α-(±)-4-dimethylamino-1,2-diphenyl-3-methyl-2-butanol (III)
(75%, crystallizes)
along with 15% of the diastereomeric β form, which is better soluble and remains in solution

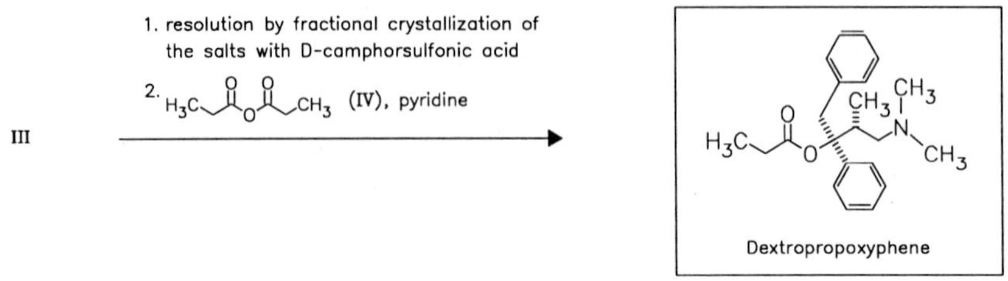

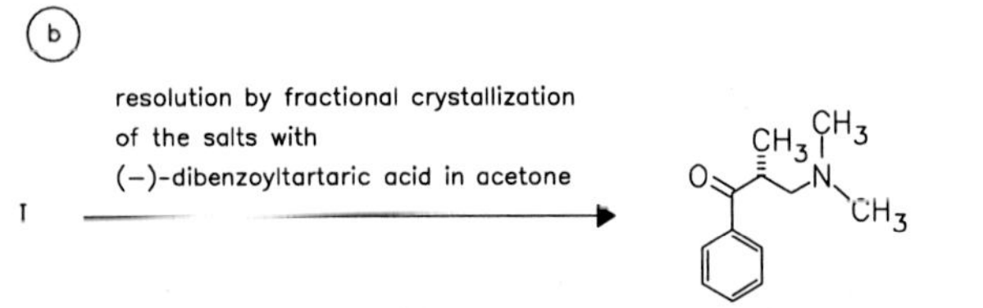

III

1. resolution by fractional crystallization of the salts with D-camphorsulfonic acid

2. H_3C—(IV), pyridine

→ Dextropropoxyphene

(b)

resolution by fractional crystallization of the salts with (–)-dibenzoyltartaric acid in acetone

I →

(–)-β-dimethylamino-isobutyrophenone (V)

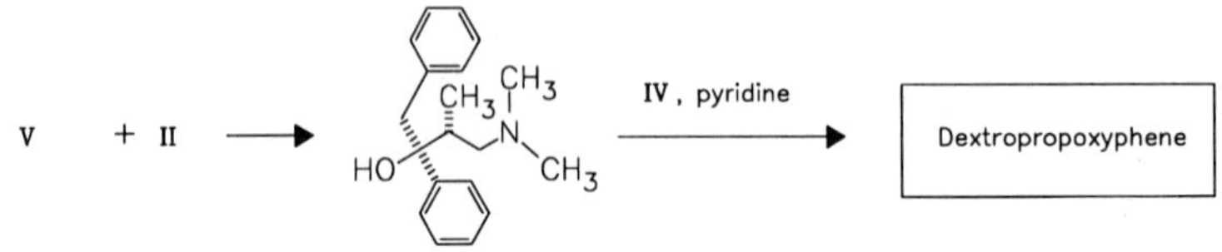

V + II →

α-(+)-4-dimethyl-amino-1,2-diphenyl-3-methyl-2-butanol
(69%)

IV, pyridine → Dextropropoxyphene

Reference(s):
US 2 728 779 (Lilly; 1955; prior. 1952).
Pohland, A.; Sullivan, H.R.: J. Am. Chem. Soc. (JACSAT) **75**, 4458 (1953); **77**, 3400 (1955).
Pohland, A. et al.: J. Org. Chem. (JOCEAH) **28**, 2483 (1963).

Formulation(s):   cps. 150 mg (hydrochloride; s. r. formulation); cps. 65 mg (hydrochloride); tabl. 100 mg (napsylate); susp. 50 mg/5 ml (napsylate)

Trade Name(s):
D: Develin retard (Gödecke)
F: Antalvic (Hoechst Houdé; as hydrochloride)

Di-Antalvic (Hoechst Houdé; as hydrochloride)-comb.
Propofan (Marion Merrell)-comb.

GB: Cosalgesic (Lox)-comb.
Distalgesic (Dista)-comb.
Doloxene (Lilly)
Doloxene Co. (Lilly)-comb.
I: Liberen (Lisapharma)

Dextrothyroxine

(D-Thyroxine)

ATC: C10AX01
Use: cholesterol depressant,
        antihyperlipidemic

RN: 51-49-0   MF: $C_{15}H_{11}I_4NO_4$ MW: 776.87   EINECS: 200-102-7
CN: O-(4-hydroxy-3,5-diiodophenyl)-3,5-diiodo-D-tyrosine

sodium salt
RN: 137-53-1   MF: $C_{15}H_{10}I_4NNaO_4$ MW: 798.85   EINECS: 205-301-2

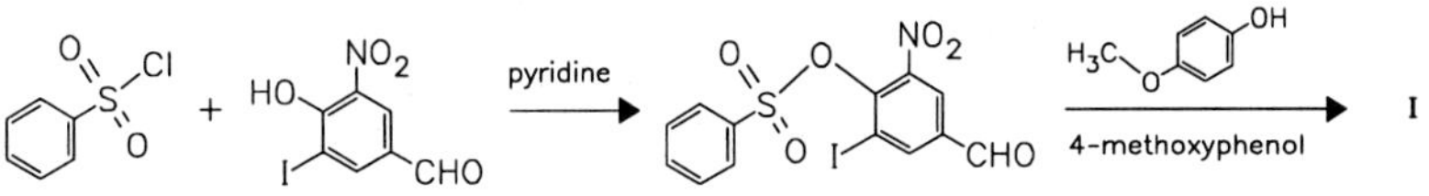

benzene-
sulfochloride

4-hydroxy-
3-iodo-5-nitro-
benzaldehyde

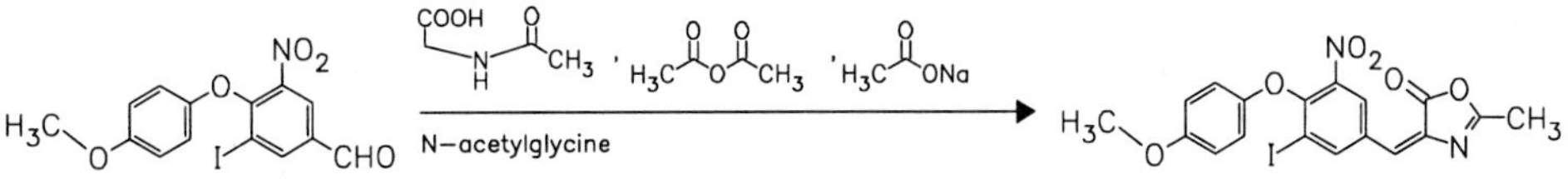

4-methoxyphenol

3-iodo-4-(4-methoxy-
phenoxy)-5-nitro-
benzaldehyde (I)

N-acetylglycine

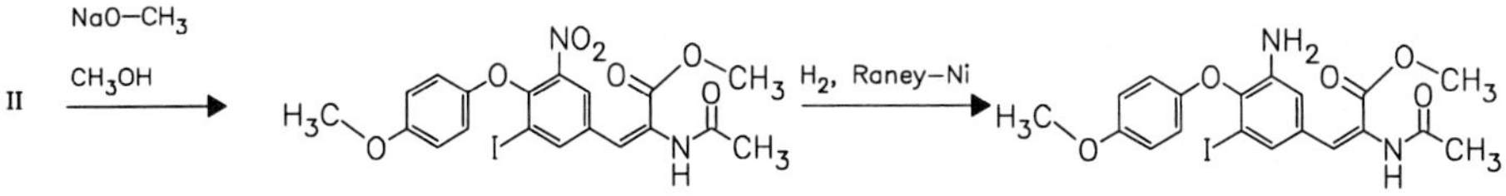

(II)

NaO–CH₃

CH₃OH

H₂, Raney–Ni

(III)

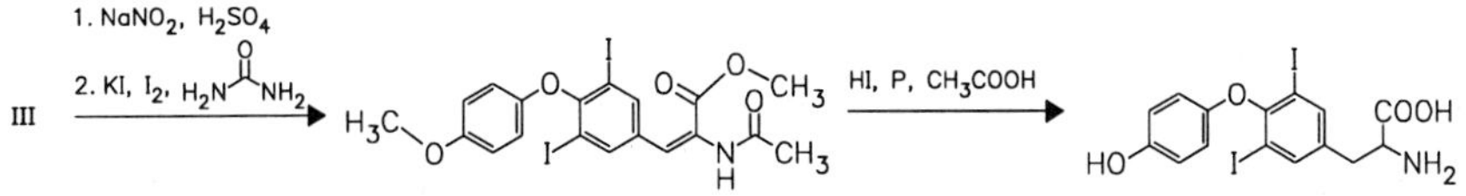

1. NaNO₂, H₂SO₄
2. KI, I₂, H₂N–NH₂

HI, P, CH₃COOH

methyl α-acetamido-3,5-diiodo-4-
(1-methoxyphenoxy)cinnamate

DL-3,5-diiodothyronine (IV)

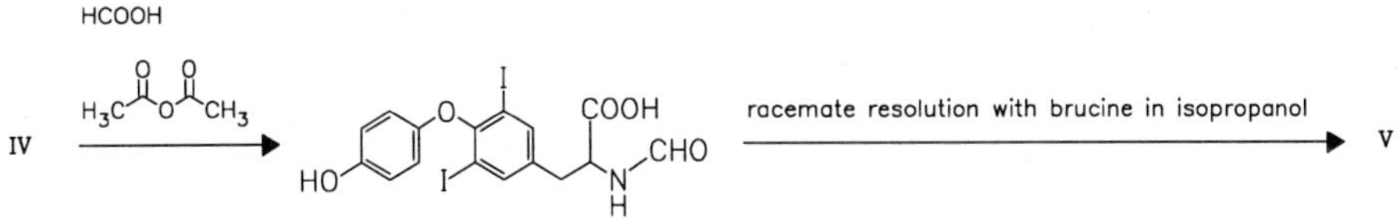

HCOOH

DL-N-formyl-3,5-
diiodothyronine

racemate resolution with brucine in isopropanol   V

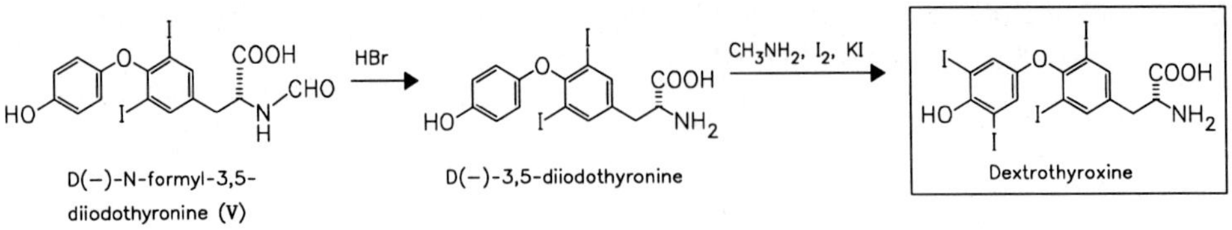

D(−)-N-formyl-3,5-
diiodothyronine (V)

D(−)-3,5-diiodothyronine

Dextrothyroxine

Reference(s):
Nahm, H.; Siedel, W.: Chem. Ber. (CHBEAM) **96**, 1 (1963).
DE 1 067 826 (Hoechst; appl. 24.12.1955).
DE 1 077 673 (Hoechst; appl. 19.8.1958).

Formulation(s):   tabl. 2 mg

Trade Name(s):
D: Dynothel (Henning Berlin)   Nadrothyron-D (Nadrol)   USA: Choloxin (Flint); wfm
      Eulipos (Boehringer   F: Biotirmone (Solac); wfm
      Mannh.)   Débétrol (Choay); wfm

Dezocine

(Wy-16225)

ATC: N02AX03
Use: central acting analgesic, mixed opioid
        agonist antagonist related to
        pentazocine

RN: 53648-55-8   MF: C$_{16}$H$_{23}$NO   MW: 245.37
LD$_{50}$:   129 mg/kg (M, i.m.); 313 mg/kg (M, p.o.);
        270 mg/kg (R, i.m.); 232 mg/kg (R, p.o.)
CN: [5R-(5α,11α,13S*)]-13-amino-5,6,7,8,9,10,11,12-octahydro-5-methyl-5,11-methanobenzocyclodecen-3-
       ol

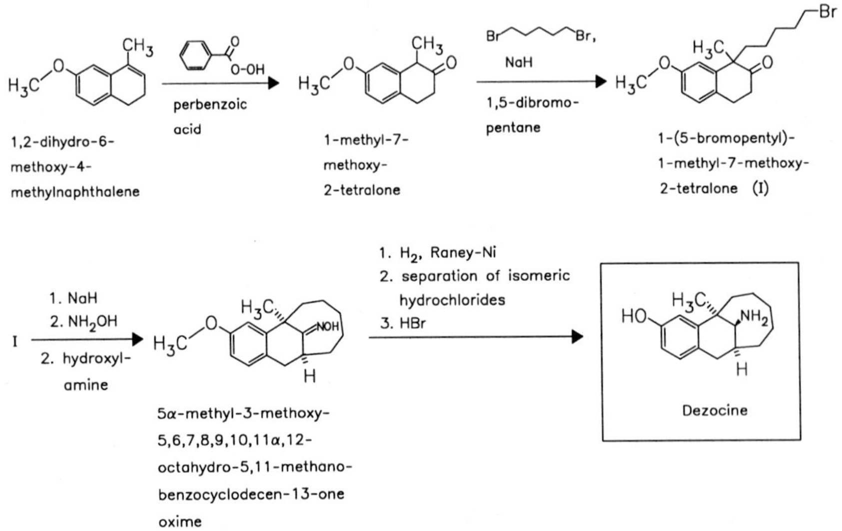

1,2-dihydro-6-
methoxy-4-
methylnaphthalene

perbenzoic
acid

1-methyl-7-
methoxy-
2-tetralone

1,5-dibromo-
pentane

1-(5-bromopentyl)-
1-methyl-7-methoxy-
2-tetralone (I)

5α-methyl-3-methoxy-
5,6,7,8,9,10,11α,12-
octahydro-5,11-methano-
benzocyclodecen-13-one
oxime

Dezocine

Reference(s):
BE 776 173 (American Home; appl. 2.12.1971; USA-prior. 4.12.1970).
DE 2 159 324 (American Home; appl. 30.11.1971; USA-prior. 3.12.1970).
Freed, M.E. et al.: J. Med. Chem. (JMCMAR) **19**, 560 (1976); **16**, 595 (1973).

synthesis of 1-methyl-7-methoxy-2-tetralone:
Howele, F.H.; Taylor, D.A.H.: J. Chem. Soc. (JCSOA9) **1958**, 1248.

pharmaceutical formulations:
US 4 605 671 (American Home; 12.8.1986; appl. 23.7.1985; prior. 28.9.1984).
WO 9 000 390 (American Home; appl. 19.6.1989; S-prior. 8.7.1988).
EP 180 303 (American Home; appl. 27.8.1985; USA-prior. 28.9.1984, 23.7.1985).

Formulation(s):   vial and Tubex syringe 5 mg/2 ml, 10 mg/2 ml, 15 mg/2 ml

Trade Name(s):
USA: Dalgan (Wyeth-Ayerst;
          1990)

Diacerein

ATC: M01AX21
Use: anti-inflammatory

RN: 13739-02-1   MF: $C_{19}H_{12}O_8$ MW: 368.30   EINECS: 237-310-2
LD_{50}: 7500 mg/kg (R, route unreported)
CN: 4,5-bis(acetyloxy)-9,10-dihydro-9,10-dioxo-2-anthracenecarboxylic acid

1,8-dihydroxy-3-
hydroxymethyl-
anthraquinone
(e.g. from
barbaloin)

1,8-dihydroxy-3-
anthraquinone-
carboxylic acid

Diacerein

Reference(s):
DOS 2 711 493 (C. A. Friedmann; appl. 16.3.1977; SA-prior. 16.3.1976).
Oesterle, O.A.: Arch. Pharm. (Weinheim, Ger.) (ARPMAS) **241**, 604 (1903).
Robinson, R.; Simonsen, J.L.: J. Chem. Soc. (JCSOA9) **1909**, 1085.
Cahn, R.S.; Simonsen, J.L.: J. Chem. Soc. (JCSOA9) **1932**, 2573.

Formulation(s):   cps. 50 mg

Trade Name(s):
I: Artrodar (Proter)   Fisiodar (Gentili)

Diazepam

ATC: N05BA01
Use: tranquilizer, hypnotic

RN: 439-14-5   MF: $C_{16}H_{13}ClN_2O$ MW: 284.75   EINECS: 207-122-5
CN: 7-chloro-1,3-dihydro-1-methyl-5-phenyl-2H-1,4-benzodiazepin-2-one

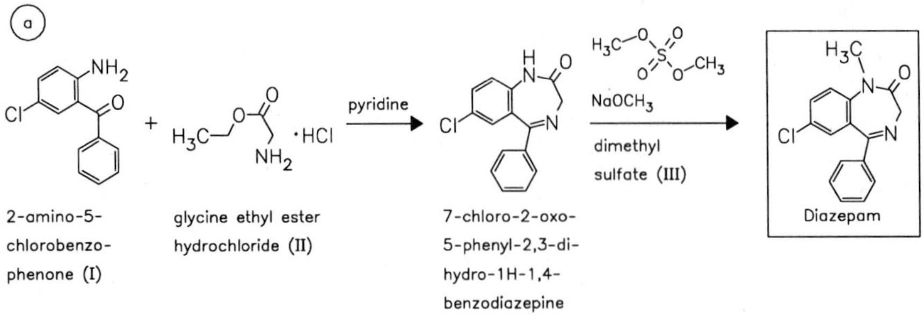
a
2-amino-5-chlorobenzophenone (I)
glycine ethyl ester hydrochloride (II)
pyridine
7-chloro-2-oxo-5-phenyl-2,3-dihydro-1H-1,4-benzodiazepine
$NaOCH_3$
dimethyl sulfate (III)
Diazepam

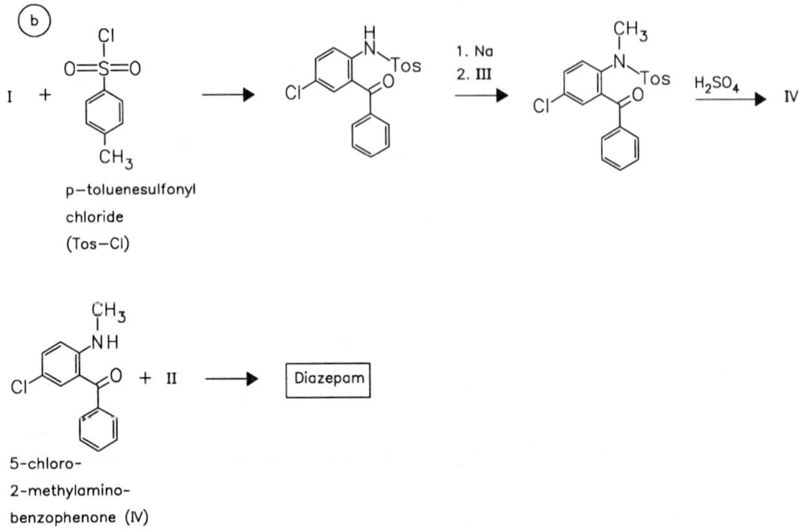
b
I +
p-toluenesulfonyl chloride (Tos–Cl)
1. Na
2. III
H_2SO_4
IV
5-chloro-2-methylaminobenzophenone (IV)
+ II
Diazepam

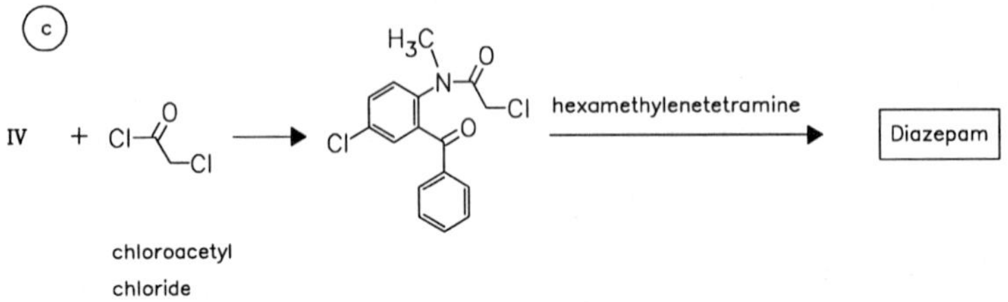
c
IV +
chloroacetyl chloride
hexamethylenetetramine
Diazepam

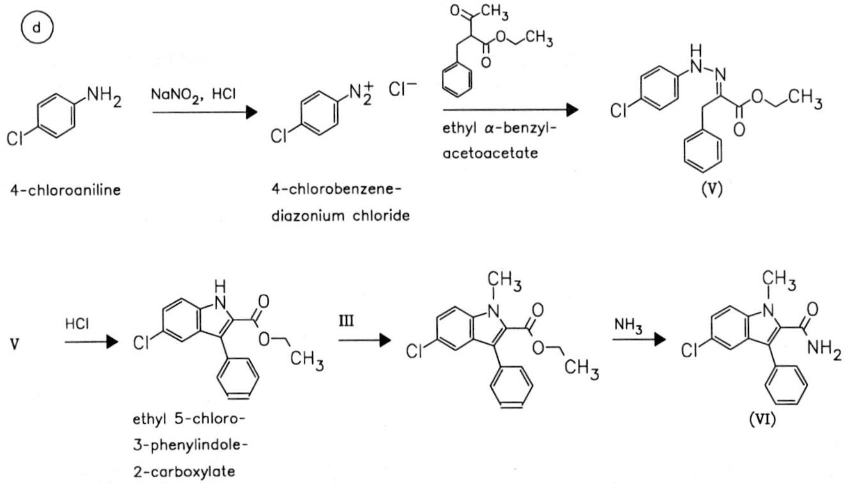
d
4-chloroaniline
$NaNO_2$, HCl
4-chlorobenzenediazonium chloride
ethyl α-benzylacetoacetate
(V)
V
HCl
ethyl 5-chloro-3-phenylindole-2-carboxylate
III
NH_3
(VI)

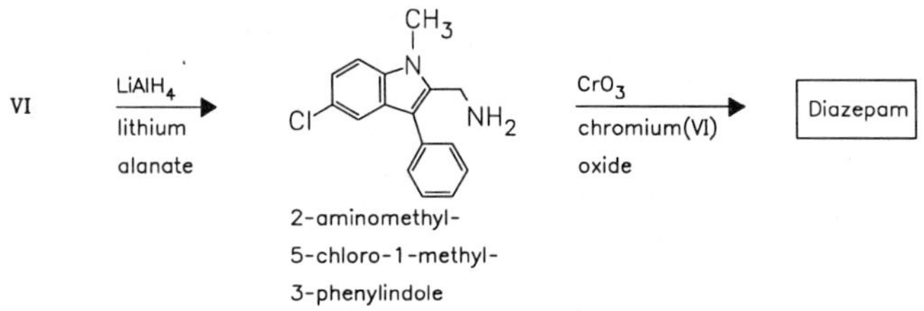

VI $\xrightarrow[\substack{\text{lithium} \\ \text{alanate}}]{\text{LiAlH}_4}$ 2-aminomethyl-5-chloro-1-methyl-3-phenylindole $\xrightarrow[\substack{\text{chromium(VI)} \\ \text{oxide}}]{\text{CrO}_3}$ Diazepam

[Alternatively to the Japp-Klingemann reaction phenylpyruvic acid or ethyl phenylpyruvate can be condensed with 4-chlorophenylhydrazine.]

Reference(s):

a,b US 3 109 843 (Hoffmann-La Roche; 5.11.1963; appl. 21.6.1962; prior. 28.7.1961).
    US 3 136 815 (Hoffmann-La Roche; 9.6.1964; USA-prior. 10.12.1959).
    DE 1 136 709 (Hoffmann-La Roche; appl. 7.12.1960; USA-prior. 10.12.1959).
    DE 1 145 626 (Hoffmann-La Roche; appl. 7.12.1960; USA Prior. 10.12.1959).
    DE 1 290 143 (Hoffmann-La Roche; prior. 7.12.1960).
    US 3 371 085 (Roche; 27.2.1968; CH-prior. 2.10.1960).
c DAS 2 016 084 (Hoffmann-La Roche; appl. 3.4.1970; CH-prior. 16.10.1969).
    DOS 2 233 482 (Hoffmann-La Roche; appl. 7.7.1972; GB-prior. 8.7.1971).
d Yamamoto, H. et al.: Chem. Ber. (CHBEAM) **101**, 4245 (1968).
    US 3 632 573 (Sumitomo; 4.1.1972; J-prior. 9.10.1967).

variant with α-benzylcyanoacetic acid ester:
US 4 069 230 (Sumitomo; 17.1.1978; J-prior. 4.6.1975, 9.6.1975).

alternative syntheses:
DAS 1 545 724 (Delmar Chemicals; appl. 14.1.1965; GB-prior. 14.1.1964).
DAS 1 695 789 (Sumitomo; appl. 2.11.1967; J-prior. 2.11.1966, 16.11.1966, 6.9.1967).
DAS 1 944 404 (Takeda; appl. 2.9.1969; J-prior. 3.9.1968).
DOS 2 252 378 (Roche; appl. 25.10.1972; CH-prior. 18.11.1971).
Sugasawa, T. et al.: J. Heterocycl. Chem. (JHTCAD) **16**, 445 (1979).

purification:
US 3 102 116 (Hoffmann-La Roche; 27.8.1963; prior. 12.3.1962).
DAS 1 906 262 (Sumitomo; appl. 7.2.1969; J-prior. 21.2.1968).

Formulation(s):   amp. (i.v. or i.m.) 10 mg/2 ml; tabl. 2 mg, 5 mg, 10 mg

Trade Name(s):

D: Faustan (ASTA Medica AWD)	I: Aliseum (Zoja)	J: Cercine (Takeda)
Lamra (Merckle)	Ansiolin (Roussel)	Horizon (Yamanouchi)
Stesolid (Dumex)	Eridan (SIT)	Sedaril (Kodama)
Tranquase (Azuchemie)	Noan (Ravizza)	Serenamin (Toyo Jozo)
Valiquid (Roche; 1985)	Spasmeridan (UCB)-comb.	Serenzin (Sumitomo; 1968)
Valium (Roche; 1963)	Spasmomen (Menarini)-comb.	Sonacon (Chugai)
F: Novazam (Génévrier)	Tranquirit (Rhône-Poulenc	USA: Diastat (Athena)
Valium (Roche)	Rorer)	Dizac (Ohmeda)
GB: Diazemuls (Dumex)	Valium (Roche; 1965)	Valium (Roche Products; 1963)
Stesolid (Dumex)	Valpinax (Crinos)-comb.	Valrelease (Roche)
Valclair (Sinclair)	Valtrax (Valeas)-comb.	
Valium (Roche; 1963)	Vatran (Valeas)	

Diazoxide

ATC: C02DA01; V03AH01
Use: antihypertensive, hyperglycemic

RN: 364-98-7   MF: $C_8H_7ClN_2O_2S$ MW: 230.68   EINECS: 206-668-1
LD$_{50}$:   228 mg/kg (M, i.v.); 444 mg/kg (M, p.o.);
        980 mg/kg (R, p.o.)
CN: 7-chloro-3-methyl-2H-1,2,4-benzothiadiazine 1,1-dioxide

5-chloro-2-
aminobenzene-
sulfamide

triethyl
orthoacetate

Diazoxide

Reference(s):
US 2 986 573 (Schering Corp.; 30.5.1961; prior. 18.1.1961).
US 3 345 365 (Schering Corp.; 3.10.1967; prior. 19.9.1960, 18.1.1961, 31.3.1964).

Formulation(s):   amp. 300 mg/20 ml (i.v. inj.); cps. 25 mg, 100 mg

Trade Name(s):
D: Hypertonalum (Essex
        Pharma)
        Proglicem (Essex Pharma)
F: Hyperstat (Schering-
        Plough)

Proglicem (Schering-
Plough)
GB: Eudemine (Fink)
I: Hyperstat (Schering-
        Plough)

Proglicem (Schering-
Plough)
USA: Hyperstat (Schering)
        Proglycem (Baker Norton)

Dibekacin

ATC: J01GB09; J01KD
Use: aminoglycoside antibiotic

RN: 34493-98-6   MF: $C_{18}H_{37}N_5O_8$ MW: 451.52   EINECS: 252-064-6
LD$_{50}$:   373-380 mg/kg (M, i.p.); 61-68 mg/kg (M, i.v.)
CN: O-3-amino-3-deoxy-α-D-glucopyranosyl-(1→6)-O-[2,6-diamino-2,3,4,6-tetradeoxy-α-D-*erythro*-
        hexopyranosyl-(1→4)]-2-deoxy-D-streptamine

sulfate
RN: 58580-55-5   MF: $C_{18}H_{37}N_5O_8 \cdot xH_2SO_4$ MW: unspecified   EINECS: 261-341-0
LD$_{50}$:   62.6 mg/kg (M, i.v.); >6950 mg/kg (M, p.o.);
        140 mg/kg (R, i.v.); 6950 mg/kg (R, p.o.)

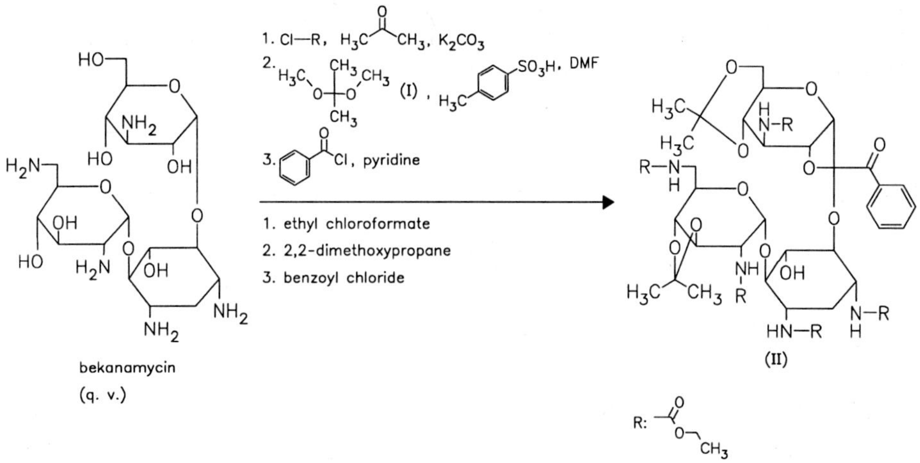

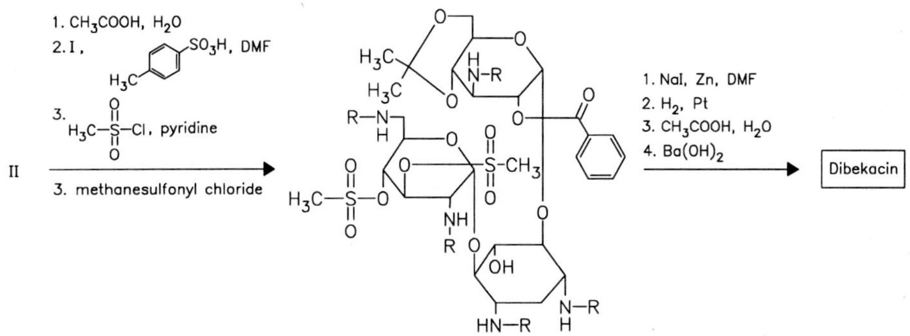

bekanamycin
(q. v.)

R: (acetyl carbonate group)

1. CH₃COOH, H₂O
2. I,
3. H₃C-S-Cl, pyridine
3. methanesulfonyl chloride

II →

1. NaI, Zn, DMF
2. H₂, Pt
3. CH₃COOH, H₂O
4. Ba(OH)₂

→ Dibekacin

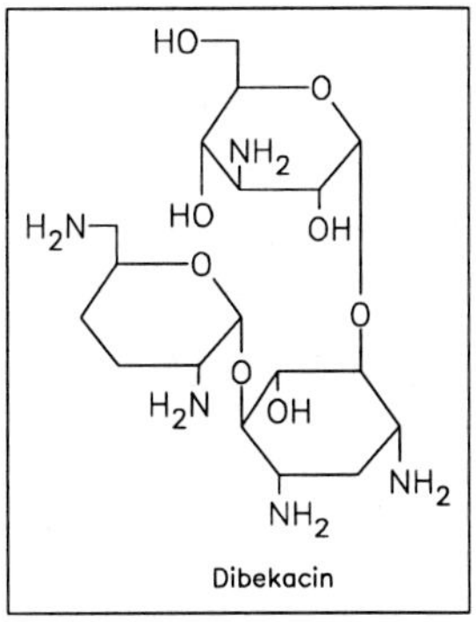

Dibekacin

Reference(s):

DE 2 135 191 (Zaidan Hojin Biseibutsu Kagaku Kenkyu Kai; appl. 14.7.1971; J-prior. 29.7.1970, 11.5.1971).
Umezawa, S. et al.: Bull. Chem. Soc. Jpn. (BCSJA8) **45**, 3624 (1972).
Umezawa, H. et al.: J. Antibiot. (JANTAJ) **24**, 485 (1971).

alternative syntheses:

DOS 2 414 416 (Hoechst; appl. 26.3.1974).
DOS 2 654 764 (Zaidan Hojin Biseibutsu Kagaku Kenkyu Kai; appl. 3.12.1976; J-prior. 10.12.1975, 9.12.1975).
DOS 2 655 731 (Zaidan Hojin Biseibutsu Kagaku Kenkyu Kai; appl. 9.12.1976; J-prior. 11.12.1975).
DOS 2 756 057 (Zaidan Hojin Biseibutsu Kagaku Kenkyu Kai; appl. 15.12.1977; J-prior. 16.12.1976).
US 4 169 939 (Zaidan Hojin Biseibutsu Kagaku Kenkyu Kai; 2.10.1979, J-prior. 16.12.1976).
Migake, T. et al.: Carbohydr. Res. (CRBRAT) **49**, 141 (1976).

Formulation(s):   amp. 50 mg/ml, 75 mg/1.5 ml

Dibenzepine

ATC: N06AA08
Use: antidepressant

RN: 4498-32-2   MF: $C_{18}H_{21}N_3O$ MW: 295.39
LD_{50}:   22 mg/kg (M, i.v.); 194 mg/kg (M, p.o.);
       22 mg/kg (R, i.v.); 220 mg/kg (R, p.o.)
CN: 10-[2-(dimethylamino)ethyl]-5,10-dihydro-5-methyl-11*H*-dibenzo[*b,e*][1,4]diazepin-11-one

monohydrochloride
RN: 315-80-0   MF: $C_{18}H_{21}N_3O \cdot HCl$ MW: 331.85   EINECS: 206-255-6
LD_{50}:   22 mg/kg (M, i.v.); 174 mg/kg (M, p.o.);
       22 mg/kg (R, i.v.); 220 mg/kg (R, p.o.)

1-bromo-
2-nitrobenzene

methyl N-methyl-
anthranilate

(I)

1. $NaNH_2$
2. 2-(dimethylamino)-
   ethyl chloride

11-oxo-5-methyl-
10,11-dihydro-5H-
dibenzo[b,e][1,4]-
diazepine

Dibenzepine

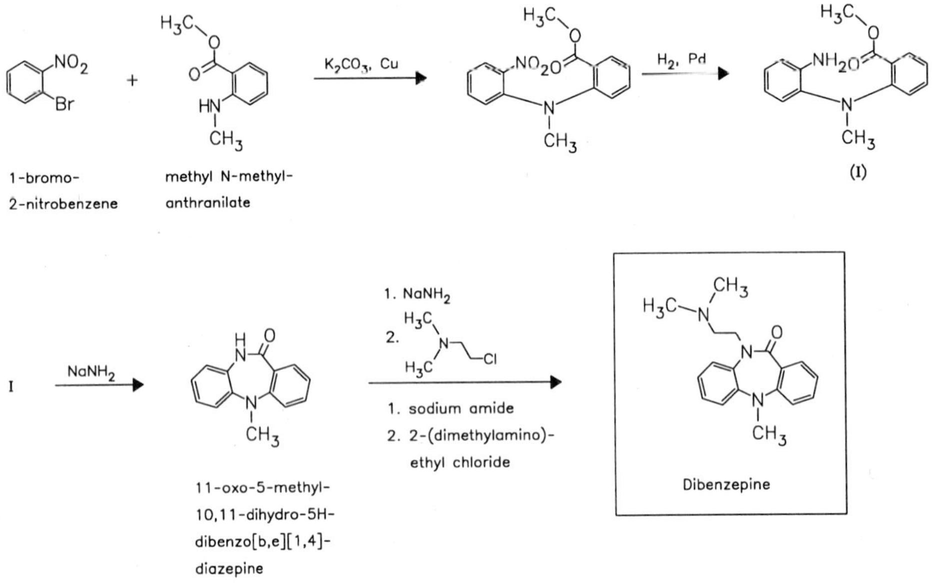

Reference(s):
DE 1 263 774 (Wander; appl. 13.9.1960; CH-prior. 22.9.1959).
US 3 419 547 (Wander; 31.12.1968; CH-prior. 22.9.1959).
GB 961 106 (Wander; appl. 22.9.1960; CH-prior. 22.9.1959).
FR 1 295 371 (Wander; appl. 20.9.1960; CH-prior. 22.9.1959).
Hunziker, F. et al.: Arzneim.-Forsch. (ARZNAD) **13**, 324 (1963).

Formulation(s):   amp. 20 mg/ml; drg. 40 mg, 80 mg; s. r. tabl. 240 mg; tabl. 80 mg

Dibrompropamidine

ATC: D08AC01; S01AX14
Use: chemotherapeutic

RN: 496-00-4   MF: $C_{17}H_{18}Br_2N_4O_2$ MW: 470.17
CN: 4,4'-[1,3-propanediylbis(oxy)]bis[3-bromobenzenecarboximidamide]

diisethionate (1:2)
RN: 614-87-9   MF: $C_{17}H_{18}Br_2N_4O_2 \cdot 2C_2H_6O_4S$ MW: 722.43   EINECS: 210-399-5

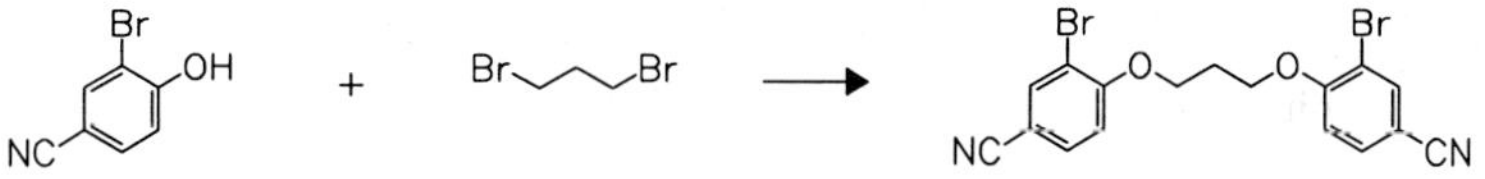

3-bromo-4-
hydroxybenzonitrile

1,3-dibromo-
propane

(I)

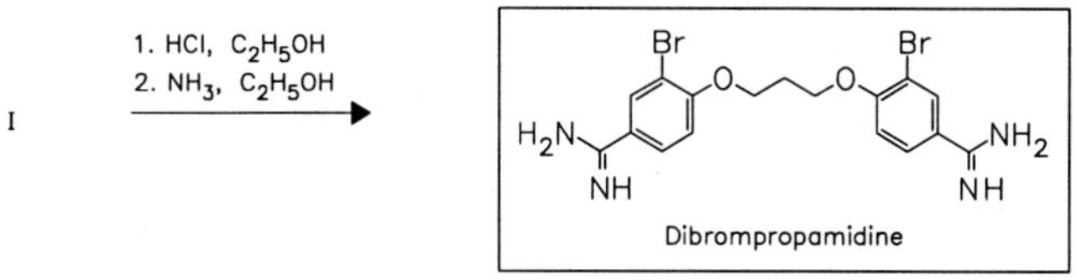

Dibrompropamidine

Reference(s):
GB 598 911 (May & Baker; appl. 1945).

Formulation(s):   eye drops 0.1 %; eye ointment 0.15 %

Trade Name(s):
GB: Brolene (May & Baker)   Golden Eye Ointment   Phenergan (May & Baker;
      Brulidine (May & Baker)   (Typharm)   as isethionate)-comb.
                                     Otamidyl (May & Baker)

Dichlorisone

ATC: D07AA
Use: topical glucocorticoid

RN: 7008-26-6   MF: $C_{21}H_{26}Cl_2O_4$ MW: 413.34   EINECS: 230-283-8
CN: (11β)-9,11-dichloro-17,21-dihydroxypregna-1,4-diene-3,20-dione

acetate
RN: 79-61-8   MF: $C_{23}H_{28}Cl_2O_5$ MW: 455.38   EINECS: 201-213-3

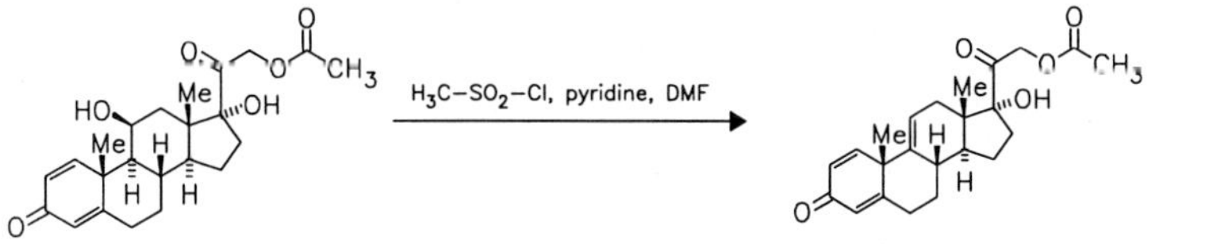

prednisolone-21-acetate

21-acetoxy-17-hydroxy-
1,4,9(11)-pregnatrien-3,20-dione (I)

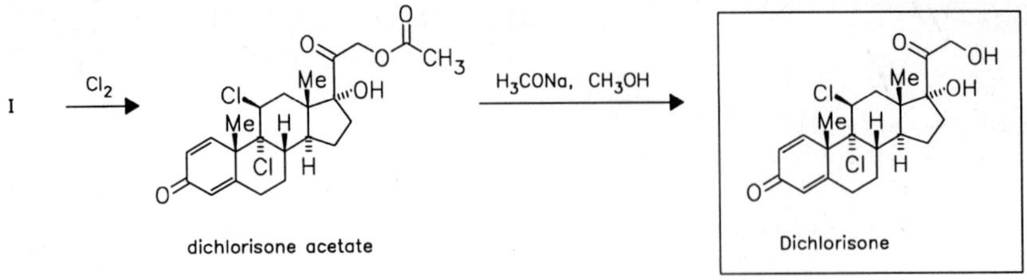

dichlorisone acetate

Dichlorisone

Reference(s):
US 2 894 963 (Schering Corp.; 1959; prior. 1958).
Robinson, C.H. et al.: J. Am. Chem. Soc. (JACSAT) **81**, 2191 (1959).

Formulation(s):   cream 0.25 %

Trade Name(s):
I: Astroderm (Lagap); wfm   J: Diloderm Cream   Neo-Diloderm Cream
                                      (Schering-Shionogi)   (Schering-Shionogi)-comb.

Dichlorophen

ATC: P02DX
Use: antifungal, antiseptic, anthelmintic

RN: 97-23-4   MF: $C_{13}H_{10}Cl_2O_2$ MW: 269.13   EINECS: 202-567-1
LD$_{50}$:   1 g/kg (M, p.o.);
      17 mg/kg (R, i.v.); 1506 mg/kg (R, p.o.);
      2 g/kg (dog, p.o.)
CN: 2,2'-methylenebis(4-chlorophenol)

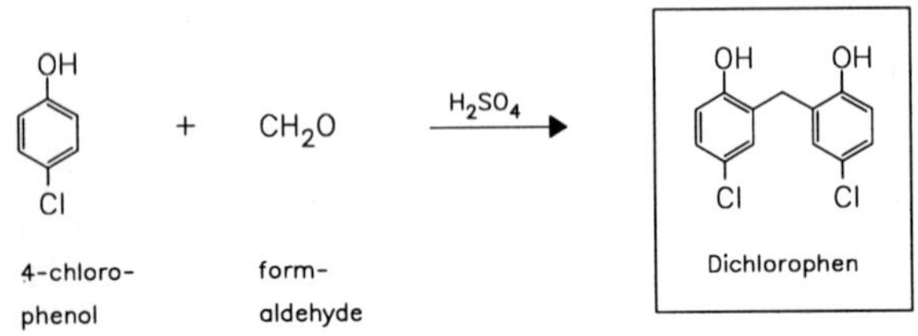

4-chloro-   form-
phenol   aldehyde

Dichlorophen

Reference(s):
DRP 530 219 (I. G. Farben; appl. 1927).
GB 1 208 325 (BDH; appl. 22.4.1968; valid from 15.4.1969).
US 2 334 408 (B.T. Bush; 1943; appl. 1941).
DAS 2 551 498 (Bayer; appl. 17.11.1975).

Formulation(s):   cream 10 mg; powder 50 mg; sol. 10 mg; spray 10 mg

Trade Name(s):
D: Fissan Brustwarzensalbe   Ovis Fußbad/Spray   GB: Anthiphen (May & Baker);
      (Fink)-comb.; wfm   (Warner); wfm   wfm
      Onychofissan (Fink)-   Wespuril (Spitzner)-comb.;
      comb.; wfm   wfm
      Ovis Flüssigkeit/salbe   F: Plath-Lyse (Génévrier);
      (Warner)-comb.; wfm   wfm

Diclofenac

ATC: M01AB05; M02AA15; S01BC03
Use: anti-inflammatory, antirheumatic

RN: 15307-86-5   MF: $C_{14}H_{11}Cl_2NO_2$ MW: 296.15   EINECS: 239-348-5
LD_{50}:   170 mg/kg (M, p.o.);
        62.5 mg/kg (R, p.o.)
CN: 2-[(2,6-dichlorophenyl)amino]benzeneacetic acid

monosodium salt
RN: 15307-79-6   MF: $C_{14}H_{10}Cl_2NNaO_2$ MW: 318.14   EINECS: 239-346-4
LD_{50}:   116 mg/kg (M, i.v.); 390 mg/kg (M, p.o.);
        117 mg/kg (R, i.v.); 150 mg/kg (R, p.o.)

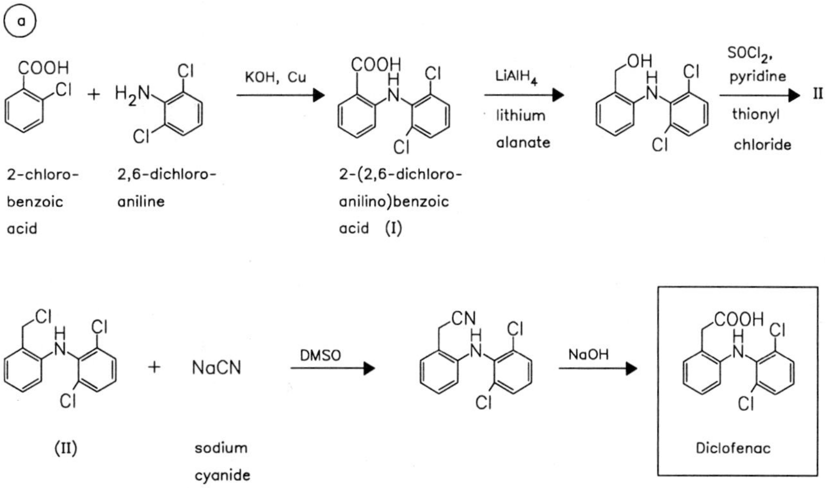

2-chloro-
benzoic
acid

2,6-dichloro-
aniline

2-(2,6-dichloro-
anilino)benzoic
acid (I)

(II)   sodium
              cyanide

Diclofenac

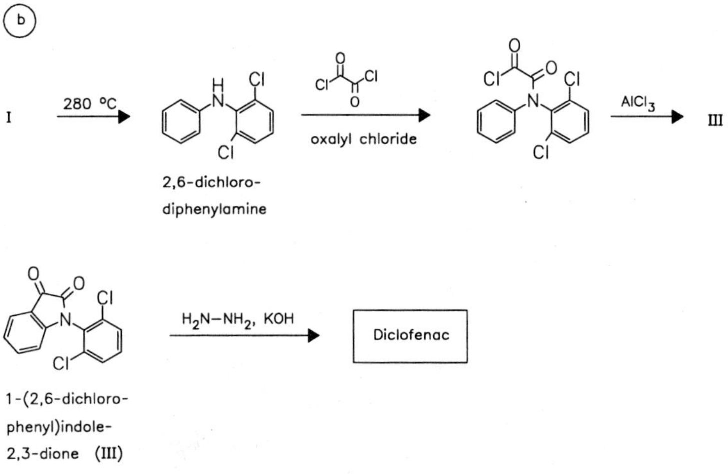

2,6-dichloro-
diphenylamine

oxalyl chloride

1-(2,6-dichloro-
phenyl)indole-
2,3-dione (III)

Diclofenac

Reference(s):
US 3 558 690 (Geigy; 26.1.1971; CH-prior. 8.4.1965, 25.2.1966, 30.3.1966, 20.12.1967).
DAS 1 543 639 (Ciba-Geigy; appl. 7.4.1966; CH-prior. 8.4.1965).
DAS 1 793 592 (Ciba-Geigy; appl. 7.4.1966; CH-prior. 8.4.1965).
US 3 652 762 (Ciba-Geigy; 28.3.1972; prior. 9.12.1968, 29.9.1969, 14.4.1970).
US 3 778 470 (Geigy; 11.12.1973; appl. 2.10.1970; prior. 4.4.1966).
CH 492 679 (Geigy; appl. 30.3.1966).

alternative synthesis:
DOS 2 613 838 (Ikeda Mohando; appl. 31.3.1976; J-prior. 31.3.1975).

Formulation(s):   amp. 75 mg; cps. and drg. 25 mg, 50 mg, 100 mg, 140 mg; eye drops 1 mg, 0.3 mg, 5 mg/5 ml; gel 11.6 mg, 1 %; inj. sol. 75 mg/3 ml; suppos. 12.5 mg, 25 mg, 50 mg, 100 mg; tabl. 25 mg, 50 mg, 75 mg

Trade Name(s):

D:	Allvoran (TAD)	Dolgit-Diclo (Dolorgiet)
	arthrex (BASF Generics)	Dolobasan (Sagitta)
	Benfofen (Sanofi	Duravolten (durachemie)
	Winthrop)	Effekton (Brenner-Efeka/
	Delphimix (Cyanamid)	Law)
	Delphinac (Lederle)	Jenafenac (Jenapharm)
	Diclac (Hexal)	Lexobene (Merckle)
	diclo (ct-Arzneimittel)	Monoflam (Lichtenstein)
	Diclofenbeta (betapharm)	Myogit (Pfleger)
	Diclo KD (Kade)	Rehumavincin (Owege)
	Diclophlogont	Rewodina (ASTA Medica
	(Azupharma)	AWD)
	Diclo-Phlogont	Sigafenac (Kytta-Siegfried)
	(Azuchemie)	Toryxil (Baer)
	Diclo-Puren (Isis Puren)	Voltaren (Novartis Pharma;
	Diclo-rectal (Beiersdorf)	1976)
	Diclo-saar (Chephasaar)	
	Diclo-Spondyril (Dorsch)	
	Diclo-Tablinen (Beiersdorf-	
	Tablinen)	
	Diclo-Wolff (Wolff)	

F: Flector (Génévrier)
       Voltaréne (Novartis; 1976)
       Xenid (Biogalénique)
GB: Voltarol (Novartis; 1979)
       numerous generics

I: Dicloreum (Alfa
        Wassermann)
        Flogofenac (Ecobi)
        Forgenac (Zoja)
        Novapirina (Zyma)
        Voltaren (Ciba-Geigy;
        1975)
J: Adefuronic (Taiyo)
        Dichronic (San-a)
        Docell (Nippon Kayaku)
        Irinatolon (Tatumi)
        Neriodin (Teikoku)
        Nifleriel (Mohan)
        Sofarin (Nippon
        Chemiphar)
        Tsudohmin (Toho)
        Voltaren (Fujisawa; 1974)
USA: Voltaren (Novartis; as
        sodium salt)

Diclofenamide
(Dichlorphenamide)

ATC: S01EC02
Use: carboanhydrase inhibitor (against glaucoma)

RN: 120-97-8   MF: $C_6H_6Cl_2N_2O_4S_2$ MW: 305.16   EINECS: 204-440-6
LD_{50}:   643 mg/kg (M, i.v.); 1710 mg/kg (M, p.o.)
CN: 4,5-dichloro-1,3-benzenedisulfonamide

2-chloro-phenol → (Cl—SO$_3$H, chlorosulfonic acid) → (PCl$_5$, phosphorus(V) chloride) → (NH$_3$) → Diclofenamide

Reference(s):
US 2 835 702 (Merck & Co.; 20.5.1958; prior. 2.5.1956).

Formulation(s):   tabl. 50 mg

Trade Name(s):
D: Diclofenamid (Mann)   F: Oratrol (Alcon); wfm

GB:	Daranide (Merck Sharp & Dohme); wfm	I:	Antidrasi (SmithKline Beecham) Fenamide (Farmigea)	J:	Glaumid (SIFI) Daranide (Merck-Banyu)
				USA:	Daranide (Merck)

Dicloxacillin

ATC: J01CF01
Use: antibiotic

RN: 3116-76-5   MF: $C_{19}H_{17}Cl_2N_3O_5S$ MW: 470.33   EINECS: 221-488-3
CN: [2S-(2α,5α,6β)]-6-[[[3-(2,6-dichlorophenyl)-5-methyl-4-isoxazolyl]carbonyl]amino]-3,3-dimethyl-7-oxo-4-thia-1-azabicyclo[3.2.0]heptane-2-carboxylic acid

monosodium salt monohydrate
RN: 13412-64-1   MF: $C_{19}H_{16}Cl_2N_3NaO_5S \cdot H_2O$ MW: 510.33
LD$_{50}$: 875 mg/kg (M, i.v.); 4560 mg/kg (M, p.o.);
520 mg/kg (R, i.v.); 3579 mg/kg (R, p.o.);
>3 g/kg (dog, p.o.)

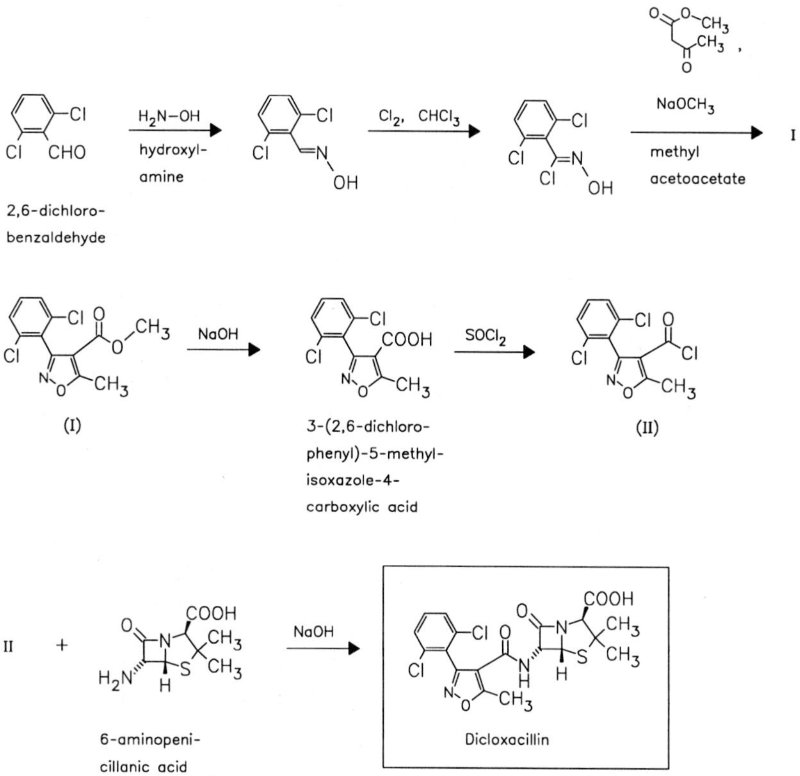

2,6-dichloro-benzaldehyde

(I)

3-(2,6-dichloro-phenyl)-5-methyl-isoxazole-4-carboxylic acid

(II)

6-aminopeni-cillanic acid

Dicloxacillin

Reference(s):
US 3 239 507 (Beecham; 8.3.1966; GB-prior. 17.10.1962).
GB 978 299 (Beecham; appl. 17.10.1962; addition to GB 905 778 from 14.3.1961).
BE 657 504 (Bayer; appl. 23.12.1964; D-prior. 24.12.1963).

Formulation(s):   cps. 500 mg (as sodium salt)

Trade Name(s):
D:	Dichlor-Stapenor (Bayer)	F:	Cefaplus (Labif)-comb.; wfm		Diclocil (Bristol); wfm
				I:	Diclo (Firma)

		J:	Clocil (Bristre-Banyu)
Diclocil (Bristol); wfm	Dicloxapen (Magis); wfm		Combipenix (Toyo Jozo)-
Diclocillin (Aristochimica);	Diflor (Coli); wfm		comb.
wfm	Etadipen (Ghimas)-comb.;		Diclex (Meiji)
Diclocillin (Lagap); wfm	wfm		Staphcillin (Banyu)
Diclocta (Lusofarmaco)-	Novapen (IBP); wfm	USA:	Dycill (Beecham); wfm
comb.; wfm	Versaclox (Bristol)-comb.;		Dynapen (Bristol); wfm
Dicloeta (Lusofarmaco)-	wfm		Pathocil (Wyeth); wfm
comb.; wfm	numerous combination		Veracillin (Ayerst); wfm
Diclomax (Pulitzer); wfm	preparations		

Dicycloverine
(Dicyclomine)

ATC: A03AA07
Use: antispasmodic, anticholinergic

RN: 77-19-0   MF: $C_{19}H_{35}NO_2$ MW: 309.49   EINECS: 201-009-4
CN: [1,1'-bicyclohexyl]-1-carboxylic acid 2-(diethylamino)ethyl ester

hydrochloride
RN: 67-92-5   MF: $C_{19}H_{35}NO_2 \cdot HCl$ MW: 345.96   EINECS: 200-671-1
LD$_{50}$:   31.5 mg/kg (M, i.v.); 625 mg/kg (M, p.o.);
1290 mg/kg (R, p.o.)

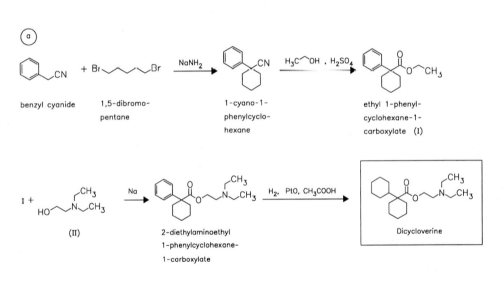

benzyl cyanide   1,5-dibromo-   1-cyano-1-   ethyl 1-phenyl-
pentane   phenylcyclo-   cyclohexane-1-
hexane   carboxylate (I)

(II)   2-diethylaminoethyl   Dicycloverine
1-phenylcyclohexane-
1-carboxylate

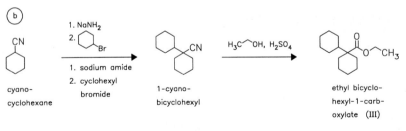

cyano-   1-cyano-   ethyl bicyclo-
cyclohexane   bicyclohexyl   hexyl-1-carb-
oxylate (III)

1. sodium amide
2. cyclohexyl
bromide

III   +   II   →   Dicycloverine

Reference(s):
US 2 474 796 (Merrell Comp.; 1949; prior. 1946).
Tilford, C.H. et al.: J. Am. Chem. Soc. (JACSAT) **69**, 2903 (1947).

Formulation(s):   cps. 10 mg

Trade Name(s):

D:	Atumin (Merrell); wfm	I:	Bentyl (Merrell)
	Spasmo-Rhoival (Tosse)-		Merankol (Lepetit)-comb.
	comb.; wfm	J:	Bentyl (Shionogi)
GB:	Diarrest (Galen)-comb.		Bentyl/Phenobarbital
	Kolanticon (Hoechst)-		(Shionogi)-comb.
	comb.		Incron (Seiko Eiyo)-comb.
	Merbentyl (Florizel)		Kolantyl (Shionogi)

Mamiesan (Kyowa
Yakuhin-Hoei)
USA: Bentyl (Hoechst Marion
Roussel; as hydrochloride)
generics

Didanosine
(DDI; Dideoxyinosine)

ATC: J05AF02
Use: anti-AIDS therapeutic, symptomatic
oral treatment

RN: 69655-05-6   MF: C$_{10}$H$_{12}$N$_4$O$_3$ MW: 236.23
LD$_{50}$: >2 g/kg (M, p.o.);
>2 g/kg (R, p.o.);
>2 g/kg (dog, p.o.)
CN: 2',3'-dideoxyinosine

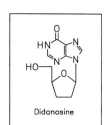

2',3'-dideoxy-
adenosine

fermentation with Acinetobacter Iwoffi (ATCC 9036)

Didanosine

Reference(s):
US 4 970 148 (Ajinomoto; 13.11.1990; appl. 7.10.1988; J-prior. 24.12.1987, 7.10.1987, 13.9.1988).
Plunkett, W.; Cohen, S.S.: Cancer Res. (CNREA8) **35**, 1547 (1975).

alternative synthesis:
US 5 011 774 (Bristol-Myers Squibb; 30.4.1991; appl. 28.2.1990; prior. 17.7.1987).
Prisbe, E.J.; Martin, J.C.: Synth. Commun. (SYNCAV) **15**, 401 (1985).
Horwitz, J.P. et al.: J. Org. Chem. (JOCEAH) **32**, 817 (1967).

purification:
US 4 962 193 (Ajinomoto; 9.10.1990; appl. 28.12.1988; J-prior. 22.12.1987).
JP 1 175 991 (Ajinomoto; appl. 29.12.1987).
JP 1 165 390 (Ajinomoto; appl. 22.12.1987).

medical use for treatment of AIDS:
EP 206 497 (Wellcome; appl. 14.5.1986; GB-prior. 15.5.1985, 20.2.1986).
EP 216 510 (US Department of Health; appl. 21.8.1986; USA-prior. 26.8.1985).
US 4 861 759 (US Department of Health; 29.8.1989; appl. 15.5.1989; prior. 11.8.1987, 26.8.1985, 4.12.1986).
US 5 026 687 (National Institute of Health; 25.6.1991; appl. 3.1.1990).

medical use for treatment of hepatitis B virus infections:
WO 9 014 091 (US Department of Health; appl. 15.5.1990; USA-prior. 15.5.1989, 4.12.1986, 11.8.1987).

synthesis of dideoxyadenosine:
The Merck Index, 3091 (Rahway 1989).

Formulation(s):   chewable tabl. 10 mg, 25 mg, 50 mg, 100 mg, 150 mg; powder 2 g, 4 g

Dienestrol

(Dienoestrol)

ATC: G03CB01; G03CC02
Use: estrogen

RN: 84-17-3   MF: $C_{18}H_{18}O_2$ MW: 266.34   EINECS: 201-519-7
CN: 4,4'-(1,2-diethylidene-1,2-ethanediyl)bis[phenol]

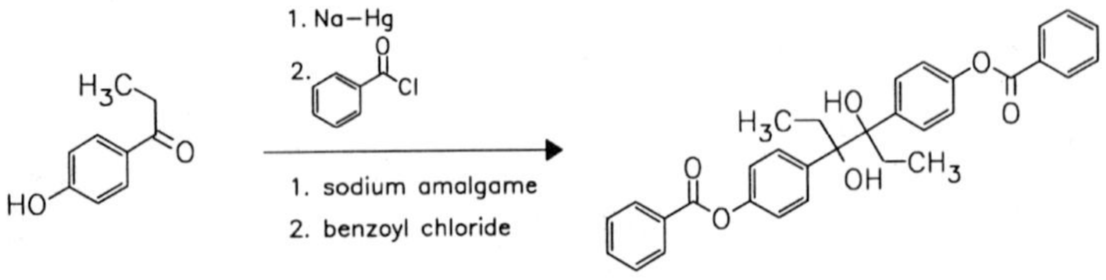

4'-hydroxy-
propiophenone

(I)

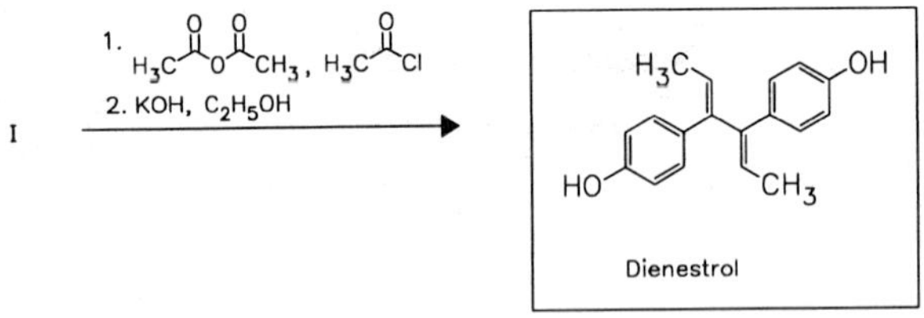

Dienestrol

Reference(s):

GB 566 881 (Boots Pure Drug; appl. 1943).
US 2 464 203 (Boots; 1949; GB-prior. 1943).
US 2 465 505 (Roche; 1949; CH-prior. 1944).

alternative synthesis:
Hobday, G.I.; Short, W.F.: J. Chem. Soc. (JCSOA9) **1943**, 609.

review:
Ehrhart, Ruschig **III**, 330.
Dodds, E.C. et al.: Proc. R. Soc. London, Ser. B (PRLBA4) **127**, 162 (1939).

Formulation(s):   cream 0.01 %; tabl. 5 mg, 25 mg

Diethylcarbamazine

ATC: P02CB02
Use: anthelmintic

RN: 90-89-1   MF: $C_{10}H_{21}N_3O$ MW: 199.30   EINECS: 202-023-3
LD_{50}:   240 mg/kg (M, i.p.)
CN: *N,N*-diethyl-4-methyl-l-piperazinecarboxamide

citrate (1:1)
RN: 1642-54-2   MF: $C_{10}H_{21}N_3O \cdot C_6H_8O_7$ MW: 391.42   EINECS: 216-696-6
LD_{50}:   180 mg/kg (M, i.v.); 660 mg/kg (M, p.o.);
        1400 mg/kg (R, p.o.)
phosphate (1:1)
RN: 16289-41-1   MF: $C_{10}H_{21}N_3O \cdot H_3PO_4$ MW: 297.29

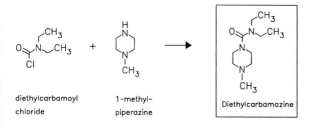

diethylcarbamoyl chloride   1-methyl-piperazine   Diethylcarbamazine

Reference(s):
US 2 467 893 (American Cyanamid; 1949; prior. 1946).
US 2 467 895 (American Cyanamid; 1949; prior. 1946).

Formulation(s):   tabl. 50 mg

Trade Name(s):
D: Hetrazan (Lederle); wfm   GB: Banocide (Wellcome); wfm   USA: Hetrazan (Lederle); wfm
F: Notézine (Specia); wfm   J: Hetrazan (Lederle)

Diethylstilbestrol
(Diäthylstilböstrol)

ATC: G03CB02; G03CC05; L02AA01
Use: formerly in estrogenic hormone
        therapy, listed as a known carcinogen

RN: 56-53-1   MF: $C_{18}H_{20}O_2$ MW: 268.36   EINECS: 200-278-5
LD_{50}:   300 mg/kg (M, i.v.); >3 g/kg (M, p.o.);
        >3 g/kg (R, p.o.)
CN: (*E*)-4,4'-(1,2-diethyl-1,2-ethenediyl)bis[phenol]

(a)

anethole

HBr, toluene
hydrogen bromide

1. NaNH$_2$, NH$_3$, – 80 °C
2. KOH, glycol, 224 °C
1. sodium amide

Diethylstilbestrol

b)

deoxyanisoin   ethyl
                    iodide   (I)

II   KOH, 220 °C   ┌─────────────────┐
      ───────────►   │ Diethylstilbestrol │
                     └─────────────────┘

c)

4'−hydroxy−
propiophenone   (III)

III   Na, amyl alcohol   HCl   ┌─────────────────┐
       ───────────────►   ───►   │ Diethylstilbestrol │
                                                   └─────────────────┘

Reference(s):
a   US 2 392 852 (Lilly; 1946; prior. 1941).
    US 2 402 054 (Lilly; 1946; prior. 1941).
b   Dodds, E.C.: Nature (London) (NATUAS) **141**, 247 (1938).
c   US 2 421 401 (Hoffmann-La Roche; 1947; S-prior. 1943).
    DRP 715 542 (Schering AG; appl. 1939).

alternative syntheses:
Ullmanns Encykl. Tech. Chem., 3. Aufl., Vol. **8**, 327.
GB 526 927 (Richter Gedeon; appl. 1939; H-prior. 1938).
BE 665 818 (Miles Lab.; appl. 23.6.1965; USA-prior. 24.6.1964).

review:
Solmssen, U.V.: Chem. Rev. (Washington, D. C.) (CHREAY) **37**, 481 (1945).
Ehrhart, Ruschig, **III**, 327.

Formulation(s):   tabl. 1 mg, 5 mg

Trade Name(s):
D: Cyren A (Bayer); wfm   Stilboestrol and Lactid   USA: Diethylstilbestrol (Lilly);
F: Distilbène (Gerda)   Acid (Norgine)-comb.;   wfm
GB: Menopax Cream   wfm   Tylosterone (Lilly)-comb.;
      (Nicholas); wfm   wfm

Diethylstilbestrol dipropionate

(Diethylstilboestrol-dipropionat; Diäthylstilboestrol-
   dipropionat)

ATC: G03CB
Use: estrogen

RN: 130-80-3   MF: $C_{24}H_{28}O_4$ MW: 380.48   EINECS: 204-995-4
CN: (E)-4,4'-(1,2-diethyl-1,2-ethenediyl)bis[phenol] dipropanoate

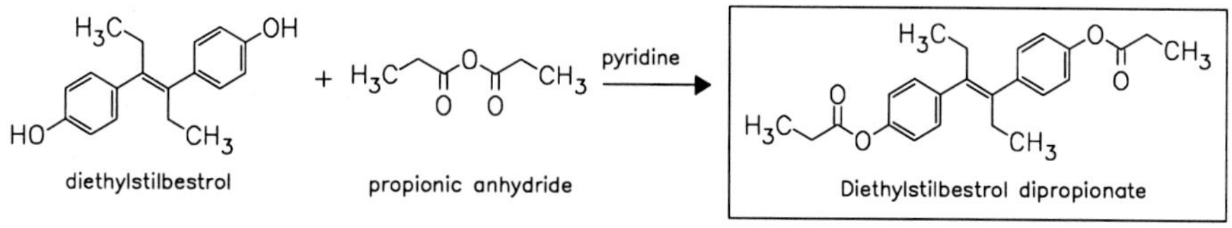

diethylstilbestrol   propionic anhydride   Diethylstilbestrol dipropionate

Reference(s):
Dodds, E.C. et al.: Proc. R. Soc. London, Ser. B (PRLBA4) **127**, 140 (1939).

Formulation(s):   amp.

Trade Name(s):
D: Klimax "Taeschner"   USA: Dibestil (Breon); wfm
   (Taeschner); wfm

Diethylstilbestrol disulfate

(Diethylstilboestroldisulfat; Diäthylstilboestroldisulfat)

ATC: G03CB
Use: estrogen

RN: 316-23-4   MF: $C_{18}H_{20}O_8S_2$ MW: 428.48   EINECS: 206-257-7
CN: (E)-4,4'-(1,2-diethyl-1,2-ethenediyl)bis[phenol]bis(hydrogen sulfate)

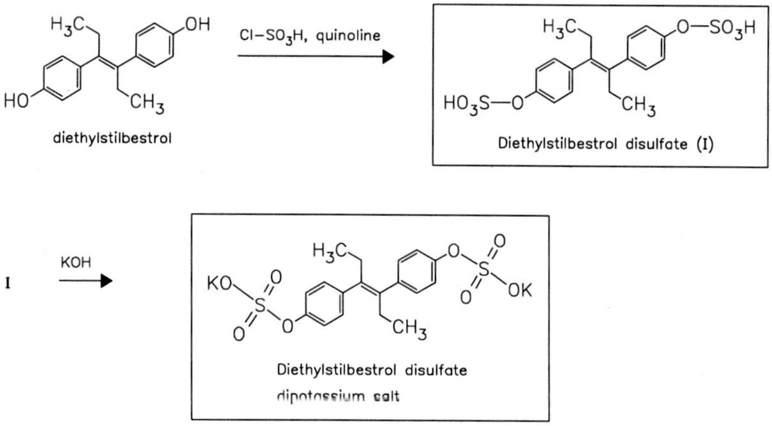

diethylstilbestrol   Diethylstilbestrol disulfate (I)

Diethylstilbestrol disulfate
dipotassium salt

Reference(s):
US 2 234 311 (Ciba; 1941; CH-prior. 1938).

Formulation(s):   ointment

Trade Name(s):
I: Idroestril (Maggioni); wfm   Pappy (Kanto)   Stilbestohormon (Tokyo
J: Estiol (Hokuriku)   Hosei)

Difenidol
(Diphenidol)

ATC: A04; D04
Use: anti-emetic, antihistaminic

RN: 972-02-1   MF: $C_{21}H_{27}NO$ MW: 309.45   EINECS: 213-540-9
LD_{50}:   32 mg/kg (M, i.v.); 450 mg/kg (M, p.o.);
       815 mg/kg (R, p.o.)
CN: α,α-diphenyl-1-piperidinebutanol

hydrochloride
RN: 3254-89-5   MF: $C_{21}H_{27}NO \cdot HCl$ MW: 345.91   EINECS: 221-850-0
LD_{50}:   37 mg/kg (M, i.v.); 400 mg/kg (M, p.o.);
       29 mg/kg (R, i.v.); 515 mg/kg (R, p.o.)
pamoate (2:1)
RN: 26363-46-2   MF: $C_{21}H_{27}NO \cdot 1/2C_{23}H_{16}O_6$ MW: 1007.28

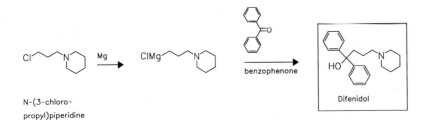

N-(3-chloro-
propyl)piperidine

benzophenone

Difenidol

Reference(s):
US 2 411 664 (Ciba; 1946; CH-prior. 1941).

Formulation(s):   tabl. 25 mg, 50 mg

Trade Name(s):
J: Ansumin (SS Seiyaku)   Meniedolin (Toyo   Verterge (Nippon
      Antiul (Tokyo Hosei)   Shinyaku)   Chemiphar)
      Cephadol (Nippon S.)   Meranom (Hokuriku)   Wansar (Hoei)
      Cerachidol (Ono)   Midnighton (Takata)   Yophadol (Yoshindo/
      Cerrosa (Toyo Pharmar)   Pineroro (Maruko)   Horita)
      Degidole (Nihon Yakuhin)   Promodor (Torii)   USA: Vontrol (Smith Kline &
      Gipsydol (Nihon Yakuhin)   Satanolon (Tatsumi)   French); wfm
      Maniol (Morishita)   Sofalead (Nikken)
      Mecalmin (Yoshitomi-   Solnomin (Zensei)
      Takeda)   Tatimil (Mohan)

Difenoxin

ATC: A07DA04
Use: antidiarrheal, antiperistaltic

RN: 28782-42-5   MF: $C_{28}H_{28}N_2O_2$ MW: 424.54
CN: 1-(3-cyano-3,3-diphenylpropyl)-4-phenyl-4-piperidinecarboxylic acid

monohydrochloride
RN: 35607-36-4   MF: $C_{28}H_{28}N_2O_2 \cdot HCl$ MW: 461.01   EINECS: 252-640-7
LD_{50}:   149 mg/kg (R, p.o.)

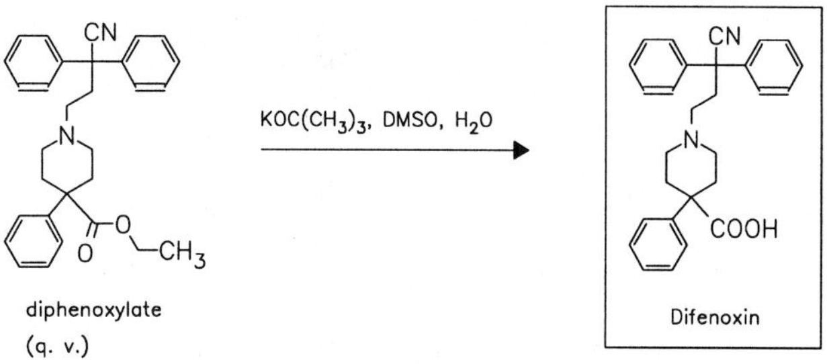

diphenoxylate
(q. v.)

Difenoxin

Reference(s):
DAS 1 953 342 (Janssen; appl. 23.10.1969; USA-prior. 4.11.1968).
US 3 646 207 (Janssen; 29.2.1972; appl. 4.11.1968).

Formulation(s):   tabl. 0.5 mg

Trade Name(s):

D:	Lyspafena (Cilag-Chemie)-comb.; wfm	I:	Motofen (Cilag)-comb.; wfm	USA: Motofen (Carnrick; as hydrochloride)

Diflorasone diacetate

ATC: D07AC10
Use: topical glucocorticoid

RN: 33564-31-7   MF: $C_{26}H_{32}F_2O_7$ MW: 494.53   EINECS: 251-575-1
LD$_{50}$:   >3 g/kg (M, p.o.);
>3 g/kg (R, p.o.)
CN: (6α,11β,16β)-17,21-bis(acetyloxy)-6,9-difluoro-11-hydroxy-16-methylpregna-1,4-diene-3,20-dione

diflorasone
RN: 2557-49-5   MF: $C_{22}H_{28}F_2O_5$ MW: 410.46   EINECS: 219-875-7

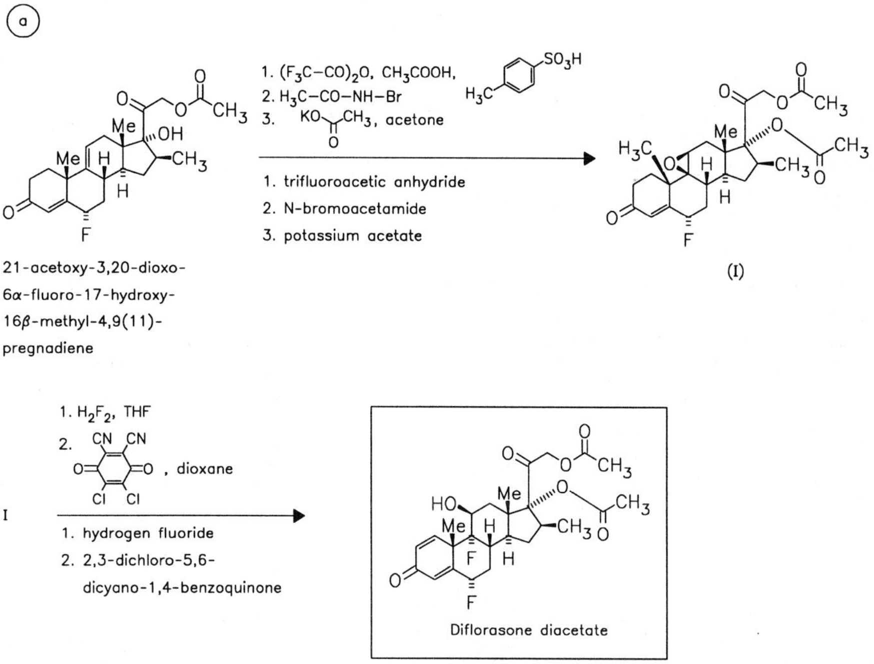

1. $(F_3C-CO)_2O$, CH_3COOH,
2. $H_3C-CO-NH-Br$
3. $KO-CH_3$, acetone

1. trifluoroacetic anhydride
2. N-bromoacetamide
3. potassium acetate

21-acetoxy-3,20-dioxo-
6α-fluoro-17-hydroxy-
16β-methyl-4,9(11)-
pregnadiene

(I)

1. H_2F_2, THF
2. CN CN / O=⬡=O , dioxane / CI CI

I

1. hydrogen fluoride
2. 2,3-dichloro-5,6-dicyano-1,4-benzoquinone

Diflorasone diacetate

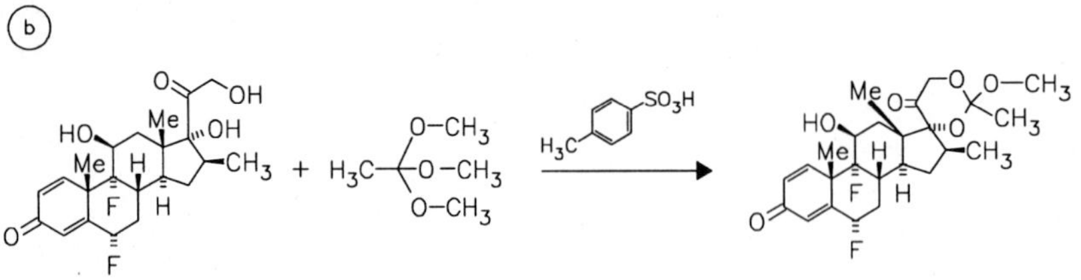

(b)

6α,9α-difluoro-3,20-
dioxo-16β-methyl-
11β,17,21-trihydroxy-
1,4-pregnadiene

trimethyl
orthoacetate

(II)

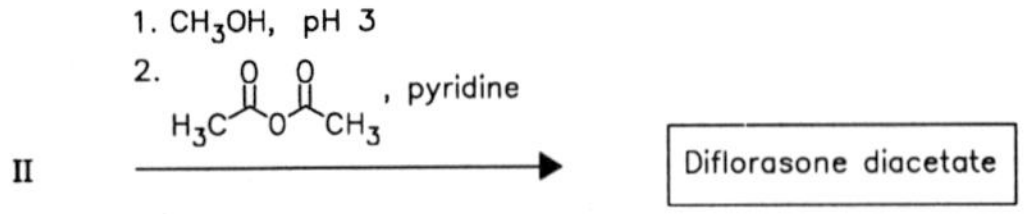

1. CH₃OH, pH 3

2. , pyridine

II ⟶ Diflorasone diacetate

Reference(s):
DE 2 308 731 (Upjohn; appl. 22.2.1973; USA-prior. 9.3.1972).
US 3 980 778 (Upjohn; 14.9.1976; appl. 20.5.1975; prior. 25.10.1973, 20.12.1972, 9.3.1972).
NL 7 303 262 (Upjohn; appl. 8.3.1973; USA-prior. 9.3.1972, 20.12.1972).

starting material:
US 3 557 158 (Upjohn; 19.1.1971; prior. 22.1.1962, 18.3.1959, 4.8.1958).

Formulation(s):   cream 0.05 %; ointment 0.05 %

Trade Name(s):

D:	Florone (Galderma; 1982)	Sterodelta crema	Diflal (Yamanouchi; 1985)
I:	Dermaflor (Brocchieri)	(Gibipharma)	USA: Florone (Dermik; 1978)
	Sterodelta (Metapharma) J:	Diacort (Upjohn-Sumitomo; 1985)	Maxiflor (Allergan; 1981)

Diflucortolone valerate

ATC: D07AC06
Use: glucocorticoid

RN: 59198-70-8   MF: C₂₇H₃₆F₂O₅   MW: 478.58   EINECS: 261-655-8
LD₅₀: 450 mg/kg (M, i.p.); >4 g/kg (M, p.o.); 180 mg/kg (M, s.c.);
     98 mg/kg (R, i.p.); 3.1 g/kg (R, p.o.); 13 mg/kg (R, s.c.)
CN: (6α,11β,16α)-6,9-difluoro-11-hydroxy-16-methyl-21-[(1-oxopentyl)oxy]pregna-1,4-diene-3,20-dione

diflucortolone
RN: 2607-06-9   MF: C₂₂H₂₈F₂O₄   MW: 394.46   EINECS: 220-022-6

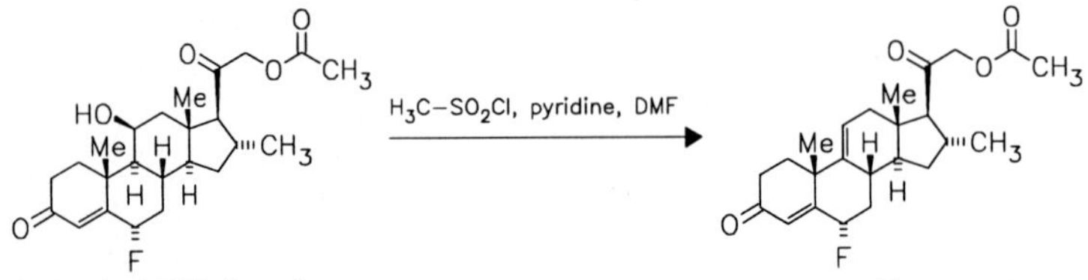

21-acetoxy-3,20-dioxo-6α-
fluoro-11β-hydroxy-16α-
methyl-4-pregnene
(cf. fluocortolone synthesis)

H₃C–SO₂Cl, pyridine, DMF

(I)

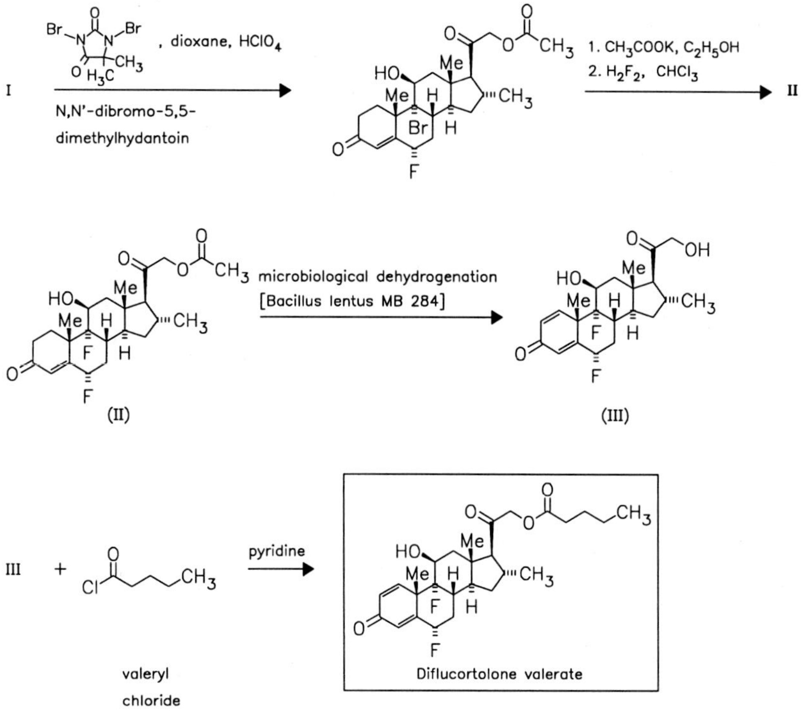

Reference(s):

DE 1 211 194 (Schering; 27.7.1963) continuation of DE 1 169 444.
DE 1 169 444 (Schering; 22.2.1961).
Kieslich, K. et al.: Arzneim.-Forsch. (ARZNAD) **26**, 1462 (1976).
(alternative syntheses described)

Formulation(s): cream 0.1 %; ointment 0.1 %

Trade Name(s):

D: Neribas (Schering)
    Nerisona (Schering)
    Nerisona C (Schering)-
    comb.
    Travocort (Schering)-comb.
F: Nerisone (Schering; as
    valerate)
    Nerisone C (Schering; as
    valerate)-comb.
GB: Nerisone (Schering)
I: Cortical (Caber)
    Cortifluoral (Schering)-
    comb.

Dermaflogil (Nuovo Cons.
Sanit. Naz.)-comb.
Dermeval (Firma)
Dermobios (Biotekfarma)-
comb.
Dervin (Boniscontro &
Gazzone)
Dicortal (Medici)
Flu-Cortanest (Piam)
Impetex (Roche)-comb.
Nerisona (Schering)
Nerisona C (Schering)-
comb.
Temetex (Roche)

several combination
preparations
J: Afusona (Toyama)
    Arusona (Hotta)
    Dertron (Sankyo)
    Lizatlone (Kaken)
    Lorizon (Shinshin)
    Neridalon (Taiyo)
    Nerisona (Nippon
    Schering)
    Sawatolone (Sawai)
    Texmeten (Roche)
    Youtolon (Tatsumi)

Diflunisal

ATC: N02BA11
Use: anti-inflammatory, analgesic,
        antipyretic

RN: 22494-42-4   MF: $C_{13}H_8F_2O_3$ MW: 250.20   EINECS: 245-034-9
LD_{50}:   439 mg/kg (M, p.o.);
        392 mg/kg (R, p.o.)
CN: 2',4'-difluoro-4-hydroxy[1,1'-biphenyl]-3-carboxylic acid

2,4-difluoro-   anisole   4-(2,4-difluoro-
aniline   phenyl)anisole

4-(2,4-difluoro-   Diflunisal
phenyl)phenol (I)

Reference(s):
Hannah, J. et al.: J. Med. Chem. (JMCMAR) **21**, 1093 (1978).
DE 1 618 663 (Merck & Co.; appl. 3.3.1967; USA-prior. 8.9.1966).
DAS 2 532 559 (Merck & Co.; appl. 21.7.1975; USA-prior. 22.7.1974, 16.4.1975, 1.5.1975).
US 3 674 870 (Merck & Co.; 4.7.1972; appl. 9.6.1970; USA-prior. 23.12.1964, 8.9.1966, 19.1.1968).
US 3 681 445 (Merck & Co.; 1.8.1972; appl. 19.1.1968; USA-prior. 23.12.1964, 8.9.1966).
US 3 714 226 (Merck & Co.; 30.1.1973; appl. 9.6.1970; prior. 23.12.1964, 8.9.1966, 19.1.1968).

alternative syntheses:
US 3 992 459 (Merck & Co.; 16.11.1976; appl. 1.5.1975).
US 4 131 618 (Merck & Co.; 26.12.1978; appl. 29.12.1977).
US 4 225 730 (Merck & Co.; 30.9.1980; appl. 11.5.1978; prior. 22.7.1974, 16.4.1975)

Formulation(s):   tabl. 250 mg, 500 mg

Trade Name(s):
D: Fluniget (Ferlux; 1982);
      wfm
F: Dolobis (Merck Sharp &
      Dohme-Chibret; 1981)
GB: Dolobid (Morson; 1978)
I: Adomal (Malesci)
      Aflogos (Biomedica
      Foscama; as arginine salt)

Artrodol (AGIPS)
Difludol (Edmond)
Diflusan (Leben's)
Dolisal (Guidotti)
Dolobid (Merck Sharp &
Dohme; 1979)
Fluodonil (Biologici Italia)
Flustar (Firma)

Reuflos (Roussel)
J: Dolobid (Merck-Banyu;
      1984)
USA: Dolobid (Merck; 1982)

Difluprednate

ATC: D07AC19
Use: topical glucocorticoid

RN: 23674-86-4 MF: $C_{27}H_{34}F_2O_7$ MW: 508.56 EINECS: 245-815-4
LD$_{50}$: >4 g/kg (M, p.o.);
       >4 g/kg (R, p.o.)
CN: (6α,11β)-21-(acetyloxy)-6,9-difluoro-11-hydroxy-17-(1-oxobutoxy)pregna-1,4-diene-3,20-dione

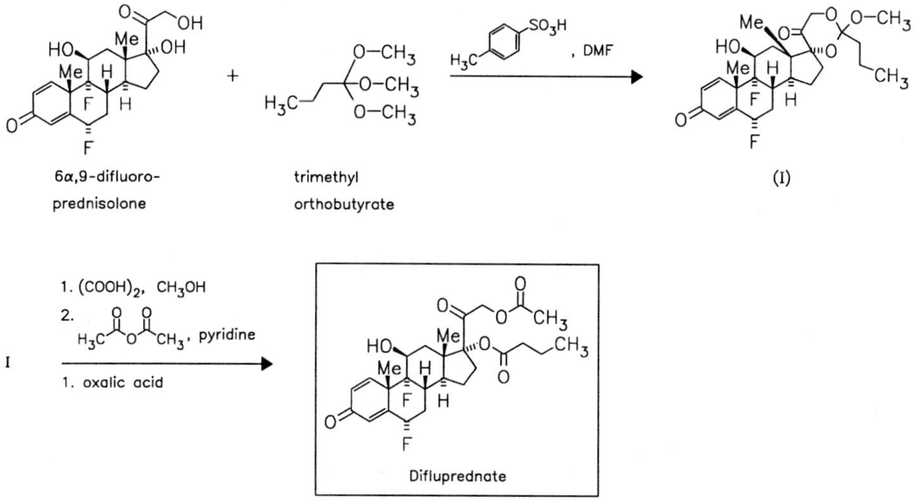

6α,9-difluoro-
prednisolone

trimethyl
orthobutyrate

(I)

Difluprednate

Reference(s):
US 3 780 177 (Warner-Lambert; 18.12.1973; I-prior. 6.6.1967).
ZA 6 803 686 (Warner-Lambert; appl. 21.5.1968; I-prior. 16.6.1967).
Gardi, R. et al.: J. Med. Chem. (JMCMAR) **15**, 556 (1972).

Formulation(s):   gel 0.05 %

Trade Name(s):
F: Epitopic (Gerda)   J: Myser (Mitsubishi Chem.-
                                             Tokyo Tanabe)

Digitoxin

ATC: C01AA04
Use: cardiac glycoside, cardiotonic

RN: 71-63-6 MF: $C_{41}H_{64}O_{13}$ MW: 764.95 EINECS: 200-760-5
LD$_{50}$: 0.18 mg/kg (cat, p.o.)
CN: (3β,5β)-3-[(O-2,6-dideoxy-β-D-*ribo*-hexopyranosyl-(1→4)-O-2,6-dideoxy-β-D-*ribo*-hexopyranosyl-
      (1→4)-2,6-dideoxy-β-D-*ribo*-hexopyranosyl)oxy]-14-hydroxycard-20(22)-enolide

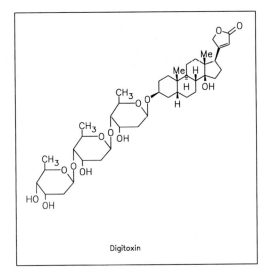

(a)

Digitalis purpurea →(extraction) purpureaglycoside A (I)

I →(enzymatic hydrolysis [Digilanidase]) Digitoxin

(b)   from Digitalis lanata

Reference(s):
a   DRP 646 930 (Sandoz; appl. 1933).
    US 2 449 673 (Wyeth; 1948; prior. 1944).
    US 2 557 916 (Wyeth; 1951; appl. 1948).
    US 2 615 884 (Wyeth; 1952; prior. 1948).
    HU 155 252 (Richter Gedeon; appl. 14.12.1966).
b   HU 156 753 (Richter Gedeon; appl. 7.1.1968).
    IN 62 497 (Council of Scientifique & Industrial Research; appl. 17.9.1958).

alternative syntheses:
Ullmanns Encykl. Tech. Chem., 3. Aufl., Vol. **8**, 229.
DOS 2 006 926 (Deutsche Akademie der Wissenschaften; appl. 16.2.1970; DDR-prior. 15.8.1969).

Formulation(s):   amp. 0.1 mg/ml, 0.25 mg/ml; eye drops 0.02 mg/ml; lotion; tabl. 0.05 mg, 0.07 mg, 0.1 mg, 0.25 mg

Trade Name(s):

D: Digicor (Henning)
     Digimed (Hormosan)
     Digimed (Trommsdorff)
     Digimerck (Merck)
     Digipural (Schaper &
     Brümmer)
     Digitoxin Didier
     (Hormosan)
     Digitoxin Hameln
     (Hameln)
     Digophton-Augentropfen
     (ankerpharm)

F: Dirautheon (Robugen)-
     comb.
     Ditaven (Cascan)-comb.
     Ditaven Lot. (Cascan)
     mono-glycocard (R.A.N.)
     Recorsan-Herzdragees
     (Recorsan)
     Tardigal (Beiersdorf)
     Targital (Beiersdorf)
     generics
     Digitaline Nativelle
     (Procter & Gamble)

     Ditavène
     (Pharmadéveloppement)-
     comb.

GB: Nativelle Digitaline
     (Wilcox); wfm

I: Digifar (Farmila)-comb.
     Digitalina (Procter &
     Gamble)
     Digitos (Formulario Naz.)

J: Digitoxin (Shionogi)
     Digitoxin-Sandoz (Sankyo)

USA: Crystodigin (Lilly)

Digoxin

ATC: C01AA05
Use: cardiac glycoside, cardiotonic

RN: 20830-75-5   MF: $C_{41}H_{64}O_{14}$ MW: 780.95   EINECS: 244-068-1
LD$_{50}$:   7.670 mg/kg (M, i.v.); 17.78 mg/kg (M, p.o.);
     25 mg/kg (R, i.v.); 28.27 mg/kg (R, p.o.)
CN: (3β,5β,12β)-3-[(*O*-2,6-dideoxy-β-D-*ribo*-hexopyranosyl-(1→4)-*O*-2,6-dideoxy-β-D-*ribo*-hexopyranosyl-
     (1→4)-2,6-dideoxy-β-D-*ribo*-hexopyranosyl)oxy]-12,14-dihydroxycard-20(22)-enolide

ⓐ   from Digitalis lanata by extraction

ⓑ

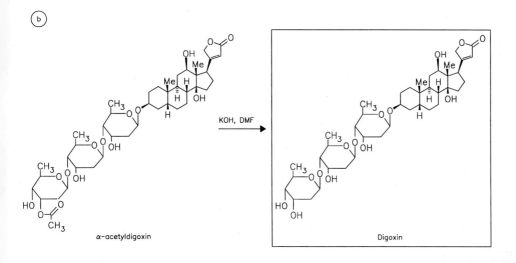

KOH, DMF

α-acetyldigoxin   Digoxin

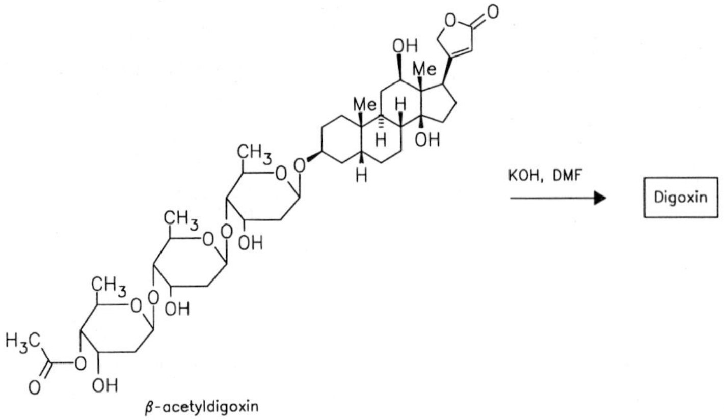

β-acetyldigoxin

Reference(s):

a GB 337 091 (Wellcome Found.; appl. 1929).
Smith, S.: J. Chem. Soc. (JCSOA9) **1930**, 508.
IN 62 497 (Council of Scientific & Industrial Research; appl. 17.9.1958).
HU 149 778 (Richter Gedeon; appl. 8.12.1959).
HU 151 897 (Richter Gedeon; appl. 29.2.1964).
HU 156 753 (Richter Gedeon; appl. 7.6.1968).
DAS 2 225 039 (VEB Arzneimittelwerke Dresden; appl. 23.5.1972; DDR-prior. 24.1.1972).
Ullmanns Encykl. Tech. Chem., 3. Aufl., Vol. **8**, 231.
b DD 70 088 (C. Lindig, K. Repke; appl. 1.11.1968).

alternative synthesis:
DD 134 644 (VEB Arzneimittelwerke Dresden; appl. 6.10.1977).

Formulation(s):   amp. 0.1 mg/ml, 0.25 mg/ml, 0.2 mg/ml, 0.5 mg/2 ml; cps. 0.1 mg, 0.2 mg; drops;
tabl. 0.125 mg, 0.25 mg

Trade Name(s):

D:	Digacin (Beiersdorf-Lilly)	GB: Lanoxin (Glaxo Wellcome)
	Dilanacin (ASTA Medica	I: Digomal (Malesci)
	AWD)	Digos (Biologici Italia)
	Lanicor (Boehringer	Digoss (Formulario Naz.;
	Mannh.)	Sifra)
	Lenoxin (Glaxo Wellcome)	Digossina (Scfm)
	Novodigal Amp.	Eudigox (Astra-Simes)
	(Beiersdorf)	Lanicor (Boehringer
F:	Digoxine Nativelle (Procter	Mannh.)
	& Gamble)	Lanoxin (Wellcome)

J: Digosin (Chugai)
Digosin Erixir (Chugai)
Digoxin (Yamanouchi)
Lanoxin (Wellcome-Tanabe)

USA: Lanoxicaps (Glaxo
Wellcome)
Lanoxin (Glaxo Wellcome)
generic

Dihydralazine

(Dihydrallazine)

ATC: C02DB01
Use: antihypertensive

RN: 484-23-1   MF: $C_8H_{10}N_6$ MW: 190.21   EINECS: 207-605-0
LD_{50}:   300 mg/kg (M, i.v.)
CN: 2,3-dihydro-1,4-phthalazinedione dihydrazone

sulfate (1:1)
RN: 7327-87-9   MF: $C_8H_{10}N_6 \cdot H_2SO_4$ MW: 288.29   EINECS: 230-808-0
LD_{50}:   400 mg/kg (M, p.o.);
400 mg/kg (R, p.o.)

phthalo-   carbon
nitrile   disulfide

Dihydralazine

Reference(s):
DE 845 200 (Cassella; appl. 1951).
DE 847 748 (Ciba; appl. 1949; CH-prior. 1947).

Formulation(s):   tabl. 25 mg, 50 mg (as sulfate)

Trade Name(s):

D:	Adelphan-Esidrix (Novartis Pharma)-comb.	Obsilazin (Isis Pharma)-comb.	Trasipressol (Novartis; as hydrogen sulfate)-comb.
	Depressan (OPW)	Triniton (Apogepha)	I: Adelfan (Novartis)-comb.
	Dihyzin (Henning Berlin)	Tri-Torrat (Boehringer Mannh.)-comb.	Ipogen (Gentili)-comb.
	Nepresol (Novartis Pharma)-comb.		Nepresol (Novartis)
		F: Népressol (Novartis; as hydrogen sulfate)	

Dihydrocodeine
(Drocode)

ATC: N02AA08
Use: antitussive, analgesic

RN: 125-28-0   MF: $C_{18}H_{23}NO_3$ MW: 301.39   EINECS: 204-732-3
LD_{50}:   80 mg/kg (M, i.v.)
CN: (5α,6α)-4,5-epoxy-3-methoxy-17-methylmorphinan-6-ol

hydrogen tartrate (1:1)
RN: 5965-13-9   MF: $C_{18}H_{23}NO_3 \cdot C_4H_6O_6$ MW: 451.47   EINECS: 227-747-7
LD_{50}:   359 mg/kg (R, p.o.)

codeine   Dihydrocodeine

Reference(s):
Ehrhart, Ruschig **I**, 118.
Stein, A.: Pharmazie (PHARAT) **10**, 180 (1955).

Formulation(s):   cps. 20 mg; sol. 10 mg/g; s. r. tabl. 60 mg, 90 mg, 120 mg; syrup 12.1 mg/5 ml; tabl. 10 mg (as hydrogen tartrate)

Trade Name(s):

D:	Antibex forte (Lappe)-comb.	Antitussivum (Ysatfabrik)-comb.	Makatussin (Roland)-comb.
		DHC (Mundipharma)	Paracodin (Knoll)

	Remedacen (Rhône-Poulenc Rorer) Tiamon (Temmler)-comb.	I:	Remedeine (Napp)-comb. Alla Paracodina (Knoll)-comb.	USA:	DHC plus (Purdue Frederick; as bitartrate)
F:	Dicodin (ASTA Medica; as tartrate)		Paracodina (Knoll) Sciroppo Knoll paracodina (Knoll)-comb.		Synalgos-DC (Wyeth-Ayerst; as bitartrate)
GB:	DF-118 forte (Napp) DHC Contiums (Napp)		Tavolette (Knoll)		

Dihydroergocristine

ATC: C04AE04
Use: adrenolytic, sympatholytic

RN: 17479-19-5   MF: $C_{35}H_{41}N_5O_5$ MW: 611.74   EINECS: 241-493-4
CN: (5'α,10α)-9,10-dihydro-12'-hydroxy-2'-(1-methylethyl)-5'-(phenylmethyl)ergotaman-3',6',18-trione

monomesylate
RN: 24730-10-7   MF: $C_{35}H_{41}N_5O_5 \cdot CH_4O_3S$ MW: 707.85   EINECS: 246-434-6
LD_{50}:   70 mg/kg (M, i.v.); >2500 mg/kg (M, p.o.);
        91 mg/kg (R, i.v.); 2643 mg/kg (R, p.o.);
        >50 mg/kg (dog, i.v.); >1250 mg/kg (dog, p.o.)

ergocristine $\xrightarrow{H_2, Pd}$ Dihydroergocristine

constituent of dihydroergotoxine (q. v.)

Reference(s):
Stoll, A.; Hofmann, A.: Helv. Chim. Acta (HCACAV) **26**, 2070 (1943).

combination with pentoxifylline:
BE 865 891 (Roussel-Uclaf; appl. 11.4.1978; F-prior. 12.4.1977).

Formulation(s):   amp. 0.3 mg/ml; sol. 1 mg/ml; tabl. 1.5 mg (as mesylate)

Trade Name(s):
D:	Bellaserp (Atmos)-comb.; wfm		Panthesin-Hydergin (Sandoz)-comb.; wfm		Decril (Damor)
	Briserin (Sandoz)-comb.; wfm		Rexiluven (Sandoz)-comb.; wfm		Defluina (Teofarma)
	Card-Hydergin (Sandoz)-comb.; wfm		Sinedyston (Steiner); wfm		Diertina (Poli)
	Decme (Spitzner); wfm		Vertebran N (Rentschler); wfm		Difluid (Bioprogress)
	Decme (Zyma); wfm				Ergo (Foletto)
	Defluina (Natrapharm)-comb.; wfm		Wallerox (Sandoz); wfm		Ergotina (Ist. Chim. Inter.)
	Enirant Tropflösung (Desitin); wfm	F:	Cervilane (Cassenne)-comb.		Gral (Boniscontro & Gazzone)
			Iskédyl (Pierre Fabre; as mesylate)-comb.		Sandoven (Sandoz)-comb. Unergol (Poli)
	Nehydrin (TAD); wfm	I:	Brinerdina (Sandoz)-comb.		

Dihydroergotamine

ATC: N02CA01
Use: sympatholytic, antimigraine agent

RN: 511-12-6 MF: $C_{33}H_{37}N_5O_5$ MW: 583.69 EINECS: 208-123-3
LD_{50}: 118 mg/kg (M, i.v.)
CN: (5'α,10α)-9,10-dihydro-12'-hydroxy-2'-methyl-5'-(phenylmethyl)ergotaman-3',6',18-trione

monomesylate
RN: 6190-39-2 MF: $C_{33}H_{37}N_5O_5 \cdot CH_4O_3S$ MW: 679.80 EINECS: 228-235-6
LD_{50}: >2 g/kg (M, p.o.);
      >2 g/kg (R, p.o.)
tartrate (2:1)
RN: 5989-77-5 MF: $C_{33}H_{37}N_5O_5 \cdot 1/2C_4H_6O_6$ MW: 1317.46 EINECS: 227-816-1
LD_{50}: 118 mg/kg (M, i.v.);
      110 mg/kg (R, i.v.)

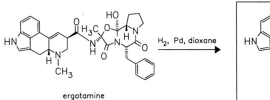

ergotamine

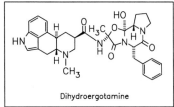

Dihydroergotamine

Reference(s):
Stoll, A.; Hoffmann, A.: Helv. Chim. Acta (HCACAV) **26**, 2070 (1943).
DE 883 153 (Sandoz; appl. 1941; CH-prior. 1940).

nasal formulation:
DOS 2 802 113 (Sandoz; appl. 19.1.1978).

Formulation(s):   amp. 1 mg, 2 mg; s. r. cps. 2.5 mg; 5.0 mg; tabl. 1 mg, 2.5 mg (as mesylate)

Trade Name(s):

D: Agit (Sanofi Winthrop)
     Agit (Sanofi Winthrop)-
     comb.
     Angionorm (Farmasan)
     clavigrenin (Hormosan)
     DET-MS (Rentschler)
     DHE-Puren (Isis Puren)
     Dihydergot-forte/retard
     (Novartis Pharma)
     Dihydergot-plus (Novartis
     Pharma)-comb.
     Dihytamin (ASTA Medica
     AWD)

Effortil (Boehringer Ing.)-
comb.
Embolex (Novartis
Pharma)-comb.
Ergo-Lonarid (Boehringer
Ing.)-comb.
Ergomimet (Klinge)
Ergont (Desitin)
ergotam (ct-Arzneimittel)
Optalidon (Novartis
Pharma)
Tonopres-forte (Boehringer
Ing.)
Venelbin (Hoechst)-comb.

          Verladyn (Verla)
F: Ikaran (Pierre Fabre; as
     mesylate)
     Séglor (Sanofi Winthrop; as
     mesylate)
     Tamik (EG Labo; as
     mesylate)
GB: Dihydergot (Sandoz); wfm
I: Diidergot (Sandoz)
     Ikaran (Formenti)
     Seglor (Synthelabo)
J: Dihydergot (Sandoz-
     Sankyo)
USA: DHE 45 (Novartis)

Dihydroergotoxine
(Dihydroergocornine Dihydro-α-ergocryptine)

ATC: C04AE
Use: sympatholytic, cognition adjuvant

RN: 11032-41-0   MF: unspecified   MW: unspecified
LD$_{50}$:   71 mg/kg (M, s.c.)
CN: dihydroergotoxine

mesylate
RN: 8067-24-1   MF: unspecified   MW: unspecified

"ergotoxine"
(ergocornine + ergocristine + ergocryptine A 1:1:1)

"Dihydroergotoxine"

9,10−Dihydroergocornine   R:

9,10−Dihydroergocristine   R:

9,10−Dihydroergocryptine A   R:

Reference(s):
Stoll, A.; Hofmann, A.: Helv. Chim. Acta (HCACAV) **26**, 2070 (1943).
DE 883 153 (Sandoz; appl. 1941; CH-prior. 1940).

nasal formulation:
DOS 2 802 113 (Sandoz; appl. 19.1.1978).

Formulation(s):   amp. 0.3 mg/ml, 1.5 mg/5 ml; sol. 1 mg/ml, 2 mg/ml; tabl. 2 mg

Trade Name(s):
D: Circanol (3M Medica)
      Dacoren (Nattermann)
      DCCK (Rentschler)
      Defluina (Nattermann)
      Enirant (gepepharm)
      Ergodesit (Desitin)
      ergoplus (Hormosan)
      ergotux (ct-Arzneimittel)
      Hydergin/-forte (Novartis
      Pharma)
      Hydro-Cebral-ratiopharm
      (ratiopharm)
      Nehydrin (TAD)
      Orphol (Opfermann)
      Sponsin (Farmasan)

F: Capergyl (Thérica)
      Ergodose (Murat; as
      mesylate)
      Hydergine (Novartis; as
      mesylate)
      Optamine (Théraplix)
      Pérénan (Sanofi Winthrop)
GB: Hydergine (Novartis)
I: Coristin (San Carlo)
      Hydergina (Sandoz)
      Ischelium (Polifarma)
      Ischelium Papaverina
      (Polifarma)-comb. with
      papaverine hydrochloride
      Progeril (Midy)

      Progeril Papaverina
      (Midy)-comb. with
      papaverine
      Trelidat (Coop. Farm.)
      Visergil (Sandoz)-comb.
J: Hydergine (Sandoz-
      Sankyo)
USA: Circanol (Riker); wfm
      Deapril (Mead Johnson);
      wfm
      Hydergine (Sandoz); wfm
      Hydro-Ergoloid (Schein);
      wfm
      Hydro-Ergot (Interstate);
      wfm

Dihydrostreptomycin sulfate

ATC: S01AA15
Use: antibiotic

RN: 5490-27-7   MF: $C_{21}H_{41}N_7O_{12} \cdot 3/2H_2SO_4$ MW: 1461.43   EINECS: 226-823-7
LD$_{50}$:   186 mg/kg (M, i.v.)
CN: O-2-deoxy-2-(methylamino)-α-L-glucopyranosyl-(1$\rightarrow$2)-O-5-deoxy-3-C-(hydroxymethyl)-α-L-lyxofuranosyl-(1$\rightarrow$4)-N,N'-bis(aminoiminomethyl)-D-streptamine sulfate (2:3) (salt)

dihydrostreptomycin
RN: 128-46-1   MF: $C_{21}H_{41}N_7O_{12}$ MW: 583.60   EINECS: 204-888-2
LD$_{50}$:   200 mg/kg (M, i.v.);
           200 mg/kg (R, i.v.)

streptomycin sulfate
(q. v.)

Dihydrostreptomycin sulfate

Reference(s):
US 2 498 574 (Merck & Co.; 1950; prior. 1946).
GB 642 249 (Squibb; appl. 1947; USA-prior. 1946).

review:
Ehrhart, Ruschig **IV**, 317.

Formulation(s):   amp. 1 g/2 ml; vial 1 g

Trade Name(s):
D: Didrothenat (Grünenthal);
       wfm
       Dihydrostreptomycin
       "Heyl" (Heyl); wfm
       Dihydrostreptomycin
       "Heyl" Double-mycin
       (Heyl)-comb.; wfm
       Entera-strept (Heyl)-comb.;
       wfm
       Penimycin (Winger)-
       comb.; wfm
       Solvo-strept (Heyl); wfm
F: Abiocine (Lepetit); wfm

       Dihydromycine (Specia);
       wfm
       Dihydrostreptomycine
       Diamant (Diamant); wfm
       Entercine (Robapharm)-
       comb.; wfm
       Tri-antibiotique Chibret
       (Chibret)-comb.; wfm
       numerous combination
       preparations; wfm
GB: Guanimycin (Allen &
       Hanburys)-comb.; wfm

I: Dihydrostreptomicina Icar
       (ISF); wfm
       Streptoguanidin
       (Lisapharma)-comb.; wfm
       Streptomagna (Wyeth)-
       comb.; wfm
       Streptomicina Morgan
       (Morgan); wfm
       Trimicina (Farmitalia)-
       comb.; wfm
       combination preparations;
       wfm

Dihydrotachysterol
(Dihydrotachysterin)

ATC: A11CC02
Use: calcium regulator, vitamin D-analog

RN: 67-96-9   MF: C_{28}H_{46}O   MW: 398.68   EINECS: 200-672-7
LD_{50}:   288 mg/kg (M, p.o.)
CN: (3β,5E,7E,10α,22E)-9,10-secoergosta-5,7,22-trien-3-ol

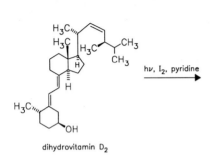

dihydrovitamin D_2

hν, I_2, pyridine

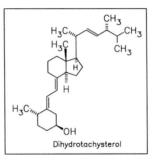

Dihydrotachysterol

Reference(s):
DE 1 108 215 (Merck AG; appl. 22.12.1959).

synthesis of dihydrovitamin D_2:
Schubert, K.: Biochem. Z. (BIZEA2) **327**, 507 (1956).

alternative syntheses:
DE 730 017 (IG Farben; appl. 1938).
DE 1 026 748 (Philips Gloilampenfabrieken; appl. 1956; NL-prior. 1955).
US 2 228 491 (Winthrop; 1941; D-prior. 1938).

medical use:
DE 1 492 177 (A. Schumacher; appl. 3.11.1965).

Formulation(s):   cps. 0.125 mg, 0.5 mg; drops 0.1 %; syrup 0.25 mg/ml

Trade Name(s):
D: A.T.10 (Bayer)   F: Calcamine (Wander); wfm   J: A.T.10 (Bayer)
      Tachystin (Chauvin   GB: A.T.10 (Sanofi Winthrop)   Hytakerol (Torii)
      ankerpharm)   I: A.T.10 (Bayer-Yoshitomi)   USA: DHT (Roxane)

Dihydroxydibutyl ether

ATC: A03
Use: choleretic, antispasmodic

RN: 821-33-0   MF: C_8H_{18}O_3   MW: 162.23   EINECS: 212-475-3
CN: 4,4'-oxybis[2-butanol]

acrylonitrile   H_2O, NaOH →   3,3'-oxydipropionitrile   H_3C—MgBr / methylmagnesium bromide →   I

4,4'-oxydi(2-butanone) (I)   H_2, Raney–Ni →   Dihydroxydibutyl ether

Reference(s):
FR 1 267 084 (M. A. Joulty; appl. 1960).

Formulation(s):   cps. 500 mg; sol. 0.35 g/ml

Trade Name(s):
F: Dyskinébyl (Novartis)   Discinil Complex   Fluidobil (Lifepharma)-
I: Discinil (Lusofarmaco)   (Lusofarmaco)-comb.   comb.
                                     Diskin (Benedetti)

Diiodohydroxyquinoline
(Diiodohydroxyquin; Iodoquinol)

ATC: G01AC01
Use: intestinal antiseptic, antiamebic

RN: 83-73-8   MF: $C_9H_5I_2NO$ MW: 396.95   EINECS: 201-497-9
LD_{50}: 56 mg/kg (M, i.v.)
CN: 5,7-diiodo-8-quinolinol

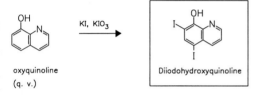

oxyquinoline
(q. v.)

Diiodohydroxyquinoline

Reference(s):
DRP 411 050 (F. Passek; 1925).

Formulation(s):   cream 1 % (comb. with hydrocortisone); tabl. 210 mg, 650 mg

Trade Name(s):
D: Entero-sediv (Grünenthal)-   Ioquin (Abbott); wfm   USA: Vytone (Dermik)
      comb.; wfm   GB: Diodoquin (Searle); wfm   Yodoxin (Glenwood)
F: Direxiode (Delalande);   I: Diiodoidrossichina (Tariff.
      wfm   Integrativo)

Diisopromine

ATC: A03AX02
Use: choleretic, antispasmodic

RN: 5966-41-6   MF: $C_{21}H_{29}N$ MW: 295.47   EINECS: 227-752-4
CN: N,N-bis(1-methylethyl)-γ-phenylbenzenepropanamine

hydrochloride
RN: 24358-65-4   MF: $C_{21}H_{29}N \cdot HCl$ MW: 331.93   EINECS: 246-201-9

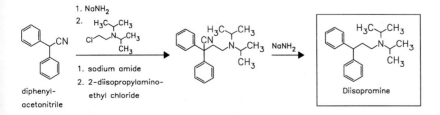

diphenyl-
acetonitrile

1. NaNH₂
2. H₃C CH₃
   Cl N CH₃
       CH₃
1. sodium amide
2. 2-diisopropylamino-
   ethyl chloride

Diisopromine

Reference(s):
GB 808 158 (Janssen; appl. 1956; NL-prior. 1955).

Formulation(s):   tabl. (comb. with 2 mg diisopromine)

Trade Name(s):
D: Agofell (Janssen)   Ulcolind (Lindopharm)-   F: Mégabyl (LeBrun); wfm
                         comb.; wfm   I: Do-Bil (Dompé); wfm

Dilazep

ATC: C01DX10
Use: coronary vasodilator

RN: 35898-87-4   MF: $C_{31}H_{44}N_2O_{10}$ MW: 604.70
CN: 3,4,5-trimethoxybenzoic acid (tetrahydro-1*H*-1,4-diazepine-1,4(5*H*)-diyl)di-3,1-propanediyl ester

dihydrochloride
RN: 20153-98-4   MF: $C_{31}H_{44}N_2O_{10} \cdot 2HCl$ MW: 677.62   EINECS: 243-548-8
LD$_{50}$:   16.8 mg/kg (M, i.v.); 2860 mg/kg (M, p.o.);
          13.7 mg/kg (R, i.v.); >2150 mg/kg (R, p.o.);
          11.2 mg/kg (dog, i.v.); >316 mg/kg (dog, p.o.)

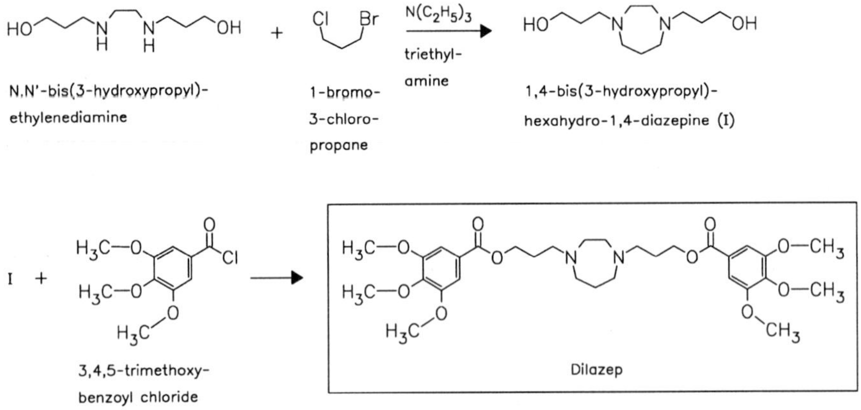

N,N'-bis(3-hydroxypropyl)-   1-bromo-   1,4-bis(3-hydroxypropyl)-
ethylenediamine   3-chloro-   hexahydro-1,4-diazepine (I)
                   propane

3,4,5-trimethoxy-   Dilazep
benzoyl chloride

Reference(s):
GB 1 107 470 (ASTA-Werke; appl. 2.12.1966; D-prior. 16.12.1965).
DE 1 545 575 (ASTA-Werke; appl. 16.12.1965).
US 3 532 685 (ASTA-Werke; 6.10.1970; D-prior. 16.12.1965).

Formulation(s):   drg. 56 mg (as dihydrochloride)

Trade Name(s):
D: Cormelian (ASTA Medica);   I: Cormelian (Schering)
      wfm   J: Comelian (Kowa)

Dilevalol

((R,R)-Labetalol)

ATC: C02CB
Use: α- and β-adrenoceptor antagonist,
isomer of labetalol, antihypertensive

RN: 75659-07-3 MF: $C_{19}H_{24}N_2O_3$ MW: 328.41
LD$_{50}$: 1719 mg/kg (M, p.o.);
1228 mg/kg (R, p.o.)
CN: [R-(R*,R*)]-2-hydroxy-5-[1-hydroxy-2-[(1-methyl-3-phenylpropyl)amino]ethyl]benzamide

monohydrochloride
RN: 75659-08-4 MF: $C_{19}H_{24}N_2O_3 \cdot HCl$ MW: 364.87
LD$_{50}$: 1079 mg/kg (M, p.o.);
82 mg/kg (R, i.v.); 1026 mg/kg (R, p.o.)

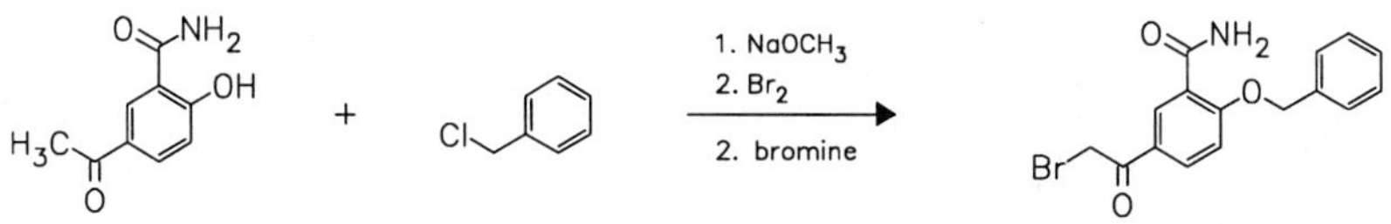

5-acetyl-
salicylamide

benzyl
chloride

1. NaOCH$_3$
2. Br$_2$
2. bromine

4'-benzyloxy-2-bromo-
3'-carbamoylacetophenone (I)

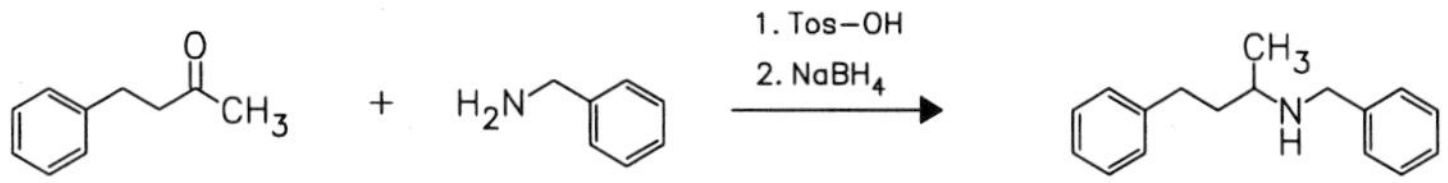

4-phenyl-
2-butanone

benzylamine

1. Tos—OH
2. NaBH$_4$

(±)-N-benzyl-1-methyl-
3-phenylpropylamine (II)

II

resolution with
N—(p—toluenesulfonyl)—L—leucine

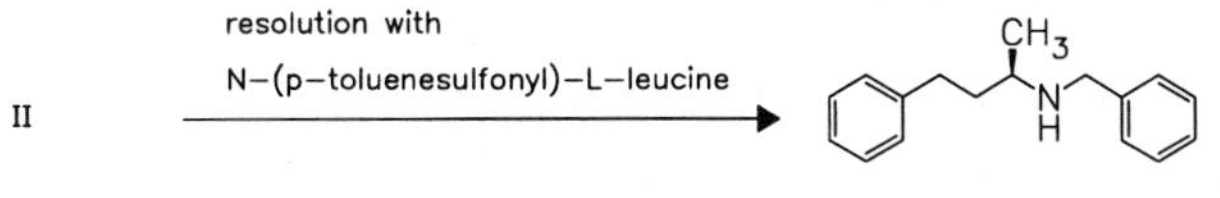

(R)-(+)-N-benzyl-1-methyl-
3-phenylpropylamine (III)

III + I

K$_2$CO$_3$

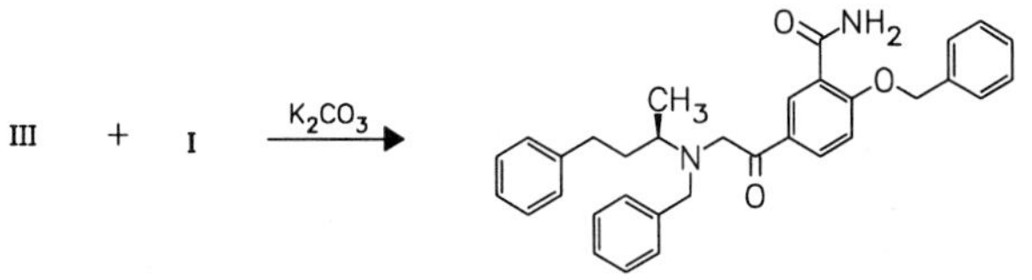

O-benzyl-5-[N-benzyl-N-
[(R)-1-methyl-3-phenylpropyl]-
glycyl]salicylamide (IV)

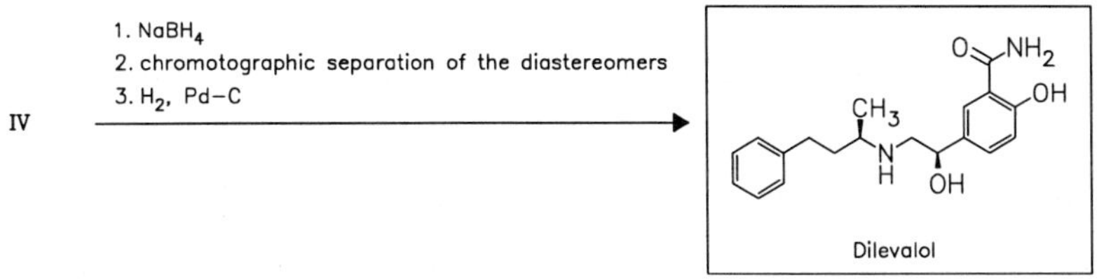

IV

1. NaBH$_4$
2. chromotographic separation of the diastereomers
3. H$_2$, Pd—C

Dilevalol

Reference(s):
EP 9 702 (Schering Corp.; appl. 17.9.1979; USA-prior. 20.9.1978).

improvement of diastereomer separation:
DOS 2 616 403 (Scherico; appl. 14.4.1976; USA-prior. 17.4.1975).
US 4 173 583 (Schering Corp.; 6.11.1979; appl. 21.9.1978; prior. 17.4.1975).

synthesis without chromatographic purification:
EP 92 787 (Schering Corp.; appl. 20.4.1983; USA-prior. 26.4.1982).

chiral reduction of IV:
EP 382 157 (Schering Corp.; appl. 6.2.1990; USA-prior. 10.2.1989, 26.9.1989).
US 4 948 732 (Schering Corp.; 14.8.1990; prior. 26.9.1989, 10.2.1989).
Clifton, J.E. et al.: J. Med. Chem. (JMCMAR) **25**, 670 (1982).
Gold, E.H. et al.: J. Med. Chem. (JMCMAR) **25**, 1363 (1982).

Formulation(s):   tabl. 50 mg, 100 mg

Trade Name(s):

I:			J:	
Abetol (CT)	Lolum (Lifepharma)		Dilevalon (Shionogi; 1989	
Alfabetal (Mitim)	Pressalolo (Locatelli)		as hydrochloride); wfm	
Amipress (Salus Research)	Pressalolo (Locatelli)-		Levadil (Schering Corp.;	
Biotens (Kemyos)-comb.	comb.		1990 as hydrochloride);	
Diurolab (Leben's)	Trandate (Glaxo)		wfm	
Ipolab (Leben's)	Trandiur (Teofarma)			

Diloxanide

ATC: P01AC01
Use: antiamebic, antiprotozoal

RN.   579-38-4   MF: $C_9H_9Cl_2NO_2$ MW: 234.08   EINECS: 209-439-4
LD$_{50}$: 2 g/kg (M, p.o.)
CN: 2,2-dichloro-*N*-(4-hydroxyphenyl)-*N*-methylacetamide

dichloroacetyl chloride   4-(methylamino)-phenol (I)   Diloxanide

chloral hydrate   chloral cyanohydrin   (II)

I + II   →   Diloxanide   (pyridine, N(C₂H₅)₃)

Reference(s):
a GB 767 148 (Boots; appl. 1954).
b GB 786 806 (Boots; appl. 22.7.1955; Compl. 3.7.1956).

Formulation(s):   1.5 g/day

Trade Name(s):
J: Entamide (Boots)

Diloxanide furoate

ATC: P01AX
Use: antiamoebic, antiprotozoal

RN: 3736-81-0   MF: C$_{14}$H$_{11}$Cl$_2$NO$_4$ MW: 328.15   EINECS: 223-108-1
CN: 2-furancarboxylic acid 4-[(dichloroacetyl)methylamino]phenyl ester

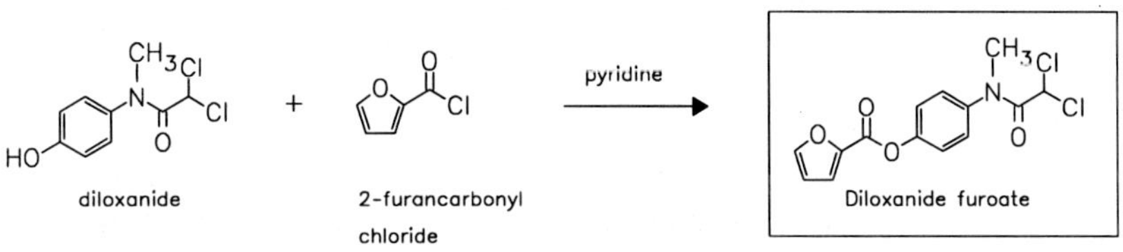

diloxanide   2-furancarbonyl   Diloxanide furoate
                  chloride

Reference(s):
GB 855 556 (Boots; prior. 6.5.1958, 4.6.1958, 14.4.1959).

Formulation(s):   tabl. 500 mg

Trade Name(s):
GB: Furamide (Knoll)

Diltiazem

ATC: C08DB01
Use: coronary therapeutic (calcium antagonist)

RN: 42399-41-7   MF: C$_{22}$H$_{26}$N$_2$O$_4$S   MW: 414.53   EINECS: 255-796-4
LD$_{50}$:   61 mg/kg (M, i.v.); 740 mg/kg (M, p.o.)
CN: (2S-*cis*)-3-(acetyloxy)-5-[2-(dimethylamino)ethyl]-2,3-dihydro-2-(4-methoxyphenyl)-1,5-benzothiazepin-4(5H)-one

monohydrochloride
RN: 33286-22-5   MF: C$_{22}$H$_{26}$N$_2$O$_4$S · HCl   MW: 450.99   EINECS: 251-443-3
LD$_{50}$:   58 mg/kg (M, i.v.); 508 mg/kg (M, p.o.);
         38 mg/kg (R, i.v.); 560 mg/kg (R, p.o.)

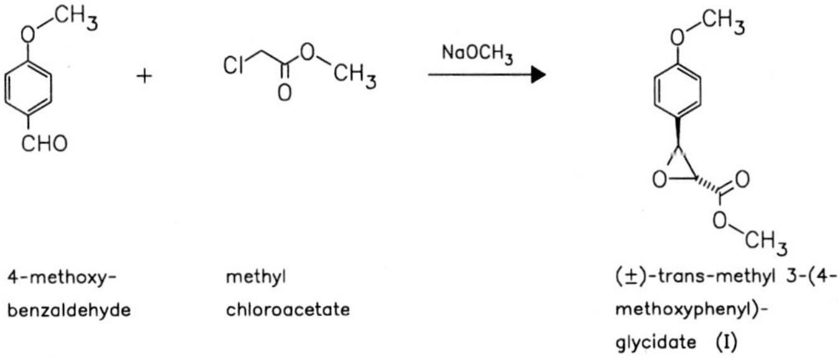

4-methoxy-   methyl   (±)-trans-methyl 3-(4-
benzaldehyde   chloroacetate   methoxyphenyl)-
                                    glycidate (I)

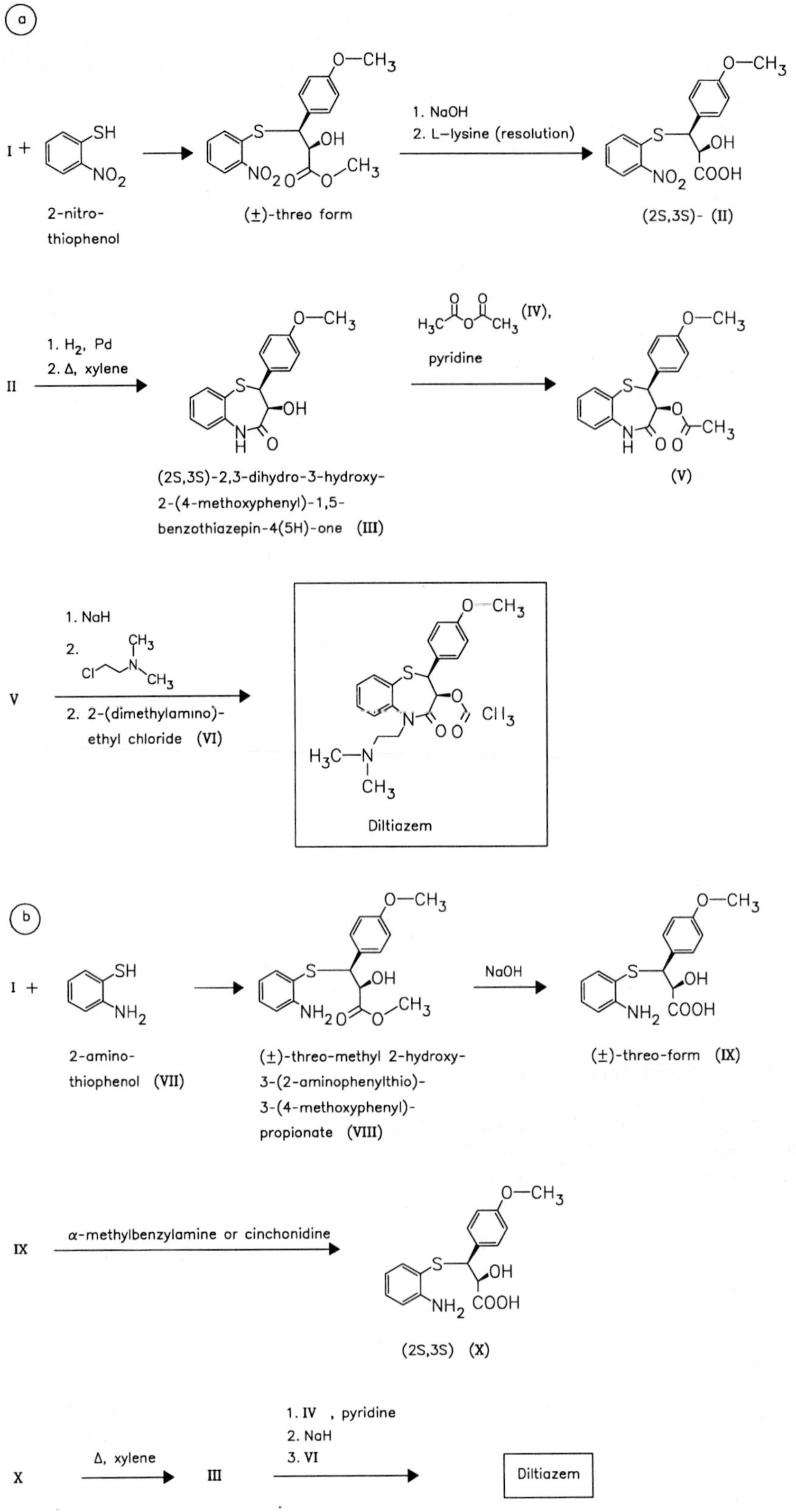

ⓐ

I + 2-nitro-thiophenol → (±)-threo form →
1. NaOH
2. L–lysine (resolution)
→ (2S,3S)- (II)

II →
1. H₂, Pd
2. Δ, xylene
→ (2S,3S)-2,3-dihydro-3-hydroxy-2-(4-methoxyphenyl)-1,5-benzothiazepin-4(5H)-one (III)
→ (IV), pyridine → (V)

V →
1. NaH
2. 2-(dimethylamino)-ethyl chloride (VI)
→ Diltiazem

ⓑ

I + 2-amino-thiophenol (VII) → (±)-threo-methyl 2-hydroxy-3-(2-aminophenylthio)-3-(4-methoxyphenyl)-propionate (VIII)
→ NaOH → (±)-threo-form (IX)

IX → α-methylbenzylamine or cinchonidine → (2S,3S) (X)

X → Δ, xylene → III →
1. IV , pyridine
2. NaH
3. VI
→ Diltiazem

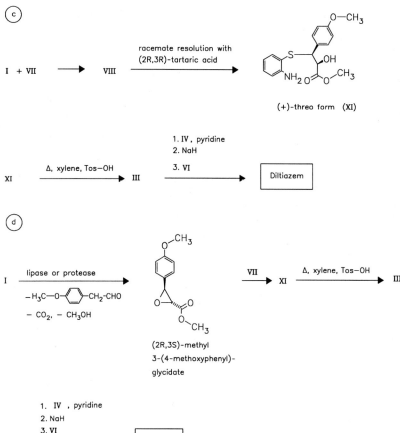

Ⓒ

I + VII ⟶ VIII ⟶ (racemate resolution with (2R,3R)-tartaric acid)

(+)-threo form (**XI**)

XI → (Δ, xylene, Tos—OH) → III → (1. IV, pyridine / 2. NaH / 3. VI) → Diltiazem

ⓓ

I → (lipase or protease / − H₃C—O—⟨⟩—CH₂-CHO / − CO₂, − CH₃OH) → (2R,3S)-methyl 3-(4-methoxyphenyl)-glycidate → VII → XI → (Δ, xylene, Tos—OH) → III

III → (1. IV, pyridine / 2. NaH / 3. VI) → Diltiazem

Reference(s):

Kugita, H. et al.: Chem. Pharm. Bull. (CPBTAL) (Tokyo) **18**, 2028, 2284 (1970); **19**, 595 (1971).
DE 1 805 714 (Tanabe; appl. 28.10.1968; J-prior. 28.10.1967, 17.6.1968).
US 3 562 257 (Tanabe; 9.2.1971; J-prior. 28.10.1967, 17.6.1968).
a US 4 420 628 (Tanabe Seiaku; appl. 13.12.1983; J-prior. 27.2.1981, 22.5.1981).
   Inoue, H. et al.: J. Chem. Soc., Perkin Trans. 1 (JCPRB4), **1984**, 1725.
   Inoue, H. et al.: J. Chem. Soc., Perkin Trans. 1 (JCPRB4), **1985**, 421.
b US 4 416 819 (Tanabe; 22.11.1983; appl. 9.7.1982).
   US 4 438 035 (Tanabe; 20.3.1984; appl. 1.12.1982; J-prior. 7.12.1981).
c US 5 144 025 (Zambon Group S.p.A.; 1.9.1992; I-prior. 2.4.1990).
   EP 669 327 (Zambon Group A.p.A.; appl. 12.4.1990; I-prior. 13.4.1989).
d US 5 274 300 (Sepracor; 28.12.1993; appl. 10.2.1089; prior. 26.10.1988).
   US 5 244 803 (Tanabe; 14.9.1993; appl. 7.9.1990; J-prior. 13.9.1989).
   Gentile, A ; Giordano, C.: J. Org. Chem. (JOCEAH) **57**, 6635 (1992).
   Rossy, G. et al. (Synthelabo): Manuf. Chem. (MCHMDI) **1993** (4), 20.

alternative synthesis:
glycidic ester via chlorohydrin route:
US 5 081 240 (Sanofi; 14.1.1992; F-prior. 18.7.1989).

enantioselective Darzens condensation:
Schwartz, A. et al.: J. Org. Chem. (JOCEAH) **57**, 851 (1992).

condensation of cyclic sulfite with 2-aminothiophenol:
Lohray, B.B. et al.: J. Org. Chem. (JOCEAH) **60**, 5983 (1995).

further routes:
DOS 3 337 176 (Ist Lusofarmaco; appl. 12.10.1983; I-prior. 15.10.1982).
DOS 3 415 035 (Shionogi; appl. 19.4.1984; J-prior. 21.4.1983).
EP 158 303 (Abic; appl. 5.4.1985; IL-prior. 13.4.1984).

combination with dihydropyridines:
US 4 504 476 (A. Schwartz et al.; 12.3.1985; appl. 16.9.1983).

inhalative formulation:
EP 133 252 (Gödecke AG; appl. 19.7.1984; D-prior. 20.7.1983).

slow and controlled release formulations:
EP 315 197 (Gödecke AG; appl. 4.11.1988; D-prior. 6.11.1987).
EP 318 398 (Ethypharm; appl. 25.11.1988; F-prior. 26.11.1987).
EP 320 097 (Elan Corp.; appl. 14.10.1988; IE-prior. 16.10.1987, 20.11.1987, 18.3.1988).
EP 340 105 (Sanofi; appl. 25.4.1989; F-prior. 27.4.1988).
US 4 859 470 (Alza Corp.; 22.8.1989; appl. 2.6.1988).
US 5 000 962 (Schering Corp.; 19.3.1991; appl. 25.8.1989).

Formulation(s): amp. 10 mg, 100 mg; cps. 60 mg, 90 mg, 120 mg, 180 mg, 240 mg; lyo. 25 mg; s. r. cps. 90 mg, 120 mg, 180 mg, 240 mg; s. r. tabl. 120 mg, 180 mg; tabl. 30 mg, 60 mg, 90 mg

Trade Name(s):

D: Corazet (Mundipharma)
     Dilicardin (Azupharma)
     Dilsal (TAD)
     Dil-Sanorania (Sanorania)
     Dilta (AbZ-Pharma)
     Diltahexal (Hexal)
     Diltaretard (betapharm)
     Dilti (ct-Arzneimittel)
     Diltiuc (durachemie)
     Dilzem (Gödecke; 1981)
     dilzereal (realpharma)
F: Bi-Tildiem (Labs.
     Synthélabo)
     Deltrazen (Pharmacia &
     Upjohn SA)
     Diacor (Labs. Houdé)
     Dilrène (Dakota)

     Mono-Tildiem (Labs.
     Synthélabo)
     Tildiem (Labs. Synthélabo;
     1980)
GB: Adizem XL (Napp)
     Angitil SR (Trinity)
     Britiazim (Thames)
     Dilzem SR (Elan)
     Slozem (Lipha)
     Tildiem (Lorex; 1984)
     Viazem XL (Du Pont)
I: Altiazem (Lusofarmaco;
     1984)
     Angizem (Inverni della
     Beffa)
     Carzem (Rottapharm)
     Citizem (CT)

     Diladel (Delalande Isnardi)
     Dilem (Ist. Chim. Inter.)
     Dilzene (Sigma-Tau)
     Tildiem (Synthelabo)
     Zilden (Schiapparelli)
J: Clarute (Santen)
     Gadoserin (Toho)
     Helsibon (Tobishi)
     Herbesser (Tanabe; 1987)
     Pazeadin (Taiyo)
     Tiaves (Rorer)
     Ziruvate (Choseido-
     Kayaku)
USA: Cardizem (Hoechst Marion
     Roussel; 1982)
     Dilacor XR (Watson)
     Tiazec (Forest)

Dimazole

(Diamthazole)

ATC: D01AE17
Use: antifungal

RN: 95-27-2   MF: $C_{15}H_{23}N_3OS$ MW: 293.44   EINECS: 202-406-5
CN: 6-[2-(diethylamino)ethoxy]-N,N-dimethyl-2-benzothiazolamine

dihydrochloride
RN: 136-96-9   MF: $C_{15}H_{23}N_3OS \cdot 2HCl$ MW: 366.36   EINECS: 205-270-5
LD$_{50}$: 98 mg/kg (M, i.v.); 430 mg/kg (M, p.o.);
     880 mg/kg (R, p.o.)

4-ethoxyphenyl   dimethyl-   6-ethoxy-2-dimethyl-
isothiocyanate   amine   aminobenzothiazole (I)

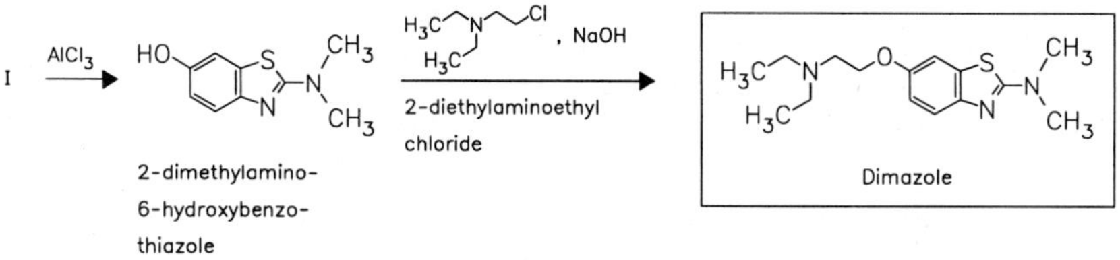

I → 2-dimethylamino-6-hydroxybenzo-thiazole

2-diethylaminoethyl chloride

Dimazole

Reference(s):

US 2 578 757 (Hoffmann-La Roche; 1951; prior. 1949).

Formulation(s):   topical 5 %

Trade Name(s):
GB: Asterol (Roche); wfm
I: Asterol (Roche); wfm

J: Asterol "Roche" (Roche-Shionogi)

Dimemorfan

ATC: R05DA11
Use: antitussive, sedative

RN: 36309-01-0   MF: $C_{18}H_{25}N$ MW: 255.41   EINECS: 252-963-3
CN: (9α,13α,14α)-3,17-dimethylmorphinan

phosphate
RN: 40678-33-9   MF: $C_{18}H_{25}N \cdot xH_3PO_4$ MW: unspecified

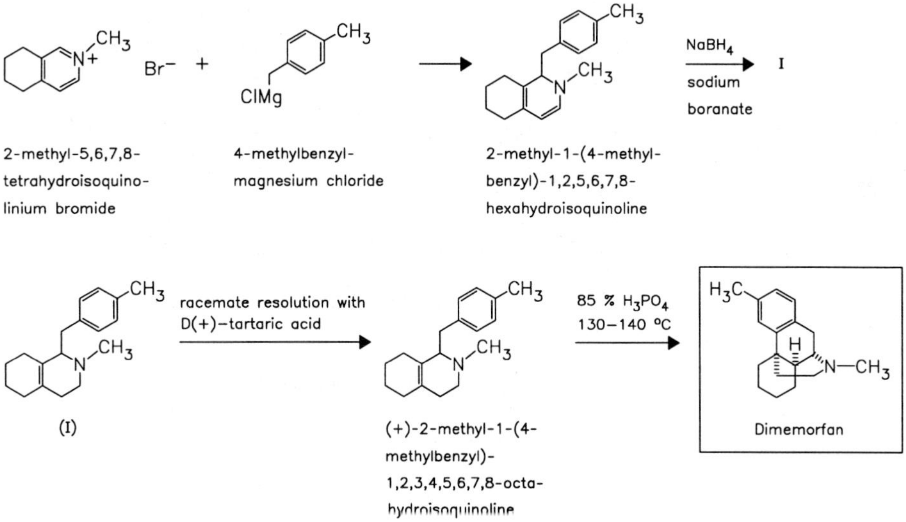

2-methyl-5,6,7,8-tetrahydroisoquino-linium bromide

4-methylbenzyl-magnesium chloride

2-methyl-1-(4-methyl-benzyl)-1,2,5,6,7,8-hexahydroisoquinoline

NaBH₄
sodium boranate → I

(I)

racemate resolution with D(+)-tartaric acid

(+)-2-methyl-1-(4-methylbenzyl)-1,2,3,4,5,6,7,8-octa-hydroisoquinoline

85 % H_3PO_4
130–140 °C

Dimemorfan

Reference(s):

DOS 2 128 607 (Yamanouchi; appl. 9.6.1971; J-prior. 20.6.1970, 2.2.1971, 9.2.1971).

Formulation(s):   drops 5 mg/ml; cps. 10 mg; syrup 2.5 mg/ml

Trade Name(s):
I: Gentus (Gentili)

Tusben (Benedetti)

J: Astomin (Yamanouchi)

Dimenhydrinate

ATC: A04AD49
Use: anti-emetic

RN: 523-87-5   MF: $C_{17}H_{21}NO \cdot C_7H_7ClN_4O_2$ MW: 469.97   EINECS: 208-350-8
LD_{50}:   203 mg/kg (M, p.o.);
        200 mg/kg (R, i.v.); 1320 mg/kg (R, p.o.)
CN: 2-(diphenylmethoxy)-N,N-dimethylethanamine, compd. with 8-chlorotheophylline

diphenhydramine   8-chlorotheophylline   Dimenhydrinate
(q. v.)

Reference(s):
US 2 499 058 (Searle; 1950; prior. 1949).
US 2 534 813 (Searle; 1950; appl. 1950).

Formulation(s):   drg. 10 mg, 20 mg, 50 mg, 150 mg, 200 mg (s. r. cps.); sol. for inj. 62 mg/10 ml (i.v.),
        100 mg/2 ml (i.m.); suppos.40 mg, 70 mg, 80 mg, 150 mg; tabl. 50 mg

Trade Name(s):
D: Arlevert (Hennig)-comb.
    Dimen (Heumann)
    Mandros (Dolorgiet)
    Migraeflux (Henning)-
    comb.
    Reisetabletten ratiopharm
    (ratiopharm)
    RubieMen (RubiePharm)
    Superprep/-forte (Hermes)
    Vertigo-Vomex
    (Yamanouchi)

    Vomacur (Hexal)
    Vomex (Yamanouchi)
F: Dramamine (Monsanto)
    Mercalm (Lab. Phygiène)-
    comb.
    Nausicalm (Lab. Brother
    SA)
GB: Dramamine (Searle)
I: Lomarin (Geymonat)
    Motozina (Biomedica
    Foscama)

    Travelgum (ASTA Medica)
    Valontan (Recordati)
    Xamamina (SmithKline
    Beecham)
    generics
J: Dramamine (Dainippon)
    Vomiles (Fujisawa)
USA: Dimenhydrinate (Wyeth-
    Ayerst)

Dimercaprol

(Dithioglycerin)

ATC: V03AB09
Use: antidote (heavy metal poisonings)

RN: 59-52-9   MF: $C_3H_8OS_2$ MW: 124.23   EINECS: 200-433-7
LD_{50}:   56 mg/kg (M, i.v.); 217 mg/kg (M, p.o.)
CN: 2,3-dimercapto-1-propanol

allyl alcohol   2,3-dibromo-   Dimercaprol
            1-propanol

Reference(s):
US 2 402 665 (Du Pont; 1946; appl. 1942).
US 2 432 797 (Minister of Supply of the United Kingdom; 1947; GB-prior. 1942).
US 2 436 137 (Du Pont; 1948; appl. 1944).
Stocken, L.A.; Thompson, R.H.S.: Biochem. J. (BIJOAK) **40**, 529, 535, 548 (1946).

synthesis via 2,3-dichloropropanol:
Ing, H.R.: J. Chem. Soc. (JCSOA9) **1948**, 1393.

Formulation(s):   amp. 100 mg

Trade Name(s):

D: Sulfactin Homburg   I: B.A.L. Boots (Boots Italia)   USA: BAL (Hynson Westcott &
    (Homburg); wfm   J: Bal (Daiichi)   Dunning); wfm
F: B.A.L (L'Arguenon)

Dimestrol
(Dimethoxydiethylstilbene)

ATC: G03
Use: estrogen

RN: 130-79-0   MF: $C_{20}H_{24}O_2$ MW: 296.41   EINECS: 204-994-9
CN: (*E*)-1,1'-(1,2-diethyl-1,2-ethenediyl)bis[4-methoxybenzene]

(a)

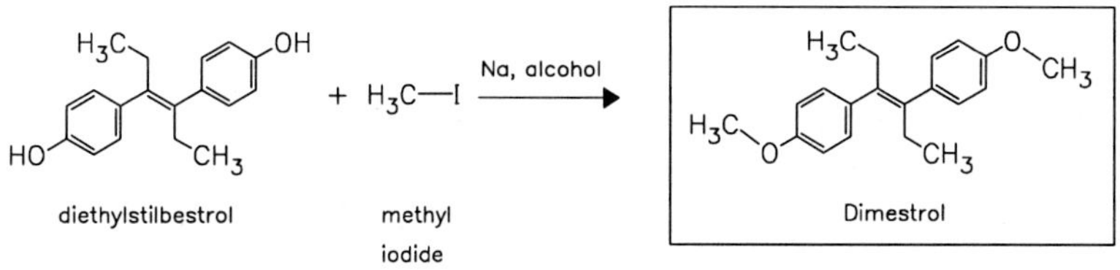

diethylstilbestrol   methyl   Dimestrol
                          iodide

(b)

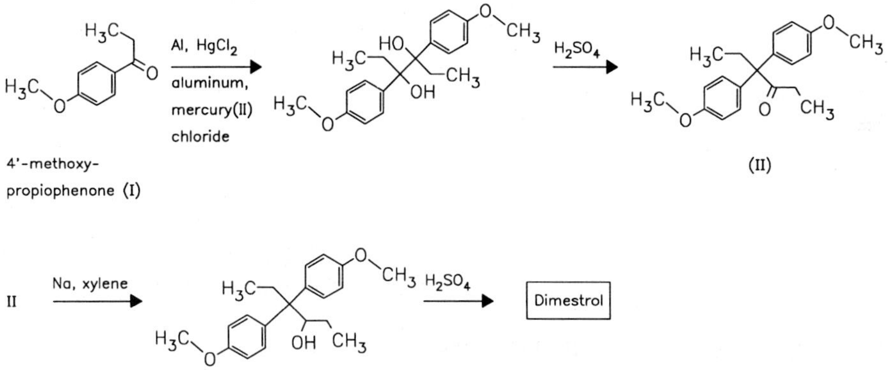

4'-methoxy-
propiophenone (I)   (II)

(c)

1-bromo-1-(4-methoxy-   (III)
phenyl)propane

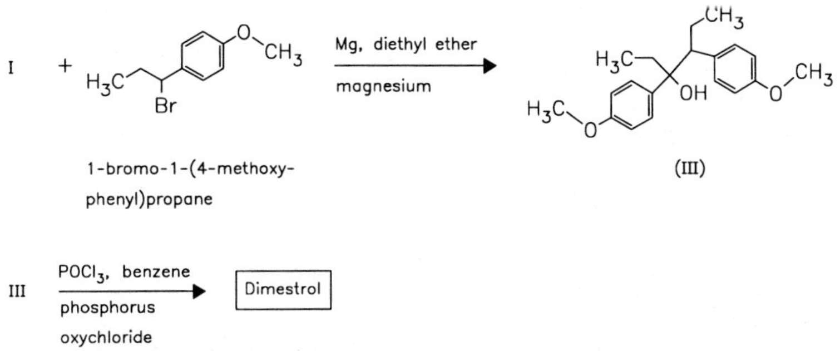

Reference(s):
a   Reid, E.E.; Wilson, E.: J. Am. Chem. Soc. (JACSAT) **64**, 1625 (1942).
b   Sisido, K.; Nozaki, H.: J. Am. Chem. Soc. (JACSAT) **70**, 776 (1948).
c   GB 584 253 (B.T. Bush; appl. 1943; USA-prior. 1941).
     GB 584 705 (B.T. Bush; appl. 1943; USA-prior. 1941).
     DE 897 559 (Bayer; appl. 1938).

alternative synthesis:
DE 824 043 (Boehringer Ing.; appl. 1949).

review:
Solmssen, U.V.: Chem. Rev. (Washington, D. C.) (CHREAY) **36**, 481 (1945).

Trade Name(s):
D: Depot-Oestromon (Merck);
      wfm

Dimetacrine

ATC: N06AA18
Use: antidepressant, thymoleptic

RN: 4757-55-5   MF: $C_{20}H_{26}N_2$ MW: 294.44
LD_{50}:   39600 µg/kg (M, i.v.); 1293 mg/kg (M, p.o.);
       1850 mg/kg (R, p.o.)
CN: *N,N*,9,9-tetramethyl-10(9*H*)-acridinepropanamine

tartrate (1:1)
RN: 3759-07-7   MF: $C_{20}H_{26}N_2 \cdot C_4H_6O_6$ MW: 444.53   EINECS: 223-166-8
LD_{50}:   40.9 mg/kg (M, i.v.); 755 mg/kg (M, p.o.);
       38 mg/kg (R, i.v.); 1671 mg/kg (R, p.o.)

diphenylamine   9-methyl-   9,9-dimethyl-
                       acridine   9,10-dihydro-
                                          acridine (I)

1. sodium amide
2. 3-dimethylamino-
   propyl chloride

Dimetacrine

Reference(s):
DE 1 224 315 (Kefalas; appl. 7.9.1961; GB-prior. 16.9.1960).
GB 933 875 (Kefalas S/A; appl. 16.9.1960; valid from 13.9.1961).
US 3 284 454 (Siegfried; 8.11.1966; CH-prior. 18.12.1961, 3.8.1962).
Molnar, I.; Wagner-Jauregg, T.: Helv. Chim. Acta (HCACAV) **48**, 1782 (1965).

Formulation(s):   drg. 100 mg (as hydrochloride)

Trade Name(s):

D: Istonil (Siegfried); wfm
J: Istonyl (Nippon Chemiphar)

Dimethadione

Use: anticonvulsant

RN: 695-53-4 MF: $C_5H_7NO_3$ MW: 129.12 EINECS: 211-781-4
LD_{50}: 850 mg/kg (M, i.p.)
CN: 5,5-dimethyl-2,4-oxazolidinedione

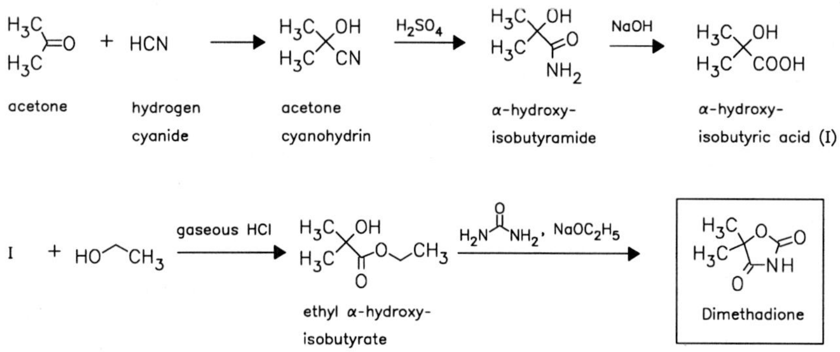

acetone
hydrogen cyanide
acetone cyanohydrin
H_2SO_4
α-hydroxy-isobutyramide
NaOH
α-hydroxy-isobutyric acid (I)
gaseous HCl
ethyl α-hydroxy-isobutyrate
$NaOC_2H_5$
Dimethadione

Reference(s):

Stoughton, R.W.: J. Am. Chem. Soc. (JACSAT) **63**, 2376 (1941).

Trade Name(s):

USA: Eupractone (Travenol); wfm

Dimethicone

(Dimethylpolysiloxane; Simethicone)

ATC: A09A
Use: antacid, antiflatulant

RN: 8050-81-5 MF: unspecified MW: unspecified
LD_{50}: 900 mg/kg (dog, i.v.)
CN: simethicone

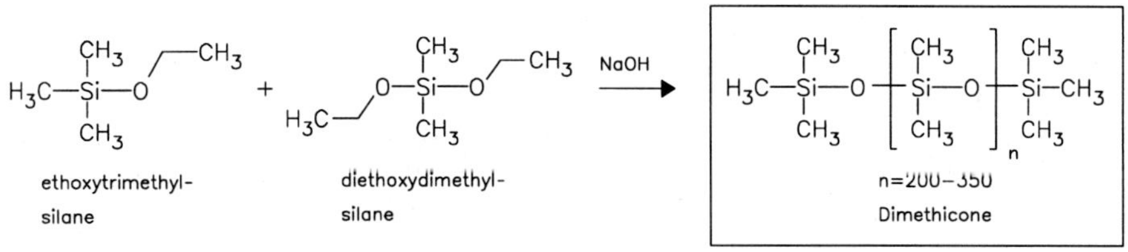

ethoxytrimethyl-silane
diethoxydimethyl-silane
NaOH
n=200–350
Dimethicone

Reference(s):

US 2 441 098 (Corning Glass; 1948; appl. 1946).

from dimethyldichlorosilane, e. *g.:*
DE 1 007 063 (General Electric; appl. 1956; USA-prior. 1955).
DOS 2 148 669 (Wacker-Chemie; appl. 29.9.1971).
DOS 2 521 742 (Wacker-Chemie; appl. 15.5.1975).

Formulation(s):　　cream 4 oz.; lotion 4 oz.

Trade Name(s):

D:　Absorber HFV (Arteva
　　Pharma)
　　Aegrosan (Opfermann)
　　Busala (Pharma Selz)
　　Ceolat (Solvay
　　Arzneimittel)
　　Dimeticon-ratiopharm
　　(ratiopharm)
　　Espumisan (Berlin-
　　Chemie)
　　ILIO-Funktion
　　Kautabletten (Robugen)
　　Meteosan (Novartis)
　　sab simplex (Parke Davis)
　　Symadal (Chauvin
　　ankerpharm)
F:　Gastrobul (Lab. Guerbet)-
　　comb.

　　Gel de polysilane (Labs.
　　UPSA)
　　Rennie Deflantine (Labs.
　　Roche Nicholas SA)-comb.
　　numerous combination
　　preparations
GB:　Infacol (Pharmax)
　　numerous combination
　　preparations
I:　Mylicon (Parke Davis)
　　Olio Silic (Tariff.
　　Integrativo)
　　Polisilon (Midy)-comb.
　　Silisan (Lipha)-comb.
J:　Aeropax (Green Cross)
　　Ganatone (Hokuriku)
　　Gasace (Kanto)
　　Gascon (Kissei)

　　Gasless (Hishiyama)
　　Gaspanon (Kotani)
　　Gasteel (Fuso)
　　Gaszeron (Nichiiko)
　　Gersmin (Kowa)
　　Harop (Toyo Pharmar)
　　Magarte (Mohan)
　　Polysilo (Toa)
　　Silies (Nippon Shoji)
　　Sili-Met-San S (Nippon
　　Shoji)
　　Spalilin (Maruishi)
USA:　Eucerin (Beiersdorf)
　　Moisturel (Westwood-
　　Squibb)

Dimethisterone

ATC:　G03D
Use:　progestogen

RN:　79-64-1　MF: $C_{23}H_{32}O_2$　MW: 340.51　EINECS: 201-215-4
CN:　(6α,17β)-17-hydroxy-6-methyl-17-(1-propynyl)androst-4-en-3-one

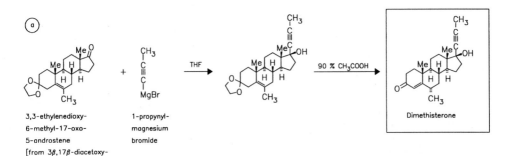

ⓐ

3,3-ethylenedioxy-
6-methyl-17-oxo-
5-androstene
[from 3β,17β-diacetoxy-
5-androstene (I)]

1-propynyl-
magnesium
bromide

Dimethisterone

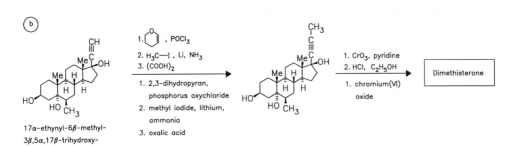

ⓑ

17α-ethynyl-6β-methyl-
3β,5α,17β-trihydroxy-
androstane
[from 3β-hydroxy-17-
oxo-5-androstene (II)]

1. 2,3-dihydropyran,
　phosphorus oxychloride
2. methyl iodide, lithium,
　ammonia
3. oxalic acid

1. chromium(VI)
　oxide

Dimethisterone

Reference(s):
a　US 2 927 119 (British Drug Houses; 1.3.1960; appl. 15.5.1958; GB-prior. 21.5.1957).

synthesis of I:
Petrov, V. et al.: J. Chem. Soc. (JCSOA9) **1957**, 4105; **1960**, 3676.
b US 2 939 819 (British Drug Houses; 7.6.1960; GB-prior. 25.1.1957).
Petrov, V. et al.: J. Chem. Soc. (JCSOA9) **1959**, 1957.
synthesis of II:
Petrov, V. et al.: J. Chem. Soc. (JCSOA9) **1957**, 4099.
Ruzicka, L.; Hofman, K.: Helv. Chim. Acta (HCACAV) **20**, 1280 (1937).

Formulation(s):   tabl. 5-15 mg

Trade Name(s):
GB: Secrosteron (Duncan,   J: Secrosteron (Santen-   USA: Oracon (Mead Johnson);
       Flockhart); wfm   Yamanouchi)   wfm

Dimethoxanate

Use: antitussive

RN: 477-93-0   MF: $C_{19}H_{22}N_2O_3S$ MW: 358.46   EINECS: 207-520-9
CN: 10*H*-phenothiazine-10-carboxylic acid 2-[2-(dimethylamino)ethoxy]ethyl ester

monohydrochloride
RN: 518-63-8   MF: $C_{19}H_{22}N_2O_3S \cdot HCl$ MW: 394.92   EINECS: 208-255-1
LD_{50}:   580 mg/kg (M, p.o.);
       1500 mg/kg (R, p.o.)

phenothiazine   phenothiazine-   2-(2-dimethylamino-   Dimethoxanate
                       10-carbonyl   ethoxy)ethanol
                       chloride

Reference(s):
DE 1 036 259 (Ayerst; appl. 1955; USA-prior. 1955).

Formulation(s):   syrup 12.5 mg/5 ml

Trade Name(s):
F: Cotrane (Clin-Midy); wfm   I: Cothera (Ayerst); wfm   Tussizid (Beolet); wfm
       Cotrane (Midyfarm); wfm   Perlatos (Farm. Mil.); wfm

Dimethyltubocurarinium chloride
(Metocurine chloride)

ATC: M03AA04
Use: muscle relaxant

RN: 33335-58-9   MF: $C_{40}H_{48}Cl_2N_2O_6$ MW: 723.74   EINECS: 251-461-1
CN: 6,6',7',12'-tetramethoxy-2,2,2',2'-tetramethyltubocuraranium dichloride

iodide
RN: 7601-55-0   MF: $C_{40}H_{48}I_2N_2O_6$ MW: 906.64   EINECS: 231-510-3
LD_{50}:   230 µg/kg (M, i.v.);
       35 µg/kg (R, i.v.)

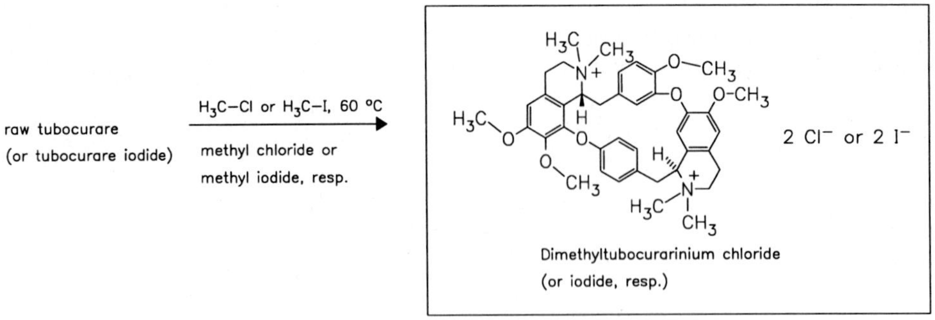

raw tubocurare
(or tubocurare iodide)

H₃C–Cl or H₃C–I, 60 °C

methyl chloride or
methyl iodide, resp.

2 Cl⁻ or 2 I⁻

Dimethyltubocurarinium chloride
(or iodide, resp.)

Reference(s):
US 2 581 903 (Eli Lilly; 1952; prior. 1949).

Formulation(s):   vial 2 mg/ml (20 ml)

Trade Name(s):
D: Methyl Curarin HAF   USA: Mecostrin (Squibb); wfm
      (Ethicon); wfm   Metubine Jodide (Lilly)

Dimetindene
(Dimethindene)

ATC: D04AA13; R06AB03
Use: antihistaminic, antipruritic

RN: 5636-83-9   MF: C₂₀H₂₄N₂   MW: 292.43   EINECS: 227-083-8
LD₅₀: 27 mg/kg (R, i.v.); 618 mg/kg (R, p.o.);
         45 mg/kg (dog, i.v.)
CN: *N,N*-dimethyl-3-[1-(2-pyridinyl)ethyl]-1*H*-indene-2-ethanamine

maleate (1:1)
RN: 3614-69-5   MF: C₂₀H₂₄N₂ · C₄H₄O₄   MW: 408.50   EINECS: 222-789-2
LD₅₀: 26.8 mg/kg (R, i.v.); 618 mg/kg (R, p.o.)

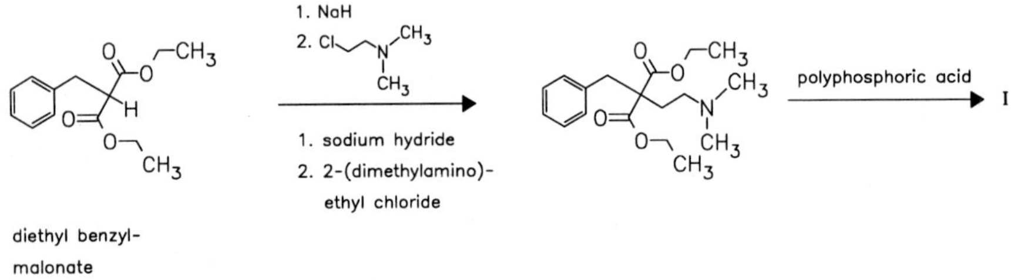

1. NaH
2. Cl–CH₂–N(CH₃)CH₃

1. sodium hydride
2. 2-(dimethylamino)-
   ethyl chloride

polyphosphoric acid ⟶ I

diethyl benzyl-
malonate

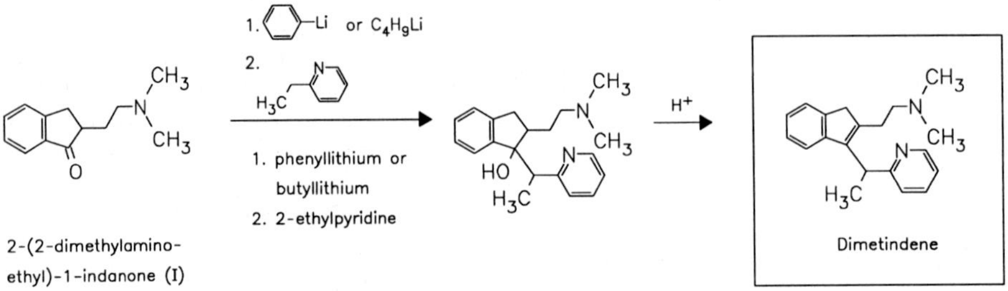

1. ⟨⟩–Li or C₄H₉Li
2.

1. phenyllithium or
   butyllithium
2. 2-ethylpyridine

H⁺

Dimetindene

2-(2-dimethylamino-
ethyl)-1-indanone (I)

Reference(s):
US 2 947 756 (Ciba; 2.8.1960; appl. 5.5.1959; prior. 12.8.1958, 3.11.1958, 10.2.1959).
US 2 970 149 (Ciba; 31.1.1961; appl. 3.11.1958).

Formulation(s):   amp. 4 mg; drg. 1 mg; drops 1 mg/ml; gel 1 mg/g; s. r. drg. 2.5 mg; s. r. tabl. 2.5 mg; syrup
0.122 mg/ml

Trade Name(s):

D:	Fenistil (Zyma-Blaes) Vibrocil (Zyma)-comb.		Vibrocil (Zyma)-comb.; wfm	J:	Foristal (Ciba-Geigy-Takeda)
GB:	Fenostil (Zyma); wfm Fenostil-Retard (Zyma); wfm	I:	Fengel (Zyma) Fenistil (Zyma) Vibrocil (Zyma)-comb.	USA:	Forhistal (Ciba); wfm Triten (Marion); wfm

---

Dimetotiazine
(Dimethothiazine; Fonazine)

ATC: N02CX05
Use: antiallergic, antihistaminic, antimigraine agent

RN: 7456-24-8   MF: $C_{19}H_{25}N_3O_2S_2$ MW: 391.56   EINECS: 231-229-6
LD_{50}:   100 mg/kg (M, i.v.); 740 mg/kg (M, p.o.)
CN: 10-[2-(dimethylamino)propyl]-*N,N*-dimethyl-10*H*-phenothiazine-2-sulfonamide

mesylate
RN: 7455-39-2   MF: $C_{19}H_{25}N_3O_2S_2 \cdot CH_4O_3S$ MW: 487.67

2-dimethylamino-
sulfonylphenothiazine

1. sodium amide
2. 2-dimethylamino-
   propyl chloride

Dimetotiazine

Reference(s):
GB 814 512 (Rhône-Poulenc; appl. 15.7.1957; F-prior. 1.8.1956, 18.12.1956).
FR 1 179 968 (Rhône-Poulenc; appl. 1.8.1956).

Formulation(s):   cps. 20 mg (base); suppos. 50 mg; tabl. 25 mg (mesylate)

Trade Name(s):

D:	Migristene (Rhodia Pharma); wfm	I:	Alius (Roussel)		Neomestin (Taiyo)
F:	Migristène (Specia); wfm	J:	Bistermin (Toyo Shinyaku)		Serevirol (Fuji Zoki)
GB:	Banistyl (May & Baker); wfm		Calsekin (Kanto-Isci) Demethotiazine (Mohan) Migristene (Shionogi)		

Dimoxyline
(Dioxyline)

ATC: A03
Use: antispasmodic, vasodilator

RN: 147-27-3   MF: $C_{22}H_{25}NO_4$ MW: 367.45
CN: 1-[(4-ethoxy-3-methoxyphenyl)methyl]-6,7-dimethoxy-3-methylisoquinoline

phosphate (1:1)
RN: 5667-46-9 MF: $C_{22}H_{25}NO_4 \cdot H_3PO_3$ MW: 449.44 EINECS: 227-126-0

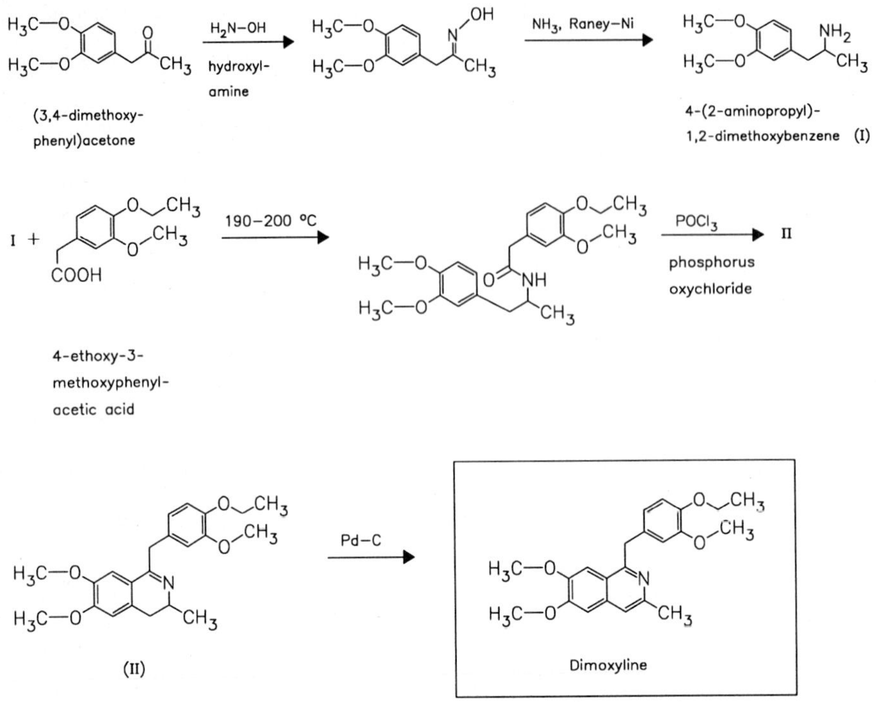

hydroxylamine
(3,4-dimethoxyphenyl)acetone
NH_3, Raney-Ni
4-(2-aminopropyl)-1,2-dimethoxybenzene (I)
I +
190-200 °C
$POCl_3$
phosphorus oxychloride
II
4-ethoxy-3-methoxyphenylacetic acid
Pd-C
(II)
Dimoxyline

Reference(s):
US 2 728 769 (Eli Lilly; 1955; prior. 1949).

Formulation(s): tabl. 100 mg

Trade Name(s):
I: Paverona (Lilly); wfm USA: Paveril (Lilly); wfm

Dinoprost
(Prostagladin $F_{2\alpha}$)

ATC: G02AD01
Use: oxytocic, abortifacient

RN: 551-11-1 MF: $C_{20}H_{34}O_5$ MW: 354.49
LD_{50}: 56 mg/kg (M, i.v.); 1300 mg/kg (M, p.o.);
106 mg/kg (R, i.v.); 1170 mg/kg (R, p.o.);
2.5-5.0 mg/kg (rabbit, i.m., i.v.)
CN: (5Z,9α,11α,13*E*,15*S*)-9,11,15-trihydroxyprosta-5,13-dien-1-oic acid

tromethamine salt (1:1)
RN: 38562-01-5 MF: $C_{20}H_{34}O_5 \cdot C_4H_{11}NO_3$ MW: 475.62 EINECS: 254-002-3
LD_{50}: 331 mg/kg (M, i.v.); 711 mg/kg (M, p.o.);
101 mg/kg (R, i.v.); 665 mg/kg (R, p.o.)

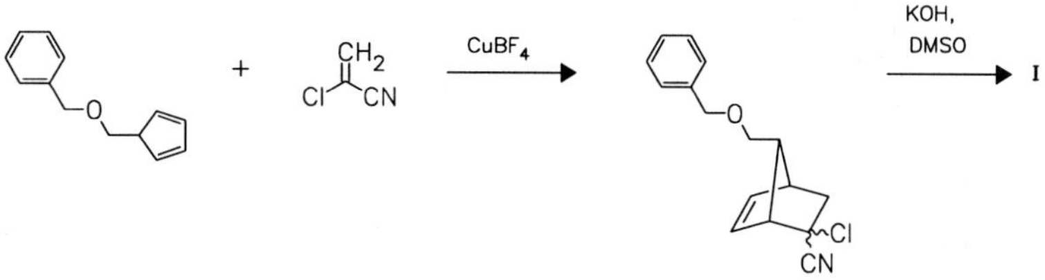

2,4-cyclopenta-
dienylmethyl
benzyl ether

2-chloro-
acrylonitrile

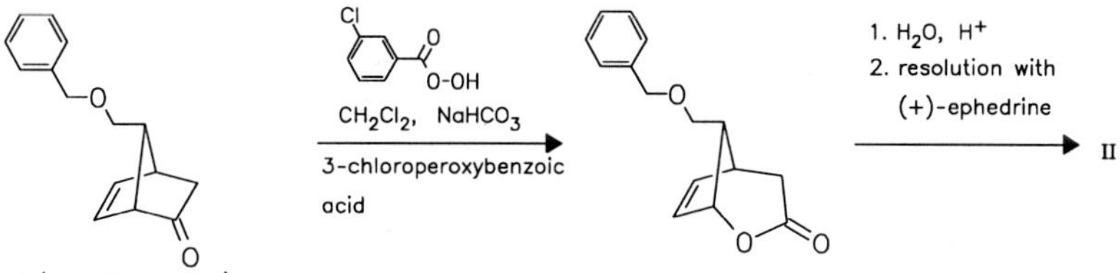

CuBF$_4$

KOH,
DMSO

→ I

7-(benzyloxymethyl)-
3-oxo-5-norbornene (I)

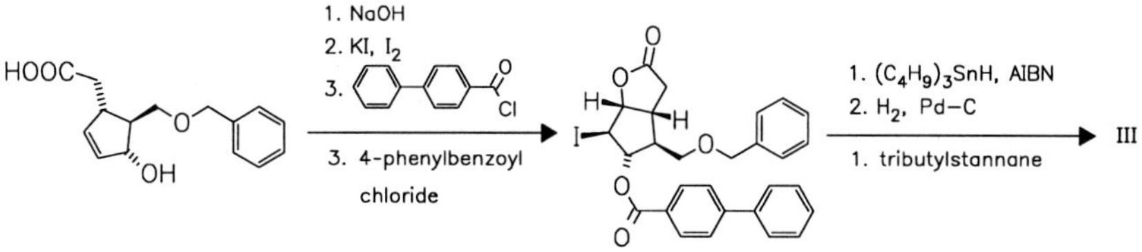

3-chloroperoxybenzoic
acid

CH$_2$Cl$_2$, NaHCO$_3$

1. H$_2$O, H$^+$
2. resolution with
   (+)-ephedrine

→ II

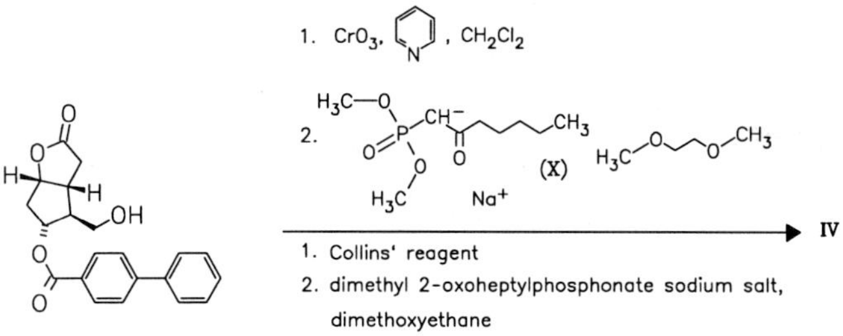

(II)

HOOC

OH

1. NaOH
2. KI, I$_2$
3.

Cl

3. 4-phenylbenzoyl
chloride

1. (C$_4$H$_9$)$_3$SnH, AIBN
2. H$_2$, Pd–C

1. tributylstannane

→ III

(−)-Corey lactone (III)

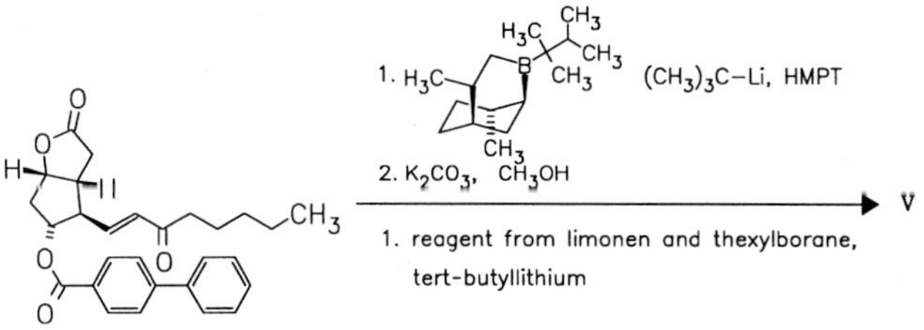

OH

1. CrO$_3$, ⟨N⟩, CH$_2$Cl$_2$

2.

H$_3$C—O

P—CH$^-$ CH$_3$

O

H$_3$C—O

H$_3$C   Na$^+$

(X)

H$_3$C—O   O—CH$_3$

1. Collins' reagent
2. dimethyl 2-oxoheptylphosphonate sodium salt,
   dimethoxyethane

→ IV

(IV)

CH$_3$

1. H$_3$C

H$_3$C   CH$_3$

B   CH$_3$

CH$_3$

(CH$_3$)$_3$C–Li, HMPT

2. K$_2$CO$_3$, CH$_3$OH

1. reagent from limonen and thexylborane,
   tert-butyllithium

→ V

(V)

(VI)

VI +

5-triphenylphosphonio-
pentanecarboxylate

Dinoprost

Reference(s):
Corey, E.J. et al.: J. Am. Chem. Soc. (JACSAT) **91**, 5675 (1969).
Corey, E.J. et al.: J. Am. Chem. Soc. (JACSAT) **92**, 397 (1970).
Corey, E.J. et al.: J. Am. Chem. Soc. (JACSAT) **92**, 2586 (1970).
Corey, E.J. et al.: J. Am. Chem. Soc. (JACSAT) **93**, 1491 (1971).

alternative syntheses:
Fried, J. et al.: J. Am. Chem. Soc. (JACSAT) **94**, 4342, 4343 (1972).
Corey, E.J. et al.: Tetrahedron Lett. (TELEAY) **1970**, 307.
Bundy, G.L. et al.: J. Am. Chem. Soc. (JACSAT) **94**, 2123 (1972).
Corey, E.J.; Varma, R.K.: J. Am. Chem. Soc. (JACSAT) **93**, 7319 (1971).
Schneider, W.P.; Murray, H.C.: J. Org. Chem. (JOCEAH) **38**, 397 (1973).
Tanouchi, T. et al.: Chem. Lett. (CMLTAG) **1976**, 739.
NL 6 505 799 (Unilever; 6.5.1965).
DOS 2 145 125 (Upjohn; 9.9.1971; USA-prior. 11.9.1970, 2.7.1971).
DOS 2 328 131 (Schering AG; 30.5.1973).
US 3 933 892 (Hoffmann-La Roche; 20.1.1976; prior. 18.1.1974, 12.2.1973).

isolation:
GB 1 040 544 (Karolinska Inst.; valid from 21.2.1963; prior. 19.3.1962).

racemic prostaglandin F$_{2\alpha}$:
US 3 933 891 (Upjohn; 20.1.1976; prior. 8.7.1974, 3.10.1973, 17.6.1975, 2.7.1971, 11.11.1970).
US 3 987 083 (Upjohn; 19.10.1976; prior. 6.12.1974, 14.3.1969).
US 3 983 155 (Upjohn; 28.9.1976; prior. 6.12.1974, 14.3.1969).
US 3 983 154 (Upjohn; 28.9.1976; prior. 6.12.1974, 14.3.1969).
US 3 983 153 (Upjohn; 28.9.1976; prior. 6.12.1974, 14.3.1969).
US 3 981 880 (Upjohn; 21.9.1976; prior. 6.12.1974, 14.3.1969).
US 3 980 691 (Upjohn; 14.9.1976; prior. 6.12.1974, 14.3.1969).
US 3 959 346 (Upjohn; 25.5.1976; prior. 6.12.1974, 14.3.1969).

tromethamine salt:
US 3 657 327 (Upjohn; 18.4.1972; prior. 1.6.1970).

use for control of conception cyclus:
DOS 1 943 492 (Upjohn; appl. 27.8.1969; USA-prior. 29.8.1968).

review:
Prostaglandin Research (Ed. P. Crabbé) p. 1, 121 New York, San Francisco, London 1977.

Formulation(s):   amp. 5 mg/ml

Trade Name(s):

D:	Minprostin F$_{2\alpha}$ (Pharmacia & Upjohn)	I:	Prostin F$_{2\alpha}$ (Upjohn); wfm	Prostarmon-F (Ono)
		J:	Glandinon (Mochida)	Zinoprost (Ono)
F:	Prostine F$_{2\alpha}$ (Pharmacia & Upjohn SA)		Penacelan-F (Glaxo-Fuji)	USA: Prostin F$_{2\alpha}$ (Upjohn); wfm
			Pronalgon (Upjohn)	
GB:	Prostin F2 (Pharmacia & Upjohn)		Prosmon (Fuji)	
			Prostamodin (Kanebo)	

Dinoprostone

(Prostaglandin E$_2$)

ATC: G02AD02
Use: oxytocic, abortifacient

RN: 363-24-6   MF: C$_{20}$H$_{32}$O$_5$ MW: 352.47   EINECS: 206-656-6
LD$_{50}$:   23.2 mg/kg (M, i.v.); 750 mg/kg (M, p.o.);
    59.5 mg/kg (R, i.v.); 500 mg/kg (R, p.o.)
CN: (5Z,11α,13E,15S)-11,15-dihydroxy-9-oxoprosta-5,13-dien-1-oic acid

prostaglandin F$_2$ 9,11-bis-
(tetrahydropyranyl ether)
(cf. dinoprost synthesis)

1. Jones oxidation
    (CrO$_3$, H$_2$SO$_4$, acetone, −10 °C)
2. CH$_3$COOH, H$_2$O

1. chromium(VI) oxide

Dinoprostone

Reference(s):
Corey, E.J. et al.: J. Am. Chem. Soc. (JACSAT) **92**, 397 (1970).

alternative syntheses:
US 3 948 981 (Upjohn; 6.4.1976; prior. 18.12.1974, 3.10.1973, 2.7.1971, 11.9.1970).
Schneider, W.P. et al.: J. Chem. Soc., Chem. Commun. (JCCCAT) **1973**, 254.
Heather, J.B. et al.: Tetrahedron Lett. (TELEAY) **1973**, 2313.

isolation:
GB 1 040 544 (Karlinska Inst.; valid from 21.2.1963; prior. 29.3.1962).

further literature:
cf. dinoprost synthesis

medical use as broncholytic:
ZA 681 055 (American Home Products; appl. 31.1.1968; USA-prior. 20.2.1967).

review:
Prostaglandin Research (Ed. P. Crabbé) p. 1, 121, New York, San Francisco, London 1977.

Formulation(s):   amp. 0.5 g/0.5 ml, 0.75 mg/0.75 ml; rectangular tabl. 0.5 mg; syringe with gel 0.5 mg; vaginal gel 0.5 mg/3 g, 1 mg/3 g, 2 mg/3 g; vaginal tabl. 3 mg

Trade Name(s):

D:	Minprostin E$_2$ (Pharmacia & Upjohn)	F:	Prépidil gel (Pharmacia & Upjohn SA)	GB: Prepidil (Pharmacia & Upjohn)
	Prepidil Gel (Pharmacia & Upjohn)		Prostine E$_2$ (Pharmacia & Upjohn SA)	Propress RS (Ferring)

	Prostin E2 (Pharmacia & Upjohn)	J: Prostadiel-E (Taiyo) Prostaglandin E2 (Kaken) Prostarmon-E (Ono)	Prepidil (Pharmacia & Upjohn) Prostin E2 (Pharmacia & Upjohn)
I:	Prepidil gel (Upjohn) Prostin E2 (Upjohn)	USA: Cervidil (Forest)	

Diodone
(Jodopyracet)

ATC: V08AA10
Use: X-ray contrast medium

RN: 101-29-1   MF: C$_7$H$_5$I$_2$NO$_3$ MW: 404.93   EINECS: 202-932-5
CN: 3,5-diiodo-4-oxo-1(4H)-pyridineacetic acid

meglumine salt
RN: 3736-90-1   MF: C$_7$H$_5$I$_2$NO$_3$ · C$_7$H$_{17}$NO$_5$ MW: 600.14
LD$_{50}$: 5900 mg/kg (R, i.v.)
diethanolamine salt (1:1)
RN: 300-37-8   MF: C$_7$H$_5$I$_2$NO$_3$ · C$_4$H$_{11}$NO$_2$ MW: 510.07   EINECS: 206-089-4
LD$_{50}$: 6400 mg/kg (M, i.v.);
      5400 mg/kg (R, i.v.)
morpholine salt (1:1)
RN: 3737-08-4   MF: C$_7$H$_5$I$_2$NO$_3$ · C$_4$H$_9$NO   MW: 492.05

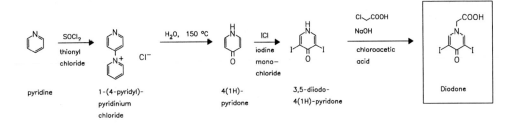

Reference(s):
DRP 554 702 (E. Koenigs, H. Greiner; 1929).
DRP 579 224 (I. G. Farben; 1930).
US 1 993 039 (I. G. Farben; 1935; D-prior. 1931).
GB 517 382 (ICI; appl. 1938).

Formulation(s):   amp. 35 %, 50 %, 70 %

Trade Name(s):
D:	Broncho-Abrodil (Schering); wfm	GB:	Umbradil (Astra)-comb.; wfm	I:	Joduron (Bracco); wfm
				J:	Pyraceton (Daiichi)

Diosmin

ATC: C05CA03
Use: antihemorrhagic, vein tonic

RN: 520-27-4   MF: C$_{28}$H$_{32}$O$_{15}$ MW: 608.55   EINECS: 208-289-7
CN: 7-[[6-O-(6-deoxy-α-L-mannopyranosyl)-β-D-glucopyranosyl]oxy]-5-hydroxy-2-(3-hydroxy-4-methoxyphenyl)-4H-1-benzopyran-4-one

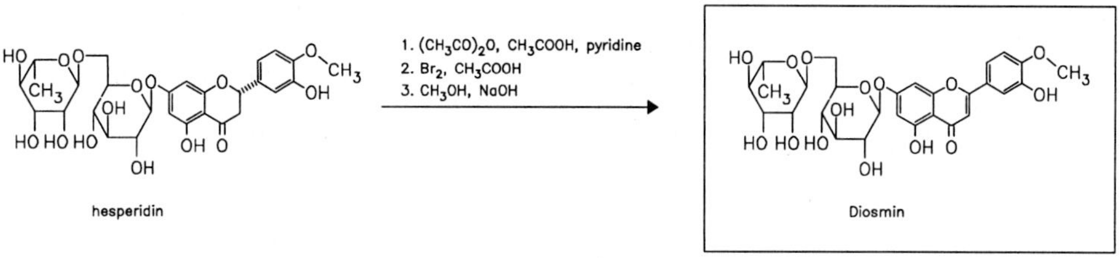

hesperidin

1. (CH₃CO)₂O, CH₃COOH, pyridine
2. Br₂, CH₃COOH
3. CH₃OH, NaOH

Diosmin

Reference(s):

Zemplén, G.; Bognár, R.: Ber. Dtsch. Chem. Ges. (BDCGAS) **76**, 452 (1943).
Lorette, N.B. et al.: J. Org. Chem. (JOCEAH) **16**, 930 (1951).
Horowitz, R.M.: J. Org. Chem. (JOCEAH) **21**, 1184 (1956).

technical method:
DOS 2 602 314 (Hommel; appl. 22.1.1976; CH-prior. 16.5.1975).

Formulation(s):   cps. 300 mg; cream 4 g/100 g; tabl. 150 mg

Trade Name(s):
D:	Tovene (Solvay Arzneimittel)	Endium (Labs. Europhta) Flebosmil (Socopharm)		Daflon (Servier) Diosven (CT)
F:	Daflon (Servier)	Litosmil (Evans Medical)		Doven (Prophin)
	Dio (Labs. Scienex)	Médiveine (Elerté)		Venosmine (Geymonat)
	Diosmil (Rhône-Poulenc Rorer)	Préparation H Veinotonic (Whitehall)		
	Diovenor (Innothéra)	I:	Arvenum (Stroder)	

Diperodon

(Diperocainum)

ATC: D04AB
Use: local anesthetic

RN: 101-08-6   MF: C₂₂H₂₇N₃O₄   MW: 397.48   EINECS: 202-913-1
CN: 3-(1-piperidinyl)-1,2-propanediol bis(phenylcarbamate) (ester)

monohydrochloride
RN: 537-12-2   MF: C₂₂H₂₇N₃O₄ · HCl   MW: 433.94   EINECS: 208-659-8
LD₅₀: 890 mg/kg (M, s.c.)

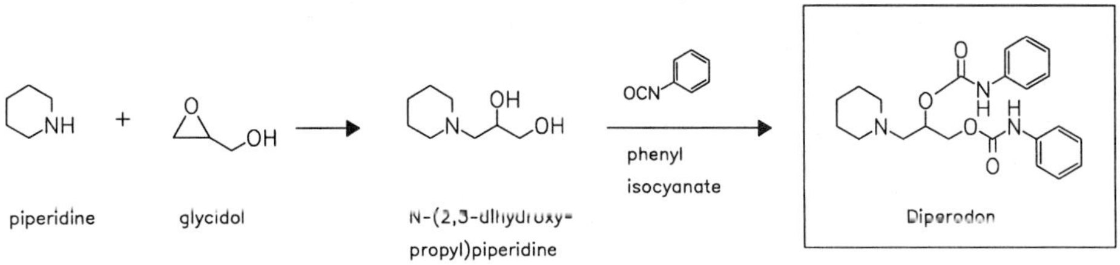

piperidine   +   glycidol   →   N-(2,3-dihydroxy-propyl)piperidine   OCN phenyl isocyanate   →   Diperodon

Reference(s):
US 2 004 132 (T. H. Rider; 1935; prior. 1931).
Rider, T.H.: J. Am. Chem. Soc. (JACSAT) **52**, 1528, 2115 (1930).

Formulation(s):   ointment (comb.)

Diphemanil metilsulfate

ATC: D11AA
Use: anticholinergic, antispasmodic

RN: 62-97-5   MF: $C_{20}H_{24}N \cdot CH_4O_3S$ MW: 374.53   EINECS: 200-552-4
LD$_{50}$:   4012 µg/kg (M, i.v.); 317 mg/kg (M, p.o.);
         5 mg/kg (R, i.v.); 1107 mg/kg (R, p.o.)
CN: 4-diphenylmethylene-1,1-dimethylpiperidinium methyl sulfate

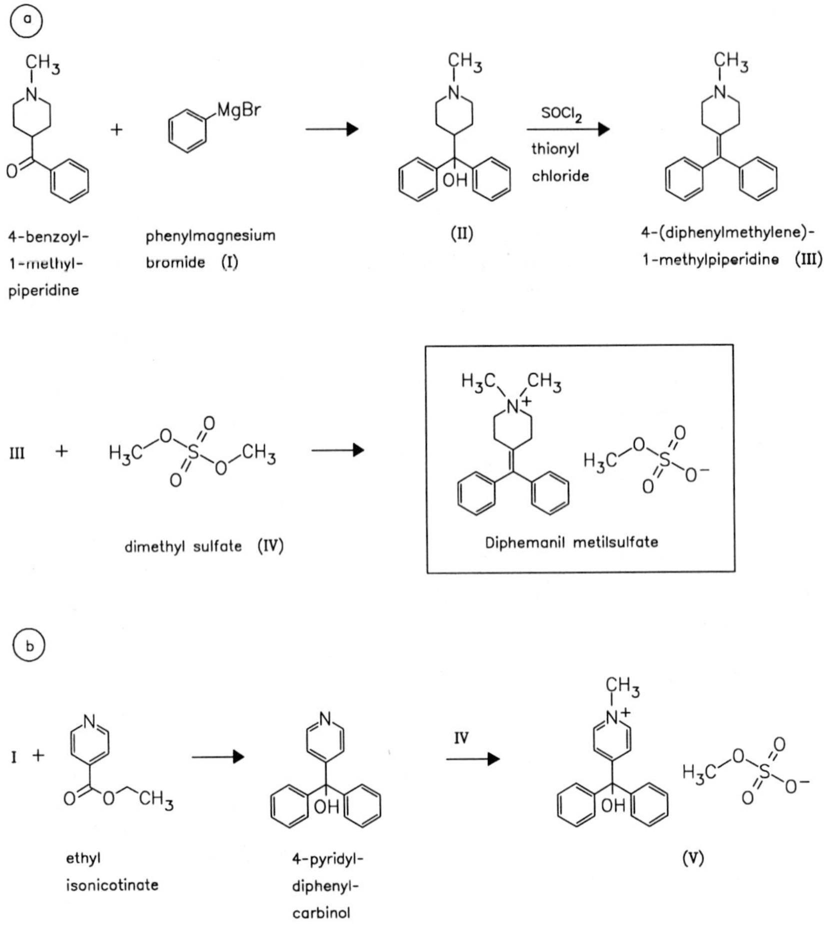

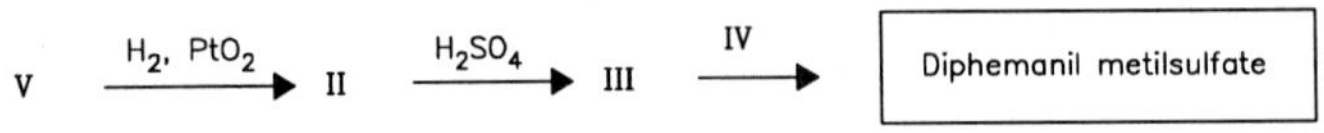

4-benzoyl-   phenylmagnesium   (II)   4-(diphenylmethylene)-
1-methyl-   bromide (I)   1-methylpiperidine (III)
piperidine

dimethyl sulfate (IV)   Diphemanil metilsulfate

ethyl   4-pyridyl-   (V)
isonicotinate   diphenyl-
                     carbinol

$V \xrightarrow{H_2, PtO_2} II \xrightarrow{H_2SO_4} III \xrightarrow{IV}$ Diphemanil metilsulfate

Reference(s):
US 2 739 968 (Schering Corp.; 1956; prior. 1951).

Formulation(s):   cream 2 %; tabl. 50 mg, 100 mg

J: Prantal (Schering-   USA: Prantal (Schering); wfm
      Shionogi)

Diphenadione

ATC: B01AA10
Use: anticoagulant, rodenticide

RN: 82-66-6   MF: $C_{23}H_{16}O_3$ MW: 340.38   EINECS: 201-434-5
LD_{50}:   28.3 mg/kg (M, p.o.);
      1500 μg/kg (R, p.o.);
      3 mg/kg (dog, p.o.)
CN: 2-(diphenylacetyl)-1H-indene-1,3(2H)-dione

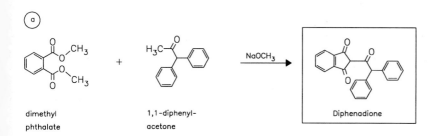

dimethyl   1,1-diphenyl-   Diphenadione
phthalate   acetone

1,3-indanedione

Reference(s):
US 2 672 483 (Upjohn; 1954; prior. 1951).

Formulation(s):   tabl. 20 mg, 50 mg

Trade Name(s):
USA: Dipaxin (Upjohn); wfm

Diphenhydramine

ATC: D04AA32; R06AA02
Use: antihistaminic, anti-emetic, sedative,
       antitussive

RN: 58-73-1   MF: $C_{17}H_{21}NO$ MW: 255.36   EINECS: 200-396-7
LD_{50}:   29 mg/kg (M, i.v.); 160 mg/kg (M, p.o.);
      42 mg/kg (R, i.v.); 390 mg/kg (R, p.o.)
CN: 2-(diphenylmethoxy)-N,N-dimethylethanamine

hydrochloride
RN: 147-24-0   MF: $C_{17}H_{21}NO \cdot HCl$ MW: 291.82   EINECS: 205-687-2
LD_{50}:   20 mg/kg (M, i.v.); 64 mg/kg (M, p.o.);
      35 mg/kg (R, i.v.); 500 mg/kg (R, p.o.);
      24 mg/kg (dog, i.v.)

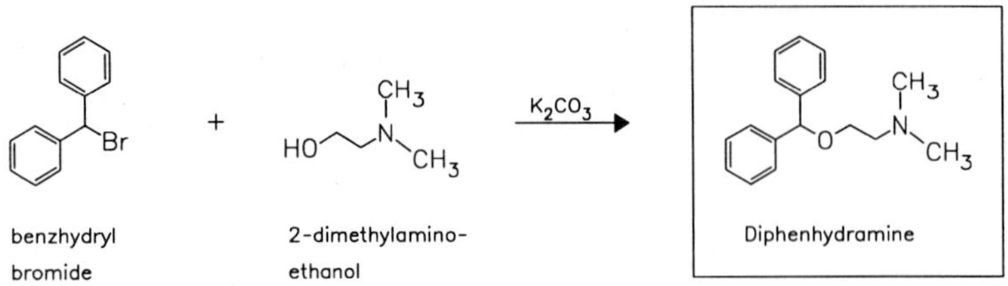

benzhydryl bromide   +   2-dimethylamino-ethanol   → (K_2CO_3) →   Diphenhydramine

Reference(s):

US 2 421 714 (Parke Davis & Co.; 1947; prior. 1944).
US 2 427 878 (Parke Davis; 1947; appl. 1947).

alternative synthesis:

US 2 397 799 (Geigy; 1946; CH-prior. 1942).

Formulation(s):   drops 12.5 mg; s. r. cps. 30 mg; suppos. 10 mg, 20 mg, 50 mg; syrup 2.67 mg/ml, 12.5 mg/ml; tabl. 25 mg, 50 mg (as hydrochloride)

Trade Name(s):

D: Anaestecomp (Ritsert)-comb.
Benadryl (Warner-Lambert)
Betadorm (Woelm)-comb.
Dibenzyl-Rhenix (Pharma Wernigerode)-comb.
Dolestan (Whitehall-Much)
Dolestan (Whitehall-Much)-comb.
Dormigoa (Scheurich)
Dormutil (Isis Pharma)
Emesan (Lindopharm)
Halbmond-Tabletten (Whitehall-Much)
Hevert-Dorm (Hevert)
Lupovalin (Pharma Selz)
Moradorm (Bouhon)-comb.
nervo OPT (Optimed)
Nytol (Block Drug Company)
Palacril (Warner-Lambert)-comb.
Palmicol (RIAM)
Pheramin (Kanoldt)
Praesidin (Medopharm)-comb.
Reisegold (Whitehall-Much)-comb.
Reisegold tabs (Whitehall-Much)
S.8 Tabletten (Chefaro)
Sediat (Pfleger)

Sedopretten (Schöning-Berlin)
Sedovegan (Wolff)
Valeriana comb. Hevert (Hevert)-comb.
Visano Cor (Kade)-comb.

F: Actifed Jour et unit (Warner-Lambert; as hydrochloride)-comb.
Butix gel (Labs. Pierre Fabre Santé; as hydrochloride)
Nautamine (Synthélabo)
Onctose hydrocortisone (Monot; as methyl sulfate)-comb.

GB: Medinex (Whitehall)
Nytol (Stafford-Miller)
numerous combination preparations

I: Allergan (Bouty)
Asmarectal (Serpero)-comb.
Benadryl (Parke Davis)-comb.
Benylin (Parke Davis)-comb.
Difeni (Formulario Naz.)
Fluvaleas (Valeas)-comb.
combination preparations

J: Benadin Salicylate (Kongo; as salicylate)
Benadol (Taisho)

Benadozol (Hokuriku; as tannate)
Benadozol-S (Hokuriku; as salicylate)
Benapon (Dainippon)
Benasin (Kanto)
Neo-Restar (Ohta; as maleate)
Restamin (Kowa)
Restar (Ohta; as salicylate)
Restin (Mohan; as salicylate)
Reston (Kowa Yakuhin)
Salibena (Fuso; as salicylate)
Vena (Tanabe)
Venerlon (Sanwa; as tannate)
Zeresmin (Juzen-Yamanouchi; as salicylate)

USA: Actifed (Warner-Lambert; as hydrochloride)
Benadryl (Parke Davis; as hydrochloride)
Dytuss (Lunsco; as hydrochloride)
Maximum (Pfizer Consumer; as hydrochloride)
Tylenol (McNeil; as hydrochloride)
Unisom (Pfizer Consumer; as hydrochloride)
generics

Diphenoxylate

ATC: A07DA01
Use: antidiarrheal, antiperistaltic

RN: 915-30-0 MF: $C_{30}H_{32}N_2O_2$ MW: 452.60 EINECS: 213-020-1
LD$_{50}$: 337 mg/kg (M, p.o.);
221 mg/kg (R, p.o.)
CN: 1-(3-cyano-3,3-diphenylpropyl)-4-phenylpiperidine-4-carboxylic acid ethyl ester

hydrochloride
RN: 3810-80-8 MF: $C_{30}H_{32}N_2O_2 \cdot HCl$ MW: 489.06 EINECS: 223-287-6
LD$_{50}$: 221 mg/kg (R, p.o.)

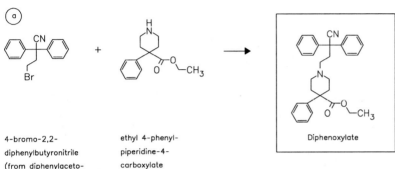

ⓐ

4-bromo-2,2-
diphenylbutyronitrile
(from diphenylaceto-
nitrile and 1,2-
dibromoethane)

ethyl 4-phenyl-
piperidine-4-
carboxylate

Diphenoxylate

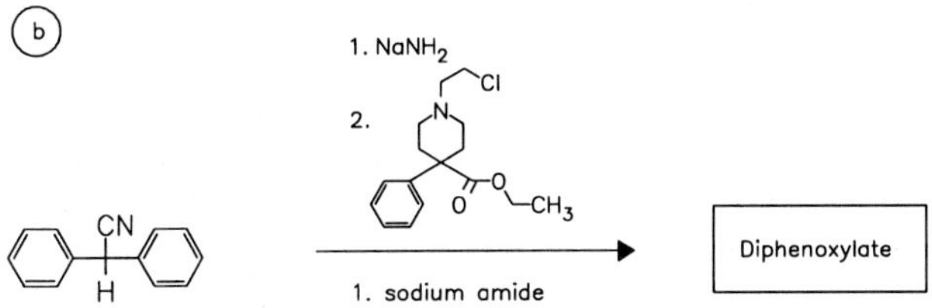

ⓑ

diphenylaceto-
nitrile

1. NaNH$_2$
2.

Diphenoxylate

1. sodium amide
2. ethyl 1-(2-chloroethyl)-
4-phenyl-piperidine-
4-carboxylate

Reference(s):
US 2 898 340 (Janssen; 4.8.1959; NL-prior. 5.7.1957).
US 4 086 234 (Searle; 25.4.1978; appl. 7.11.1975).

Formulation(s): tabl. 2.5 mg (comb. with 0.025 mg atropine sulfate)

Trade Name(s):
D: Reasec (Janssen-Cilag)-
comb. with atropine sulfate
F: Diarsed (Sanofi Winthrop;
as hydrochloride)

GB: Lomotil (Searle)-comb.
Tropergen (Norgine)-comb.
I: Reasec (Cilag)-comb.

USA: Lomotil (Searle; as
hydrochloride)
Lonox (Geneva; as
hydrochloride)

Diphenylpyraline

ATC: R06AA07
Use: antiallergic, antihistaminic

RN: 147-20-6   MF: $C_{19}H_{23}NO$ MW: 281.40   EINECS: 205-686-7
LD_{50}:   42 mg/kg (M, i.v.); 250 mg/kg (M, p.o.)
CN: 4-(diphenylmethoxy)-1-methylpiperidine

hydrochloride
RN: 132-18-3   MF: $C_{19}H_{23}NO \cdot HCl$ MW: 317.86   EINECS: 205-049-3
LD_{50}:   52 mg/kg (M, i.v.); 210 mg/kg (M, p.o.);
        28.8 mg/kg (R, i.v.); 698 mg/kg (R, p.o.)

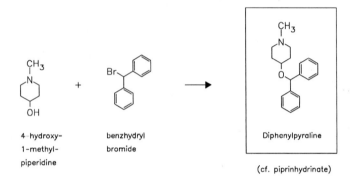

4-hydroxy-   benzhydryl   Diphenylpyraline
1-methyl-   bromide
piperidine

(cf. piprinhydrinate)

Reference(s):
US 2 479 843 (Nopco Chem. Comp.; 1949; prior. 1948).
DE 934 890 (Promonta; appl. 1951).

Formulation(s):   gel 15 mg/g

Trade Name(s):
D: Arbid (Bayer Vital)-comb.   F: Belfène (Roger Bellon);   I: Ipercron (Maggioni)-
     Perdiphen (Schwabe;   wfm   comb.; wfm
     Spitzner)-comb.   GB: Escornade Spansule (Smith   Pirazone Smit (UCB-
     Proctospre (Hennig)   Kline & French)-comb.;   Smith); wfm
     Tempil N (Temmler)-comb.   wfm   J: Plokon (Nippon Shinyaku)
     Topoderm (gepepharm)-   Histryl (Smith Kline &   USA: Diafen (Riker); wfm
     comb.   French); wfm   Hispril (Smith Kline &
                                         Lergoban (Riker); wfm   French); wfm

Dipivefrine

ATC: S01EA02
Use: antiglaucoma

RN: 52365-63-6   MF: $C_{19}H_{29}NO_5$ MW: 351.44
CN: (±)-2,2-dimethylpropanoic acid 4-[1-hydroxy-2-(methylamino)ethyl]-1,2-phenylene ester

hydrochloride
RN: 64019-93-8   MF: $C_{19}H_{29}NO_5 \cdot HCl$ MW: 387.90

a

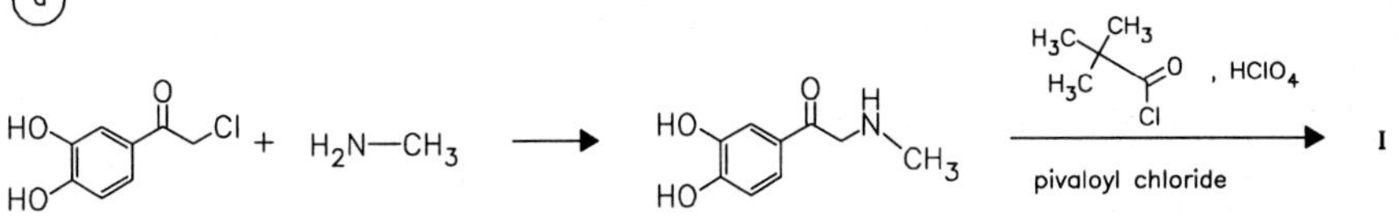

HO— ... —CCl + H₂N—CH₃ ⟶ HO— ... —N—CH₃

2-chloro-3',4'-   methylamine
dihydroxy-
acetophenone

H₃C—C(CH₃)—C(=O)—O— ... , HClO₄
pivaloyl chloride   → I

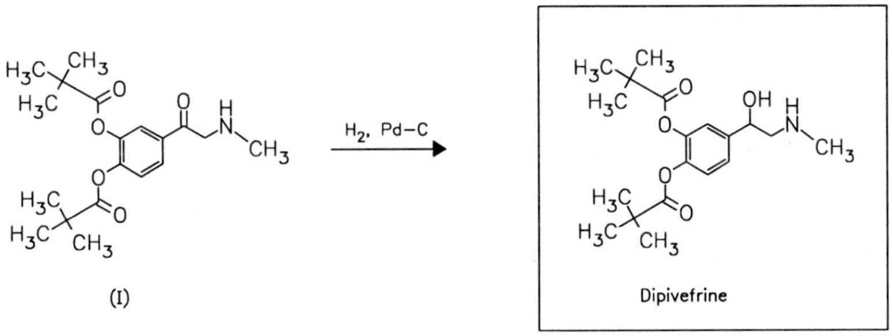

H₂, Pd–C

(I)

Dipivefrine

b

H₂, Pd–C   Dipivefrine

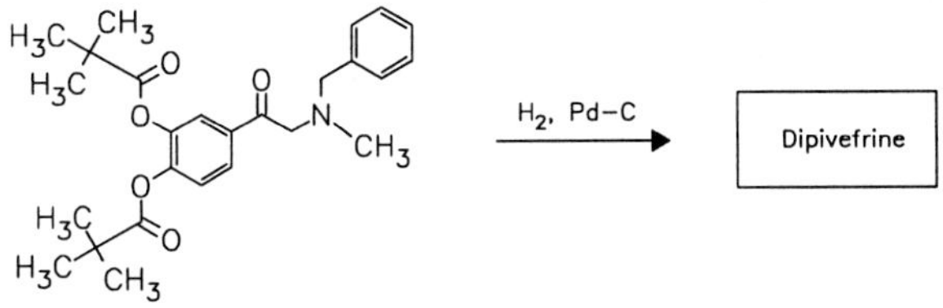

1-(3,4-dipivaloyloxy-
phenyl)-2-(benzyl-
methylamino)ethan-1-one

Reference(s):
a   DOS 2 343 657 (Interx Res. Corp.; appl. 30.8.1973; USA-prior. 31.8.1972).
    US 3 809 714 (Interx; 7.5.1974; prior. 31.8.1972) also racemate resolution.
    Hussain, A.; Truelove, J.E.: J. Pharm. Sci. (JPMSAE) 65, 1510 (1976).
b   DOS 2 152 058 (Klinge; appl. 19.10.1971).
    US 3 839 584.

Formulation(s):   eye drops 1 mg/ml (as hydrochloride)

Trade Name(s):
D: D-Epifrin (Pharm-Allergan;   Thilodigon (Alcon; 1985)-   J: Pivalephrine (Okami-
        1978)   comb.   Santen)
        Glaucothil (Alcon; 1978)   F: Propine (Allergan)
        Thiloadren (Alcon; 1980)-   GB: Propine (Allergan; 1984)
        comb.   I: Propine (Allergan)

Dipotassium clorazepate
(Clorazepate dipotassium)

ATC: N05BA05
Use: tranquilizer

RN: 57109-90-7 MF: C$_{16}$H$_{10}$ClKN$_2$O$_3$ · KOH MW: 408.92 EINECS: 260-565-6
LD$_{50}$: 173 mg/kg (M, i.v.); 700 mg/kg (M, p.o.);
279 mg/kg (R, i.v.); 880 mg/kg (R, p.o.)
CN: 7-chloro-2,3-dihydro-2-oxo-5-phenyl-1*H*-1,4-benzodiazepine-3-carboxylic acid monopotassium salt
compd. with potassium hydroxide

free acid
RN: 23887-31-2 MF: C$_{16}$H$_{11}$ClN$_2$O$_3$ MW: 314.73 EINECS: 245-926-8

2-amino-5-
chlorobenzo-
nitrile

phenyl-
magnesium
bromide

diethyl aminomalonate
hydrochloride

(I)

Dipotassium clorazepate

Reference(s):
US 3 516 988 (J. Schmitt; 23.6.1970; F-prior. 15.6.1964, 12.4.1965).
DE 1 518 764 (C. M. Industries S.A.; appl. 14.6.1965; F-prior. 15.6.1964, 12.4.1965).
DE 1 795 690 (C. M. Industries S.A.; appl. 14.6.1965).

precursors:
DOS 1 795 832 (C. M. Industries S.A.; appl. 14.6.1965; F-prior. 15.6.1964, 12.4.1965).

Formulation(s): cps. 5 mg, 10 mg, 20 mg; drops 5 mg; f. c. tabl. 20 mg, 50 mg; lyo. 50 mg, 100 mg

Trade Name(s):
D: Tranxilium 50 (Sanofi Winthrop)
F: Noctran 10 (Menarini)- comb.
GB: Tranxene (Boehringer Ing.)
I: Transene (Sanofi Winthrop)
Tranxéne (Sanofi Winthrop)
J: Cephadol (Nippon Shinyaku)
USA: Gen-XENE (Alra)
Tranxene (Abbott)

Diprophylline
(Diprofyllin; Dyphylline)

ATC: R03DA01
Use: expectorant, bronchodilator

RN: 479-18-5 MF: C$_{10}$H$_{14}$N$_4$O$_4$ MW: 254.25 EINECS: 207-526-1
LD$_{50}$: 1080 mg/kg (M, i.v.); 1954 mg/kg (M, p.o.);
860 mg/kg (R, i.v.)
CN: 7-(2,3-dihydroxypropyl)-3,7-dihydro-1,3-dimethyl-1*H*-purine-2,6-dione

ⓐ

theophylline (I) + 3−chloro-1,2-propanediol —NaOH→ Diprophylline

ⓑ

I + glycidol ⟶ Diprophylline

Reference(s):
a US 2 575 344 (State Univ. Iowa; 1951; prior. 1946).
b Roth, H.J.: Arch. Pharm. Ber. Dtsch. Pharm. Ges. (APBDAJ) **292/64**, 234 (1959).

Formulation(s):   drg. 150 mg; suppos. 200 mg, 400 mg; tabl. 200 mg, 400 mg

Trade Name(s):
D: Neobiphyllin-Clys
       (Trommsdorff)-comb.
       Ozothin (SmithKline
       Beecham)
F: Ozothine Diprophylline
       (SCAT)-comb.
GB: Silbephylline (Berk); wfm

I: Cortinal Aerosol
       (Teofarma)-comb.
       Katasma (Bruschettini)
J: Astmamasit (Showa)
       Corphyllin (Nippon
       Shinyaku)
       Dihydrophylline (Tokyo
       Hosei)

       Neophyllin-M (Eisai)
       Prophyline (Shionogi)
       Rominophyllin (Grelan)
       Theourin (Kanto)
USA: Dilor (Savage)
       Dyline (Seatrace)
       Dylix (Lunsco)
       Lufyllin (Wallace)

Dipyridamole

ATC: B01AC07
Use: coronary vasodilator

RN: 58-32-2   MF: $C_{24}H_{40}N_8O_4$ MW: 504.64   EINECS: 200-374-7
LD_{50}: 150 mg/kg (M, i.v.); 2150 mg/kg (M, p.o.);
       195 mg/kg (R, i.v.); 8400 mg/kg (R, p.o.)
CN: 2,2',2'',2'''-[(4,8-di-1-piperidinylpyrimido[5,4-d]pyrimidine-2,6-diyl)dinitrilo]tetrakis[ethanol]

urea (I) + ethyl acetoacetate (II) ⟶ 2,4-dihydroxy-6−methylpyrimidine —HNO₃→ 5-nitroorotic acid (III)

thiourea + II ⟶ 4-hydroxy-2-mercapto-6-methylpyrimidine —HNO₃→ III —SnCl₂ or Na₂S₂O₄ or H₂/Pd−C→ IV

5-aminoorotic acid (IV) → (I or KOCN, potassium cyanate) → 2,4,6,8-tetra-hydroxypyrimido-[5,4-d]pyrimidine → (POCl₃, PCl₅) → 2,4,6,8-tetra-chloropyrimido-[5,4-d]pyrimidine (V)

V + piperidine → 2,6-dichloro-4,8-dipiperidinopyri-mido[5,4-d]pyri-midine → (diethanolamine) → Dipyridamole

Reference(s):
DE 1 116 676 (Thomae; appl. 1955).
GB 807 826 (Thomae; appl. 1956; D-prior. 1955).
US 3 031 450 (Thomae; 24.4.1962; D-prior. 1959).

2,4,6,8-tetrahydroxypyrimido[5,4-d]pyrimidine:
DE 845 940 (Dr. G. F. Fischer; appl. 1950).
Fischer, F.G.; Roch, J.: Justus Liebigs Ann. Chem. (JLACBF) **572**, 216 (1951).

catalytic hydrogenation of 5-nitroorotic acid with Pd-C:
DOS 2 600 542 (Lonza; appl. 8.1.1976; CH-prior. 13.1.1975).

2,4,6,8-tetrachloropyrimido[5,4-d]pyrimidine:
Fischer, F.G.; Roch, J.; Neumann, W.P.: Justus Liebigs Ann. Chem. (JLACBF) **631**, 147 (1960).

alternative syntheses:
GB 799 177 (Thomae; appl. 1955; D-prior. 1954).
DE 1 093 801 (Thomae; appl. 1954).
DE 1 151 806 (Thomae; appl. 30.4.1959).
DAS 1 962 261 (Yamanouchi; appl. 12.12.1969; J-prior. 25.1.1969).

combination with acetylsalicylic acid (thrombocyte aggregation inhibitor):
FR-appl. 2 368 280 (Théramex; appl. 20.10.1976).
FR-appl. 2 368 272 (Théramex; appl. 20.10.1976).

Formulation(s):   amp. 10 mg/2 ml; cps. 75 mg; drg. 25 mg, 75 mg; f. c. tabl. 75 mg

Trade Name(s):
D: Asasantin (Boehringer Ing.)
      Curantyl (Berlin-Chemie)
      Persantin (Boehringer Ing.)
F: Cleridium 150 (Euro
      Generics)
      Perkod (Biogalénique)
      Persantine (Boehringer
      Ing.)

      Protangix (Expanpharm)
GB: Persantin retard
      (Boehringer Ing.)
I: Corosan (Farmacologico
      Milanese)
      Coroxin (Malesci)
      Novodil (OFF)
      Persantin (Boehringer Ing.)

      Persumbrax (Boehringer
      Ing.)-comb.
J: Anginal (Yamanouchi)
      Permiltin (Zensei)
      Persantine (Boehringer-
      Takeda)
USA: Persantine (Boehringer
      Ing.)

Disopyramide

ATC: C01BA03
Use: antiarrhythmic

RN: 3737-09-5 MF: $C_{21}H_{29}N_3O$ MW: 339.48 EINECS: 223-110-2
LD$_{50}$: 30 mg/kg (M, i.v.); 352 mg/kg (M, p.o.);
39.1 mg/kg (R, i.v.); 333 mg/kg (R, p.o.)
CN: α-[2-[bis(1-methylethyl)amino]ethyl]-α-phenyl-2-pyridineacetamide

phosphate (1:1)
RN: 22059-60-5 MF: $C_{21}H_{29}N_3O \cdot H_3PO_4$ MW: 437.48 EINECS: 244-756-1
LD$_{50}$: 81 mg/kg (M, i.v.); 820 mg/kg (M, p.o.);
41 mg/kg (R, i.v.); 880 mg/kg (R, p.o.)

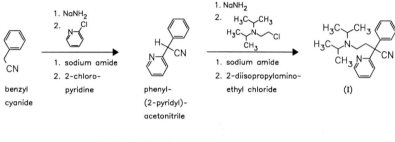

Reference(s):
US 3 225 054 (Searle; 21.12.1965; appl. 3.7.1962; prior. 17.5.1961).
DE 1 470 216 (Searle; appl. 16.5.1962; USA-prior. 17.5.1961).

Formulation(s): cps. 128.8 mg, 193.2 mg, 257.6 mg; s. r. cps. 161.25 mg, 193.2 mg, 322.5 mg (as dihydrogen phosphate)

Trade Name(s):
D:	Diso-Duriles (Astra)	F:	Isorythm (Lipha Santé)	J:	Rythmodan (MR)
	Disonorm (Solvay		Rythmodan (Roussel)	USA:	Norpace (Searle; as
	Arzneimittel)	GB:	Dirythmin (Astra)		phosphate)
	Norpace (Heumann)		Norpace (Searle)		
	Rythmodul (Albert-		Rythmodan (Roussel)		
	Roussel, Hoechst)	I:	Ritmodan (Roussel)		

Distigmine bromide
(Hexamarium bromide)

ATC: N07AA03
Use: parasympathomimetic

RN: 15876-67-2 MF: $C_{22}H_{32}Br_2N_4O_4$ MW: 576.33 EINECS: 240-013-0
LD$_{50}$: 300 μg/kg (M, i.v.); 10.5 mg/kg (M, p.o.);
740 μg/kg (R, i.v.); 10 mg/kg (R, p.o.)
CN: 3,3'-[1,6-hexanediylbis[(methylimino)carbonyl]oxy]bis[1-methylpyridinium] dibromide

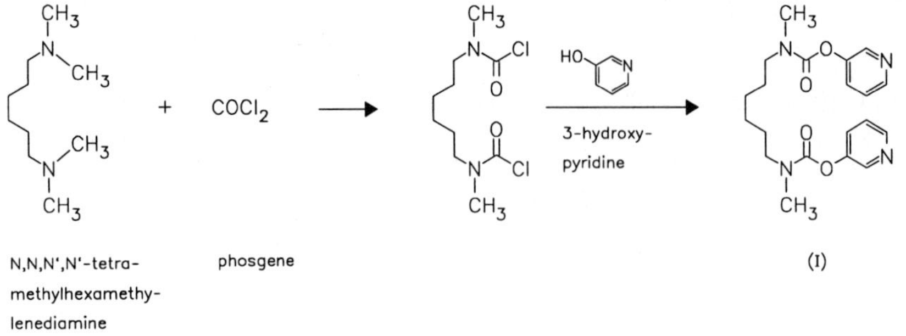

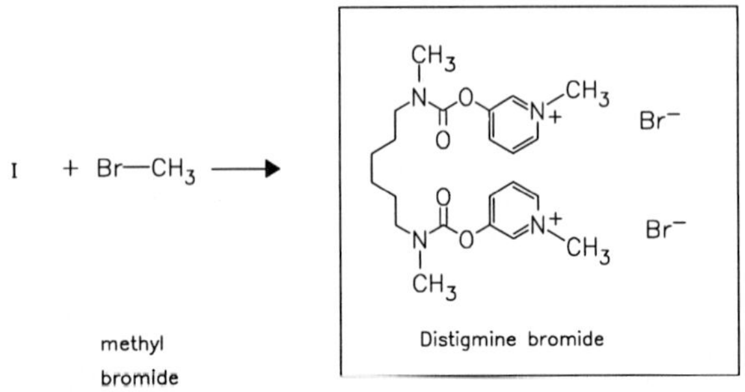

N,N,N',N'-tetra-
methylhexamethy-
lenediamine

phosgene

3-hydroxy-
pyridine

(I)

I + Br—CH₃ ⟶

methyl
bromide

Distigmine bromide

Reference(s):

US 2 789 981 (Österr. Stickstoffwerke; 1957; A-prior. 1954).

Formulation(s): amp. 0.5 mg; tabl. 5 mg

Trade Name(s):

D: Ubretid (Nycomed)   GB: Ubretid (Rhône-Poulenc   J: Ubretid (Torii)
                      Rorer)

Disulfiram

ATC: P03AA04; V03AA01
Use: alcohol deterrent

RN: 97-77-8   MF: $C_{10}H_{20}N_2S_4$ MW: 296.55   EINECS: 202-607-8
LD$_{50}$: 1980 mg/kg (M, p.o.);
       500 mg/kg (R, p.o.)
CN: tetraethylthioperoxydicarbonic diamide ([(H₂N)C(S)]₂S₂)

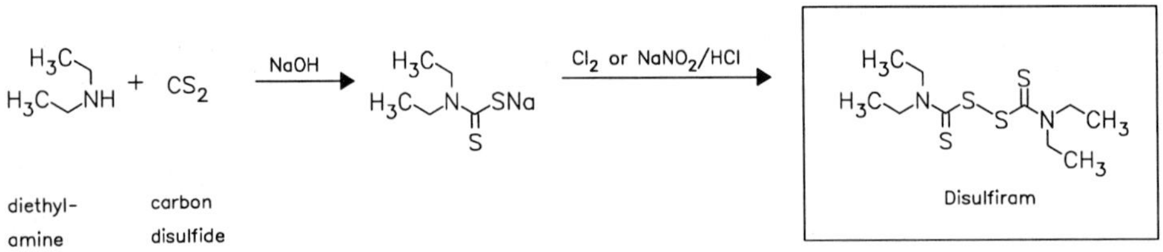

diethyl-
amine

carbon
disulfide

Disulfiram

Reference(s):

US 1 782 111 (Naugatuck; 1930; appl. 1925).
US 1 796 977 (Roessler & Hasslacher; 1931; appl. 1928).
US 2 375 083 (Monsanto; 1945; prior. 1943).
US 2 464 799 (Sharples Chemicals; 1949; prior. 1945).

Formulation(s):   tabl. 100 mg, 200 mg, 500 mg

Trade Name(s):

D:	Antabus (Byk Gulden; Byk Tosse)	GB: I:	Antabuse (Dumex) Antabuse (Crinos)		Nocbin (Tokyo Tanabe)
F:	Esperal (Sanofi Winthrop) T.T.D.-B₃-B₄ (AJC Pharma)-comb.	J:	Exiltox (Candioli) Antabuse "D" (Tokyo Tanabe)	USA:	Antabuse (Wyeth-Ayerst)

Dithranol
(Anthralin)

ATC: D05AC01
Use: antipsoriatic

RN: 1143-38-0   MF: $C_{14}H_{10}O_3$ MW: 226.23   EINECS: 214-538-0
CN: 1,8-dihydroxy-9(10H)-anthracenone

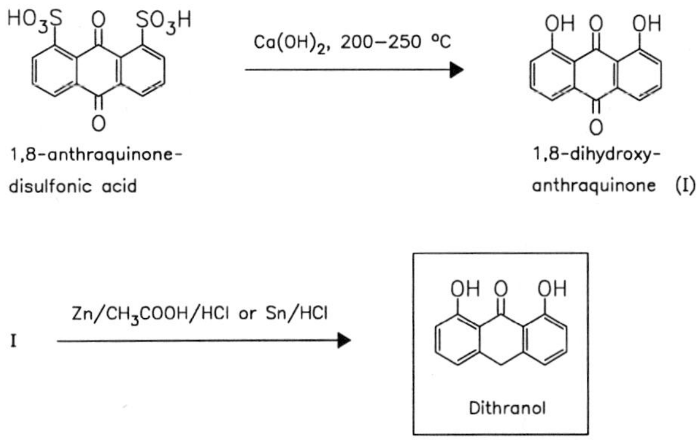

1,8-anthraquinone-
disulfonic acid

1,8-dihydroxy-
anthraquinone (I)

Dithranol

Reference(s):
DRP 296 091 (Bayer; appl. 1915).
Zahn, K.; Koch, H.: Ber. Dtsch. Chem. Ges. (BDCGAS) **71**, 172 (1938).

Formulation(s):   cream 0.5 mg/g, 1 mg/g, 2 mg/g; ointment 0.5 %, 1 %, 2 %, 3 %; pencils sticks 0.2 g/10 g, 0.5 g/10 g

Trade Name(s):

D:	Psoradexan (Hermal)-comb. Psoralon (Hermal)-comb.	I:	Micanol (Evans) Psorin (Thames)-comb. Pentagamma (IBP)-comb.;		Drithocreme (Dermik) Micanol (Bioglan)
F:	Anaxeryl (Bailly)-comb.		wfm		
GB:	Dithrocream (Dermal)	USA:	Dritho-Scalp (Dermik)		

Ditophal

ATC: D08
Use: chemotherapeutic (leprosy)

RN: 584-69-0   MF: $C_{12}H_{14}O_2S_2$ MW: 254.37
CN: 1,3-benzenedicarbothioic acid S,S-diethyl ester

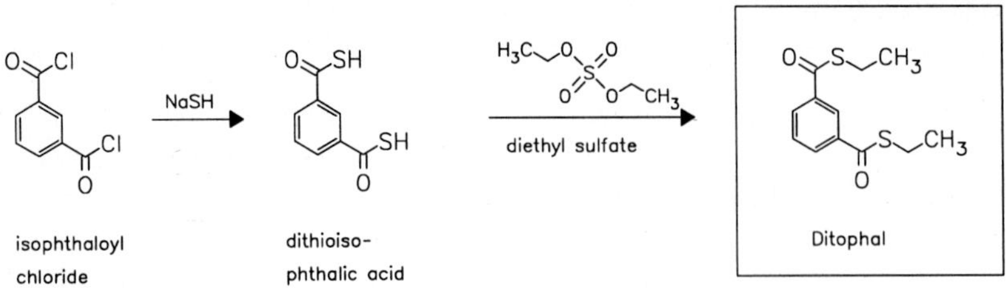

NaSH
diethyl sulfate
isophthaloyl chloride
dithioiso-phthalic acid
Ditophal

Reference(s):
GB 791 734 (ICI; appl. 1954).

Formulation(s): cream 96 %

Trade Name(s):
GB: Etisul (ICI); wfm

Dixyrazine

ATC: N05AB01
Use: neuroleptic, antihistaminic

RN: 2470-73-7 MF: $C_{24}H_{33}N_3O_2S$ MW: 427.61 EINECS: 219-591-3
LD_{50}: 37.5 mg/kg (R, i.v.); 400 mg/kg (R, p.o.)
CN: 2-[2-[4-[2-methyl-3-(10*H*-phenothiazin-10-yl)-propyl]-1-piperazinyl]ethoxy]ethanol

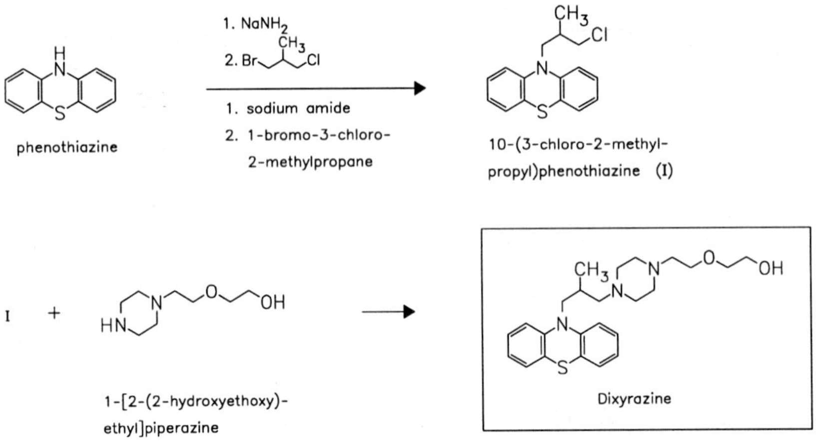

1. NaNH2
2. Br Cl
1. sodium amide
2. 1-bromo-3-chloro-2-methylpropane
phenothiazine
10-(3-chloro-2-methyl-propyl)phenothiazine (I)
I +
1-[2-(2-hydroxyethoxy)-ethyl]piperazine
Dixyrazine

Reference(s):
GB 861 420 (UCB; appl. 17.4.1959; B-prior. 19.4.1958).

Formulation(s): amp. 10 mg; drops 22 mg; tabl. 10 mg, 25 mg

Trade Name(s):
D: Esuco (UCB); wfm F: Esucos (Ucépha); wfm I: Esucos (SIT)

Dobutamine

ATC: C01CA07
Use: cardiotonic

RN: 34368-04-2 MF: $C_{18}H_{23}NO_3$ MW: 301.39
CN: (±)-4-[2-[[3-(4-hydroxyphenyl)-1-methylpropyl]amino]ethyl]-1,2-benzenediol

hydrochloride
RN: 49745-95-1   MF: C$_{18}$H$_{23}$NO$_3$ · HCl   MW: 337.85   EINECS: 256-464-1

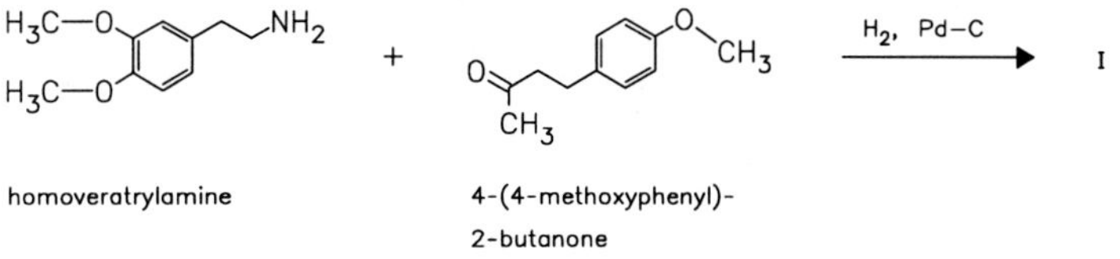

homoveratrylamine

4-(4-methoxyphenyl)-
2-butanone

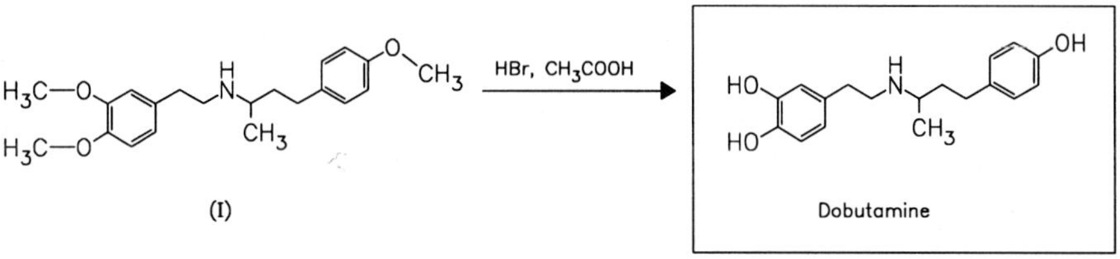

(I)

Dobutamine

Reference(s):
DOS 2 317 710 (Lilly; appl. 9.4.1973; USA-prior. 12.4.1972).
US 3 987 200 (Lilly; 19.10.1976; prior. 12.4.1972, 15.1.1975).

Formulation(s):   vial (lyo.) 280 mg (as hydrochloride)

Trade Name(s):
D: Dobutamin (ASTA Medica
     AWD; Fresenius; Hexal;   F: Dobutrex (Eli Lilly)
     Parke Davis)   GB: Dobutrex (Lilly; 1977)

I: Dobutrex (Lilly)
J: Dobutrex (Shionogi; 1982)
USA: Dobutrex (Lilly; 1978)

Dobutrex (Lilly; 1978)

Docarpamine
(TA-870; TA-8704)

ATC: C02LX
Use: cardiotonic, diuretic

RN: 74639-40-0   MF: C$_{21}$H$_{30}$N$_2$O$_8$S   MW: 470.54
LD$_{50}$: 2800 mg/kg (M, i.v.)
CN: (S)-carbonic acid 4-[2-[[2-(acetylamino)-4-(methylthio)-1-oxobutyl]amino]ethyl]-1,2-phenylene diethyl
      ester

(RS)-form
RN: 143289-50-3   MF: C$_{21}$H$_{30}$N$_2$O$_8$S   MW: 470.54

(S)-N-acetyl-
methionine (I)

1. N-hydroxysuccinimide,
   dicyclohexylcarbodiimide
2. dopamine hydrochloride,
   N-methylmorpholine

(II)

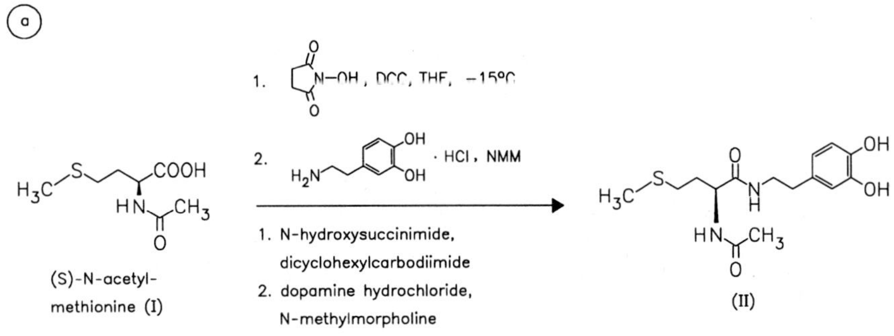

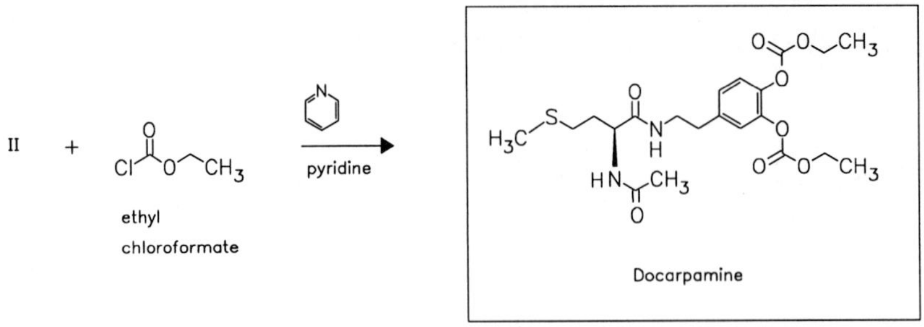

II + ethyl chloroformate →(pyridine) Docarpamine

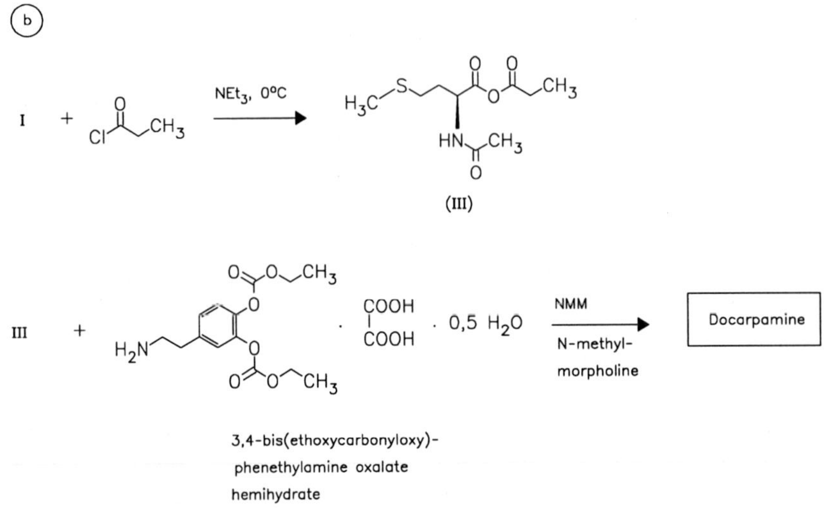

(b)

I + →(NEt₃, 0°C) (III)

III + 3,4-bis(ethoxycarbonyloxy)- phenethylamine oxalate hemihydrate →(NMM, N-methyl-morpholine) Docarpamine

Reference(s):

EP 7 441 (Tanabe Seiyaku; appl. 6.2.1980; J-prior. 30.6.1978).
JP 4 112 858 (Tanabe Seiyaku; appl. 14.4.1992; J-prior. 30.8.1990).
JP 7 165 684 (Tanabe Seiaku; appl. 27.6.1995; J-prior. 22.7.1994).

oral pharmaceuticals containing dopamine derivatives:
JP 06 183 964 (Tanabe Seiyaku; appl. 5.7.1994; J-prior. 24.12.1992).

Formulation(s): gran. 75 mg/g

Trade Name(s):
J: Tanadopa (Tanabe Seiyaku)

Docetaxel
(NSC-628503; RP-56976)

ATC: L01CD02
Use: antineoplastic, microtubule inhibitor

RN: 114977-28-5 MF: $C_{43}H_{53}NO_{14}$ MW: 807.89
CN: [2aR-[2aα,4β,4aβ,6β,9α(αR*,βS*),11α,12α,12aα,12bα]]-β-[[(1,1-dimethylethoxy)carbonyl]amino]-α-hydroxybenzenepropanoic acid 12b-(acetyloxy)-12-(benzoyloxy)-2a,3,4,4a,5,6,9,10,11,12,12a,-12b-dodecahydro-4,6,11-trihydroxy-4a,8,13,13-tetramethyl-5-oxo-7,11-methano-1H-cyclodeca[3,4]benz[1,2-b]oxet-9-yl ester

a

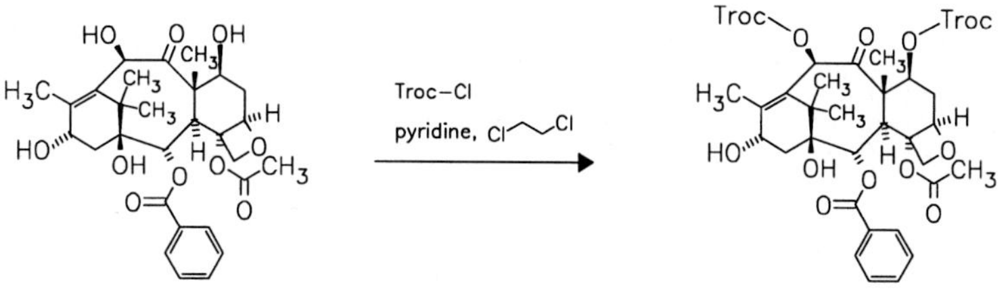

10-deacetylbaccatin III
(extracted and purified
from leaves of
Taxus baccata L.)

Troc−Cl
pyridine, Cl〜Cl

(I)

Troc:

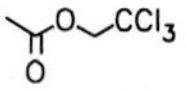

I + Boc−N−H−COOH−O−R

DCC, DMAP,
toluene
4-dimethylamino-
pyridine

(III)

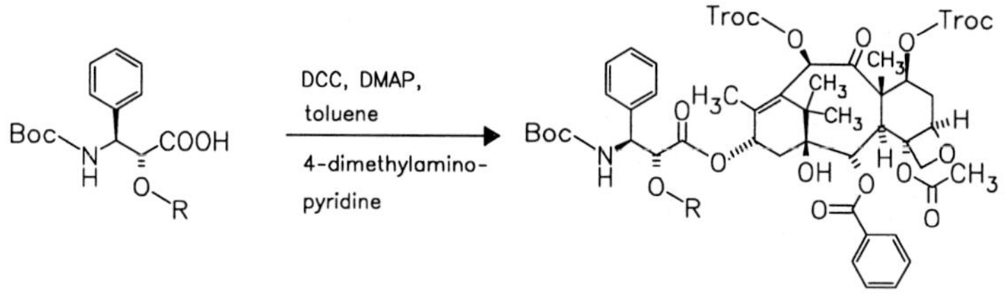

threo-2-(1-ethoxyeth-
oxy)-3-(tert-butoxy-
carbonylamino)-3-
phenylpropionic acid (II)

Boc:

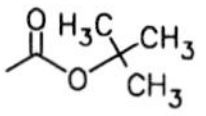

R:

Zn, CH₃COOH,
ethyl acetate

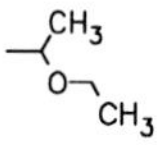

III

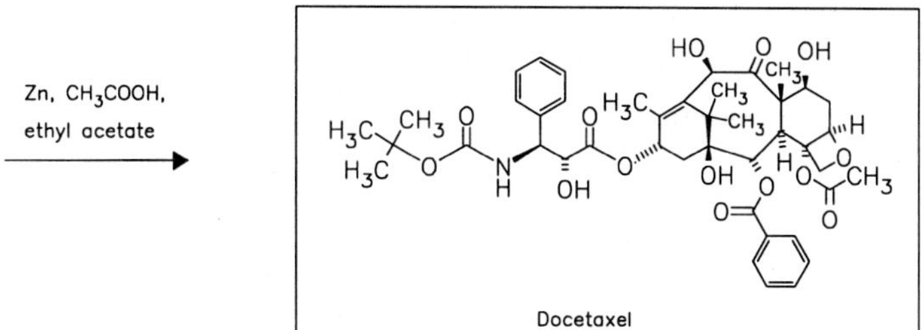

Docetaxel

aa synthesis of intermediate II

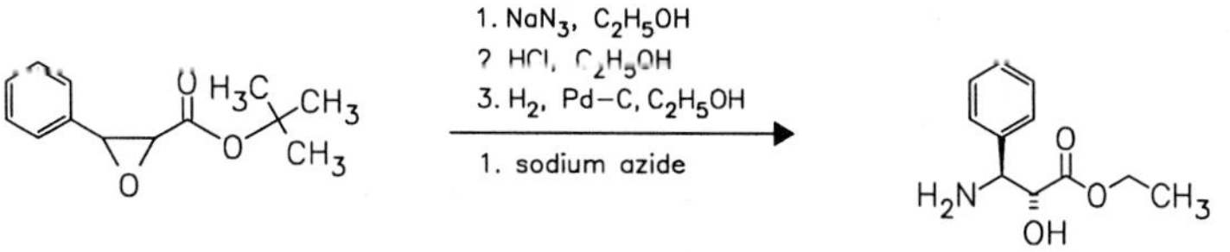

tert-butyl
3-phenylglycidate

1. NaN₃, C₂H₅OH
2. HCl, C₂H₅OH
3. H₂, Pd−C, C₂H₅OH

1. sodium azide

ethyl threo-3-amino-
2-hydroxy-3-phenyl-
propionate (IV)

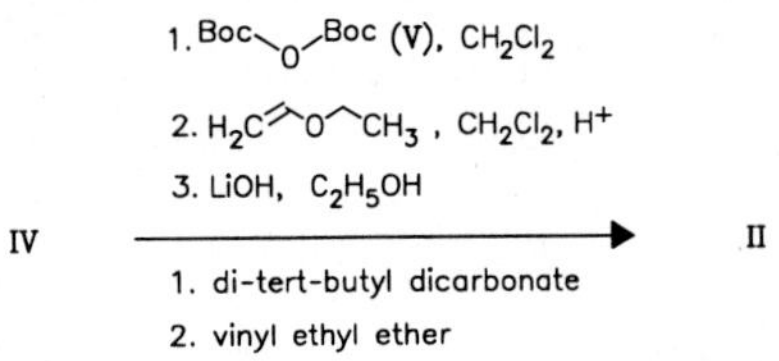

1. Boc—O—Boc (V), CH_2Cl_2

2. H_2C—O—CH_3 , CH_2Cl_2, H^+

3. LiOH, C_2H_5OH

IV $\xrightarrow{\hspace{3cm}}$ II

1. di-tert-butyl dicarbonate

2. vinyl ethyl ether

(b)

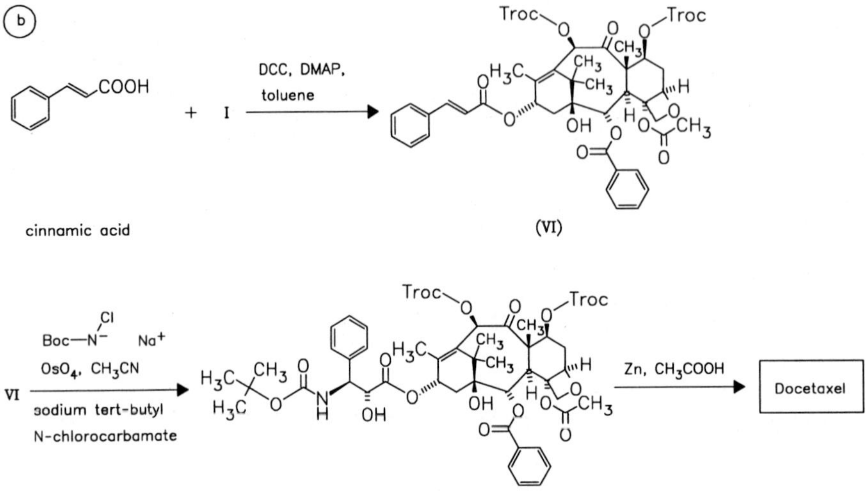

cinnamic acid ....COOH + I $\xrightarrow[\text{toluene}]{\text{DCC, DMAP,}}$ (VI)

cinnamic acid   (VI)

Boc—N$^-$—Cl Na$^+$

OsO$_4$, CH$_3$CN

VI $\xrightarrow[\text{N-chlorocarbamate}]{\text{sodium tert-butyl}}$ $\xrightarrow{\text{Zn, CH}_3\text{COOH}}$ Docetaxel

(c)

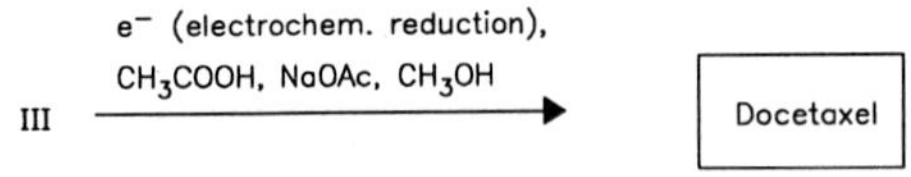

III $\xrightarrow[\text{CH}_3\text{COOH, NaOAc, CH}_3\text{OH}]{e^- \text{ (electrochem. reduction),}}$ Docetaxel

(d)

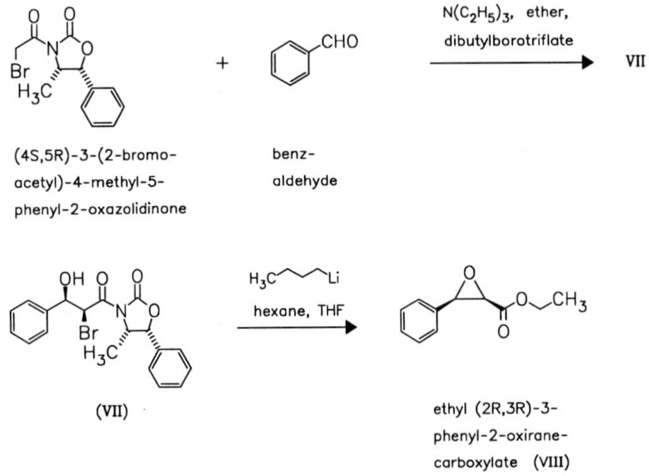

(4S,5R)-3-(2-bromo-
acetyl)-4-methyl-5-
phenyl-2-oxazolidinone

+ benz-
aldehyde

$\xrightarrow[\text{dibutylborotriflate}]{\text{N(C}_2\text{H}_5)_3, \text{ ether,}}$ VII

(VII) $\xrightarrow[\text{hexane, THF}]{\text{H}_3\text{C} \diagdown \diagup \text{Li}}$ ethyl (2R,3R)-3-
phenyl-2-oxirane-
carboxylate (VIII)

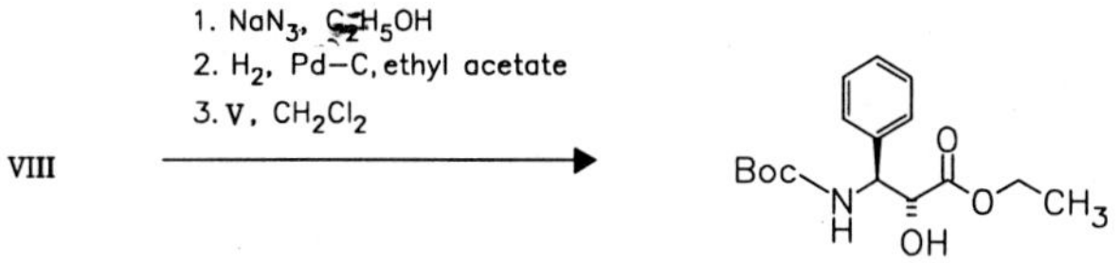

VIII →
1. NaN₃, C₂H₅OH
2. H₂, Pd–C, ethyl acetate
3. V, CH₂Cl₂

ethyl (2R,3S)-3-tert-butoxy-
carbonylamino-2-hydroxy-
3-phenylpropionate (IX)

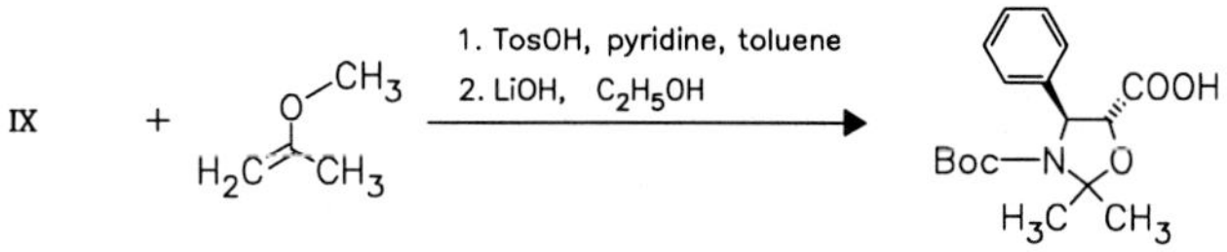

IX +

2-methoxy-
propene

1. TosOH, pyridine, toluene
2. LiOH, C₂H₅OH →

(4S,5R)-3-tert-butoxy-
carbonyl-2,2-dimethyl-
4-phenyl-5-oxazolidine-
carboxylic acid (X)

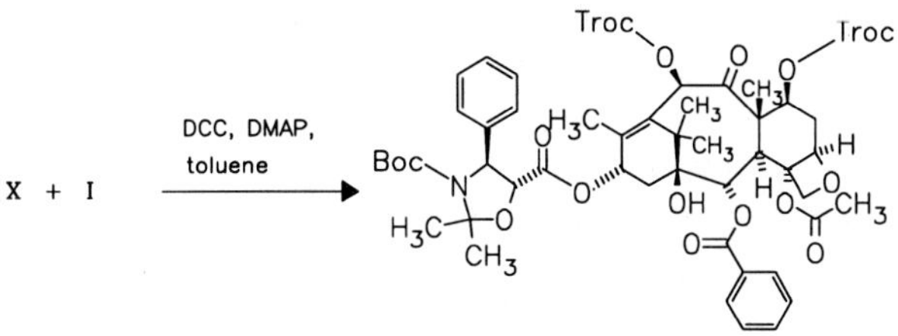

X + I →
DCC, DMAP,
toluene

(XI)

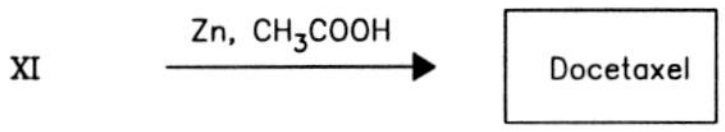

XI →
Zn, CH₃COOH

Docetaxel

(da) alternative synthesis of intermediate X:

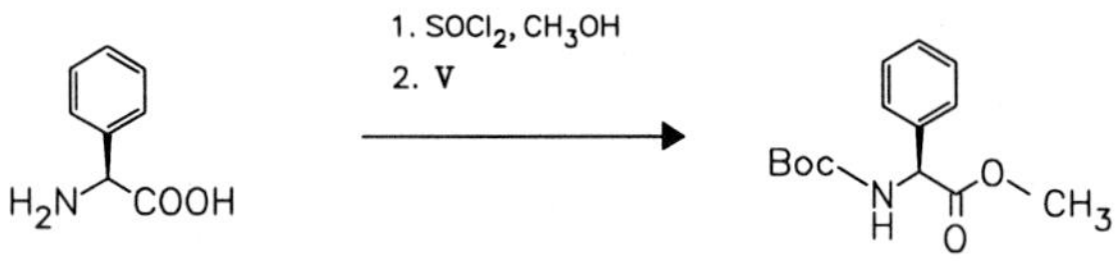

(S)-phenylglycine

1. SOCl₂, CH₃OH
2. V →

(XII)

XII →
1. DIBAL, toluene
2.

(H₃C)₃Si

tetrabutyl ammonium fluoride

1. dibutylaluminum hydride
2. 2-(trimethylsilyl)thiazole

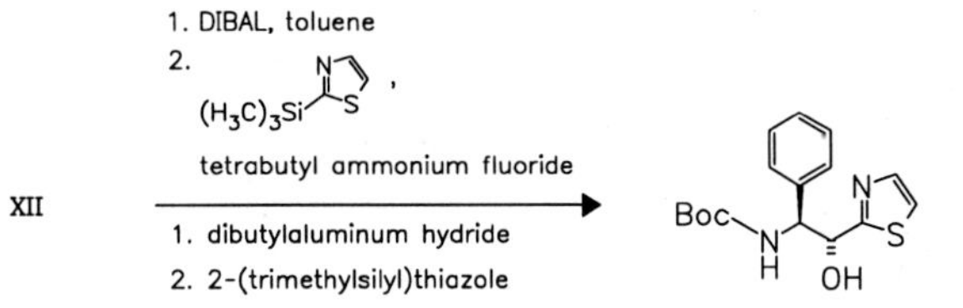

tert-butyl (1S,2R)-2-hy-
droxy-1-phenyl-2-(2-thia-
zolyl)ethylcarbamate (XIII)

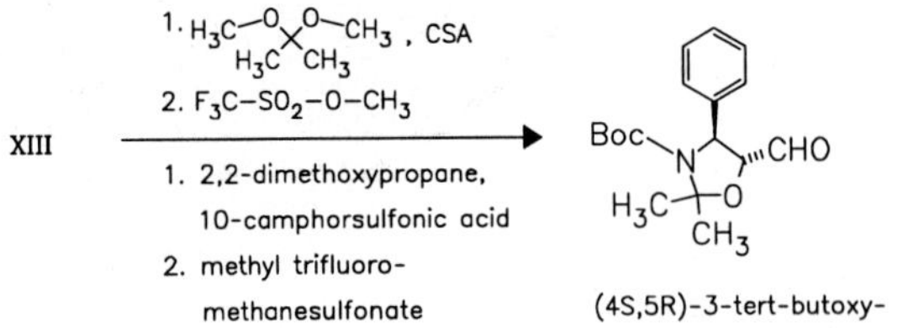

XIII

1. $H_3C-O\overset{O}{\underset{H_3C}{\underset{|}{\overset{|}{C}}}}CH_3$, CSA

2. $F_3C-SO_2-O-CH_3$

1. 2,2-dimethoxypropane,
   10-camphorsulfonic acid
2. methyl trifluoro-
   methanesulfonate

(4S,5R)-3-tert-butoxy-
carbonyl-2,2-dimethyl-4-
phenyl-5-oxazolidine-
carboxaldehyde (XIV)

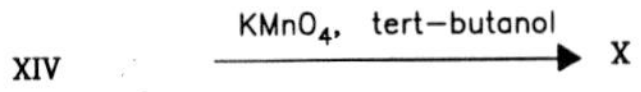

XIV

KMnO$_4$, tert−butanol

X

e

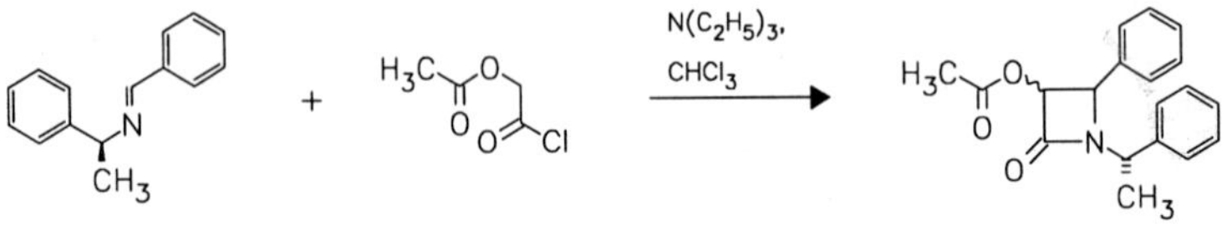

(S)-N-benzylidene-
1-phenylethylamine

+

2-acetoxy-
acetyl chloride

N(C$_2$H$_5$)$_3$,
CHCl$_3$

3-acetoxy-4-phenyl-1-
[(S)-1-phenylethyl]-2-
azetidinone (XV)

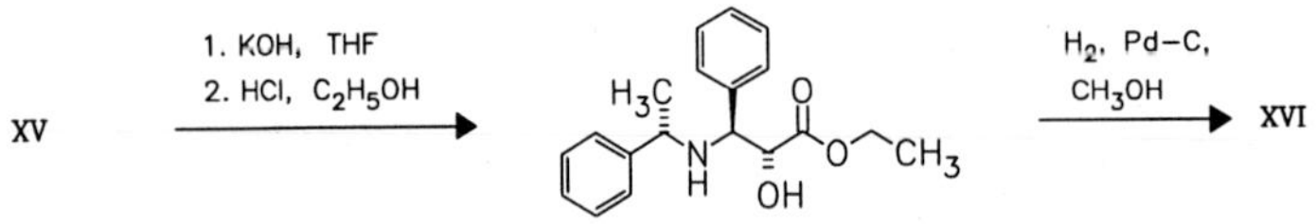

XV

1. KOH, THF
2. HCl, C$_2$H$_5$OH

ethyl (2R,3S)-2-hydroxy-
3-[(S)-1-phenylethylamino]-
3-phenylpropionate

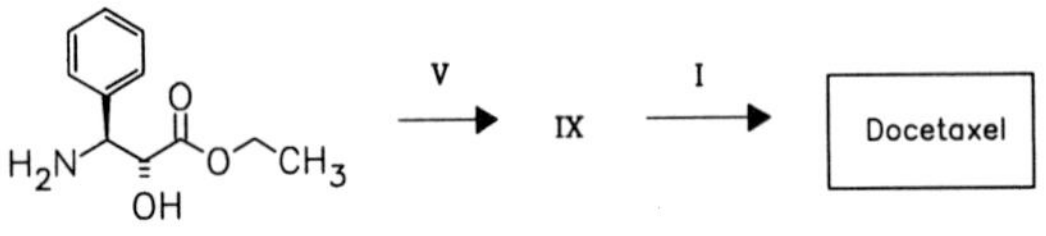

H$_2$, Pd−C,
CH$_3$OH

XVI

ethyl (2R,3S)-3-amino-
2-hydroxy-3-phenyl-
propionate (XVI)

V

IX

I

Docetaxel

f

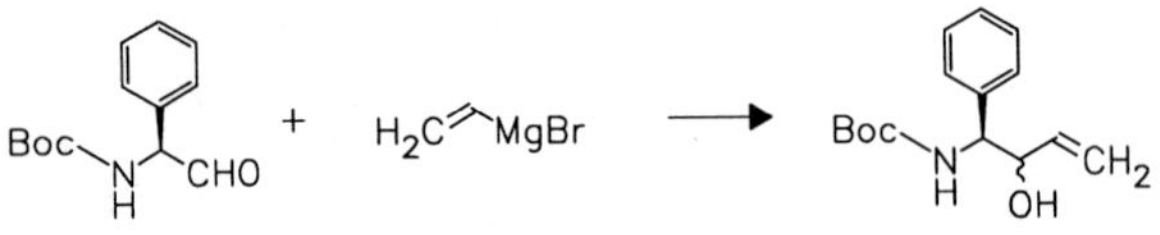

Boc-(S)-
phenylglycinal

+

vinylmagnesium
bromide

(±)-1-phenyl-1-
(tert-butoxycarbonyl-
amino)-2-hydroxy-
3-butene (XVII)

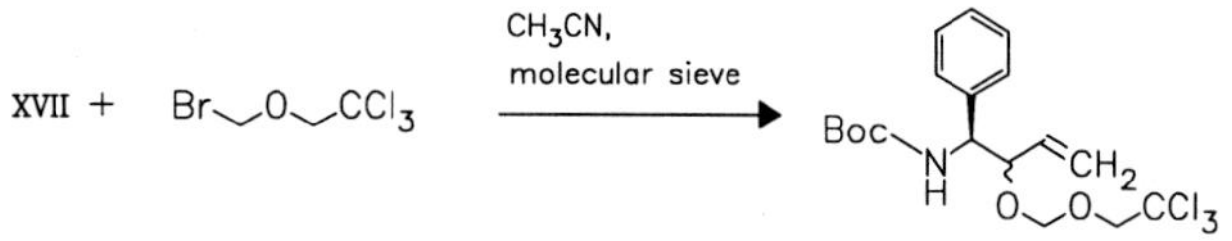

XVII + Br~O~CCl₃   →   (over arrow: CH₃CN, molecular sieve)

(2,2,2-trichloro-
ethoxy)methyl
bromide

Boc–NH–C(phenyl)(=CH₂)–O–O–CCl₃

syn-(±)-1-phenyl-1-(tert-
butoxycarbonylamino)-2-
(2,2,2-trichloroethoxymeth-
oxy)-3-butene (XVIII)

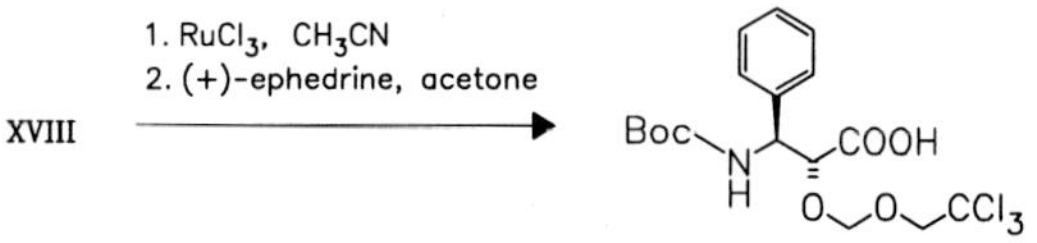

XVIII   →   (over arrow: 1. RuCl₃, CH₃CN 2. (+)-ephedrine, acetone)

Boc–NH–CH(phenyl)–CH(COOH)–O–O–CCl₃

(2R,3S)-3-tert-butoxy-
carbonylamino-3-phenyl-
2-(2,2,2-trichloroethoxy-
methoxy)propionic acid (XIX)

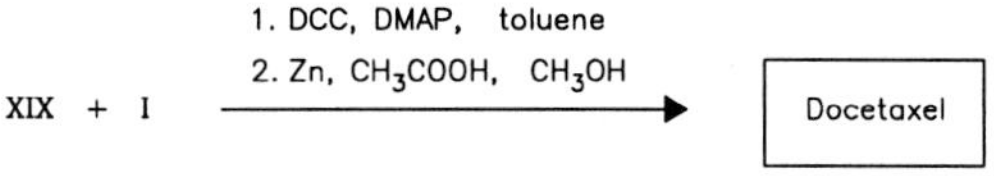

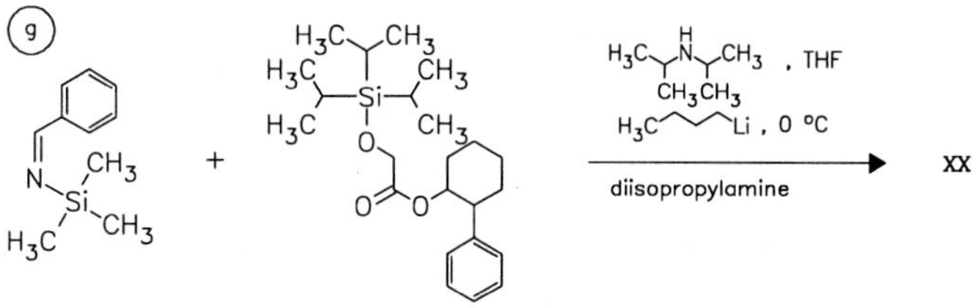

XIX + I   →   (over arrow: 1. DCC, DMAP, toluene 2. Zn, CH₃COOH, CH₃OH)   [Docetaxel]

⑨

N-(trimethyl-
silyl)benz-
aldehyde imine

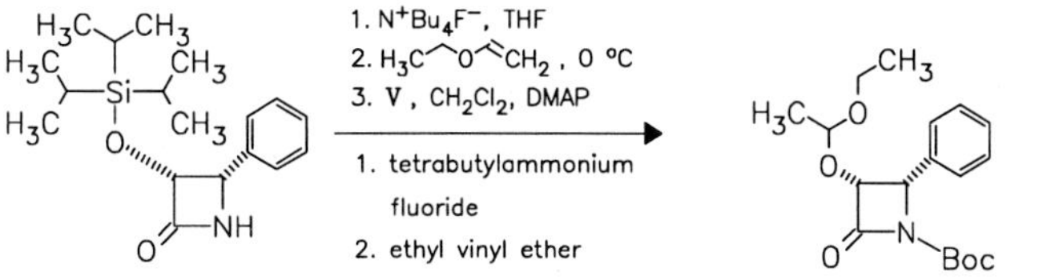

+ (TIPS glycolate cyclohexyl ester structure)   →   **XX**

(over arrow: H₃C–N(CH₃)–CH₃ ... CH₃CH₃, H₃C~~Li, 0 °C, THF / diisopropylamine)

(3R,4S)-3-(triisopropyl-
silyloxy)-4-phenyl-2-
azetidinone (XX)

  →   (over arrow: 1. N⁺Bu₄F⁻, THF 2. H₃C~O~CH₂, 0 °C 3. V, CH₂Cl₂, DMAP / 1. tetrabutylammonium fluoride 2. ethyl vinyl ether)

(3R,4S)-1-tert-butoxy-
carbonyl-3-(1-ethoxy-
ethoxy)-4-phenyl-2-
azetidinone (XXI)

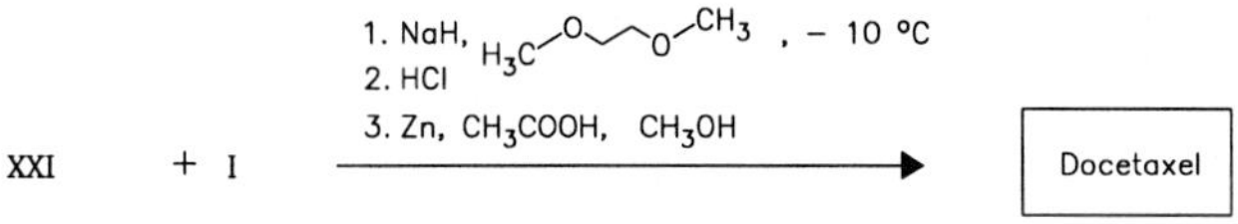

XXI + I   →   (over arrow: 1. NaH, H₃C~O~O~CH₃, –10 °C 2. HCl 3. Zn, CH₃COOH, CH₃OH)   [Docetaxel]

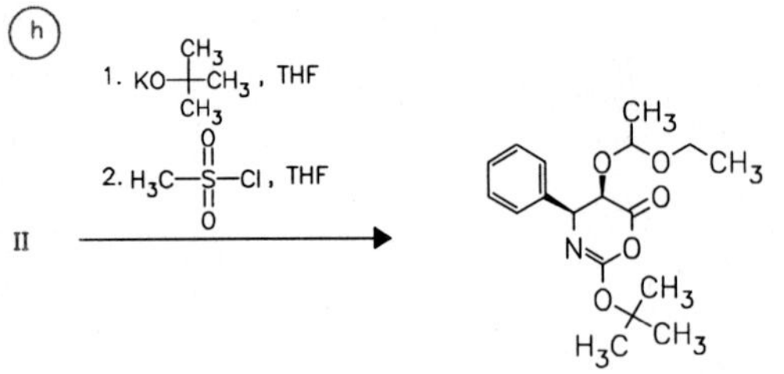

h
1. KO–C(CH3)3, THF
2. H3C–SO2–Cl, THF
II

(4S,5R)-2-tert-butoxy-4-phenyl-5-(1-ethoxyethoxy)-4,5-dihydro-1,3-oxazin-6-one (XXII)

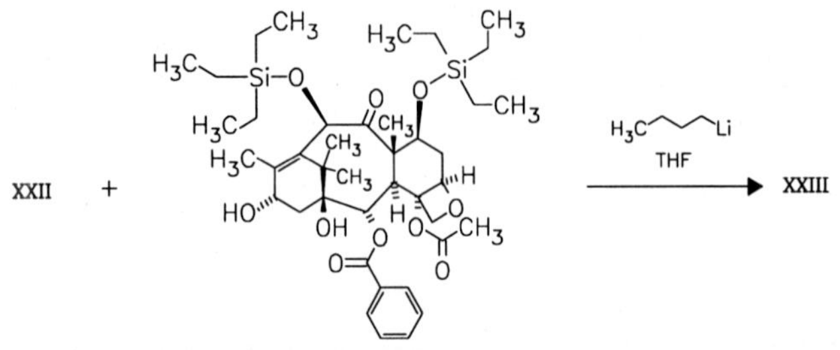

XXII
+
H3C–CH2CH2CH2–Li
THF
XXIII

7,10-bis(triethylsilyl)-10-deacetylbaccatin III

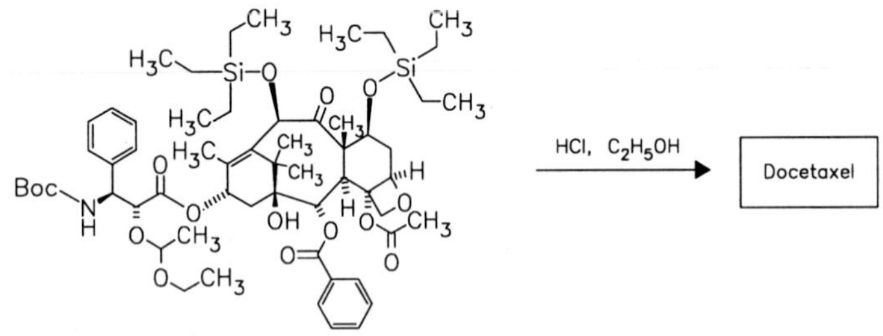

HCl, C2H5OH
Docetaxel

(2R,3S)-N-debenzoyl-N-tert-butoxycarbonyl-10-deacetyl-2-(1-ethoxyethyl)-7,10-bis-(triethylsilyl)taxol (XXIII)

Reference(s):
a EP 336 841 (Rhône-Poulenc Sante; appl. 5.4.1989; F-prior. 6.4.1988).
   Gueritte-Voegelein, F. et al.: J. Med. Chem. (JMCMAR) **34**, 992-998 (1992).
   EP 522 958 (Rhône-Poulenc Rorer; appl. 8.7.1992; F-prior. 10.7.1991).
aa Denis, J.N. et al.: J. Org. Chem. (JOCEAH) **51**, 46-50 (1986).
b EP 253 738 (Rhône-Poulenc Sante; appl. 16.7.1987; F-prior. 17.7.1986).
c WO 9 318 210 (Rhône-Poulenc Rorer; appl. 11.3.1993; F-prior. 13.3.1992).
d WO 9 209 589 (Rhône-Poulenc Rorer; appl. 22.11.1991; F-prior. 23.11.1990, 25.7.1991).
   WO 9 407 879 (Rhône-Poulenc Rorer; appl. 4.10.1993; F-prior. 5.10.1992).
   WO 9 410 169 (Rhône-Poulenc Rorer; appl. 28.10.1993; F-prior. 30.10.1992).
   WO 9 407 876 (Rhône-Poulenc Rorer; appl. 4.10.1993; F-prior. 5.10.1992).
da Dondoen, A. et al.: Synthesis (SYNTBF) **2**, 181 (1995).
e WO 9 317 997 (Rhône-Poulenc Rorer; appl. 8.3.1993; F-prior. 10.3.1992).
   WO 9 407 847 (Rhône-Poulenc Rorer; appl. 4.10.1993; F-prior. 5.10.1992).
f EP 528 729 (Rhône-Poulenc Rorer; appl. 17.8.1992; F-prior. 19.8.1991).
g WO 9 418 164 (Res. Found. SUNY; appl. 28.1.1994; USA-prior. 1.2.1993).
   WO 9 306 094 (Univ. Florida State; appl. 22.9.1992; USA-prior. 3.4.1992, 23.9.1991).
h US 5 254 703 (Univ. Florida State; appl. 6.4.1992; USA-prior. 6.4.1992).

purification of 10-deacetylbaccatin III *by partition chromatography:*
WO 9 421 622 (Rhône-Poulenc Rorer; appl. 18.3.1994; F-prior. 22.3.1993).

callus cell induction and the preparation of taxanes:
EP 568 821 (Squibb; appl. 6.4.1993; USA-prior. 7.4.1992).

purification of (2R,3R)-cis-β-phenylglycidic acid by crystallization with α-methylbenzylamine:
WO 9 113 066 (Rhône-Poulenc Rorer; appl. 20.2.1991; F-prior. 21.2.1990).

total synthesis of taxanes:
Nicolaou, K.C. et al.: Angew. Chem. (ANCEAD) **107**, 2247-2259 (1995).

composition comprising taxanes in solution with ethanol:
EP 522 936 (Rhône-Poulenc Rorer; appl. 3.7.1992; F-prior. 8.7.1991).

composition with phospholipids/surfactants:
WO 9 528 923 (Rhône-Poulenc Rorer; appl. 24.4.1995; F-prior. 25.4.1994).
US 5 415 869 (Res. Found. SUNY; appl. 12.11.1993; USA-prior. 12.11.1993).
WO 9 412 171 (Rhône-Poulenc Rorer; appl. 26.11.1993; F-prior. 2.12.1992).

formulations with cyclodextrin:
WO 9 519 994 (CNRS; appl. 24.1.1995; F-prior. 25.1.1994).

combinations with antineoplastic agents:
FR 2 697 752 (Rhône-Poulenc Rorer; appl. 10.11.1992; F-prior. 10.11.1992).

use to treat malaria:
FR 2 707 165 (Rhône-Poulenc Rorer; appl. 6.7.1993; F-prior. 6.7.1993).
WO 9 412 172 (Th. Jefferson Univ.; appl. 2.12.1993; USA-prior. 2.12.1992, 26.1.1993).

Formulation(s):   vial 20 mg, 80 mg

Trade Name(s):
D:	Taxotere (Rhône-Poulenc Rorer)	GB:	Taxotere (Rhône-Poulenc Rorer)	USA:	Taxotere (Rhône-Poulenc Rorer)
F:	Taxotère (Bellon)	J:	Taxotere (Rhône-Poulenc Rorer)		

Dodeclonium bromide

ATC: D08AJ
Use: antiseptic

RN: 15687-13-5   MF: $C_{22}H_{39}BrClNO$ MW: 448.92   EINECS: 239-779-9
CN: N-[2-(4-chlorophenoxy)ethyl]-N,N-dimethyl-1-dodecanaminium bromide

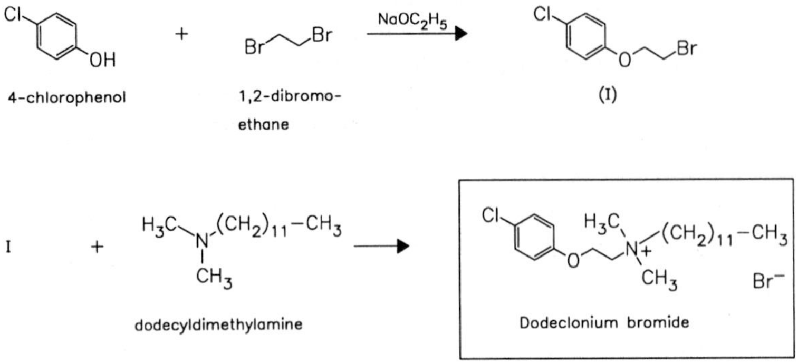

4-chlorophenol   1,2-dibromo-   (I)
                         ethane

dodecyldimethylamine   Dodeclonium bromide

Reference(s):
Gautier, J.A. et al.: Bull. Soc. Chim. Fr. (BSCFAS) **1957**, 1014.
Gautier, J.A. et al.: C. R. Hebd. Seances Acad. Sci. (COREAF) **240**, 2154 (1955).

germicidal aerosol combination:
FR 2 616 065 (J. Y. Pabst; appl. 2-6-1987).

Formulation(s):   cream 100 g/0.4 g; suppos. 1.3 mg

Trade Name(s):
F: Derméol (RPR Cooper)-   Sedorrhoide (RPR
      comb.   Cooper)-comb.

Dofetilide

Use: class III antiarrhythmic agent

(UK-68798)

RN: 115256-11-6   MF: $C_{19}H_{27}N_3O_5S_2$ MW: 441.57
CN: N-[4-[2-[Methyl[2-[4-[(methylsulfonyl)amino]phenoxy]ethyl]amino]ethyl]phenyl]methanesulfonamide

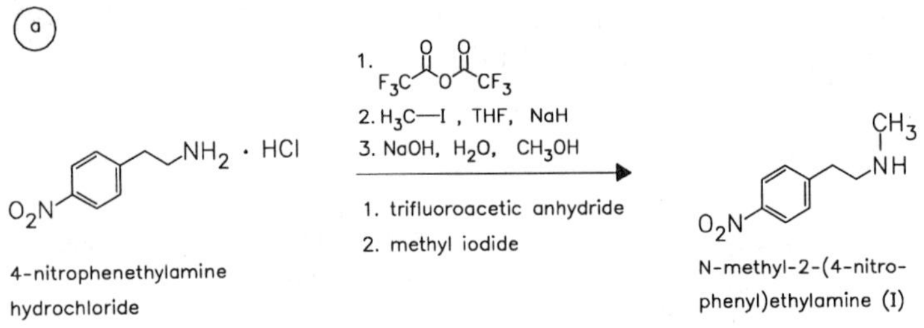

1. trifluoroacetic anhydride
2. H₃C—I , THF, NaH
3. NaOH, H₂O, CH₃OH

1. trifluoroacetic anhydride
2. methyl iodide

4-nitrophenethylamine
hydrochloride

N-methyl-2-(4-nitro-
phenyl)ethylamine (I)

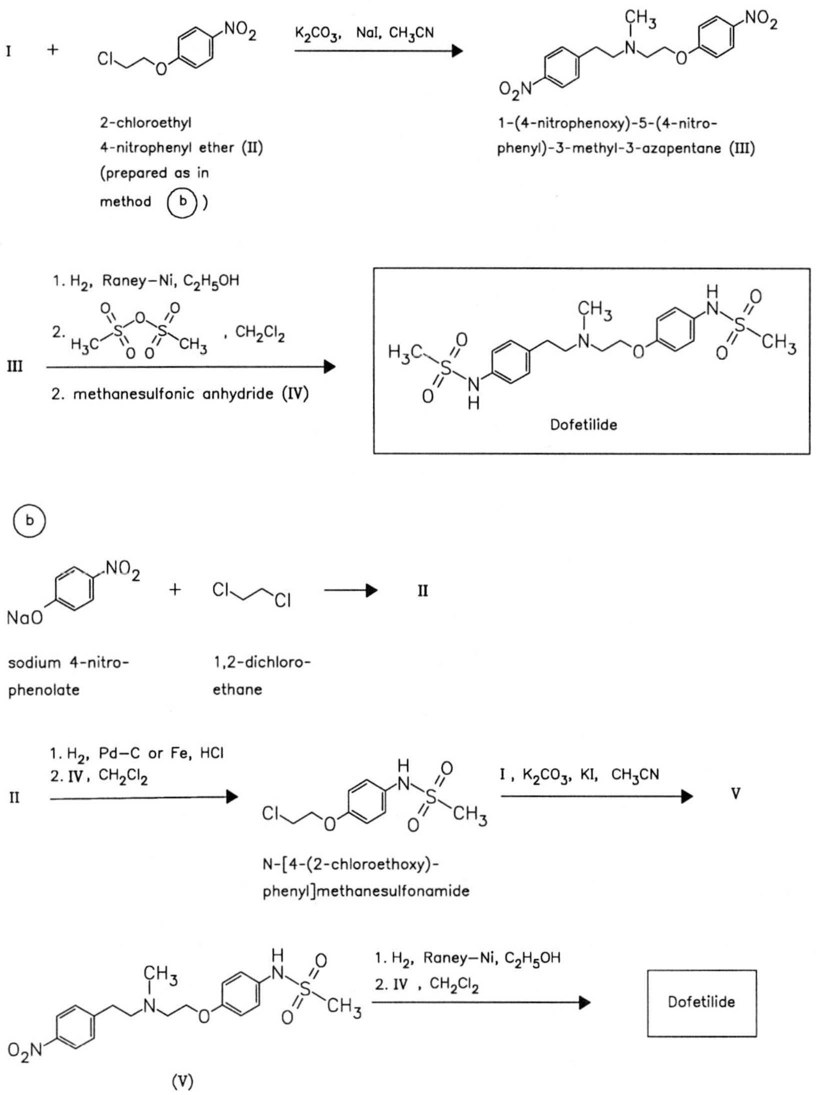

I + 2-chloroethyl 4-nitrophenyl ether (II) (prepared as in method (b)) → K₂CO₃, NaI, CH₃CN → 1-(4-nitrophenoxy)-5-(4-nitrophenyl)-3-methyl-3-azapentane (III)

III

1. H₂, Raney–Ni, C₂H₅OH
2. methanesulfonic anhydride (IV)

→ Dofetilide

(b)

sodium 4-nitrophenolate + 1,2-dichloroethane → II

II

1. H₂, Pd–C or Fe, HCl
2. IV, CH₂Cl₂

→ N-[4-(2-chloroethoxy)phenyl]methanesulfonamide

I, K₂CO₃, KI, CH₃CN → V

(V)

1. H₂, Raney–Ni, C₂H₅OH
2. IV, CH₂Cl₂

→ Dofetilide

Reference(s):

EP 245 997 (Pfizer; appl. 29.4.1987; GB-prior. 1.5.1986).
Cross, P.E.; Arrowsmith, J.E.; Geoffrey, N.; Gwilt, M.; Burges, R.A..; Higgins, A.J.: J. Med. Chem. (JMCMAR) **33**, 1151 (1990).

dofetilide *polymorphs:*
WO 9 921 829 (Pfizer; appl. 9.10.1998; GB-prior. 27.10.1997).

alternative preparation of N-methyl-(4-nitrophenethyl)amine:
Dale, W.J.; Buell, C.J.: J. Org. Chem. (JOCEAH) **21**, 45 (1956).
Theodore, L.J.; Nelson, W.L.; Dave, B.; Giacomini, J.: J. Med. Chem. (JMCMAR) **33** (2), 873 (1990).

alternative preparation of 2-chloroethyl 4-nitrophenyl ether:
Katrak: J. Indian. Chem. Soc. (JICSAH) **13**, 334 (1936).
McMahon, R.E. et al.: J. Med. Chem. (JMCMAR) **6**, 343 (1963).
US 3 937 726 (Hoechst A. G.; 10.2.1976; D-prior. 22.4.1966).

Trade Name(s):
D: Tikosyn (Pfizer; 2000)   GB: Tikosyn (Pfizer; 2000)   USA: Tikosyn (Pfizer; 1999)

Dolasetron mesilate
(MDL-73147EF)

ATC: A04AA04
Use: antiemetic agent (5-HT$_3$-receptor antagonist)

RN: 115956-13-3   MF: $C_{19}H_{20}N_2O_3 \cdot CH_4O_3S$ MW: 420.49
CN: (2α,6α,8α,9aα)-1*H*-Indole-3-carboxylic acid octahydro-3-oxo-2,6-methano-2*H*-quinolizin-8-yl ester

base
RN: 115956-12-2   MF: $C_{19}H_{20}N_2O_3 \cdot CH_4O_3S$ MW: 420.49

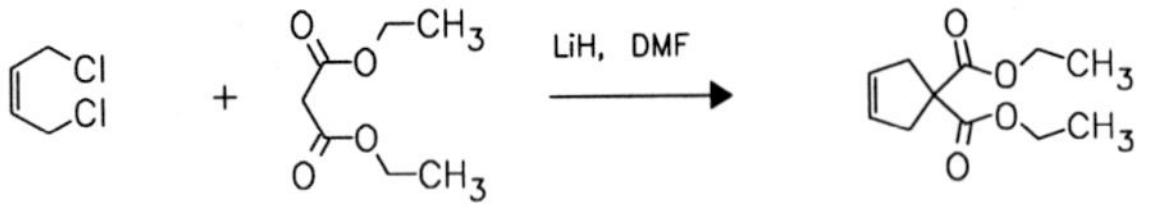

cis-1,4-
dichloro-
2-butene (I)

diethyl
malonate

LiH, DMF

diethyl 3-
cyclopentene-
1,1-dicarboxylate (II)

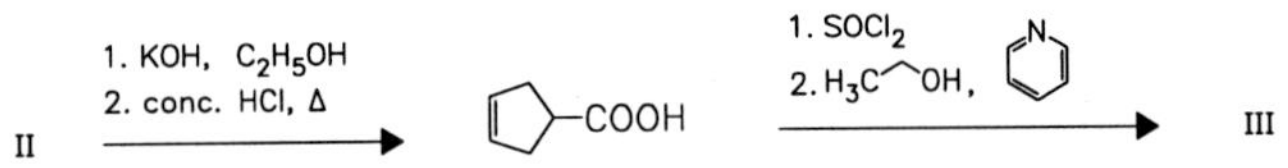

II

1. KOH, C$_2$H$_5$OH
2. conc. HCl, Δ

3-cyclopentene-
1-carboxylic acid

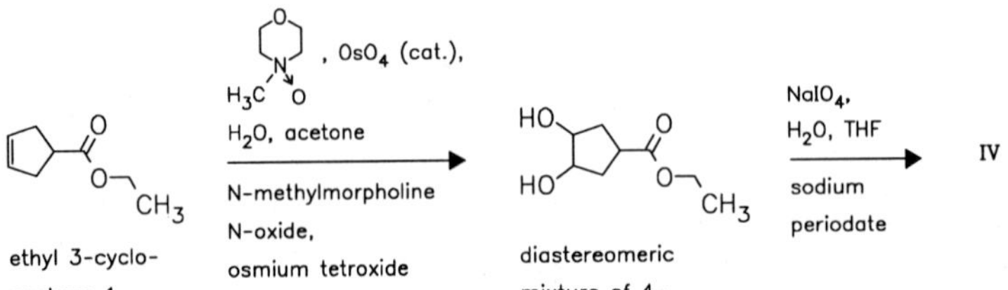

1. SOCl$_2$
2. H$_3$C^OH,

III

ethyl 3-cyclo-
pentene-1-
carboxylate (III)

, OsO$_4$ (cat.),
H$_2$O, acetone
N-methylmorpholine
N-oxide,
osmium tetroxide

diastereomeric
mixture of 4-
ethoxycarbonyl-
1,2-cyclopentanediol

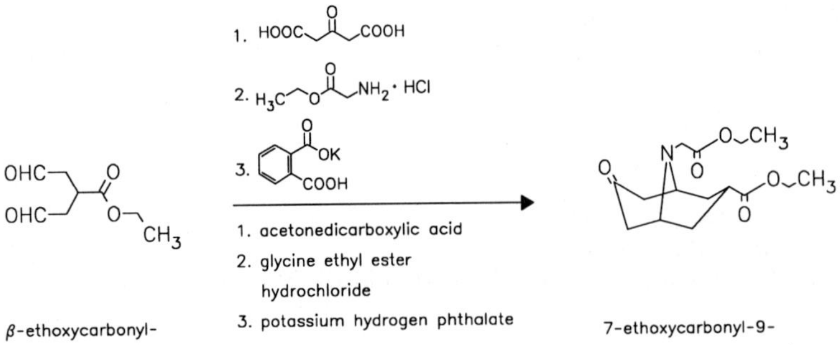

NaIO$_4$,
H$_2$O, THF
sodium
periodate

IV

β-ethoxycarbonyl-
glutaraldehyde (IV)

1. HOOC^^COOH
2. H$_3$C^O^^NH$_2$ · HCl
3. ^OK ^COOH

1. acetonedicarboxylic acid
2. glycine ethyl ester
   hydrochloride
3. potassium hydrogen phthalate

7-ethoxycarbonyl-9-
(ethoxycarbonylmethyl)-
9-azabicyclo[3.3.1]nonan-
3-one (V)

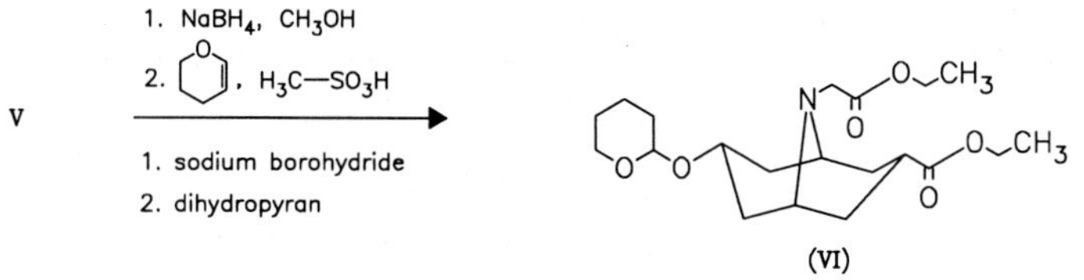

1. NaBH$_4$, CH$_3$OH

2. [structure], H$_3$C—SO$_3$H

V ⟶

1. sodium borohydride
2. dihydropyran

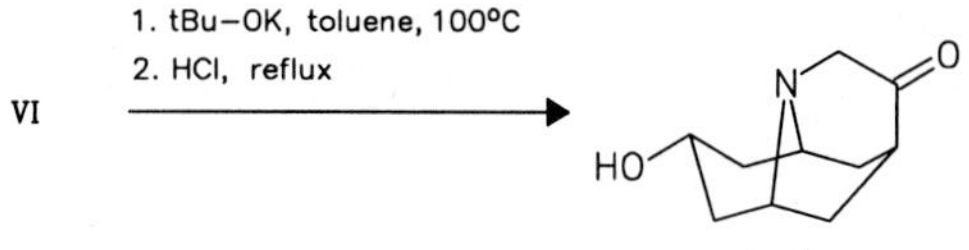

(VI)

1. tBu—OK, toluene, 100°C
2. HCl, reflux

VI ⟶

(VII)

1. [indole structure] (VIII) , CH$_2$Cl$_2$, DMAP

2. H$_3$C—SO$_3$H

VII ⟶

Dolasetron mesilate · H$_3$C—SO$_3$H

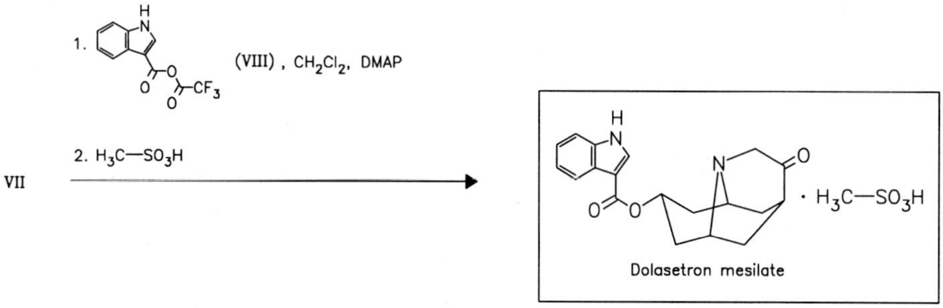

synthesis of intermediate VIII

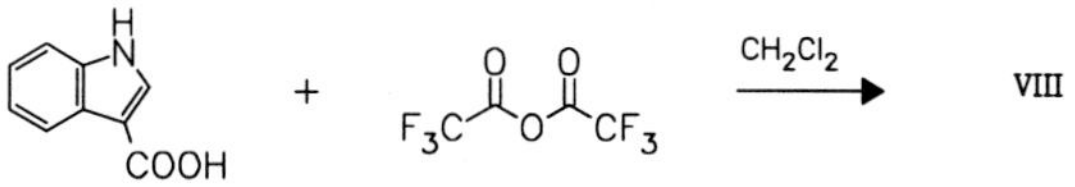

indole-3-carboxylic acid   +   trifluoroacetic anhydride $\xrightarrow{CH_2Cl_2}$ **VIII**

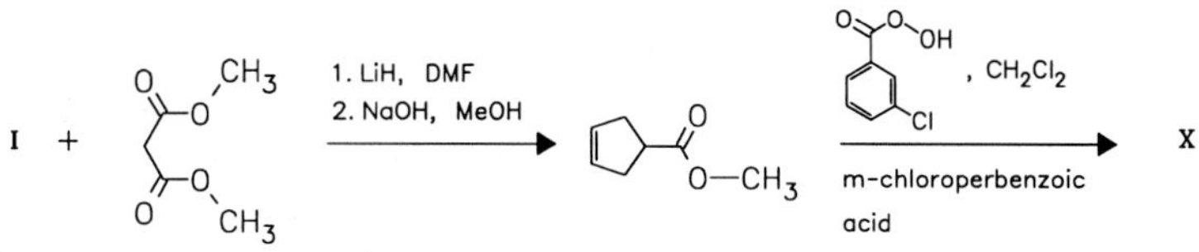

alternative preparation via β-methoxycarbonylglutaraldehyde

I +   dimethyl malonate $\xrightarrow{\substack{\text{1. LiH, DMF} \\ \text{2. NaOH, MeOH}}}$ methyl 3-cyclopentene-1-carboxylate (IX) $\xrightarrow{\substack{\text{m-chloroperbenzoic} \\ \text{acid}}}$, CH$_2Cl_2$ **X**

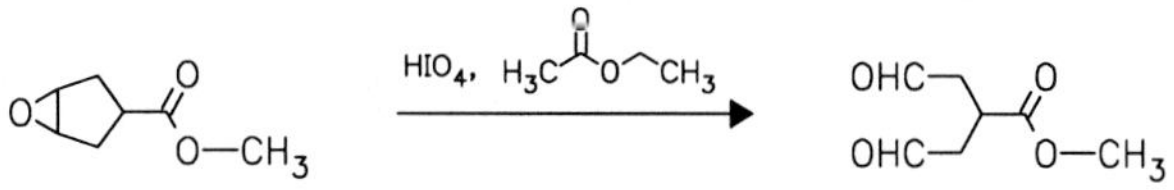

diastereomeric mixture of 1-methoxycarbonyl-3-cyclopentene oxide (trans:cis ~ 3:1) (**X**) $\xrightarrow{\text{HIO}_4,\ \text{H}_3\text{C}...}$ β-methoxycarbonyl-glutaraldehyde (**XI**)

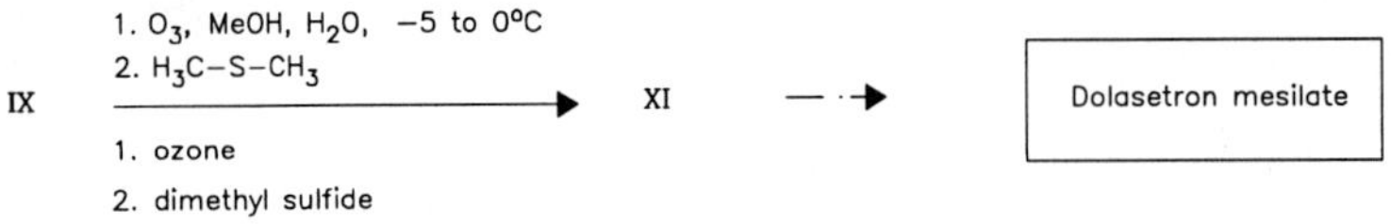

IX XI — · →

1. O_3, MeOH, H_2O, −5 to 0°C
2. $H_3C-S-CH_3$

1. ozone
2. dimethyl sulfide

Dolasetron mesilate

Reference(s):
EP 339 669 (Merrell Dow Pharm. Inc.; appl. 28.4.1989; USA-prior. 29.4.1988).
EP 266 730 (Merrell Dow Pharm. Inc.; appl. 2.11.1987; USA-prior. 3.11.1986).

Formulation(s):   amp. 12.5 mg/0.625 ml, 100 mg/5 ml; f. c. tabl. 50 mg, 200 mg

Trade Name(s):
D: Anemet (Hoechst Marion   USA: Anzemet (Hoechst Marion
      Roussel; 1997)   Roussel; 1997)

Domiodol

ATC: R05CB08
Use: mucolytic agent

RN: 61869-07-6   MF: $C_5H_9IO_3$ MW: 244.03
LD_{50}:   79-89 mg/kg (M, i.v.); 140-145 mg/kg (M, p.o.)
CN: 2-(iodomethyl)-1,3-dioxolane-4-methanol

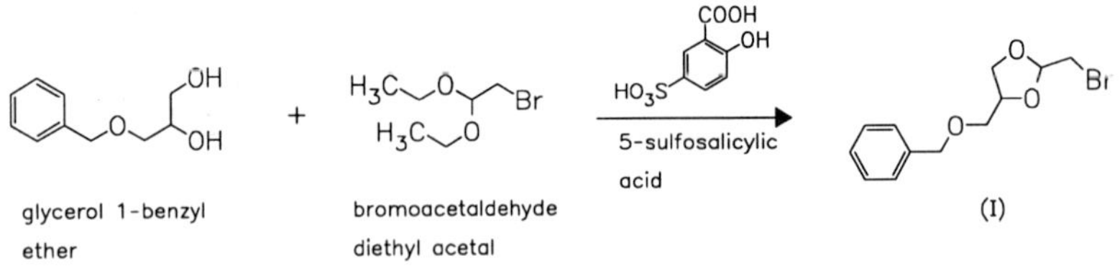

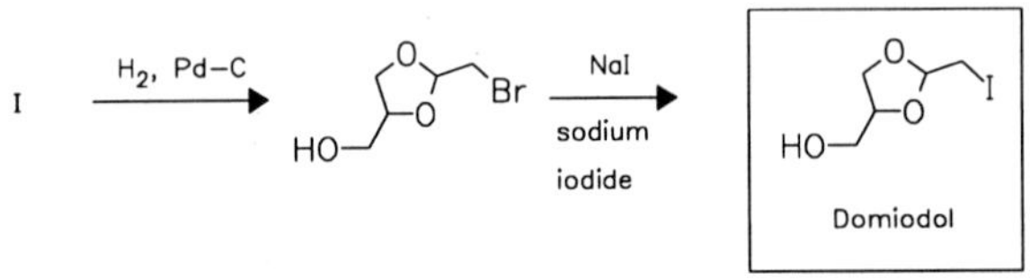

glycerol 1-benzyl   bromoacetaldehyde   (I)
ether   diethyl acetal

5-sulfosalicylic acid

Domiodol

Reference(s):
DOS 2 610 704 (Maggioni; appl. 13.3.1976; I-prior. 2.4.1975).

Formulation(s):   sachet 60 mg; sugar coated tabl. 60 mg; syrup 0.6 %

Trade Name(s):
I: Mucolitico (Maggioni-
      Winthrop)

Domiphen bromide
(Phenododecinium bromide)

ATC: A01AB06
Use: disinfectant

RN: 538-71-6   MF: $C_{22}H_{40}BrNO$ MW: 414.47   EINECS: 208-702-0
LD_{50}:   31 mg/kg (M, i.v.);
      18 mg/kg (R, i.v.)
CN: *N,N*-dimethyl-*N*-(2-phenoxyethyl)-1-dodecanaminium bromide

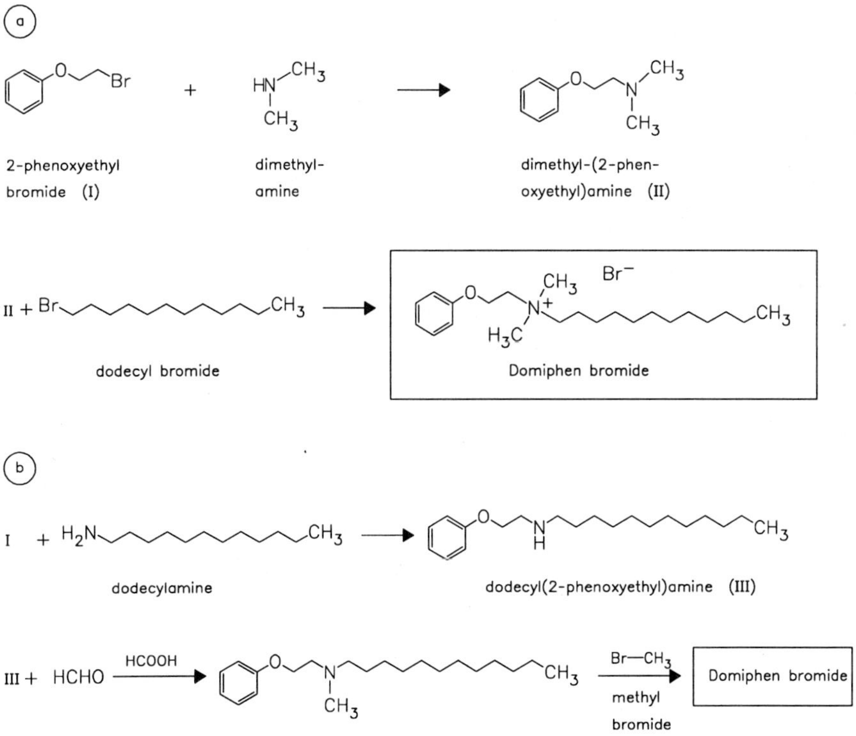

(a)

2-phenoxyethyl
bromide (I)

dimethyl-
amine

dimethyl-(2-phen-
oxyethyl)amine (II)

II + Br~~~~~~CH₃

dodecyl bromide

Domiphen bromide

(b)

I + H₂N~~~~~CH₃

dodecylamine

dodecyl(2-phenoxyethyl)amine (III)

III + HCHO → (HCOOH) → ~~~CH₃ → (Br—CH₃, methyl bromide) → Domiphen bromide

Reference(s):
US 2 581 336 (Ciba; 1952; CH-prior. 1944).

Formulation(s):   tabl. 0.5 mg

Trade Name(s):
GB: Bradosol Plus (Novartis
       Consumer)-comb.
I: Bradoral (Zyma)

J: Iodosan Nasale
       (SmithKline Beecham)-
       comb.

       Brado (Novartis-Takeda)-
       comb.

       Oradol (Novartis-Takeda)
USA: Bradosol Bromide (Ciba-
       Geigy); wfm

Domperidone

ATC: A03FA03
Use: anti-emetic

RN: 57808-66-9   MF: $C_{22}H_{24}ClN_5O_2$ MW: 425.92   EINECS: 260-968-7
LD$_{50}$:   46500 µg/kg (M, i.v.); >8 g/kg (M, p.o.);
       41700 µg/kg (R, i.v.); 5243 mg/kg (R, p.o.);
       42700 µg/kg (dog, i.v.); >160 mg/kg (dog, p.o.)
CN: 5-chloro-1-[1-[3-(2,3-dihydro-2-oxo-1H-benzimidazol-1-yl)propyl]-4-piperidinyl]-1,3-dihydro-2H-
       benzimidazol-2-one

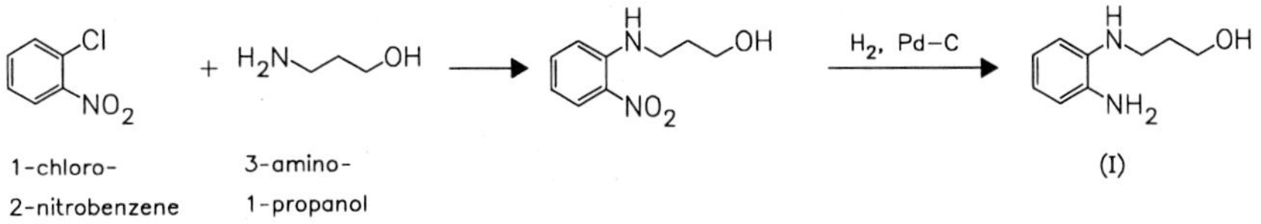

1-chloro-
2-nitrobenzene

3-amino-
1-propanol

(I)

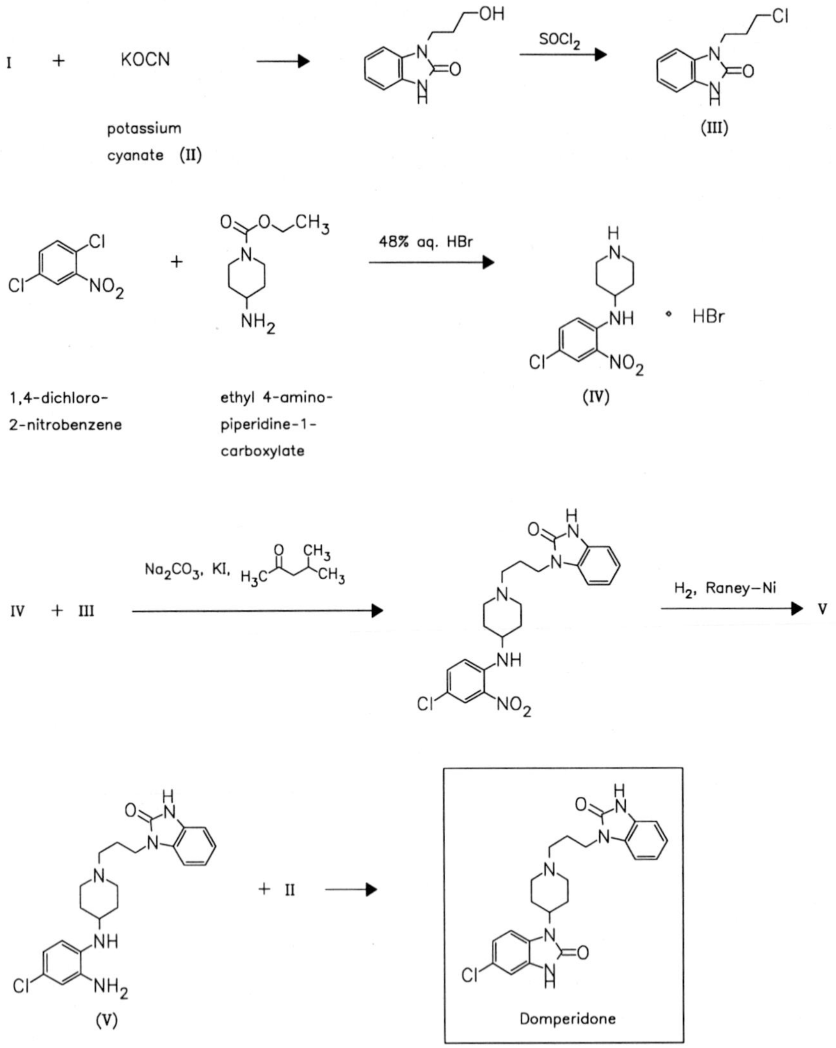

Reference(s):

US 4 066 772 (Janssen; 3.1.1978; prior. 21.7.1975, 17.5.1976).
DE 2 632 870 (Janssen; appl. 21.7.1976; USA-prior. 21.7.1975).

Formulation(s):   eff. gran. 10 mg; f. c. tabl. 10 mg; suppos. 30 mg; susp.10 mg/ml; tabl. 10 mg

Trade Name(s):

D: Motilium (Byk Gulden; 1979)

F: Motilium (Janssen-Cilag; 1983)
     Péridys (Robapharm)

GB: Domperamol (Servier)-comb.

     Motilium (Sanofi Winthrop; 1982)

I: Fobidon (Biomedica Foscama)
     Gastronorm (Janssen)
     Mod (Irbi)
     Motilium (Janssen; 1982)

     Peridon (Fisons; Italchimici)

J: Nauzelin (Kyowa Hakko; 1982)

Donepezil hydrochloride
(E-2020)

ATC: N06DA02
Use: cognition disorders,
      acetylcholinesterase inhibitor

RN: 120011-70-3   MF: $C_{24}H_{29}NO_3 \cdot HCl$ MW: 415.96
CN: 2,3-dihydro-5,6-dimethoxy-2-[[1-(phenylmethyl)-4-piperidinyl]methyl]-1H-inden-1-one hydrochloride

base
RN: 120014-06-4   MF: $C_{24}H_{29}NO_3$ MW: 379.50

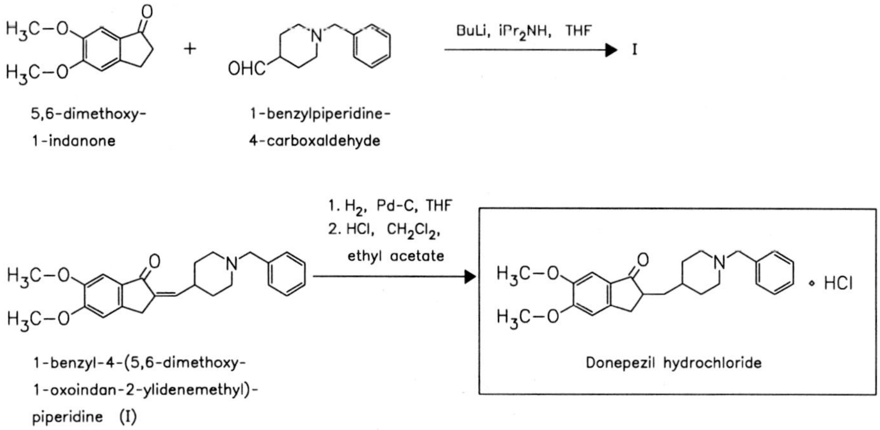

5,6-dimethoxy-
1-indanone

1-benzylpiperidine-
4-carboxaldehyde

BuLi, iPr$_2$NH, THF

I

1. H$_2$, Pd-C, THF
2. HCl, CH$_2$Cl$_2$,
   ethyl acetate

1-benzyl-4-(5,6-dimethoxy-
1-oxoindan-2-ylidenemethyl)-
piperidine (I)

Donepezil hydrochloride

Reference(s):
EP 296 560 (Eisai Co.; appl. 22.6.1988; J-prior. 22.6.1987).
Imura, J. et al.: J. Labelled Compd. Radiopharm. (JLCRD4) **27**, 835-839 (1989).

Formulation(s):   tabl. 5 mg, 10 mg

Trade Name(s):
D: Aricept (Eisai/Pfizer)   GB: Aricept (Eisai/Pfizer)   USA: Aricept (Eisai/Pfizer)

Dopamine

ATC: C01CA04
Use: sympathomimetic

RN: 51-61-6   MF: $C_8H_{11}NO_2$ MW: 153.18   EINECS: 200-110-0
LD$_{50}$: 59 mg/kg (M, i.v.)
CN: 4-(2-aminoethyl)-1,2-benzenediol

hydrochloride
RN: 62-31-7   MF: $C_8H_{11}NO_2 \cdot HCl$ MW: 189.64   EINECS: 200-527-8
LD$_{50}$: 156 mg/kg (M, i.v.); 4361 mg/kg (M, p.o.);
       4800 µg/kg (R, i.v.); 2859 mg/kg (R, p.o.);
       79 mg/kg (dog, i.v.)

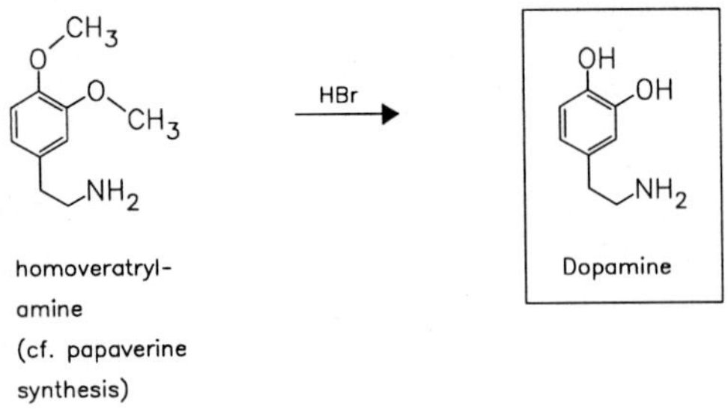

HBr
Dopamine
homoveratryl-
amine
(cf. papaverine
synthesis)

Reference(s):
Schöpf; Bayerle: Justus Liebigs Ann. Chem. (JLACBF) **513**, 196 (1934).

alternative with HCl:
DE 247 906 (K. W. Rosenmund et al.; 1909).
Hahn, G.; Stiehl, K.: Ber. Dtsch. Chem. Ges. (BDCGAS) **69**, 2640 (1936).
FR-appl. 2 332 748 (P. Fabre; appl. 28.11.1975).

combination with "nitro"-preparations (for treatment of cardiogenic shock):
DOS 2 649 162 (Nattermann; appl. 28.10.1976).

Formulation(s): vial 50 mg, 200 mg, 250 mg, 500 mg (for inf. sol.)

Trade Name(s):

D:	Dopamin AWD (ASTA Medica AWD)	
	Dopamin Fresenius (Fresenius-Klinik)	
	Dopamin ratiopharm (ratiopharm)	
	Dopamin Solvay (Solvay Arzneimittel)	
F:	Dopamine 200 Lucien (Lucien)	
	Dopamine Nativelle (Procter & Gamble)	
	Dopamine Pierre Fabre (Pierre Fabre)	
GB:	Intropin (Arnar-Stone);	wfm
I:	Revivan (Astra-Simes)	
J:	Inovan (Kyowa Hakko)	
USA:	generics	

Dopexamine

ATC: C01CA14
Use: cardiotonic

RN: 86197-47-9 MF: $C_{22}H_{32}N_2O_2$ MW: 356.51
CN: 4-[2-[[6-[(2-phenylethyl)amino]hexyl]amino]ethyl]-1,2-benzenediol

dihydrochloride
RN: 86484-91-5 MF: $C_{22}H_{32}N_2O_2 \cdot 2HCl$ MW: 429.43

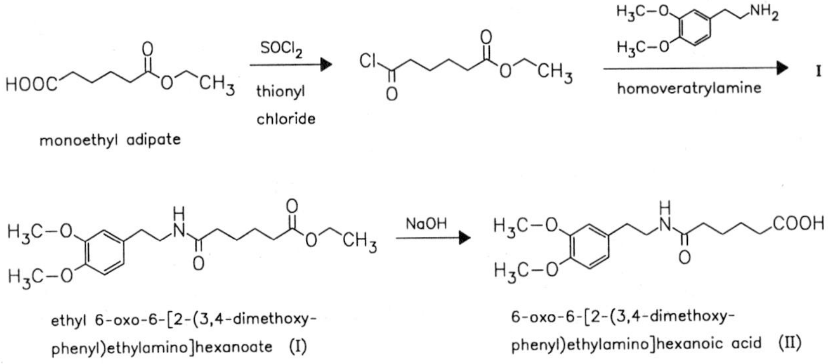

HOOC
O
O
CH3
monoethyl adipate
SOCl2
thionyl
chloride
Cl
O
H3C–O
H3C–O
NH2
homoveratrylamine
I
H
N
NaOH
COOH
ethyl 6-oxo-6-[2-(3,4-dimethoxy-
phenyl)ethylamino]hexanoate (I)
6-oxo-6-[2-(3,4-dimethoxy-
phenyl)ethylamino]hexanoic acid (II)

II + [2-phenylethylamine] (H₂N–) → N,N'-carbonyldiimidazole → N-[2-(3,4-dimethoxyphenyl)ethyl]-N'-(2-phenylethyl)hexanediamide (III)

III → 1. B₂H₆ 2. HBr (1. diborane) → Dopexamine

Reference(s):
EP 72 061 (Fisons, appl. 22.7.1982; GB-prior. 5.8.1981, 9.10.1981, 17.11.1981).

synthesis of II:
Kametani, T. et al.: Yakugaku Kenkyu (YKKKA8) **37**, 23 (1966); C.A. (CHABA8) **65**, 15320 (1966).

Formulation(s):   amp. 50 mg/5 ml for inf.

Trade Name(s):
D: Dopacard (Ipsen Pharma; as hydrochloride)   F: Dopacard (Ipsen/Biotech)   GB: Dopacard (Speywood; as hydrochloride)

Dornase alfa

(rhDNase)

ATC: R05CB13
Use: cystic fibrosis therapeutic

RN: 143831-71-4   MF: unspecified   MW: unspecified
CN: deoxyribonuclease (human clone 18-1 protein moiety reduced)

Dornase alfa is produced by genetically engineered Chinese Hamster ovary cells containing DNA encoding for the native human protein deoxyribunuclease I. It is purified by tangential flow filtration and column chromatography.

Reference(s):
WO 9 007 572 (Genentech; appl. 12.7.1990; USA-prior. 23.12.1988, 8.12.1989).
Shak, S. et al.: Proc. Natl. Acad. Sci. USA (PNASA6) **87**(23), 9188 (1990).

Formulation(s):   amp. 2.5 mg/2.5 ml

Trade Name(s):
D: Pulmozyme (Roche)   GB: Pulmozyme (Roche)   USA: Pulmozyme (Genentech)

Dorzolamide

(L-671152; MK-507)

ATC: S01EC03
Use: antiglaucoma, topical carbonic anhydrase inhibitor

RN: 120279-96-1   MF: $C_{10}H_{16}N_2O_4S_3$ MW: 324.45
CN: (4S-trans)-4-(ethylamino)-5,6-dihydro-6-methyl-4H-thieno[2,3-b]thiopyran-2-sulfonamide 7,7-dioxide

trans-base
RN: 120279-89-2   MF: $C_{10}H_{16}N_2O_4S_3$ MW: 324.45

monohydrochloride
RN: 130693-82-2   MF: $C_{10}H_{16}N_2O_4S_3 \cdot HCl$ MW: 360.91
maleate (1:1)
RN: 147600-19-9   MF: $C_{10}H_{16}N_2O_4S_3 \cdot C_4H_4O_4$ MW: 440.52

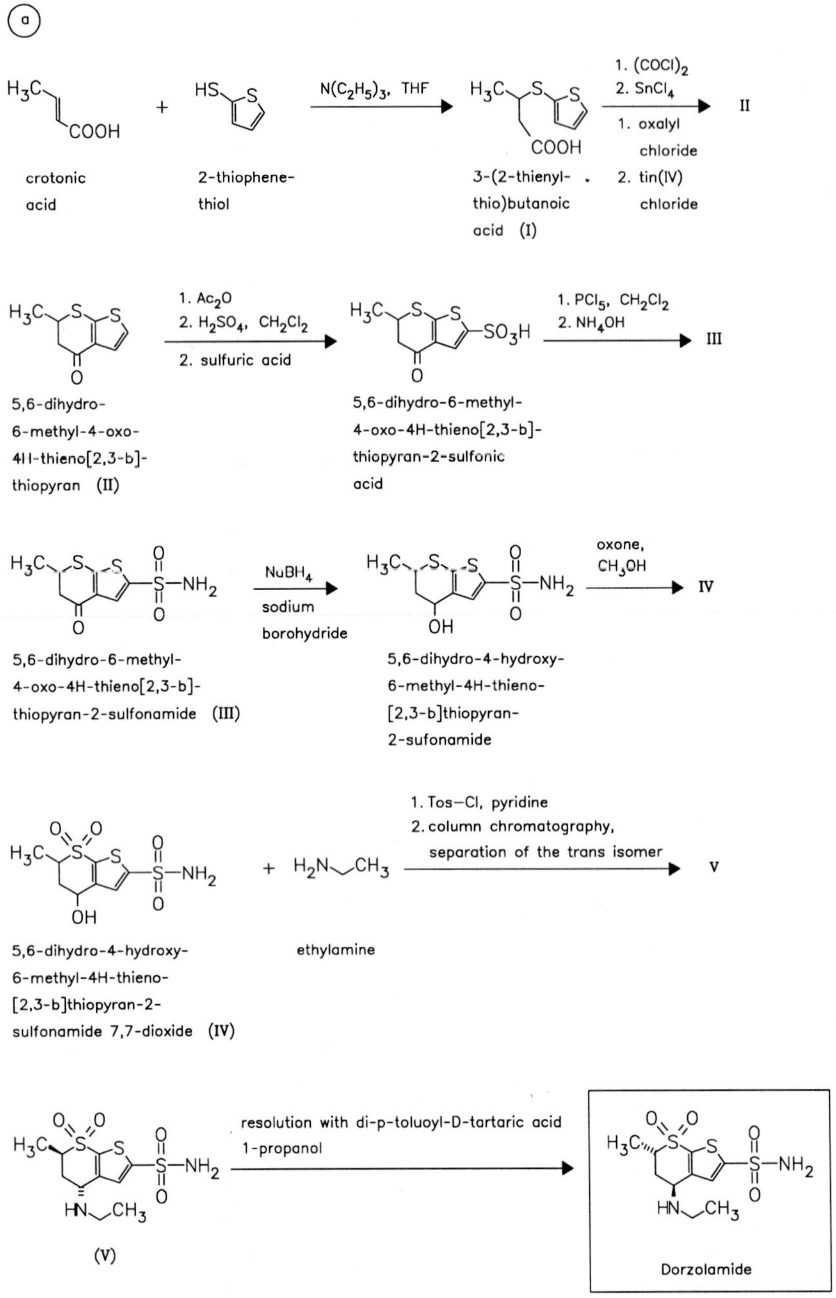

(a)

crotonic acid

2-thiophene-thiol

N(C_2H_5)$_3$, THF

3-(2-thienyl-thio)butanoic acid **(I)**

1. (COCl)$_2$
2. SnCl$_4$

1. oxalyl chloride
2. tin(IV) chloride

II

5,6-dihydro-6-methyl-4-oxo-4H-thieno[2,3-b]-thiopyran **(II)**

1. Ac$_2$O
2. H$_2$SO$_4$, CH$_2$Cl$_2$
2. sulfuric acid

5,6-dihydro-6-methyl-4-oxo-4H-thieno[2,3-b]-thiopyran-2-sulfonic acid

1. PCl$_5$, CH$_2$Cl$_2$
2. NH$_4$OH

III

5,6-dihydro-6-methyl-4-oxo-4H-thieno[2,3-b]-thiopyran-2-sulfonamide **(III)**

NuBH$_4$
sodium borohydride

5,6-dihydro-4-hydroxy-6-methyl-4H-thieno-[2,3-b]thiopyran-2-sufonamide

oxone, CH$_3$OH

IV

5,6-dihydro-4-hydroxy-6-methyl-4H-thieno-[2,3-b]thiopyran-2-sulfonamide 7,7-dioxide **(IV)**

+ H$_2$N⌒CH$_3$

ethylamine

1. Tos–Cl, pyridine
2. column chromatography, separation of the trans isomer

V

(V)

resolution with di-p-toluoyl-D-tartaric acid
1-propanol

Dorzolamide

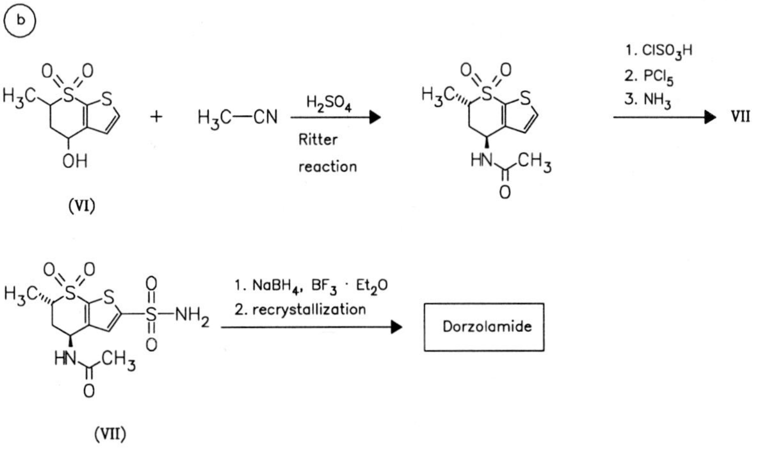

(b)

(VI)

(VII)

(c) preparation of the optically active thienothiopyran intermediate

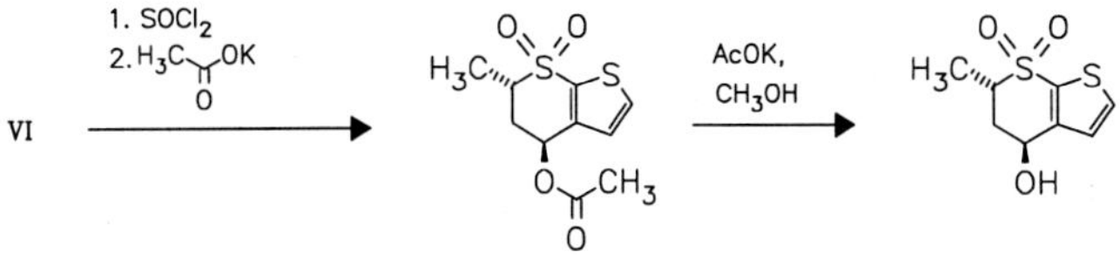

(d) stereoselective synthesis of intermediate I

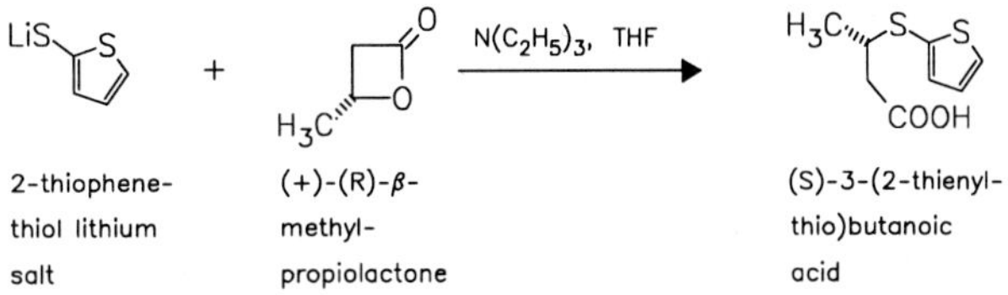

| 2-thiophene-thiol lithium salt | (+)-(R)-β-methyl-propiolactone | (S)-3-(2-thienyl-thio)butanoic acid |

Reference(s):

a US 4 797 413 (Merck & Co.; appl. 10.1.1989; USA-prior. 12.12.1984, 19.9.1985, 14.5.1986).
b EP 617 037 (Merck & Co.; appl. 17.3.1994; USA-prior. 22.3.1993, 10.2.1994).
c JP 06 107 666 (Kanegafuchi Chem.; appl. 19.4.1994; J-prior. 28.9.1992).
d US 4 968 815 (Merck & Co.; appl. 6.11.1990; USA-prior. 16.4.1990).
   US 4 968 814 (Merck & Co.; appl. 6.11.1990; USA-prior. 18.4.1990).

combination with calcium antagonists:
WO 9 323 082 (Alcon Lab.; appl. 12.5.1993; USA-prior. 13.5.1992).

combination with β-adrenergic antagonists:
EP 509 752 (Merck & Co.; appl. 14.4.1992; USA-prior. 17.4.1991, 13.2.1992).
EP 457 586 (Merck & Co.; appl. 16.5.1991; USA-prior. 17.5.1990).
EP 375 319 (Merck & Co.; appl. 18.12.1989; USA-prior. 19.12.1988).

Formulation(s):   eye drops 22.3 mg/ml (as hydrochloride)

Trade Name(s):

| D: | Trusopt (Chibret) | GB: | Trusopt (Merck Sharp & Dohme; as hydrochloride) | USA: | Trusopt (Merck; 1995 as hydrochloride) |

Dosulepin
(Dothiepin)

ATC: N06AA16
Use: antidepressant, thymoleptic

RN: 113-53-1   MF: $C_{19}H_{21}NS$ MW: 295.45   EINECS: 204-031-2
LD_{50}:   31 mg/kg (M, i.v.)
CN: 3-dibenzo[b,e]thiepin-11(6H)-ylidene-N,N-dimethyl-1-propanamine

hydrochloride
RN: 897-15-4   MF: $C_{19}H_{21}NS \cdot HCl$ MW: 331.91   EINECS: 212-978-8
LD_{50}:   29.2 mg/kg (M, i.v.); 209 mg/kg (M, p.o.);
        24 mg/kg (R, i.v.); 260 mg/kg (R, p.o.)

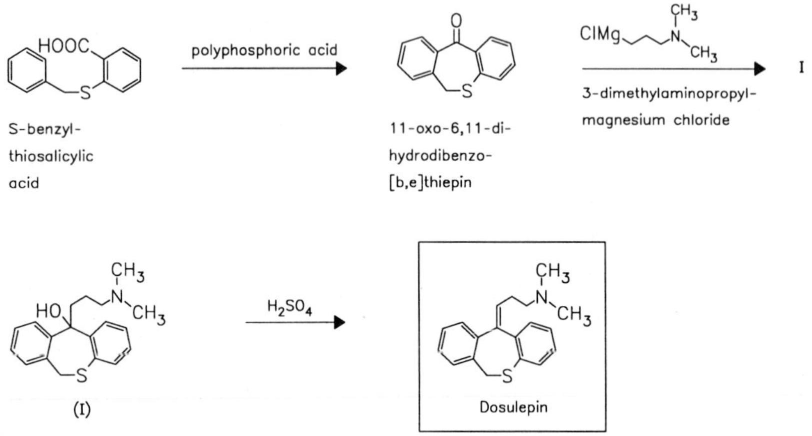

S-benzyl-
thiosalicylic
acid

11-oxo-6,11-di-
hydrodibenzo-
[b,e]thiepin

3-dimethylaminopropyl-
magnesium chloride

(I)

Dosulepin

Reference(s):
BE 618 591 (Spofa; appl. 6.6.1962; CS-prior. 8.6.1961).

Formulation(s):   cps. 25 mg, 50 mg, 75 mg.; susp. 25 mg

Trade Name(s):
D: Idom (Kanoldt)
F: Prothiaden (Knoll; as
      hydrochloride)
GB: Prothiaden (Knoll; as
       hydrochloride)
I: Protiaden (Boots Italia)

Doxapram

ATC: R07AB01
Use: central respiratory stimulant

RN: 309-29-5   MF: $C_{24}H_{30}N_2O_2$ MW: 378.52   EINECS: 206-216-3
LD_{50}:   268 mg/kg (M, i.p.)
CN: 1-ethyl-4-[2-(4-morpholinyl)ethyl]-3,3-diphenyl-2-pyrrolidinone

monohydrochloride monohydrate
RN: 7081-53-0   MF: $C_{24}H_{30}N_2O_2 \cdot HCl \cdot H_2O$ MW: 432.99
LD_{50}:   85 mg/kg (M, i.v.); 270 mg/kg (M, p.o.);
        72 mg/kg (R, i.v.); 261 mg/kg (R, p.o.);
        40 mg/kg (dog, i.v.); 150 mg/kg (dog, p.o.)

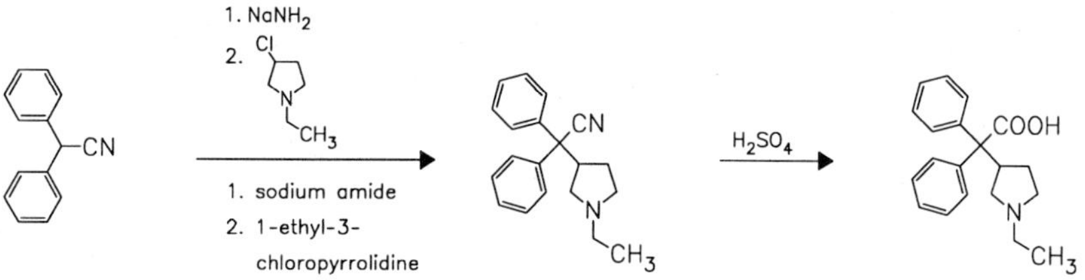
1. NaNH2
2.
Cl
N
CH3
CN
1. sodium amide
2. 1-ethyl-3-
chloropyrrolidine
CN
N
CH3
H2SO4
COOH
N
CH3

diphenyl-acetonitrile (I)

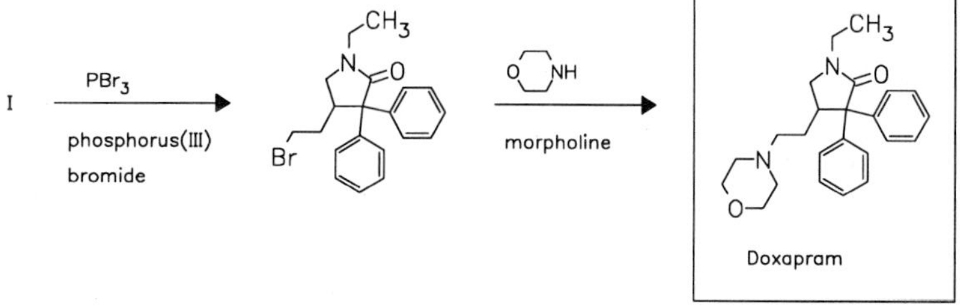
I
PBr3
phosphorus(III)
bromide
CH3
N
O
Br
O NH
morpholine
CH3
N
O
N
O
Doxapram

Reference(s):

US 3 192 230 (A. H. Robins; 29.6.1965; prior. 9.2.1961).
Lunsford, C.D. et al.: J. Med. Chem. (JMCMAR) **7**, 302 (1964).

Formulation(s): amp. 20 mg/ml

Trade Name(s):

D:	Dopram (Brenner); wfm	I:	Doxapril (Carlo Erba); wfm	J:	Dopram (Kissei)
F:	Dopram (Martinet); wfm		Doxapril (Farmalabor); wfm	USA:	Dopram (Robins; as hydrochloride)
GB:	Dopram (Anpharm)				

Doxazosin

ATC: C02CA04
Use: α_1-receptor antagonist, antihypertensive

RN: 74191-85-8 MF: $C_{23}H_{25}N_5O_5$ MW: 451.48
LD_{50}: >1000 mg/kg (M, R, p.o.)
CN: 1-(4-amino-6,7-dimethoxy-2-quinazolinyl)-4-[(2,3-dihydro-1,4-benzodioxin-2-yl)carbonyl]piperazine

hydrochloride
RN: 70918-01-3 MF: $C_{23}H_{25}N_5O_5 \cdot HCl$ MW: 487.94
mesylate
RN: 77883-43-3 MF: $C_{23}H_{25}N_5O_5 \cdot CH_4O_3S$ MW: 547.59
LD_{50}: 2935 mg/kg (M, p.o.);
>5 g/kg (R, p.o.);
>1 g/kg (dog, p.o.)

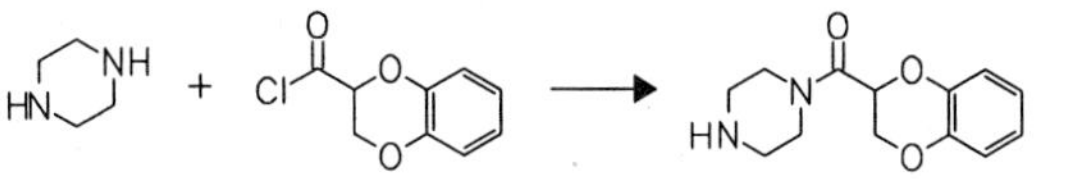
NH
HN
+
O
Cl
O
O
O
N
HN
O
O

piperazine 1,4-benzodioxan-2-ylcarbonyl chloride N-(1,4-benzodioxan-2-ylcarbonyl)piperazine (I)

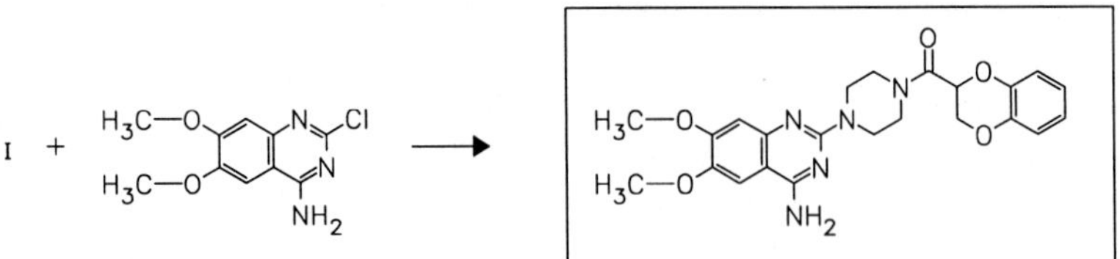

4-amino-2-chloro-6,7-
dimethoxyquinazoline
(cf. prazosine synthesis)

Doxazosin

Reference(s):
DE 2 847 623 (Pfizer; appl. 2.11.1978; GB-prior. 5.11.1977).
US 4 188 390 (Pfizer; 12.2.1980; GB-prior. 5.11.1977).
EP 848 001 (Alfa Chem.; appl. 17.10.1997; I-prior. 13.12.1996).
WO 9 935 143 (Knoll; appl. 18.12.1998; D-prior. 6.1.1998)

medical use for treatment of atherosclerosis:
US 4 758 569 (Pfizer; 19.7.1988; appl. 26.8.1987).

osmotic device:
US 4 837 111 (Alza; 6.6.1989; appl. 21.3.1988).

Formulation(s):   tabl. 1 mg, 2 mg, 4 mg

Trade Name(s):

D:	Cardular (Pfizer; 1989 as mesylate)	I:	Cardura (Roerig; 1989 as mesylate)	Normothen (Fisons Italchimici; 1989 as mesylate)
	Diblocin (Astra; 1989 as mesylate)		Dedralen (Lifepharma; 1989 as mesylate)	
GB:	Cardura (Invicta; 1989 as mesylate)			J: Cardenalin (Pfizer)
				USA: Cardura (Pfizer; 1990 as mesylate)

Doxefazepam

ATC: N05CD12
Use: hypnotic

RN: 40762-15-0   MF: $C_{17}H_{14}ClFN_2O_3$ MW: 348.76
LD_{50}:   >74 mg/kg (M, i.p.); 1500 mg/kg (M, p.o.);
586 mg/kg (R, i.p.); 1500 mg/kg (M, p.o.)
CN: 7-chloro-5-(2-fluorophenyl)-1,3-dihydro-3-hydroxy-1-(2-hydroxyethyl)-2*H*-1,4-benzodiazepin-2-one

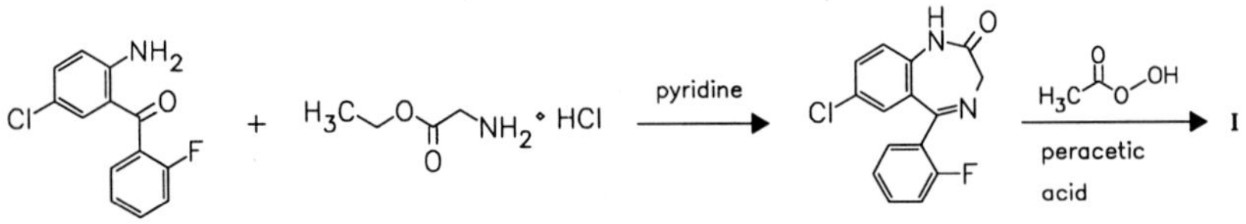

2-amino-5-chloro-
2'-fluorobenzo-
phenone
(cf. flunitrazepam
synthesis)

glycine ethyl ester
hydrochloride

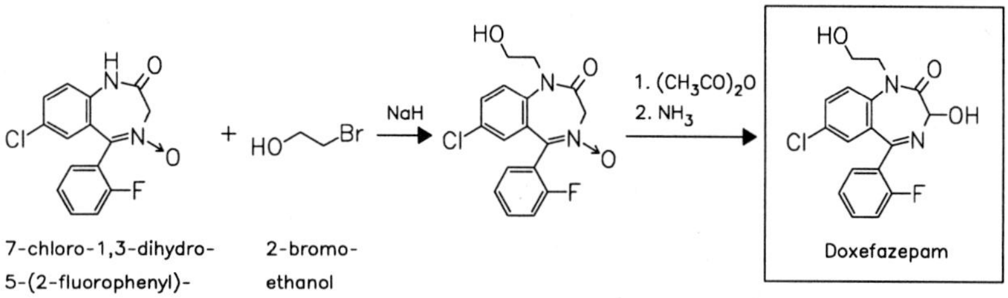

7-chloro-1,3-dihydro-   2-bromo-
5-(2-fluorophenyl)-   ethanol
2H-1,4-benzodiazepin-
2-one 4-oxide   (I)

Doxefazepam

Reference(s):

Tamagnone, G.F. et al.: Arzneim.-Forsch. (ARZNAD) **25**, 720 (1975).
DOS 2 338 058 (Schiapparelli; appl. 26.7.1973; E-prior. 28.7.1972).

synthesis of 7-chloro-1,3-dihydro-5-(2-fluorophenyl)-2H-1,4-benzodiazepin-2-one 4-oxide:
SA 6 802 239 (Hoffmann-La Roche; USA-prior. 21.4.1967, 23.10.1967).

Trade Name(s):

I: Doxans (Schiapparelli
       Searle); wfm

Doxepin

ATC: N06AA12
Use: antidepressant, tranquilizer

RN: 1668-19-5   MF: $C_{19}H_{21}NO$ MW: 279.38
LD$_{50}$:   26 mg/kg (M, i.v.); 135 mg/kg (M, p.o.);
       16 mg/kg (R, i.v.); 147 mg/kg (R, p.o.)
CN: 3-dibenz[b,e]oxepin-11(6H)-ylidene-N,N-dimethyl-1-propanamine

hydrochloride
RN: 1229-29-4   MF: $C_{19}H_{21}NO \cdot HCl$ MW: 315.84   EINECS: 214-966-8
LD$_{50}$:   15 mg/kg (M, i.v.); 180 mg/kg (M, p.o.);
       13 mg/kg (R, i.v.); 147 mg/kg (R, p.o.);
       >27 mg/kg (dog, i.v.)

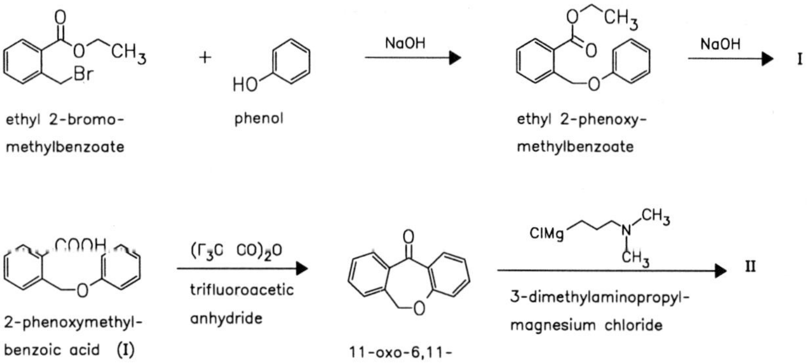

ethyl 2-bromo-   phenol   ethyl 2-phenoxy-
methylbenzoate   methylbenzoate

2-phenoxymethyl-   trifluoroacetic   11-oxo-6,11-   3-dimethylaminopropyl-
benzoic acid   (I)   anhydride   dihydrodibenz-   magnesium chloride
                                            [b,e]oxepin

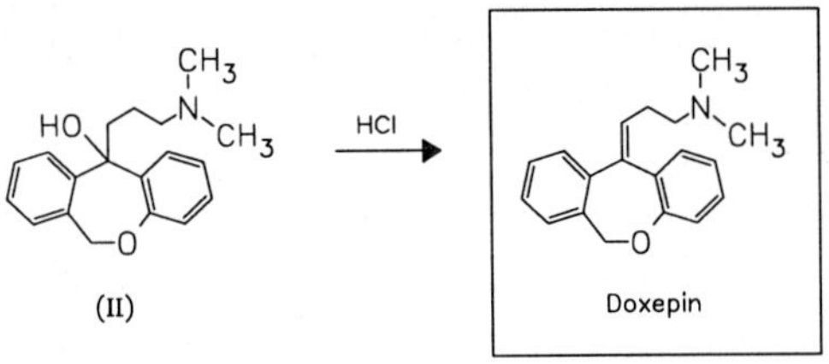

(II)   HCl →   Doxepin

Reference(s):
US 3 420 851 (Pfizer; 7.1.1969; appl. 19.12.1962; prior. 13.3.1962).
DE 1 232 161 (Boehringer Mannh.; appl. 7.10.1961).

Formulation(s):   amp. 25 mg/2 ml; coloured tabl. 50 mg, 100 mg; drg. 5 mg, 10 mg, 20 mg; drops 10 mg/ml; f.
c. tabl. 25 mg, 50 mg, 75 mg, 100 mg

Trade Name(s):

D:	Aponal (Boehringer Mannh./AWD)	F:	Quitaxon (Boehringer Mannh.; as hydrochloride)
	Maren (Krewel Meuselbach)		Sinequan (Pfizer; as hydrochloride)
	Sinquan (Pfizer)	GB:	Sinequan (Pfizer; as hydrochloride)

I: Sinequan (Pfizer); wfm
USA: Sinequan (Pfizer; as hydrochloride)
Zonalon (GenDerm)

Doxifluridine
(5'-dFUR)

ATC: L01BB
Use: antineoplastic, antimetabolite

RN: 3094-09-5   MF: $C_9H_{11}FN_2O_5$ MW: 246.19   EINECS: 221-440-1
LD$_{50}$:   >2000 mg/kg (M, i.p.); >1 g/kg (M, i.v.); >5000 mg/kg (M, p.o.);
>2000 mg/kg (R, i.p.); >1 g/kg (R, i.v.); 3390 mg/kg (R, p.o.); 3471 mg/kg (Rm, p.o.); 3390 mg/kg (Rf, p.o.)
CN: 5'-deoxy-5-fluorouridine

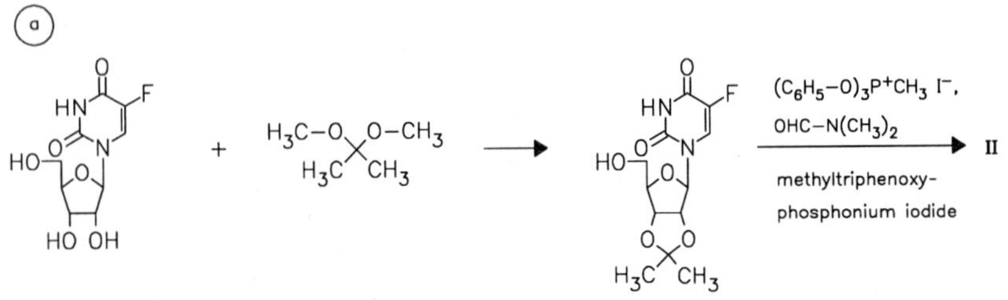

5-fluoro-
uridine (I)

2,2-dimethoxy-
propane

$(C_6H_5-O)_3P^+CH_3$ I$^-$,
OHC–N(CH_3)$_2$

methyltriphenoxy-
phosphonium iodide

II

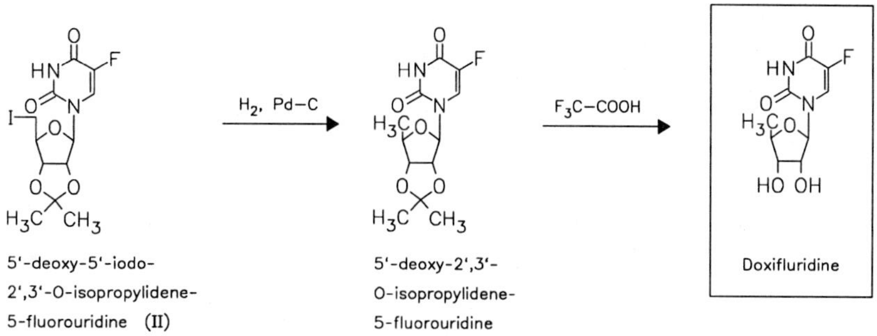

H2, Pd–C
F3C–COOH
5'-deoxy-5'-iodo-2',3'-O-isopropylidene-5-fluorouridine (II)
5'-deoxy-2',3'-O-isopropylidene-5-fluorouridine
Doxifluridine

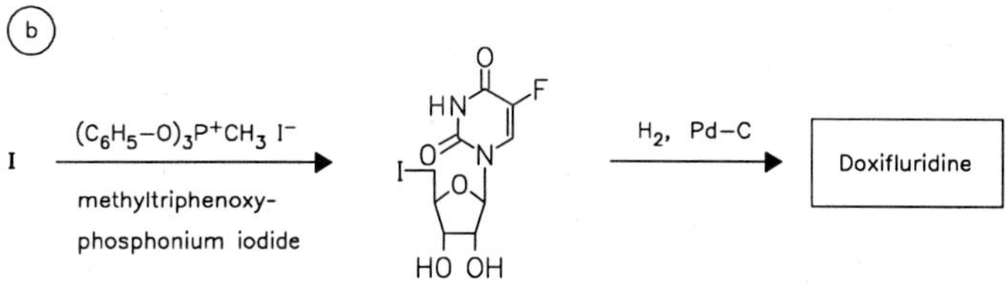

b
I
(C6H5–O)3P+CH3 I–
methyltriphenoxy-phosphonium iodide
H2, Pd–C
Doxifluridine

Reference(s):

a,b DOS 2 756 653 (Hoffmann-La Roche; appl. 19.12.1977; USA-prior. 20.12.1976).
US 4 071 680 (Hoffmann-La Roche; appl. 20.12.1976).
Cook, A.F. et al.: J. Med. Chem. (JMCMAR) **22**, 1330 (1979).

additional synthesis:

EP 21 231 (Hoffmann-La Roche; appl. 10.6.1980; CH-prior. 15.6.1979).
Hrebabecky, H.; Beranek, J.: Collect. Czech. Chem. Commun. (CCCCAK) **43**, 3268 (1978).
Kiss, J. et al.: Helv. Chim. Acta (HCACAV) **65**, 1522 (1982).
Scott, J.W. et al.: J. Carbohydr., Nucleosides, Nucleotides (JCNNAF) **8**, 171 (1981).
Ajmera, S.; Danenberg, V.: J. Med. Chem. (JMCMAR) **25**, 999 (1982).
Rosowsky, A. et al.: J. Med. Chem. (JMCMAR) **25**, 1034 (1982).

combination with purine nucleosides or nucleotides:

EP 189 755 (Hoffmann-La Roche; appl. 9.10.1985).

Formulation(s): cps. 100 mg, 200 mg

Trade Name(s):

J: Furtulon (Nippon Roche; 1987)

Doxofylline

(ABC-12/3)

ATC: R03DA11
Use: antiasthmatic, bronchodilator

RN: 69975-86-6 MF: $C_{11}H_{14}N_4O_4$ MW: 266.26 EINECS: 274-239-6
LD_{50}: 216 mg/kg (M, i.v.); 841 mg/kg (M, p.o.);
445 mg/kg (R, i.p.); 315 mg/kg (R, i.v.); 966 mg/kg (R, p.o.)
CN: 7-(1,3-dioxolan-2-ylmethyl)-3,7-dihydro-1,3-dimethyl-1*H*-purine-2,6-dione

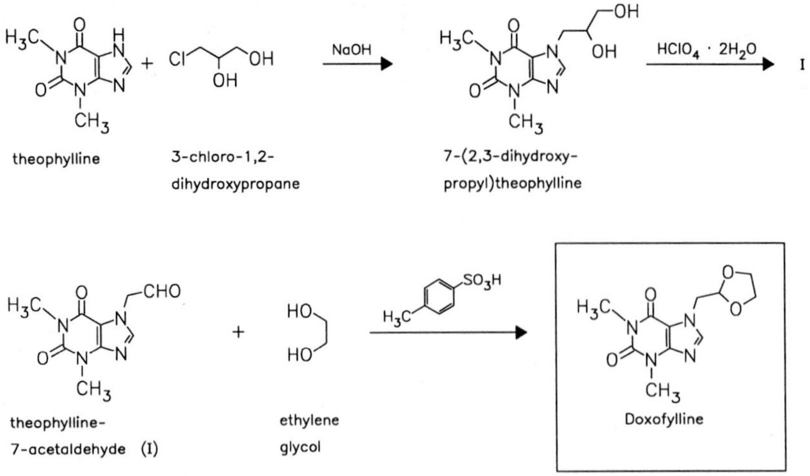

theophylline + 3-chloro-1,2-dihydroxypropane → (NaOH) → 7-(2,3-dihydroxy-propyl)theophylline → (HClO$_4$ · 2H$_2$O) → I

theophylline-7-acetaldehyde (I) + ethylene glycol → Doxofylline

Reference(s):

DE 2 827 497 (ABC; appl. 22.6.1978; I-prior. 4.6.1978).
US 4 187 308 (ABC; 5.2.1980; I-prior. 4.6.1978).
Avico, U. et al.: Farmaco, Ed. Sci. (FRPSAX) **17**, 73 (1962).

synthesis of theophylline-7-acetaldehyde:
Maney, P.V.: J. Am. Pharm. Assoc. (JPHAA3) **35**, 266 (1946).
Toffoli, F. et al.: Farmaco, Ed. Sci. (FRPSAX) **11**, 516 (1956).

Formulation(s):   amp. 100 mg/10 ml; cps. 300 mg; sachet 200 mg; s. r. tabl. 300 mg; tabl. 400 mg

Trade Name(s):
I: Ansimar (ABC; 1988)

Doxorubicin

(Adriamycin)

ATC: L01DB01
Use: antineoplastic, antibacterial

RN: 23214-92-8   MF: C$_{27}$H$_{29}$NO$_{11}$ MW: 543.53   EINECS: 245-495-6
LD$_{50}$:   10 mg/kg (M, i.v.); 570 mg/kg (M, p.o.);
10.510 mg/kg (R, i.v.);
2.4 mg/kg (dog, i.v.)
CN: (8S-*cis*)-10-[(3-amino-2,3,6-trideoxy-α-L-*lyxo*-hexopyranosyl)oxy]-7,8,9,10-tetrahydro-6,8,11-trihydroxy-8-(hydroxyacetyl)-1-methoxy-5,12-naphthacenedione

hydrochloride
RN: 25316-40-9   MF: C$_{27}$H$_{29}$NO$_{11}$ · HCl   MW: 579.99   EINECS: 246-818-3
LD$_{50}$:   1245 μg/kg (M, i.v.); 698 mg/kg (M, p.o.);
12510 μg/kg (R, i.v.)

Doxorubicin

From culture of mutant F. I. 106 of *Streptomyces peucetius var. caesius.*

Reference(s):
DE 1 770 204 (Soc. Farmaceutici Italia; prior. 13.4.1968).
US 3 590 028 (Soc. Farmaceutici Italia; 29.6.1971; appl. 18.4.1968; I-prior. 18.4.1967).
GB 1 161 278 (Soc. Farmaceutici Italia; appl. 16.4.1968; I-prior. 18.4.1967).

alternative syntheses:
partial synthesis from daunorubicin:
DOS 1 917 874 (Soc. Farmaceutici Italia; appl. 8.4.1969; I-prior. 12.4.1968).

partial synthesis from adriamycinon:
US 4 058 519 (Soc. Farmaceutici Italia; 15.11.1977; GB-prior. 22.3.1974).
US 4 098 798 (Soc. Farmaceutici Italia; 4.7.1978; GB-prior. 22.3.1974).

daunorubicin *(from cultures of Streptomyces peucetius F. I. 1762):*
GB 1 003 383 (Soc. Farmaceutici Italia; appl. 11.11.1963; I-prior. 16.11.1962).

doxorubicin-14-octanoate:
DOS 2 260 438 (Soc. Farmaceutici Italia; appl. 11.12.1972).
US 3 803 124 (Soc. Farmaceutici Italia; 9.4.1974; I-prior. 12.4.1968, 4.5.1971).

stable liposome composition:
WO 9 202 208 (Liposome Technology Inc.; appl. 2.8.1991; USA-prior. 8.8.1990).

Formulation(s):   vial (lyo.) 10 mg, 20 mg, 50 mg, 150 mg (as hydrochloride)

Trade Name(s):

D:	Adriblastin (Pharmacia & Upjohn; 1972)	GB:	Caelyx (Schering-Plough; as hydrochloride)
	Adrimedac (medac)	I:	Adriblastina (Farmitalia; 1971)
	Caelyx (Essex Pharma)		
	Ribodoxo (ribosepharm)	J:	Adriacin (Kyowa Hakko; 1975)
F:	Adriblastine (Pharmacia & Upjohn; 1974)		

USA:	Adriamycin (Pharmacia & Upjohn; 1974)
	Doxil (Sequus)
	Rubex (Bristol-Myers Squibb)

Doxycycline

ATC: J01AA02
Use: antibiotic

RN: 564-25-0   MF: $C_{22}H_{24}N_2O_8$ MW: 444.44   EINECS: 209-271-1
LD_{50}:   241 mg/kg (M, i.v.); 1870 mg/kg (M, p.o.);
228 mg/kg (R, i.v.); >2 g/kg (R, p.o.);
>100 mg/kg (dog, i.v.); >500 mg/kg (dog, p.o.)
CN: [4S-(4α,4aα,5α,5aα,6α,12aα)]-4-(dimethylamino)-1,4,4a,5,5a,6,11,12a-octahydro-3,5,10,12,12a-
pentahydroxy-6-methyl-1,11-dioxo-2-naphthacenecarboxamide

monohydrochloride
RN: 10592-13-9   MF: $C_{22}H_{24}N_2O_8 \cdot HCl$ MW: 480.90   EINECS: 234-198-7
LD_{50}:   290 mg/kg (M, i.v.); 1890 mg/kg (M, p.o.);
          137 mg/kg (R, i.v.); 1700 mg/kg (R, p.o.);
          >500 mg/kg (dog, p.o.)
monohydrate
RN: 17086-28-1   MF: $C_{22}H_{24}N_2O_8 \cdot H_2O$ MW: 462.46
hyclate
RN: 24390-14-5   MF: $C_{22}H_{24}N_2O_8 \cdot 1/2C_2H_6O \cdot HCl \cdot 1/2H_2O$ MW: 1025.89

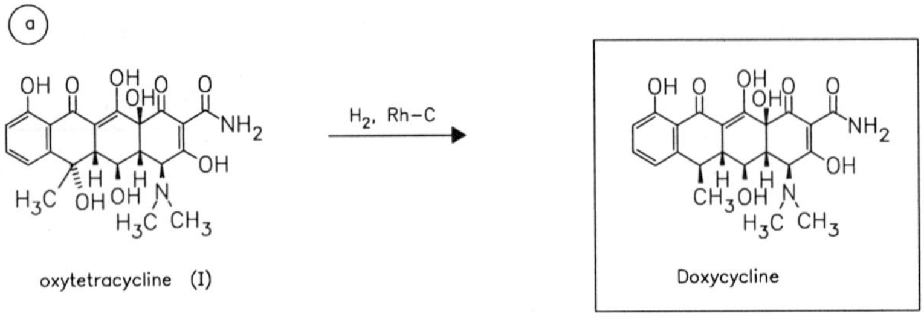

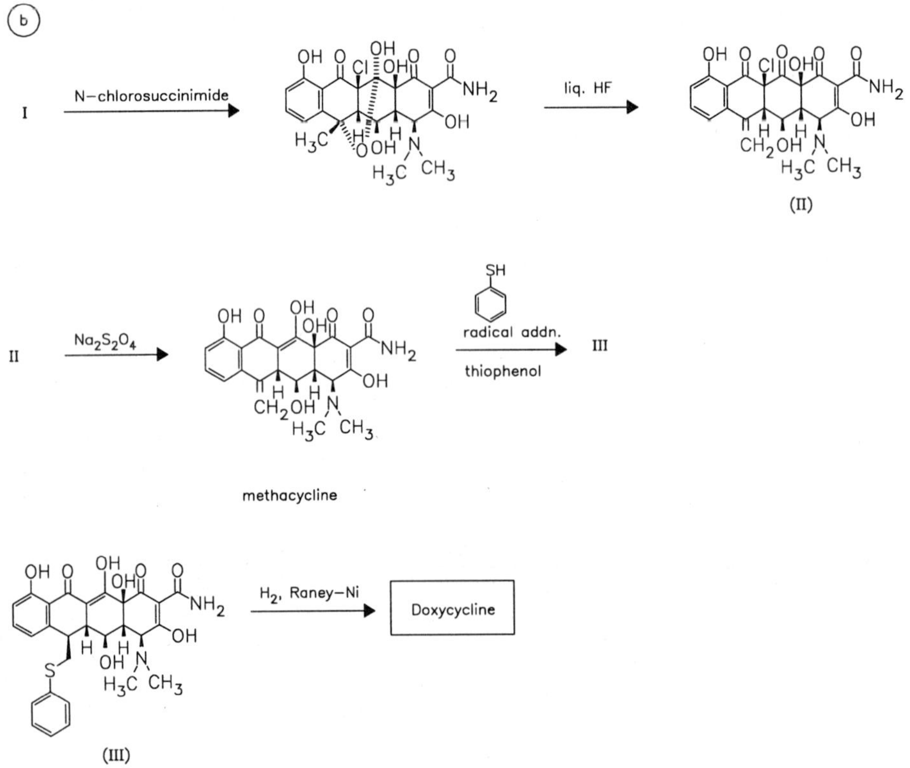

a

oxytetracycline (I) $\xrightarrow{\text{H}_2,\ \text{Rh–C}}$ Doxycycline

b

I $\xrightarrow{\text{N–chlorosuccinimide}}$ $\xrightarrow{\text{liq. HF}}$ (II)

II $\xrightarrow{\text{Na}_2\text{S}_2\text{O}_4}$ methacycline $\xrightarrow[\text{thiophenol}]{\text{radical addn.}}$ III

(III) $\xrightarrow{\text{H}_2,\ \text{Raney–Ni}}$ Doxycycline

Reference(s):
a   US 3 019 260 (American Cyanamid; 30.1.1962; prior. 13.5.1959).
    DE 1 082 905 (American Cyanamid; appl. 3.11.1958; USA-prior. 5.11.1957).
b   US 3 200 149 (Pfizer; 10.8.1965; prior. 23.5.1960).
    DAS 1 793 556 (Pfizer; appl. 19.5.1961; USA-prior. 23.5.1960).
    DE 1 298 522 (Pfizer; appl. 23.5.1961; USA-prior. 23.5.1960).
    Blackwood, R.K. et al.: J. Am. Chem. Soc. (JACSAT) **85**, 3943 (1963).

stereospecific hydrogenation of metacycline *with* diaceto(triphenylphosphine)rhodium(II) complex *to* doxycycline:
DAS 2 554 524 (Pfizer; appl. 4.12.1975; USA-prior. 28.1.1975).

Formulation(s): tabl. 50 mg, 100 mg, 200 mg

Trade Name(s):

D: Azudoxat (Azuchemie)
Clinofug (Wolff)
Mespafin 100 (Merckle)
Mucotectan (Boehringer Ing.)
Neodox (Rosen Pharma)
Sigadoxin (Kytta-Siegfried)
Supracyclin 100/200 (Grünenthal)
Vibramycin N (Pfizer)
Vibravenös (Pfizer)

F: Doxycline Plantier (ASTA Medica)
Doxygram (Pharma 2000)
Doxylets (Galephar)
Granudoxy (Pierre Fabre)
Monocline (Doms-Adrian)
Spanor (Biotherapie)

Tolexine (Biorga)
Vibramycine (Pfizer)
Vibraveineuse (Pfizer)

GB: Doxatet (Cox); wfm
Doxylar (Lagap); wfm
Nordox (Norton); wfm
Vibramycin Acne Pack (Trinity)-comb.

I: Bassado (Poli)
Doxina (Ipfi)
Farmodoxi (Lifepharma)
Gram-Val (Polifarma)
Miraclin (Farmacologico Milanese)
Monodoxin (Crosara)
Ribociclina (Puropharma)-comb.
Unacil (Firma)

J: Hydramycin (Sankyo)
Liomycin (Daiichi)
Roximycin (Kyorin)
Vibramycin (Taito Pfizer)

USA: Doryx (Warner Chilcott Professional Products; as hydrate)
Monodox (Oclassen; as monohydrate)
Vibramycin (Pfizer; as calcium salt)
Vibramycin (Pfizer; as hydrate)
Vibramycin (Pfizer; as monohydrate)

Doxylamine

ATC: R06AA09
Use: antihistaminic

RN: 469-21-6 MF: $C_{17}H_{22}N_2O$ MW: 270.38 EINECS: 207-414-2
LD$_{50}$: 62 mg/kg (M, i.v.); 470 mg/kg (M, p.o.)
CN: *N,N*-dimethyl-2-[1-phenyl-1-(2-pyridinyl)ethoxy]ethanamine

succinate (1:1)
RN: 562-10-7 MF: $C_{17}H_{22}N_2O \cdot C_4H_6O_4$ MW: 388.46 EINECS: 209-228-7
LD$_{50}$: 62 mg/kg (M, i.v.); 470 mg/kg (M, p.o.)

2-acetyl-pyridine + phenylmagnesium bromide → 1-phenyl-1-(2-pyridyl)-ethanol

1. NaNH$_2$
2. CH_3 / Cl-N-CH$_3$

1. sodium amide
2. 2-(dimethylamino)-ethyl chloride

Doxylamine

Reference(s):
Sperber, N. et al.: J. Am. Chem. Soc. (JACSAT) **71**, 887 (1949).

Formulation(s): eff. tabl. 25 mg; tabl. 25 mg (as succinate)

Trade Name(s):

D: Gittalun (Boehringer Ing.)
Hewedomir forte (Hevert)
Hoggar N (Stada)

Mereprine (Cassella-med)
Praedisup (Chephasaar)-comb.

Sedaplus (Rosen Pharma)
Wick Formel 44 S (Wick Pharma)-comb.

F: Donormyl (Oberlin)-comb.
Méréprine (Marion Merrell)
GB: Nethaprin Dospan (Merrell Dow)-comb.; wfm
Nethaprin expect (Merrell Dow)-comb.; wfm
Syndol (Merrell Dow)-comb.; wfm
I: Doxised (Corvi)
Vicks Medinait (Procter & Gamble)-comb.
USA: Unisom Nighttime Sleep-Aid (Pfizer; as succinate)

Drofenine

(Hexahydroadiphenine)

ATC: A03DA49
Use: antispasmodic

RN: 1679-76-1 MF: $C_{20}H_{31}NO_2$ MW: 317.47
LD_{50}: 37 mg/kg (R, i.v.)
CN: α-cyclohexylbenzeneacetic acid 2-(diethylamino)ethyl ester

hydrochloride
RN: 548-66-3 MF: $C_{20}H_{31}NO_2 \cdot HCl$ MW: 353.93 EINECS: 208-954-1
LD_{50}: 47 mg/kg (M, i.v.); 3700 mg/kg (M, p.o.)

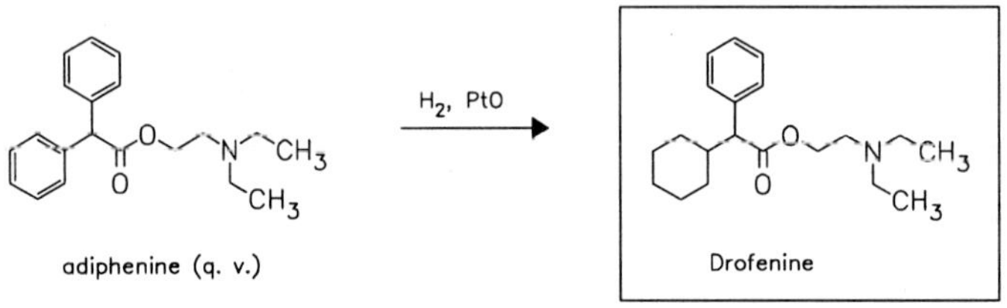

H2, PtO
adiphenine (q. v.)
Drofenine

Reference(s):
CH 219 301 (Ciba; appl. 1938).

Formulation(s): drg. 20 mg, 25 mg (comb. with 220 mg propyphenazone)

Trade Name(s):
D: Spasmo-Cibalgin/comp. (Novartis Pharma)
F: Spasmo-Cibalgine (Ciba)-comb.; wfm
I: Spasmocibalgina (Novartis)-comb.

Dronabinol

(δ-9-THC)

ATC: A04A
Use: anti-emetic, active ingredient of marijuana

RN: 1972-08-3 MF: $C_{21}H_{30}O_2$ MW: 314.47
LD_{50}: 168 mg/kg (M, i.p.); 42 mg/kg (M, i.v.); 482 mg/kg (M, p.o.);
373 mg/kg (R, i.p.); 29 mg/kg (R, i.v.); 666 mg/kg (R, p.o.)
CN: (6a*R*-*trans*)-6a,7,8,10a-tetrahydro-6,6,9-trimethyl-3-pentyl-6*H*-dibenzo[*b*,*d*]pyran-1-ol

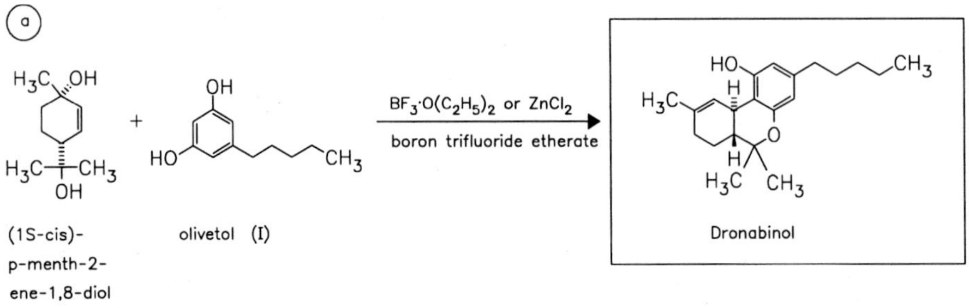

a
(1S-cis)-p-menth-2-ene-1,8-diol
olivetol (I)
BF3·O(C2H5)2 or ZnCl2
boron trifluoride etherate
Dronabinol

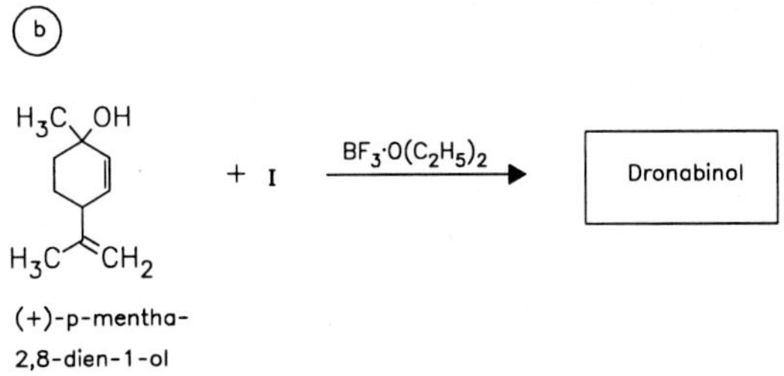

b
H3C OH
+ I
BF3·O(C2H5)2
Dronabinol
H3C CH2
(+)-p-mentha-
2,8-dien-1-ol

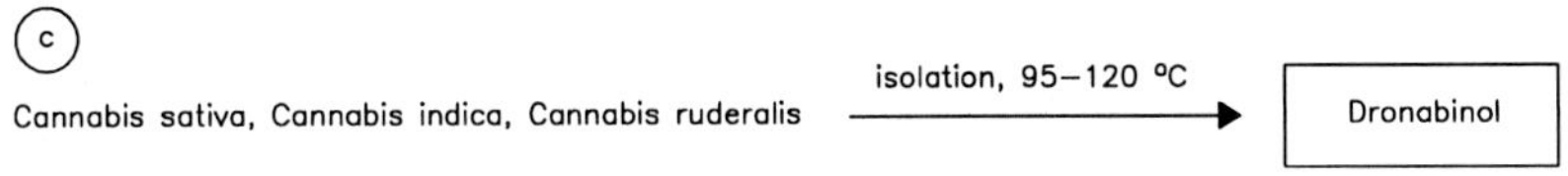

c
Cannabis sativa, Cannabis indica, Cannabis ruderalis
isolation, 95–120 °C
Dronabinol

Reference(s):

a Handrick, G.R. et al.: Tetrahedron Lett. (TELEAY) **1979**, 681.

b US 4 116 979 (Sheehan Inst. for Research; 26.9.1978; appl. 7.2.1977; prior. 28.11.1975, 24.6.1975).
US 4 381 399 (Aerojet; 26.4.1983; appl. 21.12.1981).

c US 4 279 824 (L. O. McKinney; 21.7.1981; appl. 1.11.1979).

alternative methods:

US 3 734 930 (US Dep. of Health; 22.5.1973; appl. 22.9.1971).
Straight, R. et al.: Biochem. Med. (BIMDA2) **8**, 341 (1973).
Ribi, E. et al.: Prep. Biochem. (PRBCBQ) **3**, 209 (1973).

review:

Mechoulam, R. et al.: Chem. Rev. (Washington, D. C.) (CHREAY) **76**, 75 (1976).

Formulation(s): cps. 2.5 mg, 5 mg, 10 mg

Trade Name(s):
USA: Marinol (Roxane)

Droperidol

(Dehydrobenzperidol)

ATC: N01AX01; N05AD08
Use: neuroleptic, anesthetic (neuroleptanesthesia)

RN: 548-73-2 MF: $C_{22}H_{22}FN_3O_2$ MW: 379.44 EINECS: 208-957-8
LD_{50}: 20 mg/kg (M, i.v.);
30 mg/kg (R, i.v.); 750 mg/kg (R, p.o.)
CN: 1-[1-[4-(4-fluorophenyl)-4-oxobutyl]-1,2,3,6-tetrahydro-4-pyridinyl]-1,3-dihydro-2*H*-benzimidazol-2-one

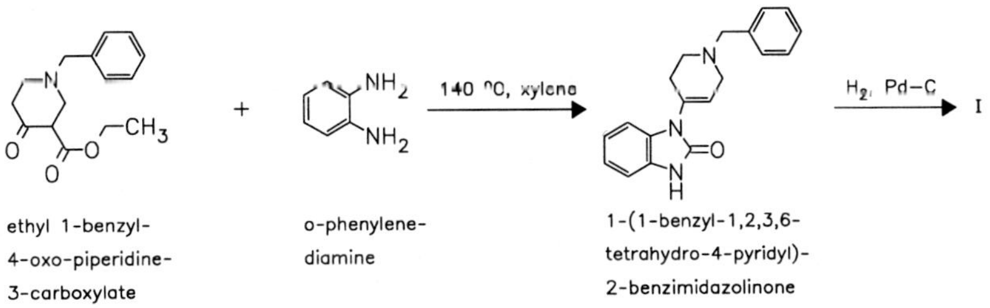

N
CH3
O
O
O
+
NH2
NH2
140 °C, xylene
N
N
O
N
H
H2, Pd–C
I
ethyl 1-benzyl-
4-oxo-piperidine-
3-carboxylate
o-phenylene-
diamine
1-(1-benzyl-1,2,3,6-
tetrahydro-4-pyridyl)-
2-benzimidazolinone

1-(1,2,3,6-tetra-
hydro-4-pyridyl)-
2-benzimidazolinone (I)

4-chloro-4'-fluoro-
butyrophenone

Droperidol

Reference(s):
GB 989 755 (Janssen; appl. 24.12.1962; USA-prior. 22.12.1961).
US 3 141 823 (Janssen; 21.7.1964; appl. 4.9.1962).
US 3 161 645 (Janssen; 15.12.1964; appl. 18.12.1962).

Formulation(s):   amp. 2.5 mg/ml, 5 mg/2 ml, 12.5 mg/5 ml, 25 mg/10 ml; tabl. 10 mg; vial 5 mg/2 ml,
12.5 mg/5 ml, 25 mg/10 ml

Trade Name(s):
D: Dehydrobenzperidol   GB: Droleptan (Janssen-Cilag)   USA;   Droperidol (Astra)
      (Jansscn-Cilag)   I: Leptofen (Carlo Erba)-   Inapsine (Janssen;
      Thalamonal (Janssen-   comb.   McNeil); wfm
      Cilag)-comb.   Sintodian (Carlo Erba)   Innovar (Janssen); wfm
F: Droleptan (Janssen-Cilag)   J: Droleptan (Sankyo)

Dropropizine

ATC: R05DB19
Use: antitussive

RN: 17692-31-8   MF: $C_{13}H_{20}N_2O_2$ MW: 236.32   EINECS: 241-683-7
LD_{50}:   200 mg/kg (R, i.v.); 750 mg/kg (R, p.o.)
CN: 3-(4-phenyl-1-piperazinyl)-1,2-propanediol

1-phenyl-
piperazine

glycide

Dropropizine

Reference(s):
DE 1 178 435 (H. Morren; appl. 13.3.1962; B-prior. 16.3.1961, 21.2.1962).

Formulation(s):   syrup 15 mg, 57 mg

Trade Name(s):
D: Dehydrobenzperidol   I: Elisir Terpina   Ribexen Espet. (Formenti)-
      (Janssen-Cilag)   (Schiapparelli Salute)-   comb.
      Thalamonal (Janssen-   comb.   Tiocalmina (Ottolenghi)-
      Cilag)-comb. with fentanyl   Guaiacalcium Complex   comb.
      hydrogen citrate   (Celsius)-comb.   Tussamag (Zilliken)-comb.
F: Catabex (Darcy)-comb.   Ribex (Formenti)

Drostanolone
(Dromostanolone)

ATC: L02AB
Use: antineoplastic

RN: 58-19-5   MF: $C_{20}H_{32}O_2$ MW: 304.47   EINECS: 200-367-9
CN: (2α,5α,17β)-17-hydroxy-2-methylandrostan-3-one

propionate
RN: 521-12-0   MF: $C_{23}H_{36}O_3$ MW: 360.54   EINECS: 208-303-1
LD_{50}: >2 g/kg (M, s.c.)

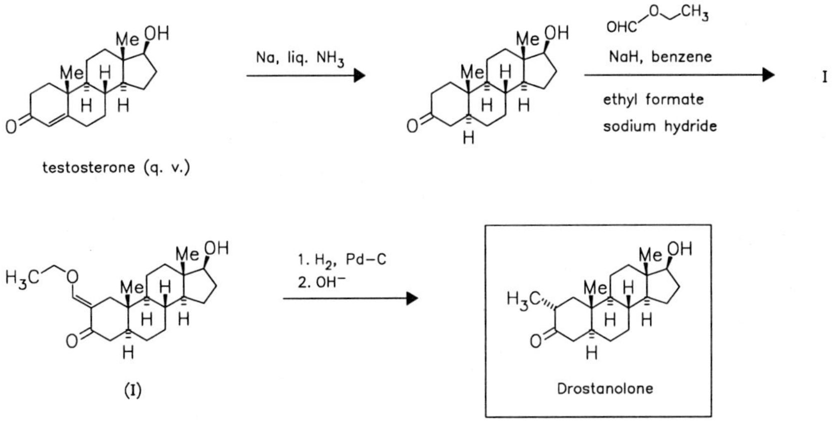

testosterone (q. v.)

(I)

Drostanolone

Reference(s):
US 2 908 693 (Syntex; 13.10.1959; MEX-prior. 17.12.1956).
Ringold, H.J. et al.: J. Am. Chem. Soc. (JACSAT) **81**, 427 (1959).

alternative syntheses:
US 3 118 915 (Syntex; 21.1.1964; MEX-prior. 24.2.1959).

Formulation(s):   amp. 100 mg (as propionate)

Trade Name(s):
D: Masterid (Grünenthal); wfm
      Masterid Spritzamp. (Grünenthal); wfm
F: Permastril (Cassenne); wfm
GB: Masteril (Syntex); wfm
I: Masteron (Recordati); wfm
J: Mastisol (Shionogi)
USA: Drolban (Lilly); wfm
      Masterone (Syntex); wfm

Droxicam

ATC: M01AC04
Use: non-steroidal anti-inflammatory, prodrug of piroxicam

RN: 90101-16-9   MF: $C_{16}H_{11}N_3O_5S$ MW: 357.35
LD_{50}: 6192 mg/kg (M, p.o.);
1434 mg/kg (Rm, p.o.); 1994 mg/kg (Rf, p.o.)
CN: 5-methyl-3-(2-pyridinyl)-2H,5H-1,3-oxazino[5,6-c][1,2]benzothiazine-2,4(3H)-dione 6,6-dioxide

3-ethoxycarbonyl-
4-hydroxy-2-methyl-
2H-1,2-benzothiazine
1,1-dioxide
(cf. piroxicam synthesis)

2-phenoxycarbonyl-
aminopyridine

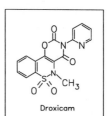

Droxicam

Reference(s):
EP 99 770 (Provesan, Esteve; appl. 8.6.1983; F-prior. 15.6.1982).
US 4 563 452 (Provesan, Esteve; 7.1.1986; appl. 8.6.1983; F-prior. 15.6.1982).

alternative synthesis:
EP 242 289 (Provesan; appl. 13.4.1987; F-prior. 15.4.1986).
EP 412 014 (Esteve; appl. 2.8.1990; F-prior. 4.8.1989).

Formulation(s):   cps. 20 mg

Trade Name(s):
I: Dobenam (Angelini)   Droxar (Upjohn)

Dyclonine

ATC: N01BX02
Use: local anesthetic (only topic)

RN: 586-60-7   MF: C$_{18}$H$_{27}$NO$_2$ MW: 289.42
CN: 1-(4-butoxyphenyl)-3-(1-piperidinyl)-1-propanone

hydrochloride
RN: 536-43-6   MF: C$_{18}$H$_{27}$NO$_2$ · HCl   MW: 325.88   EINECS: 208-633-6
LD$_{50}$:   20 mg/kg (M, i.v.);
        9500 µg/kg (dog, i.v.)

4'-butoxyacetophenone   piperidine   paraform-
                                              aldehyde

Dyclonine

Reference(s):
US 2 771 391 (Allied Laboratories; 1956; prior. 1953).
US 2 868 689 (Allied Laboratories; 1959; appl. 1956).

Formulation(s):   sol. 0.5 %, 1 %

Trade Name(s):
J: Epicain Ace (S. S. Pharm.)-   Epirocain (Eisai)   USA: Dyclone (Astra; as
      comb.   hydrochloride)

Dydrogesterone

ATC: G03DB01
Use: progestogen

RN: 152-62-5   MF: C$_{21}$H$_{28}$O$_2$ MW: 312.45   EINECS: 205-806-8
LD$_{50}$:   >7200 mg/kg (M, p.o.);
       >4600 mg/kg (R, p.o.)
CN: (9β,10α)-pregna-4,6-diene-3,20-dione

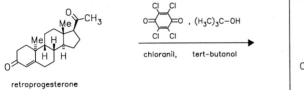

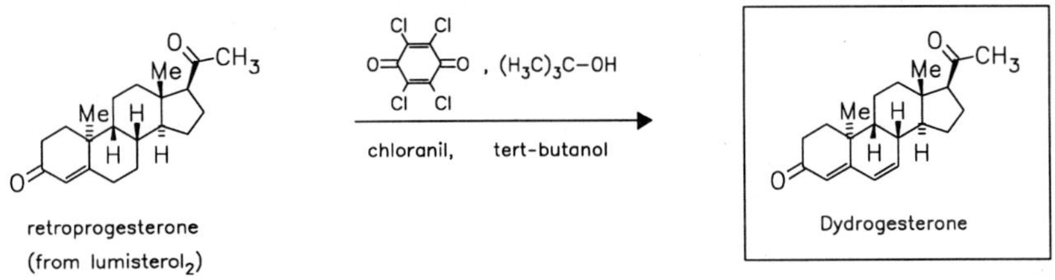

retroprogesterone
(from lumisterol$_2$)

chloranil,   tert-butanol

Dydrogesterone

Reference(s):
US 3 198 792 (North American Philips; 3.8.1965; prior. 8.4.1959, 12.6.1962).
Westerhof, P.; Reerink, E.H.: Recl. Trav. Chim. Pays-Bas (RTCPA3) **79**, 771 (1960) (also starting material).

alternative synthesis:
Rappoldt, M.P.; Westerhof, P.: Recl. Trav. Chim. Pays-Bas (RTCPA3) **80**, 43 (1961).

Formulation(s):   tabl. 10 mg

Trade Name(s):
D: Duphaston (Solvay
      Arzneimittel)
F: Duphaston (Solvay
      Pharma)
GB: Duphaston (Solvay)

Femapak 40 (Solvay)-
comb.
Femoston 1/10 (Solvay)-
comb.
I: Dufaston (UCM)

J: Duphaston (Daiichi)
USA: Duphaston (Philips
      Roxane); wfm
      Gynorest (Mead Johnson);
      wfm

Ebastine

ATC: D04AA; R06AA; R06AX22
Use: antihistaminic

RN: 90729-43-4 MF: $C_{32}H_{39}NO_2$ MW: 469.67
LD$_{50}$: 500 mg/kg (M, i.v.); >4 g/kg (M, p.o.);
>4 g/kg (R, p.o.);
>160 mg/kg (dog, p.o.)
CN: 1-[4-(1,1-dimethylethyl)phenyl]-4-[4-(diphenylmethoxy)-1-piperidinyl]-1-butanone

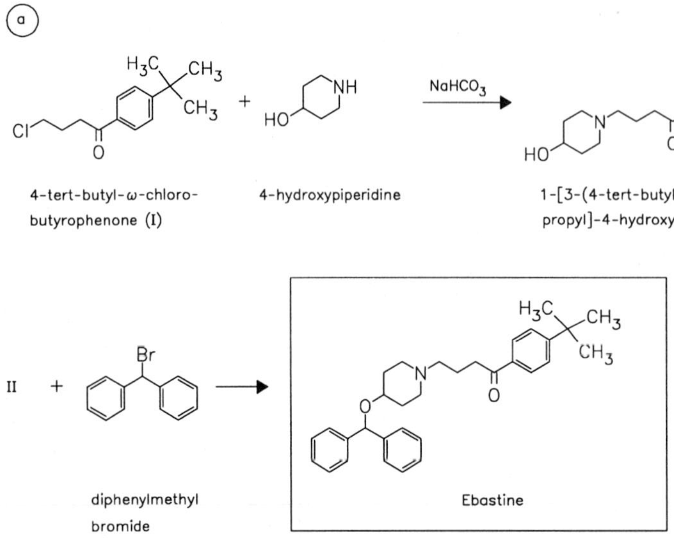

4-tert-butyl-ω-chloro-
butyrophenone (I)

4-hydroxypiperidine

1-[3-(4-tert-butylbenzoyl)-
propyl]-4-hydroxypiperidine (II)

diphenylmethyl
bromide

Ebastine

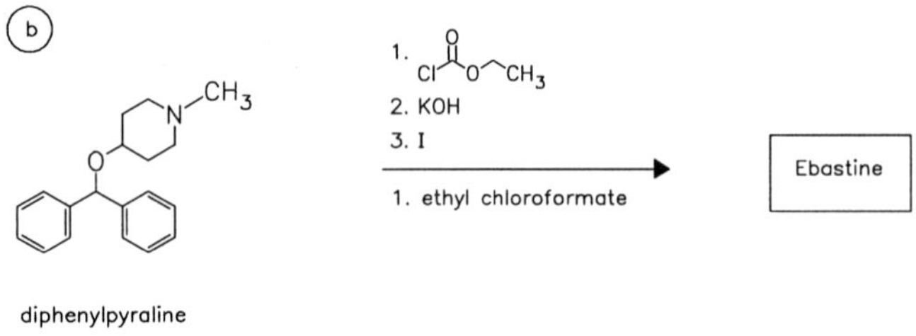

diphenylpyraline

1. ethyl chloroformate

Ebastine

Reference(s):
EP 134 124 (Fordonal; appl. 2.8.1984; GB-prior. 5.8.1983).
US 4 550 116 (Fordonal; 29.10.1985; appl. 24.7.1984; GB-prior. 5.8.1983).

Formulation(s): sol. 10 mg/10 ml; tabl. 5 mg, 10 mg

Trade Name(s):
J: Ebastel (Dainippon-Meji
Seika)

Ebrotidine

(Fl-3542)

ATC: A02B09
Use: gastric antisecretory, H_2-receptor
antagonist, gastroprotective

RN: 100981-43-9 MF: $C_{14}H_{17}BrN_6O_2S_3$ MW: 477.43
CN: [N(E)]-N-[[[2-[[[2-[(Aminoiminomethyl)amino]-4-thiazolyl]methyl]thio]ethyl]amino]-methylene]-4-bromobenzenesulfonamide

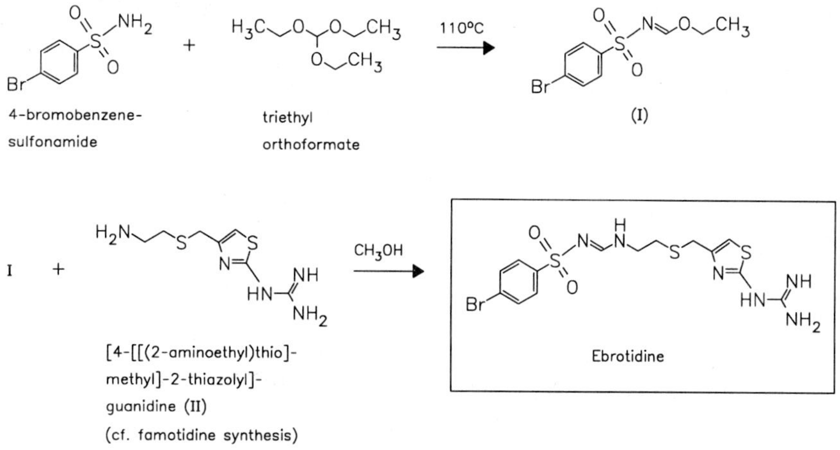

4-bromobenzene-
sulfonamide

triethyl
orthoformate

(I)

I +

[4-[[(2-aminoethyl)thio]-
methyl]-2-thiazolyl]-
guanidine (II)

(cf. famotidine synthesis)

Ebrotidine

preparation of [4-[[(2-aminoethyl)thio]methyl]-2-thiazolyl]guanidine (II):

1-amidinothiourea

1,3-dichloro-
acetone

2-aminoethanethiol

II

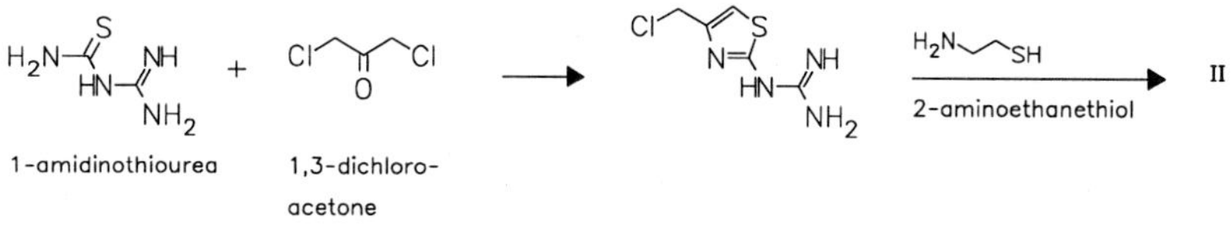

Reference(s):

EP 159 012 (Ferrer Internacional; appl. 16.4.1985; E-prior. 18.4.1984).
Anglada, L.; Marquez, M.; Sacristan, A.; Ortiz, J.A.: Eur. J. Med. Chem. (EJMCA5) **23** (1), 97 (1988).
Anglada, L.; Raga, M.; Marquez, M.; Sacristan, A.; Castello, J.M.; Ortiz, J.A.: Arzneim.-Forsch. (ARZNAD) **47** (4a), 431 (1997).

new bromobenzenesulphonamide derivatives – used as histamine receptor antagonists to inhibit acid secretion:
WO 9 614 306 (Ferrer Int.; WO-prior. 4.11.1994).

synthesis of [4-[[(2-aminoethyl)thio]methyl]-2-thiazolyl]guanidine:
DE 2 817 078 (ICI; appl. 19.4.1978; GB-prior. 20.4.1977).
Rozman, E.; Galceran, M.T.; Anglada, L.; Albet, C.: J. Pharm. Sci. (JPMSAE) **83** (2), 252 (1994).

Formulation(s):   tabl. 400 mg

Trade Name(s):
ES: Ebrocit (Ferrer; Labs.
       Robert; 1997)

Ecabet sodium

(TA-2711)

Use: ulcer therapeutic

RN: 86408-72-2   MF: $C_{20}H_{27}NaO_5S$ MW: 402.49
LD_{50}: >2 g/kg (R, p.o.)
CN: [1R-(1α,4aβ,10aα)]-1,2,3,4,4a,9,10,10a-octahydro-1,4a-dimethyl-7-(1-methylethyl)-6-sulfo-1-
       phenanthrenecarboxylic acid monosodium salt

free acid
RN: 33159-27-2   MF: $C_{20}H_{28}O_5S$ MW: 380.51

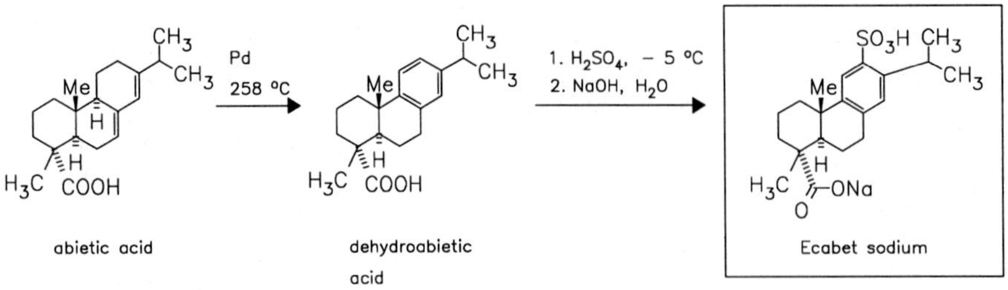

Diagram (abietic acid synthesis)

abietic acid → dehydroabietic acid → Ecabet sodium

Pd 258 °C

1. H₂SO₄, – 5 °C
2. NaOH, H₂O

Reference(s):

Fieser, L.F. et al.: J. Am. Chem. Soc. (JACSAT) **60**, 2631 (1938).
Wada, H. et al.: Chem. Pharm. Bull. (CPBTAL) **33** (4), 1472 (1985).
EP 78 152 (Tanabe Seiyaku; appl. 21.10.1982; GB-prior. 22.10.1981, 29.6.1982).

oral preparations:
JP 07 165 572 (Tanabe Seiyaku; appl. 9.12.1993; J-prior. 9.12.1993).

Formulation(s):　gran. 66.7 %

Trade Name(s):
J:　Gastrom (Tanabe Seiyaku-
　　Nippon; Boehringer Ing.)

Econazole

ATC:　D01AC03; G01AF05
Use:　fungicide, antifungal

RN:　27220-47-9　MF: $C_{18}H_{15}Cl_3N_2O$　MW: 381.69　EINECS: 248-341-6
CN:　1-[2-[(4-chlorophenyl)methoxy]-2-(2,4-dichlorophenyl)ethyl]-1*H*-imidazole

mononitrate
RN:　24169-02-6　MF: $C_{18}H_{15}Cl_3N_2O \cdot HNO_3$　MW: 444.70　EINECS: 246-053-5
LD$_{50}$:　38 mg/kg (M, i.v.); 463 mg/kg (M, p.o.);
　　50 mg/kg (R, i.v.); 668 mg/kg (R, p.o.);
　　>160 mg/kg (dog, p.o.)

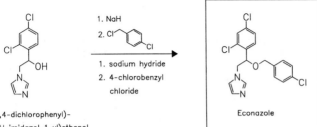

1-(2,4-dichlorophenyl)-
2-(1H-imidazol-1-yl)ethanol
(cf. miconazole synthesis)

1. NaH
2. Cl─⟨⟩─Cl

1. sodium hydride
2. 4-chlorobenzyl
　chloride

Econazole

Reference(s):

DAS 1 940 388 (Janssen; appl. 8.8.1969; USA-prior. 19.8.1968).
US 3 717 655 (Janssen; 20.2.1973; prior. 19.8.1968).
Godefroi, E.F. et al.: J. Med. Chem. (JMCMAR) **12**, 784 (1969).

Formulation(s):　cream 1 g/100 g; lotion 1 g/100 g; pastes 10 mg; powder 1 g/100 g; sol. 1 g/100 g;
　　spray 1 g/100 g (as nitrate)

D: Epi Pevaryl (Janssen-Cilag)
       Gyno-Pevaryl (Janssen-
       Cilag)
F: Dermazol (Bailleul)
       Fongéryl (L'Arguenon)
       Gyno-Pévaryl (Janssen-
       Cilag; 1976)
       Pevaryl (Janssen-Cilag;
       1976)
       Pevisone (Janssen-Cilag)-
       comb.
GB: Econacort (Bristol-Myers
       Squibb)-comb.

       Ecostatin (Bristol-Myers
       Squibb)
       Gyno Pevaryl (Janssen-
       Cilag)
       Pevaryl (Janssen-Cilag;
       1978)
I: Amicel (Salus)
       Chemionazolo (Brocchieri)
       Dermazol (CT)
       Eco Mi (Geymonat)
       Ecodergin (Von Boch)
       Ecorex (Tosi-Novara)
       Ifenec (Italfarmaco)

       Micofugal (Biopharma)
       Micogin (Crosara)
       Micos (AGIPS)
       Micosten (Bergamon)
       Pargin (Gibipharma)
       Pevaryl (Cilag; 1978)
       Pevisone (Cilag)-comb.
       Skilar (Bonomelli Farm.)
       Skilar (Italchemie)
J: Palavale (Otsuka; 1981)
USA: Spectazole (Ortho
       Dermatological; 1983)

Ecothiopate iodide
(Echothiopate iodide)

ATC: S01EB03
Use: cholinesterase inhibitor

RN: 513-10-0   MF: C9H23INO3PS   MW: 383.23   EINECS: 208-152-1
LD50: 5100 µg/kg (M, p.o.);
       174 µg/kg (R, p.o.)
CN: 2-[(diethoxyphosphinyl)thio]-N,N,N-trimethylethanaminium iodide

diethyl phospho-chloridate   2-dimethylamino-ethyl mercaptan   methyl iodide   Ecothiopate iodide

Reference(s):
US 2 911 430 (Campbell Pharmaceuticals; 3.11.1959; prior. 15.1.1958).

Formulation(s):   eye drops 1.25 mg/ml

Trade Name(s):
D: Ophtorenin (Winzer); wfm
       Pholspholinjodid
       Augentropfen (Winzer);
       wfm
F: Phospholine Iodide
       (Promedica)
GB: Phospoline Jodide (Ayerst);
       wfm
I: Phospholine Jodide
       (Chinoin); wfm
J: Phospholin Jodide
       (Tobishi)
USA: Echodide (Alcon); wfm
       Phospholine Jodide
       (Ayerst); wfm

Edetic acid
(Acide edetique; Acidum edeticum; Tetracemin)

ATC: V03AB03
Use: antidote, chelating agent

RN: 60-00-4   MF: C10H16N2O8   MW: 292.24   EINECS: 200-449-4
LD50: 28.5 mg/kg (M, i.v.); 30 mg/kg (M, p.o.)
CN: N,N'-1,2-ethanediylbis[N-(carboxymethyl)glycine]

disodium salt
RN: 139-33-3   MF: C10H14N2Na2O8   MW: 336.21   EINECS: 205-358-3

disodium salt dihydrate
RN: 6381-92-6 MF: $C_{10}H_{14}N_2Na_2O_8 \cdot 2H_2O$ MW: 372.24
calcium disodium salt
RN: 62-33-9 MF: $C_{10}H_{12}CaN_2Na_2O_8$ MW: 374.27 EINECS: 200-529-9
calcium disodium salt hydrate
RN: 23411-34-9 MF: $C_{10}H_{12}CaN_2Na_2O_8 \cdot xH_2O$ MW: unspecified
dipotassium salt
RN: 2001-94-7 MF: $C_{10}H_{14}K_2N_2O_8$ MW: 368.42 EINECS: 217-895-0
dipotassium salt monohydrate
RN: 58167-76-3 MF: $C_{10}H_{14}K_2N_2O_8 \cdot H_2O$ MW: 386.44
dipotassium salt dihydrate
RN: 25102-12-9 MF: $C_{10}H_{14}K_2N_2O_8 \cdot 2H_2O$ MW: 404.45
tetrasodium salt
RN: 64-02-8 MF: $C_{10}H_{12}N_2Na_4O_8$ MW: 380.17 EINECS: 200-573-9
LD_{50}: 330 mg/kg (M, i.p.)
trisodium salt
RN: 150-38-9 MF: $C_{10}H_{13}N_2Na_3O_8$ MW: 358.19 EINECS: 205-758-8
LD_{50}: 2150 mg/kg (M, p.o.);
2150 mg/kg (R, p.o.)
iron(III) sodium salt
RN: 15708-41-5 MF: $C_{10}H_{12}FeN_2NaO_8$ MW: 367.05 EINECS: 239-802-2
LD_{50}: 5 g/kg (M, p.o.);
5 g/kg (R, p.o.)

Reference(s):
Ullmanns Encykl. Tech. Chem., 4. Aufl., Vol. **8**, 198.
a DOS 2 150 994 (BASF; appl. 13.10.1971).
DOS 1 493 480 (BASF; appl. 30.4.1965).
DOS 2 049 223 (BASF; appl. 7.10.1970).
b DRP 694 780 (I.G. Farben; appl. 1937).

Formulation(s): inj. sol. 200 mg/ml (as calcium disodium salt)

Trade Name(s):
D: Calcium Vitis (Neopharma)
Complete all-in-one-
Lösung (Pharm-Allergan)-
comb.

Duracare (Pharm-
Allergan)-comb.
Oxysept (Pharm-Allergan)-
comb.

F: Calcitétracémate disodique
(L'Arguenon)
Chelatran (L'Arguenon)
Kélocyanor (L'Arguenon;
as cobalt salt)

Nutraflow (Alcon)-comb.   Soaclens (Alcon)-comb.
Polyclean (Alcon)-comb.   GB: Limclair (Sinclair)

Edrophonium chloride

ATC: N07A
Use: cholinergic, antidote to curare
      principles

RN: 116-38-1 MF: $C_{10}H_{16}ClNO$ MW: 201.70 EINECS: 204-138-4
LD_{50}: 8500 µg/kg (M, i.v.)
CN: N-ethyl-3-hydroxy-N,N-dimethylbenzenaminium chloride

hydroxide
RN: 473-37-0 MF: $C_{10}H_{17}NO_2$ MW: 183.25
LD_{50}: 9 mg/kg (M, i.v.); 600 mg/kg (M, p.o.)
bromide
RN: 302-83-0 MF: $C_{10}H_{16}BrNO$ MW: 246.15
LD_{50}: 9 mg/kg (M, i.v.); 600 mg/kg (M, p.o.);
        15 mg/kg (dog, i.v.)

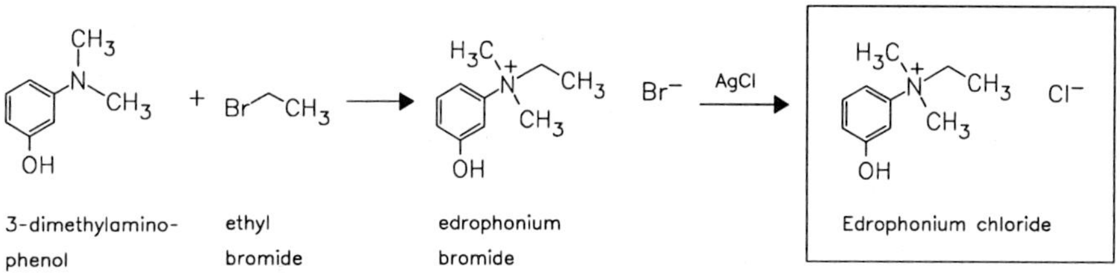

3-dimethylamino-   ethyl   edrophonium   Edrophonium chloride
phenol   bromide   bromide

Reference(s):
US 2 647 924 (Hoffmann-La Roche; 1953; prior. 1950).

Formulation(s):   amp. 10 mg/ml; vial 10 mg/10 ml

Trade Name(s):
GB: Tensilon (Roche); wfm   USA: Enlon (Ohmeda)   Tensilon (ICN)
J: Antirex (Kyorin)   Reversol (Organon)

Efavirenz
(DMP-266; L-743726)

ATC: J05AG03
Use: antiviral for AIDS, reverse
      transcriptase inhibitor

RN: 154598-52-4 MF: $C_{14}H_9ClF_3NO_2$ MW: 315.68
CN: (4S)-6-Chloro-4-(cyclopropylethynyl)-1,4-dihydro-4-(trifluoromethyl)-2H-3,1-benzoxazin-2-one

(R)-enantiomer
RN: 154801-74-8 MF: $C_{14}H_9ClF_3NO_2$ MW: 315.68
racemate
RN: 177530-93-7 MF: $C_{14}H_9ClF_3NO_2$ MW: 315.68

ⓐ

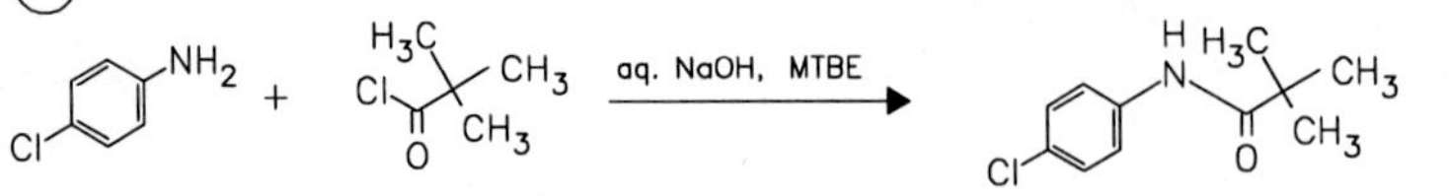

4-chloroaniline   pivaloyl
                         chloride

N-(4-chlorophenyl)-
2,2-dimethylpropanamide (I)

MTBE: methyl tert–butyl ether

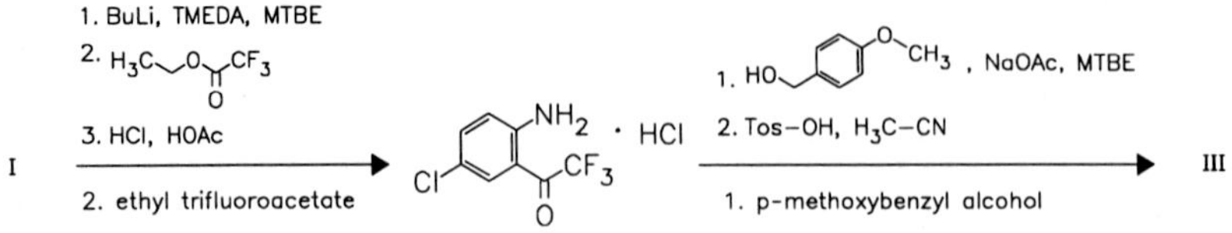

4-chloro-2-(trifluoro-
acetyl)aniline
hydrochloride (II)

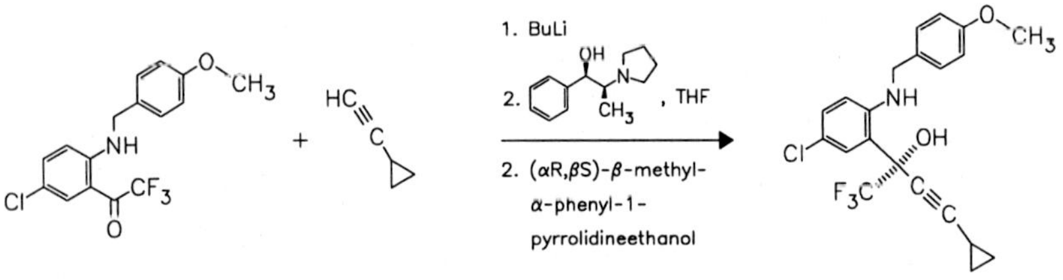

N-(4-methoxybenzyl)-   cyclopropyl-   (αS)-5-chloro-α-
4-chloro-2-(trifluoro-   acetylene (IV)   (cyclopropylethynyl)-
acetyl)aniline (III)   2-[[(4-methoxyphenyl)-
                                                    methyl]amino]-α-
                                                    (trifluoromethyl)-
                                                    benzenemethanol (V)

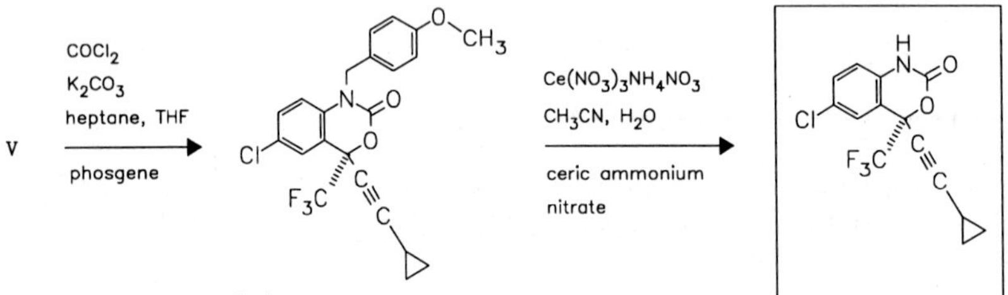

(4S)-6-chloro-4-
(cyclopropylethynyl)-
1,4-dihydro-1-[(4-meth-
oxyphenyl)methyl]-4-
trifluoromethyl-2H-
3,1-benzoxazin-2-one

Efavirenz

ⓐⓐ   preparation of cyclopropylacetylene

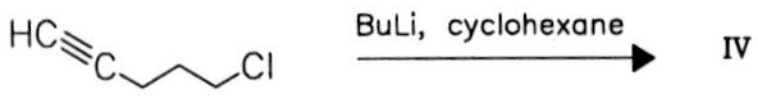

ⓑ

I + IV → BuLi, THF →

(±)-2-(2-amino-5-
chlorophenyl)-4-
cyclopropyl-
1,1,1-trifluoro-3-
butyn-2-ol (VI)

THF
carbonyl-
diimidazole
→ VII

(VII)

1. DMAP, CH₂Cl₂

2. resolution by crystallization
3. HCl
1. (S)-(−)-camphanoyl
   chloride
→ | Efavirenz |

ⓒ

1. BuLi

2.   , −60°C

II + IV →

(S)-VI

→ | Efavirenz |

Reference(s):
a Thompason, A.S. et al.: Tetrahedron Lett. (TELEAY) **36** (49), 8937-40 (1995).
   Thompason, A.S. et al.: J. Am. Chem. Soc. (JACSAT) 120, 2028-2038 (1998).
   Pierce, M.E. et al.: J. Org. Chem. (JOCEAH) **63** (23), 8536-8543 (1998).
   WO 9 637 457 (Merck + Co.; appl. 21.5.1996; USA-prior. 25.5.1995).
aa WO 9 622 955 (Merck + Co.; appl. 19.1.1996; USA-prior. 23.1.1995).
   WO 9 827 034 (Du Pont Merck; appl. 15.12.1997; USA-prior. 16.12.1996).
b EP 582 455 (Merck + Co.; appl. 3.8.1993; USA-prior. 7.8.1992, 27.4.1993).
   WO 9 520 389 (Merck + Co.; appl. 24.1.1995; USA-prior. 28.1.1994).
   WO 9 834 928 (Merck + Co.; appl. 9.2.1998; USA-prior. 12.2.1997).
   Radesca, L.A. et al.: Synth. Commun. (SYNCAV) **27** (24), 4373-4384 (1997).
   WO 9 845 278 (Du Pont; appl. 2.4.1998; USA-prior. 7.4.1997).
c Tan, L. et al.: Angew. Chem. (ANCEAD) **111** (5), 724 (1999).

process for the crystallization using an anti-solvent:
WO 9 833 782 (Merck + Co.; appl. 2.2.1998; USA-prior. 5.2.1997).

antiviral combinations:
WO 9 844 913 (Triangle Pharm.; appl. 7.4.1998; USA-prior. 7.4.1997).
WO 9 852 570 (Glaxo; appl. 14.5.1998; GB-prior. 17.5.1997).

Formulation(s):   cps. 50 mg, 100 mg, 200 mg

Trade Name(s):
D: SUSTIVA (Du Pont; 1999)   USA: Sustiva (Du Pont; 1998)

Eflornithine
(DFMO; RMI-71782)

ATC: P01CX03
Use: antineoplastic, antiprotozoal,
inhibitor of ornithine decarboxylase,
antipneumocystis

RN: 67037-37-0   MF: $C_6H_{12}F_2N_2O_2$ MW: 182.17
LD_{50}: >3000 mg/kg (M, i.p.); >5000 mg/kg (M, p.o.);
1364 µg/kg (R, intracerebral)
CN: 2-(difluoromethyl)-DL-ornithine

monohydrochloride
RN: 68278-23-9   MF: $C_6H_{12}F_2N_2O_2 \cdot HCl$ MW: 218.63   EINECS: 269-532-0
monohydrochloride monohydrate
RN: 96020-91-6   MF: $C_6H_{12}F_2N_2O_2 \cdot HCl \cdot H_2O$ MW: 236.65

N^2,N^5-dibenzylidene-ornithine methyl ester   +   chlorodifluoro-methane $\xrightarrow{\text{LiN[CH(CH}_3)_2]_2}$ lithium diiso-propylamide   Eflornithine

Reference(s):
US 4 413 141 (Merrell-Toraude; 1.11.1983; appl. 17.9.1982; prior. 11.7.1977, 2.7.1979).
US 4 330 559 (Merrell-Toraude; 18.5.1982; appl. 3.2.1981; prior. 11.7.1977, 10.4.1979).
Bey, P. et al.: J. Org. Chem. (JOCEAH) **44**, 2732 (1979).
Metcalf, B.W. et al.: J. Am. Chem. Soc. (JACSAT) **100**, 2551 (1978).

synthesis of (−)-isomer:
EP 357 029 (Merrell Dow; appl. 30.8.1989; USA-prior. 31.8.1988).

pharmaceutical composition:
BE 881 209 (Merrell-Toraude; appl. 16.5.1980; USA-prior. 10.4.1979).

combination with interferon:
US 4 499 072 (Merrell Dow; 12.2.1985; appl. 24.1.1983; prior. 29.11.1982).

Formulation(s):   vial 200 mg/ml (20 g as hydrochloride hydrate)

Trade Name(s):
USA: Ornidyl (Ilex Oncology; as hydrochloride hydrate); wfm

Ornidyl (Marion Merrell Dow; 1990); wfm

Efonidipine hydrochloride ethanol
(NZ-105)

ATC: C08CA
Use: antihypertensive, calcium channel blocker

RN: 111011-76-8   MF: $C_{34}H_{38}N_3O_7P \cdot C_2H_6O \cdot HCl$ MW: 714.20
LD$_{50}$: > 5 g/kg (R, p.o.)
CN: (±)-5-(5,5-dimethyl-1,3,2-dioxaphosphorinan-2-yl)-1,4-dihydro-2,6-dimethyl-4-(3-nitrophenyl)-3-pyridinecarboxylic acid 2-[phenyl(phenylmethyl)amino]ethyl ester *P*-oxide monohydrochloride compd. with ethanol (1:1)

efonidipine
RN: 111011-63-3   MF: $C_{34}H_{38}N_3O_7P$ MW: 631.67
hydrochloride
RN: 111011-53-1   MF: $C_{34}H_{38}N_3O_7P \cdot HCl$ MW: 668.13
(R)-base
RN: 128194-13-8   MF: $C_{34}H_{38}N_3O_7P$ MW: 631.67
(S)-base
RN: 128194-12-7   MF: $C_{34}H_{38}N_3O_7P$ MW: 631.67

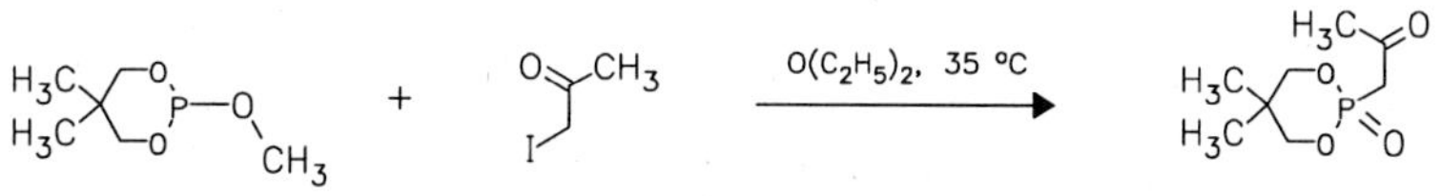

1-methoxy-4,4-dimethyl-
1-phospha-2,6-dioxa-
cyclohexane

$O(C_2H_5)_2$, 35 °C

2,2-dimethyltrimethylene
acetonylphosphonate (I)

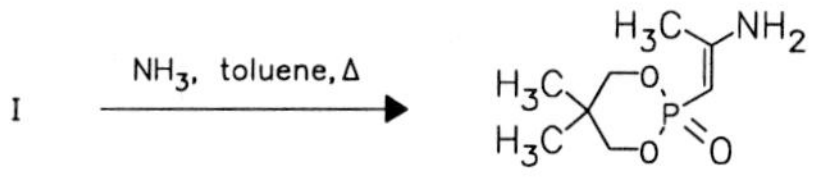

I   NH$_3$, toluene, Δ

2,2-dimethyltrimethy-
lene 2-amino-1-pro-
penylphosphonate (II)

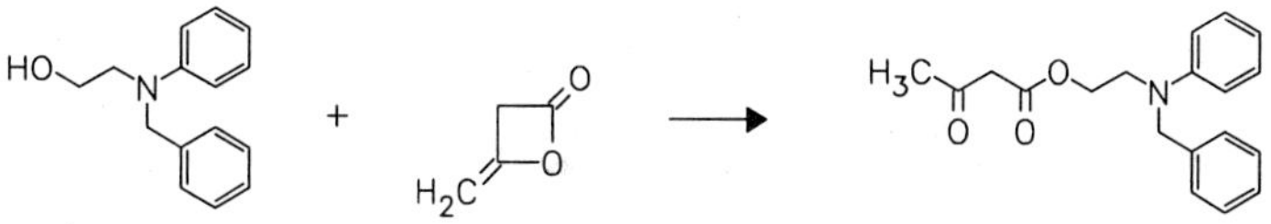

2-(benzylphenyl-
amino)ethanol

2-(benzylphenylamino)-
ethyl acetoacetate (III)

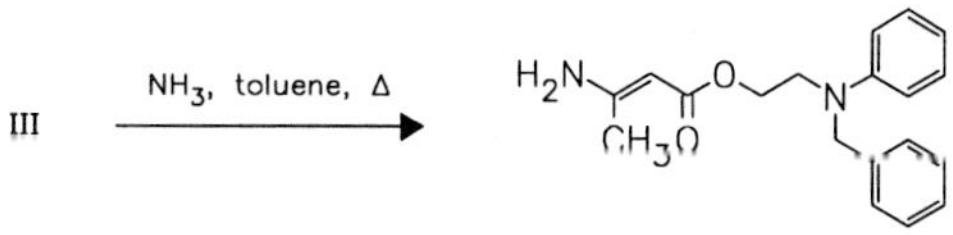

III   NH$_3$, toluene, Δ

2-(benzylphenylamino)-
ethyl 3-aminocrotonate (IV)

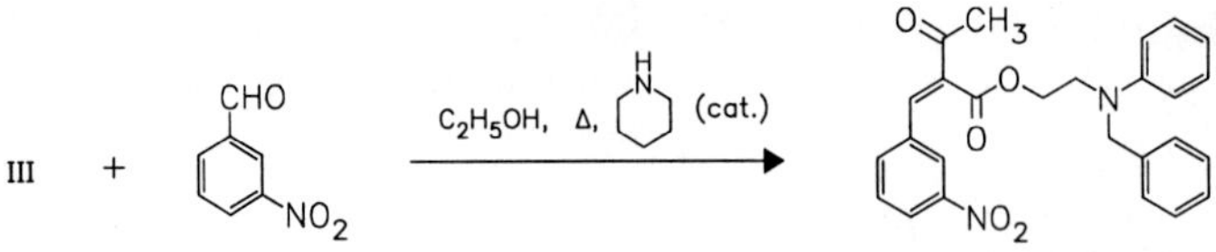

III + (V)

C₂H₅OH, Δ, piperidine (cat.) →

2-(benzylphenylamino)-
ethyl 2-(3-nitrobenzy-
lidene)acetoacetate (VI)

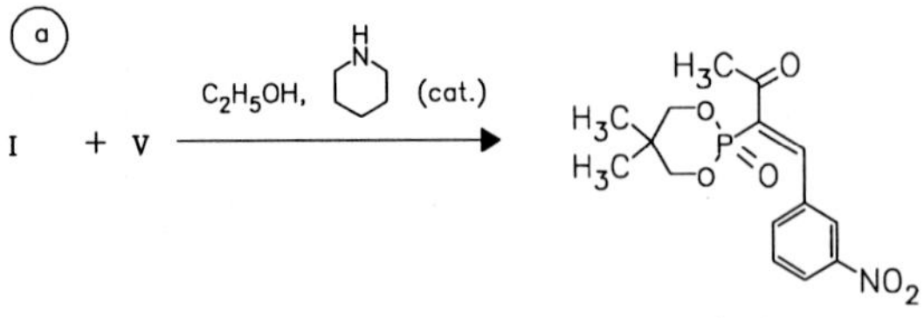

a

I + V →C₂H₅OH, piperidine (cat.)→ (VII)

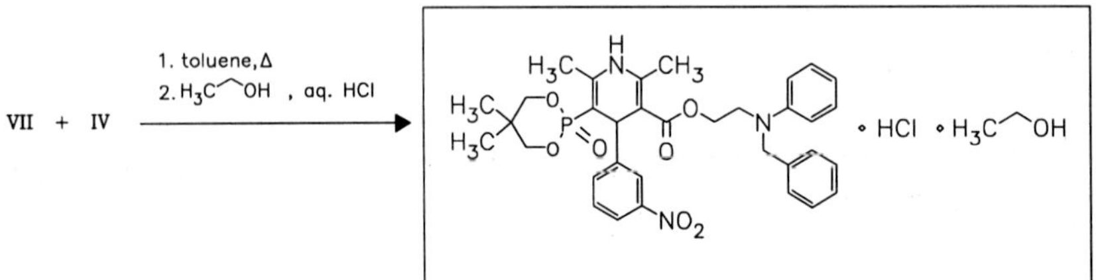

1. toluene, Δ
2. H₃C–OH, aq. HCl

VII + IV →

Efonidipine hydrochloride ethanol · HCl · H₃C–OH

b

1. toluene, Δ
2. chromatography
3. H₃C–OH, aq. HCl

I + V + IV →

Efonidipine hydrochloride ethanol

c

1. toluene, CF₃COOH, Δ
2. chromatography
3. H₃C–OH, aq. HCl

I + IV +

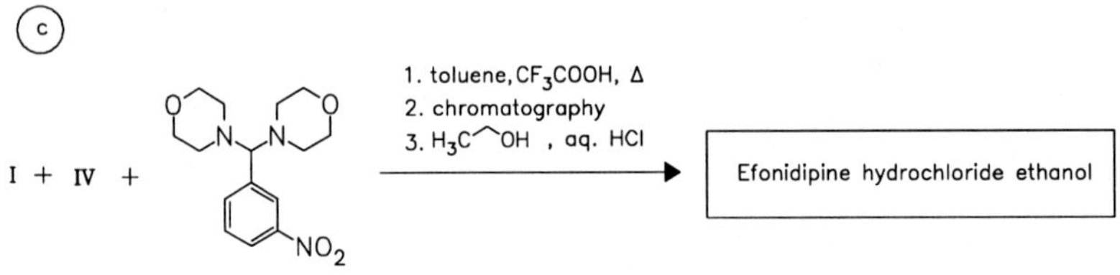

4,4'-[(3-nitro-
phenyl)methylene]-
bismorpholine

→ Efonidipine hydrochloride ethanol

d

1. toluene, Δ
2. chromatography
3. H₃C–OH, aq. HCl

II + VI →

Efonidipine hydrochloride ethanol

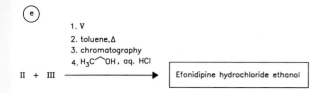

ⓔ

1. V
2. toluene,Δ
3. chromatography
4. H₃C⌒OH, aq. HCl

II + III ⟶ │ Efonidipine hydrochloride ethanol │

Reference(s):
Seto, K.; Sakoda, R.; Tanaka, S.: 10th Int. Symp. Med. Chem. (Aug. 15-19, Budapest) 1988, 301.

preparation of efonidipine hydrochloride ethanol:
WO 8 704 439 (Nissan Chemical Industries; appl. 5.8.1987; J-prior. 22.1.1986, 23.1.1986; USA-prior. 14.4.1986; J-prior. 25.11.1986).

preparation of optically active (dihydropyridyl)phosphonate esters:
JP 02 011 592 (Nissan Chemical Industries; appl. 16.1.1990; J-prior. 29.6.1988).

use of topical ophthalmic composition:
WO 9 323 082 (Alcon Laboratories; appl. 25.11.1993; USA-prior. 13.5.1992).

pharmaceutical compositions:
EP 344 603 (Zeria Pharmaceutical & Co.; Nissan Chemical Industries; appl. 6.12.1986; J-prior. 30.5.1988, 2.3.1989).

combination with immunosuppressive, cardiovascular and cerebral activity:
DE 4 430 128 (Hoechst; appl. 29.2.1996; D-prior. 25.8.1994).

Formulation(s):   tabl. 10 mg, 20 mg

Trade Name(s):
J: Landel (Nissan Chem.-
      Shionogi-Zeria)

Elliptinium acetate

ATC: L01C; L01XX
Use: antineoplastic

RN: 58337-35-2   MF: $C_{18}H_{17}N_2O \cdot C_2H_3O_2$ MW: 336.39   EINECS: 261-216-0
CN: 9-hydroxy-2,5,11-trimethyl-6*H*-pyrido[4,3-*b*]carbazolium acetate (salt)

iodide
RN: 58447-24-8   MF: $C_{18}H_{17}IN_2O$ MW: 404.25   EINECS: 261-259-5
LD$_{50}$:   5 mg/kg (M, i.p.)

9-methoxyellipticine
(extracted from
Ochrosia maculata)

pyridine
hydrochloride

9-hydroxyellipticine   (I)

I + I—CH₃ ⟶

methyl
iodide

elliptinium iodide (II)

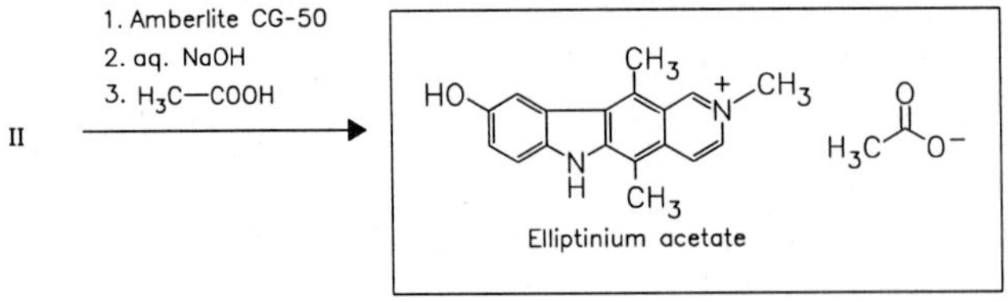

1. Amberlite CG-50
2. aq. NaOH
3. H₃C—COOH

II

Elliptinium acetate

Reference(s):
DOS 2 618 223 (Anvar; appl. 26.4.1976; F-prior. 25.4.1975).

Formulation(s):   vial (lyo.) 50 mg

Trade Name(s):
F: Celiptium (Pasteur
      Vaccins)

Emedastine

ATC: R06AE
Use: antihistaminic

RN: 87233-61-2   MF: $C_{17}H_{26}N_4O$ MW: 302.42
CN: 1-(2-ethoxyethyl)-2-(hexahydro-4-methyl-1H-1,4-diazepin-1-yl)-1H-benzimidazole

fumarate (1:2)
RN: 87233-62-3   MF: $C_{17}H_{26}N_4O \cdot 2C_4H_4O_4$ MW: 534.57
LD$_{50}$:   93 mg/kg (M, i.v.); 2206 mg/kg (M, p.o.); 609 mg/kg (M, s.c.);
        72 mg/kg (R, i.v.); 1854 mg/kg (R, p.o.); 643 mg/kg (R, s.c.);
        193 mg/kg (dog, p.o.)

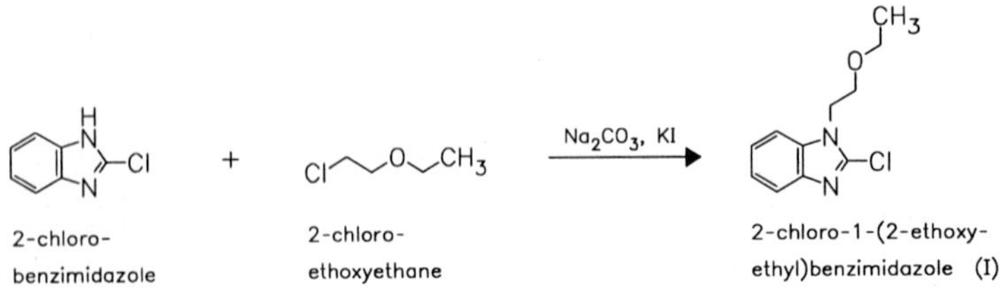

2-chloro-
benzimidazole

2-chloro-
ethoxyethane

Na_2CO_3, KI

2-chloro-1-(2-ethoxy-
ethyl)benzimidazole   (I)

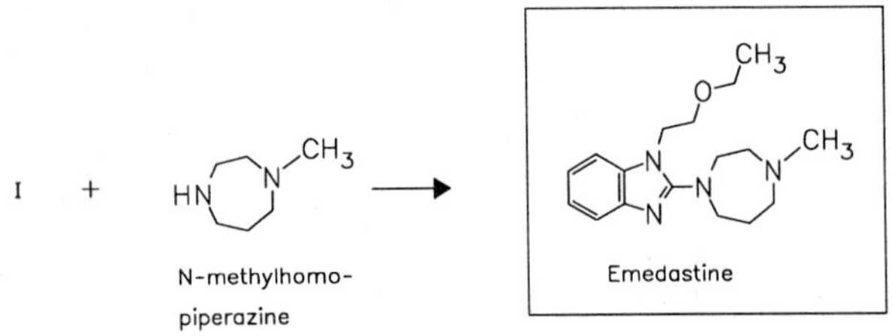

I   +   N-methylhomo-
        piperazine

Emedastine

Reference(s):
EP 79 545 (Kanebo; appl. 5.11.1982; J-prior. 6.11.1981).

percutaneous administration:
EP 440 811 (Kanebo; appl. 23.8.1990; J-prior. 28.8.1989).

Formulation(s):   cps. 1 mg, 2 mg (as difumarate)

Trade Name(s):
J: Daren (Kanebo; 1992)   Lemicut (Kowa)

Emorfazone

ATC: N02
Use: anti-inflammatory, analgesic

RN: 38957-41-4   MF: $C_{11}H_{17}N_3O_3$ MW: 239.28   EINECS: 254-220-9
LD_{50}:   700 mg/kg (M, i.p.)
CN: 4-ethoxy-2-methyl-5-(4-morpholinyl)-3-(2*H*)-pyridazinone

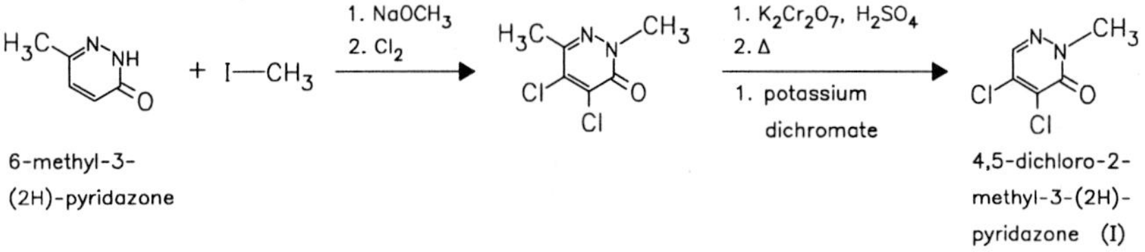

6-methyl-3-
(2H)-pyridazone

4,5-dichloro-2-
methyl-3-(2H)-
pyridazone (I)

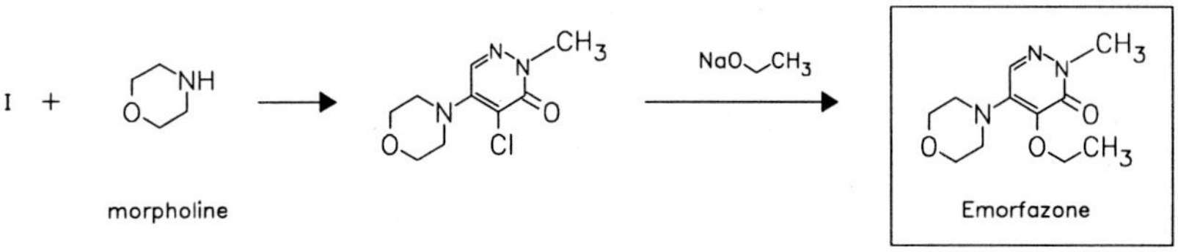

morpholine

Emorfazone

Reference(s):
DOS 2 225 218 (Morishita; appl. 24.5.1972).
GB 1 351 569 (Morishita; appl. 15.5.1972).

synthesis of 4,5-dichloro-2-methyl-3(2*H*)-pyridazone:
Homer, R.F. et al.: J. Chem. Soc. (JCSOA9) **1948**, 2191.

Formulation(s):   tabl. 100 mg, 200 mg

Trade Name(s):
J: Pentoil (Morishita; 1984)

Enalapril

ATC: C09AA02
Use: antihypertensive (ACE inhibitor)

RN: 75847-73-3   MF: $C_{20}H_{28}N_2O_5$ MW: 376.45
CN: (*S*)-1-[*N*-[1-(ethoxycarbonyl)-3-phenylpropyl]-L-alanyl]-L-proline

maleate (1:1)
RN: 76095-16-4   MF: $C_{20}H_{28}N_2O_5 \cdot C_4H_4O_4$ MW: 492.53   EINECS: 278-375-7

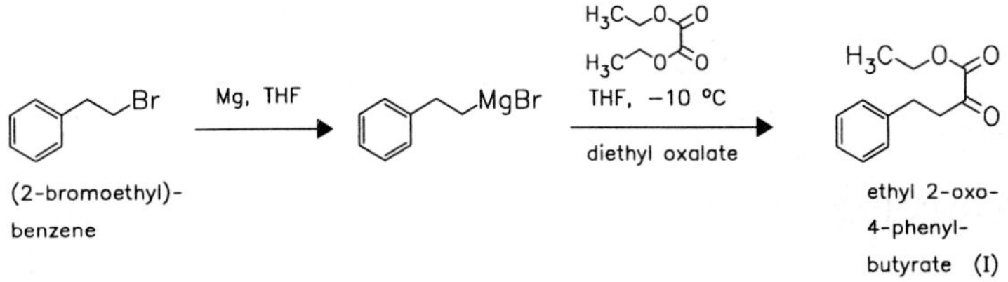

(2-bromoethyl)-
benzene

ethyl 2-oxo-
4-phenyl-
butyrate (I)

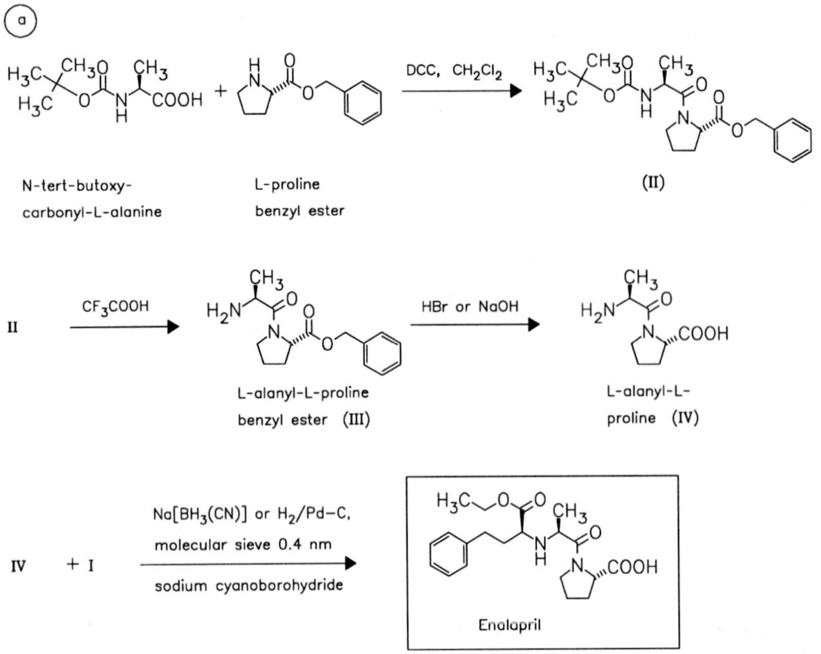

a

N-tert-butoxy-
carbonyl-L-alanine

L-proline
benzyl ester

(II)

II

CF₃COOH

L-alanyl-L-proline
benzyl ester (III)

HBr or NaOH

L-alanyl-L-
proline (IV)

IV   + I

Na[BH₃(CN)] or H₂/Pd–C,
molecular sieve 0.4 nm
sodium cyanoborohydride

Enalapril

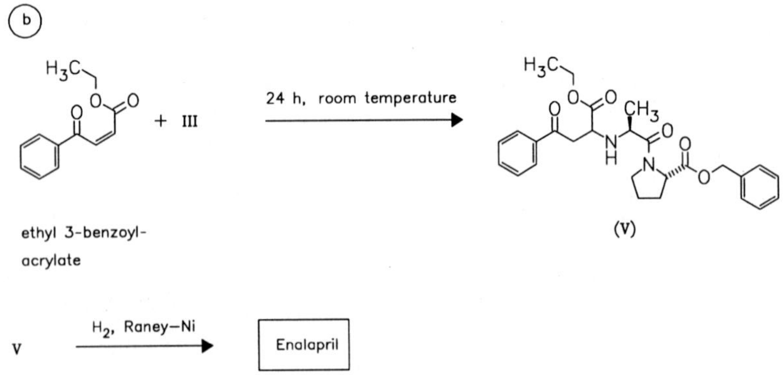

b

ethyl 3-benzoyl-
acrylate

+ III

24 h, room temperature

(V)

V

H₂, Raney–Ni

Enalapril

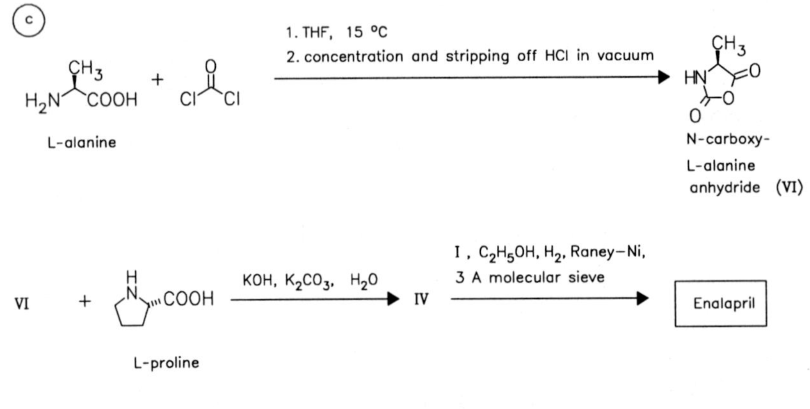

c

L-alanine

1. THF, 15 °C
2. concentration and stripping off HCl in vacuum

N-carboxy-
L-alanine
anhydride (VI)

VI   +   L-proline

KOH, K₂CO₃, H₂O

IV

I , C₂H₅OH, H₂, Raney–Ni,
3 A molecular sieve

Enalapril

Reference(s):
Patchett, A.A. et al.: Nature (London) (NATUAS) **288**, 280 (1980).

ethyl 2-oxo-4-phenylbutyrate:
Weinstock, L.M. et al.: Synth. Commun. (SYNCAV) **11**, 943 (1981).
a Wyvratt, M.J. et al.: J. Org. Chem. (JOCEAH) **49**, 2816 (1984).
  US 4 374 829 (Merck & Co.; 22.2.1983; prior. 11.12.1978).
  EP 12 401 (Merck & Co.; appl. 10.12.1979; USA-prior. 11.12.1978).
  US 4 472 380 (Merck & Co.; 18.9. 1984; prior. 11.12.1979).
  Huffmann, H.A. et al.: Tetrahedron Lett. (TELEAY) **40**, 331 (1999).
b US 4 442 030 (Merck & Co.; 10.4.1984; prior. 7.6.1982).
c Blacklock, T.J. et al.: J. Org. Chem. (JOCEAH) **53**, 836 (1988).

processes which employ reaction of activated derivatives of N-[1(S)-ethoxycarbonyl-3-phenylpropyl]-L-alanine
with L-proline:
US 4 716 235 (Kanegafuchi; 29.12.1987; J-prior. 27.8.1985).
DOS 3 542 735 (Uriach; appl. 3.12.1985; E-prior. 2.7.1985).
US 4 652 668 (Biomeasure; 24.3.1987; appl. 3.7.1985).

condensation of L-alanyl-L-proline *with* 3-phenylpropionaldehyde *and cyanides via the corresponding*
aminonitrile:
EP 79 521 (Merck & Co.; appl. 3.11.1982; USA-prior. 9.11.1981, 9.8.1982).

Formulation(s):   tabl. 2.5 mg, 5 mg, 10 mg, 20 mg (as hydrogen maleate)

Trade Name(s):

D:	Pres (Boehringer Ing.; 1984)	GB:	Innovace (Merck Sharp & Dohme; 1986)
			Naprilene (Sigma-Tau; 1985)
	Xanef (Merck Sharp & Dohme; 1984)		Innozide (Merck Sharp & Dohme)-comb.
		J:	Renivace (Banyu; 1986)
		USA:	Lexxel (Astra Merck)
F:	Co-Renitec (Merck Sharp & Dohme-Chibret)-comb.	I:	Converten (Neopharmed; 1985)
			Vaseretic (Merck; 1987)-comb. with hydrochlorothiazide
	Renitec (Merck Sharp & Dohme-Chibret; 1985)		Enapren (Merck Sharp & Dohme; 1985)
			Vasotec (Merck; 1986)

Enalaprilat
(Enalaprilic acid)

ATC: C09AA02; C09BA02
Use: angiotensin-converting enzyme inhibitor (for i.v. application as antihypertensive and in congestive heart failure, active metabolite of enalapril (q. v.))

RN: 76420-72-9   MF: $C_{18}H_{24}N_2O_5$ MW: 348.40   EINECS: 278-459-3
CN: (S)-1-[N-(1-carboxy-3-phenylpropyl)-L-alanyl]-L-proline

dihydrate
RN: 84680-54-6   MF: $C_{18}H_{24}N_2O_5 \cdot 2H_2O$ MW: 384.43

enalapril
(q. v.)

aq. NaOH

Enalaprilat

Reference(s):

Patchett, A.A. et al.: Nature (London) (NATUAS) **288**, 280 (1980).
Wyoratt, M.J. et al.: J. Org. Chem. (JOCEAH) **49**, 2816 (1984).
US 4 374 829 (Merck & Co.; 22.2.1983; USA-prior. 11.12.1978).
cf. literature cited under enalapril

Formulation(s): amp. 1.25 mg/1.25 ml

Trade Name(s):
D: Pres i.v. (Boehringer Ing.) Xanef i.v. (MSD) USA: Vasotec i.v. (Merck)

Endralazine

ATC: C02DB03
Use: antihypertensive

RN: 39715-02-1 MF: $C_{14}H_{15}N_5O$ MW: 269.31
CN: 6-benzoyl-5,6,7,8-tetrahydropyrido[4,3-*c*]pyridazin-3(2*H*)-one 3-hydrazone

monomesylate
RN: 65322-72-7 MF: $C_{14}H_{15}N_5O \cdot CH_4O_3S$ MW: 365.41
LD_{50}: 246 mg/kg (M, i.p.)

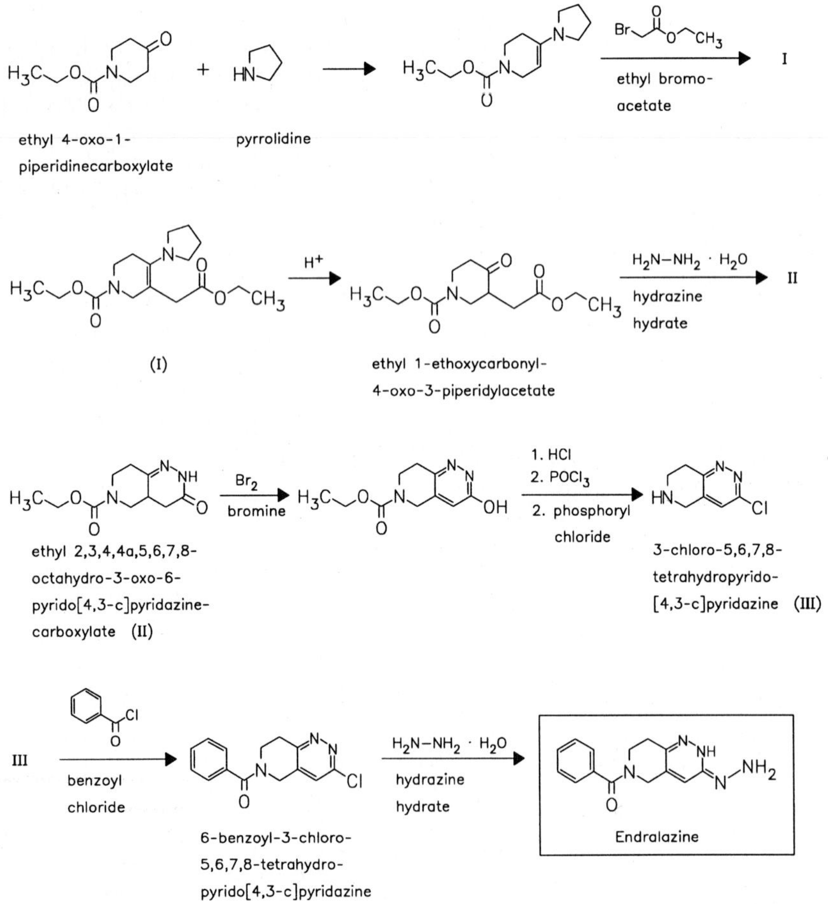

ethyl 4-oxo-1-piperidinecarboxylate
pyrrolidine
ethyl bromoacetate
I
(I)
H+
ethyl 1-ethoxycarbonyl-4-oxo-3-piperidylacetate
H2N-NH2 · H2O
hydrazine hydrate
II
Br2
bromine
ethyl 2,3,4,4a,5,6,7,8-octahydro-3-oxo-6-pyrido[4,3-c]pyridazine-carboxylate (II)
1. HCl
2. POCl3
2. phosphoryl chloride
3-chloro-5,6,7,8-tetrahydropyrido-[4,3-c]pyridazine (III)
III
benzoyl chloride
6-benzoyl-3-chloro-5,6,7,8-tetrahydro-pyrido[4,3-c]pyridazine
H2N-NH2 · H2O
hydrazine hydrate
Endralazine

Reference(s):
DOS 2 221 808 (Sandoz; appl. 4.5.1972; CH-prior. 11.5.1971, 26.5.1971, 28.5.1971, 15.10.1971).
CH 565 797 (Sandoz; appl. 16.3.1972).
Schenker, E.; Salzmann, R.: Arzneim.-Forsch. (ARZNAD) **29**, 1835 (1979).

Formulation(s):   cps. 5 mg, 10 mg (as mesylate)

Trade Name(s):
D: Miretilan (Sandoz); wfm

Enflurane

ATC: N01AB04
Use: inhalation anesthetic

RN: 13838-16-9   MF: $C_3H_2ClF_5O$ MW: 184.49   EINECS: 237-553-4
LD_{50}:   5 ml/kg (M, p.o.);
        5450 µl/kg (R, p.o.)
CN: 2-chloro-1-(difluoromethoxy)-1,1,2-trifluoroethane

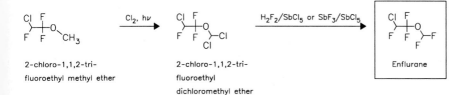

2-chloro-1,1,2-tri-
fluoroethyl methyl ether

2-chloro-1,1,2-tri-
fluoroethyl
dichloromethyl ether

Enflurane

Reference(s):
DE 1 643 591 (Air Reduction Comp.; prior. 2.10.1967).
US 3 469 011 (Air Reduction Comp.; 23.9.1969; appl. 3.10.1966).
US 3 527 813 (Air Reduction Comp.; 8.9.1970; prior. 3.10.1966, 4.9.1968).
Terrell, R.C. et al.: J. Med. Chem. (JMCMAR) **14**, 517 (1971).

Formulation(s):   liquid for inhalation 125 ml, 250 ml

Trade Name(s):
D: Enfluran-Pharmacia
      Inhalationsflüssigkeit   Ethrane (Abbott; 1976)   USA: Ethrane (Ohmeda)
      (Pharmacia & Upjohn)   GB: Ethrane (Abbott); wfm
                              J: Ethrane (Dainippon; 1981)

Enoxacin

ATC: J01MA04
Use: antibiotic (gyrase inhibitor),
        antibacterial

RN: 74011-58-8   MF: $C_{15}H_{17}FN_4O_3$ MW: 320.32
LD_{50}:   >5000 mg/kg (M, p.o.);
        >5000 mg/kg (R, p.o.)
CN: 1-ethyl-6-fluoro-1,4-dihydro-4-oxo-7-(1-piperazinyl)-1,8-naphthyridine-3-carboxylic acid

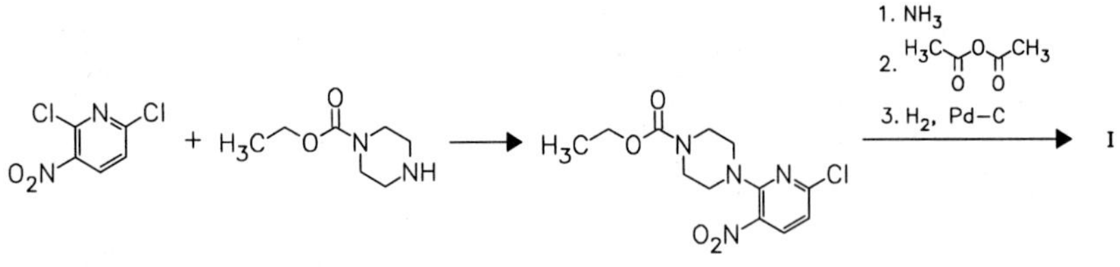

2,6-dichloro-
3-nitro-
pyridine

1-ethoxycarbonyl-
piperazine

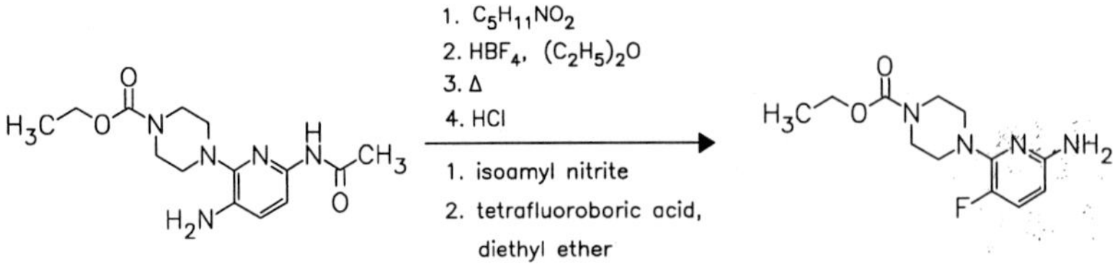

1. NH$_3$
2. H$_3$C–O–...–CH$_3$
3. H$_2$, Pd–C

I

1. C$_5$H$_{11}$NO$_2$
2. HBF$_4$, (C$_2$H$_5$)$_2$O
3. Δ
4. HCl

1. isoamyl nitrite
2. tetrafluoroboric acid,
   diethyl ether

(I)

6-amino-3-fluoro-2-
(4-ethoxycarbonyl-1-
piperazinyl)pyridine (II)

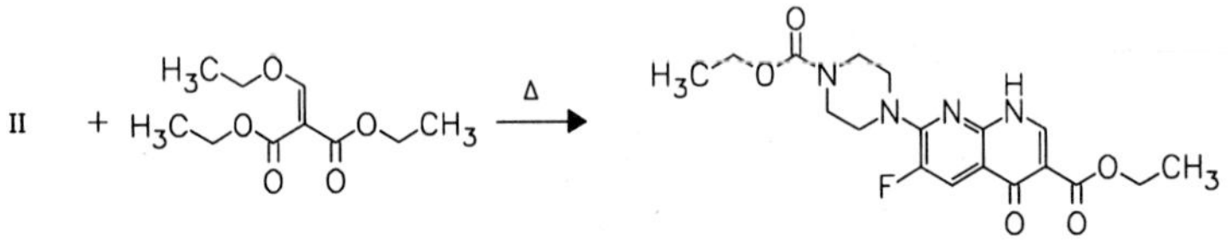

II +

diethyl ethoxy-
methylenemalonate

Δ

(III)

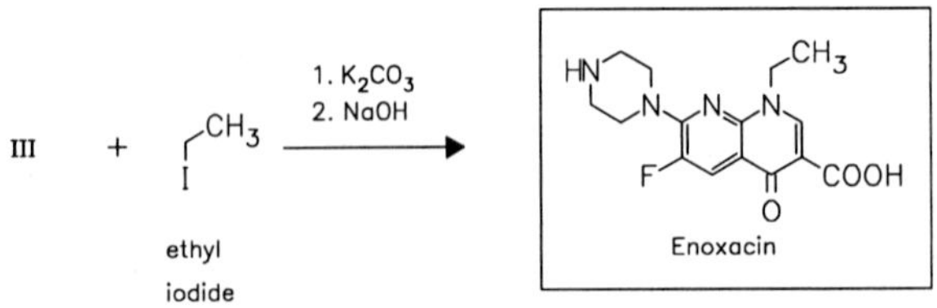

III +

ethyl
iodide

1. K$_2$CO$_3$
2. NaOH

Enoxacin

Reference(s):
EP 9 425 (Roger Bellon, Dainippon; appl. 24.8.1979; J-prior. 25.8.1978, 20.12.1978, 29.12.1978).
US 4 352 803 (Dainippon; 5.10.1982; J-prior. 25.8.1978).
US 4 359 578 (Dainippon; 5.10.1982; J-prior. 25.8.1978).

Formulation(s):   f. c. tabl. 200 mg, 300 mg, 400 mg

Trade Name(s):

D:	Enoxor (Pierre Fabre Pharma)	GB:	Comprecin (Parke Davis); wfm	J:	Flumark (Dainippon; 1986)
F:	Enoxor (Sinbio)	I:	Bactidan (Recordati)	USA:	Penetrex (Rhône-Poulenc Rorer)

Enoximone
(RMI-17043)

ATC: C01CE03
Use: cardiotonic, phosphodiesterase inhibitor

RN: 77671-31-9   MF: C$_{12}$H$_{12}$N$_2$O$_2$S   MW: 248.31
CN: 1,3-dihydro-4-methyl-5-[4-(methylthio)benzoyl]-2*H*-imidazol-2-one

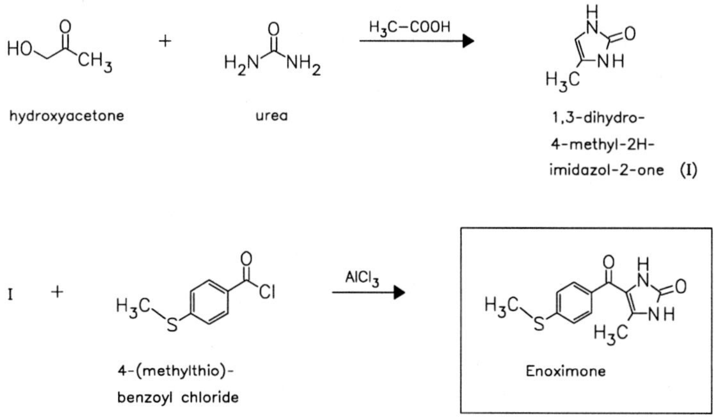

hydroxyacetone   urea

1,3-dihydro-
4-methyl-2H-
imidazol-2-one (I)

4-(methylthio)-
benzoyl chloride

Enoximone

Reference(s):
DOS 3 021 792 (Richardson-Merrell; appl. 11.6.1980; USA-prior. 18.6.1979, 7.2.1980).
GB 2 055 364 (Richardson-Merrell; appl. 18.6.1980; USA-prior. 18.6.1979, 7.2.1980).
US 4 405 635 (Richardson-Merrell; appl. 13.9.1982; prior. 18.6.1979, 7.2.1980, 13.6.1980, 18.2.1981,
30.4.1982).
EP 58 435 (Richardson-Merrell; appl. 18.2.1982; USA-prior. 18.2.1981).
Schnettler, R.A. et al.: J. Med. Chem. (JMCMAR) **25**, 1477 (1982).

synthesis of 1,3-dihydro-4-methyl-2*H*-imidazol-2-one:
WO 8 602 070 (Pfizer; appl. 26.9.1984).

Formulation(s):   amp. 100 mg/20 ml

Trade Name(s):
D: Perfan (Hoechst)   GB: Perfan (Hoechst; 1989)   I: Perfan (Lepetit)

Entacapone

ATC: N04BX02
Use: antiparkinsonian

RN: 130929-57-6   MF: $C_{14}H_{15}N_3O_5$ MW: 305.29
CN: (*E*)-2-Cyano-3-(3,4-dihydroxy-5-nitrophenyl)-*N,N*-diethyl-2-propenamide

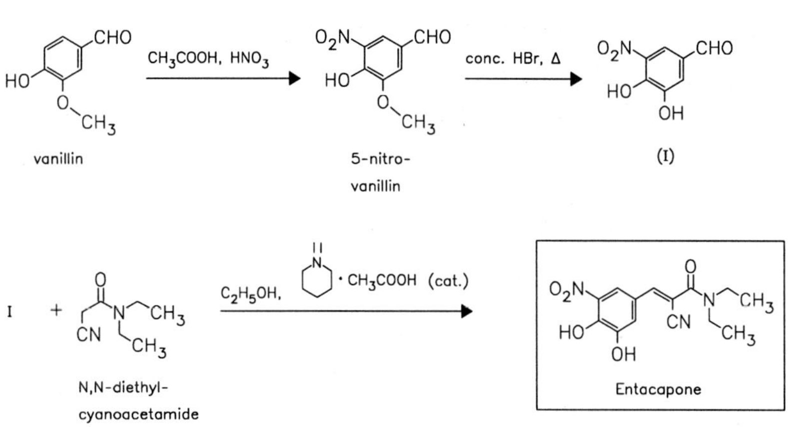

vanillin   5-nitro-
            vanillin

(I)

N,N-diethyl-
cyanoacetamide

Entacapone

Reference(s):
DE 3 740 383 (Orion Yhtymae Oy; appl. 27.11.1987; FI-prior. 28.11.1986).

preparation of 5-nitrovanillin:
Menke; Bentley: J. Am. Chem. Soc. (JACSAT) **20**, 316 (1898)

Formulation(s):   f. c. tabl. 200 mg; tabl. 200 mg

Trade Name(s):
D:	Comtess (Orion Pharma; 1998)	F:	Comtan (Novartis)	I:	Comtan (Novartis)
		GB:	Comtess (Orion)	USA:	Comtan (Orion)

Enviomycin
(Tuberactinomycin N)

ATC: J04AB
Use: antibiotic

RN: 33103-22-9   MF: $C_{25}H_{43}N_{13}O_{10}$ MW: 685.70
LD_{50}:   370 mg/kg (M, i.v.); >3 g/kg (M, p.o.);
640 mg/kg (R, i.v.); >3 g/kg (R, p.o.)
CN: (*R*)-1-(*threo*-4-hydroxy-L-3,6-diaminohexanoic acid)-6-[L-2-(2-amino-1,4,5,6-tetrahydro-4-pyrimidinyl)glycine]viomycin

sulfate (2:3)
RN: 53760-33-1   MF: $C_{25}H_{43}N_{13}O_{10} \cdot 3/2H_2SO_4$ MW: 1665.63

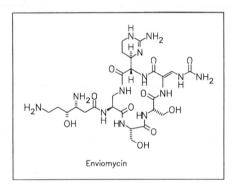

Enviomycin

From fermentation solutions of *Streptomyces griseoverticillatus var. tuberacticus* FERM P-619.

Reference(s):
DOS 2 133 181 (Toyo Jozo; appl. 30.6.1971; J-prior. 30.6.1970).
US 3 892 732 (Toyo Jozo; 1.7.1975; J-prior. 30.6.1970).
Ando, T. et al.: J. Antibiot. (JANTAJ) **24**, 680 (1971).

Formulation(s):   vial 1 g (as sulfate)

Trade Name(s):
J:	Tuberactin (Toyo Jozo)	TUM (Toyo Jozo)

Epanolol
(ICI-141292)

ATC: C07AB10
Use: β_1-adrenoceptor antagonist, antihypertensive

RN: 86880-51-5   MF: $C_{20}H_{23}N_3O_4$ MW: 369.42
CN: *N*-[2-[[3-(2-cyanophenoxy)-2-hydroxypropyl]amino]ethyl]-4-hydroxybenzeneacetamide

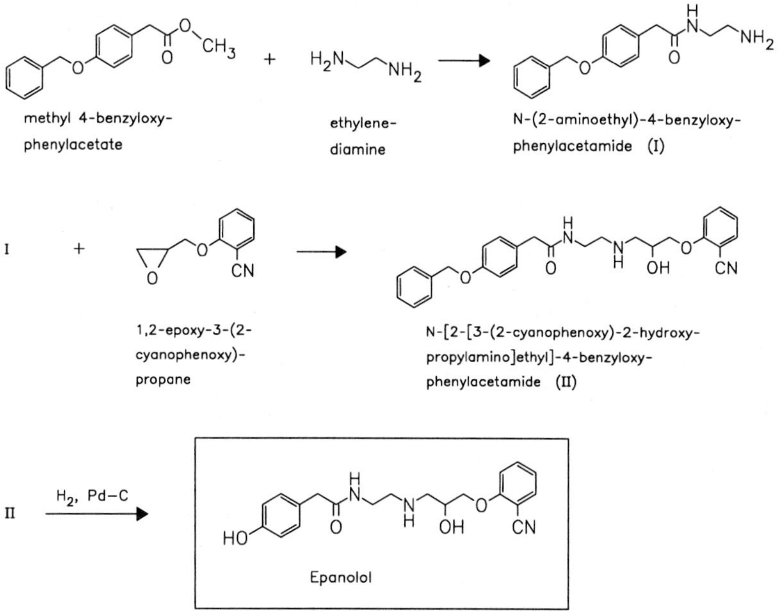

methyl 4-benzyloxy-
phenylacetate

ethylene-
diamine

N-(2-aminoethyl)-4-benzyloxy-
phenylacetamide (I)

I +

1,2-epoxy-3-(2-
cyanophenoxy)-
propane

N-[2-[3-(2-cyanophenoxy)-2-hydroxy-
propylamino]ethyl]-4-benzyloxy-
phenylacetamide (II)

II $\xrightarrow{\text{H}_2,\ \text{Pd-C}}$

Epanolol

Reference(s):
DE 2 362 568 (ICI; appl. 20.6.1974; GB-prior. 17.9.1973).
DOS 2 525 133 (ICI; appl. 5.6.1975; GB-prior. 5.6.1974).
US 4 141 987 (ICI; 27.2.1979; GB-prior. 5.6.1974).
US 4 221 807 (ICI; 9.9.1980; GB-prior. 5.6.1974).
US 4 260 632 (ICI; 7.4.1981; GB-prior. 5.6.1974).
US 4 327 113 (ICI; 27.4.1982; GB-prior. 5.6.1974).
US 4 387 099 (ICI; 7.6.1983; GB-prior. 5.6.1974).
Large, M.S.; Smith, L.H.: J. Med. Chem. (JMCMAR) **25**, 1286 (1982).

synthesis of N-(2-aminoethyl)-4-benzyloxyphenylacetamide:
DOS 2 362 568 (ICI; appl. 17.12.1973; GB-prior. 15.12.1972, 17.9.1973).

Formulation(s): tabl.

Trade Name(s):
GB: Visacor (ICI); wfm

Eperisone

ATC: A03AC
Use: skeletal muscle relaxant

RN: 64840-90-0 MF: $C_{17}H_{25}NO$ MW: 259.39
LD$_{50}$: 1024 mg/kg (M, p.o.);
1850 mg/kg (R, p.o.)
CN: 1-(4-ethylphenyl)-2-methyl-3-(1-piperidinyl)-1-propanone

hydrochloride
RN: 56839-43-1 MF: $C_{17}H_{25}NO \cdot HCl$ MW: 295.85
LD$_{50}$: 43 mg/kg (M, i.v.); 324 mg/kg (M, p.o.);
51 mg/kg (R, i.v.); 1002 mg/kg (R, p.o.);
>750 mg/kg (dog, p.o.)

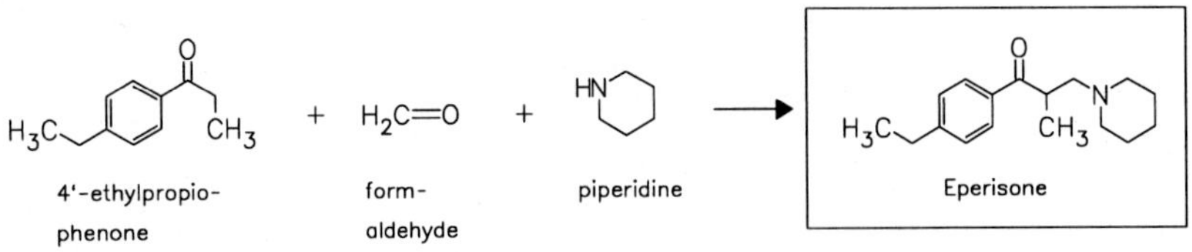

4'-ethylpropio-phenone + form-aldehyde + piperidine → Eperisone

Reference(s):
DOS 2 458 638 (Eisai; appl. 11.12.1974; J-prior. 14.12.1973).
US 4 181 803 (Eisai; 1.1.1980; J-prior. 14.12.1973).
US 39 995 047 (Eisai; 30.11.1976; J-prior. 14.12.1973).

alternative syntheses:
JP 7 930 178 (Asahi; appl. 5.8.1977).
JP 7 932 480 (Asahi; appl. 19.8.1977).
JP 7 936 274 (Asahi; appl. 24.8.1977).

Formulation(s): tabl. 50 mg (as hydrochloride)

Trade Name(s):
J: Atines (Takeda)
Dechozyl (Sawai)
Epenard (Taiyo)
Epeso (Teikoku)

Evonton (Tatsumi)
Miolease (Hotta)
Myonabase (Kotobuki)
Myonal (Eisai; 1983)

Rinpral (Nichiiko)
Sunbazon (Toyo Jozo)

L(−)-Ephedrine

ATC: R01AA03; R01AB05; R03CA02; S01FB02
Use: sympathomimetic

RN: 299-42-3 MF: $C_{10}H_{15}NO$ MW: 165.24 EINECS: 206-080-5
LD_{50}: 74 mg/kg (M, i.v.); 689 mg/kg (M, p.o.);
600 mg/kg (R, p.o.)
CN: [R-(R*,S*)]-α-[1-(methylamino)ethyl]benzenemethanol

hydrochloride
RN: 50-98-6 MF: $C_{10}H_{15}NO \cdot HCl$ MW: 201.70 EINECS: 200-074-6
LD_{50}: 95 mg/kg (M, i.v.); 400 mg/kg (M, p.o.);
69 mg/kg (R, i.v.)
sulfate (2:1)
RN: 134-72-5 MF: $C_{10}H_{15}NO \cdot 1/2H_2SO_4$ MW: 428.55 EINECS: 205-154-4
LD_{50}: 812 mg/kg (M, p.o.);
102 mg/kg (R, i.v.); 404 mg/kg (R, p.o.)

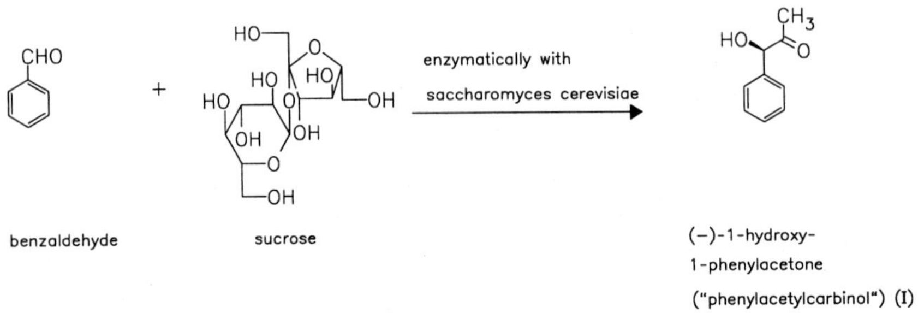

benzaldehyde + sucrose →enzymatically with saccharomyces cerevisiae→ (−)-1-hydroxy-1-phenylacetone ("phenylacetylcarbinol") (I)

Reference(s):
Budesinsky-Protiva, 24-27.
US 1 956 950 (E. Bilhuber; 1934; D-prior. 1930).
DD 51 651 (D. Gröger, H.-P. Schmauder, H. Frömmel; appl. 15.10.1965).

DL-ephedrine by hydrogenation of N-methylaminopropiophenone:
DRP 469 782 (E. Merck; appl. 1926).

Formulation(s):   amp. 10 mg, 25 mg, 50 mg (as hydrochloride); drg. 2.5 mg, 10 mg; sol. 100 mg/10 ml;
                   syrup 100 mg/100 ml, 1g/1000 ml (as hydrochloride); syrup 26.7 mg/100 ml (as sulfate);
                   tabl. 10 mg, 25 mg, 50 mg

Trade Name(s):
D: Antiföhnon (Südmedica)-   Vencipon (Artesan)-comb.   J: Ephedrine "Nagai"
        comb.   Wick MediNait (Wick   (Dainippon)
        Asthma 6-N (Hobein)-   Pharma)-comb.   numerous generic and
        comb.   generic and numerous   combination preparations
        Ephepect (Bolder)-comb.   combination preparations   USA: Broncholate (Sanofi; as
        Ephetonin (Merck); wfm   F: Ephedroides "3" (Silbert et   hydrochloride)-comb.
        Equisil (Klein)-comb.   Ripert); wfm   Kie (Laser; as
        Felsol (Roland)-comb.   generic and numerous   hydrochloride)-comb.
        Fomagrippin (Michallik)-   combination preparations   Marax (Pfizer; as sulfate)
        comb.   GB: CAM (Shire)   Pretz-D (Parnell; as
        Hevertopect (Hevert)-   numerous combination   sulfate)-comb.
        comb.   preparations   Quadrinal (Knoll Labs.; as
        Medigel (Medice)-comb.   I: Codeinol (Saba)-comb.   hydrochloride)-comb.
        Perdiphen (Schwabe/   Deltatarinolo (Lepetit)-   Rynatuss (Wallace; as
        Spitzner)-comb.   comb.   tannate)-comb.
        Pulmocordio (Hevert)-   Paidorinovit (SIT)-comb.   numerous combination
        comb.   Rinopumilene   preparations
        Rhinoguttae (Leyh)-comb.   (Montefarmaco)
        Stipo Nasenspray (Repha)-   Rinovit (SIT)-comb.
        comb.   combination preparations

Epicillin

ATC: J01CA07
Use: antibiotic

RN: 26774-90-3   MF: $C_{16}H_{21}N_3O_4S$ MW: 351.43   EINECS: 248-001-7
LD$_{50}$: 3870 mg/kg (M, i.p.)
CN: [2S-[2α,5α,6β(S*)]]-6-[(amino-1,4-cyclohexadien-1-ylacetyl)amino]-3,3-dimethyl-7-oxo-4-thia-1-
        azabicyclo[3.2.0]heptane-2-carboxylic acid

monosodium salt
RN: 34735-40-5   MF: $C_{16}H_{20}N_3NaO_4S$ MW: 373.41

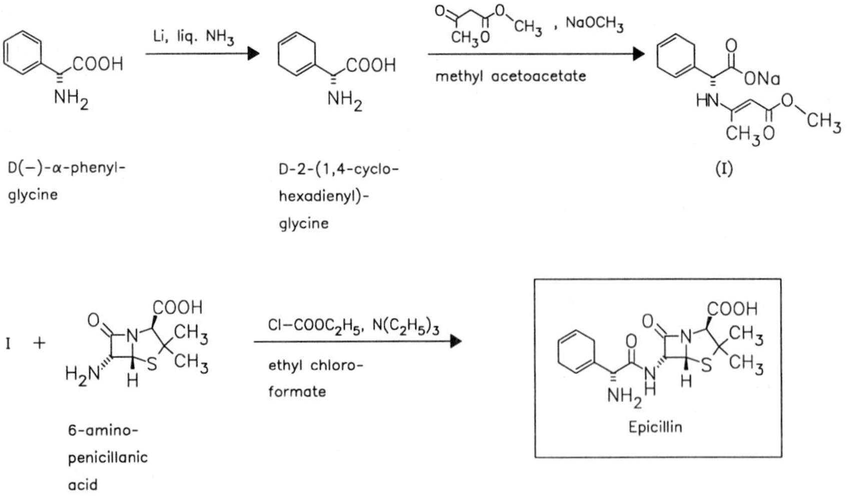

D(−)-α-phenyl-
glycine

D-2-(1,4-cyclo-
hexadienyl)-
glycine

(I)

6-amino-
penicillanic
acid

Epicillin

Reference(s):
US 3 485 819 (Squibb; 23.12.1969; USA-prior. 2.7.1968).
DAS 1 967 020 (Squibb; appl. 23.6.1969; USA-prior. 2.7.1968).

microbiological acylation by means of Aphanocladium aranearum (ATCC 20453).
US 4 073 687 (Shionogi; 14.2.1978; J-prior. 12.5.1976).

Formulation(s):   f. c. drg. 1000 mg; vial 2125.4 mg, 5313.5 mg (as sodium salt)

Trade Name(s):
D: Spectacillin (Sandoz); wfm   F: Dexacilline (Squibb); wfm   I: Dexacillin (Squibb); wfm

Epimestrol

ATC: G03GB03
Use: estrogen (ovulation stimulant),
        anterior, pituitary activator

RN: 7004-98-0   MF: $C_{19}H_{26}O_3$ MW: 302.41   EINECS: 230-278-0
CN: (16α,17α)-3-methoxyestra-1,3,5(10)-triene-16,17-diol

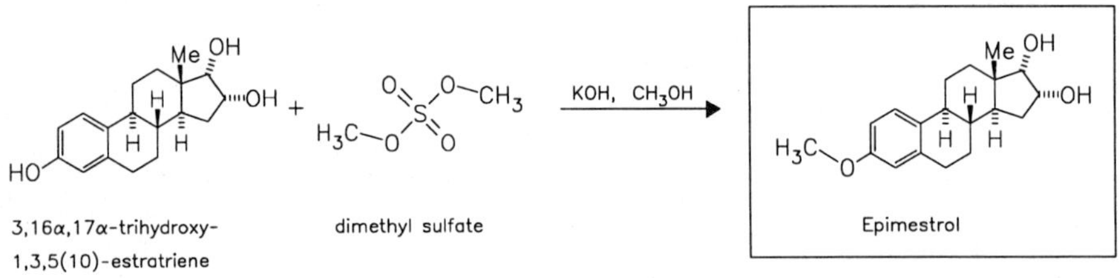

3,16α,17α-trihydroxy-
1,3,5(10)-estratriene

dimethyl sulfate

Epimestrol

Reference(s):
NL 95 257 (Organon; appl. 1958).

starting material and alternative synthesis:
US 2 584 271 (Searle; 1952; prior. 1948).
Prelog, V. et al.: Helv. Chim. Acta (HCACAV) **28**, 250 (1945).

alternative synthesis:
Caglioti, L.; Magi, M.: Tetrahedron (TETRAB) **19**, 1127 (1963).

Formulation(s):   tabl. 5 mg

Trade Name(s):
D: Stimovul (Organon); wfm   I: Stimovul (Organon Italia)   J: Stimovul (Ravasini)

Epinastine hydrochloride
(WAL-801CL)

ATC: R06AX24
Use: antihistaminic

RN: 108929-04-0   MF: $C_{16}H_{15}N_3 \cdot HCl$ MW: 285.78
LD$_{50}$:   17 mg/kg (R, i.v.); 192 mg/kg (R, p.o.)
CN: 9,13b-dihydro-1*H*-dibenz[*c,f*]imidazo[1,5-*a*]azepin-3-amine monohydrochloride

epinastine
RN: 80012-43-7   MF: $C_{16}H_{15}N_3$ MW: 249.32

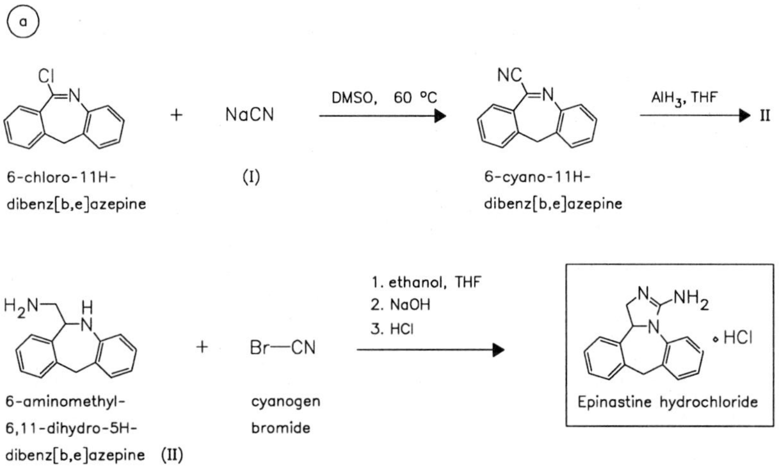

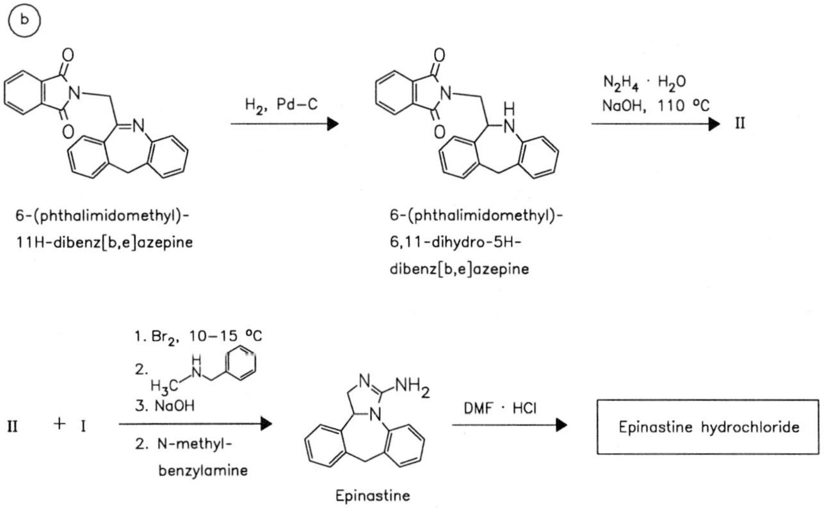

6-chloro-11H-
dibenz[b,e]azepine   (I)   6-cyano-11H-
dibenz[b,e]azepine

6-aminomethyl-
6,11-dihydro-5H-
dibenz[b,e]azepine   (II)   cyanogen
bromide   Epinastine hydrochloride

6-(phthalimidomethyl)-
11H-dibenz[b,e]azepine   6-(phthalimidomethyl)-
6,11-dihydro-5H-
dibenz[b,e]azepine

II   +   I   Epinastine   Epinastine hydrochloride

Reference(s):
a   DE 3 008 944 (Boehringer Ing.; appl. 5.3.1981; D-prior. 8.3.1980).
    starting material:
    Hunziker, E. et al.: Helv. Chim. Acta (HCACAV) **49**/II, 1433 (1966); **50**/I, 245 (1967).
b   EP 496 306 (Boehringer Ing.; appl. 18.1.1992; D-prior. 25.1.1991).

composition with PAF-antagonists:
WO 8 910 143 (Schering Corp.; appl. 24.4.1989; USA-prior. 27.4.1988).

Formulation(s):   tabl. 10 mg, 20 mg

Trade Name(s):
J: Alesion (Nippon
        Boehringer Ing./Sakyo)

Epinephrine
(Adrenaline)

ATC: A01AD01; B02BC09; C01CA24;
        R03AA01; S01EA01
Use: sympathomimetic, vasoconstrictor

RN: 51-43-4   MF: $C_9H_{13}NO_3$ MW: 183.21   EINECS: 200-098-7
LD_{50}:   217 µg/kg (M, i.v.);
        150 µg/kg (R, i.v.);
        100 µg/kg (dog, i.v.)
CN: (*R*)-4-[1-hydroxy-2-(methylamino)ethyl]-1,2-benzenediol

hydrochloride
RN: 55 31 2   MF: $C_9H_{13}NO_3 \cdot HCl$ MW: 219.67   EINECS: 200-230-3
LD_{50}:   140 µg/kg (M, i.v.);
        24 mg/kg (R, p.o.)
tartrate (1:1)
RN: 51-42-3   MF: $C_9H_{13}NO_3 \cdot C_4H_6O_6$ MW: 333.29   EINECS: 200-097-1
LD_{50}:   1780 µg/kg (M, i.v.); 4 mg/kg (M, p.o.);
        82 µg/kg (R, i.v.)

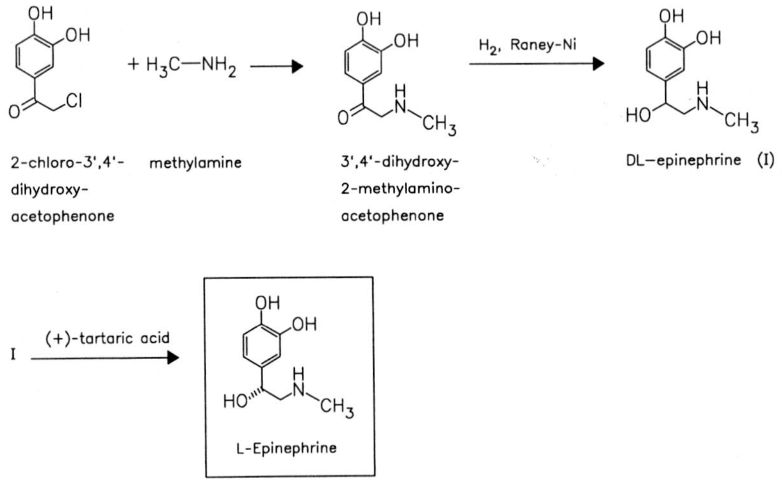

2-chloro-3',4'-   methylamine
dihydroxy-
acetophenone

3',4'-dihydroxy-
2-methylamino-
acetophenone

DL-epinephrine   (I)

L-Epinephrine

Reference(s):
DRP 152 814 (Hoechst; 1903).
DRP 157 300 (Hoechst; 1903).
DRP 222 451 (Hoechst; 1908).
Tullar, B.F.: J. Am. Chem. Soc. (JACSAT) **70**, 2067 (1948).

Formulation(s):   amp. 0.05 mg/10 ml, 1 mg/ml, 2.05 mg/2.05 ml (as hydrochloride); eye drops 2 mg/ml, 5 mg/ml; eye ointment 1 mg/g (as tartrate)

Trade Name(s):

D:	Adrenalin 1:1000 JENAPHARMA (Jenapharm)		Dyspné-Inhal (Augot) Eppy 1 % (Allergan France)
	Adrenalin Medihaler (Kettelhack-Riker)		Glaucadrine (Merck Sharp & Dohme-Chibret)-comb. numerous combination preparations and generics

D: Adrenalin 1:1000
        JENAPHARMA
        (Jenapharm)
        Adrenalin Medihaler
        (Kettelhack-Riker)
        Anaphylaxie-Besteck
        Lösung Z.J. (SmithKline
        Beecham)
        Fastjekt (Allergopharma)
        Suprarenin (Hoechst)
        numerous combination
        preparations
F: Anahelp (Stallergènes)
        Anakit (Dome-Hollister-
        Stier)

GB: Accusite (Matrix)-comb.
        Epipen (ALK)
        Eppy (Chauvin)
        Simplene (Chauvin)
        combination preparations
I: Adrenal (Lifepharma)
        Adrenalina Ism (Nuovo
        ISM)

Dyspné-Inhal (Augot)
Eppy 1 % (Allergan
France)
Glaucadrine (Merck Sharp
& Dohme-Chibret)-comb.
numerous combination
preparations and generics

J: Vaponefrin (Tokyo M.I.)
USA: Epi E-Z Pen (Dey)
        EpiPen (Dey)
        Sensorcaine with
        Epinephrine (Astra)-comb.
        Sus-Phrine (Forest)
        Xylocaine with
        Epinephrine (Astra)-comb.

Eppy (Merck Sharp &
Dohme)
Rinantipiol (Antipiol)-
comb.
Xylocaina (Astra-Simes)
Xylonor (Ogna)-comb.

Epirizole
(Mepirizole)

ATC: M01A; N02B; S01B
Use: analgesic, anti-inflammatory

RN: 18694-40-1   MF: $C_{11}H_{14}N_4O_2$ MW: 234.26   EINECS: 242-507-1
LD_{50}: 550 mg/kg (M, i.v.); 740 mg/kg (M, p.o.);
        214 mg/kg (R, i.v.); 445 mg/kg (R, p.o.)
CN: 4-methoxy-2-(5-methoxy-3-methyl-1H-pyrazol-1-yl)-6-methylpyrimidine

6-methyl-
uracil

2,4-dichloro-
6-methyl-
pyrimidine

2-chloro-4-
methoxy-6-
methylpyrimidine

2-hydrazino-4-
methoxy-2-me-
thylpyrimidine (I)

diketene

Epirizole

Reference(s):
FR-M 6 793 (Daiichi Seiyaku; appl. 31.8.1967).
DAS 2 237 632 (Daiichi Seiyaku; appl. 31.7.1972; J-prior. 31.7.1971, 5.8.1971).

intermediates:
Vanderhaeghe, H.; Claesen, M.: Bull. Soc. Chim. Belg. (BSCBAG) **68**, 30 (1959).

Formulation(s):   tabl. 50 mg, 100 mg

Trade Name(s):
I: Diacon (IBI); wfm   J: Analock (Taito Pfizer)   Mebron (Daiichi)

Epirubicin
(Pidorubicin; 4'-*epi*-Adriamycin)

ATC: L01DB03
Use: antineoplastic

RN: 56420-45-2   MF: $C_{27}H_{29}NO_{11}$ MW: 543.53
LD_{50}:   16.07 mg/kg (M, i.v.);
        14.27 mg/kg (R, i.v.);
        2 mg/kg (dog, i.v.)
CN: (8*S-cis*)-10-[(3-amino-2,3,6-trideoxy-α-L-*arabino*-hexopyranosyl)oxy]-7,8,9,10-tetrahydro-6,8,11-
        trihydroxy-8-(hydroxyacetyl)-1-methoxy-5,12-naphthacenedione

hydrochloride
RN: 56390-09-1   MF: $C_{27}H_{29}NO_{11} \cdot HCl$ MW: 579.99

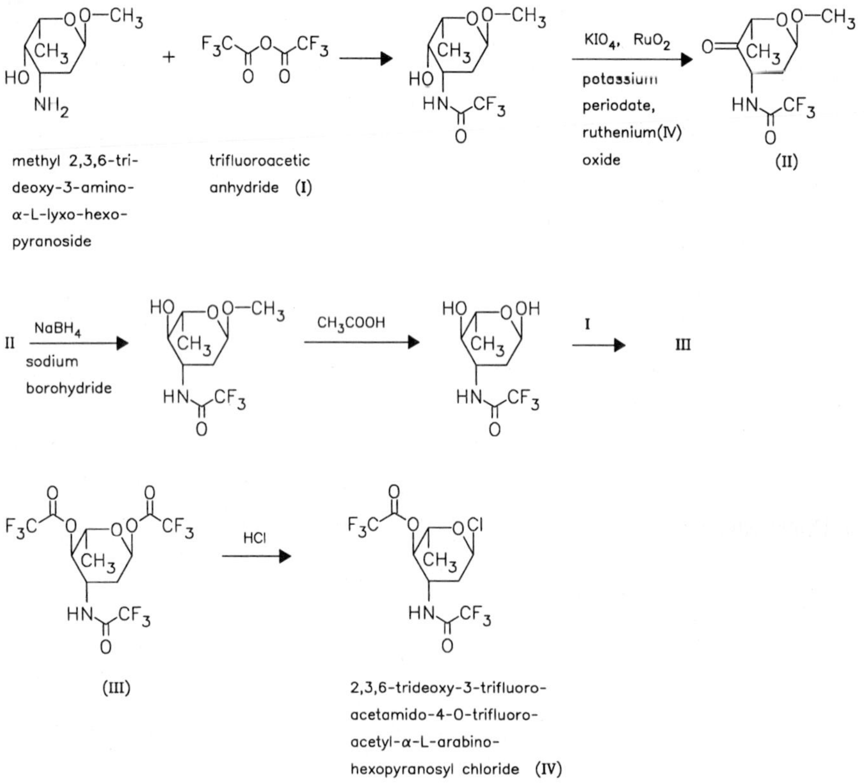

methyl 2,3,6-tri-
deoxy-3-amino-
α-L-lyxo-hexo-
pyranoside

trifluoroacetic
anhydride   (**I**)

KIO$_4$, RuO$_2$
potassium
periodate,
ruthenium(IV)
oxide

(**II**)

II   NaBH$_4$
      sodium
      borohydride

CH$_3$COOH

I

III

(**III**)

HCl

2,3,6-trideoxy-3-trifluoro-
acetamido-4-O-trifluoro-
acetyl-α-L-arabino-
hexopyranosyl chloride (**IV**)

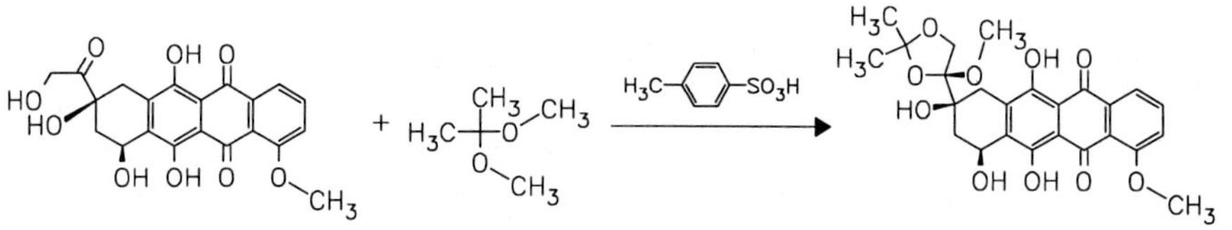

adriamycinone
2,2-dimethoxy-propane
(V)

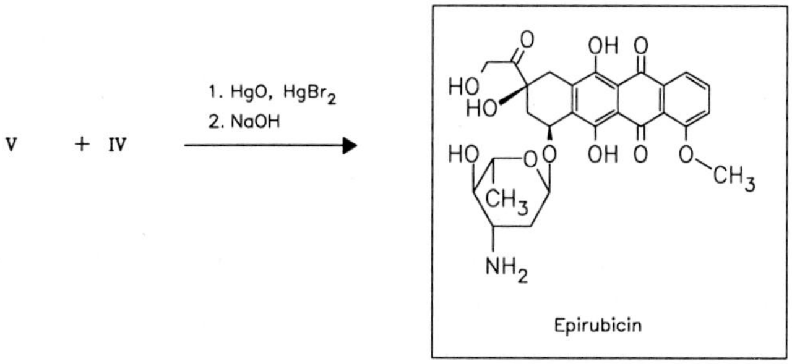

1. HgO, HgBr2
2. NaOH
V + IV
Epirubicin

Reference(s):

Arcamone, F. et al.: J. Med. Chem. (JMCMAR) **18**, 703 (1975).
DOS 2 510 866 (Farmitalia; appl. 20.3.1975; GB-prior. 22.3.1974).
US 4 058 519 (Farmitalia; 15.11.1977; appl. 19.3.1975; GB-prior. 22.3.1974).

alternative synthesis:
DOS 2 618 822 (Farmitalia; appl. 29.4.1976; GB-prior. 30.4.1975).

purification:
GB 2 133 005 (Farmitalia; appl. 16.12.1983).

Formulation(s): vial 10 mg, 20 mg, 50 mg (as hydrochloride)

Trade Name(s):

D:	Farmorubicin (Pharmacia & Upjohn; 1984)	GB:	Pharmarubicin (Pharmacia & Upjohn; 1985)	J:	Farmorubicin (Farmitalia)
F:	Farmorubicine (Pharmacia & Upjohn; 1986)	I:	Farmorubicina (Farmitalia; 1984)		

Epitiostanol

(Epithioandrostanol; Epithiostanol)

ATC: G03B; L02BA
Use: antiestrogen, antineoplastic

RN: 2363-58-8 MF: $C_{19}H_{30}OS$ MW: 306.51
LD_{50}: 1160 mg/kg (M, i.p.)
CN: (2α,3α,5α,17β)-2,3-epithioandrostan-17-ol

+ HSCN
ether
methanesulfochloride, pyridine
I

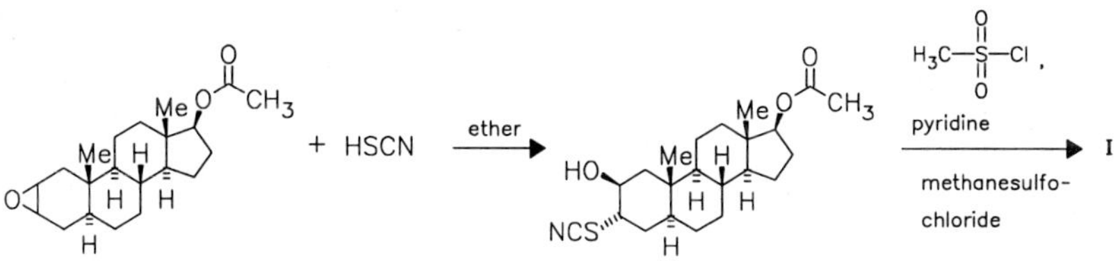

17β-acetoxy-2,3-epoxy-5α-androstane
thiocyanic acid

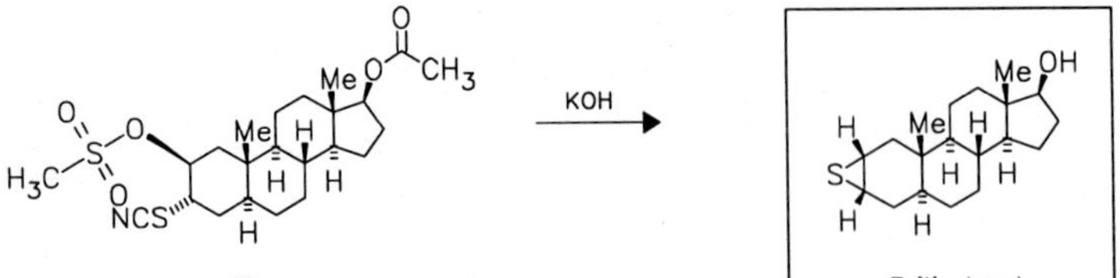

(I)

Epitiostanol

Reference(s):
GB 977 599 (Shionogi; valid from 19.12.1962; J-prior. 19.12.1961).
NL 6 400 226 (Shionogi; appl. 15.1.1964).
Takeda, K. et al.: Tetrahedron (TETRAB) **21**, 329 (1965).

Formulation(s):   vial 10 mg

Trade Name(s):
J: Thiodrol (Shionogi)

Epitizide
(Epithiazide)

ATC: C02L
Use: antihypertensive, diuretic

RN: 1764-85-8   MF: $C_{10}H_{11}ClF_3N_3O_4S_3$ MW: 425.86   EINECS: 217-181-9
CN: 6-chloro-3,4-dihydro-3-[[(2,2,2-trifluoroethyl)thio]methyl]-2H-1,2,4-benzothiadiazine-7-sulfonamide
1,1-dioxide

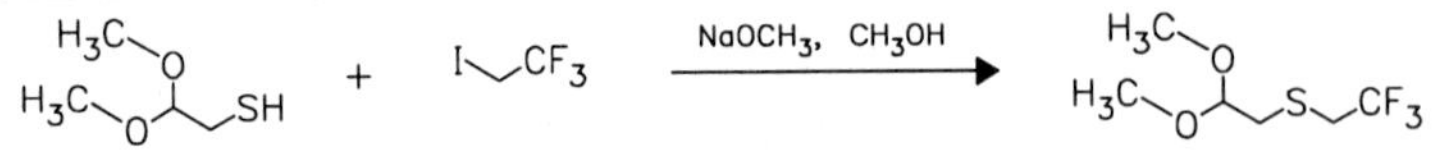

mercaptoacet-
aldehyde
dimethyl
acetal

2,2,2-tri-
fluoroethyl
iodide

(2,2,2-trifluoroethyl-
thio)acetaldehyde
dimethyl acetal   (I)

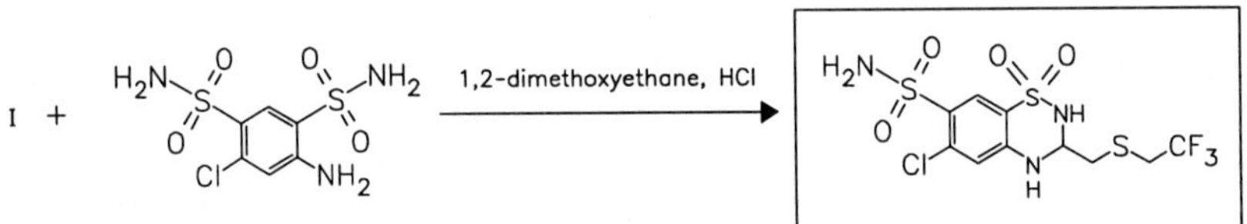

5-chloro-2,4-di-
aminosulfonyl-
aniline
(cf. chlorothiazide
synthesis)

Epitizide

Reference(s):
US 3 009 911 (Pfizer; 21.11.1961; prior. 3.6.1960, 14.9.1960, 4.1.1961).

Formulation(s):   4 mg

Trade Name(s):
GB: Thiaver (Riker); wfm

Epoprostenol

(PGI$_2$; Prostacyclin)

ATC: B01AC09
Use: anticoagulant, platelet aggregation inhibitor

RN: 35121-78-9 MF: $C_{20}H_{32}O_5$ MW: 352.47
CN: (5*Z*,9α,11α,13*E*,15*S*)-6,9-epoxy-11,15-dihydroxy-prosta-5,13-dien-1-oic acid

monosodium salt
RN: 61849-14-7 MF: $C_{20}H_{31}NaO_5$ MW: 374.45 EINECS: 263-273-7

a

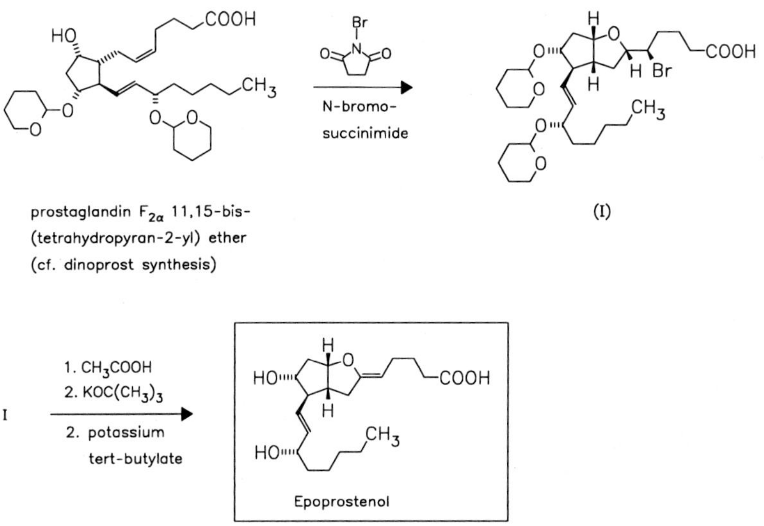
N-bromo-
succinimide
prostaglandin F2α 11,15-bis-
(tetrahydropyran-2-yl) ether
(cf. dinoprost synthesis)
(I)
I
1. CH3COOH
2. KOC(CH3)3
2. potassium
tert-butylate
Epoprostenol

b

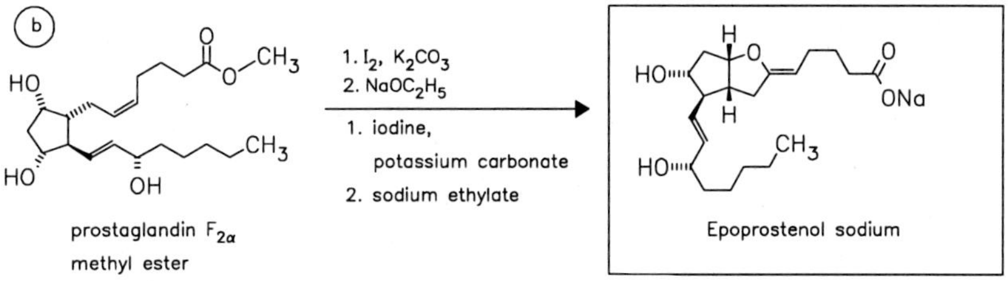
1. I2, K2CO3
2. NaOC2H5
1. iodine,
potassium carbonate
2. sodium ethylate
prostaglandin F2α
methyl ester
Epoprostenol sodium

Reference(s):
Corey, E.J. et al.: J. Am. Chem. Soc. (JACSAT) **99**, 2006 (1976).
Nicolaou, K.C. et al.: Lancet (LANCAO) **1977**, 1058.

review:
The Merck Index, 12th Ed., 1352 (Rahway 1996).

Formulation(s): vial (lyo.) 0.5 mg (as sodium salt)

Trade Name(s):
GB: Flolan (Glaxo Wellcome) USA: Flolan (Glaxo Wellcome; as sodium salt)

Eprazinone

ATC: R05CB04
Use: antitussive

RN: 10402-90-1 MF: $C_{24}H_{32}N_2O_2$ MW: 380.53 EINECS: 233-873-3
LD_{50}: 111 mg/kg (M, i.p.); 246 mg/kg (M, s.c.)
CN: 3-[4-(2-ethoxy-2-phenylethyl)-1-piperazinyl]-2-methyl-1-phenyl-1-propanone

dihydrochloride
RN: 10402-53-6 MF: $C_{24}H_{32}N_2O_2 \cdot 2HCl$ MW: 453.45 EINECS: 233-872-8
LD_{50}: 20 mg/kg (M, i.v.); 286 mg/kg (M, p.o.);
763 mg/kg (R, p.o.)

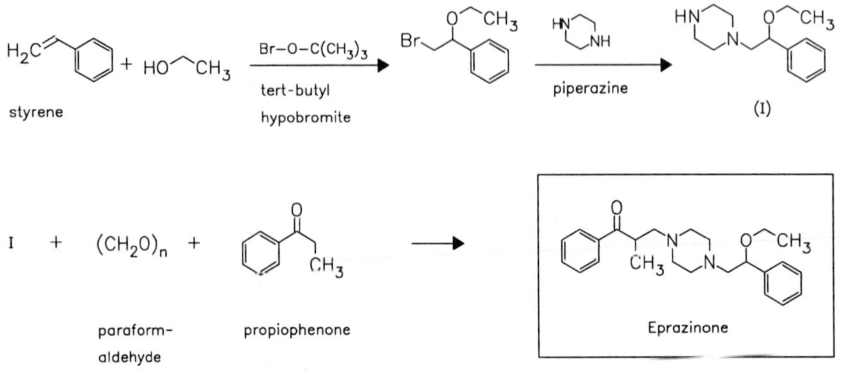

Reference(s):
DAS 1 695 431 (Mauvernay; appl. 9.6.1967; GB-prior. 27.6.1966).
JP-appl. 540 22-379 (Asahi; appl. 21.7.1977).
JP-appl. 540 22-380 (Asahi; appl. 21.7.1977).
JP-appl. 540 22-381 (Asahi; appl. 21.7.1977).
JP-appl. 540 22-382 (Asahi; appl. 21.7.1977).

Formulation(s): tabl. 5 mg, 20 mg, 50 mg; cps. 100 mg; suppos. 50 mg, 100 mg (as dihydrochloride)

Trade Name(s):
D: Eftapan (Merckle) I: Mucitux (Recordati); wfm
F: Mucitux (Riom) J: Resplen (Chugai)

Eprosartan
(SKB 108566; SKF 108566)

ATC: C09CA02
Use: antihypertensive, angiotensin II
antagonist

RN: 133040-01-4 MF: $C_{23}H_{24}N_2O_4S$ MW: 424.52
CN: (E)-α-[[2-Butyl-1-[(4-carboxyphenyl)methyl]-1H-imidazol-5-yl]methylene]-2-thiophenepropanoic acid

mesylate
RN: 144143-96-4 MF: $C_{23}H_{24}N_2O_4S \cdot CH_4O_3S$ MW: 520.63

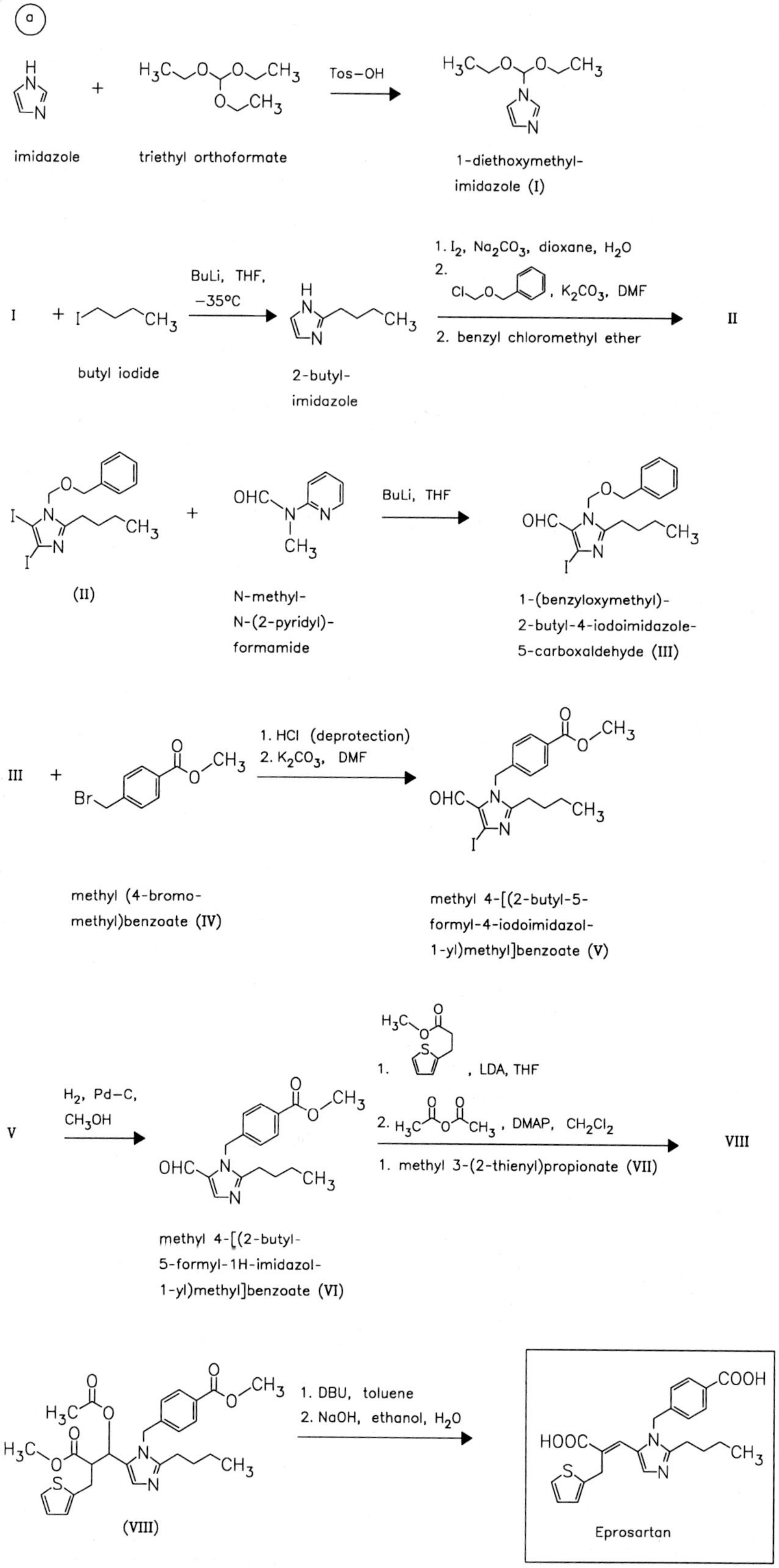

(a)

imidazole + triethyl orthoformate $\xrightarrow{\text{Tos–OH}}$ 1-diethoxymethyl-imidazole (I)

I + butyl iodide $\xrightarrow[-35°C]{\text{BuLi, THF,}}$ 2-butyl-imidazole $\xrightarrow[\text{2. benzyl chloromethyl ether}]{\substack{\text{1. } I_2, \text{ Na}_2\text{CO}_3, \text{ dioxane, H}_2\text{O} \\ \text{2. } \quad , \text{ K}_2\text{CO}_3, \text{ DMF}}}$ II

(II) + N-methyl-N-(2-pyridyl)-formamide $\xrightarrow{\text{BuLi, THF}}$ 1-(benzyloxymethyl)-2-butyl-4-iodoimidazole-5-carboxaldehyde (III)

III + methyl (4-bromo-methyl)benzoate (IV) $\xrightarrow[\text{2. K}_2\text{CO}_3, \text{ DMF}]{\text{1. HCl (deprotection)}}$ methyl 4-[(2-butyl-5-formyl-4-iodoimidazol-1-yl)methyl]benzoate (V)

V $\xrightarrow[\text{CH}_3\text{OH}]{\text{H}_2, \text{ Pd–C,}}$ methyl 4-[(2-butyl-5-formyl-1H-imidazol-1-yl)methyl]benzoate (VI) $\xrightarrow[\substack{\text{2. } \quad , \text{ DMAP, CH}_2\text{Cl}_2 \\ \text{1. methyl 3-(2-thienyl)propionate (VII)}}]{\text{1. } \quad , \text{ LDA, THF}}$ VIII

(VIII) $\xrightarrow[\text{2. NaOH, ethanol, H}_2\text{O}]{\text{1. DBU, toluene}}$ Eprosartan

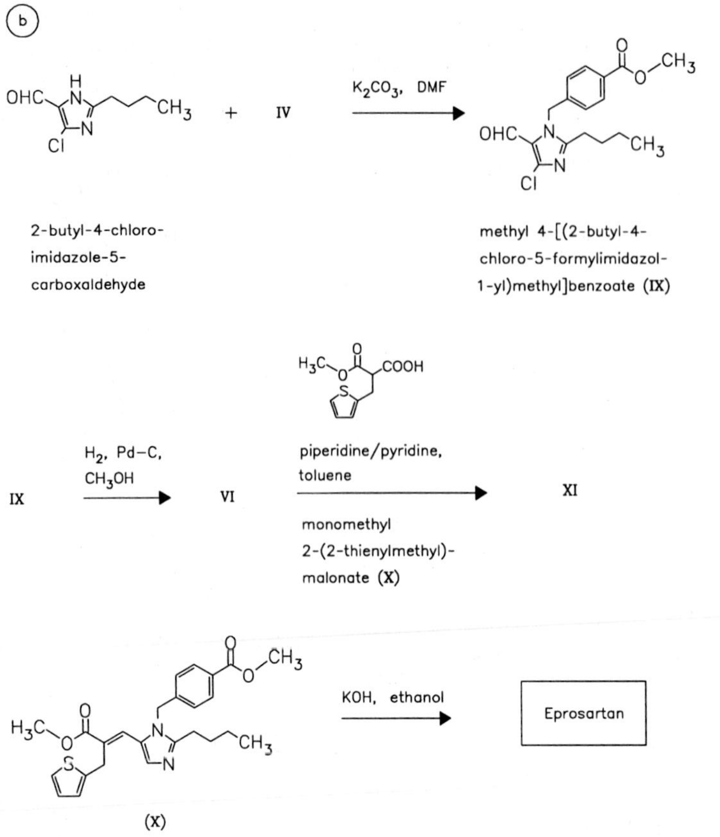

b

2-butyl-4-chloro-
imidazole-5-
carboxaldehyde

methyl 4-[(2-butyl-4-
chloro-5-formylimidazol-
1-yl)methyl]benzoate (IX)

IX $\xrightarrow{\text{H}_2,\ \text{Pd-C},\ \text{CH}_3\text{OH}}$ VI $\xrightarrow[\substack{\text{monomethyl} \\ \text{2-(2-thienylmethyl)-} \\ \text{malonate (X)}}]{\substack{\text{piperidine/pyridine,} \\ \text{toluene}}}$ XI

$\xrightarrow{\text{KOH, ethanol}}$ Eprosartan

(X)

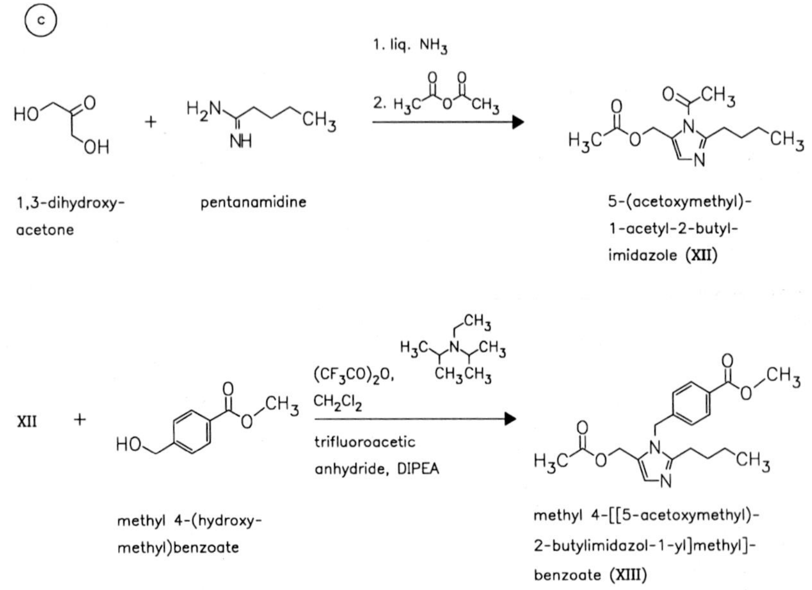

c

1,3-dihydroxy-
acetone

pentanamidine

5-(acetoxymethyl)-
1-acetyl-2-butyl-
imidazole (XII)

XII +

methyl 4-(hydroxy-
methyl)benzoate

methyl 4-[[5-acetoxymethyl)-
2-butylimidazol-1-yl]methyl]-
benzoate (XIII)

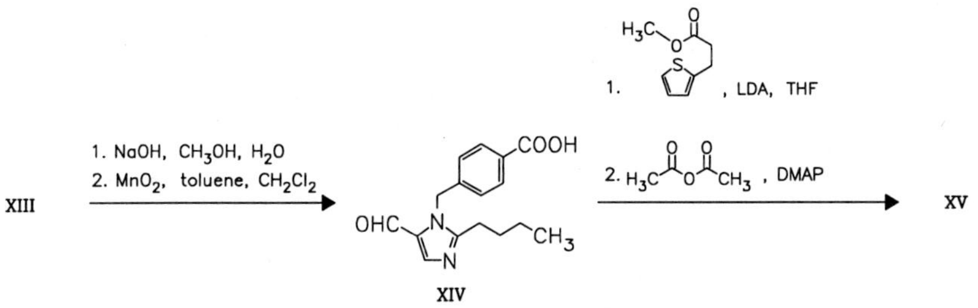

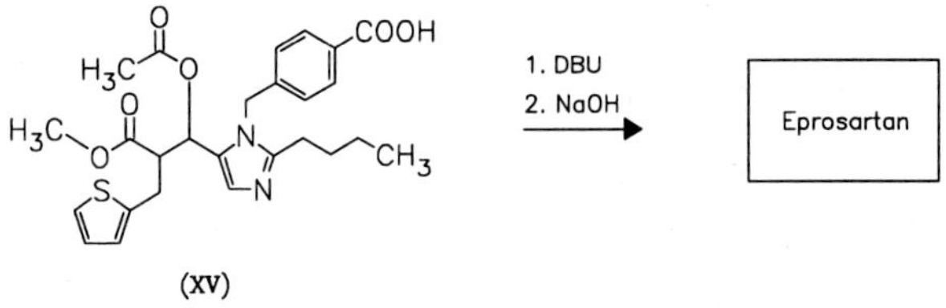

XIII

1. NaOH, CH₃OH, H₂O
2. MnO₂, toluene, CH₂Cl₂

XIV

1.
(structure with H₃C, O, S, thiophene ester), LDA, THF

2. H₃C—C(O)—O—C(O)—CH₃, DMAP

XV

(XV)

1. DBU
2. NaOH

Eprosartan

---

d

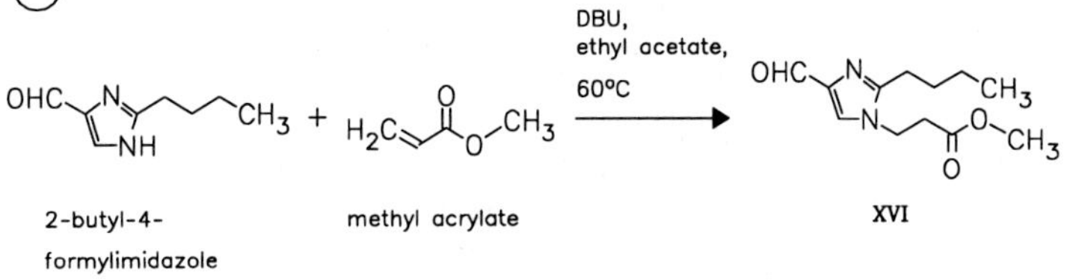

OHC (2-butyl-4-formylimidazole) + H₂C=CH—C(O)—O—CH₃ (methyl acrylate)

DBU, ethyl acetate, 60°C

XVI

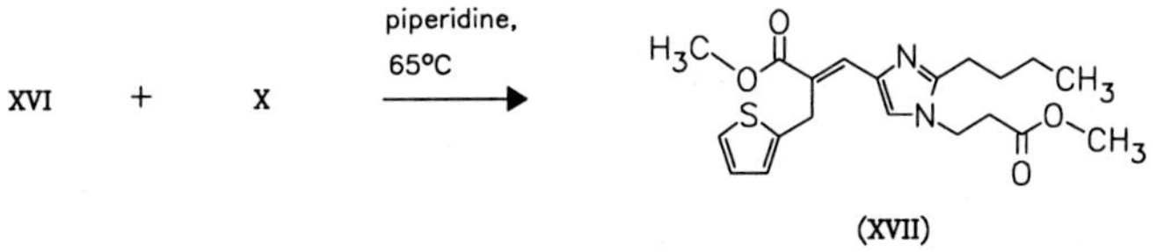

XVI + X

piperidine, 65°C

(XVII)

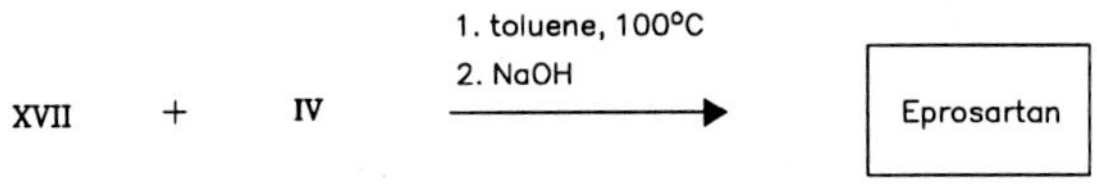

XVII + IV

1. toluene, 100°C
2. NaOH

Eprosartan

---

e

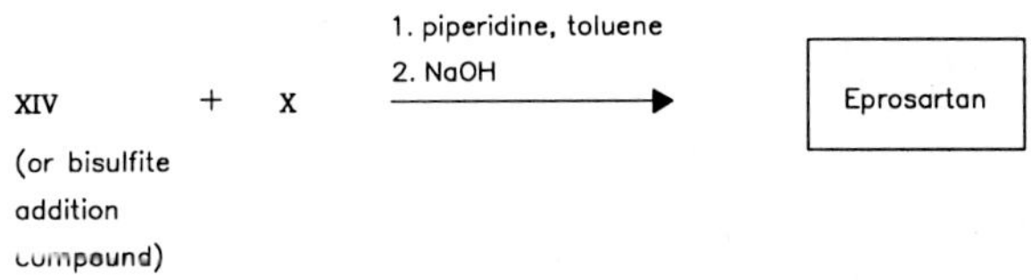

XIV (or bisulfite addition compound) + X

1. piperidine, toluene
2. NaOH

Eprosartan

---

f   alternative synthesis of VI

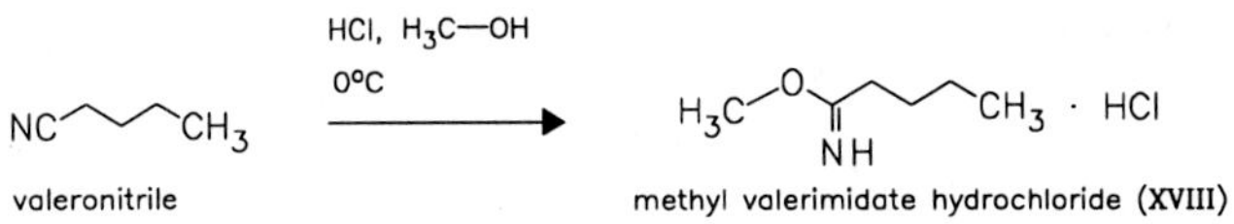

NC~~~CH₃ (valeronitrile)

HCl, H₃C—OH
0°C

H₃C—O—C(=NH)—~~~CH₃ · HCl

methyl valerimidate hydrochloride (XVIII)

methyl 4-[[N-
(1-iminopentyl)amino]-
methyl]benzoate (XIX)

2-bromo-3-(1-
methylethoxy)-
2-propenal

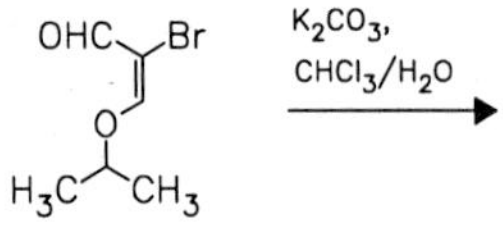

Reference(s):

a   Wittenberger, S.J. et al.: Synth. Commun. (SYNCAV) **23**, 3231 (1993).
    Keenan, R.M. et al.: J. Med. Chem. (JMCMAR) **36**, 1880 (1993).
    EP 403 159 (SmithKline Beecham; appl. 7.6.1990; USA-prior. 14.6.1989).
b   Weinstock, J. et al.: J. Med. Chem. (JMCMAR) **34**, 1514 (1991).
c   US 5 185 351 (SmithKline Beecham; 9.2.1993; USA-prior. 14.6.1989; 6.4.1990; 14.12.1990).
d   WO 9 835 962 (SmithKline Beecham; appl. 13.2.1998; USA-prior. 14.2.1997)
e   WO 9 835 963 (SmithKline Beecham; appl. 13.2.1998; USA-prior. 14.2.1997).
f   Shilera, S.C. et al.: J. Org. Chem. (JOCEAH) **62**, 8449 (1997).

Eprosartan dihydrate:
WO 9 736 874 (SmithKline Beecham; appl. 26.3.1997; USA-prior. 29.3.1996).

combination with ACE inhibitors:
EP 629 408 (MS Dohme-Chibret; appl. 14.12.1993; EP-prior. 16.6.1993).
WO 9 702 032 (Merck + Co.; appl. 26.6.1996; USA-prior. 30.6.1995).

pharmaceutical compositions and use in the treatment of macular degeneration, infarction, left ventricular hypertrophy:
WO 9 210 179 (SmithKline Beecham; appl. 12.12.1991; GB-prior. 14.12.1990).
WO 9 210 180 (SmithKline Beecham; appl. 12.12.1991; GB-prior. 14.12.1990).
WO 9 210 181 (SmithKline Beecham; appl. 12.12.1991; GB-prior. 14.12.1990).

use in the treatment of diabetic nephropathy, retinopathy, atheroma, angina pectoris, stroke or prevention of restenosis or improving cognitive function:
WO 92 101 82-88 (SmithKline Beecham; appl. 12.12.1991; GB-prior. 14.12.1990).

use to treat symptomatic heart failure:
WO 9 830 216 (Merck + Co.; appl. 7.1.1998; USA-prior. 10.1.1997).

Formulation(s):   f. c. tabl. 300 mg, 400 mg, 600 mg

Trade Name(s):

| D: | Teveten (Hoechst Marion Roussel; SmithKline Beecham) | GB: | Teveten (SmithKline Beecham; 1997) | USA: | Teveten (SmithKline Beecham) |

Eprozinol

ATC: R03DX02
Use: antiasthmatic

RN: 32665-36-4 MF: $C_{22}H_{30}N_2O_2$ MW: 354.49 EINECS: 251-146-9
LD_{50}: 350 mg/kg (M, p.o.);
640 mg/kg (R, p.o.)
CN: 4-(2-methoxy-2-phenylethyl)-α-phenyl-1-piperazinepropanol

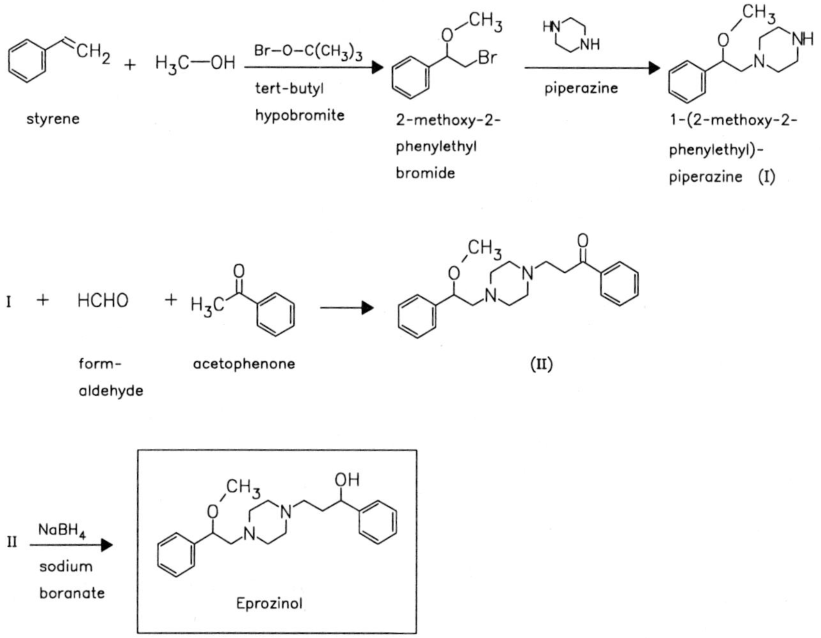

styrene + H_3C-OH →(Br-O-C(CH₃)₃ tert-butyl hypobromite) 2-methoxy-2-phenylethyl bromide →(piperazine) 1-(2-methoxy-2-phenylethyl)-piperazine (I)

I + HCHO (formaldehyde) + acetophenone → (II)

II →(NaBH₄ sodium boranate) Eprozinol

Reference(s):
GB 1 188 505 (Roland-Yves Mauvernay; appl. 27.6.1966; valid from 27.6.1967).
US 3 705 244 (Roland-Yves Mauvernay; 5.12.1972; GB-prior. 27.6.1966).

Formulation(s): suppos. 100 mg, 25 mg; syrup 15 mg/5 ml; tabl. 50 mg

Trade Name(s):
F: Eupnéron (Lyocentre) Eupnéron xantique (Lyocentre) I: Brovel (Lepetit)

Eptifibatide

(C 68 22; SB-1; Sch-60936; Intrifiban)

ATC: B01AC16
Use: platelet antiaggregatory, GPIIb/ receptor antagonist, fibrinogen receptor antagonist

RN: 188627-80-7 MF: $C_{35}H_{49}N_{11}O_9S_2$ MW: 831.98
CN: N^6-(Aminoiminomethyl)-N^2-(3-mercapto-1-oxopropyl)-L-lysylglycyl-L-α-aspartyl-L-tryptophyl-L-prolyl-L-cysteinamide cyclic (1→6)-disulfide

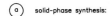

ⓐ solid-phase synthesis:

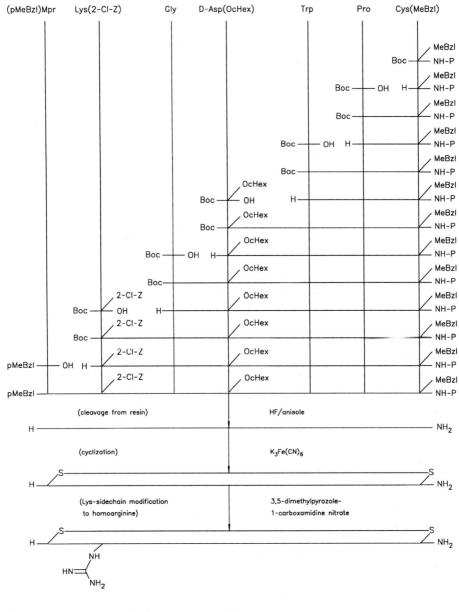

Mpr:	mercaptopropionic acid
pMeBzl:	p-methylbenzyl
MeBzl:	toluenyl
2-Cl-Z:	2-chlorobenzyloxycarbonyl
OcHex:	cyclohexyloxy
Boc:	tert-butoxycarbonyl

P: (MBHA-resin)

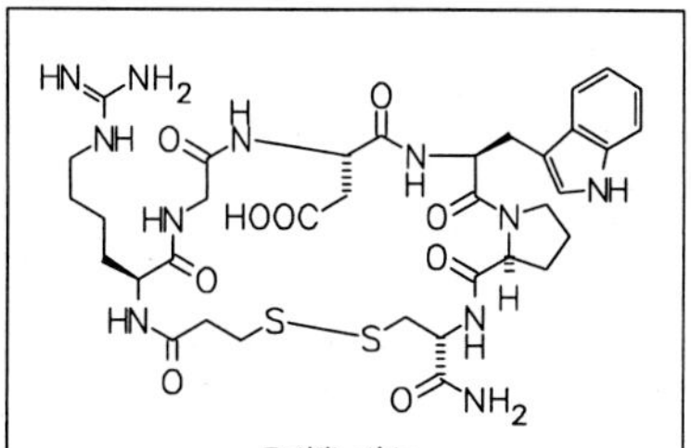

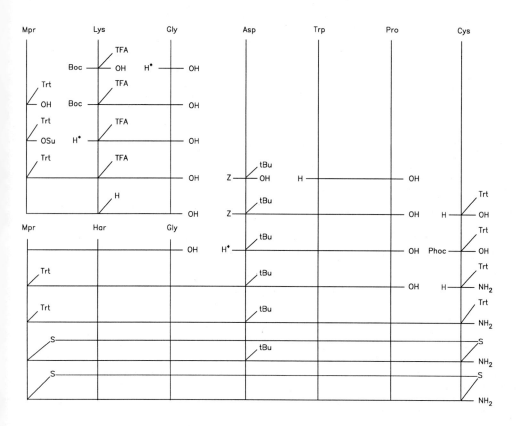

Eptifibatide

(b)   fragment synthesis in solutions:

Mpr:	mercaptopropionic acid
Har:	homoarginine
* :	persilylation
Phoc:	phenyloxycarbonyl
Trt;	trityl
OSu	succinimidooxy
TFA:	trifluoroacetyl
Boc:	tert-butoxycarbonyl
Z:	benzyloxycarbonyl
tBu:	tert-butyl

Reference(s):

a   Scarborough, R.M., et al: J. Biol. Chem. (JBCHA3) **268**, 1066-1073 (1993).
    WO 9 015 620 (Cor Therap.; appl. 15.6.1990; USA-prior. 20.2.1990).
b   Callens, R.: IBC's 2nd Internat. Conf. on Peptide Technologies San Diego 1999.

Formulation(s): vials for inj. 20 mg/10 ml, 75 mg/10 ml

Trade Name(s):
USA: Integrilin (Cor
Therapeutics/Schering-
Plough; 1998)

Erdosteine
(RV-144)

ATC: R05CB15
Use: mucolytic agent

RN: 84611-23-4 MF: $C_8H_{11}NO_4S_2$ MW: 249.31
LD_{50}: >3.5 g/kg (M, i.v.); >10 g/kg (M, p.o.);
>3.5 g/kg (R, i.v.); >10 g/kg (R, p.o.)
CN: (±)-[[2-oxo-2-[(tetrahydro-2-oxo-3-thienyl)amino]ethyl]thio]acetic acid

monopotassium salt
RN: 84611-25-6 MF: $C_8H_{10}KNO_4S_2$ MW: 287.40
monosodium salt
RN: 84611-24-5 MF: $C_8H_{10}NNaO_4S_2$ MW: 271.29
(S)-enantiomer
RN: 159701-33-4 MF: $C_8H_{11}NO_4S_2$ MW: 249.31

homocysteine   chloroacetyl   3-(chloroacetamido)-
thiolactone   chloride   2-oxotetrahydro-
                                  thiophene (I)

thioglycolic   Erdosteine
acid

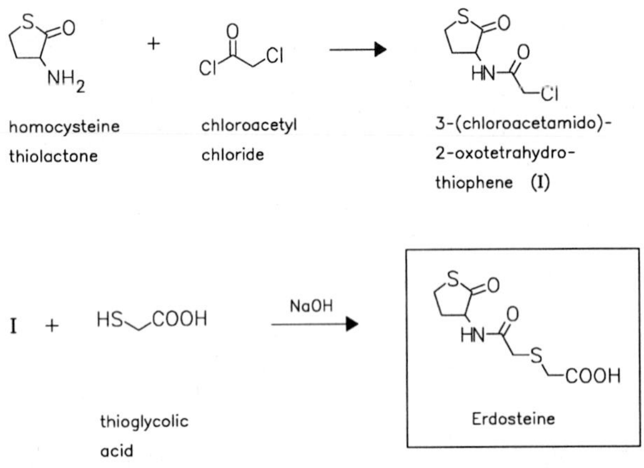

Reference(s):
EP 61 386 (Refarmed Rech. Pharm.; appl. 11.3.1982; F-prior. 19.3.1981).

combination with antibiotics:
DE 3 509 244 (Edmond Pharma; appl. 14.3.1985; I-prior. 14.3.1984).

Formulation(s): cps. 300 mg

Trade Name(s):
F: Edirel (Inava) Vectrine (Pharma 2000)

Ergocalciferol
(Vitamin D; Calciferol)

ATC: A11CC01
Use: antirachitic

RN: 50-14-6   MF: $C_{28}H_{44}O$ MW: 396.66   EINECS: 200-014-9
LD_{50}: 23.7 mg/kg (M, p.o.);
        10 mg/kg (R, p.o.)
CN: (3β,5Z,7E,22E)-9,10-secoergosta-5,7,10(19),22-tetraen-3-ol

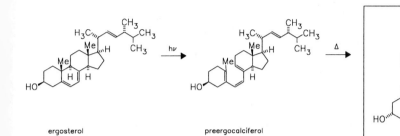

ergosterol   preergocalciferol   Ergocalciferol

Reference(s):
Kirk-Othmer Encycl. Chem. Technol., Vol. **21**, 549 ff.
Ullmanns Encykl. Tech. Chem., 3. Aufl., Vol. **18**, 236 ff.

Formulation(s):   200 iu, 400 iu in comb.

Trade Name(s):
D: Cal-C-Vita (Hoffmann-La
    Roche)-comb.
    Cobidec (Parke Davis)-
    comb.
    Frubiase (Boehringer Ing.)-
    comb.
    Geriatric (Pharmaton)-
    comb.
    Lofenalac (Lappe)-comb.
    Multiviol (Hermes)-comb.
    Natabec (Warner-Lambert)-
    comb.
    Omnival (Nordmark)-
    comb.
    Osspulvit (Madaus)-comb.
    Pregnavit (Merckle)-comb.

    Savitol (Medipharma)
    Vitalipid (Pharmacia &
    Upjohn)-comb.
    further combination
    preparations
F: Azedavit (Whitehall)-
    comb.
    Dossibil (Thérica)-comb.
    Pharmaton (Boehringer
    Ing.)-comb.
    Stérogyl (Roussel)
    Stérogyl 15 (Roussel)
    Vitalipide (Pharmacia &
    Upjohn)-comb. with
    vitamin A
    Zyma D2 (Novartis)

    numerous combination
    preparations
GB: Abidec (Warner-Lambert)-
    comb.
    Dalivit (Eastern)-comb.
    numerous combination
    preparations
I: Ostelin-800 (Teofarma)
    combination preparations
J: Chocola D (Eisai)
USA: Calciferol (Schwarz)
    further combination
    preparations

Ergometrine
(Ergobasine; Ergonovine)

ATC: G02AB03
Use: oxytocic

RN: 60-79-7   MF: $C_{19}H_{23}N_3O_2$ MW: 325.41   EINECS: 200-485-0
LD_{50}: 144 mg/kg (M, i.v.)
CN: [8β(S)]-9,10-didehydro-N-(2-hydroxy-1-methylethyl)-6-methylergoline-8-carboxamide

maleate (1:1)
RN: 129-51-1   MF: $C_{19}H_{23}N_3O_2 \cdot C_4H_4O_4$ MW: 441.48   EINECS: 204-953-5
LD_{50}: 8260 µg/kg (M, i.v.)

tartrate (2:1)

RN: 129-50-0   MF: $C_{19}H_{23}N_3O_2 \cdot 1/2C_4H_6O_6$ MW: 800.91   EINECS: 204-951-4

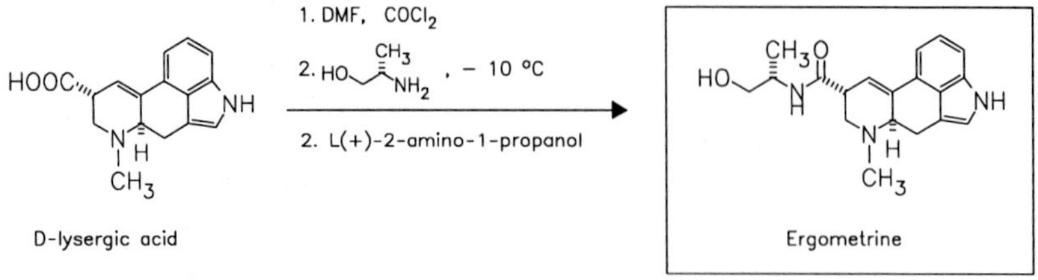

D-lysergic acid → Ergometrine

1. DMF, COCl$_2$
2. HO–CH$_3$–NH$_2$, – 10 °C
2. L(+)-2-amino-1-propanol

Reference(s):

Stoll, A.; Hofmann, A.: Helv. Chim. Acta (HCACAV) **26**, 956 (1943).
US 2 090 430 (Sandoz; 1937; CH-prior. 1936).
US 2 447 214 (Sandoz; 1948; CH-prior. 1942).
US 2 736 728 (Lilly; 1956; appl. 1954).
US 2 774 763 (Lilly; 1956; appl. 1955).
US 2 809 920 (Sandoz; 1957; CH-prior. 1953).
US 3 141 887 (Soc. Farmaceutici Italia; 21.7.1964; I-prior. 18.10.1961).

Formulation(s):   sol. 50 mg/100 ml (as maleate)

Trade Name(s):

D:	Secalysat-EM (Ysatfabrik)	Ermetrin (Takeda)	Ergotrate Maleate (Lilly);
GB:	Syntometrine (Novartis)	USA: Ergonovine Maleate	wfm
J:	Ergoton-B (Azusa-Tokyo	(Bioline; City Chem.;	
	Tanabe)	Goldline; Wyeth); wfm	
	Ergotop (Hishiyama)		

Ergotamine

ATC: N02CA02
Use: antimigraine agent

RN: 113-15-5   MF: $C_{33}H_{35}N_5O_5$ MW: 581.67   EINECS: 204-023-9
LD$_{50}$:   52 mg/kg (M, i.v.);
          80 mg/kg (R, i.v.)
CN: (5'α)-12'-hydroxy-2'-methyl-5'-(phenylmethyl)ergotaman-3',6',18-trione

tartrate (2:1)

RN: 379-79-3   MF: $C_{33}H_{35}N_5O_5 \cdot 1/2C_4H_6O_6$ MW: 1313.43   EINECS: 206-835-9
LD$_{50}$:   62 mg/kg (M, i.v.);
          80 mg/kg (R, i.v.)

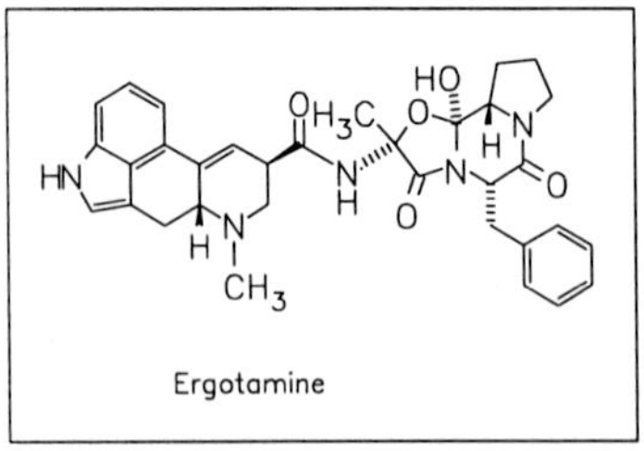

Ergotamine

By extraction of *Secale cornutum* (ergot) with e. g. benzene (1. step: extraction of the neutral substances from the slightly acidic cellular substance; 2. step: extraction of the ammonia alkaline substance).

Fermentation of *Claviceps purpurea*.

Reference(s):
Stoll, A.: Helv. Chim. Acta (HCACAV) **28**, 1283 (1945).

Formulation(s):   cps. 1 mg; suppos. 2 mg; tabl. 1 mg, 2 mg (as tartrate)

Trade Name(s):

D: Avamigran (ASTA Medica
       AWD)-comb.
       Cafergot (Novartis Pharma)
       Ergoffin (ASTA Medica
       AWD)-comb.
       Ergo-Kranit (Krewel
       Meuselbach)
       ergo-sanol (Sanol)
       Ergotamin Medihaler
       (Kettelhack-Riker)
       Gynergen (Sandoz)
       Migrätan (Berlin-Chemie)-
       comb.
       Migrexa (Sanorania)

       RubieNex (RubiePharm)-
       comb.
       numerous combination
       preparations
F: Gynergène (Novartis)-
       comb.
       Migwell (Glaxo
       Wellcome)-comb.
GB: Cafergot (Novartis)-comb.
       Lingraine (Sanofi
       Winthrop)
       Medihaler-Ergotamine (3M
       Health Care)
       Migril (Glaxo Wellcome)-
       comb.

I: Cafergot (Sandoz)-comb.
       Ergota (Sifra)
       Ergotan (Salf)
       Gynergen (Sandoz)
       Virdex (Fulton)-comb.
J: Cafergot (Sandoz-Sankyo)-
       comb.
       Ergoton A (Azusa-Tokyo
       Tanabe)
       Migretamine (Hokuriku)
USA: Ercaf (Geneva)
       Ergomar (Lotus; as tartrate)
       Wigraine (Organon)

Erythromycin

ATC: D10AF02; D10AF52; J01FA01;
       S01AA17
Use: antibiotic

RN: 114-07-8   MF: $C_{37}H_{67}NO_{13}$ MW: 733.94   EINECS: 204-040-1
LD_{50}:   426 mg/kg (M, i.v.); 2580 mg/kg (M, p.o.);
       4600 mg/kg (R, p.o.)
CN: [3R-(3R*,4S*,5S*,6R*,7R*,9R*,11R*,12R*,13S*,14R*)]-4-[(2;6-dideoxy-3-C-methyl-3-O-methyl-α-L-
       ribo-hexopyranosyl)oxy]-14-ethyl-7,12,13-trihydroxy-3,5,7,9,11,13-hexamethyl-6-[[3,4,6-trideoxy-3-
       (dimethylamino)-β-D-*xylo*-hexopyranosyl]oxy]oxacyclotetradecane-2,10-dione

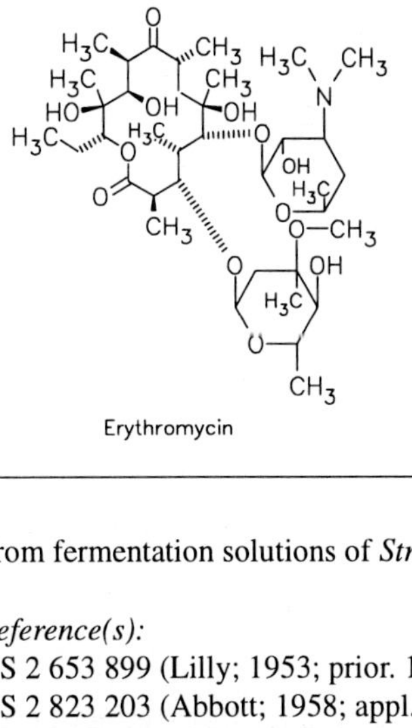

Erythromycin

From fermentation solutions of *Streptomyces erythreus*.

Reference(s):
US 2 653 899 (Lilly; 1953; prior. 1952).
US 2 823 203 (Abbott; 1958; appl. 1954).
US 2 833 696 (Abbott; 6.5.1958; prior. 1.3.1954).

Formulation(s):   cps. 250 mg; f. c. tabl. 250 mg, 500 mg; gel 0.5 mg/100 g, 1 g/100 g, 2 g/100 g (2 %),
4 g/100 g; sol. 0.2 g/10 g, 1.68 g/100 ml; spray 20 mg/ml; s. r. tabl. 250 mg, 333 mg, 500 mg;
suppos. 250 mg (as free base)

Trade Name(s):

D: Akne Cordes (Ichthyol)
Aknederm (gepepharm)
Aknefug-EL (Wolff)
Aknemago (Strathmann)
Aknemycin (Hermal)
Aknin-Winthrop (Sanofi
Winthrop)
Bisolvonat (Thomae)-
comb.
Eromerzin (Merrell)-comb.
Eryaknen (Galderma)
Erybeta (betapharm)
Erycinum (Cytochemie)
Erycinum (Schering)
Erydermec (Hexal)
Ery-Diolan (Engelhardt)
Eryhexal (Hexal)
ERY-REU (Reusch)
Erythrogenat (Azupharma)
Erythro Hefa (Hefa
Pharma)
Eupragin (Alcon)
Infectomycin
(Infectopharm)
Lederpaedit (Lederle)
Monomycin (Grünenthal)
Paediathrocin (Abbott)
Paediathrocin
Suppositorien (Abbott)
Pharyngocin (Upjohn)
Sanasepton (Pharbita)
Stiemycine (Stiefel)
Synergomycin (Abbott)-
comb.

Udima-Ery (Dermapharm)
generic
F: Ery (Bouchara; as
propionate)
Eryfluid (Pierre Fabre)
Logécine (Jacques Logeais)
Propiocine (Roussel; as
propionate)
Stimycine (Stiefel)
numerous generics
GB: Benzamycin (Bioglan)
Erymax (Elan)
Erythrocin (Abbott); wfm
Erythromid (Abbott); wfm
Erythroped (Abbott); wfm
Ilotycin (Lilly); wfm
Retcin (DDSA); wfm
I: Cicloeritrina (Proter)-
comb.; wfm
Erimec (Isola-Ibi); wfm
Eritro (Formulario Naz.)
Eritrobios (Nuovo Cons.
Sanit. Naz.); wfm
Eritrobiotic (Panther-Osfa
Chemie); wfm
Estomicina (Bergamon
Soc. It.); wfm
Ilosone (Lilly); wfm
Lauromicina (Dukron)-
comb.; wfm
Manilina (Archifar); wfm
Marocid (Lifepharma);
wfm
Mistral (Dessy); wfm

Mucolysin (Proter)-comb.;
wfm
Neobalsamocetina
Supposte (Alfa Farm.)-
comb.; wfm
Neobismocetina (Lepetit)-
comb.; wfm
Proterytrin (Proter); wfm
Proterytrin pomata
(Proter)-comb.; wfm
Stellamicina (Pierrel); wfm
J: Erythrocin (Dainippon;
Abbott)
Ilotycin (Shionogi)
USA: A/T/S (Hoechst Marion
Roussel)
Benzamycin (Dermik)
Emgel (Glaxo Wellcome)
Eryc (Warner Chilcott
Professional Products)
Erycette (Ortho
Dermatological)
Erygel (Allergan)
Erymax (Allergan)
Ery-Tab (Abbott)
Erythra-Derm (Paddock)
Ilotycin (Dista)
PCE (Abbott)
T-Stat (Westwood-Squibb)
Theramycin Z (Medicis)
generic and combination
preparations

Erythromycin estolate

ATC: D10AF02; D10AF52; J01FA01;
S01AA17
Use: antibiotic

RN: 3521-62-8   MF: $C_{40}H_{71}NO_{14} \cdot C_{12}H_{26}O_4S$ MW: 1056.40   EINECS: 222-532-4
LD$_{50}$: >6450 mg/kg (M, p.o.);
1447 mg/kg (R, p.o.)
CN: erythromycin 2'-propanoate dodecyl sulfate (salt)

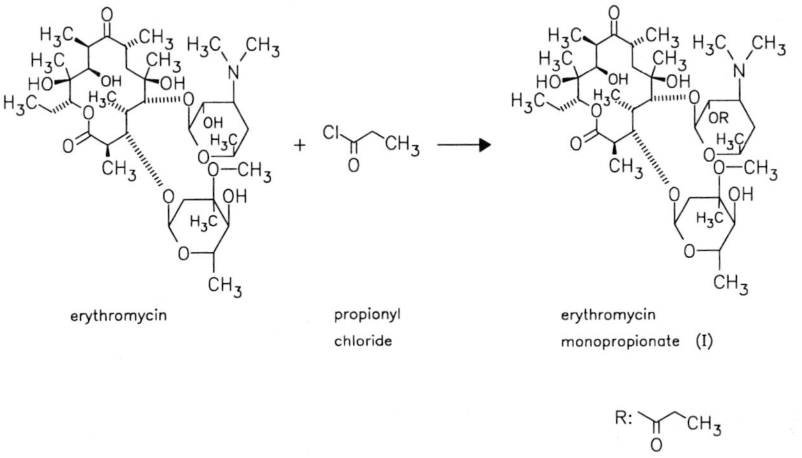

erythromycin

propionyl
chloride

erythromycin
monopropionate   (I)

R:

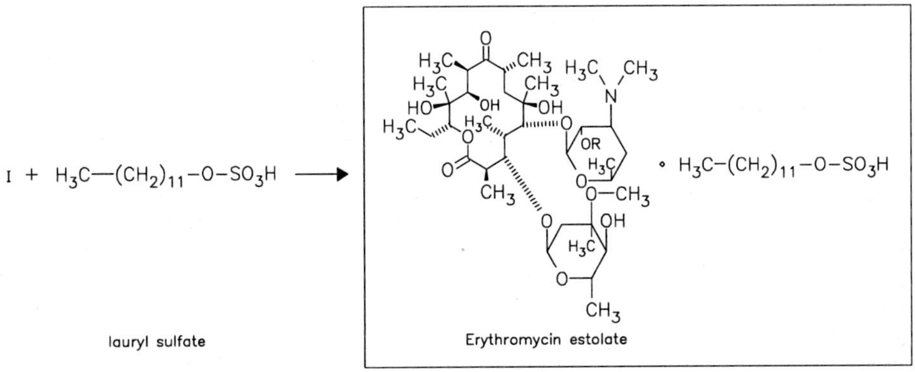

lauryl sulfate

Erythromycin estolate

Reference(s):
US 3 000 874 (Eli Lilly; 19.9.1961; prior. 8.4.1959).
DE 1 114 499 (Eli Lilly; 27.6.1959).

Formulation(s):   cleavable tabl. 125 mg, 250 mg; cps. 250 mg; susp. 125 mg/5 ml, 250 mg/5 ml; syrup 250 mg
(base equivalent)

Trade Name(s):

D:	Infectomycin (Infectopharm)		Togiren (Schwarzhaupt); wfm	I:	Ilosone (Lilly) Marocid (Lifepharma)
	Neo-Erycinum (Schering); wfm	F:	Propiocine (Roussel); wfm Rubitracine (Takeda)-		Stellamicina (Pierrel)
	Sanasepton (Pharbita)		comb.; wfm	J:	Ilosone (Shionogi)
		GB:	Ilosone (Lilly)	USA:	Ilosone (Dista)

Erythromycin ethylsuccinate

ATC: D10AF02; D10AF52; J01FA01;
S01AA17
Use: antibiotic

RN: 1264-62-6   MF: C$_{43}$H$_{75}$NO$_{16}$ MW: 862.06   EINECS: 215-033-8
LD$_{50}$: >10 g/kg (M, p.o.)
CN: erythromycin 2'-(ethyl butanedioate)

erythromycin

ethyl succinyl
chloride

Erythromycin ethylsuccinate

Reference(s):
DE 1 121 056 (Abbott; appl. 1957; USA-prior. 1956).

chewing tablets:
DOS 2 758 942 (Abbott; appl. 30.12.1977).

Formulation(s):   f. c. tabl. 400 mg; powder 1 g/4.5 g; susp. 125 mg/5 ml; syrup 100 mg/5 ml, 200 mg/5 ml, 400 mg/5 ml, 600 mg/5 ml (base equivalent)

Trade Name(s):

D: Dura Erythromycin 1000
    Granulat (durachemie)
    Durapaediat (durachemie)
    Erythrocin Granulat/-
    Ampullen (Abbott)
    Erythromycin-ratiopharm
    (ratiopharm)
    Monomycin (Grünenthal)
    Paediathrocin (Abbott)
    combination preparations
F: Abboticine (Abbott)
    Ery 125 e 250 (Bouchara)
    Erycocci (Pharmafarm)
    Erythrocine (Abbott)

    Erythrogram (Pharma
    2000)
GB: Arpimycin (Rozemont)
    Erymin (Elan)
    Erythrocin I. M. (Abbott)
    Erythroped A (Abbott)
I: Eritrocina (Abbott)
    Eritroger bustine (Isnardi);
    wfm
    Neobalsamocetina sosp.
    (Alfa Farm.)-comb.; wfm
    Proterytrin (Proter); wfm
    Proterytrin cps e i.m.
    (Proter); wfm

    Rossomicina sosp.
    (Pierrel); wfm
J: Eryromycen (Kissei)
    Erythrocin (Abbott-
    Dainippon)
    Erythro ES (Sankyo)
    Erythromycin ES (Taito
    Pfizer)
    Esinol (Toyama)
    Evesin (Torii)
USA: E.E.S. (Abbott)
    Eryped (Abbott)
    Eryzole (Alra)
    Pediazole (Ross)

Erythromycin gluceptate

(Erythromycin glucoheptonate)

ATC: J01FA01; S01AA17
Use: antibiotic

RN: 23067-13-2   MF: $C_{37}H_{67}NO_{13} \cdot C_7H_{14}O_8$ MW: 960.12   EINECS: 245-407-6
LD_{50}:   453 mg/kg (M, i.v.);
        288 mg/kg (R, i.v.)
CN: D-*glycero*-D-*gulo*-heptonic acid compd. with erythromycin (1:1)

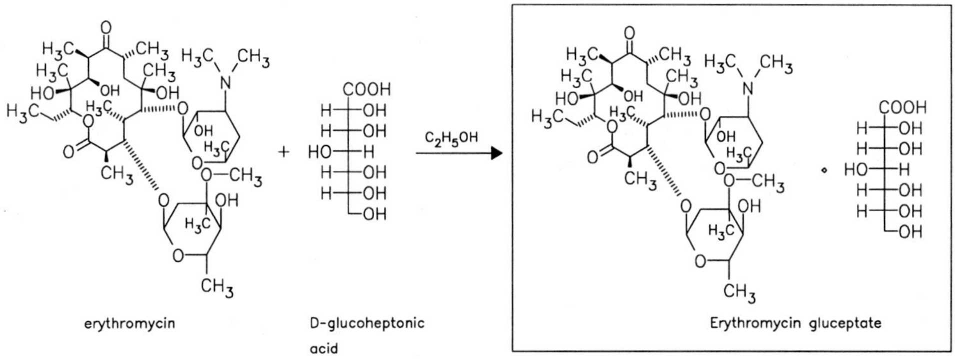

erythromycin + D-glucoheptonic acid → Erythromycin gluceptate

Reference(s):
US 2 852 429 (Lilly; 1958; appl. 1953).
DE 941 640 (Lilly; appl. 1954; USA-prior. 1953).

Formulation(s):   vial 1 g (base equivalent)

Trade Name(s):
D: Erycinum Trockensubstanz   USA: Ilotycin gluceptate (Dista)
      (Schering); wfm

Erythromycin lactobionate

ATC: J01FA01; S01AA17
Use: antibiotic

RN: 3847-29-8   MF: $C_{37}H_{67}NO_{13} \cdot C_{12}H_{22}O_{12}$ MW: 1092.23   EINECS: 223-348-7
LD$_{50}$:   735 mg/kg (M, i.p.)
CN: 4-*O*-β-D-galactopyranosyl-D-gluconic acid compd. with erythromycin (1:1)

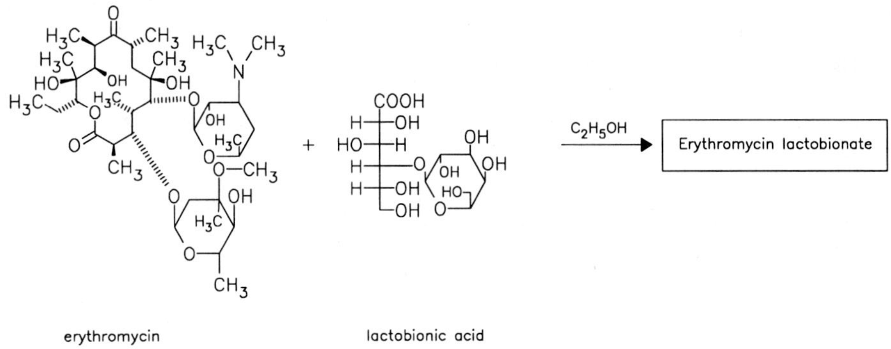

erythromycin + lactobionic acid $\xrightarrow{C_2H_5OH}$ Erythromycin lactobionate

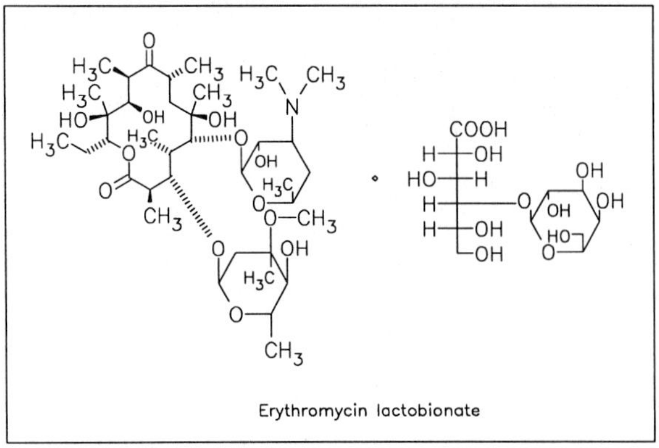

Erythromycin lactobionate

Reference(s):
US 2 761 859 (Abbott; 1956; appl. 1953).

Formulation(s): vial 500 mg, 1000 mg (base equivalent)

Trade Name(s):
D: Erythrocin I.V. (Abbott)
GB: Erythrocin I.V. Lactobionate (Abbott); wfm
I: Eritro (Formulario Naz.)
J: Erythromycin (Santen)
USA: Erythrocin Lactobionate-I.V. (Abbott); wfm

Erythromycin monopropionate mercaptosuccinate
(RV-11)

ATC: J01FA
Use: macrolide antibiotic

RN: 84252-06-2 MF: $C_{40}H_{71}NO_{14} \cdot C_4H_6O_4S$ MW: 940.16
LD$_{50}$: >3000 mg/kg (M, p.o.)
CN: erythromycin-2'-propanoate mercaptobutanedioate (1:1)

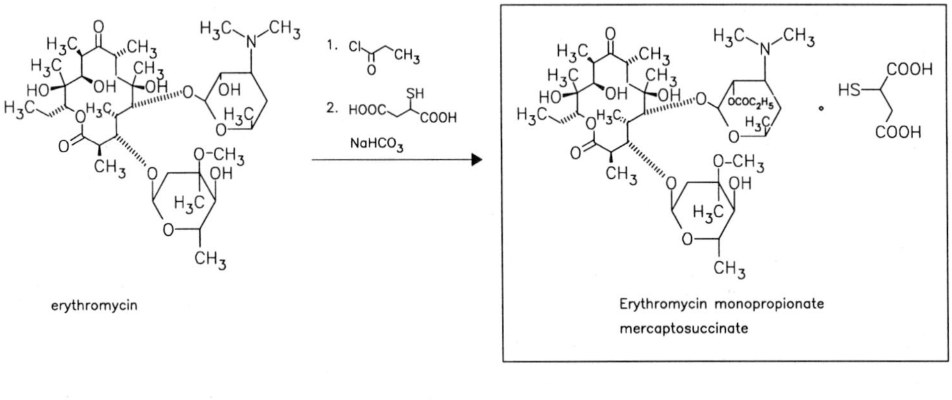

erythromycin

Erythromycin monopropionate mercaptosuccinate

Reference(s):
EP 57 489 (Pierrel; appl. 2.1.1982; F-prior. 2.2.1981).
EP 174 395 (Pierrel; appl. 2.1.1982; F-prior. 2.2.1982).
US 4 476 120 (Refarmed; 10.9.1984; appl. 2.2.1982; I-prior. 2.2.1981).

Formulation(s): gran. 200 mg; tabl. 500 mg

Trade Name(s):
I: Zalig (Pierrel; 1989)

Erythromycin stearate

ATC: D07CC02
Use: antibiotic

RN: 97327-17-8 MF: $C_{55}H_{101}NO_{14}$ MW: 1000.41
LD_{50}: 3112 mg/kg (M, p.o.)
CN: erythromycin 2'-octadecanoate

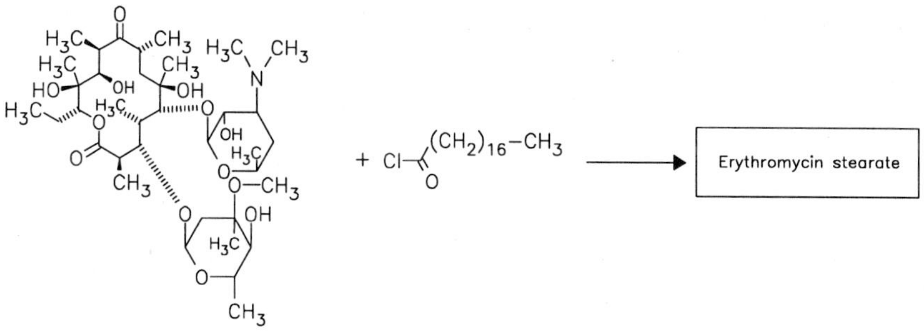

erythromycin stearoyl chloride

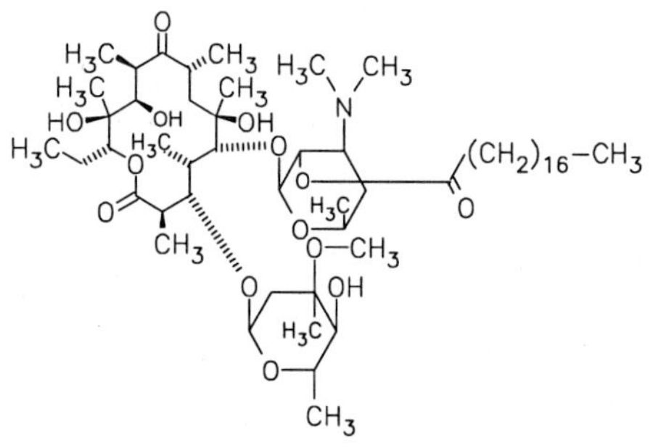

Erythromycin stearate

Reference(s):
US 2 862 921 (Upjohn; 1958; appl. 1953).

Formulation(s): f. c. tabl. 250 mg, 500 mg

Trade Name(s):
D: Dura Erythromycin (durachemie)
Erythrocin Filmtabletten (Abbott)
Emestid 500 (Abbott); wfm
F: Abboticine (Abbott); wfm
GB: Erythrocin (Abbott)
I: Eritrocina Cpr (Abbott)
Lauromicina (Lafare)
J: Erythrocin (Abbott-Dainippon)
USA: Erythrocin Stearate (Abbott)
generic

Escin
(Aescin)

ATC: C05CX; C05CX01
Use: anti-inflammatory (inhibition of edema formation and decrease of vessel fragility), vein therapeutic

RN: 6805-41-0   MF: $C_{54}H_{84}O_{23}$ MW: 1101.24   EINECS: 229-880-6
LD$_{50}$:   6.7 mg/kg (M, i.p.); 2 mg/kg (M, i.v.); 165 mg/kg (M, p.o.); 38.59 mg/kg (M, s.c.); 10.15 mg/kg (R, i.p.); 1600 g/kg (R, i.v.); 833 mg/kg (R, p.o.); 150 mg/kg (R, s.c.)
CN: 3,5-epoxypicene escin deriv.

sodium salt
RN: 20977-05-3   MF: unspecified   MW: unspecified   EINECS: 244-133-4
LD$_{50}$:   8299 µg/kg (M, i.p.); 4730 mg/kg (M, i.v.); 134 mg/kg (M, p.o.); 92.53 mg/kg (M, s.c.); 9180 µg/kg (R, i.p.); 8131 µg/kg (R, i.v.); 400 mg/kg (R, p.o.); 131 mg/kg (R, s.c.); 9130 µg/kg (g. p., i.v.); 5 mg/kg (rabbit, i.v.)

Escin

Extraction of *Aesculus hippocastanum* L. (horse-chestnut) and purification on cation-exchanger (H$^+$-form), resp. precipitation with cholesterol.

Reference(s):
extraction and purification:
DE 916 664 (Riedel-de Haen; appl. 1952).
DE 950 027 (Klinge; appl. 1951).
DAS 1 034 816 (VEB Arzneimittelwerk Dresden; appl. 1955).
DAS 1 045 597 (Dr. W. Schwabe; appl. 1953).
GB 820 787 (Klinge; appl. 1956).
GB 820 788 (Klinge; appl. 1956).
DE 1 058 208 (Klinge; appl. 1953).
DAS 1 095 989 (Madaus; appl. 11.3.1959).
US 3 163 636 (Klinge; 29.12.1964; D-prior. 14.6.1960).
DAS 1 182 385 (Chem. Fabrik Tempelhof; appl. 29.1.1962).
US 3 238 190 (Madaus; 1.3.1966; prior. 31.1.1961, 23.10.1963).
DOS 1 617 570 (J. Klosa; appl. 13.4.1967).
DOS 1 617 581 (Knoll; appl. 11.8.1967).
DAS 1 617 413 (Klinge; appl. 31.8.1967).
DE 1 667 884 (Knoll; appl. 20.1.1968).
DOS 1 902 608 (Nattermann; appl. 20.1.1969; A-prior. 31.5.1968).
DOS 2 339 760 (Klinge; appl. 6.8.1973).
DAS 2 733 204 (LEK; appl. 22.7.1977; YU-prior. 12.8.1976).

"water soluble" *(X-ray amorphous)* escin:
DE 1 282 852 (Madaus; appl. 14.12.1962).
DOS 1 902 609 (Nattermann; appl. 20.1.1969).
DOS 2 257 755 (LEK; appl. 24.11.1972; YU-prior. 6.12.1971).
GB 1 550 845 (Madaus; appl. 7.7.1976; D-prior. 11.7.1975).

separation of α- *and* β-escin:
DAS 1 125 117 (Klinge; appl. 14.6.1960).
US 3 110 711 (Klinge; 12.11.1963; D-prior. 14.6.1960).

conversion of β- *into* α-escin:
US 3 450 691 (Klinge; 17.6.1969; appl. 7.6.1967).

Formulation(s):   amp. 5 mg (as sodium salt); cps. 2 mg; drg. 10 mg, 15 mg, 20 mg; gel 1 g/100 g-comb.; s. r. drg. 40 mg

Trade Name(s):

D: Essaven (Nattermann)-comb.
Galleb forte (Hoyer)-comb.
Heweven (Hevert)-comb.
Opino, Gel (Troponwerke)-comb.
Opino retard (Troponwerke)-comb.
Opino spezial (Troponwerke)-comb.
Pe-Ce Ven (Terra-Bio-Chemie)-comb.
Proveno (Madaus)-comb.
Reparil (Madaus)
Revicain (Wiedemann)-comb.

Veno Kattwiga (Kattwiga)-comb.
Venoplant (Schwabe)-comb.
Venostasin (Klinge)-comb.
F: Flogencyl (Parke Davis)
Reparil (Madaus)
numerous combination preparations
I: Bres (Farmacologico Milanese)-comb.
Dermocinetic (Irbi)-comb.
Essaven (Nattermann)-comb.
Etascin (Rorer)-comb.

Opino (Bayropharm)-comb.
Premium (SIT)-comb.
Rectoreparil (IBI)-comb.
Reparil (IBI)-comb.
Somatoline (Manetti Roberts)-comb.
Tioscina (Inverni della Beffa)-comb.
J: Tochief (Ohta)
Tochikinon (Toho)
Yochimin (Choseido-Seiyaku)

Esmolol

ATC: C07AB09
Use: anti-arrhythmic, β-adrenoceptor antagonist, perioperative prophylactic use in supraventricular tachycardia

RN: 81147-92-4   MF: $C_{16}H_{25}NO_4$ MW: 295.38
CN: (±)-4-[2-hydroxy-3-[(1-methylethyl)amino]propoxy]benzenepropanoic acid methyl ester

hydrochloride
RN: 81161-17-3   MF: $C_{16}H_{25}NO_4 \cdot HCl$ MW: 331.84
LD_{50}:   93 mg/kg (M, i.v.);
71 mg/kg (R, i.v.);
32 mg/kg (dog, i.v.)

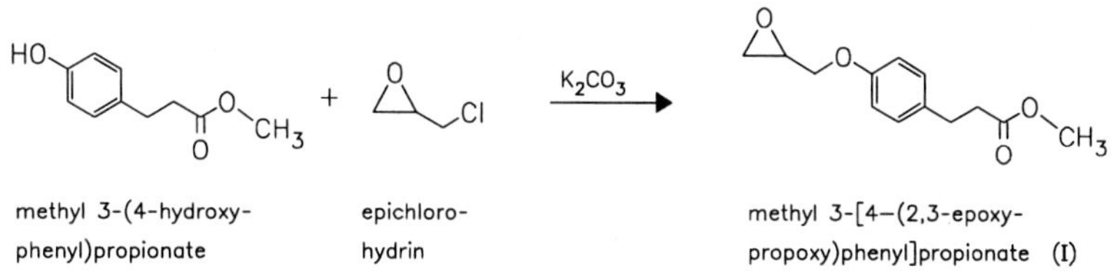

methyl 3-(4-hydroxy-phenyl)propionate   +   epichloro-hydrin $\xrightarrow{K_2CO_3}$ methyl 3-[4-(2,3-epoxy-propoxy)phenyl]propionate   (I)

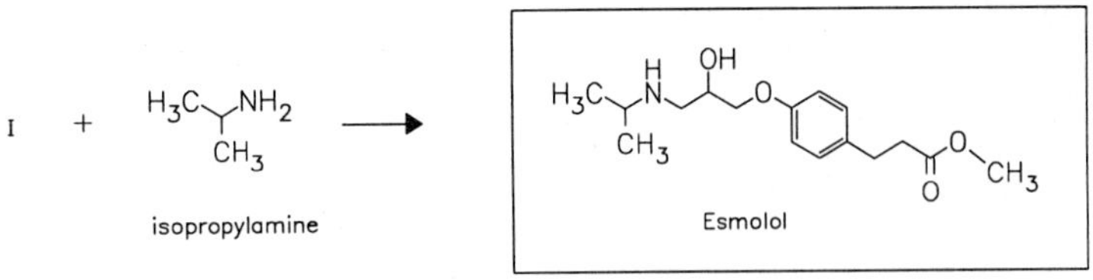

I + isopropylamine → Esmolol

Reference(s):

EP 41 491 (Hässle; appl. 27.5.1981; S-prior. 2.6.1980).

EP 53 435 (American Hospital Supply; appl. 29.10.1981; USA-prior. 28.11.1980).

Erhardt, P.W. et al.: J. Med. Chem. (JMCMAR) **25**, 1408 (1982).

US 4 387 103 (American Hospital Supply; 7.6.1983; prior. 28.11.1980).

injectable formulation:

US 4 857 552 (Du Pont; 15.8.1989; prior. 8.6.1988).

US 4 593 119 (American Hospital Supply; 3.6.1986; prior. 28.11.1980).

alternative synthesis:

ES 549 138 (Sune Coma; appl. 21.11.1985).

Formulation(s):   amp. 2.5 g/10 ml; vial 100 mg/10 ml (as hydrochloride)

Trade Name(s):

D: Brevibloc (Baxter)   F: Brevibloc (Isotec; 1989)   USA: Brevibloc (Ohmeda; 1987)

Estazolam

ATC: N05CD04
Use: hypnotic, sedative, tranquilizer

RN: 29975-16-4   MF: $C_{16}H_{11}ClN_4$ MW: 294.75   EINECS: 249-982-4
LD_{50}: 600 mg/kg (M, p.o.);
         2500 mg/kg (R, p.o.)
CN: 8-chloro-6-phenyl-4H-[1,2,4]triazolo[4,3-a][1,4]benzodiazepine

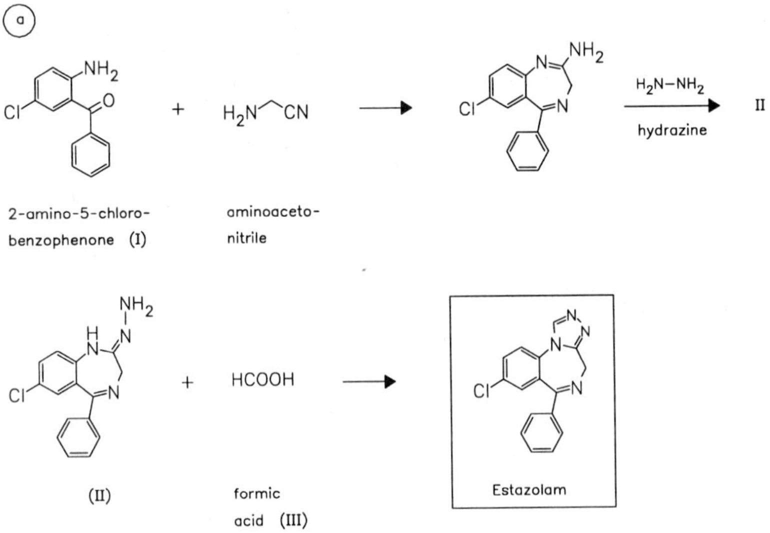

2-amino-5-chloro-   aminoaceto-
benzophenone (I)   nitrile

(II)   formic
             acid (III)

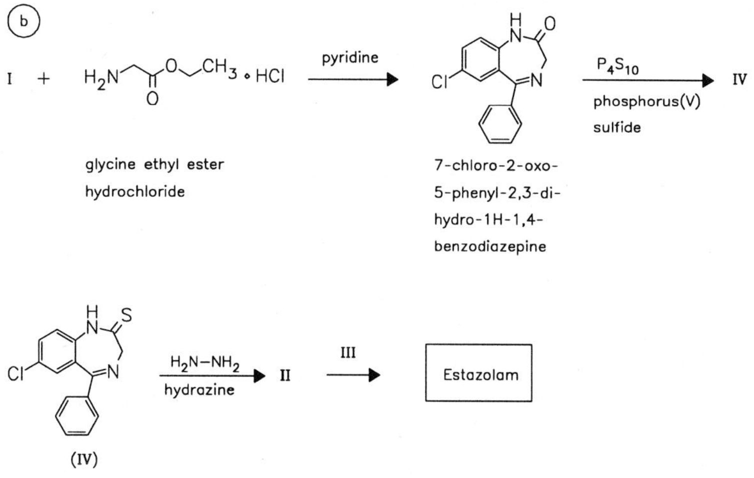

ⓑ

I + H₂N⌒O⌒CH₃ • HCl   —pyridine→

glycine ethyl ester
hydrochloride

7-chloro-2-oxo-
5-phenyl-2,3-di-
hydro-1H-1,4-
benzodiazepine

—P₄S₁₀→ IV
phosphorus(V)
sulfide

    H₂N–NH₂
    —hydrazine→ II   —III→   [Estazolam]

(IV)

Reference(s):

US 3 701 782 (Upjohn; 31.10.1972; prior. 10.2.1972).
Hester, J.B. et al.: J. Med. Chem. (JMCMAR) **14**, 1078 (1971).
US 4 116 956 (Takeda; 26.9.1978; J-prior. 5.11.1968, 17.12.1968, 25.12.1968, 13.2.1968).
DOS 1 955 349 (Takeda; appl. 4.11.1969; J-prior. 5.11.1968).
DOS 1 965 894 (Takeda; appl. 4.11.1969; J-prior. 5.11.1968, 17.12.1968, 25.12.1968, 13.2.1969).
DOS 2 012 190 (Upjohn; appl. 14.3.1970; USA-prior. 17.3.1969).
US 3 987 052 (Upjohn; 19.10.1976; prior. 17.3.1969, 29.10.1969).
DOS 2 114 441 (Takeda; appl. 25.3.1971; J-prior. 27.3.1970, 23.4.1970, 28.5.1970).
US 4 102 881 (Takeda; 25.7.1978; J-prior. 27.3.1970, 23.4.1970, 28.5.1970).
DOS 2 302 525 (Upjohn; appl. 19.1.1973; USA-prior. 31.1.1972).

review:
Schulte, E.: Dtsch. Apoth. Ztg. (DAZEA2) **115**, 1253, 1828 (1975).

Formulation(s):   tabl. 1 mg, 2 mg

Trade Name(s):
F: Nuctalon (Cassenne; 1978)   I: Esilgan (Cyanamid; 1983);   J: Eurodin (Takeda; 1975)
                                                wfm   USA: ProSom (Abbott)

Estradiol
(Oestradiol)

ATC: G03CA03
Use: estrogen

RN: 50-28-2   MF: C₁₈H₂₄O₂   MW: 272.39   EINECS: 200-023-8
CN: (17β)-estra-1,3,5(10)-triene-3,17-diol

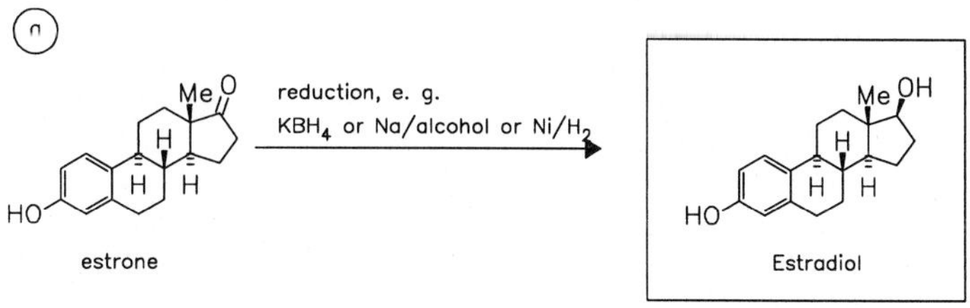

ⓝ

estrone

reduction, e. g.
KBH₄ or Na/alcohol or Ni/H₂

Estradiol

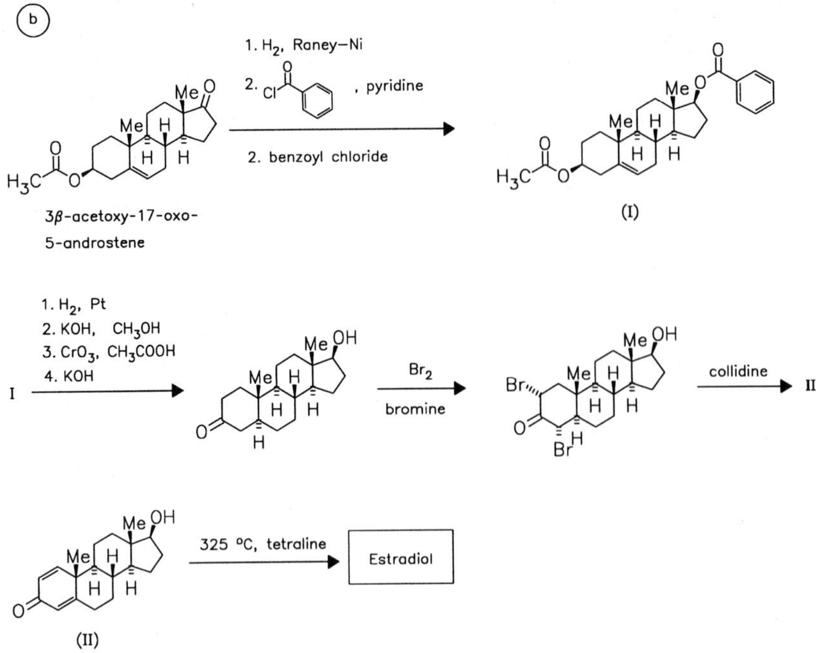

b

1. H₂, Raney–Ni

2. benzoyl chloride, pyridine

3β-acetoxy-17-oxo-5-androstene

(I)

I →
1. H₂, Pt
2. KOH, CH₃OH
3. CrO₃, CH₃COOH
4. KOH

→ Br₂ bromine → collidine → II

325 °C, tetraline → Estradiol

(II)

Reference(s):

a Ehrhart, Ruschig, **III**, 317.
US 2 096 744 (Schering Corp.; 1937; D-prior. 1932).
DRP 698 796 (Schering AG; appl. 1932).

b Inhoffen, H.H.; Zühlsdorff, G.: Ber. Dtsch. Chem. Ges. (BDCGAS) **74**, 1911 (1941).
US 2 361 847 (Schering Corp. 1944; D-prior. 1937).
Ullmanns Encykl. Tech. Chem., 3. Aufl., Vol. **8**, 657.

starting material:
The Merck Index, 12th Ed., 630 (1996).

alternative syntheses:
GB 485 388 (Lab. Franç. de Chimiothérapie; appl. 1936).
US 2 225 419 (Schering Corp.; 1940; D-prior. 1937).
US 3 128 238 (Lilly; 7.4.1974; appl. 14.9.1962).

total synthesis:
Eder, U. et al.: Chem. Ber. (CHBEAM) **109**, 2948 (1976).

Formulation(s): gel 0.5 mg/g, 1 mg/g; tabl. 2 mg, 4 mg; transdermal plaster 0.75 mg, 1.5 mg, 2 mg, 3 mg, 4 mg, 8 mg; vaginal tabl. 0.025 mg

Trade Name(s):

D: Aknefug Emulsion (Wolff)-comb.
Cerella (Asche)
Crinohermal fem. (Hermal)-comb.
Cutanum (Jenapharm)
DERMESTRIL (Opfermann)
Estracomb TTS (Novartis Pharma)-comb.
Estraderm (Novartis Pharma)

Estramon (Hexal)
Estrifam /-forte (Novo Nordisk; Rhône-Poulenc Rorer)
ESTRING (Pharmacia & Upjohn)
Evorel (Janssen-Cilag)
Fem7 (Merck)
Kliogest (Novo Nordisk; Rhône-Poulenc Rorer)-comb.
Linoladiol (Wolff)-comb.

Linoladiol-H (Wolff)-comb.
Menorest (Novo Nordisk; Rhône-Poulenc Rorer)
Osmil (Novartis Pharma)-comb.
Sandrena (Organon)
Sisare Gel (Nourypharma)
Tradelia (Sanofi Winthrop)
Trisequens (Novo Nordisk; Rhône-Poulenc Rorer)-comb.

Vagifem (Novo Nordisk;
Rhône-Poulenc Rorer)
F: Estrofem (Novo Nordisk)
Oestrogel (Besins-
Iscovesco)
Prémarin (Wyeth-Ayerst)-
comb.
Trisequens (Novo
Nordisk)-comb.

GB: Cycloprogynova (ASTA
Medica)
Hormonin (Shire)
Trisequens (Novo)-comb.
numerous generics
I: Estraderm (Ciba-Geigy)
Progynon (Schering)
J: Ovahormon Pasta (Teikoku
Zoki)

USA: Alora (Procter & Gamble)
Climara (Berlex)
Estraderm (Novartis)
Estring (Pharmacia &
Upjohn)
Estro-Plus (Rocky Mtn.)-
comb.
FemPatch (Parke Davis)
Vivelle (Novartis)

Estradiol benzoate

(Oestradiolbenzoat)

ATC: G03CA
Use: estrogen

RN: 50-50-0   MF: $C_{25}H_{28}O_3$ MW: 376.50   EINECS: 200-043-7
CN: (17β)-estra-1,3,5(10)-triene-3,17-diol 3-benzoate

estradiol   benzoyl   Estradiol benzoate
(q. v.)   chloride

Reference(s):
a   US 2 054 271 (Schering Corp.; 1933; D-prior. 1932).
GB 485 388 (Lab. Franç. de Chimiothérapie; appl. 1936).
b   DRP 641 994 (Schering AG; appl. 1932).

alternative synthesis:
US 2 225 419 (Schering Corp.; 1940; D-prior. 1937).
US 2 156 599 (Ciba; 1939; CH-prior. 1936).

Formulation(s):   amp. 2 mg/ml, 10 mg/ml, 50 mg/2 ml; sol. 5 mg/100 ml

Trade Name(s):
D: Alpicort F (Wolff)-comb.
Jephagynon (Wolff)-comb.
Ney Normin (vitOrgan)-
comb.
Syngynon (Jenapharm)
F: Benzo-Gynoestryl
(Roussel)
Dermestril (Sanofi
Winthrop)
Estraderm (Novartis)
Estreva (Théramex)
Estrofem (Specia)
Menorest (Specia)
Oesclim (Fournier)

GB: Benztrone (Paines &
Byrne); wfm
I: Benztrone (Amsa)
Duo-Ormogyn (Amsa)-
comb.
Menovis (Parke Davis)-
comb.
Progynon (Schering)
J: Estradin Susp. (Santen-
Yamanouchi)
Femihormon (Tokyo Hosei)
Follikelmon (Kyorin)
Ovahormon Benzoat Susp.
(Teikoku Zoki)

Pelanin Inj. (Mochida)
Profollior B (Schering)
Progynon B (Nihon
Schering)
USA: Gynetone Inj. (Schering)-
comb.; wfm
Testradiol (Consolidated
Midland)-comb.; wfm
Testradiol (Truxton)-comb.;
wfm
Trimonal (Vitarine)-comb.;
wfm
numerous generics; wfm

Estradiol cypionate
(Oestradiol-17-cyclopentylpropionat)

ATC: G03C
Use: estrogen

RN: 313-06-4   MF: $C_{26}H_{36}O_3$ MW: 396.57   EINECS: 206-237-8
LD$_{50}$:   >1 g/kg (M, i.p.)
CN: (17β)-estra-1,3,5(10)-triene-3,17-diol 17-cyclopentanepropanoate

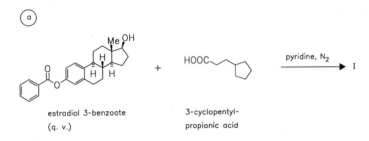

a

estradiol 3-benzoate
(q. v.)

3-cyclopentyl-
propionic acid

pyridine, N_2 I

KOH, CH$_3$OH

(I)

Estradiol cypionate

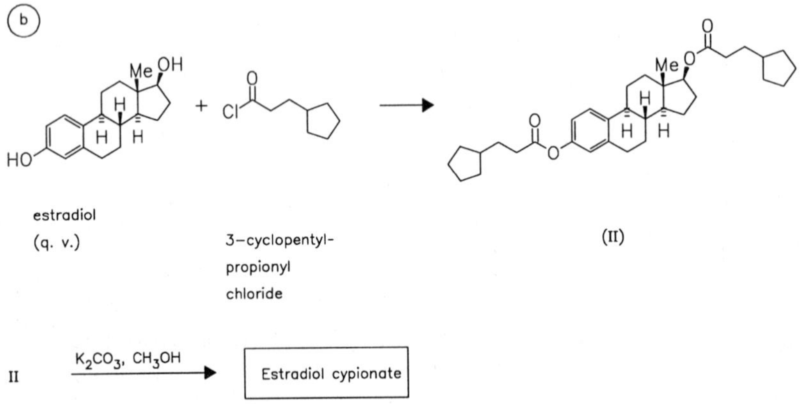

b

estradiol
(q. v.)

3-cyclopentyl-
propionyl
chloride

(II)

II $\xrightarrow{K_2CO_3,\ CH_3OH}$ Estradiol cypionate

Reference(s):
a   FR 1 215 503 (Lab. Rolland; appl. 1955).
b   US 2 611 773 (Upjohn; 1952; prior. 1951).

Formulation(s):   amp. 10 mg (1 mg/ml), 25 mg (5 mg/ml), 50 mg (5 mg/ml)

Trade Name(s):
D: Femovirin Amp. (Albert-
       Roussel)-comb.; wfm
F: Oestradiol-retard Rolland
       (L'Hépatrol); wfm

I: Cicloestradiolo (Farmigea);
      wfm
      Estradiolo Depositum
      (Orma); wfm

Neoginon Depositum
(Lusofarmaco); wfm

J: Depo-Estradiol (Upjohn-
Kodama)

USA: Depo-Estradiol (Upjohn);
wfm

Estradiol valerate

ATC: G03CA
Use: estrogen

RN: 979-32-8 MF: $C_{23}H_{32}O_3$ MW: 356.51 EINECS: 213-559-2
LD_{50}: 1224 mg/kg (M, p.o.)
CN: (17β)-estra-1,3,5(10)-triene-3,17-diol 17-pentanoate

17β—estradiol
(q. v.)

valeric anhydride

estradiol divalerate (I)

Estradiol valerate

Reference(s):
US 2 205 627 (Ciba; 1940; CH-prior. 1936).
US 2 233 025 (Ciba; 1941; CH-prior. 1936).

use as antidepressant:
DOS 2 758 549 (Schering AG; appl. 23.12.1977).

Formulation(s): amp. 5 mg/ml, 10 mg/ml; drg. 1 mg, 2 mg; drops 2 mg/0.5 ml; f. c. tabl. 2 mg; tabl. 2 mg, 4 mg

Trade Name(s):

D:	Gynokadin (Kade)	Progynova (Schering)
	Merimono (Novartis	J: Pelanin Depot (Mochida)
	Pharma)	Progynon Depot (Nihon
	Progynon Depot (Schering)	Schering)
	Progynova (Schering)	USA: Ardefem (Burgin-Arden);
	numerous combination	wfm
	preparations	Atladiol (Atlas); wfm
F:	Climène (Schering)-comb.	Delestrogen (Squibb); wfm
	Divina (Innothéra)-comb.	Depogen (Hyrex); wfm
	Progynova (Schering)	Depogen (Sig); wfm
GB:	Cyclo-Progynova (ASTA	Dioval (Keene); wfm
	Medica)-comb.	Ditate DS (Savage); wfm
	Progynova (Schering)	Dura-Estate (Ries); wfm
I:	Gravibinan (Schering)-	Dura-Estradiol (Myers-
	comb.	Carter); wfm
	Gynodian Depot	Dura-Estradiol
	(Schering)-comb.	(Ruckstuhl); wfm
	Progynon Depot (Schering)	Duratrad (Ascher); wfm

Estate (Savage); wfm
Estral-L (Pasadena Res.);
wfm
Estraval (Kay); wfm
Estravel-P.A. (Tutag); wfm
Feminate (Western Res.);
wfm
Femogen L.A. (Fellows);
wfm
Repo-Estra (Central); wfm
Repo-Estro Med (Medics);
wfm
Reposo E (Canfield); wfm
Retestrin (Rocky Mtn.);
wfm
Span-Est (Scrip); wfm
Testaval (Legere); wfm
Valergen (Hyrex); wfm

numerous combination
preparations; wfm

Estradiol undecylate
(Oestradiolundecanoat)

ATC: G03CA
Use: estrogen

RN: 3571-53-7　MF: $C_{29}H_{44}O_3$　MW: 440.67　EINECS: 222-677-3
CN: (17β)-estra-1,3,5(10)-triene-3,17-diol 17-undecanoate

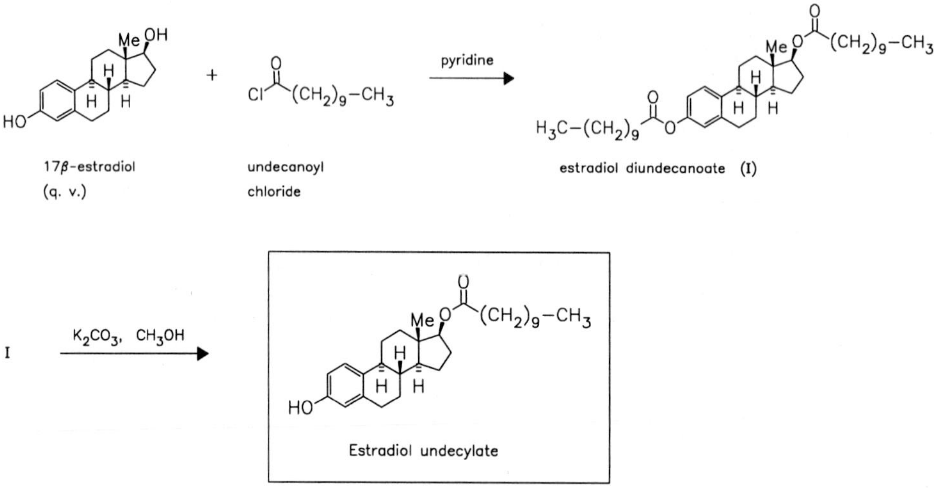

17β-estradiol
(q. v.)

undecanoyl
chloride

estradiol diundecanoate　(I)

Estradiol undecylate

Reference(s):
US 2 990 414 (Syntex; 27.1.1961; appl. 18.3.1948; MEX-prior. 26.3.1957).

Formulation(s):　amp. 100 mg/ml

Trade Name(s):
F:　Oestradiol-Retard　　　J:　　Depogin (Shionogi)
　　Théramex (Théramex)　　USA:　Delestrec (Squibb); wfm

Estramustine phosphate

ATC: L01AA
Use: antineoplastic

RN: 4891-15-0　MF: $C_{23}H_{32}Cl_2NO_6P$　MW: 520.39　EINECS: 225-512-3
CN: (17β)-estra-1,3,5(10)-triene-3,17-diol 3-[bis(2-chloroethyl)carbamate] 17-(dihydrogen phosphate)

disodium salt
RN: 52205-73-9　MF: $C_{23}H_{30}Cl_2NNa_2O_6P$　MW: 564.35

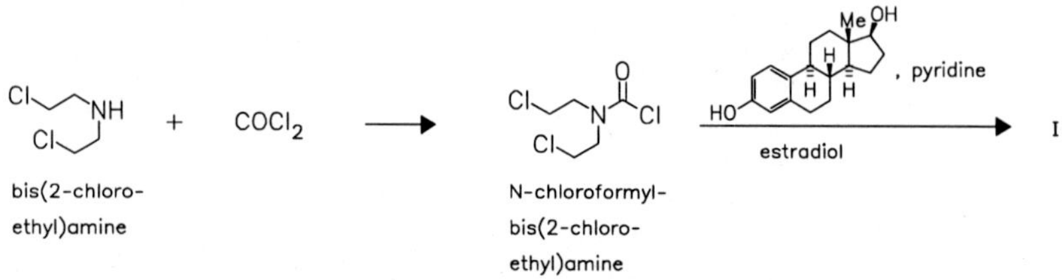

bis(2-chloro-
ethyl)amine

N-chloroformyl-
bis(2-chloro-
ethyl)amine

estradiol

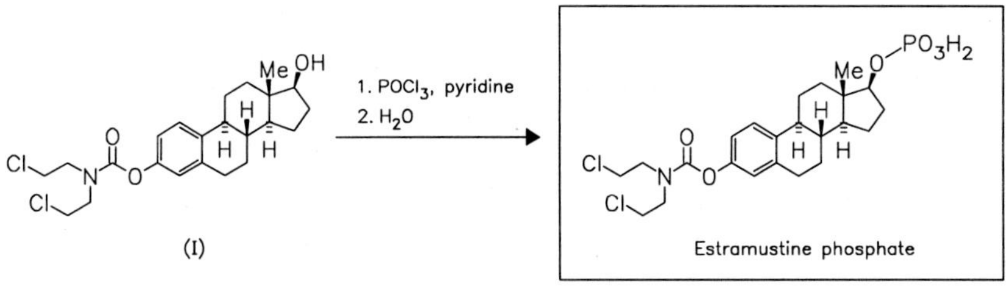

Estramustine phosphate

Reference(s):
GB 1 016 959 (Leo; appl. 9.4.1963; valid from 24.3.1964).
US 3 299 104 (Leo; 17.1.1967; GB-prior. 9.4.1963).
GB 1 523 035 (Leo; appl. 10.3.1976; valid from 8.3.1977).

complex compounds with alcohols (for purification):
US 4 115 414 (Leo; 19.9.1978; GB-prior. 10.3.1976).
DE 2 710 293 (Leo; appl. 9.3.1977; GB-prior. 10.3.1976)

Formulation(s):   cps. 151.8 mg, 303.6 mg (as disodium salt); vial 621 mg (as meglumine salt)

Trade Name(s):

D:	Cellmustin (cell pharm)	F:	Estracyt (Pharmacia & Upjohn)	J:	Estracyt (Nippon Shinyaku)
	Estracyt (Pharmacia & Upjohn)	GB:	Estracyt (Pharmacia & Upjohn)	USA:	Emcyt (Pharmacia & Upjohn; as sodium salt)
	Multosin (Takeda)				
	Prostamustin (Azupharma)	I:	Estracyt (Farmitalia)		

Estriol

ATC: G03CA04
Use: estrogen

RN: 50-27-1   MF: $C_{18}H_{24}O_3$ MW: 288.39   EINECS: 200-022-2
LD_{50}:   >2 g/kg (R, p.o.)
CN: (16α,17β)-estra-1,3,5(10)-triene-3,16,17-triol

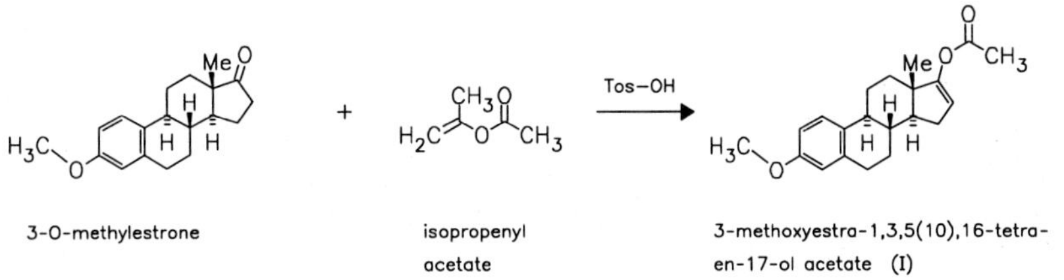

3-O-methylestrone

isopropenyl acetate

3-methoxyestra-1,3,5(10),16-tetra-en-17-ol acetate (I)

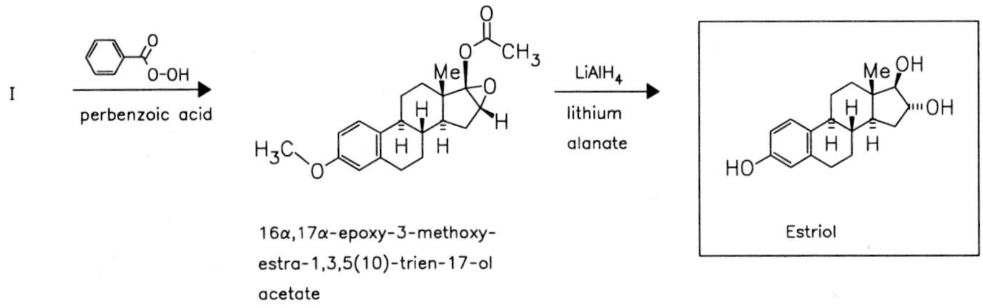

perbenzoic acid

16α,17α-epoxy-3-methoxy-estra-1,3,5(10)-trien-17-ol acetate

Estriol

Reference(s):
Gallagher, T.F.: J. Am. Chem. Soc. (JACSAT) **76**, 2943 (1954).

alternative syntheses:
Butenandt, A.;. Schäffler, E.L.: Z. Naturforsch. (ZNTFA2) **1**, 82 (1946).
Huffmann, M.N. et al.: Science (Washington, D.C.) (SCIEAS) **100**, 312 (1944).
Huffmann, M.N.; Lott, M.H.: J. Am. Chem. Soc. (JACSAT) **69**, 1835 (1947).
US 1 967 351 (Pres. and Board of Trustees of St. Louis; 1934; appl. 1930).
The Merck Index, 12th Ed., 631 (Rahway 1996).

Formulation(s):   cream 0.5 mg/g; drg. 1 mg; f. c. tabl. 2 mg; ovula 0.03 mg, 0.5 mg; tabl. 1 mg, 2 mg

Trade Name(s):
D: Cordes (Ichthyol)   F: Gydrelle (Iprad)   Ortho Gynest Depot (Cilag)
      Estiol (Jenapharm)   Physiogine (Organon)   Ovestin (Organon Italia)
      Gynäsan (Bastian-Werk)-   Trophicrème (Sanofi   J: Climatol (Santen)
      comb.   Winthrop)   Estriel (Mochida)
      Oekolp (Kade)   Trophigil (Sanofi   Molin (Teikoku Zoki)
      Ortho-Gynest (Janssen-   Winthrop)-comb.   Ovapause (Organon)
      Cilag)   GB: Hormonin (Shire)-comb.   Season (Teikoku Zoki)
      Ovestin (Organon)   Ortho-Gynest (Janssen- USA: Estro Plus Tab. (Rocky
      Ovo-Vinces 200 (Wolff)   Cilag)   Mtn.)-comb.; wfm
      Synapause (Nourypharma)   Ovestin (Organon)   Hormonin (Carnrick)-
      Xapro (Jenapharm)   Trisequens (Novo   comb.; wfm
      numerous combination   Nordisk)-comb.
      preparations   I: Colpogyn (Angelini)

Estriol succinate

ATC: G03C
Use: estrogen

RN: 514-68-1   MF: $C_{26}H_{32}O_9$ MW: 488.53   EINECS: 208-185-1
CN: (16α,17β)-estra-1,3,5(10)-triene-3,16,17-triol 16,17-bis(hydrogen butanedioate)

disodium salt
RN: 113-22-4   MF: $C_{26}H_{30}Na_2O_9$ MW: 532.50

estriol   succinic   Estriol succinate
(q. v.)   anhydride

Reference(s):
GB 879 014 (Organon; appl. 26.5.1960; NL-prior. 29.5.1959).

Formulation(s):   tabl. 2 mg, 4 mg (as disodium salt); vial 20 mg (as disodium salt)

Trade Name(s):
D: Orgastyptin (Organon   Synapause (Nourypharma);   F: Synapause (Organon); wfm
      Teknika); wfm   wfm   I: Ovestin (Organon Italia)

Estrone

(Oestron)

ATC: G03CA07; G03CC04
Use: estrogen

RN: 53-16-7 MF: $C_{18}H_{22}O_2$ MW: 270.37 EINECS: 200-164-5
CN: 3-hydroxyestra-1,3,5(10)-trien-17-one

(a)

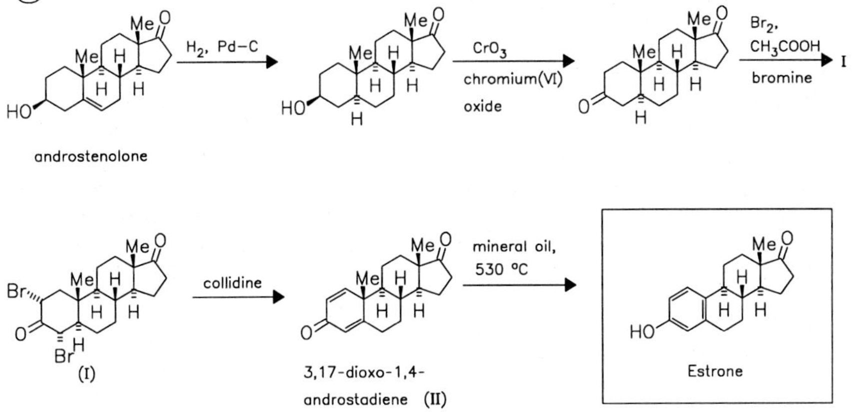

androstenolone
H2, Pd-C
CrO3
chromium(VI)
oxide
Br2,
CH3COOH
bromine
I
collidine
(I)
mineral oil,
530 °C
3,17-dioxo-1,4-
androstadiene (II)
Estrone

(b)

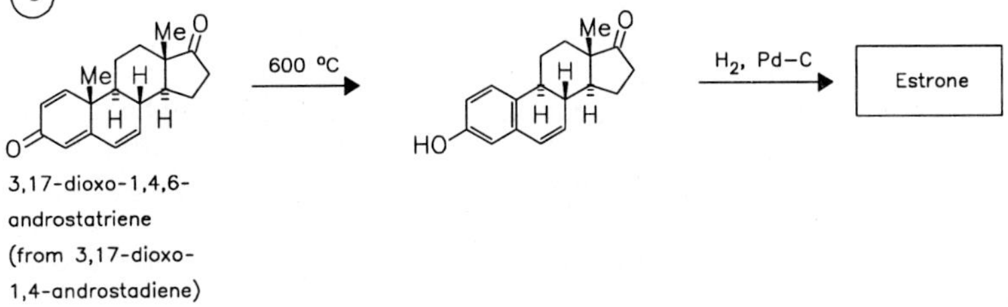

600 °C
H2, Pd-C
Estrone
3,17-dioxo-1,4,6-
androstatriene
(from 3,17-dioxo-
1,4-androstadiene)

(c)

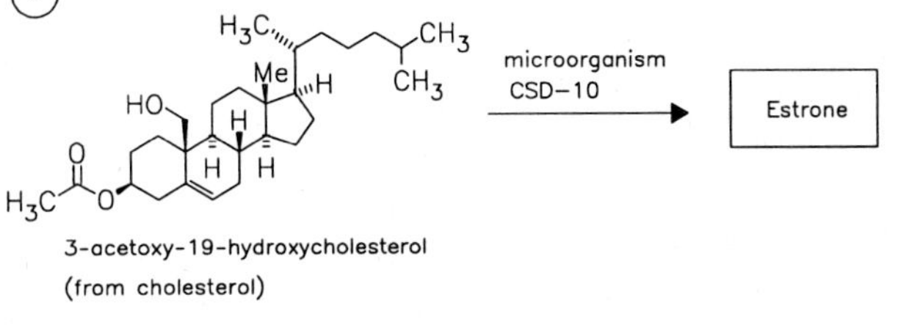

microorganism
CSD-10
Estrone
3-acetoxy-19-hydroxycholesterol
(from cholesterol)

(d)

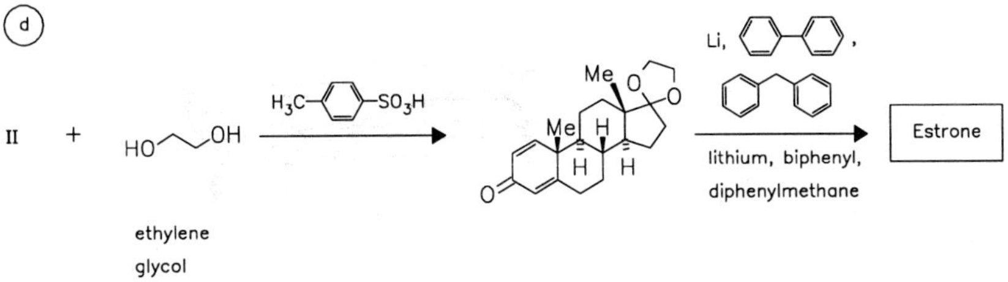

II
+
ethylene
glycol
Li,
lithium, biphenyl,
diphenylmethane
Estrone

Reference(s):

a,b Ehrhart, Ruschig, **III**, 315.
c Sih, Ch. et al.: J. Am. Chem. Soc. (JACSAT) **87**, 2765 (1965).

starting material:
Kalvoda, J. et al.: Helv. Chim. Acta (HCACAV) **46**, 1361 (1963).
d   Dryden, H.L. et al.: J. Am. Chem. Soc. (JACSAT) **86**, 742 (1964).

production of conjugated estrogens:
US 2 565 115 (Squibb; 1951; prior. 1948).
US 2 720 483 (Olin Mathieson; 1955; prior. 1951).
US 4 154 820 (Akzona; 15.5.1979; prior. 26.9.1977, 23.2.1976).

total syntheses:
EP 37 973 (Hoechst; appl. 2.4.1981; D-prior. 12.4.1980).
Morand, P.; Lyall, J.: Chem. Rev. (Washington, D. C.) (CHREAY) **68**, 85 (1968).
Velluz, L. et al.: Angew. Chem. (ANCEAD) **72**, 725 (1960).
Velluz, L. et al.: Angew. Chem. (ANCEAD) **77**, 185 (1965).
Smith, H. et al.: J. Chem. Soc. (JCSOA9)**1963**, 5072.
Smith, H. et al.: Experientia (EXPEAM) **19**, 177 (1963).
Anachenko, S.N.; Torgov, J.V.: Tetrahedron Lett. (TELEAY) **1963**, 1553; **1964**, 171.
Blickenstaff, R.T.; Ghosh, A.C.; Wolf, G.C.: Total Synthesis of Steroids (Organic Chemistry Vol. **30**) p. 58-63, 142-145, Academic Press, New York, London 1974.

Formulation(s):   e. g. 1.4 mg in comb.; amp. 20 mg; drg. 0.625 mg, 1.25 mg, 2.5 mg; vial 20 mg

Trade Name(s):

D: Coniugen (Klinge)-comb.; wfm
GT 50 B (Gewo)-comb.; wfm
Menrium (Roche)-comb.; wfm
Oestro-Feminal (Mack, Illert.; as estrogen conjugate)-comb.; wfm
Ovaribran (Thomae; as estrogen conjugate)-comb.; wfm
Ovowop (Hor-Fer-Vit)-comb.; wfm
F: Colpormon (Lipha Santé)
Prémarin (Wyeth-Lederle)
Synergon (Lipha Santé)-comb.
GB: Hormonin (Shire)-comb.
I: Emopremarin (Wyeth; as estrogen conjugate)

Premarin (Wyeth; as estrogen conjugate)
Prempak (Wyeth)-comb.
J: Estropan (Mochida)-comb.
USA: Di-Genik (Savage)-comb.; wfm
Di-Met (Organon)-comb.; wfm
Duogen (Smith, Miller & Patch)-comb.; wfm
Estro-V (Webcon); wfm
Estrusol (Smith, Miller & Patch); wfm
Follestrol (Blue Line); wfm
Foygen (Foy); wfm
Hormestrin (Smith, Miller & Patch)-comb.; wfm
Mal-O-Fem (Fellows)-comb.; wfm
Menagen (Parke Davis); wfm

Menformon (Organon); wfm
Natural Estrogenic Substance (Legere); wfm
Nestronaq (Noyes); wfm
Ogen (Abbott; as estropipate); wfm
Prinn (Scirp); wfm
Propagon-S (Spanner); wfm
Spanestrin (Savage)-comb.; wfm
Theelin (Parke Davis); wfm
Wynastron (Wyeth); wfm
further combination preparations and generic; wfm

Etacrynic acid

(Äthacrynsäure; Acide étacrynique)

ATC: C03CC01
Use: diuretic

RN: 58-54-8   MF: $C_{13}H_{12}Cl_2O_4$ MW: 303.14   EINECS: 200-384-1
LD_{50}:   176 mg/kg (M, i.v.); 600 mg/kg (M, p.o.);
1 g/kg (R, p.o.)
CN: [2,3-dichloro-4-(2-methylene-1-oxobutyl)phenoxy]acetic acid

sodium salt
RN: 6500-81-8   MF: $C_{13}H_{11}Cl_2NaO_4$ MW: 325.12

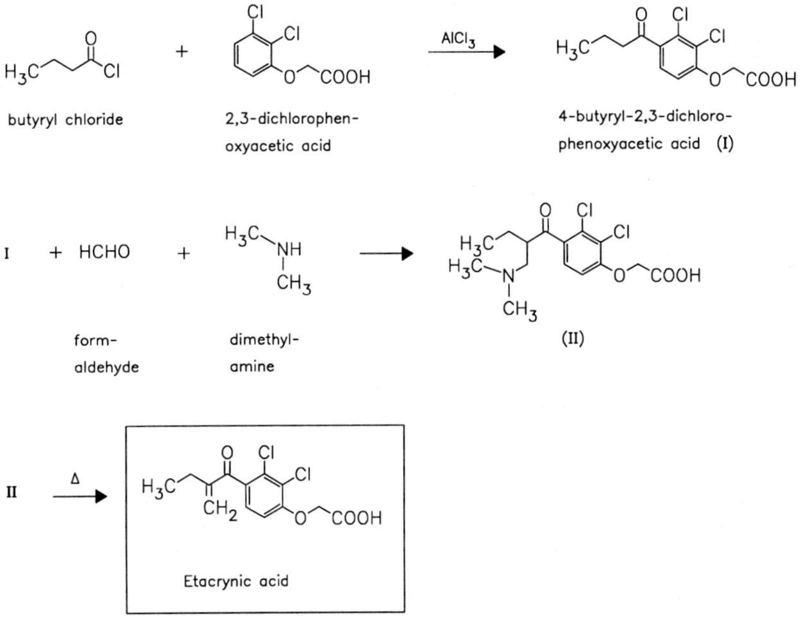

butyryl chloride + 2,3-dichlorophen-oxyacetic acid → 4-butyryl-2,3-dichloro-phenoxyacetic acid (I)

I + HCHO + dimethyl-amine → (II)

formaldehyde

II $\xrightarrow{\Delta}$

Etacrynic acid

Reference(s):
BE 612 755 (Merck & Co. appl. 17.1.1962; USA-prior. 19.1.1961).
US 3 255 241 (Merck & Co.; 7.6.1966; prior. 19.1.1961, 6.12.1961).

alternative synthesis:
DE 1 276 030 (Merck & Co.; appl. 18.12.1964; USA-prior. 23.12.1963).

Formulation(s): tabl. 25 mg, 50 mg (as free acid); vial 53.6 mg (as sodium salt)

Trade Name(s):
D: Hydromedin (Merck Sharp & Dohme)
F: Edecrine (Merck Sharp & Dohme); wfm
GB: Edecrin (Merck Sharp & Dohme)
I: Ac Etacr (Formulario Naz.) Edecrin (Merck Sharp & Dohme)
Reomax (Bioindustria)
J: Edecril (Merck-Banyu)
USA: Edecrin (Merck Sharp & Dohme)

Etafenone

ATC: C01DX07
Use: coronary vasodilator

RN: 90-54-0 MF: $C_{21}H_{27}NO_2$ MW: 325.45 EINECS: 202-002-9
CN: 1-[2-[2-(diethylamino)ethoxy]phenyl]-3-phenyl-1-propanone

hydrochloride
RN: 2192-21-4 MF: $C_{21}H_{27}NO_2 \cdot HCl$ MW: 361.91 EINECS: 218-587-9
LD_{50}: 28 mg/kg (M, i.v.); 352 mg/kg (M, p.o.);
20.8 mg/kg (R, i.v.); 716 mg/kg (R, p.o.);
50 mg/kg (dog, p.o.)

2'-hydroxy-3-
phenylpropio-
phenone

2-diethylamino-
ethyl chloride

Etafenone

Reference(s):
DAS 1 265 758 (S. p. A. Lab. Guidotti; appl. 25.5.1960).

Formulation(s):   amp. 25 mg; drg. 75 mg; s. r. cps. 50 mg (as hydrochloride)

Trade Name(s):
D: Baxacor (Helopharm); wfm   Pagano-Cor (Helopharm);   Cardilicor (Uji)
      Baxacor (Mack, Illert.);   wfm   Corodilan (Meiji)
      wfm   Seda-Baxacor   Dialicor (Kissei)
      Digi-Baxacor (Mack,   (Helopharm)-comb.; wfm   Esanthin-S (Kyoritsu
      Illert.)-comb.; wfm   Seda-Baxacor (Mack,   Yakuhin)
      Iso Baxacor (Helopharm)-   Illert.)-comb.; wfm   Etafenarin (Taiyo)
      comb.; wfm   I: Dialicor (Guidotti)   Korofenon (Nissin)
                                 J: Asamedel (Maruko)

Etamiphylline
(Dietamiphylline)

ATC: R03DA06
Use: cardiotonic, diuretic

RN: 314-35-2   MF: $C_{13}H_{21}N_5O_2$ MW: 279.34   EINECS: 206-244-6
LD_{50}:   1237 mg/kg (M, p.o.)
CN: 7-[2-(diethylamino)ethyl]-3,7-dihydro-1,3-dimethyl-1H-purine-2,6-dione

monohydrochloride
RN: 17140-68-0   MF: $C_{13}H_{21}N_5O_2 \cdot HCl$ MW: 315.81   EINECS: 241-204-1
LD_{50}:   127 mg/kg (M, i.v.)
camphersulfonate (1:1)
RN: 19326-29-5   MF: $C_{13}H_{21}N_5O_2 \cdot C_{10}H_{16}O_4S$ MW: 511.64   EINECS: 242-962-6
LD_{50}:   604 mg/kg (M, s.c.)

1. NaOH

2.

2. 2-diethylamino-
ethyl chloride

theophylline

Etamiphylline

Reference(s):
GB 669 070 (A. J.-M. Moussalli et al.; appl. 1949; F-prior. 1948).
Klosa, J.: Arch. Pharm. Ber. Dtsch. Pharm. Ges. (APBDAJ) **288/60**, 301 (1955).

Formulation(s):   75 mg in comb.; suppos. 200 mg, 500 mg; tabl. 100 mg (as camphersulfonate)

Etamivan

(Ethamivan)

ATC: R07AB04
Use: analeptic (central and respiratory
         stimulant)

RN: 304-84-7   MF: $C_{12}H_{17}NO_3$ MW: 223.27   EINECS: 206-157-3
LD$_{50}$:   15 mg/kg (M, i.v.); 67 mg/kg (M, p.o.);
         28 mg/kg (R, i.p.); 17 mg/kg (R, i.v.); 154 mg/kg (R, p.o.);
         30 mg/kg (dog, i.v.); 300 mg/kg (dog, p.o.)
CN: *N,N*-diethyl-4-hydroxy-3-methoxybenzamide

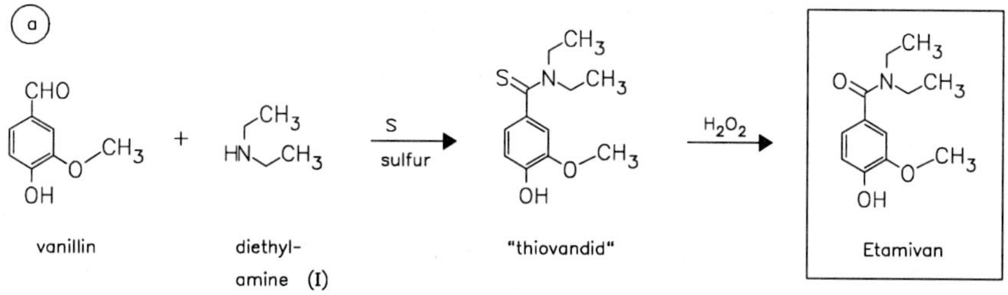

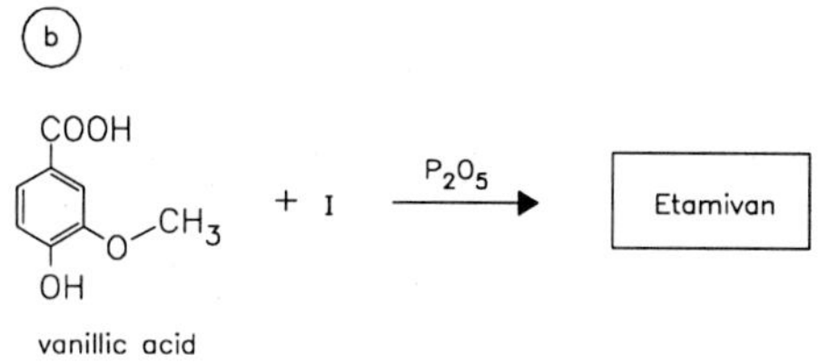

vanillin   diethyl-   "thiovandid"   Etamivan
                  amine   (I)

vanillic acid

Reference(s):
US 2 641 612 (Österr. Stickstoffwerke AG; 1953; A-prior. 1949).

Formulation(s):   drg. 10 mg in comb.; drops 20 mg/ml in comb.

Etamsylate

(Ethamsylate)

ATC: B02BX01
Use: hemostatic (capillary protective)

RN: 88-46-0   MF: $C_6H_6O_5S$ MW: 190.18   EINECS: 201-833-4
CN: 2,5-dihydroxybenzenesulfonic acid

diaethylammonium salt (1:1)
RN: 2624-44-4 MF: $C_6H_6O_5S \cdot C_4H_{11}N$ MW: 263.31 EINECS: 220-090-7
LD_{50}: 785 mg/kg (M, i.v.); 8300 mg/kg (M, p.o.);
1350 mg/kg (R, i.v.); 7500 mg/kg (R, p.o.)

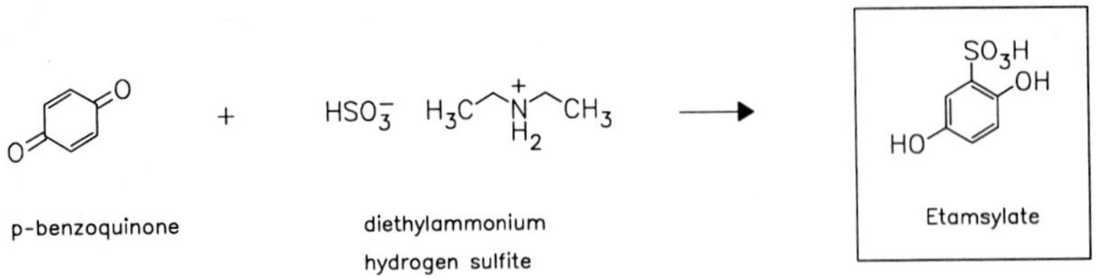

p-benzoquinone
diethylammonium hydrogen sulfite
Etamsylate

Reference(s):
GB 895 709 (Lab. OM S.A.; appl. 31.12.1959; CH-prior. 28.1.1959).

Formulation(s): amp. 250 mg/2 ml; tabl. 250 mg, 500 mg

Trade Name(s):
D: Altodor (Deutsche OM)
F: Dicynone (Synthélabo)
GB: Dicynene (Delandale)
I: Dicynone (Delalande Isnardi)
Eselin (Ravizza)
Transil (Malesci)-comb.
J: Aglumin (Eisai)
Dicynone (Torii)

Ethacridine
(Acrinol; Aethacridin)

ATC: B05CA08; D08AA01
Use: wound antiseptic, intestinal disinfectant

RN: 442-16-0 MF: $C_{15}H_{15}N_3O$ MW: 253.31 EINECS: 207-130-9
CN: 7-ethoxy-3,9-acridinediamine

lactate (1:1)
RN: 1837-57-6 MF: $C_{15}H_{15}N_3O \cdot C_3H_6O_3$ MW: 343.38 EINECS: 217-408-1
LD_{50}: 42 mg/kg (M, i.p.); 120 mg/kg (M, s.c.)

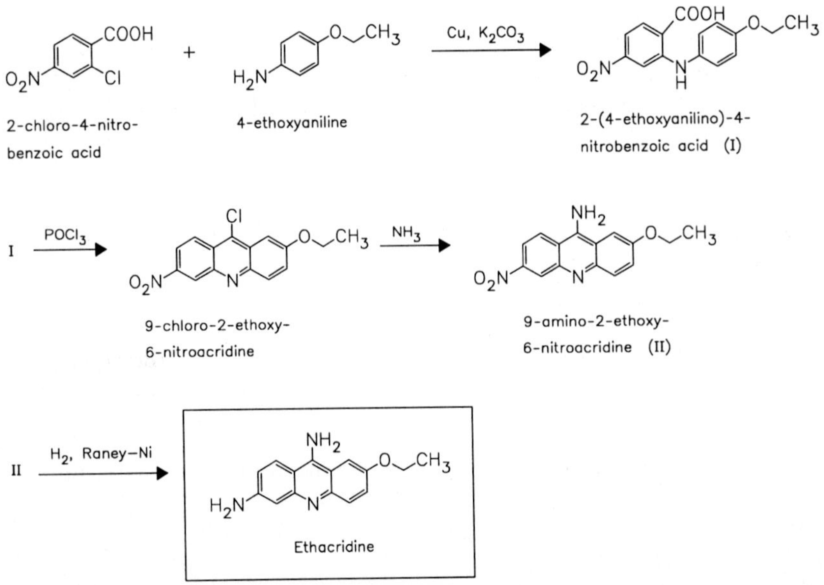

2-chloro-4-nitrobenzoic acid
4-ethoxyaniline
Cu, K2CO3
2-(4-ethoxyanilino)-4-nitrobenzoic acid (I)
POCl3
9-chloro-2-ethoxy-6-nitroacridine
NH3
9-amino-2-ethoxy-6-nitroacridine (II)
H2, Raney-Ni
Ethacridine

Reference(s):
DRP 360 421 (Hoechst; 1922).
DRP 393 411 (Hoechst; 1923).

improved method for 9-amino-2-ethoxy-6-nitroacridine:
DAS 1 952 086 (Hoechst; appl. 16.10.1969).

Formulation(s):   drg. 200 mg; eye drops 1 mg/g (as free base); gargle tabl. 25 mg (as hydrochloride); ointment
2 mg/g; sol. 0.1 %; tabl. 0.1 g

Trade Name(s):

D:	Biseptol (Winzer)		numerous combination	I:	Rivanolo (Tariff.
	Metifex (Cassella-med)		preparations		Integrativo)
	Rivanol (Chinosolfabrik)	F:	Dentinox (Pharmastra)-	J:	Hectalin (Daiichi)
	Uroseptol (Fresenius-		comb.		Rimaon (Takeda)
	Praxis; as acetate)		Pyorex (Bailly)-comb.		

Ethambutol

ATC: J04AK02
Use: tuberculostatic

RN: 74-55-5   MF: $C_{10}H_{24}N_2O_2$ MW: 204.31   EINECS: 200-810-6
LD$_{50}$:   240 mg/kg (M, i.v.); 8700 mg/kg (M, p.o.)
CN: [*S*-(*R**,*R**)]-2,2'-(1,2-ethanediyldiimino)bis[1-butanol]

dihydrochloride
RN: 1070-11-7   MF: $C_{10}H_{24}N_2O_2 \cdot 2HCl$ MW: 277.24

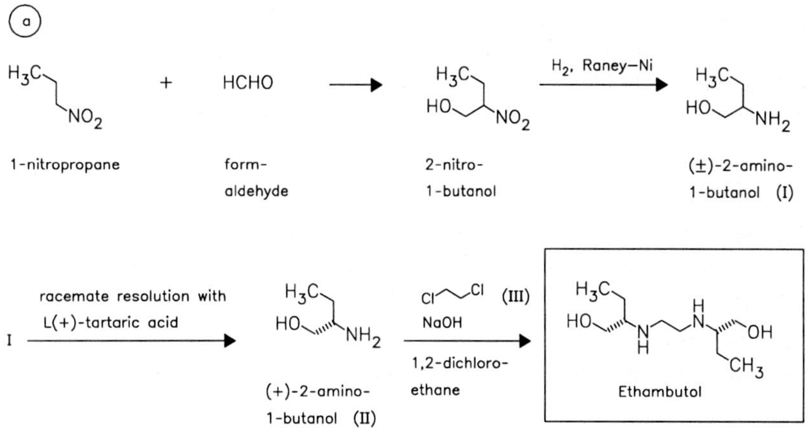

1-nitropropane   form-aldehyde   2-nitro-1-butanol   (±)-2-amino-1-butanol (I)

racemate resolution with L(+)-tartaric acid

(+)-2-amino-1-butanol (II)   1,2-dichloro-ethane   Ethambutol

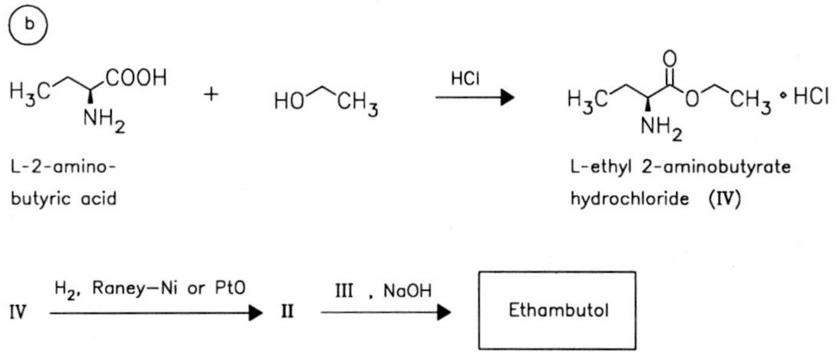

L-2-amino-butyric acid   L-ethyl 2-aminobutyrate hydrochloride (IV)

IV →[H_2, Raney–Ni or PtO] II →[III , NaOH] Ethambutol

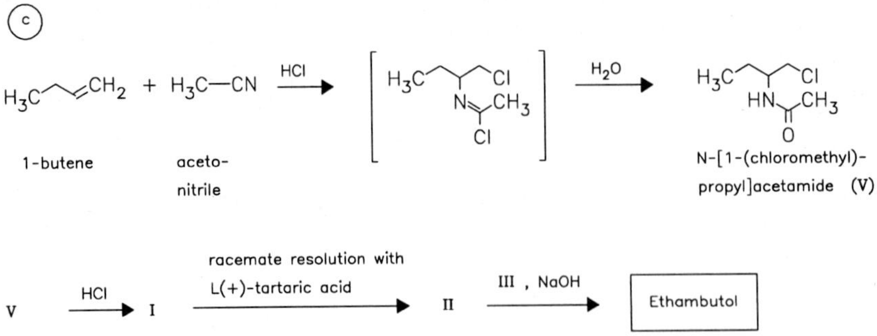

c

H₃C—CH₂ + H₃C—CN → [reaction scheme] → N-[1-(chloromethyl)-propyl]acetamide (V)

1-butene   aceto-nitrile

V →(HCl)→ I →(racemate resolution with L(+)-tartaric acid)→ II →(III, NaOH)→ Ethambutol

Reference(s):

a Wilkinson, R.G. et al.: J. Am. Chem. Soc. (JACSAT) **83**, 2212 (1961).
   Wilkinson, R.G. et al.: J. Med. Pharm. Chem. (JMPCAS) **5**, 835 (1962).
   US 3 176 040 (American Cyanamid; 30.3.1965; prior. 2.6.1960).
   BE 600 640 (American Cyanamid; appl. 24.2.1961; USA-prior. 2.6.1960, 20.12.1960).
   BE 613 545 (American Cyanamid; appl. 6.2.1962; USA-prior. 23.1.1962).
   racemate resolution of (+)-2-aminobutanol *with* tartaric acid:
   US 3 553 257 (American Cyanamid; 5.1.1971; prior. 16.9.1966).
   reaction with 1,2-dichloroethane:
   US 3 769 347 (American Cyanamid; 30.10.1973; prior. 11.2.1971).
   DOS 2 205 269 (American Cyanamid; appl. 4.2.1972; USA-prior. 11.2.1971).
   US 3 944 616 (American Cyanamid; 16.3.1976; prior. 29.10.1974).
   FR 2 351 090 (Soc. Chim. Grande Paroisse; appl. 11.5.1976).
b DAS 2 446 320 (Denki Kagaku Kogyo; appl. 27.9.1974; J-prior. 28.9.1973).
   GB 1 469 014 (Denki Kagaku Kogyo; appl. 30.9.1974; J-prior. 28.9.1973).
   reduction with sodium diethylaluminum hydride:
   JP-appl. 780 06-127 (Crc co di Ricerca; appl. 22.5.1973; CH-prior. 1.3.1973).
c US 3 944 617 (American Cyanamid; 16.3.1976; prior. 1.8.1974).
   US 3 944 618 (American Cyanamid; 16.3.1976; prior. 1.8.1974).
   US 3 944 619 (American Cyanamid; 16.3.1976; prior. 1.8.1974).
   GB 1 541 290 (American Cyanamid; appl. 9.2.1976).

alternative syntheses:
from 1,2-epoxybutane:
US 3 953 513 (Gruppo Lepetit; 27.4.1976; GB-prior. 29.11.1973).
DOS 2 454 950 (Gruppo Lepetit; appl. 20.11.1974; GB-prior. 29.11.1973).
DAS 2 410 988 (Polska Akad. Nauk Inst. Chem. Organ.; appl. 7.3.1974; PL-prior. 20.3.1973).

from 3,4-epoxybutene (butadiene monoxide):
DAS 2 263 715 (Soc. Farmaceutici Italia; appl. 28.12.1972; I-prior. 30.12.1971).

from 1-hydroxy-2-butanone:
DOS 2 547 654 (BASF; appl. 24.10.1975).

asymmetric hydrogenation of 2-acylamino-crotonic acid derivatives:
BE 862 627 (American Cyanamid; appl. 4.1.1978; USA-prior. 7.1.1977).
DOS 2 800 461 (American Cyanamid; appl. 5.1.1978; USA-prior. 7.1.1977).

racemate resolution of (±)-2-aminobutanol *with* (+)-*N*-benzoyl-*trans*-2-aminocyclohexanecarboxylic acid:
GB 1 471 838 (Nippon Soda; appl. 26.3.1975; J-prior. 4.4.1974).

Formulation(s):   amp. 400 mg/4 ml, 1000 mg/10 ml; f. c. tabl. 100 mg, 250 mg, 400 mg, 500 mg (as dihydrochloride); vial 1 g

Trade Name(s):
D: EMB-Fatol (Fatol)   Myambutol (Lederle)   F: Dexambutol (L'Arguenon)

Dexambutol-INH
(L'Arguenon)-comb. with
isoniazid
Myambutol (Wyeth-
Lederle)
GB: Myambutol (Lederle); wfm
Mynak (Lederle)-comb.
with isoniazid; wfm

I: Etambu (Formulario Naz.)
Etambu (Lifepharma)
Etanicozid (Piam)-comb.
Etapiam (Piam)
Etibi (Zoja)
Miambutol (Cyanamid)
Miazide (Cyanamid)-comb.
J: Ebutol (Kaken)

Esanbutol (Lederle)
Ethambutol (Lederle-
Takeda)
USA: Myambutol (Lederle Labs.;
as hydrochloride)

Ethaverine

ATC: A03
Use: antispasmodic

RN: 486-47-5   MF: $C_{24}H_{29}NO_4$ MW: 395.50   EINECS: 207-633-3
LD_{50}:   45600 µg/kg (M, i.v.)
CN: 1-[(3,4-diethoxyphenyl)methyl]-6,7-diethoxyisoquinoline

hydrochloride
RN: 985-13-7   MF: $C_{24}H_{29}NO_4 \cdot HCl$ MW: 431.96   EINECS: 213-573-9
LD_{50}:   86 mg/kg (M, i.v.)

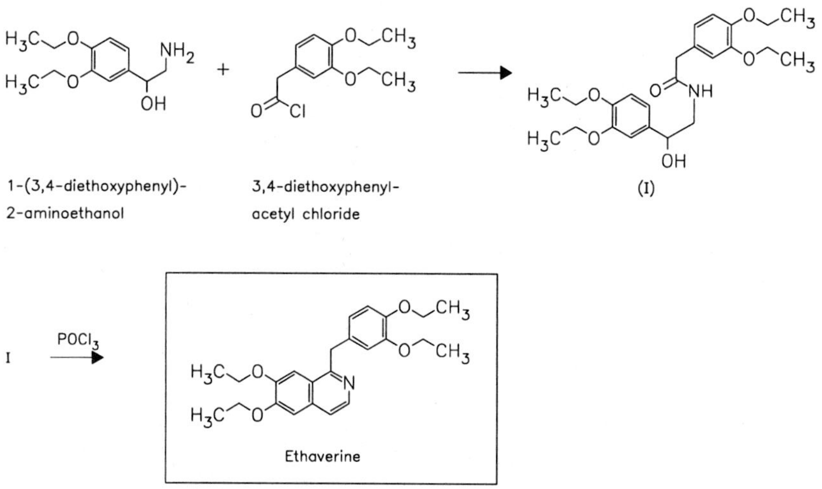

1-(3,4-diethoxyphenyl)-
2-aminoethanol

3,4-diethoxyphenyl-
acetyl chloride

(I)

Ethaverine

Reference(s):
US 1 962 224 (E. Wolf; 1934; D-prior. 1930).

Formulation(s):   suppos. 30 mg in comb. (as hydrochloride)

Trade Name(s):
D: Migräne-Kranit (Krewel
Meuselbach)
F: Etadil (Charpentier); wfm
Plaquiverine (Monal); wfm
Surparine (Licardy)-comb.;
wfm
I: Azimol ITA (ITA)-comb.;
wfm

Ceracin (Panthox &
Burck)-comb.; wfm
Etaverina (Biologici Italia);
wfm
Predem (Biologici Italia)-
comb.; wfm
USA: Ethaquin (Ascher); wfm
Ethatab (Meyer); wfm

Isovex (U.S.
Pharmaceutical); wfm
Laverin (Lemmon); wfm
Pasmol (RAM Labor); wfm
Tensodin (Knoll)-comb.;
wfm

Ethchlorvynol

ATC: N05CM08
Use: hypnotic, sedative

RN: 113-18-8 MF: C_7H_9ClO MW: 144.60
LD$_{50}$: 290 mg/kg (M, p.o.);
55 mg/kg (dog, i.v.)
CN: 1-chloro-3-ethyl-1-penten-4-yn-3-ol

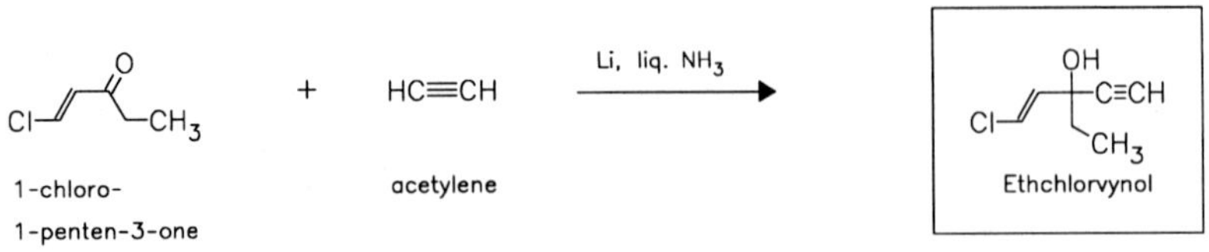

1-chloro-
1-penten-3-one

acetylene

Ethchlorvynol

Reference(s):
US 2 746 900 (Pfizer; 1956; prior. 1953).
McLamore, W.M. et al.: J. Org. Chem. (JOCEAH) **20**, 109 (1955).

Formulation(s): cps. 200 mg, 500 mg, 750 mg

Trade Name(s):
GB: Arvynol (Pfizer); wfm
Serenesil (Abbott); wfm

J: Arvynol (Taito Pfizer)
Nostel (Dainippon)

USA: Placidyl (Abbott)

Ethenzamide

(Ethoxybenzamide)

ATC: N02BA07
Use: analgesic

RN: 938-73-8 MF: $C_9H_{11}NO_2$ MW: 165.19 EINECS: 213-346-4
LD$_{50}$: 700 mg/kg (M, p.o.);
2630 mg/kg (R, p.o.)
CN: 2-ethoxybenzamide

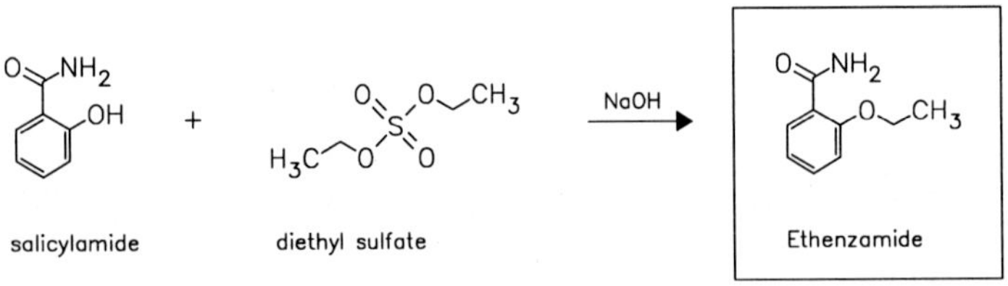

salicylamide

diethyl sulfate

Ethenzamide

Reference(s):
GB 656 746 (Lundbeck; appl. 1948; DK-prior. 1947).

Formulation(s): drg. 150 mg in comb.; tabl. 100 mg in comb.

Trade Name(s):
D: Antiföhnon (Südmedica)-
comb.
Glutisal (Ravensberg)-
comb.
Kolton grippale N (Byk
Gulden)-comb.

F: Céphil (Boiron)-comb.
I: Etocil (Biomedica
Foscama)-comb.
Etocil Pirina (Biomedica
Foscama)-comb.
J: Amisal (Daiichi)-comb.

Ethoxybenzamide (Juzen
Kagaku)
Grelan High S (Grelan)-
comb.
Grelan Shin A (Grelan)
Konjisui Soft (Tanpai)

| Pyripan A (Tanabe)-comb. | Sedes A (Shionogi) | Synpyrin F (Sumitomo) |

Ethiazide
(Aethiazidum)

ATC: C03BA
Use: diuretic

RN: 1824-58-4   MF: $C_9H_{12}ClN_3O_4S_2$ MW: 325.80   EINECS: 217-358-0
LD$_{50}$:   >310 mg/kg (M, i.v.); >2 g/kg (M, p.o.);
         >10 g/kg (R, p.o.)
CN: 6-chloro-3-ethyl-3,4-dihydro-2H-1,2,4-benzothiadiazine-7-sulfonamide 1,1-dioxide

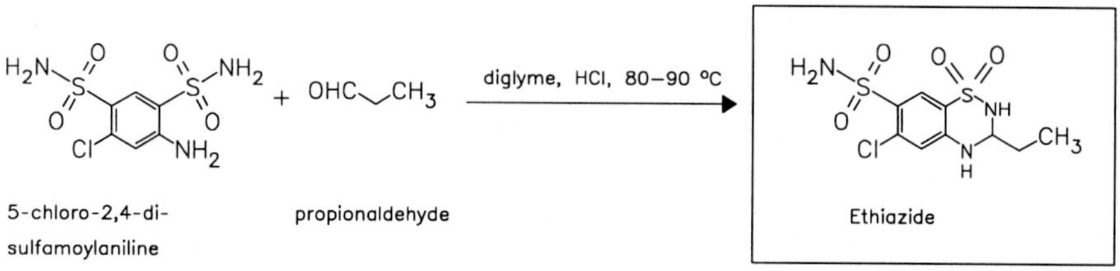

5-chloro-2,4-di-
sulfamoylaniline

propionaldehyde

Ethiazide

Reference(s):
GB 861 367 (Ciba; appl. 1959; USA-prior. 1958).

Trade Name(s):
J: Ethiazide (Tokyo Tanabe);
        wfm

Ethinamate

ATC: N05C
Use: hypnotic, sedative

RN: 126-52-3   MF: $C_9H_{13}NO_2$ MW: 167.21   EINECS: 204-789-4
LD$_{50}$:   108 mg/kg (M, i.v.); 490 mg/kg (M, p.o.);
         157 mg/kg (R, i.v.); 331 mg/kg (R, p.o.);
         144 mg/kg (dog, i.v.); 190 mg/kg (dog, p.o.)
CN: 1-ethynylcyclohexanol carbamate

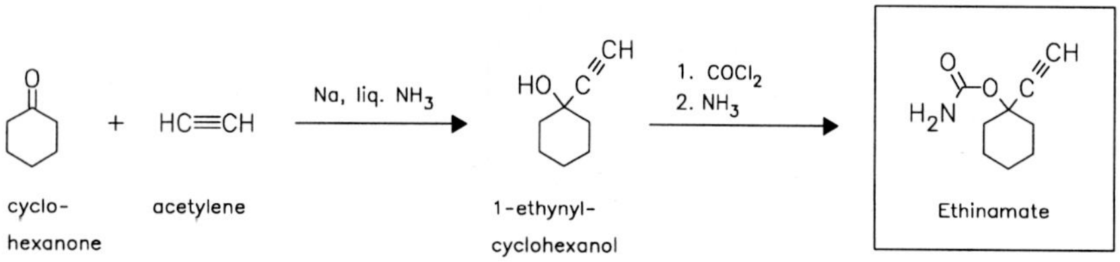

cyclo-
hexanone

acetylene

1-ethynyl-
cyclohexanol

Ethinamate

Reference(s):
US 2 816 910 (Schering; 1957; D-prior. 1953).
DE 1 021 843 (Rheinpreussen; appl. 1953).

Formulation(s):   cps. 500 mg

Trade Name(s):
D: Valamin (Asche); wfm   J: Valamin (Schering)   USA: Valamid (Dista); wfm

Ethinylestradiol
(Aethinylöstradiol; Ethinyloestradiol)

ATC: G03CA01; L02AA03
Use: estrogen (in combination with progestogen as oral contraceptive)

RN: 57-63-6   MF: $C_{20}H_{24}O_2$ MW: 296.41   EINECS: 200-342-2
LD_{50}:   1737 mg/kg (M, p.o.);
       1200 mg/kg (R, p.o.)
CN: (17α)-19-norpregna-1,3,5(10)-trien-20-yne-3,17-diol

estrone
(q. v.)

Ethinylestradiol

Reference(s):
Inhoffen, H.H. et al.: Ber. Dtsch. Chem. Ges. (BDCGAS) **71**, 1024 (1938).
DRP 702 063 (Ciba; appl. 1938; CH-prior. 1937).

Formulation(s):   tabl. 0.02 mg, 0.025 mg, 0.05 mg; drg. 1 mg

Trade Name(s):
D: Biviol (Nourypharma)-
      comb.
      Cilest (Janssen-Cilag)-
      comb.
      Concephan (Grünenthal)-
      comb.
      Cyclosan (Nourypharma)-
      comb.
      Diane 35 (Schering)-comb.
      EVE (Grünenthal)-comb.
      Femigoa (LAW)-comb.
      Femovan (Schering)-comb.
      Femranette mikro
      (Brenner-Efeka)-comb.
      Gravistat (Jenapharm)-
      comb.
      Leios (Wyeth)-comb.
      Lovelle (Organon)-comb.
      Lyndiol (Organon)-comb.
      Marvelon (Organon)-comb.
      Microgynon (Schering)-
      comb.
      Minisiston (Jenapharm)-
      comb.
      Minulet (Wyeth)-comb.
      Miranova (Schering)-comb.
      MonoStep (Asche)-comb.
      Neo-Eunomin
      (Grünenthal)-comb.
      Neogynon (Schering)-
      comb.

      Neorlest (Parke-Davis)-
      comb.
      Neo-Stedirile (Wyeth)-
      comb.
      Non-Ovlon (Jenapharm)-
      comb.
      Nuriphasic
      (Nourypharma)-comb.
      Östro-Primolut (Schering)-
      comb.
      Ovanon (Nourypharma)-
      comb.
      Ovanon (Nourypharma)-
      comb.
      Oviol (Nourypharma)-
      comb.
      Ovoresta (Organon)-comb.
      Ovysmen (Wyeth)-comb.
      Perikursal (Wyeth)-comb.
      Pramino (Janssen-Cilag)-
      comb.
      Pregnon (Schering)-comb.
      Progynon C (Schering)
      Promisiston (Schering)-
      comb.
      Prosiston (Schering)-comb.
      Sequilar (Schering)-comb.
      Sequostat (Jenapharm)-
      comb.
      Sinovula (Asche)-comb.
      Stediril (Wyeth)-comb.

      Synphasec (Grünenthal)-
      comb.
      Tetragynon (Schering)-
      comb.
      Triette (Brenner-Efeka)-
      comb.
      Trigoa (LAW)-comb.
      Triguilar (Schering)-comb.
      Trinordiol (Wyeth)-comb.
      TriNoum (Janssen-Cilag)-
      comb.
      Trisiston (Jenapharm)-
      comb.
      TriStep (Asche)-comb.
      Turisteron (Jenapharm)
      Valette (Jenapharm)-comb.
      Yermonil (Novartis
      Pharma)-comb.
      numerous combination
      preparations
F: Adepal (Wyeth-Lederle)
      Cilest (Janssen-Cilag)
      Cycleane (Monsanto)
      Diane 35 (Schering)
      Effiprev (Effik)
      Ethinyl-Estradiol Roussel
      (Roussel)
      Harmonet (Wyeth-Lederle)
      Méliane (Schering)
      Minidril (Wyeth-Lederle)
      Minulet (Wyeth-Lederle)

Tri-Minulet (Wyeth-
Lederle)
generic and numerous
combination preparations
GB: Marvelon (Schering)
numerous combination
preparations
I: Binordiol (Wyeth)-comb.
Bivlar (Schering)-comb.
Diane (Schering)-comb.
Egogyn (Schering)-comb.
Etinilestradiolo (Amsa)
Eugynon (Schering)-comb.
Evanor (Wyeth)-comb.
Ginoden (Schering)-comb.
Mercilon (Organon Italia)-
comb.
Microgynon (Schering)-
comb.
Milvane (Schering)-comb.
Minulet (Wyeth)-comb.

Novogyn (Schering)-comb.
Ovranet (Wyeth)-comb.
Planum (Menarini)-comb.
Practil (Organon Italia)-
comb.
Securgin (Menarini)-comb.
Trigynon (Schering)-comb.
Triminulet (Wyeth)-comb.
Trinordiol (Wyeth)-comb.
Trinovum (Cilag)-comb.
J: Estrogen (Nichinan Kogyo)
Ovahormon Strong
(Teikoku Zoki)
USA: Alesse (Wyeth-Ayerst)
Brevicon (Searle)
Demulen (Searle)
Desogen (Organon)
Estrostep (Parke Davis)
Ethynodiol Diacetate and
Ethinyl Estradiol (Watson)
Levlen (Berlex)

Lo/Ovral (Wyeth-Ayerst)
Modicon (Ortho-McNeil
Pharmaceutical)
Nelova (Warner Chilcott)
Nordette (Wyeth-Ayerst)
Norethin (Roberts)
Norethindrone and Ethinyl
Estradiol (Watson)
Norinyl (Searle)
Ortho-Cept (Ortho-McNeil
Pharmaceutical)
Ortho-Cyclen (Ortho-
McNeil Pharmaceutical)
Ortho Novum (Ortho-
McNeil Pharmaceutical)
Ortho-Tri-Cyclen (Ortho-
McNeil Pharmaceutical)
Ovral (Wyeth-Ayerst)
Tri-Levlen (Berlex)
Tri-Norinyl (Searle)
Triphasil (Wyeth-Ayerst)

Ethionamide

(Etionamide)

ATC: J04AD03
Use: tuberculostatic

RN: 536-33-4   MF: $C_8H_{10}N_2S$ MW: 166.25   EINECS: 208-628-9
LD_{50}:   1 g/kg (M, p.o.);
1320 mg/kg (R, p.o.)
CN: 2-ethyl-4-pyridinecarbothioamide

diethyl oxalate   +   butanone   NaOC$_2$H$_5$ →   ethyl 2,4-dioxo-hexanoate   cyanoacetamide , pyridine →   I

(I)   HCl →   6-ethyl-1,2-dihydro-2-oxo-4-pyridine-carboxylic acid   1. POCl$_3$, PCl$_5$ 2. HO CH$_3$ →   ethyl 2-chloro-6-ethyl-iso-nicotinate (II)

II   H$_2$, Pd →   ethyl 2-ethyliso-nicotinate   1. NH$_3$ 2. P$_2$O$_5$ →   2-ethyl-isonicotino-nitrile   H$_2$S →   Ethionamide

Reference(s):
GB 800 250 (Chimie et Atomistique; appl. 1957; F-prior. 1956).
Libermann, S. et al.: C. R. Hebd. Seances Acad. Sci. (COREAF) **242**, 2409, 2412 (1956).

Formulation(s):　s. c. tabl. 250 mg; tabl. 100 mg

Trade Name(s):

D:	Trécator (Théraplix); wfm	J:	Ethimide (Tanabe)		Thioniden (Kaken)
F:	Trécator (Théraplix); wfm		Ethinamin (Takeda)		Tubermin (Meiji)
GB:	Trescatyl (May & Baker);		Itiocide (Kyowa)		Tuberoid (Sankyo)
	wfm		Sertinon (Daiichi)		Tuberoson (Shionogi)
	Trescazide (May & Baker)-		Teberus (Dainippon)	USA:	Trecator-SC (Wyeth-
	comb.; wfm		Thiomid (Nikken)		Ayerst)

Ethisterone

ATC:　G03DC04
Use:　progestogen

RN:　434-03-7　MF: $C_{21}H_{28}O_2$　MW: 312.45　EINECS: 207-096-5
CN:　(17α)-17-hydroxypregn-4-en-20-yn-3-one

androstenolone　　　acetylene

(I)　　　　　Ethisterone

Reference(s):
US 2 272 131 (Ciba; 1942; CH-prior. 1937).
Ehrhart, Ruschig **III**, 343.

alternative synthesis:
US 4 041 055 (Upjohn; 9.8.1977; appl. 17.11.1975).

review:
Ullmanns Encykl. Tech. Chem., 4. Aufl., Vol. **13**, 30.

Formulation(s):　cps. 50 mg, 100 mg, 250 mg; tabl. 25 mg

Trade Name(s):

D:	Cycloestrol-A.H.		Lutogynestryl (Roussel)-		Orasecron (Schering
	Progestérone (Bruneau);		comb.; wfm		Chemicals); wfm
	wfm	GB:	Amenoren (Roussel)-	I:	Pre Ciclo (Ibis)-comb.;
F:	Cycloestrol-A.H.		comb.; wfm		wfm
	Progestérone (Bruneau)-		Menstrogen (Organon)-	J:	Estormon (Hokuriku)-
	comb.; wfm		comb.; wfm		comb.

Oophormin Luteum (Teikoku Zoki)
USA: Duosterone (Roussel)-comb.; wfm
Ora-Lutin (Parke Davis); wfm
Prodroxan (Dorsey); wfm
Progestab (Beecham); wfm
Progestoral (Organon); wfm
Syngestrotabs (Pfizer); wfm
Trosinone (Abbott); wfm

Ethoheptazine

ATC: N02A
Use: analgesic

RN: 77-15-6 MF: $C_{16}H_{23}NO_2$ MW: 261.37 EINECS: 201-007-3
LD_{50}: 65 mg/kg (M, i.v.); 318 mg/kg (M, p.o.);
34 mg/kg (R, i.v.); 355 mg/kg (R, p.o.)
CN: hexahydro-1-methyl-4-phenyl-1*H*-azepine-4-carboxylic acid ethyl ester

citrate (1:1)
RN: 6700-56-7 MF: $C_{16}H_{23}NO_2 \cdot C_6H_8O_7$ MW: 453.49 EINECS: 229-743-0
LD_{50}: 580 mg/kg (R, p.o.)

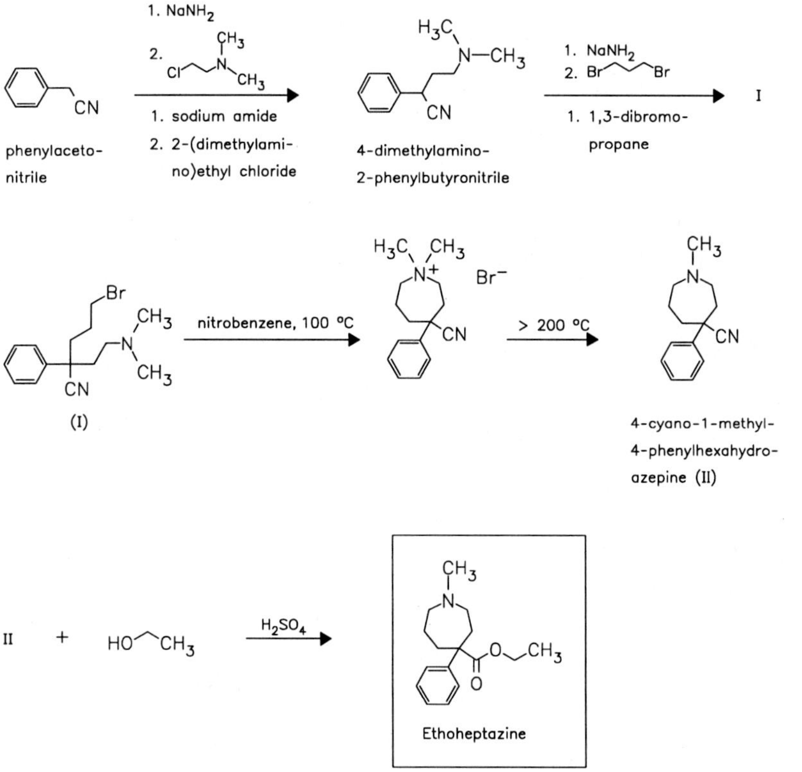

phenylaceto-nitrile
1. NaNH2
2.
1. sodium amide
2. 2-(dimethylamino)ethyl chloride
4-dimethylamino-2-phenylbutyronitrile
1. NaNH2
2.
1. 1,3-dibromo-propane
I
(I)
nitrobenzene, 100 °C
Br−
> 200 °C
4-cyano-1-methyl-4-phenylhexahydro-azepine (II)
II
H2SO4
Ethoheptazine

Reference(s):
US 2 666 050 (American Home Products; 1954; prior. 1952).

Formulation(s): tabl. 75 mg in comb.

Trade Name(s):
GB: Equagesic (Wyeth)-comb.
I: Panalgin (Padil); wfm
combination preparations; wfm
J: Zactirin (Banyu)-comb.
USA: Equagesic (Wyeth); wfm
Mepro (Schein); wfm
Zactane (Wyeth); wfm

Zactirin (Wyeth); wfm

Ethosuximide

ATC: N03AD01
Use: antiepileptic, antiparkinsonian

RN: 77-67-8   MF: $C_7H_{11}NO_2$ MW: 141.17   EINECS: 201-048-7
LD$_{50}$:   780 mg/kg (M, i.v.); 1530 mg/kg (M, p.o.)
CN: 3-ethyl-3-methyl-2,5-pyrrolidinedione

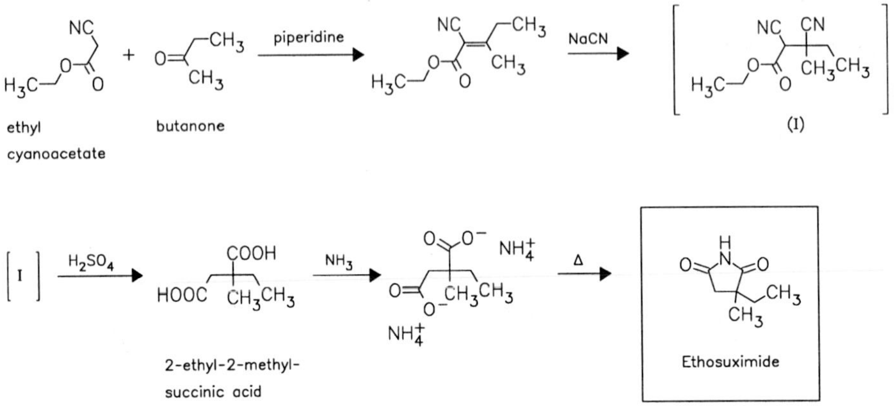

ethyl cyanoacetate   butanone   (I)

2-ethyl-2-methyl-succinic acid   Ethosuximide

Reference(s):
US 2 993 835 (Parke Davis; 25.7.1961; prior. 27.10.1958).
Sahay, S.; Sircar, G.: J. Chem. Soc. (JCSOA9) **1927**, 1252.

Formulation(s):   cps. 250 mg; sol. 50 g/100 g; syrup 250 mg/5 ml

Trade Name(s):
D: Petnidan (Desitin)
       Suxilep (Jenapharm)
       Suxinutin (Parke Davis)
F: Zarontin (Parke Davis)
GB: Emeside (Labs. for Applied
       Biology)

I: Zarontin (Parke Davis)
J: Zarontin (Parke Davis)
       Emeside (Technish-
       Kodama)
       Epileo Petitmal (Eisai)

Zarontin (Parke Davis)

Zarontin (Parke Davis-
       Sankyo)
USA: Zarontin (Parke Davis)

Ethotoin
(Aethotoin)

ATC: N03AB01
Use: antiepileptic

RN: 86-35-1   MF: $C_{11}H_{12}N_2O_2$ MW: 204.23   EINECS: 201-665-1
LD$_{50}$:   1750 mg/kg (M, p.o.);
               1500 mg/kg (R, p.o.)
CN: 3-ethyl-5-phenyl-2,4-imidazolidinedione

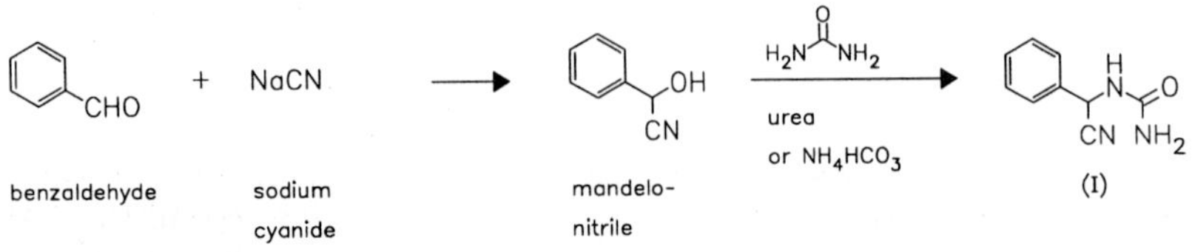

benzaldehyde   sodium cyanide   mandelo-nitrile   (I)

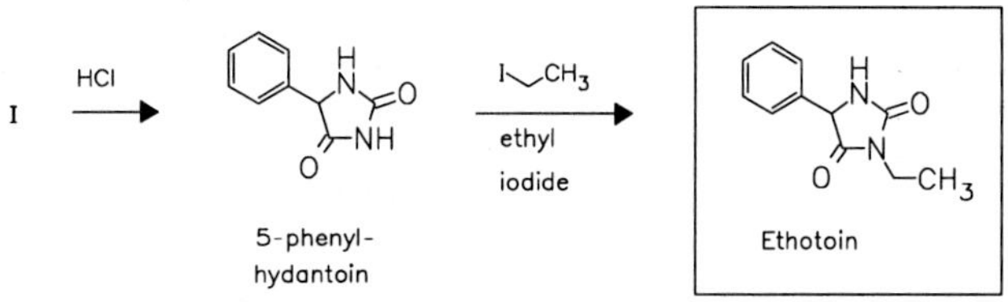

5-phenyl-
hydantoin

Ethotoin

Reference(s):
Pinner, A.: Chem. Ber. (CHBEAM) **21**, 2325 (1888).
US 2 793 157 (Abbott; 1957; appl. 1954).

Formulation(s):   tabl. 250 mg, 500 mg

Trade Name(s):
GB: Peganone (Abbott); wfm   J: Accenon (Dainippon)   USA: Peganone (Abbott)

Ethoxzolamide
(Ethoxyzolamide)

ATC: C03BA
Use: diuretic (carboanhydrase inhibitor)

RN: 452-35-7   MF: $C_9H_{10}N_2O_3S_2$ MW: 258.32   EINECS: 207-199-5
CN: 6-ethoxy-2-benzothiazolesulfonamide

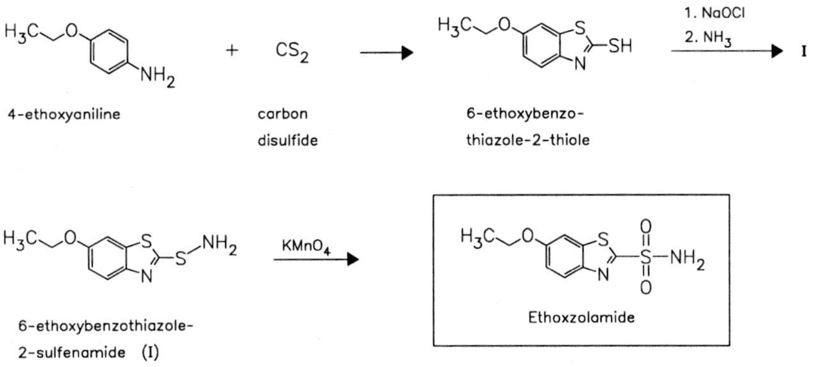

4-ethoxyaniline

carbon
disulfide

6-ethoxybenzo-
thiazole-2-thiole

6-ethoxybenzothiazole-
2-sulfenamide   (I)

Ethoxzolamide

Reference(s):
US 2 868 800 (Upjohn; 1959; appl. 1954).

Formulation(s):   tabl. 125 mg

Trade Name(s):
D: Redupresin (Thilo); wfm   I: Glaucotensil (Farmila);
                                                        wfm

Ethyl biscoumacetate

ATC: B01AA08
Use: anticoagulant, antithrombotic

RN: 548-00-5   MF: $C_{22}H_{16}O_8$ MW: 408.36   EINECS: 208-940-5
LD$_{50}$:   750 mg/kg (M, p.o.);
        840 mg/kg (R, p.o.)
CN: 4-hydroxy-α-(4-hydroxy-2-oxo-2*H*-1-benzopyran-3-yl)-2-oxo-2*H*-1-benzopyran-3-acetic acid ethyl ester

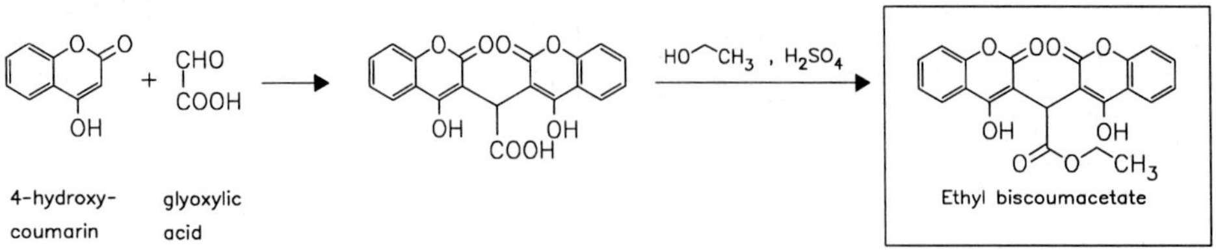

HO⌒CH3 , H2SO4
4-hydroxy-coumarin
glyoxylic acid
Ethyl biscoumacetate

Reference(s):
US 2 482 510 (Spójené farmaceutické Zovody; 1949).
US 2 482 511 (Spójené farmaceutické Zovody; 1949).
US 2 482 512 (Spójené farmaceutické Zovody; 1949).

Formulation(s): tabl. 300 mg

Trade Name(s):

D:	Tromexan (Geigy); wfm	GB:	Tromexan (Geigy); wfm
F:	Tromexane (Geigy); wfm	I:	Etilbis (Tanff. Integrativo)

Ethylestrenol

(Äthylestrenol; Äthyloestrenol)

ATC: A14AB02
Use: anabolic

RN: 965-90-2 MF: $C_{20}H_{32}O$ MW: 288.48 EINECS: 213-523-6
LD_{50}: >666.7 mg/kg (M, p.o.)
CN: (17α)-19-norpregn-4-en-17-ol

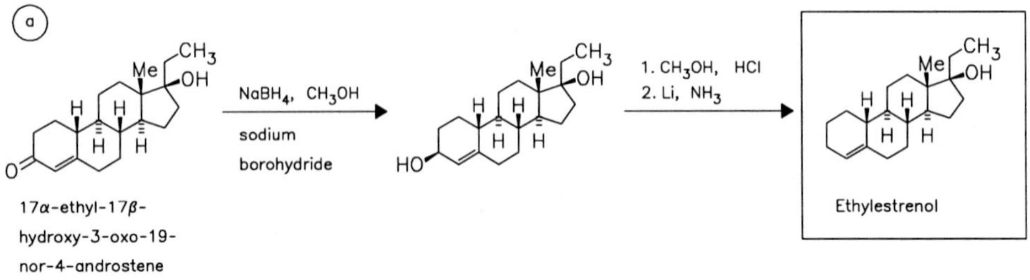

a
NaBH4, CH3OH
sodium borohydride
1. CH3OH, HCl
2. Li, NH3
17α-ethyl-17β-hydroxy-3-oxo-19-nor-4-androstene
Ethylestrenol

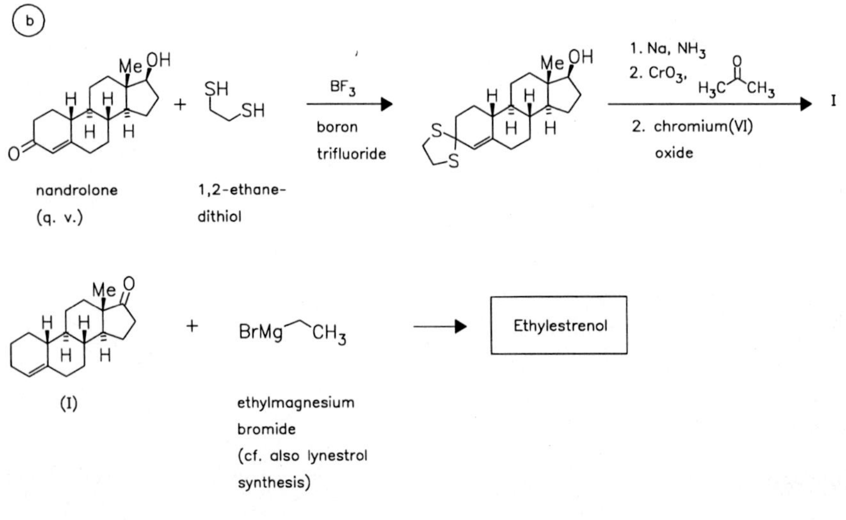

b
BF3
boron trifluoride
1. Na, NH3
2. CrO3,
2. chromium(VI) oxide
I
nandrolone (q. v.)
1,2-ethanedithiol
(I)
BrMg⌒CH3
ethylmagnesium bromide (cf. also lynestrol synthesis)
Ethylestrenol

Reference(s):
a US 2 878 267 (Organon; 1959; N-prior. 1957).
b Winter, M.S. de et al.: Chem. Ind. (London) (CHINAG) **1959**, 905.

alternative synthesis:
US 3 112 328 (Organon; 26.11.1963; NL-prior. 24.8.1956).

Formulation(s):   sol. 2 mg/5 ml; tabl. 2 mg

Trade Name(s):

F:	Orgaboline (Organon); wfm	GB:	Orabolin (Organon); wfm
	Orgaboline infantile (Organon); wfm	I:	Orgabolin (Ravasini Organon); wfm

J: Orgabolin (Organon-Sankyo)

USA: Maxibolin (Organon); wfm

Ethyl loflazepate

ATC: N05BA18
Use: tranquilizer

RN: 29177-84-2   MF: $C_{18}H_{14}ClFN_2O_3$ MW: 360.77   EINECS: 249-489-4
LD_{50}:   5506 mg/kg (M, p.o.);
         >10 g/kg (R, p.o.)
CN: 7-chloro-5-(2-fluorophenyl)-2,3-dihydro-2-oxo-1H-1,4-benzodiazepine-3-carboxylic acid ethyl ester

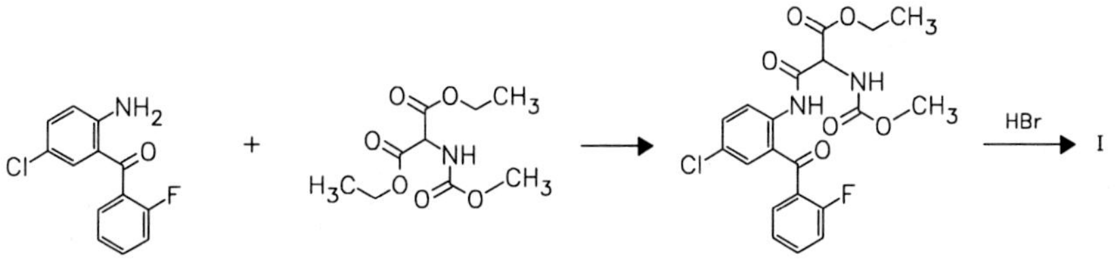

2-amino-5-chloro-2'-fluorobenzo-phenone (cf. flunitrazepam synthesis)

diethyl methoxy-carbonylamino-malonate

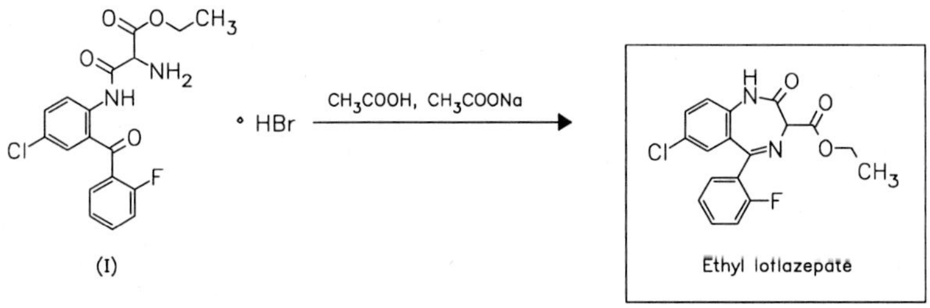

(I)

CH₃COOH, CH₃COONa

Ethyl loflazepate

Reference(s):
BE 854 249 (Clin-Midy; appl. 5.5.1977; GB-prior. 5.5.1976).
DOS 2 719 608 (Clin-Midy; appl. 2.5.1977; GB-prior. 5.5.1976).
GB 1 538 165 (Clin-Midy; appl. 5.5.1977; prior. 5.5.1976).

alternative synthesis:
EP 22 710 (Clin-Midy; appl. 8.7.1980; F-prior. 12.7.1979).

Formulation(s):   tabl. 2 mg

Ethylmorphine
(Codéthyline)

ATC: R05DA01; S01XA06
Use: antitussive, analgesic

RN: 76-58-4   MF: $C_{19}H_{23}NO_3$ MW: 313.40   EINECS: 200-970-7
LD_{50}:   120 mg/kg (M, i.p.); 520 mg/kg (M, p.o.); 136 mg/kg (M, s.c.);
       110 mg/kg (R, i.p.); 62 mg/kg (R, i.v.); 810 mg/kg (R, p.o.); 200 mg/kg (R, s.c.)
CN: (5α,6α)-7,8-didehydro-4,5-epoxy-3-ethoxy-17-methylmorphinan-6-ol

hydrochloride
RN: 125-30-4   MF: $C_{19}H_{23}NO_3 \cdot HCl$ MW: 349.86   EINECS: 204-734-4
LD_{50}:   771 mg/kg (M, p.o.); 265 mg/kg (M, s.c.);
       200 mg/kg (R, s.c.)
hydrochloride dihydrate
RN: 6746-59-4   MF: $C_{19}H_{23}NO_3 \cdot HCl \cdot 2H_2O$ MW: 385.89
LD_{50}:   200 mg/kg (M, s.c.)

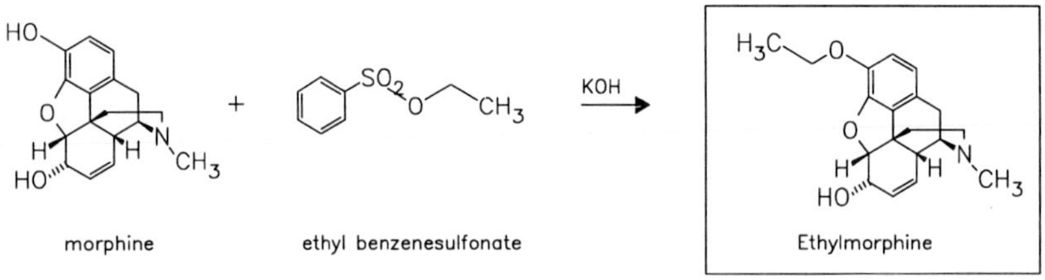

morphine   ethyl benzenesulfonate   Ethylmorphine

Reference(s):
Ehrhart, Ruschig **I**, 118
DRP 131 980 (E. Merck AG; 1902).

Formulation(s):   drg. 5 mg; tabl. 5 mg, 15 mg (as hydrochloride dihydrate)

Etidocaine

ATC: N01BB07
Use: local anesthetic

RN: 36637-18-0 MF: $C_{17}H_{28}N_2O$ MW: 276.42 EINECS: 253-143-8
LD_{50}: 47.5 mg/kg (M, i.p.)
CN: (±)-N-(2,6-dimethylphenyl)-2-(ethylpropylamino)butanamide

monohydrochloride
RN: 36637-19-1 MF: $C_{17}H_{28}N_2O \cdot HCl$ MW: 312.89 EINECS: 253-144-3
LD_{50}: 6700 µg/kg (M, i.v.)

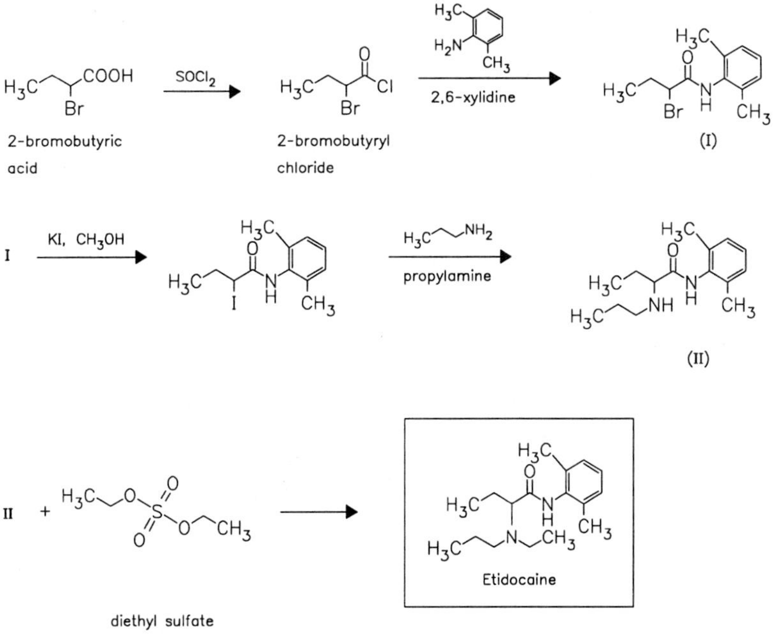

Reference(s):
US 3 812 147 (Astra; 21.5.1974; prior. 22.12.1970, 19.7.1971).
US 3 862 321 (Astra; 21.1.1975; prior. 22.12.1970, 19.7.1971, 4.3.1974).
DOS 2 162 744 (Astra; appl. 17.12.1971; USA-prior. 22.12.1970, 19.7.1971).

Formulation(s): amp. 5 mg/2 ml, 10 mg/ml, 12.5 mg/5 ml (as hydrochloride)

Trade Name(s):
D: Dur-Anest (Astra) Duranest Adrénaline USA: Duranest (Astra)
F: Duranest (Astra) (Astra)

Etidronic acid

ATC: M05BA01
Use: calcium regulator

RN: 2809-21-4 MF: $C_2H_8O_7P_2$ MW: 206.03 EINECS: 220-552-8
LD_{50}: 1800 mg/kg (M, p.o.)
CN: (1-hydroxyethylidene)bis[phosphonic acid]

disodium salt
RN: 7414-83-7   MF: $C_2H_6Na_2O_7P_2$ MW: 249.99   EINECS: 231-025-7
LD_{50}:   49 mg/kg (M, i.v.); 2050 mg/kg (M, p.o.);
        73 mg/kg (R, i.v.); 1340 mg/kg (R, p.o.)

$$P_4 + O_2 \xrightarrow{CO, \Delta} P_4O_6 \xrightarrow[\text{2. H}_2\text{O, 140 °C}]{\text{1. H}_3\text{C—COOH, 120 °C}}$$

Etidronic acid

Reference(s):
FR 1 531 913 (Procter & Gamble; appl. 19.7.1967; USA-prior. 20.7.1966).

alternative syntheses:
US 3 366 675 (Procter & Gamble; 30.1.1968; prior. 30.3.1965).
NL 6 606 548 (Procter & Gamble; appl. 12.5.1966; USA-prior. 13.5.1965).
NL 6 610 762 (Procter & Gamble; appl. 29.7.1966; USA-prior. 29.7.1965, 31.5.1966).

Formulation(s):   amp. 300 mg/6 ml; tabl. 200 mg, 400 mg (as disodium salt)

Trade Name(s):
D: Diphos (Procter & Gamble)   GB: Didronel (Procter &   USA: Didronel (MGI)
F: Didronel (Procter &   Gamble; 1992)   Didronel (Procter &
        Gamble)   I: Etidron (Gentili)   Gamble; as disodium salt)

Etifelmine

Use: antihypotensive

RN: 341-00-4   MF: $C_{17}H_{19}N$ MW: 237.35
CN: 2-(diphenylmethylene)-1-butanamine

gluconate (1:1)
RN: 28599-37-3   MF: $C_{17}H_{19}N \cdot C_6H_{12}O_7$ MW: 433.50
hydrochloride
RN: 1146-95-8   MF: $C_{17}H_{19}N \cdot HCl$ MW: 273.81
LD_{50}:   28.6 mg/kg (M, i.v.); 115 mg/kg (M, p.o.);
        17.4 mg/kg (R, i.v.); 148 mg/kg (R, p.o.)
nicotinate (1:1)
RN: 31149-45-8   MF: $C_{17}H_{19}N \cdot C_6H_5NO_2$ MW: 360.46

benzophenone   butyronitrile

(I)   Etifelmine

Reference(s):
DE 1 122 514 (Giulini; appl. 8.9.1959).

Formulation(s): drg. 11 mg in comb.

Trade Name(s):

D:	Gilutensin (Giulini)-comb.; wfm	Orthoheptamin (Giulini)-comb.; wfm	J: Tensinase-D (Nippon Chemiphar)

Etilefrine

ATC: C01CA01
Use: sympathomimetic, circulatory analeptic

RN: 709-55-7 MF: $C_{10}H_{15}NO_2$ MW: 181.24 EINECS: 211-910-4
LD_{50}: 770 mg/kg (M, p.o.);
114 mg/kg (R, p.o.)
CN: α-[(ethylamino)methyl]-3-hydroxybenzenemethanol

hydrochloride
RN: 943-17-9 MF: $C_{10}H_{15}NO_2 \cdot HCl$ MW: 217.70 EINECS: 213-398-8
LD_{50}: 860 mg/kg (M, s.c.);
>420 mg/kg (R, s.c.)

3'-hydroxy-acetophenone + benzoyl chloride → 3'-benzoyloxy-acetophenone → (I)

I → (ethylamine) → 3'-hydroxy-2-ethylamino-aceto-phenone → (H₂, Raney-Ni) → Etilefrine

Reference(s):
DRP 520 079 (H. Legerlotz; 1926).
DRP 522 790 (H. Legerlotz; 1929).

Formulation(s): amp. 10 mg/ml; drops 5 mg/ml, 7.5 mg/ml; sol. 7.5 mg/ml; s. r. cps. 20 mg, 25 mg; tabl. 5 mg, 25 mg (as hydrochloride)

Trade Name(s):

D:	Adrenam (NAM Neukönigsförder)	Circuvit (Pharma Wernigerode)	Kreislauf Katovit (Boehringer Ing.)
	Bioflutin (Südmedica)	Confidol (Medopharm)	Thomasin (Apogepha)
	Cardanat (Temmler)	Effortil Depot (Boehringer Ing.)	numerous combination preparations and generics
	Cardialgine (MIP Pharma)		
	Circupon RR-Kapseln (gegepharm)	Etilefrin (Chephasaar)	F: Effortil (Boehringer Ing.)
		Eti-Puren (Isis Puren)	I: Effortil (Boehringer Ing.)

J: Effortil (Boehringer-Tanabe)

Etiroxate

ATC: C10A
Use: antiarteriosclerotic (cholesterol depressant and antihyperlipidemic)

RN: 17365-01-4 MF: $C_{18}H_{17}I_4NO_4$ MW: 818.95
CN: O-(4-hydroxy-3,5-diiodophenyl)-3,5-diiodo-α-methyl-DL-tyrosine ethyl ester

hydrochloride
RN: 55327-22-5 MF: $C_{18}H_{17}I_4NO_4 \cdot HCl$ MW: 855.41 EINECS: 259-593-1

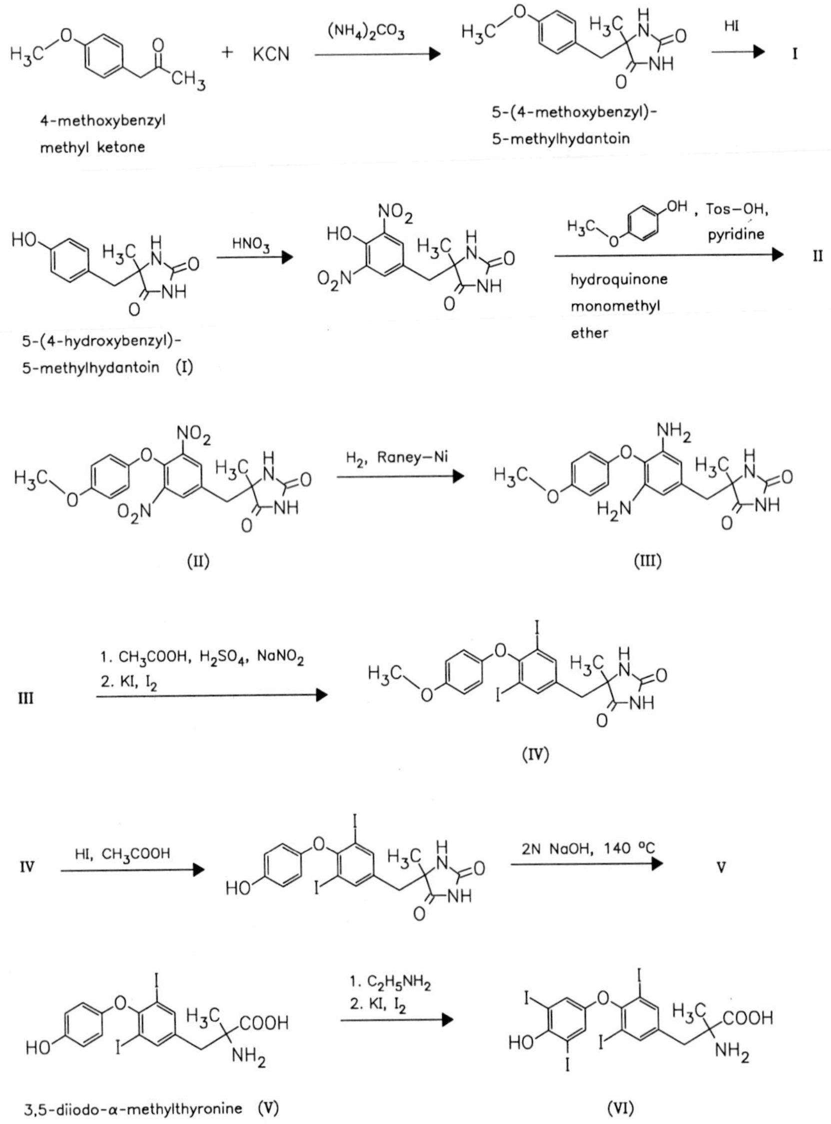

4-methoxybenzyl methyl ketone

5-(4-methoxybenzyl)-5-methylhydantoin

5-(4-hydroxybenzyl)-5-methylhydantoin (I)

hydroquinone monomethyl ether

(II)

(III)

(IV)

3,5-diiodo-α-methylthyronine (V)

(VI)

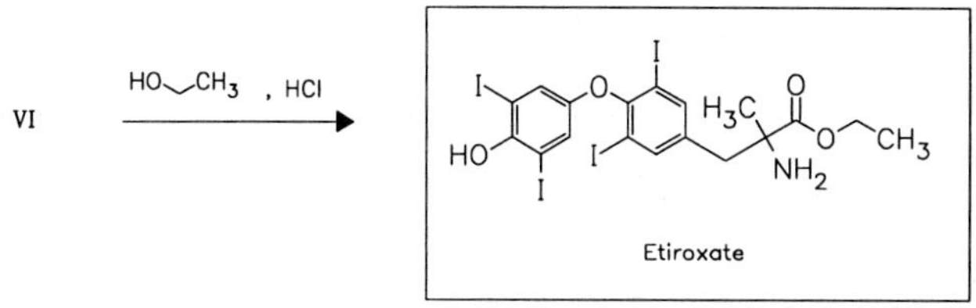
VI
HO CH3 , HCl
Etiroxate

Reference(s):
DE 1 493 533 (Chemie Grünenthal; appl. 10.4.1964).
DAS 1 493 567 (Chemie Grünenthal; appl. 7.10.1965).
US 3 930 017 (Chemie Grünenthal; 30.12.1975; D-prior. 7.10.1965).
US 4 110 470 (Chemie Grünenthal; 29.8.1978; D-prior. 7.10.1965).

Formulation(s): cps. 20 mg

Trade Name(s):
D: Skleronorm (Grünenthal);
wfm

Etizolam

ATC: N05BA19; N05CD
Use: benzodiazepine tranquilizer, anxiolytic, sedative

RN: 40054-69-1 MF: $C_{17}H_{15}ClN_4S$ MW: 342.85
LD_{50}: 4258-4358mg/kg (M, p.o.); >5000 mg/kg (M, s.c.);
3619-3509 mg/kg (R, p.o.); >5000 mg/kg (R, s.c.)
CN: 4-(2-chlorophenyl)-2-ethyl-9-methyl-6*H*-thieno[3,2-*f*][1,2,4]triazolo[4,3-*a*][1,4]diazepine

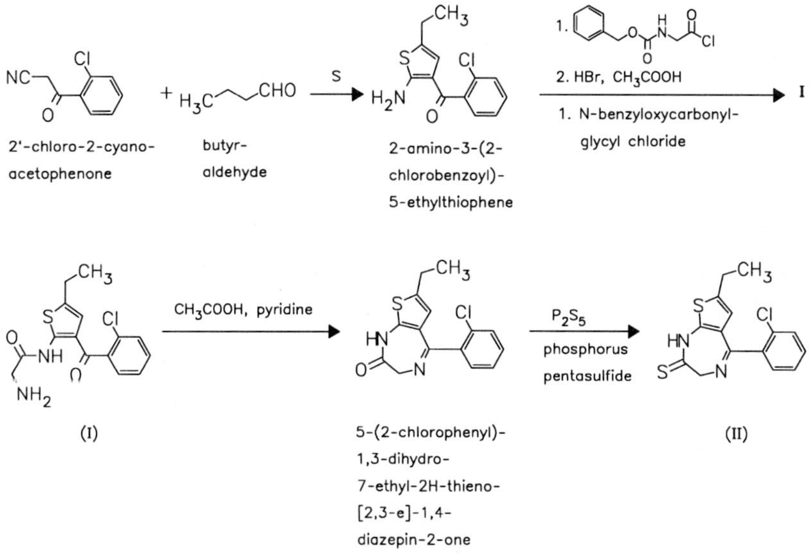
2'-chloro-2-cyano-acetophenone
butyr-aldehyde
S
2-amino-3-(2-chlorobenzoyl)-5-ethylthiophene
2. HBr, CH3COOH
1. N-benzyloxycarbonyl-glycyl chloride
I
(I)
CH3COOH, pyridine
5-(2-chlorophenyl)-1,3-dihydro-7-ethyl-2H-thieno-[2,3-e]-1,4-diazepin-2-one
P2S5
phosphorus pentasulfide
(II)

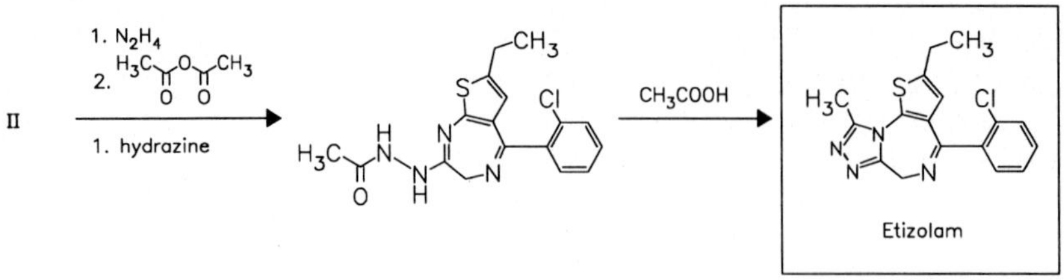

Etizolam

Reference(s):

Nakanishi, M. et al.: J. Med. Chem. (JMCMAR) **16**, 214 (1973).

Nakanishi, M. et al.: Arzneim.-Forsch. (ARZNAD) **22**, 1905 (1972).

Tahara, T. et al.: Arzneim.-Forsch. (ARZNAD) **28**, 1153 (1978).

DOS 2 229 845 (Yoshitomi; appl. 19.6.1972; J-prior. 18.6.1971, 21.6.1971, 30.6.1971, 8.7.1971, 10.7.1971, 13.7.1971).

US 3 904 641 (Yoshitomi; 9.9.1975; J-prior. 18.6.1971, 21.6.1971, 30.6.1971, 8.7.1971, 10.7.1971, 13.7.1971).

Formulation(s):   drops 0.05 %; tabl.0.5 mg, 1 mg

Trade Name(s):

I: Depas (Pierrel)   Pasaden (Farmades)   J: Depas (Yoshitomi)

Etodolac

(Etodolic acid; Etodolsäure)

ATC: M01AB08
Use: anti-inflammatory, analgesic

RN: 41340-25-4   MF: $C_{17}H_{21}NO_3$ MW: 287.36

LD$_{50}$:   593 mg/kg (M, p.o.);
          94 mg/kg (R, p.o.)

CN: 1,8-diethyl-1,3,4,9-tetrahydropyrano[3,4-*b*]indole-1-acetic acid

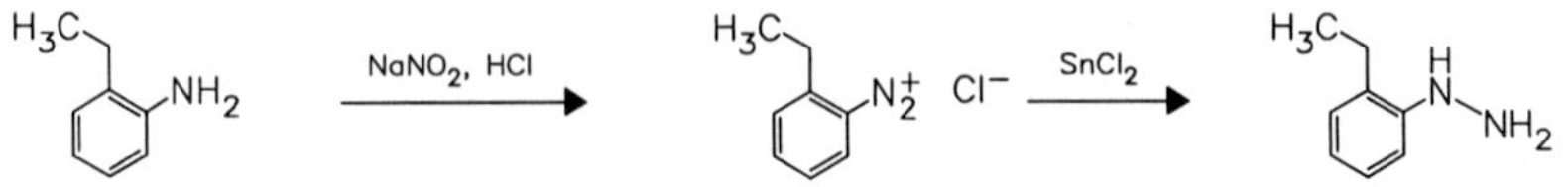

2-ethyl-
aniline

2-ethylbenzene-
diazonium
chloride

2-ethylphenyl-
hydrazine (I)

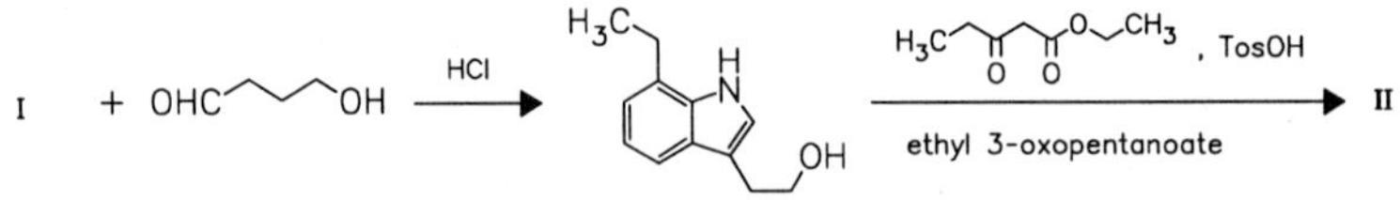

4-hydroxy-
butanal

7-ethyl-3-(2-
hydroxyethyl)-
indole

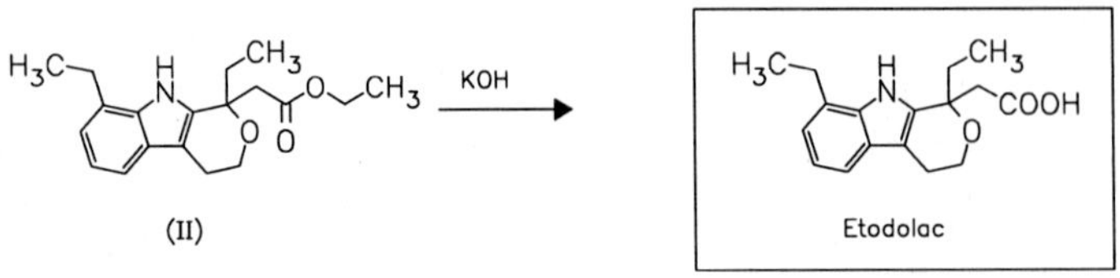

ethyl 3-oxopentanoate

(II)

Etodolac

Reference(s):
US 3 939 178 (American Home Products; 17.2.1976; appl. 15.9.1972).
US 3 843 681 (American Home Products; 22.10.1974; appl. 1.6.1971).
GB 1 391 005 (American Home Products; appl. 1.6.1972; USA-prior. 1.6.1971).
DOS 2 226 340 (American Home Products; appl. 30.5.1972; USA-prior. 1.6.1971).
FR 2 140 154 (American Home Products; appl. 1.6.1972; USA-prior. 1.6.1971).
Demerson, C.A. et al.: J. Med. Chem. (JMCMAR) **18**, 189 (1975).
Demerson, C.A. et al.: J. Med. Chem. (JMCMAR) **19**, 391 (1976).

racemate resolution:
US 4 520 203 (American Home Products; 28.5.1985; appl. 16.8.1983).
US 4 544 757 (American Home Products; 1.10.1985; appl. 16.2.1984).

Formulation(s):   cps. 200 mg, 300 mg; s. r. tabl. 400 mg, 600 mg; tabl. 100 mg, 200 mg, 400 mg, 500 mg

Trade Name(s):
F: Lodine (Wyeth)   I: Edolan (Lepetit; 1987)   Ostelac (Wyeth)
GB: Lodine SR (Monmouth;   Lodine (Wyeth; 1987)   USA: Lodine (Wyeth-Ayerst)
      1985)   J: Hypen (Nippon Shinyaku)

Etodroxizine

ATC: N05C
Use: tranquilizer, hypnotic

RN: 17692-34-1   MF: C$_{23}$H$_{31}$ClN$_2$O$_3$ MW: 418.97
LD$_{50}$:   70 mg/kg (M, i.v.); 540 mg/kg (M, p.o.);
        58 mg/kg (R, i.v.); 920 mg/kg (R, p.o.)
CN: 2-[2-[2-[4-[(4-chlorophenyl)phenylmethyl]-1-piperazinyl]ethoxy]ethoxy]ethanol

1-(4-chlorobenz-   triethylene glycol   Etodroxizine
hydryl)piperazine   monochlorohydrin
(cf. buclizine
synthesis)

Reference(s):
GB 817 231 (UCB; appl. 1957; B-prior. 1956).

Formulation(s):   tabl. 50 mg

Trade Name(s):
D: Vesparax (UCB)-comb.;   F: Drimyl (Cassenne); wfm
      wfm

Etofenamate

ATC: M02AA06
Use: anti-inflammatory

RN: 30544-47-9   MF: C$_{18}$H$_{18}$F$_3$NO$_4$ MW: 369.34   EINECS: 250-231-8
LD$_{50}$:   75 mg/kg (M, i.v.); 743 mg/kg (M, p.o.);
        139 mg/kg (R, i.v.); 292 mg/kg (R, p.o.)
CN: 2-[[3-(trifluoromethyl)phenyl]amino]benzoic acid 2-(2-hydroxyethoxy)ethyl ester

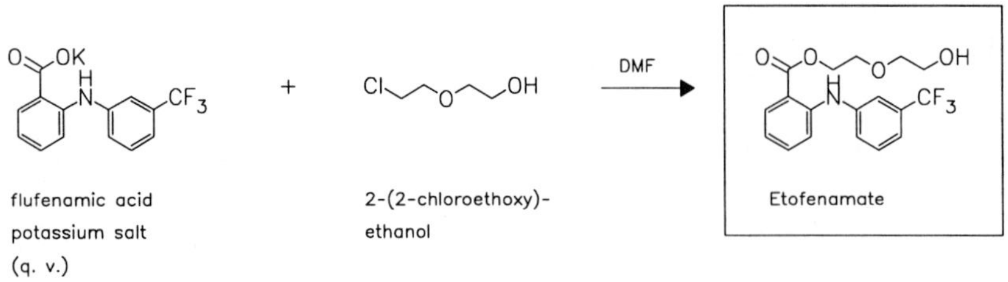

flufenamic acid
potassium salt
(q. v.)

2-(2-chloroethoxy)-
ethanol

Etofenamate

Reference(s):
DE 1 939 112 (Troponwerke; appl. 1.8.1969).
US 3 692 818 (Troponwerke; 19.9.1972; D-prior. 1.8.1969).

Formulation(s):   amp. 1 g/2 ml; cream 100 mg/g; gel 50 mg/g; lotion 100 mg/g

Trade Name(s):
D: Algesalona (Solvay
     Arzneimittel)
     Rheumon (Bayer Vital;
     1977)

     Traumon (Bayer Vital;
     1984)
I: Bayrogel (Bayropharm;
     1980)

Etofibrate

ATC: C01AB09
Use: antihyperlipidemic, cholesterol
       depressant

RN: 31637-97-5   MF: $C_{18}H_{18}ClNO_5$ MW: 363.80   EINECS: 250-743-1
CN: 3-pyridinecarboxylic acid 2-[2-(4-chlorophenoxy)-2-methyl-1-oxopropoxy]ethyl ester

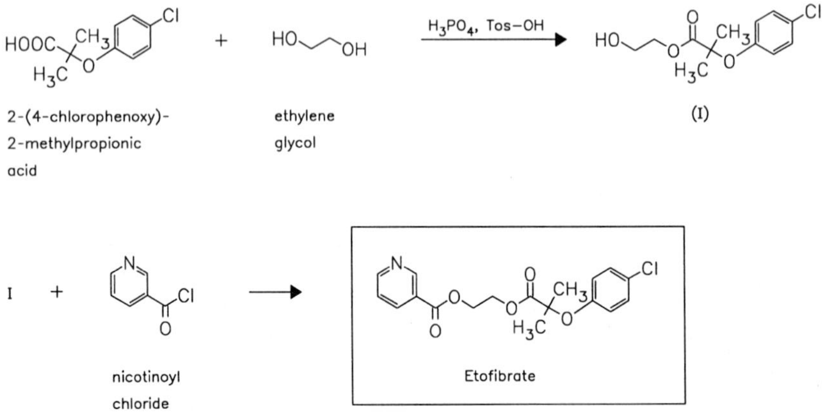

2-(4-chlorophenoxy)-
2-methylpropionic
acid

ethylene
glycol

(I)

nicotinoyl
chloride

Etofibrate

Reference(s):
DOS 1 941 217 (Merz & Co.; appl. 13.8.1969).
DOS 2 519 535 (Alter S.A.; Madrid; appl. 2.5.1975; E-prior. 29.5.1974).
DOS 2 531 254 (Merz & Co.; 12.7.1975; GB-prior. 5.9.1974).
DOS 2 542 413 (Alter S.A.; appl. 23.9.1975; E-prior. 4.6.1975).
DOS 2 542 414 (Alter S.A.; appl. 23.9.1975; E-prior. 4.6.1975).
US 3 723 446 (Merz & Co.; 27.3.1973; appl. 12.8.1970; D-prior. 13.8.1969).

Formulation(s):   s. r. cps. 500 mg

Trade Name(s):
D: Lipo-Merz (Merz & Co.;
      1974)

Etofylline

(Oxyethyltheophylline; Hydroxyäthyltheophyllin)

ATC: C03BD
Use: cardiotonic, bronchodilator

RN: 519-37-9   MF: $C_9H_{12}N_4O_3$ MW: 224.22   EINECS: 208-269-8
LD$_{50}$: 344 mg/kg (M, i.v.); 400 mg/kg (M, p.o.);
       486 mg/kg (R, i.v.); 710 mg/kg (R, p.o.)
CN: 3,7-dihydro-7-(2-hydroxyethyl)-1,3-dimethyl-1H-purine-2,6-dione

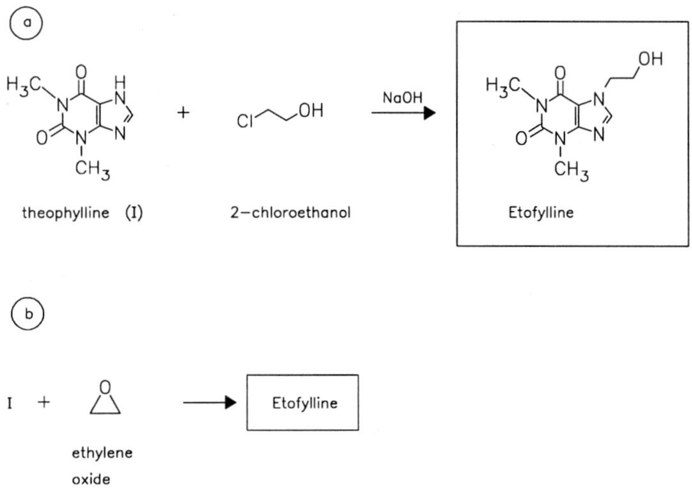

theophylline (I)   2-chloroethanol   Etofylline

I + ethylene oxide → Etofylline

Reference(s):
US 2 715 125 (Gane's Chem. Works; 1955; prior. 1953).

Formulation(s):   drg. 50 mg, 80 mg in comb.

Trade Name(s):
D: Coroverlan (Verla)   F: Oxyphylline (Amido); wfm   J: Oxyphylline (Sankyo)
      Eucebral (Südmedica)-   I: Teostallarid (SmithKline
      comb.   Beecham)

Etomidate

ATC: N01AX07
Use: anesthetic, hypnotic

RN: 33125-97-2   MF: $C_{14}H_{16}N_2O_2$ MW: 244.29   EINECS: 251-385-9
LD$_{50}$: 29.5 mg/kg (M, i.v.); 650 mg/kg (M, p.o.);
       14.8 mg/kg (R, i.v.)
CN: (R)-1-(1-phenylethyl)-1H-imidazole-5-carboxylic acid ethyl ester

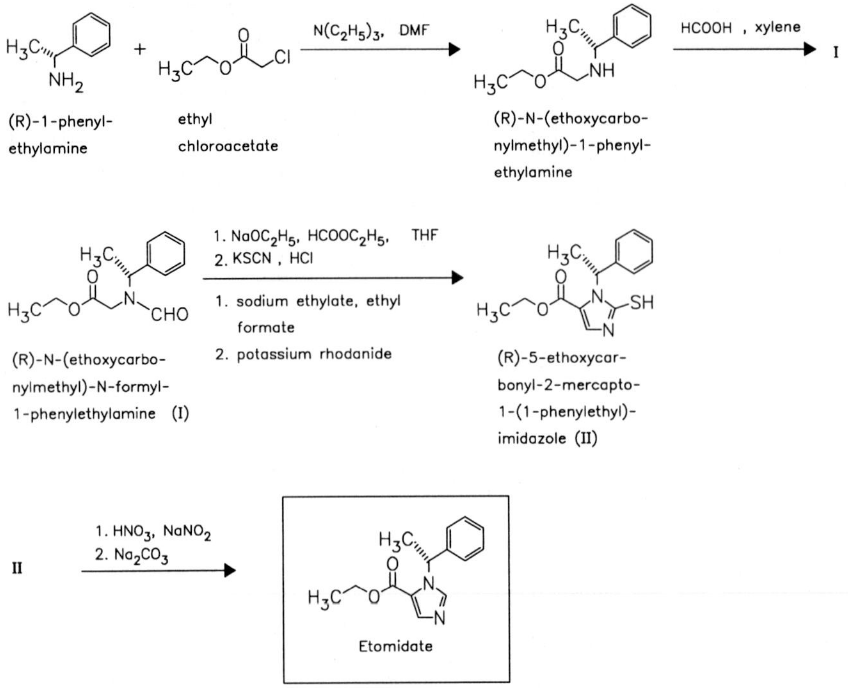

(R)-1-phenyl-ethylamine + **ethyl chloroacetate** → N(C₂H₅)₃, DMF → **(R)-N-(ethoxycarbonylmethyl)-1-phenylethylamine** → HCOOH, xylene → **I**

$N(C_2H_5)_3$, DMF

(R)-1-phenyl-ethylamine

ethyl chloroacetate

(R)-N-(ethoxycarbonylmethyl)-1-phenyl-ethylamine

HCOOH, xylene

I

(R)-N-(ethoxycarbonylmethyl)-N-formyl-1-phenylethylamine (I)

1. NaOC₂H₅, HCOOC₂H₅, THF
2. KSCN, HCl

1. sodium ethylate, ethyl formate
2. potassium rhodanide

(R)-5-ethoxycarbonyl-2-mercapto-1-(1-phenylethyl)-imidazole (II)

II

1. HNO₃, NaNO₂
2. Na₂CO₃

Etomidate

Reference(s):
US 3 354 173 (Janssen; 21.11.1967; prior. 16.4.1964).
DAS 1 545 988 (Janssen; appl. 14.4.1965; USA-prior. 16.4.1964).
Janssen, P.A.J. et al.: Arzneim.-Forsch. (ARZNAD) **21**, 1234 (1971).

injection solution:
DOS 2 937 290 (Janssen; appl. 14.9.1979; USA-prior. 14.9.1978).

Formulation(s):   amp. 2 mg/ml, 20 mg/10 ml

Trade Name(s):
D: Hypnomidate (Janssen-Cilag)
      Radenarcon (ASTA Medica AWD)
F: Hypnomidate (Janssen-Cilag)
GB: Hypnomidate (Janssen)
USA: Amidate (Abbott); wfm

Etoperidone

ATC: N06AB09
Use: antidepressant

RN: 52942-31-1   MF: C₁₉H₂₈ClN₅O   MW: 377.92
CN: 2-[3-[4-(3-chlorophenyl)-1-piperazinyl]propyl]-4,5-diethyl-2,4-dihydro-3H-1,2,4-triazol-3-one

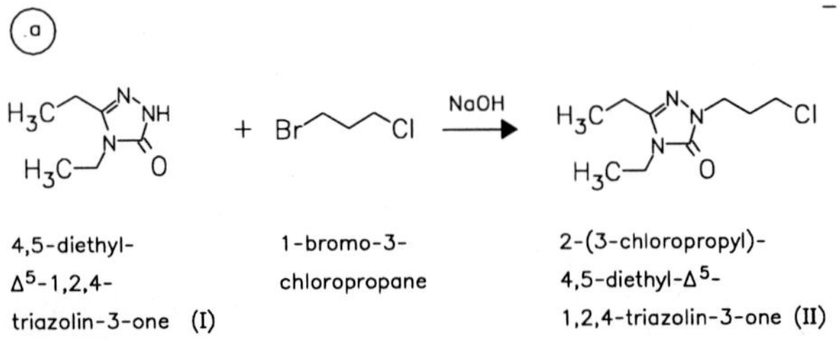

(a)

4,5-diethyl-Δ⁵-1,2,4-triazolin-3-one (I)

1-bromo-3-chloropropane

NaOH

2-(3-chloropropyl)-4,5-diethyl-Δ⁵-1,2,4-triazolin-3-one (II)

N-(3-chlorophenyl)-
piperazine

Etoperidone

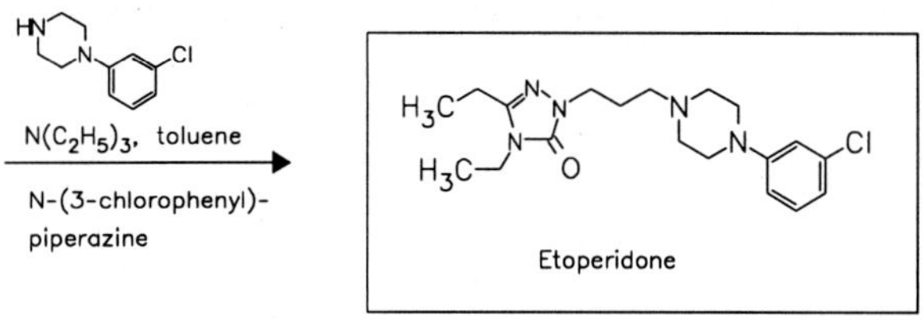

(b)

1-(3-chlorophenyl)-
4-(3-chloropropyl)piperazine

Na, dioxane

Etoperidone

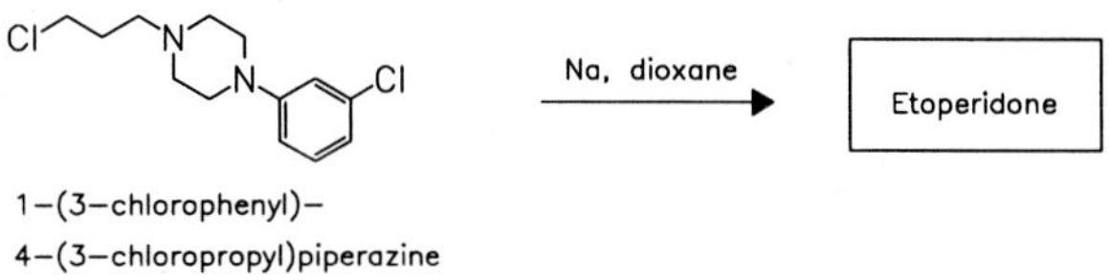

Reference(s):
DOS 2 351 739 (Angelini Francesco; appl. 15.10.1973; I-prior. 16.10.1972).
US 3 857 845 (Angelini Francesco; 31.12.1974; I-prior. 16.10.1972).

use as antiparkinsonian:
US 4 162 318 (Angelini Francesco; 24.7.1979; I-prior. 5.5.1976).
US 4 132 791 (Angelini Francesco; 2.1.1979; I-prior. 5.5.1976).

combination with L-dopa as antiparkinsonian:
US 4 131 675 (Angelini Francesco; 26.12.1978; prior. 9.2.1978).

Formulation(s):   cps. 25 mg, 50 mg

Trade Name(s):
I: Staff (Sigma-Tau); wfm

Etopophos
(BMY-40481-30)

ATC: L01CB
Use: antineoplastic (podophyllotoxin
      derivative)

RN: 122405-33-8   MF: $C_{29}H_{31}Na_2O_{16}P$ MW: 712.51
CN: [5R-[5α,5aβ,8aα,9β(R*)]]-5-[3,5-dimethoxy-4-(phosphonooxy)phenyl]-9-[(4,6-O-ethylidene-β-D-glucopyranosyl)oxy]-5,8,8a,9-tetrahydrofuro[3',4':6,7]naphtho[2,3-d]-1,3-dioxol-6(5aH)-one disodium salt

hexahydrate
RN: 151062-35-0   MF: $C_{29}H_{31}Na_2O_{16}P \cdot 6H_2O$ MW: 820.60
free acid
RN: 117091-64-2   MF: $C_{29}H_{33}O_{16}P$ MW: 668.54

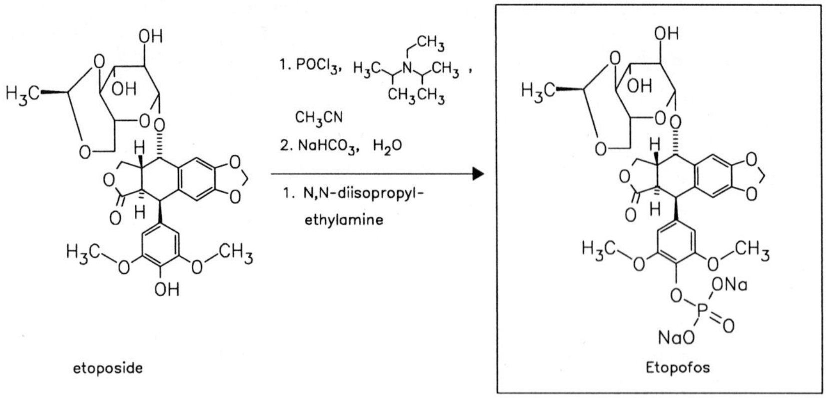

etoposide

Etopofos

Reference(s):
GB 2 207 674 (Bristol-Myers Squibb; appl. 3.8.1988; USA-prior. 27.5.1988, 4.8.1987).

synthesis of etoposide-4'-phosphate:
EP 511 563 (Bristol-Myers Squibb; appl. 16.4.1992; USA-prior. 29.4.1991, 20.2.1992).
EP 567 089 (Nippon Kayaku; appl. 21.4.1993; J-prior. 24.4.1992).

preparation of etoposide without extensive purification:
EP 652 226 (Bristol-Myers Squibb; appl. 3.11.1994; USA-prior. 4.11.1993).

stable hexahydrate with improved storage stability:
EP 548 834 (Bristol-Myers Squibb; appl. 18.12.1992; USA-prior. 23.12.1991).

Formulation(s):   vial 100 mg

Trade Name(s):
USA: Etopophos (Bristol-Myers)

Etoposide
(VP-16-213)

ATC: L01CB01
Use: antineoplastic, podophyllotoxin
           derivative

RN: 33419-42-0   MF: $C_{29}H_{32}O_{13}$ MW: 588.56   EINECS: 251-509-1
LD_{50}:   15.07 mg/kg (M, i.v.); 215 mg/kg (M, p.o.);
           75 mg/kg (R, i.v.); 1784 mg/kg (R, p.o.)
CN: [5R-[5α,5aβ,8aα,9β(R*)]]-9-[(4,6-O-ethylidene-β-D-glucopyranosyl)oxy]-5,8,8a,9-tetrahydro-5-(4-
           hydroxy-3,5-dimethoxyphenyl)furo[3',4':6,7]naphtho[2,3-d]-1,3-dioxol-6(5aH)-one

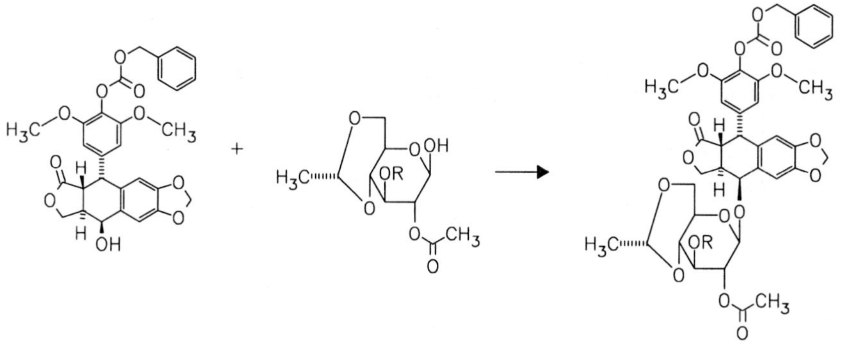

4'-benzyloxycarbonyl-
4'-demethylepipodo-
phyllotoxin
(cf. teniposide
synthesis)

4,6-O-(R)-ethylidene-
2,3-di-O-acetyl-β-D-
glucopyranose

(I)

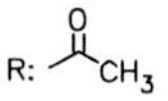

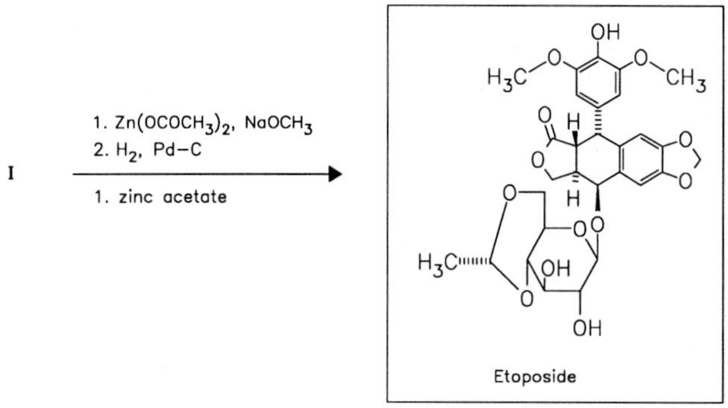

R: (acetyl group, CH₃C(=O)-)

Etoposide

Reference(s):
DE 1 643 521 (Sandoz; prior. 9.12.1967).
US 3 524 844 (Sandoz; 18.8.1970; CH-prior. 21.6.1965).
CH 514 578 (Sandoz; appl. 27.2.1968).
Keller-Juseen, C. et al.: J. Med. Chem. (JMCMAR) **14**, 936 (1971).
US 5 637 680 (Nippon Kayaku; 10.6.1997; J-prior. 24.4.1992).
EP 778 282 (Nippon Kayaku; appl. 3.12.1996; J-prior. 4.12.1995, 8.12.1995).
Allevi, P. et al.: J. Org. Chem. (JOCEAH) **58**, 4175 (1993).

Formulation(s):   cps. 50 mg, 100 mg; vial 100 mg/5 ml, 150 mg/7.5 ml, 500 mg/25 ml, 1 g/50 ml

Trade Name(s):
D: Etomedac (medac)
       Vepesid (Bristol-Myers
       Squibb; 1980)
F: Celltop (ASTA Medica)
       Etopophos (Bristol-Myers
       Squibb)

       Etoposide Pierre Fabre
       (Pierre Fabre)
       Vepeside (Novartis)
GB: Vepesid (Bristol-Myers
       Squibb; 1981)
I: Vepesid (Bristol It. Sud;
       1982); wfm

J: Lastet (Nippon Kayaku;
       1987)
       Vepesid (Bristol Squibb;
       1987)
USA: Ve Pesid (Bristol-Myers
       Squibb; 1983)

Etozolin

ATC: C03CX01
Use: diuretic

RN: 73-09-6   MF: $C_{13}H_{20}N_2O_3S$ MW: 284.38   EINECS: 200-794-0
LD_{50}:   8670 mg/kg (M, p.o.);
        10250 mg/kg (R, p.o.)
CN: [3-methyl-4-oxo-5-(1-piperidinyl)-2-thiazolidinylidene]acetic acid ethyl ester

ethyl mercapto-   ethyl cyano-   ethyl 4-oxothiazo-
acetate   acetate   lidin-2-ylideneacetate

ethyl 3-methyl-   ethyl 5-bromo-3-   Etozolin
4-oxothiazolidin-   methyl-4-oxothiazo-
2-ylideneacetate   lidin-2-ylideneacetate

Reference(s):
US 3 072 653 (Warner-Lambert; 8.1.1963; appl. 6.3.1961).
DE 1 160 441 (Warner-Lambert; appl. 21.10.1961; USA-prior. 6.3.1961).
GB 1 022 047 (Warner-Lambert; appl. 23.11.1962).
GB 1 022 048 (Warner-Lambert; appl. 23.11.1962).
Satzinger, G.: Justus Liebigs Ann. Chem. (JLACBF) **665**, 150 (1963).

Formulation(s):   tabl. 200 mg, 400 mg

Trade Name(s):
D: Elkapin (Gödecke); wfm   I: Elkapin (Parke Davis)

Etretinate

ATC: D05BB01
Use: antipsoriatic

RN: 54350-48-0   MF: $C_{23}H_{30}O_3$ MW: 354.49   EINECS: 259-119-3
LD_{50}:   1176 mg/kg (M, i.p.); >2000 mg/kg (M, p.o.);
        >2000 mg/kg (R, i.p.); >4000 (R, p.o.)
CN: (*all-E*)-9-(4-methoxy-2,3,6-trimethylphenyl)-3,7-dimethyl-2,4,6,8-nonatetraenoic acid ethyl ester

3,5-dimethyl-   2,3,5-trimethyl-   2,3,5-trimethyl-
phenol   phenol   anisole (I)

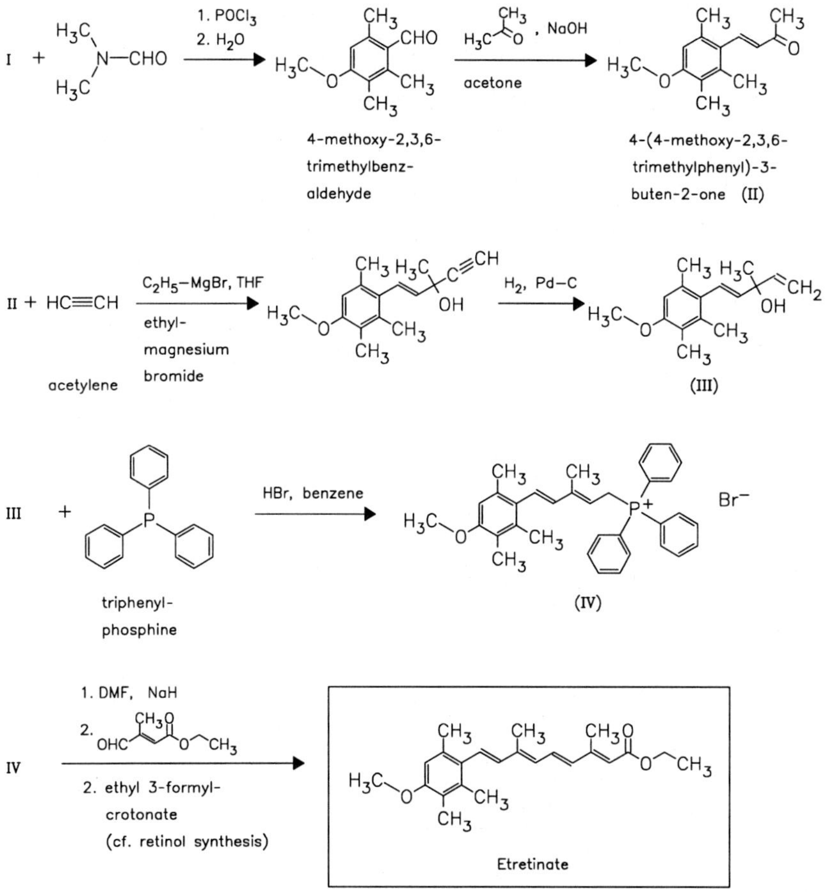
1. POCl$_3$
2. H$_2$O
acetone
4-methoxy-2,3,6-trimethylbenzaldehyde
4-(4-methoxy-2,3,6-trimethylphenyl)-3-buten-2-one (II)
acetylene
C_2H_5–MgBr, THF
ethylmagnesium bromide
H$_2$, Pd–C
(III)
HBr, benzene
triphenylphosphine
(IV)
1. DMF, NaH
2. ethyl 3-formylcrotonate
(cf. retinol synthesis)
Etretinate

Reference(s):

Mayer, H. et al.: Experientia (EXPEAM) **34**, 1105 (1978).
US 4 105 681 (Roche; 8.8.1978; prior. 22.3.1974; 1.8.1975; 13.8.1976).
DOS 2 414 619 (Roche; appl. 26.3.1974; CH-prior. 30.3.1973).
US 4 215 215 (Hoffmann-La Roche; 29.7.1980; prior. 6.7.1979).

medical use:
US 4 200 647 (Hoffmann-La Roche; 29.4.1980; appl. 12.12.1978; CH-prior. 21.12.1977).

Formulation(s): cps. 10 mg, 25 mg

Trade Name(s):

D:	Tigason (Roche; 1982); wfm	GB:	Tigason (Roche); wfm	USA:	Tegison (Roche; 1986)
F:	Tigason (Roche); wfm	I:	Tigason (Roche)		
		J:	Tigason (Roche)		

Etymemazine

(Äthylisobutrazin; Ethotrimeprazine; Ethylisobutrazine)

ATC: R06AD
Use: antihistaminic, tranquilizer, hypnotic

RN: 523-54-6 MF: $C_{20}H_{26}N_2S$ MW: 326.51
CN: 2-ethyl-*N*,*N*,β-trimethyl-10*H*-phenothiazine-10-propanamine

monohydrochloride

RN: 3737-33-5   MF: $C_{20}H_{26}N_2S \cdot HCl$ MW: 362.97   EINECS: 223-111-8

LD_{50}: 70 mg/kg (M, i.v.)

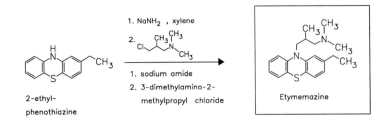

1. NaNH₂, xylene

2. CH₃ CH₃ ... N CH₃

1. sodium amide

2. 3-dimethylamino-2-
    methylpropyl chloride

2-ethyl-phenothiazine

Etymemazine

Reference(s):

DE 1 034 638 (Rhône-Poulenc; appl. 1955; F-prior. 1954).

Trade Name(s):

F: Nuital (Vaillant-Defresne);
       wfm

Etynodiol acetate

(Äthynodioldiacetat; Ethynodiol diacetate; Etynodiol diacetate)

ATC: G03AA

Use: progestogen (in combination with estrogen as oral contraceptive)

RN: 297-76-7   MF: $C_{24}H_{32}O_4$ MW: 384.52   EINECS: 206-044-9

CN: (3β,17α)-19-norpregn-4-en-20-yne-3,17-diol diacetate

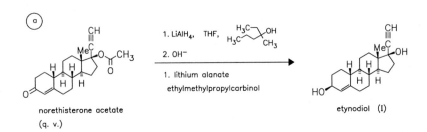

ⓐ

1. LiAlH₄, THF,

2. OH⁻

1. lithium alanate
   ethylmethylpropylcarbinol

norethisterone acetate
(q. v.)

etynodiol   (I)

I   +   acetic anhydride (II)

pyridine, I₂

Etynodiol acetate

ⓑ

I   + II

N(C₂H₅)₃,

triethylamine,
4-dimethylamino-pyridine

Etynodiol acetate

Reference(s):
a   US 3 176 013 (Searle; 30.3.1965; appl. 25.7.1963).
    DE 1 668 604 (Gedeon Richter; appl. 7.9.1967; H-prior. 7.9.1969).
    DE 2 137 557 (Gedeon Richter; appl. 27.7.1971; H-prior. 29.7.1970).
b   DE 2 137 856 (Searle; appl. 29.7.1971; USA-prior. 30.6.1970).

alternative synthesis:
DD 91 649 (G. Teichmüller et al.; appl. 2.3.1971).

Formulation(s):   tabl. 1 mg

Trade Name(s):

D:	Alfames E (Kade)-comb.; wfm	GB:	Femulen (Searle)		Ovaras (Serono)-comb.; wfm
	Ovulen (Boehringer Mannh.)-comb.; wfm	I:	Luteolas (Serono)-comb.; wfm	J:	Ovulen (Dainippon)-comb.
	Ovulen (Searle)-comb.; wfm		Luteonorm (Serono); wfm	USA:	Demulen 21/28 (Searle)-comb.
F:	Luto-métrodiol (Monsanto; as diacetate)		Metrulen (SPA)-comb.; wfm		
			Miniluteolas (Serono)-comb.; wfm		

Exalamide

ATC: D01AE
Use: topical antifungal

RN: 53370-90-4   MF: $C_{13}H_{19}NO_2$ MW: 221.30   EINECS: 258-504-3
LD$_{50}$:   13.21 g/kg (M, p.o.);
          >5 g/kg (R, p.o.)
CN: 2-(hexyloxy)benzamide

salicylamide   n-hexyl bromide   Exalamide

Reference(s):
GB 726 786 (Herts Pharm.; appl. 1952).

pharmaceutical formulation:
GB 872 891 (Smith & Nephew; appl. 1957).
Bevin, E.M. et al.: J. Pharm. Pharmacol. (JPPMAB) **4**, 872 (1952).

Formulation(s):   ointment 5 %; sol. 5 %

Trade Name(s):
J: Hyperan (S. S. Pharm.)

Exifone

ATC: N07X
Use: cognition enhancer, nootropic

RN: 52479-85-3   MF: $C_{13}H_{10}O_7$ MW: 278.22   EINECS: 257-945-9
LD$_{50}$:   355 mg/kg (R, i.p.); 1425 mg/kg (R, p.o.)
CN: (2,3,4-trihydroxyphenyl)(3,4,5-trihydroxyphenyl)methanone

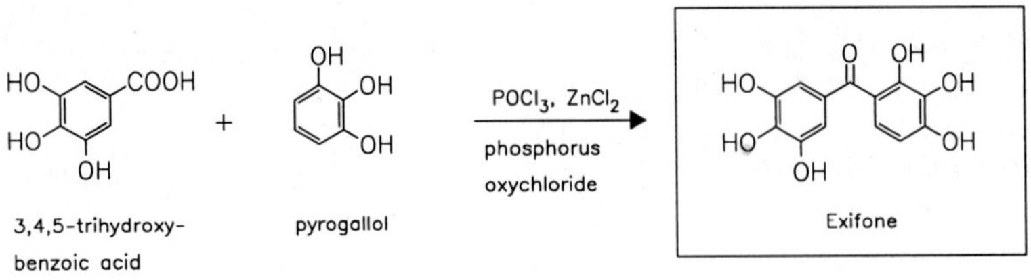
HO
COOH
HO
OH
3,4,5-trihydroxy-
benzoic acid
+
OH
OH
OH
pyrogallol
POCl3, ZnCl2
phosphorus
oxychloride
HO
O
OH
OH
HO
OH
OH
Exifone

Reference(s):
DE 2 501 443 (Lab. Pharmascience; appl. 15.1.1975; F-prior. 15.1.1974).
GB 1 495 331 (Lab. Pharmascience; appl. 15.1.1975; F-prior. 15.1.1974).

Formulation(s): tabl. 200 mg

Trade Name(s):
F: Adlone (Pharmascience; 1988); wfm

Fadrozole
(CGS-16949A)

ATC: L01
Use: antineoplastic, non-steroidal aromatase inhibitor

RN: 131833-76-6   MF: $C_{14}H_{13}N_3$ MW: 223.28
CN: (±)-4-(5,6,7,8-tetrahydroimidazo[1,5-a]pyridin-5-yl)benzonitrile

monohydrochloride
RN: 102676-96-0   MF: $C_{14}H_{13}N_3 \cdot HCl$ MW: 259.74
(S)-form
RN: 102676-86-8   MF: $C_{14}H_{13}N_3$ MW: 223.28

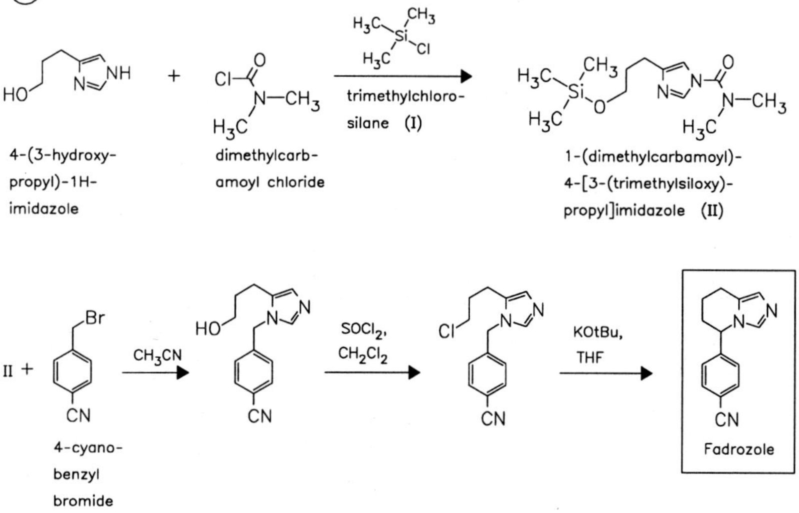

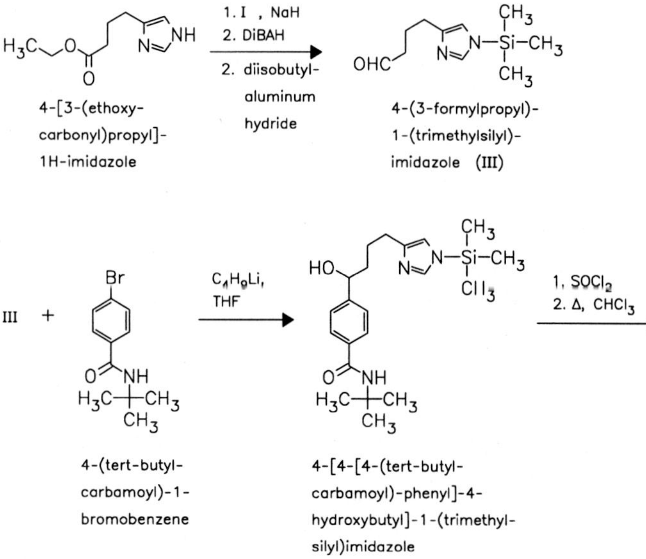

Reference(s):
EP 165 904 (Ciba-Geigy AG; appl. 17.6.1985; USA-prior. 20.6.1984; 20.6.1985).

administration of (–)-fadrozole:
WO 9 528 156 (Sepracor Inc.; appl. 11.4.1995; USA-prior. 14.4.1994).

preparation of starting materials:
Ganellin, C.R. et al.: J. Med. Chem. (JMCMAR) **39** (19), 3806 (1996).
Pasini, C.: Gazz. Chim. Ital. (GCITA9) **87**, 1464, 1473 (1957)
Akabori: Ber. Dtsch. Chem. Ges. B (BDCBAD) **66** 151, 156 (1933).

Formulation(s):   tabl. 1 mg (as hydrochloride)

Trade Name(s):
J: Afema (Ciba-Geigy)   USA: Arensin (Ciba-Geigy)

Famciclovir
(BRL-42810)

ATC: J05AB09; S01AD07
Use: antiviral

RN: 104227-87-4   MF: $C_{14}H_{19}N_5O_4$ MW: 321.34
CN: 2-[2-(2-amino-9*H*-purin-9-yl)ethyl]-1,3-propanediol diacetate (ester)

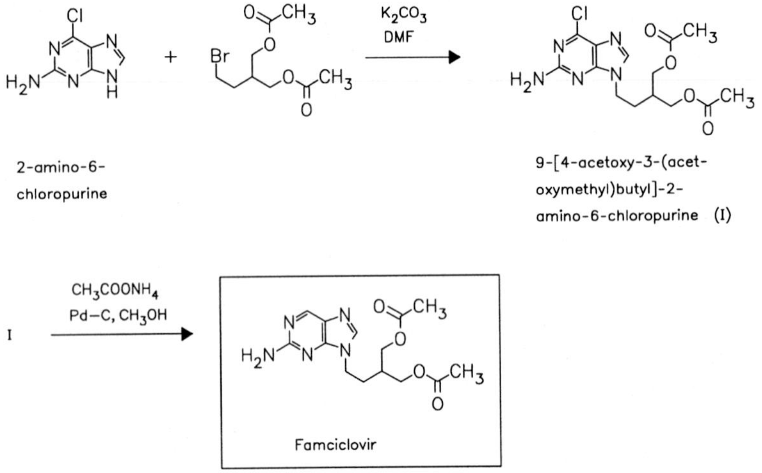

2-amino-6-
chloropurine

9-[4-acetoxy-3-(acet-
oxymethyl)butyl]-2-
amino-6-chloropurine   (I)

Famciclovir

Reference(s):
EP 182 024 (Beecham Group; appl. 9.9.1985; GB-prior. 16.8.1985).

alternative preparation of intermediate I:
WO 9 528 402 (SmithKline Beecham; appl. 19.4.1995; GB-prior. 19.4.1994).

Formulation(s):   f. c. tabl. 125 mg, 250 mg, 500 mg

Trade Name(s):
D: Famvir (SmithKline   GB: Famvir (SmithKline   USA: Famvir (SmithKline
      Beecham)   Beecham)   Beecham)

Famotidine

ATC: A02BA03
Use: ulcer therapeutic, H_2-receptor antagonist

RN: 76824-35-6   MF: $C_8H_{15}N_7O_2S_3$ MW: 337.45
LD_{50}:   244.4 mg/kg (M, i.v.)
CN: 3-[[[2-[(aminoiminomethyl)amino]-4-thiazolyl]methyl]thio]-N-(aminosulfonyl)propanimidamide

thiourea   1,3-dichloro-acetone   S-(2-amino-4-thiazolylmethyl)-isothiourea   3-chloro-propionitrile   (I)

benzoyl isothiocyanate

1. K_2CO_3
2. CH_3I
3. NH_3

2. methyl iodide

1. CH_3OH, HCl
2. $SO_2(NH_2)_2$

2. sulfamide

(II)   Famotidine

Reference(s):
DOS 2 951 675 (Yamanouchi; appl. 21.12.1979; J-prior. 2.8.1979).
DOS 3 008 056 (Yamanouchi; appl. 3.3.1980; J-prior. 6.3.1979, 23.6.1979).
GB 2 052 478 (Yamanouchi; appl. 6.3.1980; J-prior. 6.3.1979, 23.6.1979).
GB 2 055 800 (Yamanouchi; appl. 20.12.1979; J-prior. 2.8.1979).
US 4 283 408 (Yamanouchi; 11.8.1981; J-prior. 2.8.1979).

synthesis of S-[2-aminothiazol-4-ylmethyl]isothiourea:
Sprague, J.M.; Lund, A.H.; Ziegler, C.: J. Am. Chem. Soc. (JACSAT) **68**, 2155 (1946).

preparation of 4-chloromethylthiazol-2-ylamine hydrochloride:
Passarotti, C.M.; Valenti, M.; Marini, M.: Boll. Chim. Farm. (BCFAAI) **134** (11), 639-643 (1995).

Formulation(s):   f. c. tabl. 10 mg, 20 mg, 40 mg; oral susp. 40 mg/5 ml; vial (lyo.) 20 mg

Trade Name(s):

D:	Ganor (Boehringer Ing.)	GB:	Pepcid (Morson; 1987)	USA:	Mylanta (Johnson & Johnson-Merck)

D: Ganor (Boehringer Ing.)
       Pepdul (MSD Chibropharm; 1986)
F: Pepcidac (Labs. Jean-Paul Martin)
       Pepdine (Merck Sharp & Dohme-Chibret)

GB: Pepcid (Morson; 1987)
I: Famodil (Sigma-Tau)
       Gastridin (Merck Sharp & Dohme)
       Motiax (Neopharmed)
J: Gaster (Yamanouchi; 1985)

USA: Mylanta (Johnson & Johnson-Merck)
       Pepcid (Merck; 1986)
       Pepcid (Johnson & Johnson-Merck)

Faropenem sodium
(Furopenem; SUN 5555)

Use: penem antibiotic

RN: 122547-49-3   MF: $C_{12}H_{14}NNaO_5S$ MW: 307.30
CN: [5R-[3(R*),5α,6α(R*)]]-6-(1-Hydroxyethyl)-7-oxo-3-(tetrahydro-2-furanyl)-4-thia-1-azabicyclo[3.2.0]-hept-2-ene-2-carboxylic acid

sodium salt hydrate
RN: 158365-51-6   MF: $C_{12}H_{14}NNaO_5S \cdot 5/2H_2O$ MW: 704.68
acid
RN: 106560-14-9   MF: $C_{12}H_{15}NO_5S$ MW: 285.32

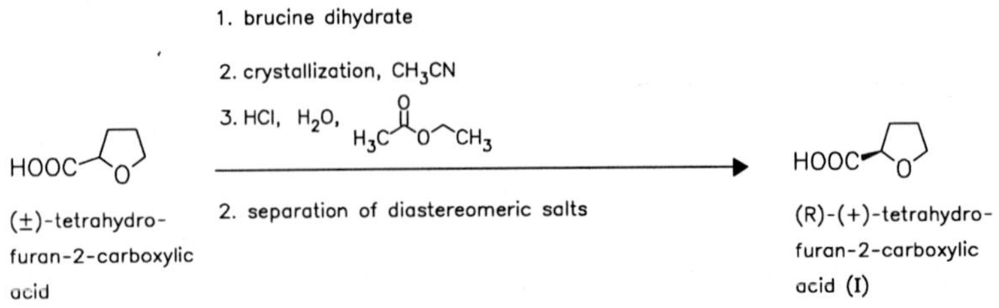

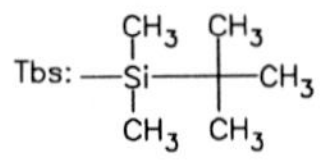

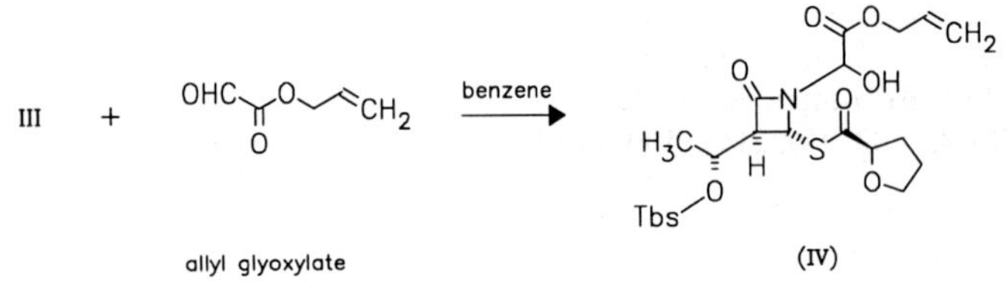

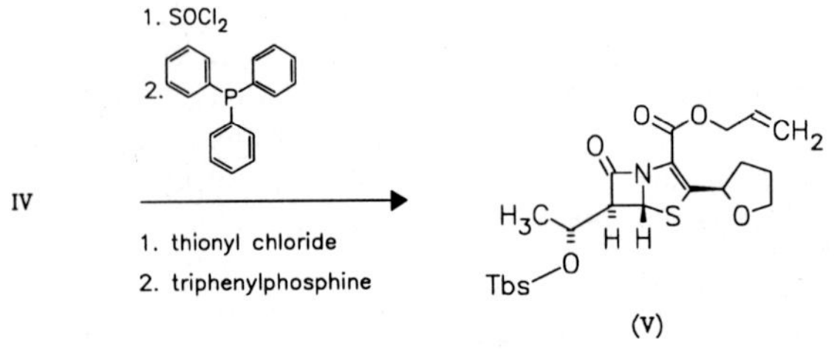

1. brucine dihydrate
2. crystallization, CH_3CN
3. HCl, H_2O, H_3C
2. separation of diastereomeric salts

(±)-tetrahydro-furan-2-carboxylic acid

(R)-(+)-tetrahydro-furan-2-carboxylic acid (I)

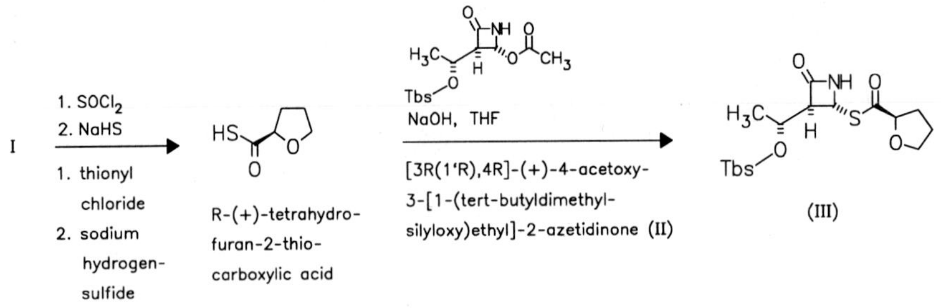

I
1. $SOCl_2$
2. NaHS

1. thionyl chloride
2. sodium hydrogen-sulfide

R-(+)-tetrahydro-furan-2-thio-carboxylic acid

[3R(1'R),4R]-(+)-4-acetoxy-3-[1-(tert-butyldimethyl-silyloxy)ethyl]-2-azetidinone (II)

Tbs, NaOH, THF

(III)

Tbs: —Si— with CH₃ groups

III   +   OHC—O—CH₂ (allyl glyoxylate)   benzene   →   (IV)

IV
1. $SOCl_2$
2. (triphenylphosphine)

1. thionyl chloride
2. triphenylphosphine

(V)

1. Bu₄N⁺ F⁻, CH₃COOH

2. H₃C, CH₃, ONa

Pd[PPh₃]₄, CH₂Cl₂

V

1. tetrabutylammonium fluoride
2. sodium 2-ethylhexanoate

Faropenem sodium

preparation of intermediate II:

methyl 2-(benzyl-
aminomethyl)-3-
oxobutanoate

1. [Ru₂Cl₄ (+)–BINAP], H₂
2. HCl

(2S,3R)-2-(aminomethyl)-
3-hydroxybutanoic acid
hydrochloride (VI)

VI

triphenylphosphine,
di-2-pyridyl disulfide

[3S(1R)]-3-
(1-hydroxyethyl)-
2-azetidinone

1. RuCl₃, HO—O
2. Cl—Si
2. TBDMS-chloride

II

Reference(s):
EP 199 446 (Suntory; appl. 7.3.1986; J-prior. 9.3.1985).
WO 9 203 443 (Suntory; appl. 16.8.1991; J-prior. 20.8.1990).

preparation of intermediate II:
EP 369 691 (Takasago Int. Corp.; appl. 10.11.1989; J-prior. 15.11.1988).
EP 371 875 (Takasago Int. Corp.; appl. 28.11.1989; J-prior. 29.11.1988).
EP 488 611 (Takasago Int. Corp.; appl. 25.11.1991; J-prior. 30.11.1990).
Murahashi, S. et al.: Tetrahedron Lett. (TELEAY) **32** (19), 2145 (1991).

alternative preparation of (3S,1'R)-3-(1'-hydroxyethyl)azetidin-2-one:
Fuganti, C. et al.: J. Chem. Soc. Perkin Trans. 1 (JCPRB4) **1** (19), 2247 (1993).
Fuganti, C. et al.: Bioorg. Med. Chem. Lett. (BMCLE8) **2** (7), 723 (1994).

preparation of racemic tetrahydrofuran-2-carboxylic acid:
Wienhaus; Sorge: Ber. Dtsch. Chem. Ges. (BDCGAS) **46**, 1929 (1913).
Kaufmann; Adams: J. Am. Chem. Soc. (JACSAT) **45**, 3041 (1923).
Wilson: J. Chem. Soc. (JCSOA9) **1945**, 58, 59.

preparation of (R)-(+)-tetrahydrofuran-2-carboxylic acid:
Ramón, A. et al.: J. Med. Chem. (JMCMAR) **38**, 2830 (1995).
Belanger, P.C., Williams, H.W.R.: Can. J. Chem. (CJCHAG) **61**, 873 (1983).

Formulation(s):　　tabl. 150 mg, 200 mg

Trade Name(s):
J: Farom (Suntory; 1999)

Fasudil

(AT-877; HA-1077)

ATC: J01CA12
Use: vasodilator, calcium channel blocker

RN: 103745-39-7   MF: $C_{14}H_{17}N_3O_2S$ MW: 291.38
CN: hexahydro-1-(5-isoquinolinylsulfonyl)-1*H*-1,4-diazepine

monohydrochloride
RN: 105628-07-7   MF: $C_{14}H_{17}N_3O_2S \cdot HCl$ MW: 327.84

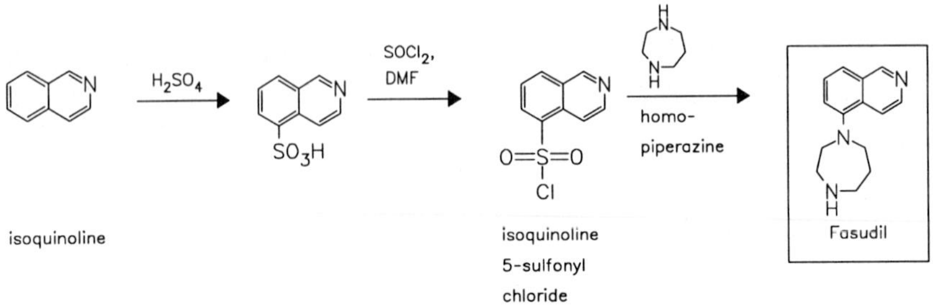

isoquinoline   isoquinoline
                                5-sulfonyl
                                chloride   Fasudil

Reference(s):
EP 187 371 (Asahi Chem.; appl. 23.12.1985; J-prior. 27.12.1984).

Formulation(s):   amp. 30 mg/2 ml (as hydrochloride)

Trade Name(s):
J: Eril (Asahi Kasei; as   Fasdil (Asahi Chem.)
      hydrochloride)

Febuprol

ATC: A05AB
Use: choleretic

RN: 3102-00-9   MF: $C_{13}H_{20}O_3$ MW: 224.30   EINECS: 221-454-8
LD_{50}:   436 mg/kg (M, i.p.); 3050 mg/kg (M, p.o.);
        400 mg/kg (R, i.p.); 2370 mg/kg (R, p.o.)
CN: 1-butoxy-3-phenoxy-2-propanol

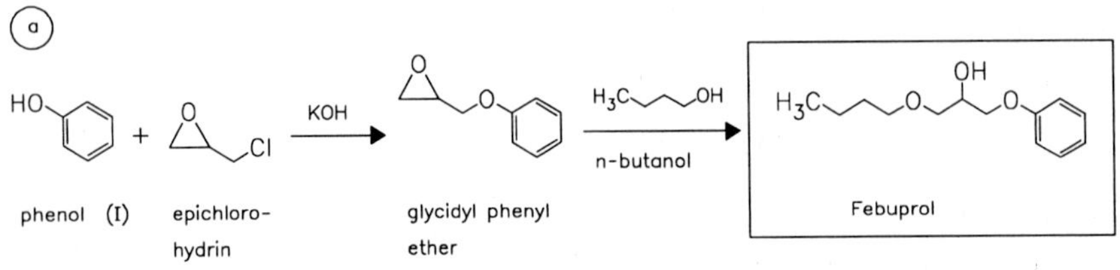

phenol (I)   epichloro-   glycidyl phenyl   Febuprol
              hydrin   ether

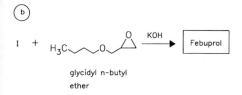

glycidyl n-butyl
ether

Reference(s):
DOS 2 207 254 (Klinge; appl. 16.2.1971).
DOS 2 120 396 (Klinge; appl. 26.4.1971).
US 3 839 587 (Klinge; 1.10.1974; D-prior. 26.4.1971, 16.2.1971).
Minor, W.F. et al.: J. Am. Chem. Soc. (JACSAT) **76**, 2993 (1954).

Formulation(s):   cps. 100 mg

Trade Name(s):
D: Valbil (Procter & Gamble)

Feclobuzone

ATC: M01A; N02B; S01B
Use: anti-inflammatory, analgesic, antipyretic

RN: 23111-34-4   MF: $C_{27}H_{25}ClN_2O_4$ MW: 476.96
CN: 4-chlorobenzoic acid (4-butyl-3,5-dioxo-1,2-diphenyl-4-pyrazolidinyl)methyl ester

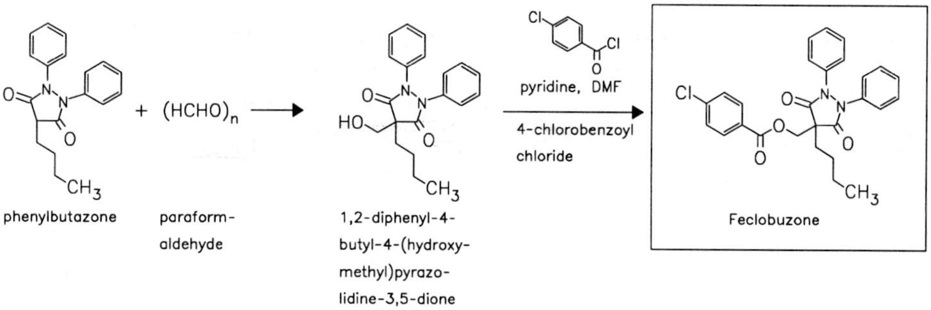

phenylbutazone   paraform-
aldehyde

+ $(HCHO)_n$ ⟶

1,2-diphenyl-4-
butyl-4-(hydroxy-
methyl)pyrazo-
lidine-3,5-dione

pyridine, DMF

4-chlorobenzoyl
chloride

Feclobuzone

Reference(s):
DE 1 809 821 (Lab. del Dr. Esteve; appl. 20.11.1968; E-prior. 23.11.1967).

Trade Name(s):
D;   Feclobuzon-Dragees
      (Atmos); wfm

Felbamate
(W-554; ADD-03055)

ATC: N03AX10
Use: anticonvulsant

RN: 25451-15-4   MF: $C_{11}H_{14}N_2O_4$ MW: 238.24   EINECS: 247-001-4
LD_{50}:   >5 g/kg (R, p. o.);
              >5 g/kg (M, p. o.)
CN: 2-Phenyl-1,3-propanediol dicarbamate

a)

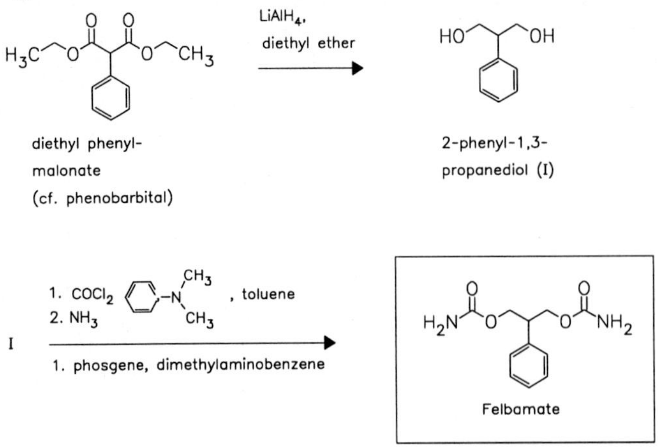

diethyl phenyl-
malonate
(cf. phenobarbital)

LiAlH₄,
diethyl ether

2-phenyl-1,3-
propanediol (I)

1. COCl₂ ... , toluene
2. NH₃

I

1. phosgene, dimethylaminobenzene

Felbamate

b)

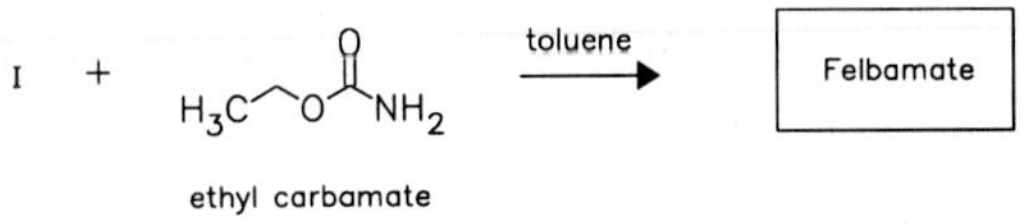

I +
ethyl carbamate

toluene

Felbamate

c)

methyl
phenylacetate

+

methyl formate

NaOMe, toluene

methyl (E)-2-formyl-
2-phenylacetate
sodium salt (II)

1. H₂SO₄
2. NaBH₄

II

2. sodium
borohydride

I

Cl–S–NCO , toluene

chlorosulfonyl
isocyanate

Felbamate

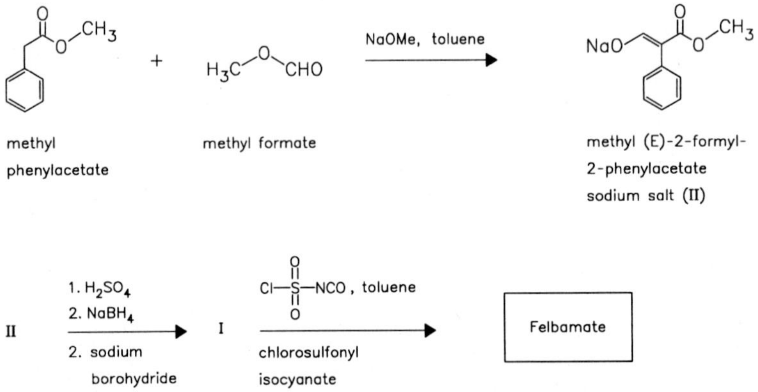

Reference(s):

a   US 4 982 016 (Carter-Wallace; 1.1.1991; USA-prior. 6.6.1989).
b   US 4 868 327 (Carter-Wallace; 20.2.1991; USA-prior. 3.6.1987).
    Ludwig, B.J. et al.: J. Med. Chem. (JMCMAR) **12**, 462 (1969).
c   WO 9 406 737 (Schering Corp./Avondale Chem.; appl. 14.9.1993; USA-prior. 18.9.1992).
    WO 9 427 941 (Avondale Chem.; appl. 18.2.1994; USA-prior. 25.5.1993).

alternative reduction of diethyl phenylmalonate *to I:*
US 5 091 595 (Choi, Y.M.; 25.2.1992; appl. 7.6.1989).

Formulation(s):   oral susp. 600 mg/5 ml; syrup 600 mg/ml; tabl. 400 mg, 600 mg

Felbinac

ATC: M01AB; M02AA08
Use: anti-inflammatory, analgesic

RN: 5728-52-9   MF: $C_{14}H_{12}O_2$ MW: 212.25   EINECS: 227-233-2
LD_{50}:   508 mg/kg (M, i.p.); 675 mg/kg (M, p.o.); 730 mg/kg (M, s.c.);
          164 mg/kg (R, p.o.); 148 mg/kg (R, s.c.);
          1280 mg/kg (rabbit, s.c.);
          320 mg/kg (dog, s.c.)
CN: [1,1'-biphenyl]-4-acetic acid

4-biphenyl-
acetonitrile

Felbinac

Reference(s):
FR-M 7 166 (R. Hurmer, J. Vernin; appl. 21.7.1967).
US 3 784 704 (American Cyanamid; 8.1.1974; prior. 13.10.1972).
Child, R.G. et al.: J. Pharm. Sci. (JPMSAE) **66**, 466 (1977).

alternative synthesis:
JP 61 036 243 (Lederle; appl. 30.7.1984).
EP 212 617 (Lederle; appl. 19.8.1986; J-prior. 23.8.1985).
JP 63 233 947 (Mitsubishi; appl. 23.3.1987).
JP 1 132 544 (Mitsubishi; appl. 18.11.1987).
JP 55 094 486 (Sumitomo; appl. 11.1.1979).
Byron, D.J.; Gray, G.W.; Wilson, R.C.: J. Chem. Soc. C (JSOOAX) **1966**, 840.

anti-inflammatory ointment:
EP 127 840 (Lederle; appl. 22.5.1984; J-prior. 1.6.1983).

analgesic patch:
JP 1 085 913 (Saitama Daiichi; appl. 26.9.1987).

cyclodextrin inclusion compound:
JP 61 030 551 (Lederle; appl. 23.7.1984).

inhibition of blood platelet aggregation:
US 3 966 978 (American Cyanamid; 29.6.1976; appl. 25.4.1975).

medical use for treatment of ocular inflammation:
US 3 991 206 (American Cyanamid; 9.11.1976; appl. 15.1.1976).

Formulation(s):   gel 30 mg/g (as 1,1'-iminobis[2-propanol] salt)

Felodipine

ATC: C02DE; C08CA02
Use: calcium antagonist, antihypertensive

RN: 72509-76-3   MF: $C_{18}H_{19}Cl_2NO_4$ MW: 384.26
LD_{50}:   3100 µg/kg (M, i.v.); 250 mg/kg (M, p.o.);
5400 µg/kg (R, i.v.); 1050 mg/kg (R, p.o.)
CN: 4-(2,3-dichlorophenyl)-1,4-dihydro-2,6-dimethyl-3,5-pyridinedicarboxylic acid ethyl methyl ester

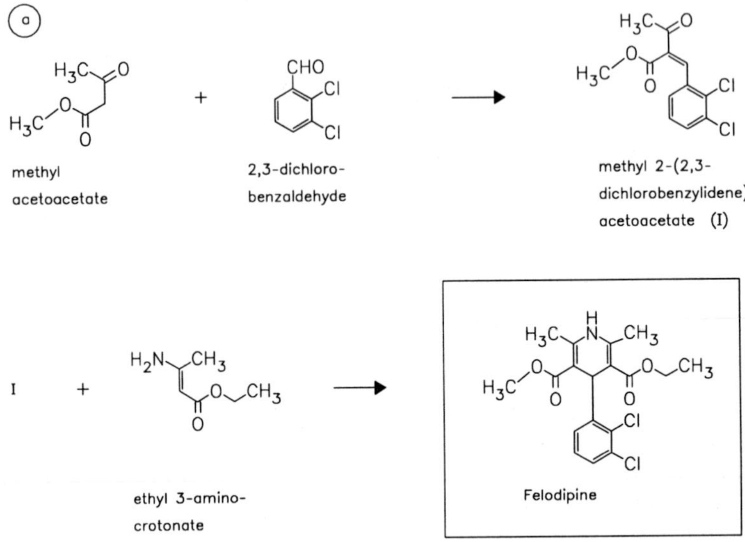

methyl
acetoacetate

2,3-dichloro-
benzaldehyde

methyl 2-(2,3-
dichlorobenzylidene)-
acetoacetate   (I)

ethyl 3-amino-
crotonate

Felodipine

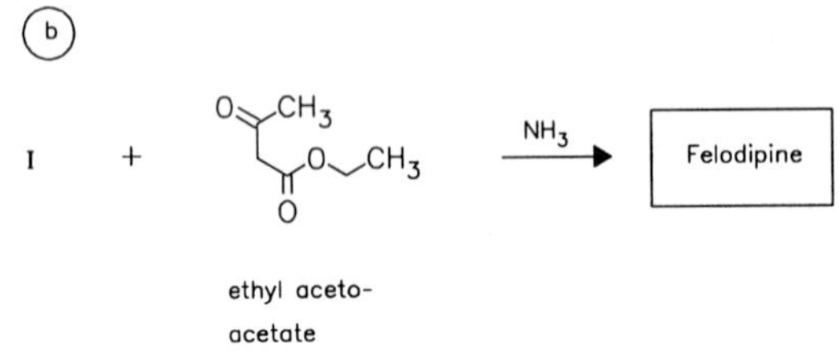

ethyl aceto-
acetate

Reference(s):
EP 7 293 (Hässle; appl. 12.6.1979; S-prior. 30.6.1978).

sustained release formulation:
EP 249 587 (Hässle; appl. 25.3.1987; S-prior. 11.4.1986).

combination with metoprolol:
EP 311 582 (Hässle; appl. 22.9.1988; S-prior. 8.10.1987).

Formulation(s):   s. r. tabl. 2.5 mg, 5 mg, 10 mg

Trade Name(s):
D: Mobloc (Astra/Promed)-
comb.
Modip (Astra/Promed)
Munobal (Hoechst)
F: Flodil (Astra)
Logimax (Astra)

GB: Plendil (Astra; 1990)
I: Feloday (Novartis)
Plendil (Sca)
Prevex (Schering-Plough)
J: Munobal (Hoechst-Nippon
HMR)

Splendil (Ciba-Geigy-
Kissei)
USA: Lexxel (Astra Merck)
Plendil (Astra Merck)

Felypressin

ATC: H01BA
Use: vasoconstrictoric effective peptide hormone

RN: 56-59-7   MF: $C_{46}H_{65}N_{13}O_{11}S_2$ MW: 1040.24   EINECS: 200-282-7
LD$_{50}$:   >10 g/kg (M, p.o.);
         5 g/kg (R, p.o.)
CN: 2-L-phenylalanine-8-L-lysinevasopressin

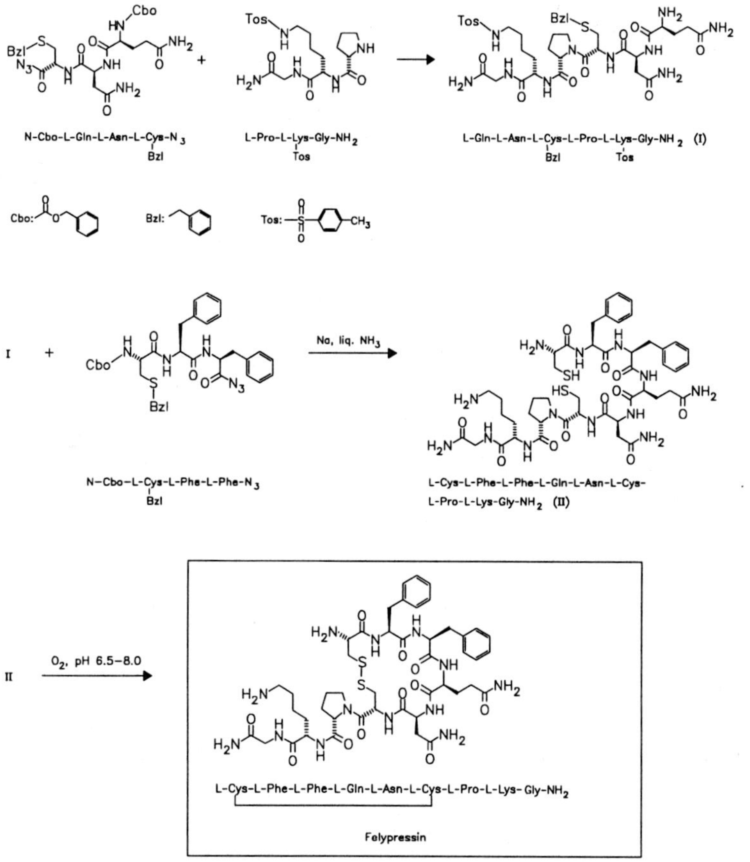

N-Cbo-L-Gln-L-Asn-L-Cys-N$_3$
         Bzl

L-Pro-L-Lys-Gly-NH$_2$
         Tos

L-Gln-L-Asn-L-Cys-L-Pro-L-Lys-Gly-NH$_2$ (I)
                    Bzl   Tos

Cbo: ...   Bzl: ...   Tos: ...

N-Cbo-L-Cys-L-Phe-L-Phe-N$_3$
         Bzl

L-Cys-L-Phe-L-Phe-L-Gln-L-Asn-L-Cys-
L-Pro-L-Lys-Gly-NH$_2$ (II)

O$_2$, pH 6.5–8.0

L-Cys-L-Phe-L-Phe-L-Gln-L-Asn-L-Cys-L-Pro-L-Lys-Gly-NH$_2$

Felypressin

Reference(s):
GB 928 607 (Sandoz; appl. 13.6.1960; CH-prior. 24.7.1959).
US 3 232 923 (Sandoz; 1.2.1966; CH-prior. 24.7.1959).

Formulation(s):   amp. 0.03 iu in comb.

Trade Name(s):
D: Xylonest mit Octapressin   F: Collupressine (Lab.   I: Citanest (Astra-Simes)-
      (Astra)-comb.   Oberlin)-comb.   comb.
                                                            J: Octapressin (Sandoz)

Fenalcomine

ATC: C01D
Use: coronary therapeutic, cardiac stimulant

RN: 34616-39-2   MF: $C_{20}H_{27}NO_2$ MW: 313.44
CN: α-ethyl-4-[2-[(1-methyl-2-phenylethyl)amino]ethoxy]benzenemethanol

hydrochloride
RN: 34535-83-6   MF: $C_{20}H_{27}NO_2 \cdot HCl$ MW: 349.90   EINECS: 252-075-6

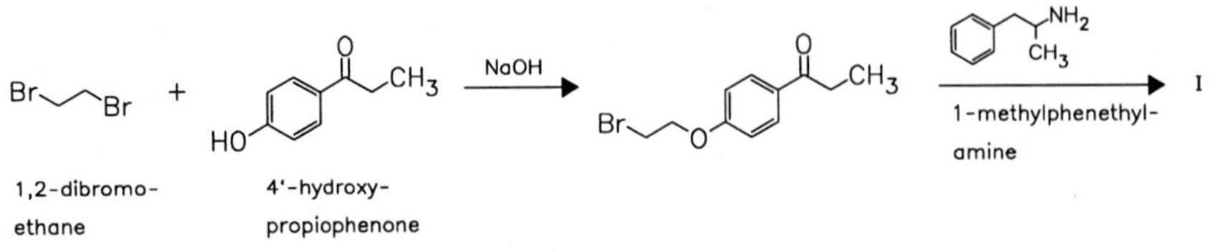

1,2-dibromo-ethane   4'-hydroxy-propiophenone

1-methylphenethyl-amine

(I)

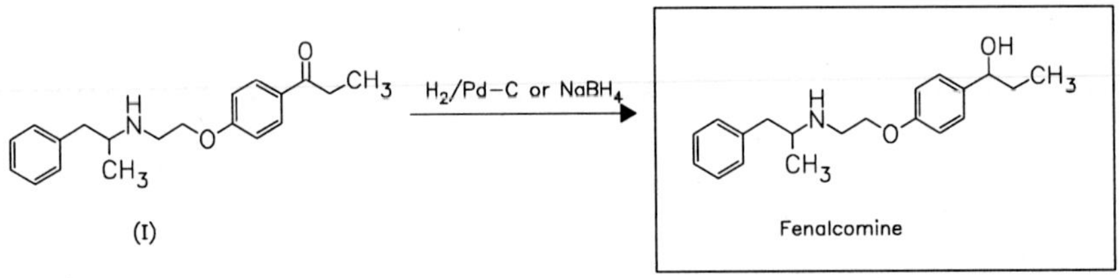

Fenalcomine

Reference(s):
FR-M 7 255 (Laroche Navarron; appl. 23.1.1968).

Formulation(s):   cps. 50 mg (as hydrochloride)

Trade Name(s):
F: Cordoxène (Laroche Navarron); wfm

Fenbufen

ATC: M01AE05
Use: anti-inflammatory, analgesic

RN: 36330-85-5   MF: $C_{16}H_{14}O_3$ MW: 254.29   EINECS: 252-979-0
LD$_{50}$:   795 mg/kg (M, p.o.);
            200 mg/kg (R, p.o.)
CN: γ-oxo[1,1'-biphenyl]-4-butanoic acid

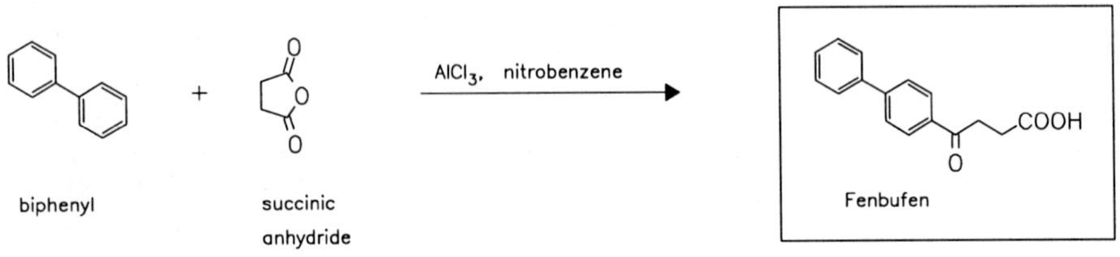

biphenyl   succinic anhydride

Fenbufen

Reference(s):

DOS 2 147 111 (American Cyanamid; appl. 21.9.1971; USA-prior. 21.9.1970).

US 3 784 701 (American Cyanamid; 8.1.1974; appl. 21.9.1970).

Child, R.G. et al.: Arzneim.-Forsch. (ARZNAD) **30** (I), 695 (1980).

Formulation(s):   cps. 300 mg; tabl. 200 mg, 300 mg, 450 mg

Trade Name(s):

D:	Lederfen (Lederle); wfm	GB:	Lederfen (Wyeth)
F:	Cinopal (Labs. Novalis)	I:	Cinopal (Cyanamid)

Fenbutrazate

(Phenbutrazate)

ATC: A08AA
Use: central stimulant, appetite depressant, anorectic

RN: 4378-36-3   MF: $C_{23}H_{29}NO_3$ MW: 367.49   EINECS: 224-480-8
CN: α-ethylbenzeneacetic acid 2-(3-methyl-2-phenyl-4-morpholinyl)ethyl ester

comb. with phenmetrazine-8-chlorotheophyllinate monohydrochloride
RN: 8004-38-4   MF: $C_{23}H_{29}NO_3 \cdot C_{18}H_{20}ClN_5O_3 \cdot HCl$ MW: 793.79

phenmetrazine   ethylene   4-(2-hydroxyethyl)-
(q. v.)   oxide   3-methyl-2-phenyl-
          morpholine (I)

2-phenyl-   Fenbutrazate
butyryl chloride

Reference(s):

US 3 018 222 (Ravensberg; 23.1.1962; D-prior. 28.8.1956).

Formulation(s):   drg. 20 mg

Trade Name(s):

D:	Cafilon (Ravensberg); wfm	F:	Cafilon (Merck-Clévenot); wfm	J:	Cafilon (Yamanouchi)-
	Cafilon (Ravensberg)-		Cafilon (Merck-Clévenot)-		comb. with phenmetrazine-
	comb. with phenmetrazine-		comb.; wfm		8-chlorotheophyllinate
	8-chlorotheophyllinate;				
	wfm				

Fencamfamin

ATC: N06BA06
Use: psychostimulant

RN: 1209-98-3   MF: $C_{15}H_{21}N$ MW: 215.34
LD$_{50}$:   83 mg/kg (R, p.o.)
CN: N-ethyl-3-phenylbicyclo[2.2.1]heptan-2-amine

hydrochloride

RN: 2240-14-4   MF: $C_{15}H_{21}N \cdot HCl$ MW: 251.80   EINECS: 218-805-2

LD$_{50}$:   15.7 mg/kg (M, i.v.); 135 mg/kg (M, p.o.);
        23.5 mg/kg (R, i.v.); 83 mg/kg (R, p.o.);
        15 mg/kg (dog, i.v.); 30 mg/kg (dog, p.o.)

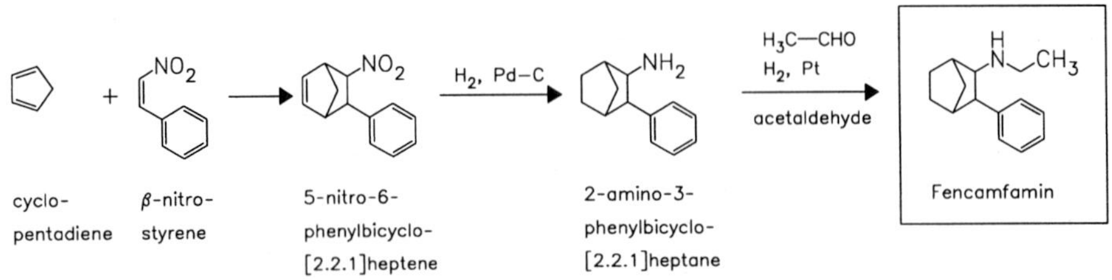

cyclo-   β-nitro-   5-nitro-6-   2-amino-3-   Fencamfamin
pentadiene   styrene   phenylbicyclo-   phenylbicyclo-
                              [2.2.1]heptene   [2.2.1]heptane

Reference(s):

DE 1 110 159 (E. Merck AG; appl. 1.8.1959).

Formulation(s):   drg. 10 mg

Trade Name(s):

D: Reactivan (Cascan)-comb.;   I: Reactivan (Bracco)-comb.;
       wfm   wfm

Fencarbamide

(Phencarbamide)

ATC: A03AC
Use: antispasmodic

RN: 3735-90-8   MF: $C_{19}H_{24}N_2OS$ MW: 328.48   EINECS: 223-103-4

LD$_{50}$:   32 mg/kg (M, i.v.);
        30 mg/kg (R, i.v.); 370 mg/kg (R, p.o.)

CN: diphenylcarbamothioic acid S-[2-(diethylamino)ethyl] ester

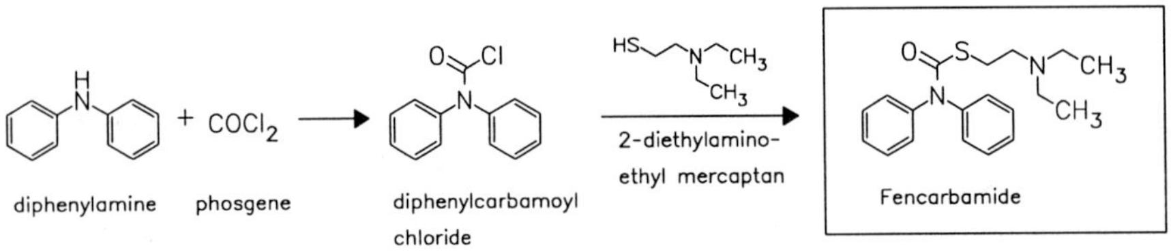

diphenylamine   phosgene   diphenylcarbamoyl   Fencarbamide
                               chloride

                               2-diethylamino-
                               ethyl mercaptan

Reference(s):

DE 1 146 693 (Bayer; appl. 18.9.1958).

Formulation(s):   suppos. 10 mg; tabl. 10 mg (as napadisilate)

Trade Name(s):

D: Spasmo-Dolviran (Bayer)-   Spasmo-Compralgyl   I: Spasmo-Dolviran (Bayer)-
       comb.; wfm   (Bayer-Pharma)-comb.;   comb.; wfm
F: Gélosédine (Bayer-   wfm
       Pharma)-comb.; wfm

Fenclofenac

ATC: M01A; N02B; S01B
Use: anti-inflammatory, analgesic

RN: 34645-84-6   MF: $C_{14}H_{10}Cl_2O_3$ MW: 297.14   EINECS: 252-126-2
LD$_{50}$:   2280 mg/kg (R, p.o.)
CN: 2-(2,4-dichlorophenoxy)benzeneacetic acid

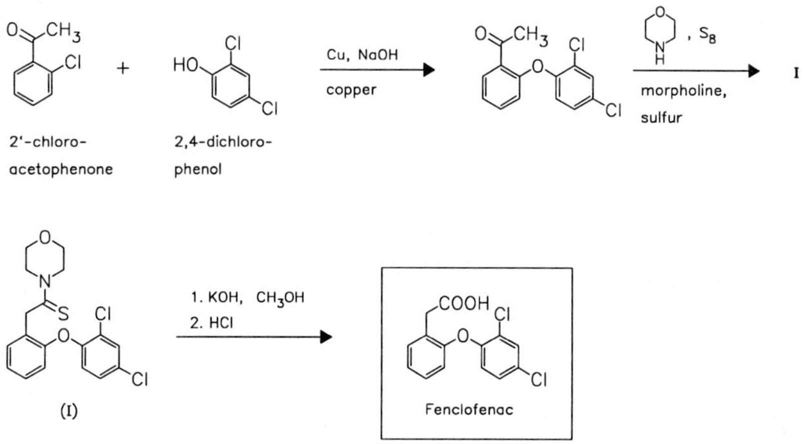

2'-chloro-
acetophenone

2,4-dichloro-
phenol

Cu, NaOH
copper

morpholine,
sulfur

I

(I)

1. KOH, CH$_3$OH
2. HCl

Fenclofenac

Reference(s):
DOS 2 117 826 (Reckitt & Colman; appl. 13.4.1971; GB-prior. 14.4.1970).
GB 1 308 327 (Reckitt & Colman; valid from 19.4.1971; prior. 14.4.1970).
US 3 766 263 (Reckitt & Colman; 16.10.1973; GB-prior. 14.4.1970).

Formulation(s):   tabl. 300 mg

Trade Name(s):
GB: Flenac (Reckitt & Colman);
       wfm

Fendiline

ATC: C08EA01
Use: coronary vasodilator

RN: 13042-18-7   MF: $C_{23}H_{25}N$ MW: 315.46   EINECS: 235-915-6
CN: γ-phenyl-N-(1-phenylethyl)benzenepropanamine

hydrochloride
RN.   13636-18-5   MF: $C_{23}H_{25}N \cdot HCl$ MW: 351.92   EINECS: 237-121-5
LD$_{50}$:   14.5 mg/kg (M, i.v.); 950 mg/kg (M, p.o.)

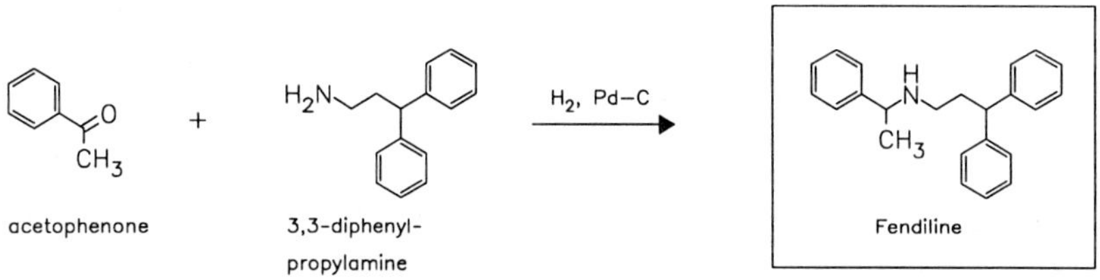

acetophenone

3,3-diphenyl-
propylamine

H$_2$, Pd–C

Fendiline

Reference(s):
DE 1 171 930 (Chinoin; appl. 24.7.1962; H-prior. 10.8.1961, 10.3.1962, 19.3.1962, 30.3.1962).
US 3 262 977 (Chinoin; 26.7.1966; H-prior. 10.3.1962, 30.3.1962).
GB 954 735 (Chinoin; appl. 10.8.1962; H-prior. 10.8.1961, 10.3.1962, 19.3.1962, 30.3.1962).

Formulation(s):   drg. 50 mg, 75 mg, 100 mg

Trade Name(s):
D: Sensit (Thiemann)   Olbiacor (Salus Research)
I: Difmecor (UCM)   Sensit-F (Organon Italia)

Fendosal

ATC: M01A
Use: anti-inflammatory

RN: 53597-27-6   MF: $C_{25}H_{19}NO_3$ MW: 381.43
LD$_{50}$:   740 mg/kg (M, p.o.);
       450 mg/kg (R, p.o.)
CN: 5-(4,5-dihydro-2-phenyl-3*H*-benz[*e*]indol-3-yl)-2-hydroxybenzoic acid

2-tetralone   pyrrolidine   2-pyrrolidino-3,4-dihydro-naphthalene   2-bromoacetophenone   I

1-phenacyl-2-tetralone (I)   5-aminosalicylic acid   Fendosal

Reference(s):
DOS 2 407 671 (Hoechst; appl. 18.2.1974; USA-prior. 1.3.1973).
Anderson, V.B. et al.: J. Med. Chem. (JMCMAR) **19**, 318 (1976).

use for thrombosis prevention:
DOS 2 502 156 (Hoechst; appl. 21.1.1975; USA-prior. 25.1.1974).

Trade Name(s):
USA: Alnovin (Hoechst-
        Roussel); wfm

Fenetylline
(Fenethylline)

ATC: N06B
Use: psychotonic, CNS stimulant

RN: 3736-08-1   MF: $C_{18}H_{23}N_5O_2$ MW: 341.42
LD$_{50}$:   347 mg/kg (M, p.o.);
       100 mg/kg (R, p.o.)
CN: 3,7-dihydro-1,3-dimethyl-7-[2-[(1-methyl-2-phenylethyl)amino]ethyl]-1*H*-purine-2,6-dione

monohydrochloride
RN: 1892-80-4 MF: $C_{18}H_{23}N_5O_2 \cdot HCl$ MW: 377.88 EINECS: 217-580-8
LD_{50}: 55 mg/kg (M, i.v.); 347 mg/kg (M, p.o.);
100 mg/kg (R, p.o.)

a

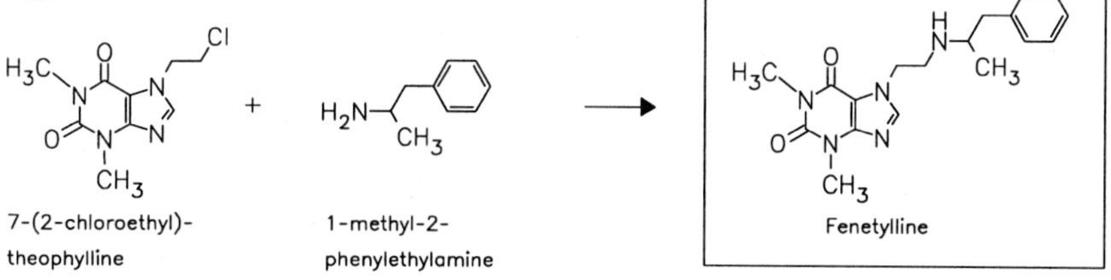

7-(2-chloroethyl)-
theophylline

1-methyl-2-
phenylethylamine

Fenetylline

b

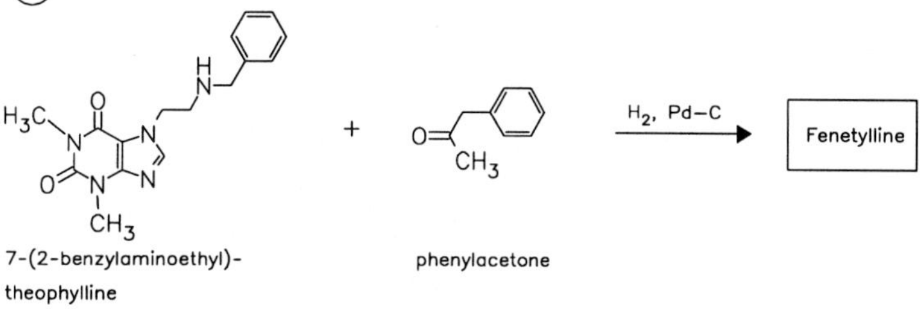

7-(2-benzylaminoethyl)-
theophylline

phenylacetone

H_2, Pd–C

Fenetylline

Reference(s):
DE 1 123 329 (Degussa; appl. 18.10.1958; addition to DE 1 095 285; appl. 25.9.1956).
US 3 029 239 (Degussa; 10.4.1962; D-prior. 17.4.1954).

Formulation(s): f. c. tabl. 50 mg (as hydrochloride)

Trade Name(s):
D: Captagon (ASTA Medica AWD)
F: Captagon (Gerda); wfm
Captagon (Promdeica); wfm

Fenfluramine

ATC: A08AA02
Use: appetite depressant, anorexic

RN: 458-24-2 MF: $C_{12}H_{16}F_3N$ MW: 231.26 EINECS: 207-276-3
LD_{50}: 145 mg/kg (M, p.o.);
130 mg/kg (R, p.o.);
100 mg/kg (dog, p.o.)
CN: *N*-ethyl-α-methyl-3-(trifluoromethyl)benzeneethanamine

hydrochloride
RN: 404-82-0 MF: $C_{12}H_{16}F_3N \cdot HCl$ MW: 267.72 EINECS: 206-968-2
LD_{50}: 90 mg/kg (M, i.v.); 170 mg/kg (M, p.o.);
69 mg/kg (R, p.o.);
23 mg/kg (dog, i.v.); 100 mg/kg (dog, p.o.)

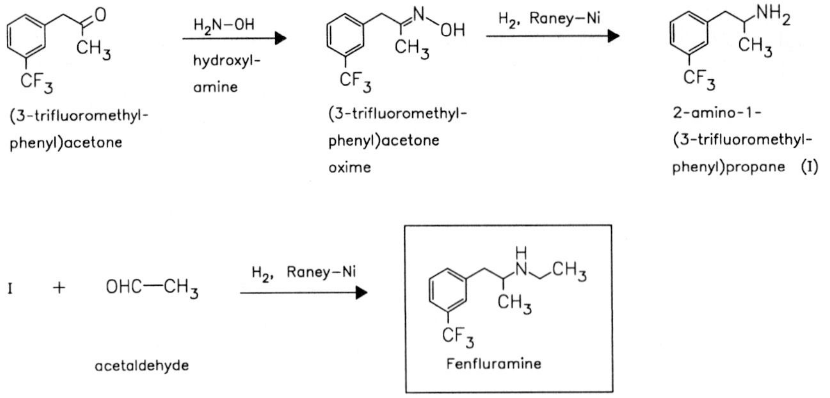

(3-trifluoromethyl-phenyl)acetone
H2N-OH
hydroxylamine
(3-trifluoromethyl-phenyl)acetone oxime
H2, Raney-Ni
2-amino-1-(3-trifluoromethyl-phenyl)propane (I)
I + OHC—CH3
acetaldehyde
H2, Raney-Ni
Fenfluramine

Reference(s):
FR-M 1 658 (Science-Union; appl. 4.4.1961; MC-prior. 5.11.1960).

Formulation(s): cps. 20 mg, 60 mg; s. r. cps. 60 mg; tabl. 20 mg, 40 mg (as hydrochloride)

Trade Name(s):

D:	Ponderax (Boehringer Ing.); wfm	F:	Pondéral (Biopharma; as hydrochloride)		Pesos (Valeas)
	Ponderax (Itherapia); wfm	GB:	Ponderax (Servier); wfm		Ponderal (Servier)
		I:	Dimafen (Stroder)	USA:	Pondimin (Robins)

Fenipentol

ATC: A05AX
Use: choleretic

RN: 583-03-9 MF: $C_{11}H_{16}O$ MW: 164.25 EINECS: 209-493-9
LD_{50}: 2900 mg/kg (M, p.o.);
5432 mg/kg (R, p.o.)
CN: α-butylbenzenemethanol

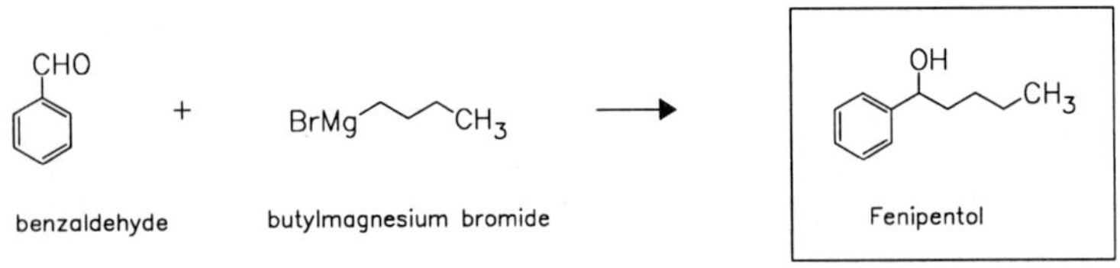

CHO
benzaldehyde
BrMg CH3
butylmagnesium bromide
OH
CH3
Fenipentol

Reference(s):
GB 915 815 (Thomae; appl. 11.4.1960; valid from 6.4.1961).
US 3 084 100 (Thomae; 2.4.1963; appl. 30.3.1961).
Adams, R.M.; Vander-Werf, C.A.: J. Am. Chem. Soc. (JACSAT) **72**, 4368 (1950).
Engelhorn, R.: Arzneim.-Forsch. (ARZNAD) **10**, 255 (1960).
Koss, F.W. et al.: Arzneim.-Forsch. (ARZNAD) **12**, 1026 (1962).

Formulation(s): cps. 100 mg

Fenofibrate
(Procetofene)

ATC: C01AB05
Use: cholesterol depressant,
       antihyperlipidemic

RN: 49562-28-9   MF: $C_{20}H_{21}ClO_4$ MW: 360.84   EINECS: 256-376-3
LD_{50}: 1600 mg/kg (M, p.o.);
       >2 g/kg (R, p.o.);
       >4 g/kg (dog, p.o.)
CN: 2-[4-(4-chlorobenzoyl)phenoxy]-2-methylpropanoic acid 1-methylethyl ester

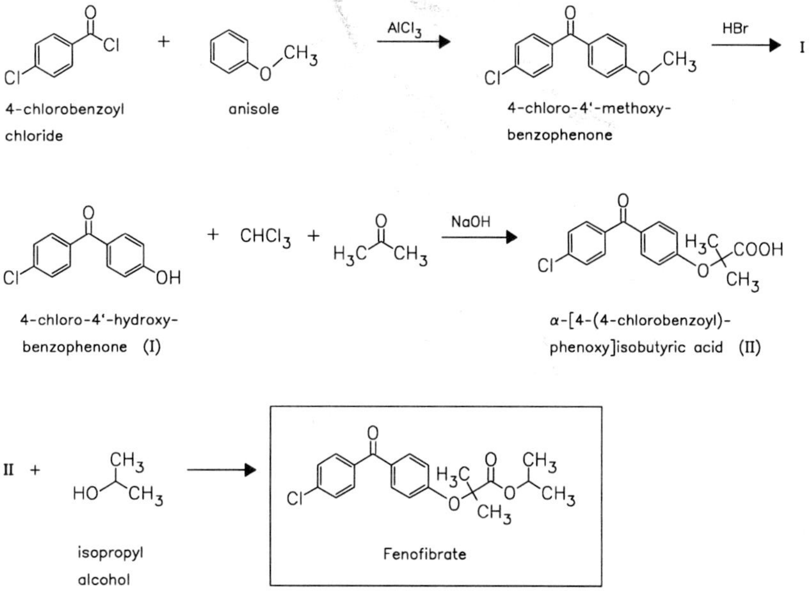

4-chlorobenzoyl chloride + anisole → (AlCl₃) 4-chloro-4'-methoxy-benzophenone → (HBr) I

4-chloro-4'-hydroxy-benzophenone (I) + CHCl₃ + acetone → (NaOH) α-[4-(4-chlorobenzoyl)-phenoxy]isobutyric acid (II)

II + isopropyl alcohol → Fenofibrate

Reference(s):
US 4 058 552 (Orchimed; 15.11.1977; CH-prior. 31.1.1969).
DOS 2 250 327 (Lab. Fournier; appl. 13.10.1972; GB-prior. 14.10.1971).
Sornay, R. et al.: Arzneim.-Forsch. (ARZNAD) **26**, 885, 889 (1976).
EP-appl. 2 151 (Devinter; appl. 10.11.1978; F-prior. 14.11.1977).

Formulation(s):   cps. 100 mg, 200 mg, 300 mg; s. r. cps. 250 mg

Fenoldopam mesilate
(SKF 82526-J)

ATC: C01CA19
Use: antihypertensive

RN: 67227-57-0   MF: $C_{16}H_{16}ClNO_3 \cdot CH_4O_3S$ MW: 401.87
CN: (±)-6-Chloro-2,3,4,5-tetrahydro-1-(4-hydroxyphenyl)-1H-3-benzazepine-7,8-diol methanesulfonate

base
RN: 67227-56-9   MF: $C_{16}H_{16}ClNO_3$ MW: 305.76
hydrochloride
RN: 181217-39-0   MF: $C_{16}H_{16}ClNO_3 \cdot HCl$ MW: 342.22

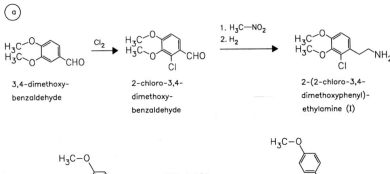

3,4-dimethoxy-
benzaldehyde

2-chloro-3,4-
dimethoxy-
benzaldehyde

2-(2-chloro-3,4-
dimethoxyphenyl)-
ethylamine (I)

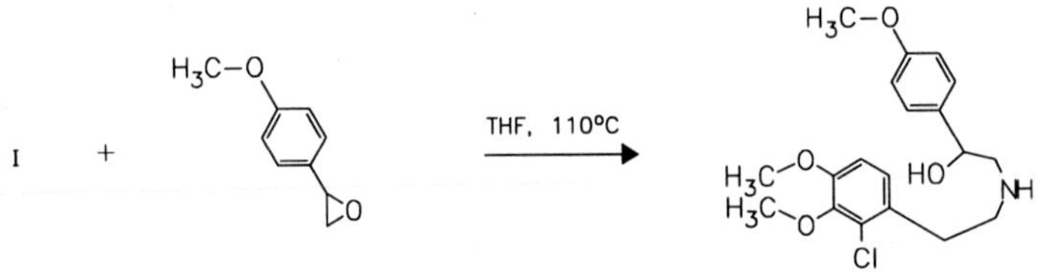

(±)-(4-methoxy-
phenyl)oxirane (II)

(III)

Fenoldopam mesilate

aa)   intermediate II

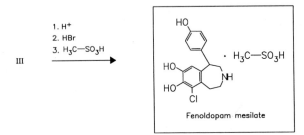

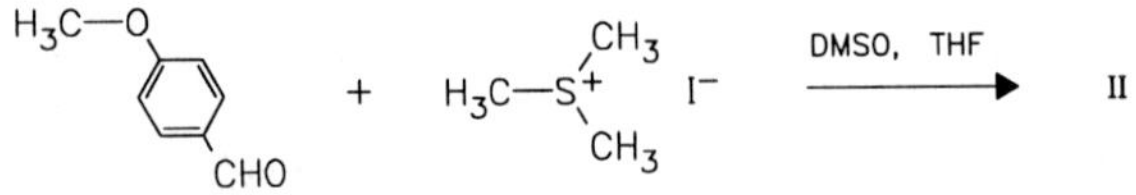

4-methoxy-
benzaldehyde

trimethylsulfonium
iodide

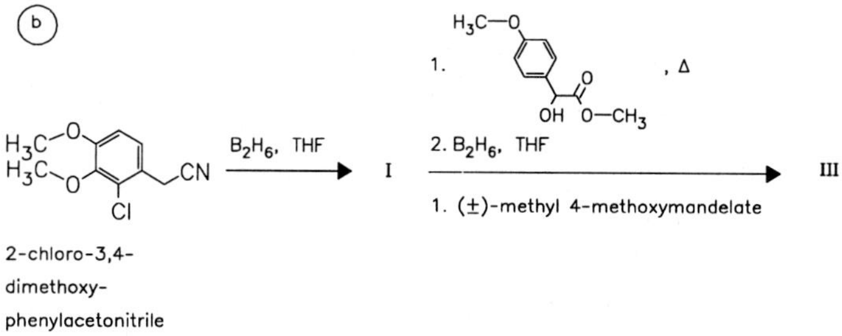

b
H_3C-O
1.
, Δ
OH O—CH_3
H_3C O
H_3C O
CN
Cl
B_2H_6, THF
I
2. B_2H_6, THF
III
1. (±)-methyl 4-methoxymandelate
2-chloro-3,4-
dimethoxy-
phenylacetonitrile

Reference(s):

a US 4 160 765 (SmithKline; 10.7.1979; USA-prior. 17.11.1976).
US 4 171 359 (SmithKline; 16.10.1979; USA-prior. 12.4.1978).
aa US 4 197 297 (SmithKline; 8.4.1980; USA-prior. 17.11.1976).
b Weinstock, J. et al.: J. Med. Chem. (JMCMAR) **23** (9), 973-975 (1980).

synergistic antihypertensive compositions:
EP 22 330 (SmithKline; appl. 26.6.1980; USA-prior. 10.7.1979).
EP 81 006 (SmithKline; appl. 8.12.1981).

controlled release dosage forms comprising separate portions of (R)- and (S)-enantiomers:
WO 9 840 053 (Darwin Discovery; appl. 11.3.1998; GB-prior. 11.3.1997).

Formulation(s): vial for inj. 10 mg/ml

Trade Name(s):
USA: Carlopam (Neurex; 1999)

Fenoprofen

ATC: M01AE04
Use: antirheumatic

RN: 31879-05-7 MF: $C_{15}H_{14}O_3$ MW: 242.27 EINECS: 250-850-3
LD_{50}: 1400 mg/kg (M, p.o.)
CN: (±)-α-methyl-3-phenoxybenzeneacetic acid

calcium salt dihydrate
RN: 53746-45-5 MF: $C_{30}H_{26}CaO_6 \cdot 2H_2O$ MW: 558.64
LD_{50}: 471 mg/kg (M, i.v.); 439 mg/kg (M, p.o.);
526 mg/kg (R, i.v.); 415 mg/kg (R, p.o.)

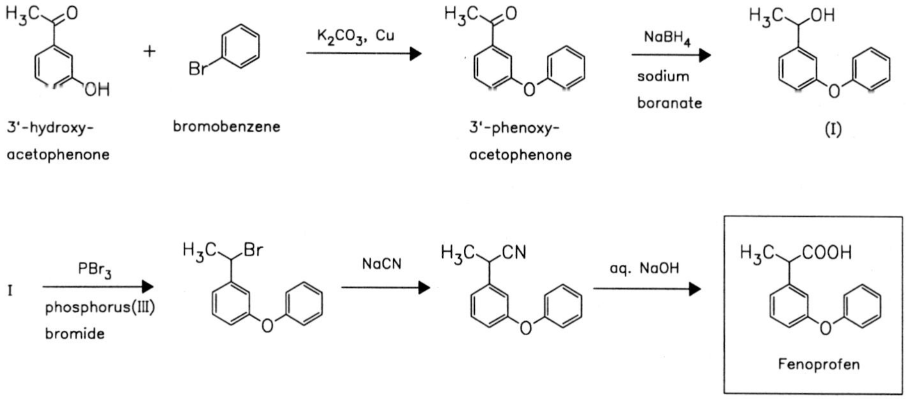

H_3C O
OH
+
Br
K_2CO_3, Cu
H_3C O
O
$NaBH_4$
sodium
boranate
H_3C OH
O
3'-hydroxy-
acetophenone
bromobenzene
3'-phenoxy-
acetophenone
(I)
I
PBr_3
phosphorus(III)
bromide
H_3C Br
NaCN
H_3C CN
aq. NaOH
H_3C COOH
Fenoprofen

Reference(s):

DOS 1 941 625 (Lilly; appl. 16.8.1969; USA-prior. 15.8.1968, 28.5.1969).
US 3 600 437 (Eli Lilly; 17.8.1971; prior. 15.8.1968, 9.5.1969, 28.5.1969).

alternative syntheses:

DOS 2 646 792 (Mitsubishi Petrochemical; appl. 16.10.1976; J-prior. 23.10.1975, 31.7.1976).
US 4 016 196 (Nisshin Flour Milling; 5.4.1977; J-prior. 27.7.1974, 29.7.1974).
DAS 2 709 504 (Sagami; appl. 4.3.1977; J-prior. 4.3.1976, 27.12.1976).

Formulation(s): powder 200 mg, 300 mg; tabl. 300 mg, 600 mg (as calcium salt dihydrate)

Trade Name(s):

D:	Feprona (Lilly; 1975); wfm		Progesic (Lilly); wfm	J:	Fenopron (Shionogi-Yamanouchi; 1982)
F:	Nalgésic (Lilly)	I:	Fepron (Lilly)		
GB:	Fenopron (Novex)			USA:	Nalfon (Dista; 1976)

Fenoterol

ATC: G02CA03; R03AC04; R03CC04
Use: bronchodilator

RN: 13392-18-2 MF: $C_{17}H_{21}NO_4$ MW: 303.36
CN: 5-[1-hydroxy-2-[[2-(4-hydroxyphenyl)-1-methylethyl]amino]ethyl]-1,3-benzenediol

hydrobromide

RN: 1944-12-3 MF: $C_{17}H_{21}NO_4 \cdot HBr$ MW: 384.27 EINECS: 217-742-8
LD_{50}: 42 mg/kg (M, i.v.); 1990 mg/kg (M, p.o.);
65 mg/kg (R, i.v.); 1600 mg/kg (R, p.o.);
150 mg/kg (dog, p.o.)

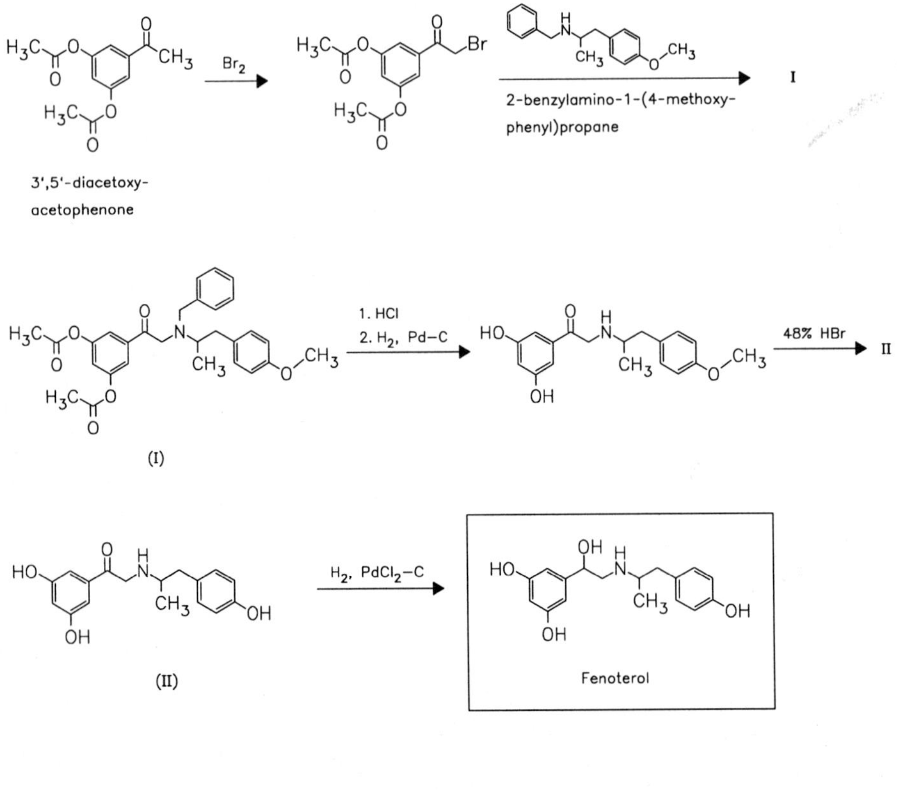

Br2
2-benzylamino-1-(4-methoxyphenyl)propane
I
3',5'-diacetoxyacetophenone
1. HCl
2. H2, Pd-C
48% HBr
II
(I)
H2, PdCl2-C
(II)
Fenoterol

Reference(s):
DE 1 286 047 (Boehringer Ing.; appl. 30.11.1962).
US 3 341 593 (Boehringer Ing.; 12.9.1967; D-prior. 30.11.1962).

alternative syntheses:
DOS 2 413 102 (Boehringer Ing.; appl. 19.3.1974).

Formulation(s):   aerosol 0.05 mg/puff in comb; amp. 0.025 mg/ml, 0.5 mg/10 ml; cps. 200 µg; sol. for
                    inhalation 0.5 mg/ml in comb., 1 mg/ml; tabl. 2.5 mg, 5 mg

Trade Name(s):

D:	Berodual Aerosol (Boehringer Ing.)	F:	Bérotec (Boehringer Ing.; as hydrobromide)	I:	Dosberotec (Boehringer Ing.)
	Berotec (Boehringer Ing.)		Bronchodual (Boehringer Ing.; as hydrobromide)		Duovent (Boehringer Ing.)-comb.
	Berotec-Dosier-Aerosol (Boehringer Ing.)	GB:	Berotec (Boehringer Ing.; as hydrobromide)		Iprafen (Chiesi)-comb.
	Ditec (Boehringer Ing.)			J:	Berotec (Boehringer Ing.; as hydrobromide)
	Partusisten (Boehringer Ing.)		Duovent (Boehringer Ing.; as hydrobromide)		

Fenoverine

ATC: A03AX05
Use: antispasmodic

RN: 37561-27-6   MF: C$_{26}$H$_{25}$N$_3$O$_3$S   MW: 459.57   EINECS: 253-552-1
LD$_{50}$:   2874 mg/kg (M, p.o.)
CN: 10-[[4-(1,3-benzodioxol-5-ylmethyl)-1-piperazinyl]acetyl]-10H-phenothiazine

phenothiazine   +   chloroacetyl chloride   →   10-(chloroacetyl)-phenothiazine   pyridine / 1-piperonyl-piperazine →   Fenoverine

Reference(s):
FR 2 092 639 (A. Buzas, R. Pierre; appl. 3.6.1970).

Formulation(s):   cps. 100 mg

Trade Name(s):

F:	Spasmopriv (Bouchard)	Spasmopriv (Vaillant-Defresne)	I:	Spasmopriv (Lusofarmaco)

Fenoxazoline

ATC: R01AA12
Use: vasoconstrictor, local anesthetic

RN: 4846-91-7   MF: C$_{13}$H$_{18}$N$_2$O   MW: 218.30   EINECS: 225-437-6
CN: 4,5-dihydro-2-[[2-(1-methylethyl)phenoxy]methyl]-1H-imidazole

monohydrochloride
RN: 23029-57-4   MF: $C_{13}H_{18}N_2O \cdot HCl$ MW: 254.76

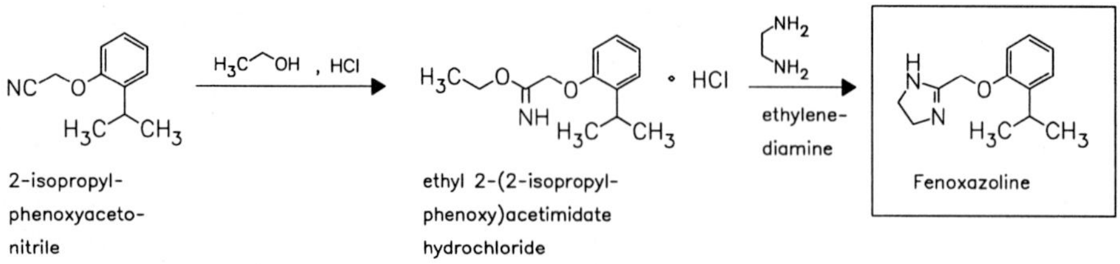

2-isopropyl-
phenoxyaceto-
nitrile

ethyl 2-(2-isopropyl-
phenoxy)acetimidate
hydrochloride

Fenoxazoline

Reference(s):
FR 1 365 971 (Lab. Dausse; appl. 19.2.1963).
US 3 198 703 (Lab. Dausse; 3.8.1965; appl. 4.5.1961).

Formulation(s):   nasal drops 0.05 %, 0.1 %; nasal spray 1 mg (as hydrochloride)

Trade Name(s):
D: Snup (Karlspharma); wfm   F: Aturgyl (Synthélabo)   Déturgylone (Synthélabo)

Fenoxedil

ATC: C01D
Use: vasodilator

RN: 54063-40-0   MF: $C_{28}H_{42}N_2O_5$ MW: 486.65
CN: 2-(4-butoxyphenoxy)-N-(2,5-diethoxyphenyl)-N-[2-(diethylamino)ethyl]acetamide

monohydrochloride
RN: 27471-60-9   MF: $C_{28}H_{42}N_2O_5 \cdot HCl$ MW: 523.11   EINECS: 248-478-1
LD_{50}:   17 mg/kg (M, i.v.); 750 mg/kg (M, p.o.);
        10 mg/kg (R, i.v.); 2400 mg/kg (R, p.o.)

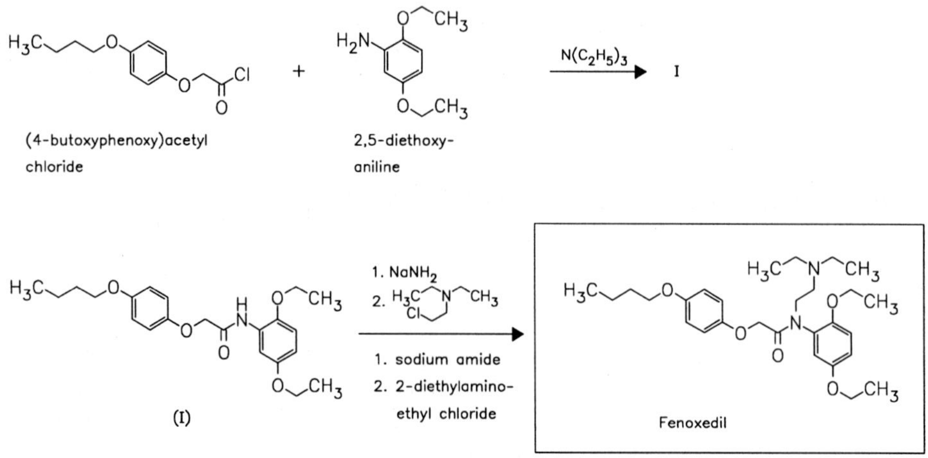

(4-butoxyphenoxy)acetyl
chloride

2,5-diethoxy-
aniline

1. NaNH$_2$
2. H$_3$C—N—CH$_3$ Cl

1. sodium amide
2. 2-diethylamino-
   ethyl chloride

(I)

Fenoxedil

Reference(s):
DE 1 964 712 (C.E.R.P.H.A.; appl. 23.12.1969; F-prior. 26.12.1968).
US 3 818 021 (C.E.R.P.H.A.; 18.6.1974; F-prior. 24.12.1968).

Formulation(s):   cps. 100 mg

Fenozolone
(Phenozolone)

ATC:　N06BA08
Use:　psychoanaleptic

RN:　15302-16-6　MF: $C_{11}H_{12}N_2O_2$　MW: 204.23　EINECS: 239-339-6
LD_{50}: 425 mg/kg (M, p.o.)
CN:　2-(ethylamino)-5-phenyl-4(5H)-oxazolone

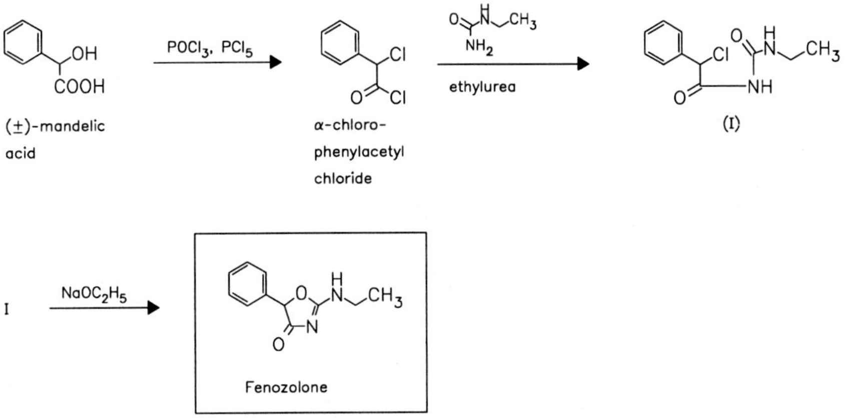

(±)-mandelic acid　α-chloro-phenylacetyl chloride　ethylurea　(I)

Fenozolone

Reference(s):
DE 1 297 108 (Lab. Dausse; appl. 20.2.1962; F-prior. 24.2.1961, 23.5.1961, 18.1.1962).

Formulation(s):　tabl. 10 mg

Fenpentadiol
(Phenpentanediol)

ATC:　N06A; N06B
Use:　antidepressant

RN:　15687-18-0　MF: $C_{12}H_{17}ClO_2$　MW: 228.72　EINECS: 239-782-5
LD_{50}: 940 mg/kg (M, p.o.);
　　1140 mg/kg (R, p.o.)
CN:　2-(4-chlorophenyl)-4-methyl-2,4-pentanediol

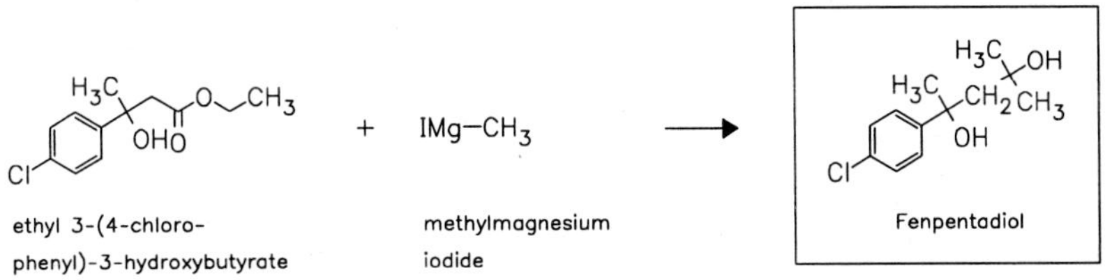

ethyl 3-(4-chloro-phenyl)-3-hydroxybutyrate　+　methylmagnesium iodide　Fenpentadiol

Reference(s):
FR-M 1 984 (Albert Rolland; appl. 26.7.1962).

Formulation(s):   cps. 100 mg

Trade Name(s):
F: Trédum (Anphar-Rolland);   Trédum (L'Hépatrol); wfm
     wfm

Fenpiverinium bromide
(Fenpipramide methylbromide)

ATC: A03AB21
Use: anticholinergic, antispasmodic

RN: 125-60-0   MF: $C_{22}H_{29}BrN_2O$ MW: 417.39   EINECS: 204-744-9
LD$_{50}$:   13.5 mg/kg (M, i.v.); 800 mg/kg (M, p.o.)
CN: 1-(4-amino-4-oxo-3,3-diphenylbutyl)-1-methylpiperidinium bromide

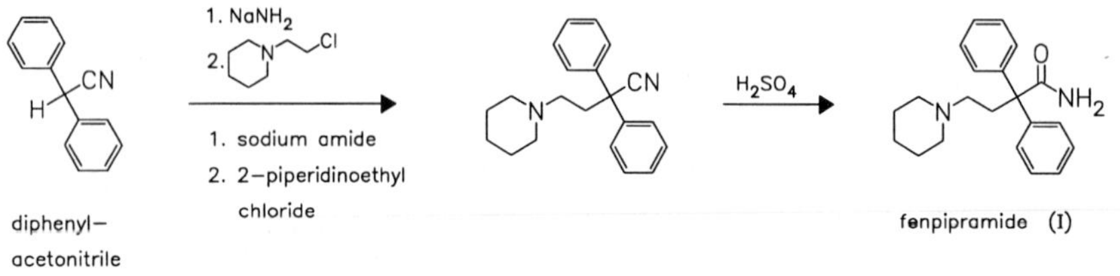

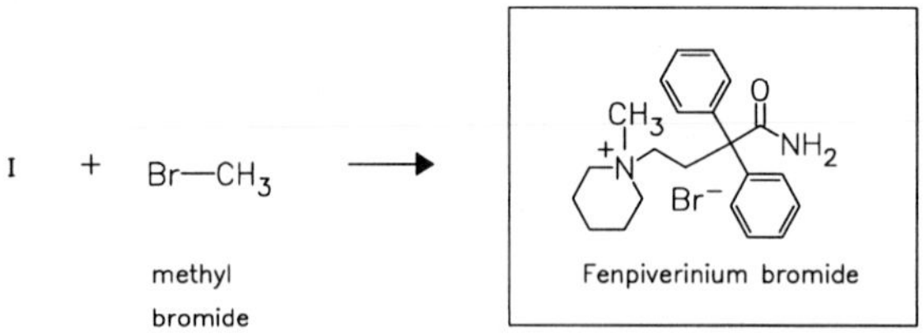

diphenyl-
acetonitrile

1. sodium amide
2. 2-piperidinoethyl
   chloride

fenpipramide   (I)

methyl
bromide

Fenpiverinium bromide

Reference(s):
DE 731 560 (Hoechst; appl. 1941).
DE 858 552 (Hoechst; appl. 1950).

Formulation(s):   amp. 0.1 mg in comb.; suppos. 0.03 mg, 0.1 mg in comb.; tabl. 0.1 mg in comb.

Trade Name(s):
D: Baralgin (Albert-Roussel)-   Baralgin compositum   F: Baralgine (Hoechst)-comb.;
     comb.; wfm   (Albert-Roussel)-comb.;   wfm
                    wfm   I: Baralgina (Hoechst Italia)-
                             comb.

Fenquizone

ATC: C03BA13
Use: diuretic

RN: 20287-37-0   MF: $C_{14}H_{12}ClN_3O_3S$ MW: 337.79   EINECS: 243-689-5
CN: 7-chloro-1,2,3,4-tetrahydro-4-oxo-2-phenyl-6-quinazolinesulfonamide

potassium salt
RN: 52246-40-9   MF: $C_{14}H_{11}ClKN_3O_3S$ MW: 375.88

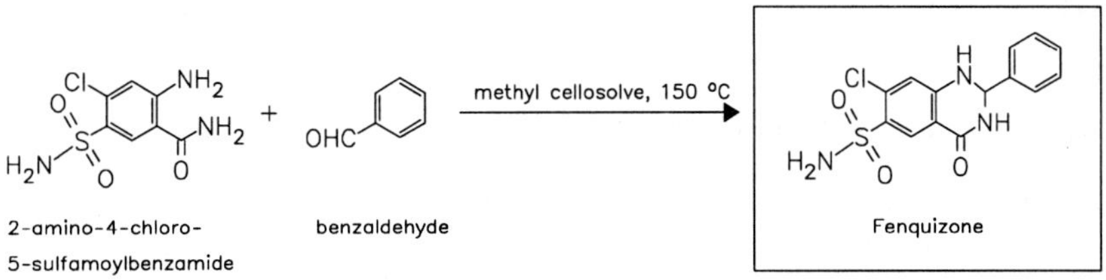

2-amino-4-chloro-
5-sulfamoylbenzamide   benzaldehyde   Fenquizone

Reference(s):
Biressi, M.E. et al.: Farmaco, Ed. Sci. (FRPSAX) **24**, 199 (1969).

Formulation(s):   cps. 11.13 mg (as potassium salt)

Trade Name(s):
I: Idrolone (Maggioni-
        Winthrop)

Fenspiride

ATC: R03BX01; R03DX03
Use: antiasthmatic, bronchodilator, α-
       adrenergic blocker

RN: 5053-06-5   MF: $C_{15}H_{20}N_2O_2$ MW: 260.34   EINECS: 225-751-3
LD_{50}:   230 mg/kg (M, i.p.)
CN: 8-(2-phenylethyl)-1-oxa-3,8-diazaspiro[4.5]decan-2-one

monohydrochloride
RN: 5053-08-7   MF: $C_{15}H_{20}N_2O_2 \cdot HCl$ MW: 296.80   EINECS: 225-752-9
LD_{50}:   106 mg/kg (M, i.v.); 250 mg/kg (M, p.o.);
       122 mg/kg (R, i.v.); 437 mg/kg (R, p.o.);
       74 mg/kg (dog, i.v.)

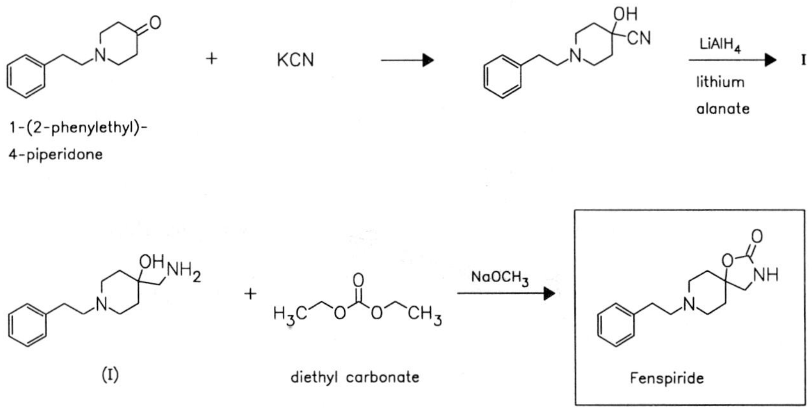

1-(2-phenylethyl)-
4-piperidone

(I)   diethyl carbonate   Fenspiride

Reference(s):
US 3 399 192 (Science Union; 27.8.1968; GB-prior. 22.4.1964).

preparation of 1-(2-phenylethyl)-4-piperidone:
Beckett et al.: J. Med. Pharm. Chem. (JMPCAS) **1**, 37, 51 (1959).
Elpern et al.: J. Am. Chem. Soc. (JACSAT) **80**, 4916 (1958).
Dutta, A.K.; Xu, C., Reith, M.F.A.: J. Med. Chem. (JMCMAR) **39** (3), 749 (1966).
Janssens, F. et al.: J. Med. Chem. (JMCMAR) **28** (12), 1925 (1985).

Formulation(s): cps. 40 mg, 80 mg; suppos. 40 mg, 80 mg

Trade Name(s):

F:	Pneumorel (Euthérapie; as hydrochloride)	I:	Espiran (ICT) Fenspir (Ibirn)		Fluiden (Lafare) Pneumorel (Stroder)

Fentanyl

ATC: N01AH01; N02AB03
Use: analgesic, narcotic

RN: 437-38-7 MF: $C_{22}H_{28}N_2O$ MW: 336.48 EINECS: 207-113-6
LD_{50}: 2900 µg/kg (M, i.v.); 368 mg/kg (M, p.o.);
2910 µg/kg (R, i.v.); 18 mg/kg (R, p.o.)
CN: *N*-phenyl-*N*-[1-(2-phenylethyl)-4-piperidinyl]propanamide

citrate (1:1)
RN: 990-73-8 MF: $C_{22}H_{28}N_2O \cdot C_6H_8O_7$ MW: 528.60 EINECS: 213-588-0
LD_{50}: 10100 µg/kg (M, i.v.); 368 mg/kg (M, p.o.);
990 µg/kg (R, i.v.); 18 mg/kg (R, p.o.)

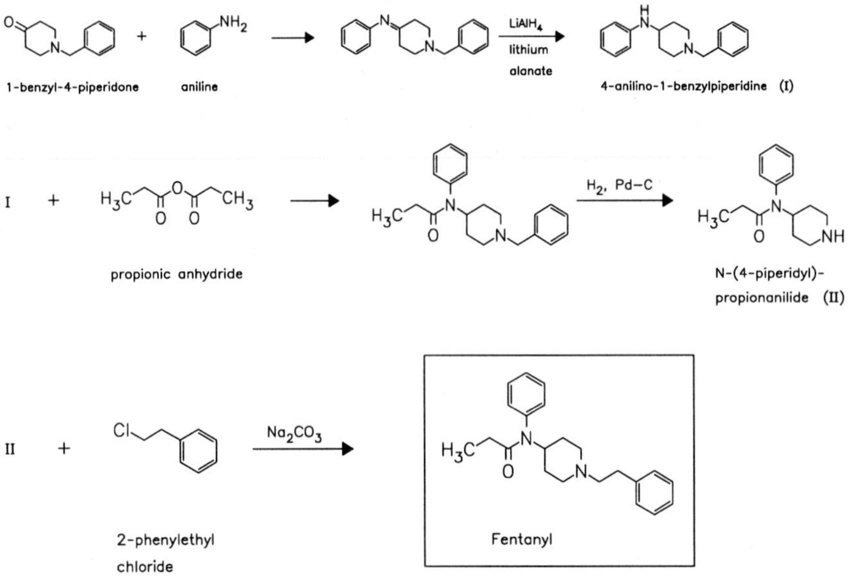

Reference(s):
FR 2 430 M (Janssen; appl. 9.10.1962; USA-prior. 10.10.1961).
US 3 141 823 (Janssen; 21.7.1964; appl. 4.9.1962).
US 3 164 600 (Janssen; 5.1.1965; appl. 10.10.1961).

Formulation(s): amp. 0.157 mg/2 ml, 0.785/10 ml (as citrate); membrane plaster

Trade Name(s):

D:	Durogesic (Janssen-Cilag) Fentanyl (Schwabe-Curamed)	Thalamonal (Janssen-Cilag)-comb. with droperidol generic	F:	Durogésic (Janssen-Cilag) generic
			GB:	Durogesic (Janssen-Cilag) Sublimaze (Janssen-Cilag)

I:	Fentanest (Carlo Erba) Leptofen (Carlo Erba)-comb.	J:	Fentanest (Sankyo; as citrate)		Thalamonal (Sankyo)-comb. with droperidol
				USA:	Duragesic (Janssen)

Fenticlor

ATC: D01A
Use: antifungal, anti-infective

RN: 97-24-5   MF: $C_{12}H_8Cl_2O_2S$ MW: 287.17   EINECS: 202-568-7
CN: 2,2'-thiobis[4-chlorophenol]

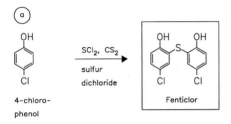

a

OH

Cl

4-chloro-phenol

SCl₂, CS₂

sulfur dichloride

OH   OH

S

Cl   Cl

Fenticlor

b

OH

phenol

SCl₂, CS₂

sulfur dichloride

OH   OH

S

bis(2-hydroxy-phenyl) sulfide

Cl₂, CH₃COOH

Fenticlor

Reference(s):
a Dunning, F. et al.: J. Am. Chem. Soc. (JACSAT) **53**, 3466 (1931).
b DRP 568 944 (I. G. Farben; appl. 1931).

Formulation(s):   ointment 5 %; sol. 5 %

Trade Name(s):
D: Antimyk (Pfleger); wfm

Fenticonazole

ATC: D01AC12; G01AF12
Use: antifungal

RN: 72479-26-6   MF: $C_{24}H_{20}Cl_2N_2OS$ MW: 455.41
LD₅₀: 1191 mg/kg (M, i.p.);
      440/309 mg/kg (R, i.p.); >3000 mg/kg (R, p.o.)
CN: 1-[2-(2,4-dichlorophenyl)-2-[[4-(phenylthio)phenyl]methoxy]ethyl]-1*H*-imidazole

mononitrate
RN: 73151-29-8   MF: $C_{24}H_{20}Cl_2N_2OS \cdot HNO_3$ MW: 518.42   EINECS: 277-302-6
LD₅₀: >3 g/kg (M, p.o.);
      >3 g/kg (R, p.o.);
      >1 g/kg (dog, p.o.)

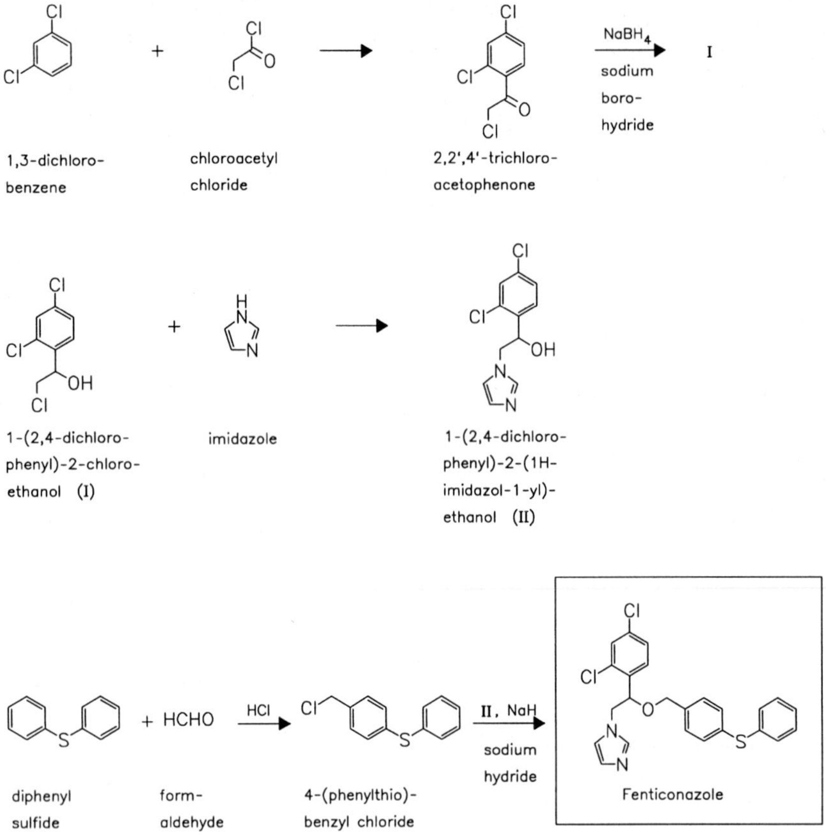

Cl
Cl
+
Cl
O
Cl
Cl
Cl
O
Cl
NaBH4
sodium
boro-
hydride
I
1,3-dichloro-
benzene
chloroacetyl
chloride
2,2',4'-trichloro-
acetophenone
Cl
Cl
OH
Cl
+
H
N
N
Cl
Cl
OH
N
N
1-(2,4-dichloro-
phenyl)-2-chloro-
ethanol (I)
imidazole
1-(2,4-dichloro-
phenyl)-2-(1H-
imidazol-1-yl)-
ethanol (II)
S
+ HCHO
HCl
Cl
S
II, NaH
sodium
hydride
Cl
Cl
O
N
N
S
Fenticonazole
diphenyl
sulfide
form-
aldehyde
4-(phenylthio)-
benzyl chloride

Reference(s):
DE 2 917 244 (Recordati; appl. 9.5.1979; I-prior. 18.5.1978).
US 4 221 803 (Recordati; 9.9.1980; appl. 9.5.1979; I-prior. 18.5.1978).

Formulation(s): cream 2 %, gel 2 %; vaginal ovules 200 mg

Trade Name(s):

F: Lomexin (Effik; as nitrate)
Terlomexin (Effik; as nitrate)

GB: Lomexin (Dominion; Pharmacia & Upjohn; as nitrate)

I: Falvin (Farmades)
Fentiderm (Zyma)
Fentigyn (Novartis)
Lomexin (Recordati; 1986)

Fentonium bromide

ATC: A03BB04
Use: anticholinergic

RN: 5868-06-4 MF: $C_{31}H_{34}BrNO_4$ MW: 564.52 EINECS: 227-520-2
LD_{50}: 12100 µg/kg (M, i.v.); >400 mg/kg (M, p.o.);
11600 µg/kg (R, i.v.)
CN: [3(*S*)-*endo,anti*]-8-(2-[1,1'-biphenyl]-4-yl-2-oxoethyl)-3-(3-hydroxy-1-oxo-2-phenylpropoxy)-8-methyl-8-azoniabicyclo[3.2.1]octane bromide

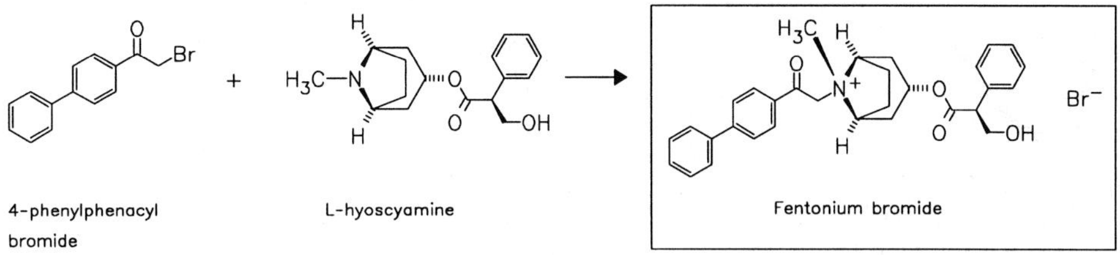

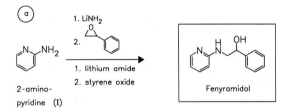

4-phenylphenacyl
bromide

L-hyoscyamine

Fentonium bromide

Reference(s):
synthesis:
US 3 356 682 (Whitefin Holding S.A.; 5.12.1967; prior. 27.10.1964).

medical use:
US 3 436 458 (Whitefin Holding S.A.; 1.4.1969; prior. 27.10.1964).

Formulation(s):   tabl. 20 mg

Trade Name(s):
D: Ulcesium (Inpharzam);   I: Duotrax (Zambon Farm.)-   Ulcesium (Zambon); wfm
      wfm   comb.; wfm

Fenyramidol
(Phenyramidol)

ATC: M03B
Use: analgesic, muscle relaxant

RN: 553-69-5   MF: $C_{13}H_{14}N_2O$ MW: 214.27   EINECS: 209-044-7
LD_{50}:   124 mg/kg (M, i.v.); 1850 mg/kg (M, p.o.);
       756 mg/kg (R, p.o.)
CN: α-[(2-pyridinylamino)methyl]benzenemethanol

monohydrochloride
RN: 326-43-2   MF: $C_{13}H_{14}N_2O \cdot HCl$ MW: 250.73   EINECS: 206-308-3
LD_{50}:   124 mg/kg (M, i.v.); 2425 mg/kg (M, p.o.)

(a)

1. $LiNH_2$

2.

1. lithium amide
2. styrene oxide

2-amino-
pyridine (I)

Fenyramidol

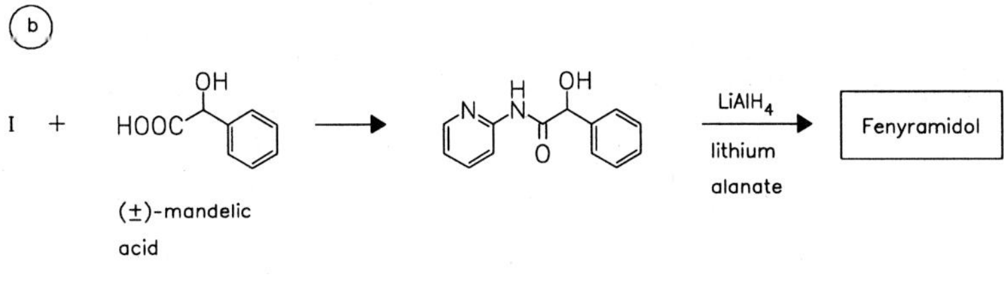

(b)

I +   HOOC

(±)-mandelic
acid

LiAlH₄

lithium
alanate

Fenyramidol

Reference(s):
DAS 1 420 056 (Neisler Labs.; appl. 14.8.1959; USA-prior. 4.11.1958).
US 3 040 050 (Lakeside Labs.; 19.6.1962; prior. 1.3.1960).
Gray, A.P. et al.: J. Am. Chem. Soc. (JACSAT) **81**, 4347, 4351 (1959).
BE 580 121 (Irwin, Neisler; appl. 26.6.1959; USA-prior. 4.11.1958).

Formulation(s):   drg. 400 mg (as hydrochloride)

Trade Name(s):

D:	Cabral (Kali-Chemie); wfm	Aramidol (ABC); wfm	Pheniramidol (Pulitzer); wfm
I:	Anabloc (Irbi); wfm	Firmalgil (Firma); wfm	
	Analexin (Biotrading); wfm	Miodar (ISM); wfm	J: Analexin-AF (Dainippon)- comb.

Fexofenadine hydrochloride
(MDL-16455A)

ATC: R06AX26
Use: antihistaminic, metabolite of
terfenadine

RN: 153439-40-8   MF: $C_{32}H_{39}NO_4 \cdot HCl$ MW: 538.13
CN: 4-[1-hydroxy-4-[4-(hydroxydiphenylmethyl)-1-piperidinyl]butyl]-α,α-dimethylbenzeneacetic acid hydrochloride

base
RN: 83799-24-0   MF: $C_{32}H_{39}NO_4$ MW: 501.67

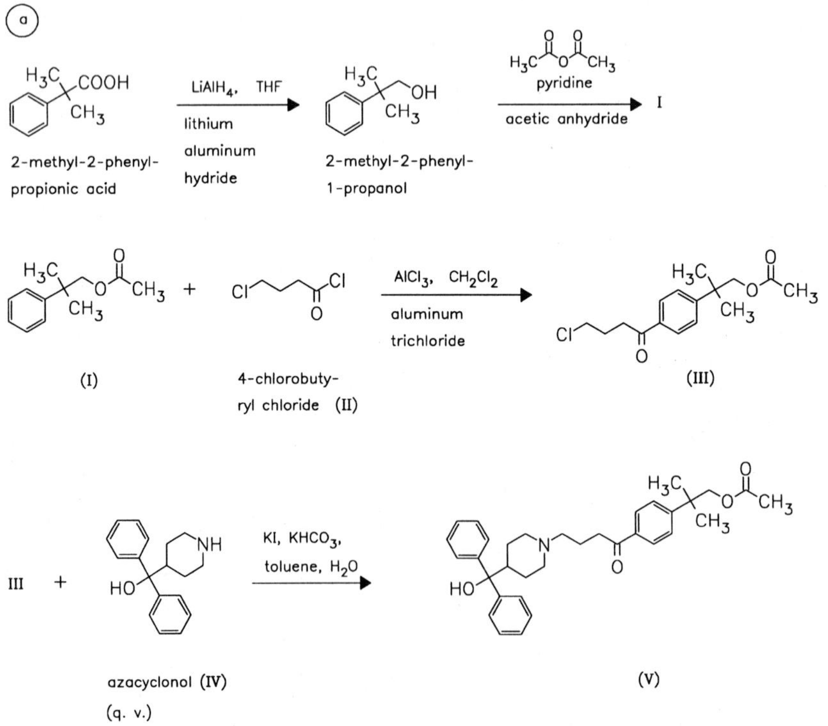

2-methyl-2-phenyl-propionic acid   →   2-methyl-2-phenyl-1-propanol   →   I

(I) + 4-chlorobutyryl chloride (II) → (III)

III + azacyclonol (IV) (q. v.) → (V)

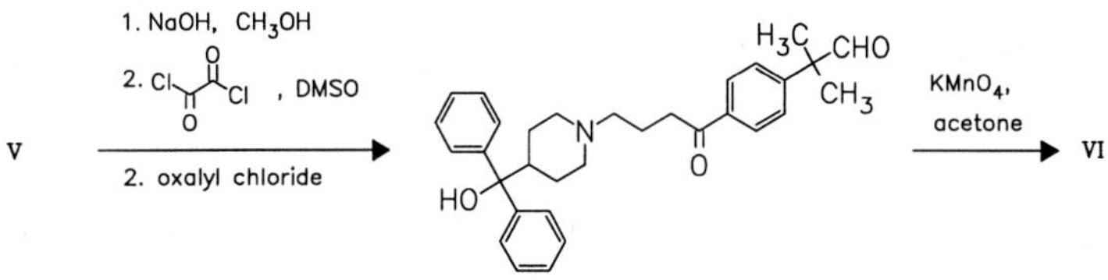

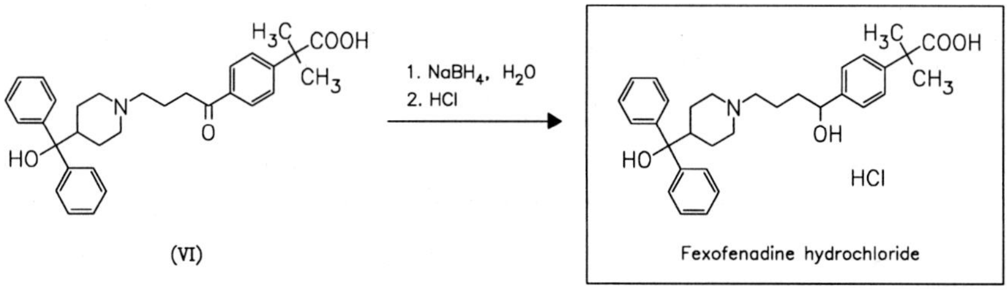

(VI)

Fexofenadine hydrochloride

b

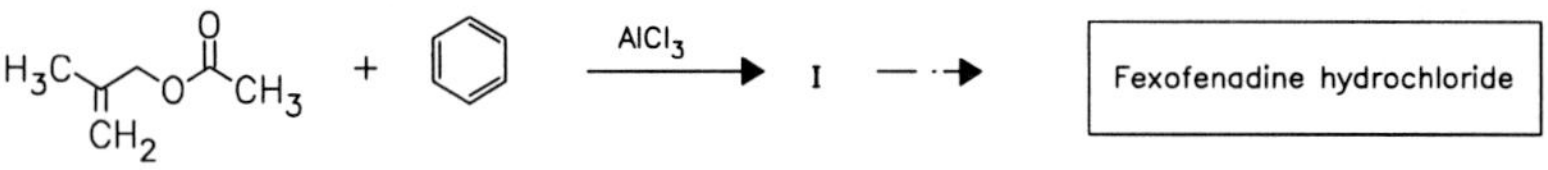

2-methyl-2-propenyl acetate

benzene

Fexofenadine hydrochloride

c

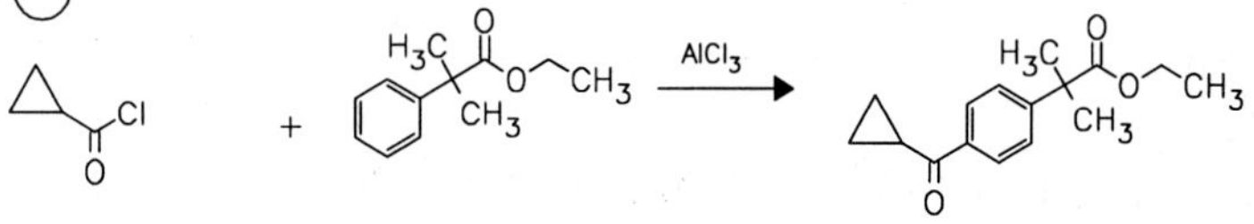

cyclopropane-carbonyl chloride

ethyl 2-methyl-2-phenyl-propionate (VII)

ethyl 2-[4-(cyclopropyl-carbonyl)phenyl]-2-methylpropionate (VIII)

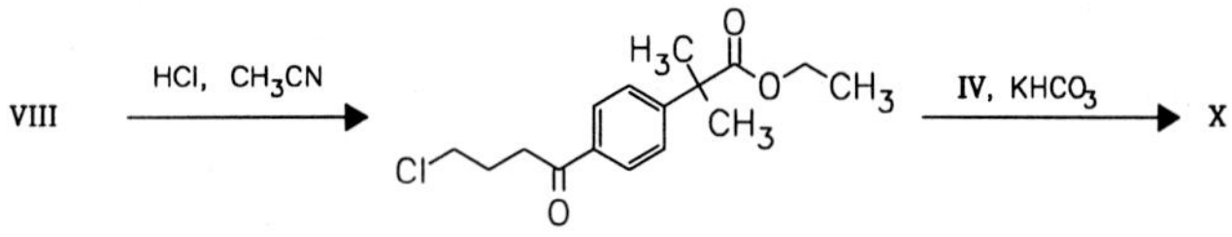

(IX)

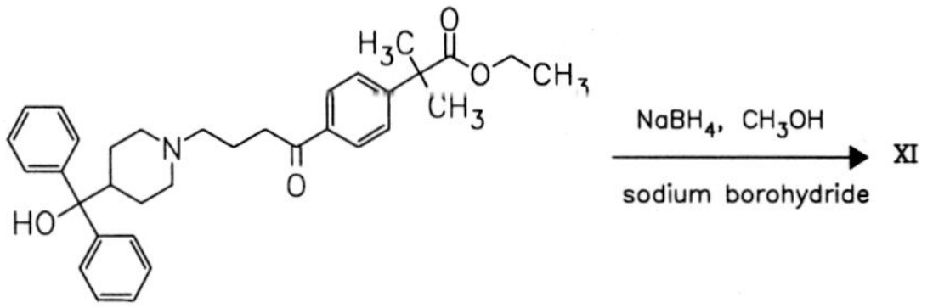

(X)

(XI)

(d)

II + VII →(AlCl₃) IX — ·→ Fexofenadine hydrochloride

Reference(s):
a,b WO 9 321 156 (Merrell Dow Pharm.; appl. 10.3.1993; USA-prior. 25.1.1993, 10.4.1992).
c,d WO 9 500 480 (Merrell Dow Pharm.; appl. 26.5.1994; USA-prior. 25.6.1993, 27.10.1993).
    WO 9 500 482 (Albany Molecular Res.; appl. 21.6.1994; USA-prior. 24.6.1993).

preparation of optically active isomers used in antihistamine treatment:
WO 9 403 170 (Sepracor Inc.; appl. 3.8.1993; USA-prior. 3.8.1992).

process for resolution using mandelic acid:
WO 9 531 436 (Merrell Pharm. Inc.; appl. 10.4.1995; USA-prior. 16.5.1994).

use in hepatic impaired patients:
WO 9 323 047 (Merrell Dow Pharm.; appl. 6.4.1993; USA-prior. 31.7.1992, 11.5.1992).
WO 9 510 278 (Marion Merrell Dow; appl. 30.9.1994; USA-prior. 15.10.1993).

anhydrous and hydrated forms:
WO 9 531 437 (Marion Merrell Dow; appl. 28.4.1995; USA-prior. 11.4.1995, 18.5.1994).

improved bioavailability with high surface area particle form:
WO 9 626 726 (Hoechst Marion Roussel; appl. 26.1.1996; USA-prior. 12.12.1995, 28.2.1995).

oral formulations in solvent comp. propylene glycol:
US 5 574 045 (Hoechst Marion Roussel; 12.11.1996; appl. 6.6.1995; USA-prior. 6.6.1995).

Formulation(s):   cps. 60 mg

Trade Name(s):
D: Telfast (Hoechst Marion   GB: Telfast (Hoechst)
    Roussel; Procter &   USA: Allegra (Hoechst Marion
    Gamble)   Roussel)

Fibrinolysin (human)
(Serum-Tryptase; Plasmin)

ATC: B01AD05
Use: thrombolytic

RN: 9004-09-5   MF: unspecified   MW: unspecified   EINECS: 232-640-3
CN: plasmin

An enzyme obtained from human plasma by conversion of profibrinolysin with streptokinase to fibrinolysin.
Proteolytic enzyme of unknown structure; molar mass ≡ 75000.
From oxalate added blood plasma by precipitation with $CaCl_2$ and purification by washing and precipitation and lyophilization.

Reference(s):
US 2 624 691 (Parke Davis; 1953; appl. 1946).
US 3 136 703 (Ortho Pharmaceutical; 9.6.1964; prior. 1.10.1957, 22.4.1958).
US 3 234 106 (Cutter Labs.; 8.2.1966; appl. 3.12.1962).

Formulation(s):   ointment 10 mg/1 g (1 %)

Trade Name(s):

D:	Fibrinolysin (Human) Lyovac (Sharp & Dohme); wfm	F:	Elase (Substantia)-comb.; wfm Thromboclase (Choay); wfm	I: USA:	Elase (Parke Davis)-comb. Elase (Fujisawa)

Finasteride

ATC: G04CA01
Use: 5α-reductase inhibitor, treatment of benign prostatic hypertrophy

RN: 98319-26-7   MF: $C_{23}H_{36}N_2O_2$ MW: 372.55
LD_{50}:   486 mg/kg (M, p.o.);
        418 mg/kg (R, p.o.);
        >1 g/kg (dog, p.o.)
CN: (5α,17β)-N-(1,1-dimethylethyl)-3-oxo-4-azaandrost-1-ene-17-carboxamide

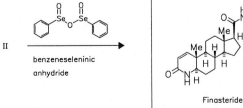

Finasteride

Reference(s):
US 155 096 (Merck & Co.; appl. 20.2.1985; USA-prior. 27.2.1984).
Rasmusson, G.H. et al.: J. Med. Chem. (JMCMAR) **29**, 2298 (1986).

medical use for treatment of androgenic alopecia:
EP 285 382 (Merck & Co.; appl. 30.3.1988; USA-prior. 3.4.1987).

medical use for treatment of prostate carcinoma:
EP 285 383 (Merck & Co.; appl. 30.3.1988; USA-prior. 3.4.1987).

Formulation(s):   f. c. tabl. 5 mg

Trade Name(s):

D:	Proscar (MSD Chibropharm)	GB:	Proscar (Merck Sharp & Dohme)		Prostide (Sigma-Tau)
				USA:	Proscar (Merck)
F:	Chibro-Proscar (Merck Sharp & Dohme-Chibret)	I:	Proscar (Merck & Co.; 1991)		

Fipexide

ATC: N06BX05
Use: antidepressant, psychotonic, nootropic

RN: 34161-24-5   MF: $C_{20}H_{21}ClN_2O_4$ MW: 388.85   EINECS: 251-857-4
CN: 1-(1,3-benzodioxol-5-ylmethyl)-4-[(4-chlorophenoxy)acetyl]piperazine

monohydrochloride
RN: 34161-23-4   MF: $C_{20}H_{21}ClN_2O_4 \cdot HCl$ MW: 425.31   EINECS: 251-856-9
LD_{50}:   4150 mg/kg (M, p.o.);
         4482 mg/kg (R, p.o.)

4-chlorophenoxy-   1-piperonyl-
acetylchloride   piperazine

Fipexide

Reference(s):
FR-M 7 524 (Lab. F. Bouchard; appl. 12.3.1968).

Formulation(s):   drg. 200 mg; tabl. 200 mg (as hydrochloride)

Trade Name(s):

F:	Vigilor (Bouchard); wfm	I:	Attentil (Lusofarmaco); wfm

Flavoxate

ATC: G04BD02
Use: antispasmodic

RN: 15301-69-6   MF: $C_{24}H_{25}NO_4$ MW: 391.47   EINECS: 239-337-5
CN: 3-methyl-4-oxo-2-phenyl-4*H*-1-benzopyran-8-carboxylic acid 2-(1-piperidinyl)ethyl ester

hydrochloride
RN: 3717-88-2 MF: $C_{24}H_{25}NO_4 \cdot HCl$ MW: 427.93 EINECS: 223-066-4
LD_{50}: 28 mg/kg (M, i.v.); 740 mg/kg (M, p.o.);
25 mg/kg (R, i.v.); 1040 mg/kg (R, p.o.)

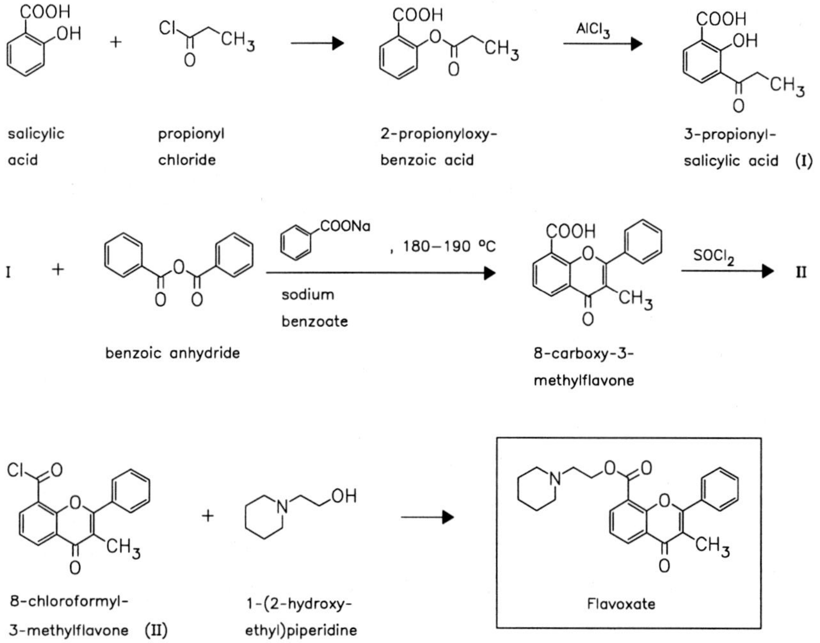

COOH
OH
Cl
CH3
O
AlCl3
salicylic acid
propionyl chloride
2-propionyloxy-benzoic acid
3-propionyl-salicylic acid (I)
I
COONa
, 180–190 °C
sodium benzoate
benzoic anhydride
8-carboxy-3-methylflavone
SOCl2
II
8-chloroformyl-3-methylflavone (II)
1-(2-hydroxy-ethyl)piperidine
N
Flavoxate

Reference(s):
US 2 921 070 (Recordati; 12.1.1960; CH-prior. 5.11.1957).

alternative synthesis:
US 3 350 411 (Seceph; 31.10.1967; I-prior. 10.10.1963).

Formulation(s): f. c. tabl. 200 mg; tabl. 100 mg (as hydrochloride)

Trade Name(s):

D:	Spasuret (Sanofi Winthrop)	I:	Cistalgan (Recordati)-comb.	J:	Bladderon (Nippon Shinyaku)
F:	Urispas (Negma)		Genurin (Recordati)	USA:	Urispas (SmithKline Beecham)
GB:	Urispas (Shire)				

Flecainide

ATC: C01BC04
Use: antiarrhythmic

RN: 54143-55-4 MF: $C_{17}H_{20}F_6N_2O_3$ MW: 414.35
CN: *N*-(2-piperidinylmethyl)-2,5-bis(2,2,2-trifluoroethoxy)benzamide

acetate
RN: 54143-56-5 MF: $C_{17}H_{20}F_6N_2O_3 \cdot C_2H_4O_2$ MW: 474.40

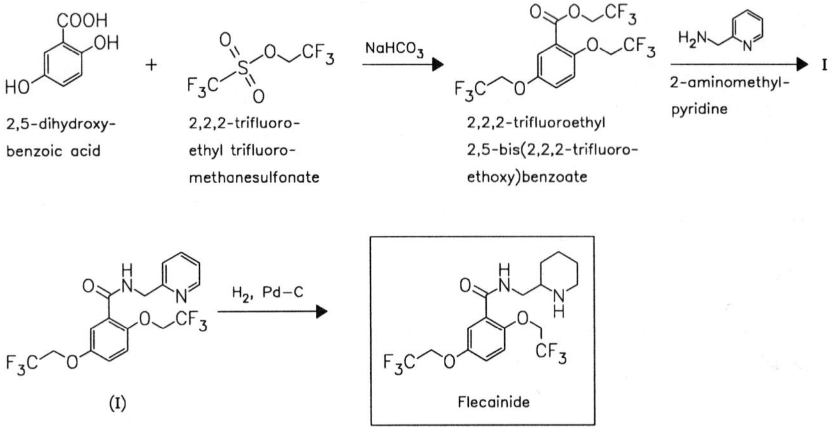

2,5-dihydroxy-
benzoic acid

2,2,2-trifluoro-
ethyl trifluoro-
methanesulfonate

2,2,2-trifluoroethyl
2,5-bis(2,2,2-trifluoro-
ethoxy)benzoate

2-aminomethyl-
pyridine

(I)

Flecainide

Reference(s):

DE 2 513 916 (Riker; prior. 27.3.1975).
US 3 900 481 (Riker; 19.8.1975; prior. 1.4.1974).
US 3 655 728 (Riker; 11.4.1972; prior. 22.7.1970).
US 4 005 209 (Riker; 25.1.1975; prior. 27.5.1975).
Bannit, E.H. et al.: J. Med. Chem. (JMCMAR) **18**, 1130 (1975); **20**, 821 (1977).

Formulation(s):   amp. 50 mg; tabl. 50 mg, 100 mg (as acetate)

Trade Name(s):

D: Tambocor (3M Medica;   GB: Tambocor (3M Health   J: Tambocor (Eisai)
      1982)   Care; 1983)   USA: Tambocor (3M; 1985)
F: Flécaïne (3M Santé; 1984)   I: Almarytm (Synthelabo;
                                         1986)

Fleroxacin

(AM 833; Ro 23-6240; Megalocin)

ATC: J01MA08
Use: antibacterial

RN: 79660-72-3   MF: $C_{17}H_{18}F_3N_3O_3$ MW: 369.34
LD$_{50}$: 20.4 mg/kg (R, i. v.); >4 g/kg (R, p. o.);
      21.7 mg/kg (M, i. v.); >4 g/kg (M, p. o.);
      >1 g/kg (dog, p. o.)
CN: 6,8-Difluoro-1-(2-fluoroethyl)-1,4-dihydro-7-(4-methyl-1-piperazinyl)-4-oxo-3-quinolinecarboxylic acid

monohydrochloride
RN: 79660-53-0   MF: $C_{17}H_{18}F_3N_3O_3 \cdot HCl$ MW: 405.80

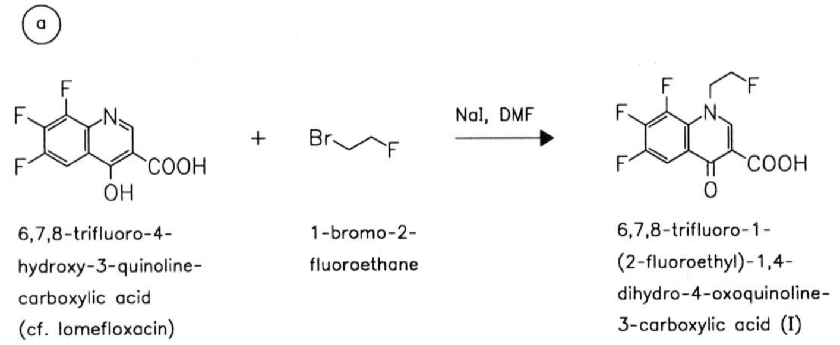

6,7,8-trifluoro-4-
hydroxy-3-quinoline-
carboxylic acid
(cf. lomefloxacin)

1-bromo-2-
fluoroethane

6,7,8-trifluoro-1-
(2-fluoroethyl)-1,4-
dihydro-4-oxoquinoline-
3-carboxylic acid (I)

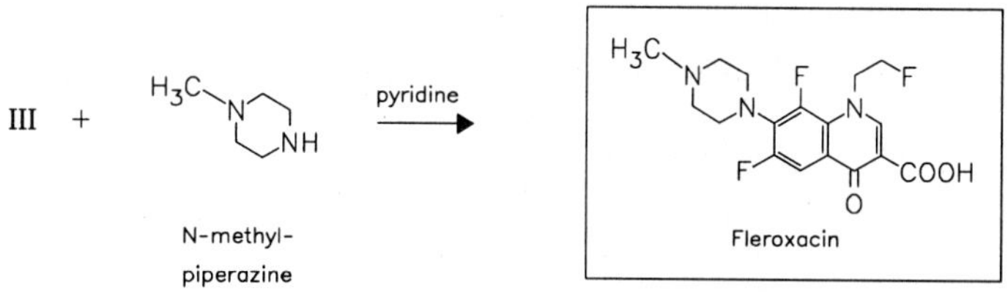

III + N-methyl-piperazine → (pyridine) → Fleroxacin

b

→ (H₂SO₄, H₂O) → Fleroxacin

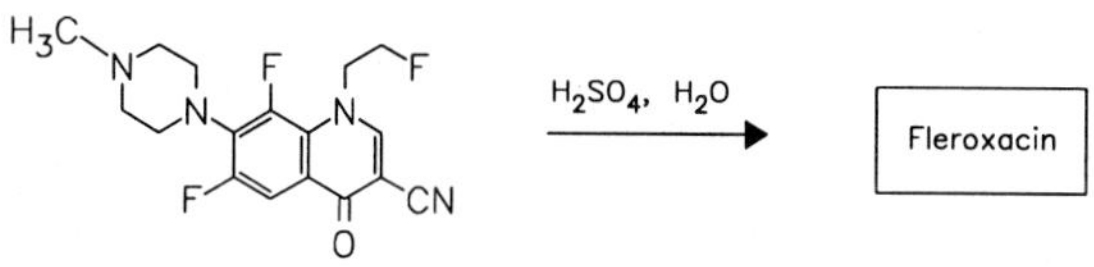

Reference(s):
a   BE 887 574 (Kyorin Pharm.; appl. 19.2.1981; BE-prior. 19.8.1980).
    ZA 8 502 065 (Kyorin Pharm.; appl. 20.3.1985; ZA-prior. 20.3.1985).
b   ES 2 010 862 (Inke S. A.; appl. 13.2.1989).

purification and recovery using porous absorbents:
JP 08 259 541 (Kyorin Seiyaku; appl. 23.3.1995).

synthesis of fluorine-labeled fleroxacin:
Livni, E. et al.: Nucl. Med. Biol. (NMBIEO) **20** (1), 883-897 (1993)

Formulation(s):   amp. for inj. 400 mg; f. c. tabl. 200 mg, 400 mg; tabl. 200 mg, 400 mg; vial 400 mg/100 ml

Trade Name(s):
D: Quinodis (Roche/
      Grünenthal)

Floctafenine

ATC: N02BG04
Use: analgesic

RN: 23779-99-9   MF: $C_{20}H_{17}F_3N_2O_4$ MW: 406.36   EINECS: 245-881-4
LD$_{50}$:   180 mg/kg (M, i.v.); 1960 mg/kg (M, p.o.);
      160 mg/kg (R, i.v.); 535 mg/kg (R, p.o.);
      >1 g/kg (dog, p.o.)
CN: 2-[[8-(trifluoromethyl)-4-quinolinyl]amino]benzoic acid 2,3-dihydroxypropyl ester

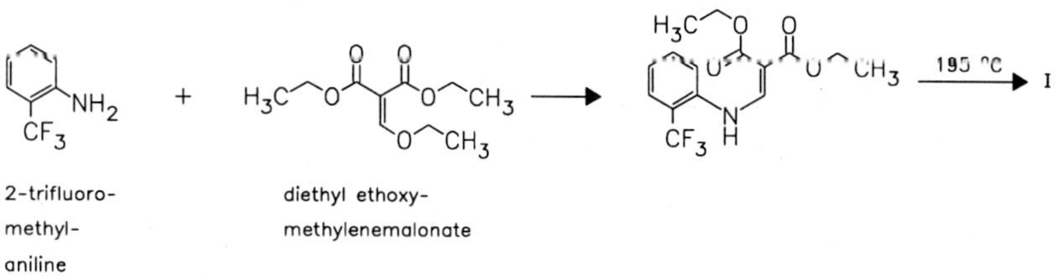

2-trifluoro-methyl-aniline   +   diethyl ethoxy-methylenemalonate   →   (195 °C) → I

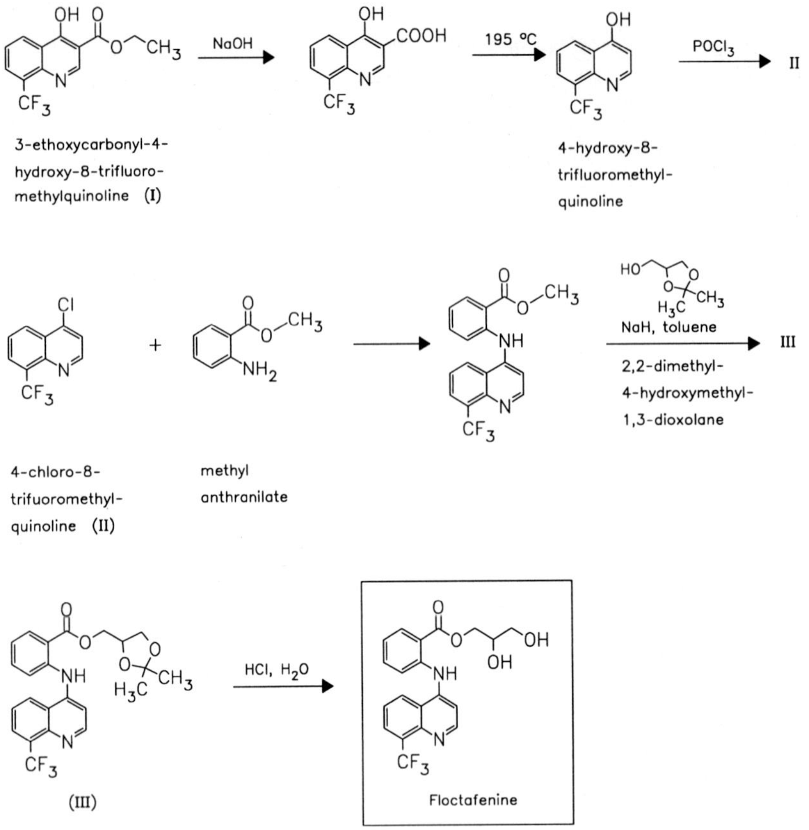
NaOH
195 °C
POCl3
II
3-ethoxycarbonyl-4-hydroxy-8-trifluoromethylquinoline (I)
4-hydroxy-8-trifluoromethyl-quinoline
NaH, toluene
III
2,2-dimethyl-4-hydroxymethyl-1,3-dioxolane
4-chloro-8-trifuoromethyl-quinoline (II)
methyl anthranilate
HCl, H2O
(III)
Floctafenine

Reference(s):

DE 1 815 467 (Roussel-Uclaf; appl. 18.12.1968; F-prior. 29.12.1967, 29.3.1968, 23.8.1968).
US 3 644 368 (Roussel-Uclaf; 22.2.1972; F-prior. 29.12.1967, 23.8.1968).
US 3 818 090 (Roussel-Uclaf; 22.2.1972; prior. 7.7.1971).

Formulation(s): tabl. 200 mg

Trade Name(s):

D:	Idarac (Roussel; 1978); wfm	F:	Idarac (Roussel Diamant; 1976)	I:	Idarac (Roussel; 1977)

Flomoxef
(6315-S)

ATC: J01C
Use: antibacterial (β-lactam antibiotic)

RN: 99665-00-6 MF: $C_{15}H_{18}F_2N_6O_7S_2$ MW: 496.47
CN: (6*R*-*cis*)-7-[[[(difluoromethyl)thio]acetyl]amino]-3-[[[1-(2-hydroxyethyl)-1*H*-tetrazol-5-yl]thio]methyl]-7-methoxy-8-oxo-5-oxa-1-azabicyclo[4.2.0]oct-2-ene-2-carboxylic acid

a   side chain I:

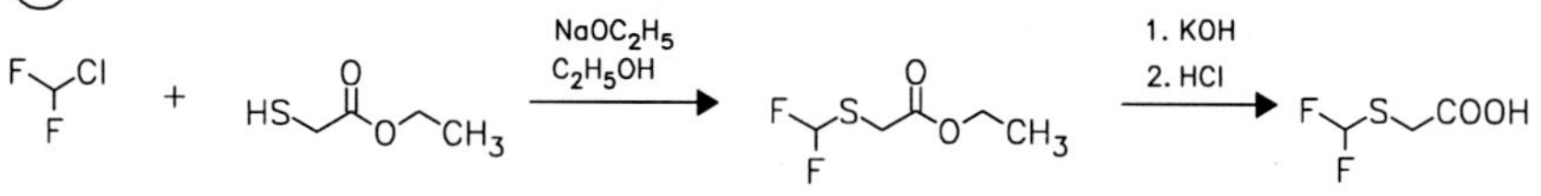

chloro-
difluoro-
methane

ethyl (difluoro-
methylthio)acetate

(difluoromethyl-
thio)acetic acid (I)

b   side chain IV:

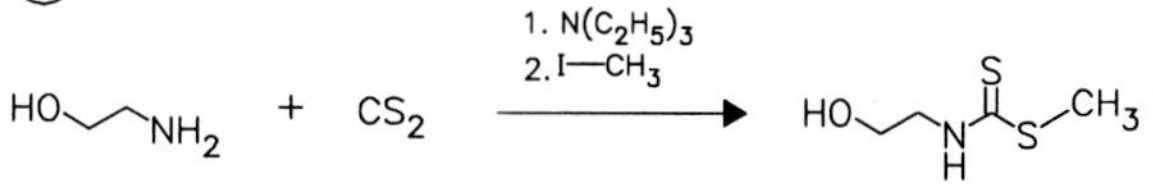

ethanolamine   carbon
                  disufide

methyl N-(2-hydroxy-
ethyl)dithiocarbamate (II)

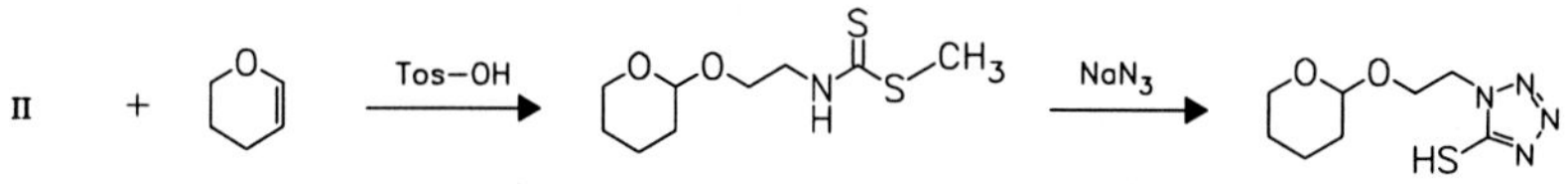

3,4-dihydro-
2H-pyran

(III)

1-(2-hydroxyethyl)-
1H-tetrazole-5-thiol (IV)

c   final product:

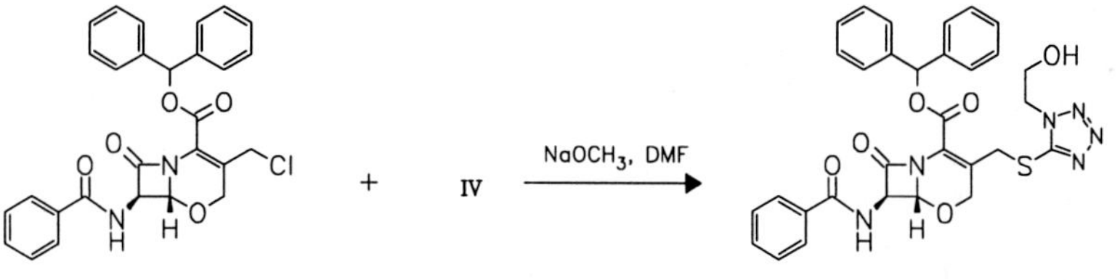

diphenylmethyl
7α-benzamido-3-
chloromethyl-1-oxa-
3-cephem-4-carboxylate
(cf. latamoxef synthesis)

(V)

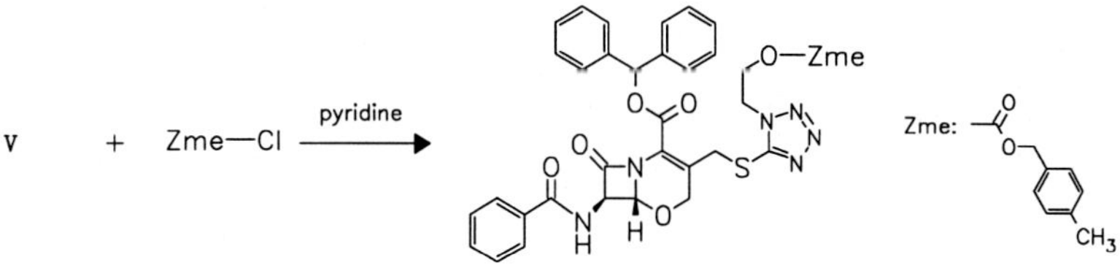

p-methyl-
benzyl
chloroformate

(VI)

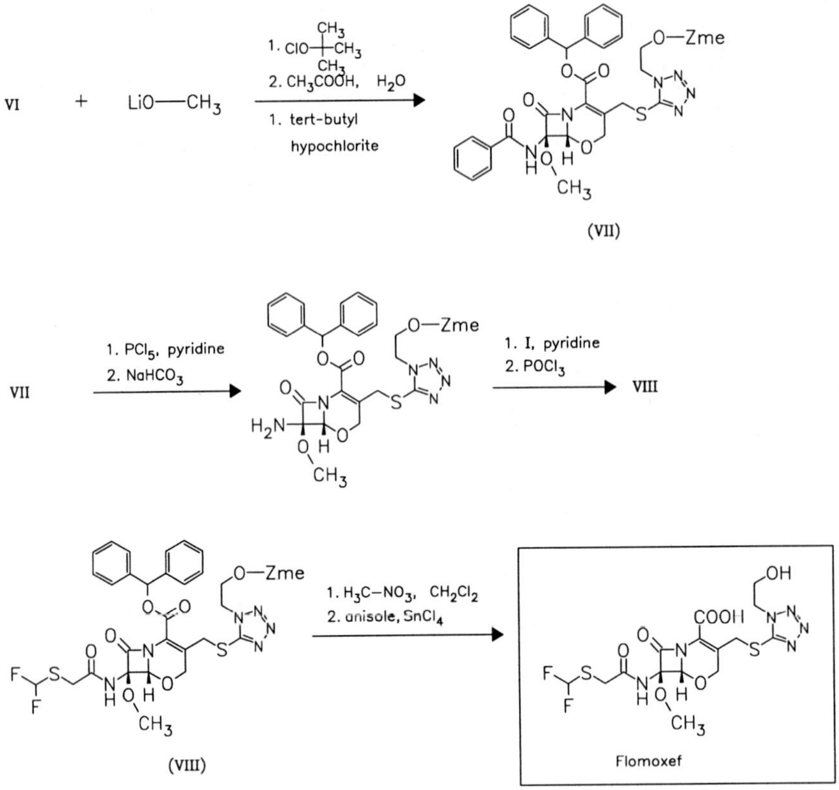

VI + LiO—CH₃

1. ClO—C(CH₃)₃
2. CH₃COOH, H₂O

1. tert-butyl hypochlorite

(VII)

VII

1. PCl₅, pyridine
2. NaHCO₃

1. I, pyridine
2. POCl₃

VIII

(VIII)

1. H₃C—NO₃, CH₂Cl₂
2. anisole, SnCl₄

Flomoxef

Reference(s):

Tsuji, T. et al.: J. Antibiot. (JANTAJ) **38**, 466 (1984).
US 4 532 233 (Shionogi; 30.7.1985; J-prior. 23.12.1982).
DOS 3 345 989 (Shionogi; appl. 20.12.1983; J-prior. 23.12.1982).
EP 128 536 (Shionogi; appl. 7.6.1984; J-prior. 14.6.1983).

purification:
DOS 3 503 303 (Shionogi; appl. 31.1.1985; J-prior. 2.2.1984).

Formulation(s):　　vial (dry substance for inj.) 500 mg, 1g

Trade Name(s):
J:　　Flumarin (Shionogi)

Flopropione

ATC:　A03A
Use:　antispasmodic

RN:　2295-58-1　MF: C₉H₁₀O₄　MW: 182.18　EINECS: 218-942-8
LD₅₀:　300 mg/kg (M, i.v.); 2780 mg/kg (M, p.o.);
　　　246 mg/kg (R, i.v.); 2380 mg/kg (R, p.o.)
CN:　1-(2,4,6-trihydroxyphenyl)-1-propanone

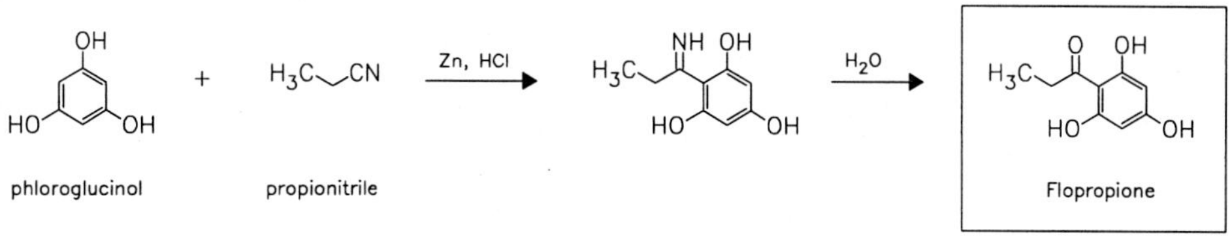

phloroglucinol   propionitrile   Flopropione

Reference(s):
Canter et al.: J. Chem. Soc. (JCSOA9) **1931**, 1245.
Shinoda, K.: Yakugaku Zasshi (YKKZAJ) **35**, 235 (1927).
Howells et al.: J. Am. Chem. Soc. (JACSAT) **54**, 2451 (1932).

pharmacology:
Cahen, R.; Boucherie, A.: C. R. Seances Soc. Biol. Ses Fil. (CRSBAW) **157**, 112 (1963).

Formulation(s):   cps. 40 mg; gran. 80 mg/g, 160 mg/g

Trade Name(s):
J: Chlonarin (Kanebo)   Cospanon (Eisai)   generic
       Colenfupan (Nichiiko)   Pasmus (Daiichi)

Florantyrone

ATC: A03A
Use: choleretic

RN: 519-95-9   MF: $C_{20}H_{14}O_3$ MW: 302.33   EINECS: 208-279-2
CN: γ-oxo-8-fluoranthenebutanoic acid

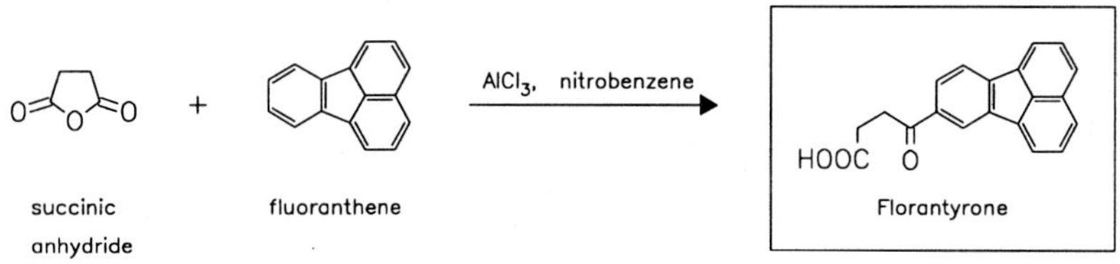

succinic anhydride   fluoranthene   Florantyrone

Reference(s):
US 2 560 425 (Miles Labs.; 1951; prior. 1948).

Formulation(s):   0.075 g, 1 g

Trade Name(s):
I: Bilyn (Janus); wfm   Idroepar (Beolet); wfm   USA: Zanchol (Searle); wfm
       Cistoplex (Borromeo); wfm   J: Zanchol (G.D.-Dainippon)

Floredil

ATC: C01DB
Use: coronary vasodilator

RN: 53731-36-5   MF: $C_{16}H_{25}NO_4$ MW: 295.38
CN: 4-[2-(3,5-diethoxyphenoxy)ethyl]morpholine

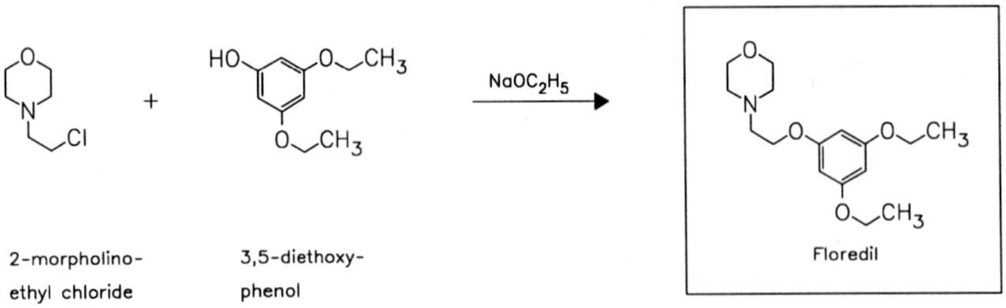
NaOC$_2$H$_5$
2-morpholino-ethyl chloride
3,5-diethoxy-phenol
Floredil

Reference(s):

DOS 2 020 464 (Orsymonde; appl. 27.4.1970; GB-prior. 29.4.1969).

Formulation(s): cps. 200 mg

Trade Name(s):

F: Carfonal (Lafon); wfm

Flosequinan

(BTS 49037; BTS 49465)

ATC: C01DB01
Use: vasodilator, antihypertensive

RN: 76568-02-0 MF: $C_{11}H_{10}FNO_2S$ MW: 239.27
CN: 7-Fluoro-1-methyl-3-(methylsulfinyl)-4(1*H*)-quinolinone

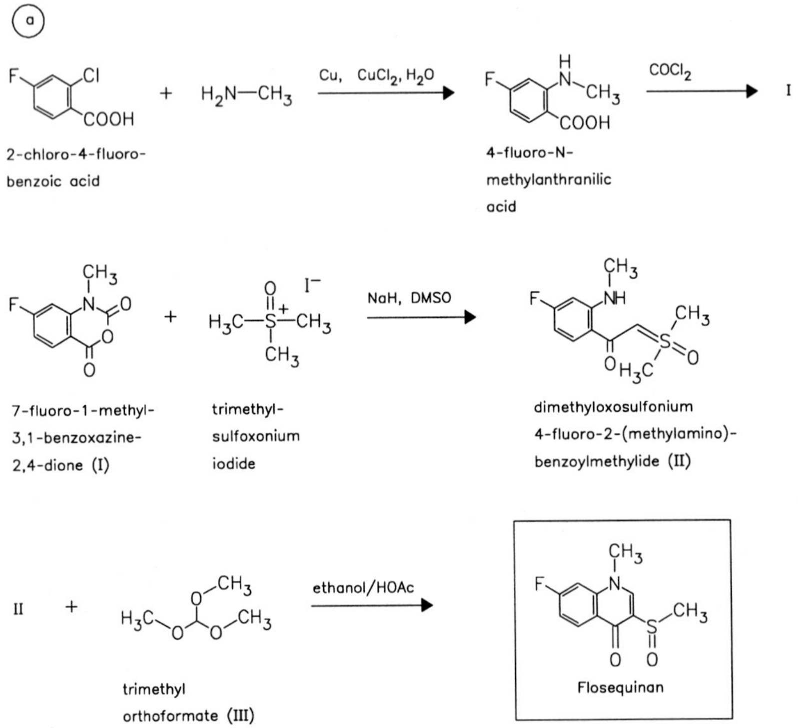
a
Cu, CuCl$_2$, H$_2$O
COCl$_2$
I
2-chloro-4-fluoro-benzoic acid
4-fluoro-N-methylanthranilic acid
NaH, DMSO
7-fluoro-1-methyl-3,1-benzoxazine-2,4-dione (I)
trimethyl-sulfoxonium iodide
dimethyloxosulfonium 4-fluoro-2-(methylamino)-benzoylmethylide (II)
II
ethanol/HOAc
trimethyl orthoformate (III)
Flosequinan

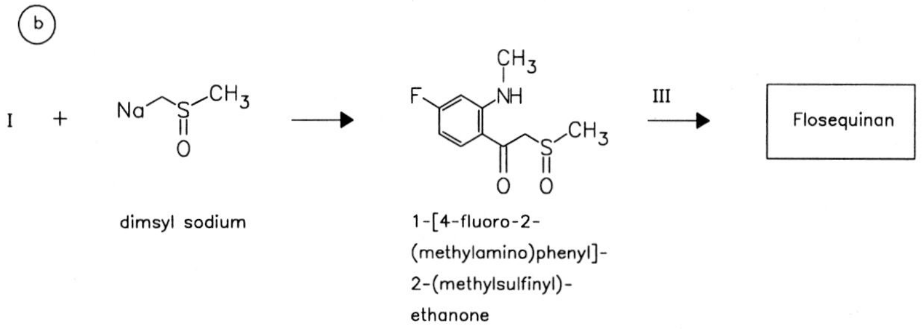

(b)

dimsyl sodium

1-[4-fluoro-2-
(methylamino)phenyl]-
2-(methylsulfinyl)-
ethanone

III → Flosequinan

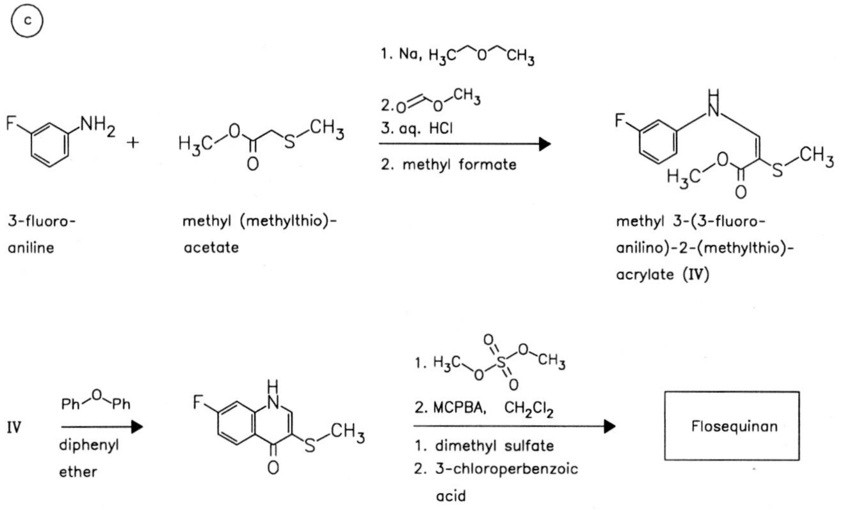

(c)

1. Na, H₃C‾O‾CH₃
2. O=⟨O‾CH₃
3. aq. HCl
2. methyl formate

3-fluoro-
aniline

methyl (methylthio)-
acetate

methyl 3-(3-fluoro-
anilino)-2-(methylthio)-
acrylate (IV)

IV

Ph‾O‾Ph
diphenyl
ether

1. H₃C‾O‾S(=O)(=O)‾O‾CH₃
2. MCPBA, CH₂Cl₂
1. dimethyl sulfate
2. 3-chloroperbenzoic
acid

→ Flosequinan

Reference(s):

a   DE 3 011 994 (Boots; appl. 27.3.1980; GB-prior. 27.3.1979).

b,c Birch, A.M. et al.: J. Chem. Soc., Perkin Trans. 1 (JCPRB4) **1994**, 387.
    EP 317 149 (Boots; appl. 7.11.1988; GB-prior. 18.11.1987).

Formulation(s):   tabl. 50 mg, 100 mg

Trade Name(s):

GB: Manoplax (Boots)

Fluanisone

ATC: N05AD09
Use: neuroleptic

RN: 1480-19-9   MF: C₂₁H₂₅FN₂O₂   MW: 356.44   EINECS: 216-038-8
LD₅₀.   25 mg/kg (M, i.v.); 550 mg/kg (M, p.o.);
        20 mg/kg (R, i.v.)
CN: 1-(4-fluorophenyl)-4-[4-(2-methoxyphenyl)-1-piperazinyl]-1-butanone

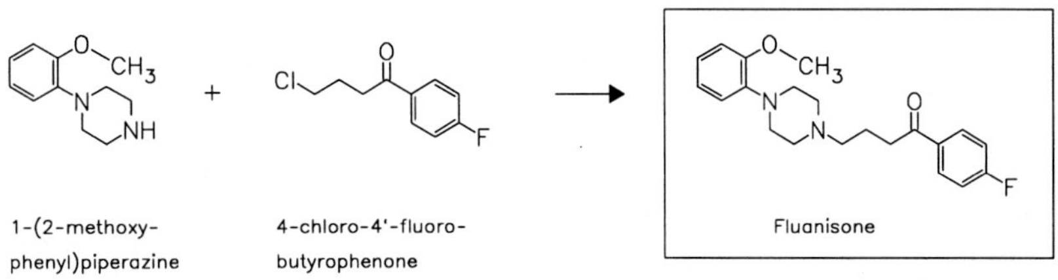

1-(2-methoxy-
phenyl)piperazine

4-chloro-4'-fluoro-
butyrophenone

Fluanisone

Reference(s):
DAS 1 185 615 (Janssen; appl. 25.3.1960; USA-prior. 26.3.1959).
US 2 997 472 (Janssen; 22.8.1961; prior. 26.3.1959).

Formulation(s):   sol. 6.25 mg/ml

Trade Name(s):
D: Sedalande (Delalande);   F: Sédalande (Delalande);
      wfm   wfm

Fluazacort

(Azacortid)

ATC: D07AB
Use: topical glucocorticoid, anti-
        inflammatory

RN: 19888-56-3   MF: $C_{25}H_{30}FNO_6$ MW: 459.51   EINECS: 243-400-2
LD$_{50}$:   54 mg/kg (M, s.c.);
       580 mg/kg (R, s.c.)
CN: (11β,16β)-21-(acetyloxy)-9-fluoro-11-hydroxy-2'-methyl-5'H-pregna-1,4-dieno[17,16-d]oxazole-3,20-dione

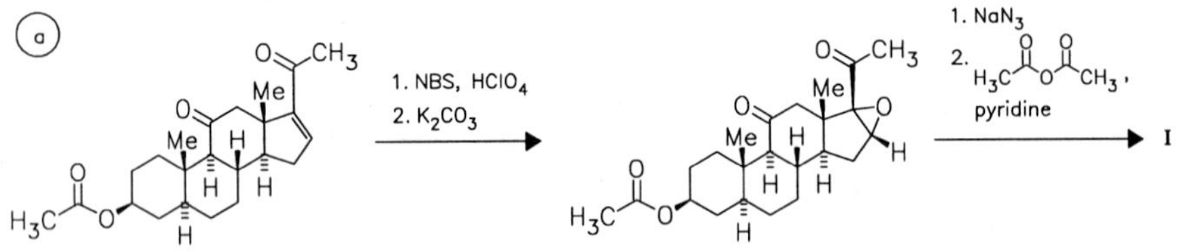

3β-acetoxy-11,20-
dioxo-16-pregnene
(from hecogenin,
cf. alfaxalone
synthesis)

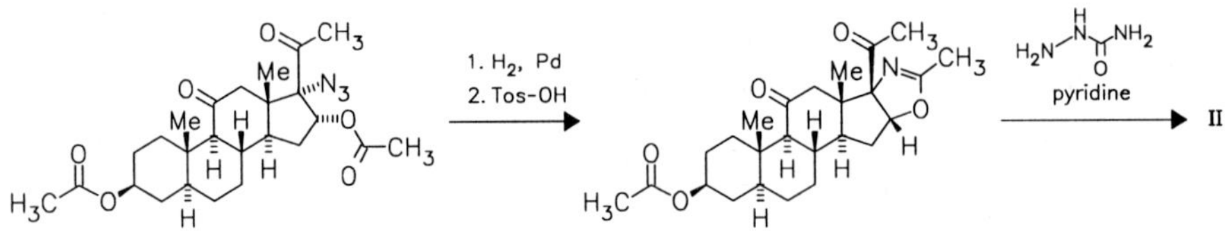

(I)

(3β,5α,5'β)-3-acetoxy-2'-
methyl-5'H-pregnano-
[17,16-d]oxazole-11,20-dione

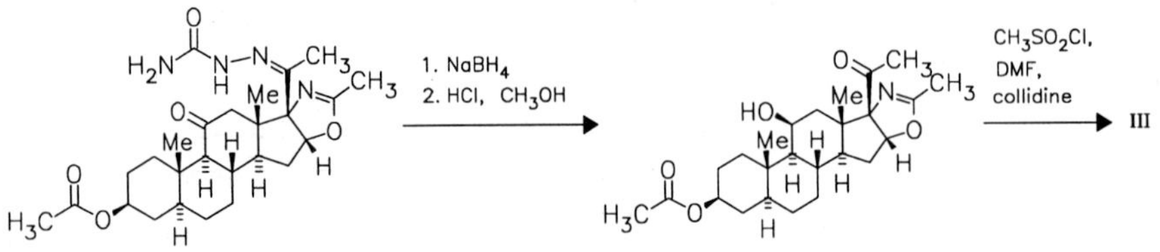

(II)

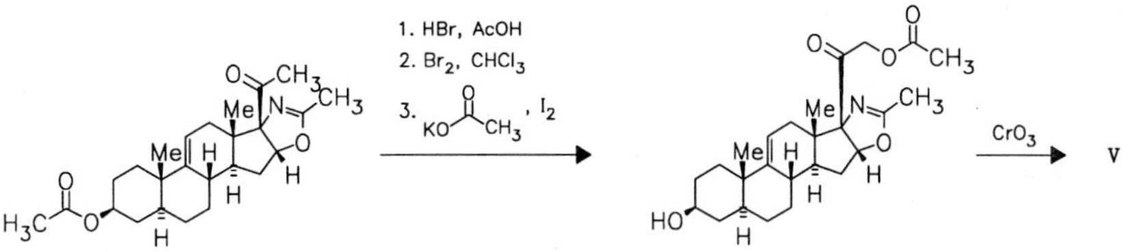

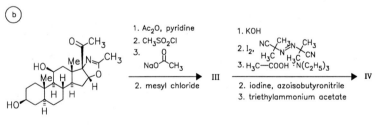

(III)

1. HBr, AcOH
2. Br₂, CHCl₃
3. [reagent], I₂

(3β,5α,5'β)-21-acetoxy-3-
hydroxy-2'-methyl-5H-
pregn-9(11)-eno[17,16-d]-
oxazol-20-one (IV)

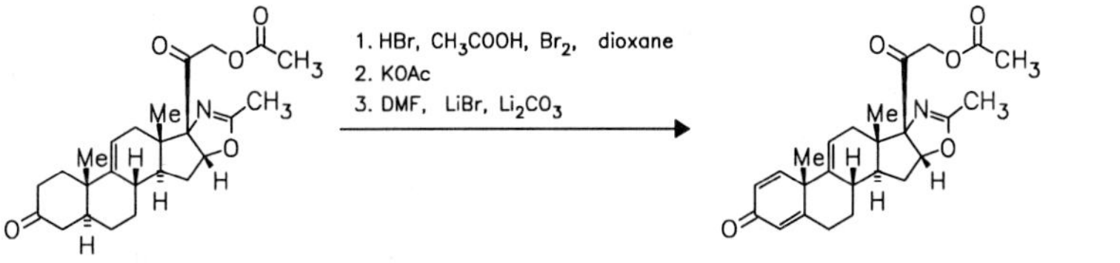

CrO₃ → V

(V)

1. HBr, CH₃COOH, Br₂, dioxane
2. KOAc
3. DMF, LiBr, Li₂CO₃

(5'β)-21-acetoxy-2'-methyl-
5'H-pregna-1,4,9(11)-trieno[17,16-d]-
oxazole-3,20-dione (VI)

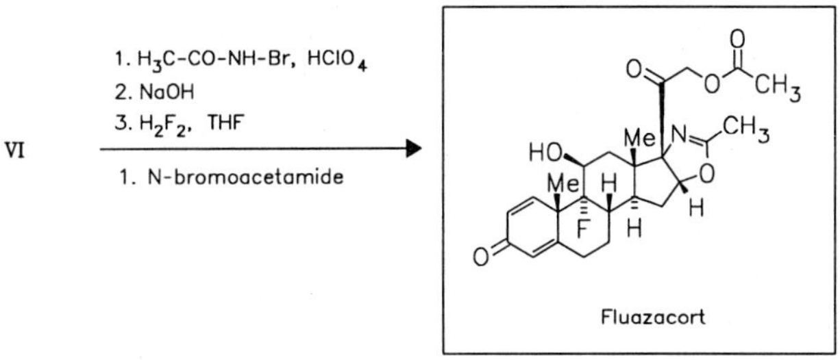

VI

1. H₃C–CO–NH–Br, HClO₄
2. NaOH
3. H₂F₂, THF

1. N-bromoacetamide

Fluazacort

(b)

(5α,5'β)-3β,11β-dihydroxy-
2'-methyl-5'H-pregnano-
[17,16-d]oxazol-20-one
(cf. Deflazacort)

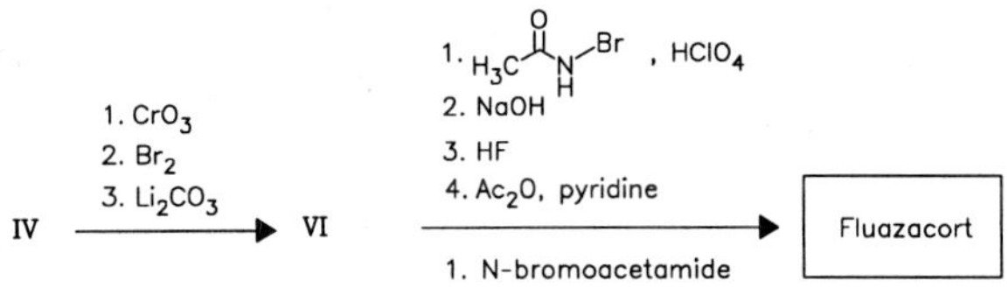

1. Ac₂O, pyridine
2. CH₃SO₂Cl
3. NaO–CO–CH₃

2. mesyl chloride

III

1. KOH
2. I₂, [reagent]
3. H₃C—COOH ·N(C₂H₅)₃

2. iodine, azoisobutyronitrile
3. triethylammonium acetate

IV

IV

1. CrO₃
2. Br₂
3. Li₂CO₃

VI

1. H₃C–CO–NH–Br , HClO₄
2. NaOH
3. HF
4. Ac₂O, pyridine

1. N-bromoacetamide

Fluazacort

Reference(s):
a DOS 1 618 613 (Lepetit; appl. 7.1.1967; GB-prior. 11.1.1966, 11.7.1966, 29.9.1966).
    US 3 461 119 (Lepetit; 12.8.1969; appl. 1967; GB-prior. 1.11.1966).

synthesis of starting compound:
DE 1 568 971 (Gruppo Lepetit; appl. 23.12.1966; GB-prior. 11.1.1966, 11.7.1966, 29.9.1966).
US 3 624 077 (Gruppo Lepetit; 30.11.1971; GB-prior. 11.1.1966).
Nathansohn, G. et al.: J. Med. Chem. (JMCMAR) **10**, 799 (1967).
Nathansohn, G. et al.: Gazz. Chim. Ital. (GCITA9) **95**, 1338 (1965).
review:
Nathanson, G. et al.: Steroids (STEDAM) **13**, 365 (1969).
alternative synthesis of VI:
DOS 1 568 971 (Lepetit; appl. 23.12.1966; GB-prior. 11.1.1966, 11.7.1966, 29.9.1966).
DOS 1 568 972 (Lepetit; appl. 23.12.1966; GB-prior. 11.1.1966, 11.7.1966, 29.9.1966).
b   Nathansohn, G. et al.: J. Med. Chem. (JMCMAR) **10**, 799 (1967).
Nathansohn, G. et al.: Steroids (STEDAM) **13**, 383 (1969).

Formulation(s):   cream 0.025 %

Trade Name(s):
I: Azacortid crema (Lepetit)

Flubendazole

ATC: P02CA05
Use: anthelmintic

RN: 31430-15-6   MF: $C_{16}H_{12}FN_3O_3$ MW: 313.29   EINECS: 250-624-4
LD$_{50}$:   >2560 mg/kg (M, p.o.);
          2560 mg/kg (R, p.o.)
CN: [5-(4-fluorobenzoyl)-1*H*-benzimidazol-2-yl]carbamic acid methyl ester

fluorobenzene   3-chloro-4-nitro-
                 benzoyl chloride

3-chloro-4-nitro-
4'-fluorobenzophenone

3,4-diamino-4'-   S-methyl-   methyl
fluorobenzophenone (I) isothiourea   chloroformate

Flubendazole

Reference(s):
DOS 2 029 637 (Janssen; appl. 16.6.1970; USA-prior. 20.6.1969).
US 3 657 267 (Janssen; 18.4.1972; appl. 20.6.1969).
Raymaekers, A.H.M. et al.: Arzneim.-Forsch. (ARZNAD) **28**, 586 (1978).

Formulation(s):   susp. 100 mg/ml; tabl. 100 mg

Trade Name(s):
D: Flubenol (Janssen); wfm   F: Fluvermal (Janssen-Cilag)

Fluclorolone acetonide
(Flucloronide)

ATC: H02AB
Use: topical glucocorticoid

RN: 3693-39-8   MF: $C_{24}H_{29}Cl_2FO_5$ MW: 487.40   EINECS: 223-010-9
CN: (6α,11β,16α)-9,11-dichloro-6-fluoro-21-hydroxy-16,17-[(1-methylethylidene)bis(oxy)]pregna-1,4-diene-3,20-dione

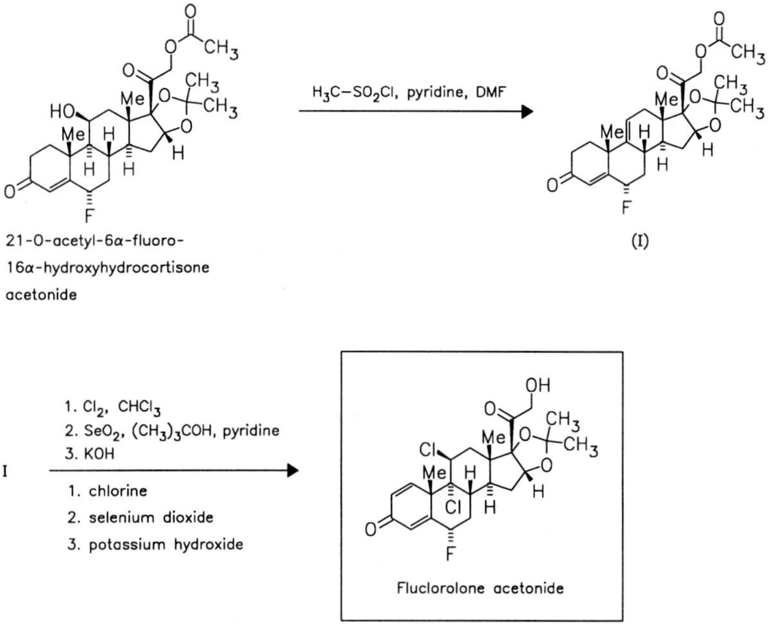

21-0-acetyl-6α-fluoro-
16α-hydroxyhydrocortisone
acetonide

(I)

Fluclorolone acetonide

Reference(s):
US 3 201 391 (Syntex; 17.8.1965; MEX-prior. 18.2.1959, 20.10.1959).

starting material:
Mills, J.S. et al.: J. Am. Chem. Soc. (JACSAT) **81**, 1264 (1959).

Formulation(s):   cream 0.025 %, 0.25 %

Trade Name(s):
F: Topilar (Syntex-Daltan);   GB: Topilar (Syntex); wfm
     wfm

Flucloxacillin
(Floxacillin)

ATC: J01CA
Use: antibiotic

RN: 5250-39-5   MF: $C_{19}H_{17}ClFN_3O_5S$ MW: 453.88   EINECS: 226-051-0
CN: [2S-(2α,5α,6β)]-6-[[[3-(2-chloro-6-fluorophenyl)-5-methyl-4-isoxazolyl]carbonyl]amino]-3,3-dimethyl-7-oxo-4-thia-1-azabicyclo[3.2.0]heptane-2-carboxylic acid

monosodium salt
RN: 1847-24-1   MF: $C_{19}H_{16}ClFN_3NaO_5S$ MW: 475.86   EINECS: 217-428-0
LD$_{50}$:   1360 mg/kg (M, i.v.); 7600 mg/kg (M, p.o.);
      680 mg/kg (R, i.v.); 11 g/kg (R, p.o.);
      670 mg/kg (dog, i.v.); >10 g/kg (dog, p.o.)

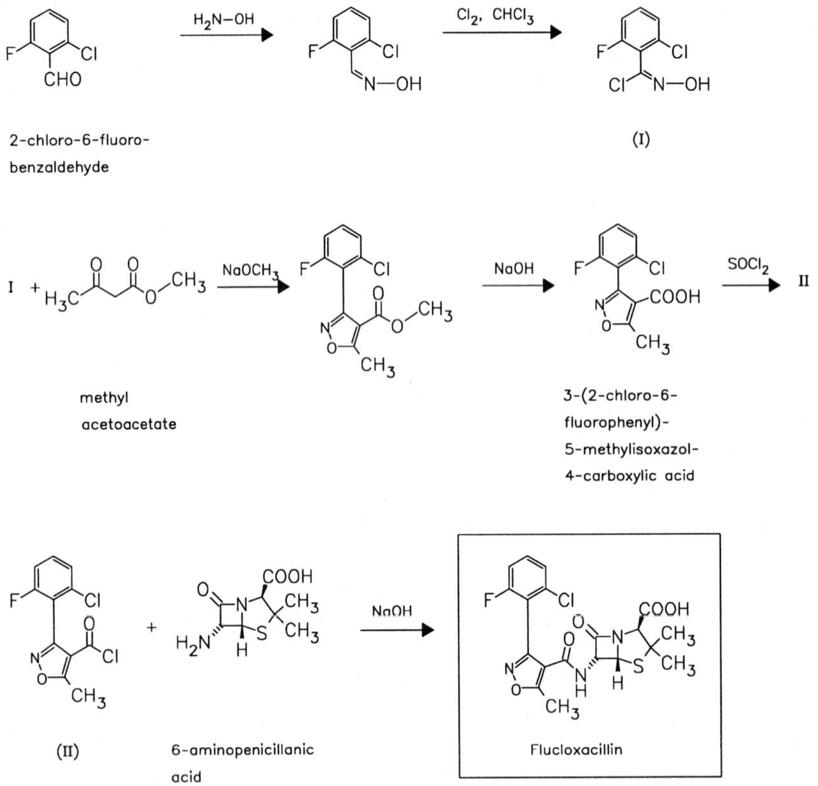

H_2N-OH
Cl_2, $CHCl_3$
2-chloro-6-fluoro-benzaldehyde
(I)
$NaOCH_3$
NaOH
$SOCl_2$
II
methyl acetoacetate
3-(2-chloro-6-fluorophenyl)-5-methylisoxazol-4-carboxylic acid
NaOH
(II)
6-aminopenicillanic acid
Flucloxacillin

Reference(s):
GB 987 299 (Beecham; appl. 17.10.1962; addition to GB 905 778 from 14.3.1961).
US 3 239 507 (Beecham; 8.3.1966; GB-prior. 17.10.1962).

Formulation(s): cps. 272 mg, 544 mg; vial 272 mg, 544 mg, 1088 mg, 2176 mg (as sodium salt hydrate)

Trade Name(s):

D:	Flanomox (Wolff)-comb. Fluxapril (Lederle)-comb. Staphylex (SmithKline Beecham)	GB:	Floxapen (SmithKline Beecham) Magnapen (SmithKline Beecham)-comb. Stafoxil (Yamanouchi)	I: J:	Infectrin (Pierrel)-comb. with ampicillin Culpen (Fujisawa) Floxapen (Beecham)

Fluconazole

(UK-49858)

ATC: J02AC01; J02AX
Use: antifungal (treatment of vaginal, oropharyngeal and atrophic oral candidiasis)

RN: 86386-73-4 MF: $C_{13}H_{12}F_2N_6O$ MW: 306.28
LD_{50}: >200 mg/kg (M, i.v.); 1408 mg/kg (M, p.o.);
>200 mg/kg (R, i.v.); 1271 mg/kg (R, p.o.);
>100 mg/kg (dog, i.v.); >300 mg/kg (dog, p.o.)
CN: α-(2,4-difluorophenyl)-α-(1*H*-1,2,4-triazol-1-ylmethyl)-1*H*-1,2,4-triazole-1-ethanol

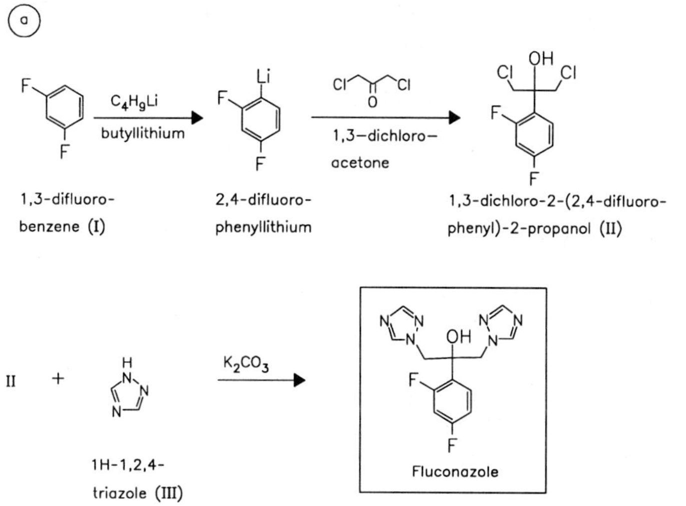

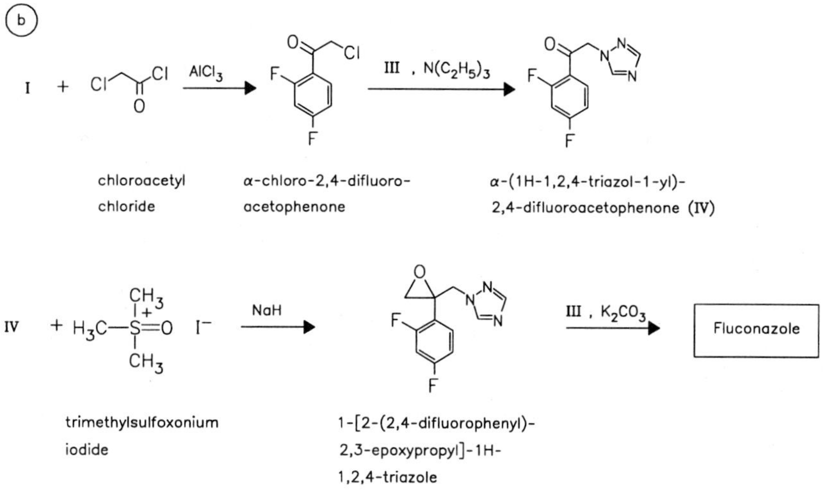

a
C_4H_9Li
butyllithium
1,3-dichloro-
acetone
1,3-difluoro-
benzene (I)
2,4-difluoro-
phenyllithium
1,3-dichloro-2-(2,4-difluoro-
phenyl)-2-propanol (II)
II
K_2CO_3
1H-1,2,4-
triazole (III)
Fluconazole

b
I
$AlCl_3$
III , $N(C_2H_5)_3$
chloroacetyl
chloride
α-chloro-2,4-difluoro-
acetophenone
α-(1H-1,2,4-triazol-1-yl)-
2,4-difluoroacetophenone (IV)
IV
NaH
III , K_2CO_3
Fluconazole
trimethylsulfoxonium
iodide
1-[2-(2,4-difluorophenyl)-
2,3-epoxypropyl]-1H-
1,2,4-triazole

Reference(s):

GB 2 099 818 (Pfizer; appl. 22.4.1982; prior. 6.6.1981, 4.3.1982).
EP 96 569 (Pfizer; appl. 6.6.1983; GB-prior. 9.6.1982, 30.7.1982).

tablet formulation:

EP 178 682 (Schering Corp.; appl. 23.4.1986; USA-prior. 19.10.1984).

alternative synthesis:

ES 549 684 (Lazlo Int.; appl. 6.12.1985).
ES 5 490 202 (Inke S. A.; appl. 19.11.1985).
US 5 710 280 (Dev. Center Biotech. Taiwan; 20.1.1998; appl. 9.7.1996).
WO 9 703 971 (Apotex; appl. 17.7.1996; NZ-prior. 17.7.1995).

Formulation(s): cps. 50 mg, 100 mg, 150 mg, 200 mg; susp. 50 mg/5 ml; syrup 50 mg/10ml; tabl. 50 mg, 100 mg, 150 mg, 200 mg; vial 100 mg, 200 mg, 400 mg

Trade Name(s):

D:	Diflucan (Pfizer)	I:	Biozolene (Bioindustria; 1989)		Elazor (Sigma-Tau)
F:	Triflucan (Pfizer; 1989)		Diflucan (Roerig)	J:	Diflucan (Pfizer Taito)
GB:	Diflucan (Pfizer; 1988)			USA:	Diflucan (Pfizer; 1990)

Flucytosine

ATC: D01AE21; J02AX01
Use: fungicide

RN: 2022-85-7   MF: $C_4H_4FN_3O$ MW: 129.09   EINECS: 217-968-7
LD_{50}:   500 mg/kg (M, i.v.); >15 g/kg (M, p.o.);
>600 mg/kg (R, i.v.); >15 g/kg (R, p.o.)
CN: 4-amino-5-fluoropyrimidin-2(1H)-one

5-fluorouracil
(q. v.)

POCl₃, dimethylaniline

2,4-dichloro-
5-fluoro-
pyrimidine

NH₃

4-amino-2-
chloro-5-fluoro-
pyrimidine

H₂O, HCl

Flucytosine

5-fluoro-2-
methylthiouracil
(cf. fluorouracil
synthesis)

PCl₅

4-chloro-5-fluoro-
2-methylthio-
pyrimidine

liq. NH₃

conc. HBr

Flucytosine

Reference(s):
a   Duschinsky, R. et al.: J. Am. Chem. Soc. (JACSAT) **79**, 4559 (1957).
Undheim, K.; Gacek, M.: Acta Chem. Scand. (ACHSE7) **23**, (1), 294 (1969).
US 3 040 026 (Roche; 19.6.1962; appl. 3.6.1959).
US 3 185 690 (Roche; 25.5.1965; prior. 3.6.1959, 14.9.1961).
b   US 2 945 038 (Roche; 12.7.1960; prior. 26.9.1956).
US 2 802 005 (Roche; 6.8.1957; prior. 26.9.1956).

medical use:
US 3 368 938 (Roche; 13.2.1968; prior. 2.3.1962).

Formulation(s):   cps. 250 mg, 500 mg; tabl. 500 mg; vial 2.5 g/250 ml

Trade Name(s):
D: Ancotil Roche (ICN)   GB: Alcobon (Roche); wfm   J: Ancotil (Roche)
F: Ancotil (Roche)   I: Ancotil (Roche)   USA: Ancobon (Roche)

Fludarabine phosphate

(2-fluoro-ara-AMP)

ATC: L01BB05
Use: antineoplastic, antimetabolite,
treatment of chronic lymphocytic
leucemia

RN: 75607-67-9   MF: $C_{10}H_{13}FN_5O_7P$ MW: 365.21
LD_{50}:   375 mg/kg (M, i.p.); 1236 mg/kg (M, i.v.)
CN: 2-fluoro-9-(5-O-phosphono-β-D-arabinofuranosyl)-9H-purin-6-amine

fludarabine
RN: 21679-14-1   MF: $C_{10}H_{12}FN_5O_4$ MW: 285.24   EINECS: 244-525-5

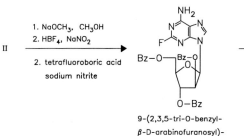

2,4,5,6-tetraamino-
pyrimidine

formamide

2-amino-
adenine (I)

1. H₃C–CO–O–CO–CH₃ , pyridine

2. Bz–O ... O–Bz ... NO₂ , HCl

I ──────────────→

2. 2,3,5-tri-O-benzyl-1-O-
p-nitrobenzoyl-β-D-
arabinofuranose

Bz: benzyl

2,6-diacetamido-9-
(2,3,5-tri-O-benzyl-
β-D-arabinofuranosyl)-
purine (II)

1. NaOCH₃, CH₃OH
2. HBF₄, NaNO₂

II ──────────────→

2. tetrafluoroboric acid
sodium nitrite

9-(2,3,5-tri-O-benzyl-
β-D-arabinofuranosyl)-
2-fluoroadenine

1. BCl₃, CH₂Cl₂
2. POCl₃, PO(OC₂H₅)₃

──────────────→

1. boron trichloride
2. phosphorus oxychloride,
triethyl phosphate

Fludarabine
phosphate

Reference(s):
US 4 357 324 (Department of Health of USA; 2.11.1982; appl. 24.2.1981).

synthesis of 9-β-D-arabinofuranosyl-2-fluoroadenine:
US 4 210 745 (Department of Health of USA; 1.7.1980; appl. 20.11.1978; prior. 10.3.1978, 4.1.1978).
Montgomery, J.A. et al.: J. Heterocycl. Chem. (JHTCAD) 16, 157 (1979).
Montgomery, J.A.; Hewson, K.: J. Med. Chem. (JMCMAR) 12, 498 (1961).

synthesis of 2-aminoadenine:
Robins, R.K. et al.: J. Am. Chem. Soc. (JACSAT) 75, 263 (1953).

Formulation(s):   vial (lyo.) 5 mg, 50 mg

Trade Name(s):
D: Fludara (meda; Schering)   GB: Fludara (Schering)
F: Fludara (Schering)   USA: Fludara (Berlex; 1991)

Fludrocortisone

ATC: H02AA02; S01CA06; S02CA07;
S03CA05
Use: glucocorticoid

RN: 127-31-1   MF: $C_{21}H_{29}FO_5$ MW: 380.46   EINECS: 204-833-2
LD$_{50}$: 170 mg/kg (M, i.p.)
CN: (11β)-9-fluoro-11,17,21-trihydroxypregn-4-ene-3,20-dione

acetate
RN: 514-36-3   MF: $C_{23}H_{31}FO_6$ MW: 422.49   EINECS: 208-180-4
LD$_{50}$: >1 g/kg (R, p.o.)

hydrocortisone 21-acetate

POCl$_3$, pyridine

1. H_3C—N—Br
2. CH$_3$COONa   → I
1. N-bromo-
acetamide

21-acetoxy-3,20-dioxo-9β,11β-
epoxy-17-hydroxy-4-pregnene (I)

H$_2$F$_2$,
anhydrous CHCl$_3$^{a,b} or CHCl$_3$/THF^c or CHCl$_3$/H$_2$O/HClO$_4$^{d} → II

fludrocortisone acetate (II)

NaOCH$_3$^{a} or
CH$_3$COOK/CH$_3$OH/N$_2$^{e}

Fludrocortisone

Reference(s):
a   Fried, J.; Sabo, E.F.: J. Am. Chem. Soc. (JACSAT) **76**, 1455 (1954).
b   GB 792 224 (Olin Mathieson; appl. 1954; USA-prior. 1954).
c   DE 1 035 133 (Merck & Co.; appl. 1956; USA-prior. 1955).
    Hirschmann, R.F. et al.: J. Am. Chem. Soc. (JACSAT) **78**, 4956 (1956).
d   US 2 894 007 (Merck & Co.; 7.7.1959).
e   DE 1 028 572 (Schering AG; appl. 21.1.1957).

synthesis of hydrocortisone acetate:
Fried, J.; Sabo, E.F.: J. Am. Chem. Soc. (JACSAT) **75**, 2273 (1953).
US 2 771 475 (Upjohn; 1956, appl. 1953).
GB 792 224 (Olin Mathieson; appl. 1954; USA-prior. 1954).

alternative syntheses:
US 2 771 475 (Upjohn; 1956; appl. 1953).
US 2 799 688 (Upjohn; 1957; appl. 1954).
US 2 852 511 (Olin Mathieson; 1958; prior. 1953).
US 4 041 055 (Upjohn; 9.8.1977; appl. 17.11.1975).

Formulation(s):   ear drops 8 mg/8 ml in comb. with polymyxin B; ointment 0.001 %; tabl. 0.1 mg (as acetate)

Trade Name(s):

D:	Astonin H (Merck)	F:	Panotile (Zambon)-comb.	USA:	Florinef (Apothecon)
	Fludrocortison (Bristol-Myers Squibb)	GB:	Florinef (Bristol-Myers Squibb; as acetate)		
	Panotile (Zambon)-comb.	J:	Florinef (Bristol Squibb)		

Fludroxycortide
(Flurandrenolide)

ATC: D07AC07
Use: glucocorticoid, anti-inflammatory

RN: 1524-88-5   MF: $C_{24}H_{33}FO_6$ MW: 436.52   EINECS: 216-196-8
CN: (6α,11β,16α)-6-fluoro-11,21-dihydroxy-16,17-[(1-methylethylidene)bis(oxy)]pregn-4-ene-3,20-dione

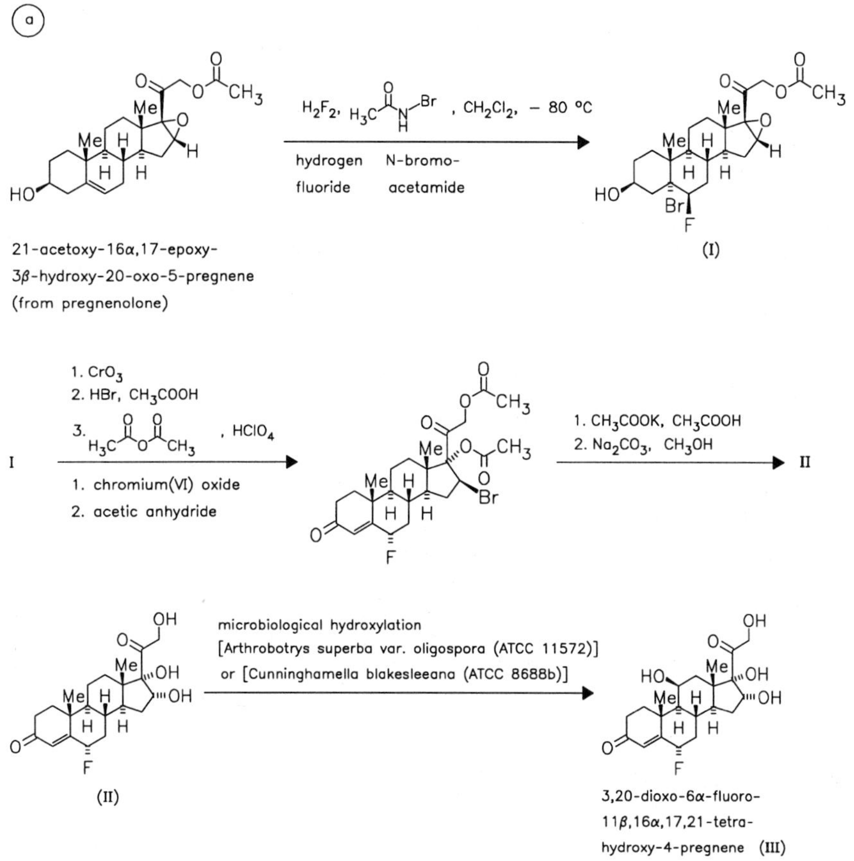

21-acetoxy-16α,17-epoxy-
3β-hydroxy-20-oxo-5-pregnene
(from pregnenolone)

(I)

microbiological hydroxylation
[Arthrobotrys superba var. oligospora (ATCC 11572)]
or [Cunninghamella blakesleeana (ATCC 8688b)]

(II)

3,20-dioxo-6α-fluoro-
11β,16α,17,21-tetra-
hydroxy-4-pregnene (III)

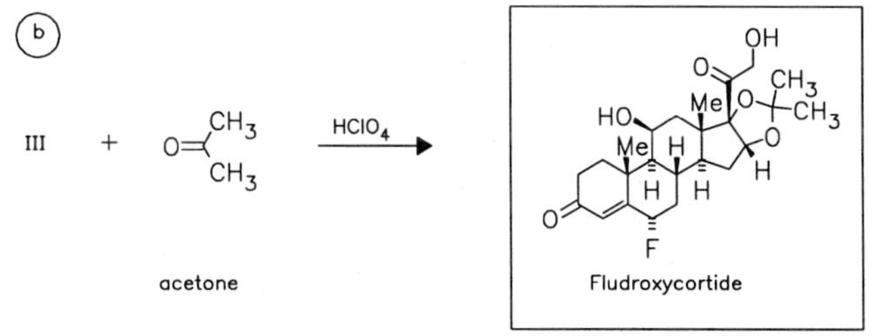

acetone

Fludroxycortide

Reference(s):
a US 3 014 938 (Syntex; 26.12.1961; appl. 23.8.1960; MEX-prior. 7.9.1959).
US 3 119 749 (Syntex; 28.1.1964; appl. 17.11.1961; MEX prior. 7.6.1961).
US 3 124 571 (Syntex; 10.3.1964; MEX-prior. 26.1.1960).
starting material:
Julian, P.L.: J. Am. Chem. Soc. (JACSAT) **72**, 5145 (1950).
US 2 678 932 (Sterling Drug; 1954; prior. 1951).
b US 3 126 375 (Syntex; 24.3.1964; appl. 11.6.1959; MEX-prior. 13.6.1958).
DE 1 131 213 (Syntex; appl. 6.6.1959; MEX-prior. 13.6.1958).

alternative syntheses:
Mills, J.S. et al.: J. Am. Chem. Soc. (JACSAT) **82**, 3399 (1960); **81**, 1264 (1959).
US 3 203 869 (Syntex; 31.8.1965; MEX-prior. 11.10.1962).

Formulation(s):   lotion 0.05 % (15 ml, 60 ml); tape 4µg/cm^2

Trade Name(s):

D:	Sermaka (Lilly)	I:	Drenison (Lilly); wfm	J:	Drenison Q (Lilly-
GB:	Drenison (Lilly); wfm		Drenison Neomicina		Dainippon)
	Haclan (Dista); wfm		(Lilly)-comb.; wfm	USA:	Cordran (Oclassen)

Flufenamic acid

(Acide flufenamique)

ATC: M01AG03
Use: anti-inflammatory, antirheumatic

RN: 530-78-9   MF: $C_{14}H_{10}F_3NO_2$ MW: 281.23   EINECS: 208-494-1
LD$_{50}$:   158 mg/kg (M, i.v.); 490 mg/kg (M, p.o.);
98 mg/kg (R, i.v.); 249 mg/kg (R, p.o.)
CN: 2-[[3-(trifluoromethyl)phenyl]amino]benzoic acid

aluminum salt
RN: 16449-54-0   MF: $C_{42}H_{27}AlF_9N_3O_6$ MW: 867.66   EINECS: 240-498-9
LD$_{50}$:   1460 mg/kg (M, p.o.);
550 mg/kg (R, p.o.)

2-chloro-
benzoic
acid

3-trifluoro-
methylaniline

Flufenamic acid

Reference(s):
FR 1 341 M (Parke Davis; appl. 11.8.1961).
Moffett, R.B.; Aspergen, B.D.: J. Am. Chem. Soc. (JACSAT) **82**, 1605 (1960).

salts with amines:
DOS 2 758 787 (T. Eckert; appl. 29.12.1977).

Formulation(s):   ointment 3 g/100 g (3 %); sol. 25 mg/g

Trade Name(s):

D:	Algesalona (Solvay)-comb.		Rheuma Lindofluid		Meralen (Merrell); wfm
	Dignodolin (Sankyo)		(Lindopharm)	I:	Mobilisin (Luitpold)-comb.
	Mobilisin (Sankyo)-comb.	F:	Arlef (Parke Davis); wfm	J:	Achless (Tatsumi)
		GB:	Arlef (Parke Davis); wfm		Arlef (Parke Davis-Sankyo)

Felunamin (Hokuriku)   Nichisedan (Nissin)   Ristogen (Kowa Yakuhin)
Flufacid (Wakamoto)   Paraflu (Dainippon)   Romazal (Tobishi)
Lanceat (Maruko)   Reumajust A (Horita)   Saal-F (Towa)

Flugestone acetate
(Flurogestone acetate)

ATC: G03
Use: progestone

RN: 2529-45-5   MF: $C_{23}H_{31}FO_5$ MW: 406.49   EINECS: 219-776-9
CN: (11β)-17-(acetyloxy)-9-fluoro-11-hydroxypregn-4-ene-3,20-dione

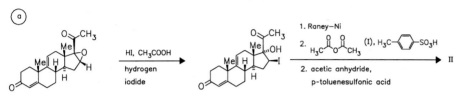

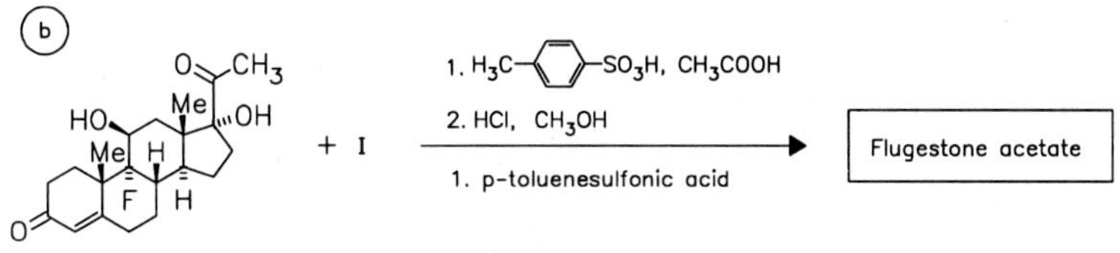

3,20-dioxo-16α,17-
epoxy-4,9(11)-pregnadiene

(II)

Flugestone acetate

11β,17-dihydroxy-3,20-
dioxo-9α-fluoro-4-pregnene

Reference(s):
a Bergstrom, C.G. et al.: J. Am. Chem. Soc. (JACSAT) **81**, 4432 (1959).
b US 2 892 851 (Searle; 30.6.1959; prior. 19.5.1958).
    US 2 963 498 (Searle; 6.12.1960; prior. 11.5.1959).

Trade Name(s):
USA: Cronolone (Searle); wfm

Flumazenil
(Ro-15-1788)

ATC: V03AB25
Use: benzodiazepine antagonist, treatment
of benzodiazepine intoxication

RN: 78755-81-4   MF: $C_{15}H_{14}FN_3O_3$ MW: 303.29
LD$_{50}$:   4000 mg/kg (M, i.p.); 143 mg/kg (M, i.v.); 1300 mg/kg (M, p.o.);
85 mg/kg (R, i.v.); 4200 mg/kg (R, p.o.)
CN: 8-fluoro-5,6-dihydro-5-methyl-6-oxo-4H-imidazo[1,5-a][1,4]benzodiazepine-3-carboxylic acid ethyl
ester

2-nitro-5-
fluorotoluene

potassium
per-
manganate

2-nitro-5-fluoro-
benzoic acid

5-fluoroisatoic
anhydride   (I)

N-methyl-
glycine

7-fluoro-4-methyl-
3,4-dihydro-2H-1,4-
benzodiazepin-2,5(1H)-
dione

ethyl isocyanato-
acetate

potassium tert-butylate

Flumazenil

Reference(s):
EP 27 214 (Hoffmann-La Roche; appl. 10.2.1980; CH-prior. 4.10.1979, 30.11.1979, 25.7.1980).
US 4 316 839 (Hoffmann-La Roche; 23.2.1982; appl. 3.10.1980; CH-prior. 4.10.1979, 30.11.1979, 25.7.1980).
US 4 346 030 (Hoffmann-La Roche; 24.8.1982; appl. 16.11.1981; CH-prior. 4.10.1979, 30.11.1979, 25.7.1980).
Hunkeler, W. et al.: Nature (London) (NATUAS) **290**, 514 (1981).

Formulation(s):   amp. 0.5 mg/5 ml, 1 mg/10 ml

Trade Name(s):
D: Anexate (Roche; 1989)   GB: Anexate (Roche)   J: Anexate (Yamanouchi)
F: Anexate (Roche)   I: Anexate (Roche; 1989)   USA: Romazicon (Roche)

Flumedroxone acetate

ATC: N02CB01
Use: antimigraine agent, progestogen

RN: 987-18-8   MF: $C_{24}H_{31}F_3O_4$ MW: 440.50   EINECS: 213-577-0
CN: (6α)-17-(acetyloxy)-6-(trifluoromethyl)pregn-4-ene-3,20-dione

17-acetoxyprogesterone

triethyl
orthoformate

p-toluenesulfonic
acid

(I)

I   +   F₃C—I   trifluoromethyl iodide

1. pyridine, hν
2. HCl

Flumedroxone acetate

Reference(s):

GB 905 694 (Lovens Kemiske Fa., valid from 14.3.1961; prior. 18.3.1960, 8.6.1960).

Godfredsen, W.O.; Vangedal, S.: Acta Chem. Scand. (ACHSE7) **15**, 1786 (1961).

Formulation(s):   drg. 1 mg in comb.

Trade Name(s):

D: Praemenstron (Nordmark)-   F: Precyclan (Leo)-comb.
      comb.; wfm

Flumequine

ATC: G04AB06
Use: chemotherapeutic, antibacterial

RN: 42835-25-6   MF: $C_{14}H_{12}FNO_3$ MW: 261.25   EINECS: 255-962-6
CN: 9-fluoro-6,7-dihydro-5-methyl-1-oxo-1H,5H-benzo[ij]quinolizine-2-carboxylic acid

6-fluoro-2-methyl-
1,2,3,4-tetrahydro-
quinoline

diethyl
ethoxymethylene-
malonate

1. polyphosphoric acid
2. NaOH, H₂O

Flumequine

Reference(s):

DOS 2 264 163 (Riker; appl. 29.12.1972; USA-prior. 30.12.1971).

US 3 896 131 (Riker; 22.7.1975; prior. 2.11.1972, 30.12.1971).

Formulation(s):   tabl. 400 mg

Trade Name(s):

F: Apurone (3M Santé)   I: Flumural (SPA)

Flumetasone
(Flumethasone)

ATC: D07AB03; D07BB01; D07CB05;
        D07XB01; S02CA02
Use: glucocorticoid, anti-inflammatory

RN: 2135-17-3   MF: $C_{22}H_{28}F_2O_5$ MW: 410.46   EINECS: 218-370-9
CN: (6α,11β,16α)-6,9-difluoro-11,17,21-trihydroxy-16-methylpregna-1,4-diene-3,20-dione

pivalate
RN: 2002-29-1   MF: $C_{27}H_{36}F_2O_6$ MW: 494.58   EINECS: 217-901-1
LD₅₀: >5 g/kg (M, p.o.);
       >2 g/kg (R, p.o.)

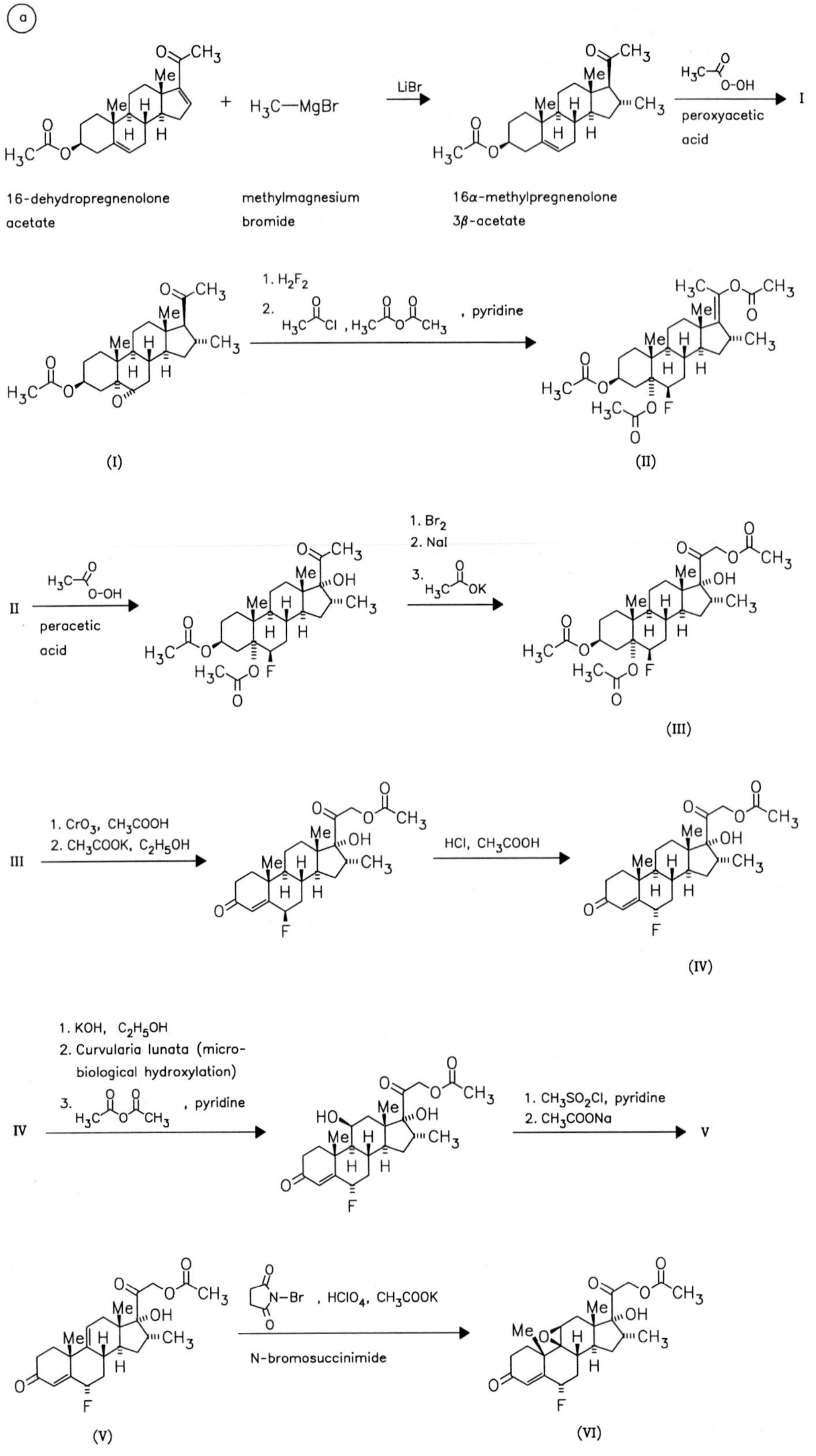

ⓐ

16-dehydropregnenolone acetate

methylmagnesium bromide

16α-methylpregnenolone 3β-acetate

LiBr

peroxyacetic acid → I

(I)

1. H₂F₂

2. H₃C–Cl , H₃C–O–CH₃ , pyridine

(II)

II

peracetic acid

1. Br₂
2. NaI
3. H₃C–OK

(III)

III

1. CrO₃, CH₃COOH
2. CH₃COOK, C₂H₅OH

HCl, CH₃COOH

(IV)

IV

1. KOH, C₂H₅OH
2. Curvularia lunata (microbiological hydroxylation)
3. H₃C–O–CH₃ , pyridine

1. CH₃SO₂Cl, pyridine
2. CH₃COONa → V

(V)

N-bromosuccinimide , HClO₄, CH₃COOK

(VI)

Flumetasone

21-(2,2-dimethylpropionyloxy)-
3,20-dioxo-6α-fluoro-17-hydroxy-
16α-methyl-1,4,9(11)-pregnatriene
(from 3,11,20-trioxo-4,16-pregnadiene)

Flumetasone pivalate

Reference(s):
a US 2 671 752 (Syntex; 1954; appl. 1951).
   Djerassi, C. et al.: J. Am. Chem. Soc. (JACSAT) **81**, 3156 (1959); **82**, 2318 (1961).
b FR 1 374 591 (Ciba; appl. 8.10.1963; CH-prior. 12.10.1962).

synthesis of 21-(2,2-dimethylpropionyloxy)-3,20-dioxo-6α-fluoro-17-hydroxy-16α-methyl-1,4,9(11)-
pregnatriene:
US 3 557 158 (Upjohn; 19.1.1971; prior. 18.3.1959).

alternative syntheses:
US 3 557 158 (Upjohn; 19.1.1971; appl. 22.1.1962; prior. 18.3.1959).
Schneider, P. et al.: J. Am. Chem. Soc. (JACSAT) **81**, 3167 (1959).
GB 902 292 (Upjohn; appl. 27.7.1959; USA-prior. 14.8.1958).
US 4 041 055 (Upjohn; 9.8.1977; appl. 17.11.1975).

Formulation(s):   sol. 0.02 g/100 g (0.02 %); cream 0.02 g/100 g (0.02 %); lotion 0.02 g/100 g (0.02 %);
          ointment 0.02 g/100 g (0.02 %)

Trade Name(s):
D:	Cerson (LAW)	Psocortène (Ciba-Geigy)-	Neolog (Zyma)-comb.
	Locacorten (Novartis	comb.; wfm	several combination
	Pharma)	GB: Locorten Vioform	preparations
F:	Locacortène (Ciba-Geigy)-	(Novartis)-comb.	J: Locorten (Ciba-Geigy)
	comb.; wfm	I: Locorten (Zyma)	Testohgen (Teisan)
	Locasalène (Ciba-Geigy)-	Locorten (Zyma)-comb.	USA: Locorten (Ciba); wfm
	comb.; wfm	Losalen (Zyma)	

Flunarizine

ATC: N07CA03
Use: cerebral and peripheral vasodilator,
     antivertigo

RN: 52468-60-7 MF: $C_{26}H_{26}F_2N_2$ MW: 404.50 EINECS: 257-937-5
LD_{50}: 960 mg/kg (M, p.o.)
CN: (*E*)-1-[bis(4-fluorophenyl)methyl]-4-(3-phenyl-2-propenyl)piperazine

dihydrochloride

RN: 30484-77-6   MF: $C_{26}H_{26}F_2N_2 \cdot 2HCl$ MW: 477.43   EINECS: 250-216-6

LD_{50}:   27 mg/kg (M, i.v.); 285 mg/kg (M, p.o.);
     35 mg/kg (R, i.v.); 503 mg/kg (R, p.o.);
     >2 g/kg (dog, p.o.)

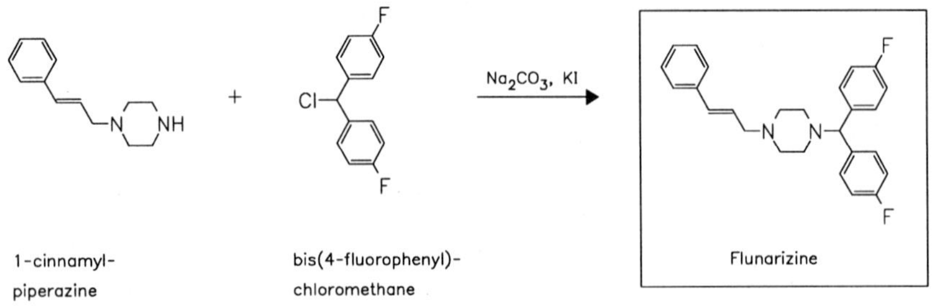

1-cinnamyl-   bis(4-fluorophenyl)-   Flunarizine
piperazine   chloromethane

Reference(s):

DAS 1 929 330 (Janssen; appl. 10.6.1969; USA-prior. 2.7.1968).
US 3 773 939 (Janssen; 20.11.1973; prior. 2.7.1968, 24.11.1971).

inhibiting effect to complementary activity:

DOS 2 254 893 (Janssen; appl. 9.11.1972; GB-prior. 9.11.1971; USA-prior. 17.10.1972).

Formulation(s):   cps. 5.9 mg, 11.8 mg (as dihydrochloride)

Trade Name(s):

D: Flunarizin (ct-Arzneimittel)   I: Flugeral (Italfarmaco;   Issium (Lifepharma)
      Flunarizin-ratiopharm   1981)   Sibelium (Janssen)
      (ratiopharm)   Flugeral mite (Italfarmaco)   Vasculene (Leben's)
      Sibelium (Janssen-Cilag;   Flunagen (Gentili)   J: Flunarl (Kyowa Hakko;
      1977)   Fluxarten (SmithKline   1984)
F: Sibélium (Janssen-Cilag;   Beecham)   USA: Sibelium (Janssen); wfm
      1986)   Gradient (Polifarma)

Flunisolide

ATC: R01AD04; R03BA03
Use: glucocorticoid, antiasthmatic

RN: 3385-03-3   MF: $C_{24}H_{31}FO_6$ MW: 434.50   EINECS: 222-193-2

LD_{50}:   >76 µg/kg (M, i.v.); >500 µg/kg (M, p.o.);
     >51 mg/kg (R, i.v.); >500 µg/kg (R, p.o.)

CN: (6α,11β,16α)-6-fluoro-11,21-dihydroxy-16,17-[(1-methylethylidene)bis(oxy)]pregna-1,4-diene-3,20-dione

hydrate (2:1)

RN: 77326-96-6   MF: $C_{24}H_{31}FO_6 \cdot 1/2H_2O$ MW: 887.02

a

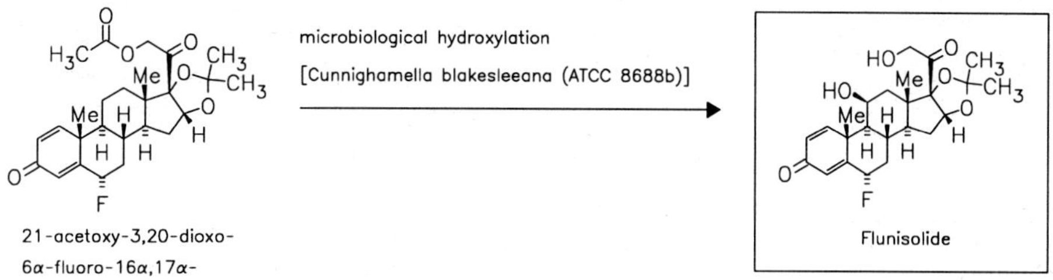

microbiological hydroxylation

[Cunninghamella blakesleeana (ATCC 8688b)]

Flunisolide

21-acetoxy-3,20-dioxo-
6α-fluoro-16α,17α-
isopropylidenedioxy-
1,4-pregnadiene

b

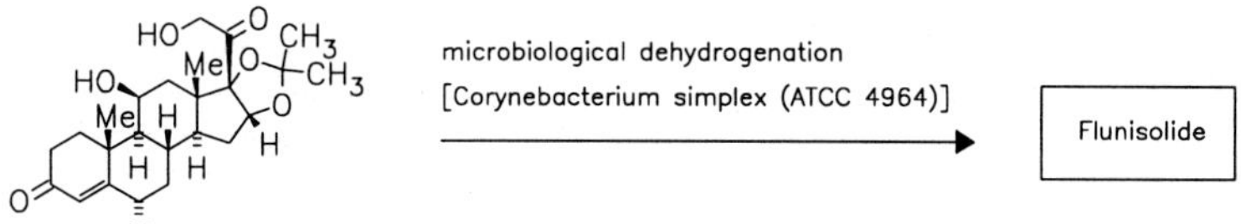

microbiological dehydrogenation

[Corynebacterium simplex (ATCC 4964)]

Flunisolide

fludroxycortide
(q. v.)

c

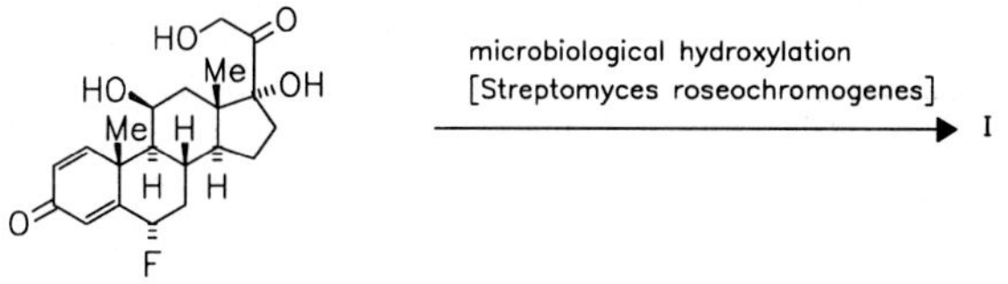

microbiological hydroxylation

[Streptomyces roseochromogenes]

I

6α-fluoroprednisolone

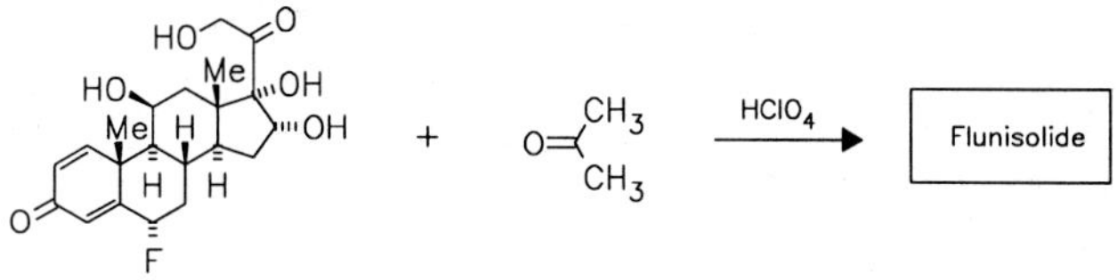

(I)   +   acetone $\xrightarrow{HClO_4}$ Flunisolide

Reference(s):
a   US 3 124 571 (Syntex; 10.3.1964; MEX-prior. 26.1.1960).
b   US 3 126 375 (Syntex; 24.3.1964; MEX-prior. 13.6.1958).
c   GB 933 867 (American Cyanamid; appl. 5.12.1959; USA-prior. 8.12.1958).

Formulation(s):   nasal spray 25 mg/metered dose inhaler with 0.25 mg/spray

Trade Name(s):
D: Inhacort (Boehringer Ing.)   F: Bronilide (Cassenne)   GB: Syntaris (Roche)
      Syntaris (Roche; Syntex)   Nasalide (Cassenne)

I: Gibiflu (Metapharma)   Syntaris (Recordati)   Nasalide (Dura)
        Lunibron-a (Valeas)   J: Synaclyn (Otsuka)   Nasarel (Dura)
        Lunis (Valeas)   USA: Aerobid (Forest)

Flunitrazepam

ATC: N05CD03
Use: anticonvulsant, hypnotic, muscle
        relaxant

RN: 1622-62-4   MF: $C_{16}H_{12}FN_3O_3$ MW: 313.29   EINECS: 216-597-8
LD$_{50}$: 1200 mg/kg (M, p.o.);
        415 mg/kg (R, p.o.)
CN: 5-(2-fluorophenyl)-1,3-dihydro-1-methyl-7-nitro-2H-1,4-benzodiazepin-2-one

4-chloroaniline   2-fluorobenzoyl   2-amino-5-chloro-   2-amino-2'-fluoro-
                   chloride   2'-fluorobenzophenone   benzophenone (I)

                   bromoacetyl   5-(2-fluorophenyl)-
                   bromide   1,3-dihydro-2H-1,4-
                                                                   benzodiazepin-2-one (II)

5-(2-fluorophenyl)-7-   methyl   Flunitrazepam
nitro-1,3-dihydro-2H-1,4-   iodide
benzodiazepin-2-one

Reference(s):
US 3 116 203 (Hoffmann-La Roche; 31.12.1963; appl. 14.3.1962).
US 3 123 529 (Hoffmann-La Roche; 3.3.1964; appl. 9.3.1962).
US 3 203 990 (Hoffmann-La Roche; 31.8.1965; prior. 27.6.1960, 20.4.1961, 21.3.1962).

Formulation(s):   amp. 2 mg; f. c. tabl. 1 mg; tabl. 1 mg, 2 mg

Trade Name(s):
D: Flunimerck (Merck)   GB: Rohypnol (Roche)   J: Rohypnol (Roche)
        Fluninoc (Neuro Hexal)   I: Darkene (Bayropharm)   Silece (Eisai)
        Rohypnol (Roche)   Roipnol (Roche)
F: Rohypnol (Roche)   Valsera (Polifarma)

Flunoxaprofen

ATC: G02CC04; M01AE15; M02AA
Use: non-steroidal anti-inflammatory, cyclooxygenase and lipoxygenase inhibitor

RN: 66934-18-7 MF: $C_{16}H_{12}FNO_3$ MW: 285.27
LD_{50}: 1275 mg/kg (M, p.o.);
521 mg/kg (R, p.o.)
CN: (*S*)-2-(4-fluorophenyl)-α-methyl-5-benzoxazoleacetic acid

DL-lysine salt (1:1)
RN: 124816-13-3 MF: $C_{16}H_{12}FNO_3 \cdot C_6H_{14}N_2O_2$ MW: 431.46
LD_{50}: 723.5 mg/kg (M, p.o.)
L-lysine salt (1:1)
RN: 124816-14-4 MF: $C_{16}H_{12}FNO_3 \cdot C_6H_{14}N_2O_2$ MW: 431.46
D-lysine salt (1:1)
RN: 124816-15-5 MF: $C_{16}H_{12}FNO_3 \cdot C_6H_{14}N_2O_2$ MW: 431.46

(a)

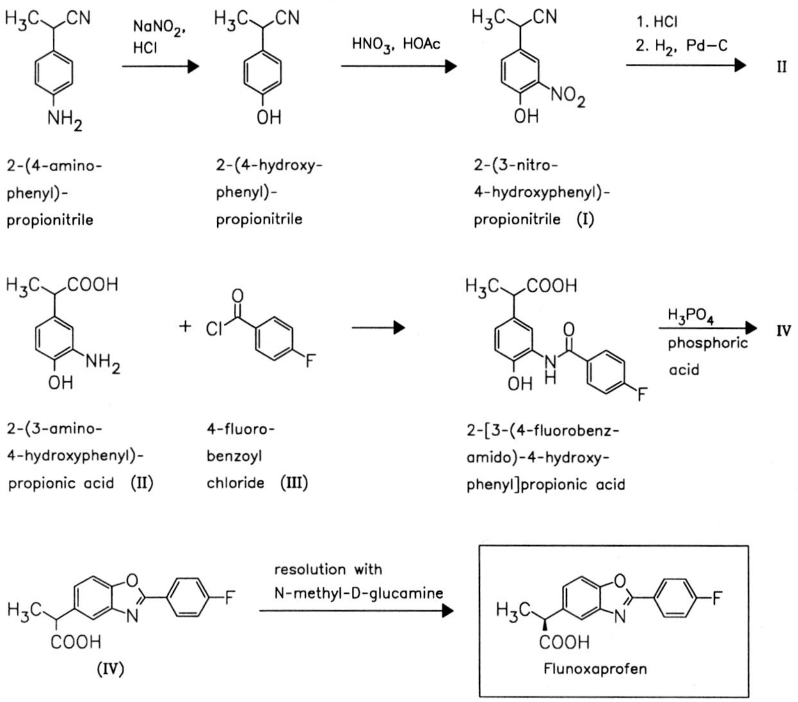
H3C CN
NH2
NaNO2, HCl
OH
HNO3, HOAc
NO2
1. HCl
2. H2, Pd–C
II
2-(4-amino-phenyl)-propionitrile
2-(4-hydroxy-phenyl)-propionitrile
2-(3-nitro-4-hydroxyphenyl)-propionitrile (I)
COOH
Cl
F
H3PO4 phosphoric acid
IV
2-(3-amino-4-hydroxyphenyl)-propionic acid (II)
4-fluoro-benzoyl chloride (III)
2-[3-(4-fluorobenz-amido)-4-hydroxy-phenyl]propionic acid
resolution with N-methyl-D-glucamine
(IV)
Flunoxaprofen

(b)

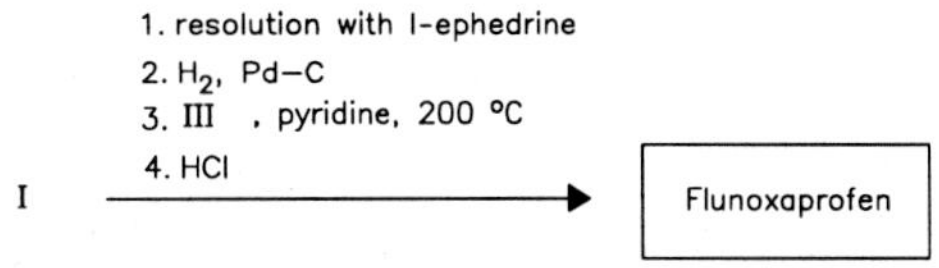
1. resolution with l-ephedrine
2. H2, Pd–C
3. III , pyridine, 200 °C
4. HCl
I
Flunoxaprofen

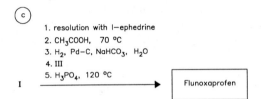

c

1. resolution with l−ephedrine
2. CH₃COOH, 70 °C
3. H₂, Pd−C, NaHCO₃, H₂O
4. III
5. H₃PO₄, 120 °C

I → Flunoxaprofen

Reference(s):

a,c DE 2 931 255 (Ravizza; appl. 1.8.1979; I-prior. 4.8.1978).
a   Dunwell, D.W. et al.: J. Med. Chem. (JMCMAR) **18**, 53 (1957).
    synthesis of 2-(4-aminophenyl)propionitrile:
    GB 1 198 212 (J. Borck et al.; appl. 1968).
    lysine salt:
    EP 324 402 (Euroresearch; appl. 9.1.1989; I-prior. 3.11.1988).
    US 4 897 408 (Euroresearch; 30.1.1990; appl. 5.1.1989; I-prior. 3.11.1988).
b   DE 2 728 323 (Ravizza; appl. 23.6.1977; GB-prior. 23.6.1976).
c   DOS 3 325 672 (Ravizza; appl. 15.7.1983; I-prior. 19.7.1982).

Formulation(s):   gel 5 %; tabl. 50 mg, 100 mg, 200 mg

Trade Name(s):
I: Priaxim (Ravizza)

Fluocinolone acetonide

ATC: C05AA10; D07AC04
Use: glucocorticoid, anti-inflammatory

RN: 67-73-2   MF: $C_{24}H_{30}F_2O_6$ MW: 452.49   EINECS: 200-668-5
LD₅₀: >4 g/kg (M, p.o.);
       >4 g/kg (R, p.o.)
CN: (6α,11β,16α)-6,9-difluoro-11,21-dihydroxy-16,17-[(1-methylethylidene)bis(oxy)]pregna-1,4-diene-3,20-dione

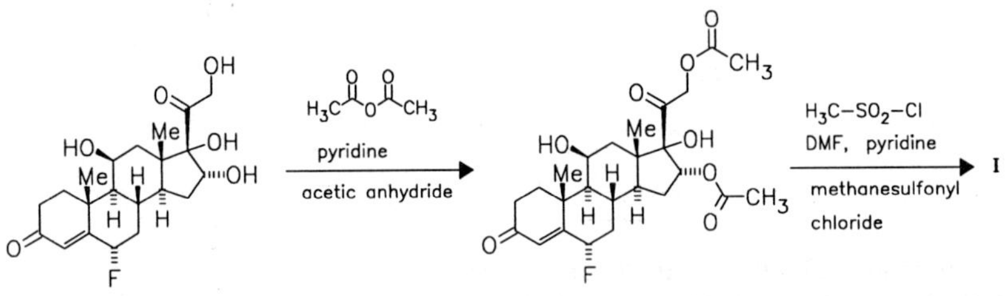

3,20-dioxo-6α-fluoro-
11β,16α,17,21-tetra-
hydroxy-4-pregnene
(cf. Fludroxycortide
synthesis)

[acetic anhydride, pyridine] → [H_3C-acetic anhydride-CH_3] → [H_3C-SO₂-Cl, DMF, pyridine, methanesulfonyl chloride] → I

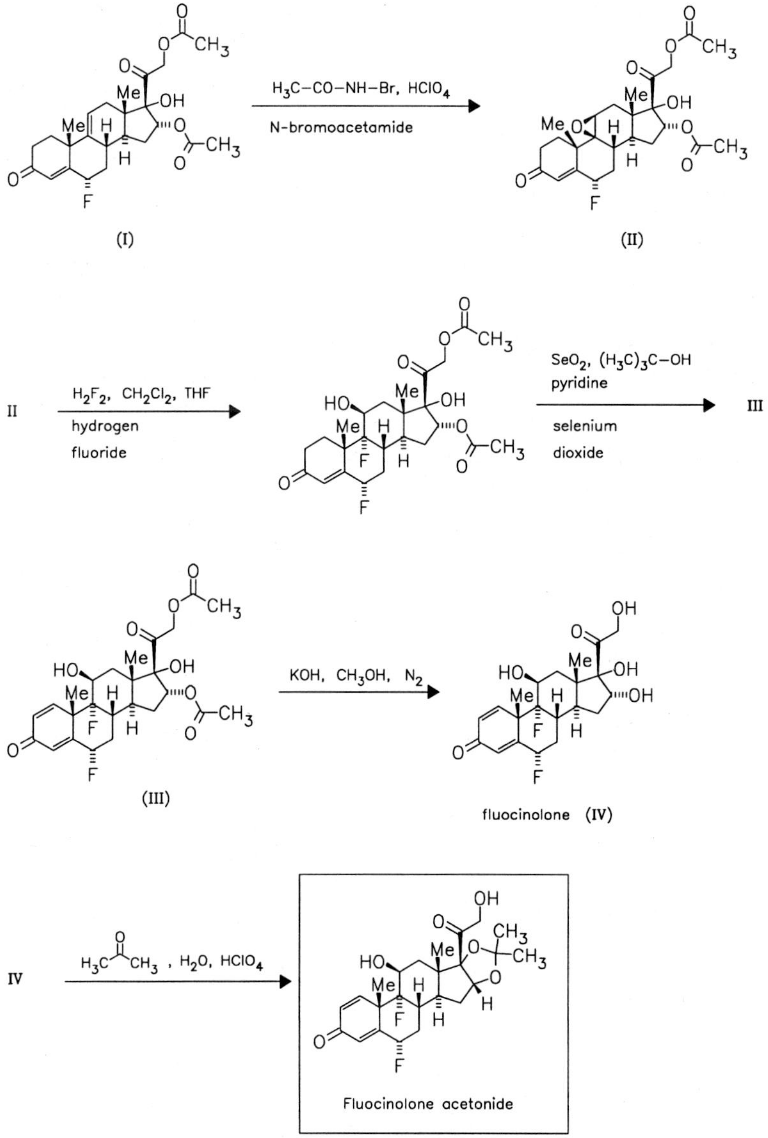

(I) → (II)

H₃C—CO—NH—Br, HClO₄
N-bromoacetamide

II

H₂F₂, CH₂Cl₂, THF
hydrogen fluoride

SeO₂, (H₃C)₃C—OH
pyridine

selenium dioxide

III

(III)

KOH, CH₃OH, N₂

fluocinolone (IV)

IV

H₃C—CO—CH₃ , H₂O, HClO₄

Fluocinolone acetonide

Reference(s):
US 3 014 938 (Syntex; 26.12.1961, appl. 23.8.1960; MEX-prior. 7.9.1959).
US 3 124 571 (Syntex; 10.3.1964; appl. 19.5.1960; MEX-prior. 26.1.1960).
Djerassi, C. et al.: J. Am. Chem. Soc. (JACSAT) **82**, 3399 (1960).

starting material:
Julian, P.L. et al.: J. Am. Chem. Soc. (JACSAT) **72**, 5145 (1950).

alternative syntheses:
GB 933 867 (American Cyanamid; appl. 5.12.1959; USA-prior. 8.12.1958).
US 3 197 469 (Pharmaceutical Research Prod.; 27.7.1965; appl. 6.8.1958).

Formulation(s):   cream 0.025 %; ointment 0.025 %; topical sol. 0.01 %

Trade Name(s):
D: Flucinar (medphano)   Jellisoft (Grünenthal)   Procto-Jellin (Grünenthal)-
      Jellin (Grünenthal)   comb.

F: Antibio-Synalar (Cassenne)-comb.
Synalar (Cassenne)
Synalar Neomycin (Cassenne)-comb.
GB: Synalar (Zeneca)
I: Alfabios (Biotekfarma)
Alfafluorone (Biotekfarma)
Boniderma (Boniscontro & Gazzone)
Coramide (Ottolenghi)
Cortanest Plus (Piam)-comb. with lidocaine
Dermobeta (Terapeutico M.R.)
Dermolin (Lafare)
Doricum (Farmila)-comb. with neomycine

Doricum Semplice (Farmila)
Esacinone (Lisapharma)
Fluocit (CT)
Fluomicetina (Zoja)-comb. with kanamycin
Fluomix Same (Savoma)
Fluovitef (Italfarmaco)
Lauromicina Pomata (Lafare)-comb. with eritromycin
Localyn (Recordati)
Mecloderm (Schwarz)-comb. with meclocycline
Meclutin (ABC-Torino)-comb. with meclocycline
Nefluan (Molteni)-comb.
Neoderm (Crosara)

Omniderm (Face)
Proctolin (Recordati)-comb. with letocaine
Sterolone (Francia Farm.)
Ultraderm (Ecobi)
J: Benamizol (Mohan Yakuhin)
Biscosal (Ohta Seiyaku)
Cortiphate (Tokyo Tanabe)
Flucort (Syntex-Tanabe)
Fluvean (Kowa)
Fluzon (Taisho Seiyaku)
USA: Derma-Smoothe (Hill)
Fluonid (Allergan)
FS Shampoo (Hill)
Synalar (Medicis)
Synemol (Medicis)

Fluocinonide

ATC: C05AA11; D07AC08
Use: glucocorticoid, anti-inflammatory

RN: 356-12-7 MF: $C_{26}H_{32}F_2O_7$ MW: 494.53 EINECS: 206-597-6
LD_{50}: >6 g/kg (M, p.o.);
14 mg/kg (R, p.o.)
CN: (6α,11β,16α)-21-(acetyloxy)-6,9-difluoro-11-hydroxy-16,17-[(1-methylethylidene)bis(oxy)]pregna-1,4-diene-3,20-dione

fluocinolone acetonide + H_3C—COOH → (pyridine) Fluocinonide

Reference(s):
GB 916 996 (Olin Mathieson; appl. 21.7.1959; USA-prior. 6.8.1958).
US 3 124 571 (Syntex; 10.3.1964; appl. 19.5.1960; MEX-prior. 26.1.1960).

Formulation(s): cream 0.5 mg/g; ointment 0.5 mg/g, sol. 0.5 mg/g

Trade Name(s):
D: Topsym (Grünenthal)
Topsym (Grünenthal)-comb. with neomycin
F: Topsyne (Cassenne)

Topsyne néomycine (Cassenne)-comb.
GB: Metosyn (Zeneca)
I: Flu 21 (Select Pharma)
Topsyn (Recordati)

J: Bestasone (Kodama)
Topsym (Syntex-Tanabe)
USA: Dermacin (Pedinol)
Lidex (Medicis)
Lidex E Cream (Medicis)

Fluocortin butyl

(Fluocortin butyl ester)

ATC: D07AB04
Use: glucocorticoid

RN: 41767-29-7 MF: $C_{26}H_{35}FO_5$ MW: 446.56 EINECS: 255-543-8
LD$_{50}$: >5 g/kg (M, p.o.);
>4 g/kg (R, p.o.);
>1 g/kg (dog, p.o.)
CN: (6α,11β,16α)-6-fluoro-11-hydroxy-16-methyl-3,20-dioxopregna-1,4-dien-21-oic acid butyl ester

fluocortin
RN: 33124-50-4 MF: $C_{22}H_{27}FO_5$ MW: 390.45 EINECS: 251-383-8

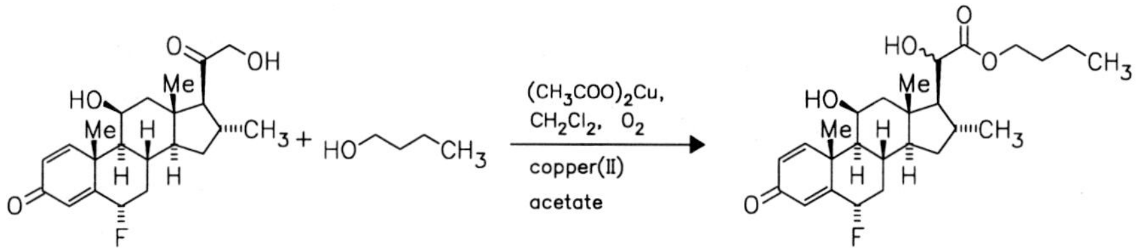

fluocortolone   1-butanol   (I)
(q. v.)

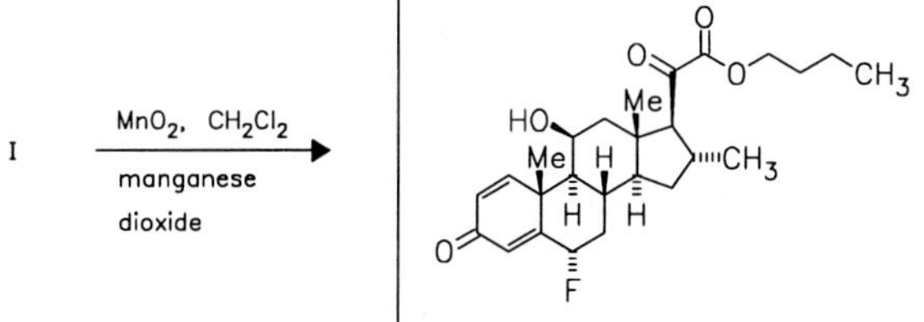

Fluocortin butyl

Reference(s):
DOS 2 150 268 (Schering AG; appl. 4.10.1971).
DOS 2 204 361 (Schering AG; appl. 27.1.1972).
DOS 2 260 303 (Schering AG; appl. 6.12.1972).
GB 1 387 911 (Schering AG; valid from 19.3.1975; D-prior. 4.10.1971, 27.1.1972).
Laurent, H. et al.: Arzneim.-Forsch. (ARZNAD) **27** (II), 2187 (1977) (also other methods).
DOS 2 441 284 (Schering AG; appl. 27.8.1974).
BE 823 682 (Schering AG; appl. 20.12.1974; D-prior. 21.12.1973, 27.8.1974, 16.9.1974).

Formulation(s):   cream 7.5 mg/g; ointment 7.5 mg/g; powder 100 mg/4 g

Trade Name(s):
D: Bi Vaspit (Asche)-comb.   Lenen (Alk-Scherax;   Vaspit (Asche)
                                       Schering)   I: Vaspit (Schering)

Fluocortolone

ATC: C05AA08; D07AC05; D07BC03;
     H02AB03; S01CA04
Use: glucocorticoid

RN: 152-97-6 MF: $C_{22}H_{29}FO_4$ MW: 376.47 EINECS: 205-811-5
CN: (6α,11β,16α)-6-fluoro-11,21-dihydroxy-16-methylpregna-1,4-diene-3,20-dione

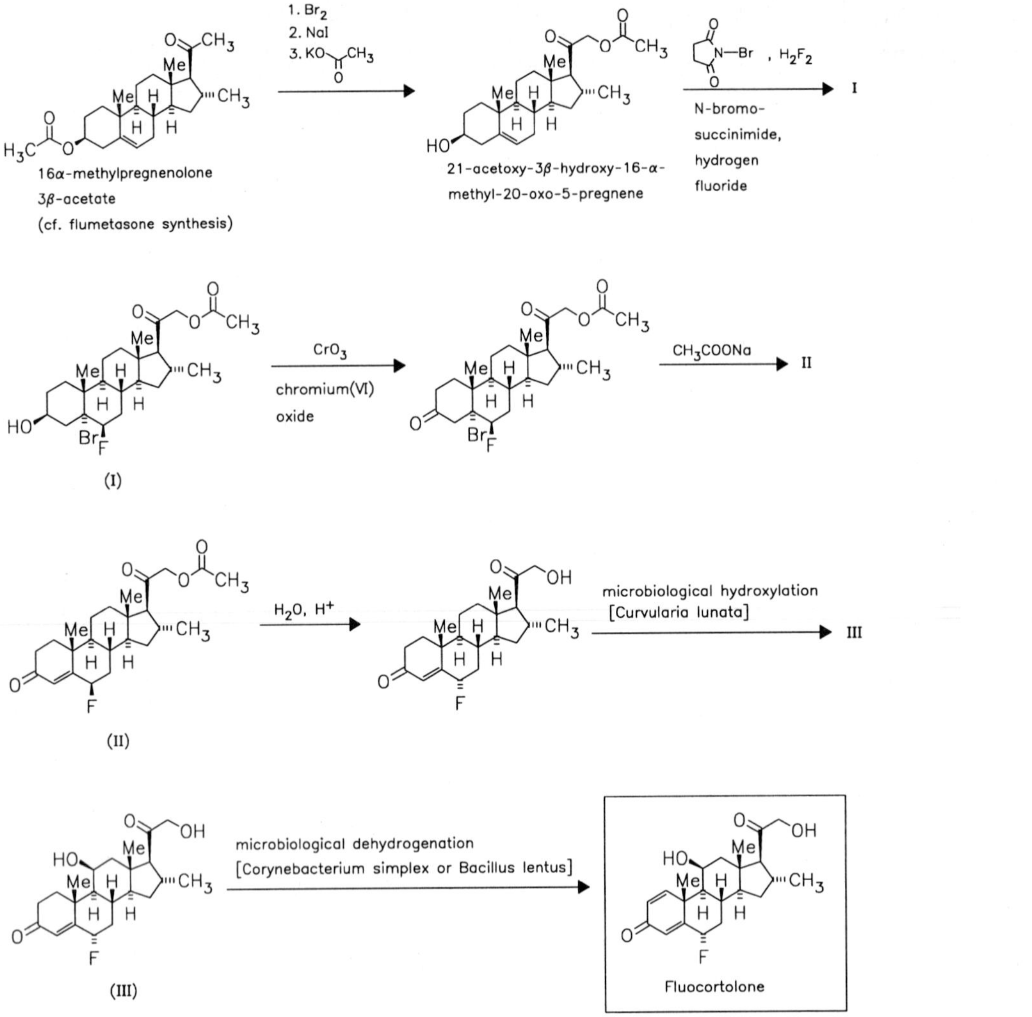

16α-methylpregnenolone
3β-acetate
(cf. flumetasone synthesis)

21-acetoxy-3β-hydroxy-16-α-
methyl-20-oxo-5-pregnene

N-bromo-
succinimide,
hydrogen
fluoride

I

(I)

CrO₃
chromium(VI)
oxide

CH₃COONa

II

(II)

H₂O, H⁺

microbiological hydroxylation
[Curvularia lunata]

III

(III)

microbiological dehydrogenation
[Corynebacterium simplex or Bacillus lentus]

Fluocortolone

Reference(s):
DE 1 135 899 (Schering AG; appl. 20.5.1960).
BE 614 196 (Schering AG; appl. 21.2.1962; D-prior. 22.2.1961).
Domenico, A. et al.: Arzneim.-Forsch. (ARZNAD) 15, 46 (1965).
DE 1 169 444 (Schering AG; appl. 22.2.1961).

synthesis of starting compound:
Petrov, V.; Williamson, D.M.: J. Chem. Soc. (JCSOA9) 1959, 3595.

alternative synthesis:
Kieslich, K. et al.: Justus Liebigs Ann. Chem. (JLACBF) 726, 168 (1969).
DOS 1 909 152 (Schering AG; appl. 19.2.1969).

review:
Akhrem, A.A. et al.: Russ. Chem. Rev. (Engl. Transl.) (RCRVAB) 34, 926 (1965).

Formulation(s):   cream 2.5 mg/g; lotion 2.5 mg/g; ointment 2.5 mg/g; tabl. 5 mg, 20 mg, 50 mg

Trade Name(s):
D: Ultralan (Schering)
      Ultrasine (Schering)
GB: Ultradil (Schering); wfm

Ultralanum (Schering)-
comb.; wfm

Ultralanum oint.
(Schering)-comb.; wfm

Ultraproct (Schering)-comb.; wfm	I:	Ultralan (Schering)-comb. Ultralan orale (Schering)	Ultraproct (Schering)-comb.

Fluocortolone caproate

ATC: C05AA08; D07AC05; H02AB03
Use: glucocorticoid

RN: 303-40-2   MF: $C_{28}H_{39}FO_5$ MW: 474.61   EINECS: 206-140-0
CN: (6α,11β,16α)-6-fluoro-11-hydroxy-16-methyl-21-[(1-oxohexyl)oxy]pregna-1,4-diene-3,20-dione

fluocortolone   caproic anhydride   Fluocortolone caproate

Reference(s):
FR 1 561 884 (Schering AG; appl. 10.5.1968; D-prior. 13.5.1967).

Formulation(s):   cream; lotion; ointment 2.5 mg/g in comb. with fluocortolone

Trade Name(s):

D:	Ultralan Creme (Schering)-comb. Ultralan Salbe (Schering)-comb. Ultraproct (Schering)-comb. F: Myco-Ultralan (Schering)-comb.		Ultralan (Schering)-comb. Ultraproct (Schering)-comb. GB: Ficoid (Fisons)-comb.; wfm Ultradil (Schering Chemicals)-comb.; wfm	Ultralanum (Schering Chemicals)-comb.; wfm Ultraproct (Schering Chemicals)-comb.; wfm I: Ultralan (Schering)-comb. Ultraproct (Schering)-comb.

Fluocortolone trimethylacetate
(Fluocortolone 21-pivalate)

ATC: C05AA08; D07AC05; H02AB03
Use: glucocorticoid

RN: 20380-10-3   MF: $C_{27}H_{37}FO_5$ MW: 460.59
CN: 6α-fluoro-17,21-dihydroxy-16α-methyl-pregna-1,4-diene-3,20-dione 21-pivalate

fluocortolone   pivalic anhydride   Fluocortolone trimethylacetate

Reference(s):
FR 1 561 884 (Schering AG; appl. 10.5.1968; D-prior. 13.5.1967).

Formulation(s):   cream 20 mg/g; cream 2.5 mg/g in comb. with fluocortolone; emulsion 2.5 mg/g in comb. with fluocortolone/-caproate; suppos. 40 mg in comb. with lidocain hydrochloride

Trade Name(s):

D:	Doloproct (Schering)- comb. with lidocaine hydrochloride Ultralan Creme (Schering)- comb.		Ultraproct (Schering)- comb.		Ultraproct (Schering Chemicals)
		F:	Ultraproct (Schering)- comb.	I:	Ultralan (Schering)-comb. Ultraproct (Schering)- comb.
		GB:	Ultralanum (Schering)- comb.		

Fluorescein

ATC: S01JA01
Use: diagnostic

RN: 2321-07-5   MF: $C_{20}H_{12}O_5$ MW: 332.31   EINECS: 219-031-8
LD_{50}:   300 mg/kg (M, i.v.)
CN: 3',6'-dihydroxyspiro[isobenzofuran-1(3H),9'-[9H]xanthen]-3-one

disodium salt
RN: 518-47-8   MF: $C_{20}H_{10}Na_2O_5$ MW: 376.28   EINECS: 208-253-0
LD_{50}:   1 g/kg (M, i.v.); 4738 mg/kg (M, p.o.);
       1 g/kg (R, i.v.); 6721 mg/kg (R, p.o.);
       1 g/kg (dog, i.v.)

phthalic anhydride   +   resorcinol   →(200 °C)   Fluorescein

Reference(s):
Ullmanns Encykl. Tech. Chem., 4. Aufl., Vol. **23**, 414.

Formulation(s):   amp. 113.2 mg/ml (as disodium salt); eye drops 1.7 mg/ml

Trade Name(s):

D:	Fluorescein-Lösung 1 0 % intravenös Inj.-Lösung (Alcon) Fluoreszein 0,15 % Thilo Augentropfen (Thilo) Pancreolauryl-Test (Temmler)		Thilorbin (Alcon)	I:	Pancreolauryl Test (Geymonat; as laurate)
		GB:	Minims Fluorescein Sodium (Chauvin) Minims lignocaine and fluorescein (Chauvin)- comb.	J:	Fluor (Tobishi-Santen) Fluores (Showa Yakuhin) Fluorescein sodium (Kobayashi) Fluorescite (Alcon)

Fluorometholone

ATC: C05AA06; D07AB06; D07XB04;
        D10AA01; S01BA07; S01CB05
Use: glucocorticoid

RN: 426-13-1   MF: $C_{22}H_{29}FO_4$ MW: 376.47   EINECS: 207-041-5
LD_{50}:   443 mg/kg (R, i.p.)
CN: (6α,11β)-9-fluoro-11,17-dihydroxy-6-methylpregna-1,4-diene-3,20-dione

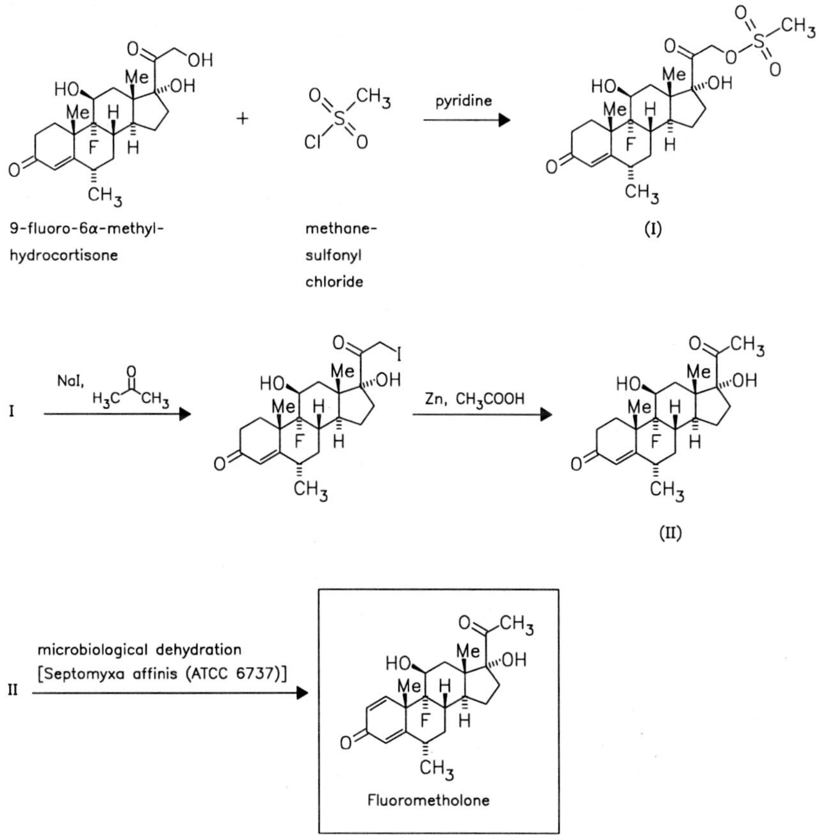

9-fluoro-6α-methyl-
hydrocortisone

methane-
sulfonyl
chloride

pyridine

(I)

I

NaI, H_3C CH_3

Zn, CH_3COOH

(II)

II

microbiological dehydration
[Septomyxa affinis (ATCC 6737)]

Fluorometholone

Reference(s):
US 2 867 638 (Upjohn; 6.1.1959; appl. 17.5.1967; prior. 10.9.1956).
DE 1 056 605 (Upjohn; appl. 6.5.1959; USA-prior. 10.9.1956).

starting material:
Spero, G.B. et al.: J. Am. Chem. Soc. (JACSAT) **79**, 1515 (1957).

Formulation(s):   eye drops 1 mg/ml

Trade Name(s):
D: Efflumidex (Pharm-
        Allergan)
        Efflumycin (Pharm-
        Allergan)-comb.
        Ehrtolan (Albert-Roussel)-
        comb.
        Ejemolin (CIBA Vision)-
        comb.
        Fluoropos (Ursapharm)

F: Isoptoflucon (Alcon)
        Flucon collyre (Alcon)
I: Efemoline (CIBA Vision)-
        comb.
        Fluaton (Allergan)
        Flumetol (Farmila)-comb.
J: Flu-Base (Kowa)
        Flumetholon (Santen)
        Okilon (Sumitomo)

        Ursnon (Nihon Yakuhin
        Kogyo)
USA: FML Liquifilm (Allergan);
        wfm
        Neo-Oxylone (Upjohn)-
        comb.; wfm
        Oxylone (Upjohn); wfm

Fluorouracil
(Fluracilum)

ATC: L01BC02
Use: antineoplastic

RN: 51-21-8   MF: $C_4H_3FN_2O_2$ MW: 130.08   EINECS: 200-085-6
LD_{50}:   81 mg/kg (M, i.v.); 115 mg/kg (M, p.o.);
245 mg/kg (R, i.v.); 230 mg/kg (R, p.o.);
30 mg/kg (dog, p.o.)
CN: 5-fluoro-2,4-pyrimidinediol or 5-fluoro-2,4(1H,3H)-pyrimidinedione

ethyl fluoroacetate   ethyl formate   (I)

S-methylthiouronium
sulfate

5-fluoro-
2-methyl-
thiouracil

Fluorouracil

uracil

Fluorouracil

fluorine
trifluoromethyl
hypofluorite

Reference(s):
a   US 2 802 005 (C. Heidelberger, R. Duschinsky; 6.8.1957; prior. 26.9.1956).
Duschinsky, R et al.: J. Am. Chem. Soc. (JACSAT) **79**, 4559 (1957).
b   US 3 682 917 (I. L. Knuniants et al.; 8.8.1972; appl. 25.3.1970).
US 3 846 429 (S. A. Giller et al.; 5.11.1974; appl. 22.9.1971).
US 3 954 758 (PCR, Inc.; 4.5.1976; prior. 7.8.1967, 1.3.1968, 27.5.1970, 4.10.1971).
DOS 2 149 504 (Research Inst. f. Med. and Chem.; appl. 4.10.1971; USA-prior. 5.10.1970).
DOS 2 719 245 (Daikin Kogyo; appl. 29.4.1977; J-prior. 29.4.1976).
DOS 2 726 258 (Daikin Kogyo; appl. 10.6.1977; J-prior. 11.6.1976).

synthesis from orotic acid by fluorination and following decarboxylation:
DOS 2 826 496 (Asahi Glass; appl. 16.6.1978; J-prior. 17.6.1977).

Formulation(s):   cream 5 %; ointment 50 mg/ml, 1 g/20 g; plaster 96 µg/1.13 cm²; vial 50 mg/ml, 250 mg/5 ml,
500 mg/10 ml, 1000 mg/20 ml

Trade Name(s):
D: Actino-Hermal Pflaster
(Hermal)
Efudix Roche (ICN)
Fluroblastin (Pharmacia &
Upjohn)
Ribofluor (ribosepharm)

Verrumal (Hermal)-comb.
numerous generics and
combination preparations
F: Efudix (Roche)
Fluoro-uracile (Roche)
generic

GB: Accusite (Matrix)
I: Efudix (Roche)
Fluoro-Uracile (Roche)
generic
J: 5-FU (Kyowa)
Arumel (SS Seiyaku)

Benton (Toyo Jozo)	Lifril (Kissei)	Ulup (Maruko)
Carzonal (Tobishi)	Timadin (Torii)	USA: Efudex (Roche)
Efudix (Roche)	Ulosagen (Kyowa Yakuhin Osaka)	Fluoroplex (Allergan)
Flacule (Nippon Kayaku)		generic

Fluoxetine

(Lilly 110140)

ATC: N06AB03
Use: antidepressant, serotonin-uptake inhibitor

RN: 54910-89-3 MF: $C_{17}H_{18}F_3NO$ MW: 309.33
LD_{50}: 464 mg/kg (M, p.o.);
825 mg/kg (R, p.o.)
CN: (±)-*N*-methyl-γ-[4-(trifluoromethyl)phenoxy]benzenepropanamine

monohydrochloride
RN: 59333-67-4 MF: $C_{17}H_{18}F_3NO \cdot HCl$ MW: 345.79
LD_{50}: 100 mg/kg (M, i.p.)

(a)

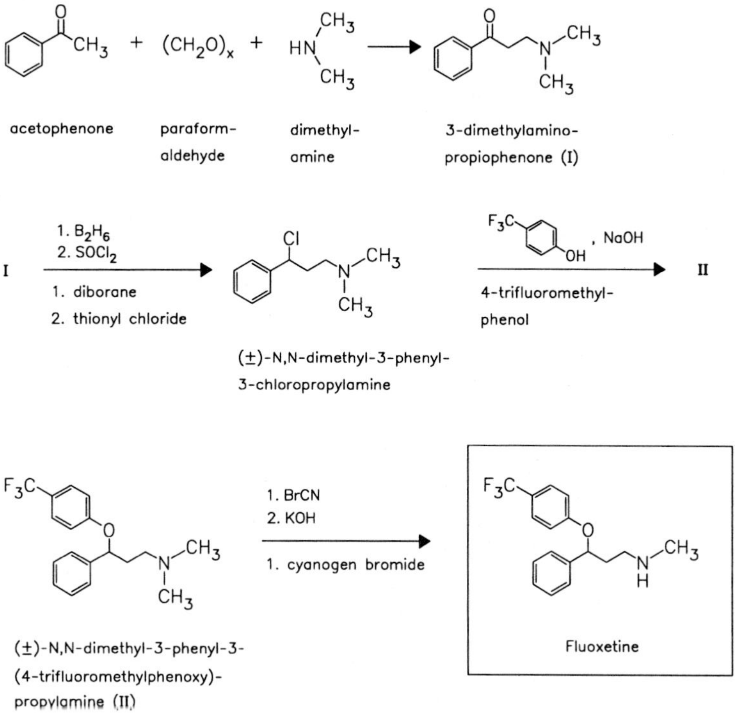
CH3
(CH2O)x
HN
CH3
CH3
N
CH3
acetophenone
paraform-
aldehyde
dimethyl-
amine
3-dimethylamino-
propiophenone (I)
I
1. B2H6
2. SOCl2
1. diborane
2. thionyl chloride
Cl
N
CH3
CH3
F3C
OH
, NaOH
4-trifluoromethyl-
phenol
II
(±)-N,N-dimethyl-3-phenyl-
3-chloropropylamine
F3C
O
N
CH3
CH3
1. BrCN
2. KOH
1. cyanogen bromide
F3C
O
N
H
CH3
Fluoxetine
(±)-N,N-dimethyl-3-phenyl-3-
(4-trifluoromethylphenoxy)-
propylamine (II)

(b)

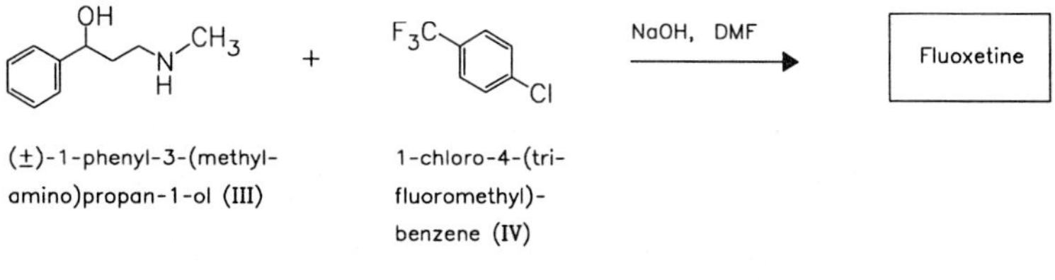
OH
N
H
CH3
F3C
Cl
NaOH, DMF
Fluoxetine
(±)-1-phenyl-3-(methyl-
amino)propan-1-ol (III)
1-chloro-4-(tri-
fluoromethyl)-
benzene (IV)

c

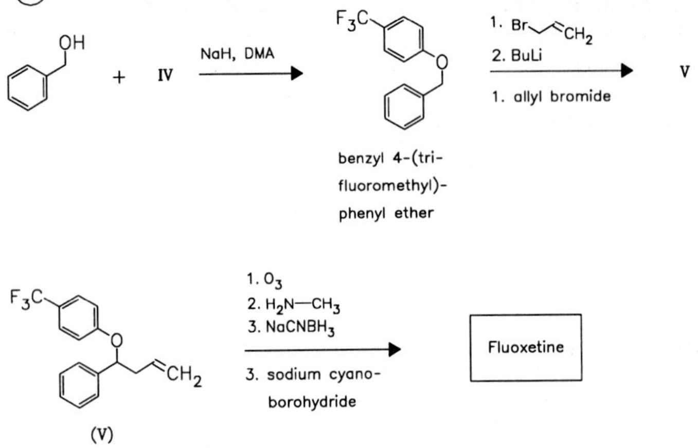
OH
+ IV
NaH, DMA
F3C
O
benzyl 4-(tri-
fluoromethyl)-
phenyl ether
1. Br CH2
2. BuLi
1. allyl bromide
V
F3C
O
CH2
(V)
1. O3
2. H2N—CH3
3. NaCNBH3
3. sodium cyano-
borohydride
Fluoxetine

d

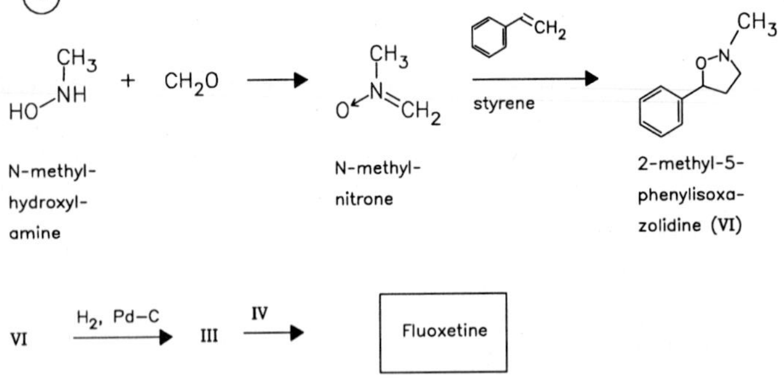
CH3
NH
HO
+ CH2O
N-methyl-
hydroxyl-
amine
CH3
N
O
CH2
N-methyl-
nitrone
CH2
styrene
CH3
O N
2-methyl-5-
phenylisoxa-
zolidine (VI)
VI
H2, Pd–C
III
IV
Fluoxetine

e

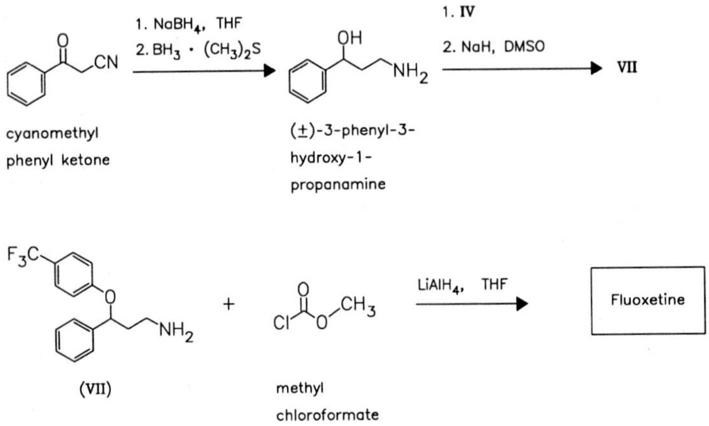
O
CN
cyanomethyl
phenyl ketone
1. NaBH4, THF
2. BH3 · (CH3)2S
OH
NH2
(±)-3-phenyl-3-
hydroxy-1-
propanamine
1. IV
2. NaH, DMSO
VII
F3C
O
NH2
(VII)
+
O
Cl O CH3
methyl
chloroformate
LiAlH4, THF
Fluoxetine

f

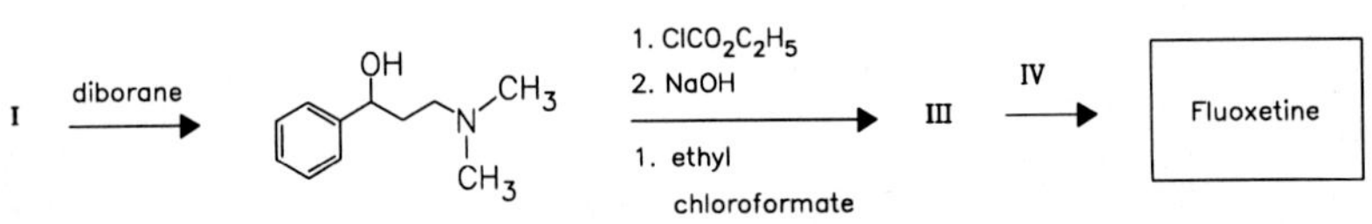
I
diborane
OH
N CH3
CH3
1. ClCO2C2H5
2. NaOH
1. ethyl
chloroformate
III
IV
Fluoxetine

Reference(s):

a   DE 2 500 110 (Lilly; appl. 3.1.1975; USA-prior. 10.1.1974).
        US 4 018 895 (Lilly; 19.4.1977; USA-prior. 10.1.1974).
        US 4 194 009 (Lilly; 18.3.1980; USA-prior. 15.9.1976).
        US 4 314 081 (Lilly; 2.2.1982; USA-prior. 10.1.1974).
        US 4 584 404 (Lilly; 22.4.1986; USA-prior. 24.10.1983, 25.1.1978, 10.1.1974).
b   US 5 847 214 (Laporte Organics; USA-prior. 7.7.1997).
c   ES 2 103 680 (Lilly; appl. 3.8.1995).
d   US 2 760 243 (Albemarle Corp.; USA-prior. 25.7.1997).
e   ES 210 654 (Lilly; appl. 24.7.1995).
f   EP 529 842 (Teva Pharm.; appl. 6.8.1992; IL-prior. 27.8.1991)

alternative synthesis:
EP 391 070 (Orion; appl. 1.3.1990; FI-prior. 3.3.1989).
EP 380 924 (E. Magnone; appl. 8.1.1990; I-prior. 10.1.1989).
WO 9 906 362 (Albemarle Corp.; appl. 4.8.1998; USA-prior. 4.8.1997).
WO 9 856 753 (Albemarle Corp.; appl. 12.6.1998; USA-prior. 12.6.1997).
ES 2 120 368 (Almirall Prodesfarma; 16.10.1998; appl. 14.6.1996).
EP 529 842 (Teva Pharm.; appl. 6.8.1992; IL-prior. 27.8.1991).
EP 617 006 (Pliva D.; appl. 4.2.1994; HR-prior. 5.2.1993).
ES 2 101 655 (Lilly; prior. 28.7.1995).
ES 2 101 654 (Lilly; prior. 24.7.1995).
ES 2 101 650 (Lilly; prior. 29.6.1995).
WO 9 811 054 (Egis Gyogyszergyar; appl. 10.9.1997; HU-prior. 10.9.1996).
US 5 760 243 (Albemarle Corp.; 2.6.1998; appl. 25.7.1996).
ES 2 103 680 (Lilly S. A.; 16.9.1997; appl. 3.8.1995).
ES 2 103 681 (Lilly; 16.9.1997; appl. 19.9.1995).

synthesis of enantiomers:
US 4 950 791 (H. C. Brown; 21.8.1990; prior. 12.6.1989, 30.3.1988).
US 4 918 242 (Aldrich; 17.4.1990; prior. 12.6.1989, 30.3.1988).
US 4 918 207 (Aldrich; 17.4.1990; prior. 12.6.1989, 30.3.1988).

fluoxetine *chiral process from* benzoylpropionic acid:
US 5 936 124 (Sepacor Inc.; appl. 22.6.1998).

treatment of nicotine withdrawal symptoms:
US 4 940 585 (W. E. Hapworth; 10.7.1990; appl. 17.2.1989).

treatment of appetite and mood disturbances:
WO 8 903 692 (MIT; appl. 21.10.1988; USA-prior. 15.9.1988, 22.10.1987).

antidiabetic combination:
EP 294 028 (Lilly; appl. 29.4.1988; USA-prior. 4.5.1987).

pharmaceutical formulation:
EP 693 281 (Lilly; appl. 17.7.1995; E-prior. 20.7.1994).

memory improvement:
US 4 647 591 (Lilly; 3.3.1987; prior. 7.10.1985, 21.6.1985).

solid oral composition:
ES 2 103 682 (Lilly; appl. 29.9.1995).

analgesic compositions:
EP 193 355 (Lilly; appl. 20.2.1986; USA-prior. 25.2.1985, 25.7.1986).
EP 193 354 (Lilly; appl. 20.2.1986; USA-prior. 25.2.1985).

treatment of anxiety:
EP 123 469 (Lilly; appl. 6.4.1984; USA-prior. 8.4.1983).

novel transdermal formulations:
WO 9 802 169 (Alza Corp.; appl. 15.7.1997; USA-prior. 15.7.1996).

pharmaceutical formulations:
EP 693 281 (Lilly; appl. 17.7.1995; E-prior. 20.7.1994).

low dose tablet:
US 5 830 500 (Pentech Pharm.; 3.11.1998; appl. 22.7.1996; USA-prior. 22.7.1996).

Formulation(s):   cps. 11.2 mg, 22.4 mg; sol. 22.4 mg/5 ml; tabl. 22.4 mg (as hydrochloride)

Trade Name(s):

D:	Fluctin (Lilly; 1990)	GB: Prozac (Lilly; 1989)	Prozac (Lilly; 1989)
F:	Prozac (Lilly; 1989)	I: Fluoxeren (Menarini; 1990)	USA: Prozac (Dista)

Fluoxymesterone

ATC: G03BA01
Use: androgen

RN: 76-43-7   MF: $C_{20}H_{29}FO_3$ MW: 336.45   EINECS: 200-961-8
LD_{50}: 2350 mg/kg (M, i.p.)
CN: (11β,17β)-9-fluoro-11,17-dihydroxy-17-methylandrost-4-en-3-one

3,17-dioxo-11β-
hydroxy-4-androstene
(from 3,17-dioxo-
4-androstene)

pyrrolidine

1. H_3C—MgBr
2. NaOH

1. methylmagnesium
   bromide

I

(I)

1. H_3C—⟨⟩—SO_2—Cl
2. base

1. p-toluenesulfonyl
   chloride

1. H_3C—CO—NH—Br
2. NaOH

1. N-bromo-
   acetamide

II

(II)

H_2F_2, CH_2Cl_2

Fluoxymesterone

Reference(s):
US 2 793 218 (Upjohn; 1957; prior. 1955).
US 2 813 881 (Upjohn; 1957; prior. 1955).
US 2 837 517 (Upjohn; 1958; prior. 1956, 1955).
US 3 029 263 (Upjohn; 10.4.1962; prior. 24.12.1959, 22.12.1958, 6.6.1958).
DAS 1 037 447 (Ciba, appl. 1955; CH-prior. 1954).
Heyl, W.F.; Herr, M.E.: J. Am. Chem. Soc. (JACSAT) **75**, 1918 (1953).
Bernstein, S. et al.: J. Org. Chem. (JOCEAH) **19**, 41 (1954).
Fried, J.; Sabo, E.F.: J. Am. Chem. Soc. (JACSAT) **75**, 2273 (1953); **76**, 1455 (1954).
Herr, M.E. et al.: J. Am. Chem. Soc. (JACSAT) **78**, 501 (1956).

alternative synthesis:
US 3 118 880 (Ciba; 21.1.1964; CH-prior. 26.5.1954).

Formulation(s):   tabl. 1 mg, 2 mg, 2.5 mg, 5 mg, 10 mg

Trade Name(s):
D:	Ultandren (Ciba); wfm	I:	Halotestin (Upjohn)	Halotestin (Upjohn); wfm
F:	Halotestin (Pharmacia &	J:	Halotestin (Kodama)	Ora-Testryl (Squibb); wfm
	Upjohn)	USA:	Halodrin (Upjohn)-comb.;	
GB:	Ultandren (Ciba); wfm		wfm	

Flupentixol
(Flupenthixol)

ATC: N05AF01
Use: neuroleptic, antipsychotic

RN: 2709-56-0   MF: $C_{23}H_{25}F_3N_2OS$ MW: 434.53   EINECS: 220-304-9
LD$_{50}$:   150 mg/kg (M, i.p.)
CN: 4-[3-[2-(trifluoromethyl)-9*H*-thioxanthen-9-ylidene]propyl]-1-piperazineethanol

dihydrochloride
RN: 2413-38-9   MF: $C_{23}H_{25}F_3N_2OS \cdot 2HCl$ MW: 507.45   EINECS: 219-321-4
LD$_{50}$:   94 mg/kg (M, i.v.); 423 mg/kg (M, p.o.);
        37 mg/kg (R, i.v.); 791 mg/kg (R, p.o.)
decanoate
RN: 30909-51-4   MF: $C_{23}H_{25}F_3N_2OS \cdot C_{10}H_{18}O$ MW: 588.78

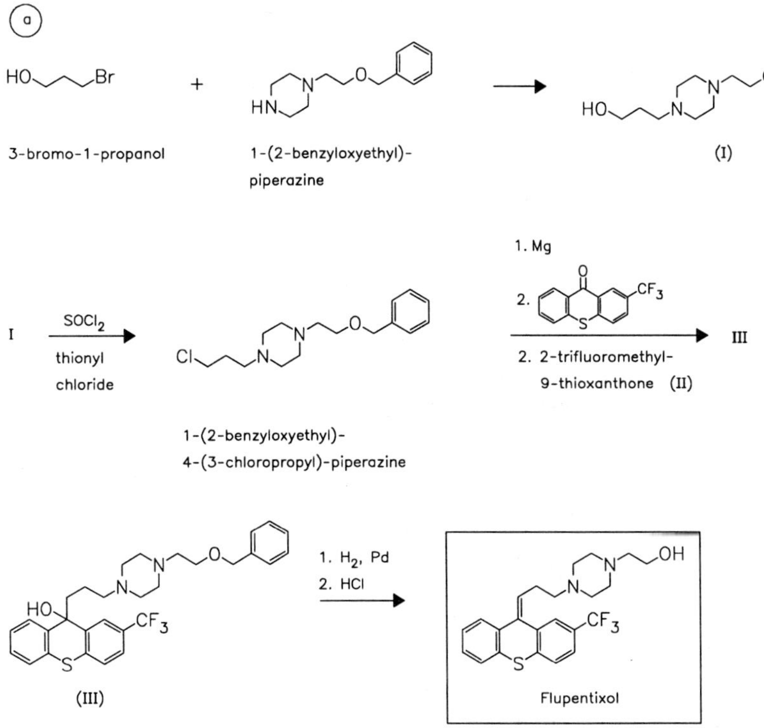

3-bromo-1-propanol   1-(2-benzyloxyethyl)-
                          piperazine   (I)

I $\xrightarrow{SOCl_2}$
     thionyl
     chloride

1-(2-benzyloxyethyl)-
4-(3-chloropropyl)-piperazine

1. Mg
2. 2-trifluoromethyl-
   9-thioxanthone   (II)

III

(III)

1. H_2, Pd
2. HCl

Flupentixol

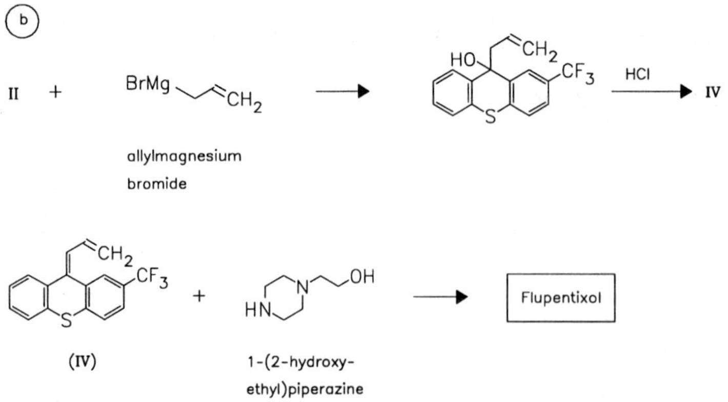

(b)

II + BrMg⌐CH₂ (allylmagnesium bromide) → [HO, CH₂, CF₃ thioxanthene structure] HCl → IV

[CH₂, CF₃ structure] (IV) + HN⌐N⌐OH 1-(2-hydroxy-ethyl)piperazine → Flupentixol

Reference(s):

a   GB 925 538 (Smith Kline & French; appl. 3.3.1961; USA-prior. 7.3.1960, 5.5.1960).
    US 3 282 930 (Smith Kline & French; 1.11.1966; prior. 7.3.1960, 5.5.1960).
b   US 3 116 291 (Kefalas; 31.12.1963; DK-prior. 4.12.1958).
    Kaiser, C. et al.: J. Med. Chem. (JMCMAR) **15**, 665 (1972).

flupentixol decanoate:
DAS 2 029 084 (Kefalas; appl. 12.6.1970; USA-prior. 20.6.1969).
US 3 681 346 (Kefalas; 1.8.1972; prior. 20.6.1969).

starting material:
GB 925 539 (Smith Kline & French; appl. 3.3.1961; USA-prior. 7.3.1960, 5.5.1960).

Formulation(s):   amp. 20 mg/ml, 100 mg/ml; drg. 0.5 mg, 1 mg, 5 mg; drops 50 mg/ml (as dihydrochloride);
                     vial 200 mg (20 mg/ml) (as decanoate)

Trade Name(s):

D:	Fluanxol (Bayer Vital)		Fluanxol (Lundbeck); wfm J:	Metamin (Takeda; 1973)
F:	Fluanxol (Lundbeck; 1976) I:		Deanxit (Lusofarmaco)-	
GB:	Depixol (Lundbeck; 1972);		comb.	
	wfm		Siplarol (Erba); wfm	

Fluperolone acetate

ATC: H02AB
Use: glucocorticoid, anti-inflammatory

RN: 2119-75-7   MF: C₂₄H₃₁FO₆   MW: 434.50   EINECS: 218-327-4
CN: [11β,17α,17(S)]-17-[2-(acetyloxy)-1-oxopropyl]-9-fluoro-11,17-dihydroxyandrosta-1,4-dien-3-one

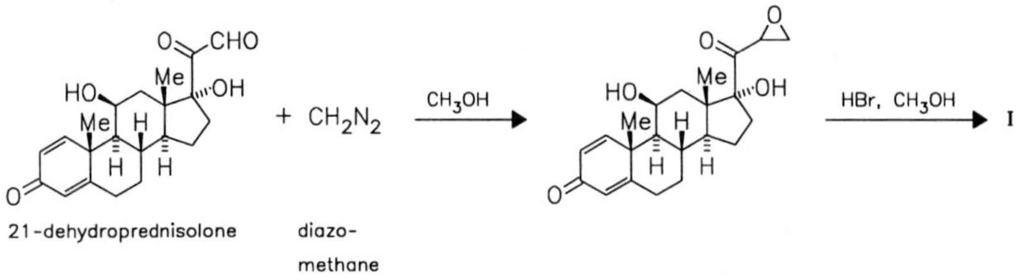

[steroid structure with O=CHO, Me, OH, HO groups] 21-dehydroprednisolone + CH₂N₂ diazomethane →(CH₃OH)→ [steroid structure with epoxide, Me, OH, HO groups] →(HBr, CH₃OH)→ I

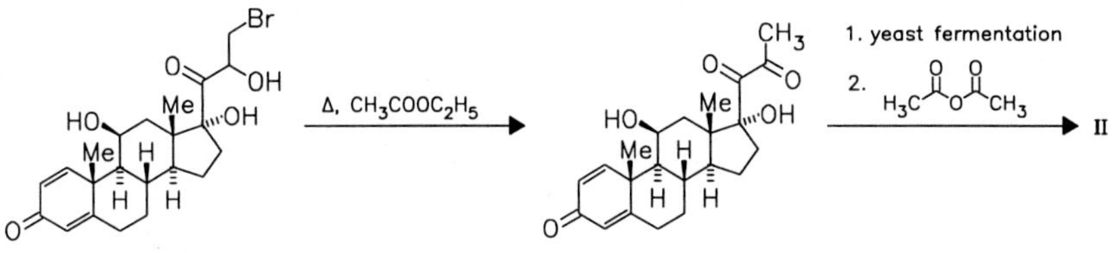
Δ, $CH_3COOC_2H_5$
1. yeast fermentation
2.
II

(I)

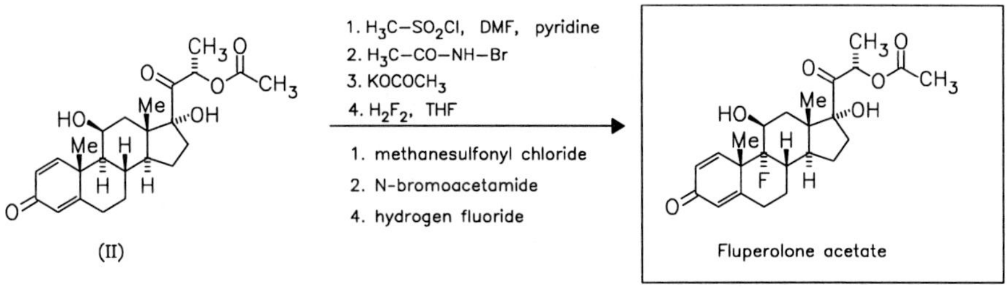
1. H_3C-SO_2Cl, DMF, pyridine
2. $H_3C-CO-NH-Br$
3. $KOCOCH_3$
4. H_2F_2, THF
1. methanesulfonyl chloride
2. N-bromoacetamide
4. hydrogen fluoride
Fluperolone acetate

(II)

Reference(s):

Agnello, E.J. et al.: J. Org. Chem. (JOCEAH) **28**, 1531 (1963).
Agnello, E.J. et al.: Experientia (EXPEAM) **16**, 357 (1960).
(also alternative syntheses)

Trade Name(s):

I: Alacortil (Pfizer); wfm USA: Methral (Pfizer); wfm

Fluphenazine

ATC: N05AB02
Use: neuroleptic, antipsychotic

RN: 69-23-8 MF: $C_{22}H_{26}F_3N_3OS$ MW: 437.53 EINECS: 200-702-9
LD_{50}: 51 mg/kg (M, i.v.); 220 mg/kg (M, p.o.)
CN: 4-[3-[2-(trifluoromethyl)-10*H*-phenothiazin-10-yl]propyl]-1-piperazineethanol

dihydrochloride
RN: 146-56-5 MF: $C_{22}H_{26}F_3N_3OS \cdot 2HCl$ MW: 510.45 EINECS: 205-674-1
LD_{50}: 56 mg/kg (M, i.v.); 220 mg/kg (M, p.o.)
decanoate
RN: 5002-47-1 MF: $C_{22}H_{26}F_3N_3OS \cdot C_{10}H_{18}O$ MW: 591.78

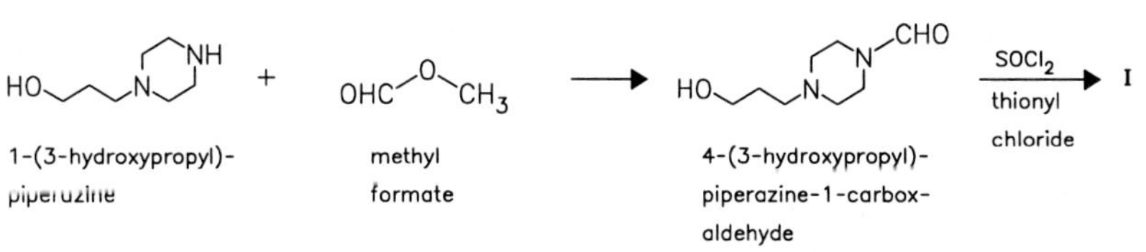
1-(3-hydroxypropyl)-piperazine
methyl formate
4-(3-hydroxypropyl)-piperazine-1-carboxaldehyde
$SOCl_2$
thionyl chloride
I

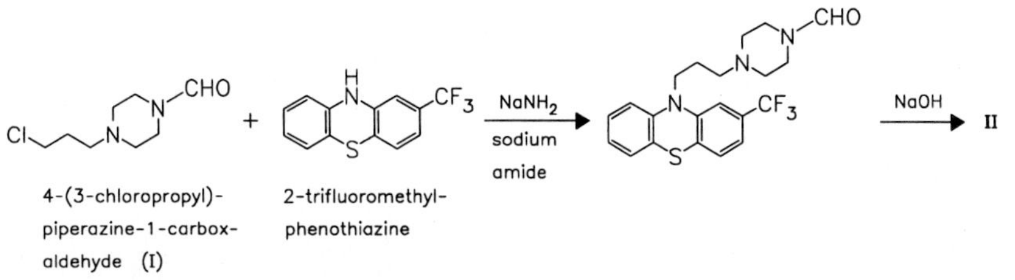
4-(3-chloropropyl)-piperazine-1-carboxaldehyde (I)
2-trifluoromethylphenothiazine
$NaNH_2$
sodium amide
NaOH
II

10-[3-(1-piperazinyl)propyl]-
2-trifluoromethyl-
phenothiazine (II)

Fluphenazine

Reference(s):
US 3 058 979 (Smith Kline & French; 16.10.1962; prior. 13.5.1957).
DE 1 095 836 (Squibb; appl. 8.12.1956; USA-prior. 23.12.1955, 12.7.1956).

alkanecarboxylic acid esters:
DE 1 165 602 (Olin Mathieson; appl. 25.4.1962; USA-prior. 26.4.1961).
US 3 194 733 (Olin Mathieson; 13.7.1965; prior. 26.4.1961, 28.1.1963).
US 3 394 131 (Squibb; 23.7.1968; prior. 26.4.1961, 28.1.1963).
Kurland, A.A. et al.: Curr. Ther. Res. (CTCEA9) **6**, 137 (1964).

Formulation(s):   amp. 2.5 mg/ml, 5 mg/ml, 12.5 mg/0.5 ml, 25 mg/ml, 25 mg/2 ml, 50 mg/0.5 ml, 100 mg/ml
                     (as decanoate); drops 4 mg/ml; f. c. drg. 3 mg, 6 mg; f. c. tabl. 0.5 mg, 1 mg, 4 mg;
                     sol. 2.5 mg/ml; tabl. 1 mg, 2.5 mg, 4 mg, 5 mg, 10 mg (as dihydrochloride)

Trade Name(s):

D:	Dapotum (Bristol-Myers Squibb; Sanofi Winthrop) Lyogen (Promonta Lundbeck) Omca (Bristol-Myers Squibb)		Moditen (Sanofi Winthrop) Motipress (Sanofi Winthrop)-comb. Motival (Sanofi Winthrop)-comb.		Sevinol (Schering-Shionogi)
F:	Modecate (Sanofi Winthrop) Moditen (Sanofi Winthrop) Motival (Sanofi Winthrop)-comb.	I:	Anatensol (Bristol-Myers Squibb) Dominans (Recordati)-comb. Moditen (Bristol-Myers Squibb)	USA:	Permitil (Schering); wfm Permitil (Schering-Plough); wfm Prolixin (Squibb); wfm Prolixin (Bristol-Myers Squibb); wfm generics
GB:	Modecate (Sanofi Winthrop)	J:	Anatensol (Showa) Fludecasine (Yoshitomi)		

Flupirtine

ATC: M03B; N02BG07
Use: analgesic

RN: 56995-20-1   MF: $C_{15}H_{17}FN_4O_2$ MW: 304.33   EINECS: 260-503-8
LD$_{50}$:   617 mg/kg (M, p.o.);
              1660 mg/kg (R, p.o.)
CN: [2-amino-6-[[(4-fluorophenyl)methyl]amino]-3-pyridinyl]carbamic acid ethyl ester

monohydrochloride
RN: 33400-45-2   MF: $C_{15}H_{17}FN_4O_2 \cdot HCl$ MW: 340.79   EINECS: 251-496-2
LD$_{50}$:   432 mg/kg (M, s.c.)
maleate (1:1)
RN: 75507-68-5   MF: $C_{15}H_{17}FN_4O_2 \cdot C_4H_4O_4$ MW: 420.40   EINECS: 278-225-0

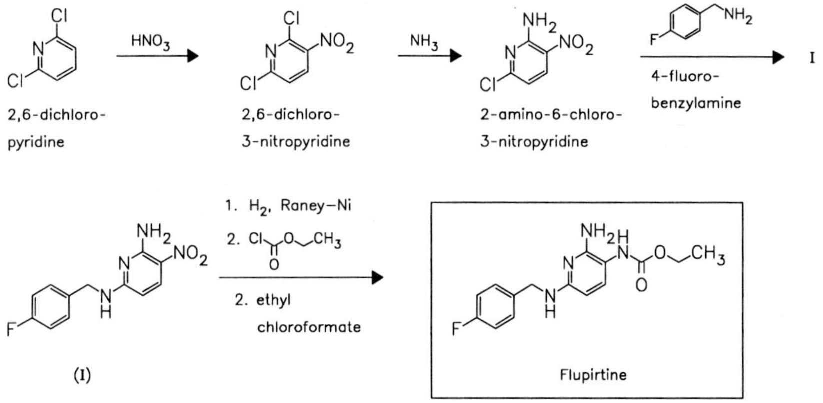

(I)

Flupirtine

Reference(s):
DE 1 670 522 (Degussa; appl. 12.5.1966).
DE 1 795 858 (Degussa; appl. 19.7.1968).
US 3 481 943 (Degussa; 2.12.1969; D-prior. 12.5.1966).
US 3 513 171 (Degussa; 19.5.1970; D-prior. 12.5.1966).
Bebenburg, W. v. et al.: Chem.-Ztg. (CMKZAT) **103**, 387 (1979).
Bebenburg, W. v. et al.: Chem.-Ztg. (CMKZAT) **105**, 217 (1981).
US 5 959 115 (ASATA Medica; 28.9.1999; appl. 23.4.1998; D-prior. 23.4.1997).

Formulation(s):   cps. 50 mg, 100 mg; suppos. 75 mg, 150 mg (as maleate)

Trade Name(s):

D:	Katadolon (ASTA Medica AWD)	Trancopal (Sanofi Winthrop)	I: Katadolon (ASTA Medica)

Fluprednidene acetate

ATC: D07AB07; D07CB02; D07XB03
Use: topical glucocorticoid

RN: 1255-35-2   MF: C$_{24}$H$_{29}$FO$_6$ MW: 432.49   EINECS: 215-013-9
CN: (11β)-21-(acetyloxy)-9-fluoro-11,17-dihydroxy-16-methylenepregna-1,4-diene-3,20-dione

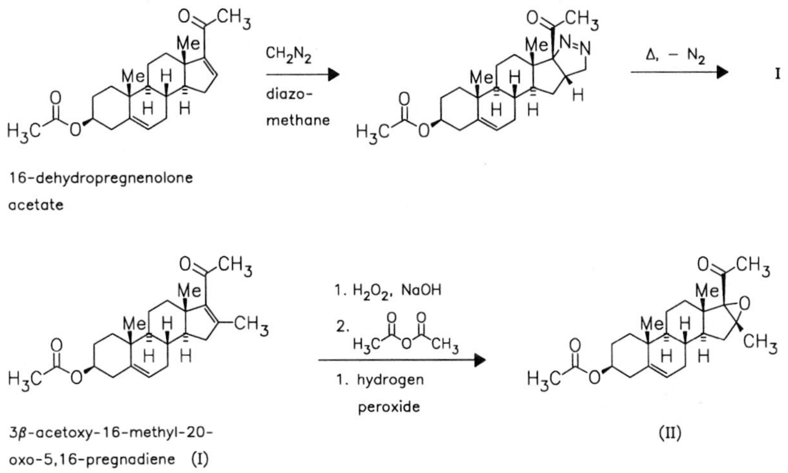

16-dehydropregnenolone acetate

3β-acetoxy-16-methyl-20-oxo-5,16-pregnadiene   (I)

(II)

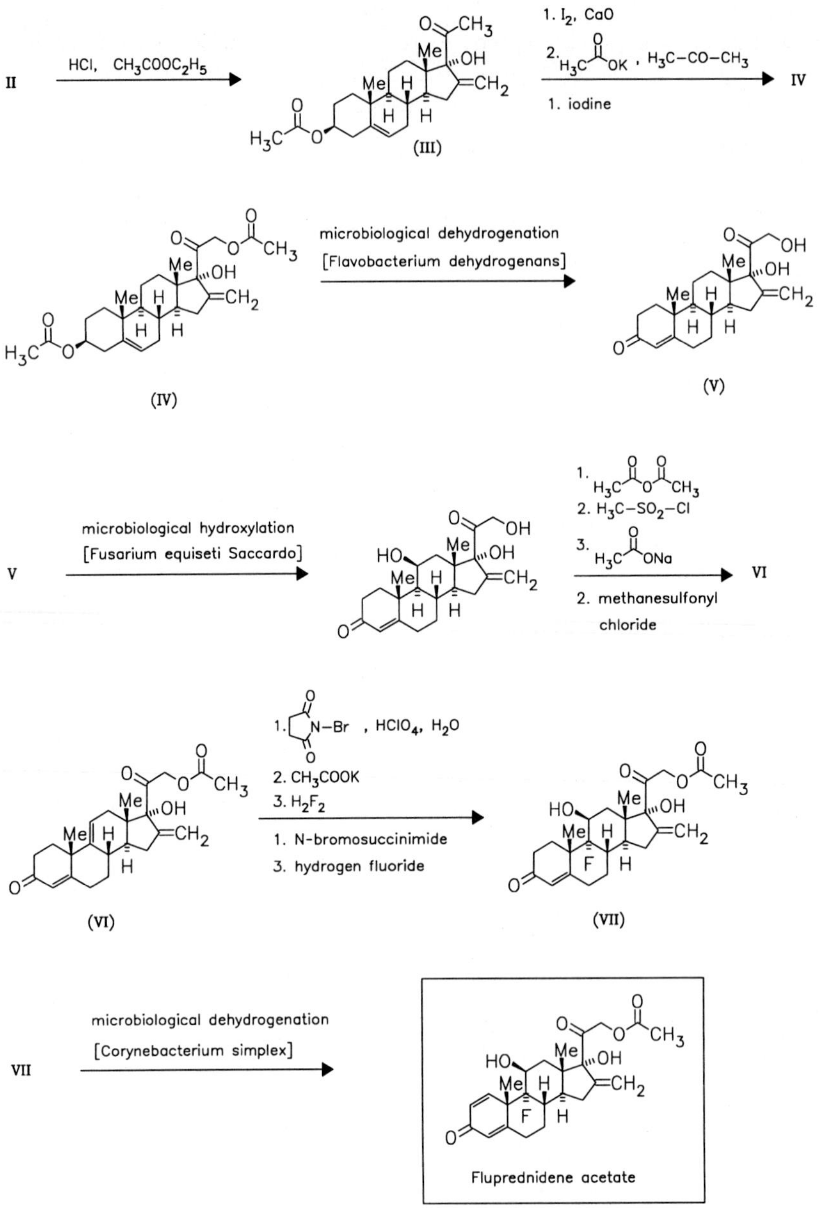

(III)

(IV)

microbiological dehydrogenation
[Flavobacterium dehydrogenans]

(V)

microbiological hydroxylation
[Fusarium equiseti Saccardo]

VI

(VI)

(VII)

microbiological dehydrogenation
[Corynebacterium simplex]

VII

Fluprednidene acetate

Reference(s):

GB 1 230 671 (Merck Patent GmbH; appl. 10.7.1969).
Irmscher, K. et al.: Arzneim.-Forsch. (ARZNAD) **18**, 7 (1968).
(also other syntheses reviewed)

synthesis of 16-dehydropregnenolone acetate:
Wettstein, A.: Helv. Chim. Acta (HCACAV) **27**, 1803 (1944).

alternative syntheses:
GB 946 860 (Merck & Co.; appl. 17.3.1960; USA-prior. 24.3.1959).
US 3 068 226 (Merck & Co.; 11.12.1962; appl. 22.12.1961; prior. 24.3.1959).
US 3 163 760 (Merck & Co.; 9.7.1964; appl. 24.3.1959).
US 3 309 272 (Merck & Co.; 14.3.1967; appl. 24.4.1961; prior. 24.3.1959).

Formulation(s):   cream 1 mg/g; ointment 0.05 g/100 g, 1 mg/g; sol. 0.025 g/100 g, 0.15 g/100 g, 1 mg/ml

Trade Name(s):

D: Candio-Hermal (Hermal)-   Decoderm (Hermal)   Sali-Decoderm (Hermal)-
      comb.   Decoderm (Hermal)-comb.   comb.

Fluprednisolone acetate

Use: glucocorticoid

RN: 570-36-5   MF: $C_{23}H_{29}FO_6$ MW: 420.48   EINECS: 209-330-1
CN: (6α,11β)-21-(acetyloxy)-6-fluoro-11,17-dihydroxypregna-1,4-diene-3,20-dione

fluprednisolone
RN: 53-34-9   MF: $C_{21}H_{27}FO_5$ MW: 378.44   EINECS: 200-170-8

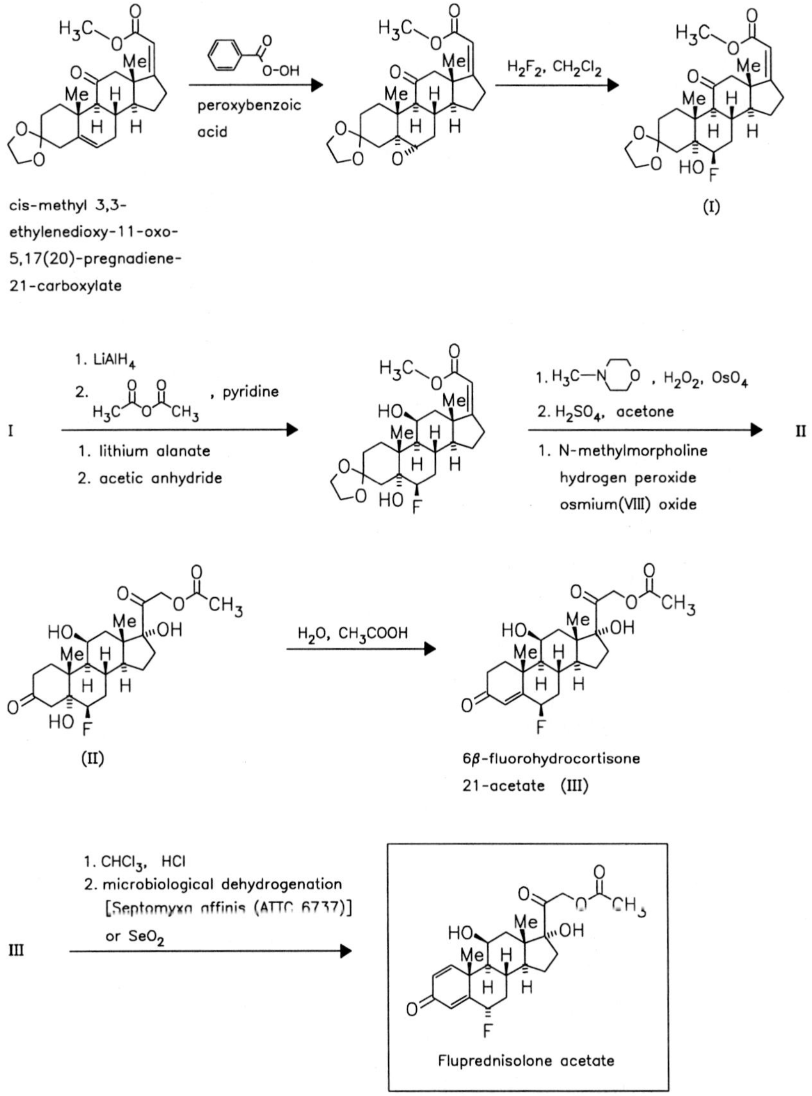

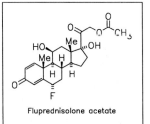

cis-methyl 3,3-
ethylenedioxy-11-oxo-
5,17(20)-pregnadiene-
21-carboxylate

(I)

I →
1. LiAlH$_4$
2. H_3C—CO—O—CO—CH_3 , pyridine

1. lithium alanate
2. acetic anhydride

1. H_3C—N(—O) , H_2O_2, OsO$_4$
2. H_2SO_4, acetone

1. N-methylmorpholine
   hydrogen peroxide
   osmium(VIII) oxide

→ II

(II)

H_2O, CH_3COOH

6β-fluorohydrocortisone
21-acetate   (III)

III →
1. CHCl$_3$, HCl
2. microbiological dehydrogenation
   [Septomyxa affinis (ATCC 6737)]
   or SeO$_2$

Fluprednisolone acetate

Reference(s):
US 2 841 600 (Upjohn; 1958; prior. 1957, 1955).
DE 1 079 042 (Syntex; appl. 1958; MEX-prior. 1957).

starting material:
US 2 707 184 (Upjohn; 1955, prior. 1953, 1952).

alternative synthesis:
US 4 041 055 (Upjohn; 9.8.1977; prior. 17.11.1975).

Formulation(s):   tabl. 1.5 mg, 2 mg, 16 mg

Trade Name(s):
D: Isopredon (Hoechst); wfm   I: Etadrol (Carlo Erba); wfm
F: Decoderme (Merck-   Etadrol (Farmitalia); wfm
      Clévenot); wfm   USA: Alphadrol (Upjohn); wfm

Flurazepam

ATC: N05CD01
Use: hypnotic, sedative

RN: 17617-23-1   MF: $C_{21}H_{23}ClFN_3O$ MW: 387.89   EINECS: 241-591-7
LD$_{50}$:   59.1 mg/kg (M, i.v.); 500 mg/kg (M, p.o.);
       38.7 mg/kg (R, i.v.); 980 mg/kg (R, p.o.)
CN: 7-chloro-1-[2-(diethylamino)ethyl]-5-(2-fluorophenyl)-1,3-dihydro-2H-1,4-benzodiazepin-2-one

dihydrochloride
RN: 1172-18-5   MF: $C_{21}H_{23}ClFN_3O \cdot 2HCl$ MW: 460.81   EINECS: 214-630-0
LD$_{50}$:   59.1 mg/kg (M, i.v.); 596 mg/kg (M, p.o.);
       38.7 mg/kg (R, i.v.); 879 mg/kg (R, p.o.)
hydrochloride
RN: 36105-20-1   MF: $C_{21}H_{23}ClFN_3O \cdot HCl$ MW: 424.35

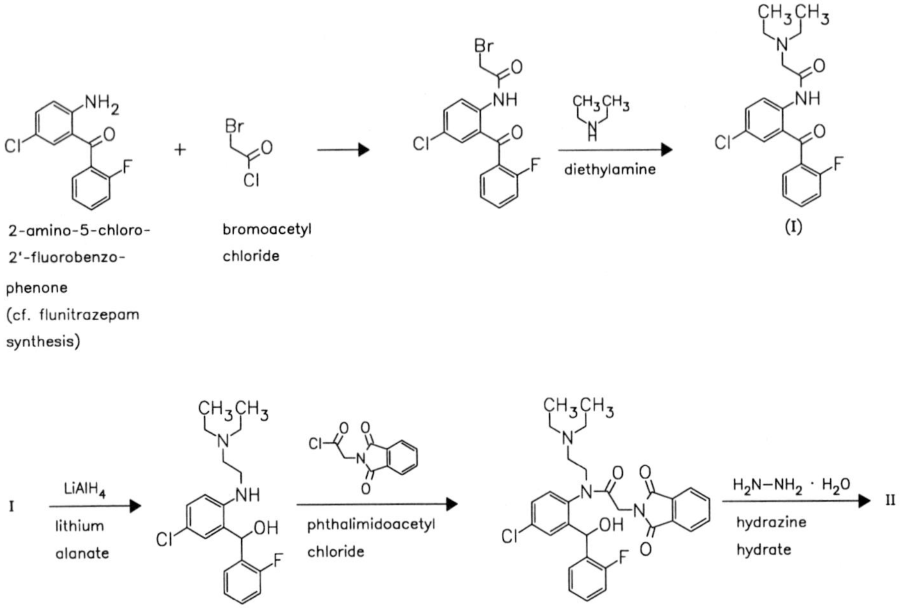

2-amino-5-chloro-
2'-fluorobenzo-
phenone
(cf. flunitrazepam
synthesis)

bromoacetyl
chloride

diethylamine

(I)

I

LiAlH$_4$
lithium
alanate

phthalimidoacetyl
chloride

$H_2N-NH_2 \cdot H_2O$
hydrazine
hydrate

II

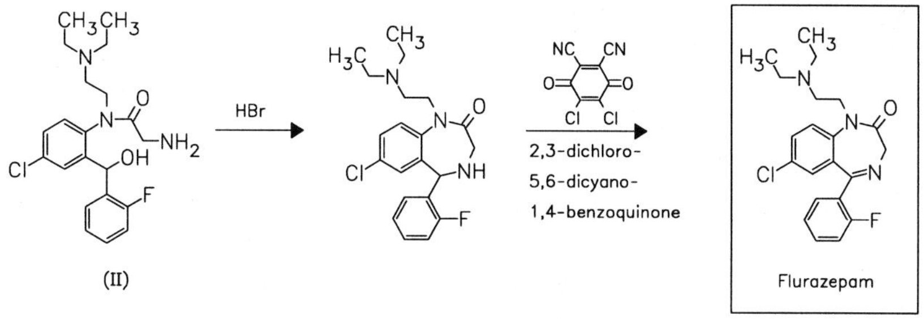

(II)

Flurazepam

Reference(s):
US 3 567 710 (Hoffmann-La Roche; 2.3.1971; prior. 3.6.1968).

alternative synthesis by reaction of 2-diethylaminoethyl chloride *with* 7-chloro-5-(2-fluorophenyl)-2-oxo-2,3-dihydro-1*H*-1,4-benzodiazepine:
GB 1 040 548 (Roche; appl. 1.3.1963; USA-prior. 2.3.1962).

Formulation(s):   cps. 10 mg, 15 mg; tabl. 27.42 mg (as base); cps. 15 mg, 30 mg; tabl. 30 mg (as monohydrochloride); cps. 15 mg, 30 mg; s. r. cps. 20 mg (as dihydrochloride)

Trade Name(s):
D: Dalmadorm (Roche)
      Flurazepam 15/30 Riker
      (3M Medica)
      Staurodorm Neu
      (Dolorgiet)
GB: Dalmane (Roche)
I: Dalmadorm (Roche)

      Felison (Bayropharm)
      Flunox (Boehringer
      Mannh.)
      Midorm A.R. (Piam)
      Remdue (Biomedica
      Foscama)
      Valdorm (Valeas)

J: Benozil (Kyowa)
      Dalmate (Nippon Roche)
      Insumin (Kyorin)
USA: Dalmane (Roche)

Flurbiprofen

ATC: M01AE09; M02AA19; S01BC04
Use: anti-inflammatory, analgesic

RN: 5104-49-4   MF: $C_{15}H_{13}FO_2$ MW: 244.27   EINECS: 225-827-6
LD$_{50}$: >385 mg/kg (M, i.v.); 640 mg/kg (M, p.o.);
       117 mg/kg (R, p.o.);
       10 mg/kg (dog, p.o.)
CN: 2-fluoro-α-methyl[1,1'-biphenyl]-4-acetic acid

sodium salt
RN: 56767-76-1   MF: $C_{15}H_{12}FNaO_2$ MW: 266.25

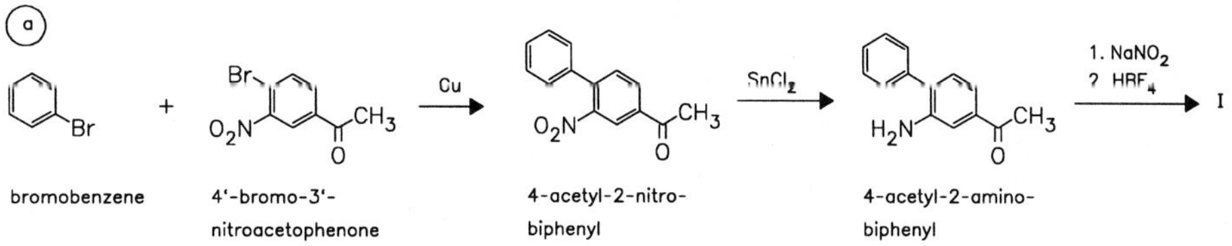

bromobenzene   4'-bromo-3'-
                nitroacetophenone

4-acetyl-2-nitro-
biphenyl

4-acetyl-2-amino-
biphenyl

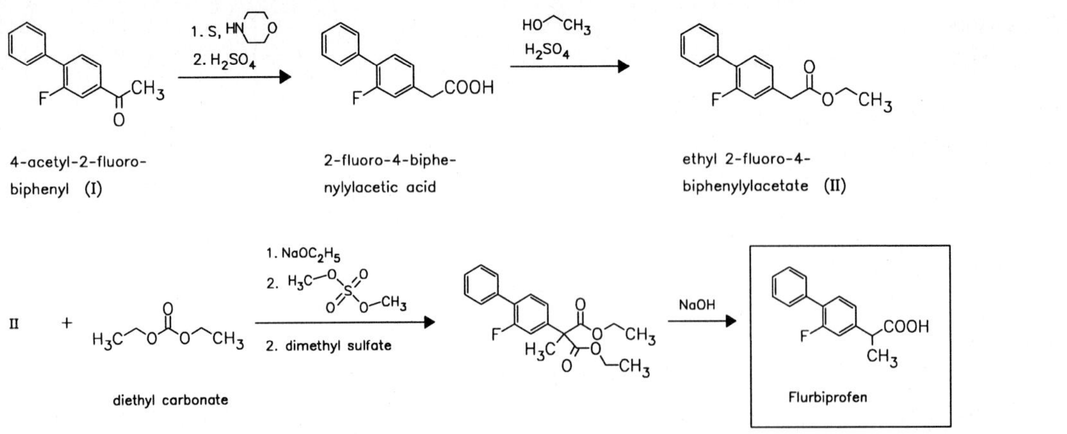
1. S, HN O
2. H_2SO_4
HO CH3
H_2SO_4
4-acetyl-2-fluoro-biphenyl (I)
2-fluoro-4-biphenylylacetic acid
ethyl 2-fluoro-4-biphenylylacetate (II)
II +
diethyl carbonate
1. $NaOC_2H_5$
2. dimethyl sulfate
NaOH
Flurbiprofen

b

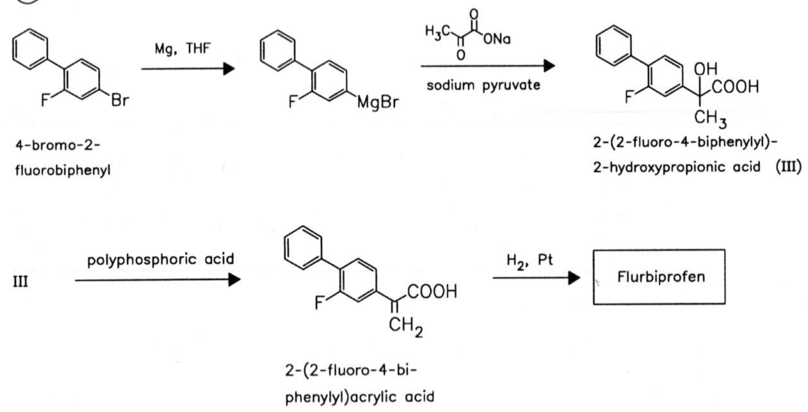
Mg, THF
sodium pyruvate
4-bromo-2-fluorobiphenyl
2-(2-fluoro-4-biphenylyl)-2-hydroxypropionic acid (III)
III
polyphosphoric acid
H_2, Pt
Flurbiprofen
2-(2-fluoro-4-biphenylyl)acrylic acid

Reference(s):

a DAS 1 518 528 (Boots; appl. 19.1.1965; GB-prior. 24.1.1964).
US 3 755 427 (Boots; 28.8.1973; GB-prior. 24.1.1964).
US 3 793 457 (Adams Sectal; 19.2.1974; GB-prior. 24.1.1964).

b GB 1 514 812 (Boots; appl. 4.4.1975; valid from 31.3.1976).

similar method:
US 3 959 364 (Boots; 25.5.1976; GB-prior. 24.5.1973).

alternative syntheses:
DOS 2 646 792 (Mitsubishi Petrochemical; appl. 16.10.1976; J-prior. 23.10.1975, 31.7.1976).

racemate resolution:
DOS 2 809 794 (Boots; appl. 7.3.1978; GB-prior. 8.3.1977, 18.1.1978).

Formulation(s): amp. 50 mg; cps. 200 mg; drg. 50 mg, 100 mg; eye drops 0.3 mg/ml (as sodium salt dihydrate); plaster 40 mg; s. r. cps. 200 mg; suppos. 100 mg; tabl. 50 mg, 100 mg

Trade Name(s):

D:	Froben (Kanoldt; 1980)	
	Ocuflur (Pharm-Allergan)	
F:	Antadys (Théramex)	
	Cebutid (Knoll; 1979)	
	Ocufen (Allergan)	
GB:	Froben (Knoll; 1977)	
	Ocufen (Allergan)	
I:	Froben (Boots Italia)	
J:	Froben (Kakenyaku)	
USA:	Ocufen (Allergan; 1987).	

Flurotyl

(Flurothyl)

Use: CNS stimulant, convulsant

RN: 333-36-8   MF: $C_4H_4F_6O$ MW: 182.06
LD$_{50}$:   46 mg/kg (M, i.v.)
CN: 1,1'-oxybis[2,2,2-trifluoroethane]

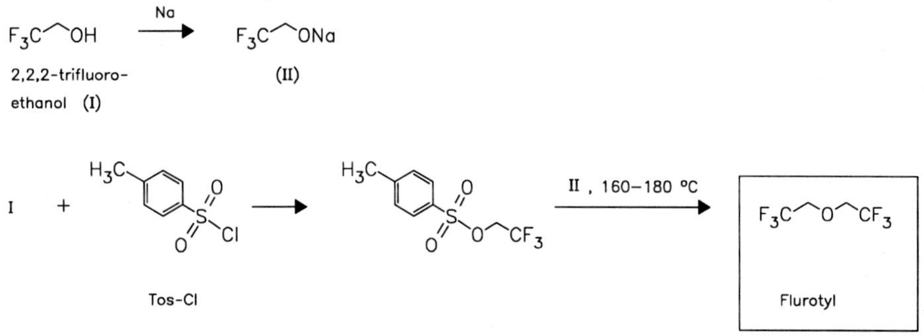

Reference(s):
US 3 363 006 (Pennwalt; 9.1.1968; prior. 29.12.1955, 20.6.1960).

Formulation(s):   2 ml in special inhalation device

Trade Name(s):
USA: Indoklon (Ohio Med.);
        wfm

Fluroxene

ATC: N01AA
Use: inhalation anesthetic

RN: 406-90-6   MF: $C_4H_5F_3O$ MW: 126.08   EINECS: 206-977-1
LD$_{50}$:   5600 mg/kg (R, i.p.)
CN: (2,2,2-trifluoroethoxy)ethene

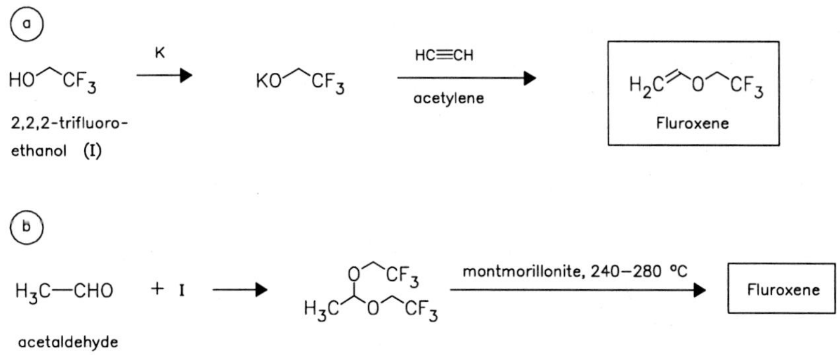

Reference(s):
a   US 2 830 007 (Air Reduction Comp.; 1958; appl. 1953).
b   US 2 870 218 (Air Reduction Comp.; 1959; appl. 1955).

Formulation(s):   liquid for inhalation 125 ml

Trade Name(s):
USA: Fluoromar (Ohio Med.);
wfm

Fluspirilene

ATC: N05AG01
Use: neuroleptic

RN: 1841-19-6 MF: $C_{29}H_{31}F_2N_3O$ MW: 475.58 EINECS: 217-418-6
LD_{50}: 106 mg/kg (M, i.m.);
>146 mg/kg (R, i.m.)
CN: 8-[4,4-bis(4-fluorophenyl)butyl]-1-phenyl-1,3,8-triazaspiro[4.5]decan-4-one

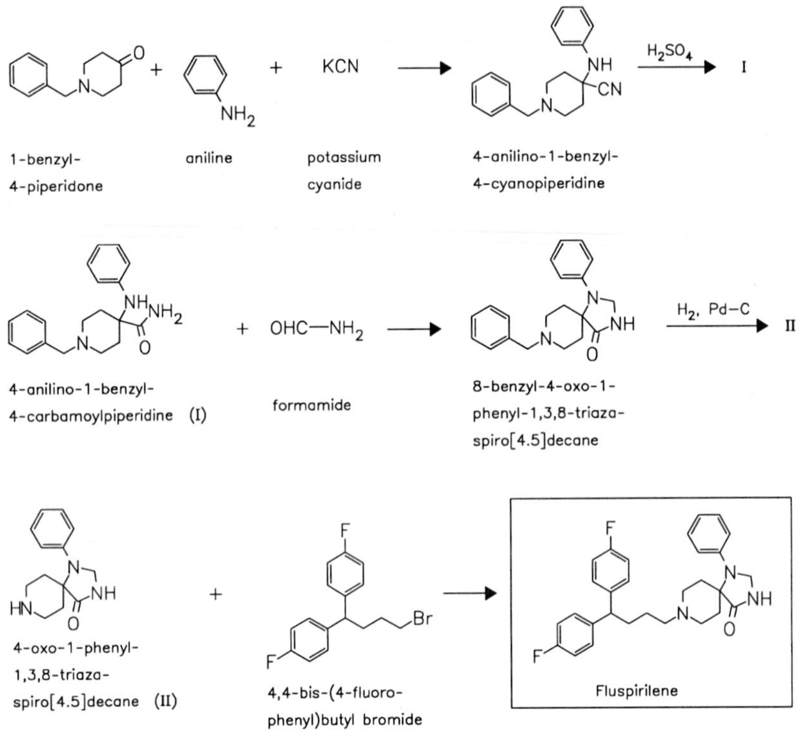

KCN
H_2SO_4
I
1-benzyl-4-piperidone
aniline
potassium cyanide
4-anilino-1-benzyl-4-cyanopiperidine
OHC—NH_2
H_2, Pd–C
II
4-anilino-1-benzyl-4-carbamoylpiperidine (I)
formamide
8-benzyl-4-oxo-1-phenyl-1,3,8-triaza-spiro[4.5]decane
4-oxo-1-phenyl-1,3,8-triaza-spiro[4.5]decane (II)
4,4-bis-(4-fluoro-phenyl)butyl bromide
Fluspirilene

Reference(s):
BE 633 914 (Janssen; appl. 20.6.1963; USA-prior. 22.6.1962).
US 3 238 216 (Janssen; 1.3.1966; prior. 22.6.1962, 20.6.1963).
DAS 1 470 125 (Janssen; appl. 21.6.1963; USA-prior. 22.6.1962).

Formulation(s): amp. 1.5 mg/0.75 ml, 2 mg/ ml, 12 mg/6 ml; vial 12 mg (2 mg/ml)

Trade Name(s):
D: Fluspi (Neuro Hexal)
Imap (Janssen)
Kivat (Hormosan)
GB: Redeptin (Smith Kline & French); wfm
USA: Imap (McNeil); wfm

Flutamide

ATC: L02BB01
Use: antiandrogen, antineoplastic
(hormonal)

RN: 13311-84-7 MF: $C_{11}H_{11}F_3N_2O_3$ MW: 276.21 EINECS: 236-341-9
LD_{50}: 787 mg/kg (R, p.o.);
>2 g/kg (dog, p.o.)
CN: 2-methyl-N-[4-nitro-3-(trifluoromethyl)phenyl]propanamide

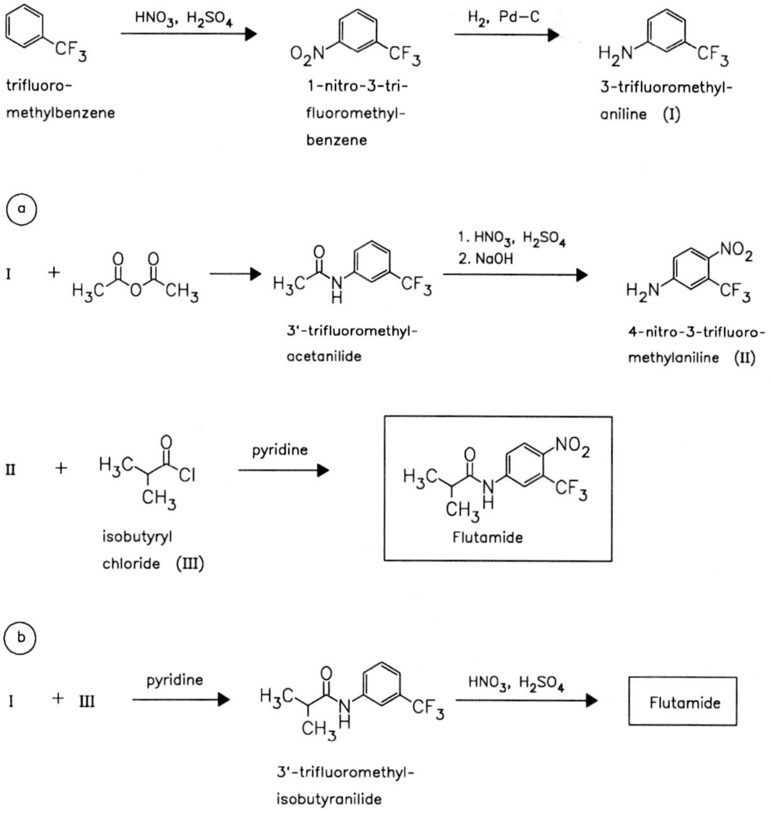

trifluoro-
methylbenzene

1-nitro-3-tri-
fluoromethyl-
benzene

3-trifluoromethyl-
aniline (I)

a

3'-trifluoromethyl-
acetanilide

4-nitro-3-trifluoro-
methylaniline (II)

isobutyryl
chloride (III)

Flutamide

b

3'-trifluoromethyl-
isobutyranilide

Flutamide

Reference(s):
J. Med. Chem. (JMCMAR) **10**, 93 (1967).
a DOS 2 130 450 (Scherico; appl. 19.6.1971).
    US 4 144 270 (Scherico; 13.3.1979; appl. 26.6.1974).
b US 4 302 599 (Schering Co.; 24.11.1981; prior. 10.9.1979).

synthesis of 4-nitro-3-trifluoromethylaniline:
Jones, R.G.: J. Am. Chem. Soc. (JACSAT) **69**, 2346 (1947).

medical use:
US 3 995 060 (Scherico; 30.11.1976; appl. 11.9.1974).
US 4 139 638 (Schering Corp.; 13.2.1979; appl. 3.10.1977).
US 4 161 540 (Schering Corp.; 13.2.1979; appl. 3.10.1977).
US 4 329 364 (Schering Corp.; 11.5.1982; appl. 23.9.1976).
US 4 474 813 (Schering Corp. 2.10.1984; appl. 24.5.1982).

Formulation(s):   cps. 125 mg; tabl. 250 mg

Trade Name(s):

D: Apimid (Apogepha)
      Cytamid (esparma)
      Flumid (Hexal)
      Fluta GRY (GRY-Pharma)
      Flutamex (Sanofi
      Winthrop)
      Fugerel (Essex Pharma;
      1984)

      Prostica (TAD)
      Prostogenat (Azupharma)
      Testac (medac)
      Testotard (Chephasaar)
F: Eulexine (Schering-Plough;
      1987)
GB: Chimax (Chiron)

      Drogenil (Schering-
      Plough)
I: Eulexin (Schering-Plough;
      1986)
J: Odyne (Nippon Kayaku)
USA: Eulexin (Schering)

Flutazolam

ATC: N05BA
Use: benzodiazepine anxiolytic,
      tranquilizer

RN: 27060-91-9   MF: $C_{19}H_{18}ClFN_2O_3$ MW: 376.82
LD_{50}: 1910 mg/kg (M, p.o.);
      >6 g/kg (R, p.o.)
CN: 10-chloro-11b-(2-fluorophenyl)-2,3,7,11b-tetrahydro-7-(2-hydroxyethyl)oxazolo[3,2-
      d][1,4]benzodiazepin-6(5H)-one

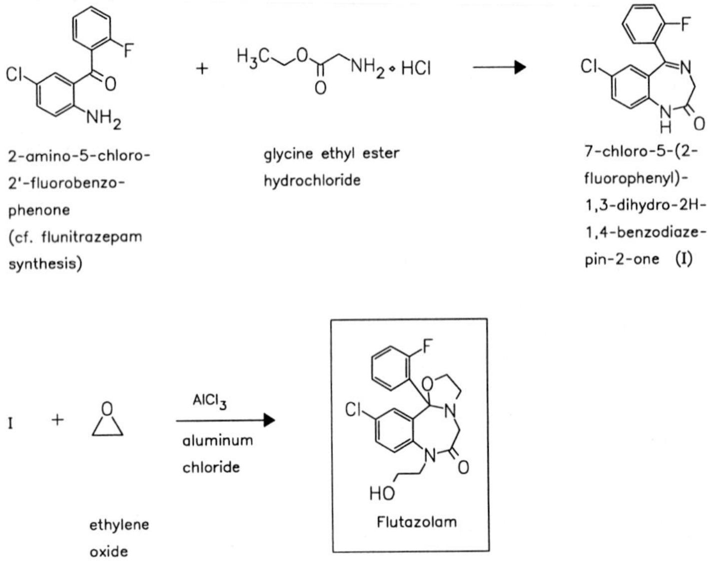

2-amino-5-chloro-
2'-fluorobenzo-
phenone
(cf. flunitrazepam
synthesis)

glycine ethyl ester
hydrochloride

7-chloro-5-(2-
fluorophenyl)-
1,3-dihydro-2H-
1,4-benzodiaze-
pin-2-one (I)

ethylene
oxide

AlCl$_3$
aluminum
chloride

Flutazolam

Reference(s):
DOS 1 952 486 (Hoffmann-La Roche; appl. 17.10.1969; USA-prior. 18.10.1968).

synthesis of 7-chloro-5-(2-fluorophenyl)-1,3-dihydro-2H-1,4-benzodiazepin-2-one:
US 3 109 843 (Hoffmann-La Roche; 5.11.1963; prior. 21.6.1962).

Formulation(s):   tabl. 4 mg

Trade Name(s):
J: Coreminal (Mitsui)

Fluticasone propionate

ATC: R01AD08; R03BA05; D07AC17
Use: locally active glucocorticosteroid

RN: 80474-14-2   MF: $C_{25}H_{31}F_3O_5S$ MW: 500.58
LD$_{50}$: >2 g/kg (R, p. o.); >1 g/kg (R, s. c.)
CN: (6α,11β,16α,17α)-6,9-Difluoro-11-hydroxy-16-methyl-3-oxo-17-(1-oxopropoxy)androsta-1,4-diene-17-carbothioic acid S-(fluoromethyl) ester

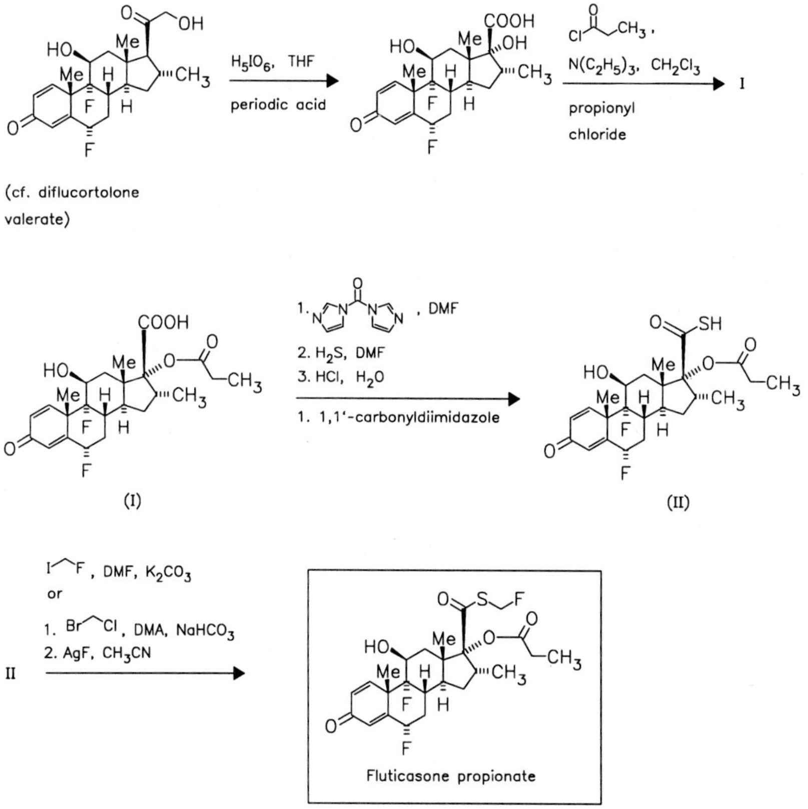

(cf. diflucortolone valerate)

Fluticasone propionate

Reference(s):
BE 887 518 (Glaxo Group; appl. 13.2.1981; GB-prior. 15.2.1980).
IL 109 656 (Chemagis LTD.; IL-prior. 15.5.1994).
Phillipps, G.H. et al.: J. Med. Chem. (JMCMAR) 37, 3717 (1994).

Formulation(s): aerosol for inh. 44 μg, 110 μg, 220 μg; cream 0.05%; ointment 0.005%; nasal spray 0.05%

Trade Name(s):
D: Atemur (ASTA Medica AWD; Glaxo Wellcome)
Flutide (Cascan; Glaxo Wellcome)
Flutivate (Cascan; Glaxo Wellcome)

Viani (Cascan; Glaxo Wellcome) comb. with Salmeterol
GB: Cutivate (Glaxo Wellcome)
Flixonase (Allen & Hanburys)
Flixotide (Allen & Hanburys)

I: Flixotide (Glaxo Wellcome)
Fluspiral (Menarini)
USA: Cutivate (Glaxo Wellcome)
Flonase (Glaxo Wellcome)
Flovent (Glaxo Wellcome)

Flutoprazepam

ATC: N05BA
Use: long acting benzodiazepine anxiolytic

RN: 25967-29-7   MF: $C_{19}H_{16}ClFN_2O$ MW: 342.80
LD$_{50}$:   2110 mg/kg (M, i.p.); 2430 mg/kg (M, p.o.);
       2230 mg/kg (R, i.p.); 10.06 g/kg (R, p.o.);
       1000 mg/kg (rabbit, p.o.);
       >10 g/kg (dog, p.o.)
CN: 7-chloro-1-(cyclopropylmethyl)-5-(2-fluorophenyl)-1,3-dihydro-2H-1,4-benzodiazepin-2-one

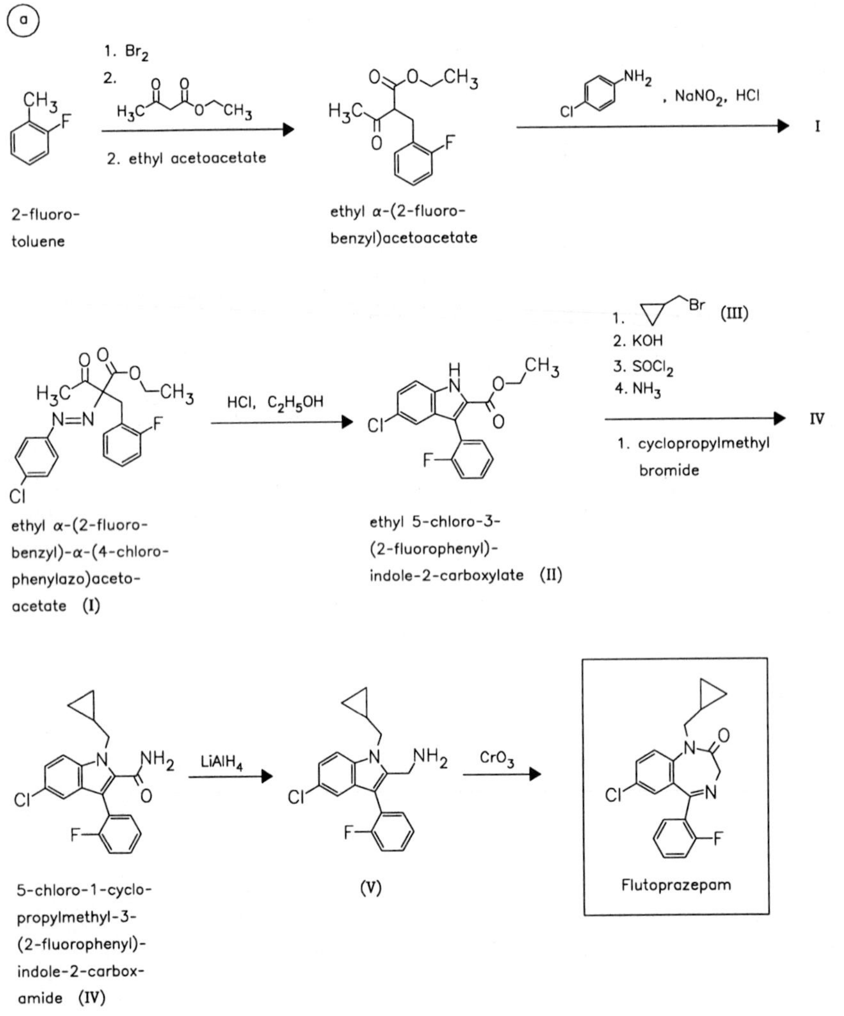

2-fluoro-
toluene

ethyl α-(2-fluoro-
benzyl)acetoacetate

ethyl α-(2-fluoro-
benzyl)-α-(4-chloro-
phenylazo)aceto-
acetate (I)

ethyl 5-chloro-3-
(2-fluorophenyl)-
indole-2-carboxylate (II)

5-chloro-1-cyclo-
propylmethyl-3-
(2-fluorophenyl)-
indole-2-carbox-
amide (IV)

(V)

Flutoprazepam

(b)

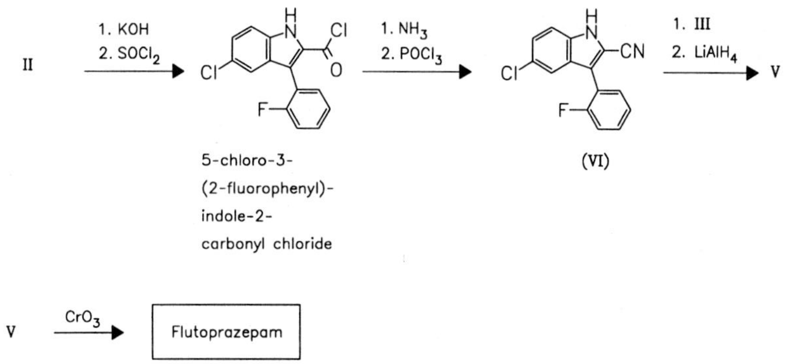

II

1. KOH
2. SOCl₂

5-chloro-3-
(2-fluorophenyl)-
indole-2-
carbonyl chloride

1. NH₃
2. POCl₃

(VI)

1. III
2. LiAlH₄

V

V

CrO₃

Flutoprazepam

(c)

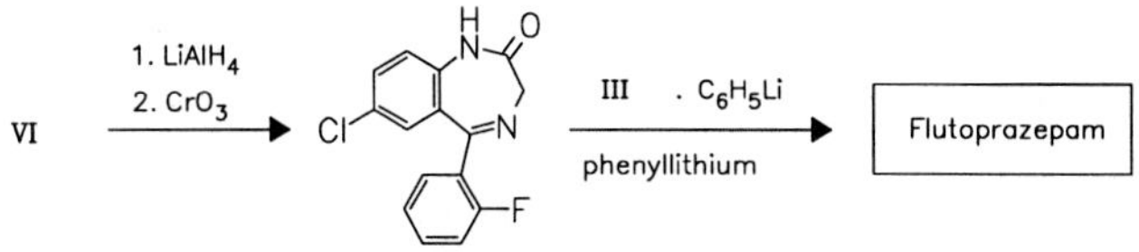

VI

1. LiAlH₄
2. CrO₃

III . C₆H₅Li

phenyllithium

Flutoprazepam

Reference(s):
DE 1 795 372 (Sumitomo; appl. 20.9.1968; J-prior. 22.9.1967).
DE 1 795 771 (Sumitomo; appl. 20.9.1968; J-prior. 2.11.1967).
US 3 925 364 (Sumitomo; 6.8.1974; appl. 16.9.1968; J-prior. 22.9.1967).

additional synthesis:
DOS 2 151 540 (Sumitomo; appl. 15.10.1971; J-prior. 17.10.1970).
DOS 2 113 122 (Sumitomo; appl. 18.3.1971; J-prior. 19.3.1970).

Formulation(s):   tabl. 2 mg

Trade Name(s):
J: Restas (Banyu; 1985)

Flutrimazole
(UR-4056)

ATC: D01A
Use: topical antifungal

RN: 119006-77-8   MF: C₂₂H₁₆F₂N₂   MW: 346.38
CN: 1-[(2-fluorophenyl)(4-fluorophenyl)phenylmethyl]-1*H*-imidazole

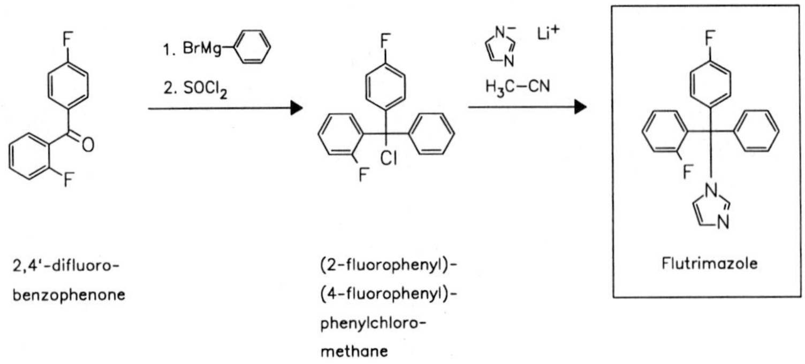

1. BrMg–
2. SOCl2
H3C–CN
Li+
2,4'-difluoro-
benzophenone
(2-fluorophenyl)-
(4-fluorophenyl)-
phenylchloro-
methane
Flutrimazole

Reference(s):
EP 352 352 (J. Uriach & Cia.; appl. 31.1.1990; prior. 28.7.1988).

Formulation(s): cream 1 %

Trade Name(s):
E: Micetal (Uriach)

Flutropium bromide
(BA-598 BR)

ATC: R03BB
Use: anticholinergic, bronchodilator

RN: 63516-07-4 MF: $C_{24}H_{29}BrFNO_3$ MW: 478.40
LD_{50}: 53 mg/kg (M, i.p.); 11 mg/kg (M, i.v.); 760 mg/kg (M, p.o.); 228 mg/kg (M, s.c.);
77 mg/kg (R, i.p.); 12.5 mg/kg (R, i.v.); 740 mg/kg (R, p.o.); 615 mg/kg (R, s.c.)
CN: (*endo,syn*)-8-(2-fluoroethyl)-3-[(hydroxydiphenylacetyl)oxy]-8-methyl-8-azoniabicyclo[3.2.1]octane bromide

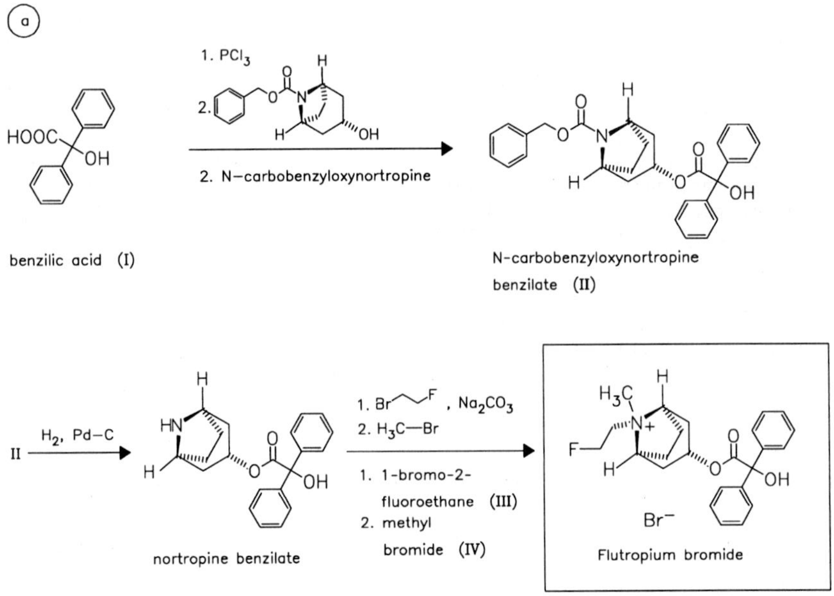

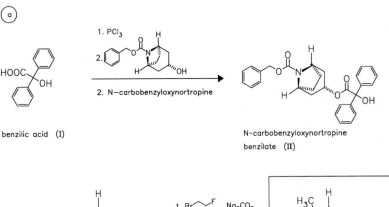

a
1. PCl3
2.
2. N-carbobenzyloxynortropine
benzilic acid (I)
N-carbobenzyloxynortropine
benzilate (II)
II
H2, Pd-C
nortropine benzilate
1. Br–CH2CH2–F, Na2CO3
2. H3C—Br
1. 1-bromo-2-
fluoroethane (III)
2. methyl
bromide (IV)
Br−
Flutropium bromide

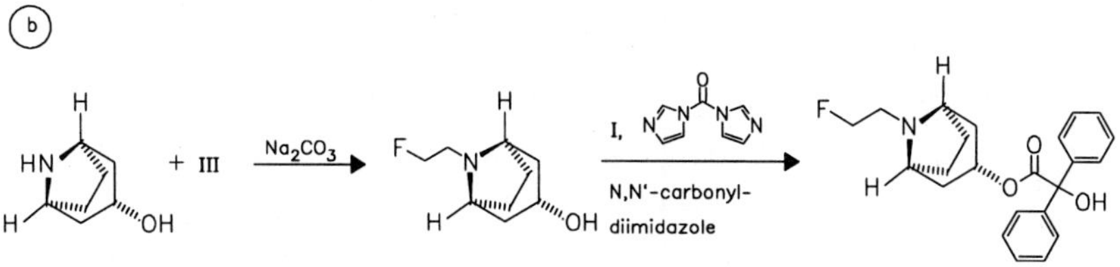
b
H
HN
H
OH
+ III
Na2CO3
F
N
H
OH
I,
N,N'-carbonyl-
diimidazole
F
N
H
O
OH

nortropine
N-(β-fluoroethyl)-nortropine
N-(β-fluoroethyl)nortropine benzilate (V)

V + IV → Flutropium bromide

Reference(s):
DE 2 540 633 (Boehringer Ing.; appl. 12.9.1976).
Banholzer, R. et al.: Arzneim.-Forsch. (ARZNAD) **36**, 1161 (1986).

synthesis of nortropine benzilate:
Bertholdt, H. et al.: Arzneim.-Forsch. (ARZNAD) **17**, 719 (1967)

Trade Name(s):
J: Flubron (S. S. Pharm.)

Fluvastatin sodium
(SRI-62320; XU-62-320; XU-620)

ATC: B04AB04
Use: hyperlipidemic, HMG-CoA-reductase inhibitor

RN: 93957-55-2 MF: $C_{24}H_{25}FNNaO_4$ MW: 433.46
CN: [*R**,*S**-(*E*)]-(±)-7-[3-(4-fluorophenyl)-1-(1-methylethyl)-1*H*-indol-2-yl]-3,5-dihydroxy-6-heptenoic acid monosodium salt

free acid
RN: 93957-54-1 MF: $C_{24}H_{26}FNO_4$ MW: 411.47

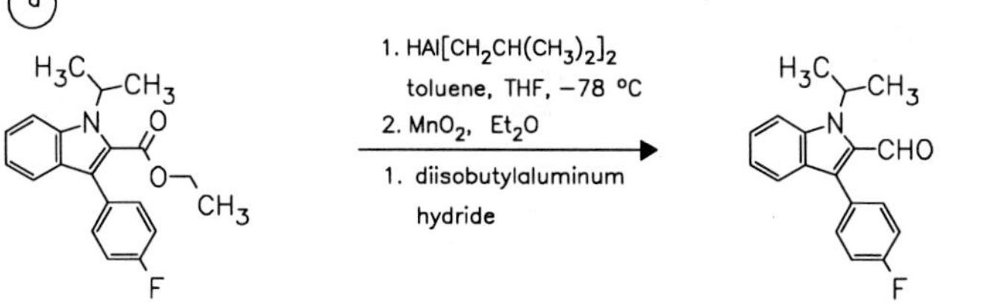
a
H3C
CH3
N
O
O
CH3
F
1. HAl[CH2CH(CH3)2]2
toluene, THF, −78 °C
2. MnO2, Et2O
1. diisobutylaluminum
hydride
H3C
CH3
N
CHO
F

ethyl 3-(4-fluorophenyl)-1-isopropylindole-2-carboxylate
3-(4-fluorophenyl)-1-isopropylindole-2-carboxaldehyde (I)

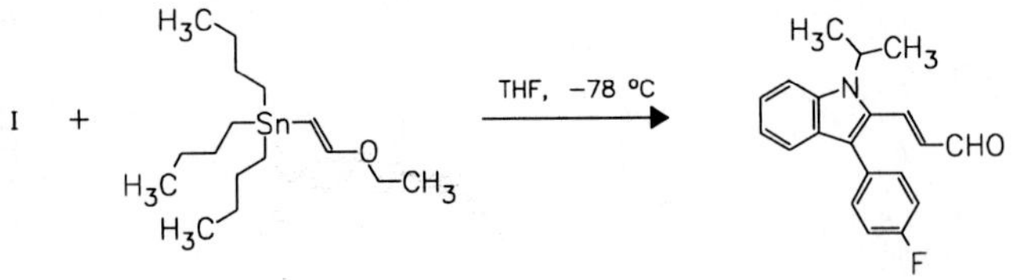

I + 2-(tributylstannyl)vinyl ethyl ether → THF, −78 °C → (E)-3-[3-(4-fluoro-phenyl)-1-isopropyl-indol-2-yl]-propenal (II)

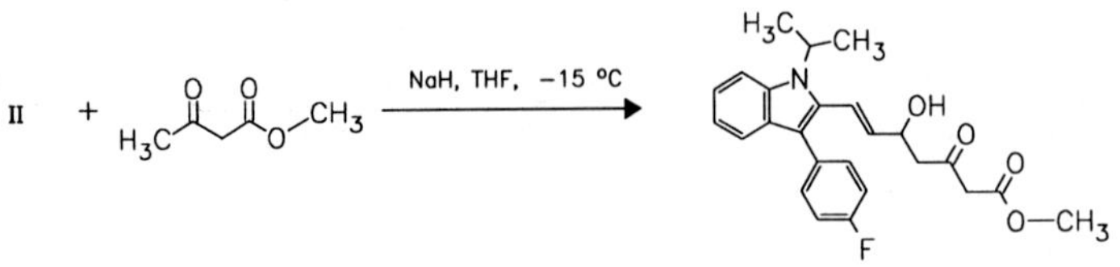

II + methyl acetoacetate → NaH, THF, −15 °C → methyl (±)-(E)-7-[3-(4-fluoro-phenyl)-1-isopropylindol-2-yl]-5-hydroxy-3-oxohept-6-enoate (III)

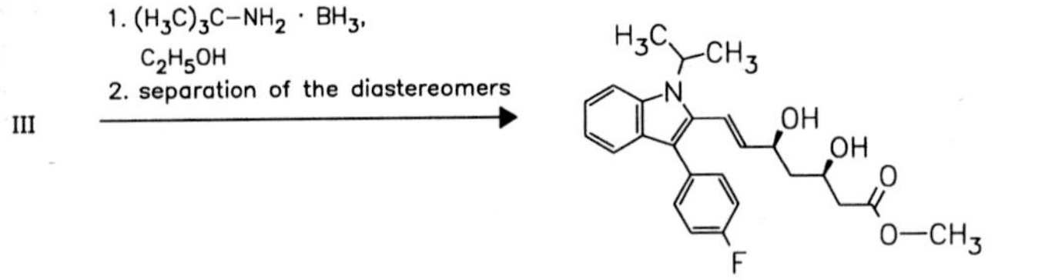

III → 1. (H₃C)₃C–NH₂ · BH₃, C₂H₅OH 2. separation of the diastereomers → methyl (±)-erythro-(E)-3,5-dihydroxy-7-[3-(4-fluorophenyl)-1-isopropyl-indol-2-yl]hept-6-enoate (IV)

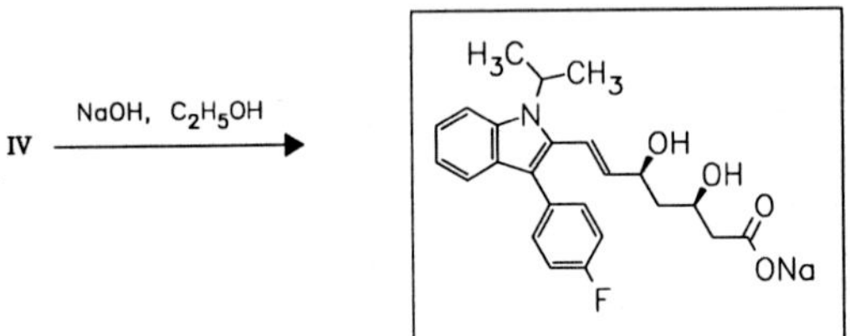

IV → NaOH, C₂H₅OH → Fluvastatin sodium

b

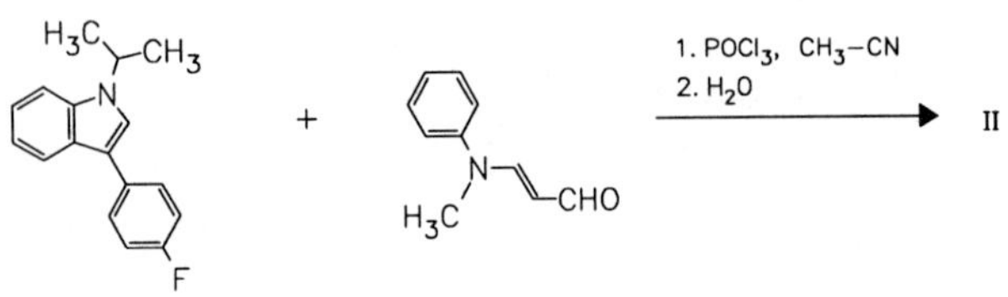

V + 3-(methylphenyl-amino)acrolein → 1. POCl₃, CH₃–CN 2. H₂O → II

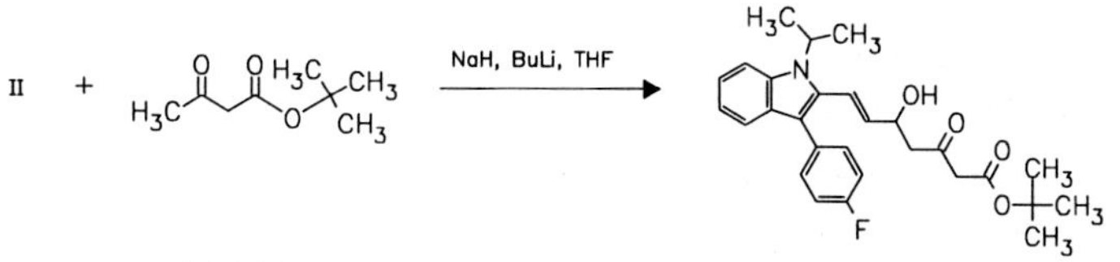

II + tert-butyl acetoacetate →(NaH, BuLi, THF)→ tert-butyl (±)-(E)-7-[3-(4-fluoro-phenyl)-1-isopropylindol-2-yl]-5-hydroxy-3-oxohept-6-enoate (VI)

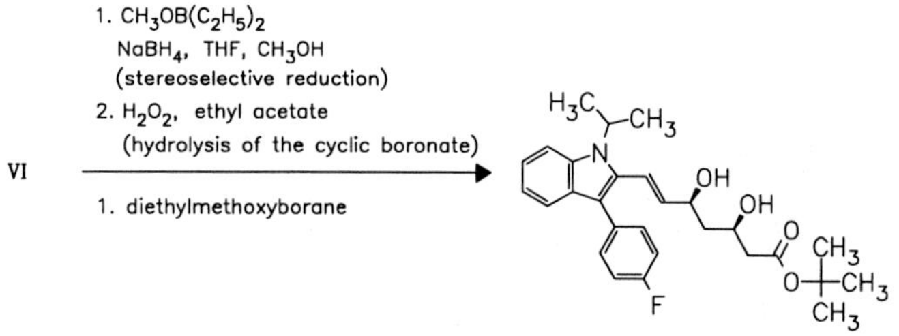

VI →
1. CH$_3$OB(C$_2$H$_5$)$_2$
NaBH$_4$, THF, CH$_3$OH
(stereoselective reduction)
2. H$_2$O$_2$, ethyl acetate
(hydrolysis of the cyclic boronate)
1. diethylmethoxyborane
→ tert-butyl (±)-erythro-(E)-7-[3-(4-fluorophenyl)-1-isopropyl-indol-2-yl]-3,5-dihydroxy-hept-6-enoate (VII)

VII →(NaOH, THF, 10 °C)→ Fluvastatin sodium

(aa) synthesis of 3-(4-fluorophenyl)-1-isopropylindole-2-carboxaldehyde (I):

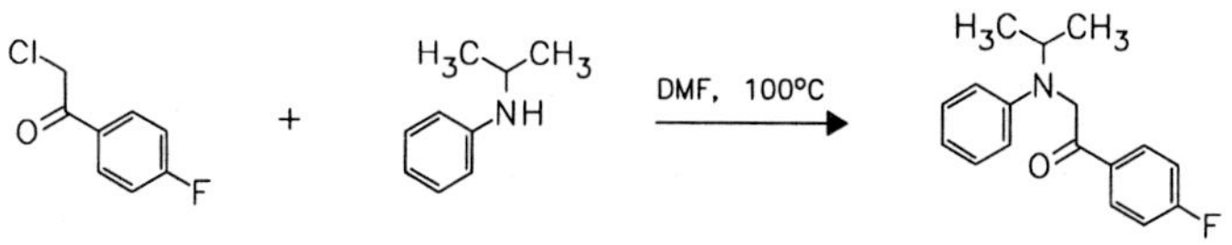

α-chloro-p-fluoroaceto-phenone + N-isopropyl-aniline →(DMF, 100°C)→ (VIII)

VIII →(ZnCl$_2$, C$_2$H$_5$OH, 100−105°C)→ V →(POCl$_3$, (H$_3$C)$_2$N−CHO)→ I

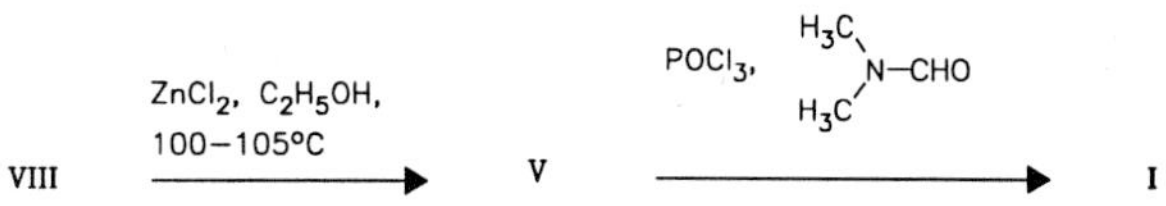

Reference(s):
a   WO 8 402 131 (Sandoz; appl. 18.11.1983; USA-prior. 22.11.1982, 1.11.1983, 4.3.1985).
aa   Walkup, R.E. et al.: Tetrahedron Lett. (TELEAY) **26** (18), 2155-2158 (1985).
b   EP 363 934 (Sandoz; appl. 11.10.1989; USA-prior. 13.10.1983, 22.5.1989).

composition with improved storage stability:
US 5 356 896 (Sandoz; 18.10.1994; appl. 22.12.1992; USA-prior. 12.12.1991).

oral pharmaceutical composition:
EP 547 000 (Sandoz; appl. 8.12.1992; USA-prior. 12.12.1991).

combination with squalene synthase inhibitors:
EP 482 498 (Squibb; appl. 16.10.1991; USA-prior. 19.10.1990).
EP 401 705 (Squibb; appl. 1.6.1990; USA-prior. 5.6.1989).

combination with ACE inhibitors:
EP 461 548 (Squibb; appl. 6.6.1991; USA-prior. 11.6.1990).
EP 457 514 (Squibb; appl. 10.5.1991; USA-prior. 15.5.1990).

combination with niacin *or* probucol:
EP 373 507 (Squibb; appl. 7.12.1982; USA-prior. 12.12.1988).

composition containing coenzyme Q10:
US 4 933 165 (Merck & Co.; 12.6.1990; appl. 18.1.1989; USA-prior. 18.1.1989).
US 4 929 437 (Merck & Co.; 29.5.1990; appl. 2.2.1989; USA-prior. 2.2.1989).

Formulation(s):　cps. 21.06 mg, 42.12 mg

Trade Name(s):

D:	Cranoc (Astra/Promed)	F:	Fractal (Sinbio)	GB:	Lescol (Novartis)
	Locol (Novartis Pharma)		Lescol (Novartis)	USA:	Lescol (Novartis)

Fluvoxamine

ATC:　N06AB08
Use:　antidepressant

RN:　54739-18-3　MF: $C_{15}H_{21}F_3N_2O_2$　MW: 318.34
CN:　(*E*)-5-methoxy-1-[4-(trifluoromethyl)phenyl]-1-pentanone *O*-(2-aminoethyl)oxime

hydrogen maleate
RN:　61718-82-9　MF: $C_{15}H_{21}F_3N_2O_2 \cdot C_4H_4O_4$　MW: 434.41

4'-trifluoromethyl-5-　　　　　O-(2-aminoethyl)-　　　　Fluvoxamine
methoxyvalerophenone　(I)　　hydroxylamine

I　+　HO—NH₂　　　　Fluvoxamine

hydroxyl-　　　2-chloro-
amine　　　　ethylamine

Reference(s):
DE 2 610 886 (Philips Gloeilampenfabrieken; appl. 7.10.1976; prior. 16.3.1976).
US 4 085 225 (Philips Corp.; 18.4.1978; appl. 13.3.1976; NL-prior. 20.3.1975).
NL 7 503 310 (Philips Gloeilampenfabrieken; appl. 20.3.1975).

Formulation(s):　f. c. tabl. 50 mg, 100 mg; tabl. 25 mg, 50 mg, 100 mg (as hydrogen maleate)

Trade Name(s):

D:	Fevarin (Solvay	GB:	Faverin (Solvay; 1987)	USA:	Luvox (Solvay)
	Arzneimittel; 1984)	I:	Dumirox (Upjohn)		
F:	Floxyfral (Solvay Pharma;		Fevarin (UCM)		
	1986)		Maveral (Farmades)		

Folescutol

ATC: C05C
Use: capillary therapeutic, capillary protectant

RN: 15687-22-6 MF: $C_{14}H_{15}NO_5$ MW: 277.28 EINECS: 239-783-0
CN: 6,7-dihydroxy-4-(4-morpholinylmethyl)-2*H*-1-benzopyran-2-one

hydrochloride
RN: 36002-19-4 MF: $C_{14}H_{15}NO_5 \cdot HCl$ MW: 313.74 EINECS: 252-831-5

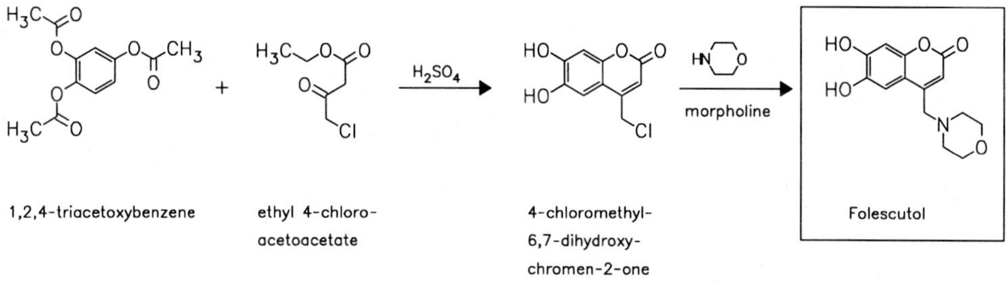

H2SO4
morpholine
1,2,4-triacetoxybenzene
ethyl 4-chloro-
acetoacetate
4-chloromethyl-
6,7-dihydroxy-
chromen-2-one
Folescutol

Reference(s):
FR-M 2 035 (Lab. Dausse; appl. 29.6.1962).

Formulation(s): drg. 20 mg in comb.

Trade Name(s):
D: Detensitral (Karlspharma)-comb.; wfm
F: Covalan (Dausse); wfm
Tensitral (Dausse)-comb.; wfm

Folic acid
(Pteroylglutamic acid)

ATC: B03BB01
Use: antianemic, growth factor

RN: 59-30-3 MF: $C_{19}H_{19}N_7O_6$ MW: 441.40 EINECS: 200-419-0
LD_{50}: 282 mg/kg (M, i.v.); 10 g/kg (M, p.o.)
CN: *N*-[4-[[(2-amino-1,4-dihydro-4-oxo-6-pteridinyl)methyl]amino]benzoyl]-L-glutamic acid

monosodium salt
RN: 6484-89-5 MF: $C_{19}H_{18}N_7NaO_6$ MW: 463.39 EINECS: 229-348-3
LD_{50}: 631 mg/kg (M, i.v.);
526 mg/kg (R, i.v.)

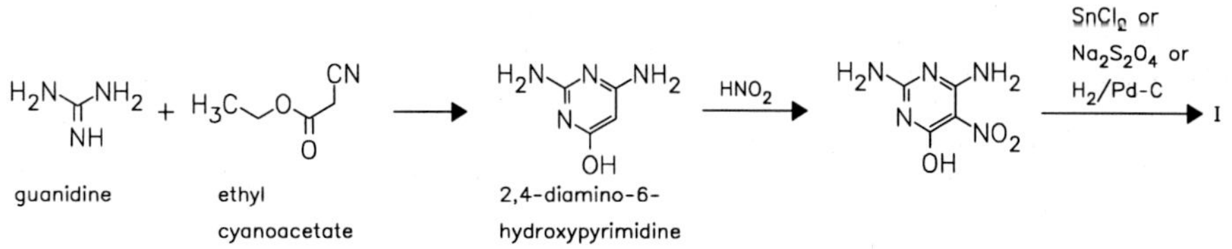

SnCl2 or
Na2S2O4 or
H2/Pd-C
HNO2
guanidine
ethyl
cyanoacetate
2,4-diamino-6-
hydroxypyrimidine

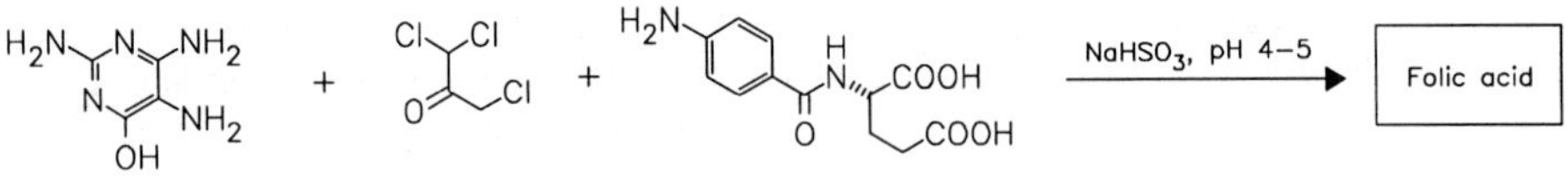

6-hydroxy-2,4,5-
triaminopyrimidine (I)

1,1,3-tri-
chloroacetone

N-(4-aminobenzoyl)-
L-glutamic acid (II)

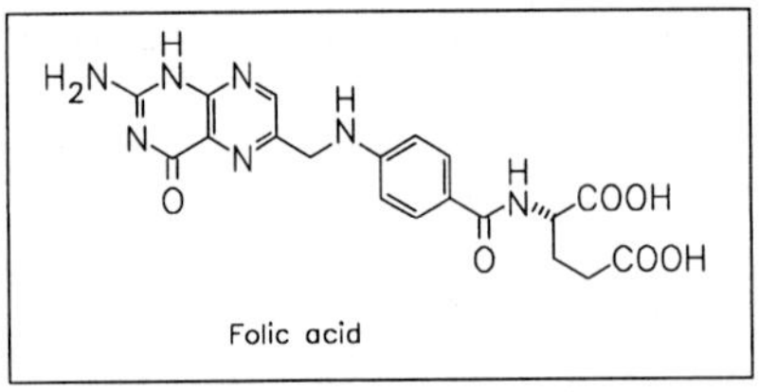

Folic acid

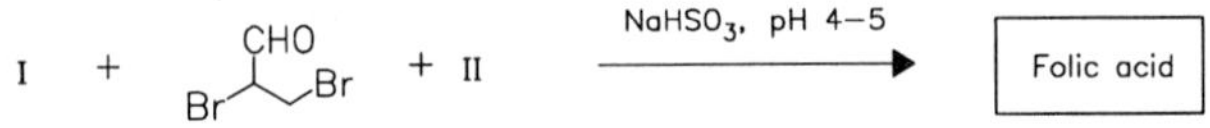

I + 2,3-dibromo-propionaldehyde + II →(NaHSO₃, pH 4–5)→ Folic acid

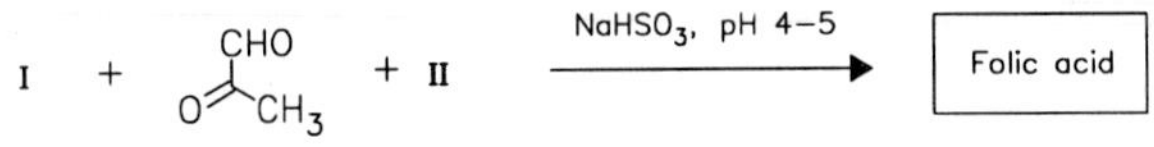

I + pyruvaldehyde + II →(NaHSO₃, pH 4–5)→ Folic acid

Reference(s):

Wailer, C.W. et al.: J. Am. Chem. Soc. (JACSAT) **70**, 19 (1948).
Hultquist, M.E. et al.: J. Am. Chem. Soc. (JACSAT) **70**, 23 (1948).
Angier, R.B. et al.: J. Am. Chem. Soc. (JACSAT) **70**, 25 (1948).
Boothe, J.H. et al.: J. Am. Chem. Soc. (JACSAT) **70**, 27 (1948).
US 2 436 073 (American Cyanamid; 1948; appl. 1945).
US 2 442 836 (American Cyanamid; 1948; appl. 1945).
US 2 443 078 (American Cyanamid; 1948; appl. 1945).
US 2 443 165 (American Cyanamid; 1948; appl. 1946).
US 2 444 002 (American Cyanamid; 1948; appl. 1946).
US 2 472 482 (American Cyanamid; 1949; appl. 1947).
US 2 477 426 (American Cyanamid; 1949; appl. 1948).
US 2 547 501 (American Cyanamid; 1951; appl. 1946).
US 2 599 526 (American Cyanamid; 1952; appl. 1951).
US 2 719 157 (Shionogi; 1955; appl. 1951).
US 2 956 057 (Kongo Kagaku Kabushiki Kaisha; 1960; J-prior. 1955).

use of propargyl aldehyde:
US 2 766 240 (Aries Labs.; 1956; appl. 1953).

condensation with α-bromoacroleine:
US 2 476 360 (Parke Davis; 1949; appl. 1946).

alternative synthesis (*via* 2-amino-6-formyl-4-hydroxy-pteridine):
US 2 786 056 (Merck & Co.; 1957; appl. 1954).
US 2 816 109 (Merck & Co.; 1957; appl. 1954).
US 2 821 527 (Merck & Co.; 1958; appl. 1954).
US 2 821 528 (Merck & Co.; 1958; appl. 1954).
US 3 067 200 (Merck & Co.; 4.12.1962; prior. 3.5.1954, 20.2.1957).
Bieri, J.H.; Viscontini, M.: Helv. Chim. Acta (HCACAV) **56**, 2905 (1973).

synthesis via 2-amino-4-hydroxy-6-halogenomethylpteridine:
US 2 547 519 (American Cyanamid; 1951; appl. 1946).
US 2 547 520 (American Cyanamid; 1951; appl. 1946).
US 2 584 538 (American Cyanamid; 1952; appl. 1948).

improved method for synthesis of 2-amino-4-hydroxy-6-methylpteridine:
GB 1 503 476 (Lonza; appl. 6.1.1977; CH-prior. 13.1.1976).
US 4 094 874 (Lonza; 13.6.1978; CH-prior. 13.1.1976).

Formulation(s):   amp. 5 mg/5 ml; tabl. 5 mg

Trade Name(s):

D: Folarell (Sanorell)
    Fol-ASmedic (Dyckerhoff)
    Folsan (Solvay
    Arzneimittel)
    Folsäure Injektionslösung
    (Hevert)
    Folverlan (Verla)
    Lafol (Brenner-Efeka;
    LAW)
    numerous combination
    preparations
F: Alvityl (Solvay Pharma)-
    comb.
    Azedavit (Whitehall)-
    comb.
    Azinc complexe
    (Arkopharma)-comb.
    Carencyl (Riom)-comb.
    Élévit Vitamine B9
    (Nicholas)-comb.
    Forvital (Whitehall)-comb.
    Lofenalac (Bristol-Myers
    Squibb)-comb.
    Plenyl (Oberlin)-comb.

    Speciafoldine (Specia)
    Vivamyne (Whitehall)-
    comb.
GB: Ferfolic SV (Sinclair)-
    comb.
    Ferrograd Folic (Abbott)-
    comb.
    Folex-350 (Shire)-comb.
    Galfer F.A. (Galen)-comb.
    Lexpec (Rosemont)
    Meterfolic (Sinclair)-comb.
    Pregaday (Evans)-comb.
    Slow-Fe folic (Novartis)-
    comb.
    numerous combination
    preparations
I: Combetasi (ISI; as calcium
    salt)-comb.
    Efargen (Teofarma)-comb.
    Epargriseovit (Farmitalia)-
    comb.
    Eparmefolin (Bracco; as
    calcium salt)-comb.

    Ferrofolin (Farmades; as
    calcium salt)-comb.
    Ferrograd Folic (Abbott)-
    comb.
    Ferrotre (Mediolanum)-
    comb.
    Folina (Astra-Simes)
    Folinemic (Firma; as
    calcium salt)-comb.
    Lederfolin (Cyanamid; as
    calcium salt)
    Oro B12 (Ripari-Gero)-
    comb.
    Tonofolin (Zyma; as
    calcium salt)
J: Foliamin (Takeda)
    Folical (Shionogi)
USA: Bevitamel (Westlake)
    Cefol (Abbott)
    Folic (Lederle)
    Materna (Lederle)
    various generic
    preparations

Folinic acid
(Citrovorum factor)

ATC: A04A; V03AB
Use: antianemic, growth factor, antidote
    (as calcium salt, at overdose of folic
    acid antagonists)

RN: 58-05-9   MF: $C_{20}H_{23}N_7O_7$ MW: 473.45   EINECS: 200-361-6
CN: *N*-[4-[[(2-amino-5-formyl-1,4,5,6,7,8-hexahydro-4-oxo-6-pteridinyl)methyl]amino]benzoyl]-L-glutamic
    acid

calcium salt (1:1) (leucovorin calcium)
RN: 1492-18-8   MF: $C_{20}H_{21}CaN_7O_7$ MW: 511.51   EINECS: 216-082-8
LD$_{50}$:   732 mg/kg (M, i.v.); >7 g/kg (M, p.o.);
    >8 g/kg (R, p.o.)
calcium salt (1:1) pentahydrate
RN: 6035-45-6   MF: $C_{20}H_{21}CaN_7O_7 \cdot 5H_2O$ MW: 601.58

folic acid (q. v.)

10-formylfolic acid (I)

Folinic acid

Reference(s):

US 2 741 608 (Research Corp.; 1956; prior. 1950).

DOS 2 836 599 (US Department of Commerce; appl. 22.8.1978; USA-prior. 22.8.1977).

Temple, C. et al.: J. Med. Chem. (JMCMAR) **22**, 731 (1979).

Formulation(s):   amp. 3 mg, 5 mg, 6 mg, 10 mg, 15 mg, 30 mg, 50 mg; cps. 5 mg; tabl. 15 mg, 25 mg (as calcium salt); amp. 1.5 mg, 3 mg, 15 mg, 30 mg, 350 mg; cps. 15 mg; powder 50 mg, 100 mg, 200 mg, 300 mg; tabl. 15 mg; vial 100 mg, 200 mg, 300 mg (as calcium salt pentahydrate)

Trade Name(s):

D:	Calciumfolinat (Rhône-Poulenc); wfm	Lederfoline (Wyeth-Lederle)	I:	Adinepar (Leben's)-comb. Hepafactor (Sigma-Tau)-comb.

D: Calciumfolinat (Rhône-
      Poulenc); wfm
      Leucovorin (Lederle); wfm
      Rescuvolin (medac); wfm
F: Elvorine (Wyeth-Lederle)
      Folinate de calcium
      (Aguettant)
      Folinoral (Therabel Lucien
      Pharma)

      Lederfoline (Wyeth-
      Lederle)
      Osfolate (ASTA Medica)
      Perfolate (ASTA Medica)
GB: Calcium Leucovorin
      (Lederle); wfm
      Refolinon (Pharmacia &
      Upjohn)
      Rescufolin (Nordic); wfm

I: Adinepar (Leben's)-comb.
      Hepafactor (Sigma-Tau)-
      comb.
      Rekord B12 complex
      (Sigma-Tau)-comb.
J: Leucovorin (Lederle)
USA: Calcium Leucovorin
      (Lederle); wfm

Fomepizole

Use: antidote for ethlene glycol, competitive inhibitor of alcohol dehydrogenase

RN: 7554-65-6   MF: C$_4$H$_6$N$_2$ MW: 82.11   EINECS: 231-445-0

CN: 4-Methyl-1*H*-pyrazole

2-methyl-2-propenal

4,5-dihydro-4-methyl-pyrazole

Fomepizole

Reference(s):
Pechmann, H.; Burkard, E.: Ber. Dtsch. Chem. Ges. (BDCGAS) **33**, 3590 (1900).
Hoyce, D.S. et al.: J. Org. Chem. (JOCEAH) **20**, 1681 (1955).
Momose, T. et al.: Heterocycles (HTCYAM) **30**, 789 (1990).
DE 4 328 228 (BASF; prior. 23.8.1993).
US 5 569 769 (BASF; 29.10.1996; D-prior. 23.8.1993).
DE 3 918 979 (BASF; prior. 10.6.1989).
EP 366 328 (Nissan Chem. Ind.; appl. 17.10.1989; J-prior. 26.10.1988).

Formulation(s):   vials, 1 g/ml; 1.5 ml

Trade Name(s):
USA: Antizol (Orphan Medical;
      1998)

Fominoben

ATC: N06
Use: antitussive, respiratory analeptic

RN: 18053-31-1   MF: $C_{21}H_{24}ClN_3O_3$ MW: 401.89   EINECS: 241-964-4
CN: N-[3-chloro-2-[[methyl[2-(4-morpholinyl)-2-oxoethyl]amino]methyl]phenyl]benzamide

monohydrochloride
RN: 24600-36-0   MF: $C_{21}H_{24}ClN_3O_3 \cdot HCl$ MW: 438.36   EINECS: 246-344-7

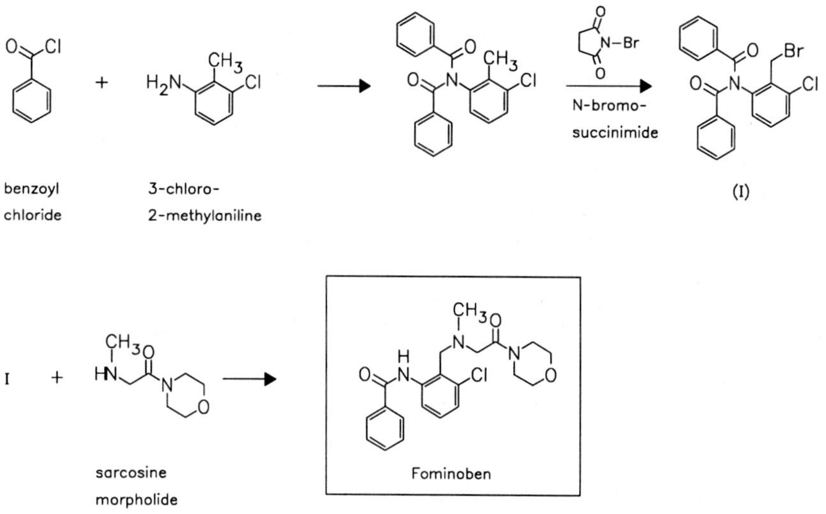

benzoyl   3-chloro-
chloride   2-methylaniline   N-bromo-succinimide   (I)

sarcosine
morpholide   Fominoben

Reference(s):
DE 1 795 259 (Thomae; appl. 13.7.1966).
Krüger, G. et al.: Arzneim.-Forsch. (ARZNAD) **23**, 290 (1973).

Formulation(s):   amp. 40 mg/5 ml; drg 160 mg (as hydrochloride)

Trade Name(s):

D:	Broncho-Noleptan (Thomae); wfm	Noleptan (Thomae); wfm	Tussirama (Serpero); wfm
	I:	Terion (Lusofarmaco); wfm	

Fomocaine

ATC: R02AD
Use: local anesthetic

RN: 17692-39-6   MF: $C_{20}H_{25}NO_2$ MW: 311.43
CN: 4-[3-[4-(phenoxymethyl)phenyl]propyl]morpholine

(a)

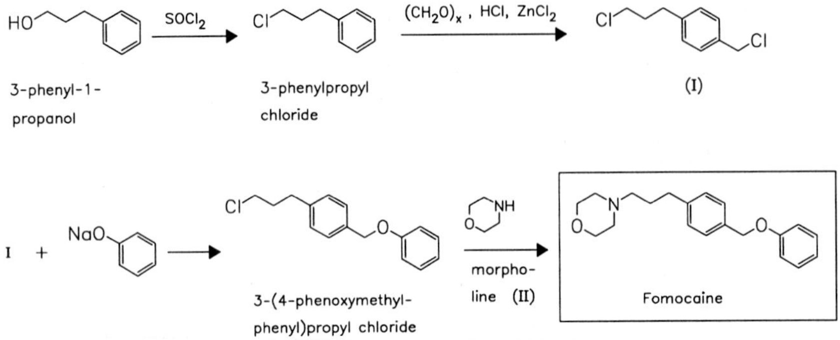

3-phenyl-1-
propanol

3-phenylpropyl
chloride

(I)

3-(4-phenoxymethyl-
phenyl)propyl chloride

morpho-
line   (II)

Fomocaine

(b)

4-bromomethyl-
benzonitrile

4-phenoxymethyl-
benzonitrile

ethylmagnesium
bromide

4'-phenoxymethyl-
propiophenone   (III)

Fomocaine

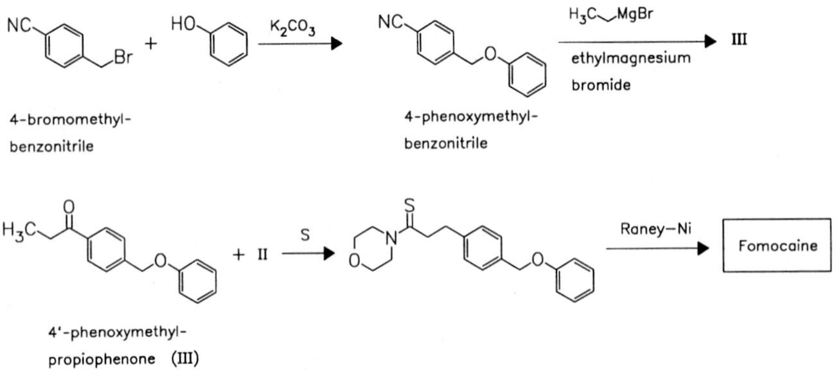

Reference(s):
a   Oelschläger, H.: Arzneim.-Forsch. (ARZNAD) **9**, 313 (1959).
     GB 786 128 (Promonta; appl. 15.11.1955; D-prior. 15.11.1954).
b   Oelschläger, H. et al.: Arzneim.-Forsch. (ARZNAD) **27**, 1625 (1977).

pharmacology:
Nieschulz, O. et al.: Arzneim.-Forsch. (ARZNAD) **8**, 539 (1958).

Formulation(s):   cream 4 g/100 g (4 %); ointment 4 g/100 g (4 %)

Trade Name(s):
D: Brand- und Wund-Gel   Erbocain (Heilit); wfm
      Herit (Engelhard)-comb.;   Erboproct (Heilit)-comb.;
      wfm   wfm

Formebolone

(Formyldienolone)

ATC: A14
Use: anabolic, anti-inflammatory

RN: 2454-11-7   MF: $C_{21}H_{28}O_4$ MW: 344.45   EINECS: 219-523-2
LD$_{50}$: 187 mg/kg (M, i.p.); 293 mg/kg (M, s.c.);
104 mg/kg (R, i.p.); 270 mg/kg (R, s.c.)
CN: (11α,17β)-11,17-dihydroxy-17-methyl-3-oxoandrosta-1,4-diene-2-carboxaldehyde

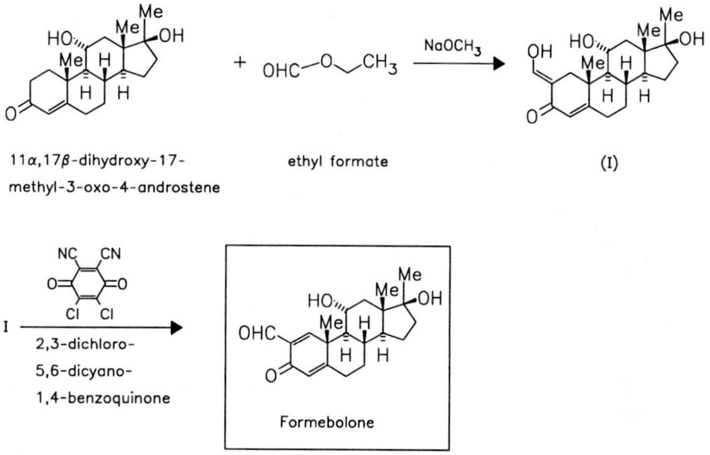

11α,17β-dihydroxy-17-
methyl-3-oxo-4-androstene

ethyl formate

(I)

2,3-dichloro-
5,6-dicyano-
1,4-benzoquinone

Formebolone

Reference(s):
DE 1 618 616 (LPB Braglia; appl. 8.2.1967).
GB 1 168 931 (LPB Braglia; valid from 20.1.1967).

Formulation(s):   amp. 2 ml/2 ml; tabl. 5 mg

Trade Name(s):
I: Esiclene (LPB)

Formestane

(4-HAD; 4-DHA)

ATC: L02B; G03BA
Use: aromatase inhibitor (for treatment of breast cancer)

RN: 566-48-3   MF: $C_{19}H_{26}O_3$ MW: 302.41
CN: 4-hydroxyandrost-4-ene-3,17-dione

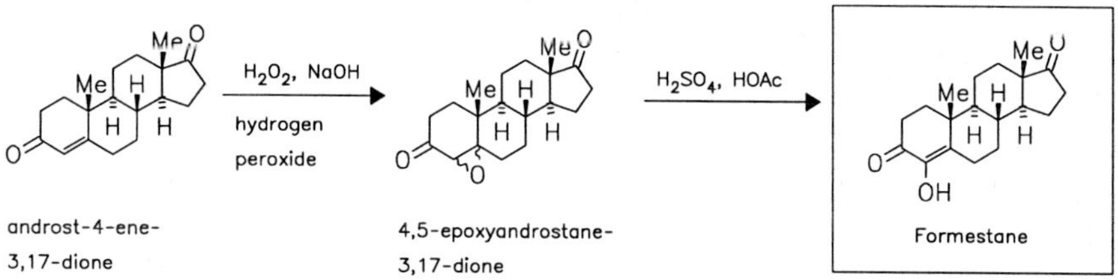

androst-4-ene-
3,17-dione

hydrogen
peroxide

4,5-epoxyandrostane-
3,17-dione

Formestane

Reference(s):
Marsh, D.A. et al.: J. Med. Chem. (JMCMAR) **28**, 788 (1985).

alternative synthesis:
Mann, J.; Pietrzak B.: J. Chem. Soc., Perkin Trans. 1 (JCPRB4) **1983**, 2681.
Burnett, R.O.; Kirk, D.N.: J. Chem. Soc., Perkin Trans. 1 (JCPRB4) **1973**, 1830.
Brodil, A.M. et al.: Endocrinology (ENDOAO) **100**, 1684 (1977).

micronised formestane*:*
EP 346 953 (Ciba-Geigy; appl. 31.10.1985; CH-prior. 6.11.1984).

stable suspension for injection:
EP 181 287 (Ciba-Geigy; appl. 31.10.1985; CH-prior. 6.11.1984).
US 5 002 940 (Ciba-Geigy; 26.3.1991; appl. 31.10.1985; CH-prior. 6.11.1984).

medical use for treatment of breast cancer:
WO 9 010 462 (Endorecherche; appl. 9.3.1990; USA-prior. 10.3.1989).

medical use for treatment of prostate hyperplasia:
DOS 3 339 295 (Schering AG; appl. 15.11.1982).
WO 9 100 731 (Endorecherche; appl. 5.7.1990; USA-prior. 7.7.1989).

medical use for treatment of gynecomastia:
US 4 895 715 (Schering Corp.; 23.1.1990; appl. 14.4.1988).

method for inhibition of estrogen biosynthesis:
US 4 235 893 (A. M. Brodic et al.; 25.11.1980; appl. 8.5.1978).

Formulation(s):   amp. 250 mg/2 ml

Trade Name(s):

D:	Lentaron (Novartis Pharma)	F: Lentaron (Novartis Pharma)	GB: Lentaron (Novartis Pharma)

Formocortal

(Formocortol)

ATC: S01BA12
Use: glucocorticoid

RN: 2825-60-7   MF: $C_{29}H_{38}ClFO_8$ MW: 569.07   EINECS: 220-584-2
LD$_{50}$:   537 mg/kg (M, i.p.); 490 mg/kg (M, s.c.)
CN: (11β,16α)-21-(acetyloxy)-3-(2-chloroethoxy)-9-fluoro-11-hydroxy-16,17-[(1-methylethylidene)bis(oxy)]-20-oxopregna-3,5-diene-6-carboxaldehyde

21-acetoxy-3,20-dioxo-
9α-fluoro-11β-hydroxy-
16α,17-isopropylidene-
dioxy-4-pregnene

ethylene
glycol

(I)

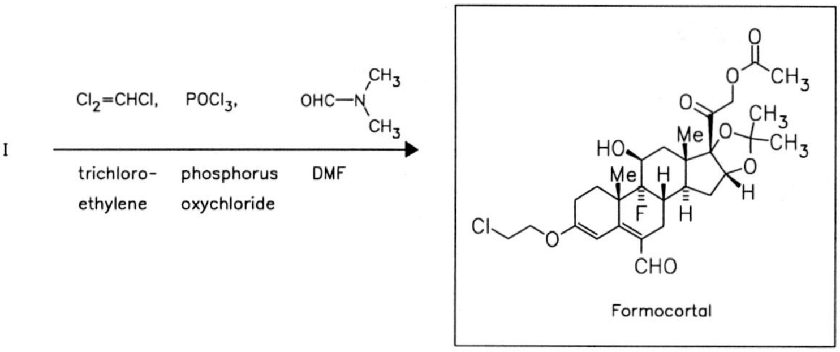

Formocortal

Reference(s):
US 3 314 945 (Societa Farmaceutici; 18.4.1967; I-prior. 15.7.1964).
Baldratti, G. et al.: Experientia (EXPEAM) **22**, 468 (1966).

starting material:
Holmund, C.E. et al.: J. Am. Chem. Soc. (JACSAT) **83**, 2586 (1961).
Bernstein, S. et al.: J. Am. Chem. Soc. (JACSAT) **81**, 1689 (1959).

Formulation(s):   eye drops 0.05 %; ointment 0.05 %; susp. 0.05 %

Trade Name(s):

D: Deidral S (Montedison)-   Deflamene (Farmitalia);   Formomicin (Farmigea)-
      comb.; wfm   wfm   comb. with gentamycin
GB: Deflamene (Carlo Erba);   I: Formoftil (Farmigea)
      wfm

Formoterol

ATC: R03AC13; R03CC
Use: selective β_2-adrenoceptor agonist

RN: 73573-87-2   MF: $C_{19}H_{24}N_2O_4$ MW: 344.41
LD$_{50}$:   71 mg/kg (M, i.v.); 8310 mg/kg (Mf, p.o.); 6700 mg/kg (Mm, p.o.);
       98-100 mg/kg (R, i.v.)
CN: (R*,R*)-(±)-N-[2-hydroxy-5-[1-hydroxy-2-[[2-(4-methoxyphenyl)-1-
       methylethyl]amino]ethyl]phenyl]formamide

fumarate (2:1)
RN: 43229-80-7   MF: $C_{19}H_{24}N_2O_4 \cdot 1/2C_4H_4O_4$ MW: 804.89
fumarate dihydrate
RN: 183814-30-4   MF: $C_{19}H_{24}N_2O_4 \cdot 1/2C_4H_4O_4 \cdot 2H_2O$ MW: 840.92

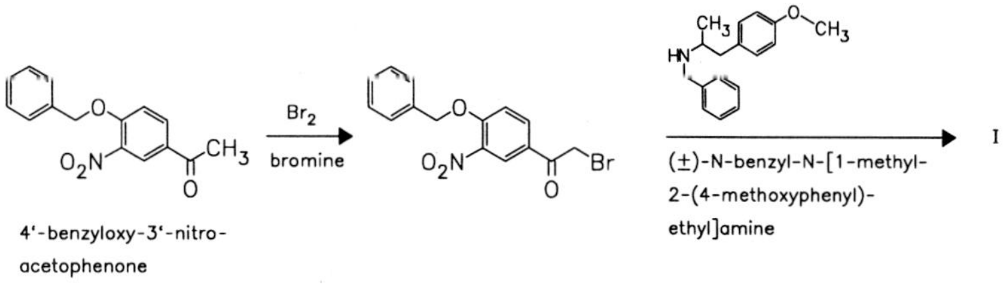

4'-benzyloxy-3'-nitro-
acetophenone

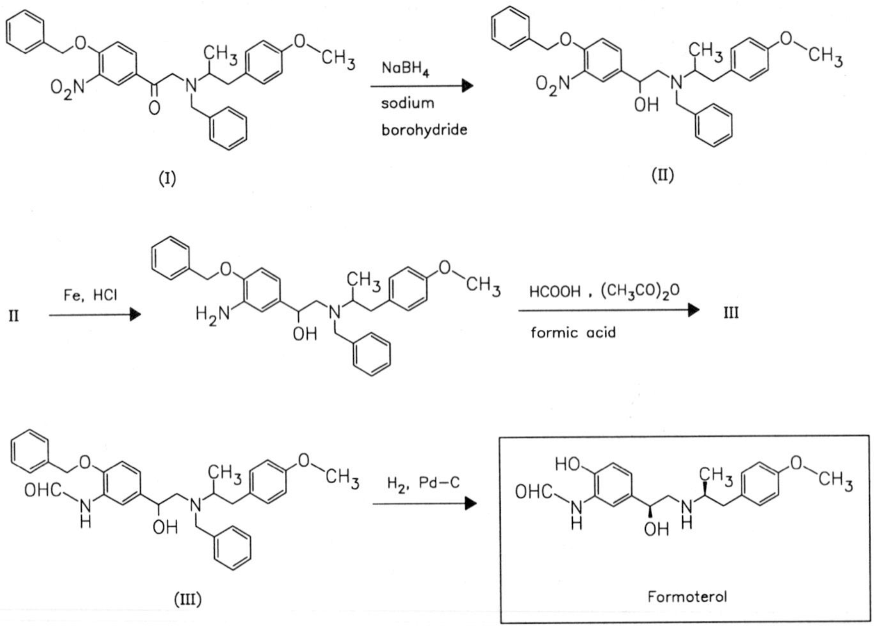

Reference(s):
US 3 994 974 (Yamanouchi; 30.11.1976; prior. 22.1.1973).
DOS 2 305 092 (Yamanouchi; appl. 2.2.1973; J-prior. 5.2.1972).
Hett, R. et al.: Org. Process Res. Dev. (OPRDFK) **2** 96 (1998).

preparation of enantiomers from (+)- or (–)-1-methyl-2-phenylethylamine:
Kibura, R.; Nakahara, Y.: Biol. Pharm. Bull. (BPBLEO) **18** (12), 1694 (1995).
Glennon, R.A.; Smith, J.D.; Ismaiel, A.M.; Ashmawy, M.; Bataglia, G.; Fisher, J.B.: J. Med. Chem. (JMCMAR)
34 (3), 1094 (1991).
Kerwin et al.: J. Am. Chem. Soc. (JACSAT) **72**, 3983, 3986 (1950).

preparation of 4'-benyloxy-3'-nitroacetophenone:
Meglio, P. de; Ravenna, F.; Gentili, P.; Manzardo, S., Riva, M.: Pharmaco, Ed. Sci. (FRPSAX) **38** (12), 998
(1983).
Oelschlaeger et al.: Arch. Pharm. Ber. Dtsch. Pharm. Ges. (APBDAJ) **296**, 107 (1963).

preparation of N-benzyl-N-[1-methyl-2-(4-methoxyphenyl)ethyl]amine:
Woodruff; Lambooy; Bust: J. Am. Chem. Soc. (JACSAT) **62**, 922 (1940).
FR 844 228 (Temmler-Werke; 1938).

alternative syntheses:
JP 7 512 040 (Yamanouchi; appl. 31.5.1973).
JP 81 115 751 (Yamanouchi; appl. 11.6.1980).

Formulation(s):   cps. for inhalation 12 µg (as fumarate dihydrate); powder inhaler 6 µg/puff, 12 µg/puff

Trade Name(s):
D: Oxis Turbohaler (Astra/   Foradil P (Novartis)   J: Atock (Yamanouchi; 1986)
      pharma-stern)   F: Foradil (Novartis)

Foscarnet sodium

ATC: J05
Use: antiviral (for treatment of CMV retinitis)

RN: 63585-09-1   MF: CNa_3O_5P MW: 191.95
LD_{50}:   384 mg/kg (M, i.p.)
CN: dihydroxyphosphinecarboxylic acid oxide trisodium salt

hexahydrate
RN: 34156-56-4   MF: $CNa_3O_5P \cdot 6H_2O$ MW: 300.04

triethyl phosphite   ethyl chloroformate   diethyl ethoxy-carbonylphosphonate   Foscarnet sodium

Reference(s):
ES 541 567 (Esp. Latinas Med. Universales; appl. 26.3.1985).
ES 556 513 (Lab. Esp. Farm. Centrum; appl. 24.6.1986).
CS 253 848 (V. Zikan, F. Roubinek; appl. 18.7.1986).
Nylen, P.: Ber. Dtsch. Chem. Ges. (BDCGAS) **57b**, 1023 (1924).

synthesis and use for regulation of plant growth:
US 4 018 854 (Du Pont; 19.4.1977; prior. 25.6.1975, 30.5.1974, 23.7.1973).
DOS 2 435 407 (Du Pont; appl. 23.7.1974; USA-prior. 23.7.1973, 17.9.1973, 30.5.1974).

medical use for treatment of virus infections:
US 4 339 445 (Astra; 13.7.1982; appl. 21.12.1978; S-prior. 1.7.1976).

Formulation(s):   cream 2 g/100 g; vial 6 g (24 mg/ml) (hexahydrate)

Trade Name(s):
D: Foscavir (Astra)   F: Foscavir (Astra)   Virudin (Bracco)
       Triapten Antiviralcreme   GB: Foscavir (Astra; 1990)   USA: Foscavir (Astra)
       (LAW/Wyeth)   I: Foscavir (Astra-Simes)

Fosfestrol

(Diethylstilbestrol diphosphate)

ATC: L02AA04
Use: antineoplastic

RN: 522-40-7   MF: $C_{18}H_{22}O_8P_2$ MW: 428.31   EINECS: 208-328-8
LD_{50}:   630 mg/kg (M, i.v.); 2 g/kg (M, p.o.);
       425 mg/kg (R, i.v.); 3 g/kg (R, p.o.)
CN: (E)-4,4'-(1,2-diethyl-1,2-ethenediyl)bis[phenol] bis(dihydrogen phosphate)

tetrasodium salt
RN: 23519-26-8   MF: $C_{18}H_{22}O_8P_2 \cdot xNa$ MW: unspecified

H₃C. OH

1. POCl₃, pyridine
2. H₂O

1. phosphorus
oxychloride

H₃C.

H_2O_3P PO_3H_2

diethylstilbestrol
(q. v.)

Fosfestrol

Reference(s):
US 2 234 311 (Ciba; 1941; CH-prior. 1938).
US 2 802 854 (ASTA-Werke; 1957; D-prior. 1952).
US 2 971 975 (Miles Labs.; 14.2.1961; appl. 30.8.1955).

Formulation(s): amp. 60 mg/5 ml; tabl. 120 mg (as tetrasodium salt)

Trade Name(s):
D: Honvan (ASTA Medica AWD)
F: ST-52 (ASTA Medica)
GB: Honvan (ASTA Medica)
I: Honvan (ASTA Medica)
J: Honvan (Kyorin)
USA: Stilphostrol (Dome); wfm
Stilphostrol (Miles Pharm.); wfm
generic

Fosfomycin

(Phosphonomycin)

ATC: J01XX01
Use: antibiotic

RN: 23155-02-4 MF: C₃H₇O₄P MW: 138.06 EINECS: 245-463-1
LD₅₀: 4 g/kg (M, i.p.)
CN: (2R-cis)-(3-methyloxiranyl)phosphonic acid

calcium salt (1:1)
RN: 26016-98-8 MF: C₃H₅CaO₄P MW: 176.12 EINECS: 247-408-7
LD₅₀: >3.5 g/kg (M, p.o.);
>7 g/kg (R, p.o.)
disodium salt
RN: 26016-99-9 MF: C₃H₅Na₂O₄P MW: 182.02
trometamol salt
RN: 78964-85-9 MF: C₃H₇O₄P · C₄H₁₁NO₃ MW: 259.20

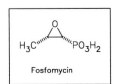

Fosfomycin

From fermentation solutions of *Streptomyces fradiae* (ATCC 21096).

Reference(s):
Hendlin et al.: Science (Washington, D.C.) (SCIEAS) **166**, 122 (1969).
BE 718 507 (Merck & Co., appl. 29.10.1968; USA-prior. 25.7.1967, 30.10.1967, 9.5.1968, 2.10.1968, 25.10.1968).

synthesis and separation of isomers:
BE 723 072 (Merck & Co., appl. 24.7.1968; USA-prior. 30.10.1967, 15.5.1968).
BE 723 073 (Merck & Co., appl. 29.10.1968; USA-prior. 30.10.1967, 15.5.1968, 30.8.1968).
Glamkowski, E.J. et al.: J. Org. Chem. (JOCEAH) **35**, 3510 (1970).

Formulation(s):   cps. 1 g; vial (lyo.) 2640 mg, 3960 mg, 6600 mg (as sodium salt); gran. 5.631 g/8 g (as trometamol); tabl. 1 g (as calcium salt); tabl. 1 g (as calcium salt monohydrate)

Trade Name(s):

D: Fosfocin pro infusione
    (Boehringer Mannh.)
    Monuril (Madaus)

F: Fosfocine (Sanofi
    Winthrop)
    Monuril (Zambon)
    Uridoz (Therabel Lucien
    Pharma)

I: Afos (Salus Research)
    Biocin (Ibirn)
    Biofos (Leben's)

Faremicin (Lafare)
Foce (Medici)-comb.
Fonofos (Pulitzer)
Fosfobiotic (Bergamon)
Fosfocin (Crinos)
Fosfogram (Firma)
Fosfolexin (Lifepharma)-
comb.
Fosforal (Farmasister)
Foximin (Caber)
Francital (Francia Farm.)

Ipamicina (IPA)
Lancetina (Farma Uno)
Lofoxin (Locatelli)
Monuril (Zambon Italia)
Neofocin (Medici)
Priomicina (San Carlo)
Ultramicina (Lisapharma)
Vastocin (Coli)

J: Fosmicin S (Meiji Seika)
USA: Fosfocina (Merck Sharp &
    Dohme); wfm

Fosinopril

(Fosenopril; SQ-28555)

ATC: C09AA09
Use: antihypertensive (ACE inhibitor)

RN: 98048-97-6   MF: $C_{30}H_{46}NO_7P$ MW: 563.67
CN: [1[$S*(R*)$],2α,4β]-4-cyclohexyl-1-[[[2-methyl-1-(1-oxopropoxy)propoxy](4-phenylbutyl)phosphinyl]acetyl]-L-proline

sodium salt
RN: 88889-14-9   MF: $C_{30}H_{45}NNaO_7P$ MW: 585.65

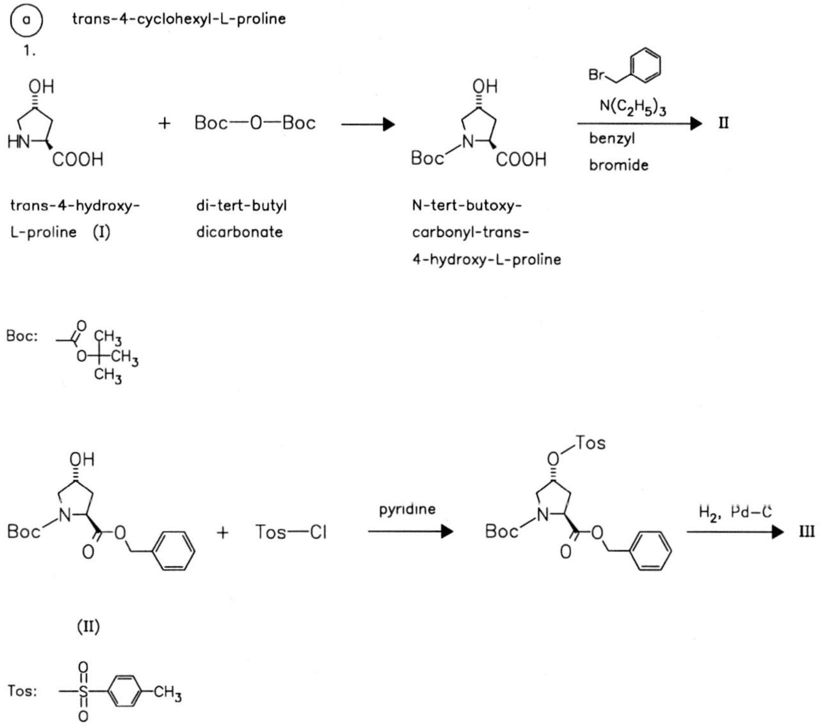

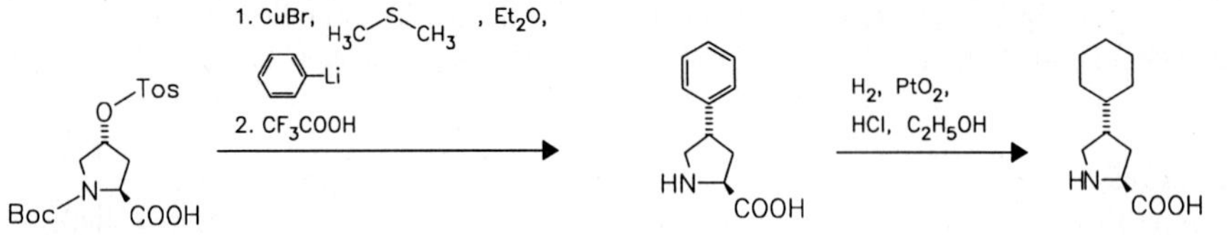

1. CuBr, $H_3C-S-CH_3$, Et$_2$O,

〈phenyl〉-Li

2. CF$_3$COOH

N-tert-butoxy-carbonyl-trans-4-tosyloxy-L-proline (III)

trans-4-phenyl-L-proline (IV)

H$_2$, PtO$_2$, HCl, C$_2$H$_5$OH

trans-4-cyclohexyl-L-proline (V)

2. alternative route

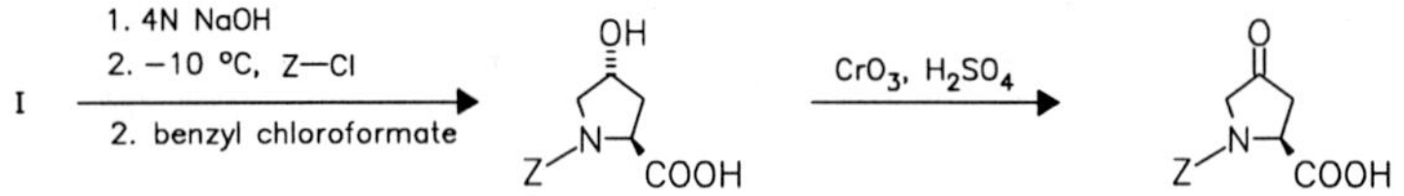

I

1. 4N NaOH
2. −10 °C, Z—Cl
2. benzyl chloroformate

CrO$_3$, H$_2$SO$_4$

N-benzyloxy-carbonyl-4-oxo-L-proline (VI)

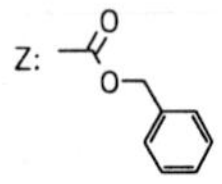

Z:

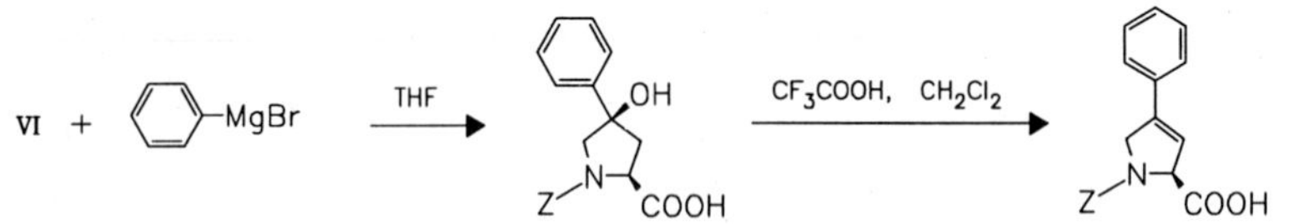

VI + 〈phenyl〉-MgBr

THF

CF$_3$COOH, CH$_2$Cl$_2$

N-benzyloxycarbonyl-3,4-didehydro-4-phenyl-L-proline (VII)

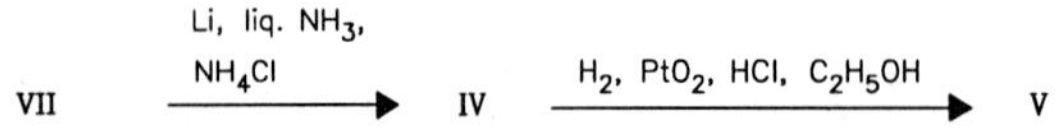

VII

Li, liq. NH$_3$, NH$_4$Cl

IV

H$_2$, PtO$_2$, HCl, C$_2$H$_5$OH

V

b

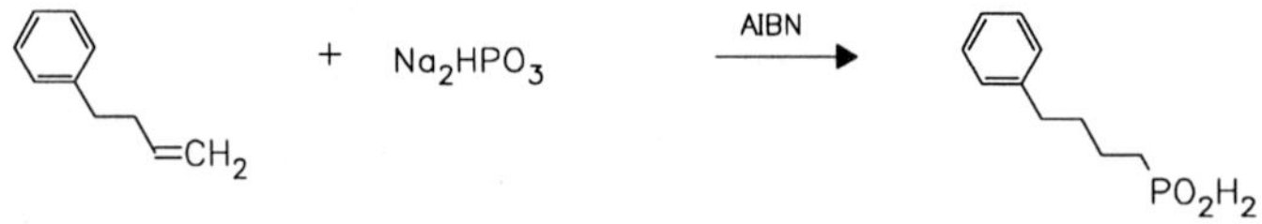

〈4-phenyl-1-butene〉 =CH$_2$

+ Na$_2$HPO$_3$

AIBN

PO$_2$H$_2$

4-phenyl-1-butene

sodium phosphite

4-phenylbutyl-phosphonous acid (VIII)

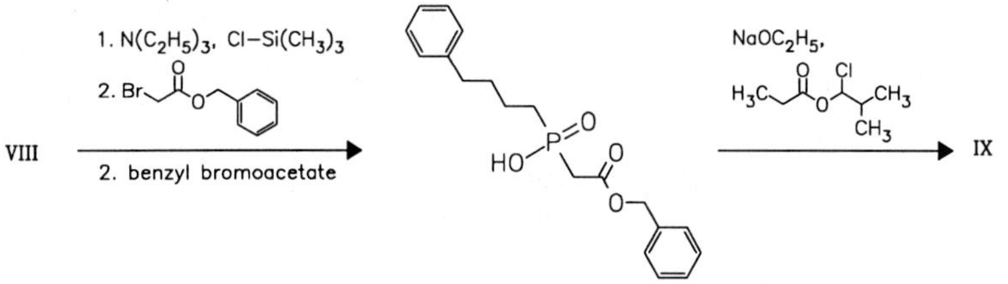

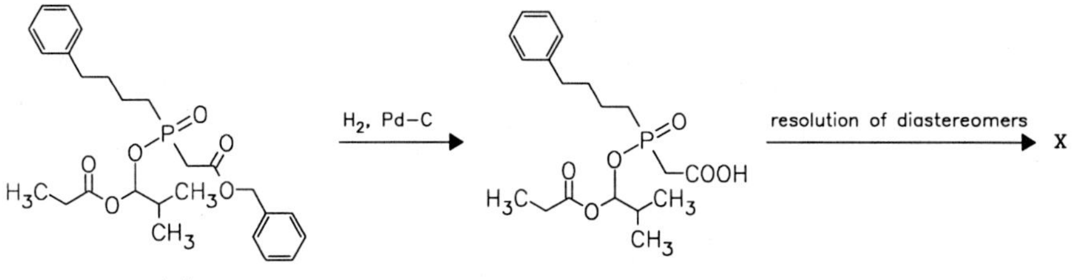

Reference(s):

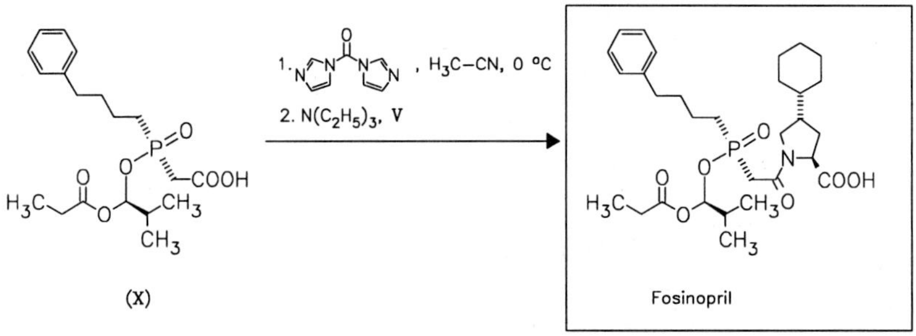

a1) DE 3 434 121 (Squibb; appl. 17.9.1984; USA-prior. 19.9.1983).
Thottathil, J.K. et al.: Tetrahedron Lett. (TELEAY) **27**, 151 (1986).
similar process:
US 4 912 231 (Squibb; 27.3.1990; prior. 15.6.1987, 17.6.1988).

a2) Krapcho, J. et al.: J. Med. Chem. (JMCMAR) **31**, 1148 (1988).
alternative route from L-pyroglutamic acid:
Thottathil, J.K. et al.: J. Org. Chem. (JOCEAH) **51**, 3140 (1986).
US 4 588 819 (Squibb; 13.5.1986; appl. 19.11.1984).
EP 183 390 (Squibb; appl. 25.10.1985; USA-prior. 19.11.1984).

b US 4 337 201 (Squibb & Sons; appl. 16.6.1982; USA-prior. 4.2.1980).
Krapcho, J. et al.: J. Med. Chem. (JMCMAR) **31**, 1148 (1988).

Formulation(s):   tabl. 5 mg, 10 mg, 20 mg, 40 mg (as sodium salt)

Trade Name(s):

D:	Dynacil (Schwarz/Sanol)	F:	Fozirétic (Lipha Santé)-comb.	I:	Eliten (Bristol-Myers Squibb)
	Fasinorm (Bristol-Myers Squibb)		Fozitec (Lipha Santé)		Fosipress (Menarini)
	Flucidine (Boehringer Ing.; Leo)	GB:	Staril (Bristol-Myers Squibb)		Tensogard (Bristol It. Sud)
	Fucithalmic (Alcon)			USA:	Monopril (Bristol-Myers Squibb)

Fosphenytoin sodium
(ACC 9653; CI-982)

ATC: N03AB
Use: anticonvulsant, prodrug of phenytoin

RN: 92134-98-0   MF: $C_{16}H_{13}N_2Na_2O_6P$ MW: 406.24
CN: 5,5-diphenyl-3-[(phosphonooxy)methyl]-2,4-imidazolidinedione disodium salt

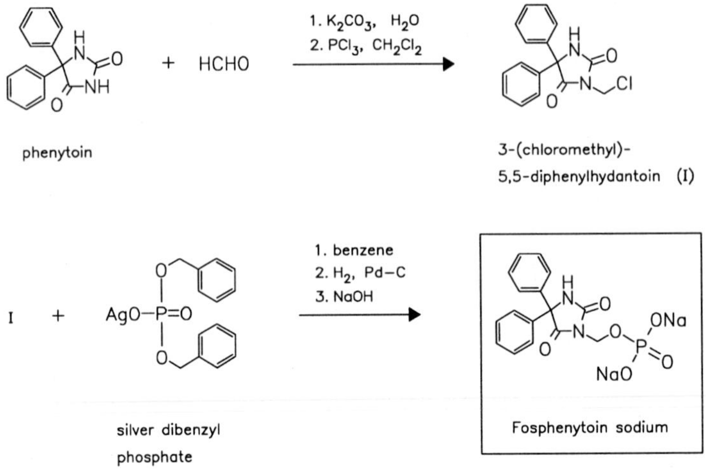

phenytoin

1. K_2CO_3, H_2O
2. PCl_3, CH_2Cl_2

3-(chloromethyl)-
5,5-diphenylhydantoin   (I)

I + AgO-P=O   silver dibenzyl phosphate

1. benzene
2. H_2, Pd–C
3. NaOH

Fosphenytoin sodium

Reference(s):
Varia, S.A. et al.: J. Pharm. Sci. (JPMSAE) **73**(8), 1068 (1984).

stable injection formulation:
US 4 925 866 (Du Pont; appl. 25.5.1989; USA-prior. 25.5.1989).

use for treatment of stroke:
EP 427 925 (Warner-Lambert; appl. 8.8.1990; USA-prior. 10.8.1989, 25.6.1990).

Formulation(s):   amp. 50 mg/ml

Trade Name(s):
USA: Cerebyx (Parke Davis)

Fotemustine

ATC: L01AD05
Use: antineoplastic, alkylating nitrosourea derivative

RN: 92118-27-9   MF: $C_9H_{19}ClN_3O_5P$ MW: 315.69
LD$_{50}$: 60 mg/kg (M, i.p.)
CN: [1-[[[(2-chloroethyl)nitrosoamino]carbonyl]amino]ethyl]phosphonic acid diethyl ester

acetylphosphonic
acid diethyl ester

NH_2OH
hydroxylamine

acetylphosphonic acid
diethyl ester oxime

1. Zn, HCOOH
2. CH_3OH, HCl

α-aminoethylphosphonic
acid diethyl ester (I)

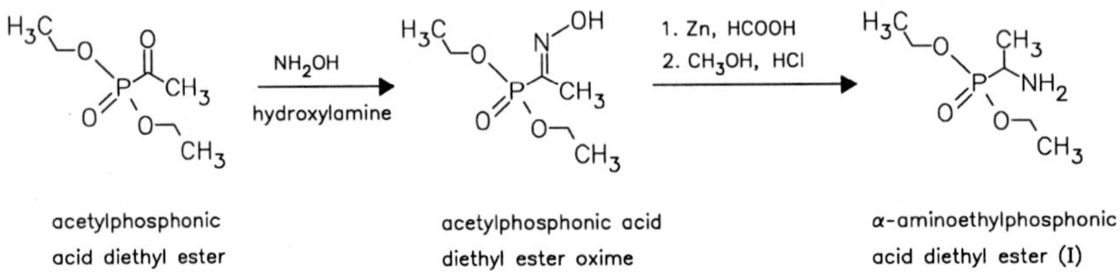

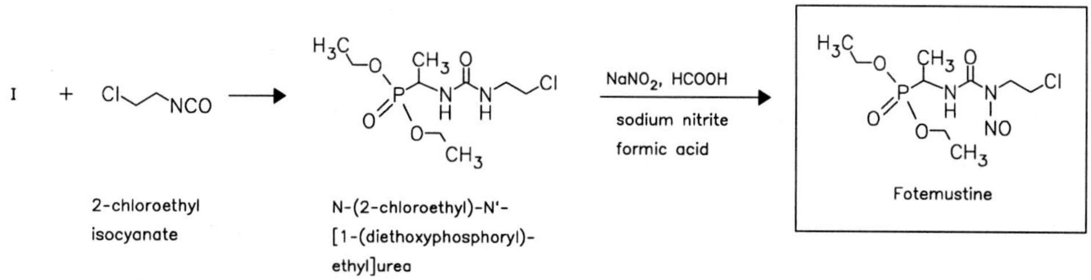

I + Cl∼NCO →

2-chloroethyl
isocyanate

N-(2-chloroethyl)-N'-
[1-(diethoxyphosphoryl)-
ethyl]urea

NaNO₂, HCOOH

sodium nitrite
formic acid

Fotemustine

Reference(s):
EP 117 959 (ADIR; appl. 16.11.1983; F-prior. 17.11.1982).
US 4 567 169 (ADIR; 28.1.1986; F-prior. 17.11.1982).

synthesis of α-aminoethylphosphonic acid diethyl ester:
Berlin, K.D. et al.: J. Org. Chem. (JOCEAH) **33**, 3090 (1968).
Kowalik, J.; Mastalerz, P.: Synthesis (SYNTBF) **1981**, 57.
Oleksyszyn, J.; Tyka, R.: Tetrahedron Lett. (TELEAY) **22**, 2823 (1977).

Formulation(s):   amp. 208 mg

Trade Name(s):
F: Muphoran (Servier; 1990)

Framycetin

(Neomycin B)

ATC: R01AX08
Use: antibiotic

RN: 119-04-0   MF: $C_{23}H_{46}N_6O_{13}$ MW: 614.65   EINECS: 204-292-2
CN: *O*-2,6-diamino-2,6-dideoxy-α-D-glucopyranosyl(1→4)-*O*-[2,6-diamino-2,6-dideoxy-β-L-
idopyranosyl-(1→3)-β-D-ribofuranosyl-(1→5)]-2-deoxy-D-streptamine

sulfate (1:3)
RN: 4146-30-9   MF: $C_{23}H_{46}N_6O_{13} \cdot 3H_2SO_4$ MW: 908.88   EINECS: 223-969-3

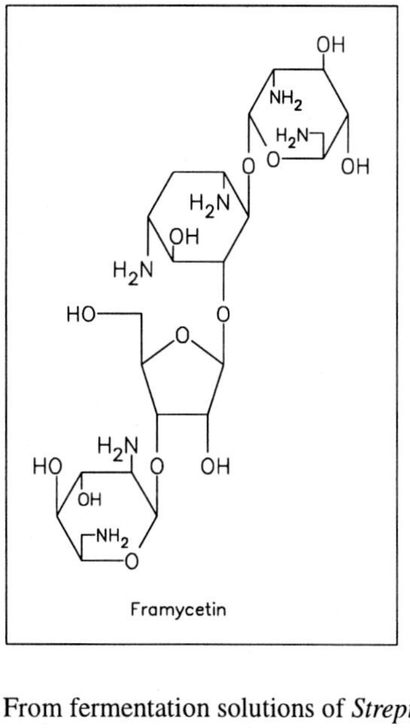

Framycetin

From fermentation solutions of *Streptomyces fradiae*.

Reference(s):
US 2 799 620 (Rutgers Res. Found.; 16.7.1957; prior. 29.6.1956).

purification:
US 2 848 365 (Upjohn; 1958; appl. 1950).
US 3 005 815 (Merck & Co.; 24.10.1961; prior. 1955, 1957).
US 3 022 228 (S. B. Penick; 20.2.1962; appl. 19.1.1960).
US 3 108 996 (Upjohn; 29.10.1963; appl. 30.7.1962).

Formulation(s):   cream 0.5 %; drops 1.25 %; eye drops 0.5 %; ointment 20 mg/g; powder 20 mg/g;
spray 500 mg/203.5 g; tabl. 250 mg (as sulfate)

Trade Name(s):

D:	Leukase (SmithKline Beecham)	I:	Cheliboldo (Terapeutico)-comb.	NeoDecadron (Merck; as sulfate)
	Sofra Tüll (Albert-Roussel, Hoechst)		Crisolax (Lifepharma)-comb.	Neomycin Sulfate (Roxane)
F:	Néomycine Diamant (Diamant)-comb. and more than 50 combination preparations		Sofra-tulle (Roussel)	Neomycin Sulfate (Teva)
		J:	Dexmy (Takeda)	Neosporin (Glaxo Wellcome; as sulfate)
			Fradio (Nippon Kayaku)	
GB:	Sofradex (Florizel)-comb.	USA:	Coly-Mycin (Parke Davis)-comb.	Neosporin (Warner-Lambert)
	Soframycin (Hoechst)-comb.		Cortisporin (Monarch; as sulfate)	
	Sofra-Tulle (Hoechst; as sulfate)		Lazersporin-C (Pedinol; as sulfate)	

Fumagillin

ATC: D06A
Use: antibiotic

RN: 23110-15-8   MF: $C_{26}H_{34}O_7$ MW: 458.55   EINECS: 245-433-8
LD$_{50}$:   2 g/kg (M, p.o.)
CN: [3R-[3α,4α(2R*,3R*),5β,6β(all-E)]]-2,4,6,8-decatetraenedioic acid mono[5-methoxy-4-[2-methyl-3-(3-methyl-2-butenyl)oxiranyl]-1-oxaspiro[2.5]oct-6-yl] ester

dicyclohexylammonium salt (1:1)
RN: 41567-78-6   MF: $C_{26}H_{34}O_7 \cdot C_{12}H_{23}N$ MW: 639.87

Fumagillin

From fermentation solutions of *Aspergillus fumigatus*.

Reference(s):
US 2 803 586 (Abbott; 1957; prior. 1953).

purification:
Tarbell, D.S. et al.: J. Am. Chem. Soc. (JACSAT) **77**, 5613 (1955).

structure and stereochemistry:
Chapman, D.D. et al.: J. Am. Chem. Soc. (JACSAT) **82**, 1009 (1960).
Chapman, D.D. et al.: J. Am. Chem. Soc. (JACSAT) **83**, 3096 (1961).
Tarbell, D.S. et al.: J. Am. Chem. Soc. (JACSAT) **77**, 5610 (1955).
McCorkindale, N.J.; Sime, J.G.: Proc. Chem. Soc., London (PCSLAW) **1961**, 331.
Turner; Tarbell, D.S.: Proc. Natl. Acad. Sci. USA (PNASA6) **48**, 733 (1962).

total synthesis:
Corey, E.J.; Snider, B.B.: J. Am. Chem. Soc. (JACSAT) **94**, 2549 (1972).

Trade Name(s):
USA: Fugillin (Upjohn); wfm Fumidil (Abbott); wfm

Furazabol

ATC: A14A
Use: anabolic

RN: 1239-29-8 MF: $C_{20}H_{30}N_2O_2$ MW: 330.47 EINECS: 214-983-0
LD_{50}: 1731 mg/kg (M, p.o.);
>4 g/kg (R, p.o.)
CN: (5α,17β)-17-methylandrostano[2,3-*c*][1,2,5]oxadiazol-17-ol

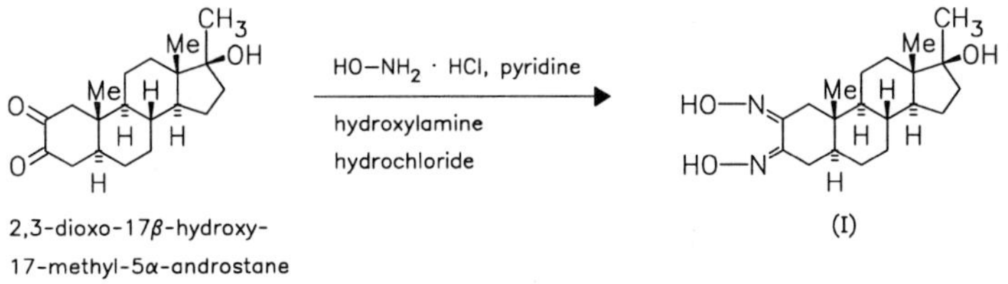

CH3
Me
OH
O
O
H
HO—NH2 · HCl, pyridine
hydroxylamine
hydrochloride
HO—N
HO—N
2,3-dioxo-17β-hydroxy-
17-methyl-5α-androstane
(I)

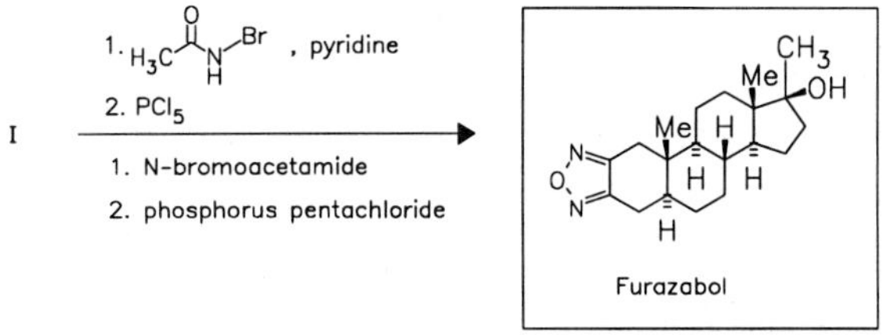

I
1. H3C N Br , pyridine
2. PCl5
1. N-bromoacetamide
2. phosphorus pentachloride
CH3
Me
OH
H
N
O
N
Furazabol

Reference(s):
US 3 415 818 (Sterling; 10.12.1968; appl. 8.7.1965).

further method:
US 3 245 988 (Daiichi; 12.4.1966; J-prior. 10.4.1963, 15.7.1963, 5.12.1963, 12.2.1964).

Formulation(s): tabl. 1 mg

Trade Name(s):
J: Miotolon (Daiichi Seiyaku)

Furazolidone

ATC: G01AX06
Use: topical anti-infective, topical antiprotozoal, chemotherapeutic (trichomonas)

RN: 67-45-8 MF: $C_8H_7N_3O_5$ MW: 225.16 EINECS: 200-653-3
LD$_{50}$: 1782 mg/kg (M, p.o.);
2336 mg/kg (R, p.o.)
CN: 3-[[(5-nitro-2-furanyl)methylene]amino]-2-oxazolidinone

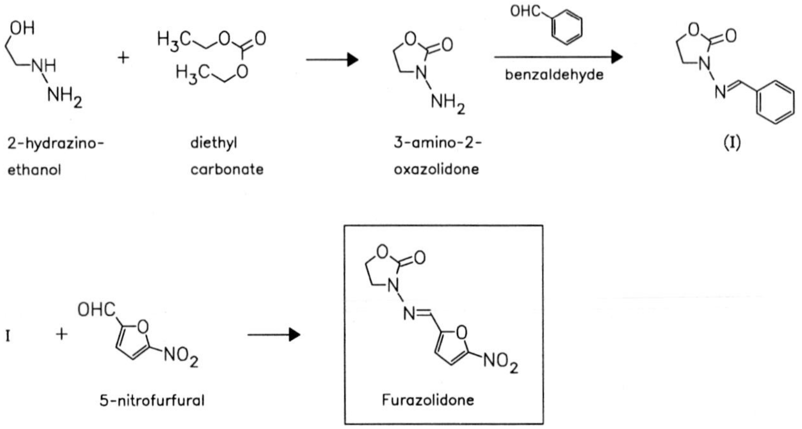

2-hydrazino-ethanol + diethyl carbonate → 3-amino-2-oxazolidone —benzaldehyde→ (I)

I + 5-nitrofurfural → Furazolidone

Reference(s):
US 2 759 931 (Norwich Pharm. Co.; 1956; prior. 1953).
US 2 927 110 (Norwich Pharm. Co.; 1.3.1960; prior. 23.1.1958).

Formulation(s): liquid 50 mg/15 ml; tabl. 100 mg

Trade Name(s):

F:	Furoxane (Oberval); wfm	Furoxone (Formenti)	J: Ginvel (Fujita)
	Tricofuron (Oberval); wfm	Ginecofuran (Crosara)-	Medaron (Yamanouchi)
GB:	Furoxone (Eaton); wfm	comb.	Purazolin T (Hokuriku)
I:	Furadone (Vebi)	Tricofur (Formenti)-comb.	USA: Furoxone (Roberts)

Furosemide
(Frusemide)

ATC: C03CA01
Use: diuretic

RN: 54-31-9 MF: $C_{12}H_{11}ClN_2O_5S$ MW: 330.75 EINECS: 200-203-6
LD$_{50}$: 308 mg/kg (M, i.v.); 2 g/kg (M, p.o.);
800 mg/kg (R, i.v.); 2600 mg/kg (R, p.o.);
>400 mg/kg (dog, i.v.); 2 g/kg (dog, p.o.)
CN: 5-(aminosulfonyl)-4-chloro-2-[(2-furanylmethyl)amino]benzoic acid

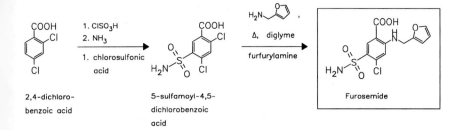

2,4-dichloro-
benzoic acid

5-sulfamoyl-4,5-
dichlorobenzoic
acid

Furosemide

Reference(s):
US 3 058 882 (Hoechst; 16.10.1962; D-prior. 28.12.1959).
DE 1 122 541 (Hoechst; appl. 28.12.1959).

alternative syntheses:
DE 1 213 846 (Hoechst; appl. 13.4.1963).
DE 1 220 436 (Hoechst; appl. 21.10.1964).
DE 1 277 860 (Hoechst; appl. 4.5.1966).
DE 1 295 566 (Hoechst; appl. 23.3.1968).
DAS 1 806 581 (Hoechst; appl. 2.11.1968).

review:
Sturm, K. et al.: Chem. Ber. (CHBEAM) **99**, 328 (1966).

Formulation(s):   amp. 80 mg, 500 mg; s. r. cps. 30 mg, 60 mg, 120 mg; sol. 20 mg, 40 mg, 50 mg, 250 mg (as
sodium salt); tabl. 20 mg, 25 mg, 40 mg

Trade Name(s):
D: Diurapid (Jenapharm)
      durafurid (durachemie)
      Furanthril (medphano)
      Furesis (Bristol-Myers
      Squibb)-comb.
      furo (ct-Arzneimittel)
      Furo-Puren (Klinge-
      Nattermann Puren)
      Furorese (Hexal)
      Furosemid (ratiopharm;
      Riker; Stadapharm)
      Fusid (GRY)
      Hydro-Rapid-Tablinen
      (Sanorania)
      Lasix (Hoechst)
      Ödemase (Azupharma)
      Osyrol (Hoechst)-comb.
      Sigasalur (Siegfried)

      generic and combination
      preparations
F: Aldalix (Monsanto)-comb.
      Furosémide (Biogalénique)
      Lasilix (Hoechst)
      Logirène (Pharmacia &
      Upjohn)-comb.
GB: Lasix (Hoechst)
      numerous combination
      preparations
I: Fluss (Roussel)-comb.
      Lasitone (Hoechst Italia
      Sud)-comb.
      Lasix (Hoechst)
      Lasix Reserpin (Hoechst)-
      comb.
      Spirofur (Lepetit)-comb.
      generics
J: Accent (Toyama)

      Arasemide (Arakawa)
      Diusemide (Nakataki)
      Diuzol (Wakamoto)
      Franyl (Seiko Eiyo)
      Fuluvamide (Kanto)
      Furfan (Nippon Roussel-
      Chugai)
      Kutrix (Kyowa)
      Lasix (Hoechst)
      Lowpston (Maruko)
      Polysquall A (Tokyo Hosei)
      Profemin (Toa Eiyo-
      Yamanouchi)
      Protargen (Ohta)
      Radonna (Nippon Kayaku)
      Rasisemid (Kodama)
      Urex (Mochida)
USA: Lasix (Hoechst Marion
      Roussel)

Fursultiamine

ATC: A11
Use: neurotropic analgesic

RN: 804-30-8   MF: $C_{17}H_{26}N_4O_3S_2$ MW: 398.55   EINECS: 212-357-1
LD_{50}: 430 mg/kg (M, i.v.); 2200 mg/kg (M, p.o.);
      2200 mg/kg (R, p.o.)
CN: *N*-[(4-amino-2-methyl-5-pyrimidinyl)methyl]-*N*-[4-hydroxy-1-methyl-2-[[(tetrahydro-2-
      furanyl)methyl]dithio]-1-butenyl]formamide

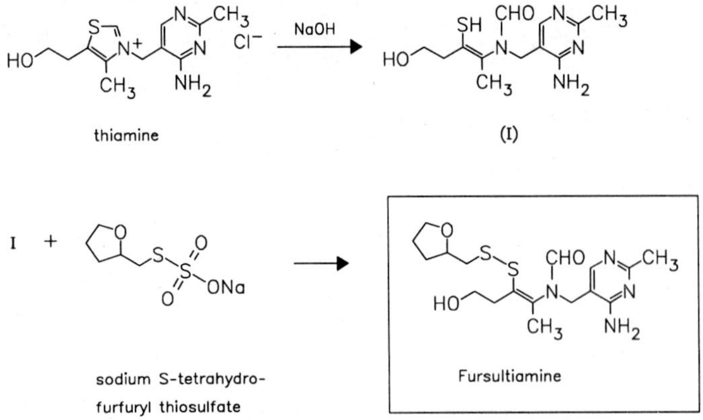

thiamine

(I)

sodium S-tetrahydro-
furfuryl thiosulfate

Fursultiamine

Reference(s):
US 3 016 380 (Takeda; 9.1.1962; J-prior. 16.8.1957).

Formulation(s):   amp. 5 mg/ml; drg. 50 mg; tabl. 5 mg, 25 mg, 50 mg

Trade Name(s):
D: Dolo-judolor (Woelm)-   judolor-Dragees (ICN);   judolor Dragees (Woelm);
      comb.; wfm   wfm   wfm
                                                                J: Alinamin F (Takeda)

Gabapentin
(GOE 2450; Go 3450; CI 945)

ATC: N03AX12
Use: anticonvulsant

RN: 60142-96-3   MF: $C_9H_{17}NO_2$ MW: 171.24   EINECS: 262-076-3
CN: 1-(Aminomethyl)cyclohexaneacetic acid

hydrochloride
RN: 60142-95-2   MF: $C_9H_{17}NO_2 \cdot HCl$ MW: 207.70   EINECS: 262-075-8

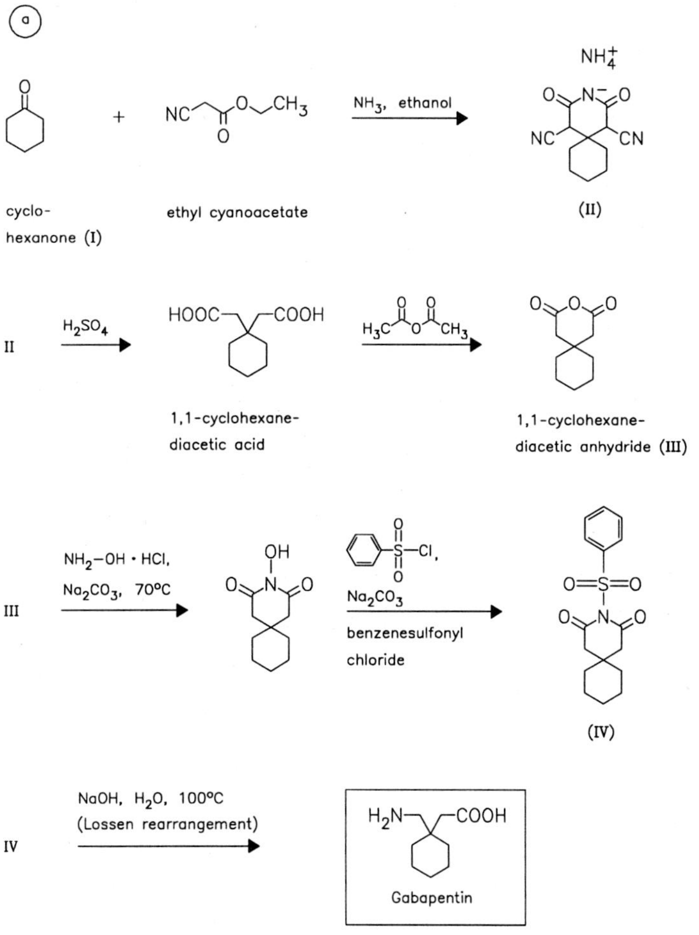

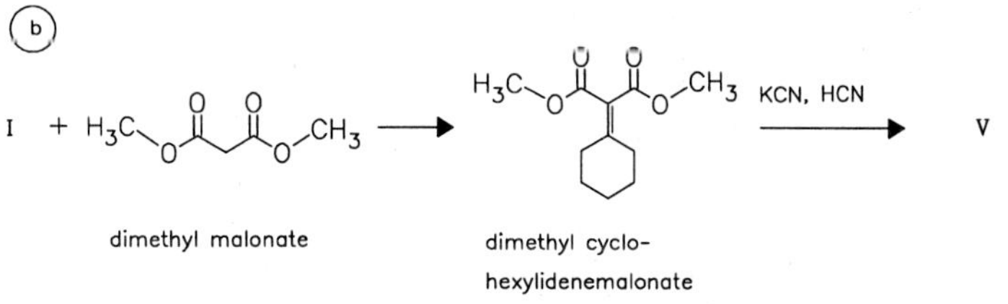

(a)

cyclo-
hexanone (I)

ethyl cyanoacetate

(II)

1,1-cyclohexane-
diacetic acid

1,1-cyclohexane-
diacetic anhydride (III)

benzenesulfonyl
chloride

(IV)

Gabapentin

(b)

dimethyl malonate

dimethyl cyclo-
hexylidenemalonate

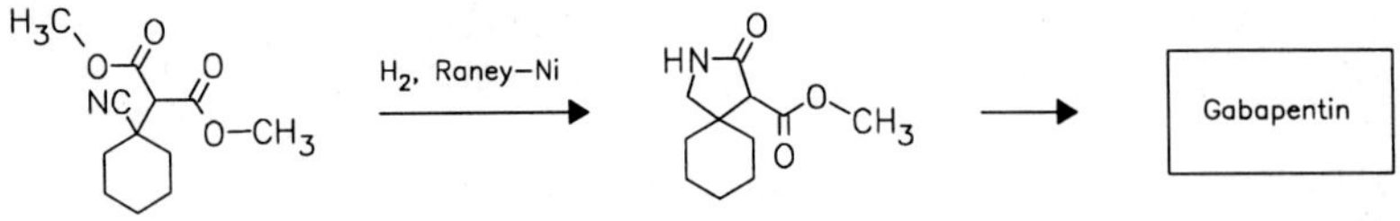

(V)

4-methoxycarbonyl-
2-azaspiro[4.5]-
decan-3-one

H₂, Raney–Ni

Gabapentin

(c)

I   +

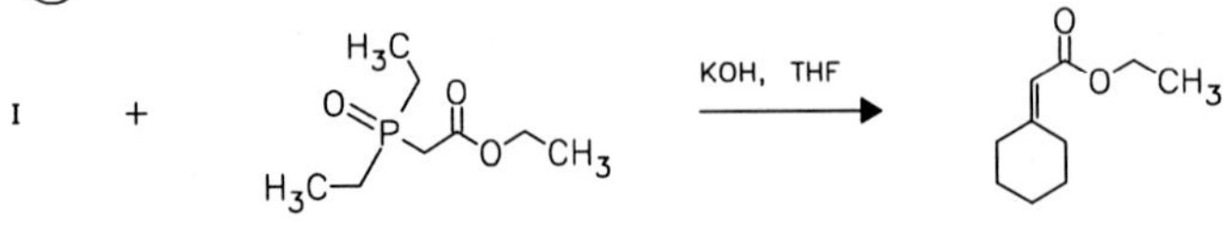

ethyl diethyl-
phosphinylacetate

KOH, THF

ethyl cyclo-
hexylidene-
acetate (VI)

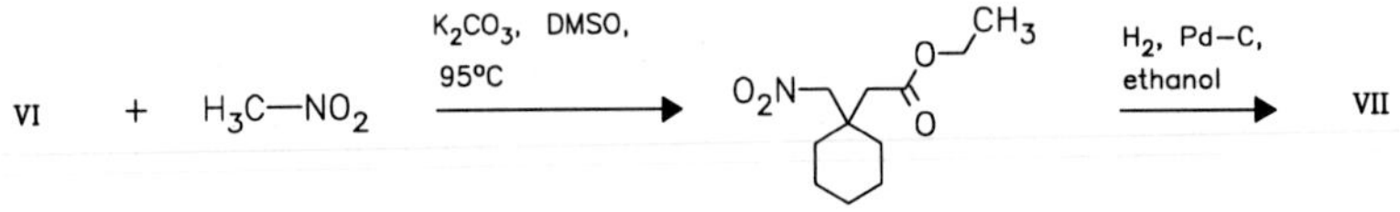

VI   +   H₃C—NO₂

K₂CO₃, DMSO,
95°C

ethyl 1-(nitro-
methyl)cyclo-
hexaneacetate

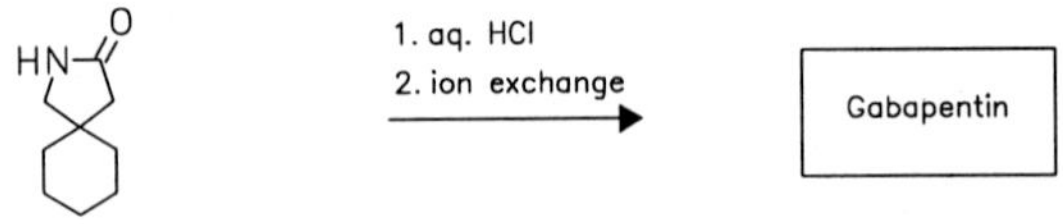

H₂, Pd–C,
ethanol

VII

2-azaspiro[4.5]-
decan-3-one (VII)

1. aq. HCl
2. ion exchange

Gabapentin

(d)

I

1. H₃C—CN
2. KCN, HCN

1. acetonitrile

1-cyanocyclo-
hexaneacetonitrile

HCl, ethanol,
toluene

(VIII)

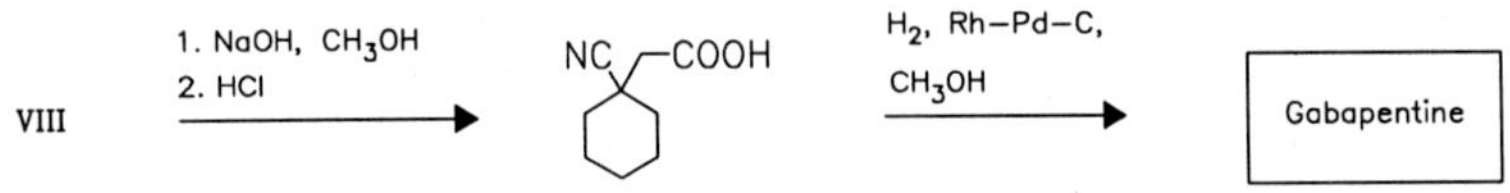

VIII

1. NaOH, CH₃OH
2. HCl

1-cyanocyclo-
hexaneacetic
acid

H₂, Rh–Pd–C,
CH₃OH

Gabapentine

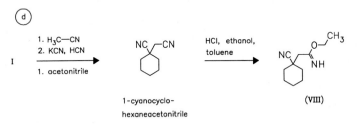

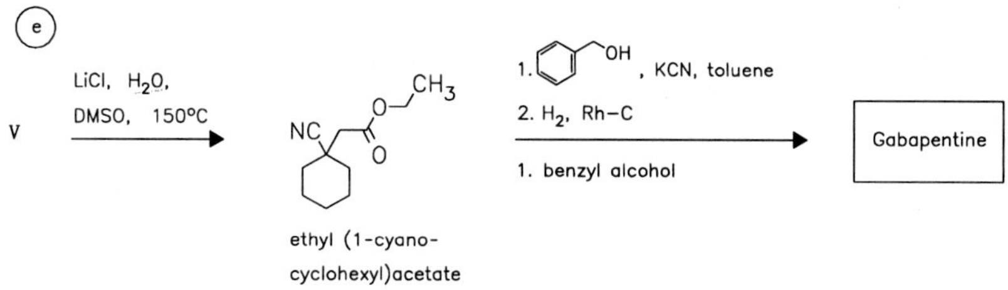

(e)

V → LiCl, H₂O, DMSO, 150°C →

ethyl (1-cyano-cyclohexyl)acetate

1. benzyl alcohol, KCN, toluene
2. H₂, Rh–C
1. benzyl alcohol
→ Gabapentine

Reference(s):

a DE 2 611 690 (Goedecke; D-prior. 19.3.1976).
US 4 152 326 (Warner-Lambert; 1.5.1979; D-prior. 19.3.1976).
b EP 358 092 (Lonza; appl. 29.8.1989; CH-prior. 1.9.1988).
c EP 414 274 (Goedecke AG; appl. 24.8.1990; D-prior. 25.8.1989).
d US 5 319 135 (Warner-Lambert; 7.6.1994; USA-prior. 25.8.1989).
EP 414 262 (Warner-lambert; appl. 24.8.1990; USA-prior. 25.8.1989).
e CA 2 030 107 (17.5.1991; appl. 15.11.1990; CH-prior. 16.11.1989).

preparation of alternate crystal forms:
WO 9 828 255 (Teva Pharm.; appl. 24.12.1997; IL-prior. 24.12.1996).

high-purity monohydrate:
EP 340 677 (Warner-Lambert; appl. 28.4.1989; USA-prior. 2.5.1988).

Formulation(s):   cps. 100 mg, 300 mg, 400 mg

Trade Name(s):
D: Neurontin (Parke Davis)   GB: Neurontin (Parke Davis)   USA: Neurontin (Parke Davis)
F: Neurontin (Parke Davis)   I: Neurontin (Parke Davis)

Gabexate

ATC: B02AB49
Use: protease inhibitor

RN: 39492-01-8   MF: C₁₆H₂₃N₃O₄   MW: 321.38
CN: 4-[[6-[(aminoiminomethyl)amino]-1-oxohexyl]oxy]benzoic acid ethyl ester

monomesylate
RN: 56974-61-9   MF: C₁₆H₂₃N₃O₄ · CH₄O₃S   MW: 417.48
LD₅₀: 218 mg/kg (M, i.v.); 8 g/kg (M, p.o.);
4020 mg/kg (R, i.v.); 6480 mg/kg (R, p.o.)

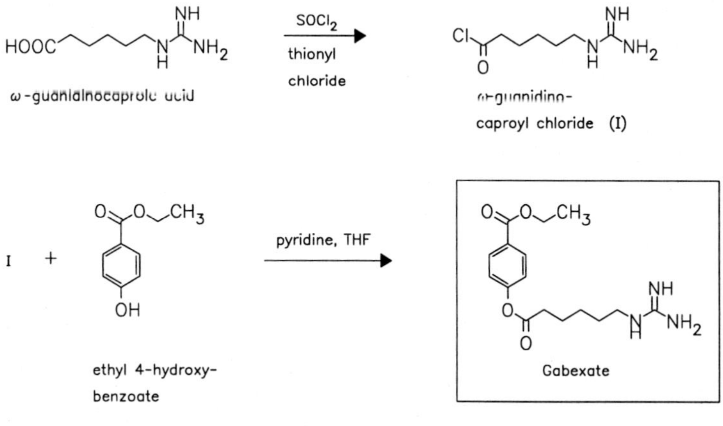

ω-guanidinocaproic acid → SOCl₂ thionyl chloride → ω-guanidino-caproyl chloride (I)

I + ethyl 4-hydroxy-benzoate → pyridine, THF → Gabexate

Reference(s):
DOS 2 050 484 (Ono; appl. 14.10.1970; J-prior. 14.10.1969).
US 3 751 447 (Ono; 7.8.1973; J-prior. 14.10.1969).

Formulation(s):   amp. 100 mg/5 ml (as mesylate)

Trade Name(s):
J: Foy (Ono; 1978)

Gallamine triethiodide

ATC: M03
Use: muscle relaxant, ganglionic blocker

RN: 65-29-2   MF: $C_{30}H_{60}I_3N_3O_3$ MW: 891.54   EINECS: 200-605-1
LD_{50}: 1800 µg/kg (M, i.v.); 425 mg/kg (M, p.o.);
        5100 µg/kg (R, i.v.); >1 g/kg (R, p.o.);
        800 µg/kg (dog, i.v.)
CN: 2,2',2''-[1,2,3-benzenetriyltris(oxy)]tris[*N,N,N*-triethylethanaminium] triiodide

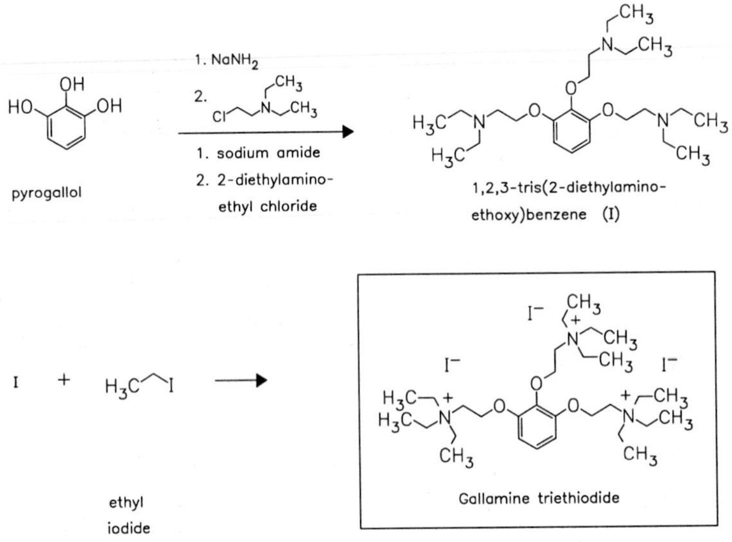

pyrogallol

1. NaNH₂ / 1. sodium amide
2. 2-diethylamino- ethyl chloride

1,2,3-tris(2-diethylamino- ethoxy)benzene (I)

ethyl iodide

Gallamine triethiodide

Reference(s):
US 2 544 076 (Rhône-Poulenc; 1951; F-prior. 1947).
DE 817 756 (Rhône-Poulenc; F-prior. 1947).

Formulation(s):   amp. 20 mg/ml, 40 mg/2 ml, 40 mg/ml, 100 mg

Trade Name(s):

D:	Flaxedil (Abbott); wfm	J: Gallamine Inj. "Teisan"
F:	Flaxédil (Specia); wfm	(Teikoku Kagaku-Nagase)
GB:	Flaxedil (Concord)	USA: Flaxedil (Lederle); wfm

Gallium nitrate

ATC: V03AG
Use: hypocalcemic agent against cancer-related hypercalcemia

RN: 13494-90-1 MF: GaN_3O_9 MW: 255.74 EINECS: 236-815-5
LD_{50}: 55 mg/kg (M, i.v.); 4360 mg/kg (M, p.o.); 600 mg/kg (M, s.c.);
46 mg/kg (R, i.v.)
CN: gallium nitrate

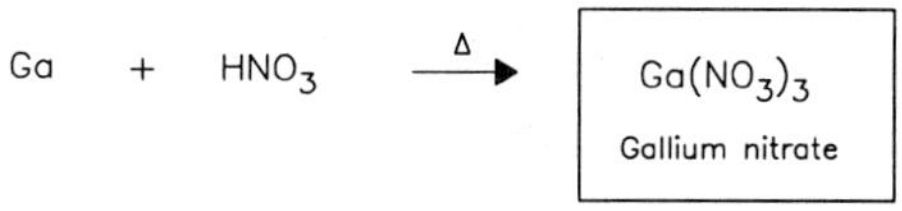

Ga + HNO₃ →Δ Ga(NO₃)₃
Gallium nitrate

Reference(s):
Dupré, A.: C. R. Hebd. Seances Acad. Sci. (COREAF) **86**, 721 (1878)

medical use as hypocalcemic agent:
EP 109 564 (Sloan Kettering Inst.; appl. 21.10.1983; USA-prior. 22.10.1982).
US 4 529 593 (Sloan Kettering Inst.; 16.7.1985; prior. 20.6.1984, 22.10.1982).

medical use as antitumor effect enhancer:
JP 1 104 016 (Taishitsu Kenkyukai; appl. 31.7.1987).

Formulation(s):   vial 500 mg

Trade Name(s):
USA: Ganite (SoloPak)

Gallopamil

ATC: C01DA; C08DA02
Use: coronary vasodilator, verapamil analog

RN: 16662-47-8 MF: $C_{28}H_{40}N_2O_5$ MW: 484.64
CN: α-[3-[[2-(3,4-dimethoxyphenyl)ethyl]methylamino]propyl]-3,4,5-trimethoxy-α-(1-methylethyl)benzeneacetonitrile

hydrochloride
RN: 16662-46-7 MF: $C_{28}H_{40}N_2O_5 \cdot HCl$ MW: 521.10

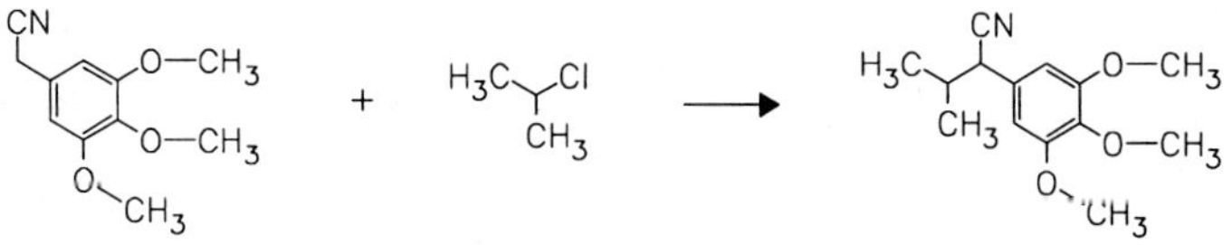

3,4,5-trimethoxy-
phenylacetonitrile

2-chloro-
propane

2-(3,4-trimethoxyphenyl)-
3-methylbutyronitrile (I)

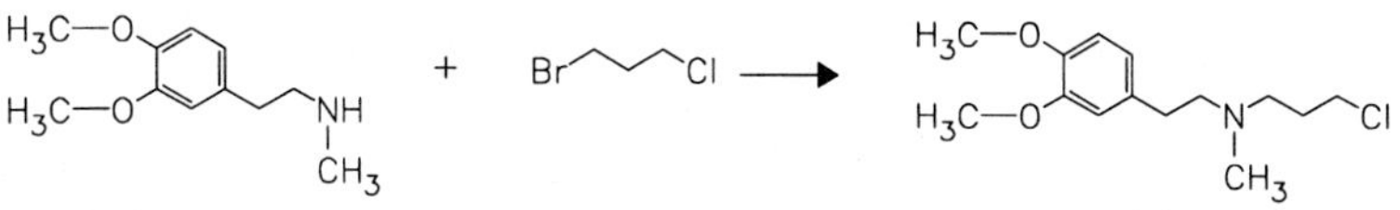

N-[2-(3,4-dimeth-
oxyphenyl)ethyl]-
methylamine

1-bromo-3-
chloropropane

(II)

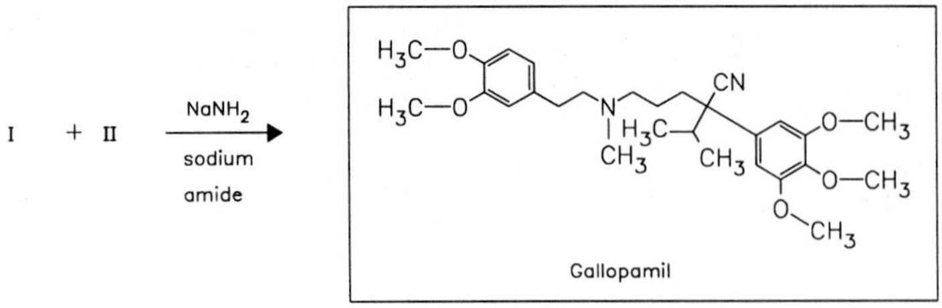

H₃C—O.

H₃C—O

CN

N
H₃C—
CH₃   CH₃

O—CH₃

O—CH₃

O
CH₃

Gallopamil

I   + II   →   NaNH₂
sodium
amide

Reference(s):
DE 1 154 810 (Knoll; appl. 28.4.1961).
DE 1 158 083 (Knoll; appl. 19.12.1962).
DE 2 059 985 (Knoll; prior. 5.12.1970).

Formulation(s):   f. c. tabl. 25 mg, 50 mg; s. r. tabl. 100 mg (as hydrochloride)

Trade Name(s):
D: Gallopamil (ct-   Procorum retard (Knoll;   I: Algocor (Ravizza)
       Arzneimittel)   1983)   Procorum (Knoll)

Gamolenic acid

ATC: D11AX02
Use: treatment of eczema

RN: 506-26-3   MF: C₁₈H₃₀O₂   MW: 278.44
CN: (Z,Z,Z)-6,9,12-octadecatrienoic acid

potassium salt
RN: 106868-38-6   MF: C₁₈H₂₉KO₂   MW: 316.53
sodium salt
RN: 86761-55-9   MF: C₁₈H₂₉NaO₂   MW: 300.42

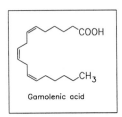

COOH

CH₃

Gamolenic acid

a From fermentation of *Mortierella.*
b From fermentation of mucor.
c Extration and isolation from natural sources (seeds of black currant, evening primrose, borage).

Reference(s):
a   JP 59 130 191 (Agency of Ind. Sciences and Techn.; appl. 12.1.1983).
    JP 60 168 391 (Agency of Ind. Sciences and Techn.; appl. 9.2.1984).
    JP 63 112 536 (Agency of Ind. Sciences and Techn.; appl. 30.10.1986).
    WO 8 604 354 (Agency of Ind. Sciences and Techn.; appl. 13.12.1985; J-prior. 22.1.1985, 21.2.1985).
    EP 155 420 (Agency of Ind. Sciences and Techn.; appl. 25.9.1984; J-prior. 9.2.1984, 5.6.1984).
    EP 253 556 (Suntory; appl. 7.7.1987; J-prior. 8.7.1986).
    EP 276 982 (Suntory; appl. 26.1.1988; J-prior. 27.1.1987).
    US 4 857 329 (Agency of Ind. Sciences and Techn.; 15.8.1989; appl. 1.8.1986; J-prior. 19.8.1985).
    Suzuki, O.: Proc. World Conf. Biotechnol. Fats Oils Ind. (56NIAQ) **1987**, p.110-116, Ed. T. H. Applewhite.
b   JP 1 132 371 (Itochu Seito; appl. 18.11.1987).
    EP 269 351 (Lion Corp.; appl. 17.11.1987; J-prior. 26.11.1986).
    Fukuda, H.; Morikawa, H.: Appl. Microbiol. Biotechnol. (AMBIDG) **27**, 15 (1987); Bioreact. Biotransform.,
    [Pap. Int. Conf.] (56GJAS), p. 386, Ed. G. W. Moody, P.B. Baker (Elsevier, London, 1987).
    Aggelis, G. et al.: Oleagineux (OLEAAF) **43**, 311 (1988).
    Hansson, L. et al.: Appl. Microbiol. Biotechnol. (AMBIDG) **31**, 223 (1989).

alternative fermentation processes:
EP 153 134 (Efamol; appl. 12.2.1985; GB-prior. 21.2.1984).
GB 2 163 424 (Nisshin Oil Mills; appl. 22.7.1985; J-prior. 31.7.1984).
JP 1 199 588 (Nitto Chem.; appl. 2.6.1988; prior. 27.10.1987).
JP 63 240 791 (Kanegafuchi; appl. 27.3.1987).
JP 62 210 995 (Nisshin Flour Milling; 12.3.1986).
JP 62 232 379 (Nisshin Oil Mills; appl. 2.4.1986).
JP 49 013 988 (A. Watanabe et al.; appl. 10.1.1969).
JP 47 022 280 (Ono; appl. 13.8.1969).
c   DOS 3 542 932 (HVG Barth, Raiser Co; appl. 4.12.1985).
    EP 271 747 (Nestle; appl. 25.11.1987; CH-prior. 17.12.1986).
    EP 178 442 (Nestle; appl. 4.9.1985; CH-prior. 10.10.1984).
    FR 1 603 383 (Ono; appl. 3.10.1968).
    US 4 703 060 (Nestec S.A.; 27.10.1987; appl. 14.10.1983; prior. 6.4.1983).
    JP 1 051 496 (Nippon Oils and Fats; appl. 21.8.1987).
    JP 63 216 845 (Agency of Ind. Sciences and Techn.; appl. 5.3.1987).
    Traitler, H. et al.: J. Am. Oil Chem. Soc. (JAOCA7) **65**, 755 (1988).
    Wille, H.; Traitler, H.: Fett Wiss. Technol. (FWTEEG) **90**, 476 (1988).

combination with calcium:
EP 261 814 (Efamol; appl. 28.8.1987; GB-prior. 10.9.1986).

modeling of androgen action in men:
EP 309 086 (Efamol; appl. 9.8.1988; GB-prior. 7.9.1987).

prevention of side effects of non-steroidal anti-inflammatories:
EP 195 570 (Efamol; appl. 7.3.1986; GB-prior. 19.3.1985).

treatment of premenstrual syndrome:
US 4 415 554 (Efamol; 15.11.1983; appl. 10.6.1981; GB-prior. 23.1.1978, 7.2.1978, 19.4.1978, 17.8.1978,
24.10.1978, 19.1.1979, 30.10.1979).
ZA 8 604 779 (Efamol; appl. 27.6.1986; GB-prior. 4.7.1985).

treatment of complications of diabetes mellitus:
EP 218 460 (Efamol; appl. 1.10.1986; GB-prior. 2.10.1985).

treatment of skin disorders:
US 4 444 755 (Efamol; 24.4.1984; appl. 10.6.1981; prior. 19.1.1979, 30.10.1979).
EP 173 478 (Efamol; appl. 5.8.1985; GB-prior. 15.8.1984).

treatment of prostatomegaly:
JP 61 207 330 (Efamol; appl. 6.3.1986; GB-prior. 8.3.1985).

treatment of endometriosis:
EP 222 483 (Efamol; appl. 1.10.1986; GB-prior. 2.10.1985).

treatment of amnesia:
EP 296 751 (Efamol; appl. 15.6.1988; GB-prior. 24.6.1987).

treatment of allergic rhinitis and asthma:
JP 61 087 621 (Nisshin Oil Mills; appl. 5.10.1984).

skin improving composition:
EP 334 507 (Efamol; appl. 7.3.1989; GB-prior. 22.3.1988).

pharmaceutical and dietary composition:
EP 3 407 (Verronmay; appl. 20.1.1979; GB-prior. 23.1.1978).
EP 4 770 (Verronmay; appl. 10.4.1979; GB-prior. 11.4.1978).
EP 19 423 (Efamol; appl. 8.5.1980; GB-prior. 18.5.1979).
US 4 273 763 (Efamol; 16.6.1981; GB-prior. 23.1.1978).

Formulation(s):   cps. 40 mg, 80 mg, 466-536 mg, 932-1073 mg extract of evening primrose seeds

Trade Name(s):
D: Epogam (Beiersdorf; 1990)   Epogam (Searle)
GB: Efamast (Searle)   I: Epogam (Whitehall)

Ganciclovir
(Biolf-62; BW-759U; DHPG; 2'-NOG)

ATC: S01AD09
Use: antiviral nucleoside for treatment of cytomegalovirus infections in AIDS patients

RN: 82410-32-0   MF: $C_9H_{13}N_5O_4$ MW: 255.23
LD_{50}:   1 g/kg (M, i.p.); 900 mg/kg (M, i.v.); >2 g/kg (M, p.o.);
      >150 mg/kg (dog, i.v.); >1 g/kg (dog, p.o.)
CN: 2-amino-1,9-dihydro-9-[[2-hydroxy-1-(hydroxymethyl)ethoxy]methyl]-6*H*-purin-6-one

monosodium salt
RN: 107910-75-8   MF: $C_9H_{12}N_5NaO_4$ MW: 277.22

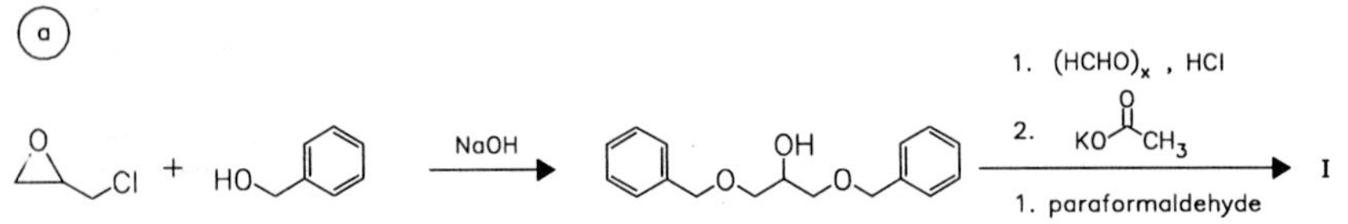

1,3-di-O-benzyl-2-O-(acetoxymethyl)glycerol (I)

N^2,9-diacetyl-guanine

Ganciclovir

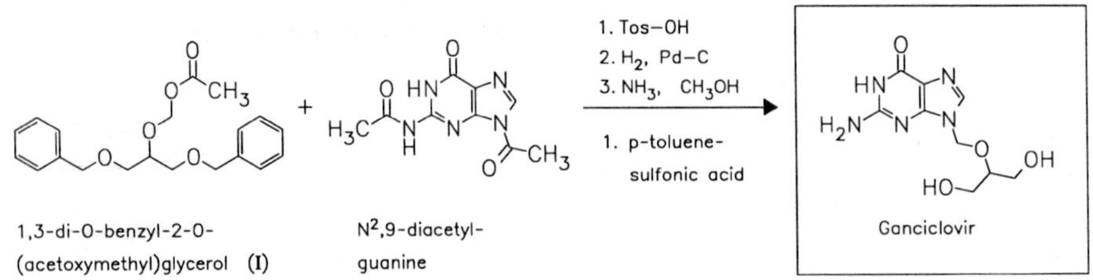

ⓑ

N²-acetylguanine

1. (NH₄)₂SO₄, HMDS
2. H₂, Pd–C
3. NH₃, CH₃OH

1. ammonium sulfate,
   hexamethyldisilazane

Ganciclovir

I +

Reference(s):
EP 85 424 (Syntex; appl. 31.1.1983; USA-prior. 1.2.1982, 22.12.1982).
EP 49 072 (Syntex; appl. 15.9.1981; USA-prior. 16.9.1980).
US 4 423 050 (Syntex; 27.12.1983; prior. 24.5.1982, 21.5.1981).
US 4 355 032 (Syntex; 14.6.1983; appl. 21.5.1981).
Martin, J.C. et al.: J. Med. Chem. (JMCMAR) **26**, 759 (1983).
Ogilvie, K.K. et al.: Can. J. Chem. (CJCHAG) **60**, 3005 (1982).
Ashton, W.A.: Biochem. Biophys. Res. Commun. (BBRCA9) **108**, 1716 (1982).

anhydrous crystalline form:
US 4 642 346 (Syntex; 10.2.1987; appl. 24.6.1985).

combination with interferon:
EP 109 234 (BioLogicals; appl. 3.11.1983; USA-prior. 4.11.1982).

alternative synthesis:
McGee, D.P. et al.: Synth. Commun. (SYNCAV) **18**, 1651 (1988).
ES 548 093 (Inke S.A.; appl. 22.10.1985).
ES 549 248 (M.J. Verde Casanova; appl. 25.11.1985).
WO 8 302 723 (BioLogicals; appl. 12.2.1982).

synthesis from guanosine:
Boryski, J.; Golankiewicz, B.: Synthesis (SYNTBF) **1999** (4), 625.

Review:
Gao, H.; Mitra, A.K.: Synthesis (SYNTBF) **2000** (3), 329.

Formulation(s):   cps. 250 mg, 500 mg; vial (lyo.) 500 mg (as sodium salt)

Trade Name(s):

D:	Cymeven (Roche)	GB:	Cymevene (Roche)
F:	Cymévan (Roche)	I:	Cymavene (Recardati)
	Virgan (Théa)	J:	Denosine (Syntex)

USA: Cytovene (Roche)
      Cytovene (Roche)

Gefarnate

ATC: A02BX07
Use: peptic ulcer therapeutic,
        antispasmodic

RN: 51-77-4   MF: $C_{27}H_{44}O_2$ MW: 400.65   EINECS. 200-121-0
LD₅₀: 2821 mg/kg (M, i.v.); >8 g/kg (M, p.o.);
       2040 mg/kg (R, i.v.); >9 g/kg (R, p.o.)
CN: (*E,E,E*)-5,9,13-trimethyl-4,8,12-tetradecatrienoic acid 3,7-dimethyl-2,6-octadienyl ester

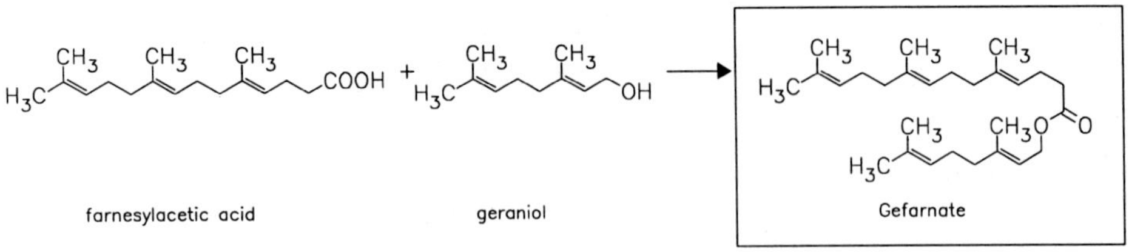

farnesylacetic acid   geraniol   Gefarnate

Reference(s):
BE 617 994 (Ist. de Angeli; appl. 23.5.1962; GB-prior. 24.5.1961).

Formulation(s):   cps. 50 mg; tabl. 50 mg; vial 50 mg

Trade Name(s):

GB: Gefarnil (Crookes); wfm
I: Famesil (AGIPS); wfm
     Farnisol (Firma); wfm
     Gefarnax/-forte (De
     Angeli)-comb.; wfm
     Gefarnil (De Angeli); wfm
     Gefarnil Compositum (De
     Angeli)-comb.; wfm
     Gefarol (Iti); wfm
     Nolesil (Geymonat); wfm
     Ulco (Elea); wfm

     Ulcofarm (Ausonia); wfm
     Ulcofarm (Iton); wfm
     Ulcotrofina (Ripari-Gero);
     wfm
     Vagogernil (Benvegna);
     wfm
J: Alsanate (Dainippon)
     Dixnalate (San-a)
     Eszyme Dental (SS
     Seiyaku)
     Gefalon (Sawai)

     Gefanil (Sumitomo)
     Gefulcer (Ohta)
     Ketonil (Mohan)
     Matorozin (Kanto-Isei)
     Polyl (Teikoku)
     Salanil (Sato)
     Terpanil (Kakenyaku)
     Zackal (SS Seiyaku)
     Zenowal (Daigo-Takeda)

Gemcitabine
(dFdC; LY-188011)

ATC: L01BC05
Use: antiviral, antineoplastic

RN: 95058-81-4   MF: $C_9H_{11}F_2N_3O_4$ MW: 263.20
CN: 2'-deoxy-2',2'-difluorocytidine

monohydrochloride
RN: 122111-03-9   MF: $C_9H_{11}F_2N_3O_4 \cdot HCl$ MW: 299.66

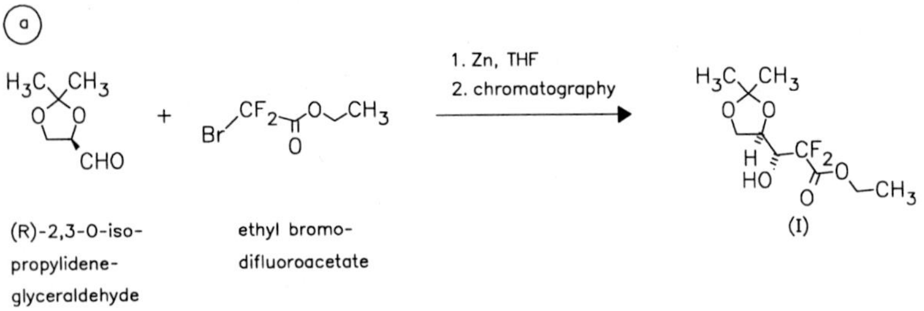

(R)-2,3-0-iso-
propylidene-
glyceraldehyde

ethyl bromo-
difluoroacetate

(I)

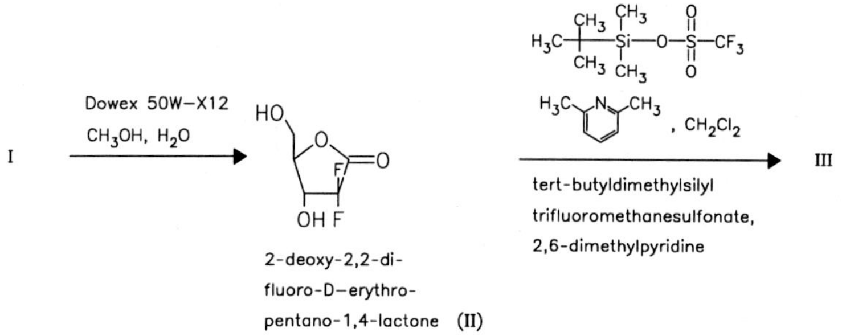
I
Dowex 50W–X12
CH3OH, H2O
HO
O
F
O
OH F
2-deoxy-2,2-di-
fluoro-D-erythro-
pentano-1,4-lactone (II)
, CH2Cl2
tert-butyldimethylsilyl
trifluoromethanesulfonate,
2,6-dimethylpyridine
III

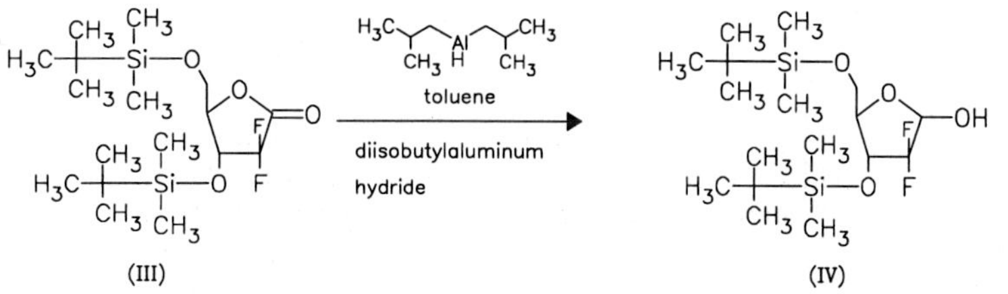
(III)
toluene
diisobutylaluminum
hydride
OH
(IV)

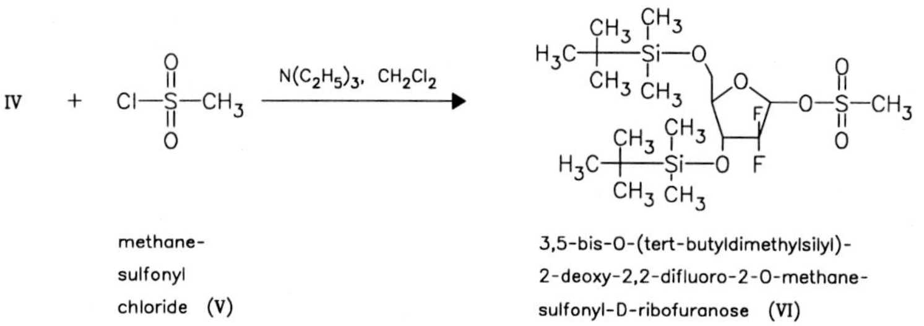
IV +
N(C2H5)3, CH2Cl2
methane-
sulfonyl
chloride (V)
3,5-bis-O-(tert-butyldimethylsilyl)-
2-deoxy-2,2-difluoro-2-O-methane-
sulfonyl-D-ribofuranose (VI)

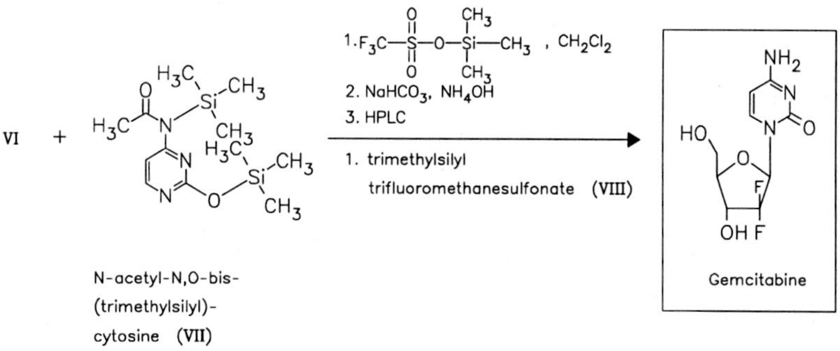
VI +
, CH2Cl2
2. NaHCO3, NH4OH
3. HPLC
1. trimethylsilyl
trifluoromethanesulfonate (VIII)
N-acetyl-N,O-bis-
(trimethylsilyl)-
cytosine (VII)
Gemcitabine

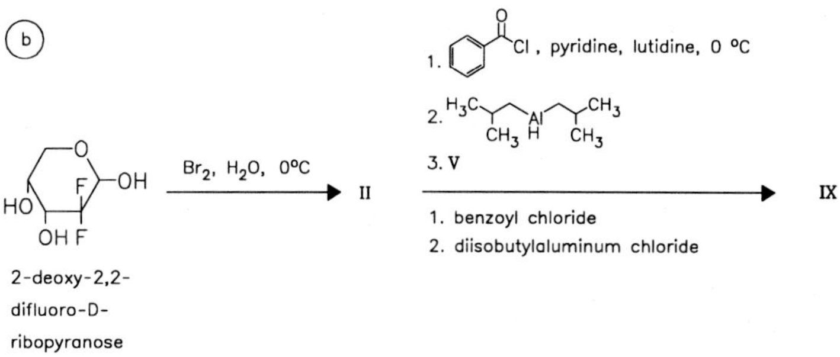
b
Br2, H2O, 0°C
II
1. Cl, pyridine, lutidine, 0 °C
3. V
1. benzoyl chloride
2. diisobutylaluminum chloride
IX
2-deoxy-2,2-
difluoro-D-
ribopyranose

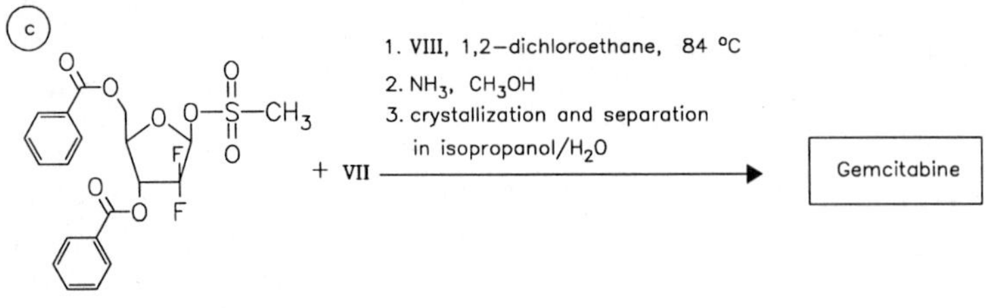

1. VIII, 1,2–dichloroethane, 84 °C
2. NH₃, CH₃OH
3. crystallization and separation in isopropanol/H₂O

+ VII ⟶ Gemcitabine

3,5-di-O-benzoyl-2-deoxy-
2,2-difluoro-1-O-methane-
sulfonyl-D-ribofuranose (IX)

Reference(s):

a   Hertel, L.W. et al.: J. Org. Chem. (JOCEAH) **53** (11), 2406 (1988).
    EP 122 707 (Eli Lilly & Co.; appl. 6.3.1984; USA-prior. 10.3.1983, 4.12.1984, 4.6.1987).
    EP 184 365 (Eli Lilly & Co.; appl. 25.11.1985; USA-prior. 10.10.1985, 4.12.1984, 3.3.1988).
b   US 4 954 623 (Eli Lilly & Co.; appl. 13.11.1989; USA-prior. 13.11.1989, 20.3.1989).
c   EP 306 190 (Eli Lilly & Co.; appl. 22.8.1988; USA-prior. 28.8.1987).
    EP 577 303 (Eli Lilly & Co.; appl. 21.6.1993; USA-prior. 7.4.1993, 22.6.1992, 30.5.1995).

Formulation(s):   vial 200 mg, 1 g (as hydrochloride)

Trade Name(s):

D: Gemzar (Lilly)	GB: Gemzar (Lilly; as	USA: Gemzar (Lilly; as
F: Gemzar (Lilly; as	hydrochloride)	hydrochloride)
hydrochloride)		

Gentamicin

(Gentamycin)

ATC: D06AX07; J01GB03; S01AA11; S03AA06
Use: antibiotic

RN: 1403-66-3   MF: unspecified   MW: unspecified   EINECS: 215-765-8
LD₅₀: 43.5 mg/kg (M, i.v.); 10 g/kg (M, p.o.);
    70 mg/kg (R, i.v.); 6600 mg/kg (R, p.o.);
    184 mg/kg (dog, i.v.)
CN: gentamicin

sulfate
RN: 1405-41-0   MF: H₂SO₄ · x unspecified   MW: unspecified   EINECS: 215-778-9
LD₅₀: 47 mg/kg (M, i.v.); >11.269 g/kg (M, p.o.);
    96 mg/kg (R, i.v.); >5 g/kg (R, p.o.)

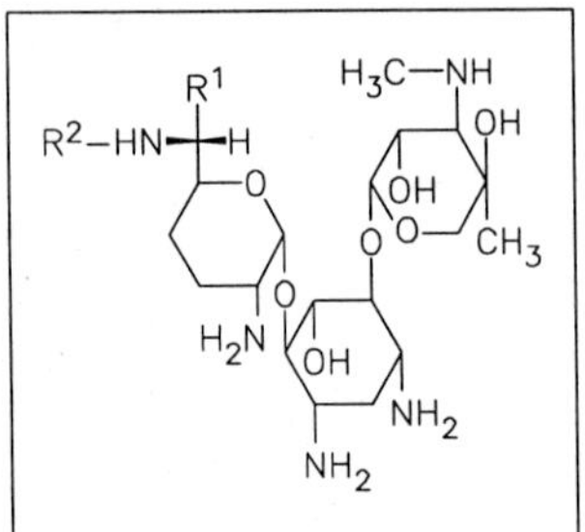

Gentamicin C₁   R¹: —CH₃ ; R²: —CH₃
Gentamicin C₂   R¹: —CH₃ ; R²: —H
Gentamicin C₁ₐ   R¹: —H ;   R²: —H

Gentamicin

From fermentation solutions of *Micromonospora purpurea; Micromonospora echinospora*.

Reference(s):
US 3 091 572 (Schering Corp.; 28.5.1963; prior. 16.7.1962).
US 3 136 704 (Schering Corp.; 9.6.1964; prior. 5.12.1962).
Weinstein, M.J. et al.: Antimicrob. Agents Chemother. (AACHAX), 1 (1963).

Formulation(s):   amp. 10 mg, 40 mg, 80 mg, 120 mg, 160 mg; eye drops 3 mg/ml; ointment 3 mg/g (as sulfate)

Trade Name(s):

D: Dispagent (CIBA Vision)
    duragentamicin 40/80/60
    (durachemie)
    Gencin (curasan)
    Genta (ct-Arzneimittel)
    Gentamicin POS
    (Ursapharm)
    Gentamytrex (Mann)
    Gent-Ophtal (Winzer)
    Ophtagram (Chauvin
    ankerpharm)
    Refobacin (Merck)
    Sulmycin (Essex Pharma)
    numerous combination
    preparations
F: Gentabilles (Schering-
    Plough)

    Gentalline (Schering-
    Plough)
    Gentamicine Chauvin
    (Chauvin)
    Gentasone (Schering-
    Plough)-comb.
    Palacos (Schering-Plough)
GB: Cidomycin (Hoechst)
    Garamycin (Schering-
    Plough)
    Genticin (Roche)
    Gentisone C (Roche)-
    comb.
    Lugacin (Lagap)
    Minims gentamicin
    (Chauvin)
I: Citrizan Antibiotico (IDI)-
    comb.

    Diprogenta (Sca)-comb.
    Farmomicin (Farmigea)-
    comb.
    Genalfa (Intes)-comb.
    Genatrop (Intes)-comb.
    Gentalyn (Schering-
    Plough)
    Gentamen (Pierrel)
    Gentibioptal (Farmila)
    Genticol (Siti)
    Ribomicin (Farmigea)
    Septopal (Bracco)
    combination preparations
J: Gentacin (Shionogi)
USA: Garamycin (Schering)
    Gentafair (Pharmafair)
    G-myticin (Pedinol)

Gentisic acid
(Acide gentisique)

ATC: M01; N02
Use: anti-inflammatory, analgesic

RN: 490-79-9   MF: $C_7H_6O_4$ MW: 154.12   EINECS: 207-718-5
LD_{50}: 374 mg/kg (M, i.v.); 4500 mg/kg (M, p.o.)
CN: 2,5-dihydroxybenzoic acid

monosodium salt
RN: 4955-90-2   MF: $C_7H_5NaO_4$ MW: 176.10   EINECS: 225-598-2
LD_{50}: 3735 mg/kg (M, i.p.); 3900 mg/kg (M, s.c.)

hydro-
quinone

Gentisic acid

Reference(s):
US 2 547 241 (Monsanto; 1951; appl. 1950).
US 2 588 336 (Monsanto; 1952; appl. 1950).
US 2 608 579 (Monsanto; 1952; appl. 1949).
US 2 816 137 (Eastman Kodak; 1957; appl. 1954).

Formulation(s):   drg. 21.1 mg (as sodium salt)

Trade Name(s):
D: Gentisinamid (Herbrand)-   Prigenta (Reiss)-comb.;   Rheumadrag (Schuck)-
      comb.; wfm   wfm   comb.; wfm

Gestodene
(SHB 331)

ATC: G03AB; G03AA
Use: progestogen, oral contraceptive

RN: 60282-87-3   MF: C$_{21}$H$_{26}$O$_2$ MW: 310.44   EINECS: 262-145-8
CN: (17α)-13-ethyl-17-hydroxy-18,19-dinorpregna-4,15-dien-20-yn-3-one

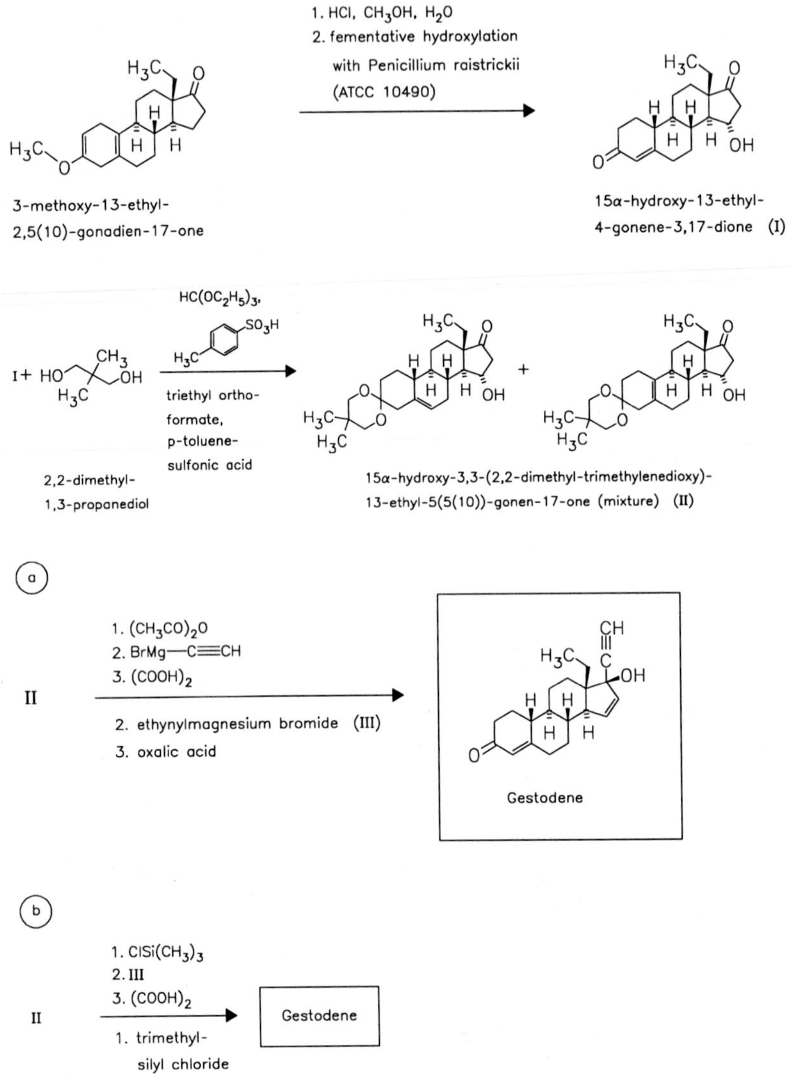

1. HCl, CH$_3$OH, H$_2$O
2. fementative hydroxylation
   with Penicillium raistrickii
   (ATCC 10490)

3-methoxy-13-ethyl-
2,5(10)-gonadien-17-one

15α-hydroxy-13-ethyl-
4-gonene-3,17-dione (I)

HC(OC$_2$H$_5$)$_3$,

triethyl ortho-
formate,
p-toluene-
sulfonic acid

2,2-dimethyl-
1,3-propanediol

15α-hydroxy-3,3-(2,2-dimethyl-trimethylenedioxy)-
13-ethyl-5(10))-gonen-17-one (mixture) (II)

(a)

1. (CH$_3$CO)$_2$O
2. BrMg—C≡CH
3. (COOH)$_2$

2. ethynylmagnesium bromide (III)
3. oxalic acid

Gestodene

(b)

1. ClSi(CH$_3$)$_3$
2. III
3. (COOH)$_2$

1. trimethyl-
   silyl chloride

Gestodene

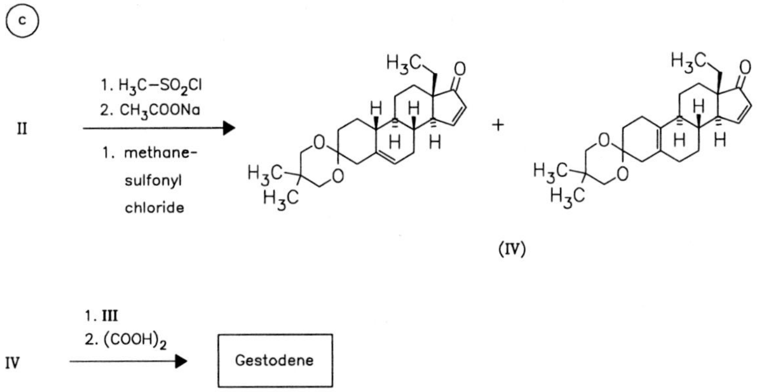

(c)

(IV)

IV → 1. III
        2. (COOH)₂ → [Gestodene]

Reference(s):

BE 847 090 (Schering AG; appl. 8.10.1976; D-prior. 10.10.1975, 12.8.1976).
DOS 2 546 062 (Schering AG; appl. 10.10.1975).
DE 2 636 404 (Schering AG; appl. 12.8.1976).
DE 2 636 405 (Schering AG; appl. 12.8.1976).
DOS 2 636 407 (Schering AG, appl. 12.8.1979).
Hofmeister, H. et al.: Arzneim.-Forsch. (ARZNAD) **36**, 781 (1986).

alternative synthesis from 3-alkoxy-18-methyl-3,5-estradien-17-one derivatives:
EP 201 452 (Schering AG; appl. 7.5.1986; D-prior. 10.5.1985).
US 4 719 054 (Schering AG; 12.1.1988; appl. 9.5.1986; D-prior. 10.5.1985).

medical use for oral contraception in combination with ethynyl-estradiol:
EP 148 724 (Schering AG; appl. 18.12.1984; D-prior. 22.12.1983).
US 4 621 079 (Schering AG; 4.11.1986; appl. 21.12.1984; D-prior. 22.12.1983).

medical use for treatment of β-TGF dependent tumors:
EP 399 631 (Schering AG; appl. 17.5.1990; D-prior. 17.5.1989).

transdermal delivery system:
EP 370 220 (Schering AG; appl. 11.10.1989; D-prior. 27.10.1988).

Formulation(s):   drg. and tabl. 75 µg in combination with 30 µg ethynylestradiol

Trade Name(s):

D: Femovan (Schering; 1987)-comb. with ethynylestradiol
     Minulet (Wyeth)-comb.
F: Harmonet (Wyeth-Lederle)-comb. with ethynylestradiol
     Meliane (Schering)-comb. with ethynylestradiol
     Minulet (Wyeth-Lederle)-comb. with ethynylestradiol

     Moneva (Schering)-comb. with ethynylestradiol
     Phaeva (Schering)-comb. with ethynylestradiol
     Tri-minulet (Wyeth)-comb. with ethynylestradiol
GB: Femodene (Schering; 1987)-comb. with ethynylestradiol
     Minulet (Wyeth; 1997)-comb. with ethynylestradiol

     Triadene (Schering)-comb. with ethynylestradiol
I: Ginoden (Schering; 1987)-comb. with ethynylestradiol
     Milvane (Schering)-comb. with ethynylestradiol
     Minulet (Wyeth; 1988)-comb. with ethynylestradiol
     Triminulet (Wyeth)-comb. with ethynylestradiol

Gestonorone caproate

(Gestronol hexanoate)

ATC: G04C
Use: progestogen (for treatment of prostate hypertrophy)

RN: 1253-28-7   MF: $C_{26}H_{38}O_4$ MW: 414.59   EINECS: 215-010-2
LD$_{50}$:   >10 g/kg (M, p.o.);
        >10 g/kg (R, p.o.)
CN: 17-[(1-oxohexyl)oxy]-19-norpregn-4-ene-3,20-dione

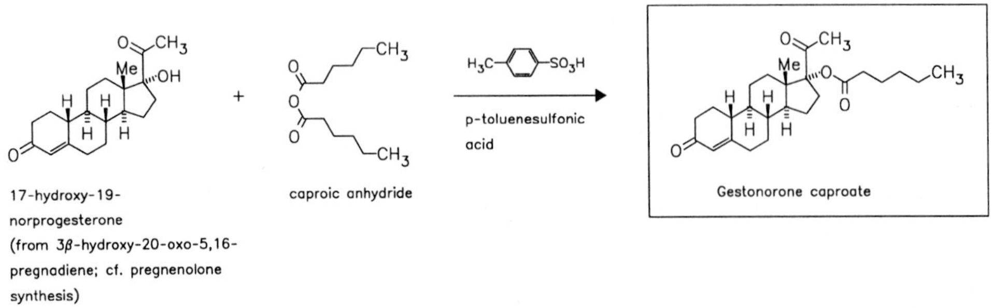

17-hydroxy-19-
norprogesterone
(from 3β-hydroxy-20-oxo-5,16-
pregnadiene; cf. pregnenolone
synthesis)

caproic anhydride

p-toluenesulfonic
acid

Gestonorone caproate

Reference(s):
DE 1 074 582 (Schering AG; appl. 24.9.1958).
Popper, A. et al.: Arzneim.-Forsch. (ARZNAD) **19**, 352 (1969).
(also starting material).

Formulation(s):   amp. 200 mg

Trade Name(s):
D: Depostat (Schering); wfm
F: Depostat (SEPPS); wfm
GB: Depostat (Schering); wfm
        generic
I: Depostat (Schering)
J: Depostat (Nihon Schering)

Gestrinone

ATC: G03D; G03XA02
Use: orally active progestogen (for treatment of endometriosis), antigonadotropin

RN: 16320-04-0   MF: $C_{21}H_{24}O_2$ MW: 308.42
CN: (17α)-13-ethyl-17-hydroxy-18,19-dinorpregna-4,9,11-trien-20-yn-3-one

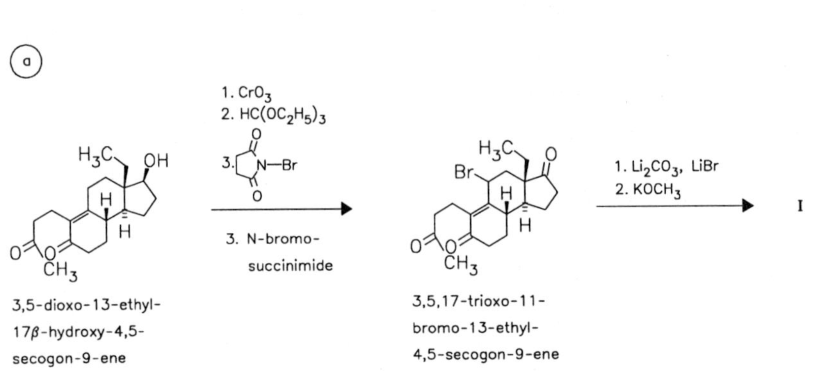

ⓐ

1. CrO$_3$
2. HC(OC$_2$H$_5$)$_3$
3. N-bromo-
succinimide

3,5-dioxo-13-ethyl-
17β-hydroxy-4,5-
secogon-9-ene

3,5,17-trioxo-11-
bromo-13-ethyl-
4,5-secogon-9-ene

1. Li$_2$CO$_3$, LiBr
2. KOCH$_3$

I

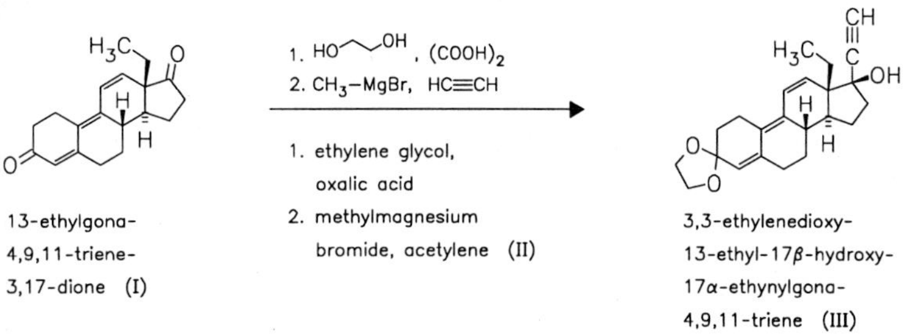

13-ethylgona-
4,9,11-triene-
3,17-dione (I)

1. HO⌒OH , (COOH)$_2$
2. CH$_3$—MgBr, HC≡CH

1. ethylene glycol,
   oxalic acid
2. methylmagnesium
   bromide, acetylene (II)

3,3-ethylenedioxy-
13-ethyl-17β-hydroxy-
17α-ethynylgona-
4,9,11-triene (III)

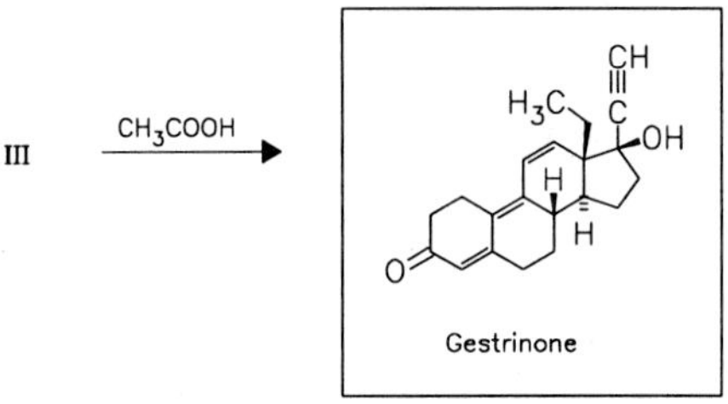

III → CH$_3$COOH

Gestrinone

b

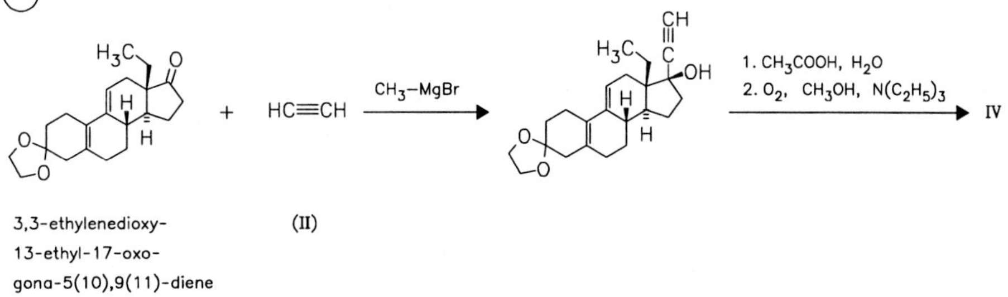

3,3-ethylenedioxy-
13-ethyl-17-oxo-
gona-5(10),9(11)-diene

(II)

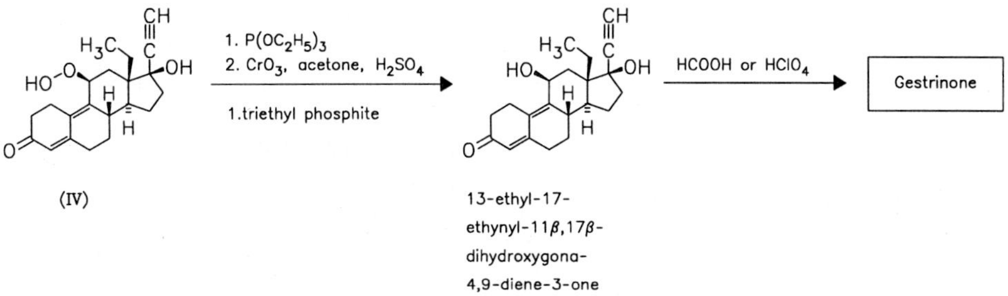

+ HC≡CH → CH$_3$—MgBr

1. CH$_3$COOH, H$_2$O
2. O$_2$, CH$_3$OH, N(C$_2$H$_5$)$_3$ IV

(IV)

1. P(OC$_2$H$_5$)$_3$
2. CrO$_3$, acetone, H$_2$SO$_4$

1.triethyl phosphite

13-ethyl-17-
ethynyl-11β,17β-
dihydroxygona-
4,9-diene-3-one

HCOOH or HClO$_4$ → Gestrinone

Reference(s):
a   DE 1 593 307 (Roussel-Uclaf; appl. 1966; F-prior. 1965).
b   DE 1 618 810 (Roussel-Uclaf; appl. 1967; F-prior. 1966).

alternative synthesis:
DOS 2 212 589 (Roussel-Uclaf; appl. 15.3.1972; F-prior. 19.3.1971).
GB 1 069 709 (Roussel-Uclaf; appl. 1966; F-prior. 1964).

synthesis of 13β-ethyl-4,9,11-gonatriene-3,17-dione:
FR 1 526 962 (Roussel-Uclaf; appl. 6.1.1967).

alternative synthesis:
NL 6 517 141 (Roussel-Uclaf; appl. 1965; F-prior. 1964).

synthesis of 3,5-dioxo-13-ethyl-17β-hydroxy-4,5-secogon-9-ene:
NL 6 414 702 (Roussel-Uclaf; appl. 1965; F-prior. 1963, 1964).
GB 1 096 761 (Roussel-Uclaf; appl. 1964; F-prior. 1963).

synthesis of 13-ethyl-17-ethynyl-11β,17β-dihydroxygona-4,9-diene-3-one:
FR-M 5 435 (Roussel-Uclaf; appl. 1966).

Formulation(s):   cps. 2.5 mg

Trade Name(s):
GB: Dimetriose (Florizel)   I: Dimetrose (Poli)

Gitaloxin
(16-Formylgitoxin)

ATC: C01AA
Use: cardiac glycoside

RN: 3261-53-8   MF: $C_{42}H_{64}O_{15}$ MW: 808.96   EINECS: 221-864-7
LD_{50}:   28.7 g/kg (M, p.o.);
       29.960 mg/kg (R, p.o.)
CN: (3β,5β,16β)-3-[(*O*-2,6-dideoxy-β-D-*ribo*-hexopyranosyl-(1→4)-*O*-2,6-dideoxy-β-D-*ribo*-hexopyranosyl-
       (1→4)-2,6-dideoxy-β-D-*ribo*-hexopyranosyl)oxy]-16-(formyloxy)-14-hydroxycard-20(22)-enolide

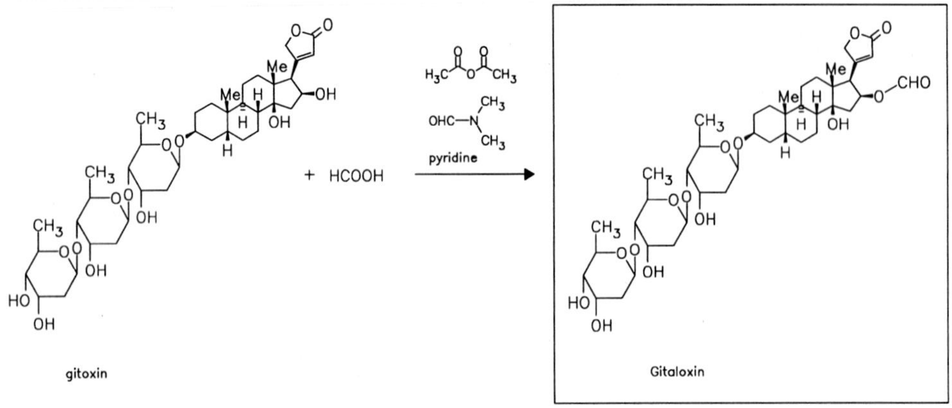

gitoxin   Gitaloxin

Reference(s):
DE 1 026 312 (Boehringer Mannh.; appl. 1955).

extraction from Digitalis purpurea:
DOS 1 042 838 (Boehringer Mannh.; appl. 1958).

mixed crystals with digitoxin:
DE 1 140 315 (Boehringer Mannh.; appl. 1961).

injection solution:
BE 618 160 (Christians; appl. 25.5.1962).

review:
Georges, A. et al.: Therapie (THERAP) **18**, 209 (1963).

Formulation(s):   tabl. 0.1 mg

Trade Name(s):
F: Cristaloxine (Sedaph); wfm

Gitoformate

(Pentaformylgitoxin)

ATC: C01AA09
Use: cardiac glycoside

RN: 10176-39-3 MF: $C_{46}H_{64}O_{19}$ MW: 921.00 EINECS: 233-450-3
LD$_{50}$: 23.9 mg/kg (M, p.o.);
39.01 mg/kg (R, p.o.)
CN: (3β,5β,16β)-3-[(O-2,6-dideoxy-3,4-di-O-formyl-β-D-*ribo*-hexopyranosyl-(1→4)-O-2,6-dideoxy-3-O-
formyl-β-D-*ribo*-hexopyranosyl-(1→4)-2,6-dideoxy-3-O-formyl-β-D-*ribo*-hexopyranosyl)oxy]-16-
(formyloxy)-14-hydroxycard-20(22)-enolide

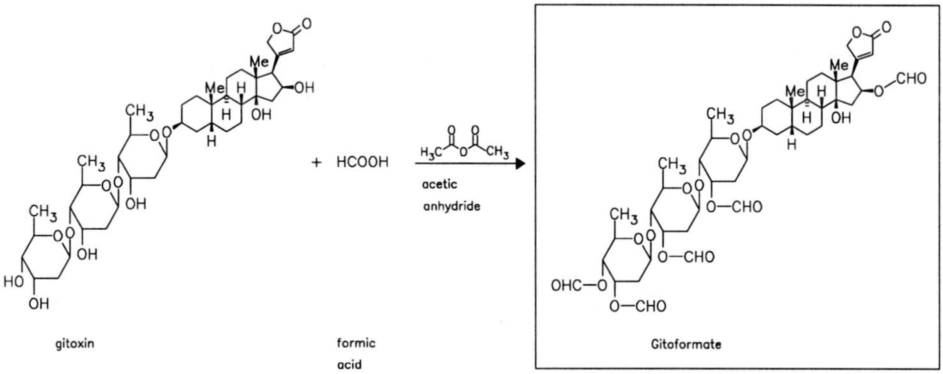

gitoxin + HCOOH formic acid acetic anhydride → Gitoformate

Reference(s):
BE 625 447 (Manufacture de Produits Pharmaceutique; appl. 28.11.1962).

Formulation(s): tabl. 0.04 mg, 0.06 mg

Trade Name(s):
D: Dynocard (Madaus); wfm I: Formiloxine (Menarini);
wfm

Glafenine

ATC: N02BG03
Use: analgesic, anti-inflammatory

RN: 3820-67-5 MF: $C_{19}H_{17}ClN_2O_4$ MW: 372.81 EINECS: 223-315-7
LD$_{50}$: 1486 mg/kg (M, p.o.);
2300 mg/kg (R, p.o.)
CN: 2-[(7-chloro-4-quinolinyl)amino]benzoic acid 2,3-dihydroxypropyl ester

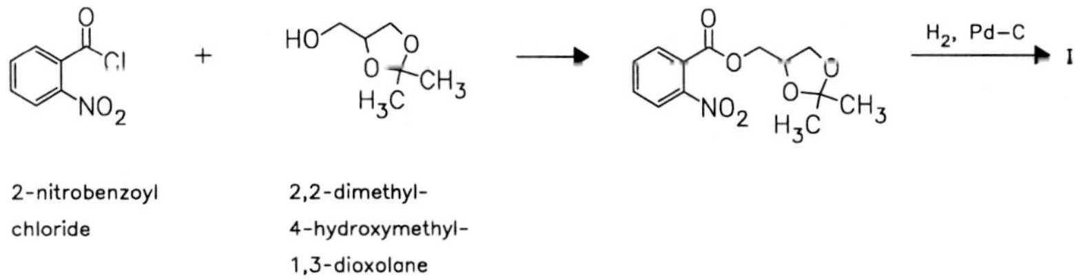

2-nitrobenzoyl
chloride

2,2-dimethyl-
4-hydroxymethyl-
1,3-dioxolane

H$_2$, Pd–C → I

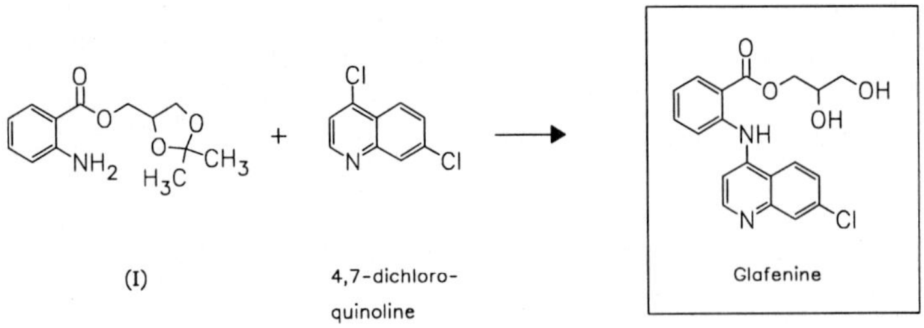

(I)　　　　4,7-dichloro-
quinoline　　　　Glafenine

Reference(s):
US 3 232 944 (Roussel-Uclaf; 1.2.1966; F-prior. 20.8.1962).
FR-M 2 413 (Roussel-Uclaf; appl. 20.11.1962; prior. 20.8.1962).

anthranilic acid monoglyceride *from* isatoic anhydride:
E-appl. 678 (Pierre Fabre; appl. 18.7.1978; F-prior. 26.7.1977).

Formulation(s):　　suppos. 500 mg; tabl. 200 mg

Trade Name(s):

D:	Glifanan (Albert-Roussel); wfm		Glifanan (Roussel); wfm Privadol (Roland-Marie); wfm	I:	Glifan (Roussel-Maestretti); wfm
F:	Adalgur (Roussel)-comb.; wfm			J:	Glifanan (Nippon Roussel)

Glaziovine

ATC:　　N05B
Use:　　tranquilizer

RN:　　17127-48-9　MF: C$_{18}$H$_{19}$NO$_3$　MW: 297.35
CN:　　(±)-2',3',8',8'a-tetrahydro-6'-hydroxy-5'-methoxy-1'-methylspiro[2,5-cyclohexadiene-1,7'(1'*H*)-cyclopent[*ij*]isoquinolin]-4-one

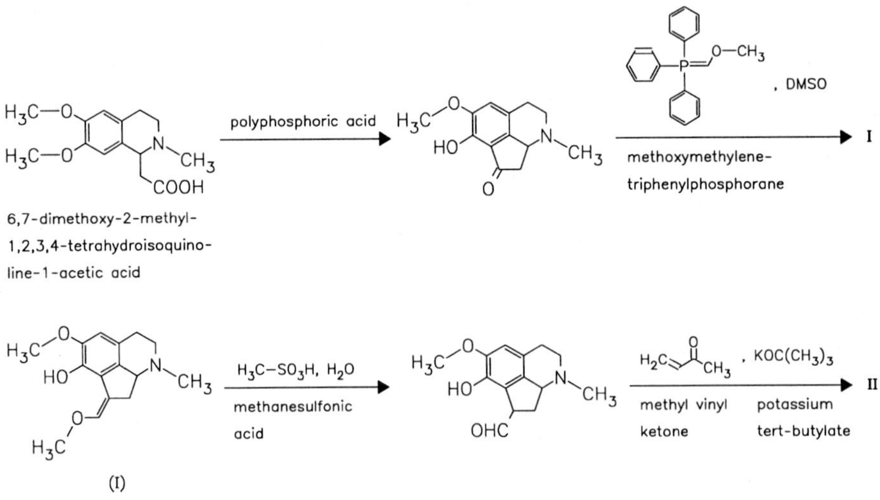

6,7-dimethoxy-2-methyl-
1,2,3,4-tetrahydroisoquino-
line-1-acetic acid

polyphosphoric acid

methoxymethylene-
triphenylphosphorane

, DMSO

(I)

H$_3$C—SO$_3$H, H$_2$O
methanesulfonic
acid

methyl vinyl
ketone

potassium
tert-butylate

. KOC(CH$_3$)$_3$

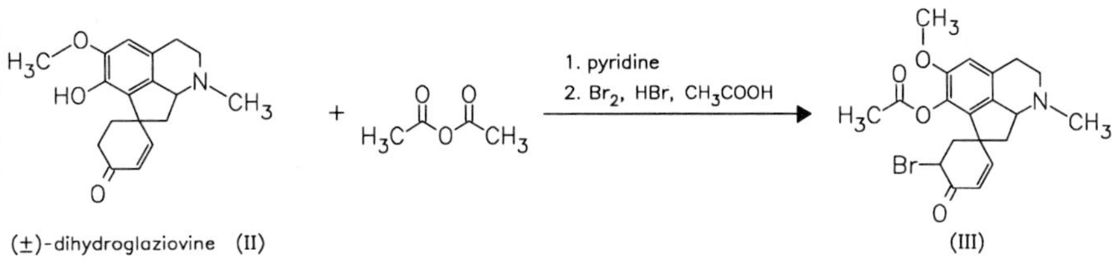

(±)-dihydroglaziovine (II)

(III)

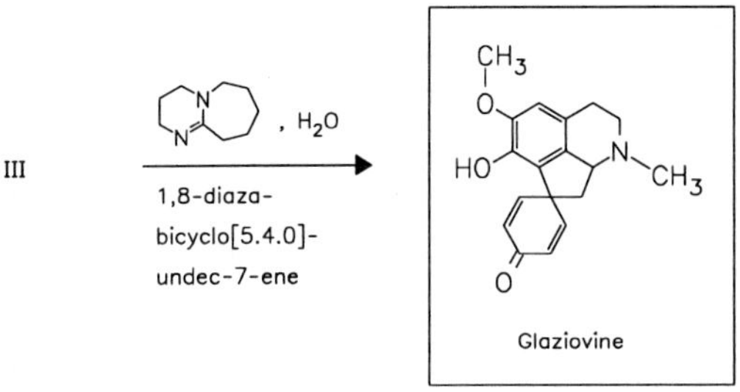

III

1,8-diaza-
bicyclo[5.4.0]-
undec-7-ene

Glaziovine

Reference(s):
DOS 2 363 531 (Siphar; appl. 20.12.1973; CH-prior. 22.12.1972).
GB 1 459 210 (Siphar; valid from 18.12.1973; CH-prior. 22.12.1972).

alternative syntheses:
US 3 886 166 (Siphar; 27.5.1975; CH-prior. 22.12.1972, 26.2.1973).
DOS 2 363 529 (Siphar; appl. 20.12.1973; CH-prior. 22.12.1972, 26.2.1973).
Kametani, T. et al.: Tetrahedron Lett. (TELEAY) **1973**, 4219.
DOS 2 363 530 (Siphar; appl. 20.12.1973; CH-prior. 22.12.1972).

isolation from the leaves of Ocotea glaziovii:
Gilbert, B. et al.: J. Am. Chem. Soc. (JACSAT) **86**, 694 (1964).

Formulation(s):   tabl. 200 mg

Trade Name(s):
I: Suavedol (Simes); wfm

Glibenclamide
(Glyburide)

ATC: A10BB01
Use: antidiabetic

RN: 10238-21-8   MF: $C_{23}H_{28}ClN_3O_5S$ MW: 494.01   EINECS: 233-570-6
LD$_{50}$:   3250 mg/kg (M, p.o.);
        >20 g/kg (R, p.o.);
        >10 g/kg (dog, p.o.)
CN: 5-chloro-*N*-[2-[4-[[[(cyclohexylamino)carbonyl]amino]sulfonyl]phenyl]ethyl]-2-methoxybenzamide

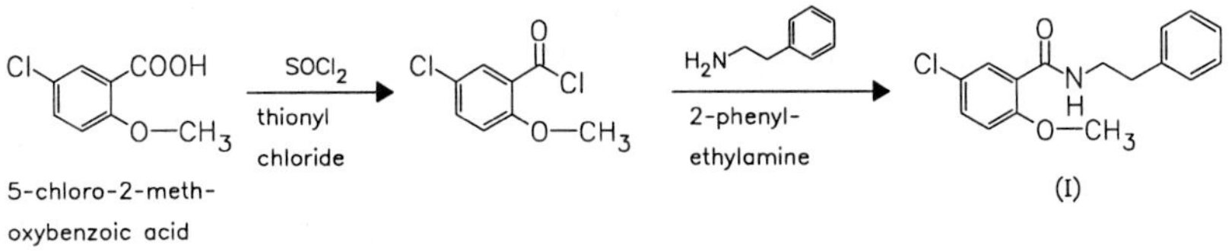

5-chloro-2-meth-
oxybenzoic acid

SOCl$_2$
thionyl
chloride

2-phenyl-
ethylamine

(I)

Glibenclamide

Reference(s):
Aumüller, W. et al.: Arzneim.-Forsch. (ARZNAD) **16**, 1640 (1966).
DE 1 283 837 (Hoechst; appl. 13.7.1967; CDN-prior. 21.7.1966)
DE 1 301 812 (Hoechst; appl. 27.7.1965).
BE 684 652 (Hoechst; appl. 27.7.1966; D-prior. 27.7.1965).
US 3 454 635 (Hoechst; 8.7.1969; appl. 13.7.1966; D-prior. 2.12.1965).

Formulation(s):   tabl. 1.75 mg, 2.5 mg, 3.5 mg, 5 mg

Trade Name(s):

D:	Azuglucon (Azuchemie)		Praeciglucon (Pfleger)	Euglucon (Boehringer
	Bastiverit (Bastian-Werk)		Semi-Euglucon N (Roche/	Mannh.)
	duraglucon (durachemie)		HMR)	Gliben (Gentili)
	Euglucon N (Roche/HMR;		Semi-Gliben-Puren N (Isis	Glibomet (Guidotti)-comb.
	1969)		Puren)	Gliboral (Guidotti)
	Glimidstada (Stada)	F:	Daonil (Hoechst)	Glucomide (Lipha)-comb.
	Gluconorm (Wolff)		Euglucan (Boehringer	combination preparations
	Glucoreduct (Sanofi		Mannh.)	J: Daonil (Hoechst)
	Winthrop)		Hemi-Daonil (Hoechst)	Euglucon (Yamanouchi)
	Glucoromed (Lichtenstein)		Miglucan (Boehringer	USA: Diabeta (Hoechst Marion
	Gluco Tablinen (Sanorania)		Mannh.)	Roussel)
	Glucovital (Wolff)	GB:	Daonil (Hoechst)	Glynase (Pharmacia &
	Glycolande (Synthelabo)		Euglucon (Hoechst)	Upjohn)
	Humedia (APS)		Semi-Daonil (Hoechst)	Micronase (Pharmacia &
	Maninil (Berlin-Chemie)	I:	Daonil (Hoechst)	Upjohn)

Glibornuride

ATC: A10BB04
Use: antidiabetic

RN: 26944-48-9   MF: $C_{18}H_{26}N_2O_4S$ MW: 366.48   EINECS: 248-124-6
LD_{50}:   >20 g/kg (M, p.o.);
          18 g/kg (R, p.o.)
CN: [1*S*-(*endo,endo*)]-*N*-[[(3-hydroxy-4,7,7-trimethylbicyclo[2.2.1]hept-2-yl)amino]carbonyl]-4-
          methylbenzenesulfonamide

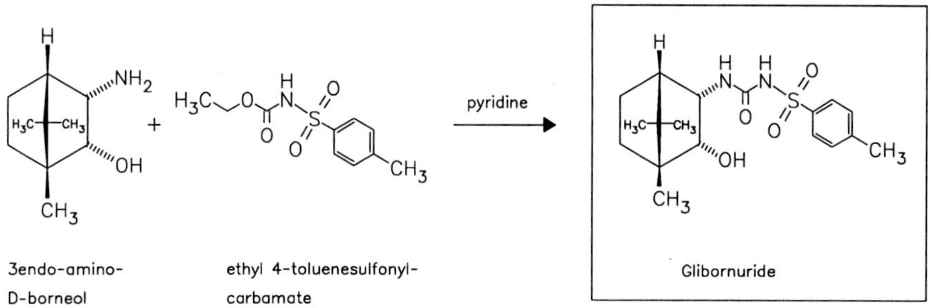

3endo-amino-
D-borneol

ethyl 4-toluenesulfonyl-
carbamate

pyridine

Glibornuride

Reference(s):
DE 1 695 201 (Hoffmann-La Roche; appl. 21.9.1967; CH-prior. 28.10.1966, 24.4.1967, 17.7.1967).
US 3 654 357 (Roche; 4.4.1972; CH-prior. 26.4.1968).
US 3 787 491 (Hoffmann-La Roche; 22.1.1974; prior. 29.9.1971).
US 3 860 724 (Hoffmann-La Roche; 14.1.1975; prior. 29.9.1971).

Formulation(s):   tabl. 25 mg

Trade Name(s):
D: Gluborid (Grünenthal)   F: Glutril (Roche; 1972)
      Glutril (ICN; 1972)   GB: Glutril (Roche; 1975); wfm

Gliclazide

ATC: A10BB09
Use: antidiabetic

RN: 21187-98-4   MF: $C_{15}H_{21}N_3O_3S$ MW: 323.42   EINECS: 244-260-5
LD$_{50}$:   295 mg/kg (M, i.v.); 3 g/kg (M, p.o.);
       382 mg/kg (R, i.v.); 3 g/kg (R, p.o.)
CN: *N*-[[(hexahydrocyclopenta[*c*]pyrrol-2(1*H*)-yl)amino]carbonyl]-4-methylbenzenesulfonamide

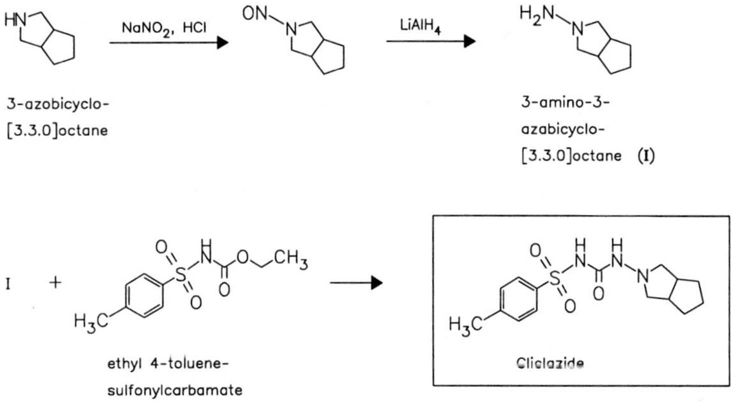

3-azobicyclo-
[3.3.0]octane

3-amino-3-
azabicyclo-
[3.3.0]octane (I)

I   +

ethyl 4-toluene-
sulfonylcarbamate

Gliclazide

Reference(s):
US 3 501 495 (Science Union; 17.3.1970; GB-prior. 10.2.1966).
FR 1 510 714 (Science Union; appl. 9.2.1967; GB-prior. 10.2.1966).

Formulation(s):   tabl. 40 mg, 80 mg

Trade Name(s):
D: Diamicron (Servier   F: Diamicron (Servier)   I: Diabrezide (Molteni)
      Deutschland)   GB: Diamicron (Servier)   Diamicron (Servier)

J: Glimicron (Dainippon)

Glimepiride
(Hoe 490)

ATC: A10BB12
Use: insulin-sparing sulfonylurea, antidiabetic

RN: 93479-97-1 MF: C₂₄H₃₄N₄O₅S MW: 490.63

RN: 93479-97-1 MF: $C_{24}H_{34}N_4O_5S$ MW: 490.63
CN: *trans*-3-ethyl-2,5-dihydro-4-methyl-*N*-[2-[4-[[[[(4-methylcyclohexyl)amino]carbonyl]amino]sulfonyl]phenyl]ethyl]-2-oxo-1*H*-pyrrole-1-carboxamide

3-ethyl-4-methyl-Δ³-pyrrolin-2-one

3-ethyl-4-methyl-Δ^3-pyrrolin-2-one   2-phenylethyl isocyanate

(I)   trans-4-methylcyclohexyl isocyanate   Glimepiride

Reference(s):
DE 2 951 135 (Hoechst AG; appl. 25.6.1981; D-prior. 19.12.1979).

preparation of 3-ethyl-4-methyl-Δ³-pyrrolin-2-one
Siedel: Justus Liebigs Ann. Chem. (JLACBF) **554**, 144, 155 (1943).
Plieninger; Decker: Justus Liebigs Ann. Chem. (JLACBF) **598**, 198, 205 (1956).
Bishop, J.E.; Nagy, J.O.; O'Connell, J.F.; Rapoport, H.: J. Am. Chem. Soc. (JACSAT) **113** (21), 8024 (1991).
Schoenleber, R.W.; Kim, Y.; Rapoport, H.: J. Am. Chem. Soc. (JACSAT) **106** (9), 2645 (1984).
Tipton, A.; Ligthner, D.A.: Monatsh. Chem. (MOCMB7) **130** (3), 425 (1999).

for treatment of arteriosclerosis:
EP 604 853 (Hoechst Japan; appl. 6.7.1994; J-prior. 28.12.1992).

formulation:
EP 649 660 (Hoechst AG; appl. 26.4.1995; D-prior. 26.1.1993).

controlled release:
DE 4 336 159 (Hoechst AG; appl. 27.4.1994; D-prior. 22.1.1993).

for treatment of obesity:
WO 9 303 724 (Upjohn Co.; appl. 4.3.1993; USA-prior. 26.8.1991).

Formulation(s): tabl. 1 mg, 2 mg, 3 mg, 4 mg

Trade Name(s):
D: Amarel (Hoechst)
F: Amarel (Hoechst Houdé)
USA: Amaryl (Hoechst Marion Roussel)

Glipizide

ATC: A10BB07
Use: antidiabetic

RN: 29094-61-9   MF: $C_{21}H_{27}N_5O_4S$ MW: 445.54   EINECS: 249-427-6
LD$_{50}$:   >3 g/kg (M, i.p.);
         1200 mg/kg (R, i.p.)
CN: N-[2-[4-[[[(cyclohexylamino)carbonyl]amino]sulfonyl]phenyl]ethyl]-5-methylpyrazinecarboxamide

Reference(s):
DAS 2 012 138 (Carlo Erba; appl. 14.3.1970; I-prior. 26.3.1969, 18.6.1969).
US 3 669 966 (Carlo Erba; 13.6.1972; I-prior. 26.3.1969, 18.6.1969).
Ambrogli, V. et al.: Arzneim.-Forsch. (ARZNAD) 21, 200 (1971).

preparation of 5-methylpyrazine-2-carboxylic acid from 2,5-dimethylpyrazine via oxidation:
Stoehr: J. Prakt. Chem. (JPCEAO) (2), 51, 464 (1895).
Stoehr: J. Prakt. Chem. (JPCEAO) (2), 47, 480 (1893).
Kiener, A.: Angew. Chem. (ANCEAD) 104 (6), 748 (1992).
Goldberg, Yu.; Shymanska, M.: Org. Prep. Proced. Int. (OPPIAK) 23 (2), 188 (1991).

electrochemical preparation of 5-methylpyrazine-2-carboxylic acid:
Borsotti, G.P.; Foà, M.; Gatti, N.: Synthesis (SYNTBF) 1990 (3), 207.
Feldman, D. et al.: Chem. Heterocycl. Compd. (N. Y.) (CHCCAL) 31 (1), 80 (1995).

Formulation(s):   tabl. 2.5 mg, 5 mg, 10 mg

Trade Name(s):

D:	Glibenese (Pfizer; 1977)		Ozidia cp à lib modifiée	I:	Minidiab (Carlo Erba;
F:	Glibénèse (Pfizer; 1974)		(CC) (Pfizer)		1972)
	Minidiab (Pharmacia &	GB:	Glibenese (Pfizer; 1975)	USA.	Glucotrol (Pfizer; 1984)
	Upjohn; 1974)		Minodiab (Pharmacia &		
			Upjohn; 1975)		

Gliquidone

ATC: A10BB08
Use: antidiabetic

RN: 33342-05-1   MF: $C_{27}H_{33}N_3O_6S$ MW: 527.64   EINECS: 251-463-2
LD$_{50}$:   234 mg/kg (M, i.v.); >2g/kg (M, p.o.)
CN: N-[(cyclohexylamino)carbonyl]-4-[2-(3,4-dihydro-7-methoxy-4,4-dimethyl-1,3-dioxo-2(1H)-
         isoquinolinyl)ethyl]benzenesulfonamide

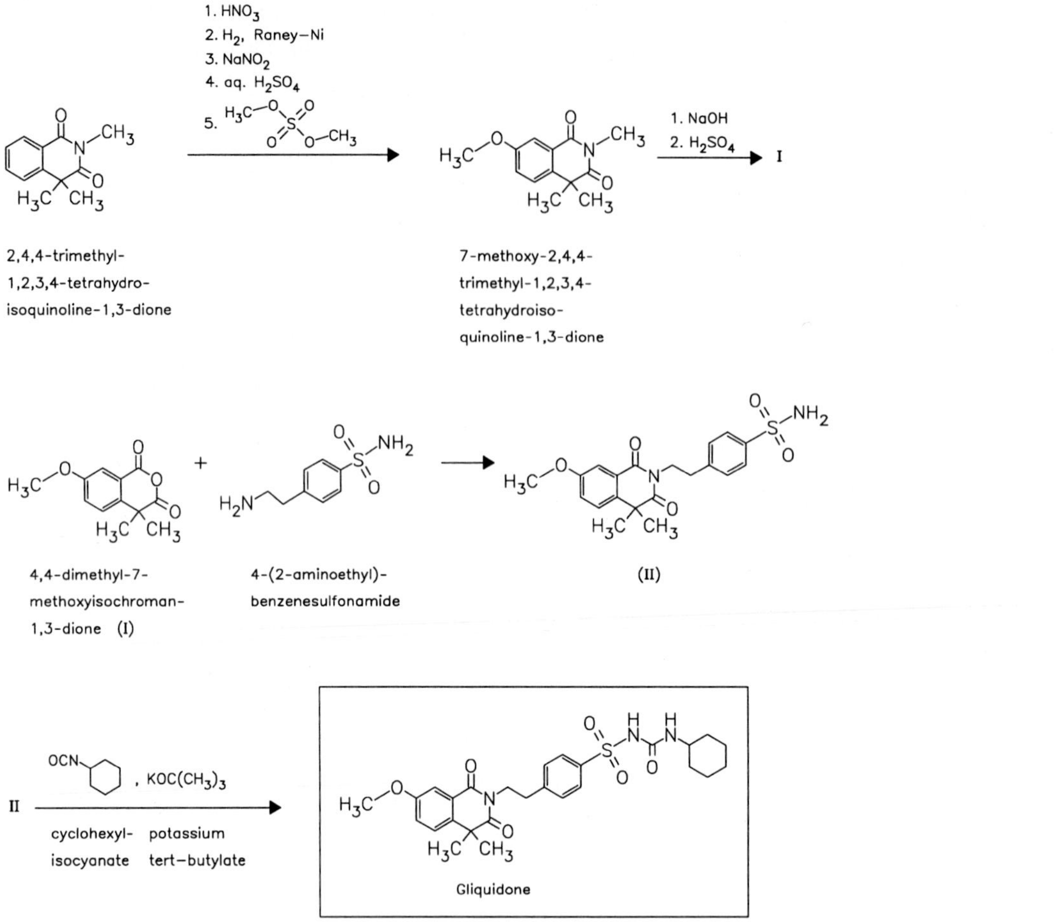

1. HNO$_3$
2. H$_2$, Raney–Ni
3. NaNO$_2$
4. aq. H$_2$SO$_4$
5. H$_3$C–O–SO$_2$–O–CH$_3$

1. NaOH
2. H$_2$SO$_4$ I

2,4,4-trimethyl-
1,2,3,4-tetrahydro-
isoquinoline-1,3-dione

7-methoxy-2,4,4-
trimethyl-1,2,3,4-
tetrahydroiso-
quinoline-1,3-dione

4,4-dimethyl-7-
methoxyisochroman-
1,3-dione   (I)

+

4-(2-aminoethyl)-
benzenesulfonamide

(II)

OCN–cyclohexyl , KOC(CH$_3$)$_3$

cyclohexyl-   potassium
isocyanate   tert–butylate

II

Gliquidone

Reference(s):
DAS 2 000 339 (Thomae; appl. 5.1.1970).
DOS 2 011 126 (Thomae; appl. 10.3.1970).
US 3 708 486 (Boehringer Ing.; 2.1.1973; D-prior. 5.1.1970 and 17.4.1969).

Formulation(s):   tabl. 30 mg

Trade Name(s):
D: Glurenorm (Yamanouchi)   GB: Glurenorm (Sanofi   I: Glurenor (Guidotti)
                                              Winthrop)

Glisoxepide

ATC: A10BB11
Use: antidiabetic

RN: 25046-79-1   MF: C$_{20}$H$_{27}$N$_5$O$_5$S   MW: 449.53   EINECS: 246-579-5
LD$_{50}$:   283 mg/kg (M, i.v.); >10 g/kg (M, p.o.);
        196 mg/kg (R, i.v.); >10 g/kg (R, p.o.);
        >2 g/kg (dog, p.o.)
CN: *N*-[2-[4-[[[[(hexahydro-1*H*-azepin-1-yl)amino]carbonyl]amino]sulfonyl]phenyl]ethyl]-5-methyl-3-
        isoxazolecarboxamide

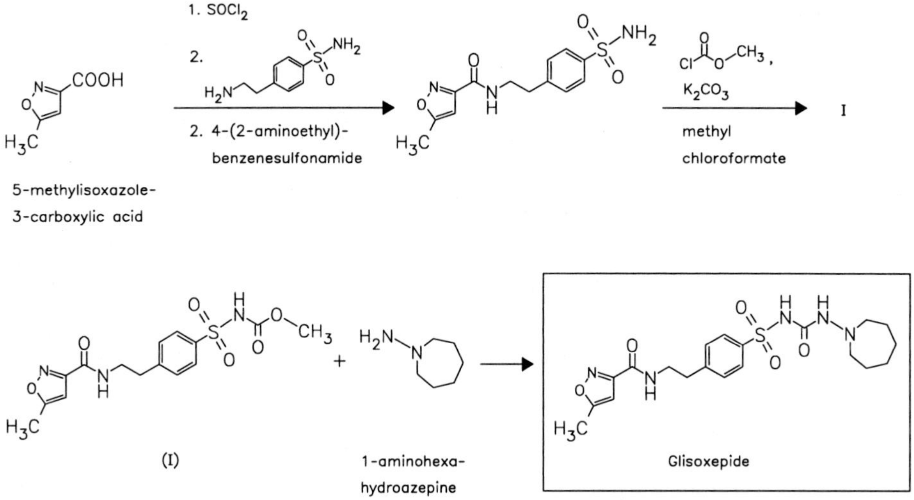

5-methylisoxazole-
3-carboxylic acid

1. SOCl₂
2.
2. 4-(2-aminoethyl)-
benzenesulfonamide

methyl
chloroformate

(I)   1-aminohexa-
                       hydroazepine

Glisoxepide

Reference(s):
DE 1 670 952 (Bayer; appl. 25.11.1967).
US 3 668 215 (Bayer; 6.6.1972; D-prior. 25.11.1967).
Plümpe, H. et al.: Arzneim.-Forsch. (ARZNAD) **24**, 363 (1974).

Formulation(s):   tabl. 4 mg

Trade Name(s):
D: Pro-Diaban (Bayer Vital)   I: Glucoben (Farmades); wfm

Glucametacin

ATC: A02A
Use: anti-inflammatory

RN: 52443-21-7   MF: C₂₅H₂₇ClN₂O₈   MW: 518.95   EINECS: 257-923-9
CN: 2-[[[1-(4-chlorobenzoyl)-5-methoxy-2-methyl-1H-indol-3-yl]acetyl]amino]-2-deoxy-D-glucose

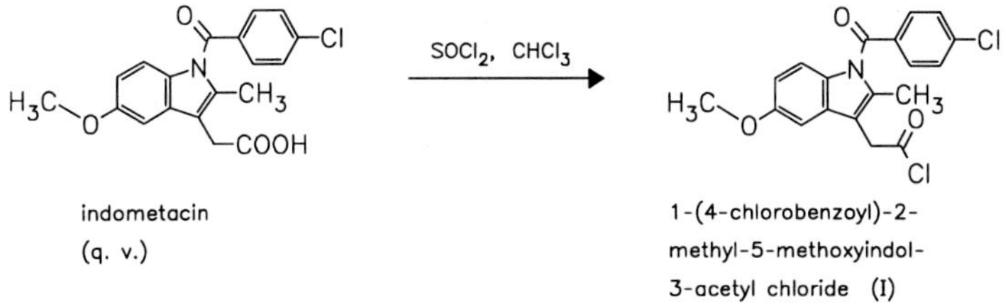

SOCl₂, CHCl₃

indometacin
(q. v.)

1-(4-chlorobenzoyl)-2-
methyl-5-methoxyindol-
3-acetyl chloride (I)

I +

α-D-glucosamine
hydrochloride

Glucametacin

Reference(s):
DOS 2 223 051 (SIR; appl. 12.5.1972; I-prior. 9.5.1972).

Formulation(s):   cps. 70 mg, 140 mg

Trade Name(s):
I: Teorema (Farmades); wfm   Teoremac (Sir); wfm

D-Glucosamine
(Chitosamine)

ATC: M01AX05
Use: antirheumatic, antiarthritic

RN: 3416-24-8   MF: $C_6H_{13}NO_5$ MW: 179.17   EINECS: 222-311-2
CN: 2-amino-2-deoxy-D-glucose

hydrochloride
RN: 66-84-2   MF: $C_6H_{13}NO_5 \cdot HCl$ MW: 215.63   EINECS: 200-638-1
sulfate (2:1)
RN: 14999-43-0   MF: $C_6H_{13}NO_5 \cdot 1/2H_2O_4S$ MW: 456.42   EINECS: 239-088-2
hydriodide
RN: 14999-44-1   MF: $C_6H_{13}NO_5 \cdot HI$ MW: 307.08   EINECS: 239-089-8

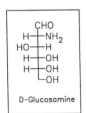

D-Glucosamine

Unit of chitin, mucoproteins and mucopolysaccharids, obtained by hydrolysis with HCl.

Reference(s):
Ledderhose, G.: Z. Physiol. Chem. (ZPCHA5) **2**, 213 (1878).

preparation of glucosamine salts:
GB 1 056 331 (Rotta Research; appl. 15.1.1964; I-prior. 18.1.1963).

stable complex from glucosamine sulfate *and* NaCl:
GB 2 101 585 (Rotta Research; appl. 26.4.1982; I-prior. 30.4.1981).

combination of glucosamine sulfate *and* hydriodide *for therapy of rheumatoid arthritis and osteoarthritis:*
US 3 683 076 (L. Rovati; 8.8.1972; I-prior. 26.10.1968).

Formulation(s):   drg. 200 mg, 250 mg (as sulfate); vial 400 mg (as sulfate)

Trade Name(s):
D: Dona-200 S (Opfermann)   I: Dona (Rottapharm; as
                                              sulfate)

Glutethimide

ATC: N05CE01
Use: hypnotic, sedative

RN: 77-21-4   MF: $C_{13}H_{15}NO_2$ MW: 217.27   EINECS: 201-012-0
LD_{50}:   360 mg/kg (M, p.o.);
         600 mg/kg (R, p.o.);
         500 mg/kg (dog, p.o.)
CN: 3-ethyl-3-phenyl-2,6-piperidinedione

methyl   2-phenyl-   Glutethimide
acrylate   butyronitrile

Reference(s):
US 2 673 205 (Ciba; 1954; CH-prior. 1951).
DE 950 193 (Ciba; appl. 1952; CH-prior. 1951).

Formulation(s):   cps.; tabl. 250 mg, 500 mg

Trade Name(s):
D: Doriden (Ciba); wfm   I: Doriden (Ciba); wfm   USA: Doriden (USV); wfm
F: Doridéne (Ciba); wfm   J: Doriden (Ciba-Geigy-   generic
GB: Doriden (Ciba); wfm   Takeda)

Glybuzole
(Desaglybuzole)

ATC: V03AH
Use: antidiabetic

RN: 1492-02-0   MF: $C_{12}H_{15}N_3O_2S_2$ MW: 297.40   EINECS: 216-081-2
LD_{50}:   193 mg/kg (M, i.v.); 550 mg/kg (M, p.o.);
         500 mg/kg (R, p.o.)
CN: N-[5-(1,1-dimethylethyl)-1,3,4-thiadiazol-2-yl]benzenesulfonamide

benzene-   2-amino-5-tert-   Glybuzole
sulfonyl   butyl-1,3,4-
chloride   thiadiazole

Reference(s):
GB 822 947 (Smith & Nephew; appl. 1957; valid from 1958).
FR-M 3 389 (Rhône-Poulenc; appl. 27.1.1964).

Formulation(s): 250 mg (oral)

Trade Name(s):
J: Gludiase (Kyowa Hakko)

Glyconiazide
(Gluconiazide)

ATC: J04A
Use: tuberculostatic

RN: 3691-74-5 MF: $C_{12}H_{13}N_3O_6$ MW: 295.25 EINECS: 223-005-1
LD$_{50}$: 641 mg/kg (M, i.v.); 748 mg/kg (M, p.o.);
1763 mg/kg (R, i.v.); 6423 mg/kg (R, p.o.)
CN: glucuronic acid γ-lactone 1-[(4-pyridinylcarbonyl)hydrazone]

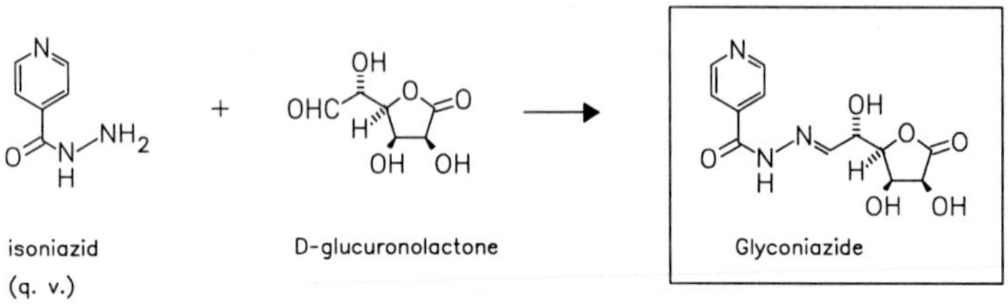

isoniazid
(q. v.)

D-glucuronolactone

Glyconiazide

Reference(s):
US 2 940 899 (Univ. of California; 14.6.1960; prior. 28.9.1953).
Sah, P.P.T.: J. Am. Chem. Soc. (JACSAT) **75**, 2512 (1953).

Formulation(s): (oral) 0.015 g/kg

Trade Name(s):
D: Gluronazid (Hormon-
Chemie); wfm

Isozidoron 444
(Saarstickstoff)-comb.;
wfm

I: Glucazide (Stholl); wfm
J: Hydronsan (Chugai)

Glycopyrronium bromide
(Glycopyrrolate)

ATC: A03AB02
Use: anticholinergic, antispasmodic

RN: 596-51-0 MF: $C_{19}H_{28}BrNO_3$ MW: 398.34 EINECS: 209-887-0
LD$_{50}$: 15 mg/kg (M, i.v.); 570 mg/kg (M, p.o.);
709 mg/kg (R, p.o.)
CN: 3-[(cyclopentylhydroxyphenylacetyl)oxy]-1,1-dimethylpyrrolidinium bromide

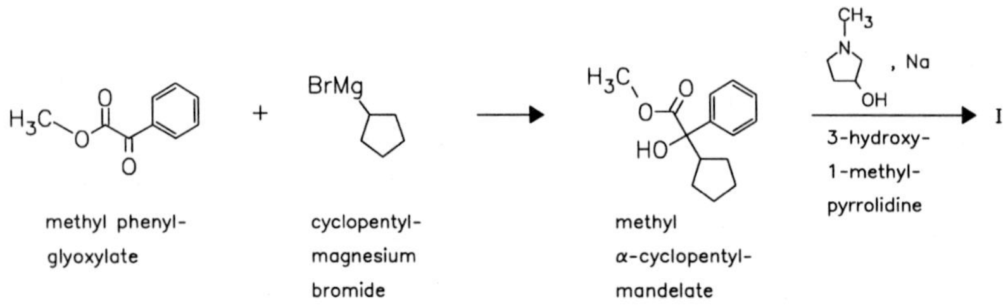

methyl phenyl-
glyoxylate

cyclopentyl-
magnesium
bromide

methyl
α-cyclopentyl-
mandelate

3-hydroxy-
1-methyl-
pyrrolidine

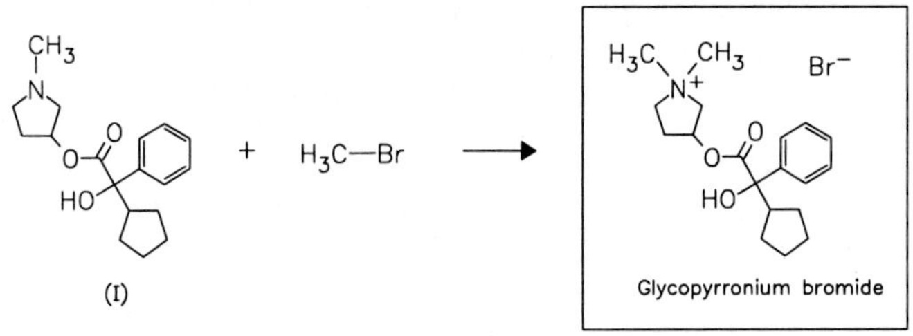

Glycopyrronium bromide

Reference(s):
US 2 956 062 (A. H. Robins; 11.10.1960; prior. 26.2.1959).

Formulation(s):   amp. 0.2 mg/ml, 500 µg/ml

Trade Name(s):
D: Robinul (Brenner-Efeka)   GB: Robinul (Anpharm)   USA: Robinul (Robins)
F: Asécryl (Martinet); wfm   J: Robinul (Kaken)

Glymidine
(Glycodiazin)

ATC: A10BC01
Use: antidiabetic

RN: 339-44-6   MF: $C_{13}H_{15}N_3O_4S$ MW: 309.35   EINECS: 206-426-5
LD_{50}:   3100 mg/kg (R, p.o.)
CN: N-[5-(2-methoxyethoxy)-2-pyrimidinyl]benzenesulfonamide

sodium salt
RN: 3459-20-9   MF: $C_{13}H_{14}N_3NaO_4S$ MW: 331.33   EINECS: 222-399-2
LD_{50}:   3100 mg/kg (R, p.o.)

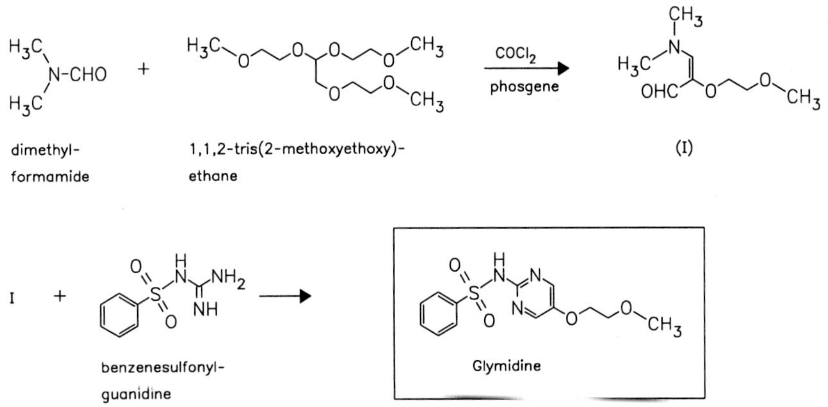

dimethyl-
formamide

1,1,2-tris(2-methoxyethoxy)-
ethane

(I)

benzenesulfonyl-
guanidine

Glymidine

Reference(s):
US 3 275 635 (Schering AG; 27.9.1966; D-prior. 18.10.1960, 22.2.1961, 23.2.1961).
DAS 1 445 142 (Schering AG; appl. 22.2.1961).
DAS 1 445 146 (Schering AG; appl. 9.9.1961; addition to DAS 1 445 142).
Gutsche, K. et al.: Arzneim.-Forsch. (ARZNAD) **14**, 373 (1964).

Formulation(s):   tabl. 0.5 g, 1 g (as sodium salt)

Trade Name(s):
D: Redul (Bayer-Schering);   wfm

| | F: | Glyconormal (Bayer- | I: | Glycanol (Bayer); wfm |

Redul plus (Bayer-		F:	Glyconormal (Bayer-		I:	Glycanol (Bayer); wfm
Schering)-comb. with			Pharma); wfm			Gondafon (Schering); wfm
buformin; wfm			Gondafon (SEPPS); wfm		J:	Lycanol (Bayer-Yoshitomi)
Redul 28 (Schering); wfm		GB:	Gondafon (Schering			
			Chemicals); wfm			

Granisetron
(BRL-43694)

ATC: A04AA02
Use: anti-emetic, 5-HT$_3$-antagonist

RN: 109889-09-0 MF: C$_{18}$H$_{24}$N$_4$O MW: 312.42
CN: *endo*-1-methyl-*N*-(9-methyl-9-azabicyclo[3.3.1]non-3-yl)-1*H*-indazole-3-carboxamide

monohydrochloride
RN: 107007-99-8 MF: C$_{18}$H$_{24}$N$_4$O · HCl MW: 348.88
LD$_{50}$: 17 mg/kg (M, i.v.); 350 mg/kg (M, p.o.);
14 mg/kg (R, i.v.); 350 mg/kg (R, p.o.)

indazole-3-carboxylic acid + methyl iodide → 1-methyl-indazole-3-carboxylic acid → (COCl)$_2$ oxalyl chloride → I

1-methyl-indazole-3-carbonyl chloride (I) + endo-9-methyl-9-azobicyclo[3.3.1]-nonan-3-amine (granataneamine) → [N(C$_2$H$_5$)$_3$] → Granisetron

Reference(s):
EP 200 444 (Beecham; appl. 21.4.1986; GB-prior. 27.4.1985, 21.10.1985).
Bermudez, J. et al.: Bioorg. Med. Chem. Lett. (BMCLE8) **4** (20), 2376 (1994).

alternative synthesis:
WO 9 730 049 (SmithKline Beecham; appl. 11.2.1992; GB-prior. 13.2.1996).

synthesis of 1-methylindazole-3-carboxylic acid:
EP 323 105 (Beecham; appl. 19.12.1988; GB-prior. 22.12.1987).
Bermudez, J. et al.: J. Med. Chem. (JMCMAR) **33**, 1924 (1990).

synthesis of granataneamine:
Jones, G.; Stanger, J.: J. Chem. Soc. C (JSOOAX) **1969**, 901.

medical use for treatment of CNS and cognitive disorders:
EP 223 385 (Beecham; appl. 7.10.1986; GB-prior. 21.10.1985).
EP 279 990 (Glaxo; appl. 16.12.1987; GB-prior. 17.12.1986, 25.3.1987).

medical use for treatment of withdrawal syndrome:
EP 278 161 (Glaxo; appl. 20.11.1987; GB-prior. 21.11.1986, 25.3.1987).
EP 279 114 (Glaxo; appl. 20.11.1987; GB-prior. 21.11.1986, 25.3.1987).

medical use for treatment of visceral pain:
EP 279 512 (Beecham; appl. 18.1.1988; GB-prior. 19.1.1987).
US 4 845 092 (Beecham; 4.7.1989; appl. 19.1.1988; GB-prior. 19.1.1987).

medical use for treatment of cough and bronchoconstriction:
EP 340 270 (Beecham; appl. 14.11.1988; GB-prior. 14.11.1987).

medical use for treatment of myocardial instability:
WO 9 109 593 (Beecham; appl. 20.12.1990; GB-prior. 21.12.1989).

Formulation(s):   amp. 1 mg, 3 mg; tabl. 1 mg (as hydrochloride)

Trade Name(s):

D:	Kevatril (Bristol-Myers Squibb/SmithKline Beecham)	GB:	Kytril (SmithKline Beecham)		Sedobex (Ecobi)-comb.
		I:	Broncosedina (Farma)-comb.	J:	Kytril (SmithKline Beecham)
F:	Kytril (SmithKline Beecham; 1991)		Kytril (SmithKline Beecham)	USA:	Kytril (SmithKline Beecham)

Grepafloxacin
(OPC-17116)

ATC: J01MA11
Use: antibacterial (gyrase inhibitor)

RN: 119914-60-2   MF: $C_{19}H_{22}FNO_3$ MW: 331.39
CN: 1-cyclopropyl-6-fluoro-1,4-dihydro-5-methyl-7-(3-methyl-1-piperazinyl)-4-oxo-3-quinolinecarboxylic acid

(±)-form
RN: 146863-02-7   MF: $C_{19}H_{22}FNO_3$ MW: 331.39
(±)-monohydrochloride
RN: 161967-81-3   MF: $C_{19}H_{22}FNO_3 \cdot HCl$ MW: 367.85
LD_{50}:   69.2 mg/kg (M, i.v.); 3900 mg/kg (M, p.o.);
        152 mg/kg (R, i.v.); 3029 mg/kg (R, p.o.)

2-methyl-3,4,6-trifluorobenzoic acid

diethyl (2-methyl-3,4,6-trifluorobenzoyl)-malonate (I)

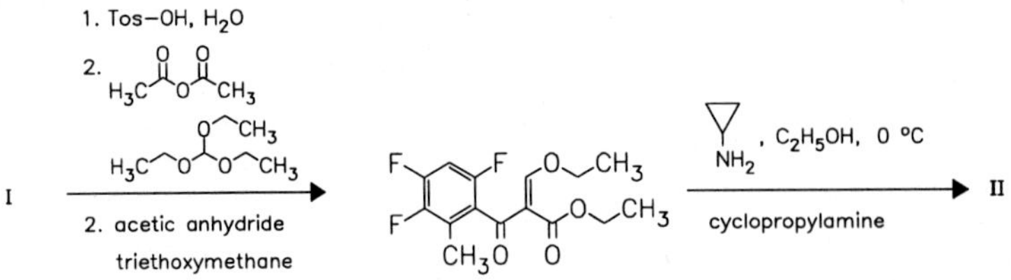

ethyl 2-(2-methyl-
3,4,6-trifluorobenzoyl)-
3-ethoxyacrylate

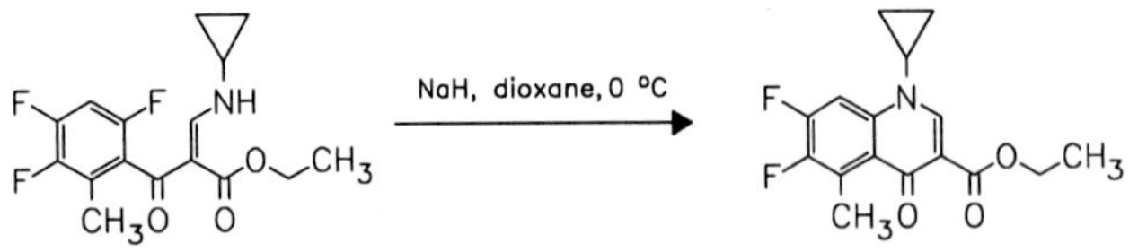

ethyl 2-(2-methyl-
3,4,6-trifluorobenzoyl)-
3-cyclopropylamino-
acrylate (II)

ethyl 1-cyclopropyl-
6,7-difluoro-5-methyl-
1,4-dihydro-4-oxo-
quinoline-3-carboxylate (III)

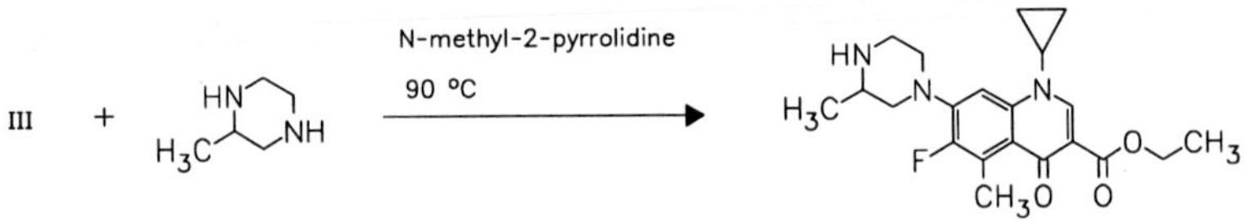

2-methyl-
piperazine

(IV)

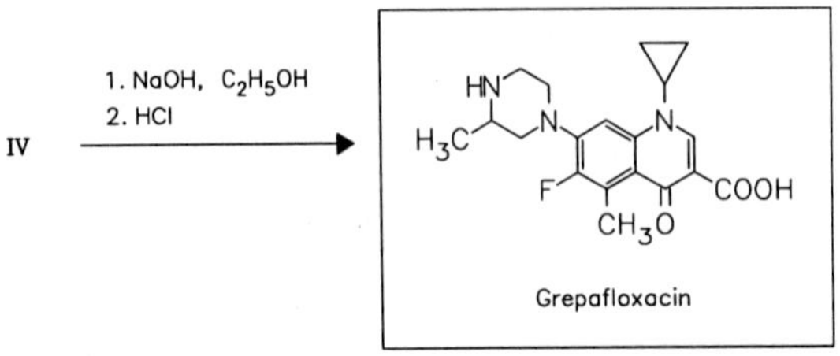

Grepafloxacin

Reference(s):
EP 287 951 (Otsuka Pharm.; appl. 14.4.1988; J-prior. 16.4.1987).
EP 364 943 (Otsuka Pharm.; appl. 17.10.1989; J-prior. 20.10.1988).

melt-extruded polymeric material:
JP 08 280 790 (Otsuka Pharm.; appl. 17.4.1995).

use as fungicide:
JP 07 149 647 (Daiichi Seiyaku; appl. 8.9.1994; USA-prior. 8.9.1993).

Formulation(s):   f. c. tabl. 400 mg, 600 mg (as hydrochloride)

Trade Name(s):
D: Vaxar (Glaxo Wellcome/   J: Lungaskin (Otsuka; as   USA: Raxar (Glaxo Wellcome;
      Cascan); wfm   hydrochloride); wfm   1997); wfm

Griseofulvin

ATC: D01AA08; D01BA01
Use: antifungal antibiotic

RN: 126-07-8   MF: $C_{17}H_{17}ClO_6$ MW: 352.77   EINECS: 204-767-4
LD$_{50}$: 280 mg/kg (M, i.v.); >50 g/kg (M, p.o.);
400 mg/kg (R, i.v.); >10 g/kg (R, p.o.)
CN: (1'S-trans)-7-chloro-2',4,6-trimethoxy-6'-methylspiro[benzofuran-2(3H),1'-[2]cyclohexene]-3,4'dione

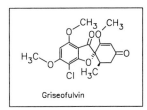

Griseofulvin

From fermentation solutions of *Penicillium patulum*.

Reference(s):
GB 784 618 (Glaxo; appl. 28.3.1955).
US 2 900 304 (ICI; 18.8.1959; GB-prior. 21.9.1956).
US 3 038 839 (Glaxo; 12.6.1962; GB-prior. 2.3.1959).
US 3 069 328 (Glaxo; 18.12.1962; GB-prior. 4.5.1960).
US 3 069 329 (Glaxo; 18.12.1962; GB-prior. 4.5.1960).

pharmaceutical formulation with polyethyleneglycol:
US 4 151 273 (Univ. of California; 24.4.1979; prior. 2.1.1970, 2.12.1970, 13.4.1972, 31.10.1974, 13.6.1978).

total syntheses:
Brossi, A. et al.: Helv. Chim. Acta (HCACAV) **43**, 1444 (1960).
Day, A.C. et al.: Proc. Chem. Soc., London (PCSLAW), **1960**, 284.
Kuo, C.H. et al.: Chem. Ind. (London) (CHINAG), **1960**, 1627.
Stork, G. et al.: J. Am. Chem. Soc. (JACSAT) **84**, 310 (1962).

Formulation(s):   cps. 125 mg, 250 mg; cream 5 g/100 g; tabl. 125 mg, 165 mg, 330 mg, 500 mg

Trade Name(s):

D:	Fulcin S (Zeneca)		Griséfuline (Sanofi		Grisetin (Nippon Kayaku)
	Gricin Creme (LAW)		Winthrop)		Grisovin (Fujisawa)
	Gricin Tabl. (ASTA Medica	GB:	Fulcin (Zeneca)		Guservin (Chugai)
	AWD)		Grisovin (Glaxo Wellcome)	USA:	Fulvicin P/G (Schering)
	griseco (ct-Arzneimittel)	I:	Fulcin (SIT)		Grifulvin V (Ortho
	Likuden M (Hoechst)		Griseofulvina (Scfm)		Dermatological)
F:	Fulcine (Zeneca Pharma)		Grisovina Fp (Teofarma)		Grisactin (Wyeth-Ayerst)
		J:	Grifulvin (Yamanouchi)		Gris-PEG (Allergan)

Guaiazulene
(Guajazulene)

ATC: S01XA01
Use: anti-inflammatory

RN: 489-84-9   MF: $C_{15}H_{18}$ MW: 198.31   EINECS: 207-701-2
LD$_{50}$: 1220 mg/kg (M, p.o.);
1550 mg/kg (R, p.o.)
CN: 1,4-dimethyl-7-(1-methylethyl)azulene

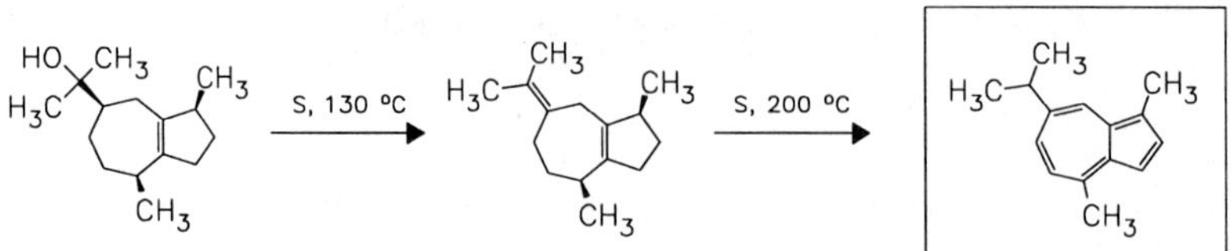

guaiol
(constituent of
guaiacum wood oil)

guaiene

Guaiazulene

Reference(s):
CH 314 487 (Dr. B. Joos; appl. 1953).

Formulation(s):   cream; drg. 20 mg; ointment (ethanolic camomile extract)

Trade Name(s):
D: Azulon Kamillen Creme   Thrombocid (bene-   I: Azulon (Armour Med.);
      (ASTA Medica AWD)   Arzneimittel)-comb.   wfm
      Azupanthenol (Parke   F: Cicatryl (Evans Medical)-   Azulon (Rorer); wfm
      Davis)-comb.   comb.   J: Azulon-Homburg (Daito)
      Garmastan (Protina)   Pepsane (Rosa-
                                     Phytopharma)-comb.

Guaifenesin
(Guajacolglycerinäther; Guaiphenesin)

ATC: R05CA03
Use: muscle relaxant, expectorant

RN: 93-14-1   MF: $C_{10}H_{14}O_4$ MW: 198.22   EINECS: 202-222-5
LD$_{50}$:   400 mg/kg (M, i.v.); 690 mg/kg (M, p.o.);
       360 mg/kg (R, i.v.); 1510 mg/kg (R, p.o.);
       335 mg/kg (dog, i.v.)
CN: 3-(2-methoxyphenoxy)-1,2-propanediol

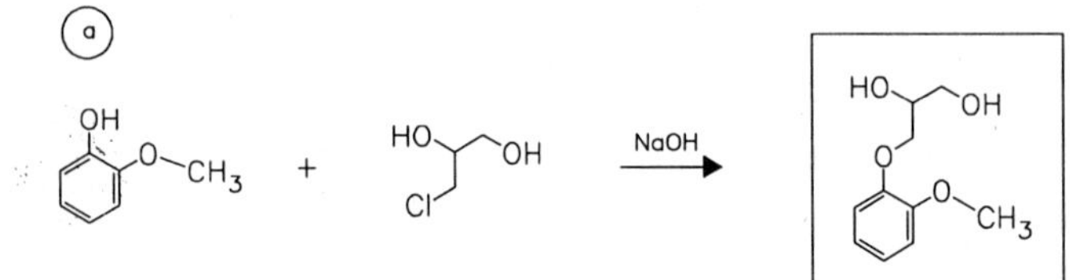

guaiacol (I)

3-chloropropane-
1,2-diol

Guaifenesin

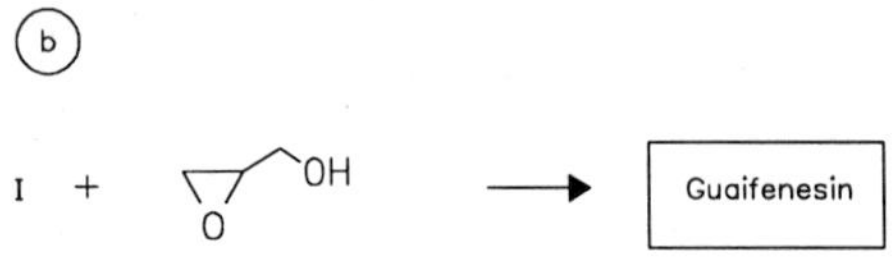

glycide

Reference(s):
GB 628 497 (British Drug Houses; appl. 1948).
Marle, E.R.: J. Chem. Soc. (JCSOA9) **101**, 305 (1912).
Yale, H.L. et al.: J. Am. Chem. Soc. (JACSAT) **72**, 3710 (1950).

Formulation(s):   cps. 200 mg; elixir 200 mg/5 ml; sol. 100 mg; syrup 100 mg, 200 mg/15 ml; tabl. 200 mg, 600 mg

Trade Name(s):

D: Anastil (Eberth)-comb.
Bricantyl (Astra)-comb.
with terbutaline sulfate
Dolestan forte (Whitehall-Much)-comb.
Fagusan (Spreewald Pharma)
Gufen (Steigerwald)
Nephulon (Redel)
Pulmotin (Serum-Werk Bernburg)
Wick Daymed (Wick Pharma)-comb.
Wick Formel 44 (Wick Pharma)
Wick Kinder Formel 44 (Wick Pharma)

F: Bronchospray (Tissot)-comb.
Catabex (Darcy)-comb.
Dimetane expectorant (Whitehall)-comb.
Hexapneumine/-composé (Doms)-comb.
Nortussine (Norgine Pharma)-comb.
Polaramine pectoral (Schering-Plough)-comb.

Pulmofluide (Phygiène)-comb.
Rectoplexil (Théraplix)-comb.
Sédophon pectoral (Mayoly-Spindler)-comb.
Toplexil (Théraplix)-comb.
numerous combination preparations

GB: Bricanyl compound (Astra)-comb.; wfm
Dimotane expect. (Robins)-comb.; wfm
Entair (Duncan, Flockhart)-comb.; wfm
Franol expect. (Winthrop)-comb.; wfm
Lotussin (Searle)-comb.; wfm
Nethaprin expect. (Merrell Dow)-comb.; wfm
Noradran (Norma)-comb.; wfm
Pholcomed expect. (Medo); wfm
Robitussin (Robins); wfm
Terpoin (Hough, Hoseason); wfm

I: Broncovanil (Scharper)
Chymoser Balsamico (Serono)-comb.
Donatiol (AGIPS)-comb.
Fepramol (Schwarz)-comb.
Idropulmina/-composta (ISI)
Lanactin scir. (Lepetit)-comb.
Polarmin Espet. scir. (Essex)-comb.
Pumilene (Montefarmaco)-comb.
Resyl (Ciba)
Rettocistin (Edmond)-comb.
Ribexen Espet. (Formenti)-comb.
Robitussin (Proter)
Torfan (Abbott)-comb.
Tuscalman Berna (Berna)-comb.
Ventolin Espet. (Glaxo)-comb.

J: Fustosil (Kyoto)

USA: numerous combination preparations

Guajacol

(Gaiacol; Guaiacolina; Guajol; Methylcatechol)

ATC: R05CA
Use: expectorant, antiseptic

RN: 90-05-1   MF: $C_7H_8O_2$ MW: 124.14   EINECS: 201-964-7
LD_{50}: 170 mg/kg (M, i.v.); 621 mg/kg (M, p.o.);
520 mg/kg (R, p.o.)
CN: 2-methoxyphenol

phenylacetate
RN: 4112-89-4   MF: $C_{15}H_{14}O_3$ MW: 242.27

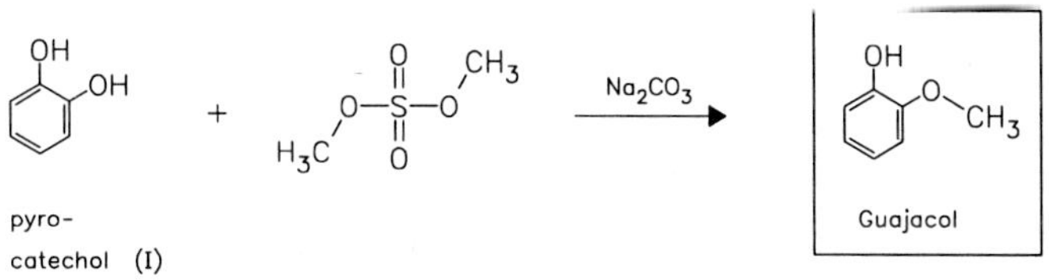

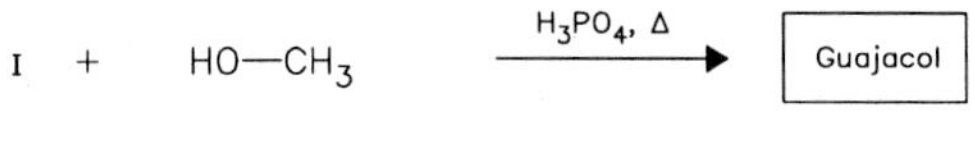

pyro-
catechol (I)

Guajacol

Reference(s):
Ullmanns Encykl. Tech. Chem., 4. Aufl., Vol. **18**, 226.

Formulation(s):   amp. 50 mg, 75 mg; drg. 100 mg; inhalation sol. 75 mg; syrup 50 mg; suppos. 500 mg (as phenylacetate)

Trade Name(s):

D:	Anastil (Eberth)	I:	Eucaliptina (Zoja)-comb.		generics
	Dalet Med Balsam		Fosfaguaiacol (Ogna)-	J:	Hustosil (Kyoto-
	(Mauermann)-comb.		comb.		Sumitomo)-comb.
	Infekt-Komplex Ho-Fu-		Lacotocol (Ogna)-comb.		
	Complex (Pharma		Lipobalsamo (Parke		
	Liebermann)-comb.		Davis)-comb.		

Guanabenz

ATC: C02
Use: antihypertensive

RN: 5051-62-7   MF: C$_8$H$_8$Cl$_2$N$_4$ MW: 231.09   EINECS: 225-750-8
CN: 2-[(2,6-dichlorophenyl)methylene]hydrazinecarboximidamide

monoacetate
RN: 23256-50-0   MF: C$_8$H$_8$Cl$_2$N$_4$ · C$_2$H$_4$O$_2$ MW: 291.14   EINECS: 245-534-7
LD$_{50}$:   260 mg/kg (M, p.o.);
        238 mg/kg (R, p.o.)

2,6-dichloro-	aminoguanidine	Guanabenz
benzaldehyde	carbonate	

Reference(s):
DOS 1 802 364 (Wyeth; appl. 10.10.1968; USA-prior. 12.10.1967).
Baum, T. et al.: Experientia (EXPEAM) **25**, 1066 (1969).

Formulation(s):   tabl. 4 mg, 8 mg, 16 mg (as acetate)

Trade Name(s):

D:	Wytensin (Wyeth); wfm		Rexitene/-plus (LPB); wfm	USA:	Wytensin (Wyeth-Ayerst;
I:	Rexitene (LPB); wfm	J:	Wytens (Nippon Shoji)		as acetate)

Guanadrel

ATC: C02
Use: antihypertensive

RN: 40580-59-4   MF: C$_{10}$H$_{19}$N$_3$O$_2$ MW: 213.28
CN: (1,4-dioxaspiro[4.5]dec-2-ylmethyl)guanidine

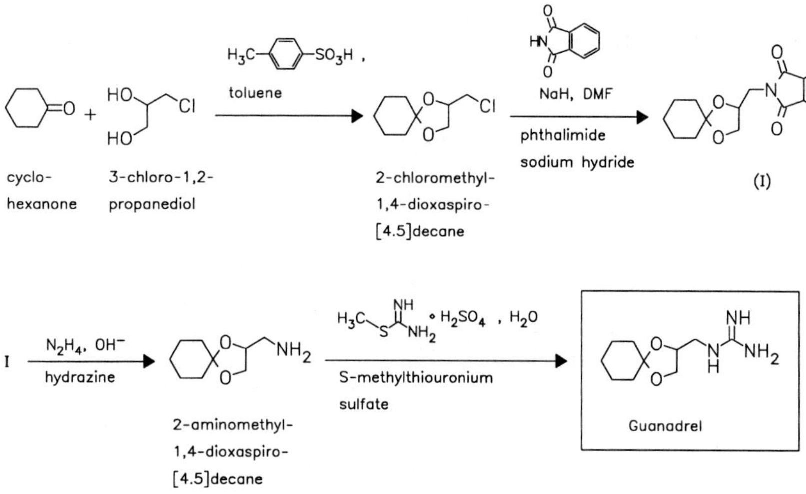

toluene
NaH, DMF
phthalimide
sodium hydride
cyclo-
hexanone
3-chloro-1,2-
propanediol
2-chloromethyl-
1,4-dioxaspiro-
[4.5]decane
(I)
hydrazine
S-methylthiouronium
sulfate
2-aminomethyl-
1,4-dioxaspiro-
[4.5]decane
Guanadrel

Reference(s):
FR 1 522 153 (Cutter Lab.; appl. 2.5.1967; USA-prior. 3.5.1966).

Formulation(s): tabl. 10 mg, 25 mg (as sulfate)

Trade Name(s):
USA: Hylorel (Medeva; as sulfate)

Guanethidine sulfate

ATC: C02CC02; S01EX01
Use: antihypertensive

RN: 60-02-6 MF: $C_{10}H_{22}N_4 \cdot 1/2H_2O_4S$ MW: 494.71 EINECS: 200-452-0
LD_{50}: 18 mg/kg (M, i.v.); 1100 mg/kg (M, p.o.);
23 mg/kg (R, i.v.); 1 g/kg (R, p.o.)
CN: [2-(hexahydro-1(2*H*)-azocinyl)ethyl]guanidine sulfate (2:1)

guanethidine
RN: 55-65-2 MF: $C_{10}H_{22}N_4$ MW: 198.31 EINECS: 200-241-3

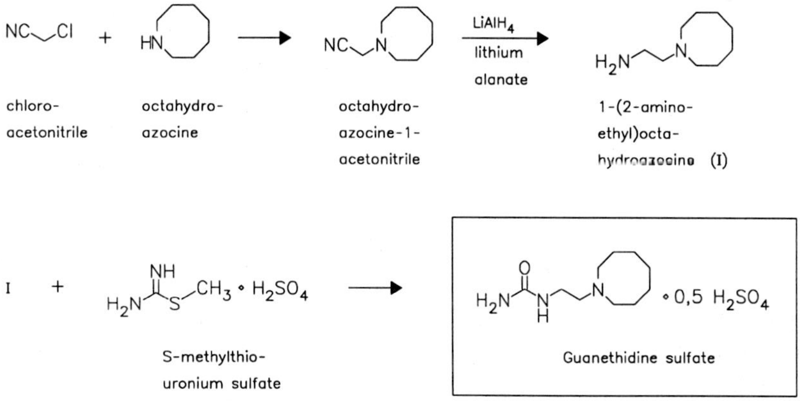

lithium
alanate
chloro-
acetonitrile
octahydro-
azocine
octahydro-
azocine-1-
acetonitrile
1-(2-amino-
ethyl)octa-
hydroazocine (I)
S-methylthio-
uronium sulfate
Guanethidine sulfate

Reference(s):
US 2 928 829 (Ciba; 15.3.1960; prior. 10.6.1958).

alternative syntheses:
US 3 006 913 (Ciba; 31.10.1961; appl. 10.6.1959).
US 3 055 882 (Ciba; 25.9.1962; appl. 10.6.1959).

Formulation(s):   eye drops 5 %, 10 %tabl. 10 mg, 25 mg

Trade Name(s):

D:	Esimil (Novartis Pharma)-comb.	F:	Isméline (CIBA Vision Ophthalmics)	J:	Ismelin (Novartis-Takeda)
	Suprexon (CIBA Vision)-comb.	GB:	Ganda (Chauvin)-comb.	USA:	Esimil (Novartis)-comb.; wfm
	Thilodigon (Alcon)-comb.		Ismelin (Novartis)		Ismelin (Novartis); wfm
		I:	Visutensil (Merck Sharp & Dohme)		

Guanfacine

ATC: C02CC
Use: antihypertensive, α-adrenoceptor agonist

RN: 29110-47-2   MF: C$_9$H$_9$Cl$_2$N$_3$O   MW: 246.10   EINECS: 249-442-8
LD$_{50}$:   165 mg/kg (M, p.o.)
CN: *N*-(aminoiminomethyl)-2,6-dichlorobenzeneacetamide

monohydrochloride
RN: 29110-48-3   MF: C$_9$H$_9$Cl$_2$N$_3$O · HCl   MW: 282.56   EINECS: 249-443-3
LD$_{50}$:   25 mg/kg (M, i.v.); 16 mg/kg (M, p.o.);
          5800 µg/kg (R, i.v.); 210 mg/kg (R, p.o.)

2,6-dichlorophenyl-acetyl chloride   guanidine (I)   Guanfacine

ethyl 2,6-dichloro-phenylacetate   + I   Guanfacine

Reference(s):
US 3 632 645 (Dr. A. Wander; 4.1.1972; appl. 23.9.1968; CH-prior. 26.9.1967).
DE 1 793 483 (Dr. A. Wander; appl. 24.9.1968; CH-prior. 26.9.1967).
Bream, J.B. et al.: Arzneim.-Forsch. (ARZNAD) **25**, 1477 (1975).

alternative syntheses:
CH 511 816 (Dr. A. Wander; appl. 26.2.1969).
CH 518 910 (Dr. A. Wander; appl. 14.11.1969).

Formulation(s):   tabl. 1 mg, 2 mg (as hydrochloride)

Trade Name(s):
D: Estulic-Wander (Novartis   F: Estulic (Novartis; 1981)   USA: Tenex (Robins; 1987)
        Pharma; 1980)   J: Estulic (Sandoz-Sankyo)

Guanoclor

ATC: C02CC05
Use: antihypertensive

RN: 5001-32-1   MF: $C_9H_{12}Cl_2N_4O$ MW: 263.13   EINECS: 225-667-7
CN: 2-[2-(2,6-dichlorophenoxy)ethyl]hydrazinecarboximidamide

sulfate (2:1)
RN: 551-48-4   MF: $C_9H_{12}Cl_2N_4O \cdot 1/2H_2SO_4$ MW: 624.33   EINECS: 208-996-0

2,6-dichloro-   1,2-dibromo-   2-(2,6-dichloro-
phenol   ethane   phenoxy)ethyl
                                            bromide

(I)   S-methylthiouronium   Guanoclor
                        sulfate

Reference(s):
BE 629 613 (Pfizer; appl. 14.3.1963; GB-prior. 15.3.1962, 20.7.1962).
US 3 271 448 (Pfizer; 6.9.1966; GB-prior. 15.3.1962, 20.7.1962).
Augstein, J. et al.: J. Med. Chem. (JMCMAR) **8**, 395 (1965).

Formulation(s):   tabl. 10 mg, 40 mg (as sulfate)

Trade Name(s):
GB: Vatensol (Pfizer); wfm   USA: Vatensol (Pfizer); wfm

Guanoxabenz

ATC: C02CC07
Use: antihypertensive

RN: 24047-25-4   MF: $C_8H_8Cl_2N_4O$ MW: 247.09
CN: 2-[(2,6-dichlorophenyl)methylene]-N-hydroxyhydrazinecarboximidamide

thiosemicarbazide   methyl   S-methylisothio-
                        iodide   semicarbazide
                                            hydriodide

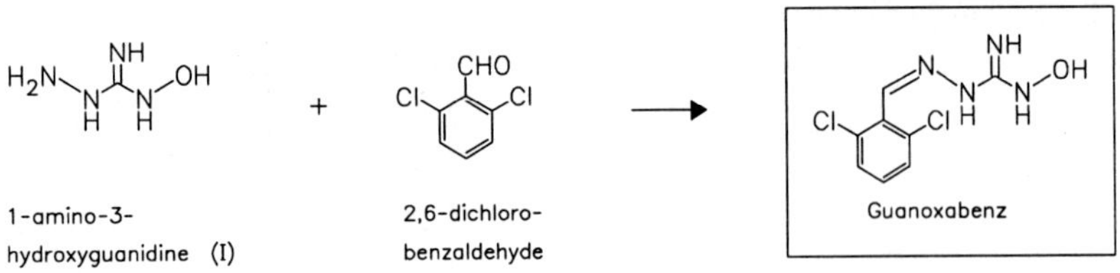
1-amino-3-hydroxyguanidine (I)
2,6-dichlorobenzaldehyde
Guanoxabenz

Reference(s):

DOS 1 902 449 (Sandoz; appl. 18.1.1969; USA-prior. 22.1.1968, 10.7.1968, 16.9.1968).
US 3 591 636 (Sandoz; 6.7.1971; prior. 22.1.1968, 10.7.1968, 16.9.1968).

Formulation(s): vial 5 mg; tabl. 25 mg

Trade Name(s):

F: Benzerial (Houdé); wfm

Guanoxan

ATC: C02CC03
Use: antihypertensive

RN: 2165-19-7 MF: $C_{10}H_{13}N_3O_2$ MW: 207.23
CN: [(2,3-dihydro-1,4-benzodioxin-2-yl)methyl]guanidine

sulfate (2:1)
RN: 5714-04-5 MF: $C_{10}H_{13}N_3O_2 \cdot 1/2H_2SO_4$ MW: 512.54
LD_{50}: 161 mg/kg (M, i.p.)

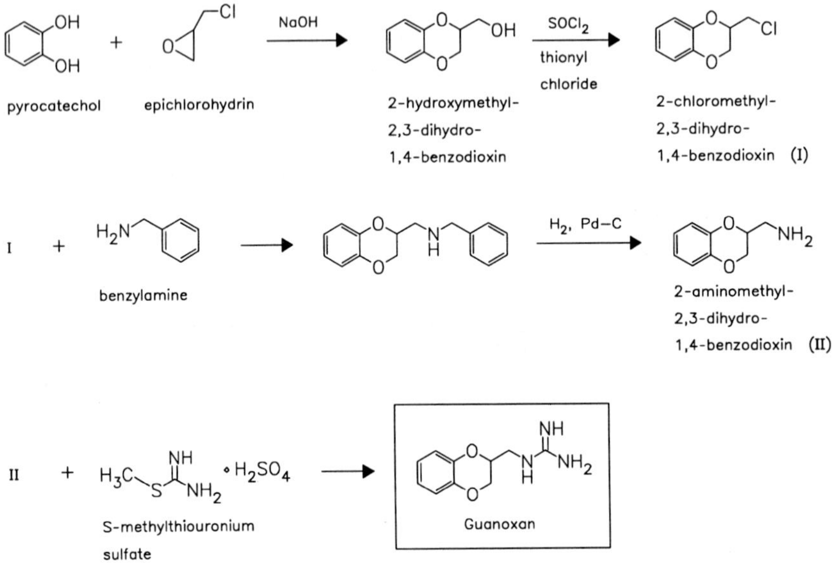
pyrocatechol
epichlorohydrin
NaOH
2-hydroxymethyl-2,3-dihydro-1,4-benzodioxin
SOCl2 thionyl chloride
2-chloromethyl-2,3-dihydro-1,4-benzodioxin (I)
benzylamine
H2, Pd-C
2-aminomethyl-2,3-dihydro-1,4-benzodioxin (II)
S-methylthiouronium sulfate
Guanoxan

Reference(s):

US 3 247 221 (Pfizer; 19.4.1966; appl. 16.5.1963; GB-prior. 22.5.1962).
Augstein, J. et al.: J. Med. Chem. (JMCMAR) **8**, 446 (1965).

Formulation(s): tabl. 10 mg

Trade Name(s):

F:	Envacar (Pfizer); wfm	GB:	Envacar (Pfizer); wfm
	Envarése (Pfizer)-comb.; wfm	USA:	Envacar (Pfizer); wfm

Gusperimus trihydrochloride

(BMY-42215-1; BMS-181173; Deoxyspergualin hydrochloride; DSG; NKT-01; NSC-356894)

ATC: L01; L04
Use: antineoplastic, immunosuppressive, multiple sclerosis therapeutic, antiangiogenic, disease modifying drug, systemic lupus erythematosus therapeutic

RN: 85468-01-5 MF: $C_{17}H_{37}N_7O_3 \cdot 3HCl$ MW: 496.91
LD_{50}: 35 mg/kg (M, i.v.)
CN: 7-[(aminoiminomethyl)amino]-*N*-[2-[[4-[(3-aminopropyl)amino]butyl]amino]-1-hydroxy-2-oxoethyl]heptanamide

S-(–)-form
RN: 84937-45-1 MF: $C_{17}H_{37}N_7O_3 \cdot 3HCl$ MW: 496.91
LD_{50}: 35 mg/kg (M, i.v.)
base (racemate)
RN: 104317-84-2 MF: $C_{17}H_{37}N_7O_3$ MW: 387.53
S-(–)-base
RN: 89149-10-0 MF: $C_{17}H_{37}N_7O_3$ MW: 387.53
R-(+)-base
RN: 114760-38-2 MF: $C_{17}H_{37}N_7O_3$ MW: 387.53
S-(–)-hydrochloride
RN: 128488-79-9 MF: $C_{17}H_{37}N_7O_3 \cdot xHCl$ MW: unspecified

intermediate I:

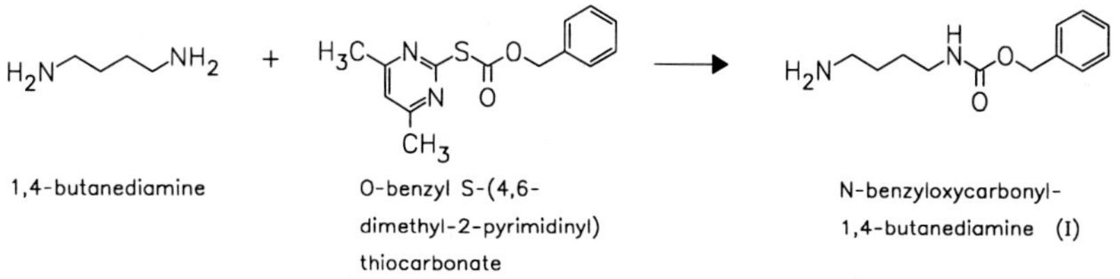

1,4-butanediamine
O-benzyl S-(4,6-dimethyl-2-pyrimidinyl) thiocarbonate
N-benzyloxycarbonyl-1,4-butanediamine (I)

intermediate V:

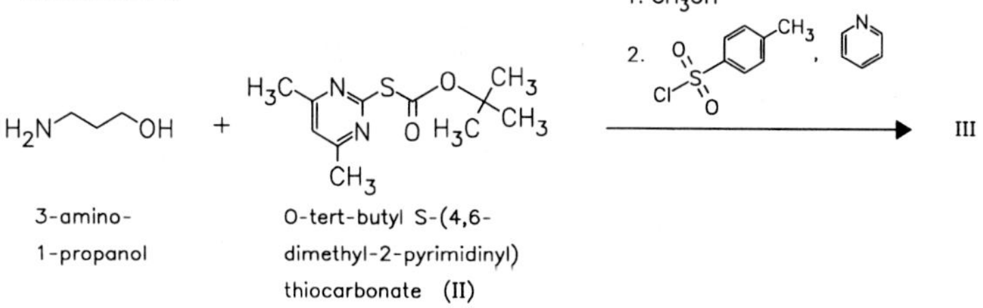

1. CH3OH
2.
III
3-amino-1-propanol
O-tert-butyl S-(4,6-dimethyl-2-pyrimidinyl) thiocarbonate (II)

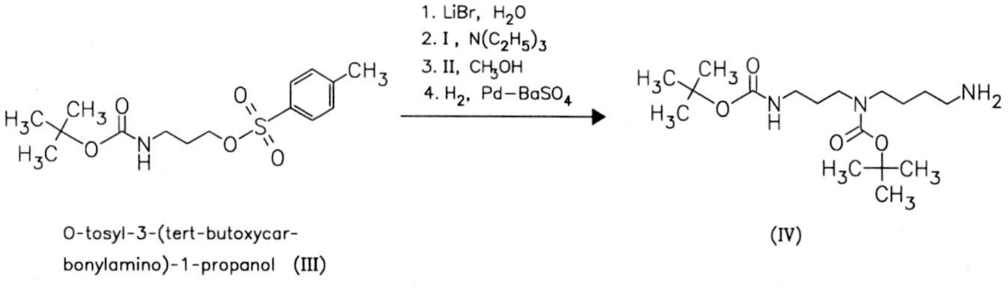

1. LiBr, H2O
2. I, N(C2H5)3
3. II, CH3OH
4. H2, Pd–BaSO4
O-tosyl-3-(tert-butoxycarbonylamino)-1-propanol (III)
(IV)

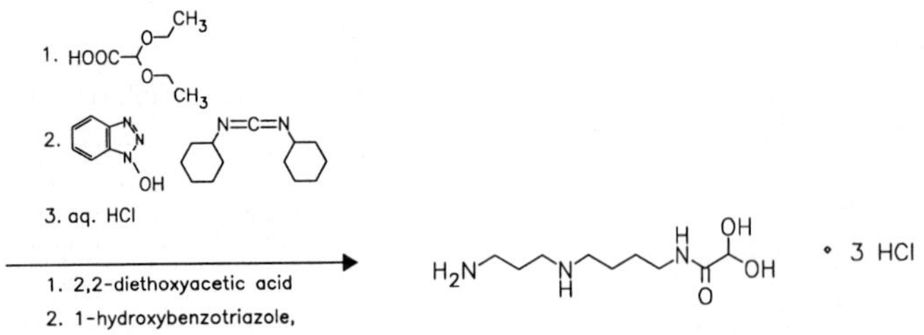

N-[4-(3-aminopropylamino)butyl]-2,2-dihydroxy-
ethanamide trihydrochloride (V)

intermediate XI:

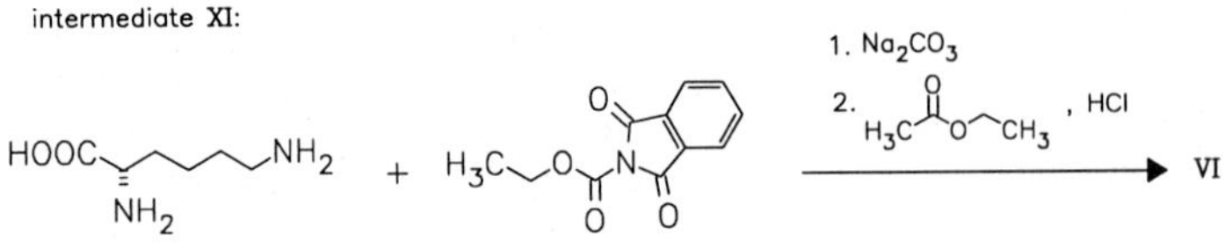

L-lysine

N-ethoxycarbonyl-
phthalimide

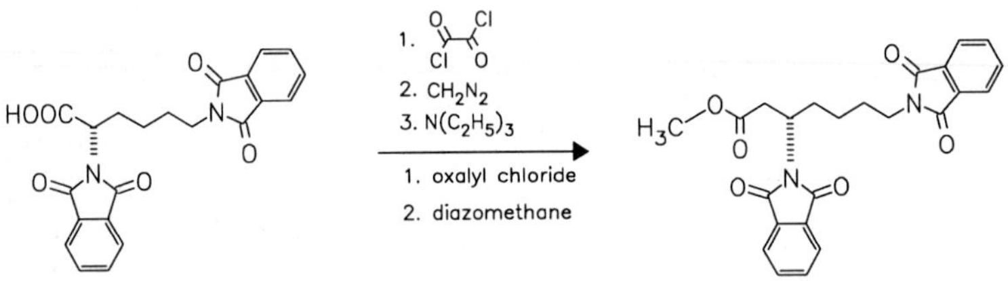

di-N-phthaloyl-
L-lysine (VI)

1. oxalyl chloride
2. diazomethane

(VII)

VII

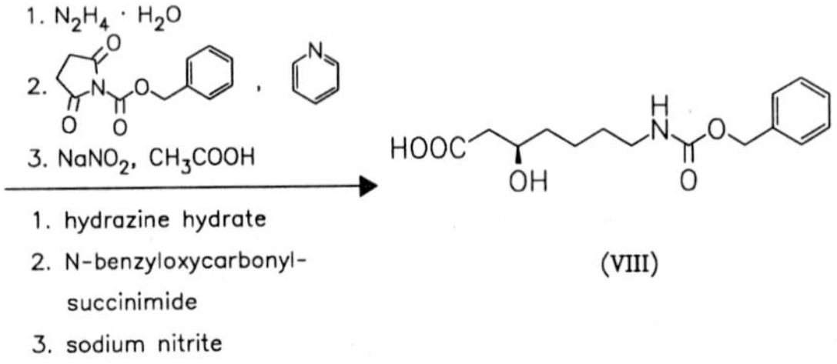

1. hydrazine hydrate
2. N-benzyloxycarbonyl-
   succinimide
3. sodium nitrite

(VIII)

VIII

1. CH_2N_2, $O(C_2H_5)_2$
2. NH_3, CH_3OH
3. CH_3OH, H_2, Pd–C

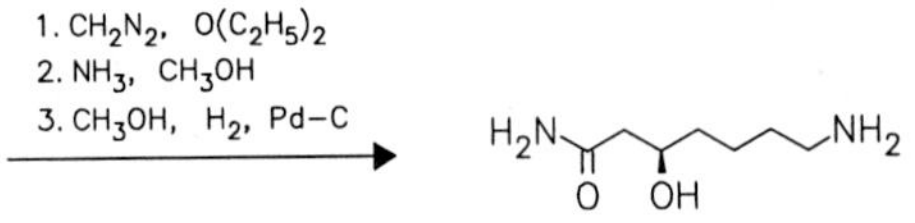

(IX)

IX   +

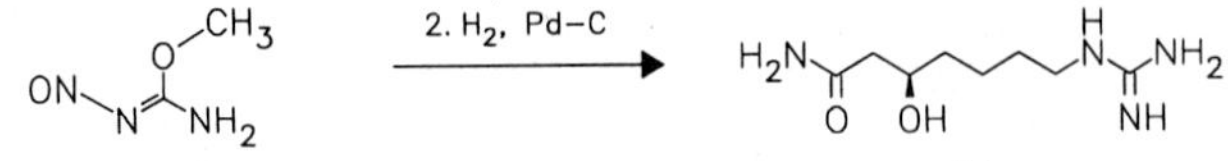

O-methyl-N-nitrosourea

1. aq. NaOH
2. H_2, Pd–C

(X)

7-guanidinoheptanamide
hydrochloride (XI)

final product:

Gusperimus trihydrochloride

Reference(s):
DE 3 626 306 (Microbiochemical Research Found.; appl. 11.2.1988; D-prior. 2.8.1986).
BE 894 651 (Microbiochemical Research Found.; appl. 31.1.1983; J-prior. 8.10.1981).
JP 08 020 533 (Nippon Kayaku; appl. 23.1.1996; J-prior. 7.7.1994).
DE 3 506 330 (Takara Skuzo Co.; Nippon Kayaku Co.; appl. 29.8.1985; J-prior. 29.2.1984).

purification:
JP 59 029 652 (Nippon Kayaku Co.; appl. 16.2.1984; J-prior. 10.8.1982).

pharmaceutical preparations:
JP 02 009 816 (Nippon Kayaku Co.; Takara Skuzo Co.; appl. 12.1.1990; J-prior. 29.6.1988).
EP 188 821 (Microbial Chemistry Research Found.; appl. 30.7.1986; J-prior. 14.1.1985).

use of gusperimus hydrochloride:
CA 2 142 376 (Bristol-Myers Squibb Co.; appl. 26.8.1995; USA-prior. 25.2.1994).
WO 9 405 323 (Jekus Hopkins Univ., School of Medicine; appl. 17.3.1994; WO-prior. 4.9.1982).
DE 3 626 306 (Behringwerke A.G.; appl. 11.2.1988; D-prior. 2.8.1986).

synthesis of O-tert-butyl *S*-(4,6-dimethyl-2-pyrimidinyl)thiocarbonate *and O*-benzyl *S*-(4,6-dimethyl-2-pyrimidinyl)thiocarbonate:
Nagasawa et al.: Bull. Chem. Soc. Jpn. (BCSJA8) **46**, 1269, 1271 (1973).
DE 2 245 392 (Nitto Boseki; appl. 12.4.1973; J-prior. 17.9.1971, 27.9.1971, 30.9.1971, 10.1.1972).
US 3 936 452 (Nitto Boseki; appl. 3.2.1976; J-prior. 8.9.1972).

Formulation(s):   vial (inj.) 100 mg (as trihydrochloride)

Trade Name(s):
J: Spanidin (Nippon Kayaku)

Halazone
(Aseptamide)

ATC:　D08A
Use:　antiseptic, chemotherapeutic

RN:　80-13-7　MF: $C_7H_5Cl_2NO_4S$　MW: 270.09　EINECS: 201-253-1
CN:　4-[(dichloroamino)sulfonyl]benzoic acid

sodium salt
RN:　5698-56-6　MF: $C_7H_4Cl_2NNaO_4S$　MW: 292.07　EINECS: 227-176-3

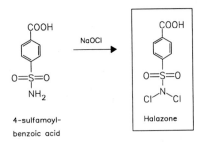

4-sulfamoyl-
benzoic acid

Halazone

Reference(s):
DE 318 899 (M. Claass; appl. 1918).

Formulation(s):　tabl. 4 mg

Trade Name(s):
F:　Gynamide (Merminod);
　　wfm

　　Théragynes (Theragynes);
　　wfm

Halcinonide

ATC:　D07AD02
Use:　glucocorticoid

RN:　3093-35-4　MF: $C_{24}H_{32}ClFO_5$　MW: 454.97　EINECS: 221-439-6
LD_{50}:　>10 g/kg (M, p.o.);
　　>5 g/kg (R, p.o.)
CN:　(11β,16α)-21-chloro-9-fluoro-11-hydroxy-16,17-[(1-methylethylidene)bis(oxy)]pregn-4-ene-3,20-dione

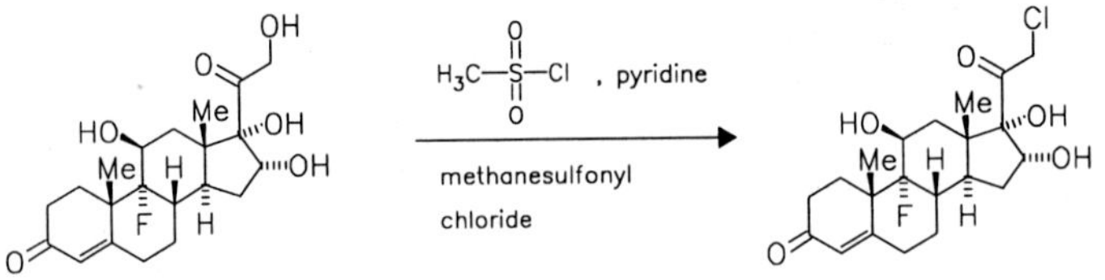

$H_3C-\overset{O}{\underset{O}{S}}-Cl$, pyridine

methanesulfonyl
chloride

9α-fluoro-16-hydroxy-
hydrocortisone
(cf. triamcinolone synthesis)

(I)

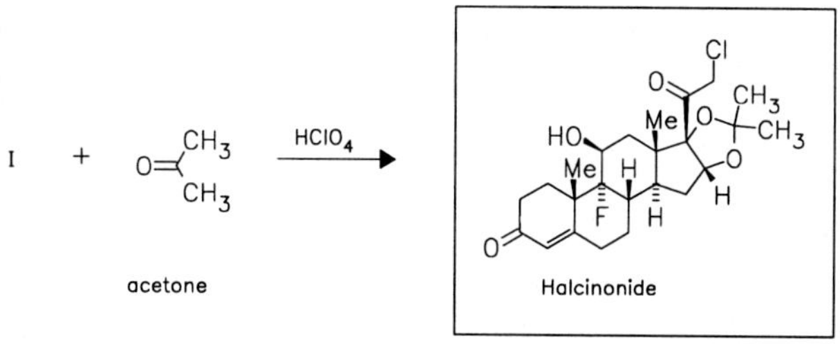

acetone   Halcinonide

Reference(s):

Bernstein, S.; Lenhard, R.H.: J. Am. Chem. Soc. (JACSAT) **82**, 3680 (1960).
Bernstein, S. et al.: J. Org. Chem. (JOCEAH) **27**, 690 (1962).

use:

DE 2 355 710 (Squibb; appl. 7.11.1973; USA-prior. 24.11.1972).

Formulation(s):   cream 0.1 %; ointment 0.1 %; sol. 0.1 %

Trade Name(s):

D: Halog (Bristol-Myers   GB: Halcicomp (F.A.I.R.)-   J: Adcortin (Sankyo)
      Squibb)   comb.   Simaderm (Bristol-Myers
F: Halog crème (Bristol-   Halciderm (Squibb)   Squibb)
      Myers Squibb)   Halcort (F.A.I.R.)   USA: Halog (Westwood-Squibb)
      Halog néomycine (Bristol-   I: Ancofort (Squibb)-comb.
      Myers Squibb)-comb.   Halciderm (Squibb)

Halofantrine
(WR-171669)

ATC: P01BX01
Use: antimalarial

RN: 69756-53-2   MF: $C_{26}H_{30}Cl_2F_3NO$ MW: 500.43   EINECS: 274-104-1
CN: 1,3-dichloro-α-[2-(dibutylamino)ethyl]-6-(trifluoromethyl)-9-phenanthrenemethanol

hydrochloride
RN: 36167-63-2   MF: $C_{26}H_{30}Cl_2F_3NO \cdot HCl$ MW: 536.89   EINECS: 252-895-4
LD_{50}:   2050 mg/kg (R, i.p.); 3400 mg/kg (R. p.o.)
(–)-enantiomer
RN: 66051-76-1   MF: $C_{26}H_{30}Cl_2F_3NO$ MW: 500.43
(+)-enantiomer
RN: 66051-74-9   MF: $C_{26}H_{30}Cl_2F_3NO$ MW: 500.43

2-nitro-4-   2,4-dichloro-   2-(2-nitro-4-tri-
trifluoromethyl-   benzaldehyde   fluoromethylphenyl)-
phenylacetic   2,4-dichloro-
acid   cinnamic acid

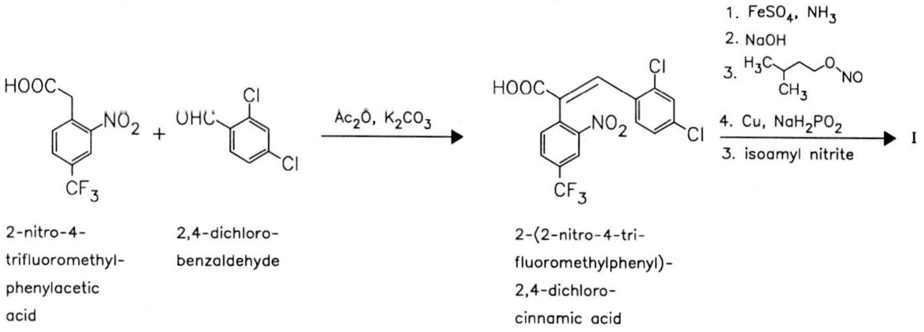

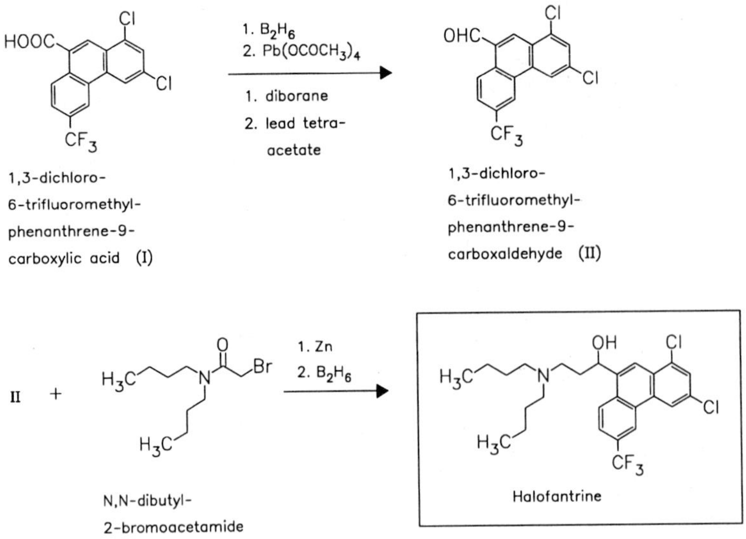

1,3-dichloro-
6-trifluoromethyl-
phenanthrene-9-
carboxylic acid (I)

1,3-dichloro-
6-trifluoromethyl-
phenanthrene-9-
carboxaldehyde (II)

N,N-dibutyl-
2-bromoacetamide

Halofantrine

Reference(s):
Colwell, W.T. et al.: J. Med. Chem. (JMCMAR) **15**, 771 (1972).

preparation of 2-nitro-4-trifluoromethylphenylacetic acid:
Simet, L.: J. Org. Chem. (JOCEAH) **28**, 358a (1963).

preparation of I:
Nodiff, E.A. et al.: J. Med. Chem. (JMCMAR) **14**, 921 (1971); **15**, 775 (1972).

resolution of the racemate:
Carroll, F.I. et al.: J. Med. Chem. (JMCMAR) **21**, 326 (1978).

glycero-phosphate, tartrate and biquinate salts:
US 4 507 288 (Smith Kline & Beckman; 20.3.1985; prior. 16.9.1983, 2.11.1983)
EP 138 374 (Smith Kline & Beckman; appl. 11.9.1984; USA-prior. 2.11.1983, 16.9.1983).

Formulation(s):   drinking amp. 2 %, 30 ml; susp. 100 mg/5 ml; tabl. 250 mg (as hydrochloride)

Trade Name(s):
D: Halfan (SmithKline   F: Halfan (SmithKline   GB: Halfan (SmithKline
      Beecham; 1991)   Beecham; 1988 as   Beecham; 1991)
                                         hydrochloride)

Halometasone

ATC: D07AB; D07AC12
Use: topical corticosteroid, anti-
        inflammatory

RN: 50629-82-8   MF: $C_{22}H_{27}ClF_2O_5$ MW: 444.90   EINECS: 256-664-9
CN: (6α,11β,16α)-2-chloro-6,9-difluoro-11,17,21-trihydroxy-16-methylpregna-1,4-diene-3,20-dione

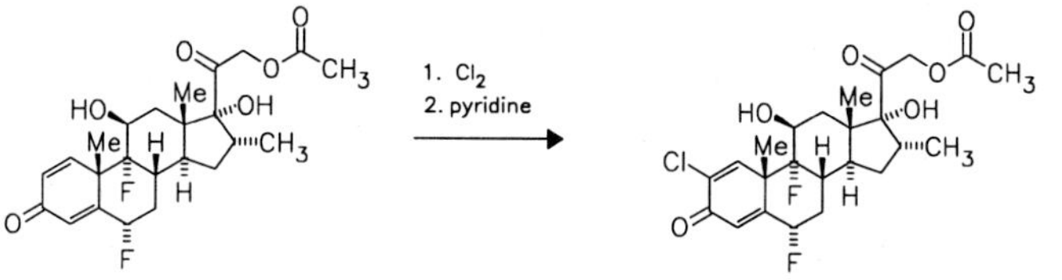

flumetasone acetate

1. Cl₂
2. pyridine

6α,9-difluoro-2-chloro-16α-methyl-
11β,17-dihydroxy-21-acetoxypregna-
1,4-diene-3,20-dione (I)

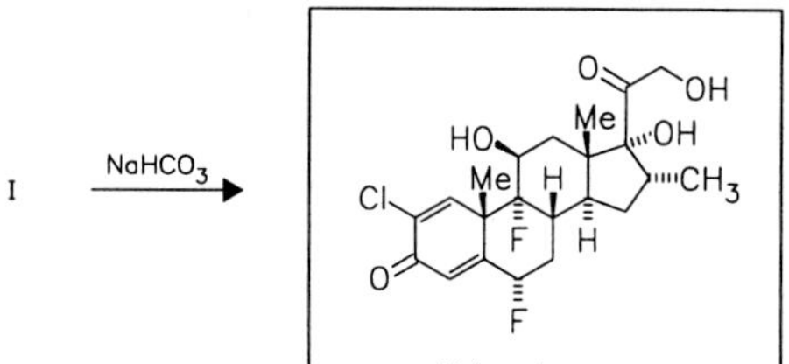

I   NaHCO₃ →

Halometasone

Reference(s):
DE 1 807 980 (Ciba-Geigy; appl. 9.11.1968; CH-prior. 17.11.1967).
CH 551 399 (Ciba-Geigy; appl. 17.10.1968).
US 3 652 554 (Ciba-Geigy; 28.3.1972; appl. 15.11.1968; CH-prior. 17.11.1967).

topical combination with triclosan:
GB 2 148 116 (Ciba-Geigy; appl. 27.10.1983).
US 4 512 987 (Ciba-Geigy; 23.4.1985; appl. 13.10.1982; GB-prior. 15.7.1982).

Formulation(s):   cream 0.5 mg/g (0.05 %),; ointment 0.5 mg/g (0.05 %)

Trade Name(s):
D: Sicorten (Novartis; 1986)   Sicorten Plus (Novartis;
                                    1986)-comb. with triclosan

Haloperidol

ATC: N05AD01
Use: neuroleptic, antidyskinetic,
        antipsychotic

RN: 52-86-8   MF: $C_{21}H_{23}ClFNO_2$ MW: 375.87   EINECS: 200-155-6
LD$_{50}$:   13 mg/kg (M, i.v.); 71 mg/kg (M, p.o.);
        15 mg/kg (R, i.v.); 128 mg/kg (R, p.o.);
        18 mg/kg (dog, i.v.); 90 mg/kg (dog, p.o.)
CN: 4-[4-(4-chlorophenyl)-4-hydroxy-1-piperidinyl]-1-(4-fluorophenyl)-1-butanone

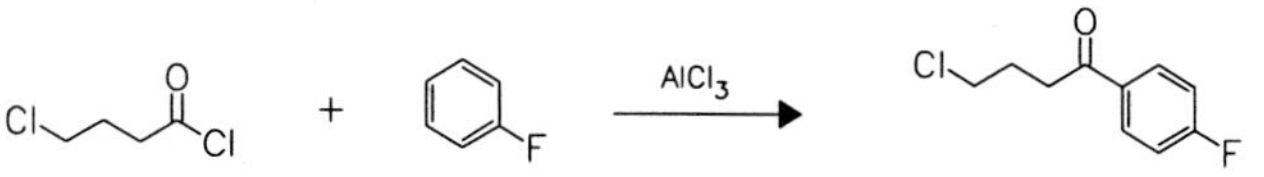

4-chlorobutyryl
chloride

+

fluoro-
benzene

AlCl₃ →

4-chloro-4'-fluoro-
buryrophenone (I)

2-(4-chlorophenyl)-
propene

4-(4-chlorophenyl)-
1,2,3,6-tetrahydro-
pyridine (II)

4-(4-chlorophenyl)-
4-hydroxypiperidine

Haloperidol

Reference(s):
Janssen, P.A.J. et al.: J. Med. Pharm. Chem. (JMPCAS) **1**, 281 (1959).
DE 1 289 845 (Janssen; appl. 18.4.1959; GB-prior. 22.4.1958).
US 3 438 991 (Janssen; 15.4.1969; GB-prior. 18.11.1959).

alternative syntheses:
GB 1 141 664 (Janssen; valid from 7.12.1966; prior. 8.12.1965, 23.9.1966).
US 4 086 234 (Searle; 25.4.1978; appl. 7.11.1975).

Formulation(s):　amp. 5 mg/ml, 100 mg/ml, 50 mg/ml; drops 2 mg, 20 mg/ml, 2 mg/ml, 0.5 mg/ml; sol. 10 mg;
　　　　　　　　oral liquid 2 mg/ml, 10 mg/ml; tabl. 1 mg, 2 mg, 5 mg, 10 mg, 20 mg

Trade Name(s):
D:	Buteridol (Promonta Lundbeck)		Serenace (Baker Norton)	Halomonth (Dainippon; as decanoate)

D:　Buteridol (Promonta
　　Lundbeck)
　　Haldol-Janssen (Janssen-
　　Cilag)
　　Sigaperidol (Kytta-
　　Siegfried)
　　generic
F:　Haldol (Janssen-Cilag)
　　Vésadol (Janssen-Cilag)-
　　comb.
GB:　Dozic (Rosemont)
　　Haldol (Janssen-Cilag; as
　　decanoate)

I:　Aloper (Sifra)
　　Aloperid (Formulario Naz.)
　　Aloperid (Biologici Italia)
　　Bioperidolo (Firma)
　　Haldol (Janssen)
　　Haldol Decanoas (Janssen;
　　as decanoate)
　　Serenase (Lusofarmaco)
J:　Brotopon (Taito Pfizer)
　　Einalon S (Maruko)
　　Halojust (Horita)

　　Serenace (Baker Norton)
　　Halomonth (Dainippon; as
　　decanoate)
　　Halosten (Shionogi)
　　Keselan (Sumitomo)
　　Linton (Yoshitomi)
　　Neoperidol (Kyowa Hakko;
　　as decanoate)
　　Peluces (Isei)
　　Serenace (Dainippon)
USA:　Haldol (Ortho-McNeil
　　Pharmaceutical)

Halopredone diacetate

ATC:　H02AB
Use:　glucocorticoid, topical anti-
　　　inflammatory

RN:　57781-14-3　MF: $C_{25}H_{29}BrF_2O_7$　MW: 559.40　EINECS: 260-951-4
LD_{50}:　>5 g/kg (M, p.o.);
　　　>5 g/kg (R, p.o.)
CN:　(6β,11β)-17,21-bis(acetyloxy)-2-bromo-6,9-difluoro-11-hydroxypregna-1,4-diene-3,20-dione

halopredone
RN:　57781-15-4　MF: $C_{21}H_{25}BrF_2O_5$　MW: 475.33　EINECS: 260-953-5

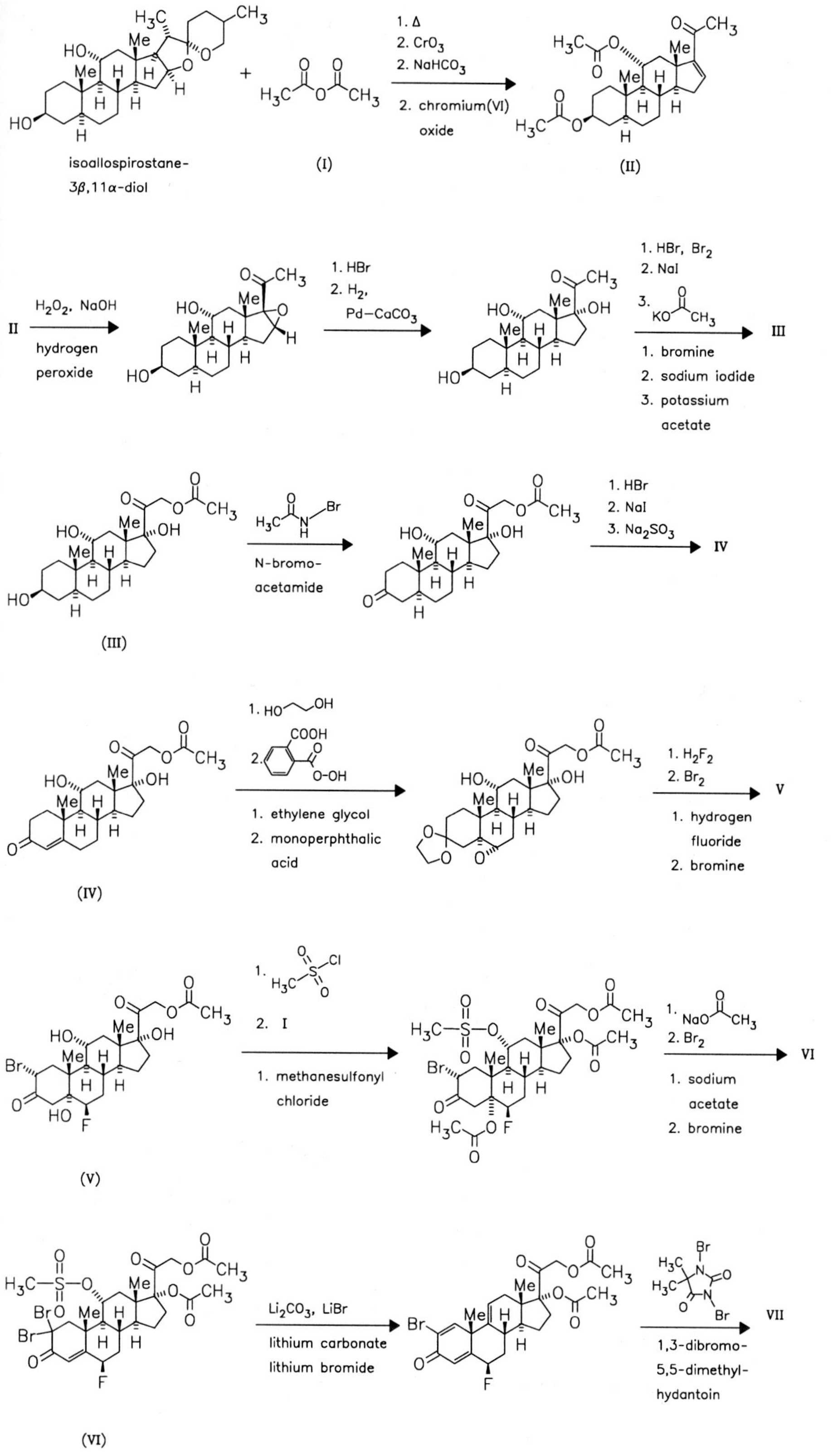

isoallospirostane-
3β,11α-diol

(I)

1. Δ
2. CrO₃
2. NaHCO₃
2. chromium(VI)
oxide

(II)

H_2O_2, NaOH

hydrogen
peroxide

II

1. HBr
2. H₂,
Pd–CaCO₃

1. HBr, Br₂
2. NaI
3. $KO \overset{O}{\overset{\|}{C}} CH_3$

1. bromine
2. sodium iodide
3. potassium
acetate

III

(III)

N-bromo-
acetamide

1. HBr
2. NaI
3. Na₂SO₃

IV

(IV)

1. HO⌒OH
2. COOH / O-OH

1. ethylene glycol
2. monoperphthalic
acid

1. H₂F₂
2. Br₂

1. hydrogen
fluoride
2. bromine

V

(V)

1. $H_3C-\overset{O}{\underset{O}{\overset{\|}{S}}}-Cl$
2. I

1. methanesulfonyl
chloride

1. $NaO \overset{O}{\overset{\|}{C}} CH_3$
2. Br₂

1. sodium
acetate
2. bromine

VI

(VI)

Li₂CO₃, LiBr

lithium carbonate
lithium bromide

1,3-dibromo-
5,5-dimethyl-
hydantoin

VII

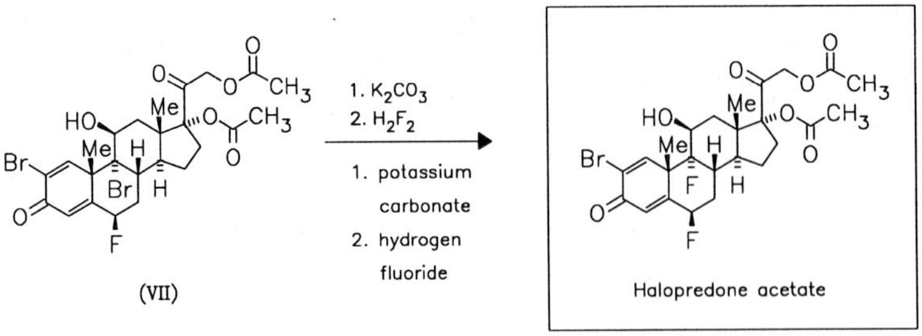

Halopredone acetate

Reference(s):
Bianchetti, A.; Riva, M.: J. Med. Chem. (JMCMAR) **20**, 213 (1977).
BE 826 030 (Pierrel; appl. 26.2.1975; GB-prior. 27.2.1974, 5.7.1974, 25.7.1974, 19.11.1974).

synthesis of 21-acetoxy-11α,17-dihydroxypregn-4-ene-3,20-dione:
Djerassi, C. et al.: J. Am. Chem. Soc. (JACSAT) **74**, 1712, 3634 (1952); **75**, 1277 (1953).

Formulation(s):   amp. 12.5 mg/ml, 25 mg/ml

Trade Name(s):
J: Haloart (Dainippon Ink-
      Taiho)

Haloprogin

ATC: D01AE11
Use: antifungal, antiseptic

RN: 777-11-7   MF: $C_9H_4Cl_3IO$ MW: 361.39   EINECS: 212-286-6
LD_{50}: >5.6 g/kg (R, p.o.);
        >3 g/kg (dog, p.o.)
CN: 1,2,4-trichloro-5-[(3-iodo-2-propynyl)oxy]benzene

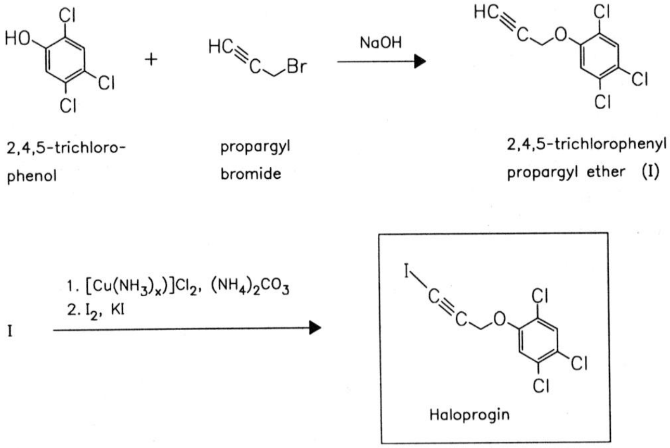

2,4,5-trichloro-   propargyl   2,4,5-trichlorophenyl
phenol   bromide   propargyl ether (I)

Haloprogin

Reference(s):
US 3 322 813 (Meiji Seika; 30.5.1967).

Formulation(s):   cream 1 %; ointment 1 %; sol. 1 %

Trade Name(s):
D: Mycanden (Asche); wfm   F: Mycilan (Théraplix); wfm   USA: Halotex (Westwood); wfm
    Mycanden (Schering); wfm   J: Polik (Meiji Seika)

Halothane

ATC: N01AB01
Use: inhalation anesthetic

RN: 151-67-7   MF: $C_2HBrClF_3$ MW: 197.38   EINECS: 205-796-5
LD_{50}: 5680 mg/kg (R, p.o.)
CN: 2-bromo-2-chloro-1,1,1-trifluoroethane

trichloroethylene	2-chloro-1,1,1-trifluoro-ethane	Halothane

Reference(s):
GB 767 779 (ICI; appl. 20.10.1954; valid from 11.10.1955).
US 2 921 098 (ICI; 12.1.1960; GB-prior. 20.10.1954).
GB 805 764 (ICI; appl. 1956; valid from 1957).

alternative syntheses:
DE 1 039 503 (Bayer; appl. 1953).
DE 1 041 937 (Hoechst; appl. 1957).
US 2 959 624 (Hoechst; 8.11.1960; D-prior. 18.7.1957).
US 3 082 263 (ICI; 19.3.1963; GB-prior. 19.9.1959).
DAS 1 161 249 (Hoechst; appl. 23.9.1960).
DAS 1 285 989 (Hoechst; appl. 9.3.1963).
DOS 2 245 372 (Biocontrol; appl. 15.9.1972).

Formulation(s):   inhalation sol. 125 ml, 250 ml

Trade Name(s):
D: Fluothane (Zeneca)   F: Fluotane (Zeneca)   J: Halothane (Hoechst)
    Halothan ASID (Rüsch   GB: Fluothane (ICI; 1957); wfm   USA: Fluothane (Wyeth-Ayerst)
    Hospital)   I: Fluothane (Zeneca)

Haloxazolam

ATC: N05BA
Use: tranquilizer, benzodiazepine
      anxiolytic, hypnotic, sedative

RN: 59128 97 1   MF: $C_{17}H_{14}BrFN_2O_2$ MW: 377.21
LD_{50}: 1850 mg/kg (M, p.o.)
CN: 10-bromo-11b-(2-fluorophenyl)-2,3,7,11b-tetrahydrooxazolo[3,2-d][1,4]benzodiazepin-6(5H)-one

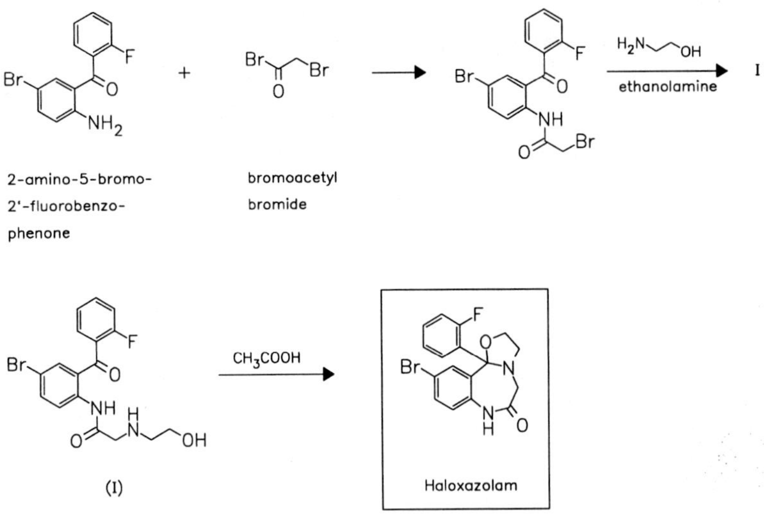
2-amino-5-bromo-2'-fluorobenzophenone
bromoacetyl bromide
ethanolamine
I
CH3COOH
(I)
Haloxazolam

Reference(s):
Miyadera, T. et al.: J. Med. Chem. (JMCMAR) **14**, 520 (1971).
JP 4 941 439 (Sankyo; appl. 21.12.1970).

Formulation(s): tabl. 5 mg, 10 mg

Trade Name(s):
J: Somelin (Sankyo)

Halquinol
(Chlorquinol)

ATC: A07
Use: intestinal antiseptic, topical anti-infective

RN: 8067-69-4 MF: unspecified MW: unspecified
CN: halquinols

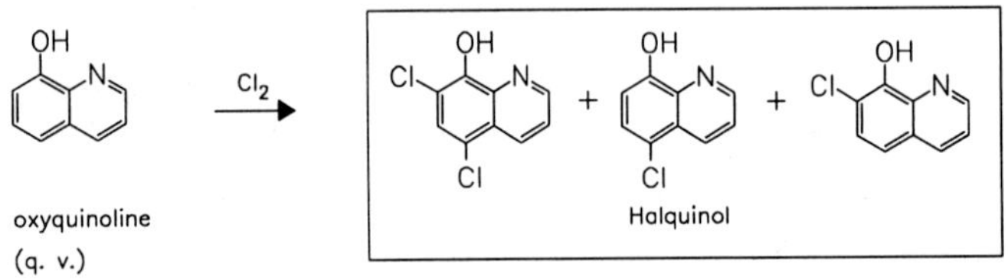
oxyquinoline (q. v.)
Halquinol

Reference(s):
FR 1 372 414 (Olin Mathieson; appl. 2.8.1961; GB-prior. 24.6.1960, 28.2.1961).

Formulation(s): cream, ointment, sol.

Trade Name(s):
D: Combiase (Luitpold)-comb.; wfm
Diarönt (Chephasaar)-comb.; wfm
Dignoquine (Luitpold)-comb.; wfm
Flamutil (Voigt)-comb.; wfm
Hyalokombun (Merckle)-comb.; wfm
Mexaform plus (Ciba)-comb.; wfm

Uzara plus (Uzara)-comb.;   GB: Quixalin (Squibb); wfm
wfm   USA: Quinolor (Squibb); wfm

Heparin

ATC: B01AB01; C05BA03
Use: anticoagulant, antithrombotic

RN: 9005-49-6   MF: [$C_{24}H_{38}N_2O_{35}S_5$]x   MW: unspecified   EINECS: 232-681-7
LD_{50}:   500 mg/kg (M, i.v.);
        1950 mg/kg (R, p.o.)
CN: mucopolysaccharide polysulfuric acid ester

sodium salt
RN: 9041-08-1   MF: unspecified   MW: unspecified
LD_{50}:   2800 mg/kg (M, i.v.); >5 g/kg (M, p.o.);
        391821 iu/kg (R, i.v.); >779000 iu/kg (R, p.o.);
        1 g/kg (dog, i.v.)
calcium salt
RN: 37270-89-6   MF: unspecified   MW: unspecified
LD_{50}:   >40 g/kg (M, i.v.);
        >40 g/kg (R, i.v.)
magnesium salt
RN: 54479-70-8   MF: unspecified   MW: unspecified

Heparin

(molar mass 6000-20000, according to origin)
From animal tissue, especially bovine lung and liver (e. g. autolysis of comminuted tissue parts, heating with ammonium sulfate in alkaline solution, filtration and acidification yield heparin as complex with protein, removal of fat with alcohol and treatment with trypsine for the purpose of decomposition of proteins, precipitation with alcohol and various purification methods).

Reference(s):
review:
Hind, H.G.: Manuf. Chem. (MACSAS) **34**, 510 (1963).

purification:
US 2 884 358 (Southerm California Gland Co; 28.4.1959; appl. 22.4.1957).
US 2 989 438 (Uclaf; 20.6.1961; appl. 29.12.1958).
DE 1 195 010 (Ormonoterapia Richter; appl. 12.5.1962).
US 3 016 331 (Ormonoterapia Richter; 9.1.1962; I-prior. 28.1.1960).
US 4 119 774 (AB Kabi; 10.10.1978; appl. 2.3.1977).
GB 1 539 332 (AB Kabi; appl. 4.3.1977; S-prior. 5.3.1976).

Formulation(s):   amp. 12500 iu, 20000 iu, 50000 iu, 60000 iu; cream, eye drops and eye ointment 30000 iu, 60000 iu, 150000 iu (as sodium salt or calcium salt); syringe 5000 iu, 7500 iu.

Trade Name(s):

D: Calciparin (Sanofi
     Winthrop)
     Heparin-Injekt (Immuno)
     Heparin-Na (Braun
     Melsungen; medac;
     Nattermann; ratiopharm)
     Heparin Novo (Novo)
     Heparin POS (Ursapharm)
     Heparin Riker (Riker)
     Liquemin (Roche)
     Thrombareduct
     (Azuchemie)
     Thrombophob (Knoll)
     Traumalitan (3M Medica;
     as sodium salt)
     Vetren (Klinge)
     numerous generic and
     combination preparations
F: Calciparin (Sanofi
     Winthrop; as calcium salt)
     Dioparine (Théa; as sodium
     salt)
     Néoparyl Framycétine
     (CIBA Vision
     Ophthalmics)-comb.
     numerous generic and
     combination preparations

GB: Canusal (CP Pharm.)
     Clexane (Rhône-Poulenc
     Rorer)
     Fragmin (Pharmacia &
     Upjohn)
     Hepsal (CP Pharm.)
     Monoparin (CP Pharm.)
     generic and combination
     preparations
I: Ateroclar (Mediolanum)
     Chemyparin (SIT)
     Clarisco (Schwarz)
     Disebrin (Allergan)
     Eparina (Tariff. Integrativo;
     Bristol-Myers Squibb;
     Manetti Roberts; Parke
     Davis)
     Eparinovis (Intes)
     Essaven Gel (Rhône-
     Poulenc Rorer)-comb.
     Flebs Crema (Pierre Fabre
     Phar.)-comb.
     Heparin Collirio
     (Farmigea)
     Idracemi Eparina
     (Farmigea)-comb.
     Lioton (Menarini)

     Liquemin (Roche; as
     sodium salt)
     Luxazone Eparina
     (Allergan)-comb.
     Normoparin (Opocrin)
     Venotrauma (Also)-comb.
     Viteparin (Teofarma)-
     comb.
     numerous generic
     preparations
J: Caprocin (Mitsui; as
     calcium salt)
     Depo-Heparin (Upjohn-
     Kodama; as sodium salt)
     Hepacarin (Eisai)
     Heparigen (Mukasa-Torii)
     Heparin Sodium (Tokyo
     Tanabe)
     Novo Heparin (Novo-
     Kodama)
     Panheprin (Nippon Abbott)
     numerous generic
     preparations
USA: Heparin-Sodium (Wyeth-
     Ayerst)
     Hep-Lock (Elkins-Sinn)

Hepronicate

ATC: C04
Use: vasodilator

RN: 7237-81-2   MF: $C_{28}H_{31}N_3O_6$ MW: 505.57
LD$_{50}$:   5 g/kg (M, p.o.)
CN: 3-pyridinecarboxylic acid 2-hexyl-2-[[(3-pyridinylcarbonyl)oxy]methyl]-1,3-propanediyl ester

octanal

2-hexyl-2-hydroxy-
methyl-1,3-
propanediol   (I)

nicotinoyl
chloride

Hepronicate

Reference(s):
US 3 384 642 (Yoshitomi; 21.5.1968; J-prior. 18.11.1964, 12.10.1965).

Formulation(s):   tabl. 100 mg

Trade Name(s):
J: Megrin (Yoshitomi)

Heptabarb
(Heptabarbital; Heptabarbitone; Heptamalum)

ATC: N05CA11
Use: hypnotic, sedative

RN: 509-86-4   MF: $C_{13}H_{18}N_2O_3$ MW: 250.30   EINECS: 208-107-6
LD$_{50}$:   180 mg/kg (M, i.v.); >800 mg/kg (M, p.o.);
     >5 g/kg (R, p.o.);
     105 mg/kg (dog, i.v.)
CN: 5-(1-cyclohepten-1-yl)-5-ethyl-2,4,6(1*H,3H,5H*)-pyrimidinetrione

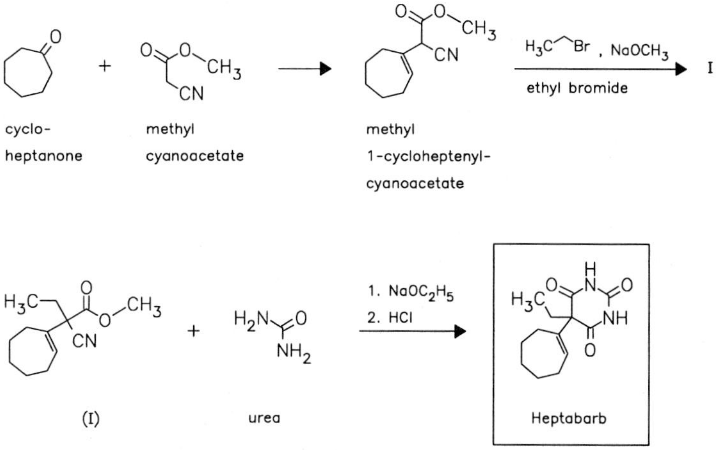

cyclo-
heptanone   methyl
cyanoacetate   methyl
1-cycloheptenyl-
cyanoacetate

(I)   urea   Heptabarb

Reference(s):
FR 870 714 (Geigy; appl. 1941; Palestine-prior. 1940).
DE 756 489 (Geigy; appl. 1941; Palestine-prior. 1940).

Formulation(s):   tabl. 200 mg

Trade Name(s):
D: Medomin (Geigy); wfm   F: Medomine (Geigy); wfm   GB: Medomin (Geigy); wfm

---

Heptaminol

ATC: C01DX08
Use: cardiotonic, sympathomimetic

RN: 372-66-7   MF: $C_8H_{19}NO$ MW: 145.25   EINECS: 206-758-0
LD$_{50}$:   1250 mg/kg (M, i.p.)
CN: 6-amino-2-methyl-2-heptanol

hydrochloride
RN: 543-15-7   MF: $C_8H_{19}NO \cdot HCl$ MW: 181.71   EINECS: 208-837-5
LD$_{50}$:   900 mg/kg (M, i.p.)

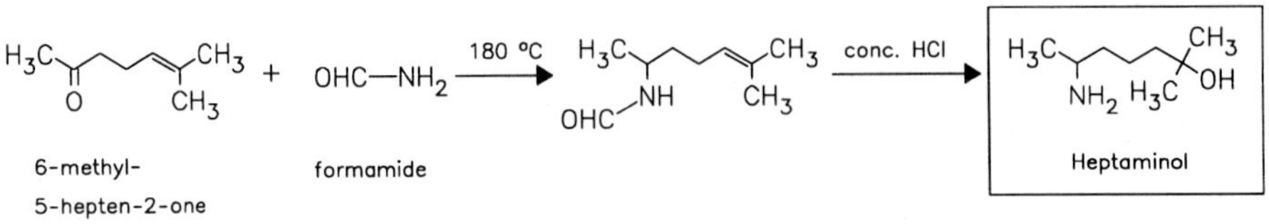

6-methyl-
5-hepten-2-one

formamide

Heptaminol

Reference(s):
Dœuvre, J.; Pozat, J.: C. R. Hebd. Seances Acad. Sci. (COREAF) **224**, 286 (1947).

Formulation(s):   amp. 250 mg, 500 mg; drg. 50 mg; drops 50 mg/ml;tabl. 150 mg (as hydrochloride)

Trade Name(s):

D: Normotin (OTW)-comb.   Débrumyl (Pierre Fabre   numerous combination
     Perivar (Intersan)-comb.   Santé)-comb.   preparations
F: Ampécyclal (Sarget)   Sureptil (Synthélabo)-   I: Coreptil (Delalande
                                  comb.   Isnardi)

Hetacillin

ATC: J01CA18
Use: antibiotic

RN: 3511-16-8   MF: $C_{19}H_{23}N_3O_4S$ MW: 389.48   EINECS: 222-512-5
CN: [2S-[2α,5α,6β(S*)]]-6-(2,2-dimethyl-5-oxo-4-phenyl-1-imidazolidinyl)-3,3-dimethyl-7-oxo-4-thia-1-
      azabicyclo[3.2.0]heptane-2-carboxylic acid

monopotassium salt
RN: 5321-32-4   MF: $C_{19}H_{22}KN_3O_4S$ MW: 427.57   EINECS: 226-182-3
LD_{50}:   650 mg/kg (M, i.v.); >15 g/kg (M, p.o.);
          >1400 mg/kg (R, i.v.); >10 g/kg (R, p.o.);
          2200 mg/kg (dog, i.v.); >4 g/kg (dog, p.o.)

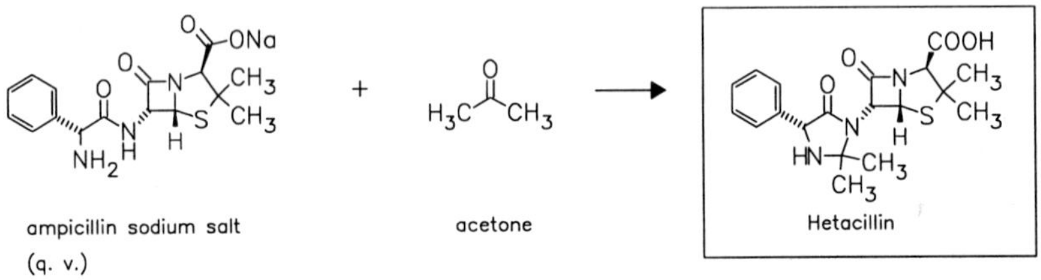

ampicillin sodium salt
(q. v.)

acetone

Hetacillin

Reference(s):
US 3 198 804 (Bristol-Myers; 3.8.1965; appl. 6.1.1965; prior. 25.1.1963).

Formulation(s):   amp. 250 mg, 500 mg, 1 g; tabl. 50 mg (as potassium salt)

Trade Name(s):

D: Penplenum (Bristol); wfm   Etadipen (Ghimas)-comb.   J: Natacillin (Banyu)
F: Versapen (Allard); wfm   with dicloxacillin; wfm   USA: Versapen (Bristol); wfm
I: Dicloeta (Lusopharma)-   Versaclox (Bristol)-comb.
     comb. with dicloxacillin;   with dicloxacillin; wfm
     wfm   Versapen (Bristol); wfm

Heteronium bromide

ATC: A03
Use: anticholinergic

RN: 7247-57-6   MF: C$_{18}$H$_{22}$BrNO$_3$S   MW: 412.35
LD$_{50}$: 3576 mg/kg (R, p.o.)
CN: 3-[(hydroxyphenyl-2-thienylacetyl)oxy]-1,1-dimethylpyrrolidinium bromide

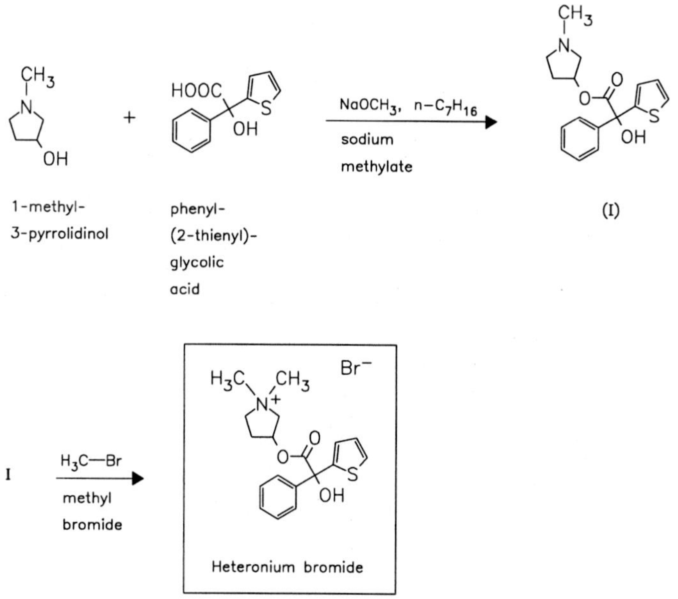

1-methyl-
3-pyrrolidinol

phenyl-
(2-thienyl)-
glycolic
acid

(I)

Reference(s):
US 3 138 614 (Eli Lilly; 23.1.1964; prior. 18.12.1961, 3.8.1960).

starting material:
US 2 830 997 (A. H. Robins; 1958; prior. 1956).

Formulation(s):   tabl. 1 mg

Trade Name(s):
I: Quentar (Ravizza)-comb.   USA: Hetrum Bromide (Lilly);
      with oxazepam; wfm   wfm

Hexachlorophene

ATC: D08AE01
Use: topical antiinfective, disinfectant,
     parasiticide

RN: 70-30-4   MF: C$_{13}$H$_6$Cl$_6$O$_2$ MW: 406.91   EINECS: 200-733-8
LD$_{50}$: 67 mg/kg (M, p.o.);
       7500 µg/kg (R, i.v.); 56 mg/kg (R, p.o.)
CN: 2,2'-methylenebis[3,4,6-trichlorophenol]

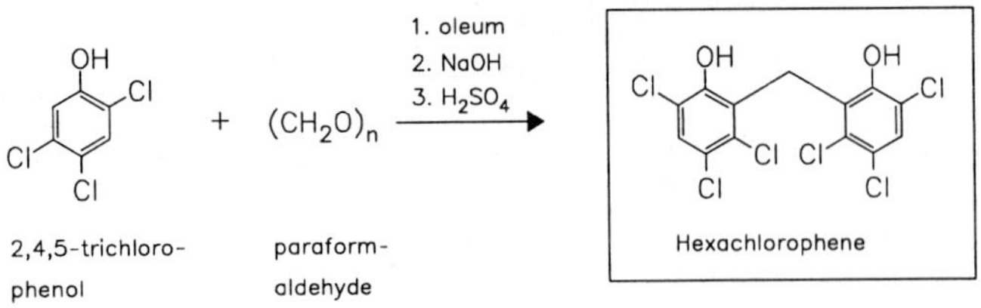
1. oleum
2. NaOH
3. H_2SO_4
$(CH_2O)_n$
2,4,5-trichloro-phenol
paraform-aldehyde
Hexachlorophene

Reference(s):
US 2 250 480 (B. T. Bush; 1941; appl. 1939).
US 2 435 593 (B. T. Bush; 1948; appl. 1945).
US 2 812 365 (Givaudan Corp.; 1957; prior. 1951, 1954).

Formulation(s): cream 0.5 g/100 g; emulsion 0.5 g/100 g; lotion 0.5 g/100 g

Trade Name(s):

D:	Aknefug (Wolff)	GB:	Dermalex (Sanofi Winthrop)-comb.		Phisohex disinf (Maggioni-Winthrop)-comb.; wfm
	Aknefug (Wolff)-comb. with estradiol		Ster-Zac D.C. (Seton)		Vestene (Eurospital); wfm
F:	Acnestrol (Poirier)-comb.	I:	Etaproctene (Angelini)-comb.; wfm	USA:	pHisoHex (Sanofi)

Hexafluronium bromide

(Hexafluorenium bromide)

ATC: M03AC05
Use: muscle relaxant

RN: 317-52-2 MF: $C_{36}H_{42}Br_2N_2$ MW: 662.55 EINECS: 206-265-0
LD_{50}: 1760 µg/kg (M, i.v.); 280 mg/kg (M, p.o.)
CN: *N,N'*-di-9*H*-fluoren-9-yl-*N,N,N',N'*-tetramethyl-1,6-hexanediaminium dibromide

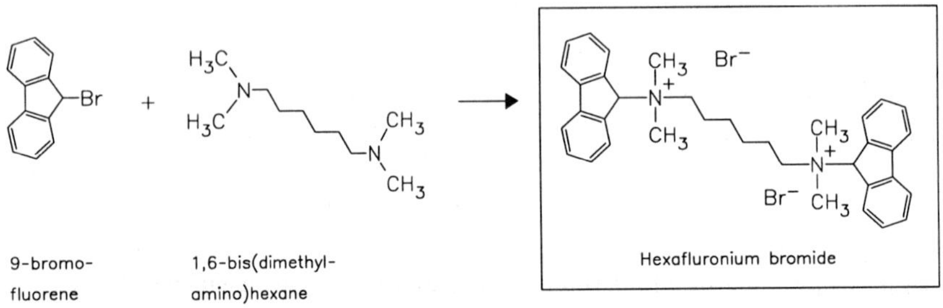
9-bromo-fluorene
1,6-bis(dimethyl-amino)hexane
Hexafluronium bromide

Reference(s):
US 2 783 237 (Irwin, Neisler & Co.; 1957; prior. 1953).

Formulation(s): amp. 20 mg/ml.

Trade Name(s):

J:	Mylaxen (Nippon Shoji); wfm	USA:	Mylaxen (Mallinckrodt); wfm	Mylaxen (Wallace); wfm

Hexamethonium chloride

ATC: C02
Use: ganglionic blocker, antihypertensive

RN: 60-25-3   MF: $C_{12}H_{30}Cl_2N_2$ MW: 273.29   EINECS: 200-465-1
LD$_{50}$:   26.7 mg/kg (M, i.v.);
        35 mg/kg (dog, i.v.)
CN: N,N,N,N',N',N'-hexamethyl-1,6-hexanediaminium dichloride

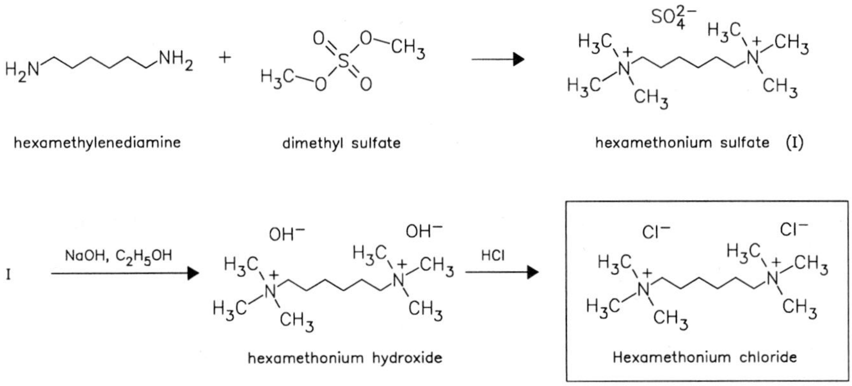

hexamethylenediamine   dimethyl sulfate   hexamethonium sulfate (I)

hexamethonium hydroxide   Hexamethonium chloride

Reference(s):
DE 900 097 (May & Baker; appl. 1951; GB-prior. 1950).

Formulation(s):   oral: 0.5 g/d

Trade Name(s):
D: Raucombin forte (Voigt);   Gastrometonio (Fabo; as
       wfm   iodide); wfm
I: Gastrometonio (Fabo);   J: Methobromin
       wfm   (Yamanouchi); wfm

Hexcarbacholine bromide
(Carbolonium bromide)

ATC: M03
Use: muscle relaxant

RN: 306-41-2   MF: $C_{18}H_{40}Br_2N_4O_4$ MW: 536.35
CN: N,N,N,N',N',N'-hexamethyl-4,13-dioxo-3,14-dioxa-5,12-diazahexadecane-1,16-diaminium dibromide

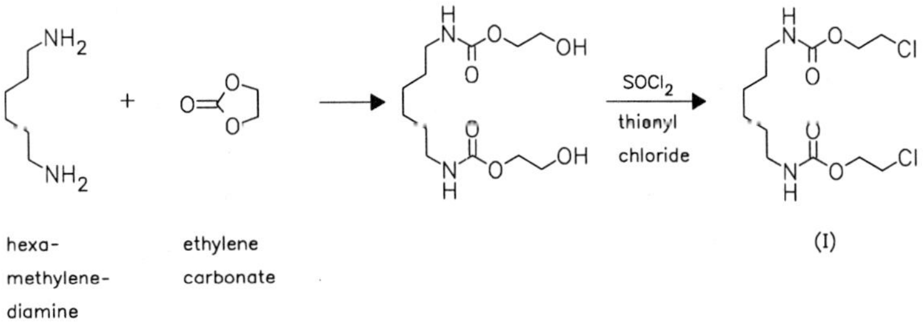

hexa-   ethylene   (I)
methylene-   carbonate
diamine

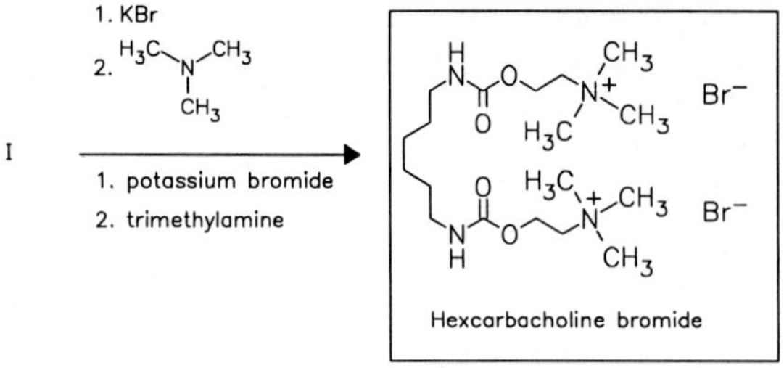

1. KBr

2. H₃C—N—CH₃
   |
   CH₃

I →

1. potassium bromide
2. trimethylamine

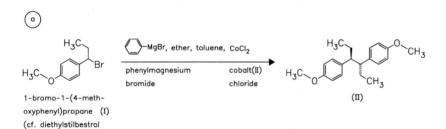

Hexcarbacholine bromide

Reference(s):
DE 1 021 842 (Österr. Stickstoffwerke; appl. 1954; A-prior. 1953).

Formulation(s):   amp. 2 mg/ml

Trade Name(s):
D: Imbretil (Hormon-
        Chemie); wfm

Hexestrol

ATC: G03
Use: estrogen, antineoplastic (hormonal)

RN: 84-16-2   MF: $C_{18}H_{22}O_2$ MW: 270.37   EINECS: 201-518-1
LD$_{50}$:   1 g/kg (M, p.o.);
        >2 g/kg (R, p.o.)
CN: (R*,S*)-4,4'-(1,2-diethyl-1,2-ethanediyl)bisphenol

a

MgBr, ether, toluene, CoCl₂

phenylmagnesium   cobalt(II)
bromide   chloride

1-bromo-1-(4-meth-
oxyphenyl)propane (I)
(cf. diethylstilbestrol
synthesis)

(II)

II →

HI

hydrogen
iodide

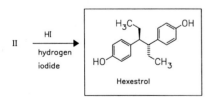

Hexestrol

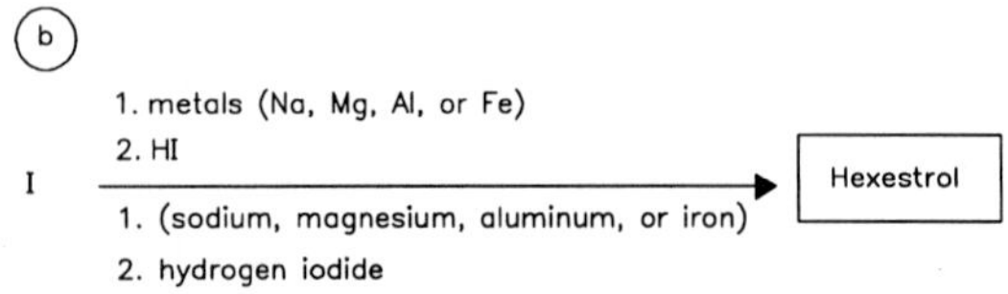

b

1. metals (Na, Mg, Al, or Fe)
2. HI

I ————————————————→   Hexestrol

1. (sodium, magnesium, aluminum, or iron)
2. hydrogen iodide

Reference(s):
a   Kharasch, M.S.; Kleimann, M.: J. Am. Chem. Soc. (JACSAT) **65**, 491 (1943).
b   Dodds, E.C. et al.: Proc. R. Soc. London, Ser. B (PRLBA4) **218**, 253 (1940).
    Bernstein, S.; Wallis, E.S.: J. Am. Chem. Soc. (JACSAT) **62**, 2871 (1940).
    Buu-Hoï, N.G.; Hoan, N.G.: J. Org. Chem. (JOCEAH) **14**, 1023 (1949).
    US 2 357 985 (Research Corp; 1944; appl. 1940).
    GB 523 320 (Boots; appl. 1938).
    FR 855 879 (Lab. Franç. de Chimiothérapie; appl. 1939).

starting material:
Bernstein, S.; Wallis, E.S.: J. Am. Chem. Soc. (JACSAT) **62**, 2871 (1940).

alternative syntheses:
Docken, A.M.;. Spielman, M.A.: J. Am. Chem. Soc. (JACSAT) **62**, 2163 (1940).
US 2 392 852 (Lilly; 1946; prior. 1941).
US 2 402 054 (Lilly; 1946; prior. 1941).
US 2 421 401 (Hoffmann-La Roche; 1947; S-prior. 1943).

review:
Solmssen, U.V.: Chem. Rev. (Washington, D. C.) (CHREAY) **36**, 481 (1945).

Formulation(s):   pessaries 10 mg

Trade Name(s):

D:	Malun (Temmler)-comb.; wfm				Hexestrol and Phenorbital (Jenkins)-comb.; wfm
F:	Cycloestrol (Bruneau); wfm	GB:	Synthrogene (Gerda); wfm		Hexestrol W/Butabarb (Bowman)-comb.; wfm
		J:	Synthovo (Boots); wfm		
			Robal (Chugai; as diacetate)		
	Micro-cristaux Cycloestrol (Bruneau); wfm	USA:	Estra-Plex (Rowell); wfm		Vagi-Plex (Rowell)-comb.; wfm

---

Hexetidine

ATC: A01AB12
Use: antiseptic antifungal

RN: 141-94-6   MF: $C_{21}H_{45}N_3$ MW: 339.61   EINECS: 205-513-5
CN: 1,3-bis(2-ethylhexyl)hexahydro-5-methyl-5-pyrimidinamine

nitro-ethane	form-aldehyde (I)	2-methyl-2-nitro-propane-1,3-diol	

(II)   Hexetidine

Reference(s):
US 2 415 047 (Commercial Solvents; 1947; appl. 1945).
US 3 054 797 (Commercial Solvents; 18.9.1962; prior. 11.10.1961).
Senkus, M.: J. Am. Chem. Soc. (JACSAT) **68**, 1611 (1946).

purification:
DE 2 011 078 (Gödecke; appl. 9.3.1970).
DAS 2 355 917 (Meditest; appl. 8.11.1973).
DOS 2 709 929 (Dolorgiet; appl. 8.3.1977).
DOS 2 310 337 (Wülfing; appl. 1.3.1973).
DAS 2 323 150 (Wülfing; appl. 8.5.1973).

salts with aromatic acids:
GB 1 538 603 (Doll; appl. 5.11.1976; D-prior. 6.11.1975, 3.6.1976, 16.6.1976).
US 4 141 968 (Doll; 27.2.1979; D-prior. 16.6.1976).
US 4 142 050 (Doll; 27.2.1979; D-prior. 6.11.1975, 3.6.1976).

nicotinate:
DOS 2 310 338 (Wülfing; appl. 1.3.1973).

Formulation(s):   sol. 100 mg/100 ml, 200 mg/100 ml; spray 0.1 g/100 g; vaginal tabl. 10 mg

Trade Name(s):

D: Anginasin Spray
      (Opfermann)-comb.
      De-menthasin (Scheurich)-
      comb.
      Doreperol (Rentschler)
      Givalex (Norgine)-comb.
      Hexetidin Gurgellösung
      (ratiopharm)

F:

Hexoral (Warner-Lambert)
Hexoral Spray (Warner-
Lambert)
generic
Collu-Hextril (Warner-
Lambert)
Givalex (Norgine)-comb.
Hextril (Warner-Lambert)

      Nifluril (UPSA)
GB: Oraldene (Warner-
      Lambert)
I: Oraseptic (Parke Davis)
USA: Sterisil (Warner Chilcott);
      wfm

Hexobarbital

ATC: N01AF02; N05CA16
Use: hypnotic, sedative

RN: 56-29-1   MF: $C_{12}H_{16}N_2O_3$ MW: 236.27   EINECS: 200-264-9
LD$_{50}$:   133 mg/kg (M, i.v.); 468 mg/kg (M, p.o.).
CN: 5-(1-cyclohexen-1-yl)-1,5-dimethyl-2,4,6($1H,3H,5H$)-pyrimidinetrione

sodium salt
RN: 50-09-9   MF: $C_{12}H_{15}N_2NaO_3$ MW: 258.25   EINECS: 200-009-1
LD$_{50}$:   165 mg/kg (M, i.v.); 1325 mg/kg (M, p.o.);
      1 g/kg (R, p.o.)

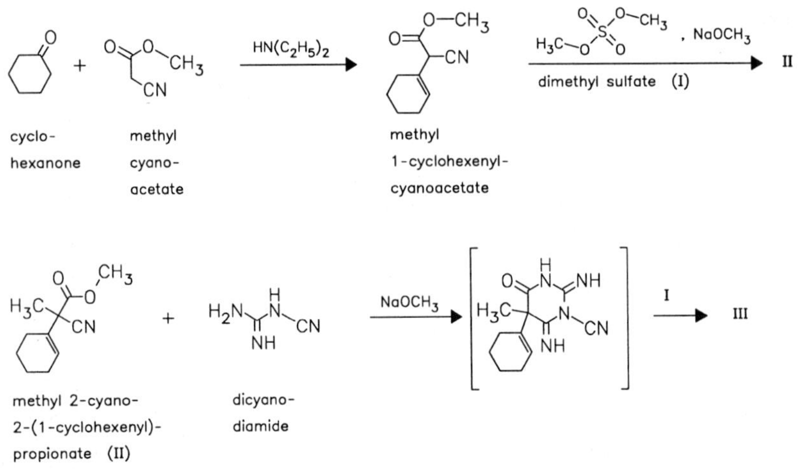

cyclo-
hexanone

methyl
cyano-
acetate

methyl
1-cyclohexenyl-
cyanoacetate

dimethyl sulfate (I)

methyl 2-cyano-
2-(1-cyclohexenyl)-
propionate (II)

dicyano-
diamide

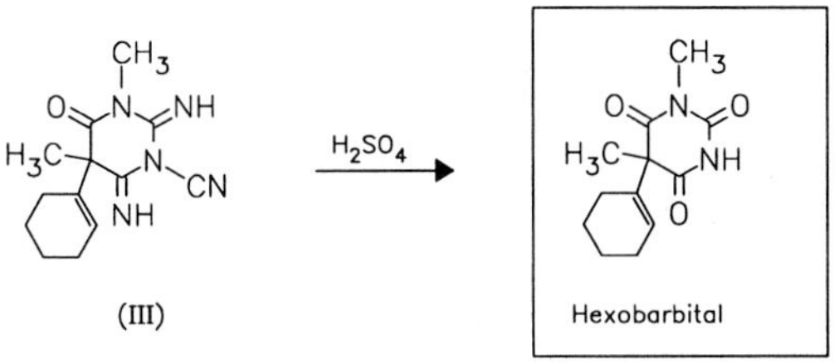

H_2SO_4
(III)
Hexobarbital

Reference(s):
DRP 595 175 (I. G. Farben; 1931).
DRP 590 175 (I. G. Farben; 1932).

Formulation(s): amp. 0.5g/5 g (10 %) (as sodium salt)

Trade Name(s):

D:	Dormopan (Bayropharm)-comb.; wfm Evipan (Bayer); wfm Stodinox (Lorenz)-comb.; wfm	F:	Dormopan (Bayer-Pharma)-comb.; wfm Noctivane (Vaillant-Defresne); wfm	J:	Cyclopan (Teikoku Kagaku-Nagase) Oltopan (Dainippon) Ouropan Soda (Shionogi)
		GB:	Evidorm (Winthrop); wfm	USA:	Sombulex (Riker); wfm

Hexobendine

ATC: C01DX06
Use: coronary vasodilator

RN: 54-03-5 MF: $C_{30}H_{44}N_2O_{10}$ MW: 592.69 EINECS: 200-189-1
LD_{50}: 34 mg/kg (R, i.v.)
CN: 3,4,5-trimethoxybenzoic acid 1,2-ethanediylbis[(methylimino)-3,1-propanediyl] ester

dihydrochloride
RN: 50-62-4 MF: $C_{30}H_{44}N_2O_{10} \cdot 2HCl$ MW: 665.61 EINECS: 200-054-7
LD_{50}: 35.2 mg/kg (M, i.v.); 682 mg/kg (M, p.o.);
52 mg/kg (R, i.v.); 2550 mg/kg (R, p.o.)

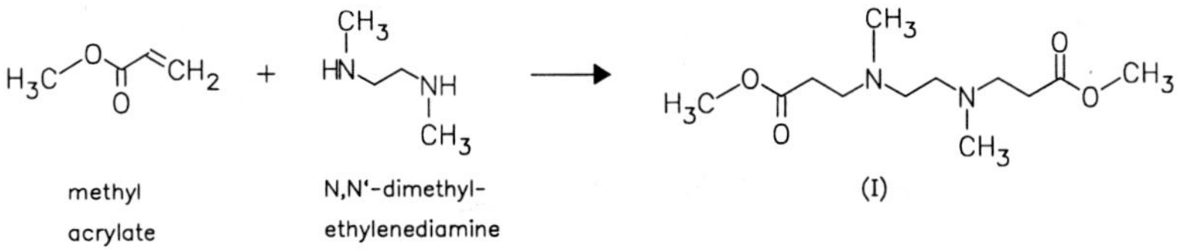

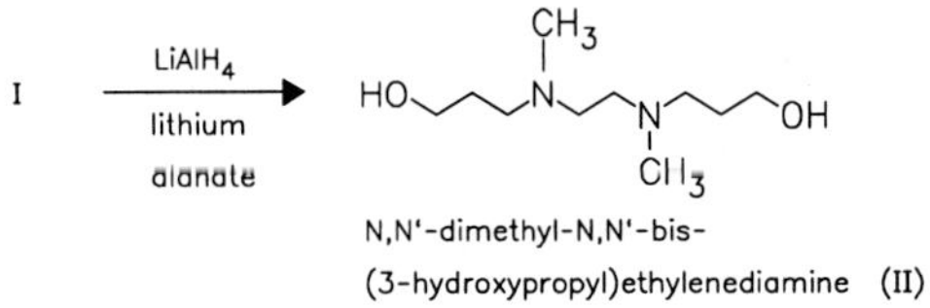

methyl acrylate
N,N'-dimethylethylenediamine
(I)

I
$LiAlH_4$
lithium alanate
N,N'-dimethyl-N,N'-bis-(3-hydroxypropyl)ethylenediamine (II)

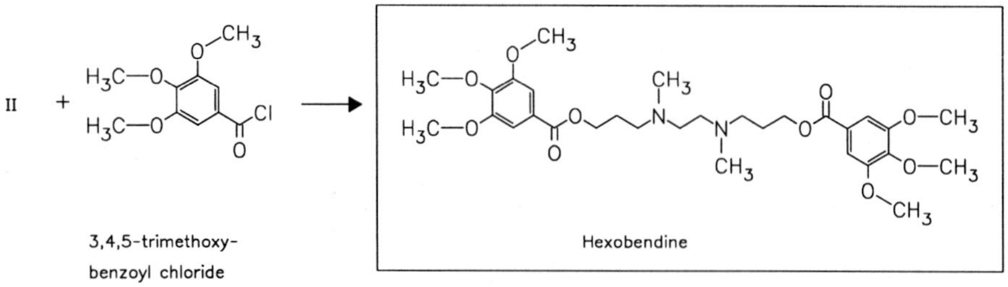

3,4,5-trimethoxy-
benzoyl chloride

Hexobendine

Reference(s):
DE 1 217 397 (Lentia; appl. 26.3.1962).

N,N'-bis-(3-hydroxypropyl)-1,2-ethylenediamine *from* 1,2-dichloroethane *and* 3-aminopropanol:
DAS 2 042 320 (Lentia; appl. 26.8.1970).

Formulation(s):   amp. 10 mg; tabl. 30 mg, 60 mg

Trade Name(s):
D: Card-Instenon (Byk   Reoxyl (Byk Gulden); wfm   Ustimon (Merck-
      Gulden)-comb.; wfm   F: Hityl (Biosedra)-comb.;   Clévenot); wfm
      Instenon (Byk Gulden)-   wfm   I: Flussicor (Farmalabor);
      comb.; wfm   wfm

Hexocyclium metilsulfate

ATC: A03AB10
Use: antispasmodic, anticholinergic

RN: 115-63-9   MF: $C_{20}H_{33}N_2O \cdot CH_3O_4S$ MW: 428.59   EINECS: 204-097-2
LD_{50}: 8900 µg/kg (M, i.v.)
CN: 4-(2-cyclohexyl-2-hydroxy-2-phenylethyl)-1,1-dimethylpiperazinium methyl sulfate (salt)

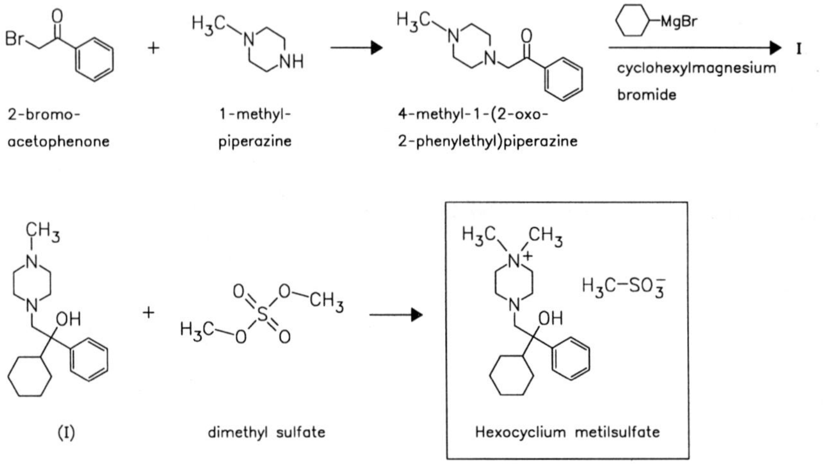

2-bromo-   1-methyl-   4-methyl-1-(2-oxo-   cyclohexylmagnesium
acetophenone   piperazine   2-phenylethyl)piperazine   bromide

(I)   dimethyl sulfate   Hexocyclium metilsulfate

Reference(s):
US 2 907 765 (Abbott; 6.10.1959; prior. 10.9.1956).

Formulation(s):   drops; f. c. tabl. 25 mg

Trade Name(s):

D: Traline retard (Abbott);   F: Traline (Abbott); wfm   USA: Tral (Abbott); wfm
     wfm   I: Tral (Abbott); wfm

Hexoprenaline

ATC: R03AC06; R03CC05
Use: bronchodilator

RN: 3215-70-1   MF: $C_{22}H_{32}N_2O_6$ MW: 420.51
CN: 4,4'-[1,6-hexanediylbis[imino(1-hydroxy-2,1-ethanediyl)]]bis[1,2-benzenediol]

dihydrochloride
RN: 4323-43-7   MF: $C_{22}H_{32}N_2O_6 \cdot 2HCl$ MW: 493.43   EINECS: 224-354-2
LD_{50}:   88 mg/kg (M, i.v.); 2036 mg/kg (M, p.o.);
     58 mg/kg (R, i.v.); 10 g/kg (R, p.o.)
sulfate
RN: 30117-45-4   MF: $C_{22}H_{32}N_2O_6 \cdot xH_2SO_4$ MW: unspecified   EINECS: 250-057-2

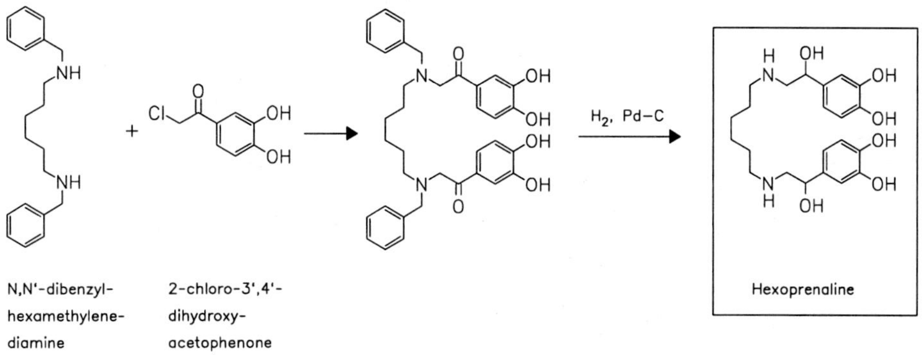

N,N'-dibenzyl-   2-chloro-3',4'-
hexamethylene-   dihydroxy-
diamine   acetophenone

Hexoprenaline

Reference(s):
DE 1 215 729 (Lentia; appl. 14.6.1963).
US 3 329 709 (Österr. Stickstoffwerke; 4.7.1967; A-prior. 11.6.1963).

Formulation(s):   aerosol 5.7 mg (as sulfate); amp. 0.005 mg, 0.025 mg; tabl. 0.5 mg (as sulfate)

Trade Name(s):
D: Etoscol (Byk Gulden); wfm   I: Tocolysan (Byk Gulden; as   J: Leanal (Yoshitomi; as
                                      sulfate)   sulfate)

Hexylcaine

ATC: N01B
Use: local anesthetic

RN: 532-77-4   MF: $C_{16}H_{23}NO_2$ MW: 261.37
CN: 1-(cyclohexylamino)-2-propanol benzoate (ester)

hydrochloride
RN: 532-76-3   MF: $C_{16}H_{23}NO_2 \cdot HCl$ MW: 297.83   EINECS: 208-544-2
LD_{50}:   23 mg/kg (M, i.v.); 1080 mg/kg (M, p.o.)

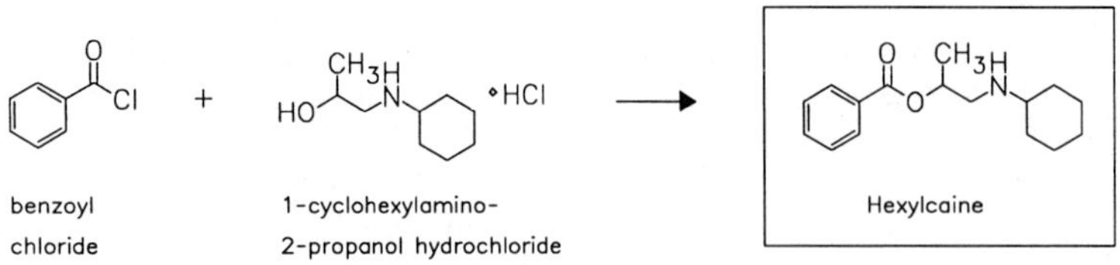

benzoyl chloride + 1-cyclohexylamino-2-propanol hydrochloride → Hexylcaine

Reference(s):
US 2 486 374 (Sharp & Dohme; 1949; prior. 1944).
Cope, A. et al.: J. Am. Chem. Soc. (JACSAT) **66**, 1453 (1944).

Formulation(s):   amp. 1 %, 2 %; sol. (as hydrochloride)

Trade Name(s):
USA: Cyclaine (Merck Sharp & Dohme); wfm

Hexylresorcinol

ATC: R02AA12
Use: anthelmintic, antiseptic

RN: 136-77-6   MF: $C_{12}H_{18}O_2$ MW: 194.27   EINECS: 205-257-4
LD_{50}: 1040 mg/kg (M, p.o.);
        550 mg/kg (R, p.o.)
CN: 4-hexyl-1,3-benzenediol

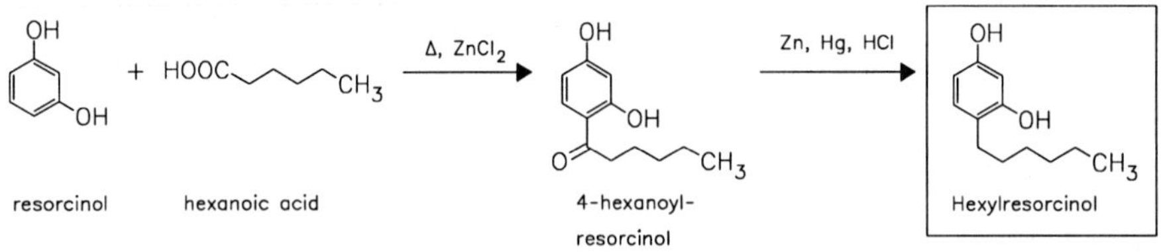

resorcinol + hexanoic acid → 4-hexanoyl-resorcinol → Hexylresorcinol

Reference(s):
DRP 488 419 (Sharp & Dohme; appl. 1923; USA-prior. 1923).
DRP 489 117 (Sharp & Dohme; appl. 1925; USA-prior. 1925).
Dohme, A.R.L. et al.: J. Am. Chem. Soc. (JACSAT) **48**, 1688 (1926).
Twiss, D.: J. Am. Chem. Soc. (JACSAT) **48**, 2206 (1926).

Formulation(s):   cream 20 mg; drg. 2.4 mg; ointment 4.3 mg

Trade Name(s):
D: Hexamon (Beiersdorf-
     Lilly)-comb.; wfm
     Jodo-Muc (Merz & Co.);
     wfm

     Mycatox (Brenner-Efeka)-
     comb.; wfm
GB: Kamillosan (Norgine)-
     comb.; wfm

I: Oxana (Biologici Italia);
     wfm

Histapyrrodine

ATC: R06AC02
Use: antihistaminic, antiallergic

RN: 493-80-1   MF: $C_{19}H_{24}N_2$ MW: 280.42   EINECS: 207-781-9
CN: N-phenyl-N-(phenylmethyl)-1-pyrrolidineethanamine

monohydrochloride
RN: 6113-17-3   MF: $C_{19}H_{24}N_2 \cdot HCl$ MW: 316.88   EINECS: 228-079-9

N-(2-chloroethyl)-   N-benzyl-   Histapyrrodine
pyrrolidine   aniline
hydrochloride

Reference(s):
US 2 623 880 (H. Hopff et al.; 1952; D-prior. 1948).
GB 659 730 (BASF; appl. 1949; D-prior. 1948).

Formulation(s):   tabl. 25 mg (as hydrochloride)

Trade Name(s):
D: Calcistin (Boehringer   Calcistin (Hestia)-comb.   Domistan (Servier); wfm
      Mannh.)-comb. with   with calcium lactate; wfm   I: Calcistin (Boehringer
      calcium lactate; wfm   F: Crème domistan vit. F   Biochemia)-comb.; wfm
                                         (Servier); wfm

L-Histidine

ATC: M01
Use: essential proteinogen amino acid (for
        infusion solutions), dietary
        supplement

RN: 71-00-1   MF: $C_6H_9N_3O_2$ MW: 155.16   EINECS: 200-745-3
LD_{50}: >2 g/kg (M, i.v.); >15 g/kg (M, p.o.);
        >2 g/kg (R, i.v.); >15 g/kg (R, p.o.)
CN: L-histidine

monohydrochloride monohydrate
RN: 5934-29-2   MF: $C_6H_9N_3O_2 \cdot HCl \cdot H_2O$ MW: 209.63
LD_{50}: >1.677 g/kg (M, i.p.)

L-Histidine

Isolation from hydrolysates of blood meal by ion-exchange chromatography.

Reference(s):
Ullmann's Encyclopedia of Industrial Chemistry, 5th Ed., Vol. **A2**, 70.
Vickery, M.B.: J. Biol. Chem. (JBCHA3) **143**, 77 (1942).

Formulation(s):   sol. 0.25 g/100 ml (as hydrochloride)

Trade Name(s):
D: Histinorm (A.S.); wfm USA: NephrAmine (R & D)

Homatropine
(Omatropina)

ATC: S01FA05
Use: anticholinergic, antispasmodic, mydriatic

RN: 87-00-3 MF: C₁₆H₂₁NO₃ MW: 275.35 EINECS: 201-716-8

RN: 87-00-3 MF: $C_{16}H_{21}NO_3$ MW: 275.35 EINECS: 201-716-8
CN: α-hydroxybenzeneacetic acid endo-(±)-8-methyl-8-azabicyclo[3.2.1]oct-3-yl ester

hydrobromide
RN: 51-56-9 MF: $C_{16}H_{21}NO_3 \cdot HBr$ MW: 356.26 EINECS: 200-105-3
LD_{50}: 107 mg/kg (M, i.v.)

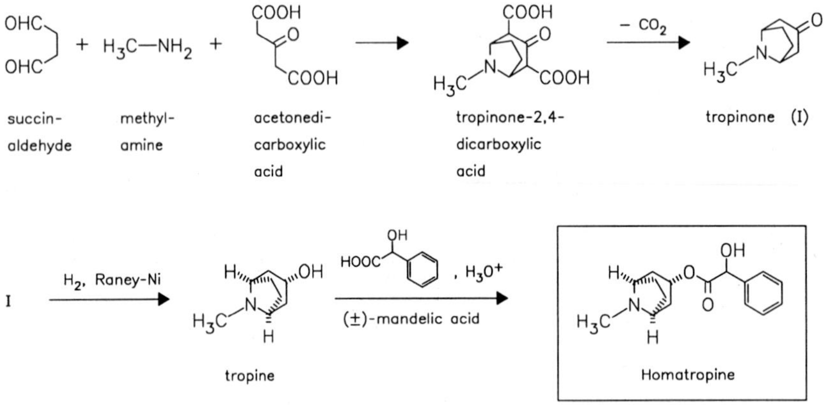

succin-aldehyde · methyl-amine · acetonedicarboxylic acid · tropinone-2,4-dicarboxylic acid · tropinone (I)

tropine · (±)-mandelic acid · Homatropine

Reference(s):
DRP 95 853 (E. Tauber; appl. 1896).
Ladenburg, A.: Justus Liebigs Ann. Chem. (JLACBF) **217**, 75 (1883).
Chemnitius, F.: J. Prakt. Chem. (JPCEAO) **117**, 142 (1927).

tropinone *synthesis:*
Robinson, R.: J. Chem. Soc. (JCSOA9) **111**, 762 (1917); **111**, 876 (1917).
Schöpf, C.: Angew. Chem. (ANCEAD) **50**, 779 (1937).

racemate resolution:
Werner, G.; Miltenberger, K.: Justus Liebigs Ann. Chem. (JLACBF) **631**, 163 (1960).

Formulation(s): eye drops 1 %; eye ointment 1 % (as hydrobromide)

Trade Name(s):
D: Homatropin POS 1 %
Augentropfen (Ursapharm)
GB: Minims Homatropine
Hydrobromide (Chauvin)

I: Omatr Br (Formulario
Naz.; Tariff. Nazionale)
Omatropina Lux coll.
(Allergan)

J: Homatropine
hydrobromide (Torii)

Homatropine methylbromide

(Methylhomatropine bromide)

ATC: A07AA54
Use: anticholinergic, antispasmodic

RN: 80-49-9   MF: $C_{17}H_{24}BrNO_3$ MW: 370.29   EINECS: 201-284-0
LD$_{50}$:   1400 mg/kg (M, p.o.);
        12 mg/kg (R, i.v.); 1200 mg/kg (R, p.o.)
CN: endo-3-[(hydroxyphenylacetyl)oxy]-8,8-dimethyl-8-azoniabicyclo[3.2.1]octane bromide

homatropine
(q. v.)

methyl
bromide

Homatropine methylbromide

Reference(s):
Cahen, R.L.; Tvede, K.: J. Pharmacol. Exp. Ther. (JPETAB) **105**, 166 (1952).

Formulation(s):   eye drops 1 %, 2 %, 5 %; syrup 1.5 mg/5 ml, 5 mg/5 ml; tabl. 1.5 mg, 5 mg

Trade Name(s):
F: Entercine (Robapharm)-
        comb.; wfm
        Supadol (Lederle)-comb.;
        wfm
        Surparine (Licardy)-comb.;
        wfm

        Ulfon (Lafon)-comb.; wfm
        Vagantyl (Robapharm)-
        comb.; wfm
GB: APP (Consolidated); wfm
        Vagantyl (Robapharm)-
        comb.; wfm

I: Novatropina (ASTA
        Medica)
USA: Hycodan (Endo)

Homofenazine

ATC: N05AK
Use: neuroleptic, psychosedative

RN: 3833-99-6   MF: $C_{23}H_{28}F_3N_3OS$ MW: 451.56
CN: hexahydro-4-[3-[2-(trifluoromethyl)-10H-phenothiazin-10-yl]propyl]-1H-1,4-diazepine-1-ethanol

dihydrochloride
RN: 1256-01-5   MF: $C_{23}H_{28}F_3N_3OS \cdot 2HCl$ MW: 524.48   EINECS: 215-017-0
LD$_{50}$:   73.5 mg/kg (M, i.v.); 790 mg/kg (M, p.o.);
        102 mg/kg (R, i.v.); 880 mg/kg (R, p.o.)

2-trifluoromethyl-
phenothiazine

1. NaNH$_2$
2. Cl ～～ N

1. sodium amide
2. 1-(3-chloropropyl)-
    hexahydro-
    1,4-diazepine

2-trifluoromethyl-
10-[3-(hexahydro-
1,4-diazepino)propyl]-
phenothiazine (I)

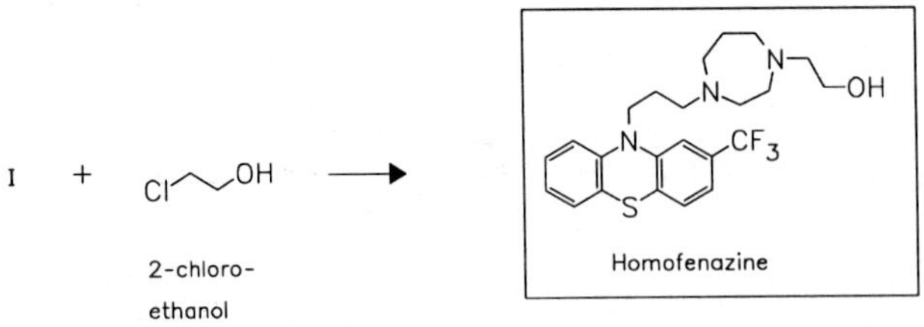

I + ClCH₂CH₂OH

2-chloro-
ethanol

Homofenazine

Reference(s):
DE 1 160 442 (Degussa; appl. 18.2.1960).
US 3 040 043 (Degussa; 19.6.1962; D-prior. 18.3.1959).

Formulation(s):   tabl. 3 mg (as hydrochloride)

Trade Name(s):
D: Pasaden (Homburg); wfm   Seda-Ildamen (Homburg)-   I: Pasaden (Farmades); wfm
                                       comb.; wfm

Hydralazine

ATC: C02DB02
Use: antihypertensive

RN: 86-54-4   MF: C₈H₈N₄   MW: 160.18   EINECS: 201-680-3
LD₅₀: 52 mg/kg (M, i.v.); 122 mg/kg (M, p.o.);
         34 mg/kg (R, i.v.); 90 mg/kg (R, p.o.);
         50 mg/kg (dog, i.v.); 100 mg/kg (dog, p.o.)
CN: 1(2H)-phthalazinone hydrazone

monohydrochloride
RN: 304-20-1   MF: C₈H₈N₄ · HCl   MW: 196.64   EINECS: 206-151-0
LD₅₀: 84 mg/kg (M, i.v.); 188 mg/kg (M, p.o.);
         34 mg/kg (R, i.v.)

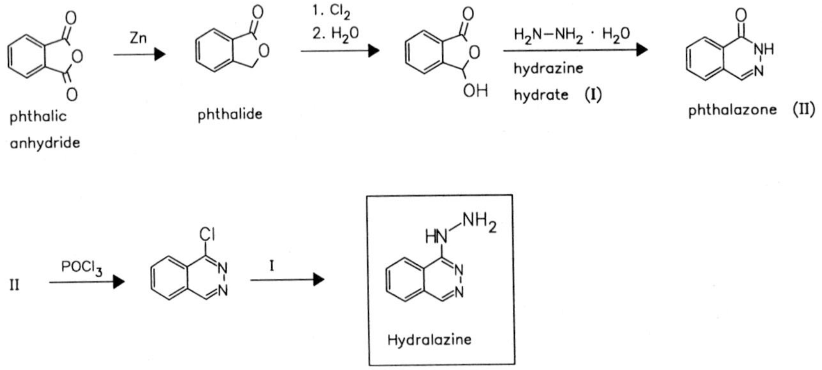

phthalic
anhydride

phthalide

1. Cl₂
2. H₂O

H₂N–NH₂ · H₂O

hydrazine
hydrate (I)

phthalazone (II)

II   POCl₃ →

I →

Hydralazine

Reference(s):
US 2 484 029 (Ciba; 1949; CH-prior. 1945).
DE 848 818 (Ciba; CH-prior. 1945).

Formulation(s):   amp. 20 mg; drg. 10 mg, 25 mg, 50 mg; tabl. 25 mg, 50 mg (as hydrochloride)

Trade Name(s):

D:	Docidrazin (Rhein-Pharma; Zeneca)-comb.	GB:	Apresoline (Novartis; as hydrochloride)	Homoton (Horii)

D: Docidrazin (Rhein-Pharma; Zeneca)-comb.
Impresso-Puren (Isis Puren)-comb.
pertenso (Fournier Pharma)-comb.
Treloc (Astra/Promed)-comb.
Trepress (Novartis Pharma)-comb.
Tri-Normin (Zeneca)-comb.

GB: Apresoline (Novartis; as hydrochloride)
I: Apresolin Retard (Novartis; as hydrochloride)
J: Anaspasmin (Vitacain)
Aprelazine (Kaigai)
Apresoline (Novartis)
Aprezine (Kanto)
Basedock D (Sawai)
Deselazine (Kobayashi Kako)
Diucholin (Toyama)

Homoton (Horii)
Hypatol (Yamanouchi)
Hypos (Nippon Shinyaku)
Pressfall (Nissin)
Propectin (Maruishi)
Solesorin (Hishiyama)
USA: Hydralazine Hydrochloride (SoloPak)
Hydra-Zide (Par)

Hydrochlorothiazide

ATC: C03AA03
Use: diuretic

RN: 58-93-5   MF: $C_7H_8ClN_3O_4S_2$ MW: 297.74   EINECS: 200-403-3
LD$_{50}$: 590 mg/kg (M, i.v.); 1175 mg/kg (M, p.o.);
990 mg/kg (R, i.v.); 2750 mg/kg (R, p.o.);
250 mg/kg (dog, i.v.)
CN: 6-chloro-3,4-dihydro-2H-1,2,4-benzothiadiazine-7-sulfonamide 1,1-dioxide

4-amino-6-chloro-benzene-1,3-disulfamide

paraform-aldehyde

Hydrochlorothiazide

chlorothiazide (q. v.)

formaldehyde

Hydrochlorothiazide

Reference(s):
a   US 3 163 645 (Ciba; 29.12.1964; appl. 25.9.1964; prior. 9.4.1958).
Stevens, G. de et al.: Experientia (EXPEAM) 14, 463 (1958).
b   US 3 164 588 (Merck & Co.; 5.1.1965; GB-prior. 19.6.1959).

alternative syntheses:
US 3 025 292 (Merck & Co.; 13.3.1962; prior. 26.11.1958).

purification:
US 3 043 840 (Merck & Co. 10.7.1962; appl. 14.10.1959).

Formulation(s):   tabl. 12.5 mg, 25 mg, 50 mg, 100 mg

Trade Name(s):

D: Disalunil (Berlin-Chemie)
diu melusin (Schwarz)-
comb.
Esidrix (Novartis Pharma)
HCT-ISIS (Isis Puren)
numerous generics and
combination preparations

F: Acuilix (Parke Davis)-
comb.
Briazide (Pierre Fabre)-
comb.
Esidrex (Novartis)
Moducren (Merck Sharp &
Dohme-Chibret)-comb.
Moduretic (Du Pont
Pharma)-comb.
Prestole (SmithKline
Beecham)-comb.
Zestoretic (Zeneca)-comb.
numerous generics and
combination preparations

GB: Hydrosaluric (Merck Sharp
& Dohme)

numerous combination
preparations

I: Accuretic (Parke Davis)-
comb.
Acediur (Guidotti)-comb.
Aceplus (Bristol-Myers
Squibb)-comb.
Acequide (Recordati)-
comb.
Acesistem (Sigma-Tau)-
comb.
Aldactazide (SPA)-comb.
Condiuren (Neopharmed)-
comb.
Esidrex (Novartis)
Indroclor (Formulario
Naz.)
Medozide (Malesci)-comb.
Moduretic (Merck Sharp &
Dohme)-comb.
Prinzide (Du Pont)-comb.
Quinazide (Malesci)-comb.
Raunova Plus (SmithKline
Beecham)-comb.

Selozide (Astra)-comb.
Spiridazide (SIT)-comb.
Vasoretic (Merck Sharp &
Dohme)-comb.
Zestoretic (Zeneca)-comb.
combination preparations

J: Chlothia (Iwaki)
Dichlotride (Merck-Banyu)
Esidrex (Novartis-Takeda)
Maschitt (Showa Shinyaku)
Mikorten (Zensei)
Newtolide (Towa)
Pantemon (Tatsumi)

USA: Hydrochlorothiazide
(Lederle)
HydroDIURIL (Merck
Sharp & Dohme)
Hydropres (Merck Sharp &
Dohme)-comb.
Oretic (Abbott)
Timolide (Merck Sharp &
Dohme)-comb.
generic and combination
preparations

Hydrocodone

ATC: R05DA03
Use: antitussive, narcotic analgesic

RN: 125-29-1 MF: $C_{18}H_{21}NO_3$ MW: 299.37 EINECS: 204-733-9
LD$_{50}$: 8.57 mg/kg (M, s.c.);
CN: (5α)-4,5-epoxy-3-methoxy-17-methylmorphinan-6-one

bitartrate hydrate
RN: 34195-34-1 MF: $C_{18}H_{21}NO_3 \cdot C_4H_6O_6 \cdot 5/2H_2O$ MW: 988.99
LD$_{50}$: 375 mg/kg (R, p.o.)
hydrochloride
RN: 25968-91-6 MF: $C_{18}H_{21}NO_3 \cdot HCl$ MW: 335.83 EINECS: 247-382-7

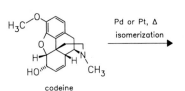

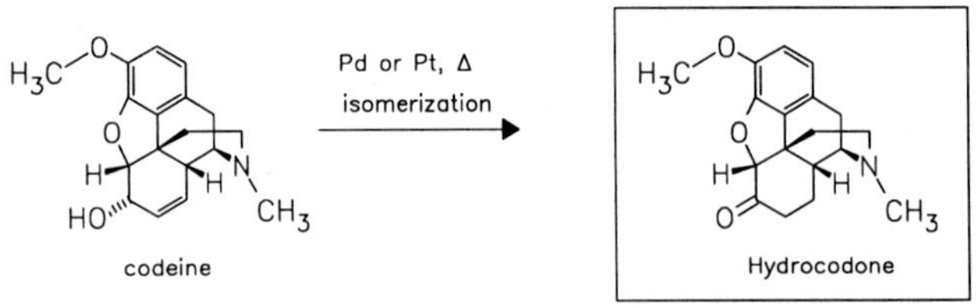

codeine → Pd or Pt, Δ isomerization → Hydrocodone

Reference(s):
Ehrhart, Ruschig **I**, 119-120.
DRP 607 931 (Knoll; 1935).
DRP 617 238 (Knoll; 1935).
DRP 623 821 (Knoll; 1935).

Formulation(s): amp. 15 mg/ml (as hydrochloride); syrup 5 mg/5 ml, 100 mg/5 ml; tabl. 5 mg, 10 mg (as
bitartrate hydrate)

Hydrocortisone

ATC: A01AC03; A07EA02; D07AA02;
C05AA01; D07XA01; H02AB09;
S01BA02; S01CB03; S02BA01

Use: glucocorticoid, anti-inflammatory

RN: 50-23-7 MF: $C_{21}H_{30}O_5$ MW: 362.47 EINECS: 200-020-1
LD_{50}: >500 mg/kg (M, s.c.);
150 mg/kg (R, i.p.); 449 mg/kg (R, s.c.)
CN: (11β)-11,17,21-trihydroxypregn-4-ene-3,20-dione

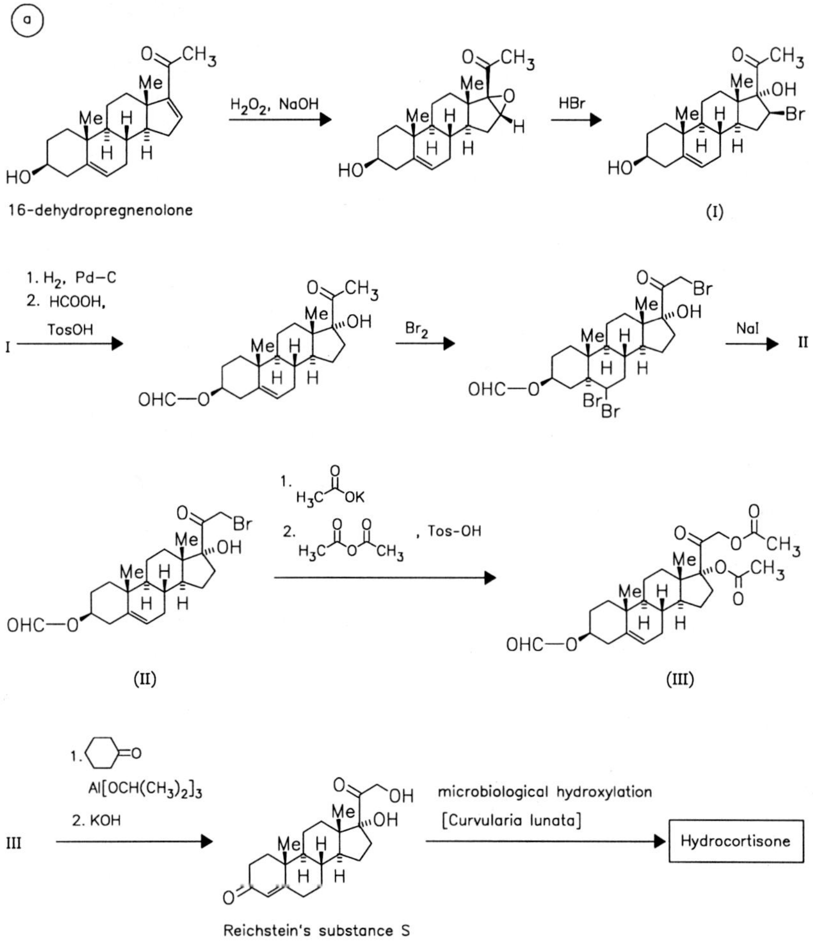

16-dehydropregnenolone

H_2O_2, NaOH

HBr

(I)

1. H_2, Pd-C
2. HCOOH,
TosOH

I

Br_2

NaI

II

OHC-O

(II)

1. $H_3C\text{—}OK$ (with C=O)
2. H_3C—(anhydride)—CH_3, Tos-OH

OHC-O

(III)

1. cyclohexanone
Al[OCH(CH$_3$)$_2$]$_3$
2. KOH

III

Reichstein's substance S

microbiological hydroxylation
[Curvularia lunata]

Hydrocortisone

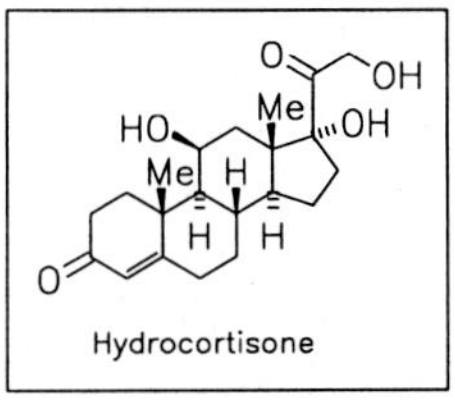

Hydrocortisone

(b)

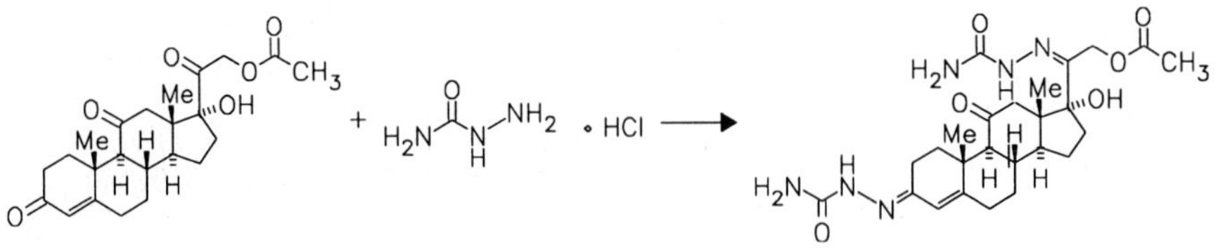

cortisone 21-acetate

semicarbazide
hydrochloride

(IV)

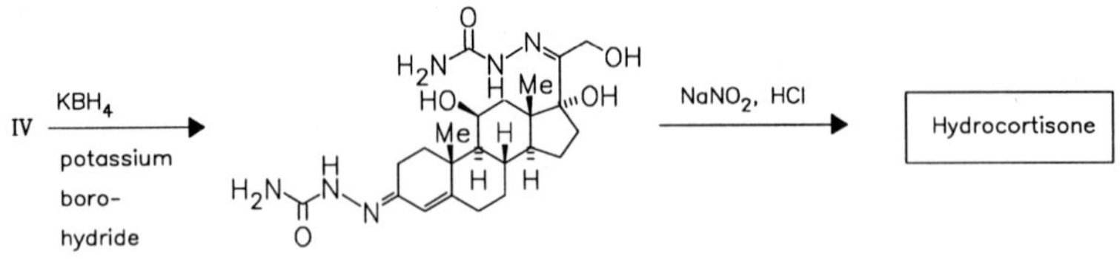

IV $\xrightarrow{\text{KBH}_4}$ potassium boro-hydride

$\xrightarrow{\text{NaNO}_2, \text{ HCl}}$ Hydrocortisone

(c)

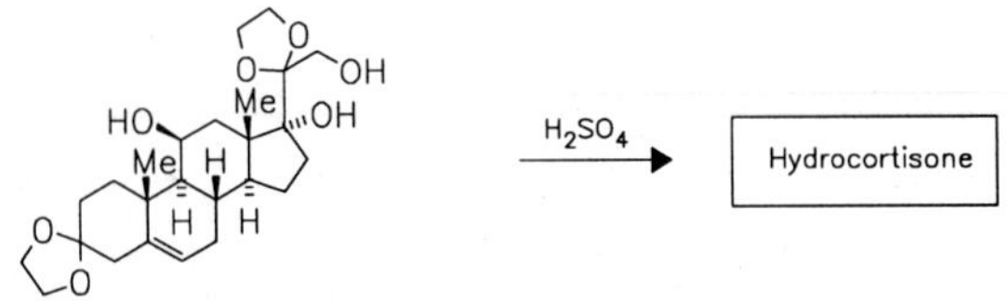

$\xrightarrow{\text{H}_2\text{SO}_4}$ Hydrocortisone

3,3:20,20-bis(ethylenedioxy)-
11β,17,21-trihydroxy-5-pregnene
(from cortisone)

(d)

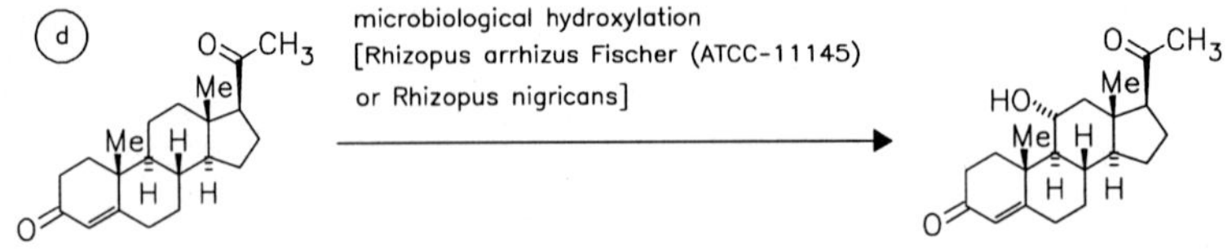

progesterone

microbiological hydroxylation
[Rhizopus arrhizus Fischer (ATCC-11145)
or Rhizopus nigricans]

11α-hydroxy-
progesterone (V)

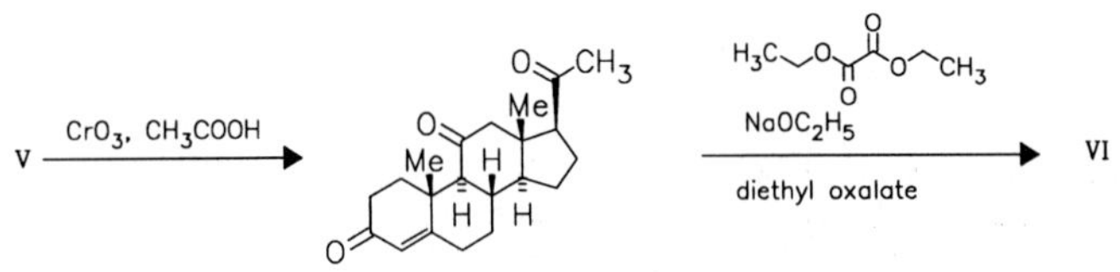

V $\xrightarrow{\text{CrO}_3, \text{ CH}_3\text{COOH}}$

11-oxoprogesterone

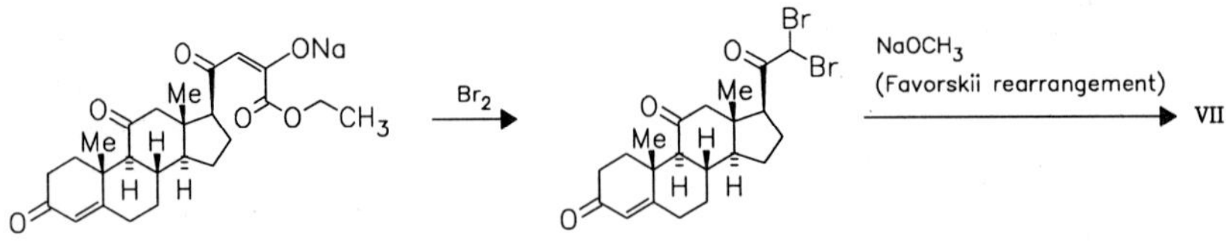

$\xrightarrow[\text{diethyl oxalate}]{\text{NaOC}_2\text{H}_5}$ VI

(VI)

$\xrightarrow{\text{Br}_2}$

$\xrightarrow[\text{(Favorskii rearrangement)}]{\text{NaOCH}_3}$ VII

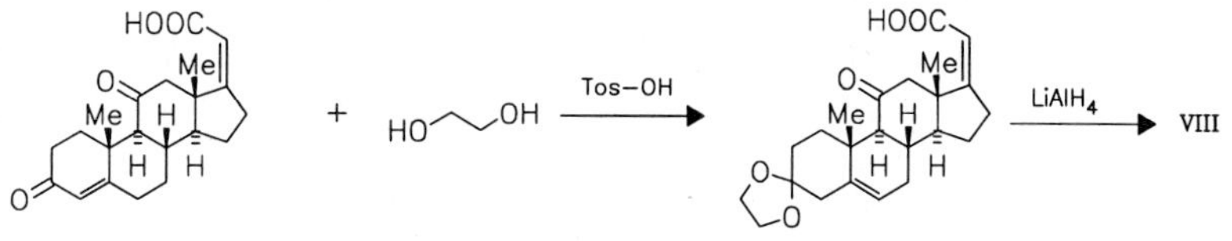

(VII)

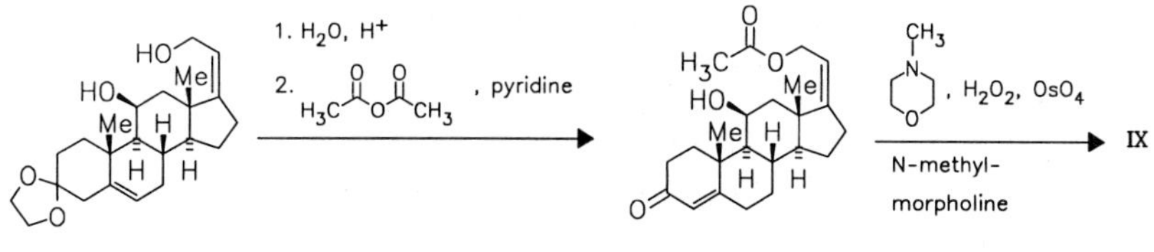

(VIII)

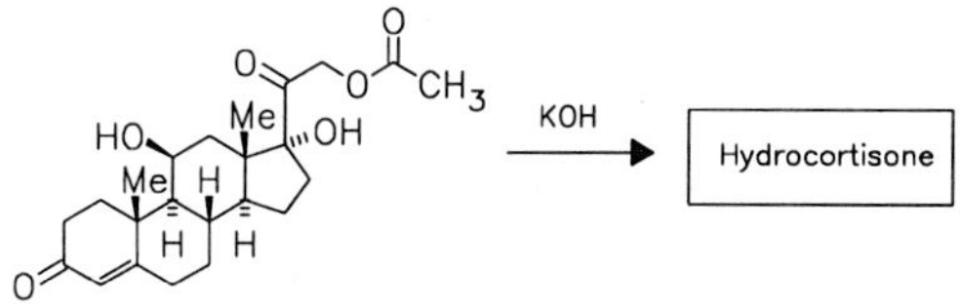

IX

hydrocortisone
21–acetate   (IX)

Reference(s):

a   US 2 649 401 (Upjohn; 1953; appl. 1950).
    US 2 658 023 (Pfizer; 1953; appl. 1952).
    US 2 794 816 (Upjohn; 1957 appl. 1954).
    synthesis of cortexolon:
    Julian, P.L.: J. Am. Chem. Soc. (JACSAT) **72**, 5145 (1950).
    Sondheimer, F. et al.: J. Am. Chem. Soc. (JACSAT) **78**, 816 (1956).
    The Merck Index, 2891 (Rahway 1976).
b   Oliveto, E. et al.: J. Am. Chem. Soc. (JACSAT) **78**, 1736 (1956).
c   US 2 666 069 (American Cyanamid; 1954; appl. 1951).
    synthesis of starting material:
    US 2 622 081 (American Cyanamid; 1952; appl. 1951).
    US 2 700 666 (American Cyanamid; 1955; appl. 1953).
d   Hogg, J.A. et al.: J. Am. Chem. Soc. (JACSAT) **77**, 4436 (1955).
    US 2 769 823 (Upjohn; 1956; appl. 1954).

alternative syntheses:
US 2 541 104 (Merck & Co.; 1951; appl. 1947).
GB 800 797 (Pfizer; appl. 1956; USA-prior. 1955).
US 4 041 055 (Upjohn; 9.8.1977; appl. 17.11.1975).
DOS 2 803 660 (Schering AG; appl. 25.1.1978).
DOS 2 803 661 (Schering AG; appl. 25.1.1978).

review:
Fieser, L.F.; Fieser, M.: Steroide, 710, 737 (Weinheim 1961).
Ullmanns Encykl. Tech. Chem., 4. Aufl., Vol. **13**, 52.

pharmaceutical formulation:
DOS 2 606 516 (Dermal; appl. 18.2.1976; GB-prior. 19.2.1975).

Formulation(s):   cream 0.5 %; lotion 0.5 %: ointment 1 %, 2.5 %; tabl. 10 mg

Hydrocortisone acetate

ATC: A07EA02; D07AA02
Use: glucocorticoid

RN: 50-03-3   MF: $C_{23}H_{32}O_6$ MW: 404.50   EINECS: 200-004-4
LD$_{50}$:   2300 mg/kg (M, i.p.); 45.05 mg/kg (M, s.c.)
CN: (11β)-21-(acetyloxy)-11,17-dihydroxypregn-4-ene-3,20-dione

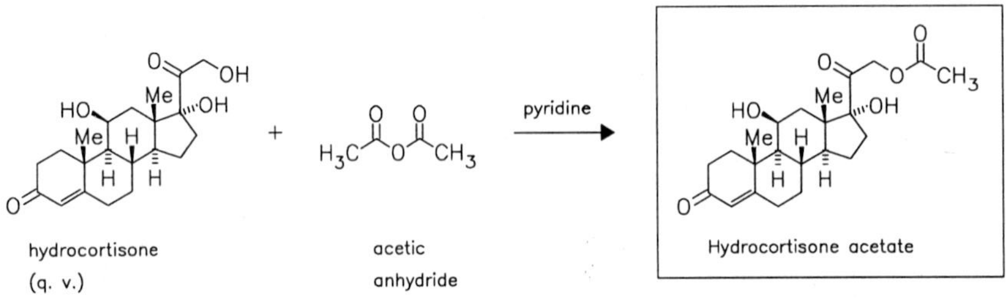

hydrocortisone   acetic   Hydrocortisone acetate
(q. v.)   anhydride

Reference(s):
US 2 183 589 (Roche-Organon; 1939; CH-prior. 1936).

alternative syntheses:
US 2 541 104 (Merck & Co.; 1951; appl. 1947).
US 2 769 823 (Upjohn; 1956; appl. 1954).

Formulation(s):   cream 3.3 mg; ointment (0.5 %, 1 %, 2 %); suppos. 3.3 mg

Colifoam (Stafford-Miller)
Fucidin H (Leo)-comb.
Neo-cortef (Dominion)-
comb.
Proctofoam (Stafford-
Miller)-comb.
Xyloproct (Astra)-comb.
numerous combination
preparations
I: Antiacne Samil (Samil)-
comb.
Antiemorroidale
Milanfarma (Milanfarma)-
comb.
Argisone (Teofarma)-comb.
Cortidro (Salus Research)
Cortinal (Teofarma)-comb.

Cortison-Chemicetina
(Carlo Erba)-comb.
Emorril (Poli)-comb.
Idrocet (Lusofarmaco)-
comb.
Idrocortisone Roussel
(Roussel)
Idroneomicil (Poli)-comb.
Lenirit (Bonomelli Farm.)
Mictasone (Zoja)-comb.
Proctosedyl (Roussel)-
comb.
Reumacort (Teofarma)-
comb.
Urecortyn (Roussel)
Vasosterone antib.
(Angelini)-comb.

Xyloproct (Byk Gulden)-
comb.
J: Dortizon Oint. (Kobayashi
Kako)
Hydrocortisone (Banyu)
Hydrocortone (Merck-
Banyu)
KC Oint. (Hokuriku)
Manosil (Sumitomo
Kagaku)
Otozon Base (Nakano)
Scheroson F (Nihon
Schering)
USA: Anusol-HC (Parke Davis)
Cortifoam (Schwarz)
Pramosone (Ferndale)
numerous combination
preparations and generic

Hydrocortisone 17-butyrate

ATC: D07AB02
Use: glucocorticoid

RN: 13609-67-1   MF: $C_{25}H_{36}O_6$ MW: 432.56   EINECS: 237-093-4
LD_{50}:   >3 g/kg (M, p.o.);
>3 g/kg (R, p.o.)
CN: (11β)-11,21-dihydroxy-17-(1-oxobutoxy)pregn-4-ene-3,20-dione

hydrocortisone
(q. v.)

N,N-dimethyl-
butyramide

Hydrocortisone 17-butyrate

Reference(s):
DAS 2 644 556 (Beiersdorf AG; appl. 2.10.1976).

alternative syntheses:
DAS 2 441 284 (Schering AG; appl. 16.9.1974).
JP 52 010 489 (Taisho; appl. 15.7.1975).
JP 52 136 157 (Taisho; appl. 14.4.1976).
JP 53 015 360 (Taisho; appl. 26.7.1976).
DOS 2 055 221 (Lab. Chimico Farma Untico; appl. 10.11.1970).
DOS 2 204 366 (Dermal; appl. 27.1.1962).

Formulation(s):   cream 1 mg/g; emulsion 1 mg/g; lotion 1 mg/g; ointment 1 mg/g

Trade Name(s):
D: Alfason (Yamanouchi)
Laticort (medphano)
F: Locoid (Yamanouchi
Pharma)

GB: Locoid (Yamanouchi)
I: Daktacort crema (Janssen)-
comb.
Locoidon (Brocades)

Molidex (Clintec)-comb.
Nasomixin (Pierrel)-comb.
J: Locoid (Torii)
USA: Locoid (Ferndale)

Hydrocortisone sodium phosphate

ATC: S01XA99
Use: glucocorticoid

RN: 6000-74-4   MF: $C_{21}H_{29}Na_2O_8P$ MW: 486.41   EINECS: 227-843-9
LD$_{50}$:   746 mg/kg (M, i.v.); 3950 mg/kg (M, p.o.);
632 mg/kg (R, i.v.); 6100 mg/kg (R, p.o.)
CN: (11β)-11,17-dihydroxy-21-(phosphonooxy)pregn-4-ene-3,20-dione disodium salt

free acid
RN: 3863-59-0   MF: $C_{21}H_{31}O_8P$ MW: 442.45   EINECS: 223-382-2

hydrocortisone
(q. v.)

methane-
sulfonyl
chloride

pyridine

NaI, DMF

(I)

1. H_3PO_4, $(H_5C_2)_3N$, H_3CCN
2. NaOH

Hydrocortisone sodium phosphate

Reference(s):
US 2 936 313 (Glaxo; 10.5.1960; appl. 18.11.1958; GB-prior. 19.11.1957).
US 2 932 657 (Merck & Co.; 12.4.1960; appl. 30.6.1957).

alternative syntheses:
US 2 870 177 (Merck & Co.; 20.1.1959; appl. 4.8.1954).
US 3 068 223 (Merck & Co.; 11.12.1962; appl. 18.11.1958; prior. 4.8.1954).
DE 1 134 075 (Merck AG; appl. 26.11.1959).

Formulation(s):   amp. 100 mg; drops 0.335 %

Trade Name(s):
D: Pantocrinale (Simons)-
comb.
GB: Efcortesol (Glaxo
Wellcome)

I: Idracemi coll. (Farmigea)
Idracemi eparina
(Farmigea)-comb.
J: Gleiton (Sankyo Zoki)

USA: Hydrocortone Phosphate
Inj. (Merck Sharp &
Dohme)

Hydroflumethiazide

ATC: C03AA02
Use: diuretic, antihypertensive

RN: 135-09-1   MF: $C_8H_8F_3N_3O_4S_2$ MW: 331.30   EINECS: 205-173-8
LD$_{50}$:   750 mg/kg (M, i.v.); >10 g/kg (M, p.o.)
CN: 3,4-dihydro-6-(trifluoromethyl)-2H-1,2,4-benzothiadiazine-7-sulfonamide 1,1-dioxide

The reaction scheme showing 3-trifluoromethyl-aniline reacting with ClSO₃H (chlorosulfonic acid) then NH₃ to form 4-amino-6-trifluoromethyl-benzene-1,3-disulfamide (I), followed by reaction with (CH₂O)n paraformaldehyde to form Hydroflumethiazide.

3-trifluoromethyl-aniline

4-amino-6-trifluoromethyl-benzene-1,3-disulfamide (I)

I + (CH₂O)ₙ → Hydroflumethiazide

paraform-aldehyde

Reference(s):
US 3 254 076 (Lovens Kemiske Fabrik; 31.5.1966; GB-prior. 13.8.1958).
Holdrege, C.T. et al.: J. Am. Chem. Soc. (JACSAT) **81**, 4807 (1959).

Formulation(s):   tabl. 25 mg, 50 mg

Trade Name(s):
F: Eusod (Leo)-comb.; wfm
      Leodrine (Leo); wfm
      Plurine (Leo)-comb. with
      KCl; wfm
GB: Aldactide 50 (Searle)-
      comb.

I: Hydrenox (Knoll)
      Diuritens (Biotrading)-
      comb.; wfm
      Rivosil (Benvegna); wfm
J: Di-Ademil (Squibb-Showa)
      Enjit (Meiji)

Robezon (Mitsu)
Rontyl (Leo-Sankyo)
USA: Diucardin (Wyeth-Ayerst)

Hydromorphone

ATC: N02AA03
Use: analgesic

RN: 466-99-9   MF: C₁₇H₁₉NO₃   MW: 285.34   EINECS: 207-383-5
LD₅₀: 104 mg/kg (M, i.v.)
CN: (5α)-4,5-epoxy-3-hydroxy-17-methylmorphinan-6-one

monohydrochloride
RN: 71-68-1   MF: C₁₇H₁₉NO₃ · HCl   MW: 321.80   EINECS: 200-762-6
LD₅₀: 55 mg/kg (M, i.v.)

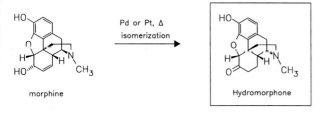

morphine   Pd or Pt, Δ
                  isomerization   Hydromorphone

Reference(s):
Ehrhart, Ruschig **I**, 120.
DRP 365 683 (Knoll; 1922).
DRP 607 931 (Knoll; 1935).
DRP 617 238 (Knoll; 1935).
DRP 623 821 (Knoll; 1935).

Formulation(s):   amp. 1 mg/ml, 2 mg/ml, 10 mg/ml, 50 mg/5 ml; tabl. 2 mg, 4 mg, 8 mg; vial 500 mg/5 ml (as hydrochloride)

Trade Name(s):

D:	Dilaudid (Knoll; as hydrochloride)	GB:	Palladone (Napp; as hydrochloride)		generics
	Dilaudid-Atropin (Knoll; as hydrochloride)-comb.	USA:	Dilaudid (Knoll; as hydrochloride)		

Hydroxocobalamin

(Aquocobalamin; Hydroxycobalamin; Vitamin B$_{12a}$)

ATC: V03AB33
Use: antipernicious vitamin (depot form: acetate)

RN: 13422-51-0   MF: C$_{62}$H$_{89}$CoN$_{13}$O$_{15}$P   MW: 1346.38   EINECS: 236-533-2
LD$_{50}$:   >50 mg/kg (M, i.v.)
CN: cobinamide dihydroxide dihydrogen phosphate (ester) mono(inner salt) 3'-ester with 5,6-dimethyl-1-α-D-ribofuranosyl-1*H*-benzimidazole

hydrate
RN: 13422-52-1   MF: C$_{62}$H$_{89}$CoN$_{13}$O$_{15}$P · H$_2$O   MW: 1364.39   EINECS: 236-534-8
acetate
RN: 22465-48-1   MF: C$_{64}$H$_{91}$CoN$_{13}$O$_{16}$P   MW: 1388.41   EINECS: 245-019-7
LD$_{50}$:   2 g/kg (M, i.v.)

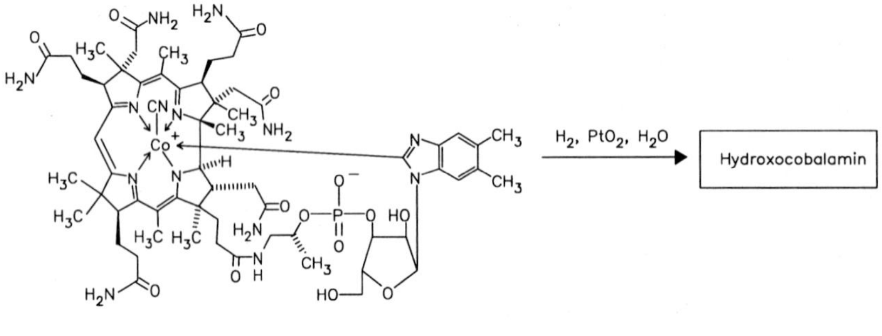

cyanocobalamin

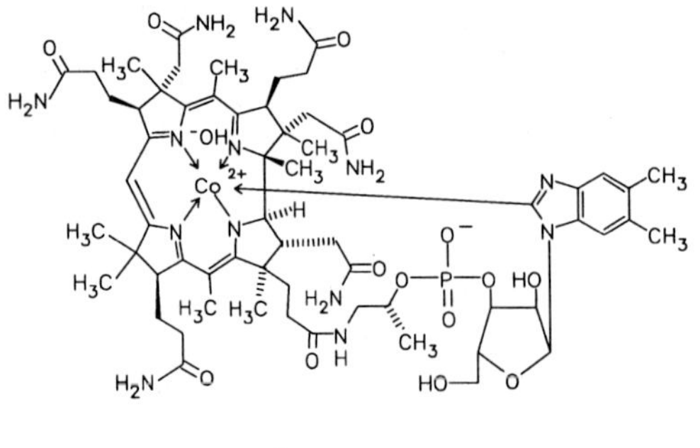

H$_2$, PtO$_2$, H$_2$O → Hydroxocobalamin

Hydroxocobalamin

Reference(s):
US 2 738 301 (Merck & Co.; 1956; appl. 1950).
US 2 738 302 (Merck & Co.; 1956; appl. 1950).

stabilized solutions:
FR 1 336 671 (Merck & Co.; appl. 28.2.1962; USA-prior. 13.3.1961).

Formulation(s):   amp. 0.5 mg/1 ml; cps. 460 mg; vial 5 mg, 10 mg (as hydrochloride)

Trade Name(s):

D: Aquo-Cytobion (Merck)
    Lophakomp-B12 Depot
    (Lomapharm)
    Novidroxin (Fatol; as
    acetate)
    numerous combination
    preparations
F: Arginotri-B (Bouchara)-
    comb.
    Dodécavit (L'Arguenon; as
    acetate)
    Hydroxo 5000 (Lipha Santé
    Division Aron-Médica)
    Inadrox (Logeais; as
    acetate)-comb.
    Néoparyl B12 (CIBA
    Vision Ophthalmics; as
    acetate)-comb.
    Terneurine H 5000 (Bristol-
    Myers Squibb; Labs.
    Allard)-comb.
    Vibalgan (Doms-Adrian; as
    acetate)-comb.
    generic and numerous
    combination preparations

GB: Cobalin-H (Link)
    Neo-Cytamen (Evans)
    numerous combination
    preparations
I: Idroxoc (Formulario Naz.)
    Idroxoc (Biologici Italia)
    Neocytamen (Teofarma)
    numerous combination
    preparations
J: Anemisol (Tobishi)
    Aquo B'av (Nippon Zoki)
    Bistin (Yamanouchi)
    B-Red S (Kyorin)
    B-Valet B_{12} (Tokyo Tanabe)
    Cobalamin H (Otsuka)
    Colsamine (Kanto)
    Dasvit H (Tanabe)
    Docelan (Nippon Roussel-
    Chugai)
    Dolevern (Seiko)
    Fresmin-S (Takeda; as
    acetate)
    Funacomin-F (Funai)
    Hicobala (Mitaka)
    Hicobalan (Maruko)

    Hydocobamin (Hishiyama)
    Hydocomin (Sanwa)
    Hydroxomin (Tokyo Hosei)
    Laseramin (Choseido)
    Masblon H (Fuso)
    Nichicoba (Nichiiko)
    OH-B_{12} (Morishita)
    Rasedon (Sawai)
    Red-B (Kowa)
    Redisol H (Merck-Banyu)
    Runova (Squibb-Sankyo)
    Solco H (Tobishi)
    Tsuerumin S (Mohan)
    Twelvmin (Mohan)
    Vigolatin (Kowa)
USA: Bevitamel (Westlake)-
    comb.
    Chromagen (Savage)-
    comb.
    Mega-B (Arco)-comb.
    numerous combination
    preparations

Hydroxycarbamide
(Hydroxyurea)

ATC: L01XX05
Use: antineoplastic

RN: 127-07-1   MF: $CH_4N_2O_2$ MW: 76.06   EINECS: 204-821-7
LD$_{50}$:   2350 mg/kg (M, i.v.); 7330 mg/kg (M, p.o.);
    4730 mg/kg (R, i.v.); 5760 mg/kg (R, p.o.);
    >1 g/kg (dog, i.v.); >2 g/kg (dog, p.o.)
CN: hydroxyurea

Reference(s):
US 2 705 727 (Du Pont; 1955; prior. 1952).

Formulation(s):   cps. 500 mg

Hydroxychloroquine

ATC:　P01BA02
Use:　antirheumatic, antimalarial

RN:　118-42-3　MF: C$_{18}$H$_{26}$ClN$_3$O　MW: 335.88　EINECS: 204-249-8
LD$_{50}$:　1240 mg/kg (M, p.o.)
CN:　2-[[4-[(7-chloro-4-quinolinyl)amino]pentyl]ethylamino]ethanol

sulfate (1:1)
RN:　747-36-4　MF: C$_{18}$H$_{26}$ClN$_3$O · H$_2$SO$_4$　MW: 433.96　EINECS: 212-019-3
phosphate (1:2)
RN:　6168-85-0　MF: C$_{18}$H$_{26}$ClN$_3$O · 2H$_3$PO$_4$　MW: 531.87

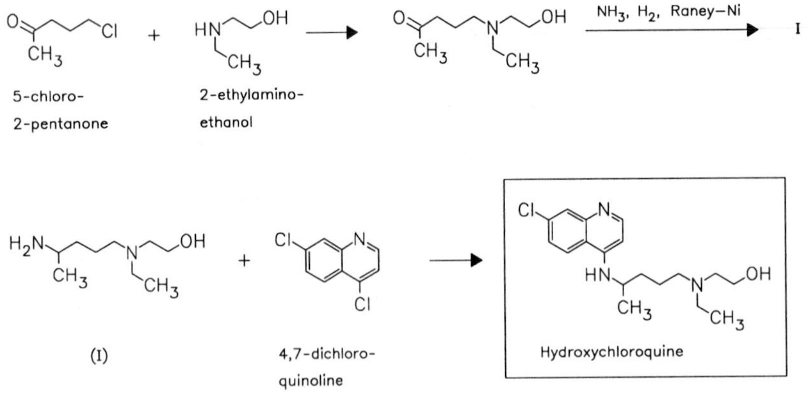

5-chloro-　　　　　2-ethylamino-
2-pentanone　　　ethanol

(I)　　　　　　　　4,7-dichloro-　　　　Hydroxychloroquine
　　　　　　　　　　quinoline

Reference(s):
US 2 546 658 (Sterling Drug; 1951; prior. 1949).

Formulation(s):　drg. 200 mg (as hydrochloride)

Hydroxyethyl salicylate
(Glycol salicylate)

ATC:　M02AC
Use:　anti-inflammatory, analgesic

RN:　87-28-5　MF: C$_9$H$_{10}$O$_4$　MW: 182.18　EINECS: 201-737-2
CN:　2-hydroxybenzoic acid 2-hydroxyethyl ester

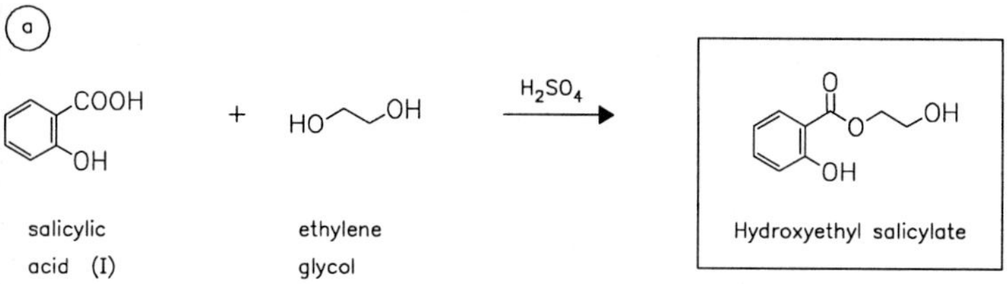

ⓐ

salicylic
acid (I)

+

ethylene
glycol

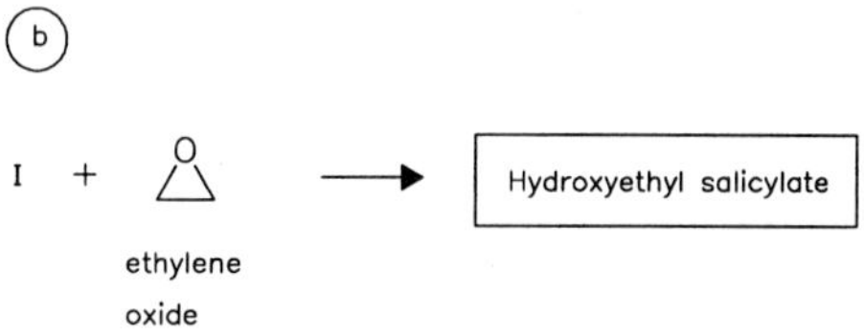

→ H₂SO₄

Hydroxyethyl salicylate

ⓑ

I + (ethylene oxide) → Hydroxyethyl salicylate

ethylene
oxide

Reference(s):

Kaufmann, H.P.: Arzneimittel-Synthese, Springer Verlag 1953, p. 79.
DD 218 616 (VEB Chem.-Pharmaz. Werk Oranienburg; appl. 18.4.1983).

Formulation(s):   cream 10.55 g/100 g, 12.5 g/100 g; gel 10.55 g/100 g, 12.5 g/100 g; ointment 10.55 g/100 g,
12.5 g/100 g

Trade Name(s):

D: Dolo-Arthrosonex
(Brenner-Efeka)
Kytta-Gel (Merck
Produkte)
Lumbinon (Lichtenstein)
Phlogont Salbe
(Azupharma)
Traumasenex (Brenner-
Efeka; LAW)

ca. 100 combination
preparations
GB: Cremalgin (Berk)-comb.;
wfm
Dubam (Norma)-comb.;
wfm
Salonair (Salonpas)-comb.;
wfm

I: combination preparations
only:
Balsamo Sifcamina (Midy)
Disalgil (Also)
Lasoreum Crema (Bayer)
Mobilisin (Luitpold)
Salonpas (Farmila)
Sloan balsamo (Parke
Davis)

Hydroxyprogesterone

ATC: G03DA03
Use: progestogen

RN: 68-96-2   MF: C₂₁H₃₀O₃   MW: 330.47   EINECS: 200-699-4
CN: 17-hydroxypregn-4-ene-3,20-dione

acetate
RN: 302-23-8   MF: C₂₃H₃₂O₄   MW: 372.51   EINECS: 206-119-6

ⓐ

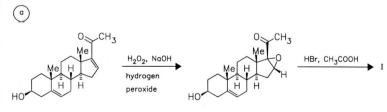

16-dehydropregnenolone
(cf. pregnenolone
synthesis)

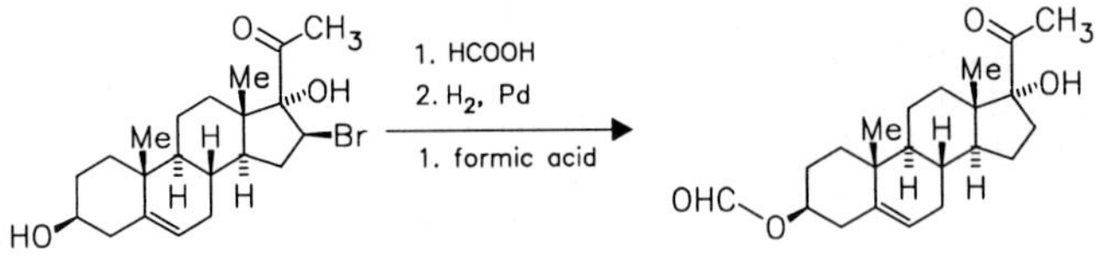

(I)   (II)

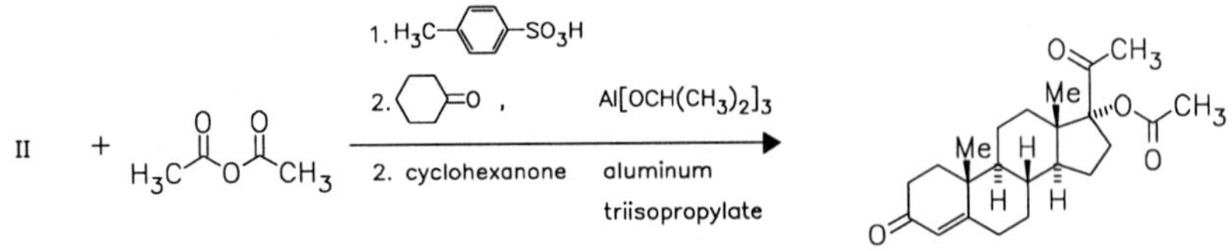

II + acetic anhydride

1. H₃C—⟨⟩—SO₃H
2. ⟨⟩=O ,   Al[OCH(CH₃)₂]₃
2. cyclohexanone   aluminum triisopropylate

Hydroxyprogesterone acetate

b

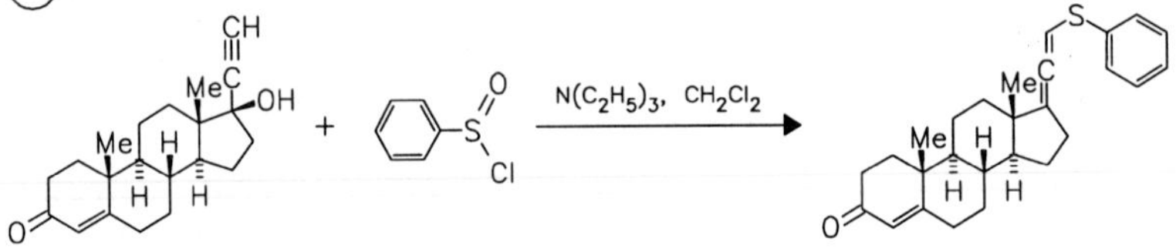

ethisterone (q. v.)   phenylsulfinyl chloride   N(C₂H₅)₃, CH₂Cl₂   (III)

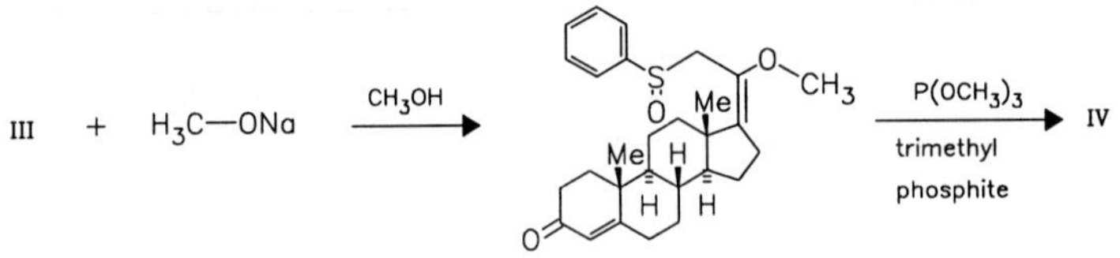

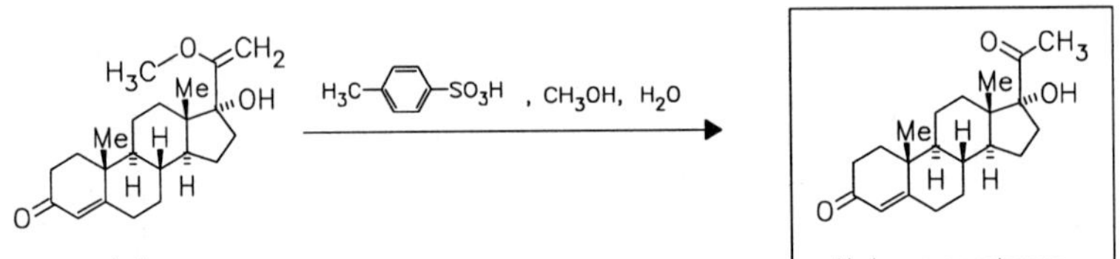

III + H₃C—ONa   CH₃OH   P(OCH₃)₃   IV
trimethyl phosphite

(IV)   H₃C—⟨⟩—SO₃H , CH₃OH, H₂O   Hydroxyprogesterone

Reference(s):
a Ringold, H.J. et al.: J. Am. Chem. Soc. (JACSAT) **78**, 816 (1956).
    US 2 802 839 (Syntex; 1957; appl. 1953; MEX-prior. 1953).
b US 4 041 055 (Upjohn; 9.8.1977; appl. 17.11.1975).

alternative syntheses:
US 2 648 662 (Glidden; 1953; appl. 1949).
US 2 777 843 (Merck & Co.; 1957; appl. 1954).
US 2 786 857 (Merck & Co.; 1957; appl. 1954; prior. 1952).
US 2 813 060 (Upjohn; 1957; appl. 1955).
US 3 000 883 (Upjohn; 1961; appl. 1957).
Cutler, F.A. et al.: J. Org. Chem. (JOCEAH) **24**, 1629 (1959).

Formulation(s):   amp. 250 mg/ml

Trade Name(s):

F: Tocogestan (Théramex; as   GB: Proluton Depot (Schering;
      17α-heptanoate)-comb.   as hexanoate)
      Trophobolène (Théramex;   USA: Prodox (Upjohn; as
      as heptanoate)-comb.   acetate); wfm

Hydroxyprogesterone caproate

ATC: G03D
Use: depot progestogen

RN: 630-56-8   MF: $C_{27}H_{40}O_4$ MW: 428.61   EINECS: 211-138-8
CN: 17-[(1-oxohexyl)oxy]pregn-4-ene-3,20-dione

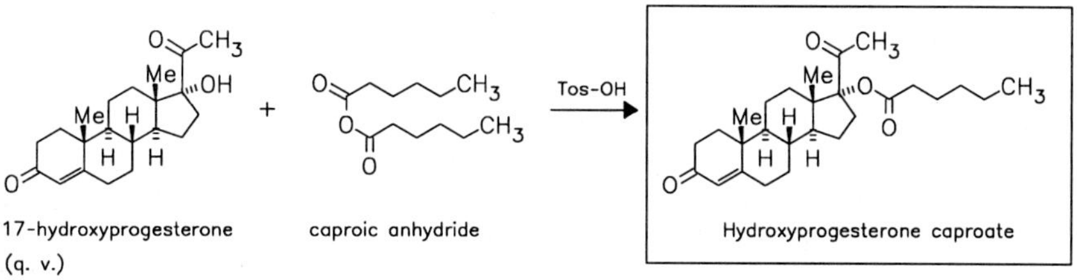

17-hydroxyprogesterone   caproic anhydride   Hydroxyprogesterone caproate
(q. v.)

Reference(s):
US 2 753 360 (Schering AG; 1956; D-prior. 1953).

alternative synthesis:
Babcock, J.C. et al.: J. Am. Chem. Soc. (JACSAT) **80**, 2904 (1958).

Formulation(s):   amp. 250 mg/ml, 500 mg/ml

Trade Name(s):

D: Gravibinon (Schering)-   GB: Primolut Depot (Schering   Oophormin Luteum Depot
      comb.   Chemicals); wfm   (Teikoku Zoki)
      Progesteron-Depot   I: Gravibinan (Schering)-   Proluton-Depot
      (Jenapharm)   comb.   (Nichidoku)
      Proluton Depot (Schering)   Lentogest (Amsa)   USA: Delalutin (Squibb); wfm
F: Progestérone-Retard-   Proluton Depot (Schering)   Deluteval (Squibb)-comb.;
      Pharlon (Schering)   J: Caprogen Depot (Kanto)   wfm
                                            Depot-Progen (Hokuriku)   Prodrox (Legere); wfm

Hydroxystilbamidine isethionate

ATC: P01C
Use: antiprotozoal (Leishmania)

RN: 533-22-2   MF: $C_{16}H_{16}N_4O \cdot 2C_2H_6O_4S$ MW: 532.60   EINECS: 208-557-3
CN: 4-[2-[4-(aminoiminomethyl)phenyl]ethenyl]-3-hydroxybenzenecarboximidamide compd. with 2-
       hydroxyethanesulfonic acid (1:2)

hydroxystilbamidine
RN: 495-99-8   MF: $C_{16}H_{16}N_4O$ MW: 280.33   EINECS: 207-811-0
LD_{50}: 27 mg/kg (M, i.v.)

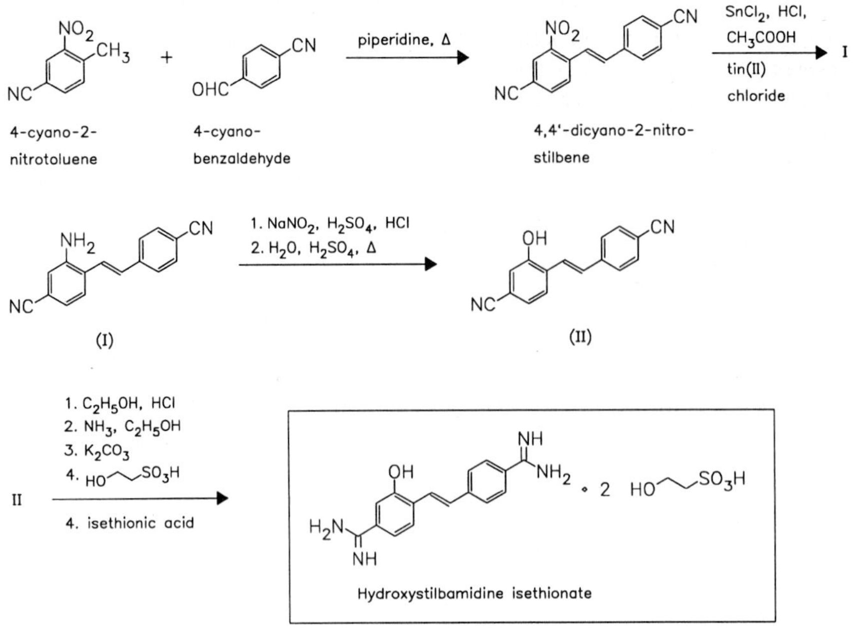

4-cyano-2-
nitrotoluene

4-cyano-
benzaldehyde

4,4'-dicyano-2-nitro-
stilbene

(I)

(II)

Hydroxystilbamidine isethionate

Reference(s):
US 2 510 047 (May & Baker; 1950; GB-prior. 1941).

Formulation(s):　vial 53.6 mg

Trade Name(s):
GB: Hydroxystilbamide (May & Baker); wfm

USA: Hydroxystilbamidine Isethionate (Merrell-National); wfm　　generic

---

Hydroxyzine

ATC:　N05BB01
Use:　tranquilizer

RN:　68-88-2　MF: $C_{21}H_{27}ClN_2O_2$　MW: 374.91　EINECS: 200-693-1
LD_{50}:　137 mg/kg (M, i.v.); 400 mg/kg (M, p.o.);
　　45 mg/kg (R, i.v.); 840 mg/kg (R, p.o.)
CN:　2-[2-[4-[(4-chlorophenyl)phenylmethyl]-1-piperazinyl]ethoxy]ethanol

dihydrochloride
RN:　2192-20-3　MF: $C_{21}H_{27}ClN_2O_2 \cdot 2HCl$　MW: 447.83　EINECS: 218-586-3
LD_{50}:　48.9 mg/kg (M, i.v.);
　　45 mg/kg (R, i.v.); 950 mg/kg (R, p.o.)
pamoate
RN:　10246-75-0　MF: $C_{23}H_{16}O_6 \cdot C_{21}H_{27}ClN_2O_2$　MW: 763.29　EINECS: 233-582-1

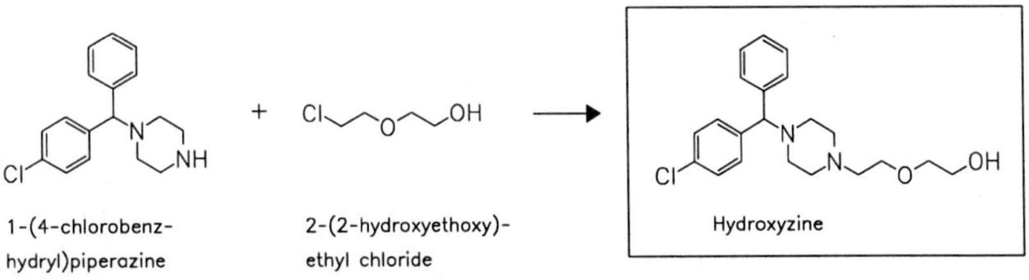

1-(4-chlorobenz-
hydryl)piperazine

2-(2-hydroxyethoxy)-
ethyl chloride

Hydroxyzine

Reference(s):
US 2 899 436 (UCB; 11.8.1959; B-prior. 30.10.1953).
DE 1 049 383 (UCB; appl. 1954; B-prior. 1953).
DE 1 061 786 (UCB; appl. 1954; B-prior. 1953).
DE 1 068 262 (UCB; appl. 1954; B-prior. 1953).
DE 1 072 624 (UCB; appl. 1954; B-prior. 1953).
DE 1 075 116 (UCB; appl. 1954; B-prior. 1953).

Formulation(s):   f. c. tabl. 10 mg, 25 mg; inj. sol. 100 mg/2 ml; syrup 10 mg/5 ml; tabl. 10 mg, 25 mg, 100 mg
(as dihydrochloride)

Trade Name(s):

D:	AH3 (Rodleben)-comb.	F:	Atarax (UCB; as	USA:	Atarax (Pfizer; as

D: AH3 (Rodleben)-comb.
     Atarax (Rodleben; UCB;
     Vedim; as hydrochloride)
     Beta-Intensain (Cassella)-
     comb.
     Diligan (Rodleben; Vedim;
     as hydrochloride)-comb.
     Elroquil (Rodleben; as
     hydrochloride)

F: Atarax (UCB; as
     dihydrochloride)
GB: Atarax (Pfizer)
     Ucerax (UCB)
I: Atarax (UCB)
J: Atarax (Lederle-Pfizer
     Taito; as hydrochloride)
     Atarax P (Pfizer Taito; as
     pamoate)

USA: Atarax (Pfizer; as
     hydrochloride)
     Marax (Pfizer; as
     hydrochloride)
     Vistaril (Pfizer; as
     hydrochloride)
     Vistaril (Pfizer; as pamoate)
     generic

Hymecromone

ATC: A05AX02
Use: choleretic

RN: 90-33-5   MF: $C_{10}H_8O_3$ MW: 176.17   EINECS: 201-986-7
LD_{50}:   2850 mg/kg (M, p.o.);
        15 mg/kg (R, i.p.); 3850 mg/kg (R, p.o.)
CN: 7-hydroxy-4-methyl-2H-1-benzopyran-2-one

resorcinol   ethyl
              acetoacetate

Hymecromone

Reference(s):
Pechmann, H. v.; Duisberg, C.: Ber. Dtsch. Chem. Ges. (BDCGAS) 16, 2119 (1883).
FR-M 1 430 (Lipha; appl. 13.7.1961).
Woods, L.L.; Sapp, J.: J. Org. Chem. (JOCEAH) 27, 3703 (1962).

Formulation(s):   amp. 200 mg; cps. 200 mg, 400 mg; tabl. 400 mg (as sodium salt)

Trade Name(s):

D: Cholspasmin (Lipha)
     Cholspasmoletten
     (Dolorgiet)
     Gallo Merz Spasmo (Merz
     & Co.)

F: Cantabiline (Lipha Santé
     Division Aron-Médica)
I: Cantabilin (Formenti)
J: Croamon (Torii)
     Crodimon (Roussel)

     Cumarote CD (Towa)
     Himecol (Kissei)
     Himecromon (Sawai)
     Paroamin (Zensei)
     generic

Ibandronate sodium monohydrate

(BM-21.0955; Ibandronic acid monosodium salt)

ATC: M05BA06
Use: bone resorption inhibitor

RN: 138926-19-9   MF: $C_9H_{22}NNaO_7P_2 \cdot H_2O$ MW: 359.23
CN: [1-hydroxy-3-(methylpentylamino)propylidene]bisphosphonic acid monosodium salt monohydrate

anhydrous
RN: 138844-81-2   MF: $C_9H_{22}NNaO_7P_2$ MW: 341.21
free acid
RN: 114084-78-5   MF: $C_9H_{23}NO_7P_2$ MW: 319.23

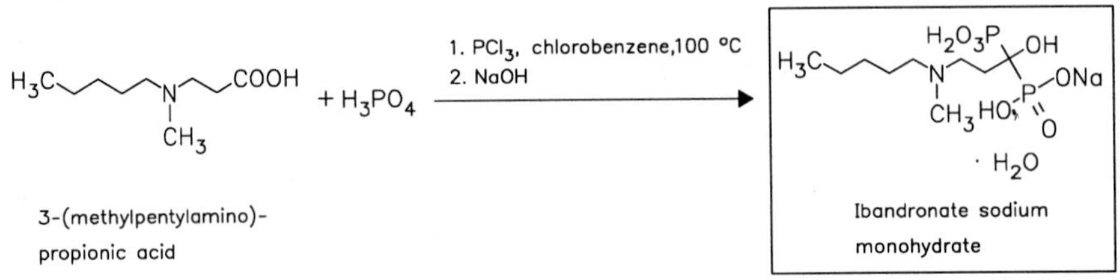

3-(methylpentylamino)-
propionic acid

Ibandronate sodium
monohydrate

Reference(s):
EP 252 504 (Boehringer Mannh.; appl. 9.7.1987; D-prior. 11.7.1986).

topical preparation:
EP 407 344 (Ciba-Geigy; appl. 28.6.1990; CH-prior. 7.7.1989).

oral formulation:
EP 566 535 (Ciba-Geigy; appl. 6.4.1993; CH-prior. 15.4.1992).

stable injection solution:
DE 4 228 552 (Boehringer Mannh.; appl. 27.8.1992; D-prior. 27.8.1992).

drymix formulation:
WO 9 412 200 (Merck & Co.; appl. 17.11.1993; USA-prior. 2.12.1992).

treatment of osteoporosis:
US 5 366 965 (Boehringer Mannh.; appl. 19.1.1993; USA-prior. 29.1.1993).

combination with growth hormone secretagones:
WO 9 511 029 (Merck & Co.; appl. 18.10.1994; USA-prior. 19.10.1993).

Formulation(s):   amp. 1 mg/ml, 2 mg/ml

Trade Name(s):
D: Bondronat (Boehringer
      Mannh.)

Ibopamine

ATC: C01CA16
Use: cardiotonic

RN: 66195-31-1   MF: $C_{17}H_{25}NO_4$ MW: 307.39   EINECS: 266-229-5
CN: 2-methylpropanoic acid 4-[2-(methylamino)ethyl]-1,2-phenylene ester

epinine   isobutyryl chloride   Ibopamine

Reference(s):
US 4 218 470 (Hal. Med. Sint. Sim; 19.8.1980; appl. 28.7.1977; I-prior. 5.8.1976).
US 4 302 471 (Hal. Med. Sint. Sim; 19.8.1980; appl. 28.7.1977; I-prior. 5.8.1976).
DOS 2 734 678 (Simes; appl. 1.8.1977; J-prior. 5.8.1976).
Casagrande, C. et al.: Arzneim.-Forsch. (ARZNAD) **36** (I), 291 (1986).

Formulation(s):   tabl. 50 mg, 100 mg, 200 mg (as hydrochloride)

Trade Name(s):
I: Inopamil (Astra-Simes;   Scandine (Zambon; 1984)
      1984)

Ibudilast

(KC-404)

ATC: R03DX04
Use: antiallergic, leukotriene antagonist

RN: 50847-11-5   MF: $C_{14}H_{18}N_2O$ MW: 230.31
LD_{50}:   146 mg/kg (M, i.v.); 1860 mg/kg (M, p.o.);
       42.5 mg/kg (R, i.v.); 1340 mg/kg (R, p.o.)
CN: 2-methyl-1-[2-(1-methylethyl)pyrazolo[1,5-*a*]pyridin-3-yl]-1-propanone

2-methyl-pyridine $H_2N-O-SO_3K$, HI
potassium
hydroxylamine
O-sulfonate

1-amino-2-methyl-pyridinium iodide

Ibudilast

Reference(s):
DE 2 315 801 (Kyorin; appl. 29.3.1973; J-prior. 30.3.1972).
US 3 850 941 (Kyorin; 26.11.1974; J-prior. 30.3.1972).

synthesis of 1-amino-2-methylpyridinium iodide:
Gösl, R.; Meuwsen, A.: Chem. Ber. (CHBEAM) **92**, 2521 (1959).

medical use for treatment of rheumatism:
EP 215 438 (Kyorin; appl. 10.9.1986; J-prior. 14.9.1985).

medical use for treatment of bronchial asthma, allergic rhinitis, urticaria:
JP 9 167 516 (Kyorin; appl. 21.9.1984; prior. 14.3.1983).

inhalant:
EP 320 002 (Kyorin; appl. 9.12.1988; J-prior. 10.12.1987).

transdermal formulation:
EP 319 902 (Kyorin; appl. 6.12.1988; J-prior. 10.12.1987).

controlled-release formulation:
EP 156 243 (Kyorin; appl. 12.3.1985; J-prior. 14.3.1984).

synthesis of potassium hydroxylamine *O*-sulfonate:
Gösl, R.; Meuwsen, A.: Chem. Ber. (CHBEAM) **92**, 2521 (1959).

Formulation(s): cps. 10 mg

Trade Name(s):
J: Ketas (Kyorin; 1989)

Ibuprofen

ATC: G02CC01; M01AE01; M02AA13
Use: anti-inflammatory, antirheumatic

RN: 15687-27-1 MF: $C_{13}H_{18}O_2$ MW: 206.29 EINECS: 239-784-6
LD_{50}: 740 mg/kg (M, p.o.);
636 mg/kg (R, p.o.)
CN: α-methyl-4-(2-methylpropyl)benzeneacetic acid

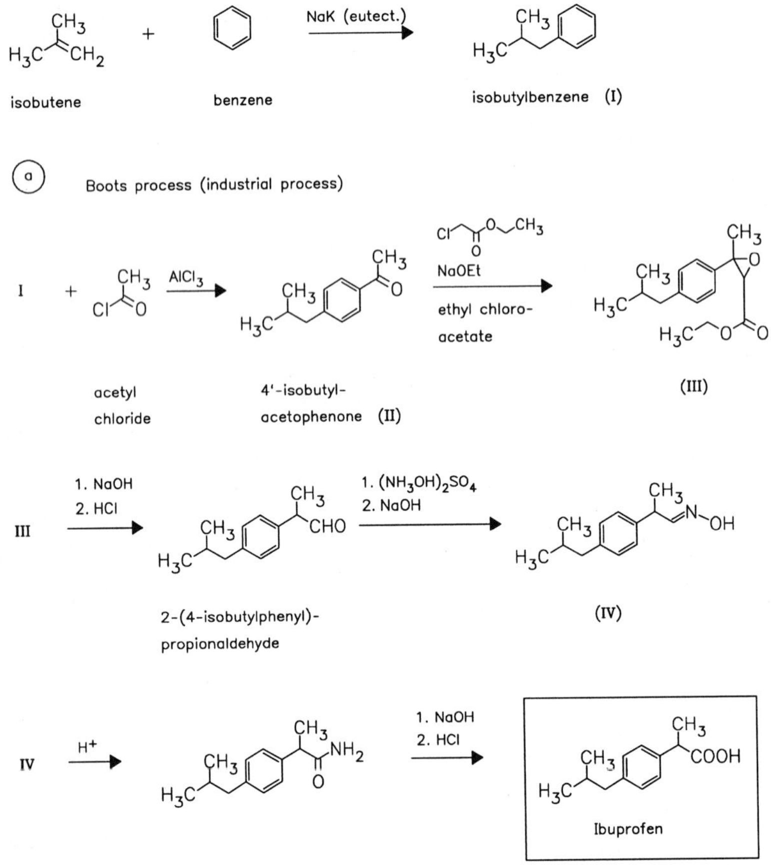
NaK (eutect.)
isobutene
benzene
isobutylbenzene (I)
a
Boots process (industrial process)
I
AlCl3
acetyl chloride
4'-isobutyl-acetophenone (II)
NaOEt
ethyl chloro-acetate
(III)
III
1. NaOH
2. HCl
2-(4-isobutylphenyl)-propionaldehyde
1. (NH3OH)2SO4
2. NaOH
(IV)
IV
H+
1. NaOH
2. HCl
Ibuprofen

b)

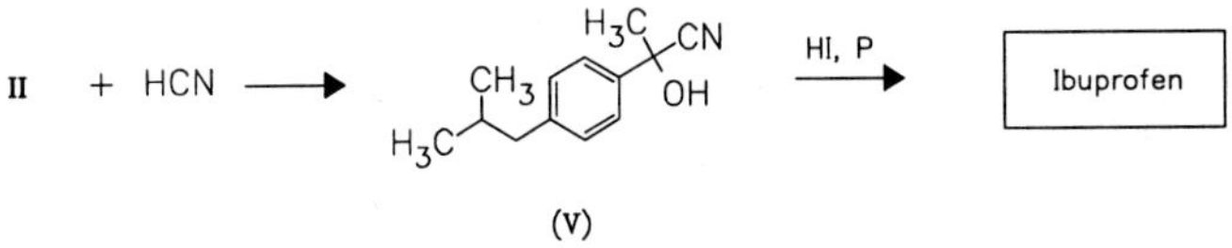

II + HCN ⟶ (structure V) $\xrightarrow{\text{HI, P}}$ [Ibuprofen]

(V)

c)

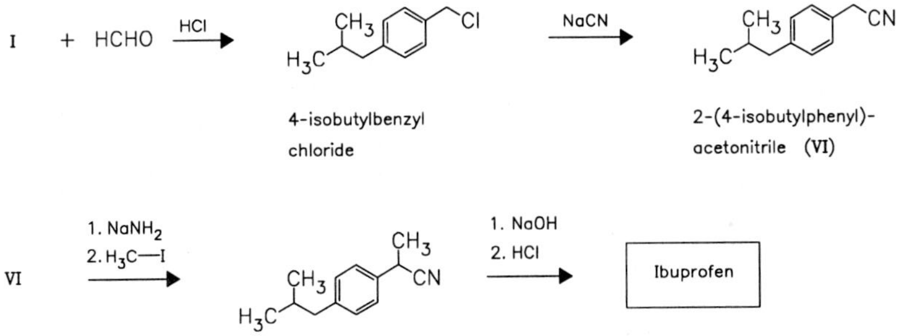

I + HCHO $\xrightarrow{\text{HCl}}$ 4-isobutylbenzyl chloride $\xrightarrow{\text{NaCN}}$ 2-(4-isobutylphenyl)-acetonitrile (VI)

VI $\xrightarrow[\text{2. H}_3\text{C—I}]{\text{1. NaNH}_2}$ (structure) $\xrightarrow[\text{2. HCl}]{\text{1. NaOH}}$ [Ibuprofen]

d) Ethyl process

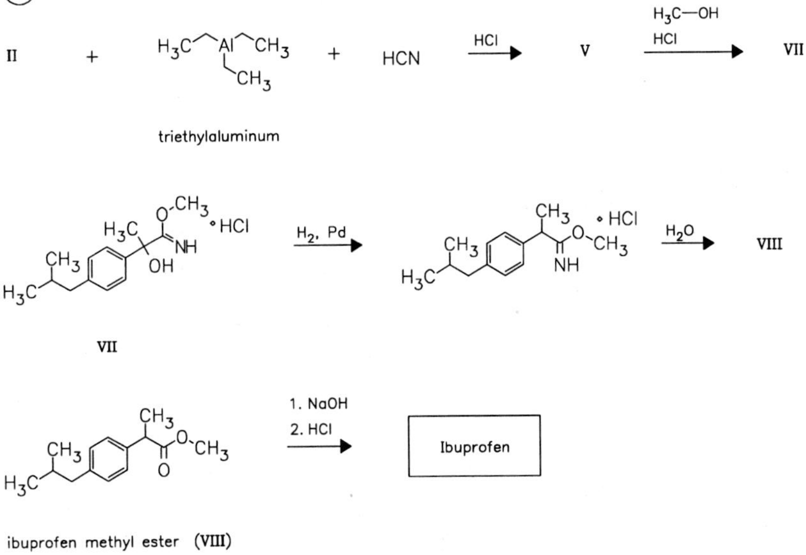

II + triethylaluminum + HCN $\xrightarrow{\text{HCl}}$ V $\xrightarrow[\text{HCl}]{\text{H}_3\text{C—OH}}$ VII

VII (structure · HCl) $\xrightarrow{\text{H}_2, \text{Pd}}$ (structure · HCl) $\xrightarrow{\text{H}_2\text{O}}$ VIII

ibuprofen methyl ester (VIII) $\xrightarrow[\text{2. HCl}]{\text{1. NaOH}}$ [Ibuprofen]

e) BHC (Boots–Hoechst–Celanese) process (industrial)

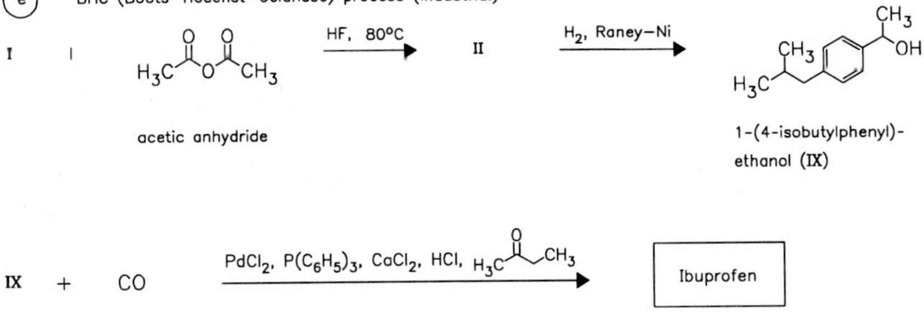

I acetic anhydride $\xrightarrow{\text{HF, 80°C}}$ II $\xrightarrow{\text{H}_2, \text{Raney–Ni}}$ 1-(4-isobutylphenyl)-ethanol (IX)

IX + CO $\xrightarrow{\text{PdCl}_2, \text{P(C}_6\text{H}_5)_3, \text{CaCl}_2, \text{HCl}, \text{H}_3\text{C-CO-CH}_3}$ [Ibuprofen]

Reference(s):
DE 1 443 429 (Boots; appl. 26.1.1962; GB-prior. 2.2.1961).
a,b GB 971700 (Boots; appl. 2.2.1961).
    US 3 228 831 (Boots; 11.1.1966; GB-prior. 2.2.1961).
    US 3 385 886 (Boots; 28.5.1968; GB-prior. 2.2.1961).
c   GB 1 514 812 (Boots; appl. 4.4.1975; valid from 31.3.1976).

similar method:
US 3 959 364 (Boots; 25.5.1976; GB-prior. 24.5.1973).

from 4'-isobutylacetophenone:
GB 1 160 725 (Boots; appl. 25.11.1966; valid from 20.11.1967).
US 4 021 478 (Upjohn; 3.5.1977; prior. 13.7.1972).

from 4'-isobutylpropiophenone *by oxidation with* thallium(III) nitrate:
GB 1 535 690 (Upjohn; appl. 20.5.1977; USA-prior. 16.6.1976).
Walker, J.A.; Pillai, M.D.: Tetrahedron Lett. (TELEAY) **42**, 3707 (1977).

from vinyl isobutyl ketone *and* diethyl 2-acetyl-3-methylsuccinate:
GB 1 265 800 (Boots; appl. 5.11.1968 and 15.11.1968; valid from 16.10.1969).
DOS 2 719 304 (Upjohn; appl. 29.4.1977; USA-prior. 24.5.1976, 15.3.1977).
DOS 2 806 424 (Upjohn; appl. 15.2.1978; USA-prior. 17.3.1977).
US 4 096 177 (L. Baiocchi; 20.6.1978; I-prior. 11.4.1974).

from 1-(4-isobutylphenyl)ethyl chloride *via the Grignard compound:*
DOS 2 605 650 (Ind. Chim. Prodotti Francis; appl. 12.2.1976; I-prior. 22.5.1975).

alternative syntheses:
DOS 2 404 159 (Nisshin Flour Milling; appl. 29.1.1974; J-prior. 29.1.1973).
DOS 2 646 792 (Mitsubishi Petrochemical; appl. 16.10.1976; J-prior. 23.10.1975, 31.7.1976).
DAS 2 709 504 (Sagami; appl. 4.3.1977; J-prior. 4.3.1976, 27.12.1976).
DOS 2 724 702 (Valles Chimica; appl. 1.6.1977; E-prior. 2.6.1976).
US 4 016 196 (Nisshin Flour Milling; 5.4.1977; J-prior. 27.7.1974, 29.7.1974).
US 4 131 747 (Ono Pharmaceutical; 26.12.1978; J-prior. 19.11.1975).
BE 859 846 (Sagami; appl. 27.12.1976; J-prior. 18.10.1976).
DOS 2 824 856 (Upjohn; appl. 6.6.1978; USA-prior. 16.6.1977).

(S)-ibuprofen:
Cleij, M. et al.: J. Org. Chem. (JOCEAH) **64**, 5029-5035 (1999)

Formulation(s):   amp. 400 mg; drg. 200 mg, 400 mg; eff. gran. 200 mg; f. c. tabl. 200 mg, 400 mg, 600 mg; s. r. tabl. 800 mg; suppos. 600 mg; syrup 100 mg/5 ml

Trade Name(s):

D:	Aktren (Bayer Vital)	Ibuhexal (Hexal)	Antalfène (Bouchara Santé
	Anco (Kanoldt)	Ibumerck (Merck | Active)
	Brufen (Kanoldt; 1971)	Generika)	Brufen (Lab. Knoll; 1972)
	Dignoflex (Sankyo)	Imbun (Merckle)	Dolgit (Lab. Merck-
	Dolgit (Dolorgiet)	Jenaprofen (Jenapharm)	Clévenot)
	DOLO PUREN (Isis Puren)	Novogent (Temmler)	Ergix (Murat)
	Dolormin (Woelm)	Opturem (Kade)	Gélufène (Lab. CPF)
	Duralbuprofen	Parsal (Brenner-Efeka)	Nurofen (Lab. Boots
	(durachemie)	Tempil (Temmler)	Healthcare)
	Esprenit (Hennig)	Togal (Togal)	Oralfène (Pierre Fabre
	Exneural (BASF Generics)	Urem (Kade)	Médicament)
	Ibu (AbZ-Pharma)	F: Advil (Whitehall)	Rhinadvil (Whitehall)-
	Ibu Beta (betapharm)	Algifène (Lab. Nicolas,	comb.
	Ibufug (Wolff)	Division de LRN SA)	

GB:	Brufen Retard (Knoll; 1969)	J:	Andran (Takata)		Landelun (Tsuruhara)

GB: Brufen Retard (Knoll; 1969)
Fenbid spansule (Goldshield)
Motrin (Pharmacia & Upjohn)
numerous generics

I: Algofen (Blue Cross)
Antagil (Janssen)
Brufen (Boots Italia; 1972)
Brufort (Lampugnani)
Dolocyl (Novartis)
Moment (Angelini)
Nurofen (Boots Italia)
generics

J: Andran (Takata)
Anflagen (Ohta)
Bluton (Morishita)
Brufanic (Teiyo; 1976)
Brufen (Kakenyaku)
Buburone (Towa)
Butylenin (Sanwa)
Daiprophen (Daito)
Donjust-B (Horita)
Epinal (Mitsubishi Yuka)
Epobron (Ono)
Eputes (Kobayashi Kako)
IB-100 (Hishiyama)
Ibuprocin (Nisshin)
Lamidon (Kowa)

Landelun (Tsuruhara)
Liptan (Kowa)
Manypren (Zensei)
Mynosedin (Toho Yakuhin)
Napacetin (Toyama)
Nobfelon (Toho)
Nobfen (Toho)
Nobgen (Kanebo)
Roidenin (Showa)
Sednafen (Taisho)
generic

USA: Motrin (McNeil)
Vicoprofen (Knoll)
generics

Ibuprofen lysinate

ATC: M01AE; M02AA
Use: anti-inflammatory, analgesic

RN: 57469-77-9   MF: $C_{13}H_{18}O_2 \cdot C_6H_{14}N_2O_2$ MW: 352.48   EINECS: 260-751-7
LD$_{50}$:   299 mg/kg (M, p.o.);
       841 mg/kg (R, p.o.)
CN: L-lysine mono[α-methyl-4-(2-methylpropyl)benzeneacetate]

ibuprofen (q. v.)   L-lysine   Ibuprofen lysinate

Reference(s):
DOS 2 419 317 (Neopharmed; appl. 22.4.1974; I-prior. 22.3.1974).
GB 1 497 044 (SpA Soc. Prodotti Antibiotici; appl. 6.3.1975; prior. 7.3.1974).

Formulation(s):   eff. tabl. 342 mg; f. c. tabl. 250 mg, 500 mg; gel 10 %; s. r. tabl. 800 mg; suppos. 500 mg; vial (lyo.) 400 mg

Trade Name(s):
D: Imbun (Merckle); wfm   I: Aciril (Delalande Isnardi)   Arfen (Lisapharma)

Ibuproxam

ATC: M01AE13; M02AA
Use: anti-inflammatory

RN: 53648-05-8   MF: $C_{13}H_{19}NO_2$ MW: 221.30   EINECS: 258-683-8
LD$_{50}$:   >2 g/kg (M, p.o.);
       >3 g/kg (R, p.o.)
CN: N-hydroxy-α-methyl-4-(2-methylpropyl)benzeneacetamide

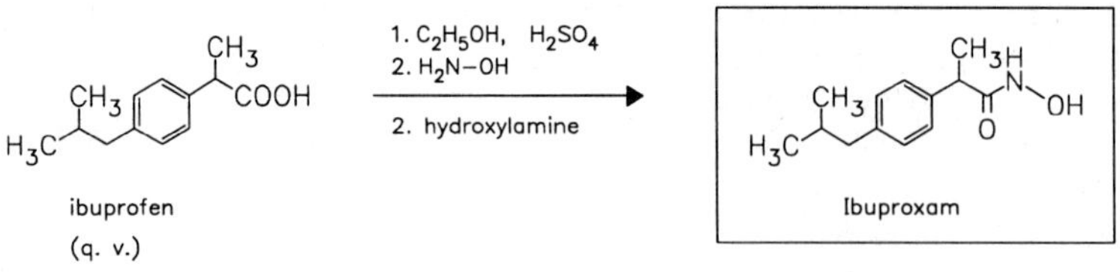

ibuprofen
(q. v.)

Ibuproxam

Reference(s):
DOS 2 400 531 (Manetti Roberts; appl. 7.1.1974; I-prior. 8.1.1973, 5.7.1973).

Formulation(s):   ointment 5 %; suppos. 600 mg; tabl. 300 mg

Trade Name(s):
I: Deflogon (Damor); wfm   Ibudros (Manetti Roberts)

Ibutilide fumarate

(U-70226E)

ATC: C01BD05
Use: antiarrhythmic

RN: 122647-32-9   MF: $C_{20}H_{36}N_2O_3S \cdot 1/2C_4H_4O_4$ MW: 885.24
LD$_{50}$: 94.2 mg/kg (R, i.v.)
CN: N-[4-[4-(ethylheptylamino)-1-hydroxybutyl]phenyl]methanesulfonamide (E)-2-butenedioate (2:1) (salt)

base
RN: 122647-31-8   MF: $C_{20}H_{36}N_2O_3S$ MW: 384.59

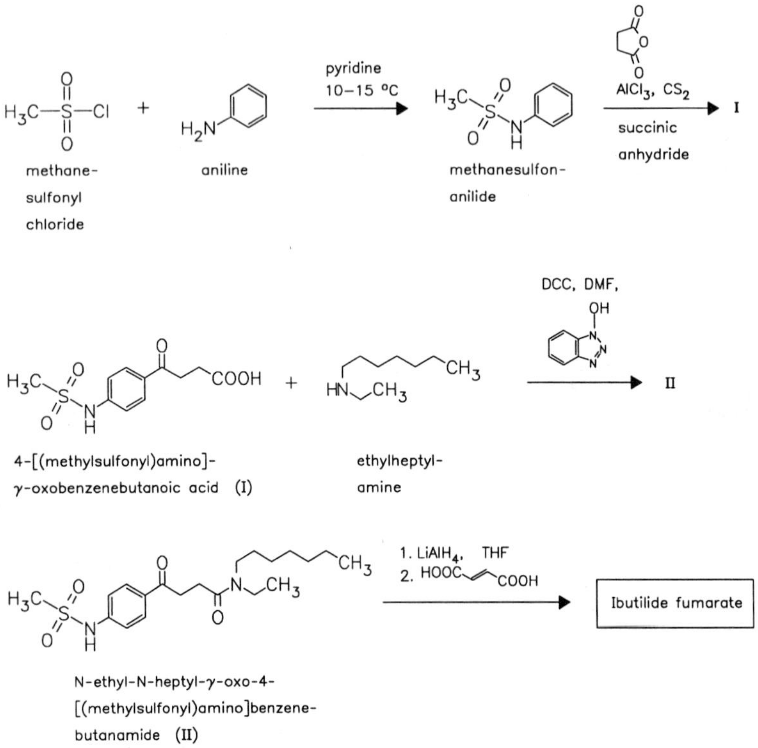

methane-
sulfonyl
chloride

aniline

methanesulfon-
anilide

4-[(methylsulfonyl)amino]-
γ-oxobenzenebutanoic acid   (I)

ethylheptyl-
amine

N-ethyl-N-heptyl-γ-oxo-4-
[(methylsulfonyl)amino]benzene-
butanamide   (II)

Ibutilide fumarate

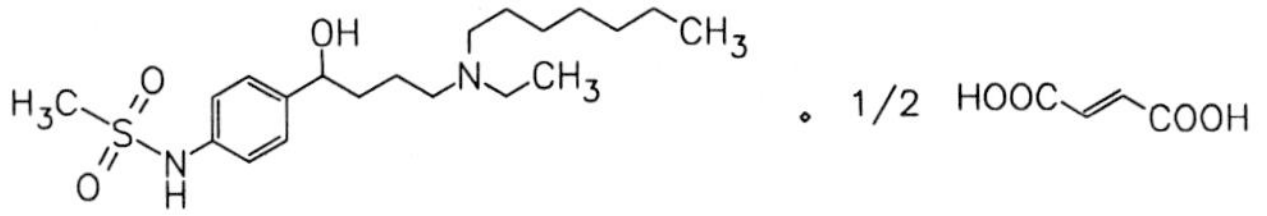

Ibutilide fumarate

Reference(s):
EP 164 865 (Upjohn; appl. 1.5.1985; USA-prior. 4.5.1984).
Hester, J.B. et al.: J. Med. Chem. (JMCMAR) **34** (1), 308 (1991).

controlled-release formulation:
WO 9 421 237 (Univ. Michigan; appl. 15.3.1994; USA-prior. 15.3.1993).

Formulation(s):   vial 1 mg/ml

Trade Name(s):
USA: Corvert (Pharmacia &
       Upjohn)

Idarubicin

ATC: L01DB06
Use: antineoplastic, anthracycline

RN: 58957-92-9   MF: $C_{26}H_{27}NO_9$ MW: 497.50
LD_{50}: 3 mg/kg (M, i.p.); 4 mg/kg (M, i.v.); 16 mg/kg (M, p.o.)
CN: (7S-cis)-9-acetyl-7-[(3-amino-2,3,6-trideoxy-α-L-lyxo-hexopyranosyl)oxy]-7,8,9,10-tetrahydro-6,9,11-trihydroxy-5,12-naphthacenedione

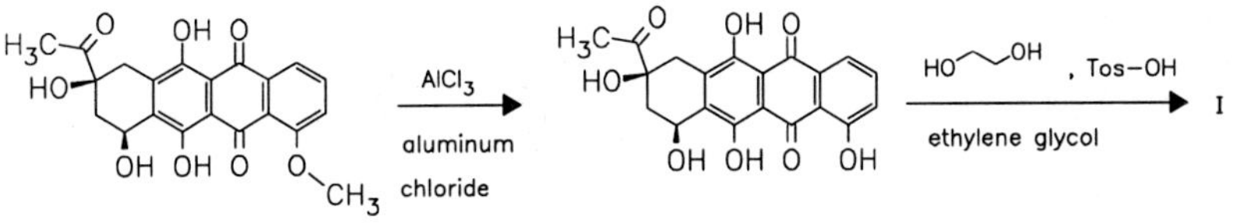

daunomycinone

4-O-demethyl-
daunomycinone

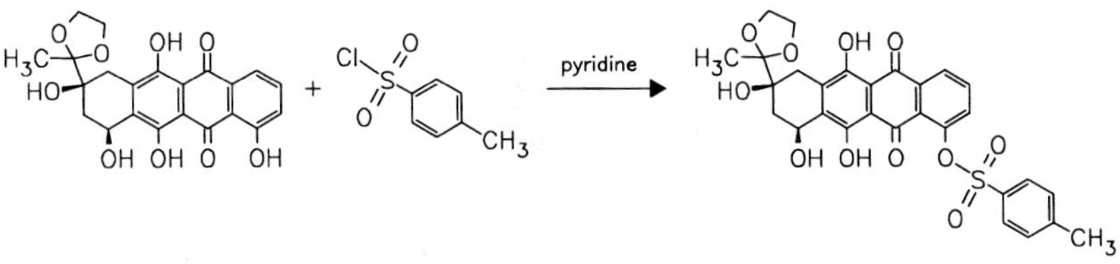

4-O-demethyldauno-
mycinone 1⁹-ethylene
acetal (I)

4-O-demethyl-4-O-(p-
toluenesulfonyl)dauno-
mycinone 1⁹-ethylene
acetal (II)

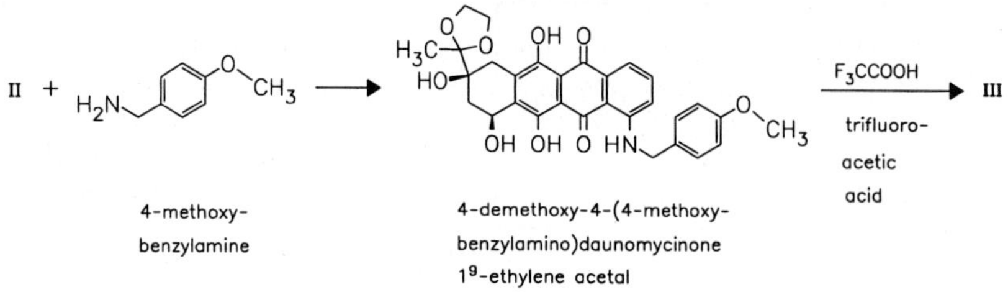

II + 4-methoxy-benzylamine → 4-demethoxy-4-(4-methoxy-benzylamino)daunomycinone 1⁹-ethylene acetal → (F₃CCOOH, trifluoro-acetic acid) → III

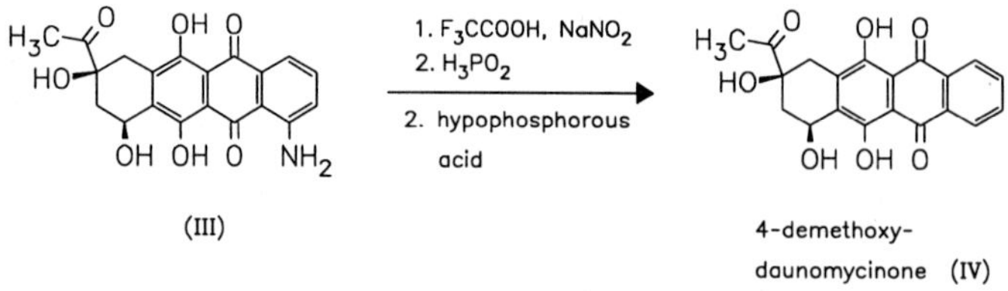

(III) — 1. F₃CCOOH, NaNO₂; 2. H₃PO₂; 2. hypophosphorous acid → 4-demethoxy-daunomycinone (IV)

a

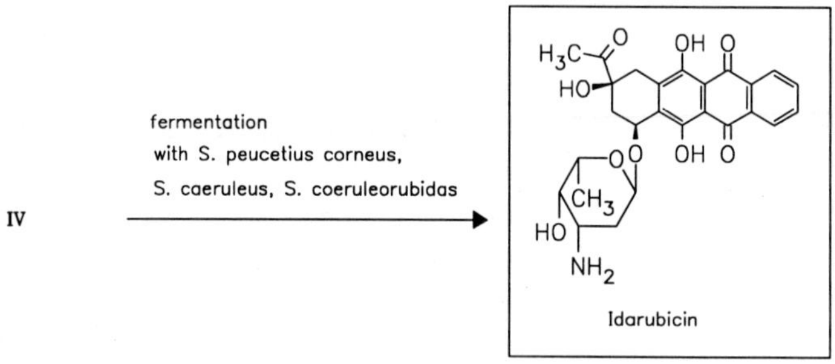

IV — fermentation with S. peucetius corneus, S. caeruleus, S. coeruleorubidas → Idarubicin

b

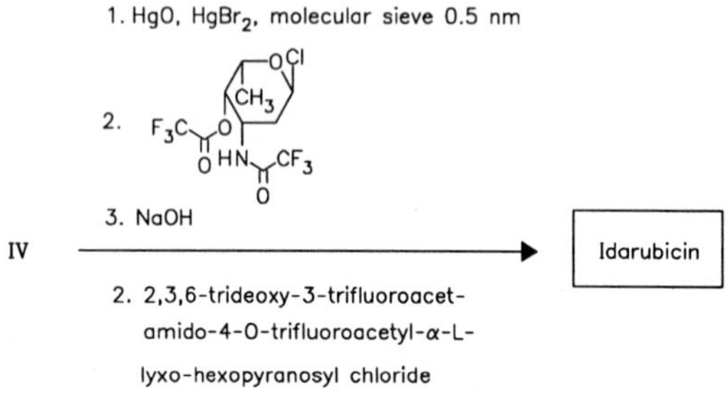

IV — 1. HgO, HgBr₂, molecular sieve 0.5 nm; 2. (F₃C...); 3. NaOH; 2. 2,3,6-trideoxy-3-trifluoroacet-amido-4-O-trifluoroacetyl-α-L-lyxo-hexopyranosyl chloride → Idarubicin

Reference(s):

a   US 4 471 052 (Adria; 9.11.1984; appl. 18.1.1982).
b   DOS 2 525 633 (Soc. Farmaceutici; appl. 6.9.1975; GB-prior. 16.12.1974).
    US 4 046 878 (Soc. Farmaceutici; 9.6.1977; appl. 22.5.1975; GB-prior. 6.12.1974).

alternative synthesis:
Arcamone, F. et al.: Experientia (EXPEAM) **34**, 1255 (1978).

synthesis of 4-demethoxydaunomycinone:
EP 328 399 (Farmitalia; appl. 2.10.1989; GB-prior. 2.12.1988).

lyophilisate:
GB 2 165 751 (Farmitalia; appl. 22.10.1984).

Formulation(s):   cps. 5 mg, 10 mg, 25 mg; vial 5 mg, 10 mg (as hydrochloride)

Idebenone

(CV-2619)

ATC: C01EB; N06BX13; N07X
Use: senile dementia therapeutic,
       coenzyme Q10 derivative, nootropic

RN: 58186-27-9   MF: $C_{19}H_{30}O_5$ MW: 338.44
LD_{50}: >10 g/kg (M, p.o.);
       10 g/kg (R, p.o.)
CN: 2-(10-hydroxydecyl)-5,6-dimethoxy-3-methyl-2,5-cyclohexadiene-1,4-dione

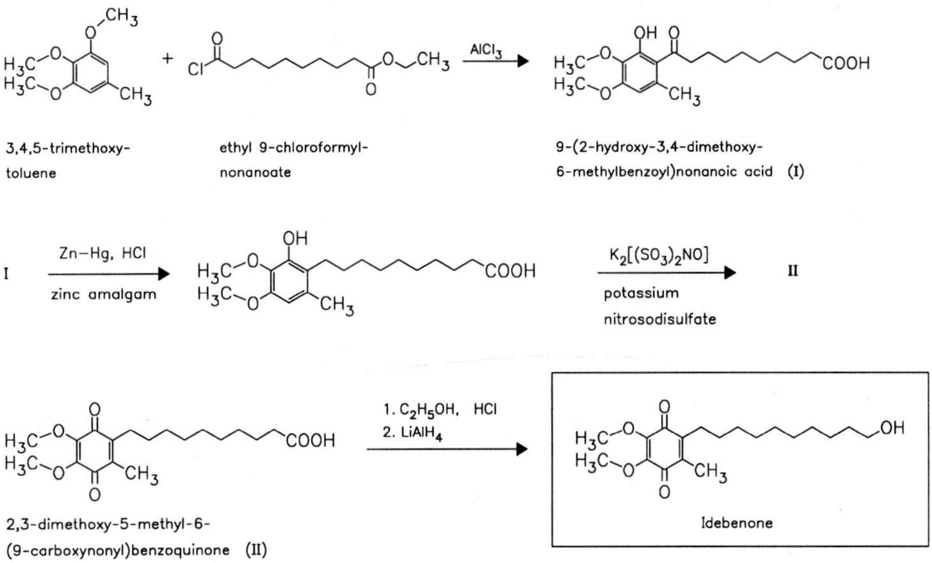

3,4,5-trimethoxy-toluene

ethyl 9-chloroformyl-nonanoate

9-(2-hydroxy-3,4-dimethoxy-6-methylbenzoyl)nonanoic acid (I)

2,3-dimethoxy-5-methyl-6-(9-carboxynonyl)benzoquinone (II)

Idebenone

Reference(s):

DOS 2 519 730 (Takeda; appl. 2.5.1975; J-prior. 2.5.1974).
JP 51 128 932 (Takeda; appl. 10.11.1976).
JP 59 039 855 (Takeda; appl. 27.8.1982).
US 4 139 545 (Takeda; 13.2.1979; J-prior. 2.5.1974).
US 4 271 083 (Takeda; 2.6.1981; J-prior. 2.5.1974).
US 4 525 361 (Takeda; 13.2.1979; appl. 3.4.1975; J-prior. 2.5.1974).

medical use for the therapy of ischemic disease:
EP 31 727 (Takeda, appl. 29.12.1980, J-prior. 30.12.1979).
DOS 3 049 039 (Takeda; appl. 24.12.1980; J-prior. 30.12.1979).

medical use for treatment of allergic disease:
EP 38 674 (Takeda; appl. 15.4.1981; J-prior. 21.4.1980).

medical use for treatment of fibrosis:
DOS 3 311 922 (Takeda; appl. 31.3.1983; J-prior. 6.4.1982).
Okamoto, K. et al.: Chem. Pharm. Bull. (CPBTAL) 33, 3745; 3756 (1985).
Goto, G. et al.: Chem. Pharm. Bull. (CPBTAL) 33, 4422 (1985).

alternative synthesis:
EP 21 841 (Takeda; appl. 27.6.1980; J-prior. 28.6.1979).
EP 58 057 (Takeda; appl. 4.2.1982; J-prior. 9.2.1981).

Formulation(s): drg. 30 mg, 45 mg; tabl. 30 mg

Trade Name(s):
I: Daruma (Cyanamid) Mnesis (Takeda) J: Avan (Takeda; 1987)

Idoxuridine

ATC: D06BB01; J05AB02; S01AD01
Use: chemotherapeutic (Herpes simplex)

RN: 54-42-2 MF: $C_9H_{11}IN_2O_5$ MW: 354.10 EINECS: 200-207-8
LD_{50}: 1 g/kg (M, i.p.);
4 g/kg (R, i.p.)
CN: 2'-deoxy-5-iodouridine

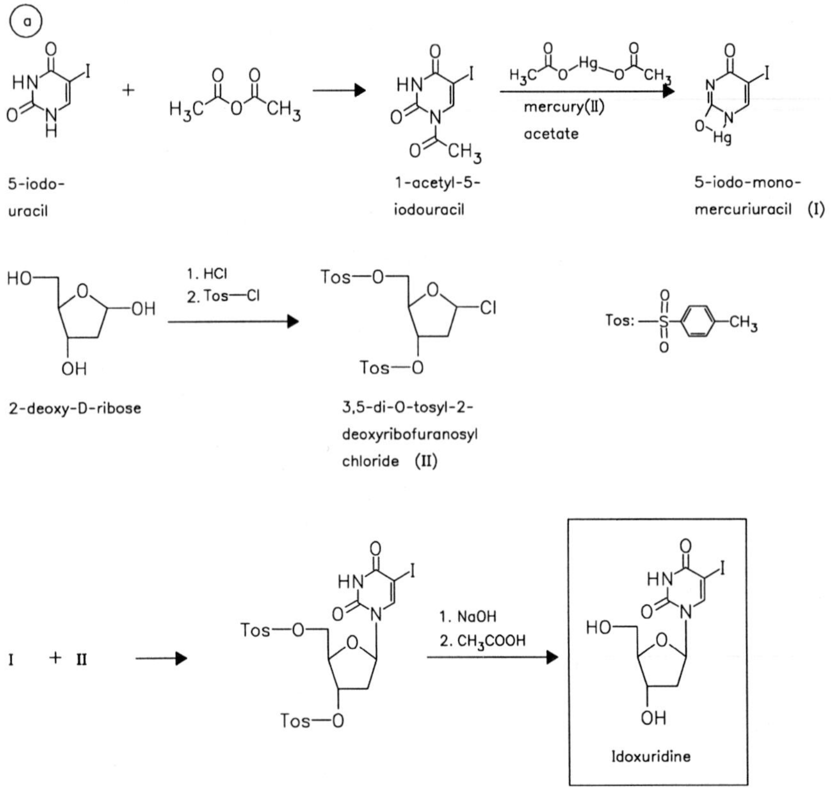
a
5-iodo-uracil
1-acetyl-5-iodouracil
mercury(II) acetate
5-iodo-mono-mercuriuracil (I)
2-deoxy-D-ribose
1. HCl
2. Tos—Cl
3,5-di-O-tosyl-2-deoxyribofuranosyl chloride (II)
Tos:
I + II
1. NaOH
2. CH3COOH
Idoxuridine

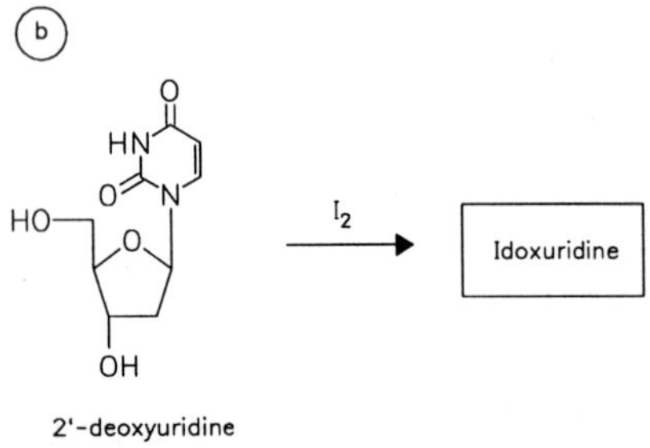
b
2'-deoxyuridine
I2
Idoxuridine

Reference(s):
Chang, P.K.; Welch, A.D.: J. Med. Chem. (JMCMAR) **6**, 428 (1963).
FR 1 336 866 (Roussel-Uclaf; appl. 27.7.1962).
GB 1 024 156 (Roussel-Uclaf; appl. 24.7.1963; F-prior. 27.7.1962).

Formulation(s):   eye drops 0.1 %; ointment 0.2 %, 0.5 %; sol. 5 %, 10 %, 40 %

Trade Name(s):

D:	Idugalen (Pharmagalen)		Iduridin (Ferring)	J:	I.D.U. (Sumitomo)
	Ophtal (Winzer)	I:	Iducher (Farmigea)	USA:	Dendrid (Alcon); wfm
	Virunguent (Hermal)		Iducol (SIFI)-comb.		Herplex (Allergan); wfm
	Zostrum (Galderma)		Iduridin (Geymonat)		Stoxil (Smith Kline &
F:	Iduviran (Chauvin)		Idustatin (Delalande		French); wfm
GB:	Herpid (Yamanouchi)		Isnardi)-comb.		

Ifenprodil

ATC: C04AX28
Use: cerebral and peripheral vasodilator

RN: 23210-56-2   MF: $C_{21}H_{27}NO_2$ MW: 325.45   EINECS: 245-491-4
LD_{50}:   320 mg/kg (M, p.o.)
CN: α-(4-hydroxyphenyl)-β-methyl-4-(phenylmethyl)-1-piperidineethanol

tartrate (2:1)
RN: 23210-58-4   MF: $C_{21}H_{27}NO_2 \cdot 1/2C_4H_6O_6$ MW: 800.99   EINECS: 245-493-5

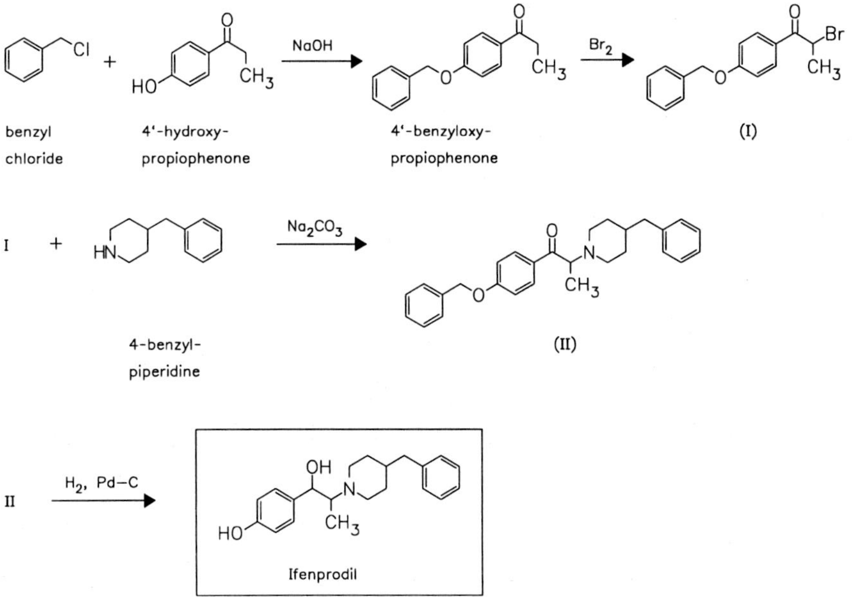

Reference(s):
DAS 1 695 772 (Lab. Robert et Carrière; appl. 7.9.1967; F-prior. 27.9.1966).
FR 5 733 M (Lab. Robert et Carrière; appl. 27.9.1966).
US 3 509 164 (Lab. Robert et Carrière; 28.4.1970; F-prior. 27.9.1966).

Formulation(s):   amp. 5 mg/2 ml; tabl. 20 mg (as tartrate)

Trade Name(s):
F: Vadilex (Synthélabo; 1972)   J: Cerocral (Funai; 1979)

Ifosfamide

ATC: L01AA06
Use: antineoplastic

RN: 3778-73-2   MF: $C_7H_{15}Cl_2N_2O_2P$ MW: 261.09   EINECS: 223-237-3
LD_{50}:   338 mg/kg (M, i.v.); 1005 mg/kg (M, p.o.);
        190 mg/kg (R, i.v.); 143 mg/kg (R, p.o.)
CN: N,3-bis(2-chloroethyl)tetrahydro-2H-1,3,2-oxazaphosphorin-2-amine 2-oxide

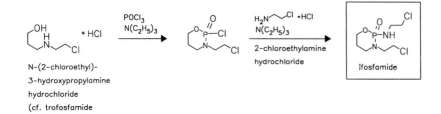

N-(2-chloroethyl)-
3-hydroxypropylamine
hydrochloride
(cf. trofosfamide
synthesis)

Reference(s):
DAS 1 645 921 (ASTA-Werke; appl. 11.7.1966).

Formulation(s):   vial 200 mg, 500 mg, 1 g, 2 g (dry powder)

Trade Name(s):
D: Holoxan (ASTA Medica   GB: Mitoxana (ASTA Medica)   USA: Ifex (Bristol-Myers
        AWD)   I: Holoxan (ASTA Medica)   Squibb)
F: Holoxan (ASTA Medica)   J: Ifomide (Shionogi)

Iloprost

(Ciloprost; E-1030; SH-401; ZK-36374)

ATC: B01AC11
Use: vasodilator, platelet aggregation
        inhibitor

RN: 78919-13-8   MF: $C_{22}H_{32}O_4$ MW: 360.49
CN: 5-[Hexahydro-5-hydroxy-4-(3-hydroxy-4-methyl-1-octen-6-ynyl)-2(1H)-pentalenylidene]pentanoic acid

(4R)
RN: 74843-30-4   MF: $C_{22}H_{32}O_4$ MW: 360.49
(4S)
RN: 74843-14-4   MF: $C_{22}H_{32}O_4$ MW: 360.49
trometamol salt
RN: 73873-87-7   MF: $C_{26}H_{43}NO_7$ MW: 481.63

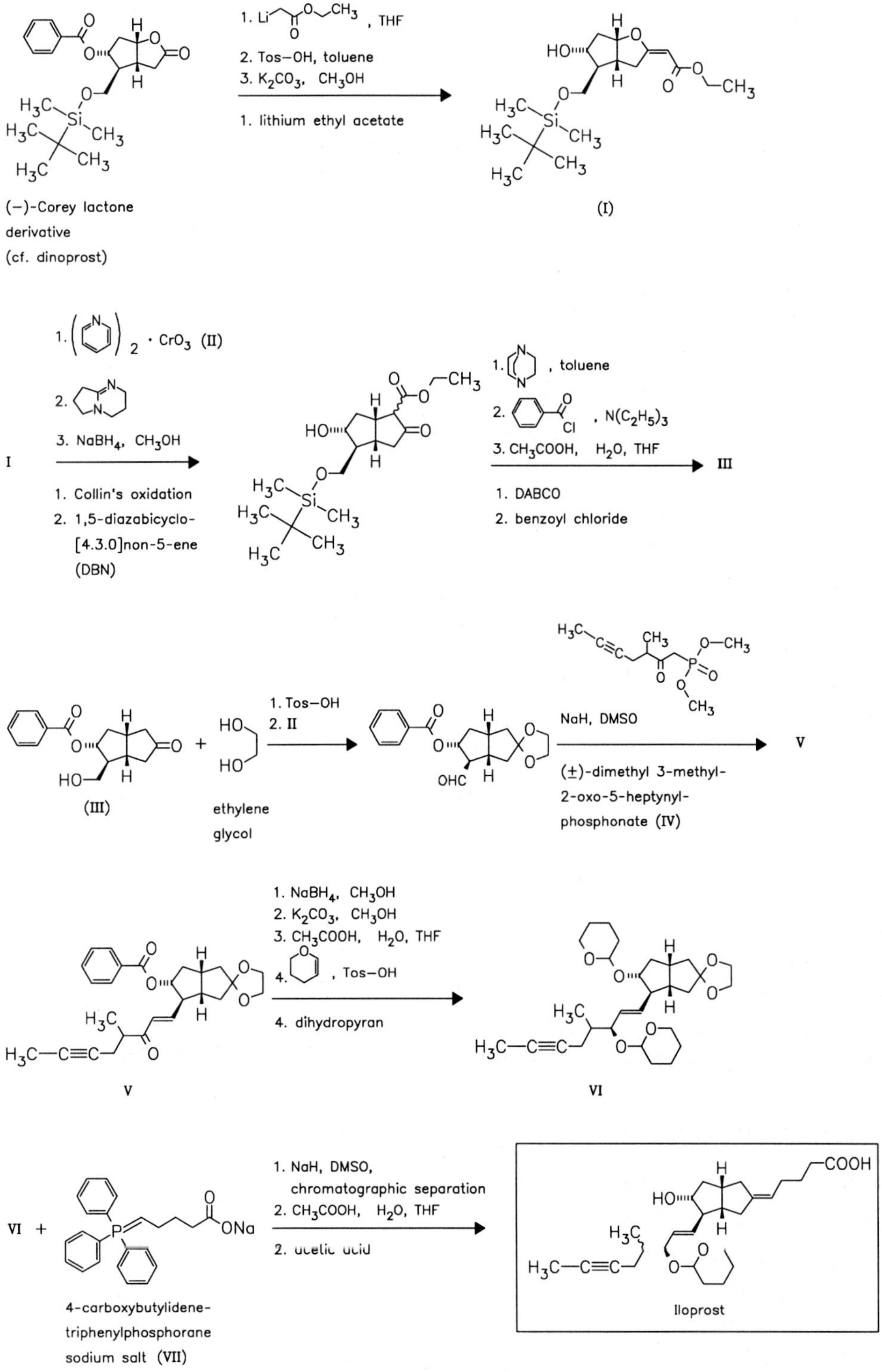

1. Li— O—CH₃ , THF
2. Tos—OH, toluene
3. K₂CO₃, CH₃OH

1. lithium ethyl acetate

(−)-Corey lactone derivative
(cf. dinoprost)

(I)

1. (pyridine)₂ · CrO₃ (II)
2. (DBN structure)
3. NaBH₄, CH₃OH

1. Collin's oxidation
2. 1,5-diazabicyclo-[4.3.0]non-5-ene (DBN)

I

1. (piperazine) , toluene
2. benzoyl chloride , N(C₂H₅)₃
3. CH₃COOH, H₂O, THF

1. DABCO
2. benzoyl chloride

III

H₃C—C≡C— CH₃ O—CH₃ P=O

CH₃

(III) + ethylene glycol

1. Tos—OH
2. II

NaH, DMSO

(±)-dimethyl 3-methyl-2-oxo-5-heptynyl-phosphonate (IV)

V

V

1. NaBH₄, CH₃OH
2. K₂CO₃, CH₃OH
3. CH₃COOH, H₂O, THF
4. (dihydropyran) , Tos—OH

4. dihydropyran

VI

VI +

4-carboxybutylidene-triphenylphosphorane sodium salt (VII)

1. NaH, DMSO, chromatographic separation
2. CH₃COOH, H₂O, THF

2. acetic acid

Iloprost

preparation of 4-carboxybutylidenetriphenylphosphorane sodium salt

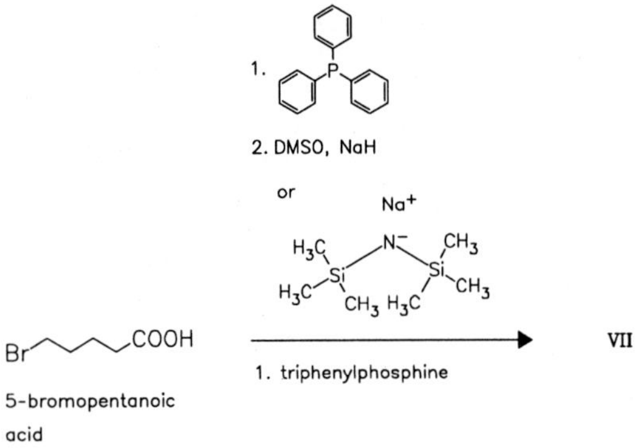

5-bromopentanoic
acid

preparation of (±)-dimethyl 3-methyl-2-oxo-5-heptynylphosphonate

1-bromo-2-butyne

diethyl
methylmalonate

(VIII)

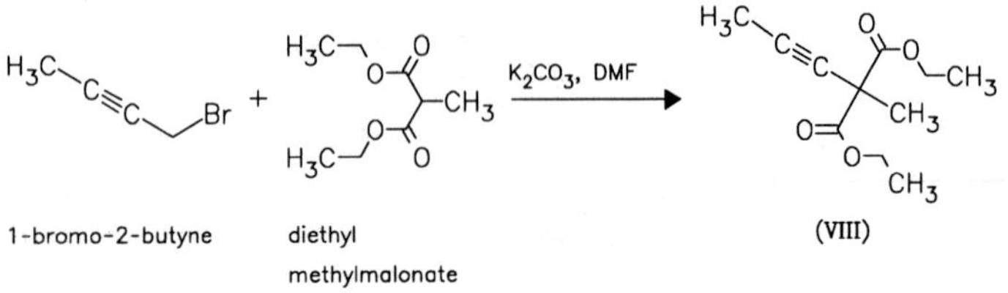

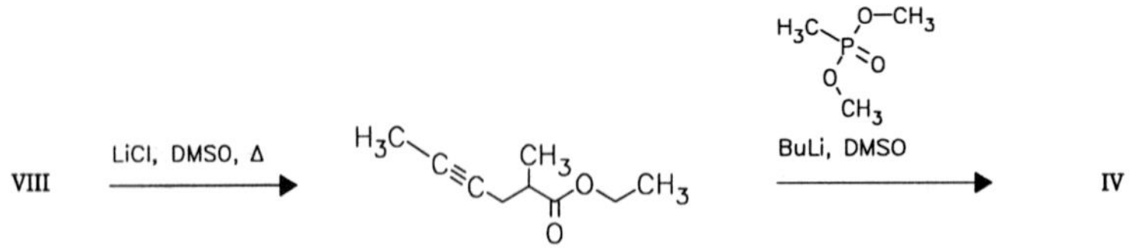

Reference(s):
DE 2 845 770 (Schering AG; D-prior. 19.10.1978)
DE 3 839 155 (Schering AG; D-prior. 17.11.1988)
Skuballa, W.; Vorbrueggen, H.: Angew. Chem. (ANCEAD) **93** (12), 1080 (1981)

preparation of 4-carboxybutylidenetriphenylphosphorane sodium salt:
Martinelli, M.J.: J. Org. Chem. (JOCEAH) **55** (17), 5065 (1990).
Johnson, F.P. et al.: J. Am. Chem. Soc. (JACSAT) **104** (8), 2190 (1982).
Niwa, H. et al.: Tetrahedron (TETRAB) **50** (25), 7385 (1994).

preparation of (±)-dimethyl 3-methyl-2-oxo-5-heptynylphosphonate:
DE 2 729 960 (Schering AG; prior. 30.6.1977).
footnote [6] in: Skuballa, W.; Vorbrueggen, H.: Angew. Chem. (ANCEAD) **93** (12), 1080 (1981).

(E)- or (Z)-selective Wittig reactions in the synthesis of carbacyclins:
Westermann, J.; Harre, M.; Hickisch, K.: Tetrahedron Lett. (TELEAY) **33** (52), 8055 (1992).

alternative synthesis of cis-bicyclo[3.3.0]octylidene derivative:
EP 153 822 (Sagami Chemical Research Center; J-prior. 10.2.1984).
Sodeoka, M.; Ogawa, Y.; Kirio, Y.; Shibasaki, M.: Chem. Pharm. Bull. (CPBTAL) **39** (2), 309 (1991)

Formulation(s):　amp. 50 μg/0.5 ml, 100 μg/1 ml (as trometamol salt)

Imidapril
(TA-6366)

ATC: C09A
Use: antihypertensive (ACE inhibitor)

RN: 89371-37-9   MF: $C_{20}H_{27}N_3O_6$ MW: 405.45
CN: [4S-[3[R*(R*)],4R*]]-3-[2-[[1-(ethoxycarbonyl)-3-phenylpropyl]amino]-1-oxopropyl]-1-methyl-2-oxo-4-imidazolidinecarboxylic acid

monohydrochloride
RN: 89396-94-1   MF: $C_{20}H_{27}N_3O_6 \cdot HCl$ MW: 441.91

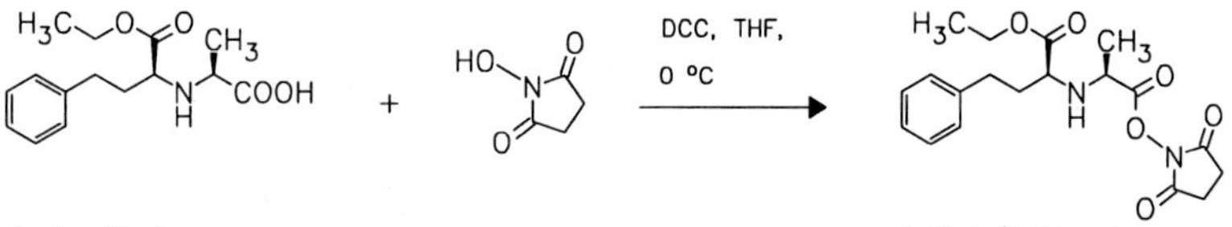

(2S)-2-[(1S)-1-ethoxy-
carbonyl-3-phenyl-
propylamino]propionic
acid

(2S)-2-[(1S)-1-ethoxy-
carbonyl-3-phenylpropyl-
amino]propionic acid
succinimido ester (I)

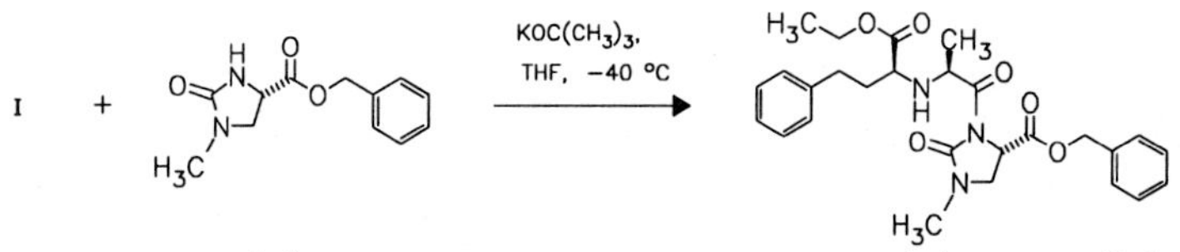

benzyl (4S)-1-methyl-
2-oxo-imidazolidine-
4-carboxylate

benzyl (4S)-1-methyl-3-[(2S)-
2-[(1S)-1-ethoxycarbonyl-3-phenyl-
propylamino]propionyl]-2-oxo-
imidazolidine-4-carboxylate (II)

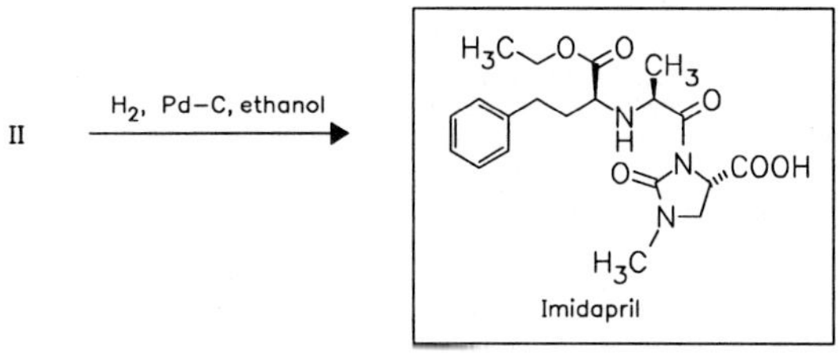

Imidapril

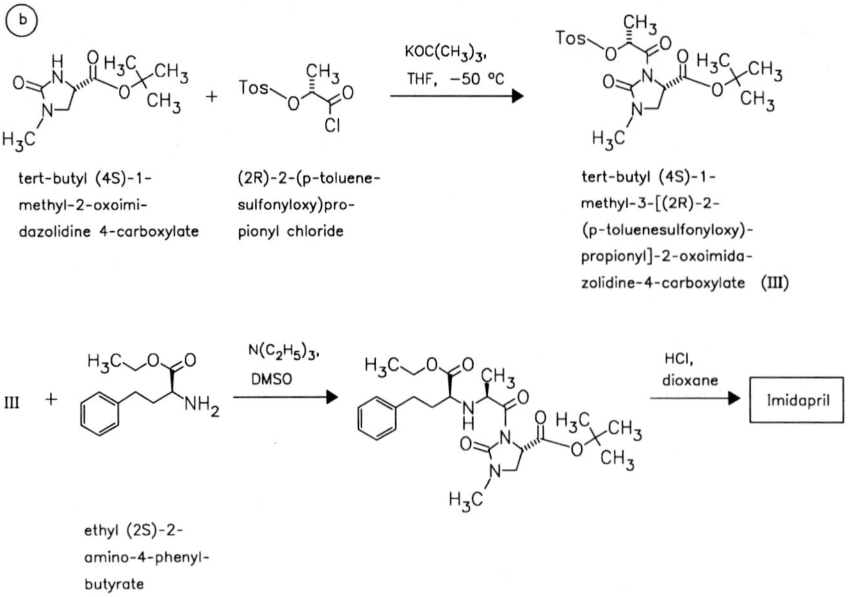

b
KOC(CH3)3,
THF, −50 °C
tert-butyl (4S)-1-methyl-2-oxoimidazolidine 4-carboxylate
(2R)-2-(p-toluenesulfonyloxy)propionyl chloride
tert-butyl (4S)-1-methyl-3-[(2R)-2-(p-toluenesulfonyloxy)-propionyl]-2-oxoimidazolidine-4-carboxylate (III)
III
N(C2H5)3,
DMSO
HCl,
dioxane
Imidapril
ethyl (2S)-2-amino-4-phenylbutyrate

Reference(s):

a EP 95 163 (Tanabe Seiyaku; appl. 20.5.1983; J-prior. 24.5.1982).
b EP 373 881 (Tanabe Seiyaku; appl. 12.12.1989; J-prior. 16.12.1988).

formulation with increased stability:
JP 06 100 447 (Tanabe Seiyaku; appl. 24.9.1992; J-prior. 24.9.1992).

composition for treatment of kidney diseases:
JP 272 849 (Tanabe Seiyaku; appl. 10.12.1987; J-prior. 12.12.1986).

composition for treatment of heart failure:
EP 274 230 (Tanabe Seiyaku; appl. 8.12.1987; J-prior. 9.12.1986).

Formulation(s): tabl. 2.5 mg, 5 mg, 10 mg (as hydrochloride)

Trade Name(s):
J: Novaroc (Nihon Schering) Tanatril (Tanabe Seiyaku)

Imipenem

(Imipemide)

ATC: J01DH51
Use: β-lactam antibiotic

RN: 64221-86-9 MF: $C_{12}H_{17}N_3O_4S$ MW: 299.35 EINECS: 264-734-5
LD_{50}: 1660 mg/kg (M, i.v.); >5 g/kg (M, p.o.);
1972 mg/kg (R, i.v.); >5 g/kg (R, p.o.)
CN: [5*R*-[5α,6α(*R**)]]-6-(1-hydroxyethyl)-3-[[2-[(iminomethyl)amino]ethyl]thio]-7-oxo-1-azabicyclo[3.2.0]hept-2-ene-2-carboxylic acid

monohydrate
RN: 74431-23-5 MF: $C_{12}H_{17}N_3O_4S \cdot H_2O$ MW: 317.37

thienamycin + methyl formimidate hydrochloride

pH 8.5 → Imipenem

Reference(s):
Leanza, W.J. et al.: J. Med. Chem. (JMCMAR) **22**, 1435 (1979).
US 4 194 047 (Merck & Co.; 18.3.1980; prior. 21.11.1975).
DOS 2 652 679 (Merck & Co.; appl. 19.11.1976; USA-prior. 21.11.1975).

production of thienamycin *(by fermentation of S. cattleya):*
US 3 950 357 (Merck & Co.; 13.4.1976; appl. 25.11.1974).
DOS 2 552 638 (Merck & Co.; appl. 24.11.1975; USA-prior. 25.11.1974).

combination with cilastatin:
EP 48 301 (Merck & Co.; appl. 24.9.1980).

Formulation(s):   amp. 250 mg, 500 mg; vial 200 mg, 500 mg

Trade Name(s):
D: Zienam (MSD; 1985)-   I: Imipem (Neopharmed)-   all combination
      comb. with cilastatin   comb.   preparations with cilastatin
F: Tiénam (Merck Sharp &   Tenacid (Sigma-Tau)-   J: Tienam (Banyu; 1987)-
      Dohme-Chibret)-comb.   comb.   comb. with cilastatin
GB: Primaxin IV (Merck Sharp   Tienam (MSD)-comb.   sodium
      & Dohme)   USA: Primaxin (Merck; 1985)

Imipramine

ATC: N06AA02
Use: antidepressant

RN: 50-49-7   MF: $C_{19}H_{24}N_2$ MW: 280.42   EINECS: 200-042-1
LD_{50}:   21 mg/kg (M, i.v.); 188 mg/kg (M, p.o.);
          9300 µg/kg (R, i.v.); 250 mg/kg (R, p.o.)
CN: 10,11-dihydro-*N,N*-dimethyl-5*H*-dibenz[*b,f*]azepine-5-propanamine

monohydrochloride
RN: 113-52-0   MF: $C_{19}H_{24}N_2 \cdot HCl$ MW: 316.88   EINECS: 204-030-7
LD_{50}:   27 mg/kg (M, i.v.); 275 mg/kg (M, p.o.);
          18 mg/kg (R, i.v.); 305 mg/kg (R, p.o.)

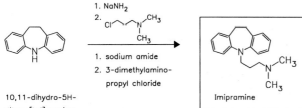

10,11-dihydro-5H-
dibenz[b,f]azepine

1. sodium amide
2. 3-dimethylamino-
   propyl chloride

Imipramine

Reference(s):
US 2 554 736 (Geigy; 1951; CH-prior. 1949).
DE 829 167 (Geigy; appl. 1950; CH-prior. 1949).

Formulation(s): amp. 25 mg/2 ml; cps. 75 mg, 100 mg, 125 mg, 150 mg; drg. 10 mg, 25 mg, 50 mg (as hydrochloride)

Trade Name(s):

D:　Imipramin (neuraxpharm; as hydrochloride)
　　Pryleugan (ASTA Medica AWD; as hydrochloride)
　　Tofranil (Novartis Pharma; as hydrochloride)
F:　Tofranil (Novartis; as hydrochloride)

GB:　Tofranil (Novartis; as hydrochloride)
I:　Impra C (Formulario Naz.)
　　Tofranil (Novartis)
　　combination preparations
J:　Depress (Toho)
　　Efuranol (Taito Pfizer)
　　Feinalmin (Sanko)

　　Imidol (Yoshitomi)
　　Imilanyle (Takata)
　　Meripramin (Kanebo Nakataki)
　　Tofranil (Novartis)
USA:　Tofranil (Novartis; as pamoate)
　　generic

Imiquimod

ATC:　D06BB10
Use:　immunomodulator, interferon alfa inducer, antiviral

RN:　99011-02-6　MF: $C_{14}H_{16}N_4$　MW: 240.31
CN:　1-(2-Methylpropyl)-1H-imidazo[4,5-c]quinolin-4-amine

hydrochloride
RN:　99011-78-6　MF: $C_{14}H_{16}N_4 \cdot HCl$　MW: 276.77

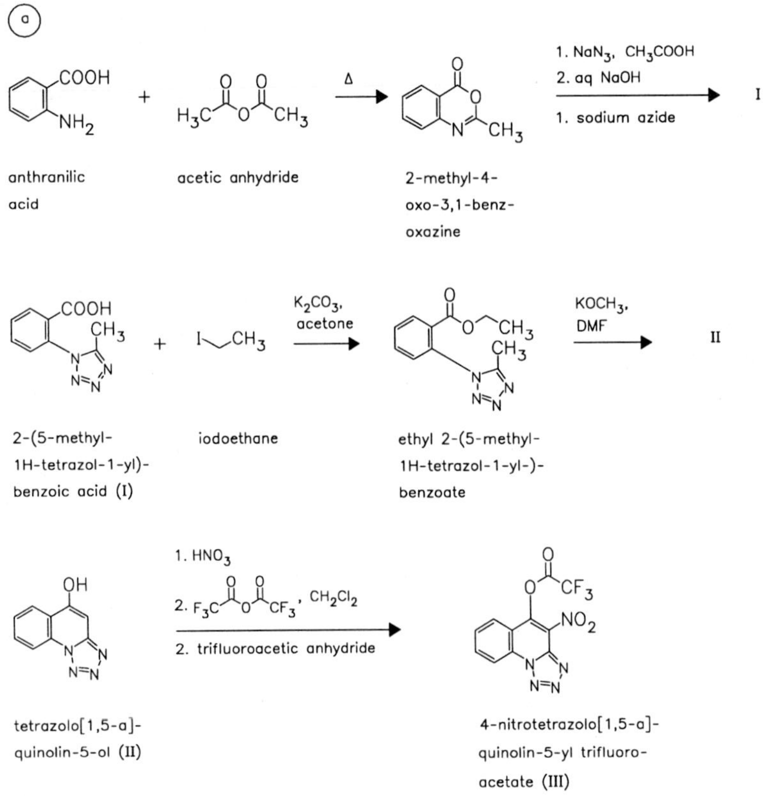

anthranilic acid
acetic anhydride
2-methyl-4-oxo-3,1-benz-oxazine

1. NaN3, CH3COOH
2. aq NaOH
1. sodium azide
I

2-(5-methyl-1H-tetrazol-1-yl)-benzoic acid (I)
iodoethane
K2CO3, acetone
ethyl 2-(5-methyl-1H-tetrazol-1-yl-)-benzoate
KOCH3, DMF
II

tetrazolo[1,5-a]-quinolin-5-ol (II)
1. HNO3
2. F3C—O—CF3, CH2Cl2
2. trifluoroacetic anhydride
4-nitrotetrazolo[1,5-a]-quinolin-5-yl trifluoro-acetate (III)

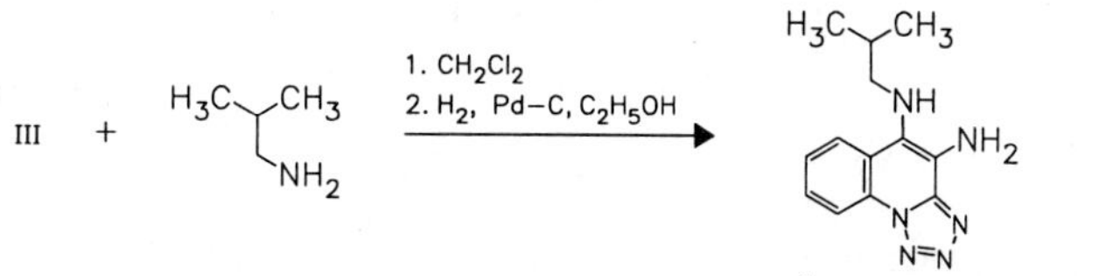

III + isobutylamine (IV)

1. CH$_2$Cl$_2$
2. H$_2$, Pd–C, C$_2$H$_5$OH

→ N^5-(2-methylpropyl)-
tetrazolo[1,5-a]quinoline-
4,5-diamine (V)

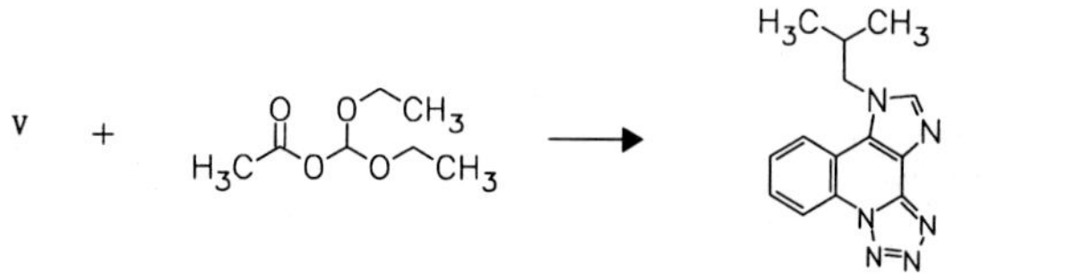

V + diethoxymethyl
acetate

→ 6-(2-methylpropyl)-6H-
imidazo[4,5-c]tetrazolo-
[1,5-a]quinoline (VI)

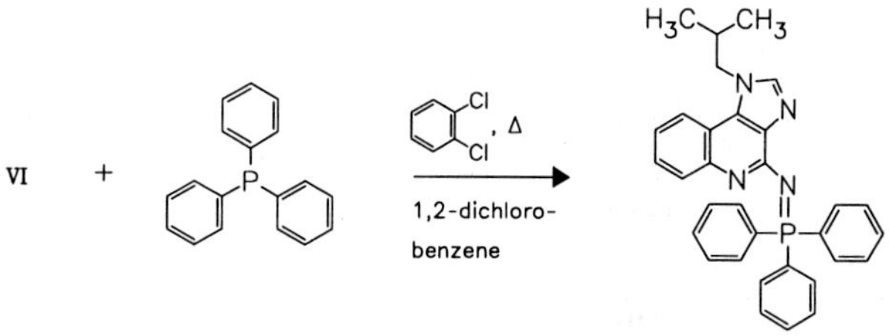

VI + triphenyl-
phosphine

1,2-dichloro-
benzene , Δ

→ (VII)

VII

1. aq. HCl, CH3OH
2. aq. NaOH

→ Imiquimod

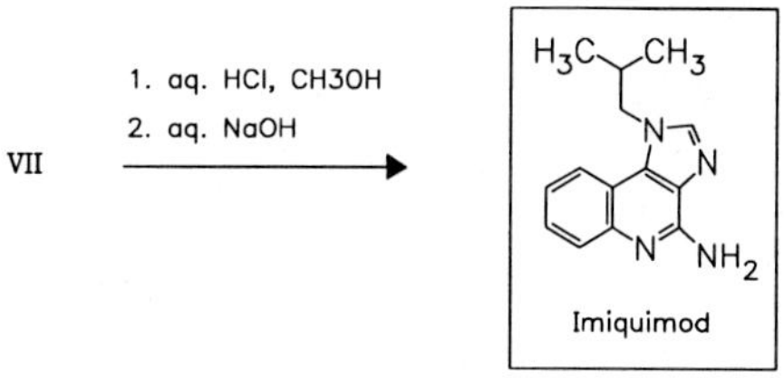

b

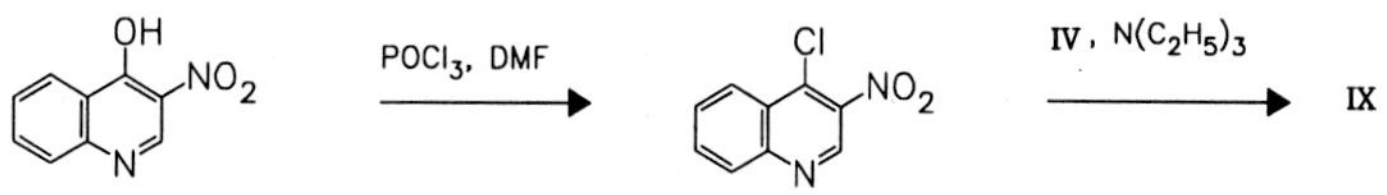

4-hydroxy-3-
nitroquinoline (VIII)

POCl$_3$, DMF →

1-chloro-3-
nitroquinoline

IV, N(C$_2$H$_5$)$_3$ → IX

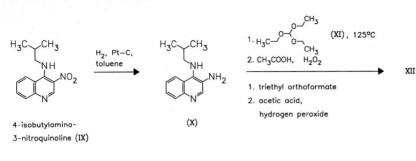

4-isobutylamino-3-nitroquinoline (IX)

(X)

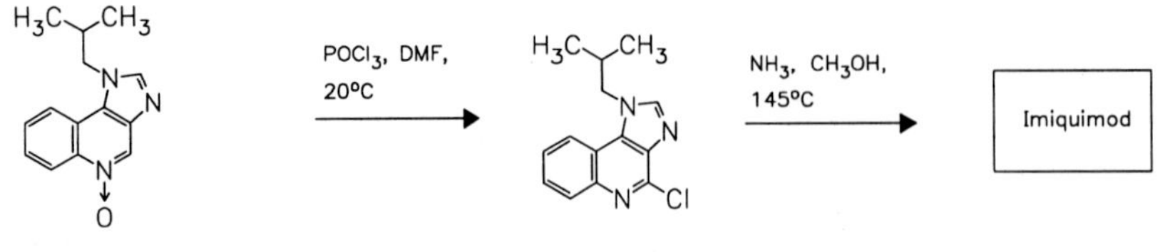

1-(2-methylpropyl)-1H-imidazo[4,5-c]-quinoline 5-oxide (XII)

POCl₃, DMF, 20°C

(XIII)

NH₃, CH₃OH, 145°C

Imiquimod

©

4-hydroxy-quinoline

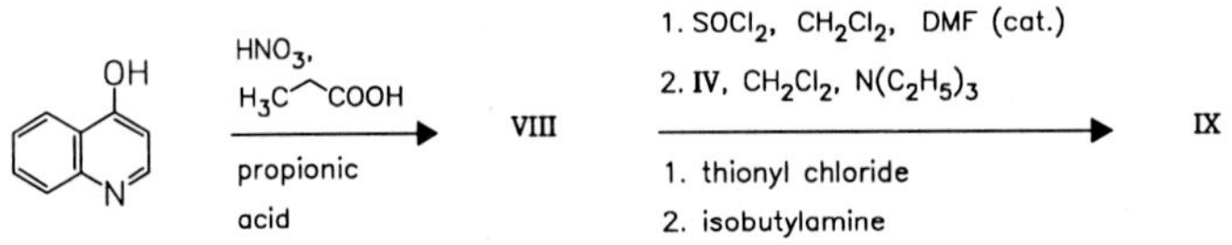

HNO₃, H₃C—COOH

propionic acid

VIII

1. SOCl₂, CH₂Cl₂, DMF (cat.)
2. IV, CH₂Cl₂, N(C₂H₅)₃

1. thionyl chloride
2. isobutylamine

IX

IX

H₂, Pt—C, MgSO₄
ethyl acetate

X

1. HCOOH, Δ
2. H₃C—CO—O—OH, CH₃COOH

1. formic acid
2. peroxyacetic acid

XII

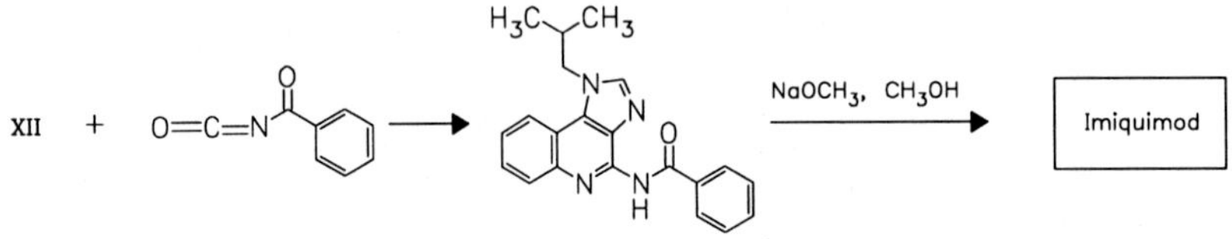

XII + O=C=N—benzoyl isocyanate

NaOCH₃, CH₃OH

Imiquimod

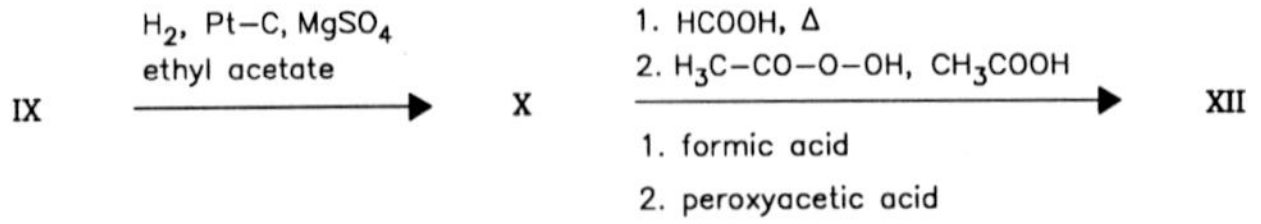

benzoyl isocyanate

ⓓ

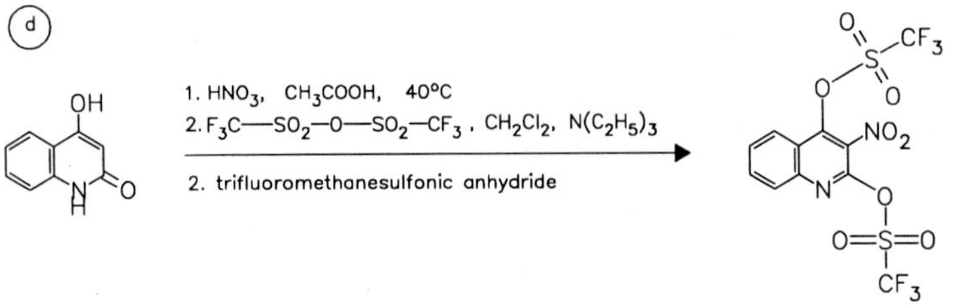

4-hydroxy-2(1H)-quinolinone (XIV)

1. HNO₃, CH₃COOH, 40°C
2. F₃C—SO₂—O—SO₂—CF₃ , CH₂Cl₂, N(C₂H₅)₃

2. trifluoromethanesulfonic anhydride

(XV)

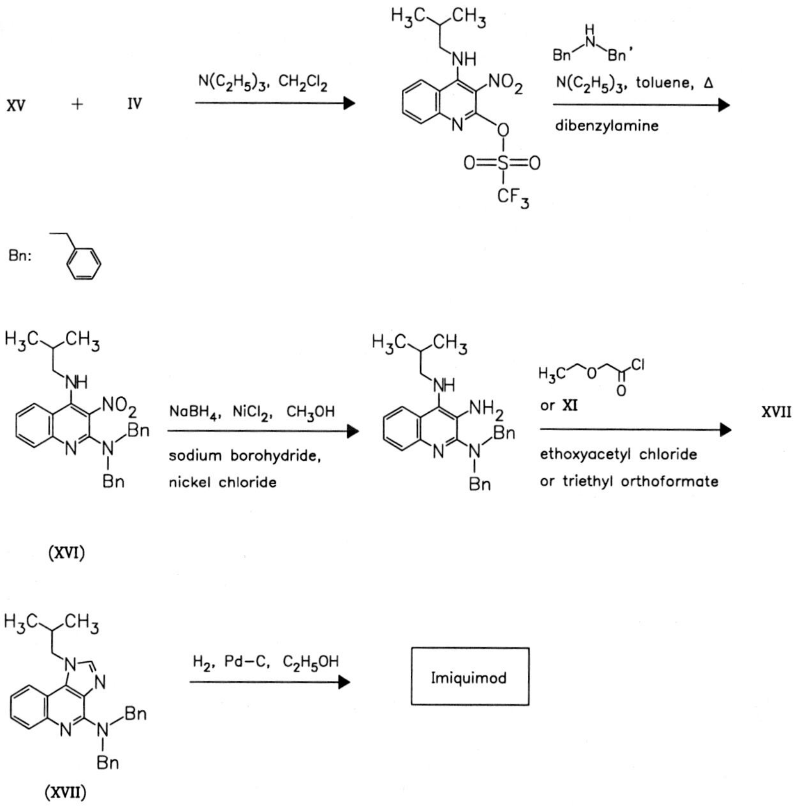

XV
+
IV
N(C2H5)3, CH2Cl2
H3C CH3
NH
NO2
N
O
O=S=O
CF3
H
N
Bn Bn'
N(C2H5)3, toluene, Δ
dibenzylamine
Bn:
H3C CH3
NH
NO2
N
N
Bn
Bn
NaBH4, NiCl2, CH3OH
sodium borohydride,
nickel chloride
NH2
H3C O Cl
O
or XI
ethoxyacetyl chloride
or triethyl orthoformate
XVII
(XVI)
H2, Pd−C, C2H5OH
Imiquimod
(XVII)

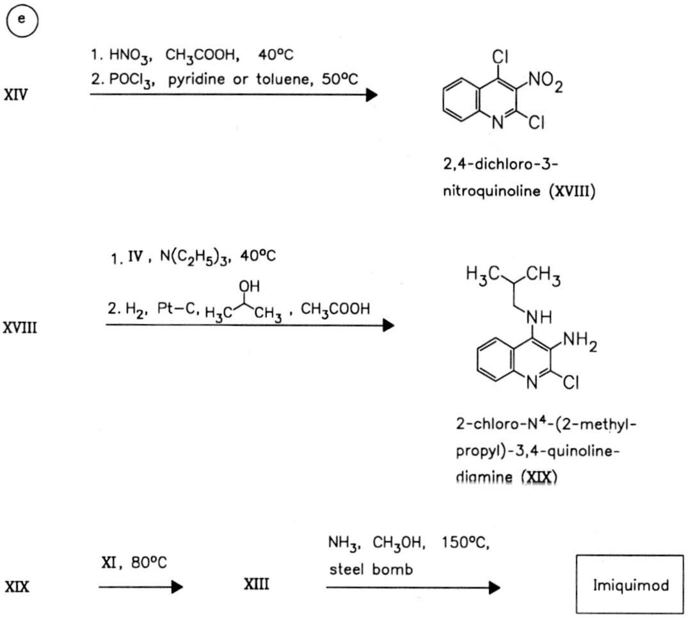

e
XIV
1. HNO3, CH3COOH, 40°C
2. POCl3, pyridine or toluene, 50°C
Cl
NO2
N
Cl
2,4-dichloro-3-
nitroquinoline (XVIII)
XVIII
1. IV, N(C2H5)3, 40°C
OH
2. H2, Pt−C, H3C CH3, CH3COOH
H3C CH3
NH
NH2
N
Cl
2-chloro-N4-(2-methyl-
propyl)-3,4-quinoline-
diamine (XIX)
XIX
XI, 80°C
XIII
NH3, CH3OH, 150°C,
steel bomb
Imiquimod

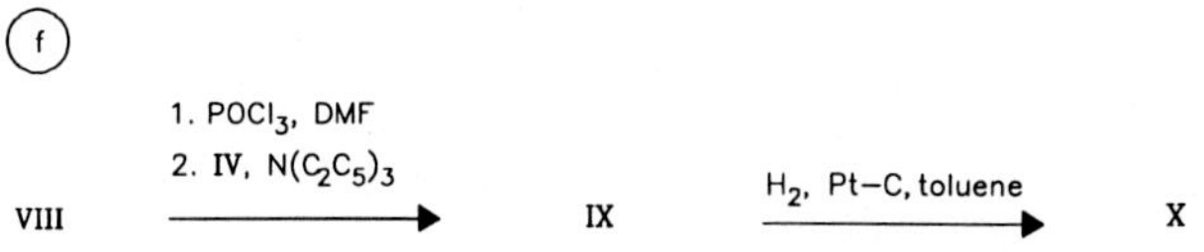

f
VIII
1. POCl3, DMF
2. IV, N(C2C5)3
IX
H2, Pt−C, toluene
X

1. XI, 185°C
2. CH₃COOH, H₂O₂

I ────────────────────→ XII

1. triethyl orthoformate
2. hydrogen peroxide

1. POCl₃, DMF, 10°C
2. NH₃, CH₃OH, 145°C

────────────────────→ Imiquimod

Reference(s):
a　WO 9 748 704 (Minnesota Mining and Manufacturing Company; appl. 22.10.1996; USA-prior. 21.6.1996).
b　EP 145 340 (Riker Lab.; appl. 16.11.1984; USA-prior. 18.11.1983).
c　WO 9 215 581 (Minnesota Mining and Manufacturing Company; appl. 13.2.1992; USA-prior. 1.3.1991).
d　WO 9 417 043 (Minnesota Mining and Manufacturing Company; appl. 25.1.1994; USA-prior. 29.1.1993).
e　US 4 988 815 (Riker Lab.; 29.1.1991; USA-prior. 26.10.1989).
f　US 4 689 338 (Riker Lab.; 25.8.1987; USA-prior. 18.11.1983).

Formulation(s):　cream 5%

Trade Name(s):
D:　Aldara (3M Medica Lab.;　　GB:　Aldara (3M Health Care)　　USA:　Aldara (3M Pharm.; 1997)
　　1998)　　　　　　　　　　 I:　Zarta (3M Medica)

Imolamine

ATC:　C01DX09
Use:　coronary vasodilator, antianginal

RN:　318-23-0　MF: C₁₄H₂₀N₄O　MW: 260.34　EINECS: 206-267-1
CN:　*N,N-diethyl-5-imino-3-phenyl-1,2,4-oxadiazol-4(5H)-ethanamine*

monohydrochloride
RN:　15823-89-9　MF: C₁₄H₂₀N₄O · HCl　MW: 296.80　EINECS: 239-920-4
LD₅₀:　25 mg/kg (M, i.v.); 475 mg/kg (M, p.o.);
　　　650 mg/kg (R, p.o.)

benzaldehyde　amine

5-amino-3-phenyl-　　2-diethylamino-　　　　Imolamine
1,2,4-oxadiazole (I)　ethyl chloride

Reference(s):
FR 2 023 M (J. Marcel, D. Aron-Samuel, J.-J. Sterne; appl. 10.10.1962; GB-prior. 12.6.1962, 11.7.1961).

Formulation(s):　inj. sol. 50 mg/5 ml; tabl. 30 mg

Trade Name(s):
D:　Irrigor (Karlspharma); wfm　　I:　Irrigor (Lipha); wfm　　　　Irrigor Aron (Spemsa);
F:　Irrigor (Aron); wfm　　　　　　　　　　　　　　　　　　　　wfm

Improsulfan

ATC: L01
Use: antineoplastic

RN: 13425-98-4   MF: $C_8H_{19}NO_6S_2$ MW: 289.37
CN: 3,3'-iminobis[1-propanol] dimethanesulfonate (ester)

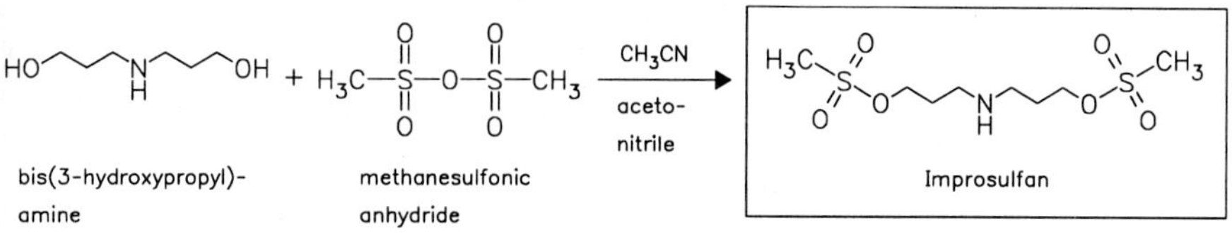

bis(3-hydroxypropyl)-   methanesulfonic   Improsulfan
amine   anhydride

Reference(s):
Sakurai, J.; El-Merzabani, M.M.: Chem. Pharm. Bull. (CPBTAL) **12**, 954 (1964).

sulfonic acid salts:
DOS 2 059 377 (Yoshitomi; appl. 2.12.1970; J-prior. 2.12.1969, 12.8.1970).
GB 1 272 497 (Yoshitomi; appl. 25.12.1970; J-prior. 2.12.1969, 12.8.1970).

Formulation(s):   tabl. 10 mg

Trade Name(s):
J: Protecton (Yoshitomi; as
        tosylate)

Incadronic acid

Use: bone resorption inhibitor, antiarthritic
        agent

RN: 124351-85-5   MF: $C_8H_{19}NO_6P_2$ MW: 287.19
CN: [(Cycloheptylamino)methylene]bis[phosphonic acid]

disodium salt monohydrate
RN: 183808-97-1   MF: $C_8H_{17}NNa_2O_6P_2 \cdot H_2O$ MW: 349.17
disodium salt
RN: 138330-18-4   MF: $C_8H_{17}NNa_2O_6P_2$ MW: 331.15

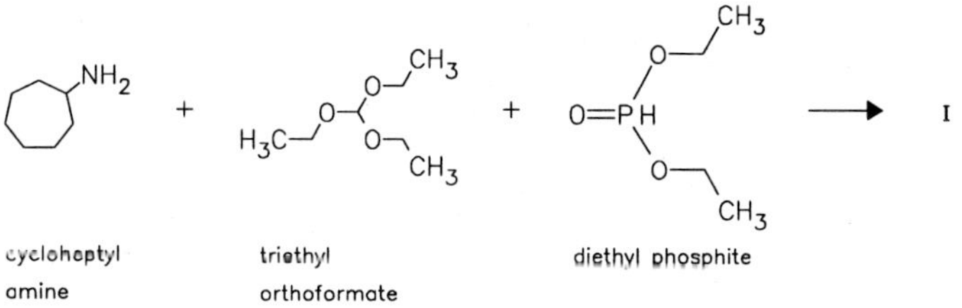

cycloheptyl   triethyl   diethyl phosphite
amine   orthoformate

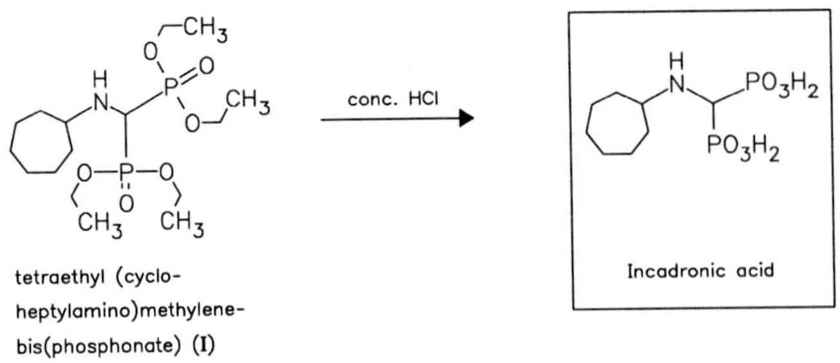

tetraethyl (cyclo-heptylamino)methylene-bis(phosphonate) (I)
conc. HCl
Incadronic acid

Reference(s):
EP 325 482 (Yamanouchi Pharm. Co.; appl. 20.1.1989; J-prior. 20.1.1988)
Takeuchi, M.; Sakamoto, S.; Yoshida, M.; Abe, T.; Isomura, Y.: Chem. Pharm. Bull. (CPBTAL) **41** (4), 688 (1993)

oral pharmaceutical compositions:
EP 550 385 (Ciba-Geigy A. G.; appl. 11.12.1992; CH-prior. 19.12.1991)

pharmaceutical compositions:
EP 693 285 (Eli Lilly & Co.; appl. 20.7.1995; USA-prior. 22.7.1994)

Formulation(s): amp. 10 mg

Trade Name(s):
J: Bisphonal (Yamanouchi; 1997)

Indalpine

ATC: N06B
Use: antidepressant, selective serotonin-uptake inhibitor

RN: 63758-79-2 MF: $C_{15}H_{20}N_2$ MW: 228.34 EINECS: 264-445-4
CN: 3-[2-(4-piperidinyl)ethyl]-1*H*-indole

monohydrochloride
RN: 63845-42-1 MF: $C_{15}H_{20}N_2 \cdot HCl$ MW: 264.80
LD_{50}: 60 mg/kg (M, i.v.); 600 mg/kg (M, p.o.)

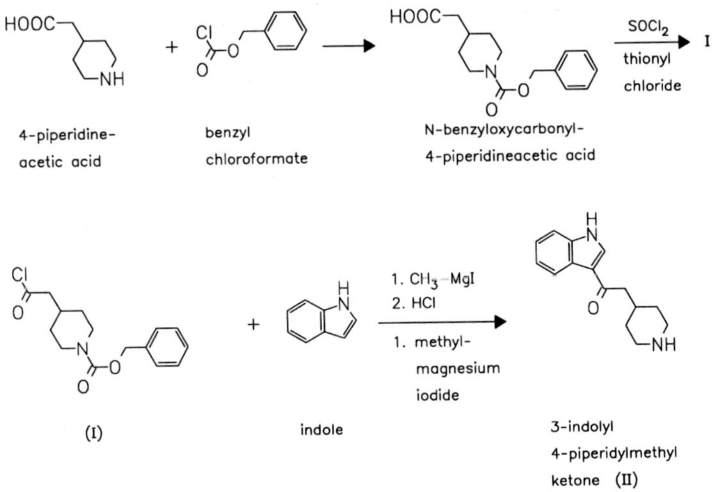

4-piperidine-acetic acid
benzyl chloroformate
N-benzyloxycarbonyl-4-piperidineacetic acid
SOCl2
thionyl chloride
I
(I)
indole
1. CH3-MgI
2. HCl
1. methyl-magnesium iodide
3-indolyl 4-piperidylmethyl ketone (II)

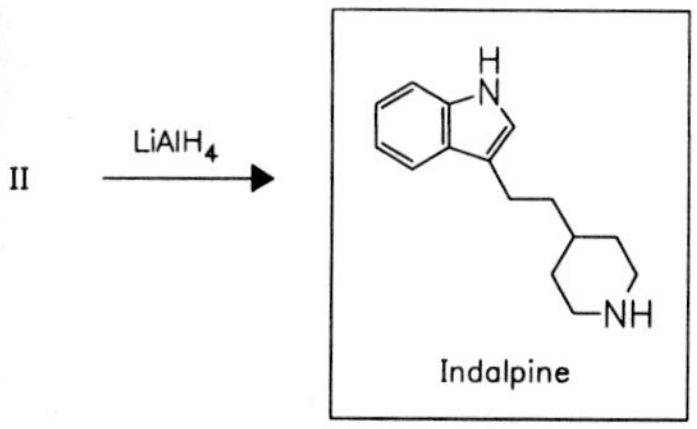

Indalpine

Reference(s):

DOS 2 618 152 (Marpha; appl. 26.4.1976; F-prior. 12.12.1975).
US 4 064 255 (Marpha; 20.12.1977; F-prior. 12.12.1975).

Formulation(s):   tabl. 50 mg

Trade Name(s):
F: Upstène (Fournier); wfm

Indanazoline

ATC: R01AA
Use: vasoconstrictor, nasal decongestant

RN: 40507-78-6   MF: $C_{12}H_{15}N_3$ MW: 201.27
LD$_{50}$:   17.6 mg/kg (Mf, i.v.); 16.3 mg/kg (Mm, i.v.); 233 mg/kg (Mf, p.o.); 179 mg/kg (Mm, p.o.)
CN: *N*-(2,3-dihydro-1*H*-inden-4-yl)-4,5-dihydro-1*H*-imidazol-2-amine

hydrochloride
RN: 56601-85-5   MF: $C_{12}H_{15}N_3 \cdot$ xHCl   MW: unspecified

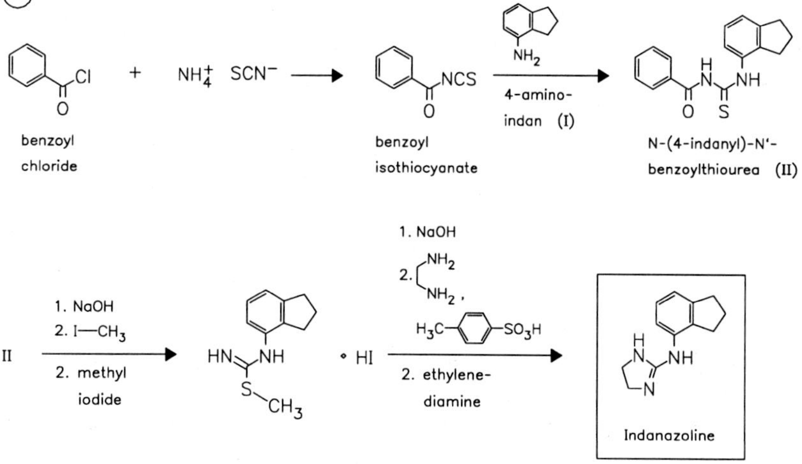

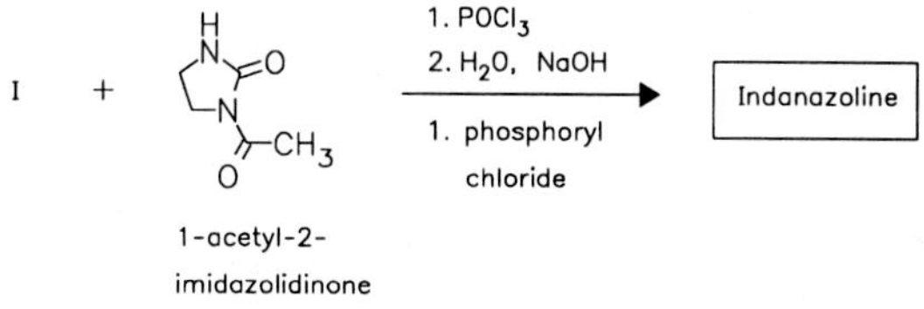

Reference(s):
May, H.J.: Arzneim.-Forsch. (ARZNAD) **30**, 1733 (1980).
a   DOS 2 136 325 (Nordmark; appl. 21.7.1971).
     US 3 882 229 (Nordmark; 6.5.1975; D-prior. 21.7.1971).
b   DOS 2 652 004 (BASF; appl. 15.11.1976).

Formulation(s):   nasal drops 1.18 mg/ml; spray 1.18 mg/ml (as hydrochloride)

Trade Name(s):
D: Farial (RIAM)

Indanorex

ATC: A08AB
Use: appetite depressant

RN: 16112-96-2   MF: $C_{12}H_{17}NO$ MW: 191.27
CN: 2-(1-aminopropyl)-2,3-dihydro-1H-inden-2-ol

hydrochloride
RN: 16112-95-1   MF: $C_{12}H_{17}NO \cdot HCl$ MW: 227.74

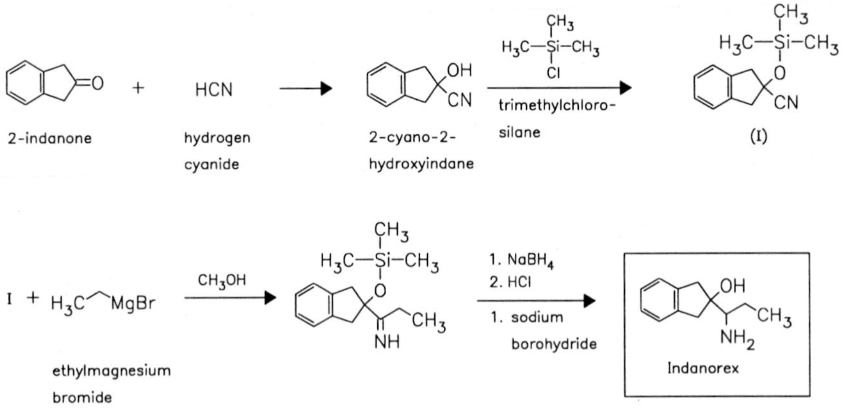

Reference(s):
DOS 2 422 879 (Lab. Logeais; appl. 11.5.1974; F-prior. 16.7.1973).

alternative synthesis:
FR 2 322 851 (Lab. Logeais; appl. 5.9.1975).

use as appetite depressant:
DOS 2 336 560 (Lab. Logeais; appl. 18.7.1973; F-prior. 20.7.1972).

Trade Name(s):
F: Dietor (Logeais); wfm

Indapamide

ATC: C03BA11
Use: diuretic, antihypertensive

RN: 26807-65-8   MF: $C_{16}H_{16}ClN_3O_3S$ MW: 365.84   EINECS: 248-012-7
CN: 3-(aminosulfonyl)-4-chloro-N-(2,3-dihydro-2-methyl-1H-indol-1-yl)benzamide

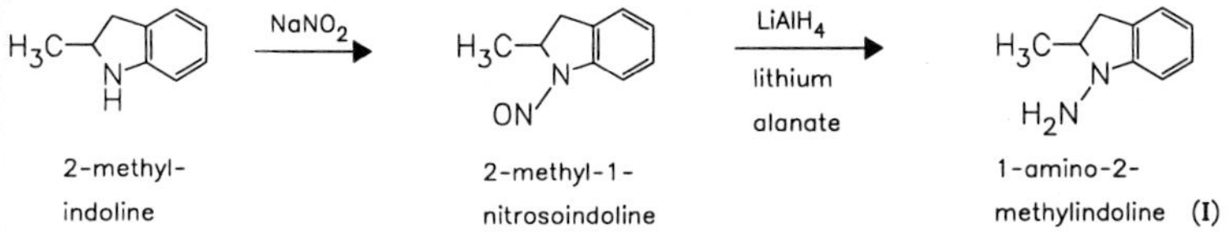

2-methyl-
indoline

2-methyl-1-
nitrosoindoline

1-amino-2-
methylindoline   (I)

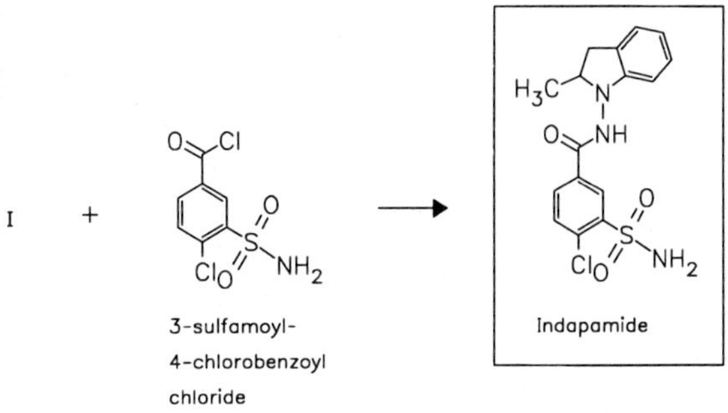

I   +

3-sulfamoyl-
4-chlorobenzoyl
chloride

Indapamide

Reference(s):

DE 1 909 180 (Science Union; appl. 2.7.1970; prior. 24.2.1969).
US 3 565 911 (Science Union; 23.2.1971; GB-prior. 6.3.1968).
FR-appl. 2 003 311 (Science Union; appl. 5.3.1969; GB-prior. 6.3.1968).

preparation of 2-methylindoline *from* 2-methyl indole:
DE 623 693 (I. G. Farbenind.; 1934).
Rogovile, V.M.; Chumale, V.T.; Dzvinka, R.T.; Shein, S.M.: J. Appl. Chem. USSR (Engl. Transl.) **54** (6), 1137 (1981).

Formulation(s):   drg. 2.5 mg, f. c. tabl. 2.5 mg

Trade Name(s):

D:	indapamid von ct (ct-Arzneimittel)	GB:	Natramid (Trinity)		Millibar (Lisapharma)
	Natrilix (Servier Deutschland; 1976)		Natrilix SR (Servier; 1978)		Veroxil (Baldacci)
	Sicco (ASTA Medica AWD)	I:	Damide (Benedetti)	J:	Natrix (Inahata-Kyoto; 1985)
			Indaflex (Lampugnani)		
			Indamol (Rhône-Poulenc Rorer)	USA:	Lozol (Rhône-Poulenc Rorer; 1983)
F:	Fludex (LBF Biopharma; Euthérapie; 1977)		Indolin (Herdel)		
			Ipamix (Gentili)		

Indecainide

(LY-135837; Ricainide)

ATC: C01B
Use: cardiac depressant (class I antiarrhythmic), therapy of life-threatening ventricular arrhythmias

RN: 74517-78-5   MF: $C_{20}H_{24}N_2O$ MW: 308.43
LD_{50}:   96 mg/kg (M, p.o.);
        10 mg/kg (R, i.v.); 82 mg/kg (R, p.o.);
        10 mg/kg (dog, i.v.); 25 mg/kg (dog, p.o.)
CN: 9-[3-[(1-methylethyl)amino]propyl]-9H-fluorene-9-carboxamide

monohydrochloride
RN: 73681-12-6   MF: $C_{20}H_{24}N_2O \cdot HCl$ MW: 344.89
LD_{50}:   96 mg/kg (M, p.o.);
        10 mg/kg (R, i.v.); 82 mg/kg (R, p.o.);
        >5 mg/kg (dog, i.v.); 25 mg/kg (dog, p.o.)

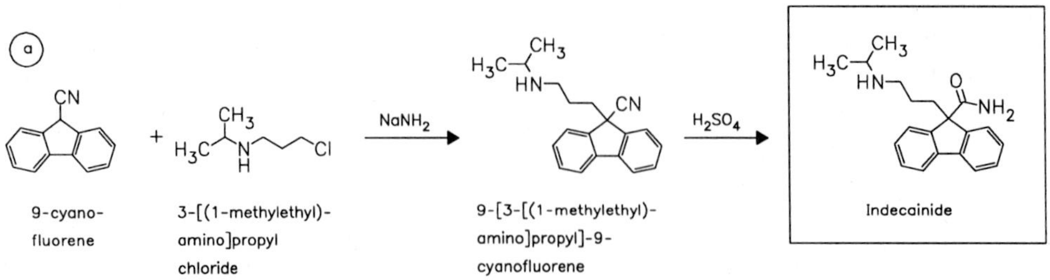

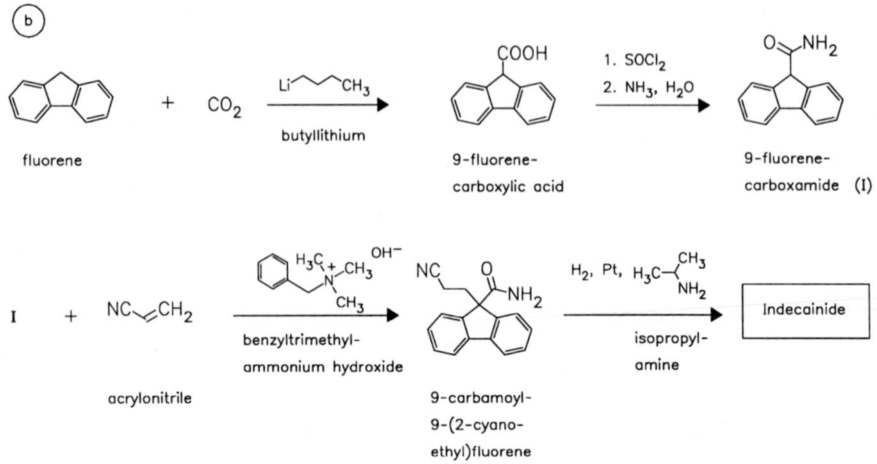

a)

9-cyano-
fluorene

3-[(1-methylethyl)-
amino]propyl
chloride

NaNH₂

9-[3-[(1-methylethyl)-
amino]propyl]-9-
cyanofluorene

H₂SO₄

Indecainide

b)

fluorene

+ CO₂

butyllithium

9-fluorene-
carboxylic acid

1. SOCl₂
2. NH₃, H₂O

9-fluorene-
carboxamide   (I)

I + NC⌒CH₂

acrylonitrile

benzyltrimethyl-
ammonium hydroxide

9-carbamoyl-
9-(2-cyano-
ethyl)fluorene

H₂, Pt, isopropyl-
amine

Indecainide

Reference(s):
EP 18 076 (Lilly; appl. 11.3.1980; USA-prior. 12.3.1978).
US 4 277 495 (Lilly; 7.7.1981; appl. 17.4.1980; prior. 12.3.1979).
US 4 282 170 (Lilly; 4.8.1981; appl. 17.4.1980).
EP 38 676 (Lilly; appl. 15.4.1981; USA-prior. 17.4.1980).

alternative synthesis:
US 4 197 313 (Lilly; 8.4.1980; appl. 12.3.1979).
US 4 552 982 (Lilly; 12.11.1985; appl. 1.8.1983).
EP 140 646 (Lilly; appl. 17.10.1984; USA-prior. 19.10.1983).

synthesis of intermediates:
US 4 486 592 (Lilly; 4.12.1984; appl. 19.10.1983).

pharmaceutical formulation:
US 4 382 093 (Lilly; 3.5.1983; appl. 29.9.1982).

Formulation(s):   s. r. tabl. 50 mg, 75 mg, 100 mg (as hydrochloride)

Trade Name(s):
USA: Decabid (Lilly; 1989); wfm

Indeloxacine

(CI-974; YM-08054)

ATC: N07X
Use: cognition activator, antidepressant, nootropic

RN: 60929-23-9   MF: C₁₄H₁₇NO₂   MW: 231.30
CN: (±)-2-[(1H-inden-7-yloxy)methyl]morpholine

hydrochloride

RN: 65043-22-3   MF: $C_{14}H_{17}NO_2 \cdot HCl$ MW: 267.76

LD$_{50}$: 47 mg/kg (M, i.v.); 444 mg/kg (M, p.o.);
77.3 mg/kg (R, i.v.); 502 mg/kg (R, p.o.);
>60 mg/kg (dog, p.o.)

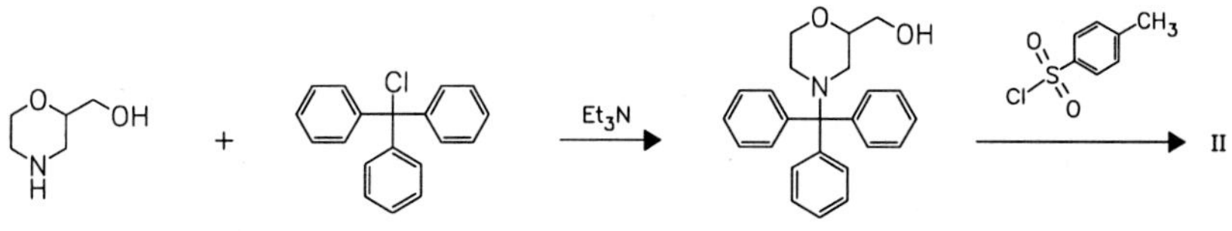

(±)-2-(hydroxy-
methyl)-
morpholine (I)

triphenyl-
chloromethane

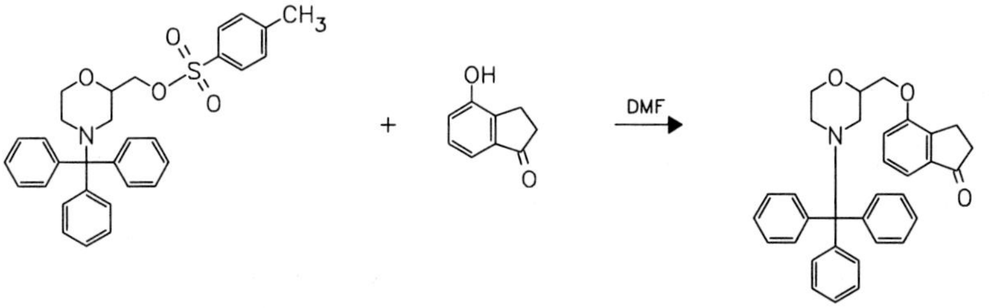

4-(triphenylmethyl)-2-
(p-toluenesulfonyloxy-
methyl)morpholine (II)

4-hydroxy-
1-indanone

2-(1-oxoindan-4-
yloxymethyl)-4-
(triphenylmethyl)-
morpholine (III)

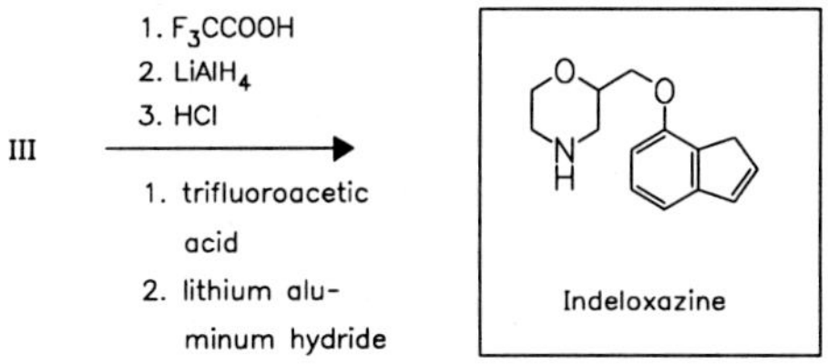

III

1. F$_3$CCOOH
2. LiAlH$_4$
3. HCl

1. trifluoroacetic
   acid
2. lithium alu-
   minum hydride

Indeloxazine

preparation of (±)-2-(hydroxymethyl)morpholine (I)

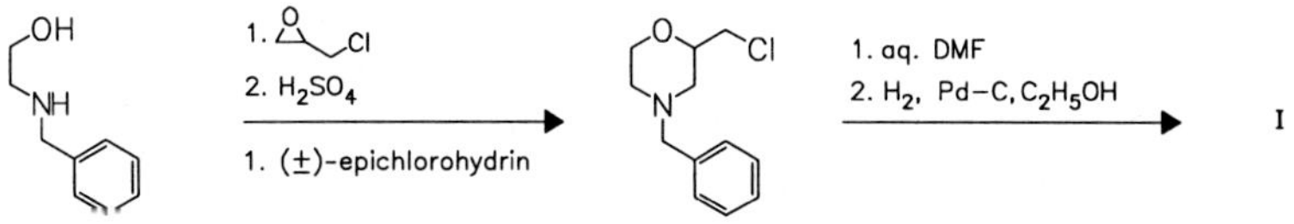

2-benzylamino-
ethanol

1. (±)-epichlorohydrin

N-benzyl-2-
(chloromethyl)-
morpholine

1. aq. DMF
2. H$_2$, Pd–C, C$_2$H$_5$OH

I

Reference(s):

US 4 109 088 (Yamanouchi; 22.8.1978; appl. 25.2.1977; prior. 5.1.1976; J-prior. 29.1.1975).
DOS 2 601 703 (Yamanouchi; appl. 19.1.1976; J-prior. 29.1.1975).
JP 52 111 580 (Yamanouchi; appl. 15.3.1976).
CA 1 103 247 (Yamanouchi; J-prior. 27.10.1976).
Kojima, T. et al.: Chem. Pharm. Bull. (CPBTAL) **33**, 3766 (1985).

synthesis of [14]*C-labeled compound:*
Arima, H.; Tamazawa, K.: J. Labelled Compd. Radiopharm. (JLCRD4) **22**, 1217 (1985).

alternative synthesis:
DE 2 707 678 (Yamanouchi; appl. 13.10.1977; J.-prior. 31.3.1976).

medical use for the treatment of mental disorders:
JP 61 145 119 (Yamanouchi; appl. 19.12.1984).
JP 56 123 915 (Yamanouchi; appl. 5.3.1980).

preparation of (±)-2-(hydroxymethyl)morpholine:
Kato, S. et al.: J. Med. Chem. (JMCMAR) **33** (5), 1406 (1990).
Berg, S. et al.: J. Med. Chem. (JMCMAR) **41** (11), 1934 (1998).
Jinbo, Y. et al.: J. Med. Chem. (JMCMAR) **37** (17), 2791 (1994).
Yanagisawa, H.; Kanazaki, T.: Heterocycles (HTCYAM) **35** (1), 105 (1993).
Loftus, F.: Synth. Commun. (SYNCAV) **10** (1), 59 (1980).

Formulation(s):   tabl. 20 mg (as hydrochloride)

Trade Name(s):
J: Elen (Yamanouchi; 1988)   Noin (Essex Nippon; 1988)

Indenolol

ATC: C07AA49
Use: beta blocking agent

RN: 60607-68-3   MF: $C_{15}H_{21}NO_2$ MW: 247.34   EINECS: 262-323-5
CN: 1-[1*H*-inden-4(or 7)-yloxy]-3-[(1-methylethyl)amino]-2-propanol

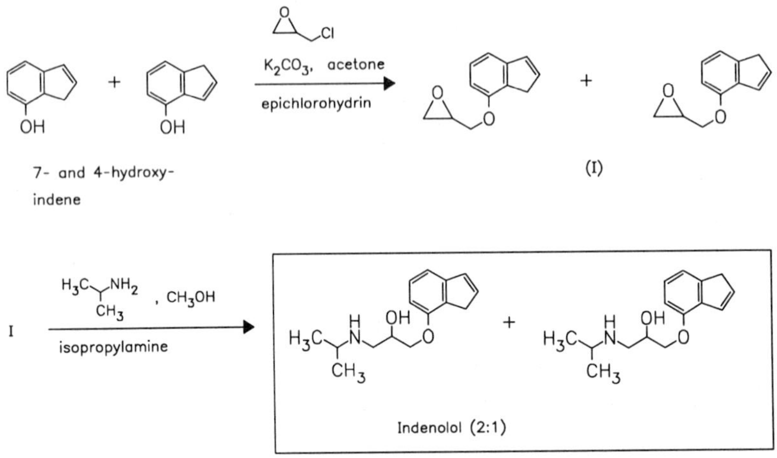

7- and 4-hydroxy-
indene

(I)

Indenolol (2:1)

Reference(s):
DOS 1 955 229 (Yamanouchi; appl. 3.11.1969; J-prior. 12.11.1968, 11.9.1969).
Murase, K. et al.: Chem. Pharm. Bull. (CPBTAL) **24**, 552 (1976).

Formulation(s):   tabl. 60 mg

Trade Name(s):
I: Myodil (Glaxo)   Securpres (Poli)   J: Pulsan (Yamanouchi)

Indinavir sulfate
(L-735524; MK-639)

ATC: J05AE02
Use: antiviral, HIV-1-protease inhibitor

RN: 157810-81-6 MF: $C_{36}H_{47}N_5O_4 \cdot H_2SO_4$ MW: 711.88
CN: [1(1S,2R),5(S)]-2,3,5-trideoxy-N-(2,3-dihydro-2-hydroxy-1H-inden-1-yl)-5-[2-[[(1,1-dimethylethyl)amino]carbonyl]-4-(3-pyridinylmethyl)-1-piperazinyl]-2-(phenylmethyl)-D-$erythro$-pentonamide sulfate (1:1) (salt)

base
RN: 150378-17-9 MF: $C_{36}H_{47}N_5O_4$ MW: 613.80

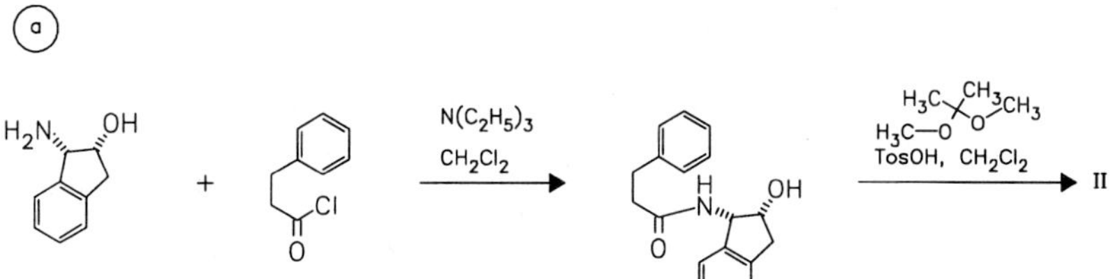

2(R)-hydroxy-
1(S)-amino-
indane (I)

dihydro-
cinnamoyl
chloride

N-[2(R)-hydroxy-
1(S)-indanyl]-3-
phenylpropan-
amide

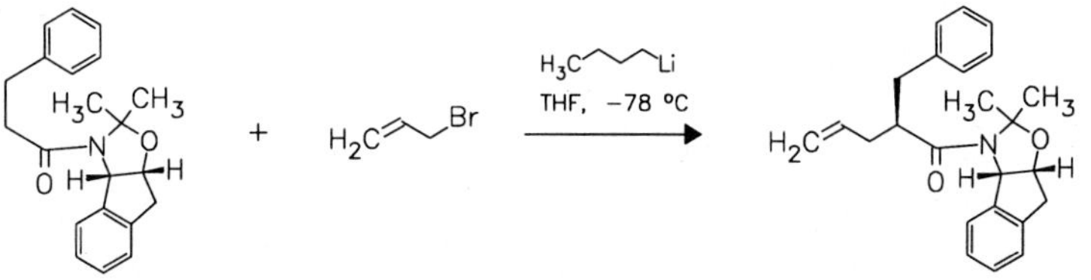

(3aS,8aR)-3,3a,8,8a-
tetrahydro-2,2-dime-
thyl-3-(3-phenylpro
pionyl)-2H-indeno-
[1,2-d]oxazole (II)

allyl
bromide

(3aS,8aR)-3,3a,8,8a-
tetrahydro-2,2-dime-
thyl-3-[(2S)-2-benzyl-
4-pentenoyl]-2H-indeno-
[1,2-d]oxazole (III)

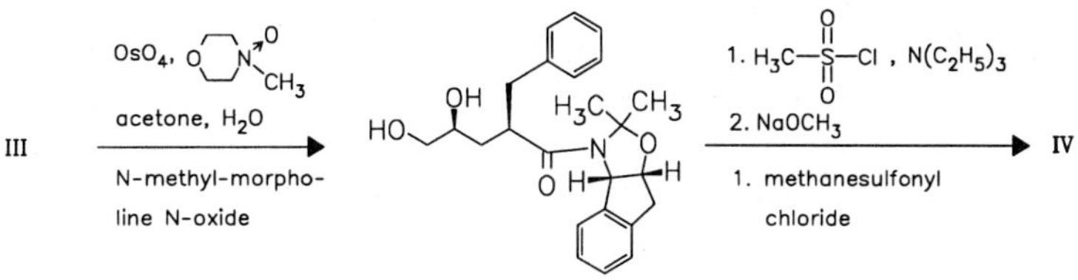

III

N-methyl-morpho-
line N-oxide

1. methanesulfonyl
chloride

IV

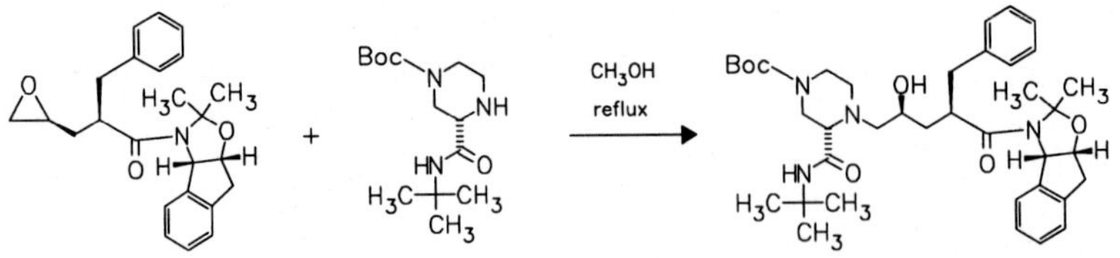
CH3OH
reflux

(IV)

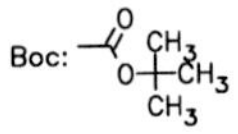
Boc:

N-tert-butyl-4-(tert-butoxycarbonyl)-piperazine-2(S)-carboxamide (V)

(3aS,8aR)-3,3a,8,8a-tetrahydro-2,2-dimethyl-3-[2(S)-benzyl-5-[4-(tert-butoxycarbonyl)-2(S)-(tert-butylcarbamoyl)piperazino]-4(R)-hydroxyvaleryl]-2H-indeno[1,2-d]oxazole (VI)

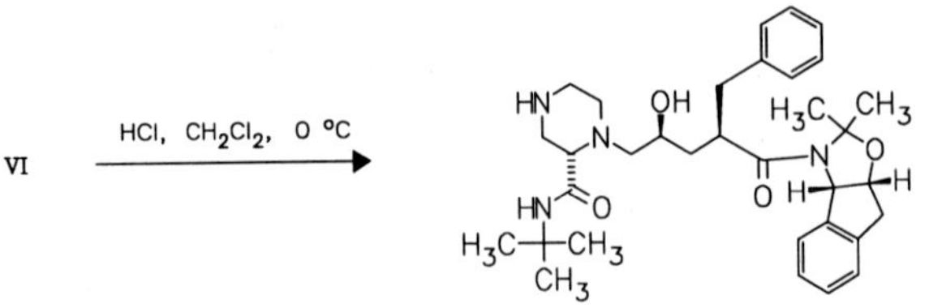
VI
HCl, CH2Cl2, 0 °C

(3a,S,8aR)-3,3a,8,8a-tetrahydro-2,2-dimethyl-3-[2(S)-benzyl-5-[2(S)-(tert-butoxycarbonyl)piperazino]-4(R)-hydroxyvaleryl]-2H-indeno[1,2-d]oxazole (VII)

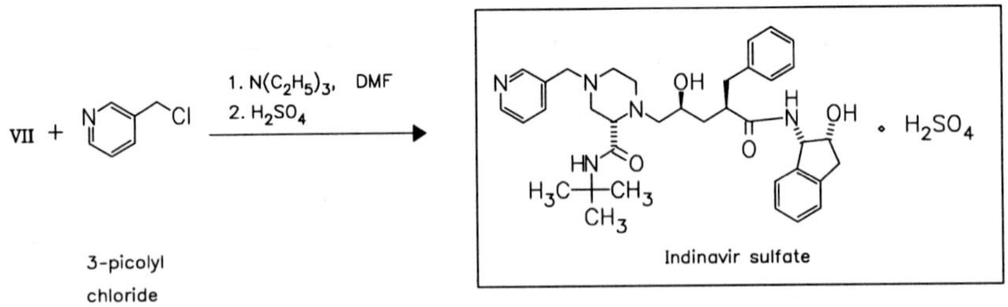
VII +
1. N(C2H5)3, DMF
2. H2SO4
• H2SO4
Indinavir sulfate

3-picolyl chloride

(b)

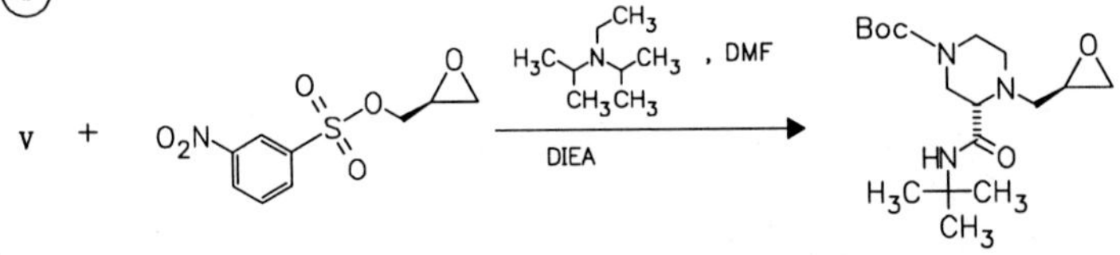
V +
, DMF
DIEA

(2S)-glycidyl 3-nitrobenzenesulfonate

N-tert-butyl-1-[(R)-oxiranylmethyl]-4-tert-butoxycarbonyl-piperazine-2(S)-carboxamide (VIII)

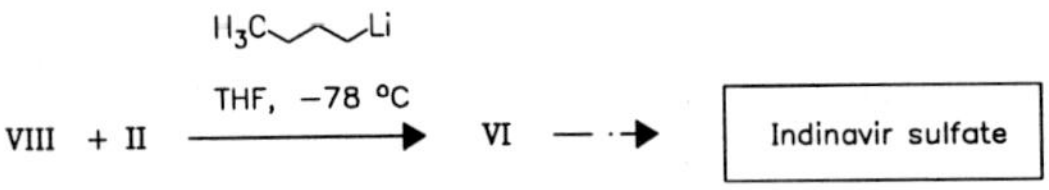
THF, −78 °C
VIII + II
VI
Indinavir sulfate

(c)

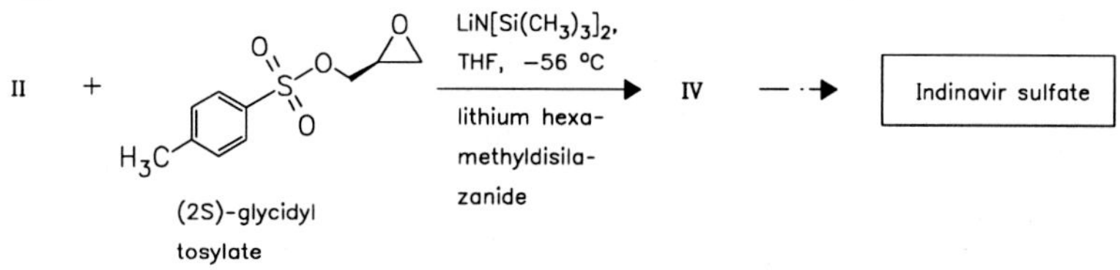

II + (2S)-glycidyl tosylate

LiN[Si(CH₃)₃]₂,
THF, −56 °C
lithium hexa-
methyldisila-
zanide

→ IV — ·→ Indinavir sulfate

(d)

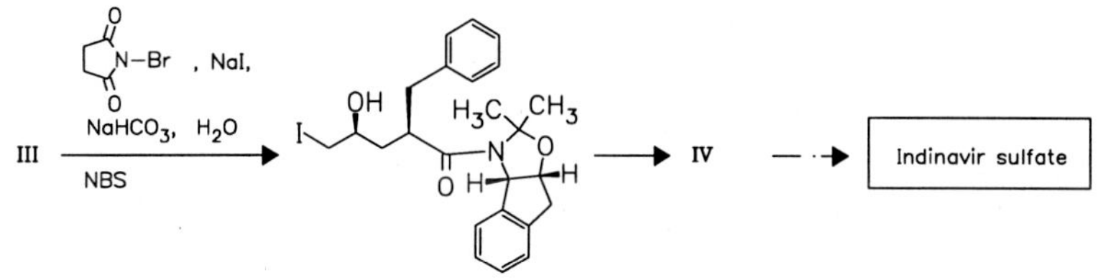

III + N–Br, NaI, NaHCO₃, H₂O
NBS

→ I → IV — ·→ Indinavir sulfate

(e)

III

e⁻, NaBr, H₂O,
CH₃CN, THF
electrochemical cell 0.2 A

→ IV — ·→ Indinavir sulfate

(f)

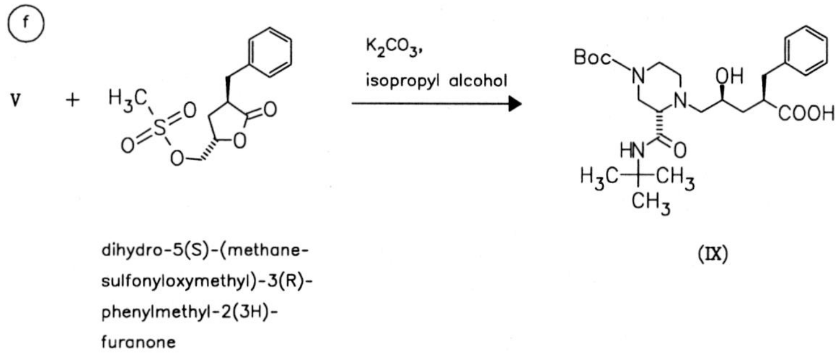

V + dihydro-5(S)-(methane-
sulfonyloxymethyl)-3(R)-
phenylmethyl-2(3H)-
furanone

K₂CO₃,
isopropyl alcohol

→ (IX)

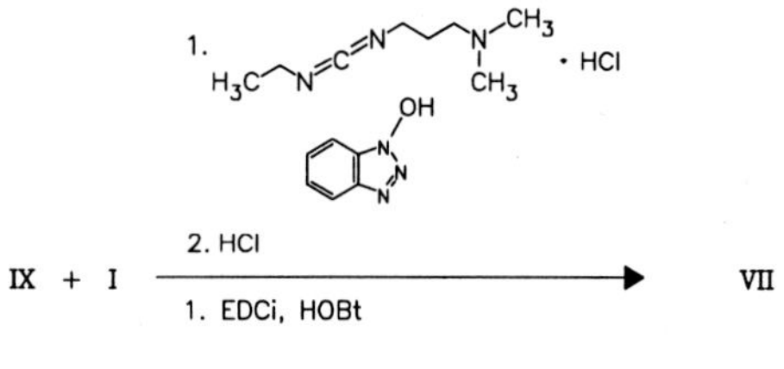

1. H₃C–N=C=N–CH₂CH₂CH₂–N(CH₃)₂ · HCl , HOBt
2. HCl
1. EDCi, HOBt

IX + I → VII

(g)

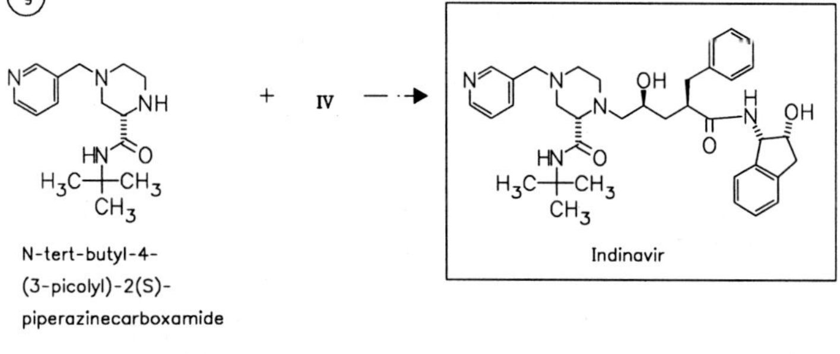

N-tert-butyl-4-
(3-picolyl)-2(S)-
piperazinecarboxamide

+ IV — ·→ Indinavir

preparation of intermediate V:

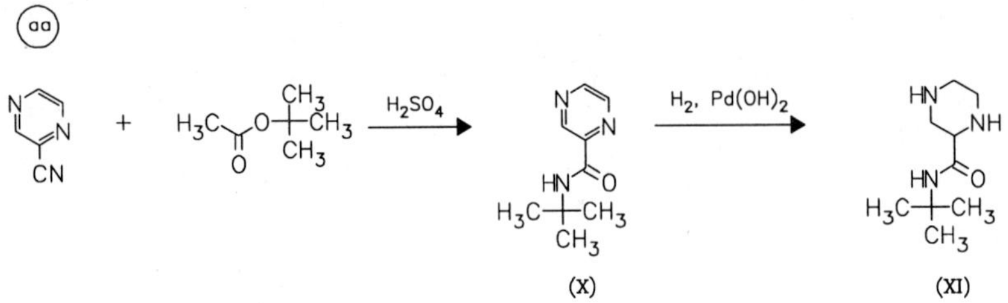

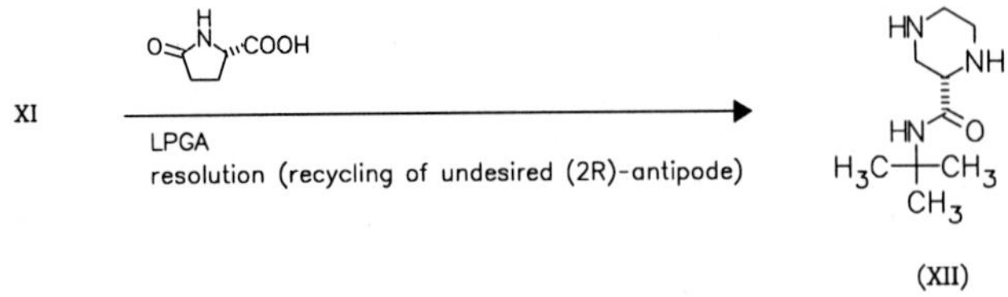

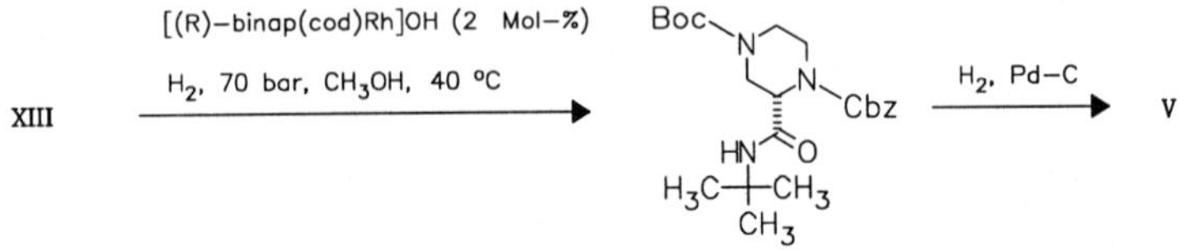

(aa)

(X)

(XI)

XI

LPGA
resolution (recycling of undesired (2R)-antipode)

(XII)

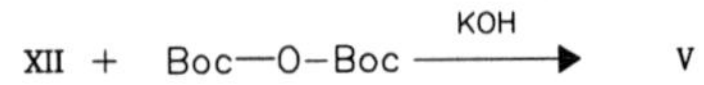

XII + Boc—O—Boc $\xrightarrow{\text{KOH}}$ V

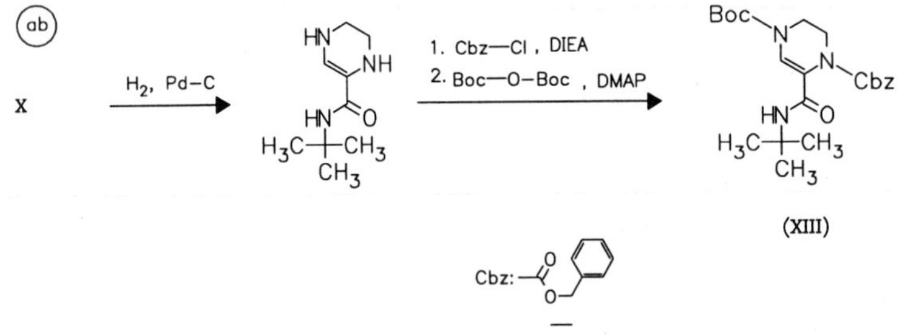

(ab)

X $\xrightarrow{\text{H}_2, \text{ Pd-C}}$

1. Cbz—Cl , DIEA
2. Boc—O—Boc , DMAP

(XIII)

Cbz: —

XIII $\xrightarrow[\text{H}_2, \text{ 70 bar, CH}_3\text{OH, 40 °C}]{\text{[(R)—binap(cod)Rh]OH (2 Mol-\%)}}$ $\xrightarrow{\text{H}_2, \text{ Pd-C}}$ V

synthesis of starting material I:

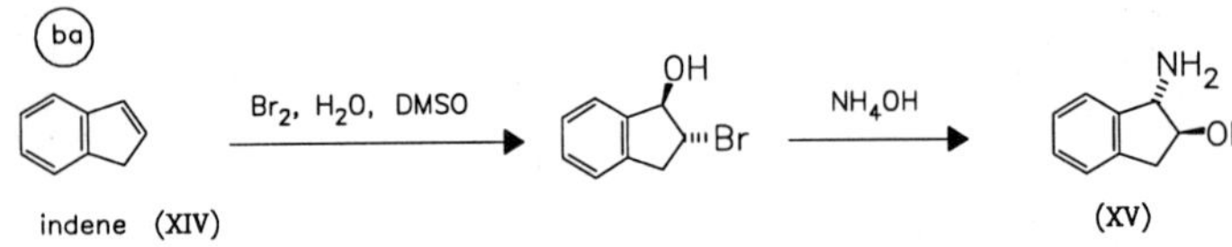

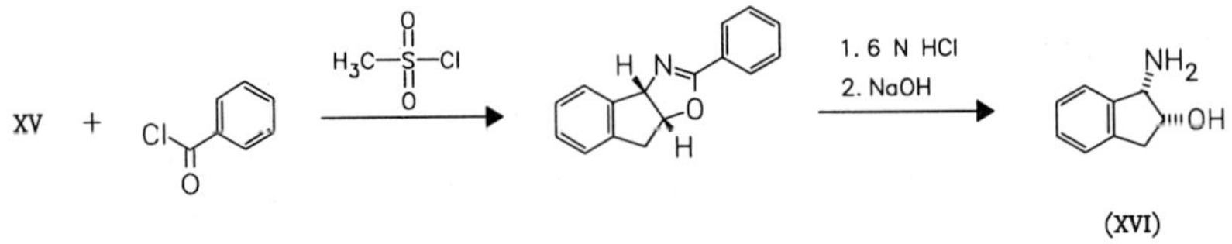

(ba)

indene (XIV)

(XV)

XV + Cl— $\xrightarrow{\text{H}_3\text{C—S—Cl}}$

1. 6 N HCl
2. NaOH

(XVI)

1. HOOC–CH(OH)–CH(OH)–COOH

2. base

XVI ———————————→ I
1. L-tartaric acid (XVII)

(bb)

0.7 % S,S–Mn(II)(salen)Cl, NaOCl,

3 %

XIV ———————————→
4-(3-phenylpropyl)pyridine N-oxide

(XVIII)

oleum, $H_3C–CN$,

H_2O, −20 °C   XVII

XVIII ———————————→ XVI ——→ I

(bc)

dioxygenase
bio-conversion

XIV ———————————→

(1S,2R)-XVIII ——→ I

Reference(s):
a   WO 9 309 096 (Merck & Co.; appl. 3.11.1992; USA-prior. 8.11.1991, 15.5.1992).
b   WO 9 502 583 (Merck & Co.; appl. 11.7.1994; USA-prior. 16.7.1993).
c   Askin, D. et al.: Tetrahedron Lett. (TELEAY) **35** (4), 673 (1994).
    WO 9 502 584 (Merck & Co.; appl. 11.7.1994; USA-prior. 16.7.1993, 26.1.1994).
d   Maligres, P.E. et al.: Tetrahedron Lett. (TELEAY) **36**, 2195 (1995).
e   WO 9 716 450 (Merck & Co.; appl. 25.10.1996; USA-prior. 30.10.1995, 22.2.1996).
f   US 5 413 999 (Merck & Co.; appl. 7.5.1993; USA-prior. 8.11.1991).
g   WO 9 628 439 (Merck & Co.; appl. 11.3.1996; USA-prior. 15.3.1995).
aa WO 9 636 629 (Merck & Co.; appl. 14.5.1996; USA-prior. 18.5.1995).
    WO 9 521 162 (Merck & Co.; appl. 30.1.1995; USA-prior. 4.2.1994).
ab GB 2 302 690 (Merck & Co.; appl. 20.6.1996; GB-prior. 13.2.1996; USA-prior. 28.6.1995).
ba WO 9 636 724 (Merck & Co.; appl. 15.5.1996; USA-prior. 19.5.1995).
    US 5 449 830 (Merck & Co., Procter/Gamble Co.; appl. 11.3.1994; USA-prior. 11.3.1994).
bb WO 9 700 966 (Merck & Co.; appl. 14.6.1996; USA-prior. 13.2.1996, 20.6.1995).
    WO 9 612 818 (Merck & Co.; appl. 17.10.1995; USA-prior. 21.10.1994).

reductive amination with pyridinecarboxaldehyde:
US 5 508 404 (Merck & Co.; appl. 15.3.1995; USA-prior. 15.3.1995).

prodrugs of indinavir:
WO 9 514 016 (Merck & Co.; appl. 14.11.1994; USA-prior. 18.11.1993).

combination with AZT:
WO 9 623 509 (Merck & Co.; appl. 29.1.1996; USA-prior. 1.2.1995).
WO 9 604 913 (Merck & Co.; appl. 7.8.1995; USA-prior. 20.7.1995, 11.8.1994, 14.11.1994).

combination with e. g. neoarapine:
WO 9 600 068 (Merck & Co.; appl. 23.6.1995; USA-prior. 27.6.1994).

combination with quinoxalines:
EP 728 481 (Bayer AG; appl. 14.2.1996; D-prior. 27.2.1995).

Formulation(s):   cps. 200 mg, 400 mg (as sulfate)

Trade Name(s):
D: Crixivan (Merck Sharp &   GB: Crixivan (Merck Sharp &   USA: Crixivan (Merck & Co.)
      Dohme)   Dohme)

Indobufen

ATC: B01AC10
Use: anti-inflammatory, antithrombocytic

RN: 63610-08-2   MF: $C_{18}H_{17}NO_3$ MW: 295.34   EINECS: 264-364-4
LD$_{50}$:   370 mg/kg (M, i.v.); 697 mg/kg (M, p.o.);
       333 mg/kg (R, i.v.); 373 mg/kg (R, p.o.)
CN: (±)-4-(1,3-dihydro-1-oxo-2H-isoindol-2-yl)-α-ethylbenzeneacetic acid

sodium salt
RN: 94135-04-3   MF: $C_{18}H_{16}NNaO_3$ MW: 317.32

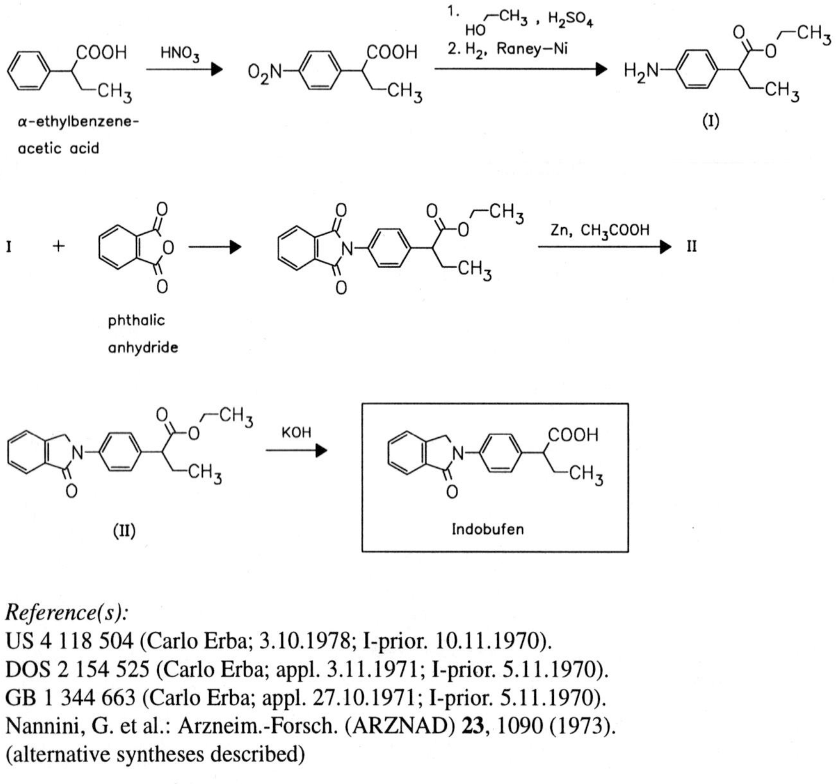

Reference(s):
US 4 118 504 (Carlo Erba; 3.10.1978; I-prior. 10.11.1970).
DOS 2 154 525 (Carlo Erba; appl. 3.11.1971; I-prior. 5.11.1970).
GB 1 344 663 (Carlo Erba; appl. 27.10.1971; I-prior. 5.11.1970).
Nannini, G. et al.: Arzneim.-Forsch. (ARZNAD) 23, 1090 (1973).
(alternative syntheses described)

synthesis of ethyl p-amino-α-ethylphenylacetate:
Wilds, A.L.; Biggerstaff, W.R.: J. Am. Chem. Soc. (JACSAT) 67, 789 (1945).

Formulation(s):   amp. 200 mg (as sodium salt); tabl. 100 mg, 200 mg

Trade Name(s):
I: Ibustrin (Pharmacia &
      Upjohn; 1984)

Indometacin
(Indomethacin)

ATC: C01EB03; M01AB01; M02AA23;
S01BC01
Use: anti-inflammatory, antipyretic,
analgesic

RN: 53-86-1   MF: $C_{19}H_{16}ClNO_4$ MW: 357.79   EINECS: 200-186-5
LD$_{50}$:   30 mg/kg (M, i.v.); 11.841 mg/kg (M, p.o.);
      21 mg/kg (R, i.v.); 2.42 mg/kg (R, p.o.);
      100 mg/kg (dog, i.v.); 160 mg/kg (dog, p.o.)
CN: 1-(4-chlorobenzoyl)-5-methoxy-2-methyl-1H-indole-3-acetic acid

sodium salt hydrate
RN: 74252-25-8   MF: $C_{19}H_{15}ClNaO_4 \cdot 3H_2O$ MW: 419.81

   ⓐ   Merck + Co.

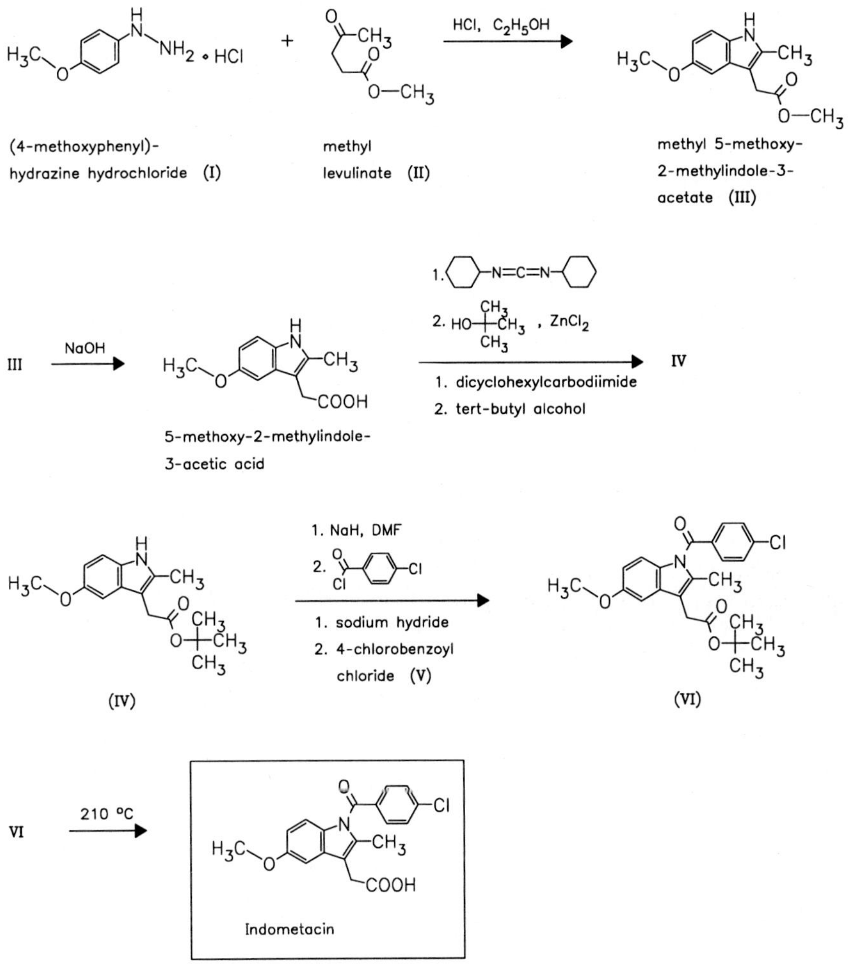

(4-methoxyphenyl)-
hydrazine hydrochloride (I)

methyl
levulinate (II)

methyl 5-methoxy-
2-methylindole-3-
acetate (III)

5-methoxy-2-methylindole-
3-acetic acid

(IV)

1. sodium hydride
2. 4-chlorobenzoyl
   chloride (V)

(VI)

Indometacin

b) Sumitomo

I + OHC—CH₃ → [structure] $\xrightarrow{\text{V, pyridine}}$ VII

acetaldehyde

[structure (VII)] $\xrightarrow{\text{HCl, C}_2\text{H}_5\text{-OH}}$ [structure] • HCl $\xrightarrow{\text{levulinic acid}}$ VIII

(VII)

[structure (VIII)] $\xrightarrow{\text{HCl}}$ Indometacin

(VIII)

c) Sumitomo

I + II → [structure] $\xrightarrow{\text{V , dioxane, 80-85 °C}}$ IX

[structure IX] $\xrightarrow{\text{NaOH}}$ Indometacin

IX

Reference(s):
a DE 1 232 150 (Merck & Co.; appl. 16.3.1962; USA-prior. 22.3.1961, 5.1.1962).
   DAS 1 620 014 (Merck & Co.; appl. 29.6.1966; USA-prior. 30.6.1965).
   DE 1 620 030 (Merck & Co.; appl. 16.3.1962; USA-prior. 22.3.1961, 5.1.1962).
   DE 1 620 031 (Merck & Co.; appl. 16.3.1962; USA-prior. 22.3.1961, 5.1.1962).
   DE 1 643 463 (Merck & Co.; appl. 12.10.1967; USA-prior. 13.10.1966, 14.8.1967).
   US 3 161 654 (Merck & Co.; 15.12.1964; appl. 11.6.1963; prior. 22.3.1961).
b Yamamoto, H.: Chem. Pharm. Bull. (CPBTAL) **16**, 17 (1968).
   Yamamoto, H. et al.: Chem. Pharm. Bull. (CPBTAL) **16**, 647 (1968).
c DAS 1 795 674 (Sumitomo; appl. 8.4.1968; J-prior. 11.4.1967, 6.5.1967, 8.5.1967, 23.5.1967, 27.5.1967, 29.5.1967, 8.11.1967, 12.12.1967, 14.12.1967).

suspension for parenteral use:
US 4 093 733 (Merck & Co.; 6.6.1978; appl. 9.9.1976).

Formulation(s):   cps. 25 mg, 50 mg; eye drops 1 %; gel 10 mg; sol. 8 mg/ml (1 %); s. r. cps. 75 mg;
                     suppos. 50 mg, 100 mg; susp. 25 mg (as sodium salt hydrate)

Trade Name(s):
D: Amuno (Merck Sharp &   Confortid (Dumex; as   Indo (ct-Arzneimittel)
      Dohme; 1965)-comb.   sodium salt)   Indometacin (Aliud
      Chibro-Amuno (Chibret)   Elmetacin (Sankyo)   Pharma)

Indometacin (Heyl)-comb.
Indomisal (Brenner-Efeka)
Indo-Top (ratiopharm)
Inflam (Lichtenstein)-
comb.
Jenatacin (Jenapharm)
various generics and
combination preparations
F: Ainscrid LP (Gerda SA)
Chrono-Indocid 75 (Merck
Sharp & Dohme)
Indocid (Merck Sharp &
Dohme; 1966)
Indocollyre (Chauvin)
GB: Artracin SR (Searle)
Flexin Continus (Napp)

Indocid (Morson)
Indomod (Pharmacia &
Upjohn)
I: Cidalgon (Ecobi)
Difmetre (UCM)-comb.
Imet (Firma)
Indocid (Merck Sharp &
Dohme)
Indoxen (Sigma-Tau)
Liometacen (Chiesi; as
megluminate)
Metacen (Chiesi)
combination preparations
J: Indacin (Merck-Banyu;
1966)
Inderapollon (Kaigai)

Indomethine (Kowa)
Inmecin (Nippon
Chemiphar)
Inmetocin (Tobishi)
Inteban (Sumitomo; 1967)
Intedarl (Choseido)
Lausit (Showa)
Methazine (Sankyo)
Mezolin (Meiji)
Salinac (Nippon Kayaku)
Taikosashin S (Taiho)
Zalbico (Toyo Pharmar)
USA: Indocin (Merck Sharp &
Dohme; 1965)

Indometacin farnesil

(Indometacin farnesyl)

ATC: M01AB
Use: non-steroidal anti-inflammatory,
indometacin prodrug

RN: 85801-02-1   MF: $C_{34}H_{40}ClNO_4$ MW: 562.15
LD$_{50}$: >4 g/kg (M, i.m.); 1305 mg/kg (M, i.p.); 6800 mg/kg (M, p.o.); 8g/kg (M, s.c.);
2000 mg/kg (R, i.m.); 3800 mg/kg (R, i.p.); 1680 mg/kg (R, p.o.); 2400 mg/kg (R, s.c.);
>3 g/kg (dog, p.o.)
CN: 1-(4-chlorobenzoyl)-5-methoxy-2-methyl-1H-indole-3-acetic acid 3,7,11-trimethyl-2,6,10-dodecatrienyl
ester

indometacin
(q. v.)

3,7,11-trimethyl-2,6,10-
dodecatrienol

Indometacin farnesil (I)

Reference(s):
DE 3 226 687 (Eisai; appl. 16.7.1982; J-prior. 23.7.1981).
US 4 455 316 (Eisai; 19.6.1984; appl. 8.7.1982; J-prior. 23.7.1981).
US 4 576 963 (Eisai; 18.3.1986; appl. 14.5.1984; prior. 8.7.1982; J-prior. 23.7.1981).

soft gelatine capsules:
EP 407 815 (Eisai; appl. 27.6.1990; J-prior. 10.7.1989).

stabilisation with tocopherol:
EP 387 655 (Eisai; appl. 5.3.1990; J-prior. 17.3.1989).

Formulation(s):   cps. 100 mg

Trade Name(s):
J: Infree (Eisai; 1991)

Indoprofen

ATC: M01AE10
Use: anti-inflammatory, analgesic

RN: 31842-01-0   MF: $C_{17}H_{15}NO_3$ MW: 281.31   EINECS: 250-833-0
LD_{50}: 700 mg/kg (M, p.o.);
         84 mg/kg (R, p.o.)
CN: 4-(1,3-dihydro-1-oxo-2*H*-isoindol-2-yl)-α-methylbenzeneacetic acid

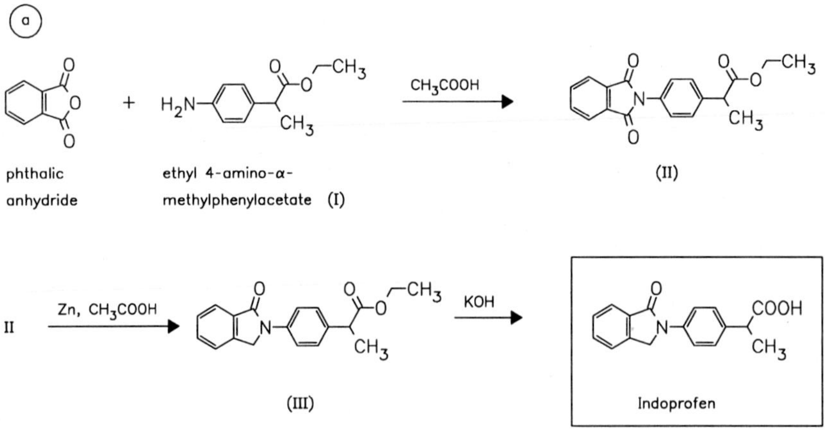

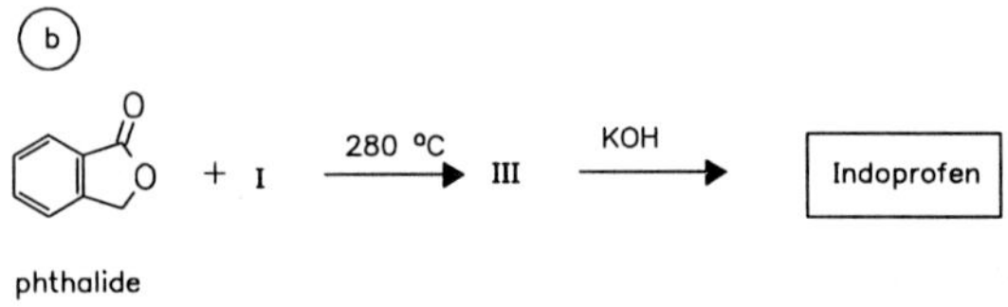

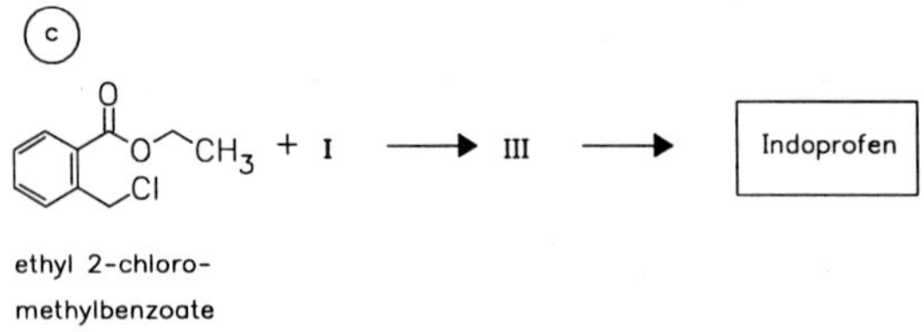

phthalic   ethyl 4-amino-α-
anhydride   methylphenylacetate (I)   (II)

phthalide

ethyl 2-chloro-
methylbenzoate

Reference(s):
DOS 2 154 525 (Carlo Erba; appl. 3.11.1971; I-prior. 5.11.1970, 10.11.1970).
US 3 767 805 (Ciba-Geigy; 23.10.1973; USA-prior. 27.3.1968, 3.9.1968, 13.1.1969, 18.3.1969, 18.7.1969, 8.9.1969, 3.2.1970).
DOS 2 034 240 (Ciba-Geigy; appl. 10.7.1970; USA-prior. 18.7.1969, 8.9.1969, 12.9.1969, 3.2.1970, 25.5.1970).
Nannini, G. et al.: Arzneim.-Forsch. (ARZNAD) **23**, 1090 (1973).

Formulation(s):   tabl. 200 mg

Trade Name(s):
J: Flosint (Carlo Erba)

Indoramin

ATC: C02CA02
Use: antihypertensive, α-adrenoceptor antagonist

RN: 26844-12-2 MF: $C_{22}H_{25}N_3O$ MW: 347.46 EINECS: 248-041-5
LD_{50}: 1800 mg/kg (R, p.o.)
CN: *N*-[1-[2-(1*H*-indol-3-yl)ethyl]-4-piperidinyl]benzamide

monohydrochloride
RN: 38821-52-2 MF: $C_{22}H_{25}N_3O \cdot HCl$ MW: 383.92 EINECS: 254-136-2

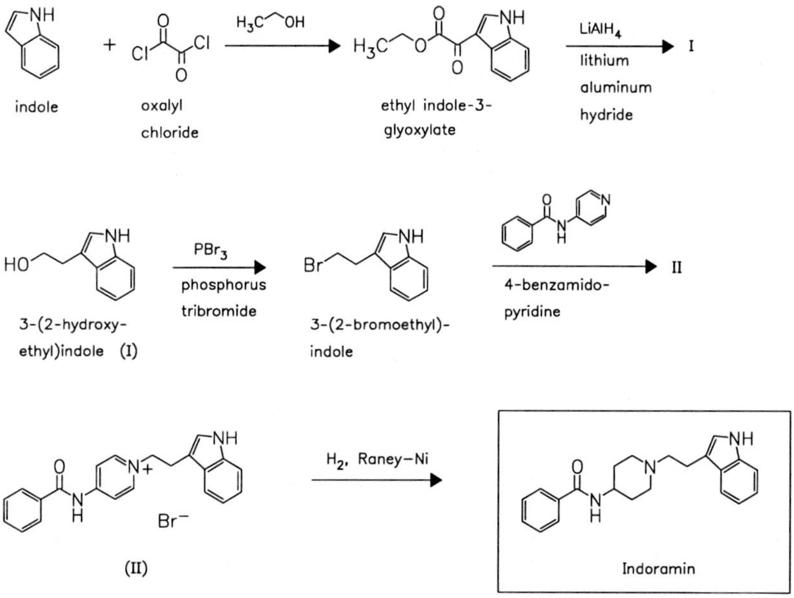

indole
oxalyl chloride
H3C OH
ethyl indole-3-glyoxylate
LiAlH4
lithium aluminum hydride
I
3-(2-hydroxyethyl)indole (I)
PBr3
phosphorus tribromide
3-(2-bromoethyl)indole
4-benzamidopyridine
II
Br−
(II)
H2, Raney-Ni
Indoramin

Reference(s):
Neumeyer, J.L. et al.: J. Med. Chem. (JMCMAR) **12**, 450 (1969).
DOS 1 770 460 (Wyeth; appl. 20.5.1968; GB-prior. 24.5.1967, 1.3.1968).

Formulation(s): tabl. 20 mg, 25 mg, 50 mg (as hydrochloride)

Trade Name(s):
D: Wydora (Brenner-Efeka)
F: Vidora (Wyeth)
GB: Baratol (Monmouth)
Doralese (Bencard)
I: Indorena (Lusofarmaco); wfm

Inositol

(Cyclohexitol; *meso*-Inositol; *myo*-Inositol)

ATC: A05
Use: liver therapeutic

RN: 87-89-8 MF: $C_6H_{12}O_6$ MW: 180.16 EINECS: 201-781-2
LD_{50}: 10 g/kg (M, p.o.);
>750 mg/kg (R, i.v.)
CN: *myo*-inositol

Inositol

Preparation by hydrolysis of phytin isolated from maize steep water [Ca- and Mg-salts of phytic acid (inositol hexa(dihydrogen phosphate))] with diluted sulfuric acid or with water under pressure.

Reference(s):
Bartow, E.B.; Walker, W.W.: Ind. Eng. Chem. (IECHAD) **30**, 300 (1938).
US 2 112 553 (E. B. Bartow, W. W. Walker; 1938; appl. 1935).
US 2 414 365 (American Cyanamid; 1947; appl. 1942).

synthesis from hexahydroxybenzene:
Wieland, H.; Wishart, R.S.: Ber. Dtsch. Chem. Ges. (BDCGAS) **47**, 2082 (1914).
Anderson, R.C.; Wallis, E.S.: J. Am. Chem. Soc. (JACSAT) **70**, 2931 (1948).

Formulation(s):   drg. 5 mg, 50 mg

Trade Name(s):
D: Geriatric Pharmaton   Enteroton (Panthox &   Lisacol Metionina
      (Pharmaton)-comb.   Burck)-comb.   (Lisapharma)-comb.; wfm
      Inosit-Zyma (Zyma); wfm   Equipar (Lampugnani)-   Neoepa (Vis)-comb.; wfm
      various generics and 50   comb.; wfm   Vitabil Composto (IBP)-
      more combination   Inobetin (Boniscontro &   comb.; wfm
      preparations   Gazzone)-comb.; wfm   USA: Amino-Cerv (Milex)-comb.
I: Colamin (UCM-Difme)   Inosital (Biomedica   Mega-B (Arco)-comb.
                                           Foscama); wfm   Megadose (Arco)-comb.

Inositol nicotinate

ATC: C04AC03
Use: vasodilator

RN: 6556-11-2   MF: $C_{42}H_{30}N_6O_{12}$ MW: 810.73   EINECS: 229-485-9
LD_{50}:   345 mg/kg (M, i.v.); >30 g/kg (M, p.o.);
       268 mg/kg (R, i.v.); >20 g/kg (R, p.o.)
CN: *myo*-inositol hexa-3-pyridinecarboxylate

nicotinic   nicotinoyl   Inositol nicotinate
acid   chloride

Reference(s):
Badgett; Woodward: J. Am. Chem. Soc. (JACSAT) **69**, 2907 (1947).
GB 1 053 689 (Bofors; appl. 19.11.1965; S-prior. 21.11.1964).

Formulation(s):   chewing tabl. 600 mg; tabl. 200 mg, 500 mg, 600 mg, 750 mg, 800 mg

Trade Name(s):

D: Hämovannad (Bastian-
    Werk)
    Hexanicit (Astra/Promed)
    Nicolip (Hennig)
    numerous combination
    preparations
F: Dilexpal (Winthrop); wfm
    Tensid (Bayer-Pharma)-
    comb.; wfm
GB: Hexopal (Sanofi Winthrop)
I: Angiokapsul (Schering)-
    comb.; wfm
    Esantene (Ibis); wfm
    Vascunicol (Boehringer
    Ing.); wfm
    Vasonicit (Ibis); wfm

J: Clevamin (Kowa)
    Cycnate (Toyo Pharmar)
    Ebelin (Samva)
    Hexalmin (Maruishi)
    Hexainosineat (Hishiyama)
    Hexanate (Nippon
    Chemiphar)
    Hexanicit (Yoshitomi)
    Hexate (Mohan)
    Hexatin (Kobayashi)
    Hexit (Toho)
    Inochinate (Nichiiko)
    Inosinit (Kanto)
    Kotanicit (Kotani)
    Mesonex (Tokyo Tanabe)
    Mesosit (Toyo Jozo)

    Nasky (Nikken)
    Neonitin (Chugai)
    Nicosamin (Toyama)
    Nicosinate (Toyo S.-Ono)
    Nicosinit (Hokuriku)
    Nicotol (Maruko)
    Nicoxatin (Fuso)
    Romanit (Kowa)
    Salex (Iwaki)
    Sannecit (Sanko)
    Secotinen (Seiko)
    Shikicit (Shiki)
    Xatolone (Showa)
    Yonomol (Sawai)

Iobenzamic acid

(Acide iobenzamique)

ATC: V08AC05
Use: X-ray contrast medium

RN: 3115-05-7   MF: $C_{16}H_{13}I_3N_2O_3$ MW: 662.00   EINECS: 221-484-1
LD_{50}: 530 mg/kg (M, i.v.); 2870 mg/kg (M, p.o.);
      500 mg/kg (R, i.v.); 2800 mg/kg (R, p.o.)
CN: *N*-(3-amino-2,4,6-triiodobenzoyl)-*N*-phenyl-β-alanine

Reference(s):
GB 870 321 (Österr. Stickstoffwerke; appl. 17.7.1959; A-prior. 23.7.1958, 2.8.1958).
DE 1 085 648 (Lentia; appl. 6.8.1958).
US 3 051 745 (Österr. Stickstoffwerke; 28.8.1962; A-prior. 23.7.1958).

Formulation(s):   tabl. 750 mg

Trade Name(s):

D: Bilibyk (Byk Gulden); wfm
Osbil (Byk Gulden); wfm
F: Osbil (Biodica); wfm
GB: Osbil (May & Baker); wfm
I: Bilibyk (Byk Gulden); wfm

Iocarmic acid

ATC: V08AA08
Use: X-ray contrast medium

RN: 10397-75-8 MF: $C_{24}H_{20}I_6N_4O_8$ MW: 1253.87 EINECS: 233-861-8
LD_{50}: 9.057 g/kg (M, i.v.); >16 g/kg (M, p.o.);
13.3 g/kg (R, i.v.); >16 g/kg (R, p.o.)
CN: 3,3'-[(1,6-dioxo-1,6-hexanediyl)diimino]bis[2,4,6-triiodo-5-[(methylamino)carbonyl]benzoic acid]

meglumine salt (1:2)
RN: 54605-45-7 MF: $C_{24}H_{20}I_6N_4O_8 \cdot 2C_7H_{17}NO_5$ MW: 1644.30 EINECS: 259-252-7
LD_{50}: 10.9 mg/kg (M, i.v.); >16 g/kg (M, p.o.);
13.3 mg/kg (R, i.v.); >16 g/kg (R, p.o.)

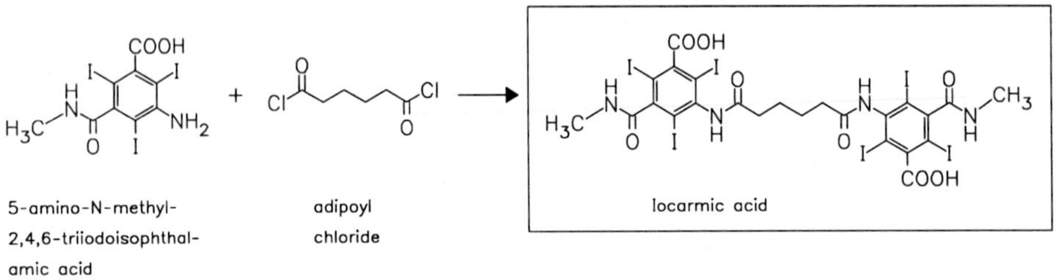
5-amino-N-methyl-2,4,6-triiodoisophthalamic acid
adipoyl chloride
Iocarmic acid

Reference(s):
US 3 290 366 (Mallinckrodt; 6.12.1966; appl. 6.3.1963).
GB 1 033 695 (Mallinckrodt; appl. 25.2.1964; USA-prior. 6.3.1963).

Formulation(s): amp. 3.02 g; inj. 60 % (as meglumine salt)

Trade Name(s):

D: Dimer-X (Byk Gulden); wfm
F: Dimer-X (Guerbet); wfm
GB: Dimer X (May & Baker); wfm
I: Osbil (Byk Gulden); wfm

Iocetamic acid

(Acide iocétamique)

ATC: V08AC07
Use: X-ray contrast medium

RN: 16034-77-8 MF: $C_{12}H_{13}I_3N_2O_3$ MW: 613.96 EINECS: 240-173-1
LD_{50}: 410 mg/kg (M, i.v.);
700 mg/kg (R, i.v.); 7100 mg/kg (R, p.o.)
CN: 3-[acetyl(3-amino-2,4,6-triiodophenyl)amino]-2-methylpropanoic acid

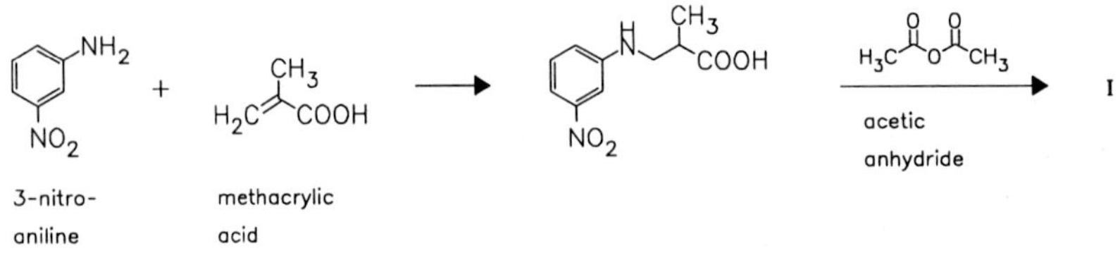
3-nitroaniline
methacrylic acid
acetic anhydride
I

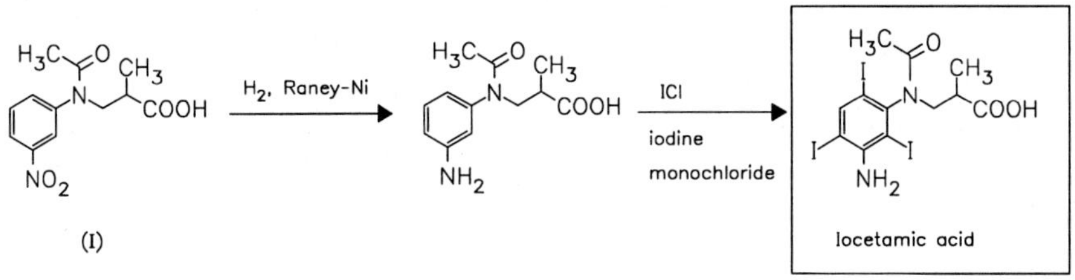

locetamic acid

Reference(s):

FR 5 997 M (Dagra; appl. 23.11.1966; prior. 25.11.1965, 26.5.1966).

Formulation(s):   tabl. 500 mg, 750 mg

Trade Name(s):

D: Cholebrine   F: Cholébrine (Schering);   USA: Cholebrine (Mallinckrodt);
     (Mundipharma); wfm   wfm   wfm

                                      I: Colebrin (Schering); wfm

---

Iodamide

(Jodamid)

ATC: V08AA03
Use: X-ray contrast medium

RN: 440-58-4   MF: $C_{12}H_{11}I_3N_2O_4$ MW: 627.94   EINECS: 207-125-1
LD$_{50}$:   >7 g/kg (M, p.o.);
        >7 g/kg (R, p.o.)
CN: 3-(acetylamino)-5-[(acetylamino)methyl]-2,4,6-triiodobenzoic acid

meglumine salt
RN: 18656-21-8   MF: $C_{12}H_{11}I_3N_2O_4 \cdot C_7H_{17}NO_5$ MW: 823.16   EINECS: 242-480-6

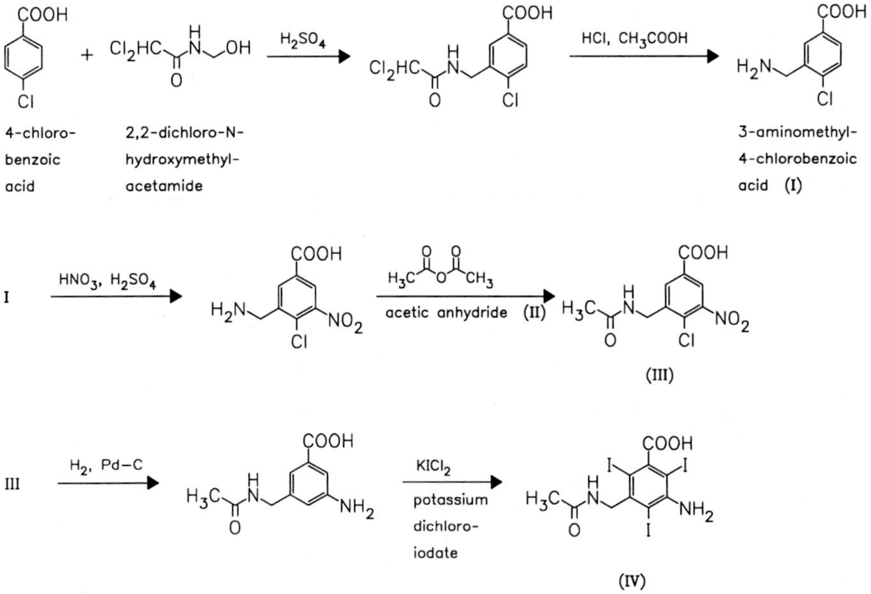

4-chloro-
benzoic
acid

2,2-dichloro-N-
hydroxymethyl-
acetamide

3-aminomethyl-
4-chlorobenzoic
acid (I)

(III)

(IV)

IV + II

COOH
I
I
H
H_3C
N
O
N
H
CH_3
O
I
Iodamide

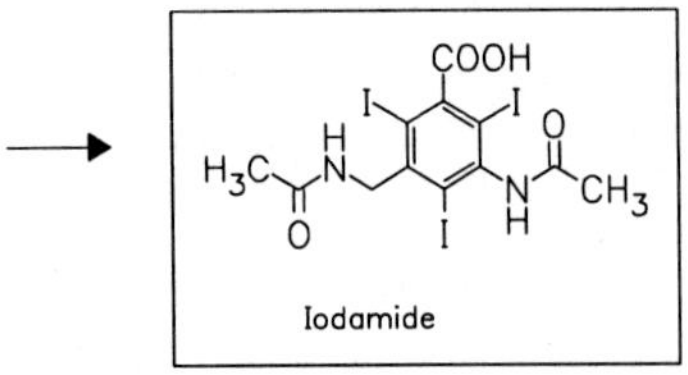

Reference(s):
US 3 360 436 (Eprova; appl. 12.11.1963; CH-prior. 23.11.1962, 9.8.1963).
GB 1 002 344 (Eprova; appl. 3.10.1963; CH-prior. 23.11.1962, 9.8.1963).
DE 1 273 747 (Eprova; appl. 24.9.1963; CH-prior. 23.11.1962).

Formulation(s): amp. 300 mg, 380 mg, 420 mg (as meglumine salt)

Trade Name(s):

D:	Uromiro (Heyden); wfm	Opacist E.R. (Bracco)	J:	Conraxin (Takeda)
I:	Isteropac E.R. (Bracco)	Uromiro (Bracco)		

Iodoxamic acid

ATC: V08AC01
Use: X-ray contrast medium (cholangiography)

RN: 31127-82-9 MF: $C_{26}H_{26}I_6N_2O_{10}$ MW: 1287.92 EINECS: 250-478-1
LD_{50}: 13.65 g/kg (M, i.v.)
CN: 3,3'-[(1,16-dioxo-4,7,10,13-tetraoxahexadecane-1,16-diyl)diimino]bis[2,4,6-triiodobenzoic acid]

meglumine salt (1:2)
RN: 51764-33-1 MF: $C_{26}H_{26}I_6N_2O_{10} \cdot 2C_7H_{17}NO_5$ MW: 1678.35 EINECS: 257-398-6

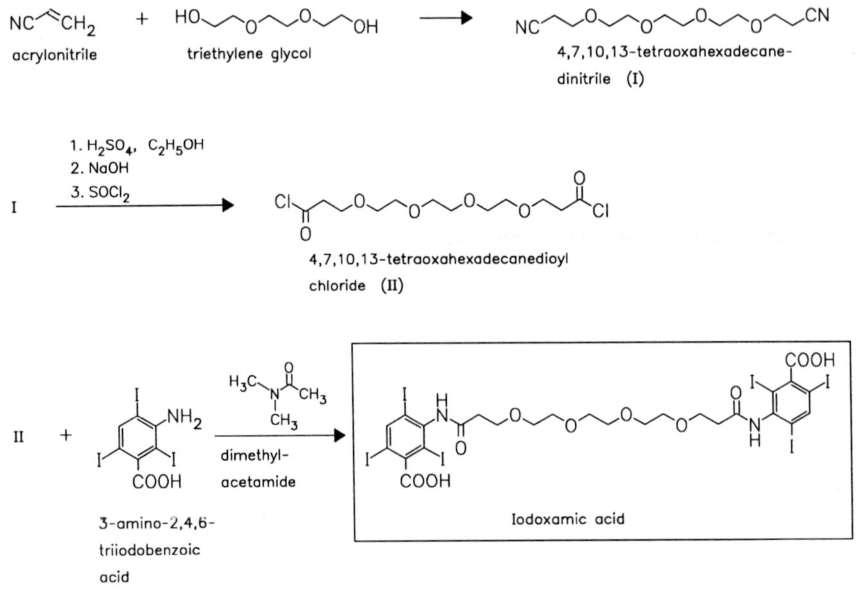

acrylonitrile
triethylene glycol
4,7,10,13-tetraoxahexadecane-dinitrile (I)
I
1. H_2SO_4, C_2H_5OH
2. NaOH
3. $SOCl_2$
4,7,10,13-tetraoxahexadecanedioyl chloride (II)
II
3-amino-2,4,6-triiodobenzoic acid
dimethyl-acetamide
Iodoxamic acid

Reference(s):
DE 1 937 211 (Bracco; appl. 22.7.1969).

Formulation(s): amp. 3.66 g, 5.49 g, 8.06 g, 9.91 g, 12.09 g (as meglumine salt)

Iofendylate
(Iophendylate)

ATC: V08AD04
Use: X-ray contrast medium
        (myelography)

RN: 99-79-6   MF: $C_{19}H_{29}IO_2$ MW: 416.34   EINECS: 202-787-8
LD_{50}:   2100 mg/kg (R, p.o.)
CN: 4-iodo-ι-methylbenzenedecanoic acid ethyl ester

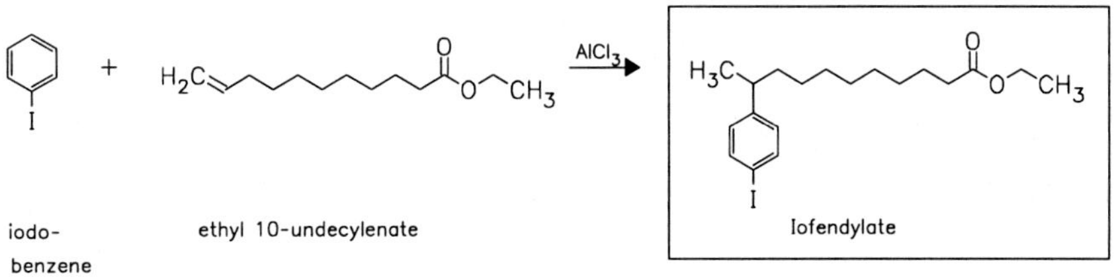

iodo-   ethyl 10-undecylenate   Iofendylate
benzene

Reference(s):
US 2 348 231 (Eastman Kodak; 1944; appl. 1940).

Formulation(s):   inj. sol.

Ioglycamic acid
(Acide ioglycamique)

ATC: V08AC03
Use: X-ray contrast medium

RN: 2618-25-9   MF: $C_{18}H_{10}I_6N_2O_7$ MW: 1127.71   EINECS: 220-048-8
CN: 3,3'-[oxybis[(1-oxo-2,1-ethanediyl)imino]]bis[2,4,6-triodobenzoic acid]

meglumine salt (1:2)
RN: 14317-18-1   MF: $C_{18}H_{10}I_6N_2O_7 \cdot 2C_7H_{17}NO_5$ MW: 1518.14

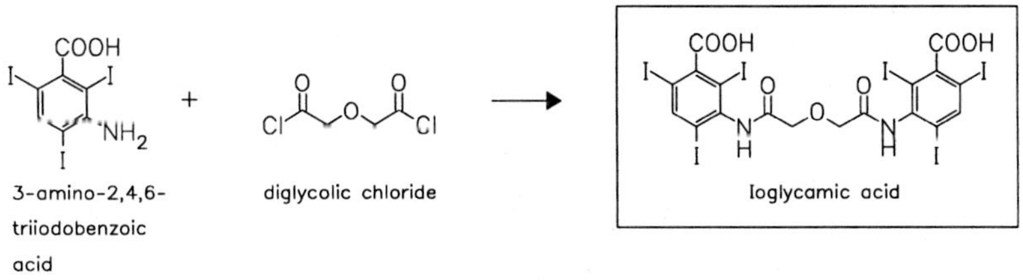

3-amino-2,4,6-   diglycolic chloride   Ioglycamic acid
triiodobenzoic
acid

Reference(s):
US 2 776 241 (Schering AG; 1957; D-prior. 1952).
US 2 853 424 (Schering AG; 1958; D-prior. 1952).
DE 936 928 (Schering AG; appl. 1952).
DE 962 698 (Schering AG; appl. 1952).
DE 962 699 (Schering AG; appl. 1953).
DE 1 006 428 (Schering AG; appl. 1955).

Formulation(s):　amp. 0.17 g/ml (as meglumine salt)

Trade Name(s):

D:	Biligram (Schering); wfm	GB:	Biligram (Schering	Bilivison (Schering); wfm
	Bilivistan (Schering); wfm		Chemicals); wfm	Bilivistan (Schering); wfm
F:	Biligram (Schering); wfm	I:	Biligram (Schering); wfm	

Iohexol

ATC:　V08AB02
Use:　X-ray contrast medium

RN:　66108-95-0　MF: $C_{19}H_{26}I_3N_3O_9$　MW: 821.14　EINECS: 266-164-2
LD$_{50}$:　50 g/kg (M, i.v.); >20 g/kg (M, p.o.);
　　　25.235 g/kg (R, i.v.); >20 g/kg (R, p.o.);
　　　>20 g/kg (dog, i.v.)
CN:　5-[acetyl(2,3-dihydroxypropyl)amino]-*N,N'*-bis(2,3-dihydroxypropyl)-2,4,6-triiodo-1,3-
　　　benzenedicarboxamide

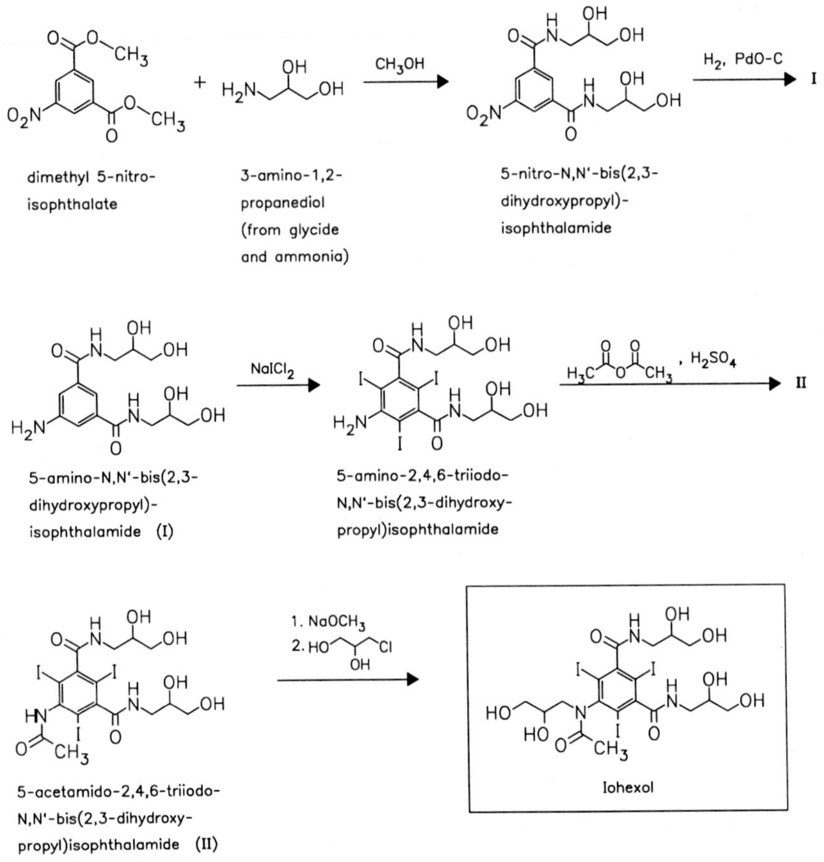

dimethyl 5-nitro-
isophthalate

3-amino-1,2-
propanediol
(from glycide
and ammonia)

5-nitro-*N,N'*-bis(2,3-
dihydroxypropyl)-
isophthalamide

5-amino-*N,N'*-bis(2,3-
dihydroxypropyl)-
isophthalamide　(I)

5-amino-2,4,6-triiodo-
N,N'-bis(2,3-dihydroxy-
propyl)isophthalamide

5-acetamido-2,4,6-triiodo-
N,N'-bis(2,3-dihydroxy-
propyl)isophthalamide　(II)

Iohexol

Reference(s):
DE 2 726 196 (Nyegaard; appl. 10.6.1977; GB-prior. 11.6.1976).
US 4 250 113 (Nyegaard; 10.2.1981; GB-prior. 11.6.1976).

Formulation(s):   amp. 240 mg, 300 mg, 350 mg

Trade Name(s):
D:	Accupaque (Nycomed) Omnipaque (Schering; 1983)	F:	Omnipaque (Nycomed; 1986)	J:	Lumopaque (Winthrop) Omnipaque (Daiichi)
		I:	Omnipaque (Schering; 1985)	USA:	Omnipaque (Winthrop-Breon); wfm

Iopamidol

ATC: V08AB04
Use: X-ray contrast medium

RN: 60166-93-0   MF: $C_{17}H_{22}I_3N_3O_8$ MW: 777.09   EINECS: 262-093-6
LD$_{50}$:   33 g/kg (M, i.v.); >49 g/kg (M, p.o.);
22.044 g/kg (R, i.v.);
35 g/kg (dog, i.v.)
CN: (S)-N,N'-bis[2-hydroxy-1-(hydroxymethyl)ethyl]-5-[(2-hydroxy-1-oxopropyl)amino]-2,4,6-triiodo-1,3-benzenedicarboxamide

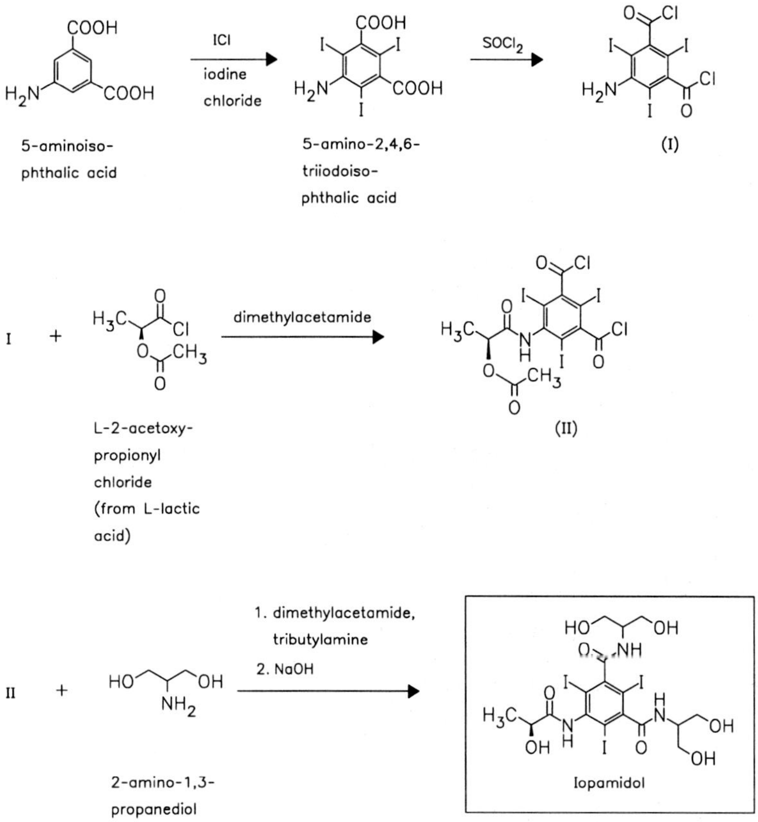

Reference(s):
DE 2 547 789 (Savac; appl. 24.10.1975; CH-prior. 13.12.1974).
US 4 001 323 (Savac; 4.1.1977; CH-prior. 13.12.1974).

Formulation(s):　　amp. 200 mg, 300 mg, 370 mg

Trade Name(s):
D:　Solutrast (Byk Gulden; 1981)　　　　GB:　Niopam (Merck; 1982); wfm　　　J:　Iopamiron (Nippon Schering)
F:　Iopamiron (Schering; 1982)　　I:　Giastromiro (Bracco)　　　　　　　USA:　Isovue (Squibb); wfm
　　　　　　　　　　　　　　　　　　　Iopamiro (Bracco)

Iopanoic acid
(Acidum iopanoicum)

ATC:　V08AC06
Use:　X-ray contrast medium

RN:　96-83-3　MF: $C_{11}H_{12}I_3NO_2$　MW: 570.93　EINECS: 202-539-9
LD$_{50}$:　320 mg/kg (M, i.v.); 6600 mg/kg (M, p.o.);
　　　280 mg/kg (R, i.v.); 1540 mg/kg (R, p.o.)
CN:　3-amino-α-ethyl-2,4,6-triiodobenzenepropanoic acid

monosodium salt
RN:　2497-78-1　MF: $C_{11}H_{11}I_3NNaO_2$　MW: 592.92　EINECS: 219-683-3
LD$_{50}$:　296 mg/kg (M, i.v.); 1602 mg/kg (M, p.o.);
　　　332 mg/kg (R, i.v.); 2986 mg/kg (R, p.o.)

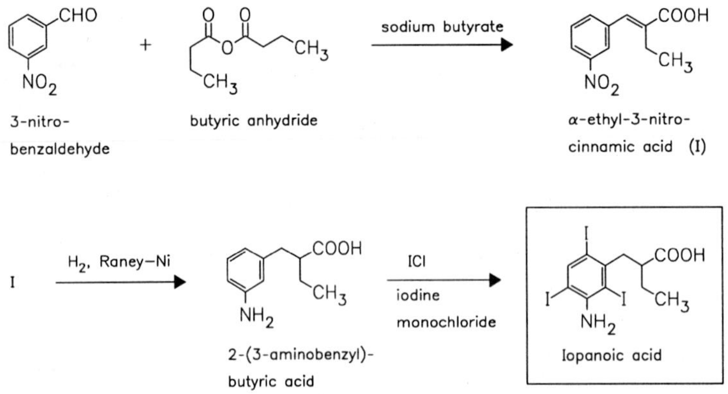

3-nitro-benzaldehyde　+　butyric anhydride　→ sodium butyrate → α-ethyl-3-nitro-cinnamic acid (I)

I → H$_2$, Raney–Ni → 2-(3-aminobenzyl)-butyric acid → ICl iodine monochloride → Iopanoic acid

Reference(s):
US 2 705 726 (Sterling Drug; 1955; prior. 1949).

Formulation(s):　　cps. 500 mg (as sodium salt); powder 375 mg/g (as calcium salt)

Trade Name(s):
D:　Telepaque (Winthrop); wfm　　I:　Cistobil (Bracco)　　　　　　　　Molpaque (Tokyo Tanabe)
F:　Télépaque (Winthrop); wfm　　　　Telepaque (Winthrop); wfm　　　　Telepaque (Kodama)
GB:　Telepaque (Winthrop); wfm　　J:　Ace-Line (Maruishi)　　　　USA:　Telepaque (Winthrop); wfm
　　　　　　　　　　　　　　　　　　Leabar (Toyo S.-Ono)

Iophenoic acid
(Acidum iophenoicum; Iophenoxic acid)

ATC: V08AD
Use: X-ray contrast medium

RN: 96-84-4   MF: $C_{11}H_{11}I_3O_3$ MW: 571.92
LD$_{50}$: 374 mg/kg (M, i.v.); 1850 mg/kg (M, p.o.);
2 g/kg (R, p.o.);
203 mg/kg (dog, i.v.)
CN: α-ethyl-3-hydroxy-2,4,6-triiodobenzenepropanoic acid

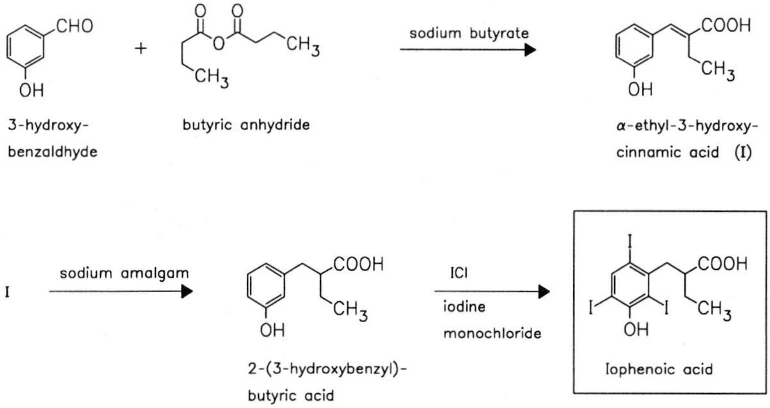

3-hydroxy-
benzaldhyde

butyric anhydride

sodium butyrate

α-ethyl-3-hydroxy-
cinnamic acid (I)

sodium amalgam

2-(3-hydroxybenzyl)-
butyric acid

ICl
iodine
monochloride

Iophenoic acid

Reference(s):
US 2 931 830 (Sterling Drug; 5.4.1960; appl. 20.3.1952).
GB 726 987 (Sterling Drug; appl. 1953; USA-prior. 1952).

Trade Name(s):
USA: Teridax (Schering); wfm

Iopydol

ATC: V08AD02
Use: X-ray contrast medium

RN: 5579-92-0   MF: $C_8H_9I_2NO_3$ MW: 420.97   EINECS: 226-968-6
CN: 1-(2,3-dihydroxypropyl)-3,5-diiodo-4(1H)-pyridinone

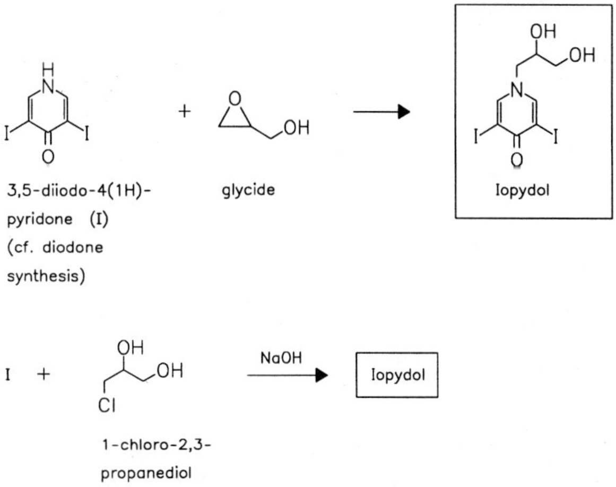

3,5-diiodo-4(1H)-
pyridone (I)
(cf. diodone
synthesis)

glycide

Iopydol

1-chloro-2,3-
propanediol

NaOH

Iopydol

Reference(s):

DRP 579 224 (I. G. Farben; appl. 1930).

Formulation(s): vial 20 mg 3,5-diiodo-4(1*H*)-pyridone/ml in comb. with iopydol

Trade Name(s):

D:	Hydrast (Byk Gulden); wfm	I:	Hytrast (Byk Gulden)-comb. with iopydone; wfm	J:	Hydrast (Guerbet-Kodama)-comb. with iopydone
F:	Hydrast (Guerbet)-comb. with iopydone; wfm				

Iotalamic acid

(Iothalamic acid; Acide iotalamique)

ATC: V08AA04
Use: X-ray contrast medium

RN: 2276-90-6 MF: $C_{11}H_9I_3N_2O_4$ MW: 613.92 EINECS: 218-897-4
CN: 3-(acetylamino)-2,4,6-triiodo-5-[(methylamino)carbonyl]benzoic acid

monosodium salt
RN: 1225-20-3 MF: $C_{11}H_8I_3N_2NaO_4$ MW: 635.90 EINECS: 214-955-8
LD_{50}: 19.2 g/kg (M, i.v.)
meglumine salt
RN: 13087-53-1 MF: $C_{11}H_9I_3N_2O_4 \cdot C_7H_{17}NO_5$ MW: 809.13 EINECS: 235-998-9
LD_{50}: 8.1 g/kg (M, i.v.);
10.5 g/kg (R, i.v.)

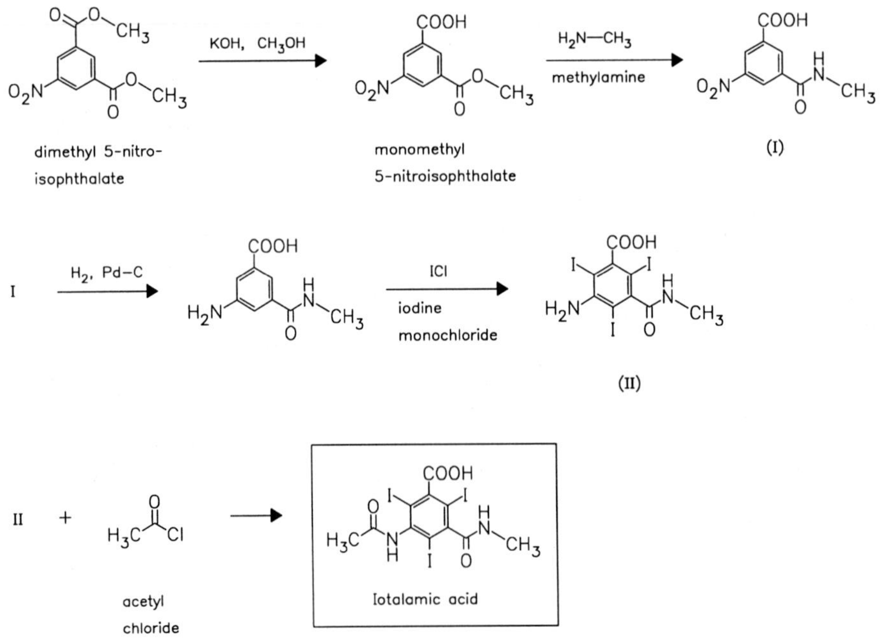

KOH, CH3OH
H2N—CH3
methylamine
dimethyl 5-nitro-isophthalate
monomethyl 5-nitroisophthalate
(I)
I
H2, Pd–C
ICl
iodine monochloride
(II)
II +
acetyl chloride
Iotalamic acid

Reference(s):

Hoey, G.B. et al.: J. Med. Chem. (JMCMAR) **6**, 24 (1963).
US 3 145 197 (Mallinckrodt; 18.8.1964; appl. 26.6.1961; prior. 25.8.1960).
GB 994 215 (Mallinckrodt; appl. 18.8.1961; USA-prior. 25.8.1960).

Formulation(s): sol. 17 %, 24 %, 36 %, 43 %, 60 %, 66.8 %, 80 % (as meglumine salt)

Trade Name(s):

D:	Conray (Byk Gulden); wfm	I:	Angio-Conray 80 (Bracco)
F:	Contrix 28 (Guerbet); wfm		Conray (Bracco)
GB:	Conray (May & Baker);	J:	Angio-Conray (Daiichi)
	wfm		Conray (Daiichi)
	Gastro-Conray (May &	USA:	Angio-Conray
	Baker); wfm		(Mallinckrodt); wfm

Conray (Mallinckrodt);
wfm
Cysto-Conray
(Mallinckrodt); wfm
Vascoray (Mallinckrodt);
wfm

Iotrolan

(DL-3117; Iotrol)

ATC: V08AB06
Use: X-ray contrast medium (water
soluble, non-ionic, for myelography
and contrast enhancement in CT)

RN: 79770-24-4   MF: $C_{37}H_{48}I_6N_6O_{18}$ MW: 1626.24
LD$_{50}$:   > 26 g/kg (M, i.v.);
12.7 g/kg (R, i.v.)
CN: 5,5'-[1,3-dioxo-1,3-propanediylbis(methylimino)]bis[*N,N'*-bis[2,3-dihydroxy-1-(hydroxymethyl)propyl]-
2,4,6-triiodo-1,3-benzenedicarbamide]

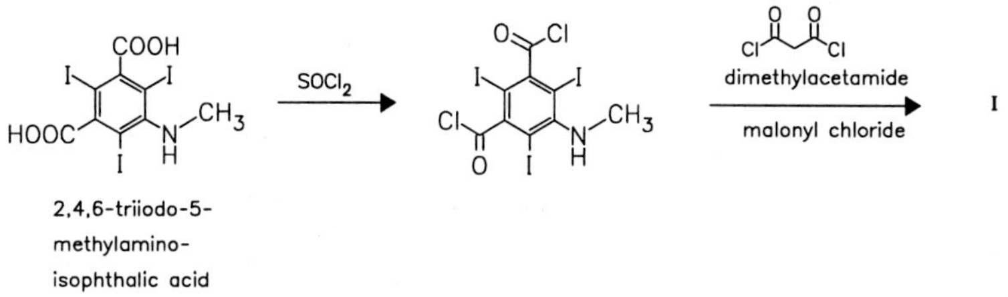

2,4,6-triiodo-5-
methylamino-
isophthalic acid

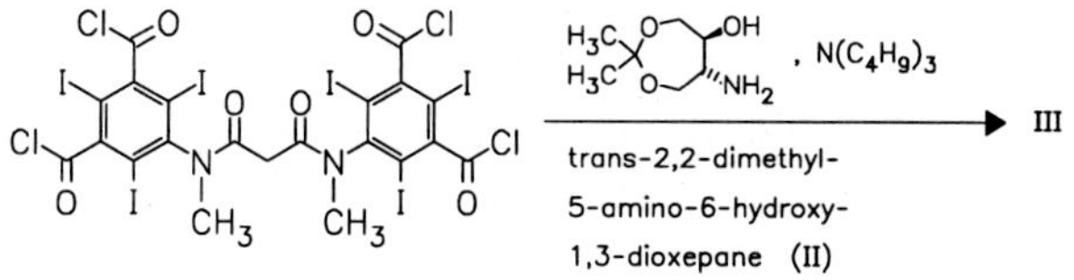

trans-2,2-dimethyl-
5-amino-6-hydroxy-
1,3-dioxepane (II)

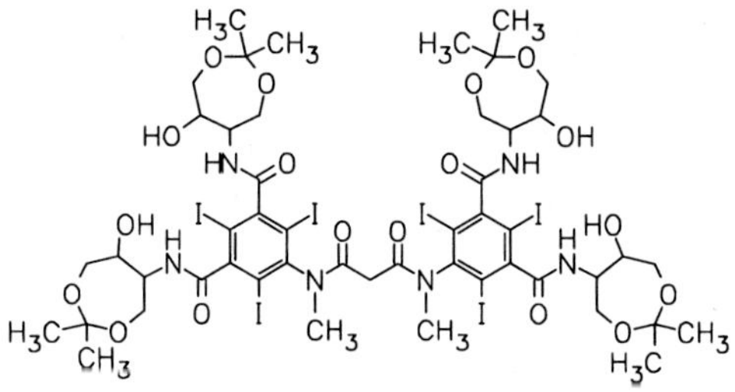

Iotrolan

90% aq. CF₃COOH

III →

preparation of II:

H₃C O—CH₃
H₃C O—CH₃   +   HO / HO

2,2-dimethoxy-
propane

1,2-dihydroxy-
2-butene

→

H₃C O / H₃C O

2,2-dimethyl-
4,7-dihydro-1,3-
dioxepin (IV)

H₃C—CN, CH₃OH, 30% H₂O₂,
40 °C, 5 h

IV ─────────────→
acetonitrile, hydrogen peroxide

H₃C O / H₃C O / O

4,4-dimethyl-
3,5,8-trioxa-
bicyclo[5.1.0]-
octane

liq.NH₃, 120 °C, 4 h
──────────────→   II

The epoxidation step can also be performed with *m*-chloroperoxybenzoic acid in CH₂Cl₂.

Reference(s):
US 4 341 756 (The Regents of the Univ. of Calif.; 27.7.1982; prior. 31.1.1980, 17.4.1980).
EP 33 426 (The Regents of the Univ. of Calif.; appl. 30.12.1980; USA-prior. 31.1.1980, 17.4.1980).

preparation of trans-2,2-dimethyl-5-amino-6-hydroxy-1,3-dioxepane:
US 4 439 613 (The Regents of the Univ. of Calif.; 27.5.1984; prior. 31.1.1980, 17.4.1980, 5.4.1982).

review:
Dawson, P.; Howell, M.: Br. J. Radiol. (BJRAAP) **59**, 987 (1986).

Formulation(s):   vial 10 ml and 20 ml (513 mg/ml), 10 ml (641 mg/ml)

Trade Name(s):
D: Isovist (Schering; 1988)   J: Isovist (Nihon Schering;
I: Isovist (Schering)   1987)

Iotroxic acid

ATC: V08AC02
Use: X-ray contrast medium

RN: 51022-74-3   MF: C₂₂H₁₈I₆N₂O₉   MW: 1215.82   EINECS: 256-917-3
LD₅₀: 2820 mg/kg (M, i.v.); >9 g/kg (M, p.o.);
4190 mg/kg (R, i.v.); >9 g/kg (R, p.o.)
CN: 3,3'-[oxybis[2,1-ethanediyloxy(1-oxo-2,1-ethanediyl)imino]]bis[2,4,6-triiodobenzoic acid]

HO ~O~O~O~ OH

HNO₃ →

HOOC~O~O~O~COOH

SOCl₂ →   I

tetraethylene glycol

3,6,9-trioxaundecanedioic acid

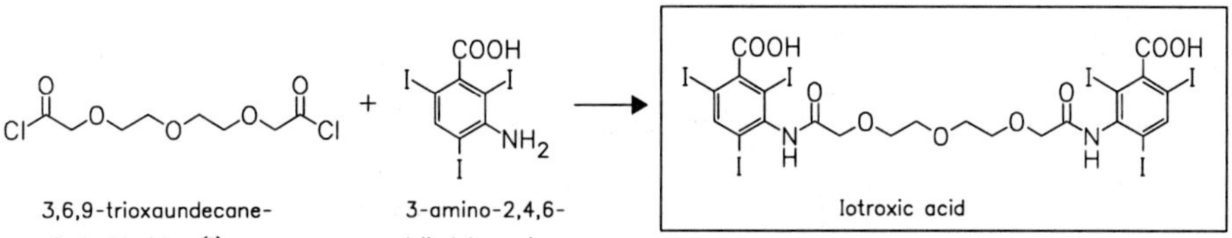

3,6,9-trioxaundecane-
dioyl chloride (I)

3-amino-2,4,6-
triiodobenzoic
acid

Iotroxic acid

Reference(s):
DAS 2 405 652 (Schering AG; appl. 4.2.1974).

Formulation(s):   vial 10.5 g/100 ml, 11.4 g/50 ml

Trade Name(s):
D: Biliscopin (Schering)   F: Biliscopine (Schering);   I: Chologram (Schering)
                                                   wfm   J: Biliscopin (Schering)

Ioxaglic acid
(Acide ioxaglique)

ATC: V08AB03
Use: X-ray contrast medium

RN: 59017-64-0   MF: C$_{24}$H$_{21}$I$_6$N$_5$O$_8$ MW: 1268.88   EINECS: 261-560-1
LD$_{50}$:   >13.3 g/kg (M, i.v.);
        13.3 g/kg (R, i.v.); 13.3 g/kg (R, p.o.);
        >10.7 g/kg (dog, i.v.)
CN: 3-[[[[3-(acetylmethylamino)-2,4,6-triiodo-5-[(methylamino)carbonyl]benzoyl]amino]acetyl]amino]-5-
        [[(2-hydroxyethyl)amino]carbonyl]-2,4,6-triiodobenzoic acid

monosodium salt
RN: 67992-58-9   MF: C$_{24}$H$_{20}$I$_6$N$_5$NaO$_8$ MW: 1290.87   EINECS: 268-060-2
meglumine salt
RN: 59018-13-2   MF: C$_{24}$H$_{21}$I$_6$N$_5$O$_8$ · C$_7$H$_{17}$NO$_5$ MW: 1464.10   EINECS: 261-561-7

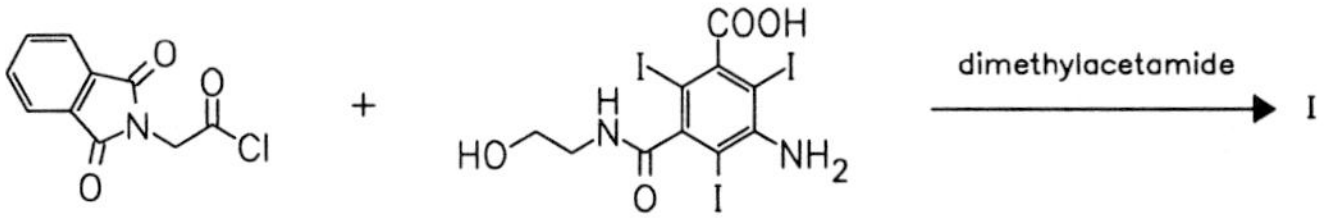

N-phthaloyl-
glycyl chloride

2,4,6-triiodo-3-(2-
hydroxyethylcarbamoyl)-
5-aminobenzoic acid
(cf. ioxitalamic acid
synthesis)

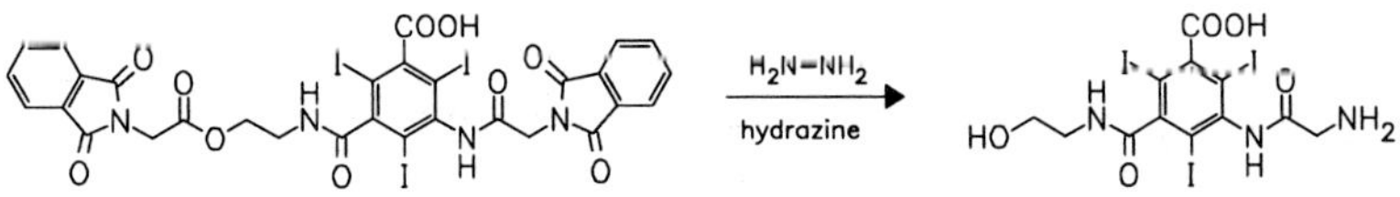

dimethylacetamide

(I)

2,4,6-triiodo-3-(2-hydroxy-
ethylcarbamoyl)-5-amino-
acetamidobenzoic acid (II)

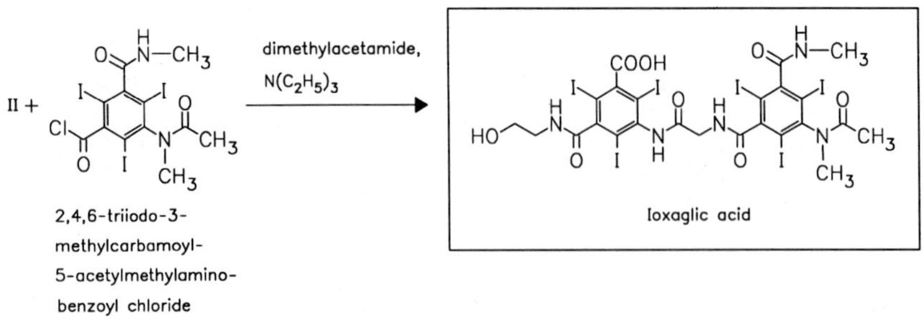

2,4,6-triiodo-3-
methylcarbamoyl-
5-acetylmethylamino-
benzoyl chloride

Ioxaglic acid

Reference(s):
US 4 014 986 (Guerbet; 29.3.1977; appl. 20.5.1975; GB-prior. 31.7.1974).
DE 2 523 567 (Guerbet; appl. 28.5.1975; GB-prior. 31.5.1974, 31.7.1974).
US 4 055 188 (Guerbet; 25.1.1977; GB-prior. 31.5.1974).

Formulation(s):   amp. 393 mg/ml (as sodium salt)

Trade Name(s):
D: Hexabrix (Guerbet; 1979)   I: Hexabrix (Byk Gulden)
F: Hexabrix (Guerbet; 1980)   J: Hexabrix (Eiken)

Ioxitalamic acid
(Acide ioxitalamique)

ATC: V08AA05
Use: X-ray contrast medium

RN: 28179-44-4   MF: $C_{12}H_{11}I_3N_2O_5$ MW: 643.94   EINECS: 248-887-5
CN: 3-(acetylamino)-5-[[(2-hydroxyethyl)amino]carbonyl]-2,4,6-triiodobenzoic acid

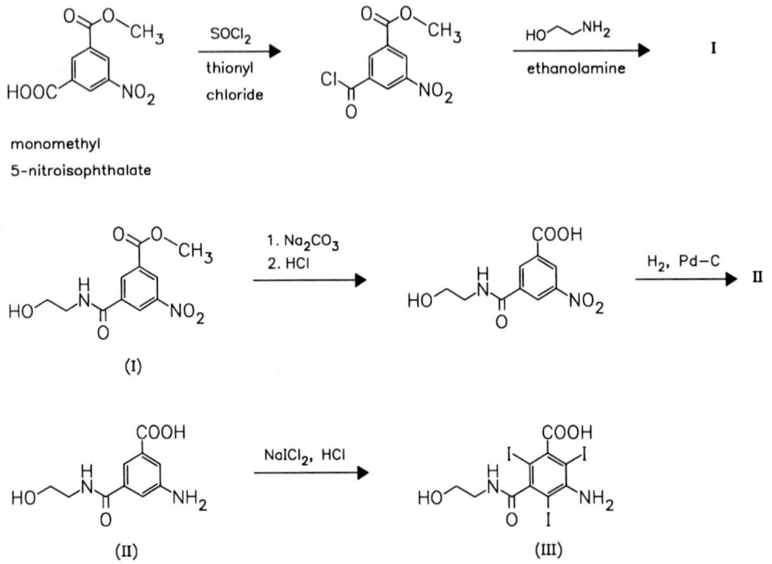

monomethyl
5-nitroisophthalate

(I)

(II)   (III)

acetyl
chloride

Ioxitalamic acid

Reference(s):
DOS 1 928 838 (Nyegaard; appl. 6.6.1969; GB-prior. 10.6.1968).

Formulation(s):   amp. 397.2 mg, 660.3 mg/ml (as meglumine salt)

Trade Name(s):
D: Telebrix (Byk Gulden)   Télébrix (Guerbet)-comb.
F: Télébrix (Guerbet)   I: Telebrix (Byk Gulden)

Ipratropium bromide

ATC: R01AX03; R03BB01
Use: bronchodilator

RN: 22254-24-6   MF: $C_{20}H_{30}BrNO_3$ MW: 412.37   EINECS: 244-873-8
LD_{50}: 12.29 mg/kg (M, i.v.); 1001 mg/kg (M, p.o.);
     15.7 mg/kg (R, i.v.); 1663 mg/kg (R, p.o.);
     1300 mg/kg (dog, p.o.)
CN: (endo,syn)-(±)-3-(3-hydroxy-1-oxo-2-phenylpropoxy)-8-methyl-8-(1-methylethyl)-8-azoniabicyclo[3.2.1]octane bromide

N-isopropylnoratropine

methyl
bromide

Ipratropium bromide

Reference(s):
DE 1 670 177 (Boehringer Ing.; prior. 28.12.1966).
US 3 505 337 (Boehringer Ing.; 7.4.1970; D-prior. 22.12.1967).

inhalation spray, also in combination with mucolytica and/or sympathomimetic effective bronchodilators:
US 3 681 500 (Boehringer Ing.; 1.8.1972; D-prior. 12.12.1969).

Formulation(s):   aerosol inhalation 20 µg/metered inhalation, 40 µg/metered inhalation; inhalation sol. 250 µg/ml; inhalation cps. 20 µg, 40 µg; doses aerosol 0.25 mg, 0.4 mg; doses aerosol susp. 0.02 mg; sol. 0.02 mg, 0.25 mg/2 ml, 0.5 mg/2 ml

Trade Name(s):
D: Atrovent (Boehringer Ing.; 1975)   Bronchodual (Boehringer Ing.)-comb.   numerous combination preparations
     Berodual (Boehringer Ing.; 1980)-comb.   Combivent (Boehringer Ing.)-comb.   I: Atem (Chiesi)
     Itrop (Boehringer Ing.)   GB: Atrovent (Boehringer Ing.)   Breva (Valeas)-comb.
F: Atrovent (Boehringer Ing.)   Duovent (Boehringer Ing.)-comb.   Duovent (Boehringer Ing.)-comb.
               Iprafen (Chiesi)-comb.

| J: | Atrovent (Tejim Phar.) | USA: | Atrovent (Boehringer Ing.; 1987) | | Combivent (Boehringer Ing.) |

Ipriflavone
(FL-113; TC-80)

ATC: M05BX01
Use: calcium regulator (for treatment of osteoporosis)

RN: 35212-22-7 MF: $C_{18}H_{16}O_3$ MW: 280.32
LD$_{50}$: >2.5 g/kg (M, i.p.); 3185 mg/kg (M, p.o.); >5 g/kg (M, s.c.);
>2.5 g/kg (R, i.p.); 2.5 g/kg (R, p.o.); >5 g/kg (R, s.c.)
CN: 7-(1-methylethoxy)-3-phenyl-4H-1-benzopyran-4-one

2,4-dihydroxyphenyl benzyl ketone + isopropyl bromide → (K$_2$CO$_3$) → 2-hydroxy-4-isopropoxy-phenyl benzyl ketone (I)

I + triethyl orthoformate → (morpholine) → Ipriflavone

Reference(s):
US 4 166 862 (Chinoin; 4.9.1979; appl. 25.5.1971; prior. 16.5.1974; H-prior. 27.5.1970).
DE 2 125 245 (Chinoin; appl. 21.5.1971; H-prior. 27.5.1970).
Szuk, G. et al.: Magy. Kem. Lapja (MGKLAL) **43**, 401, (1988) (CA **110**, 179494 d).

synthesis of I:
GB 1 374 925 (Chinoin; appl. 30.11.1972; H-prior. 2.12.1971).

pharmaceutical formulations:
JP 53 133 635 (Chinoin; appl. 20.4.1978; H-prior. 20.4.1978).
EP 129 893 (Takeda; appl. 23.6.1984; J-prior. 28.6.1983).
US 4 772 627 (Takeda; 20.9.1988; appl. 15.1.1987; J-prior. 8.6.1984, 23.6.1984, 28.6.1983).

cyclodextrin clathrates:
EP 214 647 (Chinoin; appl. 9.9.1986; J-prior. 10.9.1985).

medical use for treatment of heart and lung diseases:
JP 53 133 635 (Chinoin; appl. 20.4.1978; H-prior. 20.4.1977).

medical use for treatment of climacteric disorders:
EP 129 667 (Takeda; appl. 25.4.1984; J-prior. 26.4.1983).

dental compositions:
EP 349 535 (Reanal Finomvegyszergyar; appl. 29.1.1988; H-prior. 3.2.1987).

Formulation(s): tabl. 200 mg

Trade Name(s):
I: Iprosten (Takeda; 1991) Osteofix (Chiesi; 1991) J: Osten (Takeda; 1989)

Iproniazid

ATC: N06AF045
Use: tuberculostatic, psychoenergetic, antidepressant

RN: 54-92-2 MF: $C_9H_{13}N_3O$ MW: 179.22 EINECS: 200-218-8
LD_{50}: 719 mg/kg (M, i.v.); 681 mg/kg (M, p.o.);
365 mg/kg (R, p.o.);
95 mg/kg (dog, p.o.)
CN: 4-pyridinecarboxylic acid 2-(1-methylethyl)hydrazide

phosphate
RN: 305-33-9 MF: $C_9H_{13}N_3O \cdot H_3PO_4$ MW: 277.22

H2, Pt
isoniazid
(q. v.)
acetone
Iproniazid

Reference(s):
US 2 685 585 (Hoffmann-La Roche; 1954; prior. 1951).

Formulation(s): tabl. 50 mg (as phosphate)

Trade Name(s):
F: Marsilid (Laphal)
GB: Marsilid (Roche); wfm
I: Ellepibina (LPB)-comb.; wfm

Irbesartan

(BMS-186295; SR-47436)

ATC: C09CA04
Use: antihypertensive, angiotensin II antagonist

RN: 138402-11-6 MF: $C_{25}H_{28}N_6O$ MW: 428.54
CN: 2-butyl-3-[[2'-(1*H*-tetrazol-5-yl)[1,1'-biphenyl]-4-yl]methyl]-1,3-diazaspiro[4.4]non-1-en-4-one

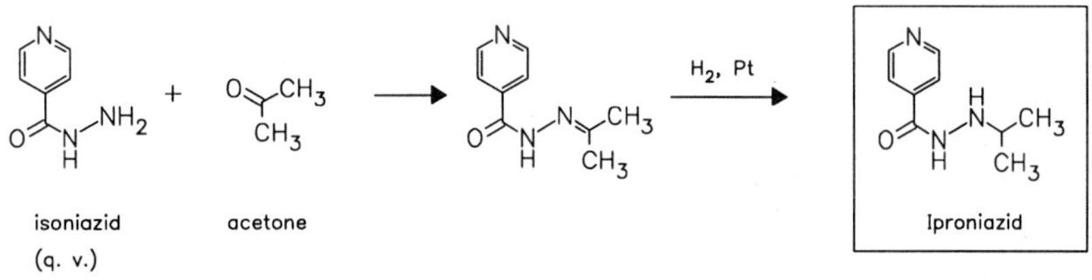

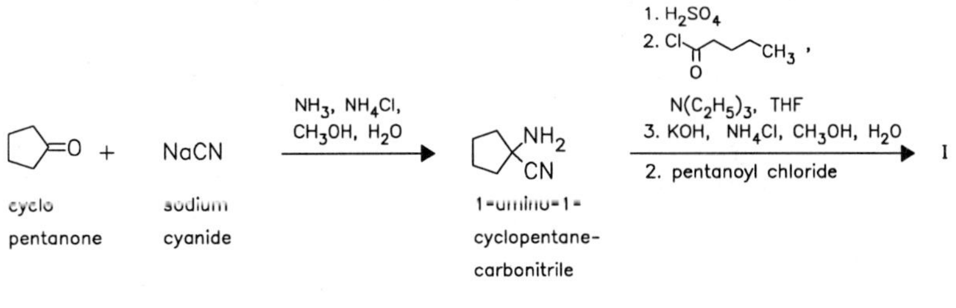

NaCN
NH3, NH4Cl, CH3OH, H2O
1. H2SO4
2. pentanoyl chloride, N(C2H5)3, THF
3. KOH, NH4Cl, CH3OH, H2O
I
cyclopentanone
sodium cyanide
1-amino-1-cyclopentanecarbonitrile

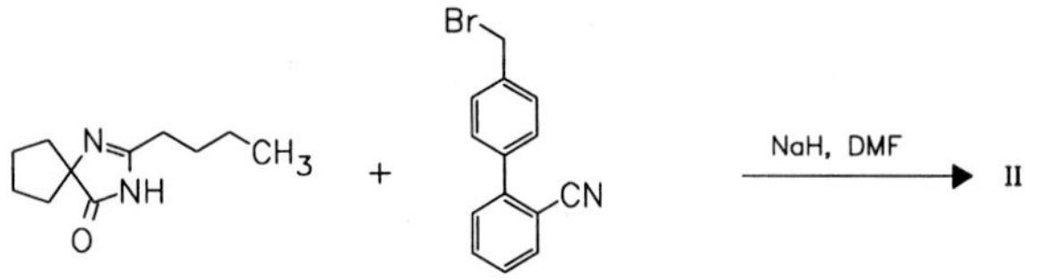

2-butyl-1,3-diaza-
spiro[4.4]non-1-en-
4-one (I)

4'-(bromomethyl)-
biphenyl-2-carbo-
nitrile
(cf. losartan
synthesis)

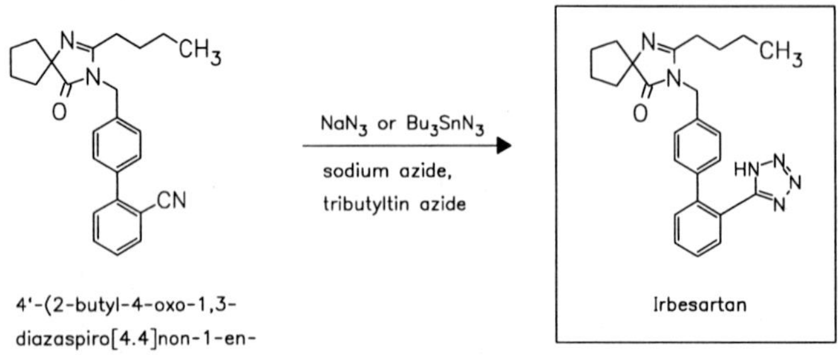

NaH, DMF
→ II

NaN₃ or Bu₃SnN₃

sodium azide,
tributyltin azide

4'-(2-butyl-4-oxo-1,3-
diazaspiro[4.4]non-1-en-
3-ylmethyl)biphenyl-2-
carbonitrile (II)

Irbesartan

alternative synthesis of intermediate I:

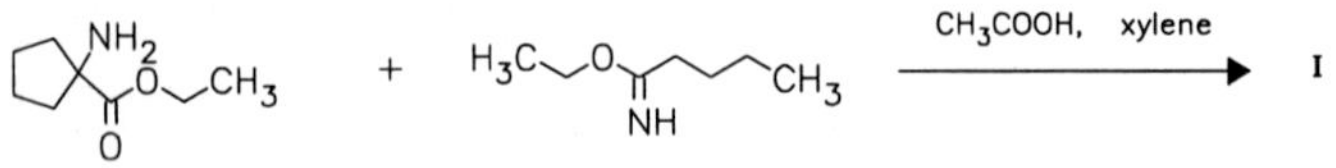

ethyl 1-amino-
1-cyclopentane-
carboxylate

ethyl pentan-
imidate

CH₃COOH, xylene
→ I

Reference(s):
US 5 270 317 (Elf Sanofi; 14.12.1993; F-prior. 20.3.1990, 1.9.1990, 10.9.1991).
Bernhart, C.A. et al.: J. Med. Chem. (JMCMAR) **36**, 3371-3380 (1993).
EP 454 511 (Sanofi; appl. 20.3.1991; F-prior. 20.3.1990, 8.8.1990).

Formulation(s):   tabl. 75 mg, 150 mg, 300 mg

Trade Name(s):

D:	Aprovel (Sanofi Winthrop; 1997)		KARVEZIDETM (BMS)-comb.	USA: Avapro (BMS)-comb.
	COAPROVEL (Sanofi Synthelabo; BMS)-comb.	F:	Aprovel (Bristol-Myers Squibb)	Avalide (Sanofi Synthelabo)
	Karvea (Bristol-Myers Squibb)	GB:	Aprovel (Bristol-Myers Squibb/Sanofi)	

Irinotecan

(CPT-11; DQ-2805; NSC-616348)

ATC: L01XX19
Use: antineoplastic, topoisomerase
inhibitor

RN: 97682-44-5   MF: C₃₃H₃₈N₄O₆   MW: 586.69
CN: (S)-[1,4'-bipiperidine]-1'-carboxylic acid 4,11-diethyl-3,4,12,14-tetrahydro-4-hydroxy-3,14-dioxo-1H-
pyrano[3',4':6,7]indolizino[1,2-b]quinolin-9-yl ester

monohydrochloride
RN: 100286-90-6   MF: $C_{33}H_{38}N_4O_6 \cdot HCl$ MW: 623.15
monohydrochloride trihydrate
RN: 136572-09-3   MF: $C_{33}H_{38}N_4O_6 \cdot HCl \cdot 3H_2O$ MW: 677.20
LD$_{50}$:   132 mg/kg (M, i.v.); 1045 mg/kg (M, p.o.);
        83.6 mg/kg (R, i.v.); 867 mg/kg (R, p.o.);
        40 mg/kg (dog, i.v.)
racemate
RN: 130144-33-1   MF: $C_{33}H_{38}N_4O_6$ MW: 586.69

7-ethyl-10-hydroxy-
camptothecin

phosgene

7-ethyl-10-(chlorocarbonyl-
oxy)camptothecin (I)

4-piperidino-
piperidine

Irinotecan

Reference(s):
EP 137 145 (Yakult Honsha; appl. 14.7.1983; J-prior. 14.7.1983).
Henegar, K.E. et al.: J. Org. Chem. (JOCEAH) **62**, 6588 (1997).

slow release formulation:
JP 07 277 981 (Daiichi Pharm.; appl. 12.4.1994; J-prior. 12.4.1994).

synergistic combinations:
JP 04 208 224 (Daiichi Pharm.; appl. 30.11.1990; J-prior. 30.11.1990).
WO 9 309 782 (SmithKline Beecham; appl. 13.11.1992; USA-prior. 15.11.1991).
WO 9 410 995 (Rhône-Poulenc Rorer; appl. 10.11.1992; F-prior. 10.11.1992).

Formulation(s):   vial 40 mg/2 ml, 100 mg/ 5 ml (as hydrochloride)

Trade Name(s):
F: Campto (Rhône-Poulenc   J: Campto (Yakult Housha; as   USA: Camptosar (Pharmacia &
       Rorer; as hydrochloride)   hydrochloride)   Upjohn)
GB: Campto (Rhône-Poulenc   Topotecin (Daiichi
       Rorer; as hydrochloride)   Seiyaku; as hydrochloride)

Isepamicin
(Sch-21420)

ATC: J01GB11; J01KD
Use: aminoglycoside antibiotic (against
        urinary and respiratory tract
        infections)

RN: 58152-03-7   MF: $C_{22}H_{43}N_5O_{12}$ MW: 569.61   EINECS: 261-143-4
LD$_{50}$:   5000 mg/kg (M, i.p.); 330 mg/kg (M, i.v.)
CN: (S)-O-6-amino-6-deoxy-α-D-glucopyranosyl-(1→4)-O-[3-deoxy-4-C-methyl-3-(methylamino)-β-L-
       arabinopyranosyl-(1→6)]-N^1-(3-amino-2-hydroxy-1-oxopropyl)-2-deoxy-D-streptamine

sulfate

RN: 67814-76-0   MF: $C_{22}H_{43}N_5O_{12} \cdot H_2SO_4$ MW: 667.69
LD_{50}: 2088 mg/kg (R, i.m.); 1591 mg/kg (R, i.p.); 476 mg/kg (R, i.v.); 3392 mg/kg (R, s.c.)

disulfate

RN: 68000-78-2   MF: $C_{22}H_{43}N_5O_{12} \cdot 2H_2SO_4$ MW: 765.77
LD_{50}: 234 mg/kg (M, i.v.); >5 g/kg (M, p.o.);
          476 mg/kg (R, i.v.); >5 g/kg (R, p.o.);
          720 mg/kg (dog, i.v.)

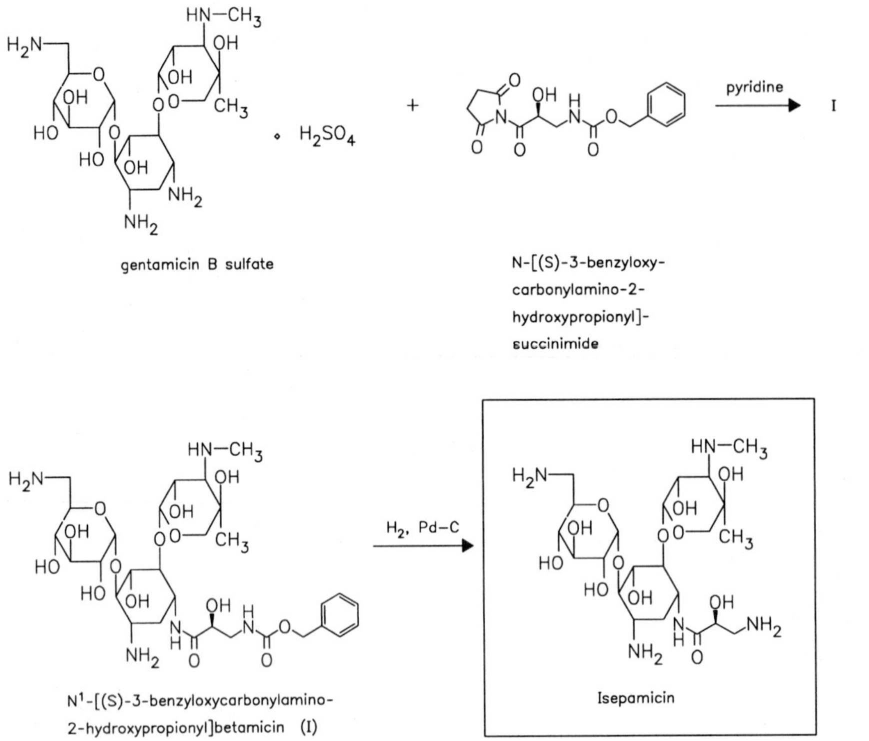

gentamicin B sulfate

N-[(S)-3-benzyloxy-
carbonylamino-2-
hydroxypropionyl]-
succinimide

N¹-[(S)-3-benzyloxycarbonylamino-
2-hydroxypropionyl]betamicin (I)

Isepamicin

Reference(s):

DOS 2 502 296 (Scherico; appl. 21.1.1975; USA-prior. 19.3.1974).
US 4 029 882 (Schering Corp.; 14.6.1977; appl. 19.3.1974).

alternative synthesis:

EP 405 820 (Schering Corp.; appl. 19.6.1990; USA-prior. 21.6.1989).
US 4 136 254 (Scherico; 23.1.1979; appl. 18.5.1978; prior. 17.6.1976).
US 4 230 847 (Schering Corp.; 28.1.1980; appl. 26.12.1979; prior. 18.5.1978, 17.6.1976).
US 4 337 335 (Schering Corp.; 29.6.1982; appl. 26.12.1979; prior. 18.5.1978, 17.6.1976).
EP 430 234 (Kanegafuchi; appl. 29.11.1990; J-prior. 29.11.1989).

stable ampoule formulation:

JP 1 268 698 (Toyo Jozo; appl. 20.4.1988).

Formulation(s):   amp. 250 mg, 500 mg

Trade Name(s):

F: Isépalline (Schering-   J: Exacin (Toyo Jozo; 1988)
      Plough)   Isepacin (Essex; 1988)

Isoaminile

ATC: R05DB04
Use: antitussive

RN: 77-51-0   MF: $C_{16}H_{24}N_2$ MW: 244.38   EINECS: 201-033-5
LD$_{50}$: 55 mg/kg (M, i.v.);
         48.4 mg/kg (dog, i.v.)
CN: α-[2-(dimethylamino)propyl]-α-(1-methylethyl)benzeneacetonitrile

citrate (1:1)
RN: 28416-66-2   MF: $C_{16}H_{24}N_2 \cdot C_6H_8O_7$ MW: 436.51   EINECS: 249-011-4
cyclamate (1:1)
RN: 10075-36-2   MF: $C_{16}H_{24}N_2 \cdot C_6H_{13}NO_3S$ MW: 423.62   EINECS: 233-207-1
LD$_{50}$: 57 mg/kg (M, i.v.); 298 mg/kg (M, p.o.);
         270 mg/kg (R, p.o.);
         84 mg/kg (dog, i.v.)

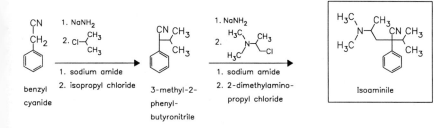

benzyl cyanide → 3-methyl-2-phenyl-butyronitrile → Isoaminile

1. NaNH$_2$
2. Cl-CH(CH$_3$)CH$_3$
1. sodium amide
2. isopropyl chloride

1. NaNH$_2$
2. (H$_3$C)$_2$N-CH(CH$_3$)-Cl
1. sodium amide
2. 2-dimethylamino-propyl chloride

Reference(s):
GB 765 510 (Kali-Chemie; appl. 1955; D-prior. 1954).
DE 960 462 (Kali-Chemie; appl. 1954).
DE 964 499 (Kali-Chemie; appl. 1954).
US 2 934 557 (Kali-Chemie; 1960; D-prior. 1957).
Krause, D.: Arzneim.-Forsch. (ARZNAD) **8**, 553 (1958).

cyclohexylsulfamic acid salt:
US 3 074 996 (Abbott; 22.1.1963; appl. 18.4.1960).

Formulation(s):   drg., tabl. 40 mg (as citrate); drops 50 mg; sol. 50 mg/15 ml (as cyclamate)

Trade Name(s):
D: Peracon (Kali-Chemie);   GB: Dimyril (Fisons); wfm   Sedotosse (Panthox &
      wfm   I: Peracon Kali-Chemie (Sir);   Burck); wfm
F: Mucalan (Delagrange);   wfm   J: Peracan (Toyo)
      wfm

Isocarboxazid

ATC: N06AF01
Use: MAO-inhibitor, antidepressant

RN: 59-63-2   MF: $C_{12}H_{13}N_3O_2$ MW: 231.26   EINECS: 200-438-4
LD$_{50}$: 193 mg/kg (M, p.o.);
         280 mg/kg (R, p.o.);
         >40 mg/kg (dog, p.o.)
CN: 5-methyl-3-isoxazolecarboxylic acid 2-(phenylmethyl)hydrazide

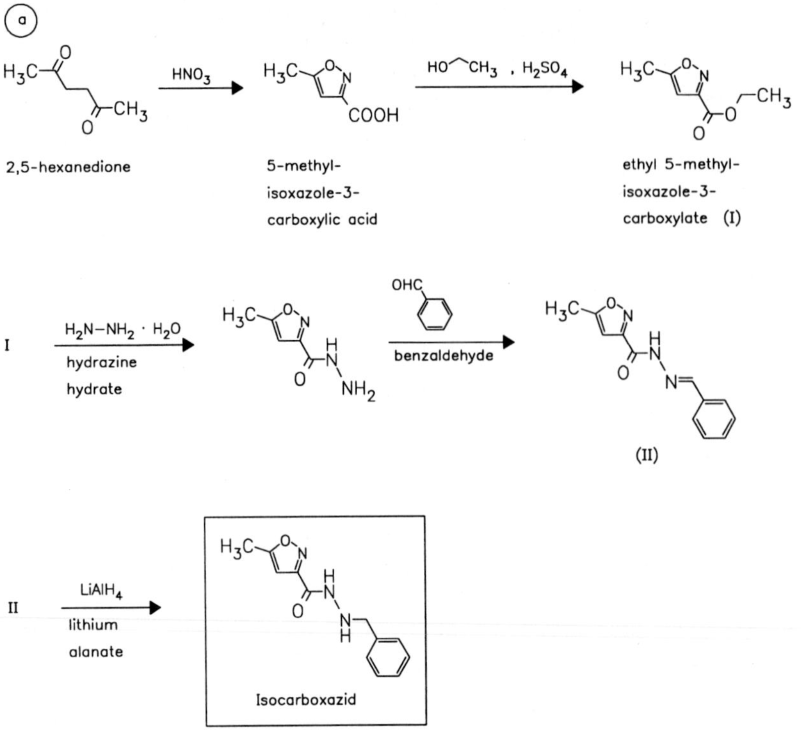

a

2,5-hexanedione → 5-methyl-isoxazole-3-carboxylic acid → ethyl 5-methyl-isoxazole-3-carboxylate (I)

reagents: HNO₃; HO–CH₃, H₂SO₄

I → (H₂N–NH₂ · H₂O, hydrazine hydrate) → (OHC–phenyl, benzaldehyde) → (II)

II → (LiAlH₄, lithium alanate) → Isocarboxazid

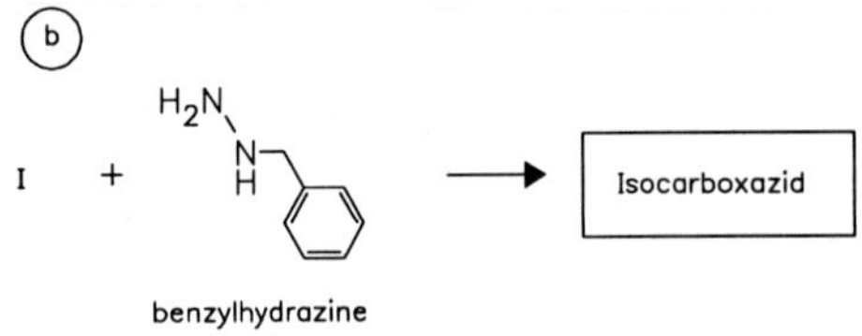

b

I + benzylhydrazine → Isocarboxazid

Reference(s):
US 2 908 688 (Hoffmann-La Roche; 13.10.1959; prior. 15.4.1958).

Formulation(s):　tabl. 10 mg

Trade Name(s):

F:	Marplan (Roche); wfm	I:	Marplan (Roche); wfm
GB:	Marplan (Roche); wfm	J:	Enerzer (Takeda)

USA:　Marplan (Roche); wfm

Isoconazole

ATC:　D01AC05; G01AF07
Use:　antifungal, antibacterial

RN:　27523-40-6　MF: $C_{18}H_{14}Cl_4N_2O$　MW: 416.14　EINECS: 248-508-3
LD$_{50}$:　189 mg/kg (M, i.p.)
CN:　1-[2-(2,4-dichlorophenyl)-2-[(2,6-dichlorophenyl)methoxy]ethyl]-1H-imidazole

mononitrate
RN:　24168-96-5　MF: $C_{18}H_{14}Cl_4N_2O \cdot HNO_3$　MW: 479.15　EINECS: 246-051-4
LD$_{50}$:　2 g/kg (M, p.o.);
　　　　5600 mg/kg (R, p.o.)

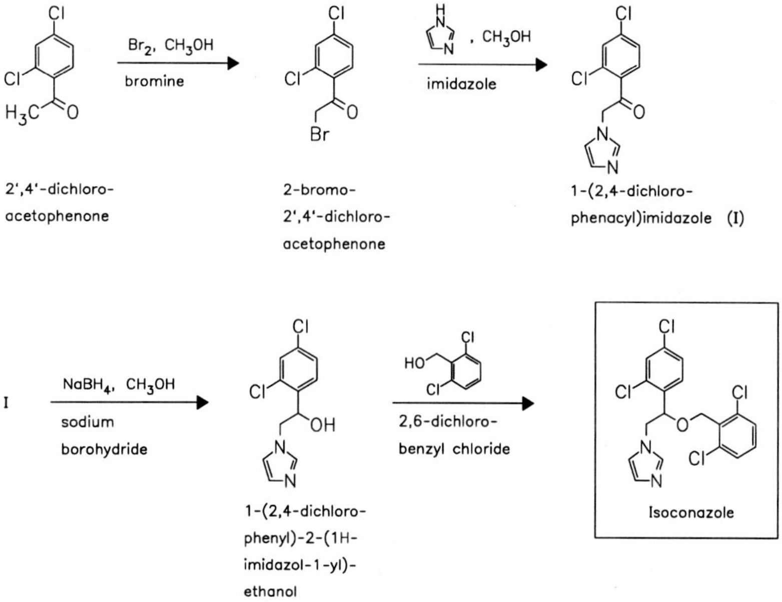

The reaction scheme for Isoconazole:

2',4'-dichloro-
acetophenone

→ (Br₂, CH₃OH / bromine) →

2-bromo-
2',4'-dichloro-
acetophenone

→ (imidazole, CH₃OH) →

1-(2,4-dichloro-
phenacyl)imidazole   (I)

I → (NaBH₄, CH₃OH / sodium borohydride) →

1-(2,4-dichloro-
phenyl)-2-(1H-
imidazol-1-yl)-
ethanol

→ (2,6-dichloro-
benzyl chloride) →

Isoconazole

Reference(s):

DOS 1 940 388 (Janssen; appl. 8.8.1969; USA-prior. 19.8.1968).
US 3 717 655 (Janssen; 20.2.1973; prior. 19.8.1968, 23.7.1969).
US 3 839 574 (Janssen; 1.10.1974; prior. 19.8.1968, 23.7.1969, 19.7.1972).
Godefroi, E.F. et al.: J. Med. Chem. (JMCMAR) **12**, 784 (1969).

Formulation(s):   cream 10 mg/g (1 %); pessaries 100 mg, 300 mg, 600 mg (as nitrate); spray 10 mg/ml

Trade Name(s):

D:	Bi-Vaspit (Asche)-comb.	F:	Fazol (Bellon; Rhône-		Travocort (Schering)-comb.
	Travocort (Schering)-comb.		Poulenc Rorer; 1979)		Travogen (Schering)
	Travogen (Schering; 1979)	GB:	Travogyn (Schering)	J:	Adestan (Schering)
		I:	Isogyn (Crosara)		

Isoetarine

(Etyprenalinum; Isoetharine)

ATC: R03AC07; R03CC06
Use: bronchodilator, sympathomimetic

RN: 530-08-5   MF: C₁₃H₂₁NO₃   MW: 239.32   EINECS: 208-472-1
CN: 4-[1-hydroxy-2-[(1-methylethyl)amino]butyl]-1,2-benzenediol

hydrochloride
RN: 50-96-4   MF: C₁₃H₂₁NO₃ · HCl   MW: 275.78
LD₅₀: 57 mg/kg (M, i.v.)

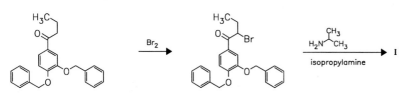

3',4'-dibenzyloxy-
butyrophenone

→ (Br₂) → (isopropylamine) → I

H₃C CH₃ ... (I) → H₂, Pd–C → Isoetarine

Reference(s):
DRP 638 650 (I. G. Farben; 1934).

Formulation(s): amp. 1 mg, 2.4 mg, 5 mg, 5.1 mg; sol. for inhalation 0.125 %, 0.61 %, 1 % (as hydrochloride)

Trade Name(s):
D: Asthmalitan (Kettelhack- GB: Bronchilator (Sterling USA: Bronkometer (Sanofi)-
    Riker); wfm   Res.); wfm   comb.
                            Numotac (Riker); wfm   Bronkosol (Sanofi)-comb.

Isoflupredone acetate

ATC: H02AB
Use: glucocorticoid

RN: 338-98-7 MF: C₂₃H₂₉FO₆ MW: 420.48 EINECS: 206-423-9
CN: (11β)-21-(acetyloxy)-9-fluoro-11,17-dihydroxypregna-1,4-diene-3,20-dione

isoflupredone
RN: 338-95-4 MF: C₂₁H₂₇FO₅ MW: 378.44 EINECS: 206-422-3

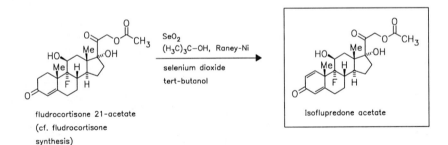

fludrocortisone 21-acetate
(cf. fludrocortisone
synthesis)

SeO₂
(H₃C)₃C—OH, Raney-Ni

selenium dioxide
tert-butanol

Isoflupredone acetate

Reference(s):
DE 1 096 900 (American Cyanamid; appl. 1959; USA-prior. 1958).

microbiological dehydrogenation with Corynebacterium simplex (A.T.C.C. 6946):
US 2 837 464 (Schering Corp.; 1958; prior. 1955).

alternative synthesis:
DE 1 159 947 (Merck & Co.; appl. 1956; USA-prior. 1955).
GB 826 364 (Merck & Co.; valid from 1956; USA-prior. 1955).

Formulation(s): amp. 10 ml

Trade Name(s):
I: Altaflor (Intes)-comb.; wfm Cortifluoral (Schering); Dermaflogil (Nuovo Cons.
    Biorinil (Farmila)-comb.;   wfm   Sanit. Naz.)-comb.; wfm
    wfm

Fluoroinil (Farmila)-comb.; wfm | Menaderm simp. (Menarini); wfm | USA: Predef 2x (Upjohn); wfm

Isoflurane

ATC: N01AB06
Use: inhalation anesthetic

RN: 26675-46-7 MF: $C_3H_2ClF_5O$ MW: 184.49 EINECS: 247-897-7
LD_{50}: 5080 µL/kg (M, p.o.);
4770 µL/kg (R, p.o.)
CN: 2-chloro-2-(difluoromethoxy)-1,1,1-trifluoroethane

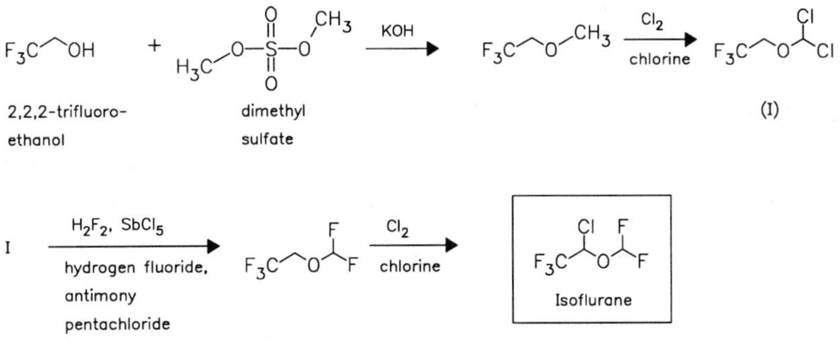

2,2,2-trifluoro-ethanol
dimethyl sulfate
KOH
Cl2
chlorine
(I)
I
H2F2, SbCl5
hydrogen fluoride, antimony pentachloride
Isoflurane

Reference(s):
DOS 1 814 962 (Air Reduction Comp.; appl. 16.12.1968; USA-prior. 15.12.1967, 22.5.1968).
US 3 535 388 (Air Reduction Comp.; 20.10.1970; prior. 15.12.1967, 21.3.1969).
US 3 535 425 (Air Reduction Comp.; 20.10.1970; prior. 15.12.1967, 18.12.1969).

alternative synthesis:
DOS 2 344 442 (Hoechst; appl. 4.9.1973).
US 3 637 477 (Air Reduction Comp.; 25.1.1972; prior. 20.2.1970, 22.5.1968).

Formulation(s): sol. for inhalation 100 ml

Trade Name(s):
D: Forene (Abbott; 1984) | J: Forane (Dainabot) | USA: Forane (Ohmeda; 1981)

Isoflurophate
(Difluorophate)

ATC: S03
Use: parasympathomimetic, miotic

RN: 55-91-4 MF: $C_6H_{14}FO_3P$ MW: 184.15 EINECS: 200-247-6
LD_{50}: 3200 µg/kg (M, i.v.); 2 mg/kg (M, p.o.);
5 mg/kg (R, p.o.);
3430 µg/kg (dog, i.v.)
CN: phosphorofluoridic acid bis(1-methylethyl) ester

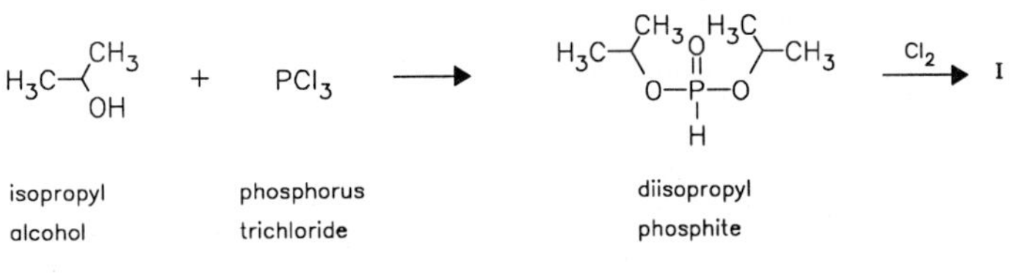

isopropyl alcohol
phosphorus trichloride
PCl3
diisopropyl phosphite
Cl2
I

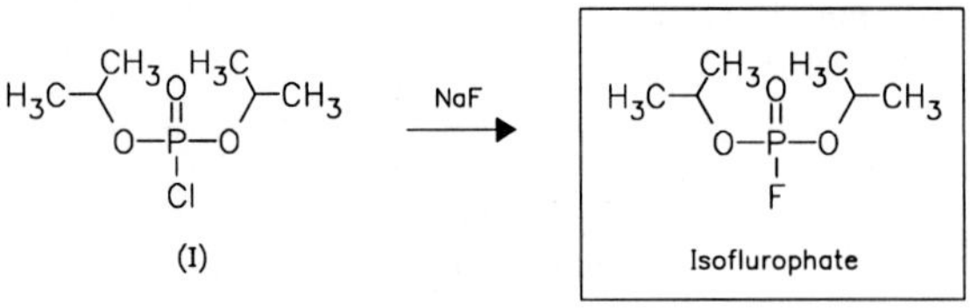

(I) → Isoflurophate

Reference(s):

US 2 409 039 (Monsanto; 1946; appl. 1944).

Formulation(s):   eye drops 0.01 %; eye ointment 0.025 %

Trade Name(s):
F: Diflupyl (Labaz); wfm   USA: Floropryl (Merck Sharp &
J: D. F. P. Inj. (Sumitomo)   Dohme); wfm

Isometheptene

ATC: A03AX10
Use: sympathetic antispasmodic for gut and urinary tract

RN: 503-01-5   MF: $C_9H_{19}N$ MW: 141.26   EINECS: 207-959-6
LD_{50}: 34 mg/kg (M, i.v.); 134 mg/kg (M, p.o.)
CN: N,6-dimethyl-5-hepten-2-amine

hydrochloride
RN: 6168-86-1   MF: $C_9H_{19}N \cdot HCl$ MW: 177.72   EINECS: 228-211-5
LD_{50}: 18 mg/kg (M, i.v.);
           26 mg/kg (dog, i.v.)
tartrate (1:1)
RN: 5984-50-9   MF: $C_9H_{19}N \cdot C_4H_6O_6$ MW: 291.34   EINECS: 227-795-9
LD_{50}: 130 mg/kg (R, i.p.)

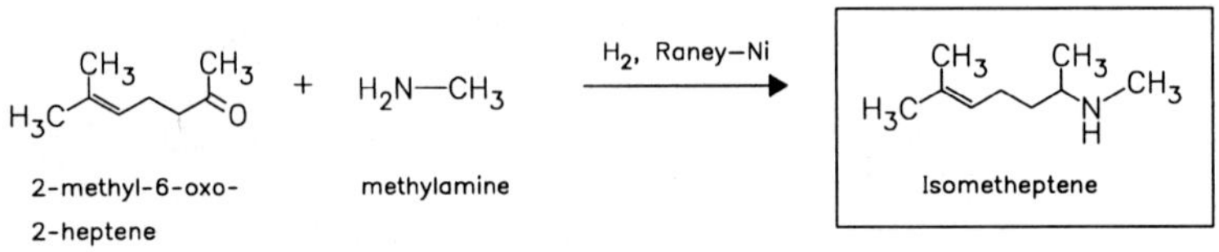

2-methyl-6-oxo-
2-heptene

methylamine

Isometheptene

Reference(s):

US 1 972 450 (Knoll; 1934; D-prior. 1931).
US 2 230 753 (E. Bilhuber; 1941; D-prior. 1937).
US 2 230 754 (E. Bilhuber; 1941; D-prior. 1937).

Formulation(s):   amp. 100 mg (as hydrochloride); drops 100 mg, 50 mg; tabl. (as tartrate)

Trade Name(s):
D: Neopyrin (Nordmark)-   GB: Midrid (Shire)-comb.   USA: Duradrin (Duramed; as
      comb.; wfm   I: Octinum (Knoll); wfm   mucate)-comb.
      Neosal (Nordmark)-comb.;   J: Cesal (Dainippon)-comb.   Midrin (Carnrick; as
      wfm   mucate)

Isoniazid

ATC: J04AC01
Use: tuberculostatic

RN: 54-85-3   MF: $C_6H_7N_3O$ MW: 137.14   EINECS: 200-214-6
LD_{50}: 149 mg/kg (M, i.v.); 133 mg/kg (M, p.o.);
365 mg/kg (R, i.v.); 1250 mg/kg (R, p.o.);
50 mg/kg (dog, p.o.)
CN: 4-pyridinecarboxylic acid hydrazide

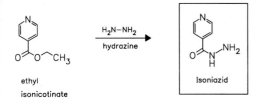

ethyl
isonicotinate

Isoniazid

Reference(s):
Meyer, H.; Mally, J.: Monatsh. Chem. (MOCMB7) **33**, 393 (1912).
US 2 596 069 (Roche; 1952; appl. 1952).
US 2 830 994 (Distillers; 15.4.1958; GB-prior. 29.6.1955).
DE 1 116 667 (BASF; appl. 3.7.1956).

combination with protionamide *and* dapsone:
DAS 2 340 515 (Saarstickstoff-Fatol; appl. 10.8.1973).

Formulation(s):   amp. 100 mg/5 ml, 250 mg/5 ml; tabl. 50 mg, 100 mg, 200 mg, 300 mg

Trade Name(s):
D: Isoprodian (Fatol)-comb.
Isozid (Fatol)
Myambutol (Lederle)-
comb.
Tebesium (Hefa Pharma)
F: Dexambutol-INH
(L'Arguenon)-comb.
Rifater (Marion Merrell)-
comb.
Rifinah (Marion Merrell)-
comb.
Rimifon (Roche)
GB: Rifater (Hoechst)-comb.
Rifinah (Hoechst)-comb.

Rimactazid (Novartis)-
comb.
Rimifon (Roche); wfm
I: Emozide B6 (Piam)-comb.
Etanicozid (Piam)-comb.
Etibi (Zoja)-comb.
Miazide (Cyanamid)-comb.
Nicazide (IFI)
Nicizina (Pharmacia &
Upjohn)
Nicozid (Piam)
Rifanicozid (Piam)-comb.
Rifinah (Lepetit)-comb.
combination preparations
J: Anteben (Dainippon)

Diazid (Nippon Shinyaku)
Hycozid (Takeda)
Hydra (Otsuka)
Iscotin (Daiichi)
Niazid (Sankyo)
Niplen (Tanabe)
Sumifon (Sumitomo)
Tuberon (Shionogi)
USA: Nydrazid (Apothecon)
Rifamate (Hoechst Marion
Roussel)
Rifater (Hoechst Marion
Roussel)
generic

Isoprenaline

(Isoproterenol)

ATC: C01CA02; R03AB02; R03CB01
Use: bronchodilator, dermatic

RN: 7683-59-2   MF: $C_{11}H_{17}NO_3$ MW: 211.26   EINECS: 231-687-7
LD_{50}: 83 mg/kg (M, i.v.); 450 mg/kg (M, p.o.);
57 mg/kg (R, i.v.); 355 mg/kg (R, p.o.);
50 mg/kg (dog, i.v.); 600 mg/kg (dog, p.o.)
CN: 4-[1-hydroxy-2-[(1-methylethyl)amino]ethyl]-1,2-benzenediol

bitartrate (1:1)
RN: 59-60-9 MF: $C_{11}H_{17}NO_3 \cdot C_4H_6O_6$ MW: 361.35 EINECS: 200-437-9
hydrochloride
RN: 51-30-9 MF: $C_{11}H_{17}NO_3 \cdot HCl$ MW: 247.72 EINECS: 200-089-8
LD$_{50}$: 77 mg/kg (M, i.v.); 1260 mg/kg (M, p.o.);
26.9 mg/kg (R, i.v.); 2221 mg/kg (R, p.o.);
50 mg/kg (dog, i.v.); 600 mg/kg (dog, p.o.)
sulfate (2:1)
RN: 299-95-6 MF: $C_{11}H_{17}NO_3 \cdot 1/2H_2SO_4$ MW: 520.60 EINECS: 206-085-2
LD$_{50}$: 188 mg/kg (M, i.v.); >3 g/kg (M, p.o.);
96 mg/kg (R, i.v.); 2230 mg/kg (R, p.o.);
50 mg/kg (dog, i.v.); 600 mg/kg (dog, p.o.)

2-chloro-3',4'-di-
hydroxyacetophenone   isopropylamine   (I)

Isoprenaline

Reference(s):
US 2 308 232 (Boehringer Ing.; 1943; D-prior. 1939).
DRP 723 278 (Boehringer Ing.; appl. 1939).

racemate resolution with (+)-tartaric acid:
US 2 715 141 (Delmar Chemicals; 1955; appl. 1952).

Formulation(s): aerosol 0.1 mg/push; amp. 0.2 mg, 2 mg (as hydrochloride); sol. for inhalation 0.25 %, 0.5 %
(as sulfate); tabl. 10 mg, 15 mg, 20 mg

Trade Name(s):
D: Ingelan (Boehringer Ing.) J: Asthpul Sol. (Nippon Sedansol "Iso" (Nippon
Kattwilon (Kattwiga) Shoji) Zoki)
F: Isuprel (Abbott) Isomenyl (Kaken) Sooner (Kaken)
generic Medihaler-Iso (Dainippon) USA: Isuprel (Sanofi; as
GB: Saventrine (Pharmax) Protemal (Nikken) hydrochloride)
I: Aleudrin (Fher); wfm Proternol (Nikken) Medihaler-Iso (3M; as
generics sulfate)

Isopropamide iodide

ATC: A03AB09
Use: anticholinergic, antispasmodic

RN: 71-81-8 MF: $C_{23}H_{33}IN_2O$ MW: 480.43 EINECS: 200-766-8
LD$_{50}$: 12.779 g/kg (M, i.v.); 1600 mg/kg (M, p.o.)
CN: γ-(aminocarbonyl)-N-methyl-N,N-bis(1-methylethyl)-γ-phenylbenzenepropanaminium iodide

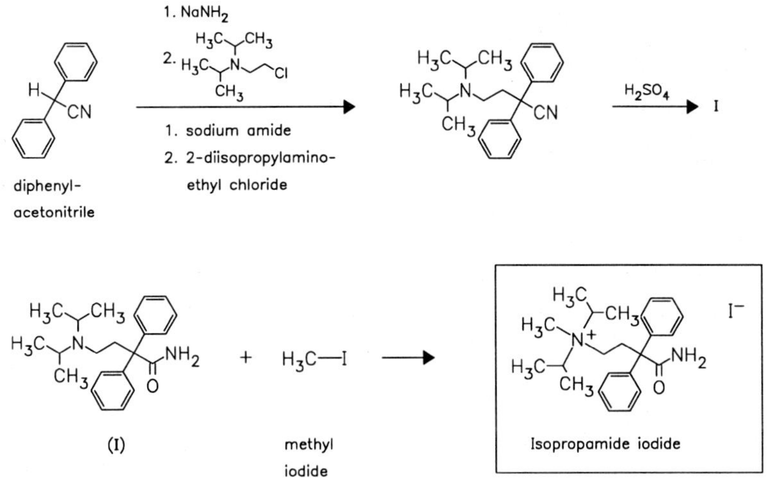

diphenyl-
acetonitrile

1. sodium amide
2. 2-diisopropylamino-
   ethyl chloride

(I)   methyl
              iodide

Isopropamide iodide

Reference(s):

GB 772 921 (Janssen; appl. 1955; NL-prior. 1954).
DE 1 003 744 (Janssen; appl. 1955; NL-prior. 1954).
Janssen, P. et al.: Arch. Int. Pharmacodyn. Ther. (AIPTAK) CIII, **82** (1955).
US 2 823 233 (Bristol; 1958; appl. 1954).

Formulation(s):   inj. sol. 3 mg/2 ml; tabl. 5 mg

Trade Name(s):

D: Ornatos (Röhm Pharma)-
      comb.; wfm
      Priamide-Eupharma
      (Janssen); wfm
      Stelabid (Röhm Pharma)-
      comb.; wfm
F: Enuretine vit. E
      isoprapamide (Le
      Marchand)-comb.; wfm
      Priamide (Delalande); wfm

GB: Stelabid (S.K.F.)-comb.;
      wfm
      Tyrimide (Smith Kline &
      French); wfm
I: Fluvaleas (Valeas)-comb.
      Iodosan (SmithKline
      Beecham)-comb.
      Valtrax (Valeas)
      combination preparations
J: Marygin M (Sumitomo)

USA: Combid (Smith Kline &
      French); wfm
      Darbid (Smith Kline &
      French); wfm
      Ornade (Smith Kline &
      French); wfm
      Prochlor-Iso (Schein); wfm
      Pro-Iso (Zenith); wfm

Isosorbide dinitrate

ATC: D03AX08
Use: coronary vasodilator

RN: 87-33-2   MF: $C_6H_8N_2O_8$ MW: 236.14   EINECS: 201-740-9
LD$_{50}$:   >40 mg/kg (M, i.v.); 1050 mg/kg (M, p.o.);
        >40 mg/kg (R, i.v.); 747 mg/kg (R, p.o.)
CN: 1,4:3,6-dianhydro-D-glucitol dinitrate

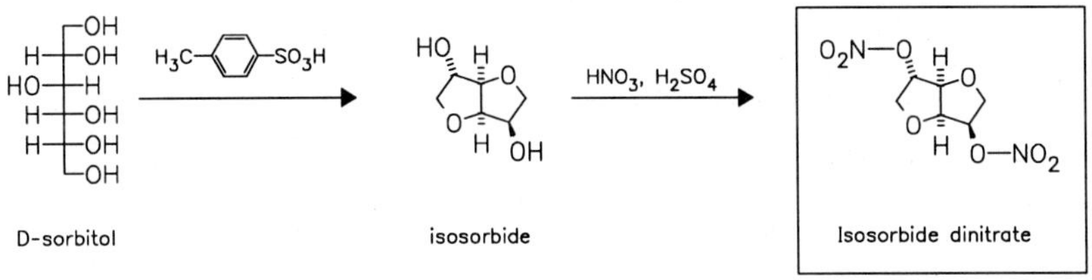

D-sorbitol   isosorbide   Isosorbide dinitrate

Reference(s):
Goldberg, L.: Acta Physiol. Scand. (APSCAX) **15**, 173 (1948).
Krantz, J.C. et al.: J. Pharmacol. Exp. Ther. (JPETAB) **67**, 187 (1939).

aqueous solutions for parenteral application:
DAS 2 623 800 (Sanol Schwarz-Monheim; appl. 28.5.1976).

Formulation(s):   r. r. cps. 20 mg, 40 mg, 60 mg, 80 mg, 100 mg; tabl. 2.5 mg, 5 mg, 10 mg, 20 mg, 30 mg,
40 mg

Trade Name(s):

D:	Corovliss (Boehringer Mannh.)	F:	Isocard (Bouchara)	Nitrosorbide (Lusofarmaco)
	Dignonitrat (Sankyo)		Langoran (Marion Merrell)	Stenodilate (Schwarz)-
	Duranitrat (durachemie)		Risordan (Specia)	comb.
	isoket (Schwarz)	GB:	Cedocard retard (Pharmacia & Upjohn)	J: Cardis (Iwaki)
	Iso Mack (Mack, Illert.)		Isocard (Eastern)	Carvanil (Banyu)
	Isostenase (Azuchemie)		Isoket retard (Schwarz)	Diretan (Ono)
	Maycor (Gödecke; Parke Davis)		Isordil (Monmouth)	Nitroret (Hishiyama)
			Sorbichew (Zeneca)	USA: Dilatrate-SR (Schwarz)
	Nitrosorbon (Pohl)		Sorbid SA (Zeneca)	Isordil (Wyeth-Ayerst)
	TD spray Iso Mack (Mack)		Sorbitrate (Zeneca)	Sorbitrate (Zeneca)
	generic and numerous	I:	Carvasin (Wyeth)	generic
	combination preparations		Diniket (Schwarz)	

Isosorbide mononitrate

(Isosorbide 5-nitrate)

ATC: C01DA14
Use: coronary vasodilator

RN: 16051-77-7   MF: $C_6H_9NO_6$ MW: 191.14   EINECS: 240-197-2
LD$_{50}$:   1820 mg/kg (M, i.v.); 1771 mg/kg (M, p.o.);
1750 mg/kg (R, i.v.); 2010 mg/kg (R, p.o.)
CN: 1,4:3,6-dianhydro-D-glucitol 5-nitrate

isosorbide (I)
(cf. isosorbide
dinitrate synthesis)

Isosorbide
mononitrate

isosorbide
2-acetate
(along with
5-acetate)

isosorbide 2-acetate
5-nitrate (II)

1. NaOH, pH 10–12
2. H$_2$SO$_4$, NaCl, pH 6.8–7.0

→ Isosorbide mononitrate

Reference(s):
a   US 3 886 186 (American Home; 27.5.1975; prior. 29.4.1971, 30.8.1973).
    EP 143 507 (Toshin Chemical; appl. 13.7.1984; J-prior. 25.11.1983).
    DOS 2 221 080 (American Home Products; appl. 28.4.1972; USA-prior. 29.4.1971).
b   DOS 2 751 934 (American Home Products; appl. 21.11.1977; USA-prior. 24.2.1977).
    US 4 065 488 (American Home Products; 27.12.1977; appl. 24.2.1977).
    EP 45 076 (Boehringer Mannh.; appl. 25.7.1981; D-prior. 30.7.1980).
    DOS 3 028 873 (Boehringer Mannh.; appl. 30.7.1980).
    US 4 431 829 (Boehringer Mannh.; 14.2.1984; D-prior. 30.7.1980).

similar methods (via 2-acyloxy derivative):
DE 2 903 927 (Sanol-Schwarz; appl. 2.2.1979).
EP 64 194 (Cassella; appl. 16.4.1982; D-prior. 5.5.1981).
EP 57 847 (H. Mack Nachf.; appl. 26.1.1982; D-prior. 29.1.1981).
DOS 3 102 947 (H. Mack Nachf.; appl. 29.1.1981).
EP 67 964 (H. Mack Nachf.; appl. 18.5.1982; D-prior. 22.6.1981).
DOS 3 124 410 (H. Mack Nachf.; appl. 22.6.1981).
US 4 417 065 (H. Mack Nachf.; 22.11.1983; D-prior. 22.6.1981).

preparation from isosorbide dinitrate:
EP 59 664 (SNPE; appl. 25.2.1982; F-prior. 27.2.1981).

formulations:
DOS 3 325 652 (Dr. Rentschler; appl. 15.7.1983).

Formulation(s):   s. r. cps. 40 mg, 60 mg; tabl. 20 mg, 40 mg

Trade Name(s):

D: Coleb (Astra/Promed)
    Corangin (Novartis
    Pharma)
    elantan (Synthelabo)
    IS 5 mono-ratiopharm
    (ratiopharm)
    Ismo (Boehringer Mannh.)
    Monit-Puren (Isis Puren)
    Monoclair (Hennig)
    Mono Mack (Mack, Illert.)
    Monostenase (Azupharma)
    Olicard (Solvay
    Arzneimittel)
    numerous generics
F: Monicor L.P. (Pierre Fabre)
GB: Elantan (Schwarz)

Imdur (Astra)
Isib XL (Ashbourne)
Ismo retard (Boehringer
Mannh.)
MCR 50 (Pharmacia &
Upjohn)
Monit (Lorex)
Mono-Cedocard
(Pharmacia & Upjohn)
Monomax (Trinity)
I: Duronitrin (Astra)
    Elan (Schwarz)
    Ismo (Boehringer Mannh.)
    Ismo Diffutab (Boehringer
    Mannh.)

Monocinque
(Lusofarmaco)
Monoket (Chiesi)
Nitralfa (Malesci)
Orasorbil (Rottapharm)
dinitrate:
Nitrosorbide
(Lusofarmaco)
Stenodilate (Schwarz)-
comb.
J: Itocol (Toa Eiyo-
    Yamanouchi)
USA: Imdur (Key Pharm.)
    Ismo (Wyeth-Ayerst)
    Monoket (Schwarz)

1110   I   Isothipendyl

Isothipendyl

ATC: D04AA22; R06AD09
Use: antiallergic, antihistaminic

RN: 482-15-5   MF: C$_{16}$H$_{19}$N$_3$S   MW: 285.42   EINECS: 207-578-5
LD$_{50}$:   222 mg/kg (M, p.o.)
CN: N,N,α-trimethyl-10H-pyrido[3,2-b][1,4]benzothiazine-10-ethanamine

monohydrochloride
RN: 1225-60-1   MF: C$_{16}$H$_{19}$N$_3$S · HCl   MW: 321.88   EINECS: 214-957-9
LD$_{50}$:   28 mg/kg (M, i.v.); 222 mg/kg (M, p.o.);
        1220 mg/kg (R, p.o.)

pyrido[3,2-b]-
[1,4]benzothiazine

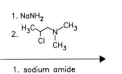
1. sodium amide
2. 1-dimethylamino-
   2-chloropropane

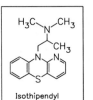
Isothipendyl

Reference(s):
DE 1 001 684 (Degussa; appl. 1954).
US 2 974 139 (Degussa; 7.3.1961; D-prior. 2.10.1954).

Formulation(s):   drg. 12 mg; gel 0.75 % (as hydrochloride)

Trade Name(s):
D: Andantol-forte/-Gelee   GB: Nilergex (ICI); wfm   J: Aczen NS (Kanebo)
      (Homburg); wfm   I: Calmogel (Rhône-Poulenc   Andantol (Sumitomo)
F: Sédermyl (RPR Cooper)   Rorer)

Isotretinoin

ATC: D10AD04; D10BA01
Use: keratolytic, acne therapeutic

RN: 4759-48-2   MF: C$_{20}$H$_{28}$O$_2$ MW: 300.44   EINECS: 225-296-0
LD$_{50}$:   3389 mg/kg (M, p.o.);
        >4 g/kg (R, p.o.)
CN: 13-cis-retinoic acid

sodium salt
RN: 13497-05-7   MF: C$_{20}$H$_{27}$NaO$_2$ MW: 322.42
potassium salt
RN: 22232-80-0   MF: C$_{20}$H$_{27}$KO$_2$ MW: 338.53

vinyl-β-ionol
(cf. retinol synthesis
and tretinoin
synthesis)

triphenyl-
phosphine

(I)

(Z)-β-formyl-
crotonic acid
(cf. retinol
synthesis)

Isotretinoin

Reference(s):
Garbers, C.F. et al.: J. Chem. Soc. C (JSOOAX) **1968**, 1982.

(Z)-β-formylcrotonic acid:
Conradie, W.J. et al.: J. Chem. Soc. C (JSOOAX) **1964**, 594.

combination with taurine:
US 4 545 977 (Searle; 8.10.1985; appl. 11.1.1985).

medical use:
DE 2 061 507 (Hoffmann-La Roche; appl. 8.7.1971; prior. 14.12.1970).
US 3 746 730 (Hoffmann-La Roche; 17.7.1973; appl. 17.12.1970; GB-prior. 13.12.1969).

Formulation(s):   cps. 2.5 mg, 5 mg, 10 mg, 20 mg; gel 0.05 %

Trade Name(s):
D:	ISOTREX (Stiefel)		Roaccutane (Roche; 1986)		Roaccutane (Roche)
	Roaccutan (Roche; 1985)	GB:	Isotrex (Stiefel)	I:	Roaccutane (Roche)
F:	Isotrex (Stiefel)		Isotrexin (Stiefel)	USA:	Accutane (Roche; 1982)

Isoxicam

ATC: M01
Use: anti-inflammatory

RN: 34552-84-6   MF: $C_{14}H_{13}N_3O_5S$ MW: 335.34   EINECS: 252-084-5
LD$_{50}$:   >5 g/kg (R, p.o.)
CN: 4-hydroxy-2-methyl-*N*-(5-methyl-3-isoxazolyl)-2*H*-1,2-benzothiazine-3-carboxamide 1,1-dioxide

3-ethoxycarbonyl-
4-hydroxy-2-methyl-
2H-1,2-benzothiazine
1,1-dioxide
(cf. piroxicam
synthesis)

3-amino-5-
methyl-
isoxazole

Isoxicam

Reference(s):
DOS 2 208 351 (Warner-Lambert; appl. 22.2.1972; USA-prior. 1.3.1971).
Lombardino, J.G.; Wiseman, E.H.: J. Med. Chem. (JMCMAR) **14**, 973 (1971).
Zinnes, H. et al.: J. Med. Chem. (JMCMAR) **25**, 12 (1982).

Formulation(s):   cps. 100 mg

Trade Name(s):
D: Pacyl (Adenylchemie);   F: Vectren (Substantia); wfm   Maxicam (Parke Davis);
     wfm   J: Floxicam (Menarini); wfm   wfm

Isoxsuprine

ATC: C04AA01
Use: vasodilator

RN: 395-28-8   MF: $C_{18}H_{23}NO_3$ MW: 301.39   EINECS: 206-898-2
LD$_{50}$:   48 mg/kg (M, i.v.); 200 mg/kg (M, p.o.)
CN: 4-hydroxy-α-[1-[(1-methyl-2-phenoxyethyl)amino]ethyl]benzenemethanol

hydrochloride
RN: 579-56-6   MF: $C_{18}H_{23}NO_3 \cdot HCl$ MW: 337.85   EINECS: 209-443-6
LD$_{50}$:   61 mg/kg (M, i.v.); 1.1 g/kg (M, p.o.);
            1.75 g/kg (R, p.o.);
            57 mg/kg (dog, i.v.); >1.2 g/kg (dog, p.o.)

4-hydroxy-   1-phenoxy-2-   Isoxsuprine
norephedrine   bromopropane

4'-benzyloxy-
propiophenone

2-amino-1-
phenoxypropane

(I)

Reference(s):
US 3 056 836 (Philips; 2.10.1962; NL-prior. 28.5.1955).
GB 832 286 (Philips; appl. 11.10.1957; NL-prior. 15.10.1956).
GB 832 287 (Philips; appl. 11.10.1957; NL-prior. 15.10.1956).

Formulation(s):   amp. 10 mg; cps. 40 mg; r. r. cps. 40 mg; tabl. 10 mg, 20 mg (as hydrochloride)

Trade Name(s):
D: Duvadilan (Thomae/   Duvadilan (Duphar); wfm   Isokulin (Toho Iyaku
     Duphar); wfm   I: Duvadilan (UCM)   Kenkyusho)
     Vasoplex (Lappe); wfm   Fenam (UCM)   Synzedrin (Teisan)
F: Duvadilan (Solvay Pharma)   Vasosuprina (Lusofarmaco)   Vahodilan (Morita)
GB: Defencin (Bristol); wfm   J: Duvadilan (Daiichi)   Vasoladin (Kanto)

| USA: | Isolait (Elder); wfm | Vasodilan (Mead Johnson); wfm | generic |

Isradipine
(Isrodipine; PN 200-110)

ATC: C02DE; C08CA03
Use: long acting calcium antagonist, antihypertensive, antianginal

RN: 75695-93-1   MF: $C_{19}H_{21}N_3O_5$ MW: 371.39
LD_{50}:   1.2 mg/kg (M, i.v.); 216 mg/kg (M, p.o.);
1.8 mg/kg (R, i.v.); >3000 mg/kg (R, p.o.);
1.2 mg/kg (rabbit, i.v.); 58 mg/kg (rabbit, p.o.)
CN: 4-(4-benzofurazanyl)-1,4-dihydro-2,6-dimethyl-3,5-pyridinedicarboxylic acid methyl 1-methylethyl ester

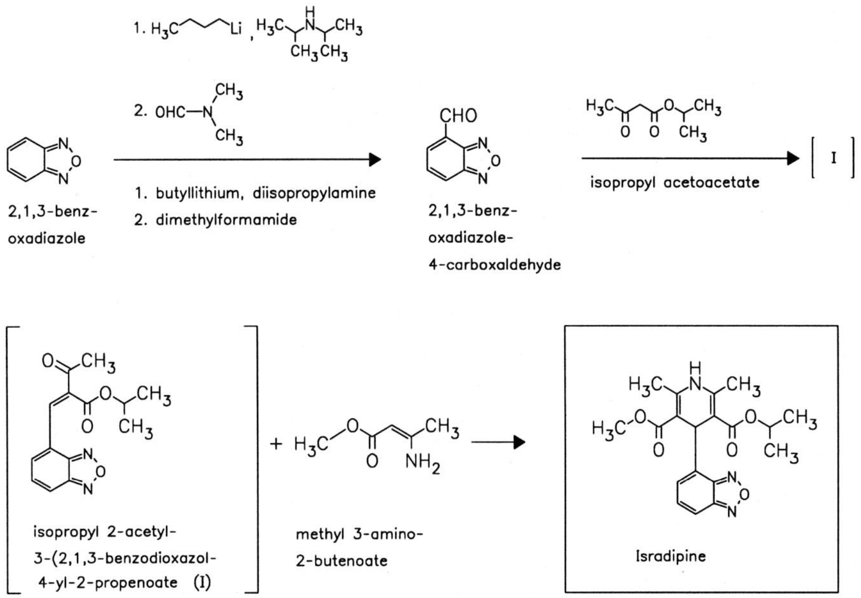

Reference(s):
DE 2 949 491 (Sandoz; appl. 8.12.1979; CH-prior. 18.12.1978).
GB 2 122 192 (Sandoz; appl. 13.6.1983; CH-prior. 15.6.1982).
US 4 466 972 (Sandoz; 21.8.1984; appl. 19.3.1982; CH-prior. 18.12.1978).

synthesis of 2,1,3-benzoxadiazole-4-carboxaldehyde:
CH 661 270 (Sandoz 15.11.1982; GB-prior. 18.11.1981).

synthesis of enantiomers:
DE 3 320 616 (Sandoz; appl. 8.6.1983; CH-prior. 15.6.1982).

sustained release formulation:
US 4 950 486 (Alza; 21.8.1990; prior. 7.11.1988, 2.10.1987).
US 4 946 687 (Alza; 7.8.1990; prior. 7.11.1988, 2.10.1987).
US 4 816 263 (Alza; 28.3.1989; prior. 7.11.1988, 2.10.1987).

hydrosol formulation:
GB 2 200 048 (Sandoz; appl. 17.12.1987; D-prior. 19.12.1986, 15.12.1987).
DE 3 742 473 (Sandoz; appl. 19.12.1986).

combination with calcitonin:
EP 202 282 (Sandoz; appl. 8.11.1985; GB-prior. 12.11.1984).

nanoparticles:
Leroueil-Le Verger, M. et al.: Eur. J. Pharm. Biopharm. (EJPBEL) **46**, 137-143 (1998)

Formulation(s):   cps. 2.5 mg, 5 mg; s. r. tabl. 5 mg, 10 mg; tabl. 2.5 mg

Trade Name(s):
D: Lomir (Novartis Pharma)   GB: Prescal (Novartis)   Lomir (Sandoz)
        Vascal (Schwarz; 1991)   I: Clivoten (Lifepharma)   USA: Dyna Circ (Novartis)
F: Icaz LP (Novartis)   Esradin (Sigma-Tau)

Itopride hydrochloride
(151235 (as hydrochloride); 149097 (as free base))

ATC: D08
Use: peristaltic stimulant, gastric
            prokinetic agent

RN: 122892-31-3   MF: $C_{20}H_{26}N_2O_4 \cdot HCl$ MW: 394.90
LD_{50}: 190.6 mg/kg (M, i.v.)
CN: N-[[4-[2-(dimethylamino)ethoxy]phenyl]methyl]-3,4-dimethoxybenzamide monohydrochloride

base
RN: 122898-67-3   MF: $C_{20}H_{26}N_2O_4$ MW: 358.44

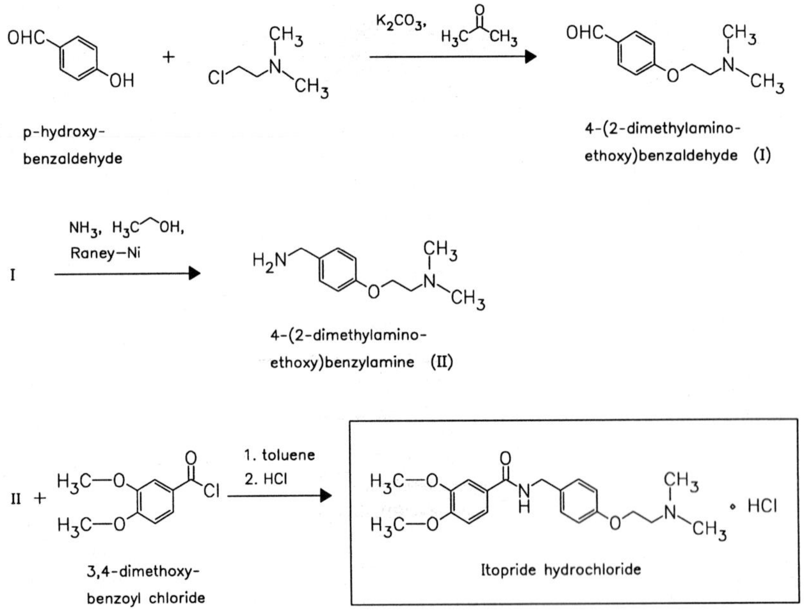

p-hydroxy-
benzaldehyde

4-(2-dimethylamino-
ethoxy)benzaldehyde (I)

4-(2-dimethylamino-
ethoxy)benzylamine (II)

3,4-dimethoxy-
benzoyl chloride

Itopride hydrochloride

Reference(s):

preparation and formulation:
EP 306 827 (Hokuriku Pharmaceutical Co.; appl. 15.3.1989; J-prior. 1.9.1988, 5.9.1987, 22.9.1987, 29.9.1987, 5.10.1987).

synthesis of intermediate II:
US 2 879 293 (Hoffmann-La Roche; 1957).

Formulation(s):   tabl. 50 mg (hydrochloride)

Trade Name(s):
J: Ganaton (Hokuriku)

Itraconazole
(R-51211)

ATC: J02AC02
Use: antifungal (treatment of vaginal
        candidiasis pityriasis versicolor,
        dermatophytes and systemic
        mycoses)

RN: 84625-61-6   MF: C$_{35}$H$_{38}$Cl$_2$N$_8$O$_4$ MW: 705.65
LD$_{50}$: 46.4 mg/kg (M, i.v.); >320 mg/kg (M, p.o.);
        40 mg/kg (R, i.v.); >320 mg/kg (R, p.o.);
        >200 mg/kg (dog, p.o.)
CN: 4-[4-[4-[4-[[2-(2,4-dichlorophenyl)-2-(1*H*-1,2,4-triazol-1-ylmethyl)-1,3-dioxolan-4-yl]methoxy]phenyl]-
        1-piperazinyl]phenyl]-2,4-dihydro-2-(1-methylpropyl)-3*H*-1,2,4-triazol-3-one

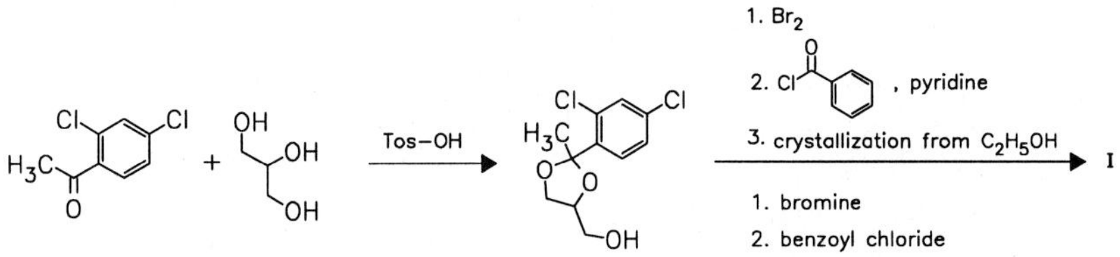

2',4'-dichloro-   glycerol
acetophenone

1. Br$_2$
2. benzoyl chloride , pyridine
3. crystallization from C$_2$H$_5$OH

1. bromine
2. benzoyl chloride

cis-[2-bromo-   1H-1,2,4-
methyl-2-(2,4-   triazole
dichlorophenyl)-
1,3-dioxolan-4-
ylmethyl] benzoate (I)

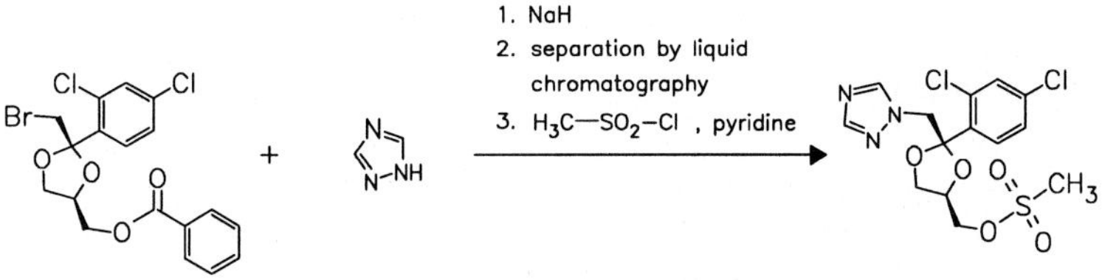

1. NaH
2. separation by liquid
   chromatography
3. H$_3$C—SO$_2$—Cl , pyridine

cis-2-(2,4-dichloro-
phenyl)-2-(1H-1,2,4-
triazol-1-ylmethyl)-
1,3-dioxolane-4-
methanol methane-
sulfonate (II)

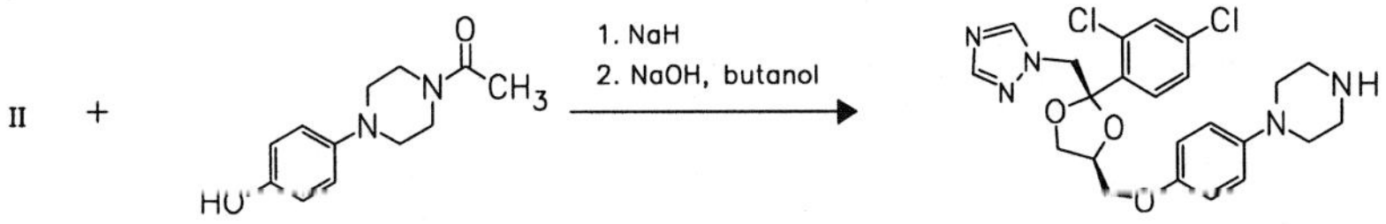

II   +

1. NaH
2. NaOH, butanol

1-(4-hydroxyphenyl)-
4-acetylpiperazine

(III)

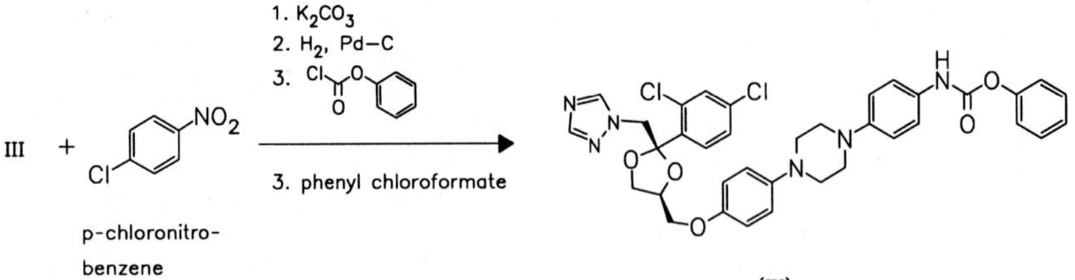

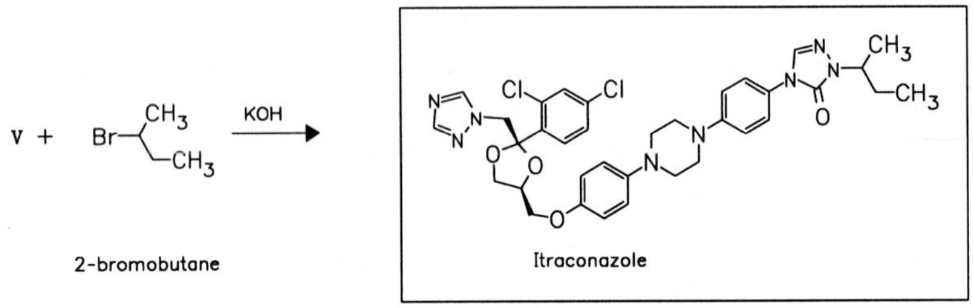

Reference(s):

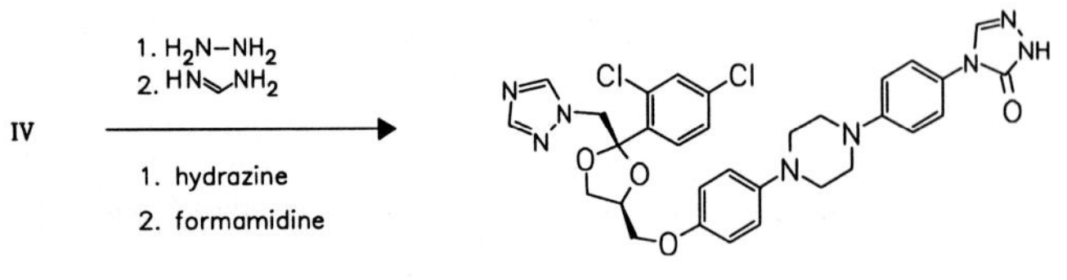

EP 118 138 (Janssen; appl. 24.1.1984; USA-prior. 28.2.1983).

alternative synthesis:

EP 6 711 (Janssen; appl. 13.6.1979; USA-prior. 23.6.1978, 14.3.1979).
US 4 267 179 (Janssen; 12.5.1981; appl. 14.3.1979; prior. 23.6.1978).
Meeres, J.; Backx, L.J.J.; Cutsem, J. van: J. Med. Chem. (JMCMAR) **27**, 894 (1984).

synthesis of cis-2-(2,4-dichlorophenyl)-2-(1H-1,2,4-triazol-1-ylmethyl)-1,3-dioxolane-4-methanol methanesulfonate:

Meeres, J. et al.: J. Med. Chem. (JMCMAR) **22**, 1003 (1979).
Meeres, J.; Hendrickx, R.; Cutsem, J. van: J. Med. Chem. (JMCMAR) **26**, 611 (1983).

topical liposomal formulation:

WO 9 315 719 (Janssen; appl. 4.2.1993; EP-prior. 12.2.1992).

pharmaceutical composition:

WO 9 416 700 (Sepracor; appl. 27.1.1994; USA-prior. 27.1.1993).

Formulation(s):   cps. 100 mg, sol. 1 %

Trade Name(s):

D:	Sempera (Glaxo Wellcome;	F:	Sporanox (Janssen-Cilag)		Triasporin (Lifepharma)
	Janssen-Cilag)	GB:	Sporanox (Janssen-Cilag)	J:	Itrizole (Janssen-Kyowa)
	Siros (Janssen-Cilag)	I:	Sporanox (Janssen)	USA:	Sporanox (Janssen; 1999)

Josamycin

ATC: J01FA07
Use: antibiotic

RN: 16846-24-5   MF: C$_{42}$H$_{69}$NO$_{15}$ MW: 828.01   EINECS: 240-871-6
CN: leucomycin V 3-acetate 4^B-(3-methylbutanoate)

propionate
RN: 51016-68-3   MF: C$_{45}$H$_{73}$NO$_{16}$ MW: 884.07

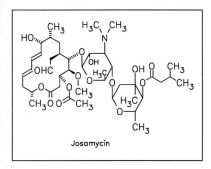

Josamycin

From fermentation solutions of *Streptomyces narbonensis var. josamyceticus* (ATTC 17835).

Reference(s):
DOS 1 492 035 (Microbial Chem. Res.; appl. 3.6.1965; J-prior. 9.6.1964).
US 3 636 197 (Yamanouchi; 18.1.1972; J-prior. 9.6.1964).

alternative syntheses:
from 10-acetyl- *and* 10,2'-diacetyljosamycin:
JP-appl. 76/41 497 (Yamanouchi; appl. 2.10.1974).

from 3-deacetyljosamycin:
JP-appl. 77/41 294 (Microb. Res. Found.; appl. 26.9.1975).

water soluble H$_2$SO$_3$-D-glucosamine *addition compound:*
JP-appl. 77/71 489 (Yamanouchi; appl. 31.10.1975).

solvent free crystals:
JP-appl. 77/51 013 (Yamanouchi; appl. 16.10.1975).
JP-appl. 76/142 519 (Yamanouchi; appl. 31.5.1975).

pharmaceutical formulation:
US 3 960 757 (Toyo Jozo; 1.6.1976; prior. 29.6.1973).

Formulation(s):   gran. 1 g; susp. 150 mg, 300 mg; tabl. 500 mg (as propionate)

Trade Name(s):
D: Wilprafen (Yamanouchi;   I: Iosalide (Schering)   J: Josamycin (Yamanouchi;
       1984)   Josaxin (UCB)   1970)
F: Josacine (Bellon; 1980)

Kanamycin (A)

ATC: A07AA08; J01GB04
Use: antibiotic

RN: 59-01-8   MF: $C_{18}H_{36}N_4O_{11}$ MW: 484.50   EINECS: 200-411-7
LD_{50}:   115 mg/kg (M, i.v.); 20.7 mg/kg (M, p.o.);
       437 mg/kg (R, i.v.); >10 g/kg (R, p.o.)
CN: *O*-3-amino-3-deoxy-α-D-glucopyranosyl-(1→6)-*O*-[6-amino-6-deoxy-α-D-glucopyranosyl-(1→4)]-2-deoxy-D-streptamine

sulfate (1:1)
RN: 25389-94-0   MF: $C_{18}H_{36}N_4O_{11} \cdot H_2SO_4$ MW: 582.58   EINECS: 246-933-9

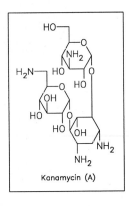

Kanamycin (A)

From fermentation solutions of *Streptomyces kanamyceticus*.

Reference(s):
US 2 931 798 (H. Umezawa, K. Maeda, M. Meda; 5.4.1960; J-prior. 5.9.1956).
US 2 936 307 (Bristol-Myers; 1960; appl. 1957).
US 2 967 177 (Bristol-Myers; 1961; appl. 1958).
US 3 032 547 (Merck & Co.; 1962; appl. 1958).

Formulation(s):   amp. 1 g (as sulfate); cream 5 mg/g; eye drops 5 mg/ml

Trade Name(s):

D: Kanamycin-POS
     (Ursapharm)
     Kanamytrex (Alcon)
     Kana-Ophthal (Winzer; as
     sulfate)
F: Kamycine (Bristol-Myers
     Squibb)
     Stérimycine (CIBA Vision
     Ophthalmics)-comb.
GB: Kannasyn (Sanofi
     Winthrop)

I: Dermaflogil (Nuovo Cons.
     Sanit. Naz.)-comb.
     Fluomicetina (Zoja)-comb.
     Kanaderm (Firma)-comb.
     Kanamicina Firma (Firma)
     Kanatrombina (Baldacci)-
     comb.
     Kanazone (SIT)-comb.
     Keimicina (Boehringer
     Mannh.)
     Roseomix (Farmigea)-
     comb.

     generic
J: Kanacillin (Banyu; Meiji)-
     comb.
     Kanacyclin (Banyu; Meiji)-
     comb.
     Kanafuracin (Fujita)-comb.
     Kanamycin (Banyu; Meiji;
     Tanabe; Yamanouchi)
     generic
USA: Kantrex (Bristol); wfm
     generic

Kawain

(Kavain; Cavain)

ATC: C04
Use: anticonvulsant, psychotonic

RN: 500-64-1   MF: $C_{14}H_{14}O_3$ MW: 230.26   EINECS: 207-907-2
LD$_{50}$:   69 mg/kg (M, i.v.); 1130 mg/kg (M, p.o.)
CN: [R-(E)]-5,6-dihydro-4-methoxy-6-(2-phenylethenyl)-2H-pyran-2-one

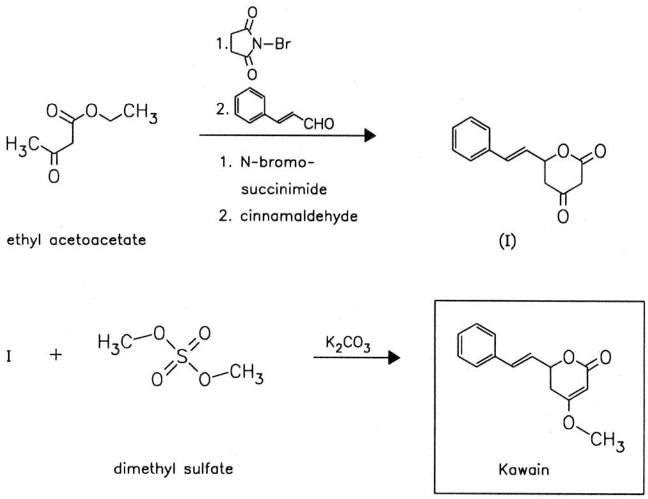

ethyl acetoacetate   (I)

1. N-bromo-
   succinimide
2. cinnamaldehyde

dimethyl sulfate   Kawain

Reference(s):
FR 1 526 596 (Spezialchemie; appl. 9.6.1967; D-prior. 29.7.1966).

alternative syntheses:
Fowler, E.M.; Henbest, H.B.: J. Chem. Soc. (JCSOA9) **1950**, 3642 (racemate).
Kostermans, D.G.F.R.: Recl. Trav. Chim. Pays-Bas (RTCPA3) **70**, 79 (1951) (racemate).

isolation of (+)-kawain:
Borsche, W.; Peitzsch, W.: Ber. Dtsch. Chem. Ges. (BDCGAS) **63**, 2414 (1930).

absolute configuration:
Snatzke, G.; Hansel, R.: Tetrahedron Lett. (TELEAY) **1968**, 1797.

preparation of an endoanesthetic effective solution:
GB 1 214 936 (Spezialchemie; valid from 5.6.1968; D-prior. 5.6.1967).

review:
Kretzschmer, R.; Teschendorf, H.J.: Chem.-Ztg. (CMKZAT) **98**, 24 (1974).

Formulation(s):   cps. 30 mg, 50 mg, 200 mg

Trade Name(s):
D: Ardeydystin
     (Ardeypharm)-comb.; wfm

     Duront (Woelm)-comb.;
     wfm

     Kavaform (Dr. Schwab)-
     comb.
     Neuronika (Klinge)

Kebuzone

(Cetophenylbutazone; Ketophenylbutazon)

ATC: M01AA06
Use: anti-inflammatory, antirheumatic

RN: 853-34-9 MF: $C_{19}H_{18}N_2O_3$ MW: 322.36 EINECS: 212-715-7
LD_{50}: 580 mg/kg (M, i.v.); 750 mg/kg (M, p.o.);
315 mg/kg (R, i.v.); 720 mg/kg (R, p.o.)
CN: 4-(3-oxobutyl)-1,2-diphenyl-3,5-pyrazolidinedione

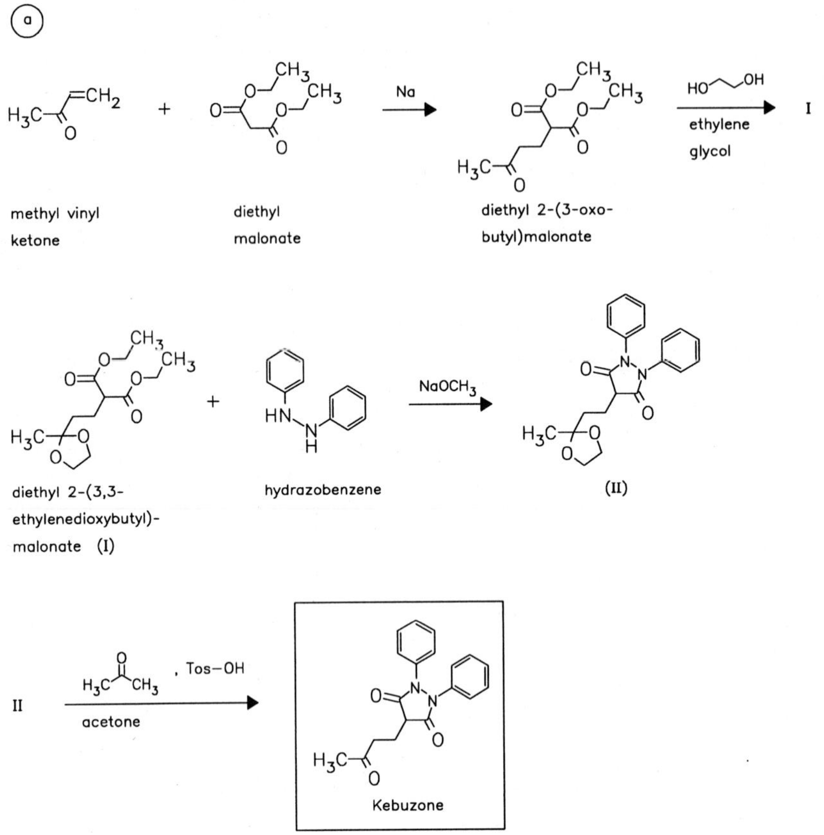
a
methyl vinyl ketone
diethyl malonate
Na
diethyl 2-(3-oxo-butyl)malonate
ethylene glycol
I
diethyl 2-(3,3-ethylenedioxybutyl)-malonate (I)
hydrazobenzene
NaOCH3
(II)
II
acetone
, Tos−OH
Kebuzone

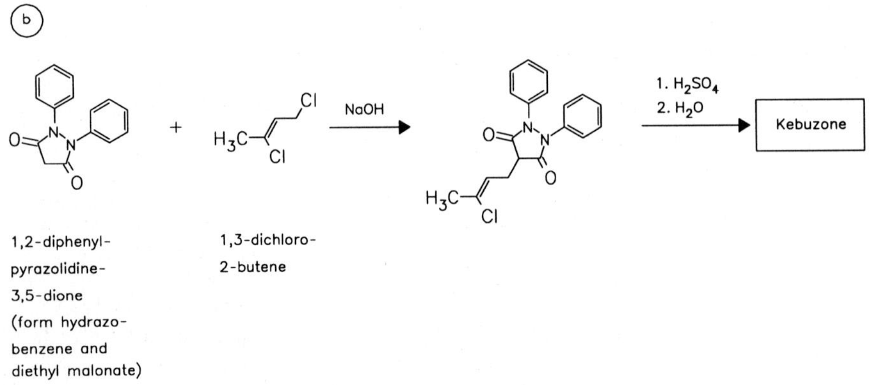
b
1,2-diphenyl-pyrazolidine-3,5-dione (form hydrazo-benzene and diethyl malonate)
1,3-dichloro-2-butene
NaOH
1. H2SO4
2. H2O
Kebuzone

Reference(s):

a Denss, R. et al.: Helv. Chim. Acta (HCACAV) **40**, 402 (1957).

starting material:
Kühn, M.: J. Prakt. Chem. (JPCPAO) **156** (II), 103 (1940).
b   AT 198 263 (Synfarma; appl. 1955).

Formulation(s):   amp. 1 g/5 ml; cps. 250 mg

Trade Name(s):
D: Kebuzon (Steiner)   Kentan-S (Sawai)   Ketobutane (Yamagata)
F: Ketazone (Beytout); wfm   Ketazon (Kyowa)   Ketobutazone (Sato; Toho)
I: Chetopir (Sarm); wfm   Ketazone (Kyowa Hakko)   Ketophezon (Kissei)
J: Hichillos (Kotani)   Ketobutan (Santen)   Vintop (Maruko)

Ketamine

ATC: N01AX03
Use: analgesic, anesthetic

RN: 6740-88-1   MF: $C_{13}H_{16}ClNO$ MW: 237.73   EINECS: 229-804-1
LD$_{50}$:   77 mg/kg (M, i.v.)
CN: (±)-2-(2-chlorophenyl)-2-(methylamino)cyclohexanone

monohydrochloride
RN: 1867-66-9   MF: $C_{13}H_{16}ClNO \cdot HCl$ MW: 274.19   EINECS: 217-484-6

2-chloro-   cyclopentyl-   (2-chloro-   (I)
benzonitrile   magnesium   benzoyl)-
                   bromide   cyclopentane

methylamine

Reference(s):
US 3 254 124 (Parke Davis; 31.5.1966; prior. 31.7.1961, 29.6.1962).
BE 634 208 (Parke Davis; appl. 27.6.1963; USA-prior. 29.6.1962).

Formulation(s):   amp. 50 mg/5 ml, 100 mg/2 ml, 500 mg/10 ml; inj. 5 mg/ml, 10 mg/ml, 25 mg/ml, 50 mg/ml
                     (as hydrochloride)

Trade Name(s):
D: Ketanest (Parke Davis; as   Kétamine Panpharma   Ketalar (Parke Davis); wfm
      hydrochloride)   (Panpharma; as   Ketalar (Parke Davis; as
      Velonarcon (ASTA Medica   hydrochloride)   hydrochloride); wfm
      AWD; as hydrochloride)   GB: Ketalar (Parke Davis)   Ketaset (Bristol-Myers
F: Kétalar (Parke Davis; as   I: Ketalar (Parke Davis)   Squibb; as hydrochloride);
      hydrochloride)   J: Ketalar (Sankyo)   wfm
                                USA: Ketaject (Bristol); wfm

Ketanserin

ATC: C02KD01
Use: antihypertensive

RN: 74050-98-9 MF: $C_{22}H_{22}FN_3O_3$ MW: 395.43 EINECS: 277-680-2
CN: 3-[2-[4-(4-fluorobenzoyl)-1-piperidinyl]ethyl]-2,4(1*H*,3*H*)-quinazolinedione

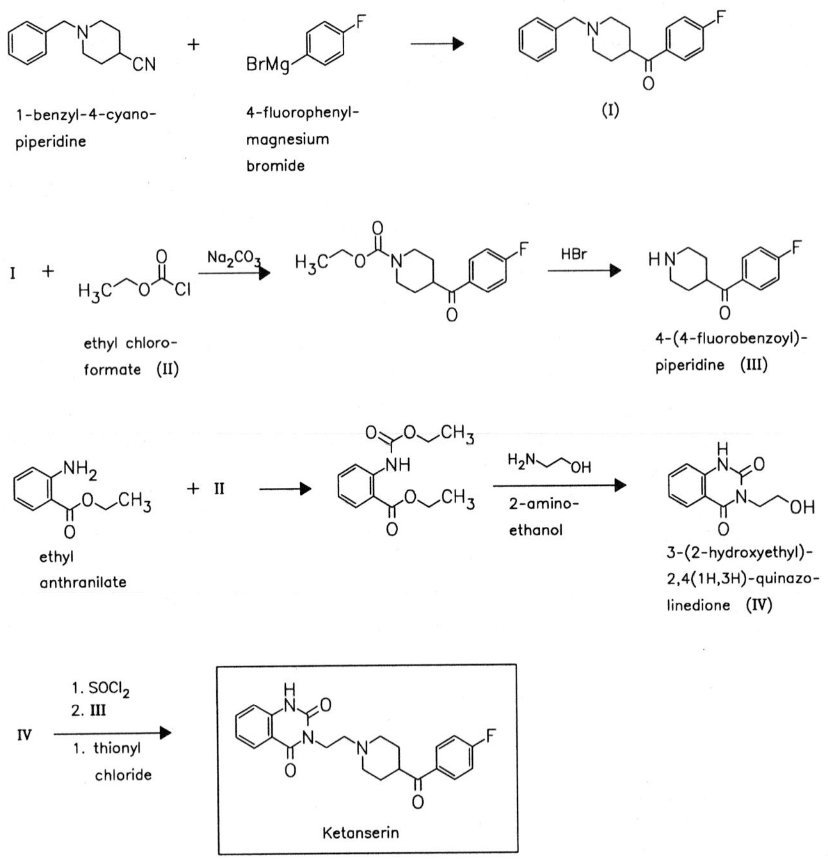
1-benzyl-4-cyano-piperidine
4-fluorophenylmagnesium bromide
(I)
I +
Na_2CO_3
ethyl chloroformate (II)
HBr
4-(4-fluorobenzoyl)-piperidine (III)
ethyl anthranilate
+ II
H_2N OH
2-aminoethanol
3-(2-hydroxyethyl)-2,4(1H,3H)-quinazolinedione (IV)
IV
1. $SOCl_2$
2. III
1. thionyl chloride
Ketanserin

Reference(s):
EP 13 612 (Janssen; appl. 7.1.1980; USA-prior. 8.1.1979, 12.10.1979).
US 4 335 127 (Janssen; 15.6.1982; prior. 8.1.1979, 12.10.1979).

alternative synthesis:
EP-appl. 98 499 (Ravizza; appl. 27.6.1983; I-prior. 6.7.1982).

Formulation(s): amp. 10 mg/2 ml, 50 mg/10 ml; tabl. 20 mg, 40 mg

Trade Name(s):
I: Perketan (Inverni della Beffa)
Serepress (Formenti)
Sufrexal (Janssen; 1987)

Ketazolam

ATC: N05BA10
Use: tranquilizer

RN: 27223-35-4   MF: $C_{20}H_{17}ClN_2O_3$ MW: 368.82   EINECS: 248-346-3
LD$_{50}$:   2 g/kg (M, p.o.);
        5 g/kg (R, p.o.)
CN: 11-chloro-8,12b-dihydro-2,8-dimethyl-12b-phenyl-4H-[1,3]oxazino[3,2-d][1,4]benzodiazepine-4,7(6H)-dione

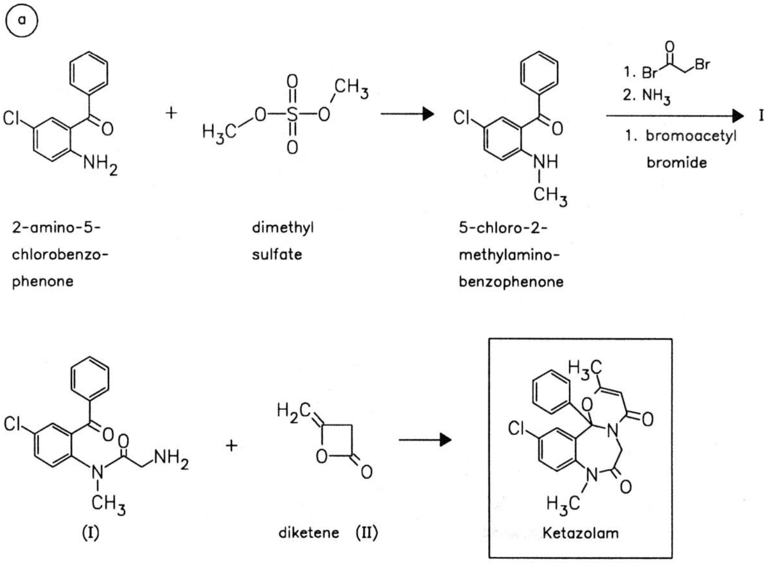

2-amino-5-chlorobenzo-phenone

dimethyl sulfate

5-chloro-2-methylamino-benzophenone

(I)

diketene (II)

Ketazolam

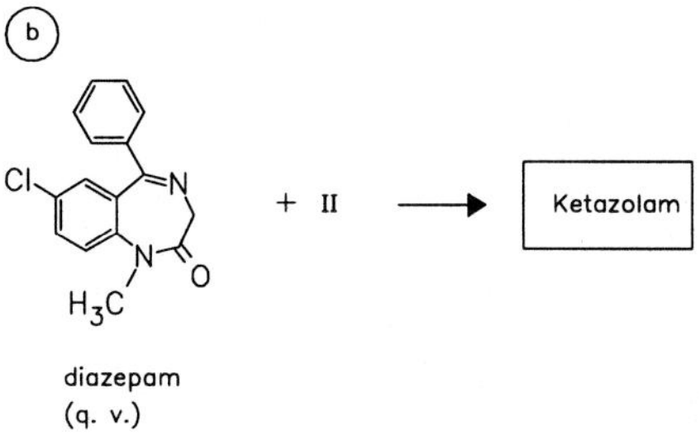

diazepam (q. v.)

Ketazolam

Reference(s):
a   US 3 575 965 (Upjohn; 20.4.1971; prior. 20.10.1969).
b   Szmuskovicz, J. et al.: Tetrahedron Lett. (TELEAY) **1971**, 3665.
        DOS 1 947 226 (Upjohn; appl. 18.9.1969; USA-prior. 19.9.1968, 27.3.1969).

Formulation(s):   cps. 15 mg, 30 mg, 45 mg

Trade Name(s):
D: Contamex (Beecham-   GB: Anxon (Beecham); wfm
        Wülfing); wfm   I: Anseren (Novartis)

Ketobemidone
(Cetobemidone)

ATC: N02AB01
Use: analgesic

RN: 469-79-4   MF: $C_{15}H_{21}NO_2$ MW: 247.34   EINECS: 207-421-0
LD$_{50}$:   14 mg/kg (M, i.v.);
10 mg/kg (R, i.v.)
CN: 1-[4-(3-hydroxyphenyl)-1-methyl-4-piperidinyl]-1-propanone

hydrochloride
RN: 5965-49-1   MF: $C_{15}H_{21}NO_2 \cdot HCl$ MW: 283.80

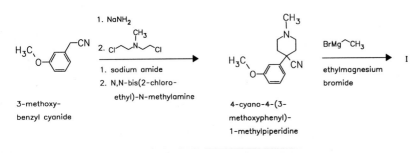

3-methoxy-benzyl cyanide

1. sodium amide
2. N,N-bis(2-chloro-ethyl)-N-methylamine

4-cyano-4-(3-methoxyphenyl)-1-methylpiperidine

ethylmagnesium bromide

4-(3-methoxyphenyl)-1-methyl-4-propionyl-piperidine (I)

Ketobemidone

Reference(s):
DRP 752 755 (I. G. Farben; appl. 1942).

alternative synthesis:
Kägi, H.; Miescher, K.: Helv. Chim. Acta (HCACAV) **32**, 2489 (1949).

Formulation(s):   amp. 2 mg/2 ml, 10 mg/10 ml, 50 mg/50 ml; tabl. 5 mg (as hydrochloride)

Trade Name(s):
D: Cliradon (Ciba); wfm

Ketoconazole

ATC: D01AC08; G01AF11; J02AB02
Use: antimycotic

RN: 65277-42-1   MF: $C_{26}H_{28}Cl_2N_4O_4$ MW: 531.44   EINECS: 265-667-4
LD$_{50}$:   32 mg/kg (M, i.v.); 618 mg/kg (M, p.o.);
86 mg/kg (R, i.v.); 166 mg/kg (R, p.o.);
23.3 mg/kg (dog, i.v.); 178 mg/kg (dog, p.o.)
CN: *cis*-1-acetyl-4-[4-[[2-(2,4-dichlorophenyl)-2-(1*H*-imidazol-1-ylmethyl)-1,3-dioxolan-4-yl]methoxy]phenyl]piperazine

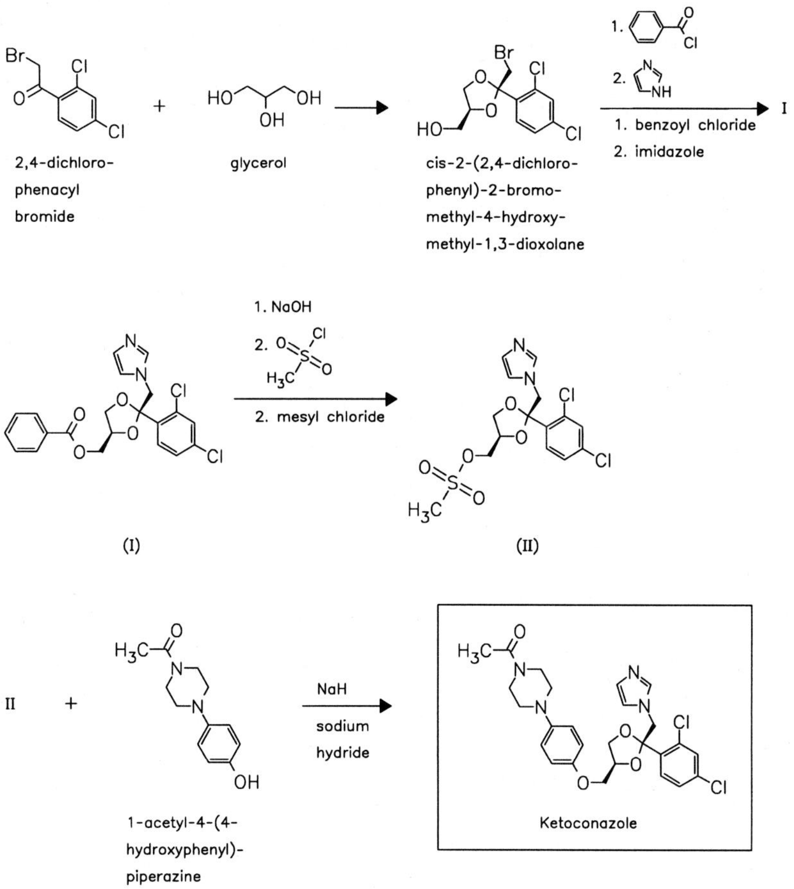

2,4-dichloro-
phenacyl
bromide

glycerol

cis-2-(2,4-dichloro-
phenyl)-2-bromo-
methyl-4-hydroxy-
methyl-1,3-dioxolane

1. benzoyl chloride
2. imidazole

1. NaOH
2. mesyl chloride

(I)   (II)

II   +

1-acetyl-4-(4-
hydroxyphenyl)-
piperazine

NaH
sodium
hydride

Ketoconazole

Reference(s):
DOS 2 804 096 (Janssen; appl. 31.1.1978; USA-prior. 31.1.1977, 21.11.1977).
US 4 335 125 (Janssen; 15.6.1982; prior. 31.1.1977).

Formulation(s):   cream 20 mg/g (2 %); shampoo 2 %; sol. 20 mg/ml; susp. 100mg; tabl. 200 mg

Trade Name(s):
D: Nizoral (Janssen-Cilag;   Nizoral (Janssen-Cilag;   I: Nizoral (Janssen; 1983)
        1981)   1983)   J: Nizoral (Kyowa Hakko)
        Terzolin (Janssen-Cilag)   GB: Nizoral (Janssen-Cilag; USA: Nizoral (Janssen; 1981)
F: Kétoderm (Janssen-Cilag)   1981)

Ketoprofen

ATC: M01AE03; M02AA10
Use: analgesic, anti-inflammatory

RN: 22071-15-4   MF: C$_{16}$H$_{14}$O$_3$ MW: 254.29   EINECS: 244-759-8
LD$_{50}$:   500 mg/kg (M, i.v.); 360 mg/kg (M, p.o.);
        350 mg/kg (R, i.v.); 62.4 mg/kg (R, p.o.)
CN: 3-benzoyl-α-methylbenzeneacetic acid

lysine salt
RN: 57469-78-0   MF: C$_{22}$H$_{28}$N$_2$O$_5$ MW: 400.48

3-methylbenzo-
phenone

3-bromomethyl-
benzophenone

(3-benzoylphenyl)-
acetonitrile (I)

diethyl carbonate

(II)

methyl
iodide

Ketoprofen

Reference(s):
US 3 641 127 (Rhône-Poulenc; 8.2.1972; F-prior. 27.1.1967).
DE 1 668 648 (Rhône-Poulenc; appl. 26.1.1968; F-prior. 27.1.1967).

alternative syntheses:
DOS 2 646 792 (Mitsubishi Petrochemical; appl. 16.10.1976; J-prior. 23.10.1975, 31.7.1976).
US 4 097 522 (Aziende Chim. Riun. Angelini Francesco; 27.6.1978; I-prior. 5.6.1975).
DOS 2 744 832 (LEK; appl. 5.10.1977; YU-prior. 18.10.1976).
DOS 2 744 833 (LEK; appl. 5.10.1977; YU-prior. 18.10.1976).
DOS 2 744 834 (LEK; appl. 5.10.1977; YU-prior. 18.10.1976).

Formulation(s):　cps. 50 mg, 100 mg; gel 2.5 %; s. r. cps. 200 mg, 320 mg (as lysine salt); suppos. 30.6 mg,
　　　　　　　　　　100 mg; vial 100 mg

Trade Name(s):
D:　Alrheumun (Bayer Vital;
　　1975)
　　Orudis (Rhône-Poulenc
　　Rorer)
F:　Bi-Profénid (Specia)
　　Kétum (Ménarini)
　　Profénid (Specia)
　　Toprec (Théraplix)
GB:　Ketocid (Trinity)
　　Orudis (Rhône-Poulenc
　　Rorer)
　　Oruvail (Rhône-Poulenc
　　Rorer)

I:　Powergel (Searle)
　　Artrosilene (Dompé; as
　　lysine salt)
　　Fastum (Menarini)
　　Flexen (Lifepharma)
　　Isok (San Carlo)
　　Ketartrium (Esseti)
　　Ketodol (Drug Research)
　　Ketofen (Delsaz &
　　Filippini; as lysine salt)
　　Meprofen (AGIPS)
　　Orudis (Rhône-Poulenc
　　Rorer)

　　Profenil (Drug Research)
　　Reuprofen (Terapeutico
　　M.R.)
　　Salient (Biomedica
　　Foscama)
　　Sinketol (Locatelli)
J:　Capisten (Kissei)
　　Orudis (Hokuriku)
USA:　Orudis (Wyeth-Ayerst;
　　1977)
　　Oruvail (Wyeth-Ayerst)

Ketorolac

ATC:　M01AB15; N02BE; S01BC05
Use:　non-steroidal anti-inflammatory,
　　　analgesic

RN:　74103-06-3　MF: $C_{15}H_{13}NO_3$　MW: 255.27
LD_{50}:　200 mg/kg (M, p.o.)
CN:　(±)-5-benzoyl-2,3-dihydro-1*H*-pyrrolizine-1-carboxylic acid

tromethamine salt
RN: 74103-07-4   MF: $C_{15}H_{13}NO_3 \cdot C_4H_{11}NO_3$ MW: 376.41
monosodium salt
RN: 110618-38-7   MF: $C_{15}H_{12}NNaO_3$ MW: 277.26

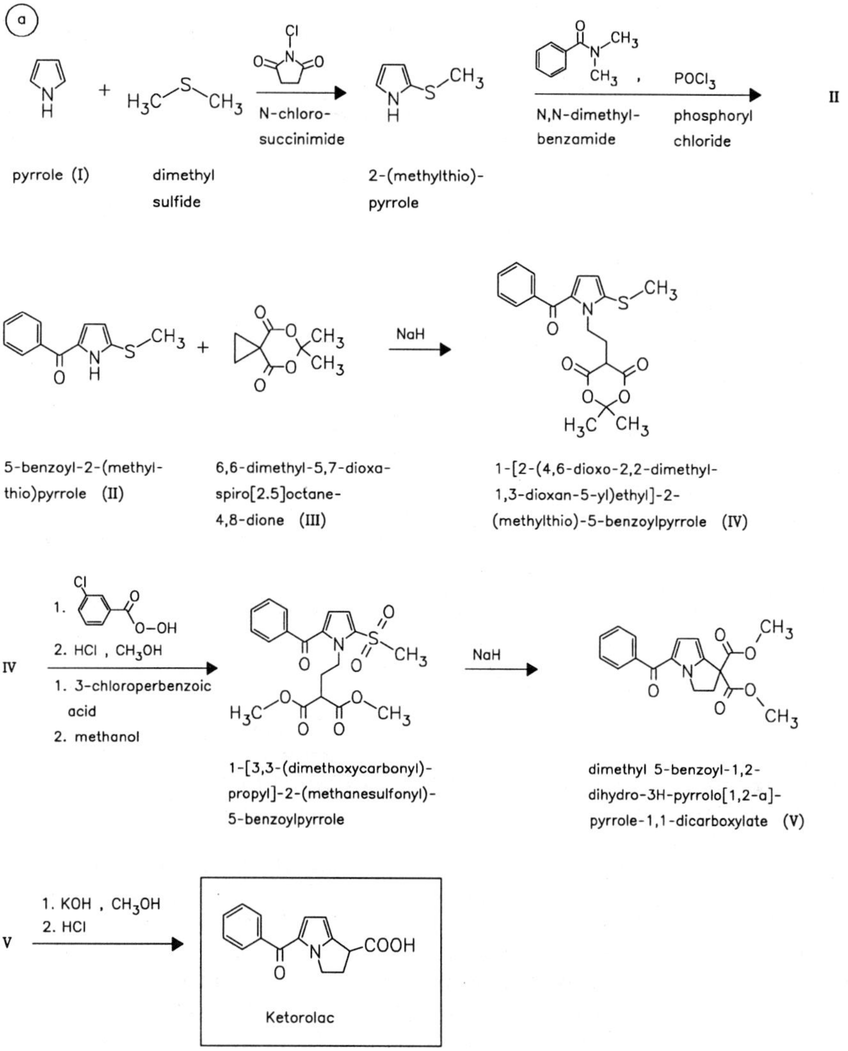

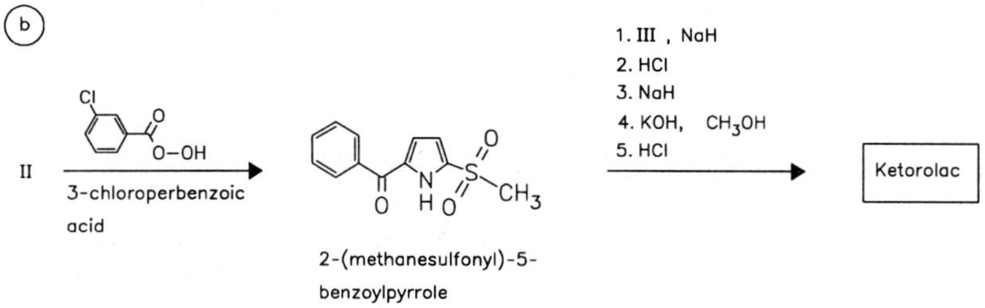

pyrrole (I)   dimethyl   2-(methylthio)-
                     sulfide   pyrrole

N-chloro-
succinimide

N,N-dimethyl-   phosphoryl
benzamide   chloride

II

5-benzoyl-2-(methyl-   6,6-dimethyl-5,7-dioxa-   1-[2-(4,6-dioxo-2,2-dimethyl-
thio)pyrrole (II)   spiro[2.5]octane-   1,3-dioxan-5-yl)ethyl]-2-
                         4,8-dione (III)   (methylthio)-5-benzoylpyrrole (IV)

IV

1.
2. HCl , CH$_3$OH
1. 3-chloroperbenzoic acid
2. methanol

1-[3,3-(dimethoxycarbonyl)-
propyl]-2-(methanesulfonyl)-
5-benzoylpyrrole

NaH

dimethyl 5-benzoyl-1,2-
dihydro-3H-pyrrolo[1,2-a]-
pyrrole-1,1-dicarboxylate (V)

V

1. KOH , CH$_3$OH
2. HCl

Ketorolac

II

3-chloroperbenzoic
acid

2-(methanesulfonyl)-5-
benzoylpyrrole

1. III , NaH
2. HCl
3. NaH
4. KOH, CH$_3$OH
5. HCl

Ketorolac

(c)

I →（N-chloro-succinimide, THF）→ 2-chloro-pyrrole →（benzmorpholide benzmorpholide, phosphoryl phosphoryl chloride, POCl₃）→ 2-chloro-5-benzoylpyrrole (VI)

VI + III →（1. NaH; 2. HCl, CH₃OH）→ 5-benzoyl-2-chloro-1-[3,3-(dimethoxycarbonyl)-propyl]pyrrole →（1. NaH, DMF; 2. KOH, CH₃OH; 3. HCl）→ Ketorolac

Reference(s):
a,b US 4 347 186 (Syntex; 31.8.1982; appl. 20.10.1980).
US 4 458 081 (Syntex; 3.7.1984; appl. 11.6.1982; prior. 20.10.1980).
c US 4 873 340 (Syntex; 10.10.1989; appl. 29.5.1986).

alternative synthesis:
DE 2 760 330 (Syntex; appl. 13.7.1977; USA-prior. 14.7.1976, 23.2.1977).
DE 2 731 678 (Syntex; appl. 13.7.1977; USA-prior. 14.7.1976, 23.2.1977).
US 4 087 539 (Syntex; 5.2.1978; appl. 23.2.1977; prior. 14.7.1976).
US 4 089 969 (Syntex; 16.5.1978; appl. 23.2.1977; prior. 14.7.1976; 23.2.1977).
US 4 097 579 (Syntex; 27.6.1978; appl. 31.3.1977).
Muchowski, J.M. et al.: J. Med. Chem. (JMCMAR) **28**, 1037 (1985).

synthesis of enantiomers:
EP 264 429 (Wisconsin Ala. Res. Found.; appl. 2.4.1987; USA-prior. 6.11.1986, 16.4.1986).
Gazman, A. et al.: J. Med. Chem. (JMCMAR) **29**, 589 (1986).

Formulation(s):   amp. 10 mg, 15 mg, 30 mg; eye drops 0.5 %; suppos. 30 mg; tabl. 10 mg (as tromethamine salt)

Trade Name(s):

D:	Acular (Pharm-Allergan)	Toradol (Recordati; 1990)	Toradol IM (Syntex;
GB:	Acular (Allergan)	Toradol IM (Recordati;	Roche; 1990)
	Toradol (Roche)	1990)	
I:	Lixidol (Farmitalia)	USA:　Acular (Allergan)	

Ketotifen

ATC:　R06AX17; S01GX08
Use:　antiasthmatic, antihistaminic

RN:　34580-13-7　MF: C₁₉H₁₉NOS　MW: 309.43　EINECS: 252-099-7
LD₅₀:　179 mg/kg (M, p.o.)
CN:　4,9-dihydro-4-(1-methyl-4-piperidinylidene)-10*H*-benzo[4,5]cyclohepta[1,2-*b*]thiophen-10-one

fumarate
RN:　34580-14-8　MF: C₂₃H₂₃NO₅S　MW: 425.51

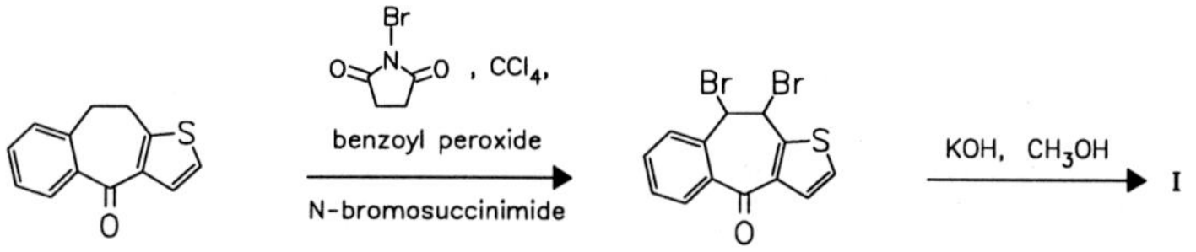

4-oxo-9,10-dihydro-
4H-benzo[4.5]cyclo-
hepta[1.2-b]thiophene
(cf. pizotifen
synthesis)

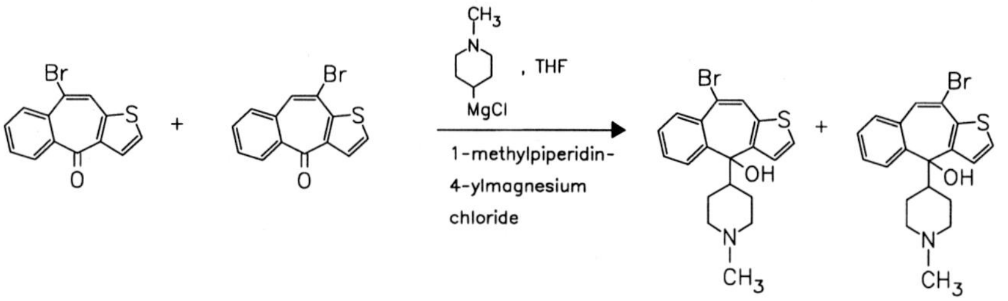

9- and 10-bromo-4-oxo-4H-
benzo[4.5]cyclohepta[1.2-b]thiophene (I)

(II)

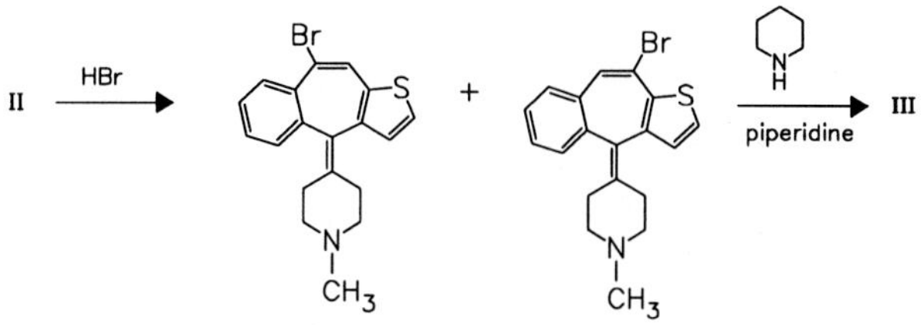

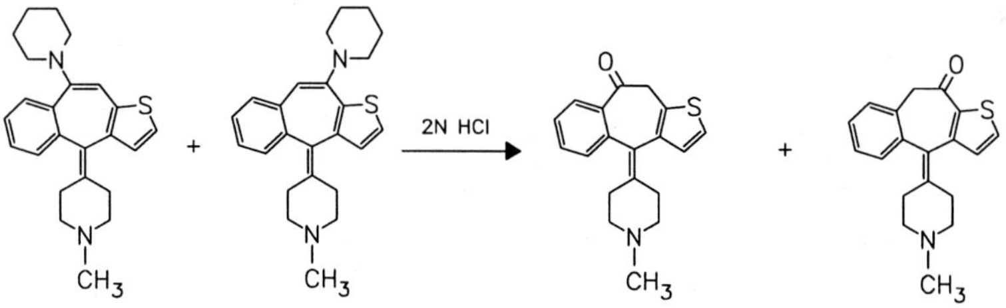

II

(III)   (IV)

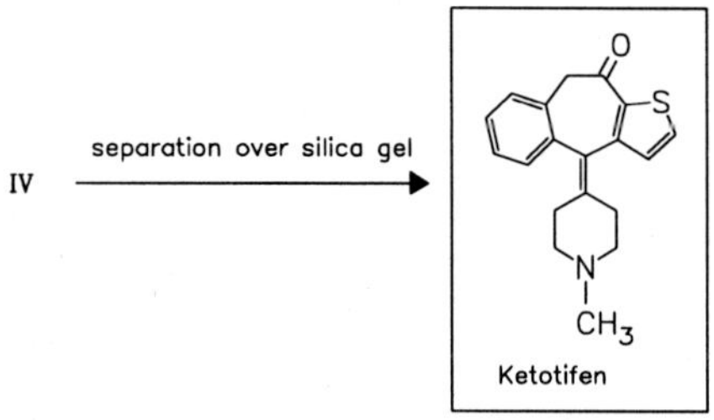

Ketotifen

Reference(s):
DAS 2 111 071 (Sandoz; appl. 9.3.1971; CH-prior. 11.3.1970, 31.7.1970).
US 3 682 930 (Sandoz; 8.8.1972; CH-prior. 11.3.1970, 31.7.1970).
DOS 2 144 490 (Sandoz; appl. 6.9.1971; CH-prior. 24.9.1970, 4.2.1971).
US 3 749 786 (Sandoz; 8.8.1972; CH-prior. 11.3.1970).

alternative syntheses:
DOS 2 302 970 (Sandoz; appl. 22.1.1973; CH-prior. 24.1.1972).
US 3 960 894 (Sandoz; 1.6.1976; CH-prior. 21.1.1972).
US 4 128 549 (Sandoz; 5.12.1978; prior. 19.1.1973, 26.7.1974, 27.2.1976, 3.2.1977).

medical use as antiasthmatic:
US 4 073 915 (Sandoz; 14.2.1978; CH-prior. 20.5.1975).

Formulation(s): cps. 1 mg; syrup 0.02 %; tabl. 1 mg, 2 mg (as fumarate)

Trade Name(s):

D:	Airvitess (Farmasan)		generics		Zaditen (Sandoz)
	Astifat (Fatol)	F:	Zaditen (Novartis; 1980)	J:	Zaditen (Sandoz-Sankyo;
	Zaditen (Novartis Pharma;	GB:	Zaditen (Novartis; 1979)		1983)
	1979)	I:	Allerkif (Edmond)		
	Zatofug (Wolff)		Totifen (Master Pharma)		

Khellin

ATC: M03
Use: vasodilator

RN: 82-02-0 MF: $C_{14}H_{12}O_5$ MW: 260.25 EINECS: 201-392-8
LD_{50}: 30.6 mg/kg (M, i.v.); 50.8 mg/kg (M, p.o.);
34.4 mg/kg (R, i.v.); 68.8 mg/kg (R, p.o.)
CN: 4,9-dimethoxy-7-methyl-5*H*-furo[3,2-*g*][1]benzopyran-5-one

Khellin

Isolation from ethanolic extracts of fruits of Umbellifera *Ammi visnaga*.

Reference(s):
Späth, E.; Gruber, W.: Ber. Dtsch. Chem. Ges. (BDCGAS) **71**, 106 (1938).
Abu-Shady, H.; Soine, T.O.: J. Am. Pharm. Assoc. (JPHAA3) **41**, 481 (1952).

total syntheses:
Baxter, R.H. et al.: J. Chem. Soc. (JCSOA9) **1949**, 30.
Clarke, J.R.; Robertson, A.: J. Am. Chem. Soc. (JACSAT) **71**, 362 (1949).
Clarke, J.R.; Robertson, A.: J. Chem. Soc. (JCSOA9) **1949**, 302.
Geissman, T.A.; Halsall, T.G.: J. Am. Chem. Soc. (JACSAT) **73**, 1280 (1951).

Formulation(s): cps. 10 mg, 12 mg; drg. 1.25 mg, 10 mg; drops 50 mg/100 ml; ointment 0.1 g/100 g,
0.01 g/100 g; sol. 0.025 g/100 ml; suppos. 2.5 mg, 5 mg; tabl. 25 mg

Trade Name(s):

D: Athmakhell (Steigerwald);
       wfm
       Bilicordan (Repha); wfm
       Broncaid (Rhône-Poulenc
       Pharma); wfm
       Cardiopax (Wider); wfm
       Coronar-Homocent (Fides);
       wfm
       Coropar (Redel); wfm

Farctil (Gewo); wfm
Hyperidyst II (Vogel &
Weber); wfm
Iosimitan (Wider); wfm
Keldrin (Thiemann); wfm
Puraeton E (Dolorgiet);
wfm
Solamin (Ardeypharm);
wfm

F: Khelline Promethazine
       Berthier (Labaz); wfm
I: Kellina (UCB); wfm
       Nefrolitin (Geymonat)-
       comb.; wfm
       Vasokellina papaverina
       (Angelini)-comb.; wfm

Labetalol

ATC:　C07AG01
Use:　antihypertensive (α- and β-receptor blocker)

RN:　36894-69-6　MF: C$_{19}$H$_{24}$N$_2$O$_3$　MW: 328.41　EINECS: 253-258-3
LD$_{50}$:　97.5 mg/kg (M, i.v.); 660 mg/kg (M, p.o.);
　　　>50 mg/kg (R, i.v.); >2 g/kg (R, p.o.)
CN:　2-hydroxy-5-[1-hydroxy-2-[(1-methyl-3-phenylpropyl)amino]ethyl]benzamide

hydrochloride
RN:　32780-64-6　MF: C$_{19}$H$_{24}$N$_2$O$_3$ · HCl　MW: 364.87　EINECS: 251-211-1
LD$_{50}$:　47 mg/kg (M, i.v.); 1.45 g/kg (M, p.o.);
　　　53 mg/kg (R, i.v.); 2.114 g/kg (R, p.o.);
　　　>1.5 g/kg (dog, p.o.)

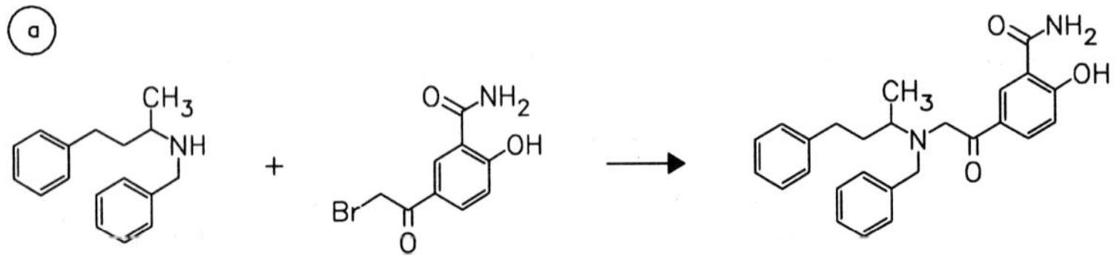

N-benzyl-1-methyl-3-phenyl-propylamine

5-bromoacetyl-salicylamide (I)

(II)

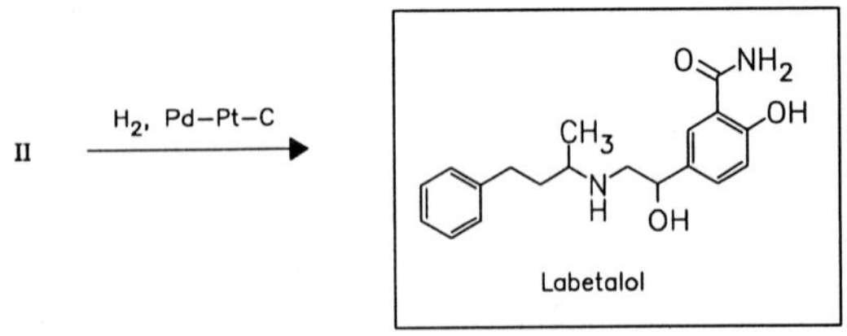

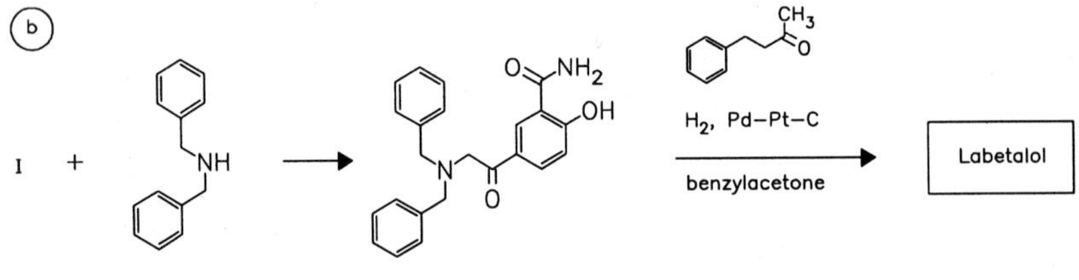

II　$\xrightarrow{\text{H}_2,\ \text{Pd–Pt–C}}$

Labetalol

dibenzyl-amine

benzylacetone

$\xrightarrow{\text{H}_2,\ \text{Pd–Pt–C}}$　Labetalol

Reference(s):
US 4 012 444 (Allen & Hanburys;.15.3.1977; prior. 29.6.1970).
US 4 066 755 (Allen & Hanburys; 3.1.1978; GB-prior. 30.11.1973).
DOS 2 032 642 (Allen & Hanburys; appl. 1.7.1970; GB-prior. 8.7.1969).
DAS 1 643 224 (Allen & Hanburys; appl. 22.9.1967; GB-prior. 23.9.1966, 21.4.1967).
US 3 705 233 (Allen & Hanburys; 5.12.1972; GB-prior. 23.9.1966).

Formulation(s):　amp. 5 mg/ml, 50 mg, 100 mg; tabl. 50 mg, 100 mg, 200 mg, 300 mg, 400 mg (as hydrochloride)

Trade Name(s):

D: Trandate (Glaxo; 1978); wfm

F: Trandate (Novartis; 1980)

GB: Trandate (Evans; 1977)

I: Abetol (CT)
Alfabetal (Mitim)
Amipres (Salus Research)
Biotens (Kemyos)-comb.

Diurolab (Leben's)-comb.
Ipolab (Leben's)
Lolum (Lifepharma)
Pressalolo (Locatelli)
Trandate (Glaxo Wellcome)
Trandiur (Glaxo
Wellcome)-comb.

J: Trandate (Shim Nihon
Jitsugyo-Graxo)

USA: Normodyne (Schering-
Plough; 1984)
Trandate (Glaxo Wellcome;
1984)

Lacidipine

(Lacipil; GR-43659X; GX-1048; SN-305)

ATC: C02DE; C08CA09

Use: once-daily calcium antagonist,
antihypertensive

RN: 103890-78-4   MF: $C_{26}H_{33}NO_6$ MW: 455.55

LD_{50}: 3150 mg/kg (M, p.o.);
880 mg/kg (R, p.o.)

CN: (E)-4-[2-[3-(1,1-dimethylethoxy)-3-oxo-1-propenyl]phenyl]-1,4-dihydro-2,6-dimethyl-3,5-
pyridinedicarboxylic acid diethyl ester

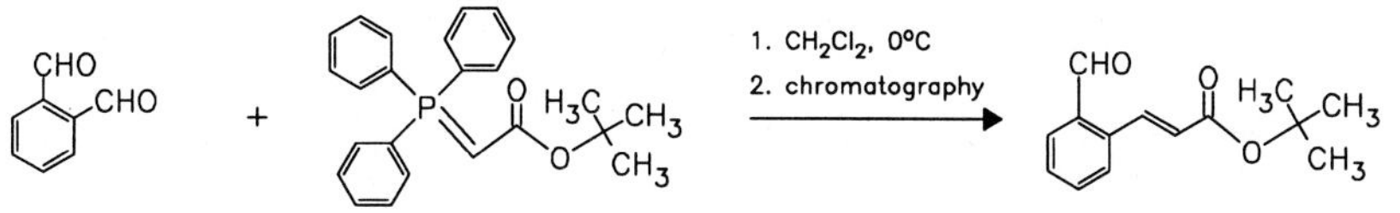

phthal-
aldehyde

triphenylphosphoranylidene-
acetic acid 1,1-dimethylethyl
ester

1. CH_2Cl_2, 0°C
2. chromatography

(E)-3-(2-formylphenyl)-
2-propenoic acid
1,1-dimethylethyl ester **(I)**

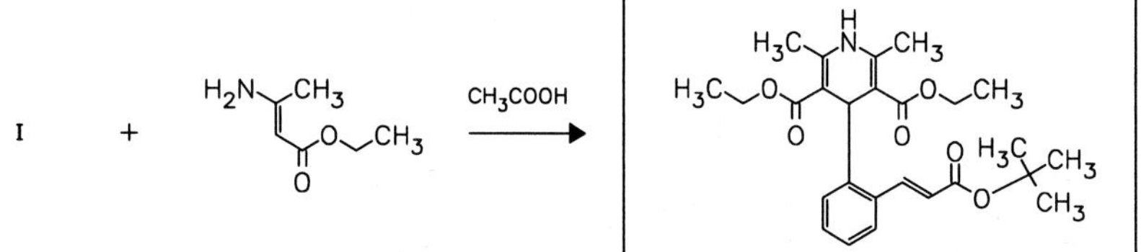

I + ethyl 3-amino-
crotonate **(II)** → CH_3COOH → Lacidipine

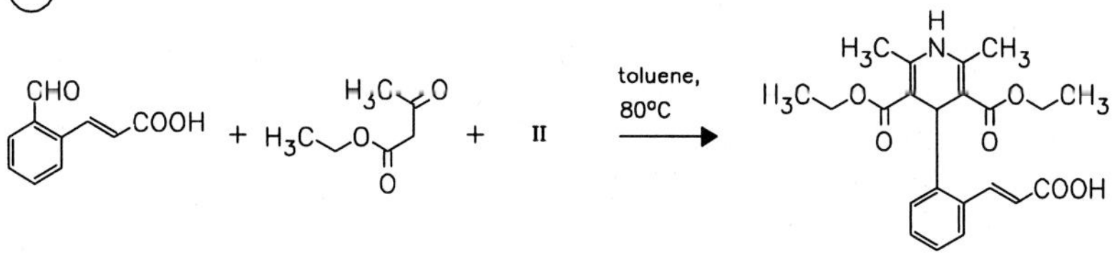

2-formyl-
cinnamic acid

ethyl aceto-
acetate

+ II

toluene,
80°C

(III)

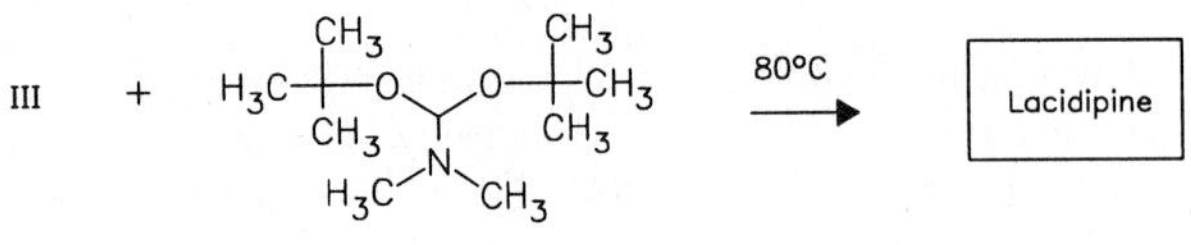

N,N-dimethylformamide
di-tert-butyl acetal

c

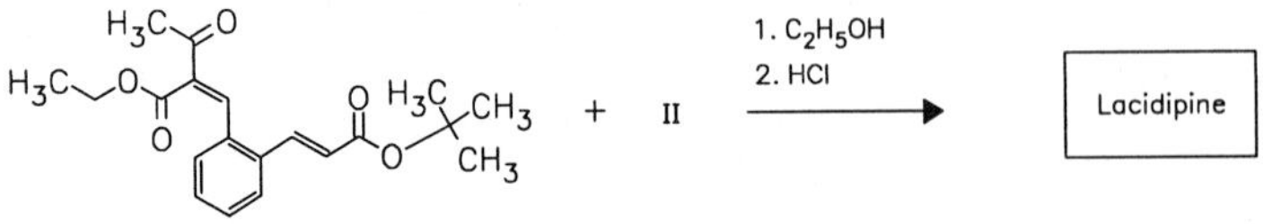

Reference(s):
a   DE 3 529 997 (Glaxo; appl. 22.8.1985; I-prior. 22.8.1984; 5.7.1985).
     US 4 801 599 (Glaxo; 31.1.1989; appl. 20.8.1985; I-prior. 22.8.1984, 5.7.1985).
     US 5 011 848 (Glaxo; 30.4.1990; appl. 23.1.1989; prior. 20.8.1985; I-prior. 22.8.1984, 5.7.1985).
b   EP 370 974 (Glaxo; appl. 21.11.1989; I-prior. 21.11.1989).
c   EP 534 520 (Merck + Co.; appl. 5.9.1992; USA-prior. 13.9.1991, 28.7.1992).

lacidipine *for treating arteriosclerosis:*
EP 499 920 (Glaxo; appl. 8.2.1992; I-prior. 13.2.1991).

long-acting formulation for dihydropyridines:
EP 301 133 (Syntex; appl. 21.12.1987; USA-prior. 26.7.1987).

prolonged-release oral pharmaceuticals:
EP 557 244 (Siegfried Pharma; appl. 3.2.1993; CH-prior. 17.2.1993).

process for preparation of solid dispersions:
WO 9 508 987 (KRKA Tovarna; appl. 26.9.1994; SI-prior. 28.9.1993).

Formulation(s):   tabl. 2 mg, 4 mg

Trade Name(s):
D:	Motens (Boehringer Ing.)	I: Apanil (Fidia) Viapres (Zambon; 1991)
F:	Caldine (Boehringer Ing.)	Lacipil (Glaxo Wellcome)
GB:	Motens (Boehringer Ing.)	Lacirex (Guidotti; 1991)

Lactulose
(Laktulose)

ATC: A06AD11
Use: laxative

RN: 4618-18-2   MF: $C_{12}H_{22}O_{11}$ MW: 342.30   EINECS: 225-027-7
CN: 4-*O*-β-D-galactopyranosyl-D-fructose

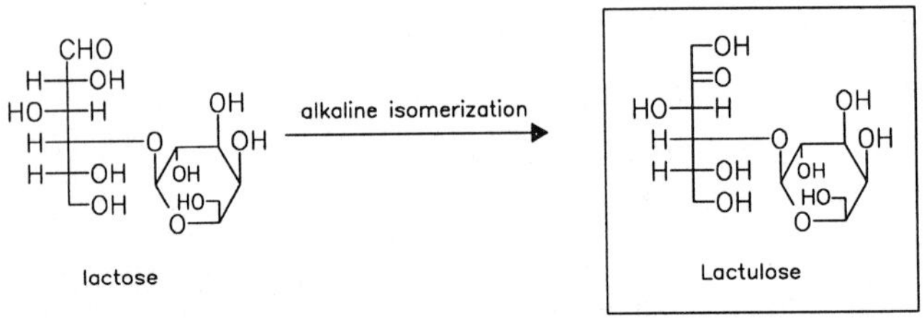

lactose   →   Lactulose

Reference(s):
Montgomery, E.M.; Hudson, C.S.: J. Am. Chem. Soc. (JACSAT) **52**, 2101 (1930).

use of alkali aluminate:
US 3 546 206 (Kraftco; 8.12.1970; prior. 20.9.1967).
US 3 850 905 (Kraftco; 26.11.1974; prior. 30.12.1972).

use of borax *and* NaOH:
JP 7 700 091 (Morinaga Milk Ind.; appl. 13.4.1971).

use of MgO, MgCO$_3$:
ES 397 810 (Jalup Jaures; appl. 26.11.1971).

use of alkali tetraborate:
US 3 505 309 (Research Corp.; 7.4.1970; prior. 25.9.1967).

use of alkali metal hydroxide or ammonia:
DOS 2 038 230 (Hayashibora; appl. 31.7.1970; J-prior. 31.7.1969, 2.8.1969).

lactulose *syrup:*
DOS 2 224 680 (Morinaga Milk Ind.; appl. 19.5.1972; J-prior. 22.5.1971).

lactulose *powder:*
DAS 1 189 839 (N.V. Tervalon; appl. 20.4.1961; N-prior. 22.4.1960).
DOS 2 153 106 (Morinaga Milk Ind.; appl. 25.10.1971; J-prior. 31.5.1971).
BE 843 777 (Morinaga Milk Ind.; appl. 5.7.1976; J-prior. 4.7.1975, 8.7.1975).
DOS 2 038 230 (Hayashibara; appl. 31.7.1970; J-prior. 31.7.1969, 2.8.1969).

crystalline lactulose:
US 5 003 061 (SIRAC; 26.3.1991; I-prior. 1.12.1987).
AT 327 224 (Laevosan; appl. 12.10.1973; valid from 15.4.1975).
Osten, B.J.: Recl. Trav. Chim. Pays-Bas (RTCPA3) **86**, 673 (1967).
CS 161 498 (M. Tadra et al.; appl. 24.5.1973).

Formulation(s): gran. 3 g, 5 g, 6 g, 10 g; syrup 3.33 g, 50 %, 66.7 %

Trade Name(s):

D:	Bifiteral (Solvay Arzneimittel)		Fitaxal (Phygiène)	Epalfen (Zambon)
	Eugalac (Töpfer)		Melaxose (Boehringer Ing.)-comb.	Lactoger (Ripari-Gero)
	Lactofalk (Falk)		Transulose (Schwarz)- comb.	Laevolac (Boehringer Mannh.)
	Lactuflor (MIP Pharma)			Lassifar (Lafare)
	Laevilac S (Fresenius- Praxis)	GB:	Duphalac (Solvay)	Normase (Molteni)
	generic		Lactugal (Galen)	Osmolac (Savio IBN)
		I:	Diacolon (Piam)	USA: Duphalac (Solvay)
F:	Duphalac (Solvay Pharma)		Duphalac (UCM)	Ixxose (ECR)

Lactylphenetidin
(Lactophenin)

ATC: N02B
Use: analgesic, antipyretic, antirheumatic

RN: 539-08-2 MF: C$_{11}$H$_{15}$NO$_3$ MW: 209.25 EINECS: 208-708-3
CN: *N*-(4-ethoxyphenyl)-2-hydroxypropanamide

DL-lactic acid + p-phenetidine → (180 °C) Lactylphenetidin

Reference(s):
DRP 70 250 (Chem. Fabr. formerly Goldenberg Geromont; 1892).

Formulation(s):   tabl. 200 mg

Trade Name(s):
D: Octadon (Thiemann)-   Quadronal (ASTA)-comb.;
      comb.; wfm   wfm

Lamivudine
(3 TC; BCH-790; GR-109714X; (–)sddc)

ATC: J05AF05
Use: antiviral, reverse transcriptase
        inhibitor

RN: 134678-17-4   MF: $C_8H_{11}N_3O_3S$ MW: 229.26
CN: (2*R-cis*)-4-amino-1-[2-(hydroxymethyl)-1,3-oxathiolan-5-yl]-2(1*H*)-pyrimidinone

(2*S-cis*)-form
RN: 134680-32-3   MF: $C_8H_{11}N_3O_3S$ MW: 229.26
racemate
RN: 136891-12-8   MF: $C_8H_{11}N_3O_3S$ MW: 229.26

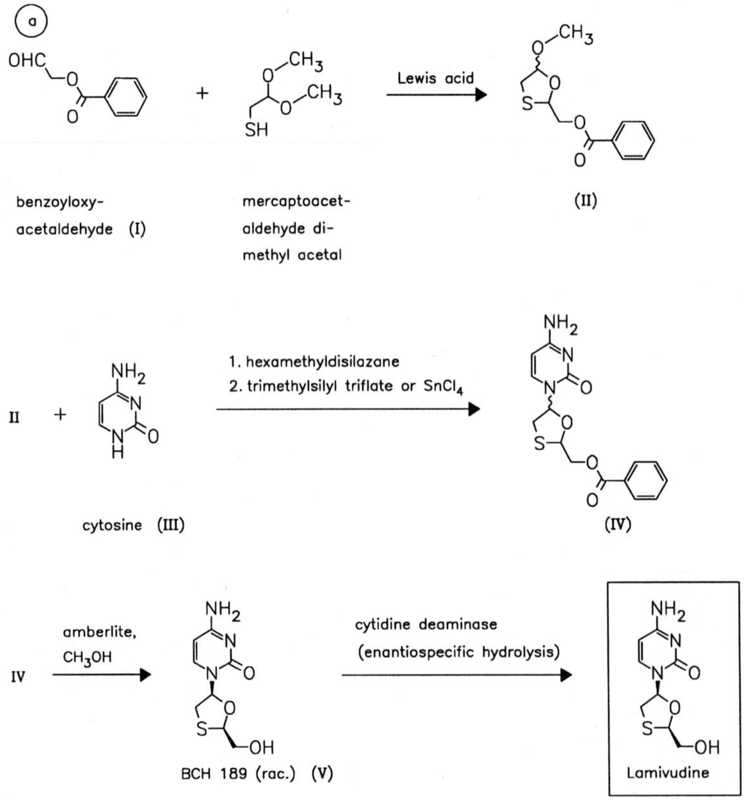

benzoyloxy-
acetaldehyde (I)

mercaptoacet-
aldehyde di-
methyl acetal

(II)

II + cytosine (III)

1. hexamethyldisilazane
2. trimethylsilyl triflate or SnCl₄

(IV)

IV → amberlite, CH₃OH

BCH 189 (rac.) (V)

cytidine deaminase
(enantiospecific hydrolysis)

Lamivudine

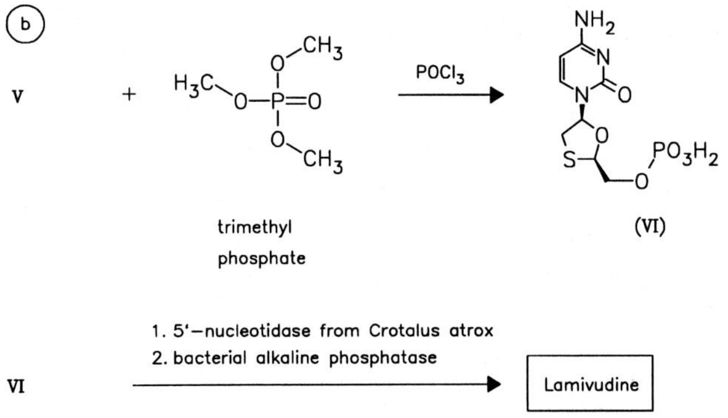

b)

V + trimethyl phosphate → (POCl₃) → (VI)

VI —[1. 5'-nucleotidase from Crotalus atrox; 2. bacterial alkaline phosphatase]→ Lamivudine

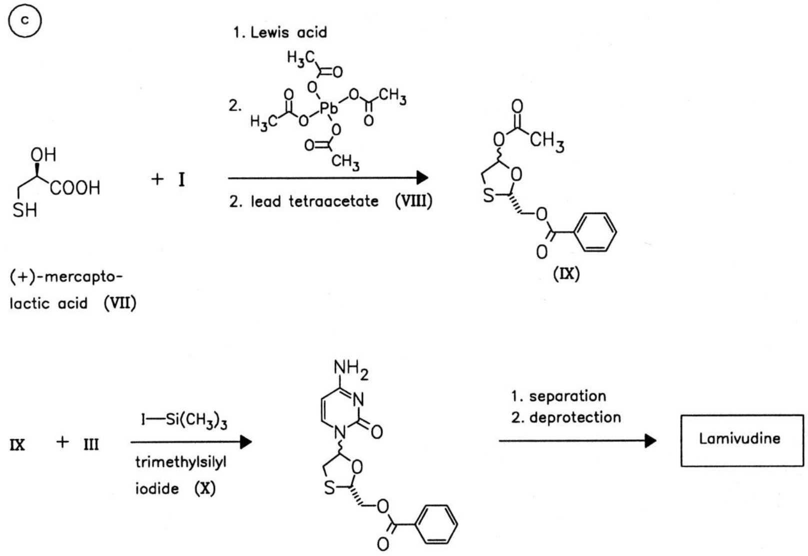

c)

(+)-mercapto-lactic acid (VII) + I —[1. Lewis acid; 2. lead tetraacetate (VIII)]→ (IX)

IX + III —[I—Si(CH₃)₃ trimethylsilyl iodide (X)]→ —[1. separation; 2. deprotection]→ Lamivudine

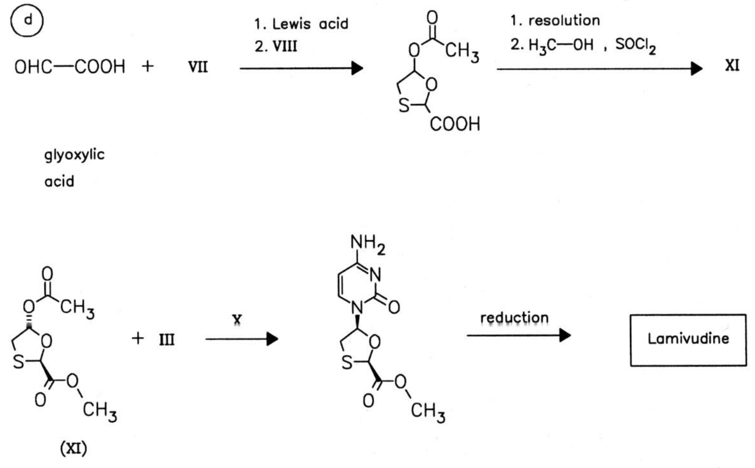

d)

glyoxylic acid (OHC—COOH) + VII —[1. Lewis acid; 2. VIII]→ —[1. resolution; 2. H₃C—OH , SOCl₂]→ XI

(XI) + III —[X]→ —[reduction]→ Lamivudine

e

1. H_3C ... Al ... CH_3 , toluene
   CH_3 CH_3

2.
   H_3C ... O ... CH_3

Bps O CHO + HOOC SH → [toluene] 2-(tert-butyldiphenyl-silyloxymethyl)-5-oxo-1,3-oxathiolane → **XII**

thioglycolic
acid

2-(tert-butyldiphenyl-
silyloxymethyl)-5-oxo-
1,3-oxathiolane

Bps: Si(CH$_3$)(CH$_3$)(CH$_3$) with two phenyl groups

[structure with CH$_3$]

+ **III**

1. Cl~~Cl , SnCl$_4$
2. N$^+$[(CH$_2$)$_3$–CH$_3$]$_4$ F$^-$, THF

→ **V**

Bps

2-(tert-butyldiphenyl-
silyloxymethyl)-5-acet-
oxy-1,3-oxathiolane (**XII**)

f

OH

S O COOH

+ H_3C O—O—CH$_3$ with CH$_3$
trimethyl
orthoformate

120 °C →

3-oxo-2,7-dioxa-
5-thiabicyclo-
[2.2.1]heptane (**XIII**)

trans-5-hydroxy-1,3-
oxathiolane-2-
carboxylic acid

XIII   + **III**

2,6–lutidine,
CH$_2$Cl$_2$
→

NH$_2$
[cytosine structure]
N
O
N
S O COOH

1. H_3C~~I , DMF
2. NaBH$_4$, CH$_3$OH
→ **V**

Reference(s):
a,b EP 382 526 (IAF Biochem. Int.; appl. 16.8.1990; USA-prior. 8.2.1989).
    WO 9 117 159 (IAF Biochem. Int.; appl. 14.11.1991; GB-prior. 2.5.1990).
    Beach, I.W. et al.: J. Org. Chem. (JOCEAH) **57** (8), 2217 (1992).
    (synthesis via L-gulose see also WO 9 210 496).
c   Humber, D.-C. et al.: Tetrahedron Lett. (TELEAY) **33** (32), 4625 (1992).
d   Drugs Future (DRFUD4) **18** (4), 319-323 (1993).
e   WO 9 111 186 (Emory Univ.; appl. 31.1.1991; USA-prior. 1.2.1990).
    US 5 210 085 (Emory Univ.; 22.2.1991; USA-prior. 1.2.1990).
f   WO 9 429 301 (Biochem. Pharma. Inc.; appl. 7.6.1994; GB-prior. 7.6.1993).

process that avoids Lewis acids:
WO 9 529 174 (Glaxo; appl. 21.4.1995; GB-prior. 23.4.1994).

crystalline new form:
EP 517 145 (Glaxo; appl. 2.6.1992; GB-prior. 3.6.1991).

use for treating and preventing hepatitis B infection:
EP 494 119 (P. Bellean; IAF Biochem. Int.; Biochem. Pharma. Inc.; appl. 3.1.1992; GB-prior. 3.1.1991).

composition for HIV infections:
EP 513 917 (Glaxo; appl. 11.5.1992; GB-prior. 16.5.1991).
WO 9 504 525 (Andrulis Pharm.; appl. 3.8.1994; USA-prior. 4.8.1993).

combination with zidovudine *and* loviride:
WO 9 601 110 (Janssen Pharm.; appl. 23.6.1995; EP-prior. 1.7.1994).

Formulation(s):   f. c. tabl. 100 mg; oral sol. 10 mg/ml; sol. 1 %; tabl. 150 mg

Trade Name(s):

D:	Combivir (Glaxo Wellcome)-comb. with Zidovudine Epivir (Glaxo Wellcome) Zeffix (Glaxo Wellcome)	GB:	Combivir (Glaxo Wellcome)-comb. with Zidovudine Epivir (Glaxo Wellcome) Zeffix (Glaxo Wellcome)	Epivir (Glaxo Wellcome)
				J: Epivir (Nippon Wellcome)
				USA: Combivir (Glaxo Wellcome)-comb. with Zidovudine
F:	Epivir (Glaxo Wellcome)	I:	Combivir (Glaxo Wellcome)-comb. with Zidovudine	Epivir (Glaxo Wellcome) 3TC (Glaxo Wellcome)

Lamotrigine
(BW-430C)

ATC: N03AX09
Use: anticonvulsant, glutamate inhibitor

RN: 84057-84-1   MF: $C_9H_7Cl_2N_5$ MW: 256.10   EINECS: 281-901-8
LD_{50}:   245 mg/kg (M, p.o.);
            205 mg/kg (R, p.o.)
CN: 6-(2,3-dichlorophenyl)-1,2,4-triazine-3,5-diamine

Reference(s):
US 4 560 687 (Wellcome; 24.12.1985; appl. 5.3.1984; prior. 15.9.1981, 29.5.1980; GB-prior. 1.6.1979).
US 4 602 017 (Wellcome; 22.7.1986; appl. 27.2.1984; prior. 15.9.1981, 29.5.1980; GB-prior. 1.6.1979).
EP 21 121 (Wellcome; appl. 30.5.1980; GB-prior. 1.6.1979).
US 5 912 345 (Glaxo Wellcome; 15.6.1999; appl. 29.12.1995; GB-prior. 30.12.1994).

Formulation(s):   tabl. 5 mg, 25 mg, 50 mg, 100 mg, 200 mg

Lanatoside C

(Lanatoside)

ATC: C01AA06
Use: cardiac glycoside

RN: 17575-22-3   MF: $C_{49}H_{76}O_{20}$ MW: 985.13   EINECS: 241-546-1
CN: (3β,5β,12β)-3-[(*O*-β-D-glucopyranosyl-(1→4)-*O*-3-*O*-acetyl-2,6-dideoxy-β-d-*ribo*-hexopyranosyl-
       (1→4)-*O*-2,6-dideoxy-β-D-*ribo*-hexopyranosyl-(1→4)-2,6-dideoxy-β-D-*ribo*-hexopyranosyl)oxy]-12,14-
       dihydroxycard-20(22)-enolide

deslanoside
RN: 17598-65-1   MF: $C_{47}H_{74}O_{19}$ MW: 943.09

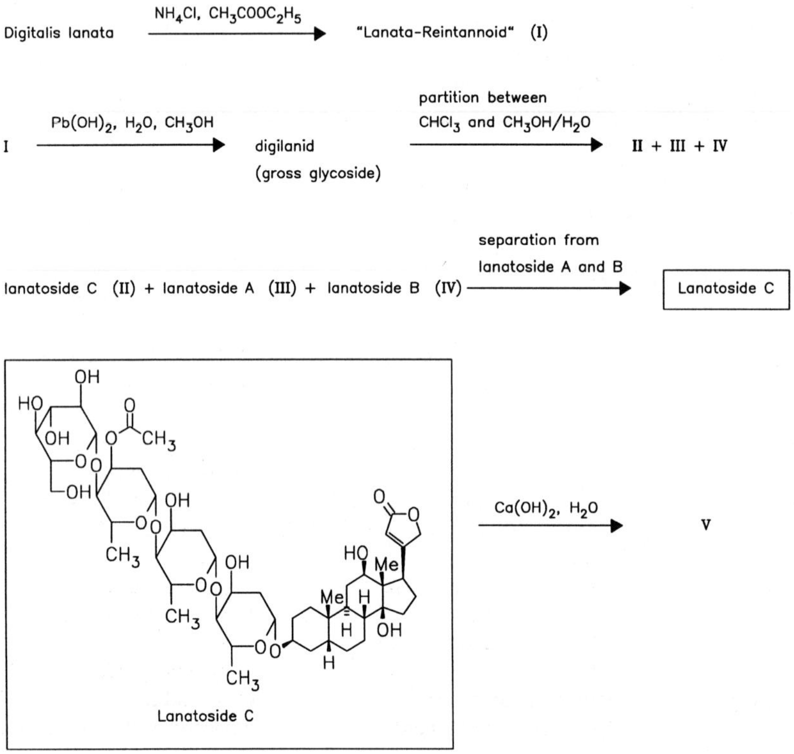

Digitalis lanata $\xrightarrow{\text{NH}_4\text{Cl, CH}_3\text{COOC}_2\text{H}_5}$ "Lanata-Reintannoid" (I)

I $\xrightarrow{\text{Pb(OH)}_2, \text{H}_2\text{O, CH}_3\text{OH}}$ digilanid $\xrightarrow[\text{CHCl}_3 \text{ and CH}_3\text{OH/H}_2\text{O}]{\text{partition between}}$ II + III + IV
(gross glycoside)

lanatoside C (II) + lanatoside A (III) + lanatoside B (IV) $\xrightarrow[\text{lanatoside A and B}]{\text{separation from}}$ Lanatoside C

Ca(OH)₂, H₂O → V

Lanatoside C

Deslanoside   (V)

Reference(s):
Stoll, A.; Kreis, W.: Helv. Chim. Acta (HCACAV) **16**, 1049 (1933).
HU 156 638 (Richter Gedeon; appl. 24.10.1967).
Ullmanns Encykl. Tech. Chem., 4. Aufl., Vol. **12**, 617.

digilanid (gross glycoside preparation):
DRP 631 790 (Sandoz; appl. 1930).
Ullmanns Encykl. Tech. Chem., 3. Aufl., Vol. **8**, 227.
CH 245 219 (Dr. Wander; appl. 1944).

deslanoside (desacetyl-lanatoside C):
Stoll, A.; Kreis, W.: Helv. Chim. Acta (HCACAV) **16**, 1049 (1933).
DD 70 088 (C. Lindig, K. Repke; appl. 1.11.1968).

Formulation(s):   amp. 0.4 mg/2 ml; drops 1 mg; tabl. 0.25 mg

Trade Name(s):

D: Cedilanid (Sandoz); wfm
    Cedilanid Amp. (Sandoz)-
    desacetyllanatoside C; wfm
    Cedilanid c. Th. (Sandoz)-
    comb.; wfm
    Cedilanid c. Th. Amp.
    (Sandoz)-
    desacetyllanatoside C; wfm
    Celadigal (Beiersdorf);
    wfm
    Ceto sanol (Sanol); wfm
    Conjunctisan-A (vitOrgan)-
    desacetyllanatoside C; wfm
    Digilanid (Sandoz)-
    glycoside total preparation;
    wfm

    Euphyllinat (Byk Gulden)-
    comb.; wfm
    Lanatorot (Heumann)-
    comb.; wfm
    Lanatosid Hameln
    (Hameln); wfm
    Lanibion (Merck)-comb.;
    wfm
    Lanimerck (Merck); wfm
    Pandigal (Beiersdorf)-
    glycoside total preparation;
    wfm
    Pulmo Frenona curn
    Digitalis (Hefa-Frenon)-
    comb.; wfm
F: Cédilanide (Novartis)

GB: Cedilanid (Sandoz); wfm
I: Cedilanid (Sandoz)
J: Cedilanid (Sandoz-Sankyo)
    Digilanogen C (Fujisawa)
    Digysid (Kanto)
    Erpasin (Kowa Yakuhin)
    Lanaside (Toyo S.-Ono)
    Lanatos (Sanko)
    Lanimerck (Doitsu)
    Ranato C (Kobayashi
    Kako)
USA: Cedilanid (Sandoz); wfm
    Cedilanid D (Sandoz)-
    desacetyllanatoside C; wfm

Lanoconazole
(NND 318; TJN-318)

ATC: D01
Use: antifungal

RN: 101530-10-3   MF: $C_{14}H_{10}ClN_3S_2$ MW: 319.84
CN: (*E*)-(±)-α-[4-(2-chlorophenyl)-1,3-dithiolan-2-ylidene]-1*H*-imidazole-1-acetonitrile

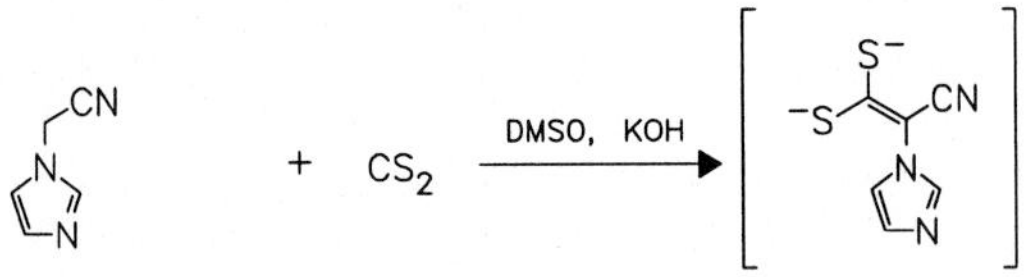

1-cyanomethyl-
imidazole

I

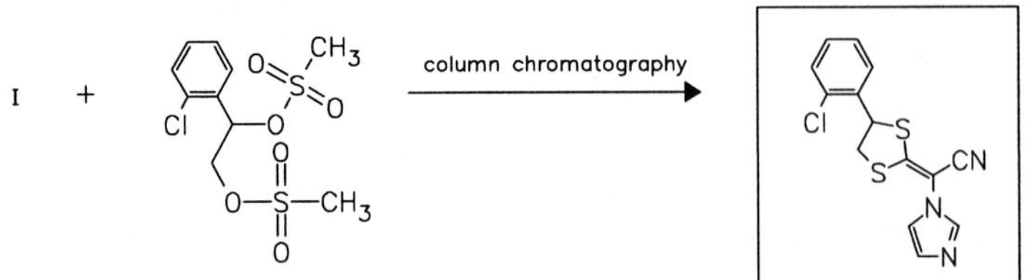

2-chloro-1-
(1,2-dimesyloxy-
ethyl)benzene

Lanoconazole

Reference(s):
EP 218 736 (Nihon Nohyaku; EP-prior. 9.10.1985).

preparation of E-isomer:
JP 02 121 983 (Nihon Noyaku; J-prior. 29.10.1988).

Formulation(s):   cream 1 %; sol. 1 %

Trade Name(s):
J: Astat (Tsumura)

Lansoprazole

ATC: A02BC03
Use: antiulcer agent H⁺/K⁺-ATPase
       inhibitor

RN: 103577-45-3   MF: $C_{16}H_{14}F_3N_3O_2S$ MW: 369.37
LD_{50}: >5 g/kg (M, p.o.);
          >5 g/kg (R, p.o.)
CN: 2-[[[3-methyl-4-(2,2,2-trifluoroethoxy)-2-pyridinyl]methyl]sulfinyl]-1*H*-benzimidazole

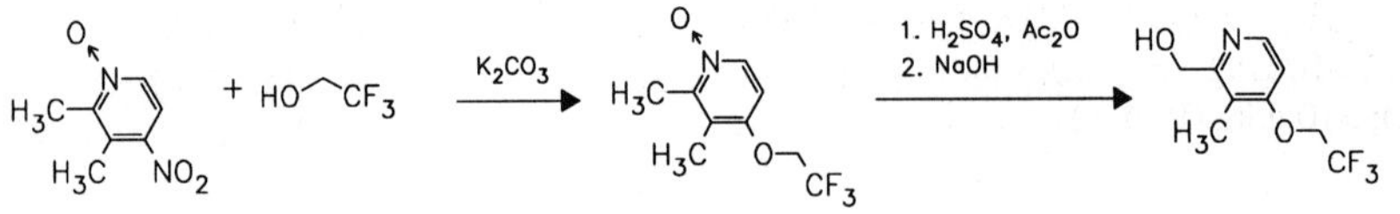

2,3-dimethyl-
4-nitropyridine
1-oxide

2,2,2-trifluoro-
ethanol

2,3-dimethyl-4-
(2,2,2-trifluoro-
ethoxy)pyridine
1-oxide

2-hydroxymethyl-
3-methyl-4-(2,2,2-
trifluoroethoxy)-
pyridine (I)

I

1. SOCl$_2$
2. [2-mercaptobenzimidazole structure] -SH

NaOCH$_3$

2. 2—mercapto-
benzimidazole

2-[3-methyl-4-(2,2,2-
trifluoroethoxy)pyrid-
2—ylmethylthio]benz-
imidazole

m-chloro-
perbenzoic
acid

Lansoprazole

Reference(s):
EP 174 726 (Takeda; appl. 31.7.1985; J-prior. 16.8.1984).

stabilized pharmaceutical formulation:
EP 237 200 (Takeda; appl. 17.10.1990; J-prior. 21.2.1986, 13.2.1986).

medical use for treatment of osteoporosis:
JP 1 203 325 (Takeda; appl. 8.2.1988).

medical use for treatment of camylobacter infections:
EP 382 489 (Takeda; appl. 6.2.1990).

Formulation(s):   cps. 15 mg, 30 mg

Trade Name(s):
D: Agopton (Takeda)   F: Lanzor (Hoechst Houdé;   I: Lansox (Takeda)
      Lanzor (Albert-Roussel,   1991)   J: Takepron (Takeda)
      Hoechst)   Ogast (Takeda; 1991)   USA: Prevacid (TAP)
                                GB: Zoton (Wyeth)

Latamoxef
(Moxalactam; S-6059)

ATC: J01DA18
Use: β-lactam antibiotic (1-oxadethia-
        cephalosporin derivative)

RN: 64952-97-2   MF: C$_{20}$H$_{20}$N$_6$O$_9$S   MW: 520.48   EINECS: 265-287-9
CN: [6R-[6α,7α,7(R*)]]-7-[[carboxy-(4-hydroxyphenyl)acetyl]amino]-7-methoxy-3-[[(1-methyl-1H-tetrazol-
       5-yl)thio]methyl]-8-oxo-5-oxa-1-azabicyclo[4.2.0]oct-2-ene-2-carboxylic acid

disodium salt
RN: 64953-12-4   MF: C$_{20}$H$_{18}$N$_6$Na$_2$O$_9$S   MW: 564.44   EINECS: 265-288-4
LD$_{50}$:   5300 mg/kg (M, i.v.); >10 g/kg (M, p.o.);
              5500 mg/kg (R, i.v.); >10 g/kg (R, p.o.)

6—amino-
penicillanic acid

benzoyl
chloride

6—benzamido-
penicillanic acid (I)

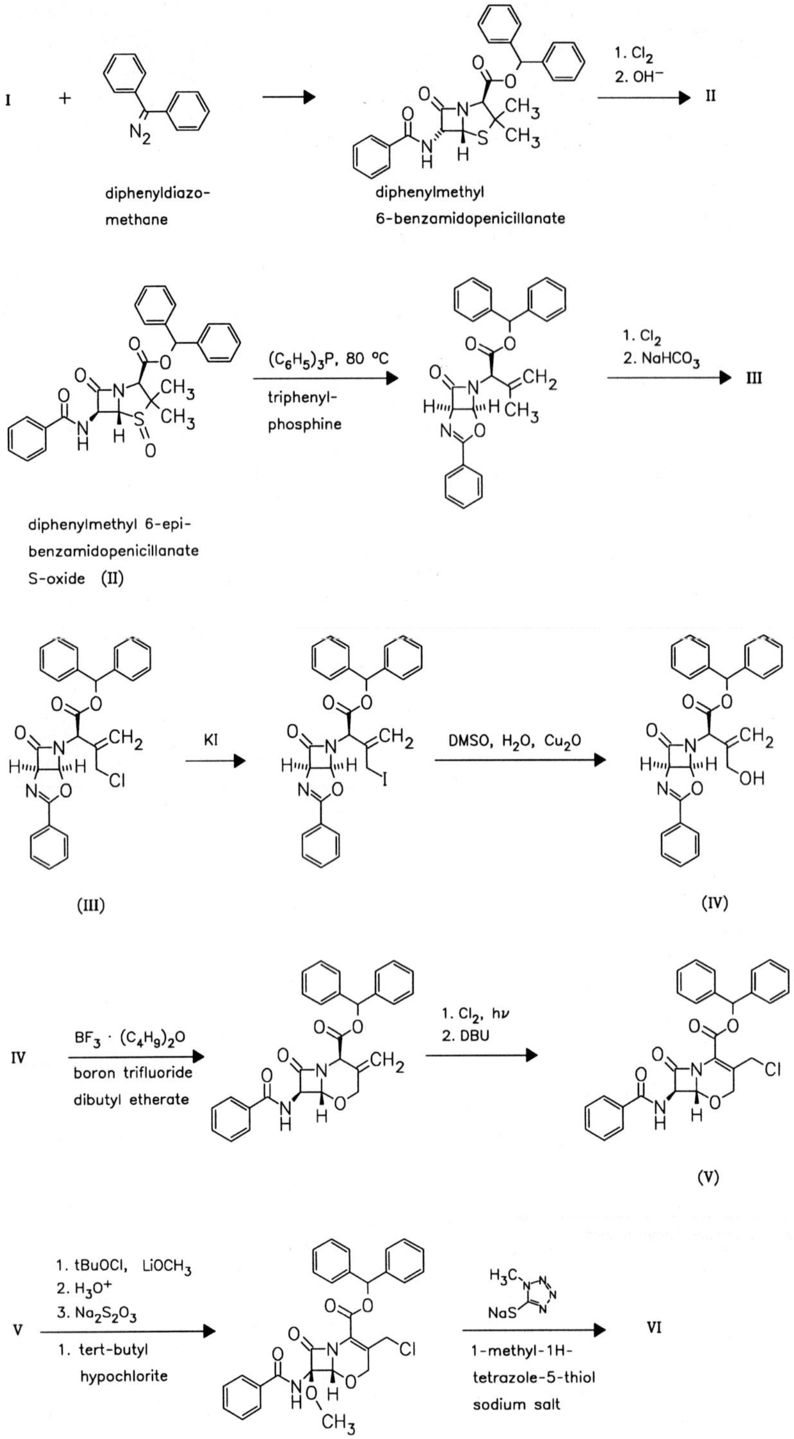

I + diphenyldiazo-methane → diphenylmethyl 6-benzamidopenicillanate

1. Cl₂
2. OH⁻ → II

diphenylmethyl 6-epi-benzamidopenicillanate S-oxide (II)

$(C_6H_5)_3P$, 80 °C
triphenylphosphine →

1. Cl₂
2. NaHCO₃ → III

(III) → KI → DMSO, H₂O, Cu₂O → (IV)

IV → $BF_3 \cdot (C_4H_9)_2O$ boron trifluoride dibutyl etherate →

1. Cl₂, hν
2. DBU → (V)

V →
1. tBuOCl, LiOCH₃
2. H₃O⁺
3. Na₂S₂O₃
1. tert-butyl hypochlorite
→

H₃C—N tetrazole
NaS
1-methyl-1H-tetrazole-5-thiol sodium salt
→ VI

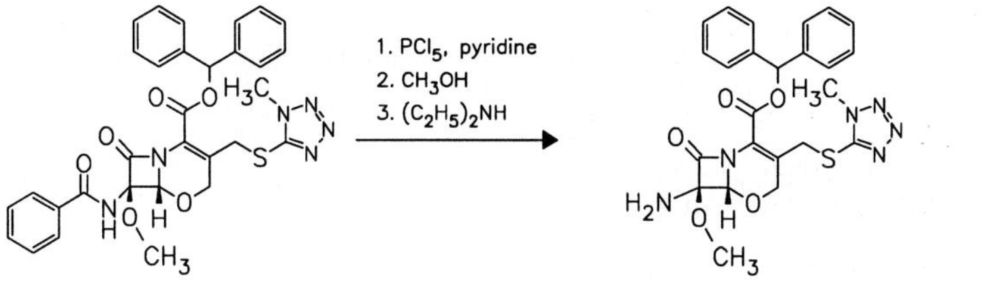

diphenylmethyl (7R)-7-amino-
7-methoxy-3-(1-methyltetrazol-
5-ylthiomethyl)-1-oxa-1-de-
thia-3-cephem-4-carboxylate (VII)

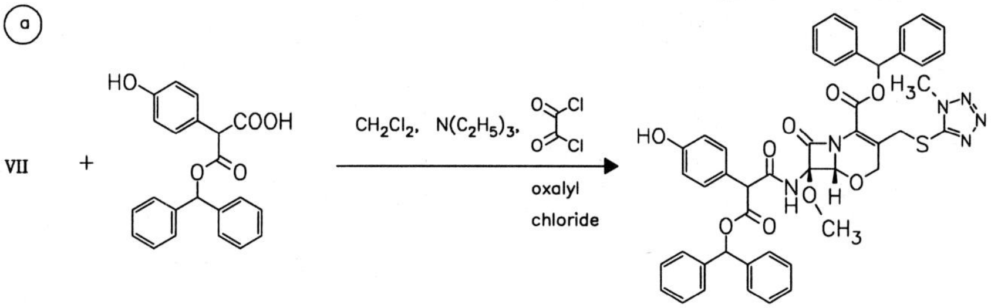

(a)

mono(diphenyl-
methyl) 4-hydroxy-
phenylmalonate

(VIII)

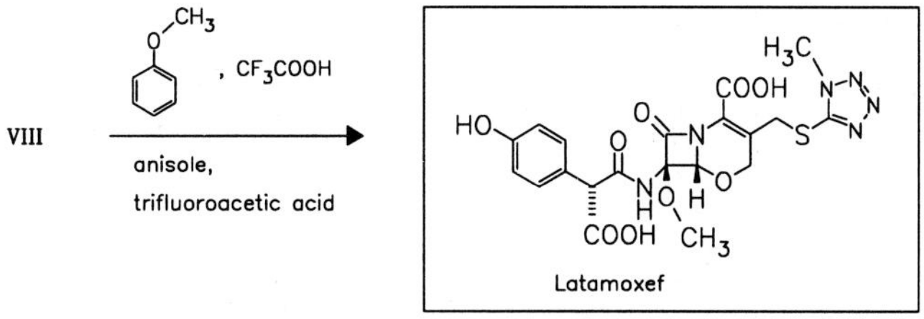

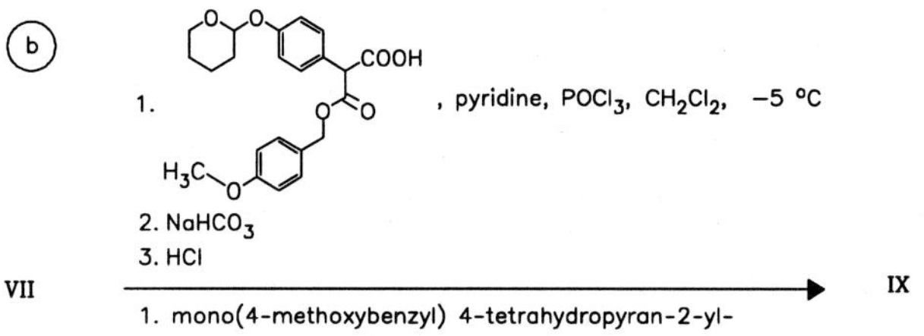

Latamoxef

(b)

1. mono(4-methoxybenzyl) 4-tetrahydropyran-2-yl-
   oxyphenylmalonate

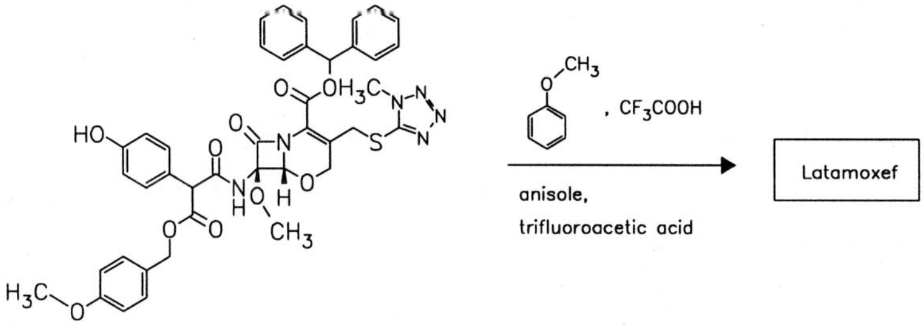

(IX)

Reference(s):

Nagata, W.: "Synthetic Aspects of 1-Oxacephem Antibiotics", in Curr. Trends Org. Synth., Proc. Int. Conf., 4th, Tokyo, 22.-27.8.1982; Ed. by H. Nozaki; Pergamon Press 1983.

DOS 2 713 370 (Shionogi; appl. 25.3.1977; J-prior. 25.3.1976, 30.4.1976).

US 4 138 486 (Shionogi; 6.2.1979; J-prior. 25.3.1976, 30.4.1976).

US 4 180 571 (Shionogi; 25.12.1979; J-prior. 25.3.1976, 30.4.1976).

alternative syntheses:

Narisada, M. et al.: J. Med. Chem. (JMCMAR) **22**, 758 (1979).

Narisada, M. et al.: Heterocycles (HTCYAM) **7**, 839 (1977).

preparation of [6R-[6α,7β,7(R*)]]-[[carboxy-(4-hydroxyphenyl)acetyl]amino] enantiomer *and epimerization methods:*

EP 98 545 (Shionogi; appl. 1.7.1983; J-prior. 2.7.1982).

US 4 504 658 (Shionogi; 12.3.1985; J-prior. 2.7.1982).

stable lyophilisates:

US 4 418 058 (Shionogi; 29.11.1983; J-prior. 23.6.1980).

combination with other antibiotics:

US 4 452 778 (Eli Lilly; 5.6.1984; appl. 4.5.1979, 31.3.1980; 21.12.1981).

Formulation(s):   amp. 250 mg, 500 mg, 1 g, 2 g, 10 g; inj. 1 g/3 ml, 1 g/20 ml (as sodium salt)

Trade Name(s):

D:	Festamoxin (Shionogi);	Mactam (Coli)	Polimoxal (Herdel)
	wfm	Moxa (Ital. Suisse)	Priolatt (Sancarlo)
I:	Baxal (Italsuisse)	Moxacef (Pulitzer)	Sectam (Locatelli)
	Betalactam (Bergamon)	Moxatres (Radiumfarma)	J: Shiomarin (Shionogi)
	Latoxacef (Magis)	Oxacef (Gibipharma)	USA: Moxam (Lilly); wfm

---

Latanoprost

(PhXA41; PhXA34 (as 15(R,S)-isomer))

ATC: S01EX03
Use: antiglaucoma, prostaglandin

RN: 130209-82-4   MF: $C_{26}H_{40}O_5$ MW: 432.60

CN: [1R-[1α(Z),2β(R*),3α,5α]]-7-[3,5-dihydroxy-2-(3-hydroxy-5-phenylpentyl)cyclopentyl]-5-heptenoic acid 1-methylethyl ester

synthesis of intermediate I:

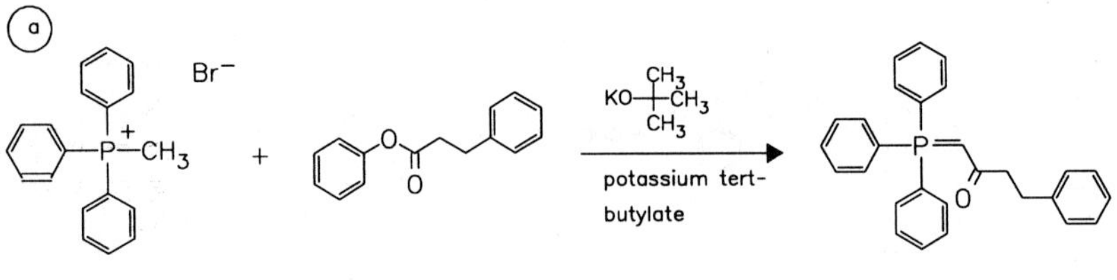

(I)

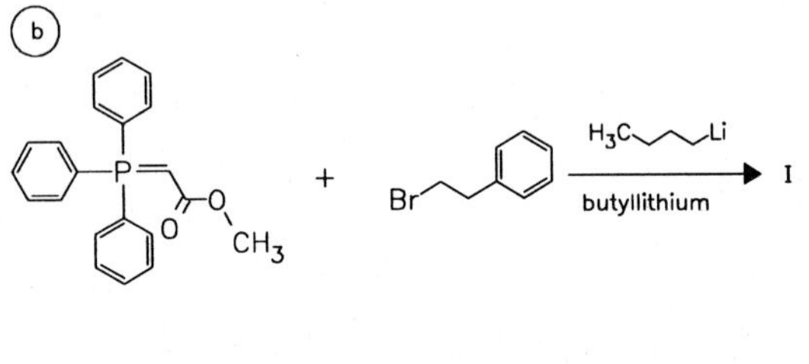

synthesis of intermediate II:

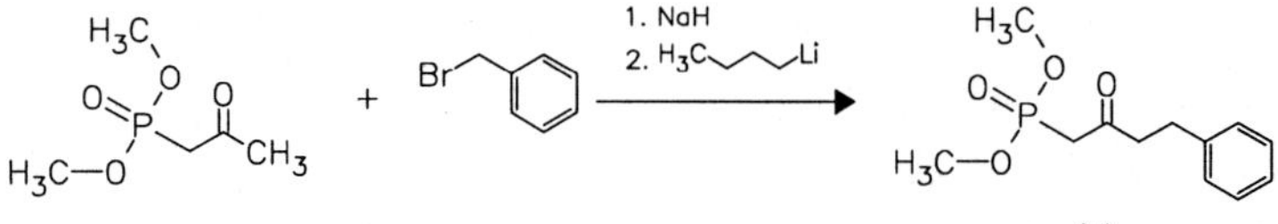

(II)

synthesis of intermediate III:

1. [cyclohexyl]–N=C=N–[cyclohexyl]

2. H₃C–S–CH₃ (with =O), NEt₃

3. H₃PO₄

1. dicyclohexylcarbodiimide
2. dimethyl sulfoxide
3. phosphoric acid

(III)

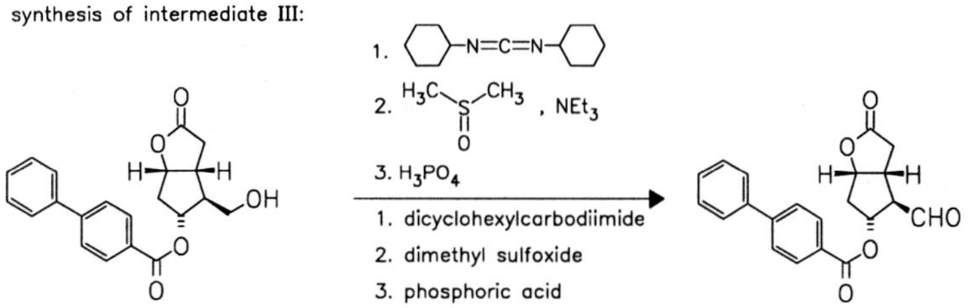

(3aα,4α,5β,6aα)-(−)-
hexahydro-4-(hydroxy-
methyl)-2-oxo-2H-
cyclopenta[b]furan-5-yl
1,1'-biphenyl-4-carboxylate

synthesis of Latanoprost:

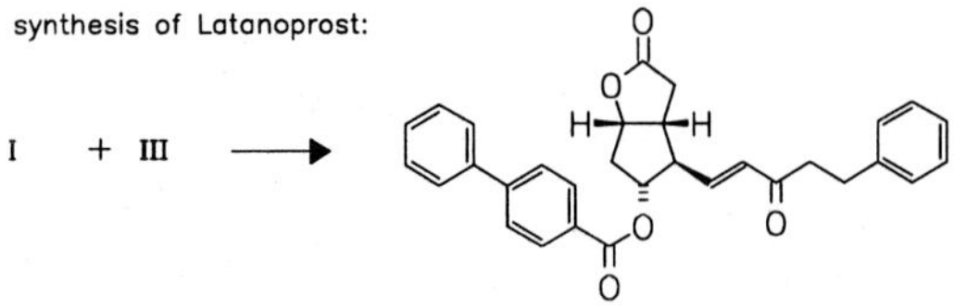

I + III ⟶

(IV)

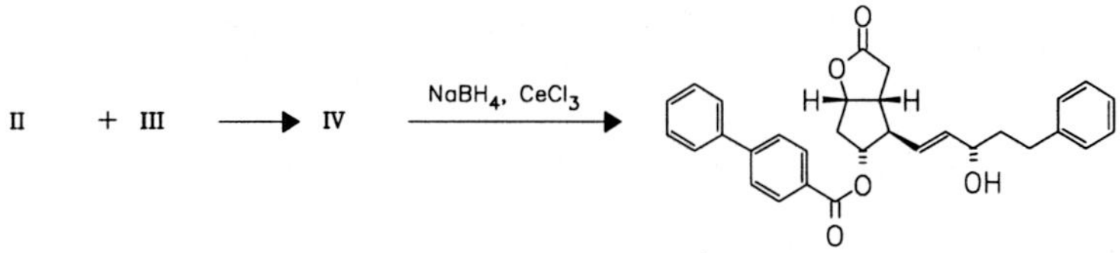

II + III ⟶ IV —NaBH₄, CeCl₃→

(V)

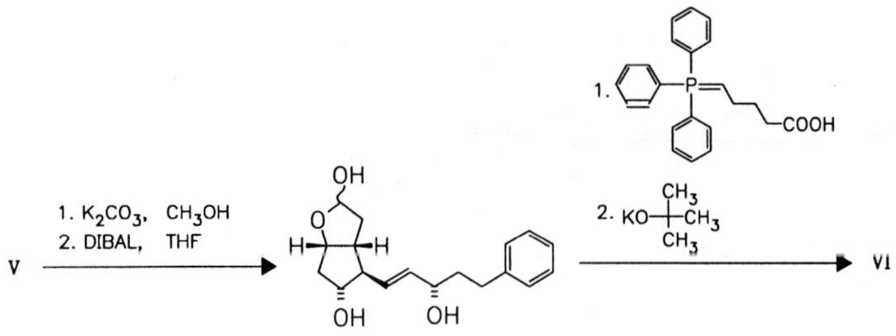

V —1. K₂CO₃, CH₃OH 2. DIBAL, THF→

1. (triphenylphosphorane)=CH–COOH

2. KO–C(CH₃)₃

⟶ VI

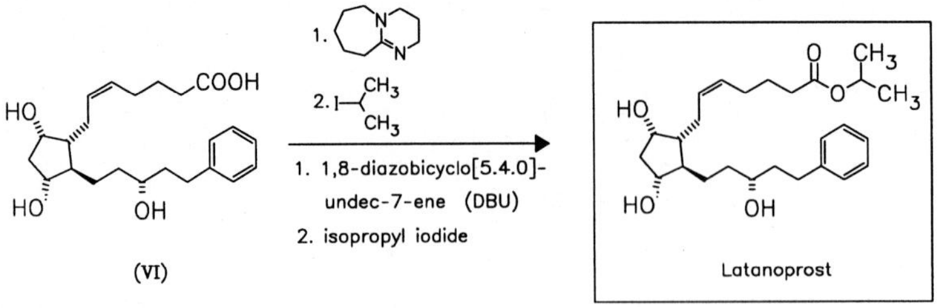

Latanoprost

Reference(s):
WO 9 002 553 (Pharmacia; appl. 22.3.1990; S-prior. 6.9.1988).
EP 364 417 (Pharmacia; appl. 18.4.1990).

synthesis of Corey lactone derivatives:
Corey, E.J. et al.: J. Am. Chem. Soc. (JACSAT) **93**, 1491 (1971).
Alm, A. et al.: Invest. Ophthalmol. Visual Sci. (IOVSDA) **1992**, Suppl. 1247.

Formulation(s):   eye drops 50 µg/ml

Trade Name(s):
D: XALATAN (Pharmacia &   GB: Xalatan (Pharmacia &
        Upjohn)   Upjohn; 1997)
F: Xalatan (Pharmacia &   USA: Xalatan (Pharmacia &
        Upjohn)   Upjohn)

Leflunomide
(HWA-486; SU 101)

ATC: L04AX
Use: antirheumatic, immunosuppressant

RN: 75706-12-6   MF: $C_{12}H_9F_3N_2O_2$ MW: 270.21
CN: 5-Methyl-*N*-[4-(trifluoromethyl)phenyl]-4-isoxazolecarboxamide

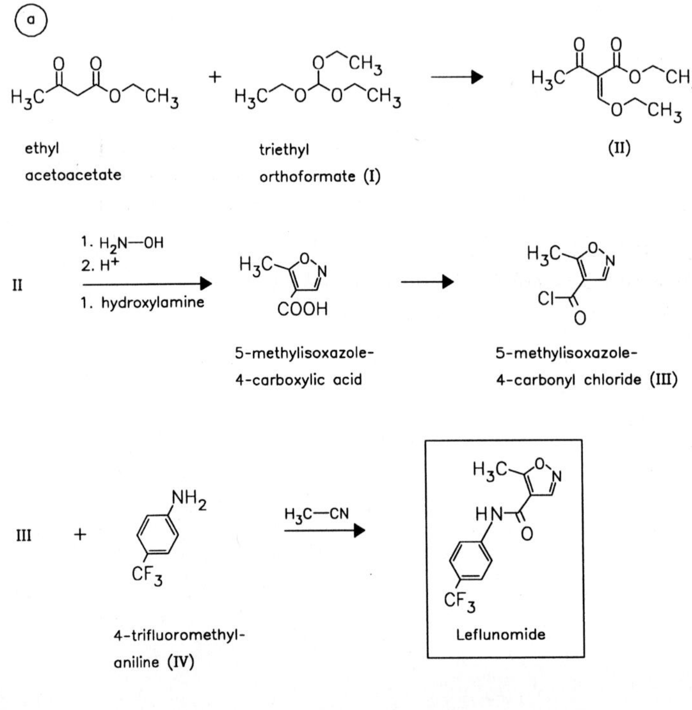

ethyl
acetoacetate

triethyl
orthoformate (I)

(II)

5-methylisoxazole-
4-carboxylic acid

5-methylisoxazole-
4-carbonyl chloride (III)

4-trifluoromethyl-
aniline (IV)

Leflunomide

(b)

IV + [diketene structure] →(H₃C—CN)→ [acetoacetic acid 4-(trifluoromethyl)-anilide structure] →(I, H₃C–C(O)–O–C(O)–CH₃ / triethyl orthoformate / acetic anhydride)→ V

diketene

acetoacetic acid
4-(trifluoromethyl)-
anilide

[structure (v)] →(H₂N—OH · HCl, NaOH, H₂O, ethanol / hydroxylamine hydrochloride)→ [Leflunomide]

(v)

Reference(s):

a,b DE 2 854 439 (Hoechst AG, D-prior. 16.12.1978).
   DE 4 127 737 (Hoechst AG; appl. 22.8.1991).

isoxazole-4-carboxamides as neoplasm inhibitors and antirheumatics:
WO 9 117 748 (Hoechst AG; appl. 24.10.1990; D-prior. 18.5.1990).

thioamide analogs with anticancer activity:
WO 9 633 179 (Sugen; appl. 19.4.1996; USA-prior. 21.4.1995).

injectable formulations:
WO 9 633 745 (Sugen; appl. 17.4.1996; USA-prior. 26.4.1995).

Formulation(s):   tabl. 10 mg, 20 mg, 100 mg

Trade Name(s):

D: Arava (Hoechst Marion   USA: Arava (Hoechst Marion
       Roussel)   Roussel)

Lenampicillin
(KBT-1585)

ATC: S01AA
Use: antibacterial, semisynthetic β-lactam
         antibiotic, derivative of ampicillin
         (prodrug for oral application)

RN: 86273-18-9   MF: $C_{21}H_{23}N_3O_7S$ MW: 461.50
CN: [2S-[2α,5α,6β(S*)]]-6-[(aminophenylacetyl)amino]-3,3-dimethyl-7-oxo-4-thia-1-
         azabicyclo[3.2.0]heptane-2-carboxylic acid (5-methyl-2-oxo-1,3-dioxol-4-yl)methyl ester

monohydrochloride
RN: 80734-02-7   MF: $C_{21}H_{23}N_3O_7S \cdot HCl$ MW: 497.96
LD$_{50}$:   >700 mg/kg (M, i.v.); >8000 mg/kg (M, p.o.);
         >800 mg/kg (R, i.v.); ca. 10000 mg/kg (R, p.o.)

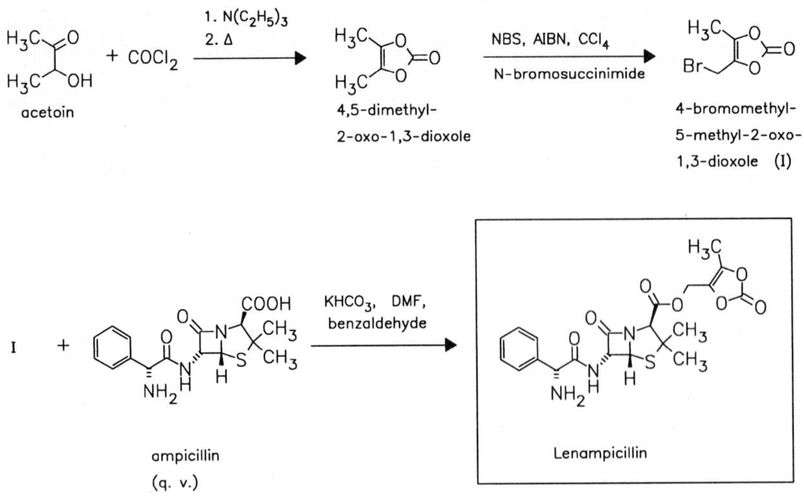

H₃C group reaction scheme: acetoin + COCl₂ → [1. N(C₂H₅)₃, 2. Δ] → 4,5-dimethyl-2-oxo-1,3-dioxole → [NBS, AIBN, CCl₄ / N-bromosuccinimide] → 4-bromomethyl-5-methyl-2-oxo-1,3-dioxole (I)

I + ampicillin (q. v.) → [KHCO₃, DMF, benzaldehyde] → Lenampicillin

Reference(s):

Sakamoto, F. et al.: Chem. Pharm. Bull. (CPBTAL) 32, 2241 (1984).
Ikeda, S. et al.: Chem. Pharm. Bull. (CPBTAL) 32, 4316 (1984).
US 4 342 693 (Kanebo; 3.8.1982; J-prior. 30.4.1980).
US 4 389 408 (Kanebo; 21.6.1983; J-prior. 30.4.1980, 22.5.1980).
EP 39 086 (Kanebo; appl. 29.4.1981; J-prior. 30.4.1980, 22.5.1980).
EP 39 477 (Kanebo; appl. 29.4.1981; J-prior. 30.4.1980).
EP 61 206 (Kanebo; appl. 29.4.1981; J-prior. 30.4.1980, 22.5.1980).
EP 90 344 (Kanebo; appl. 29.4.1981; J-prior. 30.4.1980).

Formulation(s):   vial 250 mg (as hydrochloride)

Trade Name(s):
J: Takacillin (MECT; 1987)   Varacillin (Kanebo; 1987)

Lentinan

(LC-33)

ATC: L03AX01
Use: immunostimulant, antineoplastic

RN: 37339-90-5   MF: unspecified   MW: unspecified
LD$_{50}$:   250 mg/kg (M, i.v.);
        250 mg/kg (R, i.v.);
        >100 mg/kg (dog, i.v.)
CN: lentinan

Extraction of edible fungus *Lentinus edodes* with hot water, solubilization through treatment with aqueous urea.

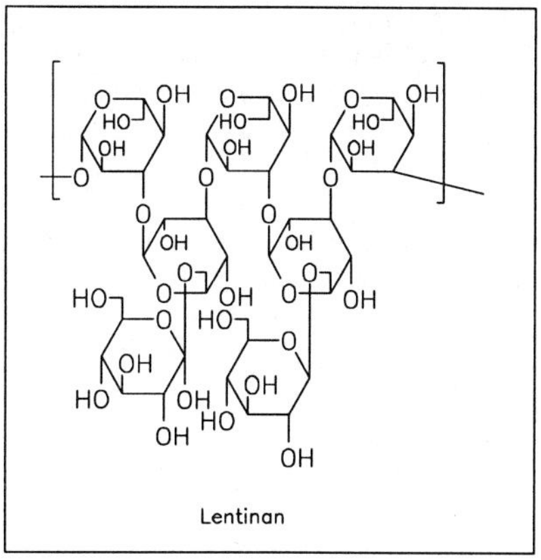

Lentinan

Reference(s):
US 3 883 505 (Ajinomoto; 13.5.1975; J-prior. 17.7.1972).
DE 2 336 378 (Ajinomoto; appl. 17.7.1973; J-prior. 17.7.1972).
Chibara, J. et al.: Cancer Res. (CNREA8) **30**, 2776 (1970).
Chihara, G. et al.: Nature (London) (NATUAS) **222**, 637 (1968).

structural study:
Sasaki, T. et al.: Carbohydr. Res. (CRBRAT) **47**, 99 (1976).

water soluble pharmaceutical formulation:
US 4 207 312 (Ajinomoto, Morishita; 10.6.1980; J-prior. 5.2.1975).

combination with CSF:
EP 326 149 (Green Cross, Morinaga; appl. 27.1.1989; J-prior. 29.1.1988).

Formulation(s):   vial 1 g

Trade Name(s):
J: Lentinan (Ajinomoto-
       Morishita; Yamanouchi;
       1986)

Lercanidipine hydrochloride

ATC: C08CA13
Use: treatment of hypertension,
        vasoselective calcium antagonist

RN: 132866-11-6   MF: $C_{36}H_{41}N_3O_6 \cdot HCl$ MW: 648.20
CN: 1,4-Dihydro-2,6-dimethyl-4-(3-nitrophenyl)-3,5-pyridinedicarboxylic acid 2-[(3,3-diphenylpropyl)methylamino]-1,1-dimethylethyl methyl ester hydrochloride

base
RN: 100427-26-7   MF: $C_{36}H_{41}N_3O_6$ MW: 611.74

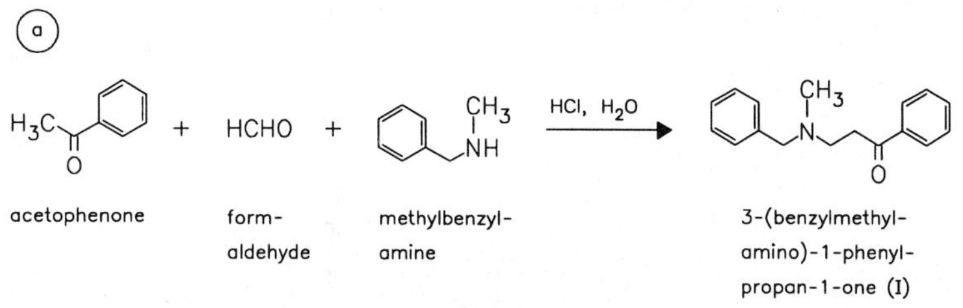

acetophenone   form-   methylbenzyl-   3-(benzylmethyl-
                    aldehyde   amine   amino)-1-phenyl-
                                                      propan-1-one (I)

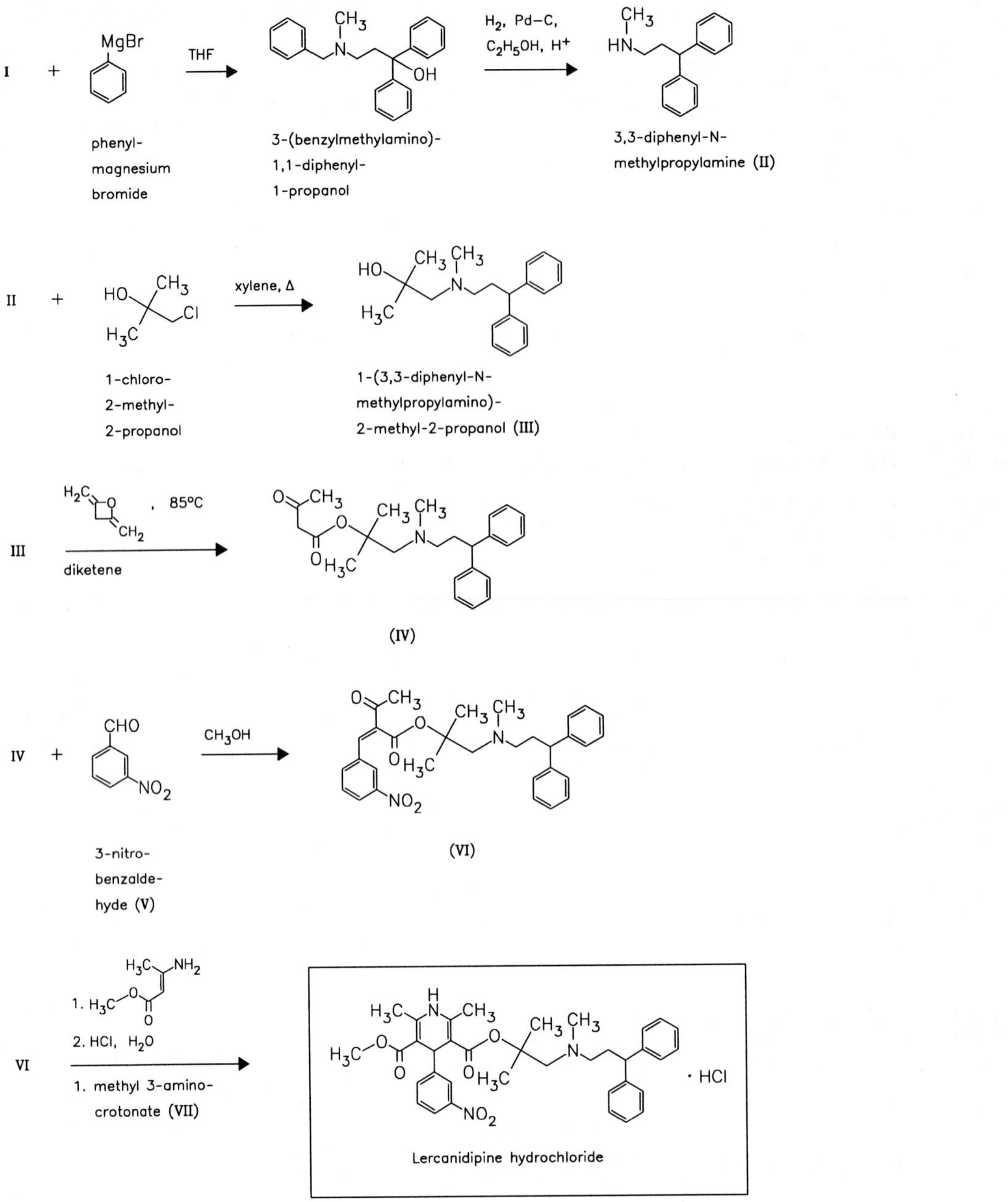

I + phenyl-
magnesium
bromide
→ (THF) 3-(benzylmethylamino)-
1,1-diphenyl-
1-propanol
→ (H₂, Pd–C, C₂H₅OH, H⁺) 3,3-diphenyl-N-
methylpropylamine (II)

II + 1-chloro-
2-methyl-
2-propanol
→ (xylene, Δ) 1-(3,3-diphenyl-N-
methylpropylamino)-
2-methyl-2-propanol (III)

III + diketene (85°C) → (IV)

IV + 3-nitro-
benzalde-
hyde (V)
→ (CH₃OH) (VI)

VI → 1. methyl 3-amino-
crotonate (VII)
2. HCl, H₂O
→ Lercanidipine hydrochloride

b

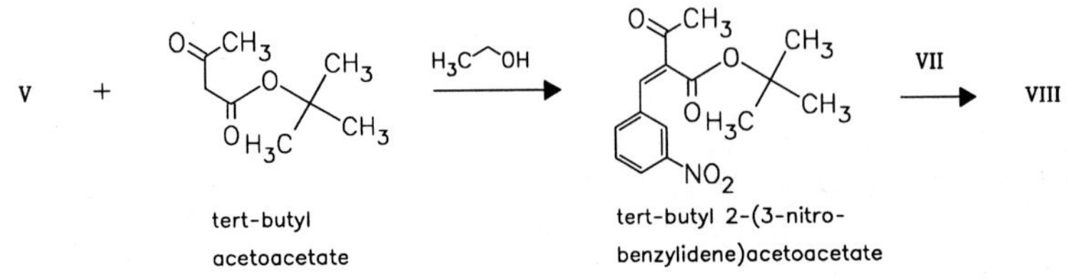

V + tert-butyl
acetoacetate
→ (H₃C–OH) tert-butyl 2-(3-nitro-
benzylidene)acetoacetate
→ (VII) VIII

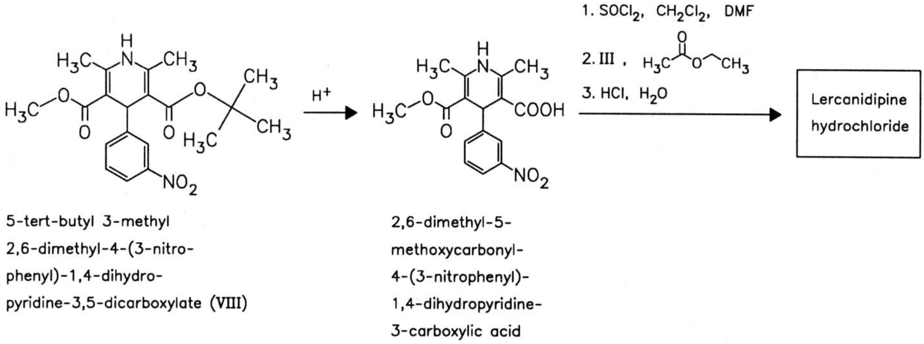

1. SOCl₂, CH₂Cl₂, DMF

2. III,

3. HCl, H₂O

Lercanidipine hydrochloride

5-tert-butyl 3-methyl
2,6-dimethyl-4-(3-nitro-
phenyl)-1,4-dihydro-
pyridine-3,5-dicarboxylate (VIII)

2,6-dimethyl-5-
methoxycarbonyl-
4-(3-nitrophenyl)-
1,4-dihydropyridine-
3-carboxylic acid

Reference(s):

Leonardi, A. et al.: Eur. J. Med. Chem. (EJMCA5) **33**, 399 (1998).

a   EP 153 016 (Recordati Chem. and Pharm.; appl. 21.1.1985; GB-prior. 14.2.1984).

b   WO 9 635 668 (Recordati Chem. and Pharm.; appl. 9.5.1996; I-prior. 12.5.1995).

preparation of 3-(benzylmethylamino)-1,1-diphenyl-1-propanol:
Morrison; Rinderknecht: J. Chem. Soc. (JOCEAH) **1950**, 1510

preparation of 3,3-diphenyl-*N*-methylpropylamine:
DE 925 468 (Farbwerke Hoechst; appl. 13.8.1941)

Formulation(s):   tabl. 10 mg (as hydrochloride)

Trade Name(s):
GB: Zanidip (Napp)   I: Zanedip (Recordati; 1999)

Letosteine

ATC: R05CB09
Use: mucolytic agent

RN: 53943-88-7   MF: C₁₀H₁₇NO₄S₂   MW: 279.38   EINECS: 258-879-3
CN: 2-[2-[(2-ethoxy-2-oxoethyl)thio]ethyl]-4-thiazolidinecarboxylic acid

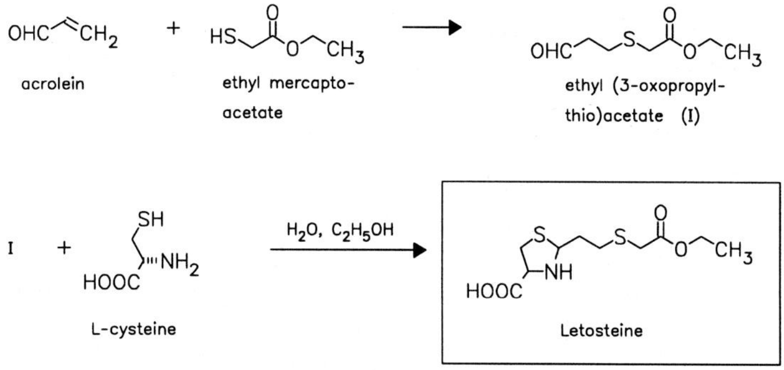

acrolein

ethyl mercapto-
acetate

ethyl (3-oxopropyl-
thio)acetate (I)

L-cysteine

H₂O, C₂H₅OH

Letosteine

Reference(s):

DOS 2 410 307 (Ferlux-Chimie; appl. 22.3.1974; F-prior. 22.3.1973).
US 4 032 534 (Ferlux-Chimie; 28.6.1977; F-prior. 22.3.1973).

Formulation(s):   cps. 50 mg; gran. 25 mg/dose, 50 mg

Trade Name(s):
F: Viscotiol (Evans Medical)   I: Letofort (Salus Research)

Viscotiol (Schiapparelli   J: Viscotiol (ISF)
Searle)

Letrozole
(CGS-20267)

ATC: L02BG04
Use: antineoplastic, aromatase inhibitor

RN: 112809-51-5   MF: $C_{17}H_{11}N_5$ MW: 285.31
CN: 4,4'-(1H-1,2,4-triazol-1-ylmethylene)bis[benzonitrile]

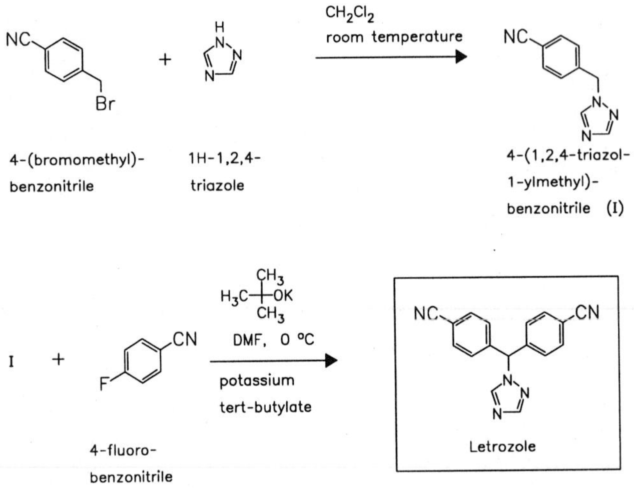

4-(bromomethyl)-   1H-1,2,4-   4-(1,2,4-triazol-
benzonitrile   triazole   1-ylmethyl)-
                                         benzonitrile   (I)

4-fluoro-
benzonitrile

Reference(s):
EP 236 940 (Ciba-Geigy; appl. 5.3.1987; USA-prior. 7.3.1986).

alternativ preparation of I with K_2CO_3/KI in acetone:
US 4 978 672 (Ciba-Geigy; appl. 6.9.1988; USA-prior. 7.3.1986, 7.3.1988).

combination with 5-α-reductase inhibitors:
WO 9 218 132 (Merck & Co.; appl. 6.4.1992; USA-prior. 17.4.1991).

use to treat androgen deficiencies:
DE 445 368 (Schering AG; appl. 22.9.1994; D-prior. 22.9.1994).

Formulation(s):   tabl. 2.5 mg

Trade Name(s):
D: Femara (Novartis Pharma)   GB: Femara (Novartis)
F: Femara (Novartis)   USA: Femara (Novartis)

Leucinocaine

ATC: N01B
Use: local anesthetic

RN: 92-23-9   MF: $C_{17}H_{28}N_2O_2$ MW: 292.42
CN: 2-(diethylamino)-4-methyl-1-pentanol 4-aminobenzoate (ester)

monomesylate
RN: 135-44-4   MF: $C_{17}H_{28}N_2O_2 \cdot CH_4O_3S$ MW: 388.53   EINECS: 205-191-6

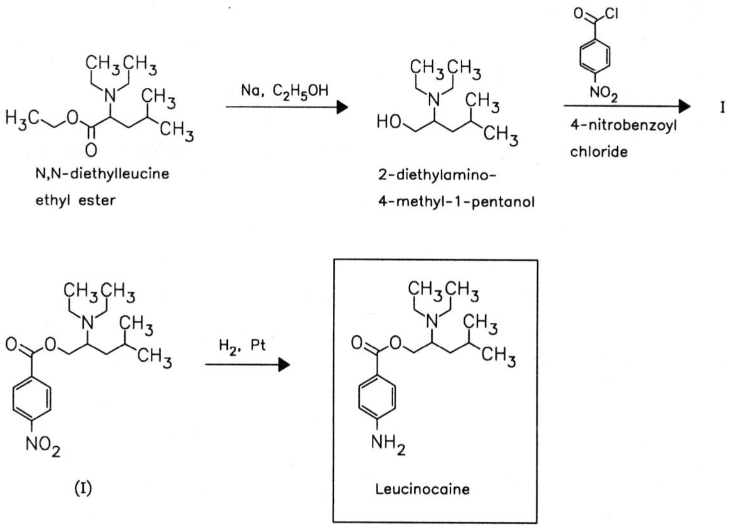

N,N-diethylleucine
ethyl ester

2-diethylamino-
4-methyl-1-pentanol

4-nitrobenzoyl
chloride

Na, C₂H₅OH

(I)

Leucinocaine

H₂, Pt

Reference(s):
DRP 464 484 (Chem. Fabr. Flora; appl. 1923; CH-prior. 1922).

Formulation(s):   amp. 200 mg/4 ml

Trade Name(s):
D: Panthesin-Balsam   Panthesin-Hydergin
      (Sandoz); wfm   (Sandoz)-comb.; wfm

Levallorphan

Use: morphine antagonist, narcotic
        antagonist

RN: 152-02-3   MF: C₁₉H₂₅NO   MW: 283.42   EINECS: 205-799-1
LD₅₀: 949 mg/kg (R, p.o.)
CN: 17-(2-propenyl)morphinan-3-ol

hydrogen tartrate
RN: 71-82-9   MF: C₁₉H₂₅NO · C₄H₆O₆   MW: 433.50

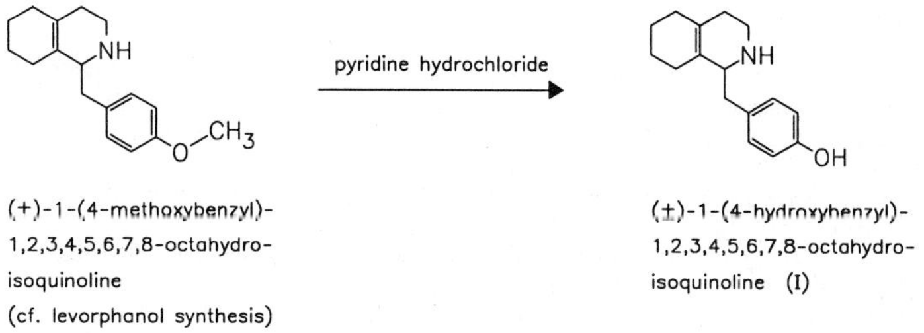

(+)-1-(4-methoxybenzyl)-
1,2,3,4,5,6,7,8-octahydro-
isoquinoline
(cf. levorphanol synthesis)

pyridine hydrochloride

(±)-1-(4-hydroxybenzyl)-
1,2,3,4,5,6,7,8-octahydro-
isoquinoline   (I)

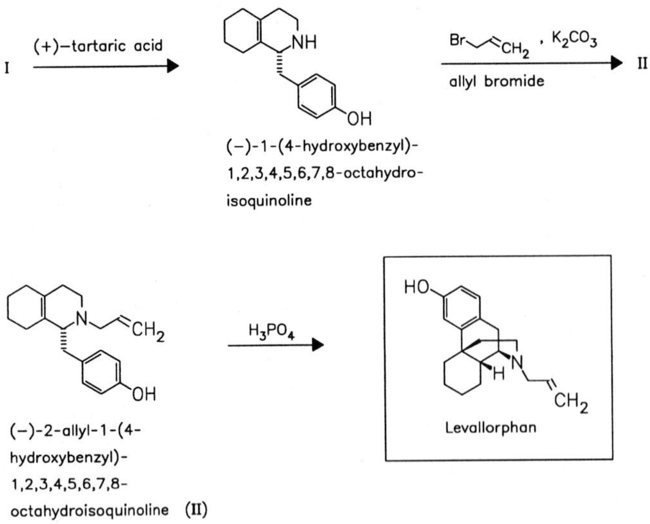

I
(+)–tartaric acid
NH
OH
(–)-1-(4-hydroxybenzyl)-1,2,3,4,5,6,7,8-octahydroisoquinoline
Br, CH2, K2CO3
allyl bromide
II
N
CH2
OH
(–)-2-allyl-1-(4-hydroxybenzyl)-1,2,3,4,5,6,7,8-octahydroisoquinoline (II)
H3PO4
HO
H
N
CH2
Levallorphan

Reference(s):

Hellerbach, J.; Grüssner, A.; Schnider, O.: Helv. Chim. Acta (HCACAV) **39**, 429 (1956).

Ehrhart, Ruschig **I**, 131-132.

Formulation(s): amp. 1 mg/ml

Trade Name(s):

D:	Lorfan (Roche); wfm	J:	Lorfan (Takeda); wfm
GB:	Lorfan (Roche); wfm	USA:	Lorfan (Roche); wfm

Levamisole

ATC: P02CE01
Use: anthelmintic, immunostimulant (tetramisole is used only in veterinary range as anthelmintic)

RN: 14769-73-4 MF: $C_{11}H_{12}N_2S$ MW: 204.30 EINECS: 238-836-5
LD_{50}: 22 mg/kg (M, i.v.); 210 mg/kg (M, p.o.);
24 mg/kg (R, i.v.); 480 mg/kg (R, p.o.)
CN: (*S*)-2,3,5,6-tetrahydro-6-phenylimidazo[2,1-*b*]thiazole

monohydrochloride
RN: 16595-80-5 MF: $C_{11}H_{12}N_2S \cdot HCl$ MW: 240.76 EINECS: 240-654-6

Tetramisole

RN: 5036-02-2 MF: $C_{11}H_{12}N_2S$ MW: 204.30 EINECS: 225-729-3
CN: (±)-2,3,5,6-tetrahydro-6-phenylimidazo[2,1-*b*]thiazole
monohydrochloride
RN: 5086-74-8 MF: $C_{11}H_{12}N_2S \cdot HCl$ MW: 240.76 EINECS: 225-799-5
LD_{50}: 22 mg/kg (M, i.v.); 210 mg/kg (M, p.o.);
24 mg/kg (R, i.v.); 480 mg/kg (R, p.o.)

a

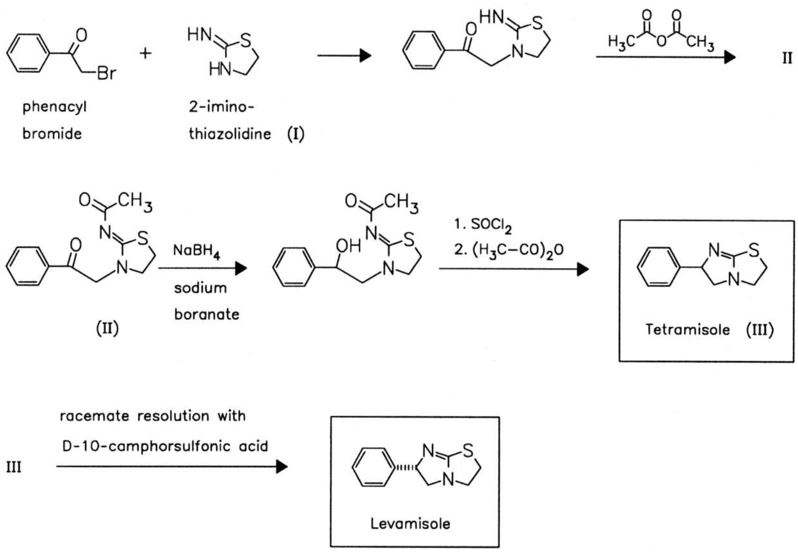

phenacyl
bromide

2-imino-
thiazolidine (I)

II

(II)

NaBH₄

sodium
boranate

1. SOCl₂
2. (H₃C–CO)₂O

Tetramisole (III)

III

racemate resolution with
D-10-camphorsulfonic acid

Levamisole

b

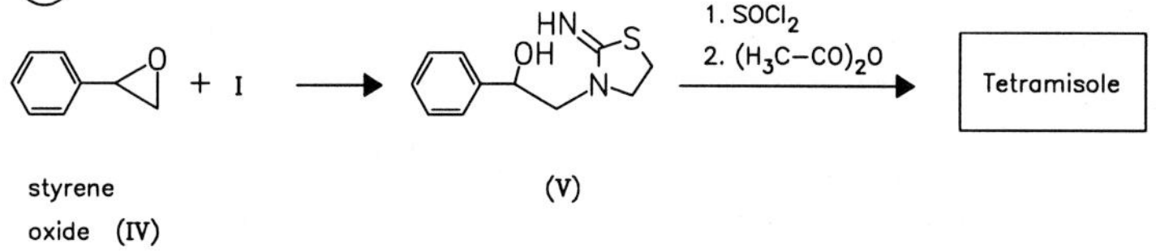

styrene
oxide (IV)

+ I

(V)

1. SOCl₂
2. (H₃C–CO)₂O

Tetramisole

c

IV +   HN◁   aziridine (VI)

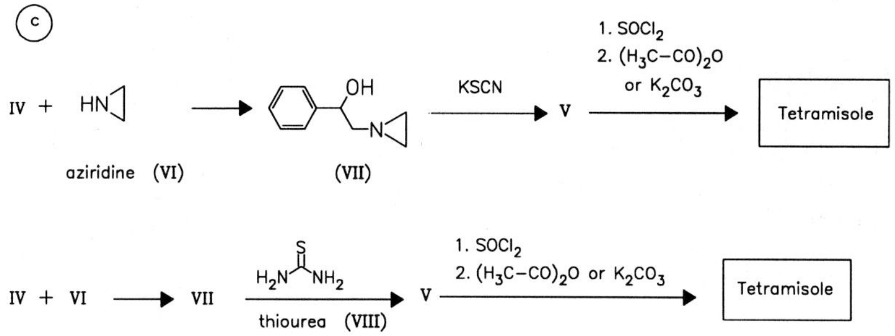

(VII)

KSCN

V

1. SOCl₂
2. (H₃C–CO)₂O
or K₂CO₃

Tetramisole

IV + VI   →   VII

H₂N–C(S)–NH₂
thiourea (VIII)

V

1. SOCl₂
2. (H₃C–CO)₂O or K₂CO₃

Tetramisole

d

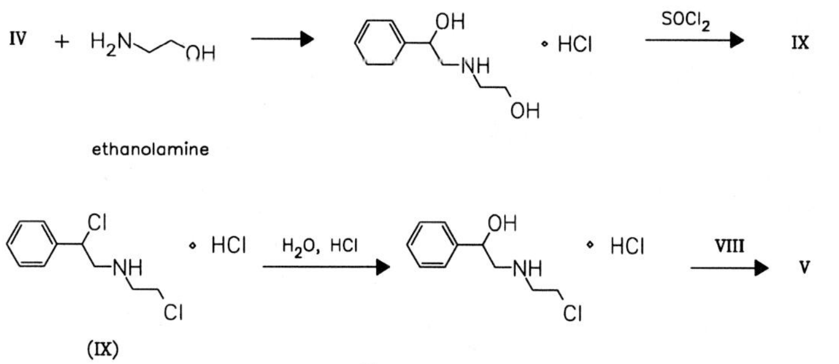

IV + H₂N⌒OH
ethanolamine

• HCl

SOCl₂

IX

• HCl   H₂O, HCl

(IX)

• HCl   VIII

V

2-amino-1-
phenylethanol

2-bromoethyl
isothiocyanate

Reference(s):
a,b US 3 274 209 (Janssen; 20.9.1966; prior. 11.5.1964, 3.8.1964, 2.10.1964, 7.4.1965).
   Raeymaekers, A.H.M. et al.: J. Med. Chem. (JMCMAR) **9**, 545 (1966).
c   Spicer, L.D. et al.: J. Org. Chem. (JOCEAH) **33**, 1350 (1968).
   DAS 1 795 651 (ICI; appl. 13.7.1966; AUS-prior. 31.8.1965, 8.9.1965).
   GB 1 076 109 (American Cyanamid; appl. 11.11.1965; USA-prior. 5.10.1965).
   US 3 679 725 (American Cyanamid; 25.7.1972; prior. 5.10.1965, 22.9.1967, 11.6.1970).
   GB 1 131 798 (ICI; appl. 4.7.1966; AUS-prior. 31.8.1965, 8.9.1965).
   DAS 1 795 651 (ICI; appl. 13.7.1966; AUS-prior. 31.8.1965, 8.9.1965).
   GB 1 131 799 (ICI; appl. 4.7.1966; AUS-prior. 19.7.1965, 26.7.1965, 31.8.1965, 8.9.1965).
   GB 1 131 800 (ICI; appl. 4.7.1966; AUS-prior. 19.7.1965, 26.7.1965, 31.8.1965, 8.9.1965).
   US 3 478 047 (ICI; 11.11.1969; GB-prior. 10.12.1965).
d   DOS 2 233 481 (ICI; appl. 7.7.1972; GB-prior. 9.7.1971, 6.4.1972).
   DOS 2 264 911 (ICI; appl. 7.7.1972; GB-prior. 9.7.1971, 6.4.1972).
   US 3 855 234 (ICI; 17.12.1974; GB-prior. 9.7.1971, 6.4.1972).
   US 4 070 363 (ICI; 24.1.1978; GB-prior. 13.4.1974).
   US 4 107 170 (American Cyanamid; 15.8.1978; prior. 18.6.1973, 24.1.1974, 29.10.1975, 14.2.1977).
e   DAS 2 034 081 (Chinoin; appl. 9.7.1970; H-prior. 1.10.1969).

other methods:
US 3 726 894 (American Cyanamid; 10.4.1973; prior. 24.6.1971).
DOS 2 326 308 (ICI; appl. 23.5.1973; GB-prior. 23.5.1973).
US 3 845 070 (ICI; 29.10.1974; GB-prior. 27.7.1971).
FR 2 224 472 (P. R. Dick, M. Rombi; appl. 5.4.1973).
US 4 090 025 (American Cyanamid; 16.5.1978; prior. 26.4.1973, 8.11.1976).
FR-appl. 2 359 844 (Propharma; appl. 28.7.1976).
FR-appl. 2 364 218 (Propharma; appl. 14.9.1976).

racemate resolution of tetramisole:
US 3 463 786 (American Cyanamid; 26.8.1969; prior. 1.6.1966, 19.12.1967).

with D-10-camphersulfonic acid:
DAS 1 645 991 (American Cyanamid; appl. 18.8.1967; USA-prior. 18.8.1966).
US 3 565 907 (American Cyanamid; 23.2.1971; prior. 18.8.1966, 23.4.1969).
Bullock, M.W. et al.: J. Med. Chem. (JMCMAR) **11**, 169 (1968).

with N-(p-toluenesulfonyl)-L-glutamic acid:
US 3 579 530 (ICI; 18.5.1971; AUS-prior. 24.8.1967, 11.1.1968, 18.1.1968).
DAS 1 795 217 (ICI; appl. 23.8.1968; AUS-prior. 24.8.1967, 11.1.1968, 18.1.1968).

with N-(p-toluenesulfonyl)-l-pyroglutamic acid *and 2,3-O,O-diaroyl-(+)-tartaric acids:*
DAS 1 907 609 (ICI; appl. 14.2.1969; GB-prior. 14.2.1968).

with di-(p-toluoyl)-(+)-tartaric acid:
DAS 2 020 142 (Rhône-Poulenc; appl. 24.4.1970; F-prior. 24.4.1969).

regioselective levamisole *synthesis by use of optical active rhodium-DIOP-complexes (asymmetric hydrogenation of 3-acyl-1-(2-methoxyethyl)-4-phenyl-4-imidazolin-2-ones):*
DOS 2 718 058 (American Cyanamid; appl. 22.4.1977; USA-prior. 26.4.1976, 8.11.1976).
DOS 2 718 059 (American Cyanamid; appl. 22.4.1977; USA-prior. 26.4.1976, 8.11.1976).
US 4 087 611 (American Cyanamid; 2.5.1978; prior. 26.4.1976, 8.11.1976).
US 4 166 824 (American Cyanamid; 4.9.1979; prior. 14.6.1977, 14.4.1978).

racemization with bases:
US 3 673 206 (American Cyanamid; 27.6.1972; prior. 14.7.1966, 2.4.1969).

via 1-vinyl-4-phenyl-2-imidazolidinthione:
US 3 726 894 (American Cyanamid; 10.4.1973; appl. 24.6.1971).

levamisole *resp.* tetramisole embonate:
DAS 1 817 509 (ICI; appl. 30.12.1968; GB-prior. 8.1.1968).

use for treatment of scabies:
DOS 2 828 200 (Johnson & Johnson; appl. 27.6.1978; USA-prior. 28.6.1977).
US 4 150 141 (Johnson & Johnson; 17.4.1979; appl. 28.6.1977).

aqueous tetramisole *preparation:*
DAS 2 036 113 (ICI; appl. 21.7.1970; AUS-prior. 21.7.1969).

Formulation(s):   tabl. 30 mg, 50 mg, 150 mg (as levamisole hydrochloride)

Trade Name(s):
D: Ergamisol (Janssen-Cilag)   I: Ergamisol (Janssen)
F: Solaskil (Specia)   USA: Ergamisol (Janssen)

Levobunolol

ATC: S01ED03
Use: beta blocking agent

RN: 47141-42-4   MF: $C_{17}H_{25}NO_3$ MW: 291.39
CN: (*S*)-5-[3-[(1,1-dimethylethyl)amino]-2-hydroxypropoxy]-3,4-dihydro-1(2*H*)-naphthalenone

hydrochloride
RN: 27912-14-7   MF: $C_{17}H_{25}NO_3 \cdot HCl$ MW: 327.85   EINECS: 248-725-3
LD$_{50}$:   78 mg/kg (M, i.v.); 1220 mg/kg (M, p.o.);
    25 mg/kg (R, i.v.); 700 mg/kg (R, p.o.);
    100 mg/kg (dog, p.o.)

5-methoxy-
1-tetralone

5-hydroxy-
1-tetralone

5-(2,3-epoxypropoxy)-
1-tetralone (I)

tert-butyl-
amine

(±)-5-(2-hydroxy-3-
tert-butylaminopropoxy)-
1-tetralone (II)

racemate resolution with l-tartaric acid

II

Levobunolol

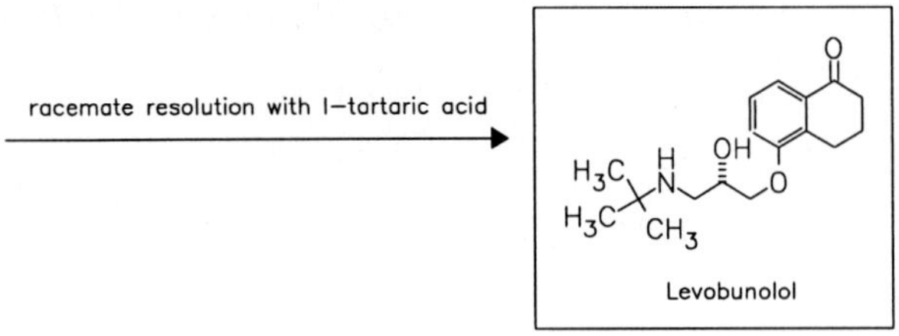

Reference(s):
DE 1 948 144 (Warner-Lambert; appl. 23.9.1969; USA-prior. 23.9.1968).
DE 1 967 162 (Warner-Lambert; appl. 23.9.1969; USA-prior. 23.9.1968).
US 3 641 152 (Warner-Lambert; 8.2.1972; prior. 23.9.1968).

racemate resolution:
DOS 2 046 043 (Warner-Lambert; appl. 17.9.1970; USA-prior. 17.9.1969).

Formulation(s):   eye drops 0.1 %, 0.25 %, 0.5 % (5 mg/ml) (as hydrochloride)

Trade Name(s):
D: Vistagan Liquifilm (Pharm-   F: Bétagan (Allergan)   I: Vistagan (Allergan; 1987)
       Allergan; 1985)   GB: Betagan (Allergan)

Levocabastine
(R-50; 547)

ATC: R01AC02; S01GX02
Use: antihistaminic (H₁-selective)

RN: 79516-68-0   MF: $C_{26}H_{29}FN_2O_2$ MW: 420.53
CN: [3S-[1(cis),3α,4β]]-1-[4-cyano-4-(4-fluorophenyl)cyclohexyl]-3-methyl-4-phenyl-4-piperidinecarboxylic acid

monohydrochloride
RN: 79547-78-7   MF: $C_{26}H_{29}FN_2O_2 \cdot HCl$ MW: 456.99

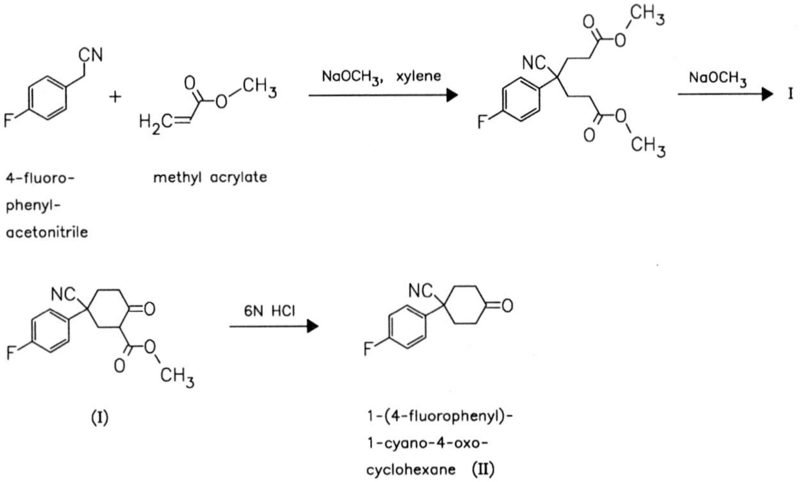

4-fluoro-
phenyl-
acetonitrile

methyl acrylate

NaOCH₃, xylene

NaOCH₃   I

(I)

6N HCl

1-(4-fluorophenyl)-
1-cyano-4-oxo-
cyclohexane (II)

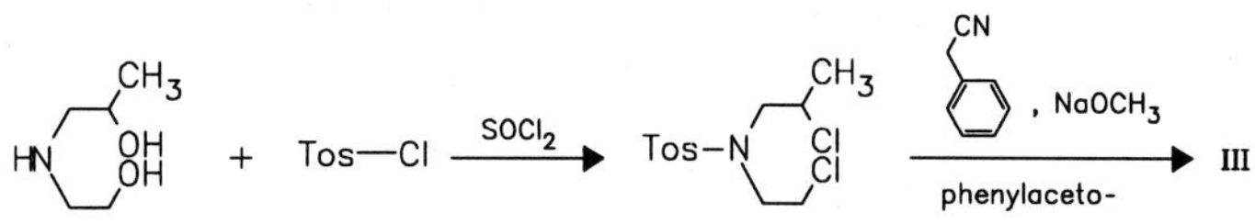

N-(2-hydroxy-ethyl)-2-hydroxy-propylamine

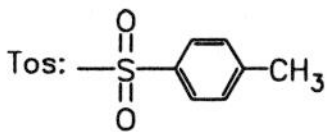

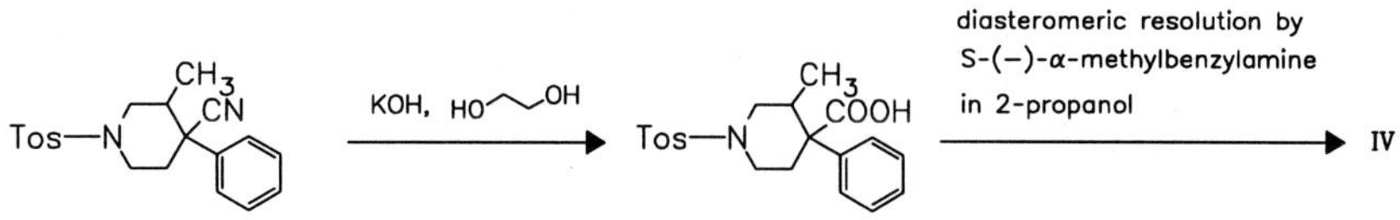

Tos: a tosyl group

(III)

diasteromeric resolution by S-(−)-α-methylbenzylamine in 2-propanol → IV

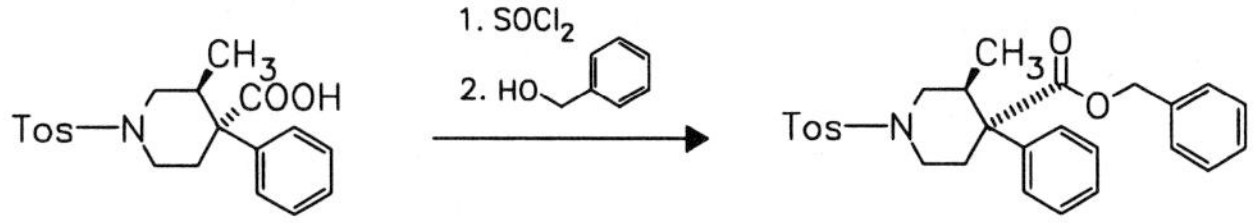

(−)-3-methyl-4-phenyl-1-tosyl-4-piperidinecarboxylic acid (IV)

(V)

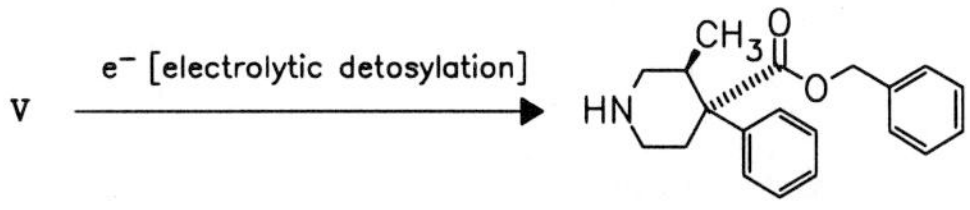

benzyl (−)-3-methyl-4-phenyl-4-piperidine-carboxylate (VI)

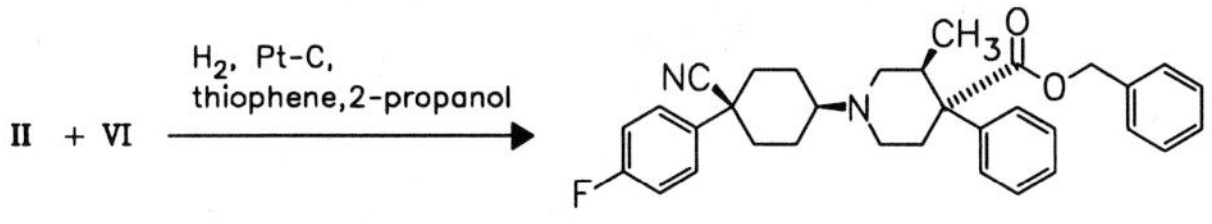

(VII)

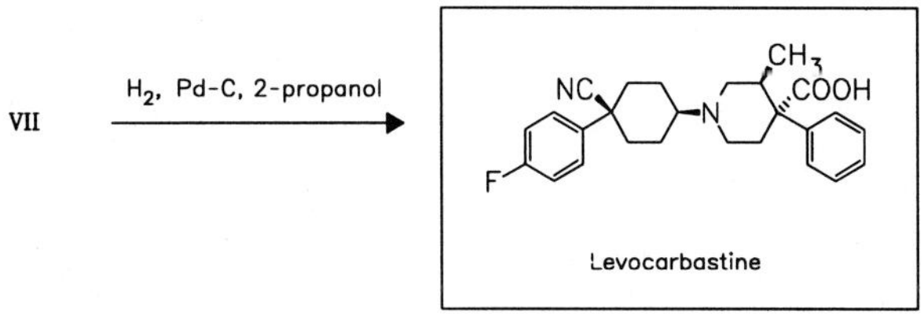

Levocarbastine

Reference(s):
US 4 369 184 (Janssen; 18.1.1983; prior. 24.1.1980, 29.9.1980).
EP 34 415 (Janssen; appl. 23.1.1981; USA-prior. 24.1.1980, 29.9.1980).

Formulation(s):   susp. 0.5 mg/ml (nasal spray, eye drops as hydrochloride)

Trade Name(s):
D: Levophta (CIBA Vision/   F: Lévophta (Chauvin)   Livostin (Janssen)
      Winzer; as hydrochloride)   GB: Livostin (CIBA Vision)
      Livocab (Janssen-Cilag)   I: Levostab (Formenti)

Levodopa

ATC: N04BA01
Use: antiparkinsonian

RN: 59-92-7   MF: C$_9$H$_{11}$NO$_4$ MW: 197.19   EINECS: 200-445-2
CN: 3-hydroxy-L-tyrosine

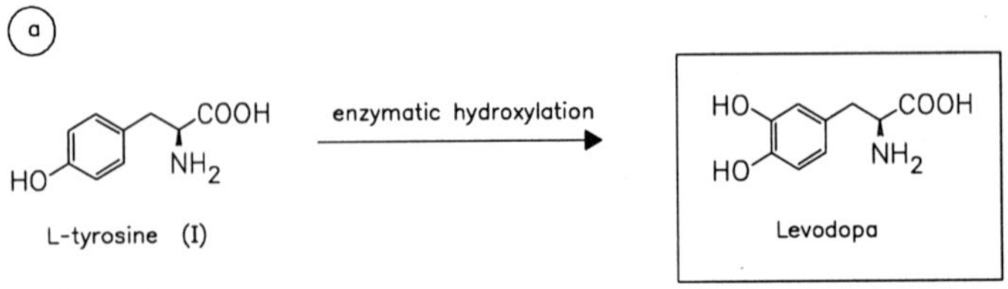

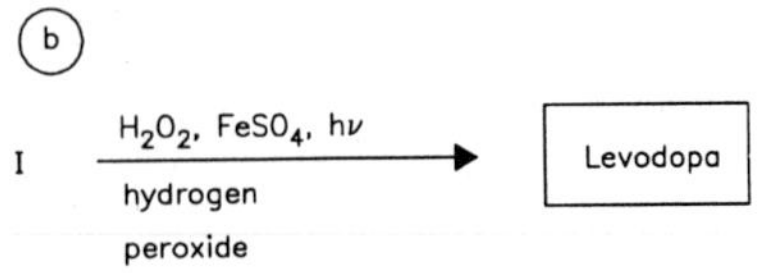

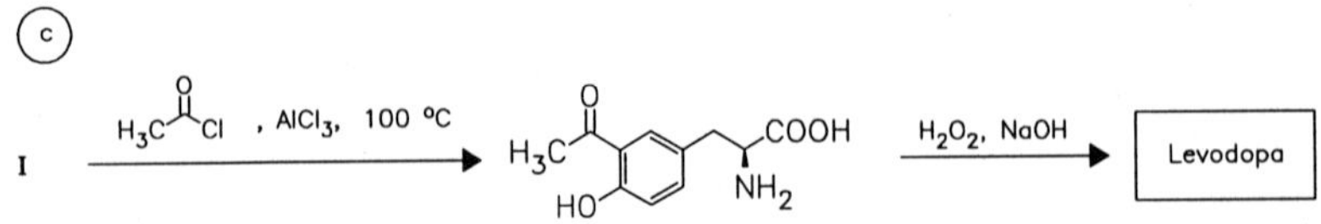

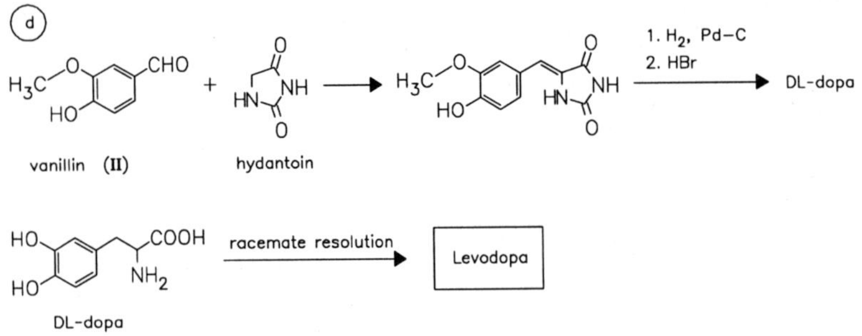

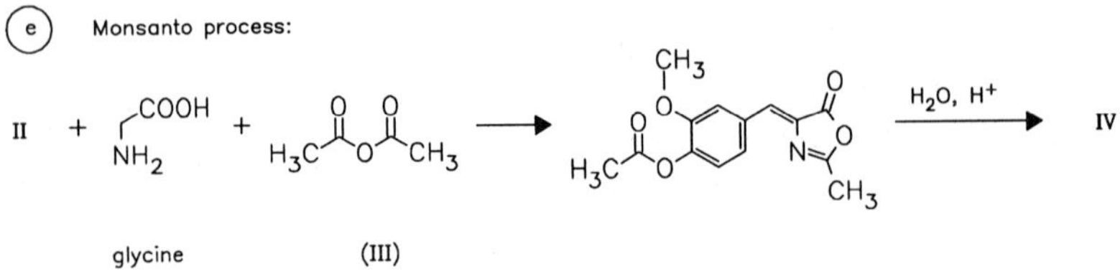

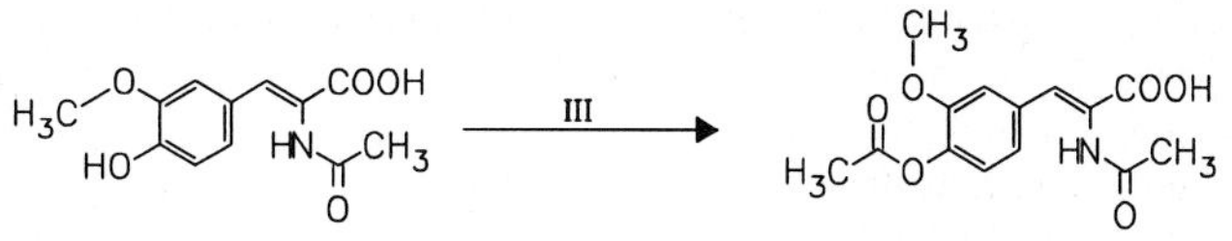

α-acetamido-4-hydroxy-
3-methoxycinnamic
acid (IV)

α-acetamido-4-acetoxy-
3-methoxycinnamic
acid (V)

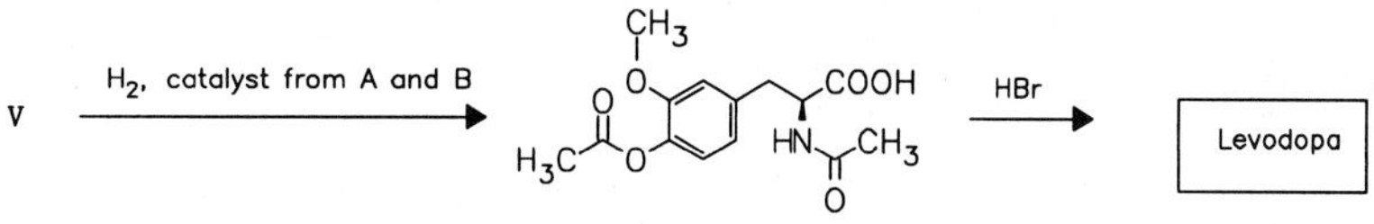

N-acetyl-3-(4-acetoxy-3-
methoxyphenyl)-L-alanine

A: 1,5-cyclooctadienylrhodium chloride
B: (+)-cyclohexylmethyl(2-methoxyphenyl)phosphine

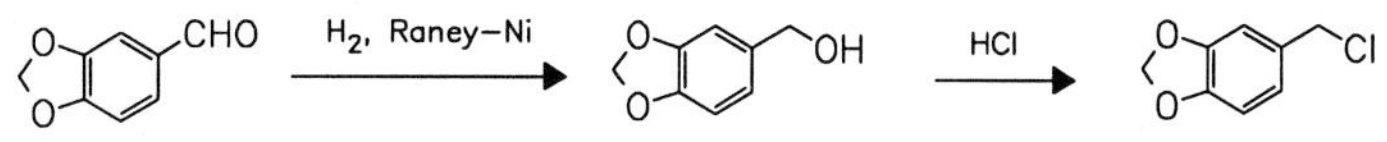

piperonal

(VI)

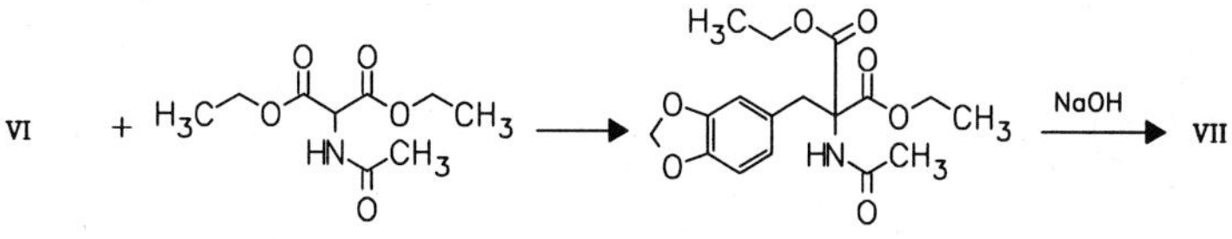

diethyl acetamidomalonate

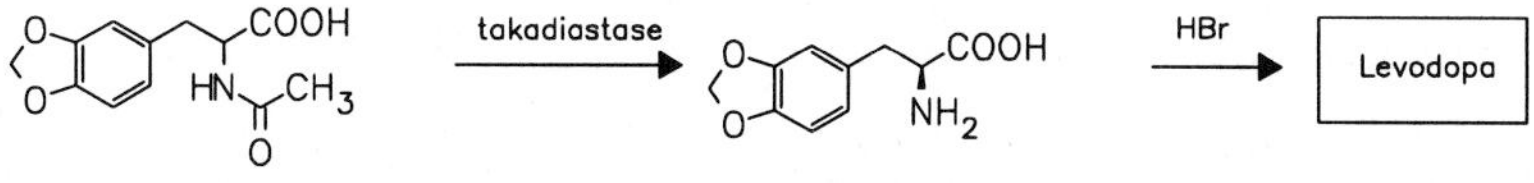

(VII)

L-3-(3,4-methylene-
dioxyphenyl)alanine

Reference(s):
a   Amao, S. et al.: Sankyo Kenkyusho Nempo (SKKNAJ) **23**, 249 (1971).
     Sih, C.J. et al.: J. Am. Chem. Soc. (JACSAT) **91**, 6204 (1969).
b   Waser, E.; Lewandowski, M.: Helv. Chim. Acta (HCACAV) **4**, 657 (1921).
     hydroxylation with benzoyl peroxide:
     DAS 2 026 952 (Schering; appl. 28.5.1970).
c   Bretschneider, H. et al.: Helv. Chim. Acta (HCACAV) **56**, 2857 (1973).
     DAS 2 023 459 (Roche; appl. 13.5.1970; CH-prior. 14.5.1969).
     DAS 2 023 460 (Roche; appl. 13.5.1970; CH-prior. 14.5.1969).
     DAS 2 023 461 (Roche; appl. 13.5.1970; CH-prior. 14.5.1969).
     similar method:
     DAS 2 026 952 (Schering AG; appl. 28.5.1970).
d   US 2 605 282 (Dow; 1952; appl. 1949).
     racemate resolution of DL-*N*-benzoyl-3-(4-hydroxy-3-methoxyphenyl)alanine *with* dehydroabietylamine:
     DOS 1 964 420 (Roche; appl. 23.12.1969; CH-prior. 27.12.1968).

racemate resolution of DL-*N*-acetyl-3-(4-acetoxy-3-methoxyphenyl)alanine *with* (−)-α-phenylethylamine:
DOS 2 052 953 (Egyt; appl. 28.10.1970; H-prior. 28.10.1969).
with (+)-*threo*-2-amino-1-(4-nitrophenyl)-1,3-propanediol:
DOS 2 052 995 (Egyt; appl. 28.10.1970; H-prior. 28.10.1969).
alternative synthesis via 2,5 dioxopiperazine:
Losse, G. et al.: J. Prakt. Chem. (JPCEAO) **21**, 32 (1963).
e   US 4 005 127 (Monsanto; 25.1.1977; prior. 8.3.1971).
DAS 2 123 063 (Monsanto; appl. 10.5.1971; USA-prior. 11.5.1970, 8.3.1971).
DAS 2 210 938 (Monsanto; appl. 7.3.1972; USA-prior. 8.3.1971).
US 4 124 533 (Monsanto; 7.11.1978; prior. 9.9.1968, 11.5.1970, 8.3.1971, 17.3.1975).
Knowles, W.S. et al.: J. Am. Chem. Soc. (JACSAT) **97**, 2567 (1975).
Vineyard, B.D. et al.: J. Am. Chem. Soc. (JACSAT) **99**, 5946 (1977).
similar method:
DOS 2 161 200 (IFR; appl. 9.12.1971; F-prior. 10.12.1970).
f   Yamada, S. et al.: Chem. Pharm. Bull. (CPBTAL) **10**, 680, 688, 693 (1963).

alternative syntheses from piperonal:
Mori, K.: Nippon Kagaku Zasshi (NPKZAZ) **81**, 464 (1960).
Barry, R.H. et al.: J. Am. Chem. Soc. (JACSAT) **70**, 693 (1948).

other methods for racemate resolution of DL-dopa *or its derivatives:*
US 3 405 159 (Merck & Co.; 8.10.1968; appl. 17.11.1964).
CH 511 774 (Ajinomoto; appl. 22.4.1970; J-prior. 23.4.1969).
Yamada, S. et al.: J. Org. Chem. (JOCEAH) **40**, 3360 (1975).

racemate resolution of DL-*N*-benzoyldopa *with* cinchonine:
DOS 1 963 992 (Dynamit Nobel; appl. 20.12.1969).

racemization of D-*N*-benzoyldopa *with* acetanhydride:
DOS 1 963 991 (Dynamit Nobel; appl. 20.12.1969).

racemization of D-dopa *by thermic treatment:*
DAS 2 126 049 (Dynamit Nobel; appl. 26.5.1971).

fermentative and enzymatic methods:
from 3,4-dihydroxyphenylpyruvic acid *by transamination by means of microorganisms:*
DOS 2 041 418 (Anm. 14.8.1970; J-prior. 16.8.1969).

by means of transaminase from Alcaligenes faecalis (IAM 1015):
DAS 2 148 953 (Nisshin Flour Milling; appl. 30.9.1971; J-prior. 30.9.1970, 1.6.1971, 19.8.1971).

from pyrocatechol, pyruvic acid *and ammonium salts by means of* β-tyrosinase:
DAS 2 152 548 (Ajinomoto; appl. 21.10.1971; J-prior. 21.10.1970, 2.11.1970, 30.12.1970).

enzymatic resolution of DL-*N*-phenylacety1-3-(3,4-methylenedioxyphenyl)alanine *or* DL-*N*-phenylacetyl-3-(3,4-dimethoxyphenyl)alanine *or* DL-*N*-phenylacety1-3-(3,4-dihydroxyphenyl)-alanine *by means of Escherichia coli-acylase:*
DOS 2 100 445 (Astra; appl. 7.1.1971; S-prior. 19.1.1970, 25.6.1970).

isolation from the seed meal of fodder beans or vetch pods:
US 3 253 023 (Dow; 24.5.1966; appl. 27.9.1963).

combination with carbidopa:
US 3 769 424 (Merck & Co.; 30.10.1973; prior. 1.10.1968, 23.6.1969, 1.10.1970).

combination with etoperidone *and* trazodone:
US 4 131 675 (Angelini Francesco; 26.12.1978; appl. 9.2.1978).

Formulation(s):   cps. 125 mg, 200 mg, 250 mg, 500 mg; tabl. 100 mg, 200 mg, 500 mg

Trade Name(s):
D: Dopaflex (medphano)

Madopar (Roche)-comb.
with benserazide

Nacom (Du Pont Pharma)-
comb. with carbidopa

numerous generics
F: Modopar (Roche)-comb. with benserazide
Sinemet (Du Pont Pharma)-comb. with carbidopa
GB: Madopar (Roche)-comb. with benserazide
Sinemet (Merck Sharp & Dohme)-comb. with carbidopa
I: Larodopa (Roche)
Madopar (Roche)-comb. with benserazide
Sinemet (Du Pont)-comb. with carbidopa
J: Doparl (Kyowa)
Dopasol (Daiichi)
Dopaston (Sankyo)
Larodopa (Roche)
Neodopaston (Sankyo)-comb. with carbidopa
USA: Larodopa (Roche)
Sinemet (Du Pont)-comb. with carbidopa
generic

Levofloxacin

((*S*)-Ofloxacin; DR-3355; HR-355; RWJ-25213)

ATC: J01MA12
Use: antibacterial

RN: 100986-85-4 MF: $C_{18}H_{20}FN_3O_4$ MW: 361.37
LD_{50}: 1803 mg/kg (M, p.o.);
1478 mg/kg (R, p.o.)
CN: (*S*)-9-fluoro-2,3-dihydro-3-methyl-10-(4-methyl-1-piperazinyl)-7-oxo-7*H*-pyrido[1,2,3-*de*]-1,4-benzoxazine-6-carboxylic acid

hydrate (2:1)
RN: 138199-71-0 MF: $C_{18}H_{20}FN_3O_4 \cdot 1/2H_2O$ MW: 740.76

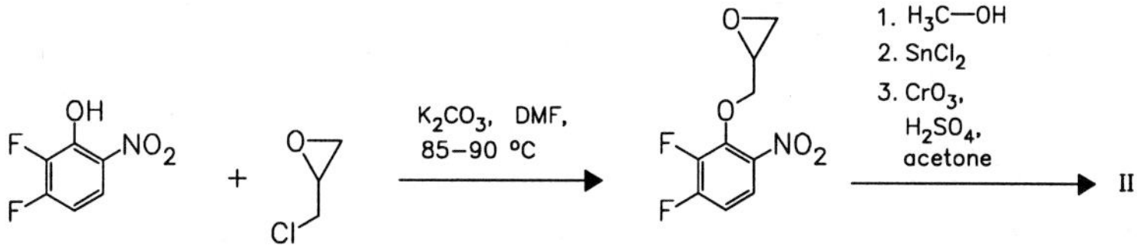

2,3-difluoro-6-nitrophenol (I)
epichlorohydrin
2,3-difluoro-6-nitrophenyl oxiranylmethyl ether

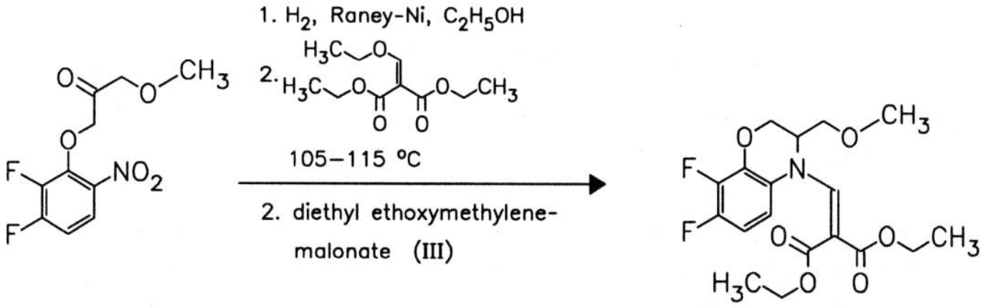

1-(2,3-difluoro-6-nitrophenoxy)-3-methoxy-2-propanone (II)
diethyl [(7,8-difluoro-3-methoxymethyl-2,3-dihydro-4H-1,4-benzoxazin-4-yl)methylene]malonate (IV)

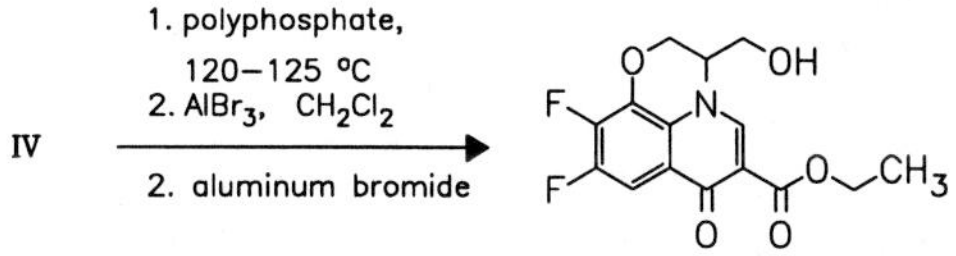

(±)-ethyl 9,10-difluoro-3-hydroxymethyl-7-oxo-2,3-dihydro-7H-pyrido-[1,2,3-de]-1,4-benzoxazine-6-carboxylate (V)

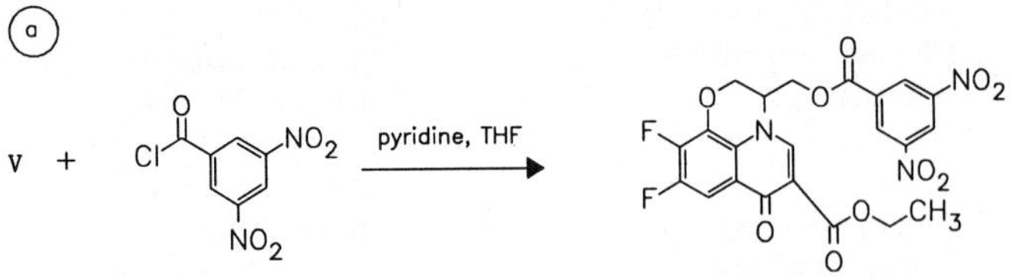

(a)

V + 3,5-dinitro-benzoyl chloride → pyridine, THF → (±)-ethyl 9,10-difluoro-3-(3,5-dinitrobenzoyl-oxymethyl)-7-oxo-2,3-dihydro-7H-pyrido-[1,2,3-de]-1,4-benzoxa-zine-6-carboxylate (VI)

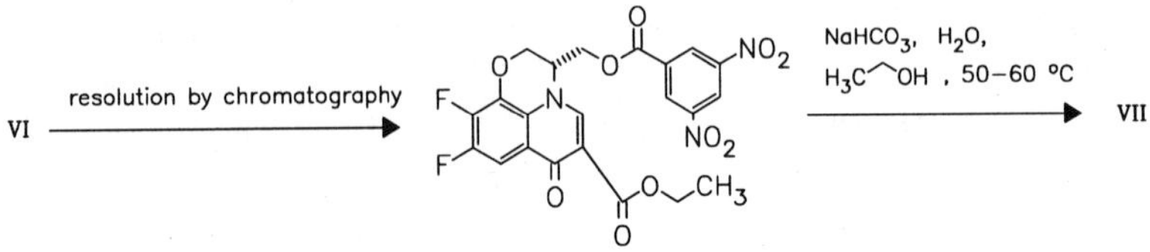

VI → resolution by chromatography → (−)-(VI) → NaHCO₃, H₂O, H₃C—OH, 50−60 °C → VII

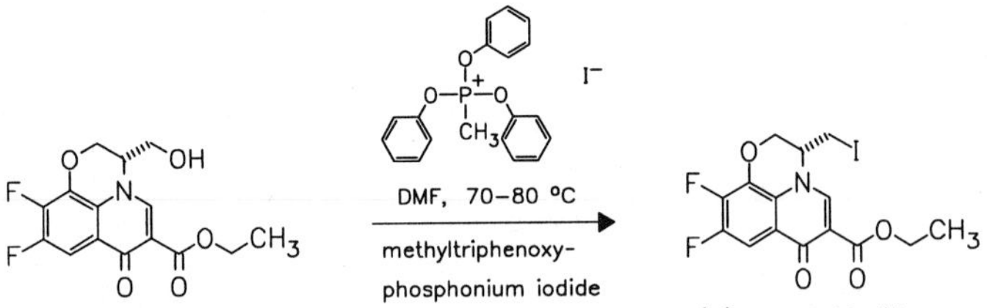

(−)-ethyl 9,10-difluoro-3-hydroxymethyl-7-oxo-2,3-dihydro-7H-pyrido-[1,2,3-de]-1,4-benzoxa-zine-6-carboxylate (VII) → DMF, 70−80 °C, methyltriphenoxy-phosphonium iodide → (−)-ethyl 9,10-difluoro-3-iodomethyl-7-oxo-2,3-dihydro-7H-pyrido-[1,2,3-de]-1,4-benzoxa-zine-6-carboxylate (VIII)

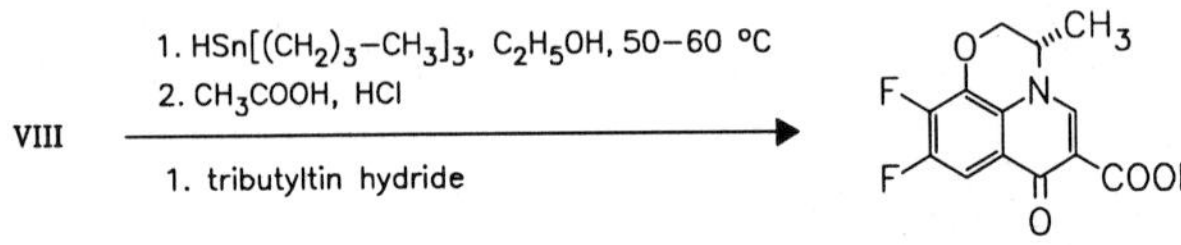

VIII → 1. HSn[(CH₂)₃−CH₃]₃, C₂H₅OH, 50−60 °C 2. CH₃COOH, HCl / 1. tributyltin hydride → 7-oxo-2,3-dihydro-7H-pyrido-(S)-9,10-difluoro-3-methyl-[1,2,3-de]-1,4-benzoxazine-6-carboxylic acid (IX)

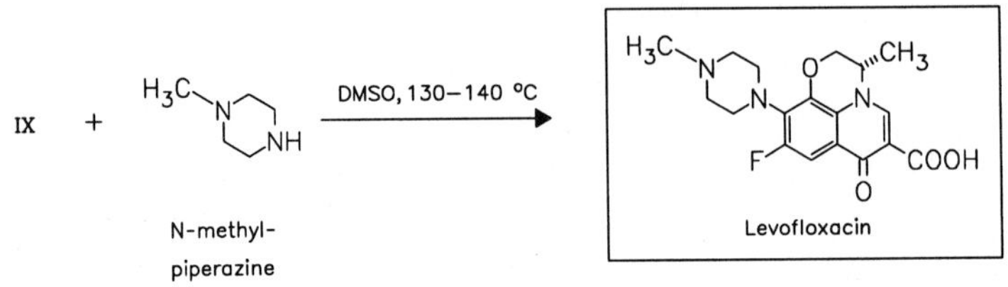

IX + N-methyl-piperazine → DMSO, 130−140 °C → Levofloxacin

(b)

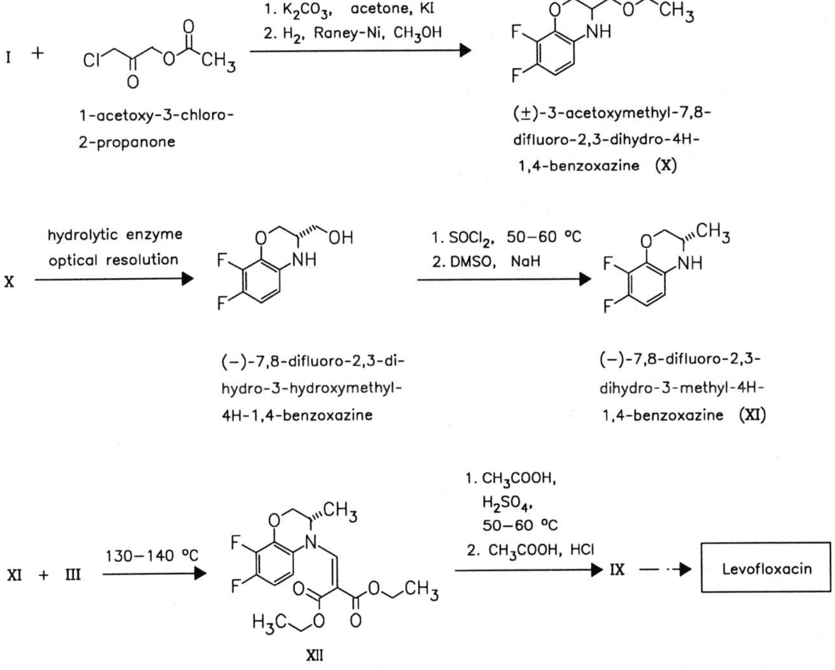

I + 1-acetoxy-3-chloro-2-propanone

1. K₂CO₃, acetone, KI
2. H₂, Raney-Ni, CH₃OH

(±)-3-acetoxymethyl-7,8-difluoro-2,3-dihydro-4H-1,4-benzoxazine (**X**)

X

hydrolytic enzyme
optical resolution

(−)-7,8-difluoro-2,3-dihydro-3-hydroxymethyl-4H-1,4-benzoxazine

1. SOCl₂, 50–60 °C
2. DMSO, NaH

(−)-7,8-difluoro-2,3-dihydro-3-methyl-4H-1,4-benzoxazine (**XI**)

XI + **III**

130–140 °C

XII

1. CH₃COOH, H₂SO₄, 50–60 °C
2. CH₃COOH, HCl

IX — ·→ Levofloxacin

(c)

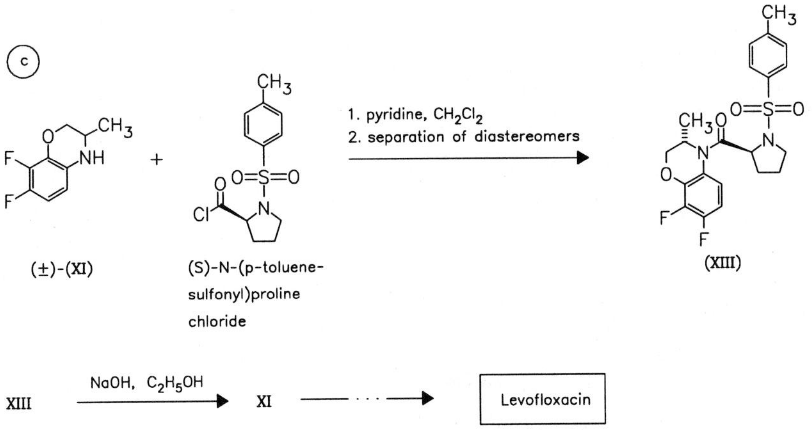

(±)-(**XI**) + (S)-N-(p-toluenesulfonyl)proline chloride

1. pyridine, CH₂Cl₂
2. separation of diastereomers

(**XIII**)

XIII

NaOH, C₂H₅OH

XI — · · ·→ Levofloxacin

(d)

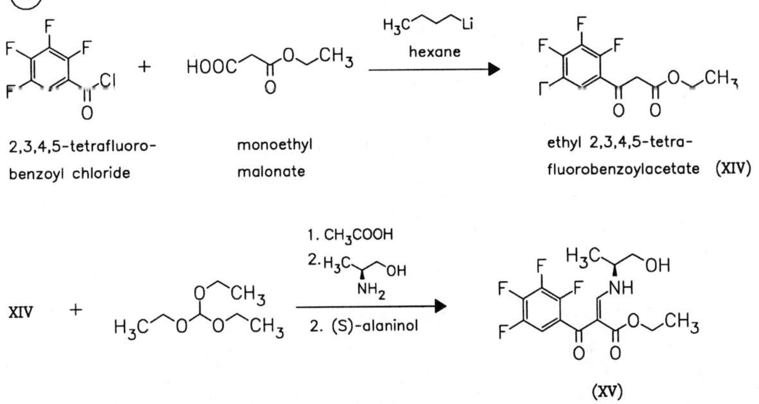

2,3,4,5-tetrafluoro-benzoyl chloride + monoethyl malonate

H₃C Li
hexane

ethyl 2,3,4,5-tetrafluorobenzoylacetate (**XIV**)

XIV +

1. CH₃COOH
2. (S)-alaninol

(**XV**)

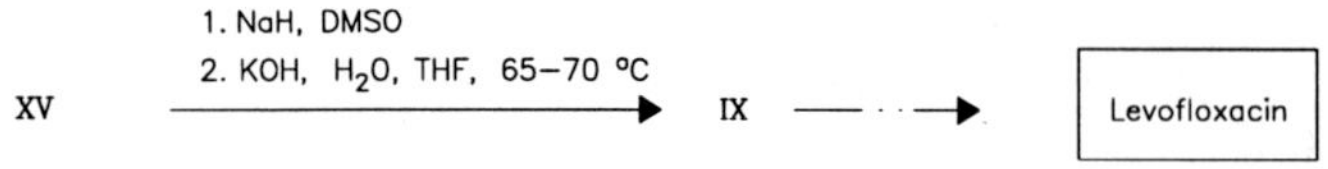

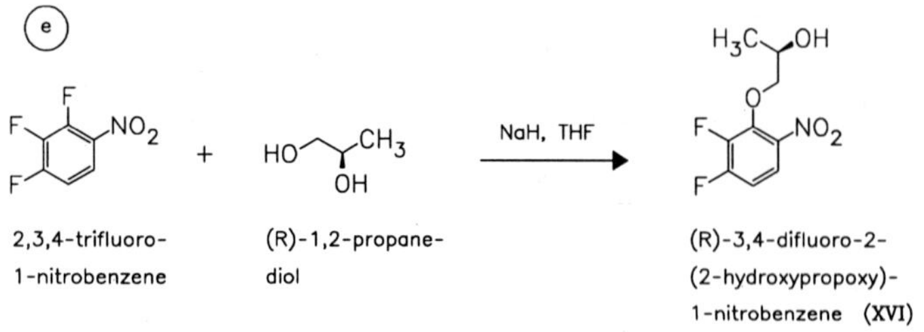

XV $\xrightarrow[\text{2. KOH, H}_2\text{O, THF, 65-70 °C}]{\text{1. NaH, DMSO}}$ IX — ·· → Levofloxacin

(e)

2,3,4-trifluoro-
1-nitrobenzene
+
(R)-1,2-propane-
diol
$\xrightarrow{\text{NaH, THF}}$
(R)-3,4-difluoro-2-
(2-hydroxypropoxy)-
1-nitrobenzene (XVI)

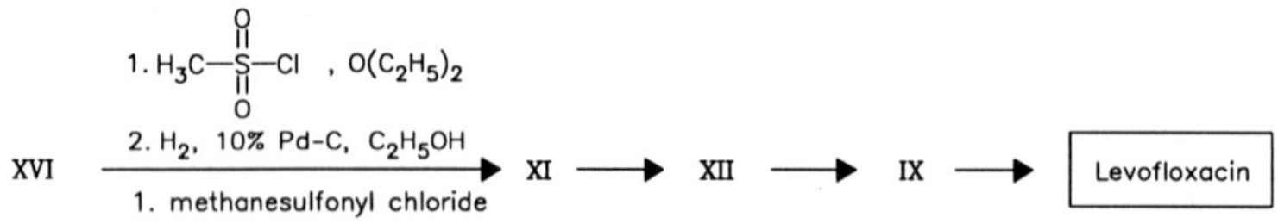

XVI $\xrightarrow[\substack{\text{2. H}_2\text{, 10% Pd-C, C}_2\text{H}_5\text{OH} \\ \text{1. methanesulfonyl chloride}}]{\substack{\text{1. H}_3\text{C-SO}_2\text{-Cl , O(C}_2\text{H}_5)_2}}$ XI → XII → IX → Levofloxacin

Reference(s):

a-c EP 206 283 (Daiichi Seiyaku; appl. 20.6.1986; J-prior. 20.6.1985, 11.10.1985, 28.1.1986).
d US 4 777 253 (Abbott Labs.; 11.10.1980; appl. 25.4.1986; USA-prior. 25.4.1986).
   DE 3 543 513 (Bayer AG; appl. 10.12.1985; D-prior. 10.12.1985).
e EP 368 410 (Gist-Brocades; appl. 6.11.1989; EP-prior. 7.11.1988).
   Atarashi, S. et al.: Chem. Pharm. Bull. (CPBTAL) **35** (5), 1896 (1987).

preparation of 2,3-difluoro-6-nitrophenol:
O'Neill, P.M et al.: J. Med. Chem. (JMCMAR) **37** (9), 1362 (1994).
Hayakawa, I.; Hiramitsu, T.; Tanaka, Y.: Chem. Pharm. Bull. (CPBTAL) **32** (12), 4907 (1984).

preparation of intermediate XI:
JP 05 068 577 (Mercian Corp.; appl. 11.12.1990; J-prior. 11.12.1990).

synergistic combination with azidothymidine:
WO 9 013 542 (Daiichi Pharm.; appl. 27.4.1990; J-prior. 23.2.1990).

topical formulation:
EP 274 714 (Daiichi Seiyaku; appl. 18.12.1987; J-prior. 18.12.1987).

liposomes with increased retention:
WO 9 526 185 (Daiichi Pharm.; appl. 27.3.1995; J-prior. 28.3.1994).

Formulation(s): gran. 100 mg/g; tabl. 100 mg, 250 mg; vial 5 mg/ml, 25 mg/ml

Trade Name(s):
D: Tavanic (Hoechst Marion   J: Cravit (Daiichi Seiyaku)
   Roussel; 1998)   USA: Levaquin (Ortho-McNeil)

Levomepromazine
(Laevomepromazine; Methotrimeprazine)

ATC: N05AA02
Use: neuroleptic

RN: 60-99-1 MF: $C_{19}H_{24}N_2OS$ MW: 328.48 EINECS: 200-495-5
LD_{50}: 39 mg/kg (M, i.v.); 370 mg/kg (M, p.o.);
      1100 mg/kg (R, p.o.)
CN: (R)-2-methoxy-N,N,β-trimethyl-10H-phenothiazine-10-propanamine

monohydrochloride
RN: 1236-99-3   MF: C$_{19}$H$_{24}$N$_2$OS · HCl   MW: 364.94   EINECS: 214-978-3
LD$_{50}$:   75 mg/kg (M, i.v.); 380 mg/kg (M, p.o.)
maleate (1:1)
RN: 7104-38-3   MF: C$_{19}$H$_{24}$N$_2$OS · C$_4$H$_4$O$_4$ MW: 444.55   EINECS: 230-412-8

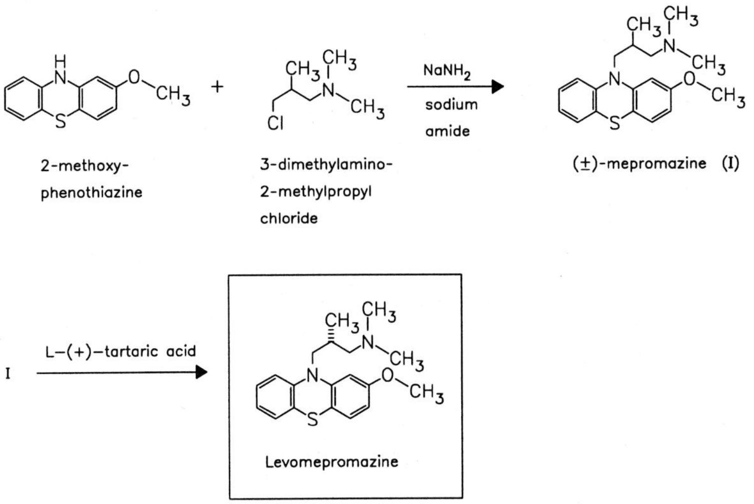

2-methoxy-
phenothiazine

3-dimethylamino-
2-methylpropyl
chloride

NaNH$_2$
sodium
amide

(±)-mepromazine   (I)

L-(+)-tartaric acid

Levomepromazine

Reference(s):
US 2 837 518 (Rhône-Poulenc; 1958; F-prior. 1954).
DE 1 034 638 (Rhône-Poulenc; appl. 1955; F-prior. 1954).

Formulation(s):   amp. 25 mg, 200 mg (as hydrochloride); sol. 40 mg/ml; tabl. 2 mg, 5 mg, 10 mg, 25 mg,
50 mg, 100 mg

Trade Name(s):
D:	Neurocil (Bayer Vital)	I:	Nozinan (Rhône-Poulenc		Levaru (Mohan)
	Tisercin (Thiemann)		Rorer)		Levomezine (Toho)
F:	Nozinan (Specia)	J:	Dedoran (Shionogi)		Levotomin (Yoshitomi)
GB:	Nozinan (Link)		Hirnamin (Shionogi; as		Sofmin (Dainippon)
			maleate)	USA:	Levoprome (Immunex)

Levonorgestrel
(D-Norgestrel; Dexnorgestrel)

ATC: G03AC03
Use: progestogen

RN: 797-63-7   MF: C$_{21}$H$_{28}$O$_2$ MW: 312.45   EINECS: 212-349-8
CN: (17α)-13-ethyl-17-hydroxy-18,19-dinorpregn-4-en-20-yn-3-one

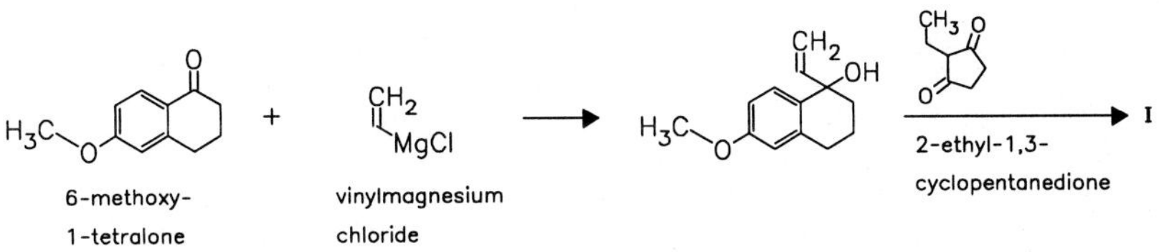

6-methoxy-
1-tetralone

vinylmagnesium
chloride

2-ethyl-1,3-
cyclopentanedione

I

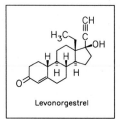

Levonorgestrel

Reference(s):
Rufer, C. et al.: Justus Liebigs Ann. Chem. (JLACBF) **702**, 141 (1967).

alternative syntheses:
DOS 1 806 410 (Hoffmann-La Roche; appl. 31.10.1968; USA-prior. 2.11.1967).
Smith, H. et al.: J. Chem. Soc. (JCSOA9) **1964**, 4472.

ethynylation methods:
DD 114 807 (Reihe, Kutz; appl. 11.10.1974).
DE 2 030 056 (Schering AG; appl. 13.7.1970).

use as contraceptive:
DOS 1 922 005 (Schering AG; appl. 24.4.1969).

combination with 17α-ethynylestradiol as contraceptive:
DOS 2 218 831 (Schering AG; appl. 14.4.1972).
DOS 2 335 265 (Schering AG; appl. 30.1.1975).
DAS 2 365 103 (Schering AG; appl. 21.12.1973).
DOS 2 431 704 (Asche; appl. 2.7.1974).

intrauterine anticonception:
DOS 2 361 206 (Schering AG; appl. 6.12.1973).

pharmaceutical formulation:
DOS 2 432 925 (Schering AG; appl. 5.7.1974).
DOS 2 449 865 (Schering AG; appl. 17.10.1974).

Formulation(s):   drg. 0.03 mg, 0.1 mg, 0.15 mg, 0.25 mg; pessaries 52 mg

Trade Name(s):
D: Micro-30-Wyeth (Wyeth)   Mirena Intrauterinpessar   numerous combination
      Microlut (Schering)   (Schering)   preparations

F: Adepal (Wyeth-Lederle)-
    comb. with ethynylestradiol
    Microval (Wyeth-Lederle)-
    comb. with ethynylestradiol
    Minidril (Wyeth-Lederle)-
    comb. with ethynylestradiol
    Trinordiol (Wyeth-
    Lederle)-comb. with
    ethynylestradiol
GB: Microval (Wyeth)
    Mirena (Schering)
    Norgeston (Schering)

numerous combination
preparations
I: Binordiol (Wyeth)-comb.
     Bivlar (Schering)-comb.
     Egogyn (Schering)-comb.
     Evanor D (Wyeth)-comb.
     Microgynon (Schering)-
     comb.
     Microlut (Schering)
     Novogyn (Schering)-comb.
     Ovranet (Wyeth)-comb.
     Trigynon (Schering)-comb.
     Trinordiol (Wyeth)-comb.

J: Microlut (Nihon Schering)
     Micro 30 (Wyeth)
     Norgeston (Nihon
     Schering)
USA: Alesse (Wyeth-Ayerst)-
     comb.
     Levlen (Berlex)-comb.
     Nordette (Wyeth-Ayerst)-
     comb.
     Tri-Levlen (Berlex)-comb.
     Triphasil (Wyeth-Ayerst)-
     comb.

Levorphanol

ATC: N02
Use: analgesic

RN: 77-07-6   MF: $C_{17}H_{23}NO$ MW: 257.38   EINECS: 201-002-6
LD$_{50}$: 41 mg/kg (M, i.v.); 285 mg/kg (M, p.o.);
      150 mg/kg (R, p.o.)
CN: 17-methylmorphinan-3-ol

tartrate (1:1)
RN: 125-72-4   MF: $C_{17}H_{23}NO \cdot C_4H_6O_6$ MW: 407.46   EINECS: 204-753-8
LD$_{50}$: 32 mg/kg (M, i.v.); 285 mg/kg (M, p.o.);
      27 mg/kg (R, i.v.); 150 mg/kg (R, p.o.);
      46 mg/kg (dog, i.v.)

cyclo-
hexanone

cyanoacetic
acid

1-cyclohexenyl-
acetonitrile (I)

2-(1-cyclo-
hexenyl)-
ethylamine

4-methoxyphenyl-
acetyl chloride

N-[2-(1-cyclohexenyl)-
ethyl]-4-methoxy-
phenylacetamide (II)

1-(4-methoxybenzyl)-
3,4,5,6,7,8-hexahydro-
isoquinoline

1-(4-methoxybenzyl)-
1,2,3,4,5,6,7,8-octa-
hydroquinoline (III)

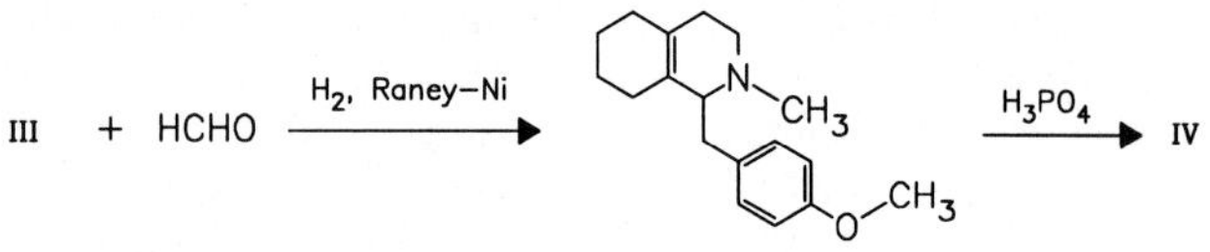

1-(4-methoxybenzyl)-
2-methyl-1,2,3,4,5,6,7,8-
octahydroisoquinoline

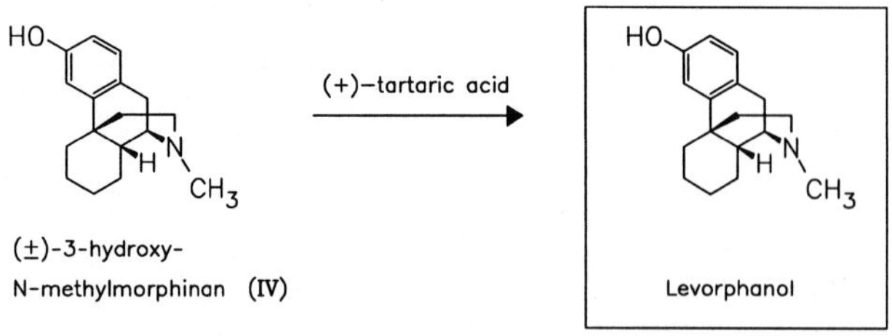

(±)-3-hydroxy-
N-methylmorphinan (IV)

Levorphanol

Reference(s):
Ehrhart, Ruschig **I**, 130-131.
Schnider, O.; Hellerbach, J.: Helv. Chim. Acta (HCACAV) **33**, 1437 (1950).
Schnider, O.; Grüssner, A.: Helv. Chim. Acta (HCACAV) **34**, 2211 (1951).

Formulation(s):   amp. 2 mg/ml; tabl. 2 mg (as tartrate)

Trade Name(s):
D: Dromoran (Roche); wfm   GB: Dromoran (Roche); wfm   USA: Levo-Dromoran (Roche)

Levothyroxine

ATC: H03AA01
Use: thyroid hormone

RN: 51-48-9   MF: $C_{15}H_{11}I_4NO_4$ MW: 776.87   EINECS: 200-101-1
CN: O-(4-hydroxy-3,5-diiodophenyl)-3,5-diiodo-L-tyrosine

monosodium salt
RN: 55-03-8   MF: $C_{15}H_{10}I_4NNaO_4$ MW: 798.85   EINECS: 200-221-4
LD_{50}: 20 mg/kg (R, i.p.); 50 mg/kg (R, s.c.)

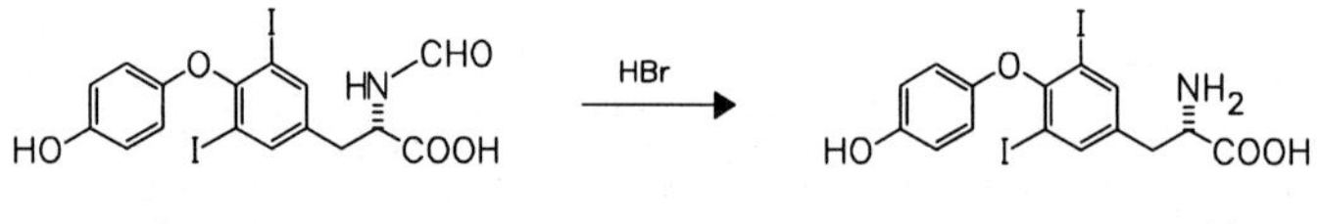

L-N-formyl-3,5-
diiodo thyronine
(cf. dextrothyronine
synthesis)

L-3,5-diiodothyronine (I)

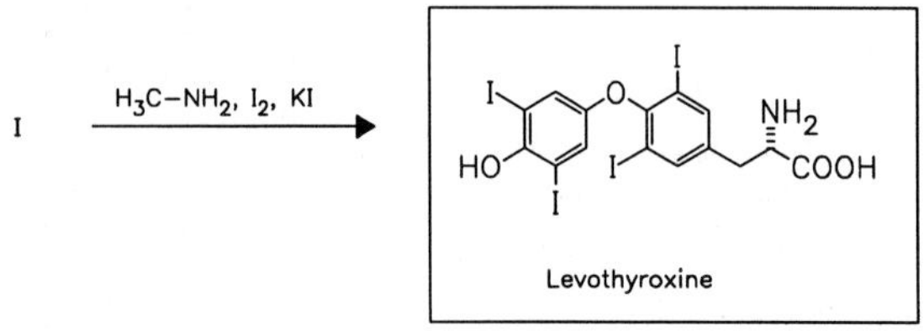

Levothyroxine

Reference(s):
Nahm, H.; Siedel, W.: Chem. Ber. (CHBEAM) **96**, 1 (1963).
DE 1 067 826 (Hoechst; appl. 1955).
DE 1 077 673 (Hoechst; appl. 1958).

alternative syntheses from L-tyrosine *via* L-*N*-acetyl-3,5-diiodotyrosine ethyl ester:
DE 1 064 529 (G. Hillmann; appl. 1956).
DE 1 065 855 (G. Hillmann; appl. 1956).
US 2 803 654 (Baxter Labs.; 1957; prior. 1953).
US 2 889 363 (Baxter Labs.; 1959; appl. 1955).
US 2 889 364 (Baxter Labs.; 1959; appl. 1957).

Formulation(s):   tabl. 0.025 mg, 0.05 mg, 0.075 mg, 0.1 mg, 0.125 mg, 0.150 mg, 0.175 mg, 0.2 mg, 0.3 mg
(as sodium salt)

Trade Name(s):

D:	Eferox (Hexal)		Lévothyrox (Lipha Santé)		Tyronamin (Takeda; as
	Euthyrox (Merck)		L-thyroxine Roche (Roche)		sodium salt)
	Thevier (Glaxo Wellcome)	GB:	Eltroxin (Goldshield)	USA:	Levothroid (Forest; as
	L-Thyroxin "Henning"	I:	Dermocinetic crema (Irbi)-		sodium salt)
	(Henning Berlin)		comb.		Levoxyl (Jones Medical
	numerous combination		Somatoline emuls. (Manetti		Industries; as sodium salt)
	preparations		Roberts)-comb.		Synthroid (Knoll; as
F:	Euthyral (Lipha Santé)-	J:	Thyradin-S (Teikoku Zoki)		sodium salt)
	comb.				

Lidocaine
(Lignocaine)

ATC: C01BB01; C05AD01; D04AB01;
        N01BB02; R02AD02; S01HA07;
        S02DA01
Use: local anesthetic, antiarrhythmic

RN: 137-58-6   MF: $C_{14}H_{22}N_2O$ MW: 234.34   EINECS: 205-302-8
LD$_{50}$:   20 mg/kg (M, i.v.); 220 mg/kg (M, p.o.);
        18 mg/kg (R, i.v.); 317 mg/kg (R, p.o.)
CN: 2-(diethylamino)-*N*-(2,6-dimethylphenyl)acetamide

monohydrochloride
RN: 73-78-9   MF: $C_{14}H_{22}N_2O \cdot HCl$ MW: 270.80   EINECS: 200-803-8
LD$_{50}$:   15 mg/kg (M, i.v.); 220 mg/kg (M, p.o.);
        21 mg/kg (R, i.v.)

chloroacetyl chloride   2,6-dimethyl-aniline   α-chloro-2,6-dimethyl-acetanilide   diethyl-amine   Lidocaine

Reference(s):
US 2 441 498 (AB Astra; 1948; S-prior. 1943).
DE 968 561 (AB Astra; appl. 1944; S-prior. 1943).

hydrochloride monohydrate:
US 2 797 241 (C.L.M. Brown, A. Poole; 1957; GB-prior. 1953).

Formulation(s):   amp. 0.5 %, 1 %, 2 %, 25 mg; gel 2 %; ointment 5 %; sol. 4 % (as hydrochloride)

Trade Name(s):

D: Gelicain (curasan)
      Heweneural (Hevert)
      Licain (curasan)
      Xylocain (Astra)
      Xyloneural (Strathmann)
      numerous generics
F: Xylocaine (Astra)
      Xylocard (Astra)
      numerous combination
      preparations
GB: Xylocaine (Astra)
      Xylocard (Astra)
      numerous combination
      preparations

I: Luan (Molteni; as
      hydrochloride)
      Neolidocaton (Dentalica)-
      comb.
      Odontalg (Giovanardi; as
      hydrochloride)
      Ortodermina (Salus
      Research; as
      hydrochloride)
      Xylocaina (Astra-Simes; as
      hydrochloride)
      Xylocaina epinefrina
      (Astra-Simes)-comb.
      Xylocaina iniett. (Astra-
      Simes; as hydrochloride)

      Xylocaina Spray (Astra-
      Simes; as hydrochloride)
      Xylonor (Ogna)-comb.
      combination preparations
J: Leostesin N (Showa)
      Xylocaine (Astra-Fujisawa)
USA: Anestacon (PolyMedica; as
      hydrochloride)
      EMLA (Astra)
      Lidocaine (Roxane)
      Lidocaine Hydrochloride
      (Elkins-Sinn)
      Xylocaine (Astra)
      Xylocaine (Astra; as
      hydrochloride)

Lidoflazine

ATC: C08EX01
Use: coronary vasodilator

RN: 3416-26-0   MF: $C_{30}H_{35}F_2N_3O$ MW: 491.63   EINECS: 222-312-8
LD_{50}:   40 mg/kg (M, i.v.); >2 g/kg (M, p.o.);
       >3.2 g/kg (R, p.o.)
CN: 4-[4,4-bis(4-fluorophenyl)butyl]-*N*-(2,6-dimethylphenyl)-1-piperazineacetamide

chloroacetyl   2,6-dimethyl-   α-chloro-2,6-
chloride   aniline   dimethyl-
                                       acetanilide (I)

1-[4,4-bis(4-fluoro-   Lidoflazine
phenyl)butyl]piperazine

Reference(s):
US 3 267 104 (Janssen; 16.8.1966; prior. 9.6.1964, 14.5.1965).
GB 1 055 100 (Janssen; appl. 8.6.1965; USA-prior. 9.6.1964, 14.5.1965).
NL-appl. 6 507 312 (Janssen; appl. 9.6.1965; USA-prior. 9.6.1964, 14.5.1965).

Formulation(s):   tabl. 60 mg

Lincomycin

ATC: J01FF02
Use: antibiotic

RN: 154-21-2   MF: $C_{18}H_{34}N_2O_6S$ MW: 406.54   EINECS: 205-824-6
LD$_{50}$: 13.9 g/kg (M, p.o.);
        1 g/kg (R, p.o.)
CN: (2S-*trans*)-methyl 6,8-dideoxy-6-[[(1-methyl-4-propyl-2-pyrrolidinyl)carbonyl]amino]-1-thio-D-*erythro*-
        α-D-*galacto*-octopyranoside

monohydrochloride
RN: 859-18-7   MF: $C_{18}H_{34}N_2O_6S \cdot HCl$ MW: 443.01   EINECS: 212-726-7

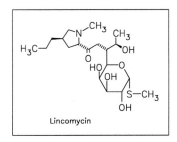

Lincomycin

From fermentation solutions of *Streptomyces lincolnensis*.

Reference(s):
US 3 086 912 (Upjohn; 23.4.1963; prior. 3.7.1961).
US 3 155 580 (Upjohn; 3.11.1964; prior. 30.8.1961).
US 4 091 204 (Upjohn; 23.5.1978; prior. 20.12.1974, 24.9.1976).

lincomycin derivatives:
US 3 380 992 (Upjohn; 30.4.1968; prior. 5.8.1964, 14.6.1965).

Formulation(s):   amp. 300 mg, 600 mg; cps. 250 mg, 500 mg; syrup 250 mg (as hydrochloride)

Liothyronine

ATC: H03AA02
Use: thyroid hormone

RN: 6893-02-3   MF: $C_{15}H_{12}I_3NO_4$ MW: 650.98   EINECS: 229-999-3
CN: O-(4-hydroxy-3-iodophenyl)-3,5-diiodo-L-tyrosine

monosodium salt
RN: 55-06-1   MF: $C_{15}H_{11}I_3NNaO_4$ MW: 672.96   EINECS: 200-223-5

L-3,5-diiodothyronine
(cf. levothyroxine synthesis)

Liothyronine

Reference(s):
US 2 823 164 (Nat. Res. Dev. Corp.; 1958; prior. 1953).
US 2 993 928 (Glaxo; 25.7.1961; GB-prior. 15.1.1957).
GB 671 070 (Glaxo; appl. 1949).

Formulation(s):   tabl. 0.005 mg, 0.02 mg, 0.025 mg, 0.05 mg, 0.1 mg; vial 0.01 mg/ml, 0.1 mg/ml

Trade Name(s):

D:	Thybon (Henning Berlin)	Euthyral (Lipha Santé)-	Cytomel (SmithKline
	Thyrotardin (Henning	comb.	Beecham; as sodium salt)
	Berlin)	GB: Tertroxin (Link)	Triostat (SmithKline
	Trijodthyronin (Berlin-	I: Titre (Teofarma)	Beecham; as sodium salt)
	Chemie)	J: Thyronamin (Takeda)	Triostat (Jones Medical
	numerous combination	Thyronine (Taisho)	Industries; as sodium salt)
	preparations	USA: Cytomel (Jones Medical	
F:	Cynomel (Marion Merrell)	Industries; as sodium salt)	

Lisinopril

(MK-521)

ATC: C09AA03
Use: angiotensin-converting enzyme
inhibitor (for use as antihypertensive
and in congestive heart failure, oral
absorption about 25 % (6-60 %), long
acting (plasma half-life 12.6 hrs),
once-daily dosing)

RN: 76547-98-3   MF: $C_{21}H_{31}N_3O_5$ MW: 405.50   EINECS: 278-488-1
CN: (S)-1-[N^2-(1-carboxy-3-phenylpropyl)-L-lysyl]-L-proline

dihydrate
RN: 83915-83-7   MF: $C_{21}H_{31}N_3O_5 \cdot 2H_2O$ MW: 441.53
LD_{50}:   >20 g/kg (M, p.o.);
      >20 g/kg (R, p.o.)

N^2-benzyloxycarbonyl-N^6-
tert-butoxycarbonyl-L-lysine

L-proline benzyl
ester hydrochloride

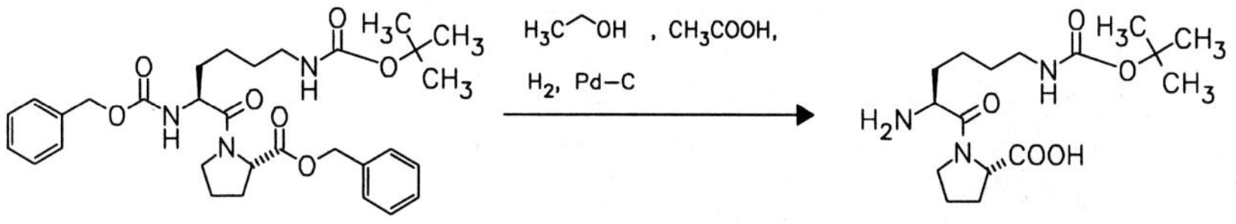

(I) → N-(N6-tert-butoxycarbonyl-L-lysyl)-L-proline (II)

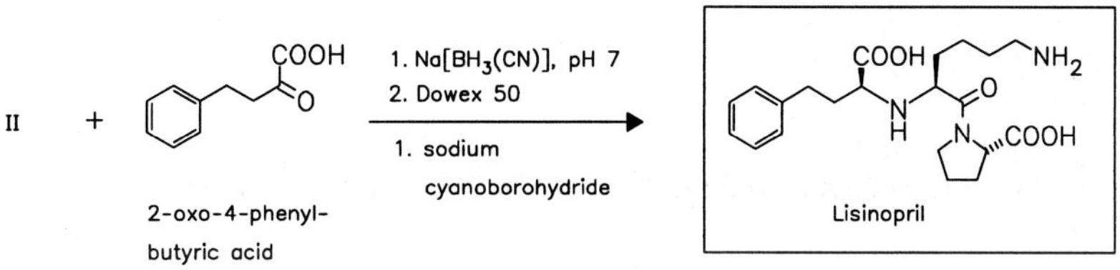

H3C–CH3, H3C–OH, CH3COOH, H2, Pd–C

II + 2-oxo-4-phenyl-butyric acid

1. Na[BH3(CN)], pH 7
2. Dowex 50
1. sodium cyanoborohydride

→ Lisinopril

(b)

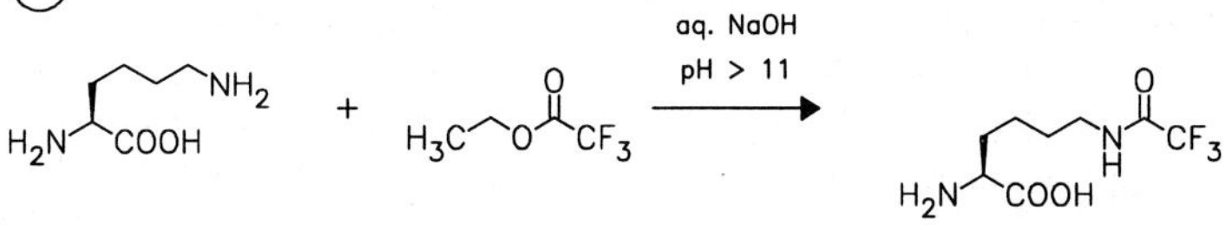

L-lysine + ethyl trifluoro-acetate

aq. NaOH pH > 11

→ N6-(trifluoroacetyl)-L-lysine (III)

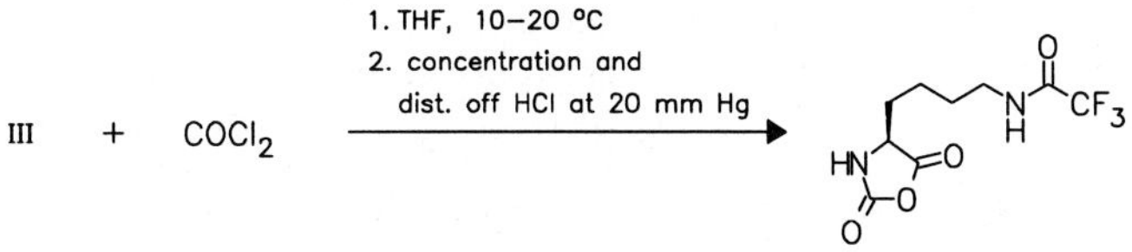

III + COCl2 (phosgene)

1. THF, 10–20 °C
2. concentration and dist. off HCl at 20 mm Hg

→ N6-(trifluoroacetyl)-N2-carboxy-L-lysine anhydride (IV)

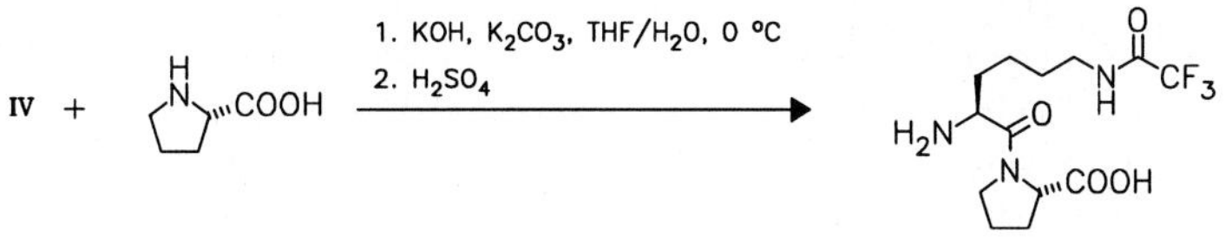

IV + L-proline

1. KOH, K2CO3, THF/H2O, 0 °C
2. H2SO4

→ N6-(trifluoroacetyl)-L-lysyl-L-proline (V)

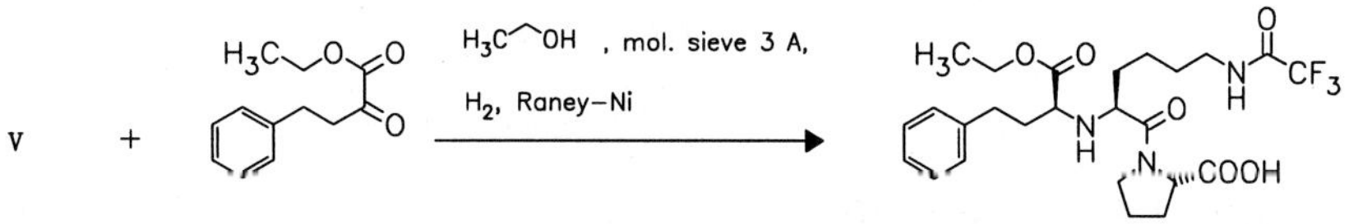

V + ethyl 2-oxo-4-phenylbutyrate

H3C–OH, mol. sieve 3 A, H2, Raney–Ni

→ (VI)

VI

1. aq. NaOH, pH 12.5, 40 °C
2. H+, [desalination on acid ion exchange resin]

→ Lisinopril

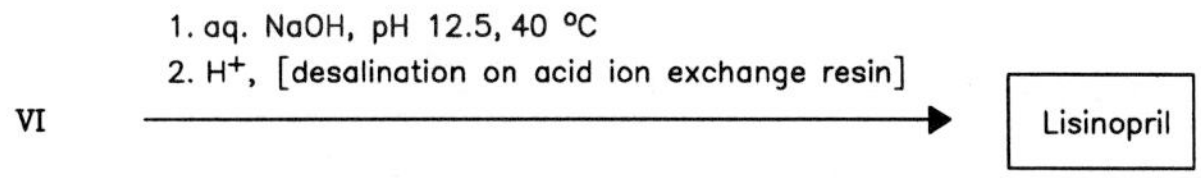

Reference(s):
Patchett, A.A. et al.: Nature (London) (NATUAS) **288**, 280 (1980).
a   US 4 374 829 (Merck & Co.; 22.2.1983; prior. 11.12.1978, 7.5.1979, 9.10.1979, 17.2.1981).
   US 4 472 380 (Merck & Co.; 18.9.1984; prior. 11.12.1978, 7.5.1979, 9.10.1978, 17.2.1981, 27.9.1982).
   EP 12 401 (Merck & Co.; appl. 10.12.1979; USA-prior. 11.12.1978).
   Wu, M.T. et al.: J. Pharm. Sci. (JPMSAE) **74**, 352 (1985).
b   Blacklock, T.J. et al.: J. Org. Chem. (JOCEAH) **53**, 836 (1988).
   EP 168 769 (Merck & Co.; appl. 11.7.1985; USA-prior. 16.7.1984).

alternative processes:
EP 79 521 (Merck & Co.; appl. 3.11.1982; USA-prior. 9.11.1981, 9.8.1982).
EP 336 368 (Kanegafuchi; appl. 4.4.1989; J-prior. 4.4.1988).

synthesis of N^6-(trifluoroacetyl)-L-lysine:
EP 279 716 (Rhône-Poulenc; appl. 18.1.1988; F-prior. 26.1.1987).

N^6-(trifluoroacetyl)-L-lysyl-L-proline, aromatic sulfonic acid salts:
US 4 720 554 (Ajinomoto; 19.1.1988; J-prior. 6.12.1985).
US 4 786 737 (Ajinomoto; 22.11.1988; J-prior. 6.12.1985).
EP 293 244 (Hamari Chemicals; appl. 27.5.1988; J-prior. 29.5.1987).

purification of N^6-(trifluoroacetyl)-L-lysyl-L-proline:
US 4 935 526 (Rhône-Poulenc; 19.6.1990; F-prior. 6.4.1988).
EP 340 056 (Rhône-Poulenc; appl. 31.3.1989; F-prior. 6.4.1988).

medical use in congestive heart failure:
EP 241 201 (Merck & Co.; appl. 31.3.1987; USA-prior. 7.4.1986).

combination with calcium antagonistic dihydropyridines:
DOS 3 437 917 (Bayer; appl. 17.10.1984).

Formulation(s):   tabl. 2.5 mg, 5mg, 10mg, 20 mg, 40 mg (USA); (as dihydrate) comb. with
            hydrochlorothiazide: tabl. 20 mg lisinopril with 12.5 mg or 25 mg hydrochlorothiazide

Trade Name(s):

D:	Acerbon (Zeneca; 1990)	I:	Alapril (Sigma-Tau)		Prinzide (Merck & Co.)-
	Acercomp (Zeneca)-comb.		Prinivil (Du Pont)		comb. with
	Coric (Du Pont; 1990)		Zestril (Zeneca)		hydrochlorothiazide
F:	Prinivil (Du Pont)	J:	Longes (Shionogi)		Zestoretic (Stuart)-comb.
	Prinzide (Du Pont)-comb.		Zestril (Zeneca-Sumitomo;		with hydrochlorothiazide
	Zestoretic (Zeneca Pharma)		ICI)		Zestril (Stuart; Zeneca;
	Zestril (Zeneca Pharma)	USA:	Prinivil (Merck & Co.;		1988)
GB:	Carace (Du Pont)		1988)		
	Zestril (Zeneca; 1988)				

Lobeline

ATC: N06
Use: respiratory analeptic, nicotine
     withdrawl agent

RN: 90-69-7   MF: $C_{22}H_{27}NO_2$ MW: 337.46   EINECS: 202-012-3
LD_{50}:   6300 µg/kg (M, i.v.)
CN: [2S-[2α,6α(R*)]]-2-[6-(2-hydroxy-2-phenylethyl)-1-methyl-2-piperidinyl]-1-phenylethanone

hydrochloride
RN: 134-63-4   MF: $C_{22}H_{27}NO_2 \cdot HCl$ MW: 373.92   EINECS: 205-150-2
LD_{50}:   7800 µg/kg (M, i.v.)
sulfate (2:1)
RN: 134-64-5   MF: $C_{22}H_{27}NO_2 \cdot 1/2H_2O_4S$ MW: 773.00   EINECS: 205-151-8
LD_{50}:   55.3 mg/kg (M, i.p.)

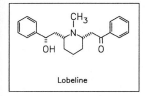

Lobeline

From *Lobelia inflata* L. by extraction of the slightly acidic extract of the drug with chloroform and subsequent purification.

Reference(s):
Wieland, H.: Ber. Dtsch. Chem. Ges. (BDCGAS) **54**, 1784 (1921).
Wieland, H.; Dragendorff, O.: Justus Liebigs Ann. Chem. (JLACBF) **473**, 83 (1929).

syntheses:
Wieland, H.; Drishaus, J.: Justus Liebigs Ann. Chem. (JLACBF) **473**, 102 (1929).
Scheuing, G.; Winterhalder, L.: Justus Liebigs Ann. Chem. (JLACBF) **473**, 126 (1929).

Formulation(s):   amp. 3 mg, 10 mg; tabl. 2 mg (as sulfate)

Trade Name(s):
D:	Citotal (Müller/		Unilobin (Rhône-Poulenc);
	Göppingen)-comb.; wfm		wfm
	Stenopressin (Efeka)-	F:	Lobatox (Sobio); wfm
	comb.; wfm	J:	Atmulatin (Dainippon)

Lobenzarit

ATC: M01
Use: anti-inflammatory

RN: 63329-53-3   MF: $C_{14}H_{10}ClNO_4$ MW: 291.69
CN: 2-[(2-carboxyphenyl)amino]-4-chlorobenzoic acid

sodium salt
RN: 64808-48-6   MF: $C_{14}H_8NNa_2O_4$ MW: 300.20

COOH   COOH K_2CO_3, Cu, I_2 ⟶   COOH COOH

Cl   NH₂   potassium carbonate,   H
                                  copper, iodine   N

Cl   Cl

2,4-dichloro-   anthranilic   Lobenzarit
benzoic acid   acid

Reference(s):
US 4 092 426 (Chugai; 30.5.1978; J-prior. 12.4.1976).
BE 842 832 (Chugai; appl. 11.6.1976; J-prior. 11.6.1975).
DE 2 526 092 (Chugai; prior. 11.6.1975).

Formulation(s):   tabl. 40 mg, 80 mg (as sodium salt)

Trade Name(s):
J: Carfenil (Chugai; 1986)

Lofepramine
(Lopramine)

ATC block

ATC: N06AA07
Use: antidepressant

RN: 23047-25-8   MF: $C_{26}H_{27}ClN_2O$ MW: 418.97   EINECS: 245-396-8
CN: 1-(4-chlorophenyl)-2-[[3-(10,11-dihydro-5H-dibenz[b,f]azepin-5-yl)propyl]methylamino]ethanone

monohydrochloride
RN: 26786-32-3   MF: $C_{26}H_{27}ClN_2O \cdot HCl$ MW: 455.43   EINECS: 248-002-2
LD$_{50}$:   >5 g/kg (M, p.o.);
      >5 g/kg (R, p.o.)

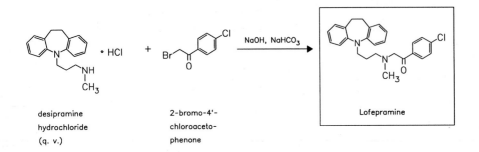

desipramine   2-bromo-4'-   Lofepramine
hydrochloride   chloroaceto-
(q. v.)   phenone

Reference(s):
GB 1 177 525 (Leo; appl. 13.4.1967; valid from 2.4.1968).
DOS 1 770 153 (Leo; appl. 8.4.1968; GB-prior. 13.4.1967).
GB 1 497 306 (Leo; appl. 3.7.1975; valid from 30.6.1976).
DOS 2 628 558 (Leo; appl. 25.6.1976; GB-prior. 3.7.1975).
US 3 637 660 (Leo; 25.1.1972; appl. 8.4.1968; GB-prior. 13.4.1967).

medical use as antidepressant:
GB 1 498 857 (Leo; appl. 3.7.1975; valid from 30.6.1976).
US 4 061 747 (Leo; 6.12.1977; GB-prior. 3.7.1975).
Eriksoo, E.; Rohte, O.: Arzneim.-Forsch. (ARZNAD) **20**, 1561 (1970).

Formulation(s):   f. c. tabl. 10 mg, 25 mg, 35 mg, 70 mg (as hydrochloride)

Trade Name(s):
D: Gamonil (Merck; 1977)   GB: Gamanil (Merck; 1983)   J: Amplit (Daiichi; 1981)

Lofexidine
(MDL-14042; BA-168; RMI-14042)

ATC: C02
Use: antihypertensive, α_2-agonist, relief of symptoms of opiate withdrawal

RN: 31036-80-3   MF: $C_{11}H_{12}Cl_2N_2O$ MW: 259.14
LD$_{50}$:   13 mg/kg (M, i.v.); 100 mg/kg (M, p.o.)
      13 mg/kg (R, i.v.); 100 mg/kg (R, p.o.)
CN: 2-[1-(2,6-dichlorophenoxy)ethyl]-4,5-dihydro-1H-imidazole

hydrochloride
RN: 21498-08-8   MF: $C_{11}H_{12}Cl_2N_2O \cdot HCl$ MW: 295.60

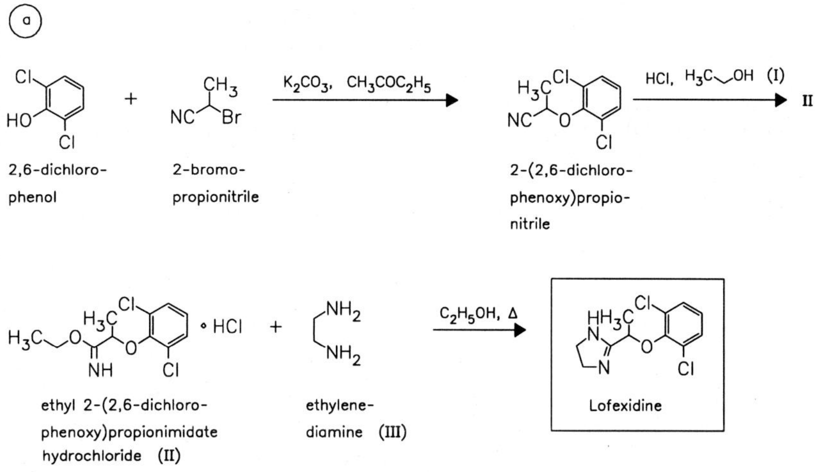
a
2,6-dichloro-
phenol
2-bromo-
propionitrile
K_2CO_3, $CH_3COC_2H_5$
2-(2,6-dichloro-
phenoxy)propio-
nitrile
HCl, $H_3C\smile OH$ (I)
II
ethyl 2-(2,6-dichloro-
phenoxy)propionimidate
hydrochloride (II)
ethylene-
diamine (III)
C_2H_5OH, Δ
Lofexidine

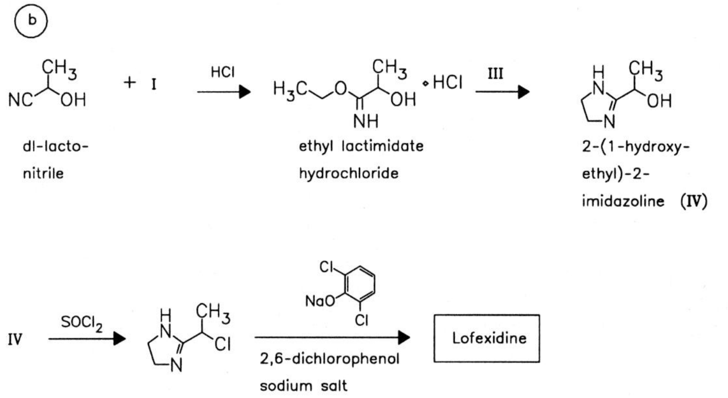
b
dl-lacto-
nitrile
+ I
HCl
ethyl lactimidate
hydrochloride
III
2-(1-hydroxy-
ethyl)-2-
imidazoline (IV)
IV
$SOCl_2$
2,6-dichlorophenol
sodium salt
Lofexidine

Reference(s):

a US 3 966 757 (Nattermann GmbH; 29.6.1976; D-prior. 23.2.1967).
DOS 1 695 555 (Nordmark-Werke; appl. 23.2.1967).
GB 1 181 356 (Nordmark-Werke; valid from 23.2.1968; D-prior. 23.2.1967).

b DE 1 935 479 (Nordmark-Werke; appl. 12.7.1967; D-prior. 12.7.1967). addition to DOS 1 695 555.

synthesis of 2-(1-chloroethyl)-2-imidazoline:
Klarer, W.; Urech, E.: Helv. Chim. Acta (HCACAV) **27**, 1762 (1944).

use as insecticides:
DE 2 818 367 (Ciba-Geigy; appl. 26.4.1978; CH-prior. 29.4.1977).

Formulation(s): tabl. 0.2 mg (as hydrochloride)

Trade Name(s):

D:	Lofetensin (Nattermann); wfm	GB:	Britlofex (Britannia Pharm.)	BritLoflex (Britannia)

Lomefloxacin
(NY-198)

ATC: J01MA07; S01AX17
Use: quinolone antibacterial, gyrase inhibitor

RN: 98079-51-7   MF: $C_{17}H_{19}F_2N_3O_3$ MW: 351.35
LD$_{50}$:   246 mg/kg (M, i.v.); >4 g/kg (M, p.o.);
         3800 mg/kg (R, p.o.)
CN: (±)-1-ethyl-6,8-difluoro-1,4-dihydro-7-(3-methyl-1-piperazinyl)-4-oxo-3-quinolinecarboxylic acid

monohydrochloride
RN: 98079-52-8   MF: $C_{17}H_{19}F_2N_3O_3 \cdot HCl$ MW: 387.81
LD$_{50}$:   253 mg/kg (M, i.v.); 1608 mg/kg (M, p.o.);
         328 mg/kg (R, i.v.); 1556 mg/kg (R, p.o.)

2,3,4-trifluoro-aniline

diethyl ethoxy-methylenemalonate

3-ethoxycarbonyl-4-hydroxy-6,7,8-trifluoroquinoline

(I)

1-ethyl-6,7,8-trifluoro-1,4-dihydro-4-oxo-3-quinolinecarboxylic acid

2-methyl-piperazine

Lomefloxacin

Reference(s):
EP 140 116 (Hokuriku; appl. 15.9.1984; J-prior. 19.9.1983, 12.3.1984, 18.6.1984).
DE 3 433 924 (Hokuriku; appl. 15.9.1984; J-prior. 19.9.1983, 12.3.1984, 18.6.1984).
US 4 528 287 (Hokuriku; 9.7.1985; appl. 17.9.1984; J-prior. 19.9.1983, 12.3.1984, 18.6.1984).

isotonic solution for i.v. administration or as ophthalmic or nasal solution:
US 4 780 465 (Hokuriku; 25.10.1988; appl. 20.5.1987; J-prior. 14.1.1987).
DOS 3 715 918 (Hokuriku; appl. 13.5.1987; J-prior. 14.1.1987).

synthesis of 1-ethyl-6,7,8-trifluoro-1,4-dihydro-4-oxo-3-quinolinecarboxylic acid:
DE 3 031 767 (Kyorin; appl. 22.8.1980; J-prior. 22.8.1979).

lyophilizate:
EP 322 892 (Kyorin; appl. 28.12.1988; J-prior. 28.12.1987).

Formulation(s):   eye drops 0.3 %; tabl. 200 mg, 400 mg (as hydrochloride)

Trade Name(s):
F: Logiflox (Monsanto)
I: Chimono (Lusofarmaco)
      Maxaquin (Schiapparelli Searle)

      Uniquin (Alfa Wassermann)
J: Bareon (Hokuriku; 1990)
      Lomefact (Shionogi; 1990)

USA: Maxaquin (Searle; Unimed)

Lomifylline

ATC: C04
Use: vasodilator (peripheral)

RN: 10226-54-7 MF: $C_{13}H_{18}N_4O_3$ MW: 278.31 EINECS: 233-547-0
CN: 3,7-dihydro-1,3-dimethyl-7-(5-oxohexyl)-1H-purine-2,6-dione

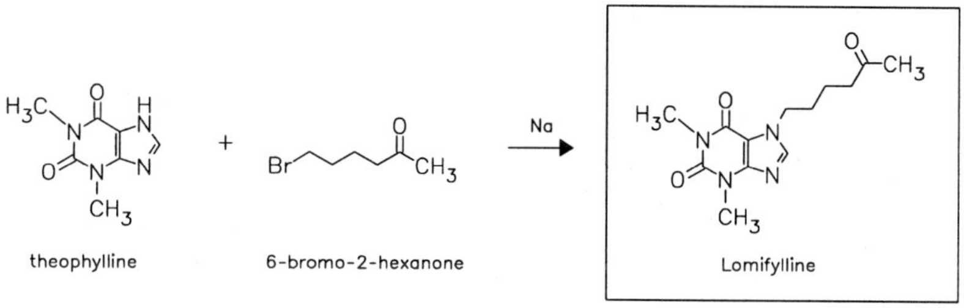

theophylline   6-bromo-2-hexanone   Lomifylline

Reference(s):
US 3 422 107 (Chemische Werke Albert; 14.1.1969; D-prior. 5.9.1964, 2.7.1965, 10.7.1965, 24.7.1965).
DE 1 233 405 (Chemische Werke Albert; appl. 5.9.1964).

alternative syntheses:
DOS 2 302 772 (Chemische Werke Albert; appl. 20.1.1973).
DOS 2 330 741 (Chemische Werke Albert; appl. 16.6.1973).

use as dissolving intermediary:
DE 1 250 968 (Chemische Werke Albert; appl. 24.7.1965).

oral pharmaceutical form:
DOS 2 520 978 (Hoechst; appl. 10.5.1975).

Formulation(s):   tabl. 80 mg

Trade Name(s):
F: Cervilane (Cassenne)-
      comb.

Lomustine
(CCNU)

ATC: L01AD02
Use: antineoplastic

RN: 13010-47-4 MF: $C_9H_{16}ClN_3O_2$ MW: 233.70 EINECS: 235-859-2
LD_{50}: 38 mg/kg (M, p.o.);
      70 mg/kg (R, p.o.)
CN: N-(2-chloroethyl)-N'-cyclohexyl-N-nitrosourea

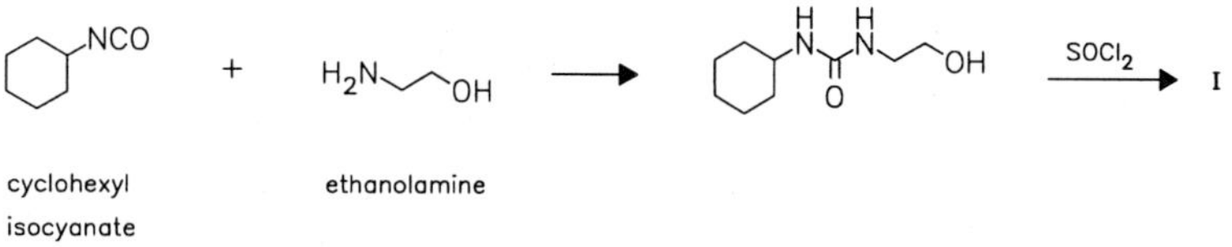

cyclohexyl   ethanolamine
isocyanate

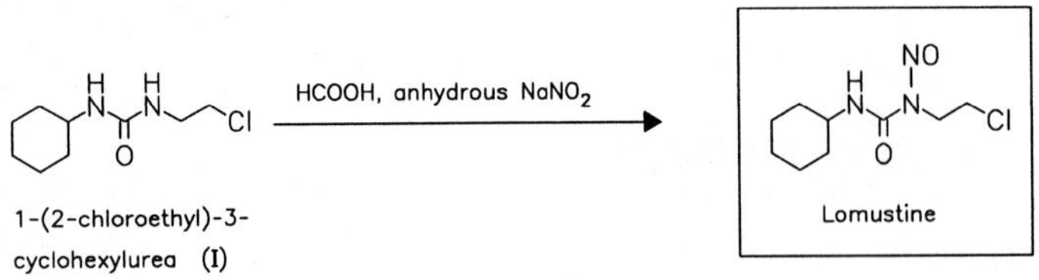
HCOOH, anhydrous $NaNO_2$
1-(2-chloroethyl)-3-cyclohexylurea (I)
Lomustine

Reference(s):
Johnston, T.P. et al.: J. Med. Chem. (JMCMAR) **9**, 892 (1966).

starting material:
Johnston, T.P. et al.: J. Med. Chem. (JMCMAR) **6**, 669 (1963).

Formulation(s): cps. 10 mg, 40 mg; tabl. 10 mg, 40 mg

Trade Name(s):

D:	Cecenu (medac)	I:	Belustine (Rhône-Poulenc Rorer)	USA:	CeeNU (Bristol-Myers Squibb)
F:	Bélustine (Roger Bellon)				

Lonazolac

ATC: M01AB09
Use: non-steroidal anti-inflammatory

RN: 53808-88-1 MF: $C_{17}H_{13}ClN_2O_2$ MW: 312.76 EINECS: 258-791-5
CN: 3-(4-chlorophenyl)-1-phenyl-1*H*-pyrazole-4-acetic acid

calcium salt
RN: 75821-71-5 MF: $C_{34}H_{24}CaCl_2N_4O_4$ MW: 663.57 EINECS: 278-322-8
LD_{50}: 670 mg/kg (M, p.o.);
845 mg/kg (R, p.o.);
790 mg/kg (g. p., p.o.);
650 mg/kg (rabbit, p.o.)

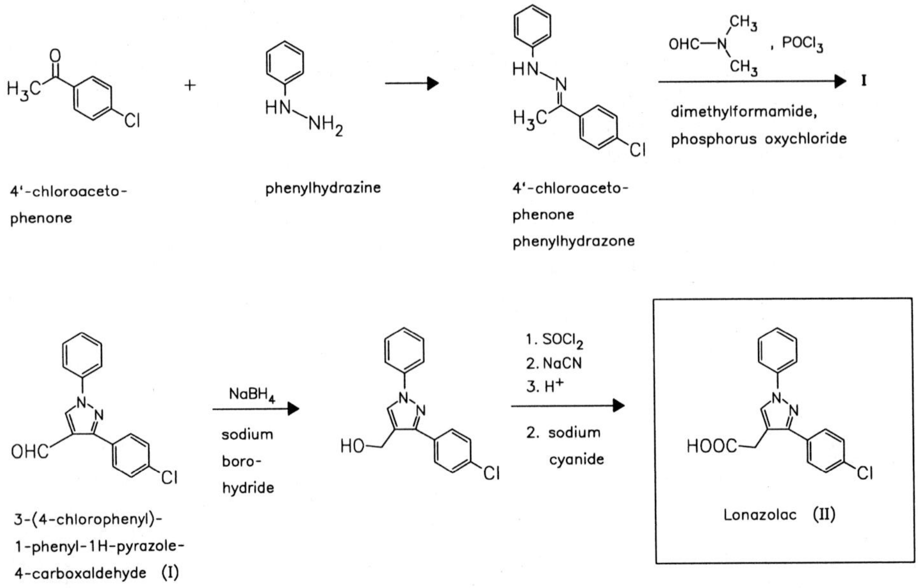
4'-chloroacetophenone
phenylhydrazine
4'-chloroacetophenone phenylhydrazone
, $POCl_3$
dimethylformamide, phosphorus oxychloride
$NaBH_4$
sodium borohydride
1. $SOCl_2$
2. NaCN
3. H^+
2. sodium cyanide
3-(4-chlorophenyl)-1-phenyl-1H-pyrazole-4-carboxaldehyde (I)
Lonazolac (II)

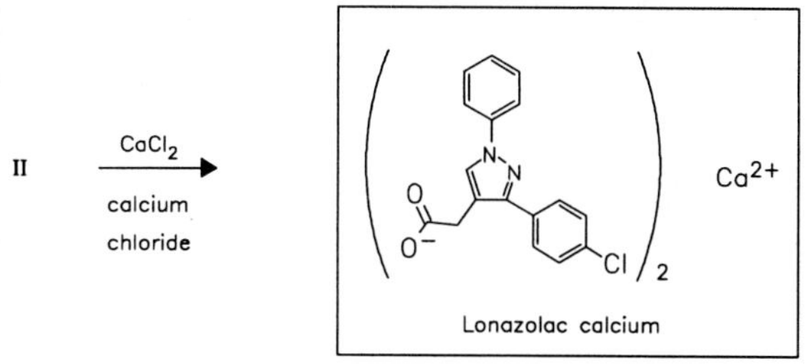

Lonazolac calcium

Reference(s):
DE 1 946 370 (Byk Gulden; appl. 12.9.1969).
US 4 325 962 (Byk Gulden; 20.4.1982; D-prior. 12.9.1969).
US 4 146 721 (Byk Gulden; 27.3.1979; D-prior. 12.9.1968).
Rainer, G. et al.: Arzneim.-Forsch. (ARZNAD) **31**, 649 (1981).

alternative synthesis of calcium salt:
EP 299 504 (Spofa; appl. 15.7.1988; CS-prior. 17.7.1987).

alternative synthesis of the free acid:
GB 1 373 212 (Wyeth; appl. 7.12.1970).

combination with analgesics:
DE 2 605 243 (Byk Gulden; appl. 11.2.1976; LUX-prior. 14.2.1975).

medical use for thrombocyte aggregation inhibition:
DE 3 444 633 (Byk Gulden; appl. 7.12.1984; CH-prior. 23.12.1983).

Formulation(s):   suppos. 400 mg; tabl. 200 mg, 300 mg

Trade Name(s):
D: Argun (Merckle)   arthro akut (Byk Gulden;
                                 Byk Tosse)

Lonidamine

ATC: L01XX07
Use: antineoplastic

RN: 50264-69-2   MF: $C_{15}H_{10}Cl_2N_2O_2$ MW: 321.16   EINECS: 256-510-0
LD$_{50}$:   435 mg/kg (M, i.p.); 900 mg/kg (M, p.o.);
        525 mg/kg (R, i.p.); 1700 mg/kg (R, p.o.)
CN: 1-[(2,4-dichlorophenyl)methyl]-1*H*-indazole-3-carboxylic acid

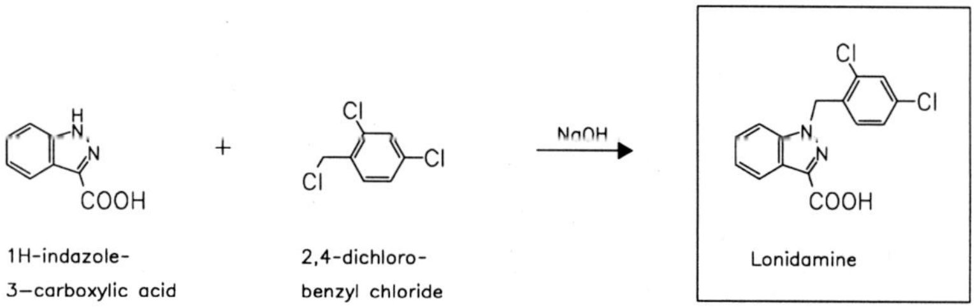

1H–indazole–   2,4–dichloro–   Lonidamine
3–carboxylic acid   benzyl chloride

Reference(s):
DE 2 310 031 (Aziende chimiche Riunite; appl. 28.2.1973; I-prior. 29.2.1972).
US 3 895 026 (Aziende chimiche Riunite; 15.7.1975; I-prior. 29.2.1972).
Corsi, G.; Palazzo, G.: J. Med. Chem. (JMCMAR) **19**, 778 (1976).

alternative synthesis:
ES 545 644 (Lab. Ausonia; appl. 29.7.1985).

medical use for treatment of cancer:
BE 894 111 (Angelini Inst.; appl. 13.8.1982; I-prior. 17.8.1981).

Formulation(s):   tabl. 150 mg

Trade Name(s):
I: Doridamina (Angelini;
      1987)

Loperamide

ATC: A07DA03
Use: antidiarrheal

RN: 53179-11-6   MF: $C_{29}H_{33}ClN_2O_2$ MW: 477.05   EINECS: 258-416-5
LD$_{50}$:   105 mg/kg (M, p.o.);
      5.1 mg/kg (R, i.v.); 98 mg/kg (R, p.o.);
      2.8 mg/kg (dog, i.v.); 40 mg/kg (dog, p.o.)
CN: 4-(4-chlorophenyl)-4-hydroxy-*N,N*-dimethyl-α,α-diphenyl-1-piperidinebutanamide

monohydrochloride
RN: 34552-83-5   MF: $C_{29}H_{33}ClN_2O_2 \cdot HCl$ MW: 513.51   EINECS: 252-082-4
LD$_{50}$:   12.64 mg/kg (M, i.v.); 105 mg/kg (M, p.o.);
      7.49 mg/kg (R, i.v.); 185 mg/kg (R, p.o.)

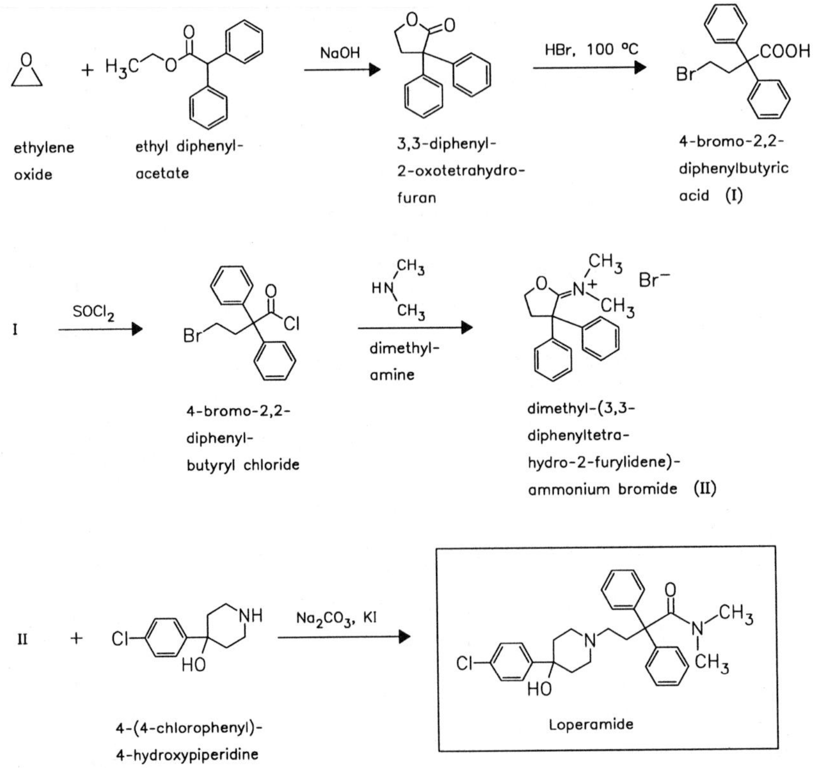

ethylene oxide   +   ethyl diphenyl-acetate   →(NaOH)   3,3-diphenyl-2-oxotetrahydro-furan   →(HBr, 100 °C)   4-bromo-2,2-diphenylbutyric acid (I)

I   →(SOCl$_2$)   4-bromo-2,2-diphenyl-butyryl chloride   + dimethyl-amine (HN(CH$_3$)CH$_3$)   →   dimethyl-(3,3-diphenyltetra-hydro-2-furylidene)-ammonium bromide (II)

II   +   4-(4-chlorophenyl)-4-hydroxypiperidine   →(Na$_2$CO$_3$, KI)   Loperamide

Reference(s):
Stokbroehx, R.A. et al.: J. Med. Chem. (JMCMAR) **16**, 782 (1973).
US 3 714 159 (Janssen; 30.1.1973; prior. 1.6.1970, 30.3.1971).
FR-appl. 2 100 711 (Janssen; appl. 28.5.1971; USA-prior. 1.6.1970, 30.3.1971).
US 3 884 916 (Janssen; 20.5.1975; prior. 1.6.1970, 30.3.1971, 7.12.1972).
DOS 2 126 559 (Janssen; appl. 28.5.1971; USA-prior. 1.6.1970, 30.3.1971).

Formulation(s): cps. 2 mg; sol. 0.2 mg; syrup 1 mg/5 ml; tabl. 2 mg (as hydrochloride)

Trade Name(s):

D:	Imodium (Janssen-Cilag; 1976)		Imodium (Janssen-Cilag; 1976)			Loperyl (SmithKline Beecham)
	Sanifug (Wolff)	GB:	Imodium (Janssen-Cilag; 1975)			Tebloc (Lafare)
	Santax (Asche)				J:	Lopemin (Dainippon; 1981)
	numerous generics		Lopergan (Norgine)			
F:	Altocel (Irex)		Novimode (Tillomed)		USA:	Imodium (Janssen; 1977)
	Arestal (Janssen-Cilag)	I:	Dissenten (SPA; 1979)			Imodium (McNeil; 1977)
	Diaretyl (RPR Cooper)		Imodium (Janssen; 1979)			
	Dyspagon (Pierre Fabre)		Lopemid (Gentili)			

Loprazolam

ATC: N05CD11
Use: tranquilizer, hypnotic

RN: 61197-73-7 MF: $C_{23}H_{21}ClN_6O_3$ MW: 464.91
LD_{50}: >1 g/kg (M, p.o.)
CN: (Z)-6-(2-chlorophenyl)-2,4-dihydro-2-[(4-methyl-1-piperazinyl)methylene]-8-nitro-1*H*-imidazo[1,2-*a*][1,4]benzodiazepin-1-one

mesylate
RN: 61197-93-1 MF: $C_{23}H_{21}ClN_6O_3 \cdot xCH_4O_3S$ MW: unspecified

| 2-amino-2'-chlorobenzophenone | glycine ethyl ester hydrochloride | 1,3-dihydro-5-(2-chlorophenyl)-2H-1,4-benzodiazepin-2-one | |

| (I) | glycine | |

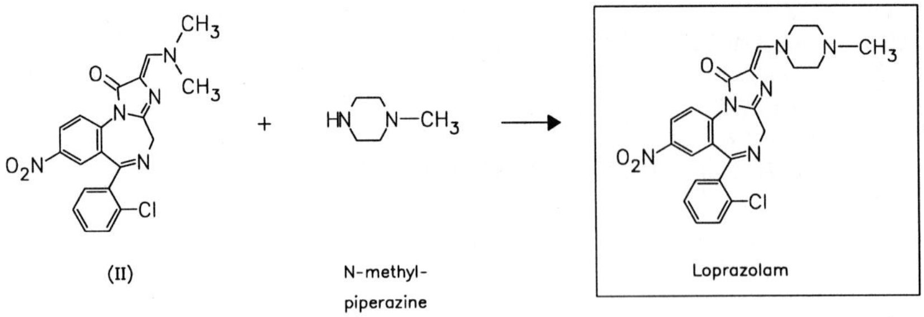

(II) + N-methyl-piperazine → Loprazolam

Reference(s):

DOS 2 605 652 (Roussel-Uclaf; appl. 12.2.1976; GB-prior. 4.11.1975, 15.2.1975).
US 4 044 142 (Roussel-Uclaf; 23.8.1977; GB-prior. 15.2.1975, 4.9.1975).

alternative synthesis:

DOS 3 211 243 (Roussel-Uclaf; appl. 26.3.1982; F-prior. 27.3.1981).

synthesis of 1,3-dihydro-7-nitro-5-(2-chlorophenyl)-2*H*-1,4-benzodiazepin-2-thione:

Sternbach, L.H. et al.: J. Med. Chem. (JMCMAR) **6**, 261 (1963).
Hester, J.B. et al.: J. Med. Chem. (JMCMAR) **14**, 1078 (1971).
DOS 2 164 777 (Upjohn; appl. 27.12.1971; USA-prior. 3.3.1971).
US 3 402 171 (Roche; 17.9.1968; CH-prior. 2.12.1960).

Formulation(s): tabl. 1 mg, 2 mg (as mesylate)

Trade Name(s):

D: Sonin (Lipha; 1987)
F: Havlane (Diamant; 1984)
GB: Dormonoct (Roussel; 1983); wfm
generics

Loratadine
(Sch-29851)

ATC: R06AX13
Use: antiallergic, non-sedating antihistaminic

RN: 79794-75-5   MF: C$_{22}$H$_{23}$ClN$_2$O$_2$ MW: 382.89
CN: 4-(8-chloro-5,6-dihydro-11*H*-benzo[5,6]cyclohepta[1,2-*b*]pyridin-11-ylidene)-1-piperidinecarboxylic acid ethyl ester

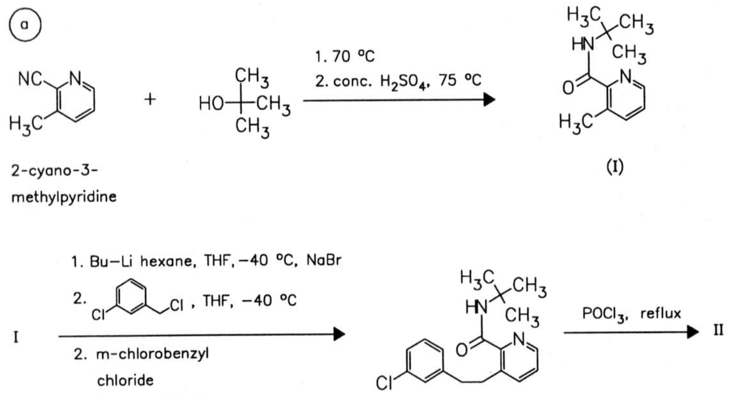

2-cyano-3-methylpyridine + HO-C(CH$_3$)$_3$ →   (I)

1. 70 °C
2. conc. H$_2$SO$_4$, 75 °C

I → II

1. Bu–Li hexane, THF, −40 °C, NaBr
2. m-chlorobenzyl chloride, THF, −40 °C

POCl$_3$, reflux

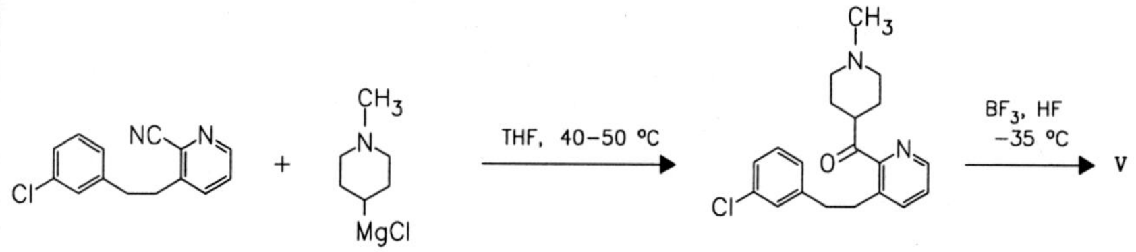

(II)

1-methyl-piperidin-4-ylmagnesium chloride (III)

(IV)

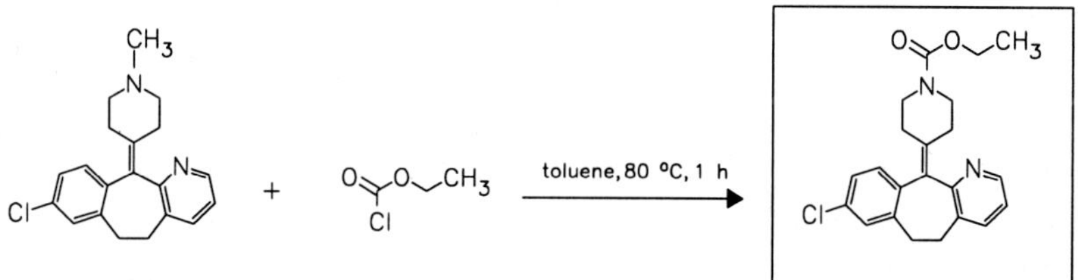

(V)

(VI)

Loratadine

(b)

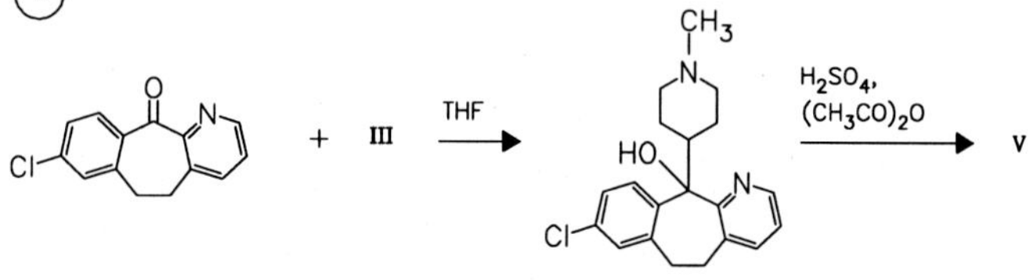

8-chloro-5,6-dihydro-11H-benzo[5,6]cyclohepta[1,2-b]pyridin-11-one (VII)

V + VI —toluene→ Loratadine

(c)

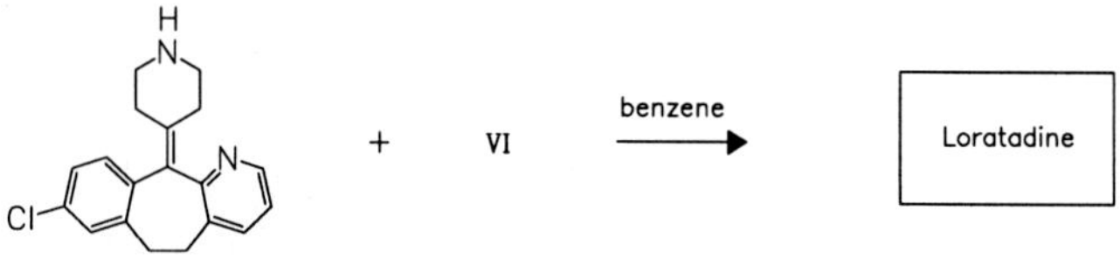

8-chloro-6,11-dihydro-11-(4-piperidylidene)-5H-benzo[5,6]cyclohepta[1,2-b]pyridine

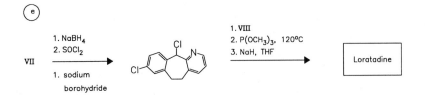

ⓓ

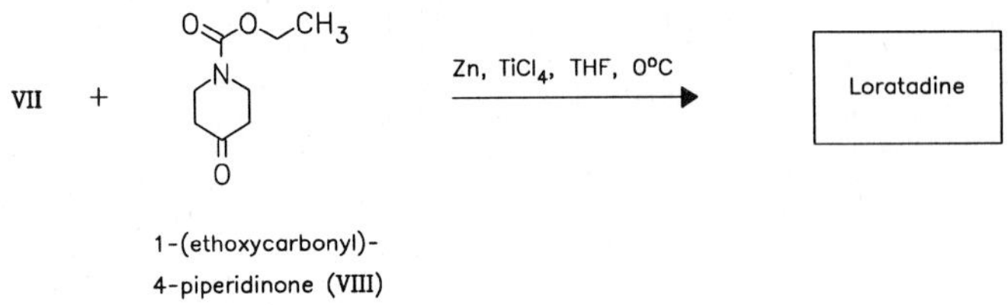

VII   +   1-(ethoxycarbonyl)-4-piperidinone (VIII)

→ Zn, TiCl₄, THF, 0°C → [Loratadine]

ⓔ

VII →
1. NaBH₄
2. SOCl₂
1. sodium borohydride

[structure with Cl]

1. VIII
2. P(OCH₃)₃, 120°C
3. NaH, THF
→ [Loratadine]

ⓕ

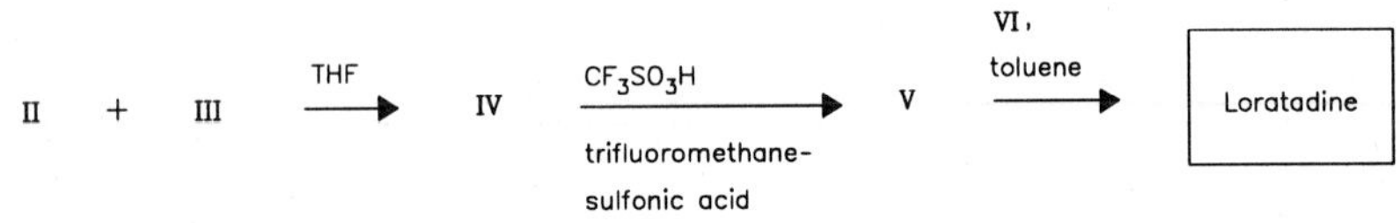

II   +   III —THF→ IV —CF₃SO₃H, trifluoromethane-sulfonic acid→ V —VI, toluene→ [Loratadine]

Reference(s):

US 4 454 143 (Schering Corp.; 12.6.1984; prior. 16.3.1981).
US 4 560 688 (Schering Corp.; 24.12.1985; prior. 16.3.1981).
US 4 282 233 (Schering Corp.; 4.8.1981; prior. 19.6.1980).
US 4 355 036 (Schering Corp.; 19.10.1982; prior. 19.6.1980, 16.3.1981).
EP 42 544 (Schering; USA-prior. 19.6.1980).

a   R Schumacher, D.P. et al.: J Org. Chem. (JOCEAH) **54**, 2242 (1989).
       US 4 731 447 (Schering Corp.; 15.3.1988; prior. 13.5.1985, 12.3.1986).
b   Villani, F.J. et al.: Arzneim.-Forsch. (ARZNAD) **36**, 1311 (1986).
       Villani, F.J. et al.: J. Med. Chem. (JMCMAR) **15** (7), 750 (1972).
c   US 4 355 036 (Schering; 19.10.1982; appl. 16.3.1981; USA-prior. 19.6.1980).
d   WO 9 840 376 (Cilag; appl. 6.3.1998; CH-prior. 11.3.1997).
       WO 9 838 166 (Jackson; appl. 26.2.1998).
e   WO 9 200 293 (Schering Corp.; appl. 21.6.1991; USA-prior. 22.6.1990).
f   WO 8 803 138 (Schering, appl. 29.10.1987; USA-prior. 31.10.1986).

Formulation(s):   eff. tabl. 10 mg; syrup 10 mg/spoon; tabl. 10 mg

Trade Name(s):

D:	Lisino (Essex Pharma; 1989)	Clarityne (Schering-Plough)	I:	Clarityn (Schering-Plough) Fristamin (Lifepharma)
F:	Clarinase Repetabs (Schering-Plough)-comb.	GB: Clarityn (Schering-Plough; 1989)	USA:	Claritin (Schering)

Lorazepam

ATC: N05BA06
Use: tranquilizer, anxiolytic

RN: 846-49-1 MF: $C_{15}H_{10}Cl_2N_2O_2$ MW: 321.16 EINECS: 212-687-6
LD_{50}: 1850 mg/kg (M, p.o.);
4500 mg/kg (R, p.o.);
>2 g/kg (dog, p.o.)
CN: 7-chloro-5-(2-chlorophenyl)-1,3-dihydro-3-hydroxy-2H-1,4-benzodiazepin-2-one

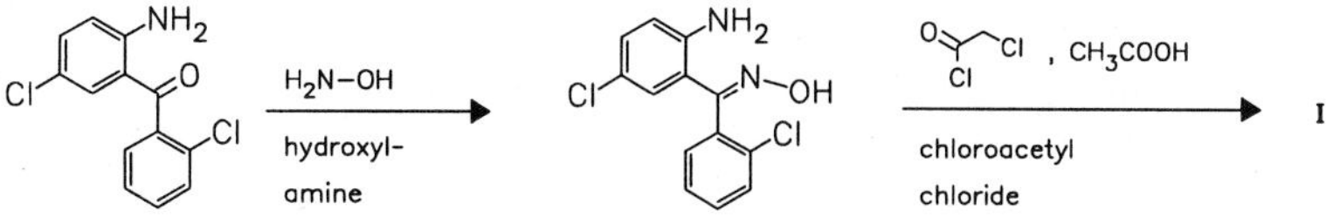

2-amino-2',5-
dichloro-
benzophenone

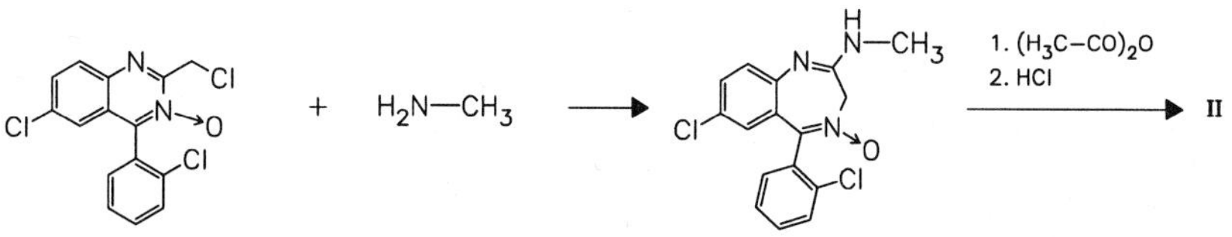

hydroxyl-
amine

chloroacetyl
chloride

6-chloro-2-
chloromethyl-4-
(2-chlorophenyl)-
quinazoline
3-oxide (I)

methyl-
amine

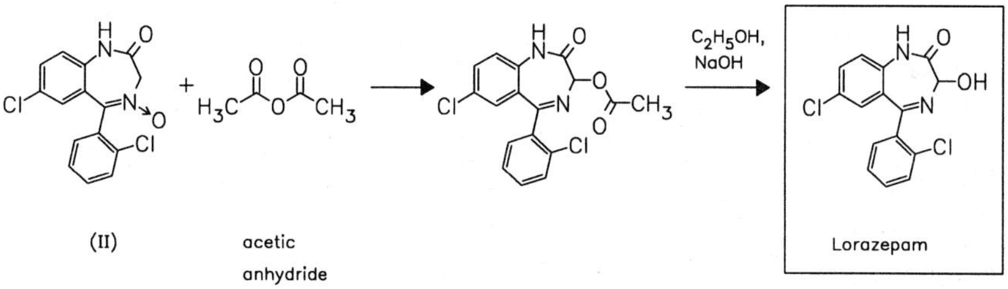

(II)

acetic
anhydride

Lorazepam

Reference(s):
US 3 296 249 (American Home Products; 3.1.1967; appl. 4.6.1963; prior. 29.8.1961, 5.3.1962).
US 3 176 009 (American Home Products; 30.3.1965; prior. 5.3.1962).
GB 1 022 642 (American Home; appl. 28.8.1962; USA-prior. 29.8.1961, 5.3.1962).
GB 1 022 644 (American Home; appl. 28.8.1962; USA-prior. 29.8.1961, 5.3.1962).
GB 1 022 645 (American Home; appl. 28.8.1962; USA-prior. 29.8.1961, 5.3.1962).
GB 1 057 492 (American Home; appl. 29.8.1968; addition to GB 1 022 642).
DE 1 445 412 (American Home; appl. 17.8.1962; USA-prior. 29.8.1961, 5.3.1962).
DE 1 645 904 (American Home; appl. 17.8.1962; USA-prior. 29.8.1961, 5.3.1962).
DE 1 795 509 (American Home; appl. 17.8.1962; USA-prior. 29.8.1961, 5.3.1962).

Formulation(s): amp. 2 mg/ml, 4 mg/ml; tabl. 0.5 mg, 1 mg, 2.5 mg

Trade Name(s):
D: Laubeel (Desitin) Punktyl (Krewel Somagerol (Brenner-Efeka)
Pro-Dorm (Synthelabo) Meuselbach) Tavor (Wyeth)

Lormetazepam

ATC: N05CD06
Use: tranquilizer, hypnotic

RN: 848-75-9   MF: $C_{16}H_{12}Cl_2N_2O_2$ MW: 335.19   EINECS: 212-700-5
LD_{50}: 1790 mg/kg (M, p.o.);
       >10 g/kg (R, p.o.)
CN: 7-chloro-5-(2-chlorophenyl)-1,3-dihydro-3-hydroxy-1-methyl-2H-1,4-benzodiazepin-2-one

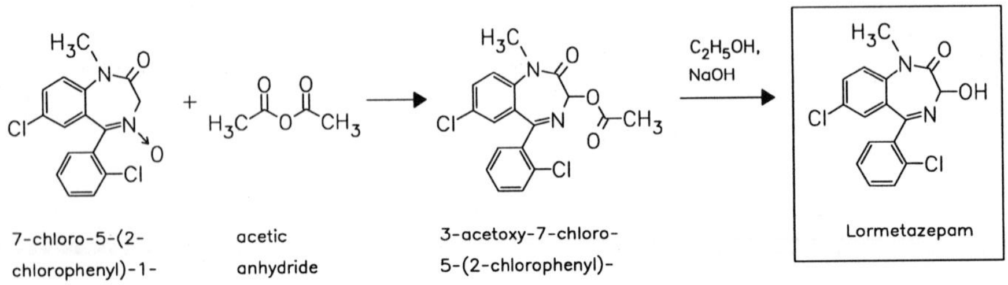

7-chloro-5-(2-
chlorophenyl)-1-
methyl-2-oxo-
1,3-dihydro-2H-
1,4-benzodiazepine
4-oxide

acetic
anhydride

3-acetoxy-7-chloro-
5-(2-chlorophenyl)-
1-methyl-1,3-dihydro-
2H-1,4-benzodiazepin-
2-one

Lormetazepam

Reference(s):
US 3 295 249 (American Home Products; 3.1.1967; prior. 4.6.1963, 5.3.1962, 29.8.1961).

Formulation(s):   cps. 0.5 mg, 1 mg, 2 mg; tabl. 0.5 mg, 1 mg, 2 mg

Trade Name(s):
D: Ergocalm (Brenner-Efeka)
       Loretam (Wyeth)
       Noctamid (Asche;
       Schering)

F: Noctamide (Schering)
GB: Loramet (Wyeth); wfm
       Noctamid (Schering); wfm
       generics

I: Minias (Farmades)
J: Evamyl (Schering)
       Loramet (Wyeth)
USA: Loramet (Wyeth); wfm

Lornoxicam

(Chlortenoxicam; Ro-13-9297)

ATC: M01AC05
Use: anti-inflammatory, nonsteroid
       antiphlogistic agent

RN: 70374-39-9   MF: $C_{13}H_{10}ClN_3O_4S_2$ MW: 371.83
CN: 6-Chloro-4-hydroxy-2-methyl-N-2-pyridinyl-2H-thieno[2,3-e]-1,2-thiazine-3-carboxamide 1,1-dioxide

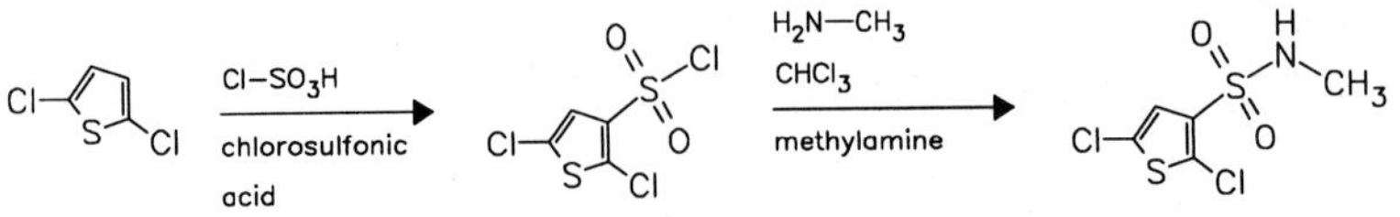

2,5-dichloro-
thiophene

chlorosulfonic
acid

2,5-dichloro-
3-thiophene-
sulfonylchloride

methylamine

(I)

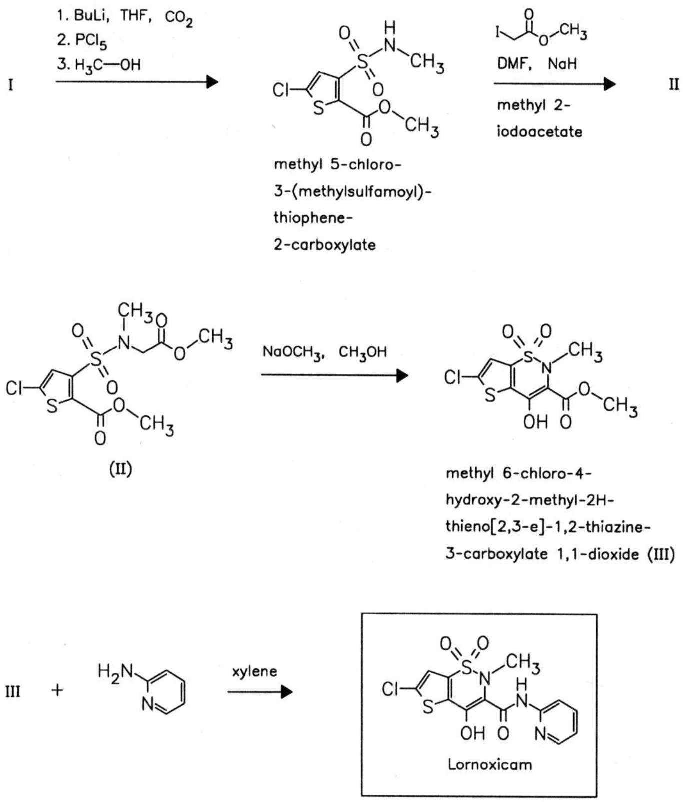

methyl 5-chloro-
3-(methylsulfamoyl)-
thiophene-
2-carboxylate

methyl 6-chloro-4-
hydroxy-2-methyl-2H-
thieno[2,3-e]-1,2-thiazine-
3-carboxylate 1,1-dioxide (III)

Lornoxicam

Reference(s):
DE 2 838 851 (Hoffmann-La Roche; appl. 6.9.1978; LU-prior. 6.9.1977).

process for the preparation of 5-chloro-3-chlorosulfonyl-2-thiophenecarboxylic esters:
EP 34 072 (CL Pharma A. G.; appl. 7.4.1989; A-prior. 2.5.1988).

pharmaceutical composition containing lornoxicam *and* disodium EDTA:
WO 9 809 654 (Nycomed Austria GmbH; appl. 1.9.1997; A-prior. 3.9.1996).

Formulation(s):   tabl. 2 mg, 8 mg

Trade Name(s):
D: Telos (Merckle; 1999)   GB: Xefo (Nycomed   I: Acabel (Formenti)
                                             Amersham)

Losartan potassium
(DuP-753; MK-954)

ATC: C02EX01; C09CA01
Use: antihypertensive, angiotensin II
        blocker

RN: 124750-99-8   MF: $C_{22}H_{22}ClKN_6O$ MW: 461.01
CN: 2-butyl-4-chloro-1-[[2'-(1H-tetrazol-5-yl)[1,1'-biphenyl]-4-yl]methyl]-1H-imidazole-5-methanol
       monopotassium salt

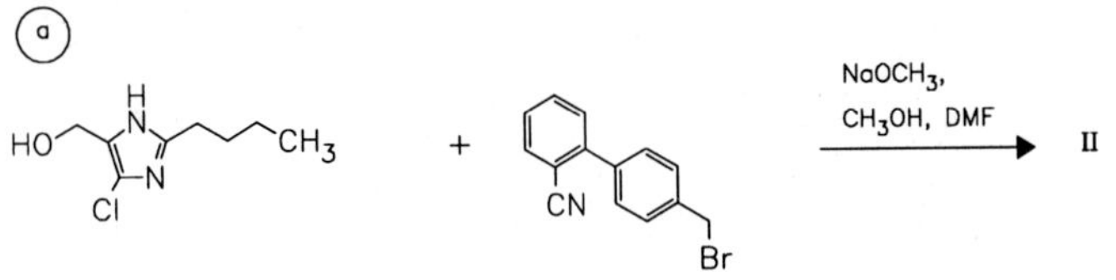

ⓐ

HO—[2-butyl-4-chloro-5-hydroxymethyl-imidazole structure] CH₃   +   [4'-bromomethyl-2-cyanobiphenyl structure with CN and Br]   → $\xrightarrow{\text{NaOCH}_3,\ \text{CH}_3\text{OH, DMF}}$ II

2-butyl-4-chloro-
5-hydroxymethyl-
imidazole

4'-bromomethyl-
2-cyanobiphenyl   (I)

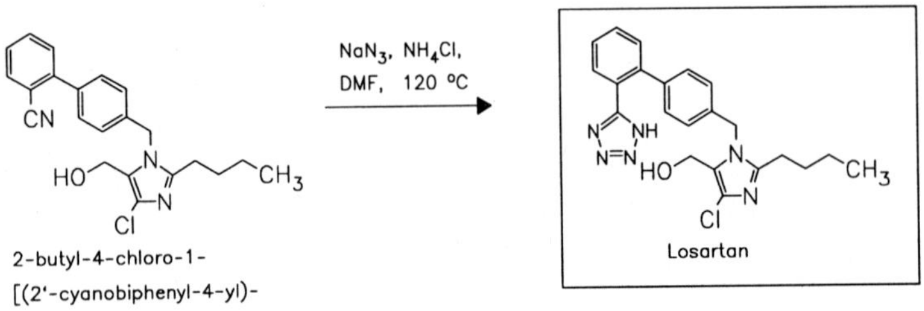

[structure of II precursor with CN, HO, CH₃, Cl] $\xrightarrow{\text{NaN}_3,\ \text{NH}_4\text{Cl},\ \text{DMF, 120 °C}}$ [Losartan structure with tetrazole, HO, CH₃, Cl]

2-butyl-4-chloro-1-
[(2'-cyanobiphenyl-4-yl)-
methyl]-5-(hydroxy-
methyl)imidazole   (II)

Losartan

preparation of I

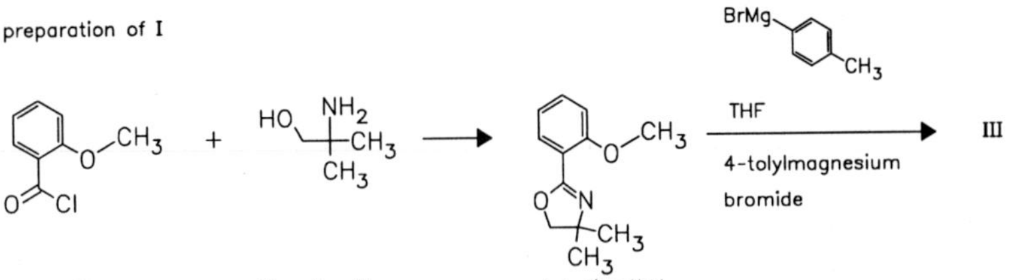

[2-methoxybenzoyl chloride structure]   +   HO—[2-amino-2-methyl-1-propanol structure, NH₂, CH₃, CH₃]   →   [4,4-dimethyl-2-(2-methoxyphenyl)-2-oxazoline structure, CH₃, CH₃] $\xrightarrow[\text{4-tolylmagnesium bromide}]{\text{BrMg}-\text{CH}_3,\ \text{THF}}$ III

2-methoxy-
benzoyl
chloride

2-amino-2-
methyl-1-
propanol

4,4-dimethyl-
2-(2-methoxy-
phenyl)-2-
oxazoline

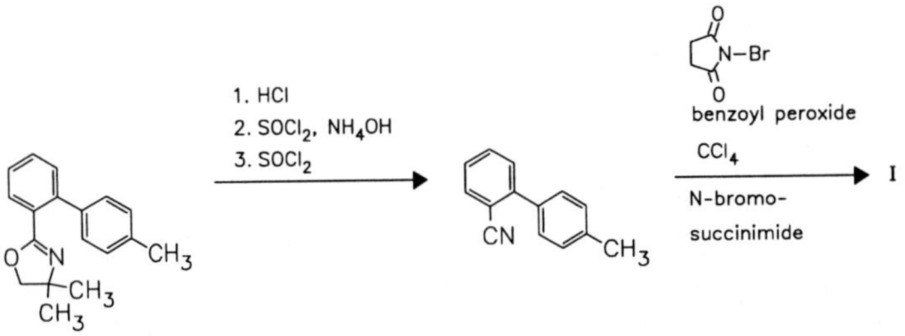

[2-(4'-methylbiphenyl-2-yl)-4,4-dimethyl-2-oxazoline structure, CH₃, CH₃, CH₃] $\xrightarrow[\text{3. SOCl}_2]{\begin{array}{l}\text{1. HCl}\\ \text{2. SOCl}_2,\ \text{NH}_4\text{OH}\end{array}}$ [4-methyl-2'-cyanobiphenyl structure, CN, CH₃] $\xrightarrow[\substack{\text{N-bromo-\\succinimide}}]{\substack{\text{N-Br succinimide}\\ \text{benzoyl peroxide}\\ \text{CCl}_4}}$ I

2-(4'-methylbiphenyl-
2-yl)-4,4-dimethyl-
2-oxazoline   (III)

4-methyl-2'-
cyanobiphenyl

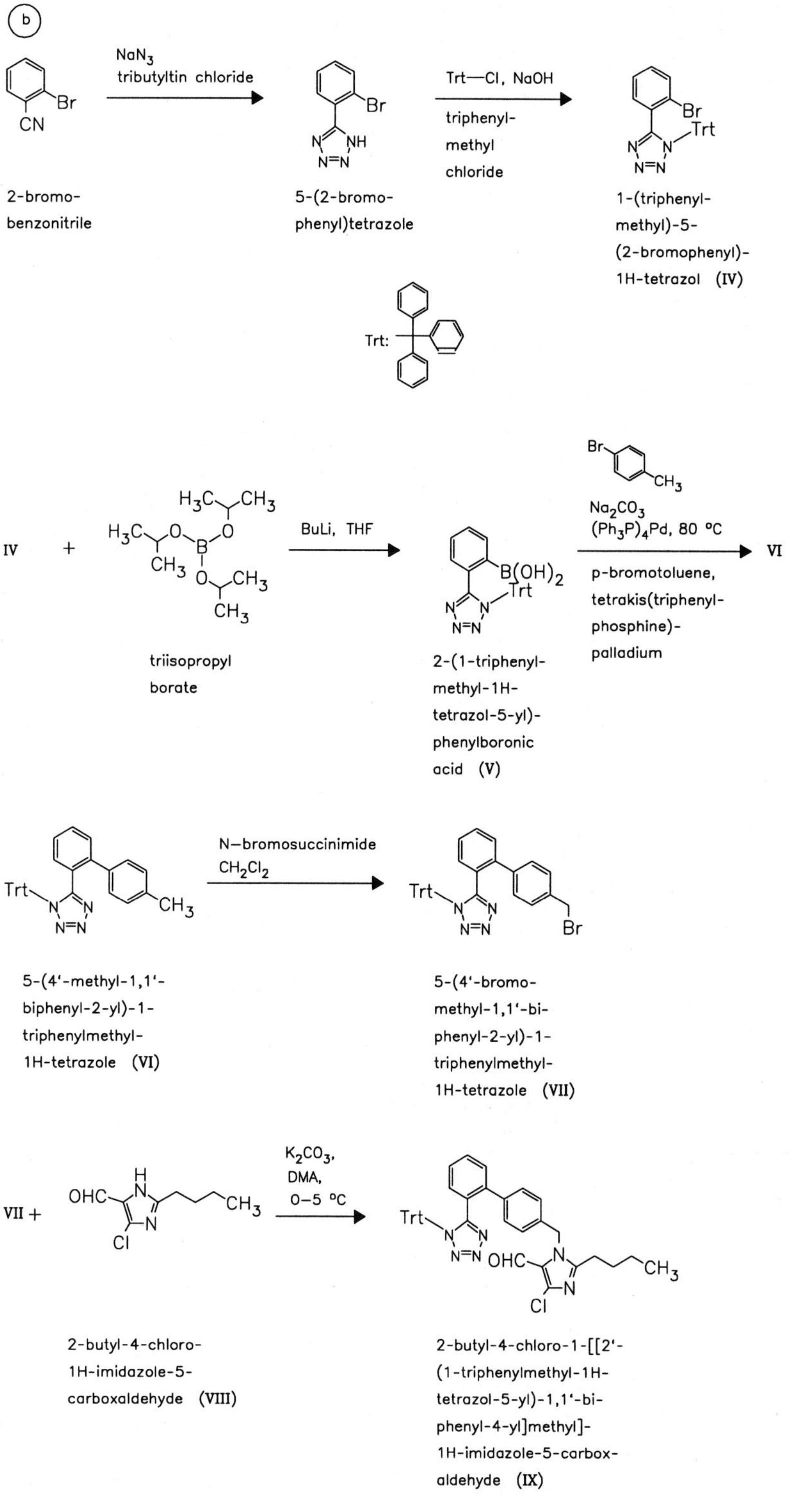

b

NaN₃
tributyltin chloride

2-bromo-
benzonitrile

5-(2-bromo-
phenyl)tetrazole

Trt—Cl, NaOH

triphenyl-
methyl
chloride

1-(triphenyl-
methyl)-5-
(2-bromophenyl)-
1H-tetrazol (IV)

Trt:

IV +

triisopropyl
borate

BuLi, THF

2-(1-triphenyl-
methyl-1H-
tetrazol-5-yl)-
phenylboronic
acid (V)

Na₂CO₃
(Ph₃P)₄Pd, 80 °C

p-bromotoluene,
tetrakis(triphenyl-
phosphine)-
palladium

VI

5-(4'-methyl-1,1'-
biphenyl-2-yl)-1-
triphenylmethyl-
1H-tetrazole (VI)

N—bromosuccinimide
CH₂Cl₂

5-(4'-bromo-
methyl-1,1'-bi-
phenyl-2-yl)-1-
triphenylmethyl-
1H-tetrazole (VII)

VII +

2-butyl-4-chloro-
1H-imidazole-5-
carboxaldehyde (VIII)

K₂CO₃,
DMA,
0–5 °C

2-butyl-4-chloro-1-[[2'-
(1-triphenylmethyl-1H-
tetrazol-5-yl)-1,1'-bi-
phenyl-4-yl]methyl]-
1H-imidazole-5-carbox-
aldehyde (IX)

Losartan potassium

p-bromobenzyl
bromide

2-butyl-4-chloro-1-
(4-bromobenzyl)-1H-
imidazole-5-carbox-
aldehyde (X)

tetrakis(triphenyl-
phosphine)palladium

Reference(s):
Larsen, R.D. et al.: J. Med. Chem. (JMCMAR) **59** (21), 6391 (1994).
a   EP 324 377 (Du Pont de Nemours; appl. 5.1.1989; USA-prior. 7.1.1988).
    Carini, D.J. et al.: J. Med. Chem. (JMCMAR) **34**, 2525 (1991).
b,c Smith, G.B. et al.: J. Org. Chem. (JOCEAH) **59**, 8151-8156 (1994).
    US 5 130 439 (Du Pont de Nemours; 14.7.1992; USA-prior. 18.11.1991).
    US 5 310 928 (Du Pont de Nemours; 10.5.1994; USA-prior. 18.11.1991).

polymorphs of losartan potassium:
WO 9 517 396 (Merck & Co.; du Pont de Nemours; appl. 21.12.1994; USA-prior. 23.12.1993).

preparation of 2-butyl-4-chloro-5-hydroxymethylimidazole:
Beoschelli, D.H.; Connor, D.T.: Heterocycles (HTCYAM) **35** (1), 121-124 (1993).
Shy, Y.-J.; Frey, L.F.; Tschaen, D.M.; Verhoeven, T.R.: Synth. Commun. (SYNCAV) **23** (18), 2623-2630 (1993).

preparation of 4'-bromomethyl-2-cyanobiphenyl *via bromination with* N-bromosuccinimide:
Tanaka, A. et al.: Bioorg. Med. Chem. (BMECEP) **6** (1), 15-30 (1998).
Huang, H.C. et al.: J. Med. Chem. (JMCMAR) **36** (15), 2172-2181 (1993).

synthesis of intermediate V:
Lo, Y.S.; Rossano, L.T.; Meloni, D.J.; Moore, J.R.; Lee, Y.-C; Arneet, J.F.: J. Heterocycl. Chem. (JHTCAD) **32** (1), 355 (1995).

synthesis of intermediate VIII:
Griffiths, G.H. et al.: J. Org. Chem. (JOCEAH) **64**, 8084 (1999).
Griffiths, G.J.: Chimia (CHIMAD) **51** (6), 283 (1997).

combination with e. g. lovastatin:
WO 9 526 188 (Merck & Co.; appl. 24.3.1995; USA-prior. 29.3.1994).

new form with specific properties:
WO 9 517 396 (Merck & Co., Du Pont; appl. 21.12.1994; USA-prior. 23.12.1993).

use for treatment of neurodegenerative processes:
WO 9 521 609 (Ciba-Geigy; appl. 26.1.1995; EP-prior. 8.2.1994).
US 5 091 390 (Du Pont de Nemours; appl. 20.9.1990; USA-prior. 20.9.1990).

composition with potassium channel activator:
EP 561 357 (Merck; appl. 16.3.1993; D-prior. 20.3.1992).

composition for direct compression tabl.:
EP 511 767 (Merck & Co.; appl. 21.4.1992; USA-prior. 29.4.1991).

use for treatment of cardiac and vascular hypertrophy:
DE 4 036 706 (Hoechst; appl. 17.11.1990; D-prior. 17.11.1990).

Formulation(s):   tabl. 12.5 mg, 25 mg, 50 mg

Trade Name(s):

D:	Lorzaar (MSD Chibropharm) Lorzaar (MSD Chibropharm)-comb. with hydrochlorothiazide	F:	Cozaar (Merck Sharp & Dohme-Chibret) Hyzaar (Merck Sharp & Dohme-Chibret)-comb.	I:	Losaprex (Merck & Co.)
				USA:	Cozaar (Merck & Co.; 1995)
		GB:	Cozaar (Merck Sharp & Dohme)		Hyzaar (Merck & Co.)-comb. with hydrochlorothiazide

---

Loteprednol etabonate

Use: ocular antiinflammatory soft corticosteroid

RN: 82034-46-6   MF: $C_{24}H_{31}ClO_7$ MW: 466.96
CN: (11β,17α)-17-[(Ethoxycarbonyl)oxy]-11-hydroxy-3-oxoandrosta-1,4-diene-17-carboxylic acid chloromethyl ester

prednisolone (q. v.)

$NaIO_4$, THF, CH_3OH

11β,17α-dihydroxy-3-oxoandrosta-1,4-diene-17β-carboxylic acid (I)

I + ethyl chloroformate

CH_2Cl_2, NEt_3

17α-(ethoxycarbonyloxy)-11β-hydroxy-3-oxoandrosta-1,4-diene-17-carboxylic acid (II)

II + chloromethyl iodide

NaOH, CH_3OH, H_2O

Loteprednol etabonate

Reference(s):
BE 889 563 (Otsuka Pharm. Co. Ltd.; appl. 9.7.1981; USA-prior. 10.7.1980).

oxidation of prednisolone *with sodium periodate:*
Hirschmann et al.: Chem. Ind. (London) (CHINAG) **1958**, 682

suspension of loteprednol etabonate:
WO 9 511 669 (Pharmos Corp.; USA-prior. 25.10.1993).
US 5 747 061 (Pharmos Corp.; 5.5.1998; USA-prior. 25.10.1993).

Formulation(s):   ophthalmic susp. 0.2% 5 ml, 10 ml, 0.5% 2.5 ml, 5 ml, 10 ml, 15 ml

Trade Name(s):
USA: Alrex (Bausch & Lomb   Lotemax (Bausch & Lomb
         Pharm.; 1998)   Pharm.)

Lovastatin
(Mevinolin; MK 803; Monakolin-K)

ATC: B04AB; C10AA02
Use: HMG-CoA-reductase inhibitor,
         antihypercholesterolemic

RN: 75330-75-5   MF: $C_{24}H_{36}O_5$ MW: 404.55
LD_{50}: >1 g/kg (M, p.o.)
CN: [1*S*-[1α(*R**),3α,7β,8β(2*S**,4*S**),8aβ]]-2-methylbutanoic acid 1,2,3,7,8,8a-hexahydro-3,7-dimethyl-8-[2-(tetrahydro-4-hydroxy-6-oxo-2*H*-pyran-2-yl)ethyl]-1-naphthalenyl ester

Fermentation of *Aspergillus terreus* (ATCC 20541).

Lovastatin

Reference(s):
US 4 294 926 (Merck & Co.; 13.10.1981, appl. 15.6.1979; prior. 23.1.1980).
US 4 342 767 (Merck & Co.; 3.8.1982; prior. 16.6.1980, 15.6.1979, 23.1.1980).
US 4 294 846 (Merck & Co.; 13.10.1981; 28.5.1980, 21.9.1979).
US 4 231 938 (Merck & Co.; 4.11.1980; prior. 15.6.1979).
EP 22 478 (Merck & Co.; 13.6.1980; USA-prior. 15.6.1979).
Alberts, A.W. et al.: Proc. Natl. Acad. Sci. USA (PNASA6) **77**, 3957 (1980).
Buckland, B. et al.: Novel Microb. Prod. Med. Agric., [Pap. Int. Conf. Biotechnol. Microb. Prod.] 1st , **1988** (56RDAV), 161, Ed. A. L. Demain (Elsevier, Amsterdam).

fermentation of Monascus ruber:
DE 3 006 216 (Sankyo; appl. 20.2.1980; J-prior. 20.2.1979).
US 4 323 648 (Sankyo; 6.4.1982; J-prior. 11.5.1979).
DOS 3 028 284 (Sankyo; appl. 25.7.1980; J-prior. 27.7.1979).

synthesis of intermediates:
JP 59 193 883 (Suntry; appl. 9.3.1983).
JP 59 186 973 (Suntry; appl. 9.3.1983).
JP 59 186 972 (Suntry; appl. 9.3.1983).

medical use for the treatment of prostatomegaly:
JP 56 115 717 (A. Endo; appl. 19.2.1980).

total synthesis:
Quinkert, G. et al.: Synform (SNFMDF) **2**, 84, 111 (1984) (review).
Majewski, M. et al.: Tetrahedron Lett. (TELEAY) **25**, 2101 (1984).
Wovkulich, P.M. et al.: J. Am. Chem. Soc. (JACSAT) **111**, 2596 (1989).
Clive, D.L. et al.: J. Am. Chem. Soc. (JACSAT) **110**, 6914 (1988).
Hirama, M.; Iwashita, M.: Tetrahedron Lett. (TELEAY) **24**, 1811 (1983).

new fermentation process:
EP 877 089 (Gist-Brocades; EP-prior. 7.5.1997).
WO 9 837 220 (Gist-Brocades; appl. 20.2.1998; EP-prior. 20.2.1997).
WO 9 837 179 (Gist-Brocades; appl. 20.2.1998; EP-prior. 20.2.1997).
WO 9 736 996 (Gist-Brocades; appl. 21.3.1997; EP-prior. 28.3.1996)

fermentation of Coniothyrium fuckelii:
US 5 409 820 (Apotex; USA-prior. 6.8.1993).

Formulation(s):   tabl. 10 mg, 20 mg, 40 mg

Trade Name(s):
D: Mevinacor (Merck Sharp &   USA: Mevacor (Merck Sharp &
       Dohme; 1989)   Dohme; 1987)

Loxapine

ATC: N05AH01
Use: neuroleptic, anxiolytic

RN: 1977-10-2   MF: $C_{18}H_{18}ClN_3O$ MW: 327.82   EINECS: 217-835-3
LD$_{50}$: 22 mg/kg (M, i.v.); 40 mg/kg (M, p.o.);
       18 mg/kg (R, i.v.); 151 mg/kg (R, p.o.)
CN: 2-chloro-11-(4-methyl-1-piperazinyl)dibenz[*b,f*][1,4]oxazepine

succinate (1:1)
RN: 27833-64-3   MF: $C_{18}H_{18}ClN_3O \cdot C_4H_6O_4$ MW: 445.90   EINECS: 248-682-0

2-(4-chlorophen-   ethyl   (I)
oxy)aniline   chloroformate

N-methyl-   Loxapine
piperazine

Reference(s):
US 3 412 193 (American Cyanamid; 19.11.1968; appl. 13.12.1965).
Schmutz, J. et al.: Helv. Chim. Acta (HCACAV) **50**, 245 (1967).

alternative syntheses:
US 3 546 226 (Dr. A. Wander; 8.12.1970, CH-prior. 30.5.1963, 27.9.1963, 13.3.1967, 22.3.1967, 9.5.1967, 14.7.1967, 3.11.1967).
DE 1 470 426 (Dr. A. Wander; appl. 25.5.1964; CH-prior. 30.5.1963, 27.9.1963).

Formulation(s): cps. 5 mg, 10 mg, 25 mg, 50 mg

Trade Name(s):

F:	Loxapac (Wyeth-Lederle; 1980)	GB:	Loxapac (Wyeth)	USA:	Loxitane (Lederle; 1975)
		I:	Loxapac (Cyanamid); wfm		generic

Loxoprofen

ATC: M01
Use: anti-inflammatory, analgesic

RN: 68767-14-6 MF: $C_{15}H_{18}O_3$ MW: 246.31
CN: α-methyl-4-[(2-oxocyclopentyl)methyl]benzeneacetic acid

sodium salt
RN: 80382-23-6 MF: $C_{15}H_{17}NaO_3$ MW: 268.29
LD_{50}: 740 mg/kg (M, i.v.); 3030 mg/kg (M, p.o.);
155 mg/kg (R, i.v.); 145 mg/kg (R, p.o.)

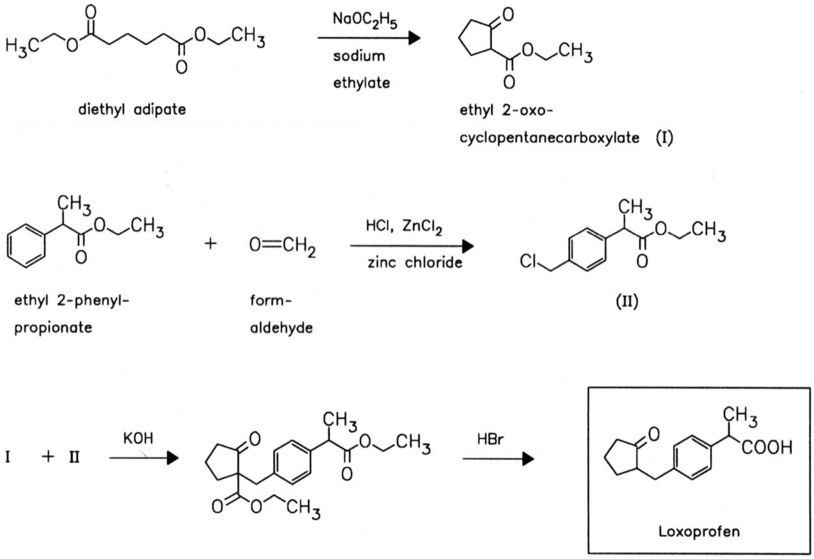

diethyl adipate
NaOC2H5
sodium ethylate
ethyl 2-oxo-cyclopentanecarboxylate (I)
ethyl 2-phenylpropionate
formaldehyde
HCl, ZnCl2
zinc chloride
(II)
I + II
KOH
HBr
Loxoprofen

Reference(s):
US 4 161 538 (Sankyo; 17.7.1979; J-prior. 5.4.1977).
DOS 2 814 556 (Sankyo; appl. 4.4.1978; J-prior. 5.4.1977).

synthesis of I:
Zupancic, B.G.; Trpin, J.: Monatsh. Chem. (MOCMB7) **98**, 369 (1967).

synthesis of II:
FR 2 134 197 (Lab. Logeais; appl. 26.4.1971).

Formulation(s): oral: 3x60 mg/d

Trade Name(s):
J: Loxonin (Sankyo; 1986)

Lymecycline

ATC: J01AA04
Use: antibiotic

RN: 992-21-2 MF: $C_{29}H_{38}N_4O_{10}$ MW: 602.64 EINECS: 213-592-2
LD_{50}: 181 mg/kg (M, i.v.)
CN: [4S-(4α,4aα,5aα,6β,12aα)]-N^6-[[[[4-(dimethylamino)-1,4,4a,5,5a,6,11,12a-octahydro-3,6,10,12,12a-pentahydroxy-6-methyl-1,11-dioxo-2-naphthacenyl]carbonyl]amino]methyl]-L-lysine

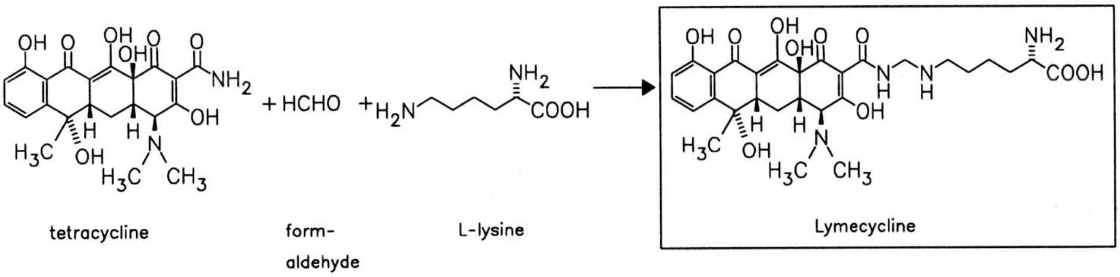

tetracycline
form-aldehyde
L-lysine
Lymecycline

Reference(s):
DE 1 134 071 (Carlo Erba; appl. 7.11.1960; I-prior. 23.11.1959).
US 3 042 716 (Pfizer; 3.7.1962; appl. 4.12.1961).

Formulation(s): cps. 150 mg, 300 mg (calculated as tetracycline)

Trade Name(s):
F: Tetralysal (Galderma)
GB: Tetralysal (Galderma)
I: Ciclisin (Fabo); wfm
Lisinbiotic (Farmacosmici); wfm
Lisinciclina (Biotrading); wfm
Stiltetra (Ellea); wfm
Tetralysal (Erba); wfm
Tralisin (Firma); wfm
J: Tetralysal (Fujisawa)

Lynestrenol

ATC: G03AC02; G03DC03
Use: progestogen (in comb. with estrogen as oral contraceptive)

RN: 52-76-6 MF: $C_{20}H_{28}O$ MW: 284.44 EINECS: 200-151-4
CN: (17α)-19-norpregn-4-en-20-yn-17-ol

BF3 etherate
Na, liq. NH3
I
nandrolone
(q. v.)
ethane-
1,2-dithiol

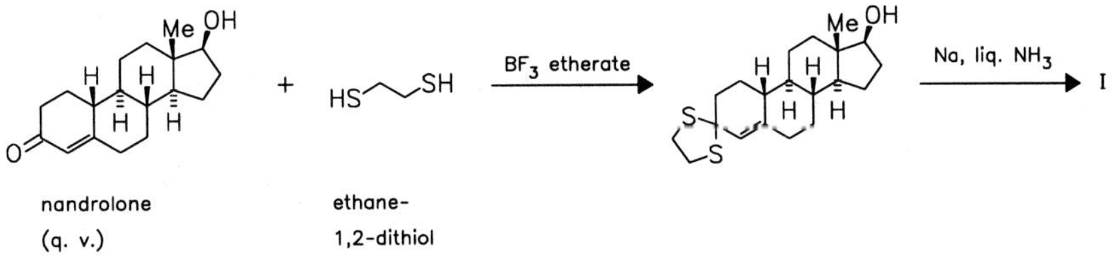

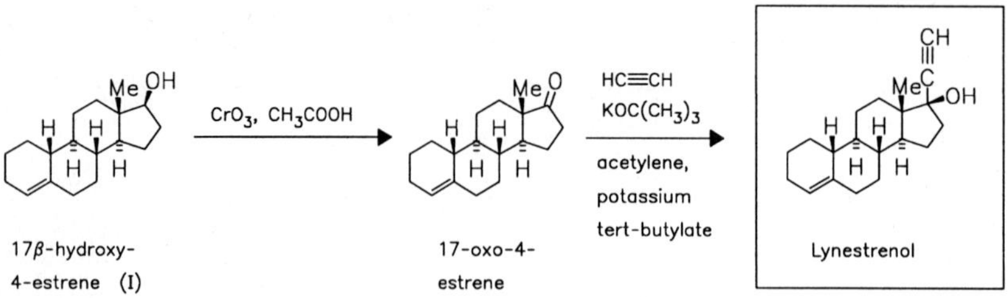

17β-hydroxy-
4-estrene (I)

17-oxo-4-
estrene

Lynestrenol

Reference(s):
GB 841 411 (Organon; appl. 2.4.1958; NL-prior. 10.4.1957).

alternative syntheses:
GB 875 549 (Organon; appl. 31.12.1959; NL-prior. 13.1.1959).
US 2 878 267 (Organon; appl. 16.4.1958; NL-prior. 1.5.1957).

Formulation(s):   tabl. 0.5 mg, 1 mg, 5 mg (in combinations)

Trade Name(s):

D: Exlutona (Organon)
     Orgametril (Organon)
     numerous generics and
     combination preparations
F: Exluton (Organon)
     Orgametril (Organon)
     Ovanon (Organon)-comb.

     Physiostat (Organon)-
     comb.
GB: Minilyn (Organon)-comb.;
     wfm
I: Franovul (Francia Farm.)-
     comb.; wfm

     Linseral (Proter)-comb.;
     wfm
     Lyndiol (Ravasini
     Organon)-comb.; wfm
J: o-Lyndiol (Organon-
     Sankyo)-comb.

Mabuprofen

(Aminoprofen; AU-7801)

ATC: M01AE; M02A
Use: topical anti-inflammatory

RN: 82821-47-4   MF: $C_{15}H_{23}NO_2$ MW: 249.35   EINECS: 280-048-9
LD_{50}:   2828 mg/kg (M, s.c.)
CN: N-(2-hydroxyethyl)-α-methyl-4-(2-methylpropyl)benzeneacetamide

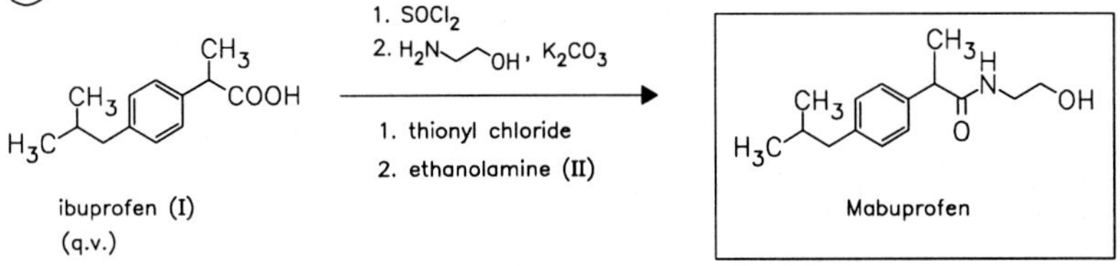

ibuprofen (I)
(q.v.)

1. SOCl₂
2. H₂N-OH, K₂CO₃

1. thionyl chloride
2. ethanolamine (II)

Mabuprofen

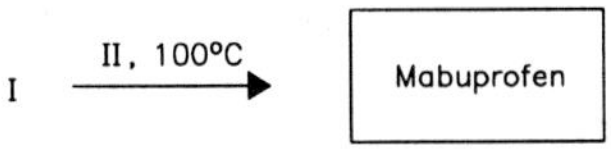

I → II, 100°C → Mabuprofen

I → 1. CH₃OH, H₂SO₄ 2. II → Mabuprofen

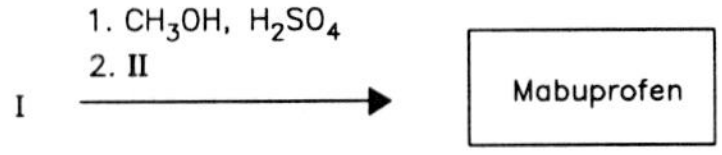

Reference(s):
a   DE 3 121 595 (Calzada; appl. 30.5.1981; E-prior. 10.3.1981).
    Zhang, D.; Ji, H., Yang, S.: Zhongguo Yiyao Gongye Zazhi (ZYGZEA) **25** (12), 535 (1994).
b   ES 2 028 601 (Prodesfarma S. A.; appl. 4.2.1991; E-prior. 4.2.1990).
    ES 2 007 236 (Laboratoio Aldo-Union S. A.; appl. 16.6.1988).
c   ES 2 023 585 (Prodesfarma S. A.; appl. 17.10.1990).

Formulation(s):   pump spray 10 %

Trade Name(s):
E: Aldospray Analgesico   Formix (Lab. Padro; 1990)
       (Aldo Union; 1989)   Sedaspray (Lusi; 1989)

Mabuterol

ATC: R03
Use: bronchodilator

RN: 56341-08-3   MF: $C_{13}H_{18}ClF_3N_2O$ MW: 310.75
CN: 4-amino-3-chloro-α-[[(1,1-dimethylethyl)amino]methyl]-5-(trifluoromethyl)benzenemethanol

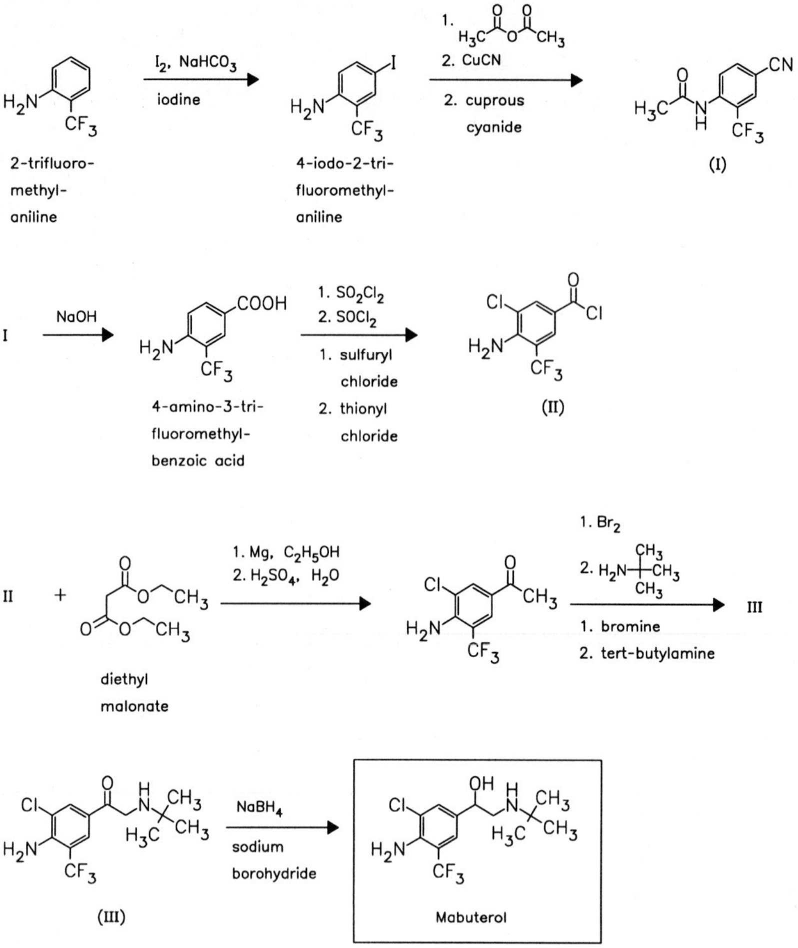

2-trifluoro-
methyl-
aniline

I_2, NaHCO$_3$
iodine

4-iodo-2-tri-
fluoromethyl-
aniline

1. H_3C acetic anhydride
2. CuCN
2. cuprous
cyanide

(I)

I $\xrightarrow{\text{NaOH}}$

4-amino-3-tri-
fluoromethyl-
benzoic acid

1. SO$_2$Cl$_2$
2. SOCl$_2$

1. sulfuryl
chloride
2. thionyl
chloride

(II)

II + diethyl
malonate

1. Mg, C$_2$H$_5$OH
2. H$_2$SO$_4$, H$_2$O

1. Br$_2$
2. H$_2$N—C(CH$_3$)$_3$

1. bromine
2. tert-butylamine

III

(III)

NaBH$_4$
sodium
borohydride

Mabuterol

Reference(s):
Keck, J. et al.: Arzneim.-Forsch. (ARZNAD) **34** (II), 1612 (1984).

Formulation(s):   tabl. 0.05 mg

Trade Name(s):
J: Broncholin (Kaken)

Mafenide

ATC: D06BA03; G01AE01
Use: chemotherapeutic

RN: 138-39-6   MF: C$_7$H$_{10}$N$_2$O$_2$S   MW: 186.24   EINECS: 205-326-9
CN: 4-(aminomethyl)benzenesulfonamide

acetate
RN: 13009-99-9   MF: C$_7$H$_{10}$N$_2$O$_2$S · C$_2$H$_4$O$_2$ MW: 246.29
hydrochloride
RN: 138-37-4   MF: C$_7$H$_{10}$N$_2$O$_2$S · HCl   MW: 222.70

Sulfatolamide

RN: 1161-88-2   MF: $C_7H_{10}N_2O_2S \cdot C_7H_9N_3O_2S_2$ MW: 417.54   EINECS: 214-600-7
CN: 4-amino-*N*-(aminothioxomethyl)benzenesulfonamide compd. with 4-(aminomethyl)benzenesulfonamide (1:1)

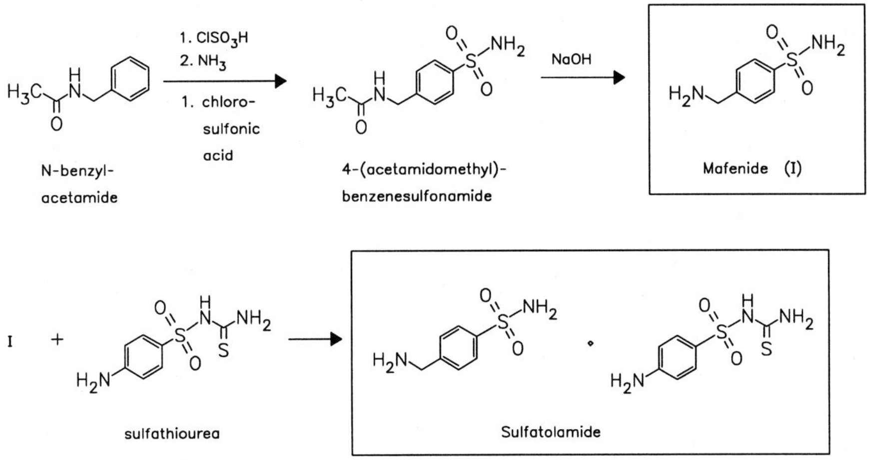

N-benzyl-acetamide → 4-(acetamidomethyl)-benzenesulfonamide → Mafenide (I)

I + sulfathiourea → Sulfatolamide

Reference(s):

mafenide:
DRP 726 386 (I. G. Farben; appl. 1939).
US 2 288 531 (Winthrop; 1942; D-prior. 1939).

sulfathiourea:
FR 913 920 (Rhône-Poulenc; appl. 1942).

sulfatolamide:
US 2 696 454 (Schenley Ind.; 1954; CH-prior. 1949).
DE 836 350 (Bayer; appl. 1944).

Formulation(s):   cream 8.5 %, 11.2 g/100 g (as acetate); eye drops 2.5 mg/g, 5 % (as propionate)

Trade Name(s):
D: Combiamid (Winzer; as hydrochloride)
     Marbaletten (Bayer; as sulfatolamide); wfm
     Napaltan (Winthrop; as mafenide); wfm
F: Anafluose (Guillaumin et Hales; as mafenide)-comb.; wfm
GB: Sulfamylon (Winthrop; as mafenide); wfm
     Sulfomyl (Winthrop); wfm
J: Paramenyl (Takeda)
USA: Sulfamylon (Dow Hickam; as acetate)

Magaldrate

ATC: A02AD02
Use: antacid

RN: 74978-16-8   MF: $Al_5H_{31}Mg_{10}O_{39}S_2 \cdot xH_2O$ MW: unspecified
CN: aluminum magnesium hydroxide sulfate ($Al_5Mg_{10}(OH)_{31}(SO_4)_2$) hydrate

$Al_5Mg_{10}(OH)_{31}(SO_4)_2$ • x H_2O

Magaldrate

$AlCl_3$ is treated with NaOH (mole ratio 1:6) to yield an aqueous sodium aluminate solution, 1.2 mole $MgSO_4$ (in aqueous solution) are added, the precipitate is washed and dried.

Reference(s):
US 2 923 660 (Byk Gulden; 2.2.1960; D-prior. 5.8.1955).

Formulation(s):   chewing tabl. 400 mg, 800 mg; gel 80 mg, 800 mg; susp. 540 mg, 800 mg; tabl. 400 mg, 480 mg

Trade Name(s):
D: Riopan (Byk Gulden;   I: Riopan (Byk Gulden)   generics; wfm
     Roland)   USA: Riopan (Ayerst); wfm

Malotilate

Use: liver therapeutic, hepatoprotectant

RN: 59937-28-9   MF: $C_{12}H_{16}O_4S_2$ MW: 288.39   EINECS: 261-987-3
LD_{50}:   729 mg/kg (M, i.v.); 3120 mg/kg (M, p.o.);
      2065 mg/kg (R, p.o.)
CN: 1,3-dithiol-2-ylidenepropanedioic acid bis(1-methylethyl) ester

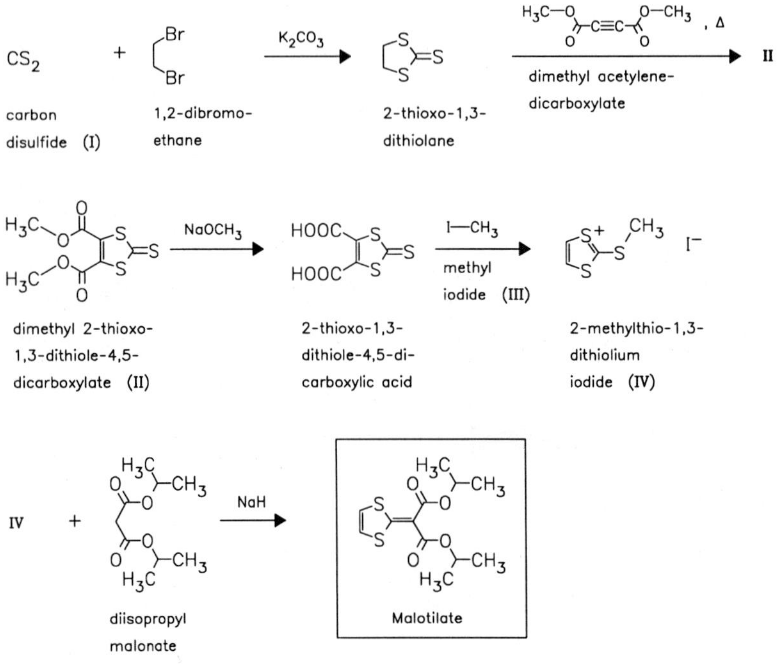

CS_2 +

carbon
disulfide (I)

1,2-dibromo-
ethane

2-thioxo-1,3-
dithiolane

dimethyl acetylene-
dicarboxylate

II

dimethyl 2-thioxo-
1,3-dithiole-4,5-
dicarboxylate (II)

2-thioxo-1,3-
dithiole-4,5-di-
carboxylic acid

methyl
iodide (III)

2-methylthio-1,3-
dithiolium
iodide (IV)

IV   +

diisopropyl
malonate

NaH

Malotilate

alternative synthesis of 2-methylthio-1,3-dithiolium iodide (IV):

acetylene

1,3-dithiole-
2-thione

Reference(s):
DOS 2 545 569 (Nihon Nohyaku; appl. 10.10.1975; J-prior. 18.10.1974, 22.10.1974).
US 4 035 387 (Nihon Nohyaku; 12.7.1977; J-prior. 18.10.1974, 22.10.1974).

medical use against liver diseases:
DOS 2 625 012 (Nihon Nohyaku; appl. 3.6.1976; USA-prior. 6.6.1975).
FR 2 313 037 (Nihon Nohyaku; appl. 4.6.1976; USA-prior. 6.6.1975).

2-thioxo-1,3-dithiolane:
Fujinami, T. et al.: Bull. Chem. Soc. Jpn. (BCSJA8) **55**, 1174 (1982).

dimethyl 2-thioxo-1,3-dithiole-4,5-dicarboxylate:
Gorgues, A. et al.: J. Chem. Soc., Chem. Commun. (JCCCAT) **1983**, 405.
O'Connor, B.R.; Jones, F.N.: J. Org. Chem. (JOCEAH) **35**, 2002 (1970).

alternative synthesis of 2-thioxo-1,3-dithiolane:
Mayer, R. et al.: Angew. Chem. (ANCEAD) **76**, 143 (1964).

Formulation(s):   tabl. 200 mg

Trade Name(s):
J: Kantec (Daiichi/Nihon
        Nohyaku)

Manidipine
(Franidipine)

ATC: C02DE; C08CA11
Use: calcium antagonist, antihypertensive

RN: 89226-50-6   MF: $C_{35}H_{38}N_4O_6$ MW: 610.71
CN: 1,4-dihydro-2,6-dimethyl-4-(3-nitrophenyl)-3,5-pyridinedicarboxylic acid 2-[4-(diphenylmethyl)-1-piperazinyl]ethyl methyl ester

dihydrochloride
RN: 89226-75-5   MF: $C_{35}H_{38}N_4O_6 \cdot 2HCl$ MW: 683.63
LD$_{50}$: 62.2 mg/kg (Mm, i.v.); 68 mg/kg (Mf, i.v.); 190 mg/kg (Mm, p.o.); 171 mg/kg (Mf, p.o.); 387 mg/kg (Mm, s.c.); 340 mg/kg (Mf, s.c.);
66.5 mg/kg (Rm, i.v.); 48.8 mg/kg (Rf, i.v.); 247 mg/kg (Rm, p.o.); 156 mg/kg (Rf, p.o.); 222 mg/kg (Rm, s.c.); 199 mg/kg (Rf, s.c.)

2-(1-piperazinyl)-
ethanol

diphenyl-
methyl bromide

2-(4-diphenylmethyl-1-
piperazinyl)ethanol (I)

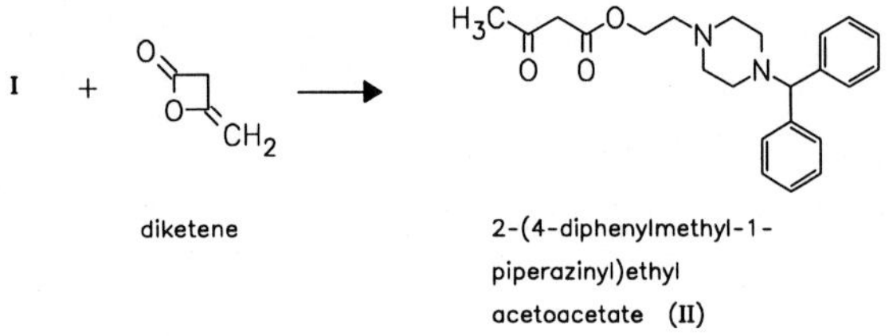

I +

diketene

2-(4-diphenylmethyl-1-
piperazinyl)ethyl
acetoacetate (II)

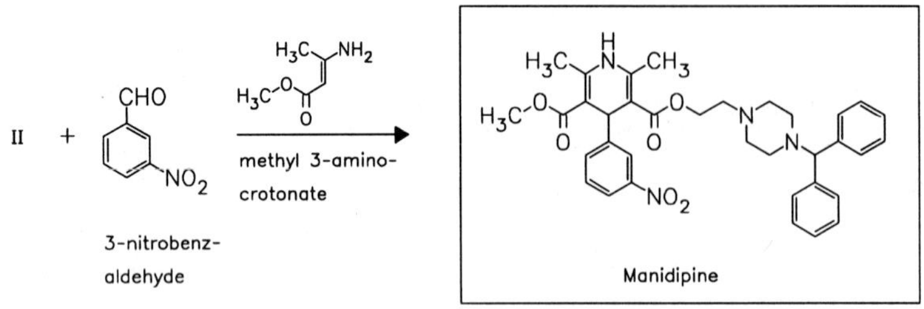

II +

CHO

NO₂

3-nitrobenz-
aldehyde

methyl 3-amino-
crotonate

Manidipine

Reference(s):
EP 94 159 (Takeda; appl. 15.4.1983; J-prior. 10.5.1982).

medical use as anti-arteriosclerotic:
JP 1 022 017 (Takeda; appl. 9.7.1984).

Formulation(s):　tabl. 5 mg, 10 mg, 20 mg (as dihydrochloride)

Trade Name(s):
J:　　Calslot (Takeda; 1991)

Maprotiline

ATC:　N06AA21
Use:　antidepressant

RN:　10262-69-8　MF: $C_{20}H_{23}N$　MW: 277.41　EINECS: 233-599-4
LD₅₀:　31 mg/kg (M, i.v.); 660 mg/kg (M, p.o.);
　　　38 mg/kg (R, i.v.); 760 mg/kg (R, p.o.)
CN:　*N*-methyl-9,10-ethanoanthracene-9(10*H*)-propanamine

hydrochloride
RN:　10347-81-6　MF: $C_{20}H_{23}N \cdot HCl$　MW: 313.87　EINECS: 233-758-8
LD₅₀:　31 mg/kg (M, i.v.); 480 mg/kg (M, p.o.);
　　　35 mg/kg (R, i.v.); 760 mg/kg (R, p.o.)

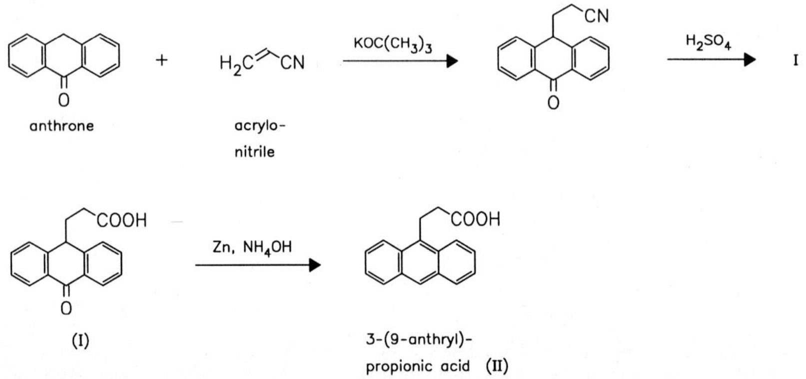

anthrone

+ H₂C CN

acrylo-
nitrile

KOC(CH₃)₃

CN

H₂SO₄

I

COOH

(I)

Zn, NH₄OH

COOH

3-(9-anthryl)-
propionic acid (II)

ⓐ

II $\xrightarrow[\text{lithium}]{\text{LiAlH}_4}$ 9-(3-hydroxy-propyl)anthracene $\xrightarrow[\text{thionyl}]{\text{SOCl}_2}$ [anthracene-CH₂CH₂CH₂Cl] $\xrightarrow[\text{methylamine (III)}]{\text{H}_3\text{C—NH}_2}$ IV

9-(3-methylamino-propyl)anthracene (IV) + H₂C=CH₂ ethylene (V) $\xrightarrow{\text{50 atm, 150 °C, 24 h}}$ Maprotiline

ⓑ

II + V $\xrightarrow{\text{DMF}}$ (COOH) $\xrightarrow[\text{2. III}]{1.}$ $\xrightarrow[\text{lithium alanate}]{\text{LiAlH}_4}$ Maprotiline

Reference(s):
DE 1 518 691 (Ciba; appl. 16.12.1965; CH-prior. 23.12.1964).
CH 467 237 (Ciba; appl. 23.12.1964).
CH 467 747 (Ciba; appl. 23.12.1964).

Formulation(s):   amp. 25 mg (as hydrochloride); f. c. tabl. 10 mg, 25 mg, 50 mg, 75 mg

Trade Name(s):
D: Deprilept (Promonta
       Lundbeck)
       Ludiomil (Novartis; as
       hydrochloride)
       Mapro-Gry (GRY)Maprolu
       (Neuro Hexal)
       Mirpan (Dolorgret)

       Psymion (Desitin)
F: Ludiomil (Novartis; as
       hydrochloride)
GB: Ludiomil (Novartis; as
       hydrochloride)
I: Ludiomil (Novartis; as
       hydrochloride)

J: Ludiomil (Novartis; as
       hydrochloride)
USA: Ludiomil (Novartis; as
       hydrochloride); wfm
       generics

Maruyama

(Z-100)

ATC: L03A
Use: immunostimulant adjuvant in
        radiation-induced leucopenia

RN: 64060-36-2   MF: unspecified   MW: unspecified
CN: Z 100 (polyester)

Extraction of *Mycobacterium tuberculosis* Aoyama B. with hot water.

Reference(s):
JP 8 094 247 (C. Maruyama; appl. 7.10.1980).
DE 3 048 699 (C. Maruyama; appl. 23.12.1980).
GB 2 088 399 (C. Maruyama; appl. 28.11.1980).
DE 3 407 823 (Zeria; appl. 2.3.1984; J-prior. 4.3.1983).
US 4 746 511 (Zeria; 24.5.1988; appl. 28.7.1986; prior. 2.3.1984; J-prior. 4.3.1983).

Formulation(s): amp. 20 µg

Trade Name(s):
J: Ancer 20 (Z-100) (Zeria; 1991)

Mazaticol

ATC: N04AA10
Use: antiparkinsonian, muscle relaxant

RN: 42024-98-6 MF: $C_{21}H_{27}NO_3S_2$ MW: 405.58
CN: *exo*-α-hydroxy-α-2-thienyl-2-thiopheneacetic acid 6,9,9-trimethyl-9-azabicyclo[3.3.1]non-3-yl ester

hydrochloride
RN: 32891-29-5 MF: $C_{21}H_{27}NO_3S_2 \cdot HCl$ MW: 442.04
LD_{50}: 20.2 mg/kg (M, i.v.); 263 mg/kg (M, p.o.);
12.9 mg/kg (R, i.v.); 1182 mg/kg (R, p.o.)

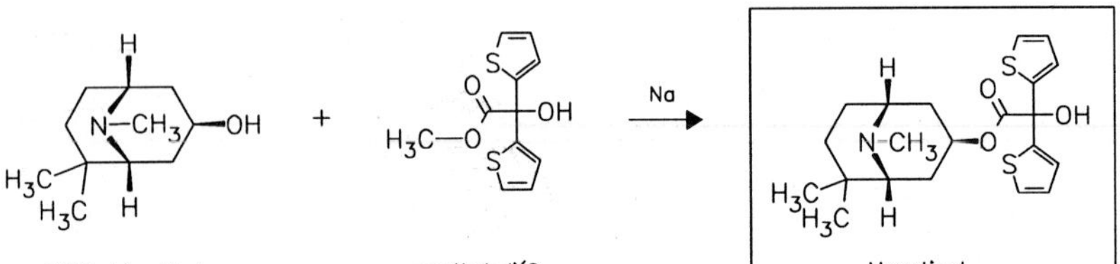

6,6,9-trimethyl-
9-azabicyclo-
[3.3.1]nonan-3β-ol

methyl di(2-
thienyl)glycolate

Mazaticol

Reference(s):
DOS 2 026 462 (Tanabe Seiyaku; appl. 29.5.1970; J-prior. 8.10.1969).
US 3 673 195 (Tanabe Seiyaku; 27.6.1972; prior. 25.5.1970).
Yoneda, N. et al.: Chem. Pharm. Bull. (CPBTAL) **20**, 476 (1972).

Formulation(s): tabl. 4 mg (as hydrochloride)

Trade Name(s):
J: Pentona (Tanabe)

Mazindol

ATC: A08AA05
Use: appetite depressant

RN: 22232-71-9 MF: $C_{16}H_{13}ClN_2O$ MW: 284.75 EINECS: 244-857-0
LD_{50}: 44.8 mg/kg (M, p.o.);
36.3 mg/kg (R, p.o.)
CN: 5-(4-chlorophenyl)-2,5-dihydro-3*H*-imidazo[2,1-*a*]isoindol-5-ol

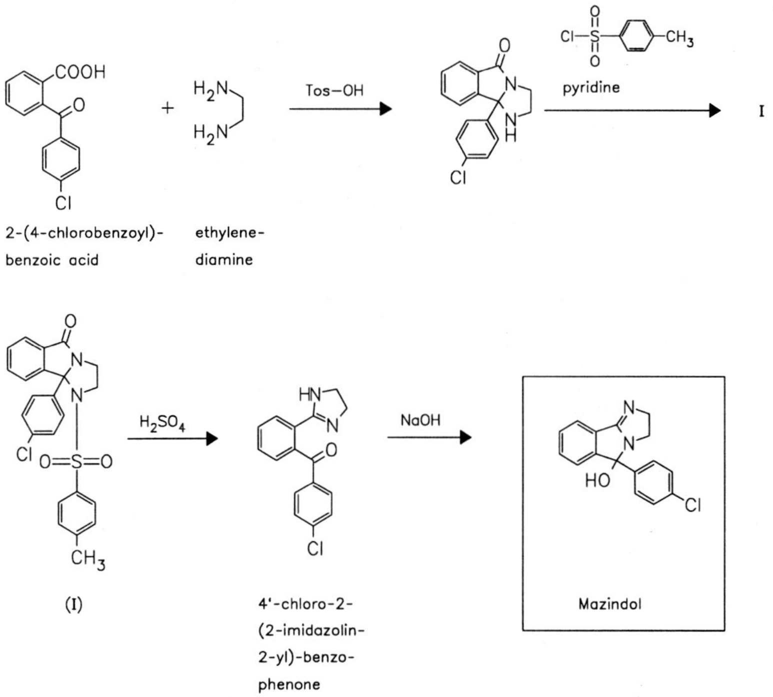

2-(4-chlorobenzoyl)-
benzoic acid

ethylene-
diamine

(I)

4'-chloro-2-
(2-imidazolin-
2-yl)-benzo-
phenone

Mazindol

Reference(s):
DOS 1 770 030 (Sandoz; appl. 22.3.1968; USA-prior. 23.3.1967).
US 3 597 445 (Sandoz-Wander; 3.8.1971; appl. 19.6.1968).
US 3 763 178 (American Home Products; 2.10.1973; appl. 5.9.1968; prior. 15.9.1965, 2.9.1966, 14.3.1967).

alternative syntheses:
DOS 1 795 105 (Sandoz; appl. 10.8.1968; USA-prior. 15.8.1967, 3.5.1968).
DOS 1 814 540 (Sandoz; appl. 12.12.1968; USA-prior. 18.12.1967, 23.7.1968).
DOS 1 930 488 (Sandoz; appl. 16.6.1969; USA-prior. 19.6.1968).

Formulation(s):   tabl. 1 mg, 2 mg

Trade Name(s):
D: Teronac (Wander); wfm   I: Mazildene (Lifepharma);   USA: Sanorex (Sandoz); wfm
GB: Teronac (Wander); wfm   wfm

Mebendazole

ATC: P02CA01
Use: anthelmintic

RN: 31431-39-7   MF: $C_{16}H_{13}N_3O_3$ MW: 295.30   EINECS: 250-635-4
LD$_{50}$:   620 mg/kg (M, p.o.);
         714 mg/kg (R, p.o.);
         1280 mg/kg (dog, p.o.)
CN: (5-benzoyl-1*H*-benzimidazol-2-yl)carbamic acid methyl ester

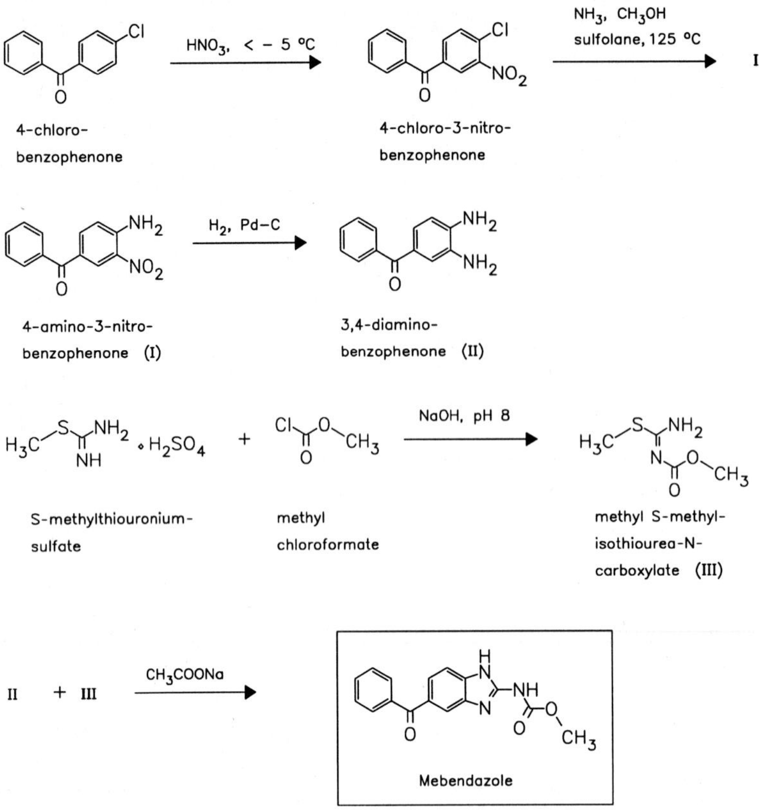

4-chloro-
benzophenone

HNO₃, < −5 °C

4-chloro-3-nitro-
benzophenone

NH₃, CH₃OH
sulfolane, 125 °C

I

4-amino-3-nitro-
benzophenone (I)

H₂, Pd−C

3,4-diamino-
benzophenone (II)

S-methylthiouronium-
sulfate

+

methyl
chloroformate

NaOH, pH 8

methyl S-methyl-
isothiourea-N-
carboxylate (III)

II   +   **III**

CH₃COONa

Mebendazole

Reference(s):

DE 2 029 637 (Janssen; appl. 16.6.1970; USA-prior. 20.6.1969).
US 3 657 267 (Janssen; 18.4.1972; prior. 20.6.1969).

Formulation(s):   chewable tabl. 100 mg; susp. 100 mg/5 ml; tabl. 100 mg, 500 mg

Trade Name(s):

D: Surfont (Ardeypharm)
     Vermox (Janssen-Cilag;
     1976)

GB: Vermox (Janssen-Cilag;
     1976)
I: Vermox (Janssen; 1979)

J: Mebendazol (Janssen
     Kyowa)
USA: Vermox (Janssen; 1975)

Mebeverine

ATC: A03AA04
Use: antispasmodic

RN: 3625-06-7   MF: C₂₅H₃₅NO₅   MW: 429.56   EINECS: 222-830-4
LD₅₀: 24 mg/kg (M, i.v.); 995 mg/kg (M, p.o.)
CN: 3,4-dimethoxybenzoic acid 4-[ethyl[2-(4-methoxyphenyl)-1-methylethyl]amino]butyl ester

hydrochloride
RN: 2753-45-9   MF: C₂₅H₃₅NO₅ · HCl   MW: 466.02   EINECS: 220-400-0
LD₅₀: 17.7 mg/kg (R, i.v.); 1540 mg/kg (R, p.o.)

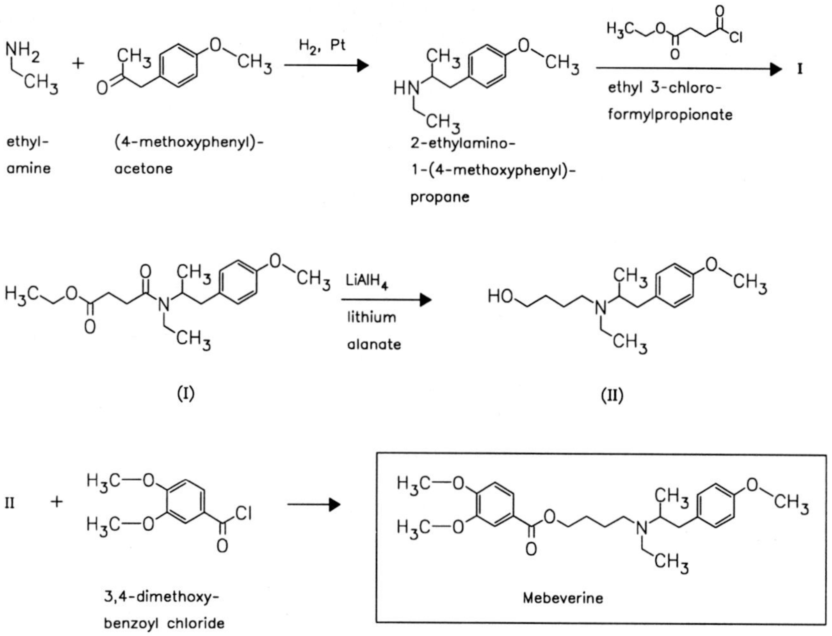

(I)   (II)

3,4-dimethoxy-
benzoyl chloride

Mebeverine

Reference(s):
DE 1 126 889 (N. V. Philips; appl. 20.11.1958; NL-prior. 23.11.1957).

alternative synthesis:
GB 1 009 082 (N. V. Philips; appl. 19.10.1961; NL-prior. 22.10.1960).

Formulation(s):   cps. 100 mg; drg. 135 mg; s. r. cps. 200 mg (as hydrochloride); susp. 10 mg

Trade Name(s):

D:	Duspatal (Solvay Arzneimittel; as hydrochloride)-comb.	Duspatalin (Solvay; as hydrochloride)	Fybogel Mebeverine (Reckitt & Colman)-comb.
F:	Colopriv (Biotherapie; as hydrochloride)	Spasmopriv (Irex; as hydrochloride)	I: Duspatal Duphar (UCM)
		GB: Colofac (Solvay)	

Mebhydrolin

ATC: R06AX15
Use: antihistaminic

RN: 524-81-2   MF: $C_{19}H_{20}N_2$ MW: 276.38   EINECS: 208-364-4
CN: 2,3,4,5-tetrahydro-2-methyl-5-(phenylmethyl)-1*H*-pyrido[4,3-*b*]indole

naphthalene-1,5-disulfonate (2:1)
RN: 6153-33-9   MF: $C_{19}H_{20}N_2 \cdot 1/2C_{10}H_8O_6S_2$ MW: 841.07   EINECS: 228-170-3
LD_{50}:   40 mg/kg (M, i.v.)

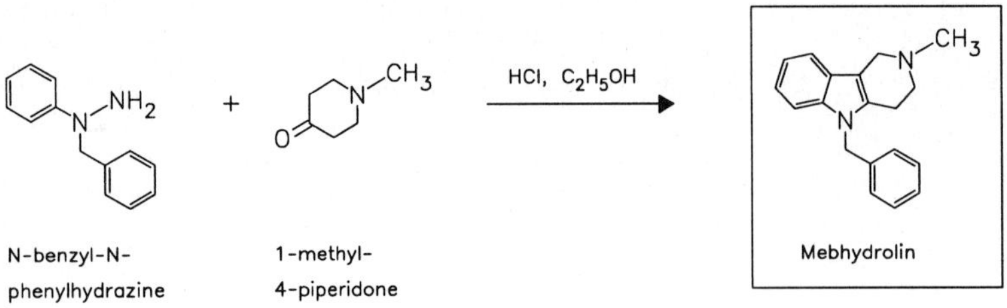

HCl, C2H5OH
N-benzyl-N-phenylhydrazine
1-methyl-4-piperidone
Mebhydrolin

Reference(s):
GB 721 171 (Bayer; appl. 1952; D-prior. 1951).

Formulation(s): drg. 50 mg, 76 mg (as napadisilate); s. r. tabl. 150 mg; susp. 50 mg; tabl. 50 mg

Trade Name(s):

D:	Omeril (Bayer); wfm	GB:	Fabahistin (Bayer); wfm
	Omeril (Tropon-Dome Hollister Stier); wfm	I:	Incidal (Bayropharm)
		J:	Incidal (Yoshitomi)

Mebutamate

ATC: N05BC04
Use: neurosedative, antihypertensive

RN: 64-55-1 MF: $C_{10}H_{20}N_2O_4$ MW: 232.28 EINECS: 200-587-5
LD_{50}: 550 mg/kg (M, p.o.);
1160 mg/kg (R, p.o.)
CN: 2-methyl-2-(1-methylpropyl)-1,3-propanediol dicarbamate

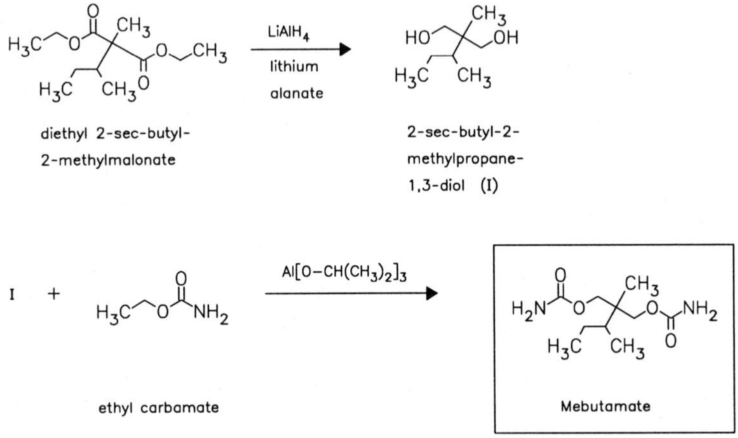

LiAlH4
lithium alanate
diethyl 2-sec-butyl-2-methylmalonate
2-sec-butyl-2-methylpropane-1,3-diol (I)
I +
Al[O-CH(CH3)2]3
ethyl carbamate
Mebutamate

Reference(s):
US 2 878 280 (Carter Products; 17.3.1959; prior. 29.11.1955).

Formulation(s): tabl. 300 mg

Trade Name(s):

F:	Dévalène (Dexo)-comb.	J:	Mega (Ono)	Dormate (Wallace); wfm
I:	Sigmafon (Lafare)	USA:	Capla (Wallace); wfm	

Mecamylamine

(Dimecamine)

ATC: C02BB01
Use: ganglionic blocker, antihypertensive

RN: 60-40-2   MF: $C_{11}H_{21}N$ MW: 167.30   EINECS: 200-476-1
LD$_{50}$:   11.9 mg/kg (M, i.v.); 90 mg/kg (M, p.o.)
CN: N,2,3,3-tetramethylbicyclo[2.2.1]heptan-2-amine

hydrochloride
RN: 826-39-1   MF: $C_{11}H_{21}N \cdot HCl$ MW: 203.76   EINECS: 212-555-8
LD$_{50}$:   14 mg/kg (M, i.v.); 92 mg/kg (M, p.o.);
        21 mg/kg (R, i.v.); 208 mg/kg (R, p.o.)

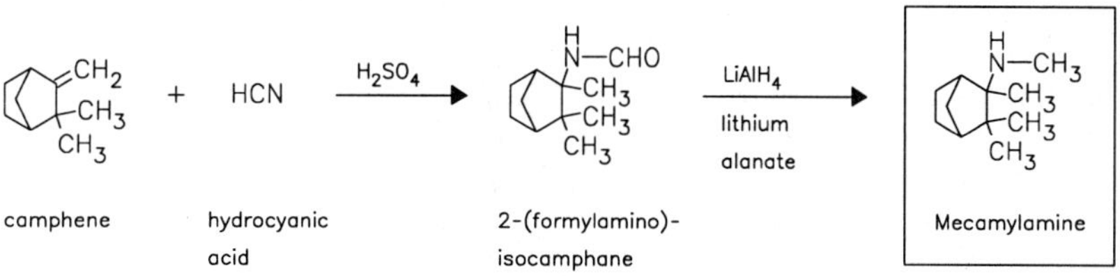

camphene   hydrocyanic   2-(formylamino)-   Mecamylamine
        acid   isocamphane

Reference(s):
US 2 831 027 (Merck & Co.; 1958, prior. 1955).
Stein, G.A. et al.: J. Am. Chem. Soc. (JACSAT) **78**, 1514 (1956).

Formulation(s):   tabl. 2.5 mg (as hydrochloride)

Trade Name(s):
D: Mevasine (Sharp &   GB: Inversine (Merck Sharp &   USA: Inversine (Merck Sharp &
        Dohme); wfm   Dohme); wfm   Dohme)
F: Inversine (Merck Sharp &   J: Mevasine (Meiji)
        Dohme); wfm

Mecillinam

ATC: J01CA11
Use: antibiotic

RN: 32887-01-7   MF: $C_{15}H_{23}N_3O_3S$ MW: 325.43   EINECS: 251-277-1
CN: [2S-(2α,5α,6β)]-6-[[(hexahydro-1H-azepin-1-yl)methylene]amino]-3,3-dimethyl-7-oxo-4-thia-1-
        azabicyclo[3.2.0]heptane-2-carboxylic acid

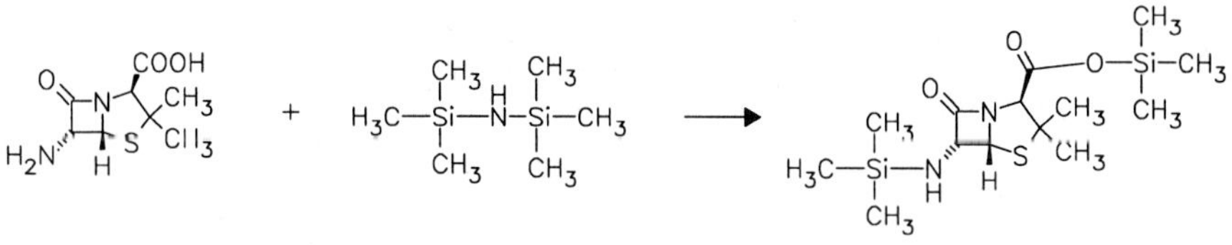

6-aminopenicillanic   hexamethyldisilazane   (I)
acid

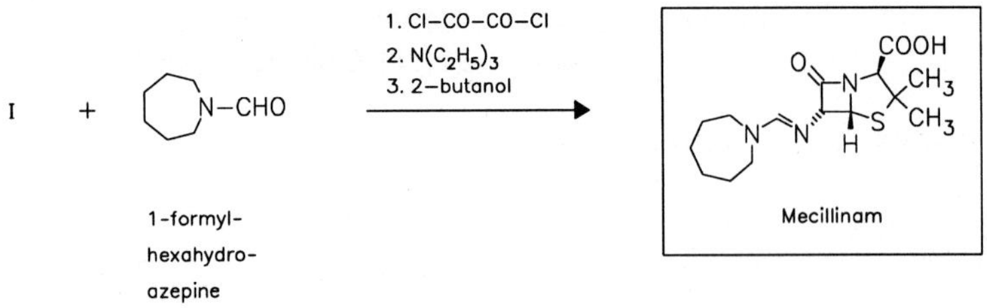

I ＋ 1-formyl-hexahydro-azepine

1. Cl—CO—CO—Cl
2. N(C₂H₅)₃
3. 2-butanol

Mecillinam

Reference(s):

DOS 2 055 531 (Loevens; appl. 11.11.1970; GB-prior. 11.11.1969, 8.7.1970).
GB 1 293 590 (Loevens; appl. 11.11.1969, 8.7.1970; valid from 10.11.1970).

Formulation(s):   amp. 0.2 g, 0.4 g, 0.5 g, 1 g

Trade Name(s):
GB: Selexidin (Burgess); wfm   Selexidin (Leo); wfm   USA: Coactin (Roche); wfm

---

Meclofenamic acid

ATC: M01AG04; M02AA18
Use: anti-inflammatory, antirheumatic, antipyretic

RN: 644-62-2   MF: $C_{14}H_{11}Cl_2NO_2$ MW: 296.15   EINECS: 211-419-5
LD_{50}:   100 mg/kg (R, p.o.)
CN: 2-[(2,6-dichloro-3-methylphenyl)amino]benzoic acid

monosodium salt
RN: 6385-02-0   MF: $C_{14}H_{10}Cl_2NNaO_2$ MW: 318.14   EINECS: 228-983-3

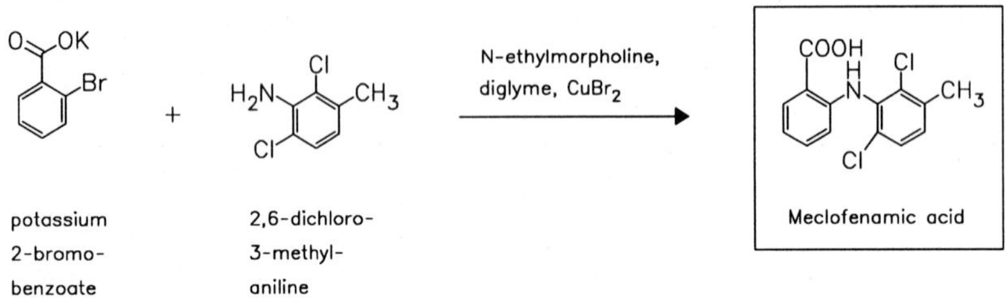

potassium 2-bromo-benzoate

＋

2,6-dichloro-3-methyl-aniline

N-ethylmorpholine, diglyme, CuBr₂

Meclofenamic acid

Reference(s):

DE 1 149 015 (Parke Davis; appl. 22.6.1961; USA-prior. 12.1.1961).
US 3 313 848 (Parke Davis; 11.4.1967; prior. 12.1.1961, 18.9.1962, 18.6.1964).

Formulation(s):   cps. 50 mg, 100 mg; suppos. 200 mg (as sodium salt)

Trade Name(s):
I: Movens (Inverni della   Meclomen (Warner-   generic
         Beffa); wfm   Lambert); wfm
USA: Meclomen (Parke Davis);   Meclomen (Parke Davis; as
         wfm   sodium salt); wfm

Meclofenoxate
(Centrophenoxine)

ATC: N06BX01
Use: neuroenergetic

RN: 51-68-3   MF: $C_{12}H_{16}ClNO_3$ MW: 257.72   EINECS: 200-116-3
LD_{50}: 1750 mg/kg (M, p.o.);
     2600 mg/kg (R, p.o.)
CN: (4-chlorophenoxy)acetic acid 2-(dimethylamino)ethyl ester

hydrochloride
RN: 3685-84-5   MF: $C_{12}H_{16}ClNO_3 \cdot HCl$ MW: 294.18   EINECS: 222-473-4
LD_{50}: 330 mg/kg (M, i.v.); 1750 mg/kg (M, p.o.);
     865 mg/kg (R, p.o.)

4-chlorophenoxy-
acetic acid

2-(dimethylamino)-
ethyl chloride

Meclofenoxate

Reference(s):
Thuillier, G.; Rumpf, P.; Thuillier, J.: C. R. Hebd. Seances Acad. Sci. (COREAF) **249**, 2081 (1959).
FR 398 M (Centre Nat'l. Recherche Sci., appl. 15.4.1959).

Formulation(s):   amp. 250 mg, 500 mg, 2 g (as hydrochloride); drg. 200 mg, 500 mg; f. c. tabl. 100 mg, 250 mg

Trade Name(s):
D: CERUTIL (Isis Pharma)
     Helfergin (Promonta
     Lundbeck; Isis Pharma; as
     hydrochloride)

F: Lucidril (Lipha Santé; as
     hydrochloride)
GB: Lucidril (Reckitt &
     Colman); wfm
I: Lucidril (Bracco); wfm

J: Lucidril (Dainippon)
     Meclon (Toho)
     Mecroeat (Hishiyama)
     Proseryl (Funai)

Mecloqualone

ATC: N05C
Use: hypnotic, sedative

RN: 340-57-8   MF: $C_{15}H_{11}ClN_2O$ MW: 270.72   EINECS: 206-432-8
LD_{50}: 470 mg/kg (M, p.o.)
CN: 3-(2-chlorophenyl)-2-methyl-4(3H)-quinazolinone

N-acetyl-
anthranilic
acid

2-chloro-
aniline

Mecloqualone

Reference(s):
Jackman, G.B. et al.: J. Pharm. Pharmacol. (JPPMAB) **12**, 528 (1960).
Klosa, J.: J. Prakt. Chem. (JPCEAO) **14** [4], 84 (1961).

Meclozine
(Histamethizine; Meclizine)

ATC: R06AE05
Use: antihistaminic

RN: 569-65-3   MF: $C_{25}H_{27}ClN_2$ MW: 390.96   EINECS: 209-323-3
LD_{50}:   1650 mg/kg (M, p.o.);
         1750 mg/kg (R, p.o.)
CN: 1-[(4-chlorophenyl)phenylmethyl]-4-[(3-methylphenyl)methyl]piperazine

dihydrochloride
RN: 1104-22-9   MF: $C_{25}H_{27}ClN_2 \cdot 2HCl$ MW: 463.88   EINECS: 214-164-8
LD_{50}:   1600 mg/kg (M, p.o.)
dihydrochloride monohydrate
RN: 31884-77-2   MF: $C_{25}H_{27}ClN_2 \cdot 2HCl \cdot H_2O$ MW: 481.90

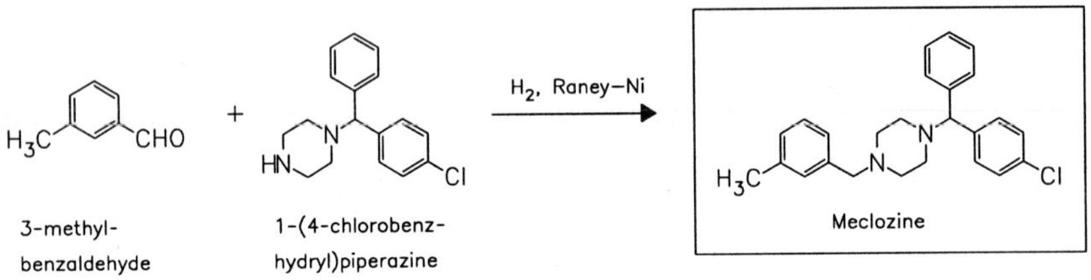

3-methyl-
benzaldehyde

1-(4-chlorobenz-
hydryl)piperazine

Meclozine

Reference(s):
US 2 709 169 (UCB; 1955; B-prior. 1951).

Formulation(s):   drg. 12.5 mg; suppos. 50 mg; tabl. 12.5 mg, 25 mg, 50 mg (as dihydrochloride)

Mecobalamin
(Methylcobalamin)

ATC: V03AB
Use: vitamin B_{12}-preparations

RN: 13422-55-4   MF: $C_{63}H_{91}CoN_{13}O_{14}P$ MW: 1344.41   EINECS: 236-535-3
CN: cobinamide *Co*-methyl deriv. hydroxide dihydrogen phosphate (ester) inner salt 3'-ester with 5,6-
       dimethyl-1-α-D-ribofuranosyl-1*H*-benzimidazole

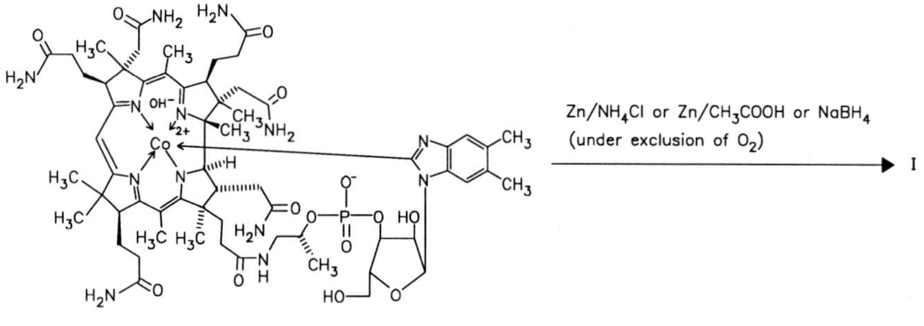

hydroxocobalamin

(q. v.)

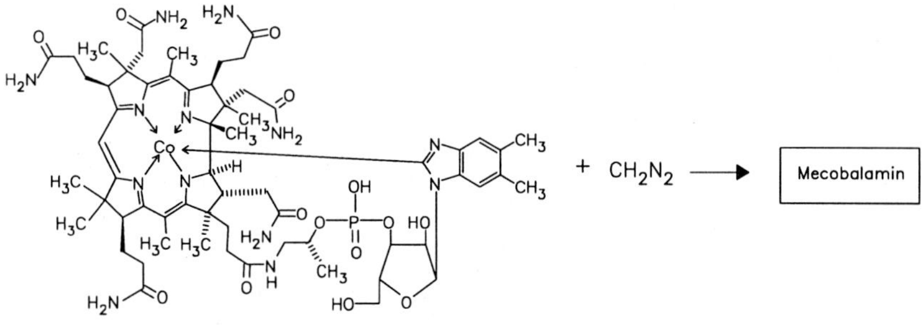

Zn/NH$_4$Cl or Zn/CH$_3$COOH or NaBH$_4$
(under exclusion of O$_2$)

$\longrightarrow$ I

hydridocobalamin   (I)

+ CH$_2$N$_2$ $\longrightarrow$ Mecobalamin

Mecobalamin

I + H$_3$C–O–S(=O)$_2$–O–CH$_3$ $\longrightarrow$ Mecobalamin

I + H$_3$C—I 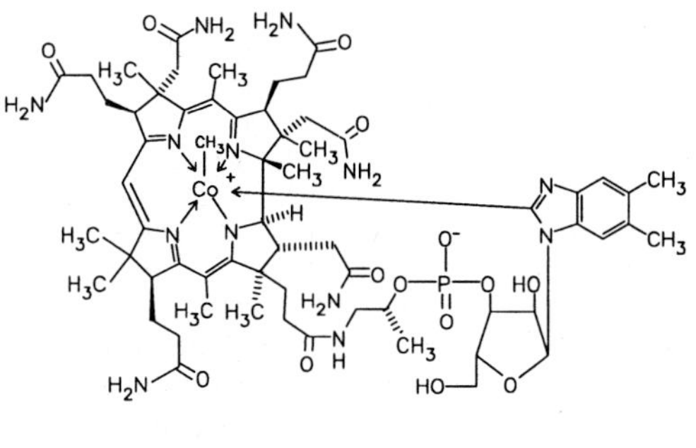 $\longrightarrow$ Mecobalamin

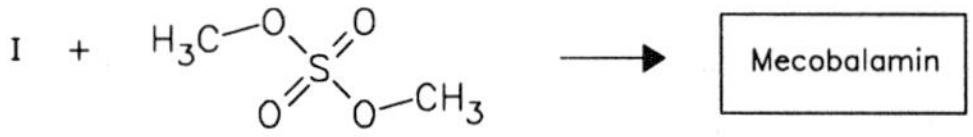

Reference(s):
Müller, O.; Müller, G.: Biochem. Z. (BIZEA2) **336**, 299 (1962).
Dolphin, D.H.; Johnson, A.W.: Proc. Chem. Soc., London (PCSLAW) **1963**, 311.
Dolphin, D.H.; Johnson, A.W.: J. Chem. Soc. (JCSOA9) **1965**, 2174.
Boos, R.N. et al.: Science (Washington, D.C.) (SCIEAS) **117**, 603 (1953).
Smith, E.L. et al.: Nature (London) (NATUAS) **194**, 1175 (1962).

review:
Bernhauer, K. et al.: Angew. Chem. (ANCEAD) **75**, 1145 (1963).

Formulation(s):   tabl. 500 µg; vial 10 µg, 500 µg

Trade Name(s):

F:	Algobaz (Labaz)	J:	Calomide-Me
	Lyométhyl (Bouchara)		(Yamanouchi)
	Méthylcobaz (Labaz)		Cobamain (Kyowa)
			Cobametin (Sankyo)

Hitocobamin-M
(Hishiyama)
Vancomin (Dainippon)

Mecysteine hydrochloride

Use: mucolytic agent

RN: 18598-63-5   MF: $C_4H_9NO_2S \cdot HCl$ MW: 171.65   EINECS: 227-208-6
LD_{50}:   2300 mg/kg (M, p.o.)
CN: L-cysteine methyl ester hydrochloride

mecysteine
RN: 2485-62-3   MF: $C_4H_9NO_2S$ MW: 135.19   EINECS: 219-625-7

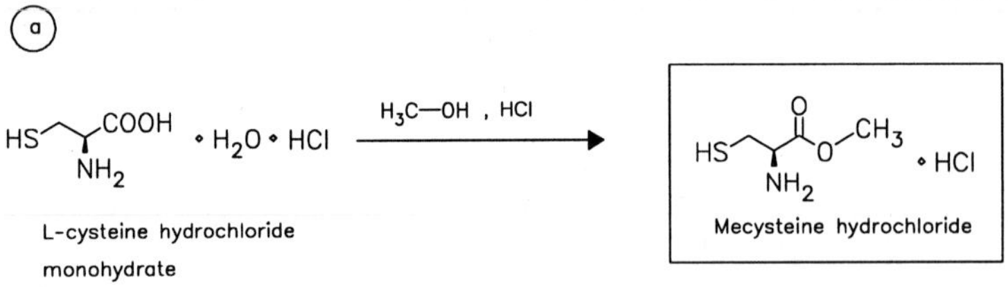

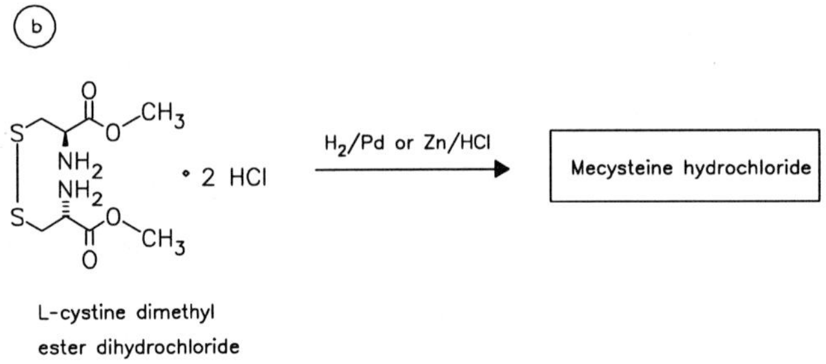

L-cysteine hydrochloride
monohydrate

Mecysteine hydrochloride

L-cystine dimethyl
ester dihydrochloride

Mecysteine hydrochloride

Reference(s):
a   Bergmann, M.; Michalis, G.: Ber. Dtsch. Chem. Ges. (BDCGAS) **63**, 987 (1930).
b   Zervas, L.; Theodoropoulos, D.M.: J. Am. Chem. Soc. (JACSAT) **78**, 1359 (1956).

Formulation(s):   drg. 100 mg; suppos. 100 mg, 200 mg; tabl. 0.05 g, 0.1 g

Trade Name(s):

F:	Acthiol J. (Joullié); wfm	Ectazis (Nichiiko)	Fuszemin S (Taiyo)
GB:	Visclair (Sinclair)	Epecoal (Beppu)	Higlomin (Wakamoto)
I:	Actiol (SIT)	Epectan (Seiko)	Jeorgen (Sanwa)
	Donatiol (AGIPS)-comb.	Equverin (Nissin)	Moltanine (Toho K.-Tokyo
J:	Aslos-C (Nissin)	Fustant (Kanto)	Tanabe)

Pectite (Kissei)
Pelmain (Sawai)

Radcol (Nippon Universal)
Sekinin (Tokyo Hosei)

Thibrin (Kyowa-Hoei)
Zeotin (Toa Eiyo)

Medazepam

ATC: N05BA03
Use: tranquilizer, anxiolytic

RN: 2898-12-6   MF: $C_{16}H_{15}ClN_2$ MW: 270.76   EINECS: 220-783-4
LD$_{50}$:   475 mg/kg (M, p.o.);
        900 mg/kg (R, p.o.)
CN: 7-chloro-2,3-dihydro-1-methyl-5-phenyl-1H-1,4-benzodiazepine

a

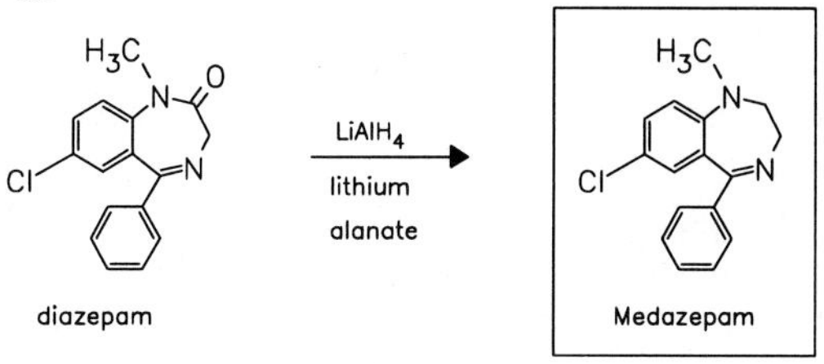

diazepam → (LiAlH₄, lithium alanate) → **Medazepam**

b

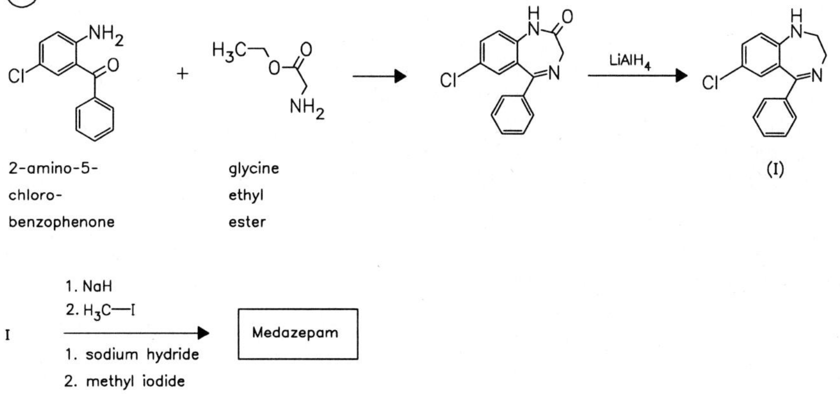

2-amino-5-chloro-benzophenone + glycine ethyl ester → (I) [via LiAlH₄]

I → (1. NaH, 2. H₃C–I / 1. sodium hydride, 2. methyl iodide) → Medazepam

c

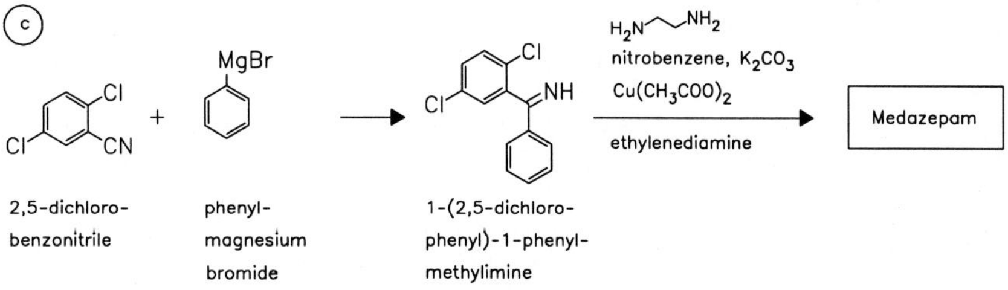

2,5-dichloro-benzonitrile + phenyl-magnesium bromide → 1-(2,5-dichloro-phenyl)-1-phenyl-methylimine → (H₂N–NH₂, nitrobenzene, K₂CO₃, Cu(CH₃COO)₂, ethylenediamine) → Medazepam

d

4-chloro-aniline + p-toluene-sulfochloride → 4-chloro-1-tosylamino-benzene → (H₃C–O–SO₂–O–CH₃, NaOCH₃, dimethyl sulfate) → II

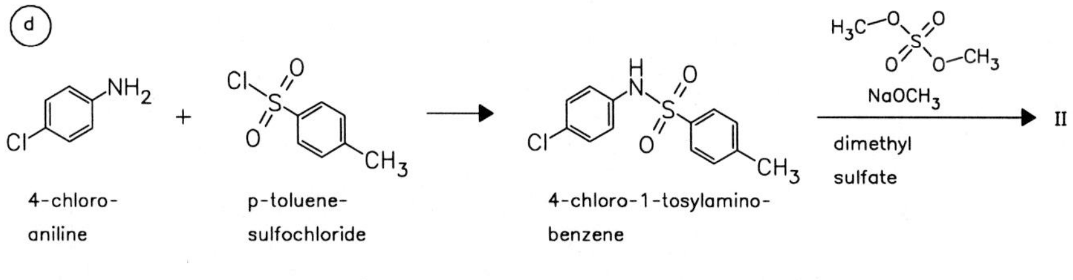

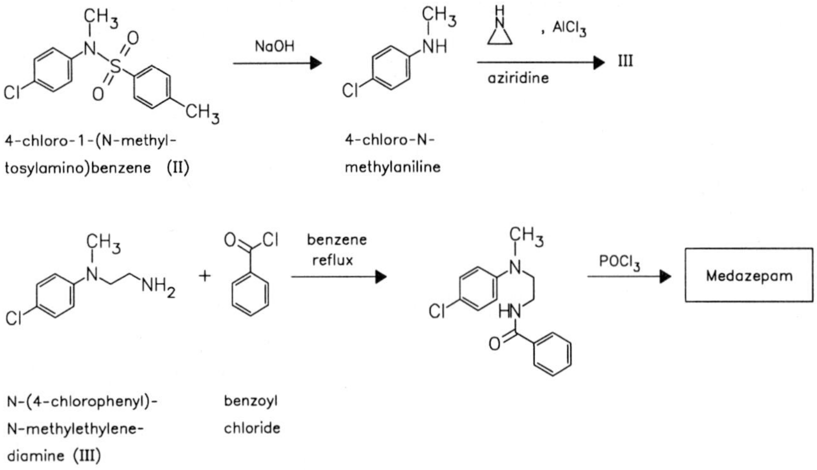

4-chloro-1-(N-methyl-tosylamino)benzene (II)

4-chloro-N-methylaniline

N-(4-chlorophenyl)-N-methylethylene-diamine (III)

benzoyl chloride

Medazepam

Reference(s):

a,b US 3 109 843 (Roche; 5.11.1963; appl. 21.6.1962; prior. 28.7.1961).
US 3 131 178 (Roche; 28.4.1964; prior. 28.7.1961, 4.12.1961, 21.6.1962).
Sternbach, L.H. et al.: J. Org. Chem. (JOCEAH) **28**, 2456 (1963).
c DAS 1 934 385 (Sumitomo; appl. 7.7.1969).
d DAS 1 695 188 (Roche; appl. 23.5.1967; USA-prior. 3.6.1966).
DAS 1 795 811 (Roche; appl. 23.5.1967).

alternative syntheses:
US 3 141 890 (Roche; 21.7.1964; prior. 28.7.1961, 4.12.1961, 21.6.1962).
US 3 144 439 (Roche; 11.8.1964; prior. 28.7.1961, 4.12.1961, 21.6.1962).
DE 1 445 864 (Roche; appl. 27.7.1962; USA-prior. 28.7.1961).
DOS 2 204 484 (Sumitomo; appl. 31.1.1972; J-prior. 9.2.1971, 6.4.1971, 28.5.1971).
DOS 2 217 301 (Sumitomo; appl. 10.4.1972; J-prior. 12.4.1971).

1-demethyl-derivative from 5-chloro-2-(2,3-dioxopiperazino)benzophenone:
DAS 1 906 254 (Sumitomo; appl. 7.2.1969; J-prior. 2.4.1968).
DAS 1 965 980 (Sumitomo; appl. 7.2.1969; J-prior. 2.4.1968).

Formulation(s):   cps. 5 mg, 10 mg; tabl. 2.5 mg, 5 mg, 10 mg

Trade Name(s):

D:	Medazepam AWD Tabletten (ASTA Medica AWD) Rudotel Tabletten (OPW)	I:	Debrum (Sigma-Tau)-comb. Nobrium (Roche)
F:	Nobrium (Roche); wfm	J:	Azepamid (Taiyo) Cerase (Torii)
GB:	Nobrium (Roche); wfm		

Kobazepam (Nihon Iyakuhin)
Metonas (Kanto)
Narsis (Sumitomo)
Nobrium (Nippon Roche)
Resmit (Shionogi)

Medibazine

ATC: C01
Use: coronary vasodilator

RN: 53-31-6   MF: $C_{25}H_{26}N_2O_2$ MW: 386.50   EINECS: 200-168-7
LD_{50}:   41 mg/kg (M, i.v.)
CN: 1-(1,3-benzodioxol-5-ylmethyl)-4-(diphenylmethyl)piperazine

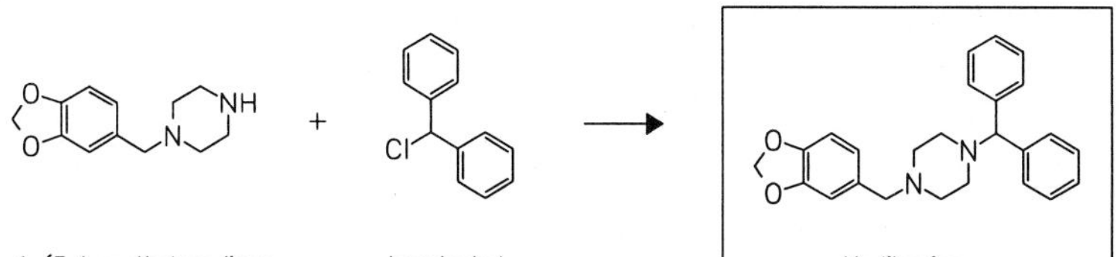

1-(3,4-methylenedioxy-benzyl)piperazine + benzhydryl chloride → Medibazine

Reference(s):
US 3 119 826 (Science Union; 28.1.1964; F-prior. 12.4.1961).

Trade Name(s):
F: Vialibran (Servier); wfm

Medifoxamine

ATC: N06A
Use: antidepressant

RN: 32359-34-5   MF: C$_{16}$H$_{19}$NO$_2$ MW: 257.33   EINECS: 251-011-4
LD$_{50}$:   750 mg/kg (M, p.o.)
CN: *N,N*-dimethyl-2,2-diphenoxyethanamine

fumarate (1:1)
RN: 16604-45-8   MF: C$_{16}$H$_{19}$NO$_2$ · C$_4$H$_4$O$_4$ MW: 373.41   EINECS: 240-657-2

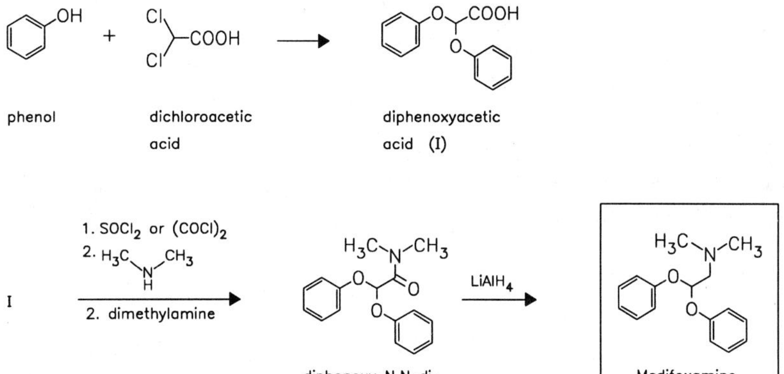

phenol + dichloroacetic acid → diphenoxyacetic acid (I)

I →
1. SOCl$_2$ or (COCl)$_2$
2. H$_3$C–N(H)–CH$_3$
2. dimethylamine
→ diphenoxy-N,N-dimethylacetamide →$_{LiAlH_4}$→ Medifoxamine

Reference(s):
FR 5 498 (Lab. Gerda; appl. 1966).
Brunet, M.A. et al.: Bull. Soc. Chim. Fr. (BSCFAS), 2000 (1967).

additional synthesis:
FR 2 645 147 (Lab. Rolland; appl. 3.4.1989).
FR 2 601 004 (Lab. Rolland; appl. 7.7.1986).
FR 2 588 553 (Lab. Rolland; appl. 16.10.1985).
EP 226 475 (Lab. Rolland; appl. 22.7.1985).

synthesis of diphenoxyacetic acid:
Alphen, J. van: Recl. Trav. Chim. Pays-Bas (RTCPA3) **46**, 144 (1927).
Scheibler, H.; Depner, M.: J. Prakt. Chem. (JPCEAO) **7**, 60 (1958).
DE 561 281 (Chem. Fabrik von Heyden; appl. 1930),
also EP 226 475, FR 2 601 004.

medical use for treatment of cerebral hypoxia and senility:
FR 2 589 357 (Lab. Rolland; appl. 5.11.1985).
FR 2 583 639 (Lab. Rolland; appl. 24.6.1985).

Formulation(s): tabl. 50 mg

Trade Name(s):
F: Clédial (Lipha Santé)

Medrogestone

ATC: G03DB03
Use: progestogen

RN: 977-79-7 MF: $C_{23}H_{32}O_2$ MW: 340.51 EINECS: 213-555-0
LD_{50}: 850 mg/kg (g.p., p.o.)
CN: 6,17-dimethylpregna-4,6-diene-3,20-dione

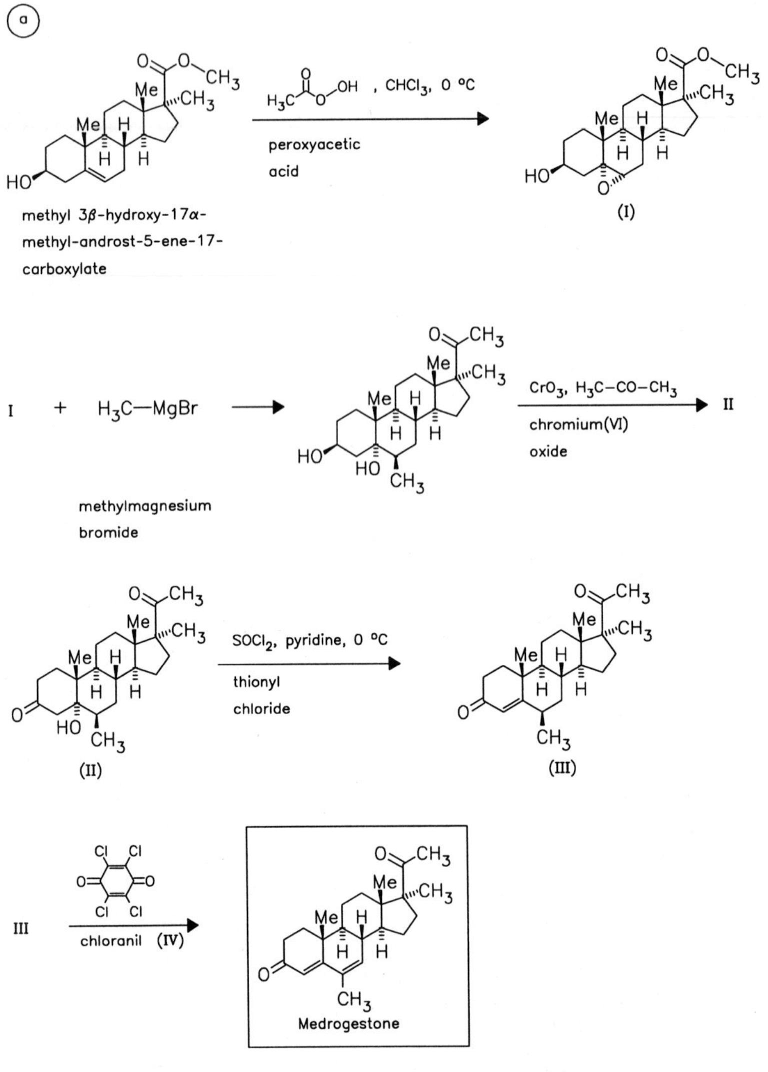
a
methyl 3β-hydroxy-17α-methyl-androst-5-ene-17-carboxylate
, CHCl3, 0 °C
peroxyacetic acid
(I)
I
H3C—MgBr
methylmagnesium bromide
CrO3, H3C–CO–CH3
chromium(VI) oxide
II
(II)
SOCl2, pyridine, 0 °C
thionyl chloride
(III)
III
chloranil (IV)
Medrogestone

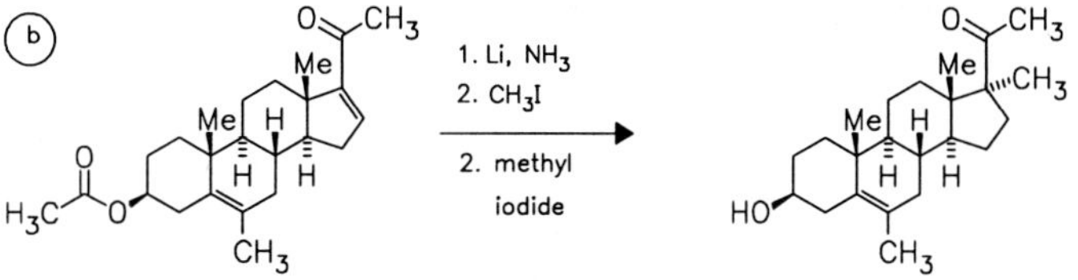

(b)

1. Li, NH₃
2. CH₃I

2. methyl
iodide

(V)

3β-acetoxy-6-methyl-
20-oxo-5,16-pregnadiene
(from dehydropregnenolone,
analogously to a)

1. Oppenauer oxidation
2. IV

V ──────────→ Medrogestone

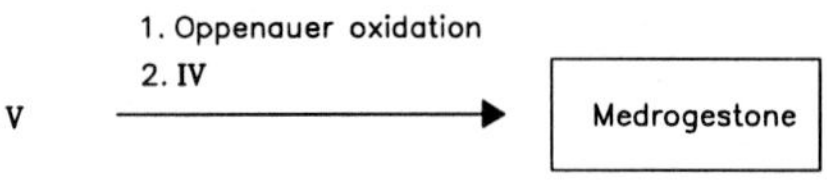

Reference(s):

a   US 3 133 913 (American Home Products, 19.5.1964; appl. 11.9.1961).
   Deghenghi, R.; Gaudry, R.: J. Am. Chem. Soc. (JACSAT) **83**, 4668 (1961).
   starting material:
   Plattner, P.A. et al.: Helv. Chim. Acta (HCACAV) **31**, 603 (1948).
b   Deghenghi, R. et al.: J. Med. Chem. (JMCMAR) **6**, 301 (1963).
   starting material:
   Burn, D. et al.: J. Chem. Soc. (JCSOA9) **1957**, 4092.

alternative syntheses:
US 3 170 936 (American Home Products; 23.2.1965; appl. 7.8.1963).
US 3 210 387 (American Home Products; 5.10.1965; appl. 6.5.1963; CDN-prior. 28.11.1962).

Formulation(s):   tabl. 5 mg, 25 mg

Trade Name(s):
D: Presomen (Solvay   Prothil (Solvay   I: Colprone (Wyeth)
     Arzneimittel)-comb.   Arzneimittel)   USA: Colprone (Ayerst); wfm
                         F: Colprone (Wyeth-Lederle)

Medroxyprogesterone acetate

ATC: G02B; G03D
Use: antineoplastic, progestogen

RN: 71-58-9   MF: C₂₄H₃₄O₄   MW: 386.53   EINECS: 200-757-9
LD₅₀: >16 g/kg (M, p.o.);
        >6.4 g/kg (R, p.o.);
        >5 g/kg (dog, p.o.)
CN: (6α)-17-(acetyloxy)-6-methylpregn-4-ene-3,20-dione

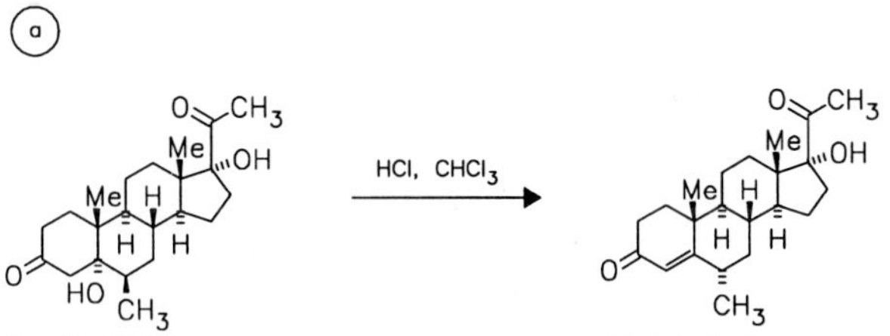

(a)

HCl, CHCl₃

5α,17α-dihydroxy-
3,20-dioxo-6β-methyl-
pregnane

17-hydroxy-6α-
methylprogesterone   (I)

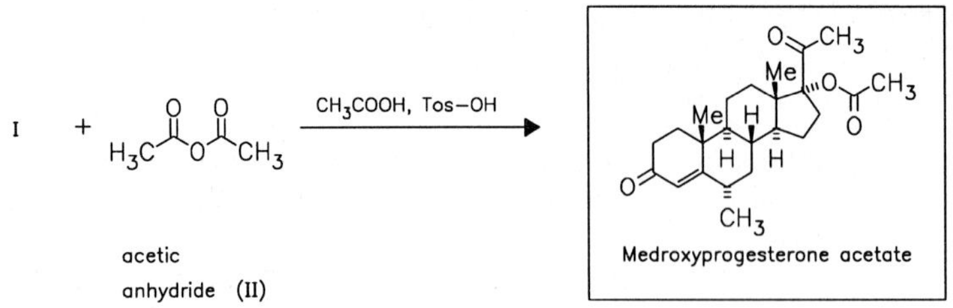

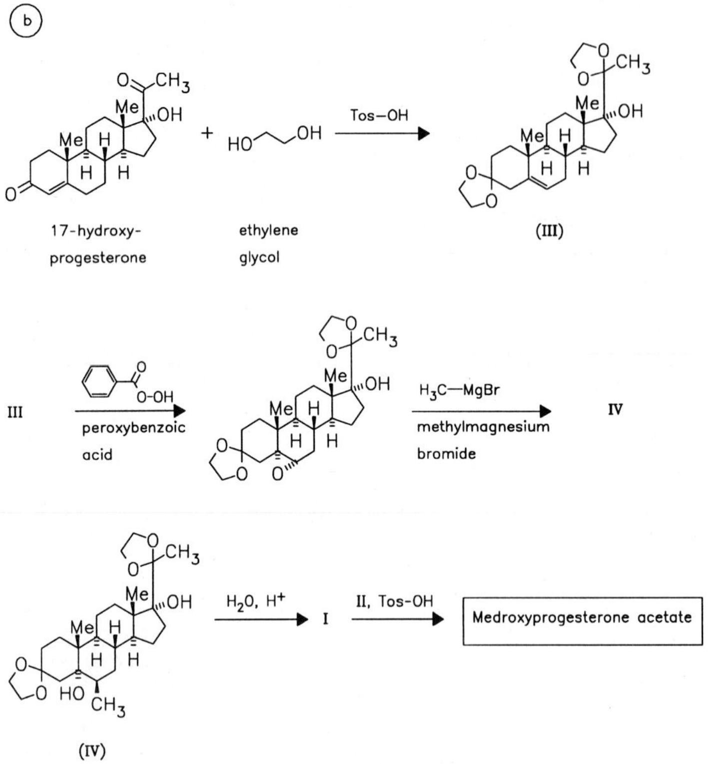

I + acetic anhydride (II) → Medroxyprogesterone acetate

(b)

17-hydroxy-progesterone + ethylene glycol → (III)

III → peroxybenzoic acid → → methylmagnesium bromide → IV

(IV) → H₂O, H⁺ → I → II, Tos-OH → Medroxyprogesterone acetate

Reference(s):
a US 3 147 290 (Upjohn; 1.9.1964; appl. 17.5.1961; prior. 23.11.1956).
Ellis, B. et al.: J. Chem. Soc. (JCSOA9) **1957**, 4092.
starting material: cf. literature cited under **a**
b US 3 061 616 (Societa Farmaceutici Italia; 30.10.1962; appl. 17.9.1958; GB-prior. 24.4.1958).
DE 1 097 986 (Syntex; appl. 29.8.1957; MEX-prior. 8.9.1956).
Babcock, J.C. et al.: J. Am. Chem. Soc. (JACSAT) **80**, 2904 (1958).
starting material:
Bernstein, S. et al.: J. Am. Chem. Soc. (JACSAT) **76**, 5674 (1954).
The Merck Index, 4756 (Rahway 1976).

alternative syntheses:
DE 1 081 456 (British Drug Houses; appl. 21.8.1958; GB-prior. 13.8.1957).
DE 1 101 415 (Searle; appl. 24.9.1958; USA-prior. 27.9.1957).
US 3 043 832 (Ormonoterapia Richter; 10.7.1962; appl. 28.4.1961; I-prior. 27.2.1961).

review:
Ehrhardt, Ruschig **III**, 352.

Formulation(s):　amp. 500 mg, 1 g; susp. 500 mg; susp. 150 mg/ml, 500 mg/ml; tabl. 2.5 mg, 5 mg, 10 mg, 100 mg, 200 mg, 250 mg, 400 mg, 500 mg

Trade Name(s):

D:　Clinofem (Pharmacia & Upjohn)-comb.
　　Clinovir (Pharmacia & Upjohn)
　　Depo-Clinovir (Pharmacia & Upjohn)
　　Farlutal (Pharmacia & Upjohn)
　　MPA (Hexal)-comb.
F:　Depo-Prodasone (Pharmacia & Upjohn)-comb.
　　Depo-Provera (Pharmacia & Upjohn)

　　　Divina (Innothéra)-comb.
　　　Farlutal (Pharmacia & Upjohn)
　　　Gestoral (Novartis)
　　　Prodasone (Pharmacia & Upjohn)
GB:　Depo-Provera (Pharmacia & Upjohn)
　　　Farlutal (Pharmacia & Upjohn)
　　　Provera (Pharmacia & Upjohn)
　　　combination preparations
I:　　Depo-Provera (Upjohn)

　　　Farlutal (Farmitalia)
　　　Lutoral (Midy)
　　　Provera (Upjohn)
J:　　Hysron (Kyowa)
　　　Provera (Upjohn)
USA:　Amen (Carnrick)
　　　Depo-Provera (Pharmacia & Upjohn)
　　　Premphase (Wyeth-Ayerst)
　　　Prempro (Wyeth-Ayerst)
　　　Provera (Pharmacia & Upjohn)

Medrylamine

ATC:　R06
Use:　topical antihistaminic

RN:　524-99-2　MF: $C_{18}H_{23}NO_2$　MW: 285.39　EINECS: 208-368-6
CN:　2-[(4-methoxyphenyl)phenylmethoxy]-*N,N*-dimethylethanamine

hydrochloride
RN:　6027-00-5　MF: $C_{18}H_{23}NO_2 \cdot HCl$　MW: 321.85　EINECS: 227-888-4
LD$_{50}$:　148 mg/kg (M, i.p.)

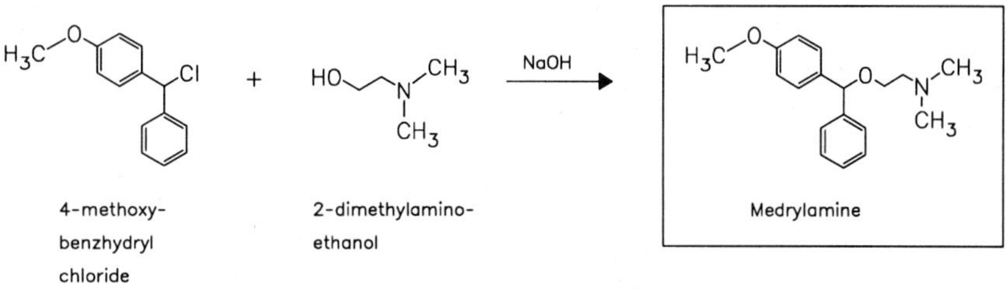

4-methoxy-
benzhydryl
chloride

2-dimethylamino-
ethanol

Medrylamine

Reference(s):
US 2 668 856 (UCB; 1954; appl. 1948).

Formulation(s):　ointment 20 mg/g (2 %) (as hydrochloride)

Trade Name(s):
D:　Corti-Postafen (UCB)-comb.; wfm

　　Postafen Salbe (UCB); wfm

Medrysone

ATC:　S01BA08
Use:　glucocorticoid

RN:　2668-66-8　MF: $C_{22}H_{32}O_3$　MW: 344.50　EINECS: 220-208-7
LD$_{50}$:　338 mg/kg (R, i.p.)
CN:　(6α,11β)-11-hydroxy-6-methylpregn-4-ene-3,20-dione

6α-methyl-11-oxo-
progesterone

1. HO⌒OH H₃C—⟨⟩—SO₃H
2. LiAlH₄, THF
3. H₂SO₄, CH₃OH

1. ethylene glycol, p-toluenesulfonic acid
2. lithium alanate

Medrysone

Reference(s):
US 2 864 837 (Upjohn; 1958; prior. 1958).
US 2 968 655 (Upjohn; 1961; prior. 1956).

starting material:
Spero, G.B. et al.: J. Am. Chem. Soc. (JACSAT) **78**, 6213 (1956).
US 2 968 655 (Upjohn; 1961; prior. 1956).

Formulation(s):   eye drops 10 mg; eye ointment 10 mg

Trade Name(s):
D: Ophtocortin (Winzer)   generic   Medrocort (Upjohn); wfm
      Spectramedryn (Pharm-   I: Medramil (Farmigea)-
      Allergan)   comb.
F: Medryson Faure (CIBA   USA: HMS Liquifilm (Allergan);
      Vision)   wfm

Mefenamic acid

ATC: M01AG01
Use: anti-inflammatory, antirheumatic,
        analgesic

RN: 61-68-7   MF: C₁₅H₁₅NO₂   MW: 241.29   EINECS: 200-513-1
LD₅₀: 96 mg/kg (M, i.v.); 525 mg/kg (M, p.o.);
       112 mg/kg (R, i.v.); 740 mg/kg (R, p.o.)
CN: 2-[(2,3-dimethylphenyl)amino]benzoic acid

potassium   2,3-dimethyl-   Mefenamic acid
2-bromo-   aniline
benzoate

Reference(s):
US 3 138 636 (Parke Davis; 23.6.1964; appl. 23.6.1960).
DE 1 163 846 (Parke Davis; appl. 22.6.1961; USA-prior. 23.6.1960).

alternative syntheses:
DAS 1 186 073 (Parke Davis; appl. 30.4.1963; CDN-prior. 18.9.1962).
DAS 1 186 074 (Parke Davis; appl. 30.4.1963; CDN-prior. 18.9.1962).
DAS 1 186 870 (Parke Davis; appl. 30.4.1963; CDN-prior. 18.9.1962).
DAS 1 186 871 (Parke Davis; appl. 30.4.1963; CDN-prior. 18.9.1962).

Formulation(s):   cps. 250 mg; powder 500 mg, 1 g; suppos. 125 mg, 500 mg; susp. 50 mg

Trade Name(s):
D: Parkemed (Parke Davis)   J: Baphameritin M   USA: Ponstan (Parke Davis);
    Ponalar (Parke Davis)   (Hishiyama)   wfm
F: Ponstyl (Parke Davis)   Bonabol (Sawai)   Ponstel (Parke Davis)
GB: Meflam (Trinity)   Pontal (Parke Davis-
    Ponstan (Elan)   Sankyo)
I: Lysalgo (SIT)   Spantac (Uji)

Mefenorex

ATC: A08AA09
Use: appetite depressant

RN: 17243-57-1   MF: C$_{12}$H$_{18}$ClN   MW: 211.74   EINECS: 241-279-0
CN: N-(3-chloropropyl)-α-methylbenzeneethanamine

hydrochloride
RN: 5586-87-8   MF: C$_{12}$H$_{18}$ClN · HCl   MW: 248.20   EINECS: 226-985-9
LD$_{50}$: 49 mg/kg (M, i.v.); 230 mg/kg (M, p.o.);
         35 mg/kg (R, i.v.); 410 mg/kg (R, p.o.)

phenylacetone   3-amino-   Mefenorex
                  1-propanol

Reference(s):
DE 1 210 873 (Hoffmann-La Roche; appl. 18.3.1959).

Formulation(s): drg. 40 mg (as hydrochloride)

Trade Name(s):
D: Rondimen (ASTA Medica   F: Incital (Pierre Fabre Santé)
    AWD; as hydrochloride)   USA: Anexate (Roche); wfm

Mefexamide

ATC: N05C
Use: psychoanaleptic, CNS stimulant

RN: 1227-61-8   MF: C$_{15}$H$_{24}$N$_2$O$_3$ MW: 280.37   EINECS: 214-963-1
LD$_{50}$: 168 mg/kg (M, i.v.); 1500 µg/kg (M, p.o.)
CN: N-[2-(diethylamino)ethyl]-2-(4-methoxyphenoxy)acetamide

monohydrochloride
RN: 3413-64-7   MF: C$_{15}$H$_{24}$N$_2$O$_3$ · HCl   MW: 316.83   EINECS: 222-304-4

hydroquinone   chloroacetic   4-methoxy-   (I)
monomethyl   acid   phenoxyacetic
ether   acid

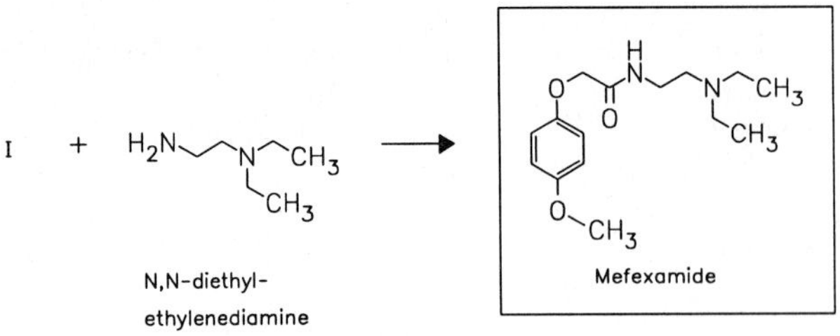

N,N–diethyl–
ethylenediamine

Mefexamide

Reference(s):
Thuillier, G.; Rumpf, P.: Bull. Soc. Chim. Fr. (BSCFAS) **1960**, 1786.

Formulation(s): amp. 150 mg; tabl. 150 mg

Trade Name(s):
F: Méféxadyne (Anphar);   Timodyne (Anphar); wfm
    wfm   I: Perneuron (Crinos); wfm

Mefloquine

ATC: P01BA05
Use: antimalarial

RN: 53230-10-7   MF: $C_{17}H_{16}F_6N_2O$ MW: 378.32
CN: (R^*,S^*)-($\pm$)-α-2-piperidinyl-2,8-bis(trifluoromethyl)-4-quinolinemethanol

monohydrochloride
RN: 51773-92-3   MF: $C_{17}H_{16}F_6N_2O \cdot HCl$ MW: 414.78   EINECS: 257-412-0
LD_{50}: 880 mg/kg (R, p.o.)

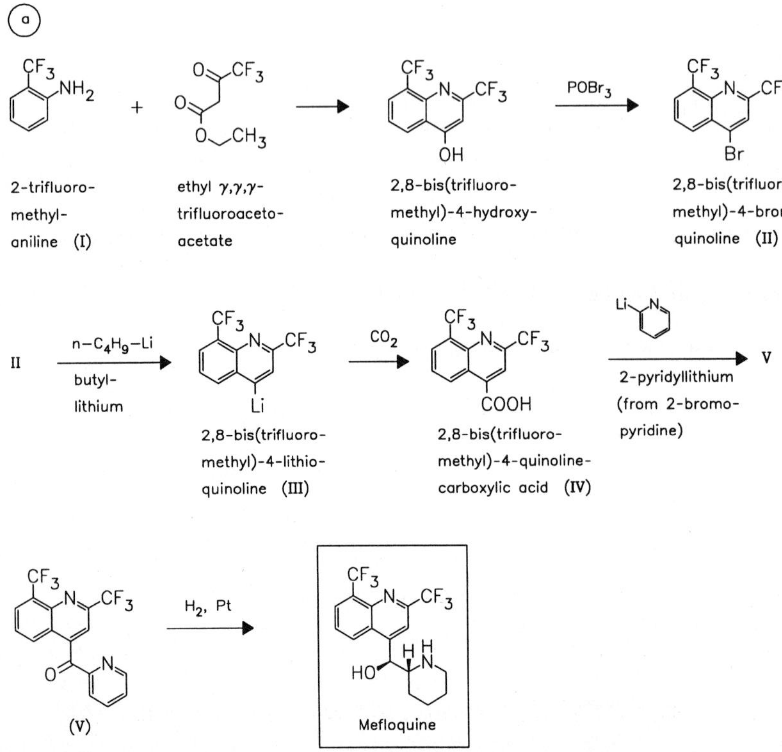

ⓐ

2-trifluoro-
methyl-
aniline (I)

ethyl γ,γ,γ-
trifluoroaceto-
acetate

2,8-bis(trifluoro-
methyl)-4-hydroxy-
quinoline

2,8-bis(trifluoro-
methyl)-4-bromo-
quinoline (II)

II   n–C₄H₉–Li
butyl-
lithium

2,8-bis(trifluoro-
methyl)-4-lithio-
quinoline (III)

CO_2

2,8-bis(trifluoro-
methyl)-4-quinoline-
carboxylic acid (IV)

2-pyridyllithium
(from 2-bromo-
pyridine)

V

(V)

H_2, Pt

Mefloquine

b

III +

pyridine-2-
carboxaldehyde

α-2-pyridyl-2,8-
bis(trifluoromethyl)-
4-quinolinemethanol

H₂, PtO → Mefloquine

c

I +

chloral
hydrate

NH₂OH · HCl
hydroxylamine
hydrochloride

H₂SO₄

7-trifluoromethyl-
isatin (VI)

VI +

1,1,1-trifluoro-
acetone

NaOH → IV

1. LiOH
   BrMg N
2.

2. 2-pyridylmagnesium
   bromide

V → H₂, Pt → Mefloquine

Reference(s):
a   Ohnmacht, C.J. et al.: J. Med. Chem. (JMCMAR) **14**, 926 (1971).
b   DOS 2 806 909 (Roche; appl. 17.2.1978; USA-prior. 17.2.1977).
c   DOS 2 940 443 (BASF; appl. 5.10.1979).

alternative synthesis:
EP 103 259 (Roche; appl. 6.9.1983; CH-prior. 10.9.1982).

preparation of pure mefloquine hydrochloride:
US 4 507 482 (Roche; 26.3.1985; CH-prior. 14.4.1982).
EP 92 185 (Roche; appl. 14.4.1983; CH-prior. 14.4.1982).

hydrochloride modification E:
EP 137 375 (Roche; appl. 20.9.1984; CH-prior. 7.10.1983).

Formulation(s):   tabl. 250 mg (as hydrochloride)

Trade Name(s):
D: Lariam (Roche)   GB: Lariam (Roche)
F: Lariam (Roche; as   I: Lariam (Roche)
       hydrochloride)   USA;   Lariam (Roche)

Mefruside

ATC: C03BA05
Use: diuretic

RN: 7195-27-9   MF: C₁₃H₁₉ClN₂O₅S₂   MW: 382.89   EINECS: 230-562-4
LD₅₀: 500 mg/kg (M, i.v.); >10 g/kg (M, p.o.);
        500 mg/kg (R, i.v.); >10 g/kg (R, p.o.);
        >5 g/kg (dog, p.o.)
CN: 4-chloro-*N*¹-methyl-*N*¹-[(tetrahydro-2-methyl-2-furanyl)methyl]-1,3-benzenedisulfonamide

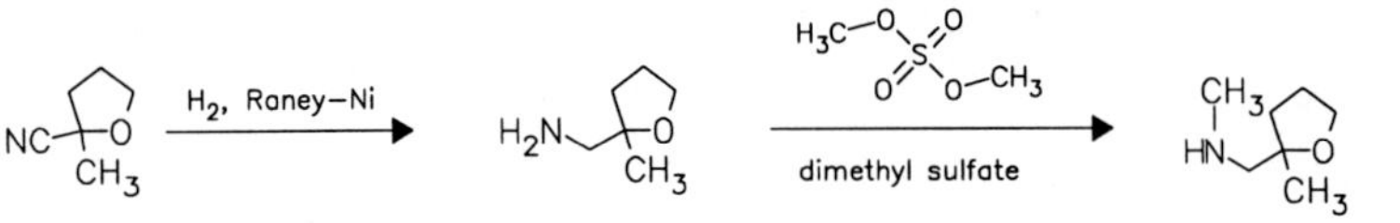

2-cyano-
2-methyl-
tetrahydro-
furan

2-aminomethyl-
2-methyltetra-
hydrofuran

2-methyl-2-
(methylamino-
methyl)tetra-
hydrofuran　(I)

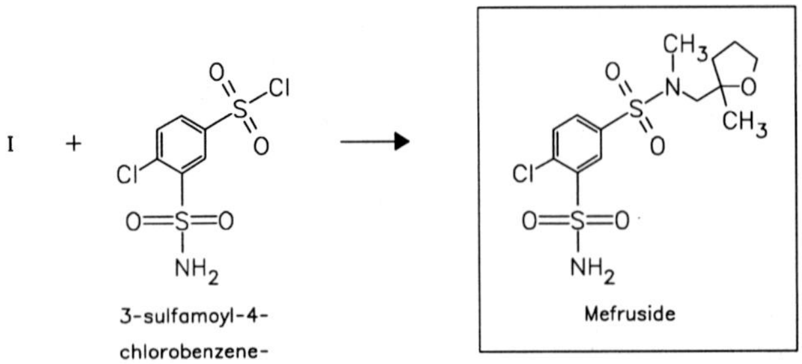

3-sulfamoyl-4-
chlorobenzene-
sulfonyl chloride

Mefruside

Reference(s):
GB 1 031 916 (Bayer; appl. 30.11.1964; D-prior. 30.11.1963).

Formulation(s):　tabl. 25 mg

Trade Name(s):

D:　Baycaron (Bayer Vital)
　　Bendigon (Bayer Vital)-
　　comb.
　　Caprinol (Bayer)-comb.;
　　wfm
　　Duranifin Sali
　　(durachemie)-comb.
　　Sali-Adalat (Bayer Vital)-
　　comb.

　　Sali-Prent (Bayer Vital)-
　　comb.
　　Sali-Presinol (Bayer)-
　　comb.
　　Thomaeamin (Thomae)-
　　comb.; wfm

F:　Tensid (Bayer Pharma)-
　　comb.; wfm

GB:　Baycaron (Bayer)

I:　Baycaron (Bayer); wfm
　　Mefrusal (Bayropharm);
　　wfm
　　Rexitene Plus (LPB)-
　　comb.; wfm

J:　Baycaron (Yoshitomi)

Megestrol acetate

ATC:　G03D
Use:　progestogen (palliative treatment of
　　　breast and endometrial carcinoma)

RN:　595-33-5　MF: $C_{24}H_{32}O_4$　MW: 384.52　EINECS: 209-864-5
LD$_{50}$: 56 mg/kg (M, i.v.)
CN:　17-(acetyloxy)-6-methylpregna-4,6-diene-3,20-dione

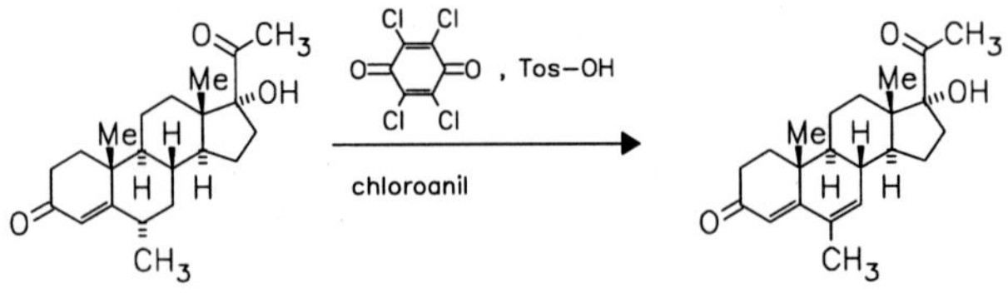

medroxyprogesterone
(q. v.)

megestrol　(I)

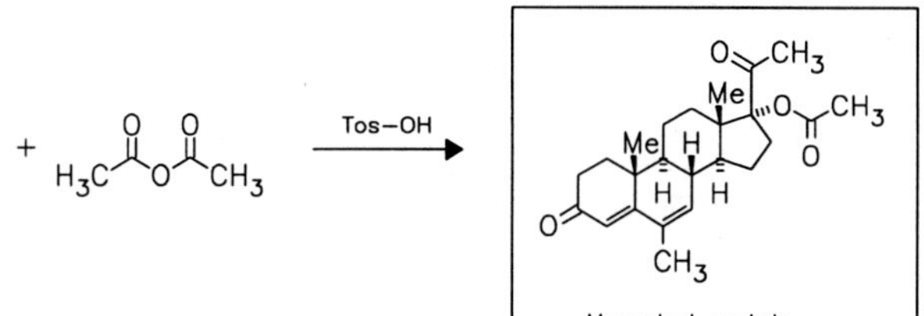

Megestrol acetate

Reference(s):

US 2 891 079 (Searle; 16.6.1959; prior. 23.1.1959).

Ringold, H.J. et al.: J. Am. Chem. Soc. (JACSAT) **81**, 3712 (1959).

Formulation(s):   oral susp. 40 mg/ml; tabl. 20 mg, 40 mg, 160 mg

Trade Name(s):

D:	Megestat (Bristol-Myers Squibb)	GB:	Megace (Bristol-Myers Squibb)		Megestil (Boehringer Mannh.)
	Niagestin (Novo); wfm	I:	Megace (Bristol-Myers Squibb)	USA:	Megace (Bristol-Myers Squibb)
F:	Megace (Bristol-Myers Squibb)				generics

Meglutol

ATC: C10AX05
Use: antihyperlipidemic

RN: 503-49-1   MF: $C_6H_{10}O_5$ MW: 162.14   EINECS: 207-971-1

LD_{50}:   7330 mg/kg (M, p.o.)

CN: 3-hydroxy-3-methylpentanedioic acid

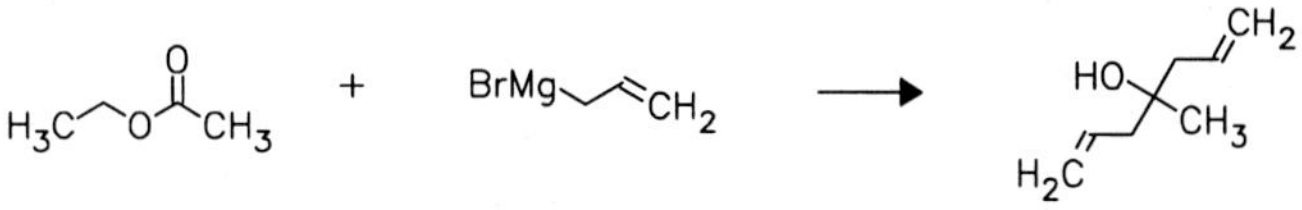

ethyl acetate	allylmagnesium bromide	(I)

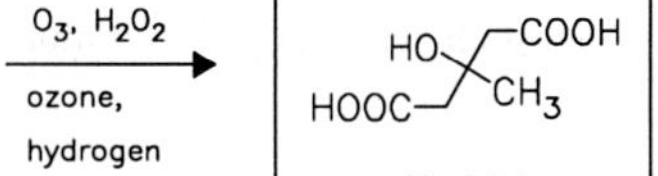

Meglutol

Reference(s):

Rabinowitz, J.L. et al.: Biochem. Prep. (BIPRAP) **6**, 25 (1958).

medical use:

US 3 629 449 (Aligarh Muslim University; 21.12.1971; prior. 22.4.1968).

Formulation(s):   cps. 500 mg; tabl. 1 g

Trade Name(s):

I: Mevalon (Guidotti)

Melengestrol acetate

ATC: G03D
Use: progestogen, antineoplastic

RN: 2919-66-6 MF: $C_{25}H_{32}O_4$ MW: 396.53 EINECS: 220-859-7
CN: 17-(acetyloxy)-6-methyl-16-methylenepregna-4.6-diene-3.20-dione

melengestrol
RN: 5633-18-1 MF: $C_{23}H_{30}O_3$ MW: 354.49 EINECS: 227-073-3

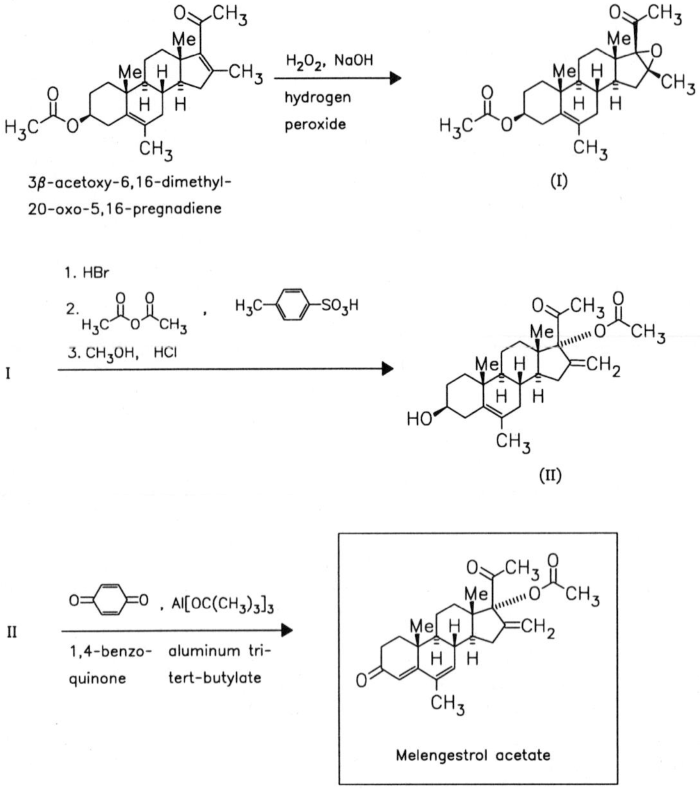

3β-acetoxy-6,16-dimethyl-
20-oxo-5,16-pregnadiene

(I)

(II)

Melengestrol acetate

Reference(s):
GB 886 619 (British Drug Houses; valid from 14.6.1960; prior. 28.12.1959).

starting material:
GB 850 423 (British Drug Houses; valid from 26.6.1959; prior. 9.7.1958).
GB 870 286 (British Drug Houses; valid from 19.10.1959; prior. 4.11.1958).
Kirk, D.N. et al.: J. Chem. Soc. (JCSOA9) **1961**, 2821.

alternative synthesis:
US 3 117 966 (British Drug Houses; 14.1.1964; prior. 27.9.1961).

Trade Name(s):
USA: MGA (Upjohn); wfm

Melitracen

ATC: N06AA14
Use: antidepressant

RN: 5118-29-6 MF: $C_{21}H_{25}N$ MW: 291.44 EINECS: 225-858-5
LD_{50}: 52 mg/kg (M, i.v.); 315 mg/kg (M, p.o.);
170 mg/kg (R, p.o.)
CN: 3-(10,10-dimethyl-9(10H)-anthracenylidene)-N,N-dimethyl-1-propanamine

hydrochloride
RN: 10563-70-9 MF: $C_{21}H_{25}N \cdot HCl$ MW: 327.90 EINECS: 234-150-5
LD_{50}: 52 mg/kg (M, i.v.); 315 mg/kg (M, p.o.);
170 mg/kg (R, p.o.)

anthrone + methyl iodide → (NaOH) 10,10-dimethyl-anthrone → (3-dimethylaminopropyl-magnesium chloride) I

(I) → (HCl) Melitracen

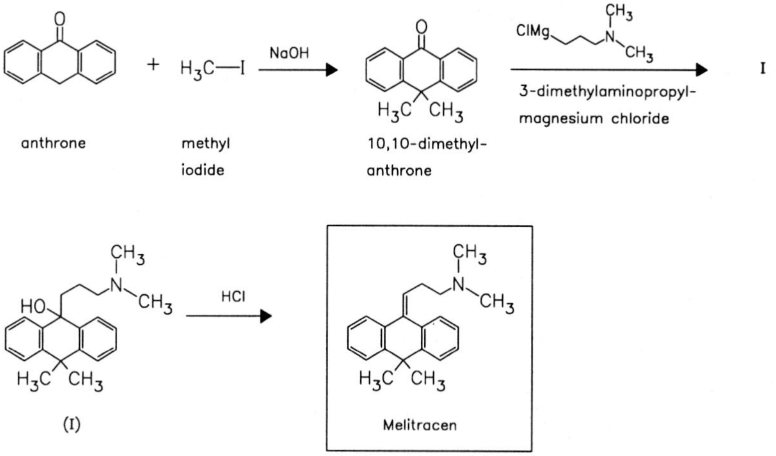

Reference(s):
US 3 177 209 (Kefalas; 6.4.1965; GB-prior. 16.9.1960, 17.2.1961).
US 3 190 893 (Kefalas; 22.6.1965; GB-prior. 17.2.1961).
DE 1 177 633 (Kefalas; appl. 14.2.1962; GB-prior. 17.2.1961).
DE 1 294 375 (Kefalas; appl. 7.9.1961; GB-prior. 16.9.1960).
Holm, T.: Acta Chem. Scand. (ACHSE7) 17, 2437 (1963).

Formulation(s): amp. 20 mg/2 ml; drg. 10 mg, 25 mg

Trade Name(s):
D: Trausabun (Byk Gulden); I: Deanxit (Lusofarmaco)-
  wfm comb.
  Trausabun (Promonta); Melixeran (Lusofarmaco)
  wfm J: Thymeol (Takeda)

Meloxicam
(UH-AC 62XX)

ATC: M01AC06
Use: anti-inflammatory, cyclooxygenase-2
  inhibitor

RN: 71125-38-7 MF: $C_{14}H_{13}N_3O_4S_2$ MW: 351.41
CN: 4-hydroxy-2-methyl-N-(5-methyl-2-thiazolyl)-2H-1,2-benzothiazine-3-carboxamide 1,1-dioxide

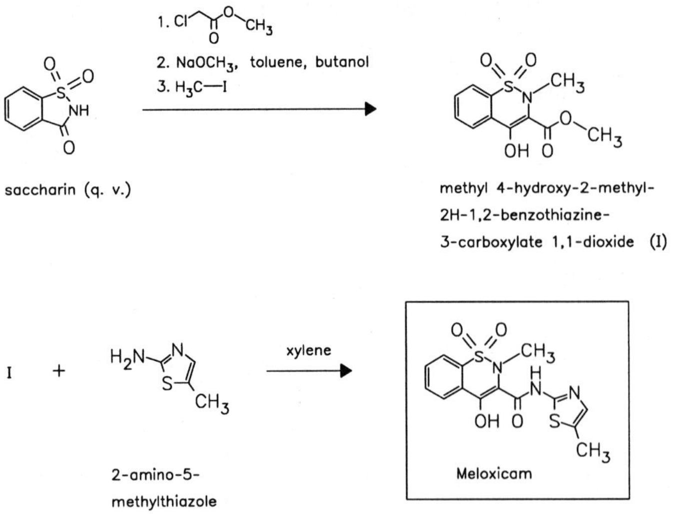

1. Cl–CH2–CO–O–CH3
2. NaOCH3, toluene, butanol
3. H3C—I
saccharin (q. v.)
methyl 4-hydroxy-2-methyl-
2H-1,2-benzothiazine-
3-carboxylate 1,1-dioxide (I)
I +
xylene
2-amino-5-
methylthiazole
Meloxicam

Reference(s):
DE 2 756 113 (Thomae GmbH; 21.6.1979; D-prior. 16.12.1977).

ophthalmic solutions:
WO 9 301 814 (Lab. Europhta; appl. 17.7.1992; F-prior. 18.7.1991).

plaster for high-bioavailability:
JP 04 321 624 (Hisamitsu Pharm.; appl. 19.4.1991; J-prior. 19.4.1991).

combination with 5-lipoxygenase inhibitors:
WO 9 641 626 (Searle & Co.; appl. 11.6.1996; USA-prior. 12.6.1995).

combination with leukotriene A hydrolase inhibitor:
WO 9 641 625 (Searle & Co.; appl. 11.6.1996; USA-prior. 12.6.1995).

combination with leukotriene inhibitors:
WO 9 641 645 (Searle & Co.; appl. 11.6.1996; USA-prior. 12.6.1995).

medical use:
WO 9 703 667 (Merck & Co.; appl. 15.7.1996; USA-prior. 19.7.1995).

Formulation(s): cps. 7.5 mg; supp. 7.5 mg, 15 mg; tabl. 7.5 mg

Trade Name(s):
D: Mobec (Boehringer Ing.) F: Mobic (Boehringer Ing.) GB: Mobic (Boehringer Ing.)

Melperone

(Methylperone; Metylperon)

ATC: N05AD03
Use: neuroleptic

RN: 3575-80-2 MF: $C_{16}H_{22}FNO$ MW: 263.36
CN: 1-(4-fluorophenyl)-4-(4-methyl-1-piperidinyl)-1-butanone

hydrochloride
RN: 1622-79-3 MF: $C_{16}H_{22}FNO \cdot HCl$ MW: 299.82 EINECS: 216-599-9
LD_{50}: 35 mg/kg (M, i.v.); 230 mg/kg (M, p.o.);
40 mg/kg (R, i.v.); 330 mg/kg (R, p.o.)

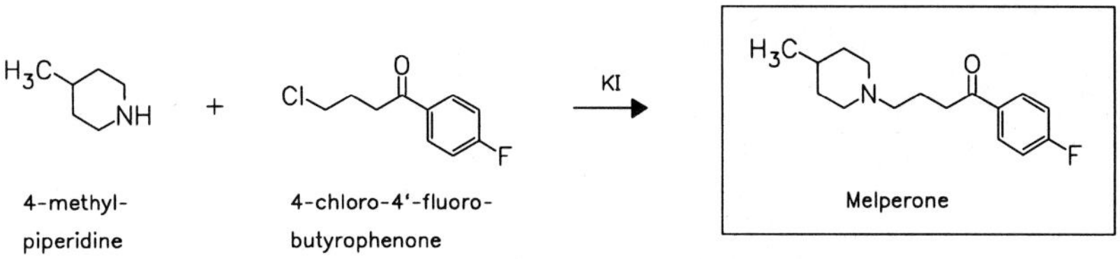

4-methyl-
piperidine

+

4-chloro-4'-fluoro-
butyrophenone

KI

Melperone

Reference(s):
US 3 816 433 (Ferrosan; 11.6.1974; prior. 24.7.1964, 22.3.1966, 29.4.1968, 5.10.1970).
DE 1 268 146 (Ferrosan; appl. 28.7.1964; GB-prior. 29.7.1963).

Formulation(s):   amp. 50 mg/2 ml; drg. 10 mg, 25 mg, 50 mg, 100 mg; sol. 25 mg/5 ml (as hydrochloride)

Trade Name(s):
D: Eunerpan (Knoll)

Melphalan

ATC: L01AA03
Use: antineoplastic

RN: 148-82-3   MF: $C_{13}H_{18}Cl_2N_2O_2$ MW: 305.21   EINECS: 205-726-3
LD$_{50}$:   20.8 mg/kg (M, i.v.);
     4.1 mg/kg (R, i.v.); 11.2 mg/kg (R, p.o.)
CN: 4-[bis(2-chloroethyl)amino]-L-phenylalanine

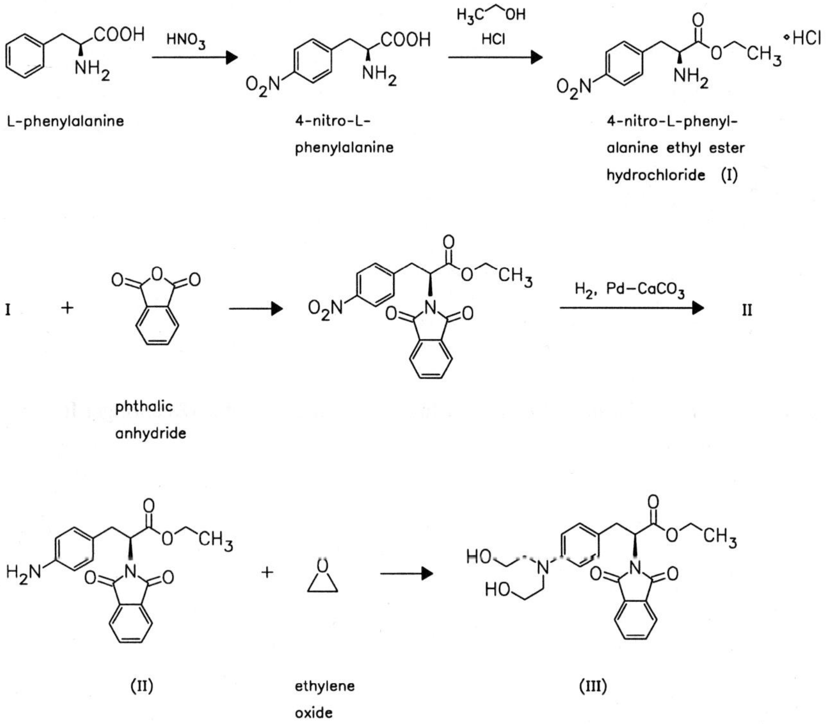

L-phenylalanine

4-nitro-L-
phenylalanine

4-nitro-L-phenyl-
alanine ethyl ester
hydrochloride   (I)

phthalic
anhydride

(II)

ethylene
oxide

(III)

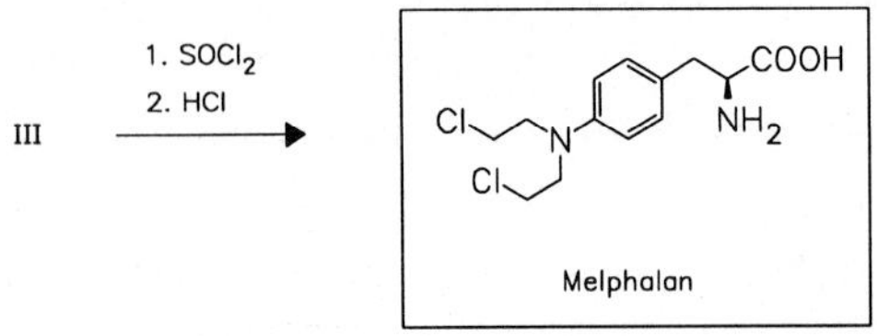

Melphalan

Reference(s):
US 3 032 584 (Nat. Res. Dev. Corp.; 1.5.1962; GB-prior. 17.3.1953).
US 3 032 585 (Nat. Res. Dev. Corp.; 1.5.1962; GB-prior. 3.12.1954).

synthesis of intermediate II:
EP 233 733 (Kureha; appl. 5.2.1987; J-prior. 19.2.1986).

Formulation(s):   amp. 50 mg/10 ml; tabl. 2 mg, 5 mg

Trade Name(s):
D: Alkeran (Glaxo Wellcome)   GB: Alkeran (Glaxo Wellcome)   I: Alkeran (Glaxo Wellcome)
F: Alkéran (Glaxo Wellcome)   Alkeran (Calmic)   USA: Alkeran (Glaxo Wellcome)

Memantine

ATC: N06DX01
Use: antispasmodic, myotonolytic,
        antiparkinsonian, muscle relaxant

RN: 19982-08-2   MF: $C_{12}H_{21}N$ MW: 179.31
CN: 3,5-dimethyltricyclo[3.3.1.1^{3,7}]decan-1-amine

hydrochloride
RN: 41100-52-1   MF: $C_{12}H_{21}N \cdot HCl$ MW: 215.77

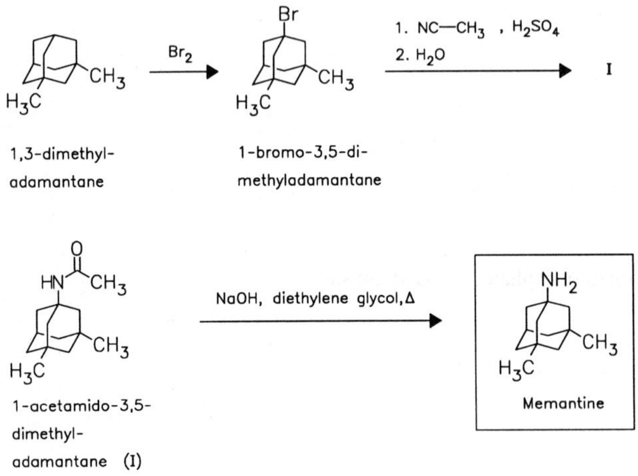

1,3-dimethyl-
adamantane

1-bromo-3,5-di-
methyladamantane

1-acetamido-3,5-
dimethyl-
adamantane   (I)

Memantine

Reference(s):
US 3 391 142 (Eli Lilly; 2.7.1968; appl. 9.2.1966).
Gerzon, K. et al.: J. Med. Chem. (JMCMAR) **6**, 760 (1963).

1,3-dimethyl-adamantane:
Schleyer, P. v. R.; Nicholas, R.D.: Tetrahedron Lett. (TELEAY) **9**, 305 (1961).

Formulation(s):   amp. 10 mg; drops 10 mg/g; f. c. tabl. 10 mg (as hydrochloride)

Menadiol diacetate

(Acetomenadione; Acetomenaftone)

ATC: A11
Use: antihemorrhagic, vitamin K-
         derivative (prothrombogenic)

RN: 573-20-6   MF: $C_{15}H_{14}O_4$ MW: 258.27   EINECS: 209-352-1
CN: 2-methyl-1,4-naphthalenediol diacetate

diphosphate dicalcium salt
RN: 74347-27-6   MF: $C_{11}H_8Ca_2O_8P_2$ MW: 410.28

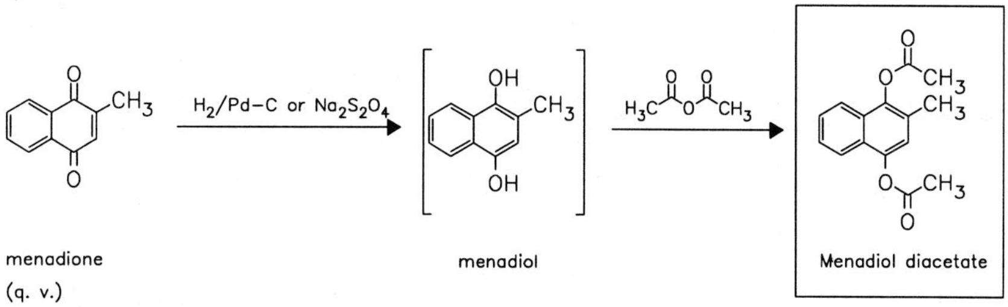

menadione
(q. v.)

menadiol

Menadiol diacetate

Reference(s):
Horii et al.: Pharm. Bull. (PHBUA9) **5**, 82 (1957).

Formulation(s):   amp. 8.86 mg/ml (as diphosphate dicalcium salt)

Menadiol sodium diphosphate

ATC: A11
Use: antihemorrhagic vitamin

RN: 131-13-5   MF: $C_{11}H_4Na_4O_8P_2$ MW: 418.05   EINECS: 205-012-1
LD_{50}:   350 mg/kg (M, s.c.);
        231 mg/kg (R, i.p.)
CN: 2-methyl-1,4-naphthalenediol bis(dihydrogen phosphate) tetrasodium salt

hexahydrate
RN: 6700-42-1   MF: $C_{11}H_8Na_4O_8P_2 \cdot 6H_2O$ MW: 530.18
menadiol diphosphate
RN: 84-98-0   MF: $C_{11}H_{12}O_8P_2$ MW: 334.16

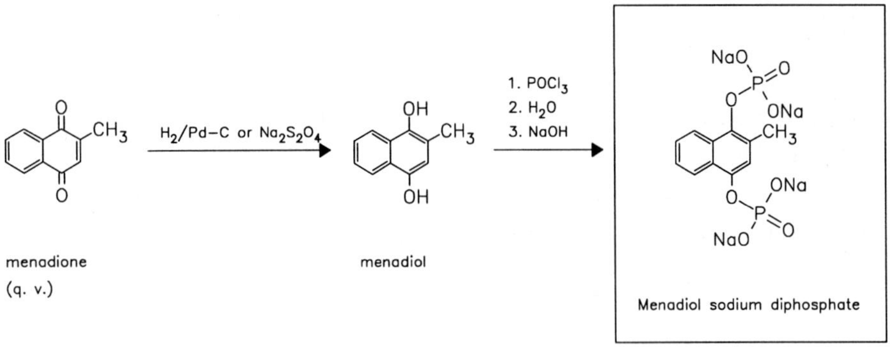

menadione
(q. v.)

menadiol

Menadiol sodium diphosphate

Reference(s):

Fieser, L.F. et al.: J. Am. Chem. Soc. (JACSAT) **62**, 228 (1940).

US 2 380 621 (Roche; 1945; CH-prior. 1942).

US 2 345 690 (Roche; 1944; appl. 1941).

US 2 354 132 (Roche; 1944; appl. 1940).

synthesis of intermediate II:

EP 233 733 (Kureha; appl. 5.2.1987; J-prior. 19.2.1986).

Formulation(s):   amp. 10 mg/1 ml, tabl. 10 mg

Trade Name(s):

D:	Styptobion (Merck)-comb.; wfm	GB:	Synkavit (Roche); wfm	USA:	Synkavite (Roche; as hexahydrate); wfm
		J:	Kativ (Takeda)		

Menadione

(Menaphthone; Menaquinone; Vitamin K₃)

ATC: B02BA02
Use: antihemorrhagic vitamin
(prothrombogenic)

RN: 58-27-5   MF: C₁₁H₈O₂   MW: 172.18   EINECS: 200-372-6

LD₅₀: 500 mg/kg (M, p.o.)

CN: 2-methyl-1,4-naphthalenedione

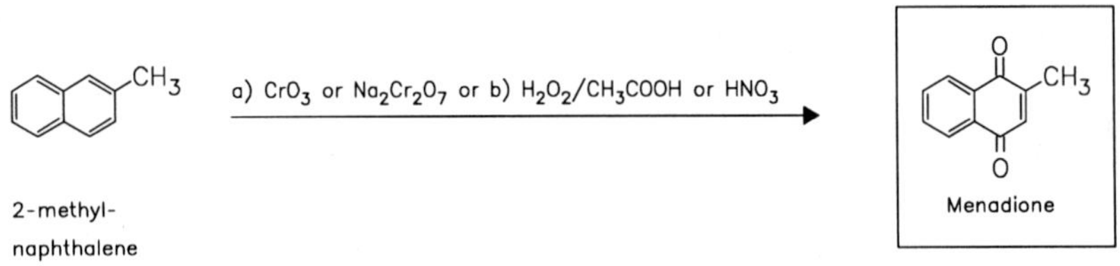

2-methyl-
naphthalene

Menadione

Reference(s):

a   *oxidation with chromic acid and derivatives:*

Fieser, L.F. et al.: J. Am. Chem. Soc. (JACSAT) **61**, 2559, 3216 (1939).

US 2 402 226 (Velsicol; 1946; appl. 1943).

b   *oxidation with hydrogen peroxide:*

Arnold; Larson: J. Org. Chem. (JOCEAH) **5**, 250 (1940).

US 2 373 003 (Univ. of Minnesota; 1945; appl. 1941).

Adam, W. et al.: Angew. Chem. (ANCEAD) **106**, 2545 (1994).

Formulation(s):   tabl. 2 mg

Trade Name(s):
F: Bilkaby (Lehning)   J: generic

Menadione sodium bisulfite

(Menaphthone sodium bisulfite)

ATC: A11
Use: antihemorrhagic vitamin

RN: 130-37-0   MF: $C_{11}H_9NaO_5S$ MW: 276.24   EINECS: 204-987-0
CN: 1,2,3,4-tetrahydro-2-methyl-1,4-dioxo-2-naphthalenesulfonic acid sodium salt

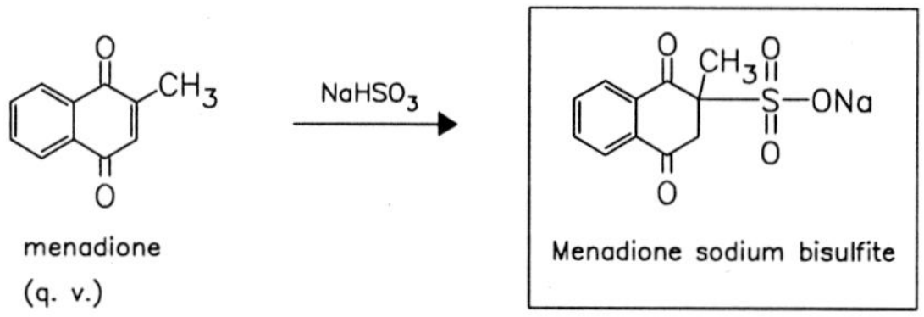

menadione
(q. v.)

Menadione sodium bisulfite

Reference(s):
Moore, M.B.: J. Am. Chem. Soc. (JACSAT) **63**, 2049 (1941).
Baker, B.R. et al.: J. Am. Chem. Soc. (JACSAT) **64**, 1096 (1942).
Menotti, A.R.: J. Am. Chem. Soc. (JACSAT) **65**, 1209 (1943).
US 2 367 302 (Abbott; 1945; appl. 1940).

Formulation(s):   amp. 1 mg, 2 mg, 3 mg, 10 mg, 50 mg

Trade Name(s):
D: Chloramsaar (Chephasaar)-
      comb.; wfm
      Geriatrie-Mulsin (Mucos)-
      comb.; wfm
      Lentinorm (Kanoldt)-
      comb.; wfm

Poly-Vitamin-Saar
(Chephasaar)-comb.; wfm
Prenatal (Cyanamid)-
comb.; wfm
Tetracycletten (Voigt); wfm

F: Arhémapectine vitaminée
      (Gallier)-comb.; wfm
      Cépévit K (UCB)-comb.;
      wfm
I: Vitamina K Salf (Salf)
J: Menadione Inj. (Nord)

(–)-Menthol

Use: anesthetic (combination ingredient in
        antitussives and expectorants)

RN: 2216-51-5   MF: $C_{10}H_{20}O$ MW: 156.27   EINECS: 218-690-9
LD$_{50}$:   3400 mg/kg (M, p.o.);
        3300 mg/kg (R, p.o.)
CN: [1*R*-(1α,2β,5)]-5-methyl-2-(1-methylethyl)cyclohexanol

a   isolation from peppermint oils, containing 70-80 % free menthol, by freezing and recrystallization

b   from (+)-citronellal (containing at 80 % in citronellol)

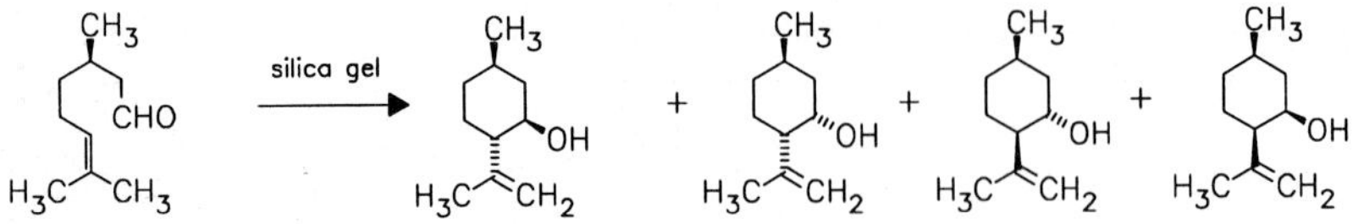

(+)-citronellal (I)   (−)-isopulegol (II)   (+)-neoiso-   (+)-isoiso-   (+)-neoisoiso-
                                                     pulegol (III)   pulegol (IV)   pulegol (V)

II —H₂, Raney–Ni→ (−)-Menthol

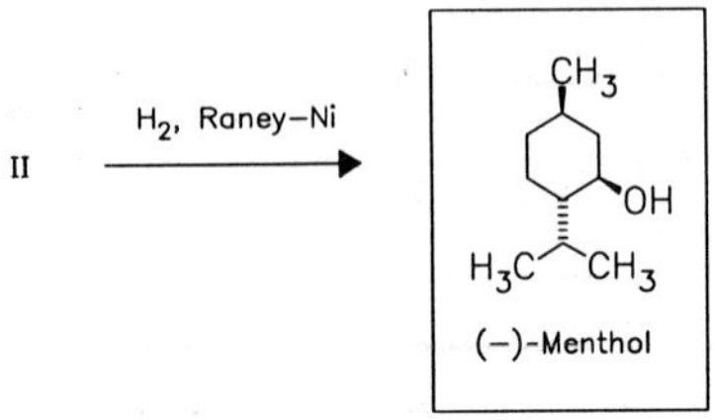

III + IV + V —pyrolysis→ I

c

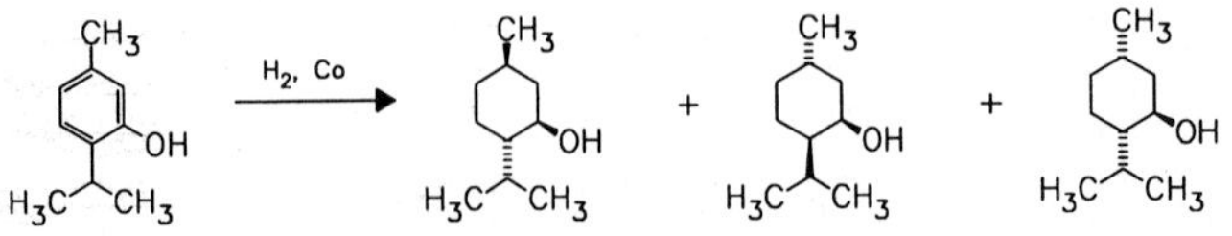

thymol   (±)-menthol (VI)   (±)-neomenthol   (±)-isomenthol

recycling: epimerization to the ratio
(±)-menthol : (±)-neomenthol : (±)-isomenthol =
6 : 3 : 1 under hydrogenation conditions;
separation of (±)-menthol by distillation

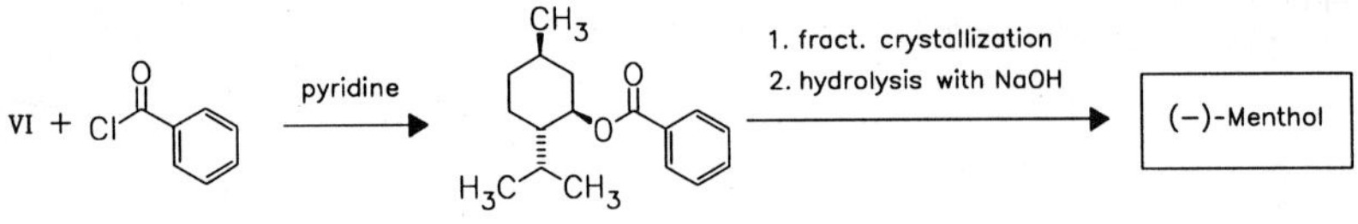

VI + Cl—(benzoyl) —pyridine→ (±)-menthyl benzoate   —1. fract. crystallization / 2. hydrolysis with NaOH→ (−)-Menthol

(+)-Menthol can be racemized under thymol hydrogenation conditions (also with Raney-Ni).

Reference(s):
review:
Ullmanns Encykl. Tech Chem., 4. Aufl., Vol. **20**, 220.
b   DAS 1 197 081 (A. Boake Roberts & Co.; appl. 31.10.1963).
c   *racemate resolution of* (±)-menthyl benzoate:
       DOS 2 109 456 (Haarmann & Reimer; appl. 27.2.1971).

Formulation(s):   cream 0.042-1 %; drg. 1 mg; ointment, sol. in numerous concentrations; powder 1 %

Trade Name(s):
D: numerous combination
    preparations
F: numerous combination
    preparations
GB: numerous combination
    preparations

I: numerous combination
    preparations
USA: Listerine (Warner-
    Lambert)-comb.
    Panalgesic Gold (ECR)-
    comb.

Thera-Gesic (Mission)-
comb.
numerous combination
preparations

Mepacrine
(Quinacrine; Atebrin)

ATC: P01AX05
Use: antimalarial

RN: 83-89-6   MF: $C_{23}H_{30}ClN_3O$ MW: 399.97   EINECS: 201-508-7
LD_{50}:   50 mg/kg (M, i.v.); 1320 mg/kg (M, p.o.)
CN: N^4-(6-chloro-2-methoxy-9-acridinyl)-N^1,N^1-diethyl-1,4-pentanediamine

dihydrochloride
RN: 69-05-6   MF: $C_{23}H_{30}ClN_3O \cdot 2HCl$ MW: 472.89   EINECS: 200-700-8
LD_{50}:   38 mg/kg (M, i.v.); 557 mg/kg (M, p.o.);
    29 mg/kg (R, i.v.); 660 mg/kg (R, p.o.)

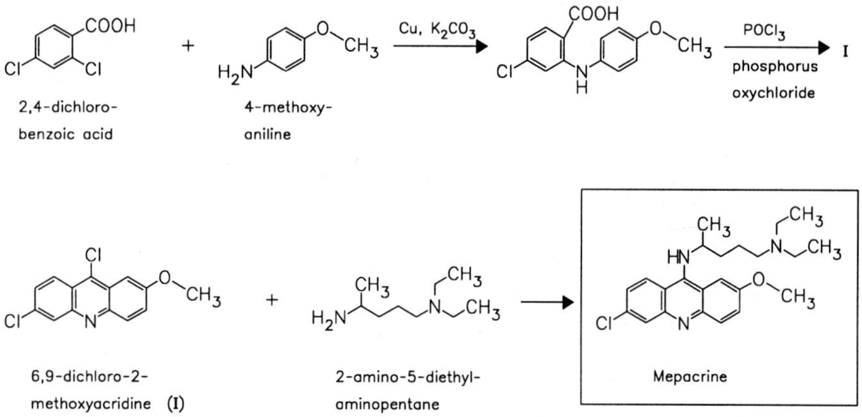

2,4-dichloro-
benzoic acid

4-methoxy-
aniline

6,9-dichloro-2-
methoxyacridine   (I)

2-amino-5-diethyl-
aminopentane

Mepacrine

Reference(s):
DRP 553 072 (I. G. Farben; appl. 1930).
Wingler, A.: Angew. Chem. (ANCEAD) **61**, 49 (1949).

Formulation(s):   tabl. 100 mg

Trade Name(s):
F: Collagenan (Sobio)-comb.;
    wfm

Tenicridine (Norgan); wfm
USA: Atabrine (Winthrop); wfm

Mepartricin
(Methylpartricin)

ATC: A01AB16; D01AA06; G01AA09
Use: polyene antibiotic (for treatment of candidal and trichomonal gynaecological infections, treatment of benign prostatic hypertrophy)

RN: 11121-32-7 MF: $C_{60}H_{88}N_2O_{19}$ MW: 1141.36
LD$_{50}$: 11.1 mg/kg (M, i.p.); 4300 µg/kg (M, i.v.); >2 g/kg (M, p.o.)
CN: partricin methyl ester

1. fermentation of Streptomyces aureofaciens
2. methylation with excess of diazomethane

Mepartricin A   R: —CH$_3$
Mepartricin B   R: —H

Reference(s):
DE 2 154 436 (Spa; appl. 2.11.1971; GB-prior. 3.11.1970).
GB 1 359 473 (Spa; appl. 3.11.1970).
GB 1 406 774 (Spa; appl. 15.2.1973).
GB 1 462 442 (Spa; appl. 29.8.1974).
DE 2 406 628 (Spa; appl. 12.2.1974; GB-prior. 15.2.1973).
US 3 773 925 (Spa; 20.11.1973; appl. 3.11.1971; GB-prior. 3.11.1970).
Bruzzese, T. et al.: Experientia (EXPEAM) 28, 1515 (1972).
Pandey, R.C. et al.: J. Antibiot. (JANTAJ) 30, 158 (1973).
Tweit, R.C. et al.: J. Antibiot. (JANTAJ) 35, 997 (1982).

water soluble formulation:
GB 1 413 256 (Spa; appl. 14.5.1973).
GB 1 463 348 (Spa; appl. 3.9.1974).

medical use for treatment of benign prostatic hypertrophy:
US 4 237 117 (Spa; 2.12.1980; prior. 6.11.1978, 5.10.1979).

liposomal formulation:
WO 89 103 677 (Board of Regents; Univ. of Texas Syst.; appl. 27.10.1988; USA-prior. 27.10.1987).

structure of partricin:
Tweit, R.C. et al.: J. Antibiot. (JANTAJ) 35, 997 (1982).

Formulation(s): tabl. 50000 iu/g, 40 mg; vaginal cream 5000 iu/g; vaginal tabl. 25000 iu.

Trade Name(s):
I: Ipertrofan (SPA; 1986)   Montricin (SPA; 1988 as sodium lauryl sulfate)   Tricandil (SPA; 1975)

Mepenzolate bromide

ATC: A03AB12
Use: anticholinergic

RN: 76-90-4 MF: $C_{21}H_{26}BrNO_3$ MW: 420.35 EINECS: 200-992-7
LD$_{50}$: 9800 µg/kg (M, i.v.); 900 mg/kg (M, p.o.);
22 mg/kg (R, i.v.); 742 mg/kg (R, p.o.)
CN: 3-[(hydroxydiphenylacetyl)oxy]-1,1-dimethylpiperidinium bromide

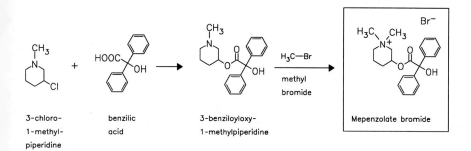

3-chloro-
1-methyl-
piperidine

benzilic
acid

3-benziloyloxy-
1-methylpiperidine

methyl
bromide

Mepenzolate bromide

Reference(s):
US 2 918 408 (Lakeside Labs.; 1959; prior. 1950).

Formulation(s): tabl. 7.5 mg, 15 mg, 25 mg

Trade Name(s):
F: Cantil (Roger Bellon); wfm
GB: Cantil (M.C.P.
Pharmaceuticals); wfm
I: Cantril Lakeside (Roger
Bellon)
Colibantil (Tosi-Novara)

Colum (Jamco)
Enterocantril (RBS
Pharma)-comb.
Enterocantril (Roger
Bellon)-comb.
Gastropidil (Fabo)

J: Eftoron (Maruko Seiyaku)
Sachicoron (Zensei)
Tendalin (Nihon Yakuhin)
Trancolon (Fujisawa)
USA: Cantil (Hoechst Marion
Roussel)

Mephenesin

ATC: M03BX06
Use: muscle relaxant

RN: 59-47-2 MF: $C_{10}H_{14}O_3$ MW: 182.22 EINECS: 200-427-4
LD$_{50}$: 175 mg/kg (M, i.v.); 720 mg/kg (M, p.o.);
133 mg/kg (R, i.v.); 625 mg/kg (R, p.o.)
CN: 3-(2-methylphenoxy)-1,2-propanediol

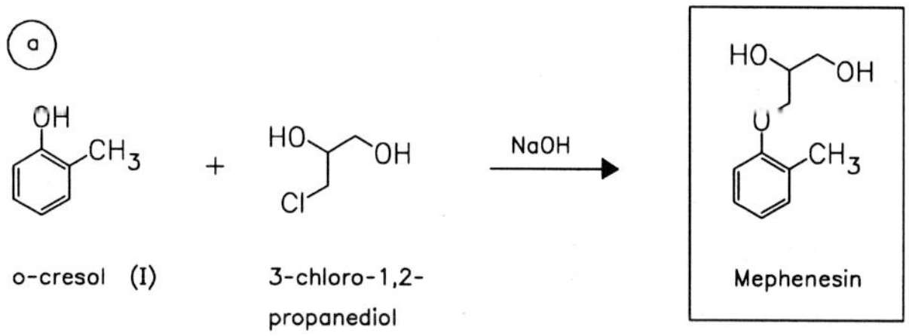

ⓐ

o-cresol (I)

3-chloro-1,2-
propanediol

NaOH

Mephenesin

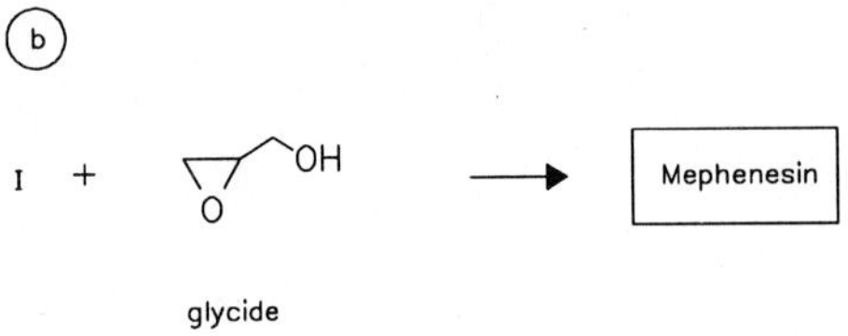

(b)

I + glycide → Mephenesin

Reference(s):

a Marie, E.R.: J. Chem. Soc. (JCSOA9) **101**, 310 (1912).
b GB 628 497 (British Drug Houses; appl. 1948).

Formulation(s):   drg. 250 mg; tabl. 500 mg

Trade Name(s):

D: Dolo Visano (Kade)
F: Algipan Baume (Darcy)-
      comb.
      Décontractyl (Synthelabo)-
      comb.

      Traumalgyl
      (Pharmadéveloppement)-
      comb.
GB: Myanesin (Duncan,
      Flockhart); wfm
I: Mefenesina (Tariff.
      Integrativo)

      Relaxar (Bouty)
      Relaxar Linimento
      (Bouty)-comb.
J: Curaresin (Kyoto)
      Myanol (Chugai)
      Myoserol (Sankyo)
USA: Tolserol (Squibb); wfm

Mephenytoin
(Methoin)

ATC: N03AB04
Use: antiepileptic, anticonvulsant

RN: 50-12-4   MF: $C_{12}H_{14}N_2O_2$ MW: 218.26   EINECS: 200-012-8
LD_{50}:   440 mg/kg (M, p.o.);
      850 mg/kg (R, p.o.)
CN: 5-ethyl-3-methyl-5-phenyl-2,4-imidazolidinedione

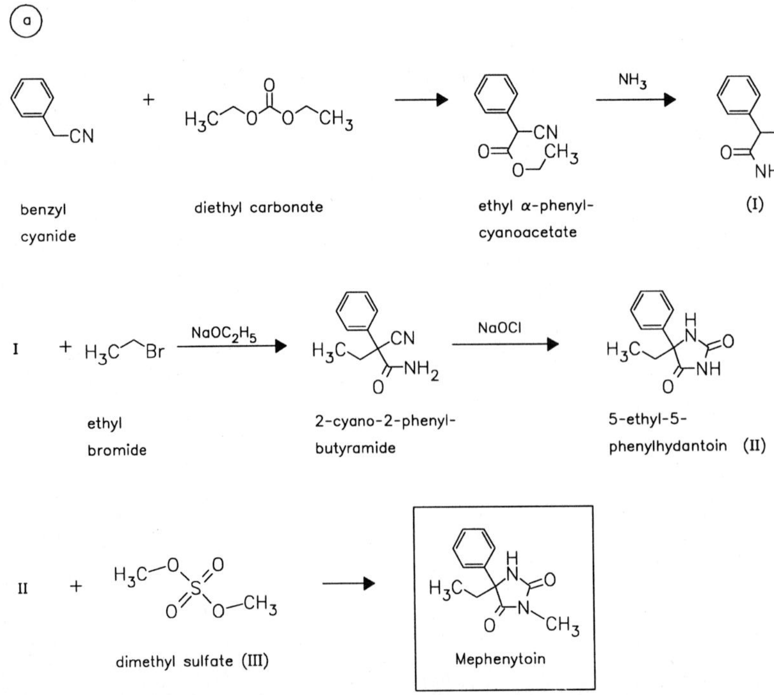

(a)

benzyl cyanide   diethyl carbonate   ethyl α-phenyl-cyanoacetate   (I)

I + ethyl bromide   2-cyano-2-phenyl-butyramide   5-ethyl-5-phenylhydantoin (II)

II + dimethyl sulfate (III)   Mephenytoin

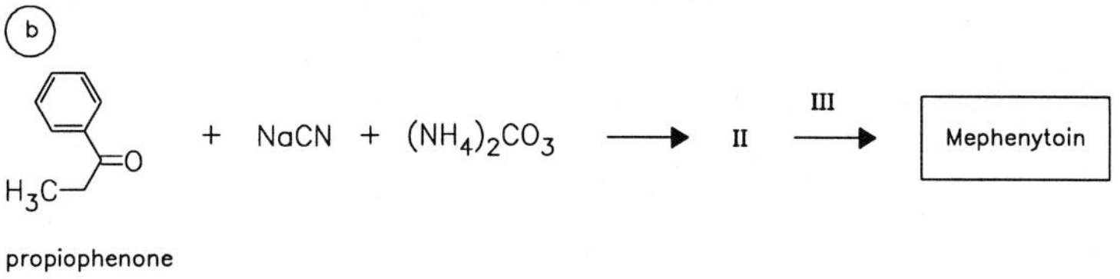

b)

propiophenone

+ NaCN + (NH₄)₂CO₃ $\longrightarrow$ II $\xrightarrow{\text{III}}$ [Mephenytoin]

Reference(s):
Ehrhart-Ruschig, Vol. **1**, 196.
DRP 309 508 (Chem. Fabrik von Heyden; appl. 1914).
FR 769 667 (Sandoz; 1934).

Formulation(s):   tabl. 100 mg

Trade Name(s):
D: Mesantoin (Sandoz); wfm   GB: Mesantoin (Sandoz); wfm   USA: Mesantoin (Sandoz); wfm
F: Sédantoinal (Sandoz); wfm   I: Mesantoina (Sandoz); wfm

Mepindolol

ATC: C07AA14
Use: beta blocking agent, antianginal

RN: 23694-81-7   MF: $C_{15}H_{22}N_2O_2$ MW: 262.35   EINECS: 245-831-1
CN: 1-[(1-methylethyl)amino]-3-[(2-methyl-1*H*-indol-4-yl)oxy]-2-propanol

sulfate
RN: 56396-94-2   MF: $C_{30}H_{44}N_4O_4 \cdot H_2SO_4$ MW: 622.78

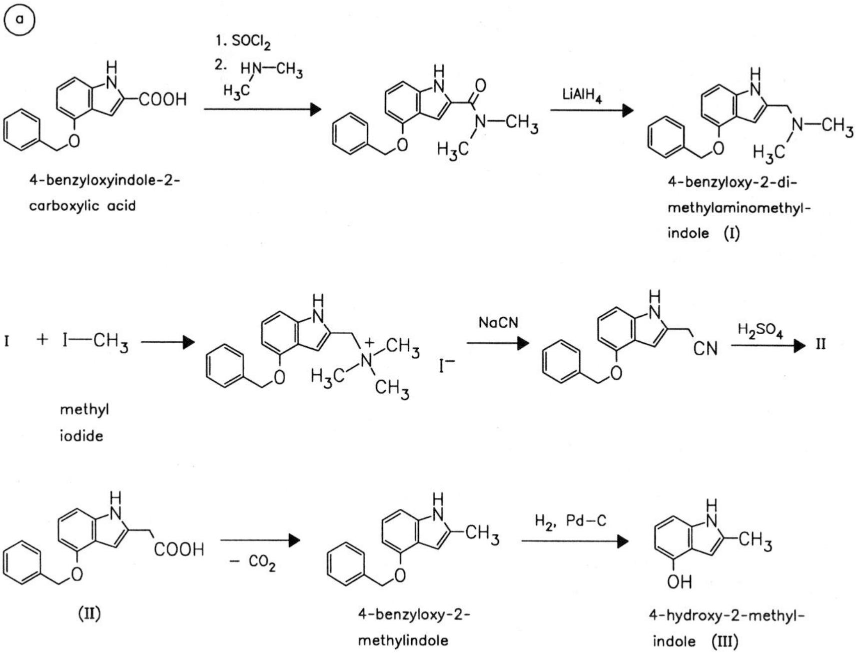

a)

4-benzyloxyindole-2-
carboxylic acid

4-benzyloxy-2-di-
methylaminomethyl-
indole (I)

I + I—CH₃ $\longrightarrow$

methyl
iodide

(II)

4-benzyloxy-2-
methylindole

4-hydroxy-2-methyl-
indole (III)

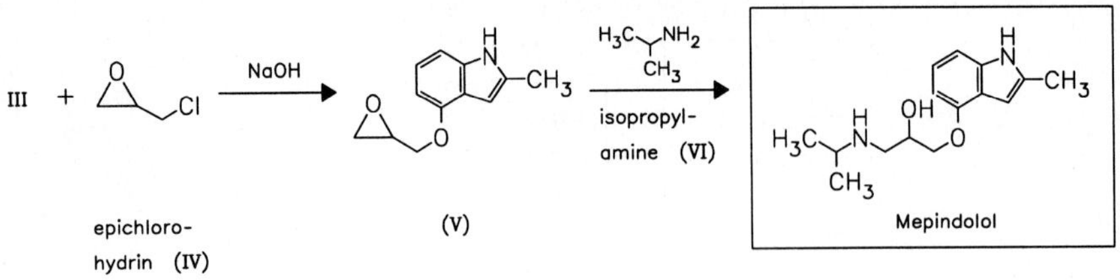

III + epichloro-hydrin (IV) → (via NaOH) → (V) → (via isopropyl-amine (VI)) → Mepindolol

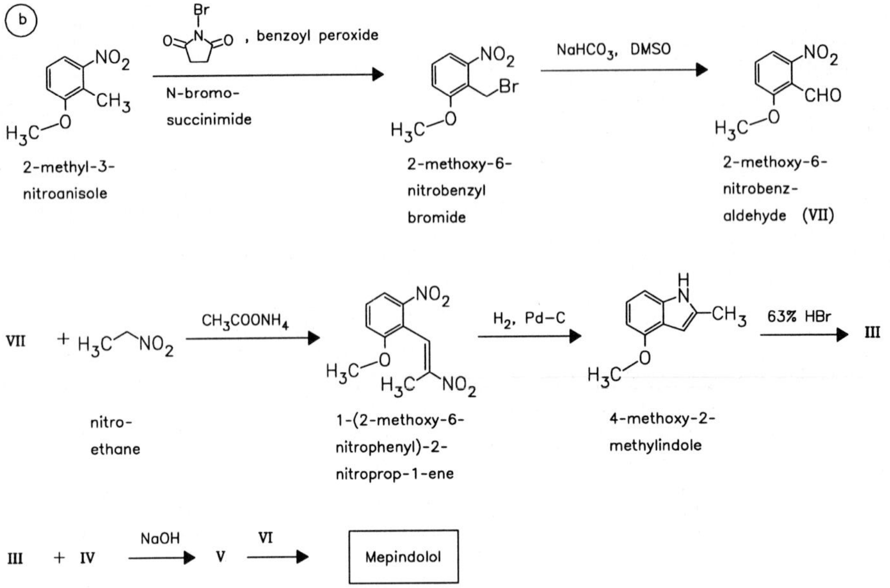

b) 2-methyl-3-nitroanisole → (N-bromo-succinimide, benzoyl peroxide) → 2-methoxy-6-nitrobenzyl bromide → (NaHCO$_3$, DMSO) → 2-methoxy-6-nitrobenz-aldehyde (VII)

VII + nitroethane (H$_3$C—NO$_2$) → (CH$_3$COONH$_4$) → 1-(2-methoxy-6-nitrophenyl)-2-nitroprop-1-ene → (H$_2$, Pd—C) → 4-methoxy-2-methylindole → (63% HBr) → III

III + IV → (NaOH) → V → (VI) → Mepindolol

Reference(s):

a GB 1 260 907 (Sandoz; appl. 23.5.1969; BR-prior. 7.6.1968).
Seemann, F. et al.: Helv. Chim. Acta (HCACAV) **54**, 2411 (1971).
b DOS 2 905 054 (Schering AG; appl. 8.2.1979).

combination with hydrochlorothiazide:
DOS 3 027 392 (Schering AG; appl. 17.7.1980).

Formulation(s): f. c. tabl. 2.5 mg, 5 mg (as sulfate)

Trade Name(s):

D: Corindocomb (Schering)-comb. with hydrochlorothiazide

Corindolan (Schering; as sulfate)

I: Betagon (Schering; as sulfate)
Mepicor (Corvi; as sulfate)

Mepitiostane

ATC: L02BA
Use: antiestrogen, antineoplastic

RN: 21362-69-6 MF: C$_{25}$H$_{40}$O$_2$S MW: 404.66
CN: (2α,3α,5α,17β)-2,3-epithio-17-[(1-methoxycyclopentyl)oxy]androstane

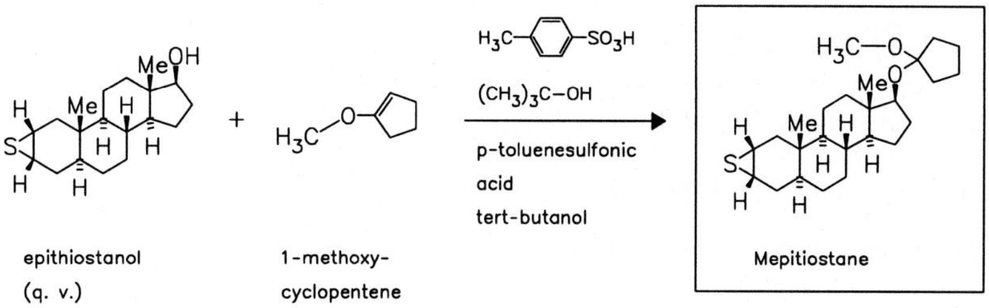

epithiostanol   1-methoxy-   Mepitiostane
(q. v.)   cyclopentene

Reference(s):
DE 1 668 659 (Shionogi; appl. 27.1.1968; J-prior. 28.1.1967).
US 3 567 713 (Shionogi; 2.3.1971; appl. 26.1.1968; J-prior. 28.1.1967).

Trade Name(s):
J: Thioderon (Shionogi;
        1979)

Mepivacaine

ATC: N01BB03
Use: local anesthetic

RN: 22801-44-1   MF: $C_{15}H_{22}N_2O$ MW: 246.35
CN: (±)-N-(2,6-dimethylphenyl)-1-methyl-2-piperidinecarboxamide

monohydrochloride
RN: 1722-62-9   MF: $C_{15}H_{22}N_2O \cdot HCl$ MW: 282.82   EINECS: 217-023-9
LD_{50}: 32 mg/kg (M, i.v.)

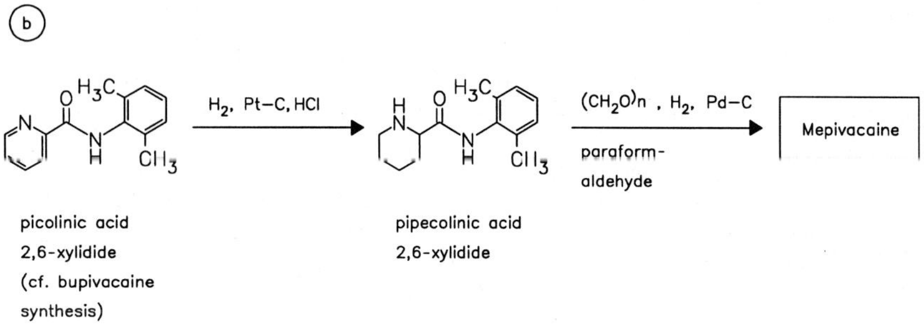

ⓐ

2,6-dimethyl-   ethyl-   ethyl 1-methyl-   Mepivacaine
aniline   magnesium   piperidine-
                   bromide   2-carboxylate

ⓑ

picolinic acid   pipecolinic acid
2,6-xylidide   2,6-xylidide
(cf. bupivacaine
synthesis)

Reference(s):

a US 2 799 679 (AB Bofors; 1957; S-prior. 1955).
  Ekenstam, B. af et al.: Acta Chem. Scand. (ACHSE7) **11**, 1183 (1957).
  Rinderknecht, H.: Helv. Chim. Acta (HCACAV) **42**, 1324 (1959).
b DOS 2 726 200 (Bofors; appl. 10.6.1977; S-prior. 22.6.1976).
  US 4 110 331 (Bofors; 29.8.1978; S-prior. 22.6.1976).

analogous method with methylation before hydrogenation of pyridine nucleus:
GB 826 668 (Crookes Labs.; appl. 1955).

D-(–)-mepivacaine:
DOS 2 259 517 (Bofors; appl. 5.12.1972; USA-prior. 6.12.1971).

Formulation(s):   amp. 0.5 %, 1 %, 2 %, 3 %, 4 % (as hydrochloride)

Trade Name(s):

D:	Meaverin (Rhône-Poulenc Rorer)	F:	Carbocaine (Astra; as hydrochloride)	Mepident (Parke Davis)

D: Meaverin (Rhône-Poulenc
      Rorer)
      Mecain (curasan)
      Mepivastesin (Espe)
      Scandicain (Astra)
      numerous combination
      preparations

F: Carbocaine (Astra; as
      hydrochloride)
GB: Estradurin (Lundbeck)-
      comb.; wfm
I: Carbocaina (Astra-Simes)
      Carbocaina adrenalina
      (Pierrel)-comb.

      Mepident (Parke Davis)
      Mepiforan (Bieffe Medital)
      Mepimynol (Molteni)
      Optocain (Bayer)
J: Carbocain (Yoshitomi)
USA: Polocaine (Astra; as
      hydrochloride)

Meprednisone

ATC: H02AB15
Use: glucocorticoid

RN: 1247-42-3   MF: $C_{22}H_{28}O_5$ MW: 372.46   EINECS: 214-996-1
CN: (16β)-17,21-dihydroxy-16-methylpregna-1,4-diene-3,11,20-trione

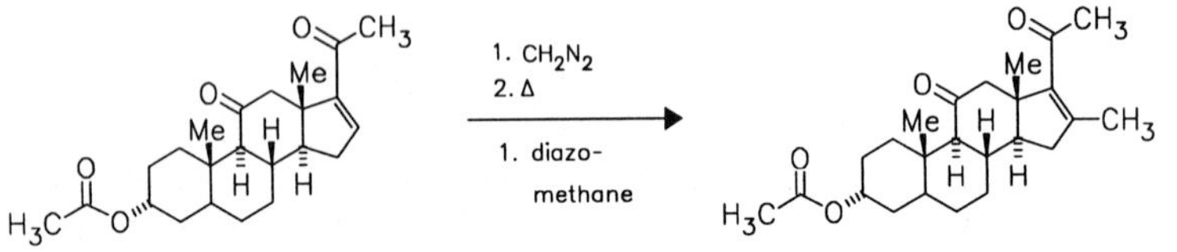

3α-acetoxy-11,20-dioxo-16-pregnene

(I)

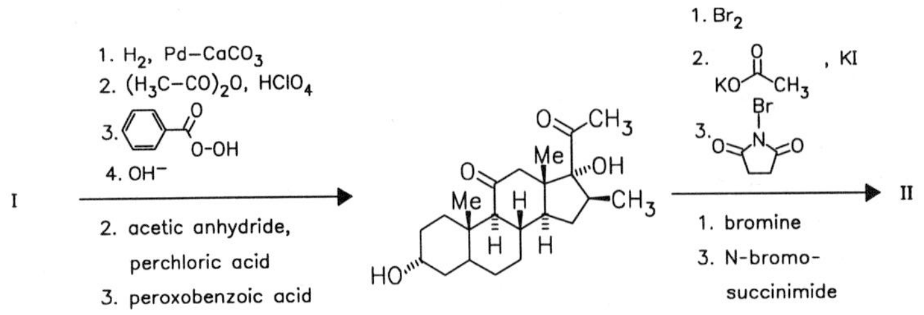

1. Br$_2$
2. OHC—N(CH$_3$)$_2$
3. KHCO$_3$, CH$_3$OH, H$_2$O

(II)

Meprednisone

b

SeO$_2$, H$_3$C—C(CH$_3$)—OH

or microbiologically with Bacillus
sphaericus var. fusiformis ATCC 7055

selenium dioxide, tert-amyl alcohol

Meprednisone

16β-methylcortisone

Reference(s):
US 3 164 618 (Schering Corp., 5.1.1965; prior. 23.7.1957, 8.5.1958).
Taub, D. et al.: J. Am. Chem. Soc. (JACSAT) **82**, 4012 (1960); **80**, 4435 (1958).
Oliveto, E.P. et al.: J. Am. Chem. Soc. (JACSAT) **80**, 4428 (1958).

starting material for **a**:
Slates, H.L.; Wandler, N.L.: J. Org. Chem. (JOCEAH) **22**, 498 (1957).
US 2 671 794 (Glidden; 1954; prior. 1950, 1949).

alternative syntheses:
from hecogenin:
Nathansohn, G.B. et al.: Experientia (EXPEAM) **17**, 448 (1961).

from sitosterin:
US 4 041 055 (Upjohn; 9.8.1977; prior. 17.11.1975).

Trade Name(s):
F: Betalone (Lepetit); wfm   Policort (Lepetit)-comb.;   USA: Betapar (Parke Davis); wfm
I: Corti-Bi (Sidus); wfm   wfm   Betapred (Schering); wfm

Meprobamate

ATC: N05BC01
Use: tranquilizer

RN: 57-53-4   MF: C$_9$H$_{18}$N$_2$O$_4$ MW: 218.25   EINECS: 200-337-5
LD$_{50}$:   230 mg/kg (M, i.v.); 750 mg/kg (M, p.o.);
        350 mg/kg (R, i.v.); 794 mg/kg (R, p.o.)
CN: 2-methyl-2-propyl-1,3-propanediol dicarbamate

H$_3$C—CH—CH$_3$ (CHO)   +   HCHO   →   HO—C(CH$_3$)—OH

1. COCl$_2$
2. NH$_3$
1. phosgene

H$_2$N—O—C(CH$_3$)—O—NH$_2$

Meprobamate

2-methyl-   form-   2-methyl-2-
valeraldehyde   aldehyde   propyl-1,3-
                                   propanediol

Reference(s):
US 2 724 720 (Carter Products; 1955; prior. 1953).
Ludwig, B.J.; Piech, E.C.: J. Am. Chem. Soc. (JACSAT) **73**, 5779 (1951).

Formulation(s):   f. c. tabl. 200 mg, 400 mg

Trade Name(s):

D:	Meprobamat Saar (Philopharm) Visano (Kade)		generics and numerous combination preparations	Harmonin (Yoshitomi) Mepron (Choseido) Mepron (Kanto)
		GB:	Equagesic (Wyeth)-comb.	Xalogen (Ono)
F:	Equanil (Sanofi Winthrop)	I:	Meprob (Tariff. Nazionale)	
	Meprobamate Richard (Richard)		Quanil (Wyeth)	USA: Equagesic (Wyeth-Ayerst) Equanil (Wyeth-Ayerst)
	Novalm (LDM Santé)	J:	Atraxin (Daiichi) Erina (Sumitomo)	Miltown (Wallace)

Meproscillarin
(Rambufaside; Meproscillaridin)

ATC: C01AB
Use: cardiac glycoside

RN: 33396-37-1 MF: $C_{31}H_{44}O_8$ MW: 544.69 EINECS: 251-493-6
LD$_{50}$: 2800 µg/kg (M, i.v.); 12.5 mg/kg (M, p.o.);
       5800 µg/kg (R, i.v.); 79 mg/kg (R, p.o.)
CN: (3β)-3-[(6-deoxy-4-O-methyl-α-L-mannopyranosyl)oxy]-14-hydroxybufa-4,20,22-trienolide

proscillaridin   triethyl   ethyl 2',3'-proscillaridin-
(q. v.)   orthoformate   orthoformate (I)

methyl   Meproscillarin
iodide

Reference(s):
DOS 2 301 382 (Knoll; appl. 12.1.1973).
DOS 2 427 976 (Knoll; appl. 10.6.1974).
Kubinyi, H.: Arzneim.-Forsch. (ARZNAD) **28** (I), 491 (1978).

alternative syntheses:
DE 1 910 207 (Knoll; appl. 28.2.1969).
Kubinyi, H.: Arch. Pharm. Ber. Dtsch. Pharm. Ges. (APBDAJ) **304**, 531 (1971).

combination with verapamil:
DOS 2 746 881 (BASF; appl. 19.10.1977).

Formulation(s):   tabl. 0.25 mg

Trade Name(s):
D: Clift (Knoll); wfm

Meprylcaine

ATC: N01B
Use: local anesthetic

RN: 495-70-5   MF: $C_{14}H_{21}NO_2$ MW: 235.33
CN: 2-methyl-2-(propylamino)-1-propanol benzoate (ester)

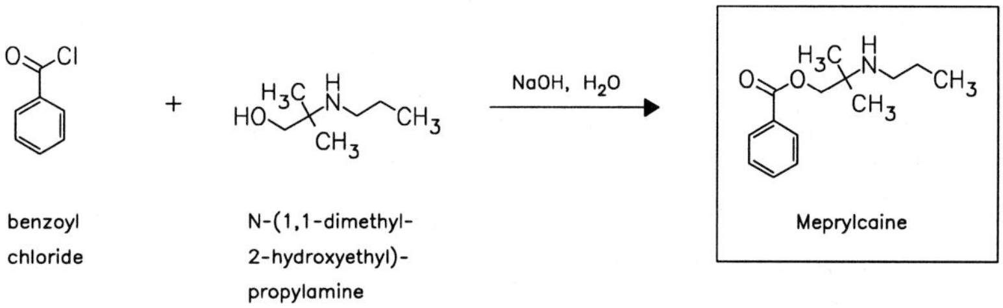

benzoyl
chloride

N-(1,1-dimethyl-
2-hydroxyethyl)-
propylamine

Meprylcaine

Reference(s):
US 2 421 129 (Oradent Chem.; 1947; prior. 1944).

local anesthetic effective injection solution:
US 2 767 207 (Mizzy Inc.; 1956; prior. 1953).

Formulation(s):   cream, gel, sol.

Trade Name(s):
J: Epirocain (Eisai); wfm   USA: Oracaine (Mizzy); wfm

Meptazinol

ATC: N02AX
Use: narcotic, analgesic

RN: 54340-58-8   MF: $C_{15}H_{23}NO$ MW: 233.36   EINECS: 259-109-9
CN: 3-(3-ethylhexahydro-1-methyl-1*H*-azepin-3-yl)phenol

hydrochloride
RN: 59263-76-2   MF: $C_{15}H_{23}NO \cdot HCl$ MW: 269.82   EINECS: 261-683-0
LD_{50}:   282 mg/kg (M, p.o.);
        1260 mg/kg (R, p.o.)

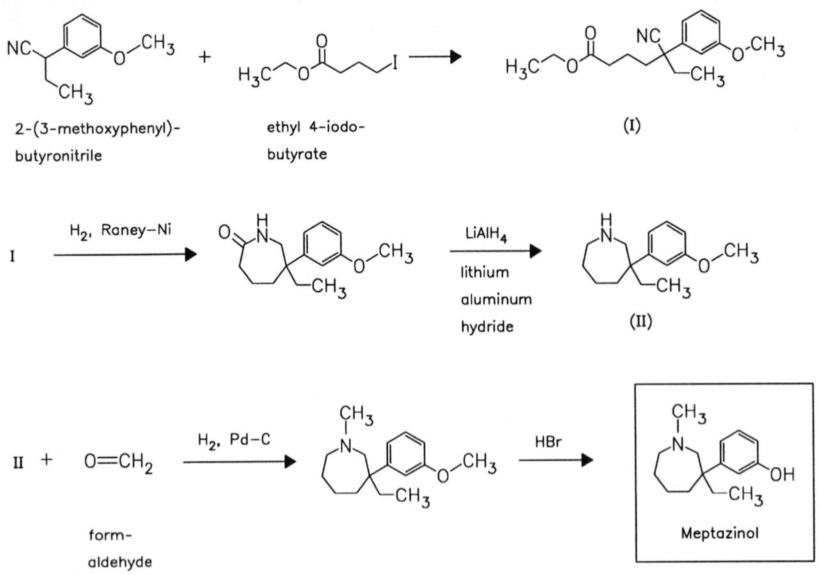

2-(3-methoxyphenyl)-
butyronitrile

ethyl 4-iodo-
butyrate

(I)

(II)

form-
aldehyde

Meptazinol

Reference(s):

DOS 1 941 534 (Wyeth; appl. 14.8.1969; GB-prior. 16.8.1968, 4.9.1968, 28.1.1969).
GB 1 285 025 (Wyeth; Complete Specification 12.8.1969; prior. 16.8.1968, 4.9.1968, 28.1.1969).

alternative syntheses:
Bradley, G. et al.: Eur. J. Med. Chem. (EJMCA5) **15**, 375 (1980).

synthesis of enantiomers:
DOS 2 105 463 (Wyeth; appl. 5.2.1971; GB-prior. 6.2.1970).

combination with ibuprofen:
EP 99 186 (Wyeth; appl. 18.6.1983; GB-prior. 8.7.1982).

Formulation(s):   amp. 100 mg/ml; tabl. 200 mg (as hydrochloride)

Trade Name(s):
D: Meptid (Wyeth; as   GB: Meptid (Monmouth)
      hydrochloride)

Mepyramine

(Pyrilamine)

ATC: D04AA02; R06AC01
Use: antihistaminic

RN: 91-84-9   MF: $C_{17}H_{23}N_3O$ MW: 285.39   EINECS: 202-102-2
LD_{50}:   23 mg/kg (M, i.v.); 220 mg/kg (M, p.o.);
       950 mg/kg (R, p.o.)
CN: N-[(4-methoxyphenyl)methyl]-N',N'-dimethyl-N-2-pyridinyl-1,2-ethanediamine

monohydrochloride
RN: 6036-95-9   MF: $C_{17}H_{23}N_3O \cdot HCl$ MW: 321.85   EINECS: 227-920-7
LD_{50}:   25 mg/kg (M, i.v.); 325 mg/kg (M, p.o.)
maleate (1:1)
RN: 59-33-6   MF: $C_{17}H_{23}N_3O \cdot C_4H_4O_4$ MW: 401.46   EINECS: 200-422-7
LD_{50}:   23 mg/kg (M, i.v.); 220 mg/kg (M, p.o.);
       513 mg/kg (R, p.o.)

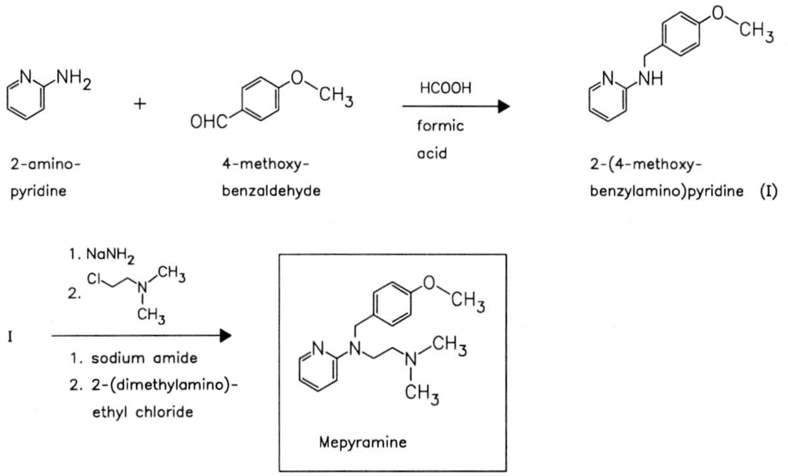

2-amino-
pyridine

4-methoxy-
benzaldehyde

2-(4-methoxy-
benzylamino)pyridine   (I)

1. NaNH₂

2.

1. sodium amide
2. 2-(dimethylamino)-
   ethyl chloride

Mepyramine

Reference(s):
US 2 502 151 (Rhône-Poulenc; 1950; F-prior. 1943).

Formulation(s):   amp. 15 mg, 25 mg; cream 2 %; tabl. 25 mg, 100 mg

Trade Name(s):

D:	Praecinal (Pfleger)-comb.; wfm	F:	Nortussine (Norgine Pharma)-comb.
	Snup (Karlspharma)-comb.; wfm		Triaminic (Novartis)-comb.
	Triaminic (Wander)-comb.; wfm	GB:	Anthisan (May & Baker); wfm
	Vistosan A (Pharm-Allergen)-comb.; wfm	USA:	Atrohist (Medeva; as tannate)-comb.; wfm

Poly-Histine-D (Sanofi; as maleate)-comb.; wfm
Rynatan (Wallace; as tannate)-comb.; wfm
Triaminic (Novartis Consumer)-comb.; wfm
Triotann (Duramed; as tannate)-comb.; wfm

Mequitazine

ATC: R06AD07
Use: antihistaminic, sedative

RN: 29216-28-2   MF: C₂₀H₂₂N₂S   MW: 322.48   EINECS: 249-521-7
LD₅₀: 210 mg/kg (M, p.o.);
        245 mg/kg (R, p.o.)
CN: 10-(1-azabicyclo[2.2.2]oct-3-ylmethyl)-10*H*-phenothiazine

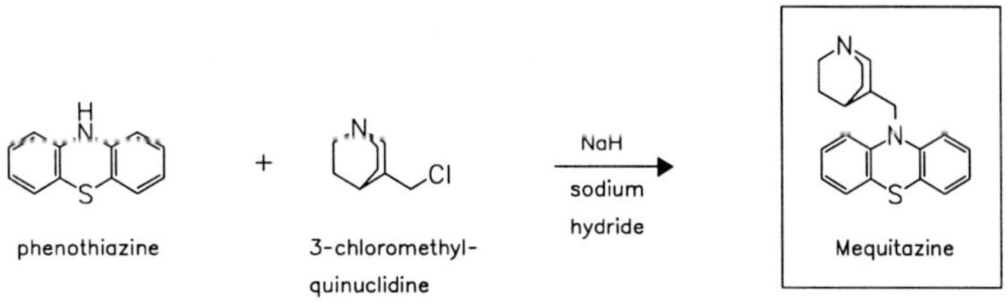

phenothiazine

3-chloromethyl-
quinuclidine

Mequitazine

Reference(s):
DOS 2 009 555 (Sogeras; appl. 28.2.1970; GB-prior. 3.3.1969).
US 3 987 042 (Auclaire; M. et al.; 19.10.1976; prior. 17.8.1973).

Formulation(s):   syrup 1.25 mg/ml, 2.5 mg/ml; tabl. 5 mg

Trade Name(s):

D:	Metaplexan (Rhône-Poulenc Rorer; 1977)	GB:	Primalan (Rhône-Poulenc Rorer; 1976)	J:	Nipdazin (Nippon Shoji; 1983)
F:	Butix (Pierre Fabre) Primalan (Inava; 1976)	I:	Primalan (Rhône-Poulenc Rorer; 1985)		

Merbromin

(Mercurochrome)

ATC: D08
Use: antiseptic

RN: 55728-51-3 MF: $C_{20}H_{10}Br_2HgO_6$ MW: 706.69 EINECS: 259-779-2
CN: (2',7'-dibromo-3',6'-dihydroxy-3-oxospiro[isobenzofuran-1(3*H*),9'-[9*H*]xanthen]-4'-yl)hydroxymercury

disodium salt
RN: 129-16-8 MF: $C_{20}H_8Br_2HgNa_2O_6$ MW: 750.66 EINECS: 204-933-6
LD_{50}: 50 mg/kg (M, i.v.)

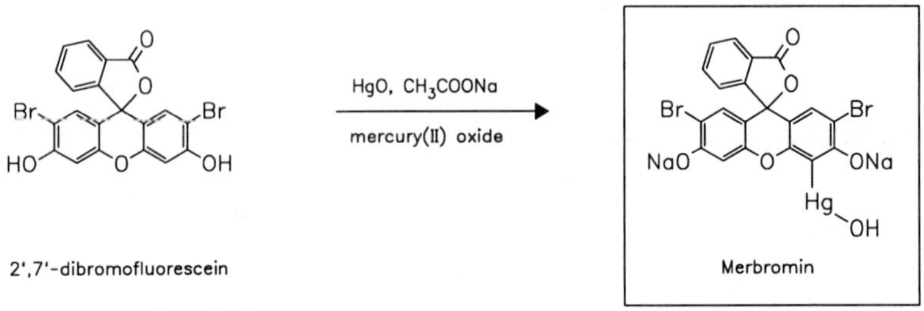

HgO, CH3COONa
mercury(II) oxide
2',7'-dibromofluorescein
Merbromin

Reference(s):
US 1 535 003 (E. C. White; 1925; prior. 1921).

Formulation(s): sol. 2 %

Trade Name(s):

D:	Mercurochrom-Lösung (Krewel Meuselbach)	F:	Pharmadose (Gilbert)
		I:	Mercurocromo (SIT)

Mercaptopurine

ATC: L01BB02
Use: antineoplastic

RN: 50-44-2 MF: $C_5H_4N_4S$ MW: 152.18 EINECS: 200-037-4
LD_{50}: 80 mg/kg (M, i.v.); 260 mg/kg (M, p.o.);
250 mg/kg (R, i.v.)
CN: 1,7-dihydro-6*H*-purine-6-thione

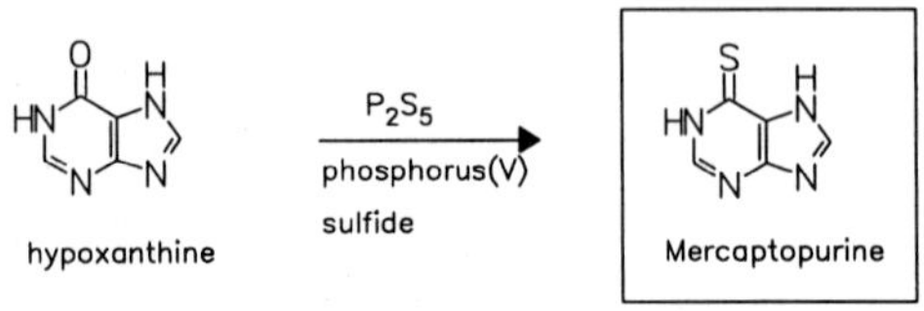

P2S5
phosphorus(V) sulfide
hypoxanthine
Mercaptopurine

Reference(s):
GB 713 286 (Wellcome Found.; appl. 1951).

alternative syntheses:
US 2 721 866 (Burroughs Wellcome; 1955; appl. 1954).
US 2 724 711 (Burroughs Wellcome; 1955; appl. 1954).
US 2 933 498 (Burroughs Wellcome; 1960; appl. 1954).

Formulation(s):   tabl. 50 mg

Trade Name(s):

D:	NERCAP (medac)	GB:	Puri-Nethol (Glaxo		Mern (Tanabe)

D: NERCAP (medac)
      Puri-Nethol (Glaxo
      Wellcome)
F: Purinéthol (Glaxo
      Wellcome)

GB: Puri-Nethol (Glaxo
      Wellcome)
I: Ismipur (Nuovo ISM)
      Purinethol (Wellcome)
J: Classen (Nippon Shoji)
      Leukerin (Takeda)

Mern (Tanabe)
6-MP (Dojin)
Thioinosie (Morishita)
USA: Purinethol (Glaxo
      Wellcome)

Meropenem
(SM-7338; ICI-194660)

ATC: J01DH02
Use: carbapenem, antibiotic

RN: 96036-03-2   MF: $C_{17}H_{25}N_3O_5S$ MW: 383.47
LD$_{50}$:   2650 mg/kg (M, i.v.); >5 g/kg (M, p.o.);
       2850 mg/kg (R, i.v.); >5 g/kg (R, p.o.)
CN: [4R-[3(3S*,5S*),4α,5β,6β(R*)]]-3-[[5-[(dimethylamino)carbonyl]-3-pyrrolidinyl]thio]-6-(1-
       hydroxyethyl)-4-methyl-7-oxo-1-azabicyclo[3.2.0]hept-2-ene-2-carboxylic acid

trihydrate
RN: 119478-56-7   MF: $C_{17}H_{25}N_3O_5S \cdot 3H_2O$ MW: 437.51

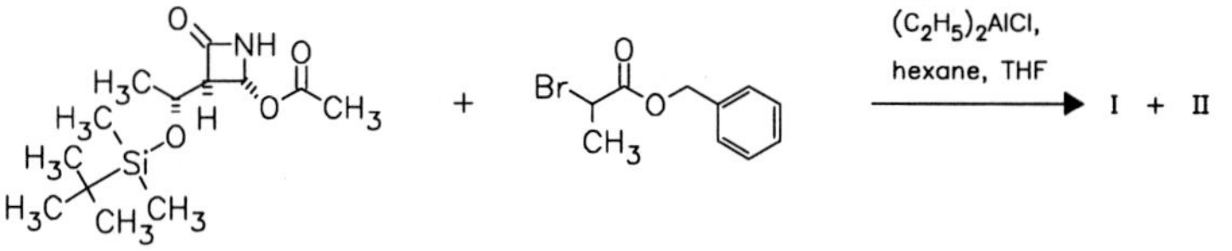

4(R)-acetoxy-3(R)-
[1(R)-(tert-butyldi-
methylsilyloxy)ethyl]-
azetidin-2-one

benzyl 2-bromo-
propionate

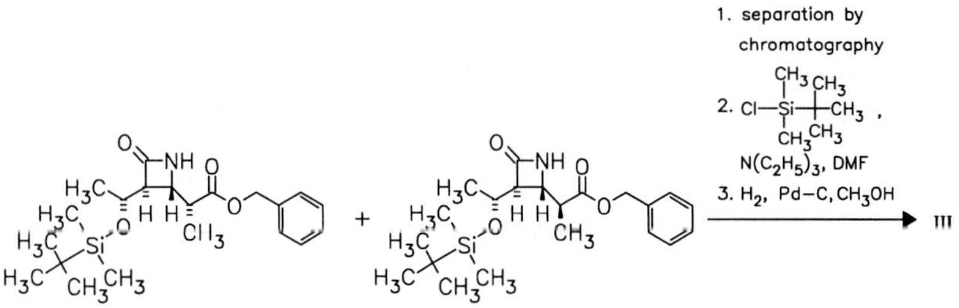

(I)   (II)

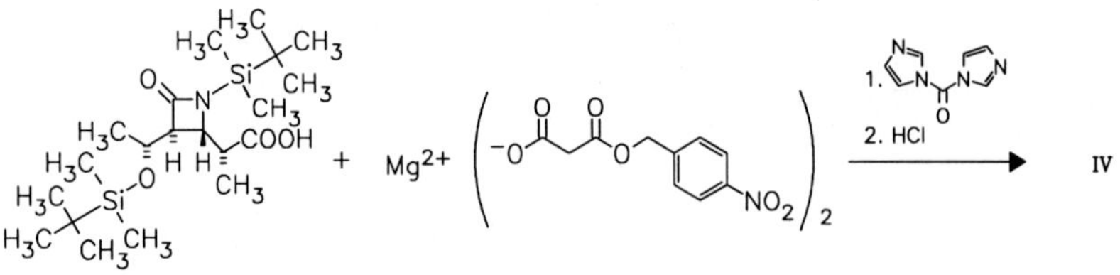

(III)

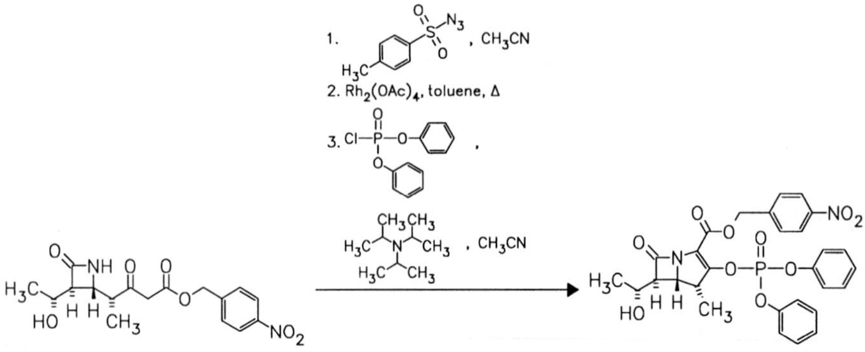

1. , CH₃CN

2. Rh₂(OAc)₄, toluene, Δ

3. Cl—P—O— (diphenyl chlorophosphate)

(IV)

CH₃ CH₃
H₃C—N—CH₃ , CH₃CN
H₃C—CH₃

(V)

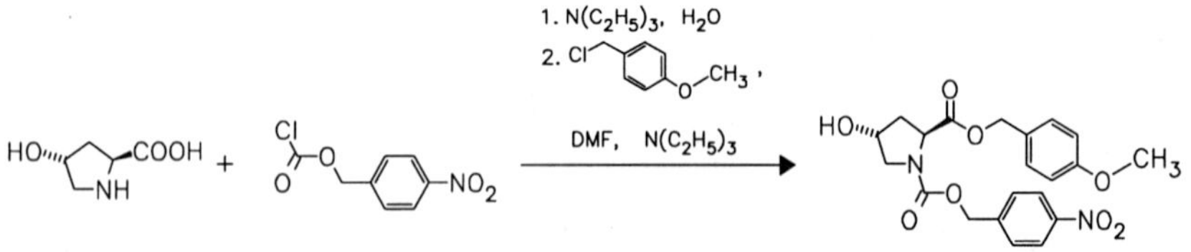

trans-4-hydroxy-
L-proline

1. N(C₂H₅)₃, H₂O
2. Cl—CH₂— —O—CH₃ ,
DMF, N(C₂H₅)₃

trans-1-(p-nitrobenzyl-
oxycarbonyl)-4-hydroxy-
L-proline p-methoxy-
benzyl ester (VI)

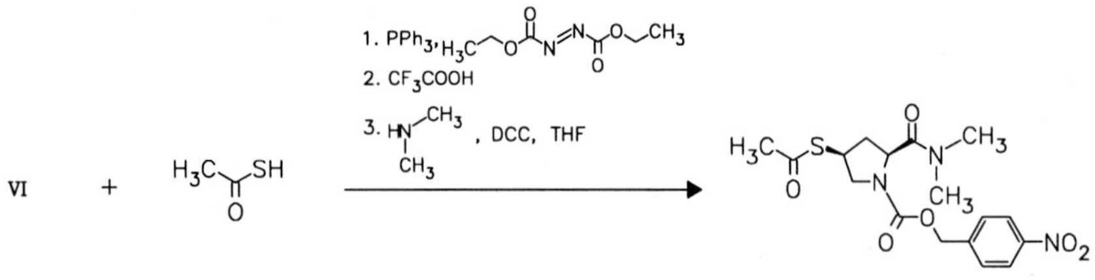

VI +

1. PPh₃, H₃C—O— N=N —O—CH₃
2. CF₃COOH
3. HN—CH₃ , DCC, THF
     CH₃

(VII)

VII   1n NaOH

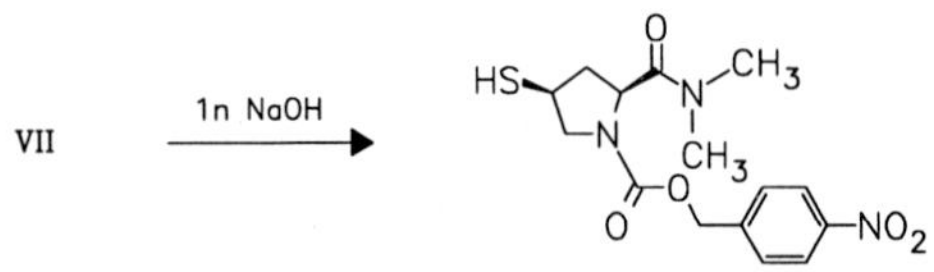

cis-4-mercapto-N,N-
dimethyl-1-(p-nitro-
benzyloxycarbonyl)-
L-prolinamide (VIII)

$$\text{VIII} + \text{V} \xrightarrow[\text{2. H}_2\text{, Pd–C, THF}]{\text{1. CH}_3\text{CN, } \Delta}$$

Meropenem

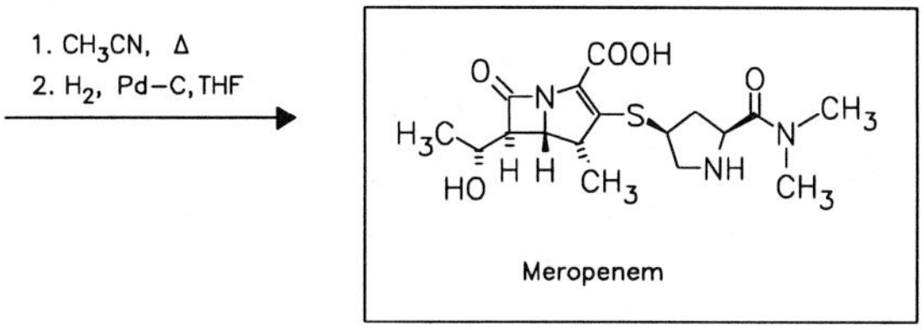

Reference(s):

EP 126 587 (Sumitomo Chemical Co. Ltd; appl. 28.11.1984; J-prior. 9.5.1983, 15.6.1983, 12.7.1983, 3.9.1983, 11.11.1983, 10.2.1984).

synthesis of 4(*R*)-acetoxy-3(*R*)-[1(*R*)-(*tert*-butyldimethylsilyloxy)ethyl]azetidin-2-one:
Reider, P.J. et al.: Tetrahedron Lett. (TELEAY) 2293 (1982).
Kobayashi, Y. et al.: Tetrahedron (TETRAB) **48**, 55 (1992).
EP 256 377 (Sumitomo Pharmaceuticals Co., Ltd; appl. 24.2.1988; J-prior. 30.7.1986, 26.6.1987) (trihydrate).

preparation of β-*lactams:*
JP 01 075 488 (Sumitomo Pharmaceuticals Co., Ltd; appl. 22.3.1989; J-prior. 17.9.1982).
JP 60 233 076 (Sumitomo Chemical Co., Ltd; appl. 19.11.1985; CA-prior. 3.5.1984).

stable ophthalmic oily suspensions containing β-*lactams:*
JP 06 340 529 (Sumitomo Pharma; Santen Pharma Co. Ltd; J-prior. 1.6.1993, 13.12.1994).

stable topical film preparations:
JP 06 001 718 (Sumitomo Pharma; appl. 11.1.1994; J-prior. 17.6.1992).

in combination with penicillin, cephalosporin, penem and carbapenem antibiotics:
EP 640 607 (Hoffmann-La Roche; appl. 1.3.1995; CH-prior. 24.8.1993; 31.5.1994).

synergistic antimicrobial pharmaceutical compositions containing carbapenem *and cephalosporins or penicillins:*
EP 384 410 (Banya Pharmaceuticals Co., Ltd; appl. 29.8.1990; J-prior. 21.2.1989, 14.4.1989).

synergistic effects with human monoclonal antibody:
EP 441 395 (Sumitomo Pharmaceuticals Co., Ltd; appl. 14.8.1991; J-prior. 8.2.1990).

manufacture of sterilized dried sodium carbonate *for pharmaceutical compounds:*
JP 04 198 137 (Sumitomo Pharmaceuticals Co., Ltd; appl. 17.7.1992; J-prior. 28.1.1990).

Formulation(s):　amp. 500 mg, 1 g; vial 250 mg, 500 mg, 1000 mg meropenem trihydrate equivalent

Trade Name(s):

D:	Meronem (Grünenthal; Zeneca)	GB:	Meronem (Zeneca)	J:	Meropen (Sumitomo)
		I:	Merrem (Zeneca); wfm	USA:	Merrem (Zeneca)

Mesalazine
(5-ASA; Fisalamine; Mesalamine)

ATC:　A07EC02
Use:　treatment of gastrointestinal disorders (ulcerative colitis, Crohn's disease)

RN:　89-57-6　MF: C$_7$H$_7$NO$_3$　MW: 153.14　EINECS: 201-919-1
LD$_{50}$:　681 mg/kg (M, i.p.); 5 g/kg (M, p.o.);
　　　　132 mg/kg (R, i.p.); 2800 mg/kg (R, p.o.)
CN:　5-amino-2-hydroxybenzoic acid

a

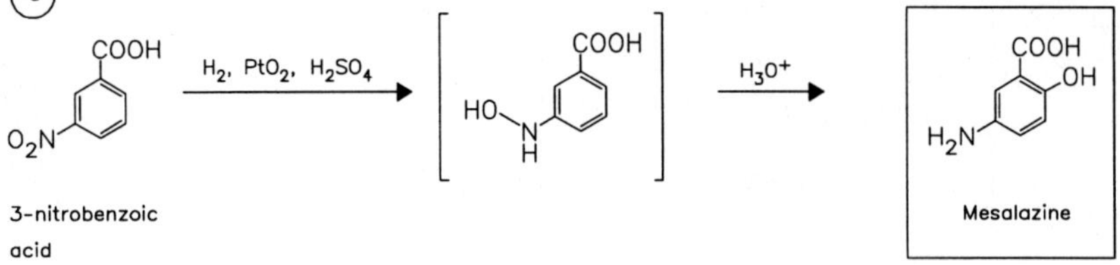
COOH
O_2N
H_2, PtO_2, H_2SO_4
COOH
HO
N
H
H_3O^+
COOH
OH
H_2N
Mesalazine
3-nitrobenzoic acid

b

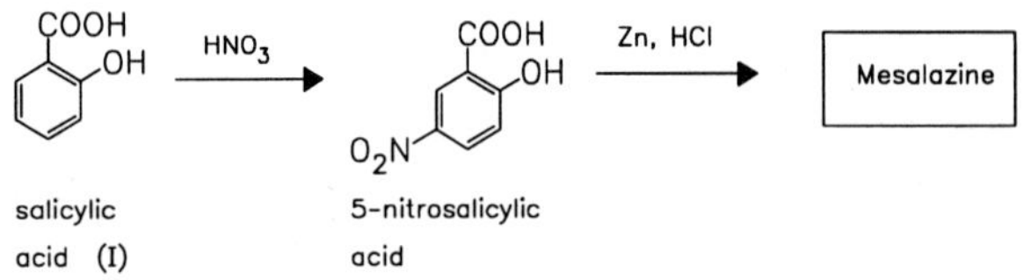
COOH
OH
HNO_3
COOH
OH
O_2N
Zn, HCl
Mesalazine
salicylic acid (I)
5-nitrosalicylic acid

c

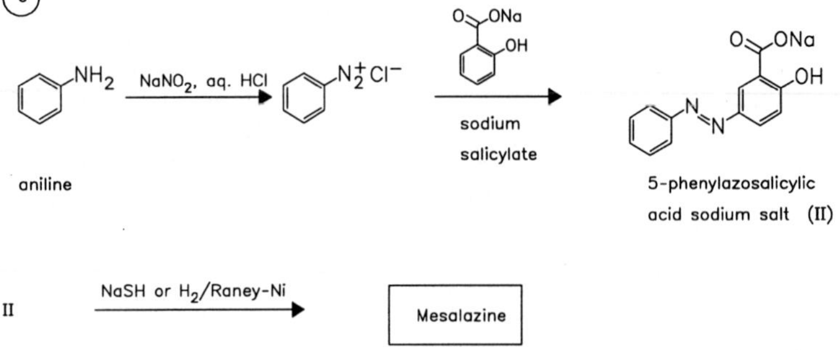
NH_2
$NaNO_2$, aq. HCl
N_2^+ Cl^-
O ONa
OH
sodium salicylate
O ONa
OH
N
N
aniline
5-phenylazosalicylic acid sodium salt (II)
II
NaSH or H_2/Raney-Ni
Mesalazine

d

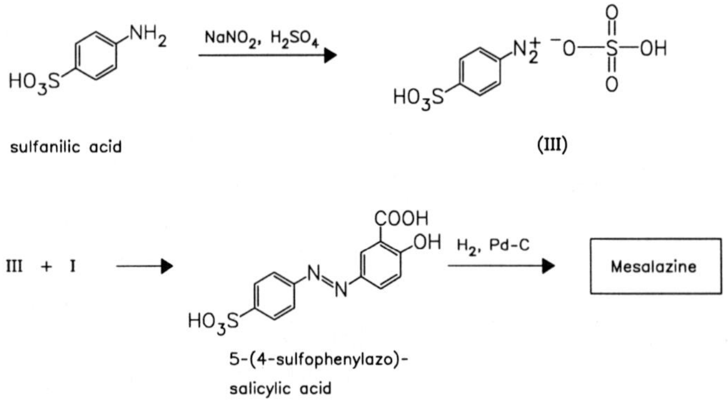
NH_2
HO_3S
$NaNO_2$, H_2SO_4
N_2^+ $^-O-SO_2-OH$
HO_3S
sulfanilic acid
(III)
III + I
COOH
OH
N
N
HO_3S
H_2, Pd-C
Mesalazine
5-(4-sulfophenylazo)-salicylic acid

Active metabolite of sulfasalazine.

Reference(s):

a US 2 198 249 (Du Pont; 1940; appl. 1938).
b Weil, H. et al.: Ber. Dtsch. Chem. Ges. (BDCGAS) **55**, 2664 (1922).
c DOS 3 638 364 (Bayer; appl. 11.11.1986).
DD 255 941 (VEB Chem. Pharm. Oranienburg; appl. 24.12.1986).
d EP 253 788 (Nobelkemi; appl. 17.6.1987; S-prior. 7.7.1986).

review:
The Merck Index, 11th Ed., 5806 (Rahway 1989).

medical use for treatment of dermatological disorders:
EP 352 826 (Gist-Brocades; appl. 1.5.1989; N-prior. 5.5.1988).

medical use for treatment of psoriasis:
EP 291 159 (Dak-Lab.; appl. 31.3.1988; GB-prior. 1.4.1987).

medical use for treatment of colitis ulcerosa and Crohn's disease:
WO 8 102 671 (Ferring; appl. 20.3.1980).

medical use for treatment of coronary circulation diseases:
WO 8 903 216 (Ferring; appl. 13.10.1988; DK-prior. 14.10.1987).

soluble pharmaceutical formulations:
DOS 3 151 196 (K. H. Bauer; appl. 23.12.1981).
US 4 664 256 (Ferring; 12.5.1987; prior. 6.9.1983).

controlled-release formulation:
EP 131 485 (Rowell Lab.; appl. 6.6.1984; USA-prior. 7.7.1983).

Formulation(s):   rectal susp. 2g/30 ml, 4 g/60 ml; suppos. 250 mg, 500 mg; tabl. 250 mg, 400 mg, 500 mg

Trade Name(s):

D:	Asacolitin (Henning Berlin)	GB:	Asacol (SmithKline Beecham)		Salofalk (Interfalk)
	Claversal (Merckle; SmithKline Beecham)		Pentasa (Yamanouchi) Salofalk (Thames)	J:	Pentasa (Kyorin; Nisshin Kyorin)
	Pentasa (Ferring)	I:	Asacol (Giuliani)	USA:	Asacol (Procter & Gamble)
	Salofalk (Falk)		Claversal (Smith Kline & French)		Pentasa (Hoechst Marion Roussel)
F:	Pentasa (Ferring)				Rowasa (Solvay)
	Rowasa (Solvay Pharma)		Pentasa (Brocades)		

Mesna

ATC: R05CB05; V03AF01
Use: detoxificant, mucolytic agent

RN: 19767-45-4   MF: $C_2H_5NaO_3S_2$ MW: 164.18   EINECS: 243-285-9
LD$_{50}$:   1720 mg/kg (M, i.v.); 6102 mg/kg (M, p.o.);
         1510 mg/kg (R, i.v.); 4440 mg/kg (R, p.o.)
CN: 2-mercaptoethanesulfonic acid

free acid
RN: 3375-50-6   MF: $C_2H_6O_3S_2$ MW: 142.20   EINECS: 222-167-0

thiourea   2-bromo-ethanesulfonic acid   Mesna

Reference(s):
US 2 695 310 (Lever Brothers; 1954; appl. 1951).
Schramm, C.H. et al.: J. Am. Chem. Soc. (JACSAT) **77**, 6231 (1955).
US 3 567 835 (UCB; 2.3.1971; GB-prior. 7.5.1965) - only medical use.

from ethylenesulfide *and* sodium hydrogen sulfite:
Reppe, W.: Justus Liebigs Ann. Chem. (JLACBF) **601**, 127 (1956).

from 2-chloroethanesulfonic acid *and* NaSH:
DRP 619 299 (Henkel; appl. 1933).

detoxicant for therapy with cyclophosphamide *and* ifosfamide:
DAS 2 756 018 (ASTA-Werke; appl. 14.12.1977).

salts with amines (mucolytics):
DAS 1 620 629 (UCB; appl. 5.5.1966; GB-prior. 7.5.1965).

Formulation(s):   amp. 100 mg/ml, 600 mg; tabl. 400 mg, 600 mg

Trade Name(s):

D:	Mistabronco (UCB)	GB:	Urimitexan (ASTA
	Uromitexan (ASTA Medica		Medica)
	AWD)	I:	Ausobronc (Biotekfarma)
F:	Mucofluid (UCB)		Mucofluid (UCB)
	Uromitexan 400 (ASTA		Mucolene (Formenti)
	Medica)		

Uromitexan (ASTA
Medica)
J: Uromitexan (Shionogi)
USA: Mesnex (Bristol-Myers
Squibb)

Mesoridazine

ATC: N05AC03
Use: psychosedative, antipsychotic

RN: 5588-33-0   MF: $C_{21}H_{26}N_2OS_2$ MW: 386.58
LD$_{50}$:   26 mg/kg (M, i.v.); 560 mg/kg (M, p.o.);
       644 mg/kg (R, p.o.)
CN: 10-[2-(1-methyl-2-piperidinyl)ethyl]-2-(methylsulfinyl)-10*H*-phenothiazine

monobenzenesulfonate
RN: 32672-69-8   MF: $C_{21}H_{26}N_2OS_2 \cdot C_6H_6O_3S$ MW: 544.76

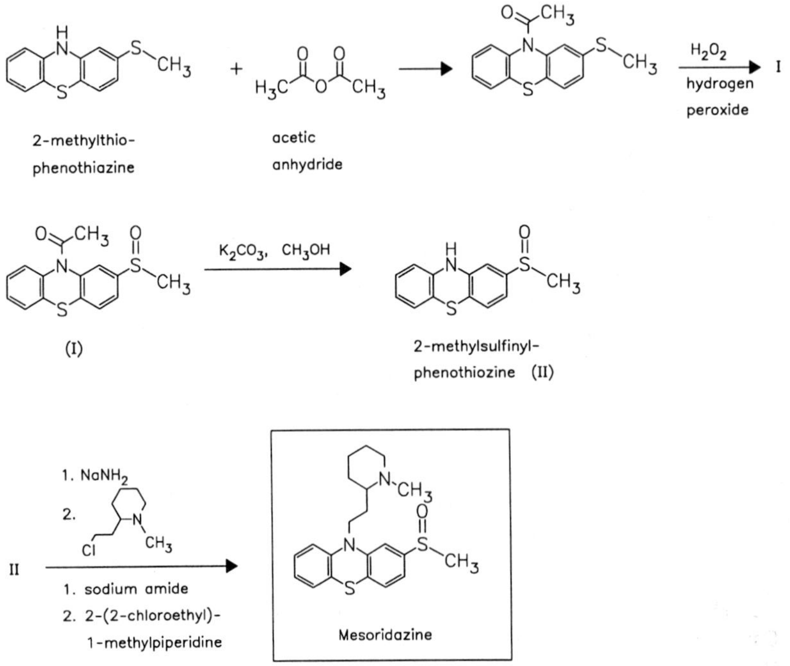

2-methylthio-
phenothiazine

acetic
anhydride

(I)

2-methylsulfinyl-
phenothiozine   (II)

1. NaNH$_2$

2.

1. sodium amide
2. 2-(2-chloroethyl)-
   1-methylpiperidine

Mesoridazine

Reference(s):
US 3 084 161 (Sandoz; 2.4.1963; CH-prior. 10.3.1960).

Formulation(s):   amp. 25 mg; drg. 5 mg; tabl. 10 mg, 25 mg, 100 mg (as monobenzenesulfonate)

Trade Name(s):
F: Lidanil (Salvoxyl-Wander);   USA: Serentil (Boehringer Ing.)
       wfm

Mestanolone

ATC: A14
Use: anabolic, androgen

RN: 521-11-9   MF: $C_{20}H_{32}O_2$ MW: 304.47   EINECS: 208-302-6
LD_{50}:   >3 g/kg (M, p.o.);
       >3 g/kg (R, p.o.)
CN: (5α,17β)-17-hydroxy-17-methylandrostan-3-one

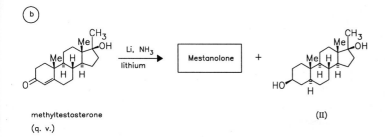

androsterone   methylmagnesium   (I)
                      bromide

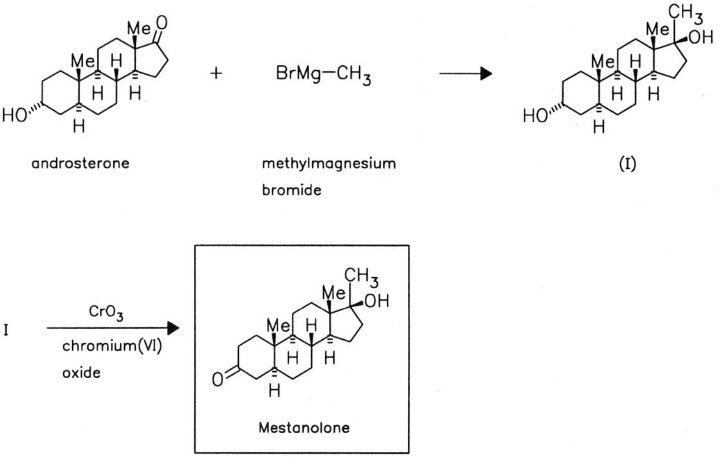

Mestanolone

methyltestosterone   (II)
(q. v.)

II   Mestanolone

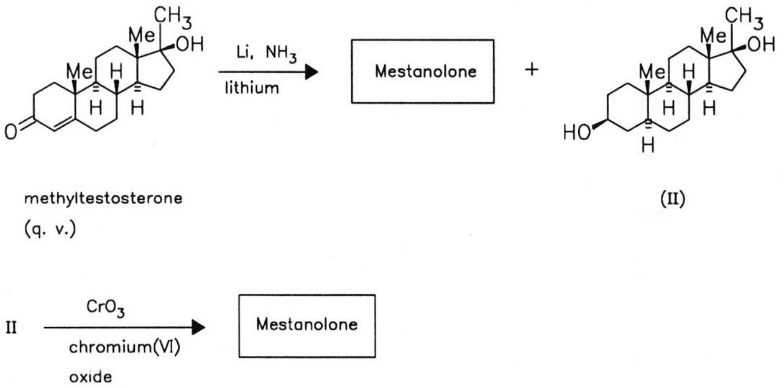

Reference(s):
a   GB 464 396 (Schering AG; appl. 1935).
    Ruzicka, L. et al.: Helv. Chim. Acta (HCACAV) **18**, 994, 1487 (1935).
    starting material:
    The Merck Index, 676 (Rahway 1976).
b   US 2 763 670 (Syntex; 1956, MEX-prior. 1954).

Formulation(s):   10 - 30 mg/day

Mesterolone

ATC: G03BB01
Use: androgen

RN: 1424-00-6   MF: $C_{20}H_{32}O_2$ MW: 304.47   EINECS: 215-836-3
CN: (1α,5α,17β)-17-hydroxy-1-methylandrostan-3-one

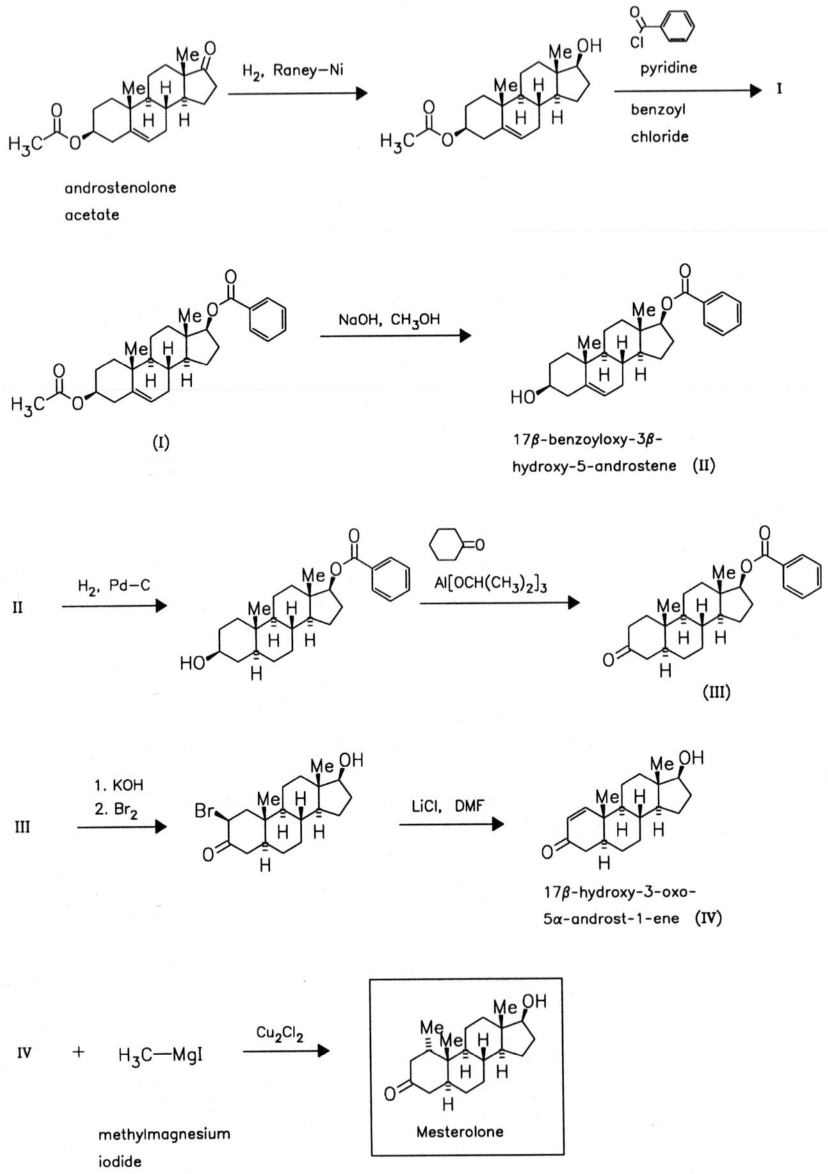

androstenolone acetate

17β-benzoyloxy-3β-hydroxy-5-androstene (II)

(III)

17β-hydroxy-3-oxo-5α-androst-1-ene (IV)

methylmagnesium iodide

Mesterolone

Reference(s):
DE 1 152 100 (Schering AG; appl. 23.12.1960).

alternative syntheses:
DE 1 122 944 (Schering AG; appl. 6.4.1960).
DE 1 131 667 (Schering AG; appl. 21.7.1960).

Formulation(s):   tabl. 25 mg, 50 mg

Trade Name(s):
D: Proviron (Schering)   F: Proviron (Schering)   I: Pro-Viron (Schering)
      Vistimon (Jenapharm)   GB: Proviron (Schering)

Mestranol

ATC: G03
Use: estrogen (in combination with
        progestogen as oral contraceptiva)

RN: 72-33-3   MF: $C_{21}H_{26}O_2$ MW: 310.44   EINECS: 200-777-8
LD_{50}: >10 g/kg (M, p.o.);
          >10 g/kg (R, p.o.)
CN: (17α)-3-methoxy-19-norpregna-1,3,5(10)-trien-20-yn-17-ol

3-O-methylestrone   acetylene   Mestranol

Reference(s):
US 2 666 769 (Searle; 1954; appl. 1952).
Colton, F.B. et al.: J. Am. Chem. Soc. (JACSAT) **79**, 1123 (1957).
DE 1 096 354 (Schering AG; appl. 1.8.1959).

Formulation(s):   drg. 0.05 mg, 0.08 mg; tabl. 0.05 mg

Trade Name(s):
D: Gestamestrol (Hermal-   Norinyl-1 (Searle)-comb.   Regovar (Recordati); wfm
      Chemie)-comb.   Ortho-Novin 1/50 (Janssen-   J: Devocin (Shionogi)
      Ortho-Novum (Janssen-   Cilag)-comb.   Enavid (Dainippon)-comb.
      Cilag)-comb.   I: Elan (Valeas)-comb.; wfm   Lutedione (Teikoku Zoki)-
      Ovosiston (Jenapharm)   Franovul (Francia Farm.)-   comb.
F: Métrulène (Searle)-comb.;   comb.; wfm   Lyndiol (Organon-Sankyo)-
      wfm   Luteolas (Serono)-comb.;   comb.
      Métrulène-test (Searle)-   wfm   Norluten D (Shionogi)-
      comb.; wfm   Lyndiol (Ravasini   comb.
      Noracycline (Ciba-Geigy)-   Organon)-comb.; wfm   USA: Nelova (Warner Chilcott)-
      comb.; wfm   Metrulen (SPA)-comb.;   comb.
      Orgaluton (Organon)-   wfm   Norethin (Roberts)-comb.
      comb.; wfm   Ortho-Novum (Cilag-   Norinyl (Searle)-comb.
      O.V. 28 (Biosedra)-comb.;   Chemie)-comb.; wfm   Ortho-Novum (Ortho-
      wfm   Ovaras (Serono)-comb.;   McNeil)-comb.
GB: Menophase (Searle)   wfm

Mesulfen

(Thianthol)

ATC: D10AB05; P03AA03
Use: topical scabicide, antipruritic

RN: 135-58-0 MF: $C_{14}H_{12}S_2$ MW: 244.38 EINECS: 205-202-4
CN: 2,7-dimeththianthrene

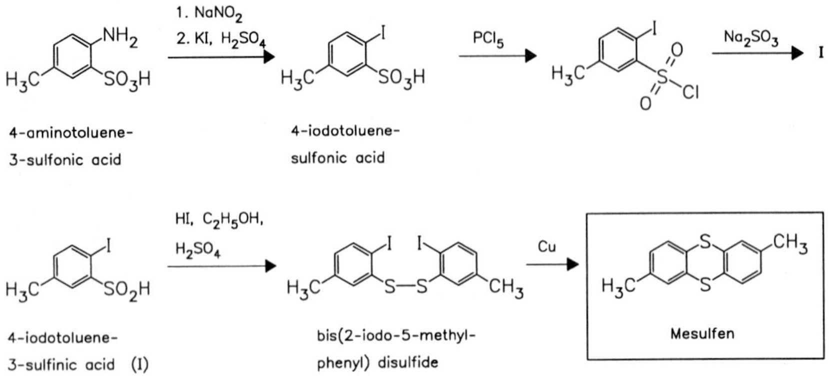
1. NaNO2
2. KI, H2SO4
PCl5
Na2SO3
I
4-aminotoluene-3-sulfonic acid
4-iodotoluenesulfonic acid
HI, C2H5OH, H2SO4
Cu
4-iodotoluene-3-sulfinic acid (I)
bis(2-iodo-5-methylphenyl) disulfide
Mesulfen

Reference(s):
Barber, H.J.; Smiles, S.: J. Chem. Soc. (JCSOA9) **1928**, 1141.

alternative synthesis (from toluene, sulfur *and* $AlCl_3$*):*
DE 365 169 (Bayer; appl. 1919).

Formulation(s): ointment 5-25 %

Trade Name(s):
D: Citemul (Medopharm)
I: Mitigal (Bayropharm); wfm
Mitigal (Sigurtà); wfm
J: Scabol (Daiichi)

Mesuximide

(Methsuximide)

ATC: N03AD03
Use: antiepileptic

RN: 77-41-8 MF: $C_{12}H_{13}NO_2$ MW: 203.24 EINECS: 201-026-7
LD_{50}: 900 mg/kg (M, p.o.)
CN: 1,3-dimethyl-3-phenyl-2,5-pyrrolidinedione

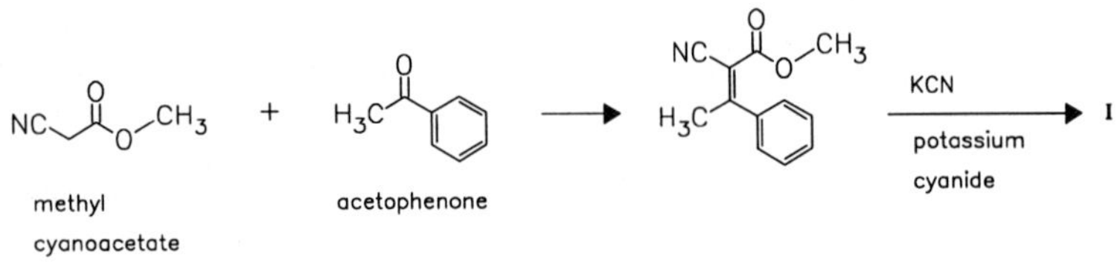
KCN
potassium cyanide
I
methyl cyanoacetate
acetophenone

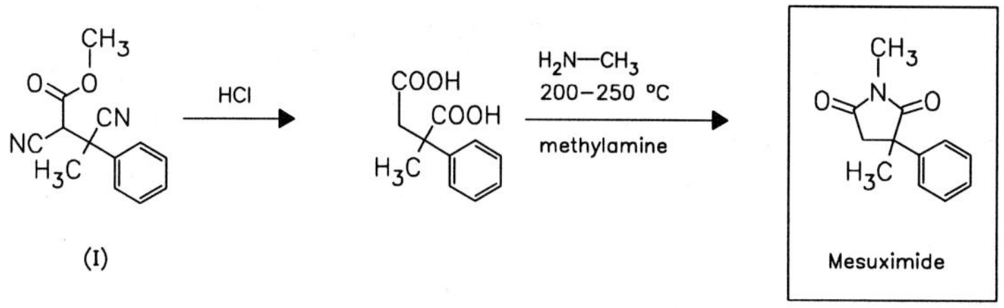

(I)

Mesuximide

Reference(s):
US 2 643 257 (Parke Davis; 1953; prior. 1950).
Miller, C.A.; Long, L.M.: J. Am. Chem. Soc. (JACSAT) **73**, 4895 (1951); **75**, 373 (1953).

Formulation(s):   cps. 150 mg, 300 mg

Trade Name(s):
D: Petinutin (Parke Davis)   GB: Celontin (Parke Davis);   USA: Celontin (Parke Davis)
                                                 wfm

Metaclazepam

(Brometazepam; Metuclazepam)

ATC: N05BA
Use: anxiolytic, benzodiazepine derivative

RN: 84031-17-4   MF: C$_{18}$H$_{18}$BrClN$_2$O   MW: 393.71
LD$_{50}$:   1578 mg/kg (M, p.o.)
CN: 7-bromo-5-(2-chlorophenyl)-2,3-dihydro-2-(methoxymethyl)-1-methyl-1H-1,4-benzodiazepine

monohydrochloride
RN: 61802-93-5   MF: C$_{18}$H$_{18}$BrClN$_2$O · HCl   MW: 430.17   EINECS: 263-234-4
LD$_{50}$:   1578 mg/kg (M, p.o.)

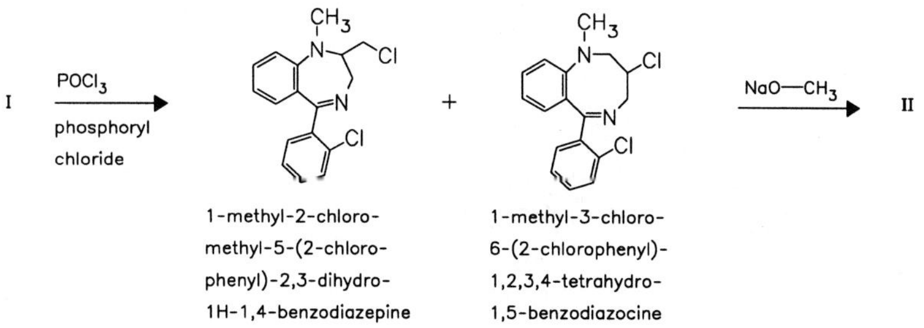

N-phenyl-N-methyl-
2-hydroxy-1,3-
diaminopropane

N-phenyl-N-methyl-N'-
(2-chlorobenzoyl)-2-hydroxy-
1,3-diaminopropane   (I)

I   POCl$_3$
      phosphoryl
      chloride

1-methyl-2-chloro-
methyl-5-(2-chloro-
phenyl)-2,3-dihydro-
1H-1,4-benzodiazepine

1-methyl-3-chloro-
6-(2-chlorophenyl)-
1,2,3,4-tetrahydro-
1,5-benzodiazocine

NaO—CH$_3$ II

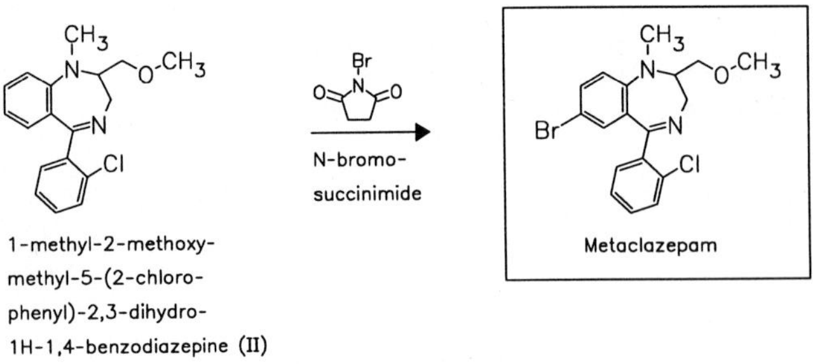

1-methyl-2-methoxy-
methyl-5-(2-chloro-
phenyl)-2,3-dihydro-
1H-1,4-benzodiazepine (II)

Metaclazepam

Reference(s):

BE 799 001 (Kali-Chemie; appl. 2.5.1973; D-prior. 3.5.1972).

DOS 2 520 937 (Kali-Chemie; appl. 10.5.1975).

US 4 098 786 (Kali-Chemie; 4.7.1978; appl. 23.9.1976; D-prior. 3.5.1972).

US 4 244 869 (Kali-Chemie; 13.1.1981; D-prior. 3.5.1972, 10.5.1975).

Liepmann, H. et al.: Eur. J. Med. Chem. (EJMCA5) **11**, 501 (1976).

medical use as analgesic:

EP 96 320 (Kali-Chemie; appl. 28.5.1983; D-prior. 5.6.1982).

Formulation(s):　drops 10 mg/ml; tabl. 5 mg, 10 mg (as hydrochloride)

Trade Name(s):

D:　Talis (Kali-Chemie; 1990);
　　wfm

Metacycline

(Methacycline; Méthylènecycline)

ATC:　J01AA05
Use:　antibiotic

RN:　914-00-1　MF: $C_{22}H_{22}N_2O_8$　MW: 442.42　EINECS: 213-017-5

LD$_{50}$:　660 mg/kg (R, i.p.)

CN:　[4S-(4α,4aα,5α,5aα,12aα)]-4-(dimethylamino)-1,4,4a,5,5a,6,11,12a-octahydro-3,5,10,12,12a-pentahydroxy-6-methylene-1,11-dioxo-2-naphthacenecarboxamide

monohydrochloride

RN:　3963-95-9　MF: $C_{22}H_{22}N_2O_8 \cdot HCl$　MW: 478.89　EINECS: 223-568-3

LD$_{50}$:　193 mg/kg (M, i.v.); 3450 mg/kg (M, p.o.);
　　202 mg/kg (R, i.v.); >2 g/kg (R, p.o.)

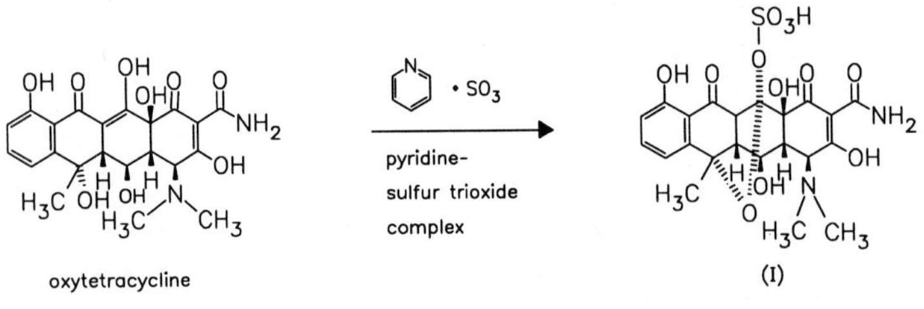

oxytetracycline

pyridine-
sulfur trioxide
complex

(I)

Metacycline

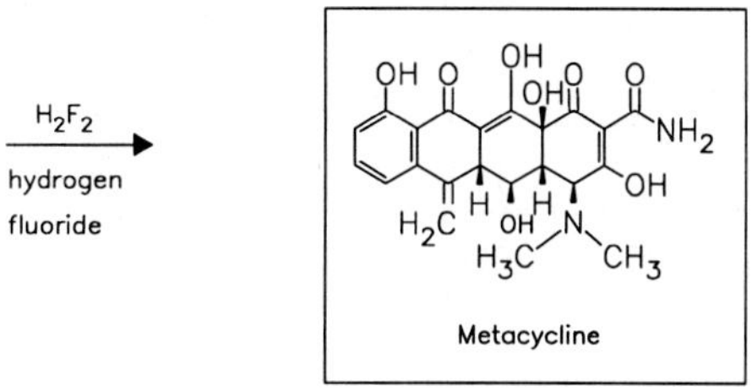

Reference(s):
US 2 984 686 (Pfizer; 16.5.1961; appl. 19.12.1960).
US 3 026 354 (Pfizer; 20.3.1962; prior. 15.12.1960).
cf. also doxycycline

Formulation(s):   cps. 150 mg, 300 mg; drops 100 mg; susp. 100 mg (as hydrochloride)

Trade Name(s):
D: Rondo-Bron (Mack)-
      comb.; wfm
      Rondo-Bron (Mack)-comb.
      with guaiphenesin; wfm
      Rondomycin (Mack); wfm
F: Lysocline (Parke Davis)

GB: Physiomycine (Laphal)
      Rondomycin (Pfizer); wfm
I: Esarondil (Terapeutico
      M.R.)
      Rotilen (Terapeutico Mil.)
      Stafilon (AGIPS)

J: Adramycin (Sanko)
      Rondomycin (Taito Pfizer)
USA: Rondomycin (Wallace);
      wfm

Metahexamide

ATC: A10BB10
Use: antidiabetic

RN: 565-33-3   MF: $C_{14}H_{21}N_3O_3S$ MW: 311.41   EINECS: 209-276-9
CN: 3-amino-*N*-[(cyclohexylamino)carbonyl]-4-methylbenzenesulfonamide

3-acetylamino-
4-methylbenzene-
sulfonamide

ethyl
chloroformate

(I)

cyclohexyl-
amine

Metahexamide

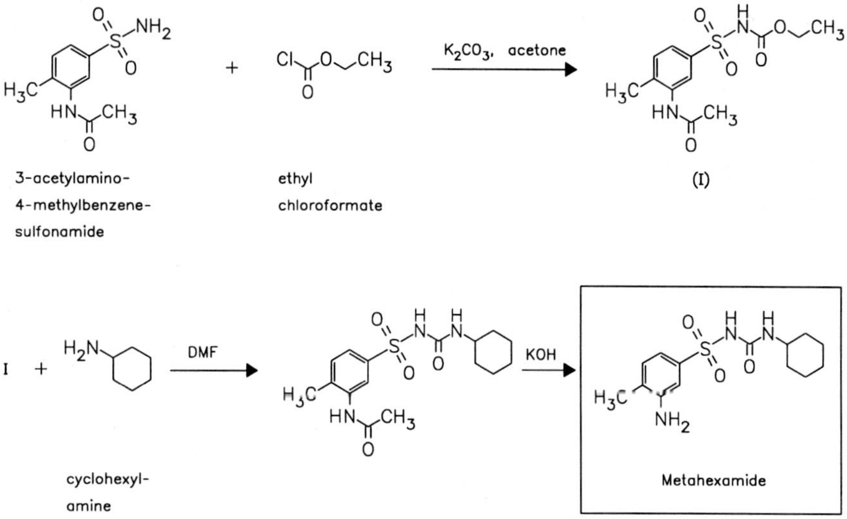

Reference(s):
GB 831 043 (Boehringer Mannh.; appl. 1958; D-prior. 1957).

Formulation(s):   tabl. 100 mg

Trade Name(s):
F: Isodiane (Servier); wfm

Metamizole sodium

ATC: N02BB02
Use: analgesic, antipyretic, anti-inflammatory

RN: 68-89-3 MF: $C_{13}H_{16}N_3NaO_4S$ MW: 333.34 EINECS: 200-694-7
LD_{50}: 2197 mg/kg (M, i.v.); 2891 mg/kg (M, p.o.);
2182 mg/kg (R, i.v.); 3 g/kg (R, p.o.)
CN: [(2,3-dihydro-1,5-dimethyl-3-oxo-2-phenyl-1*H*-pyrazol-4-yl)methylamino]methanesulfonic acid sodium salt

monohydrate
RN: 5907-38-0 MF: $C_{13}H_{16}N_3NaO_4S \cdot H_2O$ MW: 351.36
metamizole
RN: 50567-35-6 MF: $C_{13}H_{17}N_3O_4S$ MW: 311.36 EINECS: 256-627-7

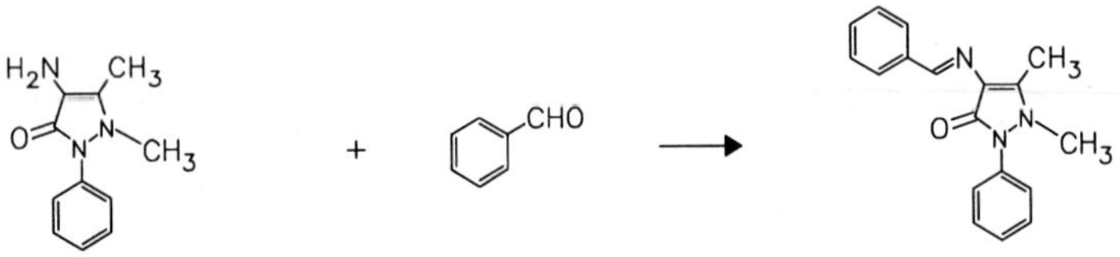

4-amino-2,3-dimethyl-1-phenyl-5-Δ^3-pyrazolone (cf. aminophenazone synthesis)

benzaldehyde

4-benzylidenamino-2,3-dimethyl-1-phenyl-5-Δ^3-pyrazolone (I)

I +

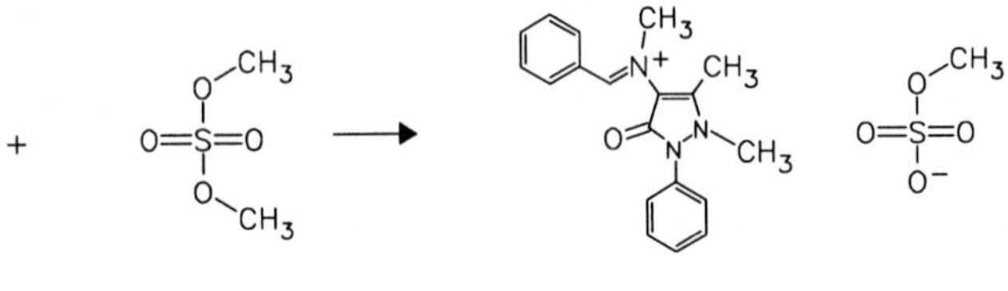

dimethyl sulfate

4-benzylidenemethylammonio-2,3-dimethyl-1-phenyl-5-Δ^3-pyrazolone methyl sulfate (II)

II

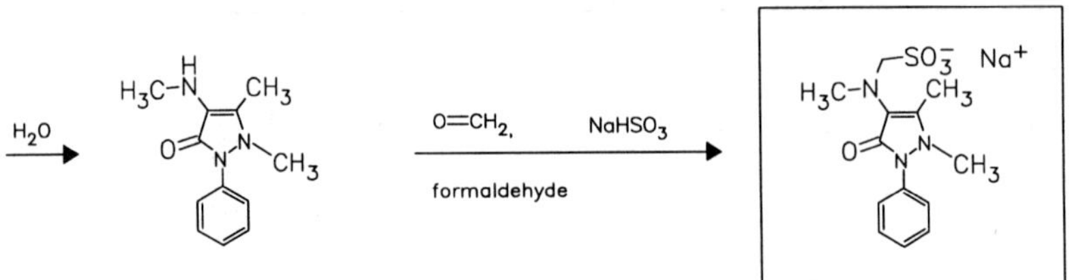

2,3-dimethyl-4-methylamino-1-phenyl-5-Δ^3-pyrazolone

Metamizole sodium

Reference(s):
Ehrhart, Ruschig, **I**, 171.
DRP 476 663 (I.G. Farben; 1922).
DRP 421 505 (I.G. Farben; appl. 1920),
DRP 467 627 (I.G. Farben; appl. 1921).
DRP 476 643 (I.G. Farben; appl. 1921).

Formulation(s): amp. 1 g, 2.5 g, 5 g; drops 500 mg; f. c. tabl. 500 mg; suppos. 300 mg, 750 mg, 1 g; syrup 250 mg

Trade Name(s):
D: Novalgin (Hoechst)
Novaminsulfon (Braun
Melsungen; Lichtenstein;
ratiopharm)
Novaminsulfon-ratiopharm
(ratiopharm)

F: Novalgine (Hoechst)
Pyréthane (Gerda)
combination preparations
I: Novalgina (Hoechst-I)-
comb.
Trisalgina (Molteni)-comb.

J: Sulpylon (Hokuriku)
Sulpyna (Kanto)
USA: Novaldin (Winthrop); wfm

Metampicillin

ATC: J01CA14
Use: antibiotic

RN: 6489-97-0 MF: $C_{17}H_{19}N_3O_4S$ MW: 361.42 EINECS: 229-365-6
CN: [2S-[2α,5α,6β(S*)]]-3,3-dimethyl-6-[[(methyleneamino)phenylacetyl]amino]-7-oxo-4-thia-1-azabicyclo[3.2.0]heptane-2-carboxylic acid

monosodium salt
RN: 6489-61-8 MF: $C_{17}H_{18}N_3NaO_4S$ MW: 383.40

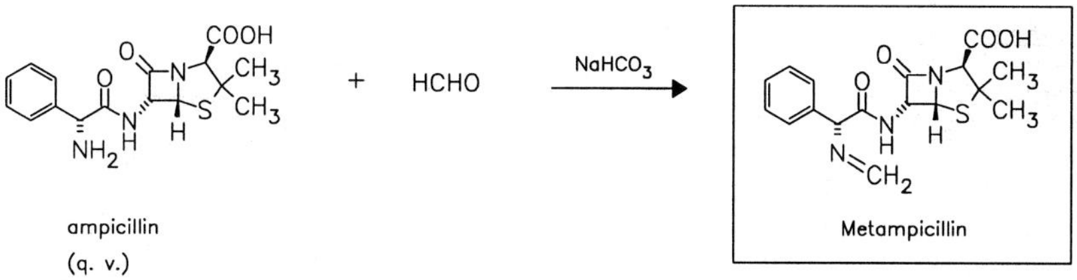

ampicillin
(q. v.)

Metampicillin

Reference(s):
GB 1 081 093 (Soc. d'Etudes de Recherche et d'Applicat. Scientifiques et Medicals; appl. 17.3.1964; valid from 12.3.1965).

Formulation(s): amp. 250 mg, 500 mg, 1 g; cps. 250 mg, 500 mg (as sodium salt)

Trade Name(s):
F: Magnipen (Clin-Midy);
wfm
I: Suvipen (Sarbach); wfm
Magnipen (Midy); wfm

Metandienone
(Methandienone; Methandrostenolone)

ATC: A14AA03; D11AE01
Use: anabolic, androgen

RN: 72 63 9 MF: $C_{20}H_{28}O_2$ MW: 300.44 EINECS: 200-787-2
LD_{50}: >1 g/kg (R, p.o.)
CN: (17β)-17-hydroxy-17-methylandrosta-1,4-dien-3-one

ⓐ

methyltestosterone (I)
(q. v.)

Metandienone

ⓑ

I   →   microbiological dehydrogenation [Didymella]   →   Metandienone

Reference(s):
a   US 2 900 398 (Ciba; 1959; CH-prior. 1956).
    Meystre, Ch. et al.: Helv. Chim. Acta (HCACAV) **39**, 734 (1956).
b   Vischer, F. et al.: Helv. Chim. Acta (HCACAV) **38**, 1502 (1955).

Formulation(s):   ointment 10 mg/g; tabl. 2 mg, 5 mg

Trade Name(s):

D:	Dianabol (Ciba); wfm	GB:	Dianabol (Ciba); wfm		Perholin (Ion); wfm
F:	Dianabol (Ciba-Geigy); wfm	I:	Dianabol (Ciba); wfm	J:	Abirol (Takeda)
	Dianavit (Ciba-Geigy)-comb.; wfm		Metabolina (Guidi); wfm		Anoredan (Kodama)
			Metastenol (Farber-Ref); wfm	USA:	Dianabol (Ciba); wfm

Metapramine

ATC: N06A
Use: antidepressant

RN: 21730-16-5   MF: $C_{16}H_{18}N_2$ MW: 238.33
CN: 10,11-dihydro-N,5-dimethyl-5H-dibenz[b,f]azepin-10-amine

fumarate
RN: 93841-84-0   MF: $C_{16}H_{18}N_2 \cdot C_4H_4O_4$ MW: 354.41
hydrochloride
RN: 21737-55-3   MF: $C_{16}H_{18}N_2 \cdot xHCl$ MW: unspecified   EINECS: 244-555-9

5-methyl-10-oxo-
10,11-dihydro-5H-
dibenz[b,f]azepine

H_2N-OH →

5-methyl-10-
hydroxyimino-
10,11-dihydro-5H-
dibenz[b,f]azepine

Na, C_4H_9OH →

(I)

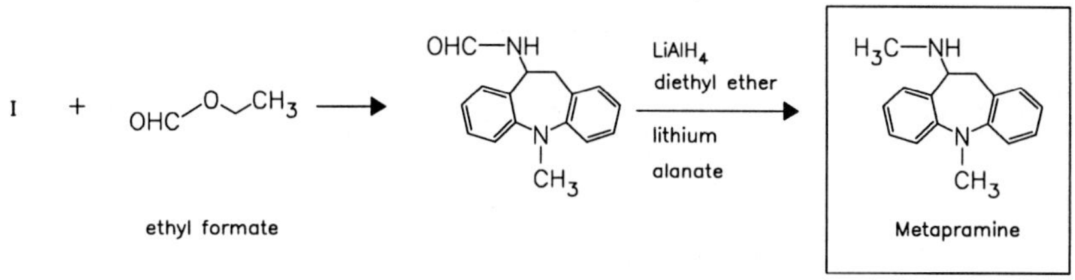

ethyl formate

Metapramine

Reference(s):

FR-M 6 616 (Rhône-Poulenc; appl. 14.4.1967).
ZA 6 800 345 (Rhône-Poulenc; appl. 19.6.1968; F-prior. 18.1.1967, 9.11.1967).

alternative synthesis:

DOS 2 159 678 (Rhône-Poulenc; appl. 1.12.1971; F-prior. 1.12.1970).

Formulation(s):   tabl. 50 mg (as fumarate)

Trade Name(s):
F: Rodostene (Rhône-   Timaxel (Specia); wfm
         Poulenc); wfm

Metaraminol

ATC: C01CA09
Use: sympathomimetic

RN: 54-49-9   MF: $C_9H_{13}NO_2$ MW: 167.21
LD$_{50}$: 51 mg/kg (M, i.v.); 99 mg/kg (M, p.o.);
         240 mg/kg (R, p.o.)
CN: [R-(R*,S*)]-α-(1-aminoethyl)-3-hydroxybenzenemethanol

hydrogen tartrate (1:1)
RN: 33402-03-8   MF: $C_9H_{13}NO_2 \cdot C_4H_6O_6$ MW: 317.29   EINECS: 251-502-3
LD$_{50}$: 39 mg/kg (M, i.v.); 99 mg/kg (M, p.o.);
         3427 µg/kg (R, i.v.); 240 mg/kg (R, p.o.)

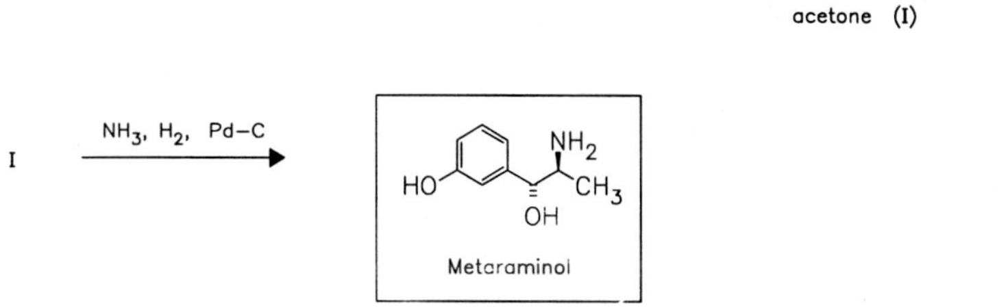

3-acetoxy-
benzaldehyde

D-glucose

(−)-1-hydroxy-1-
(3-hydroxyphenyl)-
acetone (I)

Metaraminol

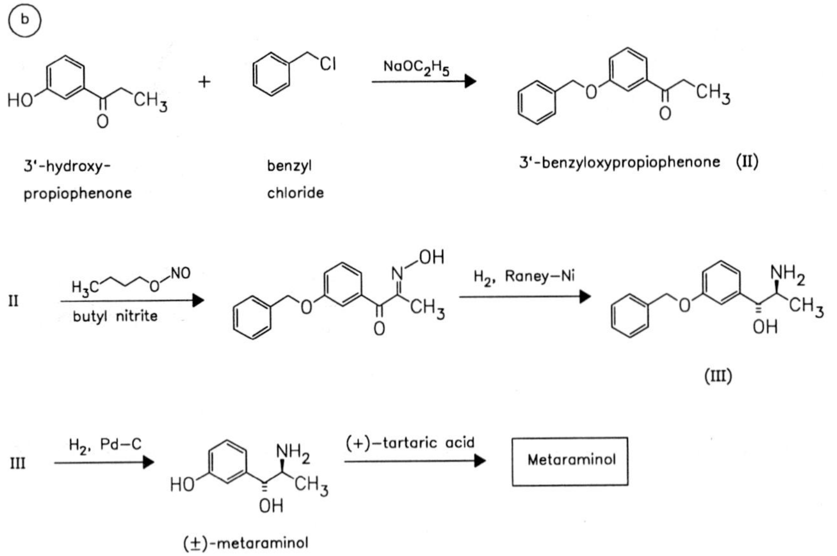

(b)

3'-hydroxy-
propiophenone

benzyl
chloride

3'-benzyloxypropiophenone (II)

II

butyl nitrite

H₂, Raney–Ni

(III)

III

H₂, Pd–C

(+)–tartaric acid

Metaraminol

(±)-metaraminol

Reference(s):

a DRP 555 404 (I. G. Farben; appl. 1930).
US 1 951 302 (Winthrop; 1934; D-prior. 1930).
b DRP 571 229 (I. G. Farben; appl. 1930).
US 1 948 162 (Winthrop; 1934; D-prior. 1930).
US 1 995 709 (Sharp & Dohme; 1935; appl. 1931).
GB 396 951 (I. G. Farben; appl. 1932; D-prior. 1931).

Formulation(s): amp. 10 mg/ml (as hydrogen tartrate)

Trade Name(s):

D:	Araminum (Sharp & Dohme); wfm	GB:	Aramine (Merck Sharp & Dohme)
F:	Aramine (Merck Sharp & Dohme); wfm	I:	Levicor (Bioindustria)
		J:	Araminon (Merck-Banyu)

USA: Aramine (Merck Sharp & Dohme; as bitartrate)

Metaxalone

ATC: M03
Use: muscle relaxant

RN: 1665-48-1 MF: C₁₂H₁₅NO₃ MW: 221.26 EINECS: 216-777-6
LD₅₀: 1690 mg/kg (M, p.o.);
775 mg/kg (R, p.o.)
CN: 5-[(3,5-dimethylphenoxy)methyl]-2-oxazolidinone

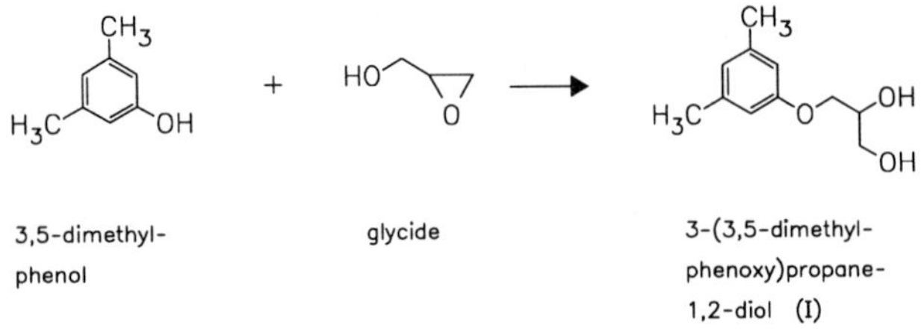

3,5-dimethyl-
phenol

glycide

3-(3,5-dimethyl-
phenoxy)propane-
1,2-diol (I)

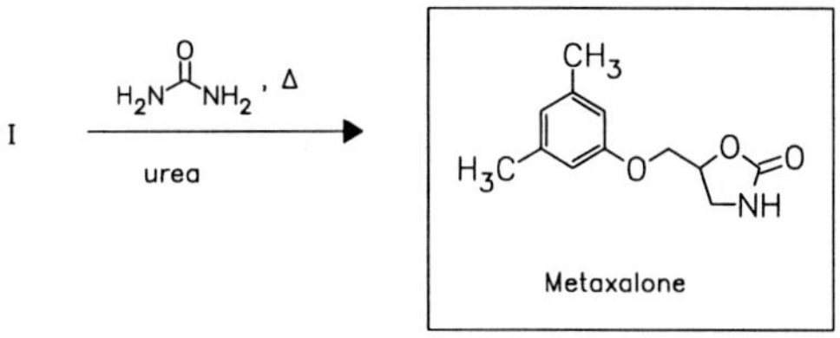
I
urea
Metaxalone

Reference(s):
US 3 062 827 (A. H. Robins; 6.11.1962; prior. 19.6.1959).

Formulation(s): tabl. 400 mg

Trade Name(s):
USA: Skelaxin (Carnrick)

Metenolone acetate

(Methenolone acetate)

ATC: A14AA04
Use: anabolic

RN: 434-05-9 MF: $C_{22}H_{32}O_3$ MW: 344.50 EINECS: 207-097-0
LD_{50}: 4 g/kg (M, p.o.);
4 g/kg (R, p.o.)
CN: (5α,17β)-17-(acetyloxy)-1-methylandrost-1-en-3-one

metenolone
RN: 153-00-4 MF: $C_{20}H_{30}O_2$ MW: 302.46 EINECS: 205-812-0
metenolone enanthate
RN: 303-42-4 MF: $C_{27}H_{42}O_3$ MW: 414.63 EINECS: 206-141-6

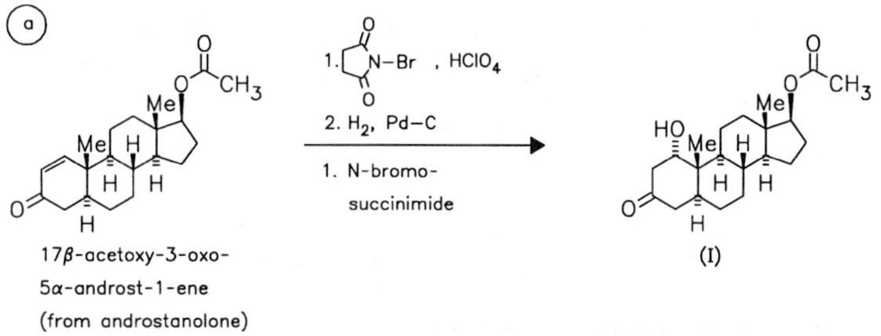
a
1. N-Br , HClO4
2. H2, Pd–C
1. N-bromo-
succinimide
17β-acetoxy-3-oxo-
5α-androst-1-ene
(from androstanolone)
(I)

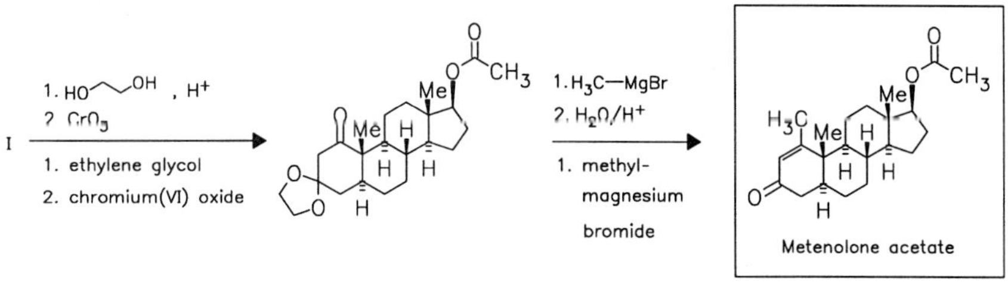
I
1. HO⌒OH , H+
2. CrO3
1. ethylene glycol
2. chromium(VI) oxide
1. H3C—MgBr
2. H2O/H+
1. methyl-
magnesium
bromide
Metenolone acetate

b

17β-hydroxy-3-oxo-5α-androst-1-ene

diazomethane

Δ, quinoline, N$_2$ → II

metenolone (II)

acetic anhydride

pyridine →

Metenolone acetate

Reference(s):

a DE 1 152 100 (Schering AG; appl. 23.11.1960).
   DE 1 154 467 (Schering AG; appl. 22.7.1961).

b DE 1 023 764 (Schering AG; appl. 6.2.1957).
   DE 1 072 991 (Schering AG; appl. 25.10.1958).
   DE 1 096 353 (Schering AG; appl. 11.7.1961).
   DE 1 117 113 (Schering AG; appl. 5.12.1959).
   DE 1 135 900 (Schering AG; appl. 27.8.1960).

starting material:
Butenandt, A.; Dannenberg, H.: Chem. Ber. (CHBEAM) **71**, 1681 (1938).

alternative syntheses:
GB 977 082 (Schering AG; valid from 17.3.1961; D-prior. 6.4.1960, 21.7.1960, 23.12.1960).
GB 977 083 (Schering AG; valid from 17.3.1961; D-prior. 6.4.1960).

review:
Wiechert, R.: Z. Naturforsch., B: Anorg. Chem., Org. Chem., Biochem., Biophys., Biol. (ZENBAX) **196**, 944 (1964).

Formulation(s):   amp. 20 mg/ml, 50 mg/ml, 100 mg/ml; tabl. 5 mg, 25 mg

Trade Name(s):

D:	Primobolan Depot (Schering; as enanthate) numerous generics as acetate	GB: Primobolan Depot (Schering Chemicals); wfm
F:	Primobolan (SEPPS); wfm Primobolan-Depot (SEPPS; as enanthate); wfm	I: Primobolan Depot (Schering) J: Primobolan Depot (Nihon Schering; as enanthate)

Primobolan Inj. (Nihon Schering)
USA: Primobolan Depot (Schering); wfm
Primobolan Depot (Schering; as enanthate); wfm

Metformin
(Dimethylbiguanide)

ATC: A10BA02
Use: antidiabetic

RN: 657-24-9   MF: C$_4$H$_{11}$N$_5$ MW: 129.17   EINECS: 211-517-8
LD$_{50}$:   247 mg/kg (M, i.p.); 230 mg/kg (M, s.c.)
CN: *N,N*-dimethylimidodicarbonimidic diamide

monohydrochloride
RN: 1115-70-4   MF: C$_4$H$_{11}$N$_5$ · HCl   MW: 165.63   EINECS: 214-230-6

dimethylamine
hydrochloride

dicyanodiamide

Metformin

Reference(s):
DE 1 023 757 (Heumann & Co.; appl. 1955) - only methods.
FR-appl. 2 322 860 (Aron S.A.R.L.; appl. 5.9.1975).

Formulation(s):   f. c. tabl. 500 mg, 850 mg; s. r. tabl. 850 mg; tabl. 500 mg, 850 mg (as hydrochloride)

Trade Name(s):
D: Diabetase (Azupharma)   I: Diabetosan (Brocchieri)   Glycoran (Nippon
     Glucophage (Lipha)   Glibomet (Guidotti)-comb.   Shinyaku)
     Mediabet (Medice)   Glucamide (Lipha)-comb.   Insuloid M (Ono)
     Mescorit (Boehringer   Glucophage (Lipha)   Langer-K (Kanto)
     Mannh.)   Glucosulfa (Lipha)-comb.   Melbin (Sumitomo)
F: Glucinan (Lipha Santé)   Metforal (Guidotti)   Metolmin (Kodama)
     Glucophage (Lipha Santé)   Pleiamide (Guidotti)-comb.   USA: Glucophage (Bristol-Myers
     Stagid (Merck-Clévenot)   J: Diabetose B (Nichiiko)   Squibb)
GB: Glucophage (Lipha)

Methadone

ATC: N02AC02
Use: analgesic, narcotic (heroin
       substitution therapy)

RN: 76-99-3   MF: $C_{21}H_{27}NO$ MW: 309.45   EINECS: 200-996-9
CN: 6-(dimethylamino)-4,4-diphenyl-3-heptanone

hydrochloride
RN: 1095-90-5   MF: $C_{21}H_{27}NO \cdot HCl$ MW: 345.91   EINECS: 214-140-7
LD$_{50}$:   16 mg/kg (M, i.v.); 124 mg/kg (M, p.o.);
       9200 µg/kg (R, i.v.); 30 mg/kg (R, p.o.)
(±)-methadone
RN: 297-88-1   MF: $C_{21}H_{27}NO$ MW: 309.45
(±)-hydrochloride
RN: 125-56-4   MF: $C_{21}H_{27}NO \cdot HCl$ MW: 345.91
(–)-methadone
RN: 125-58-6   MF: $C_{21}H_{27}NO$ MW: 309.45
(–)-hydrochloride
RN: 5967-73-7   MF: $C_{21}H_{27}NO \cdot HCl$ MW: 345.91   EINECS: 227-756-6

diphenyl-
acetonitrile

2-dimethylamino-
1-methylethyl
chloride

4-dimethylamino-
2,2-diphenyl-
valeronitrile (I)

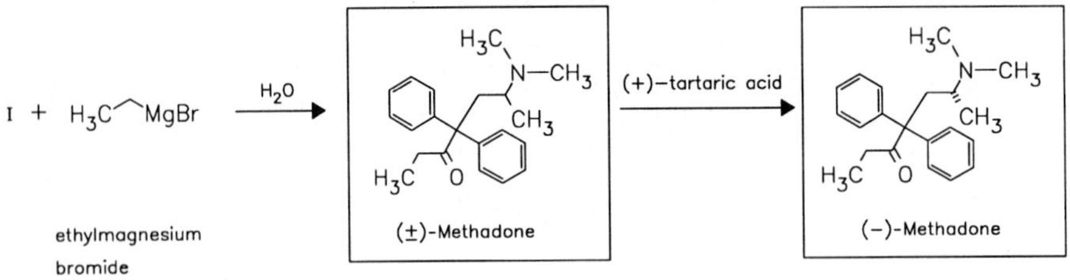

I + H₃C⌒MgBr →(H₂O) (±)-Methadone →((+)-tartaric acid) (−)-Methadone

ethylmagnesium
bromide

Reference(s):
DE 865 314 (Farbw. Hoechst; appl. 1941).
DE 870 700 (Farbw. Hoechst; appl. 1942).
DE 890 506 (Farbw. Hoechst; appl. 1944).
Ehrhart, G.; Bockmühl, M.: Justus Liebigs Ann. Chem. (JLACBF) **561**, 52 (1948).

alternative procedure for racemate resolution:
US 2 644 010 (Merck & Co.; 1953; appl. 1947).
US 2 983 757 (Abbott; 1961; appl. 1959).

Formulation(s):   amp. 5 mg/ml, 10 mg/ml; drops 5 mg; tabl. 5 mg, 10 mg, 20 mg, 40 mg

Trade Name(s):

D:	L-Polamidon (Hoechst)	GB:	Physeptone (Glaxo Wellcome)		Metadone (Molteni)
F:	Méthadone AP (Mayoly-Spindler)	I:	Eptadone (Zambon Italia) Metado (Formulario Naz.)	USA:	Dolophine Hydrochloride (Roxane) generic

---

Methallenestril
(Methallenoestril; Methallenoestrol)

ATC: G03CB03; G03CC03
Use: estrogen

RN: 517-18-0   MF: $C_{18}H_{22}O_3$ MW: 286.37   EINECS: 208-232-6
CN: β-ethyl-6-methoxy-α,α-dimethyl-2-naphthalenepropanoic acid

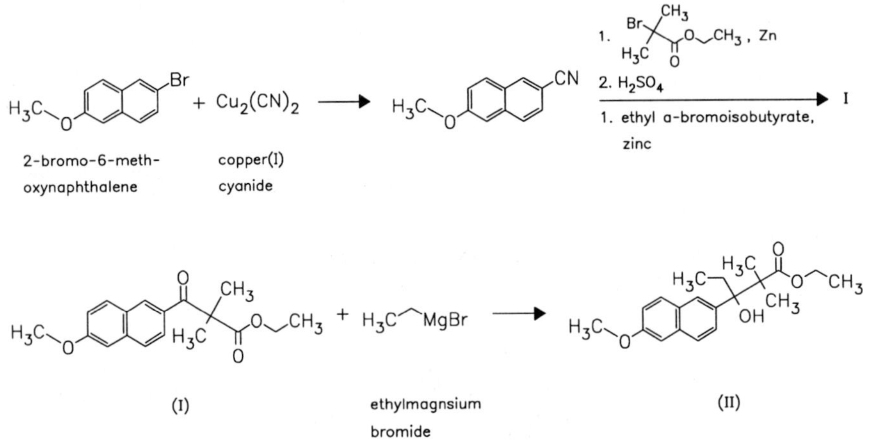

2-bromo-6-meth-
oxynaphthalene

copper(I)
cyanide

1. Br⌒CH₃ / H₃C / O⌒CH₃, Zn
2. H_2SO_4
1. ethyl α-bromoisobutyrate, zinc

→ I

(I)   + H₃C⌒MgBr → (II)

ethylmagnsium
bromide

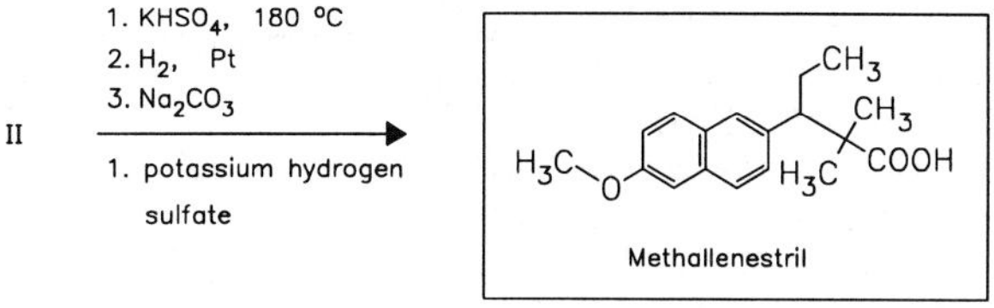

1. KHSO$_4$, 180 °C
2. H$_2$, Pt
3. Na$_2$CO$_3$

1. potassium hydrogen
   sulfate

II →

Methallenestril

Reference(s):

US 2 547 123 (A. Horeau; 1951; F-prior. 1947).

Horeau, A. et al.: C. R. Hebd. Seances Acad. Sci. (COREAF) **224**, 862 (1947).

Horeau, A. et al.: Bull. Soc. Chim. Fr. (BSCFAS) **1948**, 711; 1955, 955.

starting material:

DOS 2 619 614 (Hoechst; appl. 4.5.1976).

Formulation(s):   tabl. 3 mg

Trade Name(s):

GB: Vallestril (Searle); wfm   Vallestril (Dainippon)
J: Ercostrol (Green Cross)   USA: Vallestril (Searle); wfm

Methamphetamine

(Desoxyephedrine)

ATC: N06BA03
Use: sympathomimetic, psychostimulant,
        appetite depressant

RN: 537-46-2   MF: C$_{10}$H$_{15}$N   MW: 149.24   EINECS: 208-668-7
CN: (S)-N,α-dimethylbenzeneethanamine

hydrochloride

RN: 51-57-0   MF: C$_{10}$H$_{15}$N · HCl   MW: 185.70   EINECS: 200-106-9

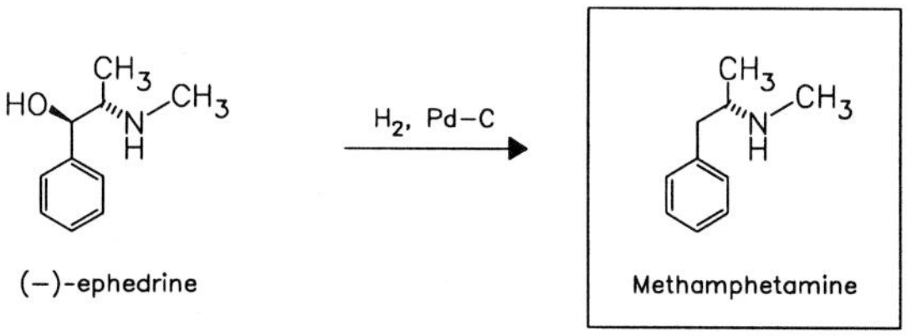

(−)-ephedrine   Methamphetamine

Reference(s):

Emde, H.: Helv. Chim. Acta (HCACAV) **12**, 365 (1929).

Formulation(s):   tabl. 5 mg, 10 mg, 15 mg (as hydrochloride)

Trade Name(s):

D: Pervitin (Temmler); wfm   USA: Desoxyn (Abbott; as
J: Philopon (Dainippon)   hydrochloride)

Methandriol

ATC: A14
Use: anabolic, androgen

RN: 521-10-8   MF: C$_{20}$H$_{32}$O$_2$ MW: 304.47   EINECS: 208-301-0
CN: (3β,17β)-17-methylandrost-5-ene-3,17-diol

dipropionate
RN: 3593-85-9 MF: $C_{26}H_{40}O_4$ MW: 416.60

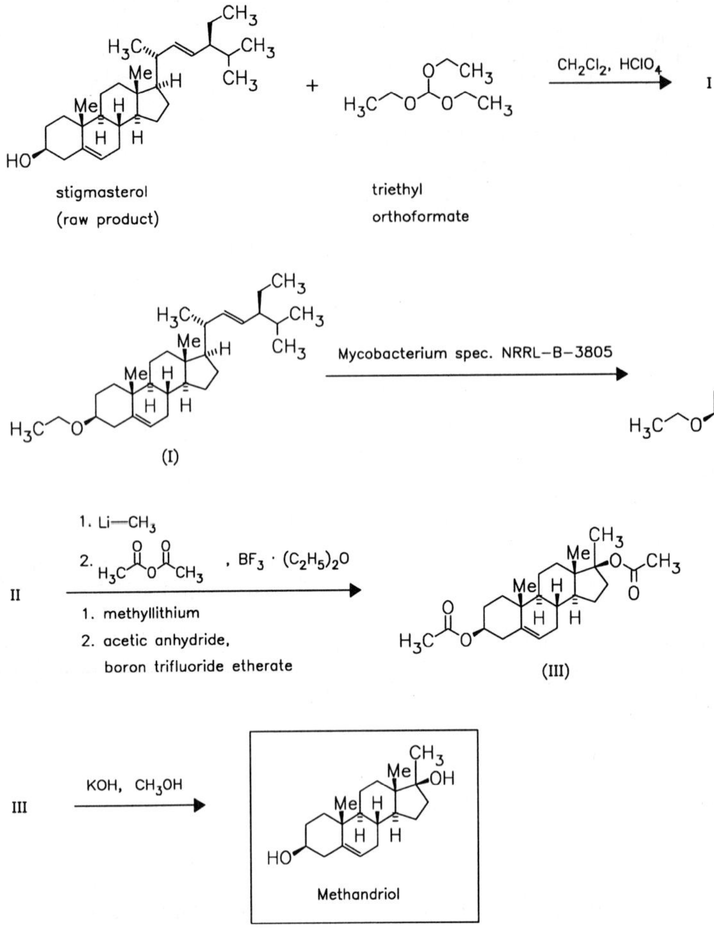

stigmasterol
(raw product)

triethyl
orthoformate

(I)

Mycobacterium spec. NRRL–B–3805

(II)

II

1. Li—CH₃

2. H_3C ... , $BF_3 \cdot (C_2H_5)_2O$

1. methyllithium
2. acetic anhydride,
   boron trifluoride etherate

(III)

III

KOH, CH₃OH

Methandriol

Reference(s):
DOS 2 534 911 (Schering AG; appl. 1.8.1975).
Ruzicka, L. et al.: Helv. Chim. Acta (HCACAV) **18**, 1487 (1935).
Miescher, K.; Klarer, W.: Helv. Chim. Acta (HCACAV) **22**, 962 (1939).

Formulation(s): amp. 50 mg/ml (as dipropionate)

Trade Name(s):
I: Anacufen (Difa
  Coopervision)-comb.; wfm
  Metilandrostendiolo
  Schering (Schering); wfm
  Metildiolo (Orma)-comb.;
  wfm

  Otormon F (Farmades).-
  comb.; wfm
  Panfaco (Difa
  Coopervision)-comb.; wfm
  Sinesex (Wells); wfm

  Troformone (Biomedica
  Foscama); wfm
USA: Methostan (Schering); wfm
  Stenediol (Organon); wfm

Methapyrilene
(Thenylpyramine)

ATC: R06AC05
Use: antihistaminic

RN: 91-80-5   MF: $C_{14}H_{19}N_3S$ MW: 261.39   EINECS: 202-099-8
LD_{50}:   20 mg/kg (M, i.v.); 182 mg/kg (M, p.o.)
CN: *N,N*-dimethyl-*N'*-2-pyridinyl-*N'*-(2-thienylmethyl)-1,2-ethanediamine

monohydrochloride
RN: 135-23-9   MF: $C_{14}H_{19}N_3S \cdot HCl$ MW: 297.85   EINECS: 205-184-8
LD_{50}:   17.5 mg/kg (M, i.v.); 182 mg/kg (M, p.o.);
       200 mg/kg (R, p.o.)
fumarate (2:3)
RN: 33032-12-1   MF: $C_{14}H_{19}N_3S \cdot 3/2C_4H_4O_4$ MW: 871.00   EINECS: 251-351-3

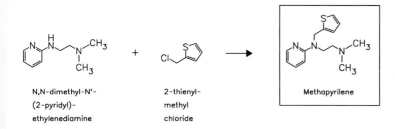

N,N-dimethyl-N'-   2-thienyl-   Methapyrilene
(2-pyridyl)-   methyl
ethylenediamine   chloride

Reference(s):
Weston, A.W.: J. Am. Chem. Soc. (JACSAT) **69**, 980 (1947).
US 2 581 868 (Monsanto; 1952; prior. 1946).

fumarate:
GB 694 805 (Monsanto; valid from 1950; USA-prior. 1949).

Formulation(s):   cps. 50 mg, 100 mg (as hydrochloride)

Trade Name(s):
D: Contac Liquid (Vonora)-
       comb.; wfm
       Copyronilum (Lilly)-
       comb.; wfm
       Sedanoct (Woelm)-comb.;
       wfm
       tiffaforte (Tiffapharm; as
       fumarate); wfm
I: Co-Pyronil (Lilly); wfm
USA: Allerest (Pharmacraft; as
       fumarate)-comb.; wfm

Brexin (Savage)-comb.;
wfm
Citra (Boyle)-comb.; wfm
Co-Pyronil (Dista)-comb.;
wfm
Ephed-Organidin
(Wallace); wfm
Excedrin P.M. (Bristol-
Myers; as fumarate)-comb.;
wfm

Hista-Clopane (Lilly)-
comb.; wfm
Histadyl E.C. (Lilly; as
fumarate)-comb.; wfm
Histadyl Fum. (Lilly; as
fumarate); wfm
Histadyl Pulvules (Lilly);
wfm

Methaqualone

ATC: N05CM01
Use: hypnotic

RN: 72-44-6   MF: $C_{16}H_{14}N_2O$ MW: 250.30   EINECS: 200-780-4
LD_{50}:   420 mg/kg (M, p.o.);
       185 mg/kg (R, p.o.)
CN: 2-methyl-3-(2-methylphenyl)-4(3*H*)-quinazolinone

monohydrochloride
RN: 340-56-7 MF: $C_{16}H_{14}N_2O \cdot HCl$ MW: 286.76 EINECS: 206-431-2
LD_{50}: 120 mg/kg (M, i.v.); 400 mg/kg (M, p.o.);
120 mg/kg (R, i.v.); 410 mg/kg (R, p.o.)

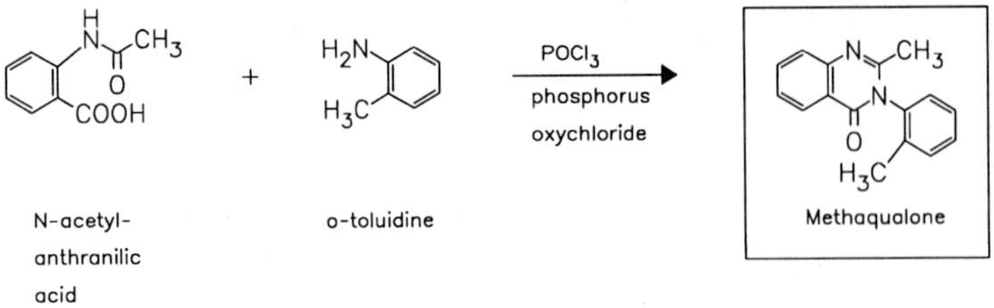

N-acetyl-
anthranilic
acid

o-toluidine

POCl₃
phosphorus
oxychloride

Methaqualone

Reference(s):
GB 843 073 (Labs. Toraude; appl. 22.8.1958; USA-prior. 9.5.1958).
Klosa, J.: J. Prakt. Chem. (JPCEAO) [4], **20**, 283 (1963).

Formulation(s): tabl. 200 mg

Trade Name(s):
D: Normi-Nox (Herbrand);
wfm
Optinoxan (Robisch); wfm
Revonal (Cascan); wfm
F: Divinoctal (I.S.H.)-comb.;
wfm
Isonox (Ucépha)-comb.;
wfm

Mandrax (Houdé)-comb.;
wfm
Mandrax (I.S.H.)-comb.;
wfm
GB: Revonal (Merck); wfm
J: Hyminal (Eisai)
Meroctan (Sanwa)
Nene (Sankyo)

Normorest (Doitsu-Aoi)
Orzolon (Kobayashi)
USA: Parest (Parke Davis); wfm
Quaalude (Rorer); wfm
Somnafac (Cooper); wfm
Sopor (Arnar-Stone); wfm

Methazolamide

ATC: C03; S01EC
Use: diuretic (carboanhydrase inhibitor)

RN: 554-57-4 MF: $C_5H_8N_4O_3S_2$ MW: 236.28 EINECS: 209-066-7
LD_{50}: >1 g/kg (M, i.v.)
CN: N-[5-(aminosulfonyl)-3-methyl-1,3,4-thiadiazol-2(3H)-ylidene]acetamide

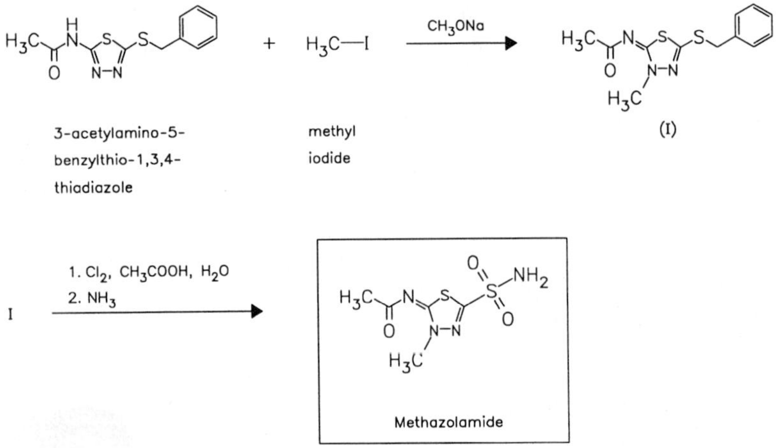

3-acetylamino-5-
benzylthio-1,3,4-
thiadiazole

methyl
iodide

CH₃ONa

(I)

1. Cl₂, CH₃COOH, H₂O
2. NH₃

I

Methazolamide

Reference(s):
US 2 783 241 (American Cyanamid; 1957; prior. 1955).

Formulation(s):   tabl. 50 mg

Trade Name(s):
F:	Neptazane (Théraplix); wfm	J:	Neptazane (Lederle)
		USA:	Neptazane (Lederle); wfm

generics

Methdilazine

ATC: R06AD04
Use: antiallergic, antihistaminic

RN: 1982-37-2   MF: $C_{18}H_{20}N_2S$ MW: 296.44   EINECS: 217-841-6
LD_{50}: 225 mg/kg (M, p.o.);
162 mg/kg (R, p.o.)
CN: 10-[(1-methyl-3-pyrrolidinyl)methyl]-10*H*-phenothiazine

monohydrochloride
RN: 1229-35-2   MF: $C_{18}H_{20}N_2S \cdot HCl$ MW: 332.90   EINECS: 214-967-3
LD_{50}: 190 mg/kg (M, p.o.);
260 mg/kg (R, p.o.)

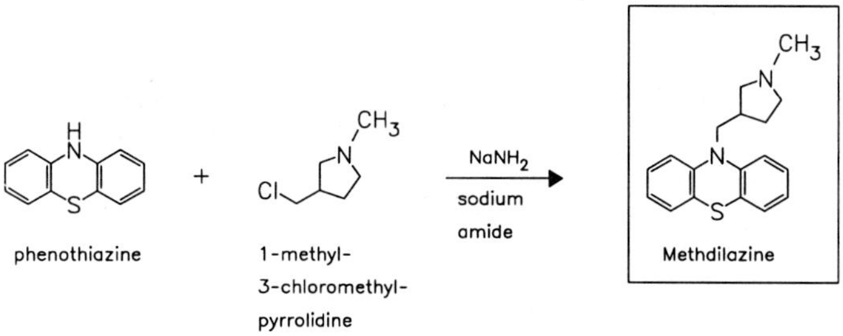

phenothiazine   +   1-methyl-3-chloromethyl-pyrrolidine   NaNH₂ sodium amide →   Methdilazine

Reference(s):
US 2 945 855 (Mead Johnson; 19.7.1960; prior. 21.10.1958).
DE 1 049 382 (Cilag; appl. 1956; CH-prior. 1955).

Formulation(s):   tabl. 8 mg (as hydrochloride)

Trade Name(s):
GB:	Dilosyn (Duncan, Flockhart); wfm	USA:	Tacaryl (Westwood); wfm

Methenamine
(Formamine; Hexamethylentetramine; HMT; HHMTA; Metenamine; Urotropin)

ATC: G04AA01
Use: antibacterial (urinary)

RN: 100-97-0   MF: $C_6H_{12}N_4$ MW: 140.19   EINECS: 202-905-8
CN: 1,3,5,7-tetraazatricyclo[3.3.1.1^{3,7}]decane

mandelate (1:1)
RN: 587-23-5   MF: $C_8H_8O_3 \cdot C_6H_{12}N_4$ MW: 292.34   EINECS: 209-597-4

hippurate (1:1)
RN: 5714-73-8   MF: $C_9H_9NO_3 \cdot C_6H_{12}N_4$ MW: 319.37   EINECS: 227-206-5
LD_{50}:   1500 mg/kg (M, i.p.); 2870 mg/kg (M, s.c.)

Methenamine

Reference(s):
US 2 762 799 (J. Meissner; 1956; D-prior. 1952).
US 2 762 800 (J. Meissner; 1956; D-prior. 1951).

Formulation(s):   f. c. tabl. 250 mg, 500 mg, 1000 mg; cream 13 g/100g; drg. 500 mg, 1000 mg

Trade Name(s):
D: Antihydral (Robugen)
     Mandelamine (Parke
     Davis)
     Urotractan (Klinge)
F: Aromalgyl (Plantes et
     Medecines)-comb.
     Mictasol (J. P. Martin)-
     comb.
     Uromil (Iprad)-comb.
GB: Hiprex (3M Health Care)
I: Cinarbile cpr. (Benvegna)-
     comb.; wfm

Elmitolo (Bayer); wfm
Esamet (Tariff. Integrativo)
Esation vitaminico
(Lafare); wfm
Etiliodina B1 (Ceccarelli)-
comb.; wfm
Jodoibs (Benvegna); wfm
Mictasol (Zoja)-comb.;
wfm
Tionamil (Ogna)-comb.;
wfm

J: Hexamine(Mohan;
     Nisshin-Yamagata)
USA: Urex (3M; as hippurate)
     Uro-Phosphate (ECR)
     Mandelamine (Warner
     Chilcott Professional
     Products; as mandelate)
     Urised (PolyMedica)-
     comb.
     Uroqid-Acid (Beach; as
     mandelate)

Methestrol dipropionate

ATC: G03C
Use: estrogen

RN: 84-13-9   MF: $C_{26}H_{34}O_4$ MW: 410.55
CN: 4,4'-(1,2-diethyl-1,2-ethanediyl)bis[2-methylphenol] dipropanoate

methestrol
RN: 130-73-4   MF: $C_{20}H_{26}O_2$ MW: 298.43

3'-methyl-4'-
propionyloxy-
propiophenone

3,4-bis(3-methyl-4-propionyl-
oxyphenyl)-3,4-hexandediol   (I)

I

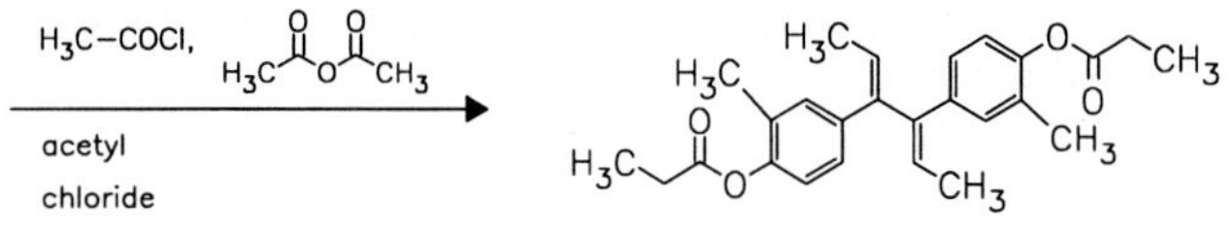

3,4-bis(3-methyl-4-propionyloxy-
phenyl)-2,4-hexadiene (II)

II

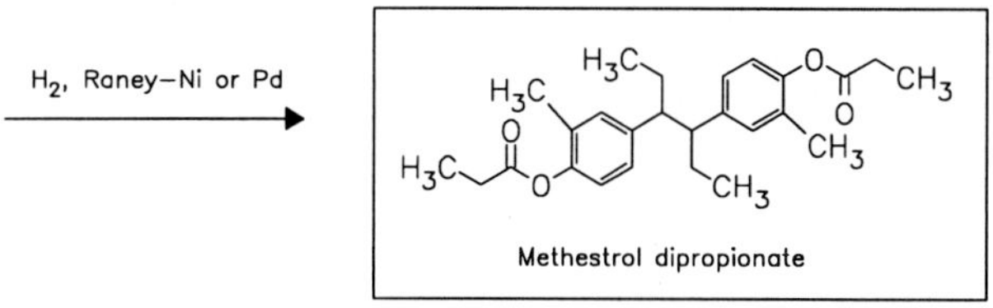

Methestrol dipropionate

Reference(s):

Niederl, V. et al.: J. Am. Chem. Soc. (JACSAT) **70**, 508 (1948).

alternative syntheses:

Marson, L.M.: Bull. Chim. Farm. (BCFAAI) **102**, 317 (1963).
Burckhalter, J.H.; Seiwald, R.J.: J. Org. Chem. (JOCEAH) **24**, 445 (1959).

Formulation(s):   4 x 1 mg/day (oral)

Trade Name(s):
USA: Meprane (Reed &
          Carnrick); wfm

Methocarbamol

ATC: M03BA03
Use: muscle relaxant

RN: 532-03-6   MF: $C_{11}H_{15}NO_5$ MW: 241.24   EINECS: 208-524-3
LD_{50}:   774 mg/kg (M, i.v.); 812 mg/kg (M, p.o.);
          1320 mg/kg (R, p.o.);
          2 g/kg (dog, p.o.)
CN: 3-(2-methoxyphenoxy)-1,2-propanediol 1-carbamate

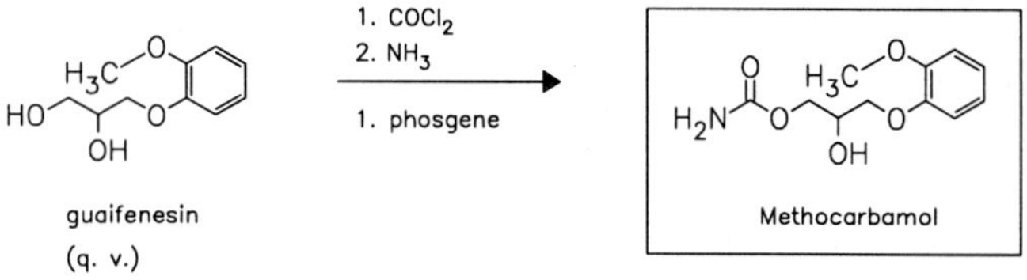

guaifenesin
(q. v.)

Methocarbamol

Reference(s):

US 2 770 649 (Robins; 1956; prior. 1955).
Yale, H.L. et al.: J. Am. Chem. Soc. (JACSAT) **72**, 3710 (1950).

Formulation(s):   amp. 100 mg/ml; tabl. 250 mg, 500 mg, 750 mg; tabl. (USA) 325 mg, 400 mg in comb. with
          aspirin

Trade Name(s):
D: Ortoton (Bastian-Werk)   GB: Robaxin (Shire)   Miowas (Wassermann);
F: Lumirelax (Jumer Sa)-   I: Miowas (IFI); wfm   wfm
      comb.   J: Carbametin (Uji)

Carxin (Kanto)
Methocabal (Zeria)
Methocal (Daiko)

Nichirakishin S (Nichiiko)
Ohlaxin (Ohta)
USA: Robaxin (Robins)

Robaxisal (Robins)-comb.
generics

Methohexital
(Methohexitone)

ATC: N01AF01; N05CA15
Use: narcotic

RN: 18652-93-2 MF: $C_{14}H_{18}N_2O_3$ MW: 262.31
CN: (±)-1-methyl-5-(1-methyl-2-pentynyl)-5-(2-propenyl)-2,4,6(1H,3H,5H)-pyrimidinetrione

sodium salt
RN: 60634-69-7 MF: $C_{14}H_{17}N_2NaO_3$ MW: 284.29

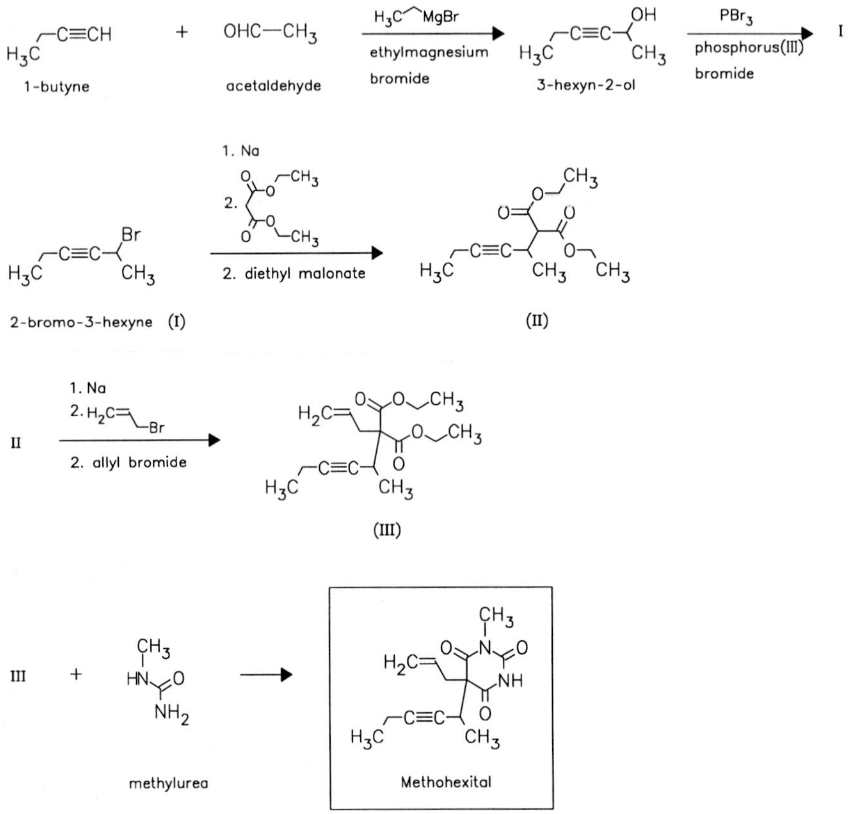

1-butyne   acetaldehyde   ethylmagnesium bromide   3-hexyn-2-ol   phosphorus(III) bromide

2-bromo-3-hexyne (I)   2. diethyl malonate   (II)

II   2. allyl bromide   (III)

III   methylurea   Methohexital

Reference(s):
US 2 872 448 (Eli Lilly; 3.2.1959; prior. 4.4.1956).

Formulation(s): vial 100 mg, 500 mg (as sodium salt)

Trade Name(s):
D: Brevimytal-Natrium (Lilly) GB: Brietal Sodium (Lilly) USA: Brevital Sodium (Jones
F: Briétal (Lilly); wfm   Medical Industries)

Methotrexate

(Amethopterin)

ATC: N01AF01; N05CA15
Use: antineoplastic (folic acid antagonist)

RN: 59-05-2   MF: $C_{20}H_{22}N_8O_5$ MW: 454.45   EINECS: 200-413-8

LD_{50}:   65 mg/kg (M, i.v.); 146 mg/kg (M, p.o.);
    14 mg/kg (R, i.v.); 135 mg/kg (R, p.o.)

CN: *N*-[4-[[(2,4-diamino-6-pteridinyl)methyl]methylamino]benzoyl]-L-glutamic acid

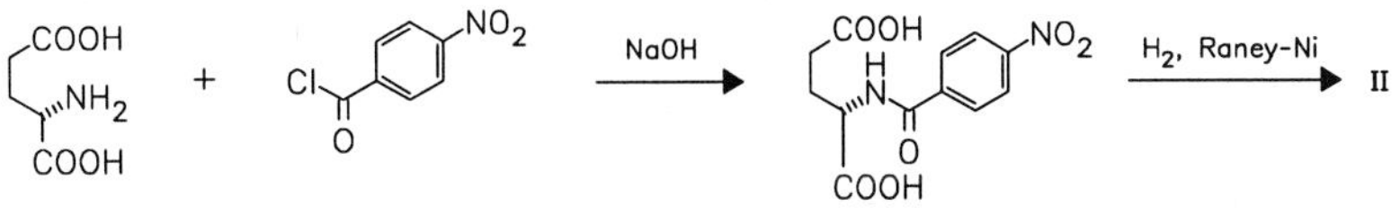

L-glutamic acid (I) + 4-nitrobenzoyl chloride → (NaOH) → → (H₂, Raney-Ni) → II

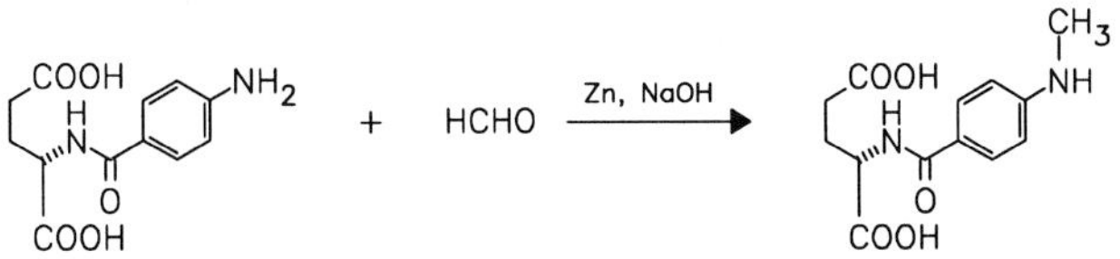

N-(4-aminobenzoyl)-L-glutamic acid (II) + HCHO → (Zn, NaOH) → N-(4-methylamino-benzoyl)-L-glutamic acid (III)

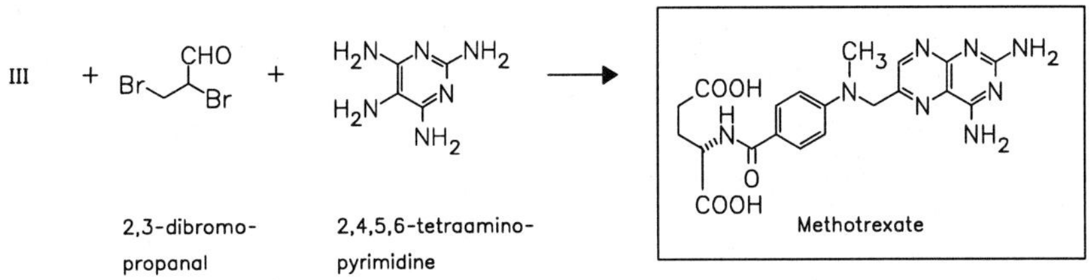

III + 2,3-dibromo-propanal + 2,4,5,6-tetraamino-pyrimidine → Methotrexate

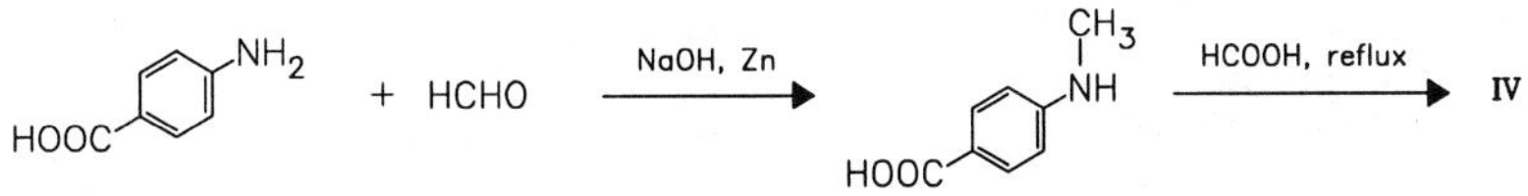

4-aminobenzoic acid + HCHO formaldehyde → (NaOH, Zn) → 4-(methylamino)-benzoic acid → (HCOOH, reflux) → IV

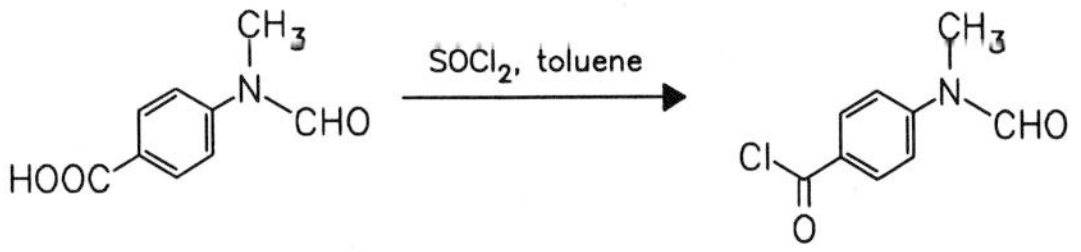

(IV) → (SOCl₂, toluene) → (V)

diethyl L-glutamate
hydrochloride

diethyl N-(4-methyl-
amino)benzoyl-L-
glutamate (VI)

2,4,5,6-tetraamino-
pyrimidine dihydrobromide

1,3-dihydroxy-
acetone

2,4-diamino-6-(hydroxy-
methyl)pteridine hydro-
bromide (VII)

(VIII)

"methotrexate ester" (IX)

Methotrexate

Reference(s):
Seeger, D.R. et al.: J. Am. Chem. Soc. (JACSAT) **71**, 1753 (1949).

alternative syntheses:
Piper, J.R.; Montgomery, J.A.: J. Heterocycl. Chem. (JHTCAD) **11**, 279 (1974).
Chaykowsky, M. et al.: J. Med. Chem. (JMCMAR) **17**, 1212 (1974).
DOS 2 741 270 (US-Secr. of Commerce Nat. Techn. Inform. Service; appl. 14.9.1977; USA-prior. 17.11.1976).
US 4 057 548 (J. Wiecko; 8.11.1977; prior. 11.11.1975, 30.3.1976).
US 4 067 867 (J. Wiecko; 10.1.1978; prior. 11.11.1975, 30.3.1976, 8.10.1976).
US 4 080 325 (US-Secr. of Health; 21.3.1978; appl. 17.11.1976).
DOS 2 741 383 (Lonza; appl. 14.9.1977; CH-prior. 12.8.1977).

various syntheses of N-(4-methylaminobenzoyl)-L-glutamic acid:
DOS 2 824 011 (Lonza; appl. 1.6.1978; CH-prior. 12.8.1977).
US 3 892 801 (American Cyanamid; 1.7.1975; appl. 11.9.1974).
US 4 136 101 (American Cyanamid; 23.1.1979; prior. 3.2.1978).

Formulation(s):   amp. 5 mg, 25 mg, 50 mg, 200 mg, 500 mg, 100 mg, 5000 mg; tabl. 2.5 g, 7.5 g, 10 g (as
disodium salt)

Trade Name(s):
D: Farmitrexat (Pharmacia &   Lantarel (Lederle)   generic
     Upjohn)   Metex (medac)   F: Ledertrexate (Lederle)

	GB:	Maxtrex (Pharmacia &	I:	generic
Méthothrexate Roger		Upjohn)	J:	generic
Bellon (Rhône-Poulenc		Methotrexate (Wyeth)	USA:	Rheumatrex (Lederle)
Rorer Bellon)		generic		generic
generic				

Methoxamine

ATC: C01CA10
Use: sympathomimetic, vasoconstrictor

RN: 390-28-3   MF: $C_{11}H_{17}NO_3$ MW: 211.26   EINECS: 206-867-3
LD_{50}:   30 mg/kg (M, p.o.)
CN: α-(1-aminoethyl)-2,5-dimethoxybenzenemethanol

hydrochloride
RN: 61-16-5   MF: $C_{11}H_{17}NO_3 \cdot HCl$ MW: 247.72   EINECS: 200-499-7
LD_{50}:   5030 μg/kg (M, i.v.)

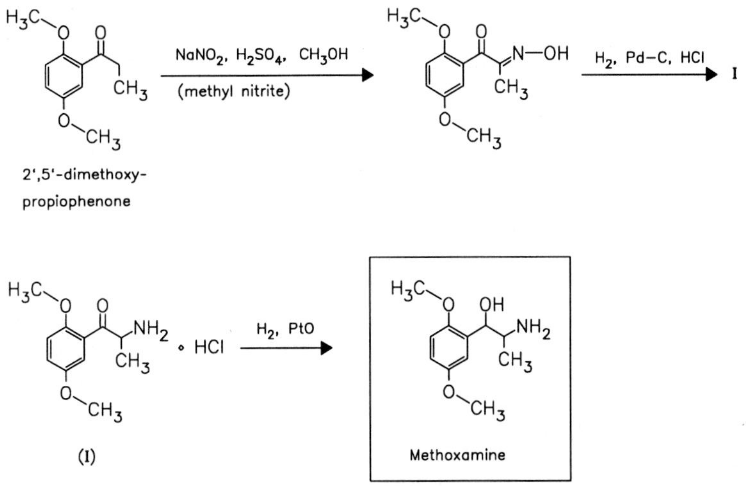

2',5'-dimethoxy-
propiophenone

(I)   Methoxamine

Reference(s):
US 2 359 707 (Burroughs Wellcome; 1944; prior. 1942).

Formulation(s):   amp. 20 mg/ml (as hydrochloride)

Trade Name(s):
D:	Rolinex (Röhm Pharma)-	GB:	Vasoxine (Glaxo	J:	Mexan (Nippon Shinyaku)
	comb.; wfm		Wellcome)	USA:	Vasoxyl (Glaxo Wellcome)
		I:	Vasoxine (Wellcome); wfm		

Methoxsalen

(Ammoidin; Methoxypsoralen; Methoxysalen;
   Xanthotoxin)

ATC: D05AD02; D05BA02
Use: radioprotector

RN: 298-81-7   MF: $C_{12}H_8O_4$ MW: 216.19   EINECS: 206-066-9
CN: 9-methoxy-7H-furo[3,2-g][1]benzopyran-7-one

(a) from plant material:

8-geranyloxypsoralen is obtained by extraction from Ammi majus with n-hexane, which is dealkylated with CH_3COOH/H_2SO_4 to 8-hydroxypsoralen and then is methylated with dimethyl sulfate

(b)

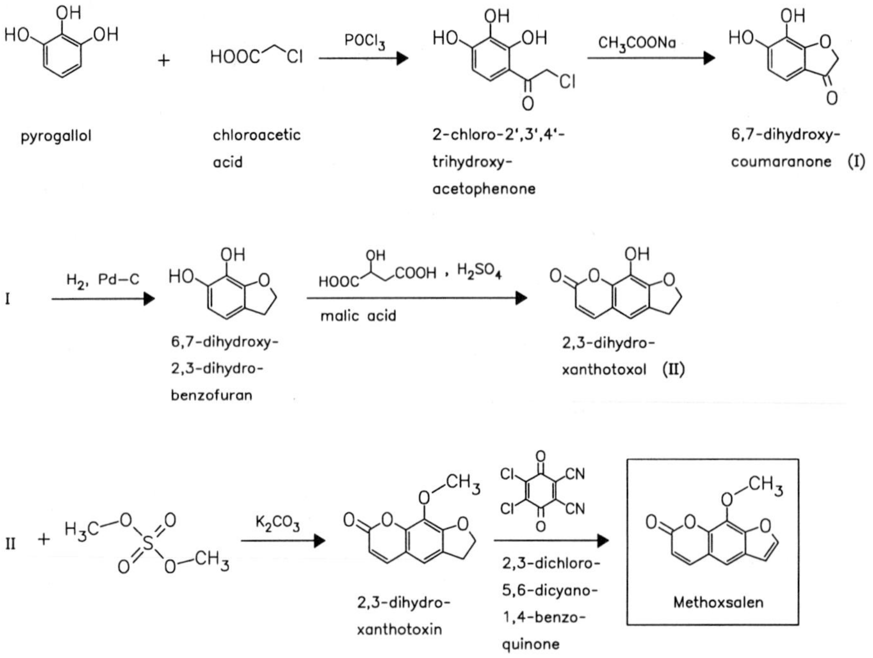

POCl3
CH3COONa
pyrogallol
chloroacetic acid
2-chloro-2',3',4'-trihydroxy-acetophenone
6,7-dihydroxy-coumaranone (I)
I
H2, Pd–C
6,7-dihydroxy-2,3-dihydro-benzofuran
COOH, H2SO4
malic acid
2,3-dihydro-xanthotoxol (II)
II
K2CO3
2,3-dihydro-xanthotoxin
2,3-dichloro-5,6-dicyano-1,4-benzo-quinone
Methoxsalen

Reference(s):

a US 2 889 337 (US-Secret. of Agriculture; 1959; appl. 1956).

b US 4 129 576 (T. C. Elder; 12.12.1978; prior. 12.4.1976, 24.6.1976).
US 4 129 575 (T. C. Elder; 12.12.1978; prior. 12.4.1976).

alternative syntheses:

US 4 150 042 (Roche; 17.4.1979; prior. 29.7.1977, 8.3.1978).
US 4 107 182 (Roche; 15.8.1978; appl. 29.7.1977).
US 4 147 703 (Roche; 3.4.1979; appl. 29.7.1977).
DOS 2 820 263 (Thomae; appl. 10.5.1978).
US 4 169 840 (Oy Star; 2.10.1979; SF-prior. 3.10.1977).

Formulation(s): sol. 0.1 g/100 ml, 0.75 g/100 ml, 1.5 mg; tabl. 10 mg

Trade Name(s):

D:	Meladinine (Galderma)	J:	Meladinine (Nippon Shoji)	Oxsoralen (ICN)
F:	Méladinine (Promedica)		Oxsoralen (Taisho)	
	Psoraderm-S (Sunlife)	USA:	8-MOP (ICN)	

Methoxyflurane

ATC: N01AB03
Use: inhalation anesthetic

RN: 76-38-0   MF: $C_3H_4Cl_2F_2O$ MW: 164.97   EINECS: 200-956-0
LD_{50}: 150 mg/kg (M, i.v.);
3600 mg/kg (R, p.o.)
CN: 2,2-dichloro-1,1-difluoro-1-methoxyethane

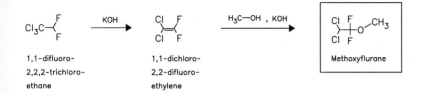

1,1-difluoro-
2,2,2-trichloro-
ethane

1,1-dichloro-
2,2-difluoro-
ethylene

Methoxyflurane

Reference(s):
GB 928 786 (Dow; appl. 9.2.1960; USA-prior. 3.4.1959, 20.7.1959).

Formulation(s):   sol. 125 mg/125 ml

Trade Name(s):
D: Penthrane (Abbott); wfm   I: Pentrane (Abbott); wfm   USA: Penthrane (Abbott); wfm
GB: Penthrane (Abbott); wfm   J: Penthrane (Abbott)

Methscopolamine bromide
(Hyoscine methobromide)

ATC: A03
Use: parasympatholytic, antispasmodic

RN: 155-41-9   MF: $C_{18}H_{24}BrNO_4$ MW: 398.30   EINECS: 205-844-5
LD_{50}: 26.806 mg/kg (M, i.v.); 619 mg/kg (M, p.o.);
42.5 mg/kg (R, i.v.); 3400 mg/kg (R, p.o.)
CN: [7(S)-(1α,2β,4β,5α,7β)]-7-(3-hydroxy-1-oxo-2-phenylpropoxy)-9,9-dimethyl-3-oxa-9-
azoniatricyclo[3.3.1.0²,⁴]nonane bromide

nitrate
RN: 6106-46-3   MF: $C_{18}H_{24}N_2O_7$ MW: 380.40   EINECS: 228-065-2
methylsulfate
RN: 18067-13-5   MF: $C_{18}H_{24}NO_4 \cdot CH_3SO_4$ MW: 429.49   EINECS: 241-975-4

scopolamine
(q. v.)

methyl
bromide

Methscopolamine bromide

methscopolamine nitrate

Reference(s):
DE 145 996 (E. Merck AG; appl. 1902).
US 2 753 288 (Upjohn; 1956; prior. 1952).

Formulation(s):   cps. 2.5 mg; syrup 1.25 mg/5 ml; tabl. 1.25 mg

Trade Name(s):
D: Holopon (Byk Gulden); wfm
      Ichtho-Spasmin (Ichthyol)-comb.; wfm
      Methyscopolamin (Upjohn); wfm

      Oragallin S (Hormon-Chemie)-comb.; wfm
      Skopyl (Pharmacia; as nitrate); wfm
      Spasmo-Bilicura (Mueller Göppingen)-comb.; wfm

GB: Pamine (Upjohn); wfm
      Skopyl (Farillon); wfm
J: Ace (Ono)
      Meporamin (Taiyo)
USA: Pamine (Upjohn); wfm

Methyclothiazide

ATC: C03AA08
Use: diuretic

RN: 135-07-9   MF: $C_9H_{11}Cl_2N_3O_4S_2$ MW: 360.24   EINECS: 205-172-2
LD$_{50}$:   400 mg/kg (M, i.v.); >10 g/kg (M, p.o.);
      >4 g/kg (R, p.o.)
CN: 6-chloro-3-(chloromethyl)-3,4-dihydro-2-methyl-2H-1,2,4-benzothiadiazine-7-sulfonamide 1,1-dioxide

5-chloro-2,4-di-sulfamoylaniline   +   urea   →(180 °C)   7-sulfamoyl-6-chloro-3-oxo-3,4-dihydro-2H-1,2,4-benzothiadiazine 1,1-dioxide   →(H_3C—I , DMF, NaH)   I

(I)   →(NaOH)   →(OHC⌐Cl chloro-acet-aldehyde)   Methyclothiazide

Reference(s):
Close, W.J. et al.: J. Am. Chem. Soc. (JACSAT) 82, 1132 (1960).

Formulation(s):   tabl. 2.5 mg, 5 mg

Trade Name(s):
F: Isobar (Jacques Logeais)-comb.
GB: Enduron (Abbott); wfm

      Enduronyl (Abbott)-comb.; wfm
I: Enduronil (Abbott)-comb.

J: D.A.II-Tablets (Dura)
      Dallergy (Laser)
      Dura-Vent (Dura)

Enduron (Dainippon)   Omnihist (We)   Enduron (Abbott)
Extendryl (Fleming)   USA: Aquatensen (Wallace)
Mescolor (Horizon)   Diutensen-R (Wallace)

Methyldopa

ATC: C02AB01
Use: antihypertensive

RN: 555-30-6   MF: $C_{10}H_{13}NO_4$ MW: 211.22   EINECS: 209-089-2
CN: 3-hydroxy-α-methyl-L-tyrosine

A

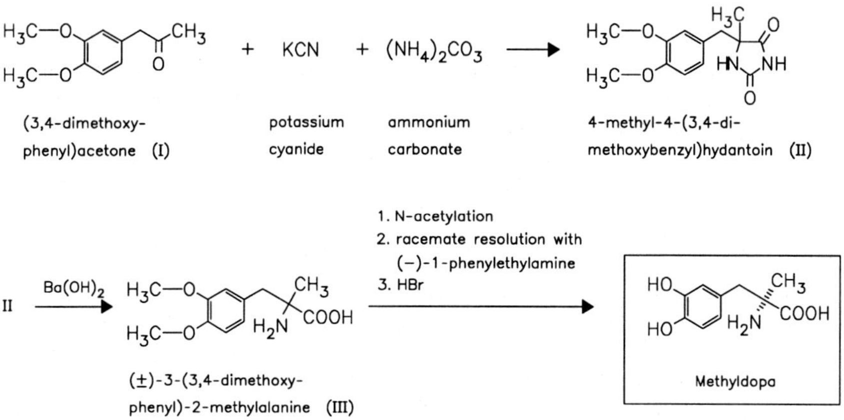

(3,4-dimethoxy-   potassium   ammonium   4-methyl-4-(3,4-di-
phenyl)acetone (I)   cyanide   carbonate   methoxybenzyl)hydantoin (II)

1. N-acetylation
2. racemate resolution with
   (−)-1-phenylethylamine
3. HBr

II $\xrightarrow{Ba(OH)_2}$

(±)-3-(3,4-dimethoxy-
phenyl)-2-methylalanine (III)

Methyldopa

B

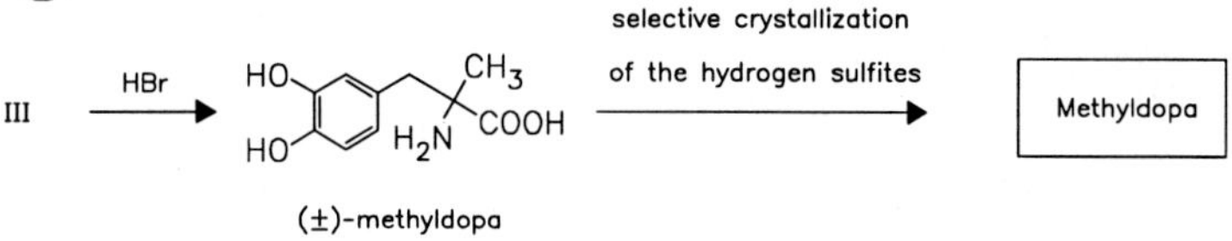

III $\xrightarrow{HBr}$

selective crystallization
of the hydrogen sulfites

Methyldopa

(±)-methyldopa

C

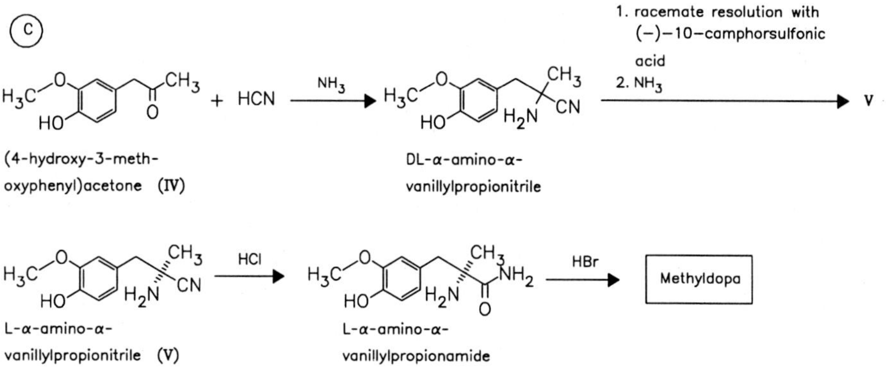

1. racemate resolution with
   (−)−10−camphorsulfonic
   acid
2. NH₃

+ HCN $\xrightarrow{NH_3}$ V

(4-hydroxy-3-meth-   DL-α-amino-α-
oxyphenyl)acetone (IV)   vanillylpropionitrile

$\xrightarrow{HCl}$ $\xrightarrow{HBr}$ Methyldopa

L-α-amino-α-   L-α-amino-α-
vanillylpropionitrile (V)   vanillylpropionamide

Ⓓ Starting products:

ⓐ

homoveratronitrile

ⓑ

isoeugenol

Reference(s):
A US 2 868 818 (Merck & Co.; 13.1.1959; prior. 15.12.1953).
   GB 936 074 (Merck & Co.; appl. 18.10.1960; USA-prior. 8.4.1960, 24.8.1960).
   DE 1 171 931 (Merck & Co.; prior. 6.10.1960).
   Tristram, E.W. et al.: J. Org. Chem. (JOCEAH) **29**, 2053 (1964).
B Stein, G.A. et al.: J. Am. Chem. Soc. (JACSAT) **77**, 700 (1955).
   Chem. Eng. from 8.11.1965; p. 247.
   US 3 158 648 (Merck & Co.; 24.11.1964; prior. 11.7.1961, 9.4.1962).
C Reinhold, D.F. et al.: J. Org. Chem. (JOCEAH) **33**, 1209 (1968).
   FR 1 492 765 (Merck & Co.; appl. 10.10.1963; USA-prior. 11.10.1962, 19.9.1963).
   similar method via L-α-acetylamino-α-vanillylpropionitrile:
   GB 1 142 595 (Merck & Co.; appl. 23.5.1967, 12.2.1969).
   alternative syntheses:
   US 3 366 679 (Merck & Co.; 30.1.1968; prior. 11.10.1962, 19.9.1963).
   DOS 2 302 937 (Tanabe; appl. 22.1.1973; J-prior. 22.1.1972).
   US 3 517 057 (Merck & Co.; 23.6.1970; appl. 21.9.1967).
   DE 1 235 946 (Boehringer Mannh.; appl. 8.8.1964).
   DE 1 235 947 (Bayer; appl. 16.1.1963).
   DE 1 258 416 (Knoll; appl. 9.10.1964).
   DE 1 269 622 (Knoll; appl. 22.12.1966).
   DOS 2 406 898 (BASF; appl. 14.2.1974).
   AT 250 936 (Egyesült; appl. 3.11.1964; HU-prior. 18.11.1963).
   FR 1 502 972 (Merck & Co.; appl. 21.10.1966; USA-prior. 22.10.1965).
   FR 1 531 877 (Sankyo; appl. 18.7.1967; J-prior. 11.8.1966, 21.2.1967).
   GB 1 321 802 (D.D.S.A.; appl. 5.2.1971).
Da Steinetal, G.A.: J. Am. Chem. Soc. (JACSAT) **77**, 700 (1955).
Db GB 2 059 955 (Merck & Co.; appl. 9.9.1980; USA-prior. 13.9.1979, 28.9.1979).

medical use:
US 3 344 023 (Merck & Co.; 12.4.1983; prior. 8.4.1960, 24.8.1960, 1.2.1963; reexamination request 21.12.1981).

Formulation(s):   drg. 250 mg, 500 mg; f. c. tabl. 125 mg, 250 mg, 500 mg; tabl. 250 mg, 500 mg

Trade Name(s):

D:	Caprinol (Bayer Vital)-comb.	Sembrina (Boehringer Mannh.)
	Dopegyt (Thiemann)	F: Aldomet (Merck Sharp & Dohme-Chibret; 1964)
	Presinol (Bayer Vital; 1963)	GB: Aldomet (Merck Sharp & Dohme; 1962); wfm
	Sali-Presinol (Bayer)-comb.	Dopamet (Berk); wfm

D: Caprinol (Bayer Vital)-
    comb.
    Dopegyt (Thiemann)
    Presinol (Bayer Vital;
    1963)
    Sali-Presinol (Bayer)-
    comb.

    Sembrina (Boehringer
    Mannh.)
F: Aldomet (Merck Sharp &
    Dohme-Chibret; 1964)
GB: Aldomet (Merck Sharp &
    Dohme; 1962); wfm
    Dopamet (Berk); wfm

    Hydromet (Merck Sharp &
    Dohme)-comb.; wfm
    Medomet (DDSA); wfm
I: Aldomet (Merck Sharp &
    Dohme)
    Medopren (Malesci)
    Medozide (Malesci)-comb.

J:	Saludopin (SIT)-comb. Aldomet (Merck-Banyu) Becanat (Kissei) Eldopane (Takata-Shionogi) Ledopan (Mochida)	Medopa (Kaigai-Nippon Kayaku) Meprin (Kyorin) Metholes (Taisho) Methoplain (Kowa Yakuhin) Polinal (Yamanouchi)	USA:	Sankaira (Hotta) Aldoclor (Merck Sharp & Dohme) Aldomet (Merck Sharp & Dohme; 1963) Aldoril (Merck Sharp & Dohme) and generics

Methyldopate

ATC: C02AB01
Use: antihypertensive

RN: 2544-09-4 MF: $C_{12}H_{17}NO_4$ MW: 239.27 EINECS: 219-821-2
CN: 3-hydroxy-α-methyl-L-tyrosine ethyl ester

hydrochloride
RN: 2508-79-4 MF: $C_{12}H_{17}NO_4 \cdot HCl$ MW: 275.73 EINECS: 219-720-3

methyldopa
(q. v.)

Methyldopate

Reference(s):
US 2 868 818 (Merck & Co.; 13.1.1959; prior. 15.12.1953).

medical use (for injection):
US 3 230 143 (Merck & Co.; 18.1.1966; appl. 22.6.1961, 28.12.1962, 14.5.1965).
FR-M 2 153 (Merck & Co.; appl. 20.9.1962; USA-prior. 22.6.1961).

Formulation(s): amp. 250 mg/ml, 500 mg/ml; f. c. tabl. 125 mg, 250 mg, 500 mg

Trade Name(s):

D:	Presinol pro inj. (Bayer Vital)		Methyldopa (Merck Sharp & Dohme-Chibret)	USA:	Aldomet Ester HC1 Inj. (Merck Sharp & Dohme)
F:	Aldomet (Merck Sharp & Dohme-Chibret)	GB:	Aldomet Inj. (Merck Sharp & Dohme)		

Methylergometrine
(Methylergonovine)

ATC: G02AB01
Use: uterotonic, oxytocic

RN: 113-42-8 MF: $C_{20}H_{25}N_3O_2$ MW: 339.44 EINECS: 204-027-0
LD_{50}: 85 mg/kg (M, i.v.); 187 mg/kg (M, p.o.);
23 mg/kg (R, i.v.); 93 mg/kg (R, p.o.)
CN: [8β(S)]-9,10-didehydro-N-[1-(hydroxymethyl)propyl]-6-methylergoline-8-carboxamide

maleate (1:1)
RN: 57432-61-8 MF: $C_{20}H_{25}N_3O_2 \cdot C_4H_4O_4$ MW: 455.51 EINECS: 260-734-4
LD_{50}: 85 mg/kg (M, i.v.); 187 mg/kg (M, p.o.);
23 mg/kg (R, i.v.); 93 mg/kg (R, p.o.)
tartrate (2:1)
RN: 6209-37-6 MF: $C_{20}H_{25}N_3O_2 \cdot 1/2C_4H_6O_6$ MW: 828.96

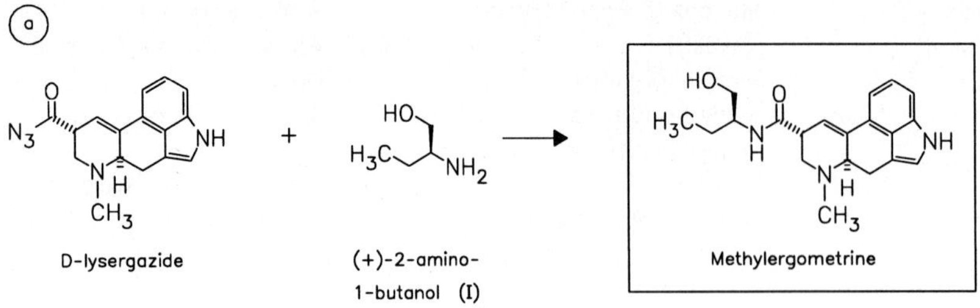

D-lysergazide + (+)-2-amino-1-butanol (I) → Methylergometrine

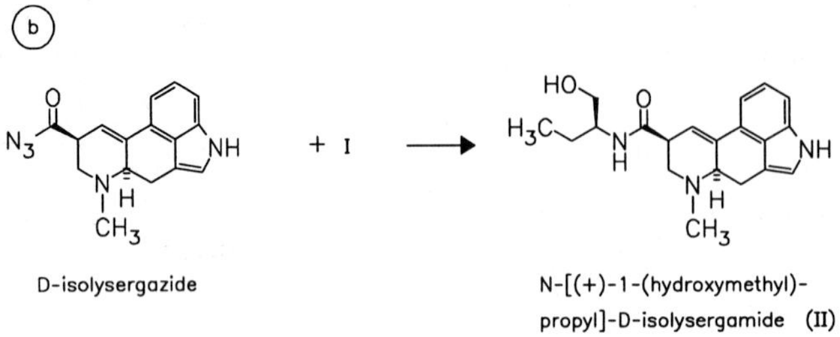

D-isolysergazide + I → N-[(+)-1-(hydroxymethyl)-propyl]-D-isolysergamide (II)

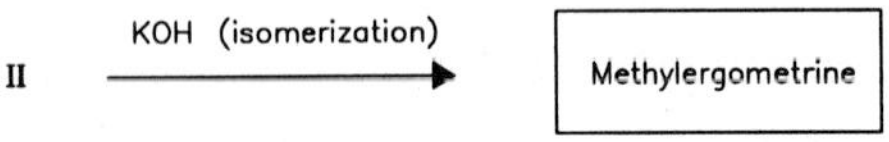

II →[KOH (isomerization)] Methylergometrine

Reference(s):
US 2 265 207 (Sandoz; 1941; CH-prior. 1939).

Formulation(s): amp. 0.2 mg/ml; drg. 0.125 mg; drops 0.25 mg/ml; sol. 0.24 mg/100 ml; tabl. 0.125 mg (as maleate)

Trade Name(s):

D:	Methergin (Novartis Pharma)	I:	Methergin (Novartis)
	Syntometrin (Novartis Pharma)-comb.	J:	Levospan (Isei)
			Metenarin (Teikoku Zoki)
F:	Methergin (Novartis)		Methergin (Sandoz-Sankyo)

Ryegonovin (Morishita)
USA: Ergotrate Maleate (Lilly); wfm
Methergine (Sandoz); wfm

Methylestrenolone
(Normethandrone; Normethandrolone)

ATC: G03DC31
Use: progestogen

RN: 514-61-4 MF: $C_{19}H_{28}O_2$ MW: 288.43 EINECS: 208-183-0
CN: (17β)-17-hydroxy-17-methylestr-4-en-3-one

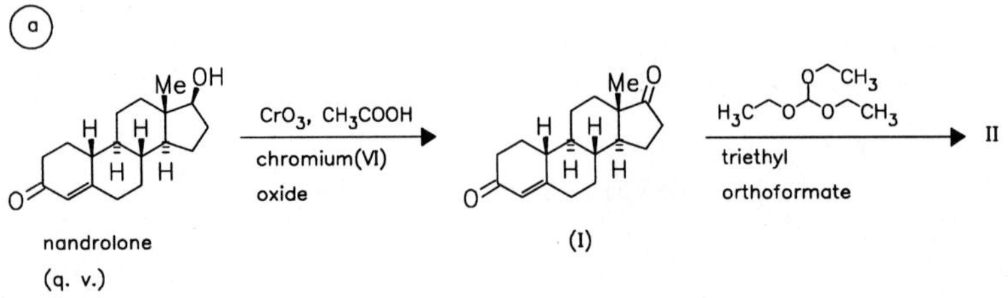

nandrolone (q. v.) →[CrO₃, CH₃COOH / chromium(VI) oxide] (I) →[triethyl orthoformate] II

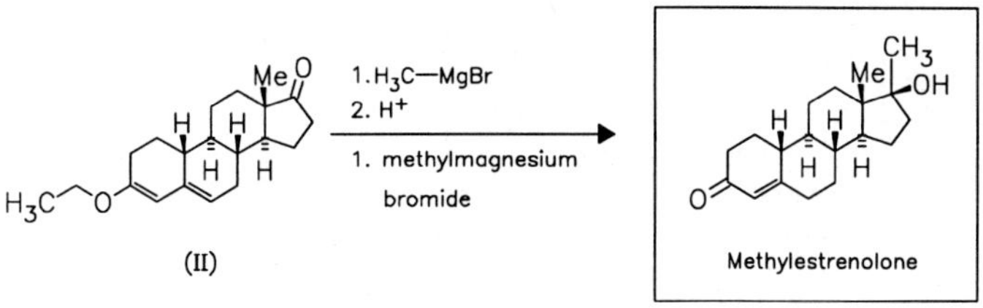

(II) → Methylestrenolone

1. H₃C—MgBr
2. H⁺

1. methylmagnesium bromide

b

3-O,17α-dimethyl-estradiol

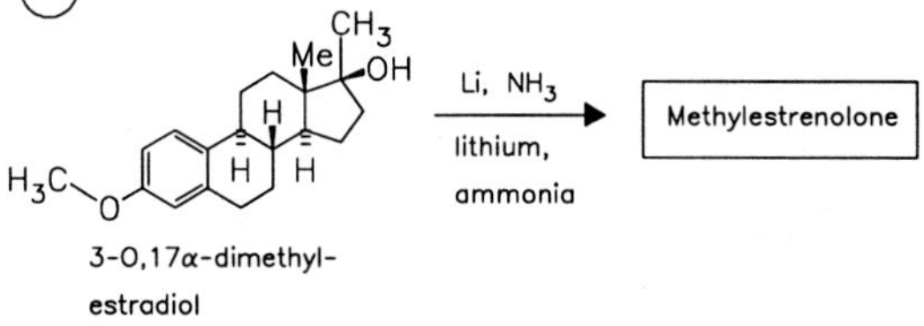

Li, NH₃

lithium, ammonia

Methylestrenolone

c

estrone 3-methyl ether

1. Li, NH₃
2. CrO₃, CH₃COOH

1. lithium, ammonia
2. chromium(VI) oxide

I contd. as under (a) Methylestrenolone

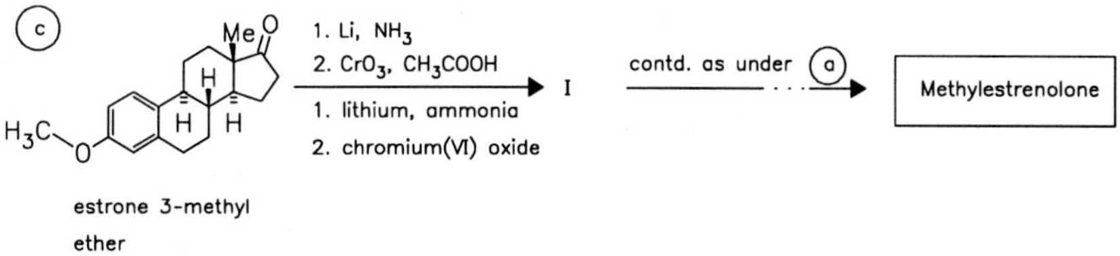

Reference(s):

a,b Djerassi, C. et al.: J. Am. Chem. Soc. (JACSAT) **76**, 4092 (1954).
c US 2 744 122 (Syntex; 1956; MEX-prior. 1951).
US 2 774 777 (Syntex; 1956; prior. 1952).
Djerassi, C. et al.: J. Am. Chem. Soc. (JACSAT) **76**, 4092 (1954).

alternative synthesis:
US 2 849 461 (P. de Ruggieri; 1958; appl. 1957).

Trade Name(s):
D: Gynäkosid (Boehringer Mannh.)-comb.; wfm
F: Orgastéron (Organon); wfm
USA: Methalutin (Parke Davis); wfm

Methylmethionine sulfonium chloride

(Methiosulfonii chloridum; MMS; Vitamin U)

ATC: A02
Use: peptic ulcer therapeutic, antidote

RN: 1115-84-0 MF: $C_6H_{14}ClNO_2S$ MW: 199.70 EINECS: 214-231-1
LD$_{50}$: 259 mg/kg (M, i.v.); >6 g/kg (M, p.o.);
432 mg/kg (R, i.v.); >6 g/kg (R, p.o.)
CN: (S)-(3-amino-3-carboxypropyl)dimethylsulfonium chloride

bromide
RN: 33515-11-6 MF: $C_6H_{14}BrNO_2S$ MW: 244.15
iodide
RN: 3493-11-6 MF: $C_6H_{14}INO_2S$ MW: 291.15

L-methionine   methyl   Methylmethionine sulfonium chloride
                      chloride

Reference(s):
DE 1 239 697 (Degussa; appl. 20.2.1963).

therapy of renal diseases:
US 4 122 189 (Kaken; 24.10.1978; J-prior. 31.3.1976).
GB 1 538 000 (Kaken; appl. 30.3.1977; J-prior. 31.3.1976).
DOS 2 714 391 (Kaken; appl. 31.3.1977; J-prior. 31.3.1976).

hyperlipidemic effect:
Seri, K. et al.: Arzneim.-Forsch. (ARZNAD) **28**, 1711 (1978).

Formulation(s):   drg. 12.5 mg; sol. 0.4 g/100 g

Trade Name(s):

D:	Medosalgon (Loges)-comb.; wfm	I:	Quamon (Neopharmed; as methylsulfate); wfm	New Edion-U (SS Seiyaku)
	Stacho-Zym (Kattwiga)-comb.; wfm	J:	Cabagin (Kowa)	Nichigreen U (Nichiiko)
			Gaston U (Tokyo Hosei)	Showa U (Showa)
F:	Ardesyl (Beytout); wfm		Kizankohl (Sanko)	U-vit. (Hamari)
	Lobarthrose (Opodex); wfm		Kizankohl U (Sanko)	Vitas U (Kaken)
			New U-TIV (Zeria)-comb.	Yucron (Daigo Eiyo)
				combination preparations

Methylpentynol
(Meparfynol)

ATC: N05CM15
Use: sedative

RN: 77-75-8   MF: $C_6H_{10}O$ MW: 98.15   EINECS: 201-055-5
LD$_{50}$:   525 mg/kg (M, p.o.)
CN: 3-methyl-1-pentyn-3-ol

acetylene   butanone   Methylpentynol

Reference(s):
DRP 285 770 (Bayer; 1913).
DRP 289 800 (Bayer; 1913).
DRP 291 185 (Bayer; 1914).

Formulation(s):   cps. 250 mg

Trade Name(s):

D:	Allotropal (Heyl); wfm		N-Oblivon (Latéma; as methylpentynol carbamate); wfm		Oblivon (British Schering); wfm
	Melval (Kattwiga)-comb.; wfm			USA:	Dormison (Schering); wfm
F:	N-Oblivon (Latema); wfm	GB:	Insomnol (Medo); wfm		

Methylphenidate

ATC: N06BA04
Use: psychotonic

RN: 113-45-1   MF: $C_{14}H_{19}NO_2$ MW: 233.31   EINECS: 204-028-6
LD_{50}:   41 mg/kg (M, i.v.); 150 mg/kg (M, p.o.);
        48 mg/kg (R, i.v.); 367 mg/kg (R, p.o.)
CN: α-phenyl-2-piperidineacetic acid methyl ester

hydrochloride
RN: 298-59-9   MF: $C_{14}H_{19}NO_2 \cdot HCl$ MW: 269.77   EINECS: 206-065-3
LD_{50}:   40 mg/kg (M, i.v.); 60 mg/kg (M, p.o.);
        50 mg/kg (R, i.v.); 350 mg/kg (R, p.o.)

benzyl   2-chloro-   phenyl-   methyl phenyl-
cyanide   pyridine   (2-pyridyl)-   (2-pyridyl)-
                                 acetonitrile   acetate (I)

Reference(s):
US 2 507 631 (Ciba; 1950; CH-prior. 1944).

separation of diastereomers:
US 2 957 880 (Ciba; 1960; CH-prior. 1953).
Panizzon, L.: Helv. Chim. Acta (HCACAV) **27**, 1748 (1948).

Formulation(s):   amp. 20 mg; tabl. 5 mg, 10 mg, 20 mg (as hydrochloride)

Trade Name(s):
D: Ritalin (Novartis Pharma)   I: Ritalin (Ciba); wfm   USA: Ritalin (Novartis)
F: Ritaline (Novartis)   J: Ritalin (Ciba-Geigy-
GB: Ritalin (Novartis)   Takeda)

Methylphenobarbital
(Mephobarbital; Methylphenobarbitone)

ATC: N03AA01
Use: anticonvulsant, sedative

RN: 115-38-8   MF: $C_{13}H_{14}N_2O_3$ MW: 246.27   EINECS: 204-085-7
LD_{50}:   300 mg/kg (M, p.o.)
CN: 5-ethyl-1-methyl-5-phenyl-2,4,6(1H,3H,5H)-pyrimidinetrione

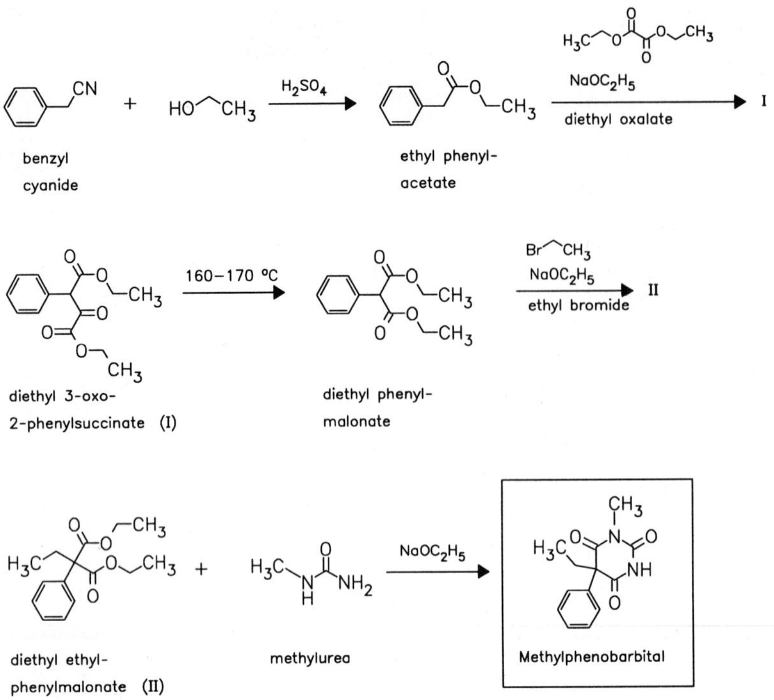

benzyl
cyanide

ethyl phenyl-
acetate

diethyl oxalate

diethyl 3-oxo-
2-phenylsuccinate (I)

diethyl phenyl-
malonate

ethyl bromide

diethyl ethyl-
phenylmalonate (II)

methylurea

Methylphenobarbital

Reference(s):
DRP 537 366 (I. G. Farben; 1929).
DRP 590 175 (I. G. Farben; 1932).

Formulation(s):   tabl. 30 mg, 60 mg, 200 mg

Trade Name(s):
D: Prominal (Bayer); wfm
GB: Prominal (Sanofi Winthrop)
I: Prominal (Bracco); wfm

Prominalette (Bracco);
wfm
J: Prominal (Bayer)

USA: Mebaral (Sanofi)

Methylprednisolone

ATC: D07AA01; D10AA02; H02AB04
Use: glucocorticoid

RN: 83-43-2   MF: $C_{22}H_{30}O_5$ MW: 374.48   EINECS: 201-476-4
LD_{50}: >4 g/kg (R, p.o.)
CN: (6α,11β)-11,17,21-trihydroxy-6-methylpregna-1,4-diene-3,20-dione

acetate
RN: 53-36-1   MF: $C_{24}H_{32}O_6$ MW: 416.51   EINECS: 200-171-3
LD_{50}: >10 g/kg (R, p.o.)
succinate
RN: 2921-57-5   MF: $C_{26}H_{34}O_8$ MW: 474.55   EINECS: 220-863-9
succinate sodium salt
RN: 2375-03-3   MF: $C_{26}H_{33}NaO_8$ MW: 496.53   EINECS: 219-156-8
LD_{50}: 750 mg/kg (M, i.v.); >5 g/kg (M, p.o.);
        640 mg/kg (R, i.v.); >5 g/kg (R, p.o.)

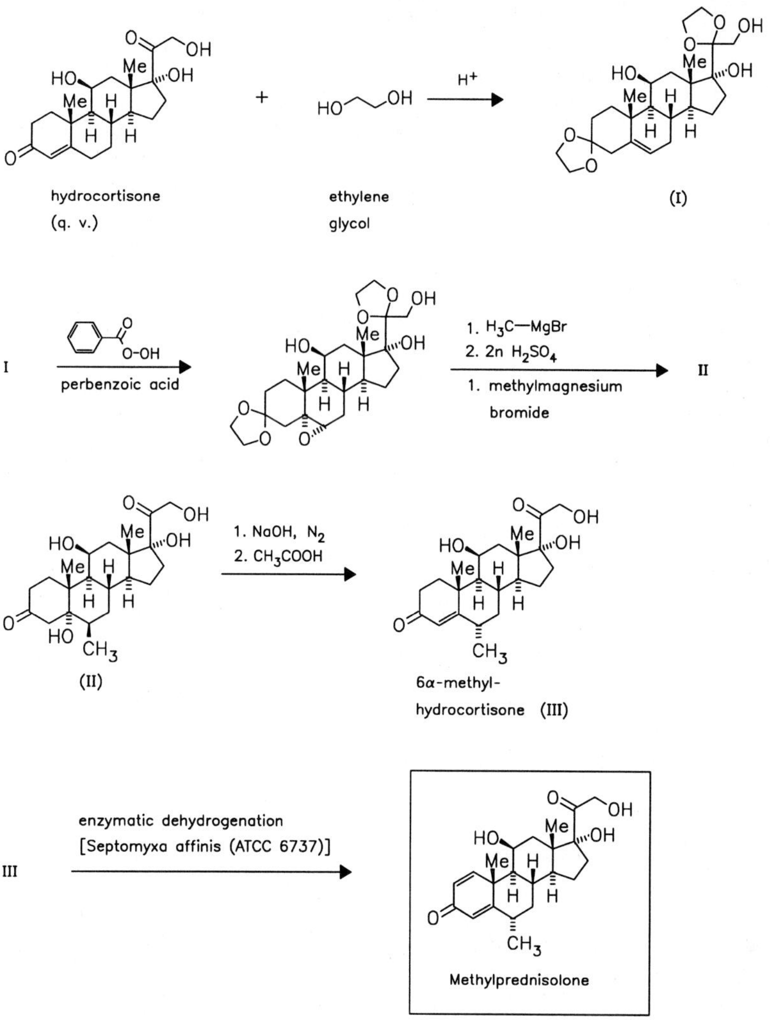

hydrocortisone
(q. v.)

ethylene
glycol

(I)

I → perbenzoic acid →

1. H₃C—MgBr
2. 2n H₂SO₄

1. methylmagnesium
bromide

II

1. NaOH, N₂
2. CH₃COOH →

(II)

6α-methyl-
hydrocortisone (III)

III

enzymatic dehydrogenation
[Septomyxa affinis (ATCC 6737)] →

Methylprednisolone

Reference(s):
US 2 897 218 (Upjohn; 28.7.1959; appl. 23.11.1956; prior. 23.4.1956).
Speero, G.G. et al.: J. Am. Chem. Soc. (JACSAT) **78**, 6213 (1956); **79**, 1515 (1957).

alternative syntheses:
US 3 053 832 (Schering Corp.; 11.9.1962; prior. 29.4.1957).
Fried, J.H. et al.: J. Am. Chem. Soc. (JACSAT) **81**, 1235 (1959).
US 4 041 055 (Upjohn; 9.8.1977; appl. 17.11.1975).

Formulation(s): amp. 20 mg, 30 mg, 40 mg, 60 mg, 80 mg; cream 0.1 %; ointment 0.1 %; tabl. 6 mg, 24 mg,
60 mg

Trade Name(s):

D:	Advantan (Schering)	F: Dépo-Medrol (Pharmacia & Upjohn)	I: Advantan (Schering)
	Depo-Medrate (Pharmacia & Upjohn)	Médrol (Pharmacia & Upjohn)	Asmacortone (Nuovo Cons. Sanit. Naz.)
	Medrate (Pharmacia & Upjohn)	Solu-Médrol (Pharmacia & Upjohn)	Avancort (Farmades)
	Metypred (Orion Pharma)		Depo Medrol (Upjohn)
	Urbason (Hoechst)	GB: Medrone (Pharmacia & Upjohn)	Emmetip (Zanoni)
			Esametone (Lisapharma)
			Firmacort (Firma)

Medrol (Upjohn)	Urbason Retard (Hoechst)	Dura-Meth (Foy); wfm
Medrol Loz. Antiance	J: Medrol (Upjohn)	Medrol (Upjohn); wfm
(Upjohn)-comb.	USA: A-methaPred (Abbott);	Neo-Medrol (Upjohn)-
Metilpre (Formulario Naz.)	wfm	comb.; wfm
Neomedrol Veriderm	Depo-Medrol (Upjohn; as	Solu-Medrol (Upjohn; as
(Upjohn)-comb.	acetate); wfm	21-hemisuccinate); wfm
Solu-medrol (Upjohn)	Depo-Predate (Legere);	generic
Urbason (Hoechst)	wfm	

Methyltestosterone

ATC: G03BA02; G03EK01
Use: androgen

RN: 58-18-4 MF: $C_{20}H_{30}O_2$ MW: 302.46 EINECS: 200-366-3
LD_{50}: 1860 mg/kg (M, p.o.);
2500 mg/kg (R, p.o.)
CN: (17β)-17-hydroxy-17-methylandrost-4-en-3-one

androstenolone   methyl-magnesium iodide   (I)

Methyltestosterone

Reference(s):
US 2 143 453 (Ciba; 1939; CH-prior. 1935).
US 2 374 369 (Ciba; 1945; CH-prior. 1939).
US 2 374 370 (Ciba; 1945; CH-prior. 1939).
Ruzicka, L.: Helv. Chim. Acta (HCACAV) **18**, 1487 (1935).

starting material:
The Merck Index, 2846 (Rahway 1976).

alternative syntheses:
US 2 384 335 (Alien Property Custodian; 1945; NL-prior. 1936).
US 2 386 331 (Ciba; 1945; CH-prior. 1938).
US 2 435 013 (Ciba; 1948; CH-prior. 1941).
Bharucha, K.R.: Experientia (EXPEAM) **14**, 5 (1958).

Formulation(s): cps. 5 mg, 10 mg, 25 mg

Trade Name(s):
D: Femoviron Dragees
(Albert-Roussel)-comb.;
wfm

Gerobion (Merck)-comb.;
wfm
Gevrabon (Cedra)-comb.;
wfm

Hormocornut B (AGM)-
comb.; wfm
Hormo-Gerobion (Merck)-
comb.; wfm

Hormovitastan (ASTA)-comb.; wfm	Testifortan (Promonta)-comb.; wfm	Veinotrope Méthyltestostéron (Lobica); wfm
Klimax Taeschner (Taeschner); wfm	Tropodil (Tropon)-comb.; wfm	GB: Mepilin (Duncan, Flockhart)-comb.; wfm
Lipogeron 300 (Nattermann)-comb.; wfm	Viracton plus (Promonta)-comb.; wfm	Perandren (Ciba); wfm
Medigeron (Medice)-comb.; wfm	F: Climatérine (Lucien)-comb.; wfm	I: Testovis (SIT)
Pasuma (Cascan)-comb.; wfm	Glosso-Stérandryl (Roussel); wfm	J: Enarmon Tab. (Teikoku Zoki)
Primodian (Schering)-comb.; wfm	Triphosadénine Methyltestostérone	Primodian (Nihon Schering)-comb.
Primogeron (Schering)-comb.; wfm	Composé (Débat)-comb.; wfm	Sanstron (Sankyo)
Reginol (Merz)-comb.; wfm		USA: Android (ICN)
		Estratest (Solvay)
		Testred (ICN)
		Virilon (Star)

Methylthioninium chloride

(Methylenblau; Methylene blue)

ATC: V03AB17; V04CG05
Use: diagnostic (for gastric function test), antidote (cyanide poisonings)

RN: 61-73-4   MF: $C_{16}H_{18}ClN_3S$ MW: 319.86   EINECS: 200-515-2
LD_{50}: 77 mg/kg (M, i.v.); 3500 mg/kg (M, p.o.);
    1250 mg/kg (R, i.v.); 1180 mg/kg (R, p.o.)
CN: 3,7-bis(dimethylamino)phenothiazin-5-ium chloride

trihydrate
RN: 7220-79-3   MF: $C_{16}H_{18}ClN_3S \cdot 3H_2O$ MW: 373.91

4-amino-N,N-di-
methylaniline

(I)   N,N-dimethyl-   Methylthioninium chloride
            aniline

Reference(s):
Ullmanns Encykl. Tech. Chem., 4. Aufl., Vol. **8**, 237.

Formulation(s):   amp. 10 mg/ml, 50 mg/ml, 100 mg/ml; tabl. 65 mg

Trade Name(s):
D: Methylenblau Vitis (Neopharma)
F: Antiseptique-Calmante(Chauvin)-comb.

Collyre Bleu Laiter (Leurquin)-comb.

Mictasol Bleu (Martin-Johnson & Johnson-MSD)-comb.

Pastilles Monléon (Toulade)-comb.
Stilla (Phygiène)-comb.

I: Blu Di Meti (Scfm)

Blu Meti (Formulario Naz.)
USA: Urised (PolyMedica)-comb.

Urolene Blue (Star Pharmaceut.)

Methylthiouracil

ATC: H03BA01
Use: thyroid therapeutic

RN: 56-04-2   MF: $C_5H_6N_2OS$ MW: 142.18   EINECS: 200-252-3
LD$_{50}$:   1500 mg/kg (R, p.o.)
CN: 2,3-dihydro-6-methyl-2-thioxo-4(1H)-pyrimidinone

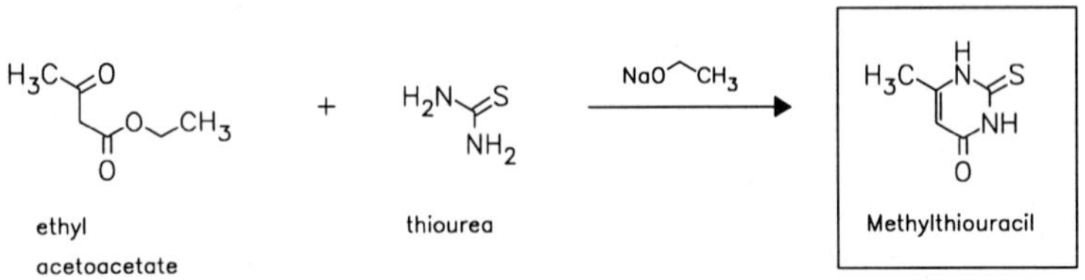

ethyl acetoacetate

thiourea

Methylthiouracil

Reference(s):

List, R.: Justus Liebigs Ann. Chem. (JLACBF) **236**, 1 (1886).
Anderson, G.W. et al.: J. Am. Chem. Soc. (JACSAT) **67**, 2197 (1945).

Formulation(s):   tabl. 0.025 g, 0.1 g

Trade Name(s):

D: Pitufren comp. (Brunnengräber)-comb.; wfm

F: Frenantol Comp. (Laroche Navarron); wfm
J: Methiocil (Chugai)

USA: Muracin (Organon); wfm

Methyprylon

ATC: N05CE02
Use: hypnotic

RN: 125-64-4   MF: $C_{10}H_{17}NO_2$ MW: 183.25   EINECS: 204-745-4
LD$_{50}$:   275 mg/kg (M, i.v.); 890 mg/kg (M, p.o.);
380 mg/kg (R, i.v.); 860 mg/kg (R, p.o.);
300 mg/kg (dog, p.o.)
CN: 3,3-diethyl-5-methyl-2,4-piperidinedione

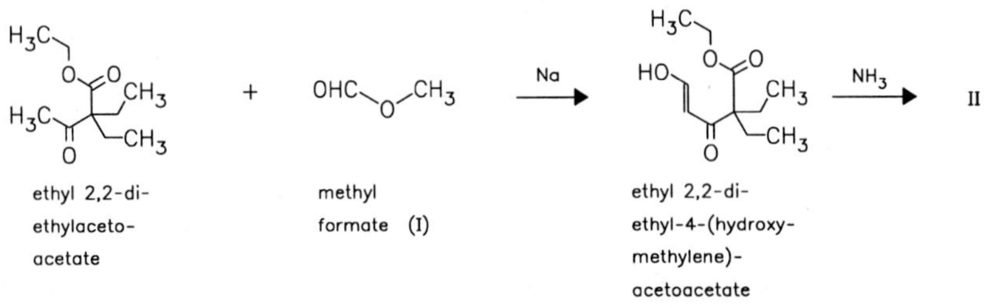

ethyl 2,2-di-ethylaceto-acetate

methyl formate (I)

ethyl 2,2-di-ethyl-4-(hydroxy-methylene)-acetoacetate

II

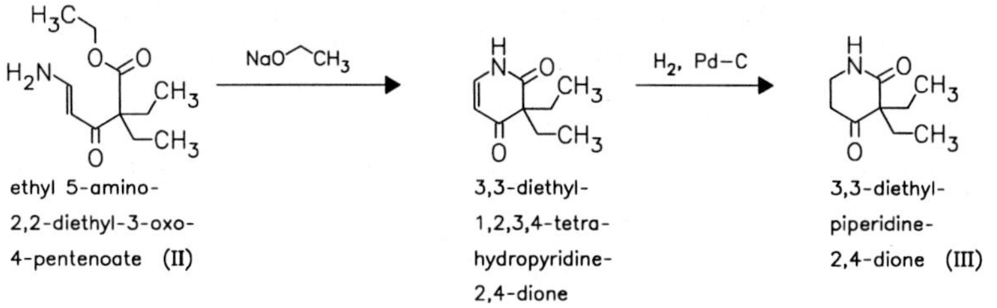

NaO⌒CH3
H2, Pd–C
ethyl 5-amino-2,2-diethyl-3-oxo-4-pentenoate (II)
3,3-diethyl-1,2,3,4-tetrahydropyridine-2,4-dione
3,3-diethyl-piperidine-2,4-dione (III)

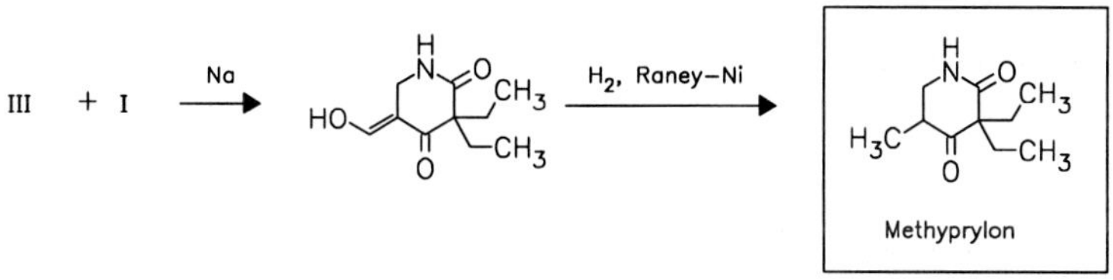

III + I
Na
H2, Raney-Ni
Methyprylon

Reference(s):
US 2 680 116 (Hoffmann-La Roche; 1954; CH-prior. 1951).
DRP 634 284 (Hoffmann-La Roche; 1935).
US 2 151 047 (Hoffmann-La Roche; prior. 1938).

Formulation(s): tabl. 200 mg

Trade Name(s):

D:	Noludar (Roche); wfm	J:	Noctan (Yamanouchi)
GB:	Noludar (Roche)	USA:	Noludar (Roche); wfm

Methysergide

ATC: N02CA04
Use: serotonin antagonist, antimigraine agent

RN: 361-37-5 MF: $C_{21}H_{27}N_3O_2$ MW: 353.47 EINECS: 206-644-0
LD_{50}: 185 mg/kg (M, i.v.); 440 mg/kg (M, p.o.)
CN: [8β(*S*)]-9,10-didehydro-*N*-[1-(hydroxymethyl)propyl]-1,6-dimethylergoline-8-carboxamide

hydrogen maleate (1:1)
RN: 129-49-7 MF: $C_{21}H_{27}N_3O_2 \cdot C_4H_4O_4$ MW: 469.54 EINECS: 204-950-9

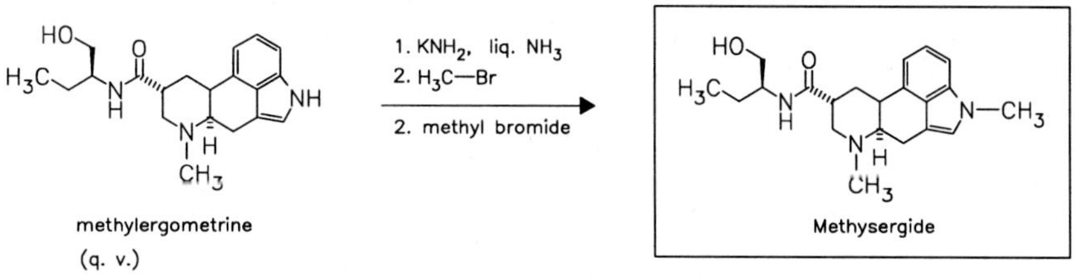

1. KNH2, liq. NH3
2. H3C—Br
2. methyl bromide
methylergometrine (q. v.)
Methysergide

Reference(s):
US 3 113 133 (Sandoz; 3.12.1963; CH-prior. 18.5.1956).
US 3 218 324 (Sandoz; 16.11.1965; CH-prior. 18.5.1956, 20.3.1957, 16.4.1957, 7.3.1958, 11.3.1960).
DE 1 076 137 (Sandoz; appl. 7.5.1957; CH-prior. 18.5.1956, 20.3.1957).

Formulation(s): s. r. tabl. 3 mg; tabl. 1 mg, 2 mg (as hydrogen maleate)

Metiazinic acid

(Acide métiazinique)

ATC: M01; N02
Use: anti-inflammatory

RN: 13993-65-2   MF: $C_{15}H_{13}NO_2S$ MW: 271.34   EINECS: 237-795-0
LD$_{50}$: 350 mg/kg (M, i.v.); 800 mg/kg (M, p.o.);
       495 mg/kg (R, p.o.);
       2 g/kg (dog, p.o.)
CN: 10-methyl-10*H*-phenothiazine-2-acetic acid

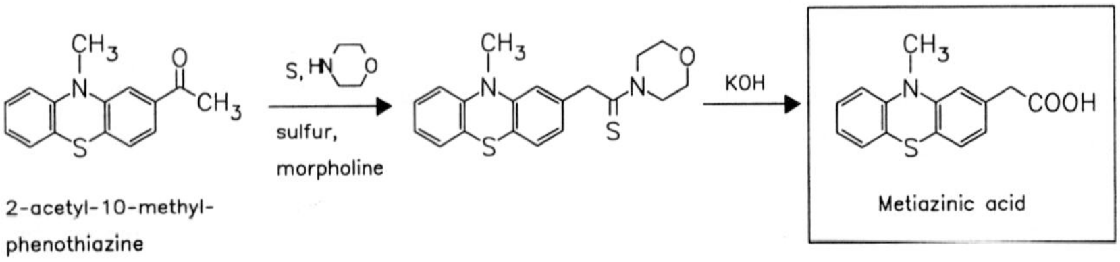

2-acetyl-10-methyl-
phenothiazine

sulfur,
morpholine

Metiazinic acid

Reference(s):
GB 1 048 680 (Rhône-Poulenc; appl. 27.10.1965; F-prior. 29.10.1964, 30.10.1964, 28.12.1964, 24.9.1965).

ester derivatives:
US 3 424 748 (Rhône-Poulenc; 28.1.1969; F-prior. 22.10.1965, 25.8.1966).

Formulation(s):   cps. 125 mg, 250 mg

Meticillin

(Methicillin)

ATC: J01CF03
Use: antibiotic

RN: 61-32-5   MF: $C_{17}H_{20}N_2O_6S$ MW: 380.42   EINECS: 200-505-8
LD$_{50}$: 3720 mg/kg (M, i.v.)
CN: [2*S*-(2α,5α,6β)]-6-[(2,6-dimethoxybenzoyl)amino]-3,3-dimethyl-7-oxo-4-thia-1-
       azabicyclo[3.2.0]heptane-2-carboxylic acid

sodium salt monohydrate
RN: 7246-14-2   MF: $C_{17}H_{19}N_2NaO_6S \cdot H_2O$ MW: 420.42

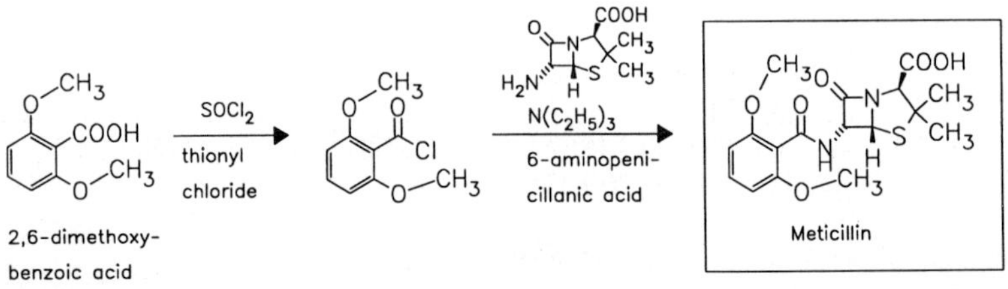

2,6-dimethoxy-
benzoic acid

thionyl
chloride

6-aminopeni-
cillanic acid

Meticillin

Reference(s):
US 2 951 839 (Beecham; 6.9.1960; GB-prior. 15.7.1959).

Formulation(s): amp. 0.5 g, 1 g/ml, 4 g, 6 g (as sodium salt)

Trade Name(s):
D: Cinopenil (Hoechst); wfm
F: Chibro-Flabelline (Merck Sharp & Dohme-Chibret)-comb.; wfm
Flabelline (Delagrange); wfm
Pénistaph (Bristol); wfm
GB: Celbenin (Beecham; 1960)
I: Staficyn (Firma)
J: Methocillin (Meiji)
Staphcillin (Banyu)
USA: Azapen (Pfizer); wfm
Celbenin (Beecham-Massengill); wfm
Staphcillin (Bristol); wfm

Meticrane

ATC: C03BA09
Use: diuretic

RN: 1084-65-7 MF: $C_{10}H_{13}NO_4S_2$ MW: 275.35 EINECS: 214-112-4
LD_{50}: 325 mg/kg (M, i.v.); >20 g/kg (M, p.o.);
445 mg/kg (R, i.v.); >16 g/kg (R, p.o.)
CN: 3,4-dihydro-6-methyl-2*H*-1-benzothiopyran-7-sulfonamide 1,1-dioxide

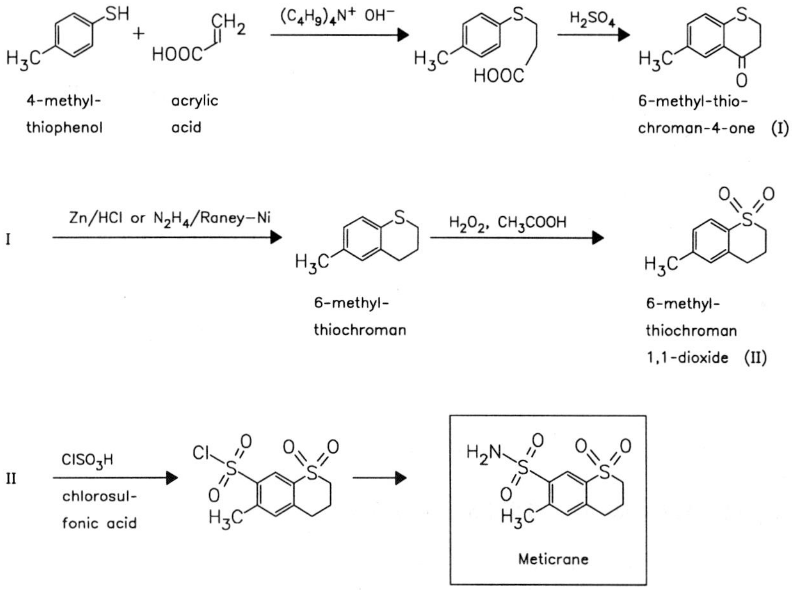

4-methyl-thiophenol
acrylic acid
(C4H9)4N+ OH−
H2SO4
6-methyl-thio-chroman-4-one (I)
Zn/HCl or N2H4/Raney-Ni
H2O2, CH3COOH
6-methyl-thiochroman
6-methyl-thiochroman 1,1-dioxide (II)
ClSO3H
chlorosul-fonic acid
Meticrane

Reference(s):
FR 1 365 504 (S.I.F.A; appl. 24.5.1963).

Formulation(s): tabl. 150 mg

Trade Name(s):
F: Fontilix (Diamant); wfm
J: Aresten (Nippon Shinyaku)

Metildigoxin
(Medigoxin; β-Methyldigoxin)

ATC: C01AA08
Use: cardiac glycoside

RN: 30685-43-9　MF: C$_{42}$H$_{66}$O$_{14}$　MW: 794.98　EINECS: 250-292-0
CN: (3β,5β,12β)-3-[(O-2,6-dideoxy-4-O-methyl-β-D-ribo-hexopyranosyl-(1→4)-O-2,6-dideoxy-β-D-ribo-hexopyranosyl-(1→4)-2,6-dideoxy-β-D-ribo-hexopyranosyl)oxy]-12,14-dihydroxycard-20(22)-enolide

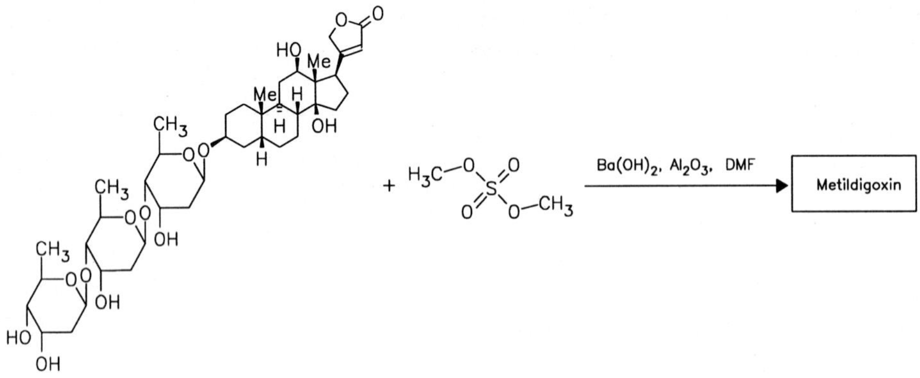

digoxin
(q. v.)

dimethyl sulfate

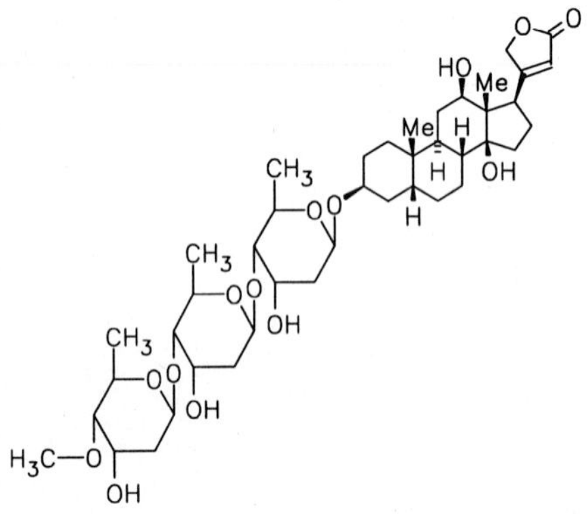

Ba(OH)$_2$, Al$_2$O$_3$, DMF

Metildigoxin

Metildigoxin

Reference(s):
DE 1 643 665 (Boehringer Mannh.; appl. 20.9.1967).
DOS 1 961 034 (Boehringer Mannh.; appl. 5.12.1969).
US 3 538 078 (Boehringer Mannh.; 3.11.1970; D-prior. 7.5.1968).

methylation with methyl mesylate:
DOS 2 734 401 (LEK; appl. 29.7.1977; YU-prior. 20.8.1976).
US 4 145 528 (LEK; 20.3.1979; YU-prior. 20.8.1976).

Formulation(s):　amp. 0.2 mg; tabl. 0.1 mg, 0.15 mg

Trade Name(s):
D: Lanitop (Boehringer   I: Cardiolan (Tosi-Novara)   Miopat (Polifarma)
    Mannh.; 1971)   Lanitop (Boehringer   J: Lanirapid (Yamanouchi;
F: Lanitop (Roussel); wfm   Mannh.; 1973)   1979)

Metipranolol

ATC: S01ED04
Use: beta blocking agent

RN: 22664-55-7   MF: $C_{17}H_{27}NO_4$ MW: 309.41   EINECS: 245-151-5
CN: 4-[2-hydroxy-3-[(1-methylethyl)amino]propoxy]-2,3,6-trimethylphenol 1-acetate

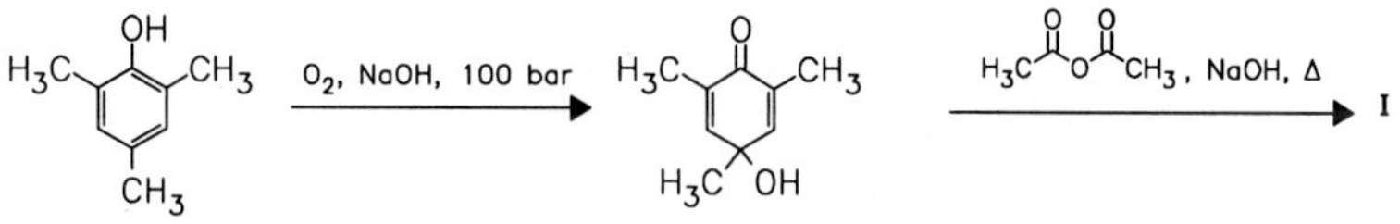

2,4,6-trimethyl-
phenol

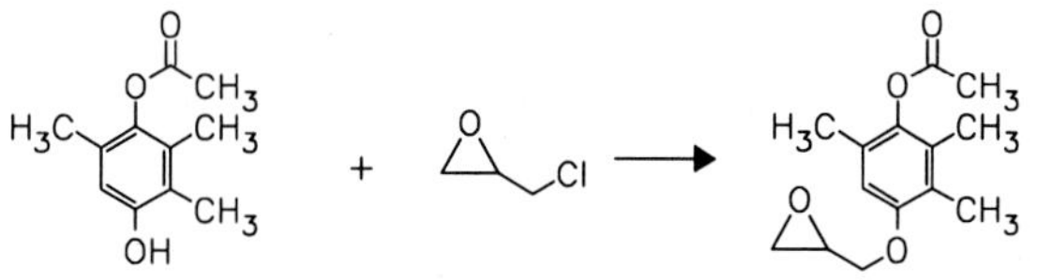

4-acetoxy-2,3,5-   epichloro-   (II)
trimethylphenol   (I)   hydrin

Metipranolol

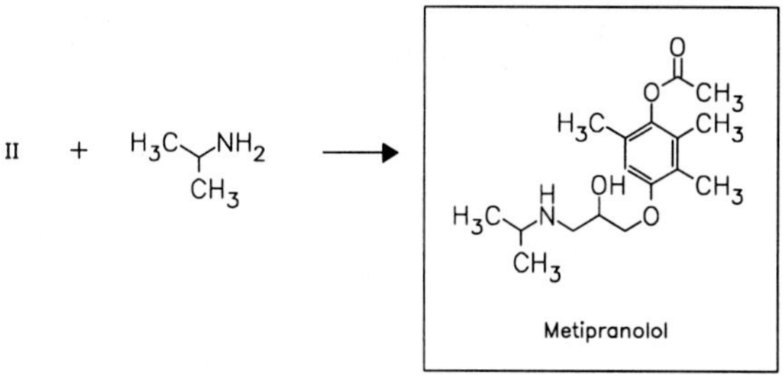

Reference(s):
DOS 1 668 964 (Spofa; appl. 1968; P-prior. 1967).
CS 1 150 020 (L. Blaha; appl. 26.11.1970).

synthesis of 4-acetoxy-2,3,5-trimethylphenol:
DOS 2 314 600 (Teijin; appl. 23.3.1973; J-prior. 25.3.1972).

Formulation(s):   drg. 20 mg; eye drops 1 mg/ml, 3 mg/ml, 6 mg/ml (0.1 %, 0.3 %, 0.6 %); tabl. 20 mg

Trade Name(s):
D: Betamann (Mann)   Torrat (Boehringer Mannh.)   F: Bétanol (Europhta)
    Normoglaucon (Mann)-   Tri-Torrat (Boehringer   I: Turoptin (CIBA Vision)-
    comb.   Mannh.)   comb.

Metirosine

(Metyrosine; α-Methyltyrosine)

ATC: C02KB01
Use: antihypertensive (at pheochromocytoma)

RN: 672-87-7 MF: $C_{10}H_{13}NO_3$ MW: 195.22 EINECS: 211-599-5
CN: α-methyl-L-tyrosine

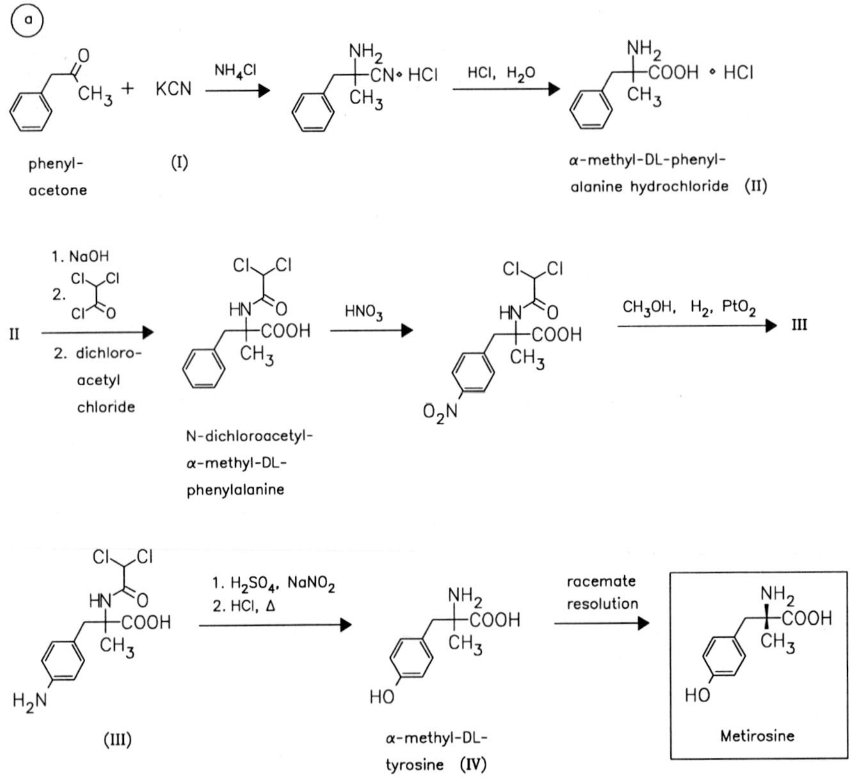
a
phenyl-acetone
KCN
(I)
NH4Cl
HCl, H2O
α-methyl-DL-phenyl-alanine hydrochloride (II)
II
1. NaOH
2. dichloro-acetyl chloride
HNO3
CH3OH, H2, PtO2
III
N-dichloroacetyl-α-methyl-DL-phenylalanine
(III)
1. H2SO4, NaNO2
2. HCl, Δ
α-methyl-DL-tyrosine (IV)
racemate resolution
Metirosine

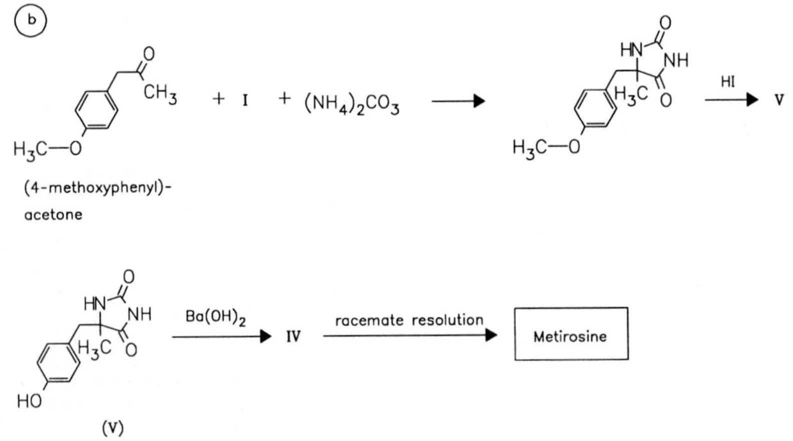
b
(4-methoxyphenyl)-acetone
+ I + (NH4)2CO3
HI
V
(V)
Ba(OH)2
IV
racemate resolution
Metirosine

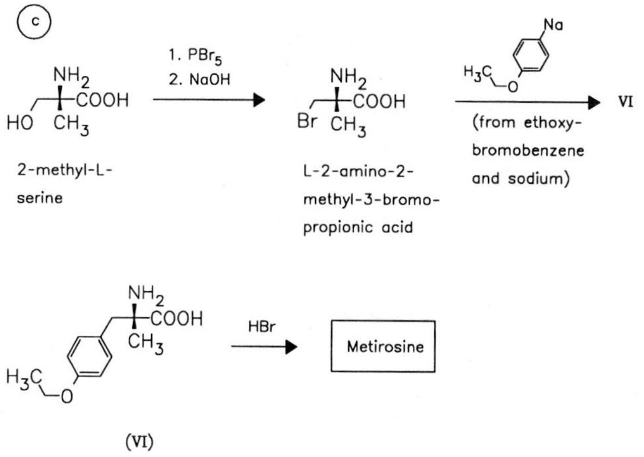

(c)

2-methyl-L-serine → L-2-amino-2-methyl-3-bromo-propionic acid (1. PBr$_5$, 2. NaOH) → (from ethoxy-bromobenzene and sodium) → VI

(VI) + HBr → Metirosine

Reference(s):

a Stein, G.A. et al.: J. Am. Chem. Soc. (JACSAT) **77**, 700 (1955).
US 2 868 818 (Merck & Co.; 13.1.1959; appl. 15.12.1953).
b Potts, K.T.: J. Chem. Soc. (JCSOA9) **1955**, 1632.
c DOS 1 543 763 (Merck & Co.; appl. 25.5.1966; USA-prior. 3.6.1965).
GB 1 105 103 (Merck & Co.; appl. 1.6.1966; USA-prior. 3.6.1965).

alternative syntheses:
enantioselective synthesis from L-tyrosine *via the reaction product from N,O*-bis(carbobenzoxy)-L-tyrosine *with* benzaldehyde *and its methylation:*
US 4 508 921 (Merck & Co.; 2.4.1985; appl. 28.6.1984).

DL-metirosine *by reaction of N,N*-dimethyl-4-hydroxybenzylamine *with* ethyl 2-nitropropionate:
Saari, W.S.: J. Org. Chem. (JOCEAH) **32**, 4074 (1967).

combination with carbidopa:
US 4 389 415 (Merck & Co.; 21.6.1983; USA-prior. 24.1.1978, 5.10.1979, 20.7.1981).
EP 3 353 (Merck & Co.; appl. 24.1.1979; USA-prior. 24.1.1978).

Formulation(s):   cps. 250 mg

Trade Name(s):
GB: Demser (Merck Sharp & Dohme); wfm
USA: Demser (Merck Sharp & Dohme)

Metixene
(Methixene)

ATC: N04AA03
Use: antiparkinsonian, antispasmodic

RN: 4969-02-2   MF: $C_{20}H_{23}NS$ MW: 309.48   EINECS: 225-610-6
LD$_{50}$:   18 mg/kg (M, i.v.); 430 mg/kg (M, p.o.)
CN: 1-methyl-3-(9H-thioxanthen-9-ylmethyl)piperidine

hydrochloride
RN: 1553-34-0   MF: $C_{20}H_{23}NS \cdot HCl$ MW: 345.94   EINECS: 216-300-1
LD$_{50}$:   18 mg/kg (M, i.v.); 346 mg/kg (M, p.o.);
24 mg/kg (R, i.v.); 1460 mg/kg (R, p.o.)
hydrochloride monohydrate
RN: 7081-40-5   MF: $C_{20}H_{23}NS \cdot HCl \cdot H_2O$ MW: 363.95

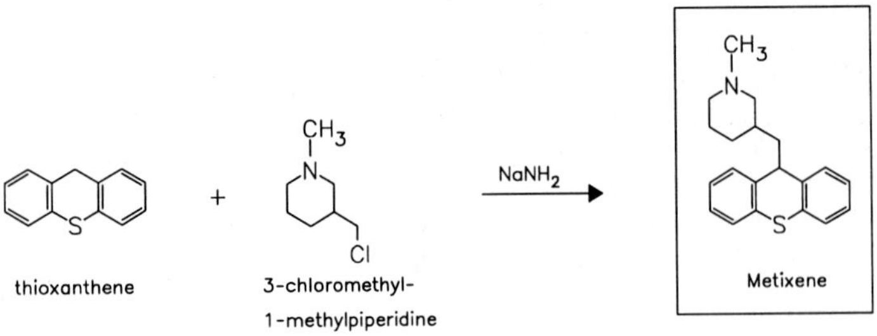

thioxanthene
3-chloromethyl-1-methylpiperidine
NaNH2
Metixene

Reference(s):
US 2 905 590 (The Wander Comp.; 22.9.1959; prior. 7.5.1958).

Formulation(s): tabl. 2.5 mg, 5 mg, 15 mg (as hydrochloride)

Trade Name(s):

D: Tremarit (Novartis Pharma)
F: Spasmenzyme (Salvoxyl-Wander)-comb.; wfm
GB: Tremonil (Wander); wfm
Tremonil (Sandoz; as hydrochloride hydrate); wfm
I: Tremaril (Sandoz)
J: Atosil (Teisan)
Cholinfall (Tokyo Tanabe)
Dalpan (Grelan)
Methyloxan (Nippon Shoji-Kodama)
Thioperkin (Hokuriku)
USA: Trest (Dorsey); wfm

Metoclopramide

ATC: A03FA01
Use: anti-emetic, gastric therapeutic

RN: 364-62-5 MF: $C_{14}H_{22}ClN_3O_2$ MW: 299.80 EINECS: 206-662-9
CN: 4-amino-5-chloro-*N*-[2-(diethylamino)ethyl]-2-methoxybenzamide

monohydrochloride
RN: 7232-21-5 MF: $C_{14}H_{22}ClN_3O_2 \cdot HCl$ MW: 336.26 EINECS: 230-634-5
monohydrochloride monohydrate
RN: 54143-57-6 MF: $C_{14}H_{22}ClN_3O_2 \cdot HCl \cdot H_2O$ MW: 354.28

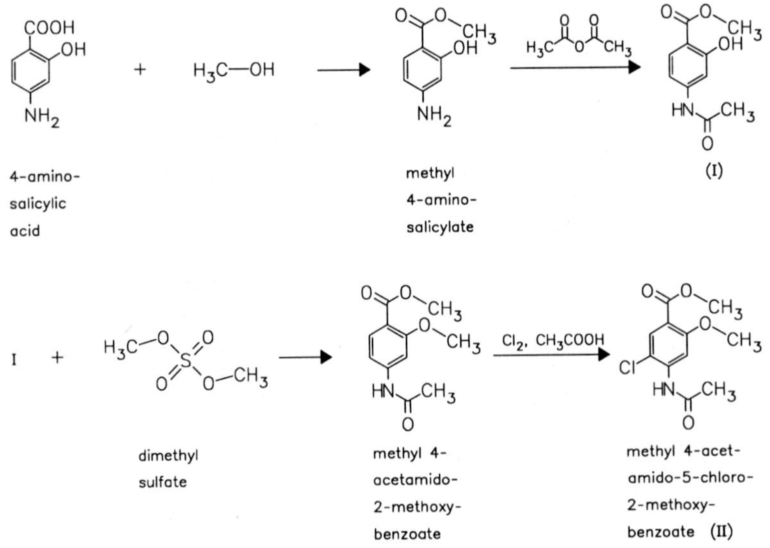

4-amino-salicylic acid
H3C—OH
methyl 4-amino-salicylate
(I)
I
dimethyl sulfate
methyl 4-acetamido-2-methoxy-benzoate
Cl2, CH3COOH
methyl 4-acetamido-5-chloro-2-methoxy-benzoate (II)

N,N-diethyl-
ethylenediamine

Metoclopramide

Reference(s):

DE 1 233 877 (Soc. d'Etudes Scientifiques et Industrielles de l'Ile-de-France; appl. 14.7.1962; F-prior. 25.7.1961).

FR 1 313 758 (Soc. d'Etudes Scientifiques et Industrielles de l'Ile-de-France; appl. 25.7.1961).

US 3 177 252 (Soc. d'Etudes Scientifiques et Industrielles de l'Ile-de-France; 6.4.1965; F-prior. 25.7.1961).

US 3 219 528 (Soc. d'Etudes Scientifiques et Industrielles de l'Ile-de-France; 23.11.1965; F-prior. 25.7.1961, 5.8.1961, 4.11.1961).

US 3 357 978 (Soc. d'Etudes Scientifiques et Industrielles de l'Ile-de-France; 12.12.1967; F-prior. 5.3.1963).

alternative syntheses:

DOS 1 932 512 (Huhtamaki; appl. 26.6.1969; SF-prior. 28.6.1968).

DAS 1 960 130 (Yamanouchi; appl. 29.11.1969; J-prior. 2.12.1968, 9.12.1968, 4.4.1969).

DAS 1 966 453 (Yamanouchi; appl. 29.11.1969; J-prior. 9.12.1968).

DAS 2 102 848 (Delmar; appl. 21.1.1971; USA-prior. 21.1.1970).

DAS 2 119 724 (Teikoku Hormone Mfg.; appl. 22.4.1971; J-prior. 24.4.1970).

DAS 2 162 917 (Soc. d'Etudes Scientifiques et Industrielles de l'Ile-de-France; appl. 17.12.1971; J-prior. 21.12.1970).

DAS 2 166 117 (Teikoku Hormone Mfg.; appl. 22.4.1971; J-prior. 24.4.1970).

DAS 2 166 118 (Teikoku Hormone Mfg.; appl. 22.4.1971; J-prior. 24.4.1970).

DAS 2 342 934 (Delmar; appl. 24.8.1973; GB-prior. 25.8.1972, 13.12.1972, 16.4.1973).

DAS 2 365 988 (Heumann & Co.; appl. 12.7.1973).

starting material:

DAS 2 335 439 (Heumann & Co.; appl. 12.7.1973).

Formulation(s): amp. 10 mg/2 ml, 50 mg/10 ml; cps. 10 mg, 30 mg; drops 4 mg, 5 mg; liquid 4 mg; s. r. cps. 30 mg; sol. 1 mg/ml, 15 mg/15 ml, 5 mg/5 ml; suppos. 10 mg, 20 mg; tabl. 10 mg (as hydrochloride hydrate)

Trade Name(s):

D:	Cerucal (ASTA Medica AWD)	Primpéran (Thera France; 1964)
	Gastronerton (Dolorgiet)	GB: Gastrobid Continus (Napp)
	Gastrosil (Heumann)	Maxolon (Monmouth;
	Gastro-Timelets (Temmler)	1967)
	Paspertase (Solvay	Paramax (Lorex)-comb.
	Arzneimittel)-comb.	I: Citroplus (Irbi)
	Paspertin (Solvay	Clopan (Firma)
	Arzneimittel; 1965)	Cronauzan (ASTA Medica)
	generics	Ede (Teofarma)-comb.
F:	Anausin Metoclopramide	Eugastran (Piam)-comb.
	(ASTA Medica)	Geffer (Boehringer
	Céphalgan (UPSA)-comb.	Mannh.)
	Metoclopramide GNR	Plasil (Lepetit; 1967)
	(GNR-pharma)	

Plasil enzimatico (Lepetit)-comb.
Randum (Roussel)
Viscal (Zoja)
J: Donopon-GP (Sana)
Peraprin (Taiyo Yakuko Takayama)
Primperan (Fujisawa; 1970)
Putoprin (Mohan)
Terperan (Teikoku Zoki)
USA: Reglan (Robins; 1979)

Metolazone

ATC: C03BA08
Use: diuretic, antihypertensive

RN: 17560-51-9 MF: $C_{16}H_{16}ClN_3O_3S$ MW: 365.84 EINECS: 241-539-3
LD_{50}: >5 g/kg (M, p.o.);
>5 g/kg (R, p.o.)
CN: 7-chloro-1,2,3,4-tetrahydro-2-methyl-3-(2-methylphenyl)-4-oxo-6-quinazolinesulfonamide

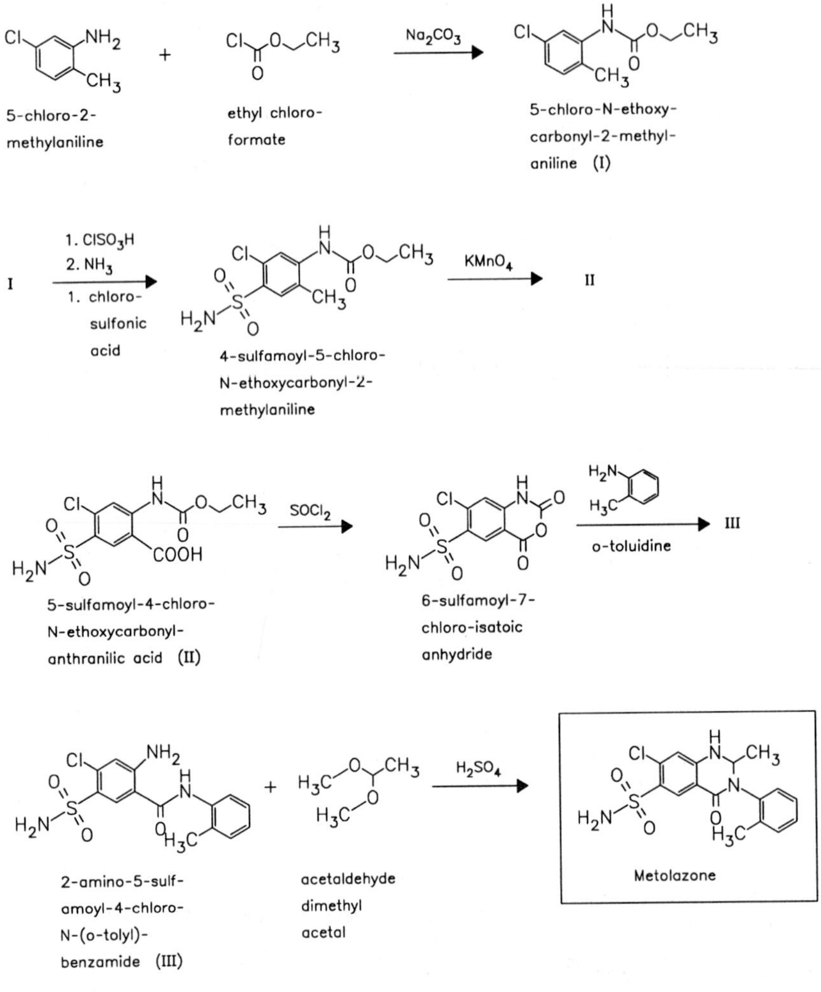

5-chloro-2-methylaniline
ethyl chloroformate
Na_2CO_3
5-chloro-N-ethoxycarbonyl-2-methylaniline (I)
I
1. $ClSO_3H$
2. NH_3
1. chlorosulfonic acid
4-sulfamoyl-5-chloro-N-ethoxycarbonyl-2-methylaniline
$KMnO_4$
II
$SOCl_2$
5-sulfamoyl-4-chloro-N-ethoxycarbonylanthranilic acid (II)
6-sulfamoyl-7-chloro-isatoic anhydride
o-toluidine
III
2-amino-5-sulfamoyl-4-chloro-N-(o-tolyl)benzamide (III)
acetaldehyde dimethyl acetal
H_2SO_4
Metolazone

Reference(s):
DAS 1 620 740 (Pennwalt; appl. 24.12.1966; USA-prior. 3.1.1966).
DOS 2 131 622 (Pennwalt; appl. 25.6.1971; USA-prior. 29.6.1970).
US 3 360 518 (Wallace & Tiernan; 26.12.1967; prior. 3.1.1966).
US 3 557 111 (Wallace & Tiernan; 19.1.1971; prior. 29.3.1968).
US 3 761 480 (Pennwalt; 25.9.1973; prior. 10.7.1968, 7.11.1969, 15.3.1972).
DOS 2 035 657 (Sumitomo; appl. 17.7.1970; J-prior. 22.7.1969, 25.2.1970, 27.3.1970).

Formulation(s): tabl. 2.5 mg, 5 mg, 10 mg

Trade Name(s):
D: Zaroxolyn (Heumann)
GB: Metenix (Hoechst)
I: Zaroxolyn (SmithKline Beecham)
J: Normelan (Sandoz-Sankyo)

USA: Mykrox (Medeva)
        Zaroxolyn (Medeva)

Metopimazine

ATC: A04AD05
Use: anti-emetic

RN: 14008-44-7   MF: $C_{22}H_{27}N_3O_3S_2$ MW: 445.61   EINECS: 237-818-4
CN: 1-[3-[2-(methylsulfonyl)-10H-phenothiazin-10-yl]propyl]-4-piperidinecarboxamide

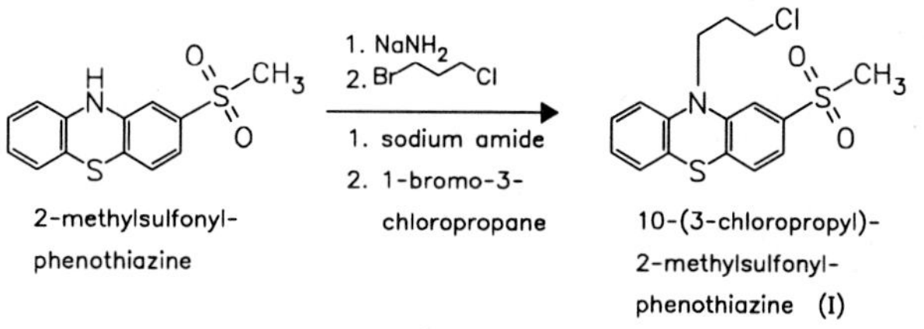

2-methylsulfonyl-
phenothiazine

1. NaNH$_2$
2. Br⌐Cl

1. sodium amide
2. 1-bromo-3-
   chloropropane

10-(3-chloropropyl)-
2-methylsulfonyl-
phenothiazine   (I)

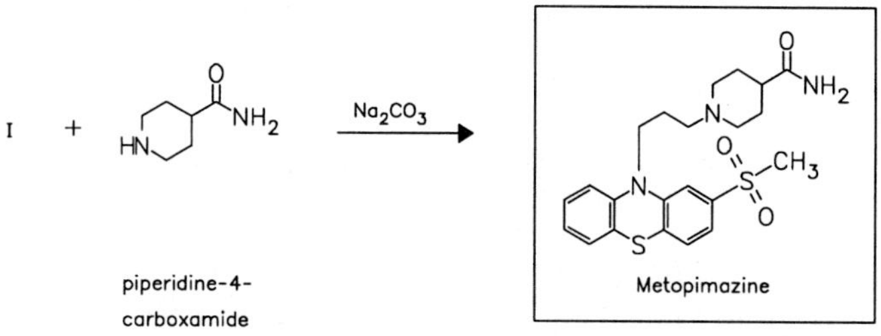

I   +   piperidine-4-
        carboxamide

Na$_2$CO$_3$

Metopimazine

Reference(s):
DE 1 092 476 (Rhône-Poulenc; appl. 14.4.1959; F-prior. 24.4.1958).

Formulation(s):   amp. 10 mg/1 ml; cps. 15 mg; drg. 2.5 mg; sol. 4 mg/ml, 5 mg/5 ml; suppos. 5 mg;
                     tabl. 2.5 mg

Trade Name(s):
F: Vogalène (Schwarz)

Metoprolol

ATC: C07AB02
Use: beta blocking agent

RN: 51384-51-1   MF: $C_{15}H_{25}NO_3$ MW: 267.37   EINECS: 253-483-7
LD$_{50}$:   62 mg/kg (M, i.v.); 1050 mg/kg (M, p.o.);
              71.9 mg/kg (R, i.v.); 3470 mg/kg (R, p.o.);
              60 mg/kg (dog, i.v.)
CN: 1-[4-(2-methoxyethyl)phenoxy]-3-[(1-methylethyl)amino]-2-propanol

tartrate (2:1)
RN: 56392-17-7   MF: $C_{15}H_{25}NO_3 \cdot 1/2C_4H_6O_6$ MW: 684.82   EINECS: 260-148-9

4-(2-methoxy-
ethyl)phenol

epichloro-
hydrin

1,2-epoxy-3-[4-(2-
methoxyethyl)phen-
oxy]propane　(I)

isopropyl-
amine

Metoprolol

Reference(s):
DAS 2 106 209 (AB Hässle; appl. 10.2.1971; S-prior. 18.2.1970).
US 3 873 600 (AB Hässle; 25.3.1975; S-prior. 18.2.1970).
US 3 998 790 (AB Hässle; 21.12.1976; appl. 15.1.1974; prior. 19.3.1973).

(S)-enantiomer:
US 5 034 535 (Astra; 23.7.1991; S-prior. 22.4.1988).
US 5 362 757 (Sepracor; 8.11.1994; appl. 16.11.1992; prior. 18.3.1991).

Formulation(s):　amp. 5 mg/5 ml; s. r. f. c. tabl. 200 mg; s. r. tabl. 200 mg; tabl. 50 mg, 100 mg (as tartrate)

Trade Name(s):

D:	Azumetop (Azupharma)	F:	Logimax (Astra)-comb.	I:	Igroton Lopresor
	Beloc (Astra; 1976)		Logroton (Novartis Pharma		(Novartis)-comb.
	Lopresor (Novartis Pharma;		SA)-comb.		Lopresor (Novartis; 1978)
	1976)		Lopressor (Novartis		Seloken (Astra; 1978)
	Prelis (Novartis Pharma;		Pharma SA; 1980)		Selozide (Astra)-comb.
	1982)		Seloken (Astra; 1980)	J:	Lopresor (Ciba-Geigy)
	Sigaprolol (Kytta-	GB:	Betaloc (Astra; 1975)		Seloken (Fujisawa; 1983)
	Siegfried)		Co-betaloc (Astra; as	USA:	Lopressor (Novartis; 1978)
	Treloc (Astra)-comb.		tartrate)-comb.		
	generics		Lopresor (Novartis; 1975)		

Metrizamide

ATC:　V08AB01
Use:　X-ray contrast medium

RN:　55134-11-7　MF: $C_{18}H_{22}I_3N_3O_8$　MW: 789.10
CN:　2-[[3-(acetylamino)-5-(acetylmethylamino)-2,4,6-triiodobenzoyl]amino]-2-deoxy-D-glucopyranose

metrizoic acid
(q. v.)

metrizoyl chloride　(I)

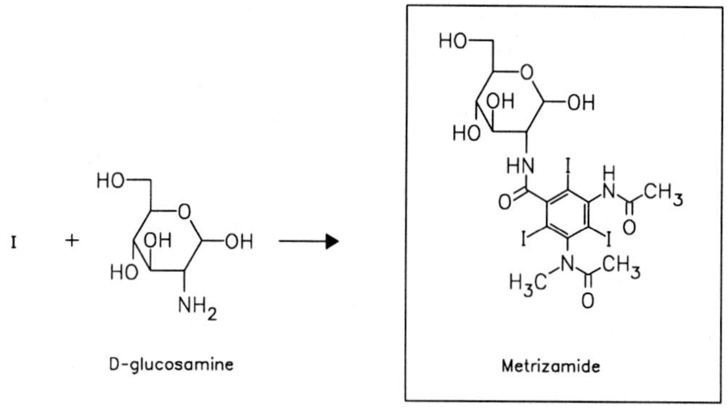

D-glucosamine   Metrizamide

Reference(s):

US 3 701 771 (Nyegaard; 31.10.1972; GB-prior. 27.6.1969, 9.2.1970).
DOS 2 031 724 (Nyegaard; appl. 26.6.1970; GB-prior. 27.6.1969, 9.2.1970).

Formulation(s):   amp. 3.75 g, 6.75 g (12.5 %, 13.5 %, 18.75 %)

Trade Name(s):

D:	Amipaque (Schering; 1977); wfm	F:	Amipaque (Sterling Winthrop; 1980); wfm	USA:	Amipaque (Winthrop-Breon; 1975); wfm
	Arnipaque (Schering); wfm	J:	Amipaque (Schering; 1981)		

Metrizoic acid

ATC: V08AA02
Use: X-ray contrast medium

RN: 1949-45-7   MF: $C_{12}H_{11}I_3N_2O_4$ MW: 627.94   EINECS: 217-761-1
LD_{50}:   10 g/kg (M, i.v.); >46.8 mg/kg (M, p.o.);
          14.3 g/kg (R, i.v.); 38.1 mg/kg (R, p.o.)
CN: 3-(acetylamino)-5-(acetylmethylamino)-2,4,6-triiodobenzoic acid

monosodium salt
RN: 7225-61-8   MF: $C_{12}H_{10}I_3N_2NaO_4$ MW: 649.92   EINECS: 230-624-0

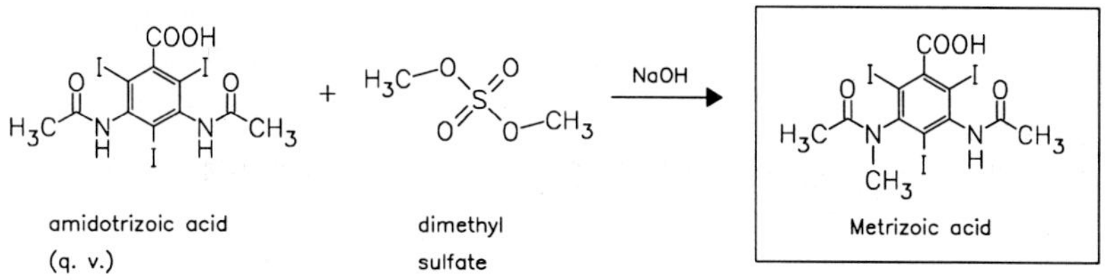

amidotrizoic acid   dimethyl   Metrizoic acid
(q. v.)   sulfate

Reference(s):

GB 973 881 (Nyegaard; appl. 5.12.1960; N-prior. 8.12.1959).
GB 987 796 (Nyegaard; appl. 26.2.1962; N-prior. 28.2.1961).
US 3 178 473 (Nyegaard; 13.4.1965; appl. 2.3.1962).

Formulation(s):   amp. 100 mg, 150 mg, 260 mg, 350 mg, 370 mg, 440 mg (as Ca-, Mg-, Na- and meglumine salt)

Metronidazole

ATC: A01AB17; D06BX01; G01AF01;
       J01XD01; P01AB01
Use: chemotherapeutic (trichomonas)

RN: 443-48-1   MF: $C_6H_9N_3O_3$ MW: 171.16   EINECS: 207-136-1
LD_{50}:   3800 mg/kg (M, p.o.);
       3 g/kg (R, p.o.)
CN: 2-methyl-5-nitro-1H-imidazole-1-ethanol

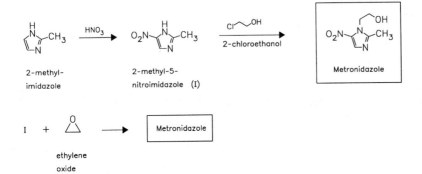

2-methyl-
imidazole

2-methyl-5-
nitroimidazole   (I)

Metronidazole

I   +   ethylene oxide $\longrightarrow$ Metronidazole

Reference(s):
US 2 944 061 (Rhône-Poulenc; 5.7.1960; F-prior. 20.9.1975).

Formulation(s):   cps. 250 mg, 375 mg; f. c. tabl. 250 mg, 400 mg; suppos. 100 mg (vaginal); tabl. 250 mg,
                    400 mg; vaginal tabl. 100 mg; vial 5 g/1000 ml

Metyrapone

ATC: V04CD01
Use: adrenocorticostatic

RN: 54-36-4   MF: $C_{14}H_{14}N_2O$ MW: 226.28   EINECS: 200-206-2
CN: 2-methyl-1,2-di-3-pyridinyl-1-propanone

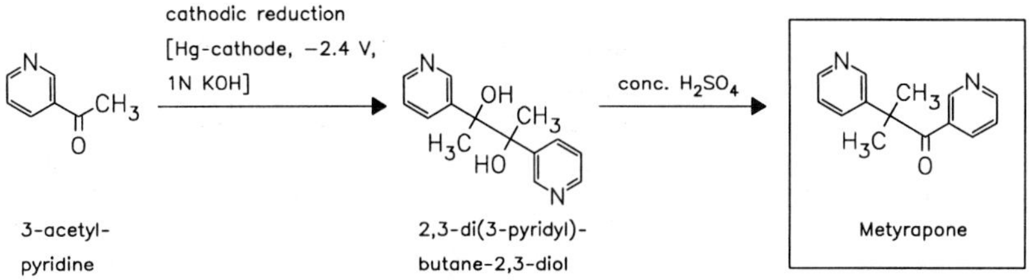
cathodic reduction
[Hg-cathode, −2.4 V, 1N KOH]
conc. H2SO4
3-acetyl-pyridine
2,3-di(3-pyridyl)-butane-2,3-diol
Metyrapone

Reference(s):

Chart, J.J. et al.: Experientia (EXPEAM) **14**, 151 (1958).
Allen, M.J.: J. Org. Chem. (JOCEAH) **15**, 435 (1950). - (pinacol-synthesis)
Bencze, W.L.; Allen, M.J.: J. Am. Chem. Soc. (JACSAT) **81**, 4015 (1959).

as intermediate mentioned in:

US 2 923 710 (Ciba; 2.2.1960, prior. 14.7.1958).
US 2 966 493 (Ciba; 27.12.1960; appl. 10.3.1958).

Formulation(s): cps. 250 mg

Trade Name(s):

D:	Metopiron (Ciba); wfm	GB:	Metopirone (Novartis)	J:	Metopiron (Ciba-Geigy-Takeda)
F:	Metopirone (Novartis Pharma SA)	I:	Metopiron (Ciba); wfm	USA:	Metopirone (Ciba); wfm

Mexazolam

ATC: N05
Use: tranquilizer, anxiolytic

RN: 31868-18-5 MF: $C_{18}H_{16}Cl_2N_2O_2$ MW: 363.24
LD_{50}: 4571 mg/kg (M, p.o.);
4500 mg/kg (R, p.o.)
CN: 10-chloro-11b-(*o*-chlorophenyl)-2,3,7,11b-tetrahydro-3-methyloxazolo[3,2-*d*][1,4]benzodiazepin-6(5*H*)-one

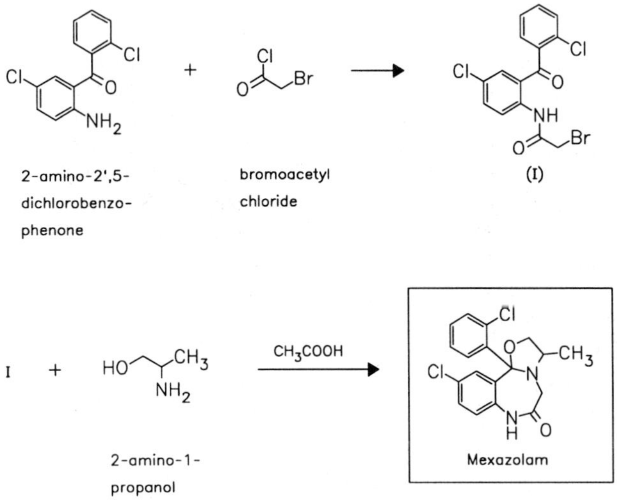
2-amino-2',5-dichlorobenzophenone
bromoacetyl chloride
(I)
I
CH3COOH
2-amino-1-propanol
Mexazolam

Reference(s):

Migadera, T. et al.: J. Med. Chem. (JMCMAR) **14**, 520 (1971).
JP 4 941 439 (Sankyo; appl. 21.12.1970).

Formulation(s):   0.5 mg, 1 mg

Trade Name(s):
J: Melex (Sankyo)

Mexenone

ATC: D02B
Use: ultraviolet screen

RN: 1641-17-4   MF: $C_{15}H_{14}O_3$ MW: 242.27   EINECS: 216-691-9
CN: (2-hydroxy-4-methoxyphenyl)(4-methylphenyl)methanone

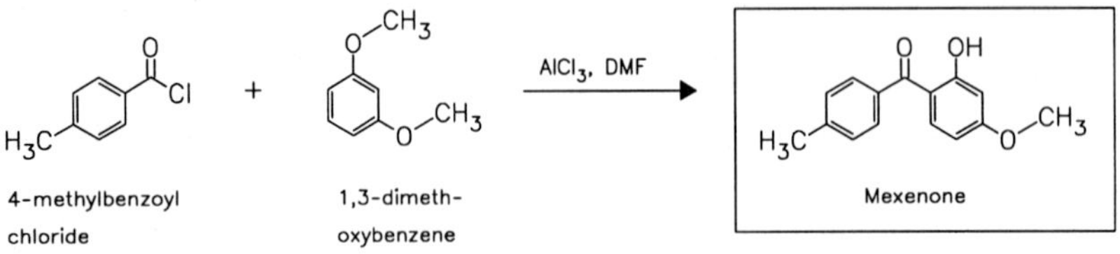

4-methylbenzoyl chloride   +   1,3-dimethoxybenzene $\xrightarrow{\text{AlCl}_3,\ \text{DMF}}$ Mexenone

Reference(s):
US 2 773 903 (American Cyanamid; 1956; prior. 1955).

Formulation(s):   cream 4 %

Trade Name(s):
GB: Uvistat (WB
        Pharmaceuticals); wfm

Mexiletine

ATC: C01BB02
Use: anticonvulsant, antiarrhythmic

RN: 31828-71-4   MF: $C_{11}H_{17}NO$ MW: 179.26   EINECS: 250-825-7
LD$_{50}$:   23 mg/kg (M, i.v.); 320 mg/kg (M, p.o.);
        41 mg/kg (R, i.v.)
CN: 1-(2,6-dimethylphenoxy)-2-propanamine

hydrochloride
RN: 5370-01-4   MF: $C_{11}H_{17}NO \cdot HCl$ MW: 215.72   EINECS: 226-362-1
LD$_{50}$:   21 mg/kg (M, i.v.); 272 mg/kg (M, p.o.);
        27 mg/kg (R, i.v.); 330 mg/kg (R, p.o.);
        19 mg/kg (dog, i.v.); 356 mg/kg (dog, p.o.)

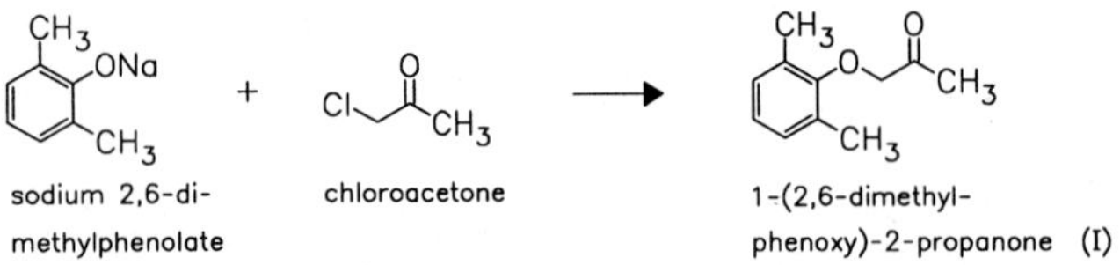

sodium 2,6-dimethylphenolate   +   chloroacetone   →   1-(2,6-dimethylphenoxy)-2-propanone   (I)

I $\xrightarrow[\text{hydroxylamine}]{\text{H}_2\text{N—OH}}$ $\xrightarrow{\text{H}_2,\ \text{Raney-Ni}}$ Mexiletine

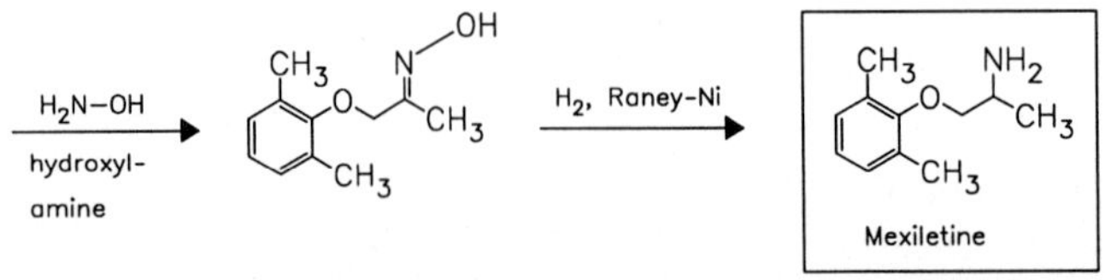

Reference(s):
US 3 954 872 (Boehringer Ing.; 4.5.1976; D-prior. 16.9.1966, 17.8.1967).
DE 1 543 369 (Boehringer Ing.; prior. 16.9.1966).

composition and use:
US 4 031 244 (Boehringer Ing.; 21.6.1977; D-prior. 17.8.1967).

Formulation(s):   amp. 25 mg/ml, 250 mg/10 ml; cps. 100 mg, 150 mg, 200 mg; s. r. cps. 360 mg (as hydrochloride)

Trade Name(s):

D:	Mexitil (Boehringer Ing.; 1979)	GB:	Mexitil (Boehringer Ing.; 1976)	J:	Mexitil (Boehringer Ing.; 1985)
F:	Mexitil (Boehringer Ing.; 1980)	I:	Mexitil (Boehringer Ing.; 1982)	USA:	Mexitil (Boehringer Ing.; 1986) generics

Mezlocillin

ATC: J01CA10
Use: antibiotic

RN: 51481-65-3   MF: C$_{21}$H$_{25}$N$_5$O$_8$S$_2$ MW: 539.59   EINECS: 257-233-8
CN: [2S-[2α,5α,6β(S*)]]-3,3-dimethyl-6-[[[[[3-(methylsulfonyl)-2-oxo-1-imidazolidinyl]carbonyl]amino]phenylacetyl]amino]-7-oxo-4-thia-1-azabicyclo[3.2.0]heptane-2-carboxylic acid

monosodium salt
RN: 42057-22-7   MF: C$_{21}$H$_{24}$N$_5$NaO$_8$S$_2$ MW: 561.57   EINECS: 255-640-5
LD$_{50}$:   6329 mg/kg (M, i.v.); >16 g/kg (M, p.o.);
2636 mg/kg (R, i.v.); >20 g/kg (R, p.o.)

methanesulfonyl chloride   2-imida-zolidinone   1-methanesulfonyl-2-imidazolidinone

1. (CH$_3$)$_3$SiCl, N(C$_2$H$_5$)$_3$
2. COCl$_2$
1. trimethylchlorosilane, triethylamine
2. phosgene
I

3-chloroformyl-1-methanesulfonyl-2-imidazolidinone   (I)   ampicillin (q v)   Mezlocillin

N(C$_2$H$_5$)$_3$
pH 7–8

Reference(s):
DOS 2 152 967 (Bayer; appl. 23.10.1971).
DOS 2 152 968 (Bayer; appl. 23.10.1971).
DOS 2 318 955 (Bayer; appl. 14.4.1973).
US 3 974 142 (Bayer; 10.8.1976; appl. 3.9.1974; D-prior. 23.10.1971).

combination with e. g. oxacillin:
DOS 2 737 673 (Bayer; appl. 20.8.1977).

Formulation(s): vial 0.5g/5 ml, 1 g/10 ml, 2 g/20 ml, 5 g/50 ml

Trade Name(s):

D:	Baypen (Bayer Vital; 1977)	F:	Baypen (Bayer-Pharma)	J:	Baypen (Yoshitomi; 1982)
	Melocin (curasan)	GB:	Baypen (Bayer; 1980);	USA:	Mezlin (Bayer; 1981)
	Optocillin (Bayer Vital;		wfm		
	1979)-comb. with oxacillin	I:	Baypen (Bayer; 1983)		

Mianserin

ATC: N06AX03
Use: antidepressant

RN: 24219-97-4 MF: $C_{18}H_{20}N_2$ MW: 264.37 EINECS: 246-088-6
LD_{50}: 365 mg/kg (M, p.o.)
CN: 1,2,3,4,10,14b-hexahydro-2-methyldibenzo[c,f]pyrazino[1,2-a]azepine

monohydrochloride
RN: 21535-47-7 MF: $C_{18}H_{20}N_2 \cdot HCl$ MW: 300.83 EINECS: 244-426-7
LD_{50}: 31 mg/kg (M, i.v.); 224 mg/kg (M, p.o.);
31.85 mg/kg (R, i.v.); 780 mg/kg (R, p.o.)

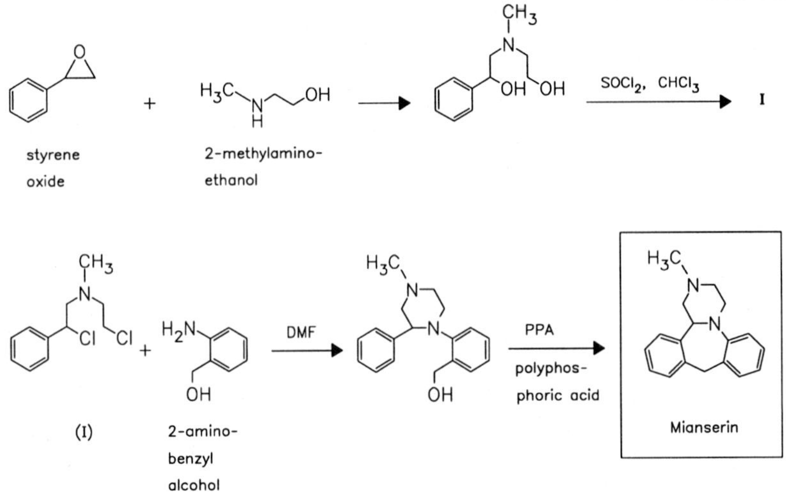

Reference(s):
DOS 2 505 239 (Akzo; appl. 7.2.1975; NL-prior. 9.2.1974).

medical use:
US 4 128 641 (HZJ Research Center; 5.12.1978; prior. 31.7.1975, 28.2.1977).

older methods:
DOS 1 695 556 (Organon; appl. 9.3.1967; NL-prior. 12.3.1966).
US 3 534 041 (Organon; 13.10.1970; NL-prior. 12.3.1966).
Burg, W.J. Van der et al.: J. Med. Chem. (JMCMAR) **13**, 35 (1970).

Formulation(s): f. c. tabl. 10 mg, 30 mg, 60 mg (as hydrochloride)

Trade Name(s):

D:	Tolvin (Organon; 1975)	F:	Athymil (Organon; 1979)	GB:	Bolvidon (Organon; 1976);
	generics				wfm

Norval (Bencard; 1976); wfm

I: Lantanon (Organon Italia; 1979)

J: Tetramide (Sankyo; 1983)

Mibefradil hydrochloride
(Ro-40-5967; Ro-40-5967/001)

ATC: C08CX01
Use: antihypertensive, calcium channel blocker

RN: 116666-63-8 MF: $C_{29}H_{38}FN_3O_3 \cdot 2HCl$ MW: 568.56
CN: (1*S-cis*)-methoxyacetic acid 2-[2-[[3-(1H-benzimidazol-2-yl)propyl]methylamino]ethyl]-6-fluoro-1,2,3,4-tetrahydro-1-(1-methylethyl)-2-naphthalenyl ester dihydrochloride

base
RN: 116644-53-2 MF: $C_{29}H_{38}FN_3O_3$ MW: 495.64

synthesis of intermediate **II**:

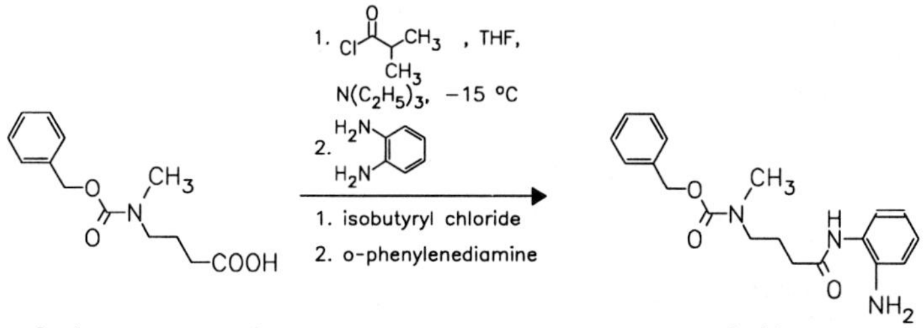

4-[N-(benzyloxycarbonyl)-methylamino]butyric acid

1. isobutyryl chloride
2. o-phenylenediamine

benzyl [3-[(2-amino-phenyl)carbamoyl]propyl]-methylcarbamate (I)

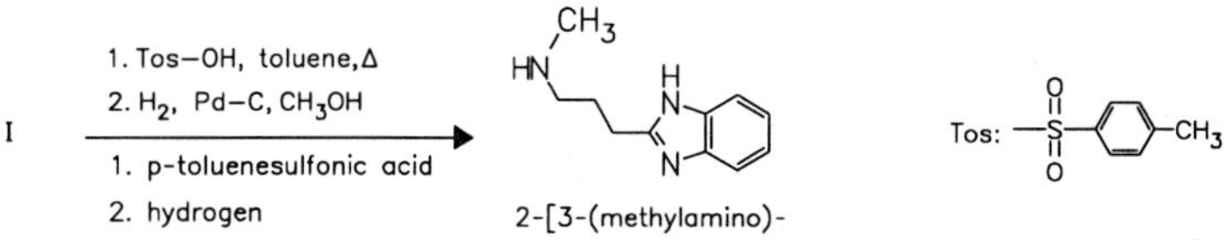

1. p-toluenesulfonic acid
2. hydrogen

2-[3-(methylamino)-propyl]benzimidazole (II)

Tos:

synthesis of intermediate **VII**:

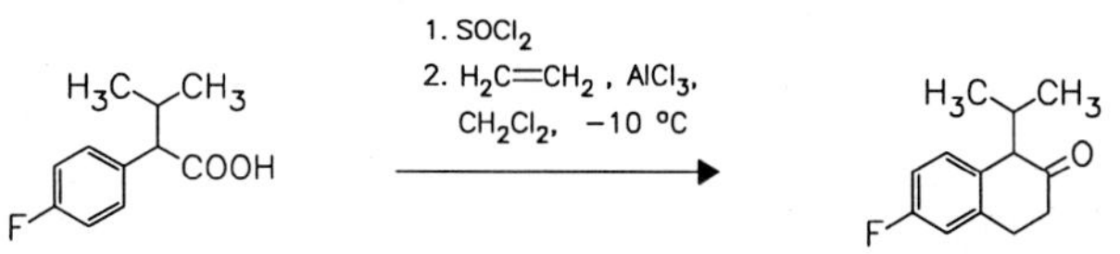

(RS)-2-(p-fluorophenyl)-3-methylbutyric acid

(RS)-6-fluoro-3,4-di-hydro-1-isopropyl-2(1H)-naphthalenone (III)

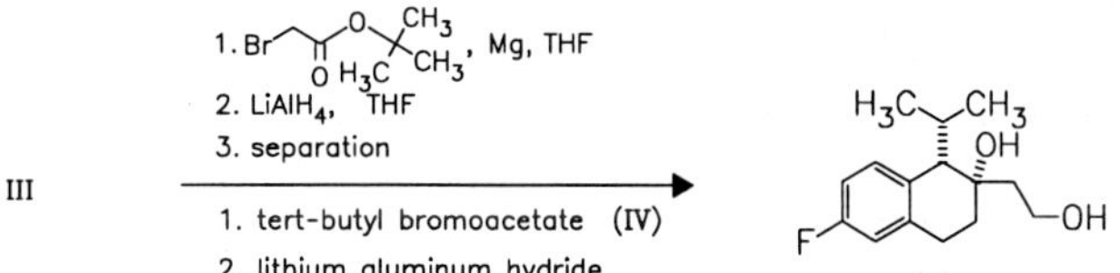

1. tert-butyl bromoacetate (IV)
2. lithium aluminum hydride

(V)

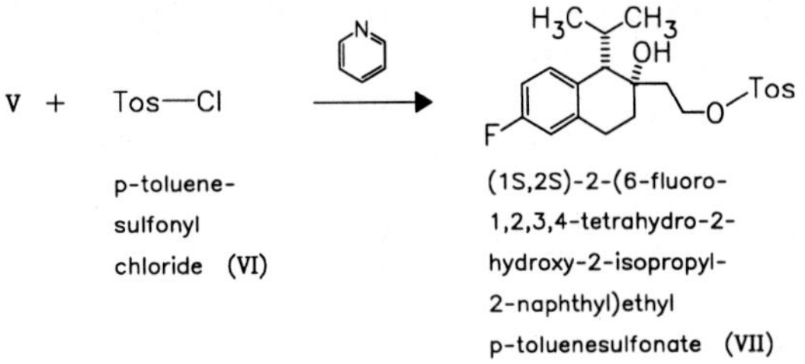

V + Tos—Cl

p-toluene-
sulfonyl
chloride (VI)

(1S,2S)-2-(6-fluoro-
1,2,3,4-tetrahydro-2-
hydroxy-2-isopropyl-
2-naphthyl)ethyl
p-toluenesulfonate (VII)

asymmetric synthesis of intermediate VII:

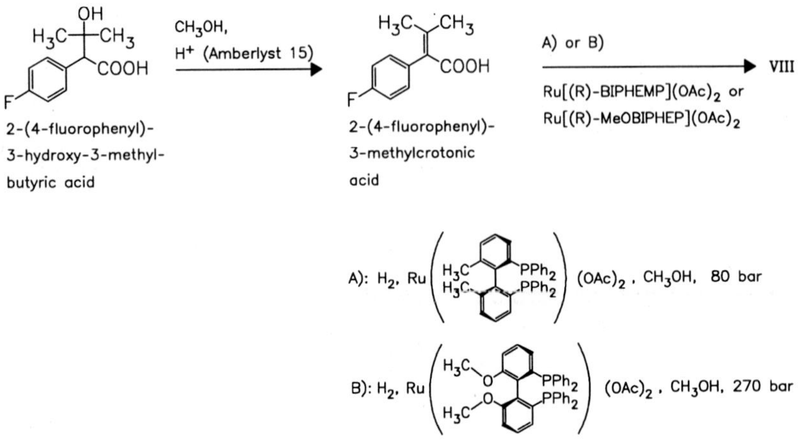

2-(4-fluorophenyl)-
3-hydroxy-3-methyl-
butyric acid

2-(4-fluorophenyl)-
3-methylcrotonic
acid

A) or B)

Ru[(R)-BIPHEMP](OAc)$_2$ or
Ru[(R)-MeOBIPHEP](OAc)$_2$

VIII

A): H$_2$, Ru (...) (OAc)$_2$. CH$_3$OH, 80 bar

B): H$_2$, Ru (...) (OAc)$_2$. CH$_3$OH, 270 bar

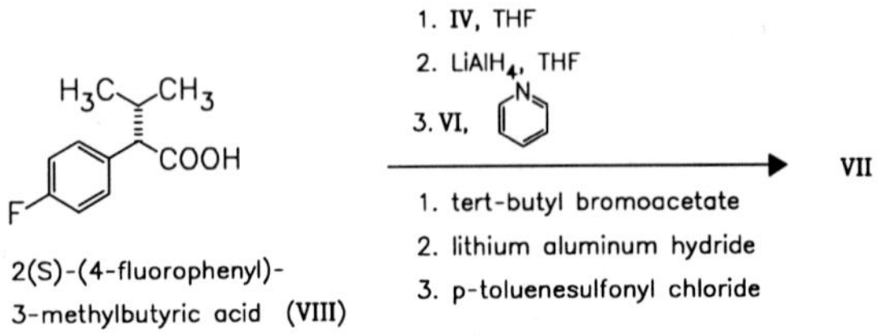

2(S)-(4-fluorophenyl)-
3-methylbutyric acid (VIII)

1. IV, THF
2. LiAlH$_4$, THF
3. VI,

1. tert-butyl bromoacetate
2. lithium aluminum hydride
3. p-toluenesulfonyl chloride

VII

synthesis of the final product:

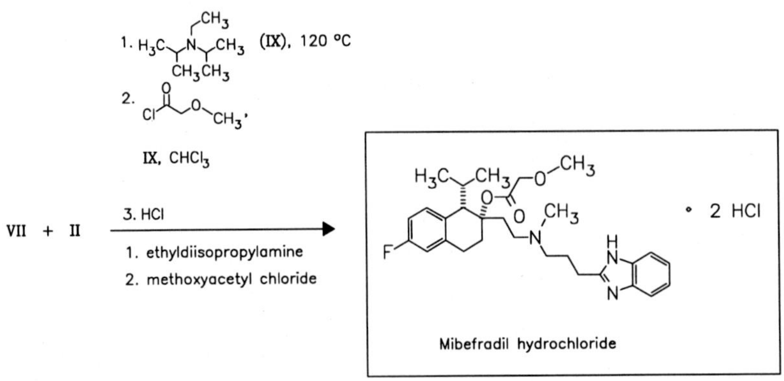

1. H$_3$C ... CH$_3$ (IX), 120 °C

2. Cl ... O ... CH$_3$,

IX, CHCl$_3$

3. HCl

VII + II

1. ethyldiisopropylamine
2. methoxyacetyl chloride

• 2 HCl

Mibefradil hydrochloride

Reference(s):
EP 268 148 (Hoffmann-La Roche AG; appl. 25.5.1988; CH-prior. 14.11.1986).

synthesis of intermediate VII:
EP 177 960 (Hoffmann-La Roche AG; appl. 16.4.1986; CH-prior. 11.10.1984).

asymmetric synthesis of 2(S)-(4-fluorophenyl)-3-methylbutyric acid:
Crameri, Y. et al.: Chimia (CHIMAD) **51** (6), 303 (1997).

synthesis of 2-(4-fluorophenyl)-3-methylcrotonic acid:
Noyori, R. et al.: J. Org. Chem. (JOCEAH) **52**, 3176 (1987).
Takaya, H. et al.: J. Org. Chem. (JOCEAH) **61**, 5510 (1996).
Schmid, R. et al.: Helv. Chim. Acta (HCACAV) **71**, 897 (1988).
Heiser, B. et al.: Tetrahedron: Asymmetry (TASYE3) **2**, 51 (1991).
EP 787 711 (Hoffmann-La Roche AG; appl. 6.8.1997; CH-prior. 31.1.1996).

Formulation(s):   f. c. tabl. 50 mg, 100 mg

Trade Name(s):
D: Cerate (ASTA Medica   Posicor (Roche); wfm
      AWD); wfm   USA: Posicor (Roche); wfm

Mibolerone

ATC: G03
Use: oral contraceptive, anabolic

RN: 3704-09-4   MF: $C_{20}H_{30}O_2$ MW: 302.46   EINECS: 223-046-5
CN: (7α,17β)-17-hydroxy-7,17-dimethylestr-4-en-3-one

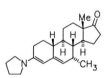

7α-methyl-3-pyrrolidino-19-
norandrosta-3,5-dien-17-one
(from 3,17-dioxo-7α-methyl-
19-nor-4-androstene)

1. H_3C—MgBr
2. NaOH

1. methylmagnesium
   bromide

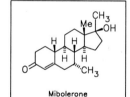

Mibolerone

Reference(s):
US 3 341 557 (Upjohn; 12.9.1967; prior. 5.6.1961, 6.11.1960, 6.6.1958).
FR-M 4 521 (Upjohn; appl. 4.6.1962; USA-prior. 5.6.1961).

alternative synthesis:
Campbell, J.A.; Babcock, J.C.: J. Am. Chem. Soc. (JACSAT) **81**, 4069 (1959).

separation of the 7α-, 7β-isomers:
NL 6 601 855 (Ciba; appl. 14.2.1966; CH-prior. 15.2.1965).

Trade Name(s):
GB: Matenon (Upjohn); wfm   USA: Cheque (Upjohn); wfm

Micinicate
(Nicotinyl cyclandelate)

ATC: C04A
Use: vasodilator (for treatment of chromic
        obliterating peripheral arteriopathy
        and microcircular disorders)

RN: 39537-99-0   MF: $C_{23}H_{27}NO_4$ MW: 381.47
CN: (±)-*cis*-3-pyridinecarboxylic acid 2-oxo-1-phenyl-2-[(3,3,5-trimethylcyclohexyl)oxy]ethyl ester

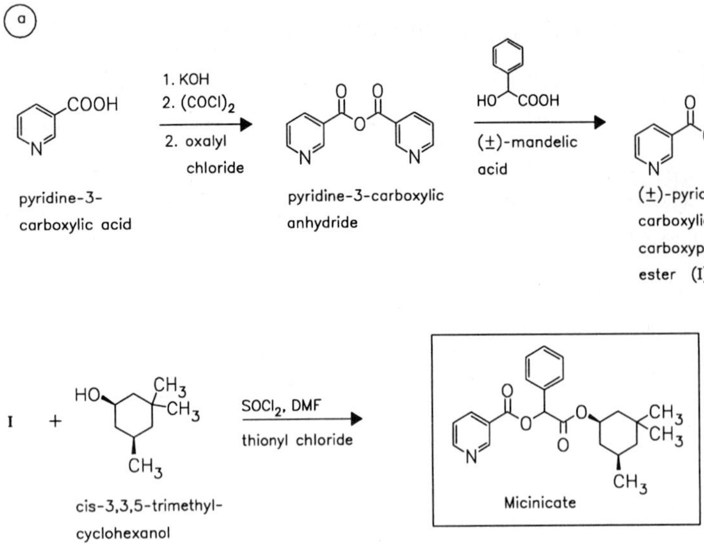
a
COOH
1. KOH
2. (COCl)2
2. oxalyl chloride
pyridine-3-carboxylic acid
pyridine-3-carboxylic anhydride
HO COOH
(±)-mandelic acid
COOH
(±)-pyridine-3-carboxylic acid carboxyphenylmethyl ester (I)
I +
HO
CH3
CH3
CH3
cis-3,3,5-trimethyl-cyclohexanol
SOCl2, DMF
thionyl chloride
Micinicate

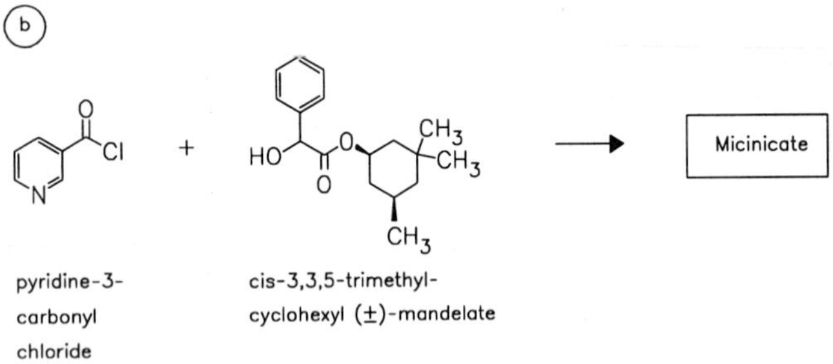
b
Cl
HO
CH3
CH3
CH3
pyridine-3-carbonyl chloride
cis-3,3,5-trimethyl-cyclohexyl (±)-mandelate
Micinicate

Reference(s):

a EP 157 151 (Ravizza; appl. 22.2.1985; I-prior. 3.7.1984).

b JP 47 034 365 (Mitsui; appl. 12.4.1971).
DOS 2 461 909 (Gaver; appl. 31.12.1974; CH-prior. 3.1.1974).

synthesis of cis-3,3,5-trimethylcyclohexyl mandelate:
Funcke, A.B.H. et al.: Arzneim.-Forsch. (ARZNAD) **3**, 505 (1953).

Formulation(s): tabl. 400 mg

Trade Name(s):
I: Micivas (Ravizza; IT); wfm

Miconazole

ATC: A01AB09; A07AC01; D01AC02; G01AF04; J02AB01; S02AA13
Use: topical antifungal

RN: 22916-47-8 MF: $C_{18}H_{14}Cl_4N_2O$ MW: 416.14 EINECS: 245-324-5
LD_{50}: 90.57 mg/kg (M, i.v.); 872 mg/kg (M, p.o.);
105 mg/kg (R, i.v.); >3 g/kg (R, p.o.);
60 mg/kg (dog, i.v.)
CN: 1-[2-(2,4-dichlorophenyl)-2-[(2,4-dichlorophenyl)methoxy]ethyl]-1*H*-imidazole

mononitrate

RN: 22832-87-7 MF: $C_{18}H_{14}Cl_4N_2O \cdot HNO_3$ MW: 479.15 EINECS: 245-256-6

LD_{50}: 28 mg/kg (M, i.v.); 578 mg/kg (M, p.o.);
14.7 mg/kg (R, i.v.); 920 mg/kg (R, p.o.);
>160 mg/kg (dog, p.o.)

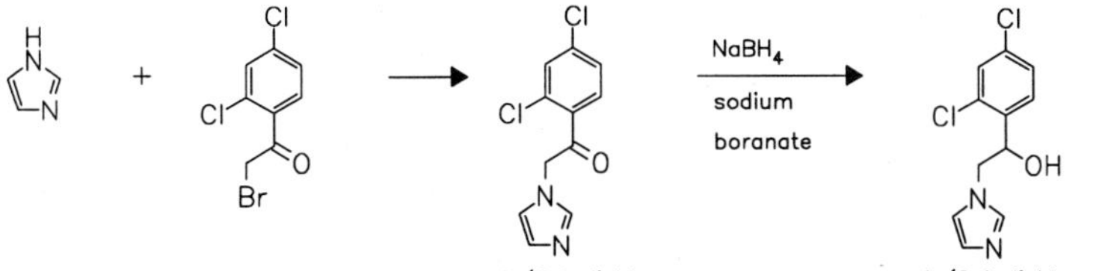

NaBH4
sodium
boranate
imidazole
2-bromo-
2',4'-dichloro-
acetophenone
1-(2,4-dichloro-
benzoylmethyl)-
imidazole
1-(2,4-dichloro-
phenyl)-2-(1H-
imidazol-1-yl)-
ethanol (I)

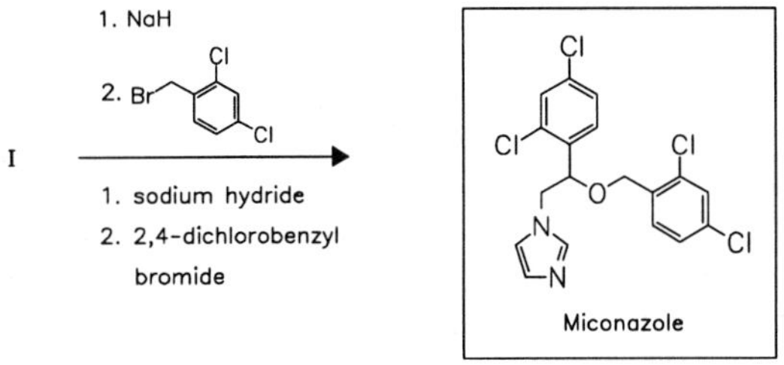

I
1. NaH
2.
1. sodium hydride
2. 2,4-dichlorobenzyl
bromide
Miconazole

Reference(s):

DAS 1 940 388 (Janssen; appl. 8.8.1969; USA-prior. 19.8.1968, 23.7.1969).
US 3 717 655 (Janssen; 20.2.1973; appl. 19.8.1968).
US 3 839 574 (Janssen; 1.10.1974; prior. 23.7.1969).

Formulation(s): cream 2 g/100 g, 20 mg/g; vaginal cream 20 mg/g; powder 2 g/100 g, 20 mg/g (as mononitrate); sol. 20 mg/ml; tabl. 250 mg (as free base)

Trade Name(s):

D: Acnidazil (Janssen-Cilag)-comb.
Daktar (Janssen-Cilag; 1974)
Epi-Monistat (Janssen-Cilag; 1974)
Gyno-Daktar (Janssen-Cilag; 1974)
Gyno-Monistat (Janssen-Cilag; 1974)

F: Britane (M. Johnson & Johnson-MSD)
Daktarin (Janssen-Cilag SA; 1974)
Gyno-Daktarin (Janssen-Cilag SA)
Miconazole GNR (GNR-pharma)

GB: Acnidazil (Janssen-Cilag)-comb.
Daktacort (Janssen-Cilag)-comb.
Daktarin (Janssen-Cilag; 1974)
Gyno-Daktarin (Janssen-Cilag)

I: Andergin (Pierrel)
Daktacort (Janssen)-comb.
Daktarin (Janssen; 1975)
Micoderm (Kemyos)
Miconal (Ecobi)
Micotef (LPB)
Prilagin (Gambar)

J: Florid (Mochida; 1980)

USA: Monistat (Ortho)

Micronomicin

ATC: S01AA22
Use: antibiotic, antibacterial

RN: 52093-21-7 MF: $C_{20}H_{41}N_5O_7$ MW: 463.58

CN: *O*-2-amino-2,3,4,6-tetradeoxy-6-(methylamino)-α-D-*erythro*-hexopyranosyl-(1→4)-*O*-[3-deoxy-4-*C*-methyl-3-(methylamino)-β-L-arabinopyranosyl(1→6)]-2-deoxy-D-streptamine

sulfate
RN: 66803-19-8 MF: $C_{20}H_{41}N_5O_7 \cdot xH_2SO_4$ MW: unspecified

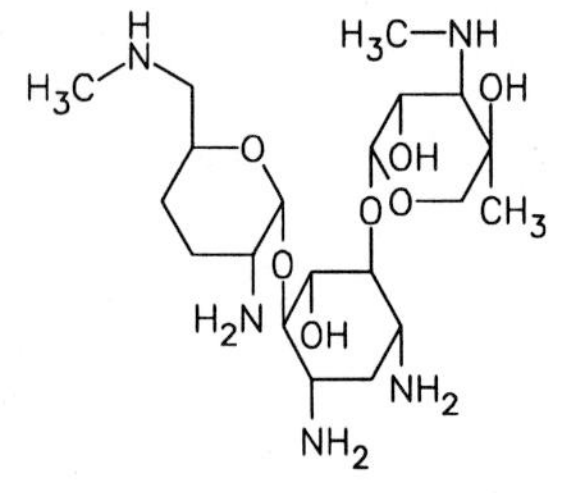

Micronomicin

Preparation by fermentation of *Micromonospora sagamiensis var. nonreductans nov. sp.* MK-65, ATCC 21826.

Reference(s):
Okachi, R. et al.: J. Antibiot. (JANTAJ) **27**, 793 (1974).
Nara, T. et al.: J. Antibiot. (JANTAJ) **28**, 21 (1975).
DOS 2 326 781 (Kyowa Ferm.; appl. 25.5.1973; J-prior. 27.5.1972).
US 4 045 298 (Kyowa Ferm.; 30.8.1977; J-prior. 27.5.1972).

structure:
Egan, R.S. et al.: J. Antibiot. (JANTAJ) **28**, 29 (1975).

total synthesis from gentamicin $C_{1\alpha}$:
JP 50 126 639 (appl. 25.3.1974).
JP 50 149 647 (appl. 28.5.1974).
JP 50 149 646 (appl. 28.5.1974).
JP 50 131 949 (appl. 9.4.1974).
JP 50 123 640 (appl. 15.5.1974).
JP 50 129 531 (appl. 29.3.1974).

Formulation(s): amp. 60 mg, 120 mg; eye drops 0.3 %; ointment 0.3 % (as sulfate)

Trade Name(s):
F: Microphta (Europhta) J: Sagamicin (Kyowa Hakko) Santemycin (Santen)

Midazolam

ATC: N05CD08
Use: hypnotic

RN: 59467-70-8 MF: $C_{18}H_{13}ClFN_3$ MW: 325.77 EINECS: 261-774-5
LD_{50}: 50 mg/kg (M, i.v.);
75 mg/kg (R, i.v.); 215 mg/kg (R, p.o.)
CN: 8-chloro-6-(2-fluorophenyl)-1-methyl-4*H*-imidazo[1,5-*a*][1,4]benzodiazepine

monohydrochloride
RN: 59467-96-8 MF: $C_{18}H_{13}ClFN_3 \cdot HCl$ MW: 362.24 EINECS: 261-776-6
maleate (1:1)
RN: 59467-94-6 MF: $C_{18}H_{13}ClFN_3 \cdot C_4H_4O_4$ MW: 441.85 EINECS: 261-775-0

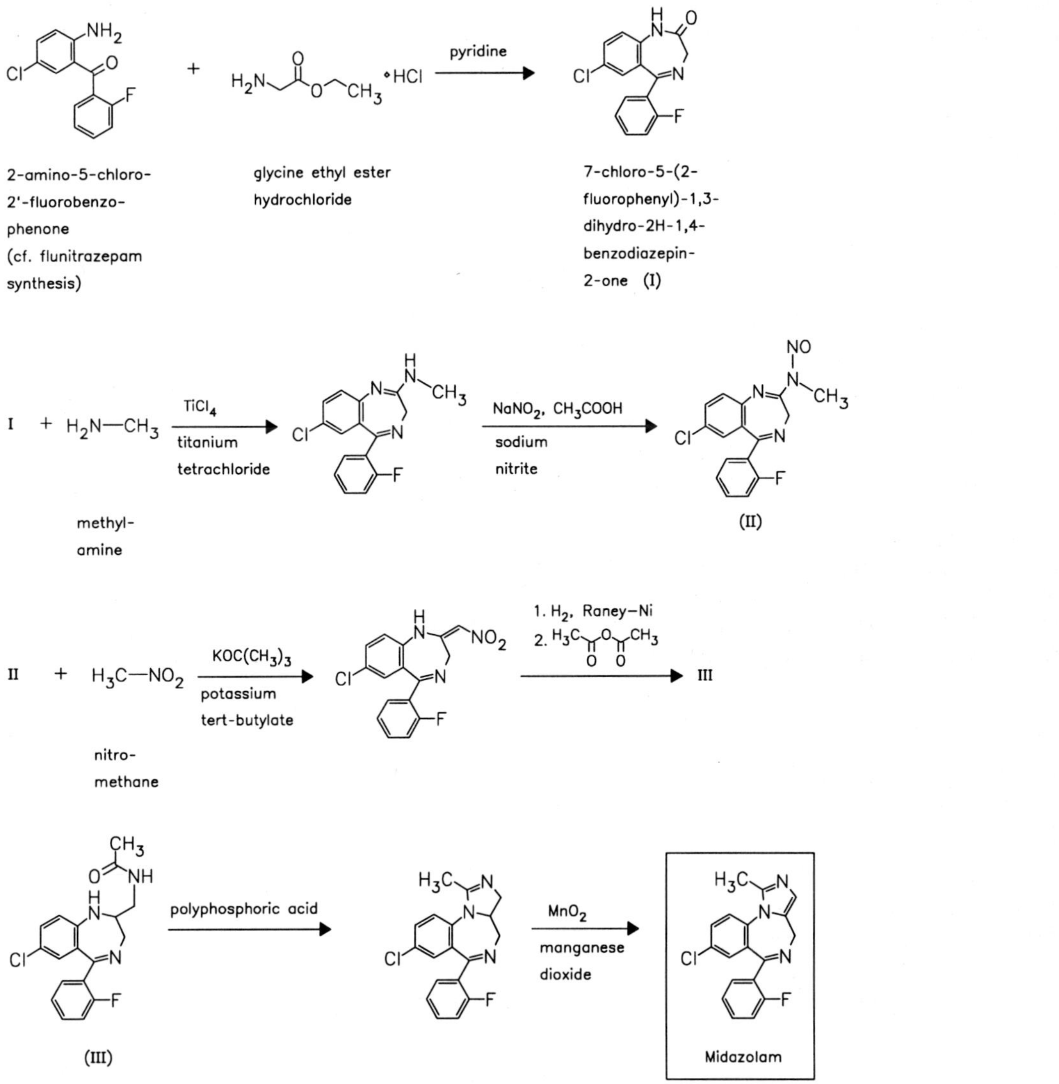

2-amino-5-chloro-
2'-fluorobenzo-
phenone
(cf. flunitrazepam
synthesis)

glycine ethyl ester
hydrochloride

7-chloro-5-(2-
fluorophenyl)-1,3-
dihydro-2H-1,4-
benzodiazepin-
2-one (I)

methyl-
amine

(II)

nitro-
methane

(III)

Midazolam

Reference(s):
DOS 2 540 522 (Hoffmann-La Roche; appl. 11.9.1975; USA-prior. 11.9.1974).
US 4 280 957 (Hoffmann-La Roche; 28.7.1981; prior. 8.2.1977).

Formulation(s):   amp. 5 mg/5 ml, 15mg/3 ml; f. c. tabl. 7.5 mg (as hydrochloride)

Trade Name(s):
D: Dormicum (Roche; 1984)   GB: Hypnovel (Hoffmann-La   J: Dormicum (Roche)
F: Hypnovel (Produits Roche)   Roche; 1983)   USA: Versed (Roche; 1986)

Midecamycin

(Espinomycin; Midecamicin; Mydecamycin)

ATC: J01FA03
Use: macrolide antibiotic

RN: 35457-80-8   MF: $C_{41}H_{67}NO_{15}$ MW: 813.98   EINECS: 252-578-0
LD_{50}:   1 g/kg (M, i.v.); 5800 mg/kg (M, p.o.);
    9600 mg/kg (R, p.o.)
CN: leucomycin V 3,4^B-dipropanoate

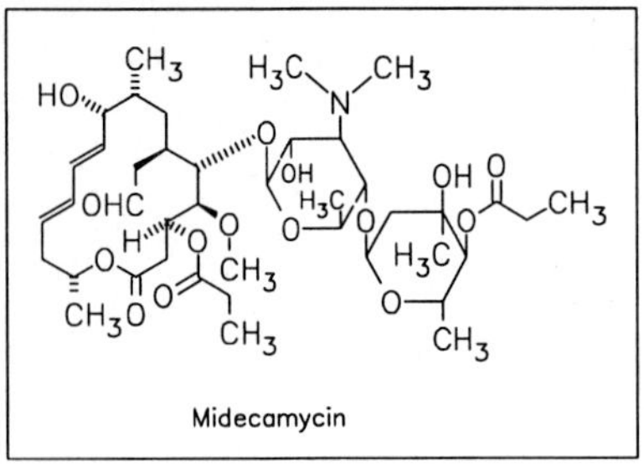

Midecamycin

Macrolide antibiotic from cultures of *Streptomyces mycarofaciens*. Midecamycin A$_2$, A$_3$ and A$_4$ exist also in small amounts in the complex beside main component Midecamycin.

Reference(s):

US 3 761 588 (Meiji Seika; 25.9.1973; J-prior. 6.2.1969, 25.9.1969).
Niida, T. et al.: J. Antibiot. (JANTAJ) **24**, 319 (1971).
Tsuruoka, T. et al.: J. Antibiot. (JANTAJ) **24**, 452 (1971).
Inouye, S. et al.: J. Antibiot. (JANTAJ) **24**, 460 (1917).

Formulation(s):   cps. 50 mg, 100 mg, 200 mg; tabl. 400 mg

Trade Name(s):

F:	Midécacine (Clin-Midy);	I:	Midecin (Farmaka)	Rubimycin (Nikken)
	wfm	J:	Medemycin (Meiji)	

Midecamycin acetate

(Miokamycin)

ATC: J01FA03
Use: antibiotic

RN: 55881-07-7   MF: C$_{45}$H$_{71}$NO$_{17}$ MW: 898.05   EINECS: 259-879-6
LD$_{50}$:   >5 g/kg (M, p.o.)
CN: leucomycin V 3^B,9-diacetate 3,4^B-dipropanoate

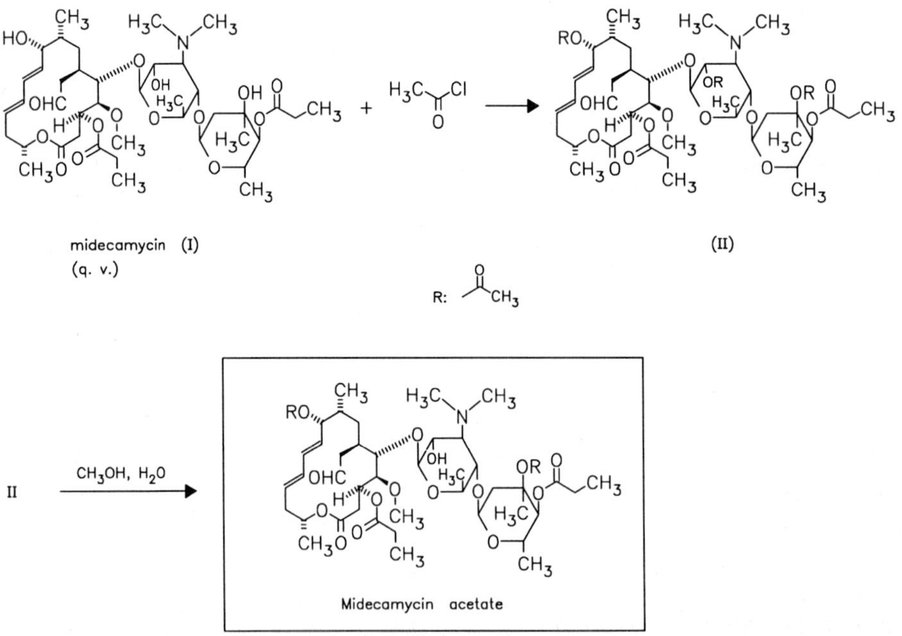

midecamycin (I)
(q. v.)

R:

(II)

Midecamycin acetate

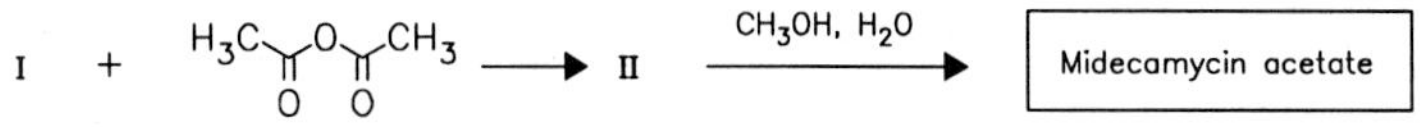

I + $H_3C\underset{O}{\overset{}{\diagup}}O\underset{O}{\overset{}{\diagdown}}CH_3$ $\longrightarrow$ II $\xrightarrow{CH_3OH,\ H_2O}$ [Midecamycin acetate]

Reference(s):

DE 2 004 686 (Meiji; prior. 3.2.1970).
US 3 761 588 (Meiji; 25.9.1973; J-prior. 25.9.1969).
DOS 2 835 547 (Meiji; appl. 14.8.1978; J-prior. 15.8.1977).
DOS 2 537 375 (Meiji; appl. 22.8.1975; J-prior. 27.8.1974).
US 4 017 607 (Meiji; 12.4.1977; J-prior. 27.8.1974).
US 4 188 480 (Meiji; 12.2.1980; J-prior. 15.8.1977).
Omoto, S. et al.: J. Antibiot. (JANTAJ) **24**, 536 (1976).
Nakamura, K. et al.: Chem. Lett. (CMLTAG) **1978**, 1293.

Formulation(s):   tabl. 400 mg

Trade Name(s):

F: Mosil (Menarini)   Miocamen (Menarini;   J: Miocamycin (Meiji Seika;
I: Macroral (Malesci; 1985)   1985)   1985)
                                       Miokacin (Firma; 1986)

Midodrine

ATC: C01CA17
Use: antihypotensive, α-adrenergic,
        vasoconstrictor

RN: 97476-58-9   MF: $C_{12}H_{18}N_2O_4$ MW: 254.29
CN: (±)-2-amino-*N*-[2-(2,5-dimethoxyphenyl)-2-hydroxyethyl]acetamide

monohydrochloride
RN: 3092-17-9   MF: $C_{12}H_{18}N_2O_4 \cdot HCl$ MW: 290.75
LD$_{50}$:   56.2 mg/kg (M, i.v.); 246 mg/kg (M, p.o.);
        18.2 mg/kg (R, i.v.); 68.8 mg/kg (R, p.o.);
        150 mg/kg (dog, p.o.)

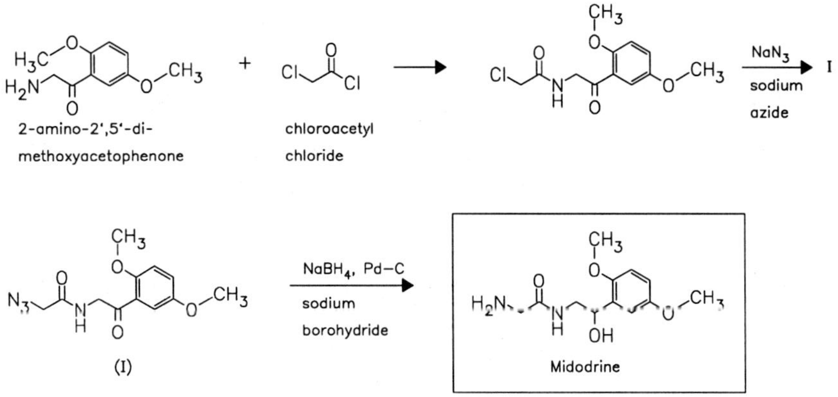

2-amino-2',5'-di-
methoxyacetophenone

chloroacetyl
chloride

NaN$_3$
sodium
azide
→ I

(I)

NaBH$_4$, Pd–C
sodium
borohydride
→

Midodrine

Reference(s):

DAS 2 523 735 (Lentia; appl. 28.5.1975; A-prior. 24.7.1974).

alternative syntheses:

AT 241 435 (Österr. Stickstoffwerke Linz; appl. 11.6.1963; valid from 15.12.1964).
DOS 2 506 110 (Lentia; appl. 13.2.1975).
BE 838 512 (Chemie Linz AG; appl. 12.8.1976; D-prior. 13.2.1975).

Formulation(s):   amp. 5 mg; drops 10 mg/ml; tabl. 2.5 mg, 5 mg (as hydrochloride)

Trade Name(s):
D: Gutron (Nycomed)   I: Gutron (Guidotti)   USA: ProAmatine (Roberts)
F: Gutron (Nycomed SA)   J: Metligine (Taisho)

Midoriamin
(Thiamine cobaltichlorophylline complex)

ATC: A02B
Use: ulcer therapeutic

RN: 87211-44-7   MF: $C_{46}H_{53}CoN_8O_9S$ MW: 952.98
LD$_{50}$:   209 mg/kg (M, i.p.); 3066 mg/kg (M, p.o.); 406 mg/kg (M, s.c.);
        82 mg/kg (R, i.p.); 3590 mg/kg (R, p.o.); 201 mg/kg (R, s.c.)
CN: [*OC*-6-24-(2*S*-*trans*)]-[*N*-[(4-amino-2-methyl-5-pyrimidinyl)methyl]-*N*-(4-hydroxy-2-mercapto-1-methyl-1-butenyl)formamide]aqua[18-carboxy-20-(carboxymethyl)-8-ethenyl-13-ethyl-2,3-dihydro-3,7,12,17-tetramethyl-21*H*,23*H*-porphine-2-propanoato(5-)-*N*^{21},*N*^{22},*N*^{23},*N*^{24}]dihydrogencobaltate(2-)

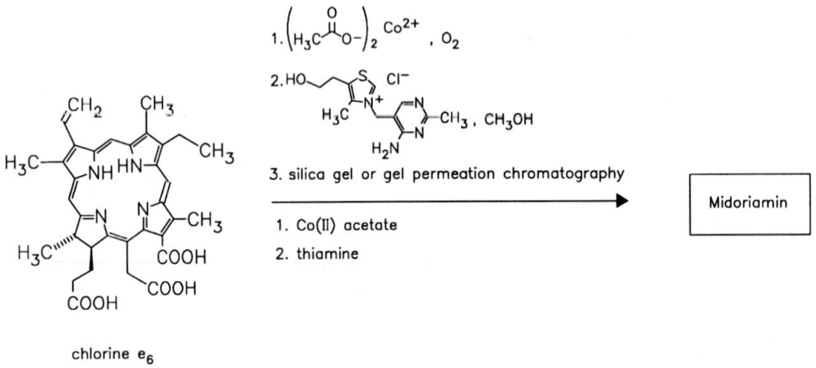

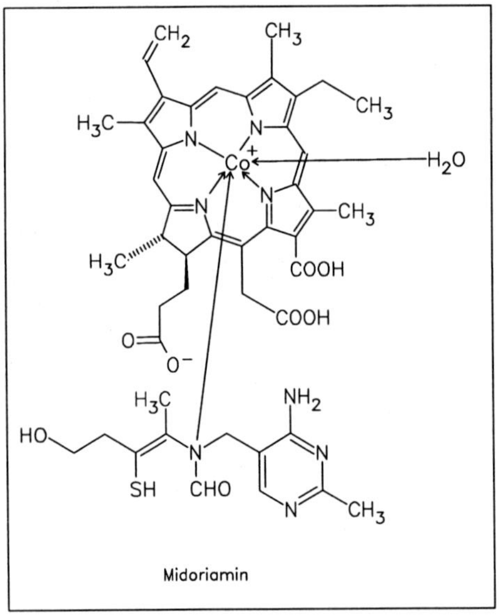

chlorine e$_6$

Midoriamin

Reference(s):

JP 1 052 779 (Green Cross; appl. 15.7.1988).
JP 63 264 483 (Green Cross; appl. 11.3.1988).
JP 63 264 420 (Green Cross; appl. 11.3.1988).
JP 57 062 281 (Green Cross; Nisshin Flour Mill; appl. 1.10.1980).
JP 58 041 885 (Green Cross, Nisshin Flour Mill; appl. 1.4.1982).

medical use for treatment of gastritis:
JP 2 149 522 (Green Cross; appl. 30.11.1988).

Formulation(s):   tabl. 5 mg

Trade Name(s):
J: Midoriamin (Green Cross;
        Nisshin Flour; 1988)

Mifepristone
(RU-486)

ATC: G03A; G03D; G03XB01
Use: abortifacient, orally active
        progesterone and glucocorticoid
        receptor antagonist, contraceptive

RN: 84371-65-3   MF: $C_{29}H_{35}NO_2$ MW: 429.60
CN: (11β,17β)-11-[4-(dimethylamino)phenyl]-17-hydroxy-17-(1-propynyl)estra-4,9-dien-3-one

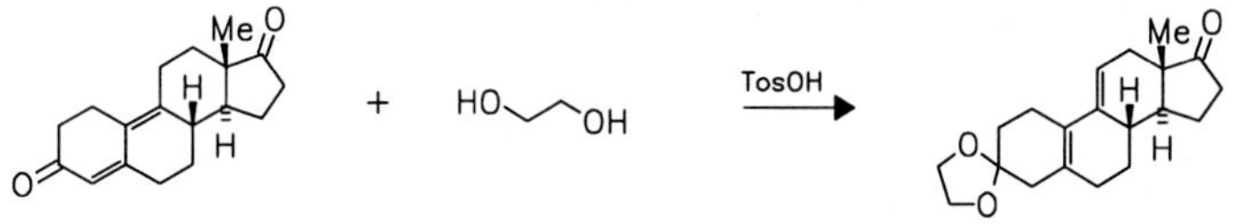

estra-4,9-diene-
3,17-dione

ethylene
glycol

3,3-(ethylenedioxy)-
estra-5(10),9(11)-
dien-17-one   (I)

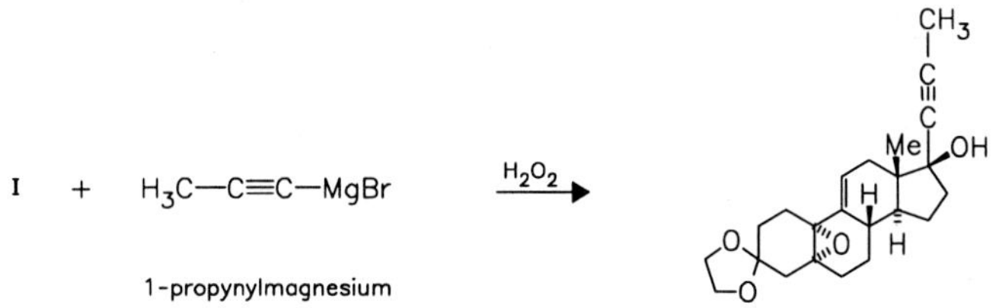

1-propynylmagnesium
bromide

3,3-(ethylenedioxy)-17α-
(1-propynyl)-5α,10α-epoxy-
estr-9(11)-en-17β-ol   (II)

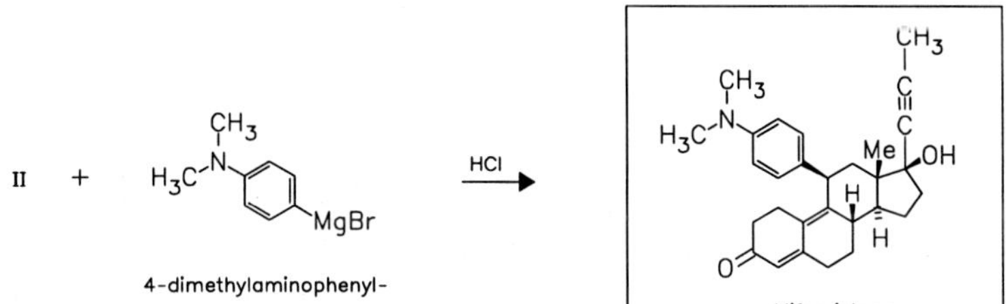

4-dimethylaminophenyl-
magnesium bromide

Mifepristone

Reference(s):
EP 57 115 (Roussel-Uclaf; appl. 8.1.1982; F-prior. 9.1.1981).
US 4 386 085 (Roussel-Uclaf; 31.5.1983; appl. 10.6.1982; F-prior. 9.1.1981).
US 4 447 424 (Roussel-Uclaf; 8.5.1984; appl. 10.6.1982; F-prior. 9.1.1981).
US 4 519 946 (Roussel-Uclaf; 28.5.1985; appl. 25.5.1984; prior. 9.1.1982, 10.6.1982, 30.3.1984; F-prior. 9.1.1981).
US 4 634 695 (Roussel-Uclaf; 6.1.1987; appl. 22.1.1985; prior. 9.1.1982, 25.5.1984, 10.6.1982, 30.3.1984; F-prior. 9.1.1981).

synthesis of 3,3-(ethylenedioxy)estra-5(10),9(11)-dien-17-one:
BE 651 813 (Merck & Co.; appl. 1964).

alternative synthesis:
FR 1 336 083 (Roussel-Uclaf; appl. 1962).
NL 6 406 712 (Roussel-Uclaf; appl. 1964; F-prior. 1963).
BE 651 812 (Merck & Co.; appl. 1964).

Formulation(s):   tabl. 200 mg

Trade Name(s):
F: Mifégyne (Exelgyn)   GB: Mifegyne (Exelgyn)

Miglitol

(Bay-m-1099)

ATC: A10BF02
Use: antidiabetic, α-glucosidase inhibitor

RN: 72432-03-2   MF: $C_8H_{17}NO_5$ MW: 207.23   EINECS: 276-661-6
CN: [2R-(2α,3β,4α,5β)]-1-(2-Hydroxyethyl)-2-(hydroxymethyl)-3,4,5-piperidinetriol

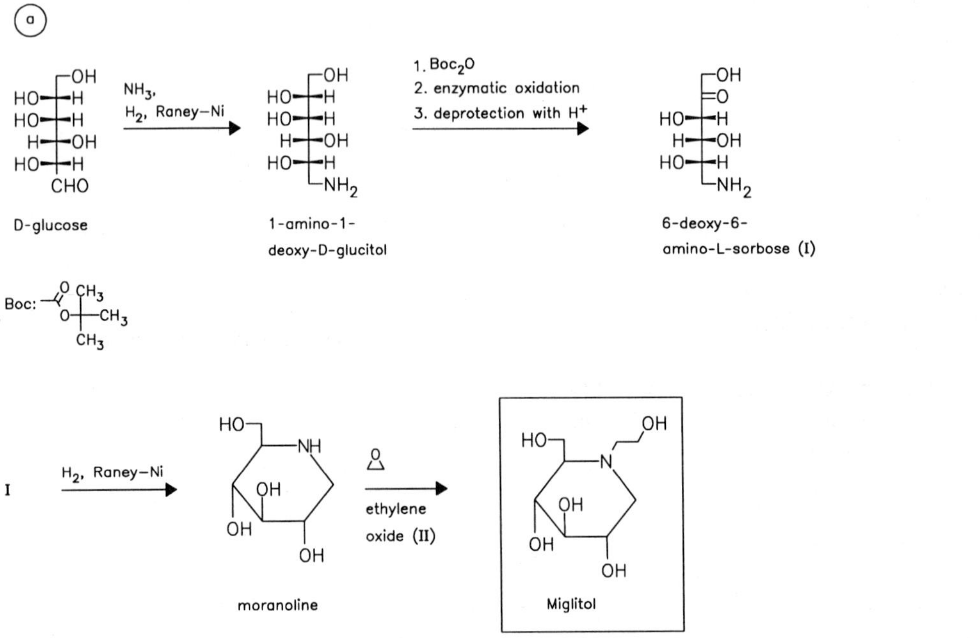

D-glucose

1-amino-1-deoxy-D-glucitol

6-deoxy-6-amino-L-sorbose (I)

Boc:

moranoline

Miglitol

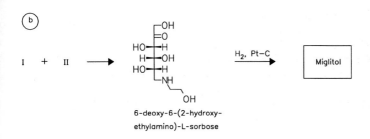

6-deoxy-6-(2-hydroxy-
ethylamino)-L-sorbose

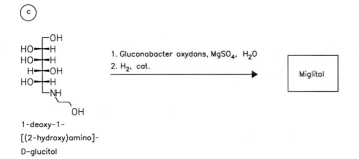

1-deoxy-1-
[(2-hydroxy)amino]-
D-glucitol

Reference(s):

preparation of moranoline *via* N-formyl-6-amino-6-deoxy-L-sorbose:
DE 3 611 841 (Bayer; appl. 9.4.1986; D-prior. 9.4.1986).
a   DE 2 758 025 (Bayer AG; 12.7.1979; appl. 24.12.1977; D-prior. 27.8.1977).
   EP 49 858 (Bayer AG; appl. 7.10.1981; D-prior. 15.10.1981).
   JP 54 106 477 (Nippon Shinyaku; appl. 3.2.1978).
b   DE 3 024 901 (Bayer AG; appl. 1.7.1980).
c   EP 477 160 (Monsanto Co.; 25.3.1992; appl. 19.9.1991; USA-prior. 20.9.1990).

Formulation(s):   tabl. 25 mg, 50 mg, 100 mg

Trade Name(s):
D: Diastabol (Sanofi-   USA: Glyset (Pharmacia &
     Synthelabo; 1998)   Upjohn; 1999)

Milnacipran hydrochloride

(Midalcipran hydrochloride)

ATC: N06AX17
Use: antidepressant, 5-HT and
       norepinephrine reuptake-inhibitor

RN: 101152-94-7   MF: $C_{15}H_{22}N_2O \cdot HCl$ MW: 282.82
CN: (±)-*cis*-2-(Aminomethyl)-*N,N*-diethyl-1-phenylcyclopropanecarboxamide monohydrochloride

(±)-*cis*-base
RN: 92623-85-3   MF: $C_{15}H_{22}N_2O$ MW: 246.35

synthesis of intermediate I: 2-oxo-1-phenyl-3- oxabicyclo[3.1.0]hexane

benzyl
cyanide

1. NaNH$_2$, benzene
2. [epichlorohydrin]
3. H$^+$

1. sodium amide
2. (±)-epichlorohydrin

2-(hydroxymethyl)-
2-phenylcyclo-
propanecarboxylic
acid ((±)-cis-trans-
mixture)

1. 150°C, 45 min
2. distillation
or
1. KOH
2. HCl

(±)-2-oxo-1-
phenyl-3-oxa-
bicyclo[3.1.0]-
hexane (I)

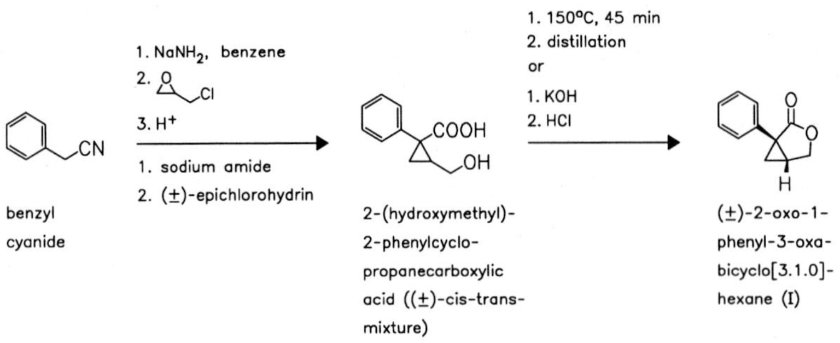

finalisation of Milnacipran hydrochloride

a

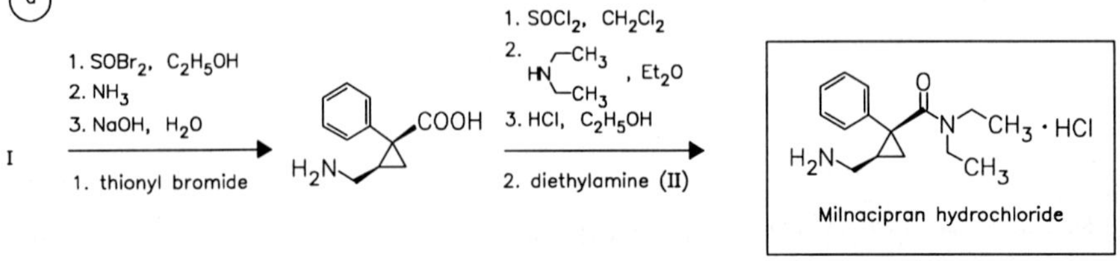

I

1. SOBr$_2$, C$_2$H$_5$OH
2. NH$_3$
3. NaOH, H$_2$O

1. thionyl bromide

1. SOCl$_2$, CH$_2$Cl$_2$
2. HN(CH$_3$)(CH$_3$), Et$_2$O
3. HCl, C$_2$H$_5$OH

2. diethylamine (II)

Milnacipran hydrochloride

b

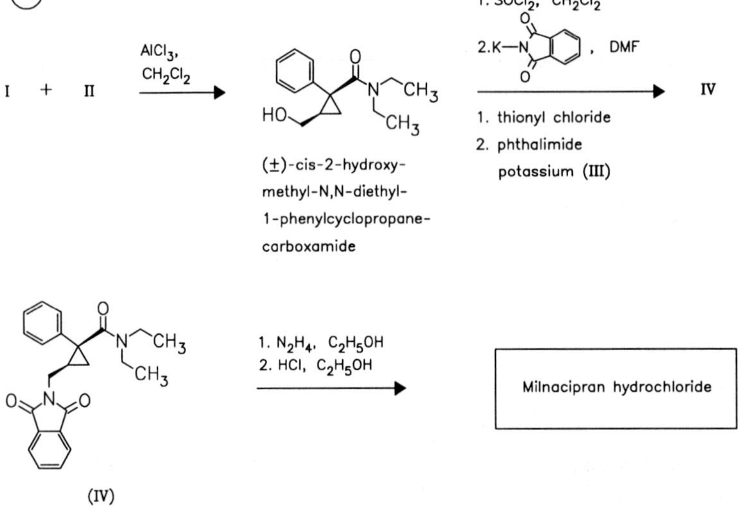

I + II

AlCl$_3$,
CH$_2$Cl$_2$

(±)-cis-2-hydroxy-
methyl-N,N-diethyl-
1-phenylcyclopropane-
carboxamide

1. SOCl$_2$, CH$_2$Cl$_2$
2. K—N(phthalimide), DMF

1. thionyl chloride
2. phthalimide
potassium (III)

IV

(IV)

1. N$_2$H$_4$, C$_2$H$_5$OH
2. HCl, C$_2$H$_5$OH

Milnacipran hydrochloride

c

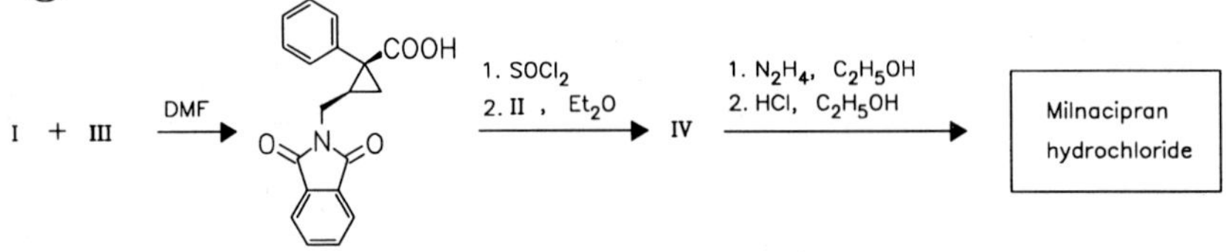

I + III

DMF

1. SOCl$_2$
2. II , Et$_2$O

IV

1. N$_2$H$_4$, C$_2$H$_5$OH
2. HCl, C$_2$H$_5$OH

Milnacipran
hydrochloride

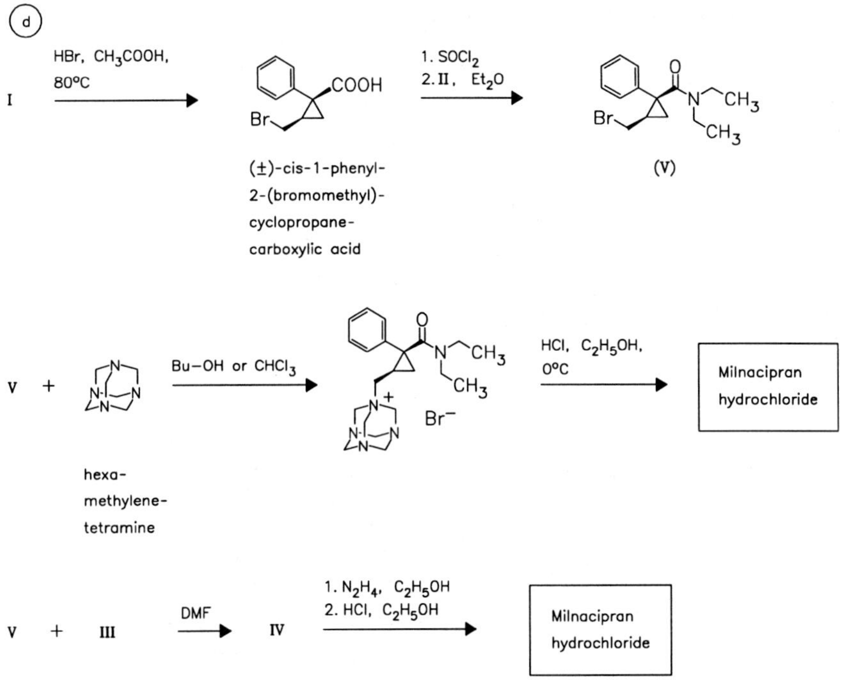

(d)

I →[HBr, CH₃COOH, 80°C] (±)-cis-1-phenyl-2-(bromomethyl)-cyclopropane-carboxylic acid →[1. SOCl₂ 2. II, Et₂O] (V)

V + hexamethylenetetramine →[Bu–OH or CHCl₃] [intermediate] →[HCl, C₂H₅OH, 0°C] Milnacipran hydrochloride

V + III →[DMF] IV →[1. N₂H₄, C₂H₅OH 2. HCl, C₂H₅OH] Milnacipran hydrochloride

Reference(s):
synthesis of intermediate I:
Mouzi, G.; Cousse, H.; Bonnaud, B: Synthesis (SYNTBF) **1978** (4), 304.
a   EP 068 999 (Pierre Fabre S. A.; appl. 21.6.1982; F-prior. 23.6.1981).
b   EP 377 381 (Pierre Fabre S. A.; appl. 27.12.1987; F-prior. 28.12.1988).
c   EP 200 638 (Pierre Fabre S. A.; appl. 22.4.1986; F-prior. 25.4.1985).
d   FR 2 581 060 (Pierre Fabre Medicament; appl. 31.10.1986; F-prior. 25.4.1985).

synthesis of 1-aryl-2-(aminomethyl)cyclopropanecarboxylic acid derivatives:
Bonnaud, B. et al.: J. Med. Chem. (JMCMAR) **30**, 318 (1987)

alternative syntheses:
Shuto, S. et al.: J. Org. Chem. (JOCEAH) **61**, 915 (1996)
Shuto, S. et al.: J. Med. Chem. (JMCMAR) **38**, 2964 (1995)

prolonged-release pharmaceuticals containing milnacipran:
WO 9 808 495 (Pierre Fabre S. A.; appl. 26.8.1997; F-prior. 28.8.1996)

compositions containing milnacipran and idazoxan:
WO 9 735 574 (Pierre Fabre S. A.; appl. 25.3.1997; F-prior. 25.3.1996)

Formulation(s):   cps. 50 mg (as hydrochloride)

Trade Name(s):
F: Ixel (Pierre Fabre; 1997)

Milrinone

ATC: C01CE02
Use: cardiotonic, phosphodiesterase III-inhibitor

RN: 78415-72-2 MF: $C_{12}H_9N_3O$ MW: 211.22 EINECS: 278-903-6
LD_{50}: 79 mg/kg (M, i.v.); 137 mg/kg (M, p.o.);
73 mg/kg (R, i.v.); 91 mg/kg (R, p.o.)
CN: 1,6-dihydro-2-methyl-6-oxo[3,4'-bipyridine]-5-carbonitrile

lactate
RN: 100286-97-3 MF: $C_{12}H_9N_3O \cdot xC_3H_6O_3$ MW: unspecified

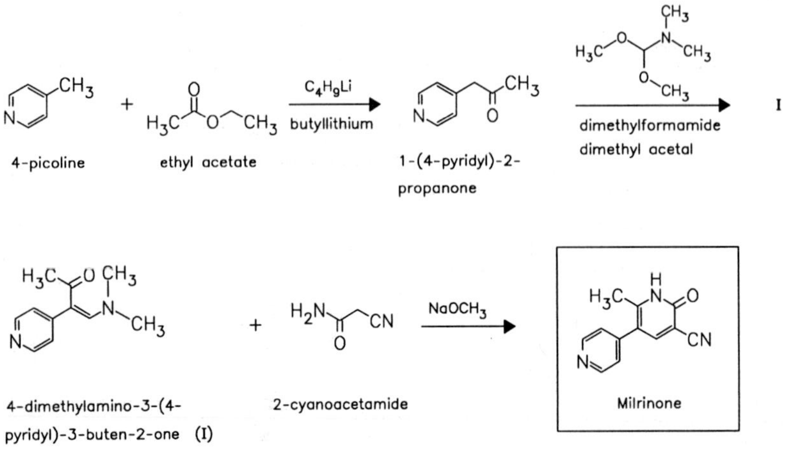

4-picoline
ethyl acetate
C_4H_9Li
butyllithium
1-(4-pyridyl)-2-propanone
dimethylformamide dimethyl acetal
I
4-dimethylamino-3-(4-pyridyl)-3-buten-2-one (I)
2-cyanoacetamide
$NaOCH_3$
Milrinone

Reference(s):
DOS 3 044 568 (Sterling Drug; appl. 26.11.1980; USA-prior. 26.11.1979, 20.10.1980, 28.3.1980, 6.11.1980).
US 4 312 875 (Sterling Drug; 26.1.1982; prior. 26.11.1979, 20.10.1980, 28.3.1980, 6.11.1980).
US 4 313 951 (Sterling Drug; 2.2.1982; prior. 26.11.1979, 20.10.1980, 28.3.1980, 6.11.1980).
Singh, B.: Heterocycles (HTCYAM) **23**, 1479 (1985).

alternative synthesis:
ES 544 504 (Inke; appl. 25.6.1985).
DD 274 620 (Arzneimittelwerk Dresden; appl. 2.8.1988).
DD 256 131 (Akademie der Wissenschaften; appl. 4.7.1986).

sustained release pharmaceutical composition:
EP 164 959 (Sterling Drug; appl. 30.5.1985; GB-prior. 4.6.1984, 30.5.1985).

Formulation(s): amp. 10 mg/10 ml (as free base); USA: bag 100 ml, 200 ml (200 µg/ml); vial 10 ml, 20 ml (1 mg/ml) (as lactate)

Trade Name(s):
D: Corotrop (Sanofi Winthrop)
F: Corotrope (Sanofi Winthrop; as lactate)
J: Milrila (Yamanouchi)
USA: Primcor (Sanofi)

Miltefosine

(D 18506; Hexadecylphosphocholine)

ATC: L01XX09
Use: antitumor (topical treatment)

RN: 58066-85-6 MF: $C_{21}H_{46}NO_4P$ MW: 407.58
LD_{50}: 246 mg/kg (R, p.o);
680 mg/kg (Mm, p.o);
603 mg/kg (Mf, p.o).
CN: 2-[[(Hexadecyloxy)hydroxyphosphinyl]oxy]-*N*,*N*,*N*-trimethylethanaminium inner salt

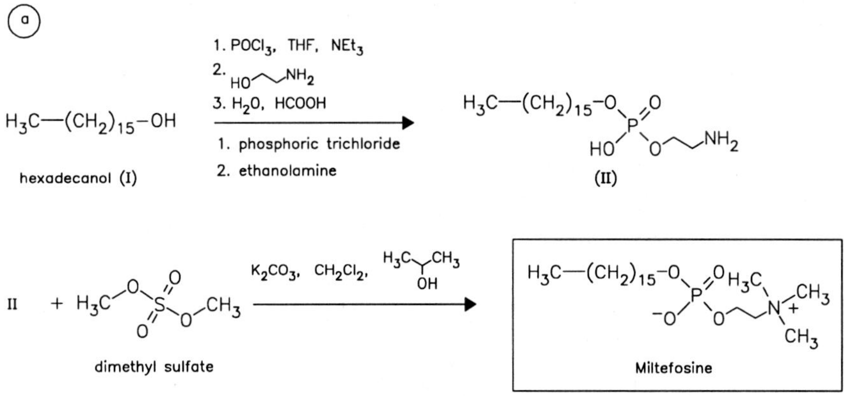
a
1. POCl3, THF, NEt3
2. HO NH2
3. H2O, HCOOH
H3C—(CH2)15—OH
1. phosphoric trichloride
2. ethanolamine
hexadecanol (I)
H3C—(CH2)15—O
O
P
HO
O
NH2
(II)
II
+ H3C
O
S
O
CH3
dimethyl sulfate
K2CO3, CH2Cl2,
H3C CH3
OH
H3C—(CH2)15—O
H3C
CH3
N+
CH3
Miltefosine

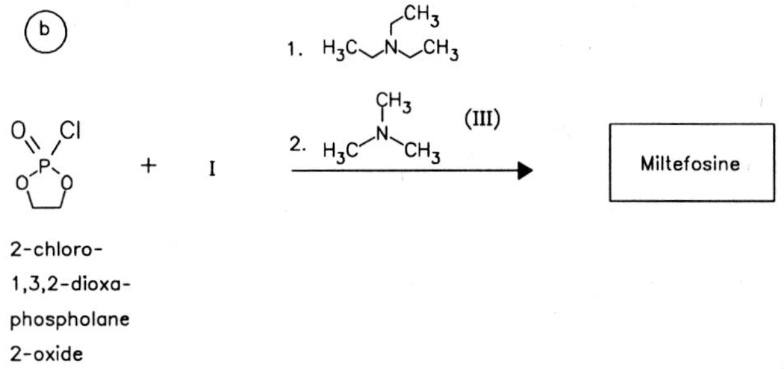
b
1. H3C N CH3
CH3
2. H3C N CH3
CH3
(III)
O Cl
P
O O
+ I
Miltefosine
2-chloro-
1,3,2-dioxa-
phospholane
2-oxide

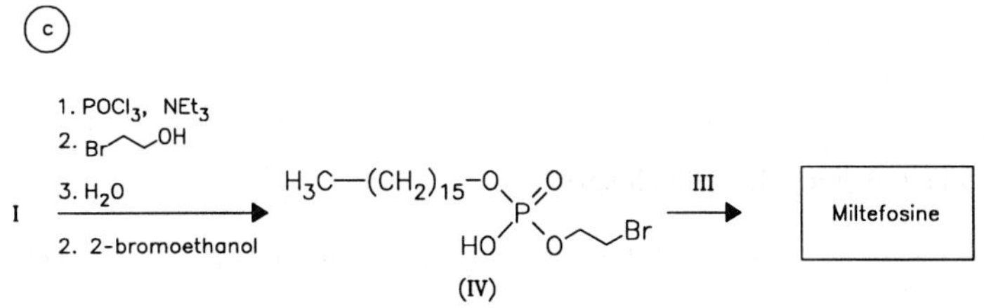
c
1. POCl3, NEt3
2. Br OH
3. H2O
I
2. 2-bromoethanol
H3C—(CH2)15—O
P
HO
O
Br
(IV)
III
Miltefosine

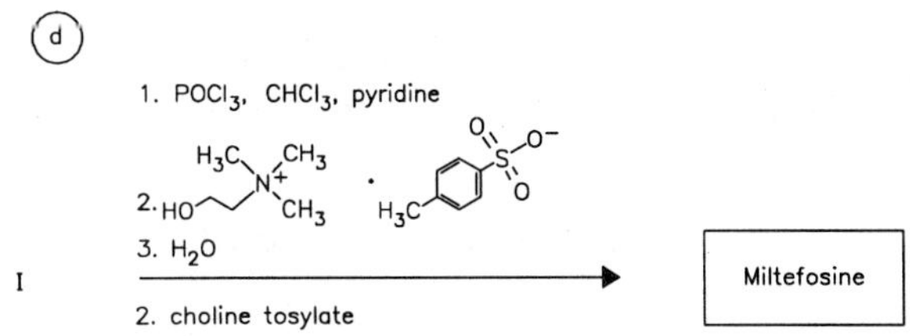
d
1. POCl3, CHCl3, pyridine
2. HO
H3C CH3
N+
CH3
H3C
O
S
O−
3. H2O
I
2. choline tosylate
Miltefosine

Reference(s):

a   Kaatze, U. et al.: Chem. Phys. Lipids (CPLIA4) **27** (3), 263-280 (1980).
    EP 225 608 (Max-Planck-Ges.; appl. 4.12.1986; D-prior. 4.12.1985).

preparation of quaternized ethanamine phosphate esters for oral or topical treatment of leishmaniasis:
EP 534 445 (Max-Planck-Ges.; appl. 24.9.1992; D-prior. 27.9.1991).
b   Eibl, H.; Engel, J.: Prog. Exp. Tumor Res. (EXPTAR) **34**, 1 (1992).
    Kametani, F. et al.: Nippon Kagaku Kaishi (NKAKB8) **9**, 1452-1458 (1984).
c   Nuhn, P. et al.: Pharmazie (PHARAT) **37** (10), 706-708 (1982).
d   EP 521 297 (ASTA Medica; appl. 26.6.1992; D-prior. 4.7.1991).

synergistic antitumor pharmaceuticals containing them and allylglycerins:
AT 393 505 (Max-Planck-Gesellschaft; appl. 27.4.1987).

Formulation(s):   sol. 60 mg/ml (10 ml bottles)

Trade Name(s):
D: Miltex (ASTA Medica
     AWD)

Minaprine

ATC: N06AX07
Use: antidepressant

RN: 25905-77-5   MF: $C_{17}H_{22}N_4O$ MW: 298.39   EINECS: 247-329-8
CN: N-(4-methyl-6-phenyl-3-pyridazinyl)-4-morpholineethanamine

dihydrochloride
RN: 25953-17-7   MF: $C_{17}H_{22}N_4O \cdot 2HCl$ MW: 371.31
LD_{50}: 63 mg/kg (M, i.p.)

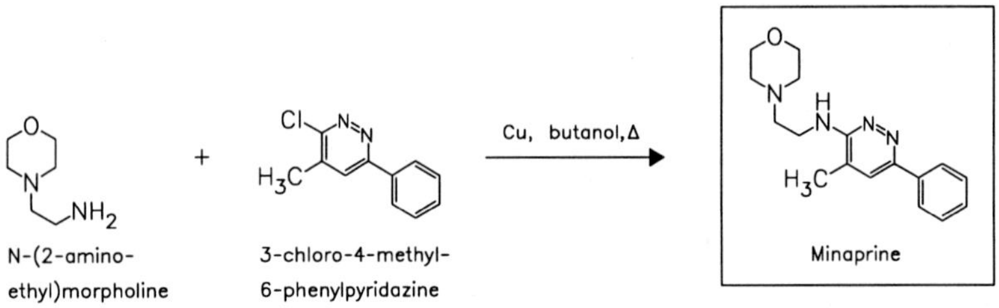

N-(2-amino-   3-chloro-4-methyl-   Minaprine
ethyl)morpholine   6-phenylpyridazine

Reference(s):

DOS 2 229 215 (CEPBEPE; appl. 15.6.1972; GB-prior. 18.6.1971).
GB 1 345 880 (CEPBEPE; valid from 16.6.1972; prior. 18.6.1971).
ZA 730 671 (CEPBEPE; appl. 3.1.1973).

medical use:
US 4 169 158 (Laborit Henri; 25.9.1979, GB-prior. 18.6.1971).

Formulation(s):   drops 5 %; tabl. 50 mg, 100 mg (as dihydrochloride)

Trade Name(s):
F: Cantor (Clin-Comar-Byla;   Cantor (Clin-Midy); wfm   J: Alcas (Taisho)
     1980); wfm   I: Cantor (Midy; 1984)

Minocycline

ATC: J01AA08
Use: antibiotic

RN: 10118-90-8 MF: $C_{23}H_{27}N_3O_7$ MW: 457.48
LD$_{50}$: 140 mg/kg (M, i.v.); 3100 mg/kg (M, p.o.)
CN: [4S-(4α,4aα,5aα,12aα)]-4,7-bis(dimethylamino)-1,4,4a,5,5a,6,11,12a-octahydro-3,10,12,12a-tetrahydroxy-1,11-dioxo-2-naphthacenecarboxamide

monohydrochloride
RN: 13614-98-7 MF: $C_{23}H_{27}N_3O_7 \cdot HCl$ MW: 493.94 EINECS: 237-099-7
LD$_{50}$: 154 mg/kg (M, i.v.); 3600 mg/kg (M, p.o.);
164 mg/kg (R, i.v.); 2380 mg/kg (R, p.o.)

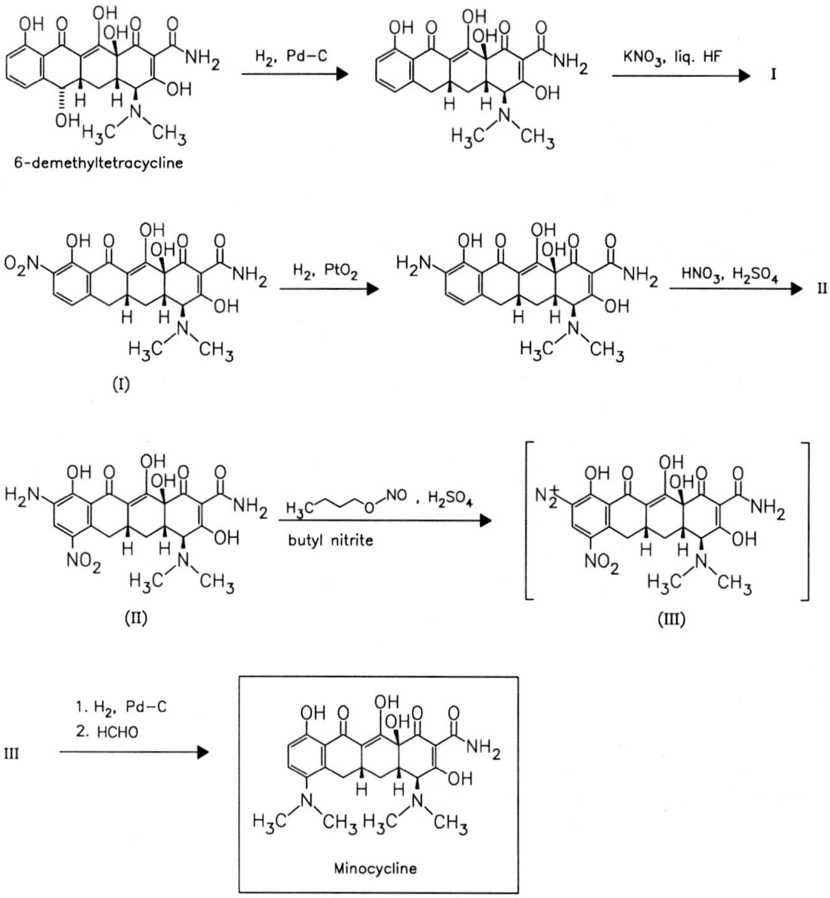

6-demethyltetracycline

(I)

(II)

(III)

Minocycline

Reference(s):
US 3 148 212 (American Cyanamid; 8.9.1964; appl. 22.12.1961).
US 3 226 436 (American Cyanamid; 28.12.1965; prior. 24.10.1961, 22.12.1961, 17.5.1963).
US 3 345 410 (American Cyanamid; 3.10.1967; prior. 14.3.1966, 1.12.1966).
Church, R.F.R. et al.: J. Org. Chem. (JOCEAH) **36**, 723 (1971).
DE 1 245 942 (American Cyanamid; appl. 15.5.1962; USA-prior. 21.10.1961, 22.12.1961, 7.2.1962).
DE 1 643 767 (American Cyanamid, prior. 16.1.1968).

intermediates:
US 3 403 179 (American Cyanamid; 24.9.1968; prior. 10.1.1967).
US 3 483 251 (American Cyanamid; 9.12.1969; prior. 3.3.1967).

purification:
DOS 2 309 582 (American Cyanamid; appl. 26.2.1973; USA-prior. 11.5.1972).

Formulation(s):　cps. 50 mg, 100 mg; f. c. tabl. 50 mg, 100 mg; susp. 50 mg/60 ml (oral); vial 100 mg (as hydrochloride)

Trade Name(s):

D:　Aknin (Sanofi Winthrop)
　　Klinomycin (Lederle; 1972)
　　generics
F:　Acneline (Wyeth-Lederle)
　　Minolis (Noviderm)
　　Mynocine (Wyeth-Lederle; 1973)

　　Zacnan (Lipha Santé)
GB:　Aknemin (Merck Sharp & Dohme)-comb.
　　Dentomycin (Wyeth)
　　Minocin (Wyeth; 1973)-comb.
I:　Minocin (Cyanamid; 1972)

J:　Minomycin (Lederle-Takeda)
USA:　Dynacin (Medicis)
　　Minocin (Lederle; 1971)
　　Vectrin (Warner Chilcott; 1973)

Minoxidil

ATC:　C02DC01; D11AX01
Use:　antihypertensive

RN:　38304-91-5　MF: $C_9H_{15}N_5O$　MW: 209.25　EINECS: 253-874-2
LD_{50}:　51 mg/kg (M, i.v.); >1 g/kg (M, p.o.);
　　　49 mg/kg (R, i.v.); 1321 mg/kg (R, p.o.)
CN:　6-(1-piperidinyl)-2,4-pyrimidinediamine 3-oxide

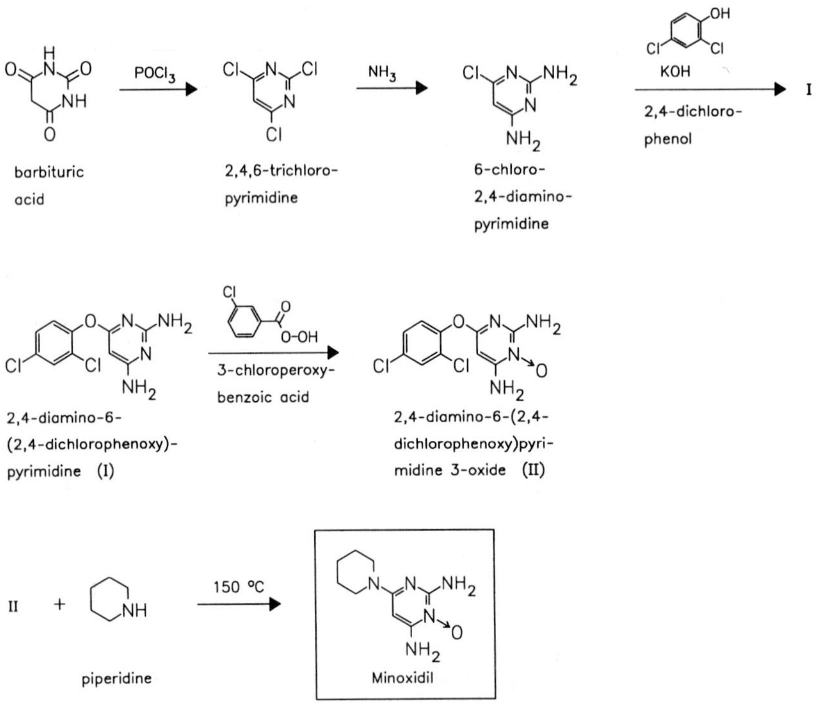

barbituric acid → 2,4,6-trichloro-pyrimidine → 6-chloro-2,4-diamino-pyrimidine → 2,4-dichloro-phenol → I

2,4-diamino-6-(2,4-dichlorophenoxy)-pyrimidine (I) → 3-chloroperoxy-benzoic acid → 2,4-diamino-6-(2,4-dichlorophenoxy)pyrimidine 3-oxide (II)

II + piperidine → 150 °C → Minoxidil

Reference(s):
DE 1 620 649 (Upjohn; prior. 28.10.1966).
US 3 382 247 (Upjohn; 7.5.1968; appl. 1.11.1965).
US 3 461 461 (Upjohn; 12.8.1969; appl. 1.1.1965).
US 3 644 364 (Upjohn; 22.2.1972; appl. 31.3.1970).
DAS 2 114 887 (Upjohn; appl. 27.3.1971; USA-prior. 31.3.1970).
DOS 2 114 887 (Upjohn; appl. 27.3.1971; USA-prior. 31.3.1970).

topical composition and use for hair growth:
US 4 139 619 (Upjohn; 13.2.1979; prior. 24.5.1976).

Formulation(s):   topical gel 2 %; topical sol. 2 %; tabl. 2.5 mg, 10 mg

Trade Name(s):

D:	Lonolox (Pharmacia & Upjohn; 1982)	GB:	Loniten (Pharmacia & Upjohn; 1980)
F:	Alostil (Sanofi Winthrop)		Regaine (Pharmacia & Upjohn; 1988)
	Lonoten (Pharmacia & Upjohn; 1984)	I:	Aloxidil (IDI)
	Néoxidil (Galderma)		Loniten (Pharmacia & Upjohn; 1983)
	Regaine (Pharmacia & Upjohn; 1987)		Minovital (Terapeutico)
			Minoximen (Menarini)

Normoxidil (Medosan)
Regaine (Pharmacia & Upjohn)
Tricoxidil (Bioindustria)
USA: Loniten (Upjohn; 1979); wfm
Rogaine (Pharmacia & Upjohn); wfm
generics

Mirtazapine

(6-Azamianserin; Mepirzepine; Org-3770)

ATC: N06AX11
Use: antidepressant, 5-HT$_{2/3}$-antagonist

RN: 61337-67-5   MF: C$_{17}$H$_{19}$N$_3$ MW: 265.36
CN: 1,2,3,4,10,14b-hexahydro-2-methylpyrazino[2,1-*a*]pyrido[2,3-*c*][2]benzazepine

racemate
RN: 85650-52-8   MF: C$_{17}$H$_{19}$N$_3$ MW: 265.36   EINECS: 288-060-6
(*R*)-enantiomer
RN: 61364-37-2   MF: C$_{17}$H$_{19}$N$_3$ MW: 265.36   EINECS: 262-735-5
(*S*)-enantiomer
RN: 61337-87-9   MF: C$_{17}$H$_{19}$N$_3$ MW: 265.36   EINECS: 262-714-0

2-chloro-
nicotino-
nitrile

1-methyl-3-
phenyl-
piperazine

KF, DMF, 140 °C

2-(4-methyl-2-
phenyl-1-piperazinyl)-
3-pyridinecarbonitrile (I)

I

1. KOH, ethanol
2. LiAlH$_4$, THF

1-(3-hydroxymethyl-
2-pyridinyl)-2-
phenyl-4-methyl-
piperazine

H$_2$SO$_4$

Mirtazapine

Reference(s):
DE 2 614 406 (AKZO; appl. 2.4.1976; NL-prior. 5.4.1975).

separation of enantiomers:
WO 9 407 814 (AKZO; appl. 1.10.1993; NL-prior. 7.10.1992).

oral formulations:
EP 436 252 (AKZO; appl. 19.12.1990; NL-prior. 30.12.1989).

combination with L-*amino acid decarboxylase inhibitors:*
WO 8 901 774 (British Technology Group; appl. 1.9.1988; GB-prior. 2.9.1987, 1.9.1988).

Formulation(s): tabl. 15 mg, 30 mg

Trade Name(s):
I: Remeron (Organon Italia) USA: Remeron (Organon)

Misoprostol

ATC: A02BB01
Use: peptic ulcer therapeutic

RN: 59122-46-2 MF: $C_{22}H_{38}O_5$ MW: 382.54
CN: (11α,13*E*)-11,16-dihydroxy-16-methyl-9-oxoprost-13-en-1-oic acid methyl ester

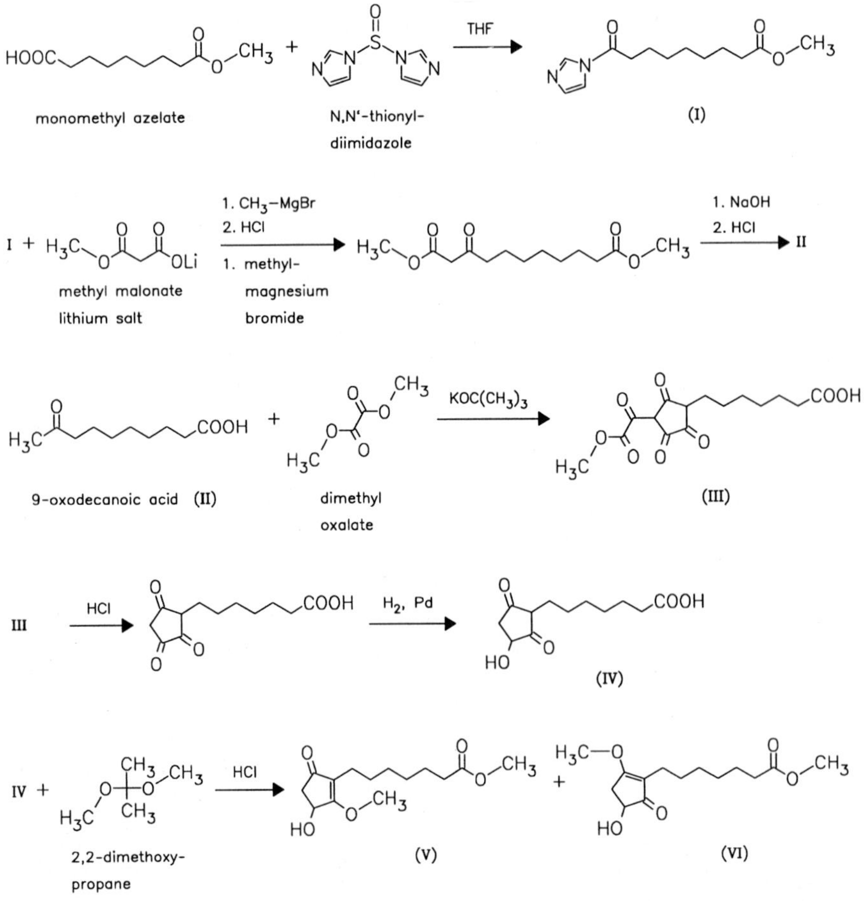
monomethyl azelate
N,N'-thionyl-diimidazole
THF
(I)
1. CH3–MgBr
2. HCl
1. methyl-magnesium bromide
methyl malonate lithium salt
1. NaOH
2. HCl
9-oxodecanoic acid (II)
dimethyl oxalate
KOC(CH3)3
(III)
HCl
H2, Pd
(IV)
2,2-dimethoxy-propane
HCl
(V)
(VI)

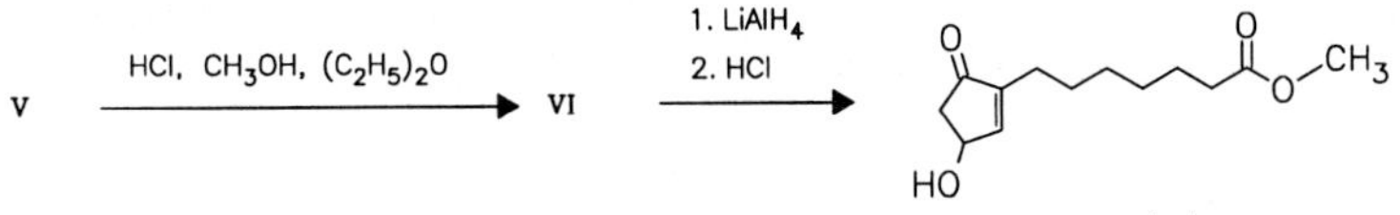

V $\xrightarrow{\text{HCl, CH}_3\text{OH, (C}_2\text{H}_5)_2\text{O}}$ VI $\xrightarrow[\text{2. HCl}]{\text{1. LiAlH}_4}$

(VII)

VII + dihydropyran $\longrightarrow$

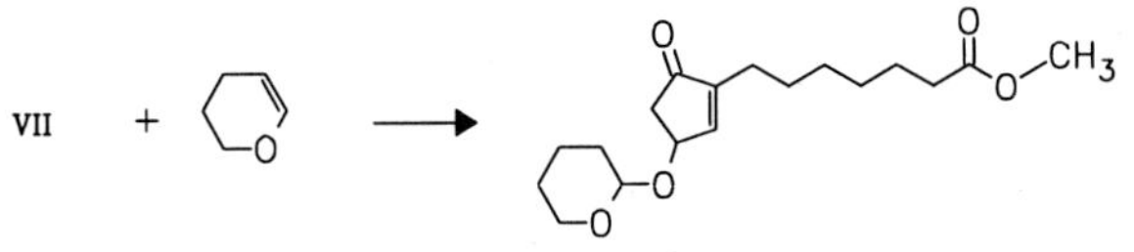

dihydro-
pyran

methyl 7-[3(RS)-tetrahydro-
pyran-2-yloxy-5-oxocyclopent-
1-en-1-yl]heptanoate (VIII)

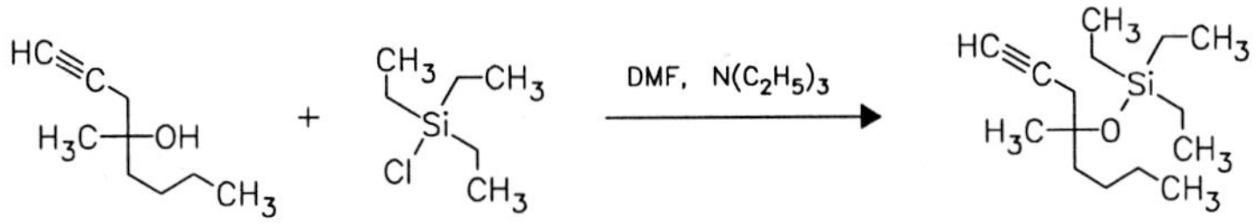

(±)-4-methyl-1-
octyn-4-ol

triethylsilyl
chloride

$\xrightarrow{\text{DMF, N(C}_2\text{H}_5)_3}$

(IX)

IX + diisobutyl-
aluminum
hydride $\xrightarrow{\text{hexane}}$ $\xrightarrow{\text{I}_2,\ \text{THF}}$ X

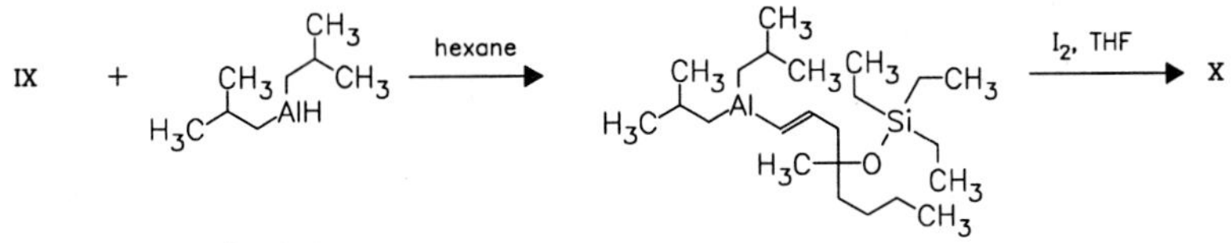

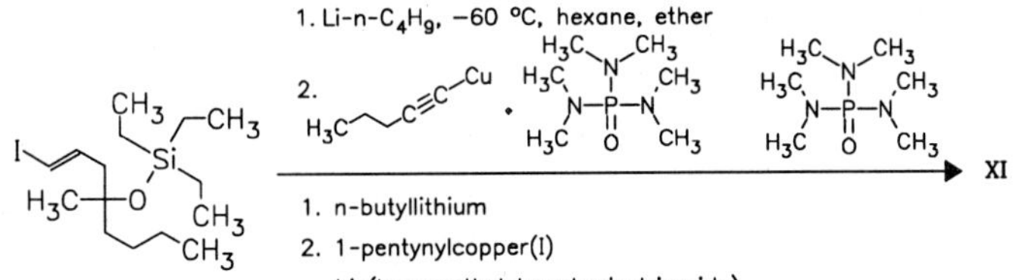

(±)-(E)-4-methyl-4-
triethylsilyloxy-1-
octenyl iodide (X)

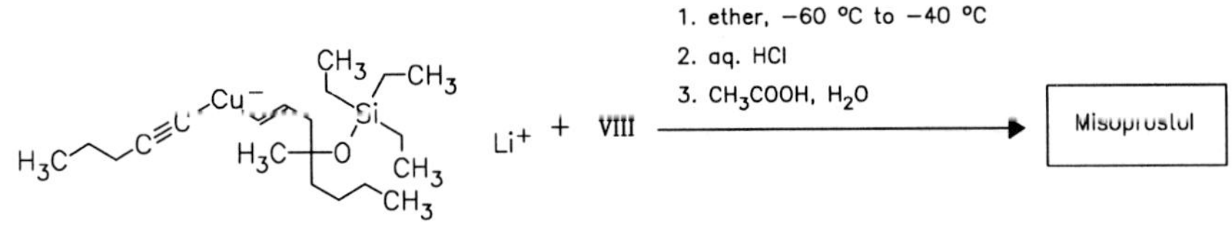

1. Li-n-C$_4$H$_9$, −60 °C, hexane, ether
2.

1. n-butyllithium
2. 1-pentynylcopper(I)
    bis(hexamethylphosphoric triamide)

$\longrightarrow$ XI

(XI)

Li$^+$ + VIII $\xrightarrow[\text{3. CH}_3\text{COOH, H}_2\text{O}]{\begin{array}{l}\text{1. ether, −60 °C to −40 °C}\\\text{2. aq. HCl}\end{array}}$ Misoprostol

Misoprostol

Reference(s):
Collins, P.W. et al.: J. Med. Chem. (JMCMAR) **20**, 1152 (1977).
DOS 2 513 212 (Searle; appl. 25.3.1975; USA-prior. 26.3.1974).
US 3 965 143 (Searle; 22.6.1976; appl. 26.3.1974).
US 4 060 691 (Searle; 29.11.1977; prior. 26.3.1974).
FR 2 274 289 (Searle; appl. 26.3.1975; USA-prior. 26.3.1974).
GB 1 492 426 (Searle; appl. 25.3.1975; USA-prior. 26.3.1974).

Formulation(s):   f. c. 0.2 mg (comb. with diclofenac sodium); tabl. 0.1 mg, 0.2 mg

Trade Name(s):

D:	Cytotec (Heumann; 1986)		Napratec (Searle)-comb.	J:	Cytotec (Nippon
F:	Artotec (Monsanto)-comb.	I:	Artotec (Monsanto)-comb.		Monsanto-Kaken)
	Cytotec (Monsanto; 1987)		Cytotec (Monsanto)	USA:	Cytotec (Searle)
GB:	Arthrotec (Searle)-comb.		Misofenac (Sefarm)-comb.		
	Cytotec (Searle)		Symbol (Sefarm)		

Mitobronitol

ATC: L01AX01
Use: antineoplastic

RN: 488-41-5   MF: $C_6H_{12}Br_2O_4$ MW: 307.97   EINECS: 207-676-8
LD_{50}:   2200 mg/kg (M, i.v.); 1380 mg/kg (M, p.o.);
       1370 mg/kg (R, i.v.); 1500 mg/kg (R, p.o.)
CN: 1,6-dibromo-1,6-dideoxy-D-mannitol

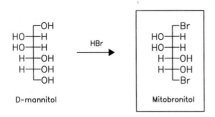

Reference(s):
GB 959 407 (Chinoin; appl. 15.9.1961; H-prior. 15.9.1960, 6.9.1961).

Formulation(s):   cps. 50 mg; tabl. 125 mg, 250 mg

Trade Name(s):

D:	Myelobromol (Hormon-	GB:	Myelobromol (Berk); wfm
	Chemie); wfm	J:	Myebrol (Kyorin)

Mitomycin

ATC: L01DC03
Use: antineoplastic

RN: 50-07-7 MF: C$_{15}$H$_{18}$N$_4$O$_5$ MW: 334.33 EINECS: 200-008-6
LD$_{50}$: 4 mg/kg (M, i.v.); 23 mg/kg (M, p.o.);
3 mg/kg (R, i.v.); 30 mg/kg (R, p.o.);
720 µg/kg (dog, i.v.)
CN: [1a*R*-(1aα,8β,8aα,8bα)]-6-amino-8-[[(aminocarbonyl)oxy]methyl]-1,1a,2,8,8a,8b-hexahydro-8a-methoxy-5-methylazirino[2',3':3,4]pyrrolo[1,2-*a*]indole-4,7-dione

Mitomycin

From culture of *Streptomyces caespitosus*; column chromatographic purification.

Reference(s):
GB 830 874 (Kyowa Hakko; appl. 8.4.1958; J-prior. 6.4.1957).
US 3 042 582 (Bristol-Myers; 3.7.1962; prior. 11.12.1958).

Formulation(s): vial 2 mg, 5 mg, 20 mg, 40 mg

Trade Name(s):
D: Mito-medac (medac)
Mitomycin (medac)

F: Amétycine (Sanofi Winthrop)
I: Mitomycin-C (Kyowa)

J: Mitomycin (Kyowa Hakko)
USA: Mutamycin (Bristol-Myers Squibb)

Mitopodozide

ATC: L01
Use: antineoplastic

RN: 1508-45-8 MF: C$_{24}$H$_{30}$N$_2$O$_8$ MW: 474.51 EINECS: 216-138-1
LD$_{50}$: 140 mg/kg (R, i.v.)
CN: [5*R*-(5α,6α,7β,8α)]-5,6,7,8-tetrahydro-8-hydroxy-7-(hydroxymethyl)-5-(3,4,5-trimethoxyphenyl)naphtho[2,3-*d*]-1,3-dioxole-6-carboxylic acid 2-ethylhydrazide

podophyllotoxin

H$_2$N—NH$_2$
hydrazine

1. H$_3$C—CHO
2. H$_2$, Raney-Ni

1. acetaldehyde

Mitopodozide

Reference(s):
US 3 054 802 (Sandoz; 18.9.1962; CH-prior. 7.10.1960).

hydrazinolysis of podophyllotoxin:
Rutschmann, J.; Renz, J.: Helv. Chim. Acta (HCACAV) **42**, 890 (1959).

Formulation(s): amp. 200 mg/ml, 1000 mg/5 ml

Trade Name(s):
D: Proresid (Sandoz); wfm

Mitotane

ATC: H02CA
Use: antineoplastic

RN: 53-19-0 MF: $C_{14}H_{10}Cl_4$ MW: 320.05 EINECS: 200-166-6
LD_{50}: >4 g/kg (M, p.o.);
>5 g/kg (R, p.o.)
CN: 1-chloro-2-[2,2-dichloro-1-(4-chlorophenyl)ethyl]benzene

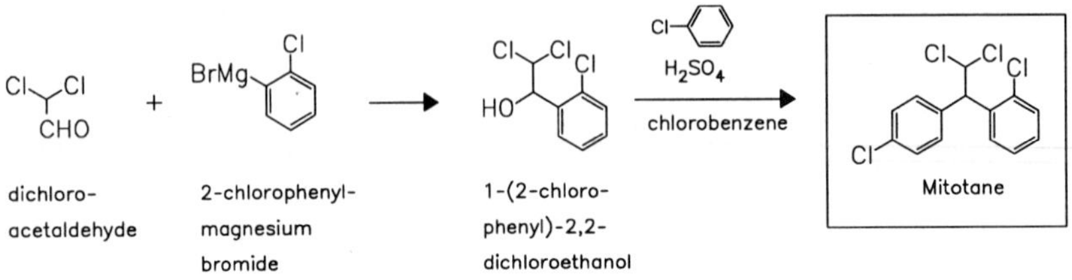

dichloro-
acetaldehyde

2-chlorophenyl-
magnesium
bromide

1-(2-chloro-
phenyl)-2,2-
dichloroethanol

Mitotane

Reference(s):
Hailer, B.L. et al.: J. Am. Chem. Soc. (JACSAT) **67**, 1591 (1945).

Formulation(s): tabl. 500 mg

Trade Name(s):
USA: Lysodren (Bristol-Myers
Squibb)

Mitoxantrone

ATC: L01DB07
Use: antineoplastic

RN: 65271-80-9 MF: $C_{22}H_{28}N_4O_6$ MW: 444.49
CN: 1,4-dihydroxy-5,8-bis[[2-[(2-hydroxyethyl)amino]ethyl]amino]-9,10-anthracenedione

dihydrochloride
RN: 70476-82-3 MF: $C_{22}H_{28}N_4O_6 \cdot 2HCl$ MW: 517.41 EINECS: 274-619-1
LD_{50}: 11300 µg/kg (M, i.v.); 502 mg/kg (M, p.o.);
4800 µg/kg (R, i.v.); 682 mg/kg (R, p.o.);
375 µg/kg (dog, i.v.)

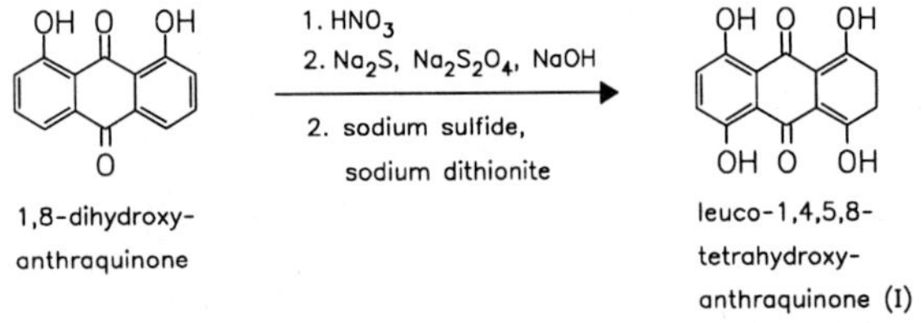

1,8-dihydroxy-
anthraquinone

leuco-1,4,5,8-
tetrahydroxy-
anthraquinone (I)

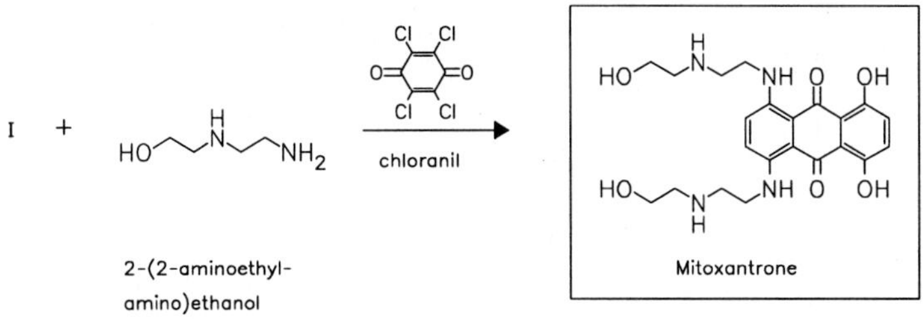

2-(2-aminoethyl-
amino)ethanol

Mitoxantrone

Reference(s):

DE 2 835 661 (American Cyanamid; prior. 14.8.1978).
US 4 197 249 (American Cyanamid; 8.4.1980; prior. 15.8.1977).
US 4 278 689 (American Cyanamid; 14.7.1981; prior. 11.7.1978).
Zee-Cheng, R.K.Y.; Cheng, C.C.: J. Med. Chem. (JMCMAR) **21**, 291 (1978).
Murdock, K.C. et al.: J. Med. Chem. (JMCMAR) **22**, 1024 (1979).

synthesis of 1,4,5,8-tetrahydroxyanthraquinone:
SU 230 188 (I. D. Belkin et al.; appl. 3.8.1967).
SU 266 777 (I. D. Belkin et al.; appl. 5.5.1968).

Formulation(s):   amp. 2 mg/ml, 10 mg/5 ml, 20 mg/10 ml, 25 mg/12.5 ml, 30 mg/15 ml (as dihydrochloride)

Trade Name(s):

D:	Novantron (Lederle; 1985)	F:	Novantrone (Wyeth-Lederle; 1986)	I:	Novantrone (Wyeth-Lederle; 1987)
	Onkotrone (ASTA Medica AWD)	GB:	Novantrone (Wyeth; 1984)	J:	Novantron (Lederle; 1987)
				USA:	Novantrone (Immunex)

Mizolastine

(SL-85.0324; MKC-431)

ATC: R06AX25
Use: antihistamine, histamine H_1-receptor antagonist

RN: 108612-45-9   MF: $C_{24}H_{25}FN_6O$ MW: 432.50
CN: 2-[[1-[1-[(4-Fluorophenyl)methyl]-1*H*-benzimidazol-2-yl]-4-piperidinyl]methylamino]-4(3*H*)-pyrimidinone

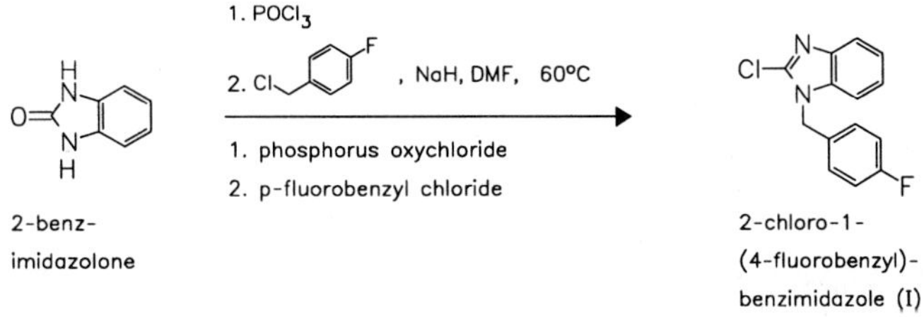

2-benz-
imidazolone

1. phosphorus oxychloride
2. p-fluorobenzyl chloride

2-chloro-1-
(4-fluorobenzyl)-
benzimidazole (I)

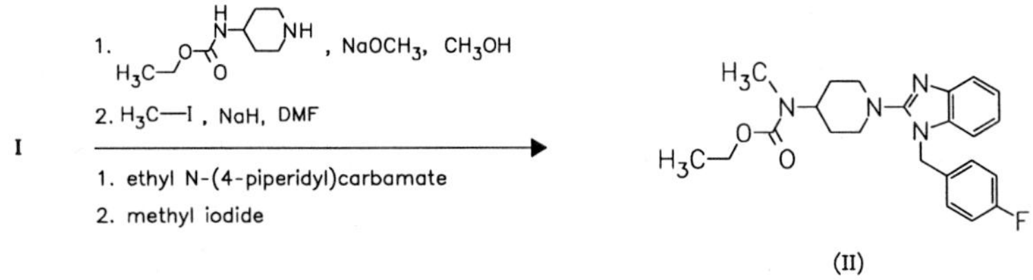

1. ethyl N-(4-piperidyl)carbamate
2. methyl iodide

(II)

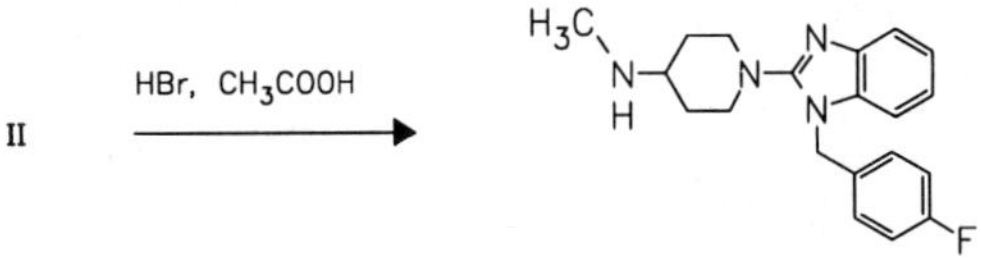

II → N-[1-[1-(4-fluorobenzyl)-
benzimidazol-2-yl]-
piperidin-4-yl]-N-methylamine (III)

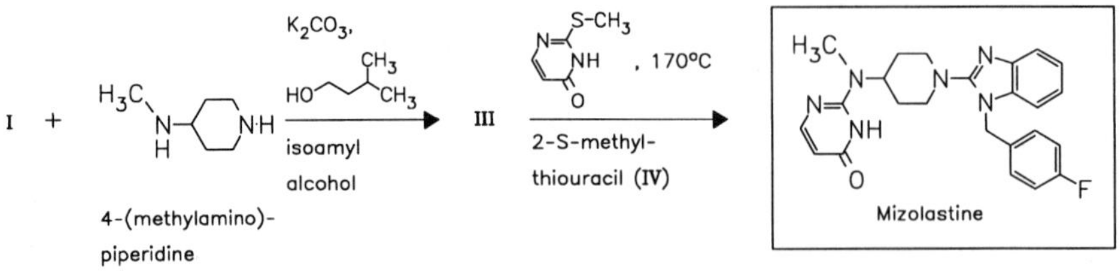

I + 4-(methylamino)-piperidine → III → Mizolastine

K₂CO₃, isoamyl alcohol

2-S-methyl-thiouracil (IV), 170°C

alternative way:

ethyl 4-(methyl-amino)piperidine-1-carboxylate

1. IV, xylene
2. HBr, CH₃COOH
1. 2-S-methyl-thiouracil
2. hydrobromic acid

→ (V)

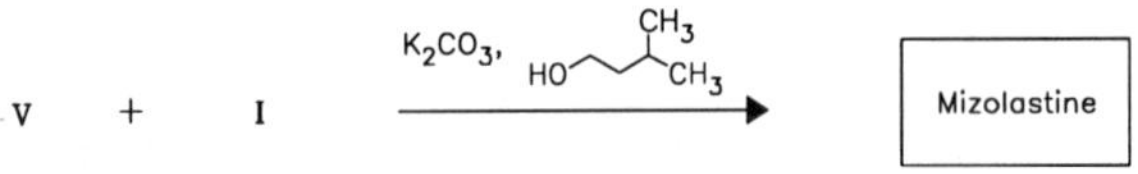

V + I → Mizolastine

K₂CO₃, HO–CH(CH₃)CH₃

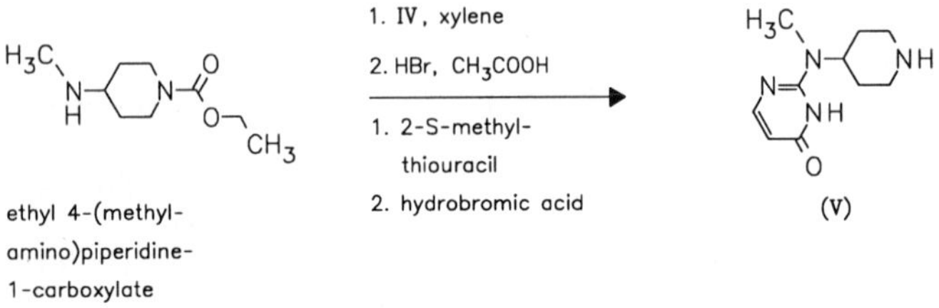

Reference(s):
synthesis of 2-chloro-1-(4-fluorobenzyl)benzimidazole:
Parrodi, C.; Quintero-Cortes, L.; Sandoval-Ramirez: Synth. Commun. (SYNCAV) **26**, 17 (1996).

synthesis:
EP 217 700 (Synthélabo; appl. 2.9.1986; F-prior. 11.9.1985).

formulation:
WO 9 732 584 (Synthélabo; appl. 28.2.1997; F-prior. 4.3.1996).

Formulation(s):   f. c. tabl. 10 mg; tabl. 10 mg

Trade Name(s):

D:	Mizollen (Synthélabo; 1998)	Zollistam (Vita)
	Zolim (Schwarz Pharma)	
GB:	Mizollen (Lorex Synthélabo)	
I:	Mizollen (Synthelabo)	

Mizoribine

ATC: L04
Use: immunosuppressive

RN: 50924-49-7   MF: C₉H₁₃N₃O₆   MW: 259.22
LD₅₀: 500 mg/kg (M, i.v.); >4.883 g/kg (M, p.o.);
1500 mg/kg (R, i.v.); 2.847 g/kg (R, p.o.)
CN: 5-hydroxy-1-β-D-ribofuranosyl-1H-imidazole-4-carboxamide

Isolation from cultures of *Eupenicillium brefeldianum* NRRL 5734.
a Amberlite IRA-411/pH 10.
b Chromatography on DEAE-Sephadex A-25.

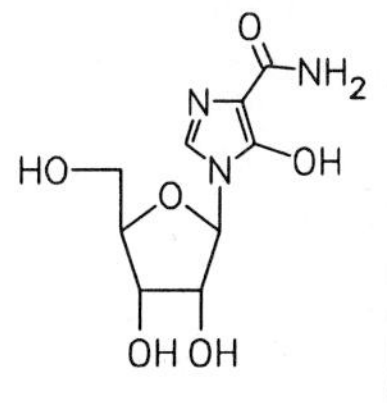

Mizoribine

Reference(s):
BE 799 805 (Toyo Jozo; appl. 31.11.1973; J-prior. 21.5.1973).
DOS 2 326 916 (Toyo Jozo; appl. 23.5.1973; J-prior. 21.5.1973).
Mizuno, K. et al.: J. Antibiot. (JANTAJ) **27**, 775 (1974).

controlled-release formulation:
JP 59 227 817 (Toyo Jozo; appl. 7.6.1983).

Formulation(s):   f. c. tabl. 10 mg; tabl. 10 mg

Trade Name(s):
J: Bredinin (Toyo Jozo)

Moclobemide
(Ro-11-1163)

ATC: N06AG02
Use: antidepressant, reversible
        nonhydrazide MAO-A-inhibitor,
        antiparkinsonian

RN: 71320-77-9   MF: $C_{13}H_{17}ClN_2O_2$ MW: 268.74
LD_{50}:   707 mg/kg (R, p.o.)
CN: 4-chloro-*N*-[2-(4-morpholinyl)ethyl]benzamide

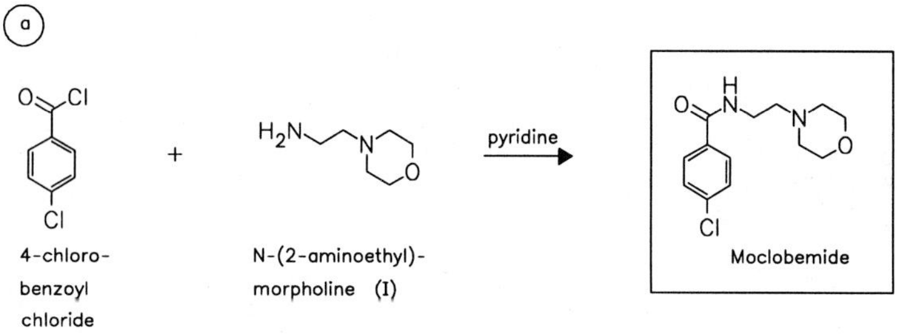

4-chloro-
benzoyl
chloride

N-(2-aminoethyl)-
morpholine (I)

pyridine

Moclobemide

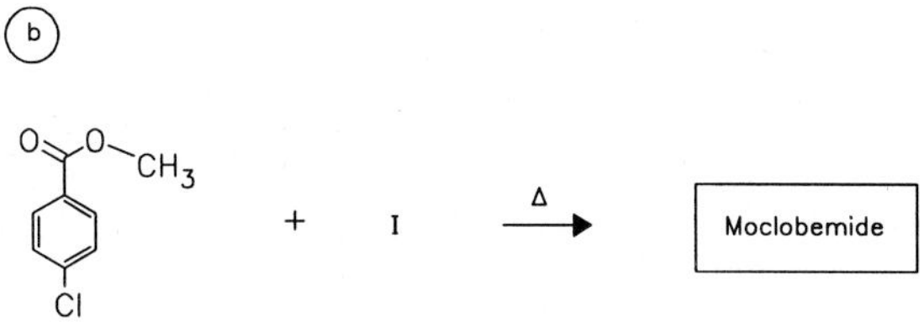

Moclobemide

Reference(s):
DE 2 706 179 (Hoffmann-La Roche; appl. 14.2.1977; A-prior. 16.2.1976).
GB 1 512 194 (Hoffmann-La Roche; appl. 24.5.1978; A-prior. 16.2.1976, 2.9.1976, 20.4.1977, 17.2.1976, 4.6.1976).

medical use for treatment of cognitive disorders:
US 4 906 626 (Hoffmann-La Roche; 6.3.1990; appl. 13.1.1989; CH-prior. 8.12.1988).

Formulation(s): f. c. tabl. 150 mg, 300 mg; tabl. 100 mg, 150 mg, 300 mg

Trade Name(s):

D:	Aurorix (Roche)	F:	Moclamine (Produits Roche)	GB:	Manerix (Roche)
				I:	Aurorix (Roche)

Modafinil
(CRL-40476)

ATC: N06BA07
Use: psychostimulant, α_1-adrenoceptor agonist (treatment of narcolepsy and idiopathic hypersomnia)

RN: 68693-11-8 MF: $C_{15}H_{15}NO_2S$ MW: 273.36
CN: 2-[(diphenylmethyl)sulfinyl]acetamide

racemate
RN: 112111-49-6 MF: $C_{15}H_{15}NO_2S$ MW: 273.36
(+)-form
RN: 112111-47-4 MF: $C_{15}H_{15}NO_2S$ MW: 273.36
(–)-form
RN: 112111-43-0 MF: $C_{15}H_{15}NO_2S$ MW: 273.36

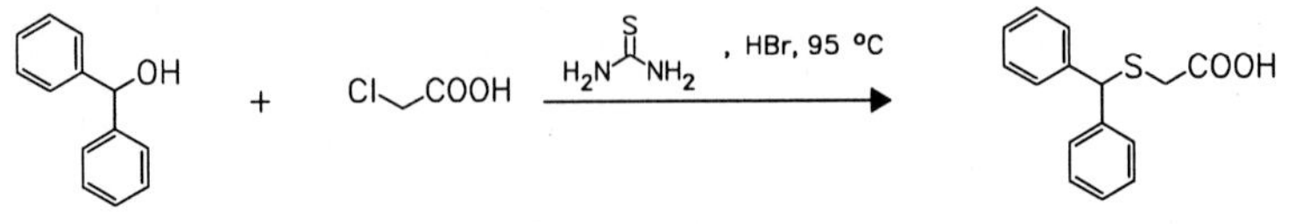
OH
+
Cl COOH
S
H2N NH2
, HBr, 95 °C
S COOH

benzhydrol

2-(diphenylmethylthio)acetic acid (I)

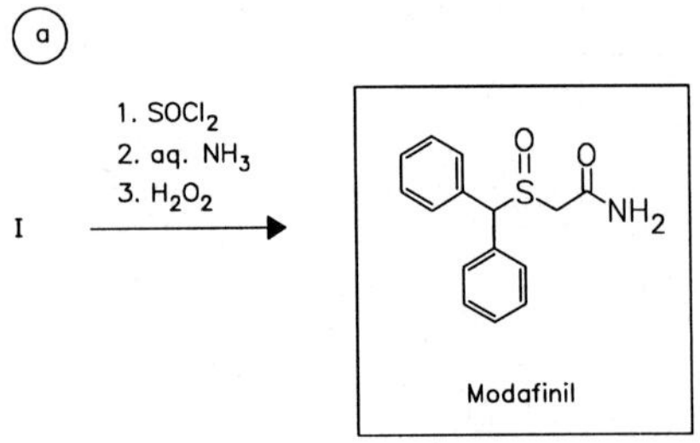
a
1. SOCl2
2. aq. NH3
3. H2O2
I
O
S
O
NH2
Modafinil

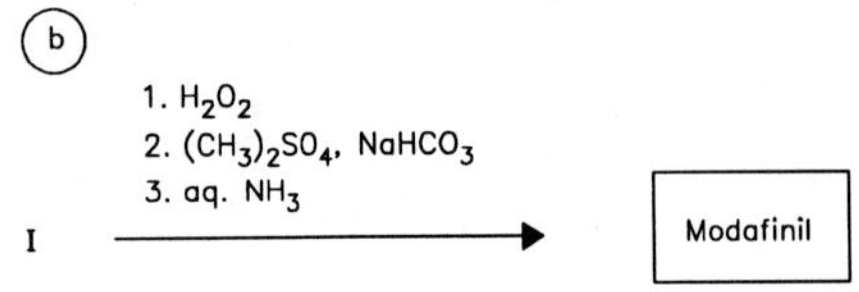
b
1. H2O2
2. (CH3)2SO4, NaHCO3
3. aq. NH3
I
Modafinil

Reference(s):
DE 2 809 625 (Laboratoire L. Lafon; appl. 5.10.1978; GB-prior. 31.3.1977).
EP 233 106 (Laboratoire L. Lafon; appl. 19.8.1987; F-prior. 31.1.1986).

use of modafinil *as neuroprotective agent:*
EP 462 004 (Laboratoire L. Lafon; appl. 18.12.1991; F-prior. 14.6.1990).

use for treatment of urinary and fecal incontinence:
EP 594 507 (Laboratoire L. Lafon; appl. 27.4.1994; F-prior. 23.1.1992).

use as a brain anti-ischemia:
EP 547 952 (Laboratoire L. Lafon; appl. 23.6.1993; F-prior. 13.12.1991).

use for treatment of sleep apnea and ventilation problems of central origin:
WO 9 500 132 (Laboratoire L. Lafon; appl. 5.1.1995; F-prior. 22.6.1993).

use for modifying feeding behavior:
WO 9 501 171 (Laboratoire L. Lafon; appl. 12.1.1995; 30.6.1993).

Formulation(s):   tabl. 100 mg

Trade Name(s):
F: Modiodal (Lafon)

Moexipril

(CI-925; RS-10085 (base); RS-10085-197; SPM-925)

ATC: C09AA13; C09AB13
Use: antihypertensive (ACE inhibitor)

RN: 103775-10-6   MF: $C_{27}H_{34}N_2O_7$ MW: 498.58
CN: [3*S*-[2[*R**(*R**)],3*R**]]-2-[2-[[1-(ethoxycarbonyl)-3-phenylpropyl]amino]-1-oxopropyl]-1,2,3,4-tetrahydro-6,7-dimethoxy-3-isoquinolinecarboxylic acid

monohydrochloride
RN: 82586-52-5   MF: $C_{27}H_{34}N_2O_7 \cdot HCl$ MW: 535.04

L-alanine
tert-butyl ester

ethyl 2-bromo-
4-phenylbutanoate

N-[1(S)-ethoxycarbonyl-
3-phenylpropyl]-L-alanine
tert-butyl ester (I)

benzyl 1,2,3,4-tetrahydro-
6,7-dimethoxy-3(S)-iso-
quinolinecarboxylate (II)

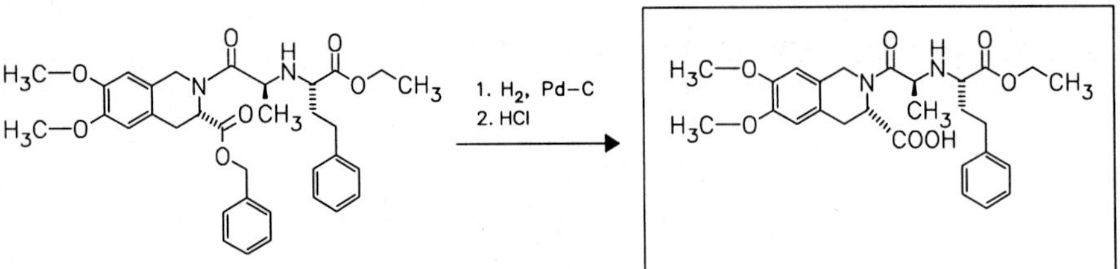

benzyl 2-[N-[1(S)-ethoxycarbonyl-
3-phenylpropyl]-L-alanyl]-1,2,3,4-
tetrahydro-6,7-dimethoxy-3(S)-
isoquinolinecarboxylate (III)

Moexipril

synthesis of the starting material II:

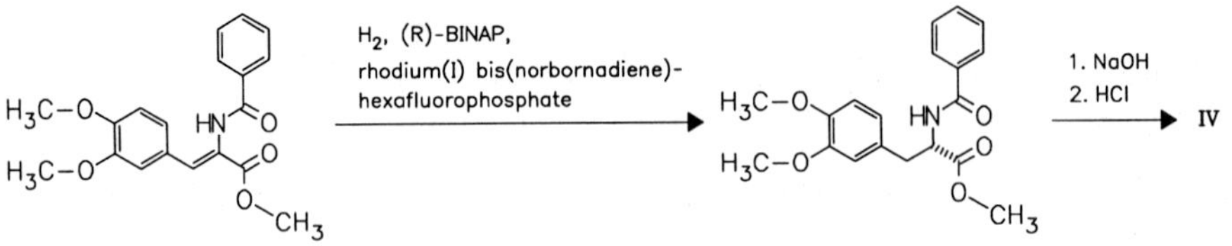

methyl N-benzoyl-
3,4-dimethoxy-L-
phenylalanine

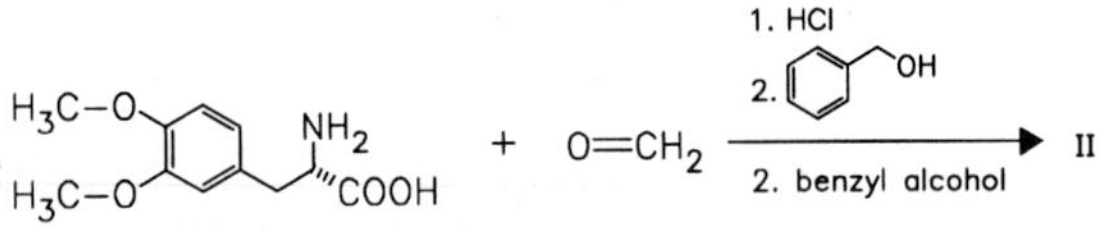

(S)-2-amino-3-(3,4-
dimethoxyphenyl)-
propanoic acid (IV)

Reference(s):

EP 96 157 (Warner-Lambert; appl. 1.10.1981; USA-prior. 3.10.1980, 20.2.1981).

EP 49 605 (Warner-Lambert; appl. 1.10.1981; USA-prior. 3.10.1980).

O'Reilly, N.J. et al.: Synthesis (SYNTBF) **7**, 550-556 (1990).

US 4 912 221 (Occidental Chemical Corp., appl. 27.10.1988).

formulation stabilized with ascorbic acid:

EP 264 887 (Warner-Lambert Co.; appl. 19.10.1987; USA-prior. 20.10.1986).

Formulation(s): f. c. tabl. 7.5 mg, 15 mg; USA: f. c. tabl. 7.5 mg, 12.5 mg, 15 mg, 25 mg (in comb. with hydrochlorothiazide) (as hydrochloride)

Trade Name(s):

D: Fempres (Isis Pharma)
GB: Perdix (Schwarz)

I: Femipres (Schwarz)
Primoxil (Bayer Italia)

USA: Uniretic (Schwarz)
Univasc (Schwarz)

Mofebutazone

ATC: M01AA02; M02AA02
Use: antirheumatic, anti-inflammatory, analgesic

RN: 2210-63-1 MF: $C_{13}H_{16}N_2O_2$ MW: 232.28 EINECS: 218-641-1

LD_{50}: 600 mg/kg (M, i.v.);
1750 mg/kg (R, p.o.)

CN: 4-butyl-1-phenyl-3,5-pyrazolidinedione

sodium salt

RN: 41468-34-2 MF: $C_{13}H_{15}N_2NaO_2$ MW: 254.27

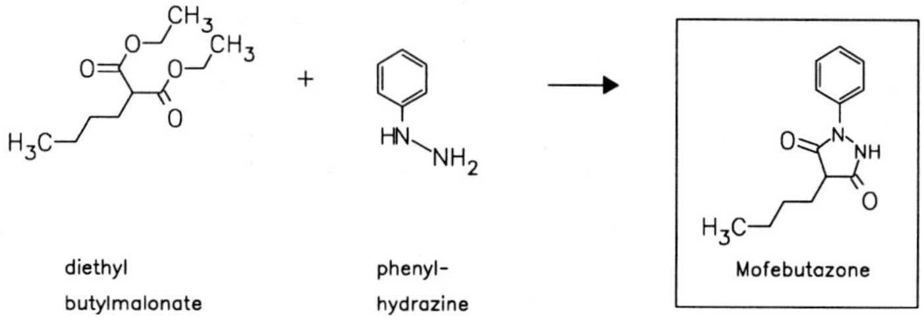

diethyl
butylmalonate

+

phenyl-
hydrazine

→

Mofebutazone

Reference(s):

GB 839 057 (Comm. Farmaceutica Milanese; appl. 27.11.1957; I-prior. 28.11.1956).

Büchi, J. et al.: Helv. Chim. Acta (HCACAV) **36**, 75 (1953).

Formulation(s): amp. 650 mg/3 ml (as sodium salt); cps. 200 mg; drg. 150 mg-comb.; f. c. tabl. 300 mg

Trade Name(s):

D:	Diadin (Diadin)	F:	Arcobutina (Silbert et		Monbutina (Lafare); wfm
	Mofesal (Medice)		Ripert)-comb.; wfm		Reumatox (Medosan); wfm
	Vasotonin (Merz)-comb.	I:	Chemiartrol (Gazzoni);		
			wfm		

Mofezolac

(N-22)

ATC: M01; N02
Use: anti-inflammatory, analgesic

RN: 78967-07-4 MF: $C_{19}H_{17}NO_5$ MW: 339.35
LD_{50}: 1528 mg/kg (M, p.o.);
887 mg/kg (R, p.o.);
800 mg/kg (dog, p.o.)
CN: 3,4-bis(4-methoxyphenyl)-5-isoxazoleacetic acid

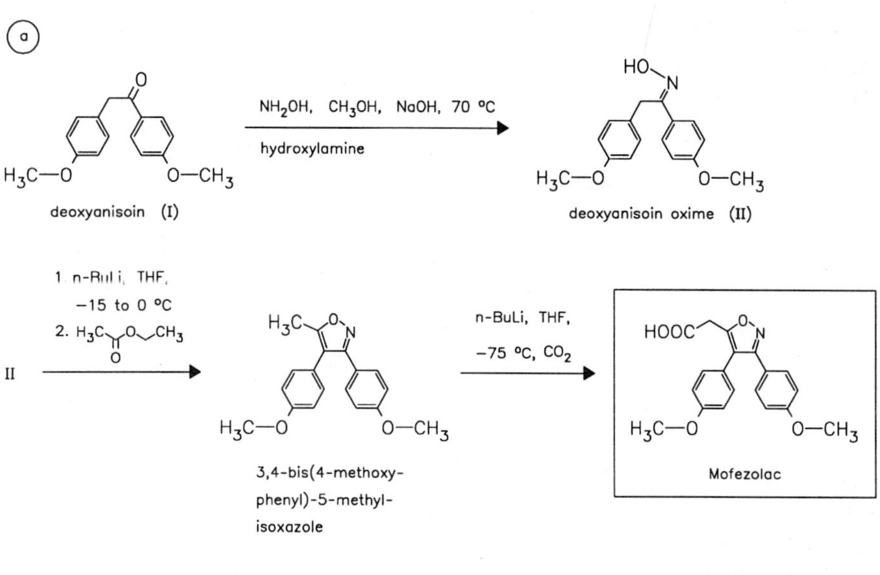

deoxyanisoin (I)

NH_2OH, CH_3OH, NaOH, 70 °C
hydroxylamine

deoxyanisoin oxime (II)

1. n-BuLi, THF,
   −15 to 0 °C
2. H_3C—O—CH_3

II →

3,4-bis(4-methoxy-
phenyl)-5-methyl-
isoxazole

n-BuLi, THF,
−75 °C, CO_2

Mofezolac

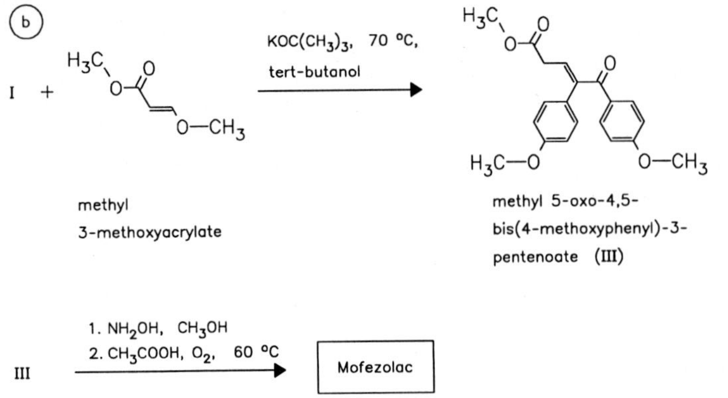

(b)

I + methyl 3-methoxyacrylate

$\xrightarrow{\text{KOC(CH}_3)_3, \ 70 \ °C, \ \text{tert-butanol}}$

methyl 5-oxo-4,5-bis(4-methoxyphenyl)-3-pentenoate (III)

III $\xrightarrow[\text{2. CH}_3\text{COOH, O}_2, \ 60 \ °C]{\text{1. NH}_2\text{OH, CH}_3\text{OH}}$ Mofezolac

Reference(s):
EP 26 928 (CDC Life Sciences Inc.; appl. 3.10.1980; CA-prior. 5.10.1979).

synthesis with ClCO$_2$Et *instead* CO$_2$:
JP 02 223 568 (Taiho Pharmaceuticals; appl. 20.11.1989; J-prior. 2.11.1988).
EP 454 871 (Taiho Pharmaceuticals; appl. 19.11.1990; J-prior. 21.11.1989).
JP 03 220 180 (Taiho Pharmaceuticals; appl. 24.1.1990; J-prior. 24.1.1990).
EP 464 218 (Taiho Pharmaceuticals; appl. 22.1.1991; J-prior. 24.1.1990).

transdermal formulations:
JP 05 017 354 (Nichiban KK; appl. 3.7.1991; J-prior. 3.7.1991).

stable injection solution:
AT 391 415 (Kwizda; appl. 26.7.1989; A-prior. 26.7.1989).

Trade Name(s):
J: Disopain (Taiho-Yoshitomi)

Molindone

ATC: N05AE02
Use: tranquilizer, sedative

RN: 7416-34-4   MF: C$_{16}$H$_{24}$N$_2$O$_2$ MW: 276.38
CN: 3-ethyl-1,5,6,7-tetrahydro-2-methyl-5-(4-morpholinylmethyl)-4H-indol-4-one

monohydrochloride
RN: 15622-65-8   MF: C$_{16}$H$_{24}$N$_2$O$_2$ · HCl   MW: 312.84
LD$_{50}$: 670 mg/kg (M, p.o.);
   261 mg/kg (R, p.o.)

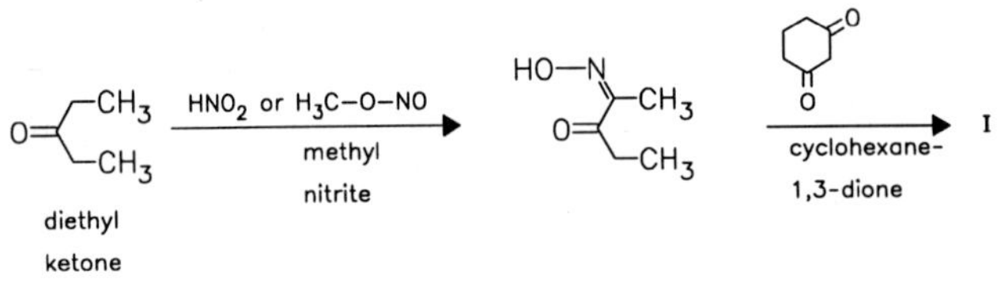

diethyl ketone $\xrightarrow[\text{methyl nitrite}]{\text{HNO}_2 \text{ or } \text{H}_3\text{C}-\text{O}-\text{NO}}$ $\xrightarrow{\text{cyclohexane-1,3-dione}}$ I

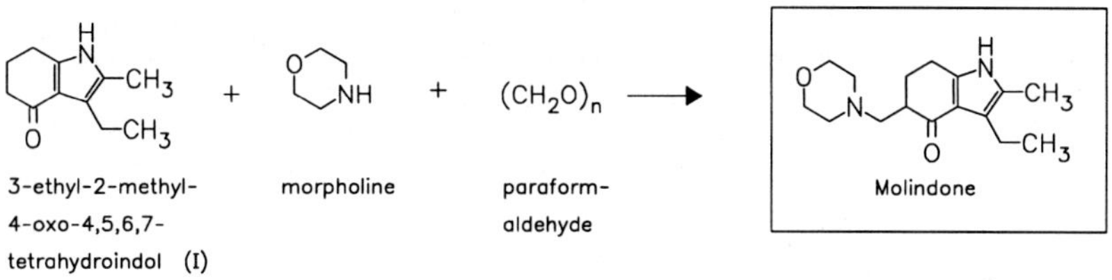
3-ethyl-2-methyl-
4-oxo-4,5,6,7-
tetrahydroindol (I)
morpholine
paraform-
aldehyde
Molindone

Reference(s):
DAS 1 545 774 (Endo Labs.; appl. 16.10.1965).
US 3 491 093 (Endo Labs.; 20.1.1970; prior. 2.3.1964, 3.4.1964, 11.5.1966, 29.11.1967).

combination with amantadine *(antidepressant):*
US 4 148 896 (Du Pont; 10.4.1979; appl. 22.2.1978).

Formulation(s): sol. 20 mg/ml; tabl. 5 mg, 10 mg, 25 mg, 50 mg, 100 mg

Trade Name(s):
USA: Moban (Gate)

Molsidomine

ATC: C01DX12
Use: coronary vasodilator

RN: 25717-80-0 MF: $C_9H_{14}N_4O_4$ MW: 242.24 EINECS: 247-207-4
LD_{50}: 800 mg/kg (M, i.v.); 830 mg/kg (M, p.o.);
760 mg/kg (R, i.v.); 1050 mg/kg (R, p.o.)
CN: 5-[(ethoxycarbonyl)amino]-3-(4-morpholinyl)-1,2,3-oxadiazolium inner salt

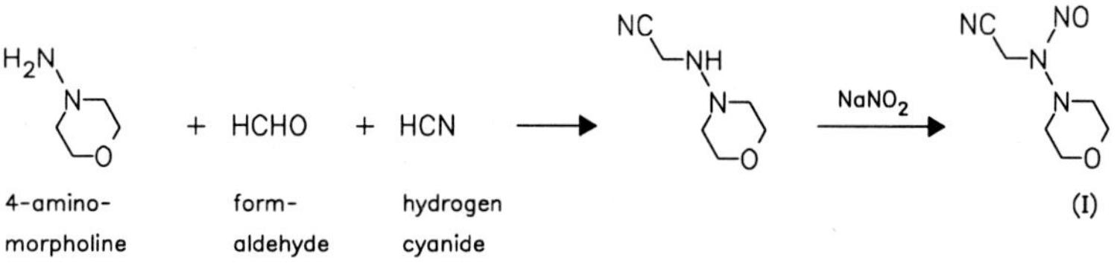
4-amino-
morpholine
form-
aldehyde
hydrogen
cyanide
NaNO2
(I)

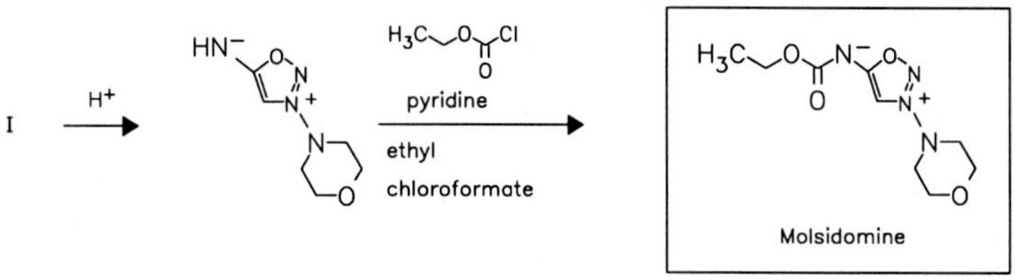
I
H+
pyridine
ethyl
chloroformate
Molsidomine

Reference(s):
DAS 1 695 897 (Takeda; appl. 1.7.1967; J-prior. 4.7.1966).

synthesis of 4-aminomorpholine:
DAS 2 532 124 (Cassella; appl. 18.7.1975).
US 3 769 283 (Takeda, 30.10.1973; J-prior. 4.7.1966).

Formulation(s): amp. 2 mg; s. r. tabl. 8 mg; tabl. 1 mg, 2 mg, 4 mg

Mometasone furoate

ATC: D07AB
Use: topical glucocorticoid, anti-
       inflammatory

RN: 83919-23-7   MF: $C_{27}H_{30}Cl_2O_6$ MW: 521.44
LD_{50}: 300 mg/kg (R, s.c.)
CN: (11β,16α)-9,21-dichloro-17-[(2-furanylcarbonyl)oxy]-11-hydroxy-16-methylpregna-1,4-diene-3,20-dione

mometasone
RN: 105102-22-5   MF: $C_{22}H_{28}Cl_2O_4$ MW: 427.37

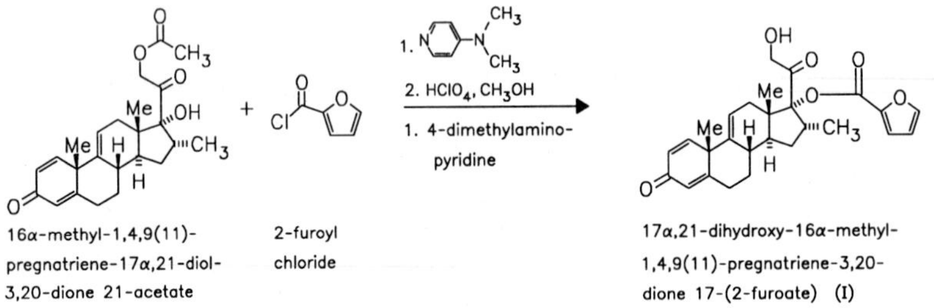

16α-methyl-1,4,9(11)-
pregnatriene-17α,21-diol-
3,20-dione 21-acetate

2-furoyl
chloride

17α,21-dihydroxy-16α-methyl-
1,4,9(11)-pregnatriene-3,20-
dione 17-(2-furoate) (I)

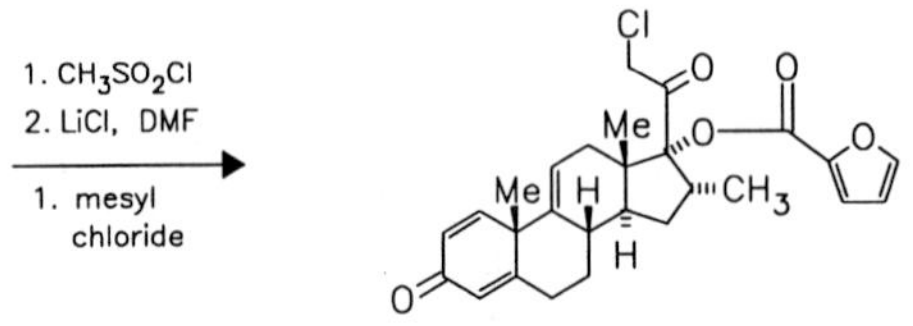

(II)

Mometasone furoate

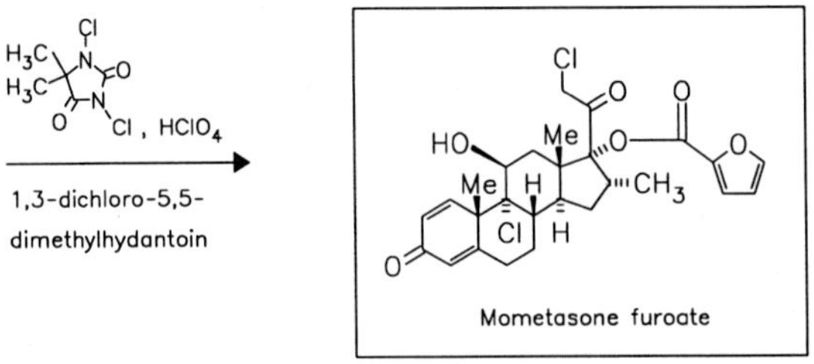

Reference(s):
EP 57 401 (Schering Corp.; appl. 25.1.1982; USA-prior. 2.2.1981).
US 4 472 393 (Schering Corp.; 18.9.1984; appl. 29.7.1982; prior. 2.2.1981).
Shapiro, E.L. et al.: J. Med. Chem. (JMCMAR) **30**, 1581 (1987).

cream:
EP 262 681 (Schering Corp.; appl. 1.10.1987; USA-prior. 2.10.1986).

lotion:
US 4 775 529 (Schering Corp.; 4.10.1988; appl. 21.5.1987).

Formulation(s): cream 0.1 %; lotion 0.1 %; ointment 0.1 % (1 mg/g)

Trade Name(s):

D:	Ecural (Essex Pharma)	I:	Altosone (Essex Italia)	USA:	Elocon (Schering-Plough;
GB:	Elocon (Schering-Plough)		Elocon (Schering-Plough)		1988)
	Nasorex (Schering-	J:	Ecotone (Schering-Plough)		
	Plough)-comb.		Flumeta (Shionogi)		

Monobenzone

Use: depigmentant (melanin inhibitor against hyperpigmentation of skin)

RN: 103-16-2 MF: $C_{13}H_{12}O_2$ MW: 200.24 EINECS: 203-083-3
LD_{50}: >600 mg/kg (M, i.p.);
4500 mg/kg (R, i.p.)
CN: 4-(phenylmethoxy)phenol

hydroquinone + benzyl bromide → KOH, C_2H_5OH → Monobenzone

Reference(s):
Schiff, H.; Pellizzari, G.: Justus Liebigs Ann. Chem. (JLACBF) **221**, 365 (1883).

Formulation(s): cream 20 %

Trade Name(s):

D:	Depigman (Hermal); wfm	I:	Dermochinona (Chinoin); wfm	USA:	Benoquin (Elder)

Montelukast sodium
(MK-476; MK-0476; L-706631)

ATC: R03DC03
Use: antiallergic, antiasthmatic, leukotriene antagonist

RN: 151767-02-1 MF: $C_{35}H_{35}ClNNaO_3S$ MW: 608.18
CN: 1-[[[(1R)-1-[3-[(1E)-2-(7-Chloro-2-quinolinyl)ethenyl]phenyl]-3-[2-(1-hydroxy-1-methylethyl)phenyl]propyl]thio]methyl]cyclopropaneacetic acid sodium salt

acid
RN: 158966-92-8 MF: $C_{35}H_{36}ClNO_3S$ MW: 586.20

ⓐ

7-chloroquinaldine + isophthalaldehyde → H_3C—acetic anhydride → 3-[2(E)-(7-chloroquinolin-2-yl)vinyl]benzaldehyde (I)

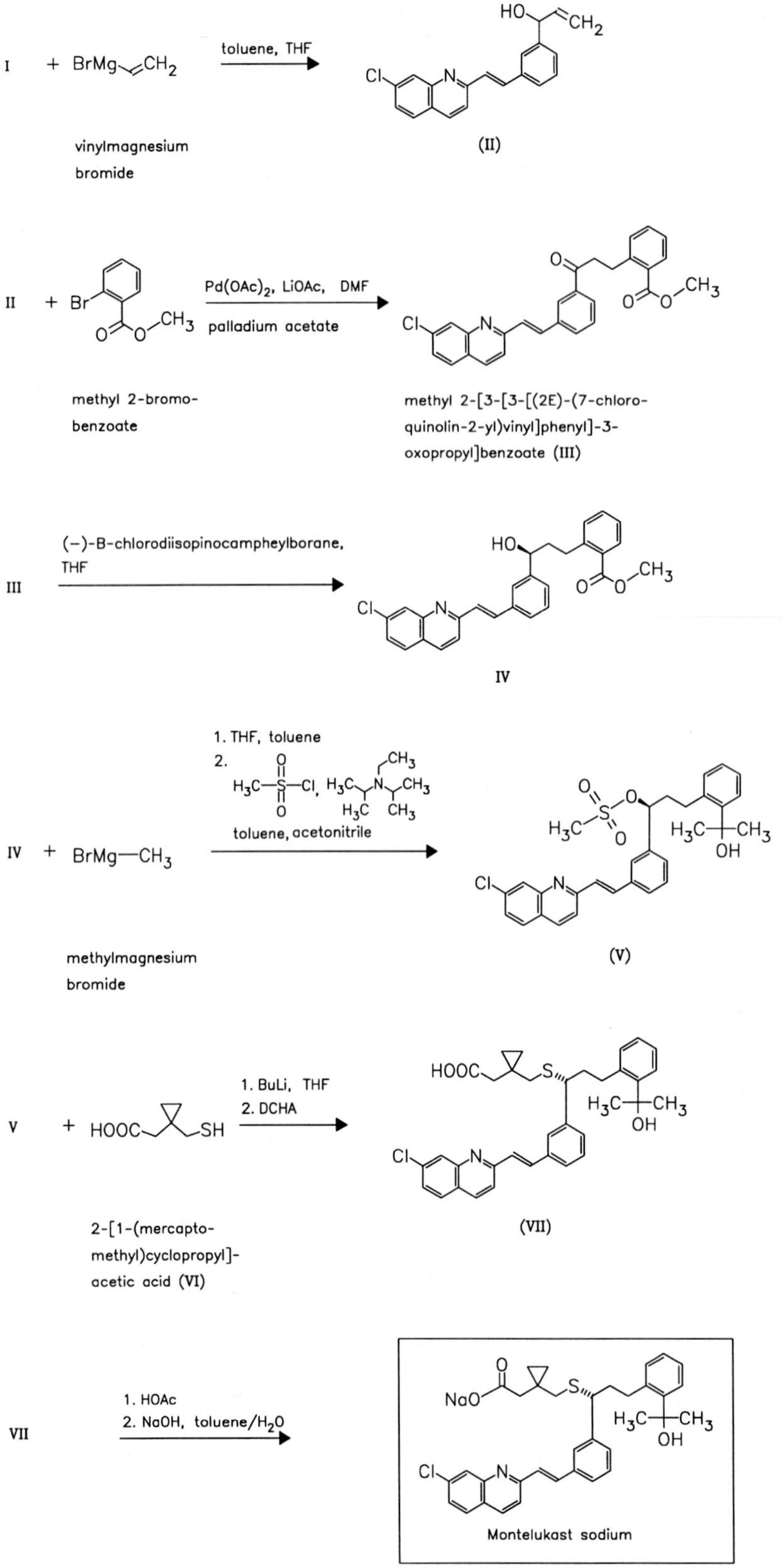

I   + BrMg—CH₂   →   toluene, THF

vinylmagnesium
bromide

(II)

II   + methyl 2-bromo-
benzoate   →   Pd(OAc)₂, LiOAc, DMF
palladium acetate

methyl 2-[3-[3-[(2E)-(7-chloro-
quinolin-2-yl)vinyl]phenyl]-3-
oxopropyl]benzoate (III)

III   →   (−)-B-chlorodiisopinocampheylborane,
THF

IV

IV   + BrMg—CH₃   →   1. THF, toluene
2. H₃C—S—Cl, diisopropylethylamine
toluene, acetonitrile

methylmagnesium
bromide

(V)

V   + 2-[1-(mercapto-
methyl)cyclopropyl]-
acetic acid (VI)   →   1. BuLi, THF
2. DCHA

(VII)

VII   →   1. HOAc
2. NaOH, toluene/H₂O

Montelukast sodium

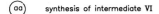

aa synthesis of intermediate VI

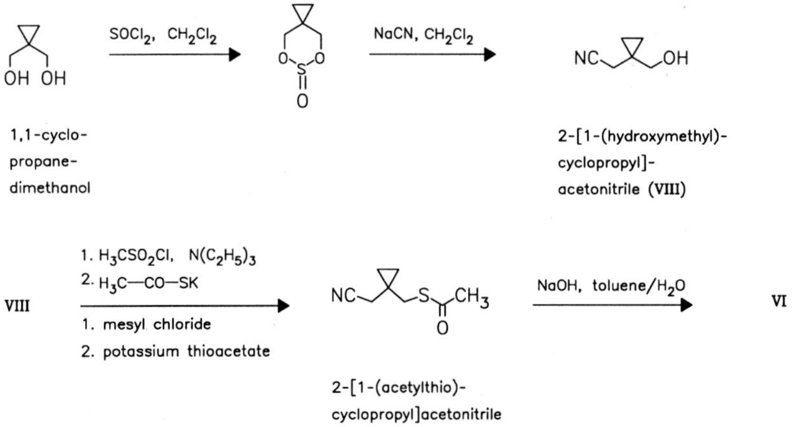

1,1-cyclo-
propane-
dimethanol

SOCl₂, CH₂Cl₂ →

NaCN, CH₂Cl₂ →

2-[1-(hydroxymethyl)-
cyclopropyl]-
acetonitrile (VIII)

VIII

1. H₃CSO₂Cl, N(C₂H₅)₃
2. H₃C—CO—SK
1. mesyl chloride
2. potassium thioacetate

→

2-[1-(acetylthio)-
cyclopropyl]acetonitrile

NaOH, toluene/H₂O →

VI

---

b

H₃C—O—C(O)—[cyclopropane]—C(O)—O—CH₃

diethyl cyclo-
propane-1,1-di-
carboxylate

LiAlH₄, THF →

HO—[cyclopropane]—OH

Cl—C(O)—C₆H₅,
pyridine, CH₂Cl₂
benzoyl chloride

→ IX

(IX)

1. H₃CSO₂Cl
2. NaCN, DMSO

→

2-[1-(benzoyloxy-
methyl)cyclopropyl]-
acetonitrile (X)

X

1. KOH, ethanol
2. CH₂N₂
3. H₃C—S(O₂)—Cl

→

H₃C—O—C(O)—CH₂—[cyclopropane]—CH₂—O—S(O₂)—CH₃

H₃C—C(O)—SCs'
CH₂Cl₂
cesium
thioacetate

→ XI

H₃C—O—C(O)—CH₂—[cyclopropane]—CH₂—S—C(O)—CH₃ +

methyl 2-[1-(acetyl-
thiomethyl)cyclopropyl]-
acetate (XI)

XII

Cs₂CO₃, acetonitrile → XIII

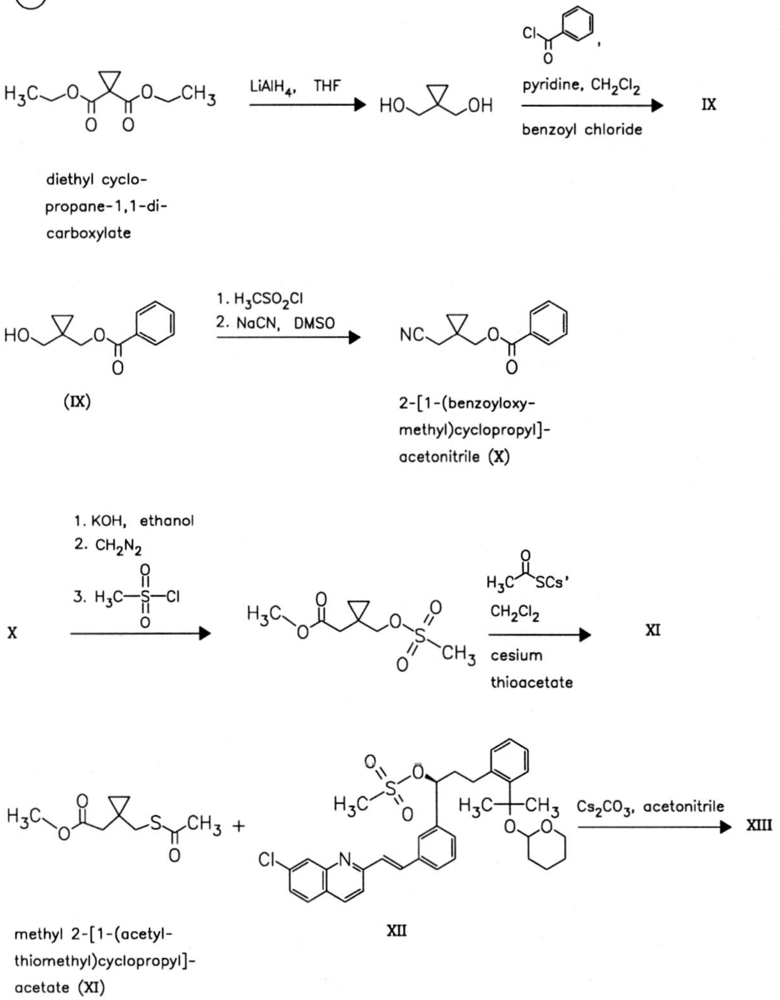

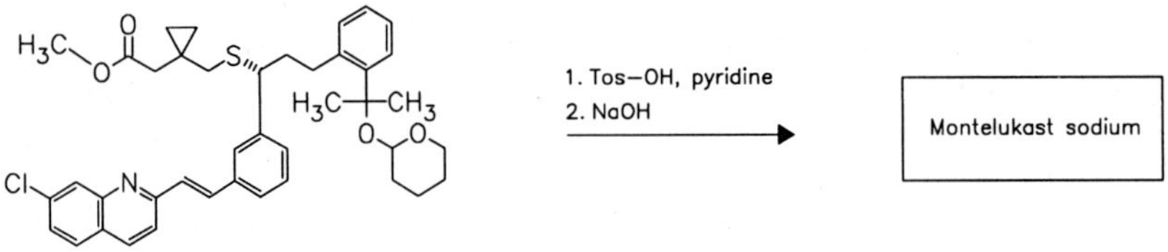

XIII

1. Tos—OH, pyridine
2. NaOH

Montelukast sodium

(ba) intermediate XII can be sythesized from

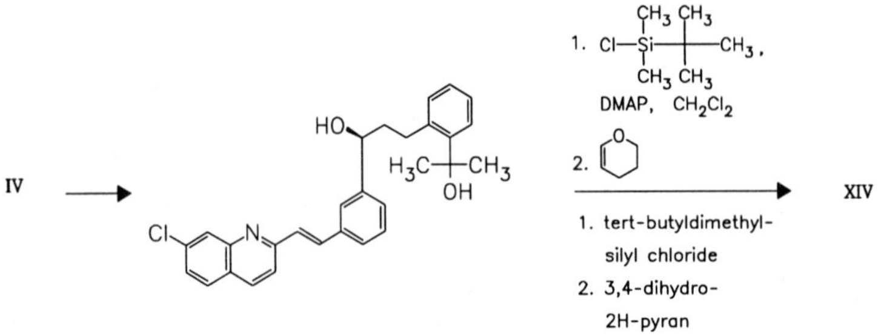

IV →

1. Cl—Si(CH₃)₂—C(CH₃)₃,
DMAP, CH₂Cl₂

2. (tetrahydropyran)

1. tert-butyldimethyl-
silyl chloride

2. 3,4-dihydro-
2H-pyran

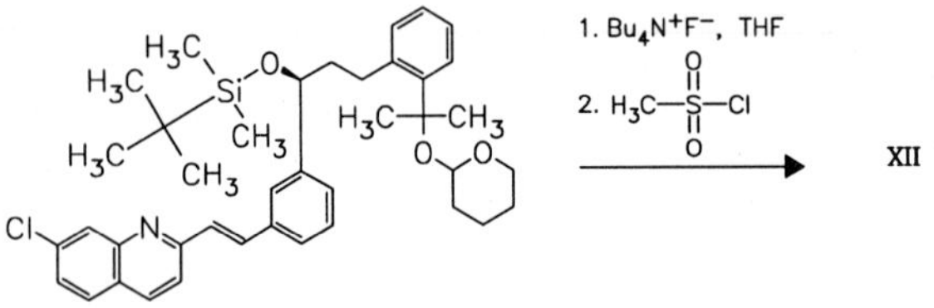

XIV

(XIV)

1. Bu₄N⁺F⁻, THF

2. H₃C—SO₂—Cl

XII

Reference(s):

a EP 480 717 (Merck Frosst; appl. 10.10.1991; USA-prior. 12.10.1990, 8.8.1991).

aa US 5 523 477 (Merck + Co.; 4.6.1996; USA-prior. 23.1.1995).

b WO 9 518 107 (Merck + Co.; appl. 22.12.1994; USA-prior. 28.12.1993, 9.12.1994).
Labelle, M. et al.: Bioorg. Med. Chem. Lett. (BMCLE8) **5**, 283 (1995).

synthesis of starting material I:
US 4 851 409 (Merck Frosst Can.; 25.7.1987; USA-prior. 14.2.1986).

pharmaceutical composition with loratidine:
WO 9 728 797 (Merck + Co.; appl. 4.2.1997; USA-prior. 8.2.1996).

Formulation(s):　chewable tabl. 5 mg, 50 mg; f. c. tabl. 10 mg; tabl. 5 mg, 10 mg (as sodium salt)

Trade Name(s):

D:	Singulair (MSD Dieckmann)	I:
GB:	Singulair (Merck Sharp & Dohme)	

I: Lukasm (Sigmatau)
Montegen (Gentili)
Singulair (Merck Sharp & Dohme)

USA: Singulair (Merck Sharp & Dohme; 1998)

Moperone

(Methylperidol; Mopiperone)

ATC: N05AD04
Use: neuroleptic, antipsychotic

RN: 1050-79-9　MF: C₂₂H₂₆FNO₂　MW: 355.45　EINECS: 213-887-6
CN: 1-(4-fluorophenyl)-4-[4-hydroxy-4-(4-methylphenyl)-1-piperidinyl]-1-butanone

hydrochloride
RN: 3871-82-7   MF: $C_{22}H_{26}FNO_2 \cdot HCl$ MW: 391.91   EINECS: 223-392-7
LD_{50}:   15.5 mg/kg (M, i.v.); 218 mg/kg (M, p.o.);
          12.1 mg/kg (R, i.v.); 152 mg/kg (R, p.o.)

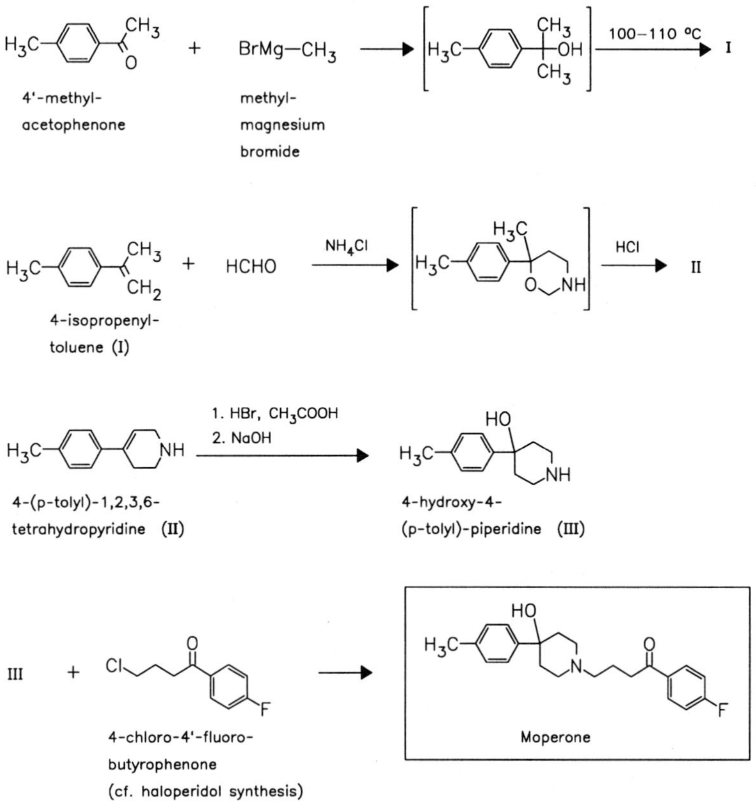

4'-methyl-
acetophenone

methyl-
magnesium
bromide

100–110 °C   I

4-isopropenyl-
toluene (I)

HCHO NH_4Cl HCl   II

4-(p-tolyl)-1,2,3,6-
tetrahydropyridine (II)

1. HBr, CH_3COOH
2. NaOH

4-hydroxy-4-
(p-tolyl)-piperidine (III)

III   +   4-chloro-4'-fluoro-
butyrophenone
(cf. haloperidol synthesis)

Moperone

Reference(s):
GB 881 893 (P. A. J. Janssen; appl. 22.4.1958; valid from 14.4.1959).

Formulation(s):   amp. 5 mg/1 ml; tabl. 5 mg, 20 mg (as hydrochloride)

Trade Name(s):
D: Luvatrena (Cilag-Chemie);   F: Sedalium (Fournier   I: Luvatren (Cilag-Chemie);
        wfm   Frères)-comb.; wfm   wfm

                                                                              J: Luvatren (Yamanouchi)

Mopidamol

ATC: B01AC
Use: thrombosis and metastasis
         prophylactic, antineoplastic

RN: 13665-88-8   MF: $C_{19}H_{31}N_7O_4$ MW: 421.50   EINECS: 237-145-6
LD_{50}:   148 mg/kg (M, i.v.); 465 mg/kg (M, p.o.);
          3 g/kg (R, p.o.)
CN: 2,2',2'',2'''-[[4-(1-piperidinyl)pyrimido[5,4-*d*]pyrimidine-2,6-diyl]dinitrilo]tetrakis[ethanol]

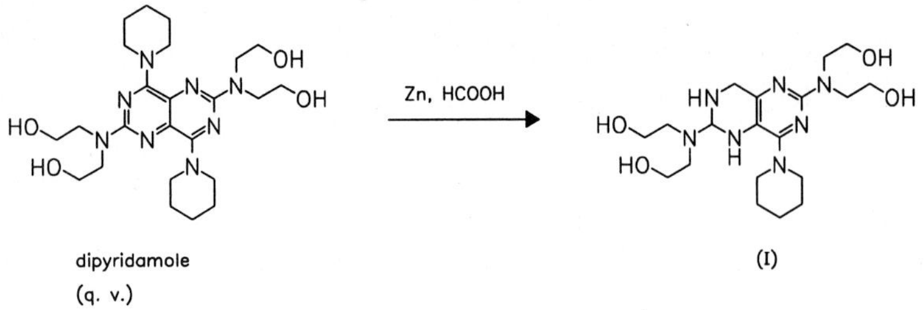
Zn, HCOOH
dipyridamole
(q. v.)
(I)

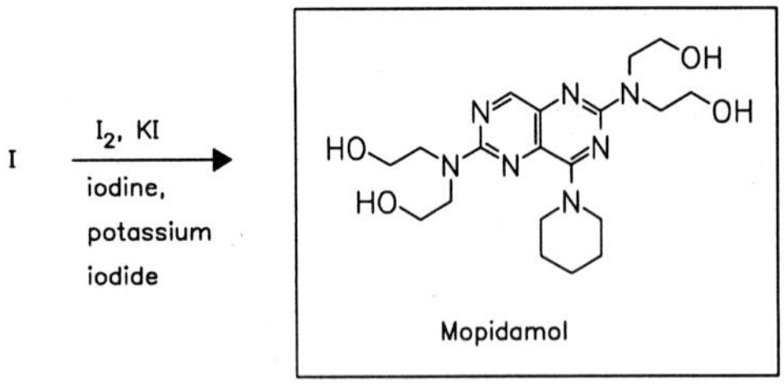
I
I_2, KI
iodine,
potassium
iodide
Mopidamol

Reference(s):
DAS 1 470 341 (Thomae; appl. 9.3.1963).

starting material:
DE 1 116 676 (Thomae; appl. 14.3.1955).
US 3 031 450 (Thomae; 24.4.1962; D-prior. 30.4.1959).

Formulation(s): amp. 150 mg/3 ml; cps. 250 mg

Trade Name(s):
D: Rapenton (Thomae); wfm

Moracizine

(Ethmosine; Etmosin; Moricizine)

ATC: C01BG01
Use: class I antiarrhythmic

RN: 31883-05-3 MF: $C_{22}H_{25}N_3O_4S$ MW: 427.53 EINECS: 250-854-5
LD_{50}: 131 mg/kg (M, i.p.); 36 mg/kg (M, i.v.);
105 mg/kg (R, i.p.); 11 mg/kg (R, i.v.)
CN: [10-[3-(4-morpholinyl)-1-oxopropyl]-10*H*-phenothiazin-2-yl]carbamic acid ethyl ester

monohydrochloride
RN: 29560-58-5 MF: $C_{22}H_{25}N_3O_4S \cdot HCl$ MW: 463.99
LD_{50}: 36 mg/kg (M, i.v.);
11 mg/kg (R, i.v.); 1 g/kg (R, p.o.)

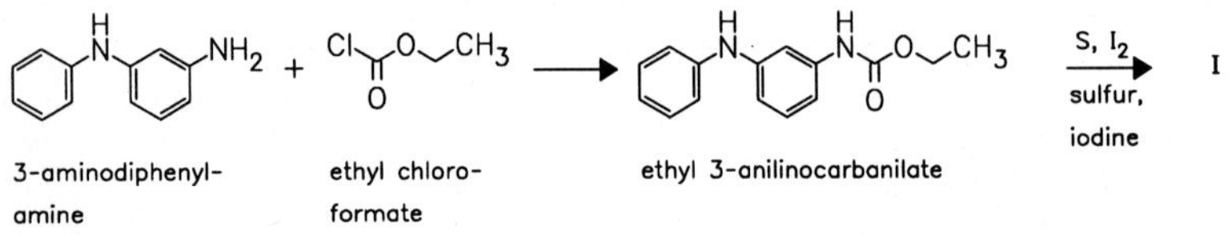
S, I_2
sulfur,
iodine
I
3-aminodiphenylamine
ethyl chloroformate
ethyl 3-anilinocarbanilate

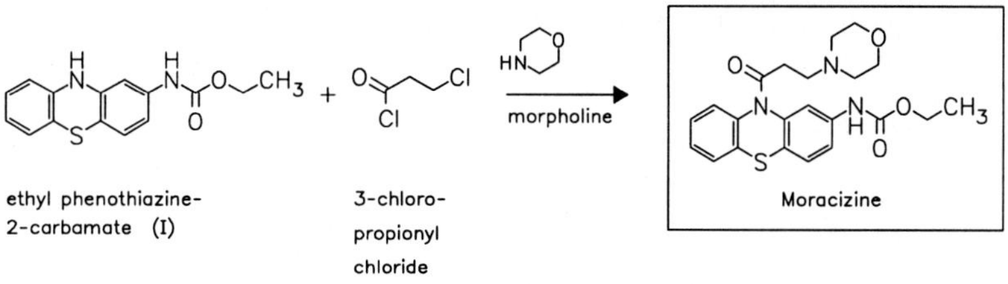

ethyl phenothiazine-
2-carbamate (I)

3-chloro-
propionyl
chloride

morpholine

Moracizine

Reference(s):
DE 2 014 201 (Academy of Medical Sci. USSR; appl. 24.3.1970).
US 3 740 395 (Academy of Medical Sci. USSR; prior. 16.8.1971; 14.10.1969).
US 3 864 487 (A.N. Gritsenko et al.; 4.2.1975; prior. 16.8.1971, 10.10.1969).
GB 1 269 969 (Academy of Medical Sci. USSR; appl. 25.9.1969).
Gritsenko, A.N. et al.: Khim. Farm. Zh. (KHFZAN) **6**, 17 (1972).
SU 332 835 (Academy of Medical Sci. USSR; appl. 15.1.1965).
SU 329 891 (Academy of Medical Sci. USSR; appl. 19.7.1965).

Formulation(s):   tabl. 200 mg, 250 mg, 300 mg (as hydrochloride)

Trade Name(s):
USA: Ethmozine (Roberts)

Morclofone

ATC: R05DB25
Use: antitussive

RN: 31848-01-8   MF: $C_{21}H_{24}ClNO_5$ MW: 405.88   EINECS: 250-838-8
LD$_{50}$:   552 mg/kg (M, p.o.)
CN: (4-chlorophenyl)[3,5-dimethoxy-4-[2-(4-morpholinyl)ethoxy]phenyl]methanone

hydrochloride
RN: 31848-02-9   MF: $C_{21}H_{24}ClNO_5 \cdot HCl$ MW: 442.34   EINECS: 250-839-3
LD$_{50}$:   609 mg/kg (M, p.o.);
       1290 mg/kg (R, p.o.)

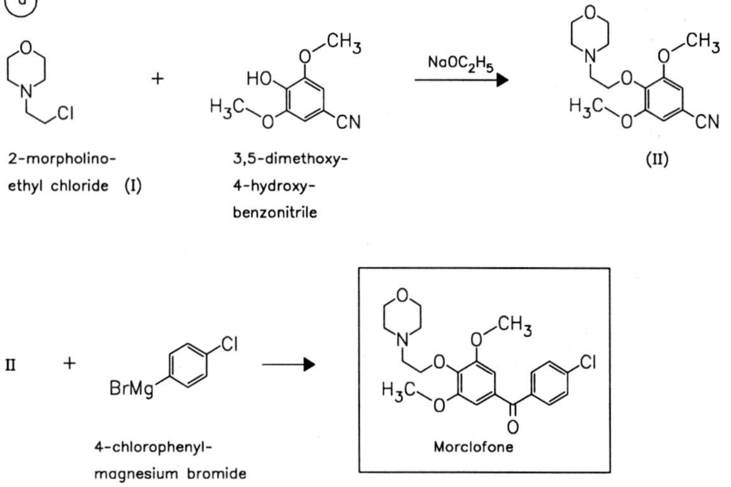

2-morpholino-
ethyl chloride (I)

3,5-dimethoxy-
4-hydroxy-
benzonitrile

NaOC$_2$H$_5$

(II)

II +

4-chlorophenyl-
magnesium bromide

Morclofone

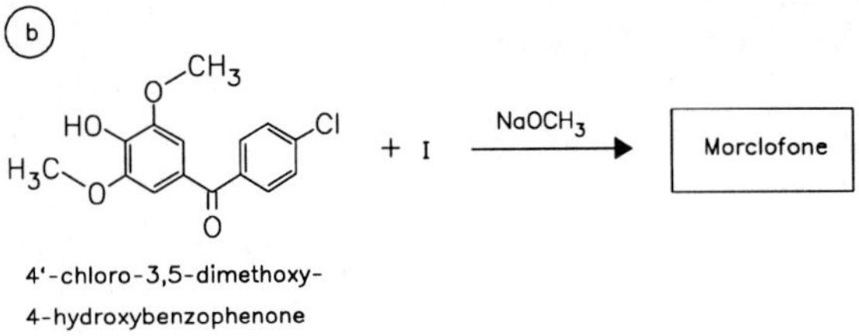

(b)

4'-chloro-3,5-dimethoxy-
4-hydroxybenzophenone

Reference(s):

DOS 2 016 707 (Carlo Erba; appl. 8.4.1970; I-prior. 15.4.1969).

Formulation(s):   syrup 50 mg (1 %) (as hydrochloride)

Trade Name(s):
I: Plausitin (Carlo Erba)

Morinamide
(Morfazinamida)

ATC: J04AK04
Use: tuberculostatic

RN: 952-54-5   MF: $C_{10}H_{14}N_4O_2$ MW: 222.25   EINECS: 213-460-4
LD_{50}:   2750 mg/kg (M, i.p.)
CN: *N*-(4-morpholinylmethyl)pyrazinecarboxamide

monohydrochloride
RN: 1473-73-0   MF: $C_{10}H_{14}N_4O_2 \cdot HCl$ MW: 258.71   EINECS: 216-013-1

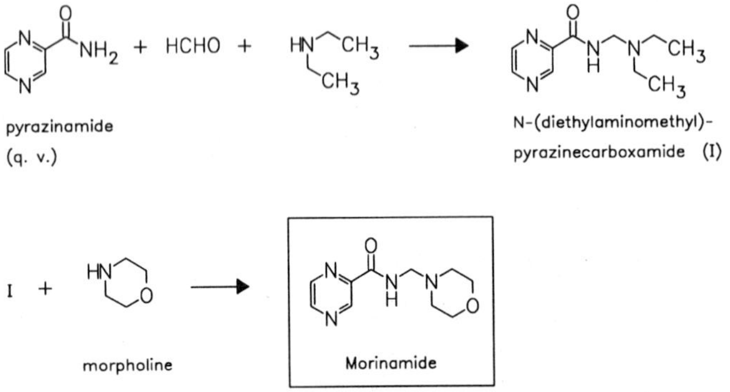

pyrazinamide
(q. v.)

N-(diethylaminomethyl)-
pyrazinecarboxamide   (I)

morpholine   Morinamide

Reference(s):
DE 1 129 492 (Bracco; appl. 23.6.1960; CH-prior. 31.7.1959).

Formulation(s):   amp. 1 g; tabl. 500 mg (as hydrochloride)

Trade Name(s):
F: Piazoline (Beytout); wfm   I: Piazofolina (Bracco)

Moroxydine

ATC: J05AX01
Use: antiviral, influenza therapeutic

RN: 3731-59-7   MF: $C_6H_{13}N_5O$ MW: 171.20   EINECS: 223-093-1
CN: *N*-(aminoiminomethyl)-4-morpholinecarboximidamide

monohydrochloride
RN: 3160-91-6   MF: $C_6H_{13}N_5O \cdot HCl$ MW: 207.67   EINECS: 221-612-6
LD_{50}:   325 mg/kg (M, i.v.); >6.25 g/kg (M, p.o.)

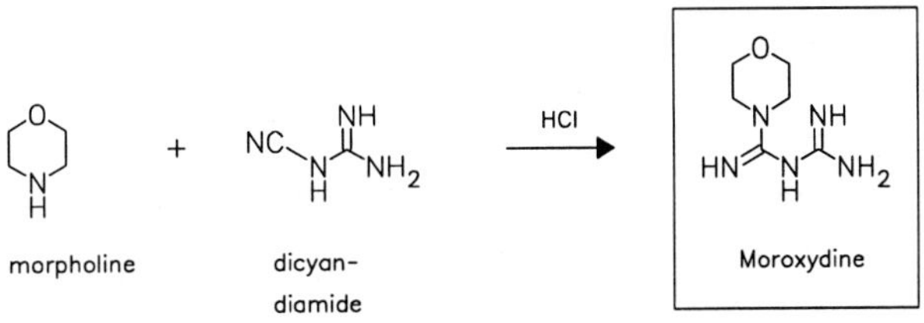

morpholine   +   dicyan-diamide $\xrightarrow{\text{HCl}}$ Moroxydine

Reference(s):
GB 776 176 (A. B. Kabi; appl. 1954; S-prior. 1953).

Formulation(s):   tabl. 100 mg, 400 mg (as hydrochloride)

Trade Name(s):

D: Flumidin (Kabi); wfm
F: Assur (Delagrange)-comb.; wfm
     Virustat (Delagrange); wfm
J: ABOB (Nichiiko; Sankyo)
     Aboryl (Taisho)

Anrus (Towa S.-Aoi)
Enless (Zeria)
Flue (Kobayashi Kako)
Infermine (Hokuriku)
Nicefull (Kyorin)
Pansil (Iwaki)

Pathin (Tokyo Tanabe)
Sanflumin (Sanwa)
Tamaxin (Sawai)
Vilusron (Kanto)
Virusmin (Sumitomo)
Virusmohin (Mohan)

Morphine
(Morphium)

ATC: N02AA01
Use: analgesic, sedative

RN: 57-27-2   MF: $C_{17}H_{19}NO_3$ MW: 285.34   EINECS: 200-320-2
LD_{50}:   135 mg/kg (M, i.v.); 524 mg/kg (M, p.o.);
      140 mg/kg (R, i.v.); 335 mg/kg (R, p.o.);
      133 mg/kg (dog, i.v.)
CN: (5α,6α)-7,8-didehydro-4,5-epoxy-17-methylmorphinan-3,6-diol

hydrochloride
RN: 52-26-6   MF: $C_{17}H_{19}NO_3 \cdot HCl$ MW: 321.80   EINECS: 200-136-2
LD_{50}:   180 mg/kg (M, i.v.); 745 mg/kg (M, p.o.);
      265 mg/kg (R, i.v.); 335 mg/kg (R, p.o.);
      175 mg/kg (dog, i.v.)

sulfate (2:1)
RN: 64-31-3   MF: $C_{17}H_{19}NO_3 \cdot 1/2H_2SO_4$ MW: 668.76   EINECS: 200-582-8
LD_{50}:   156 mg/kg (M, i.v.); 600 mg/kg (M, p.o.);
      70 mg/kg (R, i.v.); 461 mg/kg (R, p.o.);
      316 mg/kg (dog, i.v.)

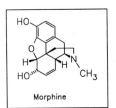

Morphine

By extraction of poppy-heads or opium with water, precipitation with aqueous Na$_2$CO$_3$-solution, washing of the precipitate with ethanol and dissolving in diluted acetic acid.

Reference(s):
Ullmanns Encykl. Tech. Chem., 3. Aufl., Vol. **3**, 233.

newer methods:
DOS 2 726 925 (Knoll; appl. 15.6.1977).

isolation on ion-exchanger:
DOS 2 905 468 (Kutnowskie Zaklady Farmaceut. Polfa; appl. 13.2.1979; PL-prior. 15.2.1978).

Formulation(s):   amp. 10 mg, 20 mg, 100 mg; s. r. cps. 10 mg, 20 mg, 30 mg, 50 mg, 60 mg, 100 mg; suppos. 10 mg, 20 mg, 30 mg (as sulfate)

Trade Name(s):

D:	Capros (Rhône-Poulenc Rorer; medac)	MXL (Napp; as sulfate)	MS Contin (Shionogi; as sulfate)
	Kapanol (Glaxo Wellcome)	Oramorph (Boehringer Ing.; as sulfate)	
	MST Mundipharma (Mundipharma)	Rapiject (Evans; as sulfate)	USA: Duramorph (Elkins-Sinn; as sulfate)
	generic	Sevredol (Napp; as sulfate)	Kadian (Zeneca; as sulfate)
F:	Colchimax (Hoechst Houdé)-comb.	Zomorph (Link; as sulfate)	MS Contin (Purdue Frederick; as sulfate)
		I: Cardiostenol (Molteni)-comb.	
	Lamaline (Solvay Pharma)-comb.	MS Contin (ASTA Medica; as sulfate)	MSIR (Purdue Frederick; as sulfate)
	Morphine Meram sans conservateur (RPR Cooper)	Skenan (Ethypharm)	OMS (Upsher-Smith; as sulfate)
	Moscontin (ASTA Medica)	numerous generics as hydrochloride	RMS Suppos. (Upsher-Smith; as sulfate)
GB:	Cyclimorph (Glaxo Wellcome)-comb.	J: Morphine Hydrochloride (Dainippon; Sankyo; Takeda; Tanabe)	Roxanol (Roxane; as sulfate)
	MST Continus (Napp)		generics

Mosapramine

ATC: N05AX10
Use: neuroleptic, metabolite of clocapramine, combined 5-HT/ dopamine receptor antagonist

RN: 89419-40-9   MF: C$_{28}$H$_{35}$ClN$_4$O   MW: 479.07
LD$_{50}$:   74 mg/kg (M, i.p.); 1008 mg/kg (M, p.o.); 1147 mg/kg (M, s.c.);
      201 mg/kg (R, i.p.); 4912 mg/kg (R, p.o.)
CN: (±)-1'-[3-(3-chloro-10,11-dihydro-5*H*-dibenz[*b,f*]azepin-5-yl)propyl]hexahydrospiro[imidazo[1,2-*a*]pyridine-3(2*H*),4'-piperidin]-2-one

dihydrochloride
RN: 98043-60-8   MF: C$_{28}$H$_{35}$ClN$_4$O · 2HCl   MW: 551.99
LD$_{50}$:   1008 mg/kg (M, p.o.);
      4912 mg/kg (R, p.o.)

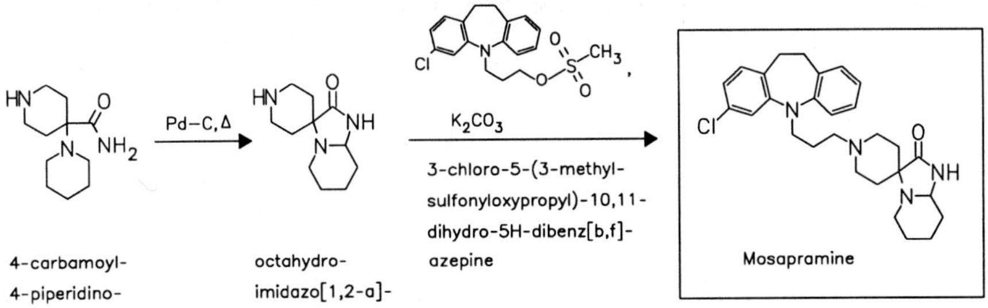

4-carbamoyl-
4-piperidino-
piperidine

octahydro-
imidazo[1,2-a]-
pyridine-3-
spiro-4'-piperi-
din-2-one

3-chloro-5-(3-methyl-
sulfonyloxypropyl)-10,11-
dihydro-5H-dibenz[b,f]-
azepine

Mosapramine

Reference(s):
EP 73 845 (Yoshitomi; appl. 3.9.1981).
US 4 337 260 (Yoshitomi; 29.6.1982; appl. 10.9.1981).
DE 3 170 724 (Yoshitomi; appl. 3.9.1981).
Tashiro, Ch. et al.: Yakugaku Zasshi (YKKZAJ) **109**, 93 (1989); C.A. (CHABA8) **112**, 35749j (1989).

sustained-release microsphere:
WO 9 410 982 (Yoshitomi Pharm.; appl. 15.11.1993; J-prior. 17.11.1992).

Formulation(s): tabl. 10 mg, 15 mg, 25 mg (as dihydrochloride)

Trade Name(s):
J: Cremin (Yoshitomi; 1989)

Mosapride citrate
(AS-4370)

Use: 5-HT$_4$-agonist, promotility agent,
treatment of gastroesophageal reflux
disease

RN: 112885-42-4 MF: C$_{21}$H$_{25}$ClFN$_3$O$_3$ MW: 421.90
CN: 4-Amino-5-chloro-2-ethoxy-*N*-[[4-[(4-fluorophenyl)methyl]-2-morpholinyl]methyl]benzamide citrate

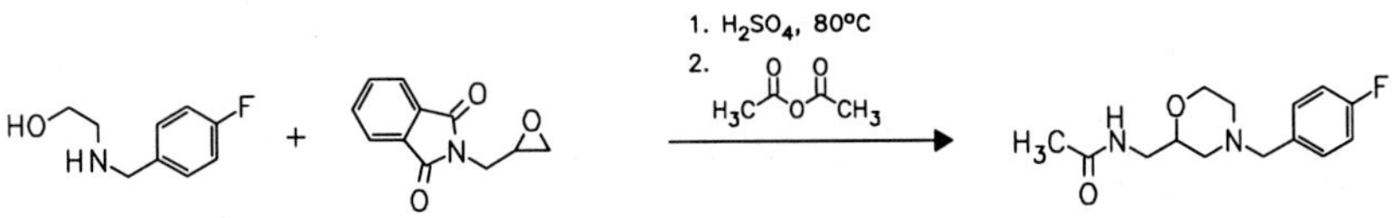

N-(4-fluorobenzyl)-
ethanolamine

(±)-N-(2,3-epoxypropyl)-
phthalimide

(±)-2-(acetylaminomethyl)-
4-(4-fluorobenzyl)-
morpholine (I)

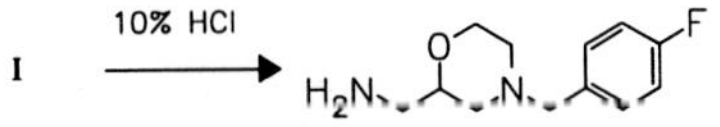

(±)-2-aminomethyl-
4-(4-fluorobenzyl)-
morpholine (II)

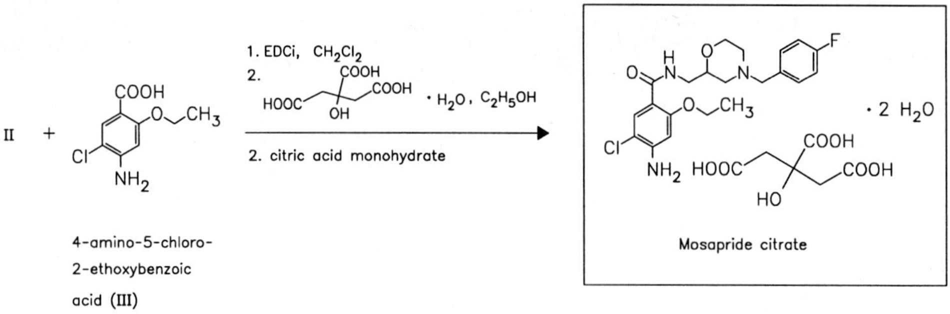

4-amino-5-chloro-
2-ethoxybenzoic
acid (III)

Mosapride citrate

intermediate III can be prepared from

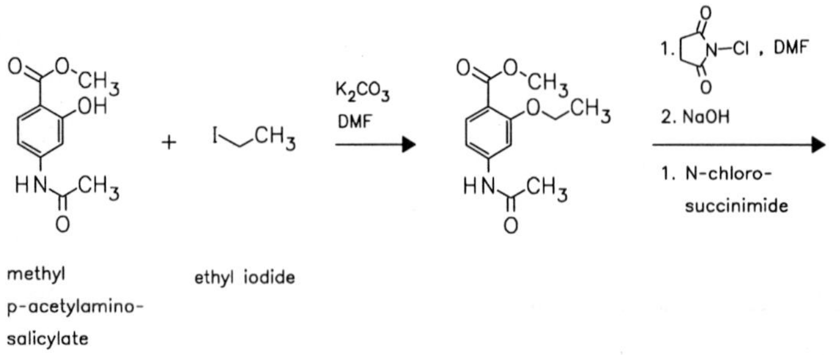

methyl
p-acetylamino-
salicylate

ethyl iodide

Reference(s):
EP 243 959 (Dainippon Pharm.; appl. 29.4.1987; J-prior. 30.4.1986).
Kato, S. et al.: J. Med. Chem. (JMCMAR) **34** (2), 616 (1991).
Kato, S. et al.: Chem. Pharm. Bull. (CPBTAL) **40** (6), 1470 (1992).

optical isomers:
Morie, T. et al.: Chem. Pharm. Bull. (CPBTAL) **42** (4), 877 (1994).
Morie, T. et al.: Heterocycles (HTCYAM) **38** (5), 1033 (1994).

oral dosage form with a proton pump inhibitor:
WO 9 725 065 (Astra; 17.7.1997; appl. 20.12.1996; S-prior. 8.1.1996)

Formulation(s):　powder 10 mg/g (as citrate); tabl. 2.5 mg, 5 mg

Trade Name(s):
J:　　Gasmotin (Dainippon;
　　　1998)

Moxaverine

ATC:　G04BE
Use:　antispasmodic

RN:　10539-19-2　MF: $C_{20}H_{21}NO_2$　MW: 307.39　EINECS: 234-117-5
CN:　3-ethyl-6,7-dimethoxy-1-(phenylmethyl)isoquinoline

hydrochloride
RN:　1163-37-7　MF: $C_{20}H_{21}NO_2 \cdot HCl$　MW: 343.85　EINECS: 214-607-5

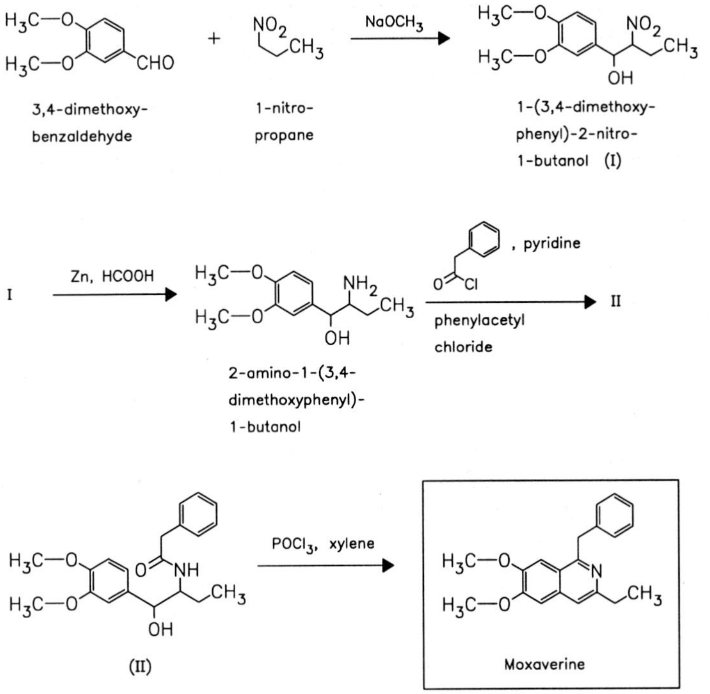

NaOCH3
3,4-dimethoxy-benzaldehyde
1-nitro-propane
1-(3,4-dimethoxy-phenyl)-2-nitro-1-butanol (I)
Zn, HCOOH
, pyridine
phenylacetyl chloride
2-amino-1-(3,4-dimethoxyphenyl)-1-butanol
POCl3, xylene
(II)
Moxaverine

Reference(s):
GB 1 030 022 (Orgamol; appl. 7.5.1963; CH-prior. 16.6.1962).

Formulation(s): amp. 150 mg/5 ml; drg. 100 mg, 150 mg (as hydrochloride)

Trade Name(s):

D: Certonal (Sertürner)
Kallaterol (Ursapharm)-comb.
Kollateral (Ursapharm)
I: Eupaverina (Bracco); wfm

Moxestrol

ATC: G03CB04
Use: estrogen

RN: 34816-55-2 MF: $C_{21}H_{26}O_3$ MW: 326.44
CN: (11β,17α)-11-methoxy-19-norpregna-1,3,5(10)-trien-20-yne-3,17-diol

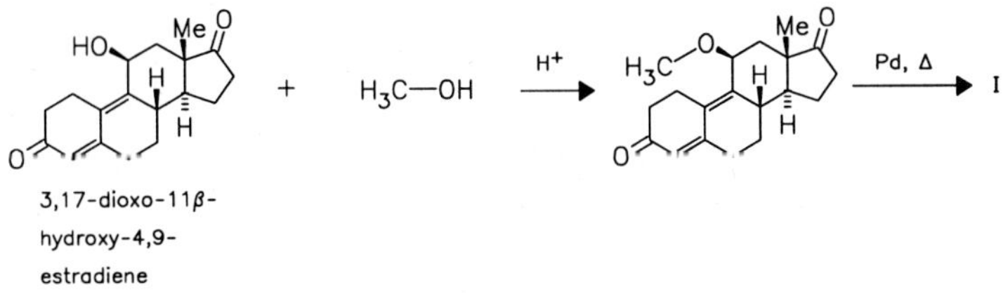

H3C—OH
H+
Pd, Δ
I
3,17-dioxo-11β-hydroxy-4,9-estradiene

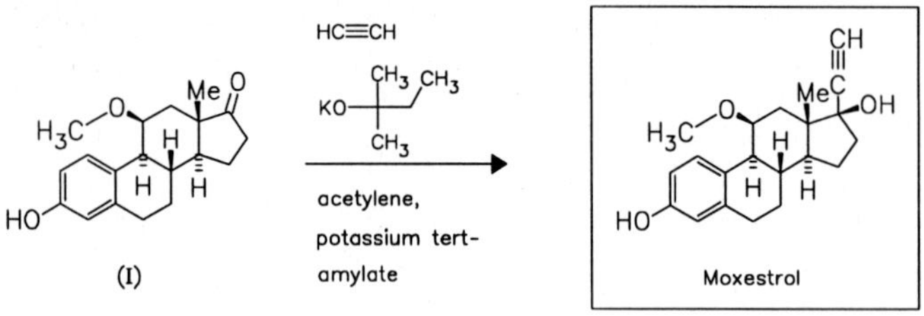

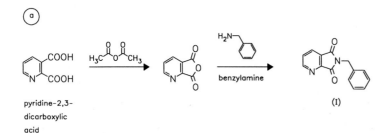

(I)

acetylene,
potassium tert-
amylate

Moxestrol

Reference(s):

US 3 579 545 (Roussel-Uclaf; 18.5.1971; F-prior. 7.9.1966, 9.12.1966, 28.2.1967, 9.3.1967).
FR-M 6 182 (Roussel-Uclaf; appl. 28.2.1967).
Azadian-Boulanger, G.; Bertin, D.: Chim. Ther. (CHTPBA) **8**, 451 (1973).

starting material:

Joly, R. et al.: C. R. Seances Acad. Sci., Ser. C (CHDCAQ) **258**, 5669 (1964) (total synthesis).

Formulation(s):　　tabl. 50 μg

Trade Name(s):

F:　　　Surestryl (Roussel-Uclaf)

Moxifloxacin hydrochloride

(Bay-12-8039)

ATC:　J01MA14
Use:　fluoroquinolone antibacterial

RN:　186826-86-8　MF: $C_{21}H_{24}FN_3O_4 \cdot HCl$　MW: 437.90
CN:　1-Cyclopropyl-6-fluoro-1,4-dihydro-8-methoxy-7-[(4a*S*,7a*S*)-octahydro-*6H*-pyrrolo[3,4-*b*]pyridin-6-yl]-4-oxo-3-quinolinecarboxylic acid

base
RN:　151096-09-2　MF: $C_{21}H_{24}FN_3O_4$　MW: 401.44

pyridine-2,3-
dicarboxylic
acid

benzylamine

(I)

1. H_2, Pd–C
2. LiAlH₄

I →

(±)-cis-8-benzyl-
2,8-diazabicyclo-
[4.3.0]nonane ((±)-II)

1. L-(+)-tartaric acid, DMF
2. crystallization
3. (mother liquors), NaOH
4. D-(−)-tartaric acid (crystallization)
5. NaOH

→ (S,S)-II

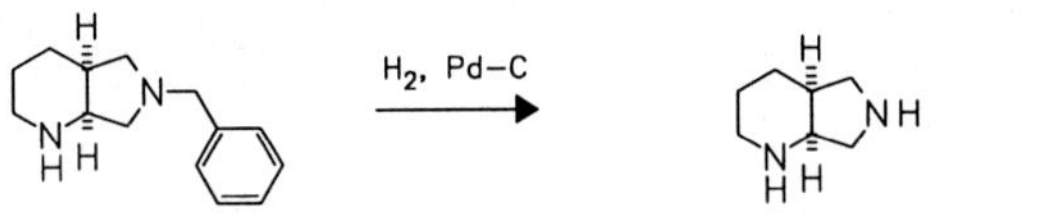

((S,S)-II)

(S,S)-2,8-diaza-
bicyclo[4.3.0]nonane (III)

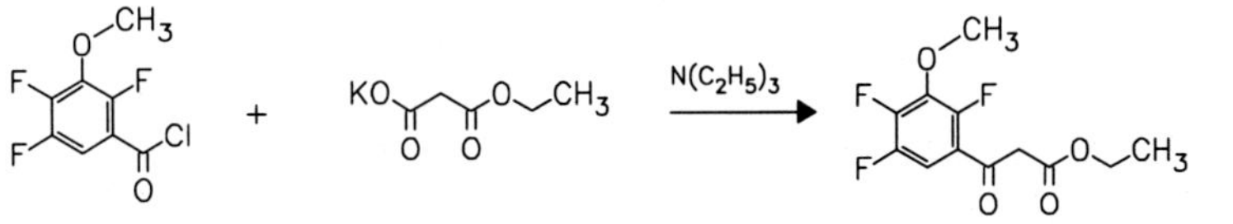

2,4,5-trifluoro-
3-methoxybenzoyl
chloride (IV)

potassium monoethyl
malonate

ethyl 2-(2,4,5-trifluoro-
3-methoxybenzoyl)acetate (V)

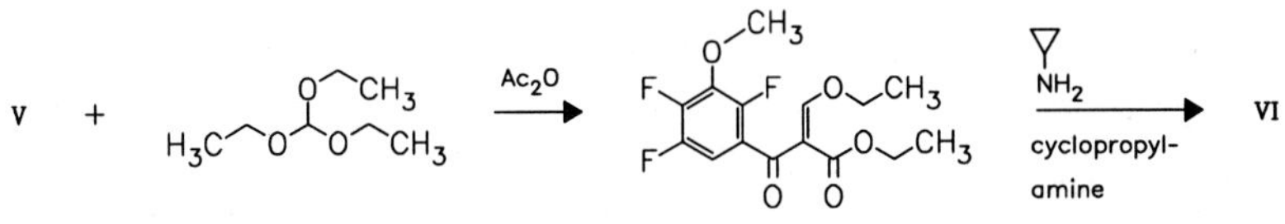

triethyl orthoformate

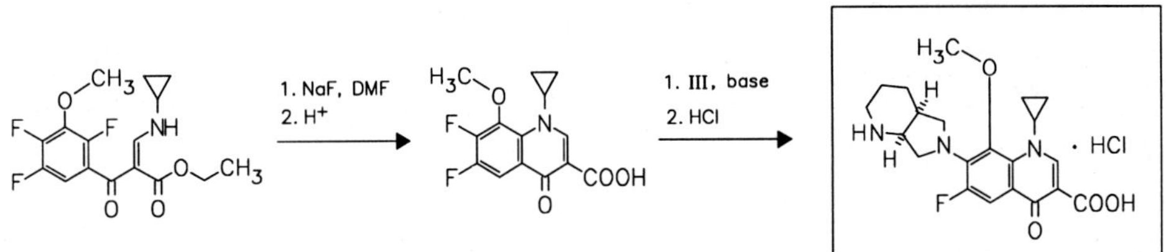

cyclopropyl-
amine

(VI)

1-cyclopropyl-6,7-
difluoro-8-methoxy-
4-oxo-1,4-dihydro-
quinoline-3-carboxylic
acid

Moxifloxacin hydrochloride

(aa)   synthesis of IV

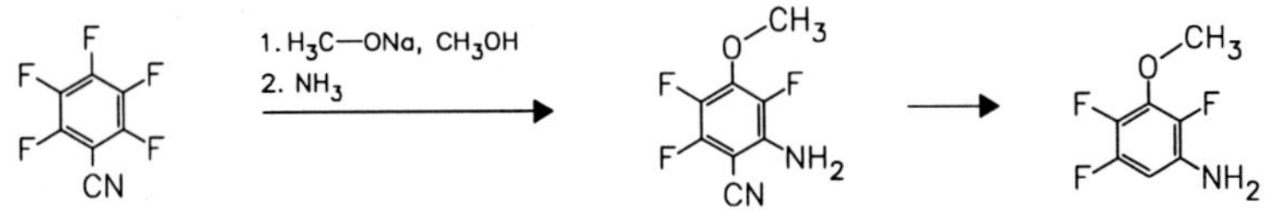

pentafluoro-
benzonitrile

(VII)

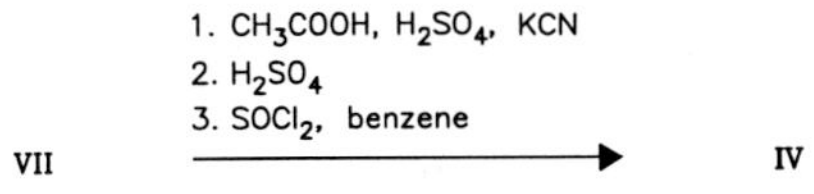

VII → IV

1. CH_3COOH, H_2SO_4, KCN
2. H_2SO_4
3. $SOCl_2$, benzene

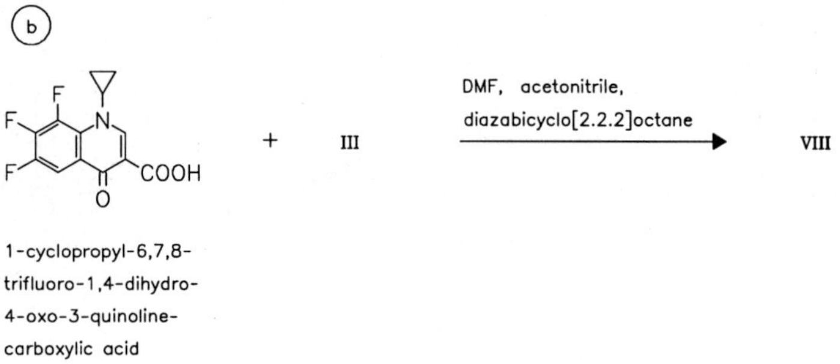

b
F
F
F
N
COOH
O
+
III
DMF, acetonitrile,
diazabicyclo[2.2.2]octane
VIII
1-cyclopropyl-6,7,8-
trifluoro-1,4-dihydro-
4-oxo-3-quinoline-
carboxylic acid

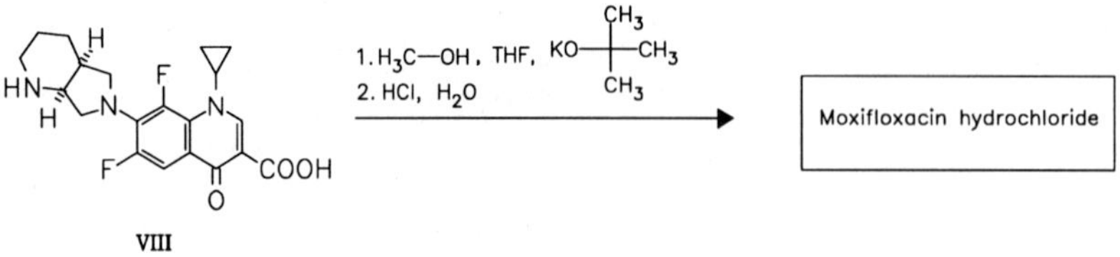

HN
H
H
N
F
N
F
COOH
O
VIII
1. H3C—OH, THF, KO—C(CH3)3
2. HCl, H2O
Moxifloxacin hydrochloride

Reference(s):

a DE 4 208 792 (Bayer AG; appl. 19.3.1992; D-prior. 19.3.1992).
aa EP 241 206 (Sankyo; appl. 31.3.1987; J-prior. 31.3.1986).
b EP 550 903 (Bayer AG, appl. 28.12.1992; D-prior. 19.3.1992).
DE 19 751 948 (Bayer AG; appl. 24.11.1997; D-prior. 24.11.1997).

formulation with controlled release of moxifloxacin:
DE 19 546 249 (Bayer AG; appl. 12.12.1995; D-prior. 12.12.1995).

Formulation(s): f. c. tabl. 400 mg

Trade Name(s):
D: Avalox (Bayer) Avelox (Bayer; 1999) Tovan (Bayer AG)

Moxisylyte
(Thymoxamine)

ATC: C04AX10
Use: vasodilator (peripheral)

RN: 54-32-0 MF: $C_{16}H_{25}NO_3$ MW: 279.38 EINECS: 200-204-1
LD_{50}: 225 mg/kg (M, p.o.)
CN: 4-[2-(dimethylamino)ethoxy]-2-methyl-5-(1-methylethyl)phenol acetate (ester)

hydrochloride
RN: 964-52-3 MF: $C_{16}H_{25}NO_3 \cdot HCl$ MW: 315.84 EINECS: 213-519-4
LD_{50}: 28 mg/kg (M, i.v.); 225 mg/kg (M, p.o.);
32 mg/kg (R, i.v.); 740 mg/kg (R, p.o.);
54.5 mg/kg (dog, i.v.)

CH3
H3C
CH3 OH
thymol
NaNO2, HCl
NO
4-nitroso-
thymol
H2S
NH2
4-amino-
thymol
acetic
anhydride (I)
II

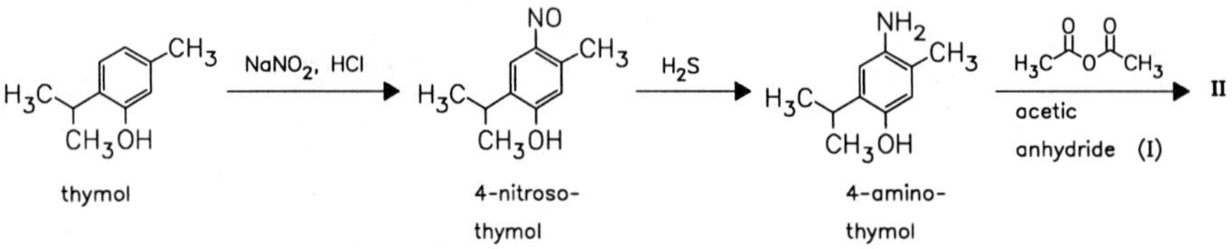

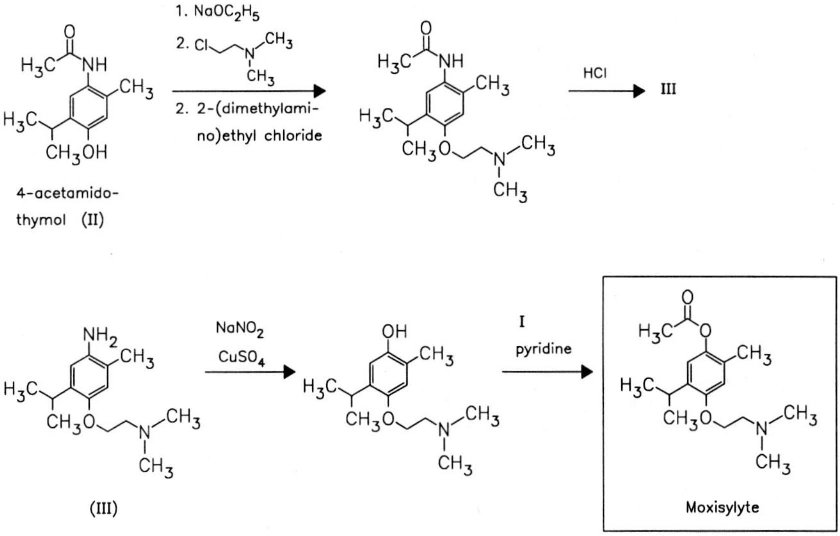

4-acetamido-
thymol (II)

(III)

Moxisylyte

Reference(s):
DE 905 738 (Diwag; appl. 1943).

combination with steroids (e. g. for treatment of asthma):
GB 1 535 531 (W. R. Warner & Co.; appl. 17.10.1975; valid from 15.10.1976).

Formulation(s):   drg. 30 mg; tabl. 30 mg, 40 mg, 60 mg, 120 mg (as hydrochloride)

Trade Name(s):
D: Vasoklin (Gödecke); wfm   Icavex (ASTA Medica)   J: Moxyl (Fujirebio)
F: Carlytène (ASTA Medica)   GB: Erecnos (Fournier)   Thimozil (Kohjin)
      Erecnos (Débat)   I: Arlitene (ASTA Medica)

Moxonidine
(BDF-5895)

ATC: C02AC05
Use: antihypertensive, presynaptic α_2-adrenoceptor agonist

RN: 75438-57-2   MF: $C_9H_{12}ClN_5O$ MW: 241.68
LD$_{50}$:   320 mg/kg (M, p.o.);
       115 mg/kg (Rf, p.o.); 143 mg/kg (Rm, p.o.)
CN: 4-chloro-N-(4,5-dihydro-1H-imidazol-2-yl)-6-methoxy-2-methyl-5-pyrimidinamine

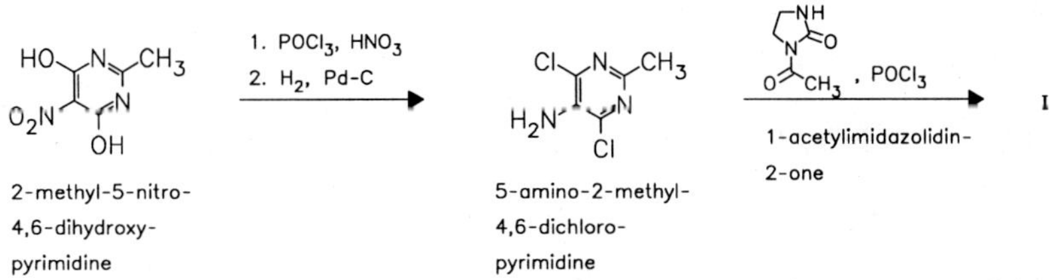

2-methyl-5-nitro-
4,6-dihydroxy-
pyrimidine

5-amino-2-methyl-
4,6-dichloro-
pyrimidine

1-acetylimidazolidin-
2-one

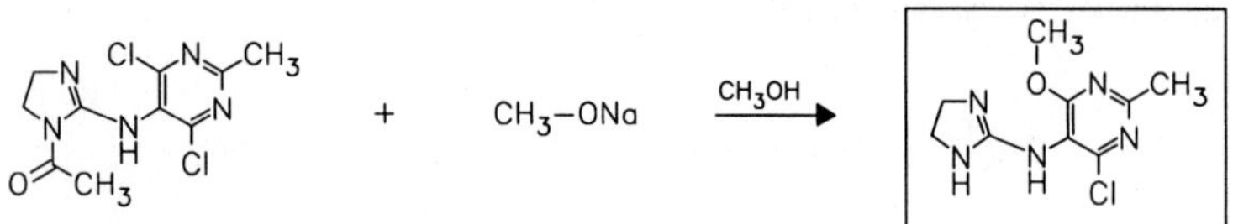

4,6-dichloro-2-methyl-
5-(1-acetyl-2-imidazolin-
2-ylamino)pyrimidine (I)

Moxonidine

Reference(s):

DOS 2 849 537 (Beiersdorf; appl. 15.11.1978).
DOS 2 937 023 (Beiersdorf; appl. 13.9.1979).

synthesis of 5 amino-2-methyl-4,6-dichloropyrimidine:
Huber, W.; Hölscher, H.A.: Chem. Ber. (CHBEAM) **71B**, 87 (1938).
Ochiai, E.; Kashida, Y.: Yakugaku Zasshi (YKKZAJ) **62**, 97 (1942).

medical use for promoting growth:
DOS 3 904 795 (Beiersdorf; appl. 17.2.1989).

combination with hydrochlorothiazide, triamterene:
EP 317 855 (Beiersdorf; appl. 12.11.1988; D-prior. 24.11.1987).

Formulation(s): f. c. tabl. 0.2 mg, 0.3 mg, 0.4 mg

Trade Name(s):

D:	Cynt (Beiersdorf-Lilly/ Lilly; 1991)	Physiotens (Solvay Arzneimittel; 1991)	F:	Physiotens (Solvay Pharma)
			GB:	Physiotens (Solvay)

Mupirocin

(Pseudomonic acid)

ATC: D06AX09; R01AX06
Use: antibiotic

RN: 12650-69-0 MF: $C_{26}H_{44}O_9$ MW: 500.63
LD_{50}: 1638 mg/kg (M, i.v.); 5 g/kg (M, p.o.);
1310 mg/kg (R, i.v.); 5 g/kg (R, p.o.)
CN: [2S-[2α(E),3β,4β,5α[2R*,3R*(1R*,2R*)]]]-9-[[3-methyl-1-oxo-4-[tetrahydro-3,4-dihydroxy-5-[[3-(2-hydroxy-1-methylpropyl)oxiranyl]methyl]-2H-pyran-2-yl]-2-butenyl]oxy]nonanoic acid

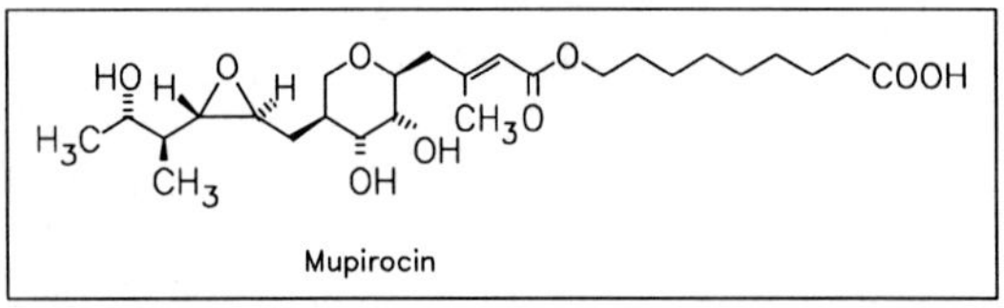

Mupirocin

Fermentation of *Pseudomonas fluorescens* NCIB 10586.

Reference(s):

DE 2 227 739 (Beecham; GB-prior. 12.6.1971).
US 3 977 943 (Beecham; 31.8.1976; appl. 7.7.1975; prior. 27.3.1974).
US 4 071 536 (Beecham; 31.1.1978; prior. 12.6.1971).
Fuller, A.T. et al.: Nature (London) (NATUAS) **234**, 416 (1971).

structure:
Chain, E.B. et al.: J. Chem. Soc., Chem. Commun. (JCCCAT) **20**, 847 (1974).

Formulation(s): cream 20 mg/g; ointment 2 % (as calcium salt)

Trade Name(s):

D:	Turixin (SmithKline Beecham)	GB:	Bactroban (SmithKline Beecham)	J:	Bactroban (SmithKline Beecham)
F:	Bactroban (SmithKline Beecham; 1985)	I:	Bactroban (SmithKline Beecham)	USA:	Bactroban (SmithKline Beecham; 1988)

Muzolimine

ATC: C03CD01
Use: diuretic

RN: 55294-15-0 MF: $C_{11}H_{11}Cl_2N_3O$ MW: 272.14 EINECS: 259-573-2
CN: 5-amino-2-[1-(3,4-dichlorophenyl)ethyl]-2,4-dihydro-3*H*-pyrazol-3-one

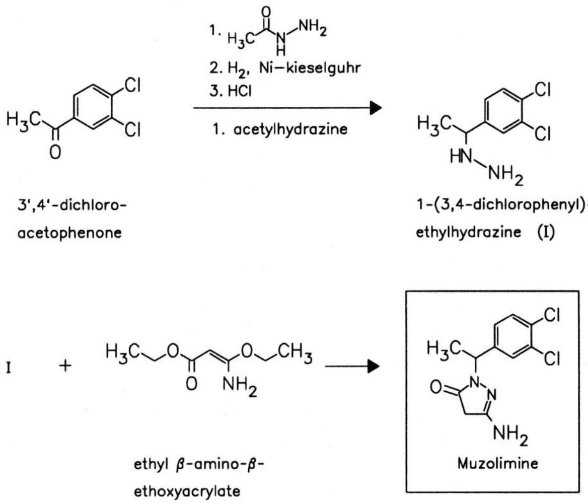

Reference(s):
DE 2 366 559 (Bayer; prior. 17.4.1973).
DOS 2 363 139 (Bayer; appl. 19.12.1973).
Möller, E.; Horstmann, H.; Meng, K.: Pharmatherapeutica (PHARDW) **1**, 540 (1977).
US 3 957 814 (Bayer; 18.5.1976; appl. 15.4.1974; D-prior. 17.3.1973).

synthesis of 1-(3,4-dichlorophenyl)ethylhydrazine:
DE 1 003 215 (Hoechst; appl. 1954).
Houben-Weyl **10/2**, 47.

Formulation(s): tabl. 20.7 mg, 31 mg, 248 mg

Trade Name(s):

D:	Edrul (Zyma/Bayer; 1985); wfm	I:	Edrul (Bayropharm; 1983); wfm

Mycophenolate mofetil
(ME-MPA; RS-61443)

ATC: L04AA06
Use: anti-inflammatory,
       immunosuppressive

RN: 128794-94-5   MF: $C_{23}H_{31}NO_7$ MW: 433.50
CN: (E)-6-(1,3-dihydro-4-hydroxy-6-methoxy-7-methyl-3-oxo-5-isobenzofuranyl)-4-methyl-4-hexenoic acid
      2-(4-morpholinyl)ethyl ester

hydrochloride
RN: 116680-01-4   MF: $C_{23}H_{31}NO_7 \cdot HCl$ MW: 469.96

(E)-6-(1,3-dihydro-4-hydroxy-
6-methoxy-7-methyl-3-oxo-
5-isobenzofuranyl)-4-methyl-
4-hexenoic acid

mycophenolic acid
chloride (I)

2-morpholino-
ethanol (II)

Mycophenolate mofetil

Reference(s):
US 4 727 069 (Syntex; appl. 30.1.1987; USA-prior. 30.1.1987).

preparation by azeotropic water removal from I and II:
US 5 247 083 (Syntex; appl. 10.7.1992; USA-prior. 10.7.1992).

treatment of allograft rejection:
US 4 786 637 (Syntex; appl. 22.1.1988; USA-prior. 30.1.1987, 4.9.1987, 22.1.1988, 17.8.1988).

immunoassays for mycophenolic acid derivatives:
WO 9 602 004 (Behringwerke AG; appl. 29.6.1995; USA-prior. 7.7.1994).

high dose oral suspension:
WO 9 509 626 (Syntex; appl. 27.9.1994; USA-prior. 1.10.1993).

high-dosage unit formulation obtained by hot-melt filling:
WO 9 426 266 (Syntex; appl. 10.5.1994; USA-prior. 13.5.1993).

intravenous formulation using crystalline anhydrous mycophenolate mofetil:
WO 9 507 902 (Syntex; appl. 12.9.1994; USA-prior. 15.9.1993).

combination of immunosuppressives:
WO 9 202 229 (SmithKline Beecham Corp.; appl. 7.8.1991; USA-prior. 10.8.1990, 5.12.1990).
WO 9 119 498 (Du Pont Merck; appl. 5.6.1991; USA-prior. 6.11.1990).

Formulation(s):   cps. 250 mg; tabl. 500 mg

Trade Name(s):
D: CellCept (Roche)   GB: CellCept (Roche)   USA: CellCept (Roche)
F: CellCept (Produits Roche) I: CellCept (Roche)

Mycophenolic acid

ATC: L04AA06
Use: antibiotic, antineoplastic, antiviral

RN: 24280-93-1 MF: $C_{17}H_{20}O_6$ MW: 320.34 EINECS: 246-119-3
LD$_{50}$: 1 g/kg (M, p.o.);
450 mg/kg (R, i.v.); 352 mg/kg (R, p.o.)
CN: (E)-6-(1,3-dihydro-4-hydroxy-6-methoxy-7-methyl-3-oxo-5-isobenzofuranyl)-4-methyl-4-hexenoic acid

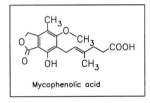

By fermentation of *Penicillium stoloniferum*.

Reference(s):
BE 862 618 (Lilly; appl. 4.1.1978; USA-prior. 21.3.1977).
GB 1 157 099 (ICI; valid from 31.7.1967; prior. 27.9.1966, 13.6.1967).

structure:
Logan, W.R.; Newbold, G.T.: J. Chem. Soc. (JCSOA9) **1957**. 1946.

use as immunosuppressive and anticancer agent:
Williams, R.H.: J. Antibiot. (JANTAJ) **21**, 463 (1968).
Suzuki, S. et al.: J. Antibiot. (JANTAJ) **22**, 297 (1969).
GB 1 157 100 (ICI; valid from 2.8.1967; prior. 27.9.1966).

antiviral activity:
Ando, K. et al.: J. Antibiot. (JANTAJ) **21**, 649 (1968).

total synthesis:
Birch, A.J.: J. Chem. Soc., Chem. Commun. (JCCCAT) **1969**, 788.
Canonica, L. et al.: Tetrahedron Lett. (TELEAY) **1971**, 2691.

review:
Carter, S.B. et al.: Nature (London) (NATUAS) **223**, 848 (1969).

Trade Name(s):
USA: Melbex (Lilly); wfm

Myrtecaine
(Nopoxamine)

ATC: N01B
Use: local anesthetic, antispasmodic

RN: 7712-50-7 MF: $C_{17}H_{31}NO$ MW: 265.44 EINECS: 231-735-7
LD$_{50}$: 48 mg/kg (M, i.v.)
CN: 2-[2-(6,6-dimethylbicyclo[3.1.1]hept-2-en-2-yl)ethoxy]-N,N-diethylethanamine

lauryl sulfate
RN: 76157-55-6 MF: $C_{17}H_{31}NO \cdot C_{12}H_{26}O_4S$ MW: 531.84

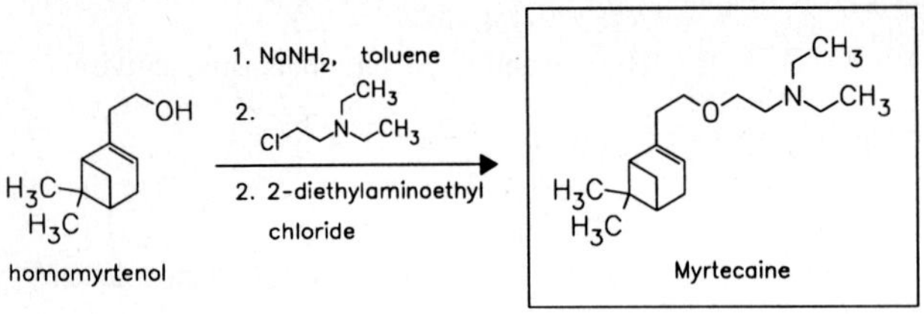

homomyrtenol

Myrtecaine

Reference(s):

GB 861 900 (O. P. Gaudin; appl. 1959).

Formulation(s):   chewing tabl. 2.5 mg (as lauryl sulfate); cream 1 g/100 g, 10 g/100 g; tabl. 2.5 mg

Trade Name(s):

D:	Acidrine (Solvay Pharma)-comb.		Algesalona (Solvay Pharma)-comb.	Algésal Suractivé (Solvay Pharma)-comb.
	Algesal (Solvay Pharma)-comb.	F:	Acidrine (Solvay Pharma)-comb.	I: Acidrine (Solvay Pharma)-comb.

List of Abbreviations

Δ	heating	DOS	Deutsche Offenlegungsschrift
A	Austria	drg.	dragees
Ac	acetyl	DRP	Deutsches Reichspatent
ACE	angiotensin converting enzyme	E	Spain
Ala	alanine	e.g.	exempli gratia (for example)
amp.	ampoules	EDCi	*N*-ethyl-*N'*-
appl.	application (patent)		(3-dimethylamino-propyl)
aq.	aqueous		carbodiimide hydrochloride
Arg	arginine	eff.	effervescent
Asn	asparagine	eff. gran.	effervescent granules
Asp	aspartic acid	eff. tabl.	effervescent tablets
ATC	ATC code (see preface)	EINECS	EINECS Registry Number
B	Belgium		(see preface)
BINAP	[1,1'-binaphthalene]-2,2'-	Encycl.	Encyclopedia
	diylbis[diphenylphosphine]	enzym.	enzymatic
Boc	*tert*-butoxycarbonyl	eq.	equivalent
BPH	benign prostatic hyperplasia	F	France
Bu	butyl	f. c. drg.	film coated dragees
BuLi	butyllithium	f. c. tabl.	film coated tablets
cat.	catalytic	ff	and the following pages
Cbo	carbobenzoxy or benzyloxycar-	FL	Liechtenstein
	bonyl	fract.	fractionated
CDN	Canada	g. p.	guinea pig
cf.	confer (compare)	GB	Great Britain
CN	chemical name (Chemical	Gln	glutamine
	Abstracts Registry name [unin-	Glu	glutamic acid
	verted])	Gly	glycine
	Note: CA Registry names are	gran.	granules
	subject to change without notice	H	Hungary
comb.	combination preparation	hv	irradiation
conc.	concentrated	His	Histidine
cps.	capsules	HMG-CoA	Hydroxymethylglutaryl-
CS	Czechoslovakia		coenzyme A
Cys	Cysteine	HMPT	hexamethylphosphoric triamide
D	Germany	HOBt	1-Hydroxybenzotriazole
DAS	Deutsche Auslegungsschrift	HPLC	high-performance liquid
DBU	1,8-diazabicyclo[5.4.0]undec-		chromatography
	7-ene	I	Italy
DEAD	diethylazodicarboxylate	i.a.	intraarterial
DCC	dicyclohexylcarbodiimide	i.m.	intramuscular
DCHA	dicyclohexylamine	i.p.	intraperitoneally
DHQ-PHAL	dihydroquinine-phthalazine	i.v.	intravenously
DIBAL	dibutylaluminum hydride	Ile	isoleucine
DIPEA	*N,N*-diisopropylethylamine	immobil.	immobilized
DK	Denmark	inf.	infusion
DMAA	*N,N*-dimethylacetamide	inh.	inhalation
DMAP	4-(dimethylamino)pyridine	inj.	injection
DMF	*N,N*-dimethylformamide	IRL	Ireland
DMSO	dimethylsulfoxide	iu	International units